MOLECULAR DETECTION OF HUMAN FUNGAL PATHOGENS

MOLECULAR DETECTION OF HUMAN FUNGAL PATHOGENS

EDITED BY
DONGYOU LIU

CRC Press
Taylor & Francis Group
Boca Raton London New York

CRC Press is an imprint of the
Taylor & Francis Group, an **informa** business

CRC Press
Taylor & Francis Group
6000 Broken Sound Parkway NW, Suite 300
Boca Raton, FL 33487-2742

Printed in the United States of America on acid-free paper
Version Date: 20110510

International Standard Book Number: 978-1-4398-1240-2 (Hardback)

Visit the Taylor & Francis Web site at
http://www.taylorandfrancis.com

and the CRC Press Web site at
http://www.crcpress.com

This volume is dedicated to a group of international mycologists, whose in-depth knowledge and technical expertise have made an all inclusive coverage of major human fungal pathogens possible.

Contents

Pezizomycotina: Sordariomycetes

Saccharomycotina and Taphrinomycotina

PART III Entomohpthoromycotina and Mucoromyotina

PART IV Microsporidia

Preface

Fungi are a diverse group of eukaryotic organisms that range from yeasts, molds, mushrooms, lichens, rusts, smuts, to microsporidia. Forming a kingdom of their own and being ubiquitously distributed in all environments, most fungi are saprophytes involved in the decomposition and recycling of organic matters as well as in the formation of symbiotic relationship with plants and animals. However, some fungi have the capacity to cause diseases in plants, animals, and humans. Often occurring as a result of trauma or underlying immunosuppression, human mycoses may manifest as superficial, cutaneous, subcutaneous, or systemic diseases. The inability to distinguish human mycoses caused by various fungal pathogens on clinical ground necessitates the development and use of laboratory diagnostic procedures in order to facilitate their treatment and prevention.

Given their complex life cycle and their tendency to produce morphologically similar structures, fungi are notoriously difficult to identify on the basis of their macroscopic and microscopic features, even for an experienced mycologist. To increase the accuracy, sensitivity, and efficiency of fungal identification, molecular techniques such as PCR and nucleotide sequencing have been increasingly adopted and applied in research and clinical laboratories worldwide. Consequently, a large number of molecular protocols have been described in the literature for the identification and detection of fungal organisms. As the saying goes, one person's medicine could easily turn into another's poison. There is certainly no exception here. The overabundance of original protocols and subsequent modifications has created a dilemma for anyone who was not directly involved in their developments to know which are most appropriate to use for streamlined identification and detection of fungal organisms of interest.

With contributions from international scientists in respective fungal pathogen research and diagnosis, this book aims to provide a reliable and comprehensive source relating the molecular detection and identification of major human fungal pathogens. Each chapter consists of a brief review on the classification, epidemiology, clinical features, and diagnosis of one or a group of related fungal species; an outline of clinical sample collection and preparation procedures; a selection of representative stepwise molecular protocols; and a discussion on additional research for further improving the diagnosis. This book represents an indispensable tool for both upcoming and experienced medical, veterinary, and industrial laboratory scientists engaged in fungus characterization and provides an essential reference for undergraduate and graduate students majoring in mycology.

An all-encompassing book such as this clearly demands a concerted team's efforts. I am fortunate and extremely honored to have had a large group of international mycologists as chapter contributors, whose in-depth knowledge and technological insights into human fungal pathogen detection have significantly enriched this book. Additionally, the professionalism and dedication of executive editor Barbara Norwitz and senior project coordinator Jill Jurgensen at CRC Press have enhanced its presentation. Finally, without the understanding and support of my family, Liling Ma, Brenda, and Cathy, the compilation of this comprehensive book would have been unimaginable.

Editor

Dongyou Liu, PhD, undertook his veterinary science education at Hunan Agricultural University, Changsha, China. Upon graduation, he received an overseas postgraduate scholarship from the Chinese Ministry of Education to pursue further training at the University of Melbourne, Melbourne, Victoria, Australia, where he worked toward improved immunological diagnosis of human hydatid disease. During the past two decades, he has crisscrossed between research and clinical laboratories in Australia and the United States, with focuses on molecular characterization and virulence determination of microbial pathogens such as ovine footrot bacterium (*Dichelobacter nodosus*), dermatophyte fungi (*Trichophyton*, *Microsporum*, and *Epidermophyton*), and listeriae (*Listeria* species). He is the first author of more than 50 original research and review articles in various international journals and the editor of the recently released *Handbook of* Listeria monocytogenes (2008), *Handbook of Nucleic Acid Purification* (2009), *Molecular Detection of Foodborne Pathogens* (2009), *Molecular Detection of Human Viral Pathogens* (2010), and *Molecular Detection of Human Bacterial Pathogens* (2011), as well as the forthcoming *Molecular Detection of Human Parasitic Pathogens* (2012), all of which are published by CRC Press.

Contributors

Ana Alastruey-Izquierdo
Centro Nacional de Microbiologia
Instituto de Salud Carlos III
Majadahonda, Spain

Priscila Costa Albuquerque
Laboratório de Micologia
Instituto de Pesquisa Clinica Evandro Chagas
Fundação Oswaldo Cruz
Rio de Janeiro, Brazil

Artur Alves
Departamento de Biologia
Centro de Estudos do Ambiente e do Mar
Universidade de Aveiro
Aveiro, Portugal

María J. Andrade
Higiene y Seguridad Alimentaria
Facultad de Veterinaria
Universidad de Extremadura
Cáceres, Spain

Zsuzsanna Antal
Laboratoire de Génomique Fonctionnelle des Champignons Pathogènes des Plantes
Université Claude Bernard Lyon 1
Villeurbanne, France

Maiken Cavling Arendrup
Unit of Mycology and Parasitology
Statens Serum Institute
Copenhagen, Denmark

S.N. Arseculeratne
Faculty of Medicine
University of Peradeniya
Peradeniya, Sri Lanka

Miguel A. Asensio
Higiene y Seguridad Alimentaria
Facultad de Veterinaria
Universidad de Extremadura
Cáceres, Spain

Frank W. Austin
College of Veterinary Medicine
Mississippi State University
Mississippi State, Mississippi

Eduardo Bagagli
Departamento de Microbiologia e Imunologia
Instituto de Biociências
Universidade Estadual Paulista-Botucatu
Sao Paulo, Brazil

Abhishek Baghela
Division of Mycology
Department of Medical Microbiology
Postgraduate Institute of Medical Education and Research
Chandigarh, India

Rosemary A. Barnes
School of Medicine
University Hospital of Wales
Cardiff University
Cardiff, United Kingdom

Alex van Belkum
Department of Medical Microbiology and Infectious Diseases
Erasmus Medical Center
Rotterdam, the Netherlands

Elena Bermúdez
Higiene y Segurided Alimentaria
Facultad de Veterinaria
Universidad de Extremadura
Cáceres, Spain

Leticia Bernal-Martinez
Centro Nacional de Microbiologia
Instituto de Salud Carlos III
Majadahonda, Spain

Antoine Berry
Department of Parasitology and Mycology
Faculty of Medicine Rangueil
Toulouse University Hospitals
Toulouse University
Toulouse, France

Sandra de Moraes Gimenes Bosco
Departamento de Microbiologia e Imunologia
Instituto de Biociências
Universidade Estadual Paulista-Botucatu
Sao Paulo, Brazil

Raimunda Sâmia Nogueira Brilhante
Specialized Medical Mycology Center
Federal University of Ceará
Ceará, Brazil

Sophie Cassaing
Department of Parasitology and Mycology
Faculty of Medicine Purpan
Toulouse University Hospitals
Toulouse University
Toulouse, France

Maria Victoria Castelli
Centro Nacional de Microbiologia
Instituto de Salud Carlos III
Majadahonda, Spain

Arunaloke Chakrabarti
Division of Mycology
Department of Medical Microbiology
Postgraduate Institute of Medical Education and Research
Chandigarh, India

Shiv Sekhar Chatterjee
Division of Mycology
Department of Medical Microbiology
Postgraduate Institute of Medical Education and Research
Chandigarh, India

Sharon C.-A. Chen
Centre for Infectious Diseases and Microbiology
Westmead Hospital
Westmead, New South Wales, Australia
and
Molecular Mycology Research Laboratory
Sydney Medical School—Western
University of Sydney at Westmead Hospital
Sydney, New South Wales, Australia

Massimo Cogliati
Laboratory of Medical Mycology
Department of Public Health – Microbiology – Virology
Università degli Studi di Milano
Milan, Italy

Susan Coloe
Microbiology Department
Melbourne Pathology
Collingwood, Australia

Rossana de Aguiar Cordeiro
Specialized Medical Mycology Center
Federal University of Ceará
Fortaleza, Brazil

Juan J. Córdoba
Higiene y Seguridad Alimentaria
Facultad de Veterinaria
Universidad de Extremadura
Cáceres, Spain

António Correia
Departamento de Biologia
Centro de Estudos do Ambiente e do Mar
Universidade de Aveiro
Aveiro, Portugal

Manuel Cuenca-Estrella
Centro Nacional de Microbiologia
Instituto de Salud Carlos III
Majadahonda, Spain

Eric Dannaoui
Unité de Mycologie Moléculaire
Centre National de Référence Mycologie et Antifongiques
Institut Pasteur
and
Faculté de Médecine
Université Paris Descartes
Unité de Parasitologie—Mycologie
Assistance Publique-Hôpitaux de Paris
Hôpital Européen Georges Pompidou
Paris, France

Silvia D'Arezzo
Unità di Microbiologia Molecolare
Azienda Policlinico Umberto I
Istituto Nazionale per le Malattie Infettive "Lazzaro Spallanzani"
Sapienza Università de Roma
Rome, Italy

Elizabeth S. Didier
Division of Microbiology
Tulane National Primate Research Center
Covington, Louisiana

Xianghong Du
College of Agronomy
Northwest A&F University
Shaanxi, China

David Ellis
Mycology Unit
South Australia Pathology
Adelaide, South Australia, Australia

Ahmed H. Fahal
Mycetoma Research Centre
University of Khartoum
Khartoum, Sudan

László Galgóczy
Faculty of Science and Informatics
Department of Microbiology
University of Szeged
Szeged, Hungary

Varghese K. George
Division of Mycology
Department of Medical Microbiology
Postgraduate Institute of Medical Education and Research
Chandigarh, India

Corrado Girmenia
Dipartimento di Ematologia, Oncologia, Anatomia Patologica e Medicina Rigenerativa
Azienda Policlinico Umberto I
Sapienza Università de Roma
Rome, Italy

Yvonne Gräser
Institute für Mikrobiologie und Hygiene
Charité Universitätsmedizin Berlin
Berlin, Germany

Joanna Gray
Royal College of Pathologists of Australasia
BioSecurity Quality Assurance Programs
Surrey Hills, New South Wales, Australia

Allan Jefferson Guimarães
Laboratório de Micologia
Instituto de Pesquisa Clinica Evandro Chagas
Fundação Oswaldo Cruz
Rio de Janeiro, Brazil

Larry Hanson
College of Veterinary Medicine
Mississippi State University
Mississippi State, Mississippi

D. Jane Hata
Division of Clinical Microbiology
Mayo Clinic College of Medicine
Jacksonville, Florida

Lóránt Hatvani
Faculty of Science and Informatics
Department of Microbiology
University of Szeged
Szeged, Hungary

Francisca Hernández-Hernández
Faculty of Medicine
Department of Microbiology and Parasitology
National Autonomous University of Mexico
Mexico City, Mexico

Kerstin Hoffmann
Jena Microbial Resource Collection (JMRC)
Institute of Microbiology
University of Jena
and
Department of Microbiology and Molecular Biology
Leibniz-Institute for Natural Product Research and Infection Biology e.V.
Hans-Knöll-Institute (HKI)
Neugasse, Jena, Germany

G. Sybren de Hoog
Central Bureau of Fungal Cultures
Fungal Biodiversity Centre
Royal Netherlands Academy of Arts and Sciences
Utrecht, the Netherlands

Po-Ren Hsueh
Departments of Laboratory Medicine and Internal Medicine
National Taiwan University Hospital
National Taiwan University College of Medicine
Taipei, Taiwan

Reiko Ikeda
Department of Microbiology
Meiji Pharmaceutical University
Tokyo, Japan

Yoshio Ishibashi
Department of Immunobiology
Meiji Pharmaceutical University
Tokyo, Japan

Jouni Issakainen
Department of Biology
University of Turku
Turku, Finland

Peter C. Iwen
Department of Pathology and Microbiology
University of Nebraska Medical Center
Omaha, Nebraska

Ayse Kalkanci
Department of Microbiology
School of Medicine
Gazi University
Ankara, Turkey

Sándor Kocsubé
Faculty of Science and Informatics
Department of Microbiology
University of Szeged
Szeged, Hungary

Laura Kovács
Faculty of Science and Informatics
Department of Microbiology
University of Szeged
Szeged, Hungary

Theerapong Krajaejun
Faculty of Medicine
Department of Pathology
Ramathibodi Hospital
Mahidol University
Bangkok, Thailand

László Kredics
Faculty of Science and Informatics
Department of Microbiology
University of Szeged
Szeged, Hungary

Semra Kustimur
Department of Microbiology
School of Medicine
Gazi University
Ankara, Turkey

Diego Libkind
Laboratorio de Microbiología Aplicada y Biotecnología
Centro Regional Universitario Bariloche
Consejo Nacional de Investigaciones Científicas y Tecnológicas
Instituto de Investigaciones en Biodiversidad y Medio Ambiente
Universidad Nacional del Comahue
Bariloche, Argentina

Marco Ligozzi
Dipartimento di Patologia
Università di Verona
Verona, Italy

N. Lima
Centre of Biological Engineering
Institute for Biotechnology and Bioengineering
University of Minho
Braga, Portugal

Marie-Denise Linas
Department of Parasitology and Mycology
Faculty of Medicine Purpan
Toulouse University Hospitals
Toulouse University
Toulouse, France

Mark D. Lindsley
Mycotic Diseases Branch
Centers for Disease Control and Prevention
Atlanta, Georgia

Thomas Lion
Division of Molecular Microbiology and Development of Genetic Diagnostics
Children's Cancer Research Institute and
LabDia Labordiagnostik
Vienna, Austria

Dongyou Liu
Royal College of Pathologists of Australasia
BioSecurity Quality Assurance Programs
Surrey Hills, New South Wales, Australia

Giuliana Lo Cascio
Departmento ad Attività Integrata di Patologia e Diagnostica
Azienda Ospedaliera Universitaria Integrata
Verona, Italy

Rubén Lopez-Martínez
Faculty of Medicine
Department of Microbiology and Parasitology
National Autonomous University of Mexico
Mexico City, Mexico

Marie Machouart
Faculté de Médecine
Centre Hospitalier Universitaire de Nancy
Hôpital Brabois
Université Henri Poincaré
Vandoeuvre-les-Nancy, France

H.N. Madhavan
L&T Microbiology Research Centre
Vision Research Foundation, Sankara Nethralaya
Chennai, India

László Manczinger
Faculty of Science and Informatics
Department of Microbiology
University of Szeged
Szeged, Hungary

Palanisamy Manikandan
Department of Microbiology
Aravind Eye Hospital and Postgraduate Institute of Ophthalmology
Coimbatore, India

Sílvio Alencar Marques
Faculdade de Medicina
Departamento de Dermatolologia e Radioterapia
Universidade Estadual Paulista-Botucatu
Sao Paulo, Brazil

Nanthawan Mekha
Department of Medical Sciences
National Institute of Health
Ministry of Public Health
Nonthaburi, Thailand

Jean Menotti
Hôpital Saint-Louis
Assistance Publique-Hôpitaux de Paris
Université Paris-Diderot
Paris, France

Steven R. Meshnick
Department of Epidemiology
Gillings School of Global Public Health
University of North Carolina
Chapel Hill, North Carolina

Wieland Meyer
Centre for Infectious Diseases and Microbiology
Westmead Hospital
Westmead, New South Wales, Australia
and
Molecular Mycology Research Laboratory
Sydney Medical School—Western
University of Sydney at Westmead Hospital
Sydney, New South Wales, Australia

Hideki Miyagi
Division of Dermatology
University of the Ryukyus
Okinawa, Japan

Mayumi Miyamoto
Department of Dermatology
Tokyo Medical University
Tokyo, Japan

Mauro de Medeiros Muniz
Laboratório de Micologia
Instituto de Pesquisa Clinica Evandro Chagas
Fundação Oswaldo Cruz
Rio de Janeiro, Brazil

László G. Nagy
Faculty of Science and Informatics
Department of Microbiology
University of Szeged
Szeged, Hungary

Venkatapathy Narendran
Department of Microbiology
Aravind Eye Hospital and Postgraduate Institute of Ophthalmology
Coimbatore, India

Akemi Nishikawa
Department of Immunobiology
Meiji Pharmaceutical University
Tokyo, Japan

Joshua D. Nosanchuk
Division of Infectious Disease
Departments of Medicine and Immunology
Albert Einstein College of Medicine
Yeshiva University
Bronx, New York

Félix Núñez
Higiene y Seguridad Alimentaria
Facultad de Veterinaria
Universidad de Extremadura
Cáceres, Spain

Ildikó Nyilasi
Faculty of Science and Informatics
Department of Microbiology
University of Szeged
Szeged, Hungary

Rosely Maria Zancopé Oliveira
Laboratório de Micologia
Instituto de Pesquisa Clinica Evandro Chagas
Fundação Oswaldo Cruz
Rio de Janeiro, Brazil

Rodrigo de Almeida Paes
Laboratório de Micologia
Instituto de Pesquisa Clinica Evandro Chagas
Fundação Oswaldo Cruz
Rio de Janeiro, Brazil

Malena P. Pantou
Molecular Immunopathology and Histocompatibility Laboratory
Onassis Cardiac Surgery Center
Athens, Greece

Tamás Papp
Faculty of Science and Informatics
Department of Microbiology
University of Szeged
Szeged, Hungary

R.R.M. Paterson
Centre for Biological Engineering
Institute for Biotechnology and Bioengineering
University of Minho
Braga, Portugal

Amelia Pérez-Mejía
Facultad de Medicina
Departamento de Microbiología y Parasitología
Universidad Nacional Autónoma de México
México City, Mexico

Armando Pérez-Torres
Facultad de Medicina
Departamento de Biología Celular y Tisular
Universidad Nacional Autónoma de México
México City, Mexico

David S. Perlin
Public Health Research Institute
New Jersey Medical School
University of Medicine and Dentistry of New Jersey
Newark, New Jersey

Michael D. Perry
Public Health Wales
Microbiology—Cardiff
University Hospital of Wales
Cardiff, United Kingdom

Tamás Petkovits
Faculty of Science and Informatics
Department of Microbiology
University of Szeged
Szeged, Hungary

Alan J.L. Phillips
Faculdade de Ciências e Tecnologia
Departamento de Ciências da Vida
Centro de Recursos Microbiológicos
Universidade Nova de Lisboa
Caparica, Portugal

Miia Pitkäranta
DNA Sequencing and Genomics Laboratory
Institute of Biotechnology
University of Helsinki
Helsinki, Finland

Natteewan Poonwan
Department of Medical Sciences
National Institute of Health
Ministry of Public Health
Nonthaburi, Thailand

Brunella Posteraro
Istituto di Microbiologia
Istituto Dermopatico dell'Immacolata
Istituto Di Ricovero e Cura a Carattere Scientifico
Ospedale San Carlo
Università Cattolica del Sacro Cuore
Rome, Italy

Patrizia Posteraro
Laboratory of Clinical Pathology and Microbiology
Istituto Dermopatico dell'Immacolata
Istituto Di Ricovero e Cura a Carattere Scientifico
Ospedale San Carlo
Università Cattolica del Sacro Cuore
Rome, Italy

Sandra Preuner
Division of Molecular Microbiology and Development of Genetic Diagnostics Children's Cancer Research Institute
and
LabDia Labordiagnostik
Vienna, Austria

María del Rocío Reyes-Montes
Facultad de Medicina
Departamento de Microbiología y Parasitología
Universidad Nacional Autónoma de México
México City, Mexico

Malcolm D. Richardson
Regional Mycology Laboratory
Education and Research Centre
Wythenshawe Hospital
University Hospital of South Manchester
University of Manchester
Manchester, United Kingdom

Virgínia Bodelão Richini-Pereira
Departamento de Microbiologia e Imunologia
Instituto de Biociências
Universidade Estadual Paulista-Botucatu
Sao Paulo, Brazil

Marcos Fábio Gadelha Rocha
Faculty of Veterinary Medicine
Postgraduate Program in Veterinary Science
State University of Ceará
and
Specialized Medical Mycology Center
Federal University of Ceará
Ceará, Brazil

Mar Rodríguez
Higiene y Seguridad Alimentaria
Facultad de Veterinaria
Universidad de Extremadura
Cáceres, Spain

Uwe H. Roesler
Institute of Animal Hygiene and Environmental Health
Freie University Berlin
Berlin, Germany

Franca Rossi
Biotechnology Department
University of Verona
Verona, Italy

Johannes E. Rothhardt
Jena Microbial Resource Collection (JMRC)
Institute of Microbiology
University of Jena
and
Department of Microbiology and Molecular Biology
Leibniz-Institute for Natural Product Research and Infection Biology e.V.
Hans-Knöll-Institute (HKI)
Neugasse, Jena, Germany

Wendy W.J. van de Sande
Department of Medical Microbiology and Infectious Diseases
Erasmus Medical Center
Rotterdam, the Netherlands

Maurizio Sanguinetti
Istituto di Microbiologia
Istituto Di Ricovero e Cura a Carattere Scientifico
Istituto Dermopatico dell'Immacolata
Ospedale San Carlo
Università Cattolica del Sacro Cuore
Rome, Italy

Ayako Sano
Medical Mycology Research Center
Chiba University
Chuo-ku, Japan

Boonmee Sathapatayavongs
Faculty of Medicine
Department of Medicine
Ramathibodi Hospital
Mahidol University
Bangkok, Thailand

Audrey N. Schuetz
Weill Cornell Medical College
NewYork-Presbyterian Hospital
New York, New York

Volker U. Schwartze
Institute of Microbiology
School of Biology and Pharmacy
University of Jena
Jena, Germany

Kanesan Panneer Selvam
Department of Microbiology
Dr. G.R. Damodaran College of Science
Coimbatore, India

Rahul Sharma
Plant Science Division
Agharkar Research Institute
Pune, India

M.R. Shivaprakash
Department of Medical Microbiology
Postgraduate Institute of Medical Education and Research
Chandigarh, India

Coimbatore Subramanian Shobana
Department of Microbiology
Dr. G.R. Damodaran College of Science
Coimbatore, India

Shoo Peng Siah
Human Genetic Signatures
North Ryde, New South Wales, Australia

José Júlio Costa Sidrim
Specialized Medical Mycology Center
Federal University of Ceará
Ceará, Brazil

Tania C. Sorrell
Centre for Infectious Diseases and Microbiology
Westmead Hospital
Westmead, New South Wales, Australia
and
Sydney Medical School
Western University of Sydney at Westmead Hospital
Sydney, New South Wales, Australia

Takashi Sugita
Department of Microbiology
Meiji Pharmaceutical University
Tokyo, Japan

Thomas D. Sullivan
Department of Pediatrics
School of Medicine and Public Health
University of Wisconsin
Madison, Wisconsin

Mami Tajima
Department of Dermatology
Tokyo Medical University
Tokyo, Japan

Masako Takashima
Japan Collection of Microorganisms, RIKEN BioResource Center, Saitama, Japan

Miklós Takó
Faculty of Science and Informatics
Department of Microbiology
University of Szeged
Szeged, Hungary

Yi-Wei Tang
Departments of Pathology and Medicine
Vanderbilt University School of Medicine
Nashville, Tennessee

Steve M. Taylor
Department of Epidemiology
Gillings School of Global Public Health
University of North Carolina
Chapel Hill, North Carolina
and
Division of Infectious Diseases and International Health
Duke University Medical Center
Durham, North Carolina

Raquel Cordeiro Theodoro
Departamento de Microbiologia e Imunologia
Instituto de Biociências
Universidade Estadual Paulista-Botucatu
Sao Paulo, Brazil

K. Lily Therese
L&T Microbiology Research Centre
Vision Research Foundation, Sankara Nethralaya
Chennai, India

Conchita Toriello
Facultad de Medicina
Departamento de Microbiología y Parasitología
Universidad Nacional Autónoma de México
México City, Mexico

Sandra Torriani
Biotechnology Department
University of Verona
Verona, Italy

Anna Maria Tortorano
Laboratory of Medical Mycology
Department of Public Health – Microbiology – Virology
Università degli Studi di Milano
Milan, Italy

Ryoji Tsuboi
Department of Dermatology
Tokyo Medical University
Tokyo, Japan

Hisae Tsubuku
Department of Dermatology
Tokyo Medical University
Tokyo, Japan

Juan Luis Rodríguez Tudela
Centro Nacional de Microbiologia
Instituto de Salud Carlos III
Majadahonda, Spain

Milton A. Typas
Faculty of Biology
Department of Genetics and Biotechnology
University of Athens
Athens, Greece

Csaba Vágvölgyi
Faculty of Science and Informatics
Department of Microbiology
University of Szeged
Szeged, Hungary

Jaco J. Verweij
Department of Parasitology
Leiden University Medical Center
Leiden, the Netherlands

Paolo Visca
Dipartimento di Biologia
Università Roma Tre
and
Azienda Policlinico Umberto I
Sapienza Università de Roma
and
Unità di Microbiologia Molecolare
Istituto Nazionale per le Malattie Infettive "Lazzaro Spallanzani"
Rome, Italy

Govinda S. Visvesvara
Department of Health and Human Services
Public Health Service
Centers for Disease Control and Prevention
Atlanta, Georgia

Maria Anna Viviani
Laboratory of Medical Mycology
Department of Public Health – Microbiology – Virology
Università degli Studi di Milano
Milan, Italy

Kerstin Voigt
Jena Microbial Resource Collection (JMRC)
Institute of Microbiology
University of Jena
and
Department of Microbiology and Molecular Biology
Leibniz-Institute for Natural Product Research and Infection Biology e.V.
Hans-Knöll-Institute (HKI)
Neugasse, Jena, Germany

Lysett Wagner
Institute of Microbiology
School of Biology and Pharmacy
University of Jena
Jena, Germany

Song Weining
College of Agronomy
Northwest A&F University
Shaanxi, China

Nancy L. Wengenack
Division of Clinical Microbiology
Mayo Clinic College of Medicine
Rochester, Minnesota

P. Lewis White
Public Health Wales
Microbiology—Cardiff
University Hospital of Wales
Cardiff, United Kingdom

Lihua Xiao
Department of Health and Human Services
Public Health Service
Centers for Disease Control and Prevention
Atlanta, Georgia

Kyoko Yarita
Medical Mycology Research Center
Chiba University
Chuo-ku, Japan

Enshi Zhang
Department of Dermatology
Tokyo Medical University
Tokyo, Japan

Yanan Zhao
Public Health Research Institute
New Jersey Medical School of New Jersey
University of Medicine and Dentistry
Newark, New Jersey

1 Introductory Remarks

Dongyou Liu

CONTENTS

1.1 PREAMBLE

Fungi (singular fungus, meaning "mushroom" in Latin) are a diverse group of eukaryotic organisms (ranging from yeasts, molds, mushrooms, lichens, rusts, smuts to microsporidia) that constitute one of the five kingdoms (i.e., Prokaryotae, Fungi, Protista, Plantae, and Animalia) in the current classification system for living organisms. Similar to other eukaryotic kingdoms (Protista, Plantae, and Animalia), fungi harbor membrane-bound nuclei with chromosomal DNA, which consists of noncoding regions (introns) and coding regions (exons), possess membrane-bound cytoplasmic organelles (e.g., mitochondria), sterol-containing membranes, and 80S ribosomes and produce a variety of soluble carbohydrates and storage compounds, including sugar alcohols, disaccharides, and polysaccharides. Furthermore, Fungi resemble Protista and Animalia by the lack of chloroplasts and thus the requirement of preformed organic compounds as energy sources. Although both Fungi and Plantae possess a cell wall and vacuoles, reproduce by sexual as well as asexual means, generate spores (as in ferns and mosses), and have haploid nuclei (as in mosses and algae), Fungi differ from Plantae by the presence of chitin (which also exists in the exoskeleton of arthropods), instead of cellulose in the cell walls, and the absence of chloroplasts. On the other hand, Fungi differ from Prokaryotae by having nuclear membrane, plasma membrane, and cell wall.

In this introductory chapter, a brief overview is presented on the classification, biology, and genetics of fungal organisms, and clinical manifestations in human hosts resulting from their infections. This is followed by a summary of laboratory approaches that are useful for phenotypic characterization of fungi, including sample collection and processing, microscopic examination, in vitro cultivation, biochemical and anti-fungal testing. The subsequent section focuses on key attributes relating to molecular characterization of fungi, such as nucleic acid extraction, target gene selection, template amplification, and amplicon detection. Finally, the importance of rational result interpretation, standardization, and quality control and assurance in the molecular fungal testing is emphasized.

1.2 CLASSIFICATION, BIOLOGY, GENETICS, AND CLINICAL PRESENTATION

1.2.1 Classification

Fungi are an extremely diverse and abundant group of eukaryotic organisms, whose sizes range from single-celled aquatic chytrids to large mushrooms and whose number has been estimated to be between 700,000 and 1.5 million species, with nearly 100,000 species being described to date [1–8]. However, fewer than 500 of the recognized fungal species (including about 200 yeast species) have been shown to cause human infections.

Based on morphological criteria, fungi are often divided into two major categories: filamentous fungi (true fungi) and yeasts. Accounting for the bulk of fungal species, filamentous fungi produce tubular, elongated, and thread-like (filamentous) cellular structures (known as hyphae), which contain multiple nuclei and extend at their tips. With about 700 known species, yeasts are single-celled organisms that reproduce by budding or binary fission. In addition, a few fungal species are able to switch between a yeast phase and a hyphal phase in response to environmental conditions and are referred to as dimorphic fungi. Despite their relatively insignificant proportion in relation to filamentous fungi, about 200 of the 700 recognized yeast species are responsible for a majority of clinical cases of human mycoses.

Using a combination of morphological characteristics and mechanisms of reproduction, fungi have been traditionally separated into five phyla: Ascomycota (sac fungi), Basidiomycota (club fungi), Mycophycophyta (lichens fungi), Zygomycota (conjugation fungi), and Deuteromycota (imperfect fungi, or mitosporic fungi, which are fungi with no known sexual cycle) [9].

Recent phylogenetic analyses of 18S rRNA, 28S rRNA, 5.8S rRNA, *rpb1*, *rpb2*, and *tef1* genes indicate that the kingdom Fungi consists of one subkingdom (Dikarya including phyla Ascomycota and Basidiomycota), seven phyla (all with the suffix -mycota except Microsporidia; i.e., Ascomycota, Basidiomycota, Chytridiomycota, Glomeromycota, Blastocladiomycota, Neocallimastigomycota, and Microsporidia, in addition to Fungi incertae sedis, which encompasses fungi with indeterminate taxonomical status) (Table 1.1), 10 subphyla (with the suffix -mycotina), 35 classes (with the suffix -mycetes), 12 subclasses (with the suffix -mycetidae), and 129 orders (with the suffix -ales) [10–13]. A most notable feature of this taxonomical scheme is the reorganization of the former phylum Zygomycota into the phylum Glomeromycota and four separate subphyla (Mucoromycotina, Entomophthoromycotina, Zoopagomycotina, and Kickxellomycotina), which may form independent phyla upon further confirmation (Table 1.1). However, this scheme does not take into account of organisms such as oomycetes and slime molds that were formerly included in the kingdom Fungi [12]. Also, the genera *Caulochytrium*, *Olpidium*, *Rozella* (formerly of Chytridiomycota), and *Basidiobolus* (formerly of Entomophthorales, Zygomycota) are not included in any higher taxa in this scheme, pending further taxonomical resolutions [12]. In addition, a clade (i.e., Symbiomycota) sharing similarity between Glomeromycota and Dikarya is not included in the current scheme as Symbiomycota may possibly constitute a rank between kingdom and subkingdom [12].

Most human pathogenic fungi are found in the phyla Ascomycota, Basidiomycota, and Microsporidia as well as Fungi incertae sedis (principally Mucoromycotina and Entomophthoromycotina of the former phylum Zygomycota). From a medical mycologist's perspective, human pathogenic fungi are conveniently separated into eight subgroups: (i) dermatophytes (represented by *Epidermophyton*, *Microsporum*, and *Trichophyton*); (ii) yeasts (represented by *Blastoschizomyces*, *Candida*, *Cryptococcus*, *Lacazia*, *Malassezia*, *Rhodotorula*, *Saccharomyces*, and *Trichosporon*); (iii) dimorphic fungi (represented by *Blastomyces*, *Coccidioides*, *Histoplasma*, and *Paracoccidioides*); (iv) hyaline hyphomycetes (hyaline molds) (represented by *Acremonium*, *Aspergillus*, *Beauveria*, *Chrysosporium*, *Cylindrocarpon*, *Fusarium*, *Geotrichum*, *Gliocladium*, *Graphium*, *Madurella*, *Malbranchea*, *Onychocola*, *Paecilomyces*, *Penicillium*, *Scedosporium*, *Scopulariopsis*, *Sepedonium*, *Trichoderma*, *Trichothecium*, and *Verticillium*); (v) dematiaceous hyphomycetes (dematiaceous molds) (represented by *Acrophialophora*, *Alternaria*, *Aureobasidium*, *Bipolaris*, *Cladophialophora*, *Cladosporium*, *Curvularia*, *Drechslera*, *Exophiala*, *Exserohilum*, *Fonsecaea*, *Hortaea*, *Lecythophora*, *Ochroconis*, *Phaeoacremonium*, *Phialophora*, *Ramichloridium*, *Rhinocladiella*, *Scedosporium*, *Sporothrix*, *Ulocladium*, and *Veronaea*); (vi) coelomycetes (represented by *Colletotrichum*, *Lasiodiplodia*, *Nattrassia*, and *Phoma*); (vii) zygomycetes (represented by *Apophysomyces*, *Basidiobolus*, *Conidiobolus*, *Cunninghamella*, *Mortierella*, *Mucor*, *Absidia*, *Rhizomucor*, *Rhizopus*, *Saksenaea*, and *Syncephalestrum*); and (viii) basidiomycetes [7].

TABLE 1.1
Classification of the Kingdom Fungi

Phylum (Subkingdom)	Subphylum (Former Classification)	Brief Description
Ascomycota (*Dikarya*)	Pezizomycotina	Ascomycota (commonly known as sac fungi) represents the largest phylum of Fungi, with over 64,000 species grouped under three subphyla (Taphrinomycotina, Saccharomycotina, and Pezizomycotina). Ascomycota produce *ascus* (from *Greek askos*, meaning "sac" or "wineskin"), in which nonmotile *spores* (a *sexual structure* also known as *ascospores*) are formed. However, some Ascomycota (formerly belonging to *Deuteromycota*) are *asexual*, do not have a sexual cycle, and thus do not form asci (or ascospores).
		Forming part of Ascomycota, Pezizomycotina consist of *Orbiliomycetes*, *Pezizomycetes*, *Dothideomycetes*, *Arthoniomycetes*, Eurotiomycetes, *Laboulbeniomycetes*, Lichinomycetes, *Lecanoromycetes*, Leotiomycetes, and *Sordariomycetes*, as well as three unassigned orders (Lahmiales, Medeolarials, and Triblidiales). Pezizomycotina cover all ascomycetes that produce *ascocarps* (fruiting bodies), except for *Neolecta* in Taphrinomycotina.
	Saccharomycotina	Forming part of Ascomycota, Saccharomycotina consist of the "true" yeast class *Saccharomycetes*
	Taphrinomycotina	Forming part of Ascomycota, Taphrinomycotina consist of four classes: *Neolectomycetes* (hyphal fungi), *Pneumocystidomycetes* (mammalian lung pathogen Pneumocystis), *Schizosaccharomycetes* (fission yeasts), and *Taphrinomycetes* (hyphal fungi).
Basidiomycota (*Dikarya*)	Pucciniomycotina (Urediniomycetes)	Basidiomycota (commonly known as club fungi) is the second largest phylum of Fungi, with 31,515 species grouped under three subphyla (Pucciniomycotina, Ustilaginomycotina, and Agaricomycotina, in addition to two separate classes Wallemiomycetes and Entorrhizomycetes).
		Forming part of *Basidiomycota*, Pucciniomycotina consist of eight classes of rust fungi (i.e., *Classiculomycetes*, *Cryptomycocolacomycetes*, *Mixiomycetes*, *Atractiellomycetes*, *Agaricostilbomycetes*, *Cystobasidiomycetes*, *Pucciniomycetes*, and *Microbotryomycetes*).
	Ustilaginomycotina (Ustilaginomycetes)	Forming part of *Basidiomycota*, Ustilaginomycotina consist of two smut fungus classes *Exobasidiomycetes* and *Ustilaginomycetes*, as well as a separate smut fungus order Malasseziales.
	Agaricomycotina (Basidiomycetes)	Forming part of *Basidiomycota*, Agaricomycotina consist of three classes: *Agaricomycetes* (hymenia-forming fungi), *Dacrymycetes* (hymenia-lacking fungi), and *Tremellomycetes* (jelly fungi).
Chytridiomycota		Consisting of two classes Chytridiomycetes (with three orders: Chytridiales, Spizellomycetales, and Rhizophydiales) and Monoblepharidomycetes (with one order Monoblephariales), *Chytridiomycota* include more than 1000 known species. Chytridion (meaning "little pot") describes the structure containing unreleased *spores*. Chytrids are mostly primitive, *aquatic*, *saprobic* fungi involved in the degradation of *chitin* and *keratin*, have *coenocytic thalli*, and usually form *rhizoids* (instead of true mycelium).
Neocallimastigomycota	(Chytridiomycota)	Consisting of Neocallimastigales, a traditional member of Chytridiomycota, *Neocallimastigomycota* include a small group of anaerobic fungi that inhabit the digestive system of larger herbivorous mammals and possibly other terrestrial and aquatic environments. Although lacking *mitochondria*, *Neocallimastigomycota* possess *hydrogenosomes* of mitochondrial origin. Similar to chrytrids, neocallimastigomycetes form zoospores with posteriorly uniflagellate or polyflagellate. However, neocallimastigomycetes are distinct from other chytrids on the basis of both morphology and molecular phylogeny.
Blastocladiomycota	(Chytridiomycota)	Consisting of Blastocladiales, also a traditional member of Chytridiomycota, *Blastocladiomycota* are *saprotrophs*, and also parasites of all eukaryotic groups. *Blastocladiomycota* undergo *sporic meiosis* in contrast to chytrids, which mostly exhibit *zygotic meiosis*.
Glomeromycota	(*Zygomycota*)	Forming part (commonly known as "sugar" and "pin" molds) of former *Zygomycota*, and consisting of one class *Glomeromycetes* (with four orders: *Glomerales*, *Diversisporales*, *Paraglomerales*, and *Archaeosporales*) with about 200 described species (all of which reproduce asexually), Glomeromycota produce *arbuscular mycorrhizas* with roots or thalli, and are obligate *biotrophs*, dependent on symbiosis with land plants for carbon and energy.
Fungi *incertae sedis*	*Mucoromycotina* (*Zygomycota*)	Fungi that were placed in Zygomycota are now being reassigned to Glomeromycota, and Fungi *incertae sedis* (including four subphyla *Mucoromycotina*, *Entomophthoromycotina*, *Zoopagomycotina*, and *Kickxellomycotina*).
		Consisting part of Fungi *incertae sedis*, *Mucoromycotina* cover three orders: *Mucorales*, *Endogonales*, and *Mortierellales*.
	Entomophthoromycotina (*Zygomycota*)	Consisting part of Fungi *incertae sedis*, *Entomophthoromycotina* (with one order Entomophthorales) are pathogens of *insects*, *nematodes*, and *tardigrades*, as well as free-living *saprotrophs*.

(*continued*)

TABLE 1.1 (continued)
Classification of the Kingdom Fungi

Phylum (Subkingdom)	Subphylum (Former Classification)	Brief Description
	Zoopagomycotina (*Zygomycota*)	Consisting part of Fungi *incertae sedis*, *Zoopagomycotina* (with one order Zoopagales) are pathogens of microscopic animals such as amoebae.
	Kickxellomycotina (*Zygomycota*)	Consisting part of Fungi *incertae sedis*, *Kickxellomycotina* include four orders: *Asellariales*, *Kickxellales*, *Dimargaritales*, and *Harpellales*.
Microsporidia		Microsporidia cover about 150 genera (containing >1200 species) that were previously considered as protozoa, of which 12 species (representing 8 genera) have been shown to cause opportunistic infections in humans. Microsporidia lack mitochondria and motile structures (e.g., flagella) and produce highly resistant spores, the morphology (oval or pyriform, occasionally rod-shaped or spherical) of which is often used for their differentiation.

Sources: James, T.Y. et al., *Nature*, 443, 818, 2006; Hibbett, D.S. et al., *Mycol. Res.*, 111, 509, 2007.

1.2.2 Biology

Filamentous fungi are characterized by the production of hyphae, which are cylindrical, thread-like structures of 2–10 μm in diameter and up to several centimeters in length. Hyphae can be either septate (with two or more compartments separated by right-angled internal cell walls called septa) or aseptate (or coenocytic, with each compartment containing one or more nuclei). Septa have pores that facilitate passage and interchange of cytoplasm, organelles, and at times nuclei. Hyphae are important for penetration/invasion into the host cells and for the uptake of nutrients from living hosts and other substrates. New hyphae typically emerge from hyphal tips (apices), arise from existing hyphae by a process called branching, or occasionally grow hyphal tips bifurcate (fork) giving rise to two parallel-growing hyphae. The combined effects of apical growth and branching/forking result in the formation of an interconnected network of hyphae (with high surface area to volume ratios) known as mycelium (plural mycelia), which is also commonly called mold. Mycelia grown on solid agar media are referred to as colonies, which may exhibit a variety of sizes, shapes, and colors (pigmentations) [9].

In general, fungi reproduce by means of microscopic propagules called spores (conidia) as a result of an asexual process. Near a third of all fungi reproduce by different modes of propagation, showing two well-differentiated stages (i.e., the teleomorph or sexual stage and the anamorph or asexual stage) within the life cycle of a species. Achieved via vegetative spores or mycelial fragmentation, asexual reproduction helps clonal populations to adapt to a specific niche and allows more efficient dispersal than sexual reproduction. Sexual reproduction through meiosis involves various sexual structures (e.g., fruiting bodies) and reproductive strategies. Compatible fungi may fuse their hyphae into an interconnected network in a process known as anastomosis, which is required for the initiation of the sexual cycle.

Yeasts commonly undergo asexual reproduction (mitosis) by budding or fission, although some have the capacity to reproduce both asexually and sexually. During the budding process, a small bud (or daughter cell) forms on the parent cell, and the nucleus of the parent cell splits into a daughter nucleus which migrates into the daughter cell. The growing bud eventually separates from the parent cell to become a new cell.

Fungi are widespread in all environments and habitats, including soil, plants, insects, animals, humans, air, deserts, and deep-sea sediments [14]. Most fungi are saprophytes that play an essential environmental role in the decomposition and recycling of organic matters, and form symbiotic relationship with plants and animals; some have the capacity to cause diseases in plants, animals, and humans. Furthermore, some fungi have other properties that can be exploited for bread/beverage making, insect pest control, medicine, and scientific research. For instance, yeasts have been employed in (i) the two-hybrid screening systems for the general detection of protein–protein interactions; (ii) the yeast artificial chromosomes (YACs) for cloning large fragments (200–800 kb) of DNA; and (iii) expression systems for heterologous proteins.

1.2.3 Genetics

Relative to other higher level eukaryotes (e.g., mammals), fungal genomes are simple and compact, with sizes ranging from 12,068 kb in *Saccharomyces cerevisiae*, 22,540 kb in *Trichophyton verrucosum* HKI 0517 (GenBank ACYE00000000), 28,467 kb in *Penicillium marneffei* ATCC 18224 (GenBank ABAR00000000), 32,228 kb in *Penicillium chrysogenum* Wisconsin 54-1255 to 51,230 kb in *Nectria haematococca* (anamorph *Fusarium solani*) (GenBank ACJF00000000).

The 12 Mb genome of baker's yeast *Saccharomyces cerevisiae* is clustered into 16 chromosomes (of 200–2200 kb in size), with a total of 6183 open-reading frames (ORFs), of which 5885 are predicated to be protein-coding genes. Its ribosomal RNA (rRNA) genes are coded by about 140 genes of a single tandem array on chromosome XII; small nuclear RNAs are coded by 40 genes; and transfer RNAs (tRNAs) are coded by 275 genes. *S. cerevisiae* mitochondrial DNA encodes components of the mitochondrial translational machinery and about 15% of the mitochondrial proteins [15].

Whereas the 22 Mb genome of *Penicillium marneffei* ATCC 18224 harbors 10,136 ORF; the 32 Mb genome of *Penicillium chrysogenum* Wisconsin 54-1255 contains 13,911 ORF, with 12,791 being protein-coding genes [16]. As a member of the "*Fusarium solani* species complex" that encompasses >50 species, *Nectria haematococca* MPVI (anamorph *Fusarium solani*) has been shown to possess a 51 Mb genome, which is organized in 17 chromosomes (of 530 kb–6.52 Mb in size) with 15,707 predicted genes [17].

On the other hand, microsporidia possess extremely reduced eukaryotic genomes, which may be as small as 2.6 Mb with 2000 genes. These organisms have remnant mitochondria and show unique morphologies related to parasitism, including polar tube to penetrate host cells and initiate infection.

1.2.4 Clinical Presentation

Although most fungal species are saprophytic organisms with a very low inherent virulence, some have the ability to take advantage of the weakened host defense (e.g., trauma and immunosuppression) and invade the host cells, causing a variety of clinical diseases, ranging from (i) superficial, (ii) cutaneous, (iii) subcutaneous to (iv) systemic mycoses [18].

Superficial mycoses: As cosmetic fungal infections of the skin or hair shaft, superficial mycoses do not invade the living tissue nor elicit cellular response from the host. Patients with superficial mycoses seeking medical advices are largely for social or cosmetic reasons. Examples of such superficial mycoses include pityriasis versicolor and seborrhoeic dermatitis due to *Malassezia furfur*, tinea nigra due to *Hortaea werneckii*, white piedra due to *Trichosporon* species, and black piedra due to *Piedraia hortae*.

Cutaneous mycoses: Being another form of superficial fungal infections of the skin, hair, or nails, cutaneous mycoses do not invade the living tissue but may cause a variety of pathological changes in the host due to the presence of the infectious agent and its metabolic products. Examples of such cutaneous mycoses comprise dermatophytosis due to *Epidermophyton*, *Microsporum*, and *Trichophyton*; candidiasis (of skin, mucous membranes, and nails) due to *Candida* species; and dermatomycosis due to non-dermatophyte molds such as *Onychocola*, *Scopulariopsis*, and *Scytalidium*.

Subcutaneous mycoses: As chronic, localized infections of the skin and subcutaneous tissue following the traumatic implantation of a soil saprophyte, subcutaneous mycoses may present a variety of clinical symptoms. These range from sporotrichosis due to *Sporothrix*; chromoblastomycosis due to *Cladosporium*, *Fonsecaea*, and *Phialophora*; phaeohyphomycosis due to *Bipolaris*, *Cladosporium*, *Curvularia*, *Exophiala*, and *Exserohilum*; eumycetoma due to *Acremonium*, *Madurella*, and *Pseudallescheria*; subcutaneous zygomycosis due to *Basidiobolus* and *Conidiobolus*; rhinosporidiosis due to *Rhinosporidium*; and lacaziosis (or lobomycosis) due to *Lacazia loboi*.

Systemic mycoses: Some fungi, especially dimorphic fungi, have the capacity to breach the physical and immunological defenses of the human host, causing pulmonary and other infections after the inhalation of conidia. Examples of such systemic mycoses include histoplasmosis due to *Histoplasma capsulatum*; coccidioidomycosis due to *Coccidioides immitis*; blastomycosis due to *Blastomyces dermatitidis*; and paracoccidioidomycosis due to *Paracoccidioides brasiliensis* [7].

1.3 PHENOTYPIC CHARACTERIZATION

Due to the fact that clinical presentations of human mycoses caused by various fungal species are nonspecific and indistinguishable, and that different fungal pathogens demonstrate varied resistance to commonly used antifungal drugs, there is a need to identify the causative agents to genus and species level in order to implement effective control and prevention strategies.

Traditionally, laboratory identification and characterization of fungi rely mainly on morphological (e.g., the size and shape of spores or fruiting structures), biochemical (e.g., the ability to metabolize certain biochemicals, or the reaction to chemical tests), biological (e.g., the ability to mate), and other phenotypic criteria. Apart from some mycotic/hyphal elements, most fungi present in the clinical samples are impossible to distinguish upon direct microscopic examination. Therefore, in vitro cultivation is vital for isolation of the fungal pathogens of interest, permitting subsequent determination on the basis of distinct colonial (macroscopic) and microscopic features [19–21].

1.3.1 Sample Collection and Processing

1.3.1.1 General Guidelines for Specimen Handling

Fungal pathogens are capable of spreading through spores and may pose danger to laboratory personnel if sufficient caution is not heeded. Therefore, when dealing with fungal specimens in laboratory, it is essential to (i) wear a protective gown or laboratory coat while in the laboratory, (ii) wear gloves when handling clinical and culture materials, (iii) transport cultures in a rack or canister, (iv) disinfect specimen containers contaminated on the outside by wiping with gauze before opening, (v) open specimens in laminar flow safety cabinet, (vi) use mechanical pipetting devices for any material or reagent, and (vii) clean the work area with a 2% amphyl solution when work is completed.

A diverse range of clinical and environmental samples can be utilized for fungal testing. In order to ensure accurate and consistent results, samples intended for mycological investigations need to be collected and processed in such a way that offers the best chance for isolation and identification of causative fungal agents.

1.3.1.2 Sputum, Bronchial Washings, and Throat Swabs

Collection: (i) Collect sputum (5–10 mL, as a result of a deep cough not saliva) in sterile container in the early morning. Patients are advised not to eat before sputum collection. Thick sputum can be emulsified by the addition of 12–20 sterile glass beads and 3–5 mL of sterile distilled water followed by shaking. (ii) Collect bronchial washings (tracheal lavage or bronchial lavage) aseptically by physicians. (iii) Obtain throat specimens by rolling a moist sterile swab over the affected area. For suspected *Candida* specimen, scrape the affected area with a sterile tongue depressor. (iv) Store samples at 4°C in case of short delays in processing. *Processing*: (i) Make wet mounts in KOH (1 drop) and Gram stained smears (1 drop) for direct microscopy. Use periodic acid-Schiff (PAS) stain if KOH preparation is unsatisfactory. (ii) Inoculate sample onto Sabouraud's dextrose agar with chloramphenicol and gentamicin and incubate duplicate cultures at 26°C and 35°C. (iii) Inoculate sample onto brain heart infusion agar (BHIA) supplemented with 5% sheep blood and incubate at 35°C. Maintain cultures for 4 weeks.

1.3.1.3 Blood, Bone Marrow, and Body Fluids

Collection: (i) Collect blood (8–10 mL aseptically using vacutainer tube [#4960, containing 1.7 mL of 0.35% sodium polyanethol sulfonate (SPS) as an anticoagulant]). Clean the collection site with a disinfectant at the time of collection. (ii) Collect bone marrow and body fluids (pleural, synovial, and peritoneal) aseptically by physicians. Add SPS or heparin as an anticoagulant. *Processing*: (i) Prepare smears for Giemsa, Gram, and PAS staining. (ii) Inoculate 0.5–1.0 mL of buffy coat (after centrifugation of 5–8 mL of blood) onto the surface of culture media (Sabouraud's dextrose agar with chloramphenicol and gentamicin, and BHIA supplemented with 5% sheep blood) with a loop, or 1 part blood to 10–20 parts brain/heart infusion broth. (iii) Inoculate bone marrow and body fluids (pleural, synovial, and peritoneal) on culture media. Maintain cultures at 26°C and 35°C for 4 weeks.

1.3.1.4 Pus, Exudate, and Drainage

Collection: (i) Aspirate material from undrained abscesses using a sterile needle and syringe. (ii) Express pus using a sterile, sharp-pointed scalpel. (iii) Place the material in a sterile container.

1.3.1.5 Vaginal Swabs

Collection: (i) Collect material from the vagina using several sterile swabs. (ii) Insert swabs into a sterile tube. *Processing:* Smear swab onto heat-sterilized glass slide for Gram stain.

1.3.1.6 Urine

Collection: (i) Collect an early morning, mid-stream catch of >2.0 mL (do not process more than 50.0 mL) in a sterile container when aspiration or cystoscopy cannot be done. (ii) Store urine at 4°C for up to 12–14 h if necessary. *Processing*: (i) Centrifuge the urine for 10–15 min at 2000 rpm. Decant the supernatant and pool the sediment if necessary. (ii) Prepare a direct smear of the sediment in KOH for direct microscopy. PAS, Gram, or India ink preparations may also be helpful.

1.3.1.7 Cerebrospinal Fluid

Collection: (i) Collect 2–5 mL CSF aseptically by clinician. (ii) Leave CSF at room temperature or incubate at 30°C if there is a delay in processing. (iii) Centrifuge CSF and process the sediment as follows. *Processing*: (i) Use 1 drop of the sediment to make an India ink mount. (ii) Resuspend the remaining sediment in 1–2 mL of CSF and inoculate onto Sabouraud's dextrose agar with chloramphenicol and gentamicin and incubate at 26°C and 35°C. (iii) Inoculate sediment onto BHIA supplemented with 5% sheep blood and incubate at 35°C. Maintain cultures for at least 4 weeks.

1.3.1.8 Tissue Biopsies from Visceral Organs

Collection: (i) Collect tissue from the center and edge of the lesion aseptically by clinician. Include normal tissue for comparison. (ii) Keep a portion of the tissue in formalin for rapid frozen sectioning with staining by hematoxylin and eosin (H&E), Grocott's methenamine silver (GMS), and PAS. (iii) Keep tissue samples moist with sterile water, saline, or BHI broth. Do not refrigerate at 4°C for more than 8–10 h. *Processing*: (i) Tease apart tissue specimens aseptically in a sterile Petri dish. (ii) Perform a smear for direct microscopic examination with staining by H&E, GMS, and PAS (as well as Gram stain, Ziehl Neelsen stain, and modified Ziehl Neelsen stain if necessary) and inoculate directly onto the isolation media, if areas of pus and necrosis are present. (iii) Mince tissue specimen with a sterile scalpel blade, or grind in a sterile glass tissue grinder, if no areas of pus or necrosis are present, and inoculate the minced or homogenized material onto the isolation media. (iv) Inoculate onto Sabouraud's dextrose agar with chloramphenicol and gentamicin and incubate duplicate cultures at 26°C and 35°C; (v) Inoculate onto BHIA supplemented with 5% sheep blood and incubate at 35°C. Maintain cultures for 4 weeks.

1.3.1.9 Nail, Hair, Skin Scraping, and Swabs

Collection: For nail, (i) clean nail with 70% alcohol. (ii) Scrape outer dorsal plate surface and discard. (iii) Scrape the deeper portion and remove a portion of debris from under the nail with a scalpel. (iv) Collect whole nail or nail clippings. (v) Place all material in a clean envelope labeled with the patient's data. For hair, (i) Select infected areas and with forceps, epilate at least 10 hairs. (ii) Use a scalpel or a blade knife for hairs broken off at the scalp level. (iii) Place hairs between two clean glass slides or in a clean envelope labeled with the patient's data. For skin scraping and swabs, (i) wipe lesions and interspaces between the toes with alcohol sponge or sterile water. (ii) Scrape the entire lesion and both sides of interspaces with a sterile scalpel and place scrapings between two clean glass slides or place in a clean envelope labeled with the patient's data. (iii) Swab the lesion with a sterile swab (wetted in distilled water if necessary) and place the swab into a clean tube. *Processing*: (i) For nail, hair, and skin scrapings, make a wet mount preparation in KOH for direct microscopy. Calcofluor-stained mount may also be useful. (ii) For skin swabs, smear swab onto heat-sterilized glass slide for Gram stain. (iii) Inoculate skin scraping and swab specimens onto Sabouraud's dextrose agar slopes containing chloramphenicol and gentamicin, but NO cycloheximide and incubate at 35°C. (iv) For suspected secondary bacterial infection, inoculate the swab onto a blood agar plate, followed by the Sabouraud's agar containing the antibiotics and then place into brain heart infusion broth. Incubate at 35°C. Maintain the cultures for 4 weeks.

1.3.2 Microscopic Examination

In addition to macroscopic assessment of colonial size, shape, and color, direct microscopic observation of mycotic elements in clinical specimens and subsequent examination of fungal structures of cultured isolates are critical for correct identification of causal fungal agents and accurate diagnosis of mycoses. Direct microscopy not only facilitates the selection of the proper portion of clinical specimen, the appropriate media, and inoculation techniques for enhanced recovery of the fungus, but also alerts the physician as to the likely etiology of the disease. In general, all fungal specimens of sufficient quantity are examined microscopically and inoculated on culture media. However, when the quantity of a fungal sample is insufficient, culture should take priority over direct microscopy due to its higher sensitivity.

A number of stains (e.g., KOH and its derivatives, lactophenol cotton blue, India ink, and Southgate's mucicarmine stain) can be applied for improved identification and recognition of mycotic elements in clinical samples and structural details of fungal isolates (Table 1.2) [22,23]. Ocular lens containing a micrometer disc may be employed on light microscope for the measurement of hyphae, conidia, and other fungal structures. For detection of fungal elements in tissue biopsies, PAS stain, GMS stain, Fontana-Masson stain, Gridley's stain, and H&E may be utilized. In particular, GMS stain represents an essential stain for detection of fungal elements in tissue sections. Fontana-Masson stain is indispensable for detection of melanin in the cell walls of dematiaceous fungi. Besides standard light or phase-contrast microscopy, transmission electronic microscopy (TEM) enables visualization of fine structural details (e.g., the outer wall layers of conidia and ascospores) of fungal organisms, leading to more accurate speciation.

Cellotape flag preparation may be utilized for rapid mounting and keeping intact of the reproductive structures of sporulating fungi. This is performed by (i) using clear 2 cm wide cellotape and a wooden applicator stick (orange stick) to make a small cellotape flag (2 × 2 cm), (ii) using sterile technique to gently press the sticky side of the flag onto the surface of the culture, (iii) applying a drop of 95% alcohol to the flag to act as a wetting agent and also to dissolve the adhesive glue holding the flag to the applicator stick, (iv) placing the flag onto a small drop of lactophenol cotton blue on a clean glass slide, removing the applicator stick and discarding, and (v) adding another drop of stain, covering with a coverslip, gently pressing and moping up any excess stain before microscopy [7].

For macroscopic examination of colonial morphology, attention should be paid to (i) surface texture (e.g., glabrous, suede-like, powdery, granular, fluffy, downy, cottony, and velvety); (ii) surface topography (e.g., flat, raised, heaped, folded, domed, radial, and grooved); (iii) surface pigmentation (e.g., white, cream, yellow, brown, green, grey, and black); (iv) reverse pigmentation (e.g., none, yellow, brown, red, and dark); (v) growth rate (e.g., fast, moderate, and slow); and (vi) growth temperature (e.g., 25°C, 37°C, 40°C, and 45°C). For microscopic assessment of cultured isolates, the morphologic characteristics of conidia are recognized in terms of septation, shape (e.g., spherical, pyriform, clavate, and ellipsoidal), size, color (hyaline and darkly pigmented), surface texture (e.g., smooth, rough, verrucose, and echinulate), type (microconidia and macroconidia), and arrangement (e.g., single, in mass) [7].

1.3.3 In Vitro Cultivation

In vitro cultivation remains a critical step for the phenotypic characterization of fungi. Macroscopic examination of colonial size, shape, and color followed by microscopic investigation of final structures of fungal isolates allows proper determination of fungal organisms implicated in human mycoses in most cases. Further characterization of fungal isolates is possible by using various biological and biochemical techniques as well as antifungal drug resistance testing. The composition, preparation, and application of common mycological media are summarized in Table 1.3.

Slide culture preparation allows observation of the precise arrangement of the conidiophores and conidial ontogeny (the way the conidia are produced). This is conducted by (i) using a sterile blade to cut out an agar block (7 × 7 mm) from a plate of cornmeal agar or Czapek dox agar that is small enough to fit under a coverslip, (ii) flipping the block up onto the surface of the agar plate, (iii) inoculating the four sides of the agar block with spores or mycelial fragments of the fungus, (iv) placing a flamed coverslip centrally upon the agar block, (v) incubating the plate at 26°C until growth and sporulation occur, (vi) removing the cover slip from the agar block, (vii) applying a drop of 95% alcohol as a wetting agent, (viii) gently lowering the coverslip onto a small drop of lactophenol cotton blue on a clean glass slide, (ix) leaving the slide overnight to dry and resealing later with fingernail polish, and (x) applying a coat of clear polish followed by one coat of red-colored polish before microscopy [7].

Apart from in vitro culture, in vivo animal models (e.g., rodents and rabbits) have been occasionally applied to compare diagnostic procedures for the estimation of fungal burdens in blood, bronchoalveolar lavage (BAL), and tissue samples.

TABLE 1.2
Composition, Preparation, and Application of Common Mycological Stains

Stain	Composition	Preparation	Application
15% KOH	KOH 15 g Glycerol 20 mL Distilled water 80 mL	1. Place a portion of specimen onto a clean glass microscope slide. 2. Add a drop of 15% KOH to specimen and mix. 3. Pace a cover glass over the preparation. 4. Leave slide at room temperature until the material is cleared. The slide may be warmed for speedy clearing. 5. Observe slide under microscopy. 6. Slides that appear negative for fungi may be reexamined the following day.	For direct microscopic examination of skin scrapings, hairs, nails, and other clinical specimens for fungal elements.
10% KOH with Calcofluor white	Solution A KOH 10 g Glycerin 10 mL Distilled water 90 mL Solution B Calcofluor white (fluorescent brightener 28, F6259, Sigma) 0.1 g Distilled water 100 mL	1. Mix one drop of solution A and a drop of solution B on the center of a clean microscope slide. 2. Place specimen in the solution and cover with a coverslip, squash the preparation with the butt of the inoculation needle and then blot off the excess fluid. 3. Gently heat the slide and examine microscopically for the presence of fungal elements that fluoresce a chalk-white or brilliant apple green color, depending on the filter combination.	For direct microscopic examination of skin scrapings, hairs, nails, and other clinical specimens for fungal elements. Calcofluor white (or Uvitex 2B, Blankophor) binds to cellulose and chitin and fluoresces blue-white or apple-green when exposed to ultraviolet light.
KOH with Parker Ink	KOH 10 g Glycerol 10 mL Parker Quink permanent blue ink 10 mL Distilled water 80 mL (dissolve the KOH in water; add glycerol and Parker ink. The glycerol prevents crystallization of the reagent and prevents the specimen from drying out).	1. Remove a portion of specimen with an inoculation needle and mount in a drop of KOH on a clean microscope slide. 2. Cover with a coverslip, squash the preparation with the butt of the inoculation needle, and then blot off the excess fluid. 3. Gently heat by passing through a flame two or three times. 4. When the specimen is cleared (from 20 min for skin scrapings to several hours for nail scrapings), examine microscopically for faintly blue-stained fungal elements.	For direct microscopic examination of skin scrapings, hairs, nails, and other clinical specimens for fungal elements. Keep negative specimens and reexamine the next day to avoid reporting false-negative. Preparations may be kept until culture result is known.
KOH-DMSO preparation	Dimethyl sulfoxide (DMSO) 40 mL Distilled water 60 mL KOH 10 g	1. Remove a small portion of the specimen with an inoculation needle and mount in a drop of KOH-DMSO on a clean microscope slide. 2. Cover with a coverslip, squash the preparation with the butt of the inoculation needle, and then blot off the excess fluid. Do not heat the preparation. 3. Examine the mount within 20 min microscopically for unstained refractile fungal elements.	For direct microscopic examination of skin scrapings, hairs, and nails for fungal elements. DMSO gives more rapid maceration and clearing, but preparations do not keep long.
Lactophenol cotton blue	Cotton blue (Aniline blue) 0.05 g Phenol crystals ($C_6H_5O_4$) 20 g Glycerol 40 mL Lactic acid ($CH_3CHOH\ COOH$) 20 mL Distilled water 20 mL	1. Dissolve the cotton blue in the distilled water overnight; 2. Add phenol crystals to the lactic acid in a glass beaker, mix to dissolve, and then add glycerol; 3. Filter the cotton blue and distilled water solution into the phenol/glycerol/lactic acid solution, mix, and store at room temperature.	For direct microscopic examination of skin scrapings, hairs, nails, and other clinical specimens for fungal elements.

India Ink	India Ink (colloidal carbon)	Place a drop of India Ink on the specimen, mix well with a sterilized loop, and cover with a coverslip.	For direct microscopic examination of *Cryptococcus neoformans* and other encapsulated fungi in a cell suspension (e.g., CSF sediment).
Southgate's Mucicarmine stain	Carmine 1 g Aluminium hydroxide 1 g 50% alcohol 100 mL (mix above by shaking) Aluminium chloride (anhydrous) 0.5 g (boil in water bath for 2–3 min, cool, make up to original volume with 50% alcohol and filter. The stain is stable for a few months).	1. Take sections to water. 2. Stain nuclei with hematoxylin. 3. Differentiate in acid-alcohol and blue in tap water. 4. Stain with Southgate's mucicarmine solution for 30 min and rinse in distilled water. 5. Dehydrate, clear, and mount.	For detection of capsular material in *Cryptococcus neoformans*. Mucicarmine stains acidic mucins pink.
Periodic acid-Schiff (PAS) and PAS digest stain	1% periodic acid (50%) Periodic acid 2 mL Distilled water 98 mL Schiff's reagent Basic fuchsin (C.I. 42500) 1 g Potassium metabisulfite 2 g Distilled water 200 mL HCl concentrate 2 mL Deactivated charcoal 1–2 g (add basic fuchsin slowly to boiling distilled water, mix, and cool to 50°C. Add potassium metabisulfite, mix, and cool to room temperature before adding HCl. Keep in the dark overnight. Add charcoal and filter through coarse filter paper, then fine filter paper. Store in fridge).	1. Take sections to water. 2. For PAS digest only, digest glycogen with saliva for 15 min, wash in water. 3. Treat with 1% periodic acid for 5 min, rinse in water. 4. Treat with Schiff's reagent for 10 min, wash in running tap water for 5 min (to help develop the color). 5. Counterstain nuclei lightly with Mayer's hematoxylin for 1 min, wash, and "blue up" in Li2Co3. 6. Dehydrate, clear, and mount.	For demonstration of glycogen and neutral mucins, and for detection of fungal elements in tissue sections. As a counter stain, PAS reveals the background host cellular detail, tissue architecture, and inflammatory response. PAS-positive material appears magenta; nuclei appear blue; PAS digest material appears colorless.
Gomori's methenamine silver stain (GMS) [or Grocott-Gomori silver stain]	5% aqueous chromic acid 1% aqueous sodium bisulfite 5% aqueous borax 0.1% aqueous gold chloride 2% aqueous sodium thiosulfate Stock methenamine silver solution (add 5 mL of 5% silver nitrate to 100 mL of 3% hexamine, mix, and keep at 4°C for 2 months). Working silver solution (filter before use). (stock methenamine silver solution 25 mL, distilled water 25 mL, 5% borax 1–2 mL)	1. Take sections to water. 2. Oxidize in 5% chromic acid for 1 h and wash in running tap water for 10 min. 3. Treat with sodium bisulfite for 1 min to remove any residual chromic acid, and wash in tap water then distilled water. 4. Place section in the working silver solution at 60°C in a water bath and rinse in distilled water. 5. Tone in 0.1% gold chloride for 5 min and rinse in distilled water. 6. Treat with 2% sodium thiosulfate for 1–2 min to remove unreduced silver and wash thoroughly. 7. Counterstain with light green. 8. Dehydrate, clear, and mount.	An essential stain for detection of fungal elements in tissue sections. As a counter stain, GMS removes the fine details of background host cells and tissues, but provides a more sensitive stain for detecting small fragments of cell wall. Fungi stain black; cell walls stain brown to black; background stains pale green.
Fontana-Masson stain	10% silver nitrate 0.1% gold chloride 5% hypo Nuclear-Fast Red	1. Deparaffinize and hydrate to distilled water. 2. Add 10% silver nitrate, place in 60°C oven for 1 h, and rinse in distilled water. 3. Add 0.1% gold chloride, leave for 10 min, and rinse in distilled water. 4. Add 5% hypo, leave for 5 min, wash in tap water, and rinse in distilled water. 5. Add nuclear-Fast Red, leave for 5 min, and wash in tap water. 6. Dehydrate, clear, and coverslip.	For staining cell walls of dematiaceous fungi and *Cryptococcus neoformans* in tissue sections. Melanin, argentaffin cells stain black; nuclei stain red; background stains pale pink.

(*continued*)

TABLE 1.2 (continued)
Composition, Preparation, and Application of Common Mycological Stains

Stain	Composition	Preparation	Application
Gridley's stain	Chromic acid Metabisulfite bleach Schiff's reagent Aldehyde fuchsin Metanil yellow	1. Bring sections to water via xylene and ethanol. 2. Place in chromic acid for 1 h, wash with tap water. 3. Treat with the metabisulfite bleach for 1 min, wash with tap water, and rinse with distilled water. 4. Place in Schiff's reagent for 20 min, wash with tap water, and rinse with 70% ethanol. 5. Place in aldehyde fuchsin 30 min, rinse off excess with 95% ethanol, and wash with tap water. 6. Counterstain with metanil yellow for 1 min, and rinse with distilled water. 7. Dehydrate, clear and mount in a resinous medium.	For enhanced visualization of fungi and their morphology in tissue sections. Cell walls stain purple to magenta; yeasts stain rose to purple; capsules stain deep purple; background stain yellow.
Hematoxylin and eosin	Alum hematoxylin 0.3% acid alcohol Scott's tap water substitute Eosin/phloxine	1. Bring sections to distilled water. 2. Stain nuclei with alum hematoxylin, rinse in running tap water. 3. Differentiate with 0.3% acid alcohol, rinse in running tap water, then in Scott's tap water substitute, and again in tap water. 4. Stain with eosin/phloxine for 2 min. 5. Dehydrate, clear, and mount.	For visualization of host response to fungus in tissue sections. Nuclei stain blue, cartilage; calcium deposits stain dark blue; cytoplasm and other components stain shades of red; erythrocytes stain bright red.

Sources: McGinnis, M.R., *Laboratory Handbook of Medical Mycology*, Academic Press, New York, 1980; Schwarz, J., *Human Pathol.*, 13, 519, 1982; Koneman, E.W. and Roberts, G.D., *Practical Laboratory Mycology*, The Williams and Wilkins Co., Baltimore, MD, 1985.

TABLE 1.3
Composition, Preparation, and Application of Common Mycological Media

Medium	Composition	Preparation	Application
Potato dextrose agar (PDA)	Potato infusion 200 g Dextrose 20 g Bacto agar 15 g Distilled water 1000 mL (potato infusion, dextrose, and Bacto agar may be replaced with 39 g PDA, Oxoid CM139).	1. Mix ingredients in 1000 mL water. 2. Heat with frequent stirring, bring to boil for 1 min. 3. Dispense for slopes as required. 4. Autoclave at 121°C for 15 min; slope or pour for plates as required.	For routine cultivation and identification of fungi.
Cornmeal agar (CA)	CA (BD) 8.5 g Distilled water 500 mL	1. Mix agar in 500 mL water. 2. Heat with frequent stirring, bring to boil for 1 min. 3. Dispense for slopes as required. 4. Autoclave at 121°C for 10 min; slope or pour for plates as required.	For routine cultivation and identification of fungi.
Oatmeal agar (OA)	Oatmeal (Difco) 60 g Agar 12.5 g Distilled water 1000 mL (oatmeal and agar may be replaced with 72.5 g OA, Difco 255210).	1. Mix oatmeal and agar in 1000 mL water. 2. Heat with frequent stirring, bring to boil for 1 min. 3. Dispense for slopes as required. 4. Autoclave at 121°C for 15 min; slope or pour for plates as required.	For routine cultivation and identification of fungi.
Malt extract agar (MEA)	Malt extract 20 g Peptone 1 g Dextrose 20 g Bacto agar 15 g Distilled water 1000 mL (malt extract, peptone and dextrose may be replaced with 20 g malt extract, Oxoid L39)	1. Mix malt extract, peptone and dextrose in 1000 mL water (adjust to pH 6.5 with NaOH if necessary). 2. Heat with frequent stirring, bring to boil for 1 min. 3. Dispense for slopes as required. 4. Autoclave at 121°C for 10 min, slope or pour for plates as required.	For routine cultivation and identification of fungi. MEA is a useful alternative to bread and BHI broth for recovery of zygomycetes.
Inhibitory mould agar (IMA)	Pancreatic digest of casein 3.0 g Sodium phosphate 2.0 g Peptic digest of animal tissue 2.0 g Magnesium sulfate 0.8 g Yeast extract 5.0 g Ferrous sulfate 0.04 g Dextrose 5.0 g Sodium chloride 0.04 g Starch 2.0 g Manganese sulfate 0.16 g Dextrin 1.0 g Agar 15.0 g Chloramphenicol 0.125 g Distilled water to 1000 mL	IMA is an enriched medium with inorganic salts, chloramphenicol, and gentamicin and can be obtained commercially.	Fór isolation of fungi and inhibition of bacteria.

(*continued*)

TABLE 1.3 (continued)
Composition, Preparation, and Application of Common Mycological Media

Medium	Composition	Preparation	Application
1% Peptone agar (PA)	Bacto peptone (BD) 5g Bacto agar (BD) 10g Distilled water 500mL	1. Mix agar and peptone in 500mL water. 2. Heat with frequent stirring, bring to boil for 1 min. 3. Dispense for slopes as required. 4. Autoclave at 121°C for 10min; slope or pour for plates as required.	For cultivation and differentiation of fungi.
Sabouraud's dextrose agar (SDA) with chloramphenicol and gentamicin	SDA (Oxoid CM41) 65g Chloramphenicol 1× 250mg capsule Distilled water 1000mL Gentamicin (40mg/mL) 0.65mL	1. Mix SDA and chloramphenicol in 1000mL water. 2. Heat with frequent stirring, bring to boil for 1 min. 3. Add gentamicin, mix, dispense for slopes as required. 4. Autoclave at 121°C for 10min; slope, or pour for plates as required.	For primary isolation and cultivation of yeasts and moulds.
SDA with cycloheximide, chloramphenicol, gentamicin, and yeast extract.	SDA (Oxoid CM41) 65g Cycloheximide (actidione) 0.5g Chloramphenicol 1× 250mg capsule Yeast extract 5g Distilled water 1000mL Gentamicin (40mg/mL) 0.65mL	1. Mix the first four ingredients in 1000mL water. 2. Heat with frequent stirring, bring to boil for 1 min. 3. Autoclave at 121°C for 10min and cool to 50°C. 4. Add gentamicin, dispense for slopes, or plates as required.	For primary isolation and cultivation of dermatophytes. Mycosel (BBL) and mycobiotic (Difco) agars (containing SDA, 1% glucose, chloramphenicol, and cycloheximide) are commercially available.
SDA with 5% NaCl	SDA (Oxoid CM41) 32.5g NaCl 25g Distilled water 500mL	1. Mix agar and NaCl in 500mL water. 2. Heat with frequent stirring, bring to boil for 1 min. 3. Dispense for slopes as required. 4. Autoclave at 118°C for 10min; slope or pour for plates as required.	For the cultivation and differentiation of dermatophytes especially *T. rubrum* from *T. mentagrophytes*.
Dermatophyte test medium (DTM)	Papaic digest of soybean meal 10.0g Dextrose 10.0g Phenol red 0.2g Cycloheximide 0.5g Agar 20.0g Distilled water 1000mL	1. Mix ingredients to 1000mL water. 2. Heat with frequent stirring, bring to boil for 1 min. 3. Autoclave at 121°C for 15min.	For recovery of dermatophytes from heavily contaminated clinical specimens and for presumptive indication of the presence of a dermatophyte. Dermatophytes and a few other fungi/bacteria turn medium from pink to red.
Trichophyton agar nos. 1–7	Trichophyton agar nos. 1–7 (BD) 11.8g Distilled water 200mL	1. Mix agar in water. 2. Heat with frequent stirring, bring to boil for 1 min. 3. Dispense for slopes, autoclave at 118°C for 10min, and slope.	For differentiation of *Trichophyton* species. (i) *Trichophyton mentagrophytes* grows well on agar nos. 1,2,3, and 4 (ii) *Trichophyton tonsurans* grows well on agar nos. 3 and 4; poorly on agar nos. 1 and 2. (iii) *Trichophyton verrucosum* grows well on agar nos. 2 and 3; poorly on agar nos. 1 and 4.
Lactrimel agar (LA)	Dutch Jug skimmed milk powder 7g Glycine 10g Honey 10g CA (BD) 17g Chloramphenicol 1× 250mg capsule Distilled water 1000mL	1. Mix milk powder with 150mL water, add other ingredients, and 850mL water. 2. Heat with frequent stirring, bring to boil for 1 min. 3. Dispense for slopes as required. 4. Autoclave at 115°C for 10min, slope, or pour plates as required.	For production of pigment by *Trichophyton rubrum*.

Hair perforation test (HPT)	Autoclaved pre-pubital hair (blonde if available) cut into short pieces (1 cm) Sterile distilled water 5 mL in vial	1. Place autoclaved hair in a vial containing 5 mL sterile water. 2. Inoculate with small fragments of the test fungus and incubate at room temperature. 3. Remove individual hairs at intervals up to 4 weeks and examine microscopically in lactophenol cotton blue.	For differentiation of dermatophytes. *Trichophyton mentagrophytes* and its variants produce marked localized areas of pitting and marked erosion whereas *T. rubrum* does not.
Urease agar (UA) with 0.5% glucose	Bacto agar (BD) 1.5 g Distilled water 91 mL Sterile urea solution (Oxoid SR20) 5 mL 10% sterile glucose solution 4 mL	1. Autoclave agar in 100 mL bottle at 115°C for 20 min and cool to 50°C. 2. Add aseptically 5 mL of sterile urea solution and 4 mL of 10% sterile glucose solution. 3. Dispense for slopes.	For differentiation of dermatophytes.
Rice grain slopes (RGS)	Rice 1/2 teaspoon Distilled water 8 mL	1. Add 1/2 teaspoon rice grains into wide neck 20 mL vials containing 8 mL distilled water to each vial. 2. Close lid, autoclave at 121°C for 15 min, and slope.	For induction of sporulation and for differentiation of *M. audouinii*, *M. canis*, and *M. distortum*.
Littman oxgall agar (LOA)	LOA (BD) 27.5 g Distilled water 500 mL	1. Mix agar in 500 mL water. 2. Heat with frequent stirring, bring to boil for 1 min. 3. Dispense for slopes as required. 4. Autoclave at 121°C for 10 min; slope or pour for plates as required.	For routine inoculation of specimens from skin, nails, and hair, etc.
CGB (L-canavanine, glycine, 2 bromthymol blue) agar	Solution A Glycine 10 g KH_2PO_4 1 g $MgSO_4$ 1 g Thiamine HCl 1 mg L-canavanine sulfate 30 mg Distilled water 100 mL Solution B Bromthymol blue 0.4 g 0.01 N NaOH 64 mL Distilled water 36 mL CGB agar Bacto agar 20 g Solution B 20 mL Distilled water 880 mL Solution A 100 mL	Solution A 1. Dissolve ingredients in beaker and adjust to pH 5.6. 2. Filter sterilize solution using 0.22 µm filter; store in refrigerator. Solution B 1. Dissolve the bromthymol blue in 64 mL of 0.01 N NaOH. 2. Add 36 mL water. CGB agar 1. Mix 20 g agar and 20 mL solution B, autoclave to 121°C for 15 min, and cool to 48°C. 2. Add 100 mL of the filtered solution A, mix, and dispense in plates.	For differentiation between *Cryptococcus neoformans* var. *neoformans* and *Cryptococcus neoformans* var. *gattii*.
Brain-heart infusion agar (BHIA) with 5% sheep blood	BHIA (Oxoid) 52 g Chloramphenicol 1× 250 mg capsule Distilled water 1000 mL Sheep blood 50 mL Gentamicin (40 mg/mL) 0.65 mL	1. Mix BHI agar and chloramphenicol in 1000 mL water. 2. Heat with frequent stirring, bring to boil for 1 min. 3. Autoclave at 121°C for 15 min, cool to 50°C. 4. Add sheep blood and gentamicin, dispense for slopes, or plates as required.	For recovery of *Cryptococcus neoformans* from sterile specimens such as CSF, and for yeast-mould conversions of *Sporothrix* and *Paracoccidioides*. BHIA is an alternative to chocolate agar for isolation of fungi and bacteria, and used routinely by some ophthalmologists for corneal scrapings. Ready-to-use BHIA is commercially available.
Brain heart infusion (BHI) broth	BHI agar (Oxoid) 52 g Distilled water 1000 mL Penicillin G (20 U/mL) 1 mL	Aseptic addition of sterile penicillin to BHI broth inhibits bacteria.	BHI broth with penicillin is useful for isolation of zygomycetes.

(continued)

TABLE 1.3 (continued)
Composition, Preparation, and Application of Common Mycological Media

Medium	Composition	Preparation	Application
Dixon's agar (DA)	Malt extract (Oxoid L39) 18 g Peptone (BDH 44075) 18 g Bacto agar (BD) 7.25 g Ox-bile desiccated (Oxoid L50) 10 g Tween 40 5 mL Glycerol monooleate 2.5 mL Distilled water 500 mL	1. Mix ingredients in 500 mL water. 2. Heat with frequent stirring, bring to boil for 1 min. 3. Dispense for slopes. 4. Autoclave at 121°C for 10 min and then slope.	For primary isolation and cultivation of *Malassezia furfur*.
Czapek dox agar (CDA)	CDA (Oxoid CM97) 45.4 g Distilled water 1000 mL	1. Mix agar in 1000 mL water. 2. Heat with frequent stirring, bring to boil for 1 min. 3. Dispense for slopes as required. 4. Autoclave at 121°C for 10 min, slope or pour for plates as required.	For routine cultivation of fungi, especially *Aspergillus*, *Penicillium*, and non-sporulating moulds.
Bird seed agar	Niger seed (*Guizotia abyssinica*) 50 g Glucose 1 g KH_2PO_4 1 g Creatinine 1 g Bacto agar (BD) 15 g Distilled water 1000 mL Penicillin G (20 U/mL) 0.5 mL per 500 mL Gentamicin (40 mg/mL) 0.5 mL per 500 mL	1. Grind *Guizotia abyssinica* seeds finely and add to 1000 mL distilled water. 2. Boil for 30 min, pass through filter paper and adjust volume to 1000 mL. 3. Add the remaining ingredients except Bacto agar to filtrate and dissolve. 4. Cool to room temperature, adjust to pH 5.5, and dispense into 500 mL bottles. 5. Add 7.5 g Bacto agar to each 500 mL reagent bottle and autoclave at 110°C for 20 min. 6. Cool to 48°C, add 0.5 mL Penicillin G and 0.5 mL gentamicin to each 500 mL, mix, and pour into 90 mm plastic Petri dishes.	For selective isolation of *Cryptococcus neoformans*.
Cornmeal glucose sucrose yeast extract agar	CA (BD) 17 g Dextrose (glucose) 2 g Sucrose 3 g Yeast extract (Difco) 1 g Distilled water 1000 mL	1. Mix dry ingredients into 1000 mL water. 2. Heat with frequent stirring, bring to boil for 1 min. 3. Dispense for slopes. 4. Autoclave at 120°C for 10 min and slope.	For sporulation of some zygomycetes, such as *Saksenaea* and *Apophysomyces*.
Sterile bread	Bread Glass Petri dish	1. Sterilize a piece of bread in a humidified glass Petri dish. 2. Inoculate specimens from non-contaminated sites directly; treat contaminated specimens with antibacterial agents before inoculation.	Sterile bread without preservatives is superior to other media for recovery of zygomycetes from clinical specimens.

Sources: McGinnis, M.R., *Laboratory Handbook of Medical Mycology*, Academic Press, New York, 1980; Schwarz, J., *Human Pathol.*, 13, 519, 1982; Koneman, E.W. and Roberts, G.D., *Practical Laboratory Mycology*, The Williams and Wilkins Co., Baltimore, MD, 1985.

1.3.4 Biochemical and Antifungal Testing

Many fungi demonstrate varied tolerance to temperature, which can be exploited as a complementary tool in the differentiation of dematiaceus fungi. Examination of primary metabolites such as ubiquinones (coenzyme Q) has proven useful for the taxonomy of black yeasts and filamentous fungi. Secondary metabolites (e.g., steroids, terpenes, alkaloids, cyclopeptides, and coumarins) produced by fungal organisms may be examined by thin-layer chromatography. The resulting pattern of secondary-metabolite production provides a reliable approach for identification and classification of lichens. Using pyrolysis gas chromatography, pyrolysis mass spectrometry, gas chromatography, and partition aqueous polymer two-phase systems, the cellular fatty acid composition of fungi can be determined, which represents another useful means for differentiating filamentous fungi. In addition, the structure and composition of the cell wall may be targeted for the definition of fungi. For example, chitin and glucan are present in ascomycetes and basidiomycetes whereas chitosan and polyglucuronic acid are found in zygomycetes. The presence or absence of polysaccharides (e.g., fucose, galactose, rhamnose, and xylose) in the walls of yeast cells allows their differentia on. Further, isoenzyme patterns generated by electrophoretic techniques (zymograms) enable the determination of generic relationships of fungi [9]. Various serological tests have been also described for specific detection of fungal antigens in clinical specimens. Matrix-assisted laser desorption/ionization time-of-flight intact cell mass spectrometry (MALDI-TOF ICMS) has been used to identify fungal organisms, including (i) the terverticillate penicillia, (ii) aflatoxigenic, black, and other aspergilli, (iii) *Fusarium*, (iv) *Trichoderma*, (iv) wood rotting fungi (e.g., *Serpula lacrymans*), and (v) dermatophytes [24]. Moreover, assessment of the sensitivity of fungal isolates to various antifungal drugs offers an additional way to their discrimination as well as treatment [25–27].

1.4 GENOTYPIC CHARACTERIZATION

Phenotypic characterization of fungal organisms on the basis of morphological, biological, and biochemical features suffers from the drawbacks of laborious, time-consuming, and variable, especially for poorly differentiated filamentous fungi. For improved taxonomic resolution and epidemiological study of fungal organisms, molecular techniques detecting the nucleic acids have been increasingly utilized [28–38].

1.4.1 Nucleic Acid Purification

Due to the presence of a tough cell wall in fungi, it is often necessary to undertake several steps to purify the nucleic acids before molecular testing becomes feasible. These include (i) disruption of cell walls, (ii) denaturation of nucleoprotein complexes, (iii) inactivation of endogenous DNase/RNAse, and (iv) removal of contaminating proteins, polysaccharides, polyphenolic pigments, and other compounds.

Grinding lyophilized or fresh mycelia in liquid nitrogen with a mortar and pestle represents a common way to disrupt the fungal cell walls. Because of its time-consuming and laborious nature and its potential for cross-contamination between samples, this technique is not suitable for dealing with a large number of samples. Another means to mechanically break the fungal cell walls is through the use of glass beads with a vortex mixer. In addition, sonicator may be employed for disruption of fungal cell walls. Alternative methods to disrupt the fungal cell walls include enzymatic digestion (using a combination of lyticase, zymolase, chitinase, gluculase, and proteinase K), acid, and alkali treatments. Subsequent treatment with organic solvents (e.g., phenol/chloroform) and detergents (e.g., sodium dodecyl sulfate, SDS; hexadecyltrimethylammonium bromide, CTAB; and *N*-lauroylsarcosine) denatures cytosolic proteins and lipid membranes and inactivates endogenous DNase/RNAse, facilitating their removal. Subsequent precipitation using ethanol or isopropanol results in the isolation of high-purity nucleic acids.

The development of various easy-to-use commercial kits has eliminated the use of hazardous organic solvents in the isolation of DNA/RNA from fungi. For the preparation of fungal DNA, Qiagen DNeasy Plant Kit (Qiagen), UltraClean™ Microbial DNA kit (Mo Bio Laboratories), DNAzol® (Invitrogen), and Whatman FTA cards (Whatman) are highly efficient [39]. Furthermore, automated DNA extraction systems have become increasingly sophisticated and affordable, contributing to the reduction of potential cross-contamination during manual handling.

1.4.2 Target Genes

For accurate and efficient identification of fungal organisms, a number of gene regions have been proven valuable. The most versatile and widely used target is rRNA gene [40–42]. This is followed by *rpb1*, *rpb2*, *tef1a*, and *atp6* genes. Other genes of interest include those encoding β-tubulin, actin, chitin synthase, acetyl coenzyme A synthase, glyceraldehyde-3-phosphate dehydrogenase, isoepoxydon dehydrogenase (idh), lignin peroxidase, and orotidine 5′-monophosphate decarboxylase genes [43,44].

Ribosome is an essential cellular organelle that is involved in protein synthesis in all living organisms. Being the key component of the ribosome, rRNA molecules consists of two complex folded subunits of differing sizes (small and large), whose main functions are to provide a mechanism for decoding messenger RNA (mRNA) into amino acids (at the center of small ribosomal subunit) and to interact with tRNA during translation by providing petidytransferase activity (large subunit [LSU]). The two rRNA subunits in eukaryotes including fungi have sedimentation coefficiency values of 40S (Svedberg units) and 60S. The small rRNA subunit (40S) in eukaryotes contains a single RNA species (i.e., 18S rRNA), whereas the large rRNA subunit (60S) in eukaryotes comprises three RNA species (5S, 5.8S, and 25S–28S rRNA). On the other hand, the two rRNA subunits in prokaryotes

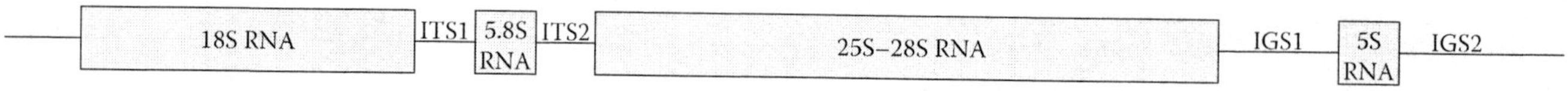

FIGURE 1.1 Organization of fungal ribosomal RNA (rRNA) genes. Considerable variations exist in the small and large subunits of rRNA genes among different fungal groups. Notably, the small subunit (SSU) of rRNA gene in filamentous fungi and yeasts is 18S in size, while that in microsporidia is 16S. Similarly, the large subunit (LSU) of rRNA gene in filamentous fungi is 28S in size, that in yeasts is 25S and that in microsporidia is 23S. ITS, internal transcribed spacer; IGS, intergenic spacer.

measure 30S and 50S, respectively. While the small rRNA subunit (30S) in prokaryotes contains a single RNA species (i.e., 16S rRNA), the large rRNA subunit (50S) in prokaryotes contains two RNA species (5S and 23S rRNA). Interestingly, despite their recent redesignation as a phylum in the kingdom Fungi, microsporidia possess rRNA subunits (30S and 50S) and RNA species (16S and 23S) that are characteristic of prokaryotes, with a notable absence of 5.8S rRNA in microsporidia.

The fungal rRNA genes exist as a family of multiple-copy genes that are arranged in a head-to-toe manner, with each copy (of 8–12 kb in size) consisting of 18S RNA (small subunit [SSU]), ITS 1 (internal transcribed spacer 1), 5.8S RNA, ITS 2, 25S–28S RNA (LSU), IGS 1 (intergenic spacer), 5S, and IGS 2 (Figure 1.1). The tandemly repeated copies of rRNA genes have been homogenized by concerted evolution and contain highly similar nucleotide sequence. Therefore, they are almost always treated as a single-locus gene. Along with ITS, the 18S, 5.8S, and 25S–28S rRNAs are transcribed as a 35S–40S precursor, with all spacers being later spliced out of the transcript. A nontranscribed or IGS region exists between the copies of the 18S, 5.8S, and 25S–28S rRNA repeats, serving to separate the repeats from one another on the chromosome. A 5S RNA gene takes a position within the IGS region and is transcribed in the opposite direction. In filamentous ascomycetes, the 5S RNA gene is absent [45–49].

Much of the 25–28S RNA (LSU) gene is conserved across widely divergent taxa, and only the first 600–900 bases contain three divergent domains (D1, D2, and D3) that are useful for phylogenetic study of fungal organisms. Typically, the first 900 bases of LSU are amplified with primers 5.8SR (located in the 5.8S RNA) and LR7 (located in the 28S RNA), and the resulting amplicon is sequenced with primers LR5, LR16, LR0R, and LR3R (see Table 1.4) [43]. In yeasts, the D1 and D2 variable regions of 25S rRNA regions are often targeted [50].

The 18S rRNA (SSU) gene also includes alternating regions of sequence conservation and heterogeneity. The conserved regions are often targeted for phylogenetic analysis of higher taxonomic orders (e.g., phylum, family, and genus), while the regions of sequence diversity are valuable for characterization of isolates to the genus or species level (with isolates showing sequence identity >97% in the 18S rRNA gene being considered as the identical species).

Compared to the rRNA genes, the rRNA ITS1 and ITS2 regions are not as essential, and thus tend to be more variable, offering extremely valuable targets for fungal speciation and identification. Frequently, the ITS1 and ITS2 regions together with 5.8S RNA are amplified with primers ITS1 (located at the end of 18S RNA) and ITS4 (located at the beginning of 28S RNA). The resulting product is sequenced with primers ITS1, ITS2, ITS3, and ITS4 (see Table 1.4) [43,51]. The identity of the testing fungus is defined by its ITS sequence similarity (%) to the type strain or control isolate: species, ≥99%; genus, 93%–99%; and inconclusive, ≤93%.

Similar to ITS1 and ITS2 regions, the rRNA IGS regions (of 2–5 kb in length, depending on fungal taxa) also experience more genetic drift and consequently are not as highly conserved. This makes the rRNA IGS regions another potential target for fungal identification. The rRNA IGS regions can be amplified with primers LR12R (located at the end of 28S RNA) and invSR1R (located at the beginning of 18S RNA); the resulting product is sequenced with primers LR12R, 5SRNA, 5SRNAR, and invSR1R for all basidiomycetes and some ascomycetous yeasts, and with primers LR12R and invSR1R plus other internal primers for filamentous ascomycetes (as 5S RNA is nonexistent in filamentous ascomycetes) (see Table 1.4) [43,52].

1.4.3 Template Amplification

Prior to the advent of polymerase chain reaction (PCR) in the mid-1980s, molecular procedures for fungal identification were insensitive and cumbersome. With the development of PCR and other nucleic acid amplification technologies (such as ligase chain reaction [LCR], nucleic acid sequence-based amplification [NASBA], transcription-mediated amplification [TMA], strand displacement amplification [SDA], rolling circle amplification [RCA], cycling probe technology [CPT], branched DNA [bDNA], and loop-mediated isothermal amplification [LAMP]), it has become possible to rapidly and specifically detect a single copy of nucleic acid in a matter of hours [38].

Due to their efficiency, simplicity, robustness, and versatility, PCR and its derivatives have been widely adopted in both research and clinical laboratories for identification and determination of fungi and other microbial organisms. From the original version using a pair of primers and gel-based detection, improvements have been made in the forms of nested PCR, multiplex PCR, reverse transcription-PCR (RT-PCR), real-time PCR, quantitative PCR (Q-PCR), and arbitrarily primer PCR (or random amplified polymorphic DNA, RAPD), etc. [38,53]. These developments have not only enhanced the assay sensitivity (nested PCR) and versatility (multiplex PCR detecting multiple genes and/or organisms and RT-PCR detecting RNA instead of DNA), but also enabled the accurate quantitation of fungal organisms (Q-PCR) and elimination of manual handling post-amplification (real-time PCR).

TABLE 1.4
Identity and Sequence of Common rRNA, ITS, and IGS Primers for PCR Amplification and Sequencing Analysis of Fungal Organisms

Gene Region	Primer	Sequence (5′–3′)	Orientation	Nucleotide Positions in *M. grisea* (Nucleotide Positions in *S. cereviseae*)
18S rRNA	NS1	GTA GTC ATA TGC TTG TCT C	Forward	413–422
18S rRNA	NS1R	GAG ACA AGC ATA TGA CTA C	Reverse	413–422
18S rRNA	NS2	GGC TGC TGG CAC CAG ACT TGC	Reverse	943–963
18S rRNA	NS3	GCAAGTCTGGTGCCAGCAGCC	Forward	943–963
18S rRNA	NS4	CTT CCG TCA ATT CCT TTA AG	Reverse	1525–1544
18S rRNA	NS5	AAC TTA AAG GAA TTG ACG GAA G	Forward	1523–1544
18S rRNA	NS6	GCA TCA CAG ACC TGT TAT TGC CTC	Reverse	1806–1829
18S rRNA	NS7	GAG GCA ATA ACA GGT CTG TGA TGC	Forward	1806–1829
18S rRNA	NS8	TCC GCA GGT TCA CCT ACG GA	Reverse	2162–2181
18S rRNA	NS17	CAT GTC TAA GTT TAA GCA A	Forward	447–465
18S rRNA	NS18	CTC ATT CCA ATT ACA AGA CC	Reverse	887–906
18S rRNA	NS19	CCG GAG AAG GAG CCT GAG AAA C	Forward	771–792
18S rRNA	NS20	CGT CCC TAT TAA TCA TTA CG	Reverse	1243–1262
18S rRNA	NS21	GAA TAA TAG AAT AGG ACG	Forward	1193–1210
18S rRNA	NS22	AAT TAA GCA GAC AAA TCA CT	Reverse	1687–1706
18S rRNA	NS23	GAC TCA ACA CGG GGA AAC TC	Forward	1579–1598
18S rRNA	NS24	AAA CCT TGT TAC GAC TTT TA	Reverse	2143–2162
18S rRNA	SR1R	TAC CTG GTT GAT TCT GC	Forward	394–410 (1–21)
18S rRNA	SR1	ATT ACC GCG GCT GCT	Reverse	(578–564)
18S rRNA	SR2	CGG CCA TGC ACC ACC	Reverse	(1277–1263)
18S rRNA	SR3	GAA AGT TGA TAG GGC T	Reverse	696–711 (318–302)
18S rRNA	SR4	AAA CCA ACA AAA TAG AA	Reverse	(838–820)
18S rRNA	SR5	GTG CCC TTC CGT CAA TT	Reverse	(1146–1130)
18S rRNA	SR6	TGT TAC GAC TTT TAC TT	Reverse	(1760–1744)
18S rRNA	SR6R	AAG WAA AAG TCG TAA CAA GG	Forward	(1744–1763); similar to ITS 1
18S rRNA	SR7	GTT CAA CTA CGA GCT TTT TAA	Reverse	(617–637)
18S rRNA	SR7R	AGT TAA AAA GCT CGT AGT TG	Forward	(637–617)
18S rRNA	SR8R	GAA CCA GGA CTT TTA CCT T	Forward	(732–749)
18S rRNA	SR9R	QAG AGG TGA AAT TCT	Forward	(896–910)
18S rRNA	SR10R	TTTG ACT CAA CAC GGG	Forward	(1181–1196)
18S rRNA	SR11R	GGA GCC TGA GAA ACG GCT AC	Forward	779–798
18S rRNA	SSU1Fd	CTG CCA GTA GTC ATA TGC TTG TCT C	Forward	407–431
18S rRNA	SSU1Rd	CTT TGA GAC AAG CAT ATG AC	Reverse	416–435
18S rRNA	SSU2Fd	GAA CAA YTR GAG GGC AAG	Forward	930–947
18S rRNA	SSU2Rd	TAT ACG CTW YTG GAG CTG	Reverse	974–991
18S rRNA	SSU3Fd	ATC AGA TAC CGT YGT AGT C	Forward	1389–1407
18S rRNA	SSU3Rd	TAY GGT TRA GAC TAC RAC GG	Reverse	1397–1416
18S rRNA	SSU4Fd	CCG TTC TTA GTT GGT GG	Forward	1670–1686
18S rRNA	SSU4Rd	CAG ACA AAT CAC TCC ACC	Reverse	1682–1699
18S rRNA	SSU5Fd	TAC TAC CGA TYG AAT GGC	Forward	2037–2054
18S rRNA	SSU5Rd	CGG AGA CCT TGT TAC GAC	Reverse	2148–2165
18S rRNA	SSU6Fm	GCT TGT CTC AAA GAT TAA GCC ATG CAT GTC	Forward	423–452
18S rRNA	SSU6Rm	GCA GGT TAA GGT CTC GTT CGT TAT CGC	Reverse	1707–1733
18S rRNA	SSU7Fm	GAG TGT TCA AAG CAG GCC TNT GCT CG	Forward	1153–1178
18S rRNA	SSU7Rm	CAA TGC TCK ATC CCC AGC ACG AC	Reverse	1921–1943
18S rRNA	SSU8Fm	GCA CGC GCG CTA CAC TGA C	Forward	1848–1866
18S rRNA	V9G	TTA CGT CCC TGC CCT TTG TA	Forward	2002–2021
ITS	ITS1	TCC GTA GGT GAA CCT GCG G	Forward	2162–2180
ITS	ITS1F	CTT GGT CAT TTA GAG GAA GTA A	Forward	2124–2145; similar to ITS 1
ITS	ITS1Fd	CGA TTG AAT GGC TCA GTG AGG C	Forward	2043–2064
ITS	ITS1Rd	GAT ATG CTT AAG TTC AGC GGG	Reverse	2671–2691

(*continued*)

TABLE 1.4 (continued)
Identity and Sequence of Common rRNA, ITS, and IGS Primers for PCR Amplification and Sequencing Analysis of Fungal Organisms

Gene Region	Primer	Sequence (5′–3′)	Orientation	Nucleotide Positions in *M. grisea* (Nucleotide Positions in *S. cereviseae*)
ITS	ITS2	GCT GCG TTC TTC ATC GAT GC	Reverse	
ITS	ITS3	GCA TCG ATG AAG AAC GCA GC	Forward	
ITS	ITS4	TCC TCC GCT TAT TGA TAT GC	Reverse	2685–2704
ITS	ITS4B	CAG GAG ACT TGT ACA CGG TCC AG	Reverse	
ITS	ITS4S	CCT CCG CTT ATT GAT ATG CTT AAG	Reverse	2680–2703
ITS	ITS5	GGA AGT AAA AGT CGT AAC AAG G	Forward	2138–2159; similar to ITS 1
ITS	ITS5R	CCT TGT TAC GAC TTT TAC TTC C	Reverse	
5.8S rRNA	5.8S	CGC TGC GTT CTT CAT CG	Forward	(51–35)
5.8S rRNA	5.8SR	TCG ATG AAG AAC GCA GCG	Reverse	(34–51)
5.8S rRNA	5.8S1Fd	CTC TTG GTT CBV GCA TCG	Forward	2333–2350
5.8S rRNA	5.8S1Rd	WAA TGA CGC TCG RAC AGG CAT G	Reverse	2451–2472
28S rRNA	F377	AGA TGA AAA GAA CTT TGA AAA GAG AA	Forward	3005–3030
28S rRNA	LR0R	GTA CCC GCT GAA CTT AAG C	Forward	2668–2686
28S rRNA	LR1	GGT TGG TTT CTT TTC CT	Reverse	(73–57); similar to ITS 4
28S rRNA	LR2	TTT TCA AAG TTC TTT TC	Reverse	3009–3025
28S rRNA	LR2R	AAG AAC TTT GAA AAG AG	Forward	3012–3028
28S rRNA	LR3	GGT CCG TGT TTC AAG AC	Reverse	3275–3291
28S rRNA	LR3R	GTC TTG AAA CAC GGA CC	Forward	3275–3291
28S rRNA	LR5	TCC TGA GGG AAA CTT CG	Reverse	3579–3595
28S rRNA	LR5R	GAA GTT TCC CTC AGG AT	Forward	3580–3596
28S rRNA	LR6	CGC CAG TTC TGC TTA CC	Reverse	3756–3772
28S rRNA	LR7	TAC TAC CAC CAA GAT CT	Reverse	4062–4078
28S rRNA	LR8	CAC CTT GGA GAC CTG CT	Reverse	4473–4489
28S rRNA	LR8R	AGC AGG TCT CCA AGG TG	Forward	4473–4489
28S rRNA	LR9	AGA GCA CTG GGC AGA AA	Reverse	4799–4815
28S rRNA	LR10	AGT CAA GCT CAA CAG GG	Reverse	5015–5031
28S rRNA	LR10R	GAC CCT GTT GAG CTT GA	Forward	5013–5029
28S rRNA	LR11	GCC AGT TAT CCC TGT GGT AA	Reverse	5412–5431
28S rRNA	LR12	GAC TTA GAG GCG TTC AG	Reverse	5715–5731
28S rRNA	LR12R	CTG AAC GCC TCT AAG TCA GAA	Forward	5715–5735
28S rRNA	LR13	CAT CGG AAC AAC AAT GC	Reverse	5935–5951
28S rRNA	LR14	AGC CAA ACT CCC CAC CTG	Reverse	5206–5223
28S rRNA	LR15	TAA ATT ACA ACT CGG AC	Reverse	2780–2796
28S rRNA	LR16	TTC CAC CCA AAC ACT CG	Reverse	3311–3327
28S rRNA	LR17R	TAA CCT ATT CTC AAA CTT	Forward	3664–3681
28S rRNA	LR20R	GTG AGA CAG GTT AGT TTT ACC CT	Forward	5570–5592
28S rRNA	LR21	ACT TCA AGC GTT TCC CTT T	Reverse	3054–3072
28S rRNA	LR22	CCT CAC GGT ACT TGT TCG CT	Reverse	2982–3001
28S rRNA	LSU1Fd	GRA TCA GGT AGG RAT ACC CG	Forward	2655–2674
28S rRNA	LSU1Rd	CTG TTG CCG CTT CAC TCG C	Reverse	2736–2754
28S rRNA	LSU2Fd	GAA ACA CGG ACC RAG GAG TC	Forward	3280–3299
28S rRNA	LSU2Rd	ATC CGA RAA CWT CAG GAT CGG TCG	Reverse	3379–3402
28S rRNA	LSU3Fd	GTT CAT CYA GAC AGC MGG ACG	Forward	3843–3863
28S rRNA	LSU3Rd	CAC ACT CCT TAG CGG ATT CCG AC	Reverse	3876–3898
28S rRNA	LSU4Fd	CCG CAG CAG GTC TCC AAG G	Forward	4469–4487
28S rRNA	LSU4Rd	CGG ATC TRT TTT GCC GAC TTC CC	Reverse	4523–4545
28S rRNA	LSU5Fd	AGT GGG AGC TTC GGC GC	Forward	3357–3373
28S rRNA	LSU5Rd	GGA CTA AAG GAT CGA TAG GCC ACA C	Reverse	5355–5379
28S rRNA	LSU6Fd	CCG AAG CAG AAT TCG GTA AGC G	Forward	5499–5520
28S rRNA	LSU6Rd	TCT AAA CCC AGC TCA CGT TCC C	Reverse	5543–5564
28S rRNA	LSU7Fd	GTT ACG ATC TRC TGA GGG TAA GCC	Forward	5943–5966
28S rRNA	LSU7Rd	GCA GAT CGT AAC AAC AAG GCT ACT CTA C	Reverse	5927–5954

TABLE 1.4 (continued)
Identity and Sequence of Common rRNA, ITS, and IGS Primers for PCR Amplification and Sequencing Analysis of Fungal Organisms

Gene Region	Primer	Sequence (5′–3′)	Orientation	Nucleotide Positions in *M. grisea* (Nucleotide Positions in *S. cereviseae*)
28S rRNA	LSU8Fd	CCA GAG GAA ACT CTG GTG GAG GC	Forward	3469–3491
28S rRNA	LSU8Rd	GTC AGA TTC CCC TTG TCC GTA CC	Reverse	4720–4742
28S rRNA	LSU9Fm	GGT AGC CAA ATG CCT CGT CAT C	Forward	4882–4903
28S rRNA	LSU9Rm	GAT TYT GCS AAG CCC GTT CCC	Reverse	4979–4999
28S rRNA	LSU10Fm	GGG AAC GTG AGC TGG GTT TAG A	Forward	5543–5564
28S rRNA	LSU10Rm	CGC TTA CCG AAT TCT GCT TCG G	Reverse	5499–5520
28S rRNA	LSU11Fm	TTT GGT AAG CAG AAC TGG CGA TGC	Forward	3753–3776
28S rRNA	LSU12Fd	GTG TGG CCT ATC GAT CCT TTA GTC C	Forward	5355–5379
IGS	LR12R	GAA CGC CTC TAA GTC AGA ATC C	Forward	
IGS	5SRNA	ATC AGA CGG GAT GCG GT	Reverse	
IGS	5SRNAR	ACQ GCA TCC CGT CTG AT	Forward	
IGS	invSR1R	ACT GGC AGA ATC AAC CAG GTA	Reverse	

Sources: White, T.J. et al., Amplification and direct sequencing of fungal ribosomal RNA genes for phylogenetics, in Innis, M.A. et al. (eds.), *PCR Protocols: A Guide to Methods and Applications*, Academic Press, New York, 1990, pp. 315–322; Bruns, T.D. et al., *Ann. Rev. Ecol. Syst.*, 22, 525, 1991; Bruns, T.D. et al., *Mol. Phylog. Evol.*, 1, 231, 1992; Gardes, M. and Bruns, T.D., *Mol. Ecol.*, 2, 113, 1993; Vilgalys, R. et al., *Mycol. Helvet.*, 6, 73, 1994; Gargas, A. and DePriest, P.T., *Mycologia*, 88, 745, 1996; Crous, P.W. et al., *Stud. Mycol.*, 64, S17, 2009.

Notes: Primer names with a "d" ending denote degenerate primers, whereas those with an "m" ending denote specific primers. The nucleotide positions of the primers refer to the rRNA gene sequence of *Magnaporthe grisea* (GenBank accession *AB026819*) or that of *Saccharomyces cereviseae* in the 5′–3′ direction [43,61].

1.4.4 Product Detection

In its standard form, the products generated by PCR are separated by agarose gel electrophoresis with or without modification (e.g., enzymatic digestion), stained with a DNA-binding dye (e.g., ethidium bromide or gel star), and visualized under UV light. For PCR products <100 bp or for distinction of products with minor size differences, polyacrylamide gel electrophoresis and its derivatives (e.g., single-strand conformational polymorphism [SSCP] analysis, denaturing gradient gel electrophoresis [DGGE], and temperature DGGE [TGGE]) may be used.

To improve the sensitivity of PCR product detection, enzymatic signal amplification (e.g., ELISA and flow cytometry) can be applied. In a common version of PCR-ELISA, streptavidin-coated microtiter plate is incubated with a biotinylated capture probe (or oligonucleotide) with specificity for a fungal gene. Aliquots of a PCR products generated with digoxigenin-labeled primers (derived from the same gene) are denatured in NaOH and subsequently hybridized to the capture probe. Specific hybridization products are then visualized by a colorimetric detection system based on an anti-digoxigenin horseradish peroxidase conjugate in the presence of a chromogenic substrate solution. After stopping the enzyme reaction with H_2SO_4, the absorbance is measured in an ELISA reader with a 450 nm filter.

PCR amplicons can be also detected by DNA microarray (also known as DNA chip, gene or genome chip, or gene array). Typically, a collection of microscopic DNA spots (either in the form of oligonucleotides or amplified DNA fragments with specificity for unique portions of the 18S rRNA gene) are affixed to a solid surface (e.g., glass, plastic, or silicon chip) as probes, forming an array for simultaneous identification of fungal organisms. Universal 18S rRNA gene primers (one of which contains a fluorescent label) are used in PCR to amplify all the 18S rRNA genes present in a sample. The resulting PCR products are added to the array and will only bind to the probes for which they have a complementary sequence. Pathogens are identified by the pattern of fluorescing spots in the array. Line probe assay (LiPA) is another nucleic acid hybridization test that is modified from DNA microarray. Instead of a glass, silicon, plastic chip, LiPA is conducted on a nitrocellulose strip, on which specific oligonucleotide probes are attached at known positions as parallel lines and are hybridized with biotin-labeled PCR products.

Recent advances in instrument automation and fluorescent dye chemistry permit instant monitoring of PCR amplicons (so-called real-time PCR) without additional manual handling. In one form of real-time PCR, a double-stranded DNA intercalating dye (e.g., SYBR Green) is used. SYBR Green increases its emission spectrum by 50- to 100-fold when binding to double-stranded DNA. As the double-stranded DNA is synthesized during PCR, an increase in fluorescence correlates to an increasing concentration of PCR products, which can be determined real time with reference to a standard sample. Discrimination of amplicons generated by multiplex PCR from different genes is also possible if these gene products have sufficiently different T_m values.

A melting curve analysis is performed post-PCR, using the SYBR Green as a fluorescent marker. As the melting point is reached, the DNA denatures and the fluorescence decreases sharply. The plotting of fluorescence versus temperature in a graph assists calculation of the melting temperature for each product. Other forms of real-time PCR employ specifically designed probes that target a region of amplicon and incorporate a fluorescent dye. Examples of these probes include hydrolysis dual-labeled probes (TaqMan®), hybridization probes (LightCycler), molecular beacons, peptide nucleic acid (PNA) probes, TaqMan minor groove-binding (MGB™) probes, locked nucleic acid (LNA®) primers and probes, and scorpions™ [38].

DNA sequencing analysis provides a most accurate way to ascertain the identity of PCR amplicons generated from fungi and other organisms. The classical chain termination sequencing method (or Sanger method) utilizes primers or dideoxynucleotides that are labeled with radioactive isotope or fluorescent tag, and the sequencing products are detected by exposure to x-rays or UV light. More recently, pyrosequencing, Roche 454, and Illumina Solexa platforms have been adopted for high-throughput sequencing analysis of PCR products. The nucleotide sequences of the PCR amplicon are then compared with those stored at reference databases such as GenBank, and the phylogenetic relationships of related fungi are displayed in the form of trees, constructed with distance matrix methods (resulting in phenograms) and maximum-parsimony methods (resulting in cladograms) [54].

1.5 RESULT INTERPRETATION, STANDARDIZATION, QUALITY CONTROL, AND ASSURANCE

1.5.1 Key Performance Characteristics

The performance of a diagnostic assay is often evaluated by using several key parameters, including detection limit, sensitivity, specificity, accuracy, intra-assay precision, inter-assay precision, and linearity (as in the case of a quantitative assay). Detection limit (or limit of detection) is the lowest concentration or quantity of bacteria that can be detected by a given assay. Sensitivity is the percentage of samples containing bacteria of interest that are identified by the assay as positive for the bacteria. Specificity is the percentage of samples without bacteria of interest that are identified by the assay as negative for the bacteria. Accuracy (or trueness) is the degree of conformity of an assay's measurements to the actual (true) value. It is often estimated by analyses of reference materials or comparisons of results with those obtained by a reference method. The closer an assay's measurements to the accepted value, the more accurate the assay is. Precision is the degree of mutual agreement among a series of assay's individual measurements, values, or results. Usually characterized in terms of the standard deviation of the measurements, precision can be stratified into (i) repeatability—the variation arising using the same instrument and operator in a single rune (i.e., intra-assay precision) or repeating during a short time period; and (ii) reproducibility—the variation arising using the same measurement process among different instruments and operators from one run to another (i.e., inter-assay precision) or over longer time periods. Linearity refers to the tendency of measurements by a quantitative assay to form a straight line when plotted on a graph. Data from linearity experiments may be subjected to linear regression analysis with an ideal regression coefficient of 1. In case of a nonlinear curve, other objective, statistically valid methods may be utilized.

1.5.2 Result Interpretation

A positive result by a molecular assay for a given pathogen normally confirms the etiologic relationship if the clinical syndrome is compatible with the pathogen identified. Considering the sensitive nature of the amplified methods such as PCR, it is important to rule out the possibility of a false-positive result. Occasionally, false-positive results may originate from the low diagnostic specificity of the assay, in which primers bind to irrelevant sequences and occasionally a homologous sequence that is shared among related or unknown bacteria. More often, false-positive results in the molecular testing come from contamination, which may arise during manual handling of the samples in the testing laboratory either at the pre- or post-extraction (while setting up the PCR) stages. This risk is heightened when a high copy-number polynucleotide (or plasmid) is used as a quantification standard and distributed around the laboratory, contaminating reaction source. Additionally, contamination may be attributable to samples that are referred from other laboratories, which do not utilize manipulation techniques that are mandatory for the molecular testing. These may include the use of unplugged pipette tips, infrequent changing of gloves, and using pipette for long periods without decontamination. Another cause of contamination is by amplification products from previous tests. Contamination may also occur by leakage from tubes or microtiter plates with lids not tightly closed or by breakage of glass capillaries leading to spillage of the amplification mixture. Besides the adoption of stringent laboratory practice, the risk of contamination with PCR products may be reduced by replacing nucleotide dTTP with dUTP in PCR and implementing a digestion step with uracil-DNA-glycosylase (UNG) to remove previous PCR products containing dUTP prior to each amplification reaction. Furthermore, inclusion of multiple negative controls, such as no-template controls (NTC) and no-amplification controls (NAC), may help identify the likely source of contamination and prevent false-positive results. Moreover, microbial DNA may come with PCR reagents.

Similarly, a negative result by a molecular assay for a given pathogen normally indicates the absence of the pathogen. However, it is equally important to rule out the possibility of false-negative results. One possible cause is due to the low sensitivity of the assay employed. Alternatively, insufficient amount of bacteria may be present in the sample (due to sample degradation or prior antibiotic treatment). Another may be due to the impurity of the processed sample. Enzymes (e.g., DNA polymerase, reverse transcriptase) used in PCR and RT-PCR are impeded by components in blood and feces (e.g., heme, hemoglobulin, lactoferrin, immunoglobulin G, leukocyte DNA, polysaccharides, and urea), in foods (e.g., phenolics, glycogen, calcium ions, fat, and other organic substances), in environmental specimens (e.g., phenolics, humic acids, and heavy metals), and in added anticoagulants (e.g., EDTA and heparin) as well as nucleic acid purification reagents (e.g., detergents, lysozyme, NaOH, alcohol, EDTA, EGTA, phenol, and high salt concentrations). Any impurities and contaminations present in the samples after nucleic acid isolation may contribute to false-negative results. A useful way to determine the effective of nucleic acid purification procedure for removing inhibitory substances is to spike samples with well-defined DNA or RNA prior to and after sample preparation (as process and amplification internal controls). In light of the high sensitivity of PCR, the occurrence of false-negative results is probably a truly underestimated problem [55].

Because few species-specific molecular assays are available for fungal organisms, PCR amplification and DNA sequencing analysis of the rRNA genes, ITS and other gene regions have remained a most useful tool for fungal identification. As this approach is contingent on sequence comparison, inaccuracy of data deposited in reference databases may lead to incorrect identification. For example, a lack of pigmentation in *Alternaria infectoria* may contribute to its misidentification using macroscopic characteristics. Sequence data from the incorrectly identified isolates that are stored in reference databases may result in erroneous determination of testing isolates. Indeed, de Hoog and Horré [56] demonstrated that about 14% of the *Alternaria infectoria* sequences deposited in GenBank were found to be misidentified. In a separate study, Nilsson et al. [57] reported that about 20% of the entries related to fungi in the International Nucleotide Sequence Database might have been incorrectly identified to species level.

1.5.3 Standardization and Validation

As molecular tests such as PCR and sequencing offer improved sensitivity, specificity, accuracy, precision, and result availability for fungal identification and diagnosis, they have been increasingly adopted and applied in routine diagnostic laboratories. Considering the possibility of false-positive and false-negative results that may occur in these highly sensitive tests, it is essential to properly standardize and validate them prior to their adoption, and to put in place appropriate quality control measures to ensure their consistent performance.

Standardization of molecular tests addresses the need for standardized reagents and common units, contamination control mechanisms, inhibition control mechanisms, clinically relevant dynamic ranges and internal controls, etc. Validation helps to verify the sensitivity, specificity, accuracy, repeatability (intra-assay precision), reproducibility (inter-assay precision), detection limit, and linearity (if quantitative) of molecular tests.

Before validating a method, it is important to have all instruments calibrated and maintained throughout the testing process. The validation process may involve a series of steps including (i) testing of dilution series of positive samples (or plasmid construct) to determine the limits of detection of the assay and their linearity over concentrations to be measured in quantitative test (using minimal number of reference calibrators such as previously tested patient samples or pooled sera); (ii) evaluating the sensitivity and specificity of the assay, along with the extent of cross-reactivity with other genomic material; (iii) establishing the day-to-day variation of the assay's performance; (iv) assuring the quality of assembled assays using quality control procedures that monitor the performance of reagent batches; and (v) aligning the in-house primer and probe sequences with a genome sequence databank to avoid extended specificity testing [25–27,58].

1.5.4 Quality Control and Assurance

1.5.4.1 Quality Control

Quality control strategies for nucleic acid-based tests include (i) designation of a "clean" area for reaction setup (e.g., room under negative air pressure; positive-displacement pipettes; aerosol-block pipette tips; UV-equipped PCR cabinet); (ii) use of personal protective equipment (PPE) (e.g., disposable gloves and laboratory coats to prevent the introduction of contaminating DNA or nucleases); (iii) use of uracil-*N*-glycosylase (UNG) in real-time PCR (to eliminate cross-over amplicon contamination); (iv) use of a "hot-start" method (to minimize false priming events by withholding a crucial reaction component until appropriate temperature is reached); (v) use of external positive and negative controls (to monitor reaction performance and contamination) and homologous or heterologous internal controls (to monitor presence of inhibitors).

A variety of test controls may be considered for diagnostic PCR. These include (i) internal amplification control (IAC) (negative sample spiked with sufficient pathogen and processed throughout the entire protocol); (ii) processing positive control (PPC) (negative sample spiked with sufficient closely related, but non-target, strain processed throughout the entire protocol.); (iii) reagent control (blank) (containing all reagents, but no nucleic acid apart from the primers.); (iv) premise control (tube containing the master mixture left open in the PCR setup room) to detect possible contaminating DNA in the environment (carried out at regular intervals as part of the quality assurance program); (v) standard (three to four samples containing 10-fold dilution series of known number of target DNA copies in a range) [31,59].

1.5.4.2 Quality Assurance

One way to assess preparedness of the diagnostic laboratories is through the conduct of an external quality assurance (EQA) program providing characterized specimens containing pathogens of interest. The design of a quality assurance program has the following components: (i) internal quality control (IQC) materials are distributed every month and comprising three pools of clinical samples of known pathogen status (typically one negative, one positive containing 1 log 10 over the lower limit of detection of the assay, and one low positive containing up to 1 log 10 of the lower limit of detection of the assay). These are incorporated in test runs on a weekly basis. The purpose of IQC is to provide samples of known status for repeated testing in parallel with clinical samples to ensure reproducibility of the test system in an individual laboratory. (ii) EQA distributions of panels of five unknown samples distributed quarterly. Results are returned to the QA laboratory for assessment. EQA compares the performance of different testing sites using specimens of known but undisclosed content. (iii) Aliquots of all samples sent from the reference laboratory are posted back to Site A for repeat testing to check for integrity of the pools and for transport problems. (iv) A final element of the pilot program involves Sites B, C, and D sending an aliquot of every 50th sample to Site A to check for reproducibility. (v) A detailed record of distributions is kept to provide an audit trail.

1.6 CONCLUSIONS

Fungi are a diverse group of eukaryotic organisms that are ubiquitously distributed in all environments. Although most fungi are saprophytes involved in the decomposition and recycling of organic matters as well as in the formation of symbiotic relationship with plants and animals, some have the capacity to infect plants, animals, and humans. Human mycoses often occur as a result of trauma or underlying immunosuppression, with clinical symptoms ranging from superficial, cutaneous, subcutaneous, to systemic mycoses. Since human mycoses caused by various fungal species are almost impossible to distinguish clinically, it is important to identify the causative agents to genus and species level for implementation of appropriate treatment and prevention measures.

Due to their complex life cycle and their production of morphologically similar structures, fungi are notoriously difficult to identify on the basis of macroscopic and microscopic features [60]. Use of biochemical, biological, and other phenotypic procedures to assess cultured isolates adds further delay in the laboratory speciation of fungal organisms. With recent development and application of molecular techniques, especially PCR amplification and sequencing analysis of the rRNA gene, ITS, and other gene regions, rapid, sensitive, and accurate determination of fungal organisms has become a reality [61]. Moreover, molecular characterization of fungal organisms is much less technically demanding than phenotypic procedures.

Considering that sequence-based approach for fungal identification is dependent on the reliability and accuracy of reference databases for comparison with other related organisms, it is critical to be aware that inaccuracy in reference databases that hold fungal nucleotide sequences may lead to incorrect identification. In addition, frequent taxonomical changes for many fungal organisms in the past as well as in recent times represent another potential source of misidentification if care is not taken. Therefore, future identification and characterization of novel taxon-specific gene markers will contribute to the increased accuracy in the molecular determination of fungal species/varieties implicated in human mycoses. These gene markers may come in the forms of previously uncharacterized, uniquely present genes, or of taxon-specific probes recognizing distinct portions of the shared genes. The latter category is exemplified by the development of species-specific probes from the internal transcribed spacer (ITS) regions that are common to all dermatophytes [62].

REFERENCES

1. The Index Fungorum Partnership (CABI, CBS, Landcare Research-NZ). Available at http://www.indexfungorum.org/, accessed on August 1, 2010.
2. International Mycology Association. Available at http://www.mycobank.org/, accessed on August 1, 2010.
3. The official website of the Mycoses Study Group. Available at http://www.doctorfungus.org/, accessed on August 1, 2010.
4. The Assembling the Fungal Tree of Life (AFTOL) project. Available at http://aftol.org/, accessed on August 1, 2010.
5. The UniProt Consortium. Available at http://www.uniprot.org/, accessed on August 1, 2010.
6. The Global Biodiversity Information Facility. Available at http://www.data.gbif.org/, accessed on August 1, 2010.
7. Mycology online, School of Molecular & Biomedical Science, The University of Adelaide, Australia. Available at http://www.mycology.adelaide.edu.au/, accessed on August 1, 2010.
8. Crous, P.W. et al. 2009. Phylogeny and taxonomy of obscure genera of microfungi. *Persoonia* 22:139–161.
9. Guarro, J., Gené, J., and A.M. Stchigel. 1999. Developments in fungal taxonomy. *Clin. Microbiol. Rev.* 12(3):454–500.
10. Hibbett, D.S. and R. Vilgalys. 1993. Phylogenetic relationships of the Basidiomycete genus *Lentinus* inferred from molecular and morphological characters. *Syst. Bot.* 18:409–433.
11. James, T.Y. et al. 2006. Reconstructing the early evolution of the fungi using a six gene phylogeny. *Nature* 443:818–822.
12. Hibbett, D.S. et al. 2007. A higher-level phylogenetic classification of the Fungi. *Mycol. Res.* 111:509–547
13. McLaughlin, D.J. 2009. The search for the fungal tree of life. *Trends Microbiol.* 17:488.
14. Crous, P.W. et al. (eds.) 2009. Fungal biodiversity. *CBS Laboratory Manual Series 1.* Centraalbureau voor Schimmelcultures, Utrecht, Netherlands.
15. Goffeau, A. et al. 1996. Life with 6000 genes. *Science* 274: 563–567.
16. van den Berg, M.A. et al. 2008. Genome sequencing and analysis of the filamentous fungus *Penicillium chrysogenum*. *Nat. Biotechnol.* 26(10):1161–1168.
17. Coleman, J.J. et al. 2009. The genome of *Nectria haematococca*: Contribution of supernumerary chromosomes to gene expansion. *PLoS Genet.* 5(8):e1000618.
18. Chandler, F.W. 1985. Pathology of the mycoses in patients with the acquired immunodeficiency syndrome (AIDS). *Curr. Top. Med. Mycol.* 1:1–23.
19. McGinnis, M.R. 1980. *Laboratory Handbook of Medical Mycology*. Academic Press, New York.

20. Schwarz, J. 1982. The diagnosis of deep mycoses by morphologic methods. *Human Pathol.* 13:519–533.
21. Dixon, D.M. and A. Polak-Wyss. 1991. The medically important dematiaceous fungi and their identification. *Mycoses* 34:1–18.
22. Kaufman, L. 1992. Immunohistologic diagnosis of systemic mycoses: An update. *Eur. J. Epidemiol.* 8:377–382.
23. Gonçalves, A.B. et al. 2006. FISH and Calcofluor staining techniques to detect in situ filamentous fungal biofilms in water. *Rev. Iberoam. Micol.* 23:194.
24. Santos, C. et al. 2010. Filamentous fungal characterizations by matrix-assisted laser desorption/ionization time-of-flight mass spectrometry. *J. Appl. Microbiol.* 108:375.
25. McGinnis, M.R. and M.G. Rinaldi. 1985. Antifungal drugs: Mechanisms of action, drug resistance, susceptibility testing, and assays of activity in biological fluids. In V. Lorian (ed.), *Antibiotics in Laboratory Medicine*. The Williams & Wilkins Co., Baltimore, MD, pp. 223–281.
26. Rinaldi, M.G. and A.W. Howell. 1988. Antifungal antimicrobics: Laboratory evaluation. In B. Wentworth (ed.), *Diagnostic Procedures for Mycotic and Parasitic Infections*, 7th edn. American Public Health Association, Washington, DC, pp. 325–356..
27. Pfaller, M.A. et al. 1995. Quality control guidelines for National Committee for Clinical Laboratory Standards recommended broth macrodilution testing of amphotericin B, fluconazole, and flucytosine. *J. Clin. Microbiol.* 33:1104–1107.
28. DePriest, P.T. and M.D. Been. 1992. Numerous group I introns with variable distributions in the ribosomal DNA of a lichen fungus. *J. Mol. Biol.* 228:315–321.
29. Hopple, J.S. Jr. and R. Vilgalys. 1994. Phylogenetic relationship among coprinoid taxa and allies based on data from restriction site mapping of nuclear rDNA. *Mycologia* 86:96–107.
30. Bretagne, S. et al. 1995. Detection of *Aspergillus* species DNA in bronchoalveolar lavage samples by competitive PCR. *J. Clin. Microbiol.* 33:1164–1168.
31. Bretagne, S. and J.-M. Costa. 2006. Towards a nucleic acid-based diagnosis in clinical parasitology and mycology. *Clin. Chim. Acta* 363:221.
32. de Hoog, G.S. and A.H.G. Gerrits van den Ende. 1998. Molecular diagnostics of clinical strains of filamentous Basidiomycetes. *Mycoses* 41:183–189.
33. Bougnoux, M. et al. 1999. Serum is more suitable than whole blood for diagnosis of systemic candidiasis by nested PCR. *J. Clin. Microbiol.* 37:925–930.
34. Chen, S.C., C.L. Halliday, and W. Meyer. 2002. A review of nucleic acid-based diagnostic tests for systemic mycoses with an emphasis on polymerase chain reaction-based assays. *Med. Mycol.* 40:333–357.
35. Yeo, S.F. and B. Wong. 2002. Current status of nonculture methods for diagnosis of invasive fungal infections. *Clin. Microbiol. Rev.* 15:465.
36. Balajee, S.A., L. Sigler, and M.E. Brandt. 2007. DNA and the classical way: Identification of medically important molds in the 21st century. *Med. Mycol.* 45:475.
37. Borman, A.M. et al. 2008. Molecular identification of pathogenic fungi. *J. Antimicrob. Chemother.* 61:S7.
38. Wengenack, N.L. and M.J. Binnicker. 2009. Fungal molecular diagnostics. *Clin. Chest Med.* 30:391.
39. Borman, A.M. et al. 2010. An improved protocol for the preparation of total genomic DNA from isolates of yeast and mould using Whatman FTA filter papers. *Mycopathologia* 169:445–449.
40. Bruns, T.D. et al. 1992. Evolutionary relationships within the fungi: Analyses of nuclear small subunit rRNA sequences. *Mol. Phylogenet. Evol.* 1:231–241.
41. Bruns, T.D., T.J. White, and J.W. Taylor, 1991. Fungal molecular systematics. *Ann. Rev. Ecol. Syst.* 22:525–564.
42. Borman, A.M. et al. 2009. Pyrosequencing analysis of 20 nucleotides of internal transcribed spacer 2 discriminates *Candida parapsilosis*, *Candida metapsilosis*, and *Candida orthopsilosis*. *J. Clin. Microbiol.* 47:2307–2310.
43. Vilgalys, R., J.S. Hopple Jr., and D.S. Hibbett. 1994. Phylogenetic implications of generic concepts in fungal taxonomy: The impact of molecular systematic studies. *Mycol. Helvet.* 6:73.
44. Paterson, R.R. 2007. The isoepoxydon dehydrogenase gene PCR profile is useful in fungal taxonomy. *Rev. Iberoam. Micol.* 24:289.
45. Hillis, D.M. and M.T. Dixon, 1991. Ribosomal DNA: Molecular evolution and phylogenetic inference. *Quart. Rev. Biol.* 66:411.
46. Gardes, M. and T.D. Bruns, 1993. ITS primers with enhanced specificity for basidiomycetes—Application to the identification of mycorrhizae and rusts. *Mol. Ecol.* 2:113–118.
47. Gargas, A. and J.W. Taylor. 1992. Polymerase chain reaction (PCR) primers for amplifying and sequencing 18S rDNA from lichenized fungi. *Mycologia* 84:589–592.
48. Gargas, A. and P.W. DePriest. 1996. A nomenclature for fungal PCR primers with examples from intron-containing SSU rDNA. *Mycologia* 88:745–748.
49. Hibbett, D.S. 1992. Ribosomal RNA and fungal systematics. *Trans. Mycol. Soc. Jpn.* 33:533–556.
50. Fell, J.W. et al. 2000. Biodiversity and systematics of basidiomycetous yeasts as determined by large-subunit rDNA D1/D2 domain sequence analysis. *Int. J. Syst. Evol. Microbiol.* 50:1351.
51. White, T.J. et al. 1990. Amplification and direct sequencing of fungal ribosomal RNA genes for phylogenetics. In M.A. Innis et al. (eds.), *PCR Protocols: A Guide to Methods and Applications*. Academic Press, New York, pp. 315–322.
52. Vilgalys, R. and D. Gonzalez. (1990). Organization of ribosomal DNA in the basidiomycete *Thanatephorus praticola*. *Curr. Genet.* 18:277.
53. Liu, D. et al. 2000. Application of PCR to the identification of dermatophyte fungi. *J. Med. Microbiol.* 49:493.
54. Shenoy, B.D., R. Jeewon, and K.D. Hyde. 2007. Impact of DNA sequence-data on the taxonomy of anamorphic fungi. *Fungal Divers.* 26:1–54.
55. Wilson, I.G. 1997. Inhibition and facilitation of nucleic acid amplification. *Appl. Environ. Microbiol.* 63:3741–3751.
56. de Hoog, G.S. and R. Horré. 2002. Molecular taxonomy of the *Alternaria* and *Ulocladium* species from humans and their identification in the routine laboratory. *Mycoses* 45:259.
57. Nilsson, R.H. et al. 2006. Taxonomic reliability of DNA sequences in public sequence databases: A fungal perspective, *PLoS One* 1:e59.
58. Kontoyiannis, D.P. and R.E. Lewis. 2002. Antifungal drug resistance of pathogenic fungi. *Lancet* 359:1135–1144.
59. Paterson, R.R. 2007. Internal amplification controls have not been employed in fungal PCR hence potential false negative results. *J. Appl. Microbiol.* 102:1.
60. Koneman, E.W. and G.D. Roberts. 1985. *Practical Laboratory Mycology*. The Williams and Wilkins Co., Baltimore, MD.
61. Crous, P.W. et al. 2009. Phylogenetic lineages in the *Capnodiales*. *Stud. Mycol.*, 64, S17.
62. Bergmans, A.M. et al. 2010. Evaluation of a single-tube real-time PCR for detection and identification of 11 dermatophyte species in clinical material. *Clin. Microbiol. Infect.*, 16, 704.

Part I

Ascomycota

Pezizomycotina: Dothideomycetes

2 *Alternaria*

Giuliana Lo Cascio and Marco Ligozzi

CONTENTS

2.1 INTRODUCTION

2.1.1 Classification, Morphology, and Biology

Alternaria is a genus of asexual or imperfect fungi assigned with the class hyphomycetes. Fungi in this genus are anamorphs of ascomycetes, including members of the genus *Pleospora*. *Alternaria* is a dematiaceous mold, which includes, at present, about 50 species and varieties. However, only eight species have been implicated as human pathogens: *A. alternata*, *A. brassicicola*, *A. chartarum*, *A. stemphylioides*, *A. dianthicola*, *A. infectoria*, *A. pluriseptata*, and *A. tenuissima*. *A. iridis* is mentioned only as an allergic strain.[1,2]

In 1817 Nees described this telluric fungus is characterized by chains of spores with apical beak under the name of *Alternaria tenuis*. Subsequently Fries noted in 1832 described that *Alternaria* species are characterized by very distinctive large multicellular dictyospores that have a beak and are produced in chains. Species of *Alternaria* occur as parasites on a number of crop plants, causing early blight or leaf spot diseases, or as saprobes on a wide variety of organic substrates. This genus is prominent in aerobiological literature because it is recognized as an important aeroallergen as well as plant pathogen.[3,4]

Many pathogenic species of *Alternaria* have a worldwide distribution and are infective to a variety of plants including potato, tomato, onion, and members of Brassicacee.[5] In general, *Alternaria* attacks plants under stress, especially those affected by drought, insect infestation, or senescence. *Alternaria* spores are passively dispersed from infected leaves by moderate to strong gusty wind, with velocities of 2–3 m/s required for spore release. As a component of the dry air spora, dispersal typically occurs during dry weather that immediately follows periods of rain or heavy dew.

Alternaria is also commonly isolated from soil, food, and indoor air environment, in particular on carpets and house dust.[6,7]

Alternaria spp. grow rapidly and the colony size reaches a diameter of 3–9 cm following incubation at 25°C for 7 days on potato glucose agar. The colony is flat, downy to woolly, and is covered by grayish, short, aerial hyphae in time. The surface is grayish white at the beginning, which later darkens and becomes greenish black or olive brown with a light border. The reverse side is typically brown to black due to pigment production.[8–10]

On microscopic examination, *Alternaria* spp. have septate, brown hyphae. The conidiophores arise singly or in small groups, simple or branched, straight or flexible, sometimes geniculate, pale to mid-olivaceous or golden brown, smooth, up to 50 μm long, 3–6 μm thick with one or several conidial scars. They bear simple or branched large conidia ($7–10 \times 23–34$ μm) that have both transverse and longitudinal septations. These conidia may be observed singly or in acropetal chains and may produce germ tubes. They are ovoid to obclavate, darkly pigmented, muriform, and smooth or roughened. The end of the conidium nearest to the conidiophore is round while it tapers toward the apex. This gives the typical beak or club-like appearance of the conidia.[8–10]

2.1.2 Clinical Features

The genus *Alternaria* comprises a large number of mostly saprobic or plant-pathogenic species and infects mainly immunocompromised hosts. Although infections in immunocompetent hosts have also been reported, these rarely involve invasive disease.[11] *Alternaria* can be found on normal human and animal skin[12] and conjunctiva.[13] This fungus has been associated frequently with hypersensitivity pneumonitis, bronchial asthma, and allergic sinusitis and rhinitis.[14–21] It can also cause several different types of human infections, for example, paranasal sinusitis, ocular infections, onychomycosis,[22] cutaneous, and subcutaneous infections,[23–25] and in some cases also soft palate perforation[26] and disseminated disease.[27,28]

The portal of entry of the infection is usually through corneal trauma or breakdown of the skin barrier.

2.1.2.1 Ocular infections

The incidence of *Alternaria* spp. in oculomycosis ranges from 2.3% to 10.4%[29–32] but it can vary according to the geographical location and is probably related to the risk of trauma caused by organic matter. The majority of cases are keratitis, although endophthalmitis is frequently associated.[32] Patients, in general, are farmers or gardeners who have been exposed to soil and garbage. Accidental or surgical ocular traumas are the predisposing factors.[33,34] The causative agent was generally identified as *A. alternata* or *A. infectoria*, although identification to the species level has not been performed on many occasions.

2.1.2.2 Rhinosinusitis

Sinusitis, or more accurately rhinosinusitis (RS), is a common disorder affecting approximately 20% of the population.[35] A report by the Centers for Disease Control and Prevention has estimated that the prevalence of sinusitis is up to 14.1% of the U.S. adult population, and the fact that there is no treatment for this disorder approved by the U.S. Food and Drug Administration emphasizes the profound impact of this disease.[36] Acute RS is well categorized and is often attributed to bacterial or viral causal agents. However, controversies surround chronic rhinosinusitis (CRS) and the role of fungi in this condition. CRS accounts for >90% of all cases of RS, and the correct diagnosis of each category of CRS is important for optimum therapy and predicting the course. Patients with CRS suffer from long-term nasal congestion, thick mucus production, loss of sense of smell, and intermittent acute exacerbations secondary to bacterial infections.

CRS is characterized histologically by an intense eosinophilic infiltration into the nasal mucosa.[37] The role of fungi in CRS is noninvasive, and it is not a fungal infection. It needs to be differentiated from other forms of fungal sinusitis, such as fungus balls (noninvasive) and invasive fungal sinusitis (acute, fulminant, or chronic form). Recently, striking progress has been made in the development of better techniques to detect fungi in nasal secretion or in ethmoid sinus. Polymerase chain reaction (PCR) with specific fungal primers has been used to find fungal DNA in polypoid nasal tissue,[38] in ethmoid sinus mucus, and nasal fluids, and it has provided greater sensitivity in detecting fungal elements. *Alternaria* DNA was detected in surgical specimens from the mucosa of the middle meatus and the paranasal sinuses or in ethmoid sinus mucosal specimens in almost all patients with CRS but in none of the normal controls.[38,39] It has been postulated that fungi act as non-IgE-mediated immunologic targets which initiate and maintain the inflammatory reaction in paranasal sinuses.[40] It has been shown that patients with CRS have exaggerated cytokine and humoral immune response to fungi, particularly to *Alternaria*.[41] The fungal DNA detected in the mucosa by PCR has been assumed to be related to an antigen-presenting process involving cells such as dendritic cells.[38]

2.1.2.3 Onychomycosis

Different epidemiological studies reported *Alternaria* as a causative agent of onychomycosis only in the 0.08%–2.5% of the cultured-proved clinical cases of nail infections.[42–47]

As with onychomycosis caused by other fungi, a history of contact with soil or trauma in the nail exists in most of the cases. Nails appear dystrophic and distal subungual hyperkeratosis and onycholysis are frequent manifestations. More than one nail could be affected, and in some cases both fingernails and toenails show infections.[42–47] Identification to the species level is not always reported, although *A. alternata* is the prevalent identified species, with some reports of *A. humicola*, *A. pluriseptata* and *A. chlamydospora* as causative agents.[47]

2.1.2.4 Cutaneous and Subcutaneous Infections

Most of clinical presentations concerned localized cutaneous infections resulting from direct, traumatic inoculation, even if in a compromised host a systemic spread is possible. Most of *Alternaria* infections are in patients with immunosuppression, and transplant recipients are one of the groups at risk for developing cutaneous alternariosis.[23–25] Phaeohyphomycosis caused by *Alternaria* may be difficult to recognize because lesions appear variable in size and aspect, ranging from crusted lesions, papules to erythematous macules or subcutaneous nodules, and, rarely, as cellulitis with secondary ulceration. Lesions are located on exposed areas, mainly on upper or lower limbs. Histopathologically, *Alternaria* infection is sometimes misdiagnosed as yeast infection or blastomycosis and in vitro distinction of species may be problematic, since clinical isolates may remain degenerate and sterile. An inflammatory diffuse infiltrate with mixed cellularity (lymphocytes, neutrophils, plasma cells, and histiocytes) mixed with a sarcoid-like and suppurative granulomatous inflammation is usually observed. Biopsy specimens from early lesions (<3 months evolution) are often characterized by the presence of epidermal changes (pseudoepitheliomatous hyperplasia, acanthosis) and a diffuse dermal mixed inflammatory infiltrate with lymphocytes, plasma cells, histiocytes, and neutrophils. Histiocytic cells often presented with intracytoplasmic hyphae that adopted a typical aspect of round-to-oval, thick refractile walls (spore-like structures). In biopsy specimens from more advanced lesions (>3 months evolution), the observation of an inflammatory granulomatous infiltrate is a constant feature. Round, yeast-like structures and branched hyphae are found within the granulomas and microabscesses.

Two different routes of cutaneous infection have been reported: (1) from an exogenous source either as a consequence of traumatic inoculation of fungal elements (after injury by a plant spine) or to colonization of pathologically altered skin, or (2) from an exogenous focus (inhalation of fungal conidia and systemic spread) resulting in secondary cutaneous involvement. Some authors have also defined a "dermatopathic" cutaneous alternariosis, consisting of secondary colonization by Alternaria of preexisting lesions such as steroid-treated eczema of the face.[48]

An altered host resistance appears to be the prerequisite in the majority of affected patients. The majority of the reported patients with cutaneous alternariosis are solid-organ transplant recipients, although many of these patients were receiving systemic corticosteroids. Some authors have postulated that cutaneous fragility induced by corticosteroids may increase the risk of percutaneous inoculation from the environment.[49]

Rarely, *Alternaria* species may cause a systemic infection involving other organs apart from the skin. This endogenous cutaneous variant is manifested by multilocular involvement without having a site of predilection. Visceral and mucosal infection is more common among patients with HIV infection.[50] The observation of cutaneous alternariosis progressing to systemic involvement seems to be an exceedingly rare phenomenon.[25,51]

Differentiation down to the species level is recommended, because species may have different virulence or resistance to antimycotic therapy due to the occurrence of chlamydospores.[52,53] Moreover blastomycosis, sporotrichosis, cryptococcosis, and other subcutaneous mycosis that may present similar clinical features require different therapeutical approaches.

2.1.3 Pathogenesis

Melanins play an important role in the evasion of host defence by fungal pathogens, including dematiaceous molds. Melanins are believed to be composed of polymerized phenolic and/or indolic compounds. Melanins are dark in color, insoluble in aqueous or organic fluids, resistant to concentrated acid, and susceptible to bleaching by oxidizing agents.[54] At this moment these pigments remain poorly characterized because current biochemical and biophysical techniques are unable to provide a chemical structure for this complex polymer. Melanin in dematiaceous molds such as *Alternaria alternata* is synthesized through the dihydroxynapththalene (DHN) polyketide pathway.[55,56] The production of this pigment may protect the fungi from diverse environmental insult. The microbiota at the site of the Chernobyl nuclear reactor accident provides one of the most striking associations between fungal melanogenesis and the ability of these organisms to survive in an extreme environment. The extreme conditions present in the Chernobyl environment, radiation levels over 10,000 times the lethal human dose, select a radiation tolerant *A. alternata* clone.[57]

Melanin synthesis is associated with virulence for mammals in several pathogenic fungi including *C. neoformans* and *Exophiala dermatitidis* by comparing the relative pathogenicity of wild-type strains to mutants incapable of melanization, and it seems that melanin formation occurs in vivo. Melanins contribute to virulence with different mechanisms: they give protection against oxidants because they are highly effective scavengers of free radicals and have electron transfer properties that can facilitate redox cycling. Moreover DHN melanin in *A. alternata* protects against permanganate and hypoclorite and protects against neutrophil oxidative burst. Melanization interferes with phagocytosis and protects against killing by macrophages.

Melanin induces resistance to antimicrobial compounds in *Cryptococcus neoformans*, binds amphotericin B and caspofungin, and prevents them from reaching their target sites.[58]

Melanins are immunologically active compounds; they can elicit a vigorous inflammatory reaction inducing granuloma formation. It is conceivable that cell-wall-associated melanin in tissue provides an indigestible material that serves as "foreign body"-like material that interferes with clearance of infection while at the same time stimulating intense inflammation.

Factors other than 1,8-DHN melanin contribute to the virulence of dematiaceous fungi, although most of the pathogenic virulence factors are studied in *Wangiella dermatitidis* used as a model for black molds.[59] Among the many possibilities, chitin is receiving the most attention. The cell wall content of chitin is significantly enriched in most of the alternative morphotypes of *W. dermatitidis* compared to that of its budding yeast.[59] Furthermore, additional chitin is delocalized from predominantly yeast septal regions to all cell wall locations in hyphae and sclerotic forms. Chitin could act as a virulence factor by adding additional strength to the cell walls of black molds, whenever conditions are encountered in hosts that temporarily retard rapid yeast growth or induce transition of yeast cells to other morphotypes. Many studies using chitin synthase gene disruption showed morphological changes like abnormal hyperpigmentation or pseudohyphal-like growth with defective septa. Tissue sections from patients and animals with chronic cutaneous and subcutaneous infections frequently show yeast and hyphae with abnormally thickened walls. Even brain and other tissue from systemically infected humans and animals with rapidly progressing disease often show many types of thickwalled forms, which are variously described as yeast, chain and clusters of yeast, sclerotic cells, and cells with internal septa. Whether these different in vivo tissue forms are actually enriched with chitin is not known, but chemical and cytochemical analyses of identical morphotypes produced in vitro indicate that they are.[60]

2.1.4 Laboratory Diagnosis

2.1.4.1 Conventional Diagnosis

The collection, transport, and processing of clinical specimens encompass some of the most important considerations in determining the etiology of fungal disease. Only with the appropriate handling of specimens can the recovery of fungal organisms be clearly associated with a disease process. As with all disease processes, the best specimen for determining the etiologic agent is the one taken from the active infection site. The collection details for fungal cultures are very similar to those for bacterial cultures, although the volume of material required for recovery of fungi may exceed those necessary for bacteria. The presence of more material for primary inoculum and concentration of large volumes of fluid greatly increases the likelihood of recovery of fungal species. Appropriate transport and storage of specimens are necessary for fungal elements to remain viable. Fungal viability may be affected by excessive heat and cold; room temperature transport and storage, within 2h of collection, are recommended.

The most frequently submitted specimens for the recovery of dematiaceous fungi include aspirates, scrapings, and surgical tissue. Transport media should not be used, but specimens must not be allowed to become desiccated prior to processing. Specimen portions that are necrotic, purulent, or caseous should be selected for microscopic examination and inoculation onto isolation media. Tissue specimens should be minced with scalpels into 0.5mm pieces and used to directly inoculate culture media. The use of tissue homogenizer for fungal cultures is discouraged because some molds do not have regularly septate hyphae and thus can be killed easily during homogenizing.

Media for primary isolation must support fungal growth while inhibiting bacteria, especially with specimens from nonsterile sites. The most commonly used media are brain heart infusion (BHI) plus antibiotics, Sabouraud dextrose agar alone or plus antibiotics, and selective media containing cycloheximide, a eukaryotic protein synthesis inhibitor, as suppressor of many saprobic fungi. In facilities where a large portion of the patient population may be neutropenic or immunocompromised, the rapid identification of opportunistic fungi, which may quickly disseminate in these individuals, is critical. The addition of cycloheximide and/or antibiotics to potato dextrose agar may be more appropriate in this setting, since it facilitates the identification of these organisms by providing plates for tape mounts of diagnostic structures from primary media. Sheep blood on BHI may inhibit conidiation, so is useful for primary isolation but not for identification.

Microscopic examination of a clinical specimen is essential to detect fungal elements. In laboratories that use fluorescence microscopy, bright-field examination of a positive field must be done to evaluate the amount of melanin present. Heavily melanized fungal cells are not reliably detected by using calcofluor white. Fluorescent-antibody-specific conjugates are not available for *Alternaria* sp.

Colonies of *Alternaria* grow rapidly and appear flat, downy to woolly, and are covered by grayish, short, aerial hyphae in time. The surface is grayish white at the beginning, which later darkens and becomes greenish black or olive brown with a light border. The reverse side is typically brown to black due to pigment production.

Until recently, the identification of *Alternaria* rested upon microscopic morphology, with the most significant characteristics being the morphology of the conidia and the formation of conidial chains. Morphology evaluations normally are based upon cultures that have been grown on a medium such as potato dextrose agar or cornmeal dextrose agar at 25°C–30°C for approximately 2 weeks. Slide culture preparation using cornmeal dextrose agar is ideal for conidiogenesis because these nutritionally minimal media usually stimulate the formation of spores.[61]

The morphological characteristics useful for distinction among the three species more frequently isolated are the following: *A. alternata* shows medium-brown conidia with a short, cylindrical beak, forming long and profusely branched chains, usually 10 or more conidia; in *A. tenuissima* the conidia are golden-brown, frequently tapering gradually into a beak that is up to half the length of the conidium, and usually occur in unbranched chains of three to five conidia; *A. infectoria* species group comprises more than 30 named anamorph taxa. Morphologically, the *A. infectoria* species group differs from other Alternaria species in the three-dimensional sporulation pattern and has more scarce conidia, as this species usually sporulates poorly in common media, and its small conidia (up to 70 µm in length) occur in branched chains with long, geniculate multiseptate secondary conidiophores (up to 120 µm) between conidia.[62]

2.1.4.2 Molecular Techniques

Identification of pathogenic dematiaceous fungi such as *Alternaria* spp. is typically done by morphological and physiological procedures.[63,64] However, these procedures are time-consuming, require technical expertise, and are ineffective for identification of species with poor conidia production and a wide diversity in anamorphic life cycles.[65]

The use of molecular techniques facilitates the identification of rare pathogenic fungi, such as *Alternaria* spp., and the most immediate need for nucleic acid detection methods is for the immunocompromised patient group. In this context, rapid diagnosis of mycological infection by in vitro amplification and detection of fungal DNA is a common method used in clinical laboratories.

2.1.4.2.1 Selection of Target DNA

Two avenues are available for this application; first is the selection of the target DNA using specific sequence information from databases, allowing primers to be designed across conserved and variable regions, and second is cloning and sequencing of arbitrary parts of the fungal genome.

Ribosomal RNA is an essential component of protein synthesis, thus ubiquitous. The sequence of rRNA has highly conserved as well as variable regions. Examination of this sequence reveals the relatedness between the species or the genetic distance between the organisms in question. An additional advantage of this is that these genes are not horizontally transferred, like other prokaryotic genes, for example, drug resistance genes.

Nuclear ribosomal RNA is coded by ribosomal DNA (rDNA), which is organized in ribosomal operons (usually 100–200 identical copies in fungi) located in chromosomes. The nuclear-encoded rRNA genes of fungi exist as a multiple-copy gene family consisting of highly similar DNA sequences (typically from 8 to 12 kb each) arranged in a head-to-toe manner. Each operon codes for the large subunit (LSU-rRNA; 28S rRNA), small subunit (SSU-rRNA; 18S rRNA), 5.8S rRNA and 5S rRNA. The position of 5S rRNA varies, but the organization of the rest of the genes is the same in all fungi (Figure 2.1).

The fungal identification is based on detection of conserved sequences in 5.8S and 28S rDNA that enable the amplification of the ITS region between these two regions and detection of D1/D2 domain contained in 28S rDNA. Then the tools for molecular investigation are represented by (i) DNA sequencing of the full length internal transcribed spacer (ITS) region ITS1–2 and (ii) DNA sequencing of the D1/D2 region of LSU gene in 28S rDNA.

The main area for the development of fungal diagnostics is ribosomal genes,[66] present in all organisms and at high copy numbers aiding detection and the sensitivity of PCR. The fungal nuclear rDNA consists of three genes: the large subunit gene (28S), the SSU gene (18S), and the 5.8S gene, separated by ITS regions, in a unit repeated many times. The ITS region is an area of particular importance to fungal diagnostics. It has areas of high conservation and areas of high variability and is an ideal starter for the development of specific PCR primers for identification of fungal species. Universal primers[66] are available for fungi that isolate the regions of the ITS. Once cloned these sequences can be compared to the wealth of other sequences in the sequence database and diagnostic primers developed for a particular fungus. The MicroSeq D2 large-subunit rDNA sequencing kit appears to be accurate and useful for the identification of filamentous fungi, even those that are relatively uncommon that are seen in the clinical laboratory. However, the library includes more of the common *Alternaria* spp. and other environmental flora that cause disease in immunocompromised patients.

2.2 METHODS

2.2.1 Sample Preparation

A key step in the detection of fungi using PCR is the ability to efficiently extract DNA from hyphae and/or conidia. The efficiency of this step is even more critical when attempting to detect small quantities of fungal material in biological samples such as human blood, mucus, and tissues.

The detection of fungal pathogens in clinical samples by PCR requires the use of extraction methods that efficiently lyse fungal cells and recover DNA suitable for amplification. The DNA extraction stage of the sample is critical processing because fungi have cell walls that impede lysis and the recovery of nucleic acids. Because of the structure and composition of the fungal cell wall, consisting of thick layers of chitin, (1–3)-β-D-glucan, (1,6) β-glucans, mannan, mannoproteins, lipids, and peptides, and the presence of a melanin complex in the cell wall, the release of fungal DNA usually requires additional lysis steps, such as mechanical disruption and enzymatic digestion or use of toxic chemicals.[54] In general, the sample preparation method should release intracellular DNA from the fungal cell wall and/or thick capsule; it must concentrate DNA targets that may be present in very small amounts; and it must eliminate protein debris, contaminants, potential inhibitors, and other extraneous materials without degrading the target DNA.

Alternaria molecular detection test can be used indirectly on molds isolated from a patient's specimen or directly on clinical samples. Several commercial kits are available for mold DNA extraction in a clinical microbiology laboratory.[67] Many extraction methods use a bead matrix and lysis buffer to destroy mycelial cells, followed by adsorption of DNA to a spin filter, a wash step, and the elution of DNA in buffer, prior to amplification steps.

(i) Clinical sample pretreatment—Step 1

Tissue: For DNA extraction, all tissue samples are incubated for ≥2 h in proteinase K and digestion buffer at 55°C, and the DNA is extracted using the miniMAG™ or easyMAG instrument (bioMérieux) according to the manufacturer's instructions. The DNA could be stored at −20°C prior to use.

(Note: Digestion buffer: 50 mM Tris–HCl, 100 mM EDTA, 100 mM NaCl, 1% SDS, pH 8.0. Proteinase K: 0.5 mg/mL in digestion buffer).

1. Place 20–25 mg of tissue into a polypropylene microfuge tube.

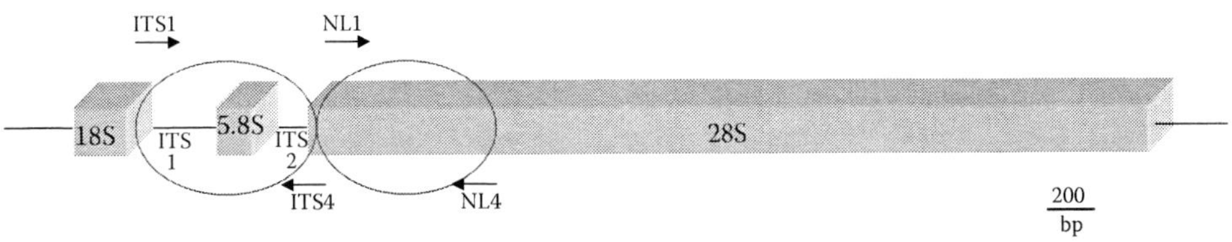

FIGURE 2.1 Map of fungal rRNA from the 3′ end of the 18S rRNA gene to the 3′ end of the 28S rRNA gene. PCR primers for sequencing and PCR assay development (arrows).

2. Add 0.5 mL of DNA digestion buffer with proteinase K.
3. Incubate for a minimum of 3 h to overnight at 55°C with gentle shaking.

Corneal scrapings are collected from all of the patients while they are under local anesthesia by use of a slit-lamp microscope and a flame-sterilized Kimura spatula.[68] One part of each scraping is directly transferred to lysis buffer for DNA extraction.

Cutaneous swab is immersed in lysis buffer in a microcentrifuge tube and incubated at room temperature for 5 min with occasional agitation.

(ii) Clinical sample pretreatment—Step 2

1. Up to 1 mL of lysis buffers each pre-treated sample containing fungal cells or conidia.
2. Add 300 mg of MicroBeads in each sample tube.
3. Continue DNA purification as described below in step 3 for Method UC (UltraClean DNA isolation kit [MoBio, Inc., Solana Beach, CA]) of DNA extraction from culture.

The supernatant solution (above the beads) is transferred to a mini-MAG for a semi-automated magnetic extraction or an easy MAG for automated magnetic extraction.

(iii) DNA extraction from culture

Method UC (UltraClean DNA isolation kit [MoBio, Inc., Solana Beach, CA]) uses a bead matrix and lysis buffer to pulverize cells by horizontal shaking on a vortex mixer, followed by adsorption of DNA to a spin filter, a wash step, and the elution of DNA in buffer.[69] Microcentrifuge tubes with sample and bead matrices were attached to a horizontal platform on a vortex mixer and agitated vigorously for 40 min. Each sample is eluted in a TE1X buffer. In detail the method:

1. Add 300 μL of MicroBead solution to a MicroBead tube. MicroBead Solution contains salts such as guanidine thiocyanate and a buffer which stabilizes and homogeneously disperses the microbial cells prior to lysis.
2. Use an inoculation loop to collect approximately half of the mycelia covering the cellophane square. Transfer mycelia into the MicroBead tube.
3. Add 50 μL of solution MD1 to the MicroBead tube. Solution MD1 contains sodium dodecyl sulfate (SDS) and other disruption agents required for cell lysis.
4. Place MicroBead tubes in a liquid nitrogen safe container (such as a Styrofoam cup) and cover with liquid nitrogen. Incubate for 1 min. Remove MicroBead tubes from the liquid nitrogen using forceps. Immediately vortex tubes for 10 min.
5. Repeat previous step two times.
6. Vortex MicroBead tubes an additional 10 min.
7. Centrifuge the MicroBead tubes for 2 min at 10,000 × *g*.
8. Add 100 μL of solution MD2 to a clean 2.0 mL microcentrifuge tube. Solution MD2 contains ammonium acetate, a reagent to precipitate non-DNA organic and inorganic material including cell debris and proteins.
9. Transfer the supernatant from step 7 to the tube containing solution MD2. Vortex briefly. Incubate at 4°C until tubes are completely chilled (at least 15 min).
10. Centrifuge tubes for 1 min at 10,000 × *g*.
11. Add 900 μL of solution MD3 to a clean 2.0 mL microcentrifuge tube. Solution MD3 contains guanidine hydrochloride/isopropanol which is a highly concentrated salt solution necessary to bind DNA to the spin filter membrane in the following step.
12. Transfer the supernatant in step 10 to the tube containing solution MD3. Vortex briefly and spin down. During transfer of supernatant, do not touch the protein pellet with the pipette tip.
13. Transfer 650 μL to a clean spin filter. Centrifuge for 30 s at 10,000 × *g*. Discard the flow through. Replace the filter into the same collection tube.
14. Repeat step 13 with remaining supernatant MD3.
15. Add 300 μL of solution MD4 to the spin filter and centrifuge for 30 s at 10,000 × *g*. Solution MD4 is an ethanol based wash solution used to further clean the DNA that is bound to the silica filter membrane in the spin filter. Discard the flow through. Replace the filter into the same collection tube.
16. Centrifuge the empty spin filter for 60 s at 10,000 × *g*. Discard the collection tube. (*Note*: Alcohol is a PCR inhibitor. Make sure no residual alcohol remains near the filter basket).
17. Place the spin filter into a clean 2.0 mL microcentrifuge tube.
18. Add 35 μL of solution MD5 onto the center of the filter. Solution MD5 is 10 mM Tris pH 8. Incubate for 2 min. Centrifuge for 30 s at 10,000 × *g*.
19. Be careful not to touch the filter with the pipette tip.
20. Discard the spin filter. Extracted DNA is contained in the flow through and is ready for use. Store DNA at −20°C.

(iv) Semi-automated extraction

DNA could be isolated using a semi-automated magnetic extraction method, NucliSens® miniMAG, and the NucliSens Magnetic Extraction Reagents (bioMérieux S.A.).

In the first step, 2 mL of NucliSens lysis buffer containing guanidine thiocyanate (5 mol/L) is added to each 1.5 mL tube containing a single pretreated sample.

The tubes are pulse vortexed for 15 s, and then incubated for 10 min at room temperature. The NucliSens lysis buffer contains a chaotropic salt for efficient lysis and inactivation of nucleases. In the second step, 50 µL of Magnetic Silica Particles (bioMérieux) are added to allow DNA binding during a 10 min incubation period at room temperature.

Next, the silica particles are washed twice with 400 µL of wash buffer 1 (5 mol/L guanidine thiocyanate, Tris/HCl, Triton X 100, EDTA), twice with 400 µL of wash buffer 2 (MES hydrate), and once with 500 µL of wash buffer 3 (component: disodium tetraborate) using the NucliSens miniMAG instrument to remove any contaminants from the biological specimen.

Removal of wash buffer is done by vacuum aspiration.

Finally, the DNA is recovered from the particles using 40 µL of wash buffer 3 during a 10 min incubation period at 70°C and under constant shaking (1400 rpm). Thirty microliter of each supernatant is used as template for PCR amplification.

(v) Automatic extraction
Extraction with the easyMAG (bioMérieux S.A.) is done according to the manufacturer's recommendations. Up to 1 mL of each pretreated sample is placed in the disposable sample vessel and it is loaded onto the extractor.

After the initial lysis incubation, 100 µL of magnetic silica particles, prepared as recommended by the manufacturer, are added to each sample, and the extractor restart.

Samples are eluted in 55–110 µL. All samples are transferred to a 1.5 mL microcentrifuge tube and stored at 4°C for 24 h, or −20°C for no more than 2 months.

The internal control is added after the initial incubation step, immediately before the magnetic silica is added.

2.2.2 Detection Procedures

PCR mixes (50 µL) are set up as follows: 10 mM Tris–HCl pH 8.3, 50 mM KCl, 2.5 mM $MgCl_2$, 200 µM (each) dATP, dCTP, dGTP, and dTTP; 1.25 U of *Taq* DNA polymerase (Amplitaq Gold (Applied Biosystems), 0.2 µM (each) of the appropriate primers targeting both the D1/D2 region NL1 (forward) 5-GCATATCAATAAGCGGAGGAAAAG-3 and NL4 (reverse) 5-GGTCCGTGTTTCAAGACGG-3′, as well as the larger 18S-ITS1-5.8SITS2-28S (ITS1/ITS4), ITS1 (forward) 5-TCCGTAGGTGAACCTGCGG-3′ and ITS4 (reverse) 5-TCCTCCGCTTAT TGATATGC-3, as previously described,[70] and 5 µL of DNA template. The reaction mixtures are subjected to the following thermal cycling parameters in a MJ PTC 20 thermocycler: 95°C for 3 min followed by 35 cycles of 94°C for 1 min, 55°C for 1 min, 72°C for 1 min, followed by a final extension at 72°C for 10 min.

To validate the presence of amplifiable DNA and absence of inhibitory substances a PCR is performed using the primer set GH20 (5′-GAAGAGCCAAGGACAGGTAC-3′) and PC04 (5′-CAACTTCATCCACGTTCACC-3′) targeting the human β globin gene.[71] The conditions are as described above except that 5 µL plasmid DNA containing partial β globin gene are used, in noncellular sample. When the internal control result is negative, DNA extraction must be repeated if enough material is available.

Following amplification, aliquots (10 µL) are removed from each reaction mixture and examined by electrophoresis (80 V, 45 min) in gels composed of 2% (w/v) agarose (Gibco, United Kingdom) in TAE buffer (40 mM Tris, 20 mM acetic acid, 1 mM EDTA pH 8.3), stained with ethidium bromide (5 µg/100 mL).

The two most important clinical species, *A. alternata* and *A. infectoria*, can easily be differentiated according to the length of the ITS1-4 amplicon, with the ITS spacer domain being ca. 570 bp in the former species and ca. 600 bp in the latter. *A. tenuissima* cannot reliably be distinguished from *A. alternata* using this method.

All PCR products are purified before DNA sequence analysis particularly to remove dNTPS, polymerases, salts, and primers using a QIAquick PCR purification kit (Qiagen) according to the manufacturer's instructions. Purified amplicons were then sequenced on both strands using the same primers as described above. BigDye terminator cycle sequencing Ready Reaction kits (Applied Biosystems) were employed as recommended by the manufacturer. All cycle sequencing reactions were performed on a MJ PTC200 thermal cycler using an initial denaturation at 96°C for 5 s, followed by 30 cycles of 96°C for 10 s, 50°C for 5 s, and 60°C for 4 min. Products are purified using a Dye-Ex spin kit (Qiagen), dried in a vacuum centrifuge, and resuspended in either template suppression reagent (Applied Biosystems; D1/D2R products) or formamide (Applied Biosystems; ITS products). Products are then analyzed on an automated capillary DNA sequencer (ABI Prism 310 genetic analyzer; Applied Biosystems). Comparative sequence analysis and GenBank searches are assisted by the Genetics Computer Group software package (FASTA, BESTFIT, STRETCHER, and PILEUP algorithms; University of Wisconsin, Madison), the Clustal W alignment program,[72] and the nucleotide–nucleotide Basic Local Alignment Search Tool (BLAST) algorithm (blastn).[73]

2.3 CONCLUSION AND FUTURE PERSPECTIVES

Alternaria is a cosmopolitan mold, found in soil or more often on living and dead plants. This mold is a well-known phytopathogen fungal genus, but numerous published cases of human alternarioses are reported. The most frequent clinical presentations are cutaneous and subcutaneous infections, followed by oculomycosis, RS, and onychomycosis. Immunosuppression is frequently associated with clinical manifestations, and solid organ transplantation is the most important risk factor. The portal of entry of this mold is usually through mechanical trauma or breakdown of the skin barrier.

The use of molecular techniques, the detection of fungal DNA by PCR has been described as an important tool in the early diagnosis of fungal infection. Moreover it facilitates the identification of rare pathogenic fungi, such as *Alternaria* spp., even by nonexperts. In fact, using BLAST, it is now easy to compare sequences of a given unidentified mold, even if it is a non-sporulating strain. Sequences deposited previously in GenBank are useful in order to find the region of similarity among different strains. However it is important to ensure that unknown sequences are compared with the sequence of reference strains that have been identified by experts.

Amplification of the ITS region with pan-fungal primers, followed by downstream sequencing, has been shown to be a useful method for identification of *Alternaria* spp. De Hoog and Horré[52] have provided an accurate procedure to distinguish between *Alternaria* spp. based on a PCR method using general primers, followed by restriction enzyme digestion of the amplicons. *A. alternata* and *A. infectoria* are easily differentiated with the length of the ITS1-4 amplicon, with 570 bp in the *A. alternata* and 600 bp in the *A. infectoria*.

There is a regular need for identification of *Alternaria* isolates, because they have acquired different abilities in nature, which affect the possibilities that the isolates could be an opportunistic human pathogen or plant pathogen or saprotrophic on cereals producing biologically active metabolites.

In medical mycology, molecular identification using ITS is fast, well-known, and often the only method to identify strains from human lesions that rapidly lose their ability to sporulate in vitro.

Concerning alternarioses in humans or animals, generally only identification to species-group level is needed, since the same medical treatment is likely to be applicable regardless of taxon identification. Otherwise the combination of real-time PCR and direct sequencing of positive PCR products obtained from clinical samples improves the diagnostic turnaround time.

Considering that molecular techniques offer the possibility of faster detection and identification of fungi in various samples, the application of PCR has to be evaluated as a supplementary method to mycological cultivation. Cultivation methods ensure the possibility to distinguish between colonization and acute disease and moreover allow to measure the antibiotic resistance.

In the modern medicine era, a critical use of molecular methods is essential in a microbiology laboratory. Application of DNA typing could provide rapid diagnosis and consequently results in the administration of specific and effective therapy. Moreover molecular methods could help in the near future to understand more on the pathogenic role of environmental fungi, even in the era of large-scale immunosuppressant usage, the most important risk factor in predisposing to mycotic disease.

REFERENCES

1. De Bièvre, C., Les *Alternaria* pathogens pour l'homme: Mycologie èpidemiologique, *J. Mycol. Med.*, 118, 50, 1991.
2. de Hoog, G.S., Guarro, J., Genè, J., and Figueras, M.J., *Atlas of Clinical Fungi*, 2nd edn., Centraalbureau voor Schimmelcultures/Universitat Rovira I Virgili, Utrecht, the Netherlands, 2000.
3. Levetin, E., Aerobiology of agricultural pathogens, in *Manual of Environmental Microbiology*. Hurst, C.J., et al. (eds.), Vol. 81, Wiley-Blackwell, Malden, MA, 2007.
4. American Academy of Allergy, Asthma, and Immunology, *1998 Pollen and Spore Report*, American Academy of Allergy, Asthma, and Immunology, Milwaukee, WI, 1999.
5. Rotem, J., *The Genus Alternaria*, APS Press, St. Paul, MN, 1994.
6. Kozak, P.P. et al., Currently available methods for home mold surveys. II. Examples of problem homes surveyed, *Ann. Allergy*, 45, 167, 1980.
7. Miller, J.D. et al., Fungi and fungal products in some Canadian houses, *Int. Biodeterior.*, 24, 103, 1988.
8. Collier, L., Balows, A., and Sussman, M., *Topley & Wilson's Microbiology and Microbial Infections*, 9th edn., Vol. 4, Arnold, London, U.K., 1998.
9. Larone, D.H., *Medically Important Fungi—A Guide to Identification*, 4th edn, ASM Press, Washington, DC, 2002.
10. St-Germain, G. and Summerbell, R., *Identifying Filamentous Fungi—A Clinical Laboratory Handbook*, 1st edn., Star Publishing Company, Belmont, CA, 1996.
11. Pastor, F.J. and Guarro, J., *Alternaria* infections: Laboratory diagnosis and relevant clinical features, *Clin. Microbiol. Infect.*, 14, 734, 2008.
12. Yu, H., Studies on fungi of the normal skin, *Acta Dermatol.* (Kyoto), 60, 126, 1965.
13. Nema, H.V. et al., Mycotic flora of the conjunctiva, *Am. J. Ophthalmol.*, 62, 968, 1966.
14. Denning, D.W. et al., The link between fungi and severe asthma: A summary of the evidence, *Eur. Respir. J.*, 27, 615, 2006.
15. Shugar, M.A., Montgomery, W.W., and Hyslop, N.E., *Alternaria* sinusitis, *Ann. Otol. Rhinol. Laryngol.*, 90, 251, 1981.
16. Ogawa, H. et al., Eosinophilic pneumonia caused by *Alternaria alternate*, *Allergy*, 52, 1005, 1997.
17. Lyke, K.E. et al., A case of cutaneous ulcerative alternariosis: Rare association with diabetes mellitus and unusual failure of itracoanzole treatment, *Clin. Infect. Dis.*, 32, 1178, 2001.
18. Bush, R.K. and Pochnau, J.J., *Alternaria*-induced asthma, *J. Allergy Clin. Immunol.*, 113, 227, 2004.
19. Cantani, A. and Ciaschi, V., Epidemiology of *Alternaria alternata* allergy: A prospective study in 6840 Italian asthmatic children, *Eur. Rev. Med. Pharmacol. Sci.*, 8, 289, 2004.
20. Nelson, H.S., Advances in upper airway diseases and allergen immunotherapy, *J. Allergy Clin. Immunol.*, 119, 872, 2007.
21. Stark, P.C. et al., Fungal levels in the home and allergic rhinitis by 5 years of age, *Environ. Health Perspect.*, 113, 1405, 2005.
22. De Berker, D., Fungal nail disease, *N. Engl. J. Med.*, 360, 2108, 2009.
23. Lo Cascio, G. et al., Utility of molecular identification in opportunistic mycotic infections: A case of cutaneous *Alternaria infectoria* infection in a cardiac transplant recipient, *J. Clin. Microbiol.*, 42, 5334, 2004.

24. Farina, C. et al., Phaehyphomycotic soft tissue disease caused by *Alternaria alternata* in a kidney transplant patient: A case report and literature review, *Trans. Proc.*, 39, 1655, 2007.
25. Gilaberte, M. et al., Cutaneous alternariosis in transplant recipients: Clinicopathologic review of 9 cases, *J. Am. Acad. Dermatol.*, 52, 653, 2005.
26. Fathallah, M. et al., Alternariose palatine et cutanèe à *Alternaria infectoria* chez une patiente immunocompetente tunisienne, *J. Mycol. Med.*, 13, 145, 2003.
27. Kucan, J.O. and Hall, S., *Alternaria* burn wound sepsis, *J. Burn Care Rehabil.*, 6, 501, 1985.
28. Revankar, S.G. et al., Disseminated phaeomycosis: Review of an emerging mycosis, *Clin. Infect. Dis.*, 34, 467, 2002.
29. Chander, J. and Sharma, A., Prevalence of fungal corneal ulcers in Northern India, *Infection*, 22, 207, 1994.
30. Panda, A. et al., Mycotic keratitis in children: Epidemiologic and microbiologic evaluation, *Cornea*, 16, 295, 1997.
31. Tanure, M.A. et al., Spectrum of fungal keratitis at Wills Eye Hospital, Philadelphia, Pennsylvania, *Cornea*, 19, 307, 2000.
32. Chowdhary, A. and Singh, K., Spectrum of fungal keratitis in North India, *Cornea*, 1, 8, 2005.
33. Ferrer, C. et al., Polymerase chain reaction diagnosis in fungal keratitis caused by *Alternaria alternate*, *Am. J. Ophthalmol.*, 133, 398, 2002.
34. Zahara, L.V. et al., Case report. Keratomycosis due to *Alternaria alternate* in a diabetic patient, *Mycoses*, 45, 512, 2002.
35. International Rhinosinusitis Advisory Board, Infectious rhinosinusitis in adults: Classification, etiology and management, *Ear Nose Throat J.*, 76, 5, 1997.
36. Pleis, J.R. and Lethbridge-Cejku, M., Summary health statistics for U.S. adults: National Health Interview survey, 2006, *Vital Health Stat.*, 235, 1, 2007.
37. Chakrabarti, A. et al., Fungal rhinosinusitis: A categorization and definitional schema addressing current controversies, *Laryngoscope*, 119, 1809, 2009.
38. Gosepath, J. et al., Fungal DNA is present in tissue specimens of patients with chronic rhinosinusitis, *Am. J. Rhinol.*, 18, 9, 2004.
39. Jiang, R.S. et al., Ethmoid sinus mycology of chronic rhinosinusitis, *Eur. J. Clin. Microbiol. Infect. Dis.*, 27, 253, 2008.
40. Ponikau, J.U. et al., The role of ubiquitous airborne fungi in chronic rhinosinusitis, *Clin. Rev. Allergy Immunol.*, 30,187, 2006.
41. Shin, S.H. et al., Chronic rhinosinusitis: An enhanced immune response to ubiquitous airborne fungi, *J. Allergy Clin. Immunol.*, 114, 369, 2004.
42. Arrese, J.E., Piérard-Franchimont, C., and Piérard, G.E., Onychomicosis and keratomicosis caused by *Alternaria* spp, *Am. J. Dermatophatol.*, 18, 611, 1996.
43. Singh, S.M., Naidu, J., and Pouranik M., Ungual and cutaneous phaeohyphomycosis caused by *Alternaria alternate* and *Alternaria chlamydospora*, *J. Med. Vet. Mycol.*, 28, 275, 1990.
44. Gianni, C., Cerri, A., and Crosti, C., Ungual phaeohyphomycosis caused by *Alternaria alternate*, *Mycoses*, 40, 219, 1996.
45. Baykal, C. et al., Simultaneous cutaneous and ungula alternariosis in a renal transplant recipient, *Br. J. Dermatol.*, 143, 910, 2000.
46. Romano, C., Paccagnini, E., and Difonzo, E.M., Onychomycosis caused by *Alternaria* spp. in Tuscani, Italy from 1985 to 1999, *Mycoses*, 44, 73, 2001.
47. Hilmioğlu-Polat, S. et al., Non-dermatophytic molds as agents of onychomycosis in Izmir, Turkey—A prospective study, *Mycopathologia*, 160, 125, 2005.
48. Male, O. and Pehamberger, H., Secondary cutaneous mucose caused by *Alternaria* species, *Hautarzt*, 37, 94, 1986.
49. Machet, L. et al., Cutaneous alternariosis: Role of corticosteroid-induced cutaneous fragility, *Dermatology*, 193, 342, 1996.
50. Wiest, P.M. et al., *Alternaria* infection in a patient with acquired immunodeficiency syndrome: A case report and review of invasive *Alternaria* infections, *Rev. Infect. Dis.*, 9, 799, 1987.
51. Lerner, L.H., Lerner, E.A., and Bello, Y.M., Co-existence of cutaneous and presumptive pulmonary alternariosis, *Int. J. Dermatol.*, 36, 276, 1997.
52. De Hoog, G.S. and Horré R., Molecular taxonomy of the *Alternaria* and *Ulocladium* species from humans and their identification in the routine laboratory, *Mycoses*, 45, 259, 2001.
53. Pujol, I. et al., In vitro antifungal susceptibility of *Alternaria* spp. and *Ulocladium* spp., *J. Antimicrob. Chemother.*, 46, 323, 2000.
54. Nosanchuk, J.D. and Casadevall, A., The contribution of melanin to microbial pathogenesis, *Cell. Microbiol.*, 5, 203, 2003.
55. Walsh, T.J. et al., Infections due to emerging and uncommon medically important fungal pathogens, *Clin. Microbiol. Infect.*, 10, 48, 2004.
56. Taylor, B., Wheeler, M.H., and Szaniszlo, P.J., Evidence for pentaketide melanin biosynthesis in dematiaceous human pathogenic fungi, *Mycologia*, 79, 320–322. 1987.
57. Mironeko, N.V. et al., Intraspecific variation in gamma-radiation resistance and genomic structure in the filamentous fungus *Alternaria alternata*: A case study of strains inhabiting Chernobyl reactor, 4, *Ecotoxicol. Environ. Saf.*, 45, 177, 2000.
58. Van Duin, D., Casadeval. A., and Nosanchuk, J.D., Melanization of *Cryptococcus neoformans* and *Histoplasma capsulatum* reduces their susceptibility to amphotericin B and caspofungin, *Antimicrob. Agents Chemother.*, 46, 3394, 2002.
59. Szaniszlo, P.J., Virulence factors in Black Molds with emphasis on melanin, chitin, and *Wangiella* as a molecularly tractable model, in *Molecular Principles of Fungal Pathogenesis*, p. 407, Heitman, J. et al. (eds.), ASM Press, Washington, DC, 2006.
60. Szaniszlo, P.J. and Momany M., Chitin, chitin synthase and chitin synthase conserved region homologues in *Wangiella dermatitidis*, in *NATO Workshop Proceedings on Molecular Biology and its Application to Medical Mycology*, p. 229, Maresca, G., Kobayashi, G., and Yamaguchi, H. (eds.), Springer-Verlag, New York, 1993.
61. Schell, W.A., Salkin, I.F., and McGinnis M.R., *Bipolaris*, *Exophiala*, *Scedosporium*, *Sporothrix*, and other dematiaceous fungi, in *Manual of Clinical Microbiology*, 8th edn., p. 1820, Murray, P.R. et al. (eds.), ASM Press, Washington, DC, 2003.
62. Simmons, E.G., *Alternaria*. An identification manual. CBS Fungal Biodiversity Centre, Utrecht, the Netherlands, 2007.
63. Steadham, J.E., Geis, P.A., and Simmank, J.L., Use of carbohydrate and nitrate assimilations in the identification of dematiaceous fungi, *Diagn. Microbiol. Infect. Dis.*, 5, 71, 1986.
64. Fothergill, A.W., Identification of dematiaceous fungi and their role in human disease, *Clin. Infect. Dis.*, 22, 179, 1996.
65. de Hoog, G.S. et al., Phaeoanamorphic life cycle of Exophiala (Wangiella) dermatitidis, *Anton. van Leeuwenh.*, 65, 143, 1994.
66. White, T.J. et al., Amplification and direct sequencing of fungal ribosomal RNA genes for phylogenetics, in *PCR Protocols: A Guide to Methods and Applications*, p. 315, Innis N., Gelfand J., White T. (eds.), Academic Press, New York, 1990.

67. Fredricks, D.N, Smith, C., and Meier, A., Comparison of six DNA extraction methods for recovery of fungal DNA as assessed by quantitative PCR, *J. Clin. Microbiol.*, 43, 5122, 2005.
68. Bharathi, M.J. et al., Microbiological diagnosis of infective keratitis: Comparative evaluation of direct microscopy and culture results, *Br. J. Ophthalmol.*, 90, 1271, 2006.
69. Montero, C.I. et al., Evaluation of pyrosequencing technology for the identification of clinically relevant non-dematiaceous yeasts and related species, *Eur. J. Clin. Microbiol. Infect. Dis.*, 27, 821, 2008.
70. Hinrikson, H.P. et al., Assessment of ribosomal large-subunit D1–D2, internal transcribed spacer 1, and internal transcribed spacer 2 regions as targets for molecular identification of medically important *Aspergillus* species, *J. Clin. Microbiol.*, 43, 2092, 2005.
71. Vahey, M.T., Wong, M.T., and Michael, N.L., A standard PCR protocol: Rapid isolation of DNA and PCR assay for beta-globin, in *PCR Primers: Laboratory Manual*, p. 17, Dieffenbacher, C.W. and Dveksler, G.S. (eds.), Cold Spring Harbor Laboratory Press, New York, 1995.
72. Thompson, J.D., Higgins, D.G., and Gibson, T.J., CLUSTAL W: Improving the sensitivity of progressive multiple sequence alignment through sequence weighting, position-specific gap penalties and weight matrix choice, *Nucleic Acids Res.*, 22, 4673, 1994.
73. Zhang, J. and Madden, T.L., PowerBLAST: A new network BLAST application for interactive or automated sequence analysis and annotation, *Genome Res.*, 7, 649, 1997.

3 *Aureobasidium*

Miia Pitkäranta and Malcolm D. Richardson

CONTENTS

3.1 INTRODUCTION

Aureobasidium is a saprophyte distributed widely throughout the environment, commonly isolated from plant debris, soil, wood, textiles, and wet areas of bathrooms. In particular, it is found in osmotic environments, such as in food and on plant leaves. It is a frequent colonizer of damp stone and glass and commonly found in the clinical laboratory as a contaminant. Only exceptionally is *Aureobasidium* involved in human disease.

3.1.1 Classification, Morphology, and Environmental Sources

3.1.1.1 Classification

Hyphomycetes form the major part of the moulds encountered in the medical mycology laboratory. They are composed of regularly septate hyphae and produce asexual propagules directly on the hyphae, without fruit bodies.[1] Included in the hyphomycetes are the black yeasts[1] that are defined as asexual fungi potentially able to produce melanized budding cells (a yeast phase) in any stage of their life cycle. Often strictly filamentous relatives of pathogenic black yeasts are included in this group. Here the term "black yeast-like fungi" is applied. *Aureobasidium pullulans*, the only well-known species in the genus, is a dematiaceous fungus, characterized by the presence of melanin pigment in the cell wall and the production of hyaline blastoconidia that develop darkly pigmented chains of arthroconidia in culture.[2]

According to present taxonomy, *Aureobasidium* belongs to Ascomycota, order Dothideales, family *Dothioraceae*.[3] More than 20 species are listed under this genus in Mycobank (http://www.mycobank.org). As such, the genus is polyphyletic and many species are better known by their synonyms (Table 3.1). Most species are saprophytes on various substrates or are plant pathogens causing leaf spots.[2] The species of *Aureobasidium* and closely related genera *Hormonema* and *Kabatiella* are difficult to distinguish in culture by

TABLE 3.1
Members of the Genus *Aureobasidium* and Availability of Molecular Data

Species	Currently Used Synonym	Pathogenicity	Sequence Data
Aureobasidium aleuritidis (Vassiljevsky) Herm.-Nijh. 1977		Plant	N/A
A. apocryptum (Ellis & Everh.) Herm.-Nijh. 1977		Plant	N/A
A. australiense McAlpine 1896		N/A	N/A
A. bolleyi (Sprague) Arx 1957	*Microdochium bolleyi* (Sprague) de Hoog & Herm.-Nijh. 1977	Plant	18S, ITS
A. caulivorum (Kirchn.) Cooke 1962	*Kabatiella caulivora* (Kirchn.) Karak. 1923	Plant	18S, 28S, ITS, ELO1
A. dalgeri (Morelet) Herm.-Nijh. 1977		Plant	N/A
A. foliicola (Oudem.) Muell. 1964	*Lecythophora hoffmannii* (Beyma) Gams & McGinnis 1983	Human[14–16]	18S, 28S, ITS, Cytb
A. harposporum (Bres. & Sacc.) Herm.-Nijh. 1977	*Kabatiella harpospora* (Bres. & Sacc.) Arx 1957	Plant	N/A
A. indicum Pande & Ghate 1985		N/A	N/A
A. lilii Crisan & Hodisan 1964		Plant	N/A
A. lini (Laff.) Herm.-Nijh. 1977	*Kabatiella lini* (Laff.) Karak. 1950	Plant	18S, 28S, ITS, EF1, TUB, ELO1
A. mansonii (Castell.) Cooke 1962	*Exophiala castellanii* Iwatsu, Nishim. & Miyaji 1999	Human[17]	18S, 28S, RPB1, Cytb
A. microstictum (Bubák) Cooke 1962	*Kabatiella microsticta* Bubák 1907	Plant	18S, 28S, ITS, EF1, TUB, ELO
A. microstromoides (Moesz) Cooke 1962		N/A	N/A
A. nigricans (Atk. & Edgerton) Cooke 1962	*Kabatiella nigricans* (Atk. & Edgerton) Karak. 1923	Plant	N/A
A. nigrum (Marpmann) Cif. & Dalla Torre 1963	*Torula dematia* Berkhout 1923	N/A	N/A
A. prunicola (Ellis & Everh.) Herm.-Nijh. 1977		Plant	N/A
A. prunorum Dennis & Buhagiar 1973	*Hormonema prunorum* (Dennis & Buhagiar) Herm.-Nijh. 1977	Saprophytic	18S, ITS
A. pullulans (de Bary) Arnaud 1918		Human, other mammalian, saprophytic	18S, 28S, ITS, EF1, TUB, ELO, other
A. ribis (Vassiljevsky) Herm.-Nijh. 1977		Plant	N/A
A. salmonis (Carmich.) Borelli 1969	*Exophiala salmonis* Carmich. 1966	Fish	18S, 28S, ITS, mt ssu rRNA, RPB1, RPB2, EF1, Cytb
A. sanguinariae (Ellis & Everh.) Herm.-Nijh. 1977		Plant	N/A
A. slovacum Svob.-Pol., Chmel & Bojan. 1966	*Chmelia slovaca* (Svob.-Pol., Chmel & Bojan.) Svob.-Pol. 1966	Human[18]	N/A
A. thujae-plicatae Morelet 1978		Plant	N/A
A. umbellulariae (Harv.) Herm.-Nijh. 1977		Plant	N/A

TABLE 3.1 (continued)
Members of the Genus *Aureobasidium* and Availability of Molecular Data

Species	Currently Used Synonym	Pathogenicity	Sequence Data
A. vaccinii Richit. & Teodoru 1989		Plant	N/A
A. zeae (Narita & Hirats.) Dingley 1973	*Kabatiella zeae* Narita & Y. Hirats. 1959	Plant	ITS

Abbreviations: 18S, nuclear small subunit rRNA gene; 28S, nuclear large subunit rRNA gene; ITS, nuclear internal transcribed spacer region; ELO1, fatty acid elongase gene; Cytb, mitochondrial gene for cytochrome b; EF1, translation elongation factor 1-α-like gene; TUB, β-tubulin gene; RPB1, RNA polymerase II largest subunit gene; RPB2, RNA polymerase II second largest subunit gene; N/A, data not available. Synonym data: http://www.mycobank.org

morphology or nutritional physiology.[4] Several studies have assessed the taxonomy of the group using molecular methods. The genera *Aureobasidium* and *Kabatiella* are overlapping based on common micromorphological, physiological, and molecular features.[2,5,6] The taxonomy of the group remains largely uncertain.[7]

A. pullulans is a clonal anamorph species without a confirmed teleomorph. It has got no close relatives among human pathogenic fungi, but certain other species, that is, plant pathogenic *Kabatiella* spp., as well as *Selenophoma mahoniae*, a pycnidial fungus with *Aureobasidium*-like cultural synanamorph, and *Discosphaerina* (*Columnosphaeria*) *fagi*, a potential teleomorph of *A. pullulans*, show high genetic affinity with it. *Hormonema dematioides* resembles *Aureobasidium pullulans* phenotypically and these species are often mixed up.[1,4,6–11] Phytopathogenic species *A. bolleyi*, *A. prunorum*, and *A. zeae*, as well as *A. salmonis*, a pathogen of fish, show considerable genetic distance to *A. pullulans* complex (see Table 3.1 for synonyms in use). Unknown *Aureobasidium*-like strains have been isolated from various environments.[7,12] DNA sequence analysis is a powerful tool for characterizing these isolates, and novel entities belonging to *Aureobasidium* will probably be described in the future.[7,13] For most species in the genus *Aureobasidium*, pheno- and genotypic data is not available or is available only from a restricted number of strains.

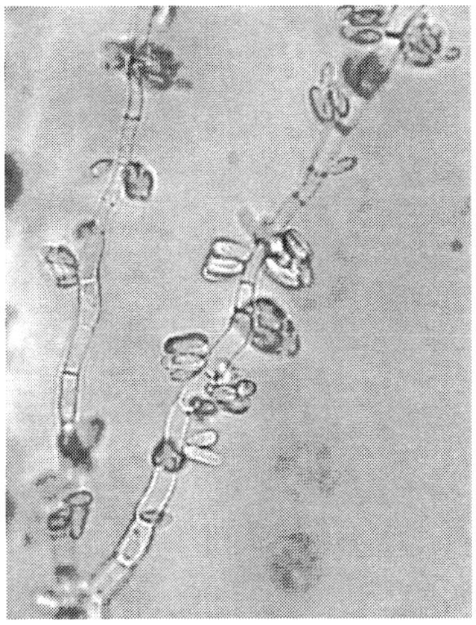

FIGURE 3.1 Septate, hyaline hyphae with groups of synchronous conidia of *Aureobasidium pullulans*. (From Andreoni, S. et al., *Medical Mycology Atlas*, 2004.)

3.1.1.2 Morphology

On malt extract agar (MEA) or glucose peptone agar, *Aureobasidium* grows rapidly when incubated at 30°C, reaching 30 mm in 1 week, appearing flat and smooth, and soon covered with a slimy exudate. The colonies are cream or pink and then later becoming brown or black. The reverse of the colony is cream, turning brown or black with age. The varieties show different pigmentation behavior, the entire colonies of var. *melanigenum* turning green to black usually within 1 week, while other varieties stay mainly light colored for at least 1–2 weeks. Bright color variants have been isolated from the tropics.[13] The microscopical appearance (Figure 3.1)[70] is vegetative hyphae 3–12 μm wide, hyaline, locally converting into blackish-brown, thick-walled chlamydospores, and expanding hyphae with irregular dichotomous branching. The conidia-bearing cells appear undifferentiated and mostly intercalary in hyaline hyphae. Conidia are initially produced synchronously in dense groups from small denticles. They are smooth, oval, and non-pigmented. Subsequently, the conidia are produced percurrently and adhere to one another in slimy heads. The conidia are hyaline, ellipsoidal, vary considerably in shape and size, 7.5–16.0 × 3.5–7.0 μm, and are one celled, often with an indistinct hilum. Polar budding of the conidia is often observed. An additional feature is the presence of endoconidia (asexual propagules formed inside a cell) in intercalary cells.

3.1.1.3 Environmental Sources

The environmental sources of *Aureobasidium* are varied. It is a ubiquitous microorganism that can be easily isolated from the phyllosphere and from plant residues, flowers, soil, wood, indoor and outdoor air, and even stone (reviewed in Cooke and Taylor et al.).[19–21]

3.1.2 Pathogenesis and Clinical Features

3.1.2.1 Pathogenesis

In *Aureobasidium*, *A. pullulans* is the most important species with respect of pathogenicity, but *A. foliicola* (*L. hoffmannii*), *A. mansonii* (*E. castellanii*), and *A. slovacum* (*C. slovaca*) have occasionally been implicated as causative agents of clinical conditions in compromised patients.[16–18] The latter three species are not related to other species of *Aureobasidia*,[8,22] and, regarding their close affinity to other genera these species are discussed in detail elsewhere. Individual cases of cutaneous phaeohyphomycosis, peritonitis, and fungemia caused by *H. dematioides* in immunocompromised patients have been described.[23–25]

Aureobasidium pullulans is an opportunistic pathogen able to cause disease in compromised patients (BSL-1). It may cause phaeohyphomycosis, keratomycosis, pulmonary mycosis with sepsis, peritonitis, and other opportunistic infections, as well as cutaneous mycoses such as eumycotic dermatitis. *Aureobasidium* may also colonize hair, skin, and nails in humans, but its ability to penetrate healthy tissues is limited. Patients with an *Aureobasidium* blood stream infection as a result of major trauma such as road traffic accidents exemplify the possibility of accidental inoculation of the pathogen into the host. Although the mechanism of infection remains unknown, it is likely that inoculation occurs as open fractures come into contact with contaminating material like soil during the accident. After entering the body, the fungus may be capable of survival or dissemination, depending on the general condition and immunological status of the patient and the virulence of the fungal strain. A similar mechanism of infection is possible in immunocompromised patients with peritoneal dialysis catheter, central venous line, or other piece of indwelling synthetic material. *A. pullulans* has affinity for synthetic materials, for example, Silastic devices and indwelling catheters which, when contaminated, may serve as the primary source and focus of infection. Removal of the contaminated device is generally needed to resolve the infection.[26] The pathogenicity in healthy subjects is low and the fungus is commonly considered as a contaminant in clinical specimens.

Aureobasidium pullulans is an agent of phaeohyphomycosis.[27] Clinically, it has been reported to cause a variety of localized infections, including peritonitis (among patients on peritoneal dialysis), cutaneous infection, pneumonia, meningitis, corneal and scleral infection, and abscesses in the spleen and jaw (reviewed in Hawkes et al. and Richardson and Warnock).[26,28] The underlying conditions of patients with reported *Aureobasidium* infection include carcinoma, leukemia, chronic renal failure, diabetes mellitus, disseminated lymphoma, multiple traumatic injuries, organ transplantation, congenital heart lesion repaired with intracardiac prosthetic material, and pregnancy. Many human infections by *Aureobasidium* have followed traumatic inoculation. Published reports have included keratitis, onychomycosis, cutaneous and subcutaneous phaeohyphomycosis, osteomyelitis of the mandible after tooth extraction, and systemic phaeohyphomycosis in both HIV-infected and non-HIV-infected individuals. Systemic infections have been reported including peritonitis and invasive disease in AIDS patients.

3.1.2.2 Hypersensitivity Pneumonitis (Extrinsic Allergic Alveolitis)

Aureobasidium pullulans is present in the environment and exposure may result in sensitization to the organism and is associated with the development of allergic diseases.[20,29,30] A study in a cohort of 405 children suggested that specific dust-borne fungi (*Aspergillus* and *Aureobasidium pullulans*) found at home in the first 3 months of life might be associated with an increased risk of developing allergic rhinitis by 5 years of age.[29] A further study showed that sensitivity to *A. pullulans* was significantly associated with more severe asthma.[31]

A. pullulans has been associated with an outbreak of hypersensitivity pneumonitis (extrinsic allergic alveolitis).[32] Hypersensitivity pneumonitis is a syndrome that is caused by a broad spectrum of inhaled organic dusts or chemical products that cause an immunologically mediated inflammatory response of the alveoli and bronchioles, frequently accompanied by systemic symptoms. This expression depends on several factors including immunological responsiveness of the host, intensity of the exposure, and antigenicity of the inhaled biological dust. Despite extensive studies, the exact immunological mechanisms are not known. Antigen exposure is associated with the presence of circulating IgG antibodies in exposed individuals. Various forms of hypersensitivity pneumonitis have been described including ventilation hypersensitivity pneumonitis caused by thermophilic actinbacteria, moulds such as *Aspergillus fumigatus*, and *Aureobasidium pullulans*. Cases of hypersensitivity pneumonitis secondary to residential exposure to *A. pullulans* and other fungi proliferating in wet, contaminated building constructions have also been described.[32–34]

3.1.2.3 Mycotic Keratitis

Rare cases of keratomycosis (corneal and scleral ulcer) caused by *Aureobasidium pullulans* have been described. An illustrative case report describes a patient who developed *Aureobasidium pullulans* keratitis following refractive laser epithelial keratomileusis (LASEK).[35] The patient was referred to a tertiary care center 1 month after LASEK for the treatment of a corneal ulcer that was unresponsive to conventional therapy. Mycology culture and direct microscopy identified *Aureobasidium* as the infectious organism. The infection responded well to treatment with topical natamycin and systemic itraconazole.

3.1.2.4 Disseminated Infections, Fungemia and Peritonitis

A. pullulans is a very rare cause of systemic infection in humans. Disseminated systemic infection has been reported in only four cases so far (reviewed in Joshi et al).[36] Of the four cases, one had acute myeloid leukemia, the second patient had ovarian carcinoma (both had Hickman catheters in situ), the third patient had met with a road traffic accident with accidental inoculation of pathogen, and the fourth was a child with congenital heart disease who had undergone closure of an atrial septal defect with a Goretex patch. Very occasionally, cutaneous involvement by *Aureobasidium* in patients with systemic infection has been reported, for example, in kidney and liver transplant patients.[36]

Nosocomial infections caused by *A. pullulans* are rare and have been described only in some cases of peritonitis involving patients undergoing peritoneal dialysis and in one case of severe infection where the fungus was isolated from a splenic abscess (reviewed in Bolignano and Criseo[37]). Even more uncommon are infections where *A. pullulans* can be isolated from blood cultures. Catheter-related fungemia has been reported.[38]

Cases of *A. pullulans* fungemia after allogeneic stem cell transplantation have been reported.[36] In one case, the blood culture initially grew yeast-like colonies suggestive of *Candida* species.[36] However, on subculture, the characteristic colony morphology (moist and creamy colonies in 2 days, which matured into shiny brownish black colonies with a gray fringe and pigment production) was suggestive of *A. pullulans* fungemia. The source of fungemia in this patient was the central venous catheter.

A rare case of disseminated nosocomial fungal infection was seen in a patient who was severely traumatized following a road traffic accident.[37] The patient was diagnosed with diffuse cerebral edema, hemoperitoneum, multiple fractures of the left femur and tibia, a suspected pelvic fracture, and multiple ruptures of the liver's right lobe, which made immediate surgery mandatory. After surgery, the patient underwent mechanical ventilation and prolonged total parenteral nutrition. *Aureobasidium pullulans* was isolated from blood and urine samples. Isolation from urine was not achieved initially and became possible only after isolation from blood, thus suggesting a progressive colonization of the patient's organs. The patient was treated with fluconazole 400 mg/day. Subsequently, all cultures became negative. The isolates from this case were further identified as *A. pullulans* var. *melanigenum* on the basis of early black pigmentation of the colonies.[2] This appears to be the first reported case of disseminated nosocomial fungal infection by *A. pullulans* var. *melanigenum*.

3.1.2.5 Antifungal Treatment

There is no standard treatment for infection caused by *A. pullulans* because of the paucity of human cases reported in literature. Amphotericin-B alone or in combination with azoles has been tried with variable success. Combination therapy is probably the treatment of choice. The duration of treatment is not certain, though most patients received antifungal treatment for 4–8 weeks. *Aureobasidium* is susceptible to the majority of antifungal drugs with the exception of 5-fluorocytosine.[17,26]

3.1.3 Diagnosis

The diagnosis of *Aureobasidium* infection is seldom suspected until the fungus is isolated from blood, cutaneous lesions, or other clinical specimens.[28] Often, however, patients have died and the infection has remained unrecognized until postmortem material obtained has been investigated.

3.1.3.1 Conventional Techniques

There are no descriptions of direct microscopy being applied to primary clinical specimens from cases of presumed *Aureobasidium* infection. *Aureobasidium* can be isolated from clinical and environmental specimens using various media (see Section 3.2). Histopathological features are similar to those seen for other agents of phaeohyphomycosis. The pigment produced by these fungi is also produced in vivo. The pigment is often a complex phenol-derived substance. Dark-walled, short, septate hyphae may be observed in H&E-stained sections. If the pigment production is not prominent, the use of periodic acid-schiff stain (PAS), Gömöri methenamine silver stain (GMS), or the Fontana-Masson silver stain may be advantageous.[39] There are no serological tests for the diagnosis of *Aureobasidium* infection.

3.1.3.2 Molecular Techniques

Diagnostics, taxonomic placement, and intraspecies variability of *Aureobasidium pullulans* have been assessed by various molecular methods in addition to traditional morphological and physiological studies. These include fingerprinting techniques like restriction enzyme analysis of polymerase chain reaction (PCR)-amplified ribosomal DNA target area (PCR-ribotyping or restriction fragment length polymorphism (RFLP)-PCR),[4,40,41] universally primed PCR (UP-PCR),[40,42] arbitrary primed PCR (ap-PCR),[43] random amplified polymorphic DNA (RAPD-PCR),[43,44] fluorescent amplified fragment length polymorphism (fAFLP),[45] and repetitive-element PCR (rep-PCR) method,[41] as well as sequence analysis of rRNA genes and internal transcribed spacer (ITS) regions, and protein coding genes.[6,7,9,41] Intraspecies variability has also been assessed using whole-cell protein analysis with sodium dodecyl sulphate-polyacrylamide gel electrophoresis (SDS-PAGE) technique.[41] Strain-specific sequence-characterized amplification region (SCAR) primers,[46,47] species-specific oligonucleotide probes,[48] and PCR primers[49] have been developed for the detection of *A. pullulans*, as well as a quantitative PCR (qPCR) assay,[50] see http://www.epa.gov/nerlcwww/moldtech.htm. The applicability of some of these methods for clinical identification and strain typing of *A. pullulans* is discussed below.

Apart from qPCR and other specific PCR protocols, the methods described here require the isolation of the microbe in pure culture. Also a monoculture (in respect of fungi) in normally sterile medium can serve as identification target. At present there are no commercial tools for molecular diagnostics of *Aureobasidium*.

3.1.3.2.1 Molecular Identification of Aureobasidium pullulans

Both DNA sequencing-independent and sequencing-based methods can be used to identify *A. pullulans*. Due to the genetic distance between *A. pullulans* and other pathogens,

DNA-based separation is usually straightforward. This applies to the separation from other melanized hyphomycetes, as well as from yeasts like *Candida albicans*, *Candida parapsilosis*, and *Cryptococcus albidus*, which resemble *A. pullulans* in young culture. Distinguishing *A. pullulans* from nonpathogenic relatives is more challenging, but several unique genetic markers are nevertheless available. RFLP analysis of a PCR amplicon containing partial large subunit (lsu or 28S) rRNA gene and ITS2 region with *DdeI* restriction endonuclease creates unique banding pattern for *A. pullulans*, distinguishing it from phytopathogenic relatives.[4,41,51] For further identification of non-*A. pullulans* strains, enzymes *RsaI*, *HhaI*, *MspI*, and *AluI* can be used according to the scheme presented by Yurlova et al.[4] Since restricted amount of reference information for this region is available, the use of RFLP in diagnostics calls for further validation and may become obsolete for identification along with the reducing prices of more unambiguous DNA sequencing-based methods. It may, however, be used for screening large numbers of isolates for epidemiologic purposes. When more precise results of the identity of studied isolates are wanted, the ITS2 region of the obtained undigested PCR products can be sequenced.

Species specific oligonucleotide probe for small subunit (ssu or 18S) rRNA gene was published for *A. pullulans* in 1996.[48] The probe has been used to detect *A. pullulans* by fluorescent in situ hybridization on microscope slides and leaf surfaces,[52] but it has not been reportedly used in clinical applications. Along with increasing amount of available sequence data, the probe seems to be to some extent unspecific and the use of this probe in clinical settings calls for validation. Other species-wide *A. pullulans*-specific probes have not been published so far in the scientific literature, yet molecular data for the species suggests that this approach would be feasible. Recently, a PCR assay targeting ITS2 region was described for species wide detection of endophytic *A. pullulans* in grapewine.[49] Based on publicly available sequence data, the primer pair is prone to amplify most strains in *A. pullulans* var. *melanigenum* with lowered efficiency compared to var. *pullulans* due to suboptimal matching of the primers.

The U.S. Environmental Protection Agency (EPA) has developed validated qPCR assays for more than 130 major indoor air fungi, including *A. pullulans*. The assays are based on TaqMan chemistry.[53] The primer-probe set targets the ITS1 region, matching well with presently available *A. pullulans* nucleotide data, and has been used to detect and enumerate *A. pullulans* among other fungi in indoor environments.[21,50] The patented assays are available through a license.

Along with lowering costs of sequencing technology and an increasing amount of reference data in public databases, DNA sequencing-based identification is becoming one of the golden standards in mycology. At present, a relatively well covering set of molecular data exists from *A. pullulans* in public DNA databases, including DNA sequence information from about 200 distinct strains or isolates of the species, as well as from related fungi. Most data is retrieved from ribosomal genes and spacers, the ITS region being most widely represented (more than 150 sequences), followed by partial lsu (more than 85 seqs) and full or partial ssu rRNA genes (more than 45 seqs). DNA sequences from the genes elongase (*ELO*, 39 seqs), translation elongation factor (*EF1α*, 29 seqs), and β-tubulin (*TUB*, 29 seqs) are available, but reference data from other fungal species is limited compared to ribosomal targets. The disadvantage of the abundance of published ribosomal sequences is that public databases harbor also misannotated *A. pullulans* sequences as well as non-*A. pullulans* sequences annotated as *A. pullulans*, and hence interpretation of database query results must always be done with deliberation. (Data from EMBL Nucleotide database in April 2009.)

Several studies point out the potential of ITS region sequencing for fungal identification in clinical laboratory.[54–59] Due to the varying degree of inter- and intraspecies variability of ITS region among different fungal clades, few general rules have been suggested concerning the interpretation of sequence comparison results. The intraspecies variation of the whole ITS region is below 3% in *A. pullulans*, whereas the distance to closest relatives is 5% or more; ITS region sequencing readily distinguishes *A. pullulans* from all known pathogenic fungi, as well as from *Hormonema* spp., *Selenophoma mahoniae*, *Pringsheimia* sp., *Dothiora* sp., and *Kabatiella* spp. apart from *K. lini*.[7] ITS2 region is more variable than ITS1 in *Aureobasidium* and relatives,[6] and can be used for the identification alone or in combination with 5.8S rRNA gene and ITS1. Recently, a detailed ITS sequencing-based protocol was presented for the wide-spectrum identification of fungi in clinical laboratory. This included fast DNA extraction from pure culture, universal PCR, and ITS sequencing followed by comparison against a novel commercial database as well as against public DNA database (GenBank).[58] In evaluation study, the protocol provided an identical or more accurate identification compared to traditional identification in case of 84% of tested 244 random clinical isolates. The study included isolates that could not be identified to species level within 5 days by standard phenotypic criteria, including two strains of *A. pullulans*. The database and data handling platform are commercially available from SmartGene (SmartGene, Zug, Switzerland). The authors represent acceptable limits for sequence similarity between analyzed and database sequences for identification to species and genus level. The species-level limits can be applied to *A. pullulans* (see Section 3.2). ITS sequence comparison has been used to identify a pathogen cultivated from septic patient as *A. pullulans*.[38] The authors report the identification of two yeast-like isolates yielded from blood cultures of two septic patients. For one isolate, an initial identification using VITEK system (Yeast Biochemical Card, bioMerieux, France) revealed 90% similarity with *Cryptococcus laurentii*. ITS sequencing revealed 100% similarity with *A. pullulans* for both of the strains.[38]

A quick, culture-independent detection of fungal pathogens directly from clinical samples is achievable by a broad-range fungal real-time PCR reaction recently described by

Vollmer et al.[60] The assay utilizes TaqMan chemistry and is designed for the variable D1/D2 region of lsu rRNA gene. Species identification is attainable by sequencing the positive PCR product, yet only genus-level identification can be achieved in case of some species belonging to genera with little genetic variation within the region. *Aureobasidium pullulans* is detected by the assay with slightly lowered sensitivity due to suboptimal primer matching. The authors report results consistent with cultivation-based identification and a diagnostic turnaround time of 9h for the identification of fungal pathogens in cervical swabs and tracheal secretion samples.[60] The method is not quantitative. Traditional PCR targeting the same region has commonly been applied to identifying numerous ascomycetous yeasts as well as medically important zygomycetes and dematiaceous fungi, including *A. pullulans*.[10,61–63]

3.1.3.2.2 Molecular Subtyping of A. pullulans

A. pullulans shows considerable variability in its vegetative morphology and physiological properties.[5] There is also a remarkable amount of genetic diversity inside *A. pullulans* that differentiates between distinct varieties, and can also be used for strain-specific monitoring.[7,41,45–47] Redefinition of *A. pullulans* and its varieties has recently been presented by Zalar et al.[7] Data from multigene analysis supports the division of *A. pullulans* into two well-known entities, *A. pullulans* var. *pullulans* and var. *melanigenum*, as well as into two novel varieties, var. *subglaciale* and var. *namibiae*.[7] Sequence comparison of variable D1/D2 region of lsu rRNA gene clusters strains with shared morphological and metabolic characteristics into the above-mentioned varieties with high confidence.[7] Even more precise separation is achieved by comparison of partial sequences of the ELO gene.[7] The genetic relationships of the four varieties as depicted by alignment of D1/D2 region of the lsu rRNA gene are shown in Figure 3.2. Potentially novel entities comprising tropical isolates of *Aureobasidium* were recently characterized from Thailand using sequence analysis of five gene loci. Combined data analysis of BT2, RPB2, EF1-1α, and ITS sequences produced 12 well-supported phylogenetic clades. The studied strains shared a good level of congruence in respect of colony characteristics and level of *pullulan* production and xylanase activity within each clade. The clades are separate from the var. *pullulans*, but their relation to other entities remains to be solved.[13] Certain strains of *A. pullulans* are of commercial interest due to their biocontrol abilities and production of specific metabolites like the water-soluble polysaccharide *pullulan* and β1.3–1.6 glucans with reported beneficial immunomodulatory properties.[13,46]

Several papers describe the design of strain-specific primer and probing systems for tracking biocontrol strains of *A. pullulans* in agricultural environment, especially in phylloplane.[46,47,52,64] Meanwhile, the genetic properties of clinical *Aureobasidium* isolates have not been assessed, and there is no information of genotypic differences between clinical strains and the rest of *A. pullulans*. According to available geno- and phenotypic data, practically all clinically significant strains of *A. pullulans* cluster to the var. *melanigenum*.[10,37,38,56,60,65] This is consistent with the thermophilic nature and strong

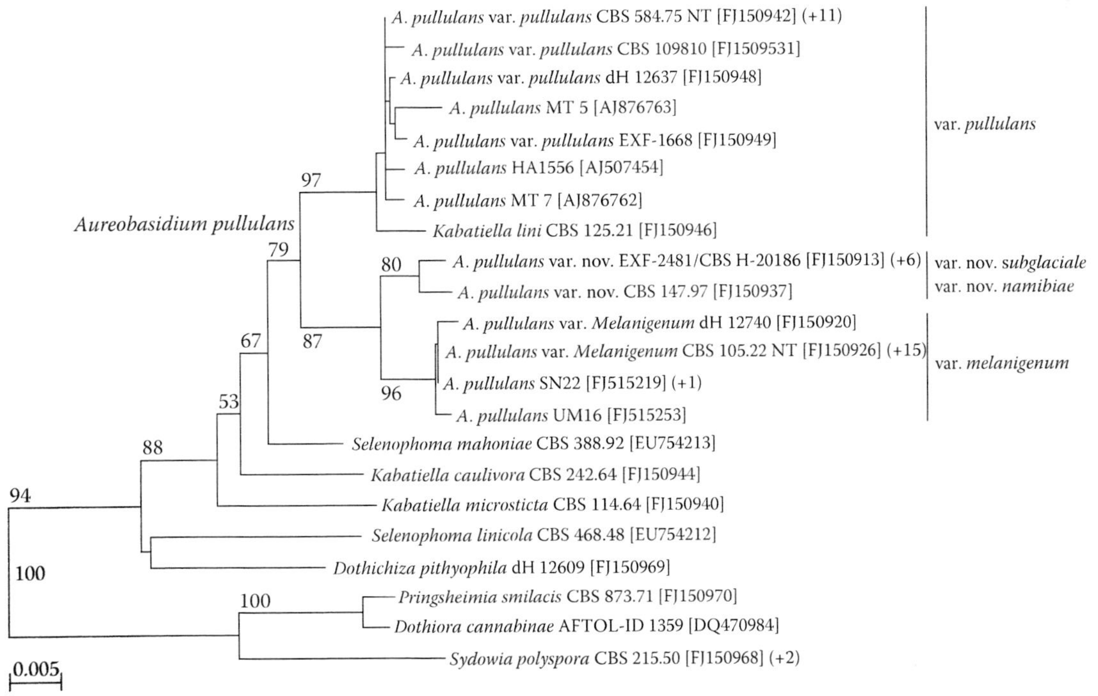

FIGURE 3.2 Neighbor joining tree of *Dothioraceae* showing relationships among varieties of *Aureobasidium pullulans* and relatives. The taxon and strain name, sequence id (in square brackets), and number of additional strains with identical sequence presently available (in brackets) are given for representative strains. Bootstrap values above 50% are shown.

pigmentation of *A. pullulans* var. *melanigenum* that may play a role in increased pathogenicity of the variety.[2,7]

3.2 METHODS

Like many fungi of low virulence, *A. pullulans* is considered a contaminant when isolated from a healthy host. It may be considered a pathogen when isolated from a normally sterile site, from multiple samples, in the presence of clinical signs of infection, or with pathological evidence of tissue-invasive disease.[26]

3.2.1 Sample Preparation

3.2.1.1 In Vitro Pure Culture

Aureobasidium can be isolated from clinical and environmental specimens using various media, for example, half-strength corn meal agar (CMA), rose bengal agar, dichloran rose bengal chloramphenicol (DRBC), MEA, sabourad dextrose agar (SDA), or brain heart infusion agar with gentamicin and chloramphenicol, and identified on yeast malt extract agar (YMA), MEA, glucose peptone agar, SDA with gentamicin and chloramphenicol, potato dextrose agar (PDA), or blakeslee's MEA.[7,13,58] On MEA or glucose peptone agar, *Aureobasidium* grows rapidly when incubated at 30°C, reaching 30 mm in 1 week, appearing flat and smooth, and soon covered with slimy exudates. The optimum temperature for growth is 25°C, maximum near 35°C. The var. *melanigenum* has higher optimum temperature for growth (30°C) than other varieties, and some strains can grow in 37°C. *Aureobasidium* does not grow on media containing cycloheximide. It tolerates salt concentrations up to 10%–15%. In the setting of fungemia, *Aureobasidium* can be isolated in blood culture systems, for example, the Bactec 9240, using pediatric aerobic culture media.[26]

3.2.1.2 DNA Extraction

Alternative DNA extraction protocols suitable for putative *Aureobasidium* and other pathogenic fungal isolates are presented here. They differ in the utilization of in-house methods and reagents, or fast but more costly commercial products and equipment. All used equipment must be DNA free. The methods below are for fungal pure cultures. For DNA extraction from clinical samples and direct PCR detection of fungi, see Vollmer et al.[60] For additional information about DNA extraction procedures, see Hebart et al.[66]

1. Cetyltrimethylammonium bromide (CTAB) extraction. Collect 1 cm² of mycelium from plate culture with disposable loop and transfer to 2 mL Eppendorf tube containing 2:1 mixture of silicagel and Celite 545 and 300 μL CTAB buffer (Tris–HCl 200 mM, pH 7.5, 200 mM Na-EDTA, 8.2% NaCl w/v, 2% CTAB w/v). Grind with micropestle for 1–2 min. Add 200 μL of CTAB buffer. Shake to homogenize. Incubate in 65°C water bath for 10 min. Add 500 μL of chloroform and vortex briefly. Centrifuge at 14,000 rpm for 5 min. Transfer the supernatant to clean Eppendorf tube, and add two volumes of cold (−20°C) 96% ethanol. Mix gently. Incubate in −20°C for 30 min to overnight. Centrifuge at 14,000 rpm for 5 min, discard supernatant, and wash the pellet twice with 500 μL cold 70% ethanol. Dry in room temperature until no liquid is visible on the pellets (~5–30 min). Resuspend in 50 μL TE buffer (10 mM Tris–HCl, pH 7.5, 0.1 mM EDTA) containing 20 mg/mL RNase A. Incubate for 5–30 min at 37°C to dissolve DNA. Adapted from Gerrits vd Ende and de Hoog.[67]
2. Fastprep extraction. Collect 1 cm² of mycelium from plate culture with a disposable loop into 1 mL of TE buffer, pH 8.0. Add about 0.5 mL volume of glass beads (0.5 mM Zirconia/Silica beads, BioSpec Products, Inc., Bartlesville, OK). Disrupt the mycelium using FastPrep instrument (QBiogene), set at number 4, for 30 s. Isolate DNA from the lysate using QIAquick PCR purification kit (Qiagen) according to manufacturer's instructions. Adapted from Manitchotpisit et al.[13]
3. FTA® filter extraction. Collect and suspend fungal mycelium by scraping from culture plate to 0.2–1 mL of sterile water yielding a mixture of fungal cell mass and water in proportion of ca. 1:10–20 v/v and vortex briefly to suspend. Transfer 125 μL of the solution directly to a FTA Microcard (Whatman International Ltd.). Place the damp cards open in a microwave and subject to two cycles of heating at 800 W for 30 s, with a pause of at least 30 s between cycles. Place a pyrex beaker containing 50 mL of sterile water to dissipate excess heat. Store the treated filters in sterile plastic bags containing desiccant. For use in PCR, remove a 2 mm punch from FTA filter with Harris micro-punch (Whatman International Ltd.) into a PCR reaction tube and wash the filter twice for 1 min with 100 μL of Whatman FTA wash reagent and twice for 1 min with 100 μL of TE buffer. Dry tubes with filters for 5 min at 55°C on a dry heat block. Add PCR reaction mixture directly to the washed and dried filter punch. The punches can be reused at least five times in a new PCR reaction with same or different primers without carryover of primers or products from the first reaction, after washing and drying the filter as described before. Adapted from Borman et al.[68]

3.2.2 Detection Procedures

3.2.2.1 Morphological Identification

The colonies are cream or pink and then later becoming brown or black (see Section 3.1.1.2). The microscopical morphology of *Aureobasidium*, particularly the synchronous blastoconidial formation, is best studied utilizing the Dalmau plate method for demonstrating chlamydospores in *Candida albicans*. *Aureobasidium* is differentiated from *Hormonema dematiodes* by synchronous blastoconidia produced only from hyaline hyphae. However, Yurlova et al. have noted that it is often difficult to distinguish between these modes of conidiogenesis and they suggested differences in the number of conidiogenesis loci and physiological tests as methods for differentiation.[40] These differences, however, are also small and variable. *A. pullulans* and *H. dematioides* are easily distinguishable by genotype, for example, by means of DNA sequencing or RFLP analysis.[6]

3.2.2.2 PCR Amplification of rDNA

Perform amplification of ribosomal DNA target in 50 μL reaction volume in 250 μL reaction tubes or plates. Prepare a PCR reaction mastermix for samples and negative control reaction, containing 1 U of DNA polymerase (e.g., DyNAzyme™ II DNA polymerase, Finnzymes, Espoo, Finland), 1× solution of matching reaction buffer, 250 μM of each dNTP, and 50 pmol of each primer per reaction. Divide into 49 μL aliquots and add 1 μL of genomic DNA extract for template. Use following thermal conditions: initial denaturation for 2 min in 95°C followed by 40 cycles of denaturation at 95°C for 30 s, annealing at T_{ann} for 45 s and elongation at 72°C for 1 min, followed by a final elongation in 72°C for 10 min in thermocycler apparatus with heated lid. For amplification of ITS region, use primers ITS1 (5′-TCCGTAGGTGAACCTGCGG-3′) and ITS4 (5′-TCCTCCGCTTATTGATATGC-3′)[69] and T_{ann} of 50°C. For the D1/D2 region of lsu rDNA gene, use primers NL 1 (5′-GCATATCAATAAGCGGAGGAAAAG-3′) and NL 4 (5′-GGTCCGTGTTTCAAGACGG-3′)[61] and T_{ann} of 52°C. For PCR-RFLP analysis, perform PCR protocol with primers 5.8SR (5′-TCGATGAAGAACGCAGCG-3′) and LR7 (5′-TACTACCACCAAGATCT-3′).[51] Use elongation time of 2 min instead of 1 min at 72°C and T_{ann} of 50°C. Visualize the products by electrophoresis in 1% agarose gel with ethidium bromide staining (0.5 μg/mL). The expected products sizes are 581–582 bp for ITS1–ITS4, 614 bp for NL1–NL4, and ~1900 bp for 5.8SR-LR7. The primers are universal fungal, that is, amplify the target region from all fungal species. Purify the products for sequencing and RFLP application with, for example, QIAquick PCR purification kit (Qiagen). In case of failed amplification, try diluting the template DNA in 1:100 and 1:1000 in sterile water, use more efficient DNA polymerase (e.g., Phusion® high fidelity DNA polymerase, Finnzymes) or repeat DNA extraction with different protocol.

3.2.2.3 DNA Sequencing

Comparison of ITS region sequence is suitable for species identification of putative *A. pullulans*. Varieties are most reliably resolved using the variable D1/D2 region of lsu rRNA gene or ELO gene.[7] The most popular method for sequencing is enzymatic Sanger sequencing, using, for example, BigDye™ 3.1 chemistry of Applied Biosystems (Life Technologies, Carlsbad, CA). Sequencing is available as commercial service. The obtained PCR products can be sequenced using one of the primers used in PCR, that is, ITS4 for ITS region and NL4 for lsu D1/D2, or with both forward and reverse primers to get sequence from both DNA strands, for best fidelity.

3.2.2.4 Phylogenetic Analysis of Obtained Sequences

Compare the obtained sequences with published DNA sequences using online Blast or Fasta alignment tools (http://blast.ncbi.nlm.nih.gov/Blast.cgi and http://www.ebi.ac.uk/Tools/fasta33/nucleotide.html) against GenBank or EMBL DNA database. Note that Blast alignment easily "drops out" the perimeters of the alignment if lower similarity occurs on the area, thus giving a partial alignment with high percentage identity. Fasta is slower compared to Blast but performs full alignment for the target sequence. High similarity percentages (>99%) with reference sequences from authenticated strains should be emphasized in interpretation of the results. See Figure 3.2 for reference strains. An ITS sequence from *A. pullulans* is supposed to have a similarity of 97%–100% with preferably two or more independent *A. pullulans* database sequences. The D1/D2 lsu rRNA sequences of isolates belonging to same variety generally show 99%–100% similarity, whereas the distance between varieties is generally 2%–3%. For ITS-sequencing-based identification of non-*A. pullulans* strains, guidelines given by Ciardo et al.[58] are recommended to be followed.

3.2.2.5 RFLP Analysis of PCR Products

Digest the obtained 5.8SR-LR7 PCR products (10 μL per digestion) with *DdeI* and optionally with *HhaI*, *MspI*, and *RsaI* restriction endonucleases according to manufacturer's instructions. Amplify and digest corresponding fragment in parallel from representative, authenticated reference strains. Separate the digests in 2% agarose gel with appropriate size standard. Stain with ethidium bromide. *DdeI* produces unique digestion pattern for *A. pullulans*, whereas other enzymes distinguish between *Hormonema* sp. and other relatives of *Aureobasidium*.[4]

3.3 CONCLUSIONS AND FUTURE PERSPECTIVES

With increasing survival of immunocompromised patients, the significance of rare and weakly pathogenic fungi becomes more obvious. During past years, several species with some pathogenicity to humans have been affiliated to the genus *Aureobasidium*, but apart from *A. pullulans*, these have later been shown to be more closely related to other genera. *Aureobasidium pullulans* represents an opportunistic fungus that only occasionally causes disease, but it has been reported to be involved in serious conditions including pneumonia, meningitis, and systemic fungemia. The diagnosis of *Aureobasidium* traditionally relies on observing characteristic macro- and micromorphology, yet these characteristics vary by strains and *Aureobasidium* is often mixed up with other black yeast-like fungi or even yeasts. *A. pullulans* is of commercial interest due to its biocontrol abilities and production of various metabolites. In this perspective, the fungus is relatively well-characterized and the taxonomy, identification, and subspecific typing of agricultural, biotechnological, and environmental strains have been assessed in numerous studies. In contrast, there are few studies that involve both clinical aspects, and phylogeny or development of molecular identification methods. No commercial products are presently available for specific detection of *A. pullulans*. Molecular methods like probing techniques, qPCR, and DNA sequence analysis are nevertheless very well suited also for the identification of *Aureobasidium* in clinical laboratory, but most methods still need further validation.

REFERENCES

1. de Hoog, G.S., Guarro, J., Gené, J., and Figuerras, M.J., *Atlas of Clinical Fungi*, Centraalbureau voor Schimmelcultures/Universitat Rovira I Virgili, Utrecht, the Netherlands, 2000.
2. Hermanides-Nijhof, E.J., *Aureobasidium* and allied genera, *Stud. Mycol.*, 15, 141, 1977.
3. Kirk, P.M. et al., *Dictionary of the Fungi*, CAB International, Wallingford, U.K., 2008.
4. Yurlova, N.A., Uijthof, J.M., and de Hoog, G.S., Distinction of species in *Aureobasidium* and related genera by PCR-ribotyping, *Anton. Van. Leeuwen.*, 69, 323, 1996.
5. de Hoog, G.S. and Yurlova, N.A., Conidiogenesis, nutritional physiology and taxonomy of *Aureobasidium* and *Hormonema*, *Anton. Van. Leeuwen.*, 65, 41, 1994.
6. Yurlova, N.A., de Hoog, G.S., and Gerrits van den Ende, A.H.G., Taxonomy of *Aureobasidium* and allied genera, *Stud. Mycol.*, 43, 63, 1999.
7. Zalar, P. et al., Redefinition of *Aureobasidium pullulans* and its varieties, *Stud. Mycol.*, 61, 21, 2008.
8. Spatafora, J.W., Mitchell, T.G., and Vilgalys, R., Analysis of genes coding for small-subunit rRNA sequences in studying phylogenetics of dematiaceous fungal pathogens, *J. Clin. Microbiol.*, 33, 1322, 1995.
9. de Hoog, G.S. et al., Relationships of dothideaceous black yeasts and meristematic fungi based on 5.8S and ITS2 rDNA sequence comparison, *Stud. Mycol.*, 43, 31, 1999.
10. Abliz, P. et al., Identification of pathogenic dematiaceous fungi and related taxa based on large subunit ribosomal DNA D1/D2 domain sequence analysis, *FEMS Immunol. Med. Microbiol.*, 40, 41, 2004.
11. Schoch, C.L. et al., A multigene phylogeny of the Dothideomycetes using four nuclear loci, *Mycologia*, 98, 1041, 2006.
12. Alvarez-Rodriguez, M.L. et al., Degradation of vanillic acid and production of guaiacol by microorganisms isolated from cork samples, *FEMS Microbiol. Lett.*, 220, 49, 2003.
13. Manitchotpisit, P. et al., Multilocus phylogenetic analyses, *pullulan* production and xylanase activity of tropical isolates of *Aureobasidium pullulans*, *Mycol. Res.*, 113, 1107, 2009.
14. Rinaldi, M.G., McCoy, E.L., and Winn, D.F., Gluteal abscess caused by *Phialophora hoffmannii* and review of the role of this organism in human mycoses, *J. Clin. Microbiol.*, 16, 181, 1982.
15. McGinnis, M.R., Human pathogenic species of *Exophiala*, *Phialophora*, and *Wangiella*. *PAHO Sci. Publ.*, 356, 37, 1978.
16. Marriott, D.J. et al., *Scytalidium dimidiatum* and *Lecythophora hoffmannii*: Unusual causes of fungal infections in a patient with AIDS, *J. Clin. Microbiol.*, 35, 2949, 1997.
17. de Hoog, G.S., Guarro, J., Gené, J., and Figueras, M.J., *Atlas of Clinical Fungi*, a pilot CD-ROM version of the 3rd edition, Centraalbureau voor Schimmelcultures, Baarn, the Netherlands, 2009.
18. Svobodova, Y., *Chmelia slovaca* gen. nov. a dematiaceous fungus, pathogenic for man and animals, *Biologia (Bratisl)*, 21, 81, 1966.
19. Cooke, W.B., An ecological life history of *Aureobasidium pullulans* (De Bary) Arnaud, *Mycopathologia*, 12, 1, 1959.
20. Taylor, P.E. et al., Identification and possible disease mechanisms of an under-recognized fungus, *Aureobasidium pullulans*, *Int. Arch. Allergy Immunol.*, 139, 45, 2006.
21. Vesper, S.J. et al., Relative moldiness index as predictor of childhood respiratory illness, *J. Expo. Sci. Environ. Epidemiol.*, 17, 88, 2007.
22. Haase, G. et al., Phylogenetic analysis of ten black yeast species using nuclear small subunit rRNA gene sequences, *Anton. Van Leeuwen.*, 68, 19, 1995.
23. Coldiron, B.M., Wiley, E.L., and Rinaldi, M.G., Cutaneous phaeohyphomycosis caused by a rare fungal pathogen, *Hormonema dematioides*: Successful treatment with ketoconazole, *J. Am. Acad. Dermatol.*, 23, 363, 1990.
24. Shin, J.H. et al., Fatal *Hormonema dematioides* peritonitis in a patient on continuous ambulatory peritoneal dialysis: Criteria for organism identification and review of other known fungal etiologic agents, *J. Clin. Microbiol.*, 36, 2157, 1998.
25. Kent, D. et al., Fungemia due to *Hormonema dematioides* following intense avian exposure, *Clin. Infect. Dis.*, 26, 759, 1998.
26. Hawkes, M. et al., *Aureobasidium pullulans* infection: Fungemia in an infant and a review of human cases, *Diagn. Microbiol. Infect. Dis.*, 51, 209, 2005.
27. Rinaldi, M.G., Phaeohyphomycosis, *Dermatol. Clin.*, 14, 147, 1996.
28. Richardson, M. D. and Warnock, D. W., *Fungal Infection: Diagnosis and Management*, Blackwell Publishing, Oxford, 2003.
29. Stark, P.C. et al., Fungal levels in the home and lower respiratory tract illnesses in the first year of life, *Am. J. Respir. Crit. Care Med.*, 168, 232, 2003.
30. Woodcock, A.A. et al., Fungal contamination of bedding, *Allergy*, 61, 140, 2006.
31. Niedoszytko, M. et al., Association between sensitization to *Aureobasidium pullulans* (*Pullularia* sp) and severity of asthma, *Ann. Allergy Asthma Immunol.*, 98, 153, 2007.
32. Woodard, E.D. et al., Outbreak of hypersensitivity pneumonitis in an industrial setting, *JAMA*, 259, 1965, 1988.
33. Temprano, J. et al., Hypersensitivity pneumonitis secondary to residential exposure to *Aureobasidium pullulans* in 2 siblings, *Ann. Allergy Asthma Immunol.*, 99, 562, 2007.
34. Engelhart, S. et al., Childhood hypersensitivity pneumonitis associated with fungal contamination of indoor hydroponics, *Int. J. Hyg. Environ. Health*, 212, 18, 2009.
35. Maverick, K.J. and Conners, M.S., *Aureobasidium pullulans* fungal keratitis following LASEK, *J. Refract. Surg.*, 23, 727, 2007.
36. Joshi, A. et al., Subcutaneous mycosis and fungemia by *Aureobasidium pullulans*: A rare pathogenic fungus in a post allogeneic BM transplant patient, *Bone Marrow Transplant.*, 45, 203, 2009.
37. Bolignano, G. and Criseo, G., Disseminated nosocomial fungal infection by *Aureobasidium pullulans* var. *melanigenum*: A case report, *J. Clin. Microbiol.*, 41, 4483, 2003.
38. Huang, Y.T. et al., Catheter-related septicemia due to *Aureobasidium pullulans*, *Int. J. Infect. Dis.*, 12, e137, 2008.

39. Evans, E.G.V. and Richardson, M.D., *Medical Mycology: A Practical Approach*, IRL Press at Oxford University Press, Oxford, 1989.
40. Yurlova, N.A., Mokrousov, I.V., and de Hoog, G.S., Intraspecific variability and exopolysaccharide production in *Aureobasidium pullulans*, *Anton. Van Leeuwen.*, 68, 57, 1995.
41. Loncaric, I. et al., Phenotypic and genotypic diversity among strains of *Aureobasidium pullulans* in comparison with related species, *Anton. Van Leeuwen.*, 95, 165, 2009.
42. Bulat, S.A. and Mironenko, N.V., Polymorphism of the yeast-like fungus *Aureobasidium pullulans* (de Bary) detected with universally primed-polymerase reaction: Species divergence, *Genetics*, 28, 19, 1992.
43. Schena, L. et al., Genetic diversity and biocontrol activity of *Aureobasidium pullulans* isolates against postharvest rots, *Postharvest Biol. Technol.*, 17, 189, 1999.
44. Urzì, C. et al., Intra-specific diversity of *Aureobasidium pullulans* strains isolated from rocks and other habitats assessed by physiological methods and by random amplified polymorphic DNA (RAPD), *J. Microbiol. Methods*, 36, 95, 1999.
45. De Curtis, F. et al., Use of fluorescent amplified fragment length polymorphism (fAFLP) to identify specific molecular markers for the biocontrol agent *Aureobasidium pullulans* strain LS30, *Postharvest Biol. Technol.*, 34, 179, 2004.
46. Schena, L., Finetti Sialer, M.M., and Gallitelli, D., Molecular detection of strain L47 of *Aureobasidium pullulans*, a biocontrol agent of postharvest diseases, *Plant Dis.*, 86, 54, 2002.
47. Loncaric, I. et al., Strain-specific detection of two *Aureobasidium pullulans* strains, fungal biocontrol agents of fire blight by new, developed multiplex-PCR, *J. Appl. Microbiol.*, 104, 1433, 2008.
48. Li, S. et al., Development of an oligonucleotide probe for *Aureobasidium pullulans* based on the small-subunit rRNA gene, *Appl. Environ. Microbiol.*, 62, 1514, 1996.
49. Martini, M. et al., DNA-dependent detection of the grapevine fungal endophytes *Aureobasidium pullulans* and *Epicoccum nigrum*, *Plant Dis.*, 93, 993, 2009.
50. Meklin, T. et al., Quantitative PCR analysis of house dust can reveal abnormal mold conditions, *J. Environ. Monit.*, 6, 615, 2004.
51. Vilgalys, R. and Hester, M., Rapid genetic identification and mapping of enzymatically amplified ribosomal DNA from several *Cryptococcus* species, *J. Bacteriol.*, 172, 4238, 1990.
52. Li, S., Spear, R.N., and Andrews, J.H., Quantitative fluorescence in situ hybridization of *Aureobasidium pullulans* on microscope slides and leaf surfaces, *Appl. Environ. Microbiol.*, 63, 3261, 1997.
53. Wittwer, C.T. et al., Continuous fluorescence monitoring of rapid cycle DNA amplification, *BioTechniques*, 22, 130, 1997.
54. Chen, Y.C. et al., Polymorphic internal transcribed spacer region 1 DNA sequences identify medically important yeasts, *J. Clin. Microbiol.*, 39, 4042, 2001.
55. Iwen, P.C., Hinrichs, S.H., and Rupp, M.E., Utilization of the internal transcribed spacer regions as molecular targets to detect and identify human fungal pathogens, *Med. Mycol.*, 40, 87, 2002.
56. Rakeman, J.L. et al., Multilocus DNA sequence comparisons rapidly identify pathogenic molds, *J. Clin. Microbiol.*, 43, 3324, 2005.
57. Leaw, S.N. et al., Identification of medically important yeast species by sequence analysis of the internal transcribed spacer regions, *J. Clin. Microbiol.*, 44, 693, 2006.
58. Ciardo, D.E. et al., Identification of moulds in the diagnostic laboratory—An algorithm implementing molecular and phenotypic methods, *Diagn. Microbiol. Infect. Dis.*, 59, 49, 2007.
59. Borman, A.M. et al., Molecular identification of pathogenic fungi, *J. Antimicrob. Chemother.*, 61, S7, 2008.
60. Vollmer, T. et al., Evaluation of novel broad-range real-time PCR assay for rapid detection of human pathogenic fungi in various clinical specimens, *J. Clin. Microbiol.*, 46, 1919, 2008.
61. Kurtzman, C.P. and Robnett, C.J., Identification of clinically important ascomycetous yeasts based on nucleotide divergence in the 5′ end of the large-subunit (26S) ribosomal DNA gene, *J. Clin. Microbiol.*, 35, 1216, 1997.
62. Kurtzman, C.P. and Robnett, C.J., Identification and phylogeny of ascomycetous yeasts from analysis of nuclear large subunit (26S) ribosomal DNA partial sequences, *Anton. Van Leeuwen.*, 73, 331, 1998.
63. Voigt, K., Cigelnik, E., and O'donnell, K., Phylogeny and PCR identification of clinically important Zygomycetes based on nuclear ribosomal-DNA sequence data, *J. Clin. Microbiol.*, 37, 3957, 1999.
64. Spear, R.N. et al., Quantitative imaging and statistical analysis of fluorescence in situ hybridization (FISH) of *Aureobasidium pullulans*, *J. Microbiol. Methods*, 35, 101, 1999.
65. Hall, L., Wohlfiel, S., and Roberts, G.D., Experience with the MicroSeq D2 large-subunit ribosomal DNA sequencing kit for identification of commonly encountered, clinically important yeast species, *J. Clin. Microbiol.*, 41, 5099, 2003.
66. Hebart, H., Loeffler, J., and Einsele, H., Molecular diagnostics: Present and future, in *Diagnosis of Fungal Infections*, 1st edn., p. 133, Maertens, J.A. and Marr, K.A. (Eds.), Informa Health Care, New York, 2007.
67. Gerrits vd Ende, H.G. and de Hoog, G.S., Variability and molecular diagnostics of the neurotropic species *Cladophialophora bantiana*, *Stud. Mycol.*, 43, 162, 1999.
68. Borman, A.M. et al., Ultra-rapid preparation of total genomic DNA from isolates of yeast and mould using Whatman FTA filter paper technology—A reusable DNA archiving system, *Med. Mycol.*, 44, 389, 2006.
69. White, T.J. et al., Amplification and direct sequencing of fungal ribosomal RNA genes for phylogenetics, in *PCR Protocols: A Guide to Methods and Applications*, p. 315, Gelfand, D.H., Sninsky, J.J., and White, T.J. (Eds.), Academic Press, New York, 1990.
70. Andreoni, S., Farina, C., and Lombardi, G., *Medical Mycology Atlas*, 2004.

4 *Bipolaris* and *Drechslera*

Dongyou Liu and Joanna Gray

CONTENTS

4.1 INTRODUCTION

The term "dematiaceous fungi" refers to a heterogeneous group of darkly pigmented fungal organisms, which derive their coloration from melanin in the cell walls, and produce brown yeast-like cells, pseudohyphae, and irregular true hyphae in tissues. Being commonly present in the soil, plants, and other environments, dematiaceous fungi are introduced into human hosts through inhalation, or after injury and other traumatic events, causing a range of superficial and deep infections that are collectively known as chromoblastomycosis, eumycetoma, and phaeohyphomycosis.

Currently, over 130 dematiaceous fungal species belonging to 70 genera have been implicated in human infections. While both chromoblastomycosis and eumycetoma are associated with a relatively small group of dark-walled fungi, the former is characterized by the formation of sclerotic bodies in tissue and is usually seen in tropical areas, and the latter is a deep tissue infection characterized by the presence of mycotic granules, usually of the lower extremities. On the other hand, phaeohyphomycosis is generally reserved for the remainder of clinical syndromes (ranging from superficial infections, allergic disease, pneumonia, cerebral infection to disseminated disease) caused by a large number of dematiaceous fungi including members of *Bipolaris* and *Drechslera* genera.

4.1.1 Classification and Morphology

The genus *Bipolaris* belongs to the mitosporic *Cochliobolus* group, family *Pleosporaceae*, order Pleosporales, class Dothideomycetes, subphylum Pezizomycotina, phylum Ascomycota, and kingdom Fungi. The family *Pleosporaceae* covers the genera of *Brachycladium*, *Cochliobolus*, *Crivellia*, *Decorospora*, *Edenia*, *Lewia*, *Macrospora*, *Macroventuria*, mitosporic *Cochliobolus*, mitosporic *Pleosporaceae*, *Pleospora*, *Pyrenophora*, *Setosphaeria*, and some unclassified *Pleosporaceae*; the mitosporic *Cochliobolus* group is made up of the genera of *Bipolaris* and *Curvularia*. The genus *Bipolaris* (teleomorph: *Cochliobolus*) is further divided into 33 recognized species and 36 unassigned species [1].

Bipolaris spp. are primarily plant pathogens; however, several (i.e., *Bipolaris spicifera* [obsolete synonyms: *Curvularia spicifera*, *Brachycladium spiciferum*, *Drechslera spicifera*, and *Helminthosporium spiciferum*], *Bipolaris hawaiiensis* [obsolete synonyms: *Drechslera hawaiiensis* and *Helminthosporium hawaiiensis*] and *Bipolaris australiensis* [obsolete synonyms: *Drechslera australiensis* and *Helminthosporium australiensis*]) have been implicated in human infections [2].

Bipolaris colonies grow moderately fast and are effuse, grey to blackish brown, suede-like to floccose with a black reverse. Pale brown pigmented, pseudoseptate conidia are produced through pores (poroconidia) in a sympodially elongating geniculate (or zigzag) conidiophore and are straight, fusiform to ellipsoidal, rounded at both ends, smooth to finely roughened, and germinating only from the ends (bipolar) [3].

At the species level, *B. spicifera* conidia are cylindrical with a broadly rounded base, have three septa and a distinctly raised hilum, which shows a zone of dark pigmentation. *B. australiensis* conidia are ellipsoidal with a more or less tapering base, which becomes truncated because of the hilum, and 20% of its conidia are with four or five septa. In addition, *B. spicifera* conidia are slightly broader than those of *B. australiensis*, and *B. spicifera* isolates have a small hyaline area just above the hilum, which is absent in *B. australiensis*. *B. hawaiiensis* conidia are approximately 6.8 by 23.3 μm and have predominantly four or five septa [2]. Moreover, *B. spicifera* is differentiated from *B. australiensis* by its thinner cell wall, from *B. hawaiiensis* by its production of swollen conidia with 3 to 7 septa, and from *B. papendorfii* by its being curved and broadest at the second cell.

The genus *Drechslera* belongs to the mitosporic *Pleosporaceae* group in the family Pleosoraceae, and contains the genera *Alternaria*, *Dendryphiella*, *Dendryphion*, *Drechslera*, *Embellisia*, *Exserohilum*, *Nimbya*, *Pithomyces*, *Pyrenochaeta*, *Stagonospora*, *Stemphylium*, *Ulocladium*, and *Unifilum*. At present, there are 20 recognized species and 11 unassigned species within the genus *Drechslera* (teleomorph: *Pyrenophora*) [1].

Drechslera spp. are generally present in soil and plants although one of its members, i.e., *Drechslera biseptata*, has been reported recently from a brain abscess. Many previous *Drechslera* or *Helminthosporium* isolates from human and animal cases have been shown to actually belong to the genera *Bipolaris* or *Exserohilum* [2].

Drechslera colonies grow rapidly and are suede-like to downy, brown to blackish brown with a black reverse. Conidia are pale to dark brown, usually cylindrical or subcylindrical, straight, smooth walled, and are formed apically through a pore (poroconidia) in a sympodially elongating geniculate conidiophore. Conidia are transversely septate (phragmoconidia), and the hilum is not protuberant [3].

As closely related members of the mitosporic *Cochliobolus* group and the mitosporic *Pleosporaceae* group within the family *Pleosporaceae*, *Bipolaris* (teleomorph: *Cochliobolus*), *Curvularia* (teleomorph: *Cochliobolus*), *Drechslera* (teleomorph: *Pyrenophora*), and *Exserohilum* (teleomorph: *Setoshaeria*) are differentiated through a combination of characters such as conidial shape, the presence or absence of a protruding hilum, the contour of the basal portion of the conidium and its hilum, the point at which the germ tube originates from the basal cell, and the sequence and location of the first three conidial septa.

Specifically, *Bipolaris* conidia are fusiform ellipsoidal and central cells are not much darker and broader than the distal ones; hilum is not protuberant and germination is bipolar. *Drechslera* conidia are cylindrical, germinating from any cell, and hilum is not protuberant. *Curvularia* conidia have two to three broader and darker central cells, often curved, with or without a prominent hilum; germination is bipolar. *Exserohilum* conidia are fusiform cylindrical to obclavate, with a protuberant hilum; germination is bipolar [4].

Furthermore, in common with *Curvularia*, some *Bipolaris* spp. also produce curved conidia with hyaline apical cells. The curvature in *Bipolaris* conidia involves a slight change in the central cell, whereas those in *Curvularia* conidia are due to the enlargement of the central cell. Thus, *Bipolaris* is distinguishable from *Curvularia* by the absence of an enlarged cell in its curved conidia.

4.1.2 Clinical Features and Pathogenesis

Bipolaris and *Drechslera* are dematiaceous (dark-walled) fungi that are occasionally involved in human infections including cutaneous infections, sinusitis, keratitis, peritonitis, allergic bronchopulmonary disease, brain abscess, meningitis, fatal fungal endarteritis, fungemia, and disseminated fungal disease [5–29].

Bipolaris spicifera is the most commonly reported human pathogenic *Bipolaris* species, especially in subtropical and tropical regions (e.g., Texas, South Carolina, Arizona, and Georgia in the United States; Brisbane, Australia; Pakistan; and India). Other less common *Bipolaris* spp. associated with human disease are *Bipolaris hawaiiensis* and *Bipolaris australiensis*.

4.1.2.1 Cutaneous infections

Bipolaris spicifera has been reported as the causative agent of cutaneous infections in both immunocompetent and immunocompromised individuals [30–34]. Bilu et al. [35] documented a case of *Bipolaris spicifera*-related cutaneous fungal infection on the left cheek of a 5-year-old boy with B-precursor-cell acute lymphoblastic leukemia.

4.1.2.2 Mycotic keratitis

Bipolaris spp. have also been shown to cause mycotic keratits [36–43]. Bashir et al. [44] described a *Bipolaris hawaiiensis*-related corneal ulcer with abscess and hypopyon in an immunocompetent male following trauma with rice stalk to the left eye. Direct examination of the corneal scrapings showed septate, branched fungal hyphae. Culture of the scrapings on blood agar and Sabouraud dextrose agar yielded *Bipolaris hawaiiensis*.

4.1.2.3 Rhinosinusitis

Allergic fungal sinusitis (AFS) is a noninvasive form of fungal rhinosinusitis accounting for 6%–9% of all rhinosinusitis requiring surgery. Patients with AFS typically show chronic rhinosinusitis with nasal polyps, inhalant atopy, elevated total serum immunoglobulin E (IgE), and sinus-obstructing inspissates of a characteristic extramucosal "peanut buttery," viscoelastic, eosinophil-rich, hyphae-containing material called "allergic mucin." Sinus computer tomography (CT) shows findings of chronic rhinosinusitis that often include central areas of increased contrast ("hyperattenuation") within abnormal paranasal sinuses due to the presence of fungal-containing allergic mucin. AFS allergic mucin cultures are often positive for either dematiaceous fungi such as *Bipolaris spicifera* or *Curvularia lunata*, or *Aspergillus* species such as *A. fumigatus*, *A. flavus* or *A. niger* [45,46]. Indeed, *Bipolaris* spp. are increasingly recognized as the cause of fungal sinusitis in humans with the sphenoid and posterior ethmoid sinuses being most often involved followed by the anterior ethmoid sinus, frontal sinus, and maxillary sinus involvement [47–58].

Buzina et al. [59] presented a detailed description of *Bipolaris spicifera*-associated rhinitis in a 19-year-old immunocompetent man after the patient presented with restricted nasal breathing. A CT scan showed massive sinusitis with significant decalcification and destruction of the bone at skull base. An endoscopic examination revealed a total obstruction of both nostrils with glassy polyps. The immunological examination of the patient's blood demonstrated a highly elevated level of total IgE. Histological examinations of formalin-fixed and paraffin-embedded tissue and mucus samples resected/collected from sinus showed inflammatory sinonasal polyps and clusters of eosinophilic granulocytes within the mucus. Gomori's methenamine silver (GMS) staining uncovered septate fungal hyphae in the fungal masses (fungus balls) as well as within the mucus, although a fungal invasion of the tissue was not observed. Culture of the mucus on Sabouraud dextrose agar at 25°C showed fast-growing colonies, reaching a diameter of 4 cm in 7 days. Colonies were velvety, brownish, and flat. Microscopic observation showed septate, pigmented hyphae and unbranched, zigzagged conidiophores with thick-walled, darkly pigmented, cylindrical conidia (predominantly with three septate). The fungus was identified as *Bipolaris spicifera*, which was further confirmed by sequencing analysis of the internal transcribed spacer (ITS) region of the ribosomal gene cluster using fungus-specific primers. Indeed, the resulting 575 bp amplicon demonstrated a 100% identity with *Cochliobolus spiciferus*, which is the teleomorph of *Bipolaris spicifera*. In a separate study, Castelnuovo et al. [53] described a phaeohyphomycotic sinusitis due to *Bipolaris hawaiiensis* in an immunologically competent patient. The patient was successfully treated by surgical drainage and amphotericin B.

Dyer et al. [60] reported a *Bipolaris australiensis*–associated allergic bronchopulmonary disease in a 40-year-old white male with past history of allergic rhinitis and asthma. The patient presented with back pain and cough. CT scan of the chest showed a mass in the lower lobe of the right lung. Culture of the bronchial biopsy and lavage yielded a pure isolate of *Bipolaris australiensis* that was confirmed by DNA technique.

4.1.2.4 Other Infections

Kobayashi et al. [61] described a *Bipolaris spicifera*-related case of disseminated infection in an immunocompetent male. Histopathological studies of lymph node, lung, and liver biopsy specimens revealed a dark pigmented, granular fungal structure inside the granuloma. Culture of lymph node specimen grew a fungal isolate that was identified as *Bipolaris spicifera* on the basis of morphology and molecular analysis. The patient responded to combination therapy with intravenous amphotericin B and voriconazole, but the infection reoccurred during oral itraconazole.

Gadallah et al. [14] documented a case of *Bipolaris hawaiiensis* peritonitis in a 73-year-old female on continuous cyclic peritoneal dialysis (CCPD) with a nonfunctioning peritoneal catheter. The catheter harbored characteristic dark gray particles consisting of a fungal ball within the lumen of the catheter. Microscopic examination confirmed the organism attached to the inner wall of the catheter. The patient was treated by removing the catheter and the administration of a 2 week course of oral itraconazole 100 mg twice daily (without using amphotericin B or ketoconazole).

In combination with surgical debridement *Bipolaris* infections are often treated with amphotericin B, which is the antifungal drug of choice for chronic invasive sinusitis [7]. Ketoconazole, itraconazole, voriconazole, and natamycin may also be utilized [8,23,41]. Itraconazole is a highly effective antifungal agent for chromoblastomycosis and subcutaneous phaeohyphomycosis; ketoconazole is useful for mycetoma. Postoperative oral corticosteroids and antiallergic inflammation therapy may also be prescribed for AFS [62,63].

4.1.3 Laboratory Diagnosis

Dematiaceous (brown- or dark-pigmented) fungi are a heterogeneous group of molds that are classified in the genera *Alternaria*, *Bipolaris*, *Cladophialophora*, *Curvularia*, *Drechslera*, *Exserohilum*, *Exophiala* (*Wangiella*), *Fonsecaea*, *Madurella*, *Phialophora*, *Scedosporium*, and *Scytalidium*. These fungi are widespread in soil, wood, and decomposing plant debris and are opportunistic human pathogens causing phaeohyphomycosis, chromoblastomycosis, and eumycotic mycetoma.

Traditionally, dematiaceous fungi are identified on the basis of morphological characteristics such as the presence of annellides (*Phaeoannellomyces*, *Exophiala*), phialides (*Phialophora*, *Wangiella*), or adelophialides (*Phialemonium* without collarettes, *Lecythophora* with collarettes); differentiation of conidiophores (*Xylohypha*, *Cladosporium*) and conidial hilum, septation and germination (*Bipolaris*, *Drechslera*, *Exserohilum*) [2]. Fontana-Masson stain is helpful for distinguishing the pigmented dematiaceous organisms from other septated fungal forms [45]. For histological examination, GMS and hematoxylin and eosin (H&E) stains may be employed. Enhancement of sporulation by growth of species of *Bipolaris*, *Curvularia*, *Drechslera*, and *Exserohilum* on cellulose substrates may facilitate their identification in culture and production of spores at relatively high concentrations [64].

Biochemical tests (e.g., 12% gelatin test, nitrate assimilation) and determination of growth temperature (e.g., 37°C for *Exophiala jeanselmei*, 40°C for *Exophiala* [*Wangiella*] *dermatitidis* and *Bipolaris spicifera*, 42°C for *Xylohypha bantiana*, and 45°C for *Dactylaria constricta* var. *gallopava* and *Scedosporium inflatum*) may also be useful for their differentiation [65]. Fluorescent halogen immunoassay (fHIA) may be utilized for examination of mycoaerosols from individuals with exposure and sensitization to fungal allergens in indoor environments. Indeed, *Bipolaris* conidia were commonly detected from these individuals by this immunostaining procedure [66].

In recent years, molecular techniques have been applied for identification of dematiaceous fungi including *Bipolaris* and *Drechslera* [58,67]. Polymerase chain reaction (PCR) amplification and sequencing analysis of rRNA internal transcribed spacer (ITS) regions allow rapid and precise diagnosis of fungal infections in humans [67–69].

4.2 METHODS

4.2.1 Sample Preparation

Specimens may be stained with lactophenol cotton blue stain or periodic acid-Schiff (PAS) stain and examined under the microscope for mycotic elements. Portions of samples are grown on inhibitory mold agar, modified Sabouraud agars, or potato dextrose agar at 25°C and 35°C.

After growth of 1–7 days on potato dextrose agar slants, mycelia are collected by scraping the slant with a sterile stick in 1 mL of sterile water. The material is transferred to a 2 mL screw-cap tube. The tubes are centrifuged for 1 min at 6000×*g*. If the mycelia do not pellet, the material is contained with a pediatric blood serum filter (Porex Corp). The supernatant is removed and the material is resuspended in 200 μL of IDI sample buffer and transferred to the lysis tube containing glass beads. The lysis tubes are vortexed on the highest setting for 5 min. The tubes are then placed in a boiling water bath for 15 min. The tubes are centrifuged for 5 min at 16000×*g*. The supernatant is stored at −20°C until amplification [70].

Alternatively, about 1 cm^2 of fungal material is transferred to a 2 mL Eppendorf tube containing a 2:1 (wt/wt) mixture of silica gel and Celite (silica gel H, Merck 7736/Kieselguhr Celite 545; Machery) and 300 μL of TES buffer (2 g Tris [hydroxymethyl]-aminomethane, 0.38 g Na-EDTA [ethylenediaminetetraacetic acid], and 2 g sodium dodecyl sulfate in 80 mL of ultrapure water [pH 8]). The fungal material is ground with a micropestle for 1–2 min. The volume is adjusted by adding 200 μL of TES buffer. After vigorous shaking and the addition of 10 μL of a 10 mg/mL concentration of proteinase K to the tube, the mixture is incubated at 65°C for 10 min. With the addition of 140 μL of 5 M NaCl solution, the mixture is combined with 1/10 volume (~65 μL) of 10% CTAB (cetyltrimethylammonium bromide) buffer, followed by incubation for another 30 min at 65°C. One volume (~700 μL) of chloroform-isoamyl alcohol (vol/vol = 24/L) is added and mixed by inversion. After incubation for 30 min at 0°C (on ice water) and centrifugation at 14,000 rpm at 4°C for 10 min, the top layer is transferred to a clean Eppendorf tube. The sample is added with 225 μL of 5 M NH_4-acetate and incubated for at least 30 min (on ice water) and centrifuged. The supernatant is transferred to a clean sterile Eppendorf tube and mixed with a 0.55 volume (~510 μL) of ice-cold isopropanol. After centrifugation for 7 min at 14,000 rpm and 4°C (or room temperature), the supernatant is decanted. The pellet is washed with ice-cold 70% ethanol twice and dried by using a vacuum dryer. The powder is resuspended in 48.5 μL of Tris-EDTA buffer with 1.5 μL of 10 mg of RNase/mL, incubated at 37°C for 15–30 min, and stored at −20°C until use [71].

4.2.2 Detection Procedures

4.2.2.1 Sequence Analysis of ITS Regions

Pounder et al. [70] described a real-time PCR with SYBR green DNA-binding dye and amplicon melting temperature analysis for fungal detection using pan-fungal primers ITS1 forward (5′-TCCGTAGGTGAACCTGCGG-3′) and ITS4 reverse (5′-TCCTCCGCTTATTGATATGC-3′) [72]. The identity of the fungi is verified by subsequent sequencing analysis.

Procedure

1. PCR mixture is composed of 1× Lightcycler FastStart DNA Master Hybridization Probes mixture (Roche Applied Science) (containing deoxynucleoside triphosphates, FastStart *Taq* DNA polymerase, and 1 mM $MgCl_2$, additional $MgCl_2$ is added to a final concentration of 4.6 mM), 0.4 μM each of ITS1 forward and ITS4 reverse primers, 1× SYBR green (Molecular Probes), and 3 μL template DNA.
2. Thermal cycling parameters include 95°C for 10 min; 50 cycles of 95°C for 5 s, 60°C for 20 s, and 76°C for 30 s; and a final extension at 72°C for 2 min.
3. The quality of the amplicon is determined using the derivative of the melt analysis curve (55°C–99°C, 45 s hold at 55°C, 5 s/°C) using the RotorGene 3000 (Corbett Robotics, Inc).
4. The amplified product is purified for bidirectional sequencing using ExoSAP-IT (USB Corp). Five microliters of Big Dye Terminator Ready Reaction Mix v. 1.1 (Applied Biosystems) is added to 4 μL of each primer (0.8 pmol/μL) and 3 μL of purified PCR product. Cycle sequencing is performed with a 9700 thermal cycler (ABI), using 25 cycles of 96°C for 10 s, 50°C for 5 s, and 60°C for 4 min. Sequencing reaction products are passed through a Sephadex G-50 fine column to remove unincorporated dye terminators. Purified sequencing reaction products are run on an ABI Prism 3100 Genetic Analyzer with a 50 cm capillary array.
5. Sequences are analyzed with the SmartGene Integrated Database Network software version 3.2.3 vr. SmartGene is a web-based software and database system with reference sequences derived from the National Center for Biological Information (NCBI) GenBank repository.

Note. In case that real time PCR instrument is not available, standard PCR may be performed with primers ITS1 and ITS4, and the resulting amplicon is sequenced with the same primers. Sequence-based identifications are defined by percent identity: species, ≥99%; genus, 93%–99%; and inconclusive, ≤93%. For strains producing discrepant identification between the methods based on phenotypic characteristics and ITS sequence analysis, the D1–D2 region of the large-subunit RNA gene is amplified with primers NL1 (5′-GCATATCAATAAGCGGAGGAAAAG-3′) and NL4 (5′-GGTCCGTGTTTCAAGACGG-3′) and sequenced for species clarification [73].

4.2.2.2 Sequence Analysis of D1–D2 Region of 28S rRNA

Kobayashi et al. [61] utilized primers NL-1 and NL-4 in the first-round PCR and primers Saga-F (5′-TTGCGCTAGTAACGGCGAGTGAAG-3′) and Saga-R (5′-CCATTACGCCAGCATCCTAGCAGA-3′) in the second-round PCR to amplify the D1–D2 region of fungal 28S rRNA gene. The resulting PCR product (of 614 bp in length) is sequenced directly by using the primers Saga-F and Saga-R. The sequence from the fungal isolate is then compared to those of dematiaceous fungi in GenBank via nucleotide–nucleotide BLAST analysis.

4.3 CONCLUSION

Bipolaris and *Drechslera* are dematiaceous (dark-walled) fungi that are widely distributed in soil, plants, and air. Several *Bipolaris* and *Drechslera* species are occasionally implicated in human phaeohyphomycosis including cutaneous and subcutaneous infections, keratitis, allergic fungal sinusitis, allergic bronchopulmonary mycosis, disseminated disease, and brain abscess, sometimes with fatal consequence [74].

Conventional diagnosis of phaeohyphomycosis relies on pathological investigation of clinical specimens with the help of Fontana–Masson stain, which is specific for melanin, along with gross and microscopic examination of cultures. As *Bipolaris* and *Drechslera* demonstrate close morphological similarity, the only distinction between them is that *Bipolaris* germinates only from polar cells and *Drechslera* spores germinate from any cell (of the spore).

Considering the difficulty in identification and differentiation of *Bipolaris* and *Drechslera* from other dematiaceous fungi based on phenotypic characterization, application of molecular techniques such as PCR amplification and sequence analysis of ribosomal internal transcribed spacer (ITS) regions is essential for their correct determination, which in turn aids the selection of appropriate control and prevention strategies against these pathogens.

REFERENCES

1. The UniProt Consortium. Available at http://www.uniprot.org/taxonomy/, accessed on August 1, 2010.
2. McGinnis, M.R., Rinaldi, M.G., and Winn, R.E. Emerging agents of phaeohyphomycosis: Pathogenic species of *Bipolaris* and *Exserohilum*. *J Clin Microbiol*. 1986;24(2):250–259.
3. Kwon-Chung, K.J. and Bennett, J.E. *Medical Mycology*. Lea & Febiger, Philadelphia, PA, 1992.
4. Domsch, K.H., Gams, W., and Anderson, T.H. *Compendium of Soil Fungi*, Vol. 1. Academic Press, London, U.K., 1980.
5. Zapater, R.C., Albesi, E.J., and Garcia, G.H. Mycotic keratitis by *Drechslera spicifera*. *Sabouraudia*. 1975;13(3):295–298.
6. Yoshimori, R.N. et al. Phaeohyphomycosis of brain: Granulomatous encephalitis caused by *Drechslera spicifera*. *Am J Clin Pathol*. 1982;77(3):363–370.
7. Rolston, K.V., Hopfer, R.L., and Larson, D.L. Infections caused by *Drechslera* species: Case report and review of the literature. *Rev Infect Dis*. 1985;7(4):525–529.
8. Adam, R.D. et al. Phaeohyphomycosis caused by the fungal genera *Bipolaris* and *Exserohilum*: A report of 9 cases and review of the literature. *Medicine (Baltimore)*. 1986;65(4):203–217.
9. Ruben, S.J., Scott, T.E. Jr., and Seltzer, H.M. Intracranial and paranasal sinus infection due to *Drechslera*. *South Med J*. 1987;80(8):1057–1058.
10. Pratt, M.F. and Burnett, J.R. Fulminant *Drechslera* sinusitis in an immunocompetent host. *Laryngoscope*. 1988;98(12):1343–1347.
11. McGinnis, M.R. et al. Phaeohyphomycosis caused by *Bipolaris spicifera*: An informative case. *Eur J Epidemiol*. 1992;8(3):383–386.
12. Ogden, P.E., Hurley, D.L., and Cain, P.T. Fatal fungal endarteritis caused by *Bipolaris spicifera* following replacement of the aortic valve. *Clin Infect Dis*. 1992;14(2):596–598.
13. Karim, M. et al. Disseminated *bipolaris* infection in an asthmatic patient: Case report. *Clin Infect Dis*. 1993;17(2):248–253.
14. Gadallah, M.F. et al. Peritoneal dialysis complicated by *Bipolaris hawaiiensis* peritonitis: Successful therapy with catheter removal and oral itraconazol without the use of amphotericin-B. *Am J Nephrol*. 1995;15(4):348–352.
15. Douer, D. et al. Human *Exserohilum* and *Bipolaris* infections: Report of *Exserohilum* nasal infection in a neutropenic patient with acute leukemia and review of the literature. *J Med Vet Mycol*. 1987;25(4):235–241.
16. Flanagan, K.L. and Bryceson, A.D. Disseminated infection due to *Bipolaris australiensis* in a young immunocompetent man: Case report and review. *Clin Infect Dis*. 1997;25(2):311–313.
17. Pauzner, R. et al. Phaeohyphomycosis following cardiac surgery: Case report and review of serious infection due to *Bipolaris* and *Exserohilum* species. *Clin Infect Dis*. 1997;25(4):921–923.
18. Walsh, T.J. and Groll, A.H. Emerging fungal pathogens: Evolving challenges to immunocompromised patients for the twenty-first century. *Transpl Infect Dis*. 1999;1:247.
19. Khan, J.A. et al. Disseminated *bipolaris* infection in an immunocompetent host: An atypical presentation. *J Pak Med Assoc*. 2000;50(2):68–71.
20. Latham, R.H. *Bipolaris spicifera* meningitis complicating a neurosurgical procedure. *Scand J Infect Dis*. 2000;32(1):102–103.
21. Groll, A.H. and Walsh, T.J. Uncommon opportunistic fungi: New nosocomial threats. *Clin Microbiol Infect*. 2001;7:S8–S24.
22. Moore, M.L. et al. Disseminated *Bipolaris spicifera* in a neonate. *J Perinatol*. 2001;21:399.
23. Saenz, R.E., Brown, W.D., and Sanders, C.V. Allergic bronchopulmonary disease caused by *Bipolaris hawaiiensis* presenting as a necrotizing pneumonia: Case report and review of literature. *Am J Med Sci*. 2001;321:209–212.

24. Fleming, R.V., Walsh, T.J., and Anaissie, E.J. Emerging and less common fungal pathogens. *Infect Dis Clin North Am.* 2002;16:915–933.
25. Tomsiková, A. Causative agents of nosocomial mycoses. *Folia Microbiol (Praha).* 2002;47(2):105–112.
26. Filizzola, M.J., Martinez, F., and Rauf, S.J. Phaeohyphomycosis of the central nervous system in immunocompetent hosts: Report of a case and review of the literature. *Int J Infect Dis.* 2003;7:282–286.
27. Chander, J. et al. Keratomycosis in and around Chandigarh: A five-year study from a north Indian tertiary care hospital. *Indian J Pathol Microbiol.* 2008;51:304–306.
28. Romano, C., Ghilardi, A., and Massai, L. Subungual hyperkeratosis of the big toe due to *Bipolaris hawaiiensis. Acta Derm Venereol.* 2004;84:476.
29. El Khizzi, N., Bakheshwain, S., and Parvez, S., *Bipolaris*: A plant pathogen causing human infections: An emerging problem in Saudi Arabia. *Res J Microbiol*, 2010;5:212.
30. Estes, S.A., Merz, W.G., and Maxwell, L.G. Primary cutaneous phaeohyphomycosis caused by *Drechslera spicifera. Arch Dermatol.* 1977;113:813–815.
31. Chalet, M. et al. Isolation of *Bipolaris australiensis* from a lesion of viral vesicular dermatitis on the scalp. *J Med Vet Mycol.* 1986;24(6):461–465.
32. Straka, B.F., Cooper, P.H., and Body, B.A. Cutaneous *Bipolaris spicifera* infection. *Arch Dermatol.* 1989;125: 1383–1386.
33. Costa, A.R. et al. Subcutaneous phaeohyphomycosis caused by *Bipolaris hawaiiensis*: A case report. *Rev Inst Med Trop Sao Paulo.* 1991;33:74–79.
34. Robb, C.W., Malouf, P.J., and Rapini, R.P. Four cases of dermatomycosis: Superficial cutaneous infection by *Alternaria* or *Bipolaris. Cutis.* 2003;72:313–316, 319.
35. Bilu, D. et al. Cutaneous *Bipolaris* infection in a neutropenic patient with acute lymphoblastic leukemia. *J Cutan Med Surg.* 2004;8:446–449.
36. Anandi, V. et al. Corneal ulcer caused by *Bipolaris hawaiiensis. J Med Vet Mycol.* 1988;26(5):301–306.
37. Hemashettar, B.M. et al. Mycotic keratitis caused by *Bipolaris spicifera. Indian J Pathol Microbiol.* 1992; 35(3):274–277.
38. Jacobson, M. et al. *Bipolaris*-induced orbital cellulitis. *J Clin Neuroophthalmol.* 1992;12:250.
39. Pavan, P.R. and Margo, C.E. Endogenous endophthalmitis caused by *Bipolaris hawaiiensis* in a patient with acquired immunodeficiency syndrome. *Am J Ophthalmol.* 1993;116:644–645.
40. Gopalakrishnan, K. et al. Bilateral *Bipolaris* keratomycosis in a borderline lepromatous patient. *Int J Lepr Other Mycobact Dis.* 2003;71:14–17.
41. Saha, R. and Das, S. *Bipolaris* keratomycosis. *Mycoses.* 2005;48:453–455.
42. Newell, C.K., Steinmetz, R.L., and Brooks, H.L. Jr. Chronic postoperative endophthalmitis caused by *Bipolaris australiensis. Retina.* 2006;26(1):109–110.
43. Durkin, S.R. et al. Successful treatment of phaeohyphomycotic keratitis caused by *Bipolaris australiensis. Clin Experiment Ophthalmol.* 2008;36(7):697–699.
44. Bashir, G., Hussain, W., and Rizvi, A. *Bipolaris hawaiiensis* keratomycosis and endophthalmitis. *Mycopathologia.* 2009;167(1):51–53.
45. Torres, C. et al. Allergic fungal sinusitis: A clinicopathologic study of 16 cases. *Hum Pathol.* 1996;27(8):793–799.
46. Schubert, M.S. Allergic fungal sinusitis: Pathophysiology, diagnosis and management. *Med Mycol.* 2009;47:S324–S330.
47. Koshi, G. et al. Nasal phaeohyphomycosis caused by *Bipolaris hawaiiensis. J Med Vet Mycol.* 1987;25:397–402.
48. Antoine, G.A. and Raterink, M.H. *Bipolaris*: A serious new fungal pathogen of the paranasal sinus. *Otolaryngol Head Neck Surg.* 1989;100:158–162.
49. Rao, A. et al. Phaeohyphomycosis of the nasal sinuses caused by *Bipolaris* species. *Pathology.* 1989;21:280–281.
50. Chang, C.Z., Hwang, S.L., and Howng, S.L. Allergical fungal sinusitis with intracranial abscess—A case report and literature review. *Kaohsiung J Med Sci.* 1997;13(11):685–689.
51. Klapper, S.R. et al. Orbital involvement in allergic fungal sinusitis. *Ophthalmology.* 1997;104(12):2094–2100.
52. Fryen, A. et al. Allergic fungal sinusitis caused by *Bipolaris (Drechslera) hawaiiensis. Eur Arch Otorhinolaryngol.* 1999;256(7):330–334.
53. Castelnuovo, P. et al. Invasive fungal sinusitis due to *Bipolaris hawaiiensis. Mycoses.* 2004;47:76–81.
54. Ambrosetti, D. et al. An expansive paranasal sinus tumour-like lesion caused by *Bipolaris spicifera* in an immunocompetent patient. *Histopathology.* 2006;49:660–662.
55. Toul, P. et al. A pseudo tumoral sinusitis caused by *Bipolaris* sp. *J Infect.* 2006;53:e235–e237.
56. Aribandi, M. and Bazan, C. 3rd. CT and MRI features in *Bipolaris* fungal sinusitis. *Australas Radiol.* 2007;51:127–132.
57. Taguchi, K. et al. Allergic fungal sinusitis caused by *Bipolaris spicifera* and *Schizophyllum commune. Med Mycol.* 2007;45(6):559–564.
58. El-Morsy, S.M. et al. Allergic fungal rhinosinusitis: Detection of fungal DNA in sinus aspirate using polymerase chain reaction. *J Laryngol Otol.* 2010;124:152.
59. Buzina, W. et al. *Bipolaris spicifera* causes fungus balls of the sinuses and triggers polypoid chronic rhinosinusitis in an immunocompetent patient. *J Clin Microbiol.* 2003;41:4885.
60. Dyer, Z.A. et al. Back pain associated with endobronchial mucus impaction due to *Bipolaris australiensis* colonization representing atypical allergic bronchopulmonary mycosis. *Med Mycol.* 2008;46(6):589–594.
61. Kobayashi, H. et al. Disseminated infection by *Bipolaris spicifera* in an immunocompetent subject. *Med Mycol.* 2008;46(4):361–365.
62. Brandt, M.E. and Warnock, D.W. Epidemiology, clinical manifestations, and therapy of infections caused by dematiaceous fungi. *J Chemother.* 2003;15:S36–S47.
63. Revankar, S.G. Phaeohyphomycosis. *Infect Dis Clin North Am.* 2006;20(3):609–620.
64. Pratt, R.G. Enhancement of sporulation in species of *Bipolaris, Curvularia, Drechslera,* and *Exserohilum* by growth on cellulose-containing substrates. *Mycopathologia*, 2006;162:133.
65. Dixon, D.M. and Polak-Wyss, A. The medically important dematiaceous fungi and their identification. *Mycoses.* 1991;34(1–2):1–18.
66. Green, B.J. et al, Surveillance of fungal allergic sensitization using the fluorescent halogen immunoassay. *J Mycol Med.* 2009;19:253–261.
67. Kumar, M., Mishra, N.K., and Shukla, P.K. Sensitive and rapid polymerase chain reaction based diagnosis of mycotic keratitis through single stranded conformation polymorphism. *Am J Ophthalmol.* 2005;140:851–857.
68. Shin, E.J. et al. Screening of middle ear effusion for the common sinus pathogen *Bipolaris. Eur Arch Otorhinolaryngol.* 2003;260:78–80.

69. Bagyalakshmi, R. et al. Newer emerging pathogens of ocular non-sporulating molds (NSM) identified by polymerase chain reaction (PCR)-based DNA sequencing technique targeting internal transcribed spacer (ITS) region. *Curr Eye Res.* 2008;33:139–147.
70. Pounder, J.I. et al. Discovering potential pathogens among fungi identified as nonsporulating molds. *J Clin Microbiol.* 2007;45:568–571.
71. Zeng, J.S. et al. Spectrum of clinically relevant *Exophiala* species in the United States. *J Clin Microbiol.* 2007;45:3713–3720.
72. White, T.J. et al. Amplification and direct sequencing of fungal ribosomal RNA genes for phylogenetics. In M.S. Innis and D.H. Gelfand (Eds.), *PCR Protocols: A Guide to Methods and Applications.* Academic Press, New York, pp. 315–322. 1990.
73. Leaw, S.N. et al. Identification of medically important yeast species by sequence analysis of the internal transcribed spacer regions. *J Clin Microbiol.* 2006;44:693–699.
74. Rosow, L. et al. Cerebral phaeohyphomycosis caused by *Bipolaris spicifera* after heart transplantation. *Transpl. Infect. Dis.* February 15, 2011: [Epub ahead of print]

5 *Botryomyces*

Dongyou Liu and R.R.M. Paterson

CONTENTS

5.1 INTRODUCTION

'Dematiaceous fungi' is a colloquial, nontaxonomic term used to describe a heterogeneous group of fungal organisms that have melanized cell walls. The presence of melanins in these fungi not only accounts for their dark-green, brown, or black color, but also enhances their survival under stressful conditions. Further morphological, physiological, and biological adaption has endowed these organisms with the ability to tolerate desiccation, temperature, and osmolarity changes. Because of their distinct morphological and biological features, dematiaceous fungi are separated into black yeasts, meristematic fungi, microcolonial fungi, and other darkly pigmented fungi, with each of these categories encompassing multiple fungal taxa of its own.

Black yeasts are characterized by the production of daughter cells through yeast-like multilateral or polar budding. Many black yeasts also demonstrate mycelial growth and generate conidia (either unseptated or containing three transversal septa) from phialides (ranging from simple phialides, phialides with collarettes, annelated phialides, to rhachides) or undifferentiated conidiogenous cells.

Meristematic fungi form aggregates of thick-walled, densely melanized cells and grow by isodiametric cell wall expansion and division. Propagules are released by breaking apart of aggregates or by endogenous conidiogenesis that subsequently disrupts the mother cell wall. Some may form blastic conidia from yeast-like budding cells and thus may be regarded as black yeasts. Meristematic fungi are found on the exposed surfaces of desert rock, outdoor statues, leathery plant leaves, and Antarctic rock, as well as in hypersaline coastal ponds. They have the potential to erode and destroy marble, sandstone, and glass.

Microcolonial fungi refer to the in situ growth pattern of the meristematic fungi and some black yeasts on mineral substrates (e.g., rock, glass, or metal), which is characterized by the formation of black cauliflower-like colonies consisting of densely aggregated thick-walled cells. Although colonies of microcolonial fungi belonging to the different taxa are indistinguishable morphologically, upon isolation on suitable growth media, microcolonial fungi readily develop into various morphologies, allowing genus- and species-specific identification.

5.1.1 Classification and Morphology

The genus *Botryomyces* (Botrys, Greek for bunch of grapes, + mykes, Greek for fungus) is a meristematic black fungus belonging to the mitosporic Dothideomycetes group, class Dothideomycetes, subphylum Pezizomycotina, phylum *Ascomycota*, and kingdom Fungi [1]. The mitosporic Dothideomycetes group is divided into nine genera: *Asteromella*, *Botryomyces*, *Cenococcum*, *Cryomyces*, *Cyclothyrium*, *Cystocoleus*, *Racodium*, *Sclerostagonospora*, and *Seifertia*. The genus *Botryomyces* contains a single species *Botryomyces caespitosus*, which was first described from human skin lesions [2]; *Botryomyces angioformans* is a doubtful species from an unconfirmed disease of which no material has been preserved [3].

B. caespitosus colonies are pink when young and form restricted, meristematic cell clumps that disarticulate irregularly. Meristematic growth is characterized by the production of swollen isodiametric cells with thick cell walls, in which melanin is deposited. The fungus produces multicelled, irregularly septate, thick-walled spores, which may be regarded by some to be vegetative structures [2].

The genus *Botryomyces* is one of the 25 genera of meristematic black fungi that inhibit cracks in marble and rock surfaces in the Mediterranean Basin (Italy, Spain) and Ukraine. They are associated with biodeterioration of monuments, sculptures, and archaeological objects. Because meristematic black fungi present characteristics comparable to those of fungi isolated from deserts rocks, they are occasionally referred to as "microcolonial fungi." In addition, the term "black yeast" is sometimes used to describe black fungi that have yeast-like stages of reproduction and a meristematic growth pattern.

Meristematic black fungi are classified under four families within the *Ascomycota*: (i) *Herpotrichellaceae* (order Chaetothyriales) covers *Exophiala* and *Sarcynomices petricola*; (ii) *Dothideaceae* (order Dothideales) consists of some epiphytic species occasionally isolated from rocks such as *Trimmatostroma abietis*, *Aureobasidium pullulans*, and *Hortaea werneckii*; (iii) *Capnodiaceae* (order Capnodiales) includes *C. renispora*, which was isolated from a tile; and (iv) *Pleosporaceae* (order Pleosporales) includes *B. caespitosus*, which is closely related to *Alternaria* (which is also frequently found on stones) based on internal transcribed spacer (ITS) sequence analysis [4,5].

Stone-inhabiting meristematic black fungi tend to show intercrystalline growth by colonizing the weakest parts along marble crystals, leading to the detachments of crystals. They also grow preferentially in cavities and in already-formed cracks and fissures, often producing a deepening of the fissures.

5.1.2 Clinical Features

Meristematic black fungi such as *B. caespitosus* are recognized agents of phaeohyphomycosis [6] as distinct from chromoblastomycosis. Readers should be aware of the term "botryomycosis," which, surprisingly, refers to a bacterial infection (affecting the skin, and sometimes the viscera due to *Staphylococcus aureus* and several other bacteria) [7] and should not be confused with the current disease. Phaeohyphomycosis is cosmopolitan although patients are usually adults and approximately half are immunocompromised; however, this figure has probably increased substantially until the present time. Lesions may occur on almost any part of the body, often on exposed areas, with the upper arm lesions being most prevalent. The most typical and common lesions are cutaneous or subcutaneous abscesses or cysts. Primary lesion is a single, discrete, asymptomatic small nodule and this evolves gradually to an encapsulated, fluctuant abscess with a liquefied center. However, the overlying epidermis is hardly affected. Occasionally, a slightly elevated, granulomatous plaque appears when the main site of the lesion is in the dermis and epidermis. Infrequently, it is observed as a small verrucous nodule or a verrucous plaque comprising a coalescent nodule, which actually resembles chromoblastomycosis. Phaeohyphomycosis may involve the central nervous system or other internal organs (e.g., liver, lungs, and pancreas) and may appear as a hematogenous metastasis from cutaneous or subcutaneous infections or with no visible lesions.

Specifically, *B. caespitosus* may gain entry into human hosts by traumatic inoculation, through prolonged contact with domestic animals, and presumably via airborne dissemination of propagules. The fungus is responsible occasionally for chromoblastomycosis-like subcutaneous infections after trauma. Clinical symptoms may range from dermatomycosis (mycoses), cutaneous phaeohyphomycosis to mycotic granuloma. Skin lesions appear on arms and legs, usually in immuno compromised patients or in patients with chronic renal failure, transplants, and immunosuppressive therapy [8,9]. Human infection of tonsils with *B. caespitosus* may exhibit recurrent tonsillitis, sore throat, dysphasia, high temperature, and enlarged tonsils. Upon examination, the tonsils may show "grains" in the crypts [10].

5.1.3 Diagnosis

As meristematic fungi lack pronounced diagnostic features, species-specific identification on the basis of microscopic morphology and reproductive structures (e.g., conidiophores, conidia, and conidial ontogeny) is often difficult. This is further exacerbated by the fact that many meristematic fungal species are highly pleomorphic, with anamorph life cycles and widely divergent methods of propagation. Some species display meristematic growth as the only type of reproduction, consisting of isodiametrically dividing cells and endoconidiation, which do not allow delimitation of taxa [11]. Thus, morphology gives only a presumptive identification at genus level, and the use of physiological characteristics (e.g., nitrogen and carbohydrate assimilation tests, growth at different temperatures and proteolytic activity) are helpful for their identification.

B. caespitosus may be detected from surfaces by tape lifts or tease mounts from bulk samples. The laboratory diagnosis of chromoblastomycosis is performed by the demonstration of sclerotic bodies upon direct microscopical examination of wet KOH mounts of aspirated pus, skin scrapings, or biopsy material. However, *B. caespitosus* causes occasionally a chromoblastomycosis-like infection; it is in fact a phaeohyphomycosis.

Colonies on malt extract agar are restricted, cauliflower-like, heaped, pale brown initially becoming brown-black with age. Microscopy reveals that hyphae and budding cells are absent. The thallus is composed of clumps of irregularly septate, thick-walled cells, which are subhyaline, becoming dark brown with age. These disarticulate into smaller cell packets. Blastic conidia are in fact occasionally present. A series of approximately 50 physiological tests (e.g., growth on glucose, arabinose, salicin) are also available [3]. Hence, some useful characters are present in the case of this species at least. *B. caespitosus* differs from *Sarcinomyces phaeomuriformis* by young colonies being pink.

Molecular techniques have been applied for the identification of meristematic black fungi including *B. caespitosus*. Polymerase chain reaction (PCR), restriction fragment length polymorphism (RFLP) and analyses of amplified small (SSU, 18S rRNA) and large (LSU, 5.8S rRNA and internal transcribed spacers ITS1, ITS2) subunit ribosomal genes are employed [4,11–14].

5.2 METHODS

5.2.1 Sample Preparation

Molds from a specimen are grown on inhibitory mold agar, modified Sabouraud agars, or potato dextrose agar. Microscopic structures are observed on tease or tape preparations and slide cultures for up to 21 days.

After growth for 1–7 days on potato dextrose agar slants, lysates are prepared from approximately 1 cm^2 of mycelia with IDI lysis kits (GeneOhm Sciences, San Diego, CA). Briefly, in a biological safety cabinet, mycelia are collected by scraping the slant with a sterile stick in 1 mL of sterile, molecular-grade H_2O. The material is transferred to a 2 mL screw-cap tube. The tubes are centrifuged for 1 min at 6000 × *g*. If the mycelia does not pellet, the material is contained using a pediatric blood serum filter (Porex Corp., Fairburn, GA). The supernatant is removed. The material is resuspended in 200 µL of IDI sample buffer and transferred to the lysis tube, which contains glass beads. Lysis tubes are vortexed on the highest setting for 5 min. The tubes are placed in a boiling water bath for 15 min and centrifuged for 5 min at 16,000 × *g*. The supernatant is stored at −20°C until amplification [13].

5.2.2 Detection Procedures

Pounder et al. [13] described a real-time PCR with SYBR green DNA-binding dye and amplicon melting temperature analysis for fungal detection using pan-fungal primers ITS1 forward (5′-TCCGTAGGTGAACCTGCGG-3′) and ITS4 reverse (5′-TCCTCCGCTTATTGATATGC-3′) that cover the internal transcribed spacer 1 (ITS1)–5.8S–ITS2 rRNA gene cluster. The identities of the fungi are verified by subsequent sequencing analyses.

Procedure

1. PCR mixture is composed of 1 × Lightcycler FastStart DNA Master Hybridization Probes mixture (Roche Applied Science) (containing deoxynucleoside triphosphates, FastStart *Taq* DNA polymerase, and 1 mM $MgCl_2$, additional $MgCl_2$ is added to a final concentration of 4.6 mM), 0.4 µM each of ITS1 forward and ITS4 reverse primers, 1 × SYBR green (Molecular Probes), and 3 µL template DNA.
2. Thermal cycling parameters include 95°C for 10 min; 50 cycles of 95°C for 5 s, 60°C for 20 s, and 76°C for 30 s; and a final extension at 72°C for 2 min.
3. The quality of the amplicon is determined using the derivative of the melt analysis curve (55°C–99°C, 45 s hold at 55°C, 5 s/°C) using the RotorGene 3000 (Corbett Robotics, Inc).
4. The amplified product is purified for bidirectional sequencing using ExoSAP-IT (USB Corp). Five µL of Big Dye Terminator Ready Reaction Mix v. 1.1 (Applied Biosystems) is added to 4 µL of each primer (0.8 pmol/µL) and 3 µL of purified PCR product. Cycle sequencing is performed with a 9700 thermal cycler (ABI), using 25 cycles of 96°C for 10 s, 50°C for 5 s, and 60°C for 4 min. Sequencing reaction products are passed through a Sephadex G-50 fine column to remove unincorporated dye terminators. Purified sequencing reaction products are run on an ABI Prism 3100 Genetic Analyzer with a 50 cm capillary array.
5. Sequences are analyzed with the SmartGene Integrated Database Network software version 3.2.3 vr. SmartGene is a web-based software and database system with reference sequences derived from the National Center for Biological Information (NCBI) GenBank repository.

Note. In case that real time PCR instrument is unavailable, standard PCR may be performed with primers ITS1 and ITS4, and the resulting amplicon is sequenced with the same primers. Sequence-based identifications are defined by percent identity: species, ≥99%; genus, 93%–99%; and inconclusive, ≤93%. For strains producing discrepant identification between the methods based on phenotypic characteristics and ITS sequence analysis, the D1–D2 region of the large-subunit RNA gene is amplified with primers NL1 (5′-GCATATCAATAAGCGGAGGAAAAG-3′) and NL4 (5′-GGTCCGTGTTTCAAGACGG-3′) and sequenced for species clarification [15].

5.3 CONCLUSIONS

B. caespitosus is a black fungus in the mitosporic Dothideomycetes group that is responsible for a phaeohyphomycosis in humans. The organism is also associated with biodeterioration of monuments. Given the close morphological similarity among black fungi, use of molecular techniques such as PCR and sequencing is critical for accurate and specific identification of the fungus.

REFERENCES

1. The UniProt Consortium. Available at http://www.uniprot.org/, accessed on January 27, 2011.
2. de Hoog, G.S. and Rubio, C., A new dematiaceous fungus from human skin. *Med. Mycol.*, 20, 15, 1982.
3. de Hoog, G.S. et al., *Atlas of Clinical Fungi*, 2nd edn., 1126pp, Centraalbureau voor Schimmelculture, Utrecht, the Netherlands, 2000.
4. Sterflinger, K., de Hoog, G.S., and Haase, G., Phylogeny and ecology of meristematic ascomycetes. *Stud. Mycol.*, 43, 5, 1999.
5. de Hoog, G.S. and Horré, R., Molecular taxonomy of the *Alternaria* and *Ulocladium* species from humans and their identification in the routine laboratory. *Mycoses*, 45, 259, 2002.
6. Matsumoto, T. and Ajello, L., Agents of phaeohyphomycosis. In: *Topley & Wilson's Microbiology and Microbial Infections*, 9th edn., *Medical Mycology* (Eds.: Ajello, L. and Hay, R.J.), pp. 503–524, Chap. 27, Arnold, London, U.K., 1998.
7. Kwon-Chung, K.J. and Bennett J.E., *Medical Mycology*, 866pp., Lea & Febiger, Philadelphia, PA, 1992.

8. Carapeto, F.J. et al., Dermatomycosis in plaques caused by *Botryomyces caespitosus*. A new causative agent. *Med. Cutan. Ibero. Lat. Am.*, 13, 71, 1985.
9. Benoldi, D. et al., *Botryomyces caespitosus* as an agent of cutaneous phaeohyphomycosis. *J. Med. Vet. Mycol.*, 29, 9, 1991.
10. Martins, R.H. et al., Actinomycosis and botryomycosis of the tonsil. *Auris Nasus Larynx.*, 18, 377, 1991.
11. de Hoog, S. et al., Relationships of dothideaceous black yeasts and meristematic fungi based on 5.8S and ITS2 rDNA sequence comparison. *Stud. Mycol.*, 43, 31, 1999.
12. White, T.J. et al., Amplification and direct sequencing of fungal ribosomal RNA genes for phylogenetics. In: *PCR Protocols: A Guide to Methods and Applications* (Eds.: Innis, M.A. et al.), pp. 315–322, Academic Press, San Diego, CA, 1990.
13. Pounder, J.I. et al., Discovering potential pathogens among fungi identified as nonsporulating molds. *J. Clin. Microbiol.*, 45, 568, 2007.
14. Bagyalakshmi, R. et al., Newer emerging pathogens of ocular non-sporulating molds (NSM) identified by polymerase chain reaction (PCR)-based DNA sequencing technique targeting internal transcribed spacer (ITS) region. *Curr. Eye Res.*, 33, 139, 2008.
15. Leaw, S.N. et al., Identification of medically important yeast species by sequence analysis of the internal transcribed spacer regions. *J. Clin. Microbiol.*, 44, 693, 2006.

6 *Botryosphaeria* and *Lasiodiplodia*

Dongyou Liu

CONTENTS

6.1 INTRODUCTION

The genus *Botryosphaeria* consists of a large number of cosmopolitan, plant-infecting fungi that are commonly associated with dieback and cankers of woody hosts in tropical and subtropical regions. Of these, *Botryosphaeria rhodina* and its anamorph *Lasiodiplodia theobromae* are occasionally involved in human phaeohyphomycosis, producing clinical diseases of varying severity (e.g., subcutaneous abscess, keratitis, pneumonia, and death).

6.1.1 CLASSIFICATION AND MORPHOLOGY

The genus *Botryosphaeria* belongs to the family *Botryosphaeriaceae*, order Botryosphaeriales, class Dothideomycetes, subphylum Pezizomycotina, phylum Ascomycota, and kingdom Fungi. The family *Botryosphaeriaceae* covers 11 recognized genera: *Barriopsis*, *Botryosphaeria*, *Guignardia*, *Melanops*, mitosporic *Botryosphaeriaceae*, *Neodeightonia*, *Otthia*, *Phaeobotryon*, *Phaeobotryosphaeria*, *Saccharata*, and *Spencermartinsia* as well as some unclassified *Botryosphaeriaceae* [1].

The genus *Botryosphaeria* consists of 16 recognized and 58 unassigned species, of which *Botryosphaeria rhodina* (synonym *Botryodiplodia theobromae*) is a teleomorph of *Lasiodiplodia theobromae* [1]. *Botryosphaeria rhodina*/*Lasiodiplodia theobromae* is a common endophyte and opportunistic pathogen affecting >500 tree species in the tropics and subtropics, leading to stem-end rot and dieback as well as rot in fruits in most species it infects [21].

Botryosphaeria spp. tend to form uni- to multilocular ascomata with multi-layered walls, occurring singly or in clusters, and intermixed with pycnidial conidiomata. Asci are bitunicate, with a thick endotunica, stalked or sessile, clavate, with a well-developed apical chamber, forming in a basal hymenial layer, intermixed among hyaline pseudoparaphyses that are constricted at the septa. Ascospores are hyaline, aseptate, fusoid to ellipsoid or ovoid, bi- to triseriate, without a mucoid sheath or appendages; ascospores turn brown and become septate and slightly verruculose upon germination.

The genus *Lasiodiplodia* is classified in the mitosporic *Botryosphaeriaceae* group, family *Botryosphaeriaceae*. The mitosporic *Botryosphaeriaceae* group contains 17 species: *Aplosporella*, *Chaetoconis*, *Diplodia*, *Dothiorella*, *Endomelanconiopsis*, *Fusicoccum*, *Lasiodiplodia*, *Macrophomina*, *Microdiplodia*, *Neofusicoccum*, *Neoscytalidium*, *Phyllosticta*, *Pseudofusicoccum*, *Sphaeropsis*, *Stenocarpella*, *Taeniolella*, and *Tiarosporella*. In turn, the genus *Lasiodiplodia* consists of nine recognized (*Lasiodiplodia crassispora*, *Lasiodiplodia gonubiensis*, *Lasiodiplodia margaritacea*, *Lasiodiplodia parva*, *Lasiodiplodia plurivora*, *Lasiodiplodia pseudotheobromae*, *Lasiodiplodia rubropurpurea*, *Lasiodiplodia theobromae*, and *Lasiodiplodia venezuelensis*) and eight unassigned species, of which anamorph *Lasiodiplodia theobromae* (obsolete synonyms *Botryodiplodia tubericola*, *Diplodia theobromae*, *Diplodia tubericola*, and *Lasiodiplodia tubericola*) has been associated with mycotic keratitis, lesions on nail and subcutaneous tissue in humans [2,3].

Lasiodiplodia theobromae colonies are grayish sepia to mouse grey to black, fluffy with abundant aerial mycelium, and fuscous black to black on reverse. Pycnidia (up to 5 mm in width) are simple or compound, often aggregated, stromatic, ostiolate, frequently setos. Conidiophores are hyaline, simple, sometimes septate, rarely branched cylindrical, arising from the inner layers of cells lining the pycnidial cavity. Conidiogenous cells are hyaline, simple, cylindrical to subobpyriform, holoblastic, annellidic. Conidia are initially unicellular, hyaline, granulose, subovoid to ellipsoid oblong, thick walled, base truncate; mature conidia (of 20–30 × 10–15 μm) are two celled (with one septate), cinnamon to fawn or dark brown, often longitudinally striated. Paraphyses when present are hyaline, cylindrical, sometimes septate, up to 50 μm long [4].

6.1.2 Clinical Features

Botryosphaeria rhodina/Lasiodiplodia theobromae is a common plant pathogen in tropical Asia (India, Cambodia, Hong Kong, Japan, and the Philippines), Australia, Central and South America (Columbia, Guyana, Jamaica, and United States), where *Lasiodiplodia theobromae* has been shown to infect the eyes, skin, and soft tissues of humans, with clinical manifestations ranging from keratitis, corneal ulcers and abscess, onychomycosis, subcutaneous abscess and phaeohyphomycosis, endophthalmitis, sinusitis, pneumonia to death [5–18].

Summerbell et al. [15] documented a case of subcutaneous phaeohyphomycosis caused by *Lasiodiplodia theobromae* in a 50-year-old Canadian woman who stumbled on an outdoor wooden staircase and sustained an injury to the right leg while visiting Jamaica. Five weeks later, the patient presented with an ulcer at the injury site. An excisional biopsy revealed broad, septate, melanized fungal filaments penetrating into the tissue and its culture generated a nonsporulating melanized mycelium. After 16 weeks cultivation on modified Leonian's agar at 25°C, the fungus developed pycnidia characteristic of *Lasiodiplodia theobromae*. The ulcer was excised during the biopsy procedure and ultimately resolved after treatment with saline compresses.

Woo et al. [19] described a case of *Lasiodiplodia theobromae* pneumonia in a 45-year-old patient with hepatitis B virus-related hepatocellular carcinoma. Analysis of a direct KOH smear of the bronchoalveolar lavage (BAL) fluid revealed septate hyphae. Culture of BAL fluid specimens on Sabouraud dextrose agar (SDA) yielded a dematiaceous mold, which grew as cottony colonies that became dark gray within 7 days at 37°C. Microscopic examination of the mold after lactophenol cotton blue staining revealed septate brown hyphae of 6 μm in width. The mold was also grown on oatmeal agar in sunlight to stimulate sporulation. Colonies produced hairy, dark brown structures (pycnidia). Conidia released from the pycnidia were hyaline and nonseptate when young but were septate and brown, with longitudinal striations, when mature. Pycnidia were sometimes produced in aggregates as shown in periodic acid-Schiff-stained histological sections of the mold on oatmeal agar. Examination under higher magnification demonstrated a pycnidium with an obvious neck and conidiogenous cells and paraphyses (sterile filaments among conidia) lining the internal wall. The fungus was identified as *Lasiodiplodia theobromae* by microscopic morphology and elongation factor 1α (EF1α) gene sequencing. The patient died 14 days after cadaveric liver transplantation. Because the conidia of this fungus are borne inside pycnidia and are released in a slimy mass, it was speculated that the patient may have acquired the infection through consumption of longan (*Dimocarpus longan*), a very popular fruit in southern China during summertime, which is often infected with *L. theobromae*. Alternatively, the fungus may have been acquired from the cadaveric transplanted liver and may have invaded the lung via the bloodstream.

6.1.3 Laboratory Diagnosis

Traditionally, laboratory identification of molds is based on detection of fungal elements in clinical specimens and further examination of macroscopic and microscopic features of cultured isolates [20]. To augment sporulation specifically for identification, mold may be grown on oatmeal agar in sunlight. Despite being time consuming, phenotypic identification of molds and other fungi requires expertise for the recognition of characteristic microscopic features of various fungi.

More recently, molecular techniques such as polymerase chain reaction (PCR) and sequencing analysis of the internal transcribed spacer (ITS) regions have been employed for more precise and objective determination of fungi [21–24]. However, for accurate identification of *L. theobromae* to the species level, partial EF1α sequencing appears to be more reliable than ITS sequencing [19]. The ITS sequence of the patient isolate differed by six bases from that of another strain of *L. theobromae* but differed by only eight bases from that of a strain of *L. gonubiensis* [19]. In addition, the *Lasiodiplodia* spp. are clearly distinguishable from *Diplodia* spp. and *Dothiorella* spp. based on morphology and ITS and EF1α nucleotide sequences [2,3].

6.2 METHODS

6.2.1 Sample Preparation

Molds from a specimen are grown on inhibitory mold agar, modified Sabouraud agars, or potato dextrose agar. Microscopic structures are observed on tease or tape preparations and slide cultures for up to 21 days.

After growth for 1–7 days on potato dextrose agar slants, lysates are prepared from approximately 1 cm^2 of mycelia with IDI lysis kits (GeneOhm Sciences, San Diego, CA). Briefly, in a biological safety cabinet, mycelia are collected by scraping the slant with a sterile stick in 1 mL of sterile, molecular-grade H_2O. The material is transferred to a 2 mL screw-cap tube. The tubes are centrifuged for 1 min at 6000 × *g*. If the mycelia does not pellet, the material is contained with a pediatric blood serum filter (Porex Corp., Fairburn, GA). The supernatant is removed. The material is resuspended in 200 μL of IDI sample buffer and transferred to the lysis tube, which contains glass beads. Lysis tubes are vortexed on the highest setting for 5 min. The tubes are placed in a boiling water bath for 15 min and centrifuged for 5 min at 16,000 × *g*. The supernatant is stored at −20°C until amplification [22].

6.2.2 Detection Procedures

6.2.2.1 ITS1–5.8S–ITS2 rRNA Gene Cluster Sequencing Analysis

Pounder et al. [22] described a real-time PCR with SYBR green DNA-binding dye and amplicon melting temperature analysis for fungal detection using pan-fungal primers ITS1 forward (5′-TCCGTAGGTGAACCTGCGG-3′) and ITS4 reverse (5′-TCCTCCGCTTATTGATATGC-3′) [24]. These primers generate a 500 bp fragment from the internal transcribed spacer 1 (ITS1)–5.8S–ITS2 rRNA gene cluster (ITS). Subsequent sequencing analysis of the amplicon helps verify the identity of the fungal isolates.

Procedure

1. PCR mixture is composed of 1×Lightcycler FastStart DNA Master Hybridization Probes mixture (Roche Applied Science) (containing deoxynucleoside triphosphates, FastStart *Taq* DNA polymerase, and 1 mM $MgCl_2$, additional $MgCl_2$ to a final concentration of 4.6 mM), 0.4 μM each of ITS1 forward and ITS4 reverse primers, 1×SYBR green (Molecular Probes), and 3 μL template DNA.
2. Thermal cycling parameters include 95°C for 10 min; 50 cycles of 95°C for 5 s, 60°C for 20 s, and 76°C for 30 s; and a final extension at 72°C for 2 min.
3. The quality of the amplicon is determined using the derivative of the melt analysis curve (55°C–99°C, 45 s hold at 55°C, 5 s/°C) using the RotorGene 3000 (Corbett Robotics, Inc).
4. The amplified product is purified for bidirectional sequencing using ExoSAP-IT (USB Corp). Five microliters of Big Dye Terminator Ready Reaction Mix v. 1.1 (Applied Biosystems) is added to 4 μL of each primer (0.8 pmol/μL) and 3 μL of purified PCR product. Cycle sequencing is performed with a 9700 thermal cycler (ABI), using 25 cycles of 96°C for 10 s, 50°C for 5 s, and 60°C for 4 min. Sequencing reaction products are passed through a Sephadex G-50 fine column to remove unincorporated dye terminators. Purified sequencing reaction products are run on an ABI Prism 3100 Genetic Analyzer with a 50 cm capillary array.
5. Sequences are analyzed with the SmartGene Integrated Database Network software version 3.2.3 vr. SmartGene is a web-based software and database system with reference sequences derived from the National Center for Biological Information (NCBI) GenBank repository.

Note: In case that real time PCR instrument is not available, standard PCR may be performed with primers ITS1 and ITS4, and the resulting amplicon is sequences with the same primers.

6.2.2.2 Partial EF1α Gene Sequencing Analysis

Woo et al. [19] utilized primers EF1–728F (5′-CATCGAGAAGTTCGAGAAGG-3′) and EF1–986R (5′-TACTTGAAGGAACCCTTACC-3′) for PCR amplification and DNA sequencing of a 289 bp fragment of the EF1α gene from *Lasiodiplodia theobromae* isolate. The sequence was compared with those of related species listed in the GenBank database by using ClustalX 1.83. Phylogenetic relationships were determined using the neighbor-joining method.

6.3 CONCLUSION

Botryosphaeria rhodina/Lasiodiplodia theobromae is a ubiquitous plant pathogen in tropical regions that may infect the eyes, skin, and soft tissues of humans, leading to keratitis, corneal ulcers and abscess, onychomycosis, subcutaneous abscess and phaeohyphomycosis, endophthalmitis, pneumonia, and death. As phenotypic techniques for mold identification are time consuming and variable, molecular methods are increasingly utilized. PCR and sequencing analysis of the ITS regions have proven valuable for determination of a variety of molds including *Lasiodiplodia theobromae*. Furthermore, examination of the EF1α gene sequence offers an additional tool for differentiation among *Lasiodiplodia* spp.

REFERENCES

1. The UniProt Consortium. Available at http://www.uniprot.org/, accessed on August 1, 2010.
2. Pavlic, D. et al., *Lasiodiplodia gonubiensis* sp. nov., a new *Botryosphaeria* anamorph from native *Syzygium cordatum* in South Africa. *Stud. Mycol.*, 50, 313, 2004.
3. Burgess, T.I. et al., Three new *Lasiodiplodia* spp. from the tropics, recognized based on DNA sequence comparisons and morphology. *Mycologia*, 98, 423, 2006.
4. Mycology online, School of Molecular & Biomedical Science, The University of Adelaide, Australia. Available at http://www.mycology.adelaide.edu.au/, accessed on August 1, 2010.
5. Laverde, S. et al., Mycotic keratitis: 5 cases caused by unusual fungi. *Sabouraudia*, 11, 119, 1973.
6. Valenton, M.J., Rinaldi, M.G., and Butler, E.E., A corneal abscess due to the fungus *Botryodiplodia theobromae*. *Can. J. Ophthalmol.*, 10, 416, 1975.
7. Rebell, G. and Forster, R.K., *Lasiodiplodia theobromae* as a cause of keratomycoses. *Sabouraudia*, 14, 155, 1976.
8. Restrepo, A. et al., The isolation of *Botryodiplodia theobromae* from a nail lesion. *Sabouraudia*, 1, 41, 1976.
9. Slomovic, A.R., Forster, R.K., and Gelender, H., *Lasiodiplodia theobromae* panophthalmitis. *Can. J. Ophthalmol.*, 20, 225, 1985.
10. Vélez, H. and Díaz, F., Onychomycosis due to saprophytic fungi. Report of 25 cases. *Mycopathologia*, 91, 87, 1985.
11. Marasas, W.F. and Van Rensburg, S.J., Mycotoxicological investigations on maize and groundnuts from the endemic area of Mseleni joint disease in Kwazulu. *S. Afr. Med. J.*, 69, 369, 1986.
12. Maslen, M.M., Collis, T., and Stuart, R., *Lasiodiplodia theobromae* isolated from a subcutaneous abscess in a Cambodian immigrant to Australia. *J. Med. Vet. Mycol.*, 34, 279, 1996.

13. Borderie, V.M. et al., Endophthalmitis after *Lasiodiplodia theobromae* corneal abscess. *Graefe's Arch. Clin. Exp. Opthalmol.*, 235, 259, 1997.
14. Thomas, P.A., Current perspectives on ophthalmic mycoses. *Clin. Microbiol. Rev.*, 16, 730, 2003.
15. Summerbell, R.C. et al., Subcutaneous phaeohyphomycosis caused by *Lasiodiplodia theobromae* and successfully treated surgically. *Med. Mycol.*, 42, 543, 2004.
16. Bhartiya, P. et al., Fungal keratitis in Melbourne. *Clin. Exp. Ophthalmol.*, 35, 124, 2007.
17. Thew, M.R. and Todd, B., Fungal keratitis in far north Queensland, Australia. *Clin. Exp. Ophthalmol.*, 36, 721, 2008.
18. Kindo, A.J. et al., Maxillary sinusitis caused by *Lasiodiplodia theobromae. Indian J. Med. Microbiol.*, 28, 167, 2010.
19. Woo, P.C.Y. et al., *Lasiodiplodia theobromae* pneumonia in a liver transplant recipient. *J. Clin. Microbiol.*, 46, 380, 2008.
20. Thomas, O.A. et al., Use of lactophenol cotton blue mounts of corneal scraping as an aid to the diagnosis of mycotic keratitis. *Diagn. Microbiol. Infect. Dis.*, 14, 219, 1991.
21. Denman, S. et al., Circumscription of *Botryosphaeria* species associated with Proteaceae based on morphology and DNA sequence data. *Mycologia*, 95, 294, 2003.
22. Pounder, J.I. et al., Discovering potential pathogens among fungi identified as nonsporulating molds. *J. Clin. Microbiol.*, 45, 568, 2007.
23. Bagyalakshmi, R. et al., Newer emerging pathogens of ocular non-sporulating molds (NSM) identified by polymerase chain reaction (PCR)-based DNA sequencing technique targeting internal transcribed spacer (ITS) region. *Curr. Eye Res.*, 33, 139, 2008.
24. White, T.J. et al., Amplification and direct sequencing of fungal ribosomal RNA genes for phylogenetics, pp. 315–322. In Innis, M.A. et al. (Eds.), *PCR Protocols: A Guide to Methods and Applications*. Academic Press, San Diego, CA, 1990.

7 Corynespora

Dongyou Liu and Po-Ren Hsueh

CONTENTS

7.1 INTRODUCTION

Dematiaceous fungi are characterized by their production of dark-pigmented hyphae or yeast-like cells. Being widely distributed in the soil, plants and other environments, these dark-walled fungi have the ability to cause opportunistic infections in humans and animals, affecting the skin, subcutaneous tissues and internal organs. Among the 130 species of 70 genera of dematiaceous fungi that have been implicated in human infections, *Corynespora cassiicola* is a common pathogen of grasses and other herbaceous plants in the tropics, and has been confirmed as an etiologic agent of subcutaneous phaeohyphomycosis in humans.

7.1.1 Classification, Morphology, and Biology

The genus *Corynespora* is a mold belonging to the mitosporic Pleosporales group, order Pleosporales, subclass Pleosporomycetidae, class Dothideomycetes, subphylum Pezizomycotina, phylum Ascomycota, kingdom Fungi [1].

The mitosporic Pleosporales group encompasses 16 genera: *Ascochyta, Camarosporium, Chaetodiplodia, Chaetophoma, Chalastospora, Cheirosporium, Coniothyrium, Corynespora, Monascostroma, Ochrocladosporium, Phialophorophoma, Plenodomus, Pleurophoma, Pseudodiplodia, Pseudorobillarda,* and *Zeloasperisporium*. In turn, the genus *Corynespora* is separated into seven recognized species: *Corynespora cassiicola, Corynespora citricola, Corynespora olivacea, Corynespora melongenae, Corynespora proliferata, Corynespora sesamum,* and *Corynespora smithii*, in addition to an unassigned species *Corynespora* sp. SXZ-01 [1].

Within the *Corynespora* genus, *Corynespora cassiicola* is a well-known plant pathogen causing leaf spot disease in more than 70 host plant species including cowpea, cucumber, papaya, pepper, rubber, soybean, and tomato, primarily found in the tropics and subtropics [2–4]. Rubber trees infected with *C. cassiicola* develop necrotic lesions on leaves, as well as chlorosis and darkening of the veins, leading to defoliation and loss of crop [5]. *C. cassiicola* has also been associated occasionally with human infections [6,7] (Figure 7.1).

On potato dextrose agar (PDA), *C. cassiicola* colonies are effuse, gray or brown in the front and gray or black on reverse. Conidiophores (measuring 4–11 µm by 110–850 µm) are pale to light brown, 3–10 septate, cylindrical, straight or curved, unbranched, and smooth. Conidia (of 5–10 µm by 27–192 µm in size) develop at the conidiophore apex and appear smoky to olivaceous, smooth, obclavate to cylindrical, subhyaline to pale olivaceous brown, 4–20 pseudoseptate, singly or in acropetal chains, with a rounded apex and truncated base [8,9].

Phylogenetic analysis in the ribosomal DNA internal transcribed spacer (ITS) regions of ribosomal RNA (rRNA) gene, random hypervariable loci (caa5 and ga4), and the actin-encoding locus act1 identified six phylogenetic lineages in *C. cassiicola*, corresponding to host of origin, pathogenicity, and growth rate but not to geographic location [10]. In addition, restriction fragment length polymorphism (RFLP) analysis of ITS regions of rRNA and random amplified polymorphic DNA (RAPD) studies revealed significant genetic variation among *C. cassiicola* isolates collected from different host plants [2–5,10,11].

7.1.2 Clinical Features and Pathogenesis

Corynespora cassiicola is a pathogenic fungus associated with leaf spot disease in plants. The involvement of this fungus in human disease was first noted in 1969, when a case of

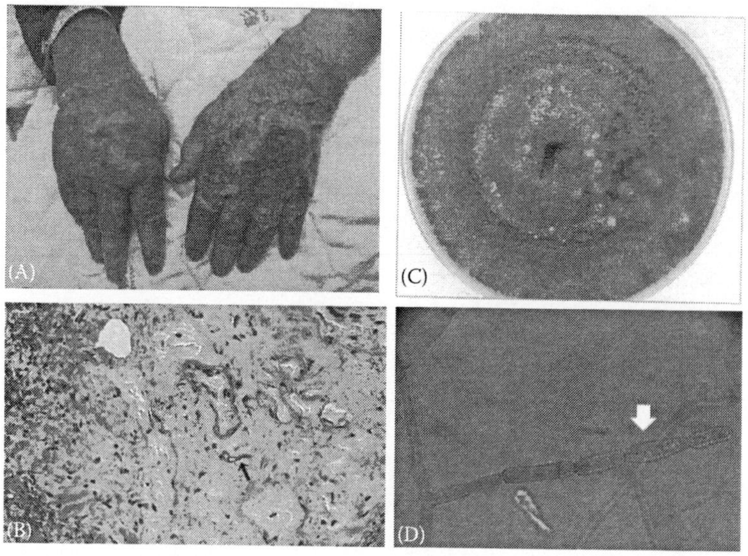

FIGURE 7.1 (A) Necrotic subcutaneous lesions over the dorsal aspects of both hands with purulent discharge in a patient with chronic subcutaneous infection caused by *Corynespora cassiicola*. (B) Hyphae with septa (black arrow) in the edematous papillary dermis (PAS stain, 400× magnification). (C) Confluent growth with velvety and gray–black colonies after 7 days of incubation on PDA at 28°C. (D) Straight and septate conidiophores and catenate conidia (white arrow) (lactophenol cotton blue stain, 400× magnification). (From Huang, H.K. et al., *J. Infect.*, 60, 188, 2010.)

maduromycetoma of the foot was attributed to *C. cassiicola* [6]. Recently, *C. cassiicola* was shown to cause chronic subcutaneous infection in a 69-year-old female farmer with diabetes mellitus and on long-term oral antihyperglycemic therapy [7]. The patient presented erythematous lesions on the bilateral forearms and dorsal aspect of both hands with purulent discharge, which was accompanied by pain, heat, swelling, and tenderness. Multiple bullae developed on the erythematous lesions, which ruptured with milk-colored, pus-like discharge. The patient underwent debridement and culture of the debrided tissue and discharged pus on PDA (Becton Dickinson) formed velvety and gray-black colonies after 7 days of incubation at 28°C. Microscopically, the organism showed long, straight, septate conidiophores and elliptic, catenate conidia using lactophenol cotton blue stain. Histological examination of biopsy using periodic acid-Schiff (PAS) stain revealed tissue necrosis with the presence of septate hyphae in the edematous papillary dermis. Upon sequence analysis of the region comprising the ITS1 and ITS2 regions of the rRNA genes using fungus-specific universal primers ITS1 and ITS4, the fungus demonstrated a nucleotide identity of 98% (527/534) with the published *C. cassiicola* sequences in the GenBank (accession number FJ852715). The mold was thus identified as *C. cassiicola* according to the characteristic phenotypes and relevant molecular methods. The patient recovered fully after amphotericin B treatment [7].

The exposure to the fungus in plants such as betel nut and banana through trivial wounds (farmer) and the immunocompromised status (old age, diabetes mellitus, and adrenal insufficiency) might have contributed partly to this case of *C. cassiicola* infection [7]. The ability of *C. cassiicola* to cause cellular damages was demonstrated recently with the isolation and characterization of a 28 kDa, 27 residue *O*-glycosylated protein (known as cassiicolin) in this fungus [12,13]. Cassiicolin induces host-selective cellular damages resembling those observed in leaf fall disease of rubber trees with *C. cassiicola* infection [12,13].

7.1.3 Laboratory Diagnosis

Traditional methods for identification of fungi are based on examination of their characteristic macroscopic and microscopic features. Typically, clinical specimens (e.g., discharge and scrapings) are examined under microscope to identify hyphae and other mycotic elements, with the help of 10% KOH or lactophenol cotton blue stain. Biopsies are examined for mycotic elements and pathological changes using PAS stain or other specific fungal stains. Portions of clinical specimens are cultured on Sabouraud dextrose agar (SDA) or potato dextrose agar (PDA) at 25°C, 35°C, respectively. The resulting fungal isolates are further characterized on their colony morphology and microscopic features as well as biochemical properties.

Given the close morphological and biochemical similarity among many pathogenic fungal species, laboratory diagnosis of fungal infections using phenotypic criteria can be challenging and time consuming. With the development and application of nucleic acid amplification techniques, rapid and precise identification of fungal organisms has become possible. In particular, PCR amplification and sequencing analysis of the small subunit (SSU) and large subunit (LSU) as well as ITS1 and ITS2 regions of the rRNA genes provide a useful tool for confirmation of fungal identity including *C. cassiicola* [7,14–17].

7.2 METHODS

7.2.1 Sample Preparation

Specimens are stained with 10% KOH, lactophenol cotton blue stain or PAS stain and examined under microscope for mycotic elements. Portions of samples are grown on inhibitory mold agar, modified Sabouraud agars, or PDA at 25°C and 35°C.

After growth for 1–7 days on PDA slants, mycelia are collected by scraping the slant with a sterile stick in 1 mL of sterile water. The material is transferred to a 2 mL screw-cap tube. The tubes are centrifuged for 1 min at 6000 × *g*. If the mycelia do not pellet, the material is contained with a pediatric blood serum filter (Porex Corp). Supernatant is removed, and the material is resuspended in 200 μL of IDI sample buffer and transferred to the lysis tube, which contains glass beads. Lysis tubes are vortexed on the highest setting for 5 min. The tubes are placed in a boiling water bath for 15 min. Tubes are centrifuged for 5 min at 16,000 × *g*. The supernatant is stored at −20°C until amplification [15].

Alternatively, about 1 cm^2 of fungal material is transferred to a 2 mL Eppendorf tube containing a 2:1 (wt/wt) mixture of silica gel and Celite (silica gel H, Merck 7736/Kieselguhr Celite 545; Machery) and 300 μL of TES buffer (2 g Tris [hydroxymethyl]-aminomethane, 0.38 g Na-EDTA [ethylenediaminetetraacetic acid], and 2 g sodium dodecyl sulfate in 80 mL of ultrapure water, pH 8). The fungal material is ground with a micropestle for 1–2 min. The volume is adjusted by adding 200 μL of TES buffer. After vigorous shaking and the addition of 10 μL of a 10 mg/mL concentration of proteinase K to the tube, the mixture is incubated at 65°C for 10 min. With the addition of 140 μL of 5 M NaCl solution, the mixture is combined with 1/10 volume (~65 μL) of 10% cetyltrimethylammonium bromide (CTAB) buffer, followed by incubation for another 30 min at 65°C. One volume (~700 μL) of chloroform-isoamyl alcohol (vol/vol = 24/1) is added and mixed by inversion. After incubation for 30 min at 0°C (on ice water) and centrifugation at 14,000 rpm at 4°C for 10 min, the top layer is transferred to a clean Eppendorf tube. The sample is added with 225 μL of 5 M NH_4–acetate and incubated for at least 30 min (on ice water) and centrifuged. The supernatant is transferred to a clean sterile Eppendorf tube and mixed with a 0.55 volume (~510 μL) of ice-cold isopropanol. After centrifugation for 7 min at 14,000 rpm and 4°C (or room temperature), the supernatant is decanted. The pellet is washed with ice-cold 70% ethanol twice and dried by using a vacuum dryer. The powder is resuspended in 48.5 μL of Tris–EDTA buffer with 1.5 μL of 10 mg of RNase/mL, incubated at 37°C for 15–30 min, and stored at −20°C until use [16].

7.2.2 Detection Procedures

7.2.2.1 Pan-Fungal Real-Time PCR and Sequencing Analysis

Pounder et al. [15] described a real-time PCR with SYBR green DNA-binding dye and amplicon melting temperature analysis for fungal detection using pan-fungal primers ITS1 forward (5′-TCCGTAGGTGAACCTGCGG-3′) and ITS4 reverse (5′-TCCTCCGCTTATTGATATGC-3′). The identity of the fungi is verified by subsequent sequencing analysis.

Procedure

1. PCR mixture is composed of 1× Lightcycler FastStart DNA Master Hybridization Probes mixture (Roche Applied Science) (containing deoxynucleoside triphosphates, FastStart *Taq* DNA polymerase, and 1 mM $MgCl_2$, additional $MgCl_2$ is added to a final concentration of 4.6 mM), 0.4 μM each of ITS1 forward and ITS4 reverse primers, 1× SYBR green (Molecular Probes), and 3 μL template DNA.
2. Thermal cycling parameters include 95°C for 10 min; 50 cycles of 95°C for 5 s, 60°C for 20 s, and 76°C for 30 s; and a final extension at 72°C for 2 min.
3. The quality of the amplicon is determined using the derivative of the melt analysis curve (55°C–99°C, 45 s hold at 55°C, 5 s/°C) using the RotorGene 3000 (Corbett Robotics, Inc).
4. The amplified product is purified for bidirectional sequencing using ExoSAP-IT (USB Corp). Five microliters of Big Dye Terminator Ready Reaction Mix v. 1.1 (Applied Biosystems) is added to 4 μL of each primer (0.8 pmol/μL) and 3 μL of purified PCR product. Cycle sequencing is performed with a 9700 thermal cycler (ABI), using 25 cycles of 96°C for 10 s, 50°C for 5 s, and 60°C for 4 min. Sequencing reaction products are passed through a Sephadex G-50 fine column to remove unincorporated dye terminators. Purified sequencing reaction products are run on an ABI Prism 3100 Genetic Analyzer with a 50 cm capillary array.
5. Sequences are analyzed with the SmartGene Integrated Database Network software version 3.2.3 vr. SmartGene is a web-based software and database system with reference sequences derived from the National Center for Biological Information (NCBI) GenBank repository.

Note. In case that real time PCR instrument is not available, standard PCR may be performed with primers ITS1 and ITS4, and the resulting amplicon is sequenced with the same primers [18]. Sequence-based identifications are defined by percent identity: species, ≥99%; genus, 93%–99%; and inconclusive, ≤93%.

7.2.2.2 Sequencing Analysis of ITS1 and ITS2

Leaw et al. [14] utilized fungus-specific universal primers ITS1 (5′-TCCGTAGGTGAACCTGCGG-3′) and ITS2 (5′-GCATCGATGAAGAACGCAGC-3′) to amplify the ITS1 region, and universal primers ITS3 (5′-GCATCGATGAAGAACGCAGC-3′) and ITS4

(5′-GCATATCAATAAGCGGAGGA-3′) to amplify the ITS2 region. Subsequent sequencing analysis of the resulting amplicons facilitated identification of a range of fungal pathogens.

Procedure

1. PCR mixture (50 μL) is made up of 10 mM Tris–HCl pH 8.3, 50 mM KCl, 1.5 mM $MgCl_2$, 0.2 mM each of deoxynucleoside triphosphates, 1.2 U of *Taq* DNA polymerase, 0.4 μM (each) of the ITS1 region primers (ITS1/ITS2) or the ITS2 region primers (ITS3/ITS4), 2 μL (1–5 ng) of DNA template. A negative control is included in each run by replacing the template DNA with sterile water in the PCR mixture.
2. Amplification is conducted at an initial 94°C for 3 min; 30 cycles of 94°C, 60°C, and 72°C for 1 min each; and a final 72°C for 3 min.
3. After checking the PCR products on 1.5% agarose gel, the amplicons are purified using a commercial PCR cleanup kit. The DNA fragments are sequenced using an ABI Prism 377 automated DNA sequencer (Applied Biosystems) with a BigDye Terminator cycle sequencing kit (version 3.1; Applied Biosystems). All amplicons are sequenced on both strands using primers ITS1 and ITS2 for the ITS1 region and primers ITS3 and ITS4 for the ITS2 region.
4. After sequencing, portions of the 18S, 5.8S, and 26S rRNA gene sequences of the PCR products are removed to obtain the exact ITS1 and ITS2 sequences.
5. For all yeasts, the sequences of the 3′ ends of the 18S and 5.8S rRNA genes are GCGGAAGGATCATTA and GTTTGAGCGTCATTT, respectively, and the sequences of the 5′ ends of the 5.8S and 26S rRNA genes are AAACTTTCAACAA and GACCTCAAATCAG, respectively.
6. The ITS1 or ITS2 sequence is compared with those available at the GenBank databases using the BLAST sequence analysis tool (http://www.ncbi.nlm.nih.gov/BLAST/) (blastn) with default settings except that sequences are not filtered for low complexity. Species identification is determined from the lowest expected value of the BLAST output.

Note. For strains producing discrepant identification between the methods based on phenotypic characteristics and ITS sequence analysis, the D1–D2 region of the LSU RNA gene is amplified with primers NL1 (5′-GCATATCAATAAGCGGAGGAAAAG-3′) and NL4 (5′-GGTCCGTGTTTCAAGACGG-3′) and sequenced for species clarification.

7.3 CONCLUSION

The genus *Corynespora* is composed of eight mold species that are distributed ubiquitously in the environment of which *C. cassiicola* causes leaf spot disease in >70 plant species, and is occasionally associated with human infections [6,7]. As shown in a recent case of *C. cassiicola*-related chronic subcutaneous infection, the fungus may be acquired through exposure to plants and it has the ability to take advantage of the weakened host immune defense such as old age, diabetes mellitus, and adrenal insufficiency to establish in human hosts [7,18].

As conventional laboratory procedures for identification of *Corynespora* and other fungal pathogens are slow and demand specialized skills, molecular techniques are increasingly utilized for rapid and precise determination of these organisms. Indeed, PCR and sequencing analysis of the SSU and LSU as well as ITS1 and ITS2 regions of the rRNA genes offer a reliable supplemental approach for accurate speciation of fungal organisms including *C. cassiicola*.

REFERENCES

1. The UniProt Consortium. Available at http://www.uniprot.org/, accessed on August 1, 2010.
2. Silva, W.P.K., Deverall, B.J., and Lyon, B.R., Molecular, physiological and pathological characterization of *Corynespora* leaf spot fungi from rubber plantations in Sri Lanka. *Plant Pathol.*, 47, 267, 2002.
3. Silva, W.P. et al., Genetic variation in *Corynespora cassiicola*: A possible relationship between host origin and virulence. *Mycol. Res.*, 107, 567, 2003.
4. Silva, W.P.K. et al., RFLP and RAPD analyses in the identification and differentiation of isolates of the leaf spot fungus *Córynespora cassiicola*. *Aust. J. Botany*, 43, 609, 1995.
5. Nghia, N.A. et al., Morphological and inter simple sequence repeat (ISSR) markers analyses of *Corynespora cassiicola* isolates from rubber plantations in Malaysia. *Mycopathologia*, 166, 189, 2008.
6. Mahgoub, E., *Corynespora cassiicola*, a new agent of madu-romycetoma. *J. Trop. Med. Hyg.*, 72, 218, 1969.
7. Huang, H.K. et al., Subcutaneous infection caused by *Corynespora cassiicola*, a plant pathogen. *J. Infect.*, 60, 188, 2010.
8. Zaher, E.A. et al., Leaf spots of ornamental foliage plants in Egypt with special reference to *Corynespora cassiicola* [(Berk. & Curt.)Wei] as a new causal. *Egypt. J. Phytopathol.*, 33, 87, 2005.
9. Jayasuriya, K.E. and Thennakoon, B.I., First report of *Corynespora cassiicola* on *Codiaeum variegatum* (Croton) in Sri Lanka. *J. Sci. (Bio. Sci.)*, 36, 138, 2007.
10. Dixon, L.J. et al., Host specialization and phylogenetic diversity of *Corynespora cassiicola*. *Phytopathology*, 99, 1015, 2009.
11. Qi, Y. et al., Molecular and pathogenic variation identified among isolates of *Corynespora cassiicola*. *Mol. Biotechnol.*, 41, 145, 2009.
12. Barthe, P. et al., Structural analysis of cassiicolin, a host-selective protein toxin from *Corynespora cassiicola*. *J. Mol. Biol.*, 367, 89, 2007.
13. de Lamotte, F. et al., Purification and characterization of cassiicolin, the toxin produced by *Corynespora cassiicola*, causal agent of the leaf fall disease of rubber tree. *J. Chromatogr. B Analyt. Technol. Biomed. Life Sci.*, 849, 357, 2007.

14. Leaw, S.N. et al., Identification of medically important yeast species by sequence analysis of the internal transcribed spacer regions. *J. Clin. Microbiol.*, 44, 693, 2006.
15. Pounder, J.I. et al., Discovering potential pathogens among fungi identified as nonsporulating molds. *J. Clin. Microbiol.*, 45, 568, 2007.
16. Zeng, J.S. et al., Spectrum of clinically relevant *Exophiala* species in the United States. *J. Clin. Microbiol.*, 45, 3713, 2007.
17. Bagyalakshmi, R. et al., Newer emerging pathogens of ocular non-sporulating molds (NSM) identified by polymerase chain reaction (PCR)-based DNA sequencing technique targeting internal transcribed spacer (ITS) region. *Curr. Eye Res.*, 33, 139, 2008.
18. Lv, G.X. et al., Phaeohyphomycosis caused by a plant pathogen, *Corynespora cassiicola*. *Med. Mycol.*, posted online on January 31, 2011.

8 *Curvularia*

Audrey N. Schuetz

CONTENTS

8.1 INTRODUCTION

8.1.1 Classification and Epidemiology

8.1.1.1 Classification

Curvularia comprises part of the Dematiaceae family of thermally monomorphic dematiaceous molds, which possess brown melanin or melanin-like pigments in their hyphal and/or conidial cell walls. Related to the sexual teleomorph ascomycete *Cochliobolus* (family Pleosporaceae, order Pleosporales), *Curvularia* is a member of the Hyphomycete class of fungi. The genus is comprised of approximately 40 species, including *C. lunata* (the most commonly isolated species in humans), *C. clavata*, *C. brachyspora*, *C. geniculata*, *C. inaequalis*, *C. pallescens*, *C. senegalensis*, and others.

Increasingly recognized as an emerging cause of human infection, particularly in immunocompromised persons, *Curvularia* causes a wide range of human diseases [1]. First reported from a mycetoma by Baylet et al. in 1959 [2], *Curvularia* spp. have been commonly associated with certain diseases, such as infections of the cornea and sinuses, as well as allergic fungal sinusitis (AFS). Other diseases associated with *Curvularia* include cutaneous and subcutaneous lesions (such as phaeohyphomycosis and chromoblastomycosis), ocular infections (such as keratitis and endophthalmitis), peritonitis, onychomycosis, and other less common clinical entities. Dissemination to the brain and to other organs rarely occurs. However, isolation in culture is not always clinically significant, as *Curvularia* may occasionally contaminate a culture. Infections due to *Curvularia* can occur in immunocompetent hosts as well. In their review of human *Curvularia* infection, Rinaldi et al. reported that only two patients were immunosuppressed in their series of 24 patients [3].

8.1.1.2 Epidemiology and Risk Factors for Infection

Most *Curvularia* spp. are saprobic soil agents and live on dead plant or grain material as weak pathogens. *Curvularia* phytopathogens cause plant diseases, ranging from seed failure to leaf blight and grass "fade-out" during hot and humid weather [4]. The spores of *Curvularia* are large (i.e., 20–30 μm × 8–12 μm) [5].

Risk factors for human infection include soil contact associated with outdoor recreational or occupational activities. Persons are usually exposed through the respiratory tract; however, skin injury is a second important route of exposure [6].

Lawn-tool injuries due to projectiles have caused ocular or cutaneous lesions [7]. *Curvularia* has been cultured from rose bush thorns on plants in the Oklahoma City area [8]. In one study from the United States of 43 cases of *Curvularia* keratitis, half of the cases were associated with trauma due to plant or dirt inoculation, or metal injury [4].

Environmental contamination has been documented in many cases of *Curvularia* infection, such as the case of a patient with *Curvularia* peritonitis who worked in his garden prior to receiving dialysis exchanges [9]. Canon et al. report a case of peritonitis in a 13-year-old boy undergoing peritoneal dialysis who played in a wooded area and wore his dialysis tubing and transfer set in his underpants, predisposing him to infection [10]. Cutaneous inoculation due to skin injury may cause severe disease, especially in an immunocompromised host. Tessari et al. report a 69-year-old heart transplant recipient who died from overwhelming sepsis due to cutaneous *C. lunata* infection, from a splinter introduced while working in his garden [11]. At autopsy, *Curvularia* was recovered in numerous organs, including the skin, lungs, mouth, upper third of the esophagus, and right lung. Rohwedder et al. report the case of an immunocompetent football player with disseminated *C. lunata* who developed cervical lymphadenopathy, pleural effusion, and a paraspinous abscess after presumptive cutaneous inoculation from skin scrapes and cuts during football practice [12]. Other cases have been documented in gardeners and children who played outdoors, including *Curvularia* phaeohyphomycosis in 22-year-old male who had overdosed on cocaine, was wandering nude in a wooded area, and subsequently developed infection of the lacerations on his extremities [3,13,14]. Finally, Grieshop et al. describe acute, localized cutaneous and subcutaneous phaeohyphomycosis due to *C. lunata*, after explosion at a chemical plant, which was believed to have inoculated organisms into body tissues due to the force of the explosion [15].

Environmental contamination with *Curvularia* spp. is not restricted to outdoor, wooded areas but has been documented in a healthcare institution as well. Kainer et al. reported a hospital-associated outbreak of *Curvularia* spp., which contaminated saline-filled silicone shell prosthetic breast implants in women who had undergone cosmetic breast augmentation surgery [16]. In their series, five patients who underwent surgery at their institution developed *Curvularia* contamination of their breast implants; one patient's breast implants contained macroscopically apparent black specks, proven to be microcolonies of *Curvularia*. An epidemiologic investigation revealed that *Curvularia* was isolated from the sterile supply room, in which saline bottles were stored underneath a water-damaged ceiling, as well as from the corridor outside of the operating room. The authors postulated that the saline used to fill the silicone implants was likely contaminated while sitting in an open bowl before or during surgery, when fungal spores circulated in air drafts introduced into the operating room, which maintained a positive air pressure differential. Risk factors associated with contamination included increased length of time spent in the operating room and longer duration of surgery.

Fungal rhinosinusitis predominates in hot, humid climates, such as the southwestern United States, India, and the Sudan [5]. The number of aerially borne *Curvularia* varies seasonally and rises just after a wet, warm month [17]. The relative prevalence of filamentous fungal keratitis appears to increase near tropical latitudes [18]. At a Texas institution [4], Wilhelmus and Jones reported that two-thirds of the *Curvularia* keratitis cases occurred during the summer, corresponding to the peak of airborne spores along the Texas coast [19]. In a retrospective ecological study of 52 episodes of dematiaceous fungal keratitis along the Gulf of Mexico spanning the years 1972 to 2004, researchers reported an increase in the diagnosis of ocular phaeohyphomycosis during the late summer, which persisted throughout the autumn months [20]. Specifically, new diagnoses of *Curvularia* keratitis clustered in warm, humid months.

8.1.2 Pathogenesis

Relatively little is known regarding the pathogenic mechanisms by which *Curvularia* and other dematiaceous fungi lead to disease, especially in immunocompetent individuals. *Curvularia* produces several mycotoxins, such as curvularins, brefeldins, and radicinin, which can act as both cytotoxins and antivirals [4,21]. *Curvularia* also metabolizes certain steroids and produces other toxins and enzymes, including anthroquinones, curvapallides, cytochalasins, neocoprogen, triticones, pyrenocenes, spirostaphylotrichins, pectinases, zaragozic acid, lipid phosphatase, galactosidase, glucosidase, endoglucanase, chloroperoxidase, and cellulases [22]. The roles of these enzymes and toxins have not yet been fully described.

There is increasing evidence that melanin, present in the cell walls of dematiaceous fungi, may act as a virulence factor [23–25]. Melanin binds to and inactivates hydrolytic enzymes and scavenges free radicals and hypochlorite [23].

Some authors have postulated that *Curvularia* shows a predilection for organic silicone or inorganic silicon substrate, which may predispose to infection in those patients who have silicone-containing foreign bodies such as catheters [26]. Reports have demonstrated that *Curvularia* may colonize silicon-containing stone churches [27] and silicone voltage insulators [28]. Likewise, *Curvularia* has been implicated in cases of silicone peritoneal dialysis catheters, silicone breast implants [16], and silicone lenses [16,29]. Silicone appears to be associated not only with the presence of *Curvularia*, but low levels of silicone also appear to enhance the growth of *Curvularia* and *Fusarium* in an alkaline medium, such as the alkaline environment of fresh peritoneal dialysis fluid [30].

Curvularia is an important inhalant allergen associated with respiratory allergic disorders, such as rhinitis. Patients with nonspecific nasobronchial symptoms demonstrate positive skin reactions to *C. lunata* extract [31]. A study of asthma and rhinitis patients in Singapore showed that patients demonstrated a high prevalence of fungal sensitization to *Curvularia* (26%–32%) [32]. In addition, approximately 60% of rural allergic patients in agricultural India showed sensitization to *Curvularia* extracts [33]. Cross-reactivity of *Curvularia* allergens has been demonstrated to *Aspergillus*, *Alternaria*, and *Epicoccum* [34]. Studies have identified that proteins of 26, 31, 38, 45, 50, and 78 kDa from *C. lunata* share IgE and IgG cross-reactivity within species [35]. Two major *C. lunata* allergens have been identified and cloned: Cur l 1 was identified as a serine protease [36], while rCur l 2 proved to be an enolase, a glycolytic enzyme that has shown allergic potential in individuals allergic to molds [37]. In addition, *C. lunata* enolase shows 90% homology with a *Cladosporium* allergen [37]. Although most research on *Curvularia* allergens is recent and has been performed at the gene recombinant level, recent mouse models have also shown that the serine protease Cur l 1 plays an important role in the exacerbation of airway inflammation [38]. Allergens of this and other fungi should be further studied to contribute toward a greater understanding of *Curvularia*-related allergy and immunotherapy of allergy-mediated disorders [39].

8.1.3 Clinical Features

The clinical manifestations are diverse and range from noninvasive disease to severe, disseminated infection. Furthermore, as molecular studies are becoming more available and easier to perform, speciation of *Curvularia* has broadened our understanding of the spectrum of disease caused by this genus. Reports of infection in various sites are described below.

8.1.3.1 Ocular Disease

Curvularia can cause various ocular infections. In their review of 32 patients with culture-proven *Curvularia* keratitis over a 30-year period, Wilhelmus and Jones demonstrated that, while dematiaceous fungi accounted for 22% of all fungal corneal isolates, one-third of the isolates were comprised of *Curvularia* spp. [4]. *Curvularia* was the fourth most common fungal isolate identified from cases of keratitis, following *Candida*, *Fusarium*, and *Aspergillus*, at a large hospital in Texas [4]. *Curvularia* spp. isolated from the Texas study included *C. senegalensis*, *C. lunata*, *C. pallescens*, and *C. prasadii*. Many of the largest case series of keratitis involving dematiaceous fungi originate from around India [40–42]. In these series, *Curvularia* caused 3.3%–7.4% of all mycotic keratitis cases [41–43], 8.2% of fungal corneal ulcers in North India [40], and 6% of all cases of suppurative keratitis in Bangladesh [44]. In a large series of mycotic keratitis reported from India, *Curvularia* spp. was the most common cause of dematiaceous fungal keratitis (approximately 23% of all dematiaceous keratitis cases) [45]. In South India, among 1352 cases of fungal keratitis reviewed, *Curvularia* represented 2.8% of all fungal etiologies and was the most common dematiaceous fungus isolated [46]. Similarly, among a hospital-based study in New Delhi of 191 patients with mycotic keratitis presenting with corneal ulcers, *Curvularia* spp. was the second most common cause of ulceration, second only to *Aspergillus* spp. [47].

Curvularia has been reported as the most frequent pigmented fungal pathogen in corneal ulcers by other investigators as well [48–51]. In Thailand, *Curvularia* was second only to *Fusarium* as a cause of severe mycotic keratitis [52]. *Curvularia* was also a fairly common cause of fungal keratitis in far north Queensland, Australia, where a retrospective review was performed of all cases of fungal keratitis over a 10-year period [53]. A higher incidence of fungal keratitis was demonstrated during the monsoon and winter seasons than during the drier summer months.

Risk factors for ocular *Curvularia* infections include soft contact lens wear [54] and laser in situ keratomileusis (LASIK) [55,56]. Tuli and Yoo report the case of a 50-year-old male who had recently undergone LASIK but developed *Curvularia* keratitis postsurgically, likely linked to a fungal skin infection of the patient's pet cat [55]. Although *Curvularia* may be isolated from unwashed skin, it is an uncommon colonizer, present on less than 1% of normal human eyelids [57]. Therefore, ocular infection is usually related to injury or manipulation of the eye. In their retrospective review of fungal keratitis, Iyer et al. reported that, in 2005, contact lens use surpassed trauma as the most common risk factor for fungal keratitis, which remains a common cause of corneal ulcers in the southern United States [58]. In their study, *Curvularia* was the most common of the filamentous fungi cultured from corneal ulcers, comprising 12%.

Ocular *Curvularia* infections may also present as conjunctivitis, dacryocystitis [59], endophthalmitis after intraocular lens implantation [60], or extension to sino-orbital cellulitis; however, the cornea is most commonly affected. One sign of probable dematiaceous keratitis is the presence of clinically pigmented corneal ulcers [45]. Keratitis due to *Curvularia* spp. commonly presents as a slowly progressing, feathery, superficial corneal infection, Endophthalmitis may present as a posterior capsule plaque [61,62].

8.1.3.2 Allergic Fungal Sinusitis

A form of noninvasive fungal rhinosinusitis, AFS is caused by a hypersensitivity response to the presence of fungal hyphae in the sinuses. Accounting for up to 7% of all sinusitis cases requiring surgery, AFS is commonly caused by dematiaceous molds [63,64]. The diagnosis of AFS is primarily histopathologic, with demonstration of scattered fungal hyphae within the viscoelastic allergic mucin viewed on biopsy. In most instances, invasion of the mucosal surface by the organism is not seen. Cultures of surgically obtained material commonly are positive for organisms such as *Curvularia* spp. [65,66]. Although the criteria for diagnosis of AFS have changed over time, the current diagnostic criteria favor the presence of characteristic "allergic mucin," accompanied by the presence of noninvasive fungal elements within the mucin, detected by either staining or culture [67]. The allergic mucin usually contains numerous eosinophils, sometimes with Charcot-Leyden crystals, and the mucin seems to be more commonly viewed in biopsies of patients from warm, humid climates [68].

A clinical history of nasal polyposis, thick mucus, and abnormal sinus computed tomography scans may be demonstrated in patients with AFS [69]. Other diagnostic criteria include a positive skin test to fungal allergens [68] or the presence of allergies to multiple aeroallergens [67]. In a study of 67 AFS patients from the southwestern United States, all patients demonstrated inhalant atopy and skin test positivity to the AFS-implicated mold [70]. The majority of patients from this study also presented with reactive airway disease, and a minority demonstrated aspirin/nonsteroidal anti-inflammatory drug hypersensitivity. Elevated total serum IgE in AFS is also common, and serial monitoring of total serum IgE over time reflects patient clinical status, as can changes in antigen-specific IgG [71].

Manifestations of AFS are due to type I immediate hypersensitivity response to the extramucosal hyphae. Although much remains unknown about the immunopathology of the disease, certain features have been consistently demonstrated in AFS patients, including an eosinophilic/lymphocytic mucosal inflammatory response [70], elevated total serum IgE, and potential roles for IgG, and perhaps *Staphylococcus aureus* superantigens, in disease development [71].

Curvularia can also cause colonization of the tracheobronchial passage and sinus cavities, leading to recurrent or persistent hypersensitivity in certain individuals. True paranasal sinus infections may occur [72]. Intrasinus mycoses may lead to extension to contiguous structures such as the orbital contents or intracranial area. Chronic invasive sinusitis is progressive and may spread to the orbit and brain. At least one case report has documented *C. lunata* pansinusitis extension to the brain in an immunocompetent host [73].

8.1.3.3 Cutaneous and Subcutaneous Infections

Curvularia has been associated with both cutaneous and deep subcutaneous infections. Hiromoto et al. describe the case of a 67-year-old immunocompetent male with *Curvularia* infection of an ulcerating lesion on his arm, after having been pricked and injured by a tree [6]. *Curvularia* has also been associated with an ecthyma gangrenosum-like lesion on the palm of a bone marrow transplant recipient [74], dermatomycosis of the toe web mimicking intertriginous tinea pedis [75], and invasive fungal dermatitis, characterized by erosive, crusting lesions, in extremely low birth weight neonates [76]. Onychomycosis can also occur; in India, 4.5% of onychomycosis was due to *Curvularia* [77–79].

Deeper infections include chromoblastomycosis, mycetoma, and phaeohyphomycosis. Chromoblastomycosis refers to subcutaneous and soft tissue infections that are characterized on histopathology by pigmented, thick-walled sclerotic bodies. Phaeohyphomycosis, on the other hand, is a broader term and includes infections ranging from localized to disseminated invasive disease. Phaeohyphomycosis is differentiated from mycetoma by the absence of sulfur granules and from chromoblastomycosis by the absence of sclerotic bodies within the biopsy [80]. *Curvularia* spp. have been associated with many types of deep infections, including eumycetoma, as illustrated by the case of a 27-year-old farmer with multiple nodules and sinuses draining black granules [81,82]. *C. lunata* also caused a deep sternal wound infection in a neonate who underwent operation for congenital heart disease, likely due to direct inoculation of aerosolized spores into the open sternal wound [83].

8.1.3.4 Cerebral Infections

The first reported case of cerebral infection due to *Curvularia* was a 13-year-old boy with multiple cerebral lesions [84]. Other reported cases include a 41-year-old male with multiple cerebral and pulmonary lesions [85] and *C. lunata* presenting as a pituitary mucocoele associated with optic atrophy [86]. Carter and Boudreaux report the case of fatal cerebral *C. lunata* infection, which presented as a cerebral mass in an immunocompetent 21-year-old male [87].

8.1.3.5 Peritonitis

In 1985, DeVault et al. reported catheter malfunction and obstruction due to *C. lunata* in a continuous ambulatory peritoneal dialysis (CAPD) patient, who presented without signs or symptoms of peritonitis [88]. Several other reports of *C. lunata* peritonitis have since been reported from peritoneal dialysis patients. Many of the peritonitis cases are reported from the southern United States, such as Georgia, North Carolina, Tennessee, and Louisiana [20,88–90]. A case of *Curvularia* peritonitis has also been reported from Brazil in a 63-year-old male with end-stage renal failure due to diabetes mellitus who presented with turbid dialysis effluent with gelatinous brown exudates discovered on exploratory laparotomy [91]. Several cases have documented the presence of black flakes in the dialysate fluid or the Tenckhoff catheter, corresponding to *Curvularia* contamination of the fluid. In one peritonitis case, black material obstructed the lumen of the catheter, preventing flow [9]. CAPD patients with peritoneal infection due to *Curvularia* may not always be symptomatic. In many reported cases, there were no problems with the catheters, such as site infection or mishandling of the tubing, leading to contamination. However, Diskin et al. suggest that the risk factor for acquisition of *Curvularia* may be the silicone dialysis tubing itself [26].

8.1.3.6 Miscellaneous Infections

Infections of internal organs are rare. A urinary tract infection (UTI) has been reported in a 5-year-old girl, who presented with recurrent UTIs and black specks in her urine [92]. Additionally, Shigemori et al. report a 10-year-old girl with acute monocytic leukemia, who presented with neutropenia and developed a hepatosplenic abscess caused by *Curvularia boedijn* [93].

8.1.3.7 Curvularial Infections, Other Than *C. lunata*

Although rarely isolated from humans, *C. clavata* has been reported to cause infections in various sites. Ebright et al. [89] described the first known case of *C. clavata* implicated in invasive sinusitis of a 46-year-old female, with bony destruction of the cribiform plate and bifrontal cerebritis. Another case of *C. clavata* causing cutaneous, red plaque-like lesions on the foot of a 64-year-old farmer with no known immunosuppression has recently been described [1]. None of the three case patients described above were known to be immunocompromised at the time of presentation. Therefore, immunosuppression does not appear to be a necessary state for infection with this species.

C. brachyspora was documented to cause cutaneous infection in an immunocompetent 58-year-old male [94]. A month prior to presentation, the patient had cryotherapy performed on some pigmented thigh lesions, which were suspected to be seborrheic keratoses. After cryotherapy, the lesions worsened, developing into necrotizing ulcers, which demonstrated necrotizing cellulitis on biopsy and *C. brachyspora* on culture.

Occasional reports of *C. geniculata* have emerged in the literature. *C. geniculata* has been reported as a cause of mycotic keratitis [95] and endocarditis associated with an aortic valve graft [96]. Disseminated disease due to *C. geniculata* was reported in a football player with B- and T-cell hypofunction, after an initial infected skin lesion disseminated to lymph nodes, vertebrae, and the pulmonary cavity [97]. The first known case of *C. geniculata* peritonitis was described by Vachharajani et al. in a 45-year-old male on CAPD, despite no macroscopic evidence of catheter infection [98]. *C. geniculata* has also been shown to cause central nervous system infection in a 35-year-old male who presented with obstructive hydrocephalus due to a large cranial base lesion, which initially appeared on imaging and biopsy to be a meningioma [99]. *C. geniculata* was isolated in culture, and speciation was confirmed by molecular sequencing. An underlying plasma cell dyscrasia was subsequently diagnosed in this patient, which likely predisposed him to development of this infection. Despite aggressive therapy with two antifungal agents, the patient died 4 months after initial biopsy from multiorgan failure.

C. inaequalis was recently reported to cause human disease in 2005 by Pimentel et al., who described the case of an 85-year-old CAPD patient with peritonitis [100]. The peritoneal fluid grew *C. inaequalis*, which was confirmed by sequencing of the internal transcribed spacer (ITS) region. As the patient reported no recent contact with soil, risk factors for acquiring infection with this organism were unknown. *C. inaequalis* was also reported to cause eosinophilic fungal rhinosinusitis in a 45-year-old male, manifesting as bony erosions of the nasal cavity [101]. Although the patient was deemed to be immunocompetent in this case report (human immunodeficiency virus negative and no evidence of malignancy), the patient did receive steroids for his long-standing nasal obstruction, which may have led to some degree of immunosuppression.

The first known cerebral case *Curvularia* infection was that of *C. pallescens* in an immunocompetent 13-year-old boy with pulmonary and cerebral abscesses [84]. The second reported case of *C. pallescens* was described in a report of cutaneous infection in a 75-year-old woman with rheumatoid arthritis who presented with an ulcerated lesion on her leg with no antecedent history of trauma [102]. Other cutaneous infections due to *C. pallescens* have been reported as well [103].

C. senegalensis has been associated with mycotic keratitis [51,104]. In at least one of these keratitis cases, there was no evidence of trauma to the eye to explain fungal inoculation; however, one of the patients was diabetic and had used steroid-containing eye drops [104].

In summary, various *Curvularia* species have been associated with a wide variety of disease presentations. As molecular techniques become more available, it is likely that newer disease states associated with certain *Curvularia* species will continue to be discovered.

8.1.4 Diagnosis

8.1.4.1 Conventional Techniques

8.1.4.1.1 Microbiologic Culture

In the laboratory, colonies of *Curvularia* spp. display a brownish-black to dark olive-green surface and a dark-colored to black reverse. The colonies are filamentous and expanding. Colonies grow rapidly on plates, generally reaching maturity within 5 days. The optimal temperature for growth is 28°C, although growth may occur from 25°C to 35°C [105,106]. Microscopically, a lactophenol cotton blue, or other preparation of the colony, will demonstrate that the conidiophores are brown, erect, and multicellular, producing conidia sympodially (bent at points of conidium formation). Conidia are dark, thin-walled, ellipsoidal, and large (up to 14 μm in width and 35 μm in length), with a variable number of true transverse septae that differ according to the species. As the central cell of the conidium is larger and darker than the others, the conidia show a characteristic curve, or boomerang-like bend, which becomes more pronounced with age (Figure 8.1).

Immature *Curvularia* spp. may be mistaken for *Bipolaris* spp., which possess straight, multicellular conidia. Conidia of some *Curvularia* species may show a prominent hilum. In the mycology laboratory, speciation can be performed with the use of conidial morphology and growth patterns [105].

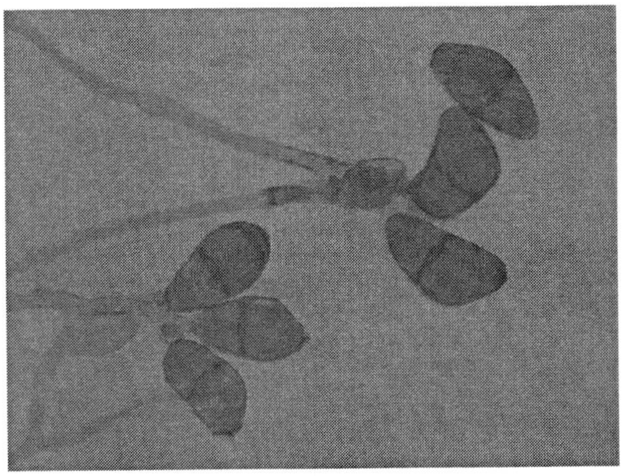

FIGURE 8.1 Slide culture preparation of *Curvularia* spp., demonstrating the characteristic curved shape of the conidia.

8.1.4.1.2 Histopathology

On a hematoxylin and eosin (H&E)-stained slide, the fungus appears as yeast-like cells that are solitary or in short chains, or as septate hyphae or pseudohyphae (Figure 8.2). Hyphae may exhibit vesicular swelling and bizarre shapes [107]. The similarity on tissue sections to *Aspergillus* spp., as well as to other organisms, such as *Fusarium* and *Scedosporium*, renders diagnosis almost impossible without microbiologic culture results. Brown pigmentation of *Curvularia* on routine H&E-stained tissue is common but not invariably present [108]. Although Fontana-Masson silver staining, which highlights the melanin in the cell walls of dematiaceous fungi, may be used to differentiate dematiaceous from nonpigmented fungi, the stain is not specific, as many nondematiaceous fungi may stain with this technique [108].

8.1.4.2 Molecular Techniques

Identification of fungal pathogens in the microbiology laboratory may be challenging, as filamentous fungi may not sporulate, or morphologic features may seem similar among species. Phenotypic assays allow differentiation among species but may take up to 12 weeks, and more rapid identification is often needed [105]. In addition, new species of fungi are being discovered, which may not be familiar enough to the laboratory worker to be included in the differential. Species differentiation may be important for appropriate therapy. Although nucleic acid probes are commercially available for a limited number of fungi, such as the thermally dimorphic fungi, molecular methods of detection are proving to increase both the reliability and the speed of diagnosis.

A common target for the molecular diagnosis of filamentous fungi is the ITS region. ITS regions are located within ribosomal DNA (rDNA). rDNA is a useful fragment for fungal identification, as it is universally present and is the most conserved region in the genome of fungi but maintains some variability to allow for species separation and identification [109]. The rRNA gene is structured as a small-subunit (SSU) 18S rRNA, a large-subunit (LSU) 28S rRNA, and a 5.8S rRNA. ITS regions I (ITSI) and ITSII show more variability than the remainder of the ribosomal gene subunits. ITSI is located between SSU rRNA and 5.8S rRNA, while ITSII is located between 5.8S rRNA and LSU rRNA. ITSI is more variable than ITSII and, thus, allows for better species differentiation [110].

Researchers have used many methods to assess sequence variation of amplified DNA such as direct sequencing of the amplified product, random fragment length polymorphism, temperature gradient gel electrophoresis, and denaturing gradient electrophoresis, as well as the single-stranded conformation polymorphism (SSCP) technique [110–113].

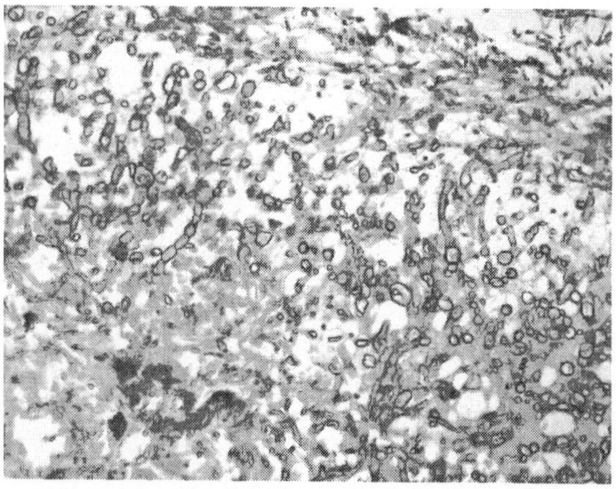

FIGURE 8.2 Histopathologic slide of *Curvularia* spp. in a sinus biopsy of a patient with paranasal disease, demonstrating the swollen, elongated, yeast-like cells viewed on biopsy.

8.2 METHODS

8.2.1 Sample Preparation

Many methods for extraction of fungal DNA have been proposed [110,114–117]. Outlined below is one of several extraction methods that may be used, which is adopted from Kumar and Shukla [114]:

1. For extraction of fungal DNA, it is suggested to begin with a strain that has been inoculated in 100 mL of Sabouraud dextrose broth under shaking conditions at 37°C, which optimizes extraction, as log-phase growth is generally achieved under these conditions.
2. The cultures should be examined microscopically to insure purity.
3. Centrifuge the fungus for 15 min at 6000 × *g* at 4°C.
4. Wash the pellets twice with 0.8% physiologic saline.
5. Transfer the pellets to 200 μL extraction buffer (0.2 M Tris–HCl pH 7.6, 0.5 M NaCl, 0.1% sodium dodecyl sulfate, 0.01 M EDTA).
6. Add glass beads to this mixture in a 1:1 ratio.
7. Vortex vigorously in a bead beater to achieve 60% lysis of the mycelial mass (approximately 30 min, with intermission every 5 min to avoid overheating).
8. Fungal DNA may then be recovered with the use of kits such as the DNeasy plant mini kit (QIAGEN).
9. In order to assess for product, electrophorese the product on a 1% agarose gel with 1× TBE buffer (8.9 mM Tris-borate, 0.2 mM EDTA).
10. Stain with ethidium bromide and assess for product.
11. Check for purity of the extracted DNA at 260 and 280 nm with spectrophotometric analysis.

8.2.2 Detection Procedures

Some detection methods that have been assessed using *Curvularia* spp. are outlined in Table 8.1. Below is a suggested method of detection using sequence analysis of the ITS rRNA gene region, as described by Sanguinetti et al. [117]:

TABLE 8.1
Published Molecular Methods for Identification of *Curvularia* spp.

Method	Target	Examined Material	No. of *Curvularia* spp. Examined	Sample Preparation and Extraction	Detection Method	Performance of System, as Relates to *Curvularia* spp.	Miscellaneous	Reference
MicroSeq D2 LSU rDNA sequencing kit	D2 LSU rDNA fragment	Filamentous fungi, including dematiaceous	2 *Curvularia* spp.	PrepMan ultra sample preparation reagent (Applied Biosystems)	Cycle sequencing followed by ABI 3100 16 capillary genetic analyzer (Applied Biosystems)	*Curvularia* identified correctly. One isolate identified as *Bipolaris hawaiiensis/Cochliobolus*	Commercially available	[116]
ITS (internal transcribed spacer)1 sequencing	ITS1-5.8S-ITS2 DNA region	Black-grain mycetoma agents	4 *Curvularia* spp.	Glass bead lysis method, snap-frozen, then DNeasy plant kit (QIAGEN)	Amplification of the rRNA gene with primers V9D and LS266 in ICycler thermocycler (Bio-Rad). Sequencing performed using BigDye Terminator Cycle Sequencing Ready Reaction kit, version 3.1 (Applied Biosystems, Foster City, CA). Analysis performed with ABI Prism 3700 automated DNA analyzer (Applied Biosystems)	Differentiated *Curvularia* well from other fungi, but some difficulty with intraspecies variation	Certain components are commercially available	[115]
Sequencing of ITSI and ITSII, confirmed by SSCP	ITSI and ITSII sequencing	Direct corneal scrapings	1 case of *Curvularia* keratitis	Glass bead lysis method, then DNeasy plant kit (QIAGEN)	Six fungus-specific primers (primers ITSI, ITS2, ITS3, ITS4, invSR1R, and LR12R) were used in amplification. Sequencing of ITSI and ITSII was performed with primer pair ITS1 and ITS4 with an ABI Prism automated DNA sequencer. Results confirmed with SSCP	One case of *C. inaequalis* keratitis confirmed by culture	Certain components are commercially available	[114]

Note: ITS, internal transcribed spacer; SSCP, single-stranded conformation polymorphism.

1. Amplify the ITS1-5.8S-ITS2 region of the rRNA gene, using a 1:50 dilution of template DNA.
2. The total reaction volume should be 50 µL (10 mM Tris–HCl pH 8.3, 50 mM KCl, 1.5 mM $MgCl_2$, 0.8 mM deoxynucleoside triphosphates at 0.2 mM each, 1.2 U *Taq* DNA polymerase, 0.5 µM each of the fungus-specific universal primers ITS1 (5′-TCCGTAGGTGAACCTGCGG-3′), and ITS4 (5′-TCCTCCGCTTATTGATATGC-3′).
3. Thermal conditions: 94°C for 3 min, 35 cycles each at 94°C for 1 min, 55°C for 1 min, 72°C for 1 min and 30 s, and the final extension step of 72°C for 10 min.
4. Purify amplicons (suggestion includes a Minielute PCR purification kit [QIAGEN]).
5. Sequence with primer ITS1 or ITS4 and BigDye terminator cycle sequencing kit (Applied Biosystems) on the ABI Prism 3100 genetic analyzer (Applied Biosystems).
6. Species may be identified by searches on the BLAST sequence analysis tool (http://www.ncbi.nlm.nih.gov/BLAST/). An isolate may be assigned to a species if there is ≥99% homology in sequence to the highest entry, and if the next entry/species shows less than 95% homology when the entire length of the sequence is examined.

Hall et al. reported on their experience with the commercially available MicroSeq D2 LSU rDNA fungal sequencing kit (Applied Biosystems) [116]. The MicroSeq kit has multiple components and is comprised of PCR and cycle sequencing modules, identification and analysis software, and a library of fungal nucleic acid sequences. The authors compared nucleic acid sequencing using the MicroSeq kit with traditional identification of cultures by phenotypic methods. A total of 234 filamentous fungi from culture were evaluated, 2 of which were *Curvularia* spp. In their study, DNA extraction was performed using PrepMan Ultra sample preparation reagent (Applied Biosystems). The D2 LSU rDNA fragment was then amplified. Once the presence of amplicon was confirmed, purification of the PCR product was performed to remove excess primers and nucleotides. Next, cycle sequencing was performed using the sequence module reagents, with the following parameters: 25 cycles of 96°C for 10 s, 50°C for 5 s, and 60°C for 4 min. Labeled amplicon was placed onto the ABI 3100 16 capillary genetic analyzer (Applied Biosystems). The organism that gave the closest match to the sequence file was considered to be the most closely related organism.

Of the two *Curvularia* spp. in the study by Hall et al. that were identified based on phenotypic methods and then sequenced, one had a distance score of ≤1% (99% similarity) to the MicroSeq D2 fungal library (which contained 788 species of filamentous fungi) (version 1.4.2, February 2002). The other *Curvularia* spp. showed a distance score of ≥1%, and its sequence was identified as *Bipolaris hawaiiensis/Cochliobolus* (the MicroSeq system was unable to distinguish between the two genera listed). However, the distance score was only mildly elevated, at 2.48. In summary, the D2 rDNA fungal library contains sequences that represent 205 genera of filamentous fungi and include type culture for many fungi, while offering rapid and accurate identification of organisms.

Desnos-Ollivier et al. described the accuracy of molecular identification of several dematiaceous fungi causing black-grain mycetoma, using the ITS1-5.8S-ITS2 DNA region [115]. The authors examined 54 strains of fungi from the Pasteur Institute and the National Reference Center for Mycoses and Antifungals; 78% of strains were clinical in origin. Four *C. lunata* stains were examined in their study. Their methods are outlined briefly below. For DNA extraction, strains cultured in RPMI 1640 medium were extracted using a glass bead lysis method and the DNeasy plant kit (QIAGEN). The rRNA gene was then amplified by PCR using primers V9D and LS266 [118]. Amplification took place in an ICycler thermocycler (Bio-Rad). PCR products were purified on P-100 Gel Fine (Bio-Rad). Strands were then sequenced using the BigDye Terminator Cycle Sequencing Ready Reaction kit, version 3.1 (Applied Biosystems), with primers V9D and LS266. Products were then analyzed using an ABI Prism 3700 automated DNA analyzer (Applied Biosystems). Sequences were edited with Chromas, version 2.24 (Technelysium, Helensvale, Queensland, Australia).

Three of the four *C. lunata* strains from the study by Desnos-Ollivier et al. showed intraspecies similarities of >98%. The single strain that showed sequence differences of more than 50 base pairs (>10% differences) was tentatively identified as *Curvularia* species, as the species could not be confirmed. In comparison to other dematiaceous black-grain mycetoma agents analyzed in this study, the four *C. lunata* strains showed sequence similarities ranging from 60% to 79%, both between species and between genera. In conclusion, although intraspecies variation was observed for *C. lunata*, the wide intergenera and interspecies variation observed between *C. lunata* and other black-grain mycetoma agents demonstrate that ITS sequencing may be a useful tool for reliable separation of *Curvularia* spp. from other common black-grain mycetoma agents examined in this study. However, more ITS sequencing studies should be performed on *Curvularia* spp. to examine the system's ability to speciate this organism.

Kumar et al. diagnosed four cases of mycotic keratitis using SSCP of the ITS1 region [110]. Clinical testing material consisted of corneal scrapings from affected patients, from whom genomic DNA was directly extracted. The authors found their method to be both sensitive as well as cost-effective, although no cases of *Curvularia* were diagnosed in this study.

In another study, Kumar and Shukla diagnosed mycotic keratitis in three patients using SSCP with six fungus-specific primers (primers ITSI, ITS2, ITS3, ITS4, invSR1R, and LR12R) [114]. All three patients were positive for fungi by this technique, one of whom had *C. inaequalis* keratitis. Molecular results correlated with culture results. Complete sequencing of the ITSI and ITSII was performed with primer pair ITS1 and ITS4 using an ABI Prism automated DNA sequencer (model 3100, version 3.0, Applied Biosystems, Warrington, United Kingdom), with the primer ITS1, and organism identifications were confirmed by SSCP. The authors noted that direct clinical material (corneal scrapings) was used, but DNA extraction with the DNeasy plant mini kit (QIAGEN) did not perform as well as the QIAamp tissue mini kit (QIAGEN). In conclusion, the authors diagnosed three cases of mycotic keratitis directly from corneal scrapings within 8 h of collection by the ITS SSCP technique.

Thus, although there have been relatively few published molecular studies of *Curvularia* detection in clinical samples, these methods have shown promise as rapid and accurate tools for diagnosis.

8.3 CONCLUSIONS

Full identification of *Curvularia* to the species level can be difficult and time-consuming, when the laboratory must rely solely on morphology and phenotypic differentiation. A successful diagnosis can be highly dependent on the experience of the microbiologist. As the immunocompromised populations of transplant recipients and other immunosuppressed individuals are growing, confirmation to the species level can be important to establish reliably and as rapidly as possible. In addition, a growing number of *Curvularia* species are being identified as causes of human disease.

Researchers should continue to conduct molecular studies in an effort to gain a better understanding of the role of *Curvularia* in disease. As the authors of one study pointed out, *C. inaequalis* and *C. lunata* appear morphologically very similar, and only molecular analysis enabled the authors to correctly identify the isolate [101]. Further work should be performed on a larger subset of *Curvularia* spp. to allow for speciation. In addition, as further differentiation between and among fungal species is made possible by molecular methods, new knowledge will likely be gained of the therapeutic efficacy of certain drugs against newly discovered fungi. Although few guidelines are available regarding the use of molecular methods, Hall et al. suggest that sequencing may prove most useful for those organisms which do not sporulate within a 48-h period on subculture [116]. It is also promising that molecular tools may be used for direct diagnosis of fungal agents from infected tissues, such as the data presented by Kumar et al. on direct detection of fungal pathogens from corneal scrapings [110, 114].

Cost–benefit analyses should be performed to compare phenotypic methods of identification to molecular methods, while weighing the advantage of rapidity which molecular methods typically offer. Although accurate, molecular approaches to diagnosis may cause potential confusion for clinicians, as new species are identified and information on treatment may be unknown. It will be necessary for microbiology laboratories and clinicians to communicate these potentially significant findings.

REFERENCES

1. Fan, Y.M. et al., Cutaneous phaeohyphomycosis of foot caused by *Curvularia clavata*. *Mycoses*, 52, 544, 2008.
2. Baylet, J., Camain, R., and Segretain, G., Identification of the agents of maduromycoses of Senegal and Mauritania. Description of a new species. *Bull. Soc. Pathol. Exot. Filiales*, 52, 448, 1959.
3. Rinaldi, M.G. et al., Human *Curvularia* infections: Report of five cases and review of the literature. *Diagn. Microbiol. Infect. Dis.*, 6, 27, 1987.
4. Wilhelmus, K.R. and Jones, D.B., *Curvularia* keratitis. *Trans. Am. Ophthalmol. Soc.*, 99, 111, 2001.
5. Revankar, S.G., Phaeohyphomycosis. *Infect. Dis. Clin. N. Am.*, 20, 609, 2006.
6. Hiromoto, A., Nagano, T., and Nishigori, C., Cutaneous infection caused by *Curvularia* species in an immunocompetent patient. *Br. J. Dermatol.*, 158, 1374, 2008.
7. Clinch, T.E. et al., Fungal keratitis from nylon line lawn trimmers. *Am. J. Ophthalmol.*, 114, 437, 1992.
8. Flournoy, D.J., Mullins, J.B., and McNeal, R.J., Isolation of fungi from rose bush thorns. *J. Okla. State Med. Assoc.*, 93, 271, 2000.
9. Ujhelyi, M.R. et al., Treatment of peritonitis due to *Curvularia* and *Trichosporon* with amphotericin B. *Rev. Infect. Dis.*, 12, 621, 1990.
10. Canon, H.L. et al., Fungal peritonitis caused by *Curvularia* species in a child undergoing peritoneal dialysis. *Pediatr. Nephrol.*, 16, 35, 2001.
11. Tessari, G. et al., Lethal systemic dissemination from a cutaneous infection due to *Curvularia lunata* in a heart transplant recipient. *J. Eur. Acad. Dermatol. Venereol.*, 17, 440, 2003.
12. Rohwedder, J.J. et al., Disseminated *Curvularia* infection in a football player. *Arch. Intern. Med.*, 139, 940, 1979.
13. Lavoie, S.R., Espinel-Ingroff, A., and Kerkering, T., Mixed cutaneous phaeohyphomycosis in a cocaine user. *Clin. Infect. Dis.*, 17, 114, 1993.
14. Fernandez, M. et al., Cutaneous phaeohyphomycosis caused by *Curvularia lunata* and a review of *Curvularia* infections in pediatrics. *Pediatr. Infect. Dis.*, 18, 727, 1999.
15. Grieshop, T.J., Yarbrough, D., and Farrar, W.E., Case report: Phaeohyphomycosis due to *Curvularia lunata* involving skin and subcutaneous tissue after an explosion at a chemical plant. *Am. J. Med. Sci.*, 305, 387, 1993.
16. Kainer, M.A. et al., Saline-filled breast implant contamination with *Curvularia* species among women who underwent cosmetic breast augmentation. *J. Infect. Dis.*, 192, 170, 2005.
17. Sandhu, D.K. and Randhawa, I.S., Studies on the air-borne fungal spores in Amristar: Their role in keratomycosis. *Mycopathologia*, 68, 47, 1979.
18. Leck, A.K. et al., Aetiology of suppurative corneal ulcers in Ghana and south India, and epidemiology of fungal keratitis. *Br. J. Ophthalmol.*, 86, 1211, 2002.
19. Dixit, A. et al., Comparison of mold capture and mold sensitivity in the United States Gulf Coast [abstract]. *J. Allergy Clin. Immunol.*, 91, 275, 1993.
20. Wilhelmus, K.R., Climatology of dematiaceous fungal keratitis. *Am. J. Ophthalmol.*, 140, 1156, 2005.

21. Rout, N., Nanda, B.K., and Gangopadhyaya, S., Experimental phaeohyphomycosis and mycotoxicosis by *Curvularia lunata* in albino rats. *Indian J. Pathol. Microbiol.*, 32, 1, 1989.
22. Ortega, J., Production of extracellular cell wall degrading enzymes by *Curvularia senegalensis*. *Texas J. Sci.*, 45, 335, 1993.
23. Jacobson, E.S., Pathogenic roles for fungal melanins. *Clin. Microbiol. Rev.*, 13, 708, 2000.
24. Butler, M.J. and Day, A.W., Fungal melanins: A review. *Can. J. Microbiol.*, 44, 1115, 1998.
25. Hamilton, A.J. and Gomez, B.L., Melanins in fungal pathogens. *J. Med. Microbiol.*, 51, 189, 2002.
26. Diskin, C.J. et al., Case report and review: Is the tendency for *Curvularia* tubular obstruction significant in pathogenesis? *Perit. Dial. Int.*, 28, 678, 2009.
27. Herrera, L.K. et al., Biodeterioration of peridotite and other constructional materials in a building of the Colombian cultural heritage. *Int. Biodeterior. Biodegrad.*, 54, 135, 2004.
28. Wallstrom, S. and Karlsson, S., Biofilms on silicone rubber insulators; microbial composition and diagnostics of removal by use of ESEM/EDS: Composition of biofilms infecting silicone rubber insulators. *Polym. Degrad. Stabil.*, 85, 841, 2004.
29. Simmons, R.B., Zhang, S.T., and Ahearn, D.H., Fungi associated with hydrogel contact lenses. *Microsc. Microanal.*, 13, S312, 2007.
30. Kaiser, C. et al., In vitro inhibition of mycelia growth of several pathogenic fungi including *Phytophthora cinnamomi* by soluble silicon. *South Afr. Growers Assoc. Yearbook*, 28, 60, 2005.
31. Chakraborty, S., Sen, S.K., and Battacharya, K., Indoor and outdoor aeromycological survery in Burdwan, West Bengal, India. *Aerobiologia*, 16, 211, 2000.
32. Chew, F.T. et al., Evaluation of the allergenicity of tropical pollen and airborne spores in Singapore. *Allergy*, 55, 340, 2000.
33. Adhikari, A. et al., Airborne viable, non-viable, and allergenic fungi in a rural agricultural area of India: A 2-year study at five outdoor sampling stations. *Sci. Total Environ.*, 326, 123, 2004.
34. Gupta, R. et al., Allergenic cross-reactivity of *Curvularia lunata* with other airborne fungal species. *Allergy*, 57, 636, 2002.
35. Gupta, R. et al., Identification of cross-reactive proteins amongst different *Curvularia* species. *Int. Arch. Allergy Immunol.*, 127, 38, 2002.
36. Gupta, R. et al., Identification of serine protease as a major allergen of *Curvularia lunata*. *Allergy*, 59, 421, 2004.
37. Sharma V. et al., Cloning, recombinant expression and activity studies of a major allergen "enolase" from the fungus *Curvularia lunata*. *J. Clin. Immunol.*, 26, 360, 2006.
38. Tripathi, P. et al., Serine protease activity of Cur 1 1 from *Curvularia lunata* augments Th2 response in mice. *J. Clin. Immunol.*, 29, 292, 2009.
39. Gupta, R. et al., Allergens of *Curvularia lunata* during cultivation in different media. *J. Allergy Clin. Immunol.*, 104, 857, 1999.
40. Chander, J. and Sharma, A., Prevalence of fungal corneal ulcers in northern India. *Infection*, 22, 207, 1994.
41. Panda, A. et al., Mycotic keratitis in children: Epidemiologic and microbiologic evaluation. *Cornea*, 16, 295, 1997.
42. Chander, J. et al., Keratomycosis in and around Chandigarh: A five-year study from a north Indian tertiary care hospital. *Indian J. Pathol. Microbiol.*, 51, 304, 2008.
43. Qiu, W.Y. et al., Fungal spectrum identified by a new slide culture and in vitro drug susceptibility using Etest in fungal keratitis. *Curr. Eye Res.*, 30, 1113, 2005.
44. Dunlop, A.A. et al., Suppurative corneal ulceration in Bangladesh: A study of 142 cases examining the microbiological diagnosis, clinical and epidemiological features of bacterial and fungal keratitis. *Aust. N. Z. J. Ophthalmol.*, 22, 105, 1994.
45. Garg, P. et al., Keratomycosis: Clinical and microbiologic experience with dematiaceous fungi. *Ophthalmol.*, 107, 574, 2000.
46. Gopinathan, U. et al., The epidemiological features and laboratory results of fungal keratitis: A 10-year review at a referral eye care center in South India. *Cornea*, 21, 555, 2002.
47. Chowdhary, A. and Singh, K., Spectrum of fungal keratitis in North India. *Cornea*, 24, 8, 2005.
48. Liesegang, T.J. and Forster, R.K., Spectrum of microbial keratitis in south Florida. *Am. J. Ophthalmol.*, 90, 38, 1980.
49. Rosa, R.H., Miller, D., and Alfonso, E.C., The changing spectrum of fungal keratitis in south Florida. *Ophthalmology*, 101, 1005, 1994.
50. Poria, V.C. et al., Study of mycotic keratitis. *Indian J. Ophthalmol.*, 33, 229, 1985.
51. Forster, R.K., Rebell, G., and Wilson, L.A., Dematiaceous fungal keratitis. Clinical isolates and management. *Br. J. Ophthalmol.*, 59, 372, 1975.
52. Boonpasart, S. et al., Infectious keratitis at King Chulalongkorn Memorial Hospital: A 12-year retrospective study of 391 cases. *J. Med. Assoc. Thai.*, 85 (Suppl. 1), S217, 2002.
53. Thew, M. and Franzco, B.T., Fungal keratitis in far north Queensland, Australia. *Clin. Exp. Ophthalmol.*, 36, 721, 2008.
54. Wilson, L.A. and Ahearn, D.G., Association of fungi with extended wear soft contact lenses. *Am. J. Ophthalmol.*, 101, 434, 1986.
55. Tuli, S. and Yoo, S.H., *Curvularia* keratitis after laser in situ keratomileusis from a feline source. *J. Cataract Refract. Surg.*, 29, 1019, 2003.
56. Chung, M.S. et al., Fungal keratitis after laser in situ keratomileusis: A case report. *Cornea*, 19, 236, 2000.
57. Wilson, L.A. et al., Fungi from the normal outer eye. *Am. J. Ophthalmol.*, 67, 52, 1969.
58. Iyer, S.A., Tuli, S.S., and Wagoner, R.C., Fungal keratitis: Emerging trends and treatment outcomes. *Eye Contact Lens*, 32, 267, 2006.
59. Sodhi, P.K. and Kaur, R., *Curvularia* dacryocystitis: Report of two cases. *Orbit*, 19, 45, 2000.
60. Kaushik, S. et al., *Curvularia lunata* endophthalmitis with secondary keratitis. *Am. J. Ophthalmol.*, 131, 140, 2001.
61. Pathengay, A. et al., *Curvularia lunata* endophthalmitis presenting with a posterior capsule plaque. *Indian J. Ophthalmol.*, 54, 65, 2009.
62. Thomas, P.A., Fungal infections of the cornea. *Eye*, 17, 852, 2003.
63. deShazo, R.D., Chapin, K., and Swain, R.E., Fungal sinusitis. *N. Engl. J. Med.*, 337, 254, 1997.
64. Corey, J.P., Delsupehe, K.G., and Ferguson, B.J., Allergic fungal sinusitis: Allergic, infectious, or both? *Otolaryngol. Head Neck Surg.*, 13, 110, 1995.
65. Corey, J.P., Allergic fungal sinusitis. *Otolaryngol. Clin. North Am.*, 25, 225, 1992.
66. Taj-Aldeen, S.J., Hilal, A.A., and Schell, W.A., Allergic fungal rhinosinusitis: A report of 8 cases. *Am. J. Otolaryngol.*, 25, 213, 2004.

67. Schubert, M.S., Allergic fungal sinusitis: Pathophysiology, diagnosis and management. *Med. Mycol.*, 47, S324, 2009.
68. Houser, S.M. and Corey, J.P., Allergic fungal rhinosinusitis. *Otolaryngol. Clin. North Am.*, 33, 251, 2000.
69. Asero, R. and Bottazi, G., Nasal polyposis: A study of its association with airborne allergen hypersensitivity. *Ann. Allergy Asthma Immunol.*, 86, 283, 2001.
70. Schubert, M.S. and Goetz, D.W., Evaluation and treatment of allergic fungal sinusitis. I. Demographics and diagnosis. *J. Allergy Clin. Immunol.*, 102, 387, 1998.
71. Schubert, M.S. and Goetz, D.W., Evaluation and treatment of allergic fungal sinusitis. II. Treatment and follow-up. *J. Allergy Clin. Immunol.*, 102, 395, 1998.
72. Chakrabarti, A. and Sharma, S.C., Paranasal sinus mycoses. *Indian J. Chest Dis. Allied Sci.*, 42, 293, 2000.
73. Ismail, Y. et al., Invasive sinusitis with intracranial extension caused by *Curvularia lunata*. *Arch. Intern. Med.*, 153, 1604, 1993.
74. Bonduel, M. et al., Atypical skin lesions caused by *Curvularia* sp. and *Pseudallescheria boydii* in two patients after allogeneic bone marrow transplantation. *Bone Marrow Transplant.*, 27, 1311, 2001.
75. Lopes, J.O. and Jobim, N.M., Dermatomycosis of the toe web caused by *Curvularia lunata*. *Rev. Inst. Med. Trop. Sao Paulo*, 40, 327, 1998.
76. Rowen, J.L. et al., Invasive fungal dermatitis in the < or = 1000-gram neonate. *Pediatrics*, 95, 682, 1995.
77. Gupta, M. et al., Onychomycosis: Clinic-mycologic study of 130 patients from Himachal Pradesh, India. *Indian J. Dermatol. Venereol. Leprol.*, 73, 389, 2007.
78. Veer, P., Patwardhan, N.S., and Damie, A.S., Study of onychomycoses: Prevailing fungi and pattern of infection. *Indian J. Med. Microbiol.*, 25, 53, 2007.
79. Ramani, R. et al., Molds in onychomycosis. *Int. J. Dermatol.*, 32, 877, 1993.
80. Brandt, M.E. and Warnock, D.W., Epidemiology, clinical manifestations, and therapy of infections caused by dematiaceous fungi. *J. Chemother.*, 15(Suppl. 2), 36, 2003.
81. Janaki, C. et al., Case report. Eumycetoma due to *Curvularia lunata*. *Mycoses*, 42, 345, 1999.
82. Garg, A. et al., Eumycetoma due to *Curvularia lunata*. *Indian J. Dermatol. Venereol. Leprol.*, 74, 515, 2008.
83. Yau, Y.C.W. et al., Fungal sternal wound infection due to *Curvularia lunata* in a neonate with congenital heart disease: Case report and review. *Clin. Infect. Dis.*, 19, 735, 1994.
84. Lampert, R.P. et al., Pulmonary and cerebral mycetoma caused by *Curvularia pallescens*. *J. Pediatr.*, 91, 603, 1977.
85. De la Monte, S.M. and Hutchins, G.M., Disseminated *Curvularia* infection. *Arch. Pathol. Lab. Med.*, 109, 872, 1985.
86. Smith, T. et al., Optic atrophy due to *Curvularia* lunata mucocoele. *Pituitary*, 10, 295, 2007.
87. Carter, E. and Boudreaux, C., Fatal cerebral phaeohyphomycosis due to *Curvularia lunata* in an immunocompetent patient. *J. Clin. Microbiol.*, 42, 5419, 2004.
88. DeVault, G.A. Jr. et al., Tenchkoff catheter obstruction resulting from invasion by *Curvularia lunata* in the absence of peritonitis. *Am. J. Kidney Dis.*, 6, 124, 1985.
89. Ebright, J.R. et al., Invasive sinusitis and cerebritis due to *Curvularia clavata* in an immunocompetent adult. *Clin. Infect. Dis.*, 28, 687, 1999.
90. Parva, P., Rojas, R., and Palacios, E., Unusual rhinosinusitis caused by *Curvularia* fungi. *Ear Nose Throat J.*, 84, 270, 2005.
91. Lopes, J.O. et al., *Cuvularia lunata* peritonitis complicating peritoneal dialysis. *Mycopathologia*, 127, 65, 1994.
92. Robson, A.M. and Craver, R.D., *Curvularia* urinary tract infection: A case report. *Pediatr. Nephrol.*, 8, 83, 1994.
93. Shigemori, M. et al., Hepatosplenic abscess caused by *Curvularia boedijn* in a patient with acute monocytic leukemia. *Pediatr. Infect. Dis. J.*, 12, 1128, 1996.
94. Torda, A.J. and Jones, P.D., Necrotizing cutaneous infection caused by *Curvularia brachyspora* in an immunocompetent host. *Australas. J. Dermatol.*, 38, 85, 1997.
95. Georg, L.K., *Curvularia geniculata*, a cause of mycotic keratitis. *J. Med. Assoc. State Ala.*, 33, 234, 1964.
96. Kaufman, S.M., *Curvularia* endocarditis following cardiac surgery. *Am. J. Clin. Pathol.*, 56, 466, 1971.
97. Harris, J.J. and Downham, T.F., Unusual fungal infections associated with immunological hyposensitivity. *Int. J. Derm.*, 17, 323, 1978.
98. Vachharajani, T.J. et al., *Curvularia geniculata* fungal peritonitis: A case report with review of literature. *Int. Urology Nephrol.*, 37, 781, 2005.
99. Singh, H. et al., *Curvularia* fungi presenting as a large cranial base meningioma: Case report. *Neurosurgery*, 63, E177, 2008.
100. Pimentel, J.D. et al., Peritonitis due to *Curvularia inaequalis* in an elderly patient undergoing peritoneal dialysis and a review of six cases of peritonitis associated with other *Curvularia* spp. *J. Clin. Microbiol.*, 43, 4288, 2005.
101. Posteraro, B. et al., Eosinophilic fungal rhinosinusitis due to the unusual pathogen *Curvularia inaequalis*. *Mycoses*, 15, 36, 2009.
102. Berg, D. et al., Cutaneous infection caused by *Curvularia pallescens*: A case report and review of the spectrum of disease. *J. Am. Acad. Dermatol.*, 32, 375, 1995.
103. Agrawal, A. and Singh, S.M., Two cases of cutaneous phaeohyphomycosis caused by *Curvularia pallescens*. *Mycoses*, 38, 301, 1995.
104. Guarro, J. et al., Mycotic keratitis due to *Curvularia senegalensis* and in vitro antifungal susceptibilities of *Curvularia* spp. *J. Clin. Microbiol.*, 37, 4170, 1999.
105. de Hoog, G.S., Guarro, J., Gene, J., and Figueras, M.J., *Atlas of Clinical Fungi*, 2nd edn., Vol. 1. Centraalbureau voor Schimmelcultures, Utrecht, the Netherlands, 2000.
106. De Luna-Alvea and Maia, L.C., Morphological, cytological, and cultural aspects of *Curvularia pallescens*. *Rev. Microbiol.*, 29, 197, 1998. From Wilhelmus, K.R. and Jones, D.B., *Curvularia* keratitis. *Trans. Am. Ophthalmol. Soc.*, 99, 111, 2001.
107. Salfelder, K., *Atlas of Fungal Pathology*. Kluwer Academic Publishers, Hingham, MA, 1990, p. 56.
108. Kimura, M. and McGinnis, M.R., Fontana-Masson—Stained tissue from culture-proven mycoses. *Arch. Pathol. Lab. Med.*, 122, 1107, 1998.
109. Iwen, P.C., Hinrichs, S.H., and Rupp, M.E., Utilization of the internal transcribed spacer region as molecular targets to detect and identify human fungal pathogens. *Med. Mycol.*, 40, 87, 2002.
110. Kumar, M., Mishra, N.K., and Shukla, P.K., Sensitive and rapid polymerase chain reaction based diagnosis of mycotic keratitis through single stranded conformation polymorphism. *Am. J. Ophthalmol.*, 140, 851, 2005.
111. Jackson, C.J. et al., Species identification and strain differentiation of dermatophyte fungi by analysis of ribosomal-DNA intergenic spacer regions. *J. Clin. Microbiol.*, 37, 931, 1999.

112. Kowalchuk, G.A., Gerards, S., and Woldendorp, J.W., Detection and characterization of fungal infections of *Ammophila arenaria* (Marram Grass) roots by denaturing gradient gel electrophoresis of specifically amplified 18S rDNA. *Appl. Environ. Microbiol.*, 63, 3858, 1997.
113. Smit, E. et al., Analysis of fungal diversity in the wheat rhizosphere by sequencing of cloned PCR-amplified genes encoding 18S rRNA and temperature gradient gel electrophoresis. *Appl. Environ. Microbiol.*, 65, 2614, 1999.
114. Kumar, M. and Shukla, P.K., Use of PCR targeting of internal transcribed spacer regions and single-stranded conformation polymorphism analysis of sequence variation in different regions of rRNA genes in fungi for rapid diagnosis of mycotic keratitis. *J. Clin. Microbiol.*, 43, 662, 2005.
115. Desnos-Ollivier, M. et al., Molecular identification of black-grain mycetoma agents. *J. Clin. Microbiol.*, 44, 3517, 2006.
116. Hall, L., Wohlfiel, S., and Roberts, G.D., Experience with the MicroSeq D2 large-subunit ribosomal DNA sequencing kit for identification of filamentous fungi encountered in the clinical laboratory. *J. Clin. Microbiol.*, 42, 622, 2004.
117. Sanguinetti, M. et al., Evaluation of VITEK 2 and RapID yeast plus systems for yeast species identification: Experience at a large clinical microbiology laboratory. *J. Clin. Microbiol.*, 45, 1343, 2007.
118. Gerrits van den Ende, A.H.G. and de Hoog, G.S., Variability and molecular diagnostics of the neurotropic species *Cladophialophora bantiana*. *Stud. Mycol.*, 43, 151, 1999.

9 *Exserohilum*

K. Lily Therese and H.N. Madhavan

CONTENTS

9.1 INTRODUCTION

Phaeohyphomycosis is the term used to denote the infections caused by more than 100 species in 60 genera belonging to dematiaceous fungal pathogens (including the common and closely related genera *Bipolaris*, *Drechslera*, and *Exserohilum*). The genus *Exserohilum* was established[1] for species having distinctly protruding hila.[2] This eliminated the inconsistency and permitted a more logical grouping of species into three anamorphic genera, *Bipolaris*, *Drechslera*, and *Exserohilum*. Alcorn et al.[3] described in detail the concept and justification for grouping them into three genera. *Exserohilum* species are ubiquitous but rarely pathogenic for human beings. They are encountered in grass or rotting wood and thrive in warm and humid climates. The genus is characterized by its conidia, which are ellipsoidal, distoseptate, and have a protruding and truncate hilum. Three species of *Exserohilum* have been recognized as human pathogens: *Exserohilum rostratum*, *Exserohilum longirostratum*, and *Exserohilum mcginnisii*. The most common infections caused by *Exserohilum* species are sinusitis and skin infections, although a few cases of cerebral abscesses, keratitis, osteomyelitis, prosthetic valve endocarditis, and disseminated infection have been described.[4–7]

9.1.1 Classification, Morphology, and Biology

9.1.1.1 Classification

The taxonomic classification is as follows: Kingdom Fungi, Phylum Ascomycota, Class Euascomycetes, Order Pleosporales, Family *Pleosporaceae*, Genus *Exserohilum*. The distinct telemorphic state of *Exserohilum* is *Setosphaeria*. *Exserohilum* is a terrestrial plant pathogen and is differentiated from the closely related genera based on the conidial formation from a protruding exserted hilum. The dark pigmentation is caused by deposition of dihydroxynaphthalene melanin.[4] The hilum is present as a scar on a conidium, at the point of attachment to the conidiophore. It is distinguished on the basis of conidial shape and size, hilar morphology, location and sequence of the conidial septa, and the origin of the germ tubes from the conidial cells. The conidium of

Exserohilum species is ellipsoidal to fusiod and germinates by germ tubes originating from either one or both of the end cells or other intermediate cells.

9.1.1.2 Description and Natural Habitats

Exserohilum is one among the three dematiaceous (*Bipolaris, Drechslera, Exserohilum*) fungi that are commonly known to cause infections in human beings. It is cosmopolitan in nature inhabiting plant material, particularly grasses, and soil. More than 20 host-specific species are known worldwide. It has been shown that human *Exserohilum* infections occur mainly in warm, tropical, and subtropical areas such as the southern United States, India, and Israel. The two recently reported new species *Exserohilum israeli*[5] and *Exserohilum sodomii*[6] are from Southern Israel. The differential identification of the common three species of *Exserohilum* is based on the macroscopic and microscopic morphology of the conidia.

9.1.1.2.1 Exserohilum rostratum (E. rostratum)

E. rostratum grows well on Sabouraud's dextrose agar (SDA) and potato dextrose agar (PDA). The colonies appear cottony to woolly in texture, raised in the central area measuring about 25–27 mm in diameter, at the end of 2 weeks of incubation at 25°C. The colonies are gray to blackish-brown, suede-like to floccose in texture, and have an olivaceous black reverse. The conidial formation on PDA may take up to 3 weeks. The fungus grows well at 37°C, but the growth is very slow at 40°C with only 5–6 mm in diameter after 2 weeks. The growth is partially inhibited when grown on SDA containing cycloheximide.[7]

Hyphae are septate and dematiaceous, with conidiophores up to 200 μm long, which are nonbranched, septate, geniculate, and are pale near the apex. Conidia are straight to slightly curved, rostrate or beaklike, and olivaceous brown, and contain typically 7–9 pseudosepta (range from 4 to 14). The conidia have prominent, dark basal, and distal septa, and a strongly protruding truncate hilum. The mature conidia range from 8 (ranging from 6 to 13 μm) to 23 μm (ranging from 14 to 34) in size, straight, slightly curved or bent, and cylindrical to rostrate. The end cells, often paler than the other cells, and the walls are often finely roughened. Conidia have a strongly protruding truncate hilum and the septum above the hilum is usually thickened and dark. The mature conidia have characteristic darkly pigmented septum at both ends of conidium that is characteristic of *E. rostratum*, useful to differentiate conidia of *E. mcginnis*.[8]

9.1.1.2.2 Exserohilum longirostratum (E. longirostratum)

E. longirostratum is considered as a variant of *E. rostratum*. Colonies on PDA are woolly to cottony and are black to dark gray with a black reverse. Growth is rapid at 25°C.[7]

Hyphae are septate and brown. Conidiophores are pale near the apex, septate, simple, geniculate, and measure up to 200 μm in length. The mature conidia are straight, ellipsoidal, and strongly protruding with a truncate hilum measuring about 13–16 × 196–260 μm in size. Conidia are broadest at the basal end with a strongly protruding truncate basal hilum, contain 13–21 distosepta, and narrow gradually towards the apex into a long beak. They resemble the conidia of *E. rostratum* with the pale end cells and finely roughened walls. *E. longistratum* has predominantly 9–11 septa. The conidia are present in two forms—long (13–21 septa) and short conidia (5–7 septa). The conidia have prominent dark basal and distal septa, frequently displaying a curvature in the middle. The size of the conidia measures between 60 and 475 μm in length × 12–26 μm in width. *E. longistratum* produce elongate conidia with a rostrate shape when periodically exposed to fluorescent light, but when grown in the dark, the septa separating the distal and basal conidial cells appears darkly pigmented and thickened. The obsolete names for anamorph of *E. longirostratum* are *Bipolaris longirostratum, Drechslera longirostratum*, and *Helminosporium longirostratum*.[8]

9.1.1.2.3 Exserohilum mcginnisii (E. mcginnisii)

The colonies on PDA are raised, wooly, and deep mouse gray. The size of the colony varies depending on the temperature at which it grows. When grown at 25°C the diameter is 35–37 mm, at 37°C the diameter is 25–27 mm, and at 40°C the diameter is only 5–6 mm at the end of 2 weeks of incubation.[7]

The hyphae are septate, branched, and pale to mid brown in color. The conidiophores are simple, arising singly, erect or flexuous, upper part geniculate, and brown to mid brown.

Conidia are straight, cylindrical to ellipsoidal, broadest in the middle with rounded apices, 64–100 × 10–15 μm, and typically with 4–13 septa, but the mature conidia have typically 9–11 septa. The conidial walls are irregular with warty projections. The hilum is distinctly protuberant with irregular warty projections. Germination is bipolar. The pale end cells of the conidia are not separated from the intercalary golden brown cells by thick walled distosepta.[7] The size of the septa also ranges from 10 to 15 × 64 to 100 μm. The characteristic features useful for differential identification of *E. mcginnisii* from *E. rostratum* are lack of darkly pigmented bands at the ends of its conidia, irregular warty projections from the outer conidial cell walls, and slightly concave depressions in the conidial cell walls with larger numbers of septa and broader basal cells.[8]

9.1.2 Epidemiology

Exserohilum species are ubiquitous but rarely pathogenic for human beings. They are encountered in grass or rotting wood and thrive in warm and humid climates.[9] The major predisposing factors reported are trauma, corticosteroid therapy, diabetes, autoimmune diseases, lymphoma, leukemia, etc. Some are true pathogens and are kept under biosafety level two containment, others are opportunistic pathogens. The clinical localization of the pathogens has been seen in nails, skin, cornea, lungs, sinuses, liver, heart, genitals, bone, kidneys, brain, lymph nodes, etc. The four major clinical manifestations are cerebral infections, subcutaneous infections, paranasal sinusitis, and disseminated form of infections.

The first review on 10 published reports on human infections caused by *Bipolaris* and *Exserohilum* species[10] indicated that these genera were able to cause infections in apparently healthy hosts. Involvement of the paranasal sinuses, central nervous system, and other tissues was common to both genera. The ability of the above-mentioned species to cause life threatening neurotropic disease is indicated by their ability to grow at 40°C. The authors concluded that *Bipolaris* or *Exserohilum* species isolates from clinical specimens should not be regarded simply as contaminants; appropriate studies should be carried out to determine if they play an etiological role. In yet another study by Padhye et al. (1986),[7] seven cases of *Bipolaris* infection and two of *Exserohilum* infection were reported, demonstrating the capability of these two genera to cause invasive as well as "allergic" diseases in humans. These two genera of fungi are to be included in the differential diagnosis of the central nervous system, disseminated fungal diseases, sinusitis, keratitis, peritonitis associated with continuous ambulatory peritoneal dialysis, and allergic bronchopulmonary diseases. Although pathologic evidence of bone invasion may not be found, there frequently is radiographic evidence of invasive disease. As the analysis did not show a correlation between the regional mold counts and the incidence of infection, the geographic difference in prevalence could be related to a climate-induced alteration in the physiology of the mold, which renders it more pathogenic.[11] The most common infections caused by *Exserohilum* species are sinusitis and skin infections, although a few cases of cerebral abscesses, keratitis, osteomyelitis, prosthetic valve endocarditis, and disseminated infection have been described in the literature.[12–15]

A PubMed search of the English-language literature using the key words *Exserohilum* and *Drechslera* from 1950 till 2006 by Adler et al.[11] resulted in 33 cases of *Exserohilum* infection, of which 23 were reported since 1993. The increase may be due to improved diagnostic techniques. The majority of the *Exserohilum* infections were reported from regions with hot climates—16 cases from southern or southwestern states of the United States, 6 cases from Israel, and 5 cases from India.[11] Impaired immunity was present in the majority of patients with invasive and skin infections. In cases with corneal infections and allergic fungal sinusitis, local trauma and atopy were the predisposing factors.

9.1.3 Pathogenesis and Clinical Features

9.1.3.1 Invasive and Noninvasive Infections

Exserohilum infections are divided into two groups: invasive and noninvasive infections. The invasive infections occurred mostly in immunocompromised patients in the respiratory tract. The possible mode of transmission could be inhalation. Of the 33 cases reported by Adler et al.,[11] only 23 (20 *E. rostratum*, 2 *E. longirostratum*, and 1 *E. mcginnisi*) were identified to species level. The details of all the 33 patients were grouped into three different categories: invasive infections (10), skin and corneal infections (14), and chronic sinusitis (9). All the details pertaining to the 33 are tabulated in their review article. Briefly, in the invasive infection category, 7 out of the 10 were children below the age of 18 and the predisposing factors were acute lymphoblastic leukemia (4), aplastic anemia (2), and acute myeloid leukemia (1). Noninvasive infections were subdivided according to the main site of infection: solitary skin infections, corneal infections, and chronic sinusitis. The corneal infections followed local trauma without any systemic infections. The majority of the skin infections with *Exserohilum* infections were immunocompromised patients (cases of sinusitis were categorized as acute-invasive or chronic).[16]

9.1.3.2 Signs and Symptoms

Sinusitis, the most common form of disease caused by *Exserohilum*, occurs in otherwise healthy patients with nasal polyps and allergic rhinitis. The most frequent findings in the immunocompromised patient are acute inflammation and with prominent vascular invasion, thrombosis and infraction. In contrast, granulomatous inflammation and leukocytoclastic vasculitis are seen in meningoencephalitis caused by these fungi. *Exserohilum* is capable of causing central nervous system disease, osteomyelitis, and sinusitis and is associated with allergic bronchopulmonary disease. In immunocompromised patients the disseminated disease is characterized by tissue necrosis and vascular invasion. These various entities have distinct histopathologic characteristics that are described as case reports in the literature.

Pathogenesis of the disease caused by *Exserohilum* is little understood, however, melanin, enzymes, and toxins are reported as virulent factors. *Exserohilum* may infect immunocompromised as well as immunocompetent patients. The clinical manifestations range from cutaneous infections to severe disseminated disease. The infiltrating inflammatory cells include lymphocytes, macrophages, focal neutrophils and eosinophils, and plasma cells. There are different types of *Exserohilum* infections associated with distinct clinical presentations and modes of therapy. Invasive infections, most commonly of the respiratory tract, affected mainly immunocompromised patients. Apparently, the mode of transmission in these cases was inhalation, sometimes with hematogenous spread to other sites (brain, bone, skin). The management of patients with invasive disease included prolonged courses of systemic antifungal agents and, if feasible, surgery. Outcome appeared to be better (attributed mortality of 30% in invasive infections) than for other mold infections and depended mainly on the underlying diseases. Surgery was required in the following cases: paranasal sinus infection, other invasive infections (bone and cardiac valve), subcutaneous infections, and in two of the corneal infections that were similar to other mold infections.[17,18]

9.1.4 Immune Response

The host's innate immune response plays an important role in protection against invading fungal pathogens. The first line of defense includes acellular microbicidal components such as products of complement activation cascade, cationic proteins, and proteolytic enzyme secreted along the mucosal lining of the respiratory tract. The principal cellular component during this early nonantigen-specific innate immune response is mediated via phagocytic activity of alveolar macrophages and recruited neutrophils; toll-like receptor (TLR) 2 and TLR 4, were recently recognized as critical in the early recognition of infection. Interestingly, TLR 4-mediated immune activation is more focused against the fungal microconidia, whereas TLR 2 is important for eliciting immune response against microconidia and the developed fungal hyphae.[19] Tumor necrosis factor (TNF) has also been shown to be an essential mediator of early immune activation following ingestion of infectious fungal spores or microconidia.[20] Improved understanding of the molecular pathogenesis of fungal infections and of the complexity of host antifungal immune responses has provided the critical information to readdress existing treatment paradigms.

9.1.5 Diagnosis

9.1.5.1 Phenotypic Identification

The clinical microbiology laboratory should provide and have experience with methods used for the rapid detection and identification of the fungi associated with infections of the immunocompromised patient. The clinician must submit the specimens to the microbiology laboratory to confirm the diagnosis of opportunistic fungal infection. Diagnosis of *Exserohilum* infections are based on direct microscopy, culture, and histopathology. Early, rapid, and accurate identification of pathogenic fungi is important for the selection of appropriate antifungal therapy. Most often stains and cultures of tissue biopsies are necessary to establish a primary diagnosis. Fungal serologic tests are not helpful in making the diagnosis for *Exserohilum* infections. The most rapid and useful of the methods available is the microscopic detection of fungal elements in clinical specimens.

The potassium hydroxide (10% KOH) preparation is the most widely used method for the detection of fungal elements directly from any clinical specimen. Basically, samples of a clinical specimen, usually body fluid and 10% KOH are placed on a microscope slide to digest the tissue debris and artifact which may interfere with the detection of fungal elements. This method may provide the earliest evidence of a fungal infection for the clinician. In addition, a fluorescent brightener calcofluor white, is recommended for use with the KOH preparation. This reagent fluoresces upon excitation with ultraviolet light and stains fungi, causing them to exhibit fluorescence which can be detected using the florescent microscope. Tissue section specimens can also be stained with calcofluor white, and results are immediately available within 15–20 min. This method has better sensitivity than the simple KOH preparation by allowing for easier and faster detection of fungal elements using a florescent microscope.

The diagnosis of *Exserohilum* infections is based primarily upon the morphologic findings. *Exserohilum*, like its related genera, is characterized by a sympodial conidiophore. It has an ellipsoidal to fusoid conidium and, more specifically, a protruding, truncate, pigmented hilum. However, current culture-based phenotypic methods are insensitive and slow, may initially be nonspecific, and require considerable expertise for correct morphological identification of less common or unusual fungi.

Recent efforts to improve the sensitivity and specificity of diagnostic tests have focused on culture-independent methods, in particular nucleic acid-based methods, such as polymerase chain reaction (PCR) assays. These can be applied to fresh and formalin-fixed, paraffin-embedded (FFPE) sections. Numerous studies have highlighted the advantages of using PCR technology to detect viable and nonviable fungal pathogens in a variety of clinical specimens. The majority of assays target multicopy genes, in particular the ribosomal RNA (rRNA) genes (18S, 28S, and 5.8S) and the intervening internal transcribed spacer (ITS) regions (ITS1 and ITS2), in order to maximize sensitivity and specificity. Sequence-based identification of PCR products is a sensitive alternative, provided that accurate sequences have been submitted to public databases like GenBank.

9.1.5.2 Genotypic Identification

The identification of fungi to species level is very important in directing the treatment strategy due to the high incidence of antifungal drug resistance and with regard to newly emerging fungal pathogens. In about 75% cases fungal identification to species level is possible by conventional methods. In the remaining, the identification is not possible because of atypical characters of certain fungi, ability to grow in different morphological forms, and emergence of new species.

Numerous targets within the fungal genome have been evaluated, with much of the current work using sequence areas within the rRNA gene complex. Multiple applications using this as target include direct detection from clinical specimens, culture identification, phylogenetic research, and molecular typing for epidemiological purposes. The breadth of applications of rRNA gene complex has great potential in characterization and identification of fungi. PCR based detection of fungal DNA sequences are proven to be specific, sensitive rapid and reliable. The coding regions of the 18S, 5.8S, and 28S nuclear rRNA genes are conserved among fungi as they are known to evolve slowly and provide a molecular basis of establishing phylogenetic relationships. Between coding regions are the ITS1 and 2 regions (ITS1 and ITS2, respectively) which evolve more rapidly and may therefore vary among different species within a genus. Thus, PCR amplification targeting ITS region may facilitate the identification of DNA sequences with sufficient polymorphism to be useful for identifying the *Exserohilum* to species level.

Identification of fungi to species level is essential to direct the antifungal treatment. Development of rapid and accurate assays to diagnose fungal infections could potentially impact care and improve outcome of affected patients.[21,22] Fungi have a rRNA gene complex with comparable characteristics. The organization of this complex in fungi includes a sequence coding for the 18S rRNA gene, an internal transcribed spacer region 1(ITS1), the 5.8 S rRNA gene coding region, another ITS region called ITS2 and the sequence coding for 28S rRNA gene. The 28S rRNA gene evolved slowly and is relatively conserved among fungi providing a molecular basis of establishing phylogenetic relationships. Early work in molecular testing using rRNA complex as target concentrated in the region of 18S rRNA also referred as the small subunit rRNA gene. The only way to identify these fungi is by using rapid molecular techniques—PCR-based RFLP and PCR-based DNA sequencing. PCR-based DNA sequencing on the ITS region is useful specially to identify non-sporulating molds (NSM).

Other genes within the molecular complex having also been used for the molecular evaluation of fungi include the 5S rRNA, 5.8S rRNA, and 28S like rRNA. The comparison of nucleotide sequences within this region has been successful for the separation of genera and species. The 18S gene of fungi is about 1800 bp in size with both conserved and variable domain sequences. Sequence variation within this region has been used to assess the taxonomic relationships and to separate genera and species based on sequence polymorphisms. However, the drawback in using this region for the identification of the species is the relative sequence homology among fungal species and the need to sequence a large number of bases in order to do comparative analysis. The 5.8S region on the other hand is only about 160 bp long and highly conserved. Owing to its small size and conserved nature, it is not used for phylogenetic studies. However this sequence is used as an attachment site for primers to amplify flanking spacer regions. The 28S rRNA which is around 3400 bp in size and the variable domains of this subunit have also been used to allow comparisons from higher taxonomic levels to the species level. Much of the 28S rRNA gene is conserved limiting the usefulness of this region for species identification.[21]

The application of PCR technology to molecular diagnostics holds great promise for the early identification of medically important pathogens. PCR has been shown to be useful for the detection of the presence of fungal DNA in clinical specimens. Considerable interest has been focused on the utility of selecting universal primers that recognize constant regions among most, if not all, medically important pathogens.

9.1.5.2.1 Target Selection

Targets that have been used in molecular diagnostic tests for fungal infections include single and multicopy nuclear and mitochondrial genes and RNA. In general, molecular diagnostic methods targeting multicopy genes have better detection thresholds than those targeting single-copy genes. The ribosomal genes contain conserved sequences that are common to all fungi and also variable domains and highly variable ITS regions. The conserved sequences can be used to screen for fungal infection, while the variable sequences can be exploited for species identification. Indeed, favorable results targeting this ribosomal gene as a tool for fungal detection and identification have been obtained.

9.1.5.2.2 Amplification and Detection Methods

Nucleic acid hybridization and amplification methods are fundamental to molecular diagnosis. Hybridization techniques employ a DNA probe to determine whether a particular organism is present. PCR is the most frequently used amplification procedure because it can readily adapt to many applications. However, when a single-copy gene is used as the target for PCR, special amplification steps such as nested PCR (nPCR) are often necessary to achieve the necessary degree of sensitivity. The PCR-generated product is usually analyzed by ethidium bromide-stained gel electrophoresis. Although gel electrophoresis is simple and inexpensive, it is much less sensitive than Southern blotting.[23] Detection of PCR products by ethidium bromide staining and by Southern blotting is time consuming, and the interpretation of results may be subjective. Hence, PCR–enzyme immunoassay (EIA)—a three part method that includes PCR amplification, hybridization with the complementary labeled probe, and detection of reaction products in an EIA that provides either colorimetric or fluorescence readout—was developed.[24] In addition, the PCR–EIA format provides further amplification without losing the species-specific binding associated with Southern blotting and multiple samples can be assayed in parallel and semiquantitation of DNA is possible.[25]

9.1.5.2.3 Species Identification

Studies have demonstrated that the most promising targets for molecular identification of fungi are the ITS1 and/or the ITS2 region, followed by the D1–D2 region of the large-subunit rRNA gene (28S rRNA). Most molecular diagnostic methods are able to screen patients in the initial stages of fungal infection, but not all protocols can identify the source of the DNA to the genus or species level. Bagyalakshmi et al.[26] applied the PCR-based DNA sequencing to identify the NSM targeting the ITS region and one of the NSM in their study was *Exserohilum*.

9.1.5.2.4 Diagnostic Considerations

The detection and identification of fungal pathogens by DNA-based methods can yield results sooner than cultivation. Although the disappearance of fungal DNA from the blood of patients correlates with successful therapy, no experimental data are available to explain these aspects of therapy.

9.1.6 Treatment and Outcome

Surgical excision is reported as the treatment of choice for localized infections with *Exserohilum* infection with or without adjunctive systemic antifungal chemotherapy in the immunocompromised patients.[11] The modes of therapy differ according to the type of infection. Invasive infections are treated with systemic antifungal agents though some of the infections require surgical procedures. The duration of systemic antifungal therapy was prolonged, from 2 to 13 months. Skin infections are managed with systemic drugs, with or without topical antifungal agents. Corneal infections were managed with topical antifungal agents in most of the patients reported in the literature and some patients required surgical intervention. Surgical debridement was the principal mode of therapy for allergic fungal sinusitis and cerebral infection with *Exserohilum* infection. Successful use of amphotericin B and itraconazole as systemic antifungal chemotherapeutic agents are reported in the literature.[7,8,11]

In case of invasive infections amphotericin B was considered as the drug of choice.[12,27] Most patients who were treated initially with surgical debridement and amphotericin B were cured, but longer follow-up were necessary in these patients. Adler et al. (2006)[11] reported in their recent review of the literature, that in three cases the triazoles (itraconazole and voriconazole) seemed to have better activity against *Exserohilum*. However, there is insufficient data to support recommendation of any single agent as initial therapy as it is impossible to deduce in vitro–in vivo correlations. The prevention of local expansion, cosmetic deformity, and possible dissemination of infection with *Exserohilum* is made possible by prompt aggressive surgical intervention. However disseminated disease has to be treated with systemic antifungal chemotherapy.[27,28] The minimum inhibitory concentrations (MICs) usually show sensitivity to amphotericin B, itraconazole, and terbinafine, and resistance to 5-fluorocytosine and ketoconazole. No specific MIC were found in the literature for the echinocandins. There is limited clinical experience reported with azoles, but voriconazole seems very promising from in vitro data and low MICs.[8]

9.2 METHODS

9.2.1 Sample Preparation

The slide culture system consists of a mini incubation chamber designed to produce optimum conditions for sporulation. It allows for examination of the colony in various stages of development, and improves the chances of observing the natural configuration of the spores and conidia on the sporulating structures.[8]

The dissecting microscope is useful for observing the gross colony characteristics and microscopic observations. The criteria for identification of molds are the morphology, culture characteristics, nutritional requirements, temperature tolerance, proteolytic activity, cycloheximide resistance, and the ability to hydrolyze urea. Characteristics like color of the colony, pigment production, or spores are dependent on the media and environmental conditions; they are not consistently to be used as the main criteria to identify molds. The important characteristic features like texture, topography of the colony, diffusible pigments, color of the colony, aerial and submerged hyphae, growth rate, and macroscopic structures such as ascocarps, pycnidia, sclerotia, sporodochia, and synnemata can be well appreciated by visual examination of the colony. However the recent classification schemes emphasize more on the conidial ontogeny rather than color, conidial septation, or the appearance of growth on natural substrates.

Accurate identification of filamentous molds is based on microscopic examination of the sporulating parts of a colony. Living cultures are necessary for preparing slide cultures and the presence of growth verifies that the slide culture has been set up appropriately. The tease preparation from cultures is also another method for identification of the fungus but the disadvantages are that if the growth is not teased apart well, or if the material is taken from a non-sporulating area, it may be difficult to observe the sporulating structures and, moreover, the conidia and spores may also get disrupted during preparation of the mount.

9.2.2 Detection Procedures

9.2.2.1 Phenotypical Criteria for Identification of Molds

Visual examination of the fungal colony based on color, texture, diffusible pigments, exudates, growth zones, aerial and submerged hyphae, growth rate, and colony topography, is useful for identification of the fungus. It is easier to identify the fungi based on conidial ontogeny than keys based on the arrangement and color of the conidiophores and conidia. The dissection microscope is very useful to observe the distinct gross colony characteristics and microscopic appearance.[29]

Mycelia Sterilia is a form that contains the filamentous fungi that remain sterile in spite of the attempts to induce the formation of conidia or spores due to lack of appropriate environmental and nutritional needs, or both. Such fungi can rarely act as opportunistic pathogens and it becomes essential to identify the fungus by inducing the spore/conidia formation. Though there are various spore inducing culture media available, there is no universal culture medium and various media and techniques must be tried until the correct combination of variables is found.

9.2.2.2 Slide Culture

Accurate identification of filamentous molds is based on microscopic examination of the sporulating parts of a colony. The slide culture system consists of a mini incubation chamber designed to produce optimum conditions for sporulation. It allows for examination of the colony in various stages of development, and improves the chances of observing the natural configuration of the spores and conidia on the sporulating structures. Specimen living cultures are necessary for preparing slide cultures.

Procedure

Sterile slide culture Petri dish preparation:

A filter paper is laid to line the bottom of a glass Petriplate.

A V-shaped piece of bent glass tubing is kept on the filter paper to keep the slide culture.

A 22×22 mm cover slip is kept on the slide.
The lid of the Petriplate is replaced and the unit is autoclaved for 25 min at 121°C.
A medium known to induce sporulation for the fungus (PDA) is used as it stimulates sporulation.
A block of sterile PDA agar 2 mm^2 and approximately 1 mm thick is cut and placed on the microscope slide.
Using aseptic technique throughout, the material from a sporulating part of the colony is removed with a firm needle and inoculated two to three areas of the block.
The coverslip is placed on top of the block using forceps and pressed gently.
Sterile water (1–2 mL) is added to the filter paper at the bottom and covered with the Petriplate lid.
The slide culture is incubated at 25°C and observed for sporulation at regular intervals by examining the preparation under a dissecting or compound microscope.
The preparation is kept moistened by adding additional water when needed.
The proximity of the glass and the agar plus the humid atmosphere in the dish encourages hyphal development along the glass.

9.2.2.3 Lactophenol Cotton Blue Mount

The fungus is mounted when it is sporulating (takes around 3–10 days):

To mount the coverslip, remove it from the agar block using forceps.

A drop of lactophenol is placed just off center on a clean microscope slide and the coverslip with the growth is placed at the edge of the lactophenol and slowly placed on the wet preparation to avoid trapping air bubbles under the coverslip.

The excess of lactophenol is removed from the edges of the coverslip by blotting with a paper towel.

The slides are labeled using a permanent marker, with the identifying name or number and other pertinent details such as the medium used, age, and the fact that it is a slide culture.

If the mount is to be preserved the edges of the coverslip are sealed with nail polish.

Sporulation of fungi can be appreciated with modes of conidiogenesis being clearly visible for study.

9.2.2.4 Lactophenol Mounts with Tease Mount Procedure

A tease mount is the common and rapid method to mount fungi for microscopic examination. The disadvantages of this method are when the growth is teased apart with dissecting needles, conidia or spores get dislodged from the conidiogenous or sporogenous cells.

Procedure

A drop of lactophenol is placed just off center on a clean microscope slide.
A small portion of fungal growth between the colony center and edge is removed and placed on the lactophenol.
The fungus is teased apart gently by using two dissecting needles, to spread out the fungus.
A coverslip is placed at the edge of the lactophenol and slowly placed on the wet preparation to avoid trapping air bubbles under the coverslip.
The excess of lactophenol is removed from the edges of the coverslip by blotting with a paper towel.
If the mount is to be preserved the edges of the coverslip are sealed with nail polish.
While tease preparations are useful, they have two main disadvantages:
If too much growth is removed, or not teased apart well, or if the material is taken from a nonsporulating area, it may be difficult to discern sporulating structures.
Spores and conidia are often disrupted during preparation of the mount.

9.2.2.5 Scotch Tape Lactophenol Mount

The Scotch tape mount is yet another easy and rapid procedure that is used for the identification of filamentous fungi since most structures are intact for observation.

As with the lactophenol mount, the organism is immersed in the solution, rendering the organism safe for handling outside of the biological safety hood.

Procedure

A strip of Scotch transparent tape is cut and the ends are placed between thumb and index finger, gummed side out.
Making a loop by closing fingers, open plate with opposite hand and press tape against the colony to identify.
A drop of lactophenol is kept on a labeled slide and the tape is pressed against slide with lactophenol.
The tape is smoothened back on the slide by opening fingers and using gauze.
Another drop of lactophenol is placed on top of the tape.
A large 20×40 mm coverslip is placed on top of the slide and examined under the microscope.
The limitations are that the tape will dissolve eventually so that it is not to be used for permanent mounts and the procedure can only be performed on molds growing from plates.

On potato glucose agar *Exserohilum* hyphae are septate, subhyaline to pale to mid brown, and 3.5–5.0 µm in diameter. The conidiophores are simple, erect or wavy, and their upper portions are fertile and geniculate. The conidia are straight and cylindrical to ellipsoidal, with rounded apices, measuring around 64–100 × 10–15 µm, and have 8–13 distosepta (having the individual cells each surrounded by a sac-like wall distinct from the outer wall). The hila of the conidia are black and distinctly protuberant germination of the conidia is bipolar.

Many criteria are considered when identifying *Exserohilum* species such as the morphology, culture characteristics, temperature tolerance, and cycloheximide sensitivity. It is distinguished on the basis of conidial shape and size, hilar morphology, location and sequence of the conidial septa, and the origin of the germ tubes from the conidial cells. The conidium of *Exserohilum* species is ellipsoidal to fusiod and germinates by germ tubes originating from either one or both of the end cells or other intermediate cells. The conidia of *Exserohilum* species are ellipsoidal to fusoid and have distinctly protruding hila with truncated bases. The conidia germinate by germ tubes originating from either one or both of the end cells or other intermediate cells. The species level identification is based on the conidial structure as described. The fungus is identified to the species level based on the characteristic features.[7,8]

9.2.2.6 Molecular Detection Method

PCR-based DNA sequencing technique for identification of the fungus to species level targeting internal transcriber spacer region (ITS region).[26]

Specimen preparation can have a significant impact on the sensitivity and reproducibility of a molecular diagnostic test. In general, the sample preparation method should release intracellular DNA from the fungal cell wall and/or thick capsule; it must concentrate DNA targets that may be present in very small amounts, and it must eliminate protein debris, contaminants, potential inhibitors, and other extraneous materials without degrading the target DNA. At present, there are many protocols for sample preparation, but no universal method for optimally extracting, purifying, and concentrating fungal DNA from clinical specimens is available. The efficiencies of the molecular diagnostic methods applied to different types of clinical samples might not be equivalent, as DNAs present in different clinical samples are probably different in origin.

DNA extraction from fungal growth:

- After growth of 7 days on SDA, approximately 1 cm^2 mycelia is cut from the fungal growth for extraction of DNA.
- The mycelia are collected by scraping the slant with a sterile stick in a biological safety cabinet (Class II).
- The material is transferred to a presterilized microfuge vial (1.5 mL) containing 1 mL of sterile water and the tube is centrifuged for 5 min at 3000 rpm in a microcentrifuge.
- The supernatant is discarded and the material is resuspended in 200 μL of lysis buffer containing 200 mM Tris, 50 mM EDTA, 10% sodium dodecyl sulphate (SDS), and 1 M sodium chloride and incubated at 56°C for 60 min.
- To this mixture 150 μL of sodium acetate is added and stored at −20°C for 20 min.
- This is followed by centrifugation at 14,000 rpm for 10 min and the supernatant is transferred into another tube to which equal volume of isopropanol is added and kept at room temperature for 5 min. The sample is centrifuged at 14,000 rpm for 10 min and the deposit is washed with 70% ethanol thrice and finally resuspended in 50 μL of tris EDTA (TE) buffer.

Amplification using ITS primers:

Primer sequence targeting ITS region[26]:

ITS1 5′ TCC GTA GGT GAA CCT GCG G 3
ITS4 5′ TCC TCC GCT TAT TAT GC 3

- In brief, for a 50 μL reaction, 8 μL of dNTPs (200 μmol), 5 μL of 1×PCR buffer (1.5 mM $MgCl_2$, 50 mM KCl, 10 mM Tris Cl, 0.001% gelatin), 6 μL of 25 mM $MgCl_2$ (1 in 10 diluted to get 1.5 mM), 10 pmol of the forward primer, reverse primers targeting ITS region, and 10 μL of template DNA is used.[26]

PCR thermal cycling profile:

- Initial denaturation at 95°C for 5 min followed by 35 cycles of denaturation at 95°C for 30 s, annealing at 55°C for 60 s, and extension at 72°C for 60 s, followed by final extension at 72°C for 6 min.

Sequencing in ABI Prism 310/3100 AVANT genetic analyzer:

- PCR-based DNA sequencing targeting ITS region is carried out in a genetic analyzer (ABI Prism 3100 AVANT Applied Biosystems).
- The sequences of PCR amplified product are subjected to BLAST analysis to identify the species and/or genera of fungi.
- Fungi with 99% homology is identified to species level, and the genus level is assigned when there is a homology of 95%–98%.[26]
- The nucleotide sequences of different isolates belonging to the same species are analyzed using Multalin analysis software.
- The variations in nucleotide sequences, single and multiple nucleotide polymorphisms can be identified by Multalin analysis.

9.3 CONCLUSION AND FUTURE PERSPECTIVES

Exserohilum is a dematiaceious fungus that may infect both immunocompromised and immunocompetent hosts. Invasive infections, most commonly of the respiratory tract, affect mainly immunocompromised patients. There is an increased number of *Exserohilum* reported in the literature within the last two decades. *Exserohilum* infections in human beings are manifested in various clinical forms starting from cutaneous infections to fulminant disseminated disease. Local trauma and atopy are reported to be the predisposing factors for corneal infections and allergic fungal sinusitis respectively. However, majority of the affected patients with invasive infections were reported to have impaired immunity. Surgical debridement is considered as the principal mode of therapy for most of the infections with *Exserohilum*. Amphotericin B is the drug of choice for most invasive diseases reported in the literature[30] though some cases improved after the additional treatment with triazole agents.[3] Outcome appeared to be better than for other mold infections and depended mainly on the underlying diseases.[32] Studies on the growth kinetics of *Exserohilum*, its susceptibility patterns to the in-use and emerging antifungal agents is a potential area of research in the near future. The ability to determine the nucleic acid sequence of the genomic DNA has revolutionized most areas of contemporary biomedical research. Further nucleic-based amplification techniques involving real-time PCR and reverse-transcriptase PCR need to be optimized to assess the transcripts produced by *Exserohilum*. Future research needs to focus on improvement in diagnostic techniques and development of new therapeutic antifungal compounds to prevent *Exserohilum* infection in humans.

REFERENCES

1. Leonard, K.J. and Suggs, E.G., *Exserohilum rostratum* (Drechsler). *Mycologia*, 66, 290, 1974.
2. Shoemaker, R.A., *Drechslera* Ito. *Can. J. Bot.*, 40, 809, 1962.
3. Alcorn, J.L., Generic concepts in *Drechslera*, *Bipolaris*, and *Exserohilum*. *Mycotaxon*, 17, 1, 1983.
4. Levy, I. et al., Ecthyma gangrenosum caused by disseminated *Exserohilum* in a child with leukemia: A case report and review of the literature. *Pediatr. Dermatol.*, 20, 495, 2003.
5. Steiman, R., *Exserohilum israeli*, a new species isolated from soil from Timma Park (Israel), and its physiological properties. *Anton. Van Leeuwen.*, 78,153, 2000.
6. Ferguson, B.J. et al., Geographic variation in allergic fungal sinusitis. *Otolaryng. Clin. North Am.*, 33, 441, 2000.
7. Padhye, A.A. et al., Phaeohyphomycosis of the nasal sinuses caused by a new species of *Exserohilum*. *J. Clin. Microbiol.*, 24, 245, 1986.
8. McGinnis, M.R., Rinaldi, M.G., and Winn, R.E., Emerging agents of phaeohyphomycosis: Pathogenic species of *Bipolaris* and *Exserohilum*. *J. Clin. Microbiol.*, 24, 250, 1986.
9. Saint-Jean, M. et al., Hospital acquired phaeohyphomycosis due to *Exserohilum rostratum* in a child with leukemia. *Can. J. Infect. Dis. Med. Microbiol.*, 18, 200, 2007.
10. Rolston, K.V.I., Hopfer, R.L., and Larson, D.L., Infections caused by *Drechslera* species: Case report and review of the literature. *Rev. Infect. Dis.*, 7, 525–529, 1985.
11. Adler, A. et al., *Exserohilum*: An emerging human pathogen. *Eur. J. Clin. Microbiol. Infect. Dis.*, 25, 254, 2006.
12. Adam, R.D. et al., Phaeohyphomycosis caused by the fungal genera *Bipolaris* and *Exserohilum*. A report of 9 cases and review of the literature. *Medicine (Baltimore)* 65, 203, 1986.
13. Aquino, V.M. et al., Fatal disseminated infection due to *Exserohilum rostratum* in a patient with aplastic anemia: Case report and review. *Clin. Infect. Dis.*, 20, 176, 1994.
14. Rinaldi, M.G., Phaeohyphomycosis. *Dermatol. Clin.*, 14, 147, 1996.
15. Lasala, P.R. et al., Invasive *Exserohilum* sinusitis in a patient with aplastic anemia. *Pediatr. Infect. Dis. J.*, 24, 939, 2005.
16. Morpeth, J.F. et al., Fungal sinusitis: An update. *Ann. Allergy Asthma Immunol.*, 76, 128, 1996.
17. Revankar, S.G. et al., Disseminated phaeohyphomycosis: Review of an emerging mycosis. *Clin. Infect. Dis.*, 34, 467, 2002.
18. Stevens, D.A. et al., Practice guidelines for diseases caused by *Aspergillus*. *Clin. Infect. Dis.*, 30, 696, 2000.
19. Netea, M. et al., *Aspergillus fumigatus* evades immune recognition during germination through loss of tolllike receptor-4-mediated signal transduction. *J. Infect. Dis.*, 188, 320, 2003.
20. Rolides, E. et al., Tumor necrosis factor alpha enhances antifungal activities of polymorphonuclear and mononuclear phagocytesagainst *Aspergillus fumigatus*. *Infect. Immunol.*, 66, 5999, 1998.
21. Iwen, P.C., Hinriche, H., and Rupp, M.E., Utilization of ITS regions as molecular targets to detect and identify fungal pathogens. *Med. Mycol.*, 40, 87, 2001.
22. Lau, A. et al., Development and clinical application of a pan-fungal PCR assay to detect and identify fungal DNA in tissue specimens. *J. Clin. Microbiol.*, 45, 380, 2007.
23. Tang, C.M. et al., The detection of *Aspergillus* spp. by the PCR and its evaluation in bronchoalveolar lavage fluid. *Am. Rev. Respir. Dis.*, 148, 1313, 1993.
24. Loeffler, J., Henke, N., and Hebart, H., Quantification of fungal DNA by using fluorescence resonance energy transfer and the Light cycler system. *J. Clin. Microbiol.*, 38, 2902, 2000.
25. Fujita, S.I. et al., Microtitration plate enzyme immunoassay to detect PCR-amplified DNA from Candida species in blood. *J. Clin. Microbiol.*, 33, 962, 1995.
26. Bagyalakshmi, R. et al., Newer emerging pathogens of ocular non-sporulating moulds (NSM) identified by polymerase chain reaction (PCR)-based DNA sequencing technique targeting internal transcribed spacer (ITS) region. *Curr. Eye Res.*, 33, 139, 2008.
27. Ajello, L. et al., *Drechslera rostrataas* an agent of phaeohyphomycosis. *Micologia*, 72, 1094, 1980.
28. Drouhet, E. and Dupont, B., Laboratory and clinical assessment of ketoconazole in deep-seated mycoses. *Am. J. Med.*, 24(Suppl 1), 30, 1983.
29. McGinnis, M.R., *Laboratory Handbook of Medical Mycology*, Academic Press, New York, 1980.
30. Moneymaker, C.S. et al., Primary cutaneous phaeohyphomycosis due to *Exserohilum rostratum* (*Drechslera rostrata*) in a child with leukemia. *Pediatr. Infect. Dis.*, 5, 380, 1986.
31. Sharkey, P.K. et al., Itraconazole treatment of phaeohyphomycosis. *J. Am. Acad. Dermatol.*, 23, 577, 1990.
32. McGinnis, M.R. and Pasarell, L., In vitro testing of susceptibilities of filamentous ascomycetes to voriconazole, itraconazole, and amphotericin B, with consideration of phylogenetic implications. *J. Clin. Microbiol.*, 36, 2353, 1998.

10 *Fusicoccum* and *Scytalidium*

Marie Machouart and Jean Menotti

CONTENTS

10.1 INTRODUCTION

Members of the genus *Scytalidium* are generally found on plants as pathogens or saprobes, or isolated from soil, with a worldwide distribution [1]. More than 15 species have been listed by referring to different sources: *S. thermophilum*, *S. japonicum*, *S. multiseptatum*, *S. album*, *S. dimidiatum*, *S. hyalinum*, *S. lignicola*, *S. infestans*, *S. indonesiacum*, *S. muscorum*, *S. uredinicola*, *S. aurantiacum*, *S. flavobrunneum*, *S. acidophilum*, *S. vaccinii*, *S. circinatum*, *S. fulvum*, and *Neoscytalidium novaehollandiae* (http://www.ncbi.nlm.nih.gov/pubmed/, http://www.cbs.knaw.nl/, www.atcc.org/).

In this chapter, we focus on two species, *S. dimidiatum* and *S. hyalinum*, frequently found in tropical or subtropical countries and responsible for superficial (or rarely deep) infections in humans, resembling dermatophytosis [2].

10.1.1 CLASSIFICATION

The taxonomy of *S. dimidiatum* and *S. hyalinum* is very confusing, with several factors having enhanced the confusion over the years. Cultural variants have been isolated with or without production of pycnidia, with distinct macroscopic morphological features and different growth rates. The nomenclature has changed several times since the original descriptions of these molds, further to the discovery of new characteristics. Actually, the nomenclature continues to evolve with molecular studies and new names are still being proposed.

S. dimidiatum was first described under the name *Hendersonula toruloidea* in 1933 by Nattrass from pome and stone fruit trees as the conidial state of *Dothiorella mangiferae* that was originally described on mangoes [3].

In 1971, Gentles and Evans first reported the fungus under the name *Hendersonula toruloidea* from patients immigrating to the United Kingdom but originating from tropical areas. Based on clinical appearance and direct microscopic examination, these patients were first suspected to have dermatophytosis, but cultures of the biological samples rapidly grew a black mold in media without cycloheximide [4].

A few years later, in 1977, Campbell and Mulder described the first case of *S. hyalinum*, in a patient having a mycosis similar to those of Gentles and Evans, resembling dermatophytosis, but caused by a fast-growing white mold [5].

Since these first descriptions, this fungus has then been renamed several times because the production of both arthroconidial and pycnidial synanamorphs (form with conidia in a cavity [pycnidia]) has led to nomenclatural controversy.

Then, in 1989, the genus *Hendersonula* was revised by Sutton and Dyko that proposed to synonymize *Dothiorella mangiferae* and *H. toruloidea* under the name *Nattrassia mangiferae*. The mycelial synanamorph was further described as *Scytalidium dimidiatum* [6].

In 2005, Farr et al. conducted a molecular study on internal transcribed spacer (ITS) and β-tubulin of *Fusicoccum* sp. They concluded from their analysis that *Nattrassia mangiferae* and *Scytalidium dimidiatum* belong to the genus *Fusicoccum* (asexual state of *Botryosphaeria*), most closely related to *Botryosphaeria mamane*. Based on these molecular sequence data and on morphological similarities of the pycnidial and arthric conidial states with *Fusicoccum*, they introduced the name *Fusicoccum dimidiatum* to replace *Scytalidium dimidiatum* and stated that the genus *Nattrassia mangiferae* should be synonymized under the name *Fusicoccum dimidiatum* [7].

Recently, Crous et al. have proposed a taxonomic revision of the *Botryosphaeriaceae* family based on the analysis of 28S rDNA sequences. In the resulting phylogenetic tree, isolates producing arthroconidia typical of *S. dimidiatum* and strains of *Nattrassia mangiferae* obtained from mangoes and lacking synanamorphs were grouped in separate clades [8].

In this study, the type species of the genus *Scytalidium*, *S. lignicola* (CBS 233.57), was associated with Leotiomycetes and clustered outside *Botryosphaeriaceae* (Dothideomycetes). These data support that *S. lignicola* and *S. dimidiatum* would be distinct species belonging to different classes. This is in agreement with the conclusion of Sigler et al. in their literature review. These authors suggested the distinction between *S. dimidiatum* and *S. lignicola* by morphological differences, leading to a confusion between both species in reports identifying the latter species as a human pathogen [9]. According to these arguments and to DNA sequence data derived from the CBS collection, Crous et al. concluded that the genus *Scytalidium* would be polyphyletic and proposed the name *Neoscytalidium* in order to accommodate *S. dimidiatum* as *Neoscytalidium dimidiatum*. Moreover, they suggested the coelomycete synanamorph *Hendersonula toruloidea* (asexual state) as a synonym of *N. dimidiatum*.

Secondarily in this study, the proposition of Farr et al. to place into synonymy the genera *Scytalidium* and *Fusicoccum* and to synonymize *Fusiccocum dimidiatum* and *Scytalidium dimidiatum* was rejected. Thus, a new genus, *Neofusicoccum* was also proposed. *Nattrassia mangiferae*, previously widely used to refer to *S. dimidiatum* isolates, should be considered now as a distinct species, which differs from the latter taxon in lacking an arthroconidial anamorph. Therefore, isolates previously described as *Nattrassia mangiferae* would be placed in this *Neofusicoccum* group and synonymized under the name *Neofusicoccum mangiferae*.

S. hyalinum has been described as a very similar fungus to *S. dimidiatum* but nonpigmented and only reported from human skin and nail infections. According to molecular data, it has been suggested that *S. dimidiatum* and *S. hyalinum* might be conspecific [10,11]. In the study of Crous et al., no synonymy has been proposed between both fungi. Very recently, a new name based on the hypothesis that *S. hyalinum* would be an albino variant of *N. dimidiatum* has been suggested: *Neoscytalidium dimidiatum* var. *hyalinum* [12].

Because the taxonomy is controversial according to the authors, in the next paragraph, both fungi still will be named as *Scytalidium dimidiatum* and *Scytalidium hyalinum*.

10.1.2 Morphology, Biology, and Epidemiology

On Sabouraud agar media, *S. dimidiatum* produces a fast growing, aerial mycelium, that is cottony and hyaline at first but soon becomes dark brown to blackish, floccose, and dense (Figure 10.1A). The reverse is black. *S. hyalinum* grows whitish colonies with a colorless to beige reverse, the mycelium remaining flushy (Figure 10.1C) [1]. These molds generally grow on cycloheximide-free media, although some *S. dimidiatum* have already been reported not to be inhibited by cycloheximide [13].

Microscopic examination demonstrates septate hyphae with cylindrical to ellipsoidal both hyaline and darkly pigmented arthroconidia in chains or disarticulated

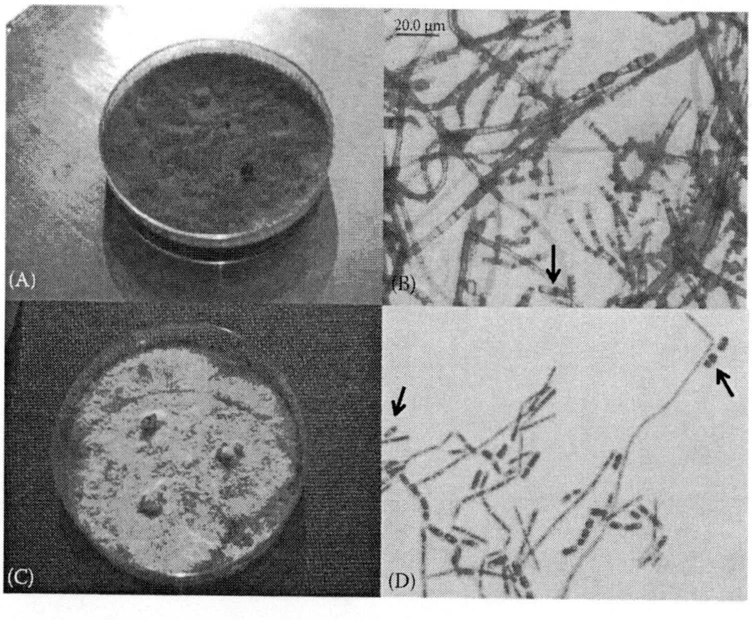

FIGURE 10.1 Macroscopy and microscopy of *Scytalidium dimidiatum* (A and B) and *Scytalidium hyalinum* (C and D). The black arrows indicate conidia with characteristic septa.

(Figure 10.1B and D). Some of these arthroconidia are characterized by the presence of two cells separated by a thick septum (arrows in Figure 10.1B and D). In old cultures, pycnidial synanamorph may be produced [8,9].

Scytalidium dimidiatum (or under the name *Nattrassia mangiferae*) has been reported as a significant cause of pathology in fruit trees or woody plants as *Ficus*, *Arbutus*, *Citrus*, *Juglans*, and others in tropical or subtropical areas [6,9,14]. This organism was first found to be responsible for gummosis and dieback of stone fruit trees in Egypt [3], drying of grape vines in Iraq and India [15,16], death on citrus trees in Iraq [17], leaf spot disease in India [18], leaf spot and dieback of mango in India and Niger [19,20], and tip rot of banana trees in Jamaica and Hawaii [21,22]. This is also the cause of the branch wilt disease on the Persian walnut tree *Juglans regia* in California [23], dieback and canker of citrus and eucalyptus trees of Arizona [24,25], and canker of almond trees [26]. More recently, *Hendersonula toruloidea* has also been found to cause a foliar disease of strawberry tree in Greece [27].

Whereas *S. dimidiatum* has been widely reported from such plants, *S. hyalinum* has never been isolated from environment.

As human pathogens, both fungi are endemic in tropical and subtropical countries comprising South America, the Caribbean, Asia, and Africa and cause diseases called scytalidioses representing approximately 40% of dermatomycoses in these areas [2,28]. Nevertheless, cases from temperate countries are increasingly reported owing to immigration and travels [12,28,29]. In a study examining data from 332 patients infected by *Scytalidium* and addressed to the dermatology department of a French hospital, the sex ratio was 2 men for 1 woman (67.5% men versus 32.5% women) with a predominance in the 36- to 55-year age group (61.5%) [13]. Most *S. hyalinum* infections seem to be related to the West Indies, South America, and West Africa, whereas patients from Asia, the Indian Ocean region, and Central Africa are mostly infected by *S. dimidiatum*. Some mixed *S. hyalinum* and *S. dimidiatum* infections have also been described. Mixed infections with dermatophytes have been reported [13,30,31].

Infection is thought to occur through contact with contaminated soil or plant material especially for barefooted individuals, or possibly through direct contact with infected nail or skin.

10.1.3 Clinical Features and Pathogenesis

Clinically, these fungi cause mostly superficial skin infections and onychomycosis indistinguishable from dermatophytosis [9,32]. Most infections are generally located on nails, toe webs, and the soles of the feet, hands being more rarely involved. Infected soles show hyperkeratosis, associated or not with desquamation. Desquamation and maceration may be observed for toe web infections. Paronychia is not rare. Toenail lesions are characterized by laterodistal invasion. Total nail dystrophy may also occur. Classically palm lesions are unilateral whereas sole affection is generally bilateral (two-foot one-hand syndrome) [33]. No clinical difference is observed between *S. hyalinum* and *S. dimidiatum* infections. According to Lacroix et al., melanonychia usually observed in people with black skin would be a result of an inflammatory process more than of the pigmentation of *S. dimidiatum* [13,32]. Asymptomatic carriage has been hypothesized but this remains to be demonstrated, because subclinical infection may be common in certain regions [2,31,32]. No studies have been carried out to test this hypothesis, and these should be conducted by experts in the field of mycological diagnosis involving dermatologists together with mycologists.

Rare cases of deep infections involving most of time *S. dimidiatum* have also been reported. Central nervous system abcesses, endophtalmitis, sinusitis, osteomyelitis, mycetoma, subcutaneous lesions, and disseminated infections have been published, mainly affecting immunocompromised patients such as renal or lung transplants, diabetic patients, or individuals receiving immunosuppressive therapy [34–48]. The underlying conditions appear to be similar to those of other invasive opportunistic fungal infections. Until now, only two cases have been reported from immunocompetent patients [45,46]: an endophtalmitis occurring after traumatic implantation and a cerebral phaeohyphomycosis. Approximately 50% of the patients concerned by deep-seated infections died. Only a few cases involving *S. hyalinum* have been described [48,49].

Virulence factors involved in the pathogenesis of *Scytalidium* sp. may be linked to a keratinase facilitating invasion of skin and production of melanin for *S. dimidiatum* mediating resistance to phagocyte-produced oxygen radicals [2,50]. The ability of *S. dimidiatum* and *S. hyalinum* to degrade nail keratin has been demonstrated, although this phenomenon is less pronounced compared to what occurs in dermatophytosis due to *T. rubrum* and *T. mentagrophytes* [50]. It is therefore possible that mechanisms operating in *Scytalidium* spp. to degrade and digest keratin may be similar to those of dermatophytes.

10.1.4 Treatment

The clinical response of *Scytalidium* spp. to antifungal treatment is poor with the available topical or systemic agents used for onychomycosis. *Scytalidium* spp. are quite resistant to drugs used in dermatology, that is, griseofulvin, ketoconazole, fluconazole, itraconazole, and terbinafine [51]. In vitro efficacy of antifungals has been evaluated by Etest methods [52,53] and by using microdilution testing with CLSI M38-A methods [54]. These studies resulted in MICs ranges of <0.03–0.5 mg/L for voriconazole, 0.06–2 mg/L for posaconazole, <0.03–16 mg/L for itraconazole, 0.06–2 mg/L for terbinafine, 0.06–8 mg/L for caspofungin, and 0.06–1 mg/L for amphotericin B. Amphotericin B, voriconazole, and posaconazole exhibit low MICs in these in vitro studies. These data are related to successful in vivo treatments with amphotericin B alone or with a regimen of liposomal amphotericin B followed by voriconazole in cases of invasive infections [29,44].

In vivo, amphotericin B dermatologic lotion has been described to be effective in curing superficial scytalidiosis in the toe webs, but this is no longer commercially available [13]. A case of *Scytalidium hyalinum* onychomycosis was also successfully treated with a 5% amorolfine nail lacquer [55]. Voriconazole has also been proposed as a new therapeutic approach for superficial infections [54]. Although the low minimal inhibitory concentration (MIC) ranges reported with voriconazole give promising results for the treatment of scytalidiosis and may be indicated in severe cases, the cost-effectiveness ratio is not in favor of such a strategy for superficial dermatomycosis.

As reported for onychomycosis in general, oral antifungals combined with topical nail lacquer may result in higher cure rate than in cases treated only per os. This combination seems also to be judicious for scytalidiosis in regard to the high level of difficulties of cure and the importance of resistance [56].

The reasons for treatment failure are not well understood. One possible explanation is that nail plate infection is often diagnosed after several years of evolution and is thus accompanied by such an extensive onycholysis that the subungual fungi are not in contact with the nail bed, where the drug presents the maximum concentration [2]. Therefore, the optimal therapy should associate a periodic toenail mechanical removal of the abnormal portions of the nail plate performed by podiatrists, dermatologists, or practitioners who are specialized in the field with an antifungal therapy [57]. Nevertheless, therapeutic success is also dependent on the good compliance of patients for this treatment, often requiring application for more than 6 months.

10.1.5 Diagnosis

10.1.5.1 Conventional Techniques

The sensitivity of diagnosis is dependent on many criteria such as the experience of the operator for sampling or reading the direct examination and the quality and quantity of the samples. For onychomycosis, high rates of false-negative reaching, respectively, 5%–15% and 30% still occur in direct examinations and cultures, even in highly experienced laboratories [58]. The optimal method to improve the accuracy of dermatomycosis diagnosis in routine practice still remains unclear. A diagnostic algorithm including validated procedures by experts is generally used and involves five main steps, each one needing to be carefully carried out: mycological sampling, direct examination, cultures, macroscopic and microscopic observation, and interpretation [59].

10.1.5.1.1 Mycological Sampling

Sampling of biological material should always be accompanied by precise examination of the clinical lesion and a complete interrogation of the patient about his or her urban or rural lifestyle, possible travel in endemic areas, antifungal treatment, duration and evolution of lesions, etc.

Each lesion should be sampled separately, after disinfection with alcohol at 70%. Samples are collected in sterile plates. Squama should be collected by scraping with a curette. For an onyxis, the nail should be cut and the under-nail zone should be scraped, just at the edge of healthy tissues. Searching for other infected sites should be systematic (fold, other foot, hands, intertoe, etc.). In case of maceration, swabs should be used.

10.1.5.1.2 Direct Examination

Squama are the most productive samples for direct examination and are directly examined in black chlorazol solution or by using staining with calcofluor. A 30% potassium hydroxide solution may be used in order to soften the nail fragments.

Microscopic examination at objectives 20× and 40× shows fungal septate hyphae indistinguishable at first from dermatophytes by an inexperienced eye. Nevertheless, the appearance of characteristic sinuous irregular hyphae should lead to the diagnosis of scytalidiosis.

10.1.5.1.3 Culture

S. dimidiatum and *S. hyalinum* grow on cycloheximide-free Sabouraud agar supplemented with antibiotics. A better sensitivity of cultures is achieved if an abundant material is cultured. Plates and slants are incubated at 30°C for up to 3 weeks in order to isolate other slow-growing potential pathogens. *Scytalidium* spp. are fast-growing molds that after 48 h give pale gray colonies soon becoming olivaceous gray to dark brown or black for *S. dimidiatum* with a black reverse. For *S. hyalinum*, cultures remain whitish with a colorless to beige reverse. The mycelium is aerial, floconnose, filling in a week petri dishes or slants for *S. dimidiatum* and remaining flushy for *S. hyalinum*.

Microscopic morphology is observed on a cellophane tape preparation with lactophenol blue. Characteristic features demonstrate septate and branched hyaline to dark brown hyphae ranging from 2 to 8 μm in width. Uni- or bicellular arthroconidia are observed that are sometimes larger than hyphae (4 × 8 μm). Hyphae remain hyaline for *S. hyalinum*. Presence of numerous arthoconidia in chains should not be confused with the *Geotrichum* species [1].

In case of invasive infection, histopathology is necessary and if positive shows pigmented or not, refractile and often sinuous, septate hyphae on biopsies as in other deep mycosis [9,60].

10.1.5.2 Molecular Techniques

Nowadays, sequencing of PCR products is widely used in order to precisely identify the genus and species of fungi that are isolated in pathogenic samples. A panfungal PCR targeting the ITS region is performed according to the procedures described by White et al. [61]. Thereafter, amplicons may be sequenced by using the same primer set. The obtained sequences are then analyzed by using bioinformatics programs such as BLAST on dedicated Web sites and databanks such as National Center for Biotechnology Information (NCBI) (http://blast.ncbi.nlm.nih.gov/Blast.cgi).

Another molecular biology-based approach has been developed in order to discriminate between dermatophytes and *Scytalidium* spp. This method combines a PCR targeting of the 18S ribosomal DNA followed by a restriction fragment length polymorphism (RFLP). This method has been evaluated experimentally on cultures and on samples from patients with dermatomycoses [62,63].

10.2 METHODS

10.2.1 Sample Preparation

DNA may be extracted from fungal cultures for the identification or confirmation of isolated species of *Scytalidium* spp., for example, in cases of atypical morphological features occurring macroscopically or microscopically (slight sporulation). Epidemiological studies may also be conducted in order to highlight interstrain polymorphisms according to their origin (geographically or biologically).

Direct manipulation of biological samples may be informative in cases of patients under treatment at the time of sampling or with an evocative lesion associated with a negative direct examination. This may also be performed retrospectively if the cultures remain negative with a direct examination that was previously positive.

Two main methods have to be considered for DNA extraction. Rapid protocol and routine use will require the use of commercial kits. For such kits, a pre-incubation with lyticase may be useful in order to efficiently digest chitin-containing fungal cell walls. Epidemiological studies or research will prefer total fungal genomic DNA extractions that may involve different techniques. The advantages of such techniques are their low cost and obtaining of a higher amount of pure DNA. The disadvantage is the delay necessary for the procedure that may take up to 1 or 2 working days. We propose here a protocol adapted from Möller et al. and slightly modified [64]. Approximately 1 cm^2 of mycelium taken from 2-week-old cultures is transferred to 2.0 mL Eppendorf tubes containing 300 µL TES buffer (Tris 1.2% [w/v], Na–ethylenediaminetetraacetic acid [EDTA] 0.38% [w/v], sodium dodecyl sulfate 2% [w/v], pH 8.0). Fungal cells are disrupted mechanically in a tight-fitting sterile pestle for approximately 1 min. Subsequently, 10 µL proteinase K is added and incubated for 10 min at 65°C. After adding 140 µL of 5 M NaCl and 1/10 vol CTAB 10% (hexadecyltrimethylammonium bromide) buffer, the mixture is vortexed and the material is incubated for 30 min at 65°C. Subsequently, 700 µL of a mixed solution of chloroform and isoamylalcohol (v/v = 24:1) is added, vortexed, incubated for 30 min on ice water, and centrifuged for 10 min at 14,000 rpm.

The supernatant is transferred to a new tube with 200 µL 5 M NH_4–acetate, incubated on ice water and centrifuged again at 4°C for 10 min at 14,000 rpm. The supernatant is transferred to a new sterile Eppendorf tube with 0.55 vol isopropanol, mixed and centrifuged at 4°C for 5 min at 14,000 rpm. Subsequently, the pellet is washed with cold 70% ethanol. Air-dried pellet is resuspended in 100 µL TE buffer (Tris 0.12% [w/v], Na–EDTA 0.04% [w/v]). DNA is used immediately for PCR amplification or stored at −80°C.

10.2.2 Detection Procedures

10.2.2.1 PCR–RFLP

According to Machouart et al. [62], PCR is performed in a reaction mixture of 25 µL containing 1.5 mM $MgCl_2$, 50 µM of each dNTP, 25 pmol of each primer, 1 U of Taq polymerase, and 25 ng of extracted DNA. A primer set, DH2L (5′-TGTACTGGTCCGGCCGGG-3′) – DH1R (5′-CGGCGGTCCTAGAAACCAAC-3′), described as preferentially targeting the 18S rDNA of *Scytalidium* spp., is used with amplification conditions consisting of 3 min at 94°C for the initial denaturation, followed by 30 cycles at 94°C for 1 min, 58°C for 1 min, and 72°C for 40 s, and one final extension step at 72°C for 5 min. Next, 5 µL of PCR product is electrophoresed in 2% agarose gel in the presence of intercalating dye and visualized under UV light. The expected size is of 184 bp.

Amplification products are then digested with *Bam*HI for 1 h at 37°C in a total volume of 20 µL containing 5 µL of the specific PCR product and 5 U of the enzyme. The digested samples are analyzed on 3.5% Nusieve ethidium bromide (BET)–agarose electrophoresis gel (Figure 10.2) [62].

10.2.2.2 Panfungal PCR

Fungal identification by molecular biology may also use panfungal primers targeting ITS regions: ITS1 (5′-TCCGTAGGTGAACCTGCGG-3′) and ITS4 (5′-TCCTCCGCTTATTGATATGC-3′) with an annealing temperature of 51°C [61].

The PCR mixture (50 µL) is composed of 2 U of FastStart Taq Polymerase (Roche Diagnostics), 5 µL of 10× PCR buffer–$MgCl_2$ (20 mM), 20 pmol of each primer, 200 µM of dATP, dGTP, dTTP, and dCTP, and 5 µL of DNA. The PCR is run in a thermal cycler under the following conditions: initial denaturation at 96°C for 1 min; 35 cycles at 94°C for 1 min, 57°C for 1 min, and 72°C for 1 min; and a final extension at 72°C for 10 min. Then 5 µL of PCR products are

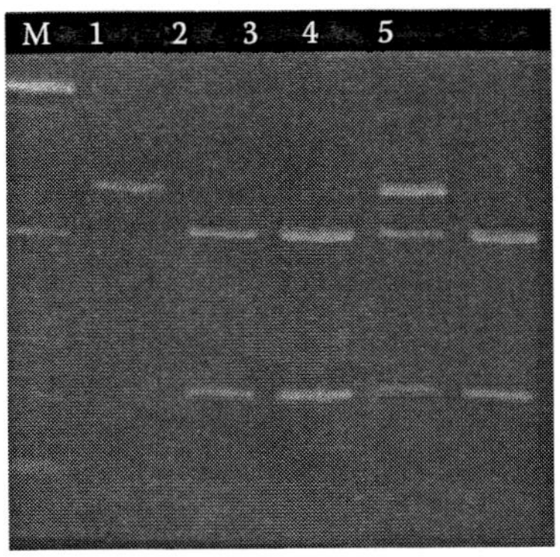

FIGURE 10.2 PCR amplification of human samples containing fungi and restriction patterns with *Bam*HI. Lane M, 100 bp ladder; lane 1, nondigested amplicon (DH2L–DH1R); lanes 2–5; *Bam*HI digested products showing only the presence of *Scytalidium* (lanes 2, 3, and 5), and showing a partial digestion (lane 4): this is characteristic of the presence of *Scytalidium* with a nondigested product corresponding to another fungus in the sample.

electrophoresed in a 2% agarose gel in the presence of BET and visualized under UV light. The expected PCR product size is approximately 600 bp.

The amplified fragments may be sequenced after purification with commercial kits. Sequencing is then performed on both strands with the same primers as those used in the initial amplification. Sequencing reactions are performed by using the Big Dye Terminator v1.1 Cycle Sequencing kit (Applied Biosystems) on an automated sequencer.

Briefly, in a final volume of 10 μL, 1 μL of PCR product is added to 2 μL of mix, 1 μL of buffer, and 1.6 pmol of primer. Sequencing reactions are performed in a thermal iCycler IQ System (Bio-Rad) with an initial denaturation at 96°C for 1 min, followed by 25 amplification cycles at 96°C for 10 s, 50°C for 5 s, and 60°C for 4 min. The resulting products are purified by using Sephadex filtration, commercial kits, or a succession of washing and centrifugation steps. For the latter procedure, the resulting product is precipitated by adding 2 μL of 3 M NaAc (Sigma Aldrich), 2 μL EDTA 125 mM, and 50 μL of 96% ethanol. After incubation on ice for 15 min and centrifugation at 13,000 rpm for 20 min, 125 μL of 70% ethanol is added. A final centrifugation is carried out at 13,000 rpm for 15 min. Each sample is run on an ABI PRISM 3100 Genetic Analyzer (Applied Biosystems).

The sequences obtained are aligned by using the Bioedit Software or the AliBee program on the Web site at http://www.genebee.msu.su/genebee.html, the consensus sequence is then analyzed with the BLASTn program (http://blast.ncbi.nlm.nih.gov/Blast.cgi).

10.3 CONCLUSIONS AND FUTURE PERSPECTIVES

Almost 40 years after the first description of *Scytalidium* spp. in humans, knowledge is still fragmented about these molds. Taxonomy of these fungi is still controversial and this has lead to an unstable nomenclature [65]. Because the genotyping approach of pathogenic fungi may clarify epidemiology and may be associated with the medical traits of fungi, some studies have been conducted on *Scytalidium* spp. [8,11,12]. This led to place *S. dimidiatum* in *Botryosphaeriaceae* and to propose a new name for *S. dimidiatum*: *Neoscytalidium dimidiatum*. Nevertheless, the position of *S. hyalinum* and other *Scytalidium* spp. not encountered in pathology has not been clearly considered.

Genetic differences have been detected between *S. hyalinum* and *S. dimidiatum* in addition to the geographical distribution and to the epidemiological sources. Introns have been found in the 18S rDNA of *S. dimidiatum* and were absent at the homologous sites in *S. hyalinum* [11,66]. Some nucleotide polymorphisms have also been highlighted in the genes coding for 18S rDNA, ITS, and tubulin. Combined analysis of polymorphic loci has thus helped define five sequence types from which two were detected exclusively in isolates from plants, two were found only in clinical isolates, and one was observed in isolates from humans and a mango tree. But this remains to be completed because the study was conducted on few strains. However, the authors proposed to subordin *S. hyalinum* as a variety of *S. dimidiatum* with this new combination of *Neoscytalidium dimidiatum* var. *hyalinum* by referring to the work of Crous et al. [8,12]. We conducted another study on ITS, 28S rDNA, actin (Act), and elongation factor (EF) on 80 isolates of *Scytalidium* species, including *S. dimidiatum*, *S. hyalinum*, and other species [67]. Our aim was to give arguments in order to specify the classification of the genus *Scytalidium* and particularly the genetic relationships between the different species. We confirmed polymorphisms in ITS and highlighted new differences on 28S, Act, and EF. We also discovered a different distribution of intronic insertions between *S. dimidiatum* and *S. hyalinum* [67] (unpublished data). These arguments let us to conclude that *S. dimidiatum* is an ancestral form of the fungus that has adapted to humans in the course of evolution. This is concomitant with genetic evolution implying nucleotidic mutations in some genes and insertion or deletion of introns. The perspective of this work is now to complete the study and to compile all already known and new data and to focus on a multilocus phylogeny in order to clarify and reclassify the species belonging to the genus *Scytalidium*. Only after this complete study will it be possible to propose new formulations for *S. dimidiatum* or *S. hyalinum* with regard to the other species and their position.

In endemic areas, *S. dimidiatum* and *S. hyalinum* are frequently isolated comparing with dermatophytes. These fungi also have the potential to cause deep and life-threatening infections mostly in patients with risk factors for developing opportunistic diseases [68]. Considering their high degree of resistance to antifungals, these molds must not be overlooked in populations of immigrants or patients having traveled in endemic zones. Because scytalidiosis is rare in temperate countries, the development of efficient treatment is obviously not a priority, even though this mycosis is very common in tropical areas. Moreover, their importance may be underestimated because of the nonconsultation of patients living in tropical and subtropical countries. Diagnosis difficulties remain a reality with the importance of false negative results in mycology even after appropriate sampling of lesions; the reasons for this are unclear. In some cases, fungal identification is also hindered by the presence of contaminants requiring subcultures, delaying the diagnosis. Correct diagnosis of scytalidiosis is necessary to avoid unnecessary lengthy and potentially toxic antifungal treatment. Thus, the ability to discriminate *Scytalidium* species from other pathogens by a PCR may ease the path to correct diagnosis in difficult cases. The PCR–RFLP rapidly discriminating among dermatophytes, *Scytalidium*, and other fungi has been evaluated in clinical practice and showed that this technique might facilitate the rapid diagnosis of onychomycosis due to dermatophytes or *Scytalidium* spp. with a similar sensitivity and specificity when compared to reference mycological techniques [46,63]. Use of such techniques may also be suitable for laboratories that are not trained to correctly sample nails or inexperienced in the field [69].

Scytalidium spp. currently remain a microbiological challenge from an epidemiological, phylogenetical, diagnostic, and therapeutic point of view.

ACKNOWLEDGMENTS

The authors thank Dr. Nicole Desbois (pôle de biologie pathologie, Laboratoire de Parasitologie-Mycologie-Immunologie, CHU de Fort-de-France—BP 632, 97261 Fort-de-France, Martinique, France) for helpful discussion on these fungi, with references to her mycological and clinical experience and for providing the strains used for microscopic illustrations of this chapter. The authors also thank Tiffany Dunham for checking the English of this chapter.

REFERENCES

1. De Hoog, G.S. et al., *Atlas of Clinical Fungi*, 2nd edn., Centraalbureau voor Schimmelcultures, Utrecht, the Netherlands, 2004.
2. Hay, R.J., *Scytalidium* infections, *Curr. Op. Infect. Dis.*, 15, 99, 2002.
3. Nattrass, R.M., A new species of *Hendersonula* (*H. toruloidea*) on deciduous trees in Egypt, *Br. Mycol. Soc. Trans.*, 18, 189, 1933.
4. Gentles, J.C. and Evans, E.G., Infection of the feet and nails with *Hendersonula totuloidea*, *Sabouraudia*, 8, 72, 1970.
5. Campbell, C.K. and Mulder, J.L., Skin and nail infection by *Scytalidium hyalinum* sp. nov., *Sabouraudia*, 15, 161, 1977.
6. Sutton, B.C. and Dyko, B.J., Revision of *Hendersonula*, *Mycol. Res.*, 93, 466, 1989.
7. Farr, D.F. et al., *Fusicoccum arbuti* sp. nov. causing cankers on Pacific madrone in western North America with notes on *Fusicoccum dimidiatum*, the correct name for *Scytalidium dimidiatum* and *Nattrassia mangiferae*, *Mycologia*, 97, 730, 2005.
8. Crous, P.W. et al., Phylogenetic lineages in the *Botryosphaeriaceae*, *Stud. Mycol.*, 55, 235, 2006.
9. Sigler, L. et al., Invasive *Nattrassia mangiferae* infections: Case report, literature review and therapeutic and taxonomic appraisal, *J. Clin. Microbiol.*, 35, 433, 1997.
10. Roeijmans, H.J. et al., Molecular taxonomy and GC/MS of metabolites of *Scytalidium hyalinum* and *Nattrassia mangiferae* (*Hendersonula toruloidea*), *J. Med. Vet. Mycol.*, 35, 81, 1997.
11. Machouart, M. et al., Polymorphisms and intronic structures in the 18S subunit ribosomal RNA gene of the fungi *Scytalidium dimidiatum* and *Scytalidium hyalinum*. Evidence of an IC1 intron with an His-Cys endonuclease gene. *FEMS Microbiol. Lett.*, 238, 455, 2004.
12. Madrid, H. et al., Genotyping and in vitro antifungal susceptibility of *Neoscytalidium dimidiatum* isolates from different origins, *Int. J. Antimicrob. Agents*, 34, 351, 2009.
13. Lacroix, C. et al., Scytalidiosis in Paris, *J. Am. Acad. Dermatol.*, 48, 852, 2002.
14. Punithalingam, E. and Waterston, J.M., *Hendersonula toruloidea*, *CMI Descript. Pathol. Fungi Bact.*, 274, 1, 1970.
15. Natour, R.M. and Ahmed, J.M., Control of branch wilt disease of grape, *Plant Dis. Rep.*, 53, 152, 1969.
16. Wangikar, P.D., Raut, J.G., and Gopalkrishna, N., Drying grape vines caused by *Hendersonula toruloidea*, *Indian Phytopathol.*, 22, 403, 1969.
17. Alizadeh, A., Heidarian, A., and Farrokhi-Nejad, R., *Citrus* branch wilt, decline, and death caused by *Nattrassia mangiferae* and its other hosts in Khuzestan province, *Iranian J. Plant Pathol.*, 36, 21, 2000.
18. Chandra, S., Some new leaf-spot diseases from Allahabad (India), *Nova Hedw Beih*, 47, 35, 1974.
19. Pandey, R.S. et al., A new leaf spot disease of mango, *Plant Dis.*, 65, 441, 1981.
20. Reckhaus, P. and Adamous, I., *Hendersonula* dieback of mango in Niger, *Plant Dis.*, 71, 1045, 1987.
21. Meredith, D.S., Tip rot of banana fruits in Jamaica, *Trans. Brit. Mycol. Soc.*, 46, 473, 1963.
22. Meredith, D.S., Fungal diseases of bananas in Hawai, *Plant Dis. Rep.*, 53, 63, 1969.
23. Wilson, E.E., The branch wilt of Persian Walnut trees and its cause, *Hilgardia*, 17, 413, 1947.
24. Calavan, E.C. and Wallace, J.M., *Hendersonula toruloidea* Nattrass on citrus in California, *Phytopathology*, 44, 635,1954.
25. Matheron, M.E. and Sigler, L. First report of *Eucalyptus* dieback caused by *Nattrassia mangiferae* in North America, *Plant Dis.*, 78, 432, 1993.
26. English, H., Davis, J.R., and deVay, J.E., Relationship of *Botryosphaeria dothidea* and *Hendersonula toruloidea* to a canker disease of almond, *Phytopathology*, 65, 114, 1975.
27. Tsahouridou, P.C. and Thanassoulopoulos, C.C., First report of *Hendersonula toruloidea* as a foliar pathogen of strawberry-tree (*Arbutus unedo*) in Europe, *Plant Dis.*, 84, 187, 2000.
28. Kane, J. et al., An autochtonous phaeohyphomycotic nail infection in Canada caused by *Hendersonula toruloidea*, *Mycoses*, 33, 37, 1990.
29. Dunn, J.J. et al., Invasive fungal sinusitis caused by *Scytalidium dimidiatum* in a lung transplant recipient, *J. Clin. Microbiol.*, 41, 5817, 2003.
30. Moore, M.K., *Hendersonula toruloidea* and *Scytalidium hyalinum* infections in London, England, *J. Met. Vet. Mycol.*, 24, 219, 1986.
31. Belloeuf, L. et al., Onychomycoses à *Scytalidium* en Martinique, *Ann. Dermatol. Venereol.*, 131, 245, 2004.
32. Hay, R.J. and Moore, M.K., Clinical features of superficial fungal infections caused by *Hendersonula toruloidea* and *Scytalidium hyalinum*, *Br. J. Dermatol.*, 110, 677, 1984.
33. Curtis, J.W., The two foot-one hand disease, *Bull. Assoc. Milit. Dermatol.*, 13, 1, 1964.
34. Levi, M.E. and Smith, J.W., Postraumatic infection due to *Scytalidium dimidiatum*, *Clin. Infect. Dis.*, 18, 127, 1994.
35. Marriott, D.J. et al., *Scytalidium dimidiatum* and *Lecythophora hoffmani*: Unusual causes of fungal infections in a patient with AIDS, *J. Clin. Microbiol.*, 35, 2949, 1997.
36. Dhindsa, M.K., Naidu, J., and Singh, S.M., A case of subcutaneous infection in a patient with discoid lupus erythematosus caused by a *Scytalidium* synanamorph of *Nattrassia mangiferae* and its treatment, *Med. Mycol.*, 36, 425, 1998.
37. Gumbo, T. et al., Case report. *Nattrassia mangiferae* endophtalmitis, *Mycoses*, 45, 118, 2002.
38. Geramishoar, M. et al., First case of cerebral phaeohyphomycosis caused by *Nattrassia mangiferae* in Iran, *Jpn. J. Infect. Dis.*, 57, 285, 2004.
39. Jabbarvand, M. et al., *Nattrassia mangiferae* keratitis after laser in situ keratomileusis, *J. Cataract. Refract. Surg.*, 30, 268, 2004.
40. Morris-Jones, R. et al., *Scytalidium dimidiatum* causing recalcitrant subcutaneous lesions produces melanin, *J. Clin. Microbiol.*, 42, 3789, 2004.

41. Willinger, B. et al., Disseminated infection with *Nattrassia mangiferae* in an immunosuppressed patient, *J. Clin. Microbiol.*, 42, 478, 2004.
42. Sadeghi Tari, A. et al., Post-traumatic fatal *Nattrassia mangiferae* orbital infection, *Int. Ophtalmol.*, 26, 247, 2005.
43. Tan, D.H.S. et al., Disseminated fungal infection in a renal transplant recipient involving *Macrophomina phaseolina* and *Scytalidium dimidiatum*: Case report and review of taxonomic changes among medically important members of the *Botryosphaeriaceae*, *Med. Mycol.*, 46, 285, 2008.
44. Benne, C.A. et al., Disseminating infection with *Scytalidium dimidiatum* in a granulocytopenic child, *Eur. J. Clin. Microbiol. Infect. Dis.*, 12, 118, 1993.
45. Mani, R.S. et al., Cerebral phaeohyphomycosis caused by *Scytalidium dimidiatum*: A case report from India, *Med. Mycol.*, 46, 705, 2008.
46. Al-Rajhi, A. et al., *Scytalidium dimidiatum* fungal endophtalmitis, *Br. J. Ophtalmol.*, 77, 388, 1993.
47. Elinav, H. et al., Invasive *Scytalidium dimidiatum* infection in an immunocompetent adult, *J. Clin. Microbiol.*, 47, 1259, 2009.
48. Zaatari, G.S., Reed, G., and Morewessel, R., Subcutaneous hyphomycosis caused by *Scytalidium hyalinum*, *Am. J. Clin. Pathol.*, 82, 252, 1984.
49. Sriaroon, C. et al., Successful treatment of subcutaneous *Scytalidium hyalinum* infection with voriconazole and topical terbinafine in a cardiac transplant patient, *Transplantation*, 85, 780, 2008.
50. Oyeka, C.A. and Gugnani, H.C., Keratin degradation by *Scytalidium* species and *Fusarium solani*, *Mycoses*, 41, 73, 1998.
51. Elewski, B.E., Onychomycosis caused by *Scytalidium dimidiatum*, *J. Am. Acad. Dermatol.*, 35, 336, 1996.
52. Dunand, J. and Paugam, A., In vitro susceptibility of isolates of *Scytalidium* spp. from superficial lesions against posaconazole, *Pathol. Biol.*, 56, 268, 2008.
53. Dunand, J. and Paugam, A., In vitro efficacy of voriconazole against clinical isolates of *Scytalidium* spp. from clinical lesions. *Int. J. Antimicrob. Agents*, 31, 176, 2008.
54. Lacroix, C. and Feuilhade de Chauvin, M., In vitro activity of amphotericin B, itraconazole, voriconazole, posaconazole, caspofungin and terbinafine against *Scytalidium dimidiatum* and *Scytalidium hyalinum* clinical isolates, *J. Antimicrob. Chemother.*, 61, 835, 2008.
55. Downs, A.M., Lear, J.T., and Archer, C.B., *Scytalidium hyalinum* onychomycosis successfully treated with 5% amorolfine nail lacquer, *Br. J. Dermatol.*, 140, 555, 1999.
56. Avner, S., Nir, N., and Henri, T., Combination of oral terbinafine and topical ciclopirox compared to oral terbinafine for the treatment of onychomycosis, *J. Dermatolog. Treat.*, 16, 327, 2005.
57. Scot Malay, D. et al., Efficacy of debridement alone versus debridement combined with topical antifungal nail lacquer for the treatment of pedal onychomycosis: A randomized controlled trial, *J. Foot Ankle Surg.*, 48, 295, 2009.
58. Hay, R., Literature review. Onychomycosis, *J. Eur. Acad. Dermatol. Venereol.*, 19, 1, 2007.
59. Fletcher, C.L., Hay, R.J., and Smeeton, N.C., Onychomycosis: The development of a clinical diagnostic aid for toenail disease. Part I. Establishing discriminating historical and clinical features, *Br. J. Dermatol.*, 150, 701, 2004.
60. Moore, M.K., The infection of human skin and nail by *Scytalidium* species, *Curr. Top. Med. Mycol.*, 4, 1, 1992.
61. White, T. J. et al., Amplification and direct sequencing of fungal ribosomal RNA genes for phylogenetics, In: *PCR Protocols: A Guide to Methods and Applications*, pp. 315–322, Innis, M.A., Gelfand, D.H., Sninsky, J.J., and White, T.J. (eds.), Academic Press, San Diego, CA, 1990.
62. Machouart, M. et al., Rapid discrimination among dermatophytes, *Scytalidium* spp., and other fungi with a PCR-restriction fragment length polymorphism ribotyping method, *J. Clin. Microbiol.*, 39, 685, 2001.
63. Menotti, J. et al., Polymerase chain reaction for diagnosis of dermatophyte and *Scytalidium* spp. onychomycosis, *Br. J. Dermatol.*, 151, 518, 2004.
64. Möller, E.M. et al., A simple and efficient protocol for isolation of high molecular weight DNA from filamentous fungi, fruit bodies, and infected plant tissues, *Nucleic Acids Res.*, 20, 6115, 1992.
65. Lacaz, C.S. et al., Onychomycosis caused by *Scytalidium dimidiatum*. Report of two cases. Review of the taxonomy of the synanamorph and anamorph forms of this coelomycete, *Rev. Inst. Med. Trop. Sao Paulo*, 41, 319, 1999.
66. Machouart, M. et al., Nucleotide structure of the *Scytalidium hyalinum* and *Scytalidium dimidiatum* 18S subunit ribosomal RNA gene: Evidence for the insertion of a group IE intron in the rDNA gene of *S. dimidiatum*, *FEMS Microbiol. Lett.*, 208, 187, 2002.
67. Machouart, M. et al., Caractéristiques moléculaires des espèces du genre *Scytalidium* spp.: Aspects phylogéniques et descriptions de structures introniques, *Congress of the French Society of Medical Mycology*, Fort-de-France, Martinique, France, January 2010.
68. Soler, C.P. et al., *Scytalidium dimidiatum* pseudodermatophyte, agent of superficial mycoses and phaehyphomycosis, *Med. Trop.*, 59, 375, 1999.
69. Savin, C. et al., Multicenter evaluation of a commercial PCR-enzyme-linked immunosorbent assay diagnostic kit (Onychodiag) for diagnosis of dermatophytic onychomycosis, *J. Clin. Microbiol.*, 45, 1205, 2007.

11 *Hortaea*

Dongyou Liu and Larry Hanson

CONTENTS

11.1 INTRODUCTION

Dematiaceous fungi cover a large group of melanised fungal organisms that may be plant-infecting pathogens or free-living saprobes. As a melanised, yeast-like fungal genus, *Hortaea* is a common plant pathogen in tropical and subtropical countries. Of the three recognized species within the *Hortaea* genus, *Hortaea werneckii* (previously incorrectly classified in the genera of *Cladosporium*, *Cryptococcus*, *Exophiala* and *Phaeoannlomyces*) is occasionally responsible for a superficial mycosis in humans known as tinea nigra. This fungus colonizes dead keratin cells, and produces hyperchromic plaques on the skin of the palms and soles, that may be mistaken for various types of pigmented birthmarks and moles (nevi) as well as melanomas.

11.1.1 Classification, Morphology, and Biology

The genus *Hortaea* is black yeast-like hyphomycete belonging to the mitosporic Dothideales group, order Dothideales, subclass Dothideomycetidae, class Dothideomycetes, subphylum Pezizomycotina, phylum Ascomycota, kingdom Fungi. The mitosporic Dothideales group consists of six genera: *Coleophoma*, *Dichomera*, *Hobsonia*, *Hortaea*, *Selenophoma*, and *Stigmina*. The genus *Hortaea* is separated into three recognized species: *Hortaea acidophila*, *Hortaea werneckii*, and *Hortaea thailandica*, in addition to an unassigned species *Hortaea* sp. F47 [1–6].

Within the genus *Hortaea*, *Hortaea werneckii* (obsolete synonyms: *Cladosporium werneckii*, *Exophiala werneckii*, and *Phaeoannellomyces werneckii*) is implicated in human diseases [7]. First described as *Cladosporium werneckii* in 1921, *Hortaea werneckii* was transferred to the genus *Exophiala* as *Exophiala werneckii* in 1970, and then moved to a new genus *Hortaea* as *Hortaea werneckii* due to its combination of sympodial and annellidic conidiogenesis [2]. To date, no teleomorph of this fungus has been identified.

Hortaea werneckii is a dematiaceous fungus possessing a yeast-like phase as well as a hyphal phase. The fungus grows on standard mycological media in 5–8 days and may take up to 21 days to mature. From the front, *H. werneckii* colonies are initially pale in color, moist/mucoid, shiny black, and yeast-like, and later become velvety, dark olivaceous in color, with abundant aerial mycelia. From the reverse, the color is black. Microscopically, *H. werneckii* hyphae (up to 6 μm wide) are septate, thick-walled, and brown to dark olivaceous in color. Its conidia (2–5 × 5–10 μm) are two-celled, hyaline to pale brown, cylindrical to spindle-shaped yeast-like cells that usually occur in aggregated masses and that taper toward the ends to form an annellide and new annelloconidia. Annelloconidia are produced at intercalary and lateral annellidic points along the hyphae. Some of the yeast-like conidia have a central septum and are bicellular. They may gradually become chlamydospore-like cells. *Hortaea werneckii* is halophilic and tolerates 3%–30% NaCl, but it does not grow at 37°C [5,8–10].

Hortaea werneckii is a common saprophytic fungus that occurs in soil, compost, humus, and on wood in humid tropical and subtropical regions. It is prevalent in highly saline water of crystallization ponds and has been isolated also from saltwater fish [11–13]. The fungus may cause superficial mycotic infection of stratum corneum (often known as tinea nigra) in humans, although tinea nigra may be also due to infection with *Stenella araguata* [14–16]. Humans often acquire *Hortaea werneckii* infection via direct inoculation of the fungus onto the skin during contact with soil, wood, decaying vegetation, or sewage.

11.1.2 Clinical Features and Pathogenesis

Hortaea werneckii causes a superficial mycosis called tinea nigra (or tinea nigra palmaris) [17]. Owing to its halophilic and lipophilic behavior, *H. werneckii* thrives in hypersaline environments such as the palm of the hand [18]. The fungus adheres to human hands through the hydrophobic feature of its yeast cells, colonizes dead keratin cells on the skin, and its branched hyphae grow between the layers of cornified epidermis, utilizing decomposed lipids in the stratum corneum as nutrient, leading to hyperkeratosis (abnormal thickening of the cornified epidermis) [19]. The incubation period typically lasts for 2–7 weeks.

Appearing as an uncommon discoloration of the skin (due to the accumulation of a melanin-like substance within the fungus), tinea nigra is characterized by the formation of a flat, painless, discrete, and light brown-to-black, noninflammatory, nonscaling, oval-shaped patch (macule) that typically occurs on the palmar aspects of hands and occasionally the plantar and other surfaces of the skin (e.g., soles of the feet and the fingers) as well as toes and nails. In some rare cases, the skin of the neck, chest wall, and penis may be affected. Occasionally, the macule is mottled or velvety and may appear irregular in shape, or there may be multiple patches. The macules range from a few millimeters to several centimeters in diameter and may show more intense pigmentations early in the morning, and fade as the day goes by. This possibly reflects the fact that the fungus is cleared from the hands as a result of daily activities. The course of *H. werneckii* infection is 1–18 months (average 3.8 months). The developing lesions may become visible within 15–30 days of infection [20].

Sometimes, *H. werneckii* behaves as an opportunistic causative agent of systemic phaeohyphomycosis. It was isolated from the serum and a splenic abscess of two leukemia patient and from the house dust of a hypersensitivity pneumonitis patient [10].

Due to its asymptomatic nature and the possibility of spontaneous cure, tinea nigra is observed rarely, accounting for about 1% of all mycoses. The disorder is frequently reported in tropical and subtropical countries such as Latin America, Asia, as well as Europe and the United States [21–25]. Predisposing factors for the illness consist of hyperhydrosis, presence in coastal sea areas, salt-marsh, river estuaria, natural, or man-made saltpans hypersaline environments, and exposure to plants and grasses or substrata of high salinity such as sea water, as well as to house dust [26]. The salty property of human palm, which has an increased concentration of eccrine sweat glands, provides an ideal environment for *H. werneckii* infection.

Most cases of tinea nigra are resolved with keratinolytic agents (e.g., urea, undecylenic acid, salicylic acid, Whitfield ointment, retinoic acid, and 4% aspirin) and epidermal tape stripping [27]. Many topical antifungals (e.g., miconazole, ketoconazole, itraconazole, bifonazole, tiabendazole, terbinafine, and ciclopirox olamine) are also effective against tinea nigra [20,28–33]. Tinea nigra does not usually recur.

11.1.3 Diagnosis

Tinea nigra is a superficial mycosis due to the melanized, polymorphic, and yeast-like fungus *Hortaea werneckii*. This fungus induces flat, pigmented, brownish, irregular and asymptomatic macules on the palm of one hand or foot, which are covered by fine scaling, with well-defined borders and without inflamed margins. Differential diagnosis includes skin lesions produced by other fungi (e.g., *Stenella araguata*, also a dematiaceous fungus) as well as viruses (e.g., herpesviruses and poxviruses). In addition, tinea nigra may be mistaken for other pigmented skin diseases (e.g., Addison's disease, melanocytic naevus, dysplastic naevus, acrolentiginous melanoma, syphilis, pinta, and yaws) or silver nitrate stain [34–36].

Morphologically, *Hortaea werneckii* is notably pleomorphic with a yeast-like phase as well as a hyphal phase. Its colonies are initially black with a creamy appearance (yeast-like phase, as a result of conidia germinating with hyphae) and later become filamentous (hyphal phase, to complete the anamorph life cycle). However, some strains with poor conidiogenesis are occasionally encountered in clinical isolates. Its relatively restricted, black primary cultures, 1 μm-wide annellated zones, and one-septate conidia are useful features for its specific identification. Direct KOH examination of cultured isolates demonstrates the short, tortuous, thick, light brown, occasionally darkened hyphae, which sometimes present short filaments and yeast-like cells [37].

Pathogenic dematiaceous fungi such as *Exophiala salmonis* (*Chaetothyriales*) also produce two-celled conidia [5]; however, the pigmentation of hyphae unambiguously distinguishes tinea nigra from other types of dermatophytoses or skin infections. In comparison with *Cladosporium* spp., *Hortaea werneckii* conidia are pigmented yeast cells with a dark central septum, and their outer wall later become thick-walled, heavily pigmented. Furthermore, whereas the chaetothyrialean black yeast-like fungus *Cladophialophora saturnica* shows invasive behavior, *Hortaea werneckii* is strictly limited to the stratum corneum. Skin biopsies of tinea nigra show fungal elements in the stratum corneum with a discrete perivascular infiltrate [37].

Use of physiobiochemical properties and antigen detection further enhances the identification of *H. werneckii* [38,39]. *Hortaea werneckii* tolerates high concentrations of sodium chloride (halotolerance) that may be used as a physiological marker for the identification of this fungus [9,40,41]. Because *H. werneckii* has a carbon-source assimilation pattern identical to that of *Exophiala dermatitidis*, *Exophiala jeanselmei*, and *Exophiala spinifera*, the commercial API 20C system does not allow differentiation among these four species [42,43].

The time-consuming and variable nature of morphological, biochemical examinations creates long delays in the identification of *H. werneckii*. Molecular techniques offer a rapid and accurate alternative for determining pathogenic fungi including *H. werneckii* [41,44]. In particular, sequencing analysis of small subunit (SSU, 18S) and large subunit (LSU, 28S) ribosomal RNA (rRNA) genes as well as rRNA internal transcribed spacer (ITS) region permits rapid and precise characterization of fungal genera and species. For example, comparison of sequences from the D1/D2 region of the LSU rRNA gene indicates that *H. werneckii* has more than 100 nucleotide differences from *Exophiala* species. Located between the 18S and 28S rRNAs, the ITS region consists of two spacers (ITS1 and ITS2) separated by a 5.8S conserved region. Nucleotide differences in the ITS1 and/or ITS2 regions provide an effective approach for species identification of pathogenic fungi from other, closely related taxa [45–47]. Sequence analysis of the ITS regions shows that *H. werneckii* has more than 200 nucleotide differences from the *Exophiala* species. Exploiting the nucleotide differences in the internal transcribed spacer (ITS) region, *H. werneckii* primers were designed for specific polymerase chain reaction (PCR) amplification of a 306bp fragment [48]. In addition, examination of restriction fragment length polymorphisms (RFLP) of mitochondrial DNA (mtDNA) enables assessment of intraspecies diversity of *H. werneckii* and subsequent separation of seven mtDNA types (populations) [44,49].

11.2 METHODS

11.2.1 Sample Preparation

Biopsy specimens are obtained from patients with tinea nigra by carefully scraping off the stratum corneum with a scalpel blade. KOH (20%) microscopic examination of cutaneous scrapings reveals one- to two-celled yeasts and thick, septate, branching hyphae with a dark pigment in their walls. Periodic acid-Schiff-positive septate hyphae are seen in the stratum corneum in biopsy specimens, along with hyperkeratosis and mild acanthosis. The pigmented cell walls of the fungus are readily observable with hematoxylin and eosin stain. Culture of the scrapings on Sabouraud glucose agar without or with antibiotics (Mycobiotic, Difco) at 25°C yields growth in a week. *H. werneckii* isolates typically shows two-celled yeast forms (annellides) that produce one- and two-celled conidia [20]. The isolates are identified on the basis of macroscopic and microscopic characteristics [25].

For DNA extraction, a small amount of fungal pellet is suspended in 100μL extraction buffer (100mM Tris–HCl, pH 7.5, 30mM EDTA, 0.5% [w/v] sodium dodecyl sulfate [SDS]) and vortexed for 15s. The mixture is then incubated at 100°C for 15min, 100μL of 2.5M potassium acetate is added and mixed, chilled on ice for 60min, and centrifuged at 12,000rpm for 5min. The supernatant is transferred to a new tube, and the DNA is precipitated with an equal volume of cold isopropanol (−20°C), washed with 500μL of 99% ethanol, air dried, and resuspended in 100μL distilled water [50].

11.2.2 Detection Procedures

11.2.2.1 Specific PCR Identification

Abliz et al. [48] designed an oligonucleotide primer set, Hor-F (5′-TGGACACCTTCATAACTCTTG-3′) and Hor-R (5′-TCACAACGCTTAGAGACGG-3′) from the ITS1-5.8S-ITS2 sequence for specific amplification of a 306bp fragment from *H. werneckii*.

Procedure

1. PCR mixture (25μL) is made up of 2.5μL template DNA, 2.5μL (2pmol) each Hor-F and Hor-R primer, 2μL (2.5mM) dNTP mixture, 0.125μL (5U/μL) *Taq* polymerase (Nippon Gene), and 2.5μL 10× reaction buffer.
2. Amplification is conducted with 1 cycle of 95°C for 4min; 30 cycles of 94°C for 1min, 58°C for 2.5min, and 72°C for 2.5min; and a final extension at 72°C for 10min.
3. Upon completion, 2μL of the amplified product is run on a 1.5% agarose gel, stained with ethidium bromide, and visualized by UV illumination.

Note. This primer set recognizes DNA from *H. werneckii* only, but not DNA from other related dematiaceous fungal taxa. Thus, the PCR assay employing primers Hor-F and Hor offers a useful tool for rapid specific identification of *H. werneckii*.

11.2.2.2 Sequencing Analysis of ITS Regions

Abliz et al. [48] employed universal primers ITS5 (5′-GGAAGTAAAAGTCGTAACAAGG-3′) and ITS4 (5′-TCCTCCGCTTATTGATATGC-3′) [51] for PCR amplification of the ITS1–5.8S–ITS2 region in *H. werneckii* DNA. The resulting PCR product is sequenced with an ABI PRISM 3100 sequencer after labeling with BigDye™ terminator cycle sequencing ready reaction (Applied Biosystems). Two external primers ITS5 and ITS4 and two internal primers ITS2 (5′-GCTGCGTTCTTCATCGATGC-3′) and ITS3 (5′-GCATCGATGAAGAACGCAGC-3′) are used for sequencing. The nucleotide sequence of the ITS region from *H. werneckii* is aligned to those from other dematiaceous fungal species and medically important yeasts using CLUSTAL W (version 1.6).

11.3 CONCLUSION

Hortaea werneckii is a causative agent of tinea nigra (superficial mycosis) in the tropical and subtropical environments. The infected patients develop asymptomatic black or brown patch with well-defined border on the palm of one hand or foot. In order to implement effective therapy, it is important to differentiate *Hortaea werneckii* from other dematiaceous fungi and pathogenic viruses that infect the skin. However, due to its pleomorphic nature and its morphological resemblances to other related dematiaceous fungal taxa, laboratory identification of *H. werneckii* on the basis of phenotypic characteristics is tedious and demands considerable technical expertise. Application of PCR and sequencing techniques facilitates rapid, specific, and sensitive identification of *H. werneckii*.

REFERENCES

1. The UniProt Consortium. Available at http://www.uniprot.org/, accessed on August 1, 2010.
2. Nishimura, K. and Miyaji, M., *Hortaea*, a new genus to accommodate *Cladosporium werneckii*. *Jpn. J. Med. Mycol.*, 25, 139, 1984.

3. Nishimura, K. and Miyaji, M., Further studies on the phylogenesis of genus *Exophiala* and *Hortaea*. *Mycophathologia*, 92, 101, 1985.
4. Rippon, J.W., *Medical Mycology*, 3rd edn. W.B. Saunders Co., Philadelphia, PA, 1988.
5. de Hoog, G.S. et al., *Atlas of Clinical Fungi*, 2nd edn. Centraalbureau voor Schimmelcultures/Universitat Rovira i Virgili, Utrecht/Reus, 2000.
6. Holker, U. et al., *Hortaea acidophila*, a new acid-tolerant black yeast from lignite. *Anton. van Leeuwen*., 86, 287, 2004.
7. Kane, J. and Summerbell, R.C., Sodium chloride as aid in identification of *Phaeoannellomyces werneckii* and other medically important dematiaceous fungi. *J. Clin. Microbiol*., 25, 944, 1987.
8. Mok, W.Y., Nature and identification of *Exophiala werneckii. J. Clin. Microbiol*., 16, 976, 1982.
9. de Hoog, G.S. and Gerrits van den Ende, A.H., Nutritional pattern and eco-physiology of *Hortaea werneckii*, agent of human tinea nigra. *Anton. van Leeuwen*., 62, 321, 1992.
10. Ng, K.P. et al., The mycological and molecular study of *Hortaea werneckii* isolated from blood and splenic abscess. *Mycopathologia*, 159, 495, 2005.
11. Iwatsu, T. and Udagawa, S., *Hortaea werneckii* isolated from sea-water. *Jpn. J. Med. Mycol*., 29, 142, 1988.
12. Petrovic, U., Role of oxidative stress in the extremely salt-tolerant yeast *Hortaea werneckii. FEMS Yeast Res*., 6, 16, 2006.
13. Plemenitaš, A. et al., Adaptation of extremely halotolerant black yeast *Hortaea werneckii* to increased osmolarity: A molecular perspective at a glance. *Stud. Mycol*., 61, 67, 2008.
14. Perez, C. et al., Tinea nigra: Report of twelve cases in Venezuela. *Mycopathologia*, 160, 235, 2005.
15. Crous, P.W., Braun, U., and Groenewald, J.Z., *Mycosphaerella* is polyphyletic. *Stud. Mycol*., 58, 1, 2007.
16. Crous, P.W. et al., Delimiting *Cladosporium* from morphologically similar genera. *Stud. Mycol*., 58, 33, 2007.
17. Hughes, J.R., Moore, M.K., and Pembroke, A.C., Tinea nigra palmaris. *Clin. Exp. Dermatol*., 18, 481, 1993.
18. Kogej, T. et al., Osmotic adaptation of the halophilic fungus *Hortaea werneckii*: Role of osmolytes and melanization. *Microbiology*, 153, 4261, 2007.
19. Göttlich, E. et al., Cell surface hydrophobicity and lipolysis as essential factors in human tinea nigra. *Mycoses*, 38, 489, 1995.
20. Schwartz, P.R.A., Superficial fungal infections. *Lancet*, 364, 1173, 2004; Kerdel-Vegas, F. and Bastardo de Albornoz, M.C., Tinea Nigra. Report of five cases seen in Venezuela. *Dermatologica*, 132, 320, 1966.
21. Uezato, H. et al., A case of tinea nigra palmaris in Okinawa, Japan. *J. Dermatol*., 33, 23, 1989.
22. Severo, L.C., Bassanesi, M.C., and Londero, A.T., Tinea nigra: Report of four cases observed in Rio Grande do Sul (Brazil) and a review of Brazilian literature. *Mycophathologia*, 126, 157, 1994.
23. Reid, B.J., *Exophiala werneckii* causing tinea nigra in Scotland. *Brit. J. Dermatol*., 139, 157, 1998.
24. Maldonado, I. et al. Tinea nigra palmaris: Presentation of a clinical case in the Argentine Republic. *Rev. Argent. Microbiol*., 39, 218, 2007.
25. Bonifaz, A. et al., Tinea nigra by *Hortaea werneckii*, a report of 22 cases from Mexico. *Stud. Mycol*., 61, 77, 2008.
26. Gunde-Cimerman, N. et al., Hypersaline waters in salterns: Natural ecological niches for halophilic black yeasts. *FEMS Microbiol. Ecol*., 32, 235, 2000.
27. Rosen, T. and Lingappan, A., Rapid treatment of tinea nigra palmaris with cilopirox olamine gel, 0.77%. *Skinmed*, 5, 201, 2006.
28. Marks, J.G., King, R.D., and Davis, B.M., Treatment of tinea nigra palmaris with miconazole topically. *Arch. Dermatol*., 116, 321, 1980.
29. Meisel, C., Treatment of tinea palmaris with Mycospor. *Dermatologica*, 169, S121, 1984.
30. Sayegh-Carreno, R., Abramovits-Ackerman, W., and Giron, G.P., Therapy of tinea nigra plantaris. *Int. J. Dermatol*., 28, 46, 1989.
31. Burke, W.A., Tinea nigra: Treatment with topical ketoconazole. *Cutis*, 52, 209, 1993.
32. Gupta, G. et al., Tinea nigra secondary to *Exophiala werneckii* responding to itraconazole. *Brit. J. Dermatol*., 137, 483, 1997.
33. Shannon, P.L. et al., Treatment of tinea nigra with terbinafine. *Cutis*, 64, 199, 1999.
34. Babel, D.E., Pelachyk, J.M., and Hurley, J.P., Tinea nigra masquerading as acral lentiginous melanoma. *J. Dermatol. Surg. Oncol*., 12, 502, 1986.
35. Hall, J. and Perry, V.E., Tinea nigra palmaris: Differentiation from malignant melanoma or junctional nevi. *Cutis*, 62, 45, 1998.
36. Tseng, S.S. et al., Bilateral tinea nigra plantaris and tinea nigra plantaris mimicking melanoma. *Cutis*, 64, 265, 1999.
37. Smith, S.B. et al., Dermoscopy in the diagnosis of tinea nigra plantaris. *Cutis*, 68, 377, 2001.
38. Standard, P.G., Kaufman, L., and Padhye, A.A., An immunological procedure for the rapid identification of *Phaeoannellomyces werneckii. Diagn. Clin. Immunol*., 5, 367, 1988.
39. Gupta, A.K., Chaudhry, M., and Elewski, B., Tinea corporis, tinea cruris, tinea nigra, and piedra. *Dermatol. Clin*., 21, 395, 2003.
40. Sterflinger, K., Temperature and NaCl-tolerance of rock-inhabiting meristematic fungi. *Anton. van Leeuwen*., 74, 271, 1998.
41. Zalar, P., de Hoog, G.S., and Gunde-Cimerman, N., Ecology of halotolerant dothideaceous black yeasts. *Stud. Mycol*., 43, 38, 1999.
42. Steadham, J.E., Geis, P.A., and Simmank, J.L., Use of carbohydrate and nitrate assimilations in the identification of dematiaceous fungi. *Diagn. Microbiol. Infect. Dis*., 5, 71, 1986.
43. Espinel-Ingroff, A. et al., Evaluation of the API 20C yeast identification system for the differentiation of some dematiaceous fungi. *J. Clin. Microbiol*., 27, 2565, 1989.
44. Uijthof, J.M. et al., Polymerase chain reaction-mediated genotyping of *Hortaea werneckii*, causative agent of tinea nigra. *Mycoses*, 37, 307, 1994.
45. Chen, Y. et al., Polymorphic internal transcribed spacer region 1 DNA sequences identify medically important yeasts. *J. Clin. Microbiol*., 39, 4042, 2001.
46. Iwen, P.C., Hinrichs, S.H., and Rupp, M.E., Utilization of the internal transcribed spacer regions as molecular targets to detect and identify human fungal pathogens. *Med. Mycol*., 40, 87, 2002.
47. Abliz, P. et al., Identification of pathogenic dematiaceous fungi and related taxa based on large subunit ribosomal DNA D1/D2 domain sequence analysis. *FEMS Immunol. Med. Microbiol*., 40, 41–49, 2004.
48. Abliz, P. et al., Specific oligonucleotide primers for identification of *Hortaea werneckii*, a causative agent of tinea nigra. *Diagn. Microbiol. Infect. Dis*., 46, 89, 2003.
49. de Cock, A.W., Population biology of *Hortaea werneckii* based on restriction patterns of mitochondrial DNA. *Anton. van Leeuwen*., 65, 21, 1994.
50. Makimura, K., Murayama, S.Y., and Yamaguchi, H., Detection of a wide range of medically important fungi by the polymerase chain reaction. *J. Med. Microbiol*., 40, 358, 1994.
51. White, T.J. et al., 1990. Amplification and direct sequencing of fungal ribosomal RNA genes for phylogenetics. In: Innis, M.A., Gelfand, J.J., Eds., *PCR Protocols: A Guide to Methods and Applications*, Academic Press, Inc., New York, pp. 315–322.

12 *Leptosphaeria*

Dongyou Liu

CONTENTS

12.1 INTRODUCTION

Leptosphaeria is a dark-walled fungal genus that grows saprophytically and also inhabits the leaves and culms of grasses, dead herbaceous stems, and driftwood. Among the multiple members within the genus, *Leptosphaeria senegalensis* and *Leptosphaeria thompkinsii* have been identified as the agents of human black-grain mycetoma in Africa, which is a highly debilitating tropical disease after a subcutaneous inoculation of the organisms by means of thorns, splinters, snake or insect bites, farm implements, knives, or scratches, resulting in subcutaneous masses with sinuses from which drain pus and grains.

12.1.1 Classification and Morphology

The genus *Leptosphaeria* is a dematiaceous (phaeoid, or dark-walled) filamentous fungus belonging to the family *Leptosphaeriaceae*, order Pleosporales, class Dothideomycetes, subphylum Pezizomycotina, phylum Ascomycota, kingdom Fungi. The family *Leptosphaeriaceae* is composed of the genera *Leptosphaeria*, mitosporic *Leptosphaeriaceae*, *Ophiobolus*, and some unassigned *Leptosphaeriaceae*. In turn, the genus *Leptosphaeria* is separated into 22 recognized and 23 unassigned species, of which *Leptosphaeria coniothyrium* (anamorph: *Coniothyrium fuckelii*), *Leptosphaeria senegalensis*, and *Leptosphaeria tompkinsii* are implicated in human diseases [1,2].

Being an ascomycetous mold, *Leptosphaeria* grows in its teleomorphic phase. On Sabouraud glucose medium containing chloramphenicol at 27°C and 37°C, *Leptosphaeria* colonies grow slowly, showing wooly texture. From the front, colonies are dark olive in color with a gray margin. From the reverse, colonies are dark olive to black in color and with a grayish margin. After 5 weeks, a brown diffusing pigment is noticeable in the medium. Microscopic examination of blue lactic-stained cultures reveals septate hyphae. The sexual form of the fungus is produced on potato-carrot agar media after 6 weeks at 30°C, with the presence of ascomata (cleistothecia), asci, and ascospores. Ascomata containing asci are globose to subglobose in shape, black in color, without ostioles. Asci are clavate to cylindrical and bitunicate, with each ascus carrying eight ascospores inside. Ascospores are four- to nine-celled, hyaline or pigmented, fusoid to curved, and have a constriction at each septum. *Leptosphaeria tompkinsii* differs from *Leptosphaeria senegalensis* in size, shape, septation, and the nature of the gelatinous sheath of the ascospores. Specifically, *Leptosphaeria tompkinsii* ascospores have six septa and pointed ends, while *Leptosphaeria senegalensis* ascospores show four septa and rounded ends [3].

12.1.2 Clinical Features and Epidemiology

Leptosphaeria spp. are present in soil and several species are associated with mycetoma and phaeohyphomycosis in humans, particularly in tropical regions [4–9].

Mycetoma is a highly debilitating, chronic infection in the tropics that is characterized by the formation of a subcutaneous mass with multiple sinuses draining pus and "grains." Depending on the color of the grains (of irregular shape and 1–2 μm in diameter), mycetoma is divided into black-grain and white-grain categories. The disease begins with a subcutaneous inoculation (through cut or splinter) of true fungi (eumycetoma) (accounting for 40% of mycetoma cases) or actinomycetes (actinomycetoma) (accounting for 60% of mycetoma cases), which remain indolent until extension to deeper tissues and bone takes place [10].

Initially described in India (Madura foot), eumycetoma is prevalent in Africa, with the area covering Sudan to Senegal delineating the "mycetoma belt." *Leptosphaeria senegalensis* and *Leptosphaeria tompkinsii* are among fungi that are associated with black-grain mycetomas in humans. Other black-grain causing fungi include *Madurella mycetomatis*, *Madurella grisea*, *Pyrenochaeta mackinnonii*, *Pyrenochaeta romeroi*, *Exophiala jeanselmei*, *Phlenodomus avramii*, and *Curvularia lunata*. On the other hand, fungi causing white or white to yellow mycetomas consist of *Acremonium (Fusarium) falciforme*, *Acremonium kiliense*, *Acremonium recifei*, *Aspergillus nidulans*, *Cylindrocarpon destructans*, *Fusarium moniliforme*, *Fusarium solani*, *Neotestudina rosatii*, and *Pseudallescheria boydii*.

Mycetoma is a chronic, suppurative infection of the subcutaneous tissue and contiguous bone. The feet are the most common site for infection and account for at least two-thirds of cases. Other less common sites comprise the lower legs, hands, head, neck, chest, shoulder, and arms. After entry through sites of local trauma, the organism causes a granulomatous reaction, producing a small, hard painless nodule, which over time begins to soften on the surface and ulcerate to discharge a viscous, purulent fluid containing grains. The infection slowly spreads to adjacent tissue including bone, often resulting in considerable deformity. Sinuses continue to discharge serosanguinous fluid containing granules (grains) of variable size, color, and degree of hardness, depending on the etiologic species. These grains are the hallmark of mycetoma.

12.1.3 Diagnosis

Several dematiaceous fungi including *Leptosphaeria* spp. are responsible for black-grain mycetoma, a chronic subcutaneous infection of humans with devastating consequence. Identification of the dematiaceous fungi causing black-grain mycetoma has traditionally relied on standard mycological procedures such as in vitro culture and microscopy. Species identification can be made on sexual reproductive structures observed on potato-carrot agar media. Because some dematiaceous fungi display poor growth or delayed sporulation, diagnosis of these organisms on the basis of macroscopic and microscopic characteristics may take up to 12 weeks [11,12].

To improve the diagnosis, epidemiological investigations, and treatment evaluation of eumycetoma, molecular techniques have been increasingly adopted in the clinical laboratories [13–15]. In particular, sequence analysis of small-subunit rRNA genes and internal transcribed spacer 1 (ITS1)-5.8S-ITS2 region allows rapid and precise determination of various fungal species causing black-grain mycetoma including *L. senegalensis* and *L. tompkinsii* [16,17].

12.2 METHODS

12.2.1 Sample Preparation

Clinical specimens are examined under microscope for mycotic elements with the help of fungal stains. Portions of the samples are cultured on standard mycology media at 28°C.

For DNA extraction, isolates are grown in 20 mL of RPMI 1640 medium with L-glutamine but without sodium bicarbonate (Sigma-Aldrich) buffered to pH 7.0 with 0.165 M morpholinepropanesulfonic acid (MOPS) (Sigma-Aldrich). After 3–14 days of growth at 30°C under agitation (100 rpm), the mycelium is transferred into a tube and washed in 40 mL of sterile distilled water. Mycelium (100 mg) is homogenized for 1 min in a tube containing 1 mL of lysis buffer (2% Triton X-100, 1% sodium dodecyl sulfate, 10 mM Tris–HCl pH 8, 100 mM NaCl, 1 mM EDTA pH 8), three 0.5 cm diameter glass beads (Sigma), and approximately 500 mg of 425–600 µm glass beads (Sigma). The homogenized mycelia are then snap-frozen in liquid nitrogen, thawed, and refrozen once. DNA extraction is then done with the DNeasy plant kit (QIAGEN) according to the manufacturer's instructions [17].

Alternatively, 4-day-old mycelium is harvested by vacuum filtration on sterile muslin, frozen at −20°C, and freeze-dried. Freeze-dried mycelium is ground with forceps in a 1.5 mL microfuge tube, so that approximately 100 µL of powder is obtained. Two hundred microliters of lysis buffer (50 mM Tris–HCl, pH 8; 50 mM EDTA; 2% sodium *N*-lauroyl sarcosinate; and 150 mM NaCl) and 200 µL of phenol (saturated with 1× Tris–HCl, pH 8) are added, and lysis takes place on ice for 1 h 30 min with occasional vortexing. The mixture is centrifuged for 5 min at 13,000 × *g* at 4°C. The aqueous phase is centrifuged an additional 15 min, the supernatant transferred to a new tube, and nucleic acids are precipitated using 0.6 volume of isopropanol at −20°C. After 70% ethanol wash, the DNA is resuspended in 1× Tris–EDTA buffer and residual RNA can be removed with RNase digestion [13].

12.2.2 Detection Procedures

12.2.2.1 Sequence Analysis of ITS Region

Desnos-Ollivier et al. [17] utilized universal primers V9D (5′-TTAAGTCCCTGCCCTTTGTA-3′) and LS266 (5′-GTAGTCATATGCTTGTCTC-3′) for polymerase chain reaction (PCR) amplification and sequence analysis of the ITS region for identification of dematiaceous fungi. In the case of negative amplification with V9D and LS266, other fungal universal primers (e.g., ITS1/ITS4, ITS4/ITS5, ITS1/ITS2, ITS3/ITS4) may be employed [18].

Procedure

1. PCR mixture (20 µL) is made up of 1 µL of genomic DNA, 1.25 U of AmpliTaq gold (Roche), 2 µL of 10× PCR buffer (Roche), 2 µL of 25 mM $MgCl_2$ (Roche), 2 µL of 2.5 mM deoxynucleoside triphosphate, and 1 µL each of 10 µM concentrated primers.
2. PCR amplification is conducted in an ICycler thermocycler (Bio-Rad) with a first cycle of denaturation for 10 min at 95°C, followed by 30 cycles of denaturation at 94°C for 30 s, annealing at 58°C for 30 s, and elongation at 72°C for 30 s, with a final extension step of 10 min at 72°C.

3. An aliquot of PCR products is separated on an agarose gel, stained with ethidium bromide, and visualized under UV light.
4. The remaining PCR products are purified on P100 Gel Fine (Bio-Rad), and sequenced on both strands by using the BigDye Terminator Cycle Sequencing Ready Reaction kit, version 3.1 (Applied Biosystems), using the primers V9D and LS266. Reaction products are analyzed using an ABI Prism 3700 automated DNA analyzer (Applied Biosystems).
5. Sequences are edited and manually corrected with Chromas, version 2.24 (Technelysium, Helensvale, Queensland, Australia). Multiple-sequence alignment is carried out using ClustalW 1.8. Phylogenetic trees are constructed by the neighbor-joining method using the Phylip package (http://www.infobiogen.fr) and visualized using Treeview.

12.2.2.2 PCR-Restriction Fragment Length Polymorphism Analysis

Balesdent et al. [13] used the 18S primer PN3 (forward) (5′-CCGTTGGTGAACCAGCGGAGGGATC-3′) and the 28S rDNA for primer PN10 (reverse) (5′-TCCGCTTATTGATATGCTTAAG-3′) for PCR amplification followed by digestion with restriction enzymes for identification and phylogenetic analysis of *Leptosphaeria* spp.

Procedure

1. PCR mixture (25 or 50 μL) is composed of 50 ng of template DNA, 200 μM of each deoxyribonucleotide triphosphate, 2.5 μM of each primer, 1.5 U (for 25 μL mixes) or 3 U (for 50 μL mixes) of *Taq* polymerase in 1× *Taq* DNA incubation mix (10 mM Tris–HCl, pH 9.0; 50 mM KCl; 1.5 mM $MgCl_2$; 0.1% Triton X-100; and 0.2 mg of bovine serum albumin per milliliter) (Appligene-Oncor).
2. The reaction is carried out for 37 cycles in a GeneAmp PCR system 9600 (Perkin-Elmer/Applied Biosystems), with each cycle consisting of 30 s at 94°C, 30 s at 58°C, and 1 min at 72°C.
3. PCR products (6–8 μL) are digested for 2–5 h with 10 U of one of the following enzymes: *Alu*I, *Bam*HI, *Eco*RII, *Hae*III, *Msp*I, *Rsa*I, or *Taq*I, (Eurogentec) according to the manufacturer's instructions.
4. PCR products and restriction fragments are analyzed upon electrophoresis in 1.4%–1.8% agarose gels with 1× TBE (45 mM Tris–borate and 1 mM EDTA, pH 8) as a buffer and staining with 0.5 μg of ethidium bromide per milliliter. As an alternative, 3%–3.5% MetaPhor agarose gels (FMC BioProducts, Rockland, ME) are used to separate restriction fragments. Gels are viewed and recorded using the Gel Doc 1000 fluorescent gel documentation system (Bio-Rad). Molecular weights are calculated with the Molecular Analyst v. 1.4 software (Bio-Rad) using the 1 kb, 100 bp, and 10 bp ladder (Life Technologies, Inc.) as references.

Note: An alternative PCR-RFLP (restriction fragment length polymorphism) protocol utilizes primer pair ITS4/ITS5 for amplification and endonuclease *Sma*I for digestion (1 h at 27°C). Restriction fragments are visualized on a 3% agarose gel after ethidium bromide staining [17].

12.3 CONCLUSION

Leptosphaeria spp. are dematiaceous fungi that are commonly distributed in soil. Several *Leptosphaeria* species such as *L. senegalensis* and *L. tompkinsii* are associated with black-grain mycetoma as well as phaeohyphomycosis in humans, particularly in tropical regions [4–10,19]. Traditional methods for differentiation of *Leptosphaeria* spp. from other mycetoma-causing dematiaceous fungi are lengthy and imprecise. PCR amplification and sequencing analysis of the ITS region provide a rapid and reliable approach for determination of *Leptosphaeria* and other dematiaceous fungi.

REFERENCES

1. The UniProt Consortium. Available at http://www.uniprot.org/, accessed on August 1, 2010.
2. El-Ani, A.S., A new species of *Leptosphaeria*, an etiologic agent of mycetoma. *Mycologia* 1966;58:406–411.
3. Kwon-Chung, K.J. and Bennett, J.E., *Medical Mycology*. Lea & Febiger, Philadelphia, PA, 1992.
4. Venugopal, P.V. and Venugopal, T.V., *Leptosphaeria tompkinsii* mycetoma. *Int J Dermatol*. 1990;29(6):432–433.
5. Venugopal, P.V. et al., Antimycotic susceptibility testing of agents of black grain eumycetoma. *J Med Vet Mycol*. 1993;31(2):161–164.
6. Mahe, A. et al., Mycetomas in Mali: Causative agents and geographic distribution. *Am J Trop Med Hyg*. 1996;54(1):77–79.
7. Khatri, M.L. et al., Mycetoma in Yemen: Clinicoepidemiologic and histopathologic study. *Int J Dermatol*. 2002;41(9):586–593.
8. Gioulekas, D. et al., Allergenic fungi spore records (15 years) and sensitization in patients with respiratory allergy in Thessaloniki-Greece. *J Investig Allergol Clin Immunol*. 2004;14(3):225–231.
9. Gonianakis, M.I. et al., Mold allergy in the Mediterranean Island of Crete, Greece: A 10-year volumetric, aerobiological study with dermal sensitization correlations. *Allergy Asthma Proc*. 2006;27(5):354–362.
10. Fahal, A.H., Mycetoma: A thorn in the flesh. *Trans R Soc Trop Med Hyg*. 2004;98:3–11.
11. de Hoog, G.S. et al., Diagnostic problems with imported cases of mycetoma in The Netherlands. *Mycoses* 1993;36:81–87.
12. de Hoog, G.S. et al., Phylogeny and typification of *Madurella mycetomatis*, with a comparison of other agents of eumycetoma. *Mycoses* 2004;47(3–4):121–130.
13. Balesdent, M.H. et al., Conidia as a substrate for internal transcribed spacer-based PCR identification of members of the *Leptosphaeria maculans* species complex. *Phytopathology* 1998;88(11):1210–1217.

14. Iwen, P.C., Hinrichs, S.H., and Rupp, M.E., Utilization of the internal transcribed spacer regions as molecular targets to detect and identify human fungal pathogens. *Med Mycol.* 2002;40:87–109.
15. Kaczmarek, J. et al., Analyses of air samples for ascospores of *Leptosphaeria maculans* and L. biglobosa by light microscopy and molecular techniques. *J Appl Genet.* 2009;50(4):411–419.
16. Ahmed, A.O. et al., Molecular detection and identification of agents of eumycetoma: Detailed report of two cases. *J Clin Microbiol.* 2003;41:5813–5816.
17. Desnos-Ollivier, M. et al., Molecular identification of black-grain mycetoma agents. *J Clin Microbiol.* 2006;44(10):3517–3523.
18. White, T.J. et al., 1990. Amplification and direct sequencing of fungal ribosomal RNA genes for phylogenetics, in M.S. Innis and D.H. Gelfand (eds.), *PCR Protocols: A Guide to Methods and Applications.* Academic Press, New York, pp. 315–322.
19. Machmachi, H. et al., Black grain mycetoma caused by *Leptosphaeria tompkinsil*, *Medi. Mycol.*, 2011:49, 186.

13 *Macrophomina*

Artur Alves, Alan J.L. Phillips, and António Correia

CONTENTS

13.1 INTRODUCTION

13.1.1 Classification, Morphology, and Biology

Macrophomina phaseolina Tassi (Goid.) is the type species of the genus *Macrophomina* Petr., and this name has also been applied, erroneously, to the coelomycete (pycnidial) synanamorph of *Rhizoctonia bataticola* (Taubenh.) E.J. Butler.[1] *Macrophomina phaseolina* is a seed and soilborne plant pathogenic fungus that causes seedling blight, root and stem rot of more than 500 cultivated and wild plant species including economically important crops, forest trees, fruit, and weed species.[2,3]

Due to the lack of knowledge about its sexual stage (teleomorph), this fungus was placed in the anamorphic Ascomycetes with no apparent affiliation to the known families of Ascomycetes. According to recent phylogenetic data,[1] *M. phaseolina* is currently recognized as a member of the family Botryosphaeriaceae even though its teleomorph is still unknown. The name of the synanamorph, and its taxonomic placement, has been the topic of much controversy (for details see Crous et al.[1]).

The asexual structures formed by the fungus are pycnidia and microsclerotia. Black sclerotia, 100–1000 μm diameter, occur in host tissue or in soil and constitute the primary inoculum source of the pathogen. They can survive several years depending on environmental conditions and whether or not the sclerotia are associated with host residues.[1–5] Secondary dispersal by pycnidiospores (conidia) is host- and isolate-dependent.[4] The pycnidial conidiomata are dark brown to black, up to 200 μm in diameter. The conidia are ellipsoid to obovoid, (16–)20–24(–32) × (6–)7–9(–11) μm. Immature conidia are hyaline, enclosed in a mucous sheath, which, upon dehiscence, encloses the top half of the conidium, transformed into two lateral tentaculiform, apical mucoid appendages (type C, Nag Raj[6]). When mature, the conidia become medium to dark brown, with a granular outer layer that in some cases appears pitted, without any mucoid appendages; conidial hilum frequently with a marginal frill.[1,4]

The genus *Tiarosporella* Höhn is characterized by having conidia formed from smooth, hyaline conidiogenous cells that lack periclinal thickenings and percurrent proliferations, and hyaline, subcylindrical to fusiform conidia that have irregular, apical mucoid appendages.[1,6] Because conidia of *M. phaseolina* have apical appendages, von Arx[7] introduced the name *Tiarosporella phaseolina* (Tassi) van der Aa for this fungus and also reduced the genus *Macrophomina* to synonymy under *Tiarosporella*, but this has not been followed by the plant pathological and mycological communities.

In their work, Crous et al.[1] studied several isolates of *M. phaseolina* and provided an updated description, as well as illustrations and photographs showing the distinctive morphological features that characterize the species and genus. The authors also pointed out the differences between *Tiarosporella* and *Macrophomina*. Thus, *Macrophomina phaseolina* has conidia with apical mucoid appendages as found in *Tiarosporella*, but it is distinguished by having percurrently proliferating conidiogenous cells and conidia that become dark brown and slightly roughened at maturity, appearing more *Diplodia*-like in morphology. Based on these differences and also on phylogenetic differences between *M. phaseolina* and *Tiarosporella* species, Crous et al.[1] retained the genus *Macrophomina* and the name *M. phaseolina*.

This fungus is known mainly as a plant pathogen on an extremely wide range of hosts and is distributed throughout the world, predominantly in regions with hot and dry conditions during the growing season. The severity of disease caused by *M. phaseolina* in various hosts is associated with environmental conditions such as high temperatures[8] or water stress.[9] However, the pathogen has also been detected in hosts without any disease symptom when growing conditions were optimal for the plant. Sudden disease outbreaks in mature plants may occur when plants are under stress and cause high yield losses in previously healthy plants.[10] This fact is a common feature among species of plant pathogenic fungi of the family Botryosphaeriaceae, which have the ability to live as endophytes inside healthy plants causing disease following the onset of stress.[11]

13.1.2 Clinical Features and Pathogenesis

The relevance of *M. phaseolina* as a human pathogen is not well understood. As with other species of Botryosphaeriaceae this fungus is most likely an opportunistic human pathogen that is capable of causing disease under appropriate conditions. The infection process (as with other coelomycetes) occurs most commonly by trauma resulting in the implantation of the fungus from plant material or soil.[12] Coelomycete fungi such as *M. phaseolina* may also be of concern to patients under immunosuppressive therapy, namely bone marrow and organ transplant recipients, cancer patients, among others.[12]

The number of reports describing this species as the etiological agent of fungal infection in humans is small. The first report of human infection by *M. phaseolina* concerns its occurrence in a patient (an adult male) who had received a renal transplant and was receiving immunosuppressive medication. The patient developed swelling, pain, and purulent discharge involving the left great toe. Swabs of the discharge and an aspirate of the metatarsophalangeal joint yielded a filamentous fungus that was later identified as *M. phaseolina*.[13] Another case of human infection (a cutaneous infection) with *Macrophomina phaseolina* was found in a child with acute myeloid leukemia who had received an unrelated hematopoietic stem cell transplant.[14] The patient was under immunosuppressive as well as prophylactic antifungal (voriconazole) and antibacterial (vancomycin, ciprofloxacin, and trimethoprim-sulfamethoxazole) therapy. The girl developed an erythematous, papular, nonulcerative, tender lesion above the right medial malleolus with no induration or inguinal lymphadenopathy. The lesion progressed and became indurated, with a central necrotic eschar and a surrounding rim of erythema, and extended into the subcutaneous tissue. A fungal culture identified as *M. phaseolina* was retrieved from a punch biopsy performed through the central ulcerative lesion.[14] The only other reported cases of *M. phaseolina* infection of humans were in a study of non-sporulating molds (NSMs) associated with ocular infections in India.[15] The species *M. phaseolina* was identified as the causal agent in seven clinical cases.

Apart from its occurrence in human infections, there is also a description of a subcutaneous granuloma on the tail of a cat that was attributed to *M. phaseolina*.[16] However, the authors' description of slow growth at 37°C and their illustration of the conidia suggest that a different fungal species was probably the etiologic agent.[13]

It is not well understood how plant pathogenic fungi like *M. phaseolina* are capable of infecting organisms from members of different biological kingdoms, namely animals. Several aspects regarding environmental conditions, population biology traits, and molecular processes may help understand how these microorganisms are capable of performing these cross-kingdom host jumps.[17] It has been speculated that pathogenic microorganisms that have a broad host range are more likely to become pathogenic on cross-kingdom hosts than highly specialized microorganisms.[17] In the case of *M. phaseolina*, being an unspecialized pathogen with a very wide host range may have enabled it to expand its pathogenic spectrum and become capable of infecting humans. A shift from an environmental state into a clinical, pathogenic state inside a host requires adaptation or tolerance to the new environment. A crucial moment occurs when fungi enter a mammalian host and experience substantially increased temperature.[17] The ability to withstand and grow at elevated temperatures (37°C) is an essential prerequisite for a microorganism to be a human pathogen. *Macrophomina phaseolina* is able to grow in vitro in a wide range of temperatures and it has been shown that optimum growth temperature is 35°C–37°C, with some isolates being capable of growth at temperatures as high as 40°C.[4] However, this characteristic is isolate dependent and some isolates have lower optimum growth temperatures. *Macrophomina phaseolina* (at least some isolates) are clearly adapted to withstand normal human body temperature, which makes this species potentially a human pathogen.

13.1.3 Diagnosis

13.1.3.1 Conventional Techniques

The identification of fungi in clinical specimens can be performed by histopathological analysis and direct microscopic examination using appropriate stains. These procedures, although useful for the detection and identification of some of the more common fungal pathogens, are not really useful for identifying unusual fungal species causing infection.[18]

Fungal identification is based largely on morphological criteria, most specifically on the morphology of the reproductive structures.[12,19] Since it is highly unlikely that fungi develop these reproductive structures on the human host tissues, it is necessary to obtain cultures of the fungus from clinical samples in order to accomplish a proper identification of the etiological agent.

Macrophomina phaseolina and coelomycetes, in general, are not particularly difficult to isolate from clinical specimens and have a moderate to rapid growth rate on a variety of routine culture media. The main problem with these fungi is related to induction of sporulation in order to obtain the diagnostic reproductive structures necessary for identification of the isolates. This process is not only time consuming but also frequently unsuccessful.[12] *Macrophomina phaseolina* is particularly recalcitrant to sporulation and commonly produces only dark sclerotia in routine cultures and these are not sufficient for identification, at least for someone who is not familiar with the species. This is well illustrated by the clinical reports concerning infections by *M. phaseolina* where the isolates of this fungus were grown on a variety of culture media and under different conditions of temperature and light. Although the fungus produced sclerotia, it invariably failed to produce pycnidia and conidia, which are fundamental diagnostic structures.[13,14]

It is known that the cultivation of fungi on sugar-rich media is prejudicial to sporulation and that culture of fungi (especially Coelomycetes) on media containing sterilized plant tissue promotes sporulation.[12,20] Our experience has shown that for fungi of the family Botryosphaeriaceae, the addition of sterilized plant material such as pine needles, oak, or poplar twigs, among others on water agar or half-strength potato dextrose agar, results in good sporulation of cultures.[1,21] Pycnidia of *M. phaseolina* have been induced on propylene oxide sterilized or autoclaved plant tissues and on groundnut meal irradiated with UV light and Whatman filter paper treated with vegetable oil.[22] Also, Crous et al.[1] were able to induce numerous strains of *M. phaseolina* to sporulate on sterile pine needles on water agar.

Considerable variation in morphology, physiology, and pathogenicity has been reported for isolates of *M. phaseolina* even when obtained from the same plant.[23] The identification of isolates of *M. phaseolina* is based on morphological criteria, but due to wide variations in the phenotype of the isolates these criteria are often not reliable and the identification process may be difficult.[24]

Selective or semi-selective culture media have been traditionally applied for detection of *M. phaseolina*.[25,26] Such methods, although effective, especially for the enumeration of microsclerotia in the soil, require exceptional expertise, are time consuming, and therefore not appropriate to use in clinical laboratories.

13.1.3.2 Molecular Techniques

Identification of fungi to genus and species level is increasingly important as the spectrum of opportunistic fungal pathogens continues to expand. Although the clinical presentation of many fungal infections may be indistinguishable, an early, fast, and accurate identification of the etiological agent is of fundamental importance for effective treatment and management of the infection.

The traditional morphological methods of fungal identification can be time consuming and laborious, may initially be nonspecific, and require considerable expertise for correct identification of less common or unusual fungi. Further shortcomings of morphological identification are the inability of some cultures to sporulate or cultures exhibiting atypical morphology making the identification difficult or impossible and resulting even in misidentifications. The incidence of NSMs is quite high in cultures derived from clinical specimens.[15,27,28]

Difficulties in the identification of *M. phaseolina* are well patent in the clinical reports regarding this fungus. For example, Tan et al.[13] reported that the fungus was initially thought to represent *Pseudallescheria boydii* based on the grayish colonies, but the identification was later revised to *Lasiodiplodia theobromae*. This was due to the fact that the isolate produced dark brown sclerotia, which were suspected to represent immature pycnidia of *L. theobromae*. A definitive identification was only accomplished when molecular methods were used. The same occurred with Srinivasan et al.[14] who were unable to induce their *M. phaseolina* isolate to sporulate and could only identify it on the basis of molecular methods.

Immunological methods, because of their high specificity and sensitivity, provide a rapid means of diagnosing fungal infections.[18,29] Several commercial enzyme-linked immunosorbent assay (ELISA) techniques are available, which reliably detect candidiasis, histoplasmosis, and coccidioidomycosis.[18]

ELISA techniques have been developed for the specific detection and quantification of *M. phaseolina* in plant tissue.[10,29] However, Srivastava and Arora[29] reported a lack of specificity of the ELISA assay since the polyclonal antisera exhibited strong cross-reactivity with many other nontarget fungal isolates tested. More recently, Afouda et al.[10] developed a double-antibody sandwich ELISA (DAS-ELISA), which revealed high specificity and sensitivity in the detection of *M. phaseolina*. This serological approach, however, has never been tested on clinical specimens and therefore its applicability and validity in the clinical setting have not been evaluated.

Molecular (DNA-based) methods are being increasingly applied in the field of medical mycology for identification of human pathogenic fungi.[15,27,28,30–34] These molecular methods present advantages over the traditional morphology-based identification. Thus, DNA-based methods are faster with a turnaround time of about 24 h from the time of DNA extraction, yield results that are objective with data portable between laboratories, and despite the higher cost could be more economical in the long run.[15,27,28,30–34] These methods, especially the ones taking advantage of the polymerase chain reaction (PCR) technology, can be applied to non-sporulating isolates or isolates for which identification could not be attained. Also, it is possible to obviate the culture step as they can be applied directly to clinical specimens.[15,27,28,30–33]

A wide array of molecular identification techniques (non-sequence-based) is available including the commercially available AccuProbe (Gen-Probe) assay, single-step PCR, RAPD-PCR, rep-PCR, nested PCR, PCR-RFLP, PCR-EIA, and microarray-based, Luminex technology-based, and real-time PCR-based methods. Many of these methods have not been widely used or rigorously validated in the clinical environment.[18,34] Some of the above-mentioned methods (e.g., RAPD, rep-PCR, PCR-RFLP) have been applied to evaluate the genetic diversity within isolates of *M. phaseolina* but were never tested regarding their potential for identification/detection within the clinical setting.[5,24,35,36]

At present, the molecular method of choice for fungal species identification is DNA sequence analysis, which constitutes PCR amplification of a selected genomic region followed by sequencing of the resulting amplicon. The obtained sequence can then be queried against a nucleotide sequence database such as GenBank.[15,31,33,34]

The genomic region to be used must be orthologous, have a high level of interspecific variability combined with low levels of intraspecific variability, and should not undergo recombination. Moreover, the target locus should be easy to amplify and sequence, and the amplicons should be within the size range obtainable with the most commonly used automated DNA sequencers (about 600–800 bp).[19,33,34]

The ITS region (noncoding sequence interspaced among highly conserved fungal rRNA genes) complies with most of these requirements since this region can be reliably amplified for most fungi, is conserved, is present as multiple copies in the fungal genome, yields sufficient taxonomic resolution for most fungi, and has the additional advantage that the GenBank (http://www.ncbi.nlm.nih.gov), European Molecular Biology Laboratory nucleotide sequence database (http://www.ebi.ac.uk/embl/), and DNA Data Bank of Japan (http://www.ddbj.nig.ac.jp/) contain a large number of sequences from this locus, enabling a ready comparison of the sequence from an unknown isolate.[19,33,34]

There is considerable consensus regarding the use of ITS sequencing in the identification of fungi. Also, the International Subcommission on Fungal Barcoding has proposed the ITS region as the prime fungal barcode or the default region for species identification. Main disadvantages regarding the use of ITS are the inability of this region for differentiation of species complexes and failure to distinguish between closely related species or cryptic species.[19,33,34] In these cases, it may be necessary to use more variable gene regions. The locus to be used will depend on the fungal group in question. Regarding the family Botryosphaeriaceae, when the ITS region fails to differentiate species complexes or cryptic species, the locus of choice has been the elongation factor 1-alpha, which has shown successful results.[37] This problem does not occur with *M. phaseolina* since it is the only species currently known within the genus *Macrophomina* and it can be easily identified on the basis of the ITS sequence. Nevertheless, an important factor that must be considered is that an accurate sequence-based identification is heavily dependent on the reliability of sequences deposited in reference databases.

The PCR-based DNA sequencing technique targeting the ITS region and other ribosomal DNA regions has been found to be a rapid and reliable tool to identify *M. phaseolina* clinical isolates.[13–15] In the clinical reports concerning *M. phaseolina*, the fungal isolates could only be definitively identified by DNA sequencing of the ITS region[13–15] and also of the 18S ribosomal DNA[13] and the D1/D2 variable region of the 28S ribosomal DNA genes.[14]

PCR-based culture-independent methods have been developed and applied for the detection of fungi on clinical specimens, thus eliminating the need for culturing.[32,38] Also, through the use of species-specific oligonucleotide primers or probes, this approach has been used to detect a particular species or group of species without the need for DNA sequencing.[39] Babu et al.[24] developed specific primers and an oligonucleotide probe targeting the ITS region for the identification/detection of *M. phaseolina* under in vitro conditions. ITS sequences of several isolates of *M. phaseolina* were aligned and two primers (MpKR1 and MpKF1) were designed from selected regions. Optimization of the PCR conditions and validation of primers yielded a specific 350 bp amplicon for *M. phaseolina* isolates. The specificity of this primer set was confirmed by the absence of an amplicon with the same size in other species of fungi, bacteria, and actinomycetes. Thus, the PCR assay with primers MpKR1 and MpKF1 could be used to rapidly identify *M. phaseolina*. Additionally, an oligonucleotide probe (MpKH1) was also designed within the ITS region and was shown to detect the target sequences at varying concentrations. The probe was also shown to be specific for *M. phaseolina* and no signals were obtained with nonspecific target ITS sequences. Before this method can be used in clinical laboratories, it first needs to be rigorously tested and validated. Moreover, the use of a specific PCR approach although fast and accurate may not be entirely feasible because in order to select the primer set to use, it is necessary to have a suspect of the fungal species responsible for the infection. Otherwise, a large set of primers needs to be tested in order to identify the etiological agent. In this respect, a panfungal PCR assay using "universal" fungal primers for amplification and sequencing of the ITS region is probably a better choice.[31,40]

13.2 METHODS

13.2.1 Sample Preparation

PCR-based methods require very small quantities of DNA. Ribosomal DNA regions are present in multiple copies in the fungal genome and therefore can be easily amplified even with minute amounts of DNA. Nevertheless, the success of any molecular detection procedure is largely dependent on the DNA extraction step, which must retrieve enough quantity of DNA and simultaneously the DNA should not be contaminated with PCR inhibitory agents. A good DNA extraction method is crucial for PCR detection to avoid the possibilities of false-negative results. Also, strict precautions must be taken throughout the whole procedure, from collection and processing of samples to DNA extraction and PCR amplification in order to avoid contaminations and consequently false-positive results.

Macrophomina phaseolina can be isolated from clinical samples using a variety of culture media. As soon as colonies start to develop, DNA can be extracted by different methods already available, including commercial kits, which are usually faster than other methods.[15,21,27,28,30,32,39] Alternatively, DNA can be extracted directly from the clinical samples using the same DNA extraction procedures used for fungal cultures or others developed specifically for human tissue samples rendering the process even faster.[31,39] By eliminating the isolation procedure, this last approach greatly reduces the time necessary for the molecular diagnosis of the fungal species responsible for the infection.

Procedure

1. Inoculate cultures onto potato dextrose agar (Difco) and incubate at 25°C. After 1 week the colony is scraped from the agar surface, frozen in liquid nitrogen, and ground in a porcelain mortar. Alternatively, cultures can be grown in potato dextrose broth.
2. Genomic DNA can be extracted from the ground mycelium using any DNA isolation protocol. In our lab, we have good results in terms of yield and purity of DNA samples following the method described by Alves et al.[21]

Note: This same procedure can be used directly on clinical samples.

13.2.2 Detection Procedures

PCR amplification of the ITS region as well as the ribosomal genes (18S and 28S rRNA) can be easily amplified using "universal" fungal primers in the vast majority of fungi. We would suggest that for identification of *M. phaseolina* the ITS region, or alternatively the D1/D2 variable region of the 28S ribosomal DNA gene, are the best options.

For *Macrophomina* the primer sets ITS1/ITS4 or ITS5/ITS4[41] give a good amplification of the target ITS region resulting in amplicons with a size of about 500–600 bp and primers NL1/NL4[42] readily amplify the D1/D2 region resulting in an amplicon with 614 bp in size. As an alternative, the forward primer (ITS1 or ITS5) can be combined with the reverse primer NL4, which anneals within the 28S rRNA gene resulting in a larger amplicon (aprox. 1200 bp)[43] and that includes the ITS region plus the D1/D2 variable region of the 28S rRNA gene. The PCR mixtures and amplification conditions have been described elsewhere.[21,43,44] In some cases where amplification is weak or unsuccessful, good results are usually obtained by performing a second PCR using as template the first PCR amplification product.[45] Another approach involving a nested or semi-nested PCR reaction can be performed. Thus, for example, an initial amplification would be performed using the primer set ITS1/NL4 and in the second PCR reaction the primer set used would be ITS1/ITS4. This kind of approach is frequently used and is known to increase the sensitivity of the PCR detection.[30] Finally, from our experience, the addition of 5% DMSO to the PCR amplification reaction is useful to help the amplification of some difficult templates and also increases reproducibility.[37,43–45]

The nucleotide sequence of the ITS region is then obtained using standard automated DNA sequencing procedures and using in the sequencing reaction the same primers used to amplify the target region. The complete sequences of the ITS region is read and edited using available software such as Chromas 1.45 (http://www.technelysium.com.au/chromas.html), FinchTV 1.4.0 (http://www.geospiza.com/finchtv), or others. The sequences must be checked manually and nucleotide arrangements at ambiguous positions clarified using both primer direction sequences. The final step in the identification procedure consists of performing a sequence similarity search of the retrieved sequence against nucleotide sequences database such as GenBank using the BLAST tool (http://www.ncbi.nlm.nih.gov/BLAST). The outcome of this similarity search must be interpreted with caution due to the well-known problems with the reliability of the ITS sequences deposited in the reference databases (e.g., GenBank/EMBL/DDBJ).[33,46] Errors in fungal sequences within GenBank, the most widely used database, have been found to be as high as 20%.[46] In order to circumvent this problem, when performing comparative sequence analyses only sequences that have been obtained from cultures thoroughly characterized and properly identified in morphological terms should be considered. Preferably, sequences from ex-type or authentic cultures should be chosen for comparison. In the case of *M. phaseolina*, there are no such sequences but nonetheless sequences from well-characterized isolates have been deposited in GenBank.

Procedure

1. Prepare a PCR reaction mixture (50 μL) containing 1× PCR buffer with $(NH_4)_2SO_4$ (MBI Fermentas, Vilnius, Lithuania), 3 mM $MgCl_2$, 200 mM of each nucleotide, 15 pmol of primer ITS1, 15 pmol of primer ITS4, 1 U of *Taq* DNA polymerase (MBI Fermentas, Vilnius, Lithuania), and 50 ng of template DNA.
   Note: The forward primer ITS5 can be used as an alternative to primer ITS1.
2. Perform the PCR amplification in a thermocycler using the following conditions: initial denaturation of 3 min at 95°C, followed by 30 cycles of 30 s at 94°C, 30 s at 50°C, and 1 min at 72°C, and a final extension period of 10 min at 72°C.
3. Purify the PCR amplicons before sequencing. In our lab, we have good results with the Jet Quick PCR Product Purification Spin Kit (Genomed, Löhne, Germany) but virtually any PCR product purification kit can be used.
4. Prepare cycle sequencing reactions (20 μL) containing 30–90 ng DNA template, 3.2 pmol of primer, and 4 μL BigDye™ Terminator from the ABI PRISM® BigDye Terminator Cycle Sequencing Ready Reaction Kit with AmpliTAQ DNA Polymerase (PE Applied Biosystems, Foster City, CA).
5. Remove excess dye terminator by mixing the reaction with 2 μL of 3 M sodium acetate (pH 4.6) and 50 μL of 100% ethanol. Precipitate for 20 min at room temperature and centrifuge for 20 min at 16,000 g. Discard the supernatant and wash the

pellet with 250 µL of 70% ethanol. Dry the pellet in a heat block at 90°C for 1 min.

6. Sequence in an automated DNA sequencer such as the ABI PRISM 310 Genetic Analyzer (PE Applied Biosystems, Foster City, CA).

13.3 CONCLUSION AND FUTURE PERSPECTIVES

Invasive fungal infections caused by opportunistic fungal pathogens have been increasing significantly in the last years. Rapid and precise identification of the infectious agent is of extreme importance for the initiation of targeted antifungal therapy. The emergence of a wide and taxonomically diverse array of unconventional fungal pathogens poses considerable diagnostic and therapeutic challenges to the field of medical mycology.

Although it is apparent that *M. phaseolina* has the ability to infect and cause disease in man, its importance as a human pathogen is not fully established yet. This species is particularly difficult to identify by the traditional morphological procedure due mainly to the difficulty of obtaining in culture the diagnostic reproductive structures. This fact is clearly highlighted by the clinical reports where unequivocal identification could only be accomplished by molecular methods.

Molecular detection methods, most specifically, PCR amplification coupled with DNA sequence analysis of the ITS regions or ribosomal DNA genes have proven extremely useful in the identification of fungal infections due to *M. phaseolina*. It is envisaged that in the future, with a wider application of these methods in the clinical laboratory, it will help reveal the real extent of fungal infections caused by this species. It is likely that many clinical cases caused by *M. phaseolina* may have previously passed unnoticed due to misidentifications or that this fungus has simply been reported as an NSM as shown by Bagyalakshmi et al.[15]

REFERENCES

1. Crous, P.W. et al., Phylogenetic lineages in the Botryosphaeriaceae, *Stud. Mycol.*, 55, 235, 2006.
2. Dhingra, O.D. and Sinclair, J.B., An annotated bibliography of *Macrophomina phaseolina* (1905–1975), Universidade Federal de Vicosa, Brazil, University of Illinois, Urbana, IL, 1977.
3. Holliday, P. and Punithalingam, E., *Macrophomina phaseolina*, CMI Descriptions of Pathogenic Fungi and Bacteria No. 275, 1988.
4. Ndiaye, M., Ecology and management of charcoal rot (*Macrophomina phaseolina*) on cowpea in the Sahel, PhD thesis, Wageningen University, the Netherlands, 2007.
5. Purkayastha, S. et al., Characterization of *Macrophomina phaseolina*, the charcoal rot pathogen of cluster bean, using conventional techniques and PCR-based molecular markers, *Plant Pathol.*, 55, 106, 2006.
6. Nag Raj, T.R., *Coelomycetous Anamorphs with Appendage-Bearing Conidia*, Mycologue Publications, Waterloo, Ontario, Canada, 1993.
7. von Arx, J.A., *The Genera of Fungi Sporulating in Pure Culture*, 3rd edn., J. Cramer, Berlin, Germany, 1981.
8. Agarwal, D.K., Gangopadhyay, S., and Sarbbhoy, A.K., Effect of temperature on the charcoal rot disease of soybean, *Indian Phytopathol.*, 26, 587, 1973.
9. Ghaffar, A. and Erwin, D.C., Effect of soil water stress on root rot of cotton caused by *Macrophomina phaseoli*, *Phytopathology*, 59, 795, 1969.
10. Afouda, L., Wolf, G., and Wydra, K., Development of a sensitive serological method for specific detection of latent infection of *Macrophomina phaseolina* in cowpea, *J. Phytopathol.*, 157, 15, 2009.
11. Slippers, B. and Wingfield, M.J., Botryosphaeriaceae as endophytes and latent pathogens of woody plants: Diversity, ecology and impact, *Fungal Biol. Rev.*, 21, 90, 2007.
12. Sutton, D.A., Coelomycetous fungi in human disease. A review: Clinical entities, pathogenesis, identification and therapy, *Rev. Iberoam. Micol.*, 16, 171, 1999.
13. Tan, D.H. et al., Disseminated fungal infection in a renal transplant recipient involving *Macrophomina phaseolina* and *Scytalidium dimidiatum*: Case report and review of taxonomic changes among medically important members of the Botryosphaeriaceae, *Med. Mycol.*, 46, 285, 2008.
14. Srinivasan, A. et al., Cutaneous infection caused by *Macrophomina phaseolina* in a child with acute myeloid leukemia, *J. Clin. Microbiol.*, 47, 1969, 2009.
15. Bagyalakshmi, R. et al., Newer emerging pathogens of ocular non-sporulating molds (NSM) identified by polymerase chain reaction (PCR)-based DNA sequencing technique targeting internal transcribed spacer (ITS) region, *Curr. Eye Res.*, 33, 139, 2008.
16. Hasegawa, T. et al., Subcutaneous granuloma associated with *Macrophomina* species infection in a cat, *Vet. Rec.*, 156, 23, 2005.
17. van Baarlen, P. et al., Molecular mechanisms of pathogenicity: How do pathogenic microorganisms develop cross-kingdom host jumps?, *FEMS Microbiol. Rev.*, 31, 239, 2007.
18. McGinnis, M.R. and Pfaller, M.A., The laboratory and clinical mycology. In: Anaissie, E.J., McGinnis, M.R., and Pfaller, M.A. (Eds), *Clinical Mycology*, pp. 55–78, 2nd edn., Chap. 4, Churchill Livingstone Inc., New York, 2009.
19. Guarro, J., Gené, J., and Stchigel, A.M., Developments in fungal taxonomy, *Clin. Microbiol. Rev.*, 12, 454, 1999.
20. Sutton, B.C., *The Coelomycetes*, CMI, Kew, England, 696pp., 1980.
21. Alves, A. et al., *Botryosphaeria corticola* sp. nov. on *Quercus* species, with notes and description of *Botryosphaeria stevensii* and its anamorph *Diplodia mutila*, *Mycologia*, 96, 598, 2004.
22. Dhingra, O.D. and Sinclair, J.B., *Basic Plant Pathology Methods*, 2nd edn., CRC Press, Boca Raton, FL, 1995.
23. Dhingra, O.D. and Sinclair, J.B., Location of *Macrophomina phaseoli* on soybean plants related to culture characteristics and virulence, *Phytopathology*, 63, 934, 1973.
24. Babu, K.B. et al., Identification and detection of *Macrophomina phaseolina* by using species-specific oligonucleotide primers and probe, *Mycologia*, 99, 797, 2007.
25. Papavizas, G.C. and Klag, N.G., Isolation and quantitative determination of *Macrophomina phaseolina* from soil, *Phytopathology*, 65, 182, 1975.
26. Cloud, G.L. and Rupe, J.C., Comparison of three media for enumeration of sclerotia of *Macrophomina phaseolina*, *Plant Disord.*, 75, 771, 1991.

27. Rakeman, J.L. et al., Multilocus DNA sequence comparisons rapidly identify pathogenic molds, *J. Clin. Microbiol.*, 43, 3324, 2005.
28. Pounder, J.I. et al., Discovering potential pathogens among fungi identified as nonsporulating molds, *J. Clin. Microbiol.*, 45, 568, 2007.
29. Srivastava, A.K. and Arora K., Evaluation of a polyclonal antibody immunoassay for detection and quantification of *Macrophomina phaseolina*, *Plant Pathol.*, 46, 785, 1997.
30. Ferrer, C. et. al., Detection and identification of fungal pathogens by PCR and by ITS2 and 5.8S ribosomal DNA typing in ocular infections, *J. Clin. Microbiol.*, 39, 2873, 2001.
31. Lau, A. et al., Development and clinical application of a panfungal PCR assay to detect and identify fungal DNA in tissue specimens, *J. Clin. Microbiol.*, 45, 380, 2007.
32. Kumar, M. and Shukla, P.K., Use of PCR targeting of internal transcribed spacer regions and single-stranded conformation polymorphism analysis of sequence variation in different regions of rRNA genes in fungi for rapid diagnosis of mycotic keratitis, *J. Clin. Microbiol.*, 43, 662, 2005.
33. Balajee, S.A. et al., Sequence-based identification of *Aspergillus*, *Fusarium*, and *Mucorales* species in the clinical mycology laboratory: Where are we and where should we go from here?, *J. Clin. Microbiol.*, 47, 877, 2009.
34. Balajee, S.A., Sigler, L., and Brandt, M.E., DNA and the classical way: Identification of medically important molds in the 21st century, *Med. Mycol.*, 45, 475, 2007.
35. Purkayastha, S. et al., Molecular genotyping of *Macrophomina phaseolina* isolates: Comparison of microsatellite primed PCR and repetitive element sequence-based PCR, *J. Phytopathol.*, 156, 372, 2008.
36. Das, I.K., Fakrudin, B., and Arora, D.K., RAPD cluster analysis and chlorate sensitivity of some Indian isolates of *Macrophomina phaseolina* from sorghum and their relationships with pathogenicity, *Microbiol. Res.*, 163, 215, 2008.
37. Alves, A. et al., Morphological and molecular data reveal cryptic speciation in *Lasiodiplodia theobromae*, *Fungal Divers.*, 28, 1, 2008.
38. Gaudio, P.A. et al., Polymerase chain reaction based detection of fungi in infected corneas, *Br. J. Ophthalmol.*, 86, 755, 2002.
39. Jaeger, E.E.M. et al., Rapid detection and identification of *Candida*, *Aspergillus*, and *Fusarium* species in ocular samples using nested PCR, *J. Clin. Microbiol.*, 38, 2902, 2000.
40. Khot, P.D., Ko, D.L., and Fredricks, D.N., Sequencing and analysis of fungal rRNA operons for development of broad-range fungal PCR assays, *Appl. Environ. Microbiol.*, 75, 1559, 2009.
41. White, T.J. et al., 1990. Amplification and direct sequencing of fungal ribosomal RNA genes for phylogenetics. In: Innis, M.A. et al. (Eds.), *PCR Protocols: A Guide to Methods and Applications*, pp. 315–322, Academic Press, San Diego, CA.
42. O'Donnell, K., 1993. *Fusarium* and its near relatives. In: Reynolds, D.R. and Taylor, J.W. (Eds), *The Fungal Holomorph: Mitotic, Meiotic and Pleomorphic Speciation in Fungal Systematics*, pp. 225–233, CAB International, Wallingford, U.K.
43. Alves, A. et al., Evaluation of amplified ribosomal DNA restriction analysis (ARDRA) as a method for the identification of *Botryosphaeria* species, *FEMS Microbiol. Lett.*, 245, 221, 2005.
44. Phillips, A.J.L. et al., Resolving the phylogenetic and taxonomic status of dark-spored teleomorph genera in the Botryosphaeriaceae, *Persoonia*, 21, 29, 2008.
45. Alves, A., Phillips, A.J.L., and Correia, A., Multi-gene genealogies and morphological data support *Diplodia cupressi* sp. nov., previously recognized as *D. pinea* f. sp. *cupressi*, as a distinct species, *Fungal Divers.*, 23, 1, 2006.
46. Nilsson, R.H. et al., Taxonomic reliability of DNA sequences in public sequence databases: A fungal perspective, *PLoS One*, 1, e59, 2006.

14 *Madurella*

Wendy W.J. van de Sande, Ahmed H. Fahal, G. Sybren de Hoog, and Alex van Belkum

CONTENTS

14.1 INTRODUCTION

14.1.1 CLASSIFICATION

The generic criteria for *Madurella* are primarily based on tissue morphology and overall sterility on mycological media, as well as an invasive potential in human and animal hosts. The genus is composed of two formally described species, namely, *Madurella mycetomatis* and *Madurella grisea*, plus a dozen poorly characterized species collectively named *Madurella* spp. [1]. *M. mycetomatis* is the generic type species. All *Madurella* species were isolated from cases of black-grain mycetoma [2]. Mycetoma is a chronic inflammatory disease, which remains localized, involves cutaneous and subcutaneous tissues, the fascia, and bones [3]. The disease is characterized by tumefaction, discharging sinuses, and the presence of fungal grains [3].

From recent molecular phylogeny studies, it appeared that *M. mycetomatis* and *M. grisea*, although both placed in the genus *Madurella*, do not share common ancestors [4]. Based on ribosomal sequences, *M. mycetomatis* has been repositioned in the order Sordariales, whereas *M. grisea* is now an acknowledged member of the order Pleosporales [4]. This explains the many differences found between the two species, as summarized in Table 14.1. In addition to *M. mycetomatis* and *M. grisea*, a wide range of closely related agents are able to cause eumycetoma. Those have presumptively been referred to as *Madurella* spp. [4–6].

14.1.1.1 *Madurella mycetomatis* (Laveran) Brumpt, 1905

M. mycetomatis has been known under many names. The species was first described in 1902 by Laveran and was named *Strepthothrix mycetomi* [3,7]. In 1905, Brumpt placed the organism in the genus *Madurella* and named it *Madurella mycetomi* [3]. This name was used for more than seven decades until it was corrected to *M. mycetomatis* in 1977 by the British Medical Research Council [3].

M. mycetomatis produces slow-growing colonies. These colonies are whitish and woolly at first, becoming olivaceous, yellow, or brown, generally producing a brownish, diffusing pigment also known as pyomelanin [2,3,8]. Next to pyomelanin, *M. mycetomatis* is also able to form another type of melanin, namely, 1,8-dihydroxynaphthalene (DHN)-melanin, which is responsible for the black color of the grain [8]. Although sporulation has not been systematically observed, two types of conidiation are known to exist *in vitro*.

TABLE 14.1
Characteristics of *Madurella mycetomatis* and *Madurella grisea*

Species	*Madurella mycetomatis*	*Madurella grisea*
Order	Sordariales	Pleosporales
Grain	Black	Black
	Cement material throughout grain	Cement material on the outer edge of grain
Optimal growth temperature	37°C	30°C
Pyomelanin	Present	Absent
Conidia	Uncommon	Uncommon
	Pyriform or spherical shaped	Chlamyospores or pycnidia
Sugar assimilation	Lactose (+)	Lactose (−)
	Sucrose (−)	Sucrose (+)
Epidemiology	Africa	America

Abbott described oval or pyriform conidia of 3.5–5.0 μm in diameter with truncate bases when he cultured *M. mycetomatis* on soil-extract agar [4,9]. These conidia were produced on simple or branched conidiophores [2,4,9]. Chalmers and Archibald also described this type of sporulation in 1916 [9]. Another type of sporulation occurs on potato-carrot agar or cornmeal agar. The spores are small, spherical, and produced on tapered tips of flask-shaped phialides [9,10].

M. mycetomatis has a distinctive character once present in the human body: it forms black grains. These grains can be firm and brittle, and the shape is usually globose, oval, or lobulated [9]. Grains are usually 0.5–1 mm in size and are composed of light brown hyphae. Terminal cells at the margin of the grains reach 12–15 μm in diameter [9]. The hyphal cells in the center of the grain are embedded in cement material, a brown matrix supposedly consisting of debris of both fungal and host cells [9,11]. When stained with hematoxylin and eosin, these grains appear uniformly rusty brown [9]. There is only one exception, in the vesicular type of grain the center is light colored and the hyphae in the periphery brown [9,11]. The grains of the latter type resemble those of *M. grisea* [9].

In the past it was difficult to determine the antifungal susceptibility of *M. mycetomatis* toward common antifungal agents. According to the CLSI M38A reference method for antifungal susceptibility testing of filamentous fungi, as starting material a conidial suspension has to be used [12]. In the case of *M. mycetomatis* this is not feasible, since conidia are only rarely obtained. Therefore, antifungal susceptibilities were determined by using hyphal fragments as starting material [13–16]. *In vitro*, *M. mycetomatis* appeared to be the most susceptible to the azoles ketoconazole, itraconazole, and voriconazole, with MIC_{90}s of 0.125, 0.064, and 0.125 μg/mL, respectively [14,17]. Fluconazole was the least effective azole, only inhibiting the fungal growth at a MIC_{90} of 16 μg/mL. Amphotericin B appeared to be less effective than ketoconazole, itraconazole, and voriconazole in inhibiting *M. mycetomatis* growth (MIC_{90} of 2 μg/mL) [14]. Interestingly, *M. mycetomatis* was also very susceptible to tea tree oil, with an MIC_{90} of 0.25% v/v [13]. *M. mycetomatis* isolates appear to be intrinsically resistant to 5-flucytosine, caspofungin, anidulafungin, and micafungin, with MIC_{90}s of >64, 128, >128, and >128 μg/mL, respectively [14,16].

14.1.1.2 *Madurella grisea* Mackinnon, Ferrade et Montemayer, 1949

M. grisea was first described as a new species in 1949 by Mackinnon [18]. This species belongs to the order Pleosporales and hence is phylogenetically closer to *Leptosphaeria* spp. than to *M. mycetomatis* [4]. Cultures of *M. grisea* are unable to grow at 37°C and grow best at 30°C. *M. grisea* forms slow-growing gray or olive-colored colonies [9]. In contrast to *M. mycetomatis*, *M. grisea* does not secrete a pigment in the culture medium. The hyphae are brown walled, septate, and 1–3 μm in diameter [9]. Isolates are usually sterile but can form chlamydospores occasionally [9]. When grown on low-nutrient media, some isolates of *M. grisea* produce pycnidia with pycnoconidia [3,9]. The pycnidia resemble those found in *Pyrenochaeta mackinnonii* and *P. romeroi*, which are also members of the order Pleosporales [3,4,9]. Since *M. grisea* is usually thought of as being sterile, most diagnostic laboratories do not attempt to induce pycnidial formation in isolates of *M. grisea*. Therefore, some *P. mackinnonii* or *P. romeroi* isolates may have been misidentified as *M. grisea* [3]. Molecular studies demonstrated that indeed many fungal mycetoma isolates have been misidentified as *M. grisea* [4–6]. Like *M. mycetomatis*, *M. grisea* forms black grains in the human body. They resemble the vesicular type of grain of *M. mycetomatis*. The size of the grain is up to 1 mm. The grain is soft at first but will become hard and brittle upon maturation [19]. The brown cement material is limited to the outer edge of the grain. The center appears to be hollow in histological sections and is composed of a loose, hyaline mycelial network of small hyphae (1–3 μm) [19]. In a couple of cases, *M. grisea* was found to grow in tissue without organization into grains [9].

For *M. grisea* very limited in vitro antifungal susceptibility data are available, and only a small number of strains of antifungal studies have been published. *M. grisea* appeared to be less susceptible to the azoles than *M. mycetomatis*, with MIC_{90}s of 5 μg/mL for ketoconazole, of 2.5 μg/mL for itraconazole and econazole, and of 10 μg/mL for miconazole [17]. Only for voriconazole a lower MIC was obtained, namely, of 0.5 μg/mL [20]. The MIC for amphotericin B was 0.25 μg/mL [20].

14.1.1.3 *Madurella* Species

Next to *Madurella grisea* and *Madurella mycetomatis*, about a dozen unidentified black-grain mycetoma causative agents have been included in the genus *Madurella* and given a number [4–6]. In older literature, numerous species have been described, but type materials are not known to exist for most of them. Not much is known about these fungi, except that they were described as causative agents of black-grain mycetoma. De Hoog et al. [21] listed numerous species as being of doubtful identity. The more recently introduced *Madurella* species which belong to the order Pleosporales could be grouped in three genetically distinct clades [6]. The *Madurella*-like species which belong to the order Sordariales are more heterogeneously distributed and are scattered within the order. Further characterization of these new *Madurella* species is needed to properly identify these agents and elucidate their phylogeny.

14.1.2 Epidemiology and Risk Factors

In addition to *M. mycetomatis* and *M. grisea*, approximately 50 species of microorganisms are capable of causing mycetoma [2]. Among these, both bacteria and fungi are found. The most prevalent causative agent of black-grain mycetoma worldwide, and in Africa in particular, is *Madurella mycetomatis* [3]. As can be seen in Figure 14.1, *M. mycetomatis* causes more than 70% of all mycetoma infections in some parts of central Africa, including the Sudan [22,23]. *M. grisea* is a less common causative agent and has been found primarily in the Americas (Figure 14.1 and Refs. [9,24]). Of the 40 cases reported worldwide, nearly 80% originated from South America [9].

Anyone living in an endemic area could become infected with *Madurella* spp. Overall, however, mycetoma is more common in herdsmen, farmers, and other field laborers who are in frequent and direct contact with the natural environment. Males are five times more often affected by mycetoma than females, even in areas where both sexes spend a lot of time outdoors [2,3,25]. Mycetoma is seen in all age groups, but it usually affects adults between 20 and 40 years old [26]. The infection is not considered to be transmissible from person to person or from animal to human. In general, mycetoma patients are considered to have a functioning immune system, although there are some reports which show impairment of the cell-mediated immune (CMI) response in patients severely infected or not responding to medical treatment [27]. This finding was supported by animal studies, since mycetoma was more successfully induced in athymic mice than in immunocompetent mice [28]. In another report,

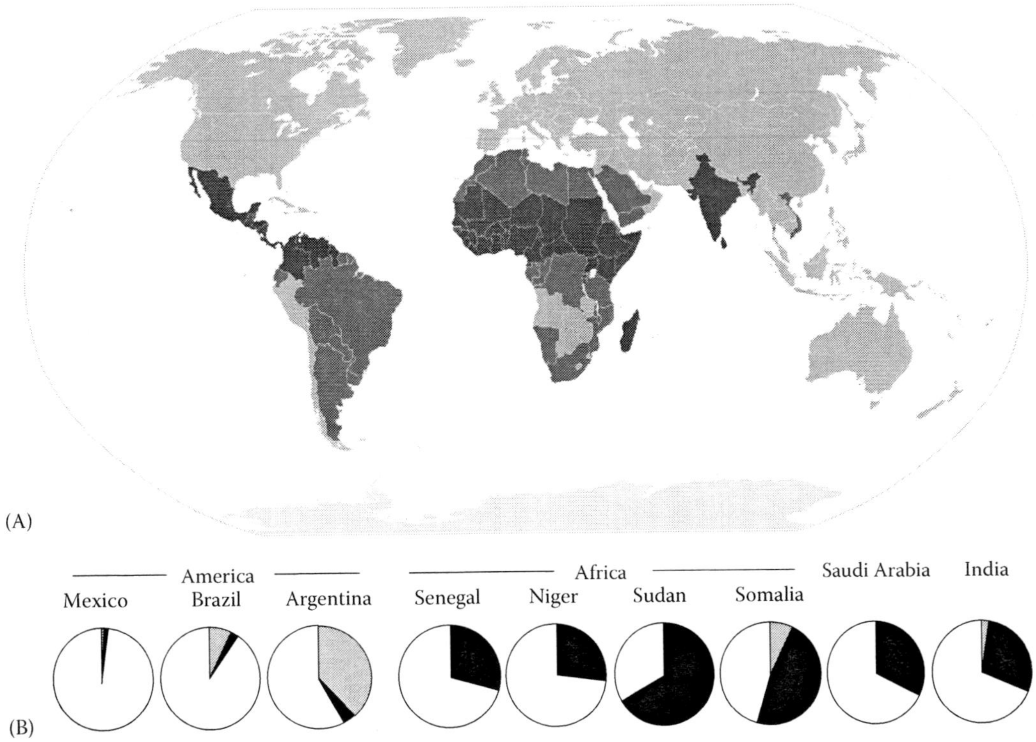

FIGURE 14.1 *Mycetoma endemic areas.* These areas include both actinomycetoma and eumycetoma regions. Overall, actinomycetoma is more often found in Middle and South America. Eumycetoma is more commonly found in Africa. (A) Dark gray colored countries are countries with common and frequent incidence. Lighter colored areas are countries with regular to moderately common incidence. (B) The incidence of *Madurella mycetomatis* (black) and *Madurella grisea* (gray) in a selection of countries. Mycetoma due to other organisms (both bacteria and fungi) are colored white. The graphs are based on data obtained from Refs. [19,74–82].

some impairment in the innate immune response was noted [29]. But overall, still no convincing evidence exists that mycetoma patients have particular immune defects.

14.1.3 Clinical Presentation

Mycetoma is primarily a subcutaneous disease. Since both *M. mycetomatis* and *M. grisea* are assumed to be soil inhabitants [30], it is believed that the etiological agent is introduced into the subcutaneous tissue by a minor trauma, such as a thorn prick [26]. That is why it is not surprising that most mycetoma lesions are usually seen in the feet (70%) (Figure 14.2). Other common sites are the hands (12%), legs, and knee joints [3,9,25]. In highly endemic areas, other parts of the body might become affected as well. After the incubation period, which can be several months or years, a small nodule will arise [3]. On the foot it can be either on the dorsum or plantar surfaces [9]. The subcutaneous nodule is usually a few centimeters in diameter, firm and rounded, but it can also be soft and lobular. It is rarely cystic [31] and is often movable. The overlying skin appears normal and may not be attached to the mass early in the course of infection. The nodule is usually painless in nature, and rarely medical advice is sought [26]. Over the ensuing months, the disease will progress. The nodule will start to grow, and multiple sinus tracts will be formed. During the active phase of the disease, the sinuses will discharge a serous, serosanguinous, or purulent discharge containing fungal grains [3,26]. The color of the grains is dependent on the causative agent and can give an indication of its identity [2,3]. Grains produced by *M. mycetomatis* or *M. grisea* are black in nature [2]. After discharge these sinuses can close transiently. Fresh adjacent sinuses may open, whereas some of the older ones may heal completely. The nodules are connected to each other through deep abscesses, and to the skin surface [2,22,25]. With time, the mycetoma granuloma will increase in size (Figure 14.2). The skin of the lesion will be stretched and become smooth and shiny. Areas of hypo- or hyper-pigmentation may develop [32]. Increased local hyperhydrosis has also been seen in some patients [25,33]. When mycetoma increases in size, it will also grow into the deeper tissues; muscles and bones will be invaded [3]. For unknown reasons, the tendons and the nerves are spared until very late in the disease process [32]. Even this late in the disease, the blood supply in the mycetoma area remains adequate and in many cases it is increased [34]. Although mycetoma usually remains a localized infection, some reports of lymphatic or bloodstream spreading have appeared in literature. Secondary nodules can arise in the affected area, sometimes with more distant lymphatic metastases in some advanced cases [32].

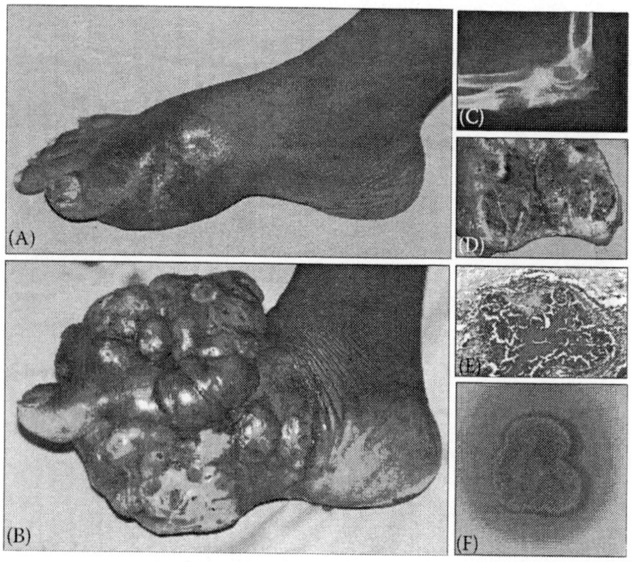

FIGURE 14.2 *Clinical presentation of mycetomaa.* (A) An early case of mycetoma of the foot. (B) An advanced case of mycetoma of the foot. (C) An x-ray showing a lateral view of the elbow joint with multiple big bone cavities with well-defined margins characteristic of eumycetoma. (D) Histological slide of *Madurella mycetomatis* surrounded by inflammatory cellular reaction (type 1 tissue reaction), H&E, 40 times enlarged. (E) Macroscopic appearance of *M. mycetomatis* in a surgical specimen. Grains are surrounded by intense fibrous capsules. (F) Culture of *Madurella mycetomatis*.

14.1.4 Disease Management and Antifungal Therapy

Disease management for mycetoma is always necessary since spontaneous cure has never been reported [2]. As mycetoma can be caused by either bacteria or fungi, the first step in proper management of mycetoma is to identify whether the causative agent is a bacterium or a fungus. In actinomycetoma, chemotherapy alone results in high cure rates (60%–90%), and surgery is only required in advanced cases and those refractory to antimicrobial treatment [35]. Unfortunately, for eumycetoma, the prospects are not as good. No fully effective antifungal agent has been discovered yet. For eumycetoma, surgery is always needed [35,36]. The aim of surgery is complete excision of the lesion. This is only possible when the mycetoma lesion is small and well-encapsulated. In larger lesions, only reduction of the amount of infected tissue is possible, and occasionally multiple surgeries are needed to excise most of the lesion [33,37]. Especially in large lesions, effective surgery needs aggressive excision or debridement under general anesthesia, which usually cripples the limb or leads to permanent disability due to mandatory amputation [37]. To prevent recurrent infections, surgery is always combined with antifungal treatment, both before and after [3,33,37]. Antifungal therapy for eumycetoma still depends mainly on the azole antifungals ketoconazole and itraconazole. Amphotericin B is generally considered ineffective [3,38,39]. Liposomal amphotericin B had somewhat better in vivo efficacy, since treatment with liposomal amphotericin B led initially to remission, but relapse occurred within 6 months after the end of therapy [40]. The azole class of antifungal agents appeared to be more suitable to treat mycetoma. The initial trials with ketoconazole and itraconazole showed a clear progress in healing of the lesions [41]. In contrast, when fluconazole was used to treat mycetoma caused by *M. mycetomatis* or *M. grisea*, only temporary improvement was achieved with quick relapses [3,42]. In one case of *M. grisea* mycetoma, itraconazole also yielded great improvement [41]. The results obtained with itraconazole in mycetoma caused by *M. mycetomatis* were more variable. According to Hay, itraconazole resulted in improvement of 42% of the *M. mycetomatis* mycetoma cases and in no response in 33% of the cases [40,43]. For both *M. mycetomatis* and *M. grisea*, favorable results have been reported with ketoconazole [44,45]. Long-term ketoconazole therapy is used in the Mycetoma Research Centre. Hay reported that in 50 patients seen in the Mycetoma Research Centre, 72% showed clinical improvement or were even cured with ketoconazole, while only 8% did not respond at all [3,40]. Also Venugopal and Venugopal noticed significant therapeutic efficacy of ketoconazole [45]. Currently, although showing varying degrees of clinical efficacy, itraconazole and ketoconazole are still the best treatment options [33,38,46,47]. Within the Mycetoma Research Centre (Khartoum, Sudan; http://www.mycetoma.edu.sd), both ketoconazole (400–800 mg daily) and itraconazole (400 mg daily) are recommended for first-line use [26,32]. The duration of treatment varies with the severity of the infection and the general health status of individual patients. Treatment may need to be continued for 18–24 months or more, and the liver function of the patients needs regular monitoring, especially with long-term ketoconazole treatment in cases of advanced mycetoma [33,37]. However, long-term treatment may lead to antifungal resistance, which may in turn complicate patient management. Both ketoconazole and itraconazole are implicated in encapsulating the mycetoma lesion. This renders final surgical treatment a likely option. But in advanced lesions, especially when bone tissue is involved, the response to chemotherapeutic treatment still remains very poor [48]. In the search for new, alternative antifungal strategies in the treatment of mycetoma, terbinafine has been tried in the treatment of eumycetoma. Twenty five percent of patients were cured and 55% showed clinical improvement [49]. Recently, the effect of posaconazole was evaluated on six eumycetoma cases, of which two were caused by *M. mycetomatis* and three by *M. grisea* [50]. In all the *M. grisea* cases, posaconazole resulted in a successful outcome with various symptomatic improvements (lesion diameter decreased, sinuses closed, secretions stopped, pain and swelling resolved, and motility increased) [50]. Resolution of bone abnormalities was also observed in osteomyelitic mycetomas, suggesting that posaconazole may penetrate this tissue after oral administration [50]. The outcome of *M. mycetomatis* mycetoma was variable, in one case a complete success was noted, in the other case no clinical improvement was noted [50].

14.1.5 Diagnosis

As is clear from the above section, early diagnosis may improve therapeutic outcome. In endemic areas, mycetoma must be considered in the differential diagnosis of all subcutaneous swellings. Grains present in the discharge of the sinuses or in biopsies taken from lesions are important for diagnosis. Biopsies without grains are not suitable for diagnosis [11]. The color of the grain can give a first impression of the causative agent but it cannot identify the causative agent in full detail. For example, not only *Madurella* spp. form black grains in mycetoma lesions, but at least 14 other fungal species including *L. senegalensis* and *P. romeroi* are known to cause black-grain mycetoma [2,9]. Therefore, for species identification, high-quality diagnostic methods have to be used, as have already been implemented in endemic areas. These include radiology, histology, serology, and culture techniques. Furthermore, new molecular diagnostic tools have been developed to identify the mycetoma etiologic agents to the species level.

14.1.5.1 Conventional Techniques

14.1.5.1.1 Imaging Methodology

To assess the extent of the mycetoma, imaging techniques can be useful. With classic radiology, multiple radiological changes are found. It can show the presence and extent of soft tissue granulomas and demonstrates when the bone is penetrated by the causative agent [32]. Magnetic resonance imaging (MRI) is also useful in visualization of soft tissue involvement and bone destruction [32,51,52]. On MRI, grains appear as conglomerates of small (2–5 mm), round hyperintense lesions, representing the granulation tissue, surrounded by a low-signal-intensity rim [52]. The central low-signal-intensity dot is caused by the grain [52]. This unique appearance was named "dot-in-circle" and is highly suggestive of mycetoma [52,53]. Still these radiological techniques are not always specific and cannot always differentiate mycetoma from chronic bacterial osteomyelitis, granulomas, bone tuberculosis, and soft tissue tumors [32]. Furthermore, in most of the endemic regions, MRI equipment is not available. A more specific imaging technique for mycetoma is ultrasonic imaging [54]. The mycetoma grains, its capsule, and the accompanying inflammatory granulomata have characteristic ultrasonic appearances [54]. Ultrasound imaging can differentiate between mycetoma and other non-mycetoma lesions and even between eumycetoma and actinomycetoma [32,54]. In eumycetoma lesions, the grains produce numerous sharp bright hyper-reflective echoes, which correspond to the black grains [32].

14.1.5.1.2 Histology and Cytology

As stated before, the mycetoma grain is a first key feature for the identification of the causative agent. In order to study the grain more closely in the tissue, histological examination is used. In stained sections, the grain of *M. mycetomatis* is rounded, oval, or trilobed [9,11]. It has a compact brown cortex and a lighter medulla. In some grains, the division into cortex and medulla is not evident. The grain filaments are usually embedded in a hard brown cement matrix [11].

In *M. mycetomatis* infection, two main morphological types of grains are identified: the filamentous and the vesicular type [9,11]. The filamentous type is the most common type and consists of brown septate and branched hyphae that may be slightly more swollen toward the periphery [11]. In the cortex, the filaments are arranged radially, while in the medulla they tend to run multidirectionally [11]. Rounded or oval cells, 7–15 μm in diameter are seen, in particular in the periphery [11]. The vesicular type is composed of unusually large cells which look like vesicles. Both types of grain can be found in the same lesion.

The inflammation reaction around the grain is variable. Three types of tissue reaction can be seen. In type I, there is a zone of neutrophils in the vicinity of the grain [11,55]. Some histocytes may also be seen among the neutrophils but they are more numerous outside the neutrophil zone [11,55]. In type II, the neutrophil zone is absent, instead a layer of histocytes and multinucleated giant cells is seen. Some of these giant cells contain fragments of grain or pigmented cement substance [11]. At this stage, the grain is usually small and fragmented. The most uncommon reaction is the type III reaction in which the grain material has largely or completely disappeared. This leaves a compact epithelioid granuloma with or without Langerhans giant cells [11].

Another technique in use to diagnose mycetoma is fine-needle aspiration cytology (FNAC). In FNAC mycetoma lesions have a distinct appearance, characterized by the presence of polymorphous inflammatory cells and grains [56]. In the smear, grains are surrounded and infiltrated by neutrophils. Outside the neutrophil zone, monocytic cells and giant cells are seen. They are surrounded by granulation tissue rich in fibroblasts and blood vessels [56]. FNAC allows morphologic identification of mycetoma and its classification into eumycetoma and actinomycetoma, in a comparable manner as with histological slides [56,57].

14.1.5.1.3 Serology

Serological assays can be used to diagnose mycetoma. Serological assays most frequently used in the endemic areas are immunodiffusion (ID) and counter-immuno-electrophoresis (CIE), although enzyme-linked immunosorbent assays (ELISAs) have been developed as well. In the past, ID was a popular test to diagnose mycetoma but since then CIE was found to be quicker and more sensitive than ID [58,59]. The ELISA used by McLaren et al. and by Taha showed that both mycetoma patients and uninfected controls living in the endemic areas had antibodies against mycetoma agents [60–62]. This makes these ELISAs less suitable for diagnostic purposes. Therefore, in the endemic areas, only CIE is still used in the diagnosis of mycetoma. In addition to its diagnostic value, CIE has been used to monitor therapeutic outcome in patients. When there is a response to treatment, both intensity and number of precipitation bands decrease and eventually disappear [59]. One of the large drawbacks of all serological techniques used to diagnose mycetoma is that they rely on crude antigens prepared in diagnostic laboratories in the endemic areas. The antigens are prepared by grinding the mycelia in physiological saline, disintegrate them ultrasonically, and collect the cytoplasmatic proteins by centrifugation [58,59,62]. This rough production of antigens may result in variation between tests. A serological assay based on a recombinant antigen would be preferred. Currently only a single recombinant antigen has been prepared for *M. mycetomatis*, none for *M. grisea*. Unfortunately the recombinant antigen, the translationally controlled tumor protein of *M. mycetomatis* was not useful for diagnostic purposes, since not all patients had produced antibodies against this protein, while in contrast some of the healthy controls did [63]. Given the experimental ease of the ELISA test, additional investigation in novel recombinant antigens is highly recommended.

14.1.5.1.4 Culture

For successful culture, a deep-seated surgical biopsy specimen is preferred to collect grains discharged with pus through the sinuses [64]. The specimen obtained is divided into two portions. One portion is used for direct microscopy, the other for culture. Direct microscopy can be performed by examination of crushed grains in 10% KOH [58]. Black grains usually are composed of hyphae of approximately 15 μm in width. For culture, grains are washed in saline containing chloramphenicol, grown on blood agar and Sabouraud dextrose agar and incubated at 37°C and 26°C for 6–8 weeks [58]. Identification of isolates is achieved by observation of the rate of growth, colony morphology, production of conidia, and assimilation patterns [58]. Culturing these fungi remains troublesome, often no growth is obtained or cultures are contaminated with bacteria. Identification of the various fungi responsible for black-grain mycetoma remains difficult with standard mycological procedures since these fungi only rarely produce conidia [65,66]. Misidentification, therefore, often occurs. Furthermore, most of these fungi have low growth rates which may delay the identification until up to 12 weeks [21].

14.1.5.2 Molecular Techniques

14.1.5.2.1 Species Identification

In order to identify the causative agents of black-grain mycetoma more accurately, molecular diagnostic tools have been developed. The identification of black-grain mycetoma agents, including *M. mycetomatis*, is based on the internal transcribed spacer (ITS) sequences [5,67]. The ITS region can be either amplified with pan-fungal primers or with *M. mycetomatis*-specific primers. The pan-fungal primers are located at the 18S ribosomal sequence and the 28S ribosomal sequence. The resulting PCR product is therefore composed of part of the 18S ribosomal sequence, the ITS 1 region, the 5.8S ribosomal sequence, the ITS 2 region, and part of the 28S ribosomal sequence. The *M. mycetomatis*-specific primers are located on the ITS 1 and ITS 2 regions. The resulting PCR products can be analyzed with either a restriction fragment length polymorphism (RFLP) digestion [67] or by sequencing [5]. The latter results in a more precise identification. The *M. mycetomatis*-specific PCR could not only be used in diagnosis, but it was even possible to detect *M. mycetomatis* DNA in soil and thorn-samples [68].

14.1.5.2.2 Molecular Typing

Three different molecular techniques have also been used so far to type individual strains of *M. mycetomatis*. The first technique was a restriction endonuclease assay (REA) [69]. Lopes et al. performed REA on 17 isolates of *M. mycetomatis* obtained from nine different countries and found two larger clusters and seven unique genotypes [69]. The two larger clusters contained isolates from Africa, the remaining genotypes mainly originated from other continents [69]. The second technique used was random amplification of polymorphic DNA (RAPD). This method has been used by two research groups, with different outcomes. The RAPD performed on the 17 isolates isolated by Lopes et al. rendered nine different genotypes [69]. In agreement with the REA data generated by Lopes, two larger clusters and seven individual genotypes were again found [69]. RAPD was also performed by Ahmed et al. [70]. In contrast to Lopes et al., they did not find genetic variation when typing 38 *M. mycetomatis* isolates from Sudan, and two from Mali [70]. The discrepancy between the study performed by Lopes et al. and the one performed by Ahmed et al. could be due to multiple factors. First of all, different primer sets and PCR conditions were used for both RAPD studies. Furthermore, Lopes et al. [69] used a worldwide collection of strains, whereas Ahmed et al. primarily used strains from Sudan [70]. van de Sande et al. later added a third typing method, namely, amplified fragment length polymorphism (AFLP) [71] analysis. AFLP proved to be more discriminatory than RAPD, and it separated the Sudanese isolates in two large and one minor cluster [71]. In cluster I mainly isolates obtained from large lesions from patients living in Central Sudan were found, while the strains from cluster II were of more heterogeneous origin [71]; Cluster III consisted only of one strain and was therefore not analyzed further.

In contrast to *M. mycetomatis*, no typing studies have been performed for *M. grisea* or the various other *Madurella* spp.

14.2 METHODS

14.2.1 Sample Preparation

Different DNA extraction methods have been used to isolate DNA from *M. mycetomatis* [5,67,68,72]. Usually DNA is isolated from fungi grown on Sabouraud dextrose agar, but DNA can be isolated from grains as well. For direct isolation of DNA from grains, the grain is frozen in liquid nitrogen and then crushed (Abdalla Ahmed, personal communication). Afterward, different DNA isolation protocols can be used, including commercial kits. Higher DNA concentrations are obtained when DNA is isolated from strains grown *in vitro*.

Procedure

1. Transfer fungal grain onto Sabouraud dextrose agar (Difco) and culture at 37°C in case of *M. mycetomatis* and at 25°C in case of *M. grisea*. For identification purposes, it is wise to split the sample to two agar plates and culture at both temperatures. After 4 weeks, the colony is excised from the agar, frozen in liquid nitrogen, and ground in a porcelain mortar.
2. DNA can be isolated from the resulting pulp with any DNA isolation protocol. In our lab we have the best results with the Promega Wizard Kit (Promega). When using this protocol, 300 μL nuclei lysis solution (Promega Wizard) is added to the ground mycelia, and the solution is mixed by pipetting gently. From this step onward, the yeast protocol from the Promega Wizard protocol is used according to the manufacturer's instructions.

14.2.2 Detection Procedures

14.2.2.1 Identification of Fungi Causing Black-Grain Mycetoma by PCR-RFLP

The molecular identification of black-grain mycetoma agents is based on the analysis of the ITS region [5,67]. Ahmed et al. developed an RFLP method to differentiate between the various black-grain mycetoma agents [67].

Procedure

1. Prepare a PCR mixture (100 μL) containing 50 ng DNA, 1× Supertaq PCR buffer 1 (HT Biotechnology, Cambridge, U.K.), 0.2 mM nucleotide mix, 25 pmol primer ITS 4 (5′-TCCTCCGCTTATTGATATGC-3′) and 25 pmol primer ITS 5 (5′-GGAAGTAAA AGTCGTAACAAGG-3′), and 1.2 U Supertaq (HT Biotechnology).
2. Perform the PCR amplification in a thermocycler using a cycling program consisting of a denaturation step at 94°C for 4–10 min; 30 cycles of 94°C for 30 s, 58°C for 30 s, and 72°C for 30 s; and a final extension step at 72°C for 10 min [5].
3. Prepare an RFLP mixture (25 μL) containing 15 μL PCR product, 2 U of either *Cfo*I, *Msp*I, *Hae*III, *Rsa*I, *Sau*3A, or *Spe*I, and 1× corresponding restriction buffer (dependent on the enzyme used and the manufacturer). Incubate overnight at the temperature indicated by the manufacturer.
4. Analyze RFLP banding patterns on 3% Nusieve GTG agarose gels. Most of the black-grain isolates are identifiable by their banding patterns [67].

Note: For most of the black-grain-producing isolates, the length of the ITS region is similar between strains of a given species, except for some *Pyrenochaeta* spp. and *Madurella grisea* isolates. In these isolates, the ITS sequences range from 951 to 1327 bp [5]. For *M. mycetomatis*, the length of the ITS region is 624 bp.

14.2.2.2 Identification of Fungi Causing Black-Grain Mycetoma by Sequencing

In order to analyze the ITS region more accurately, many scientists have chosen to sequence this region [4,5,67].

Procedure

1. Prepare a PCR mixture (50 µL) containing 50 ng DNA, 1× Supertaq PCR buffer 1 (HT Biotechnology), 0.2 mM nucleotide mix, 25 pmol primer V9D (5′-TTAAGTCCCTGCCCTTTGTA-3′) and 25 pmol primer LS266 (5′-GTAGTCATATGCTTGTCTC-3′), and 1.2 U Supertaq (HT Biotechnology).
2. Perform the PCR amplification in a thermocycler using a cycling program consisting of a denaturation step at 94°C for 4–10 min; 30 cycles of 94°C for 30 s, 58°C for 30 s, and 72°C for 30 s; and a final extension step at 72°C for 10 min [5].
3. Prepare a sequencing mixture (11 µL) containing 40 ng PCR product, 1.5 µL sequencing buffer (Applied Biosystems, Cheshire, U.K.), 1 µL sequencing reaction mix (Applied Biosystems), and 5 pmol primer ITS 4 (5′-TCCTCCGCTTATTGATATGC-3′). For the reverse reaction, replace primer ITS 4 by primer ITS 5 (5′-GGAAGTAAAAGTCGTAACAAGG-3′).
4. Perform the sequencing reaction in a thermocycler using a sequencing program consisting of 25 cycles of 96°C for 10 s, 50°C for 5 s, and 60°C for 4 min.
5. Precipitate the resulting product by adding 80 µL precipitation mix consisting of 0.11 M NaAc pH 4.8 (Sigma Aldrich) and 75% ethanol, incubate for 20 min at room temperature, and centrifuge for 30 min at 2500 × g at room temperature.
6. Wash the pellet with 150 µL 70% ethanol.
7. Add 20 µL HiDi formamide (Applied Biosystems) to each sample and sequence the product with the ABI Prism 3700 automated DNA analyzer (Applied Biosystems).

Note: By using these pan-fungal ITS primers, most of the black-grain mycetoma isolates can be amplified [5,67]. Only for four new *Madurella* species, the complete ITS region could not be amplified with primer pairs V9D/LS266 and ITS 4/ITS 5. Identification of these four isolates was based on the amplification of the ITS 2 region with primers ITS 3 (5′-GCATCGATGAAGAACGCAGC-3′) and ITS 4 [5].

14.2.2.3 Identification of *Madurella mycetomatis* by PCR

Ahmed et al. developed a *M. mycetomatis* species-specific PCR, based on the ITS region.

Procedure

1. Prepare a PCR mixture (50 µL) containing 50 ng DNA, 1× Supertaq PCR buffer 1 (HT Biotechnology), 0.2 mM nucleotide mix, 25 pmol primer 26.1A (5′-AATGAGTTGGGCTTTAACGG-3′) and 25 pmol primer 28.3A (5′-TCCCGGTAGTGTAGTGTCCCT-3′), and 1.2 U Supertaq (HT Biotechnology).
2. Perform the PCR amplification in a thermocycler using a cycling program with 40 cycles of 94°C for 1 min, 58°C for 2 min, and 72°C for 1 min [67].
3. Separate the PCR products on a 1% agarose gel and stain with ethidium bromide.

Note: With this *M. mycetomatis*-specific PCR, it is also possible to detect *M. mycetomatis* DNA in soil and thorn-samples [68].

14.2.2.4 Molecular Typing of *Madurella mycetomatis*

14.2.2.4.1 REA

Lopes et al. developed an REA for the typing of *M. mycetomatis*. In this technique, DNA is digested with restriction endonucleases, and the resulting banding patterns are used to differentiate the various genotypes [69].

Procedure

1. Prepare a restriction endonuclease mixture (20 µL) containing 100 ng of DNA, 2 U of *Sca*I (New England Biolabs), 1× NEBuffer 3 (New England Biolabs), and 5 µg of *M. mycetomatis* DNA. Incubate for 3 h at 37°C according to the manufacturer's instructions [69].
2. Electrophorese the restriction digest on a 0.7% agarose gel at 2 V/cm for 16 h and visualize the banding patterns with ethidium bromide stain [69].

Note: Next to performing an REA with *Sca*I, other restriction endonucleases can be used as well. These include *Eco*RI, *Eco*RV, *Hin*dIII, and *Hpa*I. The results obtained with *Sca*I are the most encouraging, since all 17 isolates typed by Lopes et al. give distinct, intense, and well-defined bands than those with the other restriction endonucleases [69].

14.2.2.4.2 RAPD

In the RAPD technique, usually a PCR is performed with one short primer at a low annealing temperature which will result in a banding pattern. Two different RAPD protocols have been used to type *M. mycetomatis* [69,70]. Since the RAPD protocol of Lopes et al. [69] results in different genotypes, it is presented here.

Procedure

1. Prepare an RAPD PCR mixture (50 µL) containing 25 ng isolated DNA, 0.1 mM dNTPs, 0.2 µM primer, 2.5 U Taq polymerase (HT Biotechnology), 50 mM KCl, 10 mM Tris–HCl (pH8.0), and 1.5 mM $MgCl_2$. As primer, either one of the following five primers can be used: 28289 (5′-ATGGATCCGC-3′), 28290 (5′-AAGAGCCCGT-3′), 28308 (5′-AACGCGCAAC-3′), 28309 (5′-GTTTCCGCCC-3′), and 28323 (5′-GCGATCCCCA-3′) [69].
2. Perform the RAPD in a thermocycler using a cycling program consisting of a denaturation step at

94°C for 5 min; 45 cycles of 94°C for 1 min, 35°C for 1 min, and 72°C for 2 min; and a final extension step at 72°C for 5 min [69].

3. Visualize the banding patterns on a 1.5% agarose gel with ethidium bromide strain [69].

14.2.2.4.3 AFLP

van de Sande et al. developed an AFLP method for typing *M. mycetomatis* isolates [71]. In this technique, DNA is digested with two restriction endonucleases after which the restriction fragments are amplified selectively. The resulting banding patterns are used to differentiate the various genotypes.

Procedure

1. Prepare a DNA digestion mixture (40 µL) containing 0.5 µg of DNA, 5 U *Eco*RI, 5 U *Mse*I, 10 mM Tris–HAc pH 7.5, 10 mM MgAc, 50 mM KAc, 5 mM dithiothreitol (DTT), and 50 ng/µL BSA. Incubate the mixture at 37°C for 1 h.
2. To ligate the adapters, add 10 µL of a solution containing 5 pMol *Eco*RI adapters, 50 pMol *Mse*I adapters, 1 U T4 DNA-ligase, 1 mM ATP in 10 mM Tris–HAc pH 7.5, 10 mM MgAc, 50 mM KAc, 5 mM DTT, 50 ng/µL BSA. Incubate the reaction for 3 h at 37°C [73]. The structure of the *Eco*RI adapter is

```
5′-CTCGTAGACTGCGTACC
       ||||||||||||||
       CATCTGACGCATGGTTAA-5′
```

The structure of the *Mse*I adapter is

```
5′-GACGATGAGTCCTGAG
         ||||||||||||
         TACTCAGGACTCAT-5′
```

3. Prepare a pre-amplification mixture containing 30 ng of primer E (5′-GACTGCGTACCAATTC-3′), 30 ng of primer M (5′-GACGATGAGTCCTGAGTAA-3′), 0.4 U Taq polymerase, 10 mM Tris–HCl pH 8.3, 1.5 mM $MgCl_2$, 50 mM KCl, and 0.2 mM of dNTPs [73].
4. Perform pre-amplification in a thermocycler using a cycling program consisting of 20 cycles of 94°C for 30 s, 56°C for 30 s, and 72°C for 1 min [73].
5. Dilute the pre-amplification mixture 10-fold.
6. Prepare an end-labeling mixture (50 µL) containing 500 ng *Eco*RI primer, 100 µCi $[\gamma\text{-}^{33}P]$ ATP, 10 U T4 polynucleotide kinase, 25 mM Tris–HCl pH 7.5, 10 mM $MgCl_2$, 5 mM DTT, 0.5 mM spermidine–3HCl [73]. In the method developed by van de Sande et al., two different *Eco*RI primers were used, namely, E12 (5′-GACTGCGTACCAATTCAC-3′) and E20 (5′-GACTGCGTACCAATTCGC-3′).
7. Prepare a selective amplification mixture (20 µL) containing 5 µL diluted pre-amplificated product, 5 ng primer E12, 30 ng primer M12 (5′-GACGATGAGTCCTGAGTAAAC-3′), 0.4 U *Taq* polymerase, 10 mM Tris–HCl pH 8.3, 1.5 mM $MgCl_2$, 50 mM KCl, and 0.2 mM dNTP mixture [71]. As an alternative selective amplification mixture, primer E12 can be replaced by primer E20.
8. Perform the selective amplification in a thermocycler using a cycling program consisting of 36 cycles of 94°C for 30 s, an annealing temperature for 30 s, and 72°C for 1 min. The annealing temperature in the first cycle is 65°C, is subsequently reduced each cycle by 0.7°C for the next 12 cycles, and is continued at 56°C for the remaining 23 cycles [73].
9. Following amplification, mix the reaction products with an equal volume of formamide dye consisting of 98% formamide, 10 mM EDTA pH 8.0, and bromophenol blue and xylene cyanol as tracking dyes.
10. Heat the resulting mixture at 90°C for 3 min and then cool on ice [73].
11. Load the samples on a 4.5% denaturing polyacrylamide gel (4.5% acrylamide, 0.25% methylene bisacryl, 7.5 M urea in 50 mM Tris/50 mM boric acid/1 mM EDTA, 0.05% APS, and 0.1% N,N,N′,N′-tetramethylethylenediamine [TEMED]) [73] and electrophorese for 2 h at 110 W in 100 mM Tris/100 mM boric acid/2 mM EDTA [71,73].

14.3 CONCLUSIONS AND FUTURE PERSPECTIVES

The genus *Madurella* consists of many species of which only *M. mycetomatis* and *M. grisea* are well-characterized. *M. mycetomatis* and *M. grisea* erroneously ended up in the same genus based on the fact that they both cause black-grain mycetoma and that they form sterile cultures. Furthermore, in the past, and based on these criteria many other fungi have been misidentified as either *M. mycetomatis* or *M. grisea*. This demonstrates that the identification of *Madurella* species has always been troublesome. Unfortunately, since the taxonomic description of both species, little improvement has been made in the classical diagnosis of mycetoma. With imaging techniques, histopathology, serology, and even culturing still many misidentifications occur. Until this moment, no adequate point-of-care diagnostic tools are available to accurately diagnose mycetoma.

The first real breakthrough in the differentiation of the two *Madurella* species was made when molecular tools were developed to identify these species. By characterizing the ITS region of the *Madurella* species, *M. mycetomatis* and *M. grisea* could be easily distinguished [5,6,67,68]. No misidentification could occur, since both species differ considerably in their respective ITS regions. In fact, characterizing the ITS region revealed that both species belong to completely different fungal orders [4]. Another benefit of molecular diagnostic tools is that the duration of species identification can be shortened considerably. Since DNA can be isolated directly from fungal grains, species identification can be obtained within 2–3 working days.

Next to identification of fungal species by molecular methods, the development of molecular typing tools also allowed to differentiate strains within the species. For *M. mycetomatis*, three molecular typing methods were developed: REA, RAPD, and AFLP [69,71]. AFLP proved to be the most discriminating method. Based on the results obtained from these typing methods, it was established that the species *M. mycetomatis*, although quite clonal, can still be differentiated in various genotypes that can even be found within a single geographic region [71].

The diagnostic molecular tools have gradually taken over the culture-based identification procedures for identification of black-grain mycetoma agents. Identification is now based on sequencing the ITS region of these fungi, and misidentifications are a phenomenon of the past. Unfortunately, in the endemic regions, molecular diagnostic tools have not been in use so far. Most of the endemic countries are considered to be developing countries and do not have the means to routinely implement these more reliable molecular tools in the frequently rural tropical settings. This situation may change if the expense of molecular biological reagents and equipment comes down, and alternative cost-effective molecular diagnostic procedures are developed in future.

REFERENCES

1. Ellis, D., Davis, S., Alexiou, H., Handke, R., and Bartley, R., *Descriptions of Medical Fungi*, 2nd edn., Adelaide, 2007.
2. Ahmed, A.O. et al., Mycetoma caused by *Madurella mycetomatis*: A neglected infectious burden. *Lancet Infect. Dis.*, 4, 566, 2004.
3. McGinnis, M.R., Mycetoma. *Dermatol. Clin.*, 14, 97, 1996.
4. de Hoog, G.S. et al., Phylogeny and typification of *Madurella mycetomatis*, with a comparison of other agents of eumycetoma. *Mycoses*, 47, 121, 2004.
5. Desnos-Ollivier, M. et al., Molecular identification of black-grain mycetoma agents. *J. Clin. Microbiol.*, 44, 3517, 2006.
6. Borman, A.M. et al., Molecular characterisation of the *Madurella grisea* complex reveals at least three new taxa associated with human mycetomas. In: *ISHAM 2009*, Tokyo, 2009.
7. Mackinnon, J.E., Characteristics of four cultures of Mycetomi (Laveran) Brumpt isolated in Sudan and British Somaliland. *An. Facult. Med. Montevideo*, 36, 197, 1951.
8. van de Sande, W.W. et al., Melanin biosynthesis in *Madurella mycetomatis* and its effect on susceptibility to itraconazole and ketoconazole. *Microb. Infect.*, 9, 1114, 2007.
9. Kwon-Chung, K.J. and Bennet, J.E., *Medical Mycology*. Lea and Febiger, Philadelphia, PA, 1992.
10. Rajendran, C. et al., An evaluation of straw-extract agar media for the growth and sporulation of *Madurella mycetomatis*. *Mycopathologia*, 115, 9, 1991.
11. El-hassan, A.M., Fahal, A., and El-hag, I., The pathology of mycetoma. *Sudan Med. J.*, 32(Suppl.), 23, 1994.
12. CLSI, *Reference Method for Broth Dilution Antifungal Susceptibility Testing of Filamentous Fungi: Approved Standard*. National Committee for Clinical Laboratory Standards, Wayne, Pennsylvania, PA, 2002.
13. van de Sande, W.W. et al., In vitro susceptibility of *Madurella mycetomatis*, prime agent of Madura foot, to tea tree oil and artemisinin, *J. Antimicrob. Chemother.*, 59, 553, 2007.
14. van de Sande, W.W. et al., Testing of the in vitro susceptibilities of *Madurella mycetomatis* to six antifungal agents by using the sensititre system in comparison with a viability-based 2,3-bis(2-methoxy-4-nitro-5-sulfophenyl)-5-[(phenylamino)carbonyl]-2H-tetrazolium hydroxide (XTT) assay and a modified NCCLS method. *Antimicrob. Agents Chemother.*, 49, 1364, 2005.
15. Ahmed, A.O. et al., In vitro susceptibilities of *Madurella mycetomatis* to itraconazole and amphotericin B assessed by a modified NCCLS method and a viability-based 2,3-Bis(2-methoxy-4-nitro-5-sulfophenyl)-5-[(phenylamino)carbonyl]-2H-tetrazolium hydroxide (XTT) assay. *Antimicrob. Agents Chemother.*, 48, 2742, 2004.
16. van de Sande, W.W.J. et al., *Madurella mycetomatis* is not susceptible to the echinocandin class of antifungal agents. *Antimicrob. Agents Chemother.*, 54, 2738, 2010.
17. Venugopal, P.V. et al., Antimycotic susceptibility testing of agents of black grain eumycetoma. *J. Med. Vet. Mycol.*, 31, 161, 1993.
18. Mackinnon, J.E., New species of fungus pathogenic to man in South America. *Cienc. Invest.*, 7, 77, 1951.
19. Rippon, J.W., Mycetoma. In: *Medical Mycology, the Pathogenic Fungi and the Pathogenic Actinomycetes*, 3rd edn., Chap. 5, pp. 80–84, W.B. Saunders Company, Philadelphia, PA, 1988.
20. McGinnis, M.R. and Pasarell, L., In vitro testing of susceptibilities of filamentous ascomycetes to voriconazole, itraconazole, and amphotericin B, with consideration of phylogenetic implications. *J. Clin. Microbiol.*, 36, 2353, 1998.
21. de Hoog, G.S., Guarro, J., Gene, J., and Figueras, M., *Atlas of Clinical Fungi*, 2nd edn., Centraalbureau voor Schimmelcultures/Reus. Universitat Rovira i Virgili, Utrecht, the Netherlands, 2000.
22. Boiron, P. et al., Nocardia, nocardiosis and mycetoma. *Med. Mycol.*, 36(Suppl 1), 26, 1998.
23. Gumaa, S.A., The aetiology and epidemiology of mycetoma. *Sudan Med. J.*, 32(Suppl.), 14, 1994.
24. Mahgoub, E. and Murray, I., *Mycetoma*. William Heinermann, London, U.K., 1973.
25. Fahal, A.H. et al., Clinical presentation of mycetoma. *Sudan Med. J.*, 32, S46, 1994.
26. Fahal, A.H., Mycetoma: A thorn in the flesh. *Trans. R. Soc. Trop. Med. Hyg.*, 98, 3, 2004.
27. Mahgoub, E.S., Gumaa, S.A., and El Hassan, A.M., Immunological status of mycetoma patients. *Bull. Soc. Pathol. Exot. Filiales*, 70, 48, 1977.
28. Mahgoub, E.S., Experimental infection of athymic nude New Zealand mice, nu nu strain with mycetoma agents. *Sabouraudia*, 16, 211, 1978.
29. van de Sande, W.W. et al., Polymorphisms in genes involved in innate immunity predispose toward mycetoma susceptibility. *J. Immunol.*, 179, 3065, 2007.
30. Ahmed, A. et al., Environmental occurrence of *Madurella mycetomatis*, the major agent of human eumycetoma in Sudan. *J. Clin. Microbiol.*, 40, 1031, 2002.
31. Fahal, A.H. et al., Cystic mycetoma: An unusual clinical presentation of *Madurella mycetomatis* infection. *Trans. R. Soc. Trop. Med. Hyg.*, 92, 66, 1998.
32. Fahal, A.H., *Mycetoma, Clinicopathological Monograph*, 1st edn., Khartoum University Press, Khartoum, 2006.

33. Ahmed, A. et al., Management of mycetoma: Major challenge in tropical mycoses with limited international recognition. *Curr. Opin. Infect. Dis.*, 20, 146, 2007.
34. Fahal, A.H. et al., Blood supply and vasculature of mycetoma. *J. Med. Vet. Mycol.*, 35, 101, 1997.
35. Lupi, O., Tyring, S.K., and McGinnis, M.R., Tropical dermatology: Fungal tropical diseases. *J. Am. Acad. Dermatol.*, 53, 931, quiz 952, 2005.
36. Lichon, V. and Khachemoune, A., Mycetoma: A review. *Am. J. Clin. Dermatol.*, 7, 315, 2006.
37. Fahal, A.H., Hassan, M., and Sanhouri, M., Surgical treatment of mycetoma. *Sudan Med. J.*, 32, S98, 1994.
38. Restrepo, A., Treatment of tropical mycoses. *J. Am. Acad. Dermatol.*, 31, S91, 1994.
39. Rios-Fabra, A., Moreno, A.R., and Isturiz, R.E., Fungal infection in Latin American countries. *Infect. Dis. Clin. North Am.*, 8, 129, 1994.
40. Hay, R.J. et al., Mycetoma. *J. Med. Vet. Mycol.*, 30(Suppl.), 41, 1992.
41. Borelli, D., A clinical trial of itraconazole in the treatment of deep mycoses and leishmaniasis. *Rev. Infect. Dis.*, 9(Suppl.), S57, 1987.
42. Diaz, M. et al., A Pan-American 5-year study of fluconazole therapy for deep mycoses in the immunocompetent host. Pan-American Study Group. *Clin. Infect. Dis.*, 14(Suppl.), S68, 1992.
43. Van Cutsem, J., The in-vitro antifungal spectrum of itraconazole. *Mycoses*, 32(Suppl.), 7, 1989.
44. Mahgoub, E.S. and Gumaa, S.A., Ketoconazole in the treatment of eumycetoma due to *Madurella mycetomi*. *Trans. R. Soc. Trop. Med. Hyg.*, 78, 376, 1984.
45. Venugopal, P.V. and Venugopal, T.V., Treatment of eumycetoma with ketoconazole. *Aust. J. Dermatol.*, 34, 27, 1993.
46. Develoux, M. et al., Management of mycetoma in West-Africa. *Bull. Soc. Pathol. Exot.*, 96, 376, 2003.
47. Welsh, O., Salinas, M.C., and Rodriguez, M.A., Treatment of eumycetoma and actinomycetoma. *Curr. Top. Med. Mycol.*, 6, 47, 1995.
48. Richardson, M.D. and Warnock, D.W., *Fungal Infection: Diagnosis and Management*, 3rd edn., Blackwell Publishing Ltd., Oxford, U.K., 2003.
49. N'Diaye, B. et al., Clinical efficacy and safety of oral terbinafine in fungal mycetoma. *Int. J. Dermatol.*, 45, 154, 2006.
50. Negroni, R. et al., Posaconazole treatment of refractory eumycetoma and chromoblastomycosis. *Rev. Inst. Med. Trop. Sao Paulo*, 47, 339, 2005.
51. Kumar, J. et al., The dot-in-circle sign of mycetoma on MRI. *Diag. Intervent. Radiol. (Ankara, Turkey)*, 13, 193, 2007.
52. Cherian, R.S. et al., The "dot-in-circle" sign—A characteristic MRI finding in mycetoma foot: A report of three cases. *Br. J. Radiol.*, 82, 662, 2009.
53. Sarris, I. et al., MRI of mycetoma of the foot: Two cases demonstrating the dot-in-circle sign. *Skelet. Radiol.*, 32, 179, 2003.
54. Fahal, A.H. et al., Ultrasonographic imaging of mycetoma. *Br. J. Surg.*, 84, 1120, 1997.
55. Fahal, A.H. et al., The host tissue reaction to *Madurella mycetomatis*: New classification. *J. Med. Vet. Mycol.*, 33, 15, 1995.
56. El Hag, I.A., Fine needle aspiration cytology of mycetoma. *Acta Cytol.*, 40, 461, 1996.
57. Gabhane, S.K. et al., Cytodiagnosis of eumycotic mycetoma: A case report. *Acta Cytol.*, 52, 354, 2008.
58. Suliman, E.-N., Laboratory diagnosis of mycetoma. *Sudan Med. J.*, 32, S67, 1994.
59. Mahgoub, E.-S., *Mycetoma*. In: Mahgoub E.-S., ed., *Tropical Mycoses*, pp. 58–74. Janssen Research Council, Beerse, Belgium, 1990.
60. Taha, A., A serological survey of antibodies to *Streptomyces somaliensis* and *Actinomadura madurae* in the Sudan enzyme linked immunosorbent assay (ELISA). *Trans. R. Soc. Trop. Med. Hyg.*, 77, 49, 1983.
61. McLaren, M.L., Mahgoub, E.S., and Georgakopoulos, E., Preliminary investigation of the use of the enzyme linked immunosorbent assay (ELISA) in the serodiagnosis of mycetoma. *Sabouraudia*, 16, 225, 1978.
62. Wethered, D.B. et al., Humoral immune responses to mycetoma organisms: Characterization of specific antibodies by the use of enzyme-linked immunosorbent assay and immunoblotting. *Trans. R. Soc. Trop. Med. Hyg.*, 82, 918, 1988.
63. van de Sande, W.W. et al., Translationally controlled tumor protein from *Madurella mycetomatis*, a marker for tumorous mycetoma progression. *J. Immunol.*, 177, 1997, 2006.
64. Mahgoub, E.S., Mycoses of the Sudan. *Trans. R. Soc. Trop. Med. Hyg.*, 71, 184, 1977.
65. Padhye, A.A. and McGinnis, M.R., Fungi causing eumycotic mycetoma. In: Murray P.R. et al., eds., *Manual of Clinical Microbiology*, 7th edn., pp. 1318–1326. American Society for Microbiology, Washington, DC, 1999.
66. de Hoog, G.S. et al., Diagnostic problems with imported cases of mycetoma in The Netherlands. *Mycoses*, 36, 81, 1993.
67. Ahmed, A.O. et al., Development of a species-specific PCR-restriction fragment length polymorphism analysis procedure for identification of *Madurella mycetomatis*. *J. Clin. Microbiol.*, 37, 3175, 1999.
68. Ahmed, A.O. et al., Molecular detection and identification of agents of eumycetoma: Detailed report of two cases. *J. Clin. Microbiol.*, 41, 5813, 2003.
69. Lopes, M.M., Freitas, G., and Boiron, P., Potential utility of random amplified polymorphic DNA (RAPD) and restriction endonuclease assay (REA) as typing systems for *Madurella mycetomatis*. *Curr. Microbiol.*, 40, 1, 2000.
70. Ahmed, A. et al., *Madurella mycetomatis* strains from mycetoma lesions in Sudanese patients are clonal. *J. Clin. Microbiol.*, 41, 4537, 2003.
71. van de Sande, W.W. et al., Genotyping of *Madurella mycetomatis* by selective amplification of restriction fragments (amplified fragment length polymorphism) and subtype correlation with geographical origin and lesion size. *J. Clin. Microbiol.*, 43, 4349, 2005.
72. Ahmed, A. et al., Eumycetoma caused by *Madurella mycetomatis* identified by PCR and sequencing: A report of two cases. *J. Clin. Microbiol.*, 41, 1813, 2003.
73. Vos, P. et al., AFLP: A new technique for DNA fingerprinting. *Nucleic Acids Res.*, 23, 4407, 1995.
74. Abbott, P., Mycetoma in the Sudan, *Trans. R. Soc. Trop. Med. Hyg.*, 50, 11; discussion, 50, 24, 1956.
75. Bendl, B.J. et al., Mycetoma in Saudi Arabia. *J. Trop. Med. Hyg.*, 90, 51, 1987.
76. Castro, L.G. et al., Mycetoma: A retrospective study of 41 cases seen in Sao Paulo, Brazil, from 1978 to 1989. *Mycoses*, 36, 89, 1993.
77. Destombes, P. et al., Mycetoma in Somalia—Results of a survey done from 1959 to 1964. *Acta Trop.*, 34, 355, 1977.

78. Develoux, M. et al., Mycetoma in the Republic of Niger: Clinical features and epidemiology. *Am. J. Trop. Med. Hyg.*, 38, 386, 1988.
79. Dieng, M.T. et al., Mycetoma: 130 cases. *Ann. Dermatol. Venereol.*, 130, 16, 2003.
80. Klokke, A.H. et al., The causal agents of mycetoma in South India. *Trans. R. Soc. Trop. Med. Hyg.*, 62, 509, 1968.
81. Lopez Martinez, R. et al., Epidemiology of mycetoma in Mexico: Study of 2105 cases. *Gac. Med. Mex.*, 128, 477, 1992.
82. Negroni, R. et al., Clinical and microbiological study of mycetomas at the Muniz hospital of Buenos Aires between 1989 and 2004. *Rev. Argent. Microbiol.*, 38, 13, 2006.

15 *Neodeightonia*

Artur Alves, Alan J.L. Phillips, and António Correia

CONTENTS

15.1 INTRODUCTION

15.1.1 Classification, Morphology, and Biology

The ascomycete genus *Neodeightonia* C. Booth belongs in the family Botryosphaeriaceae, order Botryosphaeriales. Members of the family Botryosphaeriaceae are best known as pathogens, saprophytes, and endophytes in a wide range of plant hosts.[1] Their occurrence and relevance as human pathogens is low in comparison to other clinical fungi, and only a few species such as *Lasiodiplodia theobromae* (Pat.) Griffon & Maubl.,[2,3] *Macrophomina phaseolina* Tassi (Goid.),[4,5] *Neoscytalidium dimidiatum* (Penz.) Crous & Slippers,[4,6] and *Neodeightonia subglobosa* C. Booth[7] have shown the potential to cause disease to man.

The genus *Neodeightonia* was introduced by Booth (in Punithalingam 1969)[8] to accommodate a single species, namely *N. subglobosa*. von Arx & Müller (1975)[9] within their broad concept of the genus transferred this species to *Botryosphaeria*. Recently, the taxonomy and systematics of the genus *Botryosphaeria* sensu lato has been subjected to considerable revision.[10,11] In their work on the dark-spored telemorph genera in the Botryosphaeriaceae, Phillips et al. (2008)[11] reinstated *Neodeightonia* on account of its phylogenetic and morphological distinction from other genera in the Botryosphaeriaceae.

Thus, currently *Neodeightonia* is a very small genus comprising only two species, namely the type species *N. subglobosa* and *N. phoenicum* A.J.L. Phillips & Crous.[11] The last species was introduced by Phillips et al. (2008)[11] so to accommodate isolates obtained from palms (*Phoenix dactylifera* and *Phoenix canariensis*) and there is no evidence to date that it may incite any kind of clinical manifestation.

Neodeightonia subglobosa is characterized by having clavate, thick-walled bitunicate, eight-spored asci typical of the members of the Botryosphaeriaceae. Ascospores are distichous, the mature ones are brown, oval to broadly ellipsoidal, becoming one-septate, measuring $20–26 \times 7–10\,\mu m$ with a finely roughened surface. Its coelomycete anamorph has spherical to subglobose conidia which are initially hyaline and become light to dark brown when mature, aseptate, and measuring $9–12 \times 6–9\,\mu m$.[8] In *N. phoenicum* conidia are ovoid to ellipsoid, with the apex and base broadly rounded, widest in middle to upper third, thick-walled, initially hyaline and aseptate, becoming dark brown and one-septate some time after discharge from pycnidia, with melanin deposits on the inner surface of wall arranged longitudinally giving a striate appearance to conidia, and measure $18.6–19.5 \times 11.2–11.8\,\mu m$.[11] The teleomorph of *N. phoenicum* has not been found. Detailed morphological descriptions as well as microphotographs and illustrations of the genus and species can be found elsewhere.[8,10,11]

Neodeightonia subglobosa is a poorly studied species and virtually nothing is known regarding its biology, geographic distribution, and ecology. *Neodeightonia subglobosa* (as *Sphaeropsis subglobosa*) was first described from dead culms of *Bambusa* in Georgetown (Guyana) in 1879 and Belgium in 1882. Both states (anamorph and telemorph) of this species were described in 1969 on dead culms of *Bambusa arundinacea* in Sierra Leone. Since then the sexual state or teleomorph was never found again and there are only two reports published concerning this species. The first one deals with its occurrence as the cause of keratomycosis on a man who had poked a bamboo cane into his eye.[7] The second and most recent one describes the occurrence of this species as an endophyte on herbaceous medicinal plants from India.[12] It is not clear if this species is a plant pathogen or merely a saprophyte that is able to grow on dead plant material. As occurs with several plant pathogens in the Botryosphaeriaceae,[1] it is likely that *N. subglobosa* is a plant pathogen that has an endophytic stage in its life cycle and that it is capable of causing diseases when the host is under stress. However, this assumption remains to be proven.

15.1.2 Clinical Features and Pathogenesis

As mentioned earlier *N. subglobosa* is a poorly known species and this applies also to the clinical environment. There is only a single case reporting this species as a human pathogen causing keratomycosis.[7] The patient was accidentally injured in the eye with a bamboo cane causing a shelving, penetrating corneal wound, which contained small slivers of bamboo. The splinters were removed and in agreement with the fungal sensitivity tests, a treatment with 2% clotrimazole eye drops was applied for 16 weeks. This resulted in the restoration of complete visual acuity. Nevertheless, 39 months after the end of the topical antifungal therapy the infection recurred and a keratoplasty had to be performed in order to remove the fungal mass that developed. The surgery was followed by topical steroid and clotrimazole therapy[7] and there was no subsequent relapse.

According to Kirkness et al.[7] fungal keratytis due to *N. subglobosa* is difficult to treat, especially when associated with a penetrating injury. These authors hypothesized that dark, thick-walled hyphae or brown conidia, which could be resistant to antifungal therapy, may have lain dormant in the cornea resulting in a recurrent infection.

Neodeightonia subglobosa is probably not a primary human pathogen and as in most coelomycetous fungi it is most likely an opportunistic human pathogen. In the coelomycetes causing opportunistic mycoses the infection occurs most frequently by implantation of the fungus from plant material (or soil) through abrasions, lacerations, puncture wounds, or other traumas rather than by inhalation of spores.[13] These fungi may also be of concern to immunocompromised patients, such as bone marrow and organ transplant recipients, cancer patients, among others.[13]

15.1.3 Diagnosis

15.1.3.1 Conventional Techniques

As for other mycoses the hispathological identification of hyphae is elemental to confirm a fungus as the causal agent of the disease. Although the pigmentation and shape of hyphae and the presence or absence of septa can give an idea of the identity, a fungal culture is required to accomplish an accurate and reliable diagnosis of the etiological agent.[13,14] The identification of fungi, unlike other important pathogens such as bacteria or viruses, relies mainly on morphological criteria. Most specifically fungal identification is based on the morphology of the reproductive structures.

Coelomycetous fungi generally display a moderate to rapid growth rate on a variety of routine culture media and are not particularly difficult to recover from excised material. The main problem lies in promoting sporulation in order to obtain the diagnostic reproductive structures necessary for identification of the isolates. Apart from the considerable amount of time required, particularly for pycnidial species (up to months in some strains), it is also necessary to employ a medium upon which these pycnidia will develop.[13]

The culture of Coelomycetes on sterilized plant tissue is deemed to produce conidiomata more representative of those in nature. It is known that culture on nutrient-rich synthetic media often results in atypical characteristics and that sporulation may be hindered or absent on sugar-rich media.[13,15] Our experience with cultures of botryosphaeriaceous fungi, such as *Neodeightonia*, has shown that the addition of sterilized plant material (e.g., pine needles, twigs of oak, poplar, or other plants) on water agar, or half-strength potato dextrose agar, encourage the production of conidiomata. Another medium that is frequently used is oatmeal agar.[11,16]

As mentioned above, the identification of fungi is based on their morphology. Despite the fact that the sexual stages (telemorphs) are the baseline of fungal taxonomy and nomenclature, medical mycologists are generally more familiar with the asexual stage or anamorph, which is the stage most frequently found in cultures of clinical isolates.[14] In the case of botryosphaeriaceous fungi the teleomorph is rarely found in nature and it is highly improbable that sexual fruiting bodies (pseudothecia) will be seen in cultures of clinical isolates. Although Punithalingam (1969)[8] reported that *N. subglobosa* is a homothallic species and its teleomorph forms in culture, we have not observed that in our work. In fact, in the closely related species *N. phoenicum* our isolates also did not form the teleomorph in culture, even after long periods of incubation (>3 months).[11]

Species identification in the Botryosphaeriaceae has relied heavily on morphological characters of the anamorph, including size, shape, color, septation, wall thickness, and texture of conidia. However, some morphological characters exhibit extensive plasticity. Thus, size ranges of conidia of different species overlap, while age and state of maturity affect conidial pigmentation and septation. Morphological characters can also be influenced by the substrate on which the fungus is growing.[1,17,18] Because species may differ in minor morphological features, identification can be a difficult task for someone who is not familiar with these fungi.

The genus *Neodeightonia* is phylogenetically related to *Diplodia* and *Lasiodiplodia* and has intermediate morphological features between both genera.[11] Conidia are initially hyaline and aseptate becoming brown and one-septate (in some cases) with age, characters that are common to both *Diplodia* and *Lasiodiplodia*. Punithalingam[8] referred to a germ slit in the conidia of *N. subglobosa*. Crous et al.[10] suggested that this is in fact a striation on the conidial wall, and that more than one could occur per conidium, a feature that was confirmed in a later study.[11] Moreover, Phillips et al.[11] described a new species in *Neodeightonia*, namely *N. phoenicum*, which has clear striations in the conidial wall. Such striate walls are a typical feature of *Lasiodiplodia* species suggesting an affinity to this genus. Nevertheless, *Neodeightonia* can be distinguished from *Lasiodiplodia* by the absence of pycnidial paraphyses which are also not present in *Diplodia*. Thus, conidial striations distinguish *Neodeightonia* from *Diplodia*, and the absence of pycnidial paraphyses distinguishes it from *Lasiodiplodia*.

Identification to species level is not as complicated as in other genera of Botryosphaeriaceae because the genus is quite small currently comprising only two species, that is, *Neodeightonia subglobosa* and *N. phoenicum*, which are easily distinguished on the size, shape, and septation of the conidia. Conidia of *N. subglobosa* are spherical to subglobose, aseptate, and smaller than the conidia of *N. phoenicum* which are oval to ellipsoid and have one transverse septum.

15.1.3.2 Molecular Techniques

Keratytis is currently recognized as the most common fungal infection of the eye, with some authors reporting up to 16%–37% of keratitis cases related to a mycotic aetiology.[19,20] Early, rapid, and accurate identification of the fungal species is important in order to guide the selection of appropriate antifungal therapy and is essential for an effective treatment of ocular infections such as fungal keratytis.[19–22]

Diagnosis of fungal infections depends on recovery of fungi from culture of clinical specimens, and their identification requires the presence of reproductive structures. Clinical diagnosis of these ocular infections is confirmed by obtaining intraocular (aqueous or vitreous) specimens or corneal scrapings. However, standard microbiological tests (Gram and Giemsa stains and culture) are positive in only 50%–80% of keratitis cases.[19,23,24] Even in the cases where culture tests are positive an identification of the fungal species can be difficult to obtain. The culture-based phenotypic methods can be time consuming and laborious, may initially be nonspecific, and require considerable expertise for correct morphological identification of less common or unusual fungi. Additional drawbacks of conventional morphological identification are the inability of some cultures to sporulate, or cultures exhibiting atypical morphology resulting in the impossibility to identify the species or even in misidentifications.[25–27] In fact, the occurrence of non-sporulating moulds seems frequent in cultures derived from ocular infections.[25,27]

For *Neodeightonia* isolation and morphological identification of cultures from clinical specimens has been successful in the single case reported to date.[7] Nevertheless, one cannot discard the chance that in some cases retrieving cultures will not be possible and some cultures/strains may not be able to sporulate thus hindering the identification. Also, even if successful isolation is accomplished the sporulation process (development of conidiomata and conidia) may take 2–3 weeks to occur,[11] a time frame that is not clinically useful.

Molecular methods for identification of pathogenic fungi have been validated for use in clinical settings in order to overcome the shortcomings of traditional morphological identification mentioned previously. The field of medical mycology has definitively embraced molecular methods of identification/detection of fungi and these methods are being increasingly applied.[19–29] Molecular methods are rapid with a turnaround time of about 24 h from the time of DNA extraction, yield results that are objective with data portable between laboratories, and despite the higher cost could be more economical in the long run.[19–29] These methods, especially the polymerase chain reaction (PCR)-based ones, can be applied to cultures for which the phenotypic approach has been found to be ineffective, and offer the additional advantage of eliminating the need for isolation of cultures as they can be applied directly to clinical samples.[19–27]

A myriad of DNA-based identification methods is available and these are being increasingly employed in clinical laboratories. These nonsequence-based molecular methods include the commercially available GenProbe assay, methods based on the polymerase chain reaction such as single-step PCR, RAPD-PCR, rep-PCR, nested PCR, PCR-RFLP, PCR-EIA, and more recent microassay-based, Luminex technology-based, and real-time PCR-based methods. Great variation in assay complexity, targets, and detection methods can be found, and many of these methods have not been widely used or rigorously validated.[29] Of the above-mentioned methods rep-PCR and MSP-PCR have proven useful for differentiating *Neodeightonia* species,[30] but their applicability in clinical settings has not been evaluated.

At present, the "gold standard" for fungal species identification is DNA sequence analysis. This method is based on PCR amplification of a selected region of genomic DNA, followed by sequencing of the resulting amplicon. The obtained sequence can then be used to perform searches against a nucleotide sequence database library.[20,22,25,28]

The success of this strategy depends largely on the choice of the appropriate locus. The gene target must be orthologous, have a high level of interspecific variability combined with low levels of intraspecific variability, and should not undergo recombination. Additionally, it must be easy to amplify and sequence and the amplicons should be within the size range obtainable with the most commonly used automated DNA sequencers (about 600–800 bp).[14,28,29]

The ITS region (noncoding sequence interspaced among highly conserved fungal rDNA genes) complies with most of these requirements since this region can be reliably amplified for most fungi, is conserved, is present as multiple copies in the fungal genome, yields sufficient taxonomic resolution for most fungi, and has the additional advantage that the GenBank (http://www.ncbi.nlm.nih.gov), European Molecular Biology Laboratory nucleotide sequence database (http://www.ebi.ac.uk/embl/), and DNA Data Bank of Japan (http://www.ddbj.nig.ac.jp/) contain a large number of sequences from this locus, enabling a ready comparison of the sequence from an unknown isolate.[14,28,29]

There is considerable consensus regarding the use of ITS sequencing in the identification of fungi. Also, the International Subcommission on Fungal Barcoding has proposed the ITS region as the prime fungal barcode or the default region for species identification. The main disadvantages regarding the use of ITS is the inability of this region for differentiation of species complexes and failure to distinguish between closely related species or cryptic species.[14,28,29] In these cases, it may be necessary to use more variable gene regions. The locus to be used will depend on the fungal group in question. For example, in the Botryosphaeriaceae when the ITS region fails to differentiate species complexes or cryptic species the locus of choice has been the elongation factor 1-alpha which has shown successful results.[31]

Regarding *Neodeightonia* the molecular identification of species can be based solely on the ITS sequence. ITS sequences of *N. subglobosa* and *N. phoenicum* are sufficiently different from one another to account for a reliable identification at species level. In the course of recent studies dealing with the genus *Neodeightonia* and other genera in Botryosphaeriaceae[10,11] sequences of the 18S and 28S rDNA (D1/D2 variable region) and ITS region, as well as the elongation factor 1-alpha and beta-tubulin became available for authentic and ex-type cultures of *N. subglobosa* and *N. phoenicum*. This fact is highly relevant since the success of the DNA sequence-based identification depends on the reliability of sequences deposited in reference databases.

PCR-based culture independent methods have been developed and applied for the detection of fungi causing keratomycosis directly on ocular specimens.[21,24] Also, through the use of species specific oligonucleotide primers or probes this approach has been used to detect a particular species or group of species without the need for DNA sequencing.[32] To date, no specific primers or oligonucleotide probes have been developed for species of *Neodeightonia* and thus if molecular detection methods are to be applied to *Neodeightonia* species these will have to depend (for now) on the PCR amplification and sequencing of the ITS region.

15.2 METHODS

15.2.1 Sample Preparation

The first and probably the most important step in any molecular detection method is the isolation of DNA from the fungal culture or directly from the clinical specimen. Although for PCR-based methods the amount of DNA required is very small it is important to use DNA extraction procedures that guarantee a sufficient quantity of DNA and simultaneously that it is not contaminated with PCR inhibitory agents. A good DNA extraction method is crucial for PCR detection to avoid the possibilities of false negative results. Also, strict precautions must be taken throughout the whole procedure, from collection and processing of samples to DNA extraction and PCR amplification in order to avoid contaminations and consequently false positive results.

Clinical diagnosis of keratomycosis is confirmed by obtaining intraocular (aqueous or vitreous) specimens or corneal scrapings.[19–21,24,25] These samples can be used to inoculate culture media in order to isolate the fungus responsible for the infection. As soon as colonies start to develop DNA can be extracted by a number of existing methods, including commercial kits which are usually faster than other approaches.[16,19,20,24–27,32] In a faster approach the DNA can be obtained directly from the ocular samples via the same DNA extraction procedures used for fungal cultures or others developed specifically for human tissue samples.[20,22,32] By eliminating the isolation and cultivation procedure this last approach greatly reduces the time necessary for the molecular diagnosis of the causal agent.

Procedure

1. Inoculate cultures onto potato dextrose agar (Difco) and incubate at 25°C. After 1 week the colony is scraped from the agar surface, frozen in liquid nitrogen and ground in a porcelain mortar. As an alternative, cultures can be grown in potato dextrose broth.
2. Genomic DNA can be extracted from the ground mycelium using any DNA isolation protocol. In our lab we have good results in terms of yield and purity of DNA samples following the method described by Alves et al. (2004).[16]

15.2.2 Detection Procedures

Once DNA is obtained it is necessary to perform a PCR amplification of the ITS region. The ITS can be amplified easily using "universal" fungal primers in the vast majority of fungi. Our experience shows that with *Neodeightonia* (and Botryospheriaceae in general) the primer sets ITS1/ITS4 or ITS5/ITS4[33] result in good amplification of the target region resulting in amplicons with a size of about 500–600 bp.[11,16,31] As an alternative the ITS1 forward primer can be combined with the primer NL4[34] which anneals within the 28S rDNA gene resulting in a larger amplicon (approx. 1200 bp) that includes the ITS region plus the D1/D2 variable region of the 28S rDNA gene.[35] The PCR mixtures and amplification conditions have been described elsewhere.[11,16,35] In some cases where amplification is weak or unsuccessful, good results are usually obtained by performing a second PCR using as template the first PCR amplification.[36] Another approach involving a nested or semi-nested PCR can be performed. Thus, for example, an initial amplification would be performed using the primer set ITS1/NL4 and in the second PCR the primer set used would be ITS1/ITS4. This kind of approach is frequently used and is known to increase the sensitivity of the PCR detection.[19] Finally, from our experience, the addition of 5% DMSO to the PCR amplification reaction is useful to help the amplification of some difficult templates and also to increase reproducibility.[11,31,35,36]

The nucleotide sequence of the ITS region is then obtained using standard automated DNA sequencing procedures. The complete sequences of the ITS region are read and edited using available software such as Chromas 1.45 (http://www.technelysium.com.au/chromas.html), FinchTV 1.4.0 (http://www.geospiza.com/products/finchtv.shtml), or others. The sequences must be checked manually and nucleotide arrangements at ambiguous positions clarified using both primer direction sequences. The final step in the identification procedure consists in performing a sequence similarity search of the retrieved sequence against nucleotide sequences database such as GenBank using the BLAST tool (http://www.ncbi.nlm.nih.gov/BLAST). The outcome of this similarity search must be interpreted with caution due to the well known problems with the reliability of the ITS sequences deposited in the reference databases (e.g., GenBank/EMBL/DDBJ).[28,37] Errors in fungal sequences within GenBank, the most widely used database, have been found to be as high as 20%.[37] However, in the case of *Neodeightonia* this task is facilitated since sequences from authentic, ex-type and well-characterized species are available in GenBank.

Procedure

1. Prepare a PCR mixture (50 μL) containing 1× PCR buffer with $(NH_4)_2SO_4$ (MBI Fermentas, Vilnius, Lithuania), 3 mM $MgCl_2$, 200 mM of each nucleotide, 15 pmol of primer ITS1, 15 pmol of primer ITS4, 1 U of *Taq* DNA polymerase (MBI Fermentas), and 50 ng of template DNA.
   Note: The forward primer ITS5 can be used as an alternative to primer ITS1.
2. Perform the PCR amplification in a thermocycler using the following conditions: initial denaturation of 3 min at 95°C, followed by 30 cycles of 30 s at 94°C, 30 s at 50°C, and 1 min at 72°C, and a final extension period of 10 min at 72°C.
3. Purify the PCR amplicons before sequencing. In our lab we have good results with the Jet Quick PCR Product Purification Spin Kit (Genomed, Löhne, Germany) but virtually any PCR product purification kit can be used.
4. Prepare cycle sequencing reactions (20 μL) containing 30–90 ng DNA template, 3.2 pmol of primer, 4 μL BigDye? Terminator from the ABI PRISM? BigDye, and Terminator Cycle Sequencing Ready Reaction Kit with AmpliTAQ DNA Polymerase (PE Applied Biosystems).
5. Remove excess dye terminator by mixing the reaction with 2 μL of 3 M sodium acetate (pH 4.6) and 50 μL of 100% ethanol. Precipitate for 20 min at room temperature and centrifuge for 20 min at 16,000× *g*. Discard the supernatant and wash the pellet with 250 μL of 70% ethanol. Dry the pellet in a heat block at 90°C for 1 min.
6. Sequence in an automated DNA sequencer such as the ABI PRISM 310 Genetic Analyzer (PE Applied Biosystems).

15.3 CONCLUSION AND FUTURE PERSPECTIVES

The incidence and prevalence of invasive mycoses due to opportunistic fungal pathogens has increased significantly over the past 2 decades. The field of medical mycology has become an extremely challenging study of infections caused by a wide and taxonomically diverse array of opportunistic fungi. These opportunistic mycoses pose considerable diagnostic and therapeutic challenges and a rapid, accurate diagnosis is essential for the initiation of targeted antifungal therapy.

Given the general lack of knowledge about *N. subglobosa* in clinical settings and the aforementioned problems with traditional morphological identification of fungi it is possible that infections caused by this species may have previously passed unnoticed due to inadequate diagnosis. The clinical significance of *N. subglobosa* is poorly understood. However, as emphasized by Kirkness et al. (1991)[7] this species should be considered in ocular infections as a result of injuries caused by plant materials. This fungus should be particularly considered in tropical and subtropical countries where incidence of keratomycosis is higher than in temperate regions.[25]

Although *Neodeightonia* species are not especially difficult to identify based on morphological characters, they require considerable expertise of the taxonomy of the Botryosphaeriaceae. Because this group of fungi is not frequent in clinical settings this may constitute a problem when trying to obtain a correct identification. In this respect, molecular methods have the advantage that no taxonomic expertise is required. Moreover, the molecular detection procedure can be performed directly on the clinical specimen without the need to isolate the fungus, which decreases the time needed for the diagnosis and consequently for the initiation of proper antifungal therapy.

REFERENCES

1. Slippers, B. and Wingfield, M.J., Botryosphaeriaceae as endophytes and latent pathogens of woody plants: Diversity, ecology and impact, *Fungal Biol. Rev.*, 21, 90, 2007.
2. Rebell, G. and Forster, R.K., *Lasiodiplodia theobromae* as a cause of keratomycoses, *Sabouraudia*, 14, 155, 1976.
3. Summerbell, R.C. et al., Subcutaneous phaeohyphomycosis caused by *Lasiodiplodia theobromae* and successfully treated surgically, *Med. Mycol.*, 42, 543, 2004.
4. Tan, D.H. et al., Disseminated fungal infection in a renal transplant recipient involving *Macrophomina phaseolina* and *Scytalidium dimidiatum*: Case report and review of taxonomic changes among medically important members of the Botryosphaeriaceae, *Med. Mycol.*, 46, 285, 2008.
5. Srinivasan, A. et al., Cutaneous infection caused by *Macrophomina phaseolina* in a child with acute myeloid leukemia, *J. Clin. Microbiol.*, 47, 1969, 2009.
6. Elinav, H. et al., Invasive *Scytalidium dimidiatum* infection in an immunocompetent adult, *J. Clin. Microbiol.*, 47, 1259, 2009.
7. Kirkness, C.M. et al., *Sphaeropsis subglobosa* keratomycosis—First reported case, *Cornea*, 10, 85, 1991.
8. Punithalingam, E., Studies on Sphaeropsidales in culture, *Mycol. Pap.*, 119, 1, 1969.
9. von Arx, J.A. and Müller, E., A re-evaluation of the bitunicate ascomycetes with keys to families and genera, *Stud. Mycol.*, 9, 1, 1975.
10. Crous, P.W. et al., Phylogenetic lineages in the Botryosphaeriaceae, *Stud. Mycol.*, 55, 235, 2006.

11. Phillips, A.J.L. et al., Resolving the phylogenetic and taxonomic status of dark-spored teleomorph genera in the Botryosphaeriaceae, *Persoonia*, 21, 29, 2008.
12. Krishnamurthy, Y.L., Naik, S.B., and Jayaram, S., Fungal communities in herbaceous medicinal plants from the Malnad Region, Southern India, *Microb. Environ.*, 23, 24, 2008.
13. Sutton, D.A., Coelomycetous fungi in human disease. A review: Clinical entities, pathogenesis, identification and therapy, *Rev. Iberoam. Micol.*, 16, 171, 1999.
14. Guarro, J., Gené, J., and Stchigel, A.M., Developments in fungal taxonomy, *Clin. Microbiol. Rev.*, 12, 454, 1999.
15. Sutton, B.C., *The Coelomycetes*, Kew, England: CMI, 1980, 696 pp.
16. Alves, A. et al., *Botryosphaeria corticola* sp. nov. on *Quercus* species, with notes and description of *Botryosphaeria stevensii* and its anamorph *Diplodia mutila*, *Mycologia*, 96, 598, 2004.
17. Denman, S. et al., An overview of the taxonomic history of *Botryosphaeria*, and a re-evaluation of its anamorphs based on morphology and ITS rDNA phylogeny, *Stud. Mycol.*, 45, 129, 2000.
18. Jacobs, K.A. and Rehner, S.A., Comparison of cultural and morphological characters and ITS sequences in anamorphs of *Botryosphaeria* and related taxa, *Mycologia*, 90, 601, 1998.
19. Ferrer, C. et al., Detection and identification of fungal pathogens by PCR and by ITS2 and 5.8S ribosomal DNA typing in ocular infections, *J. Clin. Microbiol.*, 39, 2873, 2001.
20. Mancini, N. et al., Rapid molecular identification of fungal pathogens in corneal samples from suspected keratomycosis cases, *J. Med. Microbiol.*, 55, 1505, 2006.
21. Gaudio, P.A. et al., Polymerase chain reaction based detection of fungi in infected corneas, *Br. J. Ophthalmol.*, 86, 755, 2002.
22. Lau, A. et al., Development and clinical application of a panfungal PCR assay to detect and identify fungal DNA in tissue specimens, *J. Clin. Microbiol.*, 45, 380, 2007.
23. Ferrer, C. et al., Polymerase chain reaction diagnosis in fungal keratitis caused by *Alternaria alternata*, *Am. J. Ophthalmol.*, 133, 398, 2002.
24. Kumar, M. and Shukla, P.K., Use of PCR targeting of internal transcribed spacer regions and single-stranded conformation polymorphism analysis of sequence variation in different regions of rRNA genes in fungi for rapid diagnosis of mycotic keratitis, *J. Clin. Microbiol.*, 43, 662, 2005.
25. Bagyalakshmi, R. et al., Newer emerging pathogens of ocular non-sporulating molds (NSM) identified by polymerase chain reaction (PCR)-based DNA sequencing technique targeting internal transcribed spacer (ITS) region, *Curr. Eye Res.*, 33, 139, 2008.
26. Rakeman, J.L. et al., Multilocus DNA sequence comparisons rapidly identify pathogenic molds, *J. Clin. Microbiol.*, 43, 3324, 2005.
27. Pounder, J.I. et al., Discovering potential pathogens among fungi identified as nonsporulating molds, *J. Clin. Microbiol.*, 45, 568, 2007.
28. Balajee, S.A. et al., Sequence-based identification of *Aspergillus*, *Fusarium*, and *Mucorales* species in the clinical mycology laboratory: Where are we and where should we go from here?, *J. Clin. Microbiol.*, 47, 877, 2009.
29. Balajee, S.A., Sigler, L., and Brandt, M.E., DNA and the classical way: Identification of medically important molds in the 21st century, *Med. Mycol.*, 45, 475, 2007.
30. Alves, A. et al., Rapid differentiation of species of Botryosphaeriaceae by PCR fingerprinting, *Res. Microbiol.*, 158, 112, 2007.
31. Alves, A. et al., Morphological and molecular data reveal cryptic speciation in *Lasiodiplodia theobromae*, *Fungal Divers.*, 28, 1, 2008.
32. Jaeger, E.E.M. et al., Rapid detection and identification of *Candida*, *Aspergillus*, and *Fusarium* species in ocular samples using nested PCR, *J. Clin. Microbiol.*, 38, 2902, 2000.
33. White, T.J. et al., Amplification and direct sequencing of fungal ribosomal RNA genes for phylogenetics. In Innis, M.A. et al. (eds.), *PCR Protocols: A Guide to Methods and Applications*, pp. 315–322. Academic Press, San Diego, CA, 1990.
34. O'Donnell, K., *Fusarium* and its near relatives. In Reynolds, D.R. and Taylor, J.W. (eds.), *The Fungal Holomorph: Mitotic, Meiotic and Pleomorphic Speciation in Fungal Systematics*, pp. 225–233. CAB International, Wallingford, U.K., 1993.
35. Alves, A. et al., Evaluation of amplified ribosomal DNA restriction analysis (ARDRA) as a method for the identification of *Botryosphaeria* species, *FEMS Microbiol. Lett.*, 245, 221, 2005.
36. Alves, A., Phillips, A.J.L., and Correia, A., Multi-gene genealogies and morphological data support *Diplodia cupressi* sp. nov., previously recognized as *D. pinea* f. sp. *cupressi*, as a distinct species, *Fungal Divers.*, 23, 1, 2006.
37. Nilsson, R.H. et al., Taxonomic reliability of DNA sequences in public sequence databases: A fungal perspective, *PLoS One*, 1, e59, 2006.

16 *Phoma* and *Phomopsis*

Dongyou Liu

CONTENTS

16.1 INTRODUCTION

As cosmopolitan plant-infecting pathogens or free-living saprobes, dematiaceous fungi have the capacity to take advantage of host's temporary weakness (e.g., injury and immunosuppression), and establish infections in otherwise refractory hosts. This is evidenced by the observations that along with a growing number of immunocompromized, cancer and post-operative patients in recent decades, there has been a significant increase in the incidence of unusual fungal infections. Species of *Phoma* and *Phomopsis* represent the examples of ubiquitous and plant-infective fungal organisms that have emerged as unusual pathogens in weakened human hosts, resulting in notable morbidity and mortality.

16.1.1 Classification and Morphology

16.1.1.1 The Genus *Phoma*

The genus *Phoma* is a dematiaceous, filamentous fungus belonging to the family Pleosporaceae, order Pleosporales, class Dothideomycetes, subphylum Pezizomycotina, phylum Ascomycota, kingdom Fungi. The genus *Phoma* (obsolete synonyms: *Deuterophoma*, *Leptophoma*, *Peyronellaea*, *Plenodomus*, *Polyopeus*, and *Sclerophomella*) covers over 2000 species, which produce spores of <15 μm in size as the larger spored forms are placed in the genus *Macrophoma*. The teleomorphs of the genus have been described in the genera *Didymella*, *Leptosphaeria*, *Pleospora*, and *Mycosphaerella* [1–3].

Phoma spp. are cosmopolitan in nature. Some *Phoma* spp. are saprophytes on various plants (known as plurivorous fungi, which are generally saprobic or weakly parasitic, mainly from temperate regions in Eurasia); while others are pathogens to cultivated plants and humans. For example, *Phoma beta* causes heart rot and blight in beets, and *Phoma batata* causes dry rot in sweet potato. The human pathogenic *Phoma* spp. comprise *Phoma cruris-hominis*, *Phoma denissii* var. *oculo-hominis*, *Phoma eupyrena*, *Phoma glomerata*, *Phoma herbarum*, *Phoma hibernica*, *Phoma minutella*, *Phoma minutispora*, and *Phoma sorghina*. These organisms are responsible for occasional infections in humans, often after trauma and immunosuppression, leading to cutaneous, subcutaneous, corneal, or systemic phaeohyphomycosis.

Phoma colonies grow rapidly, and are flat, spreading, grayish-brown, powdery to velvety (suede-like), and often largely submerged in the medium. From the front, colonies are initially white and later become olive grey with an occasional tint of pink. From the reverse, they are dark brown to black. Some species (e.g., *Phoma cruris-hominis* and *Phoma herbarum*) produce a reddish-purple to yellowish-brown diffusable pigment which is readily visible from the reverse. *Phoma* hyphae are septate, hyaline to brown. Pycnidia (70–100 μm in diameter) are large, round to pyriform (globose), asexual fruiting bodies. Further, pycnidia are membranous to leathery, dark in color and bear phialides at their inner lining, with one to several openings (ostioles) on their surface from which the conidia are released outside. Pycnidia may be depressed in the tissues of the host. Conidia are globose to cylindrical, unicellular, hyaline (colorless), and are produced in abundance within the pycnidia on narrow thread-like phialides, which are not easily differentiated from the inner pycnidial wall cells. Conida are usually extruded in slimy masses from the apical ostiole. Some *Phoma* species produce brown chlamydospores arranged singly or in chains, and may be unicellular or multicellular and "alternarioid" (resembling *Alternaria*) in appearance [4].

16.1.1.2 The Genus *Phomopsis*

The genus *Phomopsis* (formerly *Phoma* subgen. *Phomopsis*, and *Myxolibertella*) belongs to the mitosporic Valsaceae group, family Valsaceae, order Diaporthales, subclass Sordariomycetidae, class Sordariomycetes, subphylum Pezizomycotina, phylum Ascomycota, kingdom Fungi. The genus *Phomopsis* contains >400 described species, and the teleomorphs of the genus *Phomopsis* are identified in the genus *Diaporthe*.

Phomopsis spp. are plant pathogens, causing spots on various plant parts. For example, *Phomopsis juniperovora* infects junipers, representing an important pest of seedlings and juvenile plants in the nursery industry. *Phomopsis leptostromiformis* is infective to lupin plants, and produces the potent mycotoxins phomopsin A and phomopsin B, which are toxic to various animals following ingestion, resulting in a form of mycotoxicosis called lupinosis. *Phomopsis vexans* causes disease in aubergine, brinjal, eggplant, leading to poor seed germination and damping-off of seedlings to leaf and stem lesions and to fruit rot, both in the field and after harvest. *Phomopsis viticola* causes a plant disease called dead-arm, forming lesions on shoots, leaves, and rachises, and causing fruit rot.

Morphologically, *Phomopsis* mycelium is immersed, branched, septate, hyaline to pale brown. Conidiomata are brown to dark brown, thick-walled, immersed, stromatic (true stroma—host tissue incorporated into the pycnidium), separate or aggregated and confluent, globose, ampulliform or applanate (flattened), unilocular or multilocular (single or several cavities), with one or several circular papillate ostioles. Conidiophores are branched and septate at the base, occasionally short, more frequently multiseptate and filiform, hyaline, formed from inner cells of the locular walls. Conidiogenous cells are enteroblastic, phialidic, determinate, integrated, rarely discrete, hyaline, cylindrical, aperature apical on long or short lateral and main branches of the conidiophores, collarette, channel and periclinal thickening minute. *Phomopsis* spp. produce both alpha conidia (hyaline, fusiform, straight, usually biguttulate, and >5 μm in length) and beta conidia (hyaline, filiform to hamate or hooked, egutullate or without guttules, aseptate, usually >15 μm in length) [5].

Diaporthe vexans, the teleomorph of *Phomopsis vexans*, produces pycnidia (100–300 μm in diameter) with subepidermal, erumpent, dark, thick-walled, flattened to globose appearance and with or without a beak (to 76 μm). Phialides (10–16 μm in length) are hyaline, simple or branched, sometimes septate, arising from the innermost layer of cells lining the cavity. Alpha conidia (5–8 × 2–3 μm) are hyaline, aseptate, and sub-cylindrical. Beta conidia (18–32 μm × 0.5–2.0 μm) are filiform, curved, hyaline, septate, and non-germinating. Hyphae (2.5–4 μm wide) are hyaline, and septate. Perithecia (130–350 μm in diameter) occur in clusters in culture, with sinuous, carbonaceous, irregular beaks (80–500 μm in length). Asci (24–44 μm × 5–12 μm) are clavate, sessile, eight-spored. Ascospores (9–12 μm × 3–4.5 μm) are biseriate, hyaline, narrowly ellipsoid to bluntly fusoid, one-septate, constricted at the septum.

16.1.2 Clinical Features and Epidemiology

Members of the genera *Phoma* and *Phomopsis* are widely distributed in soil and plants, many of which are plant pathogens. Occasionally these organisms may be introduced to human hosts via traumatic implantation, resulting in subcutaneous infections, keratitis, and invasive disease [6–15].

16.1.2.1 *Phoma* Infections

Baker et al. [16] described a case of subcutaneous infection due to *Phoma minutella* on the foot of a farmer undergoing corticosteroid therapy for myasthenia gravis. Microscopic examination of stained smears and biopsy sections demonstrated dematiaceous fungal elements consistent with the mold. Culture of the specimens grew a fungus that was identified as *Phoma minutella*.

Rishi and Font [17] reported a *Phoma*-related nonhealing corneal ulcer with brownish pigmentation in a 72-year-old man. Histopathologic examination of the keratectomy specimen showed round spherules (5–30 μm in diameter) together with septate hyphae at the edges of the perforated cornea. Culture of the specimen yielded a fungus belonging to the genus *Phoma*. Polymerase chain reaction (PCR) amplification of DNA extracted from five (10 μm thick) paraffin-embedded sections using pan-fungal primers generated a single product of 360 bp.

Everett et al. [18] documented a case of subcutaneous *Phoma* infection in an immunosupressed 50-year-old female, who undertook maintenance immunosuppression with azathioprine and prednisone after bilateral nephrectomy, splenectomy, and cadaveric renal transplant for the treatment of end-stage renal disease secondary to familial nephritis. The patient presented with pain, warmth, and swelling in the dorsum of her left wrist with a decreased motion. A synovectomy of the fourth dorsal compartment and debridement of the third extensor tendon were performed. Examination of tissue sections with Grocott-Gomori methenamine-silver staining (GMS) uncovered branching filaments and pycnidia. Culture of the specimen yielded a *Phoma* sp after 2 weeks of incubation. Treatment with amphotericin B resulted in the resolution of her symptomatology and no further recurrence after 2 years of follow-up.

Balis et al. [19] reported a case of aggressive and deeply invasive hand infection due to *Phoma exigua* in a renal transplant recipient with acute myeloid leukaemia and diabetes. Treatment required surgical debridement and the use of prolonged systemic amphotericin B therapy.

16.1.2.2 *Phomopsis* Infections

Sutton et al. [5] described a case of osteomyelitis due to a *Phomopsis* species in a 61-year-old female urban gardener, who was under chronic immunosuppression (prednisone and methotrexate) for the treatment of diabetes and rheumatoid arthritis. The patient developed an erythematous, painful subcutaneous abscess on the distal phalanx of the right fourth finger complicated by osteomyelitis. Osteomyelitis was confirmed radiographically by the presence of lytic bone changes on plain x-rays and abnormally increased radiotracer uptake in the distal phalanx on bone scans. Draining of the abscess produced a large amount of purulent material that contained fungal hyphae. Culture of the specimen grew a mold that was initially white and woolly, but became slightly greenish after 4 weeks at room temperature (25°C). Subcultures of the isolate on potato flakes agar (PFA) were buff and sterile with a few black sclerotia after 4 weeks at 25°C, with chains of chlamydoconidia. Subcultures onto carnation leaf agar produced dark brown to black pycnidia after 2 weeks of incubation at 25°C. Periodic acid-Schiff (PAS) staining of carnation leaf sections containing mature pycnidia demonstrated ostiolate, thick-walled, immersed, multilocular conidiomata (101–106 μm × 196–274 μm). Crushed conidiomata showed short, ellipsoidal alpha conidia (2.0–2.5 μm × 5.8–6.0 μm) and long, filamentous beta conidia (0.4–0.5 μm × 18 to 22 μm). The fungus was identified as a *Phomopsis* species. The patient was treated with a 6-month course of itraconazole. At 16 months of follow-up, she remained free of recurrence.

Mandell and Colby [20] reported a case of *Phomopsis* fungal keratitis in a patient after a rose thorn injury that occurred 2 months earlier while gardening. The patient presented with deep stromal keratitis with extension of hyphae through Descemet's membrane. Culture of the surgical specimens yielded a *Phomopsis* organism. Treatment consisted of therapeutic keratoplasty combined with oral and topical antifungal medications.

16.1.3 Diagnosis

Species of the *Phoma* and *Phomopsis* genera are common plant pathogens that may be occasionally involved in human infections. Diagnosis of rare fungal pathogen infections is challenging and demands a rational approach that takes account of clinical manifestations, histological findings, and mycological observations.

Conventional identification of *Phoma*, *Phompsis* and other fungal organisms (e.g., *Pleurophoma*, *Pleurophomopsis*, *Pyrenochaeta*, *Chaetomium*, *Pseudallescheria*, *Coniothyrium*, *Microsphaeropsis*, and *Pseudochaetosphaeroma*) relies on the detection of distinct colony color, conidia morphology, and chlamydospore structures [21]. However, due to their apparent lack of diagnostic morphological features and their tendency to show varied phenotypic characteristics on culture media, it has been difficult and technically demanding to accurately determine their species identity.

Application of molecular techniques such as PCR and sequence analyses of the 18S rRNA (SSU) and the 28S rRNA (LSU), internal transcribed spacer (ITS), actin, and three-tubulin gene regions enables rapid and precise identification of *Phoma* and *Phomopsis* [2,3].

16.2 METHODS

16.2.1 Sample Preparation

Corneal scrapings for PCR are collected in lysis buffer (50 mM Tris–Cl pH 7.2, 50 mM EDTA, 3% SDS, and 1% β-mercaptoethanol in 400 μL). The specimens are kept at −20°C until further processing. For DNA extraction, tubes are vortexed at moderate speed for 15 s, incubated at 65°C for 1 h, and boiled for 10 min. Equal volume of phenol:chloroform:isoamylalcohol (25:24:1) is added to the tube and then revortexed. The tubes are centrifuged at 14,000 rpm for 15 min. To the aqueous phase, a 1:10 volume of 3 M sodium acetate pH 5.2 and 0.6 volume of isopropanol are added. The tubes are inverted and mixed gently, and DNA is precipitated at −20°C for 1 h. The tube is again centrifuged at 14,000 rpm for 15 min. The supernatant is discarded, and the pellet containing the DNA is washed twice with 70% alcohol. The pellet is air dried and resuspended in 30 μL of autoclaved double-distilled water [22].

16.2.2 Detection Procedures

16.2.2.1 PCR Detection of *Phoma exigua*

Balis et al. [19] described a PCR assay for specific identification of *Phoma exigua* using primers Pe1 (5′-GCTTTGCCTACCATCTCTTACCCA-3′) and Pe2 (5′-GTTATGAGTGCAAAGCGCGAGATG-3′). These primers are selected from the ITS regions 1 and 2 of the published *P. exigua* sequence (GenBank AF268192).

Procedure

1. PCR mixture (50 μL) consists of 6 μL $MgCl_2$ (25 mM), 10 μL Mg-free buffer (100 mM Tris–HCl, 500 mM KCl, and 1% Triton-X), 2.5 U Taq Polymerase (Promega), 0.04 mM of each dNTP, 100 pmol of each primer, and 5 μL of template DNA.
2. Amplification parameters include an initial 3 min at 95°C, 38 cycles of 1 min at 95°C, 1 min at 58°C, 1 min at 72°C, and a final 5 min time at 72°C.

3. The PCR product (410 bp) is separated on 2% agarose gel, stained with ethidium bromide and visualized under UV transillumination.

Note: The specific amplicon may be further verified by sequence analysis.

16.2.2.2 Real-Time PCR Detection of *Phomopsis* spp.

Hietala et al. [23] developed a real-time PCR for detection of *Phomopsis* spp. using primers and probe derived from the ITS rRNA region. The primer pair (forward: 5′-GCACCCAGAAACCCTTTGTG-3′ and reverse: 5′-AAGAGTTGACTTGGCCGCC-3′) amplifies a 116 bp product; the probe (5′-CGGTAACGAGGAGCAGCCCGC-3′) is labeled with FAM at the 5′ end and TAMRA at the 3′ end.

PCR mixture (25 µL) is composed of TaqMan Universal PCR Master Mix (P/N 4304437; Applied Biosystems), 300 nM each primer, 400 nM probe and 3 µL of the DNA.

PCR cycling parameters are 95°C for 10 min followed by 40 cycles of 95°C for 15 s and 60°C for 1 min.

Fluorescence emissions are detected with an ABI Prism 7700 (Applied Biosystems). Data acquisition and analysis are performed with the Sequence Detection System software package (1.7a; Applied Biosystems).

16.2.2.3 Real-Time PCR and Sequencing Analysis of ITS Region

Pounder et al. [24] described a real-time PCR with SYBR green DNA binding dye and amplicon melting temperature analysis for fungal detection also using pan-fungal primers ITS1 forward (5′-TCCGTAGGTGAACCTGCGG-3′) and ITS4 reverse (5′-TCCTCCGCTTATTGATATGC-3′). The identity of the fungi is verified by subsequent sequencing analysis.

Procedure

1. PCR mixture is composed of 1 × Lightcycler FastStart DNA Master Hybridization Probes mixture (Roche Applied Science) (containing deoxynucleoside triphosphates), FastStart *Taq* DNA polymerase, and 1 mM $MgCl_2$ (additional $MgCl_2$ is added to a final concentration of 4.6 mM), 0.4 µM each of ITS1 forward and ITS4 reverse primers, 1× SYBR green (Molecular Probes), and 3 µL template DNA.
2. Thermal cycling parameters include 95°C for 10 min; 50 cycles of 95°C for 5 s, 60°C for 20 s, and 76°C for 30 s; and a final extension at 72°C for 2 min.
3. The quality of the amplicon is determined using the derivative of the melt analysis curve (55°C–99°C, 45 s hold at 55°C, 5 s/°C) using the RotorGene 3000 (Corbett Robotics, Inc.).
4. The amplified product is purified for bidirectional sequencing using ExoSAP-IT (USB Corp.). Five µL of Big Dye Terminator Ready Reaction Mix v. 1.1 (Applied Biosystems) is added to 4 µL of each primer (0.8 pmol/µL) and 3 µL of purified PCR product. Cycle sequencing is performed with a 9700 thermal cycler (ABI), using 25 cycles of 96°C for 10 s, 50°C for 5 s, and 60°C for 4 min. Sequencing reaction products are passed through a Sephadex G-50 fine column to remove unincorporated dye terminators. Purified sequencing reaction products are run on an ABI Prism 3100 Genetic Analyzer with a 50 cm capillary array.
5. Sequences are analyzed with the SmartGene Integrated Database Network software version 3.2.3 vr. SmartGene is a web-based software and database system with reference sequences derived from the National Center for Biological Information (NCBI) GenBank repository.

Note: In case that real time PCR instrument is not available, standard PCR may be performed with ptimers ITS1 and ITS4, and the resulting amplicon is sequenced with the same primers. Sequence-based identifications are defined by percent identity: species, ≥99%; genus, 93%–99%; and inconclusive, ≤93%.

For strains producing discrepant identification between the methods based on phenotypic characteristics and ITS sequence analysis, the D1–D2 region of the large-subunit RNA gene is amplified with primers NL1 (5′-GCATATCAATAAGCGGAGGAAAAG-3′) and NL4 (5′-GGTCCGTGTTTCAAGACGG-3′) and sequenced for species clarification.

16.2.2.4 Sequencing Analysis of 28S rRNA Gene

Vengayil et al. [22] utilized universal primers (forward: 5′-GTG AAA TTG TTG AAA GGG AA-3′ and reverse: 5′-GAC TCC TTG GTC CGT GTT-3′) for amplification of a 28S rRNA gene region for identification of medically important fungi.

PCR mixture (25 µL) is made up of 1× PCR buffer without $MgCl_2$, 10 pmol each of forward and reverse primers, 1.5 U of *Taq* polymerase, 2 mM $MgCl_2$, 200 µM dNTPs, and 5 µL of template DNA.

Amplification is conducted with an initial denaturation at 95°C for 5 min and 50 cycles of 94°C for 1 min, 50°C for 1 min, and 72°C for 5 min in a thermocycler (GeneAmp PCR System 9700; Applied Biosystems).

The amplified product is checked in 1% agarose gel containing 0.5 µg/mL ethidium bromide and visualized in an ultraviolet transilluminator. The nucleotide sequence of the amplicon is analyzed.

16.3 CONCLUSION

Members of the genera *Phoma* and *Phomopsis* are common plant pathogens, some of which have been shown to cause human subcutaneous infections, keratitis, and invasive disease on a number of occasions. Conventional laboratory diagnosis of human infections due to these rare fungal pathogens is time-consuming and challenging, since *Phoma* and

Phomopsis are extremely complex, with hundreds of species being assigned in each of these taxa. This is further exacerbated by the fact that many of these organisms demonstrate few distinct morphological features for accurate identification and determination. Use of PCR and sequence analyses of the 18S rRNA (SSU) and the 28S rRNA (LSU), ITS, actin, and three-tubulin gene regions facilitates improved detection and differentiation of *Phoma* and *Phomopsis* from other fungal organisms.

REFERENCES

1. The UniProt Consortium. Available at http://www.uniprot.org/taxonomy/, accessed on August 1, 2010.
2. Aveskamp MM et al. DNA phylogeny reveals polyphyly of *Phoma* section *Peyronellaea* and multiple taxonomic novelties. *Mycologia* 2009;101:363–382.
3. de Gruyter J et al. Molecular phylogeny of *Phoma* and allied anamorph genera: Towards a reclassification of the *Phoma* complex. *Mycol. Res.* 2009;113:508–519.
4. http://mycology.adelaide.edu.au/, accessed on August 1, 2010.
5. Sutton DA et al. Human phaeohyphomycotic osteomyelitis caused by the coelomycete *Phomopsis saccardo* 1905: Criteria for identification, case history, and therapy. *J. Clin. Microbiol.* 1999;37(3):807–811.
6. Young NA, Kwon-Chung KJ, Freeman J. Subcutaneous abscess caused by *Phoma* sp. resembling *Pyrenochaeta romeroi*: Unique fungal infection occurring in immunosuppressed recipient of renal allograft. *Am. J. Clin. Pathol.* 1973;59:810–816.
7. Shukla N, Agarwal GP, Gupta DK. *Phoma* as a human pathogen. *Mykosen* 27, 255, 1984.
8. Rai MK. *Phoma sorghina* infection in human being. *Mycopathologia* 1989;105(3):167–170.
9. Montel E, Bridge PD, Sutton BC. An integrated approach to *Phoma* systematics. *Mycopathologia* 1991;115(2):89–103.
10. Morris JT et al. Lung mass caused by *Phoma* species. *Infect. Dis. Clin. Pract.* 1995;4:58–59.
11. Hirsh AH, Schiff TA. Subcutaneous phaeohyphomycosis caused by an unusual pathogen: *Phoma* species. *J. Am. Acad. Dermatol.* 1996;34:679–680.
12. Rosen T et al. Cutaneous lesions due to *Pleurophoma (Phoma)* complex. *South Med. J.* 1996;89(4):431–433.
13. Zaitz C et al. Subcutaneous phaeohyphomycosis caused by *Phoma* cava. Report of a case and review of the literature. *Rev. Inst. Med. Trop. Sao Paulo* 1997;39(1):43–48.
14. Oh CK et al. Subcutaneous pheohyphomycosis caused by *Phoma* species. *Int. J. Dermatol.* 1999;38(11):874–876.
15. Suh MK. Phaeohyphomycosis in Korea. *Nippon Ishinkin Gakkai Zasshi* 2005;46(2):67–70.
16. Baker J et al. First report of subcutaneous phaeohyphomycosis of the foot caused by *Phoma minutella*. *J. Clin. Microbiol.* 1987;25:2395–2397.
17. Rishi K, Font RL. Keratitis caused by an unusual fungus, *Phoma* species. *Cornea* 2003;22(2):166–168.
18. Everett JE et al. A deeply invasive *Phoma* species infection in a renal transplant recipient. *Transplant. Proc.* 2003;35:1387–1389.
19. Balis E et al. Lung mass caused by *Phoma exigua*. *Scand. J. Infect. Dis.* 2006;38:552–555.
20. Mandell KJ, Colby KA. Penetrating keratoplasty for invasive fungal keratitis resulting from a thorn injury involving Phomopsis species. *Cornea* 28, 1167, 2009.
21. Dooley DP et al. Phaeohyphomycotic cutaneous disease caused by *Pleurophoma* in a cardiac transplant patient. *J. Infect. Dis.* 159, 503, 1989.
22. Vengayil S et al. Polymerase chain reaction-guided diagnosis of mycotic keratitis: A prospective evaluation of its efficacy and limitations. *Invest. Ophthalmol. Vis. Sci.* 50, 152, 2009.
23. Hietala AM, Solheim H, Fossdal CG. Real-time PCR-based monitoring of DNA pools in the tri-trophic interaction between Norway spruce, the rust *Thekopsora areolata*, and an opportunistic ascomycetous *Phomopsis* sp. *Phytopathology* 98, 51, 2008.
24. Pounder JI et al. Discovering potential pathogens among fungi identified as nonsporulating molds. *J. Clin. Microbiol.* 45, 568, 2007.

17 *Piedraia*

Dongyou Liu

CONTENTS

17.1 INTRODUCTION

Clinical presentations of fungal infections in humans generally fall under the following four categories: (i) superficial, (ii) cutaneous, (iii) subcutaneous, and (iv) systemic mycoses. Superficial and cutaneous fungal infections are often restricted to the stratum corneum and its adnexal structures, with little or no tissue involvement.

Black piedra (piedra, meaning "stone" in Spanish) is a superficial fungal infection of humans caused by a dematiaceous fungus called *Piedraia hortae*. This infection typically affects the frontal scalp hair and facial hair, producing darkly pigmented hair-shaft nodules as large as a few millimeters in diameter.

Black piedra differs from white piedra (trichosporosis) in that white piedra is caused by basidiomycetous arthroconidial yeast *Trichosporon* spp., producing lightly pigmented, white to light-brown, loosely attached hair-shaft nodules with a soft texture on the scalp hair, pubic hair, axillary hair, beards, moustaches, eyebrows, and eyelashes. In addition, black piedra is found mainly in hot and humid tropical climate, whereas white piedra is prevalent in temperate and semitropical climates [1,2].

17.1.1 Classification and Morphology

The genus *Piedraia* is a dematiaceous (dark-walled) mold belonging to the family *Piedraiaceae*, order Capnodiales, subclass Dothideomycetidae, class Dothideomycetes, subphylum Pezizomycotina, phylum Ascomycota, kingdom Fungi. Being the only member in the family *Piedraiaceae*, the genus *Piedraia* consists of two species: *Piedraia hortae* and *Piedraia quintanilhae* [2]. While *Piedraia hortae* is a keratinolytic fungus, causing black piedra in man, *Piedraia quintanilhae* has been isolated from chimpanzees in Central Africa, and its pathogenicity in humans is unknown.

Piedraia hortae colonies are slow growing, small, folded, velvety, dark brown to black in color; reverse is black. Colonies may produce a reddish brown diffusible pigment and remain glabrous or covered with short aerial hyphae. Hyphae are septate, darkly pigmented, with intercalary chlamydoconidium-like cells. Ascostromata are pseudoparenchymatous structures of subglobose to irregular shape and black in color, with each ascostromata usually containing a single ascus. Asci (with readily dissolvable walls) are ellipsoid, solitary, or in clusters and contain eight ascospores. Ascospores (30–45 μm × 5.5–10 μm) are hyaline to darkly pigmented, one celled (aseptate), fusoid, curved, and tapering toward both ends to form the typical whip-like appendages. *P. hortae* is one of only a few pathogenic human fungi that generate sexual spores in its parasitic phase [3,4]. *Piedraia quintanilhae* differs from *Piedraia hortae* morphologically in that its ascospores do not have appendages [5].

17.1.2 Clinical Features and Pathogenesis

Piedraia is a filamentous fungus commonly present in soil as well as stagnant water and crops in tropical regions of the world. *Piedraia hortae*, the etiologic agent of black piedra, has been isolated in tropical South and Central Americas, Southeast Asia, and Africa. The hot humid environment together with the habit of using plant oil on hair facilitates *Piedraia hortae* growth [6–8].

Superficial infection of hair shafts by *Piedraia hortae* results in the formation of asymptomatic, brown to black, stone-like concretions (nodules) of up to a few millimeters in diameter on the scalp (frontal, occipital, and parietal) and facial hair but not beard, moustache, and pubic hair (commonly known as black piedra). Very firmly attached to the hair shaft, the nodules are composed of ascostromata, which are the fruiting body of the fungus containing subglobose asci and aseptate ascospores in groups of eight covered with a gelatinous sheath. The periphery of the nodule has regularly aligned hyphae and arthroconidia. The hard, darkly pigmented hair-shaft nodules of black piedra have a gritty feeling and may produce a metallic sound when hair is brushed. They rarely produce hair breakage in severe cases.

Coimbra et al. [9] conducted an epidemiological survey of black piedra among Zoró Indians in the Brazilian Amazon, who use plant oil on hair as a custom, and showed that 74 (56.9%) of the 130 individuals had the infection. Gip [10] documented a case of black piedra in a 23-year-old Swedish man after his return from 4 months' stay in India. The patient presented with typical clinical signs of black piedra of his scalp. Black nodules were found around the hair shafts, and the crushed nodules revealed numerous asci and ascospores on microscopy. *Piedraia hortae* was isolated from the concretions. After treatment with oral terbinafine 250 mg daily for 6 weeks, nodules disappeared.

Upon examination of the ultrastructural pattern of human hair infection by *Piedraia hortae* in vivo, Figueras et al. [11] demonstrated that the fungus has the capacity to destroy the cuticular layers of the hair and to penetrate deeply into the cortex. The main reasons that guarantee the long survival of the fungus, and therefore the chronic course of the disease, are possibly the slow rate of keratin degradation at the cortex together with the compacted stromatic organization of the nodules.

17.1.3 Diagnosis

Besides *Piedraia hortae*, a number of other fungi are capable of infecting human scalp and facial hair. Chief among them are white piedra (trichoporosis)-causing basidiomycetous arthroconidial yeast *Trichosporon* spp. (of which *Trichosporon ovoides*, *Trichosporon inkin*, *Trichosporon mucoides*, and *Trichosporon asahii* are linked to white piedra, with *T. inkin* likely on pubic hair and *T. ovoides* on head hair) and tinea capitis-causing dermatophytes *Trichophyton tonsurans*, *Microsporum audouinii*, and *Microsporum canis* [12]. Occasionally, *Acremonium* spp., *Brevibacterium* spp., and coryneform bacteria may also be involved in genital white piedra [13]. Other differential consideration is hair lice (pediculosis capitis).

Clinically, black piedra due to *Piedraia hortae* generates darkly pigmented, firmly attached hair-shaft nodules (concretions) on the scalp and facial hair, whereas white piedra due to *Trichosporon* spp. induces lightly pigmented, white to light-brown, loosely attached hair-shaft nodules (concretions) on the scalp hair as well as hair in other locations (including pubic hair, axillary hair, beards, moustaches, eyebrows, and eyelashes) [11,14]. The nodules of black piedra consist of ascostromata (fruiting body) with subglobose asci and aseptate ascospores in groups of eight inside the hyphae and arthroconidia in the periphery; the nodules of white piedra have darkly stained hyphae, blastoconidia (2–8 μm long), and arthroconidia fixed to the hair shaft [15,16]. Further, black piedra is distributed mainly in hot and humid tropical climate, whereas white piedra is endemic in temperate and semitropical climates [2,3,17].

On the other hand, tinea capitis due to dermatophyte fungi *Trichophyton tonsurans*, *Microsporum audouinii*, and *M. canis* does not form nodules on hair shaft. These organisms affect the base of the hair shaft and the follicle. *Trichophyton tonsurans* grows inside hair shaft (endothrix-growing), with arthroconidia forming within hair shaft; *Microsporum audouinii* and *Microsporum canis* grow outside hair shaft (ectothrix-growing), with numerous arthroconidia surrounding hair shaft and destroying its cuticle.

Direct microscopic examination of infected hair in 10% KOH allows a clear differential diagnosis of black and white piedra, as well as eggs of pediculosis. White and black piedra can also be distinguished from the color of the nodules (i.e., white to light brown, pale greenish or yellowish in white piedra and black in black piedra) and by the presence of ascospores and asci in black piedra and its absence in white piedra.

In culture, *Piedraia hortae* grows slowly at 25°C on Sabouraud's dextrose agar. As *Piedraia hortae* is uninhibited by cycloheximide, it will grow in dermatophyte test media (DTM) incorporating cycloheximide (e.g., Mycosel, DTM). *Trichosporon* spp. grow well on Sabouraud's dextrose agar at 28°C–30°C. However, they are inhibited by cycloheximide and will not grow in DTM containing cycloheximide. Because *P. hortae* rarely produces ascospores on Sabouraud agar, special techniques using transplantation and biotine may be utilized to stimulate its formation of ascospores.

Piedraia hortae and other hair invading fungal organisms can be identified in a much rapid and precise manner through polymerase chain reaction (PCR) and sequencing analysis of internal transcribed spacer (ITS) and rRNA gene regions [18,19].

Treatment options for black piedra include shaving or cutting the hair and oral terbinafine. The compact nature of black piedra nodules may compromise the effectiveness of antifungal therapy [10,20,21].

17.2 METHODS

17.2.1 Sample Preparation

Hairs with visible nodules are plucked, sectioned, and stained with toluidine blue. A 10%–15% KOH solution is used to stain hair-shaft nodules on a glass slide; a fungal stain (e.g., chlorazol black E stain or Parker blue-black ink) is added to delineate the hyphae. Nodule is crushed, taking care not to break the coverslip. Microscopic observation of septate brown hyphae, asci, and fusiform ascospores confirms the diagnosis. Culture of the organism is carried out on media with antibacterial agents, as well as media with both antibacterial agents and cycloheximide.

For DNA extraction, fungal isolates are cultured in 20 mL of RPMI 1640 medium with L-glutamine but without sodium bicarbonate (Sigma-Aldrich) buffered to pH 7.0 with 0.165 M morpholinepropanesulfonic acid (MOPS)

(Sigma-Aldrich). After 3–14 days of growth at 30°C under agitation (100 rpm), mycelium is transferred into a tube and washed in 40 mL of sterile distilled water. Mycelium is then stored at −20°C until use. DNA extraction is carried out by a glass bead lysis method. A measured quantity of 100 mg of mycelium is homogenized for 1 min in a tube containing 1 mL of lysis buffer (2% Triton X-100, 1% sodium dodecyl sulfate, 10 mM Tris–HCl pH 8, 100 mM NaCl, 1 mM ethylenediaminetetraacetic acid [EDTA] pH 8), three 0.5 cm diameter glass beads (Sigma), and approximately 500 mg of 425–600 μm glass beads (Sigma). The homogenized mycelia are then snap-frozen in liquid nitrogen, thawed, and refrozen once. DNA extraction is then done with the DNeasy plant kit (QIAGEN) [22].

17.2.2 Detection Procedures

Pounder et al. [19] described a real-time PCR with SYBR green DNA binding dye and amplicon melting temperature analysis for fungal detection also using pan-fungal primers ITS1 forward (5′-TCCGTAGGTGAACCTGCGG-3′) and ITS4 reverse (5′-TCCTCCGCTTATTGATATGC-3′). The identity of the fungi is verified by subsequent sequencing analysis.

Procedure

1. PCR mixture is composed of 1 × Lightcycler FastStart DNA Master Hybridization Probes mixture (Roche Applied Science) (containing deoxynucleoside triphosphates, FastStart *Taq* DNA polymerase, and 1 mM $MgCl_2$, additional $MgCl_2$ is added to a final concentration of 4.6 mM), 0.4 μM each of ITS1 forward and ITS4 reverse primers, 1 × SYBR green (Molecular Probes), and 3 μL template DNA.
2. Thermal cycling parameters include 95°C for 10 min; 50 cycles of 95°C for 5 s, 60°C for 20 s, and 76°C for 30 s; and a final extension at 72°C for 2 min.
3. The quality of the amplicon is determined using the derivative of the melt analysis curve (55°C–99°C, 45 s hold at 55°C, 5 s/°C) using the RotorGene 3000 (Corbett Robotics, Inc).
4. The amplified product is purified for bidirectional sequencing using ExoSAP-IT (USB Corp). Five microliters of Big Dye Terminator Ready Reaction Mix v. 1.1 (Applied Biosystems) is added to 4 μL of each primer (0.8 pmol/μL) and 3 μL of purified PCR product. Cycle sequencing is performed with a 9700 thermal cycler (ABI), using 25 cycles of 96°C for 10 s, 50°C for 5 s, and 60°C for 4 min. Sequencing reaction products are passed through a Sephadex G-50 fine column to remove unincorporated dye terminators. Purified sequencing reaction products are run on an ABI Prism 3100 Genetic Analyzer with a 50 cm capillary array.
5. Sequences are analyzed with the SmartGene Integrated Database Network software version 3.2.3 vr. SmartGene is a web-based software and database system with reference sequences derived from the National Center for Biological Information (NCBI) GenBank repository.

Note. In case that real time PCR instrument is not available, standard PCR may be performed with primers ITSI and ITS4, and the resulting amplicon is sequenced with the same primers. Sequence-based identifications are defined by percent identity: species, ≥ 99%; genus, 93%–99%; and inconclusive, ≤ 93%.

For strains producing discrepant identification between the methods based on phenotypic characteristics and ITS sequence analysis, the D1–D2 region of the large-subunit RNA gene is amplified with primers NL1 (5′-GCATATCAATAAGCGGAGGAAAAG-3′) and NL4 (5′-GGTCCGTGTTTCAAGACGG-3′) and sequenced for species clarification.

17.3 CONCLUSION

Piedraia hortae is a keratinophilic dematiaceous fungus causing black piedra, which is characterized by the formation of black, firmly adhered nodules on scalp and facial hair. Crushed nodules mounted with KOH (10%–20%) display tightly packed, thick-walled fungal cells and asci containing two to eight single-celled, fusiform, slightly curved ascospores with a single polar filament at each end. Black piedra needs to be differentiated from white piedra (trichosporosis) caused by basidiomycetous arthroconidial yeast *Trichosporon* spp. as well as tinea capitis caused by dermatophyte fungi *Trichophyton tonsurans*, *Microsporum audouinii*, and *Microsporum canis*. White piedra is characterized by the formation of white to light brown, easily detachable nodules surrounding hair shaft. Apart from causing white piedra, *Trichosporon* spp. also have the capacity to induce systemic infections (termed trichosporonosis) in immunocompromised patients. Use of molecular techniques such as PCR and sequencing analysis enables accurate and rapid identification of *Piedraia hortae* and other hair invading fungal organisms.

REFERENCES

1. Schwartz, P.R.A., Superficial fungal infections. *Lancet*. 2004;364:1173–1182.
2. Bonifaz, A. et al., Tinea versicolor, tinea nigra, white piedra, and black piedra. *Clin Dermatol*. 2010;28(2):140–145.
3. Chong, K.C., Adam, B.A., and Soo-Hoo, T.S., Morphology of *Piedra hortae*. *Sabouraudia*. 1975;13:157–160.
4. de Hoog, G.S. et al., *Atlas of Clinical Fungi*, 2nd edn., vol. 1. Centraalbureau voor Schimmelcultures, Utrecht, the Netherlands, 2000.
5. Larone, D.H., *Medically Important Fungi—A Guide to Identification*, 3rd edn. ASM Press, Washington, DC, 1995.

6. Adam, B.A., Soo-Hoo, T.S., and Chong, K.C., Black piedra in West Malaysia. *Aust J Derm.* 1977;18:45–47.
7. Venugopal, P.V. and Venugopal, T.V., Superficial mycoses in Saudi Arabia. *Australas J Dermatol.* 1992;33:45–48.
8. Kanitakis, J. et al., Black piedra: Report of a French case associated with *Trichosporon asahii. Int J Dermatol.* 2006;45:1258–1260.
9. Coimbra, C.E.A. and Santos, R.V., Black piedra among zoro Indians from Amazonia (Brazil). *Mycopathologia* 1989;107:57–60.
10. Gip, L., Black piedra: The first case treated with terbinafine (Lamisil). *Br J Dermatol.* 1994;130(Suppl 43):26–28.
11. Figueras, M.J., Guarro, J., and Zaror, L., New findings in black piedra infection. *Br J Dermatol.* 1996;135:157–158.
12. Youker, S.R. et al., White piedra: Further evidence of a synergistic infection, *J Am Acad Dermatol.* 2003;49:46–49.
13. McBride, M.E. et al., A new *Brevibacterium* species isolated from infected genital hair of patients with white piedra, *J Med Microbiol.* 1993;39:255–261.
14. Figueras, M.J., Guarro, J., and Zaror, L., Ultrastructural aspects of hair digestion in black piedra infection. *J Med Vet Mycol.* 1997a;35:1–6.
15. de Almeida, H.L. Jr., Rivitti, E.A., and Jaeger, R.G., White piedra: Ultrastructure and a new microecological aspect, *Mycoses* 1990;33:491–497.
16. de Almeida, H.L. Jr., Salebian, A., and Rivitti, E.A., Ultrastructure of black piedra, *Mycoses* 1991;34:447–451.
17. Ellner, K.M. et al., White piedra: Evidence for a synergistic infection, *Br J Dermatol.* 1990;123:355–363.
18. Abliz, P. et al., Identification of pathogenic dematiaceous fungi and related taxa based on large subunit ribosomal DNA D1/D2 domain sequence analysis. *FEMS Immunol Med Microbiol.* 2004;40:41–49.
19. Pounder, J.I. et al., Discovering potential pathogens among fungi identified as nonsporulating molds. *J Clin Microbiol.* 2007;45:568–571.
20. Gip, L., Terbinafine for black piedra. *Lancet* 1993;341:1164.
21. Drake, L.A. et al., Guidelines of care for superficial mycotic infections of the skin: Piedra—Guidelines/Outcomes Committee, *J Am Acad Dermatol.* 1996;34:122–124.
22. Desnos-Ollivier, M. et al., Molecular identification of black-grain mycetoma agents. *J Clin Microbiol.* 2006;44(10):3517–3523.

18 *Pyrenochaeta*

Dongyou Liu

CONTENTS

18.1 INTRODUCTION

Pyrenochaeta is dark-walled, coelomycete genus in the order *Dothideales* that produces asexual fruit bodies during its life cycle. Members of this genus are widely distributed in the environment, soil, wood, and plant debris and are also encountered as plant pathogens. Several *Pyrenochaeta* species such as *P. romeroi* and *P. mackinnonii* are involved as agents of chronic, suppuratives, and subcutaneous infections in immuno-competent patients after traumatic implantation, ultimately leading to mycetoma. In addition, *Pyrenochaeta* species are occasionally implicated in onychomycosis and keratitis as well as deep, non-mycetomatous infections. Although *Pyrenochaeta* is a relatively infrequent cause of human diseases in comparison with black yeast genera *Cladophialophora*, *Exophiala*, and *Fonsecaea*, in the order *Chaetothyriales*, their similarity in morphological, biological, and clinical terms make correct identification of these organisms essential in order to implement effective control and prevention strategies.

18.1.1 Classification and Morphology

The genus *Pyrenochaeta* is a dematiaceous (dark-walled) filamentous fungus belonging to the mitosporic Pleosporaceae group, family Pleosporaceae, order Pleosporales, subclass Pleosporomycetidae, class Dothideomycetes, subphylum Pezizomycotina, phylum Ascomycota, kingdom Fungi. The mitosporic Pleosporaceae group covers 13 genera: *Alternaria*, *Dendryphiella*, *Dendryphion*, *Drechslera*, *Embellisia*, *Exserohilum*, *Nimbya*, *Pithomyces*, *Pyrenochaeta*, *Stagonospora*, *Stemphylium*, *Ulocladium*, and *Unifilum*. In turn, the genus *Pyrenochaeta* contains eight recognized species: *Pyrenochaeta acicola*, *Pyrenochaeta gentianicola*, *Pyrenochaeta inflorescentiae*, *Pyrenochaeta lycopersici*, *Pyrenochaeta nobilis*, *Pyrenochaeta romeroi*, *Pyrenochaeta terrestris*, and *Pyrenochaeta unguis-hominis*, in addition to six unassigned species [1]. The teleomorphs of the *Pyrenochaeta* genus are found in the genus *Herpotrichia* (obsolete synonyms: *Herpotrichiopsis* and *Lasiophoma*) and possibly also in the genus *Leptosphaeria*, both of the family *Leptosphaeriaceae* [2,3].

Pyrenochaeta spp. are saprophytic fungal organisms that inhabit the soil and plant debris, particularly in tropical and subtropical areas. Many *Pyrenochaeta* spp. are pathogenic to plants, and several are among the rare causes of human infections. The human pathogenic *Pyrenochaeta* species include *Pyrenochaeta keratinophila*, *Pyrenochaeta mackinnonii*, *Pyrenochaeta romeroi* (obsolete synonym: *Phlenodomus avramii*), and *Pyrenochaeta unguis-hominis* [3]. *Pyrenochaeta romeroi* and *Pyrenochaeta mackinnonii* have been isolated from mycetoma, containing soft, irregular, and black grains with a subhyaline center. A number of isolates previously recognized as *Madurella grisea* have also been reidentified as *Pyrenochaeta romeroi* [4]. *Pyrenochaeta unguis-hominis* has been isolated from the infected nails of some cases. *Pyrenochaeta keratinophila* is a new species that was recently isolated from corneal scrapings of a case of keratitis in Spain [3].

Pyrenochaeta colonies grow moderately rapidly and appear flat, woolly to cottony, white initially and becoming olivaceous green to olivaceous gray with age; reverse is dark. Hyphae are septate, hyaline to subhyaline. Pycnidia (singular pycnidium, which is a round or flask-shaped fruiting body containing conidia) are globose to flask shaped, ostiolate, brown to black, with setae (rigid hair located on the pycnidia) arising from their upper portion. Phialides arise from the inner lining of the pycnidia. Conidia (2–4 × 1–2 μm) are one celled, oval to cylindrical, hyaline, and straight or slightly curved [5,6].

At the species level, *Pyrenochaeta keratinophila* colonies reach a diameter of 24–30 mm on oatmeal agar (OA) at 20°C for 10 days, with even margin, and are colorless, flat; immersed mycelium is pale gray olivaceous to greenish olivaceous, aerial mycelium is concolorous, diffuse, woolly, floccose; reverse is olivaceous gray to olivaceous black. Colonies reach a diameter of 18–23 mm on malt extract agar at 20°C for 10 days, with even margin and buff color; immersed mycelium is olivaceous black or brown vinaceous; aerial mycelium is dense, felty to woolly, pale to olivaceous gray; reverse is dark hazel. Superficial pycnidia are abundant. On OA, pycnidia (100–400 μm in diameter) are olivaceous brown to almost black, globose or flask shaped, single or confluent, with one to three ostioles (10–25 μm in diameter), displaying hyaline, thick-walled periphyses; pycnidial walls of textura angularis comprise cells (4–9 mm in diameter) with dark-brown intercellular material. The outer surface of the pycnidial wall displays scarce, brown, slightly roughened, septate setae (20–35 × 2.5–4 μm), with a blunt apex, mostly positioned near the ostiole. Conidiogenous cells (12–23 × 2–3.5 μm) arise from the entire inner surface of the pycnidial wall and are rarely discrete, ampuliform to doliform, mostly cylindrical, and integrated in conidiophores, which are branched at the base, acropleurogenous (i.e., having terminal and lateral apertures), phialidic, with a distinct periclinal thickening and sometimes with a short collarette. In older cultures, conidiogenous cells often show several distinct percurrent proliferations. Conidia (2–4 × 1–2 μm) are whitish in mass; ellipsoid, straight, or slightly curved; hyaline, continuous, smooth, with granular contents or few guttules; and rounded at both ends. Phialoconidia arising directly from solitary apertures in the aerial mycelium in 3-week-old cultures are initially indistinguishable from conidia originating from pycnidia but later become inflated unilaterally, globose or subglobose, 3–4 μm in diameter, pale brown, and slightly thick walled. Chlamydospores-like structures (8–10 μm in diameter) also form on old OA cultures and are terminal, hyaline, globose, smooth, and thick walled. Molecular data suggest that the closest sexual state of *P. keratinophila* is *Leptosphaeria* (family *Leptosphaeriaceae*) [3].

Pyrenochaeta unguis-hominis pycnidia (100–400 μm in diameter) are olivaceous brown to black, globose or flask shaped, with one to three ostioles, and with several setae positioned near the ostiole. The pycnidial wall is of textura angularis with dark brown intercellular material. The conidiophores emerging from all over the inner surface of the pycnidial wall are branched at the base, bearing terminal and lateral conidiogenous cells. Conidiogenous cells (12–18 × 2–3.5 μm) are phialidic, mostly cylindrical. Conidia (2–4 × 1–2 μm) are whitish in mass, ellipsoidal, wide, straight or slightly curved, hyaline, and smooth walled [2].

Although *P. keratinophila* and *P. unguis-hominis* have similar conidiophores and conidia, they can be distinguished by the color of the colonies and the location of the setae on the pycnidia. *P. keratinophila* colonies on OA are gray olivaceous to greenish olivaceous, while *P. unguis-hominis* colonies are brown vinaceous to fawn. The pycnidial setae of *P. keratinophila* are scarce and placed mainly near the ostiole, while those of *P. unguis-hominis* are usually more abundant and more dispersed over the entire surface of the pycnidium. *P. keratinophila* also produces conidia from the mycelium, a feature unknown in *P. unguis-hominis* or any other species of the genus *Pyrenochaeta*. *P. mackinnonii* is distinguished from *P. keratinophila* by having more restricted, raised, or wrinkled colonies. *P. romeroi* is distinguished by its grayish-sepia to fuscous-black colonies and the presence of discrete conidiogenous cells [3].

18.1.2 Clinical Features

The term "dematiaceous fungi" refers to a heterogeneous group of fungal organisms with black filaments (dematiaceae) that may cause opportunistic superficial or deep infections in humans, producing a diverse range of clinical syndromes (collectively known as phaeohyphomycosis, chromoblastomycosis, and mycetoma). To date, about 130 species belonging to 70 dematiaceous fungal genera have been shown to induce these clinical diseases.

Mycetoma is a highly debilitating disease that is characterized by a subcutaneous mass with multiple sinuses draining pus and grains (<0.5–2 mm in diameter), which may appear white (i.e., white-grain mycetoma), black (black-grain mycetoma), yellow, or red. This disease can be caused by fungi (eumycetoma) or actinomycetes (actinomycetoma). While eumycetoma usually produces either white or black grains, actinomycetoma may show white, yellow, or red grains [7].

The most common fungal species involved in white-grain eumycetoma include *Acremonium* (*Fusarium*) *falciforme*, *Acremonium kiliense*, *Acremonium recifei*, *Cylindrocarpon destructans*, *Fusarium moniliforme*, *Fusarium solani*, *Neotestudina rosatii*, and *Pseudallescheria boydii*, whereas common fungal species responsible for black-grain eumycetoma are *Exophiala jeanselmei*, *Madurella grisea*, *Madurella mycetomatis*, *Leptosphaeria tomkinsii*, *Leptosphaeria senegalensis*, *Pyrenochaeta mackinnonii*, *Pyrenochaeta romeroi*, and *Phlenodomus avramii*.

As saprophytes commonly occurring in soil and plants, *Pyrenochaeta* spp. may gain entry to human hosts through direct traumas by a plant or a soiled object. With keratomycosis, a number of factors may predispose the hosts to *Pyrenochaeta* and other fungal infections. These include chronic ocular surface disease, dry eye, contact lens wearing, atopic disease, topical steroid use, long-term treatment with broad-spectrum antibiotics, and trauma (particularly with vegetable material or soil) [8]. In addition, patients with suppressed immune functions are vulnerable to *Pyrenochaeta* and other fungal diseases [9,10].

First identified in 1959, *Pyrenochaeta romeroi* has been reported in mycetoma cases involving mainly immunosuppressed individuals [11–13]. In recent years, it has been shown that besides eumycetoma, *P. romeroi* may also be responsible for non-mycetoma phaeohyphomycosis. Girard et al. [14] reported a case of non-mycetoma deep cutaneous infection due to *Pyrenochaeta romeroi* in an immunocompromised 45-year-old man who originated from Senegal. The patient had a multibacillary leprosy treated by sequential antibiotics for more than 20 years, and presented with non-inflammatory, painful and deep lesions of both lower limbs reminiscent of cold abscesses. Clinical examination revealed a large subcutaneous, flaccid lesion of the lateral side of the left leg, with an occasional purulent white-yellow discharge through superficial sinus tracts, a 2 cm large nodule of the right tibial crest and two crusted papulo-nodules of the left foot. X-rays showed no bone alteration underlying the cutaneous lesions, while ultrasound examination of the left leg uncovered a 20 mm long, 5 mm large, and 6 mm thick cavity with multiple internal septa, extending to the subcutaneous fat but without obvious involvement of the underlying muscle or bone. A deep biopsy sample from the larger cutaneous lesion showed a chronic granulomatous dermal infiltrate with giant cells and microabscesses and the presence of septate filaments scattered within the infiltrate upon periodic acid-Schiff (PAS) staining. Direct examination of purulent material from leg sinus tract showed no granules but fungal, septate filaments. Culture of the lesion material grew a fungus, which was identified as *P. romeroi*. After surgical excision, drainage of the largest abscess, treatment with oral itraconazole (100 mg daily) and anti-bacillary antibiotics (clofazimine, rifampicine, dapsone), all cutaneous lesions slowly resolved within 1 month with minimal scarring. Itraconazole was interrupted after 4 months, and no relapse occurred during a 1-year follow-up.

In a separate study, Badali et al. [15] documented another case of non-mycetomatous infection (subcutaneous phaeohyphomycotic cyst) due to *Pyrenochaeta romeroi* in a 45-year-old Indian female. The patient presented with a verrucous plaque and a swelling (30 mm in diameter) on the right forearm that gradually increased in size over a period of 3 months. Direct microscopic examination with 10% KOH and histopathological investigation of exudates showed septate hyphae without granules. A fungus was isolated from the exudates in culture and identified as *Pyrenochaeta romeroi* on the basis of morphologic features and sequence homology in the internal transcribed spacer (ITS) regions of ribosomal RNA (rRNA). Treatment consisted of surgical excising of the cyst without any antifungal therapy, with no relapse during a 1-year follow-up. It is possible that the apparent immune dysfunction in this patient might contribute partially to the lack of granule formation characteristic of mycetoma.

Ferrer et al. [2] described a *Pyrenochaeta unguis-hominis*-related case of keratitis in a 77-year-old diabetic woman with pain and decreased vision in her left eye. Slit lamp examination showed conjunctival hyperemia, peripheral corneal pannus, corneal edema, and a central corneal ulcer (4 mm in diameter) with white stromal infiltrates inferiorly. Corneal scrapings were plated on Columbia agar plates supplemented with 5% sheep blood, Columbia chocolate agar, MacConkey agar, thioglycolate broth, and brain heart infusion broth (BioMérieux); and incubated at 37°C. Corneal scraping and biopsy taken from the ulcer edge were also inoculated onto Sabouraud dextrose agar with chloramphenicol and incubated at 30°C. Direct examination of the corneal scrapings and biopsy using Gram and calcofluor white stains showed the presence of numerous septate and branched hyphae. After 6 days, a filamentous fungus grew on chocolate agar, and a panfungal polymerase chain reaction (PCR) generated a 532 bp amplicon, which showed 86% homology to *Leptosphaeria* strain. The fungus was subcultured on malt extract agar and oatmeal agar (30 g oat flakes, 1 g $MgSO_4$, 1.5 g KH_2PO_4, 15 g agar, 1 L tap water) at 20°C in the dark. After 1 week of incubation, pycnidia typical of the coelomycetous genus *Pyrenochaeta* developed, and the fungus was identified as *Pyrenochaeta unguis-hominis*.

Recently, Verkley [3] described the first case of *Pyrenochaeta keratinophila*-associated keratitis in a 77-year-old diabetic woman in Spain. The patient presented with conjunctival hyperaemia, peripheral corneal pannus, corneal edema, and a central ulcer with white infiltrates in her left eye. Calcofluor white staining of corneal biopsy revealed fungi invading the cornea. Culture of the corneal biopsy on several media grew a fungus that was identified morphologically as a *Pyrenochaeta* species. Sequencing analysis of the ITS region in the rRNA gene showed it belonged to an unknown new species *Pyrenochaeta keratinophila*. Despite treatment with topical natamycin and oral ketoconazole for 1 month with considerable lesion improvement, the cornea could not be saved and the patient underwent an optical penetrating keratoplasty.

18.1.3 Diagnosis

Over 70 dematiaceous fungal genera containing 130 or so distinct species have been implicated in human phaeohyphomycosis, chromoblastomycosis, and mycetoma. Given that the clinical symptoms caused by these fungi are largely indistinguishable and nonspecific, and that different fungal taxa often display varied sensitivity to antifungal drugs, there is a necessity to precisely determine the species identity of the suspected fungal organisms for tailor-made, cost-effective treatment.

Together with clinical and epidemiological information, identification of dematiaceous fungi including *Pyrenochaeta* species relies on the observation of mycotic elements in biopsy and other specimens by microscopy. Subsequent isolation of suspected fungal organisms in culture media facilitates detection of characteristic colonial and microscopic features, leading to the confirmation of the species identity. Unfortunately, this testing scheme can take up to 12 weeks to complete. With cultures that are negative or contaminated with bacteria, culturing with fresh samples can further delay the diagnosis.

To improve the efficiency and specificity of laboratory diagnosis of fungal pathogens, molecular methods have been developed and applied in recent years [16–21]. PCR amplification and sequencing analysis of the ITS1–5.8S–ITS2 rRNA have enabled precise identification of the fungi responsible for black-grain mycetoma [4,22,23]. In addition, sequencing examination of the D1–D2 regions in the large subunit (LSU) rRNA D1–D2 region offers another potential avenue for taxonomical determination of dematiaceous fungi. Furthermore, combination of a modified FTA extraction protocol with pyrosequencing of ITS2 region makes it possible to identity a fungal culture within half a day [24,25].

18.2 METHODS

18.2.1 Sample Preparation

18.2.1.1 Procedure for Endophthalmitis Samples

The extraocular environment is sterilized with 5% povidone iodine solution before surgery. About 100–200 μL of aqueous fluid is withdrawn using a 30 gauge needle with a limbal paracentesis. Vitreous samples (200 μL) are taken at the time of three-port pars plana vitrectomy. The samples are divided into two aliquots and transported at 4°C. One portion is immediately examined by conventional microbiological diagnostic tests and the other is frozen at −20°C until processed by PCR.

For the microbiological diagnostic test, 50 μL of aqueous humor or 50 μL of vitreous is cultured at 30°C in Sabouraud's dextrose agar or at 37°C in thioglycolate broth, blood agar, chocolate agar, cystine lactose electrolyte deficient (CLED) agar, or MacConkey agar. Filamentous fungi are differentiated by isolation in Sabouraud dextrose agar plus chloramphenicol and morphological examination of macroscopic and microscopic characteristics. Microscopic structures may also be observed on tease or tape preparations and slide cultures for up to 21 days [18].

18.2.1.2 Procedure for Keratitis Samples

Upon completion of the ocular examination and after instillation of topical anesthetic, a sterile Kimura spatula is used to scrape the area of infection. Scrapings are inoculated into thioglycolate broth, Roiron broth, and Löwenstein-Jensen medium and are placed onto glass slides for staining with Gram and Giemsa stains. The PCR sample is obtained by scraping and stirring the spatula for a few seconds in 100 μL of sterile water in a 1.5 mL sterile Eppendorf tube. Two aliquots of 50 μL are taken from each sample and stored at −20°C [18].

18.2.1.3 DNA Extraction

Pyrenochaeta isolates are grown on potato dextrose agar on 9 cm diameter petri dishes at 25°C until mycelium completely covers the agar surface. Mycelia are collected by adding sterile distilled water containing 0.05% (v/v) Tween 80 to the surface of the culture and gently scrubbing with a sterile spatula. The mycelial suspension is transferred to a 1.5 mL tube and centrifuged at 3000 × *g* at 4°C for 5 min. The supernatant is discarded and the pellet (100 mg mycelia) is stored at −80°C until further use.

Extraction buffer (0.2 mL) (3% sodium dodecyl sulfate [SDS] [w/v] containing 0.5 mM ethylenediaminetetraacetic acid [EDTA], 1.0 M NaCl, and 0.1 mM hydroxymethylhydrochloride Tris–HCl pH 8.0) is added to 10 mg of each fungal mycelium and the suspension shaken vigorously for 15 s. Next, 0.2 mL chloroform-phenol mix (1:1) is slowly added and incubated at 65°C for 5 min. The mixture is cooled to room temperature and centrifuged at 10,000 × *g* at 4°C for 5 min. The supernatant is transferred to a new microtube, and an equal volume of cold absolute isopropanol or ethanol is added and the contents mixed thoroughly for precipitating total DNA at −20°C for 20 min; the mixture is then centrifuged at 10,000 × *g* for 10 min. The pellet is washed twice with 75% ethanol and centrifuged at 10,000 × *g* at 4°C for 5 min.

The supernatant is discarded and the pellet is resuspended in 0.03 mL MiniQuantum (deionized) water, and stored at −80°C until further use. Concentration, yield, and quality control indices based on absorbance readings at 230, 260, and 280 nm (A260/280 and A260/230 ratios) are carried out with 2 μL resuspended total DNA [26].

Alternatively, fungal strains are cultured in 20 mL of RPMI 1640 medium with L-glutamine but without sodium bicarbonate (Sigma-Aldrich) buffered to pH 7.0 with 0.165 M morpholinepropanesulfonic acid (MOPS) (Sigma-Aldrich). After 3–14 days of growth at 30°C under agitation (100 rpm), mycelium is transferred into a tube and washed in 40 mL of sterile distilled water. Mycelium is then stored at −20°C until use. DNA extraction is carried out by a glass bead lysis method. Approximately 100 mg of mycelium is homogenized for 1 min in a tube containing 1 mL of lysis buffer (2% Triton X-100, 1% SDS, 10 mM Tris–HCl pH 8, 100 mM NaCl, 1 mM EDTA pH 8), three 0.5 cm diameter glass beads (Sigma), and approximately 500 mg of 425–600 μm glass beads (Sigma). The homogenized mycelia are then snap-frozen in liquid nitrogen, thawed, and refrozen once. DNA extraction is then done with the DNeasy plant kit (QIAGEN) [4].

Furthermore, Whatman FTA filter may be used for DNA extraction. Whatman FTA filter matrices are fibrous cards pretreated with chelators and denaturants that lyse and inactivate most microorganisms on contact. The large nucleic acids released after lysis become physically entangled in the fibers of the FTA matrix, whereas cellular debris is rapidly removed by washing the inoculated card. Fungal hyphae fragments and conidia (in the case of moulds) are added to dry FTA filters, followed by brief microwave treatment, allowed liberation of amplifiable DNA with a total preparation time of ~15 min [25].

18.2.2 Detection Procedures

18.2.2.1 Standard PCR and Sequencing Analysis of ITS Region

Ferrer et al. [18] utilized universal primers ITS1 (5′-TCC GTA GGT GAA CCT GCG G-3′, which hybridizes at the end of 18S rRNA), and ITS4 (5′-TCC TCC GCT TAT TGA TAT GC-3', which hybridizes at the beginning of 28S rRNA) [27] to amplify a region covering ITS/5.8S rRNA for identification of pathogenic fungi.

Procedure

1. PCR mixture (50 μL) is made up of 10 μL of DNA template, 6 μL of 25 mM $MgCl_2$, 5 μL of PCR buffer without $MgCl_2$, 200 μM each deoxynucleoside triphosphate, 25 pmol of each primer, and 1 U of *Taq* DNA polymerase (Biotools B&M Labs).
2. Amplification is performed with 1 cycle at 95°C for 5 min; 35 cycles of 95°C for 30 s, 55°C for 1 min, and 72°C for 1 min; and 1 cycle at 72°C for 6 min.
3. Aliquots (10 μL) of each amplified product are electrophoretically separated in a 2% agarose gel in 1× Tris–borate–EDTA buffer and visualized using ethidium bromide under UV illumination.
4. PCR amplicon is purified using the GeneClean II kit (Bio 101) and directly cycle sequenced in both directions using the BigDye terminators Ready Reaction Kit (PE Applied Biosystems) on an ABI Prism automated DNA sequencer (model 377, version 2.1.1; Applied Biosystems), using primers ITS4 and ITS86.
5. The 550 bp ITS2/5.8S rRNA sequences are analyzed by using the BLAST alignment program of the GenBank database. Multiple-sequence alignment is carried out using ClustalW 1.8. Phylogenetic trees are constructed by the neighbor-joining method using the Phylip package and visualized using Treeview.

18.2.2.2 Real-Time PCR and Sequencing Analysis of ITS Region

Pounder et al. [23] described a real-time PCR with SYBR green DNA binding dye and amplicon melting temperature analysis for fungal detection also using panfungal primers ITS1 forward (5′-TCCGTAGGTGAACCTGCGG-3′) and ITS4 reverse (5′-TCCTCCGCTTATTGATATGC-3′). The identity of the fungi is verified by subsequent sequencing analysis.

Procedure

1. PCR mixture is composed of 1× Lightcycler FastStart DNA Master Hybridization Probes mixture (Roche Applied Science) (containing deoxynucleoside triphosphates, FastStart *Taq* DNA polymerase, and 1 mM $MgCl_2$, additional $MgCl_2$ is added to a final concentration of 4.6 mM), 0.4 μM each of ITS1 forward and ITS4 reverse primers, 1× SYBR green (Molecular Probes), and 3 μL template DNA.
2. Thermal cycling parameters include 95°C for 10 min; 50 cycles of 95°C for 5 s, 60°C for 20 s, and 76°C for 30 s; and a final extension at 72°C for 2 min.
3. The quality of the amplicon is determined using the derivative of the melt analysis curve (55°C–99°C, 45 s hold at 55°C, 5 s/°C) using the RotorGene 3000 (Corbett Robotics, Inc).
4. The amplified product is purified for bidirectional sequencing using ExoSAP-IT (USB Corp). Five microliters of Big Dye Terminator Ready Reaction Mix v. 1.1 (Applied Biosystems) is added to 4 μL of each primer (0.8 pmol/μL) and 3 μL of purified PCR product. Cycle sequencing is performed with a 9700 thermal cycler (ABI), using 25 cycles of 96°C for 10 s, 50°C for 5 s, and 60°C for 4 min. Sequencing reaction products are passed through a Sephadex G-50 fine column to remove unincorporated dye terminators. Purified sequencing reaction products are run on an ABI Prism 3100 Genetic Analyzer with a 50 cm capillary array.
5. Sequences are analyzed with the SmartGene Integrated Database Network software version 3.2.3 vr. SmartGene is a web-based software and database system with reference sequences derived from the National Center for Biological Information (NCBI) GenBank repository.

Note. Sequence-based identifications are defined by percent identity: species, ≥99%; genus, 93%–99%; and inconclusive, ≤93%.

For strains producing discrepant identification between the methods based on phenotypic characteristics and ITS sequence analysis, the D1–D2 region of the LSU RNA gene is amplified with primers NL1 (5′-GCATATCAATAAGCGGAGGAAAAG-3′) and NL4 (5′-GGTCCGTGTTTCAAGACGG-3′) and sequenced for species clarification.

18.3 CONCLUSION

The genus *Pyrenochaeta* consists of eight dematiaceous fungal species that are soil saprophytes with potential to cause infections in plants. Four *Pyrenochaeta* species have been associated with occasional infections in humans, producing clinical presentations ranging from mycetoma, subcutaneous infections, keratitis, and other diseases, especially in immunocompromised individuals. Because clinical diseases induced by *Pyrenochaeta* spp. are indistinguishable from those caused by other 70 dematiaceous fungal genera (and 130 distinct species), and because individual dematiaceous fungal taxa often demonstrate differing resistance to commonly used antifungal drugs, it is important to determine the species identity of the suspected fungal organisms.

Although it is possible to correctly identify dematiaceous fungi including *Pyrenochaeta* species on the basis of macroscopic and microscopic criteria, the whole process is notoriously time consuming and technically demanding. Application of molecular methods such as PCR amplification and sequencing analysis of the ITS1-5.8S-ITS2 rRNA and other gene regions offers the opportunity to vastly enhance the specificity and speed of fungal detection and determination. In addition, these techniques also enable improved phylogenetic analysis and epidemiological tracking of future outbreaks due to dematiaceous fungi, leading to the implementation of highly effective treatment and control against these emerging infective organisms.

REFERENCES

1. The UniProt Consortium. Available at http://www.uniprot.org/, accessed on August 1, 2010.
2. Ferrer, C. et al., New *Pyrenochaeta* species causing keratitis. *J Clin Microbiol.* 2009;47:1596.
3. Verkley, G.J.M., *Pyrenochaeta keratinophila* sp. nov., isolated from an ocular infection in Spain. *Rev Iberoam Micol.* 2010;27:22–24.
4. Desnos-Ollivier, M. et al., Molecular identification of black-grain mycetoma agents. *J Clin Microbiol.* 2006;44(10):3517–3523.
5. Kwon-Chung, K.J. and Bennett, J.E., *Medical Mycology*. Lea & Febiger, Philadelphia, PA, 1992.
6. de Hoog, G.S. et al., *Atlas of Clinical Fungi.* Centraalbureau voor Schimmelcultures, Utrecht, the Netherlands, 2000.
7. de Hoog, G.S. et al., Diagnostic problems with imported cases of mycetoma in The Netherlands. *Mycoses* 1993;36:81–87.
8. Thomas, P.A., Fungal infections of the cornea. *Eye* 2003;17:852–862.
9. Andre, M. et al., Fungal mycetoma with black grains due to *Pyrenochaeta romeroi* in Cambodia. *Bull Soc Pathol Exot Fil.* 1968;61:108–112.
10. English, M.P., Infection of the finger-nail by *Pyrenochaeta unguis-hominis. Br J Dermatol.* 1980;103:91–93.
11. Thammayya, A., Sanyal, M., and Basu, N., *Pyrenochaeta romeroi* causing mycetoma pedis in India. *J Indian Med Assoc.* 1979;73:66–67.
12. Rinaldi, M.G., Phaeohyphomycosis. *Dermatol Clin.* 1996;14:147–153.
13. Cerar, D. et al., Isolation, identification and susceptibility of *Pyrenochaeta romeroi* in a case of eumycetoma of the foot in the UK. *Int J Antimicrob Agents* 2009;34(6):617–618.
14. Girard, C. et al., Subcutaneous phaeohyphomycosis due to *Pyrenochaeta romeroi* in a patient with leprosy. *Acta Derm Venereol.* 2004;84(2):154–155.
15. Badali, H. et al., Subcutaneous phaeohyphomycotic cyst caused by *Pyrenochaeta romeroi. Med Mycol.* 2010;48(5):763–768.
16. Chen, Y.-C. et al., Identification of medically important yeasts using PCR-based detection of DNA sequence polymorphisms in the internal transcribed spacer region 2 of the rRNA genes. *J Clin Microbiol.* 2000;38:2302–2310.
17. Chen, Y.-C. et al., Polymorphic internal transcribed spacer region 1 DNA sequences identify medically important yeasts. *J Clin Microbiol.* 2001;39:4042–4051.
18. Ferrer, C. et al., Detection and identification of fungal pathogens by PCR and by ITS2 and 5.8S ribosomal DNA typing in ocular infections. *J Clin Microbiol.* 2001;39:2873–2879.
19. Ahmed, A.O. et al., Molecular detection and identification of agents of eumycetoma: Detailed report of two cases. *J Clin Microbiol.* 2003;41:5813–5816.
20. Schwarz, P. et al., Molecular identification of zygomycetes from culture and experimentally infected tissues. *J Clin Microbiol.* 2006;44:340–349.
21. Linton, C.J. et al., Molecular identification of unusual pathogenic yeast isolates by large ribosomal subunit gene sequencing: 2 years experience at the UK Mycology Reference Laboratory. *J Clin Microbiol.* 2007;5:1152–1158.
22. Iwen, P.C., Hinrichs, S.H., and Rupp, M.E., Utilization of the internal transcribed spacer regions as molecular targets to detect and identify human fungal pathogens. *Med Mycol.* 2002;40:87–109.
23. Pounder, J.I. et al., Discovering potential pathogens among fungi identified as nonsporulating molds. *J Clin Microbiol.* 2007;45(2):568–571.
24. Borman, A.M. et al., Molecular identification of pathogenic fungi. *J Antimicrob Chemother.* 2008;61:S7.
25. Borman, A.M. et al., An improved protocol for the preparation of total genomic DNA from isolates of yeast and mould using Whatman FTA filter papers. *Mycopathologia* 2010;169:445.
26. González-Mendoza, D. et al., A rapid method for isolation of total DNA from pathogenic filamentous plant fungi. *Genet Mol Res.* 2010;9(1):162–166.
27. White, T.J. et al., Amplification and direct sequencing of fungal ribosomal RNA genes for phylogenetics. In M.S. Innis and D.H. Gelfand (Eds.), *PCR Protocols: A Guide to Methods and Applications*. Academic Press, New York, 1990, pp. 315–322.

19 *Ramichloridium*

Dongyou Liu

CONTENTS

19.1 INTRODUCTION

Ramichloridium is an anamorph genus that comprises multiple species of saprobes, human and, plant pathogens. Morphologically, these organisms possess erect, dark, branched or unbranched conidiophores and predominantly aseptate conidia produced on a sympodially proliferating rachis. Currently, *Ramichloridium schulzeri* is the main *Ramichloridium* species involved in human disease, and other previous human pathogens within the genus, that is, *Ramichloridium mackenziei, Ramichloridium basitonum*, and *Ramichloridium musae* have been now reclassified as *Rhinocladiella mackenziei, Rhinocladiella basitonum*, and *Periconiella musae*, respectively.

19.1.1 Classification and Morphology

The genus *Ramichloridium* is a dematiaceous (dark-walled) fungus belonging to the mitosporic Capnodiales group, order Capnodiales, class Dothideomycetes, subphylum Pezizomycotina, phylum Ascomycota, kingdom Fungi. Currently, the genus *Ramichloridium* consists of nine recognized species: *Ramichloridium apiculatum* (the type species of the genus), *Ramichloridium australiense, Ramichloridium biverticillatum, Ramichloridium brasilianum, Ramichloridium epichloes, Ramichloridium indicum* (basionym: *Chloridium apiculatum*), *Ramichloridium pini, Ramichloridium schulzeri* (obsolete synonyms: *Acrotheca acuta, Chloridium schulzeri, Pleurophragmium acutum, Psilobotrys schulzeri, Rhinocladiella schulzeri* and *Rhinotrichum multisporum*), and *Ramichloridium strelitziae*, in addition to five unassigned species [1–4]. No teleomorph has been linked to the species of *Ramichloridium* so far.

Other former species in the genus include *Ramichloridium anceps* (formerly *Veronaea parvispora*) (→*Rhinocladiella anceps*), *Ramichloridium basitonum* (→*Rhinocladiella basitona*), *Ramichloridium cerophilum* (→*Rhinocladiella aquaspersa*), *Ramichloridium fasciculatum* (→*Rhinocladiella fasciculata*), *Ramichloridium mackenziei* (formerly *Ramichloridium obovoidea* and *Ramichloridium obovoiedum*) (→*Rhinocladiella mackenziei*), *Ramichloridium musae* (formerly *Veronaea musae*) (→*Periconiella musae*), and *Ramichloridium subulatum* (→*Radulidium subulatum*) (http://www.indexfungorum.org/) [5–9].

Ramichloridium colonies are flat to raised with entire margin; surface is olivaceous-green to olivaceous-black. Mycelium consists of submerged and aerial hyphae: submerged hyphae are hyaline to subhyaline and thin walled whereas aerial hyphae are smooth or verrucose. Conidiophores are straight, unbranched, rarely branched, thick walled, dark brown (darker than the subtending hyphae), and continuous or with several additional thin septa. Conidiogenous cells are integrated, terminal, polyblastic, smooth, thick walled, golden-brown, apical part subhyaline, with sympodial proliferation, straight or flexuose, geniculate or nodose, with conspicuous conidiogenous loci. Scars are crowded or scattered, unthickened, unpigmented to faintly pigmented, or slightly prominent denticles. Conidia are solitary, with 0–1 septum, subhyaline to pale brown, smooth to coarsely verrucose, thin walled, obovate, obconical or globose to ellipsoidal, fusiform, with a slightly pigmented hilum; conidial secession is schizolytic. The type species of the genus is *R. apiculatum* [7,8].

At the species level, *Ramichloridium apiculatum* (basionym: *Chloridium apiculatum*) colonies reach a diameter of 35 mm after 14 days at 24°C on malt extract plates (MEA), with minimum growth temperature of above 6°C, optimum temperature at 24°C, maximum temperature at 30°C. Colonies are raised, velvety, dense, with entire margin; surface is olivaceous-green, reverse is olivaceous-black, often with a diffusing citron-yellow pigment. Submerged hyphae (1–2.5 μm wide) are hyaline to subhyaline, thin walled; aerial hyphae are slightly darker, smooth walled. Conidiophores (up to 100 µm long) generally arise at right angles from creeping aerial hyphae, straight, unbranched, thick walled, dark brown, continuous or with 1–3 additional thin septa; intercalary cells are 10–28 µm long. Conidiogenous cells (25–47 × 2–3.5 µm) are integrated, terminal, smooth, thick walled, golden-brown, straight, cylindrical, proliferating sympodially, resulting in a straight rachis with conspicuous conidiogenous loci. Scars (less than 1 µm in diameter) are prominent, crowded, slightly pigmented. Conidia (3–7.5 × 2–4 µm) are solitary, obovate to obconical, pale brown, finely verrucose. Hilum (about 1 µm in diameter) is conspicuous, slightly pigmented [7,8].

Ramichloridium australiense colonies are slow growing, reaching a diameter of 8 mm after 14 days at 24°C on MEA, with entire, smooth margin. Mycelium is flat, olivaceous-gray, becoming granular, with gelatinous droplets at the margin developing with age; reverse is pale olivaceous-gray. Submerged hyphae (1–2 µm wide) are hyaline, smooth, thin walled; aerial hyphae are pale brown, warted. Conidiophores arise vertically and clearly differentiate from creeping aerial hyphae, up to 400 µm tall, with several additional thin septa. Intercalary cells (8–40 µm × 2–5 µm) arise from the broadest part at the base tapering toward the apex, are subhyaline, later become pale brown and warted in the lower part. Subtending hyphae are thick walled, warted. Conidiogenous cells (10–18 µm long) are integrated, terminal, proliferating sympodially, giving rise to a short rachis with conspicuous conidiogenous loci. Scars (about 1 µm in diameter) are slightly thickened and darkened. Conidia (10–23 µm × 2.5–3 µm) are solitary, aseptate, thin walled, smooth, subhyaline, subcylindrical to obclavate, with a truncated base and a slightly darkened and thickened hilum (1.5–2 µm in diameter), rarely fusing at the basal part [7,8].

Ramichloridium biverticillatum (basionym: *Periconiella musae*) is named after its biverticillate conidiophores. Colonies are slow growing, reaching a diameter of 16 mm after 14 days at 24°C on MEA, with entire, smooth, sharp margin, compact, and velvety. Surface is vinaceous-buff to olivaceous-buff; reverse is buff. Submerged hyphae (1–2 µm wide) are smooth, hyaline, thin walled; aerial hyphae are subhyaline, smooth, slightly darker. Conidiophores (2–3 µm × 250 µm) arise vertically from creeping aerial hyphae, pale brown, profusely branched, biverticillate, with up to three levels of main branches, which taper distally. Conidiogenous cells (15–50 µm long) are terminally integrated, cylindrical, variable in length, rachis straight or geniculate, pale brown, as wide as the basal part, elongating sympodially, forming a rachis with crowded, slightly darkened and thickened minute scars (<0.5 µm wide). Conidia (2–5 × 1.5–2.5 µm) are solitary, aseptate, hyaline to subhyaline, dacryoid to pyriform, smooth, thin walled, with an inconspicuous hilum. *Ramichloridium biverticillatum* is a new name based on *Periconiella musae*. It is differentiated from *Periconiella musae* by having profusely branched conidiophores, and smaller conidia (2–5 µm × 1.5–2.5 µm) than those of *Periconiella musae* (5–11 µm × 2–3 µm) [7,8].

Ramichloridium brasilianum colonies are slow-growing, reaching a diameter of 6 mm after 14 days at 24°C on MEA, and appear velvety to hairy, with entire margin; surface is dark olivaceous-gray; black gelatinous exudate droplets produced on oatmeal agar (OA). Submerged hyphae (1.5–2 µm wide) are pale olivaceous, smooth or slightly rough; aerial hyphae are olivaceous, smooth or rough, narrower and darker than the submerged hyphae. Conidiophores (2–2.5 µm × 70 µm) are unbranched, arising vertically from creeping aerial hyphae, straight or flexuose, dark brown, with up to 10 additional septa, thick walled, and cylindrical. Conidiogenous cells (10–30 µm long) are integrated, terminal, proliferating sympodially, giving rise to a long, straight rachis with crowded, slightly darkened minute scars (about 0.5 µm in diameter). Conidia (4–8.5 µm × 2–3 µm) are solitary, obovoid to fusiform with the widest part below the middle, thin walled, verruculose, aseptate, pale brown, slightly rounded at the apex, truncated at the base, with a slightly thickened and darkened hilum (1–1.5 µm in diameter) [7,8].

Ramichloridium indicum (basionym: *Chloridium indicum*; synonyms: *Veronaea indica* and *Veronaea verrucosa*) colonies reach a diameter of 35 mm after 14 days at 24°C on MEA. Colonies are velvety, rather compact, slightly elevated, with entire, smooth, whitish margin, dark olivaceous-green in the central part. Submerged hyphae (1–2.5 µm wide) are smooth, thin walled, hyaline, with thin septa; aerial hyphae (2–2.5 µm wide) are coarsely verrucose, olivaceous-green, thick walled, with thin septa. Conidiophores (250 µm × 2–4 µm) arise vertically from creeping hyphae at right angles, are straight, unbranched, thick walled, smooth, dark brown, with up to 10 thin septa, often with inflated basal cells. Conidiogenous cells (up to 165 µm long) are terminally integrated, smooth, dark brown, sympodially proliferating, rachis straight or flexuose, geniculate or nodose, subhyaline; scars (about 0.5 µm in diameter) are thickened and darkened, clustered at nodes. Conidia (5–10 µm × 4–9 µm) are solitary, with 0–1 septum, not constricted at the septum, subhyaline to pale brown, smooth or coarsely verrucose, rather thin walled, broadly ellipsoidal to globose, with truncated base; hilum (about 1 µm in diameter) is conspicuous, slightly darkened, not thickened [7,8].

Ramichloridium schulzeri colonies grow moderately rapidly, and appear compact, flat; submerged hyphae are pale orange; aerial hyphae are powdery, brownish; reverse is pink to orange. Conidiophores (up to 250 µm long) are erect, straight, unbranched, thick walled, reddish-brown, gradually becoming paler toward the apex, elongating sympodially during conidiogenesis, with scattered, pimple-shaped conidium bearing denticles, which have unpigmented scars. Conidia (6.5–37 µm × 3–4 µm) are subhyaline, smooth walled or slightly rough walled, ellipsoidal, obovoidal or fusiform, usually with an acuminate base and unpigmented scars. *Ramichloridium schulzeri*, including its varieties, is phylogenetically as well as morphologically distinct from the other genera in the *Ramichloridium* complex [7,8].

Ramichloridium strelitziae (named after its host, *Strelitzia*) colonies are slow growing, reaching a diameter of 5 mm after 14 days at 24°C on MEA, with entire margin; aerial mycelium is compact, raised, dense, olivaceous-gray; reverse is olivaceous-black. Submerged hyphae (2–2.5 μm wide) are smooth, hyaline, thin walled; aerial hyphae are pale brown, verrucose. Conidiophores (40 μm × 2 μm) arise vertically from creeping aerial hyphae, clearly differentiated from the vegetative hyphae, subhyaline, later becoming pale brown, thick walled, smooth, or verruculose, with 1–3 additional septa. Conidiogenous cells (10–35 μm long) are integrated, terminal, cylindrical, subhyaline, later turning pale brown, fertile part as wide as the basal part, proliferating sympodially, forming a straight rachis with slightly thickened and darkened, circular, somewhat protruding scars (about 0.5 μm in diameter). Conidia (3–5.5 μm × 1–2.5 μm) are solitary, aseptate, smooth or verruculose, subhyaline, oblong, ellipsoidal to clavate, with truncated base and unthickened, nonpigmented hilum [7,8].

Phylogenetically, the genus *Ramichloridium* is related to the genera *Periconiella*, *Rhinocladiella* and *Veronaea*. The main morphological feature to distinguish *Ramichloridium* from *Rhinocladiella* is the presence of exophiala-type budding cells in the species of *Rhinocladiella* [3] (see also Chapter 43 of this book). Although both *Ramichloridium* and *Veronaea* lack exophiala-type budding cells, *Veronaea* is differentiated from *Ramichloridium* by having predominantly 1-septate conidia (i.e., two-celled conidia) (in comparison with one-celled or aspetate conidia of *Ramichloridium*). The largely aseptate conidia are also found in *Rhinocladiella* [10]. *Periconiella* differs from *Veronaea* chiefly by its dark brown, apically branched conidiophores [11].

Members of the genus *Ramichloridium* are largely soil saprobes. However, *Ramichloridium pini* is a known plant pathogen, causing a needle disease on *Pinus contorta* [3], and *Ramichloridium schulzeri* may be occasionally involved in human infection.

19.1.2 Clinical Features

The genus *Ramichloridium* represents 1 of 70 dematiaceous filamentous fungal genera that are largely saprophytes, but have the potential to cause phaeohyphomycosis, chromoblastomycosis, and eumycetoma in humans, especially those with suppressed immune functions [12–16]. *Ramichloridium schulzeri* was reported in a case of "golden tongue" involving a 54-year-old woman with acute lymphocytic leukemia. The patient was neutropenic, with an erosive lesion on the left side of the tongue, which extended over the dorsum of the tongue and appeared golden orange. Microscopic examination of surface scrapings revealed branching septate mycelia, and culture of a biopsy specimen of the tongue grew a few colonies of a fungus. The fungus was golden orange on Sabouraud's glucose agar and brown-gray on corn-meal agar and was identified as *Ramichloridium schulzeri*. The lesion was resolved after the patient underwent a week therapy with amphotericin B [17].

19.1.3 Diagnosis

The genus *Ramichloridium* is characterized by the presence of erect, dark, differentiated, branched or unbranched conidiophores and predominantly aseptate conidia produced on a sympodially proliferating rachis [2].

Morphologically and phylogenetically, *Ramichloridium* demonstrates close relation to *Rhinocladiella* and *Veronaea*. *Ramichloridium* and *Rhinocladiella* are separated mainly on the basis of (i) macronematous conidiophores in *Ramichloridium* versus micronematous conidiophores in *Rhinocladiella* and (ii) presence of a yellow or an orange diffusible pigment in *Ramichloridium* and absence in *Rhinocladiella* [3].

Other useful features for differentiation among *Ramichloridium*, *Rhinocladiella*, and *Veronaea* include (i) the absence of exophiala-type budding cells in *Ramichloridium* and *Veronaea* and its presence in *Rhinocladiella*; (ii) production of largely two-celled conidia (one-septate) by *Veronaea* and one-celled conidia (aseptate) by *Ramichloridium* and *Rhinocladiella* [10].

Nonetheless, identification of *Ramichloridium* from *Rhinocladiella* and *Veronaea* on the basis of macroscopic and microscopic characteristics are not only time consuming, but also technically challenging. For these reasons, molecular techniques targeting the sequence diversity of the internal transcribed spacer (ITS) region of ribosomal RNA (rRNA), as well as small subunit (SSU) and large subunit (LSU) rRNA, have been developed and applied for improved differentiation of *Ramichloridium* from *Rhinocladiella* and *Veronaea* [18–20].

19.2 METHODS

19.2.1 Sample Preparation

Biopsy samples are examined under microscope for mycotic elements using various stains. Samples are also cultured on inhibitory mold agar, modified Sabouraud agars, or potato dextrose agar (PDA). Colony colors (surface and reverse) are assessed after 2–4 weeks on different media at 25°C in the dark. Isolates are cultured on 2% MEA by obtaining single conidial colonies. Colonies are subcultured onto fresh MEA, OA, PDA, and synthetic nutrient-poor agar (SNA) and incubated at 25°C under continuous near-ultraviolet light to promote sporulation [6,8].

Microscopic structures are observed on tease or tape preparations and slide cultures for up to 21 days. Slide cultures are set up in Petri dishes containing 2 mL of sterile water, into which a U-shaped glass rod is placed, extending above the water surface. A block of freshly growing fungal colony, about 1 cm^2, is placed onto a sterile microscope slide, covered with a somewhat larger, sterile glass cover slip and incubated in the moist chamber. Fungal sporulation is monitored over time, and when optimal, images are captured by means of a Nikon camera system (Digital Sight DS-5M, Nikon Corporation, Japan). Structures are mounted in lactic acid and 30 measurements (×1000 magnification) are determined [10].

Genomic DNA is isolated from fungal mycelium grown on MEA, using the UltraClean™ Microbial DNA Isolation Kit (Mo Bio Laboratories) according to the manufacturer's protocols. Alternatively, genomic DNA is extracted following the cetyltrimethylammonium bromide (CTAB)-based protocol [6,8].

19.2.2 Detection Procedures

Arzanlou et al. [10] utilized the universal primers ITS1 and ITS4 [21] to amplify the ITS region of the nuclear rRNA operon, including: the 3′ end of the 18S rRNA gene, the first internal transcribed spacer region (ITS1), the 5.8S rRNA gene, the second internal transcribed spacer region (ITS2) and the 5′ end of 28S rRNA gene. Subsequent sequencing analysis allows identification of fungal organisms including *Chaetomium* species.

Procedure

1. PCR mixture (25 μL) is composed of 0.5 U *Taq* polymerase (Bioline), 1× PCR buffer, 0.5 mM $MgCl_2$, 0.2 mM of each dNTP, 5 pmol of each primer, approximately 10–15 ng of fungal genomic DNA.
2. Amplification is performed on a GeneAmp PCR System 9700 (Applied Biosystems) with primary denaturation at 96°C for 5 min; 36 cycles of 96°C for 30 s, 52°C for 30 s, and 72°C for 60 s; a final extension at 72°C for 7 min.
3. The amplicons are sequenced using BigDye Terminator v. 3.1 (Applied Biosystems,) or DYEnamicET Terminator (Amersham Biosciences) Cycle Sequencing Kits and analyzed on an ABI Prism 3700 (Applied Biosystems).
4. Newly generated sequences are subjected to a Blast search of the NCBI databases, sequences with high similarity are downloaded from GenBank and comparisons are made on the basis of the alignment of the obtained sequences.
5. Phylogenetic analysis is performed with PAUP (Phylogenetic Analysis Using Parsimony) v. 4.0b10 using the neighbor-joining algorithm with the uncorrected ("p"), the Kimura 2-parameter and the HKY85 substitution models. Alignment gaps longer than 10 bases are coded as single events for the phylogenetic analyses; the remaining gaps are treated as missing data. Any ties are broken randomly when encountered.
6. Phylogenetic relationships are also inferred with the parsimony algorithm using the heuristic search option with simple taxon additions and tree bisection and reconstruction (TBR) as the branch-swapping algorithm; alignment gaps are treated as a fifth character state and all characters are unordered and of equal weight. Branches of zero length are collapsed and all multiple, equally parsimonious trees are saved.
7. Other measures calculated include tree length, consistency index, retention index, and rescaled consistency index (TL, CI, RI, and RC, respectively). The robustness of the obtained trees is evaluated by 1000 bootstrap replications. Bayesian analysis is performed. The best nucleotide substitution model is determined using MrModeltest v. 2.2. MrBayes v. 3.1.2 is used to perform phylogenetic analyses, using a general time-reversible (GTR) substitution model with inverse gamma rates, dirichlet base frequencies and the temp value set to 0.5.

Note. Part of the LSU 28S rRNA gene may be also amplified with primers LR0R and LR5 followed by sequencing analysis.

19.3 CONCLUSION

The genus *Ramichloridium* consists of nine dematiaceous fungal species of which *Ramichloridium schulzeri* is involved in human infection. Due to the fact that *Ramichloridium* spp. closely resembles *Rhinocladiella* (which contains a significant human pathogenic species *Rhinocladiella mackenziei*) and *Veronaea* morphologically, it is important to determine these organisms to species level. As phenotypical methods for identification of *Ramichloridium*, *Rhinocladiella*, and *Veronaea* lack desired speed, accuracy, and convenience, molecular techniques such as PCR and sequencing analysis of the ITS region of rRNA, as well as SSU and LSU rRNA, offer a valuable alternative for improved differentiation of these organisms.

REFERENCES

1. The UniProt Consortium. Available at http://www.uniprot.org/, accessed on August 1, 2010.
2. de Hoog GS, *Rhinocladiella* and allied genera. *Stud Mycol.* 1977;15:1–140.
3. de Hoog GS, Rahman MA, Boekhout T. *Ramichloridium*, *Veronaea*, and *Stenella*: Generic delimitation, and new combinations and two new species. *Trans Br Mycol Soc.* 1983;81:485–490.
4. de Hoog GS et al., Species diversity and polymorphism in the *Exophiala spinifera* clade containing opportunistic black yeast-like fungi. *J Clin Microbiol.* 2003;41:4767–4778.
5. McGinnis MR, Schell WA, The genus *Fonsecaea* and its relationship to the genera *Cladosporium*, *Phialophora*, *Ramichloridium*, and *Rhinocladiella*. *Pan Am Health Organ Sci Publ.* 1980;396:215–234.
6. Crous PW et al., Opportunistic, human-pathogenic species in the *Herpotrichiellaceae* are phenotypically similar to saprobic or phytopathogenic species in the *Venturiaceae*. *Stud Mycol.* 2007;58:185–217.
7. Crous PW et al., Phylogenetic lineages in the *Capnodiales*. *Stud Mycol.* 2009a;64:17–47-S7.
8. Crous PW et al., Unravelling *Mycosphaerella*: Do you believe in genera? *Persoonia* 2009b;23:99–118.

9. Taj-Aldeen SJ et al., Cerebral phaeohyphomycosis due to *Rhinocladiella mackenziei* (formerly *Ramichloridium mackenziei*): A taxonomic update and review of the literature. *Med Mycol.* 2010;48(3):546–556.
10. Arzanlou M et al., Phylogenetic and morphotaxonomic revision of *Ramichloridium* and allied genera. *Stud Mycol.* 2007;58:57–93.
11. Cheewangkoon R et al., Species of *Mycosphaerella* and related anamorphs on Eucalyptus leaves from Thailand. *Persoonia* 2008;21:77–91.
12. Fader RC, McGinnis MR, Infections caused by dematiaceous fungi: Chromoblastomycosis and phaeohyphomycosis. *Infect Dis Clin North Am.* 1988;2:925–938.
13. Rinaldi MG, Phaeohyphomycosis. *Dermatol Clin.* 1996;14:147–153.
14. Revankar SD et al., Disseminated phaeohyphomycosis: Review of an emerging mycosis. *Clin Infect Dis.* 2002;34:467–476.
15. Brandt ME, Warnock DW, Epidemiology, clinical manifestations, and therapy of infections caused by dematiaceous fungi. *J Chemother.* 2003;15(Suppl 2):36–47.
16. Chakrabarti A, Epidemiology of central nervous system mycoses. *Neurol India* 2007;55:191.
17. Rippon JW et al., "Golden tongue" syndrome caused by *Ramichloridium schulzeri*. *Arch Dermatol.* 1985;121:892–894.
18. Untereiner WA, Naveau FA, Molecular systematics of the *Herpotrichiellaceae* with an assessment of the phylogenetic positions of *Exophiala dermatitidis* and *Phialophora americana*. *Mycologia* 1999;91:67–83.
19. Caligiorne RB et al., Internal transcribed spacer rRNA gene-based phylogenetic reconstruction using algorithms with local and global sequence alignment for black yeasts and their relatives. *J Clin Microbiol.* 2005;43(6):2816–2823.
20. Zeng JS, De Hoog GS, *Exophiala spinifera* and its allies: Diagnostics from morphology to DNA barcoding. *Med Mycol.* 2008;46:193–208.
21. White TJ et al., Amplification and direct sequencing of fungal ribosomal RNA genes for phylogenetics. In: *PCR Protocols: A Guide to Methods and Applications*, M.S. Innis and D.H. Gelfand (Eds.), Academic Press, New York, 1990, pp. 315–322.

20 *Ulocladium*

Dongyou Liu

CONTENTS

20.1 INTRODUCTION

The genus *Ulocladium* covers a number of saprotrophic, darkly pigmented hyphomycetes that share similar morphological characteristics with some saprotrophic *Alternaria* species. The morphological separation between the genera *Ulocladium* and *Alternaria* hinges on whether developing conidia are ovoid or obovoid, in addition to their subtle differences in the pigmentation and verrucosity of mature conidia. Although *Ulocladium* spp. have been often considered as contaminants in medical microbiology laboratories, there is increasing evidence that supports the role of these organisms in the pathogenesis of human mycoses.

20.1.1 Classification

The genus *Ulocladium* (obsolete synonym: *Pseudostemphylium*) is a dematiaceous (dark-walled) filamentous fungus belonging to the mitosporic Pleosporaceae group, family Pleosporaceae, order Pleosporales, class Dothideomycetes, subphylum Pezizomycotina, phylum Ascomycota, kingdom Fungi. The mitosporic Pleosporaceae group consists of 13 genera: *Alternaria*, *Dendryphiella*, *Dendryphion*, *Drechslera*, *Embellisia*, *Exserohilum*, *Nimbya*, *Pithomyces*, *Pyrenochaeta*, *Stagonospora*, *Stemphylium*, *Ulocladium*, and *Unifilum*. In turn, the genus *Ulocladium* is separated into 14 recognized species: *Ulocladium alternariae*, *Ulocladium atrum*, *Ulocladium botrytis*, *Ulocladium capsicum*, *Ulocladium chartarum*, *Ulocladium consortiale*, *Ulocladium cucurbitae*, *Ulocladium dauci*, *Ulocladium multiforme*, *Ulocladium obovoideum*, *Ulocladium oudemansii*, *Ulocladium septosporum*, *Ulocladium subcucurbitae*, and *Ulocladium tuberculatum*, as well as 25 unassigned species [1].

As saprophytes inhabiting the soil and decaying herbaceous plants, *Ulocladium* spp. are widely distributed in nature and have been isolated from paper, textiles, and wood. Although often considered as contaminants, some *Ulocladium* spp. (e.g., *Ulocladium chartarum*, *Ulocladium botrytis*, and *Ulocladium atrum*) are occasionally implicated in human disease process, causing subcutaneous phaeohyphomycosis, onychomycosis, keratitis, and other infections.

Ulocladium grows moderately rapidly. Colonies are wooly (suede-like) to cottony (floccose), brown to olivaceous-black or grayish on potato dextrose agar (PDA) at 25°C. Hyphae are brown, septate. Conidiophores are simple or branched, smooth, strongly geniculate (bent at the points where the conidia are produced, leading to a zigzag or bent knee appearance) and bear the conidia. Conidia (13–30 μm × 6–19 μm) are brown to black, typically obovoid (narrowest at the base), smooth or rough and verrucous. These conidia are typically muriform with transverse and longitudinal septations, appear solitary (*Ulocladium botrytis*) or form short chains (*Ulocladium chartarum*). Solitary, multicelled conidia (dictyoconidia) are formed through a pore (poroconidia) by a sympodially elongating geniculate conidiophore. When chains are produced, a tubular, short outgrowth is formed on the conidia at the point of secondary conidium formation [2,3].

Previously, *Ulocladium chartarum* had been included in the genus *Alternaria* as *A. chartarum* and *A. stemphylioides*. *Ulocladium* differs from *Alternaria* by its strongly geniculate (zigzag) conidiophores and the absence of beak-like tapered apex of conidia. It differs from *Bipolaris*, *Curvularia*, and *Drechslera* by producing muriform conidia. *Ulocladium* is differentiated from *Stemphylium* by having geniculate, sympodial conidiophores.

20.1.2 Clinical Features

Ulocladium spp. are dematiaceous fungi that exist as saprobes on rotten plant material and soil. These organisms appear to have low pathogenicity and are occasionally associated with subcutaneous infection, keratitis, and onychomycosis in humans, especially those under immunosuppressive therapies or after a local trauma.

20.1.2.1 *Ulocladium chartarum*

Several cases of cutaneous infections due to *U. chartarum* have been reported in the literature. Because *U. chartarum* was formerly regarded as a member of the genus *Alternaria*, most of these cases were described as being alternariosis [4–6].

Duran et al. [7] documented a cutaneous mycoses caused by *Ulocladium chartarum* in a 62-year-old male heart transplant recipient under immunosuppressive therapy (consisting of tacrolimus 2 mg/day, azathioprine 100 mg/day, and prednisone 10 mg/day). The patient noticed a painless, flesh-colored cutaneous lesion on his right toe over the previous 4 weeks, with no fever, malaise, or sweating and no history of trauma. The lesion appeared as a 6×6 cm sharply demarcated plaque on the dorsal area of his right big toe, with a granular surface and a vermiculate consistency. In cutaneous biopsy tissue sections stained with hematoxylin–eosin, numerous rounded, refringent, hyaline or slightly eosinophilic thick-walled fungal structures, together with a few elongated budding yeast-like forms and branched septate hyphae, were observed in the granuloma and within the giant cells present. Fungal elements were also revealed by periodic acid-Schiff (PAS) and Grocott-Gomori methenamine-silver nitrate stains, but brown pigment was detected in fungal cell walls with Masson-Fontana stain. The lesion was completely removed by surgery, and culture of the lesion tissue homogenate on bacteriologic (blood, chocolate, and MacConkey agars and thioglycolate broth) and mycologic media (Sabouraud dextrose agar [SDA] with and without chloramphenicol and gentamicin and brain heart infusion agar with 5% of blood) grew a mold after 2 days at 30°C and 35°C, respectively. Subculture of the mold on PDA generated powdery to lanose and black to olivaceous black colonies of 6 cm in diameter after 7 days at 30°C, which grew more slowly at 35°C. Microscopic examination of slide cultures on PDA demonstrated a mycelium with pale yellow or brown septate hyphae. Conidiophores were erect, geniculate, simple or branched and golden brown. Conidia were brown and verrucose with transverse and longitudinal septa, solitary or in chains through apical production of short conidiophores (false beaks). The fungus was thus identified as *U. chartarum*. Treatment with oral itraconazole (400 mg/day) began after complete surgical removal of the lesion, with a reduced intake of tacrolimus and prednisone and switch of azathioprine to mycophenolate mofetil. After 6 months of antifungal therapy, the surgical wound had fully healed with no relapse of the lesion.

20.1.2.2 *Ulocladium botrytis*

Romano et al. [8] reported a *Ulocladium botrytis*-related case of disto-lateral onychomycosis of the third toe of the right foot in a 45-year-old man is reported. Culture of pathological material grew a mold that was identified as *Ulocladium botrytis* based on the macro and microscopic characteristics of the colonies. After 3 months of topical therapy with ciclopirox olamine, the lesion was completed resolved.

20.1.2.3 *Ulocladium atrum*

Badenoch et al. [9] described a case of *Ulocladium atrum* keratitis in a 43-year-old man. The patient presented with corneal ulcer in his right eye, showing photophobia, marked conjunctival injection, corneal edema, Descemet's membrane folding, a central stromal infiltrate with a feathery edge, and a reduced visual acuity (VA) (for hand motions only) in the right eye. Gram staining of corneal scrapings showed septate hyphae, and culture of the material on Sabouraud's dextrose agar (without antibiotics) grew a visible colony at 28°C after 4 days, which reached a diameter of 2 cm by day 8 and appeared grayish brown and powdery. Under microscope, conidia were dark brown, coarsely verrucose with transverse and longitudinal septa. Subcultures of the isolate on cornmeal agar at 23°C or 28°C revealed long, flexuous, or simple to short, geniculate, and branched conidiophores. Conidia appearing within 48 h of subculture were spherical to ellipsoidal and mostly single. By 72 h, septate conidia became dark brown, verrucose at either temperature, with the septa often intersecting at right angles. The isolate was identified as a *Ulocladium* species on morphological grounds, most closely resembling *U. atrum*, and internal transcribed spacer (ITS) sequence analysis confirmed its *U. atrum* identity. After receiving hourly eye drops of natamycin (5%) and fluconazole (0.2%; the neat intravenous preparation), the patient was comfortable, with clinical signs disappearing in 4 weeks and an improved VA (20/20).

20.1.3 Diagnosis

Ulocladium represents one of the dematiaceous mold genera that are ubiquitously distributed in the environment and that may transiently colonize the respiratory, integument, and gastrointestinal systems of human hosts, with the potential to take advantage of the weakened host defense mechanisms (or injury), causing a diverse range of clinical diseases (often referred to as phaeohyphomycosis, chromoblastomycosis, and eumycetoma) [10–12]. In the case of *Ulocladium* spp., the most common clinical presentations are subcutaneous infections in addition to keratitis and onychomycosis.

Preliminary diagnosis of phaeohyphomycoses including *Ulocladium* infections are through demonstration of brownish hyphal and/or yeast-like elements in tissue, often with melanin-specific Fontana-Masson stain. However, the dark pigmented elements of some dematiaceous fungi (e.g.,

Alternaria infectoria and *Ulocladium chartarum*) may not be visible histologically by Fontana-Masson staining [9,13]. This is clearly demonstrated by Badenoch et al. [9] in their dealing of *Ulocladium chartarum*, whose dematiaceous nature was only evident after examination of a cultured isolate.

Because of the close morphological and clinical similarities between *Ulocladium* and *Alternaria*, including the formation of brown multiseptate conidia and the induction of nonspecific cutaneous infections, *Ulocladium* needs to be distinguished from *Alternaria*. Morphologically, *Ulocladium* differs from *Alternaria* by having obovoid, coarsely verrucose conidia with tapered and narrow bases and with no or short, spindle-shaped apices (false beaks). Furthermore, *Ulocladium* conidia are in short chains or non-catenate. In addition, *Ulocladium* species should not be confused with other poroconidial genera such as *Stemphylium*, *Bipolaris*, *Exserohilum*, *Dreschlera*, and *Curvularia* [14,15].

Polymerase chain reaction (PCR) amplification and sequencing analysis of the rRNA internal transcribed spacer (ITS) regions allows clear separation of *Ulocladium* from *Alternaria* and other fungal genera [9,13,15–19].

20.2 METHODS

20.2.1 Sample Preparation

Clinical specimens are examined microscopically with fungal stains and also inoculated onto SDA for fungal culture [20]. Microscopic structures are observed on tease or tape preparations and slide cultures for up to 21 days.

Susceptibility of *Ulocladium* isolates to amphotericin B, flucytosine, fluconazole, itraconazole, ketoconazole, voriconazole is performed using the National Committee for Clinical Laboratory Standards M38-A method for molds. For *Ulocladium* susceptibility to natamycin and terbinafine, an agar disk method adapted from M38-A is carried out. Namely, Neo-Sensitabs tablets (Rosco Diagnostica, Taastrup, Denmark) containing either diffusible natamycin (50 μg) or terbinafine (30 μg) are incorporated into RPMI 1640 agar supplemented with glucose (0.2%) and buffered with MOPS (morpholinepropanesulfonic acid; 0.165 M). The inoculum is standardized between 0.4×10^4 and 5×10^4 CFU/mL using a spectrophotometer. The plates are incubated at 35°C and examined at 48 and 72 h. The zone sizes indicate the sensitivity of the fungus to natamycin and terbinafine [9,21].

For DNA extraction, the isolate is cultured on PDA at 30°C for 5 days. A suspension to a McFarland standard of 2.0 is prepared in saline (2 mL) and centrifuged. The pellet is resuspended in 200 μL of sorbitol buffer containing 200 U of lyticase (Sigma-Aldrich) and incubated at 37°C for 60 min and centrifuged (5400 × *g*; 5 min). Spheroplasts are resuspended in 180 μL of lysis solution T and 20 μL of proteinase K (GenElute Mammalian Genomic DNA Miniprep kit; Sigma-Aldrich) and then incubated at 55°C for 60 min. DNA is extracted according to the manufacturer's instructions with a final elution volume of 200 μL. Samples are stored at −20°C until use [9].

20.2.2 Detection Procedures

20.2.2.1 Standard PCR Amplification and Sequencing Analysis of ITS Region

Badenoch et al. [9] utilized universal fungal primers ITS1 (5′-TCCGTAGGTGAACCTGCGG-3′) and ITS4 (5′-TCCTCCGCTTATTGATATGC-3′) to amplify the ITS region of the rRNA gene complex, incorporating ITS1, the 5.8S gene, and ITS2 [22].

Procedure

1. PCR mixture (25 μL) is made up of 1× GeneAmp PCR buffer (Applied Biosystems), 5% glycerol, 125 μM each deoxynucleoside triphosphate, 0.5 μM each primer (ITS1 and ITS4), 1.25 U of *Taq* DNA polymerase (Applied Biosystems), and 10 μL of DNA.
2. Amplification is conducted with an initial denaturation at 94°C for 2 min; 30 cycles of 94°C for 15 s, 55°C for 30 s, and 72°C for 30 s; and a final extension of 72°C for 6 min.
3. The amplified product is purified using a GFX PCR DNA and gel band purification kit (Amersham Biosciences) and then sequenced using the ITS1 primer and a BigDye Terminator v. 3.1 cycle sequencing kit in an ABI PRISM 3100-Avant genetic analyzer (Applied Biosystems).
4. The sequence is edited using Chromas v. 2.23 software (Technelysium Pty. Ltd.), and a 520 base sequence is compared with other sequences at GenBank (02/06) using FASTA.

20.2.2.2 Real-Time PCR Amplification and Sequencing Analysis of ITS Region

Pounder et al. [18] described a real-time PCR with SYBR green DNA binding dye and amplicon melting temperature analysis for fungal detection also using pan-fungal primers ITS1 forward (5′-TCCGTAGGTGAACCTGCGG-3′) and ITS4 reverse (5′-TCCTCCGCTTATTGATATGC-3′). The identity of the fungi is verified by subsequent sequencing analysis.

Procedure

1. PCR mixture is composed of 1× Lightcycler FastStart DNA Master Hybridization Probes mixture (Roche Applied Science) (containing deoxynucleoside triphosphates, FastStart *Taq* DNA polymerase, and 1 mM $MgCl_2$, additional $MgCl_2$ is added to a final concentration of 4.6 mM), 0.4 μM each of ITS1 forward and ITS4 reverse primers, 1× SYBR green (Molecular Probes), and 3 μL template DNA.
2. Thermal cycling parameters include 95°C for 10 min; 50 cycles of 95°C for 5 s, 60°C for 20 s, and 76°C for 30 s; and a final extension at 72°C for 2 min.
3. The quality of the amplicon is determined using the derivative of the melt analysis curve (55°C–99°C, 45 s hold at 55°C, 5 s/°C) using the RotorGene 3000 (Corbett Robotics, Inc).

4. The amplified product is purified for bidirectional sequencing using ExoSAP-IT (USB Corp). Five microliters of Big Dye Terminator Ready Reaction Mix v. 1.1 (Applied Biosystems) is added to 4 μL of each primer (0.8 pmol/μL) and 3 μL of purified PCR product. Cycle sequencing is performed with a 9700 thermal cycler (ABI), using 25 cycles of 96°C for 10 s, 50°C for 5 s, and 60°C for 4 min. Sequencing reaction products are passed through a Sephadex G-50 fine column to remove unincorporated dye terminators. Purified sequencing reaction products are run on an ABI Prism 3100 Genetic Analyzer with a 50 cm capillary array.
5. Sequences are analyzed with the SmartGene Integrated Database Network software version 3.2.3 vr. SmartGene is a web-based software and database system with reference sequences derived from the National Center for Biological Information (NCBI) GenBank repository.

Note. Sequence-based identifications are defined by percent identity: species, ≥99%; genus, 93%–99%; and inconclusive, ≤93%.

For strains producing discrepant identification between the methods based on phenotypic characteristics and ITS sequence analysis, the D1–D2 region of the large-subunit RNA gene is amplified with primers NL1 (5′-GCATATCAATAAGCGGAGGAAAAG-3′) and NL4 (5′-GGTCCGTGTTTCAAGACGG-3′) and sequenced for species clarification.

20.3 CONCLUSION

The genus *Ulocladium* consists of a large number of saprotrophic, darkly pigmented hyphomycetes that occur as saprophytes in soil and decaying herbaceous plants. Several *Ulocladium* spp. (e.g., *Ulocladium chartarum*, *Ulocladium botrytis*, and *Ulocladium atrum*) are occasionally implicated in human disease process, causing subcutaneous phaeohyphomycosis, onychomycosis, keratitis, and other infections. Because *Ulocladium* spp. often appear in laboratory cultures for dermatophytes and are regarded as contaminants, confirmation of the causative role of *Ulocladium* species in human disease relies on the presence of the clinical appearance that are consistent with a fungal infection, the observation of hyphae in biopsy tissue, and the culture isolation of the organism with characteristic colonial and microscopic morphology. Application of molecular techniques especially sequence analysis of the ITS region of rRNA genes provides a rapid and precise means of identifying *Ulocladium* spp. from *Alternaria* and other fungal taxa that may potentially confuse the diagnosis based on phenotypic features.

REFERENCES

1. The UniProt Consortium. Available at http://www.uniprot.org/, accessed on August 1, 2010.
2. Simmons, E. G. 1997. Multiplex conidium morphology in species of the *Ulocladium atrum* group. *Can J Bot.* 76:1533–1539.
3. de Hoog, G. S. et al. (eds.). 2000. *Atlas of Clinical Fungi*, 2nd edn. Centraalbureau voor Schimmelcultures, Utrecht, the Netherlands.
4. Srinivasan, M. et al. 1997. Epidemiology and aetiological diagnosis of corneal ulceration in Madurai, south India. *Br J Ophthalmol.* 81:965–971.
5. Magina, S. et al. 2000. Cutaneous alternariosis by *Alternaria chartarum* in a renal transplanted patient. *Br J Dermatol.* 142:1261–1262.
6. Williamson, E. C. et al. 2000. Diagnosis of invasive aspergillosis in bone marrow transplant recipients by polymerase chain reaction. *Br J Haematol.* 108:132–139.
7. Duran, M. T. et al. 2003. Cutaneous infection caused by *Ulocladium chartarum* in a heart transplant recipient: Case report and review. *Acta Derm Venereol.* 83:218–221.
8. Romano, C. et al. 2004. Onychomycosis due to *Ulocladium botrytis*. *Mycoses* 47:346–348.
9. Badenoch, P. R. et al. 2006. *Ulocladium atrum* keratitis. *J Clin Microbiol.* 44, 1190–1193.
10. Ajello, L. 1986. Hyalohyphomycosis and phaeohyphomycosis: Two global disease entities of public health importance. *Eur J Epidemiol.* 2:243–251.
11. Rinaldi, M. G. 1996. Phaeohyphomycosis. *Dermatol Clin.* 14:147–153.
12. Thomas, P. A. 2003. Current perspectives on ophthalmic mycoses. *Clin Microbiol Rev.* 16:730–797.
13. de Hoog, G. S. and R. Horré. 2002. Molecular taxonomy of the *Alternaria* and *Ulocladium* species from humans and their identification in the routine laboratory. *Mycoses* 45:259–276.
14. Simmons, E. G. 1967. Typification of *Alternaria*, *Stemphylium*, and *Ulocladium*. *Mycologia* 59:67–92.
15. Pryor, B. M. and D. M. Bigelow. 2003. Molecular characterization of *Embellisia* and *Nimbya* species and their relationship to *Alternaria*, *Ulocladium*, and *Stemphylium*. *Mycologia* 95:1141–1154.
16. Pryor, B. M. and R. L. Gilbertson. 2000. Molecular phylogenetic relationships amongst *Alternaria* species and related fungi based upon analysis of nuclear ITS and mtSSU rDNA sequences. *Mycol Res.* 104:1312–1321.
17. Meklin, T. et al. 2004. Quantitative PCR analysis of house dust can reveal abnormal mold conditions. *J Environ Monit.* 6:615–620.
18. Pounder, J. I. et al. 2007. Discovering potential pathogens among fungi identified as nonsporulating molds. *J Clin Microbiol.* 45(2):568–571.
19. Bagyalakshmi, R. et al. 2008. Newer emerging pathogens of ocular non-sporulating molds (NSM) identified by polymerase chain reaction (PCR)-based DNA sequencing technique targeting internal transcribed spacer (ITS) region. *Curr Eye Res.* 33(2):139–147.
20. Thomas, O. A. et al. 1991. Use of lactophenol cotton blue mounts of corneal scraping as an aid to the diagnosis of mycotic keratitis. *Diagn Microbiol Infect Dis.* 14:219–224.
21. Pujol, I. et al. 2000. In vitro antifungal susceptibility of *Alternaria* spp. and *Ulocladium* spp. *J Antimicrob Chemother.* 46:337–338.
22. White, T. J. et al. 1990. Amplification and direct sequencing of fungal ribosomal RNA genes for phylogenetics. In Innis, M.A. et al. (eds.), *PCR Protocols: A Guide to Methods and Applications*. Academic Press, New York, pp. 315–322.

Pezizomycotina: Eurotiomycetes

21 *Acrophialophora*

Dongyou Liu

CONTENTS

21.1 INTRODUCTION

With an expanding spectrum of high-risk immunocompromised patients in recent years, clinical cases of human mycoses due to saprobic and opportunistic fungi have shown a dramatic increase. *Acrophialophora fusispora* in the genus *Acrophialophora* represents a poorly known, thermotolerant fungus that has been documented as the cause of human eye and lung infections as well as cerebral brain abscess.

21.1.1 Classification and Biology

The genus *Acrophialophora* belongs to the mitosporic Ascomycota group, class *Dothideomycetes*, subphylum Pezizomycotina, phylum Ascomycota, kingdom Fungi. First described in 1959, the genus currently contains three recognized species: *Acrophialophora fusispora*, *Acrophialophora levis*, and *Acrophialophora nainiana*, with *A. fusispora* (obsolete synonym: *Paecilomyces fusisporus*) being the type species. Members of the genus *Acrophialophora* are widely distributed in tropical and temperate regions and have been isolated from soil. Of the three *Acrophialophora* species, *A. fusispora* (originally placed in the genus *Paecilomyces*) is a plant pathogen that may cause occasional infections (phaeohyphomycosis) in humans in India, the Middle East, Spain, and the United States [1].

Acrophialophora fusispora grows moderately to rapidly (with optimum growth temperature of 40°C and maximum growth temperature at 50°C). *A. fusispora* colonies attain diameters of 5–8 cm at 37°C after 7 days (with the average hourly growth rate of 0.46 mm/h after 3 days, compared with 0.38 mm/h after 7 days), and diameter 3.1–5.2 cm at 30°C after 7 days. Colonies are white to buff with darker concentric circles on Sabouraud dextrose agar (SDA) and are a darker gray-brown on potato flakes agar (PFA). The reverse is brownish with centrally darker areas on SDA and a solid gray-brown on PFA. Conidiophores (15 μm × 2–5 μm) arise singly, terminally, and laterally from thin-walled, hyaline-to-pale brown septate hyphae (1.5–3.5 μm wide), and appear erect, straight or slightly flexuose, tapering toward the apex, pale brown, echinulate, with whorls of phialides on the upper part. Phialides (6–10 μm × 3.5–6 μm) are flask-shaped with a swollen base and bear long chains of limoniform-to-fusiform, one-celled, smooth to finely echinulate, pale brown conidia (4–9 μm × 2–6 μm) with indistinct spiral bands [2].

The three species of the genus *Acrophialophora* are differentiated mainly by conidial ornamentation and the degree of development of brown conidia, although they display overlapping conidial dimensions. *A. nainiana* produces mature conidia (4–10.5 μm × 2–5 μm) that are hyaline and finely echinulate; *A. fusispora* has mature conidia (4–9 μm × 2–6 μm) that are brown, thick walled, and with echinulations in spiral bands; and *A. levis* (4.5–8 μm × 2–3.5 μm) shows mature conidia that are smooth to slightly roughened and hyaline. Some considered *A. nainiana* and *A. levis* as being synonymous for *A. fusispora*.

21.1.2 Clinical Features

Acrophialophora fusispora is a thermotolerant soil fungus that grows well at 45°C or higher temperatures. The fungus is neurotropic and has been reported as the etiologic agent for a variety of human infections including brain abscess, keratitis/keratouveitis, pulmonary infection, and other invasive diseases [2–5]. In addition, *A. fusispora* may be involved in transient airway colonization without inducing apparent clinical symptoms [6,7].

Al-Mohsen et al. [2] reported a case of *A. fusispora* CNS infection in a 12-year-old Sudanese girl with acute lymphoblastic leukemia, who received four courses of induction chemotherapy with vincristine and prednisone for four weeks. The patient was febrile with no focus of infection.

A computed tomography (CT) scan of the chest revealed multiple nodular densities; and CT and magnetic resonance imaging of the brain demonstrated a brain abscess in the parieto-occipital area. Thick yellowish pus (consisting of multiple tiny pieces of grayish soft and necrotic white tissue) was drained from the abscess. Gomori methenamine silver (GMS) stain of the brain abscess aspirate revealed numerous branching septate hyphae, and culture of this material on SDA yielded several colonies of a mold that was identified initially as a *Paecilomyces* species based on its formation of conidia in short chains. Subculture of the isolate on PFA at room temperature in ambient air with alternating daylight and darkness allowed its identification as *Acrophialophora fusispora*. The fungus grew rapidly at both 35°C and 42°C. Following treatment with liposomal AMB (LAMB) and then itraconazole (ITRA), the patient demonstrated clinical and radiological improvement, with concomitant reduction in size of the brain abscess and resolution of the lung nodules.

Arthur et al. [5] described a case of *Acrophialophora fusispora* keratitis involving a 76-year-old female, which was depicted mistakenly as *Scedosporium prolificans* in the original report [8,9]. The patient presented with mild swelling of the left upper eyelid, and a necrotic mass was visible, which contained numerous, variously sized, cauliflower-like growths covered by a whitish membrane on the superior surface. Hematoxylin-and-eosin-stained sections of the mass revealed extensive areas of necrotic tissue surrounding the long-lost contact lens, and septate hyphae within the necrotic tissue were visualized with Grocott methenamine silver nitrate stain. Inoculation of a portion of the mass on SDA slant grew two colonies of a white mold with hyaline septate hyphae after 4 days of incubation at 25°C in air. Subculture of the mold on potato dextrose agar (PDA) and Mycosel agar at 25°C yielded after 5 days yielded whitish colonies that turned dark with age. Lactophenol aniline blue stain of the isolate showed hyaline septate hyphae bearing flask-shaped conidiogenous cells with elongated necks and conidia. After removal of the mass along with ciprofloxacin treatment, the clinical signs of keratitis resolved rapidly. The patient has demonstrated no further episodes of microbial keratitis, and the vision in her left eye has improved to the level that existed prior to wearing contact lenses.

Guarro et al. [10] documented three cases of *Acrophialophora fusispora* infections, with one from India and two from Europe. The first case concerned a 55-year-old Indian female agricultural worker. The patient complained of pain, discharge, watering, redness, and blurring of vision after an injury with a wood chip in her left eye a month and a half before. On examination, the patient had a central corneal ulcer of 7×4mm with grayish white plaque, a thick hypopyon, and presence of infiltration. Culture of corneal scrapings on SDA, with chloramphenicol and gentamicin for fungal isolation grew a single fungus after 5 days. Gram staining and 10% KOH preparation demonstrated some hyphae. After therapeutic keratoplasty and treatment with natamycin, fluconazole, and ciprofloxacin, the patient was discharged with the healing of the lesion.

The second case involved a 33-year-old Portuguese male with bronchiectasis after measles, who underwent corticosteroid therapy after developing transplant rejection therapy 2 years earlier. The patient presented with severe dyspnea, fever, leukocytosis, C-reactive protein elevation, hypoxemia, and bilateral interstitial infiltrate on the chest x-ray. A single fungus was isolated from several sputum and bronchoalveolar lavage (BAL) samples in routine mycological culture media. Oral treatment with voriconazole (VRC) markedly improved the existing leukocytosis and C-reactive protein normalization.

The third case was a 67-year-old Spanish male with pulmonary fibrosis and bullous emphysema who underwent immunosuppressive therapy after transplantation of the left lung previously. On day 10 posttransplantation, a filamentous fungus was isolated from respiratory secretions in routine mycological culture media. Treatment with nebulized amphotericin B (AMB) and itraconazole (ITC) resulted in negative cultures of the mucous secretions. However, the patient developed pneumonia and died on day 163 posttransplantation because of progressive worsening clinical status. Necrotic samples from lung, pleural liquid, and myocardium were positive for *E. faecalis* and the same fungus isolated before.

The isolates from the above cases were characterized by the presence of conidiogenous cells basally inflated and with an elongated neck, arising usually singly on vegetative hyphae or along the length of pale brown, verrucose, and thick-walled conidiophores. Limoniform, pale brown, echinulate conidia, often with ornamentation in spiral bands, forming long chains were also common.

In a separate study, Cimon et al. [6] reported four cases of transient or chronic airway colonization by *Acrophialophora fusispora* in patients with cystic fibrosis (CF). In each of these cases, *A. fusispora* was isolated from sputum specimens along with *Aspergillus fumigatus*. The *A. fusispora* isolates obtained were thermotolerant, showing good growth at 42°C; microscopically, they exhibited pale brown hyphae (1.5–3.5 μm wide), and basally inflated phialides with elongated necks, arising mostly singly on vegetative hyphae or along the length and near the tip of pale brown echinulate conidiophores; and they showed long chains of limoniform-to-fusiform single-celled conidia (6–10 μm × 3.5–5.0 μm) that were pale brown, and smooth or finely echinulate with ornamentations often arranged in distinct spiral bands.

21.1.3 Diagnosis

The diagnostic features for *A. fusispora* consist of the following: (1) isolates are thermotolerant, showing good growth at 42°C or higher; (2) colonies are initially buff or tan, and usually turn grayish brown with an uncolored or dark grayish brown reverse; (3) basally swollen phialides are borne mostly singly on the vegetative hyphae or along the length and near the tip of brown, echinulate conidiophores; (4) limoniform-to-fusiform or ellipsoidal conidia are borne in long chains and are smooth or finely to coarsely echinulate, sometimes in spiral bands; (5) phialides do not curve away from the main axis; and (6) phialides sometimes proliferate to form a second opening. Phialides producing more than one opening not delimited by a septum are also referred to as polyphialides. The long, dark, echinulate conidiophores that form at maturity often appear to have "fallen down" on the plate and give the colony its darker appearance.

The genus *Acrophialophora* was originally placed in the genus *Paecilomyces* due to the fact that both genera form chains of ellipsoidal to fusiform conidia from basally swollen phialides borne either on conidiophores or directly from the vegetative hyphae. However, *Acrophialophora* is differentiated from *Paecilomyces* by (1) unbranched, erect, brown, echinulate conidiophores that are fertile only near the apex; (2) a basal cell anchoring the conidiophores to the vegetative hyphae; (3) phialides with a distinct swelling at the base, not curving or bending away from the main axis; and (4) colonies that become dark. Not surprisingly, in the case reported by Al-Mohsen et al. [2], the causative agent of the brain abscess was initially identified as *Paecilomyces* species on the basis of its formation of conidia in short chains. Only after detailed examination of other characteristics of the organism, the correct identity of *A. fusispora* was made.

Other fungal organisms that may be confused with *Acrophialophora fusispora* are *Scopulariopsis chartarum* [2,11] and *Scedosporium prolificans* [5,8,9] as well as *Acremonium*. *Acrophialophora* differs from *Scopulariopsis* in forming conidia from phialides instead of annellides. Furthermore, *Scopulariopsis* conidia are spherical and usually with a wide truncate base, while *Acrophialophora* conidia are fusiform or lemon-shaped, pale brown, finely echinulate, and often with distinct spiral bands. The confusion of *A. fusispora* with *S. prolificans* is due probably to the fact that both form flask-shaped conidiogenous cells. However, the flask-shaped conidiogenous cells of *Scedosporium prolificans* are often in brush-like arrangement, not single on hyphae as of *A. fusispora*. In addition, *A. fusispora* conidia are limoniform, and their walls are usually ornamented in spiral bands and form chains, and *A. fusispora* phialides arise singly on vegetative hyphae or on conidiophores, whereas *S. prolificans* conidia are clavate, smooth, and not arranged in chains but in slimy heads [12]. *Acremonium* spp. also have single conidiogenous cells (phialides) emerging from ropes of vegetative hyphae, but they rarely produce dark colonies [8,9].

Use of molecular techniques such as PCR and sequencing analysis of rRNA genes and internal transcribed spacer (ITS) regions provides a valuable approach for accurate determination of *A. fusispora* and other morphologically confusing fungal organisms [13].

21.2 METHODS

21.2.1 Sample Preparation

Mycotic elements in clinical specimens are examined directly under microscope with the help of various stains. Cultural and microscopic features are examined on SDA, PFA, and PDA plates incubated at 30°C and 37°C. PFA slant cultures are evaluated for growth at 25°C, 35°C, and 42°C.

After growth for 1–7 days on PDA slants, approximately 1 cm^2 of mycelia are collected by scraping the slant with a sterile stick in 1 mL of sterile, molecular-grade H_2O. The material is transferred to a 2 mL screw-cap tube. The tubes are centrifuged for 1 min at 6000 × *g*. If the mycelia do not pellet, the material is contained with a pediatric blood serum filter (Porex Corp., Fairburn, GA). Supernatant is removed, and the material is resuspended in 200 μL of IDI sample buffer and transferred to the lysis tube, which contained glass (IDI lysis kits, GeneOhm Sciences). Lysis tubes are vortexed on the highest setting for 5 min. The tubes are placed in a boiling water bath for 15 min, and then centrifuged for 5 min at 16,000 × *g*. The supernatant is stored at −20°C until amplification [13].

21.2.2 Detection Procedures

Pounder et al. [13] described a real-time PCR with SYBR green DNA binding dye and amplicon melting temperature analysis for fungal detection using pan-fungal primers ITS1 forward (5′-TCCGTAGGTGAACCTGCGG-3′) and ITS4 reverse (5′-TCCTCCGCTTATTGATATGC-3′) [14]. The identity of the fungi is verified by subsequent sequencing analysis.

Procedure

1. PCR mixture is composed of 1× Lightcycler FastStart DNA Master Hybridization Probes mixture (Roche Applied Science) (containing deoxynucleoside triphosphates, FastStart *Taq* DNA polymerase, and 1 mM $MgCl_2$ (additional $MgCl_2$ is added to a final concentration of 4.6 mM), 0.4 μM each of ITS1 forward and ITS4 reverse primers, 1× SYBR green (Molecular Probes), and 3 μL template DNA.
2. Thermal cycling parameters include 95°C for 10 min; 50 cycles of 95°C for 5 s, 60°C for 20 s, and 76°C for 30 s; and a final extension at 72°C for 2 min.
3. The quality of the amplicon is determined using the derivative of the melt analysis curve (55°C–99°C, 45 s hold at 55°C, 5 s/°C) using the RotorGene 3000 (Corbett Robotics, Inc).
4. The amplified product is purified for bidirectional sequencing using ExoSAP-IT (USB Corp). Five microliters of Big Dye Terminator Ready Reaction Mix v. 1.1 (Applied Biosystems) is added to 4 μL of each primer (0.8 pmol/μL) and 3 μL of purified PCR product. Cycle sequencing is performed with a 9700 thermal cycler (ABI), using 25 cycles of 96°C for 10 s, 50°C for 5 s, and 60°C for 4 min. Sequencing reaction products are passed through a Sephadex G-50 fine column to remove unincorporated dye terminators. Purified sequencing reaction products are run on an ABI Prism 3100 Genetic Analyzer with a 50-cm capillary array.

5. Sequences are analyzed with the SmartGene Integrated Database Network software version 3.2.3 vr. SmartGene is a web-based software and database system with reference sequences derived from the National Center for Biological Information (NCBI) GenBank repository.

Note: In case that real time PCR instrument is unavailable, standard PCR may be performed with primers ITS1 and ITS4, and the resulting amplicon is sequenced with the same primers. Sequence-based identifications are defined by percent identity: species, ≥99%; genus, 93%–99%; and inconclusive, ≤93%.

For strains producing discrepant identification between the methods based on phenotypic characteristics and ITS sequence analysis, the D1–D2 region of the large-subunit RNA gene is amplified with primers NL1 (5′-GCATATCAATAAGCGGAGGAAAAG-3′) and NL4 (5′-GGTCCGTGTTTCAAGACGG-3′) and sequenced for species clarification.

21.3 CONCLUSION

Acrophialophora fusispora (obsolete synonym: *Paecilomyces fusisporus*) is a filamentous fungus that is widely distributed in tropical and temperate regions. Besides causing plant disease, this organism is occasionally associated with human phaeohyphomycosis, with clinical manifestation ranging from brain abscess, keratitis/keratouveitis, pulmonary infection, and other invasive diseases.

Morphologically, *A. fusispora* is characterized by (1) isolates are thermotolerant, showing good growth at 42°C or higher; (2) colonies are initially buff or tan, and usually turn grayish brown with an uncolored or dark grayish brown reverse; (3) basally swollen phialides are borne mostly singly on the vegetative hyphae or along the length and near the tip of brown, echinulate conidiophores; (4) Limoniform-to-fusiform or ellipsoidal conidia are borne in long chains and are smooth or finely to coarsely echinulate, sometimes in spiral bands; (5) phialides do not curve away from the main axis; (6) phialides sometimes proliferate to form a second opening.

Fungal organisms that may be confused with *Acrophialophora* include *Paecilomyces* species, *Scopulariopsis chartarum*, *Scedosporium Prolificans*, and possibly *Acremonium* species. Besides paying attention to the finer structural differences among these fungi, application of molecular methods such as PCR and sequencing analysis of rRNA genes and ITS regions should help clarify the confusion in a rapid and decisive manner.

REFERENCES

1. The UniProt Consortium. Available at http://www.uniprot.org/, accessed on August 1, 2010.
2. Al-Mohsen, I. Z. et al., 2000. *Acrophialophora fusispora* brain abscess in a child with acute lymphoblastic leukemia: Review of cases and taxonomy. *J Clin Microbiol*. 38:4569–4576.
3. Shukla, P. K. et al., 1983. Clinical and experimental keratitis caused by the Colletotrichum state of *Glomerella cingulata* and *Acrophialophora fusispora*. *Sabouraudia*. 21:137–147.
4. Sutton, D. A. et al., 1997. Pulmonary *Acrophialophora fusispora*: Case history, literature, review and mycology. In: *Proceedings of the 13th International Society for Human and Animal Mycology Congress, 1997*, Salsomaggiore Terme, Italy, Abstract 363.
5. Arthur, S. et al., 2001. *Scedosporium prolificans* keratouveitis in association with a contact lens retained intraocularly over a long term. *J Clin Microbiol*. 39(12):4579–4582.
6. Cimon, B. et al., 2005. Airway colonization by *Acrophialophora fusispora* in patients with cystic fibrosis. *J Clin Microbiol*. 43(3):1484–1487.
7. Pihet, M. et al., 2009. Occurrence and relevance of filamentous fungi in respiratory secretions of patients with cystic fibrosis—A review. *Med Mycol*. 47(4):387–397.
8. Guarro, J. and J. Gene, 2002. *Acrophialophora fusispora* misidentified as *Scedosporium prolificans*. *J Clin Microbiol*. 40:3544.
9. Sigler, L. and D. A. Sutton, 2002. *Acrophialophora fusispora* misidentified as *Scedosporium prolificans*. *J Clin Microbiol*. 40:3544–3545.
10. Guarro, J. et al., 2007. *Acrophialophora fusispora*: An emerging agent of human mycoses. A report of 3 new clinical cases. *Diagn Microbiol Infect Dis*. 59(1):85–88.
11. Welsh, R. D. and R. W. Ely, 1999. *Scopulariopsis chartarum* systemic mycosis in a dog. *J Clin Microbiol*. 37:2102–2103.
12. de Hoog, G. S. et al., 2000. *Atlas of Clinical Fungi*, 2nd edn., vol. 1. Centraalbureau voor Schimmelcultures, Utrecht, the Netherlands.
13. Pounder, J. I. et al., 2007. Discovering potential pathogens among fungi identified as nonsporulating molds. *J. Clin Microbiol*. 45(2):568–571.
14. White, T. J. et al., 1990. Amplification and direct sequencing of fungal ribosomal RNA genes for phylogenetics. In Innis, M. A. et al. (eds.), *PCR Protocols: A Guide to Methods and Applications*. Academic Press, New York, pp. 315–322.

22 *Arthrographis*

Dongyou Liu

CONTENTS

22.1 INTRODUCTION

The genus *Arthrographis* consists of several filamentous fungal species that are commonly isolated from soil and compost. Within this genus, *Arthrographis kalrae* is an uncommon human pathogen, and has been shown as a causative agent for onychomycosis, mycetoma, sinusitis, endophthalmitis, keratitis, endocarditis, and other diseases, especially in immunocompromised patients. This fungus produces cream-colored colonies that can be mistaken for those of *Candida albicans*.

22.1.1 Classification and Morphology

The genus *Arthrographis* belongs to the mitosporic Eremomycetaceae group, family Eremomycetaceae, class Dothideomycetes, subphylum Pezizomycotina, phylum Ascomycota, kingdom Fungi. Being the only member of the mitosporic Eremomycetaceae group, the genus *Arthrographis* contains five recognized species: *Arthrographis alba*, *Arthrographis cuboidea* (obsolete synonyms: *Briosia microspora*, *Coremiella cuboidea*, and *Geotrichum cuboideum*), *Arthrographis kalrae* (obsolete synonyms: *A. langeroni* and *Oidiodendron kalrae*), *Arthrographis lignicola*, and *Arthrographis pinicola*, as well as an unassigned species *Arthrographis* sp. SI-2007a [1–3]. The type species of the *Arthrographis* genus is *A. kalrae*; and the teleomorphs of the genus are found in the genus *Pithoascus* (synonym: *Eremomyces*).

Arthrographis spp. are cosmopolitan filamentous fungi, with common occurrence in soil and compost. Of the five *Arthrographis* spp., *Arthrographis kalrae* is clinically significant, causing mycetoma, photophobia in contact lens wearer, sinusitis and meningitis in an AIDS patient, and sinusitis and ophthalmitis in healthy individual after trauma to the eye.

Arthrographis grows moderately to rapidly. Some *Arthrographis* strains produce colonies of 3–9 cm in diameter, while others reach diameters of only 0.5–1 cm at 25°C for 7 days on potato glucose agar. Colonies are glabrous initially, and become downy, velvety, or powdery by maturation, and may form radial ridges or folds. The surface color is creamy white to pale yellow or tan; the reverse side is pale yellow to tan. The ability of some *Arthrographis* species to grow at 45°C is useful for identification. Colonies on potato dextrose agar (PDA) at 25°C are buff to tan initially, usually becoming darker with age; the reverse is centrally dark at maturity [4].

Hyphae are hyaline, and septate. Conidiophores are hyaline, simple or branched, short. Arthroconidia are formed either at tips of the conidiophores or at intercalary position along the hyphae. Arthroconidia formed at the tips are one-celled, cylindrical, smooth, and in chains; and those that are intercalary and arise from the undifferentiated hyphae are longer and narrower. Conidia are released by fission through double septa. *Arthrographis* may occasionally produce aleuriconidia on submerged hyphae. Primary cultures may yield yeast cells, particularly when the colony is young. By aging, intercalary arthroconidia predominate while conidiophores disappear.

At the species level, *Arthrographis kalrae* is characterized by the presence of one-celled, hyaline, smooth-walled, and cylindrical arthroconidia. These arthroconidia are formed directly by fragmentation of undifferentiated hyphae or by disjunction and segmentation of hyaline fertile branches borne at the apex of the conidiophore in fresh cultures. Mature arthroconidia become bigger and elongated. In addition, single-celled, hyaline, smooth, spherical blastoconidia occurred directly on the sides of undifferentiated hyphae or on short pedicels. *A. kalrae* grows at 42°C and is urease positive. *Pithoascus langeronii* (synonym *Eremomyces langeronii*) was initially described as the teleomorph of *A. kalrae*, but the connection between *A. kalrae* and *E. langeronii* is not supported by molecular data [5].

Among the *Arthrographis* species, *A. cuboidea* displays more rapid growth and cube-shaped arthroconidia; *A. lignicola* has broader yellow arthroconidia; *A. pinnicola* produces floccose conidiomata and fails to grow on media containing cycloheximide; and *A. alba* shows a white colony that fails to grow at 37°C and lacks a *Trichosporiella* yeast synanamorph.

Arthrographis is a genus linked to *Malbranchea*. Due to the presence of arthroconidia, *Arthrographis* resembles *Oidiodendron*, but the conidiophores of *Arthrographis* lack the characteristic pigmentation. Furthermore, the arthroconidia of *Arthrographis* are smooth walled and lack the connectives between maturing conidia that are prominent in *Oidiodendron*. *A. kalrae* is similar to *Trichosporon* and *Geotrichum* species by the formation of yeast-like colonies and arthroconidia. However, unlike *Arthographis* and *Trichosporon* species, *Geotrichum* species are urease negative; *Trichosporon* species lack conidiophores, and *Trichosporon* assimilates a lot of carbon sources, which is not the case for *Arthrographis*.

22.1.2 Clinical Features

Arthrographis is widely distributed in soil and air and has only rarely been reported as an opportunistic pathogen in humans. This fungus enters the body through the respiratory tract or by traumatic inoculation. The first case of onychomycosis due to *A. kalrae* (*A. langeronii*) was described in France in 1939. Since then, *A. kalrae* has been shown to cause cutaneous and subcutaneous infections, eumycetoma, lung infection, keratitis, corneal ulcer, panophthalmitis, sinusitis, onyxis, meningitis, and cerebral vasculitis [6].

Degavre et al. [7] described a eumycetoma caused by *Arthrographis kalrae* in an 80-year-old man in southern France. The patient was admitted to hospital with pneumonia caused by *Pseudomonas aeruginosa* and treated with ceftazidime, ciprofloxacin, and systemic corticosteroids (methylprednisolone, 120 mg daily). Within 2 days, a well-demarcated erythematous, indurated, painless tumor and scattered nodules appeared on his left hand. Microscopic examination of skin biopsy demonstrated septate, narrow (2.5 µm wide) hyphae. Culture of the biopsy sample on brain-heart infusion (BHI) agar at 25°C yielded a fungus after 3 month incubation. Subcultures in Sabouraud dextrose agar (SDA) tubes (without cycloheximide) grew round, cream-colored, and velvety colonies. Microscopically, hyphae were narrow (2.5 µm) and septate with tufted conidiophores; spores measured 1.5–2 µm × 2.5–4 µm. The fungus was identified as *A. kalrae*. Treatment with itraconazole resulted in complete clearing of lesion.

Chin-Hong et al. [8] reported a case of *Arthrographis kalrae* pansinusitis and meningitis in a 33-year-old man with AIDS after receiving an 8-week course of nafcillin for osteomyelitis. The patient presented with a 4-day history of increasing bilateral retroorbital pain, and computed tomography (CT) scan showed severe maxillary, ethmoid, and sphenoid sinusitis. Biopsy obtained by endoscopic sphenoidotomy revealed hyphal forms. Cultures of the purulent material from the left sphenoid on SDA (Emmons) and inhibitory mold agar at 30°C for 48 h grew cream-colored, glabrous colonies, which became velvety after 5 days due to formation of hyphae and developed a pale yellow reverse. Elongated oval budding yeast cells were seen under microscope. Subculture on potato flakes agar (PFA) slants at 25°C were initially cream and moist colonies with a yeast-like appearance, but colonies became beige, flat, and powdery to granular in texture after 7 days. The fungus showed growth on SDA at 25°C, 35°C, and 42°C, with moist colonies present at 42°C. Microscopically, the isolate produced hyaline, septate hyphae; unbranched, and irregularly branched (tree-like) conidiophores; chains of one-celled, rectangular arthroconidia (2 µm × 4 µm) not separated by disjunctor cells; and occasional hyaline, sessile, subglobose conidia (4 µm × 5 µm) along the sides of the hyphae. The fungus was thus identified as *Arthrographis kalrae*. The patient was treated with itraconazole for approximately 5 months, but died secondary to enterococcal sepsis and *Pneumocytis carinii* pneumonia. Postmortem examination revealed invasive fungal sinusitis that involved the sphenoid sinus and that extended through the cribiform plate into the inferior surfaces of the bilateral frontal lobes. There was also an associated fungal meningitis and vasculitis with fungal thrombosis and multiple recent infarcts that involved the frontal lobes, right caudate nucleus, and putamen. Postmortem cultures were positive for *A. kalrae*.

Xi et al. [9] documented a case of *Arthrographis kalrae*-related panophthalmitis and invasive sinusitis in 39-year-old farmer after receiving trauma to his left eye. He complained of pain, loss of visual acuity in the injured eye, bleeding and edema of the upper eyelid, and discharge of pus. A CT scan of the left eye revealed involvement of the maxillary and ethmoid sinuses. Potassium hydroxide mounts of purulent material from the left eye orbit revealed filamentous hyaline and septate hyphae, and SDA cultures at 25°C grew maize-colored, powdery colonies. The colony was initially cream colored and glabrous, gradually became maize or yellow in color and velvety, and ultimately formed a powdery, yellowish colony after 7 days. The texture of the colony was soft and frangible. Slide cultures of the isolate on SDA and potato dextrose agar (PDA) at 25°C and 37°C, produced unbranched or irregularly branched hyphae, with dendritic or tree-like conidiophores; chains of single-celled, rectangular arthroconidia not separated by disjunction cells; and occasional hyaline, sessile, subglobose conidia. Lactophenol cotton blue (LPCB)-stained mounts of cells grown on BHIA at 37°C showed mainly oval to elongated yeast cells (2.5–4 µm in diameter) and short hyphae. The isolate was capable of growth at 42°C. The sequences of regions D1 and D2 (604 bp) and the ITS (508 bp) of the case isolate were identical to those of the control *A. kalrae* UAMH 3616 type strain. The isolate was thus identified as *A. kalrae*. The patient received amphotericin B intravenously, itraconazole orally, and atomized allitridum by nebulizing allitridum therapy. The patient's wound healed following surgical removal of the maxillary, ethmoid, and paranasal sinuses and the skin fistula, but the patient lost the use of his left eye.

Pichon et al. [10] reported a case of fatal invasive fungal cerebral vasculitis due to *Arthrographis kalrae* in a 39-year-old cattle breeder. The patient presented with a 1-month history of acute sinusitis, fever (temperature up to 39°C), dyspnea, cough with purulent sputums, and weakness. Magnetic resonance imaging (MRI) demonstrated an extended right frontal stroke and showed a right frontal sinusitis. Culture of the CSF sample on chocolate agar plates supplemented with Polyvitex (BioMérieux) at 37°C with 5% CO_2 revealed a moderate growth of yeast-like colonies after 9 days. Subculture of the isolate on Sabouraud agar showed a slow growth of the isolate at 25°C, 30°C, 37°C, and 42°C. The colonies were initially mucoid, with a yeast-like appearance, and became hairy due to formation of hyphae after 3 days of incubation on Chromagar BBL and Sabouraud medium at 30°C. After 5 days of incubation at 30°C, the colonies became pale yellow (Sabouraud agar and PDA). The isolate was resistant to cycloheximide, and the urease activity was positive. Slide culture on 2% malt agar at 30°C for 6 days showed characteristic one-celled, hyaline, smooth-walled, and cylindrical, elongated arthroconidia formed by differentiation of undifferentiated hyphae. Lateral spherical blastoconidia formed directly on hyphae or on short pedicels using the description [Sigler and Carmichael]. The identification of *Arthrographis kalrae* was confirmed by sequencing the internal transcribed spacer 1 (ITS1) region of the ribosomal DNA gene using universal primers ITS1 and ITS2. Despite medical intervention, the patient died from fatal-stroke syndrome. Postmortem examination of the brain tissues revealed a necrotizing arteritis and confluent fungal hyphae, sometimes dividing into arthroconidia, massively invading the meningeal space and infiltrating the brain arteries.

Biser et al. [11] described a case of keratomycosis caused by *Arthrographis kalrae* in a 23-year-old female contact lens wearer, with photophobia and pain in her amblyopic eye. Culture of eye sample grew a fungus, which was determined as *Arthrographis kalrae*. Treatment with amphotericin 0.5% drops led to moderately rapid resolution of the active keratitis. In a recent study, Sugiura and Hironaga [12] reported a case of onychomycosis due to *Arthrographis kalrae* in a 63-year-old Japanese man with markedly dystrophic fingernails. Microscopic examination of fingernail sample showed hyaline hyphae and arthroconidia. The causal agent was identified as *A. kalrae* based on morphological characteristics and ITS region sequencing. Treatment with oral terbinafine, 125 mg/day and topical 1% miconazole for 7 months led to a complete cure of the ungual lesion. Since the patient worked with soil that harbored *A. kalrae*, it was possible that the infection was acquired from this source.

22.1.3 Diagnosis

Accurate and rapid identification of *Arthrographis* species is important for improving the clinical outcome of the infections. Conventional laboratory identification of *Arthrographis* species is based on microscopic observation of fungal elements (e.g., filamentous hyaline and septate hyphae) in clinical samples using a variety of fungal stains (e.g., KOH and LPCB). Isolation of fungal organisms from clinical specimens is conducted with several mycology media (e.g., Sabouraud agar and PDA). A slide culture on 2% malt agar may be helpful. The colonial and microscopic features of the cultured isolates are assessed. *Arthrographis kalrae* colonies are initially cream and glabrous, gradually become maize or yellow in color and velvety, and finally powdery, yellowish after 7 days. *Arthrographis kalrae* hyphae appear unbranched or irregularly branched; conidiophores are tree-like; arthroconidia are single-celled, rectangular; conidia are hyaline, sessile, and subglobose; oval to elongated yeast cells may be observed. DNA sequence analysis of LSU rRNA region D2 (604 bp) and the ITS (508 bp) makes precise and efficient determination of *Arthrographis* species possible [13,14].

22.2 METHODS

22.2.1 Sample Preparation

Besides used for direction microscopic examination, clinical samples are cultured on inhibitory mold agar, modified Sabouraud agars, or PDA. Microscopic structures of the fungal isolates are observed on tease or tape preparations and slide cultures after incubation for up to 21 days [13].

After growth for 1–7 days on PDA slants, approximately 1 cm^2 of mycelia are collected by scraping the slant with a sterile stick in 1 mL of sterile, molecular-grade H_2O. The material is transferred to a 2 mL screw-cap tube. The tubes are centrifuged for 1 min at 6000 × *g*. If the mycelia do not pellet, the material is contained with a pediatric blood serum filter (Porex Corp., Fairburn, GA). Supernatant is removed, and the material is resuspended in 200 μL of IDI sample buffer and transferred to the lysis tube, which contained glass (IDI lysis kits, GeneOhm Sciences). Lysis tubes are vortexed on the highest setting for 5 min. The tubes are placed in a boiling water bath for 15 min, and then centrifuged for 5 min at 16,000 × *g*. The supernatant is stored at −20°C until amplification [13].

22.2.2 Detection Procedures

Pounder [13] described a real-time PCR with SYBR green DNA binding dye and amplicon melting temperature analysis for fungal detection using pan-fungal primers ITS1 forward (5′-TCCGTAGGTGAACCTGCGG-3′) and ITS4 reverse (5′-TCCTCCGCTTATTGATATGC-3′) [15]. The identity of the fungi is verified by subsequent sequencing analysis.

Procedure

1. PCR mixture is composed of 1×Lightcycler FastStart DNA Master Hybridization Probes mixture (Roche Applied Science) (containing deoxynucleoside triphosphates, FastStart *Taq* DNA polymerase, and 1 mM $MgCl_2$ (additional $MgCl_2$ is added to a

final concentration of 4.6 mM), 0.4 μM each of ITS1 forward and ITS4 reverse primers, 1× SYBR green (Molecular Probes), and 3 μL template DNA.
2. Thermal cycling parameters include 95°C for 10 min; 50 cycles of 95°C for 5 s, 60°C for 20 s, and 76°C for 30 s; and a final extension at 72°C for 2 min.
3. The quality of the amplicon is determined using the derivative of the melt analysis curve (55°C–99°C, 45 s hold at 55°C, 5 s/°C) using the RotorGene 3000 (Corbett Robotics, Inc).
4. The amplified product is purified for bidirectional sequencing using ExoSAP-IT (USB Corp). Five microliters of Big Dye Terminator Ready Reaction Mix v. 1.1 (Applied Biosystems) is added to 4 μL of each primer (0.8 pmol/μL) and 3 μL of purified PCR product. Cycle sequencing is performed with a 9700 thermal cycler (ABI), using 25 cycles of 96°C for 10 s, 50°C for 5 s, and 60°C for 4 min. Sequencing reaction products are passed through a Sephadex G-50 fine column to remove unincorporated dye terminators. Purified sequencing reaction products are run on an ABI Prism 3100 Genetic Analyzer with a 50-cm capillary array.
5. Sequences are analyzed with the SmartGene Integrated Database Network software version 3.2.3 vr. SmartGene is a web-based software and database system with reference sequences derived from the National Center for Biological Information (NCBI) GenBank repository.

Note: In case that real time PCR instrument is not available, standard PCR may be performed with primers ITS1 and ITS4, and the resulting amplicon is sequenced with the same primers. Sequence-based identifications are defined by percent identity: species, ≥99%; genus, 93%–99%; and inconclusive, ≤93%.

For strains producing discrepant identification between the methods based on phenotypic characteristics and ITS sequence analysis, the D1–D2 region of the large-subunit RNA gene is amplified with primers NL1 (5′-GCATATCAATAAGCGGAGGAAAAG-3′) and NL4 (5′-GGTCCGTGTTTCAAGACGG-3′) and sequenced for species clarification.

22.3 CONCLUSION

Arthrographis spp. are cosmopolitan filamentous fungi, which are commonly isolated from soil, compost and air, etc. Of the five recognized species within the genus *Arthrographis*, *Arthrographis kalrae* is occasionally involved in human infections, leading to mycetoma, photophobia in contact lens wearer, onychomycosis, endocarditis, meningitis, sinusitis and ophthalmitis after trauma to the eye [5–12,16,17].

Morphologically, *Arthrographis kalrae* colonies are initially cream in color and glabrous, turning to maize or yellow in color and velvety, and eventually powdery, yellowish. Under microscope, *Arthrographis kalrae* hyphae are unbranched or irregularly branched; conidiophores are tree-like; arthroconidia are hyaline, single-celled, smooth-walled, and cylindrical; conidia are hyaline, sessile, and subglobose; oval to elongated yeast cells may be observed. The fungus grows at 42°C and is urease positive. DNA sequence analysis of LSU rRNA region D2 and the ITS allows rapid and accurate speciation of *Arthrographis* species.

REFERENCES

1. The UniProt Consortium. Available at http://www.uniprot.org/, accessed on August 1, 2010.
2. Sigler, L. and J. W. Carmichael. 1983. Redisposition of some fungi referred to *Oidium microspermum* and a review of *Arthrographis*. *Mycotaxon*. 18:495–507.
3. Šigler, L., Y. Yamaoka, and Y. Hiratsuka. 1990. Taxonomy and chemistry of a new fungus from bark beetle infected *Pinus contorta* var. *latifolia*. Part 1. *Arthrographis pinicola* sp. nov. *Can J Microbiol*. 36:77–82.
4. de Hoog, G. S. et al. 2000. *Atlas of Clinical Fungi*, 2nd edn., vol. 1. Centraalbureau voor Schimmelcultures, Utrecht, the Netherlands.
5. Gené, J. et al. 1996. Studies on keratinophilic fungi. X. *Arthrographis alba* sp. nov. *Can J Microbiol*. 42:1185–1189.
6. Perlman, E. M. and L. Binns. 1997. Intense photophobia caused by *Arthrographis kalrae* in a contact lens-wearing patient. *Am J Ophthalmol*. 123(4):547–549.
7. Degavre, B., J. M. Joujoux, M. Dandurand, and B. Guillot. 1997. First report of mycetoma caused by *Arthrographis kalrae*: Successful treatment with itraconazole. *J Am Acad Dermatol*. 37:318–320.
8. Chin-Hong, P. V. et al. 2001. Invasive fungal sinusitis and meningitis due to *Arthrographis kalrae* in a patient with AIDS. *J Clin Microbiol*. 39:804–807.
9. Xi, L. et al. 2004. First case of *Arthrographis kalrae* ethmoid sinusitis and ophthalmitis in the People's Republic of China. *J Clin Microbiol*. 42:4828–4831.
10. Pichon, N. et al. 2008. Fatal-stroke syndrome revealing fungal cerebral vasculitis due to *Arthrographis kalrae* in an immunocompetent patient. *J Clin Microbiol*. 46:3152.
11. Biser, S. A. et al. 2004. *Arthrographis keratitis* mimicking *Acanthamoeba keratitis*. *Cornea*. 23(3):314–317.
12. Sugiura, Y. and M. Hironaga. 2010. *Arthrographis kalrae*, a rare causal agent of onychomycosis, and its occurrence in natural and commercially available soils. *Med Mycol*. 48(2):384–389.
13. Pounder, J. I. et al. 2007. Discovering potential pathogens among fungi identified as nonsporulating molds. *J Clin Microbiol*. 45(2):568–571.
14. Bagyalakshmi, R. et al. 2008. Newer emerging pathogens of ocular non-sporulating molds (NSM) identified by polymerase chain reaction (PCR)-based DNA sequencing technique targeting internal transcribed spacer (ITS) region. *Curr Eye Res*. 33(2):139–147.
15. White, T. J. et al. 1990. Amplification and direct sequencing of fungal ribosomal RNA genes for phylogenetics. In Innis, M. A. et al. (Eds.), *PCR Protocols: A Guide to Methods and Applications*. Academic Press, New York, pp. 315–322.
16. de Diego Candelo, J. et al. 2010. Endocarditis caused by *Arthrographis kalare*. *The Annals of Thoracic surgery*. 90:e4–e5.
17. Thomas, B. C. et al. 2011. Severe *Arthrographis kalrae* Keratomycosis in an Immunocompetent Patient. *Cornea*. 30(3):364–366.

23 *Aspergillus*

Maiken Cavling Arendrup, Yanan Zhao, and David S. Perlin

CONTENTS

23.1 INTRODUCTION

23.1.1 Epidemiology and Morphology

23.1.1.1 Epidemiology

Filamentous fungi belonging to the genus *Aspergillus* are found widespread in nature. They are commonly isolated from soil, decaying plants and vegetables, and from indoor air environment.[1–5] *Aspergillus* conidia are widely dispersed and are inhaled on a daily basis unless special air filtration measures are taken. *Aspergillus* species may cause disease due to inhalation of conidia or ingestion of mycotoxins resulting in allergic or toxic disease, chronic infection, and/or acute infections.[6] Allergic aspergillosis occurs largely in patients with asthma, atopy, or cystic fibrosis (CF),[7,8] while invasive pulmonary disease usually only occurs in immunocompromised patients with inhalation being the primary route of infection. Over recent decades, the number of patients with invasive aspergillosis (IA) has increased due largely to advances in the treatment of malignant diseases with an increasing number of patients undergoing severe immunosuppression, as part of intensive chemotherapeutical regimes, hematopoietic stem cell or organ transplantation.[9–15] *A. fumigatus* is the most common species involved in infections (85%–90%), but *A. flavus*, *A. niger*, *A. terreus*, and *A. nidulans* are also regularly recovered, with varying species distribution dependent on geographical differences, antifungal selective pressure, anatomical site, and host group involved.

23.1.1.2 Morphology

Macromorphology: The growth rate of aspergilli is rapid to moderately rapid (1–9 cm in diameter after 7 days of culture at 25°C), with the exception of the slow growing species *A. nidulans* and *A. glaucus*. Colonies are powdery, granular, or cottony with a variety of colors depending on the species and the culture medium used (Table 23.1). Only *A. fumigatus* is thermo tolerant with the ability to grow up to 50°C, a feature that is helpful to separate this species from the others, including the newly defined *Aspergillus* section *Fumigati*.

Micromorphology: Characteristic features for this genus include hyphae that are hyaline and septate with dichotomous branching. Conidiophores arise from a basal foot cell and germinate into a vesicle at the apex. Conidiogenic phialides arise directly from the vesicle (uniseriate) or from intersected metulae (biseriate) depending on the species. Conidia are round (2–5 μm) and formed into radial basipetal chains (Figure 23.1). Species-specific features include the presence or absence of sclerotia, cleistothecia, Hülle cells, aleuriconidia, chlamydoconidia, and morphology of the vesicles and arrangements of phialides (Table 23.1).

TABLE 23.1
Colony and Microscopic Features of Common *Aspergillus* Species

Species	Surface	Reverse	Conidiophore[a]	Vesicle	Phialides	Sclerotia	Cleisto-thecia	Hülle cells	Aleuriconidia	Comments
A. fumigatus	Blue-green-gray	White-tan	Short, smooth, colorless/greenish	Round columnar head	U, upper half/two-thirds of the vesicle	–	–	–	–	Growth at 50°C–55°C Whitish variants have been described
A. flavus	Yellow-green	Goldish-redbrown	Long, colorless, rough	Round radiate head	U/B	+	–	–	–	Conidiophore particularly rough near the vesicle Radial heads tend to separate into columns
A. niger	Black	White-yellow	Long, smooth, colorless or brown	Round radiate head	U/B	–	–	–	–	Max. temp. 48°C Radial heads tend to separate into columns Rough conidia
A. nidulans	Green, buff-yellow	Purplish red-olive	Short, smooth, brown	Round columnar head	B, upper half of the vesicle	–	+ (red)	+	–	Slow growth. Very evident foot cell
A. terreus	Cinnamon-brown	White/pale yellow-brown	Short, smooth, colorless	Round compact columnar head	B, upper two-thirds of the vesicle	–	–	–	+ (solitary, directly on hyphae)	Clamydoconidia may be present. Elevated AmB MICs
A versicolor	White-yellow/tan/pale green or pink	White-yellow/purplish red	Long, smooth, colorless	Round loosely radiate head	B, upper three-quarters of the vesicle	–	–	+/–	–	Whitish-dark red exudate may be present
A. clavatus	Blue-green	White-brownish	Very long, smooth	Huge, elongated, clavate shaped	U	–	–	–	–	Heads tend to separate into several columns
A. glaucus	Green with yellow areas	Yellow-brown	Variable length, smooth, colorless	Round, Radiate to very loosely columnar head	U	–	+ (yellow/orange)	–	–	Phialides may appear rudimentary

U, uniseriate; B, biseriate (with metulae).
[a] Short: <300 μm.

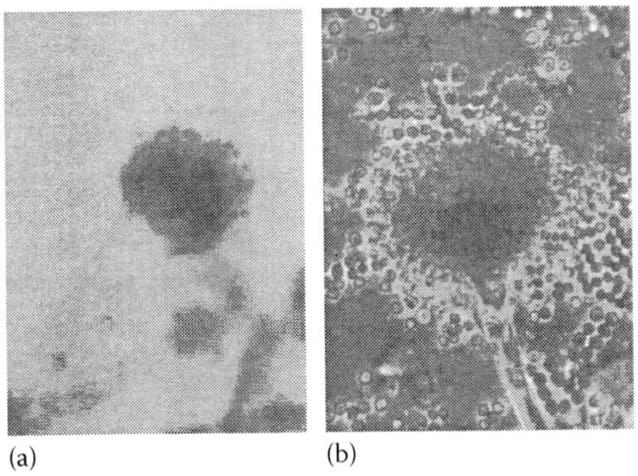

FIGURE 23.1 Wet mount of (a) *A. fumigatus* and (b) *A. flavus*. The vesicle bearing phialides on the upper half to two-thirds of the surface is clearly seen for *A. fumigatus*. The entire surface of the vesicle of *A. flavus* is covered with metulae from which conidia producing phialides arise. Also note the chains of conidia surrounding the *A. flavus* head as well as the rough appearance of the distal part of the conidiophore.

23.1.2 Clinical Features and Pathogenesis

Aspergillus is a multifaceted pathogen associated with a continuum of clinical presentations depending on the complex interaction of the fungus the host. Thus, the isolation of *Aspergillus* may represent transient or established colonization, fungal sensitization and allergy, localized infection, or acute or chronic invasive infection. The terminology used over time has changed and the disease entities are not always entirely distinct, which add to the complexity of the field. In the following, a short description of the main characteristic features concerning pathogenicity and clinical features for most acute and chronic *Aspergillus* infections will be described, although allergic manifestations are considered outside the scope of this chapter.

In the neutropenic host, *Aspergillus* infections are characterized by a hemorrhagic infection due to hyphal invasion of parenchyma and vessels, scattered inflammatory cells, and coagulative necrosis.[16,17] The lungs are the most common site of infection and depending on the site and size of vessels invaded, lesions may be spherical, nodular lesions, or wedge-shaped lesions with the base abutting the pleura.[18–20] The center of the lesion also called the sequestrum is a necrotic area surrounded by a hemorrhagic rim of hyphal invaded, congested, and hemorrhagic tissue. On computed tomography (CT) scanning, this typically present as a nodule surrounded by ground glass opacity—the so-called halo sign.[21,22] During neutrophil recovery, the sequester contracts, resulting in an air cap formation presenting on CT scanning as the air-crescent sign.[23] No clinical signs are diagnostic but rapidly progressive infection with symptoms including fever, dyspnea, cough, chest pain, and hemoptysis is characteristic. Due to the angioinvasion, hematogenous dissemination may occur involving almost any organ including the CNS, eye, heart, bone, and skin. The second most common primary sites of infection in the neutropenic host are the sinuses, potentially leading to direct invasion of the orbita and CNS, but also the skin, the valves and myocardium, and the gut may be primary sites of infection.

In the non-neutropenic host, *Aspergillus* may cause chronic forms of pulmonary aspergillosis (chronic necrotizing aspergillosis, chronic cavitary aspergillosis, and chronic fibrosing aspergillosis) and non-angioinvasive IA.[18,24] Host factors for the invasive form include GVHD, corticosteroid therapy, CGD, and solid organ transplantation, and in the majority of the chronic cases, preexisting architectural lung tissue changes are present including emphysema, prior mycobacterial infection, asthma, sarcoidosis, etc. The lesions are characterized by pyogranulomatous infection and inflammatory necrosis without angioinvasion.[16,17] Typically, the fungal burden in tissue is less, which in combination with a lack of angioinvasion and the ability of the host to mount an antibody response may be factors involved in the lower sensitivity of antigen tests.[25] Clinically, the course of infection is weeks to months in contrast to the rapidly progressive infection in the neutropenic host and symptoms include cough, shortness of breath, fatigue, weight loss, and hemoptysis. Aspergilloma, which is formation of an *Aspergillus* fungal ball of mycelium in a preexisting cavity without tissue or angioinvasion is included in the term chronic aspergillosis, but is reserved to the singularity of the infection. The disease progresses over months to years and the typical presentation on CT is a mobile fungus ball in a preexisting cavity. For recent reviews, please consult Refs. [18,26–30].

23.1.3 Diagnosis

Successful management of invasive fungal infections depends on timely and appropriate treatment.[22,31–33] Early diagnosis of IA remains a major challenge for high-risk patients because the signs and symptoms of diseases are nonspecific, blood cultures are usually negative, and biological markers can be unreliable, particularly in the non-neutropenic host or the patient receiving mold active therapy.[25,34] Diagnostic imaging is not disease-specific and often comes too late in the course of disease. Regrettably, invasive mold infections are too often positively diagnosed at autopsy, which has led to the widespread use of prophylaxis, empiric, and preemptive therapies.[35–38] Disease confirmation requires identification of hyphae in blood (rare), respiratory specimens or biopsy tissue, which can be difficult to discern as morphological features, reproductive structures, and biochemical properties can take days to weeks to properly evaluate.

Drug resistance is a confounding factor for the treatment of IA. Yet, acquired resistance in *Aspergillus* is still relatively uncommon. Thus, species identification is a valuable tool guiding choice of treatment as *Aspergillus* spp. characteristically display specific susceptibility patterns. For example, *A. terreus* is resistant to amphotericin B, while *A. lentilus* shows pleiotropic reduced susceptibility to several antifungal classes. Recently, it was reported in the United Kingdom that azole resistance in *A. fumigatus* has emerged as a significant factor in patients undergoing chronic or repeated therapy,[39] and, in the Netherlands, multi-azole resistant *A. fumigatus* with a specific resistance mechanism (L98H alteration and a tandem repeat in the promoter region) has been found in azole naïve patients and the environment.[4,40,41] On this background, microscopy and culture allowing species identification and subsequent susceptibility testing remain a corner stone in the diagnosis of aspergillosis despite associated limitations including lack of sensitivity, labor and time intensive procedures, and the difficulties in obtaining representative specimens from thrombocytopenic hematological patients, neonates, and children.

23.1.3.1 Conventional Techniques

The diagnosis of *Aspergillus* infections typically involves the combination of clinical signs and symptoms, imaging, and microbiological investigations including specimens for microscopy and culture, antigen or antibody detection. No test is perfect in the way that a positive test proves infection and a negative test rules it out. Thus, a combination of the available tools and interpretation of the results taking the clinical situation (underlying disease and other host factors) into account is necessary. Diagnostic criteria for stratification of patients into proven, probable, and possible aspergillosis has been developed and revised and represent a tool for classification of patients in clinical trials and for evaluation of new diagnostics.[42,43] In the following, the individual tests and their pros and cons are briefly summarized.

23.1.3.1.1 Imaging

IA typically involves the lung.[6,10] Chest radiographs generally show nonspecific infiltrates (alveolar syndrome, interstitial syndrome, or nodules) with lesions more characteristic for aspergillosis, such as cavitated nodules, only occasionally observed.[44] Chest CT may detect lung involvement at an early stage of infection in neutropenic patients.[22,23] In particular, the halo sign, that is, a macronodule (≥1 cm in diameter) surrounded by a perimeter of ground-glass opacity, has been shown to be an early indicator of invasive pulmonary aspergillosis (IPA), while the air-crescent sign (a pulmonary cavitation) is a later sign appearing with the bone marrow recovery.[23] In the non-immunocompromised host, CT imaging is less typical. Thus, while 71% of patients with neutropenic hematological malignancy had either halo- or air-crescent sign,[32] this was only the case in 5% or less of patients with IPA in the ICU setting.[34,45] Neuroradiological studies are helpful in the diagnosis of CNS infections.[46–48] Both CT and MRI are useful, but MRI may be more accurate and reveal more lesions than CT.[49] MRI and CT appearances of CNS aspergillosis have been studied in detail by several investigators.[49–51]

23.1.3.1.2 Microscopy

Direct microscopic examination is important for two reasons.[52] It provides rapid, easy, and cheep information about the presence of fungi and other pathogens and may allow sufficient identification to guide management and it is more sensitive than culture for a number of samples, especially if the patient has already been exposed to antifungal treatment.[53,54] Over the years, the sensitivity, specificity, and speed have been improved by the use of fluorescent brighteners such as calcofluor white or blankophor.[55–57] Important is to evaluate hyphae with respect to regularity, size, presence of septae and branching.[55] Use of special stains (Grocot Silver or PAS) increases the sensitivity and is particularly recommended for tissue biopsies.

23.1.3.1.3 Culture

Blood cultures are almost always negative in disseminated aspergillosis. Relevant sample material depends on the anatomic site involved but will in most cases be airway samples. Bronchoalveolar lavage (BAL) is superior to a tracheal aspirate, which again is superior to sputum. However, as *Aspergillus* species may be present in air, dust, etc., transient colonization of the airways or contamination of the sample is possible and should be kept in mind, especially, if growth is observed distant from the region of the agar surface that has been inoculated with the clinical material. Growth of *Aspergillus* may be suppressed by concomitant growth of bacteria. Therefore selective media inhibiting the growth of such should be included and agars incubated for a minimum of 5 days. Still sensitivity is around 25%–60%.[58–60] Species identification is traditionally performed according to macro- and micromorphology and thermotolerance (see above).

23.1.3.1.4 Susceptibility Testing

Reference methodologies using micro-broth dilution have been established[61,62] but susceptibility testing of molds is in general not performed in routine clinical microbiology laboratories. Whilst the endpoint reading is straight forward for azoles and amphotericin B due to the growth versus no growth pattern of inhibition, the endpoint determination for the echinocandins is more difficult to determine due to significant but partial growth inhibition resulting in altered hyphal morphology.[63,64] No resistance breakpoints have been established but epidemiological cut off values defining minimum inhibitory concentration (MIC) ranges for the wild-type population of *A. fumigatus* are available allowing interpretations of MICs as being within the wild-type range or above, in which case acquired resistance mechanisms should be considered.[65,66] The commercial Etest strip based upon the incubation of a plastic strip containing a concentration gradient of the antifungal compound on an inoculated agar surface has been evaluated and shows acceptable agreement if performed by experienced technicians.[64,67,68]

23.1.3.1.5 Antigen Detection

A validated antigen detection test nowadays exists for individual detection of *Aspergillus* galactomannan antigen (GM). In addition, yeast and mold infections (with the exception of *Cryptococcus* and zygomycetes) can be detected using (1,3)-β-D-glucan detection assay.

The performance of the *Aspergillus* galactomannan test has been extensively discussed in numerous publications. Best performance is seen if the test is used as a screening test twice or three times weekly in patients with neutropenic fever unresponsive to antibiotics.[36,59,69] The sensitivity is lower in patients receiving antifungal treatment and in non-neutropenic patients.[34,70] Positive results are available before a positive CT scan or culture in two-thirds of the patients.[59] In the neonatal population, false-positive results may occur more frequently in part due to gut colonization with *Bifidobacterium* spp.[71–73] Interestingly, it has recently been shown that galactomannan detection on BAL fluids and tissue biopsies may increase sensitivity.[34,74,75] In recent studies of (1,3)-β-D-glucan detection in adults, the performance has been encouraging[76–78] but direct comparisons with the galactomannan assay for detection of aspergillosis do not yield a consensus as to which test is best.[79–82] A practical drawback is that the test requires incubation and reading with a kinetic ELISA reader and that the test plates cannot be divided into separate strips, which lead to a high cost when running single samples.

23.1.3.1.6 *Antibody Detection*

While antibody detection plays an important role in the diagnosis of allergic aspergillus diseases, including allergic bronchopulmonary aspergillosis in CF patients or allergic rhinosinusitis, and in diagnosing chronic infections and aspergillomas, its role in diagnosing acute infection is limited.[18,83,84]

23.1.3.1.7 *Innate Immunity*

Innate immune molecules play a vital role in host defence against *A. fumigatus*. Mannose-binding lectin (MBL) is C-type serum lectin that acts as an opsonin and lectin complement pathway activator and inducer of proinflammatory cytokines, which helps confer innate immunity against several pathogenic organisms including *A. fumigatus*. Most individuals have low levels of circulating MBLs, and diminished levels have been linked with *Aspergillus* infections.[85,86] MBL variant alleles in patients with CF have been associated with a poor prognosis.[87]

23.1.3.2 Molecular Diagnostics

Molecular techniques are ideally suited for low abundance and difficult to culture pathogens like the *Aspergillus* spp. because they provide for faster, more sensitive, and accurate detection of infecting pathogens from a wide range of specimens. Unlike classical methodologies, they can rapidly detect in a matter of several hours a single genomic copy from blood or other specimens. Nucleic-based diagnostic assays utilizing amplification by the polymerase chain reaction (PCR) or nucleic acid sequence based assays (NASBA) are highly sensitive and can positively detect fungal infections at an early stage of infection.[88–92] The fungal ribosomal genes 18 S, 5.8 S, and 28 S, and their intervening noncoding regions, ITS1 and ITS2, provide molecular sequence signatures for panfungal, genera, and species-specific identification.[93–101] These signatures can be easily and accurately resolved by high fidelity, allele-specific molecular probes. Ribosomal genes are present in multiple (38–91) copies per genome,[102] which facilitates detection of small numbers of infecting organisms in a primary specimen. PCR primers designed to conserved regions of the rRNA targets are used to amplify the sequence-variable fragments, which are then detected by a variety of end point chemistries. Molecular diagnostic approaches with improved sensitivity and specificity are evolving rapidly, which can be used with other markers and clinical symptoms to confirm the presence of invasive disease.

The Holy Grail for molecular detection is the accurate and reproducible detection of small amounts of *Aspergillus* hyphal fragments or circulating nucleic acid in blood. This has not been easy and a wide variability has been observed in the field. An important issue is that *Aspergillus* growing in the lung is not likely to be blood-borne unless it becomes angioinvasive or cells die and nucleic acid is released into the bloodstream.

23.1.3.2.1 *PCR Detection Formats*

Standard PCR-based amplification is often insufficient for diagnostic use because of relatively high levels of false-positive results, since in the various target regions, some species vary by as little as single nucleotide.[100] Amplified DNA products can be detected in real time using intercalating dyes such as cyber green, although they provide no sequence-specific information. A wide range of endpoint detection technologies have been used to distinguish between fungal species including single-strand conformational polymorphism (SSCP),[103,104] random amplified polymorphic DNA (RAPD) PCR (RAPD-PCR),[105–107] reverse-hybridization line probe assay (LiPA),[108] restriction enzyme analysis[109,110] enzyme immunoassay (PCR-EIA) and real-time (RT) molecular probes, such as TaqMan™,[111–113] light-cycler,[114–116] and molecular beacons.[92,117–119] PCR amplification used in concert with self-reporting, high-fidelity hybridization probes has helped to improve accuracy. This approach is quantitative and has a large dynamic range, which have been used to rapidly identify fungal pathogens in both single target and multiplex formats.[92,117–119] Typically, these techniques can detect 1–10 genomic equivalents under optimal conditions. These approaches provide an opportunity to diagnose fungal infections at an early stage and in a timeframe that can influence patient management. Yet, as the technology pushes the limit of detection, it must take into account prior colonization with environmental spores of *Aspergillus* and the risk of contamination during sampling or processing. Such baseline factors require reliable quantitative data and that is linked to other clinical markers. Ultimately, validation must be correlated with clinical outcome.

23.1.3.2.2 *The Evolving PCR Experience*

The detection of fungal nucleic acid in blood and/or respiratory specimens at an early stage of invasive disease has been a major priority. PCR can be highly sensitive and predictive for detection of *Candida* and/or *Aspergillus*.[116,120] A meta-analysis on the use of PCR for the diagnosis of IA from more than 10,000 blood, serum, or plasma samples obtained from 1,618 patients showed a sensitivity and specificity of PCR of 0.75 (95% CI 0.54–0.88) and 0.87 (95% CI 0.78–0.93), respectively.[121] Yet, despite these highly promising results, home-brew PCR systems to detect *Aspergillus* are highly variable. The presence of nucleic acid, as detected by RT PCR with Taq-man or Light Cycler probes, varies from 30% to 100% when compared to circulating blood markers such as GM, diagnostic imaging, or clinical criteria.[122] Many reports indicate that PCR is superior to GM with respect to sensitivity rates.[123] Yet, some labs report more reliability detecting pathogen nucleic acid from respiratory specimens (e.g., BAL) relative to blood, suggesting that fungal burden and extraction efficiency in the specimen is important. For these reasons, the reliability of PCR as diagnostic tool for invasive fungal infections, especially those due to *Aspergillus* spp. is uncertain.

The major problems contributing to PCR variability are a lack of assay standardization and inefficient extraction of nucleic acids, especially from blood. However, significant methodological advancements in sample preparation and improved amplification/detection should greatly enhance the reliability of PCR as a diagnostic tool for invasive fungal infections. First, the lack of a standardized assay is significant as nearly every lab has a different approach. In addition, the use of positive controls for extraction and amplification are variable and quality control standards for probe and primer sets are rarely ever mentioned. The EU recognized this quandary for IA and has organized a consortium to adopt uniform standards.[121,124] Presently, only two standardized commercial PCR assays are available. The Roche Septi*Fast* RT PCR test detects and identifies 25 important bacterial and fungal species causing bloodstream infections including *C. albicans*, *C. glabrata*, *C. tropicalis*, *C. krusei*, *C. parapsilosis*, and *A. fumigatus*. The overall assay is reliable but labor intensive and complex. Myconostica has recently developed molecular assays for detecting common respiratory fungal pathogens of *Aspergillus* genus (MycAssay™ Aspergillus) and *Pneumocystis jirovecii* (*MycAssay Pneumocystis*). These are molecular beacon-based RT assays that are highly robust and easily expandable, and it has standardized internal controls for extraction and amplification. Presently, there is insufficient data with the commercial assays to fully assess their impact.

The second most important issue for PCR assays is extraction and recovery of pathogen-specific nucleic acid, especially from blood. Unlike viremias and bacteremias, which have high burdens in blood, invasive fungal infections have low circulating levels of pathogens or naked DNA. In IA, only low burdens of circulating nucleic acid or hyphal fragments are ever present. In fact, since *Aspergillus* is rarely cultured from blood, naked nucleic acid released from the respiratory tract may be most important for molecular detection. It is essential that extraction of fungal specific nucleic acid from sterile fluids is efficient, since it is critical to even the most sensitive detection modality. Fungal diagnostics require efficient extraction of nucleic acid from 1 to 50 genomic equivalents or ~40–2000 target molecules. This is more likely with respiratory samples of IA, which often contain DNA, hyphal fragments, and spores. (The value of detecting spore nucleic acid is questionable, since they are not normally associated with invasive disease.) *Aspergillus* hyphae or hyphal fragments can be extracted with reasonable efficiency with mechanical and/or enzymatic pre-lysis. For this reason, it is critical that the extraction system is highly efficient.

23.1.3.2.3 Improved Extraction of *Aspergillus* Nucleic Acid from Blood

The type of blood sample used for the molecular detection of DNA is also important, as whole blood, plasma, or serum have been analyzed. A comparison of different specimen types showed that serum is generally the preferred source for the diagnosis of IA by PCR,[121,125,126] most likely because nucleic acids are recovered more efficiently with routine extraction procedures. Recently, several promising reports have demonstrated a high degree of efficiency in isolating nucleic acid in whole blood by increasing the sample volume with an automated processing system (e.g., Roche MagnaPure™).[125] When combined with other well-established assays,[116,120,121] PCR appears poised to have an important impact on the diagnosis of invasive fungal infections. Suarez et al.,[125] in a 13 month prospective study, compared small (100 µL) and large (1000 µL) extraction volumes from blood utilizing automated extraction on a MagNaPure, and evaluated it relative to GM detection by an enzyme immunoassay. A total of 938 samples were screened for *Aspergillus* DNA following large volume extraction. The sensitivity, specificity, positive, and negative predictive values were 100%, 96.7%, 81%, and 100%, respectively, with larger serum volume. The assay sensitivity surpassed GM, produced a positive result with two cases of IA where the GM assay result remained negative, and in four IA cases, DNA was detected earlier than GM. In another newly published study,[120] a total of 2244 samples were analyzed by real-time quantitative PCR. Taking two consecutive positive results as the diagnostic criterion, PCR detected gave sensitivity, specificity, positive, and negative predictive values of 91.6%, 94.4%, 73.3%, and 98.5%, respectively. A combination of serial PCR and GM detected 100% of aspergillosis cases, with a positive predictive value of 75.1%. These new studies highlight the importance of extraction methodology and PCR standardization for improved performance of RT-PCR. There is still room for improvement in extraction efficiency, but the new approaches illustrate the value of a highly efficient RT-PCR assay.

23.1.3.2.4 Allele Discrimination and Multiplex Assays

One of the major advantages of RT-PCR assays with self-reporting probes is that they are allele discriminating, which improves accuracy and facilitates detection of single nucleotide changes that are needed to discriminate between closely related species or may result from mutations conferring drug resistance.[122,127] Molecular beacons, scorpion primers, TaqMan and several other types of commercially available probes have allele-discriminating capacities. Ultra-sensitive melt-curve analysis can also be used for allele discrimination, but polymorphisms outside the target sequence can be a limiting factor for specificity.[128] Innovations in nucleic acid amplification and detection have increased the number of targets that can be simultaneously detected in a single assay. The ability to detect multiple targets in a single multiplex assay is an important application of RT detection systems, which increases throughput and reduces assay costs. A wide range of approaches involving multiplexed probes, beads, and arrays can be employed to simultaneously detect a large number of targets facilitating rapid evaluation of multiple pathogens or drug resistance alleles. The advantage of these approaches is that they combine powerful nucleic acid amplification strategies with massive screening capability. However, a factor limiting the use of technology such as microarrays has been inadequate assay sensitivity and quantification.[129] Direct hybridization of DNA or RNA not only provides the least bias in gene detection, but also the lowest level of analytic sensitivity. This is largely due to practical limitations of efficient hybridization of linear probe-target hybrids. Ultimately, cost and limited sample throughput make conventional large planar microarrays less attractive in pathogen detection schemes. Alternative microarray formats such as bead arrays may circumvent cost and throughput limitations.

23.1.3.2.5 PanProbes Systems

A variety of panGenus probes for *Candida* and *Aspergillus* species have emerged in recent years. These probe systems utilize PCR amplification from common primers recognizing highly conserved regions of either 18S or 28S rRNA genes.[130,131] For many years, the biggest limitation of these assays was the presence of contaminating nucleic acid found in amplifying enzymes (Taq polymerase), as well as in reagents and collection tubes, which led to false-positive results. Changes in manufacturing processes, as well as new techniques to remove contaminating nucleic acid from enzyme preparations, has made the use of pan-Genera probes more reliable. Such broad fungal tests can be used to either identify or exclude fungal involvement, and in more complicated disease states, increase the certainty of invasive disease.[132]

23.1.3.2.6 Drug Resistance and Molecular Diagnostics

A wide variety of mechanisms account for triazole resistance in *Candida* spp. including mutations of the Erg11 target, overexpression of the target, and overexpression of multidrug efflux transporters of the ABC and MDS families[133,134] induced by global transcriptional regulators.[135–139] The variety and vast array of mutations makes molecular detection triazole resistance in *Candida* species difficult. In contrast, triazole resistance in *A. fumigatus* is more limited and involves largely mutations in Cyp51A (equivalent to Erg11). Triazole resistance mechanisms in *Aspergillus* can involve both overexpression of drug efflux transporters and/or modification of the target site sterol 14-α demethylase, encoded by *cyp51A* and *cyp51B*.[140,141] However, clinical resistance is mostly correlated with mutations in Cyp51A, which has been exploited to design a multiplex panel of allele-specific probes recognizing specific mutations associated with triazole resistance.[119] In a single molecular RT assay, it was possible to distinguish between triazole susceptible (wild-type sequence) and triazole resistant alleles. The new assay takes advantage of allele-discriminating probes in a multiplex assay format to evaluate drug resistance and assess differences between triazole drugs. It is suitable for primary specimens and can be incorporated into routine molecular testing regimens. Thus, molecular technology is now available for routine drug susceptibility testing of *Aspergillus* isolates to azole drugs. These targets cover a vast majority of Cyp51A mutations causing resistance. Yet, it is important to recognize that other resistance mechanisms (e.g., pump overexpression) can contribute to clinical resistance, albeit at a much lower frequency. The percentage of isolates with specific resistance mechanisms is under investigation in several large epidemiological studies.

23.1.3.2.7 Species Identification within Aspergillus fumigatus Complex

As *A. fumigatus* continues to be the most common etiological agent of IA, identification of species within the *Aspergillus* species complex (section *Fumigati*) may be important as previous studies have shown the existence of cryptic species whose antifungal susceptibility profile differs from *A. fumigates*.[142] *Aspergillus* section *Fumigati* contains 33 different species, with 10 strict anamorphs and 23 teleomorphs. Morphological observation is not sufficient to distinguish between them,[143] thus molecular typing methods mostly DNA-sequence based, is a valuable tool for species identification. As mentioned above, the ITS region was found to be useful for *Aspergillus* identification to the species complex level, but it does not always resolve very closely related phylogenetic species, whereas intron-rich protein coding genes generally do much better.[144] Therefore, the International Society for Human and Animal Mycology-sponsored *Aspergillus* Working Group has recommended the use of ITS for the species complex level identification and protein-coding locus for the identification of species within the complex.[145] Due to their prevalence in public databases, universality of application, and relative resolving power, the use of β-tubulin or calmodulin sequences has been recommended for such species identification.[144] Molecular tools other than single locus DNA sequencing, such as restriction fragment length polymorphism (RFLP), RAPD, and multilocus sequence typing (MLST) have also been applied to the same aim,[146,147] but they were either having limited discriminatory power within the complex[147–149] or low reproducibility[150] and not used as widely and effectively as β-tubulin/calmodulin sequencing.

23.2 METHODS

23.2.1 SAMPLE PREPARATION

Besides the lack of standardization and validation of PCR, another major obstacle limiting the wider use of nucleic acid detection of *Aspergillus* is the inefficient extraction of nucleic acids.[151] As for the best extraction method, a consensus is difficult to achieve because it varies with target type (DNA or RNA) and specimen source (blood, tissue, respiratory fluid). As extraction methods often vary widely from lab to lab, only the most commonly used approaches for different types of samples will be described below. Nucleic acid extraction from *Aspergillus* cells tends to be difficult and inefficient. This is largely due to the fact that the complex cell wall of filamentous fungi is hard to break. Thus, rigorous extraction methods that utilize either enzymatic digestion such as recombinant lyticase or mechanical disruption by bead-beating are required. However, the extra steps of cell lysis not only prolong the extraction procedure, but also may increase the possibility of fungal nucleic acid contamination especially when enzymatic lysis is performed,[152,153] since the sources of contamination are everywhere, including lytic enzymes, buffers, spin columns (the silica membranes), tubes, disposables, and even water. For this reason, the chosen extraction method represents a compromise between efficiency, lack of exogenous contamination, and the applicability to routine high-throughput laboratories.[83] High-speed cell disruption incorporating chaotropic reagents and lysing matrices is one combination, which provides relatively high extraction efficiency from *Aspergillus* and other filamentous fungi.[154] Either manual or automated methods can be adopted, depending on the number of samples needed to be processed and the availability of the automatic extraction instruments.

23.2.1.1 *Aspergillus* Culture

Both types of *Aspergillus* culture, conidia and mycelia (hyphae), may be the starting material for extraction, although the extractions starting with mycelia are more efficient than that with conidia.[111] Conidia are usually harvested from 3-to 5-day-old *Aspergillus* cultures grown on Sabouraud dextrose agar (SDA) or potato dextrose agar (PDA) by washing with sterile saline—0.1% Tween 20. Compared to conidia, mycelia harvest takes more efforts. The overnight *Aspergillus* culture grown in liquid media are filtrated, washed, dried, and thoroughly ground in liquid nitrogen using mortar and pestle. After collecting fungal cells, mechanical disruption takes place under the chaotropic condition of the lysis buffer (buffer AP1 or buffer RLT added with β-Mercaptoethonal if using DNeasy or RNeasy kit from QIAGEN). FastPrep instrument (MP Biomedical) with the matching lysing matrix is used in author's lab (Perlin lab) and also many other labs for the high-speed disruption. The velocity of 6 m/s for 45 s with a 5 min rest period between runs is the chosen setting. The lysis buffer is added to the conidia suspension or the fine powder of ground mycelia above, then transferred into the lysing matrix and processed on FastPrep instrument. After centrifugation, the cleared lysates (supernatants) are transferred from the lysing matrix and are ready to proceed to the rest of extraction by simply following instructions from the commercial kits.

Automated nucleic acid extraction can take place with the cleared lysates obtained from the FastPrep procedure, although the lysis buffer used for the disruption has to be replaced with that matching the automatic extraction machine. For instance, the easyMAG® system (bioMerieux) requires lysis buffer with proteinase K (20 μg/mL) added to the conidia suspension or ground mycelia powder in the lysing matrix to process the FastPrep cleared lysate. The suspension is incubated at 55°C for 1 h followed by a 10 min centrifugation at full speed using a countertop centrifuge. The cleared lysates are transferred into a sample cartridge and mixed with 2 mL of easyMAG lysis buffer. The silica solution is added, sample ID login and the program setting are performed by following the user's instructions. Other automatic extraction instrument such as MagNA Pure systems (Roche Applied Science) can also be used, except the lysis buffer should be adjusted accordingly as stated above.

23.2.1.2 Blood

The pathogenesis of IA is poorly understood and so is our knowledge of the optimal blood fraction required for the molecular detection of *Aspergillus* nucleic acid. While some researchers suggest that the yield of DNA from serum, plasma, and the white cell pellet is similar,[155] others find that whole blood is a better source for fungal DNA extraction than plasma.[156] We recommend the collection of whole blood for the nucleic acid detection, because the use of whole blood allows the detection of hyphal fragments and circulating nucleic acid while the use of serum only detects circulating nucleic acid.[151,157] The starting volume of whole blood varies from 200 μL to 10 mL in different clinical settings,[157–160] although it is expected that larger volumes produce greater yield, which promotes higher sensitivity.[161] Different extraction protocols have been widely investigated around a common principle, which is enrichment of fungal nucleic acid by removing the bulk of human nucleic acid to increase the sensitivity of the molecular assay. Currently, the hypotonic lysis of blood cells has been used very often for this purpose. Any in-house brewed or commercial blood cell lysis buffer can be used, and the main components usually consist of Tris buffer, $MgCl_2$, or with Triton X-100. By centrifugation, fungal cells are trapped in the pellet formed by blood cell debris, and the supernatant containing human nucleic acid is decanted. The pellets are resuspended in lysis buffer with proteinase K and transferred into lysing matrix as described for the extraction from *Aspergillus* culture above. Manual or automated extraction should be decided before the extraction starts since different lysis buffers will be used based on the different method.

23.2.1.3 Tissue

Histological examination and culture techniques applied to tissue are considered as the reference diagnostic standard for IA.[42] With more and more data showing that qPCR is more sensitive than culture for the detection of *Aspergillus* spp. in tissue,[162,163] the possibility of accepting a positive PCR result in tissue as the reference standard for IA deserves increasing attention. For preparation of fresh tissue samples, homogenization should be done in a sterile, nuclease-free environment. High-speed tissue homogenizer and crushing of tissues in a plastic bag are the most often used methods to prepare a tissue homogenate prior to nucleic acid extraction.[164] The tissue homogenates are then mixed with lysis buffer and proteinase K and processed on a FastPrep instrument, after which the protocols described for culture and blood are applied. For formalin-fixed, paraffin-embedded tissue samples, it has been reported that the extensive cross-linking of tissue proteins after fixation in formaldehyde can cause nucleic acid fragmentation and inhibition of the PCR process.[165] Sample preparation is more complicated than fresh tissue because of the extra dewaxing by xylene and washing by ethanol; the universal DNA extraction is applicable to formalin-fixed, paraffin-embedded tissue samples.[166] For detailed information, please consult Refs. [101,167].

23.2.2 Detection Procedures

An efficient way to detect and identify the clinically important *Aspergillus* species is using multitiered nucleic acid detection panels to recognize increasing complexity from pan-genera to species-specific and drug resistance-associated targets. Thus, different sets of primers and probes aimed at a variety of detection targets will be discussed. As representatives of DNA and RNA detection, RT-PCR and RT NASBA will be described below.

23.2.2.1 Primers, Probes, and Platforms

Among the extensive studies (>200) of detecting *Aspergillus* by PCR, most utilize ribosomal RNA genes as detection targets, as they are multicopy genes, contain highly conserved sequences as well as variable regions, which facilitate the genus and species-specific detection. Meanwhile, some studies have developed new assays using different targets such as mitochondrial DNA target.[115,155] As rapidly developing RT detection probes (e.g., TaqMan, molecular beacons) are incorporated into PCR assays, RT-PCR formats are likely to dominate the molecular diagnosis of *Aspergillus* in the near future. A summary of the most commonly used RT nucleic acid assays and the corresponding primer and probe sets for *Aspergillus* detection is listed in Table 23.2.

In addition, newly emerged nucleic acid amplification technology, NASBA offers some advantages over PCR. It is an RNA-directed isothermal amplification and as such there is no need for thermal cycling. Its robust RNA amplification is more sensitive than PCR. It hasn't been used widely for *Aspergillus* diagnosis, largely due to cost and complexity of quantification. Nevertheless, good performance has been reported from several studies using NASBA assays to detect *Aspergillus* spp.[91,92,168] For more detailed information, please consult Refs. [91,92,168,169]. Several other platforms including LightCycler with post-amplification (mostly PCR product) melting profile analysis and Luminex xMAP technology have been shown to identify different *Aspergillus* spp. simultaneously.[170,171] A LightCycler-based RT-PCR approach to detect common pathogenic *Aspergillus* species including *A. fumigatus*, *A. flavus*, *A. niger*, and *A. terreus* simultaneously was recently reported.[170] High-throughput DNA sequencing of ribosomal internal transcribed spacers (ITS1, ITS2) is ideal and is fundamentally the most accurate, but it is not as fast or as convenient as RT detection following nucleic acid amplification.

23.2.2.2 Secondary Drug Resistance

As for secondary drug resistance detection, different strategies can be applied depending on drug class and the resistance mechanisms. Our knowledge of drug resistance mechanisms in *Aspergillus* is still emerging. Secondary triazole resistance in *Aspergillus* predominantly involves mutations in genes associated with the drug target. Limited mutations in *Cyp51A* confer resistance, making them ideal targets for molecular detection. These mutations confer different susceptibility profiles: (1) resistance to itraconazole and posaconazole is associated with amino acid substitutions at Gly 54; (2) itraconazole-resistance and high MICs for voriconazole, ravuconazole, and posaconazole at Met 220; (3) azole cross-resistance with *cyp*51A overexpression produced by a tandem repeat (TR) of a 34 bp sequence in the *cyp*51A gene promoter in combination with an amino acid substitution at Leu 98 (TR-L98H); and (4) triazole cross-resistance related to a Gly 138 to cysteine substitution. They were used as the detection targets for the development of a RT-PCR assay to assess triazole resistance in *A. fumigatus*. When combined in a multiplex platform, the assay provides a rapid evaluation of drug resistance in *A. fumigatus*. By taking advantage of the single nucleotide difference detection capability of probes like molecular beacons in a high-throughput, multiplex format, a comprehensive two-tiered molecular diagnostic assay was developed for triazole-resistant *A. fumigatus* detection.[119] Molecular probes corresponding to wild-type and triazole-resistance mutations of *Aspergillus* Cyp51A (Gly54, Gly138, Met220, and L98/Tandem Promoter repeat) can be used to profile resistance in a two-tier drop out and positive identification format. Other mutations have been identified but not validated as a source of resistance. Nevertheless, the system is robust and can be expanded to include new validated targets. Table 23.3 shows the primers and beacon sequences proposed for this approach.

Resistance to echinocandin drugs remains relatively low, but it has been well recognized that genetic modifications of Fksl subunit is the universal mechanism for echinocandin resistant fungi.[172] Up to now, the widely reported amino acid substitutions in *FKS1* and *FKS2*, which are responsible for *Candida* species resistance to echinocandin drugs,[173,174] has not been observed from clinical *Aspergillus* isolates. Nonetheless, the over-expression of *FKS1* has been linked to the clinical *A. fumigatus* isolates with reduced susceptibility to caspofungin.[64] An RT TaqMan PCR assay to detect *FKS1* gene developed by Costa and colleagues demonstrated good performance of being able to detect 1 fg of *A. fumigatus* DNA

TABLE 23.2
Real-Time Nucleic Acid Assays for *Aspergillus* Detection

Target Gene	Assay Format	Primers and Probes	Sequences (5′–3′)	Specificity Determined	Sensitivity Determined	Reference
18S rRNA	TaqMan PCR	Forward primer	5′-TTGGTGGAGTGATTTGTCTGCT-3′	*Aspergillus* spp.	20 copies of *Aspergillus* DNA per assay	[158]
		Reverse primer	5′-TCTAAGGGCATCACAGACCTG-3′			
		Probe	5′-FAM-TCGGCCCTTAAATAGCCCGGTCCGC-3′-TAMRA			
18S rRNA	FRET PCR	Forward primer	5′-ATTGGAGGGCAAGTCTGGTG-3′	*Aspergillus* spp.	5 CFU/mL of blood	[175]
		Reverse primer	5′-CCGATCCCTAGTCGGCATAG-3′			
		Probe (LC)	5′-Red640-TGAGGTTCCCCAGAAGGAAAGGTGCAGC-3′		15 CFU/assay, 20 CFU/mL of blood	[159]
		Probe (FL)	5′-GTTCCCCCCACAGCCAGTGAAGGC-3′flu			
28S rRNA	TaqMan PCR	Forward primer (ASF1)	5′-GCACGTGAAATTGTTGAAAGG-3′	*Aspergillus* spp.	5 copies/assay, 10 CFU/mL of blood	Primers from [176]
		Reverse primer (ADR1)	5′-CAGGCTGGCCGCATTG-3′			
		Probe (ASP28P)	5′-FAM-CATTCGTGCCGGTGTACTTCCCCG-TAMRA-3′			Probe from [177]
28S rRNA	NASBA-Molecular Beacon	Sense primer (P2)	5′-CAGCAGTTGGACATGGGTTA-3′	*Aspergillus* spp. cross react with *Penicillium chrysogenum*	0.1–1 CFU/assay	[168]
		Antisense primer (P1)	5′-AATTCTAATACGACTCACTATAGGGGAGAATCCACATCCAGGTGC-3′			
		Probe (MB)	5′-CGCGA GTGCGCCGTGTGCCGAAA TCGCG -3′			
5.8S rRNA	TaqMan PCR	Forward primer	5′-TTGGTTCCGGCATCGA-3′	*Aspergillus* spp. also amplify *Fusarium* spp.	200 fg/assay	[178]
		Reverse primer	5′-GCAGCAATGACGCTCGG-3′			
		Probe	5′-FAM-TGAAGAACGCAGCGAAATGCGATAA-TAMRA-3′			

Mitochondrial DNA (rRNA)	FRET PCR	Antisense primer	5′-AACACCTGACCTTTCGCGTGTA-3′	*Aspergillus fumigatus*	3 fg/assay	[155]
		Sense primer/probe	5′-CTGTTAGTGCGGGAGTTCAAAXTCT-3′ (X represents internal labeling of the T with LC-Red 640)			
		Probe	5′-CTGAGCTAATTTCTTTCAACCCAAGGGA-3′			
Mitochondrial DNA (cytochrome *b* gene)	FRET PCR	Forward primer (AfLC2s)	5′-AATGCACGATACTGTAGGATCTG-3′	*Aspergillus fumigatus*, also amplify *Aspergillus clavatus*	10 copies of *Aspergillus* DNA, or 1–5 CFU/mL of blood	[115]
		Reverse primer (AfLC2as)	5′-TGCATTGGATTAGCCATAACA-3′			
		Probe (Cyt3A)	5′-Red640-TAATCTATCATAATTACCAGAAATACCTAAAGGA-3′			
		Probe (Cyt3B)	5′-AATCTTTAAATACAAAGTAAGGAGCGAAAG-3′flu			
18S rRNA	LightCycler (PCR + melting curve analysis)	Forward primer (ASP_F)	5′-ATTGGAGGGCAAGTCTGGTG-3′	*A. fumigatus*/*A. flavus* specific	5 CFU/mL of blood for *A. fumigatus*, and *A. flavus*	Primers from [157]
		Reverse primer (ASP_R)	5′-CCGATCCCTAGTCGGCAT-3′		*A. terreus* specific 50 CFU/mL of blood for *A. terreus* and *A. niger*	
		Probe (Asp.fum)	LC640Red-5′-TGAGGTTCCCCAGAAGGAAAGGTCCAGC-3′/5′-GTTCCCCCCACAGCCAGTGAAGGC-3′FL	*A. niger* specific		Probes from [170]
		Probe (Asp.nig)	LC640Red-5′-TGAGATTCCCCAGAAGGAAAGGTCCAGC-3′/5′-GTTCCCCCCACAGCCAGTGAAGGC-3′FL			
		Probe (Asp.terr)	LC705-5′-TGGGATTCCCCAGAAGGAAAGGTCCAGC-3′/5′-GTTCCCCCCACAGCCAGTGAAGGC-3′FL			

Notes: The copy numbers of ribosomal DNA were found variable, from 38 to 91 copies per genome.[102] The constant used for correlating the weight of genomic DNA of *A. fumigatus* to the number of genome was also different among studies. It was assuming 50 fg of DNA equals to 1 conidia by some studies,[157,179] and some study was calculating 33 fg of DNA as one genome.[178] For mitochondrial genes, Spiess et al. described as 10 copies of mitochondrial *A. fumigatus* cytochrome *b* gene corresponds to 13.2 fg of genomic DNA or 1–5 CFU.[115]

TABLE 23.3
Primers and Molecular Beacons for Triazole Resistance

Target region[a]	Sequence (5′–3′)[b]	Purpose
G54-F	TCATTGGGTCCCATTTCTG	Real-time PCR primer
G54-R	GCACGCAAAGAAGAACTTG	Real-time PCR primer
L98-F	GCGTTCAGGGAAACGAGT	Real-time PCR primer
L98-R	AAACGGGGGTCGTCAATG	Real-time PCR primer
G138-F	CAAGCTGATGGAGCAGAAAA	Real-time PCR primer
G138-R	CAATAAGTGGCACATGAGAC	Real-time PCR primer
M220-F	TCATGACCTGGACAAGGGC	Real-time PCR primer
M220-R	GCCGCATAACAAGAAGCGA	Real-time PCR primer
TR-F	ATCGCAGCACCACTCCAG	Real-time PCR primer
TR-R	CTTTCATTCGGCTCAGCAC	Real-time PCR primer
G54-MB	CGCGATCATCAGTTACGGGATTGATCCATCGCG	WT allele probe
G54W-MB	CGCGATCATCAGTTACTGGATTGATCCATCGCG	G54W mutant allele probe
L98-MB	CGCGACAACGGCAAGCTCAAGGATGGTCGCG	WT allele probe
G138-MB	CGCGACGTACGGCTTGACTCAGTCTGGTCGCG	WT allele probe
G138C-MB	CGCGACGTACTGCTTGACTCAGTCTGGTCGCG	G138C mutant allele probe
M220-MB	CGCGATCATCAATTTATGCTAACCGTGGGATCGCG	WT allele probe
TR-MB	CGCGTCGCTGAGCCGAATGAATCACGGACGCG	Tandem repetition probe
Gly 54-T	AAAAGGATCAATCCCGTAACTGATGAAAA	G54 allele target
G54W-T	AAAAGGATCAATCCAGTAACTGATGAAAA	G54W mutant allele target
Leu 98-T	TTTTTCATCCTTGAGCTTGCCGTTTTT	L98 allele target
Gly 138-T	AAAAACAGACTGAGTCAAGCCGTACTAAAA	G138 allele target
Gly 138-T	AAAAACAGACTGAGTCAAGCAGTACTAAAA	G138C mutant allele
Met 220-T	AAAAGCGCACGGTAGCATAAAATTGATGAAAA	M220 allele target
TR-T	AAAAACGTGATTCATTCGGCTCAGCAAAAA	Tandem repetition target

[a] F, sense; R, antisense; MB, molecular beacon; T, target.
[b] Probe and target domains are underlined.

in one PCR reaction. In this assay, the forward primer was 5′-GCCTGGTAGTGAAGCTGAGCGT-3′, the reverse primer is 5′-CGGTGAATGTAGGCA TGTTGTCC-3′, and the probe was 5′-FAM-TCACTCTCTACCCCCATGCCCGAGCC-3′-TAMRA.[112] Other RT formats can also be used for *FKS* detection, depending on the availability of different RT detecting instruments.

23.2.2.3 Assay Conditions

There is no concordant condition for *Aspergillus* nucleic acid tests described above. It varies when different platforms, primers and probes, and different commercial products are applied. It is important to make a best fit of every parameter to improve the probability for reliable detection. Regardless of the target used for detection, the working conditions always have to be optimized, especially in multiplex assays. Intricate interactions among primers, probes, and templates in multiplex format decrease the sensitivity of the assay as well as enhance the difficulty of working condition optimization. Therefore, whenever possible, consensus primers should be designed for multiplex detection. For detailed information, please consult the references listed in Table 23.2 for broad-spectrum and species-specific detection and those mentioned in Section 23.2.2.2 for drug resistance identification.

23.3 SUMMARY AND PERSPECTIVE

The challenges posed to positively and effectively diagnose IA at an early stage of disease are daunting. With current technology, no single assay, apart from direct culture, can deliver a definitive diagnosis. Rather, a combination of diagnostic approaches are required that utilize various biomarker endpoints, radiographic and clinical presentations. Molecular technology holds great promise for rapid and sensitive detection of nucleic acids from hyphal elements, and may be particularly helpful in identifying drug resistance markers against triazole drugs. As RT detection platforms become more robust, they need to detect infection at an early stage, prior to the onset of major clinical symptoms. Furthermore, they need to address common issues of spore contamination in non-sterile compartments, such as the upper respiratory tract, and the monitoring of therapeutic efficacy including emerging drug resistance.

REFERENCES

1. Hicks, J. B. et al., Fungal types and concentrations from settled dust in normal residences, *J. Occup. Environ. Hyg.*, 2, 481, 2005.
2. Horner, W. E. et al., Air- and dustborne mycoflora in houses free of water damage and fungal growth, *Appl. Environ. Microbiol.*, 70, 6394, 2004.
3. Sautour, M. et al., Profiles and seasonal distribution of airborne fungi in indoor and outdoor environments at a French hospital, *Sci. Total Environ.*, 407, 3766, 2009.
4. Snelders, E. et al., Possible environmental origin of resistance of aspergillus fumigatus to medical triazoles, *Appl. Environ. Microbiol.*, 75, 4053, 2009.
5. Zuk, J. A. et al., Locally invasive pulmonary aspergillosis occurring in a gardener: An occupational hazard? *Thorax*, 44, 678, 1989.
6. Marr, K. A. et al., Aspergillosis. Pathogenesis, clinical manifestations, and therapy, *Infect. Dis. Clin. North Am.*, 16, 875, 2002.
7. Denning, D. W. et al., The link between fungi and severe asthma: A summary of the evidence, *Eur. Respir. J.*, 27, 615, 2006.
8. Thia, L. P. and Balfour Lynn, I. M., Diagnosing allergic bronchopulmonary aspergillosis in children with cystic fibrosis, *Paediatr. Respir. Rev.*, 10, 37, 2009.
9. Pagano, L. et al., The epidemiology of fungal infections in patients with hematologic malignancies: The SEIFEM-2004 study, *Haematologica*, 91, 1068, 2006.
10. Pagano, L. et al., Fungal infections in recipients of hematopoietic stem cell transplants: Results of the SEIFEM b-2004 study—Sorveglianza epidemiologica infezioni fungine nelle emopatie maligne, *Clin. Infect. Dis.*, 45, 1161, 2007.
11. Marr, K. A., Fungal infections in hematopoietic stem cell transplant recipients, *Med. Mycol.*, 46, 293, 2008.
12. Horn, D. L. et al., Presentation of the path alliance® registry for prospective data collection and analysis of the epidemiology, therapy, and outcomes of invasive fungal infections, *Diagn. Microbiol. Infect. Dis.*, 59, 407, 2007.
13. Hammerman, K. J. et al., The incidence of hospitalized cases of systemic mycotic infections, *Sabouraudia*, 12, 33, 1974.
14. Fraser, D. W. et al., Aspergillosis and other systemic mycoses. The growing problem, *JAMA*, 242, 1631, 1979.
15. Dasbach, E. J. et al., Burden of aspergillosis-related hospitalizations in the United States, *Clin. Infect. Dis.*, 31, 1524, 2000.
16. Berenguer, J. et al., Pathogenesis of pulmonary aspergillosis. Granulocytopenia versus cyclosporine and methylprednisolone-induced immunosuppression, *Am. J. Respir. Crit. Care Med.*, 152, 1079, 1995.
17. Balloy, V. et al., Differences in patterns of infection and inflammation for corticosteroid treatment and chemotherapy in experimental invasive pulmonary aspergillosis, *Infect. Immun.*, 73, 494, 2005.
18. Hope, W. W. et al., The invasive and saprophytic syndromes due to *Aspergillus* spp., *Med. Mycol.*, 43(Suppl 1), S207, 2005.
19. Fraser, R. S., Pulmonary aspergillosis: Pathologic and pathogenetic features, *Pathol. Annu.*, 28(Pt 1), 231, 1993.
20. Denning, D. W., Invasive aspergillosis, *Clin. Infect. Dis.*, 26, 781, 1998.
21. Meziane, M. A. et al., High resolution CT of the lung parenchyma with pathologic correlation, *Radiographics*, 8, 27, 1988.
22. Caillot, D. et al., Improved management of invasive pulmonary aspergillosis in neutropenic patients using early thoracic computed tomographic scan and surgery, *J. Clin. Oncol.*, 15, 139, 1997.
23. Caillot, D. et al., Increasing volume and changing characteristics of invasive pulmonary aspergillosis on sequential thoracic computed tomography scans in patients with neutropenia, *J. Clin. Oncol.*, 19, 253, 2001.
24. Denning, D. W. et al., Chronic cavitary and fibrosing pulmonary and pleural aspergillosis: Case series, proposed nomenclature change, and review, *Clin. Infect. Dis.*, 37(Suppl 3), S265, 2003.
25. Cordonnier, C. et al., Correlation between galactomannan antigen levels in serum and neutrophil counts in haematological patients with invasive aspergillosis, *Clin. Microbiol. Infect.*, 15, 81, 2009.
26. Riscili, B. P. and Wood, K. L., Noninvasive pulmonary aspergillus infections, *Clin. Chest Med.*, 30, 315, 2009.
27. Dagenais, T. R. and Keller, N. P., Pathogenesis of aspergillus fumigatus in invasive aspergillosis, *Clin. Microbiol. Rev.*, 22, 447, 2009.
28. Bulpa, P. et al., Invasive pulmonary aspergillosis in patients with chronic obstructive pulmonary disease, *Eur. Respir. J.*, 30, 782, 2007.
29. Segal, B. H., Aspergillosis, *N. Engl. J. Med.*, 360, 1870, 2009.
30. Segal, B. H. and Romani, L. R., Invasive aspergillosis in chronic granulomatous disease, *Med. Mycol.*, 47(Suppl 1), S282, 2009.
31. Herbrecht, R. et al., Voriconazole versus amphotericin b for primary therapy of invasive aspergillosis, *N. Engl. J. Med.*, 347, 408, 2002.
32. Greene, R. E. et al., Imaging findings in acute invasive pulmonary aspergillosis: Clinical significance of the halo sign, *Clin. Infect. Dis.*, 44, 373, 2007.
33. Caillot, D. et al., Role of early diagnosis and aggressive surgery in the management of invasive pulmonary aspergillosis in neutropenic patients, *Clin. Microbiol. Infect.*, 7(Suppl 2), 54, 2001.
34. Meersseman, W. et al., Galactomannan in bronchoalveolar lavage fluid: A tool for diagnosing aspergillosis in intensive care unit patients, *Am. J. Respir. Crit. Care Med.*, 177, 27, 2008.
35. Walsh, T. J. et al., Treatment of aspergillosis: Clinical practice guidelines of the Infectious Diseases Society of America, *Clin. Infect. Dis.*, 46, 327, 2008.
36. Maertens, J. et al., Galactomannan and computed tomography-based preemptive antifungal therapy in neutropenic patients at high risk for invasive fungal infection: A prospective feasibility study, *Clin. Infect. Dis.*, 41, 1242, 2005.
37. Cordonnier, C. et al., Empirical versus preemptive antifungal therapy for high-risk, febrile, neutropenic patients: A randomized, controlled trial, *Clin. Infect. Dis.*, 48, 1042, 2009.
38. Chamilos, G. et al., Invasive fungal infections in patients with hematologic malignancies in a tertiary care cancer center: An autopsy study over a 15-year period (1989–2003), *Haematologica*, 91, 986, 2006.
39. Howard, S. J. et al., Azole resistance in aspergillus fumigatus associated with treatment failure, *Emerg. Infect. Dis.*, 15, 1068, 2009.
40. Verweij, P. E. et al., Multiple-triazole-resistant aspergillosis, *N. Engl. J. Med.*, 356, 1481, 2007.
41. Snelders, E. et al., Emergence of azole resistance in aspergillus fumigatus and spread of a single resistance mechanism, *PLoS Med.*, 5, e219, 2008.

42. Ascioglu, S. et al., Defining opportunistic invasive fungal infections in immunocompromised patients with cancer and hematopoietic stem cell transplants: An international consensus, *Clin. Infect. Dis.*, 34, 7, 2002.
43. De Pauw, B. et al., Revised definitions of invasive fungal disease from the European Organization for Research and Treatment of Cancer/Invasive Fungal Infections Cooperative Group and the National Institute of Allergy and Infectious Diseases Mycoses Study Group (EORTC/MSG) consensus group, *Clin. Infect. Dis.*, 46, 1813, 2008.
44. Cornillet, A. et al., Comparison of epidemiological, clinical, and biological features of invasive aspergillosis in neutropenic and nonneutropenic patients: A 6-year survey, *Clin. Infect. Dis.*, 43, 577, 2006.
45. Vandewoude, K. H. et al., Clinical relevance of aspergillus isolation from respiratory tract samples in critically ill patients, *Crit. Care*, 10, R31, 2006.
46. Jantunen, E. et al., Central nervous system aspergillosis in allogeneic stem cell transplant recipients, *Bone Marrow Transplant.*, 31, 191, 2003.
47. Jain, K. K. et al., Imaging features of central nervous system fungal infections, *Neurol. India*, 55, 241, 2007.
48. Almutairi, B. M. et al., Invasive aspergillosis of the brain: Radiologic-pathologic correlation, *Radiographics*, 29, 375, 2009.
49. Miaux, Y. et al., MR of cerebral aspergillosis in patients who have had bone marrow transplantation, *AJNR Am. J. Neuroradiol.*, 16, 555, 1995.
50. Dietrich, U. et al., Cerebral aspergillosis: Comparison of radiological and neuropathologic findings in patients with bone marrow transplantation, *Eur. Radiol.*, 11, 1242, 2001.
51. DeLone, D. R. et al., Disseminated aspergillosis involving the brain: Distribution and imaging characteristics, *AJNR Am. J. Neuroradiol.*, 20, 1597, 1999.
52. Denning, D. W. et al., British society for medical mycology proposed standards of care for patients with invasive fungal infections, *Lancet Infect. Dis.*, 3, 230, 2003.
53. Pagano, L. et al., The role of bronchoalveolar lavage in the microbiological diagnosis of pneumonia in patients with haematological malignancies, *Ann. Med.*, 29, 535, 1997.
54. Levy, H. et al., The value of bronchoalveolar lavage and bronchial washings in the diagnosis of invasive pulmonary aspergillosis, *Respir. Med.*, 86, 243, 1992.
55. Lass-Florl, C. et al., The value of computed tomography-guided percutaneous lung biopsy for diagnosis of invasive fungal infection in immunocompromised patients, *Clin. Infect. Dis.*, 45, e101, 2007.
56. Ruchel, R. and Schaffrinski, M., Versatile fluorescent staining of fungi in clinical specimens by using the optical brightener blankophor, *J. Clin. Microbiol.*, 37, 2694, 1999.
57. Ruchel, R. and Margraf, S., Rapid microscopical diagnosis of deep-seated mycoses following maceration of fresh specimens and staining with optical brighteners, *Mycoses*, 36, 239, 1993.
58. Reichenberger, F. et al., Diagnostic yield of bronchoscopy in histologically proven invasive pulmonary aspergillosis, *Bone Marrow Transplant.*, 24, 1195, 1999.
59. Maertens, J. et al., Use of circulating galactomannan screening for early diagnosis of invasive aspergillosis in allogeneic stem cell transplant recipients, *J. Infect. Dis.*, 186, 1297, 2002.
60. Horvath, J. A. and Dummer, S., The use of respiratory-tract cultures in the diagnosis of invasive pulmonary aspergillosis, *Am. J. Med.*, 100, 171, 1996.
61. Clinical Laboratory Standards Institute, *Reference Method for Broth Dilution Antifungal Susceptibility Testing of Filamentous Fungi; Approved Standard—Second Edition*. NCCLS document m38-a2, Clinical Laboratory Standards Institute, Wayne, PA, 2008.
62. Subcommittee on Antifungal Susceptibility Testing, o. t. E. C. f. A. S. T., EUCAST technical note on the method for the determination of broth dilution minimum inhibitory concentrations of antifungal agents for conidia-forming moulds, *Clin. Microbiol. Infect.*, 14, 982, 2008.
63. Odds, F. C. et al., Interlaboratory comparison of results of susceptibility testing with caspofungin against candida and aspergillus species, *J. Clin. Microbiol.*, 42, 3475, 2004.
64. Arendrup, M. C. et al., Establishing in vitro-in vivo correlations for aspergillus fumigatus: The challenge of azoles versus echinocandins, *Antimicrob. Agents Chemother.*, 52, 3504, 2008.
65. Rodriguez-Tudela, J. L. et al., Epidemiological cutoffs and cross-resistance to azole drugs in aspergillus fumigatus, *Antimicrob. Agents Chemother.*, 52, 2468, 2008.
66. Pfaller, M. A. et al., Wild type MIC distribution and epidemiological cutoff values for aspergillus fumigatus and three triazoles as determined by the clinical and laboratory standards institute broth microdilution methods, *J. Clin. Microbiol.*, 47(10), 3142–3146, 2009.
67. Pfaller, J. B. et al., In vitro susceptibility testing of *Aspergillus* spp.: Comparison of Etest and reference microdilution methods for determining voriconazole and itraconazole MICs, *J. Clin. Microbiol.*, 41, 1126, 2003.
68. Espinel-Ingroff, A., Comparison of three commercial assays and a modified disk diffusion assay with two broth microdilution reference assays for testing zygomycetes, *Aspergillus* spp., candida spp., and cryptococcus neoformans with posaconazole and amphotericin b, *J. Clin. Microbiol.*, 44, 3616, 2006.
69. Maertens, J. et al., Autopsy-controlled prospective evaluation of serial screening for circulating galactomannan by a sandwich enzyme-linked immunosorbent assay for hematological patients at risk for invasive aspergillosis, *J. Clin. Microbiol.*, 37, 3223, 1999.
70. Marr, K. A. et al., Antifungal therapy decreases sensitivity of the aspergillus galactomannan enzyme immunoassay, *Clin. Infect. Dis.*, 40, 1762, 2005.
71. Herbrecht, R. et al., Aspergillus galactomannan detection in the diagnosis of invasive aspergillosis in cancer patients, *J. Clin. Oncol.*, 20, 1898, 2002.
72. Mennink-Kersten, M. A. S. H. et al., Bifidobacterium lipoteichoic acid and false ELISA reactivity in aspergillus antigen detection, *Lancet*, 363, 325, 2004.
73. Mennink-Kersten, M. A. S. H. et al., Bifidobacterial lipoglycan as a new cause for false-positive platelia aspergillus enzyme-linked immunosorbent assay reactivity, *J. Clin. Microbiol.*, 43, 3925, 2005.
74. Husain, S. et al., Performance characteristics of the platelia aspergillus enzyme immunoassay for detection of aspergillus galactomannan antigen in bronchoalveolar lavage fluid, *Clin. Vac. Immunol.*, 15, 1760, 2008.
75. Clancy, C. J. et al., Bronchoalveolar lavage galactomannan in diagnosis of invasive pulmonary aspergillosis among solid-organ transplant recipients, *J. Clin. Microbiol.*, 45, 1759, 2007.
76. Senn, L. et al., 1,3-Beta-d-glucan antigenemia for early diagnosis of invasive fungal infections in neutropenic patients with acute leukemia, *Clin. Infect. Dis.*, 46, 878, 2008.

77. Obayashi, T. et al., Reappraisal of the serum (1->3)-beta-d-glucan assay for the diagnosis of invasive fungal infections—A study based on autopsy cases from 6 years, *Clin. Infect. Dis.*, 46, 1864, 2008.
78. Ellis, M. et al., Assessment of the clinical utility of serial beta-d-glucan concentrations in patients with persistent neutropenic fever, *J. Med. Microbiol.*, 57, 287, 2008.
79. Persat, F. et al., Contribution of the (1->3)-{beta}-d-glucan assay for diagnosis of invasive fungal infections, *J. Clin. Microbiol.*, 46, 1009, 2008.
80. Khan, Z. U. et al., Detection of aspergillus fumigatus-specific DNA, (1–3)-beta-d-glucan and galactomannan in serum and bronchoalveolar lavage specimens of experimentally infected rats, *Mycoses*, 51, 129, 2008.
81. Kawazu, M. et al., Prospective comparison of the diagnostic potential of real-time PCR, double-sandwich enzyme-linked immunosorbent assay for galactomannan, and a (1->3)-{beta}-d-glucan test in weekly screening for invasive aspergillosis in patients with hematological disorders, *J. Clin. Microbiol.*, 42, 2733, 2004.
82. Ahmad, S. et al., Diagnostic value of DNA, (1–3)-beta-d-glucan, and galactomannan detection in serum and bronchoalveolar lavage of mice experimentally infected with aspergillus terreus, *Diagn. Microbiol. Infect. Dis.*, 59, 165, 2007.
83. Hope, W. W. et al., Laboratory diagnosis of invasive aspergillosis, *Lancet Infect. Dis.*, 5, 609, 2005.
84. Hafen, G. M. et al., Allergic bronchopulmonary aspergillosis: The hunt for a diagnostic serological marker in cystic fibrosis patients, *Expert Rev. Mol. Diagn.*, 9, 157, 2009.
85. Lambourne, J. et al., Association of mannose-binding lectin deficiency with acute invasive aspergillosis in immunocompromised patients, *Clin. Infect. Dis.*, 49, 1486, 2009.
86. Vekemans, M. et al., Low mannose-binding lectin concentration is associated with severe infection in patients with hematological cancer who are undergoing chemotherapy, *Clin. Infect. Dis.*, 44, 1593, 2007.
87. Buranawuti, K. et al., Variants in mannose-binding lectin and tumour necrosis factor alpha affect survival in cystic fibrosis, *J. Med. Genet.*, 44, 209, 2007.
88. Van Burik, J. A. et al., Panfungal PCR assay for detection of fungal infection in human blood specimens, *J. Clin. Microbiol.*, 36, 1169, 1998.
89. Hebart, H. et al., Prospective screening by a panfungal polymerase chain reaction assay in patients at risk for fungal infections: Implications for the management of febrile neutropenia, *Br. J. Haematol.*, 111, 635, 2000.
90. Jordanides, N. E. et al., A prospective study of real-time panfungal PCR for the early diagnosis of invasive fungal infection in haemato-oncology patients, *Bone Marrow Transplant.*, 35, 389, 2005.
91. Loeffler, J. et al., Nucleic acid sequence-based amplification of aspergillus RNA in blood samples, *J. Clin. Microbiol.*, 39, 1626, 2001.
92. Zhao, Y. et al., Rapid real-time nucleic acid sequence-based amplification-molecular beacon platform to detect fungal and bacterial bloodstream infections, *J. Clin. Microbiol.*, 47, 2067, 2009.
93. Lott, T. J. et al., Nucleotide sequence analysis of the 5.8S rDNA and adjacent ITS 2 region of *Candida albicans* and related species, *Yeast*, 9, 1199, 1993.
94. Gaskell, G. J. et al., Analysis of the internal transcribed spacer regions of ribosomal DNA in common airborne allergenic fungi, *Electrophoresis*, 18, 1567, 1997.
95. Borazjani, R. N. et al., Comparison of 5.8s and ITS2 rDNA RFLP patterns among isolates of acremonium obclavatum, *A. kiliense*, and *A. strictum* from diverse sources, *Curr. Microbiol.*, 36, 70, 1998.
96. Turenne, C. Y. et al., Rapid identification of fungi by using the ITS2 genetic region and an automated fluorescent capillary electrophoresis system, *J. Clin. Microbiol.*, 37, 1846, 1999.
97. Chen, Y. C. et al., Identification of medically important yeasts using PCR-based detection of DNA sequence polymorphisms in the internal transcribed spacer 2 region of the rRNA genes, *J. Clin. Microbiol.*, 38, 2302, 2000.
98. Henry, T. et al., Identification of aspergillus species using internal transcribed spacer regions 1 and 2, *J. Clin. Microbiol.*, 38, 1510, 2000.
99. Kulik, T. et al., Development of PCR assay based on ITS2 rDNA polymorphism for the detection and differentiation of fusarium sporotrichioides, *FEMS Microbiol. Lett.*, 239, 181, 2004.
100. Hinrikson, H. P. et al., Assessment of ribosomal large-subunit d1-d2, internal transcribed spacer 1, and internal transcribed spacer 2 regions as targets for molecular identification of medically important aspergillus species, *J. Clin. Microbiol.*, 43, 2092, 2005.
101. Lau, A. et al., Development and clinical application of a panfungal PCR assay to detect and identify fungal DNA in tissue specimens, *J. Clin. Microbiol.*, 45, 380, 2007.
102. Herrera, M. L. et al., Strain-dependent variation in 18s ribosomal DNA copy numbers in aspergillus fumigatus, *J. Clin. Microbiol.*, 47, 1325, 2009.
103. Walsh, T. J. et al., PCR and single-strand conformational polymorphism for recognition of medically important opportunistic fungi, *J. Clin. Microbiol.*, 33, 3216, 1995.
104. Hui, M. et al., Rapid identification of medically important candida to species level by polymerase chain reaction and single-strand conformational polymorphism [in process citation], *Diagn. Microbiol. Infect. Dis.*, 38, 95, 2000.
105. Aufauvre-Brown, A. et al., Use of randomly amplified polymorphic DNA markers to distinguish isolates of aspergillus fumigatus, *J. Clin. Microbiol.*, 30, 2991, 1992.
106. Brandt, M. E. et al., Utility of random amplified polymorphic DNA PCR and TaqMan automated detection in molecular identification of aspergillus fumigatus, *J. Clin. Microbiol.*, 36, 2057, 1998.
107. Lass-Florl, C. et al., Aspergillus terreus infections in haematological malignancies: Molecular epidemiology suggests association with in-hospital plants, *J. Hosp. Infect.*, 46, 31, 2000.
108. Martin, C. et al., Development of a PCR-based line probe assay for identification of fungal pathogens [in process citation], *J. Clin. Microbiol.*, 38, 3735, 2000.
109. Hopfer, R. L. et al., Detection and differentiation of fungi in clinical specimens using polymerase chain reaction (PCR) amplification and restriction enzyme analysis, *J. Med. Vet. Mycol.*, 31, 65, 1993.
110. Morace, G. et al., Identification of various medically important candida species in clinical specimens by PCR-restriction enzyme analysis, *J. Clin. Microbiol.*, 35, 667, 1997.
111. Bowman, J. C. et al., Quantitative PCR assay to measure aspergillus fumigatus burden in a murine model of disseminated aspergillosis: Demonstration of efficacy of caspofungin acetate, *Antimicrob. Agents Chemother.*, 45, 3474, 2001.
112. Costa, C. et al., Development of two real-time quantitative TaqMan PCR assays to detect circulating aspergillus fumigatus DNA in serum, *J. Microbiol. Methods*, 44, 263, 2001.

113. Gonzalez-Salgado, A. et al., Specific detection of aspergillus carbonarius by SYBR green and TaqMan quantitative PCR assays based on the multicopy ITS2 region of the rRNA gene, *FEMS Microbiol. Lett.*, 295, 57, 2009.
114. Imhof, A. et al., Rapid detection of pathogenic fungi from clinical specimens using LightCycler real-time fluorescence PCR, *Eur. J. Clin. Microbiol. Infect. Dis.*, 22, 558, 2003.
115. Spiess, B. et al., Development of a LightCycler PCR assay for detection and quantification of aspergillus fumigatus DNA in clinical samples from neutropenic patients, *J. Clin. Microbiol.*, 41, 1811, 2003.
116. Klingspor, L. and Jalal, S., Molecular detection and identification of candida and *Aspergillus* spp. From clinical samples using real-time PCR, *Clin. Microbiol. Infect.*, 12, 745, 2006.
117. Park, S. et al., Rapid identification of candida dubliniensis using a species-specific molecular beacon, *J. Clin. Microbiol.*, 38, 2829, 2000.
118. Balashov, S. V. et al., Assessing resistance to the echinocandin antifungal drug caspofungin in candida albicans by profiling mutations in fks1, *Antimicrob. Agents Chemother.*, 50, 2058, 2006.
119. Garcia-Effron, G. et al., Rapid detection of triazole antifungal resistance in aspergillus fumigatus, *J. Clin. Microbiol.*, 46, 1200, 2008.
120. Cuenca-Estrella, M. et al., Value of serial quantification of fungal DNA by a real-time PCR-based technique for early diagnosis of invasive aspergillosis in patients with febrile neutropenia, *J. Clin. Microbiol.*, 47, 379, 2009.
121. Mengoli, C. et al., Use of PCR for diagnosis of invasive aspergillosis: Systematic review and meta-analysis, *Lancet Infect. Dis.*, 9, 89, 2009.
122. Perlin, D. S., Application of real-time PCR to the diagnosis of invasive fungal infections. In *Real-Time PCR: Current Technology and Applications*, J. Logan, K. Edwards, and N. Saunders (eds.), Caister Academic Press, London, U.K., 2009.
123. Buchheidt, D. et al., Prospective clinical evaluation of a LightCycler-mediated polymerase chain reaction assay, a nested-PCR assay and a galactomannan enzyme-linked immunosorbent assay for detection of invasive aspergillosis in neutropenic cancer patients and haematological stem cell transplant recipients, *Br. J. Haematol.*, 125, 196, 2004.
124. Verweij, P. E. et al., Validating PCR for detecting invasive aspergillosis, *Br. J. Haematol.*, 127, 235, 2004.
125. Suarez, F. et al., Detection of circulating aspergillus fumigatus DNA by real-time PCR assay of large serum volumes improves early diagnosis of invasive aspergillosis in high-risk adult patients under hematologic surveillance, *J. Clin. Microbiol.*, 46, 3772, 2008.
126. Bougnoux, M. et al., Serum is more suitable than whole blood for diagnosis of systemic candidiasis by nested PCR, *J. Clin. Microbiol.*, 37, 925, 1999.
127. Perlin, D. S., Multiplex detection of mutations. In *Multiplex Detection of Mutations*., Seitz., A. M. A. O. (ed.), The Humana Press Inc., Totowa, NJ, 2007.
128. Stevenson, J. et al., Effect of sequence polymorphisms on performance of two real-time PCR assays for detection of herpes simplex virus, *J. Clin. Microbiol.*, 43, 2391, 2005.
129. Vora, G. J. et al., Comparison of detection and signal amplification methods for DNA microarrays, *Mol. Cell. Probes*, 22, 294, 2008.
130. Baskova, L. et al., The Pan-AC assay: A single-reaction real-time PCR test for quantitative detection of a broad range of aspergillus and candida species, *J. Med. Microbiol.*, 56, 1167, 2007.
131. Perlin, D. S. and Zhao, Y., Molecular diagnostic platforms for detecting aspergillus, *Med. Mycol.*, 47(1), S223–232, 2008.
132. Marty, F. M. and Koo, S., Role of (1–>3)-beta-d-glucan in the diagnosis of invasive aspergillosis, *Med. Mycol.*, 47(1), S223–240, 2008.
133. Sanglard, D. and Odds, F. C., Resistance of candida species to antifungal agents: Molecular mechanisms and clinical consequences, *Lancet Infect. Dis.*, 2, 73, 2002.
134. White, T. C. et al., Resistance mechanisms in clinical isolates of candida albicans, *Antimicrob. Agents Chemother.*, 46, 1704, 2002.
135. Coste, A. et al., A mutation in Tac1p, a transcription factor regulating CDR1 and CDR2, is coupled with loss of heterozygosity at chromosome 5 to mediate antifungal resistance in candida albicans, *Genetics*, 172, 2139, 2006.
136. Coste, A. et al., Genotypic evolution of azole resistance mechanisms in sequential candida albicans isolates, *Eukaryot. Cell*, 6, 1889, 2007.
137. Selmecki, A. et al., An isochromosome confers drug resistance in vivo by amplification of two genes, Erg11 and Tac1, *Mol. Microbiol.*, 68, 624, 2008.
138. Selmecki, A. et al., Aneuploidy and isochromosome formation in drug-resistant candida albicans, *Science*, 313, 367, 2006.
139. Vermitsky, J. P. and Edlind, T. D., Azole resistance in candida glabrata: Coordinate upregulation of multidrug transporters and evidence for a Pdr1-like transcription factor, *Antimicrob. Agents Chemother.*, 48, 3773, 2004.
140. Chamilos, G. and Kontoyiannis, D. P., Update on antifungal drug resistance mechanisms of aspergillus fumigatus, *Drug Resist. Updates*, 8, 344, 2005.
141. Lupetti, A. et al., Molecular basis of resistance to azole antifungals, *Trends Mol. Med.*, 8, 76, 2002.
142. Alcazar-Fuoli, L. et al., Aspergillus section fumigati: Antifungal susceptibility patterns and sequence-based identification, *Antimicrob. Agents Chemother.*, 52, 1244, 2008.
143. Balajee, S. A. et al., Molecular studies reveal frequent misidentification of aspergillus fumigatus by morphotyping, *Eukaryot. Cell*, 5, 1705, 2006.
144. Geiser, D. M. et al., The current status of species recognition and identification in aspergillus, *Stud. Mycol.*, 59, 1, 2007.
145. Balajee, S. A. et al., Aspergillus species identification in the clinical setting, *Stud. Mycol.*, 59, 39, 2007.
146. Hong, S. B. et al., Polyphasic taxonomy of aspergillus fumigatus and related species, *Mycologia*, 97, 1316, 2005.
147. Staab, J. F. et al., Aspergillus section fumigati typing by PCR-restriction fragment polymorphism, *J. Clin. Microbiol.*, 47, 2079, 2009.
148. Bain, J. M. et al., Multilocus sequence typing of the pathogenic fungus aspergillus fumigatus, *J. Clin. Microbiol.*, 45, 1469, 2007.
149. Klaassen, C. H., MLST versus microsatellites for typing aspergillus fumigatus isolates, *Med. Mycol.*, 47(Suppl 1), S27, 2009.
150. Vanhee, L. M. et al., Comparison of multiple typing methods for aspergillus fumigatus, *Clin. Microbiol. Infect.*, 15, 643, 2009.
151. Klingspor, L. and Loeffler, J., Aspergillus PCR formidable challenges and progress, *Med. Mycol.*, 47(Suppl 1), S241, 2009.
152. Loeffler, J. et al., Contaminations occurring in fungal PCR assays, *J. Clin. Microbiol.*, 37, 1200, 1999.
153. Rimek, D. et al., Identification of contaminating fungal DNA sequences in zymolyase, *J. Clin. Microbiol.*, 37, 830, 1999.

154. Muller, F. M. et al., Rapid extraction of genomic DNA from medically important yeasts and filamentous fungi by high-speed cell disruption, *J. Clin. Microbiol.*, 36, 1625, 1998.
155. Costa, C. et al., Real-time PCR coupled with automated DNA extraction and detection of galactomannan antigen in serum by enzyme-linked immunosorbent assay for diagnosis of invasive aspergillosis, *J. Clin. Microbiol.*, 40, 2224, 2002.
156. Loeffler, J. et al., Comparison between plasma and whole blood specimens for detection of aspergillus DNA by PCR, *J. Clin. Microbiol.*, 38, 3830, 2000.
157. Einsele, H. et al., Detection and identification of fungal pathogens in blood by using molecular probes, *J. Clin. Microbiol.*, 35, 1353, 1997.
158. Kami, M. et al., Use of real-time PCR on blood samples for diagnosis of invasive aspergillosis, *Clin. Infect. Dis.*, 33, 1504, 2001.
159. Ramirez, M. et al., Molecular detection and identification of *Aspergillus* spp. From clinical samples using real-time PCR, *Mycoses*, 52, 129, 2009.
160. Lass-Florl, C. et al., Diagnosing invasive aspergillosis during antifungal therapy by PCR analysis of blood samples, *J. Clin. Microbiol.*, 42, 4154, 2004.
161. Schulz, B. et al., Effect of different sample volumes on the DNA extraction of aspergillus fumigatus from whole blood, *Clin. Microbiol. Infect.*, 15, 686, 2009.
162. Sheppard, D. C. et al., Comparison of three methodologies for the determination of pulmonary fungal burden in experimental murine aspergillosis, *Clin. Microbiol. Infect.*, 12, 376, 2006.
163. Vallor, A. C. et al., Assessment of aspergillus fumigatus burden in pulmonary tissue of guinea pigs by quantitative PCR, galactomannan enzyme immunoassay, and quantitative culture, *Antimicrob. Agents Chemother.*, 52, 2593, 2008.
164. Clemons, K. V. and Stevens, D. A., Conventional or molecular measurement of aspergillus load, *Med. Mycol.*, 47(Suppl 1), S132, 2009.
165. Bonin, S. et al., PCR analysis in archival postmortem tissues, *Mol. Pathol.*, 56, 184, 2003.
166. Sandhu, G. S. et al., Molecular probes for diagnosis of fungal infections, *J. Clin. Microbiol.*, 33, 2913, 1995.
167. Hata, D. J. et al., Real-time PCR method for detection of zygomycetes, *J. Clin. Microbiol.*, 46, 2353, 2008.
168. Zhao, Y. et al., Presented at *the 49th Interscience Conference on Antimicrobial Agents and Chemotherapy*, San Francisco, CA, 2009.
169. Yoo, J. H. et al., Comparison of the real-time nucleic acid sequence-based amplification (RTI-NASBA) with conventional NASBA, and galactomannan assay for the diagnosis of invasive aspergillosis, *J. Korean Med. Sci.*, 22, 672, 2007.
170. Faber, J. et al., Rapid detection of common pathogenic aspergillus species by a novel real-time PCR approach, *Mycoses*, 52, 228, 2008.
171. Landlinger, C. et al., Species-specific identification of a wide range of clinically relevant fungal pathogens by use of luminex xmap technology, *J. Clin. Microbiol.*, 47, 1063, 2009.
172. Perlin, D. S., Resistance to echinocandin-class antifungal drugs, *Drug Resist. Updates*, 10, 121, 2007.
173. Katiyar, S. et al., Candida albicans and candida glabrata clinical isolates exhibiting reduced echinocandin susceptibility, *Antimicrob. Agents Chemother.*, 50, 2892, 2006.
174. Garcia-Effron, G. et al., Effect of candida glabrata FKS1 and FKS2 mutations on echinocandin sensitivity and kinetics of 1,3-beta-d-glucan synthase: Implication for the existing susceptibility breakpoint, *Antimicrob. Agents Chemother.*, 53, 3690, 2009.
175. Loeffler, J. et al., Quantification of fungal DNA by using fluorescence resonance energy transfer and the light cycler system, *J. Clin. Microbiol.*, 38, 586, 2000.
176. Williamson, E. C. et al., Diagnosis of invasive aspergillosis in bone marrow transplant recipients by polymerase chain reaction, *Br. J. Haematol.*, 108, 132, 2000.
177. White, P. L. et al., The evolution and evaluation of a whole blood polymerase chain reaction assay for the detection of invasive aspergillosis in hematology patients in a routine clinical setting, *Clin. Infect. Dis.*, 42, 479, 2006.
178. Pham, A. S. et al., Diagnosis of invasive mold infection by real-time quantitative PCR, *Am. J. Clin. Pathol.*, 119, 38, 2003.
179. White, P. L. et al., A consensus on fungal polymerase chain reaction diagnosis?: A United Kingdom-Ireland evaluation of polymerase chain reaction methods for detection of systemic fungal infections, *J. Mol. Diagn.*, 8, 376, 2006.

24 *Blastomyces*

Mark D. Lindsley

CONTENTS

24.1 INTRODUCTION

Blastomycosis is a systemic fungal disease caused by the thermally dimorphic mould *Blastomyces dermatitidis* and is one of the major endemic mycoses that infect humans and animals. If left untreated, blastomycosis can spread systemically and may become fatal, regardless of immune status. *B. dermatitidis* was first described in 1894 by Thomas Caspar Gilchrist, a dermatologist, who observed numerous 10–16 μm, round, doubly contoured refractile bodies in a skin lesion of a patient thought to be suffering from chronic scrofuloderma.[1] The organism was believed to be a Blastomycete yeast by Gilchrist and others until 1896 when Gilchrist demonstrated in culture the organism's ability to produce hyphae and conidia and described its classic growth characteristics *in vitro*.[1] Observing that this organism was unique from other yeasts, he named the disease Blastomycetic Dermatitis and the organism *Blastomyces dermatitidis*.[2] The first report of invasive blastomycosis was in 1902 by Walker and Montgomery[3] in a patient believed to be suffering from disseminated miliary tuberculosis. Initially referred to as "Gilchrist's disease," blastomycosis has also been referred to as "North American blastomycosis" and "Chicago disease" after the location where a large number of early cases were observed.

24.1.1 Classification, Morphology, and Biology

24.1.1.1 Classification

B. dermatitidis is an ascomycete belonging to the family Onygenaceae in the order Onygenales. Within the genus *Blastomyces*, *B. dermatitidis* is the only known species. Other medically important members of the family Onygenaceae include the systemic pathogens *Histoplasma capsulatum*, *Paracoccidioides brasiliensis*, and *Emmonsia* sp., the latter an uncommon cause of pulmonary disease in humans.[4,5] All, including *B. dermatitidis*, are thermally dimorphic, growing as a mould at 25°C and as a yeast-like form at 37°C. Phylogenetic analysis of the members of the family Onygenaceae, comparing sequences from the 18S,[6] 28S,[7,8] and internal transcribed spacer regions[9] of ribosomal RNA (rRNA) genes, have shown that *Blastomyces* is most closely related to *Emmonsia* sp. and *Paracoccidioides brasiliensis*.

24.1.1.2 Morphology

When cultured at ambient temperature, the filamentous mould form of *B. dermatitidis* produces a white, downy to wooly colony that may become tan to brown with age. The reverse side of the culture is typically non-pigmented but may also darken with age. Microscopically, *Blastomyces dermatitidis* is a hyaline mould that produces septate hyphae which give rise to oval to round aleuriosporic conidia directly from the hyphae or on short stalks arising at right angles to the hyphae. When disturbed, the infectious conidia are aerosolized and enter the body by inhalation. After entering the body *B. dermatitidis* converts to a yeast-form, typically 8–15 μm in diameter. The yeast cell is often multinucleate and may contain up to 8–12 nuclei.[10] The yeast reproduces through budding, forming a single daughter cell from the mother cell by way of a broad-based bud with a septum which may be as wide as the daughter cell prior to separation. The yeast form of *B. dermatitidis* is not believed to be directly infectious and therefore blastomycosis cannot be spread by person-to-person transmission.

Blastomyces dermatitidis is the anamorphic or asexual form of the ascomycete *Ajellomyces dermatitidis* (teleomorph). The teleomorphic or sexual form was first described in 1968 by McDonough and Lewis[11,12] after mating different strains of *B. dermatitidis*.[13,14] Large, 200–350 μm, thick-walled cleistothecia ornamented with coiled spiral hyphae[15,16] are produced. Lateral hyphae give rise to asci that contain eight ascospores measuring 1.5–2.0 μm in diameter. *B. dermatitidis* was determined to be heterothallic, requiring two separate mating types (a+ and a– mating types) to produce sexual structures. Therefore, single ascospores from two different strains were able to produce the sexual structures whereas ascospores of the same strain could not.[12] The teleomorph of *B. dermatitidis* is sufficiently similar to the teleomorph of *Histoplasma capsulatum*[17] and *Emmonsia crescens*[4] in that all have been placed in the same teleomorphic genus *Ajellomyces* as *Ajellomyces dermatitidis*, *Ajellomyces capsulatus*, and *Ajellomyces crescens*, respectively.

24.1.1.3 Serotypes

Serologic examination of *B. dermatitidis* isolates demonstrated that *B. dermatitidis* can be divided into two serotypes.[18–20] Kaufman et al.[18] using antisera specific for the immunodominant "A" antigen of *B. dermatitidis* revealed that of 88 isolates from North American and international sources, all but 11 isolates that were recovered from Africa reacted with the antisera. The *B. dermatitidis* isolates reactive with the antisera were placed into serogroup 1. African isolates failing to react with the antisera were placed into serogroup 2. This finding was confirmed when all isolates originating from Africa failed to react with a monoclonal antibody specific for the antigen WI-1 (analogous to A antigen).[19] Further studies using Southern blot analysis demonstrated that the African strains lacked the WI-1 gene.[19]

24.1.1.4 Genotypes

B. dermatitidis can be divided into at least three genotypic groups. Using restriction fragment length polymorphism (RFLP) of genomic DNA followed by Southern blot analysis using mitochondrial DNA probes, Yates-Siilata et al.[21] placed all but 4 of 19 *B. dermatitidis* isolates from Arkansas into group 1 whereas groups 2 and 3 contained two isolates each. Group 1 could be further subdivided using random amplified polymorphic DNA (RAPD) fingerprint analysis. McCullough et al.[22] analyzed 59 isolates from 15 regions in the United States, India, Africa, and Canada using PCR-RFLP. The authors amplified the approximately 5.5 kb intergenic spacer (IGS) region of the ribosomal DNA (rDNA) with PCR primers BD28S and NS2 followed by digestion with restriction endonucleases. Three genotypic groups (A–C) were also observed. Of the 59 isolates analyzed, 17 belonged to group A, 23 to group B, and 19 to group C. Interestingly, while groups B and C contained isolates from national and international regions, 16 of 17 Group A isolates were from the upper mid-west region of the United States and Canada. RAPD fingerprinting could further subdivide the three groups into 5, 15, and 12 subtypes, respectively. Recently, Meece et al.[23] likewise performing PCR-RFLP on 106 human, animal, and environmental isolates by amplifying the IGS region and digesting with restriction endonuclease *Dde I*, found that the majority of specimens fell into one of three groups corresponding to those of McCullough et al., A (50 isolates), B (51 isolates), and C (3 isolates). However, two isolates were unable to be placed into one of the three groups and thus two new groups were formed, groups D and E, with one isolate each. Another approach for genotyping *B. dermatitidis* isolates used PCR and sequence analysis of an approximately 360 bp portion of the promoter region of the virulence factor, *Blastomyces* adhesion 1 (BAD-1).[23] Using this approach the *B. dermatitidis* isolates were placed into four haplotypes. Interestingly, a subset of the isolates contained two inserts of 35 and 251 bp, respectively. The significance of this difference remains to be determined.

PCR-RFLP genotypic analysis,[22] performed on 14 patient isolates obtained from one of the largest point source outbreaks of blastomycosis (Eagle River, WI), revealed that 10 were typed as group A, 2 were typed as group B, and 2 were typed as group C. Furthermore, the environmental isolates associated with the outbreak were also genotypically different—group B and group C, respectively. These results suggest that even within a small geographical area *B. dermatitidis* can display a range of heterogeneity. The significance of these results remains to be determined.

24.1.1.5 Virulence Factors

Some virulence factors expressed by *Blastomyces* are also shared with the other thermally dimorphic moulds. Thermal dimorphism may be considered a virulence factor in that the initial conversion from the smaller conidia, produced by the free living environmental mould form, results in larger thick-walled yeast form that may be more difficult to phagocytose, or may even be inhibitory to phagocytosis.[24]

Other virulence factors are only expressed after conversion to the yeast form. During the process of conversion to the yeast form construction of the cell wall changes from a primarily β-1,3-glucan dominant cell wall to an α-1,3-glucan dominant cell wall. Whereas the mycelial cell wall before conversion contains 60% β-1,3-glucan and 40% α-1,3-glucan, after conversion, the yeast-form cell wall contains approximately 95% α-1,3-glucan and 5% β-1,3-glucan.[25] Using a monoclonal antibody specific for α-1,3-glucan (MOPC 104e), the amount of α-1,3-glucan present in the cell wall of three different strains of *B. dermatitidis* was measured both microscopically and through flow cytometric analysis.[26,27] *B. dermatitidis* strains with increased amounts of α-1,3-glucan detected in the yeast cell wall also demonstrated an increased ability to cause disease.[26] The importance of α-1,3-glucan as a virulence factor has also been demonstrated in *Histoplasma capsulatum*[28] and *Cryptococcus neoformans*.[29] In those fungi, inhibiting the production of α-1,3-glucan by gene disruption or interference with the α-1,3-synthase (AGS1) gene substantially reduced virulence in mouse infection models and *in vitro* killing assays. A possible role of α-1,3-glucan in promoting virulence is its ability to block the innate immune response to the immune stimulating β-1,3-glucan and inhibit the production of tumor necrosis factor (TNF-α).[30]

BAD-1, a virulence factor specific for *B. dermatitidis*, is a cell surface molecule expressed only on yeast form cells.[31] Formerly known as WI-1, BAD-1 is a 120 kDa adhesion protein found both on the surface of the yeast cell and secreted in a cell-free soluble form.[32] After being secreted extracellularly, BAD-1 binds to chitin on the yeast cell wall, which facilitates binding to macrophages and lung tissue[33] through both the CD-14 receptor and complement receptor 3(CR3). This interaction, along with the secretion of soluble BAD-1, modulates the immune response by stimulating phagocytes to produce transforming growth factor-β (TGF-β), an antagonist of TNF-α expression.[34–36] Interestingly the expression of BAD-1 on the yeast cell surface is inversely correlated with virulence in that increased expression of BAD-1 is associated with decreased virulence.[32] While avirulent strains may express more BAD-1 on the cell surface, they secrete significantly less soluble form. In this situation the avirulent strains bind to phagocytes more rapidly and in greater numbers than the more virulent wild type strains.[37]

24.1.1.6 Animal Infections

B. dermatitidis has also been demonstrated as an important pathogen in animals. Other than humans, *B. dermatitidis* also infects dogs,[38,39] cats,[40] horses,[41] and sea lions.[42] Dogs are by far the most common animal infected. In the state of Louisiana, blastomycosis is the most common systemic fungal infection in dogs.[38] Dogs that spend extended time outdoors near bodies of water, such as hunting dogs, are most commonly infected. Believed to be either more susceptible or more often exposed to infected soil due to their proximity, dogs are at a 10-fold greater risk of infection than humans.[38,43] Dogs are often considered sentinels of disease in that they are often the first to be sick, before humans, when exposed to the same environmental sources of conidia. Through mating studies of human and dog *B. dermatitidis* strains, McDonough and Lewis[12] established that a single species of *Blastomyces* was responsible for infections in both dogs and humans. Whether certain subtypes of *B. dermatitidis* are responsible for canine infection compared to human infection remains to be determined. As with humans, animal infection is primarily pulmonary, but can disseminate to other organs of the body including lymph nodes, skin, bones, and eyes, and is best treated with itraconazole.[39]

Symptomatic infection in other animals is much less common. However, an outbreak of blastomycosis was observed in Chicago, IL where five suburban cats were found to have developed blastomycosis within a 3–4 month time span.[40] While all lived within 300 m of a water source, only one of the five cats was an outdoor cat. All were diagnosed by observing typical yeast forms in either cytologic smears or by histopathology. Non-domesticated cats are also susceptible to blastomycosis. Two Asian lions, an African lion, a Siberian tiger, a cheetah, and a snow leopard, all housed at a Tennesse zoo, were diagnosed with blastomycosis through cytology or a combination of serology and radiographic analysis.[44]

24.1.1.7 Epidemiology

The endemic range of blastomycosis is difficult to define due to the lack of sensitive environmental culturing methods.[45] Also an adequate skin test antigen, an important epidemiologic tool used to determine exposure and subclinical infection rates, is lacking for blastomycosis.[46,47] Endemic regions of blastomycosis have been estimated primarily through case reports and locating areas of increased incidence of symptomatic disease. Endemic areas include the regions around the Great Lakes and the Saint Lawrence Seaway in North America, and the Mississippi and Ohio River valleys in the United States. Incidence of reported disease in the endemic region averages about 1 case per 100,000 population.[48] However, hyperendemic areas of blastomycosis are found in parts of Wisconsin and parts of eastern Canada, particularly in the Province of Ontario.[49] Annual rates of infection in these areas are as high as 2–7 cases per 100,000 population in and around the town of Kenora, Ontario[49] and a mean of 5.1–41.9 cases per 100,000 population in the 10 counties that make up the northern half of Wisconsin.[50] Naturally acquired infections have been observed outside the known endemic region. Infections in two patients in Colorado who developed fungal pneumonia after working on a project to relocate prairie dogs[51] and two patients who developed osseous lesions from dermal inoculation of soil in Nebraska[52] suggest that the endemic range may be broader than once believed, or may be increasing in size. Internationally, *B. dermatitidis* is also found in localized areas of Central and South America and Africa.

The ecologic niche in which *B. dermatitidis* is most likely found is near sources of water such as rivers and lakes and in moist acidic soil enriched with decomposing organic debris.[48] In one of the largest outbreaks of blastomycosis, 48 cases of blastomycosis, with an attack rate of 51%,[48] occurred when a large party of adults and children visited a beaver dam and lodge. However, only 2 of 47 environmental samples collected on and around the beaver dam grew *B. dermatitidis*. A third specimen from the beaver dam grew *B. dermatitidis* upon retesting. Exposure to soil near waterways also has been implicated in other outbreaks.[53,54] Often, locations of endemic foci are realized only after observing an increase in incidence of blastomycosis over an extended period of time.[55] In these cases, the source of infection is often not determined.

Geographic analysis of areas associated with infection show that infections are more likely when exposures take place in areas of sandy soil near waterways located at elevations of less than 500 m.[56] In vitro, *B. dermatitidis* grows best on media containing by-products of plant and animal waste.[57] Specifically, ammonia, a by-product of animal waste, supported the growth of *B. dermatitidis*, even at concentrations that are normally toxic to most fungi.[58] Increased levels of these byproducts in the soil, allowing for selective growth of *B. dermatitidis*, may only be present for a short period of time, which may account for the difficulty of isolating *B. dermatitidis* from environmental sources.

24.1.2 Clinical Features and Treatment

24.1.2.1 Clinical Features

Blastomycosis occurs as a result of inhalation of infectious conidia released into the environment after the disruption of soil containing *B. dermatitidis*. Many of those infected will remain asymptomatic.[59] The median incubation period is 45 days from exposure to onset of symptoms.[60] Whereas symptomatic patients may present with signs and symptoms similar to those with bacterial pneumonia,[61] many acute infections often go unrecognized as a generalized influenza-like illness that resolves spontaneously.[62–64] Most commonly, however, infected individuals display a slow indolent process of cough, weight loss, and fever that may develop over weeks, months, or even years.[65] The development of this chronic disease is insidious and can mimic other more "well-known" diseases such as tuberculosis, delaying the diagnosis of blastomycosis.[61,66]

As a further manifestation of the disease, *B. dermatitidis* can spread to extrapulmonary sites. Skin[62,67] is the most common site of extrapulmonary spread and provides an accessible specimen source for culture and histopathologic diagnosis.[10] Besides skin, *B. dermatitidis* can infect any organ system of the body including the central nervous system,[68,69] bones and joints,[70,71] and the genitourinary system.[72] Because of the non-specific and generalized symptoms, the diagnosis of blastomycosis is often not attained until skin lesions or symptoms of other extrapulmonary infections occur. Skin infections have also been noted to occur through accidental needle-sticks occurring in a laboratory or veterinary setting that may introduce the organism subcutaneously resulting in localized disease.[73,74] Dermal inoculation of blastomycosis by way of a dog bite has also been reported.[75] However, dissemination from this mode of infection has not been documented.

Whereas other fungal infections such as cryptococcosis, histoplasmosis, and coccidioidomycosis have been found important causes of opportunistic infections in immunocompromised patients, blastomycosis has not. When blastomycosis does occur in this patient population, however, these patients usually have more severe disease with a much greater mortality rate than that seen in immunocompetent patients.[76,77] Transplant patients develop blastomycosis as early as 3 weeks and as late as 4 years post-transplant.[77] Most infections disseminate from an initial pulmonary focus but disseminated infection can also present at the same time as the onset of pulmonary symptoms.[77] Treatment can be effective but infection may recur after discontinuation of therapy.[77] While rare in HIV-infected persons,[78] the incidence of blastomycosis may be increasing in certain geographic areas. Pappas reported that the number of immunosuppressed patients with blastomycosis seen at the University of Alabama hospitals increased almost 10-fold over several decades.[79] Dissemination in immunosuppressed patients more often involves the central nervous system, and has a much greater mortality rate than dissemination in immunocompetent patients. Even though blastomycosis may be rare it remains important to consider *B. dermatitidis* in the differential diagnosis in immunosuppressed patients presenting with influenza-like illness.

24.1.2.2 Treatment

Guidelines and specific treatment regimens for managing blastomycosis have recently been updated by the Infectious Diseases Society of America.[80] Briefly, treatment may not be required in immunocompetent patients with mild, self-limited pulmonary infection unless there is reason to suspect that disseminated disease may occur. Mild to moderate disease can be treated with itraconazole monotherapy. Moderately severe to severe disease, whether limited to pulmonary infection or disseminated extrapulmonary infection, and including central nervous system disease, should be treated with amphotericin B (deoxycholate or lipid formulations). After improvement, the treatment may be switched to itraconazole. For immunosuppressed patients, treatment with amphotericin B may be later switched to itraconazole and continued indefinitely if the immunosuppression is irreversible. Pregnant women should be treated with amphotericin B as itraconazole is not recommended in this patient population due to the drug's teratogenic potential. Pediatric patients with severe blastomycosis should be treated with amphotericin B, followed by itraconazole after clinical improvement. Recent reports have demonstrated that voriconazole and the experimental azole isavuconazole are also active against *B. dermatitidis*.[81,82]

24.1.3 Diagnosis

24.1.3.1 Conventional Techniques

Direct specimen examination, in vitro culture, serology, and histopathology continue to be standard methods in the clinical laboratory for the diagnosis of blastomycosis. Respiratory specimens such as sputum, bronchoalveolar lavage fluid, tracheal secretions, and pleural fluid are appropriate for the diagnosis of acute localized infections. *B. dermatitidis* can disseminate to virtually any organ and may be recovered from skin,[67] blood,[83] bone,[71] CNS,[68] and the genitourinary tract.[84] When present, skin lesions are an excellent source for isolation of *B. dermatitidis* as the lesions are readily available for biopsy and cultures are frequently positive. In male patients, *B. dermatitidis* may disseminate to the prostate[72,84] and may cause dysuria. In these patients prostatic fluid is an excellent specimen for recovery of *B. dermatitidis*. In its yeast form, *B. dermatitidis* is classified a biosafety level (BSL)-2 pathogen[85] and therefore specimens from potential blastomycosis patients as well as yeast cultures grown at 37°C may be handled using BSL-2 containment. On the other hand, soil or other environmental specimens that may contain infectious conidia and sporulating mould cultures should be handled under BSL-3 conditions.[85]

24.1.3.1.1 Direct Detection

Sputum, skin scrapings, and fine needle aspirates are prepared in 10% potassium hydroxide (KOH) with or without lactophenol cotton blue stain, and examined using either light or phase contrast microscopy. Observation of the characteristic yeast form cells (8–15 μm in diameter) with daughter yeast buds with a wide septum provides presumptive diagnosis of blastomycosis. The use of the fluorescent dyes such as Calcofluor white or Blankophor increases the sensitivity of detection, and requires a fluorescence microscope to observe bright yellow to white fluorescent staining. Due to the nonspecific nature of the symptoms of blastomycosis, fungus culture is not always ordered on specimens from patients later shown to have blastomycosis. Stains used for the examination of specimens for other microbiological agents, such as the Gram stain, do not stain *B. dermatitidis* well and the yeast-forms may not be easily visualized. Therefore, when viewing specimens stained for other microorganisms it is important to be alert to the chance observation of *B. dermatitidis*. The diagnosis of a case of blastomycosis was delayed after a Gram stain-negative specimen later grew *B. dermatitidis*. Subsequent review of the Gram stain showed numerous poorly stained yeast cells resembling *B. dermatitidis* in the "background" (Dr. Glenn Roberts, personal communication).

24.1.3.1.2 Serodiagnosis

Antibody detection: The presence of antibodies in a patient suspected of having blastomycosis signifies recent or past infection. Serologic conversion from negative to positive or a fourfold or greater rise in titer between acute and convalescent sera, are necessary for a definitive serodiagnosis of acute blastomycosis. Complement fixation (CF) and immunodiffusion (ID) are the methods routinely used for the serologic diagnosis of blastomycosis. The use of a purified antigen (A antigen)[86] improved the assay over the originally unpurified culture filtrate concentrate[87] and purified A antigen is now used in most commercial kits for CF and ID. Antibody determination provides a specificity of 87%–100% and 100% for the CF and ID tests, respectively.[88,89] The sensitivity of ID and CF assays was reported to be 33% during localized infection[53] and 85% (ID) and 50% (CF), in disseminated blastomycosis.[48] Incorporation of the A antigen into an enzyme immunoassay (EIA) format improved sensitivity to between 86% and 100% with a specificity of 87%–92%.[48,53,90] A commercial version of the A antigen EIA had a reported 100% sensitivity with a specificity of 85.6%.[91] However this assay is no longer commercially available. Klein and Jones[92] described an immunodominant surface antigen of *B. dermatitidis* which they named WI-1. While immunologically similar to the A antigen,[93] WI-1 is a 120 kDa nonglycosylated protein whereas A antigen is estimated to be a 135 kDa protein with 37% carbohydrate. An EIA using the WI-1 antigen produced an assay that had a sensitivity of 77% and specificity of 92%.[94]

A Western blot assay using a culture filtrate antigen was developed and evaluated to detect serum antibodies.[95] This test increased the sensitivity to 91% in a panel of sera from blastomycosis patients, as compared to 88% using the A antigen EIA. A 98 kDa extracellular protein antigen was shown to be most highly reactive with patient sera. In an immunodiffusion assay, the 98 kDa antigen was closely related to if not identical with the A antigen and the WI-1 antigen.

Antigen detection: An EIA for the detection of *Blastomyces* antigen in urine was evaluated for the diagnosis of blastomycosis.[96] The assay has a reported sensitivity of 92.9% and specificity of 79.3%. Urine specimens from healthy volunteers and patients with coccicioidomycosis and candidiasis were negative, but cross-reactions occurred with urine from patients with histoplasmosis (96.3%), paracoccidioidomycosis (100%), and penicilliosis marneffei (70%). Likewise, reciprocal cross-reactivity was also demonstrated with specimens from blastomycosis patients when assayed using the *Histoplasma* antigen assay.[97] In a case of culture-positive blastomycosis,[98] antigen was demonstrated in the urine using the *Blastomyces* antigen EIA whereas the *Blastomyces* ID and CF antibody tests were negative. Urine antigen levels were monitored during treatment in four pediatric *Blastomyces* culture- and antigen-positive patients.[99] Elevated urine antigen levels returned to normal in three patients after treatment with itraconazole; the fourth was not compliant with the treatment regimen and elevated urine antigen levels persisted. The *Blastomyces* antigen EIA was also used to detect antigen in the urine (93.5%) and serum (87%) of 46 dogs with confirmed blastomycosis.[100] Antigen titers decreased over time during treatment with itraconazole. The *Blastomyces* antigen EIA appears to correlate well when tested with samples from confirmed cases of blastomycosis. The general usefulness of this assay for the definitive diagnosis of blastomycosis, however, remains uncertain because of its low specificity and cross-reactivity with the other major fungal pathogens.

24.1.3.1.3 Culture

Culture is still considered the gold standard for the diagnosis of blastomycosis. For the detection of *B. dermatitidis* fungemia, blood cultures should be processed using the lysis-centrifugation method for optimal sensitivity.[83] Liquid specimens (bronchoalveolar lavage fluid, cerebrospinal fluid, etc.) should be centrifuged and the resulting pellet used for culture. Tissue should be homogenized using a laboratory paddle blender or minced in sterile water or saline before plating on appropriate media. Processed specimens should be inoculated onto enriched media containing antibiotics such as chloramphenicol and/or gentamicin to inhibit bacterial growth and cycloheximide to inhibit susceptible moulds, and incubated at 25°C–30°C. Growth of *B dermatitidis* may be observed within the first 2 weeks of incubation. However, cultures should be held for 4–6 weeks before being discarded as negative.

Positive cultures show an initial growth of a glabrous (waxy) colony, becoming fluffy white in color turning beige to tan with age. Microscopic morphology demonstrates hyaline mycelia with single round to pyriform conidia, 2–10 μm in diameter, originating directly from the hyphae or on short stalks. These microscopic features may be confused with other pathogenic (e.g., *Scedosporium apiospermum*) and nonpathogenic (e.g., *Chrysosporium* species) fungi, thus requiring further confirmatory testing before making the final identification. The identification of the culture can be confirmed by (1) demonstrating the conversion of the mould form to the yeast form by growth on an enriched medium such as Kelley's at 37°C,[101,102] (2) immunological identification using the exoantigen agar gel immunodiffusion method[104], or (3) use of a commercial molecular probe, Accuprobe® Blastomyces Dermatitidis Culture Identification Test (GenProbe). The Accuprobe assay is more expensive than the exoantigen or thermal conversion methods, and requires special equipment and expertise. The decision as to which assay to perform depends on the number of cultures received by the laboratory, the technical expertise available to perform the assay, and the desired turn-around time.

24.1.3.1.4 Histopathology

Hematoxylin and eosin (H&E) staining of tissue sections from blastomycosis patients reveals an initial polymorphonuclear leukocyte infiltration in early lesions.[103] This early pyogenic reaction is not typical for fungal infections and diagnosis may be delayed if yeast forms are not present in the lesion. Later in infection, a granulomatous reaction appears with the formation of multinucleate giant cells. The yeast cells may be located free or inside giant cells. Abscess formation may be observed.

B. dermatitidis yeast forms may not stain well in the H&E stained section, but the presence of hyaline 8–15 μm in diameter, thick-walled, broad-based budding yeast cells is diagnostic. The yeast cells can be multinucleated with up to 12 nuclei per cell.[10] Periodic acid Schiff (PAS) and Grocott methenamine silver (GMS) stain the fungal cell walls a bright pink and brown to black, respectively, providing a more definitive demonstration of the typical morphology. In instances where budding is not observed, *B. dermatitidis* may resemble *Cryptococcus neoformans* and can be differentiated using the mucicarmine stain in which the capsule of *Cryptococcus* stains a bright pink color. *B. dermatitidis* is typically negative with the mucicarmine stain but the cell wall may display a faint reaction. Larger and smaller forms of *B. dermatitidis* may resemble empty *Coccidioides* spherules or *Histoplasma* yeast forms, respectively, but can be differentiated by observing the broad based budding characteristic of *B. dermatitidis*. Mycelial forms of *B. dermatitidis* in tissue are extremely rare. When present, hyphae are short, hyaline, and septate.[103]

24.1.3.2 Molecular Diagnostic Techniques

Commercially available molecular techniques for the diagnosis of blastomycosis are used for confirmation of culture isolates. The Accuprobe Blastomyces Dermatitidis Culture Identification Test (Gen-Probe) uses a chemiluminescent, acridinium ester-labeled DNA probe in a hybridization protection assay (HPA). The probe, specific for the *B. dermatitidis* rRNA molecule, binds to the rRNA target incorporating a chemiluminescent label into the double stranded molecule formed between the probe and target protecting the bound label from degradation by a hydrolyzing solution and available to emit light when the detection reagent is added. Although needing specialized equipment, the assay is quick to perform and requires a small amount of mycelium or yeast to perform the assay. The assay has excellent sensitivity (97%–100%)[104,105] and specificity (99%) with only *P. brasiliensis* the only major pathogenic fungi to cross-react in this assay. Other cross-reactions have been noted with *Gymnascella hyalinospora*.[106] The assay has also been evaluated using 72 *Blastomyces* strains isolated from a variety of non-human mammalian isolates including those from 62 dogs, 3 cats, and 1 each from a polar bear and sea lion.[107] The probe assay confirmed the identity of all isolates with 100% sensitivity. Twenty eight isolates from 21 genera other than *Blastomyces* were all negative (100% specificity). However, isolates of *P. brasiliensis* was not included for evaluation.

Another commercially available, molecular based assay for the identification of *B. dermatitidis* is based on the repetitive-sequence-based PCR (rep-PCR)[108,109] recently developed by DiversiLab and now marketed by bioMérieux Clinical Diagnostics. PCR primers bind to multiple noncoding regions in the microbial genome amplifying DNA fragments of varying sizes. Through a microfluidic electrophoresis separation step in a chip format, the DNA fragments are measured in size and intensity using laser technology and plotted in a DNA electrophoresis pattern. The resulting DNA fingerprint is compared to a library of DNA fingerprints, and a dendrogram is plotted for identification. Identification of *B. dermatitidis*[110] using rep-PCR demonstrated the ability of this technology to differentiate *B. dermatitidis* from other dimorphic fungi and placed clinical strains with the database strains. This assay also provides the user the capability to genetically compare multiple isolates which may be particularly useful in an outbreak setting. Whether rep-PCR can differentiate individual isolates will need further evaluation using greater number of *Blastomyces* isolates from more diverse, worldwide sources.

Non-commercial methods for the molecular diagnosis of *B. dermatitidis* employ the standard procedures used for most fungi and have been recently reviewed.[111,112] These methods have been used for confirmation of isolate identification, identification directly from clinical specimens, and the detection of *B. dermatitidis* in environmental samples.

Universal fungal primers that amplify the 28S[113] or the internal transcribed spacer (ITS)[114] regions of the rRNA gene have been used for confirming the identification of fungal isolates resembling *B. dermatitidis*. Species-specific probes generated against the 28S region and used in a slot blot assay were 100% specific.[113] ITS region probes were developed for the species-specific identification of yeast-like fungi found in tissue including thermally dimorphic fungi, *Cryptococcus neoformans*, and *Pneumocystis carinii*.[114] These probes were evaluated using an EIA method after PCR amplification of the ITS region using universal primers ITS1 and ITS4. All probes were 100% specific except for a slight, but not significant, cross-reactivity of the *Blastomyces* probe with DNA from *Coccidioides immitis*; the corresponding *Coccidioides immitis* probe did not react with the *B. dermatitidis* DNA.

In situ hybridization and PCR have been employed for the molecular identification of *B. dermatitidis* from formalin-fixed paraffin-embedded tissue samples. Hayden et al. designed probes to identify *B. dermatitidis* and the other yeast-like fungi in tissue using in situ hybridization.[115] Two probes, one from the 18S and one from the 28S region of the rRNA gene, were used in combination to probe fungi-containing tissue including a total of 20 tissues containing *B. dermatitidis*. In situ hybridization for the detection of *B. dermatitidis* was 90% sensitive and 97.4% specific when compared to GMS staining.

Tissue from dogs with histopathologically confirmed blastomycosis was used to evaluate the extraction and PCR amplification of *B dermatitidis* DNA in paraffin-embedded tissue.[116] Two sets of nested primers were evaluated; one set was directed to a portion of the 18S region of the rRNA gene and the other to a portion of the promoter region of the BAD-1 gene. Primers in the 18S region amplified DNA from 10 of 25 tissues from *B. dermatitidis* infected dogs but also amplified DNA from 20 of 20 *Histoplasma* positive human tissues and 9 of 33 tissues from uninfected dogs. Amplification of an approximately 360-bp portion of the BAD-1 promoter region using primers Blasto I and Blasto II was demonstrated to be as sensitive but more specific in that only *B. dermatitidis* DNA was detected. A limitation in using the BAD-1 primers is that African strains of *B. dermatitidis* which do not express BAD-1 have also been demonstrated not to have the BAD-1 gene[19] and would be negative using these primers. It is uncertain how frequently BAD-1 negative strains are found in cases of blastomycosis.

Molecular methods for the detection of *Blastomyces* in soil have been evaluated[117] also using primers Blasto I and Blasto II to amplify the BAD-1 virulence gene promoter. The primers are specific for *B. dermatitidis* in that they were unable to amplify DNA isolated from other onygenalean fungi (*Histoplasma capsulatum*, *Coccidioides immitis*, and *Emmonsia crescens*), other environmental fungi (*Aspergillus flavus*, *Penicillium chrysogenum*, and *Candida albicans*) as well as soil inhabiting bacteria (*Streptomyces griseus* and *Nocardia brasiliensis*). Isolation of DNA from natural soil samples not believed to contain *B. dermatitidis* were negative using Blasto I and Blasto II primers but were reactive using fungal universal primers ITS1 and ITS4 demonstrating the presence of natural soil fungi. DNA was isolated from eight soil samples obtained at a dog kennel near Lexington, KY where several dogs had developed blastomycosis. Of the eight samples processed, three were positive, producing amplicons with the *Blastomyces* specific primers and confirmed by sequence analysis.

24.2 METHODS

24.2.1 Sample Preparation

Methods for the molecular diagnosis of *B. dermatitidis* have not been standardized and are still considered experimental. As with most fungi, the rigid fungal cell wall consisting of α- and β-glucans and chitin can make extraction of DNA from *B. dermatitidis* difficult. Lysis of the fungal cell usually includes a form of mechanical disruption such as series of freeze/thaw steps,[6] freezing the sample in liquid nitrogen[118] or dry ice,[119] followed by grinding the frozen tissue in a mortar and pestle, and use of a mechanical homogenizer[120] or vigorous agitation in a tube containing glass beads or silica.[121] One protocol uses a combination of a freeze–thaw step prior to using a commercial bead beating method extending the bead beating time from 10 to 30 min.[110] Addition of lyticase[121,122] may facilitate the disruption of the fungal cell wall. Extraction of DNA after the mechanical disruption phase can be performed using one of several commercial DNA extraction kits.[6,110,117]

Non-commercial methods for the extraction of *Blastomyces* DNA include boiling in a guanidine thiocyanate/phenol extraction buffer[113] or the use of cetyl trimethylammonium bromide (CTAB) lysis buffer.[123] The latter reagent has been used extensively for the extraction of DNA from plants,[124,125] soil,[126] bacteria,[127] and fungi.[123,128] Extraction of DNA using CTAB lysis buffer provides a source of clean, high molecular weight DNA free of nucleases and inhibitors of proteinases and restriction enzymes,[125] and is also efficient in removing polysaccharides that may inhibit PCR.[129]

CTAB lysis buffer (100 mM Tris pH 8.0, 1.4 M NaCl, 20 mM EDTA, 2% cetyl trimethylammonium bromide [CTAB], 0.2% β-mercaptoethanol) has been used to extract DNA from a variety of mechanically disrupted samples from tissue ground in liquid nitrogen[130] or dry ice,[119] and mechanical homgenization.[120] CTAB lysis buffer can also be used in conjunction with glass bead disruption of fungal cells using a Bead Beater® system. A benefit of this method is that the mechanical disruption phase of the extraction process is self-contained, providing a greater margin of safety when working with pathogenic fungi.

Approximately 600–800 μL of CTAB lysis buffer is added to a 1.8 mL screw cap Bead Beater tube that is three fourths full of glass beads (0.5 mm in diameter). The CTAB lysis buffer should entirely wet and cover the glass beads to approximately 5 mm above the glass beads. Approximately 10–20 mg of fungal mycelium is added to the tube containing the glass beads and CTAB lysis buffer. The mould can be grown in either solid or liquid medium. However, mould

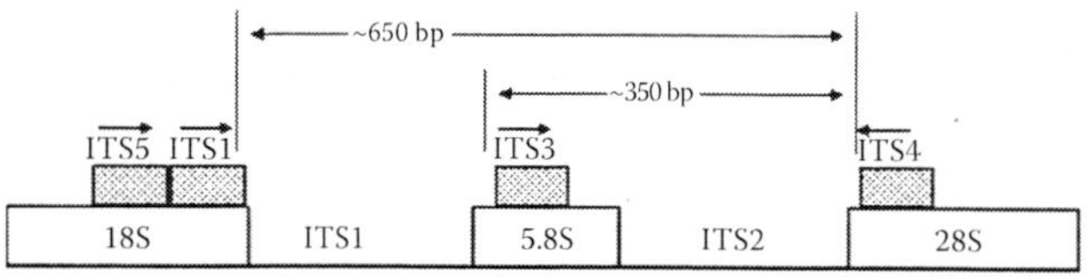

FIGURE 24.1 Diagram of the rRNA gene with the location of the binding sites of the universal fungal primers.

grown submerged in liquid medium by incubation on a shaker apparatus does not normally sporulate and therefore infectious conidia are not produced. Furthermore, the mycelium of moulds grown in liquid disperses into the CTAB lysis buffer more readily. After the mycelium is added to the Bead Beater tube, the screw cap is securely tightened and the tube is placed onto the Bead Beater shaker and the Bead Beater is run on the maximum setting for 1 min. The tubes are placed into a 65°C waterbath for 1–2 h, with inversion periodically to ensure complete digestion. Next, the tubes are centrifuged in a microcentrifuge at maximum speed for 10 min to pellet the glass beads. The supernatant is removed and placed into a 1.5 mL conical microcentrifuge tube. An equal volume of chloroform is added to the supernatant, mixed vigorously, and centrifuged at maximum speed for 10 min. The upper aqueous layer is removed without disturbing the debris layer at the interface and placed into a clean 1.5 mL conical microcentrifuge tube. The DNA is precipitated by adding an equal volume of cold isopropanol and mixed immediately with a vortex mixer. The tubes are placed on ice for at least 10 min; incubation at 4°C overnight may increase the yield of DNA. DNA is pelleted by centrifugation in a microcentrifuge at maximum speed for 15 min. The isopropanol is removed, 500 μL of cold 70% ethanol is added, and the tube is centrifuged at maximum speed for 15 min. The ethanol is removed and the DNA pellet is allowed to dry. The pellet is resuspended in 50 μL Tris/ EDTA (TE) buffer at 4°C for 1–2 h to overnight.

24.2.2 Detection Procedures

The commonly used loci for PCR amplification and identification employ universal fungal primers that are directed to portions of the rRNA gene.[112] Each of the rRNA gene regions, the 18S, 28S, and the ITS regions, has been used in the molecular identification of *B. dermatitidis* and have varying discriminating power.[113,114,116] The ITS region of the rRNA gene, particularly the ITS2 region between the 5.8S and 28S regions of the rRNA gene, demonstrates sufficient diversity to provide an accurate identification of *B. dermatitidis* and is recommended by the Clinical Laboratory Standards Institute (MM18-A) for identification by DNA sequencing.[131]

PCR amplification of the ITS region of the rRNA gene using primer pair ITS1 and ITS4 produces an approximately 650 bp product that spans the entire ITS1, 5.8S, and ITS2 regions (Figure 24.1). Another primer, ITS5, located just upstream from the ITS1 binding site, may be substituted for ITS1 and in some instances may provide better discrimination.[132,133] The amplification product is generally the same size as that using ITS1 and ITS4. ITS3 combined with ITS4 amplifies the ITS2 region of approximately 350 bp. Use of this primer set may be desirable if the extracted DNA has been damaged, containing only short stretches of amplifiable DNA, which may occur when extracting DNA from formalin fixed tissue.[134] Table 24.1 describes the primers and their sequences.

PCR is performed using 1 ng of fungal DNA added to the PCR buffer mix consisting of 10 mM Tris–HCl buffer with 50 mM KCl and 1.5 mM $MgCl_2$, pH 8.0; 0.2 mM

TABLE 24.1
Primers Used for the PCR Amplification of *B. dermatitidis* DNA

Primer	Use	Direction	Sequence (5′–3′)
ITS1	Universal fungal primer: ITS region of rRNA gene[a]	Forward	TCCGTAGGTGAACCTGCGG
ITS3	Universal fungal primer: ITS region of rRNA gene[a]	Forward	GCATCGATGAAGAACGCAGC
ITS4	Universal fungal primer: ITS region of rRNA gene[a]	Reverse	TCCTCCGCTTATTGATATGC
ITS5	Universal fungal primer: ITS region of rRNA gene[a]	Forward	GGAAGTAAAAGTCGTAACAAGG
Blasto I	Blastomyces specific primer: promoter region for the BAD-1 gene[b]	Forward	AAGTGGCTGGGTAGTTATACGCTAC
Blasto II	Blastomyces specific primer: promoter region for the BAD-1 gene[b]	Reverse	TAGGTTGCTGATTCCATAAGTCAGG

[a] From White et al.[141]
[b] From Bialek et al.[116]

deoxynucleoside triphosphate, each; 1.25 U of Taq polymerase; and 0.2 μM of each oligonucleotide primer. Conditions for thermocycling begin with an initial denaturation of 95°C for 10 min, followed by 30 cycles of 95°C for 1 min, 53°C for 1 min, 72°C for 1 min, and a final extension of 72°C for 10 min with a final hold temperature of 4°C. The PCR product can then be visualized by electrophoresis using a 1%–2% agarose gel. If amplification of the DNA produces a weak band, the annealing temperature of 53°C may be reduced to 51°C.

PCR primers Blasto I and Blasto II, specific for *B. dermatitidis* (Table 24.1), amplify an approximately 360 bp portion of the promoter region for the BAD-1 virulence gene[116,117] although some *B. dermatitidis* strains contain an insert that increases the size to approximately 660 bp. Thermocycling conditions for these primers require a denaturing step of 94°C for 5 min followed by 30 cycles of 94°C for 30 s, 50°C for 30 s, and 72°C for 1 min with final extension step of 72°C for 5 min.

Identification of the ITS generated amplicon can be accomplished through the use of *Blastomyces*-specific probes using slot blot[113] or EIA[114] methodologies or through DNA sequence analysis.[135] The original Sanger sequencing method has been continually improved upon since its original description with most of the reagents and instrumentation commercialized and automated. However, prior to sequence analysis, primers and dNTPs must be removed from the PCR product either through column- or enzyme-based methodologies.[136,137] When available, sequence analysis provides the most flexibility when identifying an unknown pathogen.

24.3 CONCLUSIONS AND FUTURE PERSPECTIVES

While generally rare even within the endemic region, blastomycosis can cause significant morbidity and mortality in infected populations. Unfortunately, the burden of infection due to *B. dermatitidis* in North America is still not fully understood. Blastomycosis is a reportable disease in only six states, all within the endemic region. However, the ease of travel to and from endemic regions makes it possible for disease to occur outside the standard endemic area. Accurate incidence rates from other states outside the endemic region are not readily available. Alternatively, reports of infections outside the known endemic region in patients without a history of travel may reflect an expansion of the endemic areas or the presence of endemic microfoci that are currently not recognized. The lack of an adequate skin test antigen to determine the rate of exposure has hindered the determination of the prevalence of infection in different areas of the country and of the true range of endemicity. Skin tests were valuable in defining the endemic ranges for both histoplasmosis[138,139] and coccidioidomycosis.[140]

Current diagnostic assays lack either sensitivity or specificity, or require weeks before obtaining a final result. Development and standardization of rapid diagnostic assays with adequate sensitivities and specificities that could determine true infection would allow for earlier treatment, reducing morbidity and mortality. Also, an accurate diagnosis of blastomycosis would aid in determining the true incidence of disease.

The environmental niche of *B. dermatitidis* remains elusive. Continued advancement in determining the nutritional needs and environmental conditions for the optimal growth of *B. dermatitidis* may aid in predicting where exposure may occur and help reduce unnecessary morbidity. Accurate and reliable methods to detect *B. dermatitidis* in the environment will help confirm the environmental requirements for *B. dermatitidis*. Molecular assays have the ability to provide sensitive and specific results but determining the accuracy of these assays is difficult when the "gold standard" assay is less sensitive than that being evaluated.

It has been over 100 years since Gilchrist first discovered *B. dermatitidis*; yet there is still much to learn and many challenges to overcome to better understand the biology, epidemiology, and environmental niche of this organism. Newer, more sensitive, and specific diagnostic technologies combined with improved molecular methods may allow for rapid advancement of our current understanding of this fascinating dimorphic fungus.

DISCLAIMER

The findings and conclusions in this chapter are those of the author and do not necessarily represent the views of the CDC.

REFERENCES

1. Gilchrist, T.C. and Stokes, W.R., The presence of an oidium in the tissues of a case of pseudo-lupus vulgaris. *Bull. Johns Hopkins Hosp.*, 7, 129, 1896.
2. Gilchrist, T.C. and Stokes, W.R., A case of pseudo-lupus vulgaris caused by a *Blastomyces*. *J. Exp. Med.*, 3, 53, 1898.
3. Walker, J.W. and Montgomery, F.H., Further report of a previously recorded case of blastomycosis of the skin; systemic infection with *Blastomyces*; death; autopsy. *J. Am. Med. Assoc.*, 32, 867, 1902.
4. Sigler, L., *Ajellomyces crescens* sp. nov., taxonomy of *Emmonsia* spp., and relatedness with *Blastomyces dermatitidis* (teleomorph *Ajellomyces dermatitidis*). *J. Med. Vet. Mycol.*, 34, 303, 1996.
5. Dot, J.M. et al., Molecular diagnosis of disseminated adiaspiromycosis due to *Emmonsia crescens*. *J. Clin. Microbiol.*, 47, 1269, 2009.
6. Bialek, R. et al., Small subunit ribosomal DNA sequence shows *Paracoccidioides brasiliensis* closely related to *Blastomyces dermatitidis*. *J. Clin. Microbiol.*, 38, 3190, 2000.
7. Leclerc, M.C., Philippe, H., and Gueho, E., Phylogeny of dermatophytes and dimorphic fungi based on large subunit ribosomal RNA sequence comparisons. *J. Med. Vet. Mycol.*, 32, 331, 1994.
8. Gueho, E. et al., Molecular taxonomy and epidemiology of *Blastomyces* and *Histoplasma* species. *Mycoses*, 40, 69, 1997.
9. Peterson, S.W. and Sigler, L., Molecular genetic variation in *Emmonsia crescens* and *Emmonsia parva*, etiologic agents of adiaspiromycosis, and their phylogenetic relationship to

Blastomyces dermatitidis (*Ajellomyces dermatitidis*) and other systemic fungal pathogens. *J. Clin. Microbiol.*, 36, 2918, 1998.
10. Bradsher, R.W. Jr., Pulmonary blastomycosis. *Semin. Respir. Crit. Care Med.*, 29, 174, 2008.
11. McDonough, E.S. and Lewis, A.L., *Blastomyces dermatitidis*: Production of the sexual stage. *Science*, 156, 528, 1967.
12. McDonough, E.S. and Lewis, A.L., The ascigerous stage of *Blastomyces dermatitidis*. *Mycologia*, 60, 76, 1968.
13. McDonough, E.S. et al., Geographic distribution of "plus" and "minus" isolates of *Blastomyces* (*Ajellomyces*) *dermatitidis* in North America. *Am. J. Epidemiol.*, 98, 63, 1973.
14. McDonough, E.S., Chan, D.M., and McNamara, W.J., Dual infection by "+" and "−" mating types of *Ajellomyces* (*Blastomyces*) *dermatitidis*. *Am. J. Epidemiol.*, 106, 67, 1977.
15. Garrison, R.G., Lane, J.W., and Johnson, D.R., Ultrastructural studies on the cleistothecium of *Ajellomyces dermatitidis*. *Sabouraudia*, 11, 131, 1973.
16. Glick, A.D. and Kwon-Chung, K.J., Ultrastructural comparison of coils and ascospores of *Emmonsiella capsulata* and *Ajellomyces dermatitidis*. *Mycologia*, 65, 216, 1973.
17. McGinnis, M.R. and Katz, B., *Ajellomyces* and its synonym *Emmonsiella*. *Mycotaxon*, 8, 157, 1979.
18. Kaufman, L. et al., Detection of two *Blastomyces dermatitidis* serotypes by exoantigen analysis. *J. Clin. Microbiol.*, 18, 110, 1983.
19. Klein, B.S., Aizenstein, B.D., and Hogan, L.H., African strains of *Blastomyces dermatitidis* that do not express surface adhesin WI-1. *Infect. Immun.*, 65, 1505, 1997.
20. Axtell, R.C. and Scalarone, G.M., Serological differences in two *Blastomyces dermatitidis* isolates from different geographical regions of North America. *Mycopathologia*, 153, 141, 2001.
21. Yates-Siilata, K.E., Sander, D.M., and Keath, E.J., Genetic diversity in clinical isolates of the dimorphic fungus *Blastomyces dermatitidis* detected by a PCR-based random amplified polymorphic DNA assay. *J. Clin. Microbiol.*, 33, 2171, 1995.
22. McCullough, M.J. et al., Molecular epidemiology of *Blastomyces dermatitidis*. *Clin. Infect. Dis.*, 30, 328, 2000.
23. Meece, J.K. et al., Genetic diversity in *Blastomyces dermatitidis*: Implications for PCR detection in clinical and environmental samples. *Med. Mycol.*, 22, 1–7, 2009.
24. Drutz, D.J. and Frey, C.L., Intracellular and extracellular defenses of human phagocytes against *Blastomyces dermatitidis* conidia and yeasts. *J. Lab. Clin. Med.*, 105, 737, 1985.
25. San-Blas, G. and San-Blas, F., Molecular aspects of fungal dimorphism. *Crit. Rev. Microbiol.*, 11, 101, 1984.
26. Hogan, L.H. and Klein, B.S., Altered expression of surface alpha-1,3-glucan in genetically related strains of *Blastomyces dermatitidis* that differ in virulence. *Infect. Immun.*, 62, 3543, 1994.
27. Klimpel, K.R. and Goldman, W.E., Cell walls from avirulent variants of *Histoplasma capsulatum* lack alpha-(1,3)-glucan. *Infect. Immun.*, 56, 2997, 1988.
28. Rappleye, C.A., Engle, J.T., and Goldman, W.E., RNA interference in *Histoplasma capsulatum* demonstrates a role for alpha-(1,3)-glucan in virulence. *Mol. Microbiol.*, 53, 153, 2004.
29. Reese, A.J. et al., Loss of cell wall alpha(1–3) glucan affects *Cryptococcus neoformans* from ultrastructure to virulence. *Mol. Microbiol.*, 63, 1385, 2007.
30. Rappleye, C.A., Eissenberg, L.G., and Goldman, W.E., *Histoplasma capsulatum* alpha-(1,3)-glucan blocks innate immune recognition by the beta-glucan receptor. *Proc. Natl. Acad. Sci. U.S.A.*, 104, 1366, 2007.
31. Rooney, P.J., Sullivan, T.D., and Klein, B.S., Selective expression of the virulence factor BAD-1 upon morphogenesis to the pathogenic yeast form of *Blastomyces dermatitidis*: Evidence for transcriptional regulation by a conserved mechanism. *Mol. Microbiol.*, 39, 875, 2001.
32. Klein, B.S. and Newman, S.L., Role of cell-surface molecules of *Blastomyces dermatitidis* in host-pathogen interactions. *Trends Microbiol.*, 4, 246, 1996.
33. Brandhorst, T.T. et al., Targeted gene disruption reveals an adhesin indispensable for pathogenicity of *Blastomyces dermatitidis*. *J. Exp. Med.*, 189, 1207, 1999.
34. Finkel-Jimenez, B. et al., The WI-1 adhesin blocks phagocyte TNF-alpha production, imparting pathogenicity on *Blastomyces dermatitidis*. *J. Immunol.*, 166, 2665, 2001.
35. Finkel-Jimenez, B., Wuthrich, M., and Klein, B.S., BAD-1, an essential virulence factor of *Blastomyces dermatitidis*, suppresses host TNF-alpha production through TGF-beta-dependent and -independent mechanisms. *J. Immunol.*, 168, 5746, 2002.
36. Rappleye, C.A. and Goldman, W.E., Defining virulence genes in the dimorphic fungi. *Ann. Rev. Microbiol.*, 60, 281, 2006.
37. Klein, B.S. et al., Altered expression of surface protein WI-1 in genetically related strains of *Blastomyces dermatitidis* that differ in virulence regulates recognition of yeasts by human macrophages. *Infect. Immun.*, 62, 3536, 1994.
38. Arceneaux, K.A., Taboada, J., and Hosgood, G., Blastomycosis in dogs: 115 cases (1980–1995). *J. Am. Vet. Med. Assoc.*, 213, 658, 1998.
39. Bromel, C. and Sykes, J.E., Epidemiology, diagnosis, and treatment of blastomycosis in dogs and cats. *Clin. Tech. Small Anim. Pract.*, 20, 233, 2005.
40. Blondin, N. et al., Blastomycosis in indoor cats: Suburban Chicago, Illinois, USA. *Mycopathologia*, 163, 59, 2007.
41. Wilson J.H. et al., Systemic blastomycosis in a horse. *J. Vet. Diagn. Invest.*, 18, 615, 2006.
42. Zwick, L.S. et al., Disseminated blastomycosis in two California sea lions (*Zalophus californianus*). *J. Zoo Wildl. Med.*, 31, 211, 2000.
43. Furcolow, M.L. et al., Prevalence and incidence studies of human and canine blastomycosis. II. Yearly incidence studies in three selected states, 1960–1967. *Am. J. Epidemiol.*, 92, 121, 1970.
44. Storms, T.N. et al., Blastomycosis in nondomestic felids. *J. Zoo Wildl. Med.*, 34, 231, 2003.
45. Denton, J.F. et al., Isolation of *Blastomyces dermatitidis* from soil. *Science*, 133, 1126, 1961.
46. Klein, B.S., Vergeront, J.M., and Davis, J.P., Epidemiologic aspects of blastomycosis, the enigmatic systemic mycosis. *Semin. Respir. Infect.*, 1, 29, 1986.
47. Chick, E.W. et al., Scientific Assembly statement. The use of skin tests and serologic tests in histoplasmosis, coccidioidomycosis, and blastomycosis, 1973. *Am. Rev. Respir. Dis.*, 108, 156, 1973.
48. Klein, B.S. et al., Isolation of *Blastomyces dermatitidis* in soil associated with a large outbreak of blastomycosis in Wisconsin. *N. Engl. J. Med.*, 314, 529, 1986.
49. Morris, S.K. et al., Blastomycosis in Ontario, 1994–2003. *Emerg. Infect. Dis.*, 12, 274, 2006.
50. Blastomycosis—Wisconsin, 1986–1995. *MMWR*, 45, 601, 1996.

51. Hannah, E.L. et al., Public health response to 2 clinical cases of blastomycosis in Colorado residents. *Clin. Infect. Dis.*, 32, E151, 2001.
52. Veligandla, S.R. et al., Delayed diagnosis of osseous blastomycosis in two patients following environmental exposure in nonendemic areas. *Am. J. Clin. Pathol.*, 118, 536, 2002.
53. Klein, B.S. et al., Two outbreaks of blastomycosis along rivers in Wisconsin. Isolation of *Blastomyces dermatitidis* from riverbank soil and evidence of its transmission along waterways. *Am. Rev. Respir. Dis.*, 136, 1333, 1987.
54. Cockerill, F.R. 3rd et al., Epidemic of pulmonary blastomycosis (Namekagon fever) in Wisconsin canoeists. *Chest*, 86, 688, 1984.
55. Baumgardner, D.J. et al., An outbreak of blastomycosis on a United States Indian reservation. *Wilderness Environ. Med.*, 13, 250, 2002.
56. Baumgardner, D.J. et al., Geographic information system analysis of blastomycosis in northern Wisconsin, USA: Waterways and soil. *Med. Mycol.*, 43, 117, 2005.
57. Baumgardner, D.J. and Laundre, B., Studies on the molecular ecology of *Blastomyces dermatitidis*. *Mycopathologia*, 152, 51, 2001.
58. Baumgardner, D.J., Microecology of *Blastomyces dermatitidis*: The ammonia hypothesis. *Med. Mycol.*, 47, 745, 2009.
59. Kauffman, C.A., Endemic mycoses: Blastomycosis, histoplasmosis, and sporotrichosis. *Infect. Dis. Clin. North Am.*, 20, 645, 2006.
60. Benenson, A.S. (ed.), *Control of Communicable Diseases Manual*. American Public Health Association, Washington, DC, 1995.
61. Wallace, J., Pulmonary blastomycosis: A great masquerader. *Chest*, 121, 677, 2002.
62. Areno, J.P., Campbell, G.D. Jr., and George, R.B., Diagnosis of blastomycosis. *Semin. Respir. Infect.*, 12, 252, 1997.
63. Tosh, F.E. et al., A common source epidemic of North American blastomycosis. *Am. Rev. Respir. Dis.*, 109, 525, 1974.
64. Recht, L.D. et al., Self-limited blastomycosis: A report of thirteen cases. *Am. Rev. Respir. Dis.*, 120, 1109, 1979.
65. Kwon-Chung, J. and Bennett, J.E., *Medical Mycology*. Lea and Febiger, Melvern, PA, 1994.
66. Bradsher, R.W., Histoplasmosis and blastomycosis. *Clin. Infect. Dis.*, 22 (Suppl 2), S102, 1996.
67. Lemos, L.B., Guo, M., and Baliga, M., Blastomycosis: Organ involvement and etiologic diagnosis. A review of 123 patients from Mississippi. *Ann. Diagn. Pathol.*, 4, 391, 2000.
68. Bariola, J.R. et al., Blastomycosis of the central nervous system: A multicenter review of diagnosis and treatment in the modern era. *Clin. Infect. Dis.*, 50, 797, 2010.
69. Sung, V.W., Blastomycosis. *Headache*, 49, 461, 2009.
70. Mahiquez, M. et al., Nonsurgical treatment of lumbosacral blastomycosis involving L2-S1: A case report. *Spine*, 33, E442, 2008.
71. Monsanto, E.H., Johnston, A.D., and Dick, H.M., Isolated blastomycotic osteomyelitis: A case simulating a malignant tumor of the distal radius. *J. Hand. Surg. Am.*, 11, 896, 1986.
72. Neal, P.M. and Nikolai, A., Systemic blastomycosis diagnosed by prostate needle biopsy. *Clin. Med. Res.*, 6, 24, 2008.
73. Graham, W.R. Jr. and Callaway, J.L., Primary inoculation blastomycosis in a veterinarian. *J. Am. Acad. Dermatol.*, 7, 785, 1982.
74. Sewell, D.L., Laboratory-associated infections and biosafety. *Clin. Microbiol. Rev.*, 8, 389, 1995.
75. Gnann, J.W. Jr. et al., Human blastomycosis after a dog bite. *Ann. Intern. Med.*, 98, 48, 1983.
76. Silveira, F.P. and Husain, S., Fungal infections in solid organ transplantation. *Med. Mycol.*, 45, 305, 2007.
77. Serody, J.S. et al., Blastomycosis in transplant recipients: Report of a case and review. *Clin. Infect. Dis.*, 16, 54, 1993.
78. Marques, S.A. et al., Mycoses associated with AIDS in the Third World. *Med. Mycol.*, 38(Suppl 1), 269, 2000.
79. Pappas, P.G., Blastomycosis in the immunocompromised patient. *Semin. Respir. Infect.*, 12, 243, 1997.
80. Chapman, S.W. et al., Clinical practice guidelines for the management of blastomycosis: 2008 update by the Infectious Diseases Society of America. *Clin. Infect. Dis.*, 46, 1801, 2008.
81. Freifeld, A. et al., Voriconazole use for endemic fungal infections. *Antimicrob. Agents Chemother.*, 53, 1648, 2009.
82. Gonzalez, G.M., In vitro activities of isavuconazole against opportunistic filamentous and dimorphic fungi. *Med. Mycol.*, 47, 71, 2009.
83. Musial, C.E. et al., Recovery of *Blastomyces dermatitidis* from blood of a patient with disseminated blastomycosis. *J. Clin. Microbiol.*, 25, 1421, 1987.
84. Seo, R., Oyasu, R., and Schaeffer, A., Blastomycosis of the epididymis and prostate. *Urology*, 50, 980, 1997.
85. Chosewood, L.C. and Wilson, D.E. (eds.), *Biosafety in Microbiological and Biomedical Laboratories*. US Government Printing Office, Washington, DC, 2007.
86. Green, J.H. et al., Isolation of an antigen from *Blastomyces dermatitidis* that is specific for the diagnosis of blastomycosis. *Curr. Mcrobiol.*, 4, 293, 1980.
87. Kaufman, L., Kovacs, J.A., and Reiss, E., Clinical immunomycology. In: *Manual of Clinical Laboratory Immunology*, Rose N.R. et al. (eds.). ASM Press, Washington, DC, p. 585, 1997.
88. Lambert, R.S. and George, R.B., Evaluation of enzyme immunoassay as a rapid screening test for histoplasmosis and blastomycosis. *Am. Rev. Respir. Dis.*, 136, 316, 1987.
89. Turner, S. and Kaufman, L., Immunodiagnosis of blastomycosis. *Semin. Respir. Infect.*, 1, 22, 1986.
90. Turner, S., Kaufman, L., and Jalbert, M., Diagnostic assessment of an enzyme-linked immunosorbent assay for human and canine blastomycosis. *J. Clin. Microbiol.*, 23, 294, 1986.
91. Sekhon, A.S. et al., The value of the Premier enzyme immunoassay for diagnosing *Blastomyces dermatitidis* infections. *J. Med. Vet. Mycol.*, 33, 123, 1995.
92. Klein, B.S. and Jones, J.M., Isolation, purification, and radiolabeling of a novel 120-kD surface protein on *Blastomyces dermatitidis* yeasts to detect antibody in infected patients. *J. Clin. Invest.*, 85, 152, 1990.
93. Klein, B.S. and Jones, J.M., Purification and characterization of the major antigen WI-1 from *Blastomyces dermatitidis* yeasts and immunological comparison with A antigen. *Infect. Immun.*, 62, 3890, 1994.
94. Klein, B.S. et al., Serological tests for blastomycosis: Assessments during a large point-source outbreak in Wisconsin. *J. Infect. Dis.*, 155, 262, 1987.
95. Hurst, S.F. and Kaufman, L., Western immunoblot analysis and serologic characterization of *Blastomyces dermatitidis* yeast form extracellular antigens. *J. Clin. Microbiol.*, 30, 3043, 1992.
96. Durkin, M. et al., Antigen assay with the potential to aid in diagnosis of blastomycosis. *J. Clin. Microbiol.*, 42, 4873, 2004.

97. Hage, C.A. et al., Diagnosis of pulmonary histoplasmosis and blastomycosis by detection of antigen in bronchoalveolar lavage fluid using an improved second-generation enzyme-linked immunoassay. *Respir. Med.*, 101, 43, 2007.
98. Tarr, M. et al., *Blastomyces* antigen detection for monitoring progression of blastomycosis in a pregnant adolescent. *Infect. Dis. Obstet. Gynecol.*, 2007, 89059, 2007.
99. Mongkolrattanothai, K. et al., Urine antigen detection of blastomycosis in pediatric patients. *Pediatr. Infect. Dis. J.*, 25, 1076, 2006.
100. Spector, D. et al., Antigen and antibody testing for the diagnosis of blastomycosis in dogs. *J. Vet. Intern. Med.*, 22, 839, 2008.
101. Kelley, W.H., A study of the cell and colony variations of *Blastomyces dermititidis, J. Infect. Med.*, 64, 293, 1939.
102. Weeks, R.J., A rapid, simplified medium for converting the mycelial phase of *Blastomyces dermatitidis* to the yeast phase. *Mycopathol. Mycol. Appl.*, 22, 153, 1964.
103. Chandler, F.W. and Watts, J.C., *Pathologic Diagnosis of Fungal Infections*. ASCP Press, Chicago, IL, 1987.
104. Padhye, A.A. et al., Comparative evaluation of chemiluminescent DNA probe assays and exoantigen tests for rapid identification of *Blastomyces dermatitidis* and *Coccidioides immitis*. *J. Clin. Microbiol.*, 32, 867, 1994.
105. Stockman, L. et al., Evaluation of commercially available acridinium ester-labeled chemiluminescent DNA probes for culture identification of *Blastomyces dermatitidis*, *Coccidioides immitis*, *Cryptococcus neoformans*, and *Histoplasma capsulatum*. *J. Clin. Microbiol.*, 31, 845, 1993.
106. Iwen, P.C. et al., Pulmonary infection caused by *Gymnascella hyalinospora* in a patient with acute myelogenous leukemia. *J. Clin. Microbiol.*, 38, 375, 2000.
107. Scalarone, G.M. et al., Evaluation of a commercial DNA probe assay for the identification of clinical isolates of *Blastomyces dermatitidis* from dogs. *J. Med. Vet. Mycol.*, 30, 43, 1992.
108. Woods, C.R. Jr. et al., Analysis of relationships among isolates of *Citrobacter diversus* by using DNA fingerprints generated by repetitive sequence-based primers in the polymerase chain reaction. *J. Clin. Microbiol.*, 30, 2921, 1992.
109. Redkar, R.J. et al., DNA fingerprinting of *Candida rugosa* via repetitive sequence-based PCR. *J. Clin. Microbiol.*, 34, 1677, 1996.
110. Pounder, J.I., Hansen, D., and Woods, G.L., Identification of *Histoplasma capsulatum*, *Blastomyces dermatitidis*, and *Coccidioides* species by repetitive-sequence-based PCR. *J. Clin. Microbiol.*, 44, 2977, 2006.
111. Wengenack, N.L. and Binnicker, M.J., Fungal molecular diagnostics. *Clin. Chest Med.*, 30, 391, 2009.
112. Lau, A. et al., Current status and future perspectives on molecular and serological methods in diagnostic mycology. *Future Microbiol.*, 4, 1185, 2009.
113. Sandhu, G.S. et al., Molecular probes for diagnosis of fungal infections. *J. Clin. Microbiol.*, 33, 2913, 1995.
114. Lindsley, M.D. et al., Rapid identification of dimorphic and yeast-like fungal pathogens using specific DNA probes. *J. Clin. Microbiol.*, 39, 3505, 2001.
115. Hayden, R.T. et al., *In situ* hybridization for the identification of yeastlike organisms in tissue section. *Diagn. Mol. Pathol.*, 10, 15, 2001.
116. Bialek, R. et al., Nested PCR assays for detection of *Blastomyces dermatitidis* DNA in paraffin-embedded canine tissue. *J. Clin. Microbiol.*, 41, 205, 2003.
117. Burgess, J.W., Schwan, W.R., and Volk, T.J., PCR-based detection of DNA from the human pathogen *Blastomyces dermatitidis* from natural soil samples. *Med. Mycol.*, 44, 741, 2006.
118. Karakousis, A. et al., An assessment of the efficiency of fungal DNA extraction methods for maximizing the detection of medically important fungi using PCR. *J. Microbiol. Methods*, 65, 38, 2006.
119. Rogers, S.O. and Bendich, A.J., Extraction of DNA from milligram amounts of fresh, herbarium and mummified plant tissues. *Plant Mol. Biol.*, 5, 69, 1985.
120. Li, R. et al., A reliable and inexpensive method of nucleic acid extraction for the PCR-based detection of diverse plant pathogens. *J. Virol. Methods*, 154, 48, 2008.
121. Saad, D.S. et al., Extraction of genomic DNA from filamentous fungi in biofilms of water-based paint coatings. *Int. Biodete. Biodegrad.*, 54, 99, 2004.
122. Paterson, P.J. et al., Development of molecular methods for the identification of aspergillus and emerging moulds in paraffin wax embedded tissue sections. *Mol. Pathol.*, 56, 368, 2003.
123. Gardes, M. and Bruns, T.D., ITS primers with enhanced specificity for basidiomycetes—Application to the identification of mycorrhizae and rusts. *Mol. Ecol.*, 2, 113, 1993.
124. Allen, G.C. et al., A modified protocol for rapid DNA isolation from plant tissues using cetyltrimethylammonium bromide. *Nat. Protoc.*, 1, 2320, 2006.
125. Murray, M.G. and Thompson, W.F., Rapid isolation of high molecular weight plant DNA. *Nucleic Acids Res*, 8, 4321, 1980.
126. Zhou, J., Bruns, M.A., and Tiedje, J.M., DNA recovery from soils of diverse composition. *Appl. Environ. Microbiol.*, 62, 316, 1996.
127. Park, D., Genomic DNA isolation from different biological materials. *Methods Mol. Biol.*, 353, 3, 2007.
128. Brandfass, C. and Karlovsky, P., Upscaled CTAB-based DNA extraction and real-time PCR assays for *Fusarium culmorum* and *F. graminearum* DNA in plant material with reduced sampling error. *Int. J. Mol. Sci.*, 9, 2306, 2008.
129. Clarke, J.D., Cetyltrimethyl ammonium bromide (CTAB) DNA miniprep for plant DNA isolation. *CSH Protoc*, 2009, pdb prot5177, 2009.
130. Filion, M., St-Arnaud, M., and Jabaji-Hare, S.H., Direct quantification of fungal DNA from soil substrate using real-time PCR. *J. Microbiol. Methods*, 53, 67, 2003.
131. Interpretive criteria for identification of bacteria and fungi by DNA target sequencing: Approved guideline. Document MM18-A. Clinical and Laboratory Standards Institute, Wayne, PA, 2008.
132. Larena, I. et al., Design of a primer for ribosomal DNA internal transcribed spacer with enhanced specificity for ascomycetes. *J. Biotechnol.*, 75, 187, 1999.
133. Bertini, L. et al., A new pair of primers designed for amplification of the ITS region in Tuber species. *FEMS Microbiol. Lett.*, 173, 239, 1999.
134. Greer, C.E. et al., PCR amplification from paraffin-embedded tissues. Effects of fixative and fixation time. *Am. J. Clin. Pathol.*, 95, 117, 1991.
135. Sanger, F., Nicklen, S., and Coulson, A.R., DNA sequencing with chain-terminating inhibitors. *Proc. Natl. Acad. Sci. U.S.A.*, 74, 5463, 1977.
136. Olsson. M. et al., A rapid and simple nested PCR assay for the detection of *Pneumocystis carinii* in sputum samples. *Scand. J. Infect. Dis.*, 28, 597, 1996.

137. Bell, J., A simple way to treat PCR products prior to sequencing using ExoSAP-IT. *Biotechniques*, 44, 834, 2008.
138. Edwards, L.B., et al., An atlas of sensitivity to tuberculin, PPD-B, and histoplasmin in the United States. *Am. Rev. Respir. Dis.*, 99, S132, 1969.
139. Furcolow, M.L., Tests of immunity in histoplasmosis. *N. Engl. J. Med.*, 268, 357, 1963.
140. Emmett, J., Coccidioidin sensitivity among school children in Phoenix (skin test and x-ray survey). *Am. J. Public Health Nations Health*, 42, 241, 1952.
141. White, T.J. et al., Amplification and direct sequencing of fungal ribosomal RNA genes for phylogenetics. In: *PCR Protocols. A Guide to Methods and Applications*, Innis, M.A., Gelfand, D.H., Sninsky, J.J., and White, T.J. (eds.). Academic Press, San Diego, CA, p. 315, 1990.

25 *Chrysosporium*

Dongyou Liu and R.R.M. Paterson

CONTENTS

25.1 INTRODUCTION

Chrysosporium is an anamorph genus whose members are common soil saprobes, with sexual states (teleomorphs) found in the ascomycete order *Onygenales*. Of the 28 recognized species within the genus *Chrysosporium*, two species (i.e., *C. zonatum* and *C. tropicum*) have been reported as causal agents for skin disorder, sinusitis, and other diseases in both immunocompromised and immunocompetent hosts.

25.1.1 Classification and Morphology

The genus *Chrysosporium* (obsolete synonym: *Glenosporella*) belongs to the mitosporic Onygenales group, order Onygenales, class Eurotiomycetes, subphylum Pezizomycotina, phylum Ascomycota, and kingdom Fungi. The group consists of nine genera: *Blastomyces*, *Chrysosporium*, *Coccidioides*, *Emmonsia*, *Geomyces*, *Locazia*, *Malbranchea*, *Myriodontium*, and *Paracoccidioides*. In turn, the genus *Chrysosporium* consists of 28 recognized species: *Chrysosporium articulatum*, *Chrysosporium carmichaelii*, *Chrysosporium chiropterorum*, *Chrysosporium europa*, *Chrysosporium evolceanui*, *Chrysosporium filiforme*, *Chrysosporium fluviale*, *Chrysosporium fluviale*, *Chrysosporium indicum*, *Chrysosporium keratinophilum*, *Chrysosporium lobatum*, *Chrysosporium lucknowense*, *Chrysosporium mephiticum*, *Chrysosporium merdarium*, *Chrysosporium minutisporosum*, *Chrysosporium ophiodiicola*, *Chrysosporium pannicola*, *Chrysosporium pilosum*, *Chrysosporium pseudomerdarium*, *Chrysosporium queenslandicum*, *Chrysosporium siglerae*, *Chrysosporium submersum*, *Chrysosporium sulfureum*, *Chrysosporium synchronum*, *Chrysosporium tropicum*, *Chrysosporium undulatum*, *Chrysosporium vallenarense*, *Chrysosporium xerophilum*, and *Chrysosporium zonatum*, and 18 unassigned species [1–3]. The teleomorphs of *Chrysosporium* spp. are found in the genera *Aphanoascus*, *Nannizziopsis*, and *Uncinocarpus*, family Onygenaceae, and order Onygenales [4,5].

The obsolete species in the genus include *Chrysosporium dermatitidis* (→ *Blastomyces dermatitidis*), *Chrysosporium pannorum* (→ *Geomyces pannorus*), *Chrysosporium pruinosum* (→ *Sporotrichum pruinosum*), and *Chrysosporium thermophilum* (→ *Myceliophthora thermophila*) [6]. Not discussed here are *Chrysosporium parvum* (or *Chrysosporium parvum* var. *parvum*) and *Chrysosporium parvum* var. *crescens*, which are considered as synonymous for *Emmonsia parva* and *Emmonsia crescens*, respectively. To simplify the presentation and reduce confusion, this chapter focuses on *Chrysosporium* spp. that are keratinophilic, causing subcutaneous infections, whereas Chapter 29 deals with *Emmonsia* spp. that mainly cause pulmonary adiaspiromycosis.

Chrysosporium colonies grow moderately at 25°C, and may appear granular, woolly, or cottony, flat, or raised, and folded. Colonies are white cream, yellow, or tan to pale brown on the top, and white to brown on the reverse. Hyphae are septate. Conidia (aleuriconidia) are hyaline, broad based, one celled, and smooth or rough walled. Conidia are broader than vegetative hyphae and occur terminally on pedicels, along the sides of the hyphae, or in intercalary positions. Conidia usually have an annular frill that is the remnant of the hyphal wall that remains after detachment from the hypha. Arthroconidia are abundant and larger than parent hyphae in diameter.

C. tropicum colonies are moderately fast growing, flat, white to tan to beige, often with a powdery or granular surface texture. Reverse pigment is absent or pale brownish yellow with age. Ameroconidia are hyaline, one celled, and produced directly on vegetative hyphae by nonspecialized conidiogenous cells. Conidia (6–7 μm × 3.5–4 μm) are typically pyriform to clavate with truncate bases and are formed either intercalary (arthroconidia), laterally (often on pedicels), or terminally.

C. zonatum (synonym: *Chrysosporium gourii*) colonies on potato dextrose agar (PDA) at 37°C are flat and coarsely powdery, and appear yellowish white initially but darken by 14–21 days to buff (grayish or brownish orange) with a light brown reverse. Colony topography and color on PDA are similar at 37°C and 25°C, but darkening of the colony obverse and growth rate are slightly faster at 37°C (73 mm in diameter after 14 days) than at 25°C (66 mm in diameter). Microscopically, the fungus forms solitary aleurioconidia that are borne at the ends of short, typically curved stalks or that are sessile (borne on the sides of the hyphae). Conidia (3.5–13 μm × 2.5–5 μm) are single celled, rarely two celled, smooth to slightly roughened, and clavate (club shaped) to broadly obovoid (egg shaped) and have a rounded tip and a broad, flat basal scar. Intercalary arthroconidia may be formed but are uncommon. Racquet hyphae (hyphae showing swellings near the septa) are common [4]. The teleomorph of *C. zonatum* is a heterothallic ascomycete *Uncinocarpus orissi* (synonyms, *Pseudoarachniotus orissi* and *Gymnoascus arxii*) (family Onygenaceae). *U. orissi* ascocarps are solitary, globose, and reddish brown and are composed of pale reddish brown ascospores surrounded by thin-walled hyaline racket hyphae and conidia. Pairing of *C. zonatum* with *U. orissi* produces ascoma-containing ascospores that are oblate (like flattened disks) with truncate ends and appear smooth to slightly pitted (punctate) [4].

Chrysosporium ophiodiicola colonies attain diameters of 27–29 mm in 14 days at 25°C on potato carrot agar (PCA; 20 g potato, 20 g carrot, 15 g agar, 1 L water). Colonies are white from top and uncolored on reverse, felty, plane, and fimbriate, with a poorly defined margin. Sparse tufts of aerial mycelium are present on the submarginal zone. Vegetative hyphae (1.5–2.5 μm wide) are hyaline, branched, septate, smooth, and thin walled. They are often disarticulated at maturity to form cylindrical arthroconidia (7.5–10 μm × 2–3 μm) adjacent to each other. Fertile hyphae arise as lateral branches. Terminal and lateral conidia are borne on straight or flexuous side branches of variable length (4.5–16 μm) or are sometimes sessile. Conidia (4–9 μm × 2–3 μm) are unicellular, solitary, thin walled, smooth, hyaline to pale yellow, and cylindrical to slightly clavate. They are released by rhexolytic dehiscence, leaving broad and long basal scars. Intercalary, solitary conidia are often present, similar to the terminal and lateral ones. Racquet hyphae are scarce, and chlamydospores are not observed. On potato flake agar at 23°C, colonies are white to pale yellow, with a similarly colored reverse side, velvety to granular with age. They produce a strong, pungent odor. Conidia borne on stalks are arthroconidia. On PDA (Difco Laboratories), *C. ophiodiicola* grows more quickly and produces denser colonies of 31–35 mm in diameter in 14 days at 25°C. Colonies are white to pale yellow, buff after 1 month, and powdery, with droplets of colorless or light yellow exudates at the periphery. On phytone-yeast extract agar (BBL), colonies measure 32–39 mm in diameter in 14 days at 25°C and are white and light yellow at the center, powdery, and dense, with the presence of droplets of colorless exudate at the center. They are light brown on the reverse side. On oatmeal agar (30 g oat flakes, 1 g $MgSO_4{\cdot}7H_2O$, 1.5 g KH_2PO_4, 15 g agar, 1 L water), colonies are similar to those on PCA, with a very restricted growth at 15°C (5 mm in diameter in 14 days). At 37°C, there is no growth. Colonies produce a strong, pungent (skunk-like) odor after 1 month of incubation in all the media tested. The fungus shows a strong keratinolytic activity. Phenotypically, *C. ophiodiicola* is separated from the *Chrysosporium* anamorph of *Nannizziopsis vriesii* by the absence of asperulate fertile hyphae and globose-to-pyriform conidia sometimes grouped in clusters and the presence of an odor from *C. ophiodiicola* [3].

Chrysosporium species are differentiated from each other by the texture of the colony and the morphology, location, and size of the conidia. *C. zonatum* differs from other species by its faster growth at 37°C than at 25°C and by forming darken to buff colonies, and clavate, broadly truncate aleurioconidia typically borne on short, curved stalks. *Chrysosporium queenslandicum* (teleomorph: *Uncinocarpus queenslandicus*, or *Apinisia queenslandica* and *Brunneospora reticulata*) is similar, but its colonies do not darken, and intercalary arthroconidia are common [4].

C. ophiodiicola is separated from *C. mephiticum* by the narrow, cylindrical-to-slightly clavate conidia of *C. ophiodiicola*, and the pyriform-to-subglobose conidia of *C. mephiticum*; *C. ophiodiicola* is differentiated from *Aphanoascus mephitalis* by the production of the telemorphic form or during the culture of *A. mephitalis*; and *C. ophiodiicola* is distinguished from *Chrysosporium europae* by the characteristic vinaceous, buff-pigmented colonies on phytone-yeast extract agar and the absence of a strong, pungent odor from *C. europae*. Some species such as *Chrysosporium pannicola* do not grow at 37°C. *Chrysosporium* differs from *Blastomyces* by being nondimorphic, from *Microsporum* and *Trichophyton* by lacking macroconidia, from *Geomyces* by lacking branched, fertile hyphae on erect conidiophores, and from *Sepedonium* by having hyaline conidia.

25.1.2 Clinical Features

Chrysosporium spp. are soil saprophytes with broad distribution. They have been isolated in soil, plant material, dung, and birds [6–8]. These organisms may enter hosts through airborne conidia and exposure to soil. Many *Chrysosporium* spp. are keratinophilic filamentous fungi involved in the breakdown of shed keratinous substrates, and may cause skin infections and onychomycosis in humans [9–12]. In addition, *Chrysosporium* species have been occasionally associated with disseminated human mycosis, affecting the brain, lungs, sinuses, liver, and kidneys and leading to sinusitis, pneumonia, pleuritis, pericarditis, and osteomyelitis [13–15].

Roilides et al. [4] reported the first case of *Chrysosporium zonatum* infection in a 15-year-old boy with X-linked chronic granulomatous disease. He developed a lobar pneumonia and tibia osteomyelitis after a 2 month prophylactic therapy with γ-interferon. He presented with pain in his right shoulder and distal part of his right tibia, infrequent cough, and fever. A computed tomography (CT) scan of the chest showed a well-demarcated large mass in the right lower lobe, enlarged left hilar lymph nodes, and lingular pneumonitis. A tibia x ray was diagnostic for osteomyelitis. A biopsy of the tibia lesion revealed granulomatous tissue and a few short and thick hyphae. *Chrysosporium zonatum* was grown from sputum and biopsy specimens, which was identified by its thermotolerance, darkening colonies (yellowish white to buff) and club-shaped terminal aleurioconidia borne at the ends of short, curved stalks. Therapy with amphotericin B, itraconazole, and then liposomal amphotericin B subdued the osteomyelitis, pneumonia, pericarditis, and pleuritis.

Guerrero Palma et al. [15] described a case of invasive sinusal mycosis caused by *Chrysosporium tropicum* in a patient with acquired immunodeficiency. Levy et al. [16] also reported a case of aggressive fungal rhinosinusitis caused by *Chrysosporium* sp. in a patient with acute lymphocytic leukemia. Histopathological and microbiological studies permitted the identification of the culprit organism. In addition, Warwick et al. [17] presented an invasive *Chrysosporium* infection in an 18-year-old patient after allogeneic sibling bone marrow transplant for T lineage acute lymphoblastic leukemia. The infection started as a facial swelling and extended into the central nervous system. Despite antifungal treatment, the disease spread rapidly, and an autopsy revealed fungal invasion of brain, lungs, liver, and kidneys.

25.1.3 Diagnosis

Conventional techniques for the identification of *Chrysosporium* spp. and other fungi rely on microscopic observation of mycotic elements in clinical specimens followed by in vitro culture. *Chrysosporium* colonial morphologies and growth rates are assessed on PDA at 25°C and 37°C. Tolerance to cycloheximide is tested by measuring the growth rate at 25°C on Mycosel agar (Becton Dickinson). Microscopic features of *Chrysosporium* isolates are examined in slide culture preparations with pablum cereal agar (pablum precooked mixed cereal 10%, agar 1.5%). The ability of *Chrysosporium* isolates to digest hairs in the hair perforation assay is diagnostic.

Molecular methods for the determination of *Chrysosporium* identity include sequence analysis of small subunit (SSU) and large subunit (LSU) of ribosomal RNA genes as well as internal transcribed spacer (ITS) regions [3,18,19]. PCR amplification of ITS2 using primers ITS86 (5′-GTGAATCATCGAATCTTTGAAC-3′) and ITS4 (5′-TCCTCCGCTTATTGATATGC-3′), restriction digestion with *Bst*UI (CG/CG), and restriction fragment length polymorphism (RFLP) assessment by capillary electrophoresis permit discrimination of *Chrysosporium* from dermatophyte fungi such as *Arthroderma*, *Epidermophyton*, *Microsporum*, and *Trichophyton* [20].

25.2 METHODS

25.2.1 Sample Preparation

Chrysosporium and other fungal isolates are revived from either freeze-dried or frozen (vapor phase of liquid nitrogen) stock and grown at 25°C on petri plates containing pablum cereal agar for 14–21 days. Blocks (1 × 1 cm) of mycelium and agar from cultures of *Chrysosporium* species are excised from the culture plates and transferred to sterile snap-cap polypropylene tubes (12 mm × 75 mm; Fisher Scientific). The mycelial blocks are freeze-dried by using an Edwards Moduylo freeze-dryer. Freeze-dried blocks of agar and *Chrysosporium* mycelium (100 mg) are placed in 1.5 mL tubes and ground to a fine powder by using a 200 μL capacity pipettor tip. The fungal material is rehydrated with 500 μL of DNA extraction buffer (50 mM Tris, 10 mM EDTA, 1% sarcosyl, pH 8.0) with gentle agitation for 10 min. An equal volume of 1:1 chloroform–phenol is added to each tube, and mixed by shaking for 20 min. Then the aqueous and organic phases are separated in a microcentrifuge for 5 min at 14,000 × *g*. The aqueous phase is pipetted into a clean tube, and 0.1 volume of 3 M sodium acetate (pH 6.0) and 1.3 volumes of ethanol are added. The tube is sealed, and the contents are mixed by inverting the tube several times. Precipitated nucleic acids are pelleted by centrifugation at 14,000 × *g* for 1 min. Ethanol is decanted, and the pellet is dried by inverting the tube over absorbent paper for 5 min. Nucleic acids are dissolved in 100 μL of TE buffer (10 mM Tris, 1 mM EDTA, pH 8.0), and 250 μL of a saturated NaI solution and 10 μL of glass milk (Gene Clean kit, Bio 101) are added to the tube. The tube is inverted periodically, and DNA is adsorbed to the glass milk for 20 min. The glass milk is pelleted and rinsed, and the genomic DNA is eluted into 50 μL of 1/10-strength TE. DNA is stored at −20°C until used [18].

Alternatively, a loopful of *Chrysosporium* mycelium is transferred from SDA to a 1.5 mL tube containing 2.5 mg zymolase (Sigma–Aldrich) in 250 μL zymolyase lysis buffer [10 mM Tris/HCl (pH 8.0), 1 mM EDTA] and incubated for 45 min at 37°C. Subsequently, 200 μL is transferred to another tube for DNA extraction using a Qiagen kit, resulting in 100 μL of DNA extract [20].

25.2.2 Detection Procedures

Pounder et al. [19] described a real-time PCR with SYBR green DNA-binding dye and amplicon melting temperature analysis for fungal detection using pan-fungal primers ITS1 forward (5′-TCCGTAGGTGAACCTGCGG-3′) and ITS4 reverse (5′-TCCTCCGCTTATTGATATGC-3′). The identity of the fungi is verified by subsequent sequence analysis.

Procedure

1. PCR mixture is composed of 1× Lightcycler FastStart DNA Master Hybridization Probes mixture (Roche Applied Science) (containing deoxynucleoside triphosphates), FastStart *Taq* DNA

polymerase, and 1 mM $MgCl_2$ (additional $MgCl_2$ is added to a final concentration of 4.6 mM), 0.4 μM each of ITS1 forward and ITS4 reverse primers, 1× SYBR green (Molecular Probes), and 3 μL template DNA.

2. Thermal cycling parameters are 95°C for 10 min; 50 cycles of 95°C for 5 s, 60°C for 20 s, and 76°C for 30 s; and a final extension at 72°C for 2 min.
3. The quality of the amplicon is determined using the derivative of the melt analysis curve (55°C–99°C, 45 s hold at 55°C, 5 s/°C) using the RotorGene 3000 (Corbett Robotics, Inc.).
4. The amplified product is purified for bidirectional sequencing using ExoSAP-IT (USB Corp.). Five μL of Big Dye Terminator Ready Reaction Mix v. 1.1 (Applied Biosystems) is added to 4 μL of each primer (0.8 pmol/μL) and 3 μL of purified PCR product. Cycle sequencing is performed with a 9700 thermal cycler (ABI), using 25 cycles of 96°C for 10 s, 50°C for 5 s, and 60°C for 4 min. Sequencing reaction products are passed through a Sephadex G-50 fine column to remove unincorporated dye terminators. Purified sequencing reaction products are run on an ABI Prism 3100 Genetic Analyzer with a 50 cm capillary array.
5. Sequences are analyzed with the SmartGene Integrated Database Network software version 3.2.3 vr. SmartGene is a web-based software and database system with reference sequences derived from the National Center for Biological Information (NCBI) GenBank repository.

Note: In case that real time PCR instrument is not available, standard PCR may be performed with primers ITS1 and ITS4, and the resulting amplicon is sequenced with the same primers. Sequence-based identifications are defined by percent identity: species, ≥99%; genus, 93%–99%; and inconclusive, ≤93%.

For strains producing discrepant identification based on phenotypic characteristics and ITS sequence analysis, the D1–D2 region of the large-subunit RNA gene is amplified with primers NL1 (5′-GCATATCAATAAGCGGAGGAAAAG-3′) and NL4 (5′-GGTCCGTGTTTCAAGACGG-3′) and sequenced for species clarification [3].

25.3 CONCLUSIONS

The genus *Chrysosporium* contains a large number of filamentous saprophytic fungi classified in the mitosporic Onygenales group, order Onygenales. The teleomorphs of the *Chrysosporium* genus are found in the genera *Aphanoascus*, *Nannizziopsis*, and *Uncinocarpus*, family Onygenaceae, and order Onygenales. Many members of the genus *Chrysosporium* are thermotolerant and keratinophilic, and occasionally implicated in human diseases, producing clinical symptoms ranging from skin infections, onychomycosis, sinusitis, pneumonia, pleuritis, pericarditis, osteomyelitis, to other disseminate infections.

Microscopic observation of mycotic elements in clinical specimens and in vitro isolation of fungal strains have been the main approaches for identification and diagnosis of *Chrysosporium* infections. More recently, molecular techniques such as PCR, sequencing, and RFLP analysis have been developed and applied for improved determinations. Through examination of nucleotide sequences in the SSU and LSU of ribosomal RNA genes as well as ITS regions, the identity of *Chrysosporium* spp. can be rapidly and unequivocally ascertained.

REFERENCES

1. The UniProt Consortium. Available at http://www.uniprot.org/, accessed on August 1, 2010.
2. Al-Musallam A, Tan CS, *Chrysosporium zonatum*, a new keratinophilic fungus. *Persoonia*. 1989;14:69–71.
3. Rajeev S et al., Isolation and characterization of a new fungal species, *Chrysosporium ophiodiicola*, from a mycotic granuloma of a black rat snake (Elaphe obsoleta obsoleta). *J Clin Microbiol*. 2009;47(4):1264–1268.
4. Roilides E. et al., Disseminated infection due to *Chrysosporium zonatum* in a patient with chronic granulomatous disease and review of non-*Aspergillus* fungal infections with this disease. *J Clin Microbiol*. 1999;37:18–25.
5. Abarca ML et al., Cutaneous hyalohyphomycosis caused by a *Chrysosporium* species related to *Nannizziopsis vriesii* in two green iguanas (*Iguana iguana*). *Med Mycol*. 2008;46:349–354.
6. Chabasse D, De Gentile L, Bouchara JP, Pathogenicity of some *Chrysosporium* species isolated in France. *Mycopathologia*. 1989;106(3):171–177.
7. Qiu WY et al., Fungal spectrum identified by a new slide culture and in vitro drug susceptibility using Etest in fungal keratitis. *Curr Eye Res*. 2005;30(12):1113–1120.
8. Singh I, Mishra A, Kushwaha R, Dermatophytes, related keratinophilic and opportunistic fungi in indoor dust of houses and hospitals. *Indian J Med Microbiol*. 2009;27(3):242–246.
9. Gan GG et al., Non-sporulating *Chrysosporium*: An opportunistic fungal infection in a neutropenic patient. *Med J Malaysia*. 2002;57(1):118–122.
10. Maghazy S, Incidence of dermatophytes and cyclohexamide resistant fungi on healthy children hairs and nails in nurseries. *Mycopathologia*. 2002;154(4):171–175.
11. Stebbins WG, Cutaneous adiaspiromycosis: A distinct dermatologic entity associated with *Chrysosporium* species. *J Am Acad Dermatol*. 2004;51(5 Suppl):S185–S189. Erratum in: *J Am Acad Dermatol*. 2004;51(6):1040.
12. Manzano-Gayosso P et al., Onychomycosis incidence in type 2 diabetes mellitus patients. *Mycopathologia*. 2008;166(1):41–45.
13. Stillwell WT, Rubin BD, Axelrod JL. *Chrysosporium*, a new causative agent in osteomyelitis. A case report. *Clin Orthop Relat Res*. 1984;(184):190–192.
14. Tomsiková A. Causative agents of nosocomial mycoses. *Folia Microbiol (Praha)*. 2002;47(2):105–112.
15. Guerrero Palma MA et al., Invasive sinusal mycosis due to *Chrysosporium tropicum*. *Acta Otorrinolaringol Esp*. 2007;58(4):164–166.

16. Levy FE et al., Invasive *Chrysosporium* infection of the nose and paranasal sinuses in an immunocompromised host. *Otolaryngol Head Neck Surg*. 1991;104(3):384–388.
17. Warwick A et al., Presumptive invasive *Chrysosporium* infection in a bone marrow transplant recipient. *Bone Marrow Transplant*. 1991;8(4):319–322.
18. Peterson SW, Sigler L. Molecular genetic variation in *Emmonsia crescens* and *Emmonsia parva*, etiologic agents of adiaspiromycosis, and their phylogenetic relationship to *Blastomyces dermatitidis* (*Ajellomyces dermatitidis*) and other systemic fungal pathogens. *J Clin Microbiol*. 1998;36(10):2918–2925.
19. Pounder JI et al., Discovering potential pathogens among fungi identified as nonsporulating molds. *J Clin Microbiol*. 2007;45(2):568–571.
20. De Baere T et al., Evaluation of internal transcribed spacer 2-RFLP analysis for the identification of dermatophytes. *J Med Microbiol*. 2010;59:48–54.

26 *Cladophialophora*

Rubén Lopez-Martínez and Francisca Hernández-Hernández

CONTENTS

26.1 INTRODUCTION

26.1.1 Classification, Morphology, and Biology

The genus *Cladophialophora* has recently attracted great interest in the scientific community. Because of the many studies performed on its morphological, taxonomic, physiological, and molecular characteristics, today there is increased knowledge regarding its biodiversity, pathogenicity, and natural behavior. Thus far, around 16 different species have been identified, many of which have the capacity to produce infectious diseases in humans and other mammals. The most recent studies published concerning this most interesting genus of fungi have described their behavior as pathogens of various clinical mycoses, such as chromoblastomycosis, phaeohyphomycosis, eumycetoma, deep tissue abscess formation, skin lesions, among others [1].

26.1.1.1 Classification

The teleomorphic state has not yet been described; however, phylogenetic studies performed on its small subunit rDNA suggest that it possibly belongs to the phylum Ascomycota of the genus *Capronia*, of the Chaetothyriales order [2]. Other references indicate that *Cladophialophora* as well as other fungi that produce the clinical picture of chromoblastomycosis belong to the Herpotrichiellaceae family, order Ophiostomatales, phylum Ascomycota [3]. The most frequent *Cladophialophora* species as human pathogen is *C. carrionii* (de Hoog, Kwon-Chung, and McGinnis) firstly known as *Cladosporium carrionii* (Trejos).

26.1.1.2 Morphology

Cladophialophora colonies grown on Sabouraud dextrose agar medium are seen as compact, smooth, or folded colonies with a dark olive green color, which measure around 4–6 cm and around their border show hyphae submerged in the agar medium. These colonies can be grown in a wide temperature range, which varies from 25°C to 35°C. Its microscopic morphology is characterized by a *Cladosporium* type of conidiation, which shows long conidiophores that hold the shape of a shield, and some may show disjunction scars. From the conidiophores, elliptical chains of conidia that are 1.5–3.0×2.5–7.0 μm long are formed. These chains of conidia have a branched, flexible appearance and can be long or short.

C. carrionii colonies grown on Sabouraud dextrose agar have a dark green olive color, with a velvety appearance. On microscopic examination, long conidiophores can be seen, with long, branched, flexible conidia chains, 1.5–3.0×2.0–7.5 μm long.

C. bantiana shows olive green to gray colonies with a slow growth rate and fluffy appearance that on its hind surface can show a gray to black color. Through a microscope, dematiaceous hyphae that are smooth and branched can be seen, along with conidiophores with long elliptical conidia chains, 2.5–4.0×5.0–9.0 μm long and acropetal pigmentation.

26.1.1.3 Biology

The different species of *Cladophialophora* have a great diversity of habitats in nature, and can be isolated either from organic or from inorganic substrates including soil, wood, vegetable detritus, eucalypt trees [4], sports drinks, iced tea, air filters, and concrete floors [5]. Table 26.1 shows the pathogenic species for human and their habitats in nature. There are reports from almost every country in the world, ranging from semiarid to humid and warm climates that lead to the conclusion that it belongs to one of the most ubiquitous and cosmopolitan genus of fungi [6,7]. Some of the species in this genus survive many sterilization methods and can resist temperatures from below 0°C up to 40°C. Due to its high transmissibility and its thermotolerance (up to 43°C), some species of *Cladophialophora*, such as *C. bantiana*, have been catalogued as having a Biosafety Level 3 containment requirement [8].

26.1.2 Clinical Features and Pathogenesis

Approximately 16 species have been described in this group, 10 of which have been isolated as a cause of different diseases in humans and some others as agents of mycosis in animals, such as dogs and cats; the rest of these are considered free-living, nonpathogenic species.

Two of the most important *Cladophialophora* species that can cause clinical disease are *C. carrionii* as an etiological agent of chromoblastomycosis and *C. bantiana* as an agent of brain abscess formation.

26.1.2.1 Chromoblastomycosis

Chromoblastomycosis is the most frequent mycosis caused by the *Cladophialophora* species. It is a chronic mycosis that is acquired from the inoculation by the agent through a wound in the skin, a frequent occurrence in field workers. Chromoblastomycosis often arise on a lower extremity; lesions start as papules, nodules, and scaly. In chronic forms, they become large tumoral lesions with cauliflower-like

TABLE 26.1
***Cladophialophora* Species Associated with Human Pathologies and Their Habitats**

Species	Clinical Manifestations	Sources
C. arxii	Tracheal abscess	
C. australiensis		Sports drink
C. bantiana	Brain abscess, eumycetoma	
C. boppii	Cutaneous phaeohyphomycosis	
C. carrionii	Chromoblastomycosis, cutaneous phaeohyphomycosis	Soil, pine, eucalyptus fence posts
C. chaetospira		Roots of fir tree (*Picea abies*)
C. devriesii	Disseminated lesion	Litter, vegetable cover/soil
C. emmonsii	Subcutaneous phaeohyphomycosis	
C. immunda	Subcutaneous phaeohyphomycosis	Decaying vegetable matter, biofilter, soil
C. minourae		Decaying wood, rotting wood
C. mycetomatis	Eumycetoma	Cultural contaminant
C. potulenturum		Sports drink
C. samöensis	Chromoblastomycosis	
C. saturnica	Cutaneous phaeohyphomycosis	Litter, vegetable cover/soil, biofilter, trunk, cut tree
C. subtilis		Iced tea
C. yegresii		Pitaya (*Stenocereus griseus*)

Source: Modified from Badali, H. et al., *Stud. Mycol.*, 61, 175, 2008; Padhye, A.A., Identification of the etiologic agents of chromoblastomycosis, Scientific Publication No. 479, p. 87, Pan American Health Organization, Washington, DC, 1986; Guía técnica para la evaluación y prevención de los riesgos relacionados con la exposición a agentes biológicos. Instituto Nacional de Seguridad e Higiene en el Trabajo. España, Real Decreto 664/1997 del 12 de mayo 1994, B.O.E. No. 124 y adaptación contenida en la Orden de 25 de marzo de 1998, 1994; Queiroz-Telles, F. et al., *Med. Mycol.*, 47, 3, 2009.

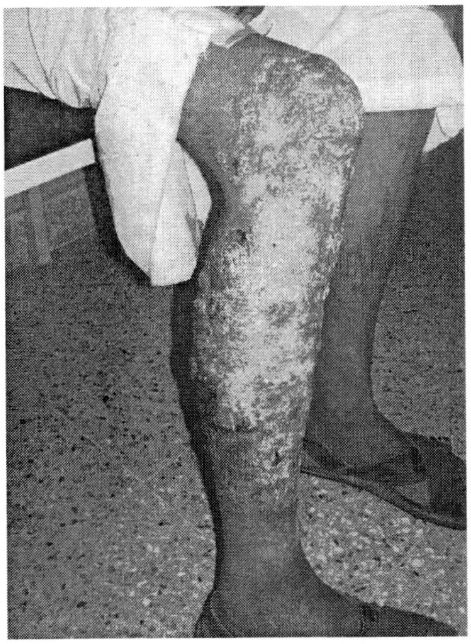

FIGURE 26.1 Chromoblastomycosis in leg showing extensive cauliflower-like lesions with some black dots and ulcers.

appearance (Figure 26.1). In some cases, ulcers are present with exudates composed of blood-serous and caseous material. Fungal cells are expelled through the epidermis expressed by small dark aggregations, known as black dots. Other cases of chromoblastomycosis resemble the clinical aspect of mycetoma, but the differential diagnosis is determined for the presence of muriform cells or for the actinomycetes or eumycetes grains (Figure 26.2). Usually, the infection occurs in individuals with no apparent factors for immunosuppression [9]; there apparently exists the possibility that

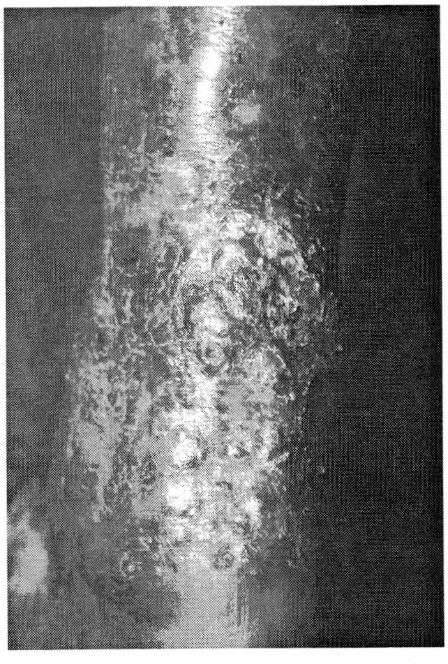

FIGURE 26.2 Chromoblastomycosis in leg with a mycetoma-like appearance. Note many nodules and some ulcers with scarring areas.

infected individuals may even have a genetic predisposition for the development of this mycosis. Its geographical distribution is very broad. Compared with the cases of chromoblastomycosis produced by *Fonsecaea pedrosoi* and *Phialophora verrucosa*, which are found in damp tropical climates, *Cladophialophora carrionii* (*Cladosporium carrionii*) is found in arid or subtropical climates with annual average rainfall of 500–600 mm, such as Madagascar, where the majority of cases are attributed to this species [10]. Other species that produce chromoblastomycosis, with a much lower frequency, are *C. arxii* and *C. samöensis* [1].

26.1.2.2 Phaeohyphomycosis

Phaeohyphomycosis is an infection generated by many pigmented fungi, and is classified as dematiaceous. These can be observed inside the tissues in the form of hyphae, yeast cells, and vesicular cells. According to the site of presentation, phaeohyphomycosis can be classified as cutaneous, subcutaneous, and systemic, the later usually manifested as lung or central nervous system (CNS) infection [11]. The clinical manifestations are very diverse, and vary according to the organ or tissue affected; in most cases, the superficial and subcutaneous forms have a benign course and are observed in immunocompetent individuals; on the other hand, lung and CNS infections are usually very severe, frequently fatal, and are characterized for their presentation as pyogenic abscesses.

26.1.2.3 Superficial and Subcutaneous Infections

The main etiological agents are *C. carrionii*, *C. emmonsii*, *C. boppii*, *C. saturnica*, and *C. immunda*. Traumatic inoculation of the skin is the mechanism of transmission most common to these infections. These fungi grow at an ideal temperature of 27°C, even when they can tolerate temperatures up to 37°C. It seems that superficial and subcutaneous phaeohyphomycosis is a disease most frequently found in immunocompetent individuals.

26.1.2.4 Systemic Forms

Systemic forms are produced by three species of *Cladophialophora*: *C. arxii*, *C. devriesii*, and *C. bantiana*; the latter is a neurotrophic fungus, which causes brain abscesses that are usually fatal in immunocompetent patients. The mechanism of transmission is through inhalation of conidia and it is able to grow in temperatures of up to 40°C. Therefore, this species is classified as an organism that should be handled under Biosafety Level 3 containment [8].

26.1.2.5 Mycetoma

Mycetoma is a syndrome caused by exogenous sources, belonging to at least one or two groups of microorganisms: fungus (Eumycetoma) or bacteria (Actinomycetoma). Mycetoma results from the traumatic implantation of soil organisms into the dermis, followed by tumefaction, suppurating abscesses, fibrosis, granulomata, and sinuses that drain the parasitic structures of the agents, so called "grains." *Nocardia*, *Actinomadura*, and *Streptomyces* are the three genera causing actinomycetoma; the principal etiological agents are *N. brasiliensis* and *A. madurae*. Eumycetoma is caused by numerous species belonging to the genera *Madurella*, *Pyrenochaeta*, *Leptosphaeria*, *Scedosporium*, and others. The most frequent species are *M. mycetomatis* and *M. grisea*. Nevertheless, it has been reported that exceptional cases of other mycoses are caused by other fungi, such as dermatophytes, including *Aspergillus* and agents of chromoblastomycosis [12,13]. Some species of *Cladophialophora* have been discovered as agents of eumycetoma, among them *C. mycetomatis* [1]. Recently, a case located in the dorsum of the foot of a 57-year-old man, manifested with edema and the formation of fistulas, was reported. Direct microscopy of the exudate revealed the presence of black granules. Both a culture and a test of the rDNA sequence determined the etiological agent as being *C. bantiana* [14].

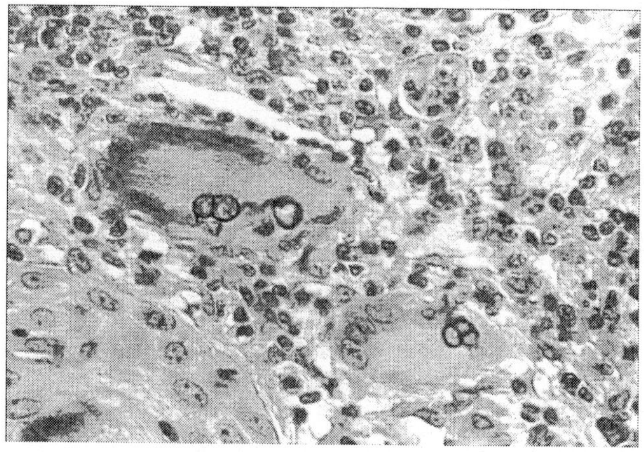

FIGURE 26.4 Histopathology of infected tissue showing muriform bodies into giant cells. H & E. ×1000.

26.1.3 Diagnosis

26.1.3.1 Conventional Techniques

Direct examination. In chromoblastomycosis, the parasitic morphology of *Cladophialophora* is known as muriform bodies or fumagoid cells, which are pigmented forms that can be observed by a direct microscopic examination (Figure 26.3). In phaeohyphomycosis, hyphae, yeast cells, and vesicular cells can be observed in the tissues.

Histopathology. This procedure is also important to confirm the diagnosis of *Cladophialophora* infections. In chromoblastomycosis, it is common to observe dermal hyperkeratosis and pseudoepitheliomatous hyperplasia, and abscesses containing muriform cells are frequent [15] (Figure 26.4).

The growth of colonies inoculated in culture in Sabouraud dextrose agar, incubated in the dark at 28°C–30°C or in many cases at room temperature, can be seen in an average of 10–15 days. Cultures can be maintained on slants of 2% malt extract agar and on oatmeal agar [16]. The colonies show a dark olive green color and velvety texture (Figure 26.5). Microscopically, single conidiophores can be observed, from which chains of elliptical conidia are shed, which are usually small (Figure 26.6).

In some cases, to differentiate the species, it is necessary to implement certain physiological techniques [3]. The effects of temperature (25°C or 37°C) and calcium (0.1 mM Ca^{2+}) under acid conditions (pH 2.5) determine the transformation

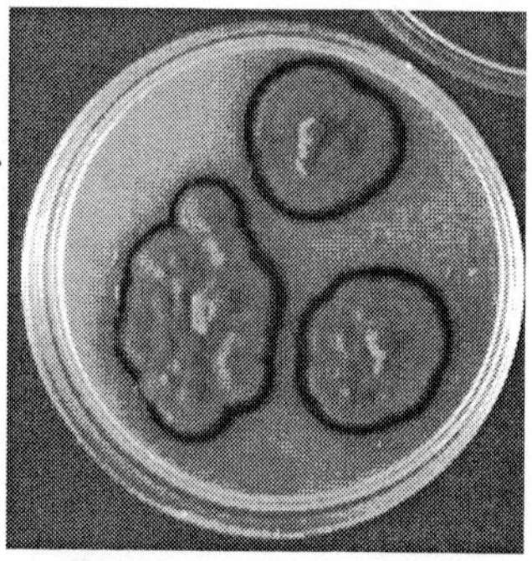

FIGURE 26.5 Colonies of *Cladophialophora carrionii* grown on Sabouraud dextrose agar at 26°C during 2 weeks.

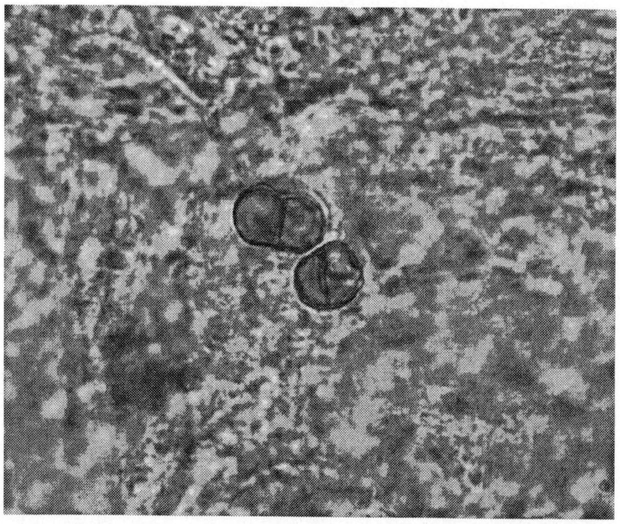

FIGURE 26.3 Direct examination of scales from a chromoblastomycosis patient, showing dematiaceous, septated muriform cells.

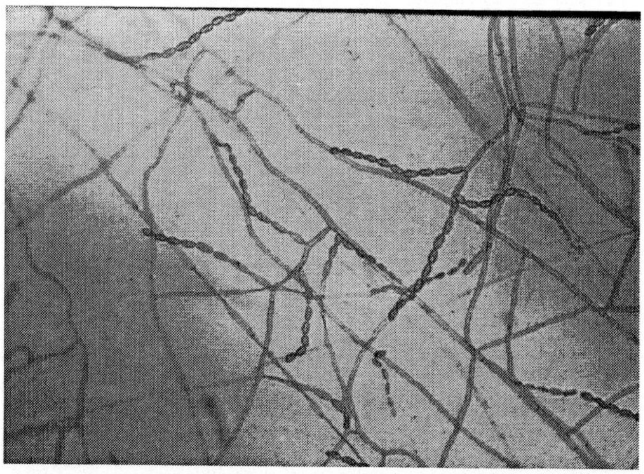

FIGURE 26.6 Microscopic examination of *Cladophialophora carrionii* forming conidia branched chains in acropetal disposition.

of the hyphae at muriform cells. Under these conditions, *C. carrionii* and *C. subtilis* produce abundant muriforms forms. On the other hand, no muriform cells are produced by *C. australiensis*, *C. immunda*, *C. mycetomatis*, *C. humicola*, or others [1].

Immune response is not constant in these infections; serological tests such as enzyme-linked immunosorbent assay (ELISA) using the antigen of *C. carrionii* (AgSPP) have been employed for the detection of antibodies in patients with chromoblastomycosis [17–20]. In healthy patients, this serological test is negative, and the criteria of remission are based on this test. The cell-mediated immune reaction is not standardized [21].

Although chromoblastomycosis is a difficult-to-treat disease because of its chronicity and the presence of melanine-like fungal pigment, there are some therapeutic approaches such as physical therapy, surgery, and several antifungal drugs (5-fluorocytosine, amphotericin B, and some triazols as voriconazol and posaconazol) that have shown efficiency [22].

26.1.3.2 Molecular Techniques

Molecular tests are now the optimal methods for the identification and differentiation of *Cladophialophora* species. By contrast with other more common fungal pathogens such as the opportunistic fungi *Candida* spp. or *Aspergillus fumigatus*, few molecular studies of *Cladophialophora* spp. have been developed. Most of these studies have focused on defining the phylogenetic relationship among highly related species, a very laborious but determinant work for reliable identification of human pathogenic species.

Thus, Caligiorne et al. [23] performed an analysis of ribosomal DNA by polymerase chain reaction (PCR)—restriction fragment length polymorphism. By this method, it was possible to differentiate *Fonsecaea* spp. from other dematiaceous fungi, but it does not differentiate *Cladophialophora* species.

Among the earliest and most important studies applying molecular approaches to the genus is that of Masclaux et al. [24], who based them on partial LS rRNA sequences that established the phylogenetic relationships of human pathogenic *Cladosporium* species. This study was crucial to future works regarding the *Cladophialophora* genus, and unambiguously established that the melanized human pathogens belong to a single ascomycete family, the Herpotrichiellaceae.

In spite of the advances in molecular biology to define fungal species, the morphological approaches and the physiological tests are valuable tools supporting the identification and taxonomy of many fungi as demonstrated by De Hoog et al. [3], who based them on nutritional physiology tests that reclassified the human pathogen *Cladosporium* species in the *Cladophialophora* genus, as well as other taxonomic changes in related species.

The complexity of *C. carrionii* strains has been explored by sequencing of three loci, internal transcribed spacers (ITS), β-tubulin (BT2), and translation elongation factor 1-α (EF1) supported by morphological studies [25].

Ecological information with phylogenetic and taxonomic data in addition to morphological description have been organized to describe the presently known *Cladophialophora* species [1].

As for many other fungal pathogens, the PCR-based amplification and sequencing of specific DNA regions until now appears to be the most promising method to identify *Cladophialophora* species. Thus, two recent studies have been performed by Abliz et al. [26,27]. The first one identifies *C. carrionii* by PCR and the second one is an approach to differentiate some *Cladophialophora* species by PCR amplification and sequencing of amplified fragments.

26.2 METHODS

This section begins with the general management of biological specimens or cultures, whose required conditions are similar to those required for other potentially infectious materials. Preparation of paraffin-embedded (PE) tissue is added in cases where this will be the material to be studied. DNA extraction procedures are those used in original papers, but we added some footnotes for those who need to make some changes according to their individual laboratory conditions. Finally, we describe two molecular protocols.

26.2.1 Sample Preparation

26.2.1.1 Sample Collection

All specimens should be presumed to contain transmissible agents and therefore should be collected and handled using standard precautions including the use of gloves, gown, mask, and protective eyewear when there is a risk of coming in contact with the specimen. A special area must be designated for processing clinical samples. Ideally, all specimens submitted for the detection of fungi should be performed in a class II biological safety cabinet.

Concerning fungal cultures from the skin and subcutaneous lesions, ideally the infected material is aspirated with a needle and syringe, and the material is expelled and transported into a sterile container that is tightly capped and promptly delivered to the laboratory. If an aspirate cannot be obtained, swab specimens of exudates collected from the deep portion of the lesion are acceptable.

For optimal evaluation of tissues, enough material should be collected to allow the culture, the histopathological examination, and eventually the molecular processing of the specimen. After collection, tissues should be placed in a sterile container and transported rapidly to the mycological laboratory to prevent drying.

26.2.1.2 Culture of Clinical Specimens

As for other purposes in medical mycology, we recommend the growth of clinical specimens in triplicate in isolation media such as Sabouraud dextrose agar with and without cycloheximide and with chloramphenicol.

The *Cladophialophora* isolates can be grown in or on YMPG broth or agar (yeast extract 0.3% [w/v], malt extract

0.3% [w/v], peptone 0.5% [w/v], glucose 1% [w/v] with or without agar 1.5% [w/v]) at 27°C for 1–2 days, as recommended by Makimura et al. [28] for other medically important fungal strains.

Other solid or liquid media can be used, depending on availability of starting materials in the mycology laboratory. We use potato dextrose agar for filamentous fungi growth with highly satisfactory results. The most important goal is to obtain a sufficient fungal mass to be submitted at the next step consisting of DNA extraction.

26.2.1.3 Treatment of PE Sections

For each PE tissue sample, 10 sections (thickness, 10 μm) are cut using a sterile microtome blade and transferred to a 10 mL centrifuge tube. To remove paraffin wax, 5 mL of histolene is added, mixed by inversion, incubated at room temperature overnight, and centrifuged at 3838 × *g* for 15 min. The supernatant is removed, and the pellet is washed with 5 mL of 100% ethanol, mixed by inversion, and centrifuged at 3838 × *g* for 15 min. The tissue pellet is transferred to a 2 mL centrifuge tube and washed in 1 mL of 100% ethanol, followed by 1 mL of 70% ethanol. The ethanol is removed by centrifugation at 5900 × *g* for 10 min, and the pellet is air dried at room temperature in preparation for DNA extraction. For DNA extraction, all tissue samples are incubated for at least 3 h in proteinase K and lysis buffer at 55°C, and the DNA is extracted using the MagNAPure LC DNA isolation kit II (for tissue) (Roche Diagnostics, Mannheim, Germany) according to the manufacturer's instructions. The DNA is stored at −20°C prior to use [29].

26.2.1.4 DNA Extraction

The following DNA extraction protocol includes the culture conditions. It is a faster procedure and less prone to contamination than culturing the fungus directly in a tube as developed by Cenis [30], and avoids the first phenol-treatment step suggested in other protocols for plant DNA extraction [31].

A 1.5 mL tube is filled with 500 μL of liquid potato dextrose medium, inoculated with some hyphal threads, and incubated for 72 h at 25°C. The mycelial mat is pelleted by centrifugation for 5 min at 13,000 rpm, washed with 500 μL of TE buffer and repelleted. The TE is decanted and 300 μL of extraction buffer is added (200 mM Tris–HCl (pH 8.5), 250 mM NaCl, 25 mM EDTA, 0.5% sodium dodecyl sulfate [SDS]). The mycelia are crushed with a conical grinder, precisely fitting the tube and manipulated by hand or electric potter at 200 rpm for some minutes. Subsequently, 150 μL of 3 M sodium acetate (pH 5.2) is added and the tube is placed at −20°C for about 10 min. The tube is then centrifuged and the supernatant transferred to another tube. An equal volume of isopropanol is added, and, after at least 5 min at room temperature, the precipitated DNA is pelleted by centrifugation. After a wash with 70% ethanol, the pellet is vacuum dried for some minutes and resuspended in 50 μL of TE. The amount of DNA obtained in this way is around 3–6 μg. This is enough for at least 50 PCR reactions, as determined by titration of the DNA concentration in several reactions. In case more material is needed, the fungal culture can be made in a 15 mL centrifuge tube. After the mycelia are pelleted and washed in a benchtop centrifuge, they are transferred to a 1.5 mL tube and crushed in the manner described above. This culture yields about 25–35 μg of DNA, enough for at least 500 PCR reactions.

In modifications made by Abliz et al., approximately 50 mg of fungal elements are suspended in 600 μL of extraction buffer (200 mM Tris–HCl [pH 7.5], 25 mM EDTA, 0.5% [w/v] SDS, 250 mM NaCl) and then incubated at 100°C for 15 min. The solution is extracted with phenol–chloroform–isoamyl alcohol (25:24:1 [v/v/v]).

26.2.1.5 Commercial Kits for DNA Extraction

Other alternative easier and faster DNA extraction methods consist of the use of several available commercial kits, most of which could possibly be used for fungal culture or clinical samples. Several authors have proposed the use of particular kits [32], methods using minimal equipment (generally a hot block and microcentrifuge), and obtaining results within 1–4 h. Fredricks et al. [33] found the MPY (MasterPure yeast DNA purification kit, Epicenter, Madison, WI) and YL GNOME (yeast cell lysis preparation kit plus GNOME kit, Qbiogene, Irvine, CA) as the best kits for DNA extraction from yeast (*Candida*) and filamentous fungi (*A. fumigatus*), respectively; the latter method being the most costly because it uses two kits to lyse fungal cells and purify DNA.

We have experience with fungal DNA extraction methods using three commercial kits: the UltraClean Fecal DNA kit (MoBio Laboratories), QIAamp DNA mini kit (Qiagen), and High Pure PCR Template Preparation kit (Roche Diagnostics), all of them giving good-quality DNA for PCR protocols including qPCR. For any of these kits, the mycelial mass is collected from a culture on agar with a scalpel in order to obtain about a 0.25 g sample, which is put in a 2 mL tube. Then the manufacturer's instructions are followed for all methods. The final volume is from 50 to 200 μL depending on the kit used, enough to quantify and run several PCR reactions.

Other interesting methods for DNA extraction have been proposed by several authors and consist of using FTA filters giving a molecular grade DNA [34–36].

26.2.2 Detection Procedures

26.2.2.1 PCR Identification of *C. carrionii*

The PCR mixture (25 μL) consists of 2.5 μL of template DNA, 2.5 μL (2 mol) of each primer (5′-TAA ACC TCA TGT TGC TTC G-3′ [Car-F] and 5′-TCG AGA M(A/C) CAC TCG ACC AA-3′ [Car-R]), 2 μL (2.5 mM) of dNTP mixture, 0.125 μL (5 U/μL) of *Taq* polymerase, and 2.5 μL of 10× reaction buffer. Amplification is performed under the following conditions: one cycle at 95°C for 4 min, followed by 30 cycles at 94°C for 1 min, at 62°C for 1.5 min, and at 72°C for 1.5 min, with a final extension at 72°C for 10 min. After thermal cycling, 2 μL of each amplified product is separated by electrophoresis on a 1.5% agarose gel, stained with ethidium bromide, and visualized with UV light. This primer set is designed based on the sequence of the ITS regions and to amplify DNAs from *C. carrionii* giving a 362-bp fragment. PCR bands are obtained within 7 h from initiating the DNA extractions: 3 h for the preparation of the template DNAs; 3 h for PCR; and 1 h for electrophoresis, staining, and detecting the band profiles. The sequences determined by this study were registered in the DNA Data Bank of Japan, where they can be consulted under the accession numbers shown in the original work [26].

26.2.2.2 Sequencing Analysis of *Cladophialophora* spp.

This protocol is based on nucleotide sequences of the D1/D2 domains of large subunit (26S) ribosomal DNA developed by Abliz et al. [27].

PCR is carried out in a 50 μL volume containing 5 μL of template DNA, 5 μL (2 pmol) of each primer (NL-1, 5′-GCA TAT CAA TAA GCG GAG GAA AAG-3′ and NL-4m, 5′-GGT CCG TGT TTC AAG ACG-3′), 4 μL (2.5 mM) dNTP mixture, 0.25 μL (5 U/μL) Taq polymerase, and 5 μL 10× reaction buffer. Amplification is performed under the following conditions: one cycle of 95°C for 4 min followed by 30 cycles of 94°C for 1 min, 55°C for 2.5 min, and 72°C for 2.5 min, with a final extension at 72°C for 10 min. The amplified products are purified and subjected to direct sequencing. The external primers, NL-1 and NL-4m, and the internal primers, N L-2m, 5′-TTG TGC GCT ATC GGT CTC-3′, and NL-3m, 5′-GAG ACC GAT AGC GCA CAA G-3, are used to sequence each DNA sample.

The length of the nucleotide sequence for *Cladophialophora* spp. is 613 bp. The number of nucleotide differences among members of *Cladophialophora* species range from 5 between *C. devriesii*, *C. minourae*, and *C. arxii* to 26 between *C. devriesii*, *C. minourae*, and *C. bantiana*. These data support the sequences of D1/D2 domains as useful for the identification of *Cladophialophora* species.

26.3 CONCLUSIONS AND FUTURE PERSPECTIVES

The identification, taxonomy, and epidemiological analyses of fungal pathogens have become increasingly dependent on molecular techniques [37]. However, for dematiaceous species such as *Cladophialophora*, relatively few molecular studies are available for their rapid, sensitive, and specific identification or diagnosis. One of the bigger problems that has limited advances in this area is that most studies on species of medical interest have until now examined regions (ITS or D1/D2) highly similar in sequence, in conditions that do not differentiate them. A second very important aspect to be considered is the presence of the melanin-like pigment in dematiaceous fungi that potentially inhibits PCR amplification. This pigment and the DNA have similar solubility and in consequence are frequently extracted together, particularly when the phenol–chloroform method is used. The approaches using the FTA filters [34–36] are particularly interesting on this topic in order to obtain a high-quality DNA.

Therefore, the challenge for future research in the molecular diagnosis of infection or in fungal identification will be to find genetic regions that can be highly discriminative, to differentiate at least the most frequent pathogenic *Cladophialophora* species. It is desirable that this region be amplified by PCR as only identification method and be sufficiently discriminative without resorting to sequencing procedures. For researchers identifying these species, it will be imperative to select the best DNA extraction method, particularly if real-time PCR is considered for performing the identification in the future because this method is highly sensitive to inhibitors associated with DNA [38].

Until now, molecular identification has been an expensive procedure. However, as the number of steps in the different procedures is reduced, costs and reliable results timing will be also reduced. Considering that molecular methods for the identification and differentiation of *Cladophialophora* species are very limited, morphological study remains a very useful tool to support identification.

REFERENCES

1. Badali, H. et al., Biodiversity of the genus *Cladophialophora*. *Stud. Mycol.*, 61, 175, 2008.
2. Haase, G. et al., Phylogenetic inference by SSU gene analysis of members of the *Herpotrichiellaceae*, with special reference to human pathogenic species. *Stud. Mycol.*, 43, 80, 1999.
3. De Hoog, G.S. et al., Nutritional physiology and taxonomy of human-pathogenic *Cladosporium-Xylohypha* species. *J. Med. Vet. Mycol.*, 33, 339, 1995.
4. Ridley, M.F., *Soil as a Source of Pathogenic Fungi*. Recent Advances in Botany, p. 312, University of Toronto, Toronto, Ontario, Canada, 1961.
5. Esterre, P., Chromoblastomycosis, in *Microbiology and Microbial Infections*, 10th edn., p. 356, Merz, W.G. and Hay, R.J. Eds., Medical Mycology, Topley and Wilson's, London, U.K., 2005.
6. Padhye, A.A., Identification of the etiologic agents of chromoblastomycosis. Scientific Publication No. 479, p. 87, Pan American Health Organization, Washington, DC, 1986.
7. Okeke, C.N. and Gugnani, H.C., Studies of pathogenic dematiaceous fungi: Isolation from natural sources. *Mycopathologia*, 94, 19, 1986.
8. Guía técnica para la evaluación y prevención de los riesgos relacionados con la exposición a agentes biológicos. Instituto Nacional de Seguridad e Higiene en el Trabajo. España. Real Decreto 664/1997 del 12 de mayo 1994. B.O.E. No. 124 y adaptación contenida en la Orden de 25 de marzo de 1998, 1994.
9. Yegres, F. et al., Virulence and pathogenicity of human and environmental isolates of *Cladosporium carrionii* in new born ddY mice. *Mycopathologia*, 114, 71, 1991.
10. Queiroz-Telles, F. et al., Chromoblastomycosis: An overview of clinical manifestations, diagnosis and treatment. *Med. Mycol.*, 47, 3, 2009.
11. Díaz, J.M. et al., *Pleurophomopsis lignicola* y *Cladophialophora bantiana*. Dos nuevos agentes de micosis oportunistas en Chile. *Bol. Micol.*, 17, 95, 2002.

12. Mehta, B. et al., *Aspergillus nidulans*. A rare etiological agent of eumycetoma. *Int. J. Infect. Dis.*, 12, e282, 2008.
13. Zoutman, D.E. and Sigler, L., Mycetoma of the foot caused by *Cylindrocarpon destructans*. *J. Clin. Microbiol.*, 29, 1855, 1991.
14. Bonifaz, A. et al., Eumycetomas caused by *Cladophialophora bantiana* successfully treated with itraconazole. *Med. Mycol.*, 47, 111, 2009.
15. Lee, M.W. et al., Spores and mycelia in cutaneous chromomycosis. *J. Am. Acad. Dermatol.*, 39, 850, 1998.
16. Gams, W. et al., *CBS Course of Mycology*, 4th edn. Centraalbureau voor Schimmelcultures, Utrecht, the Netherlands, 1998.
17. Oberto-Perdigon, L. et al., An ELISA test for the study of the therapeutic evolution of chromoblastomycosis by *Cladophialophora carrionii* in the endemic area of Falcon State, Venezuela. *Rev. Iberoam. Micol.*, 22, 39, 2005.
18. Esterre, P. et al., Humoral immune response in chromoblastomycosis during and after therapy. *Clin. Diagn. Lab. Immunol.*, 7, 497, 2000.
19. Vidal, M.S. et al., Immunoprecipitation techniques and ELISA in the detection of anti-*Fonsecaea pedrosoi* antibodies in chromoblastomycosis. *Rev. Inst. Med. Trop. Sao Paulo*, 45, 315, 2003.
20. Vidal, M.S. et al., Highly specific and sensitive, immunoblot-detected 54 kDa antigen from *Fonsecaea pedrosoi*. *Med. Mycol.*, 42, 511, 2004.
21. Correa, G.P. et al., The cell-mediated immune reaction in the cutaneous lesion of chromoblastomycosis and their correlation with different clinical form the disease. *Mycopathologia*, 156, 51, 2002.
22. López-Martínez, R. and Méndez-Tovar, L.J., Chromoblastomycosis. *Clin. Dermatol.*, 25, 188, 2007.
23. Caligiorne, R.B. et al., Dematiaceous fungal pathogens: Analysis of ribosomal DNA gene polymorphism by polymerase chain reaction–restriction fragment length polymorphism. *Mycoses*, 42, 609, 1999.
24. Masclaux, F. et al., Phylogenetic relationships of human-pathogenic *Cladosporium* (*Xylohypha*) species inferred from partial LS rRNA sequences. *J. Med. Vet. Mycol.*, 33, 327, 1995.
25. De Hoog, G.S. et al., Molecular analysis and pathogenicity of the *Cladophialophora carrionii* complex, with the description of a novel species. *Stud. Mycol.*, 58, 219, 2007.
26. Abliz, P. et al., Specific oligonucleotide primers for identification of *Cladophialophora carrionii*, a causative agent of chromoblastomycosis. *J. Clin. Microbiol.*, 42, 404, 2004.
27. Abliz, P. et al., Identification of pathogenic dematiaceous fungi and related taxa based on large subunit ribosomal DNA D1/D2 domain sequence analysis. *FEMS Immunol. Med. Microbiol.*, 40, 41, 2004.
28. Makimura, K., Murayama, S.Y., and Yamaguchi, H., Detection of a wide range of medically important fungi by the polymerase chain reaction. *J. Med. Microbiol.*, 40, 358, 1994.
29. Lau, A. et al., Development and clinical application of a panfungal PCR assay to detect and identify fungal DNA in tissue specimens. *J. Clin. Microbiol.*, 45, 380, 2007.
30. Cenis, J.L., Rapid extraction of fungal DNA for PCR amplification. *Nucleic Acids Res.*, 20, 2380, 1992.
31. Edwards, K., Johnstone, C., and Thompson, C., A simple and rapid method for the preparation of plant genomic DNA for PCR analysis. *Nucleic Acids Res.*, 19, 1349, 1991.
32. Maaroufi, Y., Comparison of different methods of isolation of DNA of commonly encountered *Candida* species and its quantitation by using a real-time PCR-based assay. *J. Clin. Microbiol.*, 42, 3159, 2004.
33. Fredricks, D.N., Smith, C., and Meier, A., Comparison of six DNA extraction methods for recovery of fungal DNA as assessed by quantitative PCR. *J. Clin. Microbiol.*, 43, 5122, 2005.
34. Borman, A.M. et al., Ultra-rapid preparation of total genomic DNA from isolates of yeast and mould using Whatman FTA filter paper technology—A reusable DNA archiving system. *Med. Mycol.*, 44, 389, 2006.
35. Borman, A.M. et al., Molecular identification of pathogenic fungi. *J. Antimicrob. Chemother.*, 61 (Suppl. 1), 7, 2008.
36. Nuchprayoon, S. et al., Flinders Technology Associates (FTA) filter paper-based DNA extraction with polymerase chain reaction (PCR) for detection of *Pneumocystis jirovecii* from respiratory specimens of immunocompromised patients. *J. Clin. Lab. Anal.*, 21, 382, 2007.
37. Atkins, S.D. and Clark, I.M., Fungal molecular diagnostics: A mini review. *J. Appl. Genet.*, 45, 3, 2004.
38. Moreira, M., Efficient removal of PCR inhibitors using agarose-embedded DNA preparations. *Nucleic Acids Res.*, 26, 3309, 1998.

27 Coccidioides

Rossana de Aguiar Cordeiro, Raimunda Sâmia Nogueira Brilhante, Marcos Fábio Gadelha Rocha, and José Júlio Costa Sidrim

CONTENTS

27.1 INTRODUCTION

27.1.1 The Genus *Coccidioides*

The genus *Coccidioides* includes the species *C. immitis* and *C. posadasii*, which exhibit very similar morphologic characteristics but have molecular polymorphism and different ecological preferences.[1,2] Although phylogenetically different,[2] *C. immitis* and *C. posadasii* cause coccidioidomycosis in both people and animals, which to date are apparently indistinct diseases when considering clinical manifestations and host immune response. The species are classified as biosafety level 3 agents and are on the Select Agent List, according to the U.S. Departments of Health and Human Services (HHS) and Agriculture (USDA).

They are telluric organisms that form hyaline hyphae during the filamentous phase of the biological cycle, which throughout the maturation originate arthroconidia interspaced with cells lacking cytoplasmatic material, called disjunctor cells. The arthroconidia are easily detached from the vegetative mycelium and contain in their extremities cell wall remnants of disjunctor cells, facilitating their air dispersion.[4] Within 72 h after inhalation by a susceptible host, each arthroconidium is converted into a multinucleated cell that undergoes enlargement of the cell wall and cytoplasmatic compartmentalization during its isotropic growth, originating a multinucleated spherule with a diameter of 60–100 μm.[3] After reaching cell maturity, the spherule may release more than 800 endospores[4] and each of them, after germinating, will prompt the development of a new spherule, thus resulting in an exponential reproduction of the fungal population. When the endospores reach the soil under adequate conditions, they grow into the filamentous form, thus guaranteeing the continuity of the microorganism's biological cycle. Even though they are classified as thermally dimorphic species, their transition from the filamentous to the yeast form under laboratory conditions also depends on nutritional factors and increased CO_2 concentration (10%–20%). Experimental observations showed that the addition of Tamol N® detergent also stimulates spherulation.[3]

Coccidioides is an ascomycete fungus that belongs to the order Onygenales, which includes other dimorphic fungi such as *Blastomyces dermatitidis*, *Paracoccidioides brasiliensis*, and *Histoplasma capsulatum*. Although the sexual phase of *Coccidioides* is not known in nature, previous experimental data have revealed the presence of a recombinant genetic structure, thus suggesting the existence of a sexual phase.[5] Recent experimental data confirm that *C. immitis* and *C. posadasii* are heterothallic species that have *MAT1-1* and *MAT1-2* idiomorphs with conserved sequences.[6] Therefore, the possibility of the occurrence of sexual reproduction within this genus may be important for the epidemiology of coccidioidomycosis, since ascospores and conidia may differ, considering survivability in the soil.[7]

27.1.2 Ecology

Knowledge on the ecological preferences of *Coccidioides* comes from studies with microorganism and/or genetic material isolation from soil. Although *Coccidioides* does not present nutritional requirements in vitro, it is rarely isolated from the environment.[7] Apparently, the distribution of *Coccidioides* in these environments is irregular and sporadic.[7] A study performed by Elconin et al.,[8] who analyzed more than 5000 soil samples from endemic areas for the disease during 8 years, showed that the isolation rates for *Coccidioides* ranged from 0% to 43%. Greene et al.[7] reported an isolation rate lower than 1% after analyzing 720 soil samples from hyperendemic areas for the disease through the double-pour method, followed by specific PCR.

Several biotic and abiotic factors, still not completely known, influence the discontinuous distribution of *Coccidioides* in soil. Apparently, the most important factors are temperature, amount and timing of rainfall, and conditions that are directly related to soil physicochemical characteristics, such as low humidity, texture, alkaline pH, and salinity. In addition, levels of exposure to sun and/or ultraviolet light, competition with other telluric microorganisms and/or vegetal species, and the presence of rodent burrows seem to influence the maintenance of *Coccidioides* in the soil.[9] However, recent studies have questioned the dominant role of these environmental variables in disease incidence in endemic areas of California and have proposed the role of the anthropogenic actions in the occurrence of coccidioidomycosis in these areas.[10]

Studies of the correlation between the natural fungal distribution and local biota have shown irreproducible results. Apparently, small mammals play an important role in the dispersion of *Coccidioides* throughout the environment and they can be implicated as temporary reservoirs of the fungus in soil.[11] Nevertheless, the exact role of animals in the ecology of *Coccidioides* still remains unclear. Emmons[12] suggested small rodents to be sentinel animals for coccidioidomycosis in the United States and considered them to be the main reservoirs of the fungus in nature, since carcasses, urine, and feces of these animals were the main source of soil contamination. Undoubtedly, small mammals such as kangaroo rats, squirrels, shrews, and rabbits have an important role for fungal dissemination in North America.[1] In Brazil, armadillos (*Dasypus novemcinctus*) have been implicated as important agents for environmental dispersion of *C. posadasii*, since the microorganism has been isolated from naturally infected animals[13] and their burrows.[14]

Until now, little is known about the ecological characteristics of *Coccidioides* outside the endemic areas of North America. The environmental isolation of the fungus has already been performed in Argentina,[15,16] but few ecological data on the fungus are available in this country.

27.1.3 Epidemiology

The occurrence of coccidioidomycosis in human beings has been determined through detection of clinical cases, environmental analysis during outbreaks, and performance of epidemiological inquiries. Such studies have shown that autochthonous cases of coccidioidomycosis come exclusively from the Americas, occurring in arid or semiarid regions located between 40°N 120°W in Northern California and 40°S 65°W in the Patagonia, Argentina.[17] In these areas, the distribution of the etiologic agents of coccidioidomycosis is very peculiar: *C. immitis* is limited to San Joaquin Valley in California, while *C. posadasii* presents wider distribution, being found in the United States, Mexico, and South America.[1,2] So far, endemic areas of the disease are known in 10 countries: the United States, Mexico, Honduras, Guatemala, Colombia, Venezuela, Bolivia, Paraguay, Argentina, and Brazil.[17] However, the most endemic regions are located in the southwestern United States and along the border with Northern Mexico. Cases of coccidioidomycosis have also been reported outside these areas, in tourists who were contaminated in endemic areas of the United States or Mexico.[18–21]

Since coccidioidomycosis is mainly acquired through inhalation,[22] activities that involve exposure to dust particles in endemic areas increase the risks of acquiring the disease.[23] Even though farming activities are considered a risk factor for coccidioidomycosis,[24] the disease is not frequent among individuals who perform only farm work, since *Coccidioides* is not a common saprophyte of agriculturable soils. Increases in the number of immunocompromised patients in risk areas, as well as increases in the number of immigrants from nonendemic areas to endemic regions, have been considered important issues for the rising incidence of coccidioidomycosis in the United States.[23,25]

In the United States, approximately 150,000 new infections are estimated per year,[22] with time-related variations in endemic areas. In that country, besides anthropogenic actions, the incidence of coccidioidomycosis is also influenced by environmental conditions, such as earthquakes[26] and sand storms[27] Possibly, outside the southwestern United States, the areas with the greatest numbers of disease cases are northern Mexico and northeastern Brazil.[17,28,29]

International demographic data show that coccidioidomycosis occurs at any age, but age extremes present greater risk of severe complications, including chronic and disseminated pulmonary diseases.[23,30,31] Pulmonary coccidioidomycosis occurs more frequently in men, which probably is related to the occupational character of the disease.[30,31] Also, men seem to be more resistant to disseminated disease, which suggests the existence of hormonal and/or genetic protective factors in women.[23] Although the risk of acquiring coccidioidomycosis is not related to any racial group, it is known that the disseminated form of the disease occurs 10–175 times more frequently in the Philippines and in African Americans.[30,31] Apparently, host genetic characteristics, in particular HLA class II and the ABO blood group genes, influence the individual susceptibility to the occurrence of grave forms of the disease.[32]

Underlying diseases that affect cellular response function, such as AIDS, especially in patients with CD4<100 cells/mL, and cancer, particularly in Hodgkin's disease, increase the risk of disease dissemination.[31] Due to the use of immunosuppressant medication, transplant patients are considered at increased risk of coccidioidomycosis, which is the most prevalent endemic mycosis among solid-organ transplant patients in the United States.[33] It is estimated that half of coccidioidomycosis cases in these patients may originate from reactivation of previously acquired coccidioidal infection, which occurs within the first year after the transplant. In spite of the opportunistic character of coccidioidomycosis in HIV-positive individuals, the disease incidence in these patients has declined in recent years due to the advent of potent antiretroviral therapy.[23] Additionally, the gestational age, especially the third trimester, and the postpartum period are also considered risk factors for the development of disseminated coccidioidomycosis.[23,31]

Through epidemiological surveys with coccidioidin and/or spherulin, important epidemiological data on coccidioidomycosis have been obtained in the past years. Several studies have shown that the incidence of coccidioidomycosis infection in endemic areas varies greatly depending on the studied population. The estimated coccidioidomycosis infection through intradermic tests have also shown variable indexes in Mexico (10%–93%), Honduras (2.5%–50%), Guatemala (0.5%–42%), Paraguay (0.9%–50%), and Argentina (7.3%–40%).[17] In the other endemic areas, the few studies that have been performed to date show positivity of 12% in Brazil[34] and 46% in Venezuela.[17] Although the reactivity indexes to the cutaneous tests show great variation, these studies indicate higher positivity among males. The reactivity increases with age until the age of 65, when it declines due to physiological alterations in cellular response. Similarly to the epidemiological characteristics of the groups affected by the disease, the percentage of reactivity to *Coccidioides* antigens also varies depending on the investigated population group and environmental alterations, such as deforestation, earthquakes, and sand storms.

As for data regarding animals, apparently, domestic dogs are the most frequently affected species. Although precise epidemiological data are unavailable, it is estimated that in endemic areas the disease incidence in dogs is similar to that in humans.[35] Coccidioidomycosis has also been described in other animal species such as cats, llamas, bats, sea lions, dolphins, wallabies and kangaroos, tapirs, Przewalski's horses,[35] and reptiles.[36] Even though *Coccidioides* has been visualized in biopsies from cattle, goats, and hogs, these animals rarely develop the disease.[35] Symptomatic infections are also infrequent in horses, and apparently, they are at greater risk of developing disseminated disease with poor outcome. Primates of different species are susceptible to *Coccidioides*.[35]

27.1.4 Pathogeny, Immunity, and Virulence

Although *Coccidioides* infection is more commonly acquired through respiratory route, the disease can also occur after traumatic inoculation. The occurrence of the disease has also been reported after pulmonary transplant[37] and following intrauterine infection,[38] even though microorganism transmission to the newborn is not frequent in some pregnant women with disseminated disease.[39]

The risk of acquiring the disease from animals is practically absent. Indeed, there are only two reports describing this form of contamination: inhalation of fungal parasitic structures during necropsy of a horse[40] and after being bit by a cat with disseminated coccidioidomycosis.[41]

After being inhaled by a susceptible host, the arthroconidia rapidly reach terminal bronchi, since their reduced size, usually smaller than 10 μm, allows them to evade certain innate defense mechanisms of the host respiratory tract, such as filtration, mucocilliary transport, and chemical inhibitors.[3,31] From that moment, these structures can be phagocytized by alveolar macrophages and/or neutrophils. However, experimental data show that only 50% of infective arthroconidia are phagocytized by neutrophils, probably due to antiphagocytic properties of the outer wall layer.[42] After arthroconidia phagocytosis, neutrophil respiratory burst occurs, and even though *Coccidioides* conidia are sensitive to reactive oxygen species and defensins, it is estimated that fewer than 20% of the phagocytized arthroconidia are eliminated through this mechanism.[3] Then, the initial response caused by *Coccidioides* infective structures in the lungs is characterized by influx of neutrophils, which seem to respond to chemotaxins formed after the activation of the complement system.[3,30]

Following the cellular transformation, endospore release also causes the influx of neutrophils, which are possibly attracted to substances released by fungal parasitic structures.[30] Nevertheless, intracellular death of these structures is even less efficient than that observed for arthroconidia.[3] The complete parasitic cycle lasts for 4–6 days and depends on the specific characteristics of each strain.[42] As for other intracellular pathogens, once phagocytized, *Coccidioides* can inhibit phagosome–lysosome fusion and survive within the phagocyte.[42] Even though they are capable of inducing a potent host inflammatory response,[4] *Coccidioides* spherules cannot be phagocytized because of their dimensions.[30,43] Hence, the resulting inflammatory response is characterized by cellular infiltration that persists throughout the infectious process, which leads to granuloma formation.[3,30]

The clinical course of coccidioidomycosis can be modified if the host response is directed to Th1- or Th2-specific cytokine pattern. Processing and presentation of important antigens by macrophages and/or dendritic cells induce production of interferon-γ, tumor necrosis factor-α, and interleukin (IL)-12 by Th1 lymphocytes and these cytokines induce macrophage recruiting and activation, which may inhibit or kill fungal structures, contributing to infection control.[30] On the other hand, IL-4, IL-5, IL-10, and IL-13, produced by Th2 lymphocytes, activate humoral response and induce downregulation of functions induced by Th1 lymphocytes,[44] increasing individual susceptibility to *Coccidioides* infection.[45] Thus, asymptomatic patients or those with benign forms of the disease present a strong intradermic reaction when stimulated by fungal antigens, and low levels of serum-specific antibodies. On the other hand, patients with chronic pulmonary or disseminated coccidioidomycosis present T-cell anergy and high titers of anti-*Coccidioides* IgE and IgG.[3] Cell immunity, manifested by delayed hypersensitivity, is long lasting and protects individuals against fungal reinfections. Thus, reactivation of coccidioidomycosis after clinical resolution has only been described in individuals who had previously suffered serious cell immunity suppression.[45] Currently, it is clear that not only the host cellular immune response pattern but also fungal virulence factors are determinants of establishment of the disease.[43]

Even though *Coccidioides* pathogenicity is multifactorial, some specific products from fungal metabolism have been associated with virulence. Proteases secreted by the fungus, which are able to degrade antibodies and other opsonins, have been suggested as important virulence factors.[43] Other proteases with collagenasic and elastasic activity have also been implicated as virulence factors, considering their potential to degrade the connective tissue matrix of the lungs, which facilitates the destruction of pulmonary parenchyma and subsequent fungal dissemination.[46] Recent studies show the role of the enzyme urease in the virulence of *Coccidioides*. The enzyme causes alkalinization of the microenvironment where the pathogen is found and possibly contributes to pulmonary parenchyma damage and exacerbation of disease symptoms.[43]

Laboratory studies have shown the presence of melanin or melanin-like compounds in arthroconidia, spherules, and endospores of *C. posadasii* from in vitro cultivation, as well as in spherules produced during experimental infection.[47] Since melanin is an important virulence factor for many fungal species, it is also likely to contribute significantly to the virulence of *Coccidioides*.

Besides the direct action of products from metabolism, several virulence mechanisms have been recognized for *Coccidioides*. Previous experimental data have shown that the fungus is capable of immunomodulating host response through the deposition of a spherule outer wall (SOW) during the growth of parasitic spherules. This material is also released to the extracellular environment as membranous sheets and is phagocytized by macrophages in infection sites. Subsequent studies have shown that a fraction of this material that is rich in glycoproteins (SOWgp) presented an immunodominant antigen, which is recognized by specific antibodies from sera of coccidioidomycosis patients, who usually presented elevated titers of anti-SOWgp IgG in severe forms of the disease.[43] Some studies show that *Coccidioides* has developed specific mechanisms to evade host immune response. Expression of SOWgp reaches its peak during isotropic growth of spherules, which are poorly phagocytized because of their size, and decreases during endospore formation, thus guaranteeing that this structure will not be opsonized by anti-SOWgp antibodies.[43] In addition, experimental data show that the fungus secretes a specific metalloproteinase (Mep1), responsible for degrading the SOWgp antigen, during endospore differentiation, which prevents host recognition of endospores.[48]

Another virulence mechanism described for *Coccidioides* involves the modulation of arginase I synthesis by macrophages.[43] Arginase I catalyzes L-arginine hydrolysis into L-ornithine, which serves as a substrate for the enzyme ornithine decarboxylase (ODC) of *Coccidioides*, considered to be a key enzyme for the metabolism of polyamine and a putative regulator of fungal parasitic differentiation.[49] In macrophages, L-arginine is the substrate for another enzyme, inducible nitric oxide synthase (iNOS), whose activity is regulated by cytokines secreted by T lymphocytes. Experimental studies suggest that host exposure to *Coccidioides* SOWgp antigen induces a Th2 response, which leads to an increase in arginase I expression and a reduction in iNOS expression. This process decreases nitric oxide production by macrophages, contributing to fungal intracellular survival.[43] In addition, increase in arginase I expression also leads to a higher urea production within infection sites.

27.1.5 Clinical Forms

Coccidioidomycosis can be classified as an asymptomatic, acute pulmonary, chronic pulmonary, disseminated, or primary cutaneous disease.[22,30] It is estimated that up to 60% of individuals exposed to infective arthroconidia are asymptomatic, being detected only through serological inquiry or cutaneous tests. Practically all of these individuals develop permanent immunity and are therefore protected against second primary infections.[22] In approximately 40% of individuals, a symptomatic disease can occur within 1–3 weeks after infection.[22] However, during epidemic outbreaks, the index of symptomatic individuals can reach up to 90%.[50,51]

The pulmonary forms of the disease are the most frequent, causing several clinical manifestations such as pneumonias of different radiological patterns, pleural effusion, hilar lymphoadenopathy, and pulmonary nodules. The most common clinical syndrome is pneumonia,[52] which usually begins within 1–3 weeks after arthroconidia inhalation and is characterized by cough, dyspnea, and thoracic pain, besides constitutional symptoms such as fever, fatigue, anorexia, and arthralgia.[53] In endemic areas, coccidioidomycosis can be considered a common cause of community-acquired pneumonia.[52] The acute pulmonary primary form may regress spontaneously, after a few months, even without specific therapy.[31] Among patients with pulmonary disease, approximately 5% develop specific manifestations like nodous erythema or multiform erythema, resulting from delayed hypersensitivity,[3] which are more frequent in females.[3] However, other cutaneous manifestations, such as acute exanthema, Sweet's syndrome, and interstitial granulomatous dermatitis, can be observed.

Approximately 5% of patients with primary pneumonia do not present spontaneous cure and may evolve to chronic pulmonary infection, which is characterized by nodular lesions or fibrocavitary pulmonary disease.[31] Clinically, patient presents nocturnal sudoresis, muscular fatigue, weight loss, chronic cough, and hemoptysis.[30,31] Due to clinical, radiographic, and histopathological resemblances, progressive pulmonary coccidioidomycosis may be mistaken for pulmonary tuberculosis. In Mexico and Brazil, endemic areas for coccidioidomycosis and tuberculosis overlap, which makes clinical diagnosis of these diseases extremely difficult.[23,54] Particularly in Brazil, where tuberculosis is the most prevalent infectious disease, our experience has shown that some cases of coccidioidomycosis were mistaken for pulmonary tuberculosis and antituberculosis therapy was instituted.[54] The incorrect diagnosis influences coccidioidomycosis epidemiology in this country, since it is believed that drugs used to treat pulmonary tuberculosis may somehow promote the recovery of patients with coccidioidomycosis, compromising detection of disease cases.[54]

The disseminated form of coccidioidomycosis can affect 1%–5% of infected individuals[31] and it can occur even when there are no clinical or radiological signs of pulmonary infection.[3] This form of the disease most frequently affects adult immunocompromised individuals.[22] Fungal dissemination occurs through hematogenic or lymphatic viae and can reach several sites, such as lymph nodes, skin, bones, central nervous system, joints, and genitourinary apparatus.[22,31] Coccidioidal meningitis is the more severe form of the disseminated disease and has a high mortality index when it is not diagnosed and properly treated.[22] The most common complications associated with coccidioidal meningitis are hydrocephalus, vasculitis, and parenchymal abscesses.[55]

Although not common, pericardic involvement in coccidioidomycosis can result from contiguous pleural-pulmonary lesions or from lympho-hematogenic dissemination. In these patients, clinical complaints are not very different from those of patients with pulmonary involvement, which are thoracic pain, cough, and dyspnea. However, some clinical findings are more characteristic for this clinical presentation, such as orthopnoea, paradoxical pulse, and pericardic attrition. Coccidioidal pericarditis usually presents unfavorable prognosis, with high indexes of morbidity and mortality.[56,57]

Primary cutaneous forms are rare and result from traumatic inoculation of the microorganism from an environmental source or through laboratory manipulation.[3,58] Dermatologic manifestations include papules, nodules, and verrucose plaques that can evolve to the formation of ulcers and abscesses.[31] Due to the clinical diversity of the lesions, cutaneous coccidioidomycosis can be mistaken for several other diseases, and therefore, laboratory procedures are necessary to confirm diagnosis.[31]

27.1.6 Treatment and In Vitro Antifungal Susceptibility

The therapeutic approach for the treatment of coccidioidomycosis basically depends on three factors: severity of the pulmonary infection, presence of disseminated disease, and the patient's individual risk factors.[22] The majority of cases of pulmonary coccidioidomycosis progress as a self-limited disease, and according to some authors, there is no need for therapeutic intervention, but the patients should be clinically and radiographically monitored for up to 2 years to document infection resolution.[30] In contrast, experts recommend the institution of systemic antifungal therapy for all symptomatic patients, with fluconazole or itraconazole v.o. for up to 6 months.[58]

More severe, chronic or disseminated clinical cases require a specific therapeutic approach. The drug choice depends on clinical manifestations and host immunological status.[59] Currently, amphotericin B and its lipid formulations are saved for severe cases of the disease. Azole derivatives, especially fluconazole and itraconazole, are used as sequential or combined treatment. In cases of meningitis, fluconazole is preferred because of its good penetration in the cerebrospinal fluid (CSF). However, due to its fungistatic activity, it demands long-lasting therapy, possibly throughout life, especially in immunocompromised patients.[60]

Recently obtained data show that voriconazole and posaconazole are efficient for the treatment of chronic pulmonary coccidioidomycosis and coccidioidal meningitis[61–63] respectively. The clinical efficacy of echinocandin antifungal caspofungin was reported in two cases of disseminated coccidioidomycosis.[64,65] The therapeutic efficacy of this drug was analyzed in murine models. The data showed an increase in survival of infected animals and a reduction of fungal parasitism in the treated animals, when compared to the control group. In addition, a synergistic effect of the combination of caspofungin and amphotericin B was observed.[66] Until now, the available data suggest the potential of caspofungin in the therapy of coccidioidomycosis. Azole chemoprophylaxis must be considered for organ transplant recipients and HIV patients who live in endemic areas, due to high risk of acquiring the disease.[67]

Even though there are no data that allow establishing a precise correlation between therapeutic response and antifungal susceptibility profile,[68] several studies have shown the in vitro response of *Coccidioides* strains to antifungal drugs. Nearly all of the results have shown that *Coccidioides* is susceptible, in vitro, to most available antifungal agents, including amphotericin B, azole derivatives, and echinocandins,[68–70] despite the absence of official endpoints. Until now, only one report describes the resistance of *Coccidioides* to azole derivatives.[71] Antifungal susceptibility tests must be performed in reference laboratories and are appropriate for surveillance studies and possibly for follow-up of patients exhibiting therapeutic refractoriness.[72] The same approach has also been applied to the prospection of drugs with antifungal activity. For example, studies have shown inhibitory in vitro effect of antituberculosis drugs against *C.* posadasii.[54,73]

27.1.7 Laboratory Diagnosis

As for other systemic mycoses, coccidioidomycosis does not have pathognomonic clinical-radiological findings, and it can be mistaken for other infectious diseases.[28,72] Thus, laboratory confirmation is essential for the definite diagnosis of coccidioidomycosis. The success of laboratory diagnosis, however, depends on the use of adequate techniques during the pre-analytical phase, which involves clinical specimen collection, transportation, and processing.

27.1.7.1 Mycological Diagnosis

Direct examination of the clinical specimen must be performed through optical microscopy with slide preparations and smears. Depending on the clinical presentation, several biological specimens may be sent to laboratories for the analysis of *Coccidioides*, such as respiratory samples, CSF, aspirates of osteoarticular lesions, pleural fluid, and biopsies.[28,72]

Under direct examination, using wet mounts on glass slides, clinical specimens are clarified by KOH 10%–40% or calcofluor white fluorescent stain (CFW) to facilitate spherule visualization; smears and imprints can also be stained by Grocott-methenamine silver (GMS). Microscopic examination shows spherules of up to 100 μm diameter, with birefringent wall, containing endospores of 2–5 μm.[43,72] This image is considered pathognomonic for the parasitic phase of *Coccidioides*. Some authors describe the morphological similarity between coccidioidic spherules and those produced in vivo by *Rhinosporidium seeberi*, *Emmonsia* spp.,[42] and *Prototheca* spp.[74] Free endospores can be mistaken for conidia of *Histoplasma capsulatum*, *Cryptococcus* spp., or *Candida* spp.[68,72] However, these findings should not represent obstacles for the establishment of mycological diagnosis, since infections caused by these agents have clinical-epidemiological characteristics that differ from coccidioidomycosis. Although cheap and feasible when performed by experienced mycologists, direct examination has low sensitivity. In patients with acute pulmonary disease, positive rates ranging from 30% to 40% have been described.[75]

Rarely, arthroconidiated filamentous structures can be formed in CSF, during CNS infections[68,72] and in pleural fluid.[74] Recently, it was demonstrated that the presentation of atypical forms in respiratory specimens can be four times more frequent in patients with cavitary lesions who have type-2 diabetes when compared with patients who are not affected by this comorbidity.[76] This finding is of great importance since these atypical parasitic structures can cause infection and thus require more caution when manipulating clinical specimens.

Despite the advances provided by molecular and immunological techniques in recent years, the isolation and growth of *Coccidioides* is still the gold standard for laboratory diagnosis of coccidioidomycosis.[68] Primary cultures obtained from clinical specimens develop quickly, within approximately 5 days with incubation at 25°C–30°C. The microorganism grows on most of the media routinely used in mycology laboratories, such as 2% Sabouraud dextrose agar, brain heart infusion (BHI) agar, and potato dextrose agar. Fungal recovery from clinical specimens can sometimes take 3 weeks or longer, especially when the patient is using antimicrobial drugs.[57,68] Usually, young colonies, with up to 5 days, are glabrous, white-colored and do not exhibit arthroconidia. As the colonies mature, they become velvety or cottony, white, cream, or gray colored and, under microscopy, exhibit septate hyaline hyphae having arthroconidia interspaced by disjunctor cells lacking cytoplasmatic material, which are 5–7.5 μm and 2.5–5 μm in length and width, respectively.[14]

27.1.7.2 Histopathological Diagnosis

In histopathological examinations, *Coccidioides* can be identified through several staining techniques, such as hematoxylin–eosin (HE), which, although not allowing easy visualization of fungal structures, continues to be the most frequent technique used in several laboratories. Other options include periodic acid-Schiff (PAS), which allows better spherule, endospore, and hyphal individualization, and Grocott-methenamine silver, which is considered the most sensitive technique.[72] Other options such as Giemsa, Papanicolau, mucicarmine, and Gram present low efficiency and should not be employed as primary stains for laboratory diagnosis of coccidioidomycosis.[72]

27.1.7.3 Immunological Diagnosis

Cellular immune response during coccidioidomycosis can be demonstrated through positive results for the cutaneous test with coccidioidin or spherulin. In this test, the fungal antigen is intradermally inoculated in the forearm of the individual. The observation within 72 h of induration of an area of at least 5 mm is considered a positive response.[77] The use of these tests to establish diagnosis and prognosis of coccidioidomycosis is very limited, because positive results can be associated with current or past infections or even cross reactions. However, the conversion of a negative result can be useful for identifying recent fungal infections. Patients with positive intradermal reaction for coccidioidin and/or spherulin have a better prognosis, while cases that cutaneous tests are negative during the infectious process generally evolve to disseminated disease.

In coccidioidomycosis, produced antibodies do not have protective activity, but serve as tools for diagnosis, prognosis, and follow-up of patients.[78] The search for specific antibodies can be performed in various clinical specimens, such as serum, CSF, pleural, peritoneal, and synovial fluids,[78] depending on the standardization of the technique used. Tests can also be used in the immunocompromised patients, though in some cases it is necessary to combine more than one procedure.[79]

After symptom appearance, early detection of IgM can be performed within the first and the third weeks, with positivity of 50% and 90% of the cases, respectively.[80] Approximately 90% of the individuals show reactivity to tests aimed at detecting IgG, around the fourth week, after the appearance of symptoms.[72,81] For IgM detection, the most frequently used methods in routine laboratories are gel immunodiffusion

(IDTP), latex agglutination, and immunoenzymatic tests. IgG detection is performed through gel immunodiffusion (IDCF), complement fixation reaction, and immunoenzymatic tests. Recently, a quantitative enzyme immune assay for antigen detection in urine became available.[82]

27.1.7.4 Molecular Diagnosis

Diagnosis of coccidioidomycosis based on molecular techniques is very promising, since it allows fungal detection from contaminated clinical samples and from mixed colonies.[4,83,84] In fact, it is the only strategy so far that allows distinguishing between the species *C. immitis* and *C. posadasii*. Besides this, the use of molecular techniques for diagnosis reduces the inherent risk of handling *Coccidioides* cultures in their filamentous phase. However, few studies have investigated the potential of molecular techniques for the diagnosis of coccidioidomycosis. Although the first techniques used for diagnosing the disease were based on hybridization reactions, currently, PCR-based protocols are preferred because they are easily executed.

The first tests for the detection of *Coccidioides* were based on hybridization reactions of nucleic acids and were performed by Stockman et al.,[85] who used an acridinium ester-labeled chemiluminescent DNA probe against rDNA sequences. Currently, commercial chemiluminescent probes ("AccuProbe") for the identification of *C. immitis* are produced by GenProbe and are commercially available. Although having high sensitivity and specificity, the detection of *Coccidioides* based on such hybridization probes is subject to false-negative results when formaldehyde-killed cultures are used.[86] However, to date, there are molecular probes for the detection of *Coccidioides* directly from clinical specimens, as well as for the differentiation between the species *C. immitis* and *C. posadasii*.

The first studies involving molecular detection of *Coccidioides* based on PCR were performed by Pan and Cole,[87] who established a protocol for the amplification of the CSA gene, directly from fungal cultures. Subsequently, several PCR-based tests were developed to confirm *Coccidioides* from cultures and clinical specimens. In this context, the gene clusters from ribosomal DNA and partial sequences from internal transcribed spacers (ITS) of rDNA have been used to detect *Coccidioides* in cultures,[7] as well as in blood samples from patients and experimentally infected animals.[83]

The detection of *Coccidioides* in filamentous cultures or in biopsies can also be performed through amplification of the partial sequence of the codifying gene for the molecule "antigen 2" or "PRA" following protocols based on nested PCR.[4] The same target sequence was used for the identification of cultures isolated from pleural fluid,[57] as well as for the diagnosis of coccidioidomycosis directly from sputum samples.[84] Through real-time PCR, partial sequences from the ITS regions can also be used as targets for *Coccidioides* detection in several clinical specimens, such as respiratory samples and fresh biopsy material fixed in formaldehyde or paraffinized.[88]

Although the distinction of the species *C. immitis* and *C. posadasii* was originally performed through analysis of microsatellite loci *GAC2* and *621*[2] and single-nucleotide polymorphisms,[1] these methods are of limited value for routine diagnosis since they are expensive and require laborious techniques.[89] In addition, apparently, the oligonucleotide primers used for microsatellite amplification are not specific for the recognition of *Coccidioides* directly from clinical samples.[4] Currently, the identification of the etiological agent can be performed through a single PCR reaction[89] or amplification of the ITS region and posterior sequencing or restriction fragment length polymorphism of the complete sequence.[90] The main techniques used for the detection of *Coccidioides* in cultures and clinical specimens, as well as for molecular identification of *C. immitis* and *C. posadasii*, are described as follows.

27.2 METHODS

Fungal DNA extraction from cultures and clinical specimens is considered to be a critical step for molecular diagnosis. Some techniques for *Coccidioides* DNA extraction from cultures and clinical specimens are described in the following sections.

27.2.1 Sample Preparation

27.2.1.1 DNA Extraction from Cultures

27.2.1.1.1 Protocol 1

Most methods of DNA preparation from fungi are time-consuming or require special equipment, such as fast prep homogenizers and liquid nitrogen suppliers. We have performed a simple method for total DNA preparation, yielding a product of good quality, suitable for PCR reactions and restriction digestion analysis. The procedure has been tested for suspicious *Coccidioides* cultures grown on mycological agar at least for 10 days.[84]

1. Add 5–7 mL of sterile phosphate-buffered saline to the agar slant.
2. Scrape the culture surface gently with a swab or microbiological loops to remove hyphae and arthroconidia without removing agar. Handle the tube carefully to prevent spillage.
3. Transfer the suspension to a plastic screw-cap tube and autoclave at 100°C for 15 min.
4. After cooling, seed a loopful of mycelial debris into BHI broth and incubate at 37°C for 1 week. Store the remaining suspension at −20°C and continue the DNA extraction procedure if no fungal growth is detected in the BHI broth.
5. Let the cultures thaw at room temperature and then collect fragments of mycelia and arthroconidia by centrifuging at 10,000 × *g* for 10 min; wash the pellet twice with phosphate-buffered saline.

6. Incubate the pellet with 1 mL of lyticase 3 mg/mL at 37°C for 2 h, and then wash with phosphate-buffered saline.
7. Resuspend cellular debris in lysis buffer (10 mM Tris–HCl [pH 8.0], 1 mM EDTA, Triton X-100 2% [v/v]), and then expose it to three cycles of heating at 100°C for 10 min and freezing at −80°C for 10 min.
8. Centrifuge at 8000 × *g* for 15 s and store the supernatant at −20°C.
9. Check DNA degradation by 0.7%–1.0% agarose gel electrophoresis.
10. Verify DNA purity from the A260/A280 ratio in a spectrophotometer.

This is a very simple protocol for DNA extraction that can be performed even in small laboratories. The DNA yield varies depending on the amount of mycelium harvested. The quality of the DNA is good and usually our results have shown an A260/A280 ratio of circa 1.8.

27.2.1.1.2 Protocol 2

A safe DNA extraction protocol previously described by Burt et al.[91] is as follows:

1. Grow *Coccidioides* cultures at 37°C in 100 mL of 2× GYE medium (20 g/L glucose; 10 g/L yeast extract) in 500 mL plastic Nalgene flasks with plastic foam stoppers.
2. When there has been sufficient growth, transfer the whole container to an autoclave and steam at 1 atm (100°C) for 15 min.
3. Allow the culture to cool, plate 1 ml on 2× GYE or potato dextrose agar, and freeze the rest at −20°C.
4. If there is no growth on the plate after 1 week, the culture can be removed from the containment room to proceed with the DNA extraction.
5. In order to prepare the samples for DNA extraction, the cultures must be filtered, frozen in liquid nitrogen, lyophilized, and ground with a mortar and pestle.
6. Mix about 10 mg of dry ground tissue with 400 μL of lysis buffer (50 mM Tris, 100 μM NaCl, 5 mM EDTA, 1% sodium dodecyl sulfate), incubate at 80°C for 10 min and cool to 40°C.
7. Add 10 μL of 10 mg/mL proteinase K for 2–3 h, and then incubate again at 80°C for 10 min.
8. Proceed with the DNA extraction with chloroform–phenol and precipitated with isopropanol and resuspend the pellet in 100 μL TE.
9. Check the yield by testing 5 μL of this material by agarose gel electrophoresis.

Many commercial extraction kits are available for fungal DNA extraction and have been tested for *Coccidioides* detection. Regardless of the method chosen, it should be borne in mind that direct manipulation of *Coccidioides* cultures is a serious hazard and should be performed only by trained technicians in laboratories with class 3 facilities.

27.2.1.2 DNA Extraction from Clinical Samples

27.2.1.2.1 Protocol 1

Although many protocols for *Coccidioides* detection in suspicious cultures have been standardized before, few procedures have focused on fungal detection directly on clinical samples. Here we describe an easy protocol for *Coccidioides* detection directly on sputum samples.[84] After the DNA extraction, specific PCR reaction must be performed to detect any sequence of diagnostic value.

1. Mix a 1 mL aliquot of sputum sample with 200 μL of NaOH 5% (v/v) and incubate at 37°C for 1 h.
2. Centrifuge at 14,000 × *g* for 5 min and wash the resulting pellet twice with phosphate-buffered saline.
3. Incubate the pellet with 100 μL of lyticase (3 mg/mL) at 37°C for 2 h and then incubate again at 56°C for 1 h with lysis buffer (10 mM Tris–HCl [pH 8.0], 1 mM EDTA, SDS 2% [w/v]) containing proteinase K (150 mg/L).
4. Centrifuge the lysed cells at 8000 × *g* for 15 s and use the supernatant for PCR reactions.

Small amounts of DNA are expected, as the quantity of parasitic structures in the clinical sample is frequently small. However, by this method, we have succeeded in obtaining DNA suitable for rapid diagnosis of coccidioidomycosis even in sputum samples heavily contaminated with commensal bacteria from the mouth and pharynx and which presented few coccidioidal spherules.[84]

27.2.1.2.2 Protocol 2

Usually clinical samples need to be prepared for DNA extraction in order to increase the amount of DNA yield, to remove PCR inhibitors, and to provide higher quality material. We have chosen the methods described by Binnicker et al.[88] for respiratory specimens and biopsies for MagNA Pure Compact extraction (Roche Applied Sciences, USA) utilizing the DNA isolation kit (large volume). We believe that these protocols can be useful as preliminary steps for other commercial kits.

1. Respiratory specimens (BAL fluid, bronchial washings, pleural fluid, and sputum)
   a. Pipette 500 μL of raw specimen and 100 μL proteinase K into a 1.5 mL tube containing 0.1 mm silica glass beads and 2.4 mm zirconia beads.
   b. Incubate samples at 55°C for 15 min in a thermomixer at 1400 rpm and subsequently place on a 95°C heat block for 5 min.
   c. Place the samples on a Disruptor Genie (Scientific Industries, Bohemia, NY) for 2 min and then centrifuge briefly at 1000 × *g* to 5000 × *g* to collect the sample at the bottom of the tube.
   d. The lysate obtained can be used for further DNA extraction/purification.

2. Fresh tissue
   a. Place a small piece of tissue, approximately 0.5 cm^2, in a sterile 1.5 mL tube containing 400 μL of 1× Tris–EDTA, 100 μL of proteinase K, and 50 μL of 10% sodium dodecyl sulfate.
   b. Vortex the samples briefly and then placed them in a thermomixer overnight at 55°C with a mixing speed of 500 rpm; digested tissue can be used for further DNA extraction/purification.

TABLE 27.1
Summary of the Most Important PCR-Based Protocols for *Coccidioides* Detection

Target	Primers (5′–3′)	Cycling Program	Amplicon Size (bp)	DNA Source	Reference
CSA antigen gene	AAGTTCTCACTCCTCAGCGCTATCG ACATTAAGGTTCCTCCCCTTCAACC	1. Initial denaturation: 94°C, 4 min 2. 94°C, 1 min; 50°C, 1 min; 72°C, 1 min (30 cycles) 3. Final extension: 72°C, 10 min	519	Cultures	[87]
ITS	CATCATAGCAAAAATCAAAC AGGCCCGTCCACACAAG	1. Initial denaturation: 94°C, 2 min 2. 94°C, 1 min; 53°C, 1 min; 72°C, 1 min (35 cycles) 3. Final extension: 72°C, 1 min	223	Cultures	[7]
PRA antigen gene	First reaction GTACTATTAGGGAGGATAATCGTT GGTGTCAACTGGTGGGATGTCAAT Second reaction ATCCCACCTTGCGCTGTATGTTCGA GGAGACGGCTGGATTTTTAACATG Hybridization probes[a] CCAAATTCTTGCATCTCGCCCA ATGGGATAAGATGAGAAGATGGAAAG	First reaction 1. Initial denaturation: 94°C, 5 min 2. 94°C, 30 s; 5°C, 30 s; 72°C, 1 min (35 cycles) 3. Final extension 72°C, 5 min Second reaction 1. Initial denaturation: 94°C, 5 min 2. 94°C, 30 s; 60°C, 30 s; 72°C, 1 min (30 cycles) 3. Final extension: 72°C, 5 min	First reaction: 526 Second reaction: 342	Cultures, lung and skin biopsy material	[4]
ITS	CATCATAGCAAAAATCAAAC AGGCCCGTCCACACAAG	1. Initial denaturation: 94°C, 10 min 2. 94°C, 1 min; 53°C, 1 min; 72°C, 1 min (35 cycles) 3. Final extension: 72°C, 10 min	239	Human and animal sera	[83]
C. immitis contig 2.2 AAEC02000002	TACGGTGTAATCCCGATACA GGTCTGAATGATCTGACGCA	1. Initial denaturation: 94°C, 3 min 2. 94°C, 30 s; 60°C, 30 s, 72°C, 45 sec (30 cycles) 3. Final extension: 72°C, 3 min	720	Cultures	[89]
ITS	CGAGGTCAAACCGGATA CCTTCAAGCACGGCTT Hybridization probes[a] GAGCGATGAAGTGATTTCCC TACACTCAGACACCAGGAACTCG	1. Initial denaturation: 95°C, 10 min 2. 95°C, 10 s; 60°C, 15 s; 72°C, 20 s (45 cycles)	170	Cultures, respiratory specimens, biopsies	[88]
ITS[b]	GCATCGATGAAGAACGCAGC TCCTCCGCTTATTGATATGC	1. Initial denaturation: 94°C, 4 min 2. 94°C, 30 s; 55°C, 30 s; 72°C, 90 s (38 cycles) 3. Final extension: 72°C, 10 min	ca. 169	Cultures	[90]

[a] Real-time PCR.
[b] Identification requires sequencing of the ITS region or restriction fragment length polymorphism (RFLP) of the entire ITS region with *Bsr*I and *Xcm*I enzymes.

3. Paraffined samples
   a. Place thin tissue sections into a 1.5 mL tube containing 500 μL xylene and incubate for 5 min at room temperature.
   b. Centrifuge for 30 s at 20,800 × *g* and remove the xylene with a disposable pipette.
   c. Repeat incubation with xylene and then centrifuge again to remove xylene.
   d. Add 500 μL of 95% ethanol in the tubes, vortex the samples gently and then incubate them for 5 min at room temperature.
   e. Centrifuge for 3 min at 20,800 × *g* and remove the alcohol with a disposable pipette.
   f. Add 400 μL of 1× Tris–EDTA, 100 μL of proteinase K, and 50 μL of 10% sodium dodecyl sulfate.
   g. Vortex the tubes and place them in a thermomixer overnight at 55°C with a mixing speed of 500 rpm and use the digested material for further DNA extraction.

Note: DNA extraction from clinical samples may be performed with commercial kits manufactured by QIAGENiagen[4,83] and Roche Applied Sciences.[88]

27.2.2 Detection Methods

27.2.2.1 PCR Amplification

Many PCR-based protocols for *Coccidioides* detection have been described in recent years. Although only a few protocols have been tested directly on clinical specimens, all of them can be performed on DNA samples obtained from cultures. PCR-based strategies are the most widely used tools for molecular detection.[92,93] Formerly, the distinction between *C. immitis* and *C. posadasii* could be achieved only by microsatellites or single-nucleotide polymorphisms (SNPs) analysis, but now a single PCR reaction can be employed for this purpose. Recently, RT-PCR has emerged as a robust method for *Coccidioides* investigation in clinical specimens, although its use is still limited as a research tool. Here we present some protocols for *Coccidioides* detection (Table 27.1) and species identification (Tables 27.2 and 27.3).

TABLE 27.2
Primer and Probe Sequences Used in SNPs Analysis for *Coccidioides* Distinction into Species, according to Castañón-Olivares et al.

Target	Primer Sequences (5′–3′)	Amplicon Size (bp)
Glucose synthase 192 gene	CCGACGGCTGGCCATAT	72
	CCAACGCCCAGTAGTATGGT	
	Hybridization probes[a]	
	CCTGGAAAACCAC	
	CCTGGGAAACCAC	
Proline-rich antigen 157 gene	GTCGTTGACCAGTGCTCCAA	63
	CGGCGGTGGTGTCAACT	
	Hybridization probes[a]	
	CCCAATTGAGATCCCA	
	CCCAATTGACATCCCA	
Proline-rich antigen 174 gene	CCGCTGAGCCGACTCA	50
	CGGTTGGGACGGCAGT	
	Hybridization probes[a]	
	TCCTCCGTAGGCTCA	
	CTCCGTGGGCTCA	

Source: Castañón-Olivares, L.R. et al., *Ann. N.Y. Acad. Sci.*, 1111, 326, 2007.
[a] Real-time PCR.

27.2.2.2 Hybridization

A hybridization system showing high levels of sensitivity and specificity for *Coccidioides* identification is currently available commercially from Gen-Probe Incorporated (AccuProbe®, USA). The test can be applied to primary cultures and, according to the manufacturer, it provides rapid and accurate results in as little as 50 min. However, the performance of this test has not yet been determined on direct

TABLE 27.3
Primer Sequences in Microsatellite Analysis for *Coccidioides* Distinction into Species

Target	Primer Sequences (5′–3′)	Allele Size (bp)	References
GAC locus	ATG TCTCGC CTG GAG CAC GAC	*C. immitis*: 216–228	2, 94
	GCT GTT CCA CGT TTC GAC	*C. posadasii*: 206	
621 locus	ACA ATG AAC GAG CAG CAA GG	*C. immitis*: 416–426	2, 94
	TGA AAG ATG TGT AGA CCC GA	*C. posadasii*: 397–400	
KO3 locus	ACCTCAAAAGGCGAGACTAC	*C. immitis*: 243–255	2, 94
	TGCCGAGTGTTTGACCACAG	*C. posadasii*: 239–241	
GA1 locus	ACCTCCCGGAATATCATGTG	Not informed	94
	GTGCGTTGAAAGCAGAAAAT		

clinical specimens. A detailed description of the procedure can be obtained from technical documents provided by the company. Here we summarize the general steps for the test.

1. Remove mycelia fragments from solid media with a disposable plastic loop or needle without taking large amounts of the solid media. Transfer the material to a tube containing 100 μL of lysis reagent and vortex briefly. In case of growth from broth media, pipette a 100 μL sample from the well-mixed broth into the lysis reagent and vortex.
2. Sonicate the material for 15 min and then incubate for 10 min at 95°C ± 5°C in a water bath or on a heating block.
3. Transfer the lysed material into probe reagent tubes and perform the hybridization by incubation for 15 min at 60°C ± 1°C in a water bath or on a heating block.
4. Pipette 300 μL of the selection reagent into each tube to remove non-hybridized probe and then incubate for 5 min at 60°C ± 1°C in a water bath or on a heating block.
5. Leave the tubes resting at room temperature for at least 5 min and then read the results in the luminometer.

27.3 CONCLUSIONS

Knowledge of coccidioidomycosis and its etiological agents has advanced rapidly in recent years. Since the description of the first case of the disease in 1892, many aspects related to the pathogeny and virulence of its etiological agents have been clarified. It is now known that the disease is caused by two genetically distinct species, but it is not known for sure if there are differences in the expression of the virulence factors of these microorganisms in vivo.

Coccidioidomycosis has been carefully studied in the United States, revealing important data on epidemiology and the ecology of its etiological agents. Nevertheless, although it can be considered an emerging disease in some Latin American countries, data on the clinical manifestations of coccidioidomycosis as well as information on the ecological preferences of *C. posadasii* in these regions are still lacking or fragmented. This means there is a need for further studies on the theme in Latin America, to consolidate information of the biology of the fungus and the clinical-epidemiological characteristics of the disease.

In endemic areas, coccidioidomycosis is often confused with other infectious diseases, since it does not present pathognomonic clinical findings. Therefore, various researchers have worked to improve laboratory diagnostic methods, through the different approaches discussed in this chapter. The use of molecular techniques is considered an excellent strategy since it does not expose researchers and lab technicians to the infection risks present in mycological diagnosis. Furthermore, these techniques enable detection of the DNA of *Coccidioides* from contaminated clinical samples and/or mixed cultures. The big challenge at present is to make these techniques more accessible, to ensure quick, safe, and unambiguous diagnosis. In the near future, the integration of the data generated by genomics and proteomics can bring great advances in the molecular diagnosis of coccidioidomycosis. Finally, it is expected that the analysis of the data generated by sequencing the *C. immitis* and *C. posadasii* genomes will reveal molecular targets with great power to detect these species.

REFERENCES

1. Fisher, M.C. et al., Biogeographic range expansion into South America by *Coccidioides immitis* mirrors New World patterns of human migration, *Proc. Natl. Acad. Sci. U.S.A.*, 98, 4558, 2001.
2. Fisher, M.C. et al., Molecular and phenotype description of *Coccidioides posadasii* sp. nov., previously recognized as the non-Californian population of *Coccidioides immitis*, *Mycologia*, 94, 73, 2002.
3. Cox, R.A. and Magee, D.M., Coccidioidomycosis: Host response and vaccine development, *Clin. Microbiol. Rev.*, 17, 804, 2004.
4. Bialek, R. et al., PCR assays for identification of *Coccidioides posadasii* based on the nucleotide sequence of the antigen 2/proline-rich antigen, *J. Clin. Microbiol.*, 42, 778, 2004.
5. Fisher, M.C. et al., Pathogenic clones versus environmentally driven population increase: Analysis of an epidemic of the human fungal pathogen *Coccidioides immitis*, *J. Clin. Microbiol.*, 38, 807, 2000.
6. Mandel, M.A. et al., Genomic and population analyses of the mating type loci in *Coccidioides* species reveal evidence for sexual reproduction and gene acquisition, *Eukaryot. Cell.*, 6, 1189, 2007.
7. Greene, D.R. et al., Soil isolation and molecular identification of *Coccidioides immitis*, *Mycologia*, 92, 406, 2000.
8. Elconin, A.F., Egberg, R.O., and Egberg, M.C., Significance of soil salinity on the ecology of *Coccidioides immitis*, *J. Bacteriol.*, 87, 500, 1964.
9. Fisher, F.S. et al., Coccidioides niches and habitat parameters in the southwestern United States: A matter of scale, *Ann. N.Y. Acad. Sci.*, 1111, 47, 2007.
10. Talamantes, J., Behseta, S., and Zender, C.S., Fluctuations in climate and incidence of coccidioidomycosis in Kern County, California: A review, *Ann. N.Y. Acad. Sci.*, 1111, 73, 2007.
11. Lacy, G.H. and Swatek, F.E., Soil ecology of *Coccidioides immitis* at Amerindian middens in California, *Appl. Microbiol.*, 27, 379, 1974.
12. Emmons, C.W., Isolation of *Coccidioides* from soil and rodents, *Public Health Rep. (Washington)*, 57, 109, 1942.
13. Eulalio, K.D. et al., *Coccidioides immitis* isolated from armadillos (*Dasypus novemcinctus*) in the state of Piaui, northeast Brazil, *Mycopathologia*, 149, 57, 2001.
14. Cordeiro, R.A. et al., Phenotypic characterization and ecological features of *Coccidioides* spp. from Northeast Brazil, *Med. Mycol.*, 44, 631, 2006.
15. Borghi, A.L. et al., *Coccidioides immitis*: Isolation from soil samples in San Luis and Mendoza provinces, *Sabouraudia*, 15, 51, 1977.
16. Rubinstein, H. et al., Isolation of human fungi from soil and identification of two endemic areas of *Cryptococcus neoformans* and *Coccidioides immitis*, *Rev. Inst. Med. Trop. Sao Paulo*, 31, 1, 1989.

17. Negroni, R., Historical evolution of some clinical and epidemiological knowledge of coccidioidomycosis in the Americas, *Rev. Argent. Microbiol.*, 40, 246, 2008.
18. Baird, R.W. et al., A fluffy white traveller: Imported *Coccidiodes immitis* infection in an Australian tourist, *Pathology*, 31, 47, 1999.
19. Osaki, T. et al., Pulmonary coccidioidomycosis that formed a fungus ball with 8-years duration, *Intern. Med.*, 44, 141, 2005.
20. Verghese, S. et al., Coccidioidomycosis in India: Report of a second imported case, *Med. Mycol.*, 40, 307, 2002.
21. Chandesris, M.O. et al., Coccidioidomycosis: An imported invasive fungal disease in France, *Med. Mal. Infect.*, 38, 336, 2008.
22. Galgiani, J.N. et al., Coccidioidomycosis, *Clin. Infect. Dis.*, 41, 1217–1223, 2005.
23. Laniado-Laborin, R., Expanding understanding of epidemiology of coccidioidomycosis in the Western hemisphere, *Ann. N.Y. Acad. Sci.*, 1111, 19, 2007.
24. McCurdy, S., Agricultural respiratory infections, *Am. J. Respir. Crit. Care Med.*, 158, S46, 1998.
25. Bissell, S.R. and Weiss, E.C., Increase in Coccidioidomycosis—California, 2000–2007, *MMWR Morb. Mortal. Wkly. Rep.*, 58, 105, 2009.
26. Schneider, E. et al., A coccidioidomycosis outbreak following the Northridge, Calif, earthquake, *JAMA*, 277, 904, 1997.
27. Pappagianis, D. and Einstein, H., Tempest from Tehachapi takes toll or *Coccidioides* conveyed aloft and afar, *West. J. Med.*, 129, 527, 1978.
28. Cordeiro, R.A. et al., Twelve years of coccidioidomycosis in Ceará State, Northeast Brazil: Epidemiologic and diagnostic aspects, *Diagn. Microbiol. Infect. Dis.*, 66, 65, 2010.
29. Cordeiro, R.A. et al., Serologic detection of coccidioidomycosis antibodies in northeast Brazil, *Mycopathologia*, 167, 187, 2009.
30. Chiller, T.M., Galgiani, J.N., and Stevens, D.A., Coccidioidomycosis, *Infect. Dis. Clin. North Am.*, 17, 41, 2003.
31. Crum, N.F. et al., Coccidioidomycosis: A descriptive survey of a reemerging disease. Clinical characteristics and current controversies, *Medicine (Baltimore)*, 83, 149, 2004.
32. Louie, L. et al., Influence of host genetics on the severity of coccidioidomycosis, *Emerg. Infect. Dis.*, 5, 672, 1999.
33. Logan, J.L., Blair, J.E., and Galgiani, J.N., Coccidioidomycosis complicating solid organ transplantation, *Semin. Respir. Infect.*, 16, 251, 2001.
34. Diógenes, M.J.N. et al., Inquérito epidemiológico com esferulina em Jaguaribara-CE, Brasil, 1993. *An. Bras. Dermatol.*, 70, 525, 1995.
35. Shubitz, L.F., Comparative aspects of coccidioidomycosis in animals and humans, *Ann. N.Y. Acad. Sci.*, 1111, 395, 2007.
36. Timm, K.I., Sonn, R.J., and Hultgren, B.D., Coccidioidomycosis in a Sonoran gopher snake, *Pituophis melanoleucus affinis*, *J. Med. Vet. Mycol.*, 26, 101, 1988.
37. Blair, J.E., Coccidioidomycosis in patients who have undergone transplantation, *Ann. N.Y. Acad. Sci.*, 1111, 365, 2007.
38. Charlton, V., Ramsdell, K., and Sehring, S., Intrauterine transmission of coccidioidomycosis, *Pediatr. Infect. Dis. J.*, 18, 561, 1999.
39. Arnold, C.A. et al., Unsuspected, disseminated coccidioidomycosis without maternofetal morbidity diagnosed by placental examination: Case report and review of the literature, *Clin. Infect. Dis.*, 46, e119, 2008.
40. Kohn, G.J. et al., Acquisition of coccidioidomycosis at necropsy by inhalation of coccidioidal endospores, *Diagn. Microbiol. Infect. Dis.*, 15, 527, 1992.
41. Gaidici, A. and Saubolle, M.A., Transmission of coccidioidomycosis to a human via a cat bite, *J. Clin. Microbiol.*, 47, 505, 2009.
42. Drutz, D.J. and Huppert, M., Coccidioidomycosis: Factors affecting the host-parasite interaction, *J. Infect. Dis.*, 147, 372, 1983.
43. Hung, C.Y., Xue, J., and Cole, G.T., Virulence mechanisms of coccidioides, *Ann. N.Y. Acad. Sci.*, 1111, 225, 2007.
44. Cutler, J.E. et al., Advances in combating fungal diseases: Vaccines on the threshold, *Nat. Rev. Microbiol.*, 5, 13, 2007.
45. Ampel, N.M., The complex immunology of human coccidioidomycosis, *Ann. N.Y. Acad. Sci.*, 1111, 245, 2007.
46. Yuan, L. and Cole, G.T., Isolation and characterization of an extracellular proteinase of *Coccidioides immitis*, *Infect. Immun.*, 55, 1970, 1987.
47. Nosanchuk, J.D. et al., *Coccidioides posadasii* produces melanin in vitro and during infection, *Fungal Genet. Biol.*, 44, 517, 2007.
48. Hung, C.Y., A metalloproteinase of *Coccidioides posadasii* contributes to evasion of host detection, *Infect. Immun.*, 73, 6689, 2005.
49. Guevara-Olvera, L. et al., Sequence, expression and functional analysis of the *Coccidioides immitis* ODC (ornithine decarboxylase) gene, *Gene*, 25, 242, 2000.
50. Schneider, E. et al., A coccidioidomycosis outbreak following the Northridge, Calif, earthquake, *JAMA*, 19, 277, 1997.
51. Cairns, L., Outbreak of coccidioidomycosis in Washington state residents returning from Mexico, *Clin. Infect. Dis.*, 30, 61, 2000.
52. Valdivia, L. et al., Coccidioidomycosis as a common cause of community-acquired pneumonia, *Emerg. Infect. Dis.*, 12, 958, 2006.
53. DiCaudo, D.J., Coccidioidomycosis: A review and update, *J. Am. Acad. Dermatol.*, 55, 929, 2006.
54. de Aguiar Cordeiro, R. et al., *In vitro* inhibitory effect of antituberculosis drugs on clinical and environmental strains of *Coccidioides posadasii*, *J. Antimicrob. Chemother.*, 58, 575, 2006.
55. Blair, J.E., Coccidioidal meningitis: Update on epidemiology, clinical features, diagnosis, and management, *Curr. Infect. Dis. Rep.*, 11, 289, 2009.
56. Arsura, E.L., Bobba, R.K., and Reddy, C.M., Coccidioidal pericarditis: A case presentation and review of the literature, *Int. J. Infect. Dis.*, 9, 104, 2005.
57. Brilhante, R.S. et al., Coccidioidal pericarditis: A rapid presumptive diagnosis by an in-house antigen confirmed by mycological and molecular methods, *J. Med. Microbiol.*, 57, 1288, 2008.
58. DiCaudo, D.J. et al., Sweet syndrome (acute febrile neutrophilic dermatosis) associated with pulmonary coccidioidomycosis, *Arch. Dermatol.*, 141, 881, 2005.
59. Parish, J.M. and Blair, J.E., Coccidioidomycosis, *Mayo Clin. Proc.*, 83, 343, 2008.
60. Galgiani, J.N., Practice guideline for the treatment of coccidioidomycosis. Infectious Diseases Society of America, *Clin. Infect. Dis.*, 30, 658, 2000.
61. Cortez, K.J., Walsh, T.J., and Bennett, J.E., Successful treatment of coccidioidal meningitis with voriconazole, *Clin. Infect. Dis.*, 36, 1619, 2003.
62. Anstead, G.M., Refractory coccidioidomycosis treated with posaconazole, *Clin. Infect. Dis.*, 40, 1770, 2005.

63. Catanzaro, A. et al., Safety, tolerance, and efficacy of posaconazole therapy in patients with nonmeningeal disseminated or chronic pulmonary coccidioidomycosis, *Clin. Infect. Dis.*, 45, 562, 2007.
64. Antony, S., Use of the echinocandins (caspofungin) in the treatment of disseminated coccidioidomycosis in a renal transplant recipient, *Clin. Infect. Dis.*, 39, 879, 2004.
65. Hsue, G., Treatment of meningeal coccidioidomycosis with caspofungin, *J. Antimicrob. Chemother.*, 54, 292, 2004.
66. González, G.M. et al., Therapeutic efficacy of caspofungin alone and in combination with amphotericin B deoxycholate for coccidioidomycosis in a mouse model, *J. Antimicrob. Chemother.*, 60, 1341, 2007.
67. Blair, J.E., Douglas, D.D., and Mulligan, D.C., Early results of targeted prophylaxis for coccidioidomycosis in patients undergoing orthotopic liver transplantation within an endemic area, *Transpl. Infect. Dis.*, 5, 3, 2003.
68. Sutton, D.A., Diagnosis of coccidioidomycosis by culture: Safety considerations, traditional methods, and susceptibility testing, *Ann. N.Y. Acad. Sci.*, 1111, 315, 2007.
69. Cordeiro, R.A. et al., *In vitro* activities of caspofungin, amphotericin B and azoles against *Coccidioides posadasii* strains from Northeast, Brazil, *Mycopathologia*, 161, 21, 2006.
70. Ramani, R. and Chaturvedi, V., Antifungal susceptibility profiles of *Coccidioides* immitis and *Coccidioides posadasii* from endemic and non-endemic areas, *Mycopathologia*, 163, 315, 2007.
71. Kriesel, J.D. et al., Persistent pulmonary infection with an azole-resistant *Coccidioides* species, *Med. Mycol.*, 46, 607, 2008.
72. Saubolle, M.A., Laboratory aspects in the diagnosis of coccidioidomycosis, *Ann. N.Y. Acad. Sci.*, 1111, 301, 2007.
73. Cordeiro, R.A. et al., *In vitro* synergistic effects of antituberculous drugs plus antifungals against *Coccidioides posadasii*, *Int. J. Antimicrob. Agents*, 34, 278, 2009.
74. Kaufman, L., Valero, G., and Padhye, A.A., Misleading manifestations of *Coccidioides immitis in vivo*, *J. Clin. Microbiol.*, 36, 3721, 1998.
75. DiTomasso, J.P. et al., Bronchoscopic diagnosis of pulmonary coccidioidomycosis. Comparison of cytology, culture, and transbronchial biopsy, *Diagn. Microbiol. Infect. Dis.*, 18, 83, 1994.
76. Muñoz, B. et al., Parasitic mycelial forms of *Coccidioides* species in Mexican patients, *J. Clin. Microbiol.*, 42, 1247, 2004.
77. Ampel, N.M. et al., An archived lot of coccidioidin induces specific coccidioidal delayed-type hypersensitivity and correlates with in vitro assays of coccidioidal cellular immune response, *Mycopathologia*, 161, 67, 2006.
78. Papagianis, D., Serologic studies in coccidioidomycosis, *Semin. Respir. Infect.*, 16, 242, 2001.
79. Blair, J.E. et al., Serologic testing for symptomatic coccidioidomycosis in immunocompetent and immunosuppressed hosts, *Mycopathologia*, 162, 317, 2006.
80. Murthy, M.H. and Blair, J.E., Coccidioidomycosis, *Curr. Fungal Infect. Rep.*, 3, 7, 2009.
81. Pappagianis, D. and Zimmer, B.L., Serology of coccidioidomycosis, *Clin. Microbiol. Rev.*, 3, 247, 1990.
82. Durkin, M. et al., Diagnosis of coccidioidomycosis with use of the *Coccidioides* antigen enzyme immunoassay, *Clin. Infect. Dis.*, 15, 47, 2008.
83. Johnson, S.M., Simmons, K.A., and Pappagianis, D. Amplification of coccidioidal DNA in clinical specimens by PCR, *J. Clin. Microbiol.*, 42, 1982, 2004.
84. de Aguiar, C.R. et al., Rapid diagnosis of coccidioidomycosis by nested PCR assay of sputum, *Clin. Microbiol. Infect.*, 13, 449, 2007.
85. Stockman, L. et al., Evaluation of commercially available acridinium ester-labeled chemiluminescent DNA probes for culture identification of *Blastomyces dermatitidis*, *Coccidioides immitis*, *Cryptococcus neoformans*, and *Histoplasma capsulatum*, *J. Clin. Microbiol.*, 31, 845, 1993.
86. Gromadzki, S.G. and Chaturvedi, V., Limitation of the AccuProbe *Coccidioides immitis* culture identification test: False-negative results with formaldehyde-killed cultures, *J. Clin. Microbiol.*, 38, 2427, 2000.
87. Pan, S and Cole, G.T., Molecular and biochemical characterization of a *Coccidioides immitis*-specific antigen, *Infect. Immun.*, 63, 3994, 1995.
88. Binnicker, M.J. et al., Detection of *Coccidioides* species in clinical specimens by real-time PCR, *J. Clin. Microbiol.*, 45, 173, 2007.
89. Umeyama, T. et al., Novel approach to designing primers for identification and distinction of the human pathogenic fungi *Coccidioides immitis* and *Coccidioides posadasii* by PCR amplification, *J. Clin. Microbiol.*, 44, 1859, 2006.
90. Tintelnot, K., Taxonomic and diagnostic markers for identification of *Coccidioides immitis* and *Coccidioides posadasii*, *Med. Mycol.*, 45, 385, 2007.
91. Burt, A. et al., A safe method of extracting DNA from *Coccidioides immitis*, *Fungal Genet. Newsl.*, 42, 23, 1995.
92. Castañón-Olivares, L.R. et al. Molecular identification of *Coccidioides* isolates from Mexican patients, *Ann. N.Y. Acad. Sci.*, 1111, 326, 2007.
93. Fisher, M.C., White, T.J., and Taylor, J.W., Primers for genotyping single nucleotide polymorphisms and microsatellites in the pathogenic fungus *Coccidioides immitis*. *Mol. Ecol.*, 8, 1082, 1999.
94. Jewell, K., Cheshier, R., and Cage, G.D., Genetic diversity among clinical *Coccidioides* spp. isolates in Arizona, *Med. Mycol.*, 46, 449, 2008.

28 *Cyphellophora*

Dongyou Liu

CONTENTS

28.1 INTRODUCTION

Cyphellophora is a dematiaceous fungal genus that is commonly present in soil, plants, and the environment. Of the 11 *Cyphellophora* species identified to date, *Cyphellophora pluriseptata* has been demonstrated to be an unusual cause of subcutaneous phaeohyphomycosis involving the ear of a healthy individual.

28.1.1 Classification and Morphology

The genus *Cyphellophora* is a dematiaceous (dark-walled) fungus belonging to the mitosporic Herpotrichiellaceae, family Herpotrichiellaceae, order Chaetothyriales, class Eurotiomycetes, subphylum Pezizomycotina, phylum Ascomycota, kingdom Fungi. The mitosporic Herpotrichiellaceae group covers 13 genera: *Blastacervulus*, *Brycekendrickomyces*, *Cladophialophora*, *Cladoriella*, *Cyphellophora*, *Exophiala*, *Heteroconium*, *Melanchlenus*, *Metulocladosporiella*, *Phaeococcomyces*, *Rhinocladiella*, *Thysanorea*, and *Veronaea*. Currently, the genus *Cyphellophora* consists of 11 recognized species: *Cyphellophora eucalypti*, *Cyphellophora eugeniae*, *Cyphellophora fusarioides* (obsolete synonym: *Pseudomicrodochium fusarioides*), *Cyphellophora guyanensis*, *Cyphellophora hylomeconis*, *Cyphellophora indica* (obsolete synonym: *Kumbhamaya indica*), *Cyphellophora laciniata*, *Cyphellophora pluriseptata*, *Cyphellophora suttonii* (obsolete synonym: *Pseudomicrodochium suttonii*), *Cyphellophora taiwanensis*, and *Cyphellophora vermispora* [1–7]. The type species of the genus is *Cyphellophora laciniata* [4].

Cyphellophora colonies are dark; hyphae (1.5–3 μm wide) are fertile, pale brown, at times constricted at septa. Conidiogenous cells are phialidic, intercalary, at times on short lateral branches, with a prominent to indistinct collarette. Conidia are sickle shaped, brown, smooth walled, 1- to 3-septate, adhering in bundles [4]. The genus *Cyphellophora* (incorporating the hyaline genus *Kumbhamaya*) is distinguished phenetically from *Pseudomicrodochium* by melanized versus hyaline thalli [8].

At the species level, *Cyphellophora guyanensis* (originally isolated from leaf litter in French Guyana) is characterized by its ampulliform to flask-shaped phialides, with a conspicuous, funnel-shaped collarette, and nearly straight to falcate or slightly sigmoid, 2- to 6-septate conidia [4].

Cyphellophora hylomeconis (named after its host genus *Hylomecon*) colonies are slow growing, reaching a diameter of 20 mm after 4 weeks at 25°C in the dark on potato dextrose agar (PDA), but 12 mm on synthetic nutrient-poor agar (SNA); colonies are fertile and slimy, with smooth, catenate margins, and without aerial mycelium; surface is crumpled and olivaceous black to iron gray. Hyphae (3–5 μm wide) are branched, greenish brown, septate, smooth, and constricted at septa. Conidiogenous cells (1 μm × 1.5–2 μm) are phialidic, intercalary, appearing denticulate, with minute collarettes (at times proliferating percurrently). Conidia (15–55 μm × 2.5–4 μm) are sickle shaped, smooth, medium brown, guttulate, 1- to 5-septate, and constricted at septa. Apex is subacutely rounded, base subtruncate, or with a slight constriction, giving rise to a subacutely rounded foot cell (1 μm × 0.5–1 μm).

Cyphellophora eugeniae (named after the host *Eugenia*) colonies are fertile, sporulating in slimy sporodochial masses, attaining a diameter of 15 mm after 4 weeks at 25°C. Colonies on PDA are erumpent, with sparse aerial mycelium and even margins; surface is olivaceous gray, with patches of iron gray; reverse is iron gray. Hyphae (3–5 μm wide) are branched, greenish brown, septate, smooth, and constricted at septa. Conidiogenous cells (1 μm wide) are phialidic, intercalary, inconspicuous to subdenticulate, with minute collarettes, with several loci aggregated at hyphal swellings. Conidia (40–90 μm × 2–3 μm) are subcylindrical, tapering toward obtuse ends, curved, smooth, hyaline to olivaceous, finely guttulate, 4- to 10-septate, and prominently constricted at septa.

Cyphellophora eucalypti (named after the host genus *Eucalyptus*) colonies attain a diameter of 4 cm after 2 weeks at 25°C in the dark, and are circular, flat, medium to dark brown; margin entire, consisting of dense, immersed mycelium. Aerial mycelium is loose, cottony, pale gray-brown (surface), appearing minutely orange-brown due to slimy conidial masses on mycelium, medium yellowish brown (reverse). Mycelium is dense, superficial, and partly immersed; hyphae (1.5–2.5 μm wide) are smooth, loosely septate, predominantly thin walled, branched, and hyaline to pale brown. Conidiogenous cells (5–12 μm × 3–5 μm) are intercalary or terminal on erect hyphal branches, solitary, subcylindrical to pyriform or lageniform, straight to slightly curved, pale to medium brown, thick walled, smooth, proliferating percurrently, with one to two annellations, and funnel-shaped collarettes, constricted and somewhat darkened and thickened below the collarette. Conidia (8–25 μm × 2–3 μm) are clavate and 1-septate when young, becoming slightly sigmoid fusiform, 1- to 3-septate, hyaline to pale brown, apex obtusely rounded, base minutely truncate or slightly protruding, thin walled, slightly thickened along the rim, refractive, and aggregating in a slimy mass.

Among the species of the genus *Cyphellophora*, conidia of *C. fusarioides* and *C. laciniata* are 1- to 3-septate, while those of other *Cyphellophora* species are >3 septate; *C. fusarioides* is distinguished from *C. laciniata* by having narrower conidia (2.5 μm versus 5 μm). *C. vermispora* is closely related to *C. suttonii* and *C. fusarioides*, and has vermiform, mostly curved conidia. *C. guyanensis* is related to the type species *C. laciniata*. *Cyphellophora hylomeconis* is the first species of the genus infecting a living plant host. *C. hylomeconis* is distinguished on the basis of its conidial dimensions and septation, with its conidia (15–55 μm × 2.5–4 μm) being larger than those of *C. fusarioides* (11–20 μm × 2–2.5 μm, 1- to 2-septate) and those of *C. laciniata* (11–25 μm × 2–5 μm, 1- to 3-septate) [5]. *C. eugeniae* conidiogenous loci are similar to those of *C. taiwanensis*; however, *C. eugeniae* is distinguished by having much longer conidia [6]. *C. eucalypti* is similar to *C. indica* and *C. pluriseptata*, as well as *C. laciniata*; but it is distinguished by having 1- to 3-septate conidia, with an average length of 15–20 μm [4].

28.1.2 Clinical Features

The genus *Cyphellophora* is one of the 70 genera (containing about 130 species) of dematiaceous fungi linked to the etiology of phaeohyphomycosis in humans. Characterized by the ability of etiologic agents to develop in their host's tissues as dark walled, septate mycelial elements, phaeohyphomycosis represents a class of mycoses that produce four distinct clinical diseases: (1) superficial; (2) cutaneous/corneal including dermatomycosis, onychomycosis, and mycotic keratitis; (3) subcutaneous; and (4) systemic [9,10]. Occurring in the upper and lower limbs, as well the buttocks, face, neck, and scalp, subcutaneous phaeohyphomycosis (improperly named phaeohyphomycotic cyst) often appears as solitary subcutaneous cysts or abscesses, firm to fluctuant [11,12]. However, it may induce solid subcutaneous masses or red indurated lesions in patients with debilitating diseases or immunosuppressive status. Other less common clinical forms of subcutaneous phaeohyphomycosis include keratotic plaques, pustules, sinus tracts, ulcers, crusts, eczematous rash, papulonodular, exophytic, or verrucous [9]. Histopathologically, subcutaneous phaeohyphomycosis may show (1) suppurative and granulomatous processes in the dermis associated with epidermal hyperplasia and epidermal abscesses; (2) intradermal multiloculated cysts lined by granulomas, with abscesses, and a normal epidermis; and (3) dermal unilocular cysts containing neutrophils and lined by granulomas, with normal epidermis [13,14]. Many *Cyphellophora* species are involved in human cutaneous/subcutaneous phaeohyphomycoses.

Bittencourt et al. [14] documented the first case of subcutaneous phaeohyphomycosis due to *Cyphellophora pluriseptata* in a 38-year-old Brazilian male. He presented a 10-year history of lesions over the left ear. Hematoxylin–eosin-stained sections of biopsy samples showed brown elongated or ovoid structures surrounded by a halo inside giant cells, and the Grocott stain revealed hyphae of different lengths, irregularly swollen to toruloid and infrequently branched. Culture of the biopsy samples on Sabouraud agar yielded rapid-growing, velvety, and dark gray colonies. Microscopic examination of the isolate demonstrated smooth-walled, pale-brown hyphae with terminal, intercalary, or lateral phialides, with a conspicuous collarette at the tip. Conidia were pale brown, cylindrical to fusiform, with rounded ends, 1- to 5-septate, straight, or slightly curved. The fungus was identified as *Cyphellophora pluriseptata*. After treatment with itraconazole and amphotericin B, the lesion showed remarkable regression.

28.1.3 Diagnosis

Microscopic examination of biopsy specimens for fungal elements (e.g., septated hyphae, irregularly swollen to toruloid, branched or unbranched; yeast-like cells that bud singly or in chains; or spherical to oval cells [sclerotic cells] that reproduce by septation in only one plane) using various stains represents a valuable approach for the diagnosis of phaeohyphomycosis and chromoblastomycosis. However, species-specific identification of the organisms involved is not possible in most cases based on direct microscopy.

Therefore, clinical samples are routinely inoculated on mycological media for the isolation of dematiaceous fungi. Subsequent assessment of macro- and microscopic characteristics of fungal isolates allows determination of their precise species identity. Because culture-based fungal identification is notoriously time consuming and technically challenging, molecular methods such as PCR and sequencing analysis have been increasingly adopted in clinical laboratories for rapid, specific, and sensitive detection and identification of dematiaceous fungi from cultured isolates as well as noncultured specimens [5,6,15,16].

28.2 METHODS

28.2.1 Sample Preparation

Biopsy samples are first examined under microscope for mycotic elements using various stains. Samples are also cultured on inhibitory mold agar, modified Sabouraud agars, or PDA. Colony colors (surface and reverse) are assessed after 2–4 weeks on different media at 25°C in the dark. Isolates are cultured on 2% malt extract agar plates (MEA) by obtaining single conidial colonies. Colonies are subcultured onto fresh MEA, oatmeal agar (OA), PDA, and synthetic nutrient-poor agar (SNA), and incubated at 25°C under continuous near-ultraviolet light to promote sporulation. Microscopic structures are observed on tease or tape preparations and slide cultures for up to 21 days [5,6].

Genomic DNA is isolated from fungal mycelium grown on MEA, using the UltraClean? Microbial DNA Isolation Kit (Mo Bio Laboratories) according to the manufacturer's protocols. Alternatively, genomic DNA is extracted following the CTAB-based protocol [5,6].

28.2.2 Detection Procedures

28.2.2.1 Real-Time PCR and Sequencing Analysis

Pounder et al. [15] described a real-time PCR with SYBR green DNA-binding dye and amplicon melting temperature analysis for fungal detection using pan-fungal primers internal transcribed spacer 1 (ITS1) forward (5′-TCCGTAGGTGAACCTGCGG-3′) and ITS4 reverse (5′-TCCTCCGCTTATTGATATGC-3′). The identity of the fungi is verified by subsequent sequencing analysis.

Procedure

1. PCR mixture is composed of 1× Lightcycler FastStart DNA Master Hybridization Probes mixture (Roche Applied Science) (containing deoxynucleoside triphosphates, FastStart *Taq* DNA polymerase, and 1 mM $MgCl_2$ (additional $MgCl_2$ is added to a final concentration of 4.6 mM), 0.4 μM each of ITS1 forward and ITS4 reverse primers, 1× SYBR green (Molecular Probes), and 3 μL template DNA.
2. Thermal cycling parameters include 95°C for 10 min; 50 cycles of 95°C for 5 s, 60°C for 20 s, and 76°C for 30 s; and a final extension at 72°C for 2 min.
3. The quality of the amplicon is determined using the derivative of the melt analysis curve (55°C–99°C, 45 s hold at 55°C, 5 s/°C) using the RotorGene 3000 (Corbett Robotics, Inc.).
4. The amplified product is purified for bidirectional sequencing using ExoSAP-IT (USB Corp.). Five microliters of Big Dye Terminator Ready Reaction Mix v. 1.1 (Applied Biosystems) is added to 4 μL of each primer (0.8 pmol/μL) and 3 μL of the purified PCR product. Cycle sequencing is performed with a 9700 thermal cycler (ABI) using 25 cycles of 96°C for 10 s, 50°C for 5 s, and 60°C for 4 min. Sequencing reaction products are passed through a Sephadex G-50 fine column to remove unincorporated dye terminators. Purified sequencing reaction products are run on an ABI Prism 3100 Genetic Analyzer with a 50 cm capillary array.
5. Sequences are analyzed with the SmartGene Integrated Database Network software version 3.2.3 vr. SmartGene is a web-based software and database system with reference sequences derived from the National Center for Biological Information (NCBI) GenBank repository.

Note: In case that real time PCR instrument is unavailable, standard PCR may be performed with primers ITS1 and ITS4, and the resulting amplicon is sequenced with the same primers. Sequence-based identifications are defined by percent identity: species, ≥99%; genus, 93%–99%; and inconclusive, ≤93%.

28.2.2.2 Standard PCR and Sequencing Analysis

Crous et al. [5] utilized primers V9G and LR5 (5′-TCCTGAGGGAAACTTCG-3′) to amplify a 1700 bp fragment covering part (ITS) of the nuclear rDNA operon spanning the 3′ end of the 18S rRNA gene (small subunit), the first ITS1, the 5.8S rRNA gene, the second ITS region, and the first 900 bases at the 5′ end of the 28S rRNA gene (large subunit). ITS4 (5′-TCCTCCGCTTATTGATATGC-3′) and LR0R (5′-ACCCGCTGAACTTAAGC-3′) are employed as internal primers for sequencing to ensure good-quality overlapping sequences. Alignment gaps are treated as new character states. The ITS sequences are compared with those sequences available in NCBI's GenBank nucleotide database using a megablast search.

28.3 CONCLUSION

The genus *Cyphellophora* consists of 11 dematiaceous fungal species, many of which are causative agents for cutaneous/subcutaneous phaeohyphomycosis in human hosts. Currently, diagnosis of phaeohyphomycosis relies on microscopic observation of fungal elements in biopsy specimens with the help of various stains. Further species-specific determination is contingent on the isolation of the fungal organisms for subsequent assessment of their distinct macroscopic and microscopic features. Due to the time-consuming and technically challenging nature of culture-based procedures, PCR and sequencing analysis of small and large subunit rRNA genes as well as ITS regions provides a rapid, specific, and sensitive alternative for identification of dematiaceous fungi including *Cyphellophora* spp. from both cultured isolates and noncultured specimens.

REFERENCES

1. The UniProt Consortium. Available at http://www.uniprot.org/, accessed on August 1, 2010.
2. De Vries, G.A., Elders, M.C.C., and Luykx, M.H.F., Description of *Cyphellophora pluriseptata* sp. nov. *Anton Leeuwen*. 1986;52:141–143.
3. Walz, A. and de Hoog, G.S., A new species of *Cyphellophora*. *Anton Leeuwen*. 1987;53:143–146.
4. Decock, C. et al., A new species and three new combinations in *Cyphellophora*, with a note on the taxonomic affinities of the genus, and its relation to *Kumbhamaya* and *Pseudomicrodochium*. *Anton Leeuwen*. 2003;84(3):209–216.
5. Crous, P.W. et al., Opportunistic, human-pathogenic species in the *Herpotrichiellaceae* are phenotypically similar to saprobic or phytopathogenic species in the *Venturiaceae*. *Stud Mycol*. 2007;58:185–217.
6. Crous, P.W. et al., Phylogeny and taxonomy of obscure genera of microfungi. *Persoonia*. 2009;22:139–161.
7. Cheewangkoon, R. et al., Myrtaceae, a cache of fungal biodiversity. *Persoonia*. 2009;23:55–85.
8. Kwon-Chung, K.J. and Bennett, J.E., *Medical Mycology*, Lea & Febiger, Philadelphia, PA, 1992.
9. McGinnis, M.R., Chromoblastomycosis and phaeohyphomycosis: New concepts, diagnosis, and mycology. *J Am Acad Dermatol*. 1983;8:1–16.
10. Rinaldi, M.G., Phaeohyphomycosis. *Dermatol Clin*. 1996;14:147–153.
11. Ziefer, A. and Connor, D.H., Phaeohyphomycotic cyst. A clinicopathologic study of twenty-five patients. *Am J Trop Med Hyg*. 1980;29:901–911.
12. O'Donnell, P.J. and Hutt, M.S.R., Subcutaneous phaeohyphomycosis: A histopathological study of nine cases from Malawi. *J Clin Pathol*. 1985;38:288–292.
13. Ronan, S.G. et al., Primary cutaneous phaeohyphomycosis: Report of seven cases. *J Cutan Pathol*. 1993;20:223–228.
14. Bittencourt, A.L. et al., Discovering potential pathogens among fungi identified as nonsporulating subcutaneous phaeohyphomycosis caused by *Cyphellophora pluriseptata*. *Eur J Dermatol*. 2002;12:103–106.
15. Pounder, J.I. et al., Discovering potential pathogens among fungi identified as nonsporulating molds. *J Clin Microbiol*. 2007;45:568–571.
16. Bagyalakshmi, R. et al., Newer emerging pathogens of ocular non-sporulating molds (NSM) identified by polymerase chain reaction (PCR)-based DNA sequencing technique targeting internal transcribed spacer (ITS) region. *Curr Eye Res*. 2008;33:139–147.

29 Emmonsia

Dongyou Liu and R.R.M. Paterson

CONTENTS

29.1 INTRODUCTION

Members of the genus *Emmonsia* are filamentous soil saprophytes that have been linked to a pulmonary fungal infection termed adiaspiromycosis in small mammals and occasionally in humans including both immunocompetent and immunocompromised individuals. *Emmonsia* conidia gain entry into the lung via inhalation, where they develop without replication into a type of spherule termed an adiaspore, with clinical symptoms ranging from cough, dyspnea, low-grade fever, to negligible manifestations. Given their morphologic similarities, *Emmonsia* can be easily misidentified as *Coccidioides*, *Blastomyces*, *Histoplasma*, or *Cryptococcus*.

29.1.1 Classification, Morphology, and Biology

The genus *Emmonsia* (obsolete synonym: *Haplosporangium*) belongs to the mitosporic Onygenales group, order Onygenales, class *Eurotiomycetes*, subphylum Pezizomycotina, phylum Ascomycota, kingdom Fungi. The group consists of nine genera: *Blastomyces*, *Chrysosporium*, *Coccidioides*, *Emmonsia*, *Geomyces*, *Locazia*, *Malbranchea*, *Myriodontium*, and *Paracoccidioides*. Currently, *Emmonsia* consists of three recognized species, *Emmonsia crescens* (synonyms: *Emmonsia parva* var. *crescens* and *Chrysosporium parvum* var. *crescens*), *Emmonsia parva* (synonyms: *Emmonsia parva* var. *parva*, *Chrysosporium parvum*, and *Haplosporangium parvum*), and *Emmonsia pasteuriana*, and nine unassigned species [1]. The type species of the genus is *E. parva* [2].

Emmonsia crescens (an agent of adiaspiromycosis), *Blastomyces dermatitidis* (the agent of blastomycosis), and *Histoplasma capsulatum* (the agent of histoplasmosis) are known to form teleomorphs, associated with the ascomycete genus *Ajellomyces* (Onygenaceae, Onygenales). However, no sexual stage is known for *Emmonsia parva* [3,4]. A recent phylogenetic study indicated that the genus *Emmonsia* forms part of a clade together with the genera *Ajellomyces*, *Blastomyces*, *Histoplasma*, and *Paracoccidioides* [2]. While *E. crescens* isolates are placed into two phylogenetic groups that correlate with their continents of origin (Eurasia and North America), *E. parva* isolates are more diverse: *B. dermatitidis* strains form a sister species of *E. parva* and *Histoplasma capsulatum* is also a close relative. *Paracoccidioides brasiliensis* and *Histoplasma capsulatum* are ancestral to most *Emmonsia* isolates [4].

Emmonsia grows moderately. Colonies are glabrous to velvety in texture, and are white, with buff to pale brown centers from the top, and cream to pale brown on the reverse. Hyphae are hyaline and septate; conidiophores are simple or occasionally branched; unicellular, thin-walled, subspherical aleurioconidia (2–5 μm × 2–4 μm) are sessile or located on slender stalks, occurring in a solitary manner or in two- to three-celled chains. On blood or brain–heart infusion agar at 37°C–40°C and in vivo, thick-walled, large, liberated conidia are produced. These are adiaspores, which are clamydospore-like and nonreproducing.

E. crescens adiaspores are multinucleate (up to a few hundred nuclei), reaching diameter of 70 μm in vitro at 37°C and 700 μm in vivo; *Emmonsia parva* adiaspores are uninucleate, reaching diameters of 10–25 μm in vitro at 40°C and 40 μm in vivo. In contrast, *Emmonsia pasteuriana* does not produce adiaspores, but forms budding cell-like structures on brain–heart infusion agar at 37°C.

In infected lung tissues, inhaled conidia (2–4 μm) enlarge in the alveoli to become adiaconidia (up to 700 μm in diameter), which are surrounded by small granulomes, with concentric layers of fibrous tissue in later stages of the infection. Adiaspores do not replicate, and become calcified at the primary implantation site, leading to a minimal localized reaction in the host tissue. However, the regular pulmonary functions may be hampered by the presence of adiaspores.

Emmonsia is a cosmopolitan filamentous fungus that is present in soil and small mammals such as rodents that act as the reservoir host of the fungus. *Emmonsia* spp. are endemic in southwestern United States, Australia, and eastern Europe, and are responsible for a pulmonary infection known as adiaspiromycosis in rodents, other small mammals, and occasionally in humans [5–9]. Human aspiromycosis typically occurs after inhalation of conidia of *Emmonsia*. While *Emmonsia crescens* is isolated primarily from humans, *Emmonsia parva* is often isolated from animals. *Emmonsia pasteuriana* was reported in an HIV-infected patient with a cutaneous disseminated infection.

29.1.2 Clinical Features and Epidemiology

Adiaspiromycosis is a lung disease of humans and small animals resulting from the inhalation of *Emmonsia* conidia, which elicit a robust, multicellular immunologic response in tissue against the growing conidia, leading to noncaseating granulomas and disturbance of normal lung function. The first case of human adiaspiromycosis due to *Emmonsia* was described in France in 1964. Subsequently, cases of human pulmonary adiaspiromycosis have been reported in Russia, the Czech Republic, Brazil, Honduras, Guatemala, and the United States, with disseminated infection occurring in immunocompromised persons [10–21].

The establishment of adiaspiromycosis appears to be dose dependent, as a small inoculum often produces no or mild clinical symptoms and scattered radiological pathology, while a heavy or repeated inoculum may develop an acute, sometimes severe, pulmonary disease (often referred to as primary progressive pulmonary adiaspiromycosis, disseminated pulmonary adiaspiromycosis, or acute pulmonary adiaspiromycosis), with diffuse granulomatous lesions in both lungs. Cleaning or playing in closed environments inhabited only by bats and mice, where conidia of *Emmonsia* spp. are present in soil and dust, represents a risk factor for human adiaspiromycosis [22].

Although most human infections are attributable to *E. crescens*, a disseminated disease caused by *E. parva* in an AIDS patient has been described. *E. crescens* tends to form larger adiaspores and has a broader host range and geographic distribution than *E. parva*.

de Almeida Barbosa et al. [22] documented three Brazilian cases of acute pulmonary adiaspiromycosis due to *Emmonsia crescens*. The first case involved a 52-year-old man, who presented with cough, mucoid expectoration, fever, and weight loss over the previous 5 months. A chest roentgenogram showed diffuse micronodular infiltrates. Hematoxylin and eosin (H&E) staining of the transbronchial biopsy sections revealed helminth-like structures. Sections of lung biopsy revealed granulomas and characteristic adiaconidia as stained with fungal stains (periodic acid-Schiff [PAS] and Gomori-Grocott metheanamine-silver). The patient improved symptomatically without treatment. The second case concerned a 47-year-old man with nonproductive cough, dyspnea, fever, and weight loss. A chest roentgenogram revealed diffuse interstitial infiltrates in both lungs. H&E-stained sections of a transbronchial biopsy showed two granulomas, one of which contained an adiaconidium; sections stained by Gomori methenamine silver demonstrated a fungal structure of 100 μm in diameter. The patient recovered without treatment. The third case involved a 43-year-old man with a 1-month history of fever (39.5°C), cough with scant expectoration, weight loss (3 kg), dyspnea, and thoracic pain. A chest roentgenogram revealed bilateral micronodular infiltrates. H&E-stained sections of transbronchial biopsy demonstrated granulomas with characteristic adiaconidia. The patient had disrupted two mice nests, in a farm, 14 days before becoming ill. Treatment with itraconazole (200 mg/day) and prednisone (30 mg/day) alleviated the patient's symptoms in a week, possibly reflecting the fact that enlargement of *Emmonsia* conidia causes a localized inflammatory response and *Emmonsia* conidia do not replicate and disseminate.

Nuorva et al. [23] described a case of disseminated bilateral pulmonary adiaspiromycosis in a 2-year-old Finnish girl. Electron microscopy showed that the three layers of the spore wall were not typical; rather, there seemed to be a gradual transition between the main wall zones, possibly split into thin layers. Varying numbers and thicknesses were seen with different staining methods and in different spores. This patient had become infected as a result of contact with soil dust containing the spores in the yard surrounding her home, and as a result of her mother's work in a large garden shop. She recovered from this rare infection after treatment with amphotericin B.

Dot et al. [24] presented a case of pulmonary adiaspiromycosis in a 30-year-old man with a 2-week history of unproductive cough, fever at 39°C, progressive dyspnea, thoracic pain, generalized weakness, and weight loss of 10 kg in 1 month. The patient became sick a few weeks after having played in the surroundings of an animal burrow, which may have functioned as a reservoir for *Emmonsia*. The chest radiography showed diffuse bilateral interstitial pneumonia with a micronodular pattern, and computerized tomography of the chest demonstrated disseminated pulmonary nodules. Formalin-fixed, paraffin-embedded transbronchial biopsy sections (4 μm) stained with PAS revealed round or ovoid, occasionally elliptical adiaspores (50–100 μm) with a thick wall in the center of some granulomas. Culture of the bronchoalveolar fluid (BAL) and transbronchial biopsy samples remained negative after 4 weeks. Sequence analysis of PCR amplicon from BAL specimen using panfungal primers indicated *Emmonsia crescens* as the culprit. Treatment with oral itraconazole (200 mg/day) for 40 days improved his respiratory status and stabilized the pulmonary lesions.

Besides pulmonary adiaspiromycosis, Mendes et al. [25] described an outbreak of conjunctivitis involving 99 children in Brazil, with symptoms ranging from photophobia (57%), ocular pain (42%), to blurred vision (40%). Unlike conjunctivitis caused by bacterial or viral pathogens, neither purulent conjunctival discharge nor hemorrhage was observed, and family members of case-patient households were not commonly affected. Furthermore, the disease was characterized by unusual, single or multiple, white, opaque scleral nodules, often with hyperemia or local edema, and in some cases with opacification extending to the limbus, or angular corneal opacities and anterior uveitis with granulomas in the anterior chamber. Microscopy of scleral biopsy samples identified spherical shapes, thick walls, and vacuous central areas with lack of organized, internal structures consistent with adiaconidia of *Emmonsia* sp. (in contrast to *Coccidiodes immitis* spores that contain internal microsporules). Conjunctival irritation was likely due to exposure to *Emmonsia* conidia in dust caused in part by dry environmental conditions. Symptomatic children responded to corticosteroid treatment.

29.1.3 Diagnosis

The diagnosis of acute pulmonary adiaspiromycosis is based on the following criteria: (1) an acute onset of respiratory and systemic symptoms, (2) a radiological picture of diffuse micronodular lesions in both lungs, and (3) histopathological demonstration of the adiaconidia in the granulomas [26]. In particular, microscopic observation of large, round or oval, thick-walled adiaspores in histopathological sections of tissue using PAS or other stains provides a key diagnostic feature for adiaspiromycosis. Adiaspores are thick walled, enlarged conidia that may be uninucleate (*E. parva*) or multinucleate (*E. crescens*). Culture of *Emmonisia* spp. from sputum or BAL is not possible because adiaconidia do not multiply and remain trapped within lung granulomas.

Emmonisia spp. are differentiated from *Blastomyces dermatitidis*, *Histoplasma capsulatum*, and *Paracoccidioides brasiliensis* by their inability to convert to a yeast phase at 37°C. They are differentiated from *Sporotrichum* by their inability to produce large chlamydospores at 25°C. *Emmonisia* spores in tissue (i.e., adiaspores or adiaconidia) resemble the parasitic spherules of *Coccidioides immitis* but differ by lacking internal spores.

The variable domain D2 located in the 5′ end of large subunit (LSU) nuclear rRNA contains sufficient nucleotide differences for discrimination between *Emmonisia* and its sibling *Blastomyces* species [27]. The other region of diagnostic interest is the internal transcribed spacer (ITS) region of rRNA (covering the ITS1, ITS2, and 5.8S rRNA), which offers a suitable target for examining *Emmonsia* species variability and for assessing evolutionary relationships among several related taxa.

29.2 METHODS

29.2.1 Sample Preparation

Biopsy specimens are stained with PAS stain and examined microscopically for adiaconidia-containing granulomas. Portions of the samples are cultured on Sabouraud agar with cycloheximide (0.5 mg/mL) at 30°C [24].

Emmonsia isolates are revived from either freeze dried or frozen (vapor phase of liquid nitrogen) stock and grown at 25°C on petri plates containing pablum cereal agar for 14–21 days. Blocks (1 cm × 1 cm) of mycelium and agar from cultures of *Emmonsia* species are excised from the culture plates and transferred to sterile snap-cap polypropylene tubes (12 mm × 75 mm; Fisher Scientific). The mycelial blocks are freeze dried using an Edwards Moduylo freeze-dryer [4].

Freeze-dried blocks of agar and *Emmonsia* mycelium (100 mg) are placed in 1.5 mL tubes and ground to a fine powder by using a 200 μL capacity pipette tip. The fungal material is rehydrated with 500 μL of DNA extraction buffer (50 mM Tris, 10 mM EDTA, 1% sarcosyl, pH 8.0) with gentle agitation for 10 min. An equal volume of 1:1 chloroform–phenol is added to each tube, and mixed by shaking for 20 min. Then, the aqueous and organic phases are separated in a microcentrifuge for 5 min at 14,000 × *g*. The aqueous phase is pipetted into a clean tube, and 0.1 volume of 3 M sodium acetate pH 6.0 and 1.3 volumes of ethanol are added. The tube is sealed, and the contents are mixed by inverting the tube several times. Precipitated nucleic acids are pelleted by centrifugation at 14,000 × *g* for 1 min. Ethanol is decanted, and the pellet is dried by inverting the tube over an absorbent paper for 5 min. Nucleic acids are dissolved in 100 μL of TE buffer (10 mM Tris, 1 mM EDTA, pH 8.0), and 250 μL of a saturated NaI solution and 10 μL of glass milk (GENECLEAN kit, Bio 101) are added to the tube. The tube is inverted periodically, and DNA is adsorbed onto the glass milk for 20 min. The glass milk is pelleted and rinsed, and the genomic DNA is eluted into 50 μL of 1/10-strength TE. DNA is stored at −20°C until used [4]. Alternatively, *Emmonsia* genomic DNA can be prepared by using Whatman FTA filter papers [28].

29.2.2 Detection Procedures

29.2.2.1 Sequencing Analysis of ITS Regions

Peterson and Sigler [4] used primers ITS1 (5′-TCCGTAGGTGAACCTGCGG-3′) and D2R to amplify a 1200-nucleotide fragment including ITS1, ITS2, 5.8S rRNA, and part of the 28S rRNA for sequencing analysis of *Emmonsia* and related organisms.

Procedure

1. PCR mixture (100 μL) is composed of 1 μL of genomic DNA, 10 μL of 10× buffer, 1 μL of 50 μM primer ITS1, 1 μL of 50 μM primer D2R, 0.5 μL (2.5 U) of *Taq* polymerase, 10 μL of dNTP (1 mM concentration), and 76.5 μL of sterile deionized water.

2. Amplification is conducted with 30 cycles of 96°C for 30 s, 53°C for 30 s, and 72°C for 2.5 min followed by 10 min at 72°C. The amplified fragment is purified by using a GENECLEAN kit (Bio 101) and eluted into 1/10-strength TE. The purified fragment is stored at −20°C until used in sequencing.
3. Sequencing is performed by using primers ITS1, ITS2 (5′-GCTGCGTTCTTCATCGATGC-3′), ITS3 (5′-GCATCGATGAAGAACGCAGC-3′), ITS4 (5′-TCCTCCGCTTATTGATATGC-3′), D1, D1R, D2, and D2R and Applied Biosystems DyeDeoxy sequencing kits. The sequencing reaction mixture is prepared with 1.5 pmol of primer, 200–400 ng of the purified DNA fragment, and sequencing reagents. DNA sequences are analyzed on an ABI 373-automated DNA sequencer.
4. Alignment of overlapping and reverse complement strands to form a consensus sequence for each isolate is accomplished with an ASCII text editor. Alignment of sequences from different isolates is performed visually also by using an ASCII text editor. Phylogenetic relationships of isolates are determined by using PAUP 3.1.1. or programs from the PHYLIP package.

29.2.2.2 Panfungal PCR and Sequencing Analysis

Dot et al. [24] used panfungal primers His3 and His4 for PCR amplification and sequencing identification of *Emmonsia* and related genera. The primer His3 (5′-GTCGTAACAAGGTTTCCGTAG-3′) is located at the end of 18S rRNA at positions 9–29, and His4 (5′-AGCGGGTATCCCTACCTGAT-3′) is located at the beginning of 28S at positions 601–620 according to the sequence of *Histoplasma capsulatum* U18363.

Procedure

Genomic DNA is extracted from the BAL by using the High Pure PCR Template Preparation Kit (Roche Diagnostics).

1. PCR mixture (50 μL) is composed of 2 U of FastStart *Taq* DNA polymerase (Roche Diagnostics), 5 μL of 10× PCR buffer-$MgCl_2$ (20 mM), 20 pmol of each primer, 200 μM of dATP, dGTP, dTTP, and dCTP, and 5 μL of DNA.
2. Amplification is conducted in an iCycler IQ Thermal Cycler (Bio-Rad) with initial denaturation at 94°C for 3 min; 30 cycles at 94°C for 1 min, 58°C for 40 s, and 72°C for 1 min; and final extension at 72°C for 5 min.
3. PCR product (5 μL) is electrophoresed in a 2% agarose gel in the presence of ethidium bromide and visualized under UV light. The expected PCR product size is 613 bp. If the band is faint, reamplification of the product may be performed by using the same primers.
4. After purification, the PCR product is sequenced with the same PCR primers using the BigDye Terminator v1.1 Cycle Sequencing Kit (Applied Biosystems). Sequencing products are analyzed on an automated sequencer ABI Prism 3100 genetic analyzer. The BLAST analysis is used to compare the sequence with those at GenBank database.

29.2.2.3 Sequencing Analysis of LSU rRNA Gene

Untereiner et al. [2] utilized primers WNS9 and LR5 (5′-ATCCTGAGGGAAACTTC-3′) to amplify a DNA fragment that extends from the 3′ end of the nuclear small subunit rRNA gene to approximately 1000 bp positions downstream from the 5′ end of the nuclear LSU rRNA gene. The amplicon is purified using a QIAquick PCR Purification Kit (Qiagen). Sequencing reactions are performed using a Prism Dye Terminator Cycle Sequencing Ready Reaction Kit (Applied Biosystems) and primers 5.8SR, LR1, WITS3, and WNS9. Excess dye terminators are removed by centrifugation using Centrisep columns (Princeton Separations) before analysis employing an Applied Biosystems 373A or 377 DNA sequencer.

Sequences are edited and assembled into larger consensus sequences using Sequencher 3.0 software (Gene Codes Corp.). Multiple alignments are produced using Clustal X version 1.7. The final multiple alignments are adjusted manually after visual inspection and areas of sequence ambiguity are eliminated. Phylogenetic relationships are inferred from aligned sequences using the maximum parsimony (MP) method found in PAUP (beta version 4.0b10). Gaps are treated as missing in all analyses. Heuristic searches are performed employing tree bisection–reconstruction (TBR) branch swapping with the MulTrees and steepest descent options activated, or using 1000 random addition sequence replicates.

29.3 CONCLUSIONS

Members of the genus *Emmonsia* are saprophytic fungi that are commonly present in soil and small mammals (e.g., rodents, otters, ground squirrels, goats, dogs, hedgehogs, raccoons, horses, and beavers). The teleomorphs of the genus *Emmonsia* genera with the *Blastomyces* and *Histoplasma* are found in the genus *Ajellomyces*.

E. crescens and *E. parva* spores are capable of entering the alveoli of mammalian hosts through inhalation, which expand to become adiaspores (adiacanidia), and cause a granulomatous inflammatory reaction, leading to a pulmonary disease known as adiaspiromycosis in animals and occasionally in humans. Most human adiaspiromycosis cases are due to *E. crescens*. Although human adiaspiromycosis often regresses spontaneously, sometimes the disease may persist, resulting in pneumonia or fatal respiratory failure [29,30].

Because *Emmonsia* spp. are not easily cultured, diagnosis of adiaspiromycosis has relied on histological examination of biopsies for large, round or oval, thick-walled adiaspores using PAS or other stains. Use of molecular techniques, especially sequencing analysis of the D2 region of the LSU rRNA gene and the ITS regions of rRNA biology, provides a valuable approach for improving the detection and diagnosis of human adiaspiromycosis [31].

REFERENCES

1. The UniProt Consortium. Available at http://www.uniprot.org/, accessed on August 1, 2010.
2. Untereiner, W.A. et al., The *Ajellomycetaceae*, a new family of vertebrate-associated *Onygenales*. *Mycologia*. 2004; 96:812–821.
3. Sigler, L., *Ajellomyces crescens* sp. nov., taxonomy of *Emmonsia* spp., and relatedness with *Blastomyces dermatitidis* (teleomorph *Ajellomyces dermatitidis*). *J Med Vet Mycol*. 1996;34(5):303–314.
4. Peterson, S.W., Sigler, L., Molecular genetic variation in *Emmonsia crescens* and *Emmonsia parva*, etiologic agents of adiaspiromycosis, and their phylogenetic relationship to *Blastomyces dermatitidis* (*Ajellomyces dermatitidis*) and other systemic fungal pathogens. *J Clin Microbiol*. 1998;36(10):2918–2925.
5. Krivanec, K., Otcenásek, M., Importance of free living mustelid carnivores in circulation of adiaspiromycosis. *Mycopathologia*. 1977;60(3):139–144.
6. Drouhet, E., Huerre, M., Yeast tissue phase of *Emmonsia pasteuriana* inoculated in golden hamster by intratesticular way. *Mycoses*. 1999;42(Suppl 2):11–18.
7. Hubálek, Z., Emmonsiosis of wild rodents and insectivores in Czechland. *J Wildl Dis*. 1999;35(2):243–249.
8. Chantrey, J.C. et al., *Emmonsia crescens* infection in a British water vole (*Arvicola terrestris*). *Med Mycol*. 2006;44:375–378.
9. Borman, A.M. et al., Adiaspiromycosis due to *Emmonsia crescens* is widespread in native British mammals. *Mycopathologia*. 2009;168(4):153–163.
10. Cueva, J.A., Little MD. Emmonsia crescens infection (adiaspiromycosis) in man in Honduras. Report of a case. *Am J Trop Med Hyg*. 1971;20(2):282–287.
11. Kodousek, R. et al., Pulmonary adiaspiromycosis in man caused by *Emmonsia crescens*. Report of a unique case. *Am J Clin Pathol*. 1971;56:394–399.
12. Kamalam, A., Thambiah, A.S., Adiaspiromycosis of human skin caused by *Emmonsia crescens*. *Sabouraudia*. 1979;17(4):377–381.
13. Bambirra, E.A., Nogueira, A.M., Human pulmonary granulomas caused by *Chrysosporium parvum* var. crescens (*Emmonsia crescens*). *Am J Trop Med Hyg*. 1983;32:1184–1185.
14. Barbas Filho, J.V. et al., Respiratory failure caused by adiaspiromycosis. *Chest*. 1990;97:1171–1175.
15. Watts, J.C., Chandler, F.W., Adiaspiromycosis. An uncommon disease caused by an unusual pathogen. *Chest*. 1990;97:1030–1031.
16. England, D.M., Hochholzer, L., Adiaspiromycosis: An unusual fungal infection of the lung. Report of 11 cases. *Am J Surg Pathol*. 1993;17:876–886.
17. Turner, D. et al., Pulmonary adiaspiromycosis in a patient with acquired immunodeficiency syndrome. *Eur J Clin Microbiol Infect Dis*. 1999;18(12):893–895.
18. dos Santos, V.M. et al., Pulmonary adiaspiromycosis: Report of two cases. *Rev Soc Bras Med Trop*. 2000;33(5):483–488.
19. Wellinghausen, N. et al., Chronic granulomatous lung infection caused by the dimorphic fungus *Emmonsia* sp. *Int J Med Microbiol*. 2003;293(6):441–445.
20. Sun, Y. et al., Fine needle aspiration of pulmonary adiaspiromycosis: A case report. *Acta Cytol*. 2007;51(2):217–221.
21. Denson, J.L. et al., Adiaspiromycosis mimicking widespread malignancy in a patient with pulmonary adenocarcinoma. *J Clin Pathol*. 2009;62(9):837–839.
22. de Almeida Barbosa, A. et al., Acute pulmonary adiaspiromycosis. Report of three cases and a review of 16 other cases collected from the literature. *Rev Iberoam Micol*. 1997;14(4):177–180.
23. Nuorva, K. et al., Pulmonary adiaspiromycosis in a two year old girl. *J Clin Pathol*. 1997;50:82–85.
24. Dot, J.M. et al., Molecular diagnosis of disseminated adiaspiromycosis due to *Emmonsia crescens*. *J Clin Microbiol*. 2009;47(4):1269–1273.
25. Mendes, M.O. et al., Acute conjunctivitis with episcleritis and anterior uveitis linked to adiaspiromycosis and freshwater sponges, Amazon region, Brazil, 2005. *Emerg Infect Dis*. 2009;15(4):633–639.
26. Barbosa, A.d.A. et al., Acute pulmonary adiaspiromycosis. Report of three cases and a review of 16 other cases collected from the literature. *Rev Iberoam Micol*. 1997;14:177–180.
27. Peterson, S.W., Kurtzman, C.P., Ribosomal RNA sequence divergence among sibling species of yeasts. *Syst Appl Microbiol*. 1991;14:124–129.
28. Borman, A.M. et al., An improved protocol for the preparation of total genomic DNA from isolates of yeast and mould using Whatman FTA filter papers. *Mycopathologia*. 2010 February 18;169(6):445–449.
29. Peres, L.C. et al., Fulminant disseminated pulmonary adiaspiromycosis in humans. *Am J Trop Med Hyg*. 1992;46:146–150.
30. Echevarria, E., Cano, E.L., Restrepo, A., Disseminated adiaspiromycosis in a patient with AIDS. *J Med Vet Mycol*. 1993;31:91–97.
31. Pelegrín, I. et al., Disseminated adiaspiromycosis: A case report of a liver transplant patient with human immunodeficiency infection and literature review. *Transpl. Infect. Dis.*, 2011, Feb 15 [Epub ahead of print].

30 Epidermophyton

Dongyou Liu and Susan Coloe

CONTENTS

30.1 INTRODUCTION

Dermatophytes are keratinophilic fungi belonging to the anamorphic (asexual or imperfect) genera *Trichophyton*, *Microsporum*, and *Epidermophyton*. These organisms have the capacity to invade keratinized tissues (skin, nails, and hair), resulting in dermatophytosis (ringworm or tinea). Unlike *Trichophyton* and *Microsporum*, which all encompass a large number of anthropophilic, zoophilic and geophilic species, *Epidermophton* consists of a single anthropophilic species (*E. floccosum*). In addition, while members of *Trichophyton* and *Microsporum* infect skin, nails, and hair, *E. floccosum* affects only skin and nails but not hair. In immunocompromised patients, dermatophyte infections may assume insidious and severe forms such as disseminated disease and dermatophytic granulomas. As clinical disease syndromes due to *Epidermophton* are indistinguishable from those caused by *Trichophyton* and *Microsporum*, accurate identification of these organisms is critical for the control and prevention of dermatophytosis.

30.1.1 Classification, Morphology, and Biology

The genus *Epidermophyton* is a filamentous fungus belonging to the mitosporic (anamorphic) Arthrodermataceae group, family Arthrodermataceae, order Onygenales, class Eurotiomycetes, subphylum Pezizomycotina, phylum Ascomycota, kingdom Fungi [1]. The mitosporic Arthrodermataceae group consists of three genera: *Epidermophyton*, *Microsporum*, and *Trichophyton*, which are collectively known as dermatophytes. While both the genera *Trichophyton* and *Microsporum* encompass a large number of closely related species that are infective to the skin, nail, and hair of human and animal hosts, the genus *Epidermophyton* consists of a single species: *Epidermophyton floccosum* (obsolete synonyms: *Acrothecium floccosum*, *Epidermophyton clypeiforme*, *E. inguinale*, *E. plicarum*, *Trichophyton cruris*, *T. floccosum*, *T. inguinale*, *T. intertriginis*, and *T. floccosum*). A former *Epidermophyton* species, that is, *Epidermophyton stockdaleae* is now considered as a variant of *Trichophyton ajelloi*.

E. floccosum colonies are generally slow growing, mustard yellow to khaki green with a suede-like surface initially appearing flat but with age becoming raised and folded in the center and having a submerged fringe. The reverse is a yellowish tan color. *E. floccosum* rapidly becomes white, downy, and sterile with age [2,3].

Microscopically, septate, hyaline hyphae, macroconidia, and chlamydospore-like cells are observed, but microconidia are always absent. Occurring singly or in clusters (of 2–3), macroconidia (measuring 20–40 μm × 6–12 μm, with 1–9 septa) are smooth- and thin walled, clavate shaped with rounded ends. Chlamydospore-like cells as well as arthroconidia are common in older cultures [2,3].

E. floccosum is differentiated from *Trichophyton ajelloi* (obsolete synonym: *E. stockdaleae*) by the former's inability to perforate hair, and by the latter's tolerance to 7% NaCl. In addition, macroconidia of *E. floccosum* are shorter than those *T. ajelloi*. On the other hand, *E. floccosum* is differentiated from *Microsporum* and other *Trichophyton* species by its absence of microconidia [4].

E. floccosum is relatively sensitive to antifungal drugs such as terbinafine, itraconazole, ketoconazole, voriconazole, amorolfine, and naftifine, but somewhat resistant to fluconazole.

30.1.2 Clinical Features and Treatment

Dermatophyte fungi *Trichophyton*, *Microsporum*, and *Epidermophyton* have the capacity to invade keratinized tissues, such as hair, skin, or nails of humans and animals, and produce a variety of superficial diseases including tinea capitis, tinea corporis, tinea cruris, tinea manuum, tinea pedis, and tinea unguium, which are often referred to as dermatophytosis (ringworm, or tinea) (Table 30.1) [5–7].

Typically acquired directly from contact with infected humans (anthropophilic organisms) or animals (zoophilic organisms) or indirectly from exposure to contaminated soil or fomites (geophilic organisms), *Trichophyton*, *Microsporum*, and *Epidermophyton* are the only causes of superficial mycoses in children [8,9]. The clinical presentations of dermatophytosis in immunocompromised hosts may be atypical and variable.

Epidermophyton floccosum is one of the common causes of dermatophytosis in otherwise healthy individuals. It primarily colonizes and infects skin (tinea corporis, tinea cruris, tinea manuum, and tinea pedis) and nails (tinea unguium), and rarely cornea (keratitis) [10]. As the fungus is incapable of penetrating the viable tissues of the immunocompetent host, its infection is limited to the nonliving cornified layers of epidermis. However, invasive *E. floccosum* infection has been described in an immunocompromised patient with Behcet's syndrome [11]. Typical of all forms of dermatophytosis, *E. floccosum* infections are communicable and transmitted by contact, particularly in common showers and gym facilities.

In a recent report, a 40-year-old man developed pain, decreased vision, and multifocal stromal infiltrates in the left eye 2 weeks after laser-assisted subepithelial keratectomy. The corneal infiltrate findings were suggestive of fungal keratitis, and corneal smears were positive for septate fungal hyphae. Mycology culture and molecular analysis revealed *Epidermophyton floccosum* as the infectious organism.

TABLE 30.1
Clinical Features, Common Causative Species, Differential Diagnosis, and Treatment of Human Dermatophytosis

Dermatophytosis	Clinical Features	Common Causative Species	Differential Diagnosis	Treatment
Tinea capitis	Superficial infection of the head and invasion of hair shafts; scaling of the scalp, circumscribed alopecia with broken hair at the scalp; adenopathy, pruritus	*Trichophyton tonsurans*, *Microsporum audouinii*, *Microsporum canis*	Alopecia areata, trichotillomania, traction alopecia, and seborrheic dermatitis	Terbinafine (Lamisil) for *Trichophyton*; griseofulvin (Grifulvin) for *Trichophyton* and *Microsporum*
Tinea corporis	Superficial infection of trunk, arms, and legs; annular patch or plaque with an advancing, raised, scaling border, and central clearing, diffuse erythema	*Trichophyton rubrum*, *Epidermophyton floccosum*	Other annular skin lesions	Butenafine (Mentax) or terbinafine (Lamisil)
Tinea cruris (jock itch)	Superficial infection of the groin predominantly in adolescent and young adults; active lesion border with pustules or vesicles; red to reddish brown background rash; symmetric macule with well-demarcated borders; pruritus of burning quality	*Trichophyton rubrum*, *Trichophyton mentagrophytes*, *Epidermophyton floccosum*	Candidal intertrigo (more uniformly red with no central clearing) and erythrasma (more uniformly brown with slight scaling and no active border)	Terbinafine (Lamisil) cream/spray; or butenafine (Mentax) 1% cream
Tinea manuum	Superficial infection of the hand; inflammatory rash with raised border and clearing in the middle; slowly extending area of peeling, dryness, and mild itching on the palm of one hand	*Trichophyton rubrum*, *Trichophyton mentagrophytes*, *Epidermophyton floccosum*	Contact and allergic dermatitis; occasionally atopic dermatitis	Terbinafine (Lamisil) or butenafine (Mentax) 1%
Tinea pedis (athlete's foot)	Superficial infection of foot; common in adolescents; white macerated area between the toes; moccasin type with a more diffuse dry scaling process often caused by *T. rubrum*; inflammatory vesiculobullous eruption on the soles of the feet; moderate scaling and deep vesicular changes	*Trichophyton rubrum*, *Trichophyton mentagrophytes*, *Epidermophyton floccosum*	Contact and allergic dermatitis; occasionally atopic dermatitis (dermatitis generally does not affect the intertriginous areas)	Terbinafine (Lamisil) or butenafine (Mentax) 1%
Tinea unguium	Infection of the finger- or toenails (a subset of onychomycosis); nail dystrophy	*Trichophyton rubrum*, *Trichophyton mentagrophytes*, *Epidermophyton floccosum*	Other acquired and congenital conditions	Terbinafine for adolescents; griseofulvin for children; itraconazole or fluconazole less effective but less costly

After treatment with natamycin, the patient's visual acuity improved [10].

Most dermatophyte infections are treated with topical therapies, although tinea capitis, severe tinea pedis, and tinea unguium may require oral doses of griseofulvin, terbinafine, itraconazole, and fluconazole (Table 30.1) [12–15].

30.1.3 Laboratory Diagnosis

Considering that dermatophytosis caused by *Trichophyton*, *Microsporum*, and *Epidermophyton* spp. is indistinguishable from each other and is easily confused with dermatitis and other superficial infections caused by nondermatophyte fungi, it is imperative that an etiologic agent involved in dermatophytosis is correctly identified to the species level for effective treatment. In addition, identification to the strain level is valuable for epidemiological tracking of dermatophytes including *E. floccosum* and for selection of tailored chemotherapy against drug-resistant dermatophyte strains [16].

30.1.3.1 Conventional Techniques

Traditionally, clinical diagnosis of dermatophytosis is made with a focused history, physical examination. Wood's lamp examination, fluorescent microscopy, histological tissue examination, and fungal culture further aid in the diagnosis.

Wood's lamp examination exploits the ability of certain dermatophyte species and other microbial agents to fluoresce under ultraviolet light. For example, zoophilic *Microsporum canis* and *Microsporum audouinii* causing tinea capitis fluoresce blue-green; *Malassezia furfur* causing tinea (pityriasis) versicolor fluoresces pale yellow to white; and *Corynebacterium minutissimum* causing erythrasma fluoresces bright coral red. On the other hand, *Trichophyton* and *Epidermophyton* causing tinea cruris and *Candida* spp. do not fluoresce. Histological examination of nail clippings using a periodic acid-Schiff stain offers an accurate means to confirm tinea unguium.

Fungal culture is useful in cases that the infection is resistant to standard topical therapy, the microscopic diagnosis is unclear, or there is a need for long-term oral therapy. Culture can be carried out with skin scraping and hair shafts. In addition, a clinical sample may be obtained with a toothbrush or cotton swab and then inoculated onto a fungal culture medium. The cotton-swab method is performed by moistening a sterile cotton swab and vigorously rubbing it over the affected areas of the scalp [17,18].

Among the key features for the differentiation of *E. floccosum* from *Microsporum* and *Trichophyton* are the former's absence of microconidia and their inability to invade hair.

30.1.3.2 Molecular Techniques

Due to their relative simplicity and efficiency, molecular techniques have been increasingly applied for the detection and identification of dermatophyte fungi. These include polymerase chain reaction (PCR), restriction fragment length polymorphism (RFLP), random amplification of polymorphic DNA (RAPD), arbitrarily primed PCR, PCR fingerprinting, and Southern blotting [18–45]. In addition to random and repetitive genomic sequences, several conserved gene regions have proven valuable for dermatophyte identification and phylogenetic analysis, including small subunit (SSU) 18S and large subunit (LSU) 28S rRNA genes, internal transcribed spacer (ITS) 1 and ITS2 regions, DNA topoisomerase II gene, and chitin synthase 1 (CHS1) gene [25,46–56].

For example, De Baere et al. [56] utilized primers ITS86 (5′-GTGAATCATCGAATCTTTGAAC-3′) and ITS4 (5′-TCCTCCGCTTATTGATATGC-3′) to amplify the ITS2 region followed by digestion with restriction enzyme *Bst*UI (ITS2-RFLP). Subsequent electrophoretic examination of digested PCR products facilitated the differentiation of 12 closely related species, including *Arthroderma cajetani*, *Arthroderma grubyi*, *Arthroderma gypseum*, *Arthroderma incurvatum*, *Arthroderma racemosum*, *Arthroderma simii*, *Arthroderma uncinatum*, *Chrysosporium georgiae*, *Chrysosporium merdarium*, *Epidermophyton floccosum*, *Microsporum praecox*, and *Trichophyton balcaneum*.

30.2 METHODS

30.2.1 Sample Preparation

A portion of skin scrapings and nail clippings from patients suspected of dermatophytosis is examined by microscopy upon treatment with 10% KOH. Another portion of the samples is inoculated onto Sabouraud dextrose agar (1% peptone, 1% glucose, and 1.5% agar) and incubated at 27°C for 5 days. Further examination of micro- and macroscopic features is undertaken on the resultant fungal isolates.

For DNA extraction, a small amount (50 mg) of mycelium from Sabouraud dextrose agar is placed in lysis buffer (200 mM Tris–HCl [pH 8.0], 0.5% [w/v] sodium dodecyl sulfate [SDS], 250 mM NaCl, 25 mM EDTA) and crushed with a conical grinder. The sample is then heated at 100°C for 15 min and mixed with 150 μL of 3.0 M sodium acetate, kept at −20°C for 10 min, and then centrifuged at 10,000 × *g* for 5 min. The supernatant is extracted once with phenol–chloroform–isoamyl alcohol (25:24:1) and is subsequently extracted once with chloroform. The DNA is precipitated with an equal volume of isopropanol at −20°C for 10 min, washed with 0.5 mL of 99% ethanol, dried, and suspended in 50 μL of ultrapure water. One microliter of solution is used as the template for PCR. The total time required to prepare the DNA was 80 min [48].

Alternatively, dermatophyte strains are cultured in 100 mL of Sabouraud liquid medium (Oxoid) and incubated with shaking for up to 7 days at 27°C. Hyphal growth is harvested by filtration and washed twice with 100 mL of sterile saline. Strains that are not processed immediately are frozen at −80°C prior to extraction. Liquid nitrogen is added to 2–3 g of frozen hyphae in a prechilled mortar, and the cells are ground finely with a pestle. Approximately 200 mg of frozen, ground mycelium is placed in a 1.5 mL microcentrifuge tube, and 600 μL of lysis buffer (400 mM Tris–HCl [pH 8.0], 60 mM EDTA, 150 mM NaCl, 1% SDS; 40 mg/mL proteinase K) is added. Samples are incubated for 1 h at 60°C with occasional mixing, and then 100 μL of 5 M sodium perchlorate is added and incubated for a further 15 min at 60°C. Tubes are cooled on ice, and extraction is performed with 500 μL of ice-cold chloroform and then with equal volumes of phenol–chloroform–isoamyl alcohol (25:24:1 [pH 8.0]) and finally with chloroform. Purified nucleic acids are precipitated with 2 volumes of ice-cold 95% ethanol, washed twice in 500 μL of 70% ethanol, air dried, and resuspended in 150–200 μL of sterile water [25].

In addition, DNA can be prepared by using a genomic DNA extraction kit (Qiagen) according to the manufacturer's instructions, preceded by lyticase digestion. Lyticase digestion is carried out by transferring a loopful of mycelium to a 1.5 mL tube containing 2.5 mg zymolase (Sigma–Aldrich) in 250 μL zymolase lysis buffer (10 mM Tris–HCl [pH 8.0], 1 mM EDTA) and incubating for 45 min at 37°C. Subsequently, 200 μL is transferred to another tube to carry out the DNA extraction, with a final elution volume of 100 μL [56].

30.2.2 Detection Procedures

30.2.2.1 PCR–RFLP Protocol

Universal primers ITS1 (5′-TCCGTAGGTGAACCTGCGG-3′) and ITS4 (5′-TCCTCCGCTTATTGATATGC-3′) can be used for PCR amplification of the ITS1–5.8S rRNA–ITS2 gene sequence. Subsequent digestion with restriction enzymes *Hin*fI, *Hha*I, and *Mvn*I generates distinct band patterns for dermatophytes including *Epidermophyton floccosum* [25,33].

Procedure

1. PCR mixture (50 μL) is made up of 1 μL of 50 pmol of each primer, 25 μL of HotStarTaq Polymerase (Qiagen), 2 μL (10 ng) of extracted DNA, and 21 μL of distilled water.
2. The reaction mixture is subjected to an initial 95°C for 6 min, 35 cycles of 95°C for 1 min, 55°C for 1 min, and 72°C for 1 min; and a final 72°C for 10 min.
3. A portion of PCR products (5 μL) is separated on 1% agarose gel, stained with ethidium bromide and visualized by UV light. The expected sizes of PCR products are between 600 and 900 bp for all dermatophyte strains.
4. Ten microliters of the PCR product is digested with 10 U of restriction enzymes *Hin*fI, *Hha*I, or *Mvn*I in 20 μL volume at 37°C for 2 h, electrophoresed on 8% polyacrylamide gels, stained, and visualized as above.

Note: Restriction enzymes *Hin*fI or *Hha*I digestion of the PCR products from dermatophyte DNA yields five characteristic, species-specific restriction patterns A–E. Profile A corresponds to *Trichophyton rubrum*, profile B corresponds to *Trichophyton verrucosum*, profile C corresponds to *Trichophyton mentagrophytes* var. *interdigitale*, profile D corresponds to *Epidermophyton floccosum*, and profile E corresponds to *Microsporum canis* [33] On the other hand, *Mvn*I digestion displays three genus-specific restriction patterns I–III, with *Epidermophyton floccosum* showing three bands of approximately 150, 250, and 350 bp [25]. The identity of the dermatophytes of interest can be further confirmed by sequencing analysis of the PCR product.

30.2.2.2 Real-Time PCR Protocol

Bergmans et al. [40] developed a single-tube Lightcycler (LC) real-time PCR assay for rapid detection and identification of 11 clinically relevant *Trichophyton*, *Microsporum*, and *Epidermophyton* species in nail, skin, and hair samples. For specific detection of *Epidermophyton floccosum*, dermatophyte ITS primer DERMF3 (5′-GGTTGCCTCGGCGGGCC-3′) and dermatophyte 5.8S rRNA primer B-DERMR2 (5′-CGGAATTCTGCAATTCACATTACT-3′) are combined with two *E. floccosum*-specific ITS LC hybridization probes Efl3S (5′-LC705-CTACGAAATCTCCATAG GTGGTTCAGTCT-3′) and Efl1A (5′-GCGCTCGCCGAAG GAGTGATTCTC-FL-3′) in a single tube.

Procedure

1. PCR mixture (20 μL) is composed of 4 μL of LC Multiplex Master Hyb Probes (Roche), 400 nM primer DERMF3 and 1300 nM primer DERMR2, 330 nM each of LC hybridization probes Efl3S and Efl1A, 1.3 mM $MgCl_2$, 0.5 μL of dimethylsulfoxide, and 0.2 U of uracil DNA glycosylase (Roche), and 5 μL of extracted DNA.
2. Amplification is conducted on a Lightcycler 2.0 (Roche) with initial denaturation for 10 min at 95°C, and 50 cycles of 20 s at 95°C, 20 s at 55°C, and 20 s at 72°C. This is followed by a melting curve analysis (1 min at 95°C, cooling for 1 min at 50°C, and ramp to 85°C, at a ramp rate of 0.1°C/s). Color compensation is performed according to the manufacturer's instructions, and fluorescence signals are corrected by the "back 530" signal.
3. The presence of an amplification curve in the 705 nm channel of the Lightcycler, in conjunction with a melting curve with the melting temperature (T_m) of 67.50°C–69.50°C and a T_m peak height of ≥0.01 is considered to be a positive result for *E. floccosum*.

Note. This single-tube Lightcycler real-time PCR assay enables sensitive and specific detection of *E. floccosum* in nail, hair, and scales in 4 h (after overnight lysis of the samples).

30.3 CONCLUSION

Dermatophytes classified in the genera *Epidermophyton*, *Microsporum*, and *Trichophyton* are keratinophilic fungi with the capacity to invade the keratinized tissues (e.g., skin, nail, and hair) of humans and animals, resulting in a diverse range of superficial infections known as dermatophytosis, tinea, or ringworm. As the clinical manifestations of dermatophytosis caused by *Trichophyton*, *Microsporum*, and *Epidermophyton* spp. are indistinguishable and dermatophyte fungal spp. demonstrate varied sensitivity to commonly used antifungal drugs, correct identification of dermatophyte species involved is critical for the effective control of the disease. Due to the slowness and impreciseness of conventional diagnostic methods, molecular techniques have been increasingly applied for identification and typing of dermatophyte fungi including *Epidermophyton floccosum*. This has not only led to a dramatic reduction in testing time from 3–4 weeks to 3–4 days in most cases but also contributed to the implementation of effective, tailor-made antifungal therapy against human dermatophytosis.

REFERENCES

1. The UniProt Consortium. Available at http://www.uniprot.org/, accessed on August 1, 2010.
2. Larone, D.H., 1995. *Medically Important Fungi—A Guide to Identification*, 3rd edn., ASM Press, Washington, DC.
3. De Hoog, G.S. et al., 2000. *Atlas of Clinical Fungi*, 2nd edn., vol. 1. Centraalbureau voor Schimmelcultures, Utrecht, the Netherlands.
4. Weitzman, I. and R.C. Summerbell, 1995. The dermatophytes. *Clin Microbiol Rev*. 8:240–259.
5. Ogawa, H. et al., 1998. Dermatophytes and host defence in cutaneous mycoses. *Med Mycol*. 36:166–173.
6. Aman, S. et al., 2001. Tinea unguium in Lahore, Pakistan. *Med Mycol*. 39:177–180.
7. Hay, R.J. et al., 2001. Tinea capitis in Europe: New perspective on an old problem. *J Eur Acad Dermatol Venereol*. 15:229–233.
8. Aly, R., 1994. Ecology and epidemiology of dermatophyte infections. *J Am Acad Dermatol*. 31:S21–S25.
9. Andrews, M.D. and M. Burns, 2008. Common tinea infections in children. *Am Fam Physician*. 77:1415.
10. Jung, S.W. et al., 2009. Epidermophyton fungal keratitis following laser-assisted subepithelial keratectomy. *J Cataract Refract Surg*. 35(12):2157–2160.
11. Seddon, M.E. and M.G. Thomas, 1997. Invasive disease due to Epidermophyton floccosum in an immunocompromised patient with Behçet's Syndrome. *Clin Infect Dis*. 25:153–154.
12. Macura, A.B., 1993. In vitro susceptibility of dermatophytes to antifungal drugs: A comparison of two methods. *Int J Dermatol*. 32:533–536.
13. Degreef, J.H. and P.R.G. DeDoncker, 1994. Current therapy of dermatophytosis. *J Am Acad Dermatol*. 31:S25–S30.
14. Venugopal, P.V. and T.V. Venugopal, 1994. Disk diffusion susceptibility testing of dermatophytes with allylamines. *Int J Dermatol*. 33:730–732.
15. Boonk, W. et al., 1998. Itraconazole in the treatment of tinea corporis and tinea cruris: Comparison of two treatment schedules. *Mycoses*. 41:509–514.
16. Weinberg, J.M. et al., 2003. Comparison of diagnostic methods in the evaluation of onychomycosis. *J Am Acad Dermatol*. 49(2):193–197.
17. Karimzadegan-Nia, M. et al., 2007. Comparison of direct smear, culture and histology for the diagnosis of onychomycosis. *Aust J Dermatol*. 48(1):18–21.
18. Uchida, T. et al., 2009. Comparative study of direct polymerase chain reaction, microscopic examination and culture-based morphological methods for detection and identification of dermatophytes in nail and skin samples. *J Dermatol*. 36(4):202–208.
19. Liu, D. et al., 1997. Molecular determination of dermatophyte fungi using the arbitrarily primed polymerase chain reaction. *Br J Dermatol*. 137:351–355.
20. Liu, D. et al., 2000 Jun. Application of PCR to the identification of dermatophyte fungi. *J Med Microbiol*. 49(6):493–497.
21. Mochizuki, T., N. Sugie, and M. Uehara, 1997. Random amplification of polymorphic DNA is useful for the differentiation of several anthropophilic dermatophytes. *Mycoses*. 40:405–409.
22. Baek, S.C. et al., 1998. Detection and differentiation of causative fungi of onychomycosis using PCR amplification and restriction enzyme analysis. *Int J Dermatol*. 37(9):682–686.
23. De Hoog, G.S. et al., 1998. Molecular phylogeny and taxonomy of medically important fungi. *Med Mycol*. 36:52–56.
24. Graser, Y. et al., 1998. Identification of common dermatophytes (*Trichophyton*, *Microsporum Epidermophyton*) using polymerase chain reactions. *Br J Dermatol*. 138:576.
25. Jackson, C.J., R.C. Barton, and E.G.V. Evans, 1999. Species identification and strain differentiation of dermatophyte fungi by analysis of ribosomal-DNA intergenic spacer regions. *J Clin Microbiol*. 37:931–936.
26. Howell, S.A., R.J. Barnard, and F. Humphreys, 1999. Application of molecular typing methods to dermatophyte species that cause skin and nail infections. *J Med Microbiol*. 48(1):33–40.
27. Faggi, E. et al., 2001. Application of PCR to distinguish common species of dermatophytes. *J Clin Microbiol*. 39:3382–3385.
28. Machouart-Dubach, M. et al., 2001. Rapid discrimination among dermatophytes, Scytalidium spp., and other fungi with a PCR-restriction fragment length polymorphism ribotyping method. *J Clin Microbiol*. 39(2):685–690.
29. Ding, J. et al., 2004. Clinical identification of common species of dermatophytes by PCR and PCR-RFLP. *J Huazhong Univ Sci Technol Med Sci*. 24(6):642–644.
30. Gutzmer, R. et al., 2004. Rapid identification and differentiation of fungal DNA in dermatological specimens by LightCycler PCR. *J Med Microbiol*. 53:1207–1214.
31. He, G. et al., 2005. Identification of common species of dermatophytes by PCR-RFLP. *J Huazhong Univ Sci Technol Med Sci*. 25(4):458–460.
32. Pounder, J.I. et al., 2005. Repetitive-sequence-PCR-based DNA fingerprinting using the Diversilab system for identification of commonly encountered dermatophytes. *J Clin Microbiol*. 43:2141–2147.
33. Dobrowolska, A. et al., 2006. PCR–RFLP analysis of the dermatophytes isolated from patients in Central Poland. *J Dermatol Sci*. 42:71.
34. Monod, M. et al., 2006. Fast and reliable PCR/sequencing/RFLP assay for identification of fungi in onychomycoses. *J Med Microbiol*. 55:1211–1216.
35. Arabatzis, M. et al., 2007. Diagnosis of common dermatophyte infections by a novel multiplex real-time polymerase chain reaction detection/identification scheme. *Br J Dermatol*. 157(4):681–689.

36. Garg, J. et al., 2007. Evaluation of pan-dermatophyte nested PCR in diagnosis of onychomycosis. *J Clin Microbiol.* 45(10):3443–3445.
37. Garg, J. et al., 2009. Rapid detection of dermatophytes from skin and hair. *BMC Res Notes.* 2:60.
38. Li, H.C. et al., 2007. Identification of dermatophytes by an oligonucleotide array. *J Clin Microbiol.* 45(10):3160–3166.
39. Bergmans, A.M. et al., 2008. Validation of PCR-reverse line blot, a method for rapid detection and identification of nine dermatophyte species in nail, skin and hair samples. *Clin Microbiol Infect.* 14(8):778–788.
40. Bergmans, A.M. et al., 2010. Evaluation of a single-tube real-time PCR for detection and identification of 11 dermatophyte species in clinical material. *Clin Microbiol Infect.* 16:704.
41. Erhard, M. et al., 2008. Identification of dermatophyte species causing onychomycosis and tinea pedis by MALDI-TOF mass spectrometry. *Exp Dermatol.* 17:356–361.
42. Shehata, A.S. et al., 2008. Single-step PCR using (GACA)4 primer: Utility for rapid identification of dermatophyte species and strains. *J Clin Microbiol.* 46(8):2641–2645.
43. Yang, G. et al., 2008. Direct species identification of common pathogenic dermatophyte fungi in clinical specimens by semi-nested PCR and restriction fragment length polymorphism. *Mycopathologia.* 166(4):203–208.
44. Beifuss, B. et al., 2011. Direct detection of five common dermatophyte species in clinical samples using a rapid and sensitive 24-h PCR-ELISA technique open to protocol transfer. *Mycoses.* 54:137.
45. Bontems, O., P.M. Hauser, and M. Monod, 2009. Evaluation of a polymerase chain reaction-restriction fragment length polymorphism assay for dermatophyte and nondermatophyte identification in onychomycosis. *Br J Dermatol.* 161(4):791–796.
46. Kano, R. et al., 1997. Phylogenetic analysis of 8 dermatophyte species using chitin synthase 1 gene sequences. *Mycoses.* 40(11–12):411–414.
47. Kano, R. et al., 1999. Phylogenetic relation of Epidermophyton floccosum to the species of Microsporum and Trichophyton in chitin synthase 1 (CHS1) gene sequences. *Mycopathologia.* 146(3):111–113.
48. Makimura, K. et al., 1999. Phylogenetic classification and species identification of dermatophyte strains based on DNA sequences of nuclear ribosomal internal transcribed spacer 1 regions. *J Clin Microbiol.* 37(4):920–924.
49. Hirai, A. et al., 2003. Molecular taxonomy of dermatophytes and related fungi by chitin synthase 1 (CHS1) gene sequences. *Anton Leeuwen.* 83(1):11–20.
50. Kanbe, T. et al., 2003a. PCR-based identification of common dermatophyte species using primer sets specific for the DNA topoisomerase II genes. *J Dermatol Sci.* 32:151–161.
51. Kanbe, T. et al., 2003b. Species-identification of dermatophytes *Trichophyton, Microsporum, Epidermophyton* by PCR and PCR–RFLP targeting of the DNA topoisomerase II genes. *J Dermatol Sci.* 33:41–54.
52. Ninet, B. et al., 2003. Identification of dermatophyte species by 28S ribosomal DNA sequencing with a commercial kit. *J Clin Microbiol.* 41(2):826–830.
53. Kamiya, A. et al., 2004. PCR and PCR–RFLP techniques targeting the DNA topoisomerase II gene for rapid clinical diagnosis of the etiologic agent of dermatophytosis. *J Dermatol Sci.* 34:35.
54. Li, H.C. et al., 2008. Identification of dermatophytes by sequence analysis of the rRNA gene internal transcribed spacer regions. *J Med Microbiol.* 57(Pt 5):592–600.
55. Ebihara, M. et al., 2009. Molecular detection of dermatophytes and nondermatophytes in onychomycosis by nested polymerase chain reaction based on 28S ribosomal RNA gene sequences. *Br J Dermatol.* 61(5):1038–1044.
56. De Baere, T. et al., 2010. Evaluation of internal transcribed spacer 2-RFLP analysis for the identification of dermatophytes. *J Med Microbiol.* 59:48–54.

31 *Exophiala*

Dongyou Liu and Shoo Peng Siah

CONTENTS

31.1 INTRODUCTION

31.1.1 Classification, Morphology, and Biology

The genus *Exophiala* is a dematiaceous (dark-walled) fungus belonging to the mitosporic Herpotrichiellaceae group, family Herpotrichiellaceae, order Chaetothyriales, class Eurotiomycetes, subphylum Pezizomycotina, phylum Ascomycota, and kingdom Fungi. The mitosporic Herpotrichiellaceae group encompasses 13 genera: *Blastacervulus*, *Brycekendrickomyces*, *Cladophialophora*, *Cladoriella*, *Cyphellophora*, *Exophiala*, *Heteroconium*, *Melanchlenus*, *Metulocladosporiella*, *Phaeococcomyces*, *Rhinocladiella*, *Thysanorea*, and *Veronaea* [1].

In turn, the genus *Exophiala* consists of at least 29 species, the most notable of which are *Exophiala castellanii* (obsolete synonyms: *Aureobasidium mansonii*, *Cladosporium mansonii*, *Dematium mansonii*, *E. jeanselmei* var. *castellanii*, *E. mansonii*, *Foxia mansonii*, *Malassezia mansonii*, *Microsporum mansonii*, *Rhinocladiella mansonii*, *Torula mansonii*, and *Wangiella mansonii*), *Exophiala dermatitidis* (synonym: *Wangiella dermatitis*), *Exophiala heteromorpha* (synonym: *Wangiella heteromorpha*), *Exophiala jeanselmei* var. *heteromorpha*, *Exophiala jeanselmei* var. *jeanselmei* (synonym: *E. jeanselmei*; obsolete synonyms: *Phialophora jeanselmei*, *Pullularia jeanselmei*, and *Torula jeanselmei*), *Exophiala jeanselmei var. lecanii-corni* (obsolete synonym: *Pullularia fermentans* and *Torula lecanii-corni*), *Exophiala moniliae*, *Exophiala pisciphila*, *Exophiala salmonis*, *Exophiala spinifera* (obsolete synonyms: *Phialophora spinifera* and *Rhinocladiella spinifera*) [2]. The teleomorphic state of the genus *Exophiala* is probably the genus *Capronia* [3].

The composition of the genus *Exophiala* has undergone significant revisions following recent application of molecular techniques [4–6]. For example, *E. jeanselmei* is a well-known heterogeneous, cryptic species, which contains three varieties (*E. jeanselmei* var. *lecanii-corni*, *E. jeanselmei* var. *heteromorpha*, and *E. jeanselmei*) on the basis of its morphological features. Sequence analysis of *E. jeanselmei* rRNA internal transcribed spacer (ITS) and other gene regions revealed that the former *E. jeanselmei* species can be separated into *E. heteromorpha*, *E. lecanii-corni*, *E. oligosperma*, and *E. xenobiotica* plus *E. jeanselmei* [7]. Furthermore, molecular examination of the *Exophiala spinifera*–*E. jeanselmei* clade led to the differentiation of *E. spinifera*, *E. jeanselmei*, *E. attenuata*, and *E.* (*Phaeococcomyces*) *exophialae* as well as an unidentified *Exophiala* sp. (represented by strain CBS 725.88) [8,9].

Morphologically, *Exophiala* colonies are slow growing and change in texture and color with age. Initially, the culture is yeast-like, moist (slimy), and brownish to greenish black in color. As the colony ages, it becomes mold-like, suede-like (velvety), or downy due to development of short, aerial grayish hyphae and varies from pale brown to black in color. The colony is olivaceous-black on the front surface and black in the reverse surface in mature colonies [12].

The *Exophiala* conidiophores are hyaline to subhyaline and hyphae-like and these vary between the different species. The conidium formation occurs on terminal or lateral conidiogenous cells (annellides), which are tubular and rocket shaped and typically taper to form a narrow elongated tip. The conidia, which ranged from 1–3 × 3–6 μm, are one celled, hyaline to pale brown, elongate, and accumulate in balls at apices of annellides [10–12]. At the species level,

E. spinifera annellides arise on spine-like conidiophores from the hyphae and have erect, multicellular, dark-brown stalks producing conidia from terminal and intercalary cells. Unlike *E. spinifera*, *E. exophialae* does not have well-differentiated conidiophores. *E. jeanselmei* possesses non-spine-like conidiophores and nonseptate, rocket-shaped, slightly darkened conidiogenous cells, with cylindrical to lageniform annellides. Like *E. jeanselmei*, the *E. oligosperma* conidiogenous cells are also nonseptated, rocket shaped, and slightly darkened. *E. attenuate* also has highly differentiated conidiophores and is situated outside the *E. spinifera* clade. *E. moniliae* has non-spine-like conidiophores and moniliform annellides with long stalks [10–12]. Although *Wangiella dermatitis* is often considered as a synonym of *Exophiala dermatitis*, both *Wangiella* and *Phialophora* differs from *Exophiala* by forming phialides instead of annellides. *Wangiella* phialides are brown and flask shaped to cylindrical, without collarettes [13–18].

Besides microscopic features, growth at different temperatures (on potato dextrose agar [PDA] at 25°C, 37°C, and 40°C for 2 weeks), hydrolysis of casein and tyrosine, ability to grow in 15% NaCl, and KNO_3 assimilation are useful for differentiation of *Exophiala* spp. as well as *Hortaea werneckii* (Table 31.1) [19,20]. Most *Exophiala* spp. are able to grow at 37°C, but *Exophiala* (*Wangiella*) *dermatitis* and *Hortaea werneckii*, in particular, are able to tolerate temperature of 40°C. The teleomorphic genus *Capronia* species of *E. psychrophila* and *E. mesophila* from cold water and/or fish are psychrophilic and do not grow at temperatures of 35°C or higher. In contrast, *Capronia epimyces* and *Capronia munkii*, which are clustered in the *E. dermatitidis* clade, are thermophilic and grow well at 37°C [3,21].

Exophiala spp. are saprophytes commonly distributed in environments including soil, plants, water, and decaying wood material [22–24]. Of the 29 known species, 18 have been isolated from humans and animals hosts, some (e.g., *E. jeanselmei*, *E. moniliae*, and *E. spinifera*) of which play a key role in human phaeohyphomycosis, mycetoma, and chromoblastomycosis [25]. A recent study reexamined 188 clinical *Exophiala* isolates from the United States. Using comparative sequencing of the ITS region in the rRNA gene, the frequency of *Exophiala* occurring in human hosts were as follows: 55 *E. dermatitidis* (29.3%), 37 *E. xenobiotica* (19.7%), 35 *E. oligosperma* (18.6%), 13 *E. lecanii-corni* (6.9%), 12 *E. phaeomuriformis* (6.4%), 7 *E. jeanselmei* (3.7%), 7 *E. bergeri* (3.7%), 6 *E. mesophila* (3.2%), 5 *E. spinifera* (2.7%), 3 *Exophiala* sp. 1 (1.6%), 3 *E. attenuata* (1.6%), 3 *Phialophora europaea* (1.3%), 1 *E. heteromorpha* (0.5%), and 1 *Exophiala* sp. 2 (0.5%) strains [25].

31.1.2 Clinical Features and Pathogenesis

As dematiaceous fungi, members of the genus *Exophiala* often cause three types of mycotic infections in humans and animals: phaeohyphomycosis, chromoblastomycosis, and mycetoma [26–29]. The infections due to *Exophiala* spp. have been reported in both immunocompetent and immunosuppressed patients [30,31]. Often the infection takes hold following traumatic implantation of the fungus through the skin or may occur as an opportunistic infection in patients with chronic or acute local or systemic immunosuppression, such as organ transplantation [32–37]. Phaeohyphomycosis usually presents as single cyst or abscess on exposed area while chromoblastomycosis and mycetoma tend to display muriform cells or grains. Clinical manifestations of *Exophiala* infections include localized cutaneous infections, subcutaneous cysts, keratitis, otitis, osteomyelitis, endocarditis, prosthetic valvular vegetations, pneumonia, nosocomial outbreaks of fungemia, stomatitis, and other cerebral and disseminated infections [38–62].

A recent study surveyed 188 *Exophiala* strains isolated in the United States and found that approximately 38%–40% of the reported infections involved deep tissues. Specifically, 39.9% (73) of isolates came from deep infections involving the lung, pleural fluid, sputum, stomach, intestines, bile, heart, brain, spleen, bone marrow, blood, dialysis fluid, lymph node, joint, breast, middle ear, throat, and intraocular tissues. Notably, the lung, pleural fluid, and sputum accounted for half of these cases. Another 38.3% (72) of the cases surveyed were cutaneous infections involving the skin, mucous membranes, nail, and corneal epithelium lesions. Subcutaneous infections including paranasal sinusitis, mycetoma, and subcutaneous cyst accounted for 12.0% (22) of the cases while a remaining 0.5% (1) was isolated from superficial infections (including hair) [25]. Interestingly, the predominant *Exophiala* spp. causing both the systemic, cutaneous, and subcutaneous infections were *E. dermatitidis* and *E. oligosperma*. Specifically, the most frequently isolated *Exophiala* spp. in systemic infections were (in decreasing order of frequency): *E. dermatitidis* (36 of 73), *E. oligosperma* (16 of 73), *E. phaeomuriformis* (9 of 73), *E. xenobiotica* (5 of 73),

TABLE 31.1
Physiological and Biochemical Characteristics of Some *Exophiala* spp.

Species	Growth at 25°C	Growth at 37°C	Growth at 40°C	Hydrolysis of Casein	Hydrolysis of Tyrosine
Exophiala jeanselmei	+	+	–	–	+
Exophiala castellanii	+	+	–	–	+
Exophiala (*Wangiella*) *dermatitis*	+	+	+	–	+
Hortaea werneckii	+	+	+	+	–

and *E. lecanii-corni* (4 of 73). In cutaneous and subcutaneous infections, the following species were reported: *E. xenobiotica* (25 of 92), *E. dermatitidis* (16 of 92), *E. oligosperma* (16 of 92), and *E. jeanselmei* (7 of 92) as well as *E. bergeri*, *E. spinifera*, *E. mesophila*, and *E. attenuate* [25].

As a localized, chronic infection of skin and subcutaneous tissues (often following penetrating trauma on the foot), mycetoma is characterized by tumefaction, abscess formation, painless slow-growing nodules or plaques, draining sinuses, and sclerotia (grains) within the abscesses and fistulae. Pain and disfigurement may eventuate if the infection spreads to underlying bones. Mycetoma can be caused by either filamentous bacteria (actinomycetoma, 60%) or fungi (eumycetoma, 40%) [63] and as such should be properly investigated as the treatment regime may differ. Over 30 fungal species are associated with eumycetoma in humans, among which *Madurella* spp. are most common (causing so-called Madura foot, which was named after the district of India where it was first described), while *Exophiala jeanselmei* has been reported as infrequent cause of eumycetoma [64–69].

Some case studies have been reported detailing the process of investigating mycetoma. In one case of *Exophiala*-related mycetoma, a 44-year-old man presented with a mass on his right foot that had multiple, recurrent sinuses that drained serosanguinous fluid within the area of tumefaction. When the patient had sustained minor trauma to the medial aspect of the right foot while outdoors about 20 years earlier, a painless mass occurred within weeks and slowly developed in size over the years. Occasionally, minute dark granules (formed by aggregates of fungal hyphae) were noticeable in the sinus drainage. A potassium hydroxide preparation of the lesion material revealed hyphal elements and in vitro culture of tissue biopsy grew *Exophiala jeanselmei*. Histopathological examination showed a deeply seated abscess surrounded by an infiltrate of lymphocytes, plasma cells, and fibrosing granulation tissue. Several grains of pigmented filamentous fungi arranged radially at the periphery were stained within the abscess using Gomori's methenamine silver stain [70].

In another case of *Exophiala* infection, a 48-year-old man developed a nodule on the right lower leg after trauma. The patient had undergone cardiac transplantation 8 months earlier with immunosuppressive therapy. Microscopic examination of the pus from lesion showed branched pale-brown hyphae and rounded thick-walled vesicles. Histopathological examination of tissue sections of a skin biopsy from the lesion displayed mild hyperplasia of the epidermis and suppurative granulomatous inflammation in dermis with budding yeasts and dematiaceous hyphae. On Sabouraud dextrose agar, dark, moist, olive to black yeast-like colony of *E. jeanselmei* was grown from the lesion material, which showed the long mycelium, thick-walled septate conidiophores with a ball of conidia (annelophores) at the tip under the microscope. After 4 months of treatment with amphotericin B, the lesion showed marked reduction in the swelling and closure of sinuses and residual scaring of the tissue [71].

Besides mycetoma, *Exophiala* spp. have been implicated in other types of human infections [72–74]. Nosocomial infections by *E. dermatitidis* have been reported following intravascular or intra-articular injection of contaminated drugs. *E. dermatitidis* has been shown to cause pulmonary infections in humans, especially those suffering from cystic fibrosis (CF). Glucocorticoid treatment, antibiotic therapy, and malnutrition are known risk factors for the colonization of the respiratory tract by *E. dermatitidis* in CF patients [75–80].

Despite being a predominantly environmental agent and with only some *Exophiala* spp. being reported as human pathogens, colonization of medical equipment and consumables have been described and may be a risk factor for opportunistic infections. There have been studies where *Exophiala* spp. have been isolated from biofilms collected from sterilization equipment and chemicals used, and indeed some infections with the black yeast had resulted in fatality [51]. For example, one study reported on the isolation of *E. mesophila* from chemical disinfectants used to reduce bacterial load in dental unit waterlines, whereas another study reported on the isolation of *Phialophora* spp. in the biofilm collected from a water pipeline that supplied an automated endoscope washer disinfectors [81,82]. The isolated Phialophora strain was further shown to be resistant to UV irradiation, peracetic acid, and hydrogen peroxide treatment [82]. While the pathogenicity of these strains may not be known, it is possible that strains such as *E. mesophila*, which is unable to grow at temperatures of greater than 35°C, may have cell walls that are capable of interfering with chemical disinfectants and subsequently facilitate colonization by bacteria [81].

31.1.3 Laboratory Diagnosis

Apart from *Exophiala* spp., phaeohyphomycosis, chromoblastomycosis, and mycetoma may be caused by other black yeasts in the teleomorph family Herpotrichiellaceae and their filamentous relatives, anamorphs of members of the order Chaetothyriales. In addition, mycetoma may be due to a number of actinobacteria. There is a need to identify these fungi for effective treatment and prevention purposes.

Chaetothyrialean fungi including *Exophiala* often exhibit a high degree of morphological variability and molecular diversity and encompass a wide array of clinically relevant species. As such, their identification and diagnosis by microscopic and histopathological examination of tissue specimens and culture can be challenging and may take up to 12 weeks [25,83,84]. Cultures are frequently negative and when positive, they may display poorly differentiated or delayed sporulation. These fungi may also appear microscopically similar to phylogenetically remote species, thus leading to misidentification [25].

In recent years, molecular approaches have been increasingly employed for improved identification of dematiaceous fungi. In particular, polymerase chain reaction (PCR) amplification and sequence analysis of the ITS1-5.8S-ITS2 gene region with universal fungal primers provides a reliable tool for routine species distinction in the genus *Exophiala*

[85–87]. Furthermore, examination of the elongation factor 1-α and β-tubulin and mitochondrial cytochrome *b* gene sequences has also proven valuable for confirmation of the identity of some of the main clinically relevant species [88].

The following section presents recent protocols for preparation of DNA from clinical samples and use of the ITS1-5.8S rRNA-ITS2 gene region to aid in the *Exophiala* spp. identification.

31.2 METHODS

31.2.1 Sample Preparation

Clinical specimens suspected of dematiaceous fungi are treated with 10% KOH or stained with calcofluor white and observed under the microscope for septate hyphal elements. Fungal isolation is carried out on inhibitory mold agar, modified Sabouraud agar, or PDA at 25°C, 35°C, and 50°C. Preliminary identification is based on macro- and microscopic features of the colonies. Halotolerance is tested in a liquid medium at 2.5%, 5%, and 10% (w/v) NaCl and $MgCl_2$ [89].

The clinical specimen is grown on PDA slants for 1–7 days after which approximately 1 cm^2 of mycelia is collected by scraping the slant with a sterile stick and emulsified in 1 mL of sterile H_2O. The material is transferred to a 2 mL screw-cap tube, and the tube is centrifuged for 1 min at 6000 × *g*. If the mycelia do not pellet, the material is concentrated with a pediatric blood serum filter. After removal of supernatant, the material is resuspended in 200 μL of GenOhm IDI sample buffer (BD diagnostics) and transferred to the lysis tube containing glass beads. The lysis tube is vortexed on the highest setting for 5 min then boiled for 15 min. The sample is then centrifuged for 5 min at 16,000 × *g* and the supernatant is stored at −20°C until PCR.

Alternatively, about 1 cm^2 of fungal material is transferred to a 2 mL Eppendorf tube containing a 2:1 (wt/wt) mixture of silica gel and Celite (silica gel H, Merck 7736/Kieselguhr Celite 545; Machery) and 300 μL of TES buffer (2 g Tris (hydroxymethyl)-aminomethane, 0.38 g Na-EDTA [ethylenediaminetetraacetic acid], and 2 g sodium dodecyl sulfate prepared in 80 mL of ultrapure water [pH 8]). The fungal material is grounded with a micropestle for 1–2 min then the volume is adjusted by adding 200 μL of TES buffer. The sample is shaken vigorously before 10 μL of proteinase K (10 mg/mL) is added and the mixture incubated at 65°C for 10 min. A measured quantity of 140 μL of 5 M NaCl solution is added and the mixture is combined with 0.1 volume (~65 μL) of CTAB (cetyltrimethylammonium bromide) buffer 10%. The sample is then incubated for another 30 min at 65°C. One volume (~700 μL) of chloroform-isoamyl alcohol (24:1 vol/vol) is added and the sample is mixed by inversion. After incubation for 30 min at 0°C (on ice water) and centrifugation at 14,000 rpm at 4°C for 10 min, the top layer is transferred to a clean Eppendorf tube. Ammonium acetate (5 M, 225 μL) is added and the sample returned to the ice bath for another 30 min before centrifugation. The supernatant is transferred to a clean sterile Eppendorf tube and mixed with a 0.55 volume (~510 μL) of ice-cold isopropanol. After centrifugation for 7 min at 14,000 rpm at 4°C (or room temperature), the supernatant is decanted and the pellet is washed twice with ice-cold 70% ethanol. The pellet is dried using a vacuum dryer, resuspended with 48.5 μL of Tris-EDTA buffer and 1.5 μL of RNase (10 mg/mL), incubated at 37°C for 15–30 min, and stored at −20°C until use [25].

In addition, fungal strains can be cultured in 20 mL of RPMI 1640 medium supplemented with L-glutamine but not sodium bicarbonate (Sigma-Aldrich). The RPMI 1640 medium is buffered to pH 7.0 with 0.165 M morpholinepropanesulfonic acid (MOPS) (Sigma-Aldrich). After 3–14 days of growth at 30°C with gentle agitation (100 rpm), the mycelium is transferred into a tube and washed in 40 mL of sterile distilled water. The mycelium is then stored at −20°C until use. For DNA extraction, 100 mg of mycelium is homogenized for 1 min in a tube containing 1 mL of lysis buffer (2% Triton X-100, 1% sodium dodecyl sulfate, 10 mM Tris–HCl pH 8, 100 mM NaCl, 1 mM EDTA pH 8), three 0.5 cm diameter glass beads (Sigma), and approximately 500 mg of 425 to 600 μm glass beads (Sigma). The homogenized mycelium is then snap-frozen in liquid nitrogen, thawed, and refrozen once. DNA is then extracted with the DNeasy plant kit (QIAGEN) [70].

31.2.2 Detection Procedures

31.2.2.1 PCR Identification of *Exophiala dermatitidis*

Nagano et al. [80] developed a novel species-specific PCR assay for *E. dermatitidis* using primers specific for the partial ITS1 region, the complete 5.8S rRNA region, and partial ITS2 region. The primers ExdF (5′-CCG CCT ATT CAG GTC C-3′) and ExdR (5′-TCT CTC CCA CTC CCG C-3′) amplify a 455 bp sequence from genomic DNA of *E. dermatitidis* only. This method facilitates the investigation on the occurrence, etiology, and epidemiology of *E. dermatitidis*.

31.2.2.2 Pan-Fungal Real-Time PCR and Sequencing Analysis

Pounder [89] described a real-time PCR assay utilizing SYBR green DNA-binding dye and high-resolution melting temperature analysis for fungal detection. Amplicons were generated using pan-fungal primers ITS1 forward (5′-TCCGTAGGTGAACCTGCGG-3′) and ITS4 reverse (5′-TCCTCCGCTTATTGATATGC-3′) [90]. The identity of the fungi is then verified by subsequent sequence analysis.

Procedure

1. A PCR mixture is prepared comprising of 1 × Lightcycler FastStart DNA Master Hybridization Probes mixture (Roche Applied Science) (containing deoxynucleoside triphosphates, FastStart *Taq* DNA polymerase, and 1 mM $MgCl_2$), 0.4 μM each of ITS1 forward and ITS4 reverse primers, 1 × SYBR green

(Molecular Probes), and 3 μL template DNA. The final $MgCl_2$ concentration is adjusted to 4.6 mM.

2. The thermal cycling parameters used is 95°C for 10 min; 50 cycles of 95°C for 5 s, 60°C for 20 s, and 76°C for 30 s; and a final extension at 72°C for 2 min.
3. The quality of the amplicon is determined using the derivative of the melt analysis curve (55°C–99°C, 45 s hold at 55°C, 5 s/°C) on the RotorGene 3000 (Corbett Robotics, Inc.).
4. The amplified product is purified for bidirectional sequencing using ExoSAP-IT (USB Corp.). Five microliters of BigDye Terminator Ready Reaction Mix v. 1.1 (Applied Biosystems) is added to 4 μL of each primer (0.8 pmol/μL) and 3 μL of purified PCR product. Cycle sequencing is performed with a 9700 thermal cycler (ABI), using 25 cycles of 96°C for 10 s, 50°C for 5 s, and 60°C for 4 min. Sequencing reaction products are purified using a Sephadex G-50 fine column to remove unincorporated dye terminators then electrophoresed on an ABI Prism 3100 Genetic Analyzer with a 50 cm capillary array.
5. Sequences are analyzed with the SmartGene Integrated Database Network software version 3.2.3. SmartGene is a web-based software and database system with reference sequences derived from the National Center for Biological Information (NCBI) GenBank repository.

Note: In case that real time PCR instrument is not available, standard PCR may be performed with primers ITS1 and ITS4, and the resulting amplicon is sequenced with the same primers. Sequence-based identifications are defined by percent identity: An isolate is placed within a species if there is greater than 99% sequence identity or within a genus if there is 93%–99% identity, and classified as inconclusive when there is less than 93% sequence identity.

31.2.2.3 Sequencing Analysis of the ITS1-5.8S-ITS2 Region

Zeng et al. [25] utilized primers V9G (5′-TTA CGT CCC TGC CCT TTG TA-3′) and LS266 (5′-GCAT TCC CAA ACA ACT CGA CTC-3′) for PCR amplification and sequence analysis of the ITS1-5.8S-ITS2 region [90].

Procedure

1. A 50 μL PCR reaction mixture comprising 10 mM Tris–HCl pH 8.3, 50 mM KCl, 1.5 mM $MgCl_2$, 0.01% gelatin, 200 mM of each deoxynucleotide triphosphate, 25 pmol of each primer, and 0.5 U of *Taq* DNA polymerase (Sigma) is prepared. Between 10 and 100 ng of rRNA is amplified.
2. Amplification is performed in a GenAmp PCR System 9700 thermocycler (Applied Biosystems) with an initial 95°C for 4 min, 35 cycles of 94°C for 45 s, 52°C for 30 s, and 72°C for 2 min, followed by a final extension step of 72°C for 7 min.
3. Amplicons are purified using a GFX column (GE Healthcare) and sequenced separately with the V9G and LS266 primers. The sequencing reaction consists of 1 μL of purified amplicon (1–10 ng), 3 μL of dilution buffer, 1 μL of BigDye v3.1, and 1 μL of primer (4 pmol/μL) and sterile MilliQ water to a final volume of 10 μL. The sequencing reaction cycling conditions consists of an initial 95°C for 1 min, 30 cycles at 95°C for 10 s, 50°C for 5 s, and final extension of 60°C for 2 min. The sequencing reactions are purified using Sephadex G-50 fine gel filtration medium (GE Healthcare Bio-Sciences) and analyzed on an ABI Prism 3730xl DNA analyzer (Applied Biosystems).
4. The sequences are adjusted using the Lasergene's SeqMan II software (DNASTAR, Inc.) and aligned using Ward's averaging in the Bionumerics package v4.0 (Applied Maths). The nearest neighbors are identified by local BLAST searches. The distance trees are based on a realigned file using the DCSE program and calculated by the neighbor-joining method of the Treecon package with Kimura-2 correction; only unambiguously aligned positions are taken into account. A total of 100 bootstrap replicates are used for analysis.

Note: If the similarity of sequences at the ITS region is greater than 99% between a studied strain and its nearest neighbor and they are distributed in the same branch of the phylogenetic tree, the strain is regarded as belonging to the same species as its nearest neighbor.

31.3 CONCLUSION

The genus *Exophiala* covers multiple species of dematiaceous (dark-walled) fungi that are widely distributed in natural environments. Of the 29 recognized *Exophiala* species, 18 have been isolated from human and animal hosts, either as colonizers or as disease-causing agents. Three types of mycosis are attributable to *Exophiala* spp., namely, phaeohyphomycosis, chromoblastomycosis, and mycetoma, with clinical manifestations ranging from localized cutaneous/subcutaneous infections, keratitis, otitis, endocarditis, pneumonia, nosocomial outbreaks of fungemia, and disseminated infections. Due to the fact that *Exophiala* spp. share morphological and biochemical characteristics with each other and also with other related fungi that cause similar clinical diseases and that they frequently fail to produce characteristic diagnostic structures in culture, it is necessary to apply molecular techniques to supplement microscopic examination and culture for their differentiation. To date, PCR amplification and sequencing analysis of the ITS1-5.8S-ITS2 gene, as well as the elongation factor 1-α and β-tubulin genes, provide a useful way to identify these fungi. It is envisaged that future identification of primers specific for individual *Exophiala* spp. will further enhance the specificity of the fungi classification and diagnosis.

REFERENCES

1. UniProt Consortium. Exophiala http://www.uniprot.org/5583, viewed June 27, 2010.
2. Li, D.M. et al. *Exophiala asiatica*, a new species from a fatal case in China. *Med. Mycol.*, 47, 101, 2009.
3. Untereiner, W.A. Capronia and its anamorphs: Exploring the value of morphological and molecular characters in the systematics of the Herpotrichiellaceae. *Stud. Mycol.*, 45, 141, 2000.
4. Matsuda, M. et al. Deoxyribonucleic acid hybridization studies of *Exophiala dermatitidis* and *Exophiala jeanselmei*. *Microbiol. Immunol.*, 33, 631, 1989
5. Kawasaki, M. et al. Mitochondrial DNA analysis of *Exophiala jeanselmei* var. *lecanii-corni* and *Exophiala castellanii*. *Mycopathologia*, 146, 75, 1999.
6. Zeng, J.S. et al. *Exophiala spinifera* and its allies: Diagnostics from morphology to DNA barcoding. *Med. Mycol.*, 46, 193, 2008.
7. De Hoog, G.S. et al. *Exophiala xenobiotica* sp. nov., an opportunistic black yeast inhabiting environments rich in hydrocarbons. *Antonie Leeuwenhoek*, 90, 257, 2006.
8. De Hoog, G.S. et al. Taxonomy of *Exophiala spinifera* and its relationship to *E. jeanselmei*. *Stud. Mycol.*, 43, 133, 1999.
9. De Hoog, G.S. et al. Species diversity and polymorphism in the *Exophiala spinifera* clade containing opportunistic black yeast-like fungi. *J. Clin. Microbiol.*, 41, 4767, 2003.
10. De Hoog G.S. et al. (eds.). *Atlas of Clinical Fungi*, 2nd edn. Centraalbureau voor Schimmelcultures/Universitat Rovira i Virgili, Utrecht, the Netherlands, Reus, Spain, p. 51, 2000.
11. Takeo, K. et al. Karyology and hyphal characters as taxonomic criteria in ascomycetous black yeasts and related fungi. *Antonie Leeuwenhoek*, 60, 35, 1991.
12. Kwon-Chung, K.J. et al. *Medical Mycology*. Lea & Febiger, Philadelphia, PA, 1992.
13. Matsumoto, T. et al. Critical review of human isolates of *Wangiella dermatitidis*. *Mycologia*, 76, 232, 1984.
14. Matsumoto, T. et al. Clinical and mycological spectra of *Wangiella dermatitidis* infections. *Mycoses*, 36, 145, 1993.
15. Kawasaki, M. et al. Mitochondrial DNA analysis of *Exophiala jeanselmei* and *Exophiala dermatitidis*. *Mycopathologia*, 110, 107, 1990.
16. McKemy, J.M. et al. Emendation of the genus *Wangiella* and a new combination, *W. heteromorpha*. *Mycologia*, 91, 200, 1999.
17. De Hoog, G.S. et al. Pleoanamorphic life cycle of *Exophiala* (*Wangiella*) dermatitidis. *Antonie Leeuwenhoek*, 65, 143, 1994.
18. De Hoog, G.S. et al. Intestinal prevalence of the neurotropic black yeast *Exophiala* (*Wangiella*) *dermatitidis* in healthy and impaired individuals. *Mycoses*, 48, 142, 2005.
19. De Hoog, G.S. et al. Nutritional physiology and selective isolation of *Exophiala dermatitidis*. *Antonie Leeuwenhoek*, 64, 17, 1993.
20. De Hoog, G.S. et al. Nutritional physiology of type isolates of currently accepted species of *Exophiala* and *Phaeococcomyces*. *Antonie Leeuwenhoek*, 68, 43, 1995.
21. Untereiner, W.A. et al. Nutritional physiology of species of Capronia. *Stud. Mycol.*, 43, 98, 1999.
22. Matos, T. et al. High prevalence of the neurotropic *Exophiala dermatitidis* and related oligotrophic black yeasts in sauna facilities. *Mycoses*, 45, 373, 2002.
23. Prenaefeta-Boldu, F.X. et al. Fungi growing on aromatic hydrocarbons: Biotechnology's unexpected encounter with biohazard? *FEMS Microbiol. Rev.*, 30, 109, 2006.
24. Vicente, V.A. et al. Environmental isolation of black yeast-like fungi involved in human infection. *Stud. Mycol.*, 61, 137, 2008.
25. Zeng, J.S. et al. Spectrum of clinically relevant *Exophiala* species in the United States. *J. Clin. Microbiol.*, 45, 3713, 2007.
26. Padhye, A.A. et al. Chromoblastomycosis caused by *Exophiala spinifera*. *Clin. Infect. Dis.*, 22, 331, 1996.
27. Rajendran, C. et al. Phaeohyphomycosis caused by *Exophiala spinifera* in India. *Med. Mycol.*, 41, 437, 2003.
28. Tomson, N. et al. Chromomycosis caused by *Exophiala spinifera*. *Clin. Exp. Dermatol.*, 31, 239, 2006.
29. Martínez-González, M.C. et al. Three cases of cutaneous phaeohyphomycosis by *Exophiala jeanselmei*. *Eur. J. Dermatol.*, 18, 313, 2008.
31. Parente, J.N. et al. Subcutaneous phaeohyphomycosis in immunocompetent patients: Two new cases caused by *Exophiala jeanselmei* and *Cladophialophora carrionii*. *Mycoses*, 2009 Webaddress: http://onlinelibrary.wiley.com/doi/10.1111/j.1439-0507.2009.01795.x/pdf, Last accessed on June 14, 2010.
32. Hague, J. et al. Subcutaneous phaeohyphomycosis caused by *Exophiala jeanselmei* in an immunocompromised host. *Cutis*, 72, 132, 2003.
33. Agger, W.A. et al. *Exophiala jeanselmei* infection in a heart transplant recipient successfully treated with oral terbinafine. *Clin. Infect. Dis.*, 38, e112, 2004.
34. De Monbrison, F. et al., Two cases of subcutaneous phaeohyphomycosis due to *Exophiala jeanselmei*, in cardiac transplant and renal transplant patients. *Br. J. Dermatol.*, 150, 597, 2004.
35. Silva Mdo, R. et al. Subcutaneous phaeohyphomycosis by *Exophiala jeanselmei* in a cardiac transplant recipient. *Rev. Inst. Med. Trop. Sao Paulo*, 47, 55, 2005.
36. González-López, M.A. et al. Subcutaneous phaeohyphomycosis caused by *Exophiala oligosperma* in a renal transplant recipient. *Br. J. Dermatol.*, 156, 762, 2007.
37. Galor, A. et al. Subconjunctival mycetoma after sub-Tenon's corticosteroid injection. *Cornea*, 28, 933, 2009.
38. Hohl, P.E. et al. Infections due to *Wangiella dermatitidis* in humans: Report of the first documented case from the United States and a review of the literature. *Rev. Infect. Dis.*, 5, 854, 1983.
39. Naka, W. et al. A case of chromoblastomycosis: With special reference to the mycology of the isolated *Exophiala jeanselmei*. *Mykosen*, 29, 445, 1986.
40. Tintelnot, K. et al. Cerebral phaeohyphomycosis caused by an *Exophiala* species. *Mycoses*, 34, 239, 1991.
41. Kusenbach, G. et al. *Exophiala dermatitidis* pneumonia in cystic fibrosis. *Eur. J. Pediatr.*, 151, 344, 1992.
42. Hiruma, M. et al. Systemic phaeohyphomycosis caused by *Exophiala dermatitidis*. *Mycoses*, 36, 1, 1993.
43. Campos-Takaki, G.M. et al. Report of chronic subcutaneous abscesses caused by *Exophiala spinifera*. *Mycopathologia*, 127, 73, 1994.
44. Kabel, P.J. et al. Nosocomial intravascular infection with *Exophiala dermatitidis*. *Lancet*, 344, 1167, 1994.
45. Horré, R. et al. Primary cerebral infections by melanized fungi: A review. *Stud. Mycol.*, 45, 176, 1999.
46. Chang, C.L. et al. Acute cerebral phaeohyphomycosis due to *Wangiella dermatitidis* accompanied by cerebrospinal fluid eosinophilia. *J. Clin. Microbiol.* 38, 1965, 2000.
47. Nucci, M. et al. Nosocomial fungemia due to *Exophiala jeanselmei* var. *jeanselmei* and a *Rhinocladiella* species: Newly described causes of bloodstream infection. *J. Clin. Microbiol.*, 39, 514, 2001.

48. Nucci, M. et al. Nosocomial outbreak of *Exophiala jeanselmei* fungemia associated with contamination of hospital water. *Clin. Infect. Dis.*, 34, 1475, 2002.
49. Ben-Simon, G.J. et al. More than tears in your eyes (*Exophiala jeanselmei* keratitis). *Cornea*, 21, 230, 2002.
50. Boisseau-Garsaud, A.M. et al. Onychomycosis due to *Exophiala jeanselmei*. *Dermatology*, 204, 150, 2002.
51. Engemann, J. et al. Exophiala infection from contaminated injectable steroids prepared by a compounding pharmacy—United States, July–November 2002. *MMWR*, 51, 1109, 2002.
52. Calista, D. et al. Subcutaneous *Exophiala jeanselmei* infection in a heart transplant patient. *Eur. J. Dermatol.*, 13, 489, 2003.
53. Greig, J. et al. Peritonitis due to the dematiaceous mold *Exophiala dermatitidis* complicating continuous ambulatory peritoneal dialysis. *Clin. Microbiol. Infect.*, 9, 713, 2003.
54. Murayama, N. et al. A case of subcutaneous phaeohyphomycotic cyst due to *Exophiala jeanselmei* complicated with systemic lupus erythematosus. *Mycoses*, 46, 145, 2003.
55. Myoken, Y. et al. Successful treatment of invasive stomatitis due to *Exophiala dermatitidis* in a patient with acute myeloid leukemia. *J. Oral Pathol. Med.*, 32, 51, 2003.
56. Negroni, R. et al. Case study: Posaconazole treatment of disseminated phaeohyphomycosis due to *Exophiala spinifera*. *Clin. Infect. Dis.*, 38, 15, 2004.
57. Tseng, P.H. et al. Central venous catheter-associated fungemia due to *Wangiella dermatitidis*. *J. Formos. Med. Assoc.*, 104, 123, 2005.
58. Al-Obaid, I. et al. Catheter-associated fungemia due to Exophiala oligosperma in a leukemic child and review of fungemia cases caused by Exophiala species. *Eur. J. Clin. Microbiol. Infect. Dis.*, 25, 729, 2006.
59. Patel, S.R. et al. *Exophiala dermatitidis* keratitis after laser in situ keratomileusis. *J. Cataract Refract. Surg.*, 32, 681, 2006.
60. Khan, S.A. Calcaneal osteomyelitis caused by *Exophiala jeanselmei* in an immunocompetent child. *J. Bone Joint Surg. Am.*, 89, 2547, 2007.
61. Leung, E.H. et al. *Exophiala jeanselmei* keratitis after laser in situ keratomileusis. *J. Cataract Refract. Surg.*, 34, 1809, 2008.
62. Harris, J.E. et al. *Exophiala spinifera* as a cause of cutaneous phaeohyphomycosis: Case study and review of the literature. *Med. Mycol.*, 47, 87, 2009.
63. Fahal, A.H. Mycetoma: A thorn in the flesh. *Trans. R. Soc. Trop. Med. Hyg.*, 98, 3, 2004.
64. Thammayya, A. et al. *Exophiala jeanselmei* causing mycetoma pedis in India. *Sabouraudia*, 18, 91, 1980.
65. Hemashettar, B.M. et al. Mycetoma due to *Exophiala jeanselmei* (a case report with description of the fungus). *Indian J. Pathol. Microbiol.*, 29, 75, 1986.
66. Neumeister, B. et al. Mycetoma due to *Exophiala jeanselmei* and Mycobacterium chelonae in a 73-year-old man with idiopathic CD^{4+} T lymphocytopenia. *Mycoses*, 38, 271, 1995.
67. Brownell, I. et al. Eumycetoma. *Dermatol. Online J.*, 11, 10, 2005.
68. Capoor, M.R. et al., Eumycetoma pedis due to *Exophiala jeanselmei*. *Indian J. Med. Microbiol.*, 25, 155, 2007.
69. Al-Tawfiq, J.A. et al. Madura leg due to *Exophiala jeanselmei* successfully treated with surgery and itraconazole therapy. *Med. Mycol.*, 47, 648, 2009.
70. Desnos-Olliver, M. et al. Molecular identification of black-grain mycetoma agents. *J. Clin. Microbiol.*, 44, 3517, 2006.
71. Ahmed, A.O. et al. Molecular detection and identification of agents of eumycetoma: Detailed report of two cases. *J. Clin. Microbiol.*, 41, 5813, 2003.
72. Bossler, A.D. et al. Exophiala oligosperma causing olecranon bursitis. *J. Clin. Microbiol.*, 41, 4779, 2003.
73. Aoyama, Y. et al. Subcutaneous phaeohyphomycosis caused by Exophiala xenobiotica in a non-Hodgkin lymphoma patient. *Med. Mycol.*, 47, 95, 2009.
74. Badali, H. et al. The clinical spectrum of *Exophiala jeanselmei*, with a case report and in vitro antifungal susceptibility of the species. *Med. Mycol.*, 29, 1, 2009.
75. Haase, G. et al. *Exophiala dermatitidis* infection in cystic fibrosis. *Lancet*, 336, 188, 1990.
76. Rath, P.M. et al. A comparison of methods of phenotypic and genotypic fingerprinting of *Exophiala dermatitidis* isolated from sputum samples of patients with cystic fibrosis. *J. Med. Microbiol.*, 46, 757, 1997.
77. Diemert, D. et al. Sputum isolation of *Wangiella dermatitidis* in patients with cystic fibrosis. *Scand. J. Infect.*, 33, 777, 2001.
78. Horré, R. et al. Isolation of fungi, especially *Exophiala dermatitidis*, in patients suffering from cystic fibrosis. A prospective study. *Respiration*, 71, 360, 2004.
79. Mukaino, T. et al. *Exophiala dermatitidis* infection in non-cystic fibrosis bronchiectasis. *Respir. Med.*, 100, 2069, 2006.
80. Nagano, Y. et al. Development of a novel PCR assay for the identification of the black yeast, *Exophiala* (*Wangiella*) *dermatitidis* from adult patients with cystic fibrosis (CF). *J. Cyst. Fibros.*, 7, 576, 2008.
81. Porteous, N.B. et al. Identification of Exophiala mesophila isolated from treated dental unit waterlines. *J. Clin. Microbiol.*, 41, 3885, 2003.
82. Phillips, G. et al. Black pigmented fungi in the water pipework supplying endoscope washer disinfectors. *J. Hosp. Infect.*, 40, 250, 1998.
83. Cox, H.H. J. et al. Growth of the black yeast *Exophiala jeanselmei* on styrene and styrene-related compounds. *Appl. Microbiol. Biotechnol.*, 39, 372, 1993.
84. Umemoto, N. et al. Two cases of cutaneous phaeohyphomycosis due to *Exophiala jeanselmei*: Diagnostic significance of direct microscopical examination of the purulent discharge. *Clin. Exp. Dermatol.*, 34, e351, 2009.
85. De Hoog, G.S. et al. Comparative rDNA diversity in medically significant fungi. *Microbiol. Cult. Collect.*, 13, 39, 1997.
86. Iwen, P. C. et al. Utilization of the internal transcribed spacer regions as molecular targets to detect and identify human fungal pathogens. *Med. Mycol.*, 40, 87, 2002.
87. Abliz, P. et al. Identification of pathogenic dematiaceous fungi and related taxa based on large subunit ribosomal DNA D1/D2 domain sequence analysis. *FEMS Immunol. Med. Microbiol.*, 40, 41, 2004.
88. Wang, L. et al. Identification, classification, and phylogeny of the pathogenic species *Exophiala jeanselmei* and related species by mitochondrial cytochrome b gene analysis. *J. Clin. Microbiol.*, 39, 4462, 2001.
89. Pounder, J.I. Discovering potential pathogens among fungi identified as nonsporulating molds. *J. Clin. Microbiol.*, 45, 568, 2007.
90. White, T.J. et al. Amplification and direct sequencing of fungal ribosomal RNA genes for phylogenetics. In M.S. Innis and D.H. Gelfand (eds.), *PCR Protocols: A Guide to Methods and Applications*, Academic Press, New York, 1990, p. 315.

32 *Fonsecaea*

Hideki Miyagi

CONTENTS

32.1 INTRODUCTION

The *Fonsecaea* infections such as chromoblastomycosis and phaeohyphomycosis affect cutaneous and subcutaneous area and have been treated mainly in dermatological field. Although these diseases occur chiefly in tropical and subtropical regions, they have also been reported in many other parts of the world. Similar to other dematiaceous fungi, *Fonsecaea* exists in environment as a saprophyte and is not an anthropophile. Generally, *Fonsecaea* infection is not lethal. However, once it invades systemically, mortality rate is high. Following recent global climate changes, this infection may extend from tropical and subtropical areas to the temperate zone as well. With limited treatment options for this fungal pathogen, it is important to identify it early and correctly for control and prevention purposes.

32.1.1 Classification, Morphology, and Biology

The genus *Fonsecaea* consists of ascomycetous fungi that are classified within order *Chaetothyriales*. Previously, two species, *F. pedrosoi* and *F. compacta*, are included in this genus on the basis of morphological characteristics. Several molecular investigations such as restriction fragment length polymorphism (RFLP) of mitochondrial DNA,[1] ribosomal RNA (rRNA) internal transcribed spacer (ITS) sequence,[2] random amplified polymorphic DNA (RAPD),[3] large subunit (LSU) rRNA D1/D2 domain sequence,[4] and RFLP of small subunit (SSU) rRNA and ITS regions[5,6] have revealed that *F. pedrosoi* and *F. compacta* have few distinctions at the molecular level, and *F. compacta* is considered as a morphological variant of *F. pedrosoi* now. A novel species *F. monophora* was suggested from sequencing of rRNA region by de Hoog et al.[2] in 2004. Thus, the genus *Fonsecaea* contains *F. pedrosoi* and *F. monophora* at present, and additional *Fonsecaea* species may exist as mentioned in literature such as group C.[7]

Morphologically, *F. monophora* forms only slightly longer conidial chains and slightly shorter denticles than *F. pedrosoi*.[2] Like other dematiaceous fungi, *Fonsecaea* sp. has melanin in its cell walls and this melanin seems to correlate with pathogenicity. Colonies of *Fonsecaea* sp. look brown with the naked eye. The slide culture of the *Fonsecaea* under light microscopy also appears brown due to the presence of melanin. Three types of conidial formations are seen on micro-slide cultures of this polymorphic fungus, i.e., *Cladosporium* forms, denticulate forms, and phialides. The conidiogenesis is important for morphological identification. However, some strains with poor conidiogenesis exist and additionally some morphological mutants such as meristematic mutant of *F. monophora* have been recovered. Xi et al.[8] postulated that the order *Chaetothyriales* may have a plesiomorphic potential to undertake extremotolerant growth.

F. pedrosoi and *F. monophora* are nonanthropophilic but saprophytes in the environment, rotten plant, soil, wood, etc. *F. monophora* was sometimes recovered from the environment, while *F. pedrosoi* was rarely isolated from the environment without mammal bait.[2,7,9,10] There were reports on the isolation of *F. pedrosoi* from thorns of the plant *Mimosa pudica*,[11] as well as from the shell of the babassu coconut (*Orbignya phalerata Martius*) in the Amazon.[12] However, these isolates were identified only morphologically but not yet molecularly. There is a need to evaluate whether clinical isolates and environmental isolates of *F. pedrosoi* are exactly same or not in terms of their pathogenicity. Molecular techniques have made it easier to examine environmental isolates in detail and also previously recovered isolates. Indeed, genetic diversity between pathogens of *F. pedrosoi* and environmental encounters of *F. pedrosoi* has been noted. In addition, many morphologically *Fonsecaea*-like fungi from environment are verified as different species by molecular examinations. de Hoog et al.[13] demonstrated that *Fonsecaea*-like strains are frequently found on rotten plant material and that fungal molecular diversity in the environment is higher than that in the patient. This complicates the precise tracking of the etiological agents of the disease and necessitates further elucidation of the correlation between pathogenic agents and environmental encounters.

F. pedrosoi is usually found in tropical and subtropical regions, particularly in humid areas. *F. pedrosoi* causes chromoblastomycosis as the principal agent, whereas *F. monophora* causes not only chromoblastomycosis but also phaeohyphomycosis, which may occasionally lead to possible invasive fungal infection. Indeed, there are a few reports of encephalitis cased by *F. monophora*.[14]

32.1.2 Clinical Features and Pathogenesis

Chromoblastomycosis is a cutaneous and subcutaneous fungal infection caused by several fungi belonging to the order *Chaetothyriales* such as *F. pedrosoi*, *F. monophora*, *Cladophialophora carrionii*, *Phialophora verrucosa*, *Rhinocladiella aquaspersa*, *Exophiala dermatitidis*, *Exophiala jeanselmei*, and *Exophiala spinifera*. *F. pedrosoi* is a leading causative agent of chromoblastomycosis among these fungi. However, each geographical area has a primary causative agent, e.g., in North America *Phialophora verrucosa* is the most common. In China, chromoblastomycosis is primarily caused by *C. carrionii* in the north and by *F. monophora* and *F. pedrosoi* in the southern and eastern parts of the country.[8,15] *Fonsecaea* sp. is not anthropophilic but saprophytic and traumatic inoculation is needed to establish infection. These traumas are often microtrauma that patients are unaware of. Mainly immunocompetent rural workers are infected with this disease but infection of immunosuppressive patients exists. Exposed sites like the lower limbs are the most affected areas because traumatic inoculations are likely to take place in these areas, although the buttocks, trunk, and face are affected as well.

Clinical manifestations of chromoblastomycosis are wide-ranging and have been classified into several types including nodular type, tumorous type, verrucous type, cicatricial type, and plaque type. Verrucous type is one of the most manifested types. As yet it is unknown whether these types are associated with specific etiologic agents or are dependent on host responses.[16] The clinical course tends to be chronic and oligosymptomatic; occasionally it could last for years. The detection of muriform cells (fumagoid, sclerotic, Medlar body), which are spherical, thick walled and brown, is needed for diagnosis of chromoblastomycosis in direct microscopic wet mount examination and histopathological sections. Fungal elements can be detected on hematoxylin-eosin-stained histopathological sections, and granulomatous reactions consist of histiocytes, macrophages, and multinucleated giant cells. Lymphocytes may also be seen in dermis with fibrosis. Pseudoepitheliomatous hyperplasia is often observed in the epidermis.

Chromoblastomycosis is recalcitrant and difficult to treat. Indeed, some therapeutic approaches such as excision, chemotherapy with antifungal drugs, cryotherapy, and topical thermotherapy have been applied to this disease but outcome is relatively variable. When the affected lesions are localized, excision is recommended. But once the lesions are expanded, chemotherapies with antifungal drugs including itraconazole and terbinafine can be chosen. Chemotherapeutic effects depend on species of causative agent and the severity of the disease. Each species has different sensitivity and *F. pedrosoi* is less sensitive than *C. carrionii*.

Phaeohyphomycosis is used to describe a broad group of infections ranging from localized cutaneous or subcutaneous disease to disseminated invasive disease[17] and is an umbrella term for the fungal infections caused by a diversity of dematiaceous fungi such as *Cladophialophora bantiana*, *Curvularia* sp., *Bipolaris* sp., *Exserohilum* sp., *Exophiala jeanselmei*, *Scedosporium prolificans*, *Ochroconis gallopava*, *Coniothyrium fuckelii*, *Phialophora parasitica*, *P. repens*, *Wangiella dermatitidis*, *Lasiodiplodia theobromae* principally, although *Fonsecaea* sp. is one of the pathogenic agents.

Phaeohyphomycosis is characterized pathologically by septate hyphae, toruloid hyphae, pseudohyphae, and no muriform cells. Subcutaneous phaeohyphomycosis presents an asymptomatic mass or nodule mainly at the extremities and head. It progresses chronically and the first-line regimen to localized lesion is surgical excision. Antifungal drugs will be applied to widely spread lesion, but later relapse is common. In systemic phaeohyphomycosis, not only brain but also joints, endocardium, and peritoneum through peritoneal dialysis are affected. Disseminated phaeohyphomycosis is uncommon, but its incidence seems to increase gradually with the rising of immunocompromised individuals such as leukemia patients, hematopoietic stem cell transplant recipient, and solid organ transplant recipients. The treatment includes excision, antifungal or a combination of these, but outcome is poor. The overall mortality rate amounts up to about 84% in immunocompromised patients. On the other hand, especially cerebral phaeohyphomycosis is most common in immunocompetent individuals with no obvious risk factors although cases have occurred in immunocompromised persons.[17] The clinical outcome of cerebral phaeohyphomycosis is dismal too, with long-term survival being reported only when complete surgical resection of discrete lesions is possible.[17] The clinical potential of *F. monophora* differs from that of *F. pedrosoi* because *F. monophora* is predominantly neurotropic in the human host.[18]

32.1.3 Diagnosis

Infectious diseases caused by the genus *Fonsecaea* such as chromoblastomycosis and phaeohyphomycosis are slow progressive and are refractory to treatment. If the diagnosis is done while the lesions are still small, somewhat effective choices of treatment including excision, antifungal, topical cryotherapy etc. are there. Once the lesions expand chronically, treatment effect will be limited and relapse is more common. Therefore, the disease should be diagnosed as early as possible. Pathogenicity, virulence, and sensitivity of *Chaetothyriales* black yeasts may differ between closely related species.[9] The causative agents have to be isolated and identified for precise diagnosis, treatment, and prevention. These efforts make an impact on prognosis of the disease. The cultures of causative agents are required for identification although actually there are some attempts to identify it directly from clinical samples by molecular procedures. Separation of *F. pedrosoi* and *F. monophora* is also significant clinically because the spectrum of infection due to *F. monophora* has been shown to be more variable than that of *F. pedrosoi*[15] and occasionally invasive. We have to know these differences for correct treatment.

32.1.3.1 Conventional Techniques

The clinical pictures of the *Fonsecaea* infection are broad range. They are divided into several clinical types as mentioned above. The differential diagnosis includes sporotrichosis, protothecosis, leprosy, lupus erythematosus, sarcoidosis and so on. If some kind of fungal infection is suspected, a series of clinical and laboratory tests are started. The scales are taken for direct wet mount preparation from especially "black dot," which represents fungal elements during transepidermal elimination,[19] and then are examined for the presence of muriform cells and fungal elements. The skin biopsy is also done for detection of muriform cells, fungal elements, granulomatous reaction, etc., in pathological sections. Clinical samples such scales, crusts, and biopsied skin tissues will be submitted to a laboratory for culture isolation and identification. Morphological, physiological, and biochemical tests are available for conventional identification of the *Fonsecaea* sp.

On morphological identification, the causative agents are cultured on Sabouraud's dextrose agar and/or another appropriate agar for observation by the naked eye, and microculture is also done for observation under light microscope. Examination of microculture under light microscope is sometimes more revealing than observation of plate with the naked eye. Combining microscopic examination of microculture with naked eye observation of the culture makes morphological identification of an isolate easier. This is similar to a correlation between clinical pictures and pathological sections in clinical practice. Pathological sections reveal the details of the disease, but only sections without clinical pictures make possible a poor diagnosis and bring us some confusion occasionally. Thus, using both types of observations, one of clinical pictures at the naked eye level and another of pathological sections at light-microscopic level, makes more precise evaluation and diagnosis.

On culture, the colony of *Fonsecaea* grows relatively slowly and shows a velvety, dark colony on a slant agar and/or on a petri dish agar (Figure 32.1a). The microculture of the *Fonsecaea* shows three types of conidiogenesis as mentioned above: phialides (Figure 32.1, d1 and d2) observed sporadically with collarettes of 5–6 μm. *Cladosporium* forms (Figure 32.1, b1 and b2) observed regularly in the strains, made of branched chains of blastospores showing little variations between the strains. Denticulate (Figure 32.1, c1 and c2) forms, acrotheca or *Rhinocladiella* form, have typically a sterile basal part and a distal fertile one which is more or less developed, bearing conidia formed in a sympodial way.[20] Exceptionally, *F. compacta*, which is not a species but a morphological variant of *F. pedrosoi* today, shows that the conidial head is elongated, and the chains of globose to subglobose conidia are held compactly together.[21]

A delayed-type skin test using a prepared antigen has been used to diagnose and identify an isolate, but it seems that specificity is limited yet.[22,23] Some assimilation tests are reported for identification.[24,25]

32.1.3.2 Molecular Techniques

Identification of pathogenic dematiaceous fungi including the genus *Fonsecaea* is typically done by morphological and physiological procedures. However, these procedures are time-consuming, require technical expertise, and are ineffective for identification of species with poor conidial production. One of the problems for morphological identification is morphological similarity among dematiaceous fungi and polymorphic conidiogenesis of the *Fonsecaea* and some other species. Recently, it has been recognized that molecular procedures (Figure 32.2) are useful for identification, phylogeny, taxonomy, typing, and epidemiology. The molecular investigation is faster, less laborious, and highly reproducible and could examine the strains with poor conidiogenesis and morphological mutation. Moreover, it can detect geographical diversity, which is useful for epidemiology, and in the near future can evaluate pathogenicity of isolates, virulence, and sensitivity to treatment from molecular diversity.

Although evaluation and comparison of all genomic DNA will reveal many interesting details, this is practical. We have to target the proper DNA regions for classification and identification. It has been reported that the rRNA regions (coding for rRNA) are suitable for the purpose of classification and identification of the *Fonsecaea*. rRNA is a component of ribosome and plays a role in protein synthesis with enzyme-like function. The rRNA is divided into several parts such as SSU rRNA, 5.8S rRNA, LSU rRNA, ITS and non-transcribed spacer (NTS) (Figure 32.3). A unit of rRNA containing all these parts is arranged repeatedly and makes a repeating structure in nucleus, although species with only one unit of rRNA in one cell such a *Pneumocystis jirovecii* exist. This repeating structure of rRNA in each cell means that enough template DNA is derived for investigation from relatively small amounts of sample, which then makes the detection of nucleotide alignment easy.

Using molecular techniques, geographical genetic diversity of *F. pedrosoi* was detected. Kawasaki et al.[1] revealed with RFLP of mtDNA that *F. pedrosoi* is divided into two groups and seven types correlating with geographical distribution respectively and clarified that this molecular method is

FIGURE 32.1 Plate of *F. monophora* (a) from chromoblastomycosis. The velvety, dark colony with relative slow growth is observed. The three types of conidiogenesis are displayed. *Cladosporium* form (b, c), denticulate form (d, e) and phialides (f, g) as seen by microculture technique (b, d, f) and scanning electron microscopy (c, e, g), respectively.

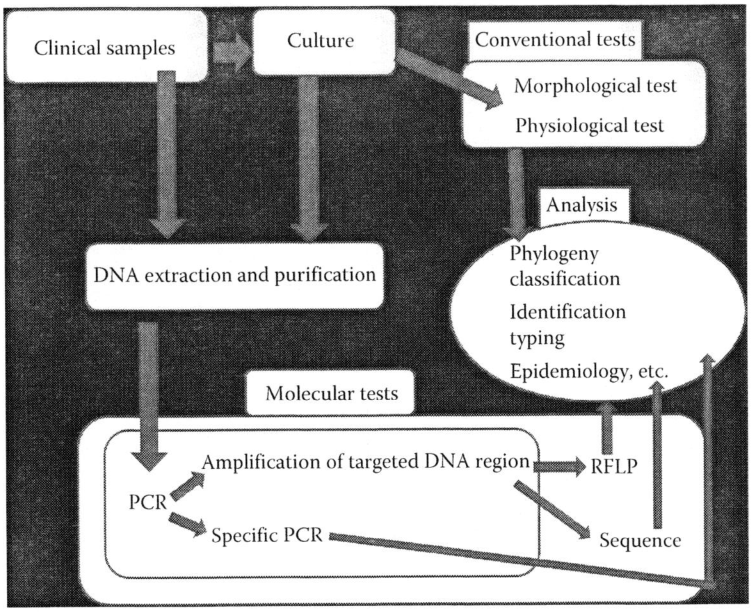

FIGURE 32.2 Course of diagnostic investigation.

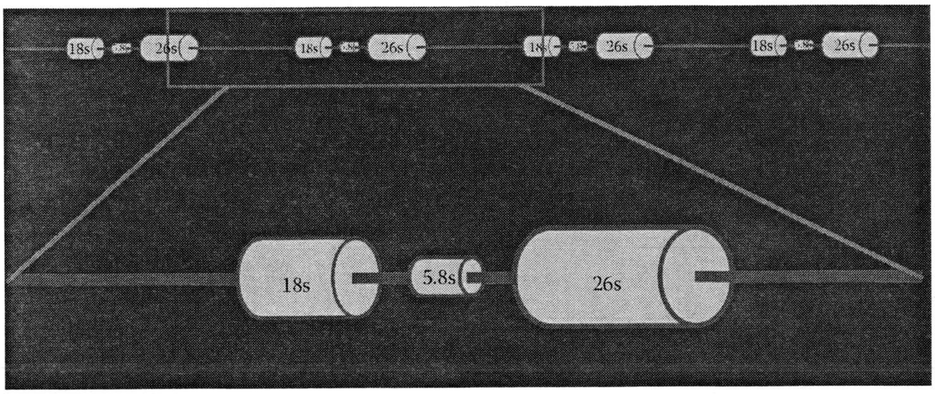

FIGURE 32.3 Schematic view of rRNA region.

useful for identification. This geographical genetic diversity was confirmed with the ITS-RFLP analysis, which classified 131 *F. pedrosoi* into five types (D1–D5) by *Dde I*, four types (M1–M4) by *Msp I*, and a combined six rRNA types corresponding to the six mtDNA types.[5] Additionally, this ITS-RFLP demonstrated that *F. pedrosoi* was discriminated from 11 other dematiaceous fungi, making ITS-RFLP a useful tool for identification and epidemiological investigation.[5]

In another report, the RFLP analysis of ribosomal gene SSU rRNA and ITS regions of 12 isolates of dematiaceous agents showed inter- and intraspecific variability in ITS and conservation in SSU rRNA region.[6] de Hoog et al.[2] suggested the existence of the novel species *F. monophora* from sequencing of ITS. *Fonsecaea* was classified into group A and group B corresponding to *F. pedrosoi* and *F. monophora*, respectively, which is unrelated to *F. compacta*. Further investigation by RAPD procedure demonstrated that group A of *F. pedrosoi* divided further into RAPD groups I and III, and group B of *F. monophora* into RAPD groups II and IV. A subsequent phylogenetic study of ITS region and cytochrome *b* gene involved 33 clinical specimens that were identified as *F. pedrosoi* morphologically and other strains of *Fonsecaea* sp. deposited already in the GenBank.[26] It was clear that *Fonsecaea* sp. is divided into two major clades, groups A and B, besides subgroups A-1 and A-2 and B-1, B-2, and B-3, respectively. The groups A and B correspond to *F. pedrosoi* and *F. monophora*, respectively. The subgroup B-1 is equal to all the *F. monophora* strains described by de Hoog et al.[2] The subgroup B-2 consisted of all the strains from Japan identified as *F. pedrosoi* morphologically and one isolated from China. Intraspecies diversities corresponding to geographic areas were detected as well. Moreover, 24 isolated strains from China and some reference strains of the *Fonsecaea* were classified by ITS rRNA sequences into *F. pedrosoi* as genotypes A-1 and A-2 and *F. monophora* as genotypes B-1, B-2, and B-3.[15] The correlation between geographical distribution and genotypes was detected.

In addition, multi-locus sequence investigation on the sequences of ITS, *TUB1*, and *ACT1*genes divided *Fonsecaea* sp. into three clades as group A for *F. pedrosoi*, group B for *F. monophora*, and group C for possible new species.[7] Not only ITS region but also LSU rRNA were investigated. Analysis of the nucleotide sequences of the D1/D2 domains of LSU ribosomal DNA of pathogenic dematiaceous fungi and related taxa revealed that intraspecies sequence diversity was extremely small, while interspecies diversity was higher. Thus, the D1/D2 domain's nucleotide alignments are valuable for identification and phylogenetic analysis.[4] Also, an oligonucleotide primer set targeting ITS region of the genus *Fonsecaea* specifically was designed for more rapid identification in 2003[27] although the novel species *F. monophora* was not yet suggested in 2003, and this valuable rapid identification method, specific PCR for *Fonsecaea*, may need reevaluation with the new species. Furthermore, the duplex PCR targeting the rRNA for rapid and specific identification was introduced.[28] This duplex PCR amplifies one band with universal primer pair ITS1 and ITS4 for broad species of fungi and the other band with specific primer pair TSA1 and TSAS for detection of *Fonsecaea* simultaneously. The band generated by universal primer functions as a positive control, making the reliable and correct identification of *Fonsecaea* possible.

Thus, these cutting-edge molecular techniques revealed the existence of the new species *F. monophora*, which couldn't be detected with conventional morphological techniques. Thanks to these molecular techniques, the phylogenetic analysis can be performed in detail, and the identification of not only the species level but also the strain level is possible. Geographical diversity in each strain and its epidemiological characteristics can also be investigated. Evaluation of pathogenicity, virulence, and sensitivity of each strain represents another useful area for molecular techniques

32.2 METHODS

32.2.1 Sample Preparation

Clinical specimens containing causative fungus elements such as scales, crusts, pus, biopsied and fragmented skin tissue, etc. are submitted for cultures and/or direct molecular tests. For fungus culture, the clinical specimens are inoculated onto a slant agar such as Sabouraud's dextrose agar and/or another proper agar containing antibiotics. Proliferated fungus elements from culture are offered to molecular tests. A pick of fungus element with a loop is taken and immersed in 400 μL resolution buffer (20 mM Tris–HCl pH 8.0, 5 mM ethylenediaminetetraacetic acid (EDTA), 400 mM NaCl, 0.3% sodium dodecyl sulfate (SDS), and 200 μg/mL proteinase K) in 1.5 mL tube. Then, this mixture is incubated for several hours at 55°C and diluted with nine volumes of distilled water. This dilution is done for reducing reaction of inhibitor to proper amounts for PCR. And then, this diluted solution becomes the template for molecular tests.

Ampdirect® Plus kit (Analytical & Measuring Instruments Division, Shimadzu Corporation) may be used in PCR mixture as it can neutralize the inhibiting materials in the sample, eliminating false negative results due to the presence of DNA polymerase inhibitors.

32.2.2 Detection Procedures

Sequencing, RFLP, and specific PCR targeting nuclear rRNA, mitochondrial DNA, and mitochondrial cytochrome b gene are useful for molecular investigation of *Fonsecaea* sp. (Figure 32.2), and to date most reports have focused on nuclear rRNA. The direct sequence of rRNA is one of the most practiced procedures in our lab.

The universal primers targeting ITS region of broad fungi include ITS1: 5′-TCCGTAGGTGAACCTGCGG-3′; ITS2: 5′-GCTGCGTTCTTCATCGATGC-3′; ITS3: 5′-GCATCGATGAAGAACGCAGC-3′; ITS4: 5′-TCCTCCGCTTATTGATATGC-3′; ITS5: 5′-GGAAGTAAAAGTCGTAACAAGG-3′.[29] The ITS region is amplified with these primers in PCR from sample's genomic DNA. The PCR mixture (20 μL) is composed of 10 μL of 2× Ampdirect Plus, 0.1 μL of Nova Taq™ Hot Start DNA Polymerase (5 U/μL), 1 μL of 10 μM forward primer, 1 μL of 10 μM reverse primer, 0.5–1 μL of template DNA, and 7.4 μL of distilled water.

Amplification is performed in Gene Amp PCR system 9700 with the following conditions: 1 cycle at 80°C for 15 min and 94°C for 4 min followed by 45 cycles at 94°C for 30 s, 55°C for 1 min, and 72°C for 1 min, with a final extension at 72°C for 7 min. The first cycle at 80°C for 15 min helps reduce activity of inhibitory substances present, And the subsequent 45 cycles are neccessary because this PCR has a relatively low amplification rate due to the existence of inhibitory substances without purification.

The amplified products are analyzed by 1% agarose gel electrophoresis. On direct sequencing, at first the amplified products are recovered from electrophoresed agarose gel with QIAquick Gel Extraction Kit (Qiagen). The concentration of amplified product contained in this recovered solution is measured using a spectrophotometer and by the comparison of electrophoresed density between this recovered solution and a solution whose concentration of DNA already known e.g., size marker. Sometimes, the concentration measured by electrophoresed density is low in spite of the high concentration measured by the spectrophotometer in the same sample. This is because the irradiation of UV light damages the structure of DNA when the amplified product is recovered. The exact amount of complete DNA is important for direct sequence reaction and the concentration value from the electrophoresed density is more reliable.

The sequencing reaction (10 μL) consists of 2 μL of premix of BigDye™ Terminator Cycle Sequencing Ready Reaction (Applied Biosystems), 2 μL of dilution buffer, 1.6 μL of primer (2 pmol/μL), 10–20 ng of template DNA, and up to 10 μL of distilled water. Sequencing reaction is performed in Gene Amp PCR system 9700 with the following conditions: 1 cycle at 96°C for 1 min followed by 25 cycles at 96°C for 30 s, 48°C for 30 s, and 60°C for 2min, with a final extension at 72°C for 7 min. Then this reaction solution is purified with ethanol and applied to a sequencer of ABI PRISM 310

Genetic Analyzer. Acquired nucleotide alignment is compared with deposited strains via internet tool BLAST offered by NCBI.

The isolated strain from our reported[30] case identified as *F. pedrosoi* was sequenced. This isolate was reidentified as *F. monophora* not *F. pedrosoi* and was related to *F. monophora* subgroup B-2[26] with four to nine nucleotide differences (similarity rate is 98.4%–99.3%). The strain differs from *F. pedrosoi* A-1 strain AY366931 and A-2 strain AB087202 with 12 (similarity 97.9%) and 13 (similarity is 97.7%) nucleotide differences, respectively; it also differs from *F. monophora* subgroup B-1 strains and B-3 strains with 6–9 (98.4%–98.9% similarity) and 6 (98.4% similarity) bp differences each. Indeed, our strain is considered near to *F. monophora* subgroup B-2, but it seems like that there is some distance to each B-2 strains. The relative distance from B-2 of each strain may be attributed to factors such as geographical location, patient's immunocompromised status, etc. Okinawa (The Ryukyu Islands) is located on the south edge of Japan in the subtropical region, several hundred miles away from the main islands of Japan and closer to the middle of south China and Taiwan. The patient has a history of Hansen's disease. The effects of geographical character and/or host's immune status on the genetic distance of the strain need to be evaluated further.

32.3 CONCLUSION AND FUTURE PERSPECTIVES

Mankind has benefited from resolving and treating infectious diseases through use of vaccines and antibiotics. Correct identification of the causative agents plays a critical role in our endeavor to overcome infections. In addition, it contributes to the understanding of causes, origins, roots, and pathogenesis of infectious diseases. Today's molecular techniques have led mycological diagnostics to a new level of sensitivity, specificity, and speed. The novel species *F. monophora* was discovered by molecular techniques, while traditional morphological and physiological techniques were unable to do so. Use of molecular techniques has also helped clarify intraspecies diversity, which may underscore geographical diversity and will be instrumental in the elucidation of *Fonsecaea* pathogenicity and virulence in future.

Dematiaceous fungi can be divided into three ecological groups: saprobes (that are not known as pathogens), low-virulent opportunists (that can directly be isolated from the environment), and highly specific pathogens (that cannot be isolated from the environment directly but require living mammal bait).[9] The correlation among the pathogen of the *Fonsecaea*, nonpathogen and/or low pathogen of the *Fonsecaea*, and other dematiaceous fungi will be unraveled through further molecular analysis. Besides rRNA, other genetic regions may be investigated for identification purpose. Multi-locus genetic analysis, DNA microarray, and the DNA chip will be also valuable for the study of *Fonsecaea* spp.

REFERENCES

1. Kawasaki, M. et al., Molecular epidemiology of *Fonsecaea pedrosoi* using mitochondrial DNA analysis, *Med. Mycol.*, 37, 435, 1999.
2. de Hoog, G.S. et al., Molecular ecology and pathogenic potential of *Fonsecaea* species, *Med. Mycol.*, 42, 405, 2004.
3. Caligiorne, R.B. et al., Use of RAPD (random amplified polymorphic DNA) to analyse genetic diversity of dematiaceous fungal pathogens, *Can. J. Microbiol.*, 45, 408, 1999.
4. Abliz, P. et al., Identification of pathogenic dematiaceous fungi and related taxa based on large subunit ribosomal DNA D1/D2 domain sequence analysis, *FEMS Immunol. Med. Microbiol.*, 40, 41, 2004.
5. Tanabe, H. et al., Species identification and strain typing of *Fonsecaea pedrosoi* using ribosomal RNA gene internal transcribed spacer regions, *Jpn. J. Med. Mycol.*, 45, 105, 2004.
6. Caligiorne, R.B. et al., Dematiaceous fungal pathogens: Analysis of ribosomal DNA gene polymorphism by polymerase chain reaction-restriction fragment polymorphism, *Mycoses*, 42, 609, 1999.
7. Najafzadeh, M.J. et al., Genetic diversity and species delimitation in the opportunistic genus *Fonsecaea*, *Med. Mycol.*, 47, 17, 2009.
8. Xi, L. et al., Chromoblastomycosis caused by a meristematic mutant of *Fonsecaea pedrosoi*, *Med. Mycol.*, 47, 77, 2009.
9. Vicente, V.A. et al., Environmental isolation of black yeast-like fungi involved in human infection, *Stud. Mycol.*, 61, 137, 2008.
10. Okeke, C.N. and Gugnani, H.C., Studies on pathogenic dematiaceous fungi, *Mycopathologia*, 94, 19, 1986.
11. Salgado, C.G. et al., Isolation of *Fonsecaea pedrosoi* from thorns of *Mimosa pudica*, a probable natural source of chromoblastomycosis, *Rev. Inst. Med. Trop. Sao Paulo*, 46, 33, 2004.
12. Marques, S.G. et al., Isolation of *Fonsecaea prdrosoi* from the shell of the babassu coconut (Orbignya phalerata Martius) in the Amazon region of Maranhao Brazil, *Jpn. J. Med. Mycol.*, 47, 305, 2006.
13. de Hoog, G.S. et al., Black fungi: Clinical and pathological approaches, *Med. Mycol.*, 38S, 243, 2000.
14. Takei, H., Goodman, J.C., and Powell, S.Z., Cerebral phaeohyphomycosis caused by *Cladophialophora bantiana* and *Fonsecaea monophora*: Report of three cases, *Clin. Neuropathol.*, 26, 21, 2007.
15. Xi, L. et al., Molecular diversity of *Fonsecaea* (*Chaetothyriales*) causing chromoblastomycosis in southern China, *Med. Mycol.*, 47S, 27, 2009.
16. Najafzadeh, M.J. et al., Successful treatment of chromoblastomycosis of 36 years duration caused by *Fonsecaea monophora*, *Med. Mycol.*, 48, 390, 2010.
17. Brandt, M.E. and Warnock, D.W., Epidemiology, clinical manifestations, and therapy of infections caused by dematiaceous fungi, *J. Chemother.*, 15S, 36, 2003.
18. Surash, S. et al., Cerebral phaeohyphomycosis caused by *Fonsecaea monophora*, *Med. Mycol.*, 43, 465, 2005.
19. Ameen, M. Chromoblastomycosis: Clinical presentation and management, *Clin. Exp. Dermatol.*, 34, 849, 2009.
20. Ibrahim-Granet, O. and de Bievre, C., Study of the conidial development and cleistothecium-like structure of some strains of *Fonsecaea pedrosoi*, *Mycopathologia*, 84, 181, 1984.
21. Wang, D.L., Li, R.Y., and Wang X.H., Observations of sporulation in *Fonsecaea* and *Phialophora* by scanning electron microscopy, *Mycopathologia*, 98, 105, 1987.

22. Iwatsu, T. et al., Evaluation of skin test for chromoblastomycosis using antigens prepared from culture filtrates of *Fonsecaea pedrosoi*, *Phialophora werrucosa*, *Wangiella dermatitidis* and *Exophiala jeanselmei*, *Mycopathologia*, 77, 59, 1982.
23. Iwatsu, T. et al., Skin test-active substance prepared from culture filtrate of *Fonsecaea pedrosoi*, *Mycopathologia*, 67, 101, 1979.
24. Espinel-ingroff, A. et al., Evaluation of the API 20C yeast identification system for the differentiation of some dematiaceous fungi, *J. Clin. Microbiol.*, 27, 2565, 1989.
25. Steadham, J.E., Geis, P.A., and Simmank, J.L., Use of carbohydrate and nitrate assimilations in the identification of dematiaceous fungi, *Diagn. Microbiol. Infect. Dis.*, 5, 71, 1986.
26. Yaguchi, T. et al., Molecular phylogenetics of strains morphologically identified as *Fonsecaea pedrosoi* from clinical specimens, *Mycoses*, 50, 255, 2007.
27. Abliz, P. et al., Rapid identification of the genus *Fonsecaea* by PCR with specific oligonucleotide primers, *J. Clin. Microbiol.*, 41, 873, 2003.
28. Andrade, T.S. et al., Rapid identification of *Fonsecaea* by duplex polymerase chain reaction in isolate from patients with chromoblastomycosis, *Diagn. Microbiol. Infect. Dis.*, 57, 267, 2007.
29. White, T.J. et al., *Amplification and Direct Sequencing of Fungal Ribosomal RNA Genes for Phylogenetics, PCR: A Guide to Methods and Applications*, Academic Press, Inc., New York, 1990, pp. 315–322.
30. Miyagi, H. et al., Case of chromoblastomycosis appearing in an Okinawa patient with a medical history of Hansen's disease, *J. Dermatol.*, 35, 354, 2008.

33 *Histoplasma*

Rosely Maria Zancopé Oliveira, Allan Jefferson Guimarães, Joshua D. Nosanchuk, Mauro de Medeiros Muniz, Priscila Costa Albuquerque, and Rodrigo de Almeida Paes

CONTENTS

33.1 INTRODUCTION

33.1.1 Classification and Morphology

33.1.1.1 Classification

Histoplasma capsulatum is the anamorphic form of *Ajellomyces capsulatum*. It belongs to the phylum Ascomycota, class Eurotiomycetes, order Onygenales, and family Onygenaceae and/or Ajellomycetaceae.[1–3] *A. capsulatum* is a heterothallic fungus with two distinct mating types: (+) and (−). They exist in a 1:1 ratio in soil; however, the (−) type is seven times more frequent in clinical samples.[4] There are three *H. capsulatum* varieties: *H. capsulatum* var. *capsulatum*, the etiological agent of classic histoplasmosis; *H. capsultaum* var. *duboisii*, the etiological agent of African histoplasmosis; and *H. capsulatum* var. *farciminosum*, the etiological agent of epizootic lymphangitis of horses and mules.[5] However, based on molecular studies, Kasuga et al.[6] suggest that *H. capsulatum* might be composed of six distinct species, instead of three varieties. This chapter focuses on *H. capsulatum* var. *capsulatum*.

33.1.1.2 Morphology

H. capsulatum is a dimorphic fungal pathogen, since it displays two distinct morphological forms, depending on the nutritional factors and temperature. The filamentous morphological form occurs at temperatures below 35°C or in the soil. This form is composed of hyaline septate hyphae that are 1–2.5 μm in diameter. Moreover, hyphae produce two different hyaline asexual reproduction structures. Microconidia are round to pear shaped, smooth, and 2–5 μm in diameter. Macroconidia are large, thick walled, round, and 7–15 μm in diameter. The macroconidia are typically tuberculate, knobby, or with short cylindrical projections, although they may occasionally be smooth. These macroconidia are similar to those of *Sepedonium*; however, this genus is unable to produce microconidia.[7]

In parasitism or when cultivated at 37°C in specific media, *H. capsulatum* var. *capsulatum* are small yeast cells (2–4 μm in length) that are ovoid with a narrow base at the smaller end. The variety *duboisii* differs from the variety *capsulatum* by producing larger yeast cells (8–15 μm in length).

The teleomorph state of *H. capsulatum* is *A. capsulatum*. When (+) and (−) *A. capsulatum* strains are paired on alphacel-yeast extract agar they produce cleistothecia. They initially appear globose or subglobose and then form irregular stellates due to the extended growth of coils beyond the outer limit of the peridium. Asci are club to pear shaped and contain eight globose ascospores with a diameter of ~1.5 μm.[4]

33.1.2 Biology and Epidemiology

33.1.2.1 Biology

H. capsulatum is a haploid ascomycete, although aneuploidy can occur in some strains, such as the reference Downs strain.[8] It is closely related to other dimorphic fungi that also cause pulmonary infections such as *Blastomyces dermatitidis* and *Coccidioides immitis* or *C. posadasii*. *H. capsulatum* has seven chromosomes and the genome size is approximately 25–30 Mb, although these values vary among different strains.[9]

The *H. capsulatum* cell wall is a rigid, largely polysaccharide structure composed of four different glycans: soluble galactomannan, α-1,3-glucan, β-1,3-glucan, and a fibrillar chitin skeleton. The amount of each glycan depends on the chemotype of the strain, although some fluctuations can occur in certain isolates from a single chemotype.[10] The α-1,3-glucan is considered relevant for *H. capsulatum* virulence, whereas the β-1,3-glucan, which is predominant in the fungal mycelial phase, is antigenic and participates in the modulation of the host immune response.[11]

The vast majority of *H. capsulatum* biomass exists in the soil as a mold, since the parasitic stage is not necessary or advantageous for the survival of the fungus. Interestingly, the parasitic stage of *H. capsulatum* may have developed in response to interactions between the fungus and other soil-living microorganisms, such as *Acanthamoeba castellanii*, since phagocytosis of the fungus can result in death of amoeba and *H. capsulatum* growth in the yeast form.[12]

In both saprophytic and parasitic stages, *H. capsulatum* must face an overabundance of diverse challenging environmental conditions. In response, *H. capsulatum* produces several biological molecules such as catalase to survive oxidative stress conditions, siderophores to survive iron starvation, and orotidine 5-monophosphate pyrophosphorylase to endure uracil limitation.[9] This fungus is also able to produce melanin, on both conidia and yeast cells.[13] Melanins are negatively charged, hydrophobic pigments of high molecular weight, formed by the oxidative polymerization of phenolic and/or indolic compounds.[14] Since *H. capsulatum* conidia synthesize melanin in the absence of exogenous phenolic substrate, it is probable that conidia are melanized in the environment. Thus, melanization may protect the conidia from environmental insults.[13] Moreover, yeast melanization appears to contribute to virulence by reducing *H. capsulatum* susceptibility to host defense mechanisms and antifungal drugs.[15,16]

Vesicular secretion of macromolecules has recently been described in *H. capsulatum*. Macromolecules transported via a trans-cell wall vesicular transport secretory mechanism by yeast cells are involved in diverse processes, including metabolism, cell recycling, signaling, and virulence. The vesicles from the yeast form of *H. capsulatum* cells react with immune serum from patients with histoplasmosis, providing an association of the vesicular products with pathogenesis and also represent a possible new target for antifungal drugs.[17]

33.1.2.2 Epidemiology

Numerous animal species can be infected by *H. capsulatum*. Contaminated soils are the infection source for both human and other animals, such as dogs, cats, horses, cattle, pigs, rodents, and marsupials.[18] Usually the fungus is associated with soils enriched with organic nitrogen sources like animal excrements, especially of bats and chickens.[5,19] In fact, *H. capsulatum* has been isolated from several protected environments, such as caves, abandoned construction sites, and chicken coops. *H. capsulatum* does not cause disease in birds, so their role in the environmental spread of *H. capsulatum* is probably due to fungal carriage on their feathers, beaks, or claws. On the other hand, bats are fungal reservoirs, since they harbor the fungus in their intestinal mucosa and release *H. capsulatum* in their feces.[20] Also, histoplasmosis can occur in the bats, including lethal disease. Recently, Julg et al. have detailed the acquisition of human histoplasmosis at the entrances of bat caves, without exposure to deposits of guano within the caves.[21] Additionally, bat migration can contribute for the generation of new *H. capsulatum* niches on the environment.[5]

H. capsulatum infection is typically acquired by microconidia inhalation after disturbances of soil or excreta. For this reason, histoplasmosis is not generally associated with person to person spread of disease. Rarely histoplasmosis can be acquired by cutaneous inoculation of the fungus and, even more unusually, vertical transmission can occur, given that there is one such case in the literature.[4]

Histoplasmosis is a cosmopolitan fungal infection with areas of high endemicity. However, patients are usually unaware of their potential exposure.[22] Endemic regions in North America are located in Midwestern and Southeastern regions of the United States of America, especially the Mississippi, Ohio, and Missouri river valleys, locations where 80% of the resident population reacts to the histoplasmin skin test.[18] Data from the United States suggests that half a million new human infections occur each year.[9] In Latin America, the most prevalent areas of endemicity are within Brazil, Venezuela, Ecuador, Paraguay, Uruguay, and Argentina. In Brazil, endemic areas are located in the Midwestern and Southeastern portions of the country.[19,23] Generally, the environmental conditions present in areas of high endemicity are a moderate climate with a relatively constant humidity level.

Multiple typing methods have been developed to study the epidemiology of *H. capsulatum*. These methodologies can be based on phenotypic characteristics such as antigenic profiles[24] and multilocus enzyme electrophoresis[25] or on DNA-based analyses. Most recently, typing has been accomplished by analysis of fatty acid profiles of *H. capsulatum*.[26] Molecular typing methods are generally considered as having advantages over phenotypic methods in terms of the stability of genomic markers and providing greater levels of typeability. Several genotype-based methods such as hybridization of target genes (probes),[27–29] restriction fragment length polymorphism (RFLP),[30] karyotyping,[31] random amplified polymorphic DNA (RAPD),[32–36] and sequencing have been described for *H. capsulatum*.[6,37–39]

33.1.3 Pathogenesis and Clinical Features

33.1.3.1 Pathogenesis

After the inhalation of *H. capsulatum* conidia or mycelial fragments by a host, the organism converts to its pathogenic yeast form that survives and replicates in the phagolysosomes or modified phagosomes of host macrophages.[40] *H. capsulatum* may be internalized after opsonization with antibody and/or complement. This process may be associated with host antimicrobial defense mechanisms, such as an oxidative burst mediated by nicotinamide adenine dinucleotide phosphate (NADPH) oxidase, phagosome–lysosome fusion, vacuolar acidification, generation of nitric oxide, and exposure to hydrolytic enzymes.[9] The fungus can also bind directly to CD18 integrins family on the surface of host mononuclear cells, and this binding occurs through a specific ligand, a heat shock 60 kDa protein (HSP60),[41] that interacts with CR3 on macrophages. The morphotypic transition from mold to yeast in *H. capsulatum* is governed predominantly by temperature and occurs after inhalation by a mammalian host. This interaction with the host involves a switch from infectious mold to pathogenic yeast that is well suited to intracellular parasitism of phagocytes and dissemination, either free in blood or lymph, or more often carried within macrophages.

The dimorphic transition and the yeast morphotype are required for the establishment of disease with *H. capsulatum*.[9] This morphological change is the manifestation of several alterations in gene expression patterns that constitute a switch to a pathogenic lifestyle. Chemical treatments of mycelia are able to prevent transition of *Histoplasma* to the yeast form when grown at 37°C; although the treated mycelia remain viable at 37°C, they are avirulent.[42] Therefore, their virulence factors relate specifically to the yeast phase of the fungus.[43] In addition to dimorphism, some virulence determinants of *H. capsulatum* have been characterized and summarized.[44] As noted, HSP60 is associated with initiation of phagocytosis, resulting in the intracellular phase of the fungus. Interestingly, vaccination with *H. capsulatum* HSP60 induces protective host responses.[45–46] The M and H antigens are immunodominant antigens of *H. capsulatum* that have long been utilized as serological markers of histoplasmosis.[19] The M antigen, also known as catalase B, is a constitutively expressed protein posited to play a role in counteracting the oxidative defense mechanisms of host phagocytic cells.[47–48] The H antigen is a secreted beta-glucosidase purportedly involved in cell wall remodeling and nutrient acquisition.[49] *H. capsulatum* secretes a calcium binding protein (CBP) during yeast-phase growth that is essential for growth in calcium limiting conditions, such as encountered in vivo, and is required for virulence during murine infection.[50] Yeast-phase-specific protein 3 (YPS3) is a cell surface and secreted protein of uncertain function, but appears as a virulence determinant, since silencing of the *YPS3* gene significantly attenuates virulence in vitro and during murine infection.[51]

33.1.3.2 Clinical Features

Histoplasmosis can affect any individual, independent of age or sex. In general, the progress of the disease depends on the amount of microconidia inhaled into the lungs. Contact with *H. capsulatum* is very common for persons living in endemic areas; however, symptomatic infection is uncommon.[52] Most primary infections in immunocompetent individuals result in mild or asymptomatic respiratory disease. In such cases, the development of antigen-specific, T-lymphocyte-mediated immunity and fungistatic activation of macrophages results in control of the infection and disease resolution within 1–2 weeks.[9] However, children less than 1 year old and adults older than 50 are at increased risk for more severe forms of the disease. Immunocompromised individuals, such as those with lymphomas, leukemia, AIDS, diabetes, and Hodgkin disease or patients on corticosteroid therapy are more likely to develop progressive opportunistic histoplasmosis, including fungal dissemination.[5] Most recently, treatment with tumor necrosis factor (TNF)-alpha inhibitors has been associated with a significant increase in risk for histoplasmosis.[53] However, histoplasmosis even in immunocompetent people can frequently result in persistent, clinically inactive infection. This latent condition can erupt into active disease at a later time when the host–pathogen balance is disrupted following a decrease in host immune function and/or due to uncharacterized contributions from the "quiescent" fungus. Many of these clinical features of histoplasmosis resemble tuberculosis.[52]

There is a broad spectrum of clinical manifestations of histoplasmosis, ranging from a self-limited pulmonary infection, which subsides without treatment to chronic pulmonary infection to more widespread disseminated disease. Acute pulmonary histoplasmosis is a usually a self-limited illness in persons exposed to the organism for the first time, especially after low inoculum exposure. In fact, over 90% of exposed individuals are asymptomatic.[54] However, symptomatic disease can occur after an incubation period of 1–3 weeks.[55] Symptoms include fever, malaise, headache, weakness, substernal chest pain, and dry cough.[56] In some patients, chest radiographs are obtained only after the pulmonary infiltrate has cleared, and hilar lymphadenopathy is the primary finding. Acute self-limited pulmonary histoplasmosis is accompanied by rheumatologic and/or dermatologic manifestations in a few patients. Disseminated histoplasmosis is the most severe form of this disease. Clinical disseminated disease most commonly occurs in individual patients receiving cytotoxic or steroid therapy, and in immunocompromised persons, especially those with AIDS.[53] Disseminated disease is also more common in the very young or very old. The manifestations of disseminated histoplasmosis are greatly variable, commonly including fever, weight loss, hepatosplenomegaly, generalized lymphadenopathy, lung infiltrates, and haematological abnormalities.[57–58]

33.1.4 Diagnosis

Although the clinical manifestations of histoplasmosis are well described, there is significant overlap of symptoms with other diseases, and the diagnosis cannot be achieved based on clinical information alone. The definitive diagnosis requires the isolation of the *H. capsulatum* on specific culture media or the visualization of the yeast form in direct examination of clinical specimens using specific fungal staining techniques. The gold standard for diagnosis is the growth of the organism; however, cultures typically take 2–4 weeks to grow and this technique lacks sensitivity. The likelihood of a positive culture depends on the clinical form, ranging from 10% to 85%[59] with the best yields in individuals with disseminated or chronic pulmonary disease. Direct microscopic examination of tissues or body fluid can facilitate diagnosis, particularly of bone marrow in acute disseminated disease where the yeast form can be identified in up to 50% of cases. However, visual inspection can result in incorrect diagnosis. Other techniques have been developed to supplement culture and microscopic examination. Serological methods usually have a rapid turnaround time and detection of either antibodies or antigens can provide information indicative of current disease. The choice of each test to be applied for diagnosis depends mainly on the clinical manifestation and host factors.[56] Non-culture-based methods are used in conjunction with culture to improve our ability to diagnosis *H. capsulatum* infection and can also guide therapy for histoplasmosis. Recently, some molecular biology techniques have been developed that may further improve the diagnosis of histoplasmosis, particularly for the detection of disease in an early stage and to improve the specificity of the diagnosis.

33.1.4.1 Conventional Techniques

33.1.4.1.1 Microscopy

Direct examination and histopathology are useful, especially as they are the most important method for alerting laboratory personnel that they may be dealing with a potential fungal pathogen.[60] Histopathologic examinations of different clinical specimens are frequently analyzed for the diagnosis of histoplasmosis.[53] Peripheral blood may show intracellular organisms in white blood cells. However, bone marrow biopsy

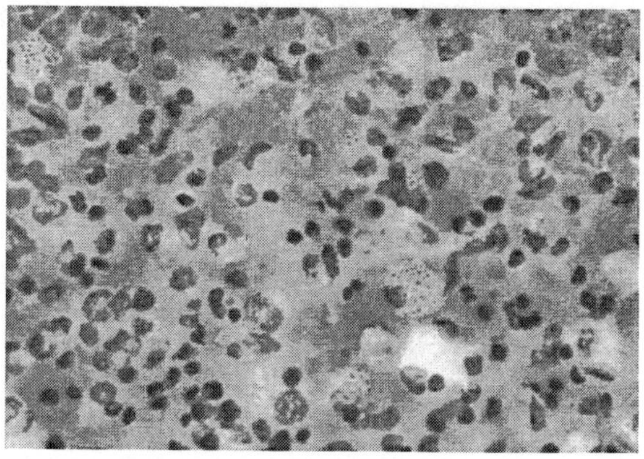

FIGURE 33.1 *Histoplasmosis.* Giemsa staining of a bone marrow smear showing intracellular *H. capsulatum* yeast phase, 40×. (Courtesy of Bodo Wanke.)

may be the most rapid method of establishing a definitive diagnosis of disseminated disease (Figure 33.1). Skin scrapings, exudates, and body fluids should be examined using 10% KOH and Parker ink or calcofluor white mounts. Tissue sections should be stained using PAS (periodic acid-Schiff) digest, Grocott's methenamine silver (GMS), or Gram stain.[61] *H. capsulatum* structures visualized microscopically can be confused with structures from other pathogens.[19]

33.1.4.1.2 Culture

Isolation of the fungus in its mycelial form can be achieved on Sabouraud agar, following incubation at 25°C for 6–12 weeks. The fungal colonies are initially smooth, becoming filamentous, cottony, and brownish with the age (Figure 33.2A). Microscopically, they are composed of septated hyaline hyphae that have tuberculated macroconidia and smooth-walled spherical, pyriform, or cigar-shaped microconidia in various developmental stages, ranging in size from 2 to 6μm in diameter[4] (Figures 33.2B and C). The conversion from the

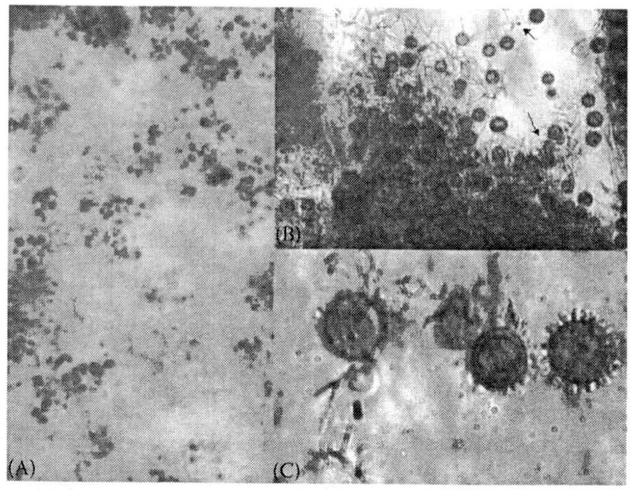

FIGURE 33.2 *Histoplasma capsulatum.* (A) Culture on Mycosel's agar for 15 days at 25°C. (B) Macroconidia from mycelia form. The arrow indicates a microconidia, 40×. (C) Globose, tuberculated macroconidia, 100×.

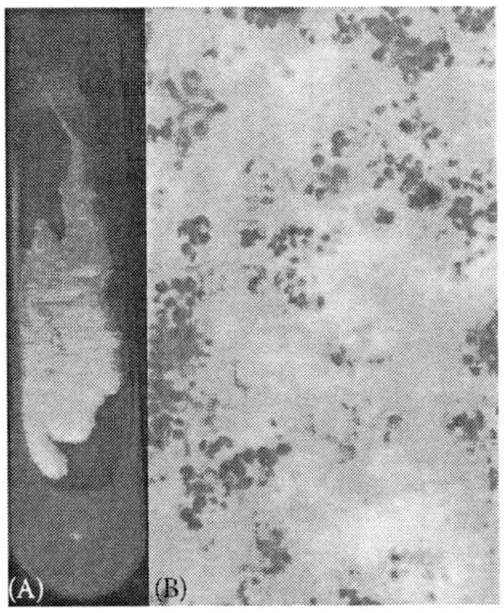

FIGURE 33.3 *Histoplasma capsulatum*. (A) Smooth yeast colony produced on YPD agar for 15 days at 37°C. (B) Globose and/or oval yeast budding cells, 40×.

mold to the yeast phase is necessary for accurate diagnosis of *H. capsulatum* and it is achieved using enriched media such as Brain-heart infusion agar (BHI) supplemented with blood and cystein and incubated at 35°C–37°C. After conversion, smooth white to brown colonies can be observed that are ovoid thick-walled yeast cells by microscopic examination (Figure 33.3). Although attempts to culture the organism should be pursued for all forms of histoplasmosis, culture-based methods are most effective when the fungal burden is high, for instance, in blood samples of patients with chronic or disseminated forms of histoplasmosis. Culture is insensitive in subacute and acute histoplasmosis.

33.1.4.1.3 Serology

Serologic diagnosis focuses on the identification of antibodies to the H[62] and/or M antigens.[47] The two routine antibody detection methodologies are complement fixation (CF) and immunodiffusion (ID), because of convenience, availability, and accuracy of these assays. In the past, CF was a popular test to diagnose histoplasmosis but ID was found to be more specific than CF.[19] The ID test qualitatively measures precipitating antibodies and is highly specific for the detection of antibodies to M and H antigens, ranging from 70% to 100%.[63] However, it has low sensitivity, principally in acute, disseminated, and opportunistic manifestation of the disease. In general, ID is useful for detecting antibodies 4–6 weeks after infection. The H band can be found in sera from patients during acute and/or progressive disease. However, the H band is present in only 7% of serum from patient with acute infection.[64] Antibodies to the H antigen may be detected 1–2 years after the resolution of the disease, but usually disappear more quickly than the antibodies to the M antigen. The M band is more frequently detected than the H precipitin and appears soon after infection and can indicate prior infection, acute disease, or a chronic progressive disease.[19] The M precipitin can persist for up to 3 years after disease resolution and can also be stimulated after skin testing with histoplasmin in people who have not been exposed to *H. capsulatum*.[65] The presence of both the H and M precipitins is considered to be conclusive for the diagnosis of histoplasmosis, though the status of disease requires clinical assessment of the patient. Antibodies to *H. capsulatum* can also be detected by immunoenzymatic methods. Western blot (WB) has been used successfully in order to diagnose histoplasmosis and epidemiologically evaluate the distribution of the disease[66–69] and should be applied in conjunction to ID in acute disease since the sensitivity can be significantly improved.[70] Several enzyme-linked immunosorbent assays (ELISA) protocols have been described for antibody detection using diverse antigenic preparations showing sensitivity among 75%–100%.[19,66] More recently, an indirect ELISA for detecting antibodies against deglycosylated histoplasmin (HMIN) was found to have a sensitivity of 92% and specificity of 96%.[71] An advantage of this method is that it can be readily applied in resource-limited laboratory settings.

Antigen detection methods are used when antibody detection is unlikely, particularly in acute disease and for immunocompromised individuals, who frequently have the disseminated form of histoplasmosis.[19,54] A radioimmunoassay for the detection of *H. capsulatum* antigen in urine and serum specimens based on the detection of a polysaccharide antigen from *H. capsulatum* (HPA) has been particularly effective, especially in patients with disseminated histoplasmosis.[72] This method has also proven especially useful for monitoring the effect of treatment.[73] ELISAs for detection of *H. capsulatum* antigens have been described and compared with the solid-phase radioimmunoassay (Table 33.1).

An inhibition ELISA has also been developed using a murine monoclonal antibody that is specific for a 70 kDa *H. capsulatum* protein.[75] The sensitivity varied from 57.1% to 88.9% depending on the clinical form of histoplasmosis, and the specificity was 85.4% when serum samples of other mycosis were used. This methodology is also useful for monitoring histoplasmosis patients during treatment and following disease in non-AIDS patients with the acute or disseminated form of the disease.[76] Recently, a capture ELISA for the detection of *H. capsulatum* using rabbit polyclonal antibodies as capture and detection agents has been described.[77] The assay had a sensitivity of 81% and specificity of 95% in the setting of testing urine samples from Guatemalan patients with AIDS who had culture-proven histoplasmosis. The authors advertise this method for its simplicity and for its capacity to facilitate the rapid diagnosis of disseminated histoplasmosis in resource limited settings.

33.1.4.2 Molecular Techniques

Molecular methods for the identification of fungal isolates can potentially reduce the time for diagnosis, while maintaining or improving the specificity, accuracy and sensitivity. Additionally, molecular methods can be safer as most methods do not require laboratory growth of the organism, limiting the possibility of laboratory infection.[78] Polymerase chain reaction (PCR) based on the amplification of fungal gene sequences is a powerful tool for identifying invasive

TABLE 33.1
Parameters of Antigen Detection Techniques for *Histoplasma capsulatum*

Antigen Detection Test		Sensitivity (%) Disseminated[a]	Non-Disseminated[b]	Specificity (%)	References
RIA		95	48	96	Wheat et al.[72]
Sandwich ELISA	AP[c]	89	25	92	Durkin et al.[74]
	HRP[d]	89	16	96	Durkin et al.[74]
Inhibition-ELISA		68	75	85	Gomez et al.[75]

[a] Disseminated includes disseminated non AIDS and opportunistic infection.
[b] Non disseminated includes acute, chronic, mediastinal and self-limited infections.
[c] Alkaline phosphatase.
[d] Horseradish peroxidase.

mycoses.[52] Several PCR-based methods have been described for *H. capsulatum* diagnosis.

A PCR was developed for rapid identification of *H. capsulatum* isolates in culture based on the sequences of the M gene.[79] The M antigen gene was amplified successfully from 31 *H. capsulatum* strains, showing 100% sensitivity. Therefore, this alternative methodology appears to offer the potential for improving the current methods for identification of this pathogenic mold, though it has mainly been applied to atypical isolates. Chemiluminescence-labeled DNA probes have been developed for the detection of specific sequences of rRNA that have been able to detect and confirm all of the 41 *H. capsulatum* clinical isolates tested.[80–81] Also, these probes could detect *H. capsulatum* directly on paraffin embedded tissue sections,[81] such as excised heart valves from patients with *Histoplasma* endocarditis.[80]

A 100 kDa *H. capsulatum* protein (HcP100) has shown promise as a diagnostic target using specific primers targeting the gene coding for HcP100. The method successfully amplified the gene in 20 of 29 histopathologically positive biopsy specimens, but, had no false-positive results.[82] Using whole blood, a nested PCR targeting a 210 bp specific segment of the HcP100 gene showed 89% sensitivity and 98% specificity.[83–84] A nested PCR based on the small-subunit (18S) rRNA gene of *H. capsulatum* has been tested in murine models of histoplasmosis and compared to quantitative cultures.[85] The nested PCR efficiently detected *H. capsulatum* DNA in tissue and blood samples from infected animals, but the assay also detected DNA of *B. dermatitidis* and *P. brasiliensis*. Using the same biopsy specimens described above, the nested PCR targeting the 18S rRNA gene was positive in 26 or 29 *H. capsulatum* infected tissues, but sequencing of the PCR products revealed that only half were identical to the *H. capsulatum* gene sequence and the assay was also positive in 18 of negative control samples[82] and nonspecific amplification of DNA from nonpathogenic fungal species had to be confirmed by sequencing. The detection of *H. capsulatum* DNA in choroidal lesions in histopathological sections from a patient with a chronic ocular histoplasmosis syndrome was also accomplished using another set of primers to the rRNA gene.[86] Detection of fungal DNA in human formalin-fixed and paraffin-embedded tissue has been described using nested PCR assays. Using the sequence of the H antigen gene, primers were selected and a semi-nested PCR developed. A comparison of a semi-nested PCR method targeting the H antigen gene to biopsy, culture, and skin scraping with May–Grunwald–Giemsa staining revealed similar results between the approaches, except that the PCR method detected *H. capsulatum* DNA in two culture-negative blood samples.[87]

A colorimetric microtiter plate PCR–enzyme immunoassay (PCR-EIA) for the detection of *H. capsulatum* in urine showed positivity in 80% of urine specimens that were positive by culture and no correlation was observed with antigenuria, limiting its use in the diagnosis of disseminated histoplasmosis.[88] Recently, real-time PCR has also shown a potential tool for the diagnosis of histoplasmosis.[89–90]

33.2 METHODS

33.2.1 Sample Preparation

The application of molecular biology techniques to the analysis of complex genomes, such as eukaryotic organisms, depends on the ability to prepare pure DNA. One of the most essential steps is the extraction of adequately pure and suitably sized nucleic acids. Several in-house methods have been described for DNA extraction from filamentous fungi and yeast,[91–94] and most of them involve disruption of the cells, and digestion with proteinase K followed by phenol and phenol/chlorophorm extractions.

In this chapter, we describe conventional microbiological approaches and selected methods of DNA preparation from *H. capsulatum* yeast phase that we have adapted from diverse techniques previously described in the specialized literature.[93] For atypical strains, which are not able to convert to the yeast phase, the DNA should be extracted from mycelia and/or spores by previously described methodologies.[95] All the work should be performed in a biosafety level two or three biological safety cabinet until cells are lysed. The methods described represent the state of the art at the present time in our laboratory and certainly will be changed as innovations occur and newer procedures are developed.

33.2.1.1 Growth of *Histoplasma capsulatum*

(i) Grow *H. capsulatum* yeast phase in HAM's F12 or BHI agar plate, and incubate at 37°C until colonies appear in order to isolate a single colony. (ii) Transfer a single colony to HAM's F12 or YPD (1% yeast extract, 2% peptone, 1% dextrose) broth, and incubate at 37°C for 72–96 h.

33.2.1.2 Extraction of *Histoplasma capsulatum* DNA from Culture

1. Harvest 5 mL *H. capsulatum* yeast culture.
2. Centrifuge at least 2 mL of the culture 10 min at 900 × *g* (3000 rpm) and discard supernatant.
3. Suspend cells in 500 μL TES buffer (100 mM Tris, pH 8.0; 50 mM EDTA, 1% sodium dodecil sulphate [SDS]).
4. Add equal volume of glass or zirconium beads 0.5 mm. Vortex 2 min.
5. Add 275 μL of 7.0 M ammonium acetate.
6. Incubate samples in tightly capped tubes 5 min at 65°C, and 5 min on ice.
7. Extract samples with 500 μL of 25:24:1 phenol/chloroform/isoamyl alcohol. Vortex for 30 s. Centrifuge 5 min at full speed (20,000 × *g*; 14,000 rpm). Transfer top layer (aqueous phase) to a new tube.
8. Add 1 mL of isoprapanol. Incubate 5 min at room temperature (RT). Centrifuge 10 min at full speed (20,000 × *g*; 14,000 rpm).
9. Wash the pellet with 70% ethanol. Mix gently several times. Centrifuge 5 min at full speed (20,000 × *g*; 14,000 rpm), air dry at RT, and suspend in 50 μL dH_2O or TE buffer.

33.2.1.3 Extraction of *Histoplasma capsulatum* DNA on Blood Samples

In order to perform a sensitive, specific, and reliable nucleic acid-based diagnostic test, availability of pure DNA-lacking inhibitors as well a rapid and easy-to-perform DNA extraction protocol is essential. Protocols for extraction of DNA of fungal cells either are time consuming or show poor release of fungal DNA.[9] In this procedure, prior to the DNA extraction with 25:24:1 phenol/chloroform/isoamyl alcohol, the whole blood sample should be prepared as follow[84]:

1. Centrifuge 2.5–4 mL of fresh blood for 10 min at 2.500 × *g*, 4°C. Do not freeze the blood samples. Better to use after 7 days at 4°C.
2. Wash the cells pellet with TE buffer. Repeat this step.
3. Suspend cells in lysing solution I (6.5 μL lyticase; 83 μL lysing enzyme from *Trichoderma harzianum* [6 μg/μL] in 230 μL sorbitol buffer). Incubate 2 h at 25°C.
4. Complete spheroplasting with lysing solution II (500 mM EDTA; 1% [w/v] *N*-laurylsarcosine; proteinase K [0.5 mg/mL]) for 2 h at 55°C, and 16 h at 37°C.
5. Adjust 1.4 M NaCl with stock solution (140 μL of 5M NaCl stock solution), and 1% hexadecyltrimethylammonium brominde (CTAB) (65 μL of 10% CTAB). Incubate 10 min at 65°C.

33.2.2 Detection Procedures

33.2.2.1 PCR for Detection and Identification of *Histoplasma capsulatum*

1. Prepare a PCR mixture (25 μL) containing 100 ng DNA, PCR buffer (10 mM Tris–HCl, 50 mM KCl), 1.5 mM $MgCl_2$, 200 μM deoxynucleoside triphosphates, 2.5 U of *Taq* DNA polymerase, and 20 pmol of each primer MsplF (5′-ACA AGA GAC GAC GGT AGC TTC ACG-3′) and Msp1R (5′-GCG TTG GGG ATC AAG CGA TGA GCC-3′).[79]
2. Carry out the PCR amplification in a thermocycler with a cycling program consisting of a denaturation step at 95°C for 5 min 35 cycles of 1 min at 95°C, 1 min at 70°C and 1 min at 72°C, followed by a single terminal extension at 72°C for 5 min.
3. Include an internal PCR control to verify the efficiency of the test and to ensure that PCR inhibition is absent. The universal fungal primers ITS1 and ITS4, derived from highly conserved regions of the fungal rRNA gene, should be used for this purpose.[96]
4. Separate the PCR product on a 1% agarose gel, and visualize with a UV transilluminator after ethidium bromide (0.5 μg/mL) staining.

33.2.2.2 Nested PCR for Detection of *Histoplasma capsulatum* from Clinical Samples

Bialek et al.[82] developed a nested PCR assay targeting the gene coding a 100 kDa-like protein described as being essential for the survival of *H. capsulatum* in human cells.[97] The outer primer set is Hc I (5′-GCG TTC CGA GCC TTC CAC CTC AAC-3′) and Hc II (5′-ATG TCC CAT CGG GCG CCG TGT AGT-3′) and delimits a 391 bp sequence. The inner primer set is Hc III (5′-GAG ATC TAG TCG CGG CCA GGT TCA-3′) and Hc IV (5′-AGG AGA GAA CTG TAT CGG TGG CTT G-3′) and the nested PCR product is 210 bp long.[82,84]

1. For the first amplification prepare a PCR mixture (25 μL) containing 2 μL of DNA, PCR buffer (10 mM Tris–HCl, 50 mM KCl), 2 mM $MgCl_2$, 200 μM deoxynucleoside triphosphates, 1 U of *Taq* DNA polymerase, 0.1% BSA Fraction V and 1 μM of each primer Hc I and Hc II. Perform the first PCR amplification using a cycling program consisting of 94°C for 5 min; 35 times at 94°C for 30 s, 65°C for 30 s, and 72°C for 1 min; and then once at 72°C for 5 min.
2. For the nested PCR the reaction mixture is identical, except that 2 μL of the first PCR product, 1 mM

$MgCl_2$, and 1 μM of each primer Hc III and Hc IV is used. The reaction mixture is thermally cycled once at 94°C for 5 min, 30 cycles of 30 s at 94°C, 30 s at 67°C and 1 min at 72°C, and then once at 72°C for 5 min.
3. Separate the PCR product on a 2% agarose gel, and visualize with a UV transilluminator after ethidium bromide (0.5 μg/mL) staining.

33.2.2.3 Identification of *Histoplasma capsulatum* by Sequencing

The sequencing of internal transcribed spacer (ITS) regions of rDNA with different primers have been chosen for identifying more precisely this species.[6,37–38,98–99] The primers ITS1 (5′-TCC GTA GGT GAA CCT GCG G-3′) and ITS4 (5′-TCCTCCGCTTATTGATATGC-3′) are the most used in our laboratory.

1. Prepare a PCR mixture consisting of 10 mM Tris–HCl buffer containing 50 mM KCl, pH 8.0, 1.5 mM $MgCl_2$, 0.2 mM deoxynucleoside triphosphate (Invitrogen), and 1.25 U of *Taq* polymerase (Invitrogen). Add primers ITS1 and ITS4 (final concentration of 0.2 mM each), and template DNA at a final concentration of 20 ng per 100 μL of the reaction mixture.
2. Carry out an initial denaturation of template DNA at 95°C for 5 min followed by 30 cycles of 30 s at 95°C, 30 s at 58°C, and 1 min at 72°C. A final extension step is conducted for 10 min at 72°C.
3. Separate the PCR product on a 1% agarose gel, and visualize with a UV transilluminator after ethidium bromide staining.
4. Purify PCR product using QIAquick PCR Purification Kit (Qiagen).
5. Prepare a sequencing mixture (10 μL) containing premix terminator (Applied Biosystems), 100 ng purified PCR product, and 3.2 pM of primers ITS1 and ITS4.
6. Direct sequencing of the PCR product is performed with a sequencing kit using the ABI PRISM 3100 sequencer (Applied Biosystems).
7. Edit and format the DNA sequences and align, by sequencing alignment programs, with those from other medically important dimorphic fungi that are available at the GenBank.

33.2.2.4 Molecular Typing of *Histoplasma capsulatum*

Muniz[100] developed a typing system composed of three methodologies, M13 DNA fingerprinting, PCR-RFLP, and sequencing of four partial gene coding protein described below.

33.2.2.4.1 M13 PCR Fingerprinting

The minisatellite-specific core sequence of the wild-type phage M13 (5′-GAGGGTGGCGGTTCT-3′)[101] must be used as a single primer.

1. Prepare the master mixture (50 μL total volume) solution containing 1×PCR buffer (10 mM Tris–HCl pH 8.3, 50 mM KCl, 1.5 mM $MgCl_2$), 0.2 mM each of the dATP, dCTP, dGTP and dTTP (Roche Diagnostics), 3 mM magnesium acetate, 30 ng of primer, and 2.5 U *Amplitaq* DNA polymerase (Applied Biosystems). Add 25 ng of genomic DNA.
2. PCR is performed for 35 cycles in a thermalcycler with 20 s of denaturation at 94°C, 1 min annealing at 50°C, and 20 s extension at 72°C, followed by a final extension cycle for 6 min at 72°C.
3. Amplification products were removed, concentrated to approximately 20 μL and separated by electrophoresis on 1.4% agarose gels (stained with ethidium bromide, 0.5 μg/mL final concentration) in 1× Tris-borate-EDTA (TBE) buffer at 60 V for 14 cm, and visualized under UV light.
4. All visualized bands on the gel were counted, independent of their intensity, and data were scored for the presence or absence of the amplified DNA bands.

33.2.2.4.2 PCR-RFLP (rRNA Gene)

PCR-RFLP analysis of the ITS region of the rDNA gene reveals species-specific variations of restriction enzymes (RE) cut sites.

1. PCR mixture (50 μL) is composed of 50 ng of DNA, 1×PCR buffer (10 mM Tris–HCl pH 8.3, 50 mM KCl, 1.5 mM $MgCl_2$)(Applied Biosystems), 0.2 mM each of dATP, dCTP, dGTP, and dTTP (Roche Diagnostics), 3 mM magnesium acetate, 1.5 U *AmpliTaq* DNA polymerase (Applied Biosystems), and 50 ng of each primer SR6R (5′-AAGTARAAGTCGTAACAAGG-3′) and LR1 (5′-GGTTGGTTTCTTTTCCT-3′).
2. PCR is performed in a thermalcycler with 1 cycle of 94°C for 2 min, 35 cycles of 45 s at 94°C, 1 min at 61°C, and 2 min at 72°C followed by a final extension cycle of 10 min at 72°C.
3. PCR products (20 μL) are incubated with *Sau*96I (10 U/μL) and *Hha*I (20 U/μL) at 37°C overnight and separated by 3% agarose gel electrophoresis at 100 V for 5 h after mixing with one fifth volume of loading buffer.

33.2.2.4.3 Sequencing

The population structure of *H. capsulatum* by DNA sequences of partial gene coding proteins *arf, H, ole, tub*, and ITS regions of rDNA has been used for phylogenetic analyses.[6,38]

1. PCR is performed with 100 ng genomic DNA template in 50 μL reaction mixture, consisting of 0.45 μM of each primer, 1.0 U of *AmpliTaq* DNA polymerase (Perkin-Elmer), 10 mM Tris–HCl (pH 8.3), 1.5 mM $MgCl_2$, 50 mM KCl, and 0.2 mM

deoxynucleotidetriphosphates. The primers (5′-3′) are as follows: arf1 (AGAATATGG GGCAAAAAGGA′) and arf2 (CGCAATTCATCTTCGTTGAG); H-anti3 (CGCAGTCACCTCCATACTATC) and H-anti4 (GCGCCGACATTAACCC); ole3 (TTTAAACGAAGCCCCCACGG) and ole 4 (CACCACCTCCAACAGCAGCA); tub1 (GGTGGCCAAATCGCAAACTC) and tub2 (GGCAGCTTTCCGTTCCTCAGT); ITS4 (TCCTCCGCTTATTGATATGC) and ITS5 (GGAAGTAAAAGTCGTAACAAGG).

2. PCR temperature profile: a 15 s DNA denaturation step at 94°C, a 30 s annealing step, and a 1 min extension step at 72°C for 32 cycles, followed by a 5 min final extension step at 72°C. The annealing temperature in the first cycle was 65°C. This annealing temperature was subsequently reduced by 0.7°C/cycle for the next 12 cycles, and thereafter, the PCR was continued at an annealing temperature of 56°C for the remaining 20 cycles (Touchdown PCR).[102]
3. Purify PCR product using commercial kit following the protocol according to manufacturer instructions.
4. Quantify the purification product and use 100 ng of PCR product for sequencing generating both strands (forward and reverse).
5. After sequencing, edit and align the raw sequences. Phylogenetic analyses (both parsimony and neighbor joining) must be done with a bootstrap test of phylogeny (500 or 1000 replicates) using appropriate software.

33.3 CONCLUSIONS AND FUTURE PERSPECTIVES

The mechanisms behind the pathobiology of *H. capsulatum* are slowly being unraveled as studies on the molecular biology, epidemiology, and host–pathogen interactions press forward. However, histoplasmosis continues to cause significant morbidity and mortality, especially in the setting of increasing numbers of immunocompromised hosts. The large numbers of individuals receiving inhibitors of TNF-alpha have recently been added to the list of significantly at-risk individuals. Despite the scientific advances in the study of *H. capsulatum*, major difficulties persist in our capacity to rapidly diagnose disease due to this fungus. Newer ELISA techniques, such as utilizing HMIN or polyclonal serums, require validation, but may facilitate improved diagnosis. Molecular techniques are in development, but there are no agreed-upon standards and the methodologies are time consuming. When available, the urine *Histoplasma* antigen assay remains extremely useful, especially in diagnosing disseminated histoplasmosis. Ongoing studies focusing on advancing automated PCR processes, high throughput sequencing techniques or mass spectroscopy may lead to improved and rapid diagnostics for *H. capsulatum*. However, the gold standard in diagnosis continues to be culture, although results are usually available after presumptive therapy has been initiated.

REFERENCES

1. Hibbet, D.S. et al., A higher-level phylogenetic classification of the fungi, *Mycol. Res.*, 111, 509, 2007.
2. Hoog, G.S. and Guarro, J., *Atlas of Clinical Fungi*. Centraalbureau voor Schimmelcultures Baarn and Delft/ Universitat Rovira i Virgili, Reus. Index Fungorum. 1995.
3. James, T.Y. et al., Reconstructing the early evolution of fungi using a six-gene phylogeny, *Nature*, 443, 818, 2006.
4. Kwon-Chung, K.J. and Bennett, J.E., *Medical Mycology*. Lea & Febiger, Philadelphia, PA, 1992.
5. Zancopé-Oliveira, R.M., Muniz, M.M., and Wanke, B., Histoplasmose. In: Coura, J.R. (ed.), *Dinâmica das Doenças Infecciosas e Parasitárias*. Guanabara Koogan, Rio de Janeiro, pp. 1207–1221, 2005.
6. Kasuga, T., Taylor, J.W., and White, T.J., Phylogenetic relationships of varieties and geographical groups of the human pathogenic fungus *Histoplasma capsulatum* Darling, *J. Clin. Microbiol.*, 37, 653, 1999.
7. Larone, D.H., Medically *Important Fungi*. ASM Press, Washington, DC, 2002.
8. Romero-Martinez, R., Canteros, C., and Taylor, M.L., Variabilidad cromossómica intraespecífica em hongos patógenos de humanos, especialmente en *Histoplasma capsulatum*, *Rev. Iberoam. Micol.*, 21, 168, 2004.
9. Woods, J.P., *Histoplasma capsulatum* molecular genetics, pathogenesis and responsiveness to its environment, *Fungal Genet. Biol.*, 35, 81, 2002.
10. Reiss, E., *Molecular Immunology of Mycotic and Actinomycotic Infections*. Elsevier, New York, pp. 77–97, 1986.
11. Gorocica, P. et al., The interaction between *Histoplasma capsulatum* cell wall carbohydrates and host components: Relevance in the immunomodulatory role of histoplasmosis, *Mem. Inst. Oswaldo Cruz*, 104, 492, 2009.
12. Steenbergen, J.N. et al., Interaction of *Blastomyces dermatitidis*, *Sporothrix schenckii* and *Histoplasma capsulatum* with *Acantamoeba castellanii*, *Infect. Immun.*, 72, 3478, 2004.
13. Nosanchuk, J.D. et al., *Histoplasma capsulatum* synthesizes melanin-like pigments in vitro and during mammalian infection, *Infect. Immun.*, 70, 5124, 2002.
14. Riley, P.A., Melanin, *Int. J. Biochem. Cell Biol.*, 29, 1235, 1997.
15. Taborda, C.P. et al., Melanin as a virulence factor of *Paracoccidioides brasiliensis* and other dimorphic pathogenic fungi, *Mycopathologia*, 165, 331, 2009.
16. van Duin, D., Casadevall, A., and Nosanchuk, J.D., Melanization of *Cryptococcus neoformans* and *Histoplasma capsulatum* reduces their susceptibilities to amphotericin B and caspofungin, *Antimicrob. Agents Chemother.*, 46, 3394, 2002.
17. Albuquerque P.C. et al., Vesicular transport in *Histoplasma capsulatum*: An effective mechanism for trans-cell wall transfer of proteins and lipids in ascomycetes, *Cell. Microbiol.*, 10, 1695, 2008.
18. Cano, M.V. and Hajjeh R.A., The epidemiology of histoplasmosis: A review, *Semin. Respir. Infect.*, 16, 109, 2001.
19. Guimarães, A.J., Nosanchuk J.D., and Zancopé-Oliveira, R.M., Diagnosis of histoplasmosis, *Braz. J. Microbiol.*, 37, 1, 2006.

20. Klite, P.D. and Diercks, F.H., *Histoplasma capsulatum* in fecal contents and organs of bats in the Canal Zone, *Am. J. Trop. Med. Hyg.*, 14, 433, 1965.
21. Julg, B. et al., Bat-associated histoplasmosis can be transmitted at entrances of bat caves and not only inside the caves, *J. Travel Med.*, 15, 133, 2008.
22. Wheat L.J. et al., A large urban outbreak of histoplasmosis: Clinical features, *Ann. Intern. Med.*, 94, 331, 1981.
23. Zancope-Oliveira, R.M., Tavares, P.M.S., and Muniz, M.M., Genetic diversity of *Histoplasma capsulatum* strains in Brazil, *FEMS Immunol. Med. Microbiol.*, 45, 443, 2005.
24. Reyes-Montes, M.R. et al., Relatedness analyses of *Histoplasma capsulatum* isolates from Mexican patients with AIDS-associated histoplasmosis by using histoplasmin electrophoretic profiles and randomly amplified polymorphic DNA patterns, *J. Clin. Microbiol.*, 37, 1404, 1999.
25. Hamrick, J.L., Lichtwardt, R.W., and Lan, C., Levels of isozyme variation within and among *Histoplasma capsulatum* localities, *Trans. Kans. Acad. Sci.*, 89, 49, 1986.
26. Zarnowski, R. et al., Typing of *Histoplasma capsulatum* strains by fatty acid profile analysis, *J. Med. Microbiol.*, 56, 788, 2007.
27. Spitzer, E.D. et al., Use of mitochondrial and ribosomal DNA polymorphisms to classify clinical and soil isolates of *Histoplasma capsulatum*, *Infect. Immun.*, 57, 1409, 1989.
28. Keath, E.J. et al., DNA probe for the identification of *Histoplasma capsulatum*, *J. Clin. Microbiol.*, 27, 2369, 1989.
29. Sandhu, G.S. et al., Molecular probes for diagnosis of fungal infections, *J. Clin. Microbiol.*, 33, 2913, 1995.
30. Keath, E.J., Kobayashi G.S., and Medoff G., Typing of *Histoplasma capsulatum* by restriction fragment length polymorphisms in a nuclear gene, *J. Clin. Microbiol.*, 30, 2104, 1992.
31. Canteros, C.E. et al., Electrophoresis karyotype and chromosome-length polymorphism of *Histoplasma capsulatum* clinical isolates from Latin America, *FEMS Immunol. Med. Microbiol.*, 45, 423, 2005.
32. Kersulyte, D. et al., Diversity among clinical isolates of *Histoplasma capsulatum* detected by polymerase chain reaction with arbitrary primers, *J. Bacteriol.*, 174, 7075, 1992.
33. Poonwan, N.T. et al., Genetic analysis of *Histoplasma capsulatum* strains isolated from clinical specimens in Thailand by a PCR-based random amplified polymorphic DNA method, *J. Clin. Microbiol.*, 36, 3073, 1998.
34. Reyes-Montes, M.R. et al., Relatedness analyses of *Histoplasma capsulatum* isolates from Mexican patients with AIDS-associated histoplasmosis by using histoplasmin electrophoretic profiles and randomly amplified polymorphic DNA patterns, *J. Clin. Microbiol.*, 35, 1404, 1999.
35. Woods, J.P. et al., Fast DNA isolation from *Histoplasma capsulatum*: Methodology for arbitrary primer polymerase chain reaction-based epidemiological and clinical studies, *J. Clin. Microbiol.*, 31, 463, 2003.
36. Muniz, M.M. et al., Genetic diversity of *Histoplasma capsulatum* strains isolated from soil, animals, and clinical specimens in Rio de Janeiro State, Brazil, by a PCR-based random amplified polymorphic DNA assay, *J. Clin. Microbiol.*, 39, 4487, 2001.
37. Jiang, B. et al., Typing *Histoplasma capsulatum* isolates based on nucleotide sequence variation in the internal transcribed spacer regions of rRNA genes, *J. Clin. Microbiol.*, 38, 241, 2000.
38. Kasuga, T. et al., Phylogeography of the fungal pathogen *Histoplasma capsulatum*, *Mol. Ecol.*, 12, 3383, 2003.
39. Carter D.A. et al., Amplified single-nucleotide polymorphisms and a $(GA)_n$ microsatellite marker reveal genetic differentiation between populations of *Histoplasma capsulatum* from the Americas, *Fungal Genet. Biol.*, 34, 37, 2001.
40. Woods, J.P., Knocking on the right door and making a comfortable home: *Histoplasma capsulatum* intracellular pathogenesis, *Curr. Opin. Microbiol.*, 6, 327, 2003.
41. Long, K.H. et al., Identification of heat shock protein 60 as the ligand on *Histoplasma capsulatum* that mediates binding to CD18 receptors on human macrophages, *J. Immunol.*, 170, 487, 2003.
42. Retallack, D.M. and Woods, J.P., Molecular epidemiology, pathogenesis, and genetics of the dimorphic fungus *Histoplasma capsulatum*, *Microbes Infect.*, 1, 817, 1999.
43. Rappley, C.A. and Goldman, W.E., Defining virulence genes in the dimorphic fungi, *Annu. Rev. Microbiol.*, 60, 281, 2006.
44. Nosanchuk, J.D. and Gacser, A., *Histoplasma capsulatum* at the host–pathogen interface, *Microbes Infect.*, 10, 973, 2008.
45. Gomez, F.J., Allendoerfer, R., and Deepe, G.S. Jr., Vaccination with recombinant heat shock protein 60 from *Histoplasma capsulatum* protects mice against pulmonary histoplasmosis, *Infect. Immun.*, 63, 2587, 1995.
46. Guimarães, A.J. et al., Monoclonal antibodies to heat shock protein 60 alter the pathogenesis of *Histoplasma capsulatum*, *Infect. Immun.*, 77, 1357, 2009.
47. Zancope-Oliveira, R.M. et al., Molecular cloning, characterization, and expression of the M antigen of *Histoplasma capsulatum*, *Infect. Immun.*, 67, 1947, 1999.
48. Guimarães, A.J. et al., Biological function and molecular mapping of M antigen in yeast phase of *Histoplasma capsulatum*, *PLOS One*, 3, e3449, 2008.
49. Fisher, K.L. and Woods, J.P., Determination of beta-glucosidase enzymatic function of the *Histoplasma capsulatum* H antigen using a native expression system, *Gene*, 247, 191, 2000.
50. Sebghati, T.S. et al., Intracellular parasitism by *Histoplasma capsulatum*: Fungal virulence and calcium dependence, *Science*, 290, 1368, 2000.
51. Bohse, M.L. and Woods, J.P., RNA interference-mediated silencing of the YPS3 gene of *Histoplasma capsulatum* reveals virulence defects, *Infect. Immun.*, 75, 2811, 2007.
52. Kauffman, C.A., Histoplasmosis, *Clin. Chest Med.*, 30, 217, 2009.
53. Smithm J.A. and Kauffman, C.A., Endemic fungal infections in patients receiving tumour necrosis factor-alpha inhibitor therapy, *Drugs*, 69, 1403, 2009.
54. Kauffman, C.A., Histoplasmosis: A clinical and laboratory update, *Clin. Microbiol. Rev.*, 20, 115, 2007.
55. Kauffman, C.A., Pulmonary histoplasmosis, *Curr. Infect. Dis. Rep.*, 3, 279, 2001.
56. Leimann, B.C.Q. et al., Histoplasmosis in a Brazilian center: Clinical forms and laboratory tests, *Rev. Iberoam. Micol.*, 22, 141, 2005.
57. Pietrobon, D. et al., Disseminated histoplasmosis and AIDS in an Argentine hospital: Clinical manifestations, diagnosis and treatment, *Enferm. Infecc. Microbiol. Clin.*, 22, 156, 2004.
58. Subramanian, S. et al., Disseminated histoplasmosis, *J. Assoc. Physicians India*, 53, 185, 2005.
59. Williams, B. et al., Diagnosis of histoplasmosis by antigen detection during an outbreak in Indianapolis, Ind., *Arch. Pathol. Lab. Med.*, 118, 1205, 1994.

60. Wheat, L.J., Improvements in diagnosis of histoplasmosis, *Expert Opin. Biol. Ther.*, 6, 1207, 2006.
61. Costa, M.R.E. et al., Conventional versus molecular diagnostic tests, *Med. Mycol.*, 38, 139, 2000.
62. Deepe, G.S. Jr. and Durose, G.G., Immunobiological activity of recombinant H antigen from *Histoplasma capsulatum*, *Infect. Immun.*, 63, 3151, 1995.
63. Johnson, J.E., Jeffrey, B., and Huppert M., Evaluation of five commercially available immunodiffusion kits for detection of *Coccidioides immitis* and *Histoplasma capsulatum* antibodies, *J. Clin. Microbiol.*, 20, 530, 1984.
64. Wheat, L.J., Laboratory diagnosis of histoplasmosis: Update 2000, *Semin. Respir. Infect.*, 16, 131, 2001.
65. Kaufman, L., Laboratory methods for the diagnosis and confirmation of systemic mycoses, *Clin. Infect. Dis.*, 14, S23, 1992.
66. Torres, M. et al., Evaluation of enzyme linked immunosorbent-assay and western blot for diagnosis of histoplasmosis, *Rev. Invest. Clin.*, 45, 155, 1993.
67. Zancope-Oliveira, R.M. et al., Evaluation of cation exchange chromatography for the isolation of M glycoprotein from histoplasmin, *J. Med. Vet. Mycol.*, 31, 29, 1993.
68. Zancope-Oliveira, R.M. et al., Immunochemical analysis of the H and M glycoproteins from *Histoplasma capsulatum*, *Clin. Diagn. Lab. Immunol.*, 1, 563, 1994.
69. Zancope-Oliveira, R.M. et al., Effects of histoplasmin M antigen chemical and enzymatic deglycosylation on cross-reactivity in the enzyme-linked immunoelectrotransfer blot method, *Clin. Diagn. Lab. Immunol.*, 1, 390, 1994.
70. Pizzini, C.V. et al., Evaluation of a western blot test in an outbreak of acute pulmonary histoplasmosis, *Clin. Diagn. Lab. Immunol.*, 6, 20, 1999.
71. Guimarães, A.J. et al., ELISA for early diagnosis of histoplasmosis, *J. Med. Microbiol.*, 53, 509, 2004.
72. Wheat, J. et al., Evaluation of cross-reactions in *Histoplasma capsulatum* serologic tests, *J. Clin. Microbiol.*, 23, 493, 1986.
73. Wheat, L.J. et al., Diagnosis of histoplasmosis by antigen detection based upon experience at the histoplasmosis reference laboratory, *Diagn. Microbiol. Infect. Dis.*, 43, 29, 2002.
74. Durkin, M.M., Connolly, P.A., and Wheat, L.J., Comparison of radioimmunoassay and enzyme-linked immunoassay methods for detection of *Histoplasma capsulatum* var. *capsulatum* antigen, *J. Clin. Microbiol.*, 35, 2252, 1997.
75. Gomez, B.L. et al., Development of a novel antigen detection test for histoplasmosis, *J. Clin. Microbiol.*, 35, 2618, 1997.
76. Gomez, B.L. et al., Detection of the 70-kilodalton *Histoplasma capsulatum* antigen in serum of histoplasmosis patients: Correlation between antigenemia and therapy during follow-up, *J. Clin. Microbiol.*, 37, 675, 1999.
77. Scheel, C.M. et al., Development and evaluation of an enzyme-linked immunosorbent assay to detect *Histoplasma capsulatum* antigenuria in immunocompromised patients, *Clin. Vacc. Immunol.*, 16, 852, 2009.
78. Rickerts, V. et al., Rapid PCR-based diagnosis of disseminated histoplasmosis in an AIDS patient, *Eur. J. Clin. Microbiol. Infect. Dis.*, 21, 821, 2002.
79. Guedes, H.L.M. et al., PCR assay for identification of *Histoplasma capsulatum* based on the nucleotide sequence of the M antigen, *J. Clin. Microbiol.*, 41, 535, 2003.
80. Chemaly, R.F. et al., Rapid diagnosis of *Histoplasma capsulatum* endocarditis using the AccuProbe on an excised valve, *J. Clin. Microbiol.*, 39, 2640, 2001.
81. Abbott, J.J., Hamacher, K.L., and Ahmed I., In situ hybridization in cutaneous deep fungal infections: A valuable diagnostic adjunct to fungal morphology and tissue cultures, *J. Cutan. Pathol.*, 33, 426, 2006.
82. Bialek, R. et al., Evaluation of two nested PCR assays for detection of *Histoplasma capsulatum* DNA in human tissue, *J. Clin. Microbiol.*, 40, 1644, 2002.
83. Maubon, D., Simon, S., and Aznar, C., Histoplasmosis diagnosis using a polymerase chain reaction method. Application on human samples in French Guiana, South America, *Diagn. Microbiol. Infect. Dis.*, 58, 441, 2007.
84. Toranzo, A.I. et al., Molecular diagnosis of human histoplasmosis in whole blood samples, *Rev. Argent. Microbiol.*, 41, 20, 2009.
85. Bialek, R. et al., Diagnosis and monitoring of murine histoplasmosis by a nested PCR assay, *J. Clin. Microbiol.*, 39, 1506, 2001.
86. Spencer, W.H. et al., Detection of *Histoplasma capsulatum* DNA in lesions of chronic ocular histoplasmosis syndrome, *Arch. Ophthalmol.*, 121, 1551, 2003.
87. Bracca, A. et al., Molecular detection of *Histoplasma capsulatum* var. *capsulatum* in human clinical samples, *J. Clin. Microbiol.*, 41, 1753, 2003.
88. Tang, Y.W. et al., Urine polymerase chain reaction is not as sensitive as urine antigen for the diagnosis of disseminated histoplasmosis, *Diagn. Microbiol. Infect. Dis.*, 54, 283, 2006.
89. Martagon-Villamil, J. et al., Identification of *Histoplasma capsulatum* from culture extracts by real-time PCR, *J. Clin. Microbiol.*, 41, 1295, 2003.
90. Buitrago, M.J. et al., Detection of imported histoplasmosis in serum of HIV-infected patients using a real-time PCR-based assay, *Eur. J. Clin. Microbiol. Infect. Dis.*, 25, 665, 2006.
91. Cenis, J.L., Rapid extraction of fungal DNA for PCR amplification, *Nucleic Acids Res.*, 20, 2380, 1992.
92. Woods J.P. et al., Fast DNA isolation from *Histoplasma capsulatum*: Methodology for arbitrary primer polymerase chain reaction-based epidemiological and clinical studies, *J. Clin. Microbiol.*, 31, 463, 1993.
93. Loffler, J. et al., Comparison of different methods for extraction of fungal pathogens from culture and blood, *J. Clin. Microbiol.*, 35, 3311, 1997.
94. Borman, A.M. et al., Molecular identification of pathogenic fungi, *J. Antimicrob. Chem.*, 61, i7, 2008.
95. Pounder, J.I., Hansen, D., and Woods, G.L., Identification of *Histoplasma capsulatum*, *Blastomyces dermatitidis*, and *Coccidioides immitis* by repetitive-sequence-based-PCR, *J. Clin. Microbiol.*, 44, 2977, 2006.
96. White, T.J. et al., Amplification and direct sequencing of fungal ribosomal RNA genes for phylogenetics. In Innis, M.A. et al. (eds.), *PCR Protocols: A Guide to Methods and Applications*. Academic Press, Inc., San Diego, CA, pp. 315–322, 1990.
97. Porta, A.S. et al., A homologue of the human 100-kDa protein (p100) is differentially expressed by *Histoplasma capsulatum* during infection of murine macrophages, *Biochem. Biophys. Res. Commun.*, 254, 605, 1999.
98. Tamura, M. et al., Phylogenetic characterization of *Histoplasma capsulatum* strains based on ITS region sequences, including two new strains from Thai and Chinese patients in Japan, *Nippon Ishinkin Gakkai Zasshi*, 43, 11, 2002.
99. Goldani L.Z. et al., Two specific strains of *Histoplasma capsulatum* causing mucocutaneous manifestations of

histoplasmosis: Preliminary analysis of a frequent manifestation of histoplasmosis in Southern Brazil, *Mycopathologia*, 167, 181, 2009.

100. Muniz M.M., Caracterização fenotípica e genotípica de isolados de *Histoplasma capsulatum* em diferentes regiões brasileiras, Thesis (Doctor in Pesquisa Clinica em Doenças Infecciosas), Fundação Oswaldo Cruz, 2009.

101. Vassart G. et al., A sequence in M13 phage detects hypervariable minisatellites in human and animal DNA, *Science*, 167, 181, 2009.

102. Don, R.H. et al., 'Touchdown' PCR to circumvent spurious priming during gene amplification, *Nucleic Acids Res.*, 19, 4008, 1991.

34 Lacazia

Dongyou Liu and Yi-Wei Tang

CONTENTS

34.1 INTRODUCTION

34.1.1 Classification, Morphology, and Biology

The genus *Lacazia* belongs to the mitosporic Onygenales group (family), order Onygenales, class Eurotiomycetes, subphylum Pezizomycotina, phylum Ascomycota, and kingdom Fungi. Apart from *Locazia*, the mitosporic Onygenales group also includes the following genera: *Blastomyces*, *Chrysosporium*, *Coccidioides*, *Emmonsia*, *Geomyces*, *Malbranchea*, *Myriodontium*, and *Paracoccidioides*. The genus *Lacazia* consists of a single species *Lacazia loboi* (formerly, *Loboa loboi*, *Paracoccidioides loboi*) that is responsible for lacaziosis.

Lacaziosis (also known as lobomycosis, Lobo's disease, keloidal blastomycosis, Amazonian blastomycosis, and blastomycoid granuloma), which was first described by Jorge Lobo in 1930,[1] is a chronic, granulomatous infection of the skin and subcutaneous tissues of humans and members of the family Delphinidae (e.g., bottle-nosed dolphins).[2,3] A fungus related to *Paracoccidioides brasiliensis* was isolated using Sabouraud glucose agar, which consisted of solitary, budding yeast cells and pseudohyphae in tissue. This etiologic agent was referred to as *Blastomyces loboi* in 1952 and as *Loboa loboi* (which is now a synonym of *P. brasiliensis*) in 1956. Another generic name *Lobomyces* was used to describe the fungus in the tissue in 1968. Then, the species name *Paracoccidioides loboi* was proposed in 1971 for the etiologic agent of lobomycosis. Finally, the etiologic agent of lobomycosis was renamed as *Lacazia loboi* in 1999, with the generic name *Lacazia* acknowledging the major contributions to our understanding of lobomycosis by Carlos da Silva Lacaz and the species name *loboi* honoring the original discoverer Jorge Lobo.[4] On the basis of its yeast-like morphology in infected tissues and its sequence homology in 18S small-subunit (SSU) ribosomal RNA (rRNA) gene, chitin synthase 2, chitin synthase 4, ADP-ribosylation factor, and gp43 genes, *L. loboi* is considered as the phylogenetic sister group to, but independent from, *Paraccidioides brasiliensis*, and is linked to the other dimorphic members of the mitosporic Onygenales group.[5–7]

Morphologically, *Lacazia loboi* in vivo phenotype consists of unicellular, thick-walled yeast-like globose to subglobose cells of 7.6–7.9 µm (range 5–12 µm) in diameter that occur singly as well as in branched and unbranched chains of three or more cells connected by short tubules. Constitutive melanin is present in yeast cell walls, which can be detected by the Fontana-Masson histologic stain. Old cells show thickened cell wall, leading to the separation of adjacent cells.[8–10]

Lacazia loboi is an uncultivated fungal pathogen that causes chronic, localized, subepidermal infection called lacaziosis (lobomycosis) in humans and dolphins, occasionally in the tropical areas of the Americas.[11,12] This disease is characterized by the presence of keloidal, verrucoid, nodular lesions (containing masses of spheroidal, yeast-like organisms) or sometimes by vegetating crusty plaques and tumors, on the face, ears, or extremities. First reported in Brazil in 1930, lacaziosis appears to be restricted to the geographic area of the Amazon (Brazil, Ecuador, Venezuela, Guyana, Suriname, Bolivia, Peru, and Colombia) as well as other South and Central American countries.[13–16] Although several recent studies indicated its occurrence in Europe, Canada, the United States, and South Africa, most patients had visited areas of endemicity in Central or South America or had handled infected dolphins.[17,18]

Lacaziosis has been documented in bottlenose dolphins (*Tursiops truncatus*) and Guiana dolphins (*Sotalia guianensis*) along the Florida and Texas coasts, the South Brazilian coast, the Suriname river estuary, and the Spanish–French coast since the 1970s.[4,19–24] The infected dolphins show white to pink, verrucous lesions on the dorsal fin, head, flukes, and peduncle, which may ulcerate and form large plaques.[25]

Being phylogenetically linked to other dimorphic fungal pathogens in the mitosporic Onygenales group, *L. loboi* may have a mycelial form in nature with high tropism for soil of damp wooded areas. Humans may become infected with the fungus through skin trauma (minor scratches or insect bites) that facilitates fungal establishment from environmental propagules or through contact with propagules from diseased hosts (humans, dolphins, and experimentally infected mice). Transmission of lobomycosis among Delphinidae may occur by direct contact including transmission from mother to calf.

34.1.2 Clinical Features

Lacaziosis is a chronic cutaneous and subcutaneous disease of humans and dolphins, such as marine dolphins (*Tursiops truncatus*) and marine-freshwater dolphins (*Sotalia fluviatilis*), that is characterized by the development of nodular parakeloidal lesions (solitary or multiple) on the ears, face, arms, and legs in humans and head, back, dorsal fin, flanks, caudal peduncle, and tail in dolphins. Being well defined, smooth, shiny, and painless, developing lesions may be easily moved around as they lie free over the deeper tissues. Older lesions may be macular, gummatous, verrucoid, or ulcerative with satellite lesions resulting from autoinoculation.

Lacaziosis predominantly affects males in close contact with vegetation and aquatic environments. Many patients with lacaziosis recalled accidental trauma with plant thorns or insect bites prior to onset of the lesions. Although lacaziosis is generally regarded as a slowly evolving disease, some variations in its incubation period have been observed. For example, after handling an *L. loboi*-infected dolphin, an aquarium caretaker developed the disease 3 months later, while man acquired the infection 2½ years after his trip to the endemic area. In experimental animal model, BALB/c mice showed extensive granulomatous infiltrate and macroscopic lesions 7–8 months after inoculation at hind foot pads with *L. loboi* cells obtained from patients with the disease. On continuous passages from mice to mice, lesions developed within 4 months of *L. loboi* inoculation.[26,27]

In a recent case of human locaziosis, a 62-year-old fisherman from Venezuela presented extensive lesions on the left ear. The patient recalled that his illness began after he accidentally injured the posterior portion of the helix of that ear with a fishhook 10 years earlier. Initially, a small, solitary, and hard nodule appeared, which was followed by similar satellite lesions with confluent and harder nodules with flat and shiny surfaces, causing occasional pruritus. Silver-stained sections of the parakeloidal lesions revealed single, uniform yeast-like cells as well as branched and unbranched chains of three or more cells connected by short tubules, typical of *L. loboi* organism.[28]

In another report, a 41-year-old female veterinarian from Brazil complained of a slowly growing, painless subcutaneous nodule of 2.0 × 1.5 cm in diameter on the inner side of her left-hand middle finger, which had appeared 10 months earlier as a small hard cutaneous swelling. The patient had worked extensively with the fungus *L. loboi* in experimentally infected mice for 10 years, and her left-hand middle finger had been extensively used to manipulate biopsied tissues and to inoculated mice with live yeast-like cells. Histopathological examination of hematoxylin–eosin-stained sections of the excised nodule displayed a granulomatous infiltrate of histiocytes and giant cells together with numerous uniform thick-walled yeast-like cells, either singly or in chains, characteristic of *L. loboi*.[29]

While lacaziosis in humans is associated with a partial deficit of cell-mediated immunity that is responsible for the lack of containment of the pathogen and no alterations of humoral immunity, the disease in dolphins shows a substantial decrease in CD4+ helper T lymphocytes and CD19+ and CD21+ B cells.[30,31]

34.1.3 Diagnosis

As *Lacazia loboi* has been uncultivable, laboratory diagnosis of locaziosis (Lobo's disease) relies on microscopic examination of the infected cutaneous and subcutaneous tissue biopsies to show yeast-like cells of 5–12 μm in diameter, often forming long chains of spherical cells interconnected by tubules. Differentiation diagnosis for *Lacazia loboi* includes *Paracoccidiodes brasiliensis* or *Blastomyces dermatitidis* that shares similar morphological features.

Microscopically, *Lacazia loboi* nodules are located between the skin and subcutaneous tissue and consist of subepidermal histiocytic granulomas, with fibrous tissue dispersing between large numbers of multinucleated giant cells (of 40–80 μm in diameter) and histiocytes. Special fungal stains such as Grocott's methenamine silver (GMS)[32] and periodic acid-Schiff (PAS) are used for the histopathological diagnosis. Under GMS staining, *L. lobio* yeast-like cells are dark or have the appearance of "empty" cells. The yeast-like cells are in chains of two or more cells and from branches, similar to those yeast cells observed in the infected tissue of patients with paracoccidioidomycosis. The cells are typically connected with small tubules, a feature also observed using other fungal stains including PAS. Alternatively, cutaneous and subcutaneous tissue is mounted in 10% KOH and examined for the presence of chains of globose cells.[33]

Compared to *Paracoccidioides brasiliensis*, which displays expanding globose to subglobose solitary yeast cells with multiple daughter blastoconidia, *L. loboi* forms yeast cells of consistent size, generating branching chains of blastoconidia. In addition, while *P. brasiliensis* is a geophilic dimorphic fungus that is associated with tegumentary surfaces, acquired through inhalation and disseminated from lungs to other organs by hematogenous means, *L. loboi* only infects cutaneous–subcutaneous tissue, especially the skin of ears, elbows, the anterior aspect of legs, and ankles, with a lower body temperature, due to its apparent intolerance of 37°C temperature (refer to Chapter 40 for further details). Similarly, *Blastomyces dermatitidis* frequently infects the lungs as well as skin and genitourinary tract. In calcofluor white or KOH-stained sputum, exudates or tissue, *B. dermatitidis* shows large, spherical, and thick-walled yeast cells of 8–15 μm in diameter, which bud singly and have wide base attachment between the buds and parent cells (see Chapter 24 for details). In addition, as both *B. dermatitidis* and *P. brasiliensis* grow in vitro and *L. loboi* does not, a useful way to distinguish *B. dermatitidis* and *P. brasiliensis* from *L. loboi* is to inoculate them on Sabouraud agar.[5]

L. loboi yeast cells can be extracted from biopsies of lobomycosis skin lesions with the help of proteolytic enzyme dispase, which is known for its action against fibronectin and collagen type IV.[34] A number of studies have examined the potential of immunological methods for diagnosis of locaziosis.[30,31,35–40] Using Western blot, an immunodominant 193 kDa *Lacazia loboi* antigen was detected in dolphin, human, and mouse sera infected with lacaziosis.[41] Because *L. loboi* does not grow in vitro, BALB/c and other rodent strains have been used for its maintenance.[27,42–44] The viability of *Lacazia loboi* may be assessed by means of fluorescein diacetate–ethidium bromide (FD-EB) staining.[32]

Molecular techniques have proven instrumental in clarifying the taxonomic status of *L. loboi*.[7,9,45,46] Sequencing analysis of *L. loboi* 18S SSU ribosomal RNA (SSU rRNA) and partial chitin synthase-2 gene indicated that *L. loboi* is the sister taxon of the human dimorphic fungal pathogen *Paracoccidioides brasiliensis* within the mitosporic Onygenales group in the order Onygenales.[5,45] Further investigation on rRNA internal transcribed spacer (ITS) and chitin synthase 4, ADP-ribosylation factor, and gp43 genes confirmed that *L. loboi* is a well-supported, monophyletic group within the *Paracoccidioides* clade, but clearly distinct from *Paracoccidioides*.[7] Through comparative assessment of the gene encoding the gp43 homologous protein in *P. brasiliensis* and *L. loboi*, three sets of primers were developed for specific differentiation of these two closely related fungi.[7,47] Current MicroSeq D2 sequence database (Applied Biosystems) does not cover *L. loboi*.

34.2 METHODS

34.2.1 Sample Preparation

The clinical specimens collected in cases of lacaziosis are mostly biopsy tissues from the infected sites. Fine-needle aspiration has been shown recently as an effective alternative to skin biopsy for collection of material for experimental infection of mice with unculturable pathogens such as *Mycobacterium leprae* and *L. lobio*.[48] A portion of biopsied tissues with *Lacazia loboi* yeast cells is fixed in formalin, sectioned and stained by hematoxylin and eosin, and GMSs for microscopic examination. The remainder is used for culture, in vivo propagation, and DNA extraction. Growth in Sabouraud agar helps differentiate *Paracoccidioides brasiliensis* from *Lacazia loboi* in case of concomitant lobomycosis and paracoccidioidomycosis.[5] For in vivo propagation, several BALB/c mice are inoculated in the footpads with *L. loboi* suspension.[32] For DNA extraction, *L. loboi*-infected skin tissue is ground under liquid nitrogen. The DNA from the ground tissues is then transferred to microcentrifuge tube, treated with sodium dodecyl sulfate, subjected to proteinase K digestion, and extracted with phenol and chloroform. The extracted *L. loboi* genomic DNA is stored at −80°C until use.[46]

34.2.2 Detection Procedures

34.2.2.1 PCR Amplification of gp43 Gene

Vilela et al. designed four primers from three of the highly conserved regions of the gp43 gene sequence of *Paracoccidioides brasiliensis* for detection and differentiation of *Lacazia loboi* and *P. brasiliensis*.[46] The sequences of these primers are primer NL1 (5′-TGC TGG AGC CAT GGA TC-3′), primer NL2 (5′-AAC GGC TTC GAC AAC AGC-3′), primer NL3 (5′-GCT GTT GTC GAA GCC GTT-3′), and primer NL4 (5′-TAG ATA CAT GGC GCA GTC-3′). While primers NL1–NL4 amplify a 917 bp fragment within the 1329 bp gp43 molecule of *P. brasiliensis*, primers NL1–NL3 and NL2–NL4 amplify the 917 bp amplicon in two fragments of 431 and 486 bp, respectively. However, primers NL1–NL4 and NL1–NL3 do not amplify the predicted 917 and 431 bp fragments, respectively, from *L. loboi* DNA, and only primers NL2 and NL4 generate a 483 bp fragment from *L. loboi* DNA. This allows differentiation of *L. loboi* from *P. brasiliensis*. The PCR protocol consisted of an initial activation at 95°C for 10 min (for activation of *Taq* Gold polymerase) and 40 cycles of 1 min at 94°C, 2 min at 50°C, and 3 min at 70°C, followed by an extension at 72°C for 7 min. The amplicons are separated on 1% agarose gels, stained with EB and visualized under UV light.

34.2.2.2 Sequencing Analysis of *L. loboi* 18S SSU rRNA Gene

Herr et al. utilized forward primer NS1 (5′-GTA GTC ATA TGC TTG TCT C-3′) and reverse primer NS8 (5′-TCC GCA GGT TCA CCW ACG GA-3′, W = A or T) to amplify the 18S SSU rRNA gene from *L. loboi* for phylogenetic analysis.[5] The resulting PCR amplicon (1768 bp) is purified and sequenced; and the sequence is analyzed using Parsimony (version 4.0b.4a; D. L. Swofford, Illinois Natural History Survey, Champlain). Neighbor-joining analysis of the 18S SSU rDNA sequences uses a maximum-likelihood multiple-hit correction with an empirical transition/transversion ratio, empirical base frequencies, a gamma distribution of 0.5, and four categories of variation. Thousand bootstrap-resampled data sets analyzed by both neighbor-joining and parsimony methods (heuristic) are used to assess branch support.

34.3 CONCLUSION

Lacazia loboi (formerly *Loboa loboi*, *Paracoccidioides loboi*) is a fungal pathogen that causes a chronic, granulomatous disease of the skin and subcutaneous tissues of humans and dolphins, mainly in Central and South America. Due to its inability to grow in culture media and its morphological resemblance to *Paracoccidioides* and *Blastomyces* spp., the taxonomic status of this organism remained unsolved until recently. Following the examination of its 18S SSU rRNA, chitin synthase 2, chitin synthase 4, ADP-ribosylation factor, and gp43 gene sequences, it is apparent that *Lacazia loboi* is phylogenetically related to *Paraccidioides brasiliensis* but contains sufficient genetic difference to warrant its independent species identity within the mitosporic Onygenales group. This implies that as a dimorphic fungus, *L. loboi* may go through a mycelial stage in nature as well as a yeast-like stage in the host. In the mammalian hosts, *L. loboi* typically produces parakeloidal lesions containing uniform yeast-like cells, singly or in chains of three or more cells connected by short tubules. While the mycelial stage of *L. loboi* has yet to be identified, it is possible that future improvement in the formation of culture media will allow in vitro growth and characterization of both developmental stages of this fungus. The story of *Lacazia loboi* highlights the power and value of molecular techniques for identification and characterization of microbial pathogens that are noncultivatable and/or that demonstrate morphological and biochemical similarity to other closely related organisms.

REFERENCES

1. Lobo, J.O. Nova especies de blastomycose. *Brasil Med* 44, 1227 (1930).
2. Elsayed, S. et al. Human case of lobomycosis. *Emerg Infect Dis* 10, 715–718 (2004).
3. Mendoza, L., Vilela, R., Rosa, P.S., and Fernandes Belone, A.F. *Lacazia loboi* and *Rhinosporidium seeberi*: A genomic perspective. *Rev Iberoam Micol* 22, 213–216 (2005).
4. Taborda, P.R., Taborda, V.A., and McGinnis, M.R. *Lacazia loboi* gen. nov., comb. nov., the etiologic agent of lobomycosis. *J Clin Microbiol* 37, 2031–2033 (1999).
5. Herr, R.A. et al. Phylogenetic analysis of *Lacazia loboi* places this previously uncharacterized pathogen within the dimorphic Onygenales. *J Clin Microbiol* 39, 309–314 (2001).
6. Mendoza, L., Ajello, L., and Taylor, J.W. The taxonomic status of *Lacazia loboi* and *Rhinosporidium seeberi* has been finally resolved with the use of molecular tools. *Rev Iberoam Micol* 18, 95–98 (2001).
7. Vilela, R. et al. Molecular phylogeny of animal pathogen *Lacazia loboi* inferred from rDNA and DNA coding sequences. *Mycol Res* 113, 851–857 (2009).
8. Bhawan, J. et al. Lobomycosis. An electronmicroscopic, histochemical and immunologic study. *J Cutan Pathol* 3, 5–16 (1976).
9. Haubold, E.M., Cooper, C.R. Jr., Wen, J.W., McGinnis, M.R., and Cowan, D.F. Comparative morphology of *Lacazia loboi* (syn. *Loboa loboi*) in dolphins and humans. *Med Mycol* 38, 9–14 (2000).
10. Woodard, J.C. Electron microscopic study of lobomycosis (*Loboa loboi*). *Lab Invest* 27, 606–612 (1972).
11. Bossart, G.D. Suspected acquired immunodeficiency in an Atlantic bottlenosed dolphin with chronic-active hepatitis and lobomycosis. *J Am Vet Med Assoc* 185, 1413–1414 (1984).
12. Rodriguez-Toro, G. Lobomycosis. *Int J Dermatol* 32, 324–332 (1993).
13. Jaramillo, D., Cortes, A., Restrepo, A., Builes, M., and Robledo, M. Lobomycosis. Report of the eighth Colombian case and review of the literature. *J Cutan Pathol* 3, 180–189 (1976).
14. Paniz-Mondolfi, A.E., Reyes Jaimes, O., and Davila Jones, L. Lobomycosis in Venezuela. *Int J Dermatol* 46, 180–185 (2007).
15. Ramos, E.S.M., Aguiar-Santos-Vilela, F., Cardoso-de-Brito, A., and Coelho-Carneiro, S. Lobomycosis. Literature review and future perspectives. *Actas Dermosifiliogr* 100(Suppl 1), 92–100 (2009).
16. Rodriguez-Toro, G. and Tellez, N. Lobomycosis in Colombian Amer Indian patients. *Mycopathologia* 120, 5–9 (1992).
17. Al-Daraji, W.I., Husain, E., and Robson, A. Lobomycosis in African patients. *Br J Dermatol* 159, 234–236 (2008).
18. Burns, R.A., Roy, J.S., Woods, C., Padhye, A.A., and Warnock, D.W. Report of the first human case of lobomycosis in the United States. *J Clin Microbiol* 38, 1283–1285 (2000).
19. Caldwell, D.K. et al. Lobomycosis as a disease of the Atlantic bottle-nosed dolphin (*Tursiops truncatus* Montagu, 1821). *Am J Trop Med Hyg* 24, 105–114 (1975).
20. Cowan, D.F. Lobo's disease in a bottlenose dolphin (*Tursiops truncatus*) from Matagorda Bay, Texas. *J Wildl Dis* 29, 488–489 (1993).
21. De Vries, G.A. and Larrman, J.J. A case of Lobo's disease in the dolphin *Sotalia guianensis*. *J Aquat Mammals* 1, 26–33 (1973).
22. Reif, J.S. et al. Lobomycosis in Atlantic bottlenose dolphins from the Indian River Lagoon, Florida. *J Am Vet Med Assoc* 228, 104–108 (2006).
23. Rotstein, D.S. et al. Lobomycosis in offshore bottlenose dolphins (*Tursiops truncatus*), North Carolina. *Emerg Infect Dis* 15, 588–590 (2009).
24. Symmers, W.S. A possible case of Lobo's disease acquired in Europe from a bottle-nosed dolphin (*Tursiops truncatus*). *Bull Soc Pathol Exot Filiales* 76, 777–784 (1983).
25. Norton, S.A. Dolphin-to-human transmission of lobomycosis? *J Am Acad Dermatol* 55, 723–724 (2006).
26. Belone, A.F., Madeira, S., Rosa, P.S., and Opromolla, D.V. Experimental reproduction of the Jorge Lobo's disease in BAlb/c mice inoculated with *Lacazia loboi* obtained from a previously infected mouse. *Mycopathologia* 155, 191–194 (2002).
27. Madeira, S., Fernandes Belone Ade, F., Padovani, C.R., and Araujo Omoprolla, D.V. Comparative experimental infection of *Lacazia loboi* in BALB/c and B10.A mice. *Rev Iberoam Micol* 20, 55–59 (2003).
28. Bermudez, L., Van Bressem, M.F., Reyes-Jaimes, O., Sayegh, A.J., and Paniz-Mondolfi, A.E. Lobomycosis in man and lobomycosis-like disease in bottlenose dolphin, Venezuela. *Emerg Infect Dis* 15, 1301–1303 (2009).
29. Rosa, P.S. et al. Accidental Jorge Lobo's disease in a worker dealing with *Lacazia loboi* infected mice: A case report. *J Med Case Reports* 3, 67 (2009).
30. Vilani-Moreno, F.R., Belone, A.F., Soares, C.T., and Opromolla, D.V. Immunohistochemical characterization of the cellular infiltrate in Jorge Lobo's disease. *Rev Iberoam Micol* 22, 44–49 (2005).
31. Vilani-Moreno, F.R., Silva, L.M., and Opromolla, D.V. Evaluation of the phagocytic activity of peripheral blood monocytes of patients with Jorge Lobo's disease. *Rev Soc Bras Med Trop* 37, 165–168 (2004).

32. Vilani-Moreno, F.R., Belone Ade, F., Rosa, P.S., Madeira, S., and Opromolla, D.V. Evaluation of the vital staining method for *Lacazia loboi* through the experimental inoculation of BALB/c mice. *Med Mycol* 41, 211–216 (2003).
33. Talhari, C., Chrusciak-Talhari, A., de Souza, J.V., Araujo, J.R., and Talhari, S. Exfoliative cytology as a rapid diagnostic tool for lobomycosis. *Mycoses* 52, 187–189 (2009).
34. Salgado, C.G. et al. Enzymatic isolation of *Lacazia loboi* cells from skin lesions of lobomycosis. *Med Mycol* 47, 119–123 (2009).
35. Camargo, Z.P., Baruzzi, R.G., Maeda, S.M., and Floriano, M.C. Antigenic relationship between *Loboa loboi* and *Paracoccidioides brasiliensis* as shown by serological methods. *Med Mycol* 36, 413–417 (1998).
36. Landman, G., Velludo, M.A., Lopes, J.A., and Mendes, E. Crossed-antigenicity between the etiologic agents of lobomycosis and *paraccocidioidomycosis* evidenced by an immunoenzymatic method (PAP). *Allergol Immunopathol (Madr)* 16, 215–218 (1988).
37. Taborda, C.P. and Camargo, Z.P. Diagnosis of *paracoccidioidomycosis* by passive haemagglutination assay of antibody using a purified and specific antigen-gp43. *J Med Vet Mycol* 31, 155–160 (1993).
38. Taborda, C.P. and Camargo, Z.P. Diagnosis of *paracoccidioidomycosis* by dot immunobinding assay for antibody detection using the purified and specific antigen gp43. *J Clin Microbiol* 32, 554–556 (1994).
39. Taborda, C.P., Juliano, M.A., Puccia, R., Franco, M., and Travassos, L.R. Mapping of the T-cell epitope in the major 43-kilodalton glycoprotein of *Paracoccidioides brasiliensis* which induces a Th-1 response protective against fungal infection in BALB/c mice. *Infect Immun* 66, 786–793 (1998).
40. Vidal, M.S., Palacios, S.A., de Melo, N.T., and Lacaz Cda, S. Reactivity of anti-gp43 antibodies from *Paracoccidioides brasiliensis* antiserum with extracts from cutaneous lesions of Lobo's disease. Preliminary note. *Rev Inst Med Trop Sao Paulo* 39, 35–37 (1997).
41. Mendoza, L. et al. Use of sera from humans and dolphins with lacaziosis and sera from experimentally infected mice for Western Blot analyses of *Lacazia loboi* antigens. *Clin Vaccine Immunol* 15, 164–167 (2008).
42. Madeira, S., Opromolla, D.V., and Belone, A. Inoculation of BALB/c mice with *Lacazia loboi*. *Rev Inst Med Trop Sao Paulo* 42, 239–243 (2000).
43. Opromolla, D.V., Madeira, S., Belone, A.F., and Vilani-Moreno, F.R. Jorge Lobo's disease: Experimental inoculation in Swiss mice. *Rev Inst Med Trop Sao Paulo* 41, 359–364 (1999).
44. Opromolla, D.V. and Nogueira, M.E. Inoculation of *Lacazia loboi* into the subcutaneous tissue of the hamster cheek pouch. *Rev Inst Med Trop Sao Paulo* 42, 119–123 (2000).
45. Vilela, R., Martins, J.E., Pereira, C.N., Melo, N., and Mendoza, L. Molecular study of archival fungal strains isolated from cases of lacaziosis (Jorge Lobo's disease). *Mycoses* 50, 470–474 (2007).
46. Vilela, R. et al. Molecular model for studying the uncultivated fungal pathogen *Lacazia loboi*. *J Clin Microbiol* 43, 3657–3661 (2005).
47. Puccia, R. and Travassos, L.R. 43-kilodalton glycoprotein from *Paracoccidioides brasiliensis*: Immunochemical reactions with sera from patients with paracoccidioidomycosis, histoplasmosis, or Jorge Lobo's disease. *J Clin Microbiol* 29, 1610–1615 (1991).
48. Rosa, P.S., Belone, A.D., Lauris, J.R., and Soares, C.T. Fine-needle aspiration may replace skin biopsy for the collection of material for experimental infection of mice with *Mycobacterium leprae* and *Lacazia loboi*. *Int J Infect Dis* 14, e49–e53 (2010).

35 *Lecythophora*

Dongyou Liu

CONTENTS

35.1 INTRODUCTION

35.1.1 Classification and Morphology

The genus *Lecythophora* is a dematiaceous (dark-walled) fungus belonging to the mitosporic Coniochaetaceae group, family Coniochaetaceae, order Coniochaetales, class Sordariomycetes, subphylum Pezizomycotina, phylum Ascomycota, and kingdom Fungi. As the only member of the mitosporic Coniochaetaceae group, the genus *Lecythophora* currently consists of six recognized species, that is, *Lecythophora decumbens*, *Lecythophora fasciculata*, *Lecythophora hoffmannii* (teleomorph: *Coniochaeta ligniaria*; obsolete synonyms: *Margarinomyces hoffmannii* and *Phialophora hoffmannii*), *Lecythophora lignicola*, *Lecythophora luteoviridis*, and *Lecythophora mutabilis* (obsolete synonyms: *Margarinomyces mutabilis* and *Phialophora mutabilis*), in addition to 25 unassigned species [1]. Of these, *Lecythophora hoffmannii* and *Lecythophora mutabilis* have been associated with human phaeohyphomycosis.

Lecythophora hoffmannii colonies are flat, smooth, moist, and pink to orange, with regular and sharp margin; reverse is pink. Hyphae are narrow and hyaline, producing conidia laterally from small collarettes directly on the hyphae or from lateral cells sometimes arranged in dense groups; lateral cells are flask shaped or nearly cylindrical. Collarettes (1.5 μm wide) are unpigmented. Conidia (3.0–3.5 × 1.5–2.5 μm) are hyaline, smooth, thin walled, slightly curved, and broadly ellipsoidal to cylindrical or allantoid and tend to aggregate in clusters at the tip of the phialides. *Lecythophora mutabilis* conidia form from intercalary cells along the hyphae [2].

35.1.2 Clinical Features

Of the six recognized dematiaceous fungal species in the genus *Lecythophora*, two (*Lecythophora hoffmannii* and *Lecythophora mutabilis*) have been implicated in human phaeohyphomycosis, leading to abscess, sinusitis, mastoiditis, endocarditis, endophthalmitis, peritonitis, and invasive disease [3–6].

35.1.2.1 *Lecythophora hoffmannii* (*Phialophora hoffmannii*)

Rinaldi et al. [7] reported a case of *L. hoffmannii* (*P. hoffmannii*) infection causing a gluteal abscess after multiple intramuscular injections of antimicrobial agents. Treatment of the purulent exudate and paraffin block sections of the exudate with an *N*-acetyl cysteine–NaOH digestion–decontamination procedure demonstrated hyphal elements. Culture of the abscess material yielded a pure growth of the mold *L. hoffmannii* (*P. hoffmannii*). Marriott et al. [8] described a rare case of simultaneous infections with *Lecythophora hoffmannii* (causing chronic sinusitis) and *Scytalidium dimidiatum* (causing skin lesions, lymphangitis, and lymphadenitis) in a 44-year-old human immunodeficiency virus-infected man. Slide cultures of sinus aspirates yielded pink colonies with a tan reverse on Sabouraud's dextrose agar. On potato dextrose agar, colonies became darker olivaceous in color with age. Microscopy showed packed strands of hyphae with simple, smooth, hyaline conidiophores. Phialides were flask shaped to cylindrical, single, and without collarettes. Conidia were hyaline, smooth, and slightly curved (0.6–2.3 μm × 2.2–6.0 mm) and tended to aggregate in clusters at the tip of the phialides. The fungus was identified as *L. hoffmannii*. Chang et al. [9] also documented an extremely rare fungal mastoiditis due to *Lecythophora hoffmannii* in an immunocompetent host. Successful treatment involved a combination of radical surgical removal of all apparent infected tissue along with local treatments with polyhexamethylene biguanide, a common swimming pool biocide agent.

35.1.2.2 *Lecythophora mutabilis* (*Phialophora mutabilis*)

Slifkin and Bowers [10] described a case of fungal endocarditis due to vegetative growth of *L. mutabilis* (*P. mutabilis*) on a prosthetic mitral valve in a patient with rheumatic heart disease. The patient developed congestive failure, and a large matted obstructive fungal vegetation was found on the prosthesis. Culture of this vegetation grew *L. mutabilis* (*P. mutabilis*). Conidia form from intercalary cells along the hyphae. In a separate study, Drees et al. [11] described a case of *Lecythophora mutabilis* prosthetic valve endocarditis in a diabetic patient, who was treated surgically and with amphotericin B lipid complex and voriconazole.

Scott et al. [12] reported delayed-onset, bleb-associated endophthalmitis caused by *Lecythophora mutabilis* in a 57-year-old woman who underwent trabeculectomy with Mitomycin C in her left eye. The patient presented with a 10 day history of pain and decreased vision in her left eye. Cultures of scraping sample yielded *L. mutabilis*. After treatment with intraocular and systemic antifungal agents, three pars plana vitrectomies, partial iridectomy, and cryopexy to the bleb and angle structures, her left eye showed no residual infection, and vision corrected to 20/80.

Taniguchi et al. [13] documented an invasive fungal infection (IFI) due to *Lecythophora mutabilis* in an 18-year-old man with mitochondrial encephalomyopathy accompanied with refractory anemia and chronic renal failure. The patient developed septic shock, and *L. mutabilis* was detected from a blood culture and was identified morphologically and genetically. Despite treatment with micafungin and liposomal amphotericin B, the patient succumbed to the infection.

35.1.3 Diagnosis

Lecythophora is one of the 70 dematiaceous (melanized) fungal genera that are opportunistic human pathogens causing phaeohyphomycosis, chromoblastomycosis, and eumycetoma in humans. Because of the morphological, ecological, and biochemical diversity of the dark-walled fungi, effective treatment and control of these organisms are reliant on their accurate identification and speciation.

Given the lengthy periods it takes to grow dematiaceous fungi and the technical expertise it demands to speciate them based on macroscopic and microscopic characteristics [14], molecular methods such as PCR and sequencing have been developed and applied for their rapid, accurate, and sensitive detection and identification. The most common molecular targets include small-subunit (SSU) and large-subunit (LSU) rRNA genes, internal transcribed spacer (ITS) regions, and other genes [15–19].

35.2 METHODS

35.2.1 Sample Preparation

Clinical specimens (including smears and tissue sections) are stained by hematoxylin and eosin, methenamine silver stain, or other fungal stains for microscopic examination. Portions of samples are inoculated on Sabouraud's dextrose agar and other mycological media. Slide cultures of resulting isolates are prepared on potato dextrose agar, cornmeal agar, and 2% agar for improved identification to the species level [8].

Fungal strains are cultured in 20 mL of RPMI 1640 medium with L-glutamine but without sodium bicarbonate (Sigma-Aldrich) buffered to pH 7.0 with 0.165 M morpholinepropanesulfonic acid (MOPS) (Sigma-Aldrich). After 3–14 days of growth at 30°C under agitation (100 rpm), the mycelium is transferred into a tube and washed in 40 mL of sterile distilled water. Mycelium is then stored at −20°C until use. Hundred milligram of mycelium is homogenized for 1 min in a tube containing 1 mL of lysis buffer (2% Triton X-100, 1% sodium dodecyl sulfate, 10 mM Tris–HCl pH 8, 100 mM NaCl, and 1 mM EDTA pH 8), three 0.5-cm-diameter glass beads (Sigma), and approximately 500 mg of 425–600 μm glass beads (Sigma). The homogenized mycelia are then snap-frozen in liquid nitrogen, thawed, and refrozen once. DNA extraction is then performed with the DNeasy plant kit (QIAGEN) [18].

Alternatively, 50 mg of fungal elements is suspended in 600 μL extraction buffer (200 mM Tris–HCl, pH 7.5, 25 mM EDTA, 0.5% w/v sodium dodecyl sulfate, and 250 mM NaCl). The mixture is vortexed for 15 s, incubated at 100°C for 15 min, kept on ice for 60 min, and then centrifuged at 14,000×*g* for 15 min. Supernatants are transferred to new tubes and extracted with phenol–chloroform–isoamyl alcohol (25:24:1 v/v). Each sample DNA is precipitated with cold isopropanol (−20°C), dried, and resuspended in 100 μL distilled water [16].

35.2.2 Detection Procedures

35.2.2.1 Real-Time PCR and Sequencing Analysis of ITS Regions

Pounder [19] described a real-time PCR with SYBR green DNA-binding dye and amplicon-melting temperature analysis for fungal detection using pan-fungal primers ITS1 forward (5′-TCCGTAGGTGAACCTGCGG-3′) and ITS4 reverse (5′-TCCTCCGCTTATTGATATGC-3′). The identity of the fungi is verified by subsequent sequencing analysis.

Procedure

1. PCR mixture is composed of 1×Lightcycler FastStart DNA Master Hybridization Probes mixture (Roche Applied Science) containing

deoxynucleoside triphosphates, FastStart *Taq* DNA polymerase, and 1 mM $MgCl_2$ (additional $MgCl_2$ is added to a final concentration of 4.6 mM), 0.4 μM each of ITS1 forward and ITS4 reverse primers, 1×SYBR green (Molecular Probes), and 3 μL template DNA.
2. Thermal cycling parameters include 95°C for 10 min; 50 cycles of 95°C for 5 s, 60°C for 20 s, and 76°C for 30 s; and a final extension at 72°C for 2 min.
3. The quality of the amplicon is determined using the derivative of the melt analysis curve (55°C–99°C, 45 s hold at 55°C, and 5 s/°C) using the RotorGene 3000 (Corbett Robotics, Inc).
4. The amplified product is purified for bidirectional sequencing using ExoSAP-IT (USB Corp). Five microliter of Big Dye Terminator Ready Reaction Mix v. 1.1 (Applied Biosystems) is added to 4 μL of each primer (0.8 pmol/μL) and 3 μL of purified PCR product. Cycle sequencing is performed with a 9700 thermal cycler (ABI), using 25 cycles of 96°C for 10 s, 50°C for 5 s, and 60°C for 4 min. Sequencing reaction products are passed through a Sephadex G-50 fine column to remove unincorporated dye terminators. Purified sequencing reaction products are run on an ABI Prism 3100 Genetic Analyzer with a 50 cm capillary array.
5. Sequences are analyzed with the SmartGene Integrated Database Network software version 3.2.3 vr. SmartGene is a web-based software and database system with reference sequences derived from the National Center for Biological Information (NCBI) GenBank repository.

Note. If real time PCR instrument is not available, standard PCR may be performed with primers ITS1 and ITS4, and the resulting amplicon is sequenced with the same primers. Sequence-based identifications are defined by percent identity: species, ≥99%; genus, 93%–99%; and inconclusive, ≤93%.

35.2.2.2 Sequencing Analysis of D1/D2 Domains of LSU rRNA Gene

Abliz et al. [16] utilized primers NL-1, 5′-GCATATCAATAAGCGGAGGAAAAG-3′, and NL-4m, 5′-GGTCCGTGTTTCAAGACG-3′, for amplification of the D1/D2 domains of the LSU rRNA gene followed by sequencing for identification of melanized fungi.

Procedure

1. PCR (50 μL) is made up of 5 μL of template DNA, 5 μL (2 pmol) each primer, 4 μL (2.5 mM) dNTP mixture, 0.25 μL (5 U/μL) *Taq* polymerase, and 5 μL 10× reaction buffer.
2. Amplification is performed with 1 cycle of 95°C for 4 min; 30 cycles of 94°C for 1 min, 55°C for 2.5 min, and 72°C for 2.5 min; and 1 cycle of 72°C for 10 min.
3. The amplified products are purified and subjected to direct sequencing with an ABI Prism 3100 sequencer after labeling with BigDye™ Terminator Cycle Sequencing Ready Reaction (Applied Biosystems). The external primers, NL-1 and NL-4m, and the internal primers, NL-2m, 5′-CTTGTGCGCTATCGGTCTC-3′, and NL-3m, 5′-GAGACCGATAGCGCACAAG-3′, are used to sequence each DNA sample.
4. The sequence data are aligned with CLUSTAL W (version 1.6). Phylogenetic trees are constructed with the neighbor-joining (NJ) method.

Note. The sequences of D1/D2 domains of the medically important species are highly conserved. For the species in which the number of nucleotide differences is less than three, the sequences of D1/D2 domains are considered to be insufficient criteria to identify confidently each species.

35.2.2.3 PCR–Restriction Fragment Length Polymorphism

Destino et al. [18] described the use of universal fungal primers V9D (5′-TTAAGTCCCTGCCCTTTGTA-3′) and LS266 (5′-GTAGTCATATGCTTGTCTC-3′) for PCR amplification and restriction fragment length polymorphism analysis for identification of fungal organisms. In the case of negative amplification with V9D and LS266, other fungal universal primers (e.g., ITS4/ITS5) may be employed.

Procedure

PCR mixture (20 μL) is composed of 1 μL of genomic DNA, 1.25 U of AmpliTaq gold (Roche), 2 μL of 10×PCR buffer (Roche), 2 μL of 25 mM $MgCl_2$, 2 μL of 2.5 mM deoxynucleoside triphosphate, and 1 μL of each 10 μM concentrated primers.

Amplification is conducted with a first cycle of 95°C for 10 min, followed by 30 cycles of 94°C for 30 s, 58°C for 30 s, and 72°C for 30 s, with a final extension of 72°C for 10 min.

PCR products are digested with endonuclease SmaI for 1 h at 27°C. Restriction fragments are visualized on a 3% agarose gel after ethidium bromide staining.

Note. PCR products may be also purified and sequenced by using the BigDye Terminator Cycle Sequencing Ready Reaction kit, version 3.1 (Applied Biosystems), with the primers V9D and LS266. Reaction products are analyzed using an ABI Prism 3700 automated DNA analyzer (Applied Biosystems). Sequences are then edited and manually corrected with Chromas, version 2.24 (Technelysium, Helensvale, Queensland, Australia). Multiple-sequence alignment was carried out using ClustalW 1.8. Phylogenetic trees are constructed by the NJ method using the Phylip package (http://www.infobiogen.fr) and visualized using Treeview. *Pseudallescheria boydii* is selected as the outgroup (GenBank accession number AY228119).

35.3 CONCLUSION

The genus *Lecythophora* contains six recognized dematiaceous fungal species, two of which (*Lecythophora hoffmannii* and *Lecythophora mutabilis*) have been implicated in human phaeohyphomycosis, resulting in a range of symptoms (e.g., abscess, sinusitis, mastoiditis, endocarditis, endophthalmitis, peritonitis, and invasive disease). Considering the time-consuming nature and difficulty of using conventional laboratory procedures for correct identification of *Lecythophora* and other melanized fungi, it is critical that molecular methods (e.g., PCR and sequencing) are adopted for their rapid, accurate, and sensitive detection and identification.

REFERENCES

1. The UniProt Consortium. Available at http://www.uniprot.org/, accessed on August 1, 2010.
2. Rippon JW. *Medical Mycology*, 3rd edn. W.B. Saunders Co., Philadelphia, PA, 1988.
3. Pierach CA et al., *Phialophora mutabilis* endocarditis. *Ann Intern Med.* 1973;79(6):900–901.
4. Ahmad S et al., Fungal peritonitis caused by *Lecythophora mutabilis. J Clin Microbiol.* 1985;22(2):182–186.
5. Marcus DM et al., *Lecythophora mutabilis* endophthalmitis after long-term corneal cyanoacrylate. *Retina.* 1999;19(4):351–353.
6. Sakaeyama S et al., *Lecythophora hoffmannii* isolated from a case of canine osteomyelitis in Japan. *Med Mycol.* 2007;45(3):267–272.
7. Rinaldi MG, McCoy EL, Winn DF. Gluteal abscess caused by *Phialophora hoffmannii* and review of the role of this organism in human mycoses. *J Clin Microbiol.* 1982;16:181–185.
8. Marriott DJ et al., *Scytalidium dimidiatum* and *Lecythophora hoffmannii*: Unusual causes of fungal infections in a patient with AIDS. *J Clin Microbiol.* 1997;35(11):2949–2952.
9. Chang CY et al., Novel use of a swimming pool biocide in the treatment of a rare fungal mastoiditis. *Laryngoscope.* 2005;115(6):1065–1069.
10. Slifkin M, Bowers HM Jr., *Phialophora mutabilis* endocarditis. *Am J Clin Pathol.* 1975;63(1):120–130.
11. Drees M et al., *Lecythophora mutabilis* prosthetic valve endocarditis in a diabetic patient. *Med Mycol.* 2007;45(5):463–467.
12. Scott IU et al., Delayed-onset, bleb-associated endophthalmitis caused by *Lecythophora mutabilis. Am J Ophthalmol.* 2004;137(3):583–585.
13. Taniguchi Y et al., Septic shock induced by *Lecythophora mutabilis* in a patient with mitochondrial encephalomyopathy. *J Med Microbiol.* 2009;58:1255–1258.
14. Espinel-Ingroff A et al., Evaluation of the API 20C yeast identification system for the differentiation of some dematiaceous fungi. *J Clin Microbiol.* 1989;27(11):2565–2569.
15. Fell JW et al., Biodiversity and systematics of basidiomycetous yeasts as determined by large-subunit rDNA D1/D2 domain sequence analysis. *Int J Syst Evol Microbiol.* 2000;50:1351–1371.
16. Abliz P et al. Identification of pathogenic dematiaceous fungi and related taxa based on large subunit ribosomal DNA D1/D2 domain sequence analysis. *FEMS Immunol Med Microbiol.* 2004;40(1):41–49.
17. Bagyalakshmi R et al., Newer emerging pathogens of ocular non-sporulating molds (NSM) identified by polymerase chain reaction (PCR)-based DNA sequencing technique targeting internal transcribed spacer (ITS) region. *Curr Eye Res.* 2008;33(2):139–147.
18. Destino L et al., Severe osteomyelitis caused by *Myceliophthora thermophila* after a pitchfork injury. *Ann Clin Microbiol Antimicrob.* 2006;5:21.
19. Pounder JI, Discovering potential pathogens among fungi identified as nonsporulating molds. *J Clin Microbiol.* 2007;45(2):568–571.

36 *Microsporum**

Rahul Sharma and Yvonne Gräser

CONTENTS

36.1 INTRODUCTION

> More has been written about ringworm than any other mycotic disease of animals and man and for no other mycosis is the older literature more confused.
>
> **G.C. Ainsworth and P.K.C. Austwick, 1973[1]**

The above quote was written when polymerase chain reaction (PCR)-based molecular methods have not yet arrived at the fungal taxonomic scene and especially for mycologists' studying dermatophytes. The PCR technology as we know now was only available when Kary B. Mullis made a breakthrough of amplifying DNA in vitro,[2] and 2 years later the use of thermostable polymerase[3] made the whole process automated, which revolutionized biological investigations. This single method has undoubtedly had more application in biology today than any other technique in whole of biological sciences including diagnosing human fungal infections.

The genus *Microsporum* established by Gruby in 1843[4] was the first genus to be described among the three genera commonly referred to as dermatophytes (*dermal*—skin; *phytes*—plants); the other two are *Trichophyton* Malmsten[5] and *Epidermophyton* Sabouraud[6] first described in 1845 and 1910, respectively. The name *Microsporum* refers to numerous micro-arthrospores covering formed during its (ectothrix type) growth on hair surface unlike *Trichophyton* that grows inside hair shaft.

The similarity among these dermatophytic genera extends to their nutritional requirement, which is mainly keratin[7] that forms exoskeleton covering in three groups of higher vertebrates, i.e., birds, reptiles, and mammals. Their keratinolytic nature itself suggests that they originated when life came to land from water because aquatic vertebrates (e.g., whales) lack exoskeleton adnexes like hair, feather, hooves, which are primarily made of keratin and we don't have *Microsporum* (or other dermatophyte), which is aquatic (marine) in nature because it has been shown that growth of human pathogenic fungi (including dermatophyte *M. gypseum*) is inhibited by high salt concentrations.[8] Also, rhexolytic dehiscence mechanism is found among fungi that have dry spores as in all *Microsporum* species (and not in mucilaginous mass as in *Fusarium* spp., which form blastospores). Among the three dermatophytic genera, *Microsporum* is well adapted to soil

* The authors dedicate this chapter to Prof. Libero Ajello, the man who knew so well the natural history of dermatophytes.

as (except a few that have newly evolved as anthropophilic species) its entire species produce macroconidia, which have rough surfaces unlike their *Trichophyton* counterparts.

The member of the genus can be grouped on the basis of ecological preference as geophilic (saprophytes in soil), zoophilic (adapted to lower animals), or anthropophilic (adapted to human).[9] These ecological groupings of various species are reflected in the phylogenetic clusters obtained by mtDNA[10] and rDNA studies.[11] This suggests that they all were previously soil dwellers and used to thrive on shed keratin. Close or direct contact with live animals and later humans made them well adapted to their zoophilic and anthropophilic lifestyle, respectively. Also, the anthropophilic members (*M. ferrugineum* and *M. audouinii*) must have evolved along with *Homo sapiens* (modern human), which have sparse hairs on their body unlike its earlier hominid viz. *Australopithecus afarensis*, *Homo habilis* or related primate like ancestors. This hypothesis needs further investigation with right molecular marker and algorithm,[12,13] which could precisely date and calibrate the divergence of *M. canis* into sibling species *M. auduoinii* in Africa and *M. ferrugineum* in East Asia to superimpose precisely with the divergence of *Homo sapiens* from Africa to East Asia. One theory suggests that modern *Homo sapiens* developed relatively hairless bodies to evade parasites[14] which certainly would have included dermatophytic fungi. For adaptation to this altered trait of nakedness[15] in humans, the evolving dermatophytes (geophiles > zoophiles > anthropophiles) that produce macroconidia for transmission, slowly lost their ability by natural selection since transmission became mainly through direct contact and propagules like arthroconidia. Macroconidia cling to hairs of animals for dispersal and host can be symptomless carriers like the cats,[16] which is an indicator that cat dwelling areas are survival grounds of *M. canis.* Cats' loose hairs are shed in large numbers (an attribute familiar to cat owners), enabling the host to dislodge clinging macroconidia and avert infection; however, the macroconidia of *M. canis* in particular is well adapted to zoophilic nature as its curved and roughened apex help attachment to animal skin, and the overall shape resemble an arrow head that penetrate through the dense hairs to reach skin apart from being anti-grazing devise postulated by Summerbell.[17] Macroconidia are rare or absent in present anthropophiles because the transmission also is hardly a problem for the fungus when the host are in plenty (considerable rise in human population), and chances of not finding a suitable host are rare because urban population recently reached 3.2 billion (crossing the world rural population of 3.1 billion),[18] where more humans live per unit area, a condition favorable for anthropophilic adaptation and evolution. Therefore, the need for dispersal spores adapted to transfer through furred animals was gradually lost or greatly reduced, leaving the production of resting spores resistant to adverse conditions like arthrospores (chlamydospores). These contagious propagules are produced in large numbers and have shown successful adhesion to stratum corneum under experimental conditions,[19] thereby effectively transmitting anthropophilic dermatophytes. The gene(s) responsible for triggering or inhibiting the production of macroconidia and the extent of this inactivation in anthropophiles is yet to be investigated. It is seen in agar or hair cultures of dermatophytes that the fungus forms spores once the nutrients are depleted (this is true for the formation of sexual stages, i.e., sexual reproduction and formation of ascomata as in Takashio dilute Sabouraud agar[20]) and the phenomenon is not uncommon in wide range of fungi. Once a fungus gets human skin (which is still attached to human body), it can ideally grow over its whole body, but due to variable microenvironment and medical arrangements, the fungus is prevented from spreading and mostly controlled; however, severe cases are often reported when no medical treatment is done or due to altered microclimate (caused by drugs or reduced inherent immunity). Human skin is like a continuous culture with regular supply of fresh keratin from beneath. The fungus anchors its mycelia with the fresh integument by spreading radial in a ring-like fashion (hence, the popular term ringworm coined by Anglo-Saxon ancestors in sixteenth century),[9] and thus ensuring its survival on host.

Africa is thought to be the center of origin of *Homo sapiens* based on the available evidence.[21] Ancestral dermatophytes including young anthropophiles also might have their origin in Africa.[22] If this is true, then the separation of two lineage to become distinct species of the *M. canis* clade after early humans moved "out of Africa" to Southeast Asia[23,24] is probable, thereby causing allopatric speciation[25] resulting into two closely related anthropophilic species *M. auduoinii* and *M. ferrugineum.* A recent trend of cryptic lineages[26] is witnessed in the distribution of minus (−) mating type, which tend to be well adapted to human but a lineage originally occurring on animals.[27] This assumption gains support from the study establishing first domestication of middle eastern wild cat by people of Israel 9500 years ago based on archeological and historical records.[28] Also, examination of DNA from 1000 wild cats and domestic cats across old world showed all modern domestic cats descended from the middle eastern wild cat *Felis silvestris lytica.*[29] Could this be the time (10,000 years ago) when *M. canis* first seriously interacted with humans?

The cosmopolitan distribution of certain species of dermatophytic fungi (not associated with birds) was greatly facilitated by humans and the animals he transported (wild and domestic) in recent past, i.e., within 5000 year BP. The massive movement of human population during wars in more recent time (1–2k years) dispersed dermatophytic fungi, which otherwise were restricted continently. Unlike certain rust fungi, which infect huge number of host plants at a time that could generate unimaginable spore inoculums and due to small size and lighter spores have intercontinental dispersal through air current,[30] the dermatophytes have comparatively large and heavy spores (due to which it can hardly reach the turbulence layer)[31] and are never able to develop such a huge spore inoculum since they infect meager number of hosts at a time to have intercontinental dispersal by themselves and need animal or other aid for long-distance travel. Due to preening habit and flight, dispersal by birds is limited to fungal species that have specialized clinging structures like *Ctenomyces serratus* (that forms ctenoid appendages)[32] whose intercontinental dispersal through birds might have enabled regular gene flow among its populations across continents and prevented allopatric speciation (the species is monotypic) as noted for some of the *M. fulvum* strains infrequently isolated in Central Europe from sand-martin (*Riparia riparia* L.) whose winter migration is in East Africa (with 8000 km migratory route).[33] The species of dermatophytes and keratinophilic fungi reported by Ajello and Padhye in 1974[34] from Galapagos Islands (Darwin fame) mostly are bird associated and no *Microsporum* species was recorded. The lack of zoophilic or geophilic dermatophyte species (either *Microsporum* or *Trichophyton*) pathogenic to humans is understandable in terms of the relative lack of past human disturbance (or migration) in the Galapagos before Darwin arrived.

Since the inception of molecular methods in disease diagnostics, dermatophyte detection, and species identification of clinically relevant (almost all) *Microsporum* species has become more accurate and precise in comparison to the less consistent morphology-based methods. However, even today morphology is still important in providing preliminary data for starting the molecular analysis for species identification. In the absence of any supporting morphological details viz. clinical symptoms, source of infection or even preliminary slides prepared from clinical specimen, molecular identification becomes blind and sometimes time consuming; however, the above morphological attributes, if checked, greatly help apply molecular methods precisely.

For molecular identification of *Microsporum* species, several approaches are now available, which initially started with the use of DNA homology studies[35] or mitochondrial DNA analysis.[10] Now, we have better approaches or loci available that can utilize even fraction of the genomic DNA to correctly diagnose a *Microsporum* species. The present chapter provides account of the preferred method, which is convenient in application by even small laboratories and diagnostic facilities. Although, in recent times, excellent reviews have been published[36–38] that gives an overall view in itself of the dermatophyte species definition, strain differentiation or the molecular methods involved in their identification and mentioning them all again will be repetition of the already available text. However, we do discuss in some detail a method that is widely applied in population-based investigations, i.e., microsatellite markers for strain typing.[37] Microsatellite markers are useful in identifying origin of infection (e.g., a pet, cattle, or environment), endemic strain introduced into a new area, multiple strain infection, or even relapse or reinfection, all of which can be detected if a good polymorphic marker is available. Recently developed microsatellite markers for two *Microsporum* species were able to detect polymorphisms first in an otherwise known clonally reproducing species (*M. canis*) and in the second (*M. persicolor*), to be a common old species in India and not introduced recently.[27,39]

Here we present the practical approach of "how to" identify or detect *Microsporum* species using modern molecular tool(s). Also, we incorporate a portion of morphological detail in the conventional method section to help non-taxonomist, parataxonomist, or even pure molecular taxonomist to appreciate the elaborate morphology that *Microsporum* species represent through their millions of years of evolutionary history, something Perkins felt for *Neurospora*.[40] The principle aim of our article is to help user accurately identify known *Microsporum* species and able to recognize newer ones in their future encounters with the genus. The article also discusses the recent developments in the dermatophyte identification like the use of alternate locus to the ribosomal DNA like the chitin synthase gene for rapid detection,[41] use of multiplex RT-PCR,[42,43] oligonucleotide array[44] along with the status of genome sequencing in two of the *Microsporum* species viz. *M. canis* (a zoophile) and *M. gypseum* (a geophile) previously prioritized among five dermatophytes (include three Trichophytons—*T. rubrum*, *T. tonsurans*, and *T. equinum*) by the Dermatophyte Genome Steering Committee.[45,46]

36.1.1 Classification, Morphology, Biology, and Epidemiology

36.1.1.1 Classification

The genus *Microsporum* is a mitosporic fungus that multiplies by producing mitospores: macro- and microconidia whose sexual state lies in ascomycete genus *Arthroderma*, some species of which produces *Chrysosporium* and *Trichophyton* asexual stages.[11,47,48] Table 36.1 lists all known teleomorphs, i.e., fungi having *Microsporum* asexual states including those that lack sexual stage in their life cycle, along with their natural habitat and distribution, which has been greatly altered due to human migration in the last century. The taxonomic positioning of *Microsporum* as per current scheme of fungal classification[49,50] and the reason thereof is given below, with example of *Microsporum gypseum*, the first *Microsporum* species (in fact the first dermatophyte) to have complete genome sequenced.

Superkingdom	Eukarya	True nucleus
Kingdom	Eumycota	True fungi
Subkingdom	Dikarya	Dikaryotic hyphae
Phylum	Ascomycota	Ascus as diagnostic character
Subphylum	Pezizomycotina	Asci within fruit bodies enclosed partially or completely
Class	Eurotiomycetes	Mostly heterogenous assemblage of taxa
Subclass	Eurotiomycetidae	Stroma absent, ascomata mostly cleistothecial, often brightly colored, evanescent asci
Order	Onygenales	Ascomata from coiled initials, peridium loosely interwoven thick-walled hyphae
Family	Arthrodermataceae	Rough-walled, dumbbell-shaped ossiform peridial hyphal cells, ascospores-minute smooth-walled discoid, anamorph either *Chrysosporium*, *Microsporum*, or *Trichophyton*

Genus	*Microsporum*	Forms echinulate holothallic two to multiseptate macroconidia
Species	*Microsporum gypseum*	Macroconidia in large clusters, rather thin-walled, regularly verrucose, fusiform, 3–6(–8) celled, 25–60 × 8.5–15 μm

36.1.1.2 Morphology

Characteristic morphology of *Microsporum* is the holothallic macroconidia the species produce, which is mainly two to multi-septate transmission propagules. The number of cells in a single macroconidium represents the same number of sub-propagules, all of which can germinate to form individual thalli. *Microsporum* spp. mostly produces abundant macroconidia in Malt agar or Sabouraud agar supplemented with horse or human hair except in species in which they are seldom formed.

There exist gross similarities among these fungal forms in their macrospore morphology in that they are multiseptate and have rhexolytic dehiscence mechanism of discharge.[51] The similarity in their macrospore morphology is probably a result of adaptive evolutionary pressures for the dermatophytic lifestyle, which is supported by their phylogenetic and morphological closeness to keratin degrading *Chrysosporium* spp. whose aleurioconidia are indistinguishable from microspore morphology of *Microsporum* spp and *Trichophyton* spp suggesting *Chrysosporium* to be ancestral to both (macroconidia forming) *Microsporum* and *Trichophyton*.[52]

TABLE 36.1
Known Sexual States of *Microsporum* Species and Their Habitat/Host Preference

S. No.	Teleomorph (Sexual State)	Anamorph (Asexual State)	Natural Reservoir	Prevalent Host(s)	Distribution	Biosafety Level[a]	Reference Strain Number and Status[b]	Sequence Accession
1	*Arthroderma borelli*	*Microsporum amazonicum*	Soil	—	Brazil/Africa	BSL1	CBS 967.68, ST	AJ877220
2	*Arthroderma cajetanum*	*Microsporum cookei*	Soil	—	Global	BSL1	CBS228.58, AUT	AJ970145
3	*Arthroderma cookiellum*	*Microsporum* anamorph	Soil	—	Global (seems to be)	—	CBS101.83, HT, MT−	AM000035
4	*Arthroderma corniculatum*	*Microsporum* anamorph	Soil	—	Africa (rare)	BSL1	CBS364.81, HT, MT+	AJ970143
5	*Arthroderma fulvum*	*Microsporum fulvum*	Soil	—	Global	BSL1	CBS287.55, T	AJ000627
6	*Arthroderma grubyi*	*Microsporum gallinae*	Soil/animal	Fowl	Global	BSL2	CBS243.66, T, MT+	AJ000612
7	*Arthroderma gypseum*	*Microsporum gypseum*	Soil	—	Global	BSL1	CBS258.61, NT	AJ970141
8	*Arthroderma incurvatum*	*Microsporum gypseum*	Soil	—	Global	BSL1	CBS174.64, T	AJ970153
9	*Arthroderma obtusum*	*Microsporum nannum*	Animal	Pig	Global	BSL2	CBS322.61, T, MT−	AJ970149
10	*Arthroderma otae*	*Microsporum canis*	Animal	Cat, dog, horse, human	Global	BSL2	CBS496.86, ET, MT−	AJ000619
11	*Arthroderma racemosum*	*Microsporum racemosum*	Soil	—	Global	BSL1	CBS424.74,T,MT−	AJ970146
12	*Arthroderma persicolor*	*Microsporum persicolor*	Soil	Vole, rat	Global	BSL2	CBS468.74, MT−	AJ000615
13	—	*Microsporum audouinii*	Human	Human	Africa	BSL2	CBS545.93, −NT	AJ000623
14	—	*Microsporum ferrugenium*	Human	Human	Asia, Africa	BSL2	CBS497.48, AUT	AJ252336
15	—	*Microsporum praecox*	Soil	—	France and USA (rare)	BSL2	CBS288.55, AUT	AJ970148

Source: de Hoog, G.S. et al., *Atlas of Clinical Fungi*, 2nd edn., Centraalbureau voor Schimmelculturales, Utrecht, the Netherlands, 2000.

[a] BSL1, saprobes occupying non-vertebrate ecological niches; BSL2, species principally occupying non-vertebrate ecological niches, but with a relatively pronounced ability to survive in vertebrate tissue.

[b] AUT, authentic strain; CBS, Centraalbureau voor Schimmelcultures, Utrecht, the Netherlands, HT, holotype strain; MT, mating type; NT, neotype strain; ST, syntype strain; T, type strain.

36.1.1.3 Biology and Epidemiology

Microsporum is an asexual stage of the sexually reproducing ascomycete genus *Arthroderma*.[47] The fungus is mostly heterothallic with two mating types plus (+) and minus (−) that mates under favorable conditions in soil to produce ascomata as in *Arthroderma otae*,[53] as the phenomenon has been demonstrated in laboratory on Takashio's dilute Sabouraud agar and soil-keratin culture where compatible mating types fuse to form fertile ascomata, which bear innumerable asci and ascospores (the product of meiosis),[54] also referred as meiospores. Sexual reproduction is essential for genetic fitness in an organism as it involves recombination and production of newer genetic traits for better adaptability to the continuously changing environment. Figure 36.1 illustrates a life cycle of sexual *Microsporum* species (*Arthroderma* sp.); only the anamorphic stage is clinically active and acquires greater virulence[27] to be able to infect humans.

Infections by *Microsporum* species (and other dermatophytes) affect a considerable part of human body[55] and population worldwide.[56] The incidence of dermatophyte infection depends largely on hosts' immunity, pathogen availability, and virulence along with favorable microenvironmental conditions (e.g., traumatized skin—wet for long time as in swimming pools, sports person or soldiers during training with shoes on for long hours, or farmers or farm workers in developing world working barefoot) and most importantly reduced hygiene that brings all these conditions together. This is true for animals also, e.g., cats licking habit (at least once a day) and avoidance of water allows it to avert dermatophyte spores to reach its skin surface that remain clinging to its dense but loose body hairs and hence are mostly nonsymptomatic carriers; inability to lick their face might be the reason why cats have inconspicuous lesions around nose, eyes, and ears caused by *M. canis*.[57] However, the habit is uncommon in dogs, which have low rates of infection in pet dogs[58]; however, this does not hold true in stray dogs (with little hygiene and medical care) that have much higher rates of dermatophyte infection (personal observation of RS in India), thus helping the fungus maintain itself in nature similar to the mechanism known for obligate plant pathogens like

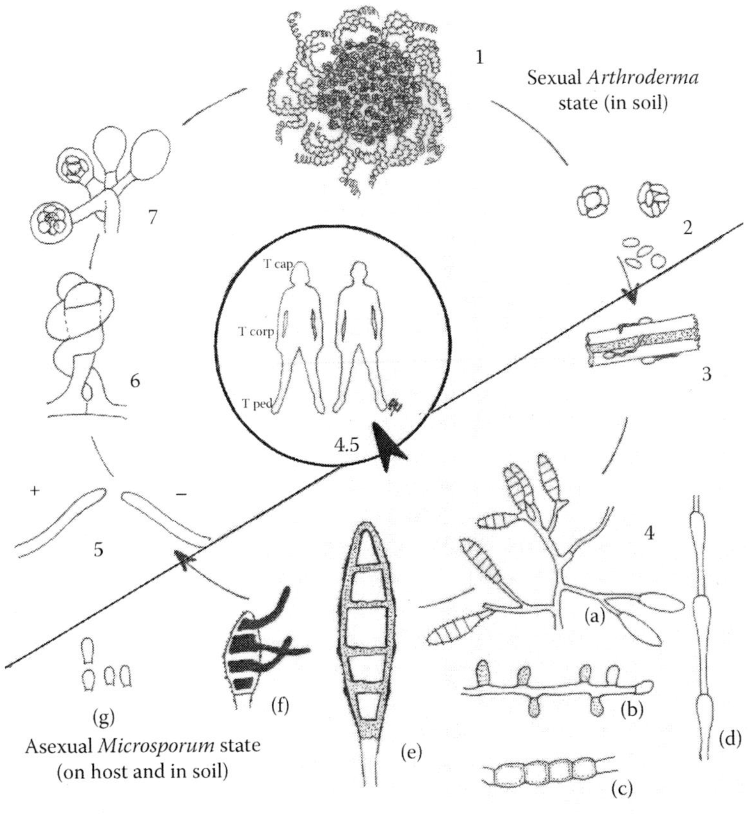

FIGURE 36.1 Life cycle of *Arthroderma* species (clockwise, not to scale) having *Microsporum* anamorph, which cause skin infections in humans. The line separating the two states are crossed when keratin is in plenty or scarcity. 1: Ascoma (peridium) bearing thick-walled dumble-shaped ascomatal hyphae with coiled appendages; 2: Evanescent asci and ascosores shed freely from ascoma after maturity; 3: Ascospores germinating and colonizing hair in soil; 4a: Fungal mycelium with conidiophores bearing young (nonseptate) and mature (septate) macroconidia, b: Clavate sessile microconidia, c: Thick-walled arthroconidia, d: Typical racquet hypha, e: Enlarged single five-celled macroconidium (four septate), f: Germinating macroconidium, g: Liberated microconidium; 4.5: Entry of *Microsporum* into human niche directly or indirectly through contact with infecting (mitotic) propagule which cause three main clinical manifestations: Tinea capitis, Tinea corporis, or Tinea pedis (first two due to adaptation to active dispersal of macroconidia by flight or through animal contact and third due to passive dispersal by arthroconidia through direct ground contact of humans); 5: Approaching receptive hyphae of opposite mating type (+) and (−); 6: Coiling of ascomatal initials; 7: Nascent asci formed after genetic recombination and meiosis.

the physiological races of wheat rusts (having alternate and collateral hosts systems)[59] for survival in absence of favored host, i.e., wheat crop. Birds are another example of hygiene maintenance as their preening habit meant for removing parasites or unwanted objects along with application of oil coating on feathers[60] prevents them from potential infections. Also, their tough covering on feet makes it difficult for a dermatophyte to colonize and digest, similar to tail hair of lions, which is also relatively unaffected compared to hairs of other animals including human when subjected to *Microsporum gypseum* treatment.[61] Birds also get rid of occasional clinging of propagules onto their feather by flight strokes; however, weak or ill ones are prone to dermatophytic infections (since they can't fly).

Tinea capitis in pre-independent India was mostly found in Europeans, Jews, and Anglo-Indians as reported in an extensive (50,000 skin samples during 5 years) study of fungal infections in Calcutta[62] possibly due to application of wide variety of hair oils viz. amla, almond, coconut, and mustard by natives, a custom followed from ages by both males and females (a similar story in nature can be found among birds that have natural coating). Garg[63] using a combination of biochemical assay and transmission electron microscopy showed the toxicity of certain hair oils used by Indians against four dermatophytes including *M. canis* and *M. gypseum* in which Amla (*Emblica officinalis*) oil was most effective.

Infections are more prevalent in persons with lower socioeconomic conditions but such cases are greatly under reported and unrecorded (reason for this is dermatophytes are never life threatening and food is a priority for survival than a mere skin infection), unlike in persons with good socioeconomic conditions and better hygiene (in such cases, 90%–95% cases are recorded). On a community scale, this holds good for developed and underdeveloped locations, cities, or even countries. Lack of basic microbiological knowledge among common people is one of the foremost causes of increased incidence of dermatophytic infections worldwide. This includes knowledge that environment is the potential reservoir of dermatophytes[64]; stray animals like cats and dogs (with or without visible symptoms); garbage (which is dumped indiscriminately in landfills areas, sewage (a big source of shed keratin),[65] etc. Also, a fact that a dermatophyte takes a day or two to cause primary athlete's foot infection in a human being after adherence of infectious propagules (viz. microconidia, macroconidia, chlamydospores) on human skin or its adnexes,[66,67] infections in other body parts may take longer due to not so congenial microenvironment like feet; not washing oneself for a day or two renders a person susceptible for infection if one is in contact with infectious propagule(s).[66] This condition can be with soldiers, hikers, mountaineers, slum dwellers, or all those persons living in water scarcity areas (a considerable portion of human population does not have enough water).[68] Human infections are generally caused when they come in contact with propagules(s), which are the overwintering structures formed in the life cycle (Figure 36.1) of a dermatophyte and when favorable environmental conditions are prolonged for some time like wet humid skin. Humans get easily infected by zoophilic dermatophyte through contact with infected or noninfected symptomatic carrier animals[57] that may cause severe symptoms in humans because naked human skin has fewer barriers compared to animals skin (which is densely covered with hair) and, in general, have high levels of immunity.

36.1.2 Clinical Features and Pathogenesis

36.1.2.1 Clinical Features

Clinical features caused by *Microsporum* apparently visible on human host are mostly characteristic for a dermatophytic infection but not always as the name suggests in the form of rings. And thus, are commonly but inappropriately referred to as *ringworm* (as no worm is involved) because of a ring-like appearance of the advancing infection, which is due to the radial growth of fungus on the skin (similar to their growth on artificial medium viz. Sabouraud Dextrose Agar [SDA]).

Principally, clinical features of dermatophytes are recognized and named after the part of the human body infected along with the word "tinea" that is being used since the time of first recorded reference of dermatophyte infection (around 30 AD).[69] It was used as a generic name for the cloth moth (Tinea sp.) that feeds on keratin of woolen garments, thus making characteristic holes in them, whose resemblance to ring-like lesion caused by fungal infection on smooth skin made Anglo-Saxon people to coin the term ringworm for such fungal infection in sixteenth century.[9] The term is now more popular for designating fungal infection types than the insect whose generic name is Tinea. The various types of dermatophyte infection or the clinical manifestations in which *Microsporum* spp. are involved (including other dermatophytes) are Tinea capitis-infection of head, scalp, eyebrows, and eyelashes (all *Microsporum* spp.); Tinea favosa infection of hair called crusty hair (*M. gypseum*, rare), Tinea barbae infection of beard (*M. canis*), and Tinea corporis infection of body or smooth skin (*M. audouinii*, *M. canis*, or any dermatophyte).[70] All *Microsporum* species grow as ectothrix, i.e., they grow outside the hair shaft and form numerous arthroconidia surrounding it while destroying the cuticle.

36.1.2.2 Pathogenesis

Pathogenesis of keratinophilic fungi including dermatophytes is due to their ability (they possess from their early origin) to break down keratin,[71] the second most abundant polymeric molecule on earth after cellulose, by releasing extraordinary amounts of enzymes often referred to as keratinases, which are in true sense proteases.[72] As studies in *M. gypseum* using radioactive tracer technique (^{35}S cystine) showed the breakdown of native keratin is by the process of sulphitolysis,[71,73] and production of keratin digesting enzymes as noted in *M. canis*.[74,75] The fungus mainly attack keratinous material by surface erosion and radial penetration by formation of boring hyphae (similar to appressoria or haustoria formed by plant pathogenic fungi to absorb nutrients from host plants[59]) where most of the secreted proteolytic enzymes are concentrated.[76] Once the mycelia formed from freshly germinated arthroconidia have penetrated keratin substrate and had established infection, extensive radial colonization under skin follows, which is seen as various clinical manifestations like the characteristic rings on smooth skin.

36.1.3 Diagnosis

Diagnosis is the key for all successful medical treatment for all types of diseases including dermatophytic infection. If diagnosis is correct, right medication can prevent further spread and agony to the patient. A dermatophytic fungal diagnosis is primarily important to successfully render appropriate therapy and eventually control of the disease. Diagnosing in early days of infection allows for immediate antifungal treatment and possible control; however, in the absence of correct diagnosis, the treatment could be false, which might help the fungus to further establish infection like in nail infections, which later becomes difficult to treat. Once infection is deep, the treatment takes long and relapses can frequently occur.

Dermatophytes can establish infection faster than we can imagine, i.e., even half day (*T. tonsurans*),[67] which prompt us to find faster and accurate diagnostic procedures. Conventional culture-based techniques are time taking and often do not always give positive results even if sample is positive for dermatophytic infection.[77] Currently, laboratory testing procedures are integrating the two techniques for better diagnoses and understanding of dermatophytes diversity, since culturing a new dermatophyte to reveal its morphology is essential in describing a novel species[78] and future drug testing for control therapy. Further section describes briefly the conventional techniques used in dermatophyte diagnoses and molecular techniques in more detail.

36.1.3.1 Conventional Techniques

Recently, an excellent review by Robert and Pihet[79] has appeared on conventional methods for diagnoses of dermatophytoses apart from a more comprehensive earlier treatment by Kane et al.[80] that greatly complement the rapid molecular techniques. Some of the conventional techniques mentioned below can be applied as preliminary tests prior to performing molecular analysis, all of which starts with the right method of collecting clinical specimen, i.e., infected material from an infected host.[81]

Wood's lamp: It is used to visualize dermatophytic infection (caused by *M. canis* and *M. audouinii*), seen as yellowish green fluorescence under dark, however not all fluoresce.

Hair brush technique: Although the method has been devised for animals,[82] it can be used for humans as well.[83] It includes use of a presterilized hair brush (having dense arrangements of hair bristles) to brush the infected portion of the hairy body (head of human or body of pet). The brush can be placed in sterile, preferably paper bags (since polythene bags tend to possess static charge, which attracts loose particles and hence can prevent retention of spores or skin fragments by the brush) and can be brought to the laboratory. The brush can then be pressed onto the SDA plate with cycloheximide.

Microscopic observation: KOH or NaOH mounts (10%–30%) plus 5% glycerol (to emulgate lipids) can be prepared directly from the clinical specimen, gently heated to observe for any hyphae or propagules, the preparation can be appropriately stained with cotton blue to visualize fungal material. Fluorescence stain with Calcoflour,[84] Kongo red or Mycetfluo®[79] can improve the visualization.

Culture preparation on Sabouraud's dextrose agar or mycotic agar: Once microscopy has revealed the presence of fungal material, a second half of the material can be planted aseptically onto SDA or Mycotic agar containing antibacterials (usually gentamicine and/or chloramphenicol plus cycloheximide).

Hair perforation tests: Using human or horse hair (sterilized by tindalization), hair perforation test[85] can be performed, which involves placing the hair on glass beads kept inside a presterilized Petri dish containing little distil water. The fungus cultured in pure form can then be inoculated aseptically onto the hair and incubated for 1–2 weeks. The fungus causing active degradation by means of boring hyphae can be visualized by cotton blue mounts of the inoculated hair. The test is particularly useful in differentiating *M. canis* (positive) from *M. audouinii* (negative).[86] However, the horse variant of *M. canis* cannot be distinguished from *M. audouinii* due to being hair-perforation negative for which molecular methods are useful.[27]

36.1.3.2 Molecular Techniques

With molecular techniques available, a diagnosis of infections caused by *Microsporum* from a clinical sample or material has become relatively rapid, exact, and authentic. In a recent study published comparing the conventional (microscopic examination and culture-based method) and molecular methods to detect dermatophytes in nail and skin samples clearly showed the contrast in positive rates of PCR versus culture (95%–99%—PCR versus culture—67% [skin] and 33% [nail]) clearly in favor of PCR method based on nuclear ribosomal internal transcribed spacer (ITS1) DNA sequences.[77]

However, morphological features are still useful like clinical symptoms on host or colony appearance on artificial media (if culture is successful, which is when the samples are collected from growing portion of the lesion) or the presence of racquet hyphae in culture; all of which even makes the molecular identification more focused by excluding most of the non-dermatophytes from picture and suggesting the use of species-specific primers (for *M. canis*[87] and *M. audouinii*[88]) or probes (for *T. rubrum*).[89] If culture is successful, and macroconidia are formed, then their surface feature indicates whether it is *Microsporum* (rough walled) or *Trichophyton* species (smooth walled).

Currently, the technique(s) most commonly applied in recognizing *Microsporum* species due to its simplicity (as it can be performed even by nonspecialists) and cost effectiveness (can be routinely applied by a small diagnostic laboratory) is—ITS-PCR-RFLP. ITS-PCR-sequencing is able to differentiate also close-related species but a method that is not so easy to perform and more expensive. Both techniques involve the targeted in vitro amplification of fungal pathogen DNA (*Microsporum* species) in thermal cycler to enhance its visibility (on a gel matrix). The amplified DNA usually from a selected site of the genome (viz. best studied is the ITS region of the ribosomal DNA) can be cut with restriction enzyme(s) to reveal a not always species-specific profile using comparison with the reference strains. The drawback of RFLP is that close-related species like *M. canis* and *M. ferrugineum* are not differentiated and a large amount of PCR product is essential. A better alternative is sequencing of the ITS region. Comparison with the reference sequences in the nucleotide databases (e.g., NCBI); better and recommended is the validated CBS database (http://www.cbs.knaw.nl) that helps in identifying the causative fungal species. Other methods, which have now become obsolete, are the mtDNA and DNA–DNA hybridization. However, many so-called "in house" PCR assays have been developed already, which mainly use nested PCR to increase the sensitivity in addition to a species-specific gene probe, which hybridizes in liquid (PCR ELISA) or firm (blot) manner to the amplified and universal target gene.[90–92] Although a few assays using real-time PCR have been developed preventing contamination of the sample during the identification process, two out of three are only able to identify species groups.[41,92–94] All of these assays involve extra cost but were developed for use in routine diagnostic laboratories. Here we describe the method that is able to recognize almost all currently known *Microsporum* species (except anthropophilic ones that can be recognized by sequencing) based on the new species concept.[36] The methods described can be applied by most laboratories and especially sequencing is a tool to identify the causative agent of any fungal infection.

36.2 METHODS

36.2.1 Sample Preparation

Direct molecular analysis of clinical sample(s) viz. hair, skin, and nail for diagnosis of fungal infectious agents makes the detection of dermatophytic species rapid in contrast to culture-dependent approach, which at least takes a couple of weeks time. In direct approach, total genomic DNA can be isolated from clinical sample using DNA extraction kits (e.g., Illustra Tissue & Cells GenomicPrep Mini Spin Kit, GE Healthcare Life Sciences; QIAamp DNA isolation Kit, Quiagen); High Pure-PCR-Template preparation Kit, Roche), and pegGOLD Fungal DNA Mini kit, Peqlab) and KOH.

Preparation of sample(s) for molecular detection of *Microsporum* does not require special care unlike systemic fungi; however, care must be taken while handling since certain animal origin strains are more virulent. Samples that include hair, nails, skin scrapings, and even floor dust or soil can be collected in fresh paper envelops or sealed plastic bags and brought to the laboratory for processing. Material (skin or hair) can be collected from growing peripheral portion of the lesions.

Most samples can be directly subjected to DNA extraction (preferably using a DNA extraction kit if the sample is less than 5–10 mg) but a part of the sample can be sent for culture (and slide preparation) in appropriate laboratory (if host laboratory lacks culturing facility) so that bulk DNA extraction can be made, which could be stored for decades for future reference studies. Several commercial kits are available for extracting DNA from clinical sample like infected nail, hair, or skin scrapings; however, it depends on the user for preferring a particular kit in terms of performance and cost (since kits tend to increase the cost of analysis).

36.2.1.1 DNA Extraction from Fungal Colony

The same kits used for clinical material can be applied when extracting DNA from a grown colony. In this case, colonies not older than 1–2 weeks should be used for the DNA preparation because pigmentation of the culture can cause problem and inhibit the following PCR reaction. The cetyl trimethyl ammonium bromide (CTAB) method described briefly, is, however, not so expensive and yields a larger amount of DNA. Mycelia were scraped from the surface of agar colony; also portions of colony are cut and placed inverted on the plain agar. The agar is then scraped off from the surface till only the mycelial matt remained. Three to four of these mycelial mats is taken in sterile pestle and grinded in liquid nitrogen using a mortar. The finely grinded mycelial powder is then added to 2 mL Eppendorf tube containing 1 mL of CTAB extraction buffer (100 mM Tris–HCl pH 8.0, 1.4 M NaCl, 20 mM EDTA, 2% cetylmethylammonium bromide, 2% β-mercaptoethanol). The suspension is then subjected to three rounds of freezing and thawing. Then, proteinase K is added to give a final concentration of 50 μg/mL and mixture is incubated for 1.5 h at 60°C. After centrifugation at 13,000 × *g* for 5 min, one volume of phenol/chloroform/isolamyl alcohol (25:24:1) is added to the supernatant and tube is vortexed and centrifuged again for 15 min at 13,000 × *g*. The upper aqueous layer is taken in a fresh tube and the procedure repeated four to five times until the middle layer is clear and shows no further protein precipitation. Finally, one volume of chloroform was added to the supernatant from previous step, mixed and centrifuged at 13,000 × *g* for 15 min. Finally, DNA was precipitated with 1/10 volume of sodium acetate (4.0 M, pH 5.2) and 1 mL of cold isopropanol. The content of tube is mixed by gentle inversion and incubating overnight at −20°C. Next day, the precipitated DNA is centrifuged at 10,000 × *g* for 10 min and pellet is washed with 70% ethanol and air-dried. DNA is dissolved in 1× TE buffer (10 mM Tris–HCl, 1 mM EDTA pH 8.0) and treated with RNase with final concentration 100 μg/mL for 15 min at 37°C. RNase was inactivated by 5 min incubation at 65°C and the pure DNA was stored at 4°C. DNA concentration was estimated by spectrophotometer reading at 280 and 260 nm, whose ratio gives the purity of DNA extracted, and also physically checked by running genomic DNA on 0.8% agarose gel along with known concentration of λ DNA.

36.2.1.2 DNA Extraction from Clinical Specimen

For extraction of DNA from clinical sample—hair, nail, or skin, 0.5–3 mg of material previously collected is cut into small pieces (2 mm) and subjected to commercially available kit extraction as per manufacturer's instruction.

Alternatively, homogenization of tissue for DNA extraction, clinical specimen can be placed in a 2 mL screw cap tube containing CTAB buffer[11] (100 mM Tris–HCl pH 8.0, 1.4 M NaCl, 20 mM EDTA, 2% cetyl methylammonium bromide, 2% β-mercaptoethanol), sterile sand and a 5 mm solid ceramic bead in an MPBio Fast Prep 24 homogenizer (Biomedicals GmbH, Germany). Further steps remain same as in the previous section on extraction from colony. A very rapid method (1 h) with slight optimization can be readily applied to bulk samples obtained by small laboratories or private clinics[94] or the one by Brillowska-Dobrowska et al.,[95] which is mainly for nail but can be applied to other clinical specimens as well.

36.2.2 Detection Procedures

36.2.2.1 Species Recognition

Identification of species of *Microsporum* using molecular means involves primarily the comparison of rDNA sequences, some regions of which are highly conserved on one hand and variable on the other to reveal interspecies differences. To visualize the interspecific variations for diagnosis, rDNA regions being multi-copy are first amplified by PCR technique and then analyzed in one or the other ways for species identification.

36.2.2.1.1 PCR Amplification

Once the genomic DNA is extracted in sufficient amount and quantified, PCR can then be performed with several universal fungal primers, e.g., ITS4 (5′TCC TCC GCT TAT TGA TAT GC3′-LSU 69-50 of *Saccharomyces cerevisiae*) and ITS5 (5′GGA AGT AAA AGT CGT AAC AA3′),[96] or LSU266[97] and V9D,[98] which were used for *Microsporum* species to amplify the ITS1, 5.8S, and ITS2 regions of rDNA giving an amplicon of nearly 1000 bp. PCR conditions-50 μL reaction mixture contained 10 mM Tris–HCl, 50 mM KCl, 3 mM $MgCl_2$, 21 pmol each primer (V9D and LSU266), 50 μM concentration of each deoxynucleotide triphosphate, 2.5 U of *Taq* polymerase, and 50 ng of template DNA. Appropriate controls viz. positive (containing template known to amplify in given conditions), negative (reaction mixture lacking any template), and inhibition (reaction mixture containing 25 ng test template and 25 ng template previously known to amplify under given conditions) are required. In case of non-amplification, even when there is no inhibition, 10% DMSO can be added.[99] The reaction mixture is finally overlaid with light mineral oil. Amplification conditions in a thermo cycler consist of 1 cycle of 95°C for 5 min; 30 cycles of 95°C for 1 min, 55°C for 1 min, 72°C for 1 min; a final extension 72°C for 10 min. Five microliter PCR product is then checked on 1.2% agarose gel for successful amplification.

36.2.2.1.2 Restriction Digestion (ITS-PCR-RFLP)

The resulting amplicon can be digested by *Mva*I restriction enzyme to give species-specific profile (Figure 36.2). For restriction digestion, 20 μL of fresh PCR product (showing single clear band on 1.2% agarose) is first transferred to an Eppendorf tube. The DNA is precipitated by 1/10 volume of sodium acetate (3 M, pH 4.8–5.2) and 2.5 volume of ethanol (95%) and incubated overnight at 4°C. Precipitated DNA is centrifuged at 13,000 × *g* for 15 min and pellet (mostly not visible) washed with ethanol (70%) and again centrifuged at 13,000 × *g* for 15 min. The pellet is dried in vacuum centrifuge for 10 min at 30°C. The dried pellet is dissolved in 18 μL of PCR grade water with the help of micropipette. 2 μL of 10× RE buffer (10 mM Tris–HCl pH 8.5, 10 mM $MgCl_2$, 100 mM KCl, and 0.1 mg/mL BSA) is added to the cleaned PCR product along with 5–10 U of *Mva*I RE (Fermentas) before incubating for 4 h at 37°C. The digested product is separated on 2% Metaphor agarose and stained with ethidium bromide and photographed for documentation.

The results of characteristic species-specific pattern considerably depend on the success of digestion reaction. Partial or incomplete digestion or even mutation (which is rare) at restriction site may give altered pattern. The method is most suitable for routine bulk samples. Figure 36.2A shows the *Mva*I restriction profile of selected species of *Microsporum* generated from nearly 1000 bp amplicon covering ITS1, 2 flanking the 5.8S region of rDNA.

36.2.2.1.3 Sequencing of PCR Products

Direct sequencing of PCR product is done by cycling sequencing kits using an automated sequencer (gel or capillary based) or first cloning them in appropriate vector and then performing cycling sequencing reaction using M13 universal primer. Direct sequencing of PCR product (obtained by primer pair V9D and LSU266) is done by performing sequencing PCR with internal primers ITS4 and ITS5.[96]

36.2.2.1.4 Sequence Alignment and Identification

Sequence obtained in the form of chromatogram from automated sequencer has to be manually checked for inconsistency and accordingly edited for accuracy. Edited sequences can then be aligned with known (reference) sequences in the database using CBS database or NCBI-BLAST search to finally give identification results.

36.2.2.2 Strain Typing

For epidemiological study including tracing the source of infection (like pet) intra-specific markers are needed like microsatellites or randomly amplified polymorphic DNA (RAPD). The choice of the former depends on the availability

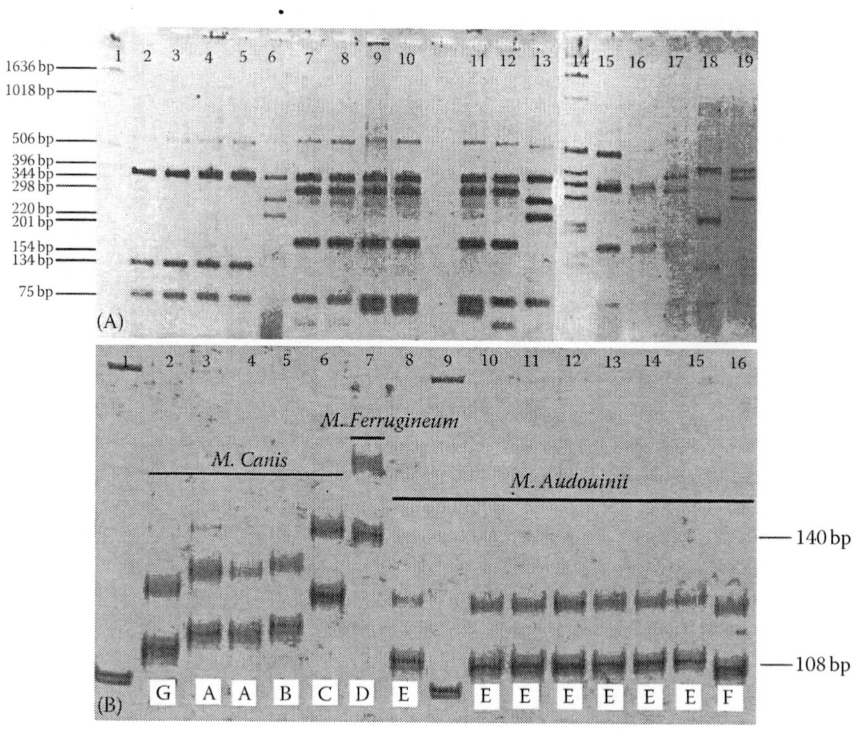

FIGURE 36.2 (A) Restriction digestion profile (*Mva*I) of reference strains (of seven *Microsporum* sp.) *Microsporum gypseum* var. *gypseum* (CBS161.69) (lane 2), same strain as previous (lane 3), *Arthroderma incurvatum* (*M. gypseum*) (CBS173.64, MT+) (lane 4), *A. incurvatum* (CBS 174.64, T) (lane 5), *A. gypsea* (CBS 258.61, NT) (lane 6), *A. fulvum* (CBS 167.64, MT+) (lane 7), *M. fulvum* (CBS 287.55, T) (lane 8), *A. persicolor* (CBS 468.74, MT–) (lane 9), *A. persicolor* (CBS 469.74, MT+) (lane 10), *M. canis* (CBS 132.88) (lane 15), *A. borelli* (*M. amazonicum*) (CBS 967.68, ST) (lane 17), *A. cookiella* (CBS 102.83) (lane 18), *Epidermophyton floccosum* (CBS 358.93) (lane 19). Test *Microsporum* species profile (lanes 11–13) corresponds to reference profile of *M. persicolor*, *M. fulvum*, and *M. gypseum* (lane 10, 8, and 6 respectively). *Abbreviations*: CBS, Centraalbureau voor Schimmelcultures, Utrecht, the Netherlands, MT, mating type; NT, neotype strain; ST, syntype strain; T, type strain. (Reproduced from Sharma, R. et al., *Med. Mycol.*, 46, 67, 2008. With permission.) (B) Polyacrylamide gel (12%) showing seven different alleles among strains of *M. canis* (G-10, A-12, B-13, and C-17 repeats), *M. ferrugenium* (D-25 repeats), and *M. audouinii* (F-7 and E-9 repeats) analyzed by marker Mc$(CA)_{13}$.

of already developed markers (e.g., *M. canis* and *M. persicolor*),[27,39] otherwise they have to be generated afresh in the species under investigation by one of the several methods available.[100,101] With the availability of the genome project of *M. canis* and *M. gypseum*, the design of such markers at least for these two species became easy.

36.2.2.2.1 Microsatellites

Microsatellites or simple sequence repeats (SSRs) are short stretches of di-, tri-, tetra-, penta-, or hexa-sequences repeated more than five times. The more the number of repetitions are present, the better the marker. The flanking regions of the microsatellite repeats are conserved (where primer binds), while variations in the number of repeat results in polymorphisms among individual of a population, which can be visually observed on PAGE gel or fragment analyzer. In the geophilic species *M. persicolor*, Sharma et al.[27] detected 26 genotypes among 56 nonclinical strains analyzed, which were isolated from various soil locations in Central India viz. public places, hair salon, burrow, cattle shed, garbage, cave, and rat carcass. The number of genotypes obtained with each of the three microsatellite marker (with 13, 9, and 5 genotypes) corresponded with the increase in repeat length of the microsatellite marker (i.e., markers 300_27ga [For: 5′-GGCTTGAGTG GCGTCTTC-3′ Rev: 5′-AGCAAACGAACCGCTGAG-3′] 196_23ca [For: 5′-TCGGCCTCCTCATCCTTC-3′ Rev: 5′-TCGGGATGTAAGTAAAGG-3′] and 300_17ct [For: 5′-GGGCAATTCTATGGGCAAG-3′3′ Rev: 5′-CTTCTTCCAAGCTCTGCCTG-3′]) had maximum 38, 26, and 18 dinucleotide repeats, respectively. The high number of multilocus genotypes (26) obtained in small population size (56 strains) suggested a fairly high level of recombination in the population and its presence in the region to be much older than its first isolation from India.[39] Misidentification of this species in earlier works on Indian soils is probable because they all were morphology based and *M. persicolor* is known to be misinterpreted by morphological features.[102,103] In zoophilic species *M. canis*, 11 multilocus genotypes were obtained with 2 microsatellite loci Mc$(GT)_{13}$ (forward: 5′-GATCGGAGCATGCCATACAG3′; reverse: 5′-TCTTCCCACCCTTCTCAATG-3′) and Mc$(GT)_{17}$ (forward: 5′-GCTCTGGGATAAGGTGTTTG-3′; reverse: 5′-GTAGCAGTAAAGCCAAGAGGG-3′) among 101 global strains analyzed.[27] The markers were also able to differentiate *M. canis* from close-related anthropophiles *M. audouinii* and *M. ferrugenium*, which represented three additional genotypes and thus can be differentiated by gel electrophoresis (Figure 36.2B).

36.2.2.2.2 RAPD Markers, AP-PCR, and PCR-Fingerprinting

A number of techniques are available like AP-PCR (arbitrary primed-PCR), PCR-fingerprinting, and RAPD, which employ a single short primer to amplify fragments scattered on the chromosomes for strain typing and population genetic studies. DNA polymorphisms, in terms of different band profile, are obtained (presence or absence of band(s) that may be of variable intensities). The reliability of these methods greatly depends on standard conditions used (including make of thermal cycler); however, reproducibility is a problem[104] among laboratories. Although the methods are for assessing intraspecific variability, they are of little use in case of dermatophytes that lack variability due to their recent adaptive radiation; and hence reveal only interspecific differences[105,106] unlike seen in other older fungal lineages.

36.3 CONCLUSIONS AND FUTURE PERSPECTIVES

Diagnosing fungal pathogen is the first step in treatment of a dermatophytic infection, which is one of the most common fungal infections worldwide infecting almost 20%–25% of population.[107] Molecular methods have proved greatly beneficial and have revolutionized detection of pathogens due to sensitivity of techniques involved (even in cases where symptoms are unapparent) and the rapidity in relation to the conventional techniques like hair perforation, mating competency tests, and even nutritional tests. Apart from the currently available molecular methods, newer improved ones are still being regularly developed like oligonucleotide array[44,108] with increased rapidity[95,109] mainly due to the reduced time required for DNA extraction.[95]

With the increase in global temperature, which is all the more evident at present like never before, there will be more cases of dermatophytoses, especially in temperate regions of the world, which already have a high rate (35%–40%) of fungal foot disease[110] due to, e.g., specialized microclimate built by footwear. The relative incidence of dermatophytic infection varies with the climatic (most important microclimate) conditions[111] prevailing in a region and on the inherent immunity of hosts involved. Warm and humid environment previously lacking in temperate regions of the world is now available for longer period during the year, which will facilitate survival of species previously unrecorded in such region. This altered environment has a direct affect on the survival of migrated species including anthropophilic dermatophytes. Also, the parasexual mechanism[112] of recombination to generate genetic diversity and fitness (virulence)[113] might be active in asexual forms like the anthropophilic dermatophytes due to its immense utility for species survival. Thus, anthropophilic nature of some dermatophytes not necessarily seems to be evolutionary dead ends for them but instead adaptation to a highly specialized life style (like rusts in plants) that will lead them to become future obligate parasites and their (evolutionary) survival will last until its host species.

Apart from detecting fungal pathogens in clinical specimens, molecular methods have proved useful in more than three ways for the study of dermatophytes, thus having direct implication on the clinicians' knowledge of dermatophytes. First, in detecting species in location where they are unrecorded,[39] second validating the taxonomic status of a closely related or cryptic species[114] or atypical isolates,[115] and third in providing phylogenetic positioning of newly found species[78,116] for nomenclatural and taxonomic considerations. We now have numerous markers and methods with us like never before; non-detection of dermatophytes in clinical specimen or the environmental samples becomes relatively rare, the only impetus that remains is how faster we can diagnose their presence for appropriate therapy. Although dermatophytes are not life threatening, they can influence the quality of life drastically.

ACKNOWLEDGMENTS

Drs. G.S. de Hoog and N.D. Sharma are greatly acknowledged for helpful discussions on dermatophytes and fungi during several years of their association with Y.G. and R.S. respectively and Dr. S.N. Kulkarni, ARI Library, for access to current literature. R.S. thanks Department of Science and Technology (DST), Govt. of India, for Fast Track grant SR/FT/L-36/2005 supporting the work on keratinophilic fungi of national parks of Central India, which led to many concepts mentioned in this chapter and Deutscher Akademischer Austausch Dienst (DAAD), Bonn, for research fellowship that enabled collaboration between the two authors to study and generate new information on *Microsporum*.

REFERENCES

1. Ainsworth, G.C. and Austwick, P.K.S., *Fungal Diseases of Animals*, Review Series 6, Commonwealth Agricultural Bureaux, England, 1973, p. 10.
2. Mullis, K.B. et al., Specific amplification of DNA in vitro: The polymerase chain reaction, *Cold Harb. Spring Symp. Quant. Biol.*, 51, 260, 1986.
3. Saiki, R.K. et al., Primer-directed enzymatic amplification of DNA with a thermostable DNA polymerase, *Science*, 239, 487, 1988.
4. Gruby, D., Recherches sur la nature, le siege et le developpment du porrigo decalvans ou phytoalopecie, *C.R. Acad. Sci.*, 17, 301, 1843.
5. Malmstem, P.H., *Trichophyton tonsurans*, harkarande Mogel, Bidrag till utredande af de ajukdomar som valla harets affal, Stockholm, translated by Creplen F.C.H., and published as *Trichophyton tonsurans* der haarscheerende Schimmel, Ein betrag zur Auseinandersetzung der Krankheiten, welche das Abfalle des Haars bewirken, *Arch. Anat. Physiol. Wiss. Med.*, 1848, 19, 1848.
6. Sabouraud, R., *Les Teignes*, Masson et Cie., Paris, 1910.
7. Fraser, R.D.B., Keratins. *Sci. Am.*, 221, 164, 1969.

8. Dzawachiszwili, N. et al., The effect of sea water and sodium chloride on the growth of fungi pathogenic to man, *J. Invest. Dermatol.*, 103, 1964.
9. Ajello, L., Natural history of the dermatophytes and related fungi, *Mycopath. Mycol. Appl.*, 53, 93, 1974.
10. Kawasaki, M., Aoki, M., and Ishizaki H., Phylogenetic relationships of some *Microsporum* and *Arthroderma* species inferred from mitochondrial DNA analysis, *Mycopathologia*, 130, 11, 1995.
11. Gräser, Y. et al., Phylogeny and taxonomy of family Arthrodermataceae (dermatophytes) using sequence analysis of the ribosomal ITS region, *Med. Mycol.*, 37, 105, 1999.
12. Magallon, S.A., Dating lineages molecular and paleontological approaches to the temporal framework of clades, *Int. J. Plant. Sci.*, 165, S7, 2004.
13. Taylor, J.W. and Berbee, M.L., Dating divergences in the fungal tree of life: Review and new analysis, *Mycologia*, 98, 838, 2006.
14. Pagel, M. and Bodmer, W., A naked ape would have fewer parasites, *Proc. R. Soc. Lond.* B, 270, S117, 2003.
15. Schwartz, G.G. and Rosenblum, L.A., Allometry of primate hair density and the evolution of human hairlessness, *Am. J. Phys. Anthropol.*, 55, 9, 1981.
16. Romano, C., Valenti, L., and Barbara, R., Dermatophytes isolated from asymptomatic stray cats, *Mycoses*, 40, 471, 1997.
17. Summerbell, R., Form and function in the evolution of dermatophytes, in *Biology of Dermatophytes and Other Keratinophilic Fungi*, Kushwaha, R.K.S. and Guarro, J., Eds., *Rev. Iberoam. Micol.*, Bilbao, Spain, 2000, p. 30.
18. Rugmer, C., The fragility of gigantism. *Research*eu*, Special issue 22, September 2008.
19. Richardson, M. and Edward, M., Model systems for the study of dermatophyte and non-dermatophyte invasion of human keratin, in *Biology of Dermatophytes and Other Keratinophilic Fungi*, Kushwaha, R.K.S. and Guarro, J., Eds., *Rev. Iberoam. Micol.*, Bilbao, Spain, 2000, p. 115.
20. Takashio, M., Sexual reproduction of some *Arthroderma* and *Nannizzia* on dilute Sabouraud agar with or without salts, *Mykosen*, 15, 11, 1972.
21. Wong, K., The human pedigree: A timeline of hominid evolution, *Sci. Am. India*, 4, 60, 2009.
22. Kaszubiak, A. et al., Population structure and evolutionary origins of *Microsporum canis*, *M. ferrugineum* and *M. audouinii*, *Infect. Genet. Evol.*, 4, 179, 2004.
23. Dixon, D. et al., *Cassell's Atlas of Evolution*, Cassell & Co. London, U.K., 2001, p. 312.
24. Bettis III, E.A., Way out of Africa: Early Pleistocene paleoenvironments inhabited by *Homo erectus* in Sangiran, Java, *J. Hum. Evol.*, 56, 11, 2009.
25. Giraud, T. et al., Speciation in fungi, *Fung. Genet. Biol.*, 45, 791, 2008.
26. Summerbell, R.C., Li, A., and Haugland, R., What constitute a functional species in the asexual dermatophytes? *Microbiol. Cult. Coll.*, 13, 29, 1997.
27. Sharma, R. et al., A virulent genotype of *Microsporum canis* is responsible for the majority of human infections, *J. Med. Microbiol.*, 56, 1377, 2007.
28. Dirscoll, C.A. et al., The near eastern origin of cat domestication, *Science*, 317, 519, 2007.
29. Dirscoll, C.A. et al., The taming of the cat, *Sci. Am. India*, 4, 54, 2009.
30. Brown, J.K.M. and Hovmoller, M.S., Epidemiology—Aerial dispersal of pathogens on the global and continental scales and its impact on plant disease, *Science*, 297, 537, 2002.
31. Burnet, J. (ed), The generation of variation-I Mutation and migration, in *Fungal Population and Species*, Burmett, J., Ed., Oxford University Press, New York, 2003, p. 121.
32. Webster, J. and Weber, R.W.S., *Introduction to Fungi*, 3rd edn., Cambridge University Press, Cambridge, U.K. 2007, p. 294.
33. Hubalek, Z. and Rush-Munro, F.M., A dermatophyte from birds: *Microsporum ripariae* sp.nov., *Sabouraudia*, 11, 287, 1973.
34. Ajello, L. and Padhye, A., Keratinophilic fungi of the Galapagos Islands, *Mykosen*, 17, 239, 1974.
35. Davidson, F.D.D., Mackenzie, D.W.R., and Owen, R.J., Deoxyribonucleic acid base composition of dermatophytes, *J. Gen. Microbiol.*, 118, 465, 1980.
36. Gräser, Y., Scott, J., and Summerbell R.C., The new species concept in dermatophytes—A polyphasic approach, *Mycopathologia*, 166, 239, 2008.
37. Abdel-Rahman, S.M., Strain differentiation of dermatophytes, *Mycopathologia*, 166, 319, 2008.
38. Kanbe, T., Molecular approaches in the diagnosis of dermatophytes, *Mycopathologia*, 166, 307, 2008.
39. Sharma, R. et al., Molecular detection of *Microsporum persicolor* in soil suggesting widespread dispersal in central India, *Med. Mycol.*, 46, 67, 2008.
40. Perkins, D.D., In praise of diversity, in *More Gene Manipulations in Fungi*, Bennet, J.W. and Lasure, L.L. Eds., Academic Press, New York, 1991, p. 3.
41. Garg, J. et al., Rapid detection of dermatophytes from skin and hair, *BMC Res. Notes*, 2, 60, 2009.
42. Arabatzis, M. et al., Diagnosis of common dermatophyte infections by a novel multiplex real time polymerase chain reaction detection/identification scheme, *Brit. J. Dermatol.*, 157, 681, 2007.
43. Vollmer, T. et al., Evaluation of novel broad-range real-time PCR assay for rapid detection of human pathogenic fungi in various clinical specimens, *J. Clin. Microbiol.*, 46, 1919, 2008.
44. Li, H.C. et al., Identification of dermatophytes by an oligonucleotide array, *J. Clin. Microbiol.*, 45, 3160, 2007.
45. White, T.C. and Henn, M.R., Genomic determinants of infection competence in dermatophyte fungi. A white paper for the Fungal Genome Initiative, http://www.genome.gov/Pages/Research/Sequencing/Seq/Proposals/Dermatophytes_WP_Seq.pdf, pp12.
46. White, T. et al., Generating and testing molecular hypothesis, *Eukar. Cell*, 7, 1238, 2008.
47. Padhye, A.A. and Carmichael, J.W., The genus *Arthroderma* Berkeley, *Can. J. Bot.*, 49, 1525, 1971.
48. Weitzman, I. et al., The genus *Arthoderma* and its later synonym *Nannizia*, *Mycotaxon*, 25, 505, 1986.
49. James, T.Y. et al., Reconstructing the early evolution of fungi using a six-gene phylogeny, *Nature*, 443, 818, 2006.
50. Hibbett, D.S. et al., A higher-level phylogenetic classification of the fungi, *Mycol. Res.*, 111, 509, 2007.
51. Cole, G.T. and Samson R.A., *Patterns of Development in Conidial Fungi*, Pitman Publishing Ltd., London, U.K., 1979.
52. Van Oorschot, C.A.N., A revision of *Chrysosporium* and allied genera, *Stud. Mycol.*, 20, 1, 1980.
53. Kubo, H., Survey of keratinophilic fungi from soils of Hokkaido prefecture of Japan—Especially on the isolation of *Nannizzia otae* from the soils, *Jpn. J. Med. Mycol.*, 27, 230, 1986.
54. Hironga, M., Nozaki, K., and Watanabe, S., Ascocarp production by *Nannizzia otae* on keratinous and non-keratinous agar media and mating behavior of *N. otae* and 123 Japanese isolates of *Microsporum canis*, *Mycopathologia*, 72, 135, 1980.

55. Degreef, H., Clinical forms of dermatophytosis (ringworm infection), *Mycopathologia*, 166, 257, 2008.
56. Havlickova, B., Czaika, V.A., and Friedrich, M., Epidemiological trends in skin mycoses worldwide, *Mycoses*, 51, 2, 2008.
57. Chermett, R., Ferreiro, L., and Guillot, J., Dermatophytoses in animals, *Mycopathologia*, 166, 385, 2008.
58. Carlotti D.N. and Bensignor, E., Dermatophytosis due to *Microsporum persicolor* (13 cases) or *Microsporum gypseum* (20 cases) in dogs, *Vet. Dermatol.*, 10, 17, 1999.
59. Agrios, G.N., *Introduction to Plant Pathology*, Elsevier Academic Press, New Delhi, 2006.
60. Moss, S., *Understanding Bird Behavior*, New Holland Publishers Ltd., U.K., 2006, p. 70.
61. Sharma, R., Taxonomic studies on keratinolytic fungal diversity of central India, PhD thesis, Rani Durgavati University, Jabalpur, India, 2002.
62. Ghosh, L.N., An analysis of 50,000 skin cases as seen in the outpatient department of school of tropical medicine, Calcutta, during 5 years from 1942–1946, *Ind. Med. Gaz.*, 83, 493, 1948.
63. Garg, A.P., Indian hair oils and protection of human hair against fungal infections, *J. Ind. Bot. Soc.*, 71, 251, 1992.
64. Katoh, T., Dermatomycosis and environment, *Jpn. J. Med. Mycol.*, 47, 63, 2006.
65. Ulfig, K., The occurrence of keratinolytic fungi in water and waste-contaminated habitats, in *Biology of Dermatophytes and Other Keratinophilic Fungi*, Kushwaha, R.K.S. and Guarro, J., Eds., *Rev. Iberoam. Micol.* Bilbao, Spain, 2000, p. 44.
66. Morishita, N. et al., Effects of temperature, humidity, minor injury and washing on penetration of dermatophytes into human stratum corneum, *Jpn. J. Med. Mycol.*, 44, 269, 2003.
67. Morishita, N. et al., Experimental studies on the penetration of dermatophytes into the human stratum corneum and the effects of washing, *Jpn. J. Med. Mycol.*, 45, 247, 2004.
68. May, R.M., Global problems and global science, *Notes Rec. R. Soc. Lond.*, 59, 105, 2005.
69. Rosenthal, T., Aulus Cornelium Celsus, *Arch. Derm.*, 84, 613, 1961.
70. Rippon, J.W., Clinical aspects of medically important conidial fungi, in *Biology of Conidial Fungi*, Vol. 2, Cole, G.T. and Kendrick, W., Ed., Academic Press Inc., New York, 1981, p. 13.
71. Kunert, J., Keratin decomposition by dermatophytes: Evidence of the sulphitolysis of protein, *Experimentia*, 28, 1025, 1972.
72. Kunert, J., Personal communication, 2001.
73. Kunert, J. and Trüper, H.G., Cystine catabolism in mycelia of *Microsporum gypseum* a dermatophyte fungus, *Arch. Microbiol.*, 145, 181, 1986.
74. Sei, Y. and Takiuchi, I., Studies on the effect of keratinase on the keratinized structure, and localization of keratinase, *Jpn. J. Med. Mycol.*, 23, 308, 1982.
75. Takiuchi, I. et al., Isolation of an extracellular proteinase (keratinase) from *Microsporum canis, Sabouraudia*, 20, 281, 1982.
76. Marchisio, V.P., Keratinophilic fungi: Their role in nature and degradation of keratinic substrates, in *Biology of Dermatophytes and Other Keratinophilic Fungi*, Kushwaha, R.K.S. and Guarro, J., Eds., *Rev. Iberoam. Micol.* Bilbao, Spain, 2000, p. 86.
77. Uchida, T. et al., Comparative study of direct polymerase chain reaction, microscopic examination and culture-based morphological methods for detection and identification of dermatophytes in nail and skin samples, *J. Dermatol.*, 36, 202, 2009.
78. Brasch, J. and Graeser, Y., *Trichophyton eboreum* sp.nov. isolated from human skin, *J. Clin. Microbiol.*, 43, 5230, 2005.
79. Robert, R. and Pihet, M., Conventional methods for the diagnosis of dermatophytosis, *Mycopathologia*, 166, 295, 2008.
80. Kane et al., *Laboratory Handbook of Dermatophytes*, Star Publishing Company, Belmont, CA, 1997.
81. Robinson, B.E. and Padhye, A.A., Collection, transport and processing of clinical specimens, in *Diagnostic Procedures for Mycotic and Parasitic Infections*, Wentworth, B.B., Ed., American Public Health Association, Washington, DC, 1988, p. 11.
82. Mackenzie, D.W.R., "Hairbrush diagnosis" in detection and eradication of non-fluorescent scalp ringworm, *Br. Med. J.*, 2, 363, 1963.
83. Klokke, A.H.A., Purushotham, A.G., and Sundaraju, D., Mass treatment of non-fluorescent tinea-capitis in an orphanage in India. Trial with a low griseofulvin dose using a modified brush sampling technique, *Trop. Georgr. Med.*, 18, 305, 1966.
84. Johnson, L., Dermatophytes-the skin eaters, *Mycologist*, 17, 147, 2003.
85. Ajello, L. and George, L.K., In vitro cultures for differentiating between atypical isolates of *Trichophyton mentagrophytes* and *Trichophyton rubrum*, *Mycopath. Mycol. Appl.*, 8, 3, 1957.
86. Padhye, A.A., Young, C.N., and Ajello, L., Hair perforation as a diagnostic criterion in the identification of *Epidermophyton*, *Microsporum* and *Trichophyton* species, *Pan. Am. Health Org. Sci. Publ.*, 396, 115, 1980.
87. Liu, D. et al., A specific PCR assay for the dermatophyte fungus *Microsporum canis*, *Med. Mycol.*, 39, 215, 2001.
88. Roque, H.D. et al., Specific primers for rapid detection of *Microsporum audouinii* by PCR in clinical samples, *J. Clin. Microbiol.*, 44, 4336, 2006.
89. El Fari, M. et al., Development of an oligonucleotide probe specific for *Trichophyton rubrum*, *Brit. J. Dermatol.*, 141, 240, 1999.
90. Savin, C. et al., Multicenter evaluation of a commercial PCR-enzyme-linked immunosorbent assay diagnostic kit (Onychodiag) for diagnosis of dermatophytic onychomycosis, *J. Clin. Microbiol.*, 45, 1205, 2007.
91. Bergmans, A.M.C. et al., Validation of PCR-reverse line blot, a method for rapid detection and identification of nine dermatophyte species in nail, skin and hair samples, *Clin. Microbiol. Infect.*, 14, 778, 2008.
92. Gutzmer, R. et al., Rapid identification and differentiation of fungal DNA in dermatological specimens by LightCycler PCR, *J. Med. Microbiol.*, 53, 1207, 2004.
93. Bergmans, A.M.C. et al., Evaluation of a single-tube real-time PCR for detection and identification of 11 dermatophyte species in clinical material, *Clin. Microbiol. Infect.*, 16, 704, 2009.
94. Liu, D. et al., Rapid mini-preparation of fungal DNA for PCR. *J. Clin. Microbiol.*, 38, 471, 2000.
95. Brillowska-Dobrowska, A., Saunte, D.M., and Arendrup, M.C., Five hour diagnosis of dermatophyte nail infection with specific detection of *Trichophyton rubrum*, *J. Clin. Microbiol.*, 45, 1200, 2007.
96. White, T.J. et al., Amplification and direct sequencing of fungal ribosomal RNA genes for phylogenetics, in *PCR Protocols: A Guide to Methods and Applications*, Innis, M.A. et al. Eds., Academic Press, New York, 1990, p. 315.
97. Masclaux, F. et al., Phylogenetic relationships of human pathogenic *Cladosporium* (*Xylohypha*) species inferred from partial LS rRNA sequences, *J. Med. Vet. Mycol.*, 33, 327, 1995.

98. de Hoog, G.S. and Gerrits van den Ende, A.H.G., Molecular diagnostics of clinical strains of filamentous Basidiomycetes, *Mycoses*, 41, 183, 1998.
99. Dieffenbach, C.W. and Dveksler, G.S., *PCR Primer—A Laboratory Manual*, 2nd edn., Cold Spring Harbor Laboratory Press, Cold Spring Harbor, New York, 2003.
100. Zane, L., Bargelloni, L., and Patarnello, T., Strategies for microsatellite isolation: A review, *Mol. Ecol.*, 11, 1, 2002.
101. Dutech, C. et al., Challenges of microsatellite isolation in fungi, *Fung, Genet. Biol.*, 44, 933, 2007.
102. Badillet, G., *Microsporum persicolor*, an easily misinterpreted dermatophyte, *Hautartz*, 29, 10, 1987.
103. Kane, J., Sigler, L., and Summerbell, R.C., Improved procedure for differentiating *Microsporum persicolor*, an easily misinterpreted dermatophyte, *J. Clin. Microbiol.*, 25, 2449, 1987.
104. Hansen, M., Hallden, C., and Sall T., Error rates and polymorphism frequencies for three RAPD protocols, *Plant Mol. Biol. Reptr.*, 16, 139, 1998.
105. Kano, R. et al., Differentiation of *Microsporum* species by random amplification of polymorphic DNA (RAPD) and southern hybridization analyses, *Mycoses*, 41, 229, 1998.
106. Graeser, Y. et al., Identification of common dermatophytes (*Trichophyton*, *Microsporum*, *Epidermophyton*) using polymerase chain reactions, *Br. J. Dermatol.*, 138, 576, 1998.
107. Male, O., The significance of mycology in medicine, in *Frontiers in Mycology*, Hawksworth, D.L. Ed., CAB International, Wallingford, 1990, p. 131.
108. Champa, D. et al., DNA microarray based on arrayed-primer extension technique for identification of pathogenic fungi responsible for invasive and superficial mycoses, *J. Clin. Microbiol.*, 46, 909, 2008.
109. Kardjeva, V. et al., Forty-eight-hour diagnosis of onychomycosis with subtyping of *Trichophyton rubrum* strains, *J. Clin. Micrbiol.*, 44, 1419, 2006.
110. Burzykowski, T. et al., High prevalence of foot diseases in Europe: Results of the Achilles Project, *Mycoses*, 46, 496, 2003.
111. Marsella, R., Fungal skin diseases. In *Current Therapy in Equine Medicine*, Robinson, E.L. and Strayberry, K.A. Eds., Saunders, Elsevier, Philadelphia, PA, 2009, p. 687.
112. Pontecorvo, G., The parasexual cycle in fungi, *Annu. Rev. Microbiol.* 10, 393, 1956.
113. Heitman, J., Sexual reproduction and the evolution of microbial pathogens, *Curr. Biol.*, 16, R711, 2006.
114. Graeser, Y. et al., Molecular and conventional taxonomy of *Microsporum canis* complex, *Med. Mycol.*, 38, 143, 2000.
115. Sharma, R. et al., Internal transcribed spacer (ITS) of rDNA of appendaged and non-appendaged strains of *Microsporum gypseum* reveals *Microsporum appendiculatum* as its synonym, *Antonie Leeuwenhoek*, 20, 197, 2006.
116. Campbell, C.K. et al., *Arthroderma olidum*, sp.nov. a new addition to the *Trichophyton terrestre* complex, *Med. Mycol.*, 44, 451, 2006.
117. de Hoog, G.S. et al., *Atlas of Clinical Fungi*, 2nd edn., Centraalbureau voor Schimmelculturales, Utrecht, the Netherlands, 2000.

37 *Myriodontium*

Dongyou Liu

CONTENTS

37.1 INTRODUCTION

37.1.1 Classification and Morphology

The genera *Myriodontium* is a mold belonging to the mitosporic Onygenales group, order Onygenales, subclass Eurotiomycetidae, class Eurotiomycetes, subphylum Pezizomycotina, phylum Ascomycota, and kingdom Fungi. The mitosporic Onygenales group encompasses nine genera: *Blastomyces*, *Chrysosporium*, *Coccidioides*, *Emmonsia*, *Geomyces*, *Locazia*, *Malbranchea*, *Myriodontium*, and *Paracoccidioides*. Currently, the genus *Myriodontium* consists of a recognized species *Myriodontium keratinophilum* and an unassigned species *Myriodontium* sp. SW150 [1].

First described in 1978 as a new species [2], *Myriodontium keratinophilum* is widespread in nature, especially where keratinous substrates are present. The fungus has been isolated from soil in Italy and California, from the hair of shrews and cats in the United Kingdom, from the penis of a bull in Germany, and also from an unknown source in Nigeria [3].

M. keratinophilum colonies are slow growing, with the appearance of a white mold. Hyphae are septate, branch non-dichotomously, and appear ribbon-like and folded. Conidia are single celled, appearing on long denticles along the sides of the fertile hyphae [3]. The teleomorph of *M. keratinophilum* is found in the genus *Neoarachnotheca*, order Onygenales. The genus *Neoarachnotheca* is characterized by its white, spherical ascomata with a wall formed by a network of hyphae and spherical, subhyaline ascospores with an irregular sheath. The type species of the genus *Neoarachnotheca*, *Nt. keratinophila*, displays wavy peridial hyphae and has been isolated from marine and river sediments [4].

37.1.2 Clinical Features

Maran et al. [3] reported the first case of *Myriodontium keratinophilum* infection in 53-year-old Nigerian man, who was not immunosuppressed nor diabetic. The patient complained of facial pain and underwent nasal polypectomy after diagnosis of sinusitis secondary to nasal polyps. Polypectomy was repeated three times during the subsequent months, with the polyps recurring often and at short intervals. Four months after the last operation, the patient developed a swelling on his forehead, which gradually enlarged and was accompanied by left proptosis. Upon further examination, brown necrotic-like material was found in the sinus, leading to the erosion of the roof of the left orbit and both ethmoid bones and blocking of both sides of the nose by polypoidal mucosa. The posterior wall of the frontal sinus was also thinned and eroded. The tissue lining the sinus consisted of floridly inflamed granulation tissue containing large numbers of plasma cells and eosinophils; most of the mucosal lining was ulcerated. There was no evidence that the tissue had been invaded by the fungus in the lumen. The contents of the sinus were mainly cellular debris and some septate hyphae, which branched non-dichotomously and appeared ribbon like and folded. These features are characteristic of mycetoma of the paranasal sinus seen in the Sudani and of allergic aspergillosis of the paranasal sinus [5,6]. Growth of the hyphae yielded a slow-growing white mold that produced a few single-celled conidia on long denticles along the sides of the fertile hyphae. This isolate was identified as *M. keratinophilum*. An antigenic extract prepared from the isolate produced a single diffuse precipitin line when tested against the patient serum by double diffusion. No reaction was observed between comparable extracts of *A. flavus*, *A. fumigatus*, and *A. versicolor* and patient serum. The patient was treated with ketoconazole 200 mg daily for 3 weeks, recovered with no recurrence.

37.1.3 Diagnosis

Molds such as *Myriodontium keratinophilum* are environmental organisms that often transiently colonize the respiratory, integument, and gastrointestinal systems of human hosts. Traditionally, molds are identified by direct microscopy in combination with culture using a variety of mycological media. To augment specific identification, in vitro sporulation using cultured isolate may also be attempted, which may take up to 3 weeks of incubation and for which success is not guaranteed, however.

For more rapid and more accurate identification of molds including *M. keratinophilum*, PCR amplification, and sequencing analysis of ribosomal RNA (rRNA) genes, internal transcribed spacer (ITS) regions as well as other genes (e.g., EF1α) can be utilized [7,8].

37.2 METHODS

37.2.1 Sample Preparation

Molds from clinical specimens are grown on inhibitory mold agar, modified Sabouraud agars, or potato dextrose agar. Microscopic structures are observed on tease or tape preparations and slide cultures for up to 21 days.

After growing for 1–7 days on potato dextrose agar slants, approximately 1 cm^2 of mycelia are collected by scraping the slant with a sterile stick in 1 mL of sterile, molecular-grade H_2O. The material is transferred to a 2 mL screw-cap tube. The tubes are centrifuged for 1 min at 6000 × *g*. If the mycelia do not pellet, the material is contained with a pediatric blood serum filter (Porex Corp., Fairburn, GA). After removal of supernatant, the material is resuspended in 200 μL of IDI sample buffer (IDI lysis kits, GeneOhm Sciences) and transferred to the lysis tubes containing glass beads. Lysis tubes are vortexed on the highest setting for 5 min. The tubes are placed in a boiling water bath for 15 min. Tubes are centrifuged for 5 min at 16,000 × *g*. The supernatant was stored at −20°C until amplification [7].

37.2.2 Detection Procedures

Pounder et al. [7] described a real-time PCR with SYBR green DNA-binding dye and amplicon-melting temperature analysis for fungal detection using pan-fungal primers ITS1 forward (5′-TCCGTAGGTGAACCTGCGG-3′) and ITS4 reverse (5′-TCCTCCGCTTATTGATATGC-3′) [9]. The identity of the fungi is verified by subsequent sequencing analysis.

Procedure

1. PCR mixture is composed of 1 × Lightcycler FastStart DNA Master Hybridization Probes mixture (Roche Applied Science) containing deoxynucleoside triphosphates, FastStart *Taq* DNA polymerase, and 1 mM $MgCl_2$ (additional $MgCl_2$ is added to a final concentration of 4.6 mM), 0.4 μM each of ITS1 forward and ITS4 reverse primers, 1× SYBR green (Molecular Probes), and 3 μL template DNA.
2. Thermal cycling parameters include 95°C for 10 min; 50 cycles of 95°C for 5 s, 60°C for 20 s, and 76°C for 30 s; and a final extension at 72°C for 2 min.
3. The quality of the amplicon is determined using the derivative of the melt analysis curve (55°C–99°C, 45 s hold at 55°C, and 5 s/°C) using the RotorGene 3000 (Corbett Robotics, Inc).
4. The amplified product is purified for bidirectional sequencing using ExoSAP-IT (USB Corp). Five microliters of Big Dye Terminator Ready Reaction Mix v. 1.1 (Applied Biosystems) is added to 4 μL of each primer (0.8 pmol/μL) and 3 μL of purified PCR product. Cycle sequencing is performed with a 9700 thermal cycler (ABI), using 25 cycles of 96°C for 10 s, 50°C for 5 s, and 60°C for 4 min. Sequencing reaction products are passed through a Sephadex G-50 fine column to remove unincorporated dye terminators. Purified sequencing reaction products are run on an ABI Prism 3100 Genetic Analyzer with a 50 cm capillary array.
5. Sequences are analyzed with the SmartGene Integrated Database Network software version 3.2.3 vr. SmartGene is a web-based software and database system with reference sequences derived from the National Center for Biological Information (NCBI) GenBank repository.

Note. If preferred or real time PCR instrument is not available, standard PCR may be performed with primers ITS1 and ITS4, and the resulting amplicon is sequenced with the same primers. Sequence-based identifications are defined by percent identity: species, ≥99%; genus, 93%–99%; and inconclusive, ≤93%.

37.3 CONCLUSION

Myriodontium keratinophilum (teleomorph *Neoarachnotheca keratinophila*) is a mold classified in the genus *Myriodontium*, the mitosporic Onygenales group. Similar to other mold organisms, *M. keratinophilum* is commonly present in the environment and has been considered as a laboratory contaminant. After the description of the first case of *M. keratinophilum*-associated sinusitis in an immunocompetent man in 1985 [3], no new cases due to this organism have bee reported.

Considering the continuing ageing of human populations and the increasing occurrence of suppressed immune functions in individuals with underlying diseases or undergoing chemotherapies, it is likely that uncommon infections due to opportunistic mold pathogens will appear with heightened frequency in future. The availability of rapid, sensitive, and specific laboratory techniques such as those based on PCR and sequencing is critical for correct determination and differentiation of molds that are difficult to identify on the basis of macroscopic and microscopic features alone.

REFERENCES

1. The UniProt Consortium. Available at http://www.uniprot.org/, accessed on August 1, 2010.
2. Samson RA, Polonelli L. *Myriodontium keratinophilum*. gen. et sp. nov. *Persoonia*. 1978;9:505–509.
3. Maran AG et al., Frontal sinusitis caused by *Myriodontium keratinophilum*. *Br Med J*. 1985;290:207.
4. Cano J et al., Studies on keratinophilic fungi. IX: *Neoarachnotheca* gen. nov. and a new species of *Nannizziopsis*. *Anton Van Leeuwen*. 1997;72(2):149–158.
5. Milosev B et al. Primary aspergilloma of the paranasal sinuses in the Sudan. *Br J Surg*. 1969;56:132–137.
6. Miller JW, Johnson A, Lamb D. Allergic aspergillosis of the maxillary sinuses. *Thorax*. 1981;36:710.
7. Pounder JI et al., Discovering potential pathogens among fungi identified as nonsporulating molds. *J Clin Microbiol*. 2007;45:568–571.
8. Bagyalakshmi R et al., Newer emerging pathogens of ocular non-sporulating molds (NSM) identified by polymerase chain reaction (PCR)-based DNA sequencing technique targeting internal transcribed spacer (ITS) region. *Curr Eye Res*. 2008;33(2):139–147.
9. White, TJ et al., 1990. Amplification and direct sequencing of fungal ribosomal RNA genes for phylogenetics. In M. A. Innis et al. (eds.), *PCR Protocols: A Guide to Methods and Applications*. Academic Press, San Diego, CA, pp. 315–322.

38 *Onychocola*

Dongyou Liu

CONTENTS

38.1 INTRODUCTION

38.1.1 Classification and Morphology

The genus *Onychocola* is the only member of the mitosporic Arachnomycetaceae group, family Arachnomycetaceae, order Onygenales, class Eurotiomycetes, subphylum Pezizomycotina, phylum Ascomycota, and kingdom Fungi. Currently, the genus *Onychocola* consists of four species: *Onychocola canadensis* (teleomorph: *Arachnomyces nodosetosus*), *Onychocola kanei* (teleomorph: *Arachnomyces kanei*), *Onychocola glareosa* (teleomorph: *Arachnomyces glareosus*), and *Onychocola sclerotica* (teleomorph: *Malbranchea sclerotica*), of which *Onychocola canadensis* and possibly *Onychocola kanei* are implicated in subungual or superficial onychomycosis in humans [1–3].

Onychocola canadensis (teleomorph: *Arachnomyces nodosetosus*) colonies grow slowly, with velvety to lanose texture and yellow to pale sandy brown color; reverse is deep brown-gray. Arthroconidia (4–16 × 2–5 μm in size) appear after 14–21 days; are broadly ellipsoidal to nearly spherical, smooth, usually single-celled (but occasionally two-celled), and hyaline to subhyaline; and often form long, more or less, upright chains that do not readily fragment into separate conidia. Old cultures may form distinctive broad, brown, thick- or rough-walled, nodose hyphae resembling peridial appendages of the *Arachnomyces* sexual state [4,5].

Onychocola kanei (teleomorph *Arachnomyces kanei*) (*kanei* is named after Canadian mycologist Julius Kane) colonies reach a diameter of 15–21 mm in 21 days on potato dextrose agar (PDA) at 25°C are downy, raised, rugose, umbonate to crateriform, initially white and often becoming yellow to grayish yellow later. Growth at 37°C is restricted, reaching a diameter of 6–13 mm in 21 days. On oatmeal agar at 25°C, colonies are white to grayish white, flat, and initially glabrous and then turn downy tufts by 21 days. Ascomata (cleistothecia) formed in paired isolates are superficial or submerged in mycelia, reddish brown, nonostiolate, and subglobose to globose with a diameter of 175–300 μm and bear 5–8 appendages (setae). Peridial wall of the ascoma is membranous of the textura angularis type. Seta (115–125 μm × 4–10 μm) is smooth to intermittently slightly nodose, sparsely septate, and uncinate to loosely coiled at the tip. Asci (7.5 μm × 6.5 μm) are evanescent, eight spored, and subglobose. Ascospores (3.5–4.5 μm × 2.5–3 μm) are smooth, oblate with a polar boss at the center, and pale to light brown. Conidia (aleurioconidia) (3.5–7 μm × 3–4.5 μm) are sessile or borne on stalks and are smooth to finely asperulate and clavate to pyriform or are intercalary and alternate and cylindrical or irregular with one or both sides swollen. Conidia detach by lytic dehiscence [2].

A. kanei is distinguished from other related species by its setal and anamorph morphology, and colony color. *A. kanei* produces cylindrical to irregularly swollen alternate arthroconidia but differs in producing solitary aleurioconidia. *A. nodosetosus* differs in producing swollen, one or two-celled arthroconidia in chains that do not easily break apart. Dehiscence occurs by rhexolysis of thin-walled cells adjacent to, or between, chains of conidia and sometimes by schizolysis of adjacent conidia. *Arachnomyces minimus* has setae that are more nodose than those of *A. kanei* and it lacks an anamorph. *Arachnomyces gracilis* differs in having setae that are straight at the tip and producing nonswollen or cylindrical, alternate arthroconidia of the *Malbranchea* type, which secede by rhexolytic dehiscence [2].

Onychocola glareosa (teleomorph: *Arachnomyces glareosus*) colonies attain a diameter of 18–25 mm in 21 days on PDA at 30°C, are initially white, later grayish brown centrally, downy, and raised to crateriform and produce abundant sclerotia giving the colony a grainy appearance. Droplets of dark reddish brown exudate occur in areas where sclerotia are produced. A pale brown diffusing pigment first appears after 7 days and turns medium dark reddish brown within 21 days. Growth at 35°C is restricted, reaching 4–5 mm diameter in 21 days. Ascomata (cleistothecia) (185–310 μm in diameter) is reddish brown, subglobose to globose, and non-ostiolate, bearing 1–5 setae. Peridial wall is membranous of textura angularis. Setae are smooth to intermittently nodose, regularly septate, and uncinate to loosely coiled or straight at the tip. Asci (6 μm × 5 μm) are evanescent, eight spored, and subglobose. Ascospores (3–3.5 μm × 1.5–2 μm) are smooth, oblate with polar boss, and light brown. Fertile hyphae are arcuate, bearing intercalary alternate arthroconidia. Arthroconidia (2.5–4.5 μm × 1.5–2 μm) are cylindrical, sometimes slightly curved, or irregular with one or both sides swollen, detaching by lytic dehiscence. Racquet hyphae are present. Sclerotia (45–60 μm in diameter) develop from swollen, thick-walled, and pale brown cells of the vegetative hyphae, enlarging and aggregating to form a solid tissue with a membranous rind of textura angularis. Sclerotia are pale brown and subglobose to globose in shape [3].

38.1.2 Clinical Features

Human onychomycosis can be separated into five types: distal and lateral subungual onychomycosis (DLSO), white superficial onychomycosis (WSO), proximal subungual onychomycosis (PSO), *Candida* onychomycosis, and total dystrophic onychomycosis (TDO) (primary or secondary type). The most common causal organisms for onychomycosis in humans are *Trichophyton mentagrophytes*, *T. rubrum*, *Acremonium* spp., *Alternaria* spp., *Aspergillus* spp., *Fusarium oxysporum*, *Scopulariopsis brevicaulis*, *Scytalidium*, and *Onychocola canadensis* [5–12].

38.1.2.1 *Onychocola canadensis*

Onychocola canadensis is a nondermatophyte first isolated from chronic infection of the great toenail in Canada [17]. The slow growth rate of *O. canadensis* and lack of resemblance to any other known nail-infecting fungus may have contributed to its delayed discovery. The fungus has since been reported in New Zealand, France, Britain, Belgium, Spain, Italy, Turkey, and other parts of the world [13–16]. As an uncommon cause of DLSO or WSO, *O. canadensis* makes the nail white or yellow in color and often hyperkeratotic and friable. *O. canadensis* onychomycosis mostly affects gardeners or farmers, suggesting its soil origin. Sometimes, *O. canadensis* may be present in an abnormal-appearing nail as an insignificant finding, not as a pathogen.

Sigler and Congly [17] reported three cases of great toenail infection in Canada, in which a slow-growing arthroconidial hyphomycete was isolated repeatedly and in pure culture. Direct microscopy revealed hyaline, round-to-barrel-shaped arthroconidia, hyaline hyphae of varying width, and broad, thick-walled brownish hyphae. The fungus is identified as *Onychocola canadensis* Sigler gen. et sp. nov. Subsequently, Sigler et al. [18] described six additional cases of toenail infection in elderly individuals and one case of glabrous skin infection due to *Onychocola canadensis* (*Arachnomyces nodosetosus*), originating from New Zealand and Canada. The development in culture of broad, brown, nodose, thick-walled hyphae suggested an affinity to the ascomycete genus *Arachnomyces*. Ascocarps were produced in six mated pairs on sterilized rice grains or rice extract agar after 7–12 month incubation. Furthermore, Gupta et al. [19] documented 10 previously unreported cases of *O. canadensis* onychomycosis in Canada, which were diagnosed on the basis of (i) the finding of compatible filaments on direct microscopy of nail and (ii) consistent culture from repeated specimens. Of these 10 cases, *O. canadensis* was associated with DLSO (six patients), WSO (one patient), and as an insignificant contaminant in the nails (three patients).

Erbagci et al. [20] described the first Turkish case of skin and nail infection due to *Onychocola canadensis* in a farmer who frequently worked barefoot on soil. The patient presented with scaly and hyperkeratotic lesions resembling tinea pedis, erythematous plaques, and Majocchi's granuloma-like dermal papulonodules. Cultures of nail plates, skin scrapings, and needle aspiration materials from papules or nodules on Sabouroud dextrose media with and without cycloheximide, trichophyton agar, and PDA at 26°C all grew a mold, which was identified as *Onychocola canadensis* on the basis of its colonial and microscopic morphology. While skin lesions responded to daily itraconazole in a dose of 200 mg for 3 months, the onychomycosis was resistant to therapy.

38.1.2.2 *Onychocola kanei*

Gibas et al. [2] reported the identification of *Onychocola kanei* from two patients. The first patient was a 78-year-old man from Toronto, Ontario, who had a 3–4 year history of an abnormally appearing left great toenail, typical of WSO. Direct examination of the nail revealed irregular filaments, and cultures yielded a slow-growing yellow fungus, which was identified as *Onychocola kanei*. The second patient was an elderly female with an abnormal left and right great toenail, which was thickened with whitish plaques on the upper surface, suggestive of WSO. Direct microscopy of nail scrapings revealed irregular filaments, and cultures of nail scrapings grew a yellow, slow-growing fungus consistent with *Onychocola kanei* as well as *Aspergillus sydowii*.

38.1.3 Diagnosis

Onychomycosis may result from infections with dermatophytes (e.g., *Trichophyton mentagrophytes* and *T. rubrum*, which invade the nail plate, resulting in tinea unguium), nondermatophytic filamentous fungi (e.g., *Acremonium*, *Alternaria*, *Aspergillus*, *Fusarium*, *Scopulariopsis*, *Scytalidium*,

Onychocola, *Pyrenochaeta*, and *Botryodiplodia*), and *Candida* yeast. Although not life threatening, onychomycosis can impact negatively on patients' emotional, social, and occupational functioning as well as health budget. Correct identification of the causative agents for onychomycosis is vital for the implementation of effective treatment and control measures.

Because nondermatophytes are opportunistic fungal pathogens with widespread distribution, isolation of these fungi from a clinical specimen alone does not prove conclusively their direct involvement in the disease process. For reliable diagnosis of opportunistic onychomycosis, direct positive microscopy and repeat isolations of the proposed organism are necessary.

To ensure that the samples are free of contaminating bacteria, thorough cleansing of the nail area with alcohol represents the first step of the sample collection. (i) For distal subungual onychomycosis (DSO) that occurs in the nail bed, the specimen is obtained from the nail bed, which contains the greatest concentration of viable fungi. The nail is clipped short with nail clippers, and the specimen is taken from the nail bed as proximally to the cuticle as possible with a small curet or a no. 15 scalpel blade. Material is also obtained from the underside of the nail plate, with emphasis placed on sampling from the advancing infected edge most proximal to the cuticle. This helps reduce contaminants. (ii) For PSO that takes place under the cuticle before settling in the proximal nail bed, the healthy nail plate is gently pared away with a no. 15 scalpel blade. A sharp curet is then used to remove material from the infected proximal nail bed as close to the lunula as possible. (iii) For WSO that affects the nail plate surface, a no. 15 scalpel blade or sharp curet is used to scrape the white area and remove the infected debris. (iv) For *Candida* onychomycosis, material is taken from the proximal and lateral nail edges; the lifted nail bed is scraped; scrapings are taken from the undersurface of the nail if insufficient debris is present in the nail bed.

The specimen is divided into two portions for direct microscopy and culture. As nail debris is thick and coarse and hyphae are usually only sparsely present, direct microscopy serves only as a screening test for the presence or absence of fungi, and its false-negative rates may range from 5% to 15%. The specimen is mounted in a solution of 20%–25% KOH (or NaOH mixed with 5% glycerol or 20% KOH and 36% dimethyl sulfoxide), heated to emulsify lipids (1 h at 51°C–54°C), and observed under ×40 magnification. The specimen may be counterstained with chitin-specific Chlorazol black E to accentuate hyphae and prevent false-positive results. Direct microscopy helps determine if the hyphae are typical of dermatophyte fungi, nondermatophyte molds, or yeasts.

Culture provides an ultimate way to identify the causative microorganism of onychomycosis. Specimens are often plated on two different media: a primary medium (e.g., Sabouraud peptone–glucose agar with cycloheximide) to select against most nondermatophytic molds and bacteria and a secondary medium (e.g., Sabouraud's glucose agar, Littman's oxgall medium, and inhibitory mold agar without cycloheximide) to facilitate such growth. A dermatophyte tends to grow on both types of media. As nails are nonsterile, fungal and bacterial contaminants may obscure the nail pathogen. In addition, approximately 40% of nail specimens may not yield fungal growth.

Because culture-based diagnosis of onychomycosis is time-consuming and difficult, molecular techniques especially PCR and sequencing of rRNA genes and internal transcribed spacer (ITS) regions have been developed for rapid and accurate identification of causal fungal agents directly from clinical specimens and cultured isolates [2,3,21].

38.2 METHODS

38.2.1 Sample Preparation

Nail specimens are divided into three portions. The first portion is mounted on a slide with 20% KOH with 40% dimethyl sulfoxide. The glass slide is heated for 1 min under a cover glass and then examined microscopically for the presence of fungal elements: spores and hyphae. The second portion is cut into small pieces by scalpel, cultured on Sabouraud's dextrose agar containing chloramphenicol (0.05%) with and without cycloheximide (0.5%), and incubated at 25°C for 4–6 weeks. Clinical isolates are identified on the basis of phenotypic characteristics of the colonies (e.g., growth morphology, colonial appearance, surface texture, shape, size, color, rate of growth, undersurface, and edge), microscopic examination of lactophenol cotton blue wet mounts, and physiological tests such as urease production, in vitro hair perforation, and nutritional requirement tests. Characteristics of conidia are examined in slide culture preparations on Pablum cereal agar (CER). The third portion is used for DNA extraction after crushing in liquid nitrogen. The crushed scrapings are suspended in 200 μL of Tris–EDTA buffer and subjected to repeated freezing and thawing. Then, 300 μL of 0.1% Triton X-100 (pH 8) and 2 μL of proteinase K solution (20 mg/mL) were added and incubated for 2 h at 65°C. The extracted DNA is purified by the phenol–chloroform–isoamyl alcohol (25:24:1) method and resuspended in 30 μL of Tris–EDTA buffer [22].

Alternatively, DNA is extracted from cultured isolate. Approximately 100 mg of fresh mycelium is scraped from the agar surface and placed in a precooled sterile porcelain mortar containing a small amount of acid-sterilized sand. Liquid nitrogen is added and the frozen mycelium is ground to a powder. 750 μL of extraction buffer (1% w/v cetyl–trimethyl ammonium bromide, 1 M NaCl, 100 mM Tris, 20 mM EDTA, and 1% w/v polyvinyl polypyrolidone) is added to the ground material. The mixture is transferred into a sterile 2 mL screw-capped microcentrifuge tube and incubated for 30 min at 65°C. An equal volume of chloroform:isoamyl alcohol (24:1 v/v) was added. The resulting complex is mixed by inverting the tube about 20 times and centrifuged for 15 min at 10 000 × g at room temperature. The upper layer, which contained crude DNA material, is collected and purified using the QIAquick DNA purification kit (QIAgen) and the purified DNA is stored at −20°C [2].

38.2.2 Detection Procedures

38.2.2.1 Standard PCR and Sequencing

Gibas et al. [2] utilized primers ITS3 and ITS4 to amplify a fragment of 271–302 bp including a portion of the 5.8S gene, the complete ITS2 region and a portion of the 28S gene followed by sequencing analysis for identification of *Onychocola/Arachnomyces.*

PCR amplification is performed with initial denaturation at 94°C for 2 min; 30 cycles of 94°C for 1 min, 55°C for 1 min, and 72°C for 2 min; and final extension 72°C for 7 min.

Sequencing of the amplicons is conducted with the DYEnamic™ ET terminator kit (Amersham Pharmacia Biotech) and run on an ABI 377 Automated sequencer (Amersham). Consensus sequences are determined using the Sequencher™ for Windows 4.0.2 (Gene Codes Corp.) and alignment is done manually using Se-Al v1.0a1 Fat, a sequence alignment program. Phylogenetic analysis is done using PAUP (Phylogenetic Analysis Using Parsimony) v. 4.0b8, and the robustness of the resultant phylogenetic tree or inferred clades is tested using bootstrap analysis of 1000 resamplings. *Ajellomyces capsulatus* (GenBank AF038354) is used as the outgroup taxon.

38.2.2.2 Real-Time PCR and Sequencing

Pounder et al. [23] described a real-time PCR with SYBR green DNA-binding dye and amplicon-melting temperature analysis for fungal detection using pan-fungal primers ITS1 forward (5′-TCCGTAGGTGAACCTGCGG-3′) and ITS4 reverse (5′-TCCTCCGCTTATTGATATGC-3′). The identity of the fungi is verified by subsequent sequencing analysis.

Procedure

1. PCR mixture is composed of 1× Lightcycler FastStart DNA Master Hybridization Probes mixture (Roche Applied Science) containing deoxynucleoside triphosphates, FastStart *Taq* DNA polymerase, and 1 mM $MgCl_2$ (additional $MgCl_2$ is added to a final concentration of 4.6 mM), 0.4 μM each of ITS1 forward and ITS4 reverse primers, 1× SYBR green (Molecular Probes), and 3 μL template DNA.
2. Thermal cycling parameters include 95°C for 10 min; 50 cycles of 95°C for 5 s, 60°C for 20 s, and 76°C for 30 s; and a final extension at 72°C for 2 min.
3. The quality of the amplicon is determined using the derivative of the melt analysis curve (55°C–99°C, 45 s hold at 55°C, and 5 s/°C) using the RotorGene 3000 (Corbett Robotics, Inc).
4. The amplified product is purified for bidirectional sequencing using ExoSAP-IT (USB Corp). Five microliters of Big Dye Terminator Ready Reaction Mix v. 1.1 (Applied Biosystems) is added to 4 μL of each primer (0.8 pmol/μL) and 3 μL of purified PCR product. Cycle sequencing is performed with a 9700 thermal cycler (ABI), using 25 cycles of 96°C for 10 s, 50°C for 5 s, and 60°C for 4 min. Sequencing reaction products are passed through a Sephadex G-50 fine column to remove unincorporated dye terminators. Purified sequencing reaction products are run on an ABI Prism 3100 Genetic Analyzer with a 50 cm capillary array.
5. Sequences are analyzed with the SmartGene Integrated Database Network software version 3.2.3 vr. SmartGene is a web-based software and database system with reference sequences derived from the National Center for Biological Information (NCBI) GenBank repository.

Note. If real time PCR instrument is not available, standard PCR may be performed with primers ITS1 and ITS4, and the resulting amplicon is sequenced with the same primers. Sequence-based identifications are defined by percent identity: species, ≥99%; genus, 93%–99%; and inconclusive, ≤93%.

38.3 CONCLUSION

The genus *Onychocola* consists of four nondematophyte fungal species: *Onychocola canadensis* (teleomorph: *Arachnomyces nodosetosus*), *Onychocola kanei* (teleomorph: *Arachnomyces kanei*), *Onychocola glareosa* (teleomorph: *Arachnomyces glareosus*), and *Onychocola sclerotica* (teleomorph: *Malbranchea sclerotica*), which are commonly present in the environment. Of these, *Onychocola canadensis* and possibly *Onychocola kanei* are implicated in subungual or superficial onychomycosis in humans.

Besides *Onychocola* spp., a number of dermatophytes (e.g., *Trichophyton mentagrophytes* and *T. rubrum*), nondermatopytic fungi (e.g., *Acremonium, Alternaria, Aspergillus, Fusarium, Scopulariopsis, Scytalidium, Onychocola, Pyrenochaeta*, and *Botryodiplodia*), and *Candida* yeast may also be responsible for onychomycosis. In order to implement effective treatment and control measures, it is important that the causative agents for onychomycosis be identified to species level.

Given their time-consuming nature and requirement of specialized skills, direct microscopy and culture for the identification of causal agents for onychomycosis are far from being satisfactory. To speed up the diagnostic process and to reduce misidentification, molecular techniques based on PCR amplification and sequencing analysis have been applied for improved determination of *Onychocola* and other fungi and *Candida* yeast. It is envisaged that future development of *Onychocola*-species-specific primers will further streamline the direct detection and identification of this fungus from clinical specimens.

REFERENCES

1. The UniProt Consortium. Available at http://www.uniprot.org/taxonomy/, accessed on August 1, 2010.
2. Gibas CF et al. *Arachnomyces kanei* (anamorph *Onychocola kanei*) sp. nov., from human nails. *Med Mycol.* 2002;40(6):573–580.
3. Gibas CF et al. Mating patterns and ITS sequences distinguish the sclerotial species *Arachnomyces glareosus* sp. nov. and *Onychocola sclerotica. Stud Mycol.* 2004;50:525.
4. de Hoog GS et al. (eds.). *Atlas of Clinical Fungi*, 2nd edn., p. 325. Centraalbureau voor Schimmelcultures, Utrecht, the Netherlands, 2004.
5. Gupta AK. Non-dermatophyte onychomycosis. *Dermatol Clin.* 2003;21(2):257–268.
6. Vélez H, Díaz F. Onychomycosis due to saprophytic fungi. Report of 25 cases. *Mycopathologia.* 1985;91(2):87–92.
7. Elewski BE. Onychomycosis: Pathogenesis, diagnosis, and management. *Clin Microbiol Rev.* 1998;11:415.
8. Gupta AK, Summerbell RC. Combined distal and lateral subungual and white superficial onychomycosis in the toenails. *J Am Acad Dermatol.* 1999;41(6):938–944.
10. Gupta AK et al. Itraconazole and terbinafine treatment of some nondermatophyte molds causing onychomycosis of the toes and a review of the literature. *J Cutan Med Surg.* 2001;5:206.
11. O'Donoghue NB, Moore MK, Creamer D. Onychomycosis due to *Onychocola canadensis. Clin Exp Dermatol.* 2003;28(3):283–284.
12. Piraccini BM, Tosti A. White superficial onychomycosis: Epidemiological, clinical, and pathological study of 79 patients. *Arch Dermatol.* 2004;140(6):696–701.
13. Campbell CK, Johnson EM, Warnock DW. Nail infection caused by *Onychocola canadensis*: Report of the first four British cases. *J Med Vet Mycol.* 1997;35(6):423–425.
14. Llovo J et al. Onychomycosis due to *Onychocola canadensis*: Report of the first two Spanish cases. *Med Mycol.* 2002;40(2):209–212.
15. Fanti F et al. First Italian report of onychomycosis caused by *Onychocola canadensis. Med Mycol.* 2003;41(5):447–450.
16. Van Esbroeck M et al. Isolation of *Onychocola canadensis* from four cases of onychomycosis in Belgium. *Acta Clin Belg.* 2003;58(3):190–192.
17. Sigler L, Congly H. Toenail infection caused by *Onychocola canadensis* gen. et sp. nov. *J Med Vet Mycol.* 1990;28(5):405–417.
18. Sigler L, Abbott SP, Woodgyer AJ. New records of nail and skin infection due to *Onychocola canadensis* and description of its teleomorph *Arachnomyces nodosetosus* sp. nov. *J Med Vet Mycol.* 1994;32(4):275–285.
19. Gupta AK, Horgan-Bell CB, Summerbell RC. Onychomycosis associated with *Onychocola canadensis*: Ten case reports and a review of the literature. *J Am Acad Dermatol.* 1998;39(3):410–417.
20. Erbagci Z et al. Cutaneous hyalohyphomycosis and onychomycosis caused by *Onychocola canadensis*: Report of the first case from Turkey. *J Dermatol.* 2002;29(8):522–528.
21. Walberg M et al. 18S rDNA polymerase chain reaction and sequencing in onychomycosis diagnostics. *Acta Derm Venereol.* 2006;86(3):223–226.
22. Garg J et al. Evaluation of pan-dermatophyte nested PCR in diagnosis of onychomycosis. *J Clin Microbiol.* 2007;45(10):3443–3445.
23. Pounder JI et al. Discovering potential pathogens among fungi identified as nonsporulating molds. *J Clin Microbiol.* 2007;45(2):568–571.

39 *Paecilomyces*

Ana Alastruey-Izquierdo, Maria Victoria Castelli, Leticia Bernal-Martinez, and Manuel Cuenca-Estrella

CONTENTS

39.1 INTRODUCTION

Paecilomyces species are widespread soil saprophytes [1]. Fungal infections caused by these species are not common despite their ubiquity in the human's environment [2,3]. These species have been mainly described as the cause of hyalohyphomycosis associated with traumatic inoculation and contamination of human skin or prosthetic devices. In the last few years, *Paecilomyces* have been reported in occasional cases of sinusitis or pulmonary infection afflicting immunosuppressant patients after inhalation of their conidia [3,4] and even as the cause of disseminated mycoses in some patients with predisposing factors [5,6]. *Paecilomyces lilacinus* and *Paecilomyces variotii* are the species most frequently involved in human infection. The differentiation between these two species is clinically important, since *P. lilacinus* and *P. variotii* seem to present marked differences in their in vitro susceptibility to the antifungal agents. *P. lilacinus* is resistant to amphotericin B and itraconazole, while other compounds such as posaconazole and voriconazole have good in vitro activity against this fungus. In contrast, *P. variotii* exhibits lower minimal inhibitory concentrations (MIC) of amphotericin B, itraconazole, and of new azoles but in vitro resistance to voriconazole has been reported [7]. In addition, the echinocandins show good in vitro activity against *P. variotii*, while *P. lilacinus* is not susceptible to this family of antifungal compounds [7–9].

39.1.1 Classification, Morphology, and Epidemiology

39.1.1.1 Classification

Paecilomyces are hyaline species of filamentous fungi whose sexual telemorphs belong to the division *Ascomycota* (*Euascomycetes*, *Eurotiales*: *Trichocomaceae*) of kingdom *Eumycota* and are classified in the genera *Byssochlamys*, *Chromocleista*, *Talaromyces*, and *Thermoascus* [10].

39.1.1.2 Morphology

Colonies of organisms belonging to these species are fast growing. Some species are termophilic; thus, *Paecilomyces crustaceus* and *P. variotii* can grow well at temperatures above 50°C [10]. Colonies are flat spreading, powdery or

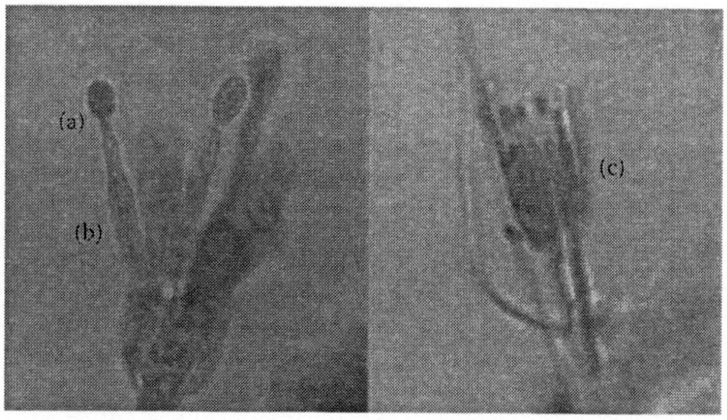

FIGURE 39.1 Microscopy of *Paecilomyces* spp. Conidiogenous cells phialidic, swollen at the base (a) and gradually narrowed into a long beak (b). Conidiophores occur as verticils (c).

velvety, and are usually white, brownish, or in bright colors, reverse off-white to brown. The most common clinical species are *P. variotii* with a yellowish brownish greenish colony and *P. lilacinus* with characteristic vinaceous to violet colonies.

Conidiophores occur solitarily, in pairs, as verticils, and in penicillate heads. Conidiogenous cells are phialidic, swollen at the base, and gradually narrowed into a long beak; they bend away from the axis of the conidiophore. Microscopically resembles *Penicillium* spp. but the phialides are more elongated (see Figure 39.1) [10].

39.1.1.3 Epidemiology and Susceptibility Profile

Paecilomyces species are saprophytic hyalohyphomycetes filamentous fungi that are found worldwide in soil, decaying vegetable matter, and as air and water contaminants causing deterioration of grain, food, and paper. Moreover, they can contaminate creams and lotions of clinical use and colonize catheters and plastic implants [11,12].

As stated above, *P. lilacinus* and *P. variotii* are the most ubiquitous species of the genus and most frequently involved in human infection. Other species of the genus such as *Paecilomyces crustaceus*, *Paecilomyces marquiandii*, *Paecilomyces viridis*, *Paecilomyces javanicus*, and *Paecilomyces puntonii* [13–18] have also been reported occasionally from human infections. Some of these species have also been described as etiology of animal diseases [19].

The identification of *Paecilomyces* spp. has clinical significance as differences in their susceptibility profile have been described. Table 39.1 displays the data of susceptibility in vitro of *P. variotii* and *P. lilacinus*, including data of 69 clinical strains tested in the Spanish Mycology Reference Laboratory following the EUCAST (European Committee on Antimicrobial Susceptibility Testing) reference procedure for susceptibility testing of filamentous fungi.

39.1.2 Clinical Features

Despite its ubiquity in the human's environment, fungal infections caused by *Paecilomyces* species are uncommon. They can cause the same clinical entities that are described for other hyaline moulds such as *Aspergillus*, *Fusarium*, and *Penicillium*. Most cases occur in patients with impaired host defences or following trauma or surgical procedures [11], but allergic bronchopulmonary diseases and pulmonary fungus ball have also been reported [3,13].

Invasion of tissues by these species causes a hyalohyphomycosis in which the invasive form of the etiologic agent is septate hyphae with no pigment in the wall. Histopathology is very similar to that observed in infections by other hyaline fungus such as *Aspergillus* spp., with infected foci that contain a central cluster of radially arranged hyphae and a margin of necrosis, with or without hemorrhagic borders. Ischemia and edema result in lesions that are grossly firm and pale. The inflammatory response is minimal in immunosuppressed patients [20,21].

39.1.2.1 *Paecilomyces lilacinus* Infections

In 2006, Pastor and Guarro published the most recent review so far on clinical importance of *P. lilacinus*. The review identified 119 reported cases of human infection by *P. lilacinus* between 1964 and 2004. Most were cases of oculomycosis (51.3%), followed by cutaneous and subcutaneous infections (35.3%), and a smaller group of miscellaneous infections (13.4%) [12].

As stated above, this species has most often associated with ocular mycosis and intraocular lens implantation has been reported as the most frequent predisposing factor for that. Other risk factors such as the use of extended-wear

TABLE 39.1
Summary of Susceptibility Results In Vitro of Antifungal Agents Tested against *Paecilomyces* species[a]

		AMB		ITC		VRC		POS		CPF		ANF		MCF	
Species	***N***	**MIC_{50}**	**MIC_{90}**	**MIC_{50}**	**MIC_{90}**	**MIC_{50}**	**MIC_{90}**	**MIC_{50}**	**MIC_{90}**	**MIC_{50}**	**MIC_{90}**	**MIC_{50}**	**MIC_{90}**	**MIC_{50}**	**MIC_{90}**
P. variotii	37	0.03	0.50	0.06	0.25	1.0	8.0	0.03	0.25	0.50	4.0	0.03	0.03	0.03	0.03
P. lilacinus	32	>16.0	>16.0	>8.0	>8.0	0.50	4.0	0.25	0.50	>16.0	>16.0	>16.0	>16.0	>16.0	>16.0

MIC_{50}, MIC causing inhibition of 50% of isolates; MIC_{90}, MIC causing inhibition of 90% of isolates; AMB, Amphotericin B; ITC, itraconazole; VRC, Voriconazole; POS, Posaconazole; CPF, Caspofungin; ANF, Anidulafungin; MCF, Micafungin.

[a] Table 39.1 displays MIC_{50} and MIC_{90}[a] in μg/mL of antifungal agents tested.

contact lenses and trauma have also been described. It can cause endophthalmitis, keratitis, corneal ulceration, infection of the lacrimal sac, and orbital granuloma. The first reported case of endophthalmitis due to *P. lilacinus* resulted from surgical procedure to treat glaucoma in Argentina [11]. An endophthalmitis epidemic caused by *P. lilacinus* occurred in the United States in 1975. A total of 13 cases were collected and it was proved that the infection resulted from the insertion of an intraocular lens that had been sterilized in sodium hydroxide and neutralized in contaminated sodium bicarbonate solution [22,23].

P. lilacinus has been involved in a number of cutaneous and subcutaneous infections. These mycoses occur mainly in solid organ and bone marrow transplant recipients, although surgery and primary or acquired immunodeficiency are also relevant predisposing factors [2,24–26]. There are also cases of histopathologically proven cutaneous infections due to *P. lilacinus* in immunocompetent patients [27,28], even one in a healthy patient without any apparent portal of entry [29].

Other entities have also been associated with *P. lilacinus*. Several cases of sinusitis have been reported in patients with predisposing causes. Diabetes mellitus and neutropenia have been reported as risk factors of these infections [4,30]. A chronic maxillary sinusitis was described in an immunocompetent patient [31], and one more infection of the paranasal sinuses was reported in a pediatric patient without predisposing causes [32]. Nasal obstruction, discharge, telecanthus, diplopia, and epiphora were the symptoms observed.

P. lilacinus has also been implicated in catheter-related fungemia in an immunocompromised pediatric patient [6], in endocarditis and subaortic aneurysm after prosthesis implantation [33], in lung abscess in a 57-year-old man without immunosuppression [34], in a report of pleural effusion [35], an in a case of proven vaginitis in an immunocompetent woman [36].

39.1.2.2 *Paecilomyces variotii* Infections

P. variotii has been mainly described as cause of endocarditis and peritonitis, although other rare infections have also been reported. Endocarditis by *P. variotii* has been observed in prosthetic heart valves [11,37] and one complicating an aortic valve allograft [38]. Other species of *Paecilomyces* have been hardly ever described as etiology of endocarditis. An infection afflicted a diabetic woman who developed endocarditis of the aortic valve 6 years after insertion of a porcine mitral valve heterograft. Infection was caused by *P. javanicus* in this case [14].

Peritonitis by *P. variotii* has been normally associated with continuous ambulatory peritoneal dialysis. Approximately half of the patients had received multiple antibiotics before the onset of the fungal peritonitis because of either bacterial peritonitis or exit site infection. No other predisposing factors seem to be related to this *Paecilomyces* infection [39–42].

Other rare mycoses by *P. variotii* have been reported. Pneumonia has been described in a patient with diabetes mellitus [43], and complicating hairy cell leukemia in a middle-aged woman [44]. Cellulitis, subcutaneous infections, and multifocal osteomyelitis have been described in renal transplant recipients and in a patient with chronic granulomatous diseases [11,45]. Finally, a case of disseminated infection in a neutropenic patient with leukemia was recently reported [7].

In addition, *P. variotii* has been isolated from deep sites after surgical procedures. It was collected from the airways of a pediatric patient with cystic fibrosis who underwent bilateral living-donor lobar lung transplantation [46]. This microorganism was found to colonize a cerebrospinal fluid shunt of a woman who underwent ventricle-peritoneal shunt placement for noncommunicating hydrocephalus [47]. It also was isolated from the renal pelvis at ureterolithotomy of a patient who had a long history of nephrolithiasis [48]. In these cases, infection was noninvasive but the isolation of that fungus further complicated the approach to postsurgical management and antimicrobial therapy.

39.1.3 Treatment of *Paecilomyces* Infections

There are no guidelines to treat these infections. But according to expert recommendations, the therapeutic strategy should be based on three points, that is, surgery, antifungal treatment, and adjuvant therapy to control predisposing causes. When possible, extensive surgical debridement of necrotic and infected tissue and replacement of infected prosthesis should be done. Antifungal treatment should be administered taking into account the susceptibility profile of those species. Infections by *P. lilacinus* should be treated with voriconazole or posaconazole as it seems to be resistant to amphotericin B and echinocandins. *P. variotii* unlike *P. lilacinus* is in vitro resistant to voriconazole and susceptible to the other licensed antifungal compounds [9].

Disseminated infections are difficult to treat and surgery cannot be performed in most cases. The granulocyte colony stimulating factor and granulocyte transfusion could play an important role in the management of these infections since they are showing some usefulness in the therapy of other life-threatening mycoses such as zygomycosis [49]. Concomitant combination antifungal therapy could be another option to treat disseminated infections. No clinical data are available on it, but results in vitro indicate that combination between triazoles and terbinafine could have synergy against both species. Other antifungal combinations have shown indifference [50].

39.1.4 Diagnosis

In general, the fungal diagnostic methods include the identification of disease symptoms as well as isolation and laboratory identification of the fungus [51]. However, in recent years, important advances in polymerase chain reaction (PCR)-based methods and other molecular tools have increased their importance in the diagnosis of fungal infections.

39.1.4.1 Conventional Techniques

Conventional techniques for the diagnosis of this mycosis rely on the isolation and culturing of the fungus and its identification including macro and micro-morphological identification. However, the two most common species, *P. variotii* and *P. lilacinus*, are frequently encountered as laboratory contaminants or as saprophytic flora, and cannot be taken as clinically significant organisms unless the presence of fungal filaments are histologically confirmed [30].

Although the morphological techniques are often enough to identify the clinically relevant species of these genera, they require high level of expertise from the laboratory staff. Another disadvantage is that these techniques are time consuming, and can often delay treatment.

39.1.4.1.1 Direct Microscopy

Material from specimens should be analyzed using 10% potassium hydroxide (KOH) or optical brighteners such as calcofluor. Septate hyaline hyphae may be observed in the direct examination of the biological samples. However, as stated above, the presence of fungal filaments in tissues is compulsory to confirm the fungal infection. Stains of fixed tissues with hematoxylin and eosin, methenamine-silver or periodic acid-Schiff show hyaline hyphae only. This technique is not specific enough to reach species identification since other genera as *Aspergillus*, *Penicillium* or *Fusarium*, produce similar structures. Lesions in tissues are caused by hyaline septate filaments with acute-angle (<45°) branching within blood vessels, causing infarction, edema, and hemorrhage [11]. Therefore morphological characteristics of the fungus are often overlooked by pathologists leading to a misidentification and a delay in the prescription of the correct treatment [30,51].

39.1.4.1.2 Isolation

These fungi grow well on conventional culture media. However, fungi are difficult to recover from some samples, particularly from biopsies of tissues. Specimens from sterile sites can be directly inoculated. Contaminated specimens should be treated with antibacterial agents before inoculation or be inoculated on assay media including antimicrobial agents in their formulation. Clinical material can be inoculated onto Sabouraud dextrose agar, blood agar, and chocolate agar and incubated at 28°C–30°C and 37°C. Alternative media such as malt extract or broths such as brain heart infusion may be used for biopsies. Recovery of the fungus from cultures allows for the identification of the organism at species level according to morphological characteristics stated earlier.

Although *Paecilomyces* are saprophytes and laboratory contaminants, the isolation of these organisms in specimens from patients with severe immunosuppression or other predisposing factor should not be disregarded as contamination and patients should be carefully evaluated.

39.1.4.1.3 Serology and Antigen Detection

There are no specific serological techniques developed for the detection of *Paecilomyces* in clinical samples. However, these species can trigger the immunological response as several reports have shown [13,52–54] and subsequently, specific antibodies could be detected.

Regarding antigen quantification, *Paecilomyces* produce beta-D-glucan [55,56], an antigen that is used for the diagnosis of invasive fungal infections and that is a criterion of probable invasive mycoses by EORTC/MSG (European Organization for Research and Treatment of Cancer/Mycoses Study Group) [57]. No data are available about the clinical utility of that technique to detect *Paecilomyces* infections since the prevalence of these mycoses is low and conclusions can not be drawn from studies published so far [58]. *Paecilomyces* species have been reported as having galactomannan as well, an antigen that is specific for the diagnosis of aspergillosis. That cross-reactivity could allow the detection of *Paecilomyces* by the galactomannan quantification in patients suffering from invasive fungal infection [59].

39.1.4.2 Molecular Techniques

An increasing number of molecular techniques for the diagnosis of fungal infections have been developed in the last few years, due to the growing prevalence of mycoses and the length of time required for diagnosis when classical microbiological methods are used.

With the PCR, it is possible to multiply a given DNA segment quickly and identify the fungus. This method has provided a rapid and powerful tool to identify organisms by amplifying specific DNA regions. PCR-based assays for the diagnosis of these species (and all mycosis) include three common steps: (i) the selection of a specific target region of DNA to identify the fungus, (ii) DNA extraction method, and (iii) a method to identify the presence of the target region.

Conventional PCR assays have been developed to detect DNA from tissues or cultured strains and are targeted mostly toward the internal transcribed spacer (ITS) region of the ribosomal DNA gene cluster. Once amplification of the target region has been performed, PCR-amplified segments must be identified. This region can also be used to detect *Paecilomyces* spp. from clinical samples or to identify theses species from cultures, but no reliable data are available so far [60]. DNA segments comprising the ITS1 and ITS2 regions can be amplified with primers ITS1 (5′-TCCGTAGGTGAACCTGCGG-3′) and ITS4 (5′-TCCTCCGCTTATTGATATGC-3′) [61].

Regarding quantitative PCR-based techniques, Atkins et al. [51] developed a real time PCR using TaqMan probes and species-specific primers targeting the ITS region of the rDNA to identify and detect *P. lilacinus* from nematode, root soil, and clinical isolates. The test showed a high specificity and authors speculated its usefulness in clinical samples.

Haugland et al. developed quantitative PCR assays of selected *Aspergillus*, *Penicillium*, and *Paecilomyces* species to analyze dust samples from different sources around the United States. The assay incorporated fluorigenic 5′ nuclease (TaqMan) chemistry and was directed at the nuclear ribosomal RNA operon, internal transcribed spacer regions (ITS1 or ITS2). The assay showed to be reliable for the detection of

Paecilomyces species, but nothing is known on its application in clinical samples [62].

39.2 METHODS

No particular procedures exist to be used in the molecular diagnosis of *Paecilomyces*. DNA extraction, amplification, sequencing, and classification follow the general principles of mycological taxonomy.

39.2.1 Sample Preparation

39.2.1.1 Clinical Specimens

Adequate specimens for detection of species of *Paecilomyces* are similar to those for other fungal species. Ocular exudates from oculomycosis, skin scrapings or biopsies from cutaneous infection, aspirates from sinuses in patients with sinusitis, bronchoalveolar lavages from pulmonary lesions, and biopsies of tissues in disseminated infections have been shown useful for the detection of these species. Blood cultures were positive in case of endocarditis. No data are available on reliability of that specimen in other sort of *Paecilomyces* infection. Sample transportation should follow the regular recommendations on biosafety, delivery, and storage.

39.2.1.2 DNA Extraction from Cultured Strains

Fungal DNA extraction is the first step in the molecular diagnosis. There are several commercial and in-house methods to perform this action. The extraction method described by Maresca et al. [63] is included below:

1. Subculture the isolated fungus in malt extract agar (2% malt extract) or potato dextrose agar until visible growth at 30°C ± 2°C.
2. Add distilled water to the subculture and scrape with a swab.
3. Inoculate plastic Petri dish containing 3 mL GYEP (0.3% yeast extract, 1% peptone, 2% glucose) with 50–100 μL of the spore suspension obtained.
4. Incubate until growth at 30°C ± 2°C.
5. Harvest mycelium with a plastic tip and put on a Whatman paper. It is important to dry mycelium completely.
6. Transfer dry mycelium to a 50 mL propylene tube with screw cap and add six glass beads (4 mm diameter).
7. Introduce the tube into liquid N_2 for 30 s and vortex for 30 s.
8. Repeat operation until all mycelial powder is obtained.
9. Add 0.8 mL of extraction buffer (0.2 M Tris–HCl, 0.5 M NaCl, 10 mM EDTA, sodium dodecyl sulfate [SDS] 1%) and 0.8 mL of phenol:chloroform:isoamilic alcohol (25:24:1) and shake gently.
10. Transfer to an eppendorf and centrifuge at 12,000 rpm for 15 min at 4°C–8°C.
11. Transfer supernatant to a new tube and add an equal volume of phenol:chloroform:isoamyl alcohol (25:24:1) and vortex for 30 s. Centrifuge 15 min a 12,000 rpm at 4°C–8°C.
12. Remove supernatant and extract with an equal volume of chloroform:isoamilic alcohol (24:1).
13. Vortex for 30 s and centrifuge for 5 min at 12,000 rpm at 4°C–8°C.
14. Take supernatant and precipitate DNA with isopropanol. Centrifuge for 15 min at 12,000 rpm at room temperature.
15. Aspirate supernatant and wash the pellet with 70% ethanol. Centrifuge at 12,000 rpm for 10 min at 4°C–8°C.
16. Aspirate supernatant and suspend the pellet again in 25–50 μL RNAase (1 mg/mL). Incubate at 37°C for 1 h.

39.2.1.3 DNA Extraction from Clinical Samples

Different DNA extraction kits can be employed depending on the clinical sample. In-house methods can be used as well. There are some examples of commercial kits available to extract DNA such as Wizard (Promega) and QiAmpTissue (Qiagen), which can be used to extract DNA from tissues, and NucleoSpinBlood (Macherey-Nagel)) and QiAmp (Qiagen) to be used with blood and other fluids (serum, bone marrow aspirate, bronchoalveolar lavage, etc.). Manufacturer's recommendations must be strictly followed if these kits are used in the clinical laboratory.

39.2.2 Detection Procedures

39.2.2.1 PCR Identification

A description of PCR procedures used in the Spanish Mycology Reference Laboratory to detect *Paecilomyces* DNA is included.

The Spanish Reference Laboratory used a panfungal PCR-based technique to detect DNA of these species by ITS fragment amplification. There are several universal primers to amplify the ITS regions. Table 39.2 displays the most commonly used.

Primers ITS1 (5′-TCCGTAGGTGAACCTGCGG-3′) and ITS4 (5′-TCCTCCGCTTATTGATATGC-3′) can be used to amplify the complete region ITS1–5.8s–ITS2 of *Paecilomyces* species. The methods used are as follows:

1. The reaction mixtures contain 0.5 μM of each primer, 0.2 mM of each deoxynucleoside triphosphate, 1.5 mM $MgCl_2$, 2.5 U *Taq* DNA polymerase, and 25 ng of DNA in a final volume of 50 μL of PCR buffer.
2. Cycling conditions include 1 initial cycle at 94°C for 2 min, followed by 35 cycles of 30 s at 94°C, 45 s at 56°C, and 2 min at 72°C, with 1 final cycle of 5 min at 72°C.
3. The reactions products must be analyzed in a 0.8% agarose gel.

TABLE 39.2
Sequences of Primers Used to Amplify ITS Regions

ITS-1-5,8s-ITS2		
Primer	**Sequence (5′–3′)**	**Size (bp)**
ITS1	TCCGTAGGTGAACCTGCGG	600–630
ITS4	TCCTCCGCTTATTGATATGC	
ITS3	GCATCGATGAAGAACGCAGC	ITS3+4=330
ITS2	GCTGCGTTCTTCATCGATGC	ITS1+2=290
ITS5	GGAAGTAAAAGTCGTAACAAGG	ITS5+4=645

Source: White, T.J. et al., *PCR Protocols: A Guide to Methods and Applications*, Innis, M. (Ed.), Academic Press, San Diego, CA, 1990, pp. 315–322.

The above conditions are examples, and each PCR-based assay should be optimized in each laboratory.

The identity of *Paecilomyces* can be further verified by sequencing analysis of the amplified product. Prior to sequencing, the PCR product is purified. There are several commercial kits available to perform this purification and are well-known for experts working in molecular procedures. For sequencing reactions, Spanish Reference Laboratory uses BigDye Terminator Cycle Sequencing Ready Reaction by Applied Biosystems Inc, typically with 1 μM of the primer and 3 μL of the purified PCR product in a final volume of 10 μL.

39.2.2.2 Quantitative PCR Detection of *Paecilomyces* Species

Two real-time PCR-based assays to specifically detect *Paecilmyces* have been developed [51,62]. Atkins et al. developed an assay for the identification and detection of *P. lilacinus* from nematode, root soil, and clinical strains. This technique could be applied in clinical laboratory as it showed good rates of reproducibility and specificity, and a low limit of detection. A summary of this assay is attached below:

1. Real-time PCR reactions were performed in Bioplastics (EU) 96 × 0.2 mL PCR plates capped with Bioplastics EU Optical thin wall eight-cap strip low background (BIOplastics BV, Landgraaf, the Netherlands).
2. Stratagene Brilliant QPCR Master Mix TaqMan Kit was used for real-time reactions (Stratagene, Amsterdam, the Netherlands).
3. Primers PLrtF and PLrtR were included at a final concentration of 900 nM each and the TaqMan probe (PLrtP) was used at 225 nM.
4. Additional $MgCl_2$ was added to increase the concentration to 3.5 mM.
5. The automated ABI Prism 7700 sequence detector (Applied Biosystems) was used for reactions.
6. Reference dye was added in accordance to the manufacture's recommendations for use with the ABI 7700. The reaction was made up to 20 μL using ultrapure dH_2O. Universal thermal cycle protocol was used for PCR amplification: Stage 1, 50°C for 2 min; stage 2, 95°C for 10 min (activation of *Taq* DNA polymerase); and stage 3, 50 cycles of 95°C for 15 s and 60°C for 1 min. The amplicon generated was 130 bp for *P. lilacinus*.

39.3 CONCLUSIONS AND FUTURE PERSPECTIVES

According to latest studies *Candida*, *Aspergillus* and, in some geographical areas, the *Zygomycetes* are the most common causes of opportunistic fungal infections. However, severe infections due to other rare and uncommon fungal species are increasingly reported. It is hard to know the real prevalence of emerging pathogens such as *Paecilomyces* but it seems that their frequency could be growing.

Several points have been signaled as causes of the rise of opportunistic infections and the emergence of rare molds. Environmental changes, antimicrobial pressure, and an expanding population of immunocompromised hosts and of patients with predisposing factors have been argued in many publications. In addition, an efficient control of more prevalent species could be extending the niche of rare and may be more resistant species. The use of voriconazole for the treatment of aspergillosis and of echinocandins as empirical treatment of invasive fungal infections have been claimed to be responsible for the increase of mycoses due to multi-resistant emerging pathogens such as zygomycosis or of infections by *Scedosporium* or *Fusarium* [64].

In addition, there has been an increase in number of experts in Medical Mycology with capabilities to identify rare fungi such as *Paecilomyces* at species level. Moreover, there is a novel interest in identifying pathogens at species level since distinct alternatives with specific spectrum of action are available to treat patients. From our point of view emergent fungal infections will continue to increase in those clinical settings where experts such as microbiologists, infectious diseases experts, hematologists, and pediatricians are interested in fungal identification at species level.

The differentiation between these *Paecilomyces* species is clinically important, since *P. lilacinus* and *P. variotii* seem to present marked differences in their in vitro susceptibility to the antifungal agents. *P. lilacinus* is resistant to amphotericin B, itraconazole, and echinocandins, while *P. variotii* is in vitro resistant to voriconazole only.

Molecular methods of identification and classification of these species can be more useful from a clinical point of view than conventional techniques. A panfungal PCR-based technique is able to identify *Paecilomyces* spp. from cultures by ITS fragment amplification. Quantitative specific PCR methods are useful to detect these species in environmental and animal samples and even to identify them from cultures, but no data are available yet about usefulness of these molecular methods in clinical samples. Direct detection of fungal DNA from clinical samples could be soon the reference procedure for the diagnosis of invasive fungal infections.

REFERENCES

1. Madsen, A.M. et al., Human exposure to airborne fungi from genera used as biocontrol agents in plant production, *Ann. Agric. Environ. Med.*, 14, 5, 2007.
2. Castro, L.G., Salebian, A., and Sotto, M.N., Hyalohyphomycosis by *Paecilomyces lilacinus* in a renal transplant patient and a review of human *Paecilomyces* species infections, *J. Med. Vet. Mycol.*, 28, 15, 1990.
3. Gutierrez, F. et al., Pulmonary mycetoma caused by an atypical isolate of *Paecilomyces* species in an immunocompetent individual: Case report and literature review of *Paecilomyces* lung infections, *Eur. J. Clin. Microbiol. Infect. Dis.*, 24, 607, 2005.
4. Gucalp, R. et al., *Paecilomyces* sinusitis in an immunocompromised adult patient: Case report and review, *Clin. Infect. Dis.*, 23, 391, 1996.
5. Antachopoulos, C., Walsh, T.J., and Roilides, E., Fungal infections in primary immunodeficiencies, *Eur. J. Pediatr.*, 166, 1099, 2007.
6. Tan, T.Q. et al., *Paecilomyces lilacinus* catheter-related fungemia in an immunocompromised pediatric patient, *J. Clin. Microbiol.*, 30, 2479, 1992.
7. Chamilos, G. and Kontoyiannis, D.P., Voriconazole-resistant disseminated *Paecilomyces variotii* infection in a neutropenic patient with leukaemia on voriconazole prophylaxis. *J. Infect.*, 51, e225, 2005.
8. Castelli, M.V. et al., Susceptibility testing and molecular classification of *Paecilomyces* spp., *Antimicrob. Agents Chemother.*, 52, 2926, 2008.
9. Cuenca-Estrella, M. et al., Activity profile in vitro of micafungin against Spanish clinical isolates of common and emerging species of yeasts and molds, *Antimicrob. Agents Chemother.*, 53, 2192, 2009.
10. de Hoog, G.S., Guarro, J., Genè, J., and Figueras, M.J., *Atlas of Clinical Fungi*. 2nd edn., Centraalbureau voor Schimmelcultures/Universitat Rovira I Virgili, 2000.
11. Kwon-Chung, K.J. and Bennett, J.E., *Medical Mycology*, Lea & Febiger, Philadelphia, PA, 1992.
12. Pastor, F.J. and Guarro, J., Clinical manifestations, treatment and outcome of *Paecilomyces lilacinus* infections, *Clin. Microbiol. Infect.*, 12, 948, 2006.
13. Akhunova, A.M., Infectious-allergic bronchopulmonary paecilomycosis, *Ter. Arkh.*, 63, 19, 1991.
14. Allevato, P.A. et al., *Paecilomyces javanicus* endocarditis of native and prosthetic aortic valve, *Am. J. Clin. Pathol.*, 82, 247, 1984.
15. Harris, L.F. et al., *Paecilomyces* cellulitis in a renal transplant patient: Successful treatment with intravenous miconazole, *South. Med. J.*, 72, 897, 1979.
16. Pore, R.S. and Larsh, H.W., First occurrence of *Thermoascus aurantiacus* from animal and human sources, *Mycologia*, 59, 927, 1967.
17. Zeltzer, E. et al., Diminished chemokine and cytokine-induced adhesion of CD4+ T cells to extracellular matrix ligands in patients with end-stage renal failure, *Isr. Med. Assoc. J.*, 2, 282, 2000.
18. Naldi, L. et al., *Paecilomyces marquandii* cellulitis in a kidney transplant patient, *Br. J. Dermatol.*, 143, 647, 2000.
19. Quance-Fitch, F.J., Schachter, S., and Christopher, M.M., Pleural effusion in a dog with discospondylitis, *Vet. Clin. Pathol.*, 31, 69, 2002.
20. Naggie, S. and Perfect, J.R., Molds: Hyalohyphomycosis, phaeohyphomycosis, and zygomycosis, *Clin. Chest. Med.*, 30, 337, 2009.
21. Nucci, M. and Anaissie, E., Emerging fungi, *Infect. Dis. Clin. North. Am.*, 20, 563, 2006.
22. Mosier, M.A. et al., Fungal endophthalmitis following intraocular lens implantation, *Am. J. Ophthalmol.*, 83, 1, 1977.
23. Pettit, T.H. et al., Fungal endophthalmitis following intraocular lens implantation: A surgical epidemic, *Arch. Ophthalmol.*, 98, 1025, 1980.
24. Diven, D.G. et al., Cutaneous hyalohyphomycosis caused by *Paecilomyces lilacinus* in a patient with lymphoma, *J. Am. Acad. Dermatol.*, 35, 779, 1996.
25. Hall, V.C. et al., Cutaneous hyalohyphomycosis caused by *Paecilomyces lilacinus*: Report of three cases and review of the literature, *Int. J. Dermatol.*, 43, 648, 2004.
26. Hilmarsdottir, I. et al., Cutaneous infection caused by *Paecilomyces lilacinus* in a renal transplant patient: Treatment with voriconazole, *Scand. J. Infect. Dis.*, 32, 331, 2000.
27. Gottlieb, T. and Atkins, B.L., Case report: Successful treatment of cutaneous *Paecilomyces lilacinus* infection with oral itraconazole in an immune competent host, *Mycoses*, 44, 513, 2001.
28. Takayasu, S., Akagi, M., and Shimizu, Y., Cutaneous mycosis caused by *Paecilomyces lilacinus*, *Arch. Dermatol.*, 113, 1687, 1977.
29. Gutierrez-Rodero, F. et al., Cutaneous hyalohyphomycosis caused by *Paecilomyces lilacinus* in an immunocompetent host successfully treated with itraconazole: Case report and review, *Eur. J. Clin. Microbiol. Infect. Dis.*, 18, 814, 1999.
30. Saberhagen, C. et al., Infection due to *Paecilomyces lilacinus*: A challenging clinical identification, *Clin. Infect. Dis.*, 25, 1411, 1997.
31. Rockhill, R.C. and Klein, M.D., *Paecilomyces lilacinus* as the cause of chronic maxillary sinusitis, *J. Clin. Microbiol.*, 11, 737, 1980.
32. Nayak, D.R. et al., *Paecilomyces* fungus infection of the paranasal sinuses, *Int. J. Pediatr. Otorhinolaryngol.*, 52, 183, 2000.
33. Muller, H., Cikirikcioglu, M., and Lerch, R., Subaortic aneurysm caused by *Paecilomyces lilacinus* endocarditis, *Arch. Cardiovasc. Dis.*, 101, 803, 2008.
34. Ono, N. et al., Lung abscess caused by *Paecilomyces lilacinus*, *Respiration*, 66, 85, 1999.
35. Fenech, F.F. and Mallia, C.P., Pleural effusion caused by *Penicillium lilacinum*, *Br. J. Dis. Chest*, 66, 284, 1972.
36. Carey, J. et al., *Paecilomyces lilacinus* vaginitis in an immunocompetent patient, *Emerg. Infect. Dis.*, 9, 1155, 2003.
37. Haldane, E.V. et al., Prosthetic valvular endocarditis due to the fungus *Paecilomyces*, *Can. Med. Assoc. J.*, 111, 963, 1974.
38. Silver, M.D., Tuffnell, P.G., and Bigelow, W.G., Endocarditis caused by *Paecilomyces varioti* affecting an aortic valve allograft, *J. Thorac. Cardiovasc. Surg.*, 61, 278, 1971.
39. Chan, T.H., Koehler, A., and Li, P.K., *Paecilomyces varioti* peritonitis in patients on continuous ambulatory peritoneal dialysis, *Am. J. Kidney Dis.*, 27, 138, 1996.
40. Crompton, C.H. et al., Peritonitis with *Paecilomyces* complicating peritoneal dialysis, *Pediatr. Infect. Dis. J.*, 10, 869, 1991.
41. Kovac, D. et al., Treatment of severe *Paecilomyces varioti* peritonitis in a patient on continuous ambulatory peritoneal dialysis, *Nephrol. Dial. Transplant.*, 13, 2943, 1998.

42. Marzec, A. et al., *Paecilomyces variotii* in peritoneal dialysate, *J. Clin. Microbiol.*, 31, 2392, 1993.
43. Byrd, R.P. Jr. et al., *Paecilomyces variotii* pneumonia in a patient with diabetes mellitus, *J. Diabetes Complications*, 6, 150, 1992.
44. Dharmasena, F.M., Davies, G.S., and Catovsky, D., *Paecilomyces variotii* pneumonia complicating hairy cell leukaemia, *Br. Med. J. (Clin. Res. Ed.)*, 290, 967, 1985.
45. Cohen-Abbo, A. and Edwards, K.M., Multifocal osteomyelitis caused by *Paecilomyces variotii* in a patient with chronic granulomatous disease, *Infection*, 23, 55, 1995.
46. Das, A. et al., *Paecilomyces variotii* in a pediatric patient with lung transplantation, *Pediatr. Transplant.*, 4, 328, 2000.
47. Fagerburg, R. et al., Cerebrospinal fluid shunt colonization and obstruction by *Paecilomyces variotii*: Case report, *J. Neurosurg.*, 54, 257, 1981.
48. Sherwood, J.A. and Dansky, A.S., *Paecilomyces* pyelonephritis complicating nephrolithiasis and review of *Paecilomyces* infections, *J. Urol.*, 130, 526, 1983.
49. Pagano, L. et al., The role of neutrophils in the development and outcome of zygomycosis in haematological patients, *Clin. Microbiol. Infect.*, 15(Suppl. 5), 33, 2009.
50. Ortoneda, M. et al., In vitro interactions of approved and novel drugs against *Paecilomyces* spp., *Antimicrob. Agents Chemother.*, 48, 2727, 2004.
51. Atkins, S.D. et al., The use of real-time PCR and species-specific primers for the identification and monitoring of *Paecilomyces lilacinus*, *FEMS Microbiol. Ecol.*, 51, 257, 2005.
52. Beezhold, D.H. et al., Prevalence of allergic sensitization to indoor fungi in West Virginia, *Allergy Asthma Proc.*, 29, 29, 2008.
53. Ujita, M. et al., Glucan-binding activity of silkworm 30-kDa apolipoprotein and its involvement in defense against fungal infection, *Biosci. Biotechnol. Biochem.*, 69, 1178, 2005.
54. Schmechel, D., Simpson, J.P., and Lewis, D.M., The production and characterization of monoclonal antibodies to the fungus *Aspergillus versicolor*, *Indoor Air*, 15(Suppl. 9), 11, 2005.
55. Yang, S. et al., Biochemical characterization of a novel thermostable beta-1,3-1,4-glucanase (lichenase) from *Paecilomyces thermophila*, *J. Agric. Food Chem.*, 56, 5345, 2008.
56. Obayashi, T. et al., Reappraisal of the serum (1->3)-beta-D-glucan assay for the diagnosis of invasive fungal infections—A study based on autopsy cases from 6 years, *Clin. Infect. Dis.*, 46, 1864, 2008.
57. de Pauw, B. et al., Revised definitions of invasive fungal disease from the European Organization for Research and Treatment of Cancer/Invasive Fungal Infections Cooperative Group and the National Institute of Allergy and Infectious Diseases Mycoses Study Group (EORTC/MSG) Consensus Group, *Clin. Infect. Dis.*, 46, 1813, 2008.
58. Koo, S. et al., Diagnostic performance of the (1->3)-beta-D-glucan assay for invasive fungal disease, *Clin. Infect. Dis.*, 49, 1650, 2009.
59. Pfeiffer, C.D., Fine, J.P., and Safdar, N., Diagnosis of invasive aspergillosis using a galactomannan assay: A meta-analysis, *Clin. Infect. Dis.*, 42, 1417, 2006.
60. Balajee, S.A. et al., Sequence-based identification of *Aspergillus*, *Fusarium*, and mucorales species in the clinical mycology laboratory: Where are we and where should we go from here? *J. Clin. Microbiol.*, 47, 877, 2009.
61. White, T.J. et al., Amplification and direct sequencing of fungal ribosomal RNA genes for phylogenetics, in: Innis, M. (Ed.), *PCR Protocols: A guide to Methods and Applications*, Academic Press, San Diego, CA, 1990, pp. 315–322.
62. Haugland, R.A. et al., Quantitative PCR analysis of selected *Aspergillus*, *Penicillium* and *Paecilomyces* species, *Syst. Appl. Microbiol.*, 27, 198, 2004.
63. Maresca, B. and Kobayashi, G.S., *Molecular Biology of Pathogenic Fungi: A Laboratory Manual*, Telos Press, New York, 1994.
64. Cuenca-Estrella, M. et al., Incidence of zygomycosis in transplant recipients. *Clin. Microbiol. Infect.*, 15, S37, 2009.

40 *Paracoccidioides*

Eduardo Bagagli, Sandra de Moraes Gimenes Bosco, Virgínia Bodelão Richini-Pereira, Raquel Cordeiro Theodoro, and Sílvio Alencar Marques

CONTENTS

40.1 INTRODUCTION

40.1.1 Paracoccidioidomycosis (PCM): PCM Infection and PCM Disease

Paracoccidioidomycosis is the most important and prevalent systemic mycosis in Latin America, mainly in Brazil, Colombia, and Venezuela. In these endemic areas, while PCM infection seems to occur at high frequencies, with no clinical symptoms, PCM disease occurs less frequently (0.1–3 cases per 10^5 individuals, approximately), affecting both immunocompromised and immunocompetent individuals [1]. In Brazil, the disease represents the eighth most common cause of death from chronic or recurrent types of infectious and parasitic diseases, with a mean annual mortality rate of 1.45 per million inhabitants [2].

Two distinct clinical forms of PCM disease have been observed: the acute–subacute form, which affects mainly young patients and evolves quickly as soon as the infection takes place, and the chronic form, which usually affects adults and has a long period of latency [3]. The acute–subacute form of the disease, also called juvenile, may be classified into moderate and severe, according to the dissemination level. It occurs in young adults under 30 years old of both sexes and represents around 5%–10% of the total casuistry of the disease. The main clinical manifestation is the involvement of the monophagocytic system showing fever, weight loss, and diffuse superficial and intraabdominal adenopathy with or without hepatosplenomegaly. Mucocutaneous involvement may be associated, however, in no more than 5% of the cases [3–6].

The chronic form of PCM disease corresponds to 90% of the cases and is often observed in men between 40 and 60 years old, whose main occupation is agricultural work [1,5]. These patients in general present some predisposing factors, such as smoking, alcoholism, and innutrition, and the main clinical manifestation is the involvement of the lungs. The pathogen can also present hematogenic dessemination to other sites such as brain [7,8], spinal cord [9], bone [10], bone marrow [11], genital tract [12,13], adrenal glands [14], oral cavity [15,16], and cutaneous tissues [17].

Autochthonous cases of PCM are observed exclusively in regions of Latin America; however, imported PCM cases have also been observed worldwide, mainly resulting from the increase in immigration of individuals from Central and South America to other regions [18–26]. In such imported cases, the diagnosis is often delayed because PCM disease frequently imitates other well-known diseases like tuberculosis or sarcoidosis [27]. It is important to call attention to the long period between the infection and the disease development in these imported cases, which in three well-documented ones were, respectively, 25, 38, and 50 years, causing many cases to remain undiagnosed without suspicion of PCM [22].

40.1.2 Mycological and Ecological Features of *Paracoccidioides brasiliensis*

Paracoccidioides brasiliensis is a thermally dimorphic fungus that grows as yeast in the host tissues or when cultured at 35°C–37°C and as mycelia under saprobic conditions or when cultured at room temperature, 18°C–23°C. The teleomorphic

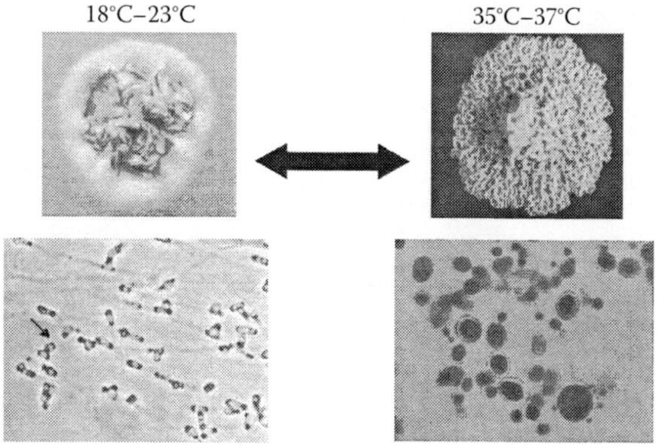

FIGURE 40.1 Thermal dimorphism of *P. brasiliensis* species. The arrow indicates the conidia produced during the mycelia phase.

or sexual phase is still unknown [28]. The exact places where the fungus occurs in nature have not been completely determined, while several pieces of evidence indicate that its saprobic form may occur in some restricted soil conditions, producing the infective propagula in the form of arthoconidia and aleuroconidia that cause primary infection in the lung by the airborne route [29,30] (Figure 40.1). Natural infection in several wild and domestic animals has been observed by intradermal, serological, and molecular tests [31–34]. The fungus has been repeatedly isolated from armadillos, in which scarce granulomas with fungal elements have been observed, suggesting that PCM disease may also occur in this wild mammals [35–39]. Naturally, PCM disease was also reported in dogs with generalized adenomegaly [40,41].

Morphological and molecular findings have confirmed that *P. brasiliensis* and the main pathogenic fungi that cause systemic mycosis are phylogenetically related, thus supporting its classification into Ajellomycetaceae, a new family of ascomycetes from the Onygenales order [42,43]. All members of this family, which also includes the species *Blastomyces dermatitidis*, *Histoplasma capsulatum*, *Emmonsia parva*, *E. crescens*, and *Lacazia loboi*, might have evolved in close association with vertebrate hosts [42,44,45]. The knowledge of the correct phylogenetic and taxonomic position of *P. brasiliensis* has contributed to a better comprehension on the fungus's lifestyle and/or strategies that the pathogen presents both to survive in nature and to interact with the hosts as well, including humans. For example, some clinical aspects of PCM disease, including its characteristic long latency period and its tendency to develop into a chronic form and to present relapses, might be greatly shaped by the long evolution of the pathogen in close association with vertebrate hosts, rather than into a free-living decomposer fungus [46,47].

Like all living organisms, *P. brasiliensis* presents a significant variability, which has been well documented by different mycological, antigenic, and virulence studies, as well as by molecular techniques [48–54]. Instead of being a unique species, *P. brasiliensis* appears to contain at least four different groups of genotypes (cryptic species), as observed by the phylogenetic studies through multi-locus sequence type (MLST) [55–57], microsatellite analysis [58], and *prp8* intein sequencing [59]. One of these four cryptic species, commonly denominated Pb01-like, differs substantially from the others (S1, PS2, and PS3), and the divergence time between it and the remaining ones was estimated as 30 mya. This phylogenetic species represented by the Pb01-like genotype has been described as a new taxon, *Paracoccidioides lutzii*, in honor of Adolpho Lutz, a physician who first recognized PCM and its agent in 1908 [57,60].

Although the occurrence of distinct genotypes in *P. brasiliensis* appears to interfere directly in the serological diagnosis, its implications for clinical and treatment responses have not yet been properly evaluated. For example, patients infected with Pb01-like genotype present negative serological results, or very low positivity, by the immunodiffusion test when its serum is challenged against antigens produced by the other genotypes of *P. brasiliensis* [61,62]. The discovery of cryptic species in *P. brasiliensis* has increased the necessity of comparative studies aiming to detect relevant phenotypic differences and molecular markers among the species in order to provide the correct diagnosis, treatment, and prophylactic measures.

40.1.3 *P. brasiliensis* Detection: Clinical and Environmental Sources

PCM diagnosis is usually made by microscopic detection of multi-budding yeast cells in direct mycological examination and/or fungal cultures from clinical specimens, histopathology of biopsied tissue, and also by serological tests, mainly with gp43, the immune dominant antigen of *P. brasiliensis* [28].

Although a definitive diagnosis can be made through fungal culture, this procedure involves high contamination risk, long incubation period, and low sensitivity [63,64]. In histological sections, the etiological agent might be undetected in some pauci-cellular samples or confused with other dimorphic fungi [65]. Several serological techniques using different *P. brasiliensis* antigen types have been employed for both diagnosis and monitoring therapy of patients; however, some problems such as anergy, poor production of antibodies, cross reactivity, and the sensitivity and specificity of the technique must be considered [66]. Besides that, patients infected with different *P. brasiliensis* genotypes can respond differently to serological assays, as already mentioned in this chapter.

Molecular approaches for detecting *P. brasiliensis*, both in several clinical specimens and in environmental sources, have been increasingly used, exploring different genomic regions as targets to develop primers and/or probes. Table 40.1 summarizes the primer/probe sequences and their applications for fungal detection.

Goldani et al. [67] designed specific primers, whose target is based on a specific sequence of β-actin gene from *P. brasiliensis*, and, therefore, they are not detected in other fungi such as *H. capsulatum*, *B. dermatitidis*, *Cryptococcus neoformans*, *Candida albicans*, *Aspergillus fumigatus*, *Saccharomyces cerevisiae*, or *Pneumocystis carinii*. Later, Goldani and Sugar

TABLE 40.1
Primers and Probes for the Molecular Detection of *P. brasiliensis*

Primers/Probes	Gene	Primer Sequences (5′–3′)	Amplicon Size (bp)	Evaluated Samples	References
Primer 1	β-actin	TCGTTATCCTCATCGAA	62	Fungal culture, serum (inoculated mice)	[67,68]
Primer 2		AAGAGTCTTCCCTCGC			
PC1	gp43	TCATCTCACGTCGCATCTCACATT	1030	Clinical (sputum)	[69]
PC5		AGCGCCAGATGGTTTGCCCGCTAGGAACGAA			
PC2		ATAGAGGGAGAGCCATATGTACAAGGT	600		
PC6		GGCTCCTCAAAGTCTGCCATGAGGAAG			
para I	gp43	AAC TAG AAT ATC TCA CTC CCA GTC C	355	Fungal culture, tissue (mice inoculated and paraffin embedded)	[65,70]
para II		TGTAGACGTTCTTGTATGTCTTGGG			
para III		GATCGCCATCCATACTCTCGCAATC	196		
para IV		GGGCAGAGAAGCATCCGAAATTGCG			
LO	p27	CAACTCTTGGCTTTGGTTGAAG	536	Tissue (armadillos, inoculated mice), soil artificially contaminated	[71,72]
UP		CTGTTGTTTCCGTCCTTGCGC			
MG2(1)F	Genomic DNA	GGGATTCCCTAGGCAAACACTTGTGTGA	285	Fungal culture, clinical (sputum, CSF)	[73]
MG2(1)R		GTGCAGTTATCCACAAGCCATATATTC			
MG2(2)F		GGAGATGATCTGACGTTAGTACGTGATG	288		
MG2(2)R		ATGCTAATTTATGTCATTCCGCGTCTG			
OL3	rRNA	CTCAGCGGGCACTT	203	Fungal culture	[74]
UNI-R		GGTCCGTGTTTCAAGACG			
PbITS1s[a]	rRNA	CCGCCGGGGACACCGTTG	418	Fungal culture	[75]
PbITS3a[a]		AAGGGTGTCGATCGAGAG			
Pb-ITS-E[a]	rRNA	GAGCTTTGACGTCTGAGACC	387	Fungal culture, tissue (armadillos and road-killed wild animals), soil	[33,76,77]
Pb-ITS-R[a]		AAGGGTGTCGATCGAGAGAG			
OliPbMB1	rRNA	ACCCTTGTCTATTCTACC	144	Fungal culture, clinical (blood, sputum, and skin biopsy)	[78]
OliPbMB2		TTACTGATTATGATAGGTCTC			
PbMB1[b]	rRNA	cgcgatCGCCGGGGACACCGTTGAAatcgcg	—		

Source: Adapted from Richini-Pereira, V.B. et al., *Mem. Inst. Oswaldo Cruz*, 104, 636, 2009.

[a] Inner primers used in nested PCR, after the first PCR with panfungal sets of primers such as ITS1-ITS4 or ITS4-ITS5 [79].

[b] Probe used in the real-time PCR to detect the OliPbMB1/2 amplicon.

[68] used the same set of primers and obtained positivity in sera from five experimentally infected mice.

Polymerase chain reaction (PCR) assays targeting *gp43* gene are widely used for molecular detection of *P. brasiliensis* DNA in several samples. Cisalpino et al. [80] cloned and characterized the entire coding region of the *gp43* gene, which is constituted by two exons interrupted by a 78 bp-intron. After this study, Gomes et al. [69] combined five primer pairs and suggested a PCR using the PC2–PC6 primer pair for direct amplification from clinical material (sputum) or nested PCR following the PC1–PC5 primer set amplification. In an experimental infection with *H. capsulatum* and *P. brasiliensis* yeasts cells, Bialek et al. [65] employed a nested PCR, also based on *gp43* gene, with high sensitivity and specificity for the sets of primers para I/II and para III/IV. Using these same sets of primers, Ricci et al. [70] demonstrated 30% positivity in pathogen detection in biopsies from PCM patients. Those authors emphasized that this low positivity might be due to the procedures used for fixation, paraffin embedding, and storage of the material, which favored DNA degradation, and the use of primers from gp43 regions that have been proven to be polymorphic.

Since *gp43* was shown to be one of the most polymorphic genes that clearly separates the four cryptic species of *P. brasiliensis* [55–57], it is now indicated for diagnostic purposes and genotyping, by PCR, of clinical samples. By using this strategy, Ricci et al. [81] detected the pathogen and differentiated it between S1 and PS2 genotypes from paraffin-embedded tissue.

As new genomic regions of the fungus became known, additional primers for molecular detection have been described. For example, the gene that codes for a 27 kDa antigen protein of *P. brasiliensis* was also cloned and sequenced [82], permitting the design of specific primers (LO and UP) that were used in ecological studies in tissue samples of armadillos, experimental infection, and contaminated soil [71,72]. After observing the common presence of a 0.72 kb fragment, generated by random amplification of polymorphic DNA (RAPD) with the arbitrary primer OPG18 (Operon Biotechnology), in a large number of different *P. brasiliensis* isolates, from distinct geographic origins [49], San-Blas et al. designed specific primers (MG2(1)F/MG2(1)R and MG2(2)F/MG2(2)R) based on this region, which were specific and highly sensitive for *P. brasiliensis*. These primers were capable of identifying the

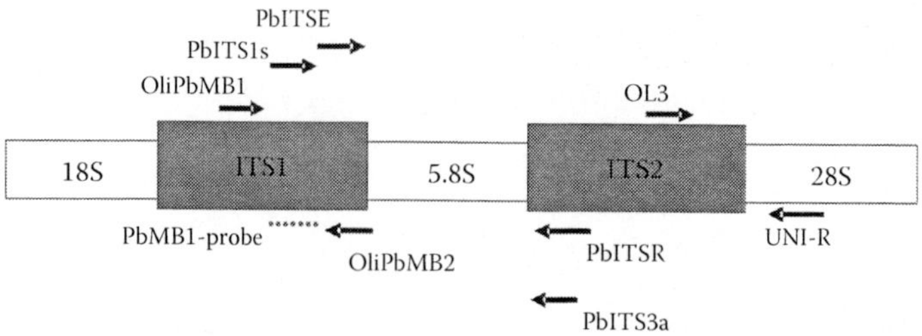

Primer/Probe	Sequence Similarity to *P. brasiliensis* Cryptic Species			
	S1	PS2	PS3	Pb01-Like
UNI-R	+	+	+	+
OL3	–	–	–	+
PbITS1s	+	+	+	+
PbITS3a	+	+	+	–
Pb-ITS-E	+	+	+	+
Pb-ITS-R	+	+	+	–
OliPbMB1	+	+	+	–
OliPbMB2	+	+	+	–
PbMB1	+	+	+	+

Note: +: Complete similarity; –: Incomplete similarity.

FIGURE 40.2 Scheme of primers and/or probe annealing in rRNA region and their similarity to the four species of *P. brasiliensis* complex.

pathogen in sputum and in cerebrospinal fluid samples from patients with different clinical forms of the disease [73].

The rRNA region, compounded by 18S, internal transcribed spacer (ITS) 1, 5.8S, ITS2, and 28S, has been widely employed for molecular detection and distinction of different fungal species, including *P. brasiliensis* (Figure 40.2). The ribosomal subunit regions (28S, 5.8S, and 18S) are known to accumulate mutations at a lower rate than the ITS regions over time. The first regions are used to establish phylogenetic relationships among taxonomic levels above genus, and the second ones are useful for the separation between genera and species, being considered the main candidate for the barcoding system of fungal identification [83]. Furthermore, this target provides higher sensitivity in PCR due to its several copies per genome (more than 100 copies) [79,84].

Using panfungal oligonucleotides, as outer primers, which anneal in 18S and 28S [79], Imai et al. [75] as well as Theodoro et al. [76] designed species-specific inner primers derived from ITS-5.8S rDNA (PbITS1s/PbITS3a and PbITSE/PbITSR, respectively) for *P. brasiliensis* detection in nested PCR. This approach proved to be useful for fungal detection in soil [76,77], road-killed wild animals [33], and also in clinical samples, such as skin biopsies and cerebrospinal fluid (data not published). Motoyama et al. [74] proposed the OL3 and UNI-R primers for PCR, also from the rRNA region (ITS2 and 28S), and obtained a 203 bp fragment when *P. brasiliensis* (Pb01 genotype) DNA was used as the template. These primers were capable of discriminating between Pb01-like isolates and *H. capsulatum*. Based on the same region, a real-time PCR, using a beacon probe, was successfully applied for *P. brasiliensis* detection in clinical samples. This approach is considered even more sensitive than nested PCR and since it is based on just one-step PCR, it minimizes the contamination risks [78].

Since *P. brasiliensis* actually consists of a species complex with distinct genetic groups, and given that new still undiscovered genotypes will probably be detected in the near future, molecular protocols for *P. brasiliensis* detection should be routinely reevaluated, both *in silico* and experimentally, in order to confirm their homology to the target sequences and their potential to discriminate among the several genotypes that occur in the endemic areas. For example, most of the rRNA primers designed until now that are supposed to be species specific for *P. brasiliensis* do not have complete similarity to all *P. brasiliensis* species (Figure 40.2).

40.2 METHODS

40.2.1 Sample Preparation

Despite the fact that *P. brasiliensis* presents hematogenic dissemination to other tissues, blood and serum from patients with chronic PCM do not represent the samples of choice for molecular diagnosis, neither for PCR nor real-time PCR, even when carried out with primers derived from the rRNA, a step that substantially increases the detection sensitivity [73,78].

The main clinical samples employed for molecular detection of *P. brasiliensis* are sputum, skin biopsies, ganglionary secretion, cerebrospinal fluid, and paraffin-embedded tissue. Special attention is given to the collection of skin biopsies, in which the tissue fragment must be collected by puncturing the most infiltrated skin lesions or on the border of ulcerated lesions. Since *P. brasiliensis* is a fungal pathogen of biosafety level 2, it is recommended that the procedures, which involve handling the fungus and clinical specimens, should be performed within a class II safety cabinet.

40.2.1.1 DNA Extraction

40.2.1.1.1 Fungal culture

1. Transfer yeast cells or fragment mycelia of *P. brasiliensis* to a 2.0 mL microcentrifuge tube (Safe-Lock type) containing 500 mg of acid-washed glass beads (0.4–0.6 mm diameter, Sigma, St. Louis, MO) and 500 µL of 125 mM EDTA/1 M sorbitol solution.
2. Homogenate in a Precellys (Precellys 24, Bertin Technologies) in three cycles of 40 s at 20 s intervals. Alternatively, vortex at the highest setting for 5 min, in a common laboratory shaker. Precautions: Tubes must be appropriate to avoid disruption.
3. Centrifuge at 13,000 rpm for 10 min at room temperature (around 22°C) and remove the supernatant.
4. Add 500 µL of extraction buffer (100 mM Tris–HCl, 20 mM EDTA, and 1% SDS) and 20 µL of proteinase K (20 mg/mL) to the pellet of cells and beads. Incubate overnight in a water bath at 54°C.
5. Add 500 µL of sodium acetate 3 M and keep the microtube on ice for 1 h.
6. Centrifuge at 13,000 rpm for 10 min at 4°C. Transfer the supernatant to a new microtube containing 1000 µL of cold absolute ethanol and mix gently by inversion.
7. Centrifuge at 13,000 rpm for 10 min at 4°C and discard the supernatant. Wash the pellet twice with 500 µL of cold ethanol 70%. Centrifuge as mentioned above.
8. Dry the pellet and then resuspend with 100 µL of ultrapure water, homogenizing carefully.
9. The purity and concentration of DNA can be determined in a drop-and-measure spectrophotometer or by 2% agarose gel electrophoresis, using a known DNA marker as standard (Low Mass Ladder, Invitrogen™).
10. Store at −20°C.

40.2.1.1.2 Tissue samples (Procedure I)

1. Grind 100 mg of frozen tissue with liquid nitrogen, using mortar and pestle, until the tissue becomes powdered.
2. Transfer the powdered tissue to a 1.5 mL microtube and add 600 µL of extraction buffer (100 mM Tris–HCl, pH 7.2, 20 mM EDTA, 1% SDS, and 0.02% 2-mercaptoethanol) and 20 µL of proteinase K (20 mg/mL).
3. Incubate in a water bath at 65°C for 2 h or at 56°C overnight.
4. Add 500 µL of phenol/chloroform/isoamyl alcohol (PHE/CHL/IAA—25:24:1) and gently homogenize the microtubes).
5. Centrifuge at 14,000 rpm for 15 min at room temperature (around 22°C).
6. The aqueous phase must be re-extracted with PHE/CHL/IAA (25:24:1) twice and with CHL/IAA (24:1) once. After each procedure, centrifuge as already described.
7. Add 1000 µL of cold isopropanol and 10 µL of sodium acetate 3M and incubate for 20–60 min at −20°C or overnight at 4°C.
8. Centrifuge at 14,000 rpm for 15 min at 4°C and discard the alcohol.
9. Wash the pellet twice with 500 µL of cold 70% alcohol.
10. Dry the pellet and then resuspend with 50 µL of ultrapure water.
11. The purity and concentration of DNA can be determined in a drop-and-measure spectrophotometer or by 2% agarose gel electrophoresis, using a known DNA marker as standard (Low Mass Ladder, Invitrogen).
12. Store at −20°C.

40.2.1.1.3 Tissue samples (Procedure II—without phenol)

1. Grind 100 mg of frozen tissue with liquid nitrogen, using mortar and pestle, until the tissue becomes powdered.
2. Transfer the powdered tissue to a 1.5 mL microtube and add 600 µL of extraction buffer (100 mM Tris–HCl, pH 7.2, 20 mM EDTA, 1% SDS, and 0.02% 2-mercaptoethanol) and 20 µL of proteinase K (20 mg/mL).
3. Incubate in water bath at 65°C for 2 h or at 56°C overnight.
4. Add 100 µL of 5 M sodium chloride solution.
5. Preheat the solution of cetyltrimethylammonium bromide (CTAB) and sodium chloride (12.5% CTAB and 5.125% NaCl) at 65°C for 10 min.
6. Add 100 µL of preheated CTAB/NaCl solution and gently homogenize the microtubes.
7. Add 750 µL of chloroform/isoamyl alcohol (24:1) and vortex for 10 s.
8. Centrifuge at 14,000 rpm for 5 min at room temperature.
9. Transfer the supernatant to a fresh tube, taking care to avoid contact with the sediment.
10. Add 450 µL of absolute ethanol or isopropanol kept at −20°C.
11. Incubate at −20°C for 10 min and then centrifuge at 14,000 rpm for 20 min at 4°C.
12. Discard the supernatant and wash the pellet twice with 500 µL of 70% ethanol.
13. Dry the pellet and resuspend with 50 µL of ultrapure water.
14. The purity and concentration of DNA can be determined in a drop-and-measure spectrophotometer or by 2% agarose gel electrophoresis, using a known DNA marker as standard (Low Mass Ladder, Invitrogen).
15. Store at −20°C.

40.2.1.1.4 Paraffin-embedded tissues

DNA extraction from paraffin-embedded tissue may also be performed by using commercial kits and with an in-house protocol.

1. Cut two or three sections of 5 μm thickness (or one of 20 μm thickness) and place into a 1.5 mL microtube. The sections must be cut in triplicate and with different blades.
2. Add 1000 μL of xylol and submit the microtube to a continuous vortexing for 30 min at room temperature (around 22°C).
3. Centrifuge at 13,000 rpm for 5 min at room temperature.
4. Remove the xylol while taking care to avoid contact with the sediment.
5. Repeat steps 2–4 twice.
6. Add 500 μL of 100% ethanol and mix by inverting the microtubes for 5 min at room temperature.
7. Centrifuge at 13,000 rpm for 3 min at room temperature. Remove the ethanol.
8. Repeat steps 6 and 7 once.
9. Dry the samples to evaporate the ethanol.
10. Proceed to the DNA extraction for tissue as described above (protocol I or II) or follow the instructions for an appropriate commercial kit.

40.2.1.2 PCR Protocols

The majority of the several PCR protocols proposed to detect *P. brasiliensis* are based on the amplification of the genes *gp43* and ribosomal (Table 40.1). While the *gp43* gene is restricted to the *P. brasiliensis* complex, it is highly polymorphic and occurs as a single copy or small number of copies per genome. However, the ribosomal genes occur as multicopies per genome and with conserved and moderately polymorphic regions that permit the design of primers, both generic and species specific.

40.2.1.2.1 Nested PCR with Ribosomal Primers

This protocol has proved to be highly sensitive for clinical and environmental sampling and for detection of the three principal genotypes of *P. brasiliensis* (S1, PS2, and PS3). However, this protocol might not work as expected for the Pb01-like genotype, which is genetically very distant from the others and has been proposed as a new species (*P. lutzii*). Therefore, a new combination of primers should be designed and properly evaluated.

Protocol step for first amplification:

1. The PCR reaction is carried out in 25 μL of reaction mixture (1 × PCR buffer) (50 mM KCl and 10 mM Tris–HCl, pH 8.0, 1.5 mM $MgCl_2$, 0.2 mM dNTP, and 1 unit of *Taq* polymerase—GE Healthcare), 10 ng of genomic DNA, and 10 pM of each primer (ITS1 and ITS4 or ITS4 and ITS5; see Table 40.1).
2. Include a positive control (DNA of the pathogen) and a negative control (ultrapure water as template).
3. Cycling conditions: initial denaturation at 94°C for 5 min, 25 cycles at 94°C for 1 min, 60°C for 2 min, and 72°C for 2 min, and a final extension at 72°C for 7 min.
4. The size of the ITS-5.8S amplicon of *P. brasiliensis* is around 634 bp and must be visualized in 1.5% agarose gel.

Protocol step for second amplification:

1. The PCR reaction is carried out in 25 μL of reaction mixture (1 × PCR buffer) (50 mM KCl and 10 mM Tris–HCl, pH 8.0, 1.5 mM $MgCl_2$, 0.2 mM dNTP, and 1 unit of *Taq* polymerase—GE Healthcare) and 10 pM of each primer (PbITSE and PbITSR; see Table 40.1).
2. Add 2.0 μL of amplified product of PCR reaction (first amplification) as DNA template. Caution: this step must be carried out in a different room from the one in which the reaction mixture is prepared to avoid contamination.
3. Include positive and negative controls from the first amplification and also include a new negative control for the nested-PCR mixture reaction (ultrapure water as template).
4. Cycling conditions: initial denaturation at 94°C for 5 min, 25 cycles at 94°C for 1 min, 62°C for 2 min, and 72°C for 2 min, and a final extension at 72°C for 10 min.
5. Amplicon of nested PCR presents a size of 387 bp and must be visualized in 1.5% agarose gel.

40.2.1.3 Molecular Strategies to Differentiate Cryptic Species of *P. brasiliensis* Complex

40.2.1.3.1 Using PRP8 Intein

The discovery of cryptic speciation in *P. brasiliensis* has highlighted the need to discriminate among these species when diagnosing PCM. To that end, Matute et al. [55] and Teixeira et al. [57] accomplished multilocus sequencing typing (MLST) with eight nuclear coding regions. The primers, PCR, and phylogenetic analysis conditions for these experiments are detailed in their articles. However, sequencing many regions for each new isolate is too much laborious for a lab that performs routine clinical diagnosis. The challenge now is to find molecular markers that easily differentiate among the four species from *P. brasiliensis* complex. For that, Theodoro et al. [59] propose to sequence the PRP8 intein, a parasitic genetic element in the *prp8* gene. An intein is transcribed and translated together with the host gene and then it performs an autocatalytic protein splicing. It has a splicing domain in its N- and C-terminus ends responsible for the protein splicing and also a homing endonuclease (in its core) that accounts for its mobility.

Protocol steps:

1. The PCR for the entire PRP8 intein is carried out in 25 µL of reaction mixture containing 20 ng of genomic DNA, 50 mM KCl, 10 mM Tris–HCl, 1.5 mM $MgCl_2$, 0.2 mM dNTP, and 0.4 µM of each primer: RCTfwd2 (5′ GCCTATCGAACCGATGTTATCC 3′) and RCTrev2 (5′ TGAACCTGGAAGCCAACGTAG 3′) and 1 unit of *Taq* polymerase (GE Healthcare) in a thermalcycler. The amplified fragment is ~1.9 kb.
2. The thermal cycling conditions are 94°C for 4 min followed by 40 cycles at 94°C for 1 min, 55°C for 1 min, 72°C for 2 min and a final cycle at 72°C for 5 min.
3. The same procedure is carried out with the primers FwdIn3 (5′ CATCGCTGACCATGCTGC 3′) and RevIn2 (5′ TGAAGCCGAACGTGTTTTCTG 3′) for core amplification of the PRP8 intein (~860 kb).
4. The PCR products are identified by 1.5% agarose gel electrophoresis stained with ethidium bromide.
5. The PCR amplicons are purified by the commercial kit GFX PCR DNA and Gel Band (GE Healthcare), whereas the sequencing reactions are carried out on both strands, using the DYEnamic ET Dye Terminator Kit, code US81090 (GE Healthcare). The MegaBACE 1000 DNA Analysis System is used for capillary electrophoresis, and the chromatogram is visualized by the Chromas program.
6. The sequences are sent to blastn for comparison with the NCBI database (http://www.ncbi.nlm.nih.gov/BLAST) and to blastn of the Broad Institute Web site for comparison with the sequences from the isolates Pb18 (S1 species), Pb01 (*P. lutzii*), and Pb03 (PS2 species), available in the *P. brasiliensis* genome database (http://www.broad.mit.edu/annotation/genome/paracoccidioides_brasiliensis). Table 40.2 [59] can be useful for species identification because it presents all the polymorphic positions in the intein PRP8 for the four cryptic species of *P. brasiliensis* complex.

40.2.1.3.2 Using HSP70

Teixeira et al. [57] revealed, by MLST, that the species *P. lutzii* is highly divergent when compared to the remaining species. Furthermore, subsequent works have shown that patients infected with this genotype present negative serological results, or very low positivity, by the immunodiffusion test when its serum is challenged with antigens produced by the other genotypes of *P. brasiliensis* [61,62]. In this sense, it is very important to distinguish between *P. lutzii* and the other species. To that end, two nuclear coding regions have been used: ITS1-5.8S-ITS2 and HSP70. The first one clearly differentiates *P. lutzii* from the remaining species by PCR with ITS4 and ITS5 primers, followed by sequencing and blast comparisons (protocol described in Section 40.2.1.2), while the second relies on only one PCR step with specific primers for *P. lutzii* isolates [57].

Protocol steps:

1. The PCR reaction is carried out in 25 µL of reaction mixture (10 ng of genomic DNA, 1×PCR buffer [50 mM KCl and 10 mM Tris–HCl], 1.5 mM $MgCl_2$, 0.2 mM dNTP, 0.4 µM of each primer, and 1 unit of *Taq* polymerase—Amersham Biosciences). The primers used in this reaction, HSPMMT1 (forward: 5′ AACCAACCCCCTCTGTCTTG 3′) and PLMMT1 (reverse: 5′ GAAATGGGTGGCAGTATGGG 3′) [57], anneal to an indel region exclusive to the *P. lutzii* species and amplify ~500 pb.
2. The reactions are performed in a thermocycler at 94°C for 4 min followed by 35 cycles at 94°C for 1 min, 55°C for 1 min, and 72°C for 2 min and a final cycle at 72°C for 5 min. The PCR products are identified by 0.8% agarose gel electrophoresis stained with ethidium bromide.

40.2.1.3.3 Using SNP

Once the MLST studies have revealed the main polymorphisms that are conserved among isolates from the different cryptic species, we can use this information to detect single nucleotide polymorphisms (SNP) that can be useful for species identification. Among the many options available for the study of SNP, we have been testing the ability of the SNaPshot® Multiplex, from Applied Biosystems, to differentiate isolates from the four cryptic species. This is a very simple technique in which we use a PCR product to perform a mini-sequencing reaction with a primer that is designed right in the upstream of the SNP, so that the first nucleotide to be incorporated by the sequencing reaction will be the one corresponding to the SNP because all the dNTPs used in this reaction are dideoxynucleotides labeled with fluorescence (different for each nucleotide). Basically, we have been using three SNP for species identification in *P. brasiliensis* complex, one from *gp43*, another from *chs2*, and the other from *arf*. The primers used for the SNaPshot reaction and SNP positions for each species are listed in Table 40.3.

Protocol steps:

1. *PCR of the genes gp43, chs2, and arf*: The primers Gp43-E2-sense2 (5′ CTAGAATATCT CACTCCCAG 3′) (not published yet) and Gp43-E2-antisense (5′ GCCCCCTCCGTCTTCCATGTCC 3′) [55] anneal in exon 2 of *gp43* gene and amplify ~700 pb. The primers CHS2 E2-4 Fwd (5′ CTTAACGGTGCCTTCTTTGCGG 3′) and CHS2 E2-4 Rev (5′ GTGAAAGTATTGTTGCCCAGCG 3′) [55] are used for *chs2* amplification of a 549 pb fragment, while the primers Arf–fwd (5′ TCTCATG GTTGGCCTCGATGCTGCC 3′) and Arf–rev

TABLE 40.2
Polymorphism in the PRP8 Intein of *P. brasiliensis* Isolates Belonging to the Four Cryptic Species, according to Theodoro et al.

Isolate or Group[a]	Block/Site																					
	A51	A78	A90	A99	A135	A159	B183	B219	B245	B246	B247	B248	B249	B250	B251	B252	B253	B254	B257	B266	B267	B279	B304
Con	A	G	C	C	T	G	T	G	C	C	C	G	T	C	C	A	T	T	C	T	A	T	T
T1F1-like	.	.	.	.	.	.	.	.	.	.	.	.	.	.	.	.	.	.	.	.	.	.	.
T10B1-like	.	.	.	.	.	.	.	.	.	.	.	.	.	.	.	.	.	.	.	.	.	.	.
Bt84-like	.	.	.	.	.	.	.	.	.	.	.	.	.	.	.	.	A	.	.	.	.	.	.
Pb262	.	.	.	.	.	.	.	.	.	.	.	.	.	.	.	.	A	.	.	.	.	.	.
Epm77-like	.	.	.	.	.	.	.	.	.	.	.	.	.	.	.	.	.	.	.	.	.	.	.
Pb01	G	T	T	G	C	C	C	A	—	—	—	—	—	—	—	—	—	A	T	C	G	C	C
	C766	C776	C825	C837	C859	C864	C879	C887	C902	C925	C937	C955	C963	C966	C970	C975	C980	C999	C1077	C1146	C1259	D1297	D1309
Con	C	A	A	G	G	G	G	T	T	A	T	A	A	C	A	G	C	C	A	T	A	A	A
T1F1-like	.	.	.	.	.	.	.	.	.	.	.	.	.	.	.	.	.	.	.	.	.	.	.
T10B1-like	.	G	.	A	.	A	.	.	.	.	.	.	.	.	.	.	.	.	.	.	.	.	C
Bt84-like	.	G	.	A	.	A	.	.	.	.	.	.	.	.	.	.	.	.	.	.	.	.	C
Pb262	.	G	.	A	.	A	.	.	.	.	.	.	.	.	.	.	.	.	.	.	.	C	C
Epm77-like	.	.	.	.	.	.	.	.	.	.	.	.	.	.	.	.	.	.	.	.	.	.	C
Pb01	T	.	T	.	T	A	C	C	A	T	C	G	C	T	C	A	T	T	T	C	G	.	C

Source: Theodoro, R.C. et al., *Fung. Genet. Biol.*, 45, 1284, 2008.
Blocks A, B, and F correspond to the splicing domain and the blocks C, D, E, and H to the homing endonuclease domain. Con, consensus sequence.
[a] T1F1-like (S1 genotype) T10B1-like; Bt84-like and Pb262 (PS2 genotype); EPM77-like (PS3 genotype) and Pb01 (*P. lutzii* genotype).

B310	B326	B337	B357	B405	B433	B441	C471	C484	C501	C577	C583	C625	C630	C638	C653	C672	C699	C713	C731	C738	C740
A	A	A	T	T	C	C	G	C	T	T	T	T	T	A	C	C	G	C	A	T	T
.	.	.	.	.	.	.	.	.	.	.	.	.	.	.	.	.	.	.	.	.	.
.	.	.	.	.	.	T	.	.	.	.	.	.	.	.	.	.	.	T	.	.	.
.	.	.	.	.	.	T	.	.	.	.	.	.	.	.	.	.	.	T	.	.	.
.	.	.	.	.	.	T	.	.	.	.	.	.	.	.	.	.	.	T	.	.	.
.	.	.	.	.	.	.	.	.	.	.	.	.	.	.	.	.	.	.	.	.	.
G	T	G	G	A	G	.	A	G	A	C	G	C	G	G	T	T	A	.	C	A	C

D1312	E1338	E1351	E1371	E1378	E1413	H1443	H1470	H1486	H1491	H1530	H1548	H1561	H1606	H1609	H1632	H1645	F1692
G	A	G	A	A	A	G	A	C	A	G	C	C	C	A	G	G	A
.	.	.	.	.	.	.	.	.	.	.	.	.	.	.	.	.	.
.	.	.	.	.	.	A	.	.	.	.	.	.	.	.	.	.	.
.	.	.	.	.	.	A	.	.	.	.	.	.	.	.	.	.	.
.	.	.	.	.	.	A	.	.	.	.	.	.	.	.	.	.	.
.	.	.	.	.	.	.	.	.	.	.	.	.	.	.	.	.	.
C	G	A	G	G	C	.	G	G	T	T	A	A	G	G	A	A	G

TABLE 40.3
SNP Primers Used for Mini-Sequencing Reactions to Differentiate the Four Species of *P. brasiliensis* Complex

Primer	Sequence 5′–3′[a]	Gene	SNP Position[b]	SNPs/Species			
				S1	PS2	PS3	*P. lutzii*
SNP3	4(C) GTATCTAAATGGGCGTGGCA	GP43-exon 2	344	T	T	T	G
SNP9	21(C) CATCCAATTTACCGTGTGGGA	ARF	298	C	T	C	C
SNP13	32(C) CTCCAGCAAAATACTCCCAGAT	CHS2 (exon 2–4)	185	C	C	T	NA

[a] The cytosines in the 5′ end of the primers were added in order to distinguish the different mini-sequencing products according to their size, in case the three reactions are performed together, in the same tube.
[b] According to the sequences deposited by Matute et al. [55] in GenBank. NA means "no annealing."

(3′GAGCCTCGACGACACGGTCACGATC 5′) [55] are used for *arf* amplification of a 407pb fragment.

PCR reactions are carried out in 25µL of reaction mixture (10ng of genomic DNA, 1×PCR buffer [50mM KCl and 10mM Tris–HCl], 1.5mM $MgCl_2$, 0.2mM dNTP, 0.4µM of each primer, and 1 unit of *Taq* polymerase—Amersham Biosciences)

The reactions are performed in a thermocycler at 94°C for 4min followed by 40 cycles of 94°C for 1min (annealing temperature 55°C for *gp43* and 60.4°C for *chs2* and *arf*) and 72°C for 1min and a final cycle of 72°C for 5min. The PCR products are identified by 1.5% agarose gel electrophoresis stained with ethidium bromide.

2. *Purification of PCR products*: 10µL of each PCR product is purified with 4µL of ExoSAP-IT (GE Healthcare) for 1h at 37°C, followed by 20min at 80°C to inactivate enzymes.
3. *Mini-sequencing reaction*: 6µL of ultrapure water, 1µL of 1/2TERM Buffer (200mM Tris–HCl, 5mM $MgCl_2$, and pH9.0), 1µL of SNaPmix (GE Healthcare), 1µL of the 1pmol/µL SNP primer (listed in Table 40.3), and 1µL of the purified PCR product were used for each mini-sequencing reaction. The dideoxynucleotides, which are left from the mini-sequencing reaction, are removed by treating the samples with 1µl of SAP (Shrimp Alcaline Phosphatase, GE Healthcare) for 1h at 37°C followed by 20min at 80°C.
4. *Electrophoresis*: The reactions are diluted to 1:1 with loading buffer containing Hi-Di™ formamide (Applied Biosystems). Before loading the samples on the gel, they are heated at 95°C for 3min and quickly chilled on ice. Then, 1µl of each sample is loaded on the 12% polyacrylamide gel using the ABI PRISM® 377 (Applied Biosystems, United States) automatic sequencer. After running the gel, the bands will have the same size of the SNP primer, and their fluorescent color will correspond to the detected SNP. See SNaPshot Multiplex Biosystems protocol for more information.

40.3 CONCLUSIONS AND FUTURE PERSPECTIVES

The dimorphic pathogenic fungus *Paracoccidioides brasiliensis* occurs in large areas of Latin America and provokes PCM, a life-threatening systemic infection that still presents serious challenges for its treatment and control. The exact place where the fungus occurs in nature, producing the infective propagula, is not completely known, which hinders the adoption of preventive measures to avoid new infections. Molecular techniques have greatly improved our capacity to address important aspects of the pathogen and the disease. Besides providing a better comprehension of the fungus taxonomy, phylogeny, and ecology, the molecular approaches have also substantially expanded the disease diagnosis. This chapter presents some recent findings on *P. brasiliensis* molecular taxonomy and the state of the art in molecular detection of the pathogen, in both environmental and clinical materials.

The discovery of cryptic speciation in *Paracoccidoides* genus has added new challenges for PCM researchers in relation to the clinical aspects, treatment, and, especially, the correct diagnosis. The structural genomes of the main genetic groups or cryptic species of *P. brasiliensis* have been established, and several molecular identification protocols have already been proposed, with promising results. Additional studies that focus on population are still necessary in order to provide a clearer picture of the pathogen biodiversity. Therefore, the use of rapid, contamination-free, and specific methodologies such as real-time PCR together with primers and/or probes specific for all species from the *Paracoccidioides* genus should be implemented as soon as possible for routine diagnosis of PCM.

REFERENCES

1. Wanke, B. and Londero, A.T., Epidemiology and paracoccidioidomycosis infection, In *Paracoccidioidomycosis*, Franco, M. et al. Eds., Boca Raton, FL: CRC Press, 1994, pp. 109–120.
2. Coutinho, Z.F. et al., Paracoccidioidomycosis mortality in Brazil (1980–1995), *Cad. Saude Publ.*, 18, 1441, 2002.
3. Franco, M., Montenegro, M.R., and Mendes, R.P., Paracoccidioidomycosis: A recently proposed classification of its clinical forms, *Rev. Soc. Bras. Med. Trop.*, 20, 32, 1987.

4. Benard, G. et al., Severe acute paracoccidioidomycosis in children, *Pediatr. Infect. Dis. J.*, 13, 510, 1994.
5. Marques, S.A., Paracoccidioidomycosis: Epidemiological, clinical and treatment up-date, *An. Bras. Dermatol.*, 78, 135, 2003.
6. Pereira, R.M. et al., Fatal disseminated paracoccidioidomycosis in a two-year-old child, *Rev. Inst. Med. Trop. Sao Paulo*, 46, 37, 2004.
7. Plá, M.P. et al., Neuroparacoccidioidomycosis: Case report and review, *Mycopathologia*, 127, 139, 1994.
8. Paniago, A.M. et al., Neuroparacoccidioidomycosis: Analysis of 13 cases observed in an endemic area in Brazil, *Trans. R. Soc. Trop. Med. Hyg.*, 101, 414, 2007.
9. Valle, A.C.F. et al., A case report of intraspinal paracoccidioidomycosis, *Rev. Inst. Med. Trop. São Paulo*, 40, 203, 1998.
10. Valera, E.T. et al., Fungal infection by *Paracoccidioides brasiliensis* mimicking bone tumor, *Pediatr. Blood Cancer*, 50, 1284, 2008.
11. Resende, L.S. et al., Bone marrow necrosis related to paracoccidioidomycosis: The first eight cases identified at autopsy, *Histopathology*, 54, 486, 2009.
12. Severo, L.C. et al., Paracoccidioidomycosis of the male genital tract. Report of eleven cases and review of Brazilian literature, *Rev. Inst. Med. Trop. São Paulo*, 42, 38, 2000.
13. Lopes, D.L. et al., Prostatic paracoccidioidomycosis: Differential diagnosis of prostate cancer, *Mem. Inst. Oswaldo Cruz*, 104, 33, 2009.
14. Agudelo, C.A. et al., Identification of *Paracoccidioides brasiliensis* in adrenal glands biopsies of two patients with paracoccidioidomycosis and adrenal insufficiency, *Rev. Inst. Med. Trop. Sao Paulo*, 51, 45, 2009.
15. Castro, L.G.M. et al., Hard palate perforation: An unusual finding in paracoccidioidomycosis, *Int. J. Dermatol.*, 40, 281, 2001.
16. Silva, C.O. et al., Gingival involvement in oral paracoccidioidomycosis, *J. Periodontol.*, 78, 1229, 2007.
17. Marques, S.A. et al., Paracoccidioidomycosis: Infiltrated, sarcoid-like cutaneous lesions misinterpreted as tuberculoid leprosy, *Rev. Inst. Med. Trop. Sao Paulo*, 50, 47, 2008.
18. Ajello, L. and Polonelli, L., Imported paracoccidioidomycosis: A public health problem in non-endemic areas, *Eur. J. Epidemiol.*, 1, 160, 1985.
19. Benoldi, D. et al., Paracoccidioidomycosis (South American blastomycosis): A report of an imported case previously diagnosed as tuberculosis, *Eur. J. Epidemiol.*, 1, 150, 1985.
20. Horre, R. et al., A case of imported paracoccidioidomycosis in a German legionnaire, *Med. Mycol.*, 40, 213, 2002.
21. Kamei, K. et al., The trend of imported mycoses in Japan, *J. Infect. Chemother.*, 9, 16, 2003.
22. Ginarte, M., Pereiro, M. Jr., and Toribio, J., Imported paracoccidioicomycosis in Spain, *Mycoses*, 46, 407, 2003.
23. Mayr, A. et al., Chronic paracoccidioidomycosis in a female patient in Austria, *Eur. J. Clin. Microbiol. Infect. Dis.*, 23, 916, 2004.
24. Van Damme, P.A. et al., A case of imported paracoccidioidomycosis: An awkward infection in the Netherlands, *Med. Mycol.*, 44, 13, 2006.
25. Mayayo, E. et al., Report of an imported cutaneous disseminated case of paracoccidioidomycosis, *Rev. Iberoam. Micol.*, 24, 44, 2007.
26. Bousquet, A. et al., Imported mycosis: A review of paracoccidioidomycosis, *Med. Mal. Infect.*, 37, S210, 2007.
27. Bräuninger, W., Hastra, K., and Rubin, R., Paracoccidioidomycosis, an imported tropical disease, *Hautarzt*, 36, 408, 1985.
28. Lacaz, C.S., Historical evolution or the knowledge on paracoccidioidomycosis and its etiologic agent, *Paracoccidioides brasiliensis*, in *Paracoccidioidomycosis*, Franco, M. et al., Eds., Boca Raton, FL: CRC Press, 1994, 410pp.
29. Restrepo, A., The ecology of *Paracoccidioides brasiliensis*: A puzzle still unsolved, *J. Med. Vet. Mycol.*, 23, 323, 1985.
30. Restrepo, A., Ecology of *Paracoccidioides brasiliensis*, in *Paracoccidioidomycosis*, Franco, M. et al. Eds. Boca Raton, FL: CRC Press, 1994.
31. Costa, E.O., Diniz, L.S.M., and Fava-Netto, C., The prevalence of positive intradermal reactions to paracoccidioidin in domestic and wild animals in São Paulo, Brazil, *Vet. Res. Commun.*, 19, 127, 1995.
32. Costa, E.O. et al., Delayed hypersensitivity test with paracoccidioidin in captive Latin American wild mammals, *J. Med. Vet. Mycol.*, 33, 39, 1995.
33. Richini-Pereira, V.B. et al., Molecular detection of *Paracoccidioides brasiliensis* in road-killed wild animals, *Med. Mycol.*, 46, 35, 2008.
34. Richini-Pereira, V.B. et al., Molecular approaches for eco-epidemiological studies of *Paracoccidioides brasiliensis*, *Mem. Inst. Oswaldo Cruz*, 104, 636, 2009.
35. Bagagli, E. et al., Isolation of *Paracoccidioides brasiliensis* from armadillos (*Dasypus novemcinctus*) captured in an endemic area of paracoccidioidomycosis, *Am. J. Trop. Med. Hyg.*, 58, 505, 1998.
36. Restrepo, A., Morphological aspects of *Paracoccidioides brasiliensis* in lymph nodes: Implications for the prolonged latency of paracoccidioidomycosis?, *Med. Mycol.*, 38, 317, 2000.
37. Silva-Vergara, M.L., Martínez, R., and Camargo, Z.P., Isolation of *Paracoccidioides brasiliensis* from armadillo (*Dasypus novemcinctus*) in an area where the fungus was recently isolated from soil, *Med. Mycol.*, 38, 185, 2000.
38. Bagagli, E. et al., High frequency of *Paracoccidioides brasiliensis* infection in armadillo (*Dasypus novemcinctus*): An ecological study, *Med. Mycol.*, 41, 217, 2003.
39. Corredor, G.G. et al., The naked-tailed armadillo *Cabassous centralis* (Miller 1899): A new host to *Paracoccidiodes brasiliensis*. Molecular identification of the isolate, *Med. Mycol.*, 43, 275, 2005.
40. Ricci, G., Mota, F.T., and Wakamatsu, A., Canine paracoccidioidomycosis, *Med. Mycol.*, 42, 379, 2004.
41. Farias, M.R. et al., Canine paracoccidioidomycosis: Case report of generalized lymphadenits [abstract], *Rev. Inst. Med. Trop. Sao Paulo*, Suppl 14, 64, 2005.
42. Untereiner, W.A. et al., The Ajellomycetaceae, a new family of vertebrate-associate Onygenales, *Mycologia*, 96, 812, 2004.
43. Wang, H. et al., A fungal phylogeny based on 82 complete genomes using the composition vector method, *BMC Evol. Biol.*, 10, 195, 2009.
44. Herr, R.A. et al., Phylogenetic analysis of *Lacazia loboi* places this previously uncharacterized pathogen within the dimorphic Onygenales, *J. Clin. Microbiol.*, 39, 309, 2001.
45. Vilela, R. et al., Molecular phylogeny of animal pathogen *Lacazia loboi* inferred from rDNA and DNA coding sequences, *Mycol. Res.*, 113, 851, 2009.
46. Bagagli, E. et al., Phylogenetic and evolutionary aspects of *Paracoccidioides brasiliensis* reveal a long coexistence with animal hosts that explain several biological features of the pathogen, *Infect. Genet. Evol.*, 6, 344, 2006.
47. Bagagli, E., *Paracoccidioides brasiliensis*: Phylogenetic and ecological aspects, *Mycopathologia*, 165, 197, 2008.

48. Franco, M. et al., *Paracoccidioides brasiliensis* antigen batches from the same isolate show immunological and biochemical differences, *Mycopathologia*, 135, 13, 1996.
49. Calgagno, A.M., Geographic discrimination of *Paracoccidioides brasiliensis* strains by randomly amplified polymorphic DNA analysis, *J. Clin. Microbiol.*, 36, 1733, 1998.
50. Sano, A. et al., Comparison between human and armadillo *Paracoccidioides brasiliensis* isolates by random amplified polymorphic DNA analysis, *Mycopathologia*, 143, 165, 1999.
51. Hebeler-Barbosa, F., Montenegro, M.R., and Bagagli, E., Virulence profiles of ten *Paracoccidioides brasiliensis* isolates obtained from armadillos (*Dasypus novemcinctus*), *Med. Mycol.*, 41, 89, 2003.
52. Hebeler-Barbosa, F. et al., Sequence comparison of the internal transcribed spacer regions and gp 43 in *Paracoccidioides brasiliensis* for patients and armadillos *Dasypus novemcinctus*, *J. Clin. Microbiol.*, 41, 5735, 2003.
53. Macoris, S.A.G. et al., Virulence attenuation and phenotypic variation of *Paracoccidioides brasiliensis* isolates obtained from armadillos and patients, *Mem. Inst. Oswaldo Cruz*, 101, 331, 2006.
54. Puccia, R., McEwen, J.G., and Cisalpino, P.S., Diversity in *Paracoccidioides brasiliensis*. The PbGP43 gene as a genetic marker, *Mycopathologia*, 165, 275, 2008.
55. Matute, D.R. et al., Cryptic speciation and recombination in the fungus *Paracoccidioides brasiliensis* as revealed by gene genealogies, *Mol. Biol. Evol.*, 23, 65, 2006.
56. Carrero, L.L. et al., New *Paracoccidioides brasiliensis* isolate reveals unexpected genomic variability in this human pathogen, *Fung. Genet. Biol.*, 45, 605, 2008.
57. Teixeira, M.M. et al., Phylogenetic analysis reveals a high level of speciation in the Paracoccidioides genus, *Mol. Phylogenet. Evol.*, 52, 273, 2009.
58. Matute, D.R. et al., Microsatellite analysis of three phylogenetic species of *Paracoccidioides brasiliensis*, *J. Clin. Microbiol.*, 44, 2153, 2006b.
59. Theodoro, R.C., Bagagli, E., and Oliveira, C., Phylogenetic analysis of PRP8 intein in *Paracoccidioides brasiliensis* species complex, *Fung. Genet. Biol.*, 45, 1284, 2008.
60. Lutz, A., Uma micose pseudo-coccidica localizada na boca e observada no Brasil: Contribuição ao conhecimento das hiphoblastomicoses americanas, *Brasil Med.*, 22, 121, 1908.
61. Batista Jr. J. et al., Is the geographical origin of a *Paracoccidioides brasiliensis* isolate important for antigen production for regional diagnosis of paracoccidioidomycosis? *Mycoses*, 53, 176, 2010.
62. Takayama, A. et al., An atypical *Paracoccidioides brasiliensis* clinical isolate based on multiple gene analysis, *Med. Mycol.*, 19, 1, 2009.
63. Salina, M.A. et al., Detection of circulating *Paracoccidioides brasiliensis* antigen in urine of paracoccidioidomycosis patients before and during treatment, *J. Clin. Microbiol.*, 36, 1723, 1998.
64. Sano, A. et al., Detection of gp43 and ITS1-5.8S-ITS2 ribosomal RNA genes of *Paracoccidioides brasiliensis* in paraffin-embedded tissue, *Nippon Ishinkin Gakkai Zasshi*, 42, 23, 2001.
65. Bialek, R. et al., Detection of *Paracoccidioides brasiliensis* in tissue samples by a Nested PCR assay, *J. Clin. Microbiol.*, 38, 2940, 2000.
66. Albuquerque, C.F., Silva, S.H.M., and Camargo, Z.P., Improvement of the specificity of an enzyme-linked immunosorbent assay for diagnosis of Paracoccidioidomycosis, *J. Clin. Microbiol.*, 43, 1944, 2005.
67. Goldani, L.Z., Maia, A.L., and Sugar, A.M., Cloning and nucleotide sequence of a specific DNA fragment from *Paracoccidioides brasiliensis*, *J. Clin. Microbiol.*, 33, 1652, 1995.
68. Goldani, L.Z. and Sugar, A.M., Short report: Use of the polymerase chain reaction to detect *Paracoccidioides brasiliensis* in murine paracoccidioidomycosis, *Am. J. Trop. Med. Hyg.*, 58, 152, 1998.
69. Gomes, G.M. et al., PCR for diagnosis of paracoccidioidomycosis, *J. Clin. Microbiol.*, 38, 3478, 2000.
70. Ricci, G. et al., Detection of *Paracoccidioides brasiliensis* by PCR in biopsies from patients with paracoccidioidomycosis: Correlation with the histopathological pattern, *Pathologica*, 99, 41, 2007.
71. Díez, S. et al., PCR with *Paracoccidioides brasiliensis* specific primers: Potential use in ecological studies, *Rev. Inst. Med. Trop. Sao Paulo*, 41, 351, 1999.
72. Corredor, G.G. et al., Isolation of *Paracoccidioides brasiliensis* from the nine-banded armadillo *Dasypus novemcinctus*, in an endemic area for paracoccidioidomycosis in Colombia, *Rev. Iberoam. Micol.*, 16, 216, 1999.
73. San-Blas, G. et al., Primer for clinical detection of *Paracoccidioides brasiliensis*, *J. Clin. Microbiol.*, 43, 4255, 2005.
74. Motoyama, A.B. et al., Molecular identification of *Paracoccidioides brasiliensis* by PCR amplification of ribosomal DNA, *J. Clin. Microbiol.*, 38, 3106, 2000.
75. Imai, T., Sano, A., and Mikami, Y., A new PCR primer for the identification of *Paracoccidioides brasiliensis* based on rRNA sequences coding the internal transcribed spacers (ITS) and 5.8 regions, *Med. Mycol.*, 38, 323, 2000.
76. Theodoro, R.C. et al., Molecular detection of *Paracoccidioides brasiliensis* in soil, *Med. Mycol.*, 43, 725, 2005.
77. Terçarioli, G.R. et al., Ecological study of *Paracoccidioides brasiliensis* in soil: Growth ability, conidia production and molecular detection, *BMC Microbiol.*, 7, 1, 2007.
78. Buitrago, M.J. et al., Utility of real-time PCR for the detection of *Paracoccidioides brasiliensis* DNA in the diagnosis of imported paracoccidioidomycosis, *Med. Mycol.*, 31, 1, 2009.
79. White, T.J. et al., Amplification and direct sequencing of fungal ribossomal RNA genes for phylogenetics, in *PCR Protocols—A Guide to Methods and Applications*, Innis, M.A. et al., Eds., San Diego, CA: Academic Press, Inc., 1990, pp. 315–322.
80. Cisalpino, P.S. et al., Cloning, characterization, and epitope expression of the major diagnostic antigen of *Paracoccidioides brasiliensis*, *J. Biol. Chem.*, 271, 4553, 1996.
81. Ricci, G. et al., Genotyping of *Paracoccidioides brasiliensis* directly from paraffin embedded tissue, *Med. Mycol.*, 43, 31, 2008.
82. McEwen, J.G. et al., Molecular cloning, nucleotide sequencing, and characterization of a 27-kDa antigenic protein from *Paracoccidioides brasiliensis*, *Fungal Genet. Biol.*, 20, 125, 1996.
83. Buckley, M., *The Fungal Kingdom Diverse and Essential Roles in Earth's Ecosystem*, Washington, DC: American Academy of Microbiology, 2008, p. 44.
84. Iwen, P.C., Hinrichs, S.H., and Rupp, M.E., Utilization of the internal transcribed spacer regions as molecular targets to detect and identify human fungal pathogens, *Med. Mycol.*, 40, 87, 2002.

41 *Penicillium*: Mycoses and Mycotoxinoses

R.R.M. Paterson and N. Lima

CONTENTS

41.1 INTRODUCTION

Fungal species are difficult to identify. The organisms are complex from a biological perspective, which contributes to the identification problem. Investigations involving human mycoses and mycotoxinoses require an improved comprehension of the structure and behavior of fungal populations. (We have employed the term mycotoxinoses as the fungal equivalent of bacterial toxinoses.) However, an understanding of the speciation, population biology, phylogeny, and evolution of fungi is not so developed as for other organisms [1].

Classification undergoes almost continual change, particularly since the introduction recently of nucleic acid-based methodology where, for example, "cryptic species" are revealed. These are species determined by "molecular biology" to have (significantly) different nucleic acid structures to other strains that are indistinguishable morphologically, although distinguishing morphology can often be found when the genetic differences are known. Additional complexity arises as systematic mycologists must work within the botanical code of nomenclature that emphasizes the naming of sexual states (teleomorphs) over asexual vegetative growth forms (anamorphs).

Identifications are time-consuming, and decisions regarding what represents a species tend to be subjective. The standard characters for identifying and classifying filamentous fungi remain morphology, and the literature provides extensive examples of problems. Unreliable morphological minutia to describe new species and variability within the morphological characters of accepted species are constant issues. *Penicillium* is difficult in these respects and the situation in the terverticillate penicillia (heavily responsible for toxin production) is particularly problematic [2,3]. In general, it is accepted that a "polyphasic" approach is optimal where morphological, physiological, biochemical, and molecular biological characters are employed to characterize fungi.

Penicillium Link is the most common of the ubiquitous fungi. It is related to other genera that produce penicilli (i.e., *Scopulariopsis*, *Geosmithia*, and *Paecilomyces*) to which it could be confused. The teleomorphic genera with anamorphs which produce penicilli are *Byssochlamys*, *Eupenicillium*, and *Talaromyces*. Furthermore, the genus can usefully be separated into subgenera where the numbers of branch points in conidiophores determine to which subgenus strains belong. In *Aspergilloides*, there is one branch point and so are monoverticillate. The biverticillate subgenus has two branch points within which there are described two subgenera: *Furcatum* and *Biverticillium*. The most complex form of the penicillus has characteristically three branch points; these are classified as terverticillate, that is, subgenus *Penicillium*. The genus contains numerous similar species that are difficult to distinguish and therein lie the problems for identifying species with human pathogenic properties [4].

There are two related areas of concern in the present chapter: (i) mycosis and (ii) "myco"toxinosis. The molecular biology of the fungi is presented in a separate section as it has assumed such importance and is, nevertheless, problematic [1].

41.1.1 Mycosis

The most important *Penicillium* species causing mycosis is the surely now-emerged fatal systemic pathogen, *Penicillium marneffei* (Figure 41.1). However, other penicillia are implicated in human mycoses (e.g., *Penicillium* spp., *P. bertai*, *P. bicolor*, *P. brevicompactum*, *P. chrysogenum*, *P. citrinum*, *P. commune*, *P. crustaceum*, *P. expansum*, *P. glaucum*, *P. purpurogenum*, and *P. spinulosum*) [5,6]. Clearly, a thorough knowledge of *Penicillium* is required even for those interested predominately in *P. marneffei*.

Patients with human immunodeficiency virus are susceptible particularly to *P. marneffei*. The disease is endemic in Asian countries such as Vietnam, Thailand, China, India, and Taiwan [7]. *P. marneffei* was first isolated from the hepatic lesions of a bamboo rat (*Rhizomys sinensis*) that was in captivity in 1956 [8]. The importance for humans was established with the HIV pandemic in Asia: infection increased in local and visitor populations from non-endemic countries [7] and only cryptococcosis and tuberculosis are of greater significance in the relevant countries [9]. Frequent manifestations are respiratory signs, anemia, fever, weight loss, lymphadenopathy, hepatosplenomegaly, and skin lesions. Primary treatment with amphotericin B and secondary prophylaxis with itraconazole are effective [10]—otherwise there is poor prognosis; infected patients who are HIV positive need suppressive therapy to prevent relapse [7]. These treatments are fairly standard for mycoses in general.

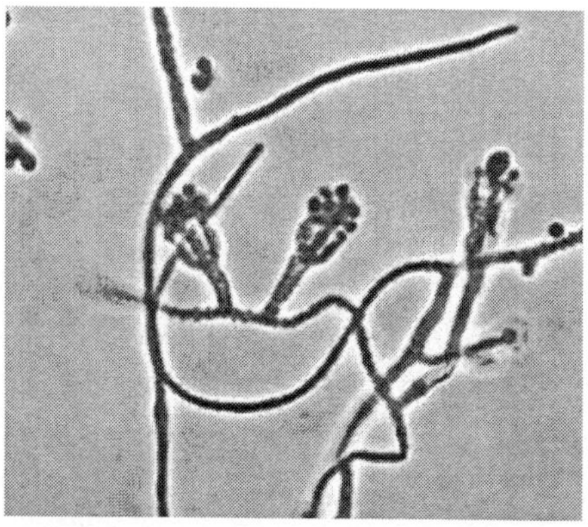

FIGURE 41.1 Hyphae and conidiophores of *Penicillium marneffei*.

41.1.1.1 Classification and Biology

41.1.1.1.1 Classification

P. marneffei was first classified into section Asymmetrica and subsection Divaricata (Raper and Thom) in 1959 [11], which is equivalent to Pitt's subgenus *Furcatum*. Frisvad and Filtenborg [12] placed the species in subgenus *Biverticillium* confirming Pitt's assignation to the subgenus. A phylogenetic analysis established that the fungus is related closely to *Penicillium* subgenus *Biverticillium* and sexual *Talaromyces* species with asexual biverticillate *Penicillium* states [13]. It is of relevance that 18S rRNA indicated that the fungus was related closely to a *Talaromyces flavus* strain (and so is the teleomorph?), second only to a near identical similarity to one of *Penicillium verruculosum* [14].

The first naturally occurring human case of "penicilliosis marneffei" was reported in 1973, of a North American minister who lived in Southeast Asia. The second (1984) was a compatriot who travelled in the Far East. Tissue sections demonstrated yeast-like cells of *P. marneffei* that were confirmed in culture. Five more cases were reported from Bangkok in the same year. Furthermore, eight cases of infection were reported from China in 1985 observed between 1964 and 1983, and 20 and 6 cases from the Guangxi region and Hong Kong, respectively, were reported shortly after [7]. However, from 1988 penicilliosis marneffei increased in Southeast Asia in patients with advanced HIV infection. The first cases were foreign AIDS patients who had visited regions of endemicity [15] and HIV patients who were native to such regions within Thailand [16]. However, the fungal infections were also observed in Cambodia, China, India, Malaysia, Taiwan, and Vietnam. External cases were detected in HIV-infected patients from Australia, Belgium, France, Germany, Japan, Sweden, Switzerland, the Netherlands, the United Kingdom, and the United States [7]. Finally, the infection is decreasing in areas where HIV infection is contracting (and vice versa, e.g., Vietnam): The welcome recent reports of a vaccine for HIV in Thailand (U.S. Military HIV research program [www01.hjf.org/apps/internet/hivnewscenter.nsf/phase3pressrelease]) and, for example, the U.K. newspaper *The Guardian* (http://www.guardian.co.uk/world/2009/sep/24/hiv-infection-vaccine-aids-breakthrough, accessed on March 24, 2011) may result ultimately in a decrease in the fungal infection, although publication in a non-scientific journal is somewhat worrying.*

41.1.1.1.2 Physiology

P. marneffei is the only *Penicillium* species where temperature-dependent dimorphic growth is reported. Dimorphism occurs in *P. chrysogenum* [17] in relation to carbon starvation and may be more widespread amongst fungi. It is well known that temperature affects, amongst other things, atmospheric CO_2 concentrations, and hence, the temperature effect may in reality be from other physical factors [18].

Conventional knowledge states that the fungus grows as (i) mycelia at temperatures below 37°C and (ii) yeast at 37°C [7]. However, the yeast transition occurs at 32°C and possibly above

30°C. One strain of those tested (9%) became yeast only at 39°C [19], again with significant ramifications for identifications.

Furthermore, *P. marneffei* produces a red pigment and a wide range of penicillia produce the same color [3,20]. Mapari et al. [21] report that the *P. marneffei* pigments are mitorubrinol (orange-red), monascorubramine (purple-red), purpactin, rubropunctatin (orange), and secalonic acid (yellow). Secalonic acid is a mycotoxin produced by, inter alia, *P. chrysogenum*: The compound is also a mutagen [22] with implications for nucleic acid analysis of producing fungi (see [1] and below). Mitorubrinol is also produced by *P. purpurogenum* (Table 41.1) with consequences for the use of "red pigment" as a character used in diagnosis. A red pigment was characterized as being (remotely) similar to herquinone by Bhardwaj et al. [23]. Color in particular is a subjective character about which an individual's perception can vary markedly. Obviously, great care is required when using these traits as characters for identification although chemical analysis will improve objectivity. A list of red or similarly colored pigments produced by penicillia is available in Table 41.1.

The enzymatic activities of *P. marneffei* have been studied [24]. However, characterization of the effects are required to determine whether secreted enzymes are linked to virulence and similarly for the pigments and mycotoxin discussed above. Wong et al. [25] demonstrated that 32 isolates of *P. marneffei* possessed the urease enzyme, which may be of use in biotyping. Some heterogeneity between isolates was observed in biochemical profiles leading to 17 biotypes being recognized. As mentioned, more work is emerging on secondary metabolite production [21,23], which will have diagnostic utility.

TABLE 41.1
List of Red or Similarly Colored Pigments from *Penicillium* (i.e., Excluding Yellow)

Taxon	Name	
Subgenus *Biverticillium*		
P. marneffei	Monascorubramine	Purple red
	Rubropunctatin	Orange
	Mitorubrinol	Orange red
	Some similarity to herqueinone (see below)	Red
P. purpurogenum	Mitorubrinol	Orange red
	PP-R	Purple red
	Purpurogenome	Yellow orange
P. islandicum	Erythroskyrin	Orange red
	Skyrin	Orange
Subgenus *Penicillium*		
P. atramentosum	Uncharacterized	Dark brown
P. atrosanguineum	Phoenicin	Red
	Uncharacterized	Red
P. atrovenetum	Norherqueinone	Red
P. cyclopium, *P. freii*	Viomellein	Red brown
	Xanthomegnin	Orange
P. herquei	Herqueinones	Red
P. oxalicum	"Arpink red"	Red
	Anthraquinone derivative	Red
P. paneum	Uncharacterized	Red
P. persicinum	Uncharacterized	Cherry red
P. viridicatum	Viomellein	Red brown

41.1.1.1.3 Ecology and Epidemiology

P. marneffei was isolated from (a) rodents *R. sinensis*, *R. pruinosus*, *R. sumatrensis*, and *Cannomys badius* and (b) soil samples from bamboo rat burrows [26,27]. Typing, including population genetic analyses, showed that genetically identical strains were isolated from humans and bamboo rats [7]. Furthermore, genes involved in asexual development, morphogenesis, and host immune response were reported [28].

The human disease may occur from zoonotic (animal) or sapronotic (environmental) transmission. The prevalence of infection in rats varies widely across Southeast Asia, which suggests that there are regional variations in the endemicity or that there are geographical variations in the predisposition to infection within different species of bamboo rats. Gugnani et al. [26] surveyed six species of sympatric rodents in north eastern India and found that only *C. badius* harbored infection; hence, host-specific factors may govern infection. All 51 animals of the grayish-black subspecies of *C. badius* were negative for *P. marneffei*, while 3 of the 10 rats in the reddish-brown group were positive [29]. These studies indicate host and geographical components.

An accurate description of the factors governing the distribution of *P. marneffei* infection in bamboo rats is necessary to understand the epidemiology in human populations. The endemic geographic range of *P. marneffei* is wide [7]. Ranges of the lesser and greater bamboo rats (*Cannomys* spp. and *Rhizomys* spp.) follow broadly the known distribution of *P. marneffei*. This may indicate that bamboo rats are an obligate stage in the life cycle of the fungus or that the ecotype favored is shared with *P. marneffei* and they are examples of sylvatic sapronotic infections. A case-control study of HIV-positive patients [30] did not implicate bamboo rats as a reservoir of infection for humans. Age and an agricultural occupation were factors associated with increased risk of infection. Temporal analyses of the incidence [7] showed extensive seasonal variation in infection rates and increased disease was associated with the rainy season. The conclusion was that soil exposure is a critical risk factor associated with infection by *P. marneffei*: The fungus must have a reservoir in the environment if it is a sapronosis.

As mentioned, a general assumption that penicillia are associated with soil has been extrapolated to *P. marneffei*. However, in *Talaromyces* and subgenus *Biverticillium*, the primary habitats are wood and its products, textiles, and "bird feathers" [31]. Hence, soil may not be the principal habitat: Vanittanakom et al. [32] demonstrated only 6% recovery from seeded non-sterilized soil, and *P. marneffei* can survive only for a few days in non-sterilized soil [7]. Furthermore, *P. marneffei* was isolated from three soil samples collected from *R. pruinosus* burrows and one out of 28 soil samples from that of *R. sumatrensis*. Definitive proof of an environmental reservoir is lacking. Many penicillia are linked to excretion, so it is perhaps surprising that examination of rat feces has not been reported.

41.1.1.2 Clinical Features and Pathogenesis

Approximately 8552 cases of *P. marneffei* infection were reported in HIV-infected patients at Chiang Mai University Hospital and Thailand from 1984 to 2004. The symptoms of 74 patients were weight loss, skin lesions, fever, hepatomegaly, and generalized lymphadenopathy respiratory signs. Observed lesions in 54 patients were papules with central necrosis, and the rest were papules or maculopapules. All patients acquired lesions on the face, neck, and other sites. Six patients had papular lesions on the palate and the average number of CD^{4+} T lymphocytes in these patients was 64 cells/mm^3. Patients (76%) were anemic, with a hemoglobin level of 10 g/dL or less. Importantly, some patients had hepatic penicilliosis without any skin lesions. Osteoarticular lesions were seen in multiple sites and arthritis was widespread. Multiple lytic bone lesions involving flat bones of the skull, long bones, and small bones of the fingers were also seen. However, attributing these signs to *P. marneffei* infection alone was equivocal [7].

P. marneffei is a primary pulmonary pathogen that disseminates to other organs by hematogenous spread. The onset of symptoms was rapid and severe in the absence of early treatment [33]. Necrotizing reactions with macrophage and histiocyte infiltrations are seen in immunocompromised patients. *P. marneffei* yeasts were observed with macrophages and histiocytes being oval or spherical cells of 2–3 μm diameter, with elongated cells of 13 μm observed extracellularly. Granulomatous and suppurative reactions were seen in tissues [7].

The mechanisms of infection are understood poorly: *P. marneffei* conidia can recognize fibronectin and bind to laminin [34,35], which are involved possibly in the attachment of conidia to bronchoalveolar epithelia. Monocyte-derived macrophages phagocytose *P. marneffei*. The fungus can be cleared within 2–3 weeks in healthy hosts, whereas the infection is fatal in nude or in T-cell-depleted mice [7]. These results demonstrated that T cells are necessary for clearing infection in mice [36]. The host immunological response to *P. marneffei* probably is activation of macrophages by T-cell-derived cytokines. Human and mouse macrophages control *P. marneffei* growth and kill intracellular yeast when activated in vitro by T-cell-derived cytokines. It was demonstrated that protective immunity follows a Th1 response, playing a crucial role in host resistance to intracellular pathogens [37]. Phosphoprotein osteopontin OPN, secreted from monocytes, is involved in the production of IL-12 from peripheral blood mononuclear cells after stimulation with *P. marneffei* [38]. Rabbit pulmonary alveolar macrophages and circulating human mononuclear phagocytes manifested antifungal activity against the fungus. Granulocyte-macrophage colony-stimulating factor enhanced the antifungal activity of human neutrophils against the yeast form of *P. marneffei* but not the conidial [7]. Youngchim et al. [24] found an expression of acid phosphatase activity by *P. marneffei*, which is one of the virulence factors for other intracellular pathogens. Finally an antigenic catalase-peroxidase protein-encoding gene (*cpeA*) was isolated recently from *P. marneffei*, by antibody screening of a cDNA yeast-phase library of this fungus [28].

41.1.1.3 Phenotypic Identification

41.1.1.3.1 Microscopy and Culture

Traditionally, identifications in clinical specimens are by microscopy and culture [9,33]. Presumptive diagnosis is made by microscopic examination of bone marrow aspirates and/or smears of skin or lymph node biopsy specimens [36]. *P. marneffei* in histopathological sections can be stained variously and appears as fission arthroconidia or roundish cells [7]. Definite diagnosis is from culture where bone marrow, blood, and skin biopsy give positive results in particular [9]. Identification of *P. marneffei* involves the morphology of the colony, mold-to-yeast conversion, and microscopic morphology (also including production of the pigment).

41.1.1.3.2 Immunological Tests

Experimental methods include identification from infected guinea pigs by monoclonal antibodies [39,40]. Immunoperoxidase staining was developed [41]: cells in skin biopsies were intracellular yeasts. Also, a specific indirect fluorescent-antibody reagent in histologic sections was devised [42]. Relevant murine immunoglobulin (IgM) monoclonal antibodies reacted in immunofluorescent staining of the yeast in tissue biopsies of patients [43]. Fine-needle aspiration cytology [44,45] was useful "potentially" to patients in whom lymphadenopathy was confined to intra-abdominal nodes. Interestingly, an exoantigen immunodiffusion test was described [46] differentiating *P. marneffei* from other *Penicillium* species, such as *P. citrinum* and *P. commune* [47], implying these could be problematic, although Ref. [47] does not mention other fungi and so from where does this information derive?

Specific antibodies against antigens of *P. marneffei* in clinical specimens [48] have low sensitivity [49]. A fluorescent-antibody test [50], an immunoblot assay reactive with serum from patients [51], and a diagnostic antigen of 38 kDa [52] have been reported. Antigens from *P. marneffei* confirmed a humoral response [53] and an enzyme-linked immunosorbent assay was developed [54]. A latex agglutination (LA) test using a monoclonal antibody [55] and a specific LA for the detection of *P. marneffei* antigens were also devised [49]. An enzyme immunoassay for the quantification of *P. marneffei* [56] was used in a dot blot ELISA and an LA test [57]. All tests were highly sensitive and specific. Finally, a monoclonal antibody-based sandwich ELISA was investigated [43,58], which requires further testing.

41.1.2 Mycotoxinosis

This section deals with chemical substances that cause disease rather than biological entities. Life expectancy has increased dramatically in the developed world at least indicating that acute poisonings from toxins are not major problems. However, this has not occurred in less technologically advanced countries. In either case, there is a vast amount of general ill health in the populations and it is required to be known to what extent mycotoxins play roles. (It has been estimated that 20,000 deaths/year occur in Indonesia from aflatoxin [an *Aspergillus* toxin]-induced liver cancer [59].)

Mycotoxins are fungal metabolites, which cause sickness or death in people when ingested, inhaled, and/or absorbed [60]. They occur and co-occur naturally in food allowing synergistic effects. Furthermore, *Penicillium* species are ubiquitous in nature and unavoidable, although they can be controlled: those that produce mycotoxins are common and are associated predominantly with the terverticillate penicillia, which are difficult taxonomically. The chemical structures of a few are provided in Figure 41.2. Mycotoxins may also be considered as weapons [61] although those from *Penicillium* are not important in this regard at present.

41.1.2.1 Human Disease Etiology

Two "named" human diseases are considered somewhat inconclusively to be caused by *Penicillium* mycotoxins [62]. Balkan endemic nephropathy (BEN) is presumed to be from cereals containing ochratoxin A, and associated urinary tract tumors are linked to *P. verrucosum* contamination. Evidence for a role of ochratoxin A in BEN has accumulated, but there is no direct epidemiological proof at the level of individual subjects.

Cardiac beriberi or "shoshin kakke" is associated with rice containing citreoviridin produced by *P. citreoviride*, a disease that occurred for centuries, including the early twentieth century in Japan and other Asian countries. The disease was characterized by palpitations, nausea, vomiting, rapid and difficult breathing, rapid pulse, abnormal heart sounds, low blood pressure, restlessness, and violent ania leading to respiratory failure and death. The causative agent was regarded as infection or avitaminosis, until a fungus was isolated and identified. An extract component was identified, its structure elucidated, and named as citreoviridin. The neurologic syndrome and respiratory failure were reproduced in laboratory animals. The Rice Act of 1921 passed by the Japanese government decreased the availability of moldy rice in the markets and resulted in rapid decline of the disease. Improved diet and inspection have made cardiac beriberi of little importance in modern times [62] but provide a useful historical example of mycotoxin-related pathogenicity and how it can be ameliorated.

Nevertheless, the *Penicillium* mycotoxins of prime importance are ochratoxin A and patulin, which have statutory limits imposed on them in foodstuff in the European Union. Table 41.2 indicates from which food penicillia are found and the associated mycotoxins. When mycotoxins are fed to animals in combination, interactive effects can be classified as additive, less than additive, synergistic, potentiative, or even antagonistic [60]. Hence, in effect, many of the studies described below provide only a partial view.

There is growing concern within medicine about mycotoxin involvement in human diseases [60]. More occult disease may occur when the mycotoxin interferes with immune processes, rendering the patient more susceptible to infectious diseases (cf. *P. marneffei*). This may explain why few data are available on mycotoxins being implicated in immunosuppressive illnesses as the presenting diseases are more obvious. The morphological effects of ochratoxin A on lymphocytes and neutrophils from healthy subjects and patients with oesophageal and breast carcinomas indicated that leukocyte death was evident from the decrease in cell survival of those exposed. The study supported that homeostasis of the immune system is compromised in cancer patients exposed to ochratoxin A, including those with breast cancer. Gliotoxin may be involved in the mycosis referred to as aspergillosis and *Penicillium bilaii* can produce this mycotoxin.

41.1.2.2 Animal Models of Human Disease

The effects of mycotoxins are usually more obvious in domesticated animals and extrapolations to humans are often based on these observations and supported by laboratory animal studies [60]. Ochratoxin A in animals (i) damages kidneys, (ii) causes intestinal necrosis and hemorrhage, (iii) suppresses immunity, and (iv) is carcinogenic. Dogs, rats, and swine are affected with kidney problems from ochratoxin A. The impaired renal function results in glucosuria and proteinuria, with casts evident in the urine, and absorption occurs in

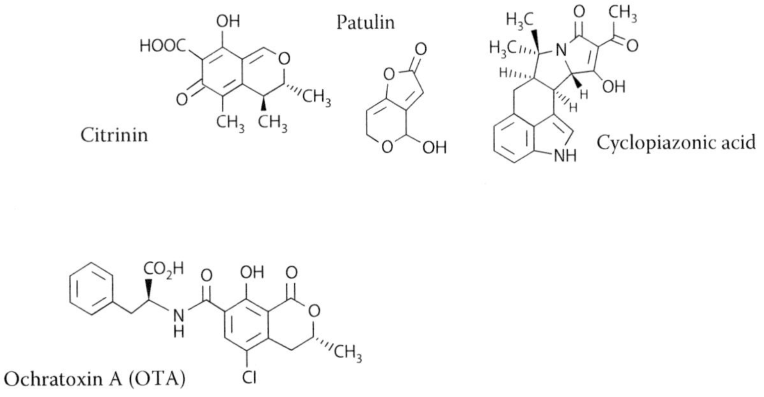

FIGURE 41.2 Chemical structure of selected mycotoxins.

TABLE 41.2
Penicillia Commonly Isolated from Food and Associated Mycotoxins

Food, Commodity	Fungi	Mycotoxins
Cereals	*P. verrucosum*	Ochratoxin A, citrinin
Dried meat products	*P. nordicum*	Ochratoxin A
Pomaceous fruit (e.g., apples)	*P. expansum*	Patulin, citrinin
Tap root plants	*P. radicicola*	Citrinin, penicillic acid
Beer, wine, rye bread, dry meat	*P. carneum*	Patulin, penicillic acid
Cereals, dry foods, indoor air	*P. griseofulvum*	Patulin, cyclopiazonic acid, secalonic acid D, F
Rye bread	*P. paneum*	Patulin
Nuts	*P. expansum*	Patulin
Yams	*P. sclerotigenum*	Patulin
Cereals	*P. aurantiogriseum, P. cyclopium, P. freii, P. melanoconidium, P. neoechinulatum, P. tulipae, P. viridicatum*	Penicillic acid
Cereals, dried meat products	*P. polonicum*	Penicillic acid
Cheese	*P. camemberti, P. commune, P. dipodomyicola, P. palitans, P. griseofulvum*	Cyclopiazonic acid

the proximal and distal tubules of the kidney. Changes in the renal function of pigs include impairment of proximal tubular function, altered urine excretion, and increased excretion of urine glucose. Carcinogenicity was evident in rats/mice. Ochratoxin A may have an immunological effect via immunoglobulins and phagocytic cells and can cause liver damage particularly at higher doses. It is reported to affect embryonic survival. Tumorigenesis/carcinogenesis was reported although mutagenicity was not established.

Cyclopiazonic acid intoxication includes weight loss, anorexia, diarrhea, dehydration, pyrexia, ataxia, immobility, and extensor spasm at death in dogs, rats, pigs, laying hens, sheep, and chickens. Patulin has an immunology effect and inhibits multiple aspects of macrophage function in vitro: neutrophilia in rats was attributed to GI inflammation. Gliotoxin has antimicrobial and immunosuppressive capabilities. The compound is produced in infected animal tissue strengthening its involvement in pathogenesis. Rubratoxin has been suspected as the cause of hepatotoxic, hemorrhagic disease of cattle and pigs. Ochratoxin A and rubratoxin B (from *P. purpurogenum*) are teratogenic in mammalian species. Finally, penitrem A elicits a tremogenic response in animals [60,63].

41.1.2.3 Mycotoxin Toxigenic Mechanisms

Mycotoxins as a unit are unclassified according to mechanisms of action because of the high diversity of chemical structures. The potential for complex toxin interactions is great given the large number and diversity of the action mechanisms. Mycotoxins with similar modes of action would be expected to have at least additive effects. Conversely, some interactions could have subtractive effects. For example, cyclopiazonic acid prevents the lipid peroxidation induced by patulin [64]. The actual combined health risk from mycotoxin exposure is unknown because many mycotoxins are undiscovered.

Mutagenic effects involving nucleic acids can be direct (e.g., changing bases) and/or indirect (e.g., inhibiting enzymes) [65]. Ochratoxin A, citrinin, patulin, penicillic acid, and penitrem have carcinogenic effects and are DNA damaging. Ochratoxin A is a potent carcinogen in rats and is classified as a possible human carcinogen by the International Agency for Research on Cancer [66]. DNA–DNA cross-links are induced by patulin: Mutations of cells might be from an indirect mutagenic mechanism (e.g., inhibition of enzymes) and direct reactivity to DNA has been demonstrated [67].

The mode of action of penitrem A is by the release of neurotransmitters from synaptosomes in the CNS and in peripheral nerves at the neuromuscular junction [68]. Mycelianamide and gliotoxin function as enzyme inhibitors: Low concentrations of gliotoxin inhibited specifically the activation of NF-κB. Hundreds of mycotoxins are known and detected frequently in plant-derived products. Various mycotoxins may occur simultaneously, depending on, amongst other things, the environmental and substrate conditions [60].

41.1.2.4 Taxonomy of Mycotoxin Fungi

The penicillia can be divided into the monoverticillate, biverticillate, and terverticillate, which refer to the number of branches in the conidiophores as mentioned. The predominant mycotoxin producers belong to the terverticillates (cf. subgenus *Penicillium*) [3]. *P. expansum* produces smooth-walled penicilli, synnemata, and rots apples: It is the predominant fungus associated with patulin production. The fungus typifies the genus although the classic features are variable. This plasticity in characters is a general feature in terverticillate fungi making the taxonomy unstable and requiring other characters to stabilize the taxonomy, hence assisting in species recognition. DNA-sequencing data have contributed, but are not suited particularly for classification and identification, although defining phylogenetics. However, Paterson et al.

have questioned the validity of such data because of "self" mutagen and inhibitor production as described in a series of papers (see [65]).

Pitt [69] reintroduced some novel characters and made rearrangements; however, overall, the concepts were similar to those in Frisvad et al. [31] and the sequence-based ribosomal DNA phylogeny of Peterson [70]. Bridge et al. [71] attempted probably the first polyphasic approach and also employed numerical taxonomy, which indicated a general similarity in most of the strains tested; some species concepts were confirmed or new groupings were made in some other cases. Indeed, few differences were found between terverticillate penicillia in the ribosomal DNA gene [70]. Nevertheless, Frisvad and Samson [3] describe, illustrate, and key out 58 species. Seifert and Lous-Seize [72] indicated that β-tubulin was more useful in delineating species, which was confirmed by Samson et al. [73].

41.1.2.5 Phenotypic Identification

Conventional methods to identify these fungi involve isolation from a substrate and microscopic observation of morphology, growth rates, and colors/morphology on growth media [3,4]. The use of secondary metabolites as characters in the group has been advocated by Frisvad and Samson [3]: Paterson found that the profiles in [3] gave relatedness to species concepts only in some cases, and there is a sense of coming full circle as the species concepts were closer to Raper and Thom in 1948 than Pitt in 1980 (see [74]). Extracellular enzymes were shown to be variable in most species by Paterson et al. [75] and have not been used subsequently with significance, although *P. brevicompactum* did appear different. A solution to identifying these penicillia was offered by Paterson et al. [2,76], which may be useful particularly in determining those strains which affect humans. This was to identify the conidiophores and analyze for the mycotoxin of interest. For patulin producing strains, an identification of subgenus *Penicillium*, patulin positive could be obtained. Finally, some of the constraints have led to an interest in matrix-assisted laser desorption/ionization time-of-flight intact cell mass spectrometry (MALDI-TOF ICMS) for rapid and reliable identifications [1]: much more work is required before its use is routine.

41.1.3 Molecular Diagnosis of Mycosis and Mycotoxinosis

Nucleic acid techniques are extremely sensitive and have been employed apparently for the successful specific detection of fungi. However, problems with the methods have been described recently by the authors of the present chapter in particular. Dupont [77] suggests that "molecular methods offer a big advantage over conventional phenotypic methods for species diagnosis in that they measure "stable" genotypic characteristics and do not rely on culture and operator interpretation." This may not be the case in what is an excellent review of molecular detection of food-borne penicillia. The author also considered that the methods of Paterson et al. [2,76] mentioned above were valid. When fungi are grown, they may produce (i) inhibitors of enzymes and (ii) mutagens of nucleic acid. Hence, the DNA polymerase of PCRs could be inhibited. Most diagnostic PCR methods lack an internal amplification control (IAC) leading to potential false-negative results from inhibitors produced in media [65]. Also, IACs are required to account for faulty thermal cyclers and reagents. This permits the possibility of false-negative results—the worst possible outcome for diagnostic microbiology, in that samples containing a pathogen may be considered as safe.

Furthermore, the nucleic acid of interest could be mutated and/or the stabilizing enzymes could be inhibited, allowing for further mutations [65]—an undesirable situation for nucleic acid analysis. Factors such as these had not been considered previously by other authors. In pure (batch) liquid culture or agar (which is how fungi are grown in molecular biology), these mutagens and inhibitors accumulate normally to a high level and will tend to overwhelm natural mechanisms to ameliorate their effects.

Penicillium mycotoxins, which are DNA damaging, include ochratoxin A, citrinin, patulin, penicillic acid, and secalonic acid produced by a wide range of penicillia. Interestingly, secalonic acid is produced by *P. marneffei* (see above). Many other secondary metabolites are likely to be mutagenic and inhibitory, which may not be mycotoxins, and so avoidance needs to be universal when analyzing fungi. However, this has not been done with potential negative consequences for the molecular biology of fungi. None of the research described below has addressed these issues and the onus is on researchers to prove these effects do not influence their results.

41.1.3.1 Molecular Identification of *P. marneffei*

Molecular diagnosis of *P. marneffei* is based on standard specific oligonucleotide primers designed from the internally transcribed spacer and 5.8S rRNA gene (ITS1-5.8S-ITS2) of *P. marneffei* [13]. First, fungal DNA was amplified with the primer pair ITS5 and ITS4 [78]. Second, nested PCR was performed successfully with primer pairs "PM2 and PM4". A 347 bp PCR product was obtained, which was used to identify *P. marneffei* from skin [79].

Furthermore, *P. marneffei* was identified using a specific PCR-enzyme immunoassay method [80]: Applicability with clinical samples required further evaluation. An 18S rRNA gene probe was specific in a PCR-hybridization reaction from human and environmental isolates [81]. However, the hybridization technique is labor- and skill-intensive; hence, single and nested PCR for the rapid identification of *P. marneffei* were developed using specific primers, also based on the 18S rRNA gene sequence [82]. Successful discrimination of a young culture of *P. marneffei* was performed. Finally, a sensitive one-tube seminested PCR assay based on the 18S rRNA sequences [83] was used in a cultural-dependent PCR (CDP) and clinical samples (or cultural-independent PCR [CIP]). The methods described in [82,83] need to be studied further to prove their utility. In none of the above studies were IACs employed allowing for the potential of the worst possible outcome of diagnostic PCR—false-negative results.

In addition to a draft sequencing of the genome [14], random sampling of genetic variation within the *P. marneffei* genome has been undertaken. Human isolates had type I (73%) and type II (27%) DNAs whereas bamboo rat isolates were type I from *R. sumatrensis* and type II from *C. badius* for 20 and 3 strains, respectively [84]. Twenty *P. marneffei* isolates from patients [85] had a similar pattern although randomly amplified polymorphic DNA (RAPD) assays yielded eight different patterns. Four genotypes were found that varied in frequency between northern and southern Thailand [86]. Isolates from Thailand [87] grouped into nine subprofiles yielding 54 genotypes. However, no correlation with geographic region or specimen source was observed [7]. In addition, sequence-specific assays of genetic variation in the *P. marneffei* genome have been investigated [88]. Some of the sequences resulted in 30 dinucleotide, 14 trinucleotide, and 5 tetranucleotide repeats [89]. Only three microsatellites were uncovered subsequently [90] and why this discrepancy between the two studies exists remains unclear. Of the (i) 49 repeats in [89], 24 were amplified to type 29 clinical and bamboo rat isolates [26,89,91], and (ii) 24 loci, 23 were amplifiable and 21 were polymorphic with between 2 and 14 alleles present at each locus, comprising 19 unique microsatellite type (MT)s. Clustering showed that isolates occurred within one of two separated clades geographically that accounted for 26% of the total observed genetic diversity. The "eastern" clade contained isolates from mainland China, Hong Kong, Indonesia, and Vietnam, while the "western" clade contained isolates from Thailand and India, showing that *P. marneffei* has a geographic component to its genetic structure. Extensive linkage disequilibria existed between loci [7], suggesting that there were either genetically differentiated subpopulations or extensive clonal reproduction occurring. A study over a smaller geographical scale in India showed that the MTs of isolates were dissimilar between plantations [26]. The population genetic structure of *P. marneffei* may be partitioned over local and geographical scales. The type isolate of *P. marneffei* [7] was characterized by a unique MT and was therefore dissimilar to human isolates [26]; a similar result was found by comparing two bamboo rat isolates with 32 clinical isolates [90]. However, 10 bamboo rat isolates collected in India demonstrated that one was identical to all 21 microsatellite loci of the human isolate CBS 101038 [92]. This is the first molecular epidemiological evidence that humans and bamboo rats share identical strains of *P. marneffei* and that host-to-host transmission may occur. However, it is possible that coinfection had occurred from a common environmental source.

41.1.3.2 Molecular Studies of Mycotoxin/Occasional Mycosis Penicillia

This work relates normally to fungi that have grown on foods, produced toxins that are then ingested by humans (or other animals), and caused disease. However, it has to be realized that these can become involved in mycosis as discussed briefly above. For example, *P. brevicompactum*, *P. chrysogenum*, and *P. expansum* can fall into both categories (cf. *P. marneffei* producing the mycotoxin secalonic acid).

As mentioned, the methods have been reviewed previously by the current authors, which can be consulted for more detail (see [65]). Also, the methods in [77] can be consulted. A novel aspect of the molecular biology of mycotoxin-producing fungi is the possibility to analyze for the genes of the metabolic pathway that creates the mycotoxin of interest, in addition to genes specific to particular taxa [93]. These have been reported for patulin, ochratoxin A, and PR-toxin.

Marek et al. [94] report their assay for *Penicillium expansum* on fruits using the polygalacturonase gene although without using an IAC. Calderon et al. [95] describe a method for determining spores of *Penicillium roqueforti* from an air sampler using PCR. There is no mention of an IAC although the possibility of inhibition of the PCR is discussed. Colombo et al. [96] provided a method that claimed to distinguish *Penicillium aurantiogriseum* from foods. Sholberg et al. [97] did not use an IAC for *Penicillium* blue rot molds although a negative control of "no DNA" in the reaction mixture was included. Again, IAC were not employed in Dupont et al. [98], and the RFLP patterns could represent inhibition rather than true differences in DNA between *Penicillium camemberti* and *Penicillium nalgiovense*. The conclusion that IACs are required in RFLP cannot be avoided, as the method is also enzyme based, although IAC will be more difficult to employ. Pedersen et al. [93] investigated subgenus *Penicillium* with two sets of primers and again an IAC was not used.

Dao et al. [99] describe a method for detecting ochratoxin A and citrinin producing fungi (see Paterson [100]). Geisen et al. [101] did not use an IAC and so the negative strains tested with the primers may have been inhibited and not specific for *Penicillium nordicum* in terms of the nucleic acid involved. Bogs et al. [102] propose a method for differentiating the ochratoxin A producing species *P. nordicum* and *Penicillium verrucosum*. As there was no IAC, the usual provisos apply concerning negative results and nonappearance of bands.

An IAC was not employed in the reports on patulin producing penicillia [100,103,104]. Nevertheless, the possibility of inhibition at least was recognized. Unpredictable results occurred on some occasions with *idh*, such as (i) presence of the gene in *Penicillium brevicompactum* and (ii) *P. expansum* lacking the gene very occasionally. One explanation for not obtaining a positive result from all members of a species was the occurrence of false-negative results from inhibitors [65]. In all cases bar one, the various media used by authors were not designed to inhibit secondary metabolites. Furthermore, Dombrink-Kurtzman [105] investigated the differences between the sequences of the *idh* of *P. griseofulvum* and *P. expansum*. Both species produced patulin in culture with *P. griseofulvum* manufacturing much more under the conditions employed.

The analysis of the sequences revealed various differences between *P. griseofulvum* strains. The three *P. expansum* strains were identical for introns "1 and 2," but there were nine additional bases in intron 1 compared to the *P. griseofulvum idh* gene. *P. expansum* had more nucleotide differences in both introns than did strains of *P. griseofulvum*. One *P. griseofulvum* had nine bp differences and another five differences compared to the standard. Dombrink-Kurtzman [106] extended the work to many other patulin-producing species and demonstrated good correlation with species concepts, although *P. griseofulvum* "3523" was different considerably from other strains of the species. A reason may have been mutations of *idh* because of mutagens in the growth medium as discussed herein. Furthermore, a discussion of various PCR methods for mycotoxigenic fungi is provided in Paterson [100] including details of the various growth media employed. It can be assumed that mutagenic compounds will be produced in these media—an unsatisfactory position for nucleic acid analysis [65].

"Bar coding" to identify fungi is being considered actively [107], in part because of problems with existing characters. The authors are also very familiar with secondary metabolites production from these penicillia. Results of a maximum parsimony analyses indicated that *P. carneum* and *P. paneum* were closely related: They had a 12 nucleotide difference in the ITS regions [108]. Peterson [70] found few differences between terverticillate penicillia in the ribosomal DNA gene. Seifert and Lous-Seize [72] indicated that β-tubulin was more useful in delineating species, which was confirmed by Samson et al. [73] who investigated the β-tubulin sequences for the phylogenetic analysis of *Penicillium* subgenus *Penicillium* in general.

41.2 METHODS

41.2.1 Sample Preparation of *P. marneffei*

As a general observation, many anatomical locations within patients have been demonstrated to contain the fungus, and so CIP may be possible from these sites [7,44,45]. Details of how the organism was isolated were not provided in [13] although these need to be known for comparative purposes. Growth of *P. marneffei* mycelium for DNA isolation was on malt extract agar at 25°C, and DNA extraction methods were described in [109]. In Vanittanakom et al. [81], filamentous-phase cultures were obtained in Sabouraud dextrose broth at 25°C. Only details were provided of the culture collection and from whom the cultures were obtained and not of how the strains were actually isolated nor from which substrate. DNA was extracted by grinding the cellular mats with a mortar and pestle in the presence of liquid nitrogen. The fungal mat was ground into a fine powder with a sterile pestle. DNA was then extracted and purified using serial applications of proteinase K and RNase.

How the human isolates were obtained was not described in [81,82]. However, the yeast form was maintained in brain heart infusion agar at 37°C and the hyphal stages on Sabouraud glucose agar at 28°C [82]. The DNA was the same in each case that was extracted from fungal cells or from fungal spheroplasts. Fungal cells from the surface of the agar medium were suspended in lysis buffer, boiled, and vortexed in the case of the direct method. Phenol–chloroform–isoamyl alcohol (25:24:1 ratio) was added and re-vortexed to extract the DNA. The upper phase was removed and DNA precipitation was carried out with cold acetone. The DNA pellet was air dried and resuspended in water.

Spheroplasts were prepared by suspending fungal cells in filter-sterilized osmotic medium, placing on ice and incubating with NovoZym 234 in the presence of bovine serum albumin at 37°C. The spheroplasts were lysed with buffer and deproteinized. The nucleic acids were collected by 2-propanol precipitation, resuspended in buffer, extracted by phenol-chloroform–isoamyl alcohol, precipitated with ethanol, and resuspended in buffer.

PCR to identify the organism in skin specimen was attempted when culturing failed [79]. DNA was extracted from the skin specimen fixed with formaldehyde and embedded in paraffin. (The use of formaldehyde warrants some concern as it is a mutagen.) The control was the unaffected part of the skin in a fixed specimen of a benign subcutaneous tumor.

41.2.2 Detection Procedures

41.2.2.1 Basic PCR for Fungi [13]

41.2.2.1.1 Reagents

10× amplification buffer

15 mM $MgCl_2$
500 mM KCl
100 mM Tris–HCl (pH 8.3) at 23°C
0.1% gelatine

10× dNTP stock solution

2 mM each of dATP, dGTP, dTTP, and dCTP

41.2.2.1.2 Procedure

1. Prepare a working reaction solution sufficient for 10 amplifications using positive displacement pipettes for steps 1–4, which consists of
   275 μL sterile distilled water
   100 μL 10× amplification buffer
   100 μL dNTP stock mixture (2 mM each dNTP)
   10 μL excess primer (50 μM stock)
   10 μL limiting primer (5 μM stock)
   5 μL *Taq* polymerase (5 units/μL)
2. Add 50 μL of the working solution to each tube.
3. Dilute the DNA samples in TE (10 mM Tris–HCl [pH 8.0] and 0.1 mM EDTA) to 0.1–10 ng 50 μL^{-1}.
4. Add 50 μL of the diluted sample DNA to each tube followed by two drops of mineral oil.
5. Spin briefly in a microcentrifuge

6. Cycling parameters

Initial denaturation	2–3 min at 95°C
Annealing	30 s at 50°–60°C
Extension	0.5–2 min (depending on product size) at 72°C
Denaturation	30 s at 95°C
Final extension	10 min at 72°C

7. Cycle numbers: Use (i) 25 for double-stranded product with 50 pM of each primer and (ii) 35 for single-stranded product using a primer ratio of 50:1 or 50:2.5 pM.
8. Briefly centrifuge the tubes when the amplification is completed and take 5 μL samples for analysis by minigels.
9. Remove the oil by extraction in chloroform and remove the unincorporated nucleotides while concentrating the DNA by centrifugal filtration before sequencing the single-stranded PCR.
10. Routinely, 7 μL of the retentate is used for sequencing reactions. Concentrate in a Speed Vac if the DNA is too dilute. Use 1 pM of the limiting primer (or an internal primer) as the sequencing primer. Anneal at 65°C, not 90°C–95°C, to avoid denaturing the double-stranded product prior to primer annealing.
11. Primers NS1–NS8 and ITS1–ITS4 [78] have been used with lower dNTP concentrations (32 μM each dNTP, instead of 200 μM in the reaction).
12. The above conditions were not optimized for enzyme and magnesium ion concentrations.

Primer sequences for fungal ribosomal RNA genes are provided in [78].

41.2.2.1.3 PCR Identification of P. marneffei

The PCR conditions in [13] were described in [91] and as modified in [109]. Primers were MS1 and MS2 for the mtSrDNA and ITS2, ITS3, ITS4, and ITS5 for the ITS1-5.8S-ITS2 rDNA region [109]. Nucleotide similarity among six isolates of *P. marneffei* was analyzed initially by restriction fragment length polymorphisms of the ITS1-5.8S-ITS2 rDNA PCR products. The ITS1-5.8S-ITS2 rDNA PCR products were digested singly with 1–3 U of the four-base-cutter restriction enzymes *Ava*I, *Ava*II, *Hha*I, *Mbo*I, and *Taq*I. The mtSrDNA and ITS1-5.8S-ITS2 rDNA regions of *P. marneffei* were sequenced from single-stranded template generated from asymmetric PCR amplifications. Nucleotide sequences from other species used in phylogenetic analyses were determined from previous studies [109]. DNA sequences were initially aligned by the use of the Genalign option in the Intelligenetics software package and then by visual optimization. Only regions without ambiguity were included in the analysis.

41.2.2.1.4 Specific PCR of P. marneffei *Protocol*

1. Amplify isolates with primers designed as general fungal primers, namely, ITS5 (GGAAGTAAAAGTCGTAACAAGG) and ITS4 (TCCTCCGCTTATTGATATGC), with an annealing temperature of 48°C.
2. Purify the ITS1-5.8S-ITS2 rDNA PCR products to eliminate primers.
3. Dilute PCR products × 1000 and then amplify with two primer combinations, PM1–PM4 and PM2–PM4, at annealing temperatures of 60°C or 65°C.
4. Test PCR amplifications from *P. marneffei* miniprep DNA diluted 1000-fold with PM1–PM4 and PM2–PM4 to ensure that the primers amplified genomic DNA.

Specific primers

Primer	Primer Location[a]	Nucleotide Sequence (59–39) bp
PM1 (59)	ITS1 (bp 125–142)	TCGCCGGGGGACGTTTGT
PM2 (59)	ITS1 (bp 187–205)	GATGGACTGTCTGAGTACC
PM4 (39)	ITS2 (bp 498–519)	ATGGTGGTGACCAACCCCCGCA

[a] Primer location is based on *P. marneffei* ITS sequence, GenBank accession no. L37406.

41.2.2.2 PCR Identification of Mycotoxin/Occasional Mycosis Penicillia

1. Marek et al. [94] report their assay for *Penicillium expansum* on fruits using the polygalacturonase gene although without using an IAC.
2. Calderon et al. [95] describe a method for determining spores of *Penicillium roqueforti* from an air sampler using PCR. There is no mention of an IAC although the possibility of inhibition of the PCR is discussed.
3. Colombo et al. [96] provided a method that claimed to distinguish *Penicillium aurantiogriseum* from foods.
4. Dupont et al. [98] detected the RFLP patterns in DNA between *Penicillium camemberti* and *Penicillium nalgiovense*.
5. Pedersen et al. [93] investigated subgenus *Penicillium* with two sets of primers.
6. Dao et al. [99] describe a method for detecting ochratoxin A and citrinin, producing fungi (see Paterson [100]).
7. Bogs et al. [102] propose a method for differentiating the ochratoxin A producing species *P. nordicum* and *Penicillium verrucosum*.

41.2.2.2.1 Protocol for Penicillium Subgenus Penicillium

In the first instance, PCR conditions can be optimized by varying the concentrations of KCl, MgCl, and Tris–HCl pH 8.3 using purified *Penicillium* subgenus *Penicillium* DNA as a template.

41.2.2.2.2 *Protocol for* Penicillium *Subgenus* Penicillium *as a Whole*

1. Wash the previously obtained DNA pellet with 80% ethanol and dry.
2. Dissolve in TE buffer (10 mM Tris–HCl pH 8.0 and 1 mM EDTA pH 8.0; 50 μL).
3. Use *Penicillium* subgenus *Penicillium* specific primers: ITS 212d (5′-AAA TAT AAA TTA TTT AAA ACT TTC-3′) and ITS 549 (5′-CTG GAT AAA AAT TTG GGT TG-3′).
4. Add 1–5 μL DNA (10 ng) to a reaction mixture containing 10 mM Tris–HCl pH 8.3, 50 mM KCl, 1 mM $MgCl_2$, 200 μM, dNTPs, 1, μM ITS 212d, and 1 μM ITS 549 and 2.5 U (Ampli)Taq (Perkin Elmer Cetus) polymerase in a final volume of 100 μL.
5. Amplify using the conditions of 45 cycles at 94°C for 30 s, 48°C for 30 s, 72°C for 30 s, and finally 72°C for 10 min using a Perkin Elmer 9600 Thermocycler.
6. Alternatively, use a Perkin Elmer 480 thermocycler for 40 cycles at 94°C for 30 s, 48°C for 60 s, and 72°C for 60 s followed by 72°C for 10 min.
7. Add 10 μL aliquots to a 2.5% agarose gel in TAE buffer and electrophorese.
8. Staining in 1 μg/mL ethidium bromide to observe PCR products.

41.2.2.2.3 *Protocol Adaption for* Penicillium *Subgenus* Penicillium *Species* P. roqueforti *and* P. carneum

1. Use ITS 183 (5′-CTG TCT GAA GAA TGC AGT CTG AGA AC-3′) and ITS 401 (5′-CCA TAC GCT CGA GGA CCG GAC-3′).
2. Add 5 μL template DNA (10 ng) to a 50 μL reaction mixture containing 10 mM Tris–HCl pH 8.3, 50 mM KCl, 2.5 mM $MgCl_2$, 0.01% gelatine, 200 μM dNTPs, 0.25% Tween 20, 10% DMSO, 0.3 μM of each of ITS 183 and ITS 401, and 2.5 U AmpliTaq polymerase.
3. Amplify at 45 cycles for 30 s at 94°C, 1 min at 60°C and 1 min at 72°C and finally 72°C for 10 min (the Perkin Elmer 9600 Thermocycler is adequate for this purpose).

41.2.2.2.4 *PCR for a Mycotoxin Metabolic Pathway Using Patulin as an Example*

Use the following *idh* primers that were designed to amplify the 600 base pairs of the *idh* gene which had the following structure: 5′CAA TGT GTC GTA CTG TGC CC and 5′ACC TTC AGT CGC TGT TCC TC (GIBO BRL, Paisley, Scotland).

1. Prepare the PCR reaction mixture consisting of 200 μM of dNTPs (Pharmacia, Herts, United Kingdom), 1.25 units of Tth polymerase enzyme, Tth buffer (both from HT Biotechnology, Cambridge, United Kingdom), 0.8 μM of primer IDH1 and IDH2, and template DNA.
2. Undertake an initial denaturing step of 94°C for 3 min. Amplify the *idh* gene in a thermal cycler (e.g., Hybaid, Omn-E) under the following conditions: 94°C for 1 min, 52°C for 1 min, 72°C for 1 min for 30 cycles and end the programme with an extension step of 72°C for 5 min.
3. Separate the amplified products by electrophoresis in 1.5% agarose gel (SeaKem LE, FMC Bioproducts) in Tris-acetate-EDTA buffer at a constant voltage of 5 V/cm.
4. Stain the gels with ethidium bromide and view under UV light.

41.3 CONCLUSIONS AND FUTURE PERSPECTIVES

There is not an absolute distinction between fungi that cause mycosis and those that produce mycotoxins as discussed above. Furthermore, standard treatments for mycosis should be considered immediately for patients with *P. marneffei* infections without waiting for a definitive identification: The rational being that treatment is not specific to any marked degree, although this needs to be confirmed by medical practitioners (see [110] for general treatments of *Penicillium* infections). Finally, analyzing samples from patients for a general fungal test such as the fungal lipid ergosterol (see [111]) may be worthwhile to facilitate immediate treatment.

It appears to be assumed that the natural habitat for *P. marneffei* is soil because "penicillia are associated with soil". However, Talaromycetes and Biverticillate penicillia are found more in wood and its products, textiles, and bird feathers. In addition, Subgenus *Penicillium* is associated with animal excretion, and the feces of rats require to be investigated (despite the fungus being biverticillate). After all, it is possible that infection of humans is via rat feces in contaminated food.

The pigments produced by *P. marneffei* could be analyzed as a means to identification. Hence, adapting Paterson et al. [2], the distinctive conidiophore could be identified by microscopy and then chromatography undertaken to determine the pigments. MALDI-TOF needs to be employed to investigate the yeast/hyphal transition in *P. marneffei* [1] and to compare this with other penicillia.

The issue of climate change is relevant to fungal diseases [112]: *P. marneffei* may become less prevalent in tropical climates as the temperature increases beyond tolerable levels for the fungus. However, this may allow increased disease in currently temperate countries (e.g., parts of Europe and United States) as the temperatures increase. Equivalent speculation can be made with *Penicillium* mycotoxins, with currently colder countries becoming more susceptible to ochratoxin A and patulin, for example. Certainly the situation merits careful monitoring.

Growing fungi in media that do not stimulate secondary metabolites production need to be investigated, and these should be employed for nucleic acid analysis. Secondary metabolites can be controlled by harvesting the fungus in trophophase (i.e., growth) rather than idiophase (i.e., secondary metabolites production). In this way, secondary metabolites can be reduced. Growing fungi for the minimum amount of time may be useful as this will limit the contact of secondary metabolites with nucleic acids. Finally, preserving the fungi in the original substrate could be undertaken. A reservoir of substrates from which a particular fungus was isolated could be maintained in freezers, freeze dried ampoules, and liquid nitrogen specifically for nucleic acid analysis. If these procedures appear laborious, then this has to be balanced with (i) the waste of financial, intellectual, and time resources, and (ii) the consequences of false diagnostic results, if the nucleic acid does not represent the taxa under analyses. There are various areas that require clarification as described above in the field of human pathogenicity by *Penicillium* and which may also apply to other fungi. Finally, it is to be hoped that recent success with a vaccine for HIV is confirmed (see above) and will result in a concomitant reduction in the incidence of *P. marneffei* infection.*

ACKNOWLEDGMENTS

RRMP is employed by the FCT framework, Commitment to Science ref. C2008-UMINHO-CEB-2.

REFERENCES

1. Santos, C. et al., Filamentous fungal characterizations by matrix-assisted laser desorption/ionization time-of-flight mass spectrometry, *J. Appl. Microbiol.*, 108, 375, 2010.
2. Paterson, R.R.M., Venâncio, A., and Lima, N., Solutions to *Penicillium* taxonomy crucial to mycotoxin research and health, *Res. Microbiol.*, 155, 507, 2004.
3. Frisvad, J.C. and Samson, R.A., Polyphasic taxonomy of *Penicillium* subgenus *Penicillium*: A guide to identification of food and air-borne terverticillate Penicillia and their mycotoxins, *Stud. Mycol.*, 49, 1, 2004.
4. Pitt, J.I. and Hocking, A.D., *Fungi and Food Spoilage*, 2nd edn., Blackie Academic & Professional, London, U.K., 1997.
5. Richardson, M.D., *Aspergillus* and *Penicillium* species, in *Medical Mycology*, Ajello, L. and Hay, R.J., Eds., Arnold, London, U.K., 1998, p. 281.
6. Hocquette, A. et al., Les champignons des genres *Acremonium, Beauveria, Chrysosporium, Fusarium, Onychocola, Paecilomyces, Penicillium, Scedosporium* et *Scopulariopsis* responsables de hyalohyphomycoses, *J. Mycol. Méd.*, 15, 136, 2005.
7. Vanittanakom, N. et al., *Penicillium marneffei* infection and recent advances in the epidemiology and molecular biology aspects, *Clin. Microbiol. Rev.*, 19, 95, 2006.
8. Cappni, M., Sureau, P., and Segretain, G., Pénicillose de *Rhizomys sinensis*, *Bull. Soc. Pathol. Exot.*, 49, 418, 1956.
9. Supparatpinyo, K. et al., Disseminated *Penicillium marneffei* infection in Southeast Asia, *Lancet*, 344, 110, 1994.
10. Supparatpinyo, K. et al., A controlled trial of itraconazole to prevent relapse of *Penicillium marneffei* infection in patients infected with the human immunodeficiency virus, *N. Engl. J. Med.*, 339, 1739, 1998.
11. Segretain, G., Description d'une nouvelle espèce de penicillium: *Penicillium marneffei* n. sp., *Bull. Soc. Mycol. Fr.*, 75, 412, 1959.
12. Frisvad, R.A. and Filtenborg, O., Revision of *Penicillium* subgenus *Furcatum* based on secondary metabolites and conventional characters, in *Modern Concepts in* Penicillium *and* Aspergillus *classification*, Samson, R.A., Pitt, J.I., Eds., Plenum Press, New York, 1990, p. 159.
13. LoBuglio, K.F. and Taylor, J., Phylogeny and PCR identification of the human pathogenic fungus *Penicillium marneffei*, *J. Clin. Microbiol.*, 33, 85, 1995.
14. Cai, J.J., Understanding the pathogenic fungus *Penicillium marneffei*: A computational genomics perspective, PhD thesis, The University of Hong Kong, 2006.
15. Piehl, M.R., Kaplan, R.L., and Haber, M.H., Disseminated penicilliosis in a patient with acquired immunodeficiency syndrome, *Arch. Pathol. Lab. Med.*, 112, 1262, 1988.
16. Sathapatayavongs, B. et al., Disseminated penicilliosis associated with HIV infection, *J. Infect.*, 19, 84, 1989.
17. Pócsi, Z. et al., Yeast-like cell formation and glutathione metabolism in autolysing cultures of *Penicillium chrysogenum*, *Acta Biol. Hungar.*, 58, 431, 2007.
18. Devlieghere, F., Debevere, J., and Van Impe, J., Concentration of carbon dioxide in the water-phase as a parameter to model the effect of a modified atmosphere on microorganisms, *Int. J. Food Microbiol.*, 43, 105, 1998.
19. Cao, C. et al., The effects of temperature, pH, and salinity on the growth and dimorphism of *Penicillium marneffei*, *Med. Mycol.*, 45, 401, 2007.
20. Viviani, M.A. and Tortorano, A.M., *Penicillium marneffei*, in *Topley & Wilson's Microbiology and Microbial Infections*, Ajello, L. and Hay, R.J., Eds., 9th edn., Medical Mycology, Arnold, London, U.K., 1989, p. 409.
21. Mapari, S.A.S. et al., Identification of potentially safe promising fungal cell factories for the production of polyketide natural food colorants using chemotaxonomic rationale, *Microb. Cell Fact.*, 8, 24, 2009.
22. Ehrlich, K.C. et al., Secalonic acid D: Natural contaminant of corn dust, *Appl. Environ. Microbiol.*, 44, 1007, 1982.
23. Bhardwaj, S. et al., Putative structure and characteristics of a red water-soluble pigment secreted by *Penicillium marneffei*, *Med. Mycol.*, 45, 419, 2007.
24. Youngchim, S., Vanittanakom, N., and Hamilton, A.J., Analysis of the enzymatic activity of mycelial and yeast phases of *Penicillium marneffei*, *Med. Mycol.*, 37, 445, 1999.
25. Wong, S.S. et al., Biotyping of *Penicillium marneffei* reveals concentration-dependent growth inhibition by galactose, *J. Clin. Microbiol.*, 39, 1416, 2001.
26. Gugnani, H.C. et al., *Cannomys badius* as a natural animal host of *Penicillium marneffei* in India, *J. Clin. Microbiol.*, 42, 5070, 2004.
27. Vanittanakom, N. and Sirisanthana, T., *Penicillium marneffei* infection in patients infected with human immunodeficiency virus, *Curr. Top. Med. Mycol.*, 8, 35, 1997.
28. Pongpom, P., Cooper, Jr. C.R., and Vanittanakom, N., Isolation and characterization of a catalase-peroxidase gene from the pathogenic fungus, *Penicillium marneffei*, *Med. Mycol.*, 43, 403, 2005.

* The trial is now published in: Rerks-Ngarm, P. et al., Vaccination with ALVAC and AIDSVAX to prevent HIV-1 infection in Thailand. *N. Engl. J. Med.*, 361, 2209, 2009.

29. Chariyalertsak, S. et al., *Rhizomys sumatrensis* and *Cannomys badius*, new natural animal hosts of *Penicillium marneffei*, *J. Med. Vet. Mycol.*, 34, 105, 1996.
30. Chariyalertsak, S. et al., Case-control study of risk factors for *Penicillium marneffei* infection in human immunodeficiency virus-infected patients in northern Thailand, *Clin. Infect. Dis.*, 24, 1080, 1997.
31. Frisvad, J.C. et al., The homogeneous species and series in subgenus *Penicillium* are related to mammal nutrition and excretion, in *Integration of Modern Taxonomic Methods for Penicillium and Aspergillus Classification*, Samson, R.A. and Pitt, J.I., Eds., Harwood Academic, the Netherlands, 2000, p. 265.
32. Vanittanakom, N. et al., Efficiency of flotation method in the isolation of *Penicillium marneffei* from seeded soil, *J. Med. Vet. Mycol.*, 33, 271, 1995.
33. Drouhet, E., Penicilliosis due to *Penicillium marneffei*: A new emerging systemic mycosis in AIDS patients travelling or living in Southeast Asia. Review of 44 cases reported in HIV infected patients during the last 5 years compared to 44 cases of non AIDS patients reported over 20 years, *J. Mycol. Med. (Paris)*, 4, 195, 1993.
34. Hamilton, A.J. et al., Recognition of fibronectin by *Penicillium marneffei* conidia via a sialic acid dependent process and its relationship to the interaction between conidia and laminin, *Infect. Immun.*, 67, 5200, 1999.
35. Hamilton, A.J. et al., Sialic-acid dependent recognition of laminin by *Penicillium marneffei* conidia, *Infect. Immun.*, 66, 6024, 1998.
36. Supparatpinyo, K. et al., *Penicillium marneffei* infection in patients infected with human immunodeficiency virus, *Clin. Infect. Dis.*, 14, 871, 1992.
37. Sisto, F. et al., Differential cytokine pattern in the spleens and livers of BALB/c mice infected with *Penicillium marneffei*: Protective role of gamma interferon, *Infect. Immun.*, 71, 465, 2003.
38. Koguchi, Y. et al., *Penicillium marneffei* causes osteopontin-mediated production of interleukin-12 by peripheral blood mononuclear cells, *Infect. Immun.*, 70, 1042, 2002.
39. Estrada, J.A. et al., Immunohistochemical identification of *Penicillium marneffei* by monoclonal antibody, *Int. J. Dermatol.*, 31, 410, 1992.
40. Notermans, S. et al., (1–5) linked beta-D-galctofuranosides are immunodominant in extracellular polysaccharides of *Penicillium* and *Aspergillus* species, *Mol. Immunol.*, 25, 975, 1988.
41. Vanittanakom, N. et al., *Penicillium marneffei*, past, present and future identification methods, in *The 15th Congress of the International Society for Human and Animal Mycology*, McGinnis, M.R., Rinaldi, M.G., and Graybill, J.R., Eds., San Antonio, TX, 2003, p. 273.
42. Kaufman, L. et al., Development of specific fluorescent-antibody test for tissue form of *Penicillium marneffei*, *J. Clin. Microbiol.*, 33, 2136, 1995.
43. Trewatcharegon, S. et al., Production and characterization of monoclonal antibodies reactive with the mycelial and yeast phases of *Penicillium marneffei*, *Med. Mycol.*, 38, 91, 2000.
44. Chaiwun, B. et al., Lymhadenopathy due to *Penicillium marneffei* infection: Diagnosis by fine needle aspiration cytology, *Mod. Pathol.*, 15, 939, 2002.
45. Ma, K.F., Tsui, M.S., and Tsang, D.N., Fine needle aspiration diagnosis of *Penicillium marneffei* infection, *Acta Cytol.*, 35, 557, 1991.
46. Sekhon, A.S., Li, J.S.K., and Garg, A.K., Penicilliosis marneffei: Serological and exoantigen studies, *Mycopathologia*, 77, 51, 1982.
47. Sekhon, A.S., Garg, A.K., and Padhye, A.A., Antigenic relationship of *Penicillium marneffei* to *P. primulinum*, *J. Med. Vet. Mycol.*, 27, 105, 1989.
48. Viviani, M.A. et al., Treatment and serological studies of an Italian case of penicilliosis marneffei contracted in Thailand by a drug addict infected with the human immunodeficiency virus, *Eur. J. Epidemiol.*, 9, 79, 1993.
49. Kaufman, L. et al., Diagnostic antigenemia tests for penicilliosis marneffei, *J. Clin. Microbiol.*, 34, 2503, 1996.
50. Yuen, K.Y. et al., Serodiagnosis of *Penicillium marneffei* infection, *Lancet*, 344, 444, 1994.
51. Vanittanakom, N. et al., Western immunoblot analysis of protein antigens of *Penicillium marneffei*, *J. Med. Vet. Mycol.*, 35, 123, 1997.
52. Chongtrakool, P. et al., Immunoreactivity of a 38-kilodalton *Penicillium marneffei* antigen with human immunodeficiency virus-positive sera, *J. Clin. Microbiol.*, 35, 2220, 1997.
53. Jeavons, L. et al., Identification and purification of specific *Penicillium marneffei* antigens and their recognition by human immune sera, *J. Clin. Microbiol.*, 36, 949, 1998.
54. Cao, L. et al., Detection of specific antibodies to an antigenic mannoprotein for diagnosis of *Penicillium marneffei* penicilliosis, *J. Clin. Microbiol.*, 36, 3028, 1998.
55. Van Cutsem, J., Fungal models in immunocompromised animals, in *Mycoses in AIDS Patients*, Van Den Bossche, H. et al., Eds., Plenum Press, New York, 1990, p. 207.
56. Desakorn, V. et al., Diagnosis of *Penicillium marneffei* infection by quantitation of urinary antigen by using an enzyme immunoassay, *J. Clin. Microbiol.*, 37, 117, 1999.
57. Desakorn, V. et al., Development and evaluation of rapid urinary antigen detection tests for diagnosis of penicilliosis marneffei, *J. Clin. Microbiol.*, 40, 3179, 2002.
58. Panichakul, T. et al., Development of a monoclonal antibody-based enzyme-linked immunosorbent assay for the diagnosis of *Penicillium marneffei* infection, *Am. J. Trop. Med. Hyg.*, 67, 443, 2002.
59. Pitt, J.I., Toxigenic fungi and mycotoxins, *Br. Med. Bull.*, 20, 184, 2000.
60. Paterson, R.R.M. and Lima, N., Toxicology of mycotoxins, in *Molecular, Clinical and Environmental Toxicology. Volume 2: Clinical Toxicology*, Luch, A., Ed., Birkhäuser Verlag, Switzerland, 2009, p. 31.
61. Paterson, R.R.M., Fungi and fungal toxins as weapons, *Mycol. Res.*, 110, 1003, 2006.
62. CAST, *Mycotoxins: Risks in Plant, Animal, and Human Systems*, Council for Agricultural Science and Technology, Ames, IA, 2003.
63. Marasas, W.F.O. and Nelson, P.E., *Mycotoxicology*, Pennsylvania State University Press, PA, 1987.
64. Riley, R.T. and Showker, J.L., The mechanism of patulin's cytotoxicity and the antioxidant activity of indole tetramic acids, *Toxicol. Appl. Pharmacol.*, 109, 108, 1991.
65. Paterson, R.R.M. et al., Mutagenic and inhibitory compounds produced by fungi affect detrimentally their PCR-based diagnosis and phylogenetic analyses. *Curr. Bioact. Comp.*, 4, 245, 2008.
66. Palma, N. et al., Ochratoxin A-induced mutagenesis in mammalian cells is consistent with the production of oxidative stress, *Chem. Res. Toxicol.*, 20, 1031, 2007.
67. Schumacher, D.M. et al., DNA–DNA cross-links contribute to the mutagenic potential of the mycotoxin patulin, *Toxicol. Lett.*, 166, 268, 2006.

68. Knaus, H.-G. et al., Tremorgenic indole alkaloids potently inhibit smooth muscle high-conductance calcium-activated potassium channels, *Biochemistry*, 33, 5819, 1994.
69. Pitt, J.I., *The genus Penicillium and Its Teleomorphic States Eupenicillium and Talaromyces*, Academic Press, London, U.K., 1979.
70. Peterson, S.W., Phylogenetic analysis of *Penicillium* species based in ITS and LSU-rDNA nucleotide sequences, in *Integration of Modern Taxonomic Methods for* Penicillium *and* Aspergillus *Classification*, Samson, R.A. and Pitt, J.I., Eds., Harwood Academic, Amsterdam, the Netherlands, 2000, p. 163.
71. Bridge, P.D. et al., A reappraisal of the terverticillate penicillia using biochemical, physiological, and morphological features. I. Numerical taxonomy, *J. Gen. Microbiol.*, 135, 2941, 1989.
72. Seifert, K.A. and Lous-Seize, G., Phylogeny and species concepts in the *Penicillium aurantiogriseum* complex as inferred from partial β-tubulin gene DNA sequences, in *Integration of Modern Taxonomic Methods for Penicillium and Aspergillus Classification*, Samson, R.A. and Pitt, J.I., Eds., Harwood Academic, Amsterdam, the Netherlands, 2000, p. 189.
73. Samson, R.A. et al., Phylogenetic analysis of *Penicillium* subgenus *Penicillium* using β-tubulin sequences, *Stud. Mycol.*, 49, 175, 2004.
74. Paterson, R.R.M. Book review, *Mycotaxon*, 91, 519–521, 2005.
75. Paterson, R.R.M. et al., A reappraisal of the terverticillate penicillia using biochemical, physiological, and morphological features. III. An evaluation of pectinases and amylase isozymes for species characterization, *J. Gen. Microbiol.*, 135, 2979, 1989.
76. Paterson, R.R.M., Venâncio, A., and Lima, N., A novel identification system based on 318 penicillia strains using the isoepoxydon dehydrogenase gene and patulin production, *Rev. Ibero. Micol.*, 23, 155, 2006.
77. Dupont, J., *Penicillium*, in *Molecular Detection of Foodborne Pathogens*, Liu, D., Ed., CRC Press, Boca Raton, FL, 2009, pp. 585–594.
78. White, T.J. et al., Amplification and direct sequencing of fungal ribosomal RNA genes for phylogenetics, in *PCR Protocols: A Guide to Methods and Applications*, Innis, M.A. et al., Eds., Academic Press Inc., New York, 1990, p. 315.
79. Tsunemi, Y., Takahashi, T., and Tamaki, T., *Penicillium marneffei* infection diagnosed by polymerase chain reaction from the skin specimen, *J. Am. Acad. Dermatol.*, 49, 344, 2003.
80. Lindsley, M.D. et al., Rapid identification of dimorphic and yeast-like fungal pathogens using specific DNA probes, *J. Clin. Microbiol.*, 39, 3505, 2001.
81. Vanittanakom, N. et al., Specific identification of *Penicillium marneffei* by a polymerase chain reaction/hybridization technique, *Med. Mycol.*, 36, 169, 1998.
82. Vanittanakom, N., Vanittanakom, P., and Hay, R.J., Rapid identification of *Penicillium marneffei* by PCR-based detection of specific sequences on the rRNA gene, *J. Clin. Microbiol.*, 40, 1739, 2002.
83. Prariyachatigul, C. et al., Development and evaluation of a one-tube seminested PCR assay for the detection and identification of *Penicillium marneffei*, *Mycoses*, 46, 447, 2003.
84. Vanittanakom, N. et al., Restriction endonuclease analysis of *Penicillium marneffei*, *J. Clin. Microbiol.*, 34, 1834, 1996.
85. Hsueh, P.R. et al., Molecular evidence for strain dissemination of *Penicillium marneffei*: An emerging pathogen in Taiwan, *J. Infect. Dis.*, 181, 1712, 2000.
86. Imwidthaya, P. et al., *Penicillium marneffei*: Types and drug susceptibility, *Mycopathologia*, 149, 109, 2001.
87. Trewatcharegon, S. et al., Molecular typing of *Penicillium marneffei* isolates from Thailand by NotI macrorestriction and pulsed-field gel electrophoresis, *J. Clin. Microbiol.*, 39, 4544, 2001.
88. Yuen, K.Y. et al., Exploring the *Penicillium marneffei* genome, *Arch. Microbiol.*, 179, 339, 2003.
89. Fisher, M.C., de Hoog, S., and Vanittanakom, N., A highly discriminatory multilocus microsatellite typing (MLMT) system for *Penicillium marneffei*, *Mol. Ecol. Notes*, 4, 515, 2004.
90. Lasker, B.A. and Ran, Y., Analysis of polymorphic microsatellite markers for typing *Penicillium marneffei* isolates, *J. Clin. Microbiol.*, 42, 1483, 2004.
91. Fisher, M.C. et al., Multilocus microsatellite typing system for *Penicillium marneffei* reveals spatially structured populations, *J. Clin. Microbiol.*, 42, 5065, 2004.
92. Ranjana, K.H. et al., Disseminated *Penicillium marneffei* infection among HIV-infected patients in Manipur state, India, *J. Infect.*, 45, 268, 2002.
93. Pedersen, L.H. et al., Detection of *Penicillium* species in complex food samples using the polymerase chain reaction, *Int. J. Food Microbiol.*, 35, 169, 1997.
94. Marek, P., Annamalai, T., and Venkitanarayanan, K., Detection of *Penicillium expansum* by polymerase chain reaction, *Int. J. Food. Microbiol.*, 89, 139, 2003.
95. Calderon, C. et al., Detection of airborne fungal spores sampled by rotating arm and Hirst-type spore traps using polymerase chain reaction assays, *J. Aerosol. Sci.*, 33, 283, 2002.
96. Colombo, F. et al., Identification of *Penicillium aurantiogriseum* species with a method of polymerase chain reaction-restriction fragment length polymorphism, *Food Contr.*, 14, 137, 2003.
97. Sholberg, P.L. et al., Benzimidazole and diphenylamine sensitivity and identity of *Penicillium* spp. that cause postharvest blue mold of apples using β-tubulin gene sequences, *Postharvest Biol. Technol.*, 36, 41, 2005.
98. Dupont, J. et al., Molecular tools for identification of *Penicillium* starter cultures used in the food industry, *Int. J. Food Microbiol.*, 49, 109, 1999.
99. Dao, H.P., Mathieu, F., and Lebrihi, A., Two primer pairs to detect OTA producers by PCR method, *Int. J. Food Microbiol.*, 104, 61, 2005.
100. Paterson, R.R.M., Identification and quantification of mycotoxigenic fungi by PCR, *Process. Biochem.*, 41, 1467, 2006.
101. Geisen, R. et al., Development of a real time PCR system for detection of *Penicillium nordicum* and for monitoring ochratoxin A production in foods by targeting the ochratoxin polyketide synthase gene system, *Appl. Microbiol.*, 27, 501, 2004.
102. Bogs, C., Battilani, P., and Geisen, R., Development of a molecular detection and differentiation system for ochratoxin A producing *Penicillium* species and its application to analyse the occurrence of *Penicillium nordicum* in cured meats, *Int. J. Food Microbiol.*, 107, 39–47, 2006.
103. Paterson, R.R.M., The isoepoxydon dehydrogenase gene of patulin biosynthesis in cultures and secondary metabolites as candidate PCR inhibitors, *Mycol. Res.*, 108, 1431, 2004.

104. Paterson, R.R.M., Primers from the isoepoxydon dehydrogenase gene of the patulin biosynthetic pathway to indicate critical control points for patulin contamination of apples, *Food Control*, 7, 741, 2006.
105. Dombrink-Kurtzman, M.A., The isoepoxydon dehydrogenase gene of the patulin metabolic pathway differs for *Penicillium griseofulvum* and *Penicillium expansum*, *Ant. Leeuwen.*, 89, 1, 2005.
106. Dombrink-Kurtzman, M.A., The sequence of the isoepoxydon dehydrogenase gene of the patulin biosynthetic pathway in *Penicillium* species, *Ant. Leeuwen.*, 91, 179, 2007.
107. Seifert, K.A. et al., Prospects for fungus identification using CO1 DNA barcodes, with *Penicillium* as a test case. *Proc. Natl. Acad. Sci. U.S.A.*, 104, 3901, 2007.
108. Boysen, M. et al., Reclassification of the *Penicillium roqueforti* group into three species on the basis of molecular genetic and biochemical profiles, *Microbiology*, 142, 541, 1996.
109. LoBuglio, K.F., Pitt, J.I., and Taylor, J.W., Phylogenetic analysis of two ribosomal DNA regions indicates multiple independent losses of a sexual *Talaromyces* state among asexual *Penicillium* species in subgenus *Biverticillium*, *Mycologia*, 85, 592, 1993.
110. Lyratzopoulos, G. et al., Invasive infection due to *Penicillium* species other than *P. marneffei*, *J. Infect.*, 45, 184, 2002.
111. Paterson, R.R.M., Fungus or bacterium and *vice versa*? *Microbiology*, 151, 641, 2005.
112. Paterson, R.R.M. and Lima, N., How will climate change affect mycotoxins? *Food Int. Res.*, 43, 1902, 2010.

42 *Phialophora*

Dongyou Liu and R.R.M. Paterson

CONTENTS

42.1 INTRODUCTION

42.1.1 Classification, Morphology, and Biology

The genus *Phialophora* is a dematiaceous (dark-walled) filamentous fungus belonging to the mitosporic Magnaporthaceae group, family Magnaporthaceae, order Magnaporthales, class Sordariomycetes, subphylum Pezizomycotina, phylum Ascomycota. The obsolete synonyms of the genus include *Cadophora* and *Margarinomyces*, and teleomorphs of this genus are found in *Chaetosphaeria*, *Gaeumannomyces*, and *Mollisia*.

Currently, the genus *Phialophora* contains 16 recognized and over 69 unassigned species, the most notable of which are *Phialophora bubakii*, *Phialophora europaea*, *Phialophora gregata*, *Phialophora repens* (obsolete synonym *Cadophora repens*), *Phialophora reptans*, *Phialophora richardsiáe* (obsolete synonyms: *Cadophora brunnescens*, *Cadophora richardsiae*, *P. brunnescens*, and *P. caliciformis*), and *Phialophora verrucosa* (synonym: *P. americana*; obsolete synonym: *Cadophora americana*; and teleomorph: *Capronia semiimmersa*) [1].

A number of former *Phialophora* species have been reclassified: *Phialophora compactum* (synonym: *Fonsecaea compacta*), *Phialophora cyanescens* (synonym: *Cylindrocarpon cyanescens*), *Phialophora dermatitidis* (synonyms: *Exophiala dermatitidis* and *Wangiella dermatitidis*), *Phialophora hoffmannii* (synonym: *Lecythophora hoffmannii*), *Phialophora jeanselmei* (synonym: *Exophiala jeanselmei var. jeanselmei*), *Phialophora mutabilis* (synonym: *Lecythophora mutabilis*), *Phialophora parasitica* (synonym: *Phaeoacremonium parasiticum*), *Phialophora pedrosoi* (synonym: *Fonsecaea pedrosoi*), and *Phialophora spinifera* (synonym: *Exophiala spinifera*).

Phialophora is a ubiquitous organism that inhabits the soil, plants, and decaying food [2,3]. Several *Phialophora* species are the causative agents of human infections such as chromoblastomycosis, phaeohyphomycosis, and eumycetoma. *P. verrucosa* and *P. richardsiae* are often involved in phaeohyphomycosis, with clinical presentations ranging from cutaneous infections, subcutaneous cysts, keratitis, endocarditis, arthritis, osteomyelitis, cerebral infection, fatal hemorrhage, and disseminated infection. *Phialophora verrucosa* is the principal causative agent of chromoblastomycosis in tropical and subtropical areas, particularly in Japan and South America. *Phialophora europaea* has been isolated from cutaneous and nail infections in North-Western Europe.

Phialophora has short conidiophores (sometimes reduced to phialides) and unicellular conidia, although conidia are sometimes absent. *Phialophora* colonies grow moderately slowly and reach a diameter of 2–3 cm after 7 days at 25°C. Colonies are gray to olivaceous black and often become dark gray-green, brown, or black with age. From the reverse, colonies appear iron-gray to black; colony texture is cottony to velvety and may be heaped and granular.

Phialophora hyphae (up to 5 μm wide) are branched and hyaline to brown. If present, condiophores are short and pale brown; conidiogenous cells (phialides) have characteristic flask-shaped or cylindrical phialides with distinctive collarettes, which are terminally or laterally located on the hyphae;

conidia are one celled, hyaline to olivaceous brown, smooth walled, ovoid to cylindrical or allantoid, and often aggregate in slimy heads at the apices of the phialides, giving the appearance of a vase of flowers; phialides may be solitary or in a brush-like arrangement. In tissues infected with *Phialophora* species, spherical or polyhedral, dark brown, thick-walled sclerotic bodies (muriform cells) and phaeoid hyphae are often present, although absence of muriform cells is noted in cases involving immunosuppressed or debilitated patients.

At the species level, *Phialophora ripens* produces intercalary phialides without basal septa or cylindrical to lageniform phialides (<20 μm in length) with narrow collarettes; *Phialophora richardsiae* conidia are globose to cylindrical and phialides with saucer- or vase-shaped collarettes; and *Phialophora verrucosa* produces flask-shaped phialides with vase-shaped collarettes; *Phialophora europaea*—a member of the *P. verrucosa* complex—shows reduced, flaring phialidic collarettes [4].

42.1.2 Clinical Features and Pathogenesis

The term dematiaceous (dark-pigmented) fungi refers to a large and heterogeneous group of molds belonging to several different genera. Some are capable of causing a diverse range of clinical diseases such as phaeohyphomycosis, chromoblastomycosis, and eumycetoma. The most notable human pathogens in the dematiaceous fungal group include *Alternaria*, *Bipolaris*, *Cladophialophora*, *Curvularia*, *Exophiala* (*Wangiella*), *Fonsecaea*, *Madurella*, *Phialophora*, *Rhinocladiella*, *Scedosporium*, and *Scytalidium*. Being widely distributed in the environment including soil, wood, and decomposing plant debris, these organisms have the capacity to take advantage of weakened defense mechanism of human hosts, inducing cutaneous, subcutaneous, and corneal infections, invasive and allergic sinusitis, arthritis, endocarditis, and systemic and disseminated infections. Traumatic implantation, injury, and inhalation represent the main routes of entry, and host immune deficiency (suppression) plays a critical role in the development of systemic and disseminated infections [5,6].

Pheohyphomycosis affects the skin and subcutis, paranasal sinuses, or central nervous system and is often categorized into four types by the region of the body in which the infection occurs: superficial, cutaneous, subcutaneous, and systemic [2]. Chromoblastomycosis is a cutaneous and subcutaneous mycotic disease (mostly of the feet and legs) with the appearance of pigmented sclerotic bodies, commonly called "copper pennies" [5]. Eumycetoma is a subcutaneous fungal infection (mostly of the feet and legs) showing a mass with multiple sinuses draining pus and "grains." The grains (0.5–1.5 mm in diameter) in eumycetoma may appear white (due to infection with *Acremonium*, *Cylindrocarpon*, *Fusarium*, *Neotestudina*, or *Pseudallescheria*) or black (due to infection with *Exophiala*, *Madurella*, *Leptosphaeria*, *Pyrenochaeta*, or *Phlenodomus*).

Phialophora spp. have been shown to cause chromoblastomycosis, mycotic keratitis, cutaneous infections, osteomyelitis, and prosthetic valve endocarditis as well as disseminated infections [7–13], with *P. verrucosa* and *P. richardsiae* being most commonly reported [9,14–17].

42.1.2.1 *Phialophora verrucosa*

A case of fatal hemorrhage due to *Phialophora verrucosa* occurred in a patient with prolonged neutropenia undergoing autologous bone marrow transplant for acute myelogenous leukemia. Autopsy uncovered granulation tissue on the tracheal wall, which showed fungal hyphae by histopathological examination and grew *P. verrucosa* in culture [18].

In a case of *Phialophora verrucosa*-related multifocal subcutaneous phaeohyphomycosis, an 85-year-old woman presented with purulent multifocal subcutaneous nodules on the right forearm and hand. The patient reported a history of gardening and recalled the occurrence of a painless, soybean-sized nodule on the dorsal site of the right forearm 3 years earlier, which gradually extended peripherally to the dorsum of the right hand. The pustules and nodules (ranging from 5×5 mm to 20×40 mm in size) showed a yellow crust on the surface and were surrounded by erythema and discharged a purulent fluid upon pressure. Examination of the biopsied specimen revealed a subcutaneous granuloma (containing faint brown septate hyphae with rare branching) with central abscess and necrosis, consistent with a phaeomycotic cyst. Culture of the discharged purulent fluid on potato dextrose agar (PDA) at room temperature grew a nearly black fungus, attaining a diameter of 20 mm in 14 days and 50 mm in 35 days. Under light microscope, flask-shaped phialides (conidiogenous cells) with cup-shaped collarettes and a bunch of elliptical conidia at the apices of the collarettes were noted, along with brown to light-brown hyphae. These morphologic features confirmed the fungus as *P. verrucosa*, which responded to oral itraconazole treatment [19].

A cutaneous fungal infection due to *P. verrucosa* was reported in a 53-year-old woman who had a history of systemic corticosteroid treatment for Evans' syndrome. The patient presented asymptomatic multiple nodules on her gluteal region. A smear from the purulent exudate of the nodules showed brown-colored hyphae, spores, and large dark-brown cells; and a dark-brown-colored colony grew on Sabouraud glucose agar, showing *Phialophora*-type conidia formation. The fungus was diagnosed as *Phialophora verrucosa* [20].

Phialophora verrucosa was also shown to be responsible for a severe, verrucous facial mycosis and sinusitis in a 12-year-old girl. The disease began with the development of verrucous, hyperkeratotic plaques, and subcutaneous violet nodules on the face and upper extremities and later evolved into severely progressive tumorous cutaneous and nasal lesions. Microscopic examination of scale samples from the upper extremities and the face showed brown, thick-walled fungal elements, which were subsequently identified as *Phialophora verrucosa*. Oral treatment with itraconazole and fluorocytosin resulted in significant improvement of the lesions [17].

A case of polymicrobial keratitis caused by *Phialophora verrucosa*, *Candida tropicalis*, and *Propionibacterium acnes* has also been described. Histopathologic evaluation revealed penetration of the crystalline lens by fungus at the site of synechiae between the intact lens capsule and iris [13].

42.1.2.2 *Phialophora richardsiae*

A case of subcutaneous phaeohyphomycosis due to *Phialophora richardsiae* was described in a renal transplant recipient. The patient developed dark-colored nodules that subsequently ulcerated after transplantation. Histopathologic examination showed dematiaceous fungal hyphae with a surrounding granulomatous reaction, and the fungus was identified as *P. richardsiae* [21].

Diseases due to dematiaceous fungi require antifungal and surgical treatments. Current antifungal therapy for chromoblastomycosis and subcutaneous phaeohyphomycosis relies on itraconazole, and for eumycetoma it is ketoconazole. Surgical debridement in combination with amphotericin B treatment is useful for chronic invasive sinusitis [6,9].

42.1.3 Laboratory Diagnosis

Phialophora represents one of the dematiaceous fungi responsible for causing phaeohyphomycosis and chromoblastomycosis in humans. There is a need to differentiate among the other dematiaceous fungi in order to implement appropriate treatment regimens as these organisms demonstrate varied sensitivity to the currently available antifungal drugs [6].

Dematiaceous fungi including *Phialophora* spp. have been traditionally identified by macroscopic and microscopic characteristics as well as biochemical properties. For example, phospholipase production, induced by egg yolk substrate, offers a useful means for the differentiation of the taxonomically related dematiaceous fungal species on the basis of their distinct enzymatic profiles [22]. Nevertheless, these methods are laborious and sometimes incapable of distinguishing species with polymorphic conidiogeneses or lacking conidia formation.

For these reasons, molecular techniques such as random amplified polymorphic DNA (RAPD), restriction fragment length polymorphism (RFLP), and DNA sequencing have been adopted increasingly for identification [23]. In particular, sequencing analysis of small subunit (SSU) 18S and large subunit (LSU) 28S rRNA genes and internal transcribed spacer regions (ITS) has proven valuable. Indeed, many *Phialophora* species can be correctly identified by examining the nucleotide sequence of the D1/D2 domains in the 28S rRNA gene [24].

42.2 METHODS

42.2.1 Sample Preparation

Clinical specimens suspected of containing dematiaceous fungi are treated with 10% KOH or stained with calcofluor white and observed by microscopy for septate hyphal elements. Samples are then inoculated onto inhibitory mold agar, modified Sabouraud agar, or PDA at 25°C, 35°C, and 50°C, respectively. Preliminary identification of the colonies is based on macroscopic and microscopic features [25].

After the growth for 1–7 days on PDA slants, approximately 1 cm^2 of mycelium is collected by scraping the slant with a sterile stick that is placed in 1 mL of sterile H_2O. The material is transferred to a 2 mL screw-cap tube, and the tube is centrifuged for 1 min at 6000×*g*. If the mycelium does not pellet, the material is passed through a pediatric blood serum filter. After removal of the supernatant, the material is resuspended in 200 μL of IDI sample buffer (IDI lysis kit, BD) and transferred to the lysis tube containing glass beads. The lysis tube is vortexed on the highest setting for 5 min, and the tube is placed in a boiling water bath for 15 min. Following centrifugation for 5 min at 16,000×*g*, the supernatant is stored at −20°C until polymerase chain reaction (PCR) is undertaken [25].

Alternatively, about 1 cm^2 of fungal material is transferred to a 2 mL Eppendorf tube containing a 2:1 (wt/wt) mixture of silica gel and Celite (silica gel H, Merck 7736/Kieselguhr Celite 545; Machery) and 300 μL of TES buffer (2 g Tris [hydroxymethyl]-aminomethane, 0.38 g Na-EDTA, and 2 g sodium dodecyl sulfate in 80 mL of ultrapure water pH 8). The fungal material is ground with a micropestle for 1–2 min. The volume is adjusted by adding 200 μL of TES buffer. After vigorous shaking and the addition of 10 μL of a 10 mg/mL concentration of proteinase K to the tube, the mixture is incubated at 65°C for 10 min. With the addition of 140 μL of 5 M NaCl solution, the mixture is combined with 1/10 volume (~65 μL) of cetyltrimethylammonium bromide (CTAB) buffer 10%, followed by incubation for another 30 min at 65°C. One volume (~700 μL) of chloroform–isoamyl alcohol (vol/vol = 24/1) is added and mixed by inversion. After incubation for 30 min at 0°C (on ice water) and centrifugation at 14,000 rpm at 4°C for 10 min, the top layer is transferred to a clean Eppendorf tube. The sample is added with 225 μL of 5 M NH_4–acetate and incubated for at least 30 min (on ice water) and centrifuged. The supernatant is transferred to a clean sterile Eppendorf tube and mixed with a 0.55 volume (~510 μL) of ice-cold isopropanol. After centrifugation for 7 min at 14,000 rpm and 4°C (or room temperature), the supernatant is decanted. The pellet is washed with ice-cold ethanol 70% twice and air dried. The powder is resuspended in 48.5 μL of Tris–EDTA buffer with 1.5 μL of 10 mg of RNase/mL, incubated at 37°C for 15–30 min, and stored at −20°C until use [26].

In addition, fungal strains can be cultured in 20 mL of RPMI 1640 medium with L-glutamine but without sodium bicarbonate (Sigma-Aldrich) buffered to pH 7.0 with 0.165 M morpholinepropanesulfonic acid (MOPS) (Sigma-Aldrich). After 3–14 days of growth with agitation (100 rpm) at 30°C, the mycelium is transferred into a tube and washed in 40 mL of sterile distilled water. Mycelium is then stored at −20°C until use. For DNA extraction, 100 mg of mycelium is homogenized for 1 min in a tube containing 1 mL of lysis buffer (2% Triton X-100, 1% sodium dodecyl sulfate, 10 mM Tris–HCl pH 8, 100 mM NaCl, and 1 mM EDTA pH 8), three 0.5 cm diameter glass beads (Sigma), and approximately 500 mg of 425–600 μm glass beads (Sigma). The homogenized mycelium is snap-frozen in liquid nitrogen, thawed, and refrozen once. DNA is then extracted using a DNeasy plant kit (Qiagen) [27].

Also, fungal DNA may be prepared by suspending 50 mg of mycelium in 600 µL extraction buffer (200 mM Tris–HCl, pH 7.5, 25 mM EDTA, 0.5% w/v sodium dodecyl sulfate, and 250 mM NaCl). The mixture is vortexed for 15 s, incubated at 100°C for 15 min, kept on ice for 60 min, and then centrifuged at 14,000 × *g* for 15 min. Supernatants are transferred to new tubes and extracted with phenol–chloroform–isoamyl alcohol (25:24:1 v/v). Each sample DNA is precipitated with cold isopropanol (−20°C), dried, and resuspended in 100 µL distilled water [24].

42.2.2 Detection Procedures

42.2.2.1 Pan-Fungal Real-Time PCR and Sequencing Analysis

Pounder [25] reported a real-time PCR with SYBR green DNA-binding dye and amplicon-melting temperature analysis for fungal detection using pan-fungal primers ITS1 forward (5′-TCCGTAGGTGAACCTGCGG-3′) and ITS4 reverse (5′-TCCTCCGCTTATTGATATGC-3′) [28]. The identity of the fungus is confirmed by sequencing.

Procedure

1. PCR mixture consists of 1 × Lightcycler FastStart DNA Master Hybridization Probes mixture (Roche Applied Science) containing deoxynucleoside triphosphates, FastStart *Taq* DNA polymerase, and 1 mM $MgCl_2$ (additional $MgCl_2$ is added to a final concentration of 4.6 mM), 0.4 µM each of ITS1 forward and ITS4 reverse primers, 1 × SYBR green (Molecular Probes), and 3 µL template DNA.
2. Thermal cycling parameters include 95°C for 10 min; 50 cycles of 95°C for 5 s, 60°C for 20 s, and 72°C for 30 s; and a final 72°C for 2 min.
3. The quality of the amplicon is determined using the derivative of the melt analysis curve (55°C–99°C, 45 s hold at 55°C, and 5 s/°C) using the RotorGene 3000 (Corbett Robotics, Inc).
4. The amplified product is purified for bidirectional sequencing using ExoSAP-IT (USB Corp). Five microliters of Big Dye Terminator Ready Reaction Mix v. 1.1 (Applied Biosystems) is added to 4 µL of each primer (0.8 pmol/µL) and 3 µL of purified PCR product. Cycle sequencing is performed with a 9700 thermal cycler (ABI), using 25 cycles of 96°C for 10 s, 50°C for 5 s, and 60°C for 4 min. Sequencing reaction products are passed through a Sephadex G-50 fine column to remove unincorporated dye terminators. Purified sequencing reaction products are run on an ABI Prism 3100 Genetic Analyzer with a 50 cm capillary array.
5. Sequences are analyzed with the SmartGene Integrated Database Network software version 3.2.3 vr. SmartGene is a web-based software and database system with reference sequences derived from the National Center for Biological Information (NCBI) GenBank repository.

Note. Sequence-based identifications are defined by percent identity: species, ≥99%; genus, 93%–99%; and inconclusive, ≤93%.

42.2.2.2 Sequencing Analysis of the ITS1-5.8S-ITS2 Region

Zeng et al. [26] utilized primers V9G (5′-TTA CGT CCC TGC CCT TTG TA-3′) and LS266 (5′-GCAT TCC CAA ACA ACT CGA CTC-3′) for PCR amplification and sequence analysis of the ITS-5.8S-ITS2 region.

Procedure

1. PCR mixture (50 µL) is composed of 10 mM Tris–HCl pH 8.3, 50 mM KCl, 1.5 mM $MgCl \cdot 6H_2O$, 0.01% gelatin, 200 mM of each dNTP, 25 pmol of each primer, 0.5 U of *Taq* DNA polymerase (Sigma), and 10–100 ng of extracted DNA.
2. The mixture is subjected to an initial 95°C for 4 min, 35 cycles of 94°C for 45 s, 52°C for 30 s, and 72°C for 2 min, and a final 72°C for 7 min in a GenAmp PCR System 9700 thermocycler (Applied Biosystems).
3. Amplicons are purified with GFX columns (GE Healthcare), and both strands are sequenced with each of the two primers separately. The sequencing PCR mixture consists of 1 µL of template DNA (1–10 ng), 3 µL of dilution buffer, 1 µL of BigDye v3.1, and 1 µL of 4 pmol primer filled with 4 µL of MilliQ water to a final volume of 10 µL. The amplification is performed with initial 95°C for 1 min, 30 cycles of 95°C for 10 s, 50°C for 5 s, and 60°C for 2 min. Reaction products are cleaned with Sephadex G-50 Fine (GE Healthcare Bio-Sciences) and analyzed by using an ABI Prism 3730xl DNA analyzer (Applied Biosystems).
4. The sequences are adjusted by using the Lasergene software program SeqMan II (DNASTAR, Inc.) and aligned iteratively using Ward's averaging in the Bionumerics package v. 4.0 (Applied Maths). Nearest neighbors are identified by local BLAST searches. The distance trees are based on a realigned file using the DCSE program and calculated by the neighbor-joining (NJ) method of the Treecon package with Kimura-2 correction; only unambiguously aligned positions are taken into account. A total of 100 bootstrap replicates are used for analysis. Bootstrap values of >90 from 100 resampled datasets are shown.

Note. If the similarity of sequences of the ITS region is more than 99% between a studied strain and its nearest neighbor and they are distributed in the same branch of the phylogenetic tree, the strain is regarded as belonging to the same species as its nearest neighbor.

42.2.2.3 Sequencing Analysis of the D1/D2 Domains of 28S rRNA Gene

Abliz et al. [24] utilized primers NL-1 (5′-GCATATCAATAAGCGGAGGAAAAG-3′) and NL-4m (5′-GGTCCGTGTTTCAAGACG-3) to amplify the conserved D1/D2 domains of the 28S rRNA from *Phialophora* and other dematiaceous fungi. Subsequent analysis of the amplicon with primers NL-1 and NL-4m and internal primers NL-2m (5′-CTTGTGCGCTATCGGTCTC-3′) and NL-3m (5′-GAGACCGATAGCGCACAAG-3′) permitted the correct identification of *Phialophora* spp.

Procedure

1. PCR mixture (50 μL) is made up of 5 μL of template DNA, 5 μL (2 pmol) each of primers NL-1 and NL-4m, 4 μL (2.5 mM) dNTP mixture, 0.25 μL (5 U/μL) *Taq* polymerase, and 5 μL 10× reaction buffer.
2. Amplification is conducted with an initial 95°C for 4 min; 30 cycles of 94°C for 1 min, 55°C for 2.5 min, and 72°C for 2.5 min; and a final 72°C for 10 min.
3. The amplified products are first checked on 1.5% agarose gel, purified with SUPREC™-02 (TaKaRa) and sequenced with an ABI Prism 3100 sequencer after labeling with BigDye™ Terminator Cycle Sequencing Ready Reaction (Applied Biosystems) using external primers, NL-1 and NL-4m, and the internal primers, NL-2m and NL-3m.
4. The sequence data are aligned with CLUSTAL W (version 1.6). Phylogenetic trees are constructed with the NJ method.

42.3 CONCLUSIONS

The genus *Phialophora* is a dematiaceous fungus that represents one of the causes for human phaeohyphomycosis and chromoblastomycosis. The most common *Phialophora* spp. identified in human infections are *P. verrucosa* and *P. richardsiae*, with clinical symptoms ranging from cutaneous and subcutaneous infections, mycotic keratitis, osteomyelitis, endocarditis, and disseminated infections.

Because different dematiaceous fungal taxa demonstrate varied sensitivity and resistance to antifungal reagents, it is important to determine their species identity with the aim of implementing the most effective treatment measures for these diseases [6]. While traditional macroscopic, microscopic, and biochemical methods provide a useful approach for the differentiation of the taxonomically related dematiaceous *Phialophora* spp., they suffer from the shortcomings of being time consuming, laborious, and incapable of distinguishing species with polymorphic conidiogeneses or lacking conidia. Exploiting the nucleotide variations within the ITS by PCR followed by sequencing, offers a valuable tool for precise determination of *Phialophora* species.

REFERENCES

1. The UniProt Consortium. Available at http://www.uniprot.org/, accessed on August 1, 2010.
2. Revankar SG et al., Disseminated phaeohyphomycosis: Review of an emerging mycosis. *Clin Infect Dis.* 2002;34:467–476.
3. Vicente VA et al., Environmental isolation of black yeast-like fungi involved in human infection. *Stud Mycol.* 2008;61:137–144.
4. de Hoog GS et al., A new species, *Phialophora europaea*, causing superficial infections in humans. *Mycoses* 2000;43:409–416.
5. McGinnis MR. Chromoblastomycosis and phaeohyphomycosis: New concepts, diagnosis, and mycology. *J Am Acad Dermatol.* 1983;8:1–16.
6. Brandt ME, Warnock DW. Epidemiology, clinical manifestations, and therapy of infections caused by dematiaceous fungi. *J Chemother.* 2003;15(Suppl. 2):36–47.
7. Duggan JM, Wolf MD, Kauffman CA. *Phialophora* infection in an AIDS patient. *Mycoses* 1995;38:215–218.
8. Hirst LW, et al., Phialophora corneal ulcer. *Aust N Z J Ophthalmol.* 1995;23(3):223–225.
9. Guerriero C, De Simone C, Tulli A. A case of chromoblastomycosis due to *Phialophora verrucosa* responding to treatment with fluconazole. *Eur J Dermatol.* 1998;8(3):167–168.
10. Sharma NL et al., Subcutaneous phaeohyphomycosis in India: A case report and review. *Int J Dermatol.* 2002;41:16–20.
11. Cornia PB, Raugi GJ, Miller RA. *Phialophora richardsiae* bursitis treated medically. *Am J Med.* 2003;115(1):77–79.
12. Eckhard M et al., Fungal foot infections in patients with diabetes mellitus—Results of two independent investigations. *Mycoses* 2007;50(Suppl. 2):14–19.
13. Banitt M et al., A case of polymicrobial keratitis violating an intact lens capsule. *Cornea* 2008;27(9):1057–1061.
14. Iwatsu T, Miyaji, M. Subcutaneous cystic granuloma caused by *Phialophora verrucosa. Mycopathologia* 1978;64:165–168.
15. Schnadig VJ et al., *Phialophora-verrucosa* induced subcutaneous phaeohyphomycosis. *Acta Cytol.* 1986;30:425–429.
16. Lieb DF et al., Case report: Fungal endophthalmitis caused by *Phialophora richardsiae. Retina* 2003;23(3):406–407.
17. Hofmann H et al., Invasive chromoblastomycosis and sinusitis due to *Phialophora verrucosa* in a child from northern Africa. *Mycoses* 2005;48(6):456–461.
18. Lundstrom TS et al., *Phialophora verrucosa* infection in a BMT patient. *Bone Marrow Transplant.* 1997;20:789–791.
19. Kimura M et al., Multifocal subcutaneous phaeohyphomycosis caused by *Phialophora verrucosa. Arch Pathol Lab Med.* 2003;127:91–93.
20. Ohira S et al., Case report: Phaeohyphomycosis caused by *Phialophora verrucosa* developed in a patient with non-HIV acquired immunodeficiency syndrome. *Mycoses* 2002;45:50–54.
21. Yehia M et al., Subcutaneous black fungus (phaeohyphomycosis) infection in renal transplant recipients: Three cases. *Transplantation* 2004;77(1):140–142.
22. Souza TF et al., Secretion of five extracellular enzymes by strains of chromoblastomycosis agents. *Rev Inst Med Trop Sao Paulo.* 2008;50(5):269–272.

23. Yamagishi Y, Kawasaki K, Ishizaki H. Mitochondrial DNA analysis of *Phialophora verrucosa*. *Mycoses* 1997;40:329–334.
24. Abliz P et al., Identification of pathogenic dematiaceous fungi and related taxa based on large subunit ribosomal DNA D1/D2 domain sequence analysis. *FEMS Immunol Med Microbiol.* 2004;40(1):41–49.
25. Pounder JI, Discovering potential pathogens among fungi identified as nonsporulating molds. *J Clin Microbiol.* 2007;45(2):568–571.
26. Zeng JS et al., Spectrum of clinically relevant *Exophiala* species in the United States. *J Clin Microbiol.* 2007; 45:3713–3720.
27. Desnos-Ollivier M et al., Molecular identification of black-grain mycetoma agents. *J Clin Microbiol.* 2006; 44(10):3517–3523.
28. White TJ et al., Amplification and direct sequencing of fungal ribosomal RNA genes for phylogenetics. In: Innis, M.A., Gelfand, J.J., Sninsky, J.J., and White, J.T., Eds., 1990. *PCR Protocols: A Guide to Methods and Applications*, Academic Press, Inc., San Diego, CA, pp. 315–322.

43 *Rhinocladiella*

Dongyou Liu

CONTENTS

43.1 INTRODUCTION

43.1.1 CLASSIFICATION AND MORPHOLOGY

The genus *Rhinocladiella* is a dematiaceous (dark-walled) fungus belonging to the mitosporic Herpotrichiellaceae group, family Herpotrichiellaceae, order Chaetothyriales, class Eurotiomycetes, subphylum Pezizomycotina, phylum Ascomycota, and kingdom Fungi. The mitosporic Herpotrichiellaceae group encompasses 13 genera: *Blastacervulus*, *Brycekendrickomyces*, *Cladophialophora*, *Cladoriella*, *Cyphellophora*, *Exophiala*, *Heteroconium*, *Melanchlenus*, *Metulocladosporiella*, *Phaeococcomyces*, *Rhinocladiella*, *Thysanorea*, and *Veronaea* [1].

Currently, the genus *Rhinocladiella* consists of eight recognized species: *Rhinocladiella anceps* (synonyms: *Ramichloridium anceps* and *Veronaea parvispora*), *Rhinocladiella aquaspersa* (obsolete synonym: *Ramichloridium cerophilum*), *Rhinocladiella atrovirens*, *Rhinocladiella basitona* (synonym: *Ramichloridium basitonum*), *Rhinocladiella fasciculata* (basionym: *Ramichloridium fasciculatum*), *Rhinocladiella mackenziei* (obsolete synonyms: *Ramichloridium mackenziei*, *Ramichloridium obovoidea*, and *Ramichloridium obovoiedum*), *Rhinocladiella phaeophora*, and *Rhinocladiella similis*, in addition to 24 unassigned species [1–4]. Of these, *Rhinocladiella aquaspersa*, *Rhinocladiella atrovirens*, and *Rhinocladiella mackenziei* have been shown to infect humans [5]. *Rhinocladiella mackenziei* is a serious human pathogen, causing cerebral phaeohyphomycosis while *Rhinocladiella aquaspersa* is often associated with localized chromoblastomycosis lesions [6–8].

A number of former *Rhinocladiella* species have been transferred to other genera. These include *Rhinocladiella apiculata* (→*Ramichloridium apiculatum*), *Rhinocladiella cellaris* (→*Zasmidium cellare*), *Rhinocladiella compactum* (→*Fonsecaea compacta*), *Rhinocladiella cristaspora* (→*Ardhachandra cristaspora*), *Rhinocladiella elatior* (→*Leptodontidium elatius* var. *elatius*), *Rhinocladiella ellisii* (→*Zasmidium cellare*), *Rhinocladiella mansonii* (synonym: *Exophiala mansonii*), *Rhinocladiella pedrosoi* (→*Fonsecaea pedrosoi*), *Rhinocladiella peruviana* (→*Dicyma peruviana*), *Rhinocladiella schulzeri* (→*Ramichloridium schulzeri*), *Rhinocladiella selenoides* (→*Ardhachandra selenoides*), and *Rhinocladiella spinifera* (→*Exophiala spinifera*) [4].

Rhinocladiella colonies are slow growing, moist in texture, and dark olivaceous brown in color. Submerged hyphae are hyaline to pale olivaceous and smooth; aerial hyphae are more darkly pigmented. Conidial apparatus (consisting of either tips of ascending hyphae or septate conidiophores) is branched and olivaceous brown. Conidiogenous cells are intercalary or terminal, polyblastic, cylindrical to acicular, with a sympodially proliferating and subdenticulate rachis; scars are unthickened and non-pigmented to darkened refractive. Conidia are solitary, hyaline to subhyaline, aseptate, thin walled, smooth, and subglobose, with a slightly pigmented hilum; conidial secession is schizolytic [4].

At the species level, *Rhinocladiella aquaspersa* (obsolete synonyms: *Acrotheca cerophila*, *Cladosporium cerophilum*, and *Ramichloridium cerophilum*) colonies grow slowly, reaching a diameter of 12 mm after 14 days at 24°C on malt extract agar (MEA). Colonies are velvety to hairy, with entire margin, and surface is dark olivaceous gray, with black gelatinous exudate droplets on oatmeal agar (OA). Submerged hyphae (1.5–3 μm wide) are pale olivaceous brown and smooth or slightly rough; aerial hyphae are olivaceous brown, smooth or slightly rough, and somewhat narrower and darker than the submerged hyphae. Conidiophores (2–3 μm × 50 μm) are unbranched, arising vertically from creeping aerial hyphae, dark brown, thick walled, smooth or verruculose, hardly tapering toward the apex, with up to three additional septa. Conidiogenous cells are integrated, terminal, proliferating sympodially, rachis short and straight, with crowded, prominent, pigmented unthickened scars, minute, about 0.5 μm in diameter. Conidia (4–11 μm × 2–3 μm) are solitary, fusiform to clavate, thin walled, smooth, 0–1 septate, and subhyaline, with a conspicuous hilum, about 0.5 μm in diameter, slightly raised with an inconspicuous marginal frill. Conidia sometimes produce one to four short secondary conidia [4].

Rhinocladiella atrovirens colonies are restricted, velvety or lanose, olivaceous, and slightly mucoid at the center; reverse is dark olivaceous green to blackish. Conidiophores are short, brown, and thick walled. Conidiogenous cells (9–19 μm × 1.6–2.2 μm) are cylindrical and intercalary or free; denticulate rachis (up to 15 μm long) shows crowded, flat or butt-shaped, unpigmented conidial denticles. Conidia (3.7–5.5 μm × 1.2–1.8 μm) are hyaline, thin and smooth walled, and short cylindrical, with truncate basal scars. Budding cells (3.0–4.3 μm × 1.7–2.5 μm) are hyaline, thin walled, and broadly ellipsoidal. Germinating cells (4.5–6.0 μm) are inflated and spherical to subspherical. An annellidic *Exophiala* synanamorph may be present [4].

Rhinocladiella basitona (synonym: *Ramichloridium basitonum*) colonies are smooth, compact, and slightly raised at the center, flat toward margin, locally with some submerged mycelium, and olivaceous black with black reverse. Hyphae (2 μm wide) are thick walled, olivaceous brown, and septate every 15–20 μm. The conidial apparatus is profusely branched with flexuose cells arising at acute angles, the lower cells often being shorter than the ultimate ones and concolourous with the hyphae. Conidiogenous cells are cylindrical, with the apical part of variable length, producing numerous conidia in sympodial sequence; denticles are truncate, with a slightly darkened scar without a hilum. Conidia (3.5–4.5 μm × 2.2 μm) are hyaline, smooth walled and thin walled, and triangular with a rounded apex and with a clearly discernible basal scar. *Rhinocladiella basitona* differs from *Rhinocladiella anceps* by its basitonously branched conidiophores and triangular conidia [9].

Rhinocladiella fasciculata (basionym: *Ramichloridium fasciculatum*) colonies attain a diameter of 8 mm after 14 days at 24°C on MEA, with entire, smooth, sharp margin; mycelium is velvety, becoming farinose in the center (due to abundant sporulation), olivaceous green to brown; reverse is dark olivaceous. Blackish droplets produced at the center contain masses of *Exophiala* conidia. Submerged hyphae (2–2.5 μm wide) are subhyaline, smooth, and thickwalled; aerial hyphae are pale brown. Conidiophores (220 μm × 2–3 μm) arise vertically from ascending hyphae in loose fascicles and are unbranched or loosely branched at acute angles, cylindrical, smooth, brown, and thick walled at the base, with 0–5 thin additional septa. Conidiogenous cells (30–100 μm long) are terminal, cylindrical, thin walled, smooth, pale brown, and fertile part as wide as the basal part, up to 2 μm wide, proliferating sympodially, giving rise to a rachis (<0.5 μm in diameter) with hardly prominent, slightly pigmented, not thickened scars. Conidia (2.5–6 μm × 2–3 μm) are solitary, smooth, thin walled, subhyaline, and ellipsoidal, with truncate, slightly pigmented hilum (0.5 μm in diameter). Synanamorph forms on torulose hyphae originating from giant cells; compact heads of densely branched hyphae form thin-walled, lateral, subglobose cells, on which conidiogenous cells are formed; conidiogenous cells (12 × 1–1.5 μm) proliferate percurrently, giving rise to tubular annellated zones with inconspicuous annellations. Conidia (2–2.5 μm in diameter) are smooth, thin walled, aseptate, subhyaline, and globose [4].

Rhinocladiella mackenziei (basionym: *Ramichloridium mackenziei*) colonies attain a diameter of 5 mm on MEA after 14 days at 24°C, with entire, smooth, sharp margin; mycelium is densely lanose and raised in the center and olivaceous green to brown; reverse is dark olivaceous. Submerged hyphae (2–3 μm wide) are subhyaline, smooth, and thin-walled; aerial hyphae are pale brown and slightly narrower. Conidiophores (10–25 μm × 2.5–3.5 μm) are slightly or not differentiated from vegetative hyphae, arising laterally from aerial hyphae, with one or two additional septa, often reduced to a discrete or intercalary conidiogenous cell, pale-brown. Conidiogenous cells (5–15 μm × 3–5 μm) are terminal or intercalary, variable in length, occasionally slightly wider than the basal part, pale brown, rachis (0.5 μm in diameter) with slightly prominent, unpigmented, non-thickened scars. Conidia (5–12 μm × 2–5 μm) are golden-brown, thin walled, smooth, ellipsoidal to obovate, and subcylindrical, with darkened, inconspicously thickened, protuberant or truncate hilum (<1 μm in diameter) [4].

Rhinocladiella similis colonies are restricted, mostly dry or initially with some black slime at the centre, velvety, and olivaceous gray with olivaceous black reverse. Budding cells (5 μm × 3 μm) are abundant, pale olivaceous, broadly ellipsoidal, and without capsule in India ink, often inflating and developing into broadly ellipsoidal brown germinating cells (5 μm × 4 μm) that often bear a clearly discernible truncate extension, which bears a very short annellated zone. Hyphae (1.5 μm wide) are pale olivaceous to brown and regularly septate every 20–40 μm. Conidiogenous cells arise at acute angles in a profusely branched conidial apparatus, which is brown, somewhat darker than the sterile hyphae; conidiogeneous cells are cylindrical, 12–20 × 2 μm apically, with an elongating sympodial part bearing conidia on small denticles mainly at the apices of the cells. Conidia (4–7 × 1.5 μm) are subhyaline, noncatenate, cylindrical, and narrowed toward the base, with a small but clearly visible scar. *Rhinocladiella similis* shows preponderantly sympodial conidiogenesis and possesses a profusely branched conidial apparatus of the same texture and pale-brown pigmentation as its mycelium. This feature is the hallmark of *Rhinocladiella* [9].

The main feature to distinguish *Rhinocladiella* from *Ramichloridium* is the presence of *exophiala*-type budding cells in species of *Rhinocladiella*. Although both *Rhinocladiella* and *Veronaea* produce sympodial conidiogenous cells, *Veronaea* is differentiated from *Rhinocladiella* by its absence of *exophiala*-type budding cells and its predominantly one-septate conidia (i.e., two-celled conidia) (in comparison with one-celled conidia of *Rhinocladiella*) [4].

43.1.2 Clinical Features and Pathogenesis

Rhinocladiella mackenziei (formerly *Ramichloridium mackenziei*, *Ramichloridium obovoideum*). *Rhinocladiella mackenziei* to the second most common cause (next is *Cladophialophora bantiana*) of central nervous system (CNS) phaeohyphomycosis due to darkly pigmented fungi. *Rhinocladiella mackenziei* infections cause high mortality and are diagnosed exclusively in patients from the Middle East, with more than one-half of the cases occurring in patients with no known underlying immunodeficiency [10–18].

Podnos et al. [19] reported a case of cerebral *Rhinocladiella mackenziei* (*R. obovoideum*) in a 58-year-old Kuwaiti woman, with a history of chronic renal failure requiring hemodialysis. The patient presented with a 3 day history of left frontal headache, blurry vision, dizziness, and right-sided clumsiness. A computed tomography-guided needle biopsy of the parieto-occipital lesion yielded 10 mL of dark caseous fluid, which contained long, branching, septate hyphae. Culture of the biopsy yielded *Rhinocladiella mackenziei* (*R. obovoideum*). Despite treatment with amphotericin B and itraconazole, the patient succumbed to the infection. In a separate study, Kanj et al. [20] described two cases of brain abscesses caused by *Rhinocladiella mackenziei* (*Ramichloridium mackenziei*) in the Middle East. One patient had chronic myelomonocytic leukemia while the other patient was a normal host. Both cases had a fatal outcome despite aggressive antifungal therapy and surgical intervention. Alkhunaizi et al. [18] also described a *Rhinocladiella mackenziei* (*Ramichloridium mackenziei*)-related CNS infection in a patient who underwent living unrelated renal transplantation (LURTX) and who developed hemiplegia and was debilitated.

Rhinocladiella aquaspersa. *Rhinocladiella aquaspersa* is occasional cause of human chromoblastomycosis, which is characterized by the slow development of polymorphic skin lesions (nodules, verrucas, tumores, plaques, and scar tissue). Inside the host, infectious propagules adhere to epithelial cells and differentiate into sclerotic forms, which effectively resist destruction by host effector cells and allow onset of chronic disease.

Arango et al. [21] documented an unusual case of chromoblastomycosis due to *R. aquaspersa* in a 60-year-old male urban resident. The patient presented with darkly pigmented, infiltrative, and crusty lesion in the ear, and the fungus was recovered in culture and identified as *R. aquaspersa* on the basis of the characteristic sporulation. Itraconazole therapy resulted in complete healing. Marques et al. [22] also reported an unusual case of chromoblastomycosis lesions localized in three different sites in a 52-year-old male farm worker from Brazil. Physical examination showed extensive plaques situated on the left leg, left arm, forehead, and left side of the face. Direct examination of biopsies revealed sclerotic cells. A fungus grew in culture and identified as *R. aquaspersa* on the basis of the characteristic conidiation. Badali et al. [23] reported an *R. aquaspersa*-related case of chromoblastomycosis in a 56-year-old male, who presented with warty nodules and lymphatic distribution on the forearm, resembling sporotrichosis. Mycological and histopathological investigation of exudates and biopsy tissue samples demonstrated a granulomatous lesion with muriform cells, the hallmark of chromoblastomycosis. The causative agent was identified as *Rhinocladiella aquaspersa*. A second case of chromoblastomycosis caused by this fungus involved a 62-year-old Brazilian female. In this case, *R. aquaspersa* hyphae instead of muriform cells were observed in tissue.

Dematiaceous (melaninized) fungi responsible for cerebral phaeohyphomycosis (e.g., *Rhinocladiella mackenziei*) are environmental organisms that may enter human host via inhalation [24–27]. These fungi may induce an initial subclinical pulmonary focus before causing CNS infection through hematogenous route [28,29]. Melanin present in their cell walls not only imparts the characteristic dark color to their conidia and hyphae but also acts as virulence factor that plays an important role in the pathogenesis of infections and possibly in their CNS localization [30]. Receptors recognizing melanin or its biochemical byproducts may allow these fungi to cross the blood–brain barrier and enter the brain parenchyma. Melanin may also confer a protective advantage by scavenging free radicals and hypochlorite that are produced by phagocytic cells in their oxidative burst and that would normally kill most organisms. Additionally, melanin may bind to hydrolytic enzymes, thereby preventing their action on the plasma membrane. Hayakawa et al. [31] showed that complement-mediated phagocytosis of the fungus by macrophages represents an important host defense mechanism against *Rhinocladiella aquaspersa*.

43.1.3 Diagnosis

Because dematiaceous fungi are common soil inhabitants and are often considered as laboratory contaminants, their isolations in culture media provide a preliminary indication of their causal roles in phaeohyphomycosis and chromoblastomycosis. Microscopic observation of moniliform hyphae or irregularly swollen hyphae with yeast-like structures (compared to septate, acutely branching, straight-walled hyphae *Aspergillus* species) or muriform cells in tissue biopsy is critical for confirming the involvement of *Rhinocladiella mackenziei* or *Rhinocladiella aquaspersa* in the disease process.

Rhinocladiella mackenziei (classified under *Eurotiomycetes*), the causative agent for severe cerebral phaeohyphomycosis in humans, may be occasionally confused with *Pleurothecium obovoideum* (classified under *Sordariomycetes*) morphologically. However, *P. obovoideum* clusters with sexual species of *Carpoligna* that have *Pleurothecium* anamorphs.

Because of the lengthy time and specialized skills that are needed for macroscopic and microscopic identification of dematiaceous (melanized) fungi, molecular techniques are increasingly utilized for improved speciation and detection of these organisms [32–34]. Spatafora et al. [35] undertook a cladistic analysis of 1050 bp of the genes coding for small-subunit ribosomal RNA (SSU rRNA) to clarify the phylogenetic relationship among species of *Exophiala*, *Fonsecaea*, *Phialophora*, *Ramichloridium*, and *Rhinocladiella*. It was noted that the conventional categories of these fungi based on asexual states are not supported by phylogenetic analysis of SSU rRNA sequences. Isolates exhibiting annellidic modes of blastic conidiogenesis (e.g., *Exophiala* spp.) are not monophyletic and are placed as sister taxa to isolates that produce phialides or sympodulae. The results indicated very close relationships between isolates of *Wangiella dermatitidis* and *Exophiala mansonii* as well as between *Rhinocladiella aquaspersa* and *Exophiala jeanselmei*. The etiological agents of chromoblastomycosis form a closely related group (clade), while the agents of phaeohyphomycosis display a broader distribution on the SSU rRNA tree. In a separate study, Zeng and de Hoog [36] exploited the sequence diversity of the internal transcribed spacer (ITS) region of rRNA for reliable identification of dematiaceous fungi.

43.2 METHODS

43.2.1 Sample Preparation

Clinical specimens are examined directly under microscope with a fungal stain. Portions of the samples are inoculated on Sabouraud glucose agar without or with antibiotics. The resulting isolates are identified on the basis of macroscopical and microscopical features. For easy microscopic observation, slide culture technique may be utilized. Slide cultures are set up in Petri dishes containing 2 mL of sterile water, into which a U-shaped glass rod is placed, extending above the water surface. A block of freshly growing fungal colony, approx. 1 cm^2, is placed onto a sterile microscope slide, covered with a somewhat larger, sterile glass cover slip, and incubated in the moist chamber. Fungal sporulation is monitored over time, and when optimal, images are captured by means of a Nikon camera system (Digital Sight DS-5M, Nikon Corporation) [4].

For fungal DNA extraction, 50 mL of Sabouraud dextrose (SAB) broth is inoculated by needle with conidia from a 7 day culture in SAB agar and incubated for 72 h at 30°C. The hyphae are recovered on a 0.45 μm-pore-size filter and washed with sterile saline. Aliquots of the fungal hyphae are stored frozen at −70°C until use. Prior to lysis, the hyphae are thawed and suspended in 400 μL of DNA extraction buffer (1 mM EDTA pH 8.0, 1% sodium dodecyl sulfate, 10 mM Tris–HCl pH 7.6, 100 mM NaCl, and 2% Triton X-100). Microcentrifuge tubes (1.5 mL) containing hyphae and buffer are sonicated in a water bath (Branson; model 2210) for 15 min, followed by heating at 100°C for 5 min. Following lysis, DNA is purified using the QIAmp blood kit (Qiagen) and protocols for crude cell lysates supplied by the manufacturer. The purified DNA is stored at 4°C until tested. Alternatively, DNA is extracted with a FastDNA kit (Qbiogene) from mycelium grown for 3–5 days in liquid complete medium or with the UltraClean™ Microbial DNA Isolation Kit (Mo Bio Laboratories) from mycelium taken from fungal colonies on MEA [37].

43.2.2 Detection Procedures

43.2.2.1 Sequence Analysis of ITS Regions

Arzanlou et al. [4] utilized the universal primers ITS1 and ITS4 to amplify the ITS region of the nuclear ribosomal RNA operon, including the 3′ end of the 18S rRNA gene, the first ITS region (ITS1), the 5.8S rRNA gene, the second ITS region (ITS2), and the 5′ end of 28S rRNA gene. Subsequent sequencing analysis allows identification of fungal organisms including *Chaetomium* species.

Procedure

1. Polymerase chain reaction (PCR) mixture (25 μL) is composed of 0.5 U *Taq* polymerase (Bioline), 1× polymerase chain reaction (PCR) buffer, 0.5 mM $MgCl_2$, 0.2 mM of each dNTP, 5 pmol of each primer, and approximately 10–15 ng of fungal genomic DNA.
2. Amplification is performed on a GeneAmp PCR System 9700 (Applied Biosystems) with primary denaturation at 96°C for 5 min; 36 cycles of 96°C for 30 s, 52°C for 30 s, and 72°C for 60 s; a final extension at 72°C for 7 min.
3. The amplicons are sequenced using BigDye Terminator v. 3.1 (Applied Biosystems,) or DYEnamicET Terminator (Amersham Biosciences) Cycle Sequencing Kits and analyzed on an ABI Prism 3700 (Applied Biosystems).
4. Newly generated sequences are subjected to a Blast search of the GenBank databases; sequences with high similarity are downloaded from GenBank, and comparisons are made on the basis of the alignment of the obtained sequences.
5. Phylogenetic analysis is performed with PAUP (Phylogenetic Analysis Using Parsimony) v. 4.0b10 using the neighbor-joining algorithm with the uncorrected ("p"), the Kimura 2-parameter and the HKY85 substitution models. Alignment gaps longer than 10 bases are coded as single events for the phylogenetic analyses; the remaining gaps are treated as missing data. Any ties are broken randomly when encountered.
6. Phylogenetic relationships are also inferred with the parsimony algorithm using the heuristic search option with simple taxon additions and tree bisection and reconstruction (TBR) as the branch-swapping algorithm; alignment gaps are treated as a fifth character state and all characters are unordered and of equal weight. Branches of zero length are collapsed and all multiple, equally parsimonious trees are saved.
7. Other measures calculated include tree length, consistency index, retention index, and rescaled consistency index (TL, CI, RI, and RC, respectively). The robustness of the obtained trees is evaluated

by 1000 bootstrap replications. Bayesian analysis is performed. The best nucleotide substitution model is determined using MrModeltest v. 2.2. MrBayes v. 3.1.2 is used to perform phylogenetic analyses, using a general time-reversible (GTR) substitution model with inverse gamma rates, dirichlet base frequencies, and the temp value set to 0.5.

Note. Part of the large-subunit 28S rRNA (LSU) gene may be also amplified with primers LR0R and LR5 followed by sequencing analysis.

43.2.2.2 Sequence Analysis of SSU rRNA Gene

Spatafora et al. [35] utilized primers NS1 and NS4 to amplify a 1150 bp of the SSU rRNA for phylogenetic analysis of dematiaceous fungi including *Exophiala*, *Fonsecaea*, *Phialophora*, *Ramichloridium*, *Rhinocladiella*, and *Wangiella*.

Procedure

1. Culture is grown in potato dextrose liquid media for 1–2 weeks, depending on growth rates. Isolates are confirmed by microscopic examination. Mycelium is collected, frozen in liquid nitrogen, and stored at −70°C prior to DNA extractions.
2. PCR amplification is conducted with 1 cycle of 94°C for 3 min; 40 cycles of 94°C for 1 min, 53°C for 30 s, and 72°C for 1 min; and 1 cycle of 72°C for 5 min.
3. PCR products are purified by using microcentrifuge ultrafiltration cartridges (UFC3 THK 00; Millipore) and concentrated to a final volume of 25 μL. Purified products are sequenced using primers NS1, NS2, and NS4 and SR11R (5′-GGAGCCTGAGAAACGGCTAC-3′) and SR7R.
4. Sequence alignments are analyzed cladistically with the software package PAUP. Informative characters are analyzed by using the tree bisection–reconnection with random sequence addition option; 25 replications are performed. Support for inferred groups is estimated by the bootstrapping technique. Five hundred bootstrap replications are performed for the informative characters alone, using the tree bisection–reconnection with random sequence addition option.

43.3 CONCLUSION

The genus *Rhinocladiella* currently comprises 12 distinct dematiaceous fungal species that are widely distributed in the environments, among which *Rhinocladiella mackenziei* is an uncommon human pathogen, causing fatal cerebral phaeohyphomycosis in the Middle East [38] while *Rhinocladiella aquaspersa* is often associated with localized chromoblastomycosis lesions in various parts of the world. Due to the morphological resemblances among *Rhinocladiella*, *Veronaea*, *Ramichloridium*, and melanized fungi, identification of *Rhinocladiella* spp. on the basis of macroscopic and microscopic characteristics is time-consuming and technically demanding. The use of molecular procedures such as PCR and sequencing analysis of ribosomal rRNA genes and internal transcribed spacer regions provides a rapid and accurate approach for the discrimination of dematiaceous fungi including *Rhinocladiella* species.

REFERENCES

1. The UniProt Consortium. Available at http://www.uniprot.org/, accessed on August 1, 2010.
2. de Hoog GS. *Rhinocladiella* and allied genera. *Stud Mycol.* 1977;15:1–140.
3. McGinnis MR, Schell WA. The genus *Fonsecaea* and its relationship to the genera *Cladosporium*, *Phialophora*, *Ramichloridium*, and *Rhinocladiella*. *Pan Am Health Organ Sci Publ.* 1980;396:215–234.
4. Arzanlou M et al. Phylogenetic and morphotaxonomic revision of *Ramichloridium* and allied genera. *Stud Mycol.* 2007;58:57–93.
5. del Palacio-Hernanz A et al. Infection of the central nervous system by *Rhinocladiella atrovirens* in a patient with acquired immunodeficiency syndrome. *J Med Vet Mycol.* 1989;27:127–130.
6. Ajello L. Hyalohyphomycosis and phaeohyphomycosis: Two global disease entities of public health importance. *Eur J Epidemiol.* 1986;2(4):243–251.
7. Campbell CK, Al-Hedaithy SSA. Phaeohyphomycosis of the brain caused by *Ramichloridium mackenziei* sp nov in Middle Eastern countries. *J Med Vet Mycol.* 1993;31:325–332.
8. Brandt ME, Warnock DW. Epidemiology, clinical manifestations, and therapy of infections caused by dematiaceous fungi. *J Chemother.* 2003;15(Suppl 2):36–47.
9. de Hoog GS et al. Species diversity and polymorphism in the *Exophiala spinifera* clade containing opportunistic black yeast-like fungi. *J Clin Microbiol.* 2003;41:4767–4778.
10. Naim UR, Mahgoub ES, Chagla AH. Fatal brain abscesses caused by *Ramichloridium obovoideum*: Report of three cases. *Acta Neurochir (Wien).* 1988;93:92–95.
11. Sutton DA et al. US case report of cerebral phaeohyphomycosis caused by *Ramichloridium obovoideum* (*R. mackenziei*): Criteria for identification, therapy, and review of other known dematiaceous neurotropic taxa. *J Clin Microbiol.* 1998;36:708–715.
12. Al Abdely HM et al. SCH 56592, amphotericin B, or itraconazole therapy of experimental murine cerebral phaeohyphomycosis due to *Ramichloridium obovoideum* ("*Ramichloridium mackenziei*"). *Antimicrob Agents Chemother.* 2000;44:1159–1162.
13. Al-Abdely HM et al. Successful therapy of cerebral phaeohyphomycosis due to *Ramichloridium mackenziei* with the new triazole posaconazole. *Med Mycol.* 2005;43(1):91–95.
14. Kashgari TQ et al. Cerebral phaeohyphomycosis caused by *Ramichloridium mackenziei* in the Eastern Province of Saudi Arabia. *Ann Saudi Med.* 2000;20:457–460.
15. Khan ZU et al. Additional case of *Ramichloridium mackenziei* cerebral phaeohyphomycosis from the Middle East. *Med Mycol.* 2002;40:429–433.
16. Revankar SD, Patterson JE, Sutton DA et al. Disseminated phaeohyphomycosis: Review of an emerging mycosis. *Clin Infect Dis.* 2002;34:467–476.
17. Revankar SG, Sutton DA, Rinaldi MG. Primary central nervous system phaeohyphomycosis: A review of 101 cases. *Clin Infect Dis.* 2004;38(2):206–216.

18. Alkhunaizi AM, Amir AA, Al-Tawfiq JA. Invasive fungal infections in living unrelated renal transplantation. *Transplant Proc*. 2005;37(7):3034–3037.
19. Podnos YD et al. Cerebral phaeohyphomycosis caused by *Ramichloridium obovoideum* (*Ramichloridium mackenziei*): Case report. *Neurosurgery* 1999;45:372–375.
20. Kanj SS, Amr SS, Roberts GD. *Ramichloridium mackenziei* brain abscess: Report of two cases and review of the literature. *Med Mycol*. 2001;39:97–102.
21. Arango M et al. Auricular chromoblastomycosis caused by *Rhinocladiella aquaspersa*. *Med Mycol*. 1998;36(1):43–45.
22. Marques SG et al. Chromoblastomycosis caused by *Rhinocladiella aquaspersa*. *Med Mycol*. 2004, 42(3):261–265.
23. Badali H et al. *Rhinocladiella aquaspersa*, proven agent of verrucous skin infection and a novel type of chromoblastomycosis. *Med Mycol*. 2010;48:696–703.
24. Dixon DM, Shadomy HJ, Shadomy S. Dematiaceous fungal pathogens isolated from nature. *Mycopathologia* 1980;70(3):153–161.
25. Dixon DM, Polak-Wyss A. The medically important dematiaceous fungi and their identification. *Mycoses* 1991;34(1–2):1–18.
26. McGinnis MR. Chromoblastomycosis and phaeohyphomycosis: New concepts, diagnosis, and mycology. *J Am Acad Dermatol*. 1983;8(1):1–16.
27. Rinaldi MG. Phaeohyphomycosis. *Dermatol Clin*. 1996;14:147–153.
28. Matsumoto T et al. Developments in hyalohyphomycosis and phaeohyphomycosis. *J Med Vet Mycol*. 1994;32(Suppl 1): 329–349.
29. Horre R, de Hoog GS. Primary cerebral infections by melanized fungi: A review. *Stud Mycol*. 1999;43:176–193.
30. Polak A. Melanin as a virulence factor in pathogenic fungi. *Mycoses* 1990;33(5):215–224.
31. Hayakawa M et al. Phagocytosis, production of nitric oxide and pro-inflammatory cytokines by macrophages in the presence of dematiaceous fungi that cause chromoblastomycosis. *Scand J Immunol*. 2006;64(4):382–387.
32. Taylor JW et al. Phylogenetic species recognition and species concepts in fungi. *Fungal Genet Biol*. 2000;31:21–32.
33. Caligiorne RB et al. Internal transcribed spacer rRNA gene-based phylogenetic reconstruction using algorithms with local and global sequence alignment for black yeasts and their relatives. *J Clin Microbiol*. 2005;43(6):2816–2823.
34. Crous PW et al. Unravelling *Mycosphaerella*: Do you believe in genera? *Persoonia* 2009;23:99–118.
35. Spatafora JW, Mitchell TG, Vilgalys R. Analysis of genes coding for small-subunit rRNA sequences in studying phylogenetics of dematiaceous fungal pathogens. *J Clin Microbiol*. 1995;33(5):1322–1326.
36. Zeng JS, de Hoog GS. *Exophiala spinifera* and its allies: Diagnostics from morphology to DNA barcoding. *Med Mycol*. 2008;46(3):193–208.
37. Cheewangkoon R et al. Species of *Mycosphaerella* and related anamorphs on Eucalyptus leaves from Thailand. *Persoonia* 2008;21:77–91.
38. Amr SS, Al-Tawfiq JA. Aspiration cytology of brain abscess from a fatal case of cerebral phaeohyphomycosis due to *Ramichloridium mackenziei*. *Diagn Cytopathol*. 2007;35(11):695–699.

44 Trichophyton

Sharon C.-A. Chen, David Ellis, Tania C. Sorrell, and Wieland Meyer

CONTENTS

44.1 INTRODUCTION

Fungi of the genus *Trichophyton* are the commonest agents of dermatophytoses. They are especially significant in nail infections but also invade as keratinized skin and hair causing ringworm infection.[1,2] Because of differences in clinical associations and geographic distribution between species,[3,4] accurate identification is important for disease management and for epidemiological purposes including the investigation of case clusters.[5,6]

44.1.1 Classification

44.1.1.1 Morphological-Based Classification

The genus *Trichophyton* was first established in 1845 with the description of *Trichophyton tonsurans*. Other species descriptions quickly followed; in the same year, *Trichophyton schoenleinii* was described as *Oidium schoenleinii* and as *Achorion schoenleinii* and, in 1847, as *Trichophyton mentagrophytes*.[1,7] The early literature is confused by the description of many new species often based on trivial differences in colony morphology. The discovery of *Arthroderma uncinatum*, the teleomorph of *Trichophyton ajelloi*, in 1959 led to the descriptions of the sexual states of many dermatophytes.[7] In 1986, the teleomorph states of *Microsporum* and *Trichophyton* were consolidated into a single teleomorph genus *Arthroderma* (family Arthrodermataceae; order Onygenales). Based on morphological characteristics, *Arthroderma* was reported to contain four anamorph genera: *Chrysosporium*, *Epidermophyton*, *Microsporum*, and *Trichophyton*.[8] A masterful review of the early literature, featuring many of the great names of dermatology/medical mycology such as David Gruby, Raymond Sabouraud, and Chester Emmons, is provided by Rippon[7]; 22 *Trichophyton* species were accepted representing the "state-of-the-art" phylogeny based on morphology. However, from this time on, molecular biology has played the more important role in delineating phylogenetic relationships (Table 44.1).

44.1.1.2 Molecular-Based Taxonomy

The application of molecular phylogenetic markers has called into question recognition of the four morphological anamorph genera of *Arthroderma* (as above). Specifically, the phylogenetic species concept has provided data, whereby the genus *Trichophyton* is split up by those of *Microsporum* and *Epidermophyton*,[9,10] with some *Chysosporium* spp. interspersed among geophilic *Trichophyton* spp. The close genetic relationship between these anamorphic genera (base genomic DNA composition 48.7–50.0 mol%)[11] indicates that species distinctions based on morphological characteristics may not be accurate. It is clear that many species previously considered unique on the basis of morphological differences are not distinct species at the molecular level. For example, the common *Trichophyton rubrum* and endemic African *Trichophyton soudanense* agents have identical internal transcribed spacer (ITS) sequences within the ribosomal DNA (rDNA) gene region, as have *Trichophyton raubitschekii*, *Trichophyton fischeri*, and *Trichophyton kane*.[12–15] Conversely, taxa formerly designated as single species are now considered as species complexes, for example, *Trichophyton mentagrophytes* sensu Emmons comprises four distinct species: *T. mentagrophytes* var. *interdigitale* → *Trichophyton interdigitale* (antropophilic), *T. mentagrophytes* var. *mentagrophytes* → *T. interdigitale* (zoophilic), *T. mentagrophytes* var. *quinckeanum* → *T. mentagrophytes*, and *T. mentogorphytes* var. *erinacei* → *Trichophyton erinacei*.[12,16]

Polymorphisms within numerous gene loci have been exploited to define phylogenetic relationships among *Trichophyton* spp. including the mitochondrial (*mt*) DNA,[17,18] small subunit (SSU),[19] ITS,[12,13,15,20–23] large subunit (LSU),[24,25] and intergenic spacer (IGS) regions of the rDNA gene,[26] actin gene,[23] chitin synthase 1 (*CHS1*) gene,[10,27] DNA topoisomerase II (*TOP2*) gene,[23,28,29] manganese superoxide dismutase gene,[16] and, most recently, whole mt genomes (Table 44.2).[30] Analyses of these loci have resulted in ongoing taxonomic refinements, with many morphologically based species now either "split" as distinct species or "synonymized" as identical species (Table 44.1).[17–36] High degrees of clonality is found in some species, for example, species within the *T. rubrum* complex, leading to the treatment of such clonal entities as separate species.[31]

Combining morphological, biological (mating, ecological, and host-specific), and phylogenetic species concepts in a polyphasic approach, Gräser et al.[9] recommend a new taxonomy for anthropophilic and zoophilic *Trichophyton* spp. based on four major species complexes (Table 44.1); three of these four have associations with the teleomorph state *Arthroderma*, while the *T. rubum* complex is exclusively asexual. Notably, phylogenetic relationships between certain species complexes, for example, those within the *Arthroderma vanbreuseghemii* and *Arthroderma simii*, vary according to the genetic locus examined.[21,23] In addition, studies of whole mt genomes have shown that *T. mentagrophytes*, *T. rubrum*, and *Trichophyton ajelloi* are highly conserved and recently diverged lineages.[30] As such, species boundaries within *Trichophyton* may require further evaluation using alternative molecular approaches.

44.1.2 Ecology and Epidemiology

Pathogenic *Trichophyton* may be anthropophilic (primarily affecting humans) or zoonotic (primarily infecting animals, e.g., cats, horses, cattle, sheep, and camels). Geophilic species, for example, *T. ajelloi* and the recently described *Arthroderma olidum/Trichophyton eboreum*, associated with keratinous hair and feather material, rarely cause disease.[37,38] The epidemiology of *Trichophyton* infections is comprehensively reviewed elsewhere[1,4,39,40] but given its continuing evolution, the major epidemiological trends are presented.

44.1.2.1 Species Distribution

Species distribution of *Trichophyton* varies with geography. Nonetheless, global trends have seen the progressive replacement of zoophilic species, for example, *Trichophyton verrucosum*, by anthropophilic species, in particular *T. rubrum*. The rise of *T. rubrum* as the commonest cause of tinea pedis (foot ringworm), nail infection (onychomycosis), tinea cruris (ringworm of the groin), and tinea corporis (body ringworm) is linked to urbanization, sociocultural practices, and immigration. Persistence of zoophilic species is associated with epidemics of infection in domestic or stray animals. *T. rubrum* is the predominant species in North and Central Europe, United States, and other developed countries, where the prevalence of tinea pedis and onychomycosis is high.[1,4] In contrast, *T. verrucosum* and *Trichophyton violaceum* are the most frequent species in the Middle East and North Africa with *T. soudanense* typically endemic to Central Africa.[4] Species presence also varies with the site of infection. Around 50%–90% of tinea capitis in the United Kingdom, United States, Brazil, and Jamaica is caused by *T. tonsurans* while, in western China, *T. violaceum* accounts for 41% of infections. *T. violaceum* is frequently isolated from hair/scalp samples of African immigrants in developed countries; infections due to *T. soudanense* likewise have been reported among migrants.[39,41,42] Tinea capitis due to *T. schoenleinii* and *Trichophyton concentricum* is endemic to South Africa and the Middle East and Southeast Asia and Polynesia, respectively.

44.1.2.2 Disease Transmission

Tinea capitis (usually in children) is acquired from infected humans or their fomites.[1,40] Many patients become carriers and intermittently shed viable organisms. Minority communities in urban areas are particularly affected through the sharing of combs, hats, and bedding. Travelers to endemic areas may develop infection after contact with stray animals. *T. tonsurans* infection (usually tinea corporis) is more common in adults, in adolescents in close contact with children or in contact with sport enthusiasts and wrestlers. Known risk factors for tinea corporis, tinea cruris, and tinea pedis (mainly

TABLE 44.1
Molecular-Based Classification of *Trichophyton* Species and Species Complexes of the Major Anthropophilic and Zoophilic Species and Selected Geophilic Species and Their Corresponding Synonymized Species Designations

Species/Species Complex	Corresponding Synonymized Species	GenBank Accession No. of ITS Reference Sequence
***Arthroderma vanbreuseghemii* complex (supported by analysis of ITS[a], LSU, and hsp[b] intron genes and PCR fingerprinting)[15,22,29,32,33]**		
Arthroderma vanbreuseghemi[c]		Z98014
Trichophyton equinum[d,e]	All varieties of *T. equinum*[d,e]	EF067316
Trichophyton interdigitale[d,f]	*Trichophyton mentagrophytes* var. *geotzii*[d,f]	AF168124
	T. mentagrophytes var. *interdigitale*[d,f]	
	T. mentagrophytes var. *nodulare*[d,f]	
	Trichophyton krajdenii[d,f]	
Trichophyton interdigitale[d,e]	*T. mentagrophytes* var. *mentagrophytes*[d,e]	AY062119
	Trichophtyon verrucosum var. *autotrophicum*[d,e]	
Trichophyton tonsurans[d,f]	All varieties of *T. tonsurans*[d,f]	EF043270
***Arthroderma simii* complex (supported by ITS[a], MLST[g], and RAPD[h] analysis)[20,21,32,33]**		
Arthroderma simii[c,e]	*Trichophyton simii*[d,e]	Z98017
T. mentagrophytes[d,e]	*T. mentagrophytes* var. *quinckeanum*[d,e]	Z97995
	Trichophyton langeronii[d,e]	
	Trichophyton sarkisovi[d,e]	
Trichophyton schoenleinii[d,f]		Z98011
***Arthroderma benhamiae* complex (supported ITS[a], IGS[i], and mt[j] DNA analysis)[15,20,21]**		
Arthroderma benhamiae[c]		Z98015
Unnamed species (*Trichophyton* anamorph of *A. benhamiae*[c])	*T. mentagrophytes* var. *granulosum*[d,e]	Z98016
Trichophyton concentricum[d,f]		Z98012
Trichophyton erinacei[d,e]	*T. mentagrophytes* var. *erinacei*[d,e]	Z97997
T. verrucosum[d,e]	All varieties of *T. verrucosum*[d,e] except	Z98003
	T. verrucosum var. *autotrophicum*[d,e]	
***Trichophyton rubrum* complex (supported by multiple gene and microsatellite analysis)[14,25–28,33–36]**		
Trichophyton rubrum[d,f]	All varieties of *T. rubrum*[d,f]	Z97993
(Population 1: worldwide)	*Trichophyton fischeri*[d,f]	
	Trichophyton kane[d,f]	
	Trichophyton raubitschenkii[d,f]	
T. rubrum[d,f]	*Trichophyton soudanense*[d,f]	AF170473
(Population 2: African)	*Trichophyton gourvillii*[d,f]	
	Trichophyton megninii[d,f]	
Trichophyton violaceum[d,f]	All varieties of *T. violaceum*[d,f]	AJ270811
	T. yaoundei[d,f]	
Selected geophilic *Trichophyton* species (soil keratinophilic fungi [supported by ITS[a]][9])		
Arthroderma flavescens[c]	*Trichophyton flavescens*[d]	AJ877219
Arthroderma gertleri[c]	*Trichophyton vanbreuseghemii*[d]	AJ877210
Arthroderma gloriae[c]	*Trichophyton gloriae*[d]	AJ877209
Arthroderma insingulare[c]	*Trichophyton terrestre*[d]	AJ000606
Arthroderma lenticulare[c]	*T. terrestre*[d]	AJ877211
Arthroderma quadrifidum[c]	*T. terrestre*[d]	AJ877214
Arthroderma uncinatum[c]	*Trichophyton ajelloi*[d]	AJ877212
Trichophyton phaseoliforme[d]	*Epidermophyton stockdaleae*[d]	AJ970152

[a] ITS, internal transcribed spacer.
[b] *hsp*, heat-shock protein.
[c] Teleomorph (sexual stage) species.
[d] Anamorph (asexual stage) species.
[e] Commonly isolated from animal infection (zoophilic).
[f] Commonly isolated from human infection (anthropophilic).
[g] MLST, multi-locus sequence typing.
[h] RAPD, random amplification of polymorphic DNA.
[i] IGS, intergenic spacer region.
[j] *mt*, mitochondrial gene.

TABLE 44.2
Characteristics of Dermatophyte-Specific PCR-Based Assays to Detect and Identify *Trichophyton* Species

Reference	Cultures/Specimens (No. Tested)	Gene Target	Primers	Primers and Primer Sequence (5′–3′)	***Trichophyton*** **Species or Genera Identified**
PCR–RFLP					
[26]	Cultures (53)	IGS(NTS) rDNA ITS1/2 rDNA	NA ITS1 ITS4	NA TCCGTAGGTGAACCTGCGG TCCTCCGCTTATTGATATGC	*Trichophyton rubrum* *Trichophyton terreste* *Trichophyton mentagrophytes*, *Trichophyton verrucosum*, *Trichophyton violaceum*, *Trichophyton soudanense/T. rubrum*, *Trichophyton tonsurans/Trichophyton equinum*, *Trichophyton quickeanum/Trichophyton schoenlenii*
[76]	Cultures (26)	ITS1/2 rDNA	ITS1 ITS4	TCCGTAGGTGAACCTGCGG TCCTCCGCTTATTGATATGC	*T. tonsurans*
[67] (review)	Cultures	*TOP2*	dPsD1 PsT PsME sets	See Kanbe et al.[67]	*T. rubrum*, *T. violaceum*, *T. erinacei*, *T. tonsurans*, *T. mentagpphytes* var. *itnerdigitale*, *T. mentagrophytes* var. *mentagrophytes*, *Arthroderma* spp.
[80]	Nails (NA)	18S rDNA	TR1 TR2 NS5 NS6	GTTTCTAGGACCGCCGTA CTCAAACTTCCATCGACTTG AACTTAAAGGAATTGACGGAAG GCATCACAGACCTGTTATTGCCTC	*Trichophyton* spp., *T. rubrum*, *Candida* spp., *Aspergillus* spp.
[73]	Nails (879)	28S rDNA	LSU1 LSU2	GATAGCGMACAAGTAGAGTG GTCCGTGTTTCAAGACGGG	*T. rubrum*, *T. mentagrophytes*[a]
[77]	Skin scales/swabs (38)	18S rDNA Real-time PCR		Multiple primer sets used[77]	*Trichophyton* spp.
Single- or multiplex PCR					
[82]	Cultures (5)	ITS1 rDNA	tonsF1 tonsR1	CGGAGGCCGGCCCCCAG GGTCCAGCGTTGAGCCGCTA	*T. tonsurans*
[74]	Nails (118)	ITS2 rDNA *CHS1* (multiplex)	panDerm1 panDerm2 Univ 1 Truburm rev	GAAGAAGATTGTCGTTTGCATCGTCTC CTCGAGGTCAAAAGCACGCCAGAG TCTTTGAACGCACATTGCGCC CGGTCCTGAGGGCCCTGAA	*T. rubrum* Pan-dermatophyte
[70]	Nails (62)	Actin ITS1 rDNA (double round)	F-actin R-actin ITS1 ITS4	CGAACCGTG AGAAGATGACC GAACCACCGATCCAGACGGAGTA TCCGTAGGTGAACCTCCGG TCCTCCGCTTATTGATATGC	*T. rubrum* (cross-reaction with *T. violaceum*)
[65]	Nails (62/75)	*CSH1* (nested PCR)	CHS11S CHS11R JF2 JR2	CATCGAGTACATGTGCTCGC CTCGAGGTCAAAAGCACGCC GCAAAGAAGCCTGGAAGAAG GGAGACCATCTGTGAGAGTTG	*Trichophyton* spp.

DNA sequencing					
[87]	Cultures (NA)	ITS/2 rDNA	SR6R LR1	AAGTATAAGTCGTAACAAGG GGTTGGTTTCTTTTCCT	*T. rubrum, T. tonsurans, T. violaceum, T. mentagrophytes, T. soudanense*
[21]	Cultures (23)	ITS1 rDNA	18SF1 58SR1	AGG TTT CCG TAG GTG AACCT TTC GCT GCG TTC TTC ATCGA	*T. rubrum, T. mentagrophytes, T. violaceum, T tonsurans, T. schoenlenii, T. terreste, T. verrucosum*
[15]	Cultures (47)	ITS1/2 rDNA	SR6R LR1	AAGTATAAGTCGTAACAAGG GGTTGGTTTCTTTTCCT	Multiple species and variation within species[15]
[25]	Cultures	18S rDNA ITS/2 rDNA	As per kit SR6R LR1	MicroSeq D2 LSU Fungal Sequencing Kit (Applied Biosystems, Foster City, CA) AAGTATAAGTCGTAACAAGG GGTTGGTTTCTTTTCCT	*T. rubrum. T. mentagrophytes, T. interdigitale, T. violaceum, T. tonsurans, T. soudanense*
[72]	Nails (195)	(GT) repeat microsatellite ITS rDNA	T1for T1rev LSU266 V9D	TGGTCTGCCCTTGACTGACC GTAAGGATGGCTAGTTAGGGGG GCATTCCCAAACAACTCGACTC TTACGTCCCTGCCCTTTGTA	*T. rubrum* *Trichophyton* spp. *T. ruburm, T. interdigitale*
[73]	Nail (879)	28S rDNA	LSU1 LSU2	GATAGCGMACAAGTAGAGTG GTCCGTGTTTTCAAGACGGG	*T. rubrum. T. mentagrophytes*[a]
[81]	Skin, hair (7)	*CHS1*	CHS11S CHS11R	CATCGAGTACATGTGCTCGC CTCGAGGTCAAAAGCACGCC	*T. mentagrophytes*
Probe-based methods					
[79]	Cultures	ITS1/2 rDNA microarray	ITS1 ITS2 ITS3 ITS4	TCCGTAGGTGAACCTCCGG GCTGCGTTCTTCATCGATGC GCATCGATGAAGAACGCAGC TCCTCCGCTTATTGATATGC	*T. rubrum, T. schoenleinii, T. violaceum, T. mentagrophytes, T. soudanense, T. verrucosum, T. terreste, T. tonsurans*
[78]	Cultures (36)	ITS1/2 rDNA padlock probes	SR6RL LR1	AAGTATAAGTCGTAACAAGG GGTTGGTTTCTTTTCCT	*T. rubrum, T. tonsurans, T. violaceum, T. mentagrophytes, T. soudanense*
[61]	Hair (32)	IGS real-time PCR	Forward Reverse	GTTTTCTAGAAATGGACCCTCTGG GCGATTTGG ACC ACT T	*T. tonsurans*
[84]	Skin (71) Hair, nails (21)	ITS1/2 rDNA real-time PCR		Multiple primers used[84]	*T. rubrum, T. mentagrophytes* complex/*T. verrucosum, T. schoenlenii, T. violaceum, T. tonsurans, T. equinum, T. soudanense*
[71]	Nails (549) Skin (230) Hair (20)	18S, 5.8S rDNA PCR–RLB, sequencing	DERMF1 BDERMP-2	GGA AAC GAC CGC CCA GG CGG AAT TCT GCA ATT CAC ATT ACT	*T. rubrum, T. mentagrophytes,* T. *interdigitale, T. verrucosum, T. violaceum, T. tonsurans, T. erinacei* (by sequencing only)

Abbreviations: CHS1, chitin synthetase 1; *hsp*, heat-shock protein; IGS, intergenic spacer region; ITS, internal transcribed spacer; *mt*, mitochondrial gene; MLST, multi-locus sequence typing; NST, non-transcribed spacer; RAPD, random amplification of polymorphic DNA; RFLP, restriction fragment length polymorphism; RLB, reverse line blot; *TOP2*, DNA topoisomerase II.

[a] *T. mentagrophytes* types I and II.

T. rubrum and *T. interdigitale*) include foot dampness and abrasion combined with exposure to high fungal inocula in communal baths and aquatic facilities, exchange of linen via communal laundering, exposure to "house dust," for example, floor carpets, and wearing of tight clothing. Immigration has led to the introduction of endemic anthropophilic species, for example *T. soudanense* to non-endemic regions. Nosocomial transmission has occurred sporadically in adult and pediatric institutions usually involving *T. tonsurans* or *T. violaceum*.[5,6,33–46] Table 44.3 presents the major case clusters. Both acute and long-term care facilities have been implicated where staff may transmit the fungus among patients. Rapid identification of the pathogen as well as strain typing is critical to inform infection control practices.

44.1.3 Specific Virulence Determinants

Infection results from destruction of keratin in conjunction with the host inflammatory response. Whereas infection may be eliminated though the host immune response, some species, for example, *T. rubrum*, reach a high degree of host adaptation and result in chronic infection with few symptoms. The almost exclusive localization of *Trichophyton* spp. in keratinized tissues and finding of in vitro keratinolytic activity have focused research on fungal secreted proteases. Relatively few data are described for other putative virulence factors.[47]

Time-dependent increase (maximal ≈ 12 h) in the number of adhering spores followed by germination and invasion of the *stratum corneum* (3 days) has been shown using *in vitro* models. Candidate mediators include carbohydrate-specific adhesions on microconidial surfaces (*T. rubrum*) and secreted proteases (e.g., the dipeptidyl-peptidase IV of *T. rubrum*.[48,49] Keratinolytic protease activity, in conjunction with reduction of disulfide bridges in skin protein to render it accessible to proteases, is important for pathogenicity. *Trichophyton* secretes multiple serine endoproteases (subtilins) and metalloendoproteases. Protease activity may be induced by restricted nutrient supply and regulated by zinc-finger transcription factors.[50] In *T. rubrum*, endoprotease expression is upregulated by the transcription factor PACC, which in turn is activated by elevated pH.[51] Disruption of the *PACC* gene hinders both protease secretion and fungal growth. Delayed and immediate hypersensitivity immune responses as well as the production of Th_2 cytokines are also important in pathogenicity. *Trichophyton* spp. counter the host immune response in several ways: lymphocyte inhibition by cell wall mannans (*T. mentagrophytes* and *T. rubrum*), alteration of macrophage function and the differential activation of keratinocytes, and secretion of proteases. *T. rubrum* proteases exhibit their effect either by their surface antigen properties or directly on immune mediators. Host–fungus interactions are detailed elsewhere.[47]

44.1.4 Clinical Manifestations

Descriptions of the clinical manifestations of *Trichophyton* infections are beyond the scope of this chapter but can be found in a number of expert texts.[1,2,52] Clinical forms of disease vary with the causative pathogen (zoophilic species cause a more intense inflammatory reaction than anthropophilic species), the body site affected, and host immune status. Infections restricted to epidermis/skin structures are typical but deep-seated infections may occur. Infections are classified according to body site affected and degree of kertinization at that site.

44.1.4.1 Tinea Corporis and Tinea Cruris

All *Trichophyton* spp. cause infection of the glabrous areas of the body, groin, and face (tinea faciei) but the prevailing fungi of a given region are most likely to be recovered. Overall, *T. rubrum* is the commonest species. Other common pathogens are the *T. mentagrophytes* complex and *T. tonsurans*. *T. concentricum* causes the tinea imbricate variant of tinea corporis (as above).[53] Tinea corporis may affect the head, neck, and arms, while the source of tinea faciei is usually a zoophilic reservoir or extension from infection elsewhere. Characteristic lesions are circular, sharply marginated, and erythematous with raised edges. Central clearing occurs with scaling of the advancing border. Crusting, vesicles, and pustules may develop. In chronic *T. rubrum* infection, central clearing is absent with little redness/scaling.[52] Lesions in acute tinea cruris are exudative and resemble eczema.[1]

TABLE 44.3
Major Reported Nosocomial Dermatophyte Outbreaks

Reference	Setting	Age (Years)	No. Infected	Location	Organism
[5]	Nursing home	26–95	7	Ottawa, Canada	*Trichophyton tonsurans*
[44]	Pediatric surgical unit	23–50	10	Illinois, USA	*T. tonsurans*
[43]	Pediatric rehabilitation unit	NA[a]	4	Illinois, USA	*T. tonsurans*
[45]	Psychiatric ward	NA	13	New Zealand	*T. tonsurans*
[46]	Neonatal ICU[b]	Neonates born at 24–28 weeks gestation	5	New York, USA	*Microsporum canis*
[6]	Pediatric medical unit	16–48	21	Missouri, USA	*T. tonsurans*

[a] NA, not available.
[b] ICU, intensive care unit.

44.1.4.2 Tinea Pedis

Three types of infection occur in the horny layer of the skin of the soles (and palms): interdigital (athelete's foot; the commonest form affecting the web spaces), vesicular (dyshidrotic), and chronic noninflammatory (moccasin). *T. rubrum* typically causes chronic dry erythematous reactions and *T. interdigitale* inflamed bullous lesions. Combined tinea pedis and hand infection (tinea manum) infrequently occurs; this is usually due to *T. rubrum*.

44.1.4.3 Tinea Capitis

Tinea capitis affects the scalp, eyebrows, or eyelashes. Infection ranges from mild patchy scales to highly inflammatory reactions with folliculitis, kerion (pus) formation, extensive scarring, and alopecia. Hair infection may be ectothrix (arthroconidia on outside of hair shaft) or endothrix (arthroconidia within the shaft). The endothrix-growing *T. tonsurans* is the predominant pathogen in the United States with *T. violaceum* and *T. soudanense* restricted to other regions (see above). Dry inflammatory patches are typical. Black dots are seen as hair break off at the opening of the follicle. Kerions often form in ectothrix infection caused by *T. verrucosum* and *T. mentagraphytes*. A unique form of the disease due to *T. schoenleinii* manifests as yellow cup-shaped crusts or scutulae around the hair shaft with scarring alopecia; crusts enlarge to form a crusting mass or favus.[1]

44.1.4.4 Tinea Unguium (Onychomycosis)

Nail plate invasion is referred as tinea unguium, although the general term "onychomycosis" is often used. Toenails are more often affected than fingernails. *T. rubrum* accounts for 70% of cases and *T. interdigitale* for 20%. There are five types of infection: invasive subungual (proximal and distal subtypes), endonyx (diffuse milky-white discoloration caused by *T. soudanense*), superficial white onychomycosis usually confined to toenails and due to *T. interdigitale*, and total dystrophic disease.[1]

Invasive Forms of Disease

Rarely *Trichophyton* spp. cause an aggressive, invasive infection, usually in patients with chronic tinea. *T. rubrum* is the usual pathogen but disease due to the *T. mentagrophytes* complex, *T. violaceum*, and *T. verrucosum* has also been reported.[54,55] There are three patterns of infection: (i) majocchi's granuloma (nodular granulomatous perifolliculits) characterized by localized dermal/subcutaneous infection due to disruption of hair follicles and spillage of fungi into the dermis with resultant granulomatous inflammation; in immunocompetent patients, infection is often restricted to the extremities following trauma but immune compromised patients may develop more generalized skin lesions; (ii) locally invasive infection with subcutaneous but without hair follicle involvement; patients are usually immunocompromised[56,57]; and (iii) very rarely, dissemination to bone, lymph nodes, liver, and the nervous system occurs (Figure 44.1).[55]

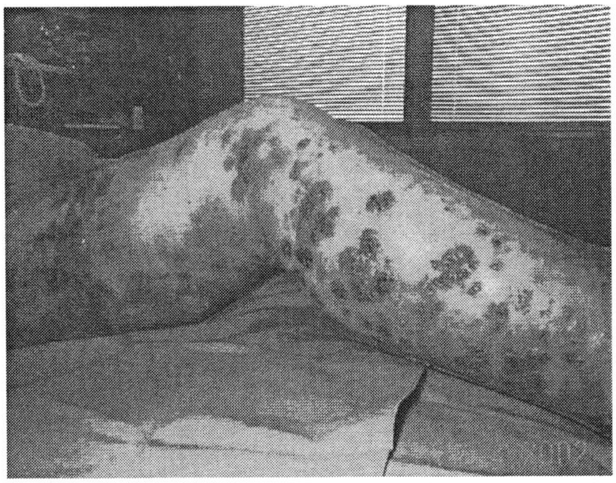

FIGURE 44.1 Disseminated *Trichophyton* infection with widespread skin lesions in an immunocompromised patient. (Matthew O' Sullivan and David Looke, Personal communication with permission.)

44.1.5 Diagnosis of *Trichophyton* Infection

Laboratory diagnosis has traditionally required the association of characteristic hyphae visualized in clinical specimens with culture isolation of the organism.[58–62] Organisms in histopathological sections stain with periodic acid-Schiff or Grocott methanamine silver (not further discussed). Standard culture-based identification methods are slow (2–4 weeks), often imprecise due to fungal phenotypic pleomorphism and require considerable expertise and the maintenance of reference "tester" strains.[63,64] Nonetheless, such methods remain widely used in the diagnosis although molecular methods are also increasingly used, both to provide rapid species identification of cultures and to directly detect *Trichophyton* DNA in clinical specimens. Proper sample collection is essential to optimize diagnostic yield.

44.1.5.1 Phenotypic-Based Methods

Phenotypic-based identification of dermatophytes is based on microscopic morphology and culture characteristics of the anamorphic state of the fungus (Table 44.4). However, the site and appearance of lesion(s), geographic location, history of travel, and animal contact and ethnicity of patient must be taken into consideration, especially in identifying rare or non-sporulating species.

Trichophyton spp. have smooth-walled macro- and microconidia. Macroconidia are mostly borne laterally directly on hyphae or on short pedicles, and are thin or thick walled, clavate to fusiform, and range from 4–8 × 8–50 mm in size; however, they may be few or absent. Microconidia (2–3 × 2–4 mm) are spherical, pyriform to clavate but may be irregular in shape. The presence of microconidia distinguishes this genus from *Epidermophyton* and the smooth-walled, mostly sessile macroconidia separate it from *Microsporum*. Differentiation of the major pathogenic species based on phenotypic criteria are summarized in Table 44.4 and illustrated by Figures 44.2 through 44.7. Good quality slide preparations and procedures to ensure stimulation of sporulation are required. Surface texture, topography, and

TABLE 44.4
Phenotypic-Based Identification of the Major Pathogenic *Trichophyton* Species (See Figures 44.2 through 44.7)

Species	Microscopy	Culture	Comments
***Trichophyton*. Macroconidia may be less distinct or absent. Microconidia are typically present. Note their shape, size, and arrangement.**			
Trichophyton rubrum (Population 1: worldwide) Anthropophilic	Most cultures show scanty to moderate numbers of slender clavate to pyriform microconidia. Macroconidia when present are smooth, thin-walled, and cylindrical.	Colonies are flat to slightly raised, white to cream, suede-like to downy, with a yellow-brown to wine-red reverse.	On primary isolation, some cultures may lack reverse pigmentation and fail to produce microconidia. Key features are culture and microscopic characteristics and failure to perforate human hair in vitro.
T. rubrum (Population 2: African) Anthropophilic.	Most cultures do not produce conidia; the hyphae may show reflexive or right-angle branching. Some strains produce pyriform microconidia; chlamydoconidia may be found in older cultures.	Colonies are often slow-growing with a flat to folded, suede-like surface. Surface mycelium and reverse pigment range from white to pink, deep apricot-orange to deep red.	Key features are clinical history, culture characteristics, and microscopic morphology and endothrix invasion of human hair.
Trichophyton interdigitale Anthropophilic form	Numerous subspherical to pyriform microconidia, occasional spiral hyphae, and spherical chlamydoconidia are present (the latter are more abundant in older cultures). Occasional slender-clavate, smooth-walled, multi-septate macroconidia are seen in some cultures.	Colonies are usually flat, white to cream in color with a powdery to suede-like surface and yellowish and pinkish-brown reverse pigment, often becoming a darker red-brown with age.	Key features include culture characteristics, microscopic morphology, and in vitro perforation of human hair.
T. interdigitale Zoophilic form (mice, guinea-pigs, kangaroos, cats, horses, sheep, and rabbits)	Numerous single-celled, spherical to subspherical microconidia are formed, often in dense clusters. Varying numbers of spherical chlamydoconidia, spiral hyphae, and smooth, thin-walled, clavate shaped, multi-celled macroconidia may also be present.	Colonies are generally flat, white to cream in color, with a powdery to granular surface. Some cultures show central folding or develop raised central tufts or pleomorphic suede-like to downy areas. Reverse pigmentation is usually a yellow-brown to reddish-brown color.	Key features include culture characteristics, microscopic morphology, and clinical disease with known animal contacts.
Trichophyton tonsurans Anthropophilic	Hyphae are broad, irregular, branched with numerous septa. Numerous microconidia varying in size and shape (long clavate to broad pyriform) are borne at right angles to the hyphae that often are unstained by lactophenol cotton blue. Very occasional smooth, thin-walled, irregular, clavate macroconidia may be present. Swollen giant forms of microconidia and chlamydoconidia are produced in older cultures.	Colonies show considerable variation in texture and color. They may be suede-like to powdery, flat with a raised center or folded, often with radial grooves. Color may vary from pale buff to yellow (the sulfureum form which resembles *Epidermophyton floccosum*), to dark brown. The reverse color varies from yellow-brown to red-brown to deep mahogany.	Key features include microscopic morphology, culture characteristics, endothrix invasion of hair, and partial thiamine requirement.
Trichophyton. equinum Zoophilic (horses)	Abundant microconidia that are clavate to pyriform and sessile or spherical, and are stalked/formed laterally along the hyphae. Macroconidia are rarely produced but when present are clavate, smooth, thin walled, and of variable size. Occasional nodular organs may be present and the microconidia often undergo a transformation to produce abundant chlamydoconidia in old cultures.	Colonies are usually flat but may develop gentle folds or radial grooves, white to buff in color, suede-like to downy in texture. Cultures usually have a deep-yellow submerged fringe and reverse, which later becomes dark red in the center.	Key features include microscopic morphology, culture characteristics, and clinical lesions in horses. Most strains require nicotinic acid for growth except those from Australia and New Zealand, which are autotrophic. Invaded hair show an ectothrix infection but do not fluoresce under Wood's ultra-violet light.

Trichophyton mentagrophytes Zoophilic (mice)	Numerous microconidia, predominantly slender clavate when young, are borne laterally along the sides of hyphae. With age, the microconidia become broader and pyriform with some subspherical forms. Occasional to moderate numbers of smooth-walled, multi-septate, clavate macroconidia may be present.	Colonies are flat or slightly raised and folded, white to cream, suede-like in texture with a pale yellow-brown to pinkish-brown reverse. A characteristic pungent "mousy" odor may be present.	Key features include microscopic morphology, culture characteristics, and a rapid urease test (within 2–3 days).
Trichophyton erinacei Zoophilic (hedgehogs)	Numerous subspherical to pyriform microconidia, occasional spiral hyphae, and spherical chlamydospores are present, the latter more abundant in older cultures. Occasional slender, clavate, smooth-walled, multi-septate macroconidia may be present.	Colonies are usually flat, white to cream in color with a powdery to suede-like surface and yellowish and pinkish-brown reverse pigment, often becoming a darker red-brown with age.	Key features include microscopic morphology and culture characteristics; slender clavate microconidia and brilliant lemon yellow reverse pigment on Sabouraud's agar; and negative hydrolysis of urea.
Nonsporulating *Trichophyton* species. No conidia are present and colonies are sterile. Chlamydoconidia or other hyphal structures may be present but are non-diagnostic. Common species include *T. verrucosum* and *T. violaceum*. Less common species include *T. concentricum* and *T. schoenleinii*.			
Trichophyton schoenleinii Anthropophilic	No macroconidia and microconidia are seen in routine cultures but numerous chlamydoconidia may be present in older cultures. Characteristic antler "nail head" hyphae or "favic chandeliers" may be observed. Sparse distorted clavate microconidia may be formed when grown on polished rice grains.	Colonies are slow growing, waxy, or suede-like with a deeply folded honey-comb-like thallus. The thallus is cream colored to yellow to orange-brown. Cultures are difficult to maintain in their typical convoluted form and rapidly become flat and downy. No reverse pigmentation is present.	Key features include clinical history, culture characteristics, and microscopic morphology showing favic chandeliers. Invaded hairs remain intact and fluoresce pale greenish-yellow under Wood's ultra-violet light.
Trichophyton violaceum Anthropophilic	Hyphae are broad, tortuous, and branched. No conidia are usually seen but occasional pyriform microconidia may be present on enriched media. Numerous chlamydoconidia are present in older cultures.	Colonies are very slow growing, glabrous or waxy, heaped and folded, and deep violet in color. Cultures often become pleomorphic, forming white sectors. Occasional non-pigmented strains may occur.	*T. violaceum* has a partial nutrient requirement for thiamine that separates this organism from *T. gourvillii*, *T. rubrum*, and other species that may produce purple pigmented colonies.
Trichophyton concentricum Anthropophilic	Cultures consist of broad, much-branched, irregular, often segmented, hyphae, which may have "antler" tips resembling *T. schoenleinii*. Chlamydospores are present in older cultures. Microconidia and macroconidia are not usually produced but some isolates will produce occasional clavate to pyriform microconidia.	Colonies are slow growing, raised, and folded, glabrous becoming suede-like, mostly white to cream colored, but sometimes orange-brown colored, often deeply folded into the agar that may produce splitting of the medium in some cultures. Reverse is buff to yellow-brown to brown in color.	Key features include association with chronic widespread noninflammatory tinea imbricata with concentric rings of scaling. The species is not known to invade human hair. NB: Hyphal segments may artificially resemble macroconidia.
Trichophyton verrucosum Zoophilic (cattle)	All strains produce chains of chlamydoconidia referred to as "chains of pearls" when grown in brain heart infusion broth containing para-aminobenzoic acid and agar at 37°C.	Colonies are slow growing, small, button, or disk-shaped, white-cream colored, with a suede to velvety surface, raised center and flat periphery with submerged growth. The reverse color varies from non-pigmented to yellow.	Key features include culture characteristics and requirements for thiamine and inositol, large ectothrix invasion of hair.

Sources: Ellis, D. et al., *Descriptions of Medical Fungi*, 2nd edn., Mycology Unit, Women's and Children's Hospital, North Adelaide, Australia, 2007; Suhonen, R.E. et al., *Fungal Infections of the Skin, Hair and Nails*, Martin Dunitz, London, U.K., 1999, p.132.

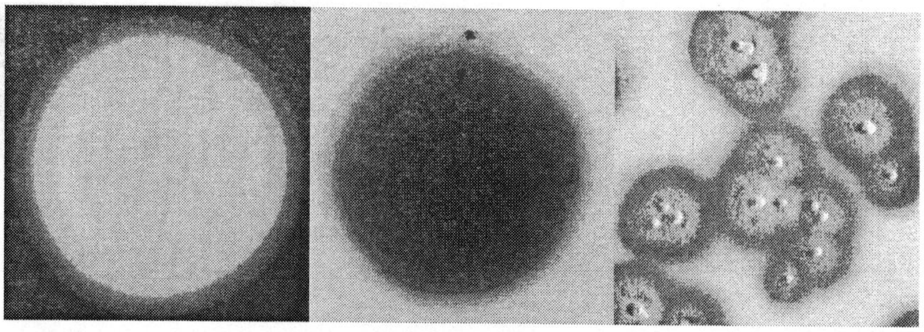

FIGURE 44.2 Culture characteristics of *T. rubrum* (population 1: worldwide form).

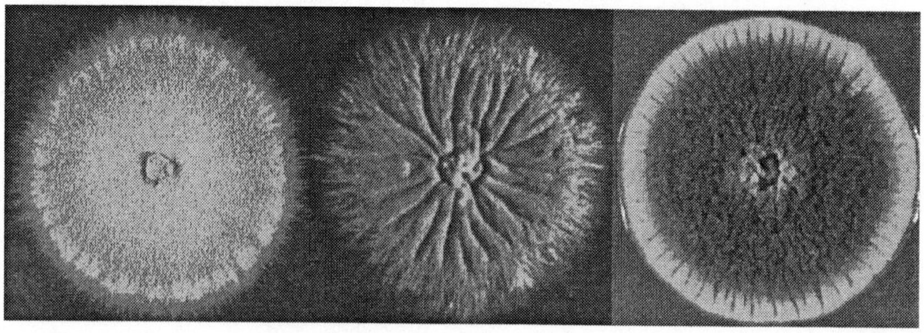

FIGURE 44.3 Culture characteristics of *T. rubrum* (population 2: African).

FIGURE 44.4 Culture characteristics of *T. interdigitale* anthropophlic (left and middle) and zoophilic (right) forms.

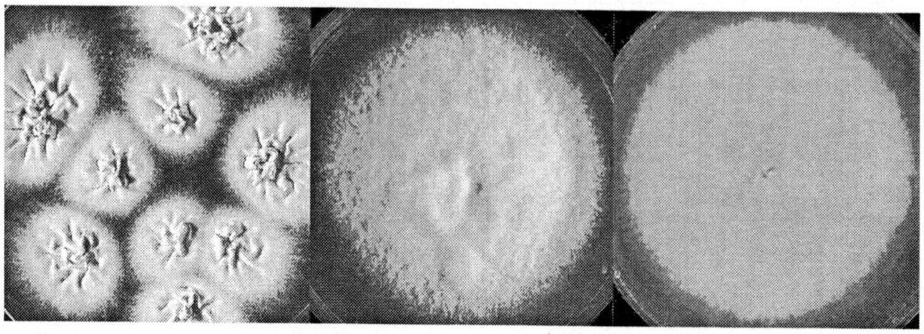

FIGURE 44.5 Culture characteristics of *T. mentagrophytes* (left and middle) and *T. erinacei* (right).

pigmentation characteristics are variable and are less reliable criteria for identification.[63]

In practice, two groups of *Trichophyton* may be recognized on microscopy (Table 44.4): (i) species that usually produce microconidia but which macroconidia may/may not be present, that is, *T. rubrum*, *T. interdigitale*, *T. mentagrophytes*, *Trichophyton equinum*, *T. erinacei*, *T. tonsurans*, *Trichophyton terrestre*, and sometimes, *T. verrucosum* (produces conidia on some media) and (ii) species that do not usually produce conidia; chlamydoconidia or other hyphal structures may be present but microscopy is often non-diagnostic, that is, *T. verrucosum*, *T. violaceum*, *T. concentricum*, *T. schoenleinii*, and *T. soudanense*. These non-sporulating species are identified by a combination of growth and colony morphology features on various media. Many *Trichophyton* have highly variable or overlapping morphological characteristics and the final identification is often based on the expert opinion of laboratory staff.

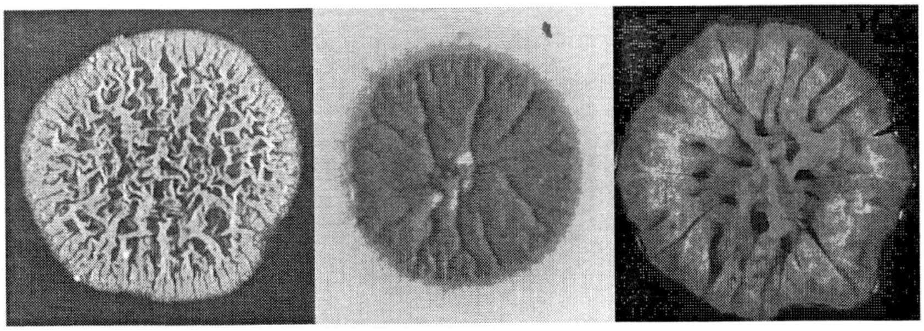

FIGURE 44.6 Culture characteristics of *T. schoenleinii*, *T. violaceum*, and *T. concentricum*.

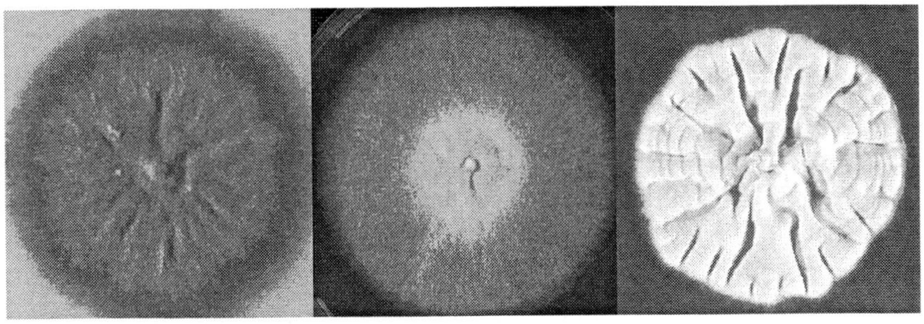

FIGURE 44.7 Culture characteristics of *T. tonsurans*, *T. equinum*, and *T. verrucosum*.

44.1.5.2 Molecular-Based Diagnostic Methods

Molecular techniques have been developed to accurately identify culture isolates, directly detect *Trichophyton* DNA in clinical specimens, and study strain variation for epidemiological purposes.[64–75] Factors influencing the detection of *Trichophyton* DNA in clinical samples include correct specimen processing, the DNA extraction protocol, and PCR design (target selection and amplicon specification method).

44.1.5.2.1 Molecular Targets

Target selection is critical for assay sensitivity and for the level of identification (to genus or species). Assays that target multi-copy, rather than single-copy, genes are generally more sensitive. The most widely used target is the rDNA gene complex. Although the variable regions of the SSU rDNA locus enables phylogenetic delineation, it is the highly variable ITS1/2 region and its intermediary 5.8S rDNA locus that are especially useful for species identification of *Trichophyton*[12,13,15,20,21,31]; the multi-copy *mt* gene has also been used.[17,18,22] Where single-copy (e.g., *CHSI*, *TOP2*, and alkaline protease 1 [*ALP1*]) genes have been targeted, PCR has often been combined with methods such as restriction fragment length polymorphism (RFLP) analysis (Table 44.2) for sufficient sensitivity and specificity.[27,65–67]

44.1.5.2.2 Overview of Methods for the Detection and/or Identification of Trichophyton spp.

Techniques developed for the identification of *Trichophyton* cultures include PCR–RFLP analysis,[26,67,76] PCR amplification of specific genetic loci followed by sequence analysis,[9,12,15,26] real-time PCR,[77] and other probe-based methods.[71,78,79] Similar techniques have been employed to detect *Trichophyton* DNA in clinical specimens, although many PCR assays are only moderately sensitive[74,80] or are unable to provide identification to species level.[65,75,77,81] Since skin, hair, and nails are not sterile sites, results of PCR tests should be interpreted in the clinical setting. The various methodology formats, gene targets, primers used, and target species of the major PCR assays are summarized in Table 44.2. Comparison of assay performance is problematic since the "gold standard" culture method is insensitive.

44.1.5.2.2.1 PCR–RFLP PCR–RFLP analyses have been used to detect and/or identify *Trichophyton* spp. (Table 44.2). Analysis of the ITS1/2 region using the restriction endonucleases *Mva*I and *HinfI* has been able to differentiate between the main *Trichophyton* species.[26,76] Other restriction enzymes useful in species identification include *Rsa*I and *Sau3A*.[67,73] PCR–RFLP analysis of the *TOP2* gene has also been employed to identify a number of *Trichophyton* as well as *Arthroderma* spp. using *HinfI* and *HincII* (as summarized in Kanbe, 2008).[67] Baek et al. used a *HaeIII* enzyme-based RFLP assay with good sensitivity and specificity to differentiate *Trichophyton* spp. including *T. rubrum* from *Candida* or *Aspergillus* spp. in onychomycotic nails.[80] RFLP analysis has also been combined with real-time PCR to detect dermatophytes in skin scales and swabs (Table 44.2).[77] Its more widespread use, however, is limited by the requirement for standard patterns for each species. Only a small number of species may be identified and certain species may not be able to be resolved (Table 44.2).

44.1.5.2.2.2 Single-Step and Multiplex PCR-Based Methods A number of single- and multiplex PCR methods, targeting the ITS1/2 region of the rDNA gene cluster as well as the single-copy actin and *CHS1* genes, have been tested for their ability to identify, and to directly detect *Trichophyton* spp. in clinical specimens (Table 44.2). Assays targeting single-copy genes may be nested.[65,70] Due to the large number of potential pathogens, multiple primers for each species are necessary and assays are limited in the number of species they are able to detect and identify (Table 44.2). Yoshida et al. evaluated an ITS1-targeted PCR assay that detected and discriminated *T. tousurans* from other dermatophytes in a single PCR amplification; however, other *Trichophyton* spp. were not able to be identified.[82] Others have reported a rapid technique whereby DNA in diseased nails was amplified using a combination of two primer sets—a pan-dermatophyte set specific to *CHS1* and a *T. rubrum*-specific set directed at the ITS2 locus. Dermatophtyes were identified to genus, and *T. rubrum* to species, level within 5 h (Table 44.2).[74] Broad range primers offer an alternative but require subsequent DNA sequencing for species identification (44.1.5.2.2.4).

44.1.5.2.2.3 Probe-Based Identification Methods Probe-based approaches have likewise been developed to assist with species identification. Early studies demonstrated the potential of SSU rDNA-directed species-specific probes to differentiate dermatophytes from *Candida* in nail and skin specimens. In one study, the sensitivity of such an approach was 92% (cf. 73% sensitivity for culture).[83] Sugita et al. reported a real-time TaqMan assay targeting the IGS rDNA region for detecting *T. tonsurans* in hair (Table 44.2).[61] Another real-time assay was tested for its ability to detect dermatophytes in 92 clinical samples. Several species could not be identified, *T. verrucosum* was not differentiated from the *T. mentagrophytes* complex and members of this complex were not distinguished from each other.[84] The assay (detection limit 2–20 cells) was limited by high costs and the need for three separate real-time reactions (two to detect all relevant species and one for the detection of an internal control).

Recently, an oligonucleotide array targeted at ITS1/2 rDNA polymorphisms was developed to detect dermatophytes including *Trichophyton* species (sensitivity 99.5% and specificity 97.8%; Table 44.2).[79] In another study, circularized ITS-targeted padlock probes were employed in a highly specific rolling circle amplification-based assay to identify five major *Trichophyton* species, including the closely related *T. rubrum* and *T. soudanense*.[78] A sensitive (detection limit 2–20 cells) PCR/reverse line blot (RLB) assay has also been reported to detect *Trichophyton* spp. in skin, hair, and nails using ITS-based probes.[71] Six species were reliably identified in 93.6% of culture- and microscopy-positive samples (Table 44.2). The assay could not differentiate between *T. soudanense* and *T. rubrum*, and a group-specific *T. mentagrophytes* probe detected both *T. mentagrophytes* and *T. schoenleinii*. The RLB assay is simple, rapid (48 h), two to three times less expensive than DNA sequencing and suited for batch testing. Within each RLB run, up to 45 samples (45 probes) can be analyzed.[71]

Finally, an ELISA-based approach (Onychodiag; Bio Advance, Bussy-Saint-Martin, France) was evaluated for the diagnosis of onychomycosis with a sensitivity of 83.6% (range 75%–100%), specificity of 100% (when compared against nails from healthy controls), and a turn-around-time of 48 h.[75] This system does not provide species identification.

44.1.5.2.2.4 Sequence-Based Identification Methods Sequence-based PCR assays have remained the backbone of molecular identification methods. Most published assays have targeted the 18S, ITS including the 5.8S, 28S rDNA regions, or the *CHS1* gene for the detection and identification of *Trichophyton* species.[9,12,27,72,73,85] ITS sequencing is the current recommended gold standard method for the identification of *Trichophyton* cultures, especially for atypical isolates or uncommon species. This gene locus is the only molecular marker for which a relatively complete dataset of sequences are available. Reviews of sequence-based identification methods that have been used for the identification of *Trichphyton* spp. are recently published.[67,86]

Monod et al. amplified template DNA of the LSU (28S) rRNA gene in nails using a universal primer set, followed by DNA sequencing to detect *T. rubrum* and *T. mentagrophytes* species complex.[73].Others have detected *T. rubrum* and *T. interdigitale* DNA in infected nails by an ITS-targeted PCR assay as well as by using the microsatellite marker T1 specific to the *T. ruburm/T. violaceum* clade (Table 44.2). There are anecdotal reports of ITS-targeted PCR assays identifying *Trichophyton* spp. in tissue specimens.[55] Sequencing of the *CHS1* gene has been successfully employed to detect *T. mentagrophytes* in the skin and hair.[81]

44.2 METHODS

44.2.1 Sample Preparation

44.2.1.1 Sample Collection for Phenotypic and Molecular-Based Identification

Specimens should be harvested by experienced staff in sufficient amounts. As a result of electrostatic attraction, plastic containers such as Petri dishes are unsuitable. Instead, sterile glass containers should be used to collect specimens.[58]

Skin: Skin lesions should be scraped from the edge or in entirety using a dermal curette but scalpel blades and glass slides may also be used. Swabs should be taken from highly inflammatory lesions. Additionally, the use of vinyl tape skin stripping improves patient compliance during the procedure as well as the diagnostic yield, especially in young children or in sensitive areas.[59]

Nails: In distal subungual infection, nails should be clipped and then scraped with a small curette or scalpel blade from the nail bed and the lower nail table at the edge of the lesion. In proximal disease, it is essential to pare the healthy upper table prior to collecting material from the infected lower table. Nail plate subungual microdrilling consistently provides better yield of fungal DNA than conventional nail clipping/subungual scraping.[60] It is essential that nails are cut into fine small pieces or mechanically ground.

Hair: Hair roots/crusts should be plucked from the infected area or its edge, and hair brushed from patients using a bristle hairbrush and impressed on culture medium.[61] Animal or human carriers can be detected by rubbing the whole scalp or hair with a sterile piece of carpet, a swab humidified with distilled water, or a toothbrush.[55] The brush/carpet square is combed through the hair and as above, impressed on culture medium.

Skin biopsy and other tissues: For histological examination, skin and tissue samples may be placed in formalin. However, specimens intended for diagnostic PCR tests should be collected into formalin-free containers as formalin is highly inhibitory to the action of *Taq* DNA polymerase. Specimens for reference facilities may be stored at −20°C or −70°C before PCR. For paraffin-embedded (PE) samples, PCR may be performed although with reduced sensitivity; several 10 μm sections should be submitted for analysis.[62]

44.2.1.2 Media and Culture Conditions

For primary isolation, specimens are usually inoculated onto Sabouraud's dextrose agar (SDA) containing cycloheximide (actidione), chloramphenicol, gentamicin, and yeast extract. Other suitable media include dermasel agar, mycobiotic agar, and dermatophyte test medium. It is useful to cut skin scales or nail pieces into the agar.[63] Cultures should be incubated at 26°C–30°C for 4 weeks and examined regularly for fungal growth. Details of media conditions and test procedures can be found in expert texts.[63,64]

When identifying individual isolates, most laboratories use additional media and employ confirmatory tests to (1) enhance sporulation and pigmentation, for example, Lactritmel and potato dextrose agar; (2) assess growth and colony morphology, for example, Sabouraud's salt agar, Littman oxgal agar, Bromcresol purple milk solids glucose agar, 1% peptone agar; (3) assess hair perforation; and (4) assess physiological characteristics (e.g., the urea hydrolysis test) and nutritional growth requirements (e.g., using the *Trichophyton* Nos. 1–7 agars).[7,63,64]

44.2.1.3 DNA Extraction Protocols

Efficient fungal cell wall lysis and DNA purification is essential for PCR sensitivity and reproducibility. There are no standardized extraction protocols for isolating *Trichophyton* DNA, either from *Trichophyton* cultures or directly from skin, hair, and nail samples.

Cultures: Treatment with glass or zirconium silicon beads, mechanical grinding, or repeated freeze-thawing using liquid nitrogen are still often used followed by chemical lysis (usually with one or more of sodium dodecyl sulfate (SDS), dithiothreitol (DTT) and proteinase K solutions). Commercial DNA extraction kits increase DNA isolation efficiency but still require fungal cell wall lysis. Various kits including the QIAmp DNA Mini kit (Qiagen), GenElute mammalian genomic DNA kit (Sigma-Aldrich), and High pure PCR template preparation kit (Roche, Mannheim, Germany) have been used with good efficiency, although comparison between kits has not been performed. Contaminating fungal DNA may be found in some commercial enzymes (e.g., lyticase), which employ yeast vector systems to release fungal DNA.[68] Automated systems (e.g., the MagNAPure LC instrument, Roche) potentially enable high throughput extraction.[69]

Skin, hair, and nail specimens: Combined mechanical and chemical disruption (usually overnight incubation) with/without commercial DNA extraction kits have also been used to extract *Trichophyton* DNA.[65,70–72] Some experts recommend an additional overnight incubation of nail pieces in a dissolvent such as sodium sulfide prior to DNA extraction.[73] Of note, a rapid extraction method from nails simply comprising a 10 min incubation at 95°C in "extraction" buffer followed by the addition of 2% bovine serum albumin (BSA) to counter inhibitors of PCR has also been successfully used.[74] One other study used only the high pure PCR template preparation kit (Roche) to extract DNA.[75]

We have selected for more detailed description three DNA extraction protocols on the basis of their practicality, performance, and potential to be incorporated into a clinical laboratory. Although commercial DNA extraction kits have been successfully used, we have not included protocols that employ these, since comparison of kit performances has not been performed and it is not possible to recommend one kit over another.

44.2.1.3.1 DNA Extraction from Trichophyton Cultures

Several DNA extraction methods have been applied to extract genomic DNA from *Trichophyton* cultures.[12,15,74,79] Two methods are presented.

44.2.1.3.1.1 Protocol of Brillowska-Dabrowska et al.[74] *Materials:* 7- to 10-day old *Trichphyton* cultures; SDA plates containing chloramphenicol (0.05%) and cycloheximide (0.5%); lysis buffers, and other solutions and solvents as specified in the procedure given below.

Procedure

1. Grow cultures on SDA containing chloramphenicol and cycloheximide (or in 2 mL of SD broth) for 7–10 days at 22°C–27°C in air.
2. Harvest cells by centrifugation and suspend the cell pellet in 500 μL lysis buffer (400 mM Tris hydrochloride [HCl] pH 8.0, 60 mM EDTA pH 8.0, 150 mM sodium chloride, 1% SDS) to incubate at room temperature for 10 min.
3. Add a small volume (150 μL) of potassium acetate (pH 4.8), vortex tubes, and then centrifuge for 1 min at 12,000 × *g*.
4. Transfer supernatant to a clean tube and add an equal volume of isopropyl alcohol.
5. Wash the DNA in 70% alcohol. Resuspend the DNA pellet in 50 μL of TE (10 mM Tris, 1 mM EDTA buffer) and store at −20°C.

44.2.1.3.1.2 Cetyltrimethylammonium Bromide (CTAB) Protocol of Gräser et al.[12,34] *Materials:* 7- to 10-day-old *Trichophyton* cultures; SDA agar plates containing chloramphenicol (0.05%) and cycloheximide (0.5%); Mortar and liquid nitrogen supply; 2% CTAB, extraction buffer, and other solutions and solvents as described below.

Procedure

1. Grow cultures on SDA containing chloramphenicol and cycloheximide for 7–10 days at 22°C–27°C in air.
2. Cut one to two fungal colonies from the agar plate, transfer to a mortar, and grind in liquid nitrogen. Suspend mycelial powder in 1 mL extraction buffer (100 mM Tris-HCl [pH 8.0], 1.4 M sodium chloride, 30 mM EDTA, 2% CTAB, 2% β-mercaptoethanol). Subject cells to three rounds of freezing and thawing.
3. To the cells, add a small volume (e.g., 50–100 μL) of proteinase K (final concentration 50 μg/mL) and incubate the mixture for 1.5 h at 60°C.
4. Centrifuge for 5 min at 13000 × *g* and add one volume of phenol: chloroform: isoamyl alcohol (25:24:1) to the supernatant, and then vortex and centrifuge again for 15 min. Remove the upper phase. Repeat the procedure four times; for the last treatment, add only one volume of chloroform
5. Precipitate the DNA with 1/10 volume of sodium acetate (4.2 M, pH 5.2) and 1 mL of cold isopropyl alcohol
6. Incubate for 16 h at −20°C, centrifuge the DNA solution for 10 min and wash the DNA in 70% alcohol. Air dry
7. Resuspend the DNA pellet in 50 μL of 1× TE buffer (10 mM Tris, 1 mM EDTA pH 8.0) and then treat with RNase (final concentration 100 μg/mL) for 15 min at 37°C. Store DNA at −20°C.

44.2.1.3.2 DNA Extraction from Nails, Skin, and Hair Samples

The method presented is adapted from Brillowska-Dabrowska et al.[74]

Materials: 7- to 10-day old *Trichphyton* cultures; SDA plates containing chloramphenicol (0.05%) and cycloheximide (0.5%); lysis buffers and other solutions and solvents as described below.

Procedure

1. Cut nails with scalpel or grind and cut hair or skin pieces into 3–5 mm fragments and place in 1.5 mL Eppendorf tubes.
2. Add 100 μL extraction buffer (60 mM sodium bicarbonate, 250 mM potassium chloride, and 50 mM Tris hydrochloride pH 9.5) and incubate at 95°C for 10 min.
3. Add 100 μL 2% BSA. Vortex gently.
4. The DNA-containing solution may be used for PCR.

44.2.2 Detection Procedures

44.2.2.1 Species Identification Using ITS Sequence Analysis

DNA sequencing is currently the most accurate method for the definitive identification of *Trichophyton* spp. A suggested protocol for ITS sequence analysis is described below.

Materials: Genomic DNA (10 ng/μL); 10× PCR buffer (100 mM Tris-hydrochloride, pH 8.3, 500 mM potassium chloride, 15 mM magnesium chloride, 0.01% [w/v] gelatine); 10× dNTPs (dinucleotide triphosphates) (2 mM dATP, dCTP, dGTP, dTTP); forward primer SR6R: 5′ AAGTATAAGTCGTAACAAGG 3′ (10 ng/μL)[87]; reverse primer LR1: 5′ GGTTGGTTTCTTTTCCT 3′ (10 ng/μL)[87]; sterile distilled water; DNA *Taq* polymerase (1 U/μL).

Procedure

1. Prepare master mix:

For one reaction	10 μL 10× PCR buffer
	10 μL 10× dNTPs
	5 μL forward primer SR6R
	5 μL reverse primer LR1
	0.5 μL DNA *Taq* polymersase

   a. To prepare 100 μL reaction/PCR tube:
      i Pipette 10 μL of 10 ng/μL genomic DNA into a PCR tube
      ii Add 53.5 μL of sterile water
      iii Add 36.5 μL of the master mix (see above)
      iv Vortex and place the PCR tubes in the thermal cycler
      v Start the reaction
2. Conduct PCR amplification with the following conditions
   First step (*initial denaturation*): 1 cycle of

		97°C	3 min
		Linked to second step	
Second step:		20 cycles of	
Denaturation	(1)	97°C	35 s
Annealing	(2)	50°C	55 s
Extension	(3)	72°C	45 s
		Auto-extension of segment (3)	
		Extension per cycle: 4 s	
		Linked to third step	
Third step:		10 cycles of:	
Denaturation	(1)	97°C	45 s
Annealing	(2)	50°C	55 s
Extension	(3)	72°C	2 min
		Auto-extension of segment (3)	
		Extension per cycle: 4 s	
		Linked to fourth step	
Fourth step (*final extension*): 1 cycle of:			
		72°C	6 min

3. Check the obtained PCR products on a 1.4% agarose gel.
4. Sequence the ITS1/2 rDNA region: (i) clean up the obtained PCR products using a commercial purification kit and prepare the sequencing template according to the instructions of the commercial sequencing service being used; (ii) sequence the PCR product in both directions using the forward (SR6R) or reverse primer (LR1); edit the sequences manually; (iii) conduct a BLASTn search against one of the two quality-controlled curated ITS databases, either at the CBS Fungal Biodiversity Center, Utrecht, the Netherlands (http://www.cbs.knaw.nl) or at the Sydney Medical School—Western of the University of Sydney, Sydney, Australia (http://www.mycologylab.org); or use the sequencing database at GenBank/European Molecular Biology Laboratory Nucleotide Sequence Database/DNA Data Bank of Japan (http://www.ncbi.nlm.nih.gov, http://www.ebi.ac.uk/embl/, http://www.ddbj.nig.ac.jp/). Note that GenBank, although the most widely used gene repository, contains unedited and nonvalidated sequences with errors of up to 20%,[88] which may lead to misidentifications.

44.2.2.2 Genotyping of *Trichoohyton* spp.

Determination of intraspecies variation is essential to answer questions in epidemiology and population genetics and in strain tracking. Strategies have targeted (1) undefined genetic markers, as in random amplified polymorphic DNA analysis (RAPD) analysis,[89–91] arbitrarily primed PCR (AP-PCR),[92,93] PCR fingerprinting,[94,95] amplified fragment length polymorphism (AFLP) analysis,[12] combined random amplified monomorphic DNA (RAMD)/single-strand conformation polymorphism (SSCP) analysis,[12] and microsatellite typing[3,34] or (2) known genetic markers as in RFLP analysis of the *mt* gene and IGS rDNA region[26,35,96–100] and single-nucleotide polymorphism (SNP) analysis of specific gene loci.[6,101]

44.2.2.2.1 Strategies Targeting Undefined Genomic Regions

Early reports indicated that AP-PCR and RAPD methods were not sufficiently sensitive to discriminate within *Trichophyton* spp., particularly species with a high degree of clonality such as *T. rubrum*.[89,92] In one study, 22 of 30 *T. rubrum* isolates were identical by RAPD analysis and the remainder exhibited minor variation[89]; similarly, AP-PCR analysis of eight strains revealed little variation.[92] A combined PCR fingerprinting (using the simple repeat primer $(AC)_{10}$), AFLP, and RAMD/SSCP analysis approach also revealed few polymorphisms.[12] However, recent analyses using two random primers identified five distinct RAPD patterns among 10 Brazilian *T. rubrum* isolates,[91] and subsequently, 12 patterns among a larger number ($n = 67$) of strains.[102]Analysis of 150 Japanese and Chinese isolates using the same two primers assigned strains into two major clusters, clusters A and B: 30% and 70% of Japanese isolates belonged to clusters A and B, respectively, while 91% of Chinese isolates belonged to cluster A.[99] RAPD analysis is also reported to demonstrate strain variation within *T. verrucosum*, including distinguishing wild type from vaccine strains,[90] and *T. interdigitale* (23 profiles among 33 isolates).[103] Prior knowledge of gene polymorphisms is required prior to analysis. RAPD and AP-PCR methods require careful standardization and may suffer from low inter-laboratory reproducibility due to the use of different assay conditions.

The high degree of clonality within *T. rubrum* is further demonstrated by PCR fingerprinting using the minisatellite primers $(AC)_{10}$ and (5′GTATTGCCCT3′).[12,94] The primer $(GACA)_4$ was used to genotype *T. mentagrophytes*, *T. rubrum*, and *T. violaceum*; few strain differences were found but different fingerprint profiles distinguished between species as well as between the varieties of *T. mentagrophytes*.[95] PCR fingerprinting has the same limitations as RAPD and AP-PCR methods even though primers are directed toward specific fungal DNA repeats. Microsatellite polymorphisms, for example, those within the marker T1 $(GT)_{8-10}$, can also be used to study intraspecies genetic variation. Within *T. rubrum* species complex, four polymorphic alleles were identified among 130 strains in one study. There was an association between genetic patterns and geographic region, suggesting an African origin for the complex but with emergence of a new genotype in Asia which spread to Europe and North America.[3] These results have been confirmed by analysis of 200 global *T. rubrum* strains using multilocus multisatellite typing (MLMT).[34] Limitations of employing microsatellite polymorphisms include hypervariability of genetic patterns. In addition, stepwise gene mutations may create alleles that manifest at identical positions on gel electrophoresis but which are homoplaseous (i.e., alleles observed are the same as a result of parallel or convergent evolution (e.g., mutation) rather than a common ancestry). This can be minimized by examining such multiple loci.

44.2.2.2.2 Strategies Targeting Defined Gene Regions

Specific genetic loci are increasingly targeted to study intraspecies genetic variation. The ITS rDNA loci are too conserved for this purpose but other loci have been useful.

Nontranscribed spacer: The IGS or non-transcribed spacer (NTS) rDNA region contains SNPs and large repetitive regions that differ between *Trichophyton* strains. In a combined RFLP/NTS-directed probe-based approach, 14 hybridization patterns among 50 *T. rubrum* isolates were found,[26] while analysis based on two tandem repeats in the NTS region (TRS-1 and TRS-2) has identified 23 genotypes among 101 strains.[35] IGS-based genetic types may vary with geography. Twenty IGS genotypes were identified among 49 *T. rubrum* isolates from three regions in China with specific hybridization patterns obtained for strains from two regions.[99,104] Correlation between genotype and increased minimal inhibitory concentration (MICs) to antifungal drugs has also been observed.[100] NTS-based RFLP analysis using three *T. mentagrophytes*-specific repetitive elements identified 19 genetic profiles within this species complex, with greater genetic diversity among *T. mentagrophytes* var. *mentagrophytes* compared with *T. mentagrophytes* var. *interdigitale*.[96] *T. tonsurans*-specific repeat elements have also been applied to examine 1048 *T. tonsurans* strains from scalp samples of preschool children—37% of children had a single predominant strain, 12% revealed a different strain each time an isolate was obtained, and symptomatic infection was found to be due to single persistent strain.[98] In sporadic *T. tonsurans* infection in Japan, NTS region-targeted RFLP analysis has identified NTS type IV to be the most prevalent but additional molecular types have been introduced from abroad.[105]

44.2.2.2.3 SNP Analysis of Intron and Coding Regions of Specific Genetic Loci

SNP analysis of the examination of *T. tonsurans* isolates from 21 individuals implicated in an outbreak by sequence analysis of the NTS region, *ALP1* and leucine aminopeptidase 1 (*LAP1*) genes demonstrated two length variations, nine SNPs, one 10 bp insertion, and a 14 bp deletion. Genetic profiles of isolates from the index patient and infected hospital personnel were identical indicating nosocomial transmission.[4] In another study, 47 genotypes were identified among 198 *T. tonsurans* strains. This study examined makers in 13 loci: NTS, ITS1, *ALP1*, *LAP1*, *LAP2*, metalloprotease-5, carboxypeptidases Y and M14, dipeptidylpeptidase IV, subtilisin-like proteases 2, 3, and 5, and ceramidase. Moderate clustering according to geographic region was observed.[101] SNP analysis lends itself to high inter-laboratory reproducibility. Construction of global databases for large-scale studies is feasible.

44.3 CONCLUSION AND PERSPECTIVES

Trichophyton infections represent a substantial health burden. Rapid accurate diagnosis is essential in guiding management, tracking infection, and identifying new pathogens. Taxonomic studies are heavily reliant on sensitive, discriminatory molecular-based techniques for species delineation. DNA sequencing of multiple genetic loci represent the "state-of-the-art" approach for phylogenetic investigations but whole genome mt analyses offer exciting insights into the genetic relationships of these complex organisms. Methods targeting specific genetic loci are particularly promising in determining intraspecies genetic variation. Many PCR-based methods have been developed to speciate *Trichophyton* and to detect *Trichophyton* DNA in clinical specimens. ITS sequencing is straightforward, accurate, and sequences can be compared against a large repository of sequences; however, where isolates can be batched, probe-based approaches also offer good potential as diagnostic assays. The selection of one or more of these techniques will vary with the population being serviced by a laboratory, its size, and function and by practical matters such as resources.

REFERENCES

1. Wietzman, L. and Summerbell, R.C. The dermatophytes. *Clin. Microbiol. Rev.*, 8, 240, 1995.
2. DeGreef, H. Clinical forms of dermatophytosis (ringworm infection). *Mycopathologia*, 166, 257, 2008.
3. Ohst, T. et al. Origins of microsatellite diversity in the *Trichophyton rubrum–T. violaceum* clade (dermatophytes). *J. Clin. Microbiol.*, 42, 4444, 2004.
4. Seebacher, C., Bouchara, J.-P., and Mignon, B. Updates on the epidemiology of dermatophyte infections. *Mycopathologia*, 166, 335, 2008.
5. Kane, J. et al. An outbreak of *Trichophyton tonsurans* dermatophytosis in a chronic care institution for the elderly. *Eur. J. Epidemiol.*, 4, 144, 1988.
6. Shroba, J. et al. A large outbreak of *Trichophyton tonsurans* among health care workers in a pediatric hospital. *Am. J. Infect. Control.*, 37, 43, 2009.
7. Rippon, J.W. *Medical Mycology*, 3rd edn., WB Saunders, Philadelphia, PA, 1988.
8. Currah, R.S. Taxonomy of the *Onygenales: Arthrodermataceae*, *Gymnoascaceae*, *Myccotrihcaceae* and *Onygenaceae*. *Mycotaxon*, 24, 1, 1985.
9. Gräser, Y., Scott, J., and Summerbell, R. The new species concept in dermatophytes—A polyphasic approach. *Mycopathologia*, 16, 239, 2008.
10. Hirai, A. et al. Molecular taxonomy of dermatophytes and related fungi by chitin synthase 1 (*CHS1*) gene sequences. *Ant. Leeuwenhoek*, 82, 11, 2003.
11. Davidson, F.D., Mackenzie, D.W.R., and Owen, R.J. Deoxyribonucleic acid base composition of dermatophytes. *J. Gen. Microbiol.*, 118, 465, 1980.
12. Gräser, Y.E. et al. Phylogeny and taxonomy of the family *Arthodermatatceae* (dermatophytes) using sequence analysis of the ribosomal ITS region. *Med. Mycol.*, 37, 105, 1999.
13. Summerbell, R.C. et al. rRNA gene internal transcribed spacer 1 and 2 sequences of asexual, anthropophilic dermatophytes related to *Trichophyton rubrum*. *J. Clin. Microbiol.*, 37, 4005, 1999.
14. Gräser, Y. et al. Molecular taxonomy of the *Trichophyton rubrum* complex. *J. Clin. Microbiol.*, 38, 3329, 2000.
15. Gräser, Y. et al. Molecular taxonomy of *Trichophyton mentagrophytes* and *T. tonsurans*. *Med. Mycol.*, 37, 315, 1999b.
16. Fréalle, E. et al. Phylogenetic analysis of *Trichophyton mentagrophytes* human and animal isolates based on MnSOD and ITS sequence comparison. *Microbiology*, 153, 3466, 2007.
17. De Bieve, C. et al. Polymorphism in mitochondrial DNA of several *Trichophyton rubrum* isolates from clinical specimens. *Ann. Inst. Pasteur Microbiol.*, 138, 719, 1987.
18. Nishio, K., Kawasaki, M., and Ishizaki, H. Phylogeny of the genera *Trichophyton* using mitochondrial DNA analysis. *Mycopathologia*, 117, 127, 1992.
19. Harmsen, D. et al. Phylogeny and dating of some pathogenic keratinophilic fungi using small subunit ribosomal RNA. *J. Med. Vet. Mycol.*, 33, 299, 1995.
20. Makimura, K. et al. Phylogenetic classification of *Trichophyton mentagrophytes* complex strains based on the DNA sequences of nuclear ribosomal internal transcribed spacer 1 regions. *J. Clin. Microbiol.*, 36, 2629, 1998.
21. Makimura, K. et al. Phylogenetic classification and species identification of dermatophyte strains based on DNA sequences of nuclear ribosomal internal transcribed spacer 1 regions. *J. Clin. Microbiol.*, 37, 920, 1999.

22. Woodgyer, A. The curious adventures of *Trichophyton equinum* in the realm of molecular biology: A modern fairy tale. *Med. Mycol.*, 42, 397, 2004.
23. Kawasaki, M. et al. Different genes can result in different phylogenetic relationships in *Trichophyton* species. *Jpn. J. Med. Mycol.*, 49, 311, 2008.
24. Leclerc, M.C., Philippe, H., and Gueho, E. Phylogeny of dermatophytes and dimorphic fungi based on large subunit ribosomal RNA sequence comparisons. *J. Med. Vet. Mycol.*, 32, 331, 1994.
25. Ninet, B. et al. Identification of dermatophyte species by 28S ribosomal DNA sequencing with a commercial kit. *J. Clin. Microbiol.*, 41, 826, 2003.
26. Jackson, C., Barton, R.C., and Evans, E.G. Species identification and strain differentiation of dermatophytes fungi by analysis of ribosomal DNA intergenic spacer regions. *J. Clin. Microbiol.*, 37, 931, 1999.
27. Kano, R. et al. Differences among chintin synthase I gene sequences in *Trichophyton rubrum* and *T. violaceum*. *Med. Mycol.*, 38, 47, 2000.
28. Kanbe, T. et al. PCR-based identification of common dermatophyte species using primer sets specific for the DNA topoisomerase II genes. *J. Dermatol. Sci.*, 32, 151, 2003.
29. Kanbe, T. et al. Species-identification of dermatophytes *Trichophyton*, *Microsporum* and *Epidermophyton* by PCR and PCR-RFP targeting of the DNA topoisomerase II genes. *J. Dermatol. Sci.*, 33, 41, 2003.
30. Wu, Y. et al. Recent dermatophyte divergence revealed by comparative phylogenetic analysis of mitochondrial genomes. *BMC Genomics*, 10, 238, 2009.
31. Gräser, Y., de Hoog, S., and Summerbell, R.C. Dermatophytes: Recognizing species of clonal fungi. *Med. Mycol.*, 44, 199, 2006.
32. Probst, S., de Hoog, G.S., and Gräser, Y. Development of DNA markers to explore host shifts in dermatophytes. *Stud. Mycol.*, 47, 57, 2003.
33. Lui, D. et al. Application of PCR to the identification of dermatophyte fungi. *J. Med. Microbiol.*, 49, 493, 2000.
34. Gräser, Y., Fröhlich, J., and Presber, W. Molecular markers reveal geographic population differentiation in *Trichophyton rubrum*. *J. Med. Microbiol.*, 56, 1058, 2007.
35. Jackson, C.J. et al. Strain identification of *Trichophyton rubrum* by specific amplification of subrepeat elements in the ribosomal DNA nontranscribed spacer. *J. Clin. Microbiol.*, 38, 4527, 2000.
36. Gupta, A.K., Kohli, Y., and Summerbell, R.C. Exploratory study of single-copy genes and ribosomal intergenic spacers for distinction of dermatophytes. *Stud. Mycol.*, 47, 87, 2002.
37. Brasch, J. and Gräser, Y. *Trichophyton eboreum* sp. nov. isolated from human skin. *J. Clin. Microbiol.*, 43, 5237, 2005.
38. Campbell, C.K. et al. *Arthroderma olidum*. sp. nov. A new addition to the *Trichophyton terrestre* complex. *Med. Mycol.*, 44, 451, 2006.
39. Borman, A.M. et al. Analysis of the dermatophyte species isolated in the British Isles between 1980 and 2005 and review of worldwide dermatophyte trends over the last three decades. *Med. Mycol.*, 45, 131, 2007.
40. Fuller, L.C. Changing face of tinea capitis in Europe. *Curr. Opin. Infect. Dis.*, 22, 115, 2009.
41. Deng, S. et al. changes in frequency of agents of tinea capitis in school children from Western China suggest slow migration rates in dermatophytes. *Med. Mycol.*, 46, 421, 2008.
42. Cetner, A. et al. Tinea capitis die to *Trichophyton soudanense* in Chicago, Illinois: Report of a case. *Pediatr. Dermatol.*, 26, 226, 2009.
43. Lewis, S. and Lewis, B. Nosocomial transmission of *Trichophyton tonsurans* tinea corporis in a rehabilitation hospital. *Infect. Control. Hosp. Epidemiol.*, 18, 322, 1997.
44. Arnow, P., Houchins, B.A., and Pugliese, G. An outbreak of tinea corporis in hospital personnel caused by a patient with *Trichophyton tonsurans* infection. *Pediatr. Infect. Dis.*, 10, 355, 1991.
45. Wood, B. and Rademaker, M. Nosocomial *Trichophyton tonsurans* in a long-stay ward. *N. Z. Med. J.*, 110, 277, 1997.
46. Drusin, L. et al. Nosocomial ringworm in a neonatal intensive care unit: A nurse and her cat. *Infect. Control. Hosp. Epidemiol.*, 21, 60, 2002.
47. Vermout, S. et al. Pathogenesis of dermatophyteosis. *Mycopathologica*, 166, 267, 2008.
48. Esquenazi, D. et al. The influence of surface carbohydrates during in vitro infection of mammalian cells by the dermatophyte *Trichophyton rubrum*. *Res. Microbiol.*, 155, 144, 2004.
49. Monod, M. et al. Aminopeptidases and dipeptidyl-peptidases secreted by the dermatophyte *Trichophyton rubrum*. *Microbiology*, 15, 145, 2005.
50. Kaufman, G. et al. Infection stages of the dermatophyte pathogen *Trichophyton*: Microscopic characterisation and proteolytic enzymes. *Med. Mycol.*, 45, 149, 2007.
51. Ferreira-Nozawa, M.S. et al. The pH signalling transcription factor PACC mediates the growth of *Trichophyton rubrum* on human nail in vitro. *Med. Mycol.*, 44, 641, 2006.
52. Bolognia, J.L. et al. *Dermatology*, Mosby, New York, 2003, pp. 1176–1178.
53. Bonifaz, A., Archer-Dubon, C., and Saul, A. Tinea imbricata or Tokelau. *Int. J. Dermatol.*, 43, 506, 2004.
54. Smith, K.J. et al. Majocchi's granuloma. *J. Cutan. Pathol.*, 18, 28, 1991.
55. Lestringent, G. et al. Deep dermatophytosis to *Trichophyton rubrum* and *T. verrucosum* in an immunosuppressed patient. *Int. J. Dermatol.*, 27, 707, 1988.
56. Nir-Paz, R. et al. Deep infection by *Trichophyton rubrum* in an immunocompromised patient. *J. Clin. Microbiol.*, 41, 5298, 2003.
57. Chastain, M.A., Reed, R.J., and Pankey, G.A. Deep dermatophytosis: Report of 2 cases and review of the literature. *Cutis*, 67, 457, 2001.
58. Robert, R. and Pihet, M. Conventional methods for the diagnosis of dermatophytes. *Mycopathologia*, 166, 295, 2008.
59. Miranda, M.F. and Silva, A.J. New uses of vinyl tape for reliable collection and diagnosis of common superficial mycoses. *Skinmed*, 2, 156, 2003.
60. Qureshi, H.S., Ornsby, H.A., and Kapadia, N. Effects of modified sample collection technique on fungal culture yield: Nail clipping/scraping versus microdrill. *J. Pak. Med. Assoc.*, 54, 301, 2004.
61. Sugita, T., Shiraki, Y., and Hiruma, M. Real time PCR TaqMan assay for detecting *Trichophyton tonsuans*, a causative agent of tinea capitis, from hairbrushes. *Med. Mycol.*, 44, 579, 2006.
62. Lau, A. et al. Development and clinical application of a panfungal PCR assay to detect and identify fungal DNA in tissue specimens. *J. Clin. Microbiol.*, 45, 380, 2007.
63. Ellis, D. et al. *Descriptions of Medical Fungi*, 2nd edn., Mycology Unit, Women's and Children's Hospital, North Adelaide, Australia. ISBN 9780959851267, 2007.

64. Suhonen, R.E., Dawber, R.P.R., and Ellis, D.H. *Fungal Infections of the Skin, Hair and Nails*, Martin Dunitz, London, U.K., 1999, p. 132.
65. Garg, J. et al. Evaluation of pan-dermatophyte nested PCR in diagnosis of onychomycosis. *J. Clin. Microbiol.*, 45, 3443, 2007.
66. Bhathena, A., Gaedigk, R., and Abdel-Rahman, S.M. Characterisation of the *ALP1* gene locus of *Trichophyton tonsurans*. *Mycopathologia*, 160, 265, 2005.
67. Kanbe, T. Molecular approaches in the diagnosis of dermatophytosis. *Mycopathologica*, 166, 307, 2008.
68. Loeffler, J. et al. Contaminations occurring in fungal PCR assays. *J. Clin. Microbiol.*, 37, 1200, 1999.
69. Loeffler, J. et al. Automated extraction of genomic DNA from medically important yeast species and filamentous fungi by using the MagNA Pure LC system. *J. Clin. Microbiol.*, 40, 2240, 2002.
70. Gupta, A.K., Zaman, M., and Singh, J. Fast and sensitive detection of *Trichophyton rubrum* DNA from the nail samples of patients with onychomycosis by a double-round polymerase chain reaction-based assay. *Br. J. Dermatol.*, 157, 698, 2007.
71. Bergmans, A.M.C. et al. Validation of PCR-reverse line blot method for rapid detection and identification of nine dermatophytes species in nail, skin and hair samples. *Clin. Microbiol. Infect.*, 14, 778, 2008.
72. Kardjeva, V. et al. Forty-eight hour diagnosis of onychomycosis with subtyping of *Trichophyton rubrum* strains. *J. Clin. Microbiol.*, 44,1419, 2006.
73. Monod, M. et al. Fast and reliable PCR/sequencing/RFLP assay for identification of fungi in onychomycoses. *J. Med. Microbiol.*, 55, 1211, 2006.
74. Brillowska-Dabrowska, A. et al. Five-hour diagnosis of dermatophyte nail infections with specific detection of *Trichophyton rubrum*. *J. Clin. Microbiol.*, 45, 1200, 2007.
75. Savin, C. et al. Multicentre evaluation of a commercial PCR-enzyme linked immunosorbent assay diagnostic kit (Onychodiag) for diagnosis of dermatophyte onychomycosis. *J. Clin. Microbiol.*, 45, 1205, 2007.
76. Mochizuki, T. et al. Rapid identification of *Trichophyton tonsurans* by PCR-RFLP analysis of ribosomal DNA regions. *J. Dermatol. Sci.*, 32, 25, 2003.
77. Gutzmer, R. et al. Rapid identification and differentiation of fungal DNA in dermatological specimens by LightCycler PCR. *J. Med. Microbiol.*, 53, 1207, 2004.
78. Kong, F. et al. Rapid identification and differentiation of *Trichophyton* species based on sequence polymorphisms of the ribosomal internal transcribed spacer regions by rolling circle amplification. *J. Clin. Microbiol.*, 46, 1192, 2008.
79. Li, H.C. et al. Identification of dermatophytes by an oligonucleotide array. *J. Clin. Microbiol.*, 45, 3160, 2007.
80. Baek, S.-C. et al. Detection and differentiation of causative fungi of onychomycosis using PCR amplification and restriction enzyme analysis. *Int. J. Dermatol.*, 37, 682, 1998.
81. Kano, R. et al. Direct detection of dermatophytes in skin samples based on sequences of the chitin synthase 1 (*CHS1*) gene. *J. Vet. Med. Sci.*, 65, 267, 2003.
82. Yoshida, E. et al. Rapid identification of *Trichophyton tonsurans* by specific PCR based on DNA sequences of nuclear ribosomal internal transcribed spacer (ITS) 1 region. *J. Dermatol. Sci.*, 42, 225, 2006.
83. Bock, M. et al. Diagnostik von dermatomykosen mit der polymerase-ketten-reaktion. *Hautarzt*, 48, 175, 1997.
84. Arabatzis, M. et al. Diagnosis of common dermatophyte infections by a novel multiplex real-time polymerase chain reaction detection/identification scheme. *Br. J. Dermatol.*, 157, 681, 2007.
85. Wallberg, M. et al. 18S rDNA polymerase chain reaction and sequencing in onychomycosis diagnostics. *Acta Derm. Venearol.*, 86, 223, 2006.
86. Kac, G. Molecular approaches to the study of dermatophytes. *Med. Mycol.*, 38, 329, 2000.
87. Vilgalys, R. and Hesters, M. Rapid genetic identification and mapping of enzymatically amplified ribosomal DNA from several *Cryptococcus* species. *J. Bacteriol.*, 172, 4238, 1990.
88. Nilsson, R.H. et al. Taxonomic reliability of DNA sequences in public sequence databases: A fungal perspective. *PLoS One*, 1, e59, 2006.
89. Zhong, Z. et al. Typing of common dermatophytes by random-amplification of polymorphic DNA. *Jpn. J. Med. Mycol.*, 38, 239, 1997.
90. Hajdúch, M. et al. Diversity among wild-type and vaccination strains of *Trichophyton verrucosum* investigated using random amplified polymorphic DNA analysis. *Folia Biol.*, 45, 151, 1999.
91. Baeza, L., Mendes, C., and Giannini, M.J.S. Strain differentiation of *Trichophyton rubrum* by random amplification pf polymorphic DNA (RAPD). *Rev. Inst. Med. Trop. S. Paulo*, 46, 339, 2004.
92. Liu, D. et al. Use of arbitrarily primed polymerase chain reaction to differentiate *Trichophyton* dermatophytes. *FEMS Microbiol. Lett.*, 136, 147, 1996.
93. Gräser, Y. et al. Identification of common dermatophytes (*Trichophyton*, *Microsporum*, *Epidermophyton*) using polymerase chain reaction. *Br. J. Dermatol.*, 138, 576, 1998.
94. Arabatzis, M. et al. First report on autochthonous urease-positive *Trichophyton rubrum* (*T. raubitschekii*) from South-east Europe. *Br. J. Dermatol.*, 153, 178, 2005.
95. Shehata, A.S. et al. Single-step PCR using $(GACA)_4$ primer: Utility for Rapid identification of dermatophyte species and strains. *J. Clin. Microbiol.*, 46, 2641, 2008.
96. Jackson, C.J., Mochizuki, T., and Barton, R.C. PCR fingerprinting of *Trichophyton mentagrophytes* var. *interdigitale* using polymorphic subrepeat loci in the rDNA non-transcribed spacer. *J. Med. Microbiol.*, 55, 1349, 2006.
97. Mochizuki, T. et al. Restriction fragment length polymorphism analysis of ribosomal DNA intergenic regions is useful for differentiating strains of *Trichophyton mentagrophytes*. *J. Clin. Microbiol.*, 41, 4583, 2003.
98. Abdel-Rahman, S.M. et al. Tracking *Trichophyton tonsurans* through a large urban child care centre: Defining infection prevalence and transmission patterns by molecular strain typing. *Paediatrics*, 118, 2365, 2006.
99. Yang, G. et al. Genotyping of *Trichophyton rubrum* by analysis of ribosomal-DNA intergenic spacer regions. *Mycopathologia*, 164, 19, 2007.
100. de Assis Santos, D. et al. Molecular typing and antifungal susceptibility of *Trichophyton rubrum* isolates from patients with onychomycosis pre- and post-treatment. *Int. J. Antimicrob. Agents*, 29, 536, 2007.
101. Abdel-Rahman, S.M. et al. Divergence among an international population of *Trichophyton tonsurans* isolates. *Mycopathologia*, 169, 1–13, doi 10.1007/s11046-009-9223-7, 2009.

102. Baeza, L.C. et al. Strain differentiation of *Trichophyton rubrum* by random amplified polymorphic DNA and analysis of rDNA non-transcribed spacer. *J. Med. Microbiol.*, 55, 429, 2006.
103. Kac, G. et al. Genetic diversity among *Trichophyton mentagrophytes isolates* using random amplification of polymorphic DNA method. *Br. J. Dermatol.*, 140, 839, 1999.
104. Yang, X. et al. Differentiation of *Trichophyton rubrum* clinical isolates from Japanese and Chinese patients by random amplified polymorphic DNA and DNA sequence analysis of the non-transcribed spacer region of the rRNA gene. *J. Dermatol. Sci.*, 54, 38, 2009.
105. Mochizuki, T. et al. Epidemiology of sporadic (non-epidemic) cases of *Trichophyton tonsurans* infections in Japan based on PCR-RFLP analysis of non-transcribed spacer region of the ribosomal RNA gene. *Jpn. J. Infect. Dis.*, 61, 219, 2008.

45 Veronaea

Dongyou Liu

CONTENTS

45.1 INTRODUCTION

45.1.1 Classification and Morphology

The genus *Veronaea* is a dematiaceous mold belonging to the mitosporic Herpotrichiellaceae group, family Herpotrichiellaceae, order Chaetothyriales, class Eurotiomycetes, subphylum Pezizomycotina, phylum Ascomycota, kingdom Fungi. The mitosporic Herpotrichiellaceae group consists of 13 genera: *Blastacervulus*, *Brycekendrickomyces*, *Cladophialophora*, *Cladoriella*, *Cyphellophora*, *Exophiala*, *Heteroconium*, *Melanchlenus*, *Metulocladosporiella*, *Phaeococcomyces*, *Rhinocladiella*, *Thysanorea*, and *Veronaea*. In turn, the genus *Veronaea* is separated into three species: *Veronaea botryosa* (obsolete synonyms: *Veronaea coprophila* and *Sympodina coprophila*), *Veronaea compacta*, and *Veronaea japonica* [1–3]. A former *Veronaea* species, that is, *Veronaea musae* (obsolete synonym: *Ramichloridium musae*), is now known as *Periconiella musae*. The type species of the genus is *Veronaea botryosa*, which was first identified in 1957 from olive slag in Italy and which is the only *Veronaea* species implicated in human infection.

Classification and identification of *Veronaea* spp. are largely based upon their conidial size, septations, shape, roughness, and conidiophore features. *Veronaea* resembles *Rhinocladiella* as both produce sympodial conidiogenous cells. However, *Veronaea* is distinguishable from *Rhinocladiella* by its absence of exophiala-type budding cells and its predominantly one-septate conidia (i.e., two-celled conidia) (in comparison with one-celled conidia of *Rhinocladiella*) and by its production of flat, barely prominent conidiogenous loci.

Veronaea colonies are moderately fast growing, velvety in texture, and pale olivaceous brown in color. Submerged hyphae are hyaline to pale olivaceous and smooth while aerial hyphae are more darkly pigmented. Conidiophores are erect, straight or flexuose, unbranched or occasionally loosely branched, geniculate, and smooth walled, with pale to medium or olivaceous brown color. Conidiogenous cells are terminally integrated, polyblastic, occasionally intercalary, cylindrical, later septate, and pale brown, with subhyaline fertile part and faintly pigmented, unthickened scars. Conidia are solitary, smooth, cylindrical to pyriform, rounded at the apex and truncate at the base, pale brown, and 1–2 septate; conidial secession is schizolytic [4].

At the species level, *Veronaea botryosa* colonies attain a diameter of 30 mm on malt extract agar (MEA) after 14 days at 24°C, with entire, sharp margin; surface is velvety and slightly raised in the center, with grayish-brown to blackish-brown color; and reverse is greenish black. Submerged hyphae are hyaline to pale olivaceous and smooth while aerial hyphae are more darkly pigmented. Conidiophores (200 μm × 2–3 μm) are erect, straight or flexuose, occasionally branched, usually geniculate (due to the sympodial development of the conidia), and smooth walled, with a pale brown to olivaceous-brown color. Conidiogenous cells (10–100 μm long) are terminal, occasionally intercalary, cylindrical, pale brown, later often becoming septate, fertile part subhyaline, rachis with crowded, flat to slightly prominent, faintly pigmented, unthickened scars. Conidia (3–12 μm × 1.5–3 μm) are solitary, smooth walled or slightly verrucose, cylindrical to pyriform, rounded at the apex and truncate at the base, pale brown, and 1–2 septate (mostly one septate or two celled), with a faintly darkened, unthickened hilum (0.5 μm in diameter) [4].

Veronaea compacta colonies are slow growing, attaining a diameter of 15 mm on MEA after 14 days at 24°C; surface is velvety to lanose, slightly raised in the center, and pale gray to pale brownish gray in color; reverse is dark gray. Submerged hyphae (1.5–3 μm wide) are subhyaline, smooth, and thin walled while aerial hyphae are thick walled and pale brown. Conidiophores (60 μm × 4 μm) are slightly differentiated from vegetative hyphae, lateral or occasionally terminal, and unbranched or branched at acute angles, with 1–3 additional septa. Conidiogenous cells (up to 10 μm in length) are terminal, occasionally intercalary, pale brown, and cylindrical to doliiform or flask shaped, with hardly prominent denticles; scars (0.5 μm in diameter) are flat, slightly pigmented, and not thickened. Conidia (4–9 μm × 2–3 μm) are solitary, pale brown, smooth, thin walled, ellipsoidal to ovoid, and 0–1 septate, often constricted at the septa, with a round apex and truncate base; hilum (0.5 μm in diameter) is prominent, slightly darkened, and unthickened [4].

Veronaea japonica colonies are slow growing, attaining a diameter of 7.5 mm on MEA after 14 days at 24°C; surface is velvety to lanose, slightly raised in the center, and olivaceous brown, with entire margin; reverse is dark olivaceous. Submerged hyphae (1.5–3 μm wide) are subhyaline, smooth, and thin walled, while aerial hyphae are slightly narrower and pale brown; hyphal cells later become swollen, thick walled, dark brown, and often aggregated. Conidiophores (65 μm × 2–3 μm) are slightly differentiated from aerial vegetative hyphae, lateral, or terminal, often wider than the supporting hypha, unbranched or occasionally branched, pale brown, thin walled, and smooth, with 1–3 additional septa. Conidiogenous cells (up to 15 μm long) are terminal, occasionally intercalary, variable in length, pale brown, and cylindrical to clavate, with hardly prominent denticles; scars (0.5 μm diameter) are flat, slightly pigmented, and not thickened. Conidia (6–10 μm × 2–4 μm) are solitary, pale brown, smooth, thin walled, ellipsoidal to ovoid, and one septate, often constricted at the septum, with a round apex and truncate base; hilum (1 μm in diameter) is unthickened but slightly darkened. Morphologically, *V. japonica* is similar to *V. compacta*; but it can be distinguished by the presence of dark brown, swollen hyphal cells in culture, which are absent in *V. compacta* [4].

45.1.2 Clinical Features

Veronaea is one of the 70 dark-walled (brown-pigmented, dematiaceous, phaeoid, or melanized) fungal genera (including 130 species) that occur as soil saprophytes, plant pathogens, and environmental contaminants and that are responsible for cutaneous, subcutaneous, corneal, and systemic diseases (often referred to as phaeohyphomycosis or chromoblastomycosis) in humans (especially immunocompromised individuals) and animals [5]. Among the more prominent dematiaceous human pathogens are members of *Alternaria*, *Bipolaris*, *Cladophialophora*, *Curvularia*, *Exophiala* (*Wangiella*), *Fonsecaea*, *Madurella*, *Phialophora*, *Scedosporium*, and *Scytalidium* genera [6]. While some thermophilic dematiaceous fungi (with growth at 40°C or higher) are neurotropic and have a predilection for central nervous system (CNS) tissue causing brain lesions and/or abscesses, other less thermotolerant fungal species (e.g., *Ramichloridium schulzeri*, *Rhinocladiella aquaspersa*, and *Veronaea botryosa*) are nonneurotropic and often induce mainly skin infections [7].

Veronaea botryosa has been isolated from soil, plant, water, and air samples in many parts of the world including China, India, Australia, New Zealand, Italy, Poland, Egypt, New Guinea, and Brazil. Human subcutaneous mycoses due to *V. botryosa* have been reported in China, the Philippines, Libya, France, and the United States [8–12].

Matsushita et al. [13] documented a case of *V. botryosa* skin infection in a 12-year-old boy from China, who received a scratch on his right elbow 6 years earlier. Later, papules (10–12 mm × 6–8 mm) with exudates appeared on his extremities and other parts of the body (i.e., face, hands, legs, scrotum, and buttocks). The lesions became nodular, forming thick, crusted areas, with adjacent lesions coalescing to form plaques. Hematoxylin- and eosin-stained biopsy tissue and KOH-treated lesion scraping revealed septate, branched, yellowish-brown-pigmented hyphae (2.5–4.0 μm wide). Culture of scrapings from the crusty lesions and biopsy tissue on Sabouraud dextrose agar containing chloramphenicol (Sab+C) and Sab+C containing cycloheximide yielded velvety, olivaceous gray colonies after 7–10 days of incubation at 25°C and 37°C in the dark. Colonies on Sab+C and potato dextrose agar were velvety to lanate and grayish brown to black, reaching diameters of 22–25 mm in 10 days at 25°C. Growth at 37°C was slow, reaching diameters of 9–11 mm after 2 weeks. No growth was evident at 40°C. Slide cultures on potato dextrose agar at 25°C for 10 days showed septate, branched, melanized hyphae (2.5–4.5 μm wide), bearing lateral and terminal conidiophores, as stained by lactophenol cotton blue. Conidiophores (200–250 μm × 2.5–4.0 μm) were erect, straight, or flexous, unbranched or occasionally branched, rarely geniculate, smooth walled, septate, and pale brown. Conidiogenous cells were terminal or lateral and cylindrical in the apical area, bearing numerous scars. Sympodially produced conidia (4.5–10 μm × 2–4 μm) were hyaline to pale brown, 0–3 septate (mostly one septate), smooth walled, and cylindrical, with rounded apices and truncate at the bases. The isolate was thus identified as *Veronaea botryosa*.

Sutton et al. [14] reported a *Veronaea botryose*-related case involving a 62-year-old orthotopic cardiac transplant recipient, who presented with chronic induration, mild erythema, and intense tenderness over the dorsum of the right hand after an attempted intravenous line insertion several months earlier. A 2 × 1.1 cm area of skin was excised along with a separate 4 × 2 × 1 cm pink–gray cyst. Histopathology of the excised lesion showed acute inflammation, microabscess formation, and brown moniliform hyphal elements with hematoxylin and eosin stain. Special fungal stains, the Gomori methenamine silver stain, and the Fontana-Masson stain for melanized hyphae were also positive. Culture of the excised tissue on Sabouraud dextrose agar plate, Sabhi agar slant, and Sabhi agar slant with chloramphenicol and gentamicin yielded dematiaceous colonies after 3 days of incubation. A potato dextrose agar slide culture revealed a "*Rhinocladiella*-like" type of conidiation. Colonies were velvety to woolly and grayish or blackish brown centrally, with a dark green periphery. The reverse was olivaceous black. At 25°C on a 60 mm-diameter potato flake agar plate, the growth rate was moderately rapid, reaching 30 mm in 2 weeks (2.1 mm/day); slower growth occurred at 35°C (suggesting its potential for invasive disease), and no growth was noted at 40°C. Conidiophores (300 µm × 2.5–4 µm) were brown, septate, smooth, straight or flexuous, and branched or unbranched, with the apical area being somewhat darker. Smooth-walled, hyaline to pale brown, mostly two-celled conidia (2–4 µm × 5–12 µm, but predominately 3.5 µm × 8 µm) with rounded apices and truncate bases were borne from the geniculate apical conidia-bearing portion of the conidiogenous cell. The organism was clearly *Veronaea botryosa*. Antifungal therapy with itraconazole and then voriconazole resolved the lesions after 10 months.

Da Cunha Filho et al. [15] described two cases of phaeohyphomycosis caused by *Veronaea bothryosa*. The first case involved a 44-year-old kidney transplant recipient, who developed a lesion on the dorsal aspect of the left foot after wooden splinter trauma several years earlier. The lesion was painful to touch, with an erythematous-violet color. Grocott–Gomori methenamine staining and direct KOH examination of exudation demonstrated dematiaceous hyphae. On Sabouraud dextrose agar, colonies were raised, velvety, initially gray, and later black. These features were typical of *Veronaea bothryosa*. Following therapy of itraconazole 200 mg daily for 10 months along with tapering dose of immunosuppressor drugs, there was reduction of pain. The patient died due to complications arising from an accidental cerebral vascular hemorrhage. The second case concerned a 64-year-old patient, who presented with an asymptomatic lesion with a small erythematous-violet papule on the dorsal aspect of the right foot, resulting from a trauma from a corn husk. Direct KOH examination of the lesion biopsy showed dematiaceous hyphae. Culture of the lesion sample yielded dark-gray colonies, with a velvety surface in the middle. These characteristics were indicative of *Veronaea bothryosa*.

Chen et al. [12] reported a *Veronaea botryose*-related case in a 76-year-old male with a history of diabetes mellitus, coronary artery disease, and Cushing's syndrome (due to his long-term use of oral corticosteroids). The patient presented with recurrent multifocal, crusted, brownish-red noduloplaques on the right forearm, left upper limb, and right knee. Grocotle's methenamine silver nitrate (GMS) staining of skin biopsy showed brownish hyphae and yeast-like cells scattered in granulomatous infiltrates. The hyphae and yeast-like cells stained black and pink with GMS and periodic acid Schiff (PAS) stains, respectively. They also stained black with Masson-Fontana stain, indicating the presence of melanin. Culture of the specimen on Sabouraud dextrose agar and potato media at room temperature yielded grayish-brown, velvety colonies after 2 weeks. Slide cultures from the colonies revealed erect and straight conidiophores with conidia borne terminally and laterally on the apices. The conidia are pale brown, smooth walled, two celled, and cylindrical with round tops and truncated bases. The fungus was identified as *Veronaea botryosa*.

Kumar and Hallikeri [16] presented a case of phaeohyphomycosis in a 29-year-old male agriculturist, who complained of painful swelling of the right supraorbital region of 1 year and pain of 6 month duration following minor trauma. Fine-needle aspiration (FNA) was performed and the smears revealed multinucleated giant cells, chronic inflammatory cells, and branching, septate hyphae. PAS stain showed bright magenta-colored hyphae with branching septae. Masson-Fontana stain demonstrated brownish black branching and septate hyphae. The occupational, clinical, and histomorphological features suggested *Veronaea botryosa* phaeohyphomycosis.

45.1.3 Diagnosis

Conventional laboratory identification of *Veronaea botryosa* is dependent on observation of the characteristic microscopic and macroscopic features of the fungus. Scrapings from skin lesions and biopsy are first examined under microscope with various stains (e.g., KOH, lactophenol cotton blue, Gomori methenamine silver stain, PAS stain, Fontana-Masson stain, and hematoxylin and eosin stain) for hyphal elements and septate hyphae. Fontana-Masson stain is particularly useful for detection of melanized hyphae. Portions of samples are cultured on mycological media (e.g., Sabouraud dextrose agar containing chloramphenicol [Sab+C] and Sab+C containing cycloheximide, and potato dextrose agar) at 25°C and 37°C in the dark for isolation of colonies with distinct morphology and structures. Slide cultures of *Veronaea* isolate on potato dextrose agar may be prepared and incubated at 25°C for identification of characteristic hyphae, conidiophores, conidiogenous cells, and conidia [13,14]. To speed up the diagnosis, molecular techniques such as polymerase chain reaction (PCR) and sequencing analysis of small and large subunit rRNA genes as well as internal transcribed spacer (ITS) regions may be employed.

45.2 METHODS

45.2.1 Sample Preparation

Clinical specimens are examined directly under microscope with a fungal stain. Portions of the samples are inoculated on Sabouraud glucose agar without or with antibiotics. The resulting isolates are identified on the basis of macroscopical and microscopical features. For easy microscopic observation, slide culture technique may be utilized. Slide cultures are set up in Petri dishes containing 2 mL of sterile water, into which a U-shaped glass rod is placed, extending above the water surface. A block of freshly growing fungal colony, approx. 1 cm^2, is placed onto a sterile microscope slide, covered with a somewhat larger, sterile glass cover slip, and incubated in the moist chamber. Fungal sporulation is monitored over time, and when optimal, images are captured by means of a Nikon camera system (Digital Sight DS-5M, Nikon Corporation) [4].

For fungal DNA extraction, 50 mL of Sabouraud dextrose (SAB) broth is inoculated by needle with conidia from a 7 day culture in SAB agar and incubated for 72 h at 30°C. The hyphae are recovered on a 0.45 µm-pore-size filter and washed with sterile saline. Aliquots of the fungal hyphae are stored frozen at −70°C until use. Prior to lysis, the hyphae are thawed and suspended in 400 μL of DNA extraction buffer (1 mM EDTA pH 8.0, 1% sodium dodecyl sulfate, 10 mM Tris–HCl pH 7.6, 100 mM NaCl, and 2% Triton X-100). Microcentrifuge tubes (1.5 mL) containing hyphae and buffer are sonicated in a water bath (Branson; model 2210) for 15 min, followed by heating at 100°C for 5 min. Following lysis, DNA is purified using the QIAmp blood kit (Qiagen) and protocols for crude cell lysates supplied by the manufacturer. The purified DNA is stored at 4°C until tested. Alternatively, DNA is extracted with a FastDNA kit (Qbiogene) from mycelium grown for 3–5 days in liquid complete medium.

45.2.2 Detection Procedures

Arzanlou et al. [4] utilized the universal primers ITS1 and ITS4 to amplify the internal transcribed spacer region (ITS) of the nuclear ribosomal RNA operon, including the 3′ end of the 18S rRNA gene, the first internal transcribed spacer region (ITS1), the 5.8S rRNA gene, the second internal transcribed spacer region (ITS2), and the 5′ end of 28S rRNA gene. Subsequent sequencing analysis allows identification of fungal organisms including *Chaetomium* species.

Procedure

1. Polymerase chain Reaction (PCR) mixture (25 μL) is composed of 0.5 U *Taq* polymerase (Bioline), 1×PCR buffer, 0.5 mM $MgCl_2$, 0.2 mM of each dNTP, 5 pmol of each primer, and approximately 10–15 ng of fungal genomic DNA.
2. Amplification is performed on a GeneAmp PCR System 9700 (Applied Biosystems) with primary denaturation at 96°C for 5 min; 36 cycles of 96°C for 30 s, 52°C for 30 s, and 72°C for 60 s; a final extension at 72°C for 7 min.
3. The amplicons are sequenced using BigDye Terminator v. 3.1 (Applied Biosystems) or DYEnamicET Terminator (Amersham Biosciences) Cycle Sequencing Kits and analyzed on an ABI Prism 3700 (Applied Biosystems).
4. Newly generated sequences are subjected to a Blast search of the GenBank databases, sequences with high similarity are downloaded from GenBank, and comparisons are made on the basis of the alignment of the obtained sequences.
5. Phylogenetic analysis is performed with PAUP (Phylogenetic Analysis Using Parsimony) v. 4.0b10 using the neighbor-joining algorithm with the uncorrected ("p"), the Kimura 2-parameter and the HKY85 substitution models. Alignment gaps longer than 10 bases are coded as single events for the phylogenetic analyses; the remaining gaps are treated as missing data. Any ties are broken randomly when encountered.
6. Phylogenetic relationships are also inferred with the parsimony algorithm using the heuristic search option with simple taxon additions and tree bisection and reconstruction (TBR) as the branch-swapping algorithm; alignment gaps are treated as a fifth character state and all characters are unordered and of equal weight. Branches of zero length are collapsed and all multiple, equally parsimonious trees are saved.
7. Other measures calculated include tree length, consistency index, retention index, and rescaled consistency index (TL, CI, RI, and RC, respectively). The robustness of the obtained trees is evaluated by 1000 bootstrap replications. Bayesian analysis is performed. The best nucleotide substitution model is determined using MrModeltest v. 2.2. MrBayes v. 3.1.2 is used to perform phylogenetic analyses, using a general time-reversible (GTR) substitution model with inverse gamma rates, dirichlet base frequencies, and the temp value set to 0.5.

Note. Part of the large subunit 28S rRNA (LSU) gene may be also amplified with primers LR0R and LR5 followed by sequencing analysis.

45.3 CONCLUSION

Veronaea is one of the 70 dematiaceous fungal genera that can take advantage of temporary weakness in the host immune system and traumatic injury, causing a variety of clinical diseases collectively referred to as phaeohyphomycosis, chromoblastomycosis, or eumycetoma. Of the eight species within the genus *Veronaea*, only *Veronaea botryosa* has been associated with human skin infections. Conventional laboratory identification of *V. botryosa* is based on detection of its characteristic phaeoid septate hyphae (3–4.5 μm wide), geniculate, unbranched, pale-brown conidiophores, scar-bearing conidiogenous cells, and sympodially produced, bicellular conidia with round tops and truncated bases through direct mycologic examination and culture microscopy. However, this diagnostic scheme is notably lengthy and requires considerable technical expertise. With the application of PCR and sequencing procedures, the identity of *V. botryosa* and other fungi can be rapidly and precisely determined, contributing to the more effective treatment and control of human phaeohyphomycosis.

REFERENCES

1. The UniProt Consortium. Available at http://www.uniprot.org/, accessed on August 1, 2010.
2. de Hoog GS, Rahman MA, Boekhout T. *Ramichloridium*, *Veronaea*, and *Stenella*: Generic delimitation, and new combinations and two new species. *Trans Br Mycol Soc.* 1983;81:485–490.
3. Kharwar RN, Singh RK. Additions to the hyphomycete genus *Veronaea* as phytoparasitic species. *Microbiol Res.* 2004;159(2):103–111.
4. Arzanlou M et al. Phylogenetic and morphotaxonomic revision of *Ramichloridium* and allied genera. *Stud Mycol.* 2007;58:57–93.

5. Chakrabarti A. Epidemiology of central nervous system mycoses. *Neurol India* 2007;55:191–197.
6. Brandt ME, Warnock DW. Epidemiology, clinical manifestations, and therapy of infections caused by dematiaceous fungi. *J Chemother*. 2003;15(Suppl. 2):36–47.
7. Kondo Y et al. Cutaneous phaeohyphomycosis caused by *Veronaea botryosa* observed as sclerotic cells in tissue. *Int J Dermatol*. 2007;46(6):625–627.
8. Zhang HE, Wang DL, Li RY. Report of the case of phaeohyphomycosis caused by *Veronaea botryosa*. *Chin J Dermatol*. 1990;23:96–98.
9. Wang DL et al. Studies on *Veronaea botryosa* agent of the first human case. *Acta Mycol Sin*. 1991;10:159–165.
10. Ayadi A, Huerre MR, de Bievre C. Phaeohyphomycosis caused by *Veronaea botryosa*. *Lancet* 1995;346:1703.
11. Foulet F et al. Cutaneous phaeohphomycosis caused by *Veronaea bothryosa* in a liver transplant recipient successfully treated with Itraconazole. *Clin Infect Dis*. 1999;29:689–690.
12. Chen YT et al. Cutaneous phaeohyphomycosis caused by an Itraconazole and Amphoterecin B resistant strain of *Veronaeae botryosa*. *Int J Dermatol*. 2006;45(4):429–432.
13. Matsushita A et al. Subcutaneous phaeohyphomycosis caused by *Veronaea botryosa* in the People's Republic of China. *J Clin Microbiol*. 2003;41:2219–2222. Erratum in: *J Clin Microbiol*. 2003;41(7):3462.
14. Sutton DA, Rinaldi MG, Kielhofner M. First U.S. report of subcutaneous phaeohyphomycosis caused by *Veronaea botryosa* in a heart transplant recipient and review of the literature. *J Clin Microbiol*. 2004;42(6):2843–2846.
15. Da Cunha Filho, RR et al. Subcutaneous phaeohyphomycosis caused by *Veronaea bothryosa*: Report of 2 cases. *An Bras Dermatol*. 2005;80:1.
16. Kumar KK, Hallikeri K. Phaeohyphomycosis. *Indian J Pathol Microbiol*. 2008;51:556–558.

Pezizomycotina: Sordariomycetes

46 Acremonium

Dongyou Liu, Xianghong Du, and Song Weining

CONTENTS

46.1 INTRODUCTION

46.1.1 Classification and Morphology

The genus *Acremonium* is classified within the mitosporic Hypocreales group, order Hypocreales, class Sordariomycetes, subphylum Pezizomycotina, phylum Ascomycota, and kingdom Fungi. The mitosporic Hypocreales group encompasses 24 genera: *Acremonium*, *Acrostalagmus*, *Cephalosporium*, *Chaetopsina*, *Cylindrocladiella*, *Escovopsis*, *Fusarium*, *Gliocladiopsis*, *Gliocladium*, *Hobsonia*, *Illosporium*, *Myrothecium*, *Parasarcopodium*, *Polycephalomyces*, *Rotiferophthora*, *Sesquicillium*, *Solheimia*, *Stachybotrys*, *Stilbella*, *Trichothecium*, *Tubercularia*, *Ustilaginoidea*, *Verticillium*, and *Xenocylindrocladium* [1].

In turn, the genus *Acremonium* consists of 33 recognized species: *Acremonium alabamense*, *A. atrogriseum*, *A. alcalophilum*, *A. alternatum*, *A. antarcticum*, *A. blochii*, *A. breve*, *A. brunnescens*, *A. butyri*, *A. collariferum*, *A. crotocinigenum*, *Acremonium cucurbitacearum*, *Acremonium curvulum*, *Acremonium exuviarum*, *A. fuci*, *A. falciforme* (obsolete synonym: *Cephalosporium falciforme*), *A. hyalinulum*, *A. kiliense* (obsolete synonyms: *Cephalosporium granulomatis*, *C. infestans*, and *C. madurae*), *A. murorum*, *A. nepalense*, *A. obclavatum*, *A. potronii*, *A. recifei* (obsolete synonym: *Cephalosporium recifei*), *A. restrictum*, *A. roseogriseum*, *A. rutilum*, *A. sclerotigenum*, *A. spinosum*, *A. strictum*, *A. stromaticum*, *A. thermophilum*, *A. zeae*, and *Cephalosporium acremonium*, in addition to over 30 unassigned species [1–3].

A recent study showed that *Acremonium falciforme* shares considerable sequencing similarity in the ribosomal large subunit (LSU) with members of the *Fusarium solani* species complex (i.e., the *F. solani* clade containing *F. solani*, *Haematonectria*, and *Neocosmospora*), and the new combination *Fusarium falciforme* was proposed for *Acremonium falciforme* [4].

Teleomorphs of *Acremonium* are found in several genera of ascomycetes such as *Emericellopsis*, *Hapsidospora*, *Nectria*, *Nectriella*, *Neocosmospora*, *Pronectria*, and *Thielavia* (Sordariales), reflecting the polyphylic nature of the *Acremonium* genus. For example, *Acremonium alabamense* is the anamorph of *Thielavia terrestris*, a member of Sordariales. It has been proposed that the name *Acremonium* should be restricted to only the anamorphs of the family Hypocreaceae, and a new genus should be established to accommodate other species of *Acremonium*, such as grass endophytes and others related to the Clavicipitaceae.

Acremonium spp. are filamentous fungi that are commonly isolated from soil, plant debris, rotting mushrooms, etc. in Europe, Asia, Egypt, and North and Central America. Several *Acremonium* species have been implicated in human infections, including *Acremonium falciforme*, *Acremonium kiliense*, *Acremonium strictum*, and *Acremonium recifei*.

Acremonium grows moderately. After 7 day incubation at 25°C on potato glucose agar, colonies measure 1–3 cm in diameter and are compact, flat or folded, and slightly raised in the center. Colonies are white to pale gray, velvety, powdery, and membrane-like initially and become cottony or fasciculate (showing spiky aggregates of hyphae) with age. On the front, colonies are white, pale gray, cottony, or pale pink in color and, on the reverse, they are either uncolored or a pink-to-rose-colored pigment production.

Acremonium spp. possess hyaline, narrow, and septate hyphae, with vegetative hyphae forming hyphal ropes. Unbranched, solitary, awl-shaped erect phialides are located on the hyphal tips, the hyphal ropes, or both. The phialides are separated from hyphae by a septum and taper toward their apices. Conidia (1–3×3–8 μm) are usually one celled (ameroconidia), hyaline or pigmented, globose to cylindrical, and mostly aggregated in slimy heads at the apex of each phialide (which may or may not possess collarettes). Bound by a gelatinous material, conidia may be single or multicellular, fusiform with a slight curve (crescent-like).

A. falciforme often produces crescentic, nonseptate conidia, sometimes being two or three celled; *A. kiliense* forms solitary hyphae, chlamydospores (especially in oatmeal agar), and short straight cylindrical conidia; *A. recifei* usually shows crescentic and nonseptate conidia; and *A. strictum* colonies may appear moist, pink, or salmon color, resembling those of *Lecythophora hoffmannii*.

46.1.2 Clinical Features and Pathogenesis

Being mainly environmental saprophytes, *Acremonium* species often gain entry to human host following penetrating injuries or surgery procedures [5,6]. Besides being one of the causative agents of eumycotic white grain mycetoma (principally by *A. falciforme*, *A. kiliense*, and *A. recifei*), *Acremonium* spp. are also responsible for rare cases of onychomycosis, keratitis, endophthalmitis, endocarditis, gastritis, fungemia, meningitis, diffuse cerebritis, peritonitis, invasive pulmonary disease, and osteomyelitis [7–13]. These infections are often diagnosed in patients with predisposing conditions (e.g., Addison's disease, neutropenia, immune suppression, burns, organ transplantation, artificial implants, intravenous drug abuse, splenectomy, bone marrow transplantation, and diabetes mellitus) and preterm infants [14,15]. Neonates who receive mechanical ventilation, umbilical vein catheterization, previous treatment with antibacterial agents, and prior use of parenteral nutrition such as intravenous lipid show increased susceptibility to *Acremonium* infections. *Acremonium* endophthalmitis may develop after penetrating keratoplasty and retinal detachment surgery, with intraocular inoculations from irrigation solutions, and display clinical symptoms of mild pain, redness, floaters, and slightly decreased visual acuity. Invasive *Acremonium* infection is difficult to treat and the outcome is generally poor. *Acremonium* species was reported to cause invasive allergic fungal sinusitis (AFS) or eosinophilic fungal rhinosinusitis (EFRS) in an immunocompetent patient, leading to unilateral visual loss [16].

Among the *Acremonium* species of clinical interest, *Acremonium kiliense* is most important and commonly described [17–19]. Examined by light microscopy, *A. kiliense* is stained Gram-positive and is characterized by the appearance of septate hyphae and chlamydospores. The fungus grows slowly in specific culture (e.g., Sabouraud agar, glucose, and malt extract). A range of clinical diseases has been attributed to *A. kiliense*, including dermatophytoses, kerion, onycomycoses, keratitis, and mycetomas in immunocompetent individuals and pneumonia, arthritis, osteomyelitis, endocarditis, peritonitis, meningitis, and sepsis in immunocompromised patients. Risk factors for *A. kiliense* infection are prosthesis, catheters, immunosuppressive therapy, and anatomic disorders. The fact that *A. kiliense* infection often presents as a cutaneous papular eruption without other clinical signs in immunocompetent patients suggests its relatively low pathogenicity.

Acremonium strictum is another species commonly identified in human diseases [13,15,20–26]. For example, a 59-year-old male who had undergone nonmyeloablative conditioning therapy and bone marrow transplantation after an initial diagnosis of acute myelogenous leukemia presented gastrointestinal graft-versus-host disease (GVHD) with severe gastrointestinal bleeding. A fungus was isolated in blood cultures and also a fungal organism was identified in multiple organs by histopathology at autopsy. Gram stains of positive blood culture bottles showed both yeast-like forms with hyphal elements and fully formed hyphal masses suggestive of a sporulating mold. On Sabouraud agar subculture at 30°C, colonies measured 2.2 cm in 7 days. Young colonies were smooth, moist, and pink, with a colorless reverse; mature colonies were velvety and raised in the center. Lactophenol aniline blue preparations revealed conidia and septate hyphae. The conidia (of 3–4 μm×1–1.5 μm in size) were one celled, cylindrical, smooth, hyaline to slightly pink, and grouped in slimy heads. The conidiophores were simple, slender, and erect phialides with basal septa arising from the vegetative hyphae. Based upon micro- and macroscopic characteristics, the fungus was identified as *A. strictum* [27].

The next common *Acremonium* species associated with human infections is *A. falciforme* [7,9,10,28]. In addition, a number of clinical cases due to *Acremonium* spp. have also been reported [29–31]. A patient with leukemia presented painful cutaneous nodules and severe myalgia indicative of pyomyositis. Culture of an aspiration grew an *Acremonium* species organism. After surgical drainage and treatment with amphotericin B and granulocyte colony-stimulating factor (G-CSF), the patient recovered fully [32]. In another case, a 64-year-old woman developed chronic uveitis in her left eye, 2 weeks after an uncomplicated cataract extraction. Potassium hydroxide (KOH)–Calcofluor staining and Gram staining of the vitreous fluid uncovered septate fungal hyphae of *Acremonium* species. Subsequent culture of the vitreous fluid grew an *Acremonium* organism. The patient was given intravenous ampotericin B daily for 5 days and oral voriconazole medication for 4 weeks. During the postoperative 18 month follow-up, she was stable without significant relapse of uveitis, and her visual acuity improved from 20/40 to 20/20 [33].

Acremonium species are generally sensitive to amphotericin B and ketoconazole. Amphotericin B therapy in combination with ketoconazole or another new azole or allylamine may be used for treatment [34]. Patients experiencing failure of amphotericin B treatment may be treated with a triazole derivative voriconazole. Voriconazole achieves a sufficient therapeutic level in aqueous and vitreous liquids by oral administration for eradication of *Acremonium* osteomyelitis [35].

Acremonium chrysogenum is the production host employed in the fermentation process of manufacturing cephalosporins with broad activity against Gram-positive and Gram-negative bacteria, while application of cephalosporins is preferred in a wide range of cases due to microbial resistance toward penicillins [36,37].

46.1.3 Laboratory Diagnosis

Acremonium is a ubiquitous, saprophytic fungus that is characterized by the formation of narrow hyphae with solitary, slender (2 µm), unbranched, awl (needle)-shaped phialides (or weakly branched conidiophores) arising from vegetative hyphae and producing clusters (slimy messes) or chains of small, one-celled conidia mostly aggregated at the apex of each phialide. In tissue sections, *Acremonium* often shows hyaline, septate hyphae and characteristic reproductive structures known as phialides and phialoconidia. The use of specific fungal stains such as Grocott methenamine-silver and Giemsa stains aids the identification of *Acremonium*, although 10% KOH and Gram stain offer alternative detection methods.

Acremonium shows a high degree of morphological similarity to *Fusarium*, *Verticillium Lecythophora*, *Phialemonium*, *Gliomastix*, and *Cylindrocarpon* as well as *Candida*. Strains of *Fusarium*, which do not produce macroconidia, is differentiated from those of *Acremonium* by their faster growth and production of deeply woolly colonies. In comparison with *Acremonium*, *Lecythophora* and *Phialemonium* phialides are not separated from hyphae by a septum; while *Gliomastix* generates olive-green to greenish-black colonies and chains or balls of dark conidia.

Acremonium is differentiated from hyaline isolates of *Phialophora* by the absence or very limited development of a collarette on the phialide and the predominant formation of awl-shaped phialides with a basal septum. Compared to *Acremonium*, microconidial *Fusarium* isolates usually grow faster and have colonies with a characteristic fluffy appearance. Potato dextrose agar and cornmeal agar are the most suitable media for their identification, and exposure to daylight maximizes their culture color characteristics.

Given the length of time for fungal culture and specialist skills for macroscopic and microscopic identification of *Acremonium* from other morphological similar fungal organisms, nucleic acid amplification and sequence analysis have been increasingly applied for their characterization and speciation. In particular, small subunit (SSU) 18S and LSU 28S ribosomal RNA (rRNA) genes as well as their internal transcribed spacer (ITS) regions provide valuable target for molecular detection of fungal organisms including *Acremonium* spp. [38,39]. For example, PCR amplification of partial 5S rRNA gene and internal transcribed spacer region 2 (ITS2) followed by digestion of the amplicon with restriction enzymes HinfI and Sau3AI divides *Acremonium* spp. into distinct restriction fragment length polymorphism (RFLP) groups, with *A. kiliense* and *A. strictum* forming similar subgroups on the basis of morphological differences and distinct RFLP patterns [40].

46.2 METHODS

46.2.1 Sample Preparation

Culture of mycetoma lesion is performed by the collection of the abscess or the fistula secretion or by tissue biopsy. The samples are cultured in media such as Sabouraud agar or mycobiotic agar to isolate fungi and/or blood agar to isolate bacteria. The etiologic agents are identified according to their macroscopic and microscopic features.

Blood culturing is performed using the BACTEC 9240 automated blood-culturing system (Becton Dickinson), with each culture consisting of one each Plus Aerobic/F, Lytic/10 Anaerobic/F, and Myco/F Lytic bottles. Aerobic and anaerobic media are held in the BACTEC cabinet for 5 days; Myco/F bottles are held for 28 days. Aerobic and anaerobic bottles positive for yeast-like fungi are subcultured to chocolate, bromcresol green, and inhibitory mold agar plates and incubated at 35°C supplemented with CO_2 to 5%. Other fungal cultures are performed using Sabouraud dextrose agar (SAB; Emmon's modification); brain heart infusion agar with blood, chloramphenicol, cycloheximide, and gentamicin; and inhibitory mold agar incubated at 30°C. Subcultures for morphological studies are made on potato dextrose agar and incubated at 30°C [27].

For DNA extraction from fungal strains, a loopful from individual colonies of each fungus culture is homogenized in 500 µL of lyticase lysis buffer (LLB; 50 mM Tris pH 7.6, 1 mM EDTA pH 8.0, 0.2% β-mercaptoethanol, and 1 U/100 µL recombinant lyticase [Sigma]) and incubated at 37°C for 1 h. After incubation, acid-washed glass beads 710–1180 µm in diameter (Sigma) are added and the solution is vortexed thoroughly for 2 min. Amounts of 400 µL of the supernatant are used for DNA extraction on a MagNA Pure compact instrument using a MagNA Pure compact nucleic acid isolation kit I (Roche Diagnostics). DNA concentrations are determined by using a PicoGreen double-stranded DNA quantification kit (Molecular Probes) on an F-2500 fluorescence spectrophotometer (Hitachi) [41].

For peripheral blood specimens, after hypotonic lysis of the erythrocytes using red blood cell lysis buffer (10 mM Tris pH 7.6, 5 mM $MgCl_2$, 10 mM NaCl), the leukocytes are pelleted and resuspended in 470 µL LLB. The subsequent steps were identical to the extraction protocol described above. For blood culture specimens, 200 µL aliquots derived from blood cultures previously shown to be fungus positive are transferred to Falcon tubes and red blood cell lysis buffer is added. The subsequent procedure is as described above. For plasma containing white blood cells, peripheral blood specimens anticoagulated with EDTA are kept at 4°C for at least 4 h to sediment the red blood cells. The entire supernatant, that is, plasma containing white blood cells, is used for DNA extraction. The samples are centrifuged at 15,000 × *g* for 10 min. Most of the plasma is removed, leaving a residual volume of 100 and 430 µL of LLB is added. The DNA extraction is performed as described above. For specimens from respiratory tract and lung biopsies, solid material is cut into small pieces and homogenized in 430 μL of LLB. The ensuing steps are as described above [41].

Alternatively, fungal DNA can be extracted from mature colonies, grown on SAB agar with chloramphenicol and gentamicin (Remel) at 30°C, using the QIAmp Mini Kit (Qiagen) following the manufacturer's tissue extraction protocol [27].

46.2.2 Detection Procedures

46.2.2.1 Sequencing Analysis of ITS and Partial 28S rRNA Gene

Novicki et al. [27] employed the ITS1 (5′-TCC GTAGGTGAACCTGCGG-3′) and ITS4 (5′-TCCTCCGCTT ATTGATATGC-3′) primers to amplify the intergenic transcribed spacer 1 (ITS1) and ITS2 regions of the rRNA operon, flanking the 5.8S rRNA gene [42]. In addition, the D1–D2 variable domain of the 28S rRNA gene is amplified with the NL-1 (5′-GCATATCAATAAGCGGAGGAAAAG-3′) and NL-4 (5′-GGTCCGTGTTTCAAGACGG-3′) primers.

PCR mixture (25 μL) is made up of GeneAmp 1× PCR buffer II (Applied Biosystems), 2.5 mM $MgCl_2$, 0.2 mM deoxynucleotide triphosphate, 400 nM each of primers ITS1 and ITS4, or primers NL-1 and NL-4, 2.5 U AmpliTaq DNA polymerase (Applied Biosystems), and 5 μL of template DNA.

PCR amplification is performed with an initial 95°C for 10 min (polymerase activation); 35 cycles of 95°C for 30 s, 55°C for 30 s, and 72°C for 1 min; and a final extension at 72°C for 10 min.

The resulting amplicons are sequenced and compared with those at the National Center for Biotechnology Information GenBank nucleotide database using the nucleotide–nucleotide BLAST program for homologous sequences. The sequences are then aligned and phylogenetic trees are drawn with Clustal X using the neighbor-joining method. The aligned sequences are edited with Jalview version 1.3b. Phylogenetic trees are displayed using Treeview version 1.6.6. This allows precise identification of fungal organisms of interest including *Acremonium* spp.

46.2.2.2 PCR-RFLP of ITS2 Region

Keynan et al. [43] showed that by using primers ITS1 and ITS4 in the first round PCR and internal primers UNF1 (5′-GCATCGATGAAGAACGCAGC-3) and UNF2 (5′-TTGATATGCTTAAGTTCAGCGG-3′) with specificity for the ITS2 region in the second round PCR, a 285 bp product is generated. Subsequent digestion of the nested PCR product with endonuclease *Hae*III results in 193 and 92 bp fragments, which are characteristic of *Acremonium* species.

The PCR mixture (50 μL) is composed of 1× HotStart Master Mix (Qiagen), 15 pmol of UNF1 mixture primers, 15 pmol of UNF2 primer, 100 μM of dUTP (Sigma), 0.5 U of uracil-DNA-gycosylase (UDG) (Sigma) for the elimination of possible cross contamination, and 5 μL of template DNA produced after standard DNA isolation.

The PCR reaction is conducted under the following conditions: (i) incubation at 37°C for 10 min for UDG treatment followed by the initial activation of DNA polymerase at 96°C for 12 min; (ii) 40 cycles consisting of a denaturation step for 10 s at 96°C, an annealing step for 10 s at 58°C, and an extension step for 30 s at 72°C; and (iii) the final extension at 72°C for 4 min.

Restriction analysis of amplified PCR products is performed by *Hae*III, endonuclease digestion (New England BioLabs). The restriction fragments are separated on a 3% (wt/vol) agarose gel, stained with ethidium bromide, visualized by UV transillumination (312 nm), and analyzed by ULTRA LUM (Ultra-Lum, Inc.) gel detection and analysis system.

46.3 CONCLUSION

The anamorphic genus *Acremonium* consists of a large number of saprophytic filamentous fungal species that have hyaline, septate hyphae and reproduce by phialidic conidiation. Several *Acremonium* spp. are involved in eumycotic mycetoma and other focal infections in otherwise healthy individuals. In addition, they have been increasingly implicated in invasive systemic mycotic diseases in immunosuppressed individuals with underlying conditions such as malignancy and other illnesses. Given their similar morphological features to a number of fungal organisms that may display varied sensitivity to antifungal agents, it is important to identify *Acremonium* to species level. Unfortunately, identification of *Acremonium* spp. by macroscopic and microscopic examination is not only time-consuming but also challenging. This is especially so with some *Acremonium* species that do not develop a teleomorphic stage in vitro. Therefore, application of molecular detection procedures such as PCR and sequencing analysis is critical for accurate determination of *Acremonium* species identity and assists effective treatment and control of *Acremonium* infections.

REFERENCES

1. The UniProt Consortium. Available at http://www.uniprot.org/, accessed on August 1, 2010.
2. Glenn AE et al. Molecular phylogeny of *Acremonium* and its taxonomic implications. *Mycologia* 1996;88:369–383.
3. Weisenborn JLF, Kirschner R, Piepenbring M. A new darkly pigmented and keratinolytic species of *Acremonium* (Hyphomycetes) with relationship to the Plectosphaerellaceae from human skin and nail lesions in Panama. *Nova Hedwigia* 2010;90:457.
4. Summerbell RC, Schroers HJ. Analysis of phylogenetic relationship of *Cylindrocarpon lichenicola* and *Acremonium falciforme* to the *Fusarium solani* species complex and a review of similarities in the spectrum of opportunistic infections caused by these fungi. *J Clin Microbiol.* 2002;40(8):2866–2875.
5. Fridkin SK, Jarvis WR. Epidemiology of nosocomial fungal infections. *Clin Microbiol Rev.* 1996a;9:499–511.
6. Geyer AS et al. *Acremonium mycetoma* in a heart transplant recipient. *J Am Acad Dermatol.* 2006;55(6):1095–1100.
7. Lau YL et al. Invasive *Acremonium falciforme* infection in a patient with severe combined immunodeficiency. *Clin Infect Dis.* 1995;20:197–198.

8. Roilides E et al. *Acremonium* fungemia in two immunocompromised children. *Pediatr Infect Dis J*. 1995;14:548–550.
9. Cameron JA et al., Chronic endophthalmitis caused by *Acremonium falciforme*. *Can J Ophthalmol*. 1996;31:367–368.
10. Koç AN, Erdem F, Patiroglu, T. Case report: *Acremonium falciforme* fungemia in a patient with acute leukaemia. *Mycoses* 2002;45:202–203.
11. Mattei D et al. Successful treatment of *Acremonium* fungemia with voriconazole. *Mycoses* 2003;46:511–514.
12. Alfonso JF et al. *Acremonium* fungal infection in 4 patients after laser in situ keratomileusis. *J Cataract Refract Surg*. 2004;30(1):262–267.
13. Miyakis S et al. Invasive *Acremonium strictum* infection in a bone marrow transplant recipient. *Pediatr Infect Dis J*. 2006;25:273–275.
14. Jeffrey WR et al. Disseminated infection due to *Acremonium* species in a patient with Addison's disease. *Clin Infect Dis*. 1993;16:170.
15. Yalaz M et al. Fatal disseminated *Acremonium strictum* infection in a preterm newborn: A very rare cause of neonatal septicaemia. *J Med Microbiol*. 2003;52:835–837.
16. Mulwafu WK et al. Allergic fungal sinusitis secondary to *Acremonium* species causing unilateral visual loss. *East Central Afr J Surg*. 2006;11:62–64.
17. Brabender W et al. *Acremonium kiliense* osteomyelitis of the calvarium. *Neurosurgery* 1985;16:554–556.
18. Fridkin SK et al. *Acremonium kiliense* endophthalmitis that occurred after cataract extraction in an ambulatory surgical care and was traced to an environmental reservoir. *Clin Infect Dis*. 1996b;22:222–227.
19. Pastorino AC et al. *Acremonium kiliense* infection in a child with chronic granulomatous disease. *Braz J Infect Dis*. 2005;9(6):529–534.
20. Schell WA, Perfect JR. Fatal, disseminated *Acremonium strictum* infection in a neutropenic host. *J Clin Microbiol*. 1996;34:1333–1336.
21. Koç AN et al. Peritonitis due to *Acremonium strictum* in a patient on continuous ambulatory peritoneal dialysis. *Nephron* 1998;79:357–358.
22. Koç AN et al. Pleuritis caused by *Acremonium strictum* in a patient with colon adenocarcinoma. *Mycoses* 2008;51(6):554–556.
23. Koç AN, Mutlu Sarigüzel F, Artiş T. Isolation of *Acremonium strictum* from pleural fluid of a patient with colon adenocarcinoma. *Mycoses* 2009;52(2):190–192.
24. Anadolu R et al. Indolent *Acremonium strictum* infection in an immunocompetent patient. *Int J Dermatol*. 2001;40:451–453.
25. Scott IU, Flynn HW Jr., Miller D. Delayed-onset endophthalmitis following cataract surgery caused by *Acremonium strictum*. *Ophthalmic Surg Lasers Imaging*. 2005;36:506–507.
26. Guarro J et al. A case of colonization of a prosthetic mitral valve by *Acremonium strictum*. *Rev Iberoam Micol*. 2009;26(2):146–148.
27. Novicki TJ et al. Genetic diversity among clinical isolates of *Acremonium strictum* determined during an investigation of a fatal mycosis. *J Clin Microbiol*. 2003;41(6):2623–2628.
28. Van Etta LL, Peterson LR, Gerding DN. *Acremonium falciforme* (*Cephalosporium falciforme*) mycetoma in a renal transplant patient. *Arch Dermatol*. 1983;119:707–708.
29. Fincher RME et al. Infection due to the fungus *Acremonium* (*Cephalosporium*). *Medicine* 1991;70:398–409.
30. Kan SF et al. Cutaneous hyalohyphomycosis caused by *Acremonium* in an immunocompetent patient. *Br J Dermatol*. 2004;150(4):789–790.
31. Yamazaki R et al. Systemic infection due to *Acremonium* after allogeneic peripheral blood stem cell transplantation. *Intern Med*. 2006;45(16):989–990.
32. Chang HY et al. *Acremonium* pyomyositis in a pediatric patient with acute leukemia. *Pediatr Blood Cancer* 2005;44:521–524.
33. Joe SG et al. Case report of *Acremonium* intraocular infection after cataract extraction. *Korean J Ophthalmol*. 2010 April;24(2):119–122.
34. Guarro J et al. *Acremonium* species: New emerging fungal opportunists—In vitro antifungal susceptibilities and review. *Clin Infect Dis*. 1997;25:1222–1229.
35. Weissgold DJ, Maguire AM, Brucker AJ. Management of postoperative *Acremonium* endophthalmitis. *Ophthalmology* 1996;103:749–756.
36. Weil J, Miramonti J, Ladisch MR. Cephalosporin-C—Mode of action and biosynthetic-pathway. *Enzyme Microb Technol*. 1995;17:85–87.
37. Barber MS, Giesecke U, Reichert A, Minas W. Industrial enzymatic production of cephalosporin-based beta-lactams. *Adv Biochem Eng Biotechnol*. 2004;88:179–215.
38. Chen YC et al. Identification of medically important yeasts using PCR-based detection of DNA sequence polymorphisms in the internal transcribed spacer 2 region of the rRNA genes. *J Clin Microbiol*. 2000;38:2302–2310.
39. Chen YC et al. Polymorphic internal transcribed spacer region 1 DNA sequences identify medically important yeasts. *J Clin Microbiol*. 2001;39:4042–4051.
40. Borazjani RN, Lott TJ, Ahearn DG. Comparison of 5.8S and ITS2 rDNA RFLP patterns among isolates of *Acremonium obclavatum*, *A. kiliense*, and *A. strictum* from diverse sources. *Curr Microbiol*. 1998;36(2):70–74.
41. Landlinger C et al. Species-specific identification of a wide range of clinically relevant fungal pathogens by use of Luminex xMAP technology. *J Clin Microbiol*. 2009;47(4):1063–1073.
42. White TJ et al. Amplification and direct sequencing of fungal ribosomal RNA genes for phylogenetics. In MA Innis, DH Gelfand, JJ Sninsky, and TJ White (eds.), *PCR Protocols: A Guide to Methods and Applications*. Academic Press, San Diego, CA, pp. 315–322, 1990.
43. Keynan Y, Sprecher H, Weber G. *Acremonium* vertebral osteomyelitis: Molecular diagnosis and response to voriconazole. *Clin Infect Dis*. 2007;45(1):e5–e6.

47 *Beauveria*

Dongyou Liu

CONTENTS

47.1 INTRODUCTION

47.1.1 Classification and Morphology

The genus *Beauveria* is a filamentous fungus belonging to the mitosporic Cordycipitaceae group, family Cordycipitaceae, Order Hypocreales, class Sordariomycetes, subphylum Pezizomycotina, phylum Ascomycota, and kingdom Fungi. The genus currently comprises 11 recognized species: *Beauveria alba*, *Beauveria amorpha*, *Beauveria bassiana*, *Beauveria brongniartii*, *Beauveria caledonica*, *Beauveria felina*, *Beauveria malawiensis*, *Beauveria tenella*, *Beauveria velata*, *Beauveria vermiconia*, and *Beauveria virella*, in addition to an unassigned species: *Beauveria* cf. *caledonica* ARSEF 2251. The teleomorphs of the genus *Beauveria* are found in *Cordyceps* (e.g., *Cordyceps bassiana* and *Cordyceps scarabaeicola*) [1,2].

Beauveria spp. are entomogenous fungi that infect insects and other arthropods. They have the ability to survive saprotrophically in the soil environment for extended time periods and to form an endophytic association with plants (e.g., corn) [3,4]. These organisms can be dispersed by sporulating cadavers as infections in migrating insect hosts or by infectious conidia on wind currents.

Beauveria bassiana was first identified in the early nineteenth century by the Italian scientist Agostino Bassi as the causative agent for white muscardine disease in silkworms (*Bombyx mori*). Subsequently, *B. bassiana* and *B. brongniartii* as well as *Metarhizium anisopliae* have been exploited as mycopesticides for biological control of insect pests (e.g., bark beetles) [5]. *Beauveria* spp. are occasionally isolated in the clinical laboratory as saprophytic contaminants; and *B. bassiana* and possibly *B. brongniartii* have been implicated in human hyalohyphomycosis.

Beauveria spp. grow moderately rapidly, producing colonies of 1–3 cm in diameter on potato glucose agar after incubation at 25°C for 7 days. Colonies are cottony to powdery or mealy in texture, initially white and later becoming yellowish white or pale pinkish in color. The reverse is white or pale. Hyphae (2–8 μm in width) are hyaline, septate, and narrow, with random branches. Conidiogenous cells on the hyphae are flask shaped with an inflation at the base and narrow zigzagging filaments at the apex. Laterally from the filament, conidia are produced from each bending point (which is termed sympodial geniculate growth). Conidia (2–4 μm in diameter) are hyaline, one celled, and globose to ovoid in shape. As conidiogenous cells tend to form dense clusters, which appear as small powdery balls in the aerial hyphae under microscope, and make it difficult to visualize the arrangement and structure of conidia, young cultures are preferred for optimal microscopic examination [6].

47.1.2 Clinical Features

Beauveria spp. are insect pathogens with ubiquitous distribution. *B. bassiana*, *B. brongniartii*, and *B. alba* have been shown to cause keratitis, deep tissue infection, disseminated infection, and other diseases in humans, especially those with suppressed immune system and/or underlying conditions (e.g., steroid treatment, prolonged granulocytopenia following chemotherapy, and hematological malignancies) as well as after accidental injury [7–11].

47.1.2.1 *Beauveria bassiana*

Tucker et al. [12] documented a case of disseminated *Beauveria bassiana* infection in a 44-year-old woman with acute lymphoblastic leukemia. After undertaking prophylactic treatment with ciprofloxacin and fluconazole, the patient developed skin lesions, sinusitis, headache, and facial pain. Histopathological examination of the skin biopsy specimen revealed areas of necrosis heavily permeated by fungal hyphae. Cultures of the tissue biopsy sample on blood agar and Sabouraud glucose agar (SGA) plates at 30°C yielded a pure growth of a white mold. On potato dextrose agar (PDA) at 30°C, colonies reached diameters of 4.5 cm in 13 days and appeared densely cottony to flocculent, yellowish white, raised, and dome shaped. Conidiogenous cells in sporodochial clusters were slightly swollen at the base and narrowed at the tip to form a zigzag rachis. Conidia (2–3 μm × 1.5–2 μm) were oval to subglobose and apiculate. However, stab inoculations of the patient isolate and a control *B. bassiana* isolate onto PDA failed to grow at 35°C. Sequence of the nuclear ribosomal internal transcribed spacer (ITS) region from the case isolate demonstrated the highest homology (98% identity) to the GenBank sequences of *B. bassiana* and its teleomorph *Cordyceps bassiana*. After treatment with liposomal amphotericin (AmBisome) in combination with itraconazole, her skin lesions healed with some scarring but no recurrence of fungal infection.

Kisla et al. [13] described a case of *Beauveria bassiana* keratitis in an 82-year-old woman, who underwent surgical repair of a graft wound dehiscence and anterior vitrectomy on her right eye following blunt trauma. Seven months later, the patient complained of decreased vision, a foreign body sensation, and mild aching in her right eye. The corneal graft was diffusely edematous; and a stromal infiltrate was noted in the area of the epithelial defect. Smears stained with Giemsa revealed the presence of septate hyphae. Samples were inoculated onto blood, chocolate, brain–heart infusion agars, and eugonic broth for anaerobes. While bacterial cultures yielded no growth, fungal cultures grew 14 colonies of a white mold. Colonies on PDA were yellowish white, moderately fast growing, dense, and from powdery to woolly in texture. Conidia (asexual spores) were produced from small pegs formed in a zigzag arrangement at the tips of the conidiogenous cells (spore-bearing cells) that are swollen at the base. These macroscopic and microscopic features confirmed the fungus as *Beauveria bassiana*. Following treatment with natamycin 5% suspension and fluconazole for 3 weeks, the infiltrate progressively improved, the epithelial defect became smaller, and the patient recovered without recurring infection.

Sonoyama et al. [14] reported a superficial keratomycosis due to *Beauveria bassiana* in an 80-year-old woman, who presented with ocular pain and hyperemia 9 days after an ocular injury by the frame of her glasses. The patient had recurrent diabetic iritis and continuously used topical antibiotics and corticosteroids. A slit-lamp examination indicated a corneal ulcer confined within the superficial stromal layer, along with a slight infiltration and edema. Direct microscopic examination of corneal scrapings revealed Gram-positive septate fungal hyphae with budding. Slide culture of the corneal scraping grew a whitish yellow colony, with the fungus exhibiting zigzag rachis and oval conidia, characteristics of *B. bassiana*. Topical voriconazole with miconazole, pimaricin, and oral itraconazole were effective, and the lesion disappeared leaving only a mild scar after 2 months. It is noteworthy that the keratomycosis caused by *B. bassiana* is restricted to the superficial cornea, as the pathogenesis for *B. bassiana* is relatively weak and does not cause severe stromal infiltration or anterior chamber inflammation. Since the lesion caused by *B. bassiana* extends horizontally and not vertically within the stroma, endothelial plaque and anterior inflammation are not evoked. On the other hand, other filamentous fungi of keratomycosis, such as *Aspergillus* and *Fusarium*, invade deeply into the corneal stroma and induce endothelial plaque and hypopyon. In another study, Pariseau et al. [15] described two contact lens-associated *Beauveria* keratitis cases involving a 55-year-old diabetic woman and a 31-year-old healthy woman in United States. Cultures of corneal scrapings from these patients grew *B. bassiana*. These isolates appeared to be unrelated to each other and also distinct from that used in *Beauveria*-based biopesticides sold in the United States as analyzed using morphological features, DNA sequencing, and random amplified polymorphic DNA (RAPD). The patients responded to natamycin, ketoconazole, amphotericin, and voriconazole treatments.

47.1.2.2 *Beauveria brongniartii*

Henke et al. [16] documented a case of human deep tissue infection due to an entomopathogenic *Beauveria* species in a 38-year-old woman receiving immunosuppressive therapy. The patient presented with pain in the right upper abdomen, severe dyspnea, dry cough, and a fever of 38.5°C. Computed tomography (CT) scan of the thorax revealed a discrete interstitial infiltrate. Allergic alveolitis was histologically confirmed by transbronchial lung biopsy. An ultrasonography of the abdomen revealed multiple lesions in the liver and spleen, suggesting systemic fungal infection. CT scans of the abdomen confirmed several focal hypodense lesions with a maximum diameter of 1.8 cm in the liver and in the spleen. Hematoxylin-eosin-stained sections of the hypodense liver specimen obtained by the CT-aided liver biopsy showed that the liver tissue was damaged in the center of the lesion with extensive focal necrosis. Grocott Gomori methenamine silver stain and calcofluor white stain from a specimen obtained from the center of the focus revealed infiltrating hyphae.

The specimen was homogenized and cultured on SGA, malt extract agar (MEA), and Sabouraud glucose broth (10 g of peptone/L and 40 g of glucose/L, pH 5.6) at 37°C and 26°C. A hyphomycete was obtained on all SGA and MEA plates incubated at 26°C, but not at 37°C. Subcultures of the Sabouraud glucose broth were performed on SGA and MEA at 37°C and 26°C. The isolate grew moderately rapidly at 26°C, forming a procumbent to flocculent mycelium with a powdery surface, which was white at first and was later pale yellow.

Conidiogenous cells of the clinical isolate were aggregated in moderately dense clusters alongside hyphae (1.5–3.0 μm wide) and were subspherical to ampulliform with peg-like, narrow, denticulate rachides that elongated in a zigzag manner. Conidia (2–3 μm in size) were smooth walled, single, and spherical to subspherical. The clinical isolate grew as an extremely floccose and velvety colony producing a low number of conidia, which is atypical for *B. bassiana*. On the basis of the snowball-like appearance of conidiogenous cells and conidia, the strain was considered as *Beauveria*, close to the generic type species, *B. bassiana*, and to its relative, *Beauveria brongniartii*. Analysis of the nucleotide sequences of the ITS regions (ITS1 and ITS2) of the nuclear ribosomal DNA showed that the patient isolate was identical, except for a single mutation in the ITS2 sequence, to three strains deposited in GenBank as *Beauveria tenella* (*B. tenella* is a synonym of *B. brongniartii*).

Following microbiological and pathological findings, antifungal therapy was initiated with 200 mg of itraconazole orally twice daily 1 week after admission. Three weeks after the start of therapy, the patient recovered, and the antifungal therapy was stopped after 3 months.

47.1.2.3 *Beauveria alba*

McDonnell et al. [17] reported a case of mycotic keratitis caused by *Beauveria alba* in a 70-year-old white man after phacoemulsification and intraocular lens implantation. A saprophytic fungus, *Beauveria alba* (also known as *Engyodontium album*), was demonstrated in the corneal button by histopathologic examination and isolated. Microscopically, conidiogenous cells formed in whorls or in aggregations are basally swollen in *B. bassiana* and are solitary and non-swollen in *B. alba*. Extensive therapeutic surgery was necessary to control the infection.

47.1.3 Diagnosis

Conventional techniques for fungal identification rely on the detection of macroscopic and microscopic features. *Beauveria* spp. are characterized by their sympodial development of single-celled conidia (ameroconidia) on a geniculate or zigzag rachis; their formation of flask-shaped, rachiform conidiogenous cells that proliferate sympodially and aggregate into sporodochia or synnemata; their production of hyaline and globose or ovoid conidia. These morphological characteristics are useful for discrimination of *Beauveria*, *Aspergillus*, *Fusarium*, and other fungi, ensuring successful treatment with antimycotic agents.

The virulence of *Beauveria* strains may be assessed in bioassays using larvae of the Colorado potato beetle (*Leptinotarsa decemlineata*) and conducted on potato leaf. Briefly, the surface of the leaf disks (diameter, 8 mm), laying on water agar, is treated with 10 μL of 10^6 conidia mL^{-1} in a first experiment and 5×10^6 conidia mL^{-1} in 0.1% Tween 80 in a second experiment. Instar larvae of *L. decemlineata* are put on the treated leaves for 2 days at 25°C (16 h light and 8 h dark each day) and then are transferred to untreated potato leaves. Control larvae are fed with leaf disks treated with 0.1% Tween 80 only. Larvae are observed for 10 days. In a control assay, the highly insect-virulent culture collection strain *B. bassiana* ARSEF No. 252 is used as Ref. [18].

The susceptibility of *Beauveria* strains to itraconazole and other antifungal drugs may be determined in a microdilution assay. A serial twofold dilution of itraconazole in RPMI broth supplemented with 2% glucose is inoculated with 10^4 conidia per milliliter. After 72 h, the minimum inhibitory concentration (MIC) is determined spectrophotometrically.

Molecular techniques are increasingly utilized for detection, identification, and phylogenetic analysis of *Beauveria* strains [19–27]. Because the nucleotide sequences of the ITS regions (ITS1 and ITS2) of the nuclear ribosomal RNA evolve relatively quickly, they provide attractive chronometers for determining the relationship of genotypically closely related species. The large- and small-subunit rRNA genes and the elongation factor 1α gene offer useful targets for *Beauveria* identification [28–30].

47.2 METHODS

47.2.1 Sample Preparation

Clinical specimens are examined under microscope with a fungal stain. Portions of the samples are inoculated on SGA without or with antibiotics. The resulting isolates are identified on the basis of macroscopical and microscopical features. For easy microscopic observation, slide culture technique may be utilized. Slide cultures are set up in Petri dishes containing 2 mL of sterile water, into which a U-shaped glass rod is placed, extending above the water surface. A block of freshly growing fungal colony, approx. 1 cm^2, is placed onto a sterile microscope slide, covered with a somewhat larger, sterile glass cover slip, and incubated in the moist chamber. Fungal sporulation is monitored over time, and when optimal, images are captured by means of a Nikon camera system (Digital Sight DS-5M, Nikon Corporation).

For DNA extraction, a small amount of fungal pellet is suspended in 600 μL extraction buffer (200 mM Tris–HCl, pH 7.5, 25 mM EDTA, 0.5% w/v sodium dodecyl sulfate [SDS], and 250 mM NaCl) and vortexed for 15 s. The mixture is then incubated at 100°C for 15 min, chilled on ice for 60 min, and centrifuged at 13,000 rpm for 15 min. The supernatant is transferred to a new tube, and the solution is extracted with phenol–chloroform–isoamyl alcohol (25:24:1 v/v). DNA is precipitated with cold isopropanol (−20°C), air dried, and resuspended in 100 μL distilled water. Alternatively, genomic DNA is extracted from colonies grown on 2% MEA, Difco, using the FastDNA kit (BIO101).

47.2.2 Detection Procedures

47.2.2.1 Standard PCR Detection

Hegedus and Khachatourians [22] reported the use of primers P1 (5′- AAGCTTCGACATGGTCTG-3′) and P3 (5′-GGAGGTGGTGAGGTTCTGTT-3′) for specific polymerase chain reaction (PCR) detection of *Beauveria bassiana*.

Procedure

1. PCR mixture (50 μL) consists of 20 mM Tris–HCl pH 8.4, 50 mM KCl, 1.5 mM $MgCl_2$, 200 μM each of dNTP, 0.75 μM of each primer, 100 ng of DNA, and 1.25 U *Taq* DNA polymerase (Gibco BRL).
2. Amplification is performed with initial denaturation at 94°C for 3 min; 40 cycles of 94°C for 30 s, 55°C for 30 s, and 72°C for 45 s; and a final 10 min extension at 72°C.
3. PCR products are separated on 1.0% (wt/vol) agarose-TBE (0.09 M Tris, 0.09 M boric acid, and 0.02 M EDTA) gels with ethidium bromide stain and visualized under UV light.

Note. A specific product of 524 bp product is expected from *B. bassiana*.

47.2.2.2 Real-Time PCR Detection

Castrillo et al. [30] designed primers and probe targeting a 445 bp sequence characterized amplified region (SCAR) fragment for real-time PCR detection of *Beauveria bassiana*. The primers GHTqF1 (5′-TTTTCAT CGAAAGGTTGTTTCTCG-3′) and GHTqR1 (5′-CTGT GCTGGGTACTGACGTG-3′) amplify a 96 bp region of the SCAR fragment; and the probe GHTqP1 (5′-TTCCGTTCCG CTCCGCCAAAAGCC-3′) is labeled at the 5′ end with fluorescent FAM reporter dye and at the 3′ end with tetramethyl-6-carboxyrhodamine (dye) (TAMRA) quencher dye.

Procedure

1. PCR mixture (25 μL) is composed of 1× iQ Supermix (BioRad), 0.5 μM of each primer, 0.125 μM of probe, 0.8 μg/μL (final concentration) of bovine serum albumin (BSA) (Ambion), and 2 μL of each of the appropriately diluted sample DNA. (BSA helps reduce PCR inhibition from co-extracted contaminants from environmental samples, especially humic acids from soil.)
2. Each assay includes a standard curve (five or six series 10-fold dilutions from 40 ng to 4 or 0.4 pg of control *B. bassiana* GHA DNA) for determining starting DNA concentration of unknowns and a non-template control.
3. Thermal cycling conditions are as follows: initial denaturation at 95°C for 10 min; 40 cycles of denaturation at 95°C for 15 s and single-step annealing and extension at 60°C for 1 min.

Note. This assay may be adapted for standard PCR detection with the following cycling parameters: initial denaturation at 94°C for 4 min and 30 cycles of 94°C for 1 min, 55°C for 1 min, and 72°C for 1 min.

47.2.2.3 Analysis of ITS Sequence

Pounder et al. [31] described a real-time PCR with SYBR green DNA-binding dye and amplicon-melting temperature analysis for fungal detection using pan-fungal primers ITS1 forward (5′-TCCGTAGGTGAACCTGCGG-3′) and ITS4 reverse (5′-TCCTCCGCTTATTGATATGC-3′). The identity of the fungi is verified by subsequent sequencing analysis.

Procedure

1. PCR mixture is composed of 1× Lightcycler FastStart DNA Master Hybridization Probes mixture (Roche Applied Science) containing deoxynucleoside triphosphates, FastStart *Taq* DNA polymerase, and 1 mM $MgCl_2$ (additional $MgCl_2$ is added to a final concentration of 4.6 mM), 0.4 μM each of ITS1 forward and ITS4 reverse primers, 1× SYBR green (Molecular Probes), and 3 μL template DNA.
2. Thermal cycling parameters include 95°C for 10 min; 50 cycles of 95°C for 5 s, 60°C for 20 s, and 76°C for 30 s; and a final extension at 72°C for 2 min.
3. The quality of the amplicon is determined using the derivative of the melt analysis curve (55°C–99°C, 45 s hold at 55°C, and 5 s/°C) using the RotorGene 3000 (Corbett Robotics, Inc).
4. The amplified product is purified for bidirectional sequencing using ExoSAP-IT (USB Corp). Five microliters of Big Dye Terminator Ready Reaction Mix v. 1.1 (Applied Biosystems) is added to 4 μL of each primer (0.8 pmol/μL) and 3 μL of purified PCR product. Cycle sequencing is performed with a 9700 thermal cycler (ABI), using 25 cycles of 96°C for 10 s, 50°C for 5 s, and 60°C for 4 min. Sequencing reaction products are passed through a Sephadex G-50 fine column to remove unincorporated dye terminators. Purified sequencing reaction products are run on an ABI Prism 3100 Genetic Analyzer with a 50 cm capillary array.
5. Sequences are analyzed with the SmartGene Integrated Database Network software version 3.2.3 vr. SmartGene is a web-based software and database system with reference sequences derived from the National Center for Biological Information (NCBI) GenBank repository.

Note. If real time PCR instrument is unavailable, standard PCR may be performed with primers ITS1 and ITS4, and the resulting amplicon is sequenced with the same primers. Sequence-based identifications are defined by percent identity: species, ≥99%; genus, 93%–99%; and inconclusive, ≤93%.

47.3 CONCLUSION

Beauveria spp. are entomopathogenic fungi that are commonly present in the environments and are utilized as biological control agents for insect pests. Several *Beauveria* species (e.g., *B. bassiana*, *B. brongniartii*, and *B. alba*) have been shown to cause hyalohyphomycosis in humans, with a variety of clinical symptoms (e.g., keratitis, deep tissue infection, and disseminated infection). The distinct morphological features of the *Beauveria* genus include (i) the sympodial development of single-celled conidia (ameroconidia) on a geniculate or zigzag rachis; (ii) the formation of flask-shaped, rachiform conidiogenous cells that proliferate sympodially and aggregate into sporodochia or synnemata; and (iii) the presence of hyaline and globose or ovoid conidia. Although in vitro culture and microscopic examination detecting these characteristics are useful for differentiation of *Beauveria* spp. from other fungi, the whole process can be lengthy and technically demanding. The use of PCR and sequencing analysis provides a much faster and more accurate alternative to the identification of *Beauveria* spp.

REFERENCES

1. The UniProt Consortium. Available at http://www.uniprot.org/, accessed on August 1, 2010.
2. Li, Z. et al. 2001. Discovery and demonstration of the teleomorph of *Beauveria bassiana* (Bals.) Vuill., an important entomogenous fungus. *Chin. Sc. Bull.*, 46:751–753.
3. Wagner, B.L. and L.C. Lewis. 2000. Colonization of corn, *Zea mays*, by the entomopathogenic fungus *Beauveria bassiana*. *Appl. Environ. Microbiol.*, 66:3468–3473.
4. Shimazu, M., H. Sato, and N. Maehara. 2002. Density of the entomopathogenic fungus, *Beauveria bassiana* Vuillemin (*Deuteromycotina*: *Hyphomycetes*) in forest air and soil. *Appl. Entomol. Zool.*, 37:19–26.
5. Strasser, H., T.M. Butt, and A. Vey. 2000. Are there any risks in using entomopathogenic fungi for pest control, with particular reference to the bioactive metabolites of *Metarhizium*, Tolypocladium and *Beauveria* species? *Biocontrol Sci. Technol.*, 10:717–735.
6. De Hoog, G.S. et al. 2000. *Atlas of Clinical Fungi*. Centraalbureau voor Schimmelcultures, Utrecht, the Netherlands.
7. Sachs, S.W., J. Baum, and C. Mies. 1985. *Beauvaria bassiana* keratitis. *Br. J. Ophthalmol.*, 69:548–550.
8. Low, C.D., P.R. Badenoch, and D.J. Coster. 1997. *Beauveria bassiana* keratitis cured by deep lamellar dissection. *Cornea*, 16:698–699.
9. Gürcan, S. et al. 2006. First case report of empyema caused by *Beauveria bassiana*. *Mycoses*, 49(3):246–248.
10. Tu, E.Y. and A.J. Park. 2007. Recalcitrant *Beauveria bassiana* keratitis: Confocal microscopy findings and treatment with posaconazole (Noxafil). *Cornea*, 26(8):1008–1010.
11. Oh, J.Y. et al. 2009. A case of necrotizing sclerokeratitis and endophthalmitis caused by *Beauveria bassiana*. *Jpn. J. Ophthalmol.*, 53(5):551–553.
12. Tucker, D.L. et al. 2004. Disseminated *Beauveria bassiana* infection in a patient with acute lymphoblastic leukemia. *J. Clin. Microbiol.*, 42(11):5412–5414.
13. Kisla, T.A. et al. 2000. Medical management of *Beauveria bassiana* keratitis. *Cornea*, 19:405–406.
14. Sonoyama, H. et al. 2008. The characteristics of keratomycosis by *Beauveria bassiana* and its successful treatment with antimycotic agents. *Clin. Ophthalmol.*, 2(3):675–678.
15. Pariseau, B. et al. 2010. *Beauveria* keratitis and biopesticides: Case histories and a random amplification of polymorphic DNA comparison. *Cornea*, 29(2):152–158.
16. Henke, M.O. et al. 2002. Human deep tissue infection with an entomopathogenic *Beauveria* species. *J. Clin. Microbiol.*, 40:2698–2702.
17. McDonnell, P.J. et al. 1984–1985. Mycotic keratitis due to *Beauveria alba*. *Cornea*, 3(3):213–216.
18. Safavi, S.A. et al. 2007. Effect of nutrition on growth and virulence of the entomopathogenic fungus *Beauveria bassiana*. *FEMS Microbiol. Lett.*, 270:116–123.
19. Kosir, J.M. and J.M. MacPherson, and G.G. Khachatourians. 1991. Genomic analysis of virulent and less virulent strain of the entomopathogenic fungus *Beauveria bassiana* using RFLP. *Can. J. Microbiol.*, 37:534–541.
20. Hegedus, D.D. and G.G. Khachatourians. 1993a. Construction of cloned DNA probes for the specific detection of the entomopathogenic fungus *Beauveria bassiana* in grasshoppers. *J. Invertebr. Pathol.*, 62:233–240.
21. Hegedus, D.D. and G.G. Khachatourians. 1993b. Identification of molecular variants in mitochondrial DNAs of members of the genera *Beauveria*, *Verticillium*, *Paecilomyces*, *Tolypocladium*, and *Metarhizium*. *Appl. Env. Microbiol.*, 59:4283–4288.
22. Hegedus, D.D. and G.G. Khachatourians. 1996. Detection of the entomopathogenic fungus *Beauveria bassiana* within infected migratory grasshoppers (*Melanoplus sanguinipes*) using polymerase chain reaction and DNA probe. *J. Invertebr. Pathol.*, 67:21–27.
23. Shih, H.L. et al. 1995. Complete nucleotide sequence of *Beauveria bassiana* 5.8s rRNA coding gene and flanking internal transcribed spacers. *DNA Seq.*, 5:381–383.
24. De Muro, M.A., S. Mehta, and D. Moore. 2003. The use of amplified fragment length polymorphism for molecular analysis of *Beauveria bassiana* isolates from Kenya and other countries and their correlation with host and geographic origin. *FEMS Microbiol. Lett.* 229:249–257.
25. Wang, C. et al. 2003. Molecular investigation on strain genetic relatedness and population structure of *Beauveria bassiana*. *Environ. Microbiol.*, 5:908–915.
26. Wang, C. et al. 2004. Molecular monitoring and evaluation of the application of the insect-pathogenic fungus *Beauveria bassiana* in southeast China. *J. Appl. Microbiol.*, 96:861–870.
27. Rehner, S.A. and Buckley, E. 2005. A *Beauveria* phylogeny inferred from nuclear ITS and EF1-α sequences: Evidence for cryptic diversification and links to *Cordyceps* teleomorphs. *Mycologia*, 97:84–98.
28. Neuvéglise, C., Y. Brygoo, and G. Riba. 1997. 28s rDNA group-I introns: A powerful tool for identifying strains of *Beauveria brongniartii*. *Mol. Ecol.*, 6:373–381.
29. Castrillo, L.A., J.D. Vandenberg, and S.P. Wraight. 2003. Strain-specific detection of introduced *Beauveria bassiana* in agricultural fields by use of sequence-characterized amplified region markers. *J. Invertebr. Pathol.*, 82:75–83.
30. Castrillo, L.A., M.H. Griggs and J.D. Vandenberg. 2008. Quantitative detection of *Beauveria bassiana* GHA (Ascomycota: Hypocreales), a potential microbial control agent of the emerald ash borer, by use of real-time PCR. *Biol. Control*, 45:163–169.
31. Pounder, J.I. et al. 2007. Discovering potential pathogens among fungi identified as nonsporulating molds. *J. Clin. Microbiol.*, 45(2):568–571.

48 *Chaetomium*

Dongyou Liu and R.R.M. Paterson

CONTENTS

48.1 INTRODUCTION

48.1.1 Classification and Morphology

The genus *Chaetomium* consists of dematiaceous (dark-walled) fungi classified in the family Chaetomiaceae, order Sordariales, class *Sordariomycetes*, subphylum Pezizomycotina, phylum Ascomycota, and kingdom Fungi. Of the 35 recognized and 69 unassigned species within the genus, the following are most common: *Chaetomium atrobrunneum* (obsolete synonym: *C. fusisporale* and *C. rectopilium*), *Chaetomium funicola*, *Chaetomium globosum* (obsolete synonyms: *C. caprophilum* and *C. cinnamomeum*), *Chaetomium perlucidum*, *Chaetomium piluliferum* (a teleomorph of *Botryotrichum piluliferum*), and *Chaetomium strumarium* (obsolete synonyms: *Achaetomium cristalliferum*, *Achaetomium strumarium*, *C. spinulosum*, and *C. sulphureum*) [1,2].

Chaetomium species are saprobic ascomycetes with worldwide distribution and are commonly found in dung, straw, paper, bird feathers, seeds, plant debris, soil, and air. Some *Chaetomium* spp. are thermophilic and neurotropic in nature and are among the fungi causing phaeohyphomycosis in humans, with clinical presentations ranging from brain abscess, peritonitis, cutaneous lesions, onychomycosis, to fatal deep mycoses.

Chaetomium species grow rapidly at between 25°C and 35°C, with some species responsible for invasive human disease growing well at 35°C–37°C, and others showing a predilection for the central nervous system growing at up to 42°C–45°C. *Chaetomium* colonies are cottony and white initially and become gray to olive with age. The colony reverse is tan to red or brown to black. Hyphae are septate. Perithecia are large, dark brown to black, fragile, and globose to flask shaped and have filamentous, hair-like, brown to black appendages (setae) on their surface. Perithecia have ostioles (small-rounded openings) and contain asci and ascospores inside. Asci are clavate to cylindrical in shape and rapidly dissolve to release their ascospores (4 to 8 in number). Ascospores are one celled, olive brown, and lemon shaped [2].

Features fundamental to the identification of *Chaetomium* species include the presence of hairs or setae covering the ascomata, the size and shape of ascospores, and the presence and position of germ pores. Among the similar *Chaetomium* species (i.e., *C. gangligerum*, *C. raii*, *C. jodhpurense*, and *C. fusisporum*), *C. gangligerum* produces broader ascospores (12–15 μm × 7.5–9.5 μm) and usually forms chlamydospores; *C. raii* has smaller ascospores (10–13 μm × 5.5–7.5 μm), and *C. jodhpurense* and *C. fusisporum* have larger ascospores (14–19 μm × 6–8 μm and 14–17 × 7–8 μm, respectively).

C. strumarium is distinguished by its formation of ascocarps covered with pale, thin-walled, flexuous hairs, a feature contributing to its original placement in the genus *Achaetomium*. Presence of pinkish exudate droplets and/or crystals associated with hyphae or ascocarps, good growth at 42°C, and production of small conidia further separate *C. strumarium* from other species [3].

Furthermore, *C. perlucidum* produces globose (spherical) to subglobose or ovoid ascomata (108–220 μm × 90–200 μm), with undulate hairs and a wide ostiole (30–50 μm in diameter) and eight-spored asci (20–40 μm × 7–18 μm). The ascospores (12.0–15.0 μm × 6–7.5 μm) are elliptical and olive brown and contain a subapical germ pore. These features help differentiate *C. perlucidum* from other invasive species of *Chaetomium* [4].

C. globosum is distinguished by its lack of growth at elevated temperatures (>35°C) and its generation of lemon-shaped ascospores.

48.1.2 Clinical Features

Despite being saprophytic ascomycetes with only occasional involvement in human disease processes, *Chaetomium* species are capable of inducing a broad spectrum of mycoses including onychomycosis, sinusitis, empyema, pneumonia, and fatal disseminated cerebral disease, especially in immunocompromised patients and intravenous drug users [5–9].

Chaetomium atrobrunneum is a notably invasive, neurotropic species, and its ability to grow at elevated temperatures may contribute to its neurotropism [10–13]. Thomas et al. [9] described a case of fatal brain abscess due to *C. atrobrunneum* in a bone marrow transplant patient. The rapid progression of cerebral infection indicates that the brain tissue provides a favorable environment for growth and proliferation of the fungus.

Chaetomium strumarium is another invasive, neurotropic species. Abbott et al. [3] reported three *C. strumarium*-related cases of fatal cerebral mycosis in males with prior histories of intravenous drug use from the United States and Australia. *C. strumarium* was detected by histopathology and isolated from the brain tissue.

Chaetomium perlucidum was recently confirmed as a neurotropic species. Barron et al. [4] documented the first two cases of invasive human mycoses caused by this phaeoid ascomycete. The first case concerned a 45-year-old female patient with acute myelogenous leukemia, who had an unrelated, 4/5 HLA-matched umbilical cord blood transplant. The patient became disoriented and febrile, and computed tomography of the chest revealed a 3 × 2 cm mass in the right lower lobe. After suffering a massive right-sided intraparenchymal hemorrhage, the patient died. Autopsy revealed disseminated invasive fungal infection in the lungs, brain, and myocardium; and cultures from the surgically obtained lung tissue yielded *C. perlucidum*. The second case involved a 78-year-old female with a history of asthma and chronic bronchiectasis. The patient underwent a lobectomy due to worsening symptoms, and cultures from the lung tissue grew *C. perlucidum*. The patient showed no further manifestations of disease after the lobectomy.

Chaetomium globosum is an occasional agent of onychomycosis [14,15]. In addition, Teixeira et al. [16] reported that *C. globosum* was responsible for a systemic infection with enlargement of the axillary and cervical lymph nodes in a chronic myeloid leukemia patient, who underwent an allogeneic sibling-matched bone marrow transplant. Culture of the aspirates from both lymph nodes resulted in the growth of *C. globosum*. The infection was successfully treated with amphotericin B.

48.1.3 Diagnosis

Considerable similarities exist between mycosis caused by *Chaetomium* and *Aspergillus* from radiographical and histopathological standpoints. A melanin-specific stain (i.e., the Masson-Fontana stain) is helpful reputedly for distinguishing the melanin-containing *Chaetomium* from most *Aspergillus* species. However, this is contradicted by reports of melanin from *Aspergillus* (e.g., *Aspergillus fumigatus*) [17], and so the information obtained may have limited value in differentiating the causal fungi.

In vitro culture techniques offer a slow but valuable way to isolate and propagate *Chaetomium* organisms for subsequent macroscopic and microscopic characterization. Inoculation of sterilized plant material with ascospore suspension may enhance induction of mature perithecia, leading to the production of well-developed ascomata on the surface of the substrate.

Molecular methods such as (i) PCR and (ii) sequence analysis of rRNA and internal transcribed spacer (ITS) regions provide an approach for the rapid and accurate identification of *Chaetomium* species from other fungi.

48.2 METHODS

48.2.1 Sample Preparation

Clinical specimens are examined by microscopy with a melanin-specific stain (i.e., the Masson-Fontana stain). However, as mentioned above, a melanin stain may not be particularly useful. Portions of the samples are inoculated on Sabouraud glucose agar with or without antibiotics. The resulting isolates are identified on the basis of macroscopical and microscopical features.

For DNA extraction, a small amount of fungal pellet is suspended in 600 μL extraction buffer (200 mM Tris–HCl, pH 7.5, 25 mM EDTA, 0.5% w/v sodium dodecyl sulfate [SDS], and 250 mM NaCl) and vortexed for 15 s. The mixture is then incubated at 100°C for 15 min, chilled on ice for 60 min, and centrifuged at 13,000 rpm for 15 min. The supernatant is transferred to a new tube, and the solution is extracted with phenol–chloroform–isoamyl alcohol (25:24:1 v/v). DNA is precipitated with cold isopropanol (−20°C), air dried, and resuspended in 100 μL distilled water.

48.2.2 Detection Procedures

Arzanlou et al. [18] utilized the universal primers ITS1 and ITS4 to amplify the ITS region of the nuclear ribosomal RNA operon, including the 3′ end of the 18S rRNA gene, the first ITS region (ITS1), the 5.8S rRNA gene, the second ITS region (ITS2), and the 5′ end of 28S rRNA gene. Subsequent sequencing allows identification of the *Chaetomium* species.

Procedure

1. PCR mixture (25 μL) is composed of 0.5 U *Taq* polymerase (Bioline), 1× PCR buffer, 0.5 mM $MgCl_2$, 0.2 mM of each dNTP, 5 pmol of each primer, and approximately 10–15 ng of fungal genomic DNA.
2. Amplification is performed on a GeneAmp PCR System 9700 (Applied Biosystems) with primary denaturation at 96°C for 5 min; 36 cycles of 96°C for

30 s, 52°C for 30 s, and 72°C for 60 s; a final extension at 72°C for 7 min.

3. The amplicons are sequenced using BigDye Terminator v. 3.1 (Applied Biosystems,) or DYEnamicET Terminator (Amersham Biosciences) Cycle Sequencing Kits and analyzed on an ABI Prism 3700 (Applied Biosystems).
4. Newly generated sequences are subjected to a Blast search of the NCBI databases, sequences with high similarity are downloaded from GenBank, and comparisons are made on the basis of the alignment of the obtained sequences.
5. Phylogenetic analysis is performed with PAUP (Phylogenetic Analysis Using Parsimony) v. 4.0b10 using the neighbor-joining algorithm with the uncorrected ("p"), the Kimura 2-parameter, and the HKY85 substitution models. Alignment gaps longer than 10 bases are coded as single events for the phylogenetic analyses; the remaining gaps are treated as missing data. Any ties are broken randomly when encountered.
6. Phylogenetic relationships are also inferred with the parsimony algorithm using the heuristic search option with simple taxon additions and tree bisection and reconstruction as the branch-swapping algorithm; alignment gaps are treated as a fifth character state and all characters are unordered and of equal weight. Branches of zero length are collapsed and all multiple, equally parsimonious trees are saved.
7. Other measures calculated include tree length, consistency index, retention index, and rescaled consistency index. The robustness of the obtained trees is evaluated by 1000 bootstrap replications. Bayesian analysis is performed. The best nucleotide substitution model is determined using MrModeltest v. 2.2. MrBayes v. 3.1.2 is used to perform phylogenetic analyses, using a general time-reversible substitution model with inverse gamma rates, dirichlet base frequencies, and the temp value set to 0.5.

Note. Part of the large-subunit 28S rRNA (LSU) gene may be also amplified with primers LR0R and LR5 followed by sequencing analysis.

48.3 CONCLUSIONS

The genus *Chaetomium* consists of a large group of dematiaceous (dark-walled) fungi that are present in a variety of environments, including dung, straw, paper, bird feathers, seeds, plant debris, soil, and air. Several *Chaetomium* spp. are opportunistic human pathogens, causing phaeohyphomycosis, with clinical manifestations of brain abscess, peritonitis, cutaneous lesions, onychomycosis, and fatal deep mycoses. Since conventional methods for identification of *Chaetomium* spp. require lengthy incubation and demand specialized skill, molecular procedures have been increasingly applied for culture-independent detection and differentiation of *Chaetomium* spp. The use of melanin stains alone may be inconclusive.

REFERENCES

1. The UniProt Consortium. Available at http://www.uniprot.org/, accessed on August 1, 2010.
2. von Arx, J. A., J. Guarro, and M. J. Figueras. 1986. The ascomycete genus *Chaetomium*. *Beih. Nova Hedwigia* 84:1–162.
3. Abbott, S. P. et al. 1995. Fatal cerebral mycoses caused by the ascomycete *Chaetomium strumarium*. *J. Clin. Microbiol.* 33:2692–2698.
4. Barron, M. A. et al. 2003. Invasive mycotic infections caused by *Chaetomium perlucidum*, a new agent of cerebral phaeohyphomycosis. *J. Clin. Microbiol.* 41(11):5302–5307.
5. Hoppin, E. C., E. L. McCoy, and M. G. Rinaldi. 1983. Opportunistic mycotic infection caused by *Chaetomium* in a patient with acute leukemia. *Cancer* 52:555–556.
6. Anandi, V. et al. 1989. Cerebral phaeohyphomycosis caused by *Chaetomium globosum* in a renal transplant recipient. *J. Clin. Microbiol.* 27:2226–2229.
7. Yeghen, T. et al. 1996. *Chaetomium* pneumonia in a patient with acute myeloid leukaemia. *J. Clin. Pathol.* 49:184–186.
8. Aru, A., L. Munk-Nielsen, and B. H. Federspiel. 1997. The soil fungus *Chaetomium* in the human paranasal sinuses. *Eur. Arch. Otorhinolaryngol.* 254:350–352.
9. Thomas, C. et al. 1999. Fatal *Chaetomium* cerebritis in a bone marrow transplant patient. *Hum. Pathol.* 30:874–879.
10. Guarro, J., L. Soler, and M. G. Rinaldi. 1995. Pathogenicity and antifungal susceptibility of *Chaetomium* species. *Eur. J. Clin. Microbiol. Infect. Dis.* 14:613–618.
11. Segal, R. and S. Kusne. 1999. Cerebral fungal infections in the immunocompromised host: A literature review and a new pathogen-*Chaetomium atrobrunneum*: Case report. *Neurosurgery*, 45:200.
12. Guppy, K. H. et al. 1998. Cerebral fungal infections in the immunocompromised host: A literature review and a new pathogen—*Chaetomium atrobrunneum*: Case report. *Neurosurgery* 43:1463–1469.
13. Rock, J. P. 1998. Cerebral fungal infections in the immunocompromised host: A literature review and a new pathogen—*Chaetomium atrobrunneum*: Case report—Comment. *Neurosurgery* 43:1469.
14. Stiller, M. J. et al. 1992. Onychomycosis of the toenails caused by *Chaetomium globosum*. *J. Am. Acad. Dermatol.* 26:775–776.
15. Lesire, V. et al. 1999. Possible role of *Chaetomium globosum* in infection after autologous bone marrow transplantation. *Intensive Care Med.* 25:124–125.
16. Teixeira, A. B. et al. 2003. Phaeohyphomycosis caused by *Chaetomium globosum* in an allogeneic bone marrow transplant recipient. *Mycopathologia* 156(4):309–312.
17. Youngchim, S. et al. 2004. Production of melanin by *Aspergillus fumigatus*. *J. Med. Microbiol.* 53:175–181.
18. Arzanlou, M. et al. 2007. Phylogenetic and morphotaxonomic revision of *Ramichloridium* and allied genera. *Stud Mycol.* 58:57–93.

49 *Colletotrichum*

M.R. Shivaprakash, Abhishek Baghela, and Arunaloke Chakrabarti

CONTENTS

49.1 INTRODUCTION

Colletotrichum species are well-known plant pathogens in tropical and subtropical countries. They are known to cause important plant disease, anthracnose, in a wide range of plants including cereals, legumes, vegetables, perennial crops, and tree fruits. They have tremendous economic impact.[1] Recently, *Colletotrichum* species have also been recognized as emerging human pathogens causing ophthalmic lesion. Occasionally, they may cause subcutaneous infections in humans.

49.1.1 Classification, Morphology, and Epidemiology

49.1.1.1 Classification

In the first description, the genus was known as *Vermicularia* (described by Tode in 1790). The name *Colletotrichum* was introduced by Corda in 1831.[2] Traditionally, these fungi are included in the Form–class *Coelomycete*, order *Melanoconiales*, and are characterized by the formation of mitosporic, asexual fruiting bodies known as acervular conidiomata (a cup-shaped fruiting body covered with the sterile pigmented and unbranched hyphae called setae). The conidiomata bearing the conidia is better seen while the fungus parasites the plant tissue bearing the conidia within the structures.[3] The sexual state of the fungi has been identified as ascomycetous sporulation and described under the Genus *Glomerella*. However, with the exception of *C. falcatum*, teleomorph stage is not reported from the field.[4]

Taxonomy and nomenclature of *Colletotrichum* is confusing, even to scientists working in this field.[5] Initially the species under the Genus *Colletotrichum* were identified based on the host specificity, location, and the morphology of the conidia. However, there was duplication in the names, as the same species were referred to by different names. Few species with curved spores were classified under *Vermicularia*. Von Arx, with his seminal works on morphological characters, could reduce the number of *Colletotrichum* species from several hundred to 11 species. A vast majority of those species was reduced as synonymy to *C. gloeosporioides*. Subsequently, based on morphological characters, species under the Genus *Colletotrichum* was classified into 22 species by Sutton (1980),[6] 40 by Sutton (1992),[7] and 60 by *Dictionary of the Fungi*.[8] Recently, due to the difficulty to differentiate the *Colletotrichum* species based on the morphological characters and host range (in plants), phylogenetic analysis on the basis of nucleic acid sequences is successfully used. The need for the polyphasic approach for the identification of the species under this genus has been highlighted by Cai et al.[9] Hyde et al. provided an overview of 66 commonly known *Colletotrichum* species along with 19 doubtfully classified species.[5] In spite of many known plant pathogenic species of *Colletotrichum*, human infections are reported due to few species only and those include *C. dematium*,[10–15] *C. coccodes* (reported as *C. atramentarium*),[16,17] *C. gloeosporioides*,[12,18,19] *C. graminicola*,[20] *C. crassipes*,[21] *C. capsici*[22] (currently named as *C. truncatum*), *C. truncatum*,[23] and an unknown species.[24]

49.1.1.2 Morphology

Traditionally, the identification of *Colletotrichum* relies on the morphological and cultural characteristics. The important morphological character that helps in speciation includes size and shape of condiomata, conidia, conidiophores, appressoria, and setae on culture or on natural substrate (Figures 49.1 and 49.2). However, the most common human pathogenic species, *Colletotrichum dematium*, is difficult to recognize on the basis of morphological characters, as different conidia sizes are described by the researchers. Colonies of putative *C. dematium* strains are also variable[7] (Table 49.1). According to Sutton,[6] the conidia of *C. dematium* are strongly curved and less than 3 μm wide and this description helps to distinguish them from *C. capsici*.[23] By multi-locus phylogenies and morphology, *Colletotrichum capsici* is synonymized with *C. truncatum* and *C. Curvatum*.[23] However, there is no consensus among researchers working in this field on this synonymy proposed by Damm et al.[5] Sutton[6,7] proposed that *C. crassipes* is similar to *C. gloeosporioides*, though the conidia of *C. crassipes* is wider and longer, and the fungus has lobed appressoria. The identification of *C. graminicola*, a grass-associated fungus and isolated from a case of human keratitis,[20] is not clear due to frequent inaccurate naming for this fungus.[4]

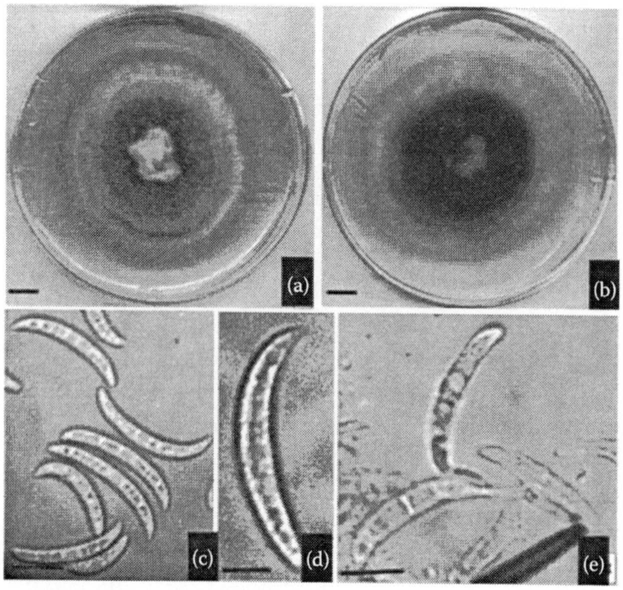

FIGURE 49.1 *Colletotrichum truncatum* (*C. capsici*): (a)–(b): Colony morphology on PDA. (c)–(d): Conidia. (e): Conidiogenous cells. Scale bars: (a)–(b) = 9 mm; (c) = 10 μm; (d)–(e) = 5 μm. (Photographs provided kindly by Shenoy, B.D. et al., *Fungal Divers*., 27, 197, 2007.)

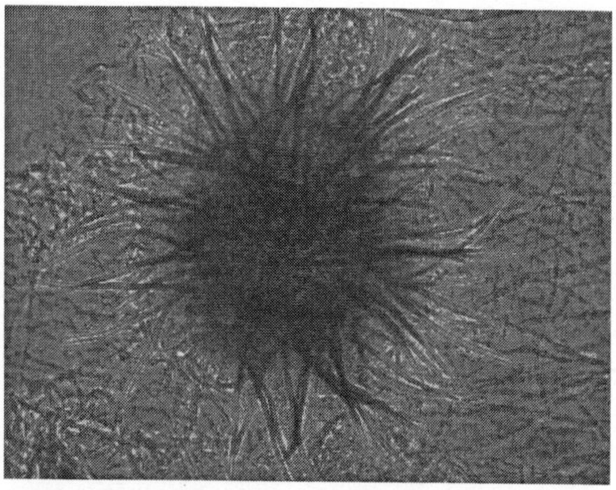

FIGURE 49.2 *Colletotrichum dematium*. Conidiomata with abundant setae and falcate conidia.

49.1.1.3 Epidemiology

Colletotrichum species are basically plant pathogens. Few species like *C. dematium*,[10–15] *C. coccodes* (reported as *C. atramentarium*),[16,17] *C. gloeosporioides*,[12,18,19] *C. graminicola*,[20] *C. crassipes*,[21] *C. capsici*[22] (currently named as *C. truncatum*), and *C. truncatum*[23] are reported to cause human infection. Among those species, *Colletotrichum capsici* (*C. truncatum*) is known to cause anthracnose on chili pepper, cotton, tomatoes, and a wide range of legume species.[27] It has also been associated with symptoms of leaf tip die-back, foliar blight, leaf spot, leaf lesions, and boll rot of various plants.[27] *Colletotrichum crassipes* was originally described from fruit of *Vitis vinifera* (grape) from Conegliano, Italy.[5] *Colletotrichum truncatum* has a wide host range and causes anthracnose diseases of many leguminous and solanaceous plants.[7,23,27] *Colletotrichum dematium* is claimed to cause several important plant diseases, such as leaf blight of Japanese radish seedlings and mulberry and cowpea anthracnose. *Colletotrichum gloeosporioides* is possibly the most common and widely distributed plant pathogen of the world and has been associated with at least 470 different host genera. *C. gloeosporioides* is prevalent in the tropics though the fungus has also been isolated from a wide range of temperate and subtropical habitats.[5–7,28] *C. graminicola* emerged as an important pathogen of corn crops during the 1970s and 1980s. However, with the development of the resistant cultivars the impact due to this fungus has reduced drastically. *C. graminicola* is the first *Colletotrichum* species to have its entire genome sequenced.[5] *C. coccodes* is known to cause potato black dot and anthracnose of tomato, chili, and mint.[2]

The burden of human infection due to *Colletotrichum* species is clearly not known due to lack of awareness of this agent in causing disease and availability of few reports. However, three series of mycotic keratitis reported by Fernandez et al.,[12] Kaliamurthy et al.,[29] and Marangon et al.[30] had shown that *Colletotrichum* was the causative agent in 2.8% (10 of 360), 1.9% (7 of 378), and 3.0% (14 of 467) of cases, respectively. Reported human infection due to *Colletotrichum* species is summarized in Table 49.2. Most patients with *Colletotrichum* keratitis are rural agricultural

TABLE 49.1
Characteristics Features of *Colletotrichum* Species Implicated in Human Infection

Species (Teleomorph)	Colony Morphology	Conidial Morphology	Conidial Dimensions (in μm)	References
C. dematium (unknown)	Extremely variable, initially white to pale mouse gray mycelium, later mouse gray, felt-like, reverse black.	Abundant setae and black conical sclerotia; conidia hyaline smooth walled, aseptate falcate, fusiform, and gradually tapered at each end; appressoria medium brown, clavate to circular, margin entire and slightly irregularly lobed aseptate, rarely septate (Figure 49.2).	Conidia 18–26 × 2.0–3.0 Appressoria 8.0–11.5 × 6.5–8.0	[6,25]
C. gloeosporioides (*Glomerella cingulata*)	Extremely variable, effuse, gray to black, with pinkish patches; reverse dark brown with vinaceous stains.	Conidia born of elongated phialides in acervular conidiomata; conidia cylindrical, obtuse end, and slightly tapered; appressoria circular with slightly irregular, pale to medium brown.	Conidia 8–11 × 3.0–4.5 Appressoria 4.5–10 × 4–7.5	[6,25,26]
C. coccodes (unknown)	Dark brown, with black sclerotia occasionally sparse white aerial mycelium; conidial mass honey colored; reverse dark brown.	Sclerotia setose, spherical; conidia straight fusiform attenuated at the ends; appressoria clavate brown.	Conidia 16–22 × 3–4 Appressoria 11.0–16.5 × 6.0–9.5	[6,25]
C. graminicola (*Glomerella graminicola*)	Colony effuse gray to brown, with salmon patches of conidial slime; reverse vinaceous to purple.	Stromata formed from dendroid brown hyphae; setae abundant; conidia fusiform to falcate; appressoria brown and irregular.	Conidia 23–29 × 3.5–5 Appressoria 17–20 × 12–14	[6,25]
C. truncatum (*C. capsici*)[a] (*Glomerella truncate*)	Colonies on oatmeal agar flat with entire margin, no aerial mycelium, surface buff, covered with olivaceous-gray to iron-gray acervuli, and reverse buff to pale olivaceous-gray (Figure 49.1).	Hyphae hyaline, septate; *Conidiomata* acervular; Setae hyaline to pale brown, smooth to verruculose, 80–150μm long, two- to five-septate, tapering toward the tip; *Conidia* hyaline, smooth-walled to verruculose, aseptate, long central part of conidia usually slightly curved with parallel walls, and ending abruptly at the round and truncate base, while tapering toward the acute and more strongly curved apex; *Appressoria* light to medium brown, entire edge to lobed, outline roundish to ellipsoidal (Figure 49.1).	Conidia 20–23.5 × 3.5–4 Appressorium 6.5–13 × 5.5–7.5	[6,23,25,27]
C. crassipes (unknown)	On potato carrot agar, cottony, olivaceous brown black reverse.	Setae acicular, septate, brown to dark brown, thick walled to dark brown; Conidia were slimy hyaline, mostly cylindrical; Appressorium-irregular, darkbrown with crenate or deeply lobed walls.	Conidia 11–18 × 6.5–8 Appressorium 18 × 10	[21]

[a] *C. capsici* is now considered as *C. truncatum*.

workers. Although precedent history of injury to the eye is obvious in the majority of cases, occasional minor unnoticed trauma may be the possible reason in a few cases.

49.1.2 Pathogenesis and Clinical Features

Like other *Coelomycetes*, *Colletotrichum* are not ubiquitous airborne organisms, and there is no report of acquisition through respiratory tract. Most infections reported are due to traumatic implantation of the fungus from plant/wood material or the soil or acquisition through abrasions, wounds, lacerations, or other trauma. This process helps the entry of the fungus in the eye and cutaneous or subcutaneous tissues. The predominant infection caused by this agent is keratitis. The major risk factor that predisposes to the *Colletotrichum* keratitis is corneal trauma. Other reported risk factors include use of corticosteroids or antiviral drugs on eye, or the presence of diabetes mellitus. The infection due to this fungus has also been reported in post-cataract surgery.[19,20] *Colletotrichum* causing endophthalmitis has been reported after subconjunctival injection of triamcinolone acetonide for treatment of anterior scerelitis.[31] The next group of infections caused by *Colletotrichum* species is subcutaneous infections in the form of phaeohyphomycosis among immunosuppressed patients.[21,32,33] Midha et al.[24] have reported a lethal case of systemic fungal infection in a

TABLE 49.2
List of Reported Human Infections due to *Colletotrichum*

S. No.	References (Nos)	Species Identification	Age/ Sex	Risk Factors	Clinical Presentation
Subcutaneous infections					
1	O'Quinn et al.[33]				
	Case 1	*C. gloeosporioides*	34/M	Acute lymphocytic leukemia on chemotherapy and relapse. Injury (fallen on cactus).	A tender erythamatous nodule measuring 1.2 × 1.2 cm with minimal superficial scaling on right forearm.
	Case 2	*C. coccodes*	47/M	Stage III non-hodgkins lymphoma of 8 years duration. Undergone autologous peripheral stem cell transplantation.	Enlarged, tender, and erythematous nodule on left arm measuring 1 × 1.5 cm with a central pustule and superficial crusting.
	Case 3	*C. coccodes*	58/M	Stage I non-hodgkins lymphoma, allogenic bone marrow transplantation.	A tender erythematous papule measuring 0.5 × 0.5 cm with a central pustule.
2	Guarro et al., 1998[32]	*C. gloeosporioides*	56/M	Diabetes mellitus. Injury by rotten wood.	Erythematous, violaceous, and tuberose nodular lesions on his left forearm and elbow measuring 1 × 3 cm in diameter.
3	Castro et al.[21]	*C. crassipes*	34/M	Renal transplant recipient.	Nodular and purulent cysts with thick walls on the right leg.
Eye infections					
1	Fernandez et al.[12]				
	Case 1	*C. gloeosporioides*	68/M	IDDM	NA
	Case 2	*Colletotrichum* spp.	34/M	IDDM, Trauma	
	Case 3	*Colletotrichum* spp.	78/F	Corneal erosion	
	Case 4	*C. gloeosporioides*	74/M	IDDM	
	Case 5	*Colletotrichum* spp.	69/M	Trauma by tree branch, topical steroids	
	Case 6	*C. dematium*	40/M	Sand in eye	
	Case 7	*Colletotrichum* spp.	35/M	Trauma by dust and prednisolone use	
	Case 8	*C. dematium*	28/M	Injury by plant liquid	
	Case 9	*C. dematium*	28/M	Injury by plant liquid	
	Case 10	*C. gloeosporioides*	78/M	Facial injury and exposure keratopathy	
2	*Kaliamurthy et al.*				
	Case 1	*C. dematium*	47/M	Topical chloramphenicol ointment	Corneal ulcer irregular edges, stromal infiltration, flare and cells in anterior chamber.
	Case 2	*C. dematium*	19/M	Injury by stick	Paracentral ulcer with dendritic pattern, stromal infiltrate.
	Case 3	*C. dematium*	30/F	Fall of insect, removed by finger	Central corneal ulcer with stromal infiltrate.
	Case 4	*C. dematium*	41/M	Acyclovir ointment	Paracentral corneal ulcer, serrated marginstromal infiltration, and hypopyon.
	Case 5	*C. dematium*	50/F	NA	Central corneal ulcer, flare.
	Case 6	*Colletotrichum* spp.	80/F	NA	Central ulcer, hypopyon, and stromal infiltratation.
	Case 7	*Not done*	70/M	Mud particles in eye	Central ulcer, hypopyon, and stromal infiltratation.
3	Yamamoto et al.[19]	*C. gloeosporioides*	82/M	Myelodysplastic syndrome, cataract surgery	NA
4	Ritterband et al.[20]	*C. graminicola*	24/M	Cataract surgery	NA
5	Upadhyay et al.[22]	*C. capsici*	NA	NA	NA
6	Matsuzaki et al.[50]				
		C. gloeosporioides	NA	Injury, topical steroid	NA
		C. gloeosporioides	NA	Injury by leaf of orange tree	NA
7	Liao et al.[14]	*C. dematium*		Injury	NA
8	Shukla et al.[18]	*G. lomerella cingulata*		Injury	NA
9	Liesegang and Forster[17]	*C. coccodes*			

TABLE 49.2 (continued)
List of Reported Human Infections due to *Colletotrichum*

S. No.	References (Nos)	Species Identification	Age/ Sex	Risk Factors	Clinical Presentation
1	Joseph et al.[51]	*C. dematium*	25/F	Trauma with stones	Central ulcer, corneal oedema, and stromal infiltratation.
2	Mendiratta et al.[15]	*C. dematium*	27/M	Injury with soybean branch	Paracentral ulcer irregular margin and leathery slough at base and stromal infilatrate.
3	Giaconi et al.[13]	*C. dematium*	59/M	Injury after work at grinding wheel, topical antiviral and steroids	NA
4	Mitani et al.[52]	*C. gloeosporioides*	80/F	Traumatic injury while working in field	Corneal ulcer with grayish stromal infiltrate, indistinct margins, and Hypopyon.
5	Singh et al.[31]	*C. dematium*	38/M	Acute nodular scerelitis, subconjunctival injection of triamcinolone acetonide	Fungal endophthalmitis.

neutropenic patient possibly caused by an unidentified species of *Colletotrichum*.

Colletotrichum keratitis may present with the symptoms similar to other fungal keratitis like ocular pain, redness, decreased vision, photophobia, and discharge. Corneal ulcer caused by this agent may be central or paracentral with irregular or serrated edges with varying amounts of stromal infiltration and edema. Hypopyon may also be present. Sometimes the ulcer on presentation may exhibit dendritic pattern resembling viral keratitis. Like other melanized fungi, *Colletotrichum* keratitis progresses slowly and less aggressively, in contrast to *Aspergillus* or *Fusarium* keratitis. Macroscopic brown pigmentation of the corneal infiltrate may also be noticed when keratitis is caused by this melanized fungus.[34]

Disease caused by *Collectotrichum* belongs to the group of infections called "phaeohyphomycosis," the name proposed by Ajello et al.[35] in 1974, as the fungus produces melanin. A case of subcutaneous infection due to *C. gloeosporioides* has been reported as hyalohyphomycosis by Guarro et al.,[32] as the hyphae were hyaline on direct microscopic examination of the tissue. In such a situation, Fontana-Masson stain may correctly identify the presence or absence of melanin. The subcutaneous infection is due to the traumatic implantation of the fungus, though the history of the trauma may not be elicited in all patients, especially in prolonged course of the progression of the disease. The presence of the fungus in the wood and soil and frequent occurrence of lesions on the exposed part of the body supports traumatic acquisition of infection. The subcutaneous infection due to *Colletotrichum* spp. is usually seen in the warm climates and particularly in immunocompromised individuals.[36] Subcutaneous *Colletotrichum* infections may present as painless or tender nodule on the extremities with or without obvious history of trauma at the site of origin. The size of the nodule may vary. The overlying skin may be normal without any signs of inflammation or may be accompanied with erythematous lesion. The nodule may be solitary or multiple satellite nodules around primary lesion.

In addition to one possible case of systemic infection reported by Midha et al.,[24] one out of three cases of phaeohyphomycosis reported by O'Quinn et al.[33] died due to fulminating infection in the lungs suggesting possible dissemination due to *C. coccodes*.

In general, most cases of keratitis caused by *Colletotrichum* species appear to respond well to the topical natamycin alone or in combination with amphotericin B. Different combinations of drugs like topical natamycin and oral itraconazole, intravitreal amphotericin B and fluconazole, and amphotericin B and topical miconazole were used successfully for the management of keratitis.[12,15,18–20,29,33] Kaliamurthy et al. reported successful resolution of corneal lesion in five cases using topical natamycin and ciprofloxacin for a mean duration of 47 ± 14 days.[29] Topical antifungal alone may not be sufficient for the resolution of the lesion and may require surgical procedure like therapeutic keratoplasty (TPK).[29] Treatment data with the use of newer triazoles are not available but one case of keratitis caused by *C dematium* was treated with topical voriconazole (1%). The lesion did not respond to the therapy though the strain was susceptible by in vitro antifungal susceptibility testing (1 μg/mL).[30] Subcutaneous phaeohyphomycosis due to *C. gloeosporioides* and *C. coccodes* was successfully treated with the combination of amphotericin B and itraconazole.[33]

Reports of antifungal susceptibility testing are few. Guarro et al. tested 16 isolates of *Colletotrichum* (*C. gloeosporioides*—7, *C. coccodes*—5, *C. dematium*—4) against azoles (fluconazole, itraconazole, ketoconazole, and miconazole), amphotericin B, and flucytosine and showed low minimum inhibitory concentration (MIC) for all antifungal agents tested except flucytosine, though high MICs were expected due to the extensive use of sublethal concentration of azoles in the agricultural practice.[32] Experimental demonstration indicated possible additive effect while combinations of azoles and caspofungin were used.[37]

49.1.3 Diagnosis

49.1.3.1 Conventional Techniques

49.1.3.1.1 Keratitis

The diagnosis of keratitis due to *Colletotrichum* species includes clinical history, clinical examination, and accurate identification of causative agent. Isolation of the causative agent is important for authoritative identification and appropriate antifungal therapy. The corneal scrapings should preferably be collected by the ophthalmologists and directly inoculated at the bedside or transported to the laboratory between two sterile glass slides (one for microscopy and the other for culture). Swabs or samples collected by laboratory loop are not good specimens and should be avoided. The direct examination of potassium hydroxide wet mount of the corneal scrapings helps in early diagnosis. The use of calcofluor white stain and examination under fluorescent microscope would increase the sensitivity and rapidity in detection of fungal elements. Isolation of the fungus can be attempted by inoculating the corneal scrapings on the blood agar/Sabourauds dextrose agar by using "C" streak. Only colonies appearing on the "C" streak should be considered as corneal pathogen.

49.1.3.1.2 Subcutaneous Phaeohyphomycosis

Excised tissue or biopsy or the aspirated fluid from the lesion are the ideal samples for the laboratory confirmation of the infection. Direct microscopic examination of the potassium hydroxide (KOH) or calcofluor wet mount, histopathology, and isolation of the fungus should be attempted from the sample. The presence of pigmented or dark-colored hyphae in the KOH wet mounts or on histopathology slides helps in the diagnosis of phaeohyphomycosis cases. However, occasionally pigment may not be noticed. Fontana-Masson stain may confirm the diagnosis in such situations. This stain specifically stains the melanin present in the fungal cell wall and appears black to brown. Various other special fungal stains like periodic acid schiff stain (PAS) and Gomori's methanamine silver stain (GMS) may be performed on histopathology sections. Materials from suspected cases should be cultured to isolate the fungus. The fungus grows within 4–6 days.

49.1.3.1.3 Identification of Fungus

Identification of *Colletotrichum* to the species level may be important for epidemiological considerations. Appearances of conidia, appressoria, and acervulus on the routine media may take longer time and this may lead to difficulty in morphological identification. Cultures are usually plated onto potato dextrose agar (PDA) and incubated at 25°C. Five 4 mm plugs may be cut from the actively sporulating areas near the growing edge of a 5-day-old culture using a sterile cork borer. Each plug may be placed on PDA plates and grown by alternating 12 h under ultraviolet light or 12 h in dark at 25°C to induce characteristic morphological characters.[6] Identification of medically important *Colletotrichum* species is usually performed on the basis of morphological features as shown in Table 49.1. It is noteworthy to mention here that some *Colletotrichum* spp. may be misidentified as *Fusarium* due to the close resemblance of the falcate conidia in both these fungi. However, the conidia of *Colletotrichum* species are nonseptate. Further, the presence of appressoria and in the later stage acervulus with setae may help in identification.

49.1.3.2 Molecular Techniques

Due to the inadequacies and plasticity of morphological characters, nucleic acid sequence analysis has been regarded as more reliable method for the identification and classification of *Colletotrichum* species.[4,7,23,26,28] Even the application of the molecular techniques directly on the clinical samples may help in rapid diagnosis.

49.1.3.2.1 Molecular Diagnosis

The utility of polymerase chain reaction (PCR) has been established for the detection of many difficult organisms like *Mycobacteria*, *Microsporidia*, and *Acanthamoeba* causing eye infections. But the same for the diagnosis of *Colletotrichum* is not well established. Kumar and Shukla developed a PCR targeting internal transcribed spacer (ITS) regions and single-stranded conformation polymorphism of rRNA genes for rapid diagnosis of mycotic keratitis.[38] One of the three fungi identified by this method is *Colletotrichum* state of *Glomerella cingulata*. In another study by Vengayil et al., PCR was proved not only as an effective rapid method, but also as a sensitive method for the diagnosis of fungal keratitis compared to KOH wet mount and Gram's smear.[39] However, the sample collection and DNA extraction procedure may interfere with the PCR test sensitivity. To overcome this problem, Menassa et al. recently developed a DNA-stabilizing Whatman (FTA) filter paper method for specimen collection. It can be used as a single-step, non-nested PCR for fungal keratitis without the need of DNA extraction.[40] Thus PCR is a promising tool for rapid diagnosis of fungal keratitis caused by *Colletotrichum* spp.

49.1.3.2.2 Molecular Identification

Due to certain limitations of the morphological identification of *Colletotrichum* species, molecular techniques seem to be an important tool for species determination. The rRNA-ITS sequencing holds the potential to identify most of the clinically important species of *Colletotrichum*. Cano et al. demonstrated the utility of ITS and D1–D2 region of rRNA gene to identify the most relevant clinical species of *Colletotrichum* causing human infections.[41] Though the ITS region is widely sequenced region, there are some concerns to use ITS sequence data for species identification. Crouch and Beirn reported a high error rate (86%) while comparing ITS sequence data within the *C. graminicola* species complex.[4] Entry of sequence data under an incorrect specific name in public domain has created more confusion. According to one analysis of 343 ITS sequences named *C. gloeosporioides* (accessed on September 6, 2009) more than 86% had considerable evolutionary divergence from the type species of *C. gloeosporioides*, and most likely the strains are other than *Colletotrichum* species.[28] However, as the spectrum of species of *Colletotrichum* causing human infection is expanding, it might not be possible to identify all the clinically relevant species of *Colletotrichum* just by rRNA-ITS sequence. Thus multigene phylogenetics is employed to systematically characterize *Colletotrichum* species.[4,23,26,42–44] Prihastuti et al. used six genes, the nuclear rDNA ITS region, partial Actin (ACT), β-tubulin (TUB2), Calmodulin (CAL), Glutamine synthetase (GS), and Glyceraldehyde 3-phosphate dehydrogenase (GPDH) to study a few closely related *Colletotrichum* species (*C. gloeosporioides sensu lato*) and established that species relationships could well be resolved by the same procedure.[26] Multigene phylogenetics is an accurate and reliable method for the diagnosis of *Colletotrichum* species, but it is neither very efficient nor economical. It is currently impractical to apply multiple gene phylogenetics to identify each clinical isolate. It is paramount that sequence data be generated from type species and used in species comparisons and phylogenetic analysis. However, the ITS region is still useful in some cases for reconstruction of interspecific relationships, although it is not ideal for inferring intraspecific relationships. Currently, ITS is the only gene region that is available from all the ex-type or ex-epitype cultures of *Colletotrichum* species.

49.1.3.2.3 Epitypification in Colletotrichum

According to article 9.7 of the *International Code of Botanical Nomenclature* (Vienna Code)

> An epitype is a specimen or illustration selected to serve as an interpretative type when the holotype, lectotype, or previously designated neotype, or all original material associated with a validly published name, is demonstrably ambiguous and cannot be critically identified for purposes of the precise application of the name of a taxon.[45–47] When an epitype is designated, the holotype, lectotype, or neotype that the epitype supports must be explicitly cited.

It has become relatively common to epitypify fungi.[27,42,46,48] One of the major reasons to epitypify is that the type material is lost or is in poor condition. Even if the type material is in relatively good condition, epitypification is needed for gene research. Another reason is that mycologists may have seen good type material and their understanding of a taxon/genus/family may be based on literature or drawings of the type or representative (possibly misidentified) collections. The individual understanding may therefore be questionable. So epitypification can solve many taxonomic problems and stabilize the understanding of species, genera, families, or orders in general and particularly in *Colletotrichum*.

49.2 METHODS

49.2.1 Sample Preparation

Whole cell DNA from the mycelia can be extracted following a slightly modified protocol of small-scale fungal DNA extraction method by Lee and Taylor.[49] Briefly, *Colletotrichum* species is allowed to grow on PDA at 37°C on a rotary shaker at 120 rpm for 3–5 days. The mycelial mat is recovered by filtration and washed with sterile normal saline. About 0.2–0.3 g of the mycelial mat is grinded in the presence of liquid nitrogen, and the resultant powder is transferred to a 1.5 mL microcentrifuge tube containing 600 μL of lysis buffer (100 mM Tris–HCl pH8.0, 50 mM EDTA, 3% sodium dodecycl sulphate (SDS)). After vortexing the tube briefly proteinase K is added to a final concentration of 20 μg/mL. The tube is incubated at 56°C for 1 h. Finally, DNA can be extracted using phenol:chloroform extraction procedure. The DNA precipitation is done with equal volume of isopropanol in the presence of 3M sodium acetate. The pellet is washed with 70% alcohol and dissolved in 100 μL of TE (10 mM Tris–HCl pH 7.5, 1 mM EDTA), and stored at −20°C for further use. Alternatively the commercial DNA isolation kit may be used for the extraction of the DNA from the culture.

49.2.2 Detection Procedures

49.2.2.1 PCR-ITS Region

The rRNA gene-ITS can be amplified by PCR in the presence of 2 mM $MgCl_2$, 200 μM of dNTP, 0.25 μM of primer (ITS1—GCATATCAATAAGCGGAGGAAAAG and ITS4—GGTCCGTGTTTCAAGACGG), 0.25 U of *Taq* polymerase, and 5–10 ng of fungal genomic DNA in a total volume of 10 μL. The amplification reaction is performed in a thermalcycler. The PCR cycling conditions consist of an initial denaturation step for 5 min at 94°C, followed by 35 cycles of denaturation at 94°C for 1 min, annealing at 55°C for 30 s and 72°C for 1 min, and a final extension step at 72°C for 5 min.

49.2.2.2 Sequencing of rRNA-ITS Region

To extract the gene fragment, agarose gel can be excised with a clean, sharp scalpel. To the weighed gel slice, three times the volume of buffer, Qiagen (QG) is added and incubated at 50°C for 10 min or until the gel slice is dissolved completely. To accelerate dissolution, the gel is vortexed every 2–3 min during incubation. One gel volume of isopropanol is added to the sample and mixed. The Q/A quick spin column (Qiagen, QIA quick) is placed in 2 mL collection tube. To bind DNA, the sample is applied to the Q/A quick column and centrifuged for 1 min. The flow-through is discarded and the Q/A quick column is placed back into the same tube. To wash 0.75 mL of buffer, Qiagen (PE) is added to Q/A quick column and centrifuged for 1 min. The flow-through is discarded. The column is centrifuged in a 2 mL collection tube for 1 min at 13,000 rpm. To elute DNA, 30–50 μL of elution buffer (10 mM Tris–HCl pH 8.5) is added to the center of the Q/A quick membrane and centrifuged for 1 min. The elute is used as the purified gene product. Sequencing reactions can be performed in any sequencing platform, but are more commonly performed with Big Dye Terminator Cycle Sequencing Kit, Version 3.1/1.1 (Applied Biosystems). The sequencing reactions may be analyzed on ABI Genetic Analyzer (Applied Biosystems) or other software depending upon the sequencing kit used. The identity of the isolate can be found by comparing the sequence with the deposited sequence in the public databases like GenBank.

49.3 CONCLUSION AND FUTURE PERSPECTIVES

Colletotrichum species, basically the plant pathogen, is recently emerging as an opportunistic agent causing eye infections and subcutaneous infections in humans. Though very few human cases are reported in the literature, exact number may be much higher due to lack of awareness and difficulty in identification of these agents in the routine clinical mycology laboratory. Among many species of *Colletotrichum*, *C. dematium*, *C. coccodes*, *C. gloeosporioides*, *C. graminicola*, *C. crassipes*, and *C. truncatum* are known to cause human infections. As this fungus takes longer time to produce conidia, appressoria, and acervulus on the routine culture media, morphological identification becomes difficult or delayed. Molecular techniques, especially the sequencing of ITS and D1–D2 region of rDNA, may help in rapid identification of the fungi. However, many sequences deposited in GenBank are wrongly labeled due to multiple names of the same species or misidentification of the fungus on the basis of morphological characters. So epitypification of all the *Colletotrichum* species may resolve many taxonomic problems and may help in accurate identification.

REFERENCES

1. Bauley, J.A. et al., Infection strategies of *Colletotrichum* species. In Bailey, J.A. and Jeger, M.J. Eds., *Colletotrichum: Biology, Pathology and Control*, CAB International, Wallingford, CT, 1992.
2. Hyde., K.D. et al., *Colletotrichum*: A catalogue of confusion. *Fungal Divers.*, 39, 1–17, 2009.
3. Sutton, D.A., Coelomycetous fungi in human disease. A review: Clinical entities, pathogenesis, identification and therapy. *Rev. Iberoam. Micol.*, 16, 171, 1999.
4. Crouch, J.A. and Beirn, L.A., Anthracnose of cereals and grasses. *Fungal Divers.*, 39, 19, 2009.
5. Hyde, K.D. et al., *Colletotrichum*—Names in current use. *Fungal Divers.*, 39, 147, 2009.
6. Sutton, B.C., Fungi imperfecti with pycnidia, acervuli and stromata. In *Coelomycetes*, Commonwealth Mycological Institute, Kew, U.K., 1980.
7. Sutton, B.C., Genus Glomerella and its anamorph *Colletotrichum*. In Bailey, J.A. and Jeger, M.J., Eds., *Colletotrichum: Biology, Pathology and Control*, CAB International, Wallingford, CT, 1992.
8. Kirk, P.M. et al., *The Dictionary of the Fungi*, 10th edn., CABI Bioscience, U.K., 2008.
9. Cai, L. et al., A Polyphasic approach for studying *colletatrichum*. *Fungal Divers.*, 39, 183, 2009.
10. Chakrabarti, A. et al., Fungal endophthalmitis: Fourteen years' experience from a center in India. *Retina*, 28, 1400, 2008.
11. Kaliamurthy, J. and Thomas, P.A., Keratitis due to *Colletotrichum dematium*. *Indian. J. Med. Microbiol.*, 23, 206, 2005.
12. Fernandez, V. et al., *Colletotrichum* keratitis. *Am. J. Ophthalmol.*, 134, 435, 2002.
13. Giaconi, J.A. et al., Voriconazole and fungal keratitis: A report of two treatment failures. *J. Ocul. Pharmacol. Ther.*, 22, 437, 2006.
14. Liao, W.Q., Shao, J.Z., and Li, S. Q., *Colletotrichum dematium* caused keratitis. *Chinese Med. J.*, 96, 391, 1983.
15. Mendiratta, D.K. et al., Keratitis due to *Colletotrichum dematium*—A case report. *Indian J. Med. Microbiol.*, 23, 56, 2005.
16. Rosa, R.H. Jr., Miller, D., and Alfonso, E.C., The changing spectrum of fungal keratitis in south Florida. *Ophthalmology*, 101, 1005, 1994.
17. Liesegang, T.J. and Forster, R.K., Spectrum of microbial keratitis in South Florida. *Am. J. Ophthalmol.*, 90, 38, 1980.
18. Shukla, P.K. et al., Clinical and experimental keratitis caused by the *Colletotrichum* state of *Glomerella cingulata* and *Acrophialophora fusispora*. *Sabouraudia*, 21, 137, 1983.
19. Yamamoto, N., Matsumoto, T., and Ishibashi, Y., Fungal keratitis caused by *Colletotrichum gloeosporioides*. *Cornea*, 20, 902, 2001.
20. Ritterband, D.C., Shah, M., and Seedor, J.A., *Colletotrichum graminicola*: A new corneal pathogen. *Cornea*, 16, 362, 1997.
21. Castro, L.G. et al., Phaeohyphomycotic cyst caused by *Colletotrichum crassipes*. *J. Clin. Microbiol.*, 39, 2321, 2001.
22. Upadhyay, M.P. et al., Epidemiologic characteristics, predisposing factors, and etiologic diagnosis of corneal ulceration in Nepal. *Am. J. Ophthalmol.*, 111, 92–99, 1991.
23. Damm, U. et al., *Colletotrichum* species with curved conidia from herbaceous hosts. *Fungal Divers.*, 39, 45, 2009.
24. Midha, N.K., Mirzanejad, Y., and Soni, M., *Colletotrichum* Sp.: Plant or human pathogen? *Antimicrob. Infect. Dis. Newsl.*, 15, 26, 1996.
25. de Hoog, G. S. et al., *Atlas of Clinical Fungi*, a pilot CD-ROM version of the 3rd edn., Centraalbureau voor Schimmelcultures, Baarn, the Netherlands, 2009.
26. Prihastuti, H. et al., Characterization of *Colletotrichum* species associated with coffee berries in northern Thailand. *Fungal Divers.*, 39, 89, 2009.
27. Shenoy, B.D. et al., Morpho-molecular characterization and epitypification of *Colletotrichum capsici* (*Glomerellaceae*, *Sordariomycetes*), the causative agent of antharacnose in chilli. *Fungal Divers.*, 27, 197, 2007.
28. Cannon, P.F., Buddie, A.G., and Bridge, P.D., The typification of *Colletotrichum gloeosporioides*. *Mycotaxon*, 104, 189, 2008.
29. Kaliamurthy, J. et al., Keratitis due to a coelomycetous fungus: Case reports and review of the literature. *Cornea*, 23, 3, 2004.
30. Marangon, F.B. et al., In vitro investigation of voriconazole susceptibility for keratitis and endophthalmitis fungal pathogens. *Am. J. Ophthalmol.*, 137, 820, 2004.
31. Singh, R. et al., Fungal endophthalmitis complicating subconjunctival injection of triamcinolone acetonide in anterior seleritis. *Ann. Ophthalmol.*, 38, 253, 2006.
32. Guarro, J. et al., Subcutaneous hyalohyphomycosis caused by *Colletotrichum gloeosporioides*. *J. Clin. Microbiol.*, 36, 3060, 1998.
33. O'Quinn, R.P., Hoffmann, J.L., and Boyd, A.S., *Colletotrichum* species as emerging opportunistic fungal pathogens: A report of 3 cases of phaeohyphomycosis and review. *J. Am. Acad. Dermatol.*, 45, 56, 2001.
34. Berger, S.T. et al., Macroscopic pigmentation in a dematiaceous fungal keratitis. *Cornea*, 10, 272, 1991.

35. Ajello, L. et al., A case of phaeohyphomycosis caused by a new species of *Phialophora*. *Mycologia*, 66, 490, 1974.
36. Sanche, S.E., Sutton, D.A., and Rinaldi, M.G., Dematiaceous fungi. In Anaissie, E.J., McGinnis, M.R., and Pfaller, M.A., Eds., *Clinical Mycology* 1st edn., Churchill Livingstone, Philadelphia, PA, 2003, p. 325.
37. Serfling, A., Wohlrab, J., and Deising, H.B., Treatment of a clinically relevant plant-pathogenic fungus with an agricultural azole causes cross-resistance to medical azoles and potentiates caspofungin efficacy. *Antimicrob. Agents Chemother.*, 51, 3672, 2007.
38. Kumar, M. and Shukla, P.K., Use of PCR targeting of internal transcribed spacer regions and single-stranded conformation polymorphism analysis of sequence variation in different regions of rRNA genes in fungi for rapid diagnosis of mycotic keratitis. *J. Clin. Microbiol.*, 43, 662, 2005.
39. Vengayil, S. et al., Polymerase chain reaction-guided diagnosis of mycotic keratitis: A prospective evaluation of its efficacy and limitations. *Invest. Ophthalmol. Vis. Sci.*, 50, 152, 2009.
40. Menassa, N. et al., Rapid detection of fungal keratitis using DNA-stabilizing FTA(R) filter paper. *Invest. Ophthalmol. Vis. Sci.*, 51, 1905, 2010.
41. Cano, J., Guarro, J., and Gene, J., Molecular and morphological identification of *Colletotrichum* species of clinical interest. *J. Clin. Microbiol.*, 42, 2450, 2004.
42. Crouch, J.A. et al., Systematic analysis of the falcate-spored graminicolous *Colletotrichum* and a description of six new species from warm-season grasses. *Mycologia*, 101, 717, 2009.
43. Farr, D.F. et al., Species of *Colletotrichum* on agavaceae. *Mycol. Res.*, 110, 1395, 2006.
44. Crouch, J.A., Clarke, B.B., and Hillman, B.I., Unraveling evolutionary relationships among the divergent lineages of *Colletotrichum* causing anthracnose disease in turfgrass and corn. *Phytopathology*, 96, 46, 2006.
45. MCNeil, J. et al., *International Code of Botanical Nomenclature* (VIENNA CODE), 2006, adapted by the seventeeth International Botanical Congress, Vienna, Austria, July 2005. Published Ruggel, Gantner, 2006. Available at http://ibot.sav.sk/icbn/main.htm
46. Hyde, K.D. and Zhang, Y., Epitypification: Should we epitypify? *J. Zhejiang. Univ. Sci.*, 9, 842, 2008.
47. Alves, A. et al., Morphological and molecular data reveal cryptic speciation in *Lasiodiplodia theobromae*. *Fungal Divers.*, 28, 1, 2008.
48. Than, P.P. et al., Epitypification and phylogeny of *Colletotrichum acutatum*. *Fungal Divers.*, 28, 97, 2008.
49. Lee, S.B. and Taylor, J.W., Isolation of DNA from fungal mycelia and single spores. In Innis, M.A., Gelfand, D.H., and Sninsky, J.J. et al., Eds., *PCR Protocols: A Guide to Methods and Applications.*, Academic Press, San Diego, CA, 1990.
50. Matsuzaki, O., Yasuda, M., and Ichinohe, M., Keratomycosis due to *Glomerella cingulata*. *Rev. Iber. Micol.*, 5, S30, 1988.
51. Joseph, J., Fernandes, M., and Sharma, S., *Colletotrichum dematium* keratitis. *J. Postgrad. Med.*, 50, 309, 2004.
52. Mitani, A. et al., In vivo and in vitro investigations of fungal keratitis caused by *Colletotrichum gloeosporioides*. *J. Ocular. Pharm. Therap.*, 25, 563, 2009.

50 *Cylindrocarpon*

Dongyou Liu

CONTENTS

50.1 INTRODUCTION

50.1.1 Classification and Morphology

The genus *Cylindrocarpon* (obsolete synonyms: *Allantospora*, *Coleomyces*, *Fusidium*, and *Moeszia*) belongs to the family Nectriaceae, order Hypocreales, class Sordariomycetes, subphylum Pezizomycotina, phylum Ascomycota, and kingdom Fungi. Depending upon the presence or absence of chlamydoconidia and microconidia, the 35 species in the genus *Cylindrocarpon* are arranged into four groups [1].

Cylindrocarpon spp. inhabit a wide range of woody and herbaceous plants and are commonly present in soil [2]. Several *Cylindrocarpon* spp. also infect humans and animals, resulting in keratitis, mycetoma, athlete's foot, peritonitis, and disseminated infection in human patients [3]. The human-infecting species include *Cylindrocarpon cyanescens* (obsolete synonym: *Phialophora cyanescens*), *Cylindrocarpon destructans*, *Cylindrocarpon lichenicola* (synonym: *Fusarium lichenicola*; obsolete synonym: *C. tonkinense*), and *Cylindrocarpon vaginae* [4–6].

Morphologically, all *Cylindrocarpon* species produce slimy macroconidia in basipetal succession (the youngest conidium at the base), which do not adhere in chains. At the species level, *Cylindrocarpon lichenicola* (synonym: *Fusarium lichenicola*) colonies grow rapidly on potato flakes agar (PFA), reaching a diameter of 35 mm in 6 days at 25°C. Colonies are velvety to floccose, initially white, then yellow, and finally pale brown at maturity. The reverse ranges from buff at the periphery to a darker brown centrally, with a brown-diffusing pigment. Hyphae are septate and hyaline, and conidiophores are long and simple or poorly branched. Subulate (slender and tapering to a point) conidiogenous cells (phialides) (of 38–50 μm × 3–5 μm) sometimes have a distinct collarette at the apices. Macroconidia (19.6–32 μm × 5 μm) are borne singly and in clusters at the apices of phialides, predominately three septate but occasionally up to five septate, hyaline, straight or curved, cylindrical to fusiform, rounded at apex, and distinctly truncate at the base with offset basal pedicels. Septations appear as rings around the macroconidia. Microconidia may be absent or one celled if present. Chlamydospores (chlamydoconidia) (of 8–15 μm in diameter) may form in old cultures, occurring singly, in chains or in clumps, intercalary, or terminal, and appear distinct hyaline to pale brown, smooth to spinulose, globose cells within the multicellular macroconidia. *C. lichenicola* grows poorly at 35°C and shows no growth at 42°C [7].

Cylindrocarpon lichenicola (*Fusarium lichenicola*) produces relatively long and narrow, several-celled, cylindrical conidia with typical rounded apices, which deviate only slightly from the macroconidia of other *Cylindrocarpon* species with straight conidia by having a conspicuously protuberant, symmetrical, truncate base. The type species of the genus *Cylindrocarpon*, *Cylindrocarpon cylindroides* (teleomorph *Neonectria neomacrospora*) is phylogenetically distinct from *C. lichenicola* (*F. lichenicola*) and other members of the genus *Fusarium*. The medically important and plant-pathogenic *Cylindrocarpon destructans* (teleomorph *Neonectria radicicola*) also clusters phylogenetically in a group of organisms well segregated from the *F. solani* and *Gibberella* clades [6]. *Cylindrocarpon* (*Phialophora*) *cyanescens* is characterized by its phialidic conidia, chlamydospores in aggregations, and intense diffusing pigment [8].

Cylindrocarpon species resemble *Fusarium* species morphologically and taxonomically, with both sharing teleomorphs in the genus *Nectria*; and *Cylindrocarpon* differs from *Fusarium* by lacking an asymmetrical foot cell on the macroconidia [9]. Through sequencing analysis of the ribosomal large subunit (LSU), it was shown recently that despite their deviating conidial morphologies, *Cylindrocarpon lichenicola* and *Acremonium falciforme* constitute members of the *Fusarium solani* species complex (i.e., the *F. solani* clade containing *F. solani*, *Haematonectria*, and *Neocosmospora*) [10]. Thus, the original name *Fusarium lichenicola* is reestablished for *Cylindrocarpon lichenicola*, and the new combination *Fusarium falciforme* is proposed for *Acremonium falciforme* [6].

While *F. lichenicola*, *F. falciforme*, and members of the *F. solani* clade all produce long, slender, cylindrical conidiophores and integrated, terminal phialides, clearly reflecting their biological unity, other *Fusarium* groups tend to form shorter, discrete monophialides (phialides with only a single fertile opening) that are subulate (awl or candle shaped, i.e., rigid looking, tapered, and narrow at the apex) or inflated, or form polyphialides (with each phialide containing several fertile openings). In addition, *F. lichenicola*, *F. falciforme*, and *F. solani* have two- to several-celled conidia that are significantly blunter at the apex and proportionately broader than the macroconidia of fusaria in the *Gibberella* clade. However, *F. solani* conidia demonstrate a classical fusarial macroconidial form (curved phragmoconidia with somewhat pointed apices and basal foot cells), whereas *F. lichenicola* conidia do not; *F. falciforme* conidia may vary from strain to strain. Furthermore, despite showing varied colony colorations, *F. lichenicola*, *F. falciforme*, and *F. solani* colonies show similar chestnut red-brown to purplish reverse pigments. The violet-purple reverse color in *F. falciforme* overlaps with that in some *F. solani* isolates [6].

50.1.2 Clinical Features

50.1.2.1 *Cylindrocarpon lichenicola* (*Fusarium lichenicola*)

Resembling *Fusarium solani*, a common agent of keratomycoses and disseminated disease in neutropenic and/or immunocompromised hosts, *C. lichenicola* (*F. lichenicola*) has been shown to cause keratitis, localized cutaneous invasive disease, disseminated infection, and peritonitis in humans [3,11–16]. Patients from areas with warm climates may develop a distinctive fusarial intertrigo caused by *Fusarium solani*, *Cylindrocarpon lichenicola* (*Fusarium lichenicola*), or *Fusarium oxysporum* [6].

Iwen et al. [7] reported a case of trauma-related invasive *Cylindrocarpon lichenicola* infection localized to the right hand in a 53-year-old male patient with acute myelogenous leukemia (AML). Histological examination of skin biopsy showed septate, branching hyphae along with globular structures (8–11 μm in diameter) resembling conidia. As invasive molds rarely produce conidia in tissues and only in lesions exposed to ambient air, the presence of the fungus on the cutaneous surface and therefore exposure to ambient air in this case may have allowed the *Cylindrocarpon* organisms to produce conidial structures. Multiple positive fungal cultures from the biopsied tissue permitted subsequent microscopic and macroscopic identification of the fungus as *C. lichenicola*. The infection resolved following marrow regeneration, aggressive debridement of the affected tissue, and treatment with amphotericin B.

Kaliamurthy et al. [17] described a fungal keratitis due to *Cylindrocarpon* (*Fusarium*) *lichenicola* in a farmer, who presented with pain, redness, and irritation of the left eye following ocular injury caused by hay 5 days earlier. Slit-lamp examination of the affected eye showed a corneal ulcer (6×5 mm) with irregular margins, raised, with necrotic slough and infiltration. The lens was cataractous and visual acuity was restricted to "hand movements" only. Scrapings from the base and edges of the corneal ulcer contained septate fungal hyphae as examined by microscopy after staining with lactophenol cotton blue and Gram stain. Histopathological examination of the infected corneal button removed at therapeutic penetrating keratoplasty (TPK) also showed septate fungal hyphae. Scrapings and corneal button are inoculated onto Sabouraud glucose-neopeptone agar, and sheep blood agar grew cottony, white-red fungal colonies within 48 h, with brown pigmentation on reverse. The fungus was identified as *Cylindrocarpon lichenicola* since it displayed branched, septate, hyaline hyphae; phialides on simple or sparsely branched conidiophores; cylindrical to fusiform, smooth-walled macroconidia, each with three to six septa, a blunt rounded apex, and distinctly truncate base; and smooth-walled chlamydoconidia on short branches.

50.1.2.2 *Cylindrocarpon destructans*

Cylindrocarpon destructans is one of the fungal species (e.g., *Acremonium* [*Fusarium*] *falciforme*, *Acremonium kiliense*, *Acremonium recifei*, *Cylindrocarpon destructans*, *Fusarium moniliforme*, *Fusarium solani*, *Neotestudina rosatii*, *Polycytella hominis*, and *Pseudallescheria boydii*) that are known to cause white grain eumycetoma. Being a slowly progressing infection, eumycetoma develops in individuals usually after traumatic implantation of fungus-contaminated plant material. Zoutman and Sigler [18] documented a *Cylindrocarpon destructans*-related eumycetoma in a 39-year-old male, originally from Antigua, West Indies. The patient presented with a 12-year history of swelling of the left foot with pus and grains expressed from a draining sinus tract, from which *Cylindrocarpon destructans* was isolated in pure culture. The fungus was identified by its microconidial morphology, the presence of chlamydospores, and an intense brown diffusible pigment.

50.1.2.3 *Cylindrocarpon cyanescens* (*Phialophora cyanescens*)

Cylindrocarpon (*Phialophora*) *cyanescens* has also been shown to cause white grain eumycetoma [5,18].

Cylindrocarpon cyanescens produces cylindrical phialides with nonflaring collarettes. Hemashettar et al. [19] reported a case of white grain eumycetoma on the foot of a 57-year-old Indian male, who sustained a trauma by a thorn 10–12 years earlier and developed a pustule at the site after thorn removal. The pustule discharged serosanguineous fluid and eventually healed. Then, the patient was admitted to the hospital with slight swelling of the left foot, pain, and draining sinuses. Histologic examination of a biopsy tissue specimen revealed oval, lobular, white granules (0.5–1.0mm in diameter) composed of hyaline, septate hyphae and thick-walled chlamydospores (up to 15µm in diameter). Culture of granules from a draining sinus on Sabouraud dextrose agar (SDA) with chloramphenicol (SDA+C) at room temperature (25°C–30°C) after 12 days produced compact, very-slow-growing, and poorly sporulating colonies with a strong reddish brown pigment that diffused into the medium. Colonies on Sabouraud dextrose agar were raised in the center, downy, and white at first, becoming grayish buff after 2 weeks, measuring 6–7.5mm in diameter at 25°C and 3.5–4.5mm in diameter at 37°C. The isolate failed to grow at 40°C. Microscopic examination of the slide culture preparations on potato dextrose agar after 3 weeks of incubation at 25°C showed sterile, septate, hyaline hyphae (2.5–5.0µm wide), producing chlamydospores but devoid of conidia. Conidiation occurred sparsely and best on Takashio and oatmeal salts agars after prolonged incubation (3–14 weeks at room temperature) and consisted of zero- to one-septate, cylindrical, sometimes slightly curved conidia produced from unbranched septate phialides without collarettes. Single-celled conidia measured 9–14µm long and 2–2.5µm wide, and two-celled conidia measured 17–25µm long × 2–3µm wide. Chlamydospores were mostly solitary, intercalary, or terminal and were subglobose and smooth to slightly roughened. The fungus was identified as a *Cylindrocarpon* sp. on the basis of the development of rare cylindrical conidia borne from solitary phialides lacking collarettes, in addition to chlamydospores formed singly or in short chains. It appeared highly similar to *C. cyanescens* in its strong production of reddish brown diffusible pigment and restricted growth. While *C. cyanescens* did not invade bone, *C. destructans* caused lytic bone destruction of the tarsals and the base of the third metatarsal.

50.1.3 Diagnosis

Like their *Fusarium* relatives, *Cylindrocarpon* species are anamorphs of ascomycetes belonging to the Hypocreaceae (teleomorph *Nectria*) and are cosmopolitan soil- and plant-associated fungi. Rapid and accurate determination of *Cylindrocarpon* species is vital for implementing appropriate treatment and control strategies for these organisms [20].

Both *Cylindrocarpon* (*Fusarium*) *lichenicola* and members of the *F. solani* clade produce elongate, filiform, often several-celled conidiophores incorporating terminal monophialides (up to 40µm in length) with a relatively broad, blunt apex. The elongate conidiophores in *F. solani* isolates are mainly associated with microconidia (which are often absent in *C. lichenicola*), and its macroconidia-forming conidiophores are somewhat shorter. On the other hand, the macroconidia-forming conidiophores in *C. lichenicola* are typically elongate. The conidiophores in *A. falciforme* may be similar in shape to the conidiophores of other members in the *F. solani* clade but are relatively strongly septate [6].

Cylindrocarpon (*Fusarium*) *lichenicola* is further differentiated macroscopically from *F. solani* by forming macroconidia, which are predominately straight rather than curved, by having apical cells that are rounded rather than tapering, by having basal cells with truncate and offset rather than attenuated pedicels (foot cells), by lacking microconidia, by having pigmented chlamydoconidia, and by having a brown color rather than cream color on reverse of SDA medium [6].

Granules produced by *Cylindrocarpon* species cannot be distinguished from each other, nor from those of *Acremonium* spp. and *Fusarium verticillioides* (*F. moniliforme*) without cultural or immunofluorescence studies. However, granules of *Cylindrocarpon* species may be differentiated from those of *P. boydii* by having less prominent, fewer, and smaller chlamydospores.

Molecular methods have offered an alternative means to identify fungal organisms including *Cylindrocarpon* species. Amplification and sequencing analysis of the internal transcribed spacer (ITS) ITS1-5.8S-ITS2 region allows differentiation of *Cylindrocarpon* species from other fungi [21–26].

50.2 METHODS

50.2.1 Sample Preparation

Clinical samples are cultured on Sabouraud dextrose agar or other mycological media. Portions of samples are examined under microscope for mycotic elements using fungal stains. Colonies are grown on synthetic nutrient agar, oatmeal agar, and malt extract agar for 7–10 days at room temperature (22°C). To induce conidiation, isolate is subcultured onto PFA plates and PFA slide cultures, both incubated at 25°C. The microscopic morphology is examined from colonies on PFA plates and slide culture mounts after 6 days of incubation. Temperature studies are performed on PFA slants to evaluate the growth rate at 25°C, 35°C, and 42°C [7].

For fungal DNA extraction, 50mL of Sabouraud dextrose (SAB) broth is inoculated by needle with conidia from a 7 day culture in SAB agar and incubated for 72h at 30°C. The hyphae are recovered on a 0.45µm-pore-size filter and washed with sterile saline. Aliquots of the fungal hyphae are stored frozen at −70°C until use. Prior to lysis, the hyphae are thawed and suspended in 400µL of DNA extraction buffer (1mM EDTA pH 8.0, 1% sodium dodecyl sulfate, 10mM Tris–HCl pH 7.6, 100mM NaCl, and 2% Triton X-100). Microcentrifuge tubes (1.5mL) containing hyphae and buffer are sonicated in a water bath (Branson; model 2210) for 15min, followed by heating at 100°C for 5min. Following lysis, DNA is purified using the QIAmp blood kit (Qiagen) and protocols for crude cell lysates supplied by the manufacturer. The purified DNA is stored at 4°C until tested [6].

Alternatively, DNA is extracted with a FastDNA kit (Qbiogene) from mycelium grown for 3–5 days in liquid complete medium.

50.2.2 Detection Procedures

50.2.2.1 Sequencing Analysis of ITS Regions

Iwen et al. [7] described the use of primers ITS1 (5′-TCC GTA GGT GAA CCT GCG G-3′) and ITS4 (5′-TCC TCC GCT TAT TGA TAT G-3′) from the conserved regions of the 18S (ITS1) and the 28S (ITS4) ribosomal RNA (rRNA) genes, respectively, to amplify the intervening 5.8S rRNA gene, ITS1 region, and ITS2 noncoding regions [27]. Subsequent sequencing analysis of the resulting amplicon enables identification of *Cylindrocarpon* species.

Procedure

1. PCR mixture (50 μL) is made up of 5 μL sample DNA, 20 mM Tris–HCl pH 8.4, and 50 mM KCl; 0.1 mM (each) dNTP, 1.5 mM $MgCl_2$, and 0.3 μM (each) primer; and 1.5 U of Platinum *Taq* high-fidelity DNA polymerase (Gibco BRL, Life Technologies).
2. Amplification is conducted in a Stratagene Robocycler model 96 thermocycler with initial denaturation at 95°C for 4.5 min; 40 cycles of 95°C for 30 s, 50°C for 30 s, and 72°C for 1 min; and a final extension at 72°C for 3 min.
3. Amplicons are separated by agarose gel electrophoresis, purified, and ligated into the PCR 2.1 plasmid vector (Invitrogen). Competent INVαF′ One Shot cells are transformed using standard protocols. Colonies are isolated and purified with a Qiagen miniprep spin kit. An aliquot of purified plasmid is digested with *Eco*RI endonuclease (New England Biolabs) and screened by agarose gel electrophoresis for a 300 bp doublet, indicating the presence of the *Eco*RI cleavage site GAATTC within the 5.8S sequence.
4. Selected plasmids are analyzed by automated dye termination sequencing procedure on a Perkin-Elmer/ABI model 373 DNA sequencer. For the sequencing of cloned fragments, both strands of the plasmid containing the fungal insert are sequenced with universal M13 forward and reverse sequencing primers. For direct sequencing of noncloned amplicons, PCR products are directly sequenced using the ITS1 and ITS4 PCR primers.
5. The resultant nucleotide sequences are aligned with the MacVector sequence analysis software, version 6.5 (Oxford Molecular Group, Inc., Campbell, CA), and the Clustal W alignment algorithm. Intraspecies sequence similarity and variation are determined by the MacVector software and are visually confirmed using pairwise nucleotide alignments. Sequences from referenced isolates are aligned to complete or partial ITS sequences available in GenBank. Comparison of sequences from referenced isolates, clinical isolates, and GenBank sequences is performed using a nongapped, advanced BLAST search. The similarities of the sequences are determined with the expectation frequency minimized to 0.0001 [28].

50.2.2.2 Sequencing Analysis of rRNA LSU

Summerbell and Schroers [6] employed primers V9G and LR5 to amplify the LSU region (encompassing domain D1 and most of domain D2, nucleotides 116–615) of rRNA followed by sequencing for identification and phylogenetic analysis of *Cylindrocarpon* and *Fusarium* species.

The PCR cycling program consists of an initial denaturation at 94°C for 1 min; 35 cycles of 94°C for 35 s, 55°C for 50 s, and 72°C for 3 min; and a final elongation at 72°C for 6 min. (final elongation). The PCR products are purified with a GFX purification kit (Amersham Pharmacia Biotech) and visualized on an electrophoresis gel after ethidium bromide staining. The rRNA is sequenced with a BigDye terminator cycle sequencing kit (Applied Biosystems) and analyzed on an ABI Prism 3700 instrument (Applied Biosystems). The primers used in the sequence reaction are ITS1 and ITS4, NL1 and NL4, and LR5.

Sequence chromatographs are assembled and edited with SeqmanII software (DNAStar) and aligned with sequences downloaded from GenBank. The alignment is initially performed with the ClustalX program (version 1.8; ftp://ftp-igbmc.u-strasbg.fr/pub/ClustalX) and adjusted manually with the Megalign program (DNAStar). The phylogenetic analysis is performed with a part of the LSU rRNA available for all accessions. A parsimony analysis is performed with the software package PAUP (version 4.0b8). Gaps are coded as missing data; characters are defined as unordered and equally weighted; characters not informative to the parsimony analysis are excluded; heuristic searches of parsimonious trees are performed for all sequences with random sequence addition and 1000 replicates, using the starting trees from the stepwise addition, tree bisection–reconnection as the swapping algorithm, and all optimal trees for the next swapping round; and branch robustness is tested by the use of 100 replications of such searches based on bootstrapped data sets with random sequence addition and 10 replicates per search.

50.3 CONCLUSION

Cylindrocarpon spp. are common plant pathogens that are widely distributed in the environment. Some *Cylindrocarpon* spp. may infect humans and animals occasionally, resulting in keratitis, mycetoma, athlete's foot, peritonitis, and disseminated infection. Among the human-infecting *Cylindrocarpon* species, *Cylindrocarpon* (*Fusarium*) *lichenicola* is the only one known to induce invasive diseases besides localized infections, whereas *Cylindrocarpon cyanescens* (obsolete synonym: *Phialophora cyanescens*) and *Cylindrocarpon destructans* are mainly responsible for white grain eumycetoma. As conventional laboratory identification of *Cylindrocarpon* species is largely based on macroscopic and microscopic characteristics of these organisms, it is not only time-consuming but also technically challenging. Application of molecular techniques such as PCR and sequencing analysis of rRNA genes and ITS regions offers a rapid, precise, and sensitive tool for *Cylindrocarpon* determination.

REFERENCES

1. Booth C. The genus *Cylindrocarpon*. Mycological Papers, no. 104. Kew, Surrey, England: Commonwealth Mycological Institute, 1996.
2. Watanabe T. *Cylindrocladium tenue* comb. nov. and two other *Cylindrocladium* species isolated from diseased seedlings of *Phellodendron amurense* in Japan. *Mycologia* 1994;86:151–156.
3. Booth C, Clayton CY, Usherwood M. *Cylindrocarpon* species associated with mycotic keratitis. *Proc Indian Acad Sci Plant Sci*. 1984;94:422–436.
4. Matsumoto T, Masaki J, Okabe T. *Cylindrocarpon tonkinense*: As a cause of keratomycosis. *Trans Br Mycol Soc*. 1979;72:503–504.
5. De Vries GA, de Hoog GS, de Bruyn HP. *Phialophora cyanescens* sp. nov. with *Phaeosclera*-like synanamorph, causing white-grain mycetoma in man. *Anton Van Leeuwen*. 1984;50:149–153.
6. Summerbell RC, Schroers HJ. Analysis of phylogenetic relationship of *Cylindrocarpon lichenicola* and *Acremonium falciforme* to the *Fusarium solani* species complex and a review of similarities in the spectrum of opportunistic infections caused by these fungi. *J Clin Microbiol*. 2002;40(8):2866–2875.
7. Iwen PC et al. Cutaneous infection caused by *Cylindrocarpon lichenicola* in a patient with acute myelogenous leukemia. *J Clin Microbiol*. 2000;38:3375–3378.
8. de Hoog GS et al. *Atlas of Clinical Fungi*. 2nd edn. Centraalbureau voor Schimmelcultures, Utrecht, the Netherlands, 2000.
9. Samuels GJ, Brayford D. Phragmosporous *Nectria* species with *Cylindrocarpon* anamorphs. *Sydowia* 1993;45:55–80.
10. O'Donnell K. Molecular phylogeny of the *Nectria haematococca-Fusarium solani* species complex. *Mycologia* 2000; 92:919–938.
11. Laney B, Blanc C, Lapalu J. *Cylindrocarpon*: A new athlete's foot agent. *Bull Soc Fr Mycol Med*. 1985;14:73–76.
12. James EA et al. Disseminated infection due to *Cylindrocarpon lichenicola* in a patient with acute myeloid leukaemia. *J Infect*. 1997;34:65–67.
13. Sharma R et al. Peritonitis in continuous ambulatory peritoneal dialysis due to *Cylindrocarpon lichenicola* infection. *Nephrol Dial Transplant*. 1998;13:2662–2664.
14. Mangiaterra M et al. Keratomycosis caused by *Cylindrocarpon lichenicola*. *Med Mycol*. 2001;39:143–145.
15. Rodríguez-Villalobos H et al. Disseminated infection due to *Cylindrocarpon* (*Fusarium*) *lichenicola* in a neutropenic patient with acute leukaemia: Report of a case and review of the literature. *Eur J Clin Microbiol Infect Dis*. 2003;22(1):62–65.
16. Chazan B et al. Mycetoma of the foot caused by *Cylindrocarpon lichenicola* in an immunocompetent traveler. *J Travel Med*. 2004;11(5):331–332.
17. Kaliamurthy J et al. Keratitis due to *Cylindrocarpon lichenicola*. *J Postgrad Med*. 2006;52(2):155–157.
18. Zoutman DE, Sigler L. Mycetoma of the foot caused by *Cylindrocarpon destructans*. *J Clin Microbiol*. 1991;29: 1855–1859.
19. Hemashettar BM et al. White grain mycetoma caused by a *Cylindrocarpon* sp. in India. *J Clin Microbiol*. 2000;38(11): 4288–4291.
20. Mitra A et al. The use of voriconazole in the treatment of *cylindrocarpon* keratomycosis. *Cornea* 2009;28(2):217–218.
21. Nelson PJ, Dignani, MC, Anaissie, EJ. Taxonomy, biology and clinical aspects of *Fusarium* species. *Clin Microbiol Rev*. 19941;7:479–504.
22. O'Donnell K, Cigelnik, E. Two divergent intragenomic rDNA ITS2 types within a monophyletic lineage of the fungus *Fusarium* are nonorthologous. *Mol Phylogenet Evol*. 1997;7:103–116.
23. de Hoog GS, Gerrits van den Ende, AHG. Molecular diagnostics of clinical strains of filamentous basidiomycetes. *Mycoses* 1998;41:183–189.
24. Hennequin C et al. Identification of *Fusarium* species involved in human infections by 28S rRNA gene sequencing. *J Clin Microbiol*. 1999;37:3586–3589.
25. Hue F-X et al. Specific identification of *Fusarium* species in blood and tissue by a PCR technique. *J Clin Microbiol*. 1999;37:2434–2438.
26. Mantiri FR et al. Phylogenetic relationships in *Neonectria* species having *Cylindrocarpon* anamorphs inferred from mitochondrial ribosomal DNA sequences. *Can J Bot*. 2001; 79:334–340.
27. White TJ et al., 1990. Amplification and direct sequencing of fungal ribosomal RNA genes for phylogenetics. In MA Innis, DH Gelfand, JJ Sninsky, and TJ White (eds.), *PCR Protocols: A Guide to Methods and Applications*. Academic Press, Inc., San Diego, CA, pp. 315–322.
28. Henry T, Iwen PC, Hinrichs SH. Identification of *Aspergillus* species using internal transcribed spacer regions 1 and 2. *J Clin Microbiol*. 2000;38:1510–1515.

51 *Fusarium*

Palanisamy Manikandan, László Galgóczy, Kanesan Panneer Selvam, Coimbatore Subramanian Shobana, Sándor Kocsubé, Csaba Vágvölgyi, Venkatapathy Narendran, and László Kredics

CONTENTS

51.1 INTRODUCTION

Fusarium species are widely distributed filamentous fungi that commonly occur in soil, water environments, and plants. The genus includes both plant pathogenic and soil saprophytic representatives.[1] During the past decades, *Fusarium* species have emerged as increasingly important causal agents for opportunistic infections in human population.[2]

51.1.1 Classification

Fusarium, first described by Link in 1809,[3] is a genus classified under the order Hypocreales, class hyphomycetes (Ascomycetes). Formerly included in the Deuteromycetes, the genus contains about 100 recognized species,[4] many of which can induce diseases in several agriculturally important crops and cause diseases in humans and domestic animals.[5–7] *Fusarium* spp. are ubiquitous filamentous fungi in all major geographic regions of the world and are routinely isolated from environmental sources such as soil, plant roots, plant debris, and water systems.[2,8] Most of the species of the genus have both an anamorphic (asexual) and teleomorphic (sexual) life cycle and exist in soils as saprophytes or "chlamydospores."[4]

As a plant pathogen, *Fusarium* has been reported as the causal agent of major devastating diseases in most economically important plants, e.g., banana,[9,10] wheat, and barley.[11] In humans, *Fusarium* species cause four patterns of invasive infections predominantly in immunocompromised patients[2,12–14]: refractory fever of unknown origin, sinopulmonary infection, disseminated infection, and a variety of focal single-organ infections[15] such as ocular infections, usually keratitis or endophthalmitis.[16–18] The mortality rate is greater than 70% in systemic infections.[9] In addition, an array of secondary metabolites is produced by *Fusarium* spp., which are associated with cancer and other growth defects in humans and animals. Interestingly, some of these secondary metabolites are used commercially either directly or as the starting material for chemical synthesis of plant and animal growth promoters in both first-world and second-world settings, and the mycotoxins produced by some of these fungi were reported to be used as biological weapons.[19–23] Few species such as *F. pallidoroseum* can be directly used as a mycoherbicide against water hyacinth.[24]

The *Fusarium* classification system is developed mainly based on the morphology of the conidia produced by the representatives of the genus. Macroscopic and microscopic features, such as color of the colony; length and shape of the macroconidia; the number, shape, and arrangement of microconidia; and the presence or absence of chlamydospores are the key features for the differentiation of *Fusarium* spp., and the main taxonomic systems have organized sections from the species sharing common morphological characteristics (Figures 51.1 through 51.3). However, speciation

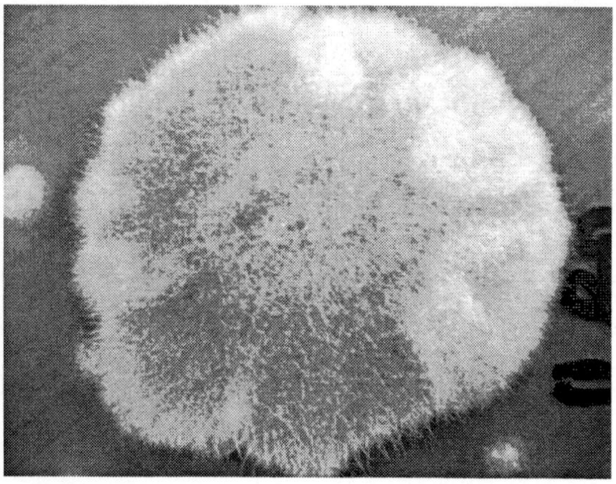

FIGURE 51.1 Colony morphology of *Fusarium* (white to violet color) from patients with *Fusarium* keratitis on PDA.

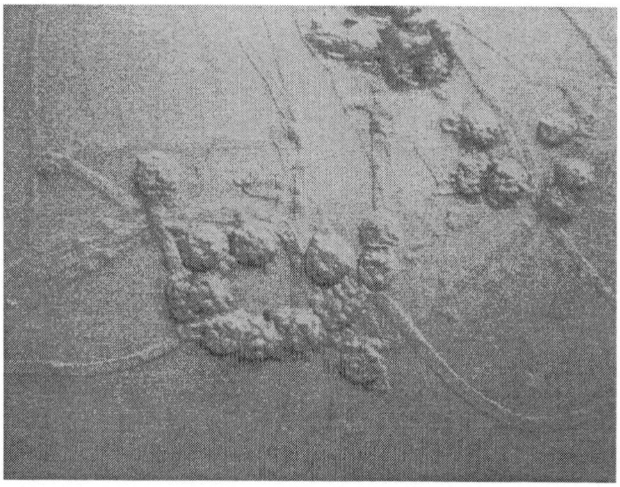

FIGURE 51.2 Differential interference contrast microphotograph showing chlamydospores produced by *Fusarium solani* (magnification 40×).

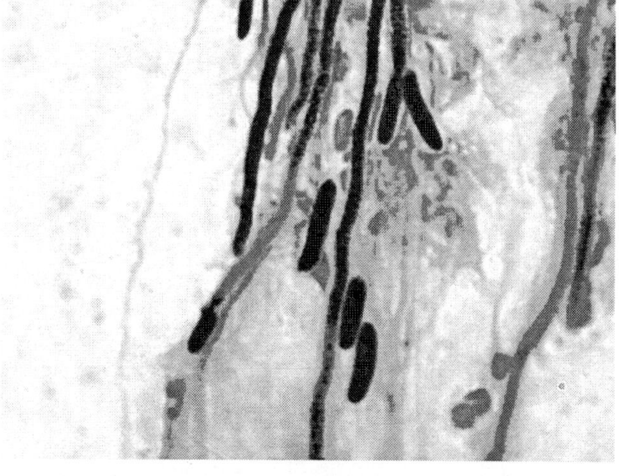

FIGURE 51.3 Gram smear showing macroconidia from corneal scraping of patients with *Fusarium* keratitis.

may be difficult due to the variability between isolates and because the features that are required are not always well developed. This would best describe why although more than 100 species have been recognized so far based on the various identification concepts, only 70 species are well described.[25]

The morphologically similar *Fusarium* species are grouped together as species complexes (SCs). For instance, the *Fusarium solani* species complex (FSSC) includes morphologically identical isolates. Further genotypic characterization of isolates from this complex is laborious and usually not routinely performed in clinical laboratories. Members of this SC are usually reported in the literature as *Fusarium solani*.[26] During the past few decades, there are controversies amongst researchers that the taxonomic system based on the sections sharing common morphological characteristics is of poor reliability compared to advanced molecular investigation methods such as random amplified polymorphic DNA (RAPD), amplified fragment length polymorphism (AFLP), restriction fragment length polymorphism (RFLP), and DNA sequencing.[9] For instance, O'Donnell et al. reported that the *Gibberella fujikuroi* species complex (GFSC) revealed 45 species using molecular methods for several gene targets, of which 23 were never reported before.[27] From the clinical point of view, identification at the species level is important for epidemiological studies and may also be very important because some new antifungal agents exhibit variable activities against *Fusarium* isolates belonging to different species.[28,29]

Together, in spite of the key work done by Wollenweber and Reinking[30] for developing a main taxonomic system, and the contributions during the last few decades by a significant number of scientists in *Fusarium* taxonomy, the systematic classification of *Fusarium* species has still a number of open question marks that need to be solved.[9]

51.1.2 Epidemiology

A variety of human and animal infections due to *Fusarium* spp. are widely reported from all over the world as the genus does not require any strict environmental conditions for survival. For instance, *Fusarium* can be recovered from diverse sources including soil, decomposing organic matter, plant roots, woody plants, trees, agricultural, and non-agricultural substrates and are found in various geographic regions with hot, temperate, or cool climates. *Fusarium* infections are being increasingly recognized as life-threatening mycotic infections worldwide.[31–33] The first documented case of disseminated invasive *Fusarium* infection was reported in 1973 in a child with acute leukaemia.[34] Infections caused by *Fusarium* species in different organs are collectively known as fusariosis. The epidemiological distribution of fusariosis is controversial,[35] as 85% of the cases until the year 2000 were reported from Western and Mediterranean countries such as United States, France, Italy, and Brazil,[35] and a large number of institutions in the world have only occasionally or have never documented invasive *Fusarium* infections.[35] The distribution of these cases only among such geographical areas cannot yet be ruled out. Among the well-described

70 *Fusarium* spp., the taxa implicated in almost half (50%) of the disease cases include *F. solani*, followed by *F. oxysporum* (14%), *F. verticillioides* (11%), *F. moniliforme* (10%), and *F. proliferatum* (5%). Further common human pathogens are *F. dimerum*, *F. chlamydosporum*, *F. nygamai*, *F. napiforme*, *F. semitectum*, and *F. equiseti*.[35]

It is likely that *Fusarium* spp. colonize patients prior to hospital admission and the subsequent immunosuppression and neutropenia could then result in a variety of infections.[31,36] Although skin, blood, lung, and sinus infections are most common, other body organ systems can also be affected. *Fusarium* species have been documented as etiological agents in localized tissue infections, including keratitis, endophthalmitis, septic arthritis, cystitis, peritonitis, brain abscesses, and breast abscess.[37–39] *F. solani* breast abscess was reported from India in an uncontrolled diabetic patient who was a paddy grower.[38] Disseminated fusariosis with heavy involvement of liver and lungs was reported in a healthy farmer.[40] Although infrequent, post-transplant invasive fusariosis is reported in Asian countries like India.[41] In systemic fusariosis, *Fusarium solani* is the predominantly isolated taxon, followed by *F. moniliforme*, *F. oxysporum*, and *F. proliferatum*.[31]

Fusarium species are isolated and reported in large numbers from eye infections in developing countries such as India.[42] Based on the reported case numbers, keratomycosis is the most frequent human infection caused by *Fusarium*. Most of the studies about fungal keratitis report the occurrence of *Fusarium* at the genus level only; however, identification of the *Fusarium* isolates at the species level would be of great importance as it is gaining consequences on clinical outcome.[43]

The literature provides a large amount of data about the epidemiology of *Fusarium* keratitis. The incidence and epidemiological pattern of *Fusarium* spp. among culture-proven cases of keratomycosis is different from country to country (Tables 51.1 and 51.2).[2,5,14,44–107] According to a series of retrospective studies, *Fusarium* spp. can be the predominant causal agents of keratomycosis[5,14,45,46,53–55,58–62,65–67,70,73,76,79,80,82,86,87,91,92,97,100] besides *Aspergillus* spp.,[47,51,56,71,83,84,88,94–96,98,99,102–104,106] *Candida* spp.,[44,68,78,83] and *Acremonium* spp.[74] Tropical and subtropical countries are the most affected, indicating that climate plays an important role in determining the predominance of certain species in fungal keratitis. Certain regions of different continents, e.g., southern Florida, Ghana, and southern India, have similar climates that are favoring the predominance of *Fusarium* spp.[55,79,86,100] The incidence of *Fusarium* spp. in keratomycosis may also vary with climatic conditions within a single country, e.g., between different parts of China,[59,60,62,63] Ghana,[55] and India (Table 51.2).[55,86,87,88,94,98–100,103,104] Keratitis caused by *Fusarium* is less frequent in regions with temperate climates like European countries: only four cases have been reported from Paris, France, in 8 years,[68] and only a single case has been diagnosed in Hungary.[108] The most frequent taxon of *Fusarium* reported from corneal infections is the FSSC.[14,58,63,64,79,86,87,109]

Fusarium keratitis may occur as a mixed infection with bacteria, mainly *Streptococcus* and *Staphylococcus* spp.[73,74,86,91,109] or herpes simplex virus.[109]

In the retrospective studies available, the proportion of farmers and agricultural workers among the affected patients was relatively high (16%–86%). Male patients were generally more frequent than females: male:female ratios were between 1.4:1 and 3.5:1, with the exception of a study from Nepal, where both sexes were equally affected.[56] The average age of the patients in the studies ranged from 35.8 to 59 years.

Among the factors predisposing for keratomycosis, corneal trauma is considered the most common, with an incidence between 31.6% and 89.9%, apart from a study from Pennsylvania, in which only 8.3% of the patients reported a recent trauma.[78] The reported injuries were caused by different traumatizing agents including plant material (corn stalks, grass, ground nuts, hay, kernel, onions, paddy, palm leaf, sugar cane, thorn, tree branch, and vegetable matter), animal matter (cat scratch, cow dung, cow's tail, hair, hen peck, and insects), chemical gas, dust, fingernails, electric welding light, glass, heat injury, metal objects, mud, soil, stones, or physical violence. In Tanzania, positive correlation was found with HIV carriage: 81.2% of patients with fungal keratitis were HIV positive.[14] Further predisposing factors include preexisting ocular diseases (e.g., atopic conjunctivitis, chronic dacryocystitis, dry eyes, lagophthalmos, recurrent corneal erosion, corneal scaring, or ulcer), the use of topical corticosteroids, previous eye surgery, systemic diseases (e.g., diabetes mellitus, leprosy, or rheumatoid arthritis), as well as contact lens wear. During the past decade, the epidemiology of contact lens solution-related *Fusarium* keratitis has been studied in detail.

Risk factors associated with microbial keratitis in contact lens wearers include continuous overnight wear, lower socioeconomic class, smoking, hypoxia, poor lens hygiene practice, blepharitis, diabetes mellitus, epithelial trauma, and steroid use, specifically in daily wear lenses.[110] Contact lens wearers have an elevated risk for fungal keratitis, but the incidence of *Fusarium* keratitis is quite rare among this group of patients. There has been no report until the beginning of 2006 on the association of multipurpose contact lens solutions with fungal keratitis. Before June 2005, the number of cases with contact lens-related fungal keratitis was about three per year in Hong Kong. In late August 2005, the Centre for Health Protection (CHP), Department of Health of Hong Kong, initiated an investigation because of the sudden increase in the number of cases of contact lens-related microbial keratitis. The initial investigation suggested that many fungal keratitis cases were caused by *Fusarium* spp. Great majority of the patients were disposable contact lens users and had reported of using a commercial multipurpose contact lens solution, namely Bausch & Lomb ReNu. Up until May 31, 2006, a total of 33 cases of contact lens-related *Fusarium* keratitis were reported to the CHP. Sixty-four percent (21/33) were female, and the age range of all cases was 16–51 years (mean 28 years). In a retrospective unmatched case–control study, Ma et al.[111] focused on the risk factors playing a role in the development of *Fusarium* keratitis among disposable soft contact lens users. They successfully interviewed 32 patients, and they choose 24 from

TABLE 51.1
Incidence of *Fusarium* spp. among Culture-Proven Cases of Fungal Keratitis (Based on Literature Search)

Country Studied	Period of Study	Number of Fungal Ulcers	Number of *Fusarium* Isolated	Reference
Australia	July 1996–May 2004	35	5 (14.3%)	[44]
Australia	1998–2008	16	8 (50.0%)	[45]
Australia	October 1999–September 2004	13	7 (53.8%)	[46]
Bangladesh	11 months	51	10 (19.6%)	[47]
Bangladesh	1987	7	1 (14.0%)	[48]
Bangladesh	1991	107	NA (28%)	[49]
Bangladesh	Unknown	63	22 (34.9%)	[50]
Brazil	1975–2003	265	137 (58.8%)	[51]
Brazil	1983–1997	49	12 (32.0%)	[52]
Brazil	January 1994–December 1999	20	12 (60.0%)	[53]
Brazil	2000–2004	66	44 (66.7%)	[54]
Ghana, Accra (south)	June 1999–May 2001	43	27 (63.0%)	[55]
Nepal	1985–1987	68	8 (11.8%)	[56]
Nepal	August 1998–July 2001	145	45 (22.0%)	[57]
Thailand (central)	January 1988–December 2000	35	12 (34.3%)	[5]
Nigeria	1974–1977	42	14 (33.3%)	[58]
Tanzania	October 1994–October 1995	32	24 (75.0%)	[14]
China, Beijing (north)	January1995–October 2000	498	321 (64.5%)	[59]
China, Zhengzhou (central)	January 1975–June 1997	615	NA (65.0%)	[60]
China, Zhengzhou (central)	January 2000–March 2009	1458	076 (73.8%)	[61]
China, Quingdao (north)	January 1996–December 1999	97	63 (64.9%)	[62]
China, North	January 1999–December 2004	596	437 (73.3%)	[63]
China, North	January 2001–December 2006	549	426 (77.6%)	[64]
China	1989–2000	775	455 (58.7%)	[65]
China	January 2001–December 2004	681	394 (57.9%)	[66]
China, Zhejiang	September 2002–July 2004	61	33 (54.1%)	[67]
France, Paris	January 1993–January 2001	19	4 (21.1%)	[68]
Iran	May 2004–March 2005	7	1 (14.3%)	[69]
Iran	1998–1999	29	10 (34.5%)	[70]
Iran	1982–2001	27	2 (7.4%)	[71]
Malaysia	January 2004–April 2005	4	2 (50.0%)	[72]
Paraguay	April 1988–April 1989	26	11 (42.3%)	[73]
Paraguay	1988–2001	136	41 (15.0%)	[74]
Paraguay (children)	1988–2002	35	5 (14.3%)	[75]
Singapore	January 1991–December 1995	29	15 (52.0%)	[76]
Thailand	January 2001–December 2004	49	13 (26.5%)	[77]
Pennsylvania, USA	January 1991–March 1999	24	6 (25.0%)	[78]
South Florida, USA	January 1982–January 1992	127	79 (62.2%)	[79]
South Florida, USA	January 1969–December 1977	133	82 (61.6%)	[80]
Florida, USA	January 2004–December 2005	122	66 (54.1%)	[81]
Florida, USA	January 1999–June 2006	59	24 (40.7%)	[82]
Minneapolis, USA	January 1971–January 1981	19	3 (15.7%)	[83]
Massachusetts, USA	January 2004–November 2007	46	19 (41.3%)	[84]

them based on their case criteria. The mean age was 30.3% and 29.2% (7/24) were male. Twenty-three of 24 used Bausch & Lomb contact lens solution and 21 could specify it as ReNu MoistureLoc. Twenty-one of the 24 stored their solutions in the bathroom or kitchen. Based on these data it can be stated that ReNu MoistureLoc was strongly connected to the cases of *Fusarium* keratitis in Hong Kong. However, poor lens hygiene practice was also characteristic among the patients. In late February 2006, the Bausch & Lomb Company voluntarily stopped sales of ReNu solution from the markets in Hong Kong. In Singapore, Wong et al.[76] conducted a retrospective case series study on fungal keratitis in the period 1991–1995 at the Singapore National Eye Centre. They identified 29 cases of fungal keratitis. Among these cases,

TABLE 51.2
Incidence of *Fusarium* spp. in India among Culture-Proven Cases of Fungal Keratitis (Based on Literature Search)

State Studied	Period of Study	Number of Fungal Ulcers	Number of *Fusarium* Isolated	Reference
Tamilnadu				
Coimbatore (children)	February 1997–January 2004	37	17 (16.8%)	[85]
Madurai (south)	January 1994–March 1994	155	73 (47.1%)	[86]
Tiruchirapalli (south)	June 1999–May 2001	353	141 (39.9%)	[55]
Tiruchirapalli (south)	July 1985–November 1985	40	19 (47.5%)	[87]
Madras (south)	1980–1982	68	8 (11.8%)	[88]
Vellore	NA	7	3 (42.8%)	[89]
Chidambaram	July 2002–June 2005	230	74 (32.0%)	[90]
Tirunelveli	September 1999–March 2001	554	254 (45.8%)	[91]
Tirunelveli	September 1999–August 2002	1100	471 (42.8%)	[92]
Tirunelveli	September 1999–August 2002	1138	511 (41.6%)	[93]
Rest of India				
New Delhi (north)	January 1999–December 2001	215	23 (10.7%)	[94]
New Delhi	January 2000–December 2004	77	6 (7.8%)	[95]
New Delhi	January 1999–June 2001	191	24 (12.5%)	[96]
New Delhi	NA	31	10 (32.3%)	[97]
Chandigarh (north)	6 years	61	10 (16.4%)	[98]
Chandigarh (north)	January 1999–December 2003	34	8 (23.5%)	[99]
Hyderabad (south)	January 1991–December 2000	1360	506 (37.2%)	[100]
Hyderabad	January 1991–December 1996	557	210 (37.6%)	[101]
Hyderabad	February 1991–June 1995	21	3 (4.1%)	[102]
Patna (east)	2 years	76	6 (7.8%)	[103]
Mumbai (west)	1988–1996	387	33 (8.5%)	[104]
Goa	February 1993–January 1994	16	2 (12.5%)	[105]
Kolkata	January 2001–December 2003	623	132 (21.2%)	[106]
Aurangabad	NA	12	4 (33.3%)	[107]

15 were due to *Fusarium* species (Table 51.1). Twelve further cases were observed in this institute from 2001 to 2004 (unpublished data). Since March 1, 2005, there was a significant increase of *Fusarium*-related eye infections among contact lens users. By May 2006, there were 66 reported cases nationwide, and 12 of these cases occurred between March and May, 2006. In a case study, Khor et al.[17] interviewed these patients. The patients were equally distributed by sex and the mean age was 27.1 years. Except from one, the patients wore disposable contact lenses. A total of 62 patients reported using multipurpose solutions from the same brand (ReNu, Bausch & Lomb, Rochester, NY). Forty-two patients (63.6%) reported using ReNu with MoistureLoc, 6 (9.1%) reported using ReNu MultiPlus, and 11 (16.7%) reported using an unspecified ReNu multipurpose cleaning solution. Based on the interview with the patients, several risk factors have been revealed. Isolates derived from these cases showed 100% identity with *Fusarium solani* CBS490.63 based on the 28S rRNA gene sequence. The main risk factors were the extended use of contact lenses after the planned replacement date (43.9%), the overnight use of daily wear contact lenses (19.7%), and swimming with contact lenses (30.3%) with or without the use of goggles. A case study—based on 61 cases and 367 controls—carried out by Saw et al.[112] showed that the risk of *Fusarium* keratitis was much higher in cases compared with control group for ReNu with MoistureLoc than for ReNu MultiPlus. Soft monthly disposable contact lenses also increased the risk of *Fusarium* keratitis. They revealed that the use of contact lenses past the replacement date increased the risk of *Fusarium* keratitis. Other factors such as washing of hands before replacing the lens, leaving the cap of the solution bottle or the lens case open, or outdoor activity were not significantly associated with *Fusarium* keratitis. On February 17, a news release warned contact lens wearers about the increasing incidence of fungal keratitis. On that same day, Bausch & Lomb voluntarily stopped all sales of ReNu products in Singapore. As a result of actions taken by the Ministry of Health and Bausch & Lomb, the number of cases appeared to be decreasing after March.

The *Fusarium* isolates responsible for the cases of contact lens-related microbial keratitis[17,112,113] showed distinct genotypes. This observation suggested that it is unlikely that common or clonal strains were the cause of the infections. The analysis of the *F. solani* and *F. oxysporum* isolates showed that these strains are capable of forming biofilms on soft contact lenses under permissive conditions, but not in the presence of, or after recommended 4h treatment with, MoistureLoc solution. These studies also concluded that there are no significant differences between the effectiveness of a freshly opened or an aged MoistureLoc solution if the conditions of usage are adequate.

Investigation carried out by both the U.S. Food and Drug Administration (FDA) and Bausch & Lomb found no point source of contamination in the factory. No contamination was found in any bottles, and all the tested products were found stable and effective. However, an analysis by Bausch & Lomb showed that the solution can induce a breach in the corneal epithelium that can lead the entry of *Fusarium* into the cornea. Bausch & Lomb concluded that ReNu with MoistureLoc's formulation could create biofilms that shield the fungus from the sterilizing agent.[114,115] In the review of Epstein,[116] the five principal elements of the *Fusarium* outbreak were specified as the decrease of the antimicrobial effect due to biocide uptake by soft contact lenses, selection of *Fusarium* due to persistence, the supported growth of microbes by MoistureLoc biofilm-forming capabilities and the nutritive properties of cellulose, chemical trauma (corneal staining) upon lens insertion due to biocide release, and the blocking of host inflammatory response by lens. In 2006, the FDA initiated an inspection against Bausch & Lomb's manufacturing site in Greenville, South Carolina. They found that the company failed to regulate the storage and transport temperatures in and beyond the plant. Bullock et al.[117] demonstrated the in vitro loss of antimicrobial activity of ReNu MoistureLoc against clinical isolates of *Fusarium* when the product was exposed for a prolonged time (4 week) to a temperature of 60°C. With this study, they simulated the possible conditions to which some of the manufacturer's bottles may have been exposed during storage and transport, or even, perhaps, after purchase. They also concluded that ReNu MoistureLoc was effective against *Fusarium* isolates after a 10min boiling. This suggests that the disinfectant alexidine inactivation is both time and temperature dependent.

Fusarium was also reported to be the cause of mycotic otitis of the external ear in Gabon, Central Africa,[118] and was also found to germinate in the middle ear of agricultural workers.[119] Four of five cases of *Fusarium* osteomyelitis were reported in healthy individuals following surgery or trauma.[120] Skin lesions are present in 80% of the cases of disseminated infection and they may be primary or metastatic. They are important for early diagnosis because of accessibility for biopsy and culture. Typical skin lesions are painful, erythematous, subcutaneous nodules and plaques, which later undergo central necrosis.[121]

Fusarium species can cause localized infections of the nails and skin.[122] *Fusarium* can also cause eumycetoma[123] and infection in burn wounds, colonization in burn wounds,[124] granulomas, ulcers, and panniculitis.[2] *Fusarium oxysporum* mainly causes infection in fingernail and toenail[125–128]; however, *F. solani* is more frequently isolated from toenails only.[35,128,129] More than 50% of the fusarial infections is involving the legs.[15] Outbreaks of nosocomial fusariosis have also been reported. *Fusarium* spp., in hospital water distribution system, may result in disseminated fusariosis in immunosuppressed patients.[130]

The possible association of the fusariosis of hospitalized patients with the colonization of a hospital water system by *Fusarium* was reported by Anaissie et al.[131] The authors applied molecular typing methods, including RAPD, RFLP, and interrepeat (IR) PCR, and demonstrated that two patients were infected by *F. solani* strains with genotypes identical to that of certain environmental isolates, while the isolates deriving from six patients matched the isolate of another patient. In a similar study, there was no identity of RAPD patterns between clinical isolates and strains isolated from water samples from the hospital environment, suggesting that the most likely source of fusariosis was the external environment, rather than the hospital water system.[132] In the study by Anaissie et al.,[131] the two clinical isolates matching environmental strains were collected several months before the environmental isolates; therefore it cannot be excluded that the patients may have contaminated the water systems.[132]

Fusarium spp. may also exist in the soil of potted plants in hospitals. These plants constitute a hazardous mycotic reservoir for nosocomial fusariosis.[133]

The reported major risk factors for invasive fusariosis include acute leukemia, bone marrow transplantation, immunosuppressed state, particularly neutropenia, and the use of corticosteroids.[134,135] The modes of entry are the respiratory tract, the skin, and injuries of the corneal epithelium in the case of mycotic keratitis. Trauma remains the major predisposing factor for the development of cutaneous and corneal infections caused by *Fusarium* in immunocompetent hosts.[37] Although many studies reported that skin (vascular catheters, periungual regions, or burns) is the main portal of entry in hematologic cancer with neutropenic patients, other portals of entries such as inhalation into the lungs or upper airways have also been documented.[136] Intake of contaminated grain foods involves gastrointestinal route of infection.[137] In particular, fusariosis in patients with cancer is predominantly a community-acquired infection, usually transmitted via the airborne route from the outdoor environment.[131]

Fusarium spp. are the only opportunistic molds that can be easily recovered from the blood stream.[128] Histopathologically, *Fusarium* infection may mimic any other mycoses exhibiting moniliaceous, septate branching, or non-branching hyphae. Because of these morphological similarities, identification of the fungus obtained from cultures is required to establish fusarial aetiology.[138]

Seasonal variation has also been observed in fusariosis. Among patients living in rural areas, most infections occurred between June and September. Fifty percent of the eye infections and pulmonary infections peaked during August[15,31] and 62% of the fusariosis occurred during summer months, June through August.[139] Fusariosis occurs more commonly in males in all age group ranging from 2 to 78.[31] *Fusarium* can also be isolated from the conjunctival sac and from pharynx as a normal flora.[140]

51.1.3 Clinical Features and Pathogenesis

The clinical features and manifestations of severe invasive fusariosis in immunosuppressed patients mimic those seen in patients with aspergillosis and are always nonspecific.[141] This can lead to inappropriate treatment regimen by the clinicians.[142] Also, nosocomial fusariosis has a longer latency period than community-acquired infection[131]; therefore the progression of disease may be unnoticed. However, the most common findings at the initial presentation are persistent fever, sinusitis, or skin lesions with a black necrotic center.[31]

The neutropenic patients, especially those with acute leukemia, allogeneic bone marrow transplant recipients, as well as the patients with extensive burns or with chronic renal failure[15] are at increased risk of disseminated fusariosis during hospitalization.[122,143–146] In particular, patients with hematological malignancies account for approximately 90% of the reported cases of disseminated fusariosis.[147] Infection typically occurs during a prolonged period of neutropenia, lasting up to 65 days.[31,128,144,148] In neutropenic cancer patients, disseminated infections by *Fusarium* spp. usually present as persistent fever, fungemia, myalgia, and unresponsiveness to a wide spectrum of antibiotics. In immunocompromised patients, *Fusarium* spp. may also cause eumycetoma, cause onychomycosis, or harmlessly colonize the ulcers.[39,138] Upon recovery from myleosuppression, the infection may either resolve completely or become chronic and localized to sinuses, lungs, eye, brain, joint, or muscle[149] with the potential for relapse and dissemination upon reinstitution of cytotoxic chemotherapy.[128] Other presentations of invasive fusariosis in compromised hosts include osteomyelitis, septic arthritis, myositis, foot abscesses, endocarditis, myocarditis, external otitis, peritonitis, brain abscesses, cystitis, meningoencephalitis, and chronic hepatic infection.[2,32,134,120,150–156]

As mentioned previously, infections caused by this fungus can mimic aspergillosis.[122] However, some distinct differences can be seen, such as an increased incidence of skin and subcutaneous lesions and a positive blood culture, mainly in the first days of fever, for patients with disseminated fusariosis.[136,143] In contrast to invasive aspergillosis, the blood cultures are positive in disseminated fusariosis in up to 50%–70% of the cases.[32,145,157] Skin lesions, the hallmarks of disseminated fusariosis, occur in 60%–90% of the cases, compared with a rare occurrence (10%) of such lesions in disseminated aspergillosis.[2,158,159] Patients with neutropenia have a higher rate of disseminated skin lesions compared with nonneutropenic immunocompromised patients.[37] Subcutaneous nodules, palpable and nonpalpable purpura, red or gray macules, red or gray papules, macules or papules with progressive central necrosis with central, flaccid pustules, vesicles, and hemorrhagic bullae are types of lesions seen in patients with disseminated fusariosis.[2,148,159] These skin lesions can involve any site, with predominance on the extremities.[37,159] The lesions, especially the subcutaneous nodules, are often tender[159] and most patients have lesions at various stages of evolution.[31,159] The number of disseminated skin lesions is variable and ranges between 4 and more than 30.[159] Skin lesions are important potential sources of diagnostic tissue in some patients with fusariosis.[37]

When four different taxa of *Fusarium* were tested in murine model, *F. solani* was found to be the most virulent species.[39,160] The production of secondary metabolites, mycotoxins are mainly associated with the pathogenesis.[2] Common mycotoxins produced by *Fusarium* spp. include moniliformin, zearalenone, trichothecenes (deoxynivalenol, nivalenol, and diacetoxyscirpenol), fumonisins, T-2 toxin, and fusaric acid.[161] Among these, trichothecenes, zearelenones, and fumonisins are the most notorious.[162] Ingestion of grains contaminated with these toxins may give rise to allergic symptoms or be carcinogenic in long-term consumption. They are known to cause myelosuppression through toxin production.[126] Furthermore, exposure to fumonisin may lead to human birth defects.[163] The best example is alimentary toxic aleukia, which has been associated with the ingestion of overwintered cereal grains colonized by the toxigenic *F. sporotrichioides* and *F. poae*.[164] The toxin produced by these organisms, called T-2 toxin, is considered as the principal component responsible for the acquisition of alimentary toxic aleukia. The first recognized trichothecene mycotoxicosis was alimentary toxic aleukia in the USSR in 1932 and the mortality rate was 60%.[165] Common manifestations of trichothecene toxicity are depression of immune responses and nausea. In vitro, they impair cellular immunity and decrease the humoral response to T-dependent antigens.[166] Experimental injection of T-2 toxin resulted in cardiomyopathy.[167] Zearalenone is mainly produced by *F. graminearum*, and it produces estrogenic effects such as infertility, vulval edema, vaginal prolapse, and mammary hypertrophy in females and feminization of males—atrophy of testes and enlargement of mammary glands. *F. moniliforme* and *F. proliferatum* produce fumonisin toxin in maize which may cause esophageal cancer.[168] In India, a single outbreak of acute foodborne disease possibly caused by fumonisin B_1 has been reported. The main features of the disease were transient abdominal pain, borborygmus, and diarrhea.[169] Studies on reduction or elimination of *Fusarium* spp. and mycotoxins from contaminated agricultural and food commodities are in progress.[170–172] As a saprophytic nature of the genus, it can easily invade in the neutropenic patient and develop invasive fusariosis. In addition, prolonged usage of corticosteroids and antibiotics in patients with organ transplantation, silastic catheters, and infections of central venous catheters (CVCs), continuous ambulatory peritoneal dialysis (CAPD) catheters, and contact lenses have been reported.[2]

Keratomycosis, one of the most frequent human infections caused by *Fusarium*, is a suppurative, usually ulcerative corneal disease, most frequently occurring as a localized infection.[173] Infection is exogenous in most of the cases, the pathogen is entering through the epithelium of the cornea. Patients usually present with photophobia and discharge from the eyes. A persistent infiltrate is often present at the site of superficial injury, which gradually increases in size and density. The cornea becomes slightly thickened, and satellite lesions may develop peripheral to the focal area of infiltrations. The signs of inflammation are minimal in comparison with bacterial keratitis.

Fusarium species are often resistant to most of the antifungal agents, and *F. solani* is the most resistant species within the genus.[29,174,175] In vitro resistance has been documented against the antifungal agents miconazole, ketoconazole, 5-fluorocytosine, fluconazole, itraconazole, and nikkomycin Z.[28,122,128,176–181] However, amphotericin B (AMB) has been shown to be the mainstay in the treatment of fusariosis.[182] In view of the inherent resistance of *Fusarium* spp. to most antifungal agents, AMB-based combination regimens have also been suggested or used for fusariosis.[128] Natamycin is also active against *Fusarium* spp. both in vitro and in vivo,[25,26,123,124] and in combination with AMB it has been the mainstay of treatment for *Fusarium* keratitis.[123] However, its toxicity precludes its systemic use in clinical practice although topical application in powder form could halt the progression of fusariosis in pediatric patients with severe burns.[125] Ketoconazole, miconazole, fluconazole, flucytosine, and itraconazole have no in vitro activity against *Fusarium* spp.[29,174] However, the newer broad-spectrum triazoles, voriconazole (VRC), posaconazole, and ravuconazole have variable in vitro activities against *Fusarium* spp. and show promise for the management of fusariosis.[29,126–129,174,175,178,183–187] Posaconazole, a broad-spectrum triazole, is active against *Fusarium* species both in vitro and in animal models.[29,174] Overall, due to the inherent resistance of *Fusarium* spp. against most of the current group of antifungal agents and the profound solid state of immunosuppression in patients who typically develop fusariosis, the currently available therapeutic strategies for invasive fusariosis particularly in heavily immunocompromised patients have less solutions.[2,5–9,25]

51.1.4 Diagnosis

51.1.4.1 Conventional Techniques

Identification of *Fusarium* species is traditionally carried out by the investigation of macro- and micromorphology. The most commonly used macromorphological characters are the colony morphology on the upper and lower surface of the medium, the linear or radial growth rates, and the pigmentation.[2,188,189] These characters are useful for describing a species under standard environmental conditions (e.g., light, temperature, and substrates)[2,190]; however, they should be applied with caution as they may vary between the isolates of a certain species, especially among clinically important ones.[2]

The micromorphological characters of the three types of spores (macroconidia, microconidia, and chlamydospores) are more suitable criteria for the differentiation of *Fusarium* species.[2,189] The absence or presence of one type of them is a key characteristic. The shape of macroconidia (which are formed in sporodochia) is the main distinguishing character, as it is relatively consistent and stable on natural substrates and under standard environmental conditions, in contrast to the size of macroconidia, which is varying within an individual species.[2,189,191] Macroconidia vary in shape from short and squat to highly curved and elongate with needle-like spores.[189] Further key macroconidial characters are the number of septa, and the shape of the apical and basal cells: the apical cell of the macroconidium may be rounded, needle-like, or whip-like and the basal cell may be barely notched or resemble an upside-down foot.[189] The presence or absence of microconidia with various shapes and sizes (which are formed in aerial mycelium) is a primary distinguishing character in *Fusarium* taxonomy. They may be produced singly, in false heads only, or in false heads and chains.[2,189] Their shapes may be napiform, oval, pyriform, clavate, fusiform, or globose.[2] The microconidiophores can be monophialidic only or both mono- and polyphialidic.[2,189] Monophialides have a single opening in the conidiogenous cell, while polyphialides have two or more.[189] Chlamydospores are more common in the older cultures, their presence or absence is a distinguishing character. These thick-walled, verrucose, light-colored spores filled with lipid-like material can be formed singly, in pairs, in clumps, or in chains in the hyphae above or below the agar surface.[2,189] Their outer wall can be smooth or rough.[2] Pseudochlamydospores were described by Marasas et al.[164] in the case of *F. andiyazi*. They are thin and smooth walled and can be found singly or in short chains in the hyphae.[164]

A lot of secondary characters can be used in *Fusarium* identification, e.g., the presence or absence of sporodochia, sclerotia, and stroma[2]; toxins and other metabolites[192]; or odor.[189] Their usage is limited, as these data are not commonly available for routine diagnoses.[189]

Beyond morphology, the studies of sexual cross-fertility can be used for the identification of *Fusarium* species.[189,193] Due to the disadvantages of this technique (that it is slow and needs adequate incubator space), it is not recommended for the purposes of clinical diagnosis.

51.1.4.2 Molecular Techniques

Molecular techniques may help us to avoid the misidentification of *Fusarium* based on their macro- and microscopic observation. The main advantage of the identification by molecular techniques is that they do not require viable organism or sporulation; therefore, they enable a rapid diagnosis.[194]

Several methods are available for molecular identification of *Fusarium* species, and they have been continuously improving by new methods. Fingerprint techniques, like RFLP, AFLP, etc., are useful to identify an unknown strain, if a set of reference strains are available.[25,189] Their advantages are that DNA sequencing is not required and these methods are also appropriate to reveal the evolutionary distance between the investigated isolates.[25,189] DNA sequences of the internal transcribed spacer (ITS) regions of the ribosomal RNA gene cluster are not very useful in distinguishing *Fusarium* species.[195] The sequencing of partial fragments from other amplified genes and their comparison with similar sequences available in the GenBank database is a commonly followed strategy for *Fusarium* identification.[25,189,196] The most frequently used targets are the β-tubulin gene, the translation elongation factor 1α gene (*tef1*), and the histone gene.[25,189,196] For rapid group- or species-level identification, specific primer pairs for PCR are described in the literature.[197,198] Real-time PCR detection protocols were evaluated for fast detection and identification based on the differences between the tef-encoding genes[199] and the genes involved in mycotoxin biosynthesis; however, this approach is not yet available for all species.[200–204] A DNA microarray was also developed for the easy and fast detection and identification of the fungal genus *Fusarium* based on the recent phylogenetic analyses of *tef1*.[205] Investigation of the intergenic spacer (IGS) regions with molecular methods is useful to resolve the identified species into various subgroups.[206]

51.2 METHODS

51.2.1 Sample Preparation

In the case of filamentous fungi, conventional DNA extraction methods are well suited for the preparation of total genomic DNA.[207] Disadvantages of these techniques are that they are slow, and relatively high amount of sample is needed. In the studies reviewed in the previous sections, commercial DNA purification kits were used for rapid and safe sample preparation.

Media and culture conditions: The following three media are preferred for morphological identification of *Fusarium* species: potato dextrose agar medium (PDA, 50% potato filtrate obtained by boiling 300g diced potato in 500mL water and filtering, 20g sucrose) for the investigation of colony morphology and pigmentation; and for measuring of the growth rates[189]; carnation leaf-piece agar medium (CLA; sterile 3–5mm carnation leaf pieces in agar) for the investigation of macroconidia[208]; and "Spezieller Nährstoffarmer agar" (SNA, 1g KH_2PO_4, 1g KNO_3, 0.5g $MgSO_4 \times 7\,H_2O$, 0.5g KCl, 0.2g glucose, 0.2g sucrose) for the investigation of microconidia and chlamydospores[209,210] (all media are given for 1L and contain 20g agar). After incubation at 20°C–25°C for 7–10 days in the presence of light, particularly some exposure to UV (black) light (λ=310–360nm), *Fusarium* isolates grow well and form conidia.[189] Artificial daylight (12h light/12h dark) is often used, although total darkness or continuous light is necessary for the evaluation of critical diagnostic characters of some species.[209,210] In the case of PDA cultures, the presence of light is not essential, but it is increasing the production and total amount of pigment. On the other hand, light is important for CLA cultures as it increases the production of sporodochia.[189] Genomic DNA can be extracted from mycelia grown either on solid or in liquid malt extract medium after incubation at 20°C–25°C for 7 days.

Microscopy sample preparation: Microscopy samples can be made from colonies grown on PDA, CLA, and SNA after incubation at 20°C–25°C for 7–10 days in the presence of light.

Genomic DNA preparation: Genomic DNA can be extracted from mycelia grown either on solid or in liquid PDA medium after incubation at 20°C–25°C for 5–7 days. Beyond conventional DNA preparation methods,[25,194] commercial DNA purification kits can be used for rapid sample preparation.

51.2.2 Detection Procedures

51.2.2.1 Morphological Identification of *Fusarium* Species

The steps of identification based on the macro- and micromorphological characters were described by Summerell et al.[189] and Leslie and Summerell.[25]

Procedure

1. Plate the strain onto PDA, CLA, and SNA medium and incubate it at 20°C–25°C for 7–10 days in the presence of light.
2. Observe the colony morphology, the growth rates, the presence of sporodochia, and the color of the culture and of the produced pigments in PDA.
3. Examine under light microscope the shape and size of macroconidia and chlamydospore production on CLA.
4. Investigate under light microscope the shape, size, and formation of microconidia, their conidiogenous cells, and chlamydospore production on SNA.
5. Examine the growth rates based on linear growth in a race tube as described by Ryan et al.,[211] and/or in a Petri-dish as described by Burgess et al.[212]
6. Organize the data with the use of a recording sheet.[25,212]
7. Make the identification using a manual. The four most commonly used manuals for morphology identification were written by Gerlach and Nirenberg,[213] Nelson et al.,[191] Burgess et al.,[212] and Leslie and Summerell.[25]

The main macromorphological characteristic features are[25,191,212,213]

a. Abundance and color of aerial mycelium
b. Pigmentation
c. Growth rate

The main micromorphological characteristic features are[25,191,212,213]

d. Macroconidia: size, shape, number of septa, shape of apical and basal cell
e. Microconidia: size, shape, number of cells, formation, nature of conidiogenous cells, and conidiophores
f. Chlamydospores: presence or absence, formation

51.2.2.2 Molecular Identification of *Fusarium* Species

51.2.2.2.1 Genus-Specific Identification

51.2.2.2.1.1 PCR Protocol of Hue et al.[197] PCR methods for rapid and reliable identification of *Fusarium* species isolated from clinical samples were evaluated by Hue et al.[197] and Hennequin et al.[214] Hue et al.[197] described a method appropriate for the preparation of PCR templates from *Fusarium*-containing blood and performed PCR with primers designed for the ribosomal RNA genes (rDNAs) based on a large number of isolates belonging to the genus *Fusarium*.

Primers P28SL (5′-ACA AAT TAC AAC TCG GGC CCG AGA-3′) and P58SL (5′-AGT ATT CTG GCG GGC ATG CCT GT-3′) amplified a fragment of 329 bp containing ITS2 and a portion of 5.8S and 28S rDNA. To avoid false-negative results, a positive internal control was used from a part of λ phage DNA. It was amplified by two primers (C1, 5′-ACA AAT TAC AAC TCG GGC CCG AGA CCA CAG CGC-3′ and C2, 5′-AGT ATT CTG GCG GGC ATG CCT GTG TAC AAC TGG-3′), whose 3′ ends correspond to the λ DNA while the 5′ ends correspond to the primers used in the PCR amplification. The PCR reaction resulted in a 517 bp fragment which contains the *Fusarium* sequence at the ends, then it was amplified with the *Fusarium* primers.

Procedure[197]:

1. Lyze the cells with a commercial genomic blood DNA purification kit. Mix it with 200 μL of enzyme buffer (0.9 M sorbitol, 0.1 M Tris, 0.1 M EDTA) and 20 μL of lyticase and incubate the samples at 37°C for 90 min. Treat the sample with proteinase K (10 μL, 20 mg/mL) and incubate it at 55°C for 30 min. Add 5 μL of RNase and incubate the sample at 37°C for 30 min. Extract the DNA with a commercial genomic blood DNA purification kit.
2. Prepare the reaction mixture according to the manufacturer's instructions. Use 1 ng of template DNA per reaction mixture.
3. Perform the first PCR in a thermocycler using the following reaction conditions: 94°C for 5 min; 40 cycles of 94°C, 68°C, and 72°C for 1 min each; and a final 72°C for 10 min. Include the positive internal control.
4. Visualize the PCR products after electrophoresis on 2% agarose gel stained with ethidium bromide and run in 1 × TAE (Tris-acetate-EDTA) buffer.

51.2.2.2.1.2 PCR Protocol of Hennequin et al.[214] Hennequin et al.[214] designed a primer pair based on the 28S rDNA sequences of *Fusarium* species associated with human infections. In contrast to the DNA from unrelated genera, the primers Fus1 (5′-TGA AAT CTG GCT CTC GGG-3′) and Fus2 (5′-CAT GCG CGA ACC TCA GTC-3′) amplified a 480 bp fragment from the DNA extracts of *Fusarium* strains and of the members of related genera *Acremonium* and *Cylindrocarpon*.

Procedure[214]:

1. Prepare the reaction mixture in a final volume of 50 μL containing 5 μL genomic DNA template, 10 mM Tris–HCl, 50 mM KCl, 2.5 mM $MgCl_2$, 0.01% (w/v) gelatin, 5 μL of DNA, 0.25 μM of each primers, 0.2 mM of each deoxynucleoside triphosphates (dNTPs), and 0.3 U of Taq polymerase.
2. Perform the PCR in a thermocycler using the following reaction conditions: 94°C for 10 min; 35 cycles of 94°C for 30 s, 64°C for 90 s, and 72°C for 90 s; and a final 72°C for 10 min.
3. Visualize the PCR products after electrophoresis on 1% agarose gels stained with ethidium bromide and run in 1 × TAE buffer.

51.2.2.2.2 Species-Specific Identification

Fusarium species can be identified at the species level by sequencing of amplified gene fragments and comparing the sequences with similar ones available in sequence databases (FUSARIUM-ID, GenBank, EMBL, and DDBJ) using a Basic Local Alignment Search Tool (BLAST) search.

51.2.2.2.2.1 β-Tubulin Gene PCR Protocol 1[206] Yli-Mattila et al.[206] used a universal forward primer (T1: 5′-ATG CGT GAG ATT GTA AGT-3′)[215] and a specific reverse primer (tub-conrev T22: 5′-TGA CCG AAA ACG AAG TTG TC-3′) for amplifying a fragment with the first two introns of the β-tubulin-encoding gene.[206]

1. Prepare the reaction mixture containing 5–50 ng genomic DNA, 2.5 mM $MgCl_2$, 10 mM Tris–HCl (pH 8.8), 50 mM KCl, 0.1% (w/v) Triton X-100, 0.4 μM of each primer, 0.12 mM of each dNTP, 3 U of Taq polymerase.
2. Perform the PCR in a thermocycler using the following reaction conditions: 1 min at 94°C; 30 cycles of 94°C, 51°C, and 74°C for 1 min each; and a final 74°C for 7 min.
3. Visualize the PCR products after electrophoresis on 1% agarose gels stained with ethidium bromide and run in 1 × TAE buffer.
4. Purify the PCR products with a commercial PCR DNA purification kit.
5. After the sequencing of the PCR product, compare it with the β-tubulin sequences available in the GenBank database using a BLAST search.

Amplification of a partial β-tubulin gene fragment was carried out by Chung et al.[216] using a specific primer pair, FU-tubulin3 (5′-CGA GCC CGG TAC CAT GGA CG-3′) and FU-tubulin2 (5′-GGT CGC CGT AAG AGG GGT TGG-3′).

51.2.2.2.2.2 β-Tubulin Gene PCR Protocol 2[216]

1. Prepare the reaction mixture in a final volume of 20 μL containing 5–50 ng genomic DNA, 0.2 μM of each primer, 0.25 mM of each dNTP, 1 U of Ex Taq polymerase and Ex Taq reaction buffer (containing 2 mM $MgCl_2$).
2. Perform the PCR in a thermocycler using the following reaction conditions: 94°C for 3 min; 35 cycles of 94°C for 1 min, 60°C for 30 s, and 72°C for 2 min; and a final 72°C for 5 min.
3. Visualize the PCR products after electrophoresis on 1% agarose gels stained with ethidium bromide and run in 1 × TAE buffer.

4. Purify the PCR products with a commercial PCR DNA purification kit.
5. After the sequencing of the PCR product, compare it with the β-tubulin sequences available in the GenBank database using a BLAST search.

51.2.2.2.2.3 Translation Elongation Factor 1α Gene (tef1) PCR The *tef1* gene has high phylogenetic utility at the species level in *Fusarium*.[196] Several universal primers have been designed that work across the phylogenetic breadth of the genus.[196,217] Geiser et al.[196] created the first generation of a database (FUSARIUM-ID, http://isolate.fusariumdb.org/intro.php), which contains more than 400 sequences representing a phylogenetically diverse selection of *tef1* sequences from the *Fusarium* genus.[217]

Procedure[196,217]

1. Perform a standard PCR protocol with an annealing temperature of 53°C to amplify the *tef1* gene region using forward primer ef1 (5′-ATG GGT AAG GA(A/G) GAC AAG AC-3′) and reverse primer ef2 (5′-GGA (G/A)GT ACC AGT (G/C)AT CAT GTT-3′).
2. Visualize the ~700 bp PCR products after electrophoresis on 1% agarose gel stained with ethidium bromide and run in 1 × TAE buffer.
3. Purify the PCR products with a commercial PCR DNA purification kit.
4. After the sequencing of the PCR product, compare it with the *tef1* sequences available in the *Fusarium* database (http://isolate.fusariumdb.org/) using a BLAST search.

51.2.2.2.2.4 Histone-Encoding Gene PCR Steenkamp et al.[218] performed PCR for amplification of a histone H3 gene fragment ranging from 519 to 527 bp with the primers H3-1a (5′-ACT AAG CAG ACC GCC CGC AGG-3′) and H3-1b (5′-GCG GGC GAG CTG GAT GTC CTT-3′).[219]

Procedure[218]

1. Prepare the reaction mixture containing 0.25 ng genomic DNA/0.05 U Taq polymerase, 2.5 mM $MgCl_2$, 1 × reaction buffer, 0.2 μM of each primer, 0.25 mM of each dNTPs.
2. Perform the PCR in a thermocycler using the following reaction conditions: 92°C for 1 min; 30 cycles of 92°C, 68°C, and 72°C for 1 min each; and a final 72°C for 5 min.
3. Visualize the PCR products after electrophoresis on 1% agarose gel stained with ethidium bromide and run in 1 × TAE buffer.
4. Purify the PCR products with a commercial PCR DNA purification kit.
5. After the sequencing of the PCR product, compare it with the histone H3 sequences available in the GenBank database using a BLAST search.

51.2.2.2.2.5 ITS Sequence PCR Oechsler et al.[220] designed primers for the end of the 18S ribosomal DNA (F18A: 5′-GCG GAG GGA TCA TTA CCG AGT T-3′) and the beginning of the 28S rRNA (F28S: 5′-CAG CGG GTA TTC CTA CCT GATC-3′) of the target *Fusarium* species, and amplified with them the ITS region comprising ITS1, 5.8S rRNA, and ITS2. Although previous ITS analyses were not very useful for distinguishing *Fusarium* species, the sequence data from this study correlated well with the morphologic classification. The authors suggested the feasibility of *Fusarium* detection and identification at the species level from ocular sources using the sequence of the ITS region.

Procedure[220]

1. Prepare the reaction mixture for a standard PCR protocol in the final volume of 25 μL containing 200 ng DNA template.
2. Perform the PCR in a thermocycler using the following reaction conditions: 95°C for 3 min; 45 cycles of 95°C for 30 s, 55°C for 30 s, and 68°C for 2 min.
3. Visualize the PCR products after electrophoresis in 1% agarose gels stained with ethidium bromide and run in 1 × TAE buffer.
4. Purify the PCR products with a commercial PCR DNA purification kit.
5. After the sequencing of the PCR product, compare it with the ITS sequences available in the GenBank database using a BLAST search.

51.2.2.2.2.6 F. oxysporum- and F. solani-Specific PCR There are a lot of species-specific primer pairs in the literature for rapid and reliable identification of *Fusarium* species.[198,214,217,221] Ghignone and Migheli[198] collected the species-specific primers for the detection of phytopathogenic fungi and created a continuously improving database from them (http://www.sppadbase.com/). This database contains primers for identification of some human pathogenic members of the *Fusarium* genus and references where the protocols are described.

Hue et al.[197] designed primer pairs for specific detection of five clinically important *Fusarium* species derived from human sources. Only the OX 31 (5′-TGA CTT GGA TGA GAC CTT GGC G-3′) and OX 32 (5′-CAG GAT TTA CCG ACA CAG CTT TTG-3′) primer pair was specific for the investigated *F. oxysporum* strains (annealing temperature = 66°C), while the SOL 31 (5′-GCT ACC GAG GCC ATC AAT TCA TG-3′) and SOL 32 (5′-TGA TGT TGT ACT TCT CCT TGC CC-3′) primer pair was specific for four of the five tested *F. solani* strains (annealing temperature = 66°C). The procedure and methodology are described in Section 51.2.2.2.1.

51.3 CONCLUSION AND FUTURE PERSPECTIVES

As human and animal diseases due to *Fusarium* spp. have become increasingly common, there is an urgent need to develop rapid, sensitive, and specific diagnostic methods. Correct identification of *Fusarium* to species/clonal level is essential for the control of human and plant fusarial diseases. Considering the fact that the morphological features of *Fusarium* at both the genus and species level are diverse and similar to those of other fungal genera/species, it has been a challenge to identify these organisms precisely. To ensure effective solutions against problems of *Fusarium* etiology, it is critical to develop common and suitable identification systems for *Fusarium* taxonomy using a combination of conventional culture, macroscopic and microscopic characteristics, and molecular methods. For clinical strains of *Fusarium*, antimicrobial susceptibility patterns of the new isolates can be compared with the confirmed patterns for additional understanding about the species and will also help select most appropriate chemotherapy. Further analysis of *Fusarium* whole-genome sequences will facilitate comparison and reconfirmation of the species/clones in future.

ACKNOWLEDGMENTS

This work was supported by the Indian Council of Medical Research (ICMR–IRIS ID: 2008-03140), New Delhi, and the Indian National Science Academy and the Hungarian Academy of Sciences within the frames of the Indo-Hungarian bilateral exchange program (No. IA/INSA-HAS Proj./2010-2012/237).

REFERENCES

1. Guy St-Germain, B.S. and Summerbell, R., *Identifying Filamentous Fungi*, Star Publishing Co., Belmont, CA, 1996.
2. Nelson, P.E., Dignani, M.C., and Anaissie, E.J., Taxonomy, biology, and clinical aspects of *Fusarium* species, *Clin. Microbiol. Rev.*, 7, 479, 1994.
3. Link, H.F., Observationes in ordines plantarum naturales, Dissertatio I, *Mag. Ges. Naturf. Freunde, Berlin*, 3, 3, 1809.
4. Verweij, P.E. and Brandt M.E., *Aspergillus*, *Fusarium* and other opportunistic moniliaceous fungi, In Murray, P.R. et al. (Eds.), *Manual of Clinical Microbiology*, 8th edn., ASM Press Washington, DC, p. 1802, 2007.
5. Boonpasart, S. et al., Infectious keratitis at King Chulalongkorn Memorial Hospital: A 12 year retrospective study of 391 cases, *J. Med. Ass. Thailand*, 85, 217, 2002.
6. Goldschmied, R.A. et al., *Fusarium* spp. isolated from non-ocular sites: A 10 year experience at an Israeli general hospital, *J. Mycol. Med.*, 3, 99, 1993.
7. Krcmery, V. et al., Fungaemia due to *Fusarium* spp. in cancer patients, *J. Hosp. Infect.*, 36, 223, 1997.
8. Burgess, L.W., General ecology, In Nelson, P.E., Toussoun, T.A., and Cook, R.J. (Eds.), *Fusarium: Diseases, Biology and Taxonomy*, Pennsylvania State University Press, University Park, PA, pp. 225–235, 1981.
9. Moretti, A.N., Taxonomy of *Fusarium* genus, a continuous fight between lumpers and splitters, *Proc. Nat. Sci. Matica Srpska Novi Sad*, 117, 7, 2009.
10. Ploetz, R.C. and Pegg, K.G., *Fusarium* wilt. In Jones, D.R. (Ed.), *Diseases of Banana, Abaca and Enset*, CABI Publishing, Wallingford, U.K., pp. 143–159, 1999.
11. Windels, C.E., Economic and social impacts of *Fusarium* head blight: Changing farms and rural communities in the Northern Great Plains, 91st Annual Meeting of The American Phytopathological Society August 8, 1999, Montreal, Quebec, Canada, *Phytopathology*, 90, 17, 2000.
12. Eljaschewitsch, J. et al., Port-a-cath-related *Fusarium oxysporum* infection in an HIV-infected patient: Treatment with liposomal amphotericin B, *Mycoses*, 39, 115, 1996.
13. Guarro, J. et al., Fungemia due to *Fusarium sacchari* in an immunosuppressed patient, *J. Clin. Microbiol.*, 38, 419, 2000.
14. Mselle, J., Fungal keratitis as an indicator of HIV infection in Africa, *Trop. Doctor*, 29, 133, 1999.
15. Nir-Paz, R. et al., Clinical and epidemiological aspects of infections caused by *Fusarium* species: A collaborative study from Israel, *J. Clin. Microbiol.*, 42, 3456, 2004.
16. Chang, D.C. et al., Multistate outbreak of *Fusarium* keratitis associated with use of a contact lens solution, *JAMA*, 296, 953, 2006.
17. Khor, W.B. et al., An outbreak of *Fusarium* keratitis associated with contact lens wear in Singapore, *JAMA*, 295, 2867, 2006.
18. Godoy, P. et al., Genotyping of 44 isolates of *Fusarium solani*, the main agent of fungal keratitis in Brazil, *J. Clin. Microbiol.*, 42, 4494, 2004.
19. Hidy, P.H. et al., Zearalenone and some derivatives: Production and biological activity, *Adv. Appl. Microbiol.*, 22, 59, 1977.
20. Shukla, R. et al., Bioprocess strategies and recovery process in gibberellic acid fermentation, *Biotech. Bioproc. Eng.*, 8, 269, 2003.
21. Tomasini, A. et al., Gibberellic acid production using different solid-state fermentation systems, *World J. Microbiol. Biotech.*, 13, 203, 1997.
22. Heyndrickx, A. et al., Detection of trichothecene mycotoxins yellow rain in blood, urine and feces of Iranian soldiers treated as victims of a gas attack, *Riv. Tossicol. Speriment. Clin.*, 19, 7, 1989.
23. Whiteside T., Annals of the cold war: The yellow rain complex, *New Yorker*, 66, 52, 1991.
24. Naseema, A. et al., *Fusarium pallidoroseum* for management of water hyacinth, *Curr. Sci.*, 86, 770, 2004.
25. Leslie, J.F. and Summerell B.A., *The Fusarium Laboratory Manual*, Blackwell Publishing, Ames, IA, 2006.
26. O'Donnell, K. et al., Phylogenetic diversity and microsphere array-based genotyping of human pathogenic Fusaria, including isolates from the multistate contact lens-associated U.S. keratitis outbreaks of 2005 and 2006, *J. Clin. Microbiol.*, 45, 2235, 48, 2007.
27. O'Donnell, K. et al., Molecular systematics and phylogeography of the *Gibberella fujikuroi* species complex, *Mycologia*, 90, 465, 1998.
28. Arikan, S. et al., In vitro susceptibility testing methods for caspofungin against *Aspergillus* and *Fusarium* isolates, *Antimicrob. Agents Chemother.*, 45, 327, 2001.
29. Arikan, S. et al., Microdilution susceptibility testing of amphotericin B, itraconazole, and voriconazole against clinical isolates of *Aspergillus* and *Fusarium* species, *J. Clin. Microbiol.*, 37, 3946, 1999.
30. Wollenweber, H.W. and Reinking, O.A., *Die Fusarien, ihre Beschreibung, Schadwirkung und Bekämpfung*, Paul Parey, Berlin, 1935.
31. Boutati, E.I. and Anassie, E.J., *Fusarium*, a significant emerging pathogen in patients with haemorrhagic malignancy: Ten years experience at a cancer centre and implications for management, *Blood*, 90, 999, 1997.

32. Walsh, T.J. and Groll, A.H., Emerging fungal pathogens: Evolving challenges to immunocompromised patients for the twenty-first century, *Transpl. Infect. Dis.*, 1, 247, 1999.
33. Marr, K.A. et al., Epidemiology and outcome of mould infections in hematopoietic stem cell transplant recipients, *Clin. Infect. Dis.*, 34, 909, 2002.
34. Cho, C.T. et al., *Fusarium solani* infection during treatment for acute leukemia, *J. Pediatr.*, 83, 1028, 1973.
35. Girmenia, C. et al., Onychomycosis as a possible origin of disseminated *Fusarium solani* infection in a patient with severe aplastic anemia, *Clin. Infect. Dis.*, 14, 1167, 1992.
36. Austen, B. et al., Fatal disseminated *Fusarium* infection in acute lymphoblastic leukaemia in complete remission, *J. Clin. Pathol.*, 54, 488, 2001.
37. Nucci, M. and Anaissie, E., Cutaneous infection by *Fusarium* species in healthy and immunocompromised hosts: Implications for diagnosis and management, *Clin. Infect. Dis.*, 35, 909, 2002.
38. Anandi, V. et al., *Fusarium solani* breast abscess, *Indian J. Med. Microbiol.*, 23, 198, 2005.
39. Guarro, J. and Gene J., Opportunistic fusarial infections in humans, *Eur. J. Microbiol. Infect. Dis.*, 14, 741, 1995.
40. Madhavan, M., Ratnakar, C., Veliath, A.J., Kanungo, R., Smile, S.R., and Bhat, S., Primary disseminated fusarial infection, *Postgrad. Med. J.*, 68, 143, 1992.
41. Rekha, A. et al., *Fusarium solani* in the post-transplant patient: An unusual fungus, *Int. J. Low. Extrem. Wounds*, 7, 38, 2008.
42. Bharathi, M.J. et al., Microbial keratitis in South India: Influence of risk factors, climate, and geographical variation, *Ophthalmic. Epidemiol.*, 14, 2007.
43. Manikandan, P. et al., Corneal ulcer due to *Neocosmospora vasinfecta* in an immunocompetent patient, *Med. Mycol.*, 46, 279, 2008.
44. Bhartiya, P. et al., Fungal keratitis in Melbourne, *Clin. Exp. Ophthalmol.*, 35, 124, 2007.
45. Michael, R.J., Thew, M.B.B.S., and Todd, B., Fungal keratitis in far north Queensland, Australia, *Clin. Exp. Ophthalmol.*, 36, 721, 2008.
46. Green, M., Apel, A., and Stapleton, F., Risk factors and causative organisms in microbial keratitis, *Cornea*, 27, 22, 2008.
47. Dunlop, A.A. et al., Suppurative corneal ulceration in Bangladesh. A study of 142 cases examining the microbiological diagnosis, clinical and epidemiological features of bacterial and fungal keratitis, *Austral. N.Z. J. Ophthalmol.*, 22, 105, 1994.
48. Williams, G. et al., Microbiological diagnosis of suppurative keratitis in Bangladesh, *Br. J. Ophthalmol.*, 71, 315, 1987.
49. Williams, G., McClellan, K., and Billson, F., Suppurative keratitis in rural Bangladesh: The value of gram stain in planning management, *Int. Ophthalmol.*, 15, 131, 1991.
50. Rahman, M.R. et al., Randomised trial of 0.2% chlorhexidine gluconate and 2.5% natamycin for fungal keratitis in Bangladesh, *Br. J. Ophthalmol.*, 82, 919, 1998.
51. Höfling-Lima, A.L. et al., Laboratory study of the mycotic infectious eye diseases and factors associated with keratitis [in Portuguese], *Arq. Bras. Oftalmol.*, 68, 21, 2005.
52. Alvarez-de-Carvalho, A.C. et al., Fungal keratitis in the State of Parana-Brazil: Clinical, epidemiological and diagnostic findings [in Portuguese], *Rev. Iberoam. Micol.*, 18, 76, 2001.
53. Salera, C.M. et al., Spectrum of fungal keratitis at the São Geraldo Hospital–Belo Horizonte–MG [in Portuguese], *Arq. Bras. Oftalmol.*, 65, 9, 2002.
54. Ibrahim, M.M. et al., Epidemiologic aspects and clinical outcome of fungal keratitis in southeastern Brazil, *Eur. J. Ophthalmol.*, 19, 355, 2009.
55. Leck, A.K. et al., Aetiology of suppurative corneal ulcers in Ghana and south India, and epidemiology of fungal keratitis, *Br. J. Ophthalmol.*, 86, 1211, 2002.
56. Upadhyay, M.P. et al., Epidemiologic characteristics, predisposing factors, and etiologic diagnosis of corneal ulceration in Nepal, *Am. J. Ophthalmol.*, 111, 92, 1991.
57. Khanal, B. et al., Laboratory diagnosis in ulcerative keratitis, *Ophthalmic Res.*, 37, 123, 2005.
58. Gugnani, H.C., Gupta, S., and Talwar, R.S., Role of opportunistic fungi in ocular infections in Nigeria, *Mycopathologia*, 65, 155, 1978.
59. Zhang, W. et al., The variance of pathogenic organisms of purulent ulcerative keratitis [in Chinese], *Zhonghua Yan Ke Za Zhi*, 38, 8, 2002.
60. Wang, L. et al., Spectrum of mycotic keratitis in China [in Chinese], *Zhonghua Yan Ke Za Zhi*, 36, 138, 2000.
61. Wang, L. et al., Spectrum of fungal keratitis in central China, *Clin. Exp. Ophthalmol.*, 37, 763, 2009.
62. Xie, L., Dong, X., and Shi, W., Treatment of fungal keratitis by penetrating keratoplasty, *Br. J. Ophthalmol.*, 85, 1070, 2001.
63. Xie, L. et al., Spectrum of fungal keratitis in North China, *Ophthalmology*, 113, 1943, 2006.
64. Xie, L. et al., Antifungal susceptibility for common pathogens of fungal keratitis in Shandong province, China. *Am. J. Ophthalmol.*, 146, 260, 2008.
65. Sun, X.G. et al., Etiological analysis on ocular fungal infection in the period of 1989–2000, *Chin. Med. J.*, 117, 598, 2004.
66. Sun, X. et al., Ocular fungal isolates and antifungal susceptibility in Northern China, *Am. J. Ophthalmol.*, 143, 131, 2007.
67. Qiu, W.-Z. et al., Fungal spectrum identified by a new slide culture and in vitro drug susceptibility using Etest in fungal keratitis, *Curr. Eye Res.*, 30, 1113, 2005.
68. Rondeau, N. et al., Fungal keratitis at the Centre Hospitalier National d'Ophtalmologie des Quinze-Vingts: Retrospective study of 19 cases [in French], *J. Fr. Ophtalmol.*, 25, 890, 2002.
69. Shokohi, T., Nowroozpoor-Dailami, K., and Moaddel-Haghighi, T., Fungal keratitis in patients with corneal ulcer in Sari, Northern Iran, *Arch. Iran. Med.*, 9, 222, 2006.
70. Mirshahi, A. et al., Fungal keratitis in patients at Farabi Hospital, Tehran (1998–1999) [in Persian], *Bina*, 5, 135, 1999.
71. Berenji, F. et al., Fungal keratitis in patients at Imam Reza Hospital, Mashhad (1982–2001) [in Persian], *Med. J. Mashhad Univ. Med. Sci.*, 45, 49, 2003.
72. Norina, T.J. et al., Microbial keratitis: Aetiological diagnosis and clinical features in patients admitted to Hospital Universiti Sains Malaysia, *Singapore Med. J.*, 49, 67, 2008.
73. Mino de Kaspar, H. et al., Mycotic keratitis in Paraguay, *Mycoses*, 34, 251, 1991.
74. Laspina, F. et al., Epidemiological characteristics of microbiological results on patients with infectious corneal ulcers: A 13-year survey in Paraguay, *Graefes Arch. Clin. Exp. Ophthalmol.*, 242, 204, 2004.
75. Maidana, E. et al., Infectious keratitis in children: An epidemiological and microbiological study in a universitary hospital in Asunción–Paraguay [in Portuguese], *Arq. Bras. Oftalmol.*, 68, 828, 2005.

76. Wong, T.Y., Fong, K.S., and Tan, D.T., Clinical and microbial spectrum of fungal keratitis in Singapore: A 5-year retrospective study, *Int. Ophthalmol.*, 21, 127, 1997.
77. Sirikul, T. et al., Predisposing factors and etiologic diagnosis of ulcerative keratitis, *Cornea*, 27, 283, 2008.
78. Tanure, M.A. et al., Spectrum of fungal keratitis at Wills Eye Hospital, Philadelphia, Pennsylvania, *Cornea*, 19, 307, 2000.
79. Rosa, R.H., Miller, D., and Alfonso, E.C., The changing spectrum of fungal keratitis in South Florida, *Ophthalmology*, 101, 1005, 1994.
80. Liesegang, T.J. and Forster, R.K., Spectrum of microbial keratitis in South Florida, *Am. J. Ophthalmol.*, 90, 38, 1980.
81. Alfonso, E.C. et al., Fungal keratitis associated with non-therapeutic soft contact lenses, *Am. J. Ophthalmol.*, 142, 154, 2006.
82. Iyer, S.A., Tuli, S.S., and Wagoner, R.C., Fungal keratitis: Emerging trends and treatment outcomes, *Eye Contact Lens*, 32, 267, 2006.
83. Doughman, D.J. et al., Fungal keratitis at the University of Minnesota: 1971–1981, *Trans. Am. Ophthalmol. Soc.*, 80, 235, 1982.
84. Jurkunas, U., Behlau, I., and Colby, K., Fungal keratitis: Changing pathogens and risk factors, *Cornea*, 28, 638, 2009.
85. Singh, G. et al., Multivariate analysis of childhood microbial keratitis in South India, *Ann. Acad. Med. Singapore*, 35, 185, 2006.
86. Srinivasan, M. et al., Epidemiology and aetiological diagnosis of corneal ulceration in Madurai, south India, *Br. J. Ophthalmol.*, 81, 965, 1997.
87. Thomas, P.A. et al., Oral itraconazole therapy for mycotic keratitis, *Mycoses*, 31, 271, 1988.
88. Sundaram, B.M., Badrinath, S., and Subramanian, S., Studies on mycotic keratitis, *Mycoses*, 32, 568, 1989.
89. Panhalkar, S. et al., Bacterial and mycotic agents of corneal ulcers in Vellore, *Indian J. Ophthalmol.*, 33, 289, 1985.
90. Vasudevan R. et al., Prevalence of bacterial and fungal keratitis in and around Chidambaram, *J. Tamil Nadu Ophthalmol. Assoc.*, 44, 25, 2006.
91. Bharathi, M.J. et al., Aetiological diagnosis of microbial keratitis in South India—A study of 1618 cases, *Indian J. Med. Microbiol.*, 20, 19, 2002.
92. Bharathi, M.J. et al., Epidemiological characteristics and laboratory diagnosis of fungal keratitis. A three-year study, *Indian J. Ophthalmol.*, 51, 315, 2003.
93. Bharathi, M.J. et al., Microbiological diagnosis of infective keratitis: Comparative evaluation of direct microscopy and culture results, *Br. J. Ophthalmol.*, 90, 1271, 2006.
94. Panda, A. et al., Mycotic keratitis in children: Epidemiologic and microbiologic evaluation, *Cornea*, 16, 295, 1997.
95. Saha, R. and Das, S., Mycological profile of infectious keratitis from Delhi, *Indian J. Med. Res.*, 123, 159, 2006.
96. Chowdhary, A. and Singh, K., Spectrum of fungal keratitis in North India, *Cornea*, 24, 8, 2005.
97. Vajpayee, R.B. et al., Ocular atopy and mycotic keratitis, *Ann. Ophthalmol.*, 22, 369, 1990.
98. Chander, J. and Sharma, A., Prevalence of fungal corneal ulcers in northern India, *Infection*, 22, 207, 1994.
99. Chander, J. et al., Keratomycosis in and around Chandigarh: A five-year study from a north Indian tertiary care hospital, *Indian J. Pathol. Microbiol.*, 51, 304, 2008.
100. Gopinathan, U. et al., The epidemiological features and laboratory results of fungal keratitis: A 10-year review at a referral eye care center in South India, *Cornea*, 21, 555, 2002.
101. Garg, P. et al., Keratomycosis: Clinical and microbiologic experience with dematiaceous fungi, *Ophthalmology*, 107, 574, 2000.
102. Kunimoto, D.Y. et al., Corneal ulceration in the elderly in Hyderabad, south India, *Br. J. Ophthalmol.*, 84, 54, 2000.
103. Kumari, N., Xess, A., and Shahi, S.K., A study of keratomycosis: Our experience, *Indian J. Pathol. Microbiol.*, 45, 299, 2002.
104. Deshpande, S.D. and Koppikar, G.V., A study of mycotic keratitis in Mumbai, *Indian J. Pathol. Microbiol.*, 42, 81, 1999.
105. Verenkar, M.P. et al., Study of mycotic keratitis in Goa, *Indian J. Med. Microbiol.*, 16, 58, 1998.
106. Basak, S.K. et al., Epidemiological and microbiological diagnosis of suppurative keratitis in Gangetic West Bengal, eastern India, *Indian J. Ophthalmol.*, 53, 17, 2005.
107. Pichare, A. et al., Bacteriological and mycological study of corneal ulcers in and around Aurangabad, *Indian J. Pathol. Microbiol.*, 47, 284, 2004.
108. Dóczi, I. et al., Involvement of *Fusarium* spp. in fungal keratitis, *Clin. Microbiol. Infect.*, 10, 773, 2004.
109. Gupta, V. et al., Polymicrobial keratitis after laser in situ keratomileusis, *J. Refract. Surg.*, 17, 147, 2001.
110. Weissman, B. and Mondino, B., Risk factors for contact lens associated microbial keratitis, *Contact Lens Anterior Eye*, 25, 3, 2002.
111. Ma, S.K. et al., A multi-country outbreak of fungal keratitis associated with a brand of contact lens solution: The Hong Kong experience, *Int. J. Infect. Dis.*, 13, 443, 2008.
112. Saw, S.M. et al., Risk factors for contact lens-related *Fusarium* keratitis: A case-control study in Singapore, *Arch. Ophthalmol.*, 125, 611, 2007.
113. Dyavaiah, M. et al., Molecular characterization, biofilm analysis and experimental biofouling study of *Fusarium* isolates from recent cases of fungal keratitis in New York State, *BMC Ophthalmol.*, 7, 1, 2007.
114. Bausch & Lomb, *Fusarium* keratitis and Bausch & Lomb's response: An issue summary in New York State, Rochester, NY, May 19, 2006.
115. http://www.bausch.com/en_US/corporate/corpcomm/news/2006_5_19_issue_summary.aspx (accessed August 23, 2006).
116. Epstein, A.B., In the aftermath of the *Fusarium* keratitis outbreak: What have we learned? *Clin. Ophthalmol.*, 1, 355, 2007.
117. Bullock, J.D. et al., Temperature instability of ReNu with MoistureLoc: A new theory to explain the worldwide *Fusarium* keratitis epidemic of 2004–2006, *Trans. Am. Ophthalmol. Soc.*, 106, 117, 2008.
118. Kombila, M. et al., Fungal otitis in Libreville, study of 83 cases, *Bull. Soc. Pathol. Exot. Filial.*, 82, 201, 1989.
119. Wadhwani, K. and Srivastava, A.K., Fungi from otitis media of agricultural field workers, *Mycopathologia*, 88, 155, 1984.
120. Nuovo, M. et al., *Fusarium solani* osteomyelitis with probable nosocomial spread, *Am. J. Clin. Pathol.*, 90, 738, 1988.
121. Yu, W.K., *Fusarium* infection, *Hong Kong Dermatol. Venereol. Bull.*, 9, 71, 2000.
122. Anaissie, E.J. et al., Emerging fungal pathogens, *Eur. J. Clin. Microbiol. Infect. Dis.*, 8, 323, 1989.
123. Kontoyiannis, D.P., Mycetoma. In Goldman, L. and Bennett, J.C. (Eds.), *Cecil Textbook of Medicine*, 22nd edn., W. B. Saunders Company, Philadelphia, PA, 2003.
124. Wheeler, M.S. et al., *Fusarium* infection in burned patients, *Am. J. Clin. Pathol.*, 75, 304, 1981.

125. Gordon, W.L., The taxonomy and habitats of *Fusarium* species from tropical and temperate regions, *Can. J. Bot.*, 38, 643, 1960.
126. Young, C.N. and Meyers, A.M., Opportunistic fungal infection by *Fusarium oxysporum* in a renal transplant patient, *Sabouraudia*, 17, 219, 1979.
127. Luque, A. et al., Aumento de la incidencia de micosis superficiales producidas por hongos del género *Fusarium*, *Rev. Iberoam. Micol.*, 12, 65, 1995.
128. Merz, W.G. et al., Diagnosis and successful treatment of fusariosis in the compromised host, *J. Infect. Dis.*, 158, 1046, 1988.
129. Robertson, M.J. et al., Successful treatment of disseminated *Fusarium* infection after autologous bone marrow transplantation for acute myeloid leukemia, *Bone Marrow Transpl.*, 8, 143, 1991.
130. Squier, C. et al., Waterborne nosocomial infections, *Curr. Infect. Dis. Rep.*, 2, 490, 2000.
131. Anaissie, E.J. et al., Fusariosis associated with pathogenic *Fusarium* species colonization of a hospital water system: A new paradigm for the epidemiology of opportunistic mold infections, *Clin. Infect. Dis.*, 33, 1871, 2001.
132. Raad, I. et al., Epidemiology, molecular mycology, and environmental sources of *Fusarium* infection in patients with cancer, *Infect. Control Hosp. Epidemiol.*, 23, 532, 2002.
133. Summerbell, R.C. et al., *Fusarium proliferatum* as an agent of disseminated infection in an immunosuppressed patient, *J. Clin. Microbiol.*, 26, 82, 1988.
134. Anaissie, E. et al., The emerging role of *Fusarium* infections in patients with cancer, *Medicine*, 67, 77, 1988.
135. Nucci, M. et al., Outcome predictors of 84 patients with hematologic malignancies and *Fusarium* infection, *Cancer*, 98, 315, 2003.
136. Fridkin, S.K. and Jarvis, W.R., Epidemiology of nosocomial fungal infections, *Clin. Microbiol. Rev.*, 9, 499, 1996.
137. Anaissie, E.J. and Rinaldi, M.G., *Fusarium* and the immunocompromised host, *N. Y. State J. Med.*, 90, 586, 1990.
138. Anaissie, E. et al., *Fusarium*, a newly recognized fungal pathogen in immunosuppressed patients, *Cancer*, 57, 141, 1986.
139. Stainbrook, T.R., Shaeffer, J., and Kusne, S., A retrospective review of *Fusarium* infections in a tertiary care hospital. Presented at the *38th Annual Meeting of the Infectious Diseases Society of America*, September 7–10, 2000; New Orleans, LA, Abstract 283.
140. Ando, N. and Takatori, K., Fungal flora of the conjunctival sac, *Am. J. Ophthalmol.*, 94, 67, 1982.
141. Lionakis, M.S. et al., Pentamidine is active in vitro against *Fusarium* species, *Antimicrob. Agents Chemother.*, 47, 3252, 2003.
142. Zhao, X.J. and Richard, A., Calderone antifungals currently used in the treatment of invasive fungal diseases. In Calderone, R.A. and Cihlar, R.L. (Eds.), *Fungal Pathogenesis, Principles and Clinical Applications*, Marcel Dekker, Inc., New York, p. 559, 2002.
143. Martino, P. et al., Clinical patterns of *Fusarium* infections in immunocompromised patients, *J. Infect.*, 1, 7, 1994.
144. Rabodonirina, M. et al., *Fusarium* infections in immunocompromised patients: Case reports and literature review, *Eur. J. Clin. Microbiol. Infect. Dis.*, 13, 152, 1994.
145. Freidank, H., Hyalohyphomycoses due to *Fusarium* spp.—Two case reports and review of the literature, *Mycoses*, 38, 69, 1995.
146. Fang, C.T. et al., *Fusarium solani* fungemia in a bone marrow transplant recipient, *J. Formos. Med. Assoc.*, 96, 129, 1997.
147. Pontón, J. et al., Emerging pathogens, *Med. Mycol.*, 38, 225, 2000.
148. Musa, M.O. et al., The spectrum of *Fusarium* infection in immunocompromised patients with haematological malignancies and in non-immunocompromised patients: A single institution experience over 10 years, *Br. J. Haematol.*, 108, 544, 2000.
149. Viscoli, C. et al., Infection with *Fusarium* species in two children with neuroblastoma, *Eur. J. Clin. Microbiol. Infect. Dis.*, 9, 773, 1990.
150. Jakle, C. et al., Septic arthritis due to *Fusarium solani*, *J. Rheumatol.*, 10, 151, 1983.
151. Young, J.B. et al., Opportunistic peritonitis in continuous ambulatory peritoneal dialysis, *Clin. Nephrol.*, 22, 268, 1984.
152. Minor, R. et al., Disseminated *Fusarium* infections in patients following bone marrow transplantation, *Bone Marrow Transpl.*, 4, 653, 1989.
153. Agamanolis, D.P. et al., *Fusarium meningoencephalitis* in a child with acute leukemia, *Neuropediatrics*, 22, 110, 1991.
154. Mohammedi, I. et al., Overwhelming myocarditis due to *Fusarium oxysporum* following bone marrow transplantation, *Scand. J. Infect. Dis.*, 27, 643, 1995.
155. Girardi, M., Glusac, E.J., and Imaeda, S., Subcutaneous *Fusarium* foot abscess in a renal transplant patient, *Cutis*, 63, 267, 1999.
156. Rolston, K.V.I., The spectrum of pulmonary infections in cancer patients, *Curr. Opin. Oncol.*, 13, 218, 2001.
157. Perfect, J.R. and Schell, W.A., The new fungal opportunists are coming, *Clin. Infect. Dis.*, 22, S112, 1996.
158. Warnock, D.W., Fungal infections in neutropenia: Current problems and chemotherapeutic control, *J. Antimicrob. Chemother.*, 41, 95, 1998.
159. Bodey, G.P. et al., Skin lesions associated with *Fusarium* infection, *J. Am. Acad. Dermatol.*, 47, 659, 2002.
160. Mayayo, E. et al., Experimental pathogenicity of four opportunist *Fusarium* species in a murine model, *J. Med. Microbiol.*, 48, 363, 1999.
161. Peraica, M. et al., Toxic effects of mycotoxins in humans, *Bull. WHO*, 77, 754, 1999.
162. Smith, S.N., An overview of ecological and habitat aspects in the genus *Fusarium* with special emphasis on the soil-borne pathogenic forms, *Plant Pathol. Bull.*, 16, 97, 2007.
163. Etzel, R.A., Mycotoxins, *JAMA*, 287, 425, 2002.
164. Marasas, W.F.O. et al., *Fusarium andiyazi* sp. nov., a new species from sorghum, *Mycologia*, 93, 1203, 2001.
165. Gajdusek, D.C., Acute infectious hemorrhagic fevers and mycotoxicoses in the Union of Soviet Socialist Republics. Medical Science Publication No. 2, Walter Reed Army Medical Center, Washington, DC, 1953.
166. Pestka, J.J. et al., Suppression of immune response in the B6C3FI mouse after dietary exposure to the *Fusarium* mycotoxins deoxynivalenol and zearalenone, *Food Chem. Toxicol.*, 25, 297, 1987.
167. Yarom, R. et al., T-2 toxin-induced pathology in the hearts of rats, *Br. J. Exp. Pathol.*, 64, 570, 1983.
168. Pitt, J.I., Toxigenic fungi and mycotoxins, *Br. Med. Bull.*, 56, 184, 2000.
169. Bhat, R.V., A foodborne disease outbreak due to the consumption of moldy sorghum and maize containing fumonisin mycotoxins, *J. Toxicol. Clin. Toxicol.*, 35, 249, 1997.

170. Visconti, A. et al., Determination of ochratoxin A in domestic and imported beers in Italy by immunoaffinity clean-up and liquid chromatography, *J. Chromatogr. A*, 888, 321, 2000.
171. Hu, Z.Q. et al., Antibody-mediated prevention of *Fusarium* mycotoxins in the field, *Int. J. Mol. Sci.*, 9, 1915, 2008.
172. Visconti, A. et al., Determination of ochratoxin A in wine and beer by immunoaffinity column cleanup and liquid chromatographic analysis with fluorometric detection: Collaborative study, *J. AOAC Int.*, 84, 1818, 2001.
173. Thomas, P.A., Current perspectives on ophthalmic mycoses, *Clin. Microbiol. Rev.*, 16, 730, 2003.
174. Anaissie, E.J. et al., Lack of activity of amphotericin B in systemic murine fusarial infection, *J. Infect. Dis.*, 165, 1155, 1992.
175. Guarro, J. et al., In vitro and in vivo experimental activities of antifungal agents against *Fusarium solani*, *Antimicrob. Agents Chemother.*, 43, 1256, 1999.
176. Richardson, S.E. et al., Disseminated fusarial infection in the immunocompromised host, *Rev. Infect. Dis.*, 10, 1171, 1988.
177. Speeleveld, E. et al., Susceptibility of clinical isolates of *Fusarium* to antifungal drugs, *Mycoses*, 39, 37, 1996.
178. Marco, F. et al., In vitro activity of a new triazole antifungal agent, SCH 56592, against clinical isolates of filamentous fungi, *Mycopathologia*, 141, 73, 1998.
179. Johnson, E.M., In vitro activity of Syn-2869, a novel triazole agent, against emerging and less common mold pathogens, *Antimicrob. Agents Chemother.*, 43, 1260, 1999.
180. Li, R.K. and Rinaldi, M.G., In vitro antifungal activity of nikkomycin Z in combination with fluconazole or itraconazole, *Antimicrob. Agents Chemother.*, 43, 1401, 1999.
181. Capilla, J. et al., In vitro antifungal activities of the new triazole UR-9825 against clinically important filamentous fungi, *Antimicrob. Agents Chemother.*, 45, 2635, 2001.
182. Pujol, I. et al., In-vitro antifungal susceptibility of clinical and environmental *Fusarium* spp. strains, *J. Antimicrob. Chemother.*, 39, 163, 1997.
183. Johnson, E.M. et al., In-vitro activity of voriconazole, itraconazole and amphotericin B against filamentous fungi, *J. Antimicrob. Chemother.*, 42, 741, 1998.
184. Espinel-Ingroff, A., Comparison of in vitro activities of the new triazole SCH56592 and the echinocandins MK-0991 (L-743,872) and LY303366 against opportunistic filamentous and dimorphic fungi and yeasts, *J. Clin. Microbiol.*, 36, 2950, 1998.
185. Espinel-Ingroff, et al., In vitro antifungal activities of voriconazole and reference agents as determined by NCCLS methods: Review of the literature, *Mycopathologia*, 150, 101, 2001.
186. Lozano-Chiu, M. et al., Treatment of murine fusariosis with SCH 56592, *Antimicrob. Agents Chemother.*, 43, 589, 1999.
187. Pfaller, M.A. et al., Antifungal activities of posaconazole, ravuconazole, and voriconazole compared to those of itraconazole and amphotericin B tested against 239 clinical isolates of *Aspergillus* spp. and other filamentous fungi: Report from the SENTRY antimicrobial surveillance program, 2000, *Antimicrob. Agents Chemother.*, 46, 1032, 2002.
188. Leslie, J.F., Genetic status of the *Gibberella fujikuroi* species complex, *Plant. Pathol. J.*, 15, 259, 1999.
189. Summerell, B.A., Salleh, B., and Leslie, J.F., A utilitarian approach to *Fusarium* identification, *Plant Dis.*, 87, 117, 2003.
190. Burgess, L.W., Liddell, C.M., and Summerell, B.A., *Laboratory Manual for Fusarium Research*, 2nd edn., University of Sydney, Sydney, Australia, 1988.
191. Nelson, P.E., Toussoun, T.A., and Marasas, W.F.O., *Fusarium species: An Illustrated Manual for Identification*, Pennsylvania State University Press, University Park, PA, 1983.
192. Thrane, U., Developments in the taxonomy of *Fusarium* species based on secondary metabolites, In Summerell, B.A., Leslie, J.F., Backhouse, D., Bryden, W.L. and Burgess, L.W. (Eds.), *Fusarium: Paul E. Nelson Memorial Symposium*, American Phytopathological Society, St. Paul, MN, pp. 29–49, 2001.
193. Klittich, C.J.R. and Leslie, J.F., Nitrate reduction mutants of *Fusarium moniliforme* (*Gibberella fujikuroi*), *Genetics*, 118, 417, 1988.
194. Petti, C.A., Detection and identification of microorganisms by gene amplification and sequencing, *Clin. Infect. Dis.*, 44, 1108, 2007.
195. O'Donnell, K., Cigelnik, E., and Casper, H.H., Molecular phylogenetic, morphological, and mycotoxin data support reidentification of the quorn mycoprotein fungus as *Fusarium venenatum*, *Fungal Genet. Biol.*, 23, 57, 1998.
196. Geiser, D.M. et al., FUSARIUM-ID v. 1.0: A DNA sequence database for identifying *Fusarium*, *Eur. J. Plant Pathol.*, 110, 473, 2004.
197. Hue, F.-X. et al., Specific detection of *Fusarium* species in blood and tissues by a PCR technique, *J. Clin. Microbiol.*, 37, 2434, 1999.
198. Ghignone, S. and Migheli, Q., The database of PCR primers for phytopathogenic fungi, *Eur. J. Plant Pathol.*, 113, 107, 2005.
199. Nicolaisen, M. et al., Real-time PCR for quantification of eleven individual *Fusarium* species in cereals, *J. Microbiol. Meth.*, 76, 234, 2009.
200. Bluhm, B.H., Cousin, M.A., and Woloshuk, C.P., Multiplex real-time PCR detection of fumonisin-producing and trichothecene-producing groups of *Fusarium* species, *J. Food Prot.*, 67, 536, 2004.
201. Strausbaugh, C.A., Overturf, K., and Koehn, A.C., Pathogenicity and real-time PCR detection of *Fusarium* spp. in wheat and barley roots, *Can. J. Plant Pathol.*, 27, 430, 2005.
202. Abd-Elsalam, K.A. et al., Molecular detection of *Fusarium oxysporum* f. sp. *vasinfectum* in cotton roots by PCR and real-time PCR assay, *J. Plant Dis. Protect.*, 113, 14, 2006.
203. Hogg, A.C., Johnston, R.H., and Dyer, A.T., Applying real-time quantitative PCR to *Fusarium* crown rot of wheat, *Plant Dis.*, 91, 1021, 2007.
204. Yli-Mattila, T. et al., Real-time PCR detection and quantification of *Fusarium poae*, *F. graminearum*, *F. sporotrichioides* and *F. langsethiae* in cereal grains in Finland and Russia, *Arch. Phytopathol. Plant Protect.*, 41, 1477, 2008.
205. Kristensen, R. et al., DNA microarray to detect and identify trichothecene- and moniliformin-producing *Fusarium* species, *J. Appl. Microbiol.*, 102, 1060, 2007.
206. Yli-Mattila, T. et al., Phylogenetic relationship of *Fusarium langsethiae* to *Fusarium poae* and *Fusarium sporotrichioides* as inferred by IGS, ITS, β-tubulin sequences and UP-PCR hybridization analysis, *Int. J. Food Microbiol.*, 95, 267, 2004.
207. Lee, S.B. and Taylor, J.W., Isolation of DNA from fungal mycelia and single spores. In Innis, M.A., Gelfand, D.H., Sninsky, J., and White, T.J. (Eds.), *PCR Protocols: A Guide to Methods and Applications*, Academic Press, New York, pp. 282–287, 1990.
208. Fisher, N.L. et al., Carnation leaves as a substrate and for preserving cultures of *Fusarium* species, *Phytopathology*, 72, 151, 1982.

209. Nirenberg, H.I., Untersuchungen über die morphologische und biologische Differenzierung in der *Fusarium* Sektion Liseola, *Mitt. Biol. Bund. Land–Forst.* (*Berlin-Dahlem*), 169, 1–117, 1976.
210. Nirenberg, H.I. and O'Donnell, K., New *Fusarium* species and combinations within the *Gibberella fujikuroi* species complex, *Mycologia*, 90, 434, 1998.
211. Ryan, F.J., Beadle, G.W., and Tatum, E.L., The tube method of measuring the growth rate of *Neurospora*, *Am. J. Bot.*, 30, 784, 1943.
212. Burgess, L.W. et al., *Laboratory Manual for Fusarium Research*, 3rd edn., University of Sydney/Royal Botanic Gardens, Sydney, Australia, 1994.
213. Gerlach, W. and Nirenberg, H.I., The genus *Fusarium*—A pictorial atlas, *Mitt. Biol. Bund. Land-Forst.* (*Berlin-Dahlem*), 209, 1–406, 1982.
214. Hennequin, C. et al., Identification of *Fusarium* species involved in human infections by 28S rRNA gene sequencing, *J. Clin. Microbiol.*, 37, 3586, 1999.
215. O'Donnell, K. and Cigelnik, E., Two different intragenomic rDNA ITS2 types within a monophyletic lineage of the fungus *Fusarium* are nonorthologous, *Mol. Phylogenet. Evol.*, 7, 103, 1997.
216. Chung, W.-H. et al., Genetic analysis and PCR-based identification of major *Fusarium* species causing head blight on wheat in Japan, *J. Gen. Plant Pathol.*, 74, 364, 2008.
217. Rahjoo, V. et al., Morphological and molecular identification of *Fusarium* isolated from maize ears in Iran, *J. Plant Pathol.*, 90, 463, 2008.
218. Steenkamp, E.T. et al., Differentiation of *Fusarium subglutinans* f. sp. *pini* by histone gene sequence data, *Appl. Environ. Microbiol.*, 65, 3401, 1999.
219. Glass, N.L. and Donaldson G.C., Development of primer sets designed for use with PCR to amplify conserved genes from filamentous ascomycetes, *Appl. Environ. Microbiol.*, 61, 1323, 1995.
220. Oechsler, R.A. et al., Utility of molecular sequence analysis of the ITS rRNA region for identification of *Fusarium* spp. from ocular sources, *Invest. Ophthalmol. Vis. Sci.*, 50, 2230, 2009.
221. Demeke, T. et al., Species-specific PCR-based assays for the detection of *Fusarium* species and a comparison with the whole seed agar plate method and trichothecene analysis, *Int. J. Food Microbiol.*, 103, 271, 2005.

52 *Microascus*, Including *Scopulariopsis*

Jouni Issakainen and Dongyou Liu

CONTENTS

52.1 INTRODUCTION

52.1.1 Classification, Morphology, and Biology

The meiosporic (i.e., sexually sporulating) genus *Microascus* includes some less common but regular nail pathogens. In rare occasions, several species have caused lethal infections in immunocompromised patients.

The genus belongs in the family Microascaceae. The same family also covers some more important human pathogenic genera, namely, the meiosporic genus *Pseudallescheria* and the mitosporic (i.e., asexually sporulating) genus *Scedosporium*. These are dealt with in other chapters of the present book. The whole family is a part of the subclass Hypocreomycetidae [1]. Due to the small number of DNA-based taxonomic studies, however, the taxonomy of these clades of fungi is far from being settled. Literature concerning the whole family as a medical mycological problem has been recently reviewed [2].

For simplicity, the name *Microascus* is used in the present chapter in a wide sense (*sensu lato, s.l.*), referring to the whole provisional "Microascoid clade" of Issakainen et al. [3]. Using this technical delimitation, the genus incorporates here practically all the few molecular clades of fungi that are known to have sexual fruiting bodies described as *Microascus*. No position is taken here whether this clade should be split into smaller genera, or whether also other, less studied meiosporic genera will be shown to be located inside the same phylogenetic clade. The definition also means that all known members of *Pseudallescheria* and *Scedosporium* remain outside *Microascus*.

More importantly, this delimitation deals the various mitosporic stages of the same molecular clade simply as members of the same large meiosporic genus *Microascus*, regardless of whether the entire life cycle between the meiosporic and mitosporic stages has been experimentally solved, or whether the name of the mitosporic stage (anamorph name) has been validly combined to any name of a meiosporic stage (teleomorph name), in terms of current taxonomic rules.

The clinically most relevant consequence of the above delimitation is that the mitosporic genus *Scopulariopsis s.l.* is handled here simply as a part of the meiosporic genus *Microascus s.l.* As the phylogenetic work of Issakainen et al. [3] shows, the morphology-based meiosporic genus and the morphology-based mitosporic genus broadly correspond with each other in molecular terms. In addition to *Scopulariopsis* anaomorphs however, some members of *Microascus s.l.* are known to produce other types of asexual sporulation. Some of these mitosporic forms are so distinctive that they have been given separate generic names such as *Doratomyces* and *Wardomyces* [3].

The sexual fruiting bodies of *Microascus* are dark and small, measuring 0.05–0.3 mm in diameter [4]. They are usually bottle-like having an opening through the neck part (then called perithecia), but more or less spherical and closed (cleistothecia) in some species, especially in those formerly classified as *Pithoascus*. The wall of the fruiting body consists of angular cells (*textura angularis*).

As the name implies, species of *Microascus* have small asci (singular: ascus, the ascospore-bearing cells, which are formed within the perithecium). The eight ascospores within each ascus are smooth, one-celled, and more or less curved. As compared with *Pseudallescheria*, the ascospores are paler brown and the single germ pore of the spore is less prominent or missing [4].

Most mitosporic forms of *Microascus* sporulate by means of annelloconidia. This type of conidiation is characterized by consecutive ring-like scars near the tip of each conidiogenous hypha. Each scar is a mark left by a detached, broad-based conidium. Of the anamorph genera of *Microascus*, annelloconidia are especially typical to *Scopulariopsis* and to its synnematal counterparts, *Doratomyces* and *Trichurus*. Annelloconidia of *Microascus* are formed in dry chains which eventually break and let the spores be dispersed by wind. This contrasts with the insect-dispersion of the slimy *Scedosporium* conidia. As compared with the oval and smooth *Scedosporium* spores, the conidia of *Microascus* species have a broader base and are more dome-shaped or bullet-shaped, sometimes conspicuously warted. Only in some clinically rare anamorphs of *Microascus*, notably the mitosporic genus *Wardomyces*, each conidium detaches by breaking and has a longitudinal germ slit [2,5].

Planning of molecular diagnostic tools for *Microascus* is currently hampered by the fact that there are many more poorly known genera and species that may prove to be parts of the natural *Microascus*. At least, such genera may be so close relatives of *Microascus* that their molecular characters must be considered in, for example, designing of primers [6]. Such potentially related genera include the meiosporic *Lophotrichus*, *Enterocarpus*, and *Canariomyces*, and the mitosporic form genera *Wardomycopsis*, *Humicola*, and *Mammaria* [4].

Because of the mentioned taxonomical problems and of limited strain collecting, the number of extant *Microascus* species is not known. The number of currently described species is most likely an underestimation but can be used as a starting point: more than 15 *Microascus* species, 20 *Scopulariopsis* species, 5 *Wardomyces* species, and 3 *Doratomyces* species are known. No position is taken, here, to the delimitation of single species. Inferring from the much more easily detected macrofungi, it is reasonable to assume that the real species number of this clade is in the magnitude of hundreds, at least.

Microascus longirostris is the taxonomical type species of the genus, described by Zukal in 1885 [7]. Thus, if some clades of the genus will be split into new genera, the genus with *M. longirostris* will retain the name *Microascus*.

The difficult task of taxonomically connecting each meiosporic species with the mitosporic stages has continued for more than a century and will do so in the future. For some species, the connection has been observed a long time ago since they produce large amounts of both sporulating stages in the same colony. This applies to, for example, *Microascus longirostris*, *M. cinereus*, *M. desmosporus*, and *M. trigonosporus* [7].

For a few species, the connection has been found only recently. This is true especially for the clinically most important species *Scopulariopsis brevicaulis*. In spite of the common cultivation of this fungus all around the world, the diligent study of Abbott et al. [8] revealed the sexual stage as late as in the 1990s. The species was thus renamed as *Microascus brevicaulis*, indicating its whole life cycle. Similar studies have been performed in some other species as well [9,10].

Despite the apparent overlap between *Scopulariopsis* and *Microascus*, these two genera cannot be treated simply as different stages of the same taxon. *Scopulariopsis* anamorphs and related synnematous forms occur in *Kernia* as well. More rarely, *Scopulariopsis* or very similar anamorphs have also been reported in other genera such as *Petriella* and *Pithoascus* [4,11]. Some extant *Microascus* species may be completely unable to form a *Scopulariopsis* stage and *vice versa*.

Microascus species can be found on a wide variety of organic substrates such as soil, rotten plants or paper. They can be termed as saprotrophs. Many species prefer more proteinous or otherwise nutrient-rich substrates such as animal remains or dung, and the fruiting bodies are often formed on these. The human body, often in various debilitated conditions, occasionally serves as a substrate for the genus, leading to various types of infection [2,7,12–15].

52.1.2 Clinical Features

Regarding their clinical significance, *Microascus* species can be broadly characterized by their two main roles: as less common but regular pathogens of the keratinous tissues and as very rare but serious opportunistic agents in deep tissues of the body. A short overview of the infections and of most relevant *Microascus* species in each body site follows.

52.1.2.1 Nails

The most common clinical phenomenon associated with *Microascus* is a chronic nail infection. Patients are typically mid-aged to older people whose big toe nail, seldom other nails, shows various symptoms of tissue damage including fragility, thickening, and discoloration of nails, nail deformations, problems in walking with shoes on, or nail loss [16]. The mechanisms of this benign but often disturbing infection are still poorly known.

M. brevicaulis is by far the most common *Microascus* species to be encountered in nails. In a Finnish material, 80% of the cases suggestive of *Microascus* nail infection concerned this species, followed by *Scopulariopsis candida* and *S. fusca* [16]. *Scopulariopsis* species have been shown to grow in symptomatic nails more often than other environmental fungi [16–20].

Genuine vegetative growth of *Microascus* in the nail (in contrast to external contamination) is often shown by typical hyphae and spores in direct nail microscopy, and by repeatedly positive culture. In a minority of the cases, common dermatophytosis (causative agents: *Trichophyton* stages of *Arthroderma*) obviously predisposes the nail for *Microascus*. Absolute exclusion of earlier dermatohyphytosis or forgotten traumas is often difficult to achieve but a great amount of indirect evidence suggests that *Microascus* often acts as the main infectious agent and at least participates in causing the symptoms [18,20,21].

M. brevicaulis is estimated to cause about 1.6% of all fungal nail infections in a Canadian survey [18]; *Scopulariopsis* spp. have been observed in 6.4% of onychomycosis cases in Italy [19], in 5.7% in Denmark [22], in 5.6% in Spain [23], in 1.9% in Colombia [24], and in 1.5% in Estonia [25], and <1% in Belgium [26].

The variety of symptoms assigned to *Microascus* may indicate that the cases do not form a homogenous entity [16]. López-Jodra and Torres-Rodríguez [21] listed *Scopulariopsis* spp. as typically responsible for white superficial onychomycosis. Gupta and Gregurek-Novak [27] referred to *Scopulariopsis* onychomycosis as reminiscent of distal and lateral subungual onychomycosis. According to Tosti et al. [18], *S. brevicaulis* can cause both distal and proximal subungual onychomycosis. On the other hand, a similar range of symptoms is also seen in dermatophytosis, suggesting variation between the human hosts.

52.1.2.2 Skin

In addition to nails, *Microascus* species have for long been known to colonize and, potentially, infect the skin. Here, it seems that the keratinous tissue needs to be previously debilitated by factors such as long-term wet conditions in the boots of soldiers [7] or deep immunosuppression. Potential predisposing factors in less extreme living conditions include primary dermatophyte infection, sweating skin folds with candidosis, and allergic lesions [16]. *M. brevicaulis* is the main agent.

52.1.2.3 Ears

Microascus species have repeatedly infected the external ear channel. Considering the thin keratinous epidermis and moist conditions in the ear, these infections may be comparable to those in the wetted skin. Details, however, are not known.

One of the earliest records of *Microascus* in the ears appears to be a case with *Microascus desmosporus* [7]. *Microascus* (*Scopulariopsis*) *brevicaulis* was documented to cause external otitis in a patient with cholestematous otitis media [28]. In an immunocompromised patient, a *Scopulariopsis* ear infection led to mastoiditis and fatal infection [29]. *Microascus* was found to be a stronger genus than *Pseudallescheria* in infecting the external ear channel. Its ecological range as an ear pathogen extended to post-operational cavities of the middle ear. In these deeper parts, however, *P. boydii* was equally common [6].

52.1.2.4 Respiratory Tract

In other than keratinous tissues, *Microascus* species are primarily seen in patients whose immune system has been suppressed due to, for example, ongoing cytostatic therapy, who have other underlying diseases or whose body has been physically injured. When these preconditions are fulfilled to such an extent that *Microascus* invades deeper tissues, the infection may turn very difficult to treat and run a lethal course.

Microascus species are known as rare causes of fungal sinusitis. In immunosuppressed individuals, sinusitis due to *Scopulariopsis* spp. may become invasive. The pale-spored species *Scopulariopsis candida* and *S. acremonium* have been involved [30]. The *Scopulariopsis candida* anamorph of *Microascus manginii* was reported as an etiologic agent in invasive sinonasal disease in a 12-year-old girl being treated for non-Hodgkin's lymphoma. Some nasal cases have become invasive in susceptible hosts [30,31].

Microascus infections of the lungs are rare. As a localized form of infection, a fungus ball in the lungs due to *Scopulariopsis* spp. was described by Endo et al. [32]. In immunocompromised patients, *Scopulariopsis* lung infections are serious and have a poor prognosis. Severe pneumonia due to *Microascus* infections has been observed a few times [33–35]. The species spectrum of *Microascus* in the lungs may differ from that of superficial infections. The ubiquitous *S. brevicaulis* has been positively recorded in only one case (pneumonia of a drug abuser). In three other cases, the species remained unidentified [32,33]. In one fatal pneumonia case [34] the causative agent was a rare species, *Microascus trigonosporus*.

52.1.2.5 Central Nervous System

At least five cases of severe CNS infections are known to be caused by *Microascus* species. Most of the infections have been in immunocompromised individuals [35]. In three patients, the infection presented as an abscess of the CNS. The fourth case was meningitis of an AIDS patient [36]. A brain infection has also been observed in an immunocompetent person [37]. In this case, a large lesion of the white matter was removed surgically. Histopathological observations were reminiscent of acute hemorrhagic leucoencephalitis. Fungal growth proceeded around the operation site. As in lungs, the rare CNS infections may have a different overall species spectrum than the keratinous tissues: *M. brevicaulis* does not have a dominant role and at least *M. cinereus* and *S. brumptii* have been seen as causative agents.

52.1.2.6 Heart and Other Organs

Serious endocarditis due to *M. brevicaulis* has occurred in a few heart surgery patients, mostly prosthetic valve recipients

[33,38]. Other types of infection include peritonitis, eye infections, and various soft tissue infections [6,33]. These have occurred in immunocompetent persons and been generally curable with surgery and/or antifungal treatments. Isolation of *Microascus* from blood is a very rare event [39].

52.1.3 Pathogenesis

Nail infection due to *Microascus* may not be actually opportunistic. However, several poorly understood factors of the patient's genotype, of his or her age or clinical history and of the keratinolytic enzyme pattern of the fungal strain may contribute to the initiation or progress of this disease [40].

Torres-Rodríguez and López-Jodra [21] and Summerbell et al. [20] showed that genuine *Scopulariopsis* growth often occurs in the nails without the help of dermatophytes. In addition to the cutaneous barrier breaks (bandage in a catheter site) [35], the nail pathogenicity has also been approached from the point of view of keratinolytic enzymes produced [41].

In some experimental models, some strains of *Microascus* were shown to be very weak utilizers of nail tissue. However, another study revealed that individual strains of a given species have significant differences in this respect. Filipello Marchisio et al. [41] showed that some strains of *M. brevicaulis* are genuinely keratinolytic, not merely keratinophilic. In other words, these fungi do enzymatically invade healthy keratinous tissues, not merely utilize secondary breakdown products of keratin. The observed enzymatic activity of *Microascus* was slower than that of dermatophytes but sufficient for giving some strains the status of pathogens [41].

Infections due to *Microascus* in non-onychomycosis cases are usually associated with some debilitating factors. These include immunosuppressive therapies due to transplantation, local barrier breaks such as catheter sites or moist bandages. Details of specific pathogenesis of *Microascus* in deep infections are not known. A few cases have been associated with just local factors, or with no observed factors at all [35,42,43].

52.1.4 Diagnosis

52.1.4.1 Conventional Techniques

Infections due to *Microascus* are mostly diagnosed by microscopy of human tissues and by isolating the causative agents on culture media. Identification of the isolated strains by conventional techniques is mainly based on morphological analysis of the sporulating structures [13,14].

52.1.4.1.1 Microscopy

Direct microscopy of *Microascus* in human tissue does not differ from that of other opportunistic fungi. In general, a portion of the sample is mounted on a glass slide, covered with a coverslip in a suitable mounting medium such as lactophenol solution, and then screened under a transmitted light microscope with a total magnification of about 200–400 times. The image may be enhanced by various methods such as phase contrast, Nomarski interference contrast or optical brighteners (e.g., Calcofluor White or Blankophor) or older chemical stains, according to facilities and the experience of the worker.

Some details apply specifically to *Microascus*. The genus includes species with brown cell walls (dematiaceous hyphae) and species with clear, transparent walls (hyaline hyphae), and intermediates in between. This may mislead a less trained microscopist since in some literature sources, these color categories are treated as separate entities (phaeohyphomycosis and hyalohyphomycosis, respectively). The most common pathogenic *Microascus* species have hyaline hypha but also some darkly pigmented species have been associated with human disease [44]. In nails, *Microascus* species may form round spores and hyphae with characteristic morphology [20].

Histopathological staining shows the position of a fungal infection in tissues and, in trained hands, gives a wealth of other information of the whole pathological process. Accordingly, it is one of the central components of the diagnosis of deep, opportunistic mycoses [30]. It is also useful in diagnosing and confirming superficial fungal infections including those of nails [45]. The periodic acid-Schiff stain (PAS) used routinely in pathological investigations also sticks to most fungal cells and stains them red. Gomori's methenamine-silver stain renders fungal cell walls black, against light green background. The latter dye gives better resolution in morphological details and should be used to confirm the PAS result especially in serious infections.

52.1.4.1.2 Fungal Cultures

A central method for diagnosing most fungal infections is to recover the living fungal strain from the infected tissue. This applies to *Microascus*, as well. An example of possible culturing procedures and growth media is illustrated as a flow chart by de Hoog et al. [13,14]. Other variations appear in numerous handbooks such as Larone [46].

In general, a clinical specimen is cultured simultaneously on several (2–5) general-purpose culture media (e.g., Sabouraud Dextrose Agar, Malt Extract Agar, and/or Brain Heart Infusion Agar) supplemented with antibacterial agents such as streptomycin and penicillin, and incubated at temperatures between 24°C and 30°C. This allows the growth of a wide variety of pathogenic fungi, including *Microascus* [13,14].

Fungal cultures are usually incubated for several weeks (up to about 1 month) and screened several times during the incubation. For most strains of pathogenic *Microascus*, the routineous culture conditions pose no problem, since they commonly form visible colonies usually in a few days or, at latest, in about 1 week. Some tissue-adapted pathogenic strains, may grow slower, however, or fail to grow at all. This phenomenon is known, for example, in *Pseudallescheria* but may apply to *Microascus* as well. The related but apathogenic genus *Kernia* is also known as difficult to culture and preceding treatments or host defense may affect the viability of any strain. Therefore, the recovery of a given pathogenic strain is never self-evident.

Regarding culture methods, there are some points which specially affect *Microascus*. A medium supplement, cycloheximide (also called actidione) is added in some culture media to suppress contaminating molds and yeasts to make the medium selective for dermatophytes. This seemingly attractive medium for fungal cultures of the skin gives unpredictable and directly misleading results when infections due to *Microascus* and related genera are concerned. For instance, the concentration of 0.5 g/L of cycloheximide allows some members of the family to grow but suppresses others [13,47]. At least *M. brevicaulis* is often found growing in cycloheximide-containing media but it is not known how consistently all species and strains of *Microascus* react to this substance. Another chemical with possible selective use in *Microascus* is benomyl. Members of the Microascaceae are usually resistant to this substance [47,48].

For colony characters and microscopic morphology of *Microascus*, special literature should be consulted [4,7,13]. In general, the colony colors of the genus vary between black, gray, whitish, and buff tones (the latter especially for *M. brevicaulis*). Green and pink colors, common in some other genera, are not encountered. The colony texture is mold-like (as opposite to yeast) but may show various types of moderately fluffy mycelium or powdery granules or masses of conidia. In some strains, perithecia are clearly visible as small black dots against the mycelium. In clinical laboratories, identification of *Microascus* is usually based on the mitosporic stage since the meiosporic stage may appear slowly, only after mating with a compatible strain, or may be totally missing [8,9].

Microscopically, the key characters of the most common *Scopulariopsis* stages include slightly to clearly branched to synnematal conidiophores with often sharp branching angles reminiscent of the common but non-related genus *Penicillium*. Using high magnifications, the densely packed annellides can be seen near the tips of the conidiogenous hyphae. The conidia are one-celled and vary according to species but are generally dome-shaped to broadly bullet-shaped. Their colors vary from hyaline to dark-walled, via brownish or gray tones. Their surface ornamentation varies from smooth to coarsely warty. Ornamentation varies gradually even in a single chain, depending on the age and position of the conidium. In some species and strains, gradual transitions occur from the normal scarcely branched *Microascus* conidiophores to large synnemata assigned to *Doratomyces*. In clinical samples, however, the latter are rarely seen [13,14].

52.1.4.1.3 Biochemical Methods

No biochemical culture methods are routineously used for the diagnosis of *Microascus*. Subcultures of the strains on different growth media, or at different temperatures may be used according to need. Fungal growth may be characteristically inhibited by the selected conditions, or sporulation may be accelerated. Members of the family show variable resistance against cycloheximide [49]. *M. brevicaulis* grows optimally in temperatures from 24°C to 30°C, but tolerates the human body temperature of 37°C. *M. brevicaulis* prefers somewhat saline or otherwise osmotically dry environments: optimal growth occurs with 3%–6% NaCl in the medium. In addition to cellulose and keratin, it can typically utilize arsenic compounds. Some *Microascus* species compete with dermatophytes by producing antifungal agents [50].

Serological tests have the capacity to identify some members of the Microascaceae directly from a clinical specimen [51,52]. However, considering that immunological tests are usually aimed at only a limited number of pathogenic species, and that there exist a large number of opportunistic fungal species [13,14], screening of all potential patients with immunological tests is currently not cost-effective.

52.1.4.2 Molecular Techniques

As in other groups of fungi, molecular techniques are useful for identification and differentiation of Microascaceae as well. The conserved rRNA coding genes and the more variable internally transcribed spacer (ITS) regions between them are most frequently targeted for phylogenetic studies [13,53]. Naturally, sequences of several protein-encoding genes also have the potential to be used with *Microascus* but have not been applied to this genus, as yet. Such study targets include β-tubulin, actin, chitin synthase, acetyl coenzyme A synthase, glyceraldehyde-3-phosphate dehydrogenase, and lignin peroxidase or orotidine-5′-monophosphate decarboxylase genes [1].

Molecular studies on this family are mainly concentrated on *Pseudallescheria* [54–56]. Molecular relationships within the genus *Microascus* have thus far been explored by only one study limited to a 350 bp gene region of the large subunit ribosomal RNA gene (LSU rDNA) [3]. Based on the analyses, the family was provisionally divided into 12 molecular lineages, seven of which were within *Microascus*. Several opportunistic human pathogenic species, including *M. brevicaulis*, were placed in one clade, together with *M. manginii*. Some deviating sporulation forms such as the synnematous *Doratomyces* and the germ-slitted *Wardomyces* were concentrated in another clade, together with *M. albonigrenscens*. An extensive molecular taxonomic revision of the genus is sorely needed.

52.2 METHODS

52.2.1 Sample Preparation

52.2.1.1 Sample Collection, Transportation, and Handling

Core criteria for diagnostic medical mycology include that the sample is representative for the infection and contains (preferably living) pieces of the causative fungus. Dying of the fungus during the transportation is usually not the problem but the increase of contaminants may be.

Tissue types and consistence of good samples vary widely, depending on the infected body site, for instance. Typical samples for diagnostic mycology include scrapings of (for culture, properly disinfected) skin, pieces taken from deep layers of nails, surgically obtained biopsies from deep tissues, as well as body liquids or excretions [57]. Also the transport system varies widely, ranging from sterile test tubes (for moist samples) to empty, sealed Petri dishes or even small paper envelopes (for dry skin and nail samples).

When a specimen arrives in the medical mycology laboratory, it is traditionally divided into at least two sections, one for direct microscopy and one for culturing. However, histopathological samples are often taken and submitted to the relevant histological laboratory separately. Separate planning can also be recommended for obtaining and handling of possible molecular samples. Procedures which prepare the specimen for microscopy, for instance, may destroy the DNA. Very little work has been done specifically on *Microascus* from the point of view of sample handling.

52.2.1.2 DNA Extraction

A number of procedures have been developed and applied for isolation of fungal DNA [58–65]. In general, chemical lysis and/or mechanical breaking (e.g., mortar, agitated glass beads, sonication and/or rapid deep-freezing using liquid nitrogen) has been utilized to disrupt the rigid cell walls of fungi. Proteinaceous cell components are eliminated enzymatically (using proteinase K). Suitable buffers such as Tris-EDTA (Tris-hydroxymethylaminomethane and ethylenediamine tetraacetic acid) are used to stabilize the conditions. In addition, several commercial kits are available for DNA extraction [60,62,65].

In DNA extraction, one must naturally consider whether the sample is taken directly from the infected human tissue which contains large amounts of human DNA and perhaps other microbes or, supposedly, a pure culture. Even in the latter case, however, pieces of human tissue debris and contaminating microbes may hide on agar under the fungal mycelium and be inadvertently taken in the sample. The toughness and other physical characters of the sample may also contribute to the outcome.

Typically, pure fungal cultures (either agar- or broth-based) are minced at room temperature (or in liquid nitrogen if RNA is to be isolated). Fungal DNA is purified by proteinase K-digestion in the presence of sodium dodecyl sulfate (SDS) followed by phenol-chloroform extraction together with 0.5 M KCl. The extract is then treated with RNase (Sigma) for 30 min at 37°C, extracted again with phenol-chloroform, precipitated by isopropanol and resuspended in a final concentration of 10 μg/mL.

To isolate DNA from nails, commercial kit based on glass beads can be utilized. Namely, 0.5–3 mg of nail material in 300 μL cell lysis solution is added with 5–10 mg glass beads (0.5 mm in diameter; BioSpec Products Inc.). The mixture is vortexed for 1 min. After incubation for 1 h in a 65°C, 3 μL proteinase K solution (20 mg/mL) is added. The tube is incubated overnight at 55°C and vortexed again. Then, 100 μL of protein precipitation solution is added. The tube is mixed a few times by inversion and centrifuged for 5 min. After incubation at −20°C overnight, the tube is centrifuged for 10 min. The DNA pellet is dissolved in 10–30 μL hydration solution [66].

52.2.2 Detection Procedures

52.2.2.1 Specific Detection of the Family Microascaceae

Principle: DNA bases or primers reliably specific to the whole family Microascaceae have not been designed. This is partly because the family itself is deficiently studied and, secondly, because the phylogenetic clade of Microascaceae is surrounded by an ocean of other, much less known taxa. Even from the clinical point of view, there is a clear need to chart the whole surrounding of the family, not to speak of the general needs of the very poorly funded basic research on fungal diversity.

Due to the current lack of information, one can identify a given strain in the natural Microascaceae only by its relative molecular phylogenetical resemblance to some known species, not based on any defined molecular borders or current definitions of taxa.

Procedure: To identify a given strain as a member of the family Microascaceae, one should sequence sufficient lengths of its genome, for example, parts of LSU, ITS, and complementing protein-coding genes. The sequences should be analyzed with proper phylogenetic methods, including parsimony analysis, with all data existing in public data banks.

Note: Naturally, all data available by traditional methods such as conidiogenesis must be evaluated together with the molecular tests. With normally sporulating strains, the typical conidia may reveal the broad affinities in a couple of days, and in some cases may be already present in human tissue a assessed by direct microscopy. Therefore, preliminary screening by microscopy provides a cost-effective, genus-level identification that can be subsequently verified by molecular procedures such as DNA sequencing.

52.2.2.2 *Microascus*-Specific Detection

Principle: LSU bases preliminarily characteristic to the main subclades of Microascaceae have been proposed by Issakainen et al. [3]. These, when continuously complemented with recent and future sequencing results, can be used for specific detection in the approximate levels of genus and subgenus. Due to lack of basic molecular mapping and delimitation, the molecular detection of individual *Microascus* species is still very unreliable. This situation will certainly improve as soon as studies based on ITS regions and on complementing protein genes will cover some hundreds of strains from *Microascus* and from all the closest genera.

Kardjeva et al. [66] described a PCR assay for *S. brevicaulis* targeting the large subunit of the rRNA gene with the SCOP primer pair (5′-GAA TGC TGC TCA AAA TGG GAG-3′ and 5′-CAT TAC GCC AGC ATC CTTG C-3′), which results in the amplification of a specific 336 bp fragment. Given that *M. brevicaulis*, *Microascus manginii* (anamorph: *Scopulariopsis candida*), and *S. koningii* share sequence homology in the region concerned, a positive result with the SCOP primers could indicate any of these three species. On the other hand, *S. brevicaulis* is more common in association with nails than either of the other two species

Procedure

1. The PCR mixture (25 μL) is comprised of 2.5 U Taq DNA polymerase and 1× PCR buffer (containing 10 mM Tris–HCl pH 9.0, 50 mM KCl, and 1.5 mM $MgCl_2$), 200 μM of each deoxynucleoside triphosphate, 15 ng of template DNA, 20 pmol each of the SCOP primers and distilled water (to 25 μL).
2. The PCR amplification is conducted in a thermocycler with the following program: 1 cycle of 94°C for 3 min; 35 cycles of 94°C for 45 sec, 69°C for 1 min, and 72°C for 1 min; and a final cycle of 72°C for 10 min.
3. Five microliters of each reaction is loaded on 1.2% agarose gels and electrophoresed for 1.5 h at 100 V in 0.5× TBE buffer (89 mM Tris-borate, 2.5 mM EDTA pH 8.3). The gels are stained with ethidium bromide and photographed.

Note: The SCOP primers are able to detect as little as 150 pg of template DNA. Given the fact that 10 pg of target DNA generally corresponds to circa 25 fungal cells or colony forming units per clinical specimens, the detection limit of the SCOP assay should correspond to 375 or more cells per clinical specimen. Sequences derived from the 336 bp PCR products shows a 100% identity with the three species that are identical in this region.

52.3 CONCLUSION AND FUTURE PERSPECTIVES

Microascus, together with some less known saprophytic genera, constitutes more than a half of the family Microascaceae as currently outlined [2,4,7,11,14,54,67,69]. A smaller part of the family consists of a separate clade which contains other human pathogenic genera, especially *Pseudallescheria* (see Chapter 58 in this book) [68].

Regarding the clinical importance and molecular detection of *Microascus*, *M. brevicaulis* stands out as a regular nail pathogen [20]. Kardjeva et al. [66] has designed a LSU rDNA-based molecular test, which seems to be specific for this clade but not for species. Tests based on ITS or other variable areas can be designed in near future for species level. However, the work of Filipello Marchisio et al. [41] point out the need for molecular or other tests which are targeted to the enzyme armamentarium of a given *Microascus* strain, since this could directly differentiate between nail pathogenic and contaminating strains.

With the advent of nucleic acid amplification technology along with vast improvement in DNA sequencing procedures, it is possible to have the identity of a fungus verified in hours, instead of days or weeks. Species of *Microascus* were successfully included in such tests of Kardjeva et al. [66] and Vollmer et al. [53]. In coming years, the most limiting factor will not be the technology but the effort put in basic mapping of the thousands of environmental fungal strains which may seem to have no importance but which form the phylogenetic framework out of which the pathogens stand out. Reliable primer design and interpretation of results can be made only after such data is at hand.

Development of routineous, affordable molecular tests even for *Microascus* species and their pathogenic strains would help early confirmation of the infecting fungal species in clinical samples [71]. This would facilitate prompt implementation of appropriate treatment regimens that would reduce suffering not only in chronic nail infections but also in the rare, severe infections of the immunocompromised patients [33,35,73–82]. Clinically, the whole family shares the feature of being resistant to many common drugs, especially in the serious infections of immunocompromised patient. In these, even *Microascus* species are lethal. A family-level diagnostic primer of Microascaceae incorporated in multimicrobial screening chips might thus be of clinical interest. Assuming that other indicators of a deep fungal infection are present, a positive result on this test would suggest, for example, that Amphotericin B is not the drug of choice.

Regarding a family-level test, Microascaceae seems to be a natural molecular taxonomic entity which has a rather long evolutionary distance from other groups of pathogenic fungi [1,83,84]. However, the evolutionary distance only applies to the pathogenic species. In the phylogenetic trees, each pathogenic species is closely surrounded by an unknown number of environmental species which have not appeared as pathogens, at least not yet. The same is true even within a single species, considering the varying pathogenities of separate *Microascus* strains [41]. In any event, molecular tests offer very effective new tools, as compared with the conventional methods.

Within *Microascus*, more basic research is definitely needed. Use of shotgun molecular methods and metagenomic approaches will probably dig out a big number of new related fungi from soils. Therefore, it would make sense to design an initial molecular test above the approximate rank of the family and then, stepwise or by nested tests, refine the study to species.

Regarding the development of test procedures, it is relevant to find out whether a given test can be performed of a pathologists' fixed microscopy slide, of a separate fresh tissue specimen, or of a pure fungal culture on a plate—or several of these. Molecular test applicable to histological stainings would elucidate the location and pathogenic role of *Microascus* in nails, for instance.

Speculating on a DNA-based clinical diagnosis of *Microascus*-related fungi within 10 years, for instance, the specimens could perhaps be divided into (a) nails and (b) all other types of specimens including upper respiratory cavities, eyes, ears, wounds, catheter sites, and skin abscesses. Due to the repeated occurrence of *M. brevicaulis* in the skin and ears, the classification of the latter specimens may be debated.

Regarding (a) *Microascus* infections of nails, the benign course and contaminated nature of such infections do actually mask a very complex question. Based on current knowledge, an ideal molecular test would simultaneously combine (i) in situ histological microscopy to confirm the nail tissue invasion; (ii) in situ identification of *M. brevicaulis* on species level; (iii) identification of other Microascaceae (including

Pseudallescheria and *Petriella*) at least on family level; (iv) presence of keratinolytic enzymes to recognize pathogenic strains; (v) simultaneous presence of dermatophytes, perhaps also of selected bacteria; (vi) perhaps unknown host immunity factors; and (vii) perhaps molecular markers of drug resistance. All this should be based on conventional laboratory procedures to ensure that test results are not based on gross contamination or other elementary pitfalls [13,14,46,85].

Regarding (b) other types of infections due to *Microascus*, especially in immunosuppressed patients: A good minimalistic molecular test for near future would be a rapid combined test which extracts the following information directly from the fresh patient tissue specimen of human tissue: (i) Family Microascaceae (including *Microascus*) positive or negative; (ii) *Pseudallescheria boydii—Scedosporium apiospermum* species complex (including the new species) positive or negative; (iii) *Scedosporium prolificans* positive or negative; and perhaps (iv) *Microascus manginii* clade positive or negative. Assuming good selectivity and sensitivity by the first three tests, already, one could in principle infer the risk of a serious *Microascus* infection from the result that the family is positive while both groups of *Scedosporium* are negative. Such a result would imply "a rare microascaceous infection," which warrants further study. The diagnosis should naturally be backed by conventional mycological and histological tests, by entirely other types of diagnosis and by more focused sequencing, according to need. In deep specimens, rapid species-level testing for *M. brevicaulis* would not be of primary importance because in the small number of serious *Microascus* infections published thus far, the causative agents have been either unidentified or other species than *M. brevicaulis*.

REFERENCES

1. Tang, A.M.C., Jeewon, R., and Hyde, K.D., Phylogenetic utility of protein (RPB2, beta-tubulin) and ribosomal (LSU, SSU) gene sequences in the systematics of Sordariomycetes (Ascomycota, Fungi). *Anton. Leeuwen.*, 91, 327, 2007.
2. Issakainen, J., Microascaceae (Ascomycota, Fungi) as a medical mycological problem. Molecular relationships and occurrence in Finnish patients (PhD thesis). *Ann. Univ. Turkuensis, Ser. A II, Tom.*, 236, 1–156, 2009.
3. Issakainen, J. et al., Relationships of *Scopulariopsis* based on LSU rDNA sequences. *Med. Mycol.*, 41, 31, 2003.
4. von Arx, J.A., Figueras, M.J., and Guarro, J., Sordariaceous Ascomycetes without ascospore ejaculation. *Beih. Nova Hedwigia*, 94, 1, 1988.
5. Dykstra, M.J., Salkin, I.F., and McGinnis, M.R., An ultrastructural comparison of conidiogenesis in *Scedosporium apiospermum*, *Scedosporium inflatum* and *Scopulariopsis brumptii*. *Mycologia*, 81, 896, 1989.
6. Issakainen, J. et al., Deep, respiratory tract and ear infections caused by *Pseudallescheria* (*Scedosporium*) and *Microascus* (*Scopulariopsis*) in Finland. A 10-year retrospective multicenter study. *Med. Mycol.*, 48, 458, 2010.
7. Morton, F.J. and Smith, G., The genera *Scopulariopsis* Bainier, *Microascus* Zukal, and *Doratomyces* Corda. *Mycol. Pap.*, 86, 1, 1963.
8. Abbott, S.P., Sigler, L., and Currah, R.S., *Microascus brevicaulis* sp. nov., the teleomorph of *Scopulariopsis brevicaulis*, supports placement of *Scopulariopsis* with the Microascaceae. *Mycologia*, 90, 297, 1998.
9. Abbott, S.P. and Sigler, L., Heterothallism in the Microascaceae demonstrated by three species in the *Scopulariopsis brevicaulis* series. *Mycologia*, 93, 1211, 2001.
10. Abbott, S.P., Lumley, T.C., and Sigler, L., Use of holomorph characters to delimit *Microascus nidicola* and *M. soppii* sp. nov., with notes on the genus *Pithoascus*. *Mycologia*, 94, 362, 2002.
11. Malloch, D. and Cain, R.F., The genus *Kernia*. *Can. J. Bot.*, 49, 855, 1971.
12. Malloch, D. and Hubart, J.-M., An undescribed species of *Microascus* from the Cave of Ramioul. *Can. J. Bot.*, 65, 2384, 1986.
13. de Hoog, G.S. et al., *Atlas of Clinical Fungi*, 2nd edn., vol. 1. Centraalbureau voor Schimmelcultures, Utrecht, the Netherlands, 2000.
14. de Hoog, G.S. et al., *Atlas of Clinical Fungi Cd-Rom (CD-ROM)*, 2nd edn., ASM Press, Washington, DC, 2007.
15. Mycology Online at http://www.mycology.adelaide.edu.au/, accessed on August 1, 2010.
16. Issakainen, J. et al., Occurrence of *Scopulariopsis* and *Scedosporium* in nails and keratinous skin. A 5-year retrospective multi-center study. *Med. Mycol.*, 45, 201, 2007.
17. André, J. and Achten, G., Onychomycosis. *Int. J. Dermatol.*, 26, 481, 1987.
18. Tosti, A. et al., Treatment of nondermatophyte mold and *Candida* onychomycosis. *Dermatol. Clin.*, 21, 491, 2003.
19. Romano, C., Gianni, C., and Difonzo, E.M., Retrospective study of onychomycosis in Italy: 1985–2000. *Mycoses*, 48, 42, 2005.
20. Summerbell, R.C. et al., Onychomycosis: A critical study of techniques and criteria for confirming the etiologic significance of nondermatophytes. *Med. Mycol.*, 43, 39, 2005.
21. López-Jodra, O. and Torres-Rodríguez, J.M., Unusual fungal species causing onychomycosis. *Rev. Iberoam. Micol.*, 16, S15, 1999.
22. Svejgaard, E.L. and Nilsson, J., Onychomycosis in Denmark: Prevalence of fungal nail infection in general practice. *Mycoses*, 47, 131, 2004.
23. García-Martos, P. et al., Onychomycoses caused by non-dermatophytic filamentous fungi in Cádiz. *Enferm. Infecc. Microbiol. Clin.*, 18, 319, 2000.
24. Escobar, M.L. and Carmona-Fonseca, J., Onychomycosis by common non-dermatophyte moulds, *Rev. Iberoam. Micol.*, 20, 6, 2003.
25. Järv, H. et al., Toenail onychomycosis in Estonia. *Mycoses*, 47, 57, 2004.
26. Lateur, N., Mortaki, A., and André, J., Two hundred ninety-six cases of onychomycosis in children and teenagers: A 10-year laboratory survey. *Pediatr. Dermatol.*, 20, 385, 2003.
27. Gupta, A.K. and Gregurek-Novak, T., Efficacy of itrakonazole, terbinafine, fluconazole, griseofulvin and ketoconazole in the treatment of *Scopulariopsis brevicaulis* causing onychomycosis of the toes. *Dermatology*, 202, 235, 2001.
28. Besbes, M. et al., Otomycose à *Scopulariopis brevicaulis*. *Rev. Laryngol. Otol. Rhinol.*, 123, 77, 2002.
29. Neglia, J.P., Invasive *Scopulariopsis* in the immunocompromised host. *Am. J. Med.*, 83, 1163, 1987.
30. Ellison, M.D. et al., Report of the first case of invasive fungal sinusitis caused by *Scopulariopsis acremonium*: Review of *Scopulariopsis* infections. *Arch. Otolaryngol. Head Neck Surg.*, 124, 1014, 1998.

31. Jabor, M.A., Greer, D.L., and Amedee, R.G., *Scopulariopsis*: An invasive nasal infection. *Am. J. Rhinol.*, 12, 367, 1998.
32. Endo, S. et al., *Scopulariopsis* fungus ball. *Ann. Thorac. Surg.*, 74, 926, 2002.
33. Migrino, R.Q., Hall, G.S., and Longworth, D.L., Deep tissue infections caused by *Scopulariopsis brevicaulis*: Report of a case of prosthetic valve endocarditis and review. *Clin. Infect. Dis.*, 21, 672, 1994.
34. Mohammedi, I. et al., Fatal *Microascus trigonosporus* (anamorph *Scopulariopsis*) pneumonia in a bone marrow transplant recipient. *Eur. J. Clin. Microbiol. Infect. Dis.*, 23, 215, 2004.
35. Steinbach, W.J. et al., Fatal *Scopulariopsis brevicaulis* infection in a paediatric stem-cell transplant patient treated with voriconazole and caspofungin and a review of *Scopulariopsis* infections in immunocompromized patients. *J. Infect.*, 48, 112, 2004.
36. Nwabuisi, C. et al., *Scopulariopsis* associated meningitis in adult Nigerian AIDS patient—A case report. *West Afr. J. Med.*, 22, 364, 2003.
37. Hart, A.P. et al., Cerebral phaeohyphomycosis caused by dematiaceous *Scopulariopsis* species. *Clin. Neuropathol.*, 20, 224, 2001.
38. Chen-Scarabelli, C. and Scarabelli, T.M. Fungal endocarditis due to *Scopulariopsis*. *Ann. Intern. Med.*, 139, W77, 2003.
39. Ng, K.P. et al., *Scopulariopsis brevicaulis* infection in a patient with acute myeloid leukemia. *Med. J. Malaysia*, 58, 608, 2003.
40. Ellis, D.H. et al., Non-dermatophytes in onychomycosis of toenails. *Br. J. Dermatol.*, 136, 490, 1997.
41. Filipello Marchisio, V., Fusconi, A., and Querio, F.L., *Scopulariopsis brevicaulis*: A keratinophilic or a keratinolytic fungus? *Mycoses*, 43, 281, 2000.
42. Bruynzeel, I. and Starink, T.M., Granulomatous skin infection caused by *Scopulariopsis brevicaulis*. *J. Am. Acad. Dermatol.*, 39, 365, 1998.
43. Sellier, P. et al., Recurrent subcutaneous infection due to *Scopulariopsis brevicaulis* in a liver transplant recipient. *Clin. Infect. Dis.*, 30, 820, 2000.
44. Krisher, K.K. et al., Disseminated *Microascus cirrosus* infection in pediatric bone marrow transplant recipient. *J. Clin. Microbiol.*, 33, 735, 1995.
45. Haneke, E., Fungal infections of the nail. *Semin. Dermatol.*, 10, 41, 1991.
46. Larone, D.H. (ed.), *Medically Important Fungi—A Guide to Identification*, 4th edn., ASM Press, Washington, DC, 2002.
47. Carillo-Munõz, A.J. et al., In vitro activity of voriconazole against dermatophytes, *Scopulariopsis brevicaulis* and other opportunistic fungi as agents of onychomycosis. *Int. J. Antimicrob. Agents*, 30, 157, 2007.
48. Summerbell, R.C., The benomyl test as a fundamental diagnostic method for medical mycology. *J. Clin. Microbiol.*, 31, 572, 1993.
49. de Hoog, G.S. et al., Ecology and physiology of the emerging opportunistic fungi *Pseudallescheria boydii* and *Scedosporium prolificans*. *Mycoses*, 37, 71, 1994. [Erratum in: *Mycoses*, 37, 223, 1994.]
50. Sage, L. et al., Description and physiological properties of two new varieties and one species of *Microascus* from Israel. *Mycotaxon*, 55, 189, 1995.
51. Kaufman, L. et al., Immunohistologic identification of Aspergillus spp. and other hyaline fungi by using polyclonal fluorescent antibodies. *J. Clin. Microbiol.*, 35, 2206, 1997.
52. Choi, J.K., Mauger, J., and McGowan, K.L., Immunohistochemical detection of *Aspergillus* species in pediatric tissue samples. *Am. J. Clin. Pathol.*, 121, 18, 2004.
53. Vollmer, T. et al., Evaluation of a novel broad-range real-time PCR assay for the rapid detection of human pathogenic fungi in various clinical specimens. *J. Clin. Microbiol.*, 46, 1919, 2008.
54. Wedde, M. et al., PCR-based identification of clinically relevant *Pseudallescheria/Scedosporium* strains. *Med. Mycol.*, 36, 61, 1998.
55. Alastruey-Izquierdo, A. et al., Prevalence and susceptibility testing of new species of *pseudallescheria* and *scedosporium* in a collection of clinical mold isolates. *Antimicrob. Agents Chemother.*, 51, 748, 2007.
56. Zeng, J.S. et al., Intraspecific diversity of species of the *Pseudallescheria boydii* complex. *Med. Mycol.*, 45, 547, 2007.
57. Hull, P.R., Gupta, A.K., and Summerbell, R.C., Onychomycosis: An evaluation of three sampling methods. *J. Am. Acad. Dermatol.*, 39, 1015, 1998.
58. Weiland, J.J., Rapid procedure for the extraction of DNA from fungal spores and mycelia. *Fungal Genet. Newsl.*, 44, 60, 1997.
59. LaMontagne, M.G. et al., Evaluation of extraction and purification methods for obtaining PCR-amplifiable DNA from compost for microbial community analysis. *J. Microbiol. Methods*, 49, 255, 2002.
60. Rantakokko-Jalava, K. and Jalava, J., Optimal DNA isolation method for detection of bacteria in clinical specimens by broad-range PCR. *J. Clin. Microbiol.*, 40, 4211, 2002.
61. Faggi, E., Pini, G., and Campisi, E., Use of magnetic beads to extract fungal DNA. *Mycoses*, 48, 3, 2005.
62. Fredricks, D.N., Smith, C., and Meier, A., Comparison of six DNA extraction methods for recovery of fungal DNA as assessed by quantitative PCR. *J. Clin. Microbiol.*, 43, 5122, 2005.
63. Cankar, K. et al., Critical points of DNA quantification by real-time PCR—Effects of DNA extraction method and sample matrix on quantification of genetically modified organisms. *BMC Biotechnol.*, 6, 37, 2006.
64. Lugert, R., Schettler, C., and Gross, U., Comparison of different protocols for DNA preparation and PCR for the detection of fungal pathogens in vitro. *Mycoses*, 49, 298, 2006.
65. Scupham, A.J., Jones, J.A., and Wesley, I.V., Comparison of DNA extraction methods for analysis of turkey cecal microbiota. *J. Appl. Microbiol.*, 102, 401, 2007.
66. Kardjeva, V. et al., Forty-eight-hour diagnosis of onychomycosis with subtyping of *Trichophyton rubrum* strains. *J. Clin. Microbiol.*, 44, 1419, 2006.
67. Baron, G.L., Gain, R.F., and Gilman, J.C., The genus *Microascus*. *Can. J. Bot.*, 39, 1609, 1961.
68. Issakainen, J. et al., Relatedness of *Pseudallescheria*, *Scedosporium* and *Graphium pro parte* based on SSU rDNA sequences. *J. Med. Vet. Mycol.*, 35, 389, 1997.
69. Issakainen, J. et al., Relationship of *S. prolificans* with *Petriella* (Microascales, Ascomycetes) confirmed by partial LSU rDNA sequences. *Mycol. Res.*, 103, 1179, 1999.
70. Monod, M. et al., Direct mycological examination in dermatology: A comparison of different methods. *Dermatologica*, 179, 183, 1989.
71. Monod, M. et al., Fast and reliable PCR/sequencing/RFLP assay for identification of fungi in onychomycoses. *J. Med. Microbiol.*, 55, 1211, 2006.

72. Baran, R. et al., A new classification of onychomycosis. *Br. J. Dermatol.*, 139, 567, 1998.
73. Agarwal, G.P. and Singh, S.M., *Microascus cinereus* infection of human nail. *Indian J. Med. Sci.*, 34, 263, 1980.
74. Aznar, C., de Bievre, C., and Guiguen, C., Maxillary sinusitis from *Microascus cinereus* and *Aspergillus repens*. *Mycopathologia*, 105, 93, 1989.
75. Aguilar, C., Pujol, I., and Guarro, J., In vitro antifungal susceptibilities of *Scopulariopsis* isolates. *Antimicrob. Agents Chemother.* 43, 1520, 1999.
76. Celard, M. et al., Early *Microascus cinereus* endocarditis of a prosthetic valve implanted after *Staphylococcus aureus* endocarditis of the native valve. *Clin. Infect. Dis.*, 29, 691, 1999.
77. Baddley, J.W. et al., *Microascus cinereus* (anamorph *Scopulariopsis*) brain abscess in a bone marrow transplant recipient. *J. Clin. Microbiol.*, 38, 395, 2000.
78. López-Jodra, O. and Torres-Rodríguez, J.M., Unusual fungal species causing onychomycosis. *Rev. Iberoam. Micol.*, 16(S), S11, 1999.
79. Heikkilä, H. and Stubb, S., The prevalence of onychomycosis in Finland. *Br. J. Dermatol.*, 133, 699, 1995.
80. Marques, A.R. et al., Suppurative cutaneous granulomata caused by *Microascus cinereus* in a patient with chronic granulomatous disease. *Clin. Infect. Dis.*, 20, 110, 1995.
81. Skou, J.P. *Microascus exsertus* sp. nov. associated with a leaf-cutting bee, with considerations on relationships of species in the genus *Microascus* Zukal. *Anton. Leeuwen.*, 39, 529, 1973.
82. Wuyts, W. et al., Fatal *Scopulariopsis* infection in a lung transplant recipient: A case report. *J. Heart Lung Transpl.*, 24, 2301, 2005.
83. Berbee, M.L. and Taylor, J.W., 18S ribosomal sequence data and dating, classifying, and ranking the fungi. In Hawksworth, D.L. (ed.), *Ascomycete Systematics: Problems and Perspectives in the Nineties*, Plenum Press, New York, 1994, pp. 213–223.
84. Tehler, A., Little, D.P., and Farris, J.S., The full-length phylogenetic tree from 1551 ribosomal sequences of chitinous fungi. *Mycol. Res.*, 107, 901, 2003.
85. Sutton, D.A., Fothergill, A.W., and Rinaldi, M.G. (ed.), *Guide to Clinically Significant Fungi*, 1st edn., Williams & Wilkins, Baltimore, MD, 1998.

53 *Myceliophthora* and *Thielavia*

Dongyou Liu

CONTENTS

53.1 INTRODUCTION

53.1.1 Classification and Morphology

The genus *Myceliophthora* is the only member of the mitosporic Chaetomiaceae group, family Chaetomiaceae, order Sordariales, class *Sordariomycetes*, subphylum Pezizomycotina, phylum Ascomycota, and kingdom Fungi. The family Chaetomiaceae covers the genera of *Achaetomium*, *Aporothielavia*, *Chaetomidium*, *Chaetomium*, *Corylomyces*, *Corynascus*, *Farrowia*, *Thielavia*, and *Zopfiella* as well as mitosporic *Chaetomiaceae* and unclassified Chaetomiaceae. Currently, the genus *Myceliophthora* consists of eight species: *Myceliophthora fergusii*, *Myceliophthora fusca*, *Myceliophthora hinnulea*, *Myceliophthora indica*, *Myceliophthora lutea*, *Myceliophthora sulphurea*, *Myceliophthora thermophila*, and *Myceliophthora vellercam*, of which *Myceliophthora thermophila* (obsolete synonyms: *Chrysosporium thermophilum* and *Sporotrichum thermophilum*) is an anamorph of *Thielavia heterothallica* [1].

The genus *Thielavia* in the family Chaetomiaceae consists of 20 recognized species: *Thielavia appendiculata*, *Thielavia arenaria*, *Thielavia australiensis*, *Thielavia basicola*, *Thielavia cephalothecoides*, *Thielavia coactilis*, *Thielavia fragilis*, *Thielavia gigaspora*, *Thielavia heterothallica*, *Thielavia hyalocarpa*, *Thielavia hyrcaniae*, *Thielavia intermedia*, *Thielavia microspora*, *Thielavia minuta*, *Thielavia ovispora*, *Thielavia rapa-nuensis*, *Thielavia subthermophila*, *Thielavia terrestris*, *Thielavia terricola*, and *Thielavia tortuosa* as well as three unassigned species (Thielavia sp. 1863, Thielavia sp. B27, and Thielavia sp. MUCL 40242) [1,2]. Of these, *Thielavia heterothallica* (the ascomycetous teleomorph or perfect state of *Myceliophthora thermophila*) and *Thielavia terrestris* are relevant to human infections.

Myceliophthora thermophila colonies reach 40 mm in diameter after 2 weeks (25°C) on potato flake agar (PFA), with more luxuriant colonies growing at the elevated temperatures. Colonies are initially white and cottony and later become pale brown and powdery, without a well-defined margin. Submerged hyphae (up to 6.0 μm in width) are wider than aerial hyphae (0.8–3.0 μm). Blastoconidia are borne terminally or laterally on the hyphae, sometimes with short or long pedicels and occasionally with a secondary blastoconidium being produced from the distal end of the first. One to four blastoconidia are borne on one hyphal cell or ampulliform swelling. Conidia (4.5–11.0 × 3.0–4.5 μm) are obovoid to pyriform, hyaline, smooth, and thick walled and become rough walled at maturity [3].

Because its hyphae lack clamp connections (which are common in fungi with basiomycetous affinities), *M. thermophila* was transferred from the genus *Sporotrichum* to *Chrysosporium*. However, given that *Chrysosponum* spp. show a maximum growth temperature of 45°C rather than 48°C and produce arthroconidia, the fungus was further transferred to the genus *Myceliophthora*.

Thielavia is a teleomorph of *Myceiophthora*, with an optimal growth temperature of 37°C. On potato dextrose agar (PDA) and Sabouroud dextrose agar (SDA), *Thielavia* forms floccose, dark gray, broadly-spreading colonies, with a distinct felt-like aerial mycelium on front and a gray to black background on reverse. Its mycelium consists of hyaline and dark olivaceous, branched, and septated hyphae, whose walls are smooth and measure 1.0–3.0 μm. Aleuriospores (conidia) appear laterally and terminally on hyphae. Being single-celled, hyaline, or light brown in color, spores are broadly clavate with a truncate base, and measure 5×3 μm in diameter. In mature cultures, ascomata (cleistothecia) develop within a mycelial mat measuring up to 180 μm. The wall of ascomata consists of tectura epidermoidea coated with dark hyphae. The broad, fusiform ascospores (16×9 μm) are single-celled, with a chacteristic superficial germ pore (measuring 1.0–1.5 μm) [7].

53.1.2 Clinical Features and Pathogenesis

Myceliophthora thermophila is a thermophilic, saprobic fungus that is commonly present in silage, fresh alfalfa grass forage, hay, wheat straw compost, wood pulp, birch chips, dry pasture soil, and animal excreta. As a disease-causing agent of cultivated mushrooms, the organism occasionally infects humans after traumatic injuries in agricultural settings or cardiac surgery. Individuals with severe immune system abrogation are also vulnerable to *M. thermophila* infection primarily through environmental contamination. Clinical manifestations range from single subcutaneous nodules, keratitis, vegetation growth in the great vessels and heart to disseminated disease and death [5,6].

Destino et al. [7] described a case of *Myceliophthora thermophila* knee and distal femur infection in a 4-year-old boy following a barnyard pitchfork injury. The patient developed osteomyelitis of the distal femur, with severe destructive osseous and cartilaginous infection. Culture of debrided bone or cartilage yielded *M. thermophila*, although fungal elements could be seen histologically in other surgical samples. Clinical improvement required the prolonged use of multiple antifungal agents.

In a separate study, Tekkok et al. [8] documented a case of combined *Clostridium perfringens* and *Myceliophthora thermophila* infections in a 21-month-old male patient, who sustained a penetrating head injury from a rusty nail in a barnyard. The patient developed a gas-containing left parietal brain abscess and presented with high fever, galeal swelling, and seizure. Initial cultures of wound aspirations grew *Clostridium perfringens*. As the wound failed to respond to antibiotic treatment, an excised specimen was obtained 6 weeks after the injury, which yielded a saprophytic and opportunistic fungus, *Myceliophthora thermophila*. The patient recovered with en bloc resection of the lesion, 6 weeks of amphotericin B, and 4 months of itraconazole.

Farina et al. [9] described the first isolation of a human pathogenic strain of *M. thermophila*, causing fatal vasculitis in a 22-year-old Italian woman affected by cystic medial necrosis. The patient developed fungal aortitis after cardiac surgery for aortic insufficiency. After experiencing two episodes of septic embolization, the patient underwent replacement of the aortic root and initial ascending aorta by a homograft. The lumina of the ascending aorta, aortic arch, and the origin of the innominate artery were completely filled with vegetation, from which the phaeoid thermophilic hyphomycete *Myceliophthora thermophila* was isolated in pure culture. The patient succumbed to the infection and its complications.

Weitzel et al. [10] reported a case of *Myceliophthora thermophila* infection in a 40-year-old man after an left ventricle assist device (LVAD) placement for dilated cardiomyopathy and inotrope-dependent New York Heart Association class IV heart failure as a bridge to transplantation. Two months later, the patient presented to the hospital for evaluation of fever, chills, and altered mental status. Computed tomography imaging of his chest revealed an abnormality located at the aortic outflow graft from the LVAD, and transesophageal echocardiography (TEE) imaging allowed visualization of a large mass or vegetation located at the aortic insertion of the LVAD outflow conduit along with a severely dilated and hypokinetic LV. The patient was taken to the operating room for exploration and replacement of the existing LVAD for presumed LVAD endocarditis. Histologic examination of the outflow graft showed extensive fungal overgrowth. Cultures of the mass yielded a rare environmental thermophilic fungus, *Myceliophthora thermophila*. Blood cultures also demonstrated fungal growth after 7 days of incubation. The patient undertook aggressive antifungal therapy with both IV liposomal amphotericin B 10 mg/kg daily for 6 weeks and voriconazole 400 mg twice daily orally. After explanation of the infected LVAD, the patient's blood cultures cleared. The patient was discharged 27 days after presentation and remained on oral voriconazole for chronic suppression until the time of his heart transplant 2 years later.

Theoulakis et al. [4] reported the first case of fungal keratitis due to *Thielavia* sp. involving a 10-year-old girl. Two weeks after suffering from ocular plant injury, the patient presented with pain and corneal stromal infiltration with central ulceration and ill-defined margins. Cultures of corneal scrapings and biopsy yielded a fungus that was confirmed as *Thielavia subthermophila* by sequence analysis of the ribosomal internal transcribed spacer (ITS) region. The organism responded to treatment of topical amphotericin B and oral voriconazole.

53.1.3 Diagnosis

Myceliophthora thermophila is a heterothallic, thermophilic fungus with spherical, black, nonostiolate cleistothecia containing ellipsoidal, evanescent asci. Asci contain eight one-celled ellipsoidal ascospores and are deep brown to black, with one germ pore. As a pigmented filamentous hyphomycete and an agent of phaeohyphomycosis, melanin in its cell wall provides a distinguishing feature. However, cell walls of phaeoid molds may appear hyaline or clear upon routine microscopy despite the presence of melanin. Staining of hyphal elements with Masson-Fontana melanin stain usually demonstrates pigment, permitting identification of a dark fungus. *M. thermophila* grows on a variety of media recommended for mold identification (e.g., potato dextrose or 2% malt agar) with optimal temperatures at 35°C–42°C, although it also grows well near 50°C, reflecting its epithet "thermophila." Considering that most clinical specimens are incubated at 30°C, the less than optimal growth temperature for *M. thermophila* may be a contributing factor for the relative infrequent isolation of this thermophilic fungus, which requires elevated temperatures for enhanced recovery.

Macroscopic and microscopic identification of *Myceliophthora thermophila* can be achieved upon observation of (i) tan to brown powdery colonies with ill-defined margins when grown on potato flakes agar at 42°C; (ii) more luxuriant growth at elevated temperatures of 35°C and 42°C than at 26°C; (iii) septate vegetative hyphae with conidial production from ampulliform swellings; and (iv) obovoid (inverted egg shaped) or pyriform (pear shaped) conidia measuring 4.5–11.0 × 3.0–4.5 μm that are hyaline and smooth when immature, becoming darker and roughened at maturity [7].

Since conventional laboratory diagnosis of fungal infections is dependent on recovery of fungi from culture of clinical specimens for subsequent identification of the presence of reproductive structures, the process can be lengthy (up to 21 days and inconclusive [some molds fail to sporulate]). The development and application of rapid, accurate diagnostic techniques are essential for the initiation of targeted antifungal therapy for a widening spectrum of emerging fungal pathogens, especially in immunocompromised individuals. Independent of mold sporulation, PCR amplification and sequencing of target regions within the ribosomal RNA gene complex (e.g., internal transcribed spacer (ITS) regions 1 and 2 located between the highly conserved small (18S) and large (28S) ribosomal subunit genes in the rRNA operon) allow identification to the species level for many fungi including *Myceliophthora thermophila* and serve as a valuable approach to detect clinically relevant pathogens [11].

53.2 METHODS

53.2.1 Sample Preparation

Clinical specimens are examined with a direct calcofluor white stain for septate hyphal elements. Portions of the samples are inoculated on inhibitory mold agar (containing cyclohexamide), modified Sabouraud agars, or potato dextrose agar. Microscopic structures are observed on tease or tape preparations and slide cultures for up to 21 days [12].

After the growth for 1–7 days on potato dextrose agar slants, lysates are prepared from approximately 1 cm^2 of mycelia with IDI lysis kits (GeneOhm Sciences, San Diego, CA). Briefly, in a biological safety cabinet, mycelia are collected by scraping the slant with a sterile stick in 1 mL of sterile, molecular-grade H_2O. The material is transferred to a 2 mL screw-cap tube. The tubes are centrifuged for 1 min at 6000 × *g*. If the mycelia do not pellet, the material is contained with a pediatric blood serum filter (Porex Corp., Fairburn, GA). Supernatant is removed. The material is resuspended in 200 μL of IDI sample buffer and transferred to the lysis tube, which contained glass beads. Lysis tubes are vortexed on the highest setting for 5 min. The tubes are placed in a boiling water bath for 15 min. Tubes are centrifuged for 5 min at 16,000 × *g*. The supernatant is stored at −20°C until amplification [12].

Alternatively, about 1 cm^2 of fungal material is transferred to a 2 mL Eppendorf tube containing a 2:1 (wt/wt) mixture of silica gel and Celite (silica gel H, Merck 7736/Kieselguhr Celite 545; Machery) and 300 μL of TES buffer (2 g Tris (hydroxymethyl)-aminomethane, 0.38 g Na-EDTA, and 2 g sodium dodecyl sulfate in 80 mL of ultrapure water (pH 8)). The fungal material is ground with a micropestle for 1–2 min. The volume is adjusted by adding 200 μL of TES buffer. After vigorous shaking and the addition of 10 μL of a 10 mg/mL concentration of proteinase K to the tube, the mixture is incubated at 65°C for 10 min. With the addition of 140 μL of 5 M NaCl solution, the mixture is combined with 1/10 volume (~65 μL) of CTAB (cetyltrimethylammonium bromide) buffer 10%, followed by incubation for another 30 min at 65°C. One volume (~700 μL) of chloroform–isoamyl alcohol (vol/vol = 24/L) is added and mixed by inversion. After incubation for 30 min at 0°C (on ice water) and centrifugation at 14,000 rpm at 4°C for 10 min, the top layer is transferred to a clean Eppendorf tube. The sample is added with 225 μL of 5 M NH_4-acetate and incubated for at least 30 min (on ice water) and centrifuged. The supernatant is transferred to a clean sterile Eppendorf tube and mixed with a 0.55 volume (~510 μL) of ice-cold isopropanol. After centrifugation for 7 min at 14,000 rpm and 4°C (or room temperature), the supernatant is decanted. The pellet is washed with ice-cold ethanol 70% two times and dried by using a vacuum dryer. The powder is resuspended in 48.5 μL of Tris-EDTA buffer with 1.5 μL of 10 mg of RNase/mL, incubated at 37°C for 15–30 min, and stored at −20°C until use [13].

53.2.2 Detection Procedures

Pounder [12] described a real-time PCR with SYBR green DNA-binding dye and amplicon-melting temperature analysis for fungal detection using pan-fungal primers ITS1 forward (5′-TCCGTAGGTGAACCTGCGG-3′) and ITS4 reverse (5′-TCCTCCGCTTATTGATATGC-3′). The identity of the fungi is verified by subsequent sequencing analysis.

Procedure

1. PCR mixture is composed of 1× Lightcycler FastStart DNA Master Hybridization Probes mixture (Roche Applied Science) containing deoxynucleoside triphosphates, FastStart *Taq* DNA polymerase, and 1 mM $MgCl_2$ (additional $MgCl_2$ is added to a final concentration of 4.6 mM), 0.4 μM each of ITS1 forward and ITS4 reverse primers, 1× SYBR green (Molecular Probes), and 3 μL template DNA.
2. Thermal cycling parameters include 95°C for 10 min; 50 cycles of 95°C for 5 s, 60°C for 20 s, and 76°C for 30 s; and a final extension at 72°C for 2 min.
3. The quality of the amplicon is determined using the derivative of the melt analysis curve (55°C–99°C, 45 s hold at 55°C, and 5 s/°C) using the RotorGene 3000 (Corbett Robotics, Inc.).
4. The amplified product is purified for bidirectional sequencing using ExoSAP-IT (USB Corp.). Five μL of Big Dye Terminator Ready Reaction Mix v. 1.1 (Applied Biosystems) is added to 4 μL of each primer (0.8 pmol/μL) and 3 μL of purified PCR product. Cycle sequencing is performed with a 9700 thermal cycler (ABI), using 25 cycles of 96°C for 10 s, 50°C for 5 s, and 60°C for 4 min. Sequencing reaction products are passed through a Sephadex

G-50 fine column to remove unincorporated dye terminators. Purified sequencing reaction products are run on an ABI Prism 3100 Genetic Analyzer with a 50 cm capillary array.

5. Sequences are analyzed with the SmartGene Integrated Database Network software version 3.2.3 vr. SmartGene is a web-based software and database system with reference sequences derived from the National Center for Biological Information (NCBI) GenBank repository.

Note. If preferred or real time PCR instrument is not available, standard PCR may be performed with primers ITS1 and ITS4, and the resulting amplicon is sequenced with the same primers. Sequence-based identifications are defined by percent identity: species, ≥99%; genus, 93%–99%; and inconclusive, ≤93%. For strains producing discrepant identification between the methods based on phenotypic characteristics and ITS sequence analysis, the D1–D2 region of the large-subunit RNA gene is amplified with primers NL1 (5′-GCATATCAATAAGCGGAGGAAAAG-3′) and NL4 (5′-GGTCCGTGTTTCAAGACGG-3′) and sequenced for species clarification [14].

53.3 CONCLUSION

Myceliophthora thermophila is a saprobic, thermophilic phaeoid mold that commonly occurs in pasture soil, wood chips, straw, moldy hay, compost piles, and other environmental settings where heat is generated. It is also present in the excreta and rumen of cattle. As a pathogen of cultivated mushrooms, *M. thermophila* may cause localized infections in immunocompetent individuals through traumatic injury and disseminated infections in immunocompromised patients, with a propensity for the vascular system. Due possibly to the fact that clinical specimens are routinely cultured at 30°C and that many saprobic fungi are discarded as contaminants without further analysis, thermophilic *M. thermophila* is infrequently reported. With the help of molecular techniques such as PCR amplification and sequencing of rRNA genes and internal transcribed spacer (ITS) regions, the true extent of opportunistic *M. thermophila* infections in humans will become clearer in future.

REFERENCES

1. The UniProt Consortium. Available at http://www.uniprot.org/, accessed on August 1, 2010.
2. Moustafa AW, Abdel-Azeem AM. *Thielavia gigaspora*, a new thermotolerant ascomycete from Egypt. *Microbiol. Res.* 2008;163(4):441–444.
3. Bourbeau P et al. Fatal disseminated infection caused by *Myceliophthora thermophila*, a new agent of mycosis: Case history and laboratory characteristics. *J. Clin. Microbiol.* 1992;30(11):3019–3023.
4. Theoulakis P et al. Keratitis resulting from *Thielavia subthermophila* Mouchacca. *Cornea* 2009;28:1067–1069.
5. Abdel-Hafez AI, el-Sharouny HM. Keratinophilic and saprophytic fungi isolated from students' nails in Egypt. *J. Basic Microbiol.* 1990;30(1):3–11.
6. Revankar SG et al. Disseminated phaeohyphomycosis: Review of an emerging mycosis. *Clin. Infect. Dis.* 2002;34:467–476.
7. Destino L et al. Severe osteomyelitis caused by *Myceliophthora thermophila* after a pitchfork injury. *Ann. Clin. Microbiol. Antimicrob.* 2006;5:21.
8. Tekkok IH, Higgins MJ, Ventureyra EC. Posttraumatic gas-containing brain abscess caused by *Clostridium perfringens* with unique simultaneous fungal suppuration by *Myceliophthora thermophila*: Case report. *Neurosurgery* 1996;39:1247–1251.
9. Farina C et al. Fatal aortic *Myceliophthora thermophila* infection in a patient affected by cystic medial necrosis. *Med. Mycol.* 1998;36:113–118.
10. Weitzel N et al. Left ventricular assist device outflow cannula obstruction by the rare environmental fungus *Myceliophthora thermophila*. *Anesth. Analg.* 2009;108:73–75.
11. Bagyalakshmi R et al. Newer emerging pathogens of ocular non-sporulating molds (NSM) identified by polymerase chain reaction (PCR)-based DNA sequencing technique targeting internal transcribed spacer (ITS) region. *Curr. Eye Res.* 2008;33:139–147.
12. Pounder JI. Discovering potential pathogens among fungi identified as nonsporulating molds. *J. Clin. Microbiol.* 2007; 45(2):568–571.
13. Zeng JS et al. Spectrum of clinically relevant *Exophiala* species in the United States. *J. Clin. Microbiol.* 2007;45:3713–3720.
14. Leaw SN et al. Identification of medically important yeast species by sequence analysis of the internal transcribed spacer regions. *J. Clin. Microbiol.* 2006;44:693–699.

54 *Neocosmosporas*

Palanisamy Manikandan, Csaba Vágvölgyi, Venkatapathy Narendran, Kanesan Panneer Selvam, and László Kredics

CONTENTS

54.1 INTRODUCTION

54.1.1 Classification

Neocosmospora is a teleomorph, homothallic, filamentous fungal genus belonging to the *Fusarium solani* species complex also known as *Fusarium* section *Martiella* (*Ascomycota*, *Pezizomycotina*, *Sordariomycetes*, *Hypocreomycetidae*, *Hypocreales*, *Nectriaceae*). The genus was introduced in 1899 by Smith.[1] He described the type species of the genus as *N. vasinfecta*, and differentiated *N. vasinfecta* var. *vasinfecta*, var. *tracheiphila*, and var. *nivea* that were isolated from diseased cotton (*Gossypium* sp.), cowpea (*Vigna unguiculata*), and watermelon (*Citrullus lanatus*), respectively. Cross-inoculation experiments revealed that none of these races could infect the host plants of the others; however, no other solid criteria for the separation of the three races could be defined.[1] Smith erroneously regarded *Fusarium vasinfectum* (later redefined as *F. oxysporum* var. *vasinfectum*) as an anamorph of *N. vasinfecta*. In 1903, von Jaczewski[2] introduced a new variety, *N. vasinfecta* var. *sesame*, from *Sesamum orientale*. A new taxon, *N. africana*, was described from Africa in 1955,[3] with cerebriform rather than rugose ascospores. This was followed by morphological and developmental investigations of the two taxa.[4–7] Both species were found later in Japan.[8] *N. ornamentata* differing in the shape and degree of rugoseness of the spores was described by Barbosa,[9] while new varieties under the names of *N. vasinfecta* var. *minor*[10] and var. *conidiifera*[11] were also introduced. Later, Udagawa and Horie,[12] Mahoney,[13] and Ueda and Udagawa[14] described *N. striata*, *N. parva*, and *N. tenuicristata*, respectively.

In 1984, Cannon and Hawksworth[15] performed a revision of the genus on the basis of scanning electron microscopical investigation of ascospore ornamentation. *N. vasinfecta* and *N. africana* were reduced to the level of varieties of *N. vasinfecta* as var. *vasinfecta* and var. *africana*, and the existence of diploid and tetraploid races corresponding to small and large ascospore morphs was postulated. Further, all previously described *N. vasinfecta* varieties (var. *tracheiphila*, var. *nivea*, var. *sesame*, var. *minor*, and var. *conidiifera*) were rejected and *N. ornamentata* was reduced to synonimity with *N. vasinfecta*. The previously described species *N. tenuicristata*, *N. parva*, *N. striata*, as well as *N. indica* Wadhwani,[16] were accepted and included in the revised taxonomy of the genus.[15] Table 54.1 shows morphological characters of these five species.

In 1989 Udagawa et al.[17] described further two species of the genus, *N. arxii* and *N. boninensis*. *N. endophytica* with a *Penicillifera* anamorph,[18] *N. diparietispora*,[19] and *N. spinulosa* were also introduced.[20]

The introduction of molecular tools for studying the taxonomic relationships of fungi revealed further information about the phylogeny of *Neocosmospora*. Using sequences of 18S and 28S ribosomal DNA it was demonstrated that *N. vasinfecta* is closely related to *Nectria haematococca*.[21,22] O'Donnell[23] suggested that the *Acremonium*-like anamorph of *N. vasinfecta* is comparable to the microconidial form of fusaria. During the revision of *Hypocreaceae* in 1999, Rossman et al.[24] excluded the species having green ascospores and *Penicillifer* anamorphs from *Neocosmospora* and replaced them to the genus *Viridispora*. This revision accepted the species *N. vasinfecta*, *N. boninensis*, *N. indica*,

TABLE 54.1
Key Morphological Characters of Five Species from the Genus *Neocosmospora*

Character	*N. vasinfecta*	*N. tenuicristata*	*N. parva*	*N. indica*	*N. striata*
Mycelium	White to pale buff, floccose	White to buff	Inconspicuous, hyaline	Very pale buff, somewhat darker toward the edges, floccose	Inconspicuous, hyaline
Hyphae	Septate, tending to aggregate in strands, there anastomosing	Tending to aggregate in strands, there anastomosing	Septate, tending to aggregate in strands	Septate, tending to aggregate in rope-like strands, there anastomosing	Septate
Ascomata	Orange-brown to red, ± globose, glabrous but for a number of rhizoidal hyphae, ostiolate, neck lined with periphyses	Orange-red to reddish brown, ± globose, ± glabrous or covered with an inconspicuous web of hyaline hyphae, ostiolate, neck lined with periphyses	Orange-brown, very fragile, ovoid to pyriform, glabrous but for a number of rhizoidal hyphae, ostiolate, neck lined with periphyses	Formed freely at edge of mycelium, orange to red, globose, glabrous but for a number of rhizoidal hyphae, ostiolate with a short neck lined with periphyses	Dark brown, pyriform, glabrous but for a number of rhizoidal hyphae, ostiolate, periphyses not observed
Ascomatal wall	Several-layered, outer pigmented wall of ± thick-walled *textura angularis*, sometimes with an inconspicuous covering hyphal web, inner layers ± hyaline, also of *textura angularis*	Several-layered, outer pigmented wall of ± thick-walled *textura angularis*, inner layers hyaline, also of *textura angularis*	Several-layered, with a pigmented outer wall of irregular *textura intricata*, inner layer almost hyaline, consisting of an ill-defined intermediate between *textura intricata* and *textura angularis*	Several-layered, with a pigmented outer layer of *textura intricata*, and thin hyaline inner layer(s) of *textura angularis*	Pigmented outer layer, almost hyaline inner layer, both of *textura epidermoidea*
Periphysoids	Present but evanescent at an early stage, consisting of vertically oriented rows of hyaline thin-walled cells	Present at maturity, composed of very thin-walled hyaline cells	Not observed	Evanescent at an early stage	Not observed
Asci	Cylindrical, thin-walled, stalk 8–15 μm long, body without discernable apical structures, not evanescent, eight-(rarely six to seven) spored	Cylindrical, thin-walled, stalk 5–12 (–20) μm long, body without discernable apical structures, not evanescent, eight-spored	Cylindrical to cylindric-clavate, thin-walled, short-stalked, without discernible apical structures, not evanescent, eight-spored	Clavate, 38–49 × 15.5–22 μm, the longer asci tending to be thinner, very thin-walled, without discernible apical structures, evanescent, eight-spored	60–85 × 6–10 μm; cylindrical to cylindric-clavate, thin-walled, with a small inconspicuous apical ring which does not stain blue in iodine, eight-spored
Ascospores	Uniseriately arranged, buff to salmon pink in mass, pale yellow individually, globose to ellipsoidal	Uniseriately arranged, yellowish-brown, globose-ellipsoidal to ellipsoidal	Obliquely uniseriately or partially biseriately arranged, orange-yellow to orange-buff en masse, golden yellow individually, ellipsoidal	Biseriately arranged, orange brown en masse, yellow-brown individually, irregularly ellipsoidal	Obliquely uniseriately or partially biseriately arranged, very pale yellow to hyaline, ellipsoidal
Ascospore wall	Germ pores absent, apical spore commonly has an irregularly thickened wall	Without germ-pores, often somewhat irregular in thickness, with an ornamentation of transversely arranged ridges, apical spore commonly has an irregularly thickened wall	Without germ-pores, with a reticulate ornamentation mostly obscured by a weakly verrucose epispore layer	Without germ-pores, inner edge of wall elliptical in section, wall very irregular in thickness, reticulate, the reticulations obscured in irregular patches by a weakly verrucose epispore layer	Without germ-pores, some spores with a median transverse straight or curved septum, then very slightly constricted at the septum with transverse hyaline flanges

Source: Based on Cannon, P.F. and Hawksworth, D.L., *Trans. Br. Mycol. Soc.*, 82, 673, 1984.

N. parva, *N. spinulosa*, *N. tenuicristata*, and *N. vasinfecta* var. *africana*.

In a subsequent study, O'Donnell[25] used sequence data from the nuclear large subunit 28S rDNA, the nuclear ribosomal internal transcribed spacer (ITS) region, and from introns and exons of the translation elongation factor 1α (*tef1*) gene for studying the molecular phylogeny of the *F. solani* species complex. The results clearly demonstrated that *Neocosmospora* is deeply nested within this complex with *Fusarium* sp. anamorphs. This molecular phylogenetic study divided the *F. solani* species complex into three clades, with *Neocosmospora* belonging to clade 3. O'Donnell[25] formally suggested synonimity of the genus *Haematonectria* Samuels and Nirenberg—a taxon proposed by Rossman et al.[24] to include *Nectria haematococca* and the closely related species *N. illudens*, *N. ipomoeae*, and *N. monilifera*—with *Neocosmospora* and noted that inclusion of morphologically similar but phylogenetically unrelated species like *N. parva* within *Neocosmospora* results in polyphyly of the genus. More recent molecular phylogenetic studies examined the *F. solani* species complex with special emphasis on its members capable of causing infections in humans.[26,27] Zhang et al.[26] first reported that all known species from clinical and veterinary sources—including *N. vasinfecta*—belong to clade 3 of the species complex. This was confirmed in a study of O'Donnell et al.[27] where the ITS, 28S rDNA and *tef1* sequence data were supplemented with sequence data from the second largest subunit of the RNA polymerase II (*rpb2*) gene in a multilocus approach. O'Donnell et al.[27] noted that besides *F. solani* f. sp. *cucurbitae* race 2,[28] *N. vasinfecta* is the only other representative of the *F. solani* species complex which has been shown to be pathogenic both to plants and to humans.

54.1.2 Epidemiology and Pathogenesis

Species of *Neocosmospora* are almost exclusively tropical or subtropical. *N. vasinfecta* var. *vasinfecta* has been isolated from different locations including South Carolina (United States), Guinea Bissau, Nigeria, Sierra Leone, Tanzania, Zaire, Guyana, India, Bangladesh, Pakistan, Hong Kong, and Australia, while *N. vasinfecta* var. *africana* is known from Egypt, Ethiopia, Gambia, Guyana, Nigeria, Sierra Leone, Zambia, South Africa, Uzbhekistan, India, Pakistan, Sri Lanka, Venezuela,[15] and Argentina.[29] *N. boninensis*, *N. parva*, *N. spinulosa*, and *N. tenuicristata* are known only from their type isolates, which derived from Japan, Galapagos Islands, Brazil, and Japan, respectively, while *N. indica* is known from two isolations from India.

Neocosmospora species can most commonly be isolated from soil, often from the rhizosphere of plants[15]; however, they were also reported to occur in special habitats. The type strain of *N. tenuicristata* has been isolated from marine sludge.[14] *N. vasinfecta* species could be recovered from hairs of rodents in Nigeria,[30] and it is also among the common genera that can be isolated from the females and cysts of nematodes.[31–34] Due to its frequent association with nematodes, *N. vasinfecta* was considered a possible myconematicide. Initial studies reported that this fungus did not parazitize eggs of the soybean cyst nematode *Heterodera glycines*[35] and also did not measurably affect egg viability.[33] In contrast, Chen et al.[36,37] found that *N. vasinfecta* is capable of penetrating the cyst wall of *H. glycines* and is moderately pathogenic to eggs with a hatch reduction ability between 21% and 56%. In a subsequent study, filtrates from *N. vasinfecta* and the closely related *F. solani* grown in malt extract broth were toxic to second-stage juveniles of *H. glycines* and also reduced egg viability.[38] The substances in the culture filtrates of the fungus may play a role themselves in colonization of eggs, or may have anti-microbial activities that enable competition with other fungi in soil and cysts.

The type species of the genus, *N. vasinfecta* is considered a weak pathogen of roots causing little crop loss, though it may have a significant effect in conjunction with other pathogens.[15] *N. vasinfecta* is known to cause root- and fruit-rot and seedling damping off in Malvaceae, Leguminosae, Piperaceae, Cucurbitaceae, etc. including pepper, peanuts, beans, pigeon pea, coconuts, *Albizziae*, and others.[39–42] The fungus is a presumed pathogen of the medicinal plant *Astragalus membranaceus*,[43] while certain strains are known to be able to infect soybean (*Glycine max*).[8,44] Nematodes can be suspected as possible vectors of *Neocosmospora* species, and may also facilitate the entering of the fungus into the host plant through nematode lesions, as suggested by Gray et al.[45] As a preventive measure, Smit and Knox-Davies[46] suggested hot-water treatment and acid scarification for the elimination of *N. vasinfecta* from tea seeds.

Temporini and VanEtten[47] provided experimental evidence for *Neocosmospora* sp. NRRL 22470 (termed by the authors as *N. boniensis*), that this species is capable of causing disease on pea (*Pisum sativum*). The experiment was performed after the discovery of the homologues for the pea pathogenicity genes (PEP cluster) of *Haematonectria haematococca* mating population VI in strain NRRL 22470. The PEP cluster involves the *pda1* gene, the product of which converts the pea phytoalexin pisatin to a less toxic compound by demethylation, and is therefore a presumed virulence factor. The PEP cluster shows a discontinuous phylogenetic distribution: it could not be found in other representatives of the *F. solani* species complex but was detected in the more distantly related pea pathogen *F. oxysporum* f. sp. *pisi*, suggesting its possible translocation via horizontal gene transfer.[47]

Watanabe et al.[48] isolated and characterized a soybean saponin hydrolase from a soil-derived isolate of *N. vasinfecta* var. *vasinfecta*, which is capable of degrading soybean saponin and generating soyasapogenol B. The gene encoding for the enzyme has also been cloned, sequenced, and successfully expressed in *T. viride* enabling further characterization. Due to their ability to form complex sterols in fungal membranes and to cause a loss of membrane integrity, saponins are supposed to take part in the protection of the producer plants from phytopathogenic fungi.[49] Deglycosylated saponins show reduced antifungal activities; saponin hydrolases appear therefore to facilitate the virulence of plant pathogenic fungi in host plants. As *N. vasinfecta* is capable of infecting soybean, its saponin hydrolase may thus represent a possible virulence factor. *N. vasinfecta* was also reported to be able to hydrolyse inulin, a polysaccharide contained in the roots of certain plants including sunflower.[50] The inulinase (2,1-beta-D-fructanohydrolase) hydrolyses inulin into pure fructose, being an excellent alternative for the production of fructose syrups. Inulin-hydrolyzing fungi might therefore be used in biotechnological processes.

Phytotoxic metabolites may also contribute to the virulence of *N. vasinfecta* as a plant pathogen. In the 1970s, phytotoxic naphtharizine derivatives were reported from *N. vasinfecta*[39,51] and supposed to be related to pathogenesis; however, Roos[52] found no correlation between their production and virulence of the fungus. *N. vasinfecta* var. *africana* NHL2298, a strain isolated from soil at Johannesburg, South Africa was found to produce an α-pyrone named neovasinone, which accelerated the root elongation of lettuce seedlings,[53] and the related metabolite neovasinin, which is phytotoxic against soybean.[54,55] Neovasinin is a fungal metabolite containing a unique bicyclic unit, 2H,5H-pyrano(4,3-*b*)pyran-2-one. To explore the biosynthetic pathway of neovasinin, a search was conducted for biogenetically related metabolites. This resulted in the identification and structure determination of further *Neocosmospora* metabolites including neovasipyrones A and B, neovasifuranones A and B[56] as well as neovasipyridones A–F.[57,58] Two phytotoxic metabolites with a benzofuranone ring, vasinfectins A and B—which are diastereomers of each other with the structure of (E)-2-(3-hydroxy-2,4-dimethyl-1-hexenyl)-2,7-dimethyl-5-propanoyl-2H-benzo[b]furan-3-one—were also isolated from *N. vasinfecta* NHL2298.[59] Vasinfectins A and B were found to cause chlorosis at 0.3 μg/leaf on young soybean leaves in the leaf spot assay. Cutler et al.[60] isolated the plant growth regulatory metabolite radicicol—5-chloro-6-(7,8-epoxy-10-hydoxy-2-oxo-3,5-undecadienyl)-β-resorcylic acid μ-lactone, also known as monorden—from the culture filtrates of *N. tenuicristata*. Radicicol significantly inhibited the growth of etiolated wheat coleoptiles, caused slight necrosis on 10-day-old corn plants, and also proved to moderately inhibit *Bacillus subtilis* and *B. cereus*. Radicicol produced by *N. tenuicristata* proved to be a cell differentiation modulator; it inhibited embryonic angiogenesis in a bioassay system involving chorioallantoic membranes of growing chick embryos.[61] The abilities of radicicol to inhibit both the proliferation of, and plasminogen activator production by vascular endothelial cells were suggested to be involved in the anti-angiogenic action. The authors concluded that radicicol might be a potential drug for treating different angiogenesis-dependent diseases, such as solid tumors, psoriasis, rheumatoid arthritis, and diabetic retinopathy.[61]

N. vasinfecta var. *vasinfecta*, var. *africana*, and *N. boninensis* were also shown to produce cyclosporins with antifungal activities against *Alternaria kikuchiana*.[62,63] Cyclosporins are fungal metabolites with powerful immunosuppressive properties, that can be applied for the prevention of rejection in organ transplant recipients. On the other hand, these metabolites might be among the possible virulence factors of *N. vasinfecta* as a human pathogen.

Neocosmospora species may carry transposable elements. Close relatives of *Fot1*, a member of the *pogo* transposon family prevalent in the *F. oxysporum* complex were found in five species from the distant section *Martiella*, including a *Neocosmospora* sp. isolate from African soil.[64] Two copies of the element, *Neo1* and *Neo2* were found in the isolate showing 98% nucleotide identity with each other. As *Fot1* elements could not be detected in other sections that are more closely related to the *F. oxysporum* complex, the authors suggested horizontal gene transfer as the possible explanation for the discontinuous distribution.

Since the 1960s, *N. vasinfecta* has been used as a model organism for research on fungal transport and metabolism of inorganic ions including ammonia,[65,66] bicarbonate,[67] potassium,[68,69] chloride,[70–74] zinc,[75–77] magnesium,[78] cadmium,[79] aluminum,[80] and mercury,[81] as well as organic compounds including acetate,[82] glucose,[83] and polyols related with osmotic adjustment.[84,85]

54.1.3 Clinical Features

The first report about the clinical involvement of the genus *Neocosmospora* dates back to 1993.[86,87] It was the description of a case of leg granuloma (a localized mycotic cyst of soft tissue in the leg) caused by *N. vasinfecta* in a renal transplant and dialysis patient from Senegal. The latency period between infection and clinical manifestation was suggested to be several years with a low level of inflammation. A circulatory source for the inoculum was suggested as there was no trauma that occurred at the infection site. Surgery and a small dose of ketoconazole (truncated prematurely because of a cyclosporin toxicity side effect) eliminated the infection. Five years later, Kac et al.[88,89] reported a case of osteoarthritis due to *Neocosmospora* in the legs of a French parachutist who suffered an open ankle dislocation in Senegal. The authors initially reported the causal agent as *N. africana*,[88] but in the final paper they modified the diagnosis to *N. vasinfecta*[89] according to the revision of the genus by Cannon and Hawksworth.[15] The fungus grew repeatedly from the wound and from the progressing osteitis despite extensive cleaning measures and treatment with amphotericin B (AMB). The minimal inhibitory concentration of AMB toward the isolate was low; however, the parenteral antifungal therapy failed. Both of the above mentioned patients were cured by radical surgical interventions that prevented dissemination of the disease, limiting it to the lower extremities. Because of the poor susceptibility of the fungus, complete surgical excision prompted by early diagnosis was the only means to influence the clinical outcome.

In 2001, a disseminated *N. vasinfecta* infection had been reported by Cornely et al.[90] in a 38-year-old patient from Germany with acute nonlymphocytic leukemia who later on died of multiorgan failure. The initial clinical manifestations included a 4-week history of fatigue, vertigo, and bone pain and subsequently the patient developed granulocytopenia with persistent fever and interstitial and patchy lung infiltrates. Investigations also revealed hepatosplenomegaly and hypodensic lesions in the liver, suggesting possible hepatosplenic involvement of fungal disease. The systemic fungal infection was confirmed as the mold grew in swab cultures of a lateral purulent ulcerous lesion of the right foot, in sputum samples, transtracheal aspirates, a second biopsy of the foot lesion, and in blood cultures and the treatment with AMB and liposomal AMB failed.

Based on the information of the NCBI Nucleotide database, a *N. vasinfecta* infection occurred in a pediatric burn patient, which was the first registered case in the American continent.[91,92] A case of corneal ulcer due to *N. vasinfecta* (initially diagnosed as *Fusarium* spp.) was reported by Manikandan et al.[93] in an immunocompetent patient from Aravind Eye Hospital, Coimbatore, South India. The patient being a farmer cultivating cotton and groundnut presented with pain, redness, and defective vision in his left eye of 5 days duration. Initially, *Fusarium* keratitis was diagnosed based on the fast growing, white colonies with cottony aerial mycelia on media and observation of the microscopic morphology of hyaline, straight or slightly curved microconidia. This diagnosis was later revised, as the isolate was identified as *N. vasinfecta* based on the teleomorph morphology and molecular methods. Further, the recipient corneal button from therapeutic keratoplasty showed fungal growth after 3 days, which was again identified as *N. vasinfecta*. Since the publication of this case report, a second case (unpublished) of *N. vasinfecta* keratitis has also been diagnosed in the same hospital, from a 77-year-old one eyed male patient who was known to be diabetic and asthmatic. The patient required therapeutic keratoplasty. Furthermore, the isolates from the *F. solani* species complex studied by O'Donnell et al.[27] included an additional clinical isolate of *N. vasinfecta* isolated from human eye in Louisiana.

A low minimal inhibitory concentration of AMB was reported by Kac et al.[89] against the isolated *N. vasinfecta* strain. Cornely et al.[90] reported the minimal inhibitory concentrations (MICs) of AMB and 5-flucytosine (5FC) as well as susceptibilities to fluconazole (FLC), itraconazole (ITC), terbinafine (TRB), and voriconazole (VRC) for their case isolate as follows: AMB: >8 μg/mL, 5FC: >128 μg/mL, FLC: >128 μg/mL, ITC: >2 μg/mL, VRC: 1 μg/mL, and TRB: 0.125 μg/mL. The overall in vitro susceptibility test results confirmed the resistance of the isolate to AMB and FLC and nearly all of the few other antifungal agents available for systemic therapy. In contrast, Manikandan et al.[93] determined that the keratitis isolate of *N. vasinfecta* was susceptible to AMB which was in congruence with the success of the postoperative AMB treatment. Nevertheless, they also reported that the strain proved to be resistant to natamycin (NAT), ketoconazole, econazole, and clotrimazole at the highest concentration evaluated, with MICs of >1024, >32, >32, and >32 μg/mL, respectively. O'Donnell et al.[27] reported the antifungal susceptibilities of AMB, NAT, 5FC, ITC, VRC, TRB, posaconazole (POS), anidulafungin (ANID) caspofungin (CAS), and micafungin (MICA) against a panel of members of the *Fusarium solani* species complex, including a clinical isolate of *N. vasinfecta*. Upon 48 h of incubation, the MICs were determined to be AMB: 8 μg/mL; 5FC: >64 μg/mL; ITC: >8 μg/mL; VRC: >8 μg/mL; POS: >8 μg/mL; NAT: 4 μg/mL; CAS: >16 μg/mL; MICA: >16 μg/mL; ANID: >16 μg/mL; and TRB: 1 μg/mL against the isolate.

54.1.4 Diagnosis

54.1.4.1 Conventional Techniques

Although their anamorphs grow easily in routine culture media, the teleomorphs of *Hypocreales* are rarely found in culture. *Neocosmospora* is an exception from this point of view, as it produces perithecia in culture enabling colony- and micromorphology-based identification. Samples can be cultivated on Sabouraud agar at 30°C and 37°C in ambient air; the fast-growing, white-to-pale-buff colonies develop without special requirements. The micromorphological features of the 3- to 4-day-old cultures resemble a *Fusarium* or *Acremonium* species. The fast growing fungal colonies on PDA are flat, thin, and appear almost transparent. Microscopic observations reveal hyaline, elongated to cylindrical conidia aggregated in slimy heads on conidiogenous cells developing on undifferentiated hyphae. The size of conidia varies from 5–10 μm in length to 2–3 μm in width and are mostly single celled or with one septum. Some conidia appear slightly curved and intercalary chlamydospores are also observed. Numerous ascomata (perithecia) form within 10–14 days that give the colony a punctate appearance (incubated at light and dark conditions at 25°C). Supported by subculturing on oatmeal agar with *Lupinus* sp., orange-brown to copper-colored fruiting bodies develop until day 8, which can be identified as perithecial ascomata (200–300 μm diameter) after 14 days of incubation.[90] Ascomata are subspherical (300–400 μm in diameter), multilayered and smooth-walled, each with an apical pore. Cylindrical asci, 90–110 × 10–12 μm in diameter, are present inside the ascomata, each containing eight ascospores in a row. The ascospores are brownish, spherical to ellipsoidal, 10–15 to 8–12 μm, with thick roughened walls and no germ pore present (Table 54.1; Figures 54.1 and 54.2).[93]

54.1.4.2 Molecular Techniques

As described above, *N. vasinfecta* can be identified by conventional methods based on morphological characters. However, a long time is needed until formation of perithecial ascomata, and the micromorphological features of *N. vasinfecta* isolates may suggest *Acremonium* or *Fusarium* species in the early days of incubation, which could lead to an inadequate diagnosis and inappropriate therapeutic regimen. This problem could be overcome by the application of fast and reliable molecular techniques for identification.

Hue et al.[94] designed a PCR primer pair that amplifies a fragment from the rDNA of the main *Fusarium* species and *N. vasinfecta*, but not of 11 other medically important fungi. The method was optimized for blood and tissue samples. In the case of a suspected fungal infection, this technique is

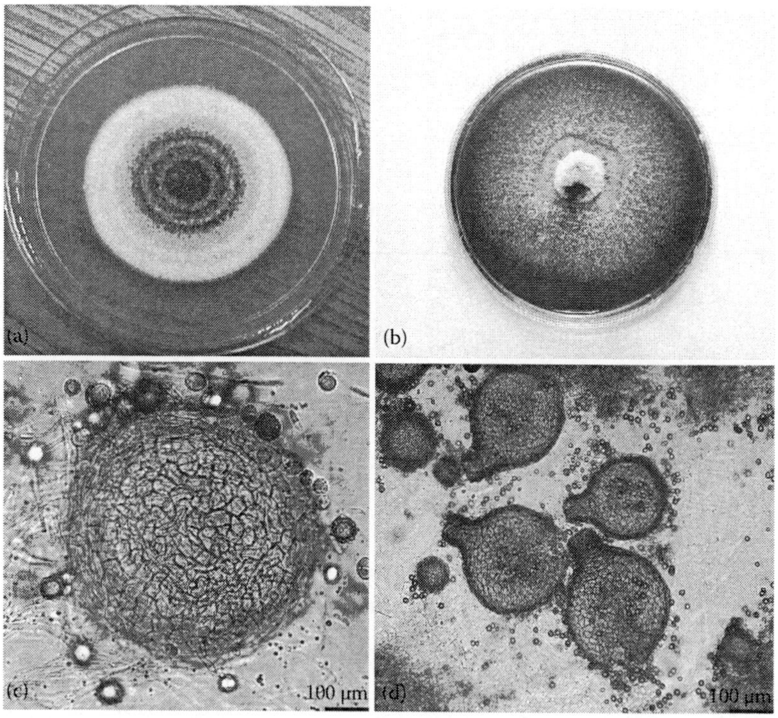

FIGURE 54.1 (a) Culture of *Neocosmospora vasinfecta* grown on PDA at 25°C for 7 days. The colony appears flat, thin, and almost transparent. Rings with punctate appearance due to the presence of numerous perithecia are apparent. (b) Culture of *N. vasinfecta* grown on recipient corneal button (5% sheep blood agar) at 37°C for 7 days. (c) Perithecia development of *N. vasinfecta*. Bar = 100 µm. (d) Perithecial ascomata of *N. vasinfecta* after 14 days on potato dextrose agar with lupine stem. Bar = 100 µm.

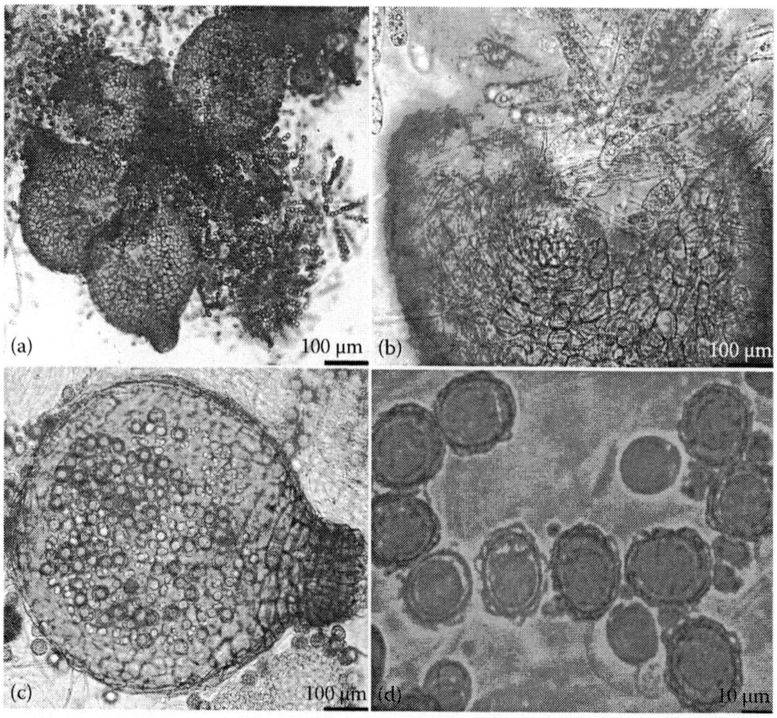

FIGURE 54.2 (a) Perithecia of *Neocosmospora vasinfecta*: orange-brown in color, subspherical (300–450 µm in diameter), smooth-walled and each has an apical pore. (b) Perithecia contain cylindrical asci (80–100 × 11–15 µm in diameter). (c) Each ascus contains eight ascospores. Bar = 100 µm. (d) Ascospores of *N. vasinfecta*: single-celled, uniseriate, brownish, spherical to ellipsoidal, 10.0–15.5 × 7.5–12.0 µm with roughened cell walls and no germ pore. Bar = 10 µm.

applicable for the restriction of the pathogen to fusaria and their teleomorphs, but not for species specific identification.

For the purposes of culture-based molecular identification of clinical *N. vasinfecta* isolates, sequence analysis of the ITS region inside the nuclear rDNA was performed both by Cornely et al.[90] and Manikandan et al.[93] O'Donnell[91] also used ITS along with a partial sequence of the 28S rRNA gene and a fragment of the *tef1* gene for the identification of a *N. vasinfecta* isolate from a pediatric burn patient. Other possibilities include the analysis of sequences from the 18S rRNA gene—which was applied by Yanai[95] for the identification of fungal isolates from arable peat soil as *N. vasinfecta*—and from the second largest subunit of the *rpb2* gene.[27]

54.2 METHODS

54.2.1 Sample Preparation

For DNA isolation, practically any protocol successfully applied for filamentous fungi could be sufficient also for *N. vasinfecta*. For the purposes of molecular identification of the *N. vasinfecta* isolate from corneal ulcer,[93] mycelia grown on yeast extract glucose medium (0.5 g yeast extract, 10 g glucose, 20 g agar in 1000 mL distilled water) at 25°C for 10 days are subjected to DNA isolation by the GeneElute™ Plant Genomic DNA Miniprep Kit (Sigma-Aldrich, St. Louis, MO) according to the manufacturers instructions. In the study of Cornely et al.[90] DNA was extracted by sonication in cetyl trimethyl ammonium bromide buffer (CTAB).[96]

54.2.2 Detection Procedures

The ITS region of the rRNA gene complex can be amplified by universal fungal primers. Subsequent analysis of the nucleotide sequence helps confirm the identity of *N. vasinfecta*.[27,90,93,97]

Manikandan et al.[93] used ITS1 (5′-TCCGTAG GTGAACCTGCGG-3′) and ITS4 (5′-TCCTCCGCTTATTG ATATGC-3′), and performed the PCR in 50 μL volumes containing 5 μL 10× PCR buffer (ZenonBio, Szeged, Hungary), 200 μM each dNTP, 2.5 mM $MgCl_2$, 0.5 μM each primer, 2.5 U of Taq DNA polymerase (ZenonBio, Szeged, Hungary), and 5 μL template DNA. PCR amplification was carried out in a T3 thermocycler (Biometra, Göttingen, Germany) with 1 cycle of 94°C for 5 min, 30 cycles of 94°C for 1 min, 48°C for 1 min, and 72°C for 1 min, and a final elongation step at 72°C for 10 min. The PCR products were purified using GenElute™ MINUS ETBr SPIN COLUMNS (Sigma-Aldrich),[93] or by the Qiagen Gel Extraction Kit (Qiagen).[90] The PCR products were sequenced using the ITS4 primer with the dye deoxy terminator chemistry on an ABI 373A DNA sequencer (Applied Biosystems).[93]

O'Donnell et al.[27] employed the following primers in their study: ITS5 (5′-GGAAGTAAAAGTCGTAACAAGG-3′) and NL4 (5v-GGTCCGTGTTTCAAGACGG-3′) amplifying a 986 bp product from the ITS region and domains D1 and D2 of 28S rRNA; EF1 (5′-ATGGGTAAGGARGACAAGAC-3′) and EF2 (5′-GGARGTACCAGTSATCATG-3′) amplifying a 716 bp product from *tef1*; as well as 5f2 (5′-GGGGWGAY CAGAAGAAGGC-3′) and 7cr (5′-CCCATRGCTTGYTTRC CCAT-3′), and 7cf (5′-ATGGGYAARCAAGCYATGGG-3′) and 11ar (5′-GCRTGGATCTTRTCRTCSACC-3′) amplifying 863 and 875 bp products from the *rpb2* gene, respectively. PCR cycling program consists of 1 cycle of 94°C for 90 s and 40 cycles of 94°C for 30 s, 55°C for 90 s, and 68°C for 2 min, followed by 1 cycle of 68°C for 5 min and a 4°C soak.[27] The PCR primers were utilized for sequencing of the nuclear ribosomal gene product and *rpb2*, while primers EF3 (5′-GTAAGGAGGASAAGACTCACC-3′) and EF22T (5′-AGGAACCCTTACCGAGCTC-3′) were used for sequencing of the *tef1* fragment on an ABI 373 or ABI 377.

Cornely et al.[90] performed PCR amplification and DNA sequencing with the primers ITS2, ITS3, ITS4, and ITS5 (5′-GGAAGTAAAAGTCGTAACAAGG-3′)[97] on an ABI 377 automated sequencer (Applied Biosystems).

Sequence analysis can be performed by BLASTN similarity search at the website of the National Center for Biotechnology Information (http://www.ncbi.nlm.nih.gov/BLAST).[98] GenBank accession numbers of some reference sequences of ITS, *tef1* and 28S rRNA for *N. vasinfecta* are AY381138-AY381143, AY381144-AY381149, and AY381150-AY381155, respectively.[91] Reference sequences of the *rpb2* gene (EU329497, EU329512, and EU329636)[27] and 18S rRNA (AB302197-AB302200, AB302205, AB302207, AB302208, AB302210, and AB302212)[95] are also available for *N. vasinfecta* in the GenBank database.

54.3 CONCLUSION AND FUTURE PERSPECTIVES

Although little is known about *N. vasinfecta* as a pathogen in humans, its clinical significance is gaining momentum as an emerging human fungal pathogen. In particular, human infections due to *N. vasinfecta* have caused serious predicaments in the recent past and have been reported from various parts of the world highlighting organ specific or systemic manifestations. The basic methods used to diagnose infections caused by *N. vasinfecta* in humans have been recognized to be complex and may also be inadequate. In the early days of incubation, the micromorphological features of *N. vasinfecta* isolates may be misleading, suggesting *Fusarium* or *Acremonium* species, which in turn could lead to an inadequate therapeutic regimen because of the different AMB susceptibilities of these genera. This problem could be overcome by the application of molecular techniques for fast and reliable identification. Furthermore, being aware of the varying ranges of MICs among different isolates of the species, each clinical isolate of *N. vasinfecta* should be examined for its minimal inhibitory concentrations to antifungals.

N. vasinfecta is one of the two species within the *F. solani* species complex which have been shown to be pathogenic both to plants and to humans.[27] Such organisms possess the potential to be developed to model organisms to investigate whether common virulence factors are involved in plant and human pathogenesis.

REFERENCES

1. Smith, E.F., Wilt disease of cotton, watermelon and cowpea, *USDA Bull.*, 17, 1, 1899.
2. von Jaczewski, V.A., Über das Vorkommen von *Neocosmospora vasinfecta* E. Smith auf *Sesamum orientale*, *Ann. Mycol.*, 1, 31, 1903.
3. von Arx, J.A., Ein neuer Ascomycet aus Afrika, *Ant. Leeuw.*, 21, 161, 1955.
4. Doguet, G., Morphologie et organogenie du *Neocosmospora vasinfecta* E. F. Smith et du *Neocosmospora africana* von Arx, *Ann. Sci. Nat. Ser. Bot. 11*, 17, 353, 1956.
5. van Warmelo, K.T., Scanning electron microscopy of *Neocosmospora vasinfecta* ascospores, *Mycologia*, 68, 1181, 1976.
6. van Warmelo, K.T., Asexual nuclear division in *Neocosmospora*, *Bothalia*, 12, 247, 1977.
7. van Warmelo, K.T., Sexual nuclear division in *Neocosmospora*, *Bothalia*, 13, 415, 1981.
8. Udagawa, S., *Neocosmospora* in Japan, *Trans. Mycol. Soc. Jpn.*, 4, 121, 1963.
9. Barbosa, M.A., A new species of *Neocosmospora* found in stored peanut, *Garcia de Orto*, 13, 17, 1965.
10. Rama Rao, P., Studies on soil fungi. 1. Ascomycetes from soils of Hyderabad (India), *Mycopathol. Mycol. Appl.*, 21, 217, 1963.
11. Kamyschko, O.P., Ascomycetes novi e terris Asiae mediae, *Novitat. Syst. Plant. Vascul.*, 1965, 115, 1965.
12. Udagawa, S. and Horie, Y., Notes on some Japanese ascomycetes XIV, *Trans. Mycol. Soc. Jpn.*, 16, 337, 1975.
13. Mahoney, D.P., A new *Neocosmospora* from Galapagos Islan soil, *Mycologia*, 68, 1111, 1976.
14. Ueda, S. and Udagawa, S., A new Japanese species of *Neocosmospora* from marine sludges, *Mycotaxon*, 16, 387, 1983.
15. Cannon, P.F. and Hawksworth, D.L., A revision of the genus *Neocosmospora* (*Hypocreales*), *Trans. Br. Mycol. Soc.*, 82, 673, 1984.
16. Wadhwani, Neocosmospora indica Wadhwani, *Indian Bot. Rep.* (1983), 2, 158, 1984.
17. Udagawa, S.-I., Horie, Y., and Cannon, P.F., Two new species of *Neocosmospora* from Japan, with a key to the currently accepted species, *Sydowia*, 41, 349, 1989.
18. Polishook, J.D., Bills, G.F., and Rossman, A.Y., A new species of *Neocosmospora* with a *Penicillifera* anamorph, *Mycologia*, 83, 797, 1991.
19. Rossman, A.Y., Samuels, G.J., and Lowen R., *Leuconectria clusiae* gen. nov. and its anamorph *Gliocephalotrichium bulbilium* with notes on *Pseudonectria*, *Mycologia*, 85, 685, 1993.
20. Pfenning, L., Neocosmospora spinulosa, *Sydowia*, 47, 66, 1995.
21. Rehner, S.A. and Samuels, G.J., Molecular systematics of the *Hypocreales*: A teleomorph gene phylogeny and the status of their anamorph, *Can. J. Bot.*, 73(S1), S816, 1995.
22. Spatafora, J.W. and Blackwell, M., Molecular systematics of unitunicate perithecial ascomycetes. I. The *Clavicipitales—Hypocreales* connection, *Mycologia*, 85, 912, 1993.
23. O'Donnell, K., Progress towards a phylogenetic classification of *Fusarium*, *Sydowia*, 48, 57, 1996.
24. Rossman, A.Y. et al., Genera of *Bionectriaceae*, *Hypocreaceae* and *Nectriaceae* (Hypocreales, Ascomycetes), *Stud. Mycol.*, 42, 1, 1999.
25. O'Donnell, K., Molecular phylogeny of the *Nectria haematococca–Fusarium solani* species complex, *Mycologia*, 92, 919, 2000.
26. Zhang, N. et al., Members of the *Fusarium solani* species complex that cause infections in both humans and plants are common in the environment, *J. Clin. Microbiol.*, 44, 2186, 2006.
27. O'Donnell, K. et al., Molecular phylogenetic diversity, multilocus haplotype nomenclature, and in vitro antifungal resistance within the *Fusarium solani* species complex, *J. Clin. Microbiol.*, 46, 2477, 2008.
28. Mehl, H.L. and Epstein, L., *Fusarium solani* species complex isolates conspecific with *Fusarium solani* f. sp. *cucurbitae* race 2 from naturally infected human and plant tissue and environmental sources are equally virulent on plants, grow at 37°C and are interfertile, *Environ. Microbiol.*, 9, 2189, 2007.
29. Godeas, A.M., Micoflora del suelo de la argentina. I algunas formas ascosporicas de la region chaqueña, *Mycopathol. Mycol. Appl.*, 46, 189, 1972.
30. Okafor, J.I. and Gugnani, H.C., Dermatophytes and other keratinophilic fungi associated with hairs of rodents in Nigeria, *Mykosen*, 24, 616, 1981.
31. Rodriguez-Kabana, R. and Morgan-Jones, G., Potential for nematode control by mycofloras endemic in the tropics, *J. Nematol.*, 20, 191, 1988.
32. Carris, L.M. and Glawe, D.A., Fungi colonizing cysts of *Heterodera glycines*, Bulletin 786, United States Department of Agriculture, University of Illinois at Urbana-Champaign, College of Agriculture, Agricultural Experiment Station, 1989.
33. Meyer, S.L.F., Huettel, R.N., and Sayre, R.M., Isolation of fungi from *Heterodera glycines* and *in vitro* bioassay for their antagonism to eggs, *J. Nematol.*, 22, 532, 1990.
34. Chen, S.Y. et al., Fungi associated with females and cysts of *Heterodera glycines* in a Florida soybean field, *J. Nematol.*, 26, 296, 1994.
35. Godoy, G., Rodriguez-Kabana, R., and Morgan-Jones, G., Parasitism of eggs of *Heterodera glycines* and *Meloidogyne arenaria* by fungi isolated from cysts of *H. glycines*, *Nematropica*, 12, 111, 1982.
36. Chen, S.Y., Dickson, D.W., and Mitchell, D.J., Pathogenicity of fungi to *Heterodera glycines*, *J. Nematol.*, 28, 145, 1996.
37. Chen, S.Y. and Dickson, D.W., Fungal penetration of the cyst wall of *Heterodera glycines*, *Phytopathology*, 86, 319, 1996.
38. Chen, S.Y., Dickson, D.W., and Mitchell, D.J., Viability of *Heterodera glycines* exposed to fungal filtrates, *J. Nematol.*, 32, 190, 2000.
39. Kern, H., Naef-Roth, S., and Défago, G., Zur Toxinbildung von *Neocosmospora vasinfecta* und *Neocosmospora africana*, *Phytopathol. Zeitschr.*, 72, 327, 1971.
40. Thakur, R.N. and Sastry, K.S.M., Studies on wilt disease of *Crotalaria mucronata*, *Indian J. Mycol. Plant Pathol.*, 1, 84, 1971.
41. Domsh, K.H., Gams, W., and Anderson, T.-H., *Compendium of Soil Fungi*, Vols. 1 and 2, Academic Press, London, U.K., 1980.
42. Vishwa, D. et al., Occurrence of pigeon pea wilt caused by *Neocosmospora vasinfecta*, *Indian J. Pulses Res.*, 9, 254, 2005.
43. Sato, T., Matsuhashi, M., and Iida, O., Fungi isolated from diseased medicinal plants, *Bull. Nat. Inst. Hyg. Sci.*, 110, 60–66, 1992.

44. Phillips, D.V., A disease of soybean caused by *Neocosmospora vasinfecta*, *Phytopathology*, 62, 612, 1972.
45. Gray, F.A., Rodriguez-Kaban, R., and Adams, J., *Neocosmospora* stem rot of soybeans in Alabama, *Plant Dis.*, 64, 321, 1980.
46. Smit, W.A. and Knox-Davies, P.S., Elimination of *Diaporthe phaseolorum* and *Neocosmospora vasinfecta* from rooibos tea seeds by hot-water treatment and acid scarification, *Phytophylactica*, 21, 297, 1989.
47. Temporini, E.D. and VanEtten, H.D., An analysis of the phylogenetic distribution of the pea pathogenicity genes of *Nectria haematococca* MPVI supports the hypothesis of their origin by horizontal transfer and uncovers a potentially new pathogen of garden pea: *Neocosmospora boniensis*, *Curr. Genet.*, 46, 29, 2004.
48. Watanabe, M. et al., A novel saponin hydrolase from *Neocosmospora vasinfecta* var. *vasinfecta*, *Appl. Environ. Microbiol.*, 70, 865, 2004.
49. Morrissey, J.P. and Osbourn, A.E., Fungal resistance to plant antibiotics as a mechanism of pathogenesis, *Microbiol. Mol. Biol. Rev.*, 63, 708, 1999.
50. de Souza-Motta, C.M. et al., Identification and characterization of filamentous fungi isolated from the sunflower (*Helianthus annus* L.) rhizosphere according to their capacity to hydrolyse inulin, *Braz. J. Microbiol.*, 34, 273, 2003.
51. Kern, H., Les naphtarizines des *Fusarium*, *Ann. Phytopathol.*, 10, 327, 1978.
52. Roos, A., Zur Physiologie und Pathologie von *Neocosmospora vasinfecta* E. F. Smith, *Phytopathol. Zeitschr.*, 88, 238, 1977.
53. Nakajima, H. et al., Structure of neovasinone, a new α-pyrine plant growth regulator produced by the fungus, *Neocosmospora vasinfecta* E.F. Smith, *Agric. Biol. Chem.*, 51, 1221, 1987.
54. Nakajima, H. et al., Structure of neovasinin, a new metabolite produced by the fungus, *Neocosmospora vasinfecta* E.F. Smith, and its biological activity to lettuce seedlings, *Agric. Biol. Chem.*, 51, 2831, 1987.
55. Nakajima, H. et al., Absolute stereochemistry of neovasinin, a phytotoxin produced by the fungus, *Neocosmospora vasinfecta*, *Biosci. Biotechnol. Biochem.*, 56, 1148, 1992.
56. Furumoto, T. et al., Neovasipyrones and neovasifuranones: four new metabolites related to neovasinin, a phytotoxin of the fungus *Neocosmospora vasinfecta*, *Phytochemistry*, 40, 745, 1995.
57. Nakajima, H. et al., Neovasipyridones A, B and C: metabolites related to neovasinin, a phytotoxin of the fungus, *Neocosmospora vasinfecta*, *Phytochemistry*, 40, 1643, 1995.
58. Nakajima, H. et al., Neovasipyridones related to neovasinin, a phytotoxin of the fungus, *Neocosmospora vasinfecta*, *Phytochemistry*, 43, 1015, 1996.
59. Furumoto, T., Hamasaki, T. and Nakajima, H., Vasinfectins A and B: new phytotoxins from *Neocosmospora vasinfecta*, *Tetrahedron Lett.*, 38, 5523, 1997.
60. Cutler, H.G. et al., Monorden from a novel source, *Neocosmospora tenuicristata*: Stereochemistry and plant growth regulatory properties, *Agric. Biol. Chem.*, 51, 3331, 1987.
61. Oikawa, T. et al., Radicicol, a microbial cell differentiation modulator, inhibits *in vivo* angiogenesis, *Eur. J. Pharmacol.*, 241, 221, 1993.
62. Nakajima, H. et al., Isolation of 2-acetylamino-3-hydroxy-4-methyloct-6-enoic acid, a derivative of the "C9-amino acid" residue of cyclosporins, produced by the fungus *Neocosmospora vasinfecta* E. F. Smith, *Agric. Biol. Chem.*, 52, 1621, 1988.
63. Nakajima, H. et al., Production of cyclosporin by fungi belonging to the genus *Neocosmospora*, *Agric. Biol. Chem.*, 53, 2291, 1989.
64. Daboussi, M.-J. et al., Evolution of the *Fot1* transposons in the genus *Fusarium*: discontinuous distribution and epigenetic inactivation, *Mol. Biol. Evol.*, 19, 510, 2002.
65. Budd, K. and Harley, J.L., The uptake and assimilation of ammonia by *Neocosmospora vasinfecta*, *New Phytol.*, 61, 138, 1962.
66. Budd, K. and Harley, J.L., The uptake and assimilation of ammonia by *Neocosmospora vasinfecta* II: Increases in the ammonia level in the mycelium during the uptake of ammonia, *New Phytol.*, 61, 244, 1962.
67. Budd, K., The assimilation of bicarbonate by *Neocosmospora vasinfecta*, *Can. J. Microbiol.*, 15, 389, 1969.
68. Budd, K., Potassium transport in non-growing mycelium of *Neocosmòspora vasinfecta*, *J. Gen. Microbiol.*, 59, 229, 1969.
69. Budd, K., Potassium fluxes in *Neocosmospora vasinfecta*, *J. Gen. Microbiol.*, 90, 311, 1975.
70. Miller, A.G. and Budd, K., Chloride uptake by mycelium of *Neocosmospora vasinfecta* and its inhibition by glucose, *J. Gen. Microbiol.*, 66, 243, 1971.
71. Miller, A.G. and Budd, K., Halide uptake by the filamentous ascomycete *Neocosmospora vasinfecta*, *J. Bacteriol.*, 121, 91, 1975.
72. Miller, A.G. and Budd, K., The development of an increased rate of Cl-uptake in the ascomycete *Neocosmospora vasinfecta*, *Can. J. Microbiol.*, 21, 1211, 1975.
73. Miller, A.G. and Budd, K., Evidence for a negative membrane potential and for movement of C1-against its electrochemical gradient in the ascomycete *Neocosmospora vasinfecta*, *J. Bacteriol.*, 128, 741, 1976.
74. Miller, A.G. and Budd, K., Influence of exogenous sugars and polyols on C1-influx and efflux by the ascomycete *Neocosmospora vasinfecta*, *J. Bacteriol.*, 126, 690, 1976.
75. Paton, W.H.N. and Budd, K., Zinc uptake in *Neocosmospora vasinfecta*, *J. Gen. Microbiol.*, 72, 173, 1972.
76. Budd, K., A high-affinity system for the transport of zinc in *Neocosmospora vasinfecta*, *Exp. Mycol.*, 12, 195, 1988.
77. Budd, K., Role of membrane potential in the transport of zinc by *Neocosmospora vasinfecta*, *Exp. Mycol.*, 13, 356, 1989.
78. Budd, K., Magnesium uptake in *Neocosmospora vasinfecta*, *Can. J. Bot.*, 57, 491, 1979.
79. Budd, K., A cadmium-tolerant strain of *Neocosmospora vasinfecta* shows reduced cadmium influx, *Can. J. Bot.*, 69, 1296, 1991.
80. Oluyedun, O.A. and VanLoon, G.W., Factors affecting uptake of aluminum by the fungus *Neocosmospora vasinfecta*, *15th World Congress of Soil Science*, Vol. 4A, 1994, p. 45.
81. Budd, K., Lefebvre, D.D., and Kelly, D.J.A., The biotransformation of mercury in pH-stat cultures of microfungi, *Can. J. Bot.*, 84, 254, 2006.
82. Budd, K., The utilization of acetate by *Neocosmospora vasinfecta*, *New Phytol.*, 65, 32, 1966.
83. Budd, K., Uptake and metabolism of D-glucose by *Neocosmospora vasinfecta* E. F. Smith., *Plant Physiol.*, 58, 193, 1976.
84. Kelly, D.J.A. and Budd, K., Polyol metabolism and osmotic adjustment in the mycelial ascomycete *Neocosmospora vasinfecta* (E. F. Smith), *Exp. Mycol.*, 13, 356, 1989.
85. Kelly, D.J.A. and Budd, K., Osmotic adjustment in the mycelia ascomycete *Neocosmospora vasinfecta* (E.F. Smith), *Exp. Mycol.*, 14, 136, 1990.

86. Ben Hamida, F. et al., Leg granuloma due to *Neocosmospora vasinfecta* in a renal graft recipient, *Transpl. Proc.*, 25, 2292, 1993.
87. Chandenier, J. et al., Leg tumefaction with *Neocosmospora vasinfecta* in a renal transplant patient. First human case report, *J. Mycol. Med.*, 3, 165, 1993.
88. Kac, G. et al., Osteo-arthritis due to *Neocosmospora africana* successfully treated by radical surgery, *Abstracts of the 4th Congress of the European Confederation of Medical Mycology*, 1998, p. 31.
89. Kac, G. et al., Osteoarthritis caused by *Neocosmospora vasinfecta*, *Med. Mycol.*, 37, 213, 1999.
90. Cornely, O.A. et al., Disseminated *Neocosmospora vasinfecta* infection in a patient with acute nonlymphocytic leukemia, *Emerg. Infect. Dis.*, 7, 149, 2001.
91. O'Donnell, K., *Neocosmospora vasinfecta* infection in a pediatric burn patient: The registry of the first case in the American continent, a case report. Direct NCBI submission, Accession numbers AY381138_AY381155. Available from http://www.ncbi.nlm.nih.gov, 2003.
92. Elfar, M.S., Heggers, J.P., Sutton, D.A., Rinaldi, M.G., O'Donnell, K., and Herndon, D.N., Neocosmospora vasinfecta infection in a pediatric burn patient: registry of the first case in North America, a case report; 2004 March, *American Burn Association 36th Annual Meeting*, Vancouver, BC, 2004.
93. Manikandan, P. et al., Corneal ulcer due to *Neocosmospora vasinfecta* in an immunocompetent patient, *Med. Mycol.*, 46, 279, 2008.
94. Hue, F.-X. et al., Specific detection of *Fusarium* species in blood and tissues by a PCR technique, *J. Clin. Microbiol.*, 37, 2434, 1999.
95. Yanai, Y., Fungal N_2O production in an arable peat soil in Central Kalimantan, Indonesia, *Soil Sci. Plant Nutr.*, 53, 806, 2007.
96. Wedde, M. et al., PCR-based identification of clinically relevant *Pseudallescheria/Scedosporium* strains, *Med. Mycol.*, 36, 61, 1998.
97. White, T.J. et al., Amplification and direct sequencing of fungal ribosomal RNA genes for phylogenetics, In Innes, M.H. et al. (eds.), *PCR Protocols*, Academic Press, San Diego, CA, 1990, pp. 315–322.
98. Altschul, S.F. et al., Basic local alignment search tool, *J. Mol. Biol.*, 215, 403, 1990.

55 Ochroconis

Ayako Sano and Kyoko Yarita

CONTENTS

55.1 INTRODUCTION

Ochroconis gallopava is a dematiaceous and thermo-tolerant fungal species that causes pulmonary, cerebral, and systemic infections in humans and animals. In avian outbreaks, *O. gallopava* induces clinical symptoms that resemble those of highly pathogenic avian flu. Therefore, correct identification of *O. gallopava* is critical for the control and prevention of this emerging zoonotic pathogen. Because of their obvious advantages, molecular tests have been increasingly applied for rapid detection of *O. gallopava*. This chapter outlines protocols based on polymerase chain reaction (PCR) and loop-mediated isothermal amplification (LAMP) for species-specific determination of *O. gallopava*.

55.1.1 Taxonomy and Morphology

Ochroconis gallopava (W.B. Cooke) de Hoog 1983 is the valid anamorphic name of a dematiaceous fungus whose synonyms include *Diplorhinotrichum gallopavum* W.B. Cooke, 1964; *Dactylaria gallopava* (W.B. Cooke) Bhatt and Kendrick, 1968; *Ochroconis gallopava* (W.B. Cooke) de Hoog, 1983; *Dactylaria constricta* (Abbott) Dixon and Salkin var. *gallopava* (W.B. Cooke) Salkin and Dixon, 1987; and *Ochroconis gallopavum* (W.B. Cooke) de Hoog, 1985 [1].

Ochroconis gallopava is related closely to *Ochroconis calidifluminalis*, *O. constricta*, *O. gamsii*, *O. humicola*, *O. tshawytschae*, *Scolecobasidium terreum*, *S. cateniphorum*, *S. avellanea*, *S. verruculosum*, *S. verruculosum*, *Chalara* spp., *Fusicladium pini*, *F. rhodense*, *F. ramoconidii*, *F. amoenum*, and *S. tricladiatum* based on nucleotide sequence analysis of the D1/D2 domain of the large-subunit (LSU) ribosomal RNA sequence [2]. As an independent species, *O. gallopava* demonstrates less than 96% in sequence identity to the closest species *O. calidifluminalis* in this gene region [3].

The colonies of *O. gallopava* are felty, flat, and brownish-black, with a reddish-brown pigment exuding into the agar [1]; however, the colors and textures may vary depending on culture media and duration of subculturing after isolation. Colonies on Sabouraud dextrose agar (SDA) at 25°C, 37°C, and 42°C are uniformly floccose or felty and dark olive green on the surface and dark brown to reddish-brown pigment on the reverse. Those on potato dextrose agar (PDA) are floccose, dark olive green on the surface, and dark brown on the reverse. Those on oat meal agar (OMA) are floccose, reddish-brown to ashy green on the surface, and lacy on the reverse, with a slight reddish-brown pigment. The fresh isolates of *O. gallopava* from the nature are extremely floccose; however, they show the same appearance as the clinical isolates after several times of subculturing [4].

Hyphae are pale or yellowish brown to brown. Conidiophores are mostly cylindrical to acicular, sometimes poorly differentiated. A few conidia attach at the tip of conidiophores. Conidia are two celled, subhyaline to pale brown, smooth walled to verrucose, cylindrical to clavate, and constricted at the septum. Typical conidia size of *O. gallopava* shown in the textbook ranged as 11–18 × 2.5–4.5 μm having wide apical cells [1]. According to the morphological studies on *O. gallopava* by Dixon and Salkin [5], the conidia sizes of six strains of this fungal species range from 3.3 to 16.6 μm in length and from 2.1 to 5.8 μm in width. The surface of the conidia is covered with mucoid-like components that stain with alcian blue (Ohori et al., Personal communication).

55.1.2 Biology and Epidemiology

Tansey (1973) reported that *O. gallopava* is one of thermo-tolerant fungi [6]. The maximum growth temperature is 48°C–50°C. It shows excellent growth at 42°C as the same as 35°C or 37°C, while slightly inhibited at 25°C [2,3]. According to Fukushiro et al. [7], the maximum growth temperature of *O. gallopava* is 50°C, while those of the related species such as *O. constricta*, *O. humicola*, and *O. tshawytschae* are less than 37°C [1,5]. Therefore, thermo-tolerance test may be important to differentiate *O. gallopava* from related species forming clavate conidia.

Distribution of *O. gallopava* might be worldwide. The natural habitat of *O. gallopava* is thought to be places with very low pH and extreme temperatures such as in thermal soils [6,8–11], hot spring effluents [4,6,10], sewage from nuclear power plants [12,13], a pulp sample [2], and broiler-house litters [14,15].

Up to the end of September 2009, more than 40 human cases have been reported [5,7,16–38]. The patients are located in the United States [2,5,16–18,20–23,25,28,30,31,36–38], South Africa [17], Australia [24–26], Canada [2,27,35], China [32,33], New Zealand [2], the Netherlands and Germany (referred from the database of CBS: Centraalbureau voor Schimmelcultures, the Netherlands database), and Japan [2,7,34] (Table 55.1). The pathogen also causes outbreaks or infections in birds and mammals [2,5,14,15,25,39–47]. The host animals include turkey [39,41], chick [14,15,25,40,42], Japanese quail chick [43], gray-winged trumpeter [44], snowy owl [45], cat [5,46], and dog [47] (Table 55.2). Animal infections, especially in birds, have raised a serious problem for differentiation from SARS and highly pathogenic avian influenza [2].

55.1.3 Pathogenesis

The main clinical symptom of *O. gallopava* infection in humans is pneumonia [2,5,17,18,24–28,31,32,37,38, and CBS database]; however, some cases develop into systemic infections [2,16,19,25,33–35], cerebral involvements [2,17,20–23,25, and CGB database], and subcutaneous [7,30] and ocular lesions [29]. Many of the patients have underlying diseases: organ transplantation as the highest risk factor [2,18–17,30,31,33,37,38], hematological malignancy [7,16,17,29,34], diabetes [16,21,24,25,30,31], AIDS [25,26,35], cancer [36], database of CBS: Centraalbureau voor Schimmelcultures, the Netherlands), immune disorder [2,17], cardiovascular disorder [17], emphigus [32], pneumoconiosis [2]. Two apparently healthy individuals, a wood pulp worker [28] and a coal mine worker [17], also became infected after frequent inhalation of or exposure to the pathogen (Table 55.1).

On the other hand, infected birds and mammals showed cerebral involvements predominantly (Table 55.2). There is a possibility that the differences of body temperatures between human and other animals influence the predominant symptoms of *O. gallopava* infection.

According to Dixon et al. and Walsh et al., clinical and animal isolates of *O. gallopava* are lethal to mice *via* intravenous route [48,49]. Furthermore, through intratracheal inoculation of *O. gallopava* conidia obtained from an outbreak case in turkeys, Blalock and Derieux succeeded reproduction of the disease caused by *O. gallopava* on turkey poultry [41].

Based on experimental studies involving intravenous infection for mice with *O. gallopava* conidia, the target organ might be the brain, although the fungal species also affected the lung and other visceral organs such as liver, kidney, and spleen [2,4]. The circulating movements shown by mice inoculated with conidia of *O. gallopava* pointed to the likely damage in the brains. The hemorrhage and marked lesions in the brains also supported the strong neurotropism [4].

These findings suggested that the central nervous system should be carefully examined in *O. gallopava*-infected patients and animals. In fact, the cerebral involvements were also recorded in the human cases [2,16,17,19–23,25,33–35, and CBS database] followed by pulmonary disorders. Similar changes were also noted in the animal cases [14,15,25,39–47].

In addition, *O. gallopava* isolates from hot springs had the same ability to elicit cerebral symptoms from experimentally infected mice as clinical or animal isolates [4]. By contrast, a clinical isolate reported by Ohori et al. [2] showed much lower virulence and no mouse died in spite of treatment with corticosteroid. The difference in virulence between *O. gallopava* strains might also depend on histories of storage and/or condition of maintenance [50].

The fungal cells have the capability to overcome the host defense mechanism in mice, although the degree of migrations of polymorphonuclear leukocytes and macrophages to mycelial cells in host tissues appear to depend on each fungal isolate. Fresh and wild strains of *O. gallopava* are highly virulent even in healthy hosts. The fact that a wood pulp worker without any underlying disease developed multiple lung abscesses after repeated inhalation of *O. gallopava* in composts provides evidence that constant exposure to a huge number of *O. gallopava* spores may cause disease in healthy subjects [17,28].

55.1.4 Antifungal Susceptibilities

Minimum inhibitory concentration (MIC) for *O. gallopava* is relatively low; MIC of amphotericin B ranges from 0.015 to 2.0 μg/mL [20,21,29,30,31,33,51–53], that of flucytosine ranges from 0.5 to 80.7 μg/mL [1,20,21,30,31], that of fluconazole ranges from 8 to 32 μg/mL [1,20,29,31], that of itraconazole ranges from 0.02 to 0.5 μg/mL [20,21,29,30,31,33,51–53], that of miconazole ranges from 0.5 to 2.0 μg/mL [52], that of voriconazole ranges from 0.03 to 1.0 μg/mL [33,51–53], that of ketoconazole ranges from 0.25 to 3.2 μg/mL [1,20], and that of terbinafine ranges from 0.03 to 0.06 μg/mL [52]. These data suggest that amphotericin B, with or without itraconazole, can be recommended as the first choice drug against *O. gallopava* infection [20,24,28].

Our recent results on the antifungal susceptibility testing of *O. gallopava* isolated from hot springs showed that the

TABLE 55.1
Human Cases of *Ochroconis gallopava* Infection

Case	Authors, Published Year, and Reference Number	Country	Sex	Age	Affected Organ	Remarks[a]	Outcome
1	Dixon and Salkin, 1986 [5]	USA	ND	ND	Lung	ND	ND
2	Fukushiro et al., 1986 [7]	Japan	F	58	Subcutaneous	Leukemia	Survived
3	Terreni et al., 1990 [16]	USA	M	62	Systemic	Leukemia, diabetic	Died
4	Sides et al., 1991 [17]	ND	M	ND	Lung	Immunocompromised	ND
5	Sides et al., 1991 [17]	South Africa	ND	ND	Lung	Coal mine worker	ND
6	Sides et al., 1991 [17]	South Africa	ND	ND	Lung	Coal mine worker	ND
7	Sides et al., 1991 [17]	USA	ND	ND	Lung	Immunocompromised	Died
8	Sides et al., 1991 [17]	USA	ND	ND	Brain	Immunocompromised	ND
9	Sides et al., 1991 [17]	USA	M	47	Lung	Cardiovascular disease	ND
10	Sides et al., 1991 [17]	USA	M	60	Brain	Lymphoma, nocardiosis	Died
11	Mancini and McGinnis, 1992 [18]	USA	M	30	Lung	Heart transplant recipient	Survived
12	Prevost-Smith et al., 1993 [19]	ND	M	46	Systemic	Heart transplant recipient	Died
13	Vukmir et al., 1994 [20]	USA	M	68	Brain	Liver transplant recipient	Survived
14	Kralovic and Rhodes 1995 [21]	USA	M	63	Brain	Liver transplant recipient, diabetic	Died
15	Rossmann et al., 1996 [22]	USA	M	59	Brain	Liver transplant recipient, nocardiosis	Died
16	Bonham et al., 1996 [23]	USA	ND	ND	Brain	Liver transplant recipient	Survived
17	Jenney et al., 1998 [24]	Australia	M	58	Lung	Heart transplant recipient, diabetic	Survived
18	Horré and de Hoog, 1999 [25]	UK	ND	ND	Systemic	AIDS	ND
19	Horré and de Hoog, 1999 [25]	Australia	ND	ND	Systemic	ND	ND
20	Horré and de Hoog, 1999 [25]	USA	ND	ND	Brain	Diabetes mellitus	ND
21	Horré and de Hoog et al., 1999 [25,26]	ND	M	48	Lung	HIV-positive transplant recipient	ND
22	Horré and de Hoog et al., 1999 [25,26]	Australia	ND	ND	Lung	ND	ND
23	Horré and de Hoog, 1999 [25]	USA	ND	ND	Lung	Transplant recipient	ND
24	Burns et al., 2000, [27]	Canada	F	58	Lung	Lung transplant recipient with pulmonary nodule	Survived
25	Odell et al., 2000 [28]	USA	M	38	Lung	Wood pulp worker with pulmonary abscess	Survived
26	Bowyer et al., 2000 [29]	UK	M	69	Eye	Chronic lymphocytic lymphoma	Died
27	Mazur and Judson, 2001 [30]	USA	F	32	Subcutaneous	Lung transplant recipient with diabetic	Survived
28	Malani et al., 2001 [31]	USA	M	32	Lung	Renal transplant recipient with diabetic	Died
29	Zhao et al., 2002 [32]	China	M	68	Lung	Pemphigus	Survived
30	Wang et al., 2003 [33]	China	M	13	Systemic	Renal transplant recipient	Died
31	Fukushima et al., 2005 [34]	Japan	F	66	Systemic	Chronic lymphocytic lymphoma	Died
32	Ohori et al., 2006 [2]	USA	M	54	Systemic	Heart transplant recipient	Died
33	Ohori et al., 2006 [2]	Japan	M	79	Lung	Pneumoconiosis	ND
34	Ohori et al., 2006 [2]	Canada	F	68	Lung	ND	ND
35	Ohori et al., 2006 [2]	USA	ND	ND	Lung	ND	ND
36	Ohori et al., 2006 [2]	New Zealand	M	83	Lung	ND	ND
37	Boggild et al., 2006 [35]	Canada	M	28	Systemic	AIDS	Died
40	Hollingsworth et al., 2007 [36]	USA	F	78	Lung	Hypothyroidism, basal cell skin cancer	Survived
41	Shoham et al., 2008 [37]	USA	F	64	Lung	Renal transplant recipient	Survived
42	Shoham et al., 2008 [37]	USA	M	60	Lung	Renal transplant recipient	Survived
43	Shoham et al., 2008 [37]	USA	M	50	Lung	Liver transplant recipient	Survived
44	Mayer and Bastani, 2009 [38]	USA	M	71	Lung	Renal transplant recipient	Died
38	Database of CBS	Netherlands	ND	ND	CSF	Spondylodiscitis	ND
39	Database of CBS	Germany	M	50	Lung	Bronchial carcinoma	Died

CBS, Centraalbureau voor Schimmelcultures, the Netherlands, CSF, cerebrospinal fluid, ND, no data.
[a] Underlying disease or occupation.

TABLE 55.2
Animal Cases of *Ochroconis gallopava* Infection

Authors, Published Year, and Reference Number	Country	Animal Species	Affected Organ(s)
Georg et al., 1964 [39]	USA	Turkey[a]	Brain
Connole, 1967 [40]	Australia	Chick[a]	Brain
Blalock et al., 1973 [41]	USA	Turkey[a]	Brain
Ranck et al., 1973 [42]	USA	Chick[a]	Brain
Waldrip et al., 1974 [14]	USA	Chick, six cases[a]	Brain
Randall and Owen, 1981 [15]	UK	Chick, two cases[a]	Brain and lung
Shane et al., 1985 [43]	USA	Japanese quail chicks[a]	Brain and lung
Dixon and Salkin, 1986 [5]	USA	Cat	Lung
Karesh et al., 1987 [44]	USA	Gray-winged trumpeter chicks	Brain
Salkin et al., 1990 [45]	USA	Snowy owl chick	Brain
Padhye et al., 1994 [46]	USA	Cat	Systemic
Horré and de Hoog, 1999 [25]	India	Chick[a]	Brain
Ohori et al., 2006 [2]	New Zealand	Antipodean parakeet	Lung
Singh et al., 2006 [47]	USA	Dog	Systemic

[a] Epidemic outbreaks at poultry farms.

MIC values range from 0.25 to 1 µg/mL for amphotericin B, 0.25–4 µg/mL for flucytosine, 0.25–1 µg/mL for itraconazole, 0.5–2 µg/mL for miconazole, 1–2 µg/mL for voriconazole, 8–64 µg/mL for fluconazole, and ≤0.03–0.125 µg/mL for micafungin [3,54].

55.1.5 Diagnosis

As opportunistic fungal infections in immunocompromised patients, aspergillosis, candidiasis, and/or cryptococcosis are well recognized by their morbidity and mortality [38]. *O. gallopava* is listed as a causative agent for emerging fungal infection, causing fatal infections in immunocompromised hosts, especially patients with hematological malignancy [34]. On the other hand, frequent exposure to fungal elements might cause the infection to humans without immune disorder [28]. We should take into consideration *O. gallopava* infection to the patients showing respiratory symptoms, systemic disorder, and/or intractable cutaneous massive lesions and to not only immunocompromised patients but also individuals working at dusty environments.

55.2 METHODS

Culture of the causative agent, *O. gallopava*, from clinical materials such as sputum, BAL (bronchoalveolar lavage), and/or biopsy is the gold standard when the isolate corresponds to cytological and histopathological findings containing fungal components. However, it is impossible to diagnose *O. gallopava* infection on the basis of cytology or histopathology alone.

It takes a longer period until mycological identification. Culture, isolation, and identification of the causative agent require at least a few weeks. To determine whether fungal culture is positive or negative, it should be kept under observation up to 4 weeks. In our experience, it took 1 week to recover from visceral organs of experimentally *O. gallopava*-inoculated mice [2,4].

Furthermore, observations of mycological profiles may take a few more weeks. When the isolate was dematiaceous fungi at a glance of the colony, having excellent growth ability at 42°C and producing clavate and two-celled conidia, the case was able to be suspected as fungal infection caused by black fungi including *O. gallopava*.

On the other hand, the hematological and blood chemistry data of the hosts may have been influenced by the underlying diseases in immunocompromised host. Chest x-ray and computed tomography also help to detect pulmonary lesions.

Therefore, rapid and accurate diagnostic methods except for culturing and DNA sequencing are useful at clinical laboratories to differentiate *O. gallopava* infection from SARS and highly pathogenic bird flu.

55.2.1 Sample Preparation

Isolation of the causative agent is the gold standard for diagnosis. We strongly recommend trying isolation of the causative fungal pathogen from clinical materials. Consistency of the findings of culture, cytology and/or histopathology, and molecular biological data may bring a valid diagnosis.

The sputum or BAL sample is directly spread on chloramphenicol (100 mg/L)-added PDA plates at least in duplicate. When bacterial contaminations are expected, autoclaved-chloramphenicol (100 mg/L)-added saline is poured on the sputum or sediment of BAL at 10 times volume, is agitated strongly, is stored at 35°C–37°C for at least 2 h, and is spread on the culture plate. Culture condition is recommended to use at 42°C and continue to observe up to 14 days. Then, suspicious colonies are picked up for morphological and molecular biological studies. Using incubators adjusted at 35°C and 42°C simultaneously might be a convenient way to isolate both *O. gallopava* and other fungal pathogen.

While waiting for the fungal colony recovering from clinical materials, some parts of molecular biological diagnosis may proceed simultaneously.

DNA samples for molecular biological studies such as species-specific PCR system and the LAMP method are prepared as follows:

DNA from purified fungal colonies is extracted with commercial DNA extraction kits (i.e., DEXPAT® kit [TaKaRa, Ohtsu, Japan]). Although DEXPAT kit is designed for extracting DNA from paraffin-embedded tissue samples, we routinely use it for the isolation of genes from fungal cultures. Our procedure for purified fungi is as follows: Approximately 100 μL of fungal cells are placed in a sterilized microtube (1.5 mL size); 0.5 mL of DEXPAT solution is then added, and the mixture is homogenized with a plastic pestle. It is then incubated at 100°C for 10 min and centrifuged at 12,000 rpm (13,201 × g) for 10 min. The supernatants are used as DNA samples. The original procedure of DEXPAT kits for paraffin-embedded tissue samples is as follows: Three sections of 10 μm-thick paraffin tissue are placed into a microtube (1.5 mL), and 0.5 mL of DEXPAT solution is added. After boiling for 10 min, the tube is centrifuged at 12,000 rpm (13,201 × g) for 10 min. The supernatant is then processed for molecular biological studies. Biopsy tissue, blood, and BAL samples are fixed with ethanol at 70% final concentration at least for 10 min. Before fixation by the ethanol, the tissue and coagulated blood are ground out using a plastic homogenizer. The fixed sample is centrifuged, washed three times with sterile water by centrifuging at 12,000 rpm (13,201 × g) for 5 min, and 0.5 mL of DEXPAT solution is added. The mixture is incubated at 100°C for 10 min and centrifuged at 12,000 rpm for 10 min. The pus samples are collected with sterile cotton buds. The cotton tip is cut off and placed in a sterilized microtube (1.5 mL), and 0.5 mL of 70% ethanol is added. After vortexing and centrifuging, the sample is washed three times with sterile water by centrifuging at 12,000 rpm (13,201 × g) for 5 min, and 0.5 mL of DEXPAT solution is added. The cotton tip and the solution are incubated together at 100°C for 10 min and centrifuged at 12,000 rpm (13,201 × g) for 10 min.

55.2.2 Detection Procedures

55.2.2.1 Species-Specific PCR

Principle. The species-specific primer set for *O. gallopava* was designed from the sequence of D1/D2 LSU rDNA of *O. gallopava* (accession number AB161059 in GenBank) after comparison with those from 14 dematiaceous fungal species (i.e., closely related species: *O. constricta*, *O. gamsii*, *O. humicola*, and *O. tshawytschae*, and other dematiaceous fungal species: *Alternaria alternata*, *Arthrobotrys javanica*, *Bipolaris spicifera*, *Cladophialophora carrionii*, *Curvularia geniculata*, *Curvularia lunata*, *Curvularia senegalensis*, *Exophiala alcalophila*, *Exophiala dermatitidis*, and *Exophiala moniliae*). The primer set was also tested on eight clinically important fungal species: *Aspergillus fumigatus*, *Blastomyces dermatitidis*, *Candida albicans*, *Coccidioides posadasii*, *Cryptococcus neoformans*, *Histoplasma capsulatum*, *Penicillium marneffei*, and *Sporothrix schenckii* for confirmation of the specificity [2].

The sequences of the forward and reverse primers are OgF3: 5′-AGG GAG TCT CGG GTT AAG GG-3′ position 119–138th, and OgB3: 5′-CAT TCC CTT CGT CTT TGT CC-3′ position 493–474th of AB161059 consisting of 615th. The D1/D2 domain of the LSU ribosomal RNA gene sequences of *O. gallopava* in the GenBank database was identical except for two wobble bases at 421st and 435th.

Procedure

1. The PCR mixture (25 μL) contains 20–40 ng of template DNA (2.5 μL), one Ready-To-Go™ PCR bead (Amersham Pharmacia), and 10 pmol each of primers OgF3 and OgB3.
2. The PCR mixture is subjected to an initial denaturation at 95°C for 4 min; 30 cycles of 94°C for 1 min, 58°C for 90 s, and 72°C for 2 min; a final extension at 72°C for 10 min.

Note. The primer set also amplified the homologous part of *O. calidifluminalis* which is the most recently discovered *Ochroconis* sp. demonstrating closest phylogenetic and morphological relationships to *O. gallopava* [3]. Furthermore, the primer set may amplify other *O. gallopava*–related species, such as *Chalara* sp., *Scolecobasidium terreum*, *S. cateniphorum*, and *S. tricladiatum*, although they have not been confirmed.

55.2.2.2 Loop-Mediated Isothermal Amplification Method

Principle. LAMP method is a simple, rapid, specific, and cost-effective nucleic acid amplification method solely developed by Eiken Chemical Co., Ltd (http://loopamp.eiken.co.jp/e/lamp/index.html). It is characterized by the use of four different primers specifically designed to recognize six distinct regions on the target gene, and the reaction process proceeds at a constant temperature using strand displacement reaction (Figure 55.1a) [55].

Amplification and detection of genes can be completed in a single step, by incubating the mixture of samples, primers, DNA polymerase with strand displacement activity and substrates at a constant temperature (around 65°C). It provides high-amplification efficiency, with DNA being amplified 10^9–10^{10} times in 15–60 min. Because of its high specificity, the presence of amplified product can indicate the presence of the target gene.

LAMP method uses four primer sets: F (Forward) 3, B (Backward) 3, FIP (Forward Inner Primer), and BIP (Backward Inner Primer) selected from six distinct

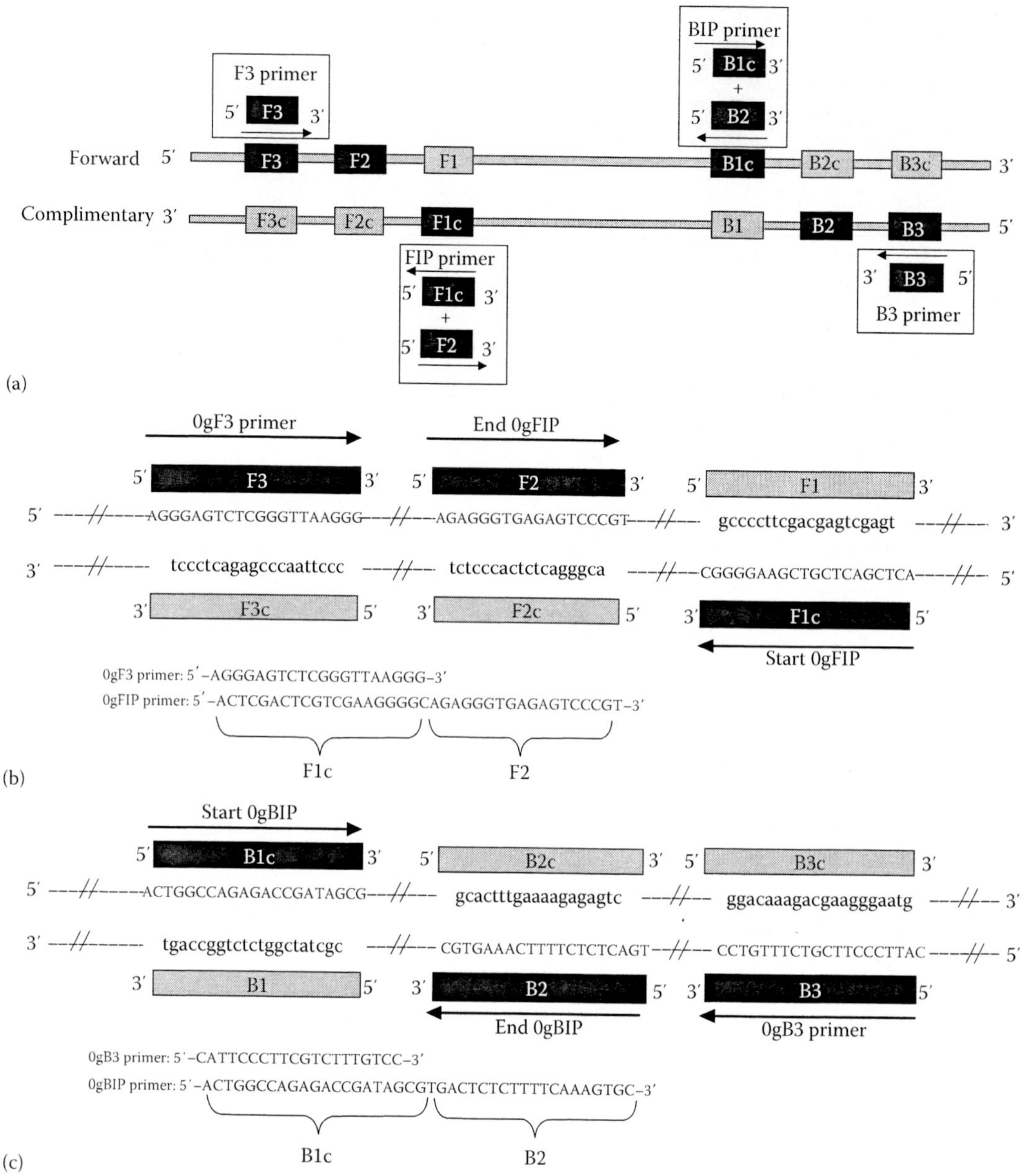

FIGURE 55.1 (a) Design four types of primers based on the following six distinct regions of the target gene: The F3c, F2c, and F1c regions at the 3′ side and the B1, B2, and B3 regions at the 5′ side. FIP: Forward inner primer consists of the F2 region (at the 3′ end) that is complementary to the F2c region, and the same sequence as the F1c region at the 5′ end. F3 primer: Forward outer primer consists of the F3 region that is complementary to the F3c region. BIP: Backward inner primer consists of the B2 region (at the 3′ end) that is complementary to the B2c region, and the same sequence as the B1c region at the 5′ end. B3 primer: Backward outer primer consists of the B3 region that is complementary to the B3c region (http://loopamp.eiken.co.jp/e/lamp/primer.html). (b) Positions and sequences of the forward primers for the species-specific LAMP for *Ochroconis gallopava*, and (c) those of the backward primers.

regions of the target gene (http://loopamp.eiken.co.jp/e/lamp/primer.html). The most important primer sets are F3 and B3. The primers should be selected from specific genes or gene sequences based on species-specific PCR and were confirmed after testing with intraspecies diversity and a huge numbers of related pathogenic fungal species. Therefore, enormous numbers of trials and errors are required until the final primers are confirmed. Further, the primers should completely differentiate the fungal genes from host ones.

Although PrimerExplore, special software to design LAMP primers, is available in the Web site (http://primerexplorer.jp/e/), it seems to be useful as reference hints for base composition, GC contents, secondary structures, and Tm value on designing primers based on our experience.

The target gene (e.g., DNA template) and the reagents are incubated at a constant temperature between 60°C and 65°C (Figure 55.2a and b). The reaction steps were available at the Web site (http://loopamp.eiken.co.jp/e/lamp/principle.html).

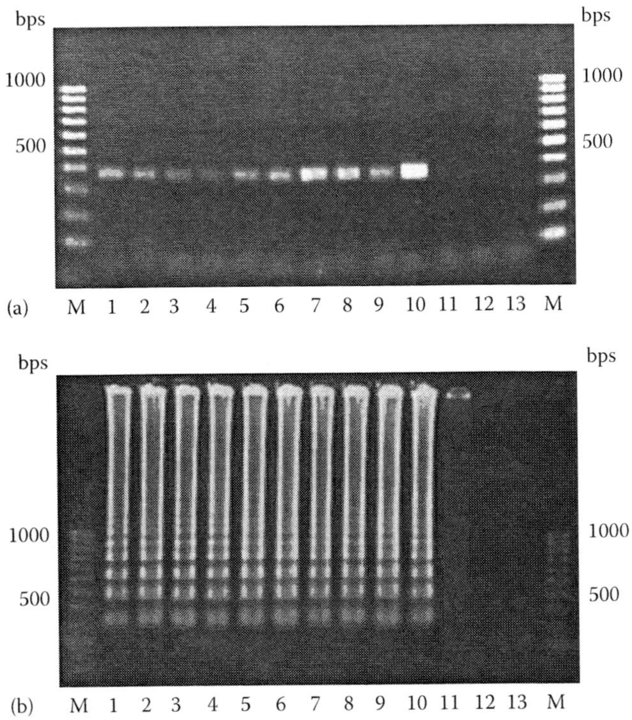

FIGURE 55.2 Images of the species-specific PCR and the LAMP for *Ochroconis gallopava*. (a) Species-specific PCR for *Ochroconis gallopava*. M: Marker, 1–10: *O. gallopava*, 11: *O. gamsii*, 12 and 13: *O. tsawytschae*. The related species such as *O. constricta*, *O. humicola*, *Alternaria alternata*, *Arthrobotrys javanica*, *Bipolaris sp.*, *Bipolaris specifera*, *Cladophialophora bantiana*, *C. carrionii*, *Curvularia geniculata*, Cu. lunata var. lunata, *Cu. senegalensis*, *Exophiala alcalophiala*, *E. dermatitidis*, *E. jeanselmei*, *E. moniliae*, *E. spinifera*, *Fonsecaea pedrosoi*, *Phialophora verrucosa*, and *Rhinocladiella atrovirens*, were negative. (b) Species-specific LAMP for *Ochroconis gallopava*. M: Marker, 1–10: *O. gallopava*, 11: *O. gamsii*, 12 and 13: *O. tsawytschae*. The related species listed above were also negative.

LAMP method is highly sensitive. We experienced many faults of contamination of the genes. Once contamination of the target gene occurs, all reactions become positive, even in a negative control using distilled water as template. Therefore, extremely careful procedures are required: reagents, pipettes, plastic pipette tips, safety cabinet, and hands. The samples should be handled separately from the reagents. The positive control for the target gene should be done separately.

This is one of the reasons why some fungal species common in normal human flora or in environments are not recommended for the LAMP method. Selection of the target fungal species is also important. The fungal species should be rare in laboratorial environment. *O. gallopava* is treated as suitable for the LAMP method because it is one of the rare species in clinical laboratory and an uncommon fungal species in our environments. The primer sequences of F3 and B3 are as the same as the species-specific primer sets for PCR: OgF3 and OgB3. The primer OgFIP (5′-ACT CGA CTC GTC GAA GGG GCA GAG GGT GAG AGT CCC GT-3′) is designed for the reverse complementary sequence of 235th–216th and the forward sequence of 170th–187th. The primer OgBIP (5′-ACT GGC CAG AGA CCG ATA GCG TGA CTC TCT TTT CAA AGT GC-3′) is designed for the forward sequence of 287th–307th and the complementary sequence of 355th–336th in AB161059 (Figure 55.1b and c).

Procedure

1. One μL of fungal DNA template and 5 pmol each of OgF3 and OgB3 and 40 pmol each of FIP and BIP primers are mixed with 12.5 μL of a reaction mixture in a kit (Loop AMP, Eiken, Tokyo, Japan) in a final volume of 25.0 μL.
2. The DNA mixtures are incubated at 63°C for 60 min and 80°C for 2 min.
3. The amplified DNAs are detected by electrophoresis in a 1.0% agarose gel stained with ethidium bromide.

Note. Detection of the fungal gene from infected tissue samples by a LAMP method requires a longer incubation period [56]. We recommend 120 min of incubation for detection of the gene from clinical samples [2].

The LAMP method for detection of species-specific gene of *O. gallopava* also amplified the closest species, *O. calidifluminalis*, as the same band pattern as *O. gallopava* [3]. The LAMP system may also amplify the partial gene of the D1/D2 domain of the LSU ribosomal RNA gene on some *O. gallopava*-related fungal species: *Chalara* sp., *Scolecobasidium terreum*, *S. cateniphorum*, and *S. tricladiatum*. However, we speculated that the band patterns might be different from those of *O. gallopava* based on our studies about genus *Paracoccidioides; P. lutzii* [57], a closely related species to *P. brasiliensis* having 89% identity in the target gene, was amplified as a different ladderlike band pattern from those of *P. brasiliensis* by a LAMP method [58]. Nevertheless, the pathogenicity of *Ochroconis calidifluminalis* is lower than that of *O. gallopava* [3], and there has been no case report involving *O. calidifluminalis*. Moreover, the pathogenicities of other new species, *Chalara* sp., *Scolecobasidium terreum*, *S. cateniphorum*, and *S. tricladiatum*, are also unknown because they are environmental isolates with no case report. Therefore, molecular biological identification of unknown dematiaceous fungi and detection of *O. gallopava* and *Ochroconis* sp. genes from clinical materials by the LAMP may remain useful (Figure 55.2).

55.3 CONCLUSION

O. gallopava is a species of dematiaceous fungi and affects central nervous and respiratory systems in humans, animals, and birds. Outbreaks of *O. gallopava* infection in birds resemble highly pathogenic bird flu and SARS. The present *O. gallopava* species-specific primer sets for PCR and LAMP methods may help differentiate *O. gallopava* infection from not only highly pathogenic bird flu and SARS but also other fungal infections.

REFERENCES

1. de Hoog, G.S., *Ochroconis gallopava*, in *Atlas of Clinical Fungi*, 2nd edn., de Hoog, G.S, Guarro, J., and Gen, J., Eds., Centraalbureau voor Schimmelcultures, Utrecht, the Netherlands, 2000, p. 782.
2. Ohori, A. et al., Rapid identification of *Ochroconis gallopava* by a loop-mediated isothermal amplification (LAMP) method, *Vet. Microbiol.*, 114, 359, 2006.
3. Yarita, K. et al., *Ochroconis calidifluminalis*, a sibling of the neurotropic pathogen *O. gallopava*, isolated from hot spring, *Mycopathologia*, 170, 21, 2010.
4. Yarita, K. et al., Pathogenicity of *Ochroconis gallopava* isolated from hot springs in Japan and a review of published reports, *Mycopathologia*, 164, 135, 2007.
5. Dixon, D.M. and Salkin, I., Morphologic and physiologic studies of three dematiaceous pathogens, *J. Clin. Microbiol.*, 24, 12, 1986.
6. Tansey, M.R. and Brock, T.D., *Dactylaria gallopava*, a cause of avian encephalitis, in hot spring effluents, thermal soils and self-heated coal waste piles, *Nature*, 242, 202, 1973.
7. Fukushiro, R. et al., Subcutaneous abscesses caused by *Ochroconis gallopavum*, *J. Med. Vet. Mycol.*, 24, 175, 1986.
8. Evans, H.C., Thermophilous fungi of coal spoil tips. I. Taxonomy, *Trans. Br. Mycol. Soc.*, 57, 241, 1971.
9. Evans, H. C., Thermophilous fungi of coal spoil tips. II. Occurrence, distribution and temperature, *Trans. Br. Mycol. Soc.*, 57, 255, 1971.
10. Weitzman, I., Rosenthal, S.A., and Shupack, J.L., A comparison between *Dactylaria gallopava* and *Scolecobasidium humicola*: First report of an infection in a tortoise caused by *S. humicola*, *Med. Vet. Mycol.*, 23, 287, 1985.
11. Redman, R.S. et al., Fungi from geothermal soils in Yellowstone National Park, *Appl. Environ. Microbiol.*, 65, 5193, 1999.
12. Tansey, M.R., Fliermans, C.B., and Kern, C.D., Aerosol dissemination of veterinary pathogenic and human opportunistic thermophilic and thermotolerant fungi from thermal effluents of nuclear production reactors, *Mycopathologia*, 69, 91, 1979.
13. Rippon, J.W., Gerholdm, R., and Heath, M., Thermophilic and thermotolerant fungi isolated from the thermal effluent of nuclear power generating reactors: Dispersal of human opportunistic and veterinary pathogenic fungi, *Mycopathologia*, 70, 169, 1980.
14. Waldrip, D.W. et al., Isolation of *Dactylaria gallopava* from broiler-house litter, *Avian Dis.*, 18, 445, 1974.
15. Randall, C.J. and Owen, D.M., Encephalytis in broiler chickens caused by a hyphomycete resembling *Dactylaria gallopava*, *Avian Pathol.*, 10, 31, 1981.
16. Terreni, A.A. et al. Disseminated *Dactylaria gallopava* infection in a diabetic patient with chronic lymphocytic leukemia of the T-cell type, *Am. J. Clin. Pathol.*, 94, 104, 1990.
17. Sides, E.H. 3rd, Benson, J.D., and Padhye, A.A., Phaeohyphomycotic brain abscess due to *Ochroconis gallopavum* in a patient with malignant lymphoma of a large cell type, *J. Med. Vet. Mycol.*, 29, 317, 1991.
18. Mancini, M.C. and McGinnis, M.R., *Dactylaria* infection of a human being: Pulmonary disease in a heart transplant recipient, *J. Heart Lung Transpl.*, 11, 827, 1992.
19. Prevost-Smith, E. et al., Fatal phaeohyphomycotic infection due to *Dactylaria gallopava* and *Scedosporium prolificans* in a cardiac transplant, in *Proceedings and Abstracts of the 93rd Annual Meeting of the American Society for Microbiology*, Atlanta, GA: ASM Press, F-35, 1993.
20. Vukmir, R.B. et al., Successful therapy for cerebral phaeohyphomycosis due to *Dactylaria gallopava* in a liver transplant recipient, *Clin. Infect. Dis.*, 19, 714, 1994.
21. Kralovic, S.M. and Rhodes, J.C., Phaeohyphomycosis caused by *Dactylaria* (human dactylariosis): Report of a case with review of the literature, *J. Infect.*, 31, 107, 1995.
22. Rossmann, S.N., Cernoch, P.L., and Davis, J.R., Dematiaceous fungi are an increasing cause of human disease, *Clin. Infect. Dis.*, 22, 73, 1996.
23. Bonham, A. et al., Central nerves system lesions in liver transplant recipients: Prospective assessment of indications for biopsy and implications for management, in *15th Annual Scientific Meeting of the American Society of Transplantation Physicians*, Dallas, TX, 66, 1596, 1998.
24. Jenney, A. et al., Pulmonary infection due to *Ochroconis gallopavum* treated successfully after orthotopic heart transplantation, *Clin. Infect. Dis.*, 26, 236, 1988.
25. Horré, R. and de Hoog, G.S., Primary cerebral infections by melanized fungi: A review, *Stud. Mycol.*, 43, 176, 1999.
26. Horré, R. et al., rDNA diversity and physiology of *Ochroconis* and *Scolecobasidium* species reported from humans and other vertebrates, *Stud. Mycol.*, 43, 194, 1999.
27. Burns, K.E., Ohori, N.P., and Iacono, A.T., *Dactylaria gallopava* infection presenting as a pulmonary nodule in a single-lung transplant recipient, *J. Heart. Lung Transpl.*, 19, 900, 2000.
28. Odell, J.A. et al., Multiple lung abscesses due to *Ochroconis gallopavum*, a dematiaceous fungus, in a nonimmunocompromised wood pulp worker, *Chest*, 118, 1503, 2000.
29. Bowyer, J.D. et al., *Ochroconis gallopava* endophthalmitis in fludarabine treated chronic lymphocytic leukemia, *Br. J. Ophthalmol.*, 84, 117, 2000.
30. Mazur, J.E. and Judson, M.A., A case report of a dactylaria fungal infection in a lung transplant patient, *Chest*, 119, 651, 2001.
31. Malani, P.N. et al., Disseminated *Dactylaria constricta* infection in a renal transplant recipient, *Transpl. Infect. Dis.*, 3, 40, 2001.
32. Zhao, J. et al., Pemphigus patient with pulmonary fungal infection caused by *Ochroconis gallopava*: The first case report in China, *Zhonghua Yi Xue Za Zhi*, 82, 1310, 2002.
33. Wang, T.K. et al., Disseminated *Ochroconis gallopavum* infection in a renal transplant recipient: The first reported case and a review of the literature, *Clin. Nephrol.*, 60, 415, 2003.
34. Fukushima, N. et al., Disseminated *Ochroconis gallopavum* infection in a chronic lymphocytic leukemia: A case report and review of the literature on hematological malignancies, *Intern. Med.*, 44, 879, 2005.
35. Boggild, A.K. et al., Disseminated phaeohyphomycosis due to *Ochroconis gallopavum* in the setting of advanced HIV infection, *Med. Mycol.*, 44, 777, 2006.
36. Hollingsworth, J.W., Shofer, S., and Zaas, A., Successful treatment of *Ochroconis gallopavum* infection in an immunocompetent host. *Infection*, 35, 367, 2007.
37. Shoham, S. et al., Transplant-associated *Ochroconis gallopava* infections, *Transpl. Infect. Dis.*, 10, 442, 2008.
38. Mayer, N. and Bastani, B., *Dactylaria constricta* var. *gallopava* in a renal transplant patient, *Nephrology*, 14, 262, 2009.
39. Georg, L.K., Bierer, B.W., and Cooke, W.B., Encephalytis in turkey poults due to a new fungus species, *Sabouraudia*, 3, 239, 1964.
40. Connole, M.D., Some aspects of animal mycoses in Australia, in *Fourth Meeting of the International Society for Human and Animal Mycology*, New Orleans, LA, 1967.

41. Blalock, H.G., Georg, L.K., and Derieux, W.T., Encephalitis in turkey poults due to *Dactylaria* (*Diplorhinotrichum*) *gallopava* case report and its experimental reproduction, *Avian Dis.*, 17, 197, 1973.
42. Ranck, F.M. Jr., Georg, L.K., and Wallace, D.H., Dactylariosis: A newly recognized fungus disease of chickens, *Avian Dis.*, 18, 4, 1973.
43. Shane, S.M. et al., Encephalitis attributed to dactylariosis in Japanese quail chicks (*Coturnix coturnix japonica*), *Avian Dis.*, 29, 822, 1985.
44. Karesh, W.B., Russell, R., and Gribble, D., *Dactylaria gallopava* encephalitis in two gray-winged trumpeters (*Psophia crepitans*), *Avian Dis.*, 31, 685, 1987.
45. Salkin, I.F. et al., Fatal encephalitis caused by *Dactylaria constricta* var. *gallopava* in a snowy owl chick (*Nyctea sandiaca*), *J. Clin. Microbiol.*, 28, 2845, 1990.
46. Padhye, A.A. et al., Fatal encephalitis caused by *Ochroconis gallopavum* in a domestic cat (*Felis domesticus*), *J. Med. Vet. Mycol.*, 32, 141, 1994.
47. Singh, K. et al., Fatal systemic phaeohyphomycosis caused by *Ochroconis gallopavum* in a dog (*Canis familaris*), *Vet. Pathol.*, 43, 988, 2006.
48. Dixon, D.M. et al., *Dactylaria constricta*: Another dematiaceous fungus with neurotropic potential in mammals, *J. Med. Vet. Mycol.*, 25, 55, 1987
49. Walsh, T.J. et al., Comparative histopathology of *Dactylaria constricta*, *Fonsecaea pedrosoi*, *Wangiella dermatitidis*, and *Xylohypha bantiana* in experimental phaeohyphomycosis of the central nervous system, *Mykosen*, 30, 215, 1987.
50. Sano, A. et al., The effects of storage at –135°C with a programmed freezing method on the virulence and morphology of *Paracoccidioides brasiliensis* yeast form cells, *Nippon Ishinkin Gakkai Zassi*, 35, 161, 1994.
51. McGinnis, M.R. and Pasarell, L., In vitro testing of susceptibilities of filamentous ascomycetes to voriconazole, itraconazole, and amphotericin B, with consideration of phylogenetic implications, *J. Clin. Microbiol.*, 36, 2353, 1998.
52. Meletiadis, J. et al., Short communication: *In vitro* antifungal activity of six drugs against 13 clinical isolates of *Ochroconis gallopava*, *Stud. Mycol.*, 43, 206, 1999.
53. Espinel-Ingroff, A., In vitro fungicidal activities of voriconazole, itraconazole, and amphotericin B against opportunistic moniliaceous and dematiaceous fungi, *J. Clin. Microbiol.*, 39, 954, 2001.
54. Method for broth dilution antifungal susceptibility testing of filamentous fungi—Second edition: Approved standard M38-A2, clinical and laboratory standards institute, Wayne, Pennsylvania, USA, 2008.
55. Sano, A. and Itano, E.N., Applications of loop-mediated isothermal amplification methods (LAMP) for identification and diagnosis of mycotic diseases: Paracoccidioidomycosis and *Ochroconis gallopava* infection, in *Molecular Identification of Fungi*, Gherbawy, Y. and Voigt, K., Eds., Springer Verlag, Berlin, Heidelberg, 2010, pp. 417–437.
56. Endo, S. et al., Detection of gp43 of *Paracoccidioides brasiliensis* by the loop-mediated isothermal amplification (LAMP) method, *FEMS Microbiol. Lett.*, 234, 93, 2004.
57. Teixeira, M.M. et al., Phylogenetic analysis reveals a high level of speciation in the *Paracoccidioides* genus, *Mol. Phylogenet. Evol.*, 52, 273, 2009.
58. Takayama, A. et al., An atypical *Paracoccidioides brasiliensis* clinical isolate based on multiple gene analysis, *Med. Mycol.*, 48, 64, 2010.

56 *Phaeoacremonium*

László Galgóczy, Laura Kovács, Tamás Papp, and Csaba Vágvölgyi

CONTENTS

56.1 INTRODUCTION

The hyphomycete genus *Phaeoacremonium*, the anamorph of *Togninia*, includes ecologically, plant pathologically, and human medically important species.[1] The majority of these fungi have been isolated from infected woody host plants, especially from grapevines (*Vitis vinifera* L.) as causative agents of Petri and Esca diseases,[2] and also from larvae of bark beetles.[3] The incidence of human infections caused by *Phaeoacremonium* spp. has increased during recent years especially among imunocompromised patients.[2] These data indicate that the diseases caused by different *Phaeoacremonium* species could be more frequent in the future.

56.1.1 Classification, Morphology, and Epidemiology

The genus *Phaeoacremonium* includes 30 species,[3,4] but their number has been increasing continuously in consequence of the reidentification and the improved molecular techniques of fungal taxonomy.

This genus was described by Crous et al.[5] in 1996 with six species: *P. parasiticum*, *P. aleophilum*, *P. angustius*, and *P. chlamydosporum* isolated from grapevines and *P. inflatipes* and *P. rubrigenum* isolated from humans and plants. Later, *P. chlamydosporum* was renamed *Phaeomoniella chlamydospora* and placed in a different genus.[6] During the past decade, further 25 species of the genus *Phaeoacremonium* have been described: *P. mortoniae*,[7] *P. viticola*,[8] *P. alvesii*, *P. amstelodamense*, *P. australiense*, *P. griseorubrum*, *P. krajdenii*, *P. scolyti*, *P. subulatum*, *P. tardicrescens*, *P. venezuelense*,[2] *P. argentinense*, *P. austroafricanum*, *P. novae-zealandiae*, *P. sphinctrophorum*, *P. theobromatis*,[1] *P. pallidum*, *P. prunicolum*, *P. fuscum*,[9] *P. croatiense*, *P. hungaricum*, *P. sicilianum*, *P. tuscanum*,[10] *P. cinereum*, and *P. hispanicum*.[4]

The genus *Phaeoacremonium* contains the anamorphs of *Togninia* species; they can be considered morphologically as intermediate between *Acremonium* and *Phialophora*. The distinctive morphological characters of *Phaeoacremonium* from *Phialophora* are its aculeate phialides and inconspicuous, nonflaring collarettes, and of *Acremonium* is its pigmented vegetative hyphae.[5] General description of the genus extended with the new species derived from Mostert et al.[1,2] Colonies on malt extract agar (MEA) medium are flat with entire (nonragged) margins, moderately dense, predominantly felty, or woolly textured. Generally, they are medium-brown becoming paler toward the conidia-forming region; but orange-brown, gray-brown, and dark-brown colonies are also common. If the colonies are paler, their color may range from pale yellow to beige. Pink-colored colonies, from pale to dark-pink, also occur. Mycelium is branched, and hyphae are septate and occur singly or bundled (4–27 hyphae). Mycelia usually have warts in their surface, but their presence depends on the area of the colony, the age of the culture, and the medium used. Warts differ in their density and their size among different species. Additional mycelial ornamentation is a useful character for distinguishing taxa, it could be smooth, verruculose, and verrucose. The mostly pale brown conidiospores could be long and branched, branched, or short and unbranched depending on the species. They differ in their number of septum and their ornamentation. Three types of phialides were identified in the genus *Phaeoacremonium*, and they differ in their size, form, and presence of the basal septum. They could be monophialidic or polyphialidic. The aseptate and smooth-walled conidia are hyaline to subhyaline aggregating into round, slimy heads at the apices of phialides. There is size difference between the aerial and agar conidia. The aerial conidia vary in shape from oblong-ellipsoidal to obovate, cylindrical, allantoids, or reniform, in contrast to agar conidia, which are relatively long and are mostly allantoid or oblongellipsoidal. Conidia have homogenous cellular contents when they are young, but they can become two-guttulate after 7–14 days, especially on agar.

Phaeoacremonium species are widely distributed throughout the world. *P. parasiticum*, *P. aleophilum*, and *P. krajdenii* are the most frequent of them. They have been mostly isolated from a range of woody hosts as causative agents of plant diseases, but *P. parasiticum*, *P. aleophilum*, and *P. argentinense* have been isolated from soil also.[1] One species, *P. scolyti* (originally identified as *P. rubrigenum*), has been isolated from bark beetles,[2,11] and only one case has been described yet, where an animal infection in dog is caused by *P. parasiticum*.[1] Nine *Phaeoacremonium* species have been confirmed as human pathogens causing mycoses in both immunocompromised and immunocompetent patients in various countries: *P. alvesii*, *P. amstelodamense*, *P. griseorubrum*, *P. krajdenii*, *P. parasiticum*, *P. rubrigenum*, *P. sphinctrophorum*, *P. tardicrescens*, and *P. venezuelense*.[1,2] From these nine species, only *P. alvesii*, *P. krajdenii*, *P. parasiticum*, and *P. venezuelense* have been isolated from woody hosts.[1] It is hypothesized by Mostert et al.[2] that species causing plant diseases, which can grow at 40°C, could be human pathogens too. According to this assumption, the following species can be expected to be detected from human infections in the future: *P. australiense*, *P. subulatum*, and *P. aleophilum*.[2]

The suspected sources of *Phaeoacremonium* infections in humans are woody splinters, soil, and air. Presumably, the major sources of human infections are the woody splinters, which cause phaeohyphomycotic cyst or mycetoma if they are passed under the skin.[1] The disease caused by *Phaeoacremonium* species is variable in their clinical manifestation including subcutaneous infections, fungemia, disseminated disease,[12] endocarditis,[13] eumycetoma,[14] onychmycosis,[15] endophthalmitis,[16] osteomyelitis,[17] and acute or chronic arthritis.[18,19] The number of human cases caused by *Phaeoacremonium* species has been continuously increasing since 1990, especially among immunocompromised hosts.[12]

56.1.2 Clinical Features and Pathogenesis

56.1.2.1 *P. alvesii*

Human infections caused by *P. alvesii* have been described in two case reports in the literature. In both cases, this fungus caused subcutaneous infection in an immunocompetent host.[20,21] Isolates from these cases were originally identified as *P. inflatipes*[20] and *P aleophilum*[21] based on their macro- and microscopic morphology only. Further investigations supplemented with molecular techniques revealed that both fungi represent a new *Phaeoacremonium* species, which was named *P. alvesii*.[2] The infected areas were surgically excised with[21] or without[20] antifungal treatment [itraconazole (ITRZ), 100 mg/day], which led to uncomplicated healing in both cases. Only one undescribed case has been found in the literature, where this fungus was isolated from synovial fluid of a patient with keratitis.[2] *P. alvesii* was proved to be sensitive to seven investigated antifungal agents with the following MICs: amphotericin B (AMB) = 2 μg/mL, fluconazole (FCZ) = 8 μg/mL, ITRZ= 16 μg/mL, ketoconazole (KCZ) = 4 μg/mL, ravuconazole (RCZ) = 1 μg/mL, terbinafine (TBF) = 2 μg/mL, and voriconazole (VCZ) = 1 μg/mL.[21]

56.1.2.2 *P. amstelodamense*

P. amstelodamense human infection was deduced from only one case, in which the fungus was isolated from elbow joint interior. It was originally misidentified as *P. inflatipes*, but Mostert et al.[2] reidentified it as a different, new species and introduced it as *P. amstelodamense* into the *Phaeoacremonium* genus.

56.1.2.3 *P. griseorubrum*

This species from human infection was originally described as *P. rubrigneum* from human infection by Matsui et al. in 1999.[22] A further study revealed that it represents a new species named *P. griseorubrum*.[2] This species was isolated as an opportunistic pathogen from a subcutaneous tumor of an immunosuppressed patient.[22] One undescribed case was reported by Mostert et al.[2], when this species was isolated from human blood.

56.1.2.4 *P. krajdenii*

The first published case where this fungus caused human infection was ascribed to *Phialophora repens*,[23] but it was reidentified as *P. krajdenii* by Mostert et al.[2] In this case, the skin lesion treated successfully by excision of the nodule from the patient, who suffered from advanced lepromatous leprosy.[23] In the second case, the fungus, misidentified as *P. repens*, was isolated from mycetoma on hand from a patient with mild diabetes mellitus.[24] The nodule was surgically excised several times with administration of flucytosine (5FC).[24] The MIC value for 5FC of this *P. krajdenii* isolate was 60 µg/mL.[24] The last described case in the literature when *P. krajdenii* caused human infection was reported as pedal white grain eumycetoma in an immunocompetent patient.[14] It was treated with 400 mg/day oral dose of ITRZ with unknown outcome.[14] Besides the above-mentioned ones, there are six cases collected by Mostert et al.[2] in which this fungus was identified as *P. inflatipes* originally in five of them. Two of them derived from unknown human source, and the remaining caused cutaneous infection in foot.

56.1.2.5 *P. parasiticum*

P. parasiticum (formerly *Phialophora parasitica*[5]) is the most commonly isolated *Phaeoacremonium* species from humans; it is able to cause infections in both immunocompetent and immunodeficient patients. Typical manifestation of the *P. parasiticum* infection is the subcutaneous phaeohyphomycosis.[12,21,25–33] In some cases, this fungus could cause sepsis or was isolated from disseminated disease with fatal outcome among immunocompromised host.[12,13,34–37] Beyond these main manifestation forms, *P. parasiticum* was isolated from endophtalmitis,[16] onychomycosis,[15] arthritis,[18,19] eumycetoma,[38] and chronic granulomatous disease.[39] Surgical excision without[25,30] or with antifungal treatment, such as AMB+ITRZ or only FCZ, was successful therapy in *P. parasiticum* cutaneous infections.[27,29,31] Application of AMB+VCZ was reported in endophthalmitis,[16] topical administration of sulconazole in onychomycosis,[15] KCZ[19] in arthritis, ITRZ+5FC in eumycetoma,[38] and AMB+VCZ with or without capsofungin in chronic granulomatous disease.[39] The MIC break- and MIC endpoints of the different drugs have not yet been established for this fungus. The MIC values documented in the literature depend on the used susceptibility methods, media, and incubation times; these are the followings:[12,21,40–42] AMB = 0.125–16 µg/mL, FCZ = 8 µg/mL, ITRZ = 0.03–32 µg/mL, KCZ = 2 µg/mL, miconazole = 2.5–10 µg/mL, posoconazole = 0.03–0.25 µg/mL, RCZ = 0.03–1 µg/mL, TBF = 2 µg/mL, and VCZ = 0.03–2 µg/mL. It is suggested that surgical excision with the administration of AMB and an azole antifungal agent would be a successful therapy for the treatment of *P. parasiticum* cutaneous infection.[12]

56.1.2.6 *P. rubrigenum*

There is no documented case report in the literature when really *P. rubrigenum* caused human infection.[2,22] Only two cases are mentioned by Crous et al.[5] and Mostert et al.[2] when this fungus was isolated from human sources. One of them was isolated from a patient with pneumonia,[5] and the other was derived from infected eye.[2]

56.1.2.7 *P. sphinctrophorum*

Two misidentified *Phialophora repens* have been caused human infection with phaeohyphomycotic cyst and subcutaneous cysts. Reidentification by morphological investigation completed with molecular techniques by Mostert et al.[1] revealed that these isolates represent a new *Phaeoacremonium* species, *P. sphinctrophorum*.

56.1.2.8 *P. tardicrescens*

Mostert et al.[2] reidentified a *P. inflatipes* isolate derived from human source as *P. tardicrescens*, and introduced it as a new species. Clinical features and pathogenesis of this fungus have not been revealed yet.

56.1.2.9 *P. venezuelense*

Till date, this species has been detected only in three human cases causing cutaneous infections.[1,2,43,44] Among them, one isolate was originally misidentified as *Cephalosporium serrae*[43]. Surgical excision of the infected area was the suggested treatment of the infection in case of a patient with hepatic and renal dysfunctions.[44]

56.1.2.10 Unidentified *Phaeoacremonium* Species

One case was reported in the literature, when olecranon osteomyelitis was caused by an unidentified *Phaeoacremonium* species in a patient who suffered traumatic olecranon bursitis.[17] This disease was cured with surgical bursal excision and antifungal therapy with ITRZ.[17] In another case report, *P. inflatipes* caused fungemia with fatal outcome (nevertheless, AMB administration) in a child suffering from severe aplastic anemia.[45] Identification of this isolate as *P. inflatipes* remained doubtful because homology of the partial rRNA gene sequences did not exceed 95%. Furthermore, all isolates from human infections, which were originally determinated as *P. infaltipes*, proved to be misidentifications in the comparative work by Mostert et al.[2] suggesting that *P. inflatipes* has not been isolated from human infection yet.

Tables 56.1 and 56.2 summarize the observed human cases and antifungal susceptibilities of *Phaeoacremonium* species.

TABLE 56.1
Reported Human Cases of *Phaeoacremonium* Infections

Species	Age (Year)	Sex	Underlying Disease	Infection Site(s)	AF Therapy	Surgical Excision	Outcome	References
P. alvesii	83	F	None	Left foot	None	Y	A (cure)	[2,20]
	19	M	None	Left ankle	ITC	Y	A (cure	[2,21]
P. amstelodamense	n.d.	n.d.	n.d.	Elbow joint int.	n.d.	n.d.	n.d.	[2]
P. griseorubrum	61	F	Malignant rheumatoid arthritis	Left foot	FCZ	Y	A (failed)	[2,22]
	n.d.	n.d.	n.d.	Blood	n.d.	n.d.	n.d.	[2]
P. krajdenii	n.d.	M	Lepromatous leprosy	Scalp	Prednisolon	Y	A (cure)	[2,23]
	63	M	Mild diabetes mellitus	Left hand	5FC	Y	A (cure)	[2,24]
	41	M	None	Right foot	ITRZ	Y	n.d.	[14]
	31	M	n.d.	Foot	n.d.	n.d.	n.d.	[2]
	n.d.	F	n.d.	Foot	n.d.	n.d.	n.d.	[2]
	n.d.	F	n.d.	Foot	n.d.	n.d.	n.d.	[2]
	n.d.	n.d.	n.d.	n.d.	n.d.	n.d.	n.d.	[2]
	n.d.	n.d.	n.d.	n.d.	n.d.	n.d.	n.d.	[2]
P. parasiticum	45	M	Renal transplant	Skin, soft tissue	None	Y	A (cure)	[26]
	41–44	M	Renal transplant	Leg, later also knee	AMB, 5FC, KCZ, ITR	Y	A (failed)	[32–34]
	45	M	Liver transplant	Disseminated	AMB, ITR	N	D	[34]
	49	M	None	Knee	AMB, 5FC	N	A (cure)	[18]
	30	F	None	Joint	KCZ	Y	A (cure)	[19]
	61	M	Renal transplant	Leg	FLC	Y	A (cure)	[27]
	92	F	Unknown	Disseminated	AMB, TBF	N	D	[35,36]
	30	M	Renal transplant	Leg	None	Y	A (cure)	[25]
	30	F	None	Foot	ITR, 5FC	Y	A (cure)	[38]
	45	M	Liver transplant	Left hand, endocarditis	AMB, FCZ, ITR	N	D	[13]
	9–17	M	Chronic granulomatous disease	Lung	CPF, ABLC, LAMB VCZ, PSZ	Y,N	A (cure)	[39]
	55	M	Renal transplant	Left ankle, foot	ITRZ, TBF, FCZ	Y	A (failed)	[2,21]
	30	F	Aplastic anemia	Disseminated	ABLC	N	D	[12]
	40	M	Cardiac transplant	Hip, forearm	ABLC, ITRZ	Y	D	[12]
	49	M	Renal transplant	Left foot	AMB, ITRZ	N	A (cure)	[29]
	n.d.	n.d.	n.d.	Eye, endophthalmitis	AMB, VCZ	N	A (cure)	[16]
	n.d.	M	Kidney transplant	Forefinger	None	Y	A (cure)	[30]
	26	F	None	Forearm	AMB, ITRZ	Y	A (cure)	[31]
	55	M	None	Onychomycosis	SCZ	Y	A (cure)	[15]
P. rubrigenum	n.d.	n.d.	Pneumonia	n.d.	n.d.	n.d.	n.d.	[5]
	n.d.	n.d.	n.d.	Eye	n.d.	n.d.	n.d.	[2]
P. sphinctrophorum	n.d.	n.d.	n.d.	Phaeohyphomycotic cyst	n.d.	n.d.	n.d.	[1]
	n.d.	M	n.d.	Subcutaneous cyst	n.d.	n.d.	n.d.	[1]
P. tardicrescens	n.d.	n.d.	n.d.	n.d.	n.d.	n.d.	n.d.	[2]
P. venezuelense	n.d.	n.d.	n.d.	Foot	n.d.	n.d.	n.d.	[2,43]
	28	M	Bone marrow transplant	Right arm	None	Y	n.d.	[44]
	n.d.	n.d.	n.d.	Right ankle	n.d.	n.d.	n.d.	[2]
Unknown	54	F	Myelodysplastic syndrome, IgA deficiency	Elbow, osteomyelitis	ITRZ	Y	A (cure)	[17]
P. infaltipes?	2	M	Aplastic anemia	Disseminated	AMB	N	D	[45]

Abbreviations: M, male; F, female; AF, antifungal; 5FC, flucytosine; AMB, amphotericin B; ABLC, amphotericin B lipid complex; CPF, caspofungin; FCZ, fluconazole; ITRZ, itraconazole; KCZ, ketoconazole; LAMB, liposomal amphotericin B; PSZ, posaconazole; SCZ, sulconazole; TBF, terbinafine; VCZ, voriconazole; Y, yes; N, no; A, alive; D, dead; n.d., no data.

TABLE 56.2
Antifungal Susceptibilities of the Clinically Important *Phaeoacremonium* spp.

	MIC (μg/mL)										
Isolate	AMB	FCZ	ITRZ	KCZ	MCZ	PSZ	RCZ	TBF	VCZ	5FC	References
P. alvesii	2	8	16	4	n.d.	n.d.	1	2	1	n.d.	[21]
P. krajdenii	n.d.	n.d.	n.d.	n.d.	n.d.	n.d.	n.d.	n.d.	n.d.	60	[24]
P. parasiticum	0.125–16	8	0.03–32	2	2.5–10	0.03–0.25	0.03–1	2	0.03–2	n.d.	[12,21,40–42]

Abbreviations: AMB, amphotericin B, FCZ, fluconazole, ITRZ, itraconazole; KCZ, ketokonazole; PSZ, posoconazole; RCZ, ravuconazole; TBF, terbinafine; VCZ, voriconazole; 5FC, flucytosine; n.d., no data.

56.1.3 Diagnosis

56.1.3.1 Conventional Techniques

Phaeoacremonium cases have been reported most frequently as subcutaneous infections and usually involving phaeohyphomycosis,[1,2] which means the invasion of the human tissue by fungi with dark, melanized cell wall. The symptoms of this disease are the dark phaeohyphomycotic cyst, a painless or painful nodule, and a cavity filled with pus under the skin.[1,2] There have been also other clinical manifestations of the *Phaeoacremonium* infections as described in Section 56.1.2.

The histological traits of the *Phaeoacremonium* infection are the mixed cell granuloma; the presence of septate, branched, and intertwined hyphal elements; and conidiogenous cells (phyalides) and conidia.[14,20,22,28,44]

Identification of *Phaeoacremonium* species is traditionally carried out by macro- and microscopic observation of the morphology and investigation of growth conditions. The identification by macroscopic examination of the *Phaeoacremonium* species is based on the colony color on MEA medium, the radial growth on MEA, and the yellow pigmentation on oatmeal agar (OA) and potato dextrose agar (PDA) medium.[1,2] Key features of clinically important *Phaeoacremonium* species for their macromorphological identification are indicated in Tables 56.3 and 56.4 based on the studies of Mostert et al.[1,2] Discriminating characters for the microscopic identification are the mycelium texture, structure of conidiophore, length of conidiospore, diameters of warts, the predominant phialide type, and the predominant Type II phialide shape.[1,2] These characters are summarized for clinically important species in Table 56.4.

TABLE 56.3
Radial Growth of Clinically Important *Phaeoacremonium* Species at Various Temperatures after 8 Days of Incubation at 25°C on MEA Medium by Mostert et al.

	Growth Temperatures (°C)			
Species	Minimum	Optimum	Maximum	Radial Growth (mm)
P. alvesii	15	30	37	9.5–11
P. amstelodamense	15	30	40	11.5–12.5
P. griseorubrum	10	30	40	6–7.5
P. krajdenii	15	30	37	9–14
P. parasiticum	15	30	40	10.5–11.5
P. rubrigenum	10	30	37	9.5–10
P. sphinctrophorum	15	25–30	30–37	6–15
P. tardicrescens	15	30	40	8–9
P. venezuelense	15	30	40	9–16

Source: Mostert, L. et al., *Stud. Mycol.*, 54, 1, 2006.

56.1.3.2 Molecular Techniques

Molecular techniques may help us to avoid the misidentification of the hypothesized species based on its macro- and microscopic observation. The main advantage of the identification by gene sequencing is that it does not require viable organism or sporulation (molds), thus enabling a rapid diagnosis,[46] This last mentioned reason is important in case of *Phaeoacremonium* species because these fungi are slow growing and their taxonomy is difficult.[3] One of the most frequently used targets for molecular identification of fungi including *Phaeoacremonium* is the internal transcribed spacer regions (ITS1 and ITS2) of the ribosomal RNA gene cluster; these regions are located between conserved genes encoding the 18S, 5.8S, and 28S rRNA.[46] However, in case of *Phaeoacremonium*, ITS region can be used only with limited success for species-level identification because of the high level of homology and thus, alternative gene targets are needed. Such alternatives may be the β-tubulin-, the actin-, or the calmodulin-encoding genes.[1,2,46]

Several methods are available for molecular identification of *Phaeoacremonium* species, and they have been continuously improved by new techniques. Restriction fragment length polymorphisms (RFLP) patterns of the ITS region and the β-tubulin gene proved to be suitable to distinguish *P. aleophilum*, *P. inflatipes*, *P. rubrigenum*, and *P. parasiticum*.[47,48] New species of *Phaeoacremonium* could successfully determined by DNA phylogenies inferred from ITS1 and ITS2, 5.8S rRNA, β-tubulin, actin, and calmodulin gene sequences.[2,7,8,49] For rapid and clear species identification and detection, several polymerase chain reaction (PCR) primer sets and combinations have been evaluated for ITS1 and ITS2, β-tubulin, and calmodulin.[1,2,12,44,50] Based on these studies, several *Phaeoacremonium* species have been identified successfully from human

TABLE 56.4

Key Features for Macroscopic and Microscopic Identification of Clinically Important *Phaeoacremonium* Species by Mostert et al.

Species	Colony Color on MEA	Yellow Pigmentation on OA and PDA	Mycelium Textures	Conidiophore Structure	Conidiophore Length (μm)	Maximum Diameter of Warts (μm)	Predominant Phialide Type	Predominant Type II Phialide Shape
P. alvesii	Medium pink or beige	Variable	Verruculose	Mostly short and unbranched	(14–)17–43(–50) av. 27	0.5	Type III	Subcylindrical or navicular
P. amstelodamense	Beige to pale brown	No	Verruculose	Mostly short and usually unbranched	(15–)16–61(–90) av. 36	1	Type II	Elongate-ampulliform, constricted at the base
P. griseorubrum	Dark pink	No	Verruculose	Mostly short and occasionally branched	(21–)23–70(–85) av. 38	1.5	Type II and III	Elongate-ampulliform or navicular
P. krajdenii	Medium brown to dark brown	No	Verrucose	Short and usually unbranched	(16–)20–45(–76) av. 28	1	Type II	Elongate-ampulliform, attenuated at the base
P. parasiticum	Brown with medium brown center	No	Verrucose	Mostly long and branched	(24–)27–80(–130) av. 47	3	Type III	Subcylindrical
P. rubrigenum	Medium to purple pink	No	Verruculose	Mostly short and unbranched	(20–)23–51(–70) av. 34	1	Type III	Elongate-ampulliform, attenuated at the base
P. sphinctrophorum	Brown to orange gray	No	Verrucose	Mostly short and often branched	(11–)13–39(–50) av. 23	No warts	Type II	Elongate-ampulliform attenuated or constricted at the base
P. tardicrescens	Medium brown to olivaceous-brown	No	Verrucose	Mostly short and unbranched	(13–)16–52(–67) av. 31	0.5	Type I and III	Subcylindrical to subulate
P. venezuelense	Beige to orange-brown	No	Verruculose	Mostly short and occasionally branched	(20–)28–48(–52) av. 31	1	Type III	Subcylindrical or navicular

Sources: Mostert, L. et al., *Stud. Mycol.*, 54, 1, 2006; Mostert, L. et al., *J. Clin. Microbiol.*, 43, 1752, 2005.

Note: Type of phialides: Type I, shortest, (2–)3–11(–17) μm, and have no basal septum, become polyphialidic in case of *P. krajdenii*; Type II, medium-sized, (5–)9–14(–16) μm, and are elongate-ampulliform or navicular in shape, become polyphialidic in case of *P. krajdenii*; Type III, the longest, (10–)15–23(–34) μm, and are subcylindrical, navicular or subulate, monophialidic.

infections.[1,2,12,14,15,44] The more rapid real-time PCR detection protocols were evaluated for detection and identification based on the differences between the ITS1 and ITS2 regions and the β-tubulin-encoding genes; however, this approach is not yet available for all clinically important species.[3,51]

56.2 METHODS

56.2.1 Sample Preparation

For filamentous fungi, conventional DNA extraction methods are well suited to total genomic sample preparation.[52] The disadvantages of these techniques are that they are slow and relatively high amount of samples are needed. In the studies reviewed in the previous sections, commercial DNA purification kits were used for rapid and safe sample preparation.

56.2.1.1 Media and Culturing Conditions

The following three media are preferred for morphological identification of *Phaeoacremonium* species: MEA (5 g malt extract, 5 g yeast extract, 5 g glucose, 10 g KH_2PO_4) medium for macro- and microscopic morphology and for measuring of the radial growth; OA (30 g of oat flakes, 1 g of $MgSO_4 \times 7H_2O$, 1.5 g of KH_2PO_4) and PDA (50% potato filtrate obtained by boiling 300 g diced potato in 500 mL water and filtering, 20 g sucrose) medium for the detection of yellow pigmentation (all media are given for 1 liter and contain 15 g agar).[1,2,21] After incubation at 25°C for 5–14 days, *Phaeoacremonium* isolates grow well.[1,2] The presence of near-UV light enhances the sporulation.[1] Genomic DNA can be extracted from mycelia grown either on solid or in liquid malt extract medium after incubation at 25°C for 5–7 days.

56.2.1.2 Light Microscopy Sample Preparation

Light microscopy sample preparation was describes by Mostert et al.[1]

Procedure

1. Plate the strain onto MEA and incubate it at 25°C for 2–3 weeks in presence of near-UV light, which enhances the sporulation.
2. Collect aerial mycelium with fine needle 2–3 cm from the colony margin to avoid contact with the agar, and mount in lactic acid on glass slide.
3. For investigating the structures formed in and on the agar surface, cut an agar block 3–5 mm from the margin of the colony with very little to no aerial mycelium, and mount it in lactic acid on glass slide.

56.2.1.3 Electron Microscopy Sample Preparation

Low-temperature scanning electron microscopy (SEM) protocol derived from Mostert et al.[1]

Procedure

1. Plate the strain onto small agar blocks (<25 mm^2) of MEA and incubate it at 25°C in the dark.
2. Mount it in a specimen holder with a mixture of cryoblock and colloidal graphite.
3. Freeze the specimens in nitrogen slush (−212°C) and transfer to the cryostage where a thin gold layer was splattered over the sample (75 s, 1.2 kV).
4. Transfer the sample to a cooled SEM stage chamber.

56.2.1.4 Genomic DNA Preparation

Beyond the conventional DNA preparation methods,[9,12,48,52] commercial DNA purification kits can be used for rapid sample preparation. Based on the literature, the following kits are suited for this: FastDNA Kit (Bio101),[1,50] DNeasy plant mini kit (QIAGEN),[3] Maxwell® 16 Tissue DNA Purification Kit from Promega®,[15] UltraClean™ Microbial DNA Kit (MO Bio).[10]

56.2.2 Detection Procedures

56.2.2.1 Morphological Identification of *Phaeoacremonium* Species

The steps of identification based on the morphological characters described by Mostert et al.[1]

Procedure

1. Plate the strain onto MEA and incubate it at 25°C in the dark for 14 days.
2. Cut mycelial plugs (2 mm diameter) from the margin of the colony and transfer onto MEA, OA, and PDA media. Incubate the plates at 25°C and note the colony characters and pigment production after 8 and 16 days.
3. For determination of the cardinal growth temperatures, plate the plugs onto MEA in three replicates for each temperature and incubate the dishes in dark at temperatures ranging from 5°C to 40°C in 5°C intervals, including 37°C to simulate human body temperature.
4. To determine radial growth, measure the radius of the colony, after incubation at 25°C for 8 days.
5. Use Tables 56.3 and 56.4 to identify the investigated strain. In the microscopic observation, use Table 56.4 for the identification of the investigated strain based on its microscopic morphology.

56.2.2.2 Molecular Identification of *Phaeoacremonium* Species

56.2.2.2.1 Identification on Genus Level

A nested PCR processed by Aroca and Raposo is appropriate to confirm that the investigated strain is actually a *Phaeoacremonium* species.[50] PCR amplification using primers ITS1F (5′-CTT GGT CAT TTA GAG GAA GTA A-3′)[53] and ITS4 (5′-TCC TCC GCT TAT TGA TAT GC-3′)[54] on DNA extracts in the first reaction and then primers Pm1 (5′-CTC CAA ACC CTT TGT GAA CAT-3′, forward) and Pm2 (5′-CGA GCC CGC CAC TGA CTT-3′, reverse) on the product of the first reaction results in a 415 bp fragments only in case of *Phaeoacremonium* species.[50] Pm1 and Pm2 were designed in the ribosomal DNA ITS regions ITS1 and ITS2, respectively.

Procedure

1. Prepare the reaction mixture for the primary PCR in a final volume of 25 µL containing 2 µL of 10× buffer, 0.2 µM (ITS1F and ITS4) primer, 2 mM $MgCl_2$, 0.2 mM deoxynucleoside triphosphates (dNTPs), 0.75 U of Taq polymerase, 2.5 µL of BSA, and 1 µL of DNA as template (approximately 10 ng of DNA).
2. Perform the first PCR in a thermocycler using the following reaction conditions: 2.5 min at 94°C; 35 cycles, with 1 cycle consisting of 15 s at 94°C, 30 s at 53°C, and 90 s at 72°C, and a final extension step at 72°C for 7 min.
3. Dilute (1:200) the PCR products from the first reaction and use it in the secondary PCR.
4. Prepare the reaction mixture for the secondary PCR in a final volume of 25 µL: 2.5 µL of 10× buffer, 0.5 µM (Pm1 and Pm2) primer, 4 mM $MgCl_2$, 0.8 mM dNTPs, and 1.25 U Taq polymerase.
5. Perform the second PCR in a thermocycler using the following reaction conditions: 5 min at 94°C; 30 cycles, with 1 cycle consisting of 30 s at 94°C, 30 s at 57°C, and 50 s at 72°C, and a final extension step at 72°C for 7 min.
6. Visualize the PCR products after electrophoresis on 1.5% agarose gels stained with ethidium bromide and run in 1× Tris-borate-EDTA buffer.

56.2.2.2.2 Identification on Species Level Using Species-Specific Primers

Molecular identification of *Phaeoacremonium* species including all clinically important taxa was carried out by Mostert et al.[1] using universal forward primer T1 (5′-AAC ATG CGT GAG ATT GTA AGT-3′)[55] with specific reverse primers developed in β-tubulin-encoding gene introns, and a reverse primer ACT-783R (5′-TAC GAG TCC TTC TGG CCC AT-3′)[56] with specific forward primers both developed in actin-encoding gene introns. The primers for the identification of all clinically important *Phaeoacremonium* species and the size of the amplified fragments are listed in Table 56.5.

Procedure

1. Prepare the reaction mixture containing 0.5 µL of diluted sample (ranging from 10 to 100 ng/µL), 1× PCR buffer, 2.5 pM of each primer, 200 µm of each of the dNTPs, 0.3 U of Taq DNA polymerase, and 1.5 mM of $MgCl_2$ and it made up to a final volume of 12.5 µL with sterile water.
2. Perform the touchdown PCR in a thermocycler using the following reaction conditions: 5 min at 94°C; 5 cycles, with 1 cycle consisting of 30 s at 94°C, 30 s at 66°C, and 60 s at 72°C; 5 cycles, with 1 cycle consisting of 30 s at 94°C, 30 s at 64°C, and 60 s at 72°C; and a final 25 cycles, with 1 cycle consisting 30 s at 94°C, 30 s at 62°C, and 60 s at 72°C, and a final extension step at 72°C for 6 min.

TABLE 56.5
Primers for the Identification of Clinically Important *Phaeoacremonium* Species by Mostert et al.

Species	Primer Pair	Species-Specific Primer (Pbr and/or Paf) Sequence (5′–3′)	Fragment Size (bp)
P. alvesii	T1 + Pbr5_1	ACG AGC TGA AGG TAA AAR GGA TC	548
	ACT-783R + Paf2	GCC AAT CTG AGG CTA TGG AA	192
P. amstelodamense	T1 + Pbr9	CGG TGA ACA TCA CGG GGG AG	456
P. griseorubrum	T1 + Pbr9	CGG TGA ACA TCA CGG GGG AG	456
	ACT-783R+Paf3	TCC GCC AAT TGA GGC TAC AA	194
P. krajdenii	T1 + Pbr13	AGA TCG TTA GAC GTG TCC CG	272
P. parasiticum	T1 + Pbr2_2	CGG TAG AGG TTT GGC GAC	446
P. rubrigenum	T1 + Pbr5_1	ACG AGC TGA AGG TAA AAR GGA TC	548
	ACT-783R + Paf1	GCC AAT CGA GGC TAT GGA G	191
P. sphinctrophorum	T1 + Pbr2_1	AGC RCC TGT AGC TTT GCA G	611
P. tardicrescens	T1 + Pbr1_2	TCC CGC TGA AGG AAA GGA AG	449
P. venezuelense	T1 + Pbr3_1	ATC TCG AGA CAG AGC GGA TG	568

Source: Mostert, L. et al., *Stud. Mycol.*, 54, 1, 2006.

Note: Pbr primers designed for introns of β-tubulin-encoding gene and Paf primers on introns of actin-encoding gene. Primer Pbr5_1 not distinguishes *P. rubrigenum* and *Pm. alvesii*, while primer Pbr9 not distinguishes *Pm. griseorubrum* and *Pm. amstelodamense*. Use additional species-specific primers (developed in actin gene introns) in these cases. *P. rubrigenum* and *P. alvesii* are distinguished by the ACT primers Paf1 and Paf2. A specific primer, Paf3, developed for *P. griseorubrum*. *P. amstelodamense* gives a positive reaction with the primer Pbr9 and a negative one with Paf3.

3. Visualize the PCR products after electrophoresis on 1% agarose gels stained with ethidium bromide and run in 1× Tris-acetate-EDTA (TAE) buffer.

56.2.2.2.3 Identification of Species Based on Partially Amplified β-Tubulin Gene Sequences

Alternatively, *Phaeoacremonium* species can be identified by sequencing of the partial amplified β-tubulin-encoding gene, and its comparison with the β-tubulin sequences available from genomic databases (GenBank, EMBL, and DDBJ) using a Basic Local Alignment Search Tool (BLAST) search. Baddley et al.[12] used T1 primer (5′-AAC ATG CGT GAG ATT GTA AGT-3′)[55] with Bt2B primer (5′-ACC CTC AGT GTA ACC CTT GGC-3′)[57] based on the study of Mostert et al.[2]

Procedure 1

1. Prepare the reaction mixture containing 5 μL of diluted sample, 1× PCR buffer 2.5 pM of each primer, 200 μM each of the dNTPs, 1.5 mM $MgCl_2$, and 0.5 U of Taq DNA polymerase; reaction mixture is made up to a final volume of 25 μL with sterile water.
2. Perform the PCR in a thermocycler using the following reaction conditions: 5 min at 96°C; 36 cycles, with 1 cycle consisting of 30 s at 94°C, 30 s at 58°C, 90 s at 72°C, and a final extension step at 72°C for 7 min.
3. Visualize the PCR products after electrophoresis on 1% agarose gels stained with ethidium bromide and run in 1× TAE buffer.
4. Purify the PCR products with a commercial PCR DNA purification kit.
5. After the sequencing of the PCR product, compare it with the β-tubulin sequences available from GenBank using a BLAST search.

Guarro et al.[44] used BT2-F (5′-GG(CT) AAC CA(AG) AT(ATC) GGT GC(CT) GC(CT)-3′) and BT2-R (5′-ACC CTC (AG)GT GTA GTG ACC CTT GGC-3′) primers based on the study of Gilgado et al.[58]

Procedure 2

1. Prepare the reaction mixture in the final volume of 25 μL containing 20–60 ng of fungal DNA template, 10 mM Tris-HCl pH 8.3, 50 mM KCl, and 2.5 mM $MgCl_2$ (10× Perkin-Elmer buffer II plus $MgCl_2$ solution), 100 μM of each dNTPs, 1 μM of each primer, and 1.5 U of Taq DNA polymerase.
2. Perform the PCR in a thermocycler using the following reaction conditions: 5 min at 94°C; 35 cycles, with 1 cycle consisting of 30 s at 95°C, 1 min at 60°C, 1 min at 72°C, and a final extension step at 72°C for 7 min.
3. Visualize the PCR products after electrophoresis on 1% agarose gels stained with ethidium bromide and run in 1× TAE buffer.
4. Purify the PCR products with a commercial PCR DNA purification kit.
5. After the sequencing of the PCR product, compare it with the β-tubulin sequences available from GenBank using a BLAST search.

The PCR product can be purified with commercial PCR DNA purification kits, for example, GFXTM PCR DNA purification kit (Pharmacia Biotech, Cerdanyola, Spain).[58]

A *Phaeoacremonium* database containing all of the known *Phaeoacremonium* species can be accessed from the website of the CBS-KNAW Fungal Biodiversity Centre (http://www.cbs.knaw.nl/databases/).

56.3 CONCLUSIONS AND FUTURE PERSPECTIVES

The incidence of fungal infections has increased dramatically over the past several years, especially among immunocompromised patient. Furthermore, more fungi have emerged as opportunistic human pathogen.[59] The data of the literature indicates that the human infections caused by different *Phaeoacremonium* species have been more frequent in the past two decades and this number could be increased in the future.[2] The treatment of *Phaeoacremonium* infections is difficult because available data concerning the MIC break- and endpoints of the generally used antifungal agents are very limited for the different species in the literature. The suggested therapy is the surgical excision with administration of AMB and extended-spectrum triazole antifungal agenst.[12]

Taxonomy, detection, and identification of the members of the *Phaeoacremonium* genus have been improved continuously; and numerous new species have been described and reidentified as new species by improving molecular techniques.[1,2] Identification of *Phaeoacremonium* isolates by morphological examination is well established and described by Mostert et al.[1,2] The disadvantages of this technique are that it is slow and sometimes lead to misidentification by an inexperienced researcher. Application of different molecular methods allows fast and reliable identification. Detection of *Phaeoacremonium* spp. in genus level was evaluated by Aroca and Raposo[50] based on the differences of the ribosomal DNA ITS regions among different fungal genus. This nested PCR method uses general fungal primers ITS1F and ITS4 in the primary reaction and specific primers (Pm1 and Pm2) designed to the ribosomal DNA ITS regions ITS1 and ITS2 of some *Phaeoacremonium* species, respectively. Because of the limited applicability of the ITS region at the species level, other molecular markers were used to process a method for distinguishing the species[46]. A touchdown PCR method appropriate to identify *Phaeoacremonium* isolates in species level was evaluated by Mostert et al.[1] This method uses specific primers designed to the different regions of the β-tubulin and the actin genes. More rapid real-time PCR detection protocols were evaluated using similar molecular markers for detection and identification of *Phaeoacremonium* species,[3,51] but these protocols have not been processed for all clinically important species.

Species belonging to the *Phaeoacremonium* genus are known principally as plant pathogenic fungi, but 9 (*P. alvesii*, *P. amstelodamense*, *P. griseorubrum*, *P. krajdenii*, *P. parasiticum*, *P. rubrigenum*, *P. sphinctrophorum*, *P. Tardicrescens*, and *P. venezuelense*) of the 30 species have been identified from human infections. This number could be increased in the future with further species (e.g., *P. australiense*, *P. subulatum*, and *P. aleophilum*)[2]. Further studies are needed to investigate their virulence factors in aspect of human side and to determine their MIC break- and MIC endpoints for the evaluation of an optimal therapy for infections caused by the members of this genus.

ACKNOWLEDGMENTS

This project was supported by a bilateral grant between Hungarian Scientific Research Fund (OTKA-NN75255) and German Research Foundation (DFG; project no. Vo 772/9-1).

REFERENCES

1. Mostert, L. et al., Taxonomy and pathology of *Togninia* (*Diaporthales*) and its *Phaeoacremonium* anamorphs. *Stud. Mycol.*, 54, 1, 2006.
2. Mostert, L. et al., Species of *Phaeoacremonium* associated with infections in humans and environmental reservoirs in infected woody plants. *J. Clin. Microbiol.*, 43, 1752, 2005.
3. Aroca, A., Raposo, R., and Lunello, P., A biomarker for the identification of four *Phaeoacremonium* species using the β-tubulin gene as the target sequence. *Appl. Microbiol. Biotechnol.*, 80, 1131, 2008.
4. Gramaje, D. et al., Novel *Phaeoacremonium* species associated with Petri disease and esca of grapevine in Iran and Spain. *Mycologia*, 101, 920, 2009.
5. Crous, P.W et al., *Phaeoacremonium* gen. nov. associated with wilt and declining diseases of woody host and human infections. *Mycologia*, 88, 786, 1996.
6. Crous, P.W. and Gams, W., *Phaeomoniella chlamydospora* gen. et. comb. nov., a causal organism of Petri grapevine decline and esca. *Phytopathol. Mediterr.*, 39, 112, 2000.
7. Groenewald, M., Kang, J.C., and Crous, P.W., ITS and β-tubulin phylogeny of *Phaeoacremonium* and *Phaeomoniella* species. *Mycol. Res.*, 105, 651, 2001.
8. Dupont, J. et al., *Phaeoacremonium viticola*, a new species associated with Esca disease of grapevine in France. *Mycologia*, 92, 499, 2000.
9. Damm, U. et al., Novel *Phaeoacremonium* species associated with necrotic wood of *Prunus* trees. *Persoonia*, 20, 87, 2008.
10. Essakhi, S. et al., Molecular and phenotypic characterisation of novel *Phaeoacremonium* species isolated from esca diseased grapevines. *Persoonia*, 21, 119, 2008.
11. Kubátová, A., Kolařík, M., and Pažoutová, S., *Phaeoacremonium rubrigenum*—Hyphomycete associated with bark beetles found in Czechia. *Folia Microbiol.*, 49, 99, 2004.
12. Baddley, J.W. et al., *Phaeoacremonium parasiticum* infections confirmed by β-Tubulin sequence analysis of case isolates. *J. Clin. Microbiol.*, 44, 2207, 2006.
13. Heath, C.H. et al., *Phaeoacremonium parasiticum* infective endocarditis following liver transplantation. *Clin. Infect. Dis.*, 25, 1251, 1997.
14. Hemashettar, B.M. et al., *Phaeoacremonium krajdenii*, a cause of white grain eumycetoma. *J. Clin. Microbiol.*, 44, 4619, 2006.
15. Sun, P.-L. and Ju, Y.-M., Onychomycosis caused by *Phaeoacremonium parasiticum*: First case report. *Mycoses*, doi:10.1111/j.1439-0507.2009.01789.x., 2009.
16. Huynh, T.K., Lee, L.R., and Ellis, D., Late-onset post-traumatic *Phaeoacremonium parasiticum* endophthalmitis. *Clin. Experiment. Ophthalmol.*, 35:366, 2007.
17. Llinas, L. et al., Osteomyelitis resulting from chronic filamentous fungus olecranon bursitis. *J. Clin. Rheumatol.*, 11, 280, 2005.
18. Kaell, A.T. and Weitzman, I., Acute monoarticular arthritis due to *Phialophora parasitica*. *Am. J. Med.*, 74, 519, 1983.
19. Rowland, M.D. and Ferrar, W.E., Case report: Thorn-induced *Phialophora parasitica* arthritis treated successfully with synovectomy and ketaconazole. *Am. J. Med. Sci.*, 293, 393, 1987.
20. Padhye, A.A. et al., Phaeohyphomycosis caused by *Phaeoacremonium inflatipes* *J. Clin. Microbiol.*, 36, 2763, 1998.
21. Guarro, J. et al., Two cases of subcutaneous infection due to *Phaeoacremonium* spp. *J. Clin. Microbiol.*, 41, 1332, 2003.
22. Matsui, T. et al., Subcutaneous phaeohyphomycosis caused by *Phaeoacremonium rubrigenum* in an immunosuppressed patient. *Jpn. J. Med. Mycol.*, 40, 99, 1999.
23. Meyers, W.M., Dooley, J.R., and Kwon-Chung, K.J., Mycotic granuloma caused by *Phialophora repens*. *Am. J. Clin. Pathol.*, 64, 549, 1975.
24. Hironaga, M. et al., *Phialophora repens*, an emerging agent of subcutaneous Phaeohyphomycosis in humans *J. Clin. Microbiol.*, 27, 394, 1989.
25. Jha, V. et al., Subcutaneous phaeohyphomycosis in a renal transplant recipient: A case report and review of the literature. *Am. J. Kidney Dis.*, 28, 137, 1996.
26. Ajello, L. et al., A case of phaeohyphomycosis caused by a new species of *Phialophora*. *Mycologia*, 66, 490, 1974.
27. Fincher, R.M. et al., Subcutaneous phaeohyphomycotic abscess caused by *Phialophora parasitica* in a renal allograft recipient. *J. Med. Vet. Mycol.*, 26, 311, 1988.
28. Kitamura, K. et al., Phaeomycotic cyst caused by *Phaeoacremonium parasiticum*. *Jpn. J. Med. Mycol.*, 41, 89, 2000.
29. Marques, S.A. et al., Subcutaneous phaeohyphomycosis caused by *Phaeoacremonium parasiticum* in a renal transplant patient. *Med. Mycol.*, 44, 671, 2006.
30. Farina, C. et al., *Phaeoacremonium parasiticum* subcutaneous infection in a kidney-transplanted patient successfully treated by surgery. *Transpl. Infect. Dis.*, 9, 253, 2007.
31. Baradkar, V.P., Mathur, M., and Kumar, S., Phaeohyphomycosis of subcutaneous tissue caused by *Phaeoacremonium parasiticum*. *Ind. J. Med. Microbiol.*, 27, 66, 2009.
32. Lavarde, V. et al., Un cas de phaeomycose a` *Phialophora parasitica* chez un transplanté. Deuxie`me observation mondiale. *Bull. Soc. Fr. Mycol. Med.*, 11, 273, 1982.
33. Ziza, J.M. et al., Ostéoarthrites a` champignons noirs (dématiés). A propos de 3 observations. *Ann. Med. Interne*, 136, 393, 1985.
34. Dupont, B. and Drouhet, E. Early experience with itraconazole in vitro and in patients: Pharmacokinetic studies and clinical results. *Rev. Infect. Dis.*, 9, S71, 1987.

35. Wong, P.K. et al., Disseminated *Phialophora parasitica* infection in humans: Case report and review. *Rev. Infect. Dis.*, 11, 770, 1989.
36. Ferraro, F.A. and Morgan, M.A., A case of disseminated *Phialophora parasitica* infection. *Arch. Pathol. Lab. Med.*, 113, 1379, 1989.
37. Revankar, S.G. et al., Disseminated phaeohyphomycosis: Review of an emerging mycosis. *Clin. Infect. Dis.*, 34, 467, 2002.
38. Hood, S.V. et al., Atypical eumycetoma caused by *Phialophora parasitica* successfully treated with itraconazole and flucytosine. *Br. J. Dermatol.*, 136, 953, 1997.
39. Segal, B.H. et al., Posaconazole as salvage therapy in patients with chronic granulomatous disease and invasive filamentous fungal infection. *Clin. Infect. Dis.*, 40, 1684, 2005.
40. Weitzman, I. et al., *Phialophora parasitica*, an emerging pathogen. *Sabouraudia*, 22, 331, 1984.
41. Espinel-Ingroff, A., Boyle, K., and Sheehan, D.J., In vitro antifungal activities of voriconazole and reference agents as determined by NCCLS methods: Review of the literature. *Mycopathologia*, 150, 101, 2001.
42. McGinnis, M.R. and Pasarell, L., In vitro testing of susceptibilities of filamentous ascomycetes to voriconazole, itraconazole, and amphotericin B, with consideration of phylogenetic implications. *J. Clin. Microbiol.*, 36, 2353, 1998.
43. De Albornoz, M.B., *Cephalosporium serrae*, agente etiológico de micetomas. *Mycopathol. Mycol. Appl.*, 54, 485, 1974.
44. Guarro, J. et al., Limitations of DNA sequencing for diagnosis of a mixed infection by two fungi, *Phaeoacremonium venezuelense* and a *Plectophomella* sp., in a transplant recipient. *J. Clin. Microbiol.*, 44, 4279, 2006.
45. Wang, S-H. et al., Fatal fungemia due to *Phaeoacremonium inflatipes* in a child with severe aplastic anemia. *Clin. Infect. Dis.*, 40, 1067, 2005.
46. Petti, C.A., Detection and identification of microorganisms by gene amplification and sequencing. *Clin. Infect. Dis.*, 44, 1108, 2007.
47. Tegli, S. et al., Genetic variation within *Phaeoacremonium aleophilum* and *P. chlamydosporum* in Italy. *Phytopathol. Mediterr.*, 39, 125, 2000.
48. Dupont, J. et al., ITS and β-tubulin markers help delineate *Phaeoacremonium* species, and the occurrence of *P. parasiticum* in grapevine disease in Argentina. *Mycol. Res.*, 106, 1143, 2002.
49. Mostert, L. et al., *Togninia* (Calosphaeriales) is confirmed as teleomorph of *Phaeoacremonium* by means of morphology, sexual compatibility, and DNA phylogeny. *Mycologia*, 95, 646, 2003.
50. Aroca, A. and Raposo, R., PCR-based strategy to detect and identify species of *Phaeoacremonium* causing grapevine diseases. *Appl. Environ. Microbiol.*, 73, 2911, 2007.
51. Overton, B.E., Stewartm, E.L., and Wenner, N.G., Molecular phylogenetics of grapevine decline fungi from Pennsylvania and New York. Paper presented at *Proceedings of the 4th International Workshop on Grapevine Trunk Diseases*, Stellenbosch, South Africa, p. 27, 2005.
52. Lee, S.B. and Taylor, J.W., Isolation of DNA from fungal mycelia and single spores, in *PCR Protocols: A Guide to Methods and Applications*, Innis, M.A., Gelfand, D. H., Sninsky, J. and White, T. J., Eds., pp. 282–287, Academic Press, New York, 1990.
53. Gardes, M. and Bruns, T.D., ITS primers with enhanced specificity for basidiomycetes—Application to the identification of mycorrhizae and rusts. *Mol. Ecol.*, 2, 113, 1993.
54. White, T. et al., Amplification and direct sequencing of fungal ribosomal RNA genes for phylogenetics, in *PCR Protocols. A Guide to Methods and Applications*, Innis, M.A., Gelfand, D. H., Sninsky, J. and White, T. J., Eds., pp. 315–322, Academic Press, San Diego, CA, 1990.
55. O'Donnell, K. and Cigelnik, E., Two different intragenomic rDNA ITS2 types within a monophyletic lineage of the fungus *Fusarium* are nonorthologous. *Mol. Phylogenet. Evol.*, 7, 103, 1997.
56. Carbone, I. and Kohn, L.M., A method for designing primer sets for speciation studies in filamentous ascomycetes. *Mycologia*, 91, 553, 1999.
57. Glass, N.L. and Donaldson, G.C., Development of primer sets designed for use with the PCR to amplify conserved genes from filamentous ascomycetes. *Appl. Environ. Microbiol.*, 61, 1323, 1995.
58. Gilgado, F. et al., Molecular phylogeny of the *Pseudallescheria boydii* species complex: Proposal of two new species. *J. Clin. Microbiol.*, 43, 4930, 2005.
59. Wengenack, N.L. and Binnicker, M.J., Fungal molecular diagnostics. *Clin. Chest. Med.*, 30, 391, 2009.

57 *Phialemonium*

Dongyou Liu and R.R.M. Paterson

CONTENTS

57.1 INTRODUCTION

57.1.1 Classification, Morphology, and Biology

The genus *Phialemonium* consists of dematiaceous (i.e., dark-walled) fungi belonging to the mitosporic Ascomycota group, class Sordariomycetes, subphylum Pezizomycotina, phylum Ascomycota, kingdom Fungi. Currently, the genus comprises two recognized species (*Phialemonium curvatum* and *Phialemonium obovatum*) and nine unassigned species. A former species in the genus *Phialemonium dimorphosporum* is synonymized with *P. curvatum* [1]. Anamorphs of the genus are found in Cephalothecaceae.

First created in 1983 by Gams and McGinnis to cover fungal isolates with morphological features intermediate between *Acremonium* (which is hyaline or nonpigmented) and *Phialophora* (which is pigmented), *Phialemonium* appears pale in culture and the innate hyphal pigmentation is observed in tissue only by histological stains with specificity for melanin (e.g., the Fontana-Masson silver stain) [2].

Phialemonium colonies grow rapidly, reaching a diameter of 25–30 mm in 7 days on Sabouraud dextrose agar (SDA). Initially, colonies appear flat, moist, and white, and later become yellowish, with yellow-green pigments diffusing into media (in the case of *P. obovatum*). Conidiophores are hyaline and resemble hyphae; intercalary and distinct septate phialides (10–16 µm in length) without collarettes arise from hyphae; conidia (2–3 µm by 4–5 µm in size) are one-celled and hyaline, smooth walled, aggregating into slimy obovate or allantoid balls [3].

P. curvatum conidia are subglobose to ellipsoidal or curved. Phialoconidia formed in tapering phialides are located in short branches of the superficial and submersed hyphae. These short prolongations are cylindrical and thin and lack a collarette. Adelophialides and some phialides are often without a basal septum. Conidia are unicellular, hyaline, and allantoid. On the other hand, *P. obovatum* produces a distinctive pale green pigment in culture, in contrast to other medically important dematiaceous fungi which are olive green or brown to grey-black. *P. obovatum* conidia are obovoid or narrowest at the base.

Phialemonium species are saprophytic in nature and represent a rare cause of invasive mold infections in humans, especially immunocompromised hosts and transplant recipients. *Phialemonium* spp. have the ability to sporulate within the matrix of the agar medium, without the requirement of an air space above the hyphae, which is also noted in hyaline filamentous molds such as *Acremonium* and *Fusarium*. The production of unicellular yeast-like forms or propagulation in vivo by *Acremonium*, *Fusarium*, and *Phialemonium* enables their ready circulation in the bloodstream (resulting in fungemia) in comparison to hyphal structures of angioinvasive fungi such as *Aspergillus* and *Mucor*. In addition, *Phialemonium* species are able to grow in culture at temperatures as high as 40°C, suggesting their potential for infection of the central nervous system.

57.1.2 Clinical Features and Pathogenesis

Once thought to possess low pathogenic potential, the dematiaceous (or melanized) fungi have increasingly become common causes of severe and disseminated phaeohyphomycosis, affecting cutaneous and subcutaneous tissues, the ocular region, frontal and maxillary sinuses, lungs, bones, and the nervous system [4]. The genus *Phialemonium* is considered as a phaeoid fungus, which may cause the lesions observed in phaeohyphomycosis and hyalohyphomycosis [5–8]. Predisposing factors for fungal infections are antibiotic therapy, post-transplant immunosuppressive therapy, and HIV infections. Another important risk factor for *Phialemonium* infection is the use of arteriovenous (AV) grafts or fistulas for hemodialysis, as many infected patients had AV grafts or fistulas rather than catheters. It is possible that implanted synthetic grafts or surgically created fistulas may act as a nidus for subclinical *Phialemonium* infection, allowing small inocula to produce overt infection more readily in a graft or fistula than in a catheter. Prior to dialysis access, thorough skin antisepsis is required for grafts and fistulas. Any suboptimal skin preparation may facilitate viable molds on the skin to be inoculated into the bloodstream or the AV graft. In addition, the use of prefilled syringes may also act as the source of *P. curvatum* infection [9].

P. curvatum infections have been reported in patients with bone marrow transplantation, and hemodialysis patients, who developed overwhelming infection associated with fungemia and endocarditis [9–15]. A case of cutaneous infection caused by *P. curvatum* was reported in a patient with multiple myeloma who underwent bone marrow transplantation. Two months after transplantation, the patient developed purplish and painful nodular lesions on the right ankle. Lesion material contained septate hyphae and some apparently hyaline dilated cells. On potato agar, an initially white colony later became chraceous with whitish borders forming moist and plane colonies. The colony became grayish on Sabouraud agar with sparse white mycelium. The fungus was identified as *P. curvatum* on the basis of its unicellular, hyaline, cylindrical, and allantoid conidia, in addition to other characteristic macroscopic and microscopic characteristics [16].

In the four cases of *P. curvatum* infections reported by Proia et al. [11], the first involved a 35-year-old man with chronic renal failure. The patient presented right ocular pain, swelling, and vision loss, 1 year after cadaveric renal transplantation surgery and oral prednisone and cyclosporine treatment through hemodialysis via AV fistula. Physical inspection of the right eye revealed periorbital edema, scleral injection with hemorrhagic foci, and vitreal clouding. Gram staining of vitreal aspiration showed both hyphal and yeast-like forms. A transesophageal echocardiogram (TEE) uncovered large vegetations on both mitral and aortic valves. Primary blood cultures on BacT/ALERT aerobic and anaerobic bottles (bioMérieux) at 30°C and 35°C became positive after 4 days; subculture showed heavy growth of *P. curvatum* mold after 3 days. The patient underwent combined aortic and mitral valve replacement and right eye vitrectomy. Gram staining of mitral valve tissue revealed septate hyphae, and cultures of mitral and aortic valve tissue specimens grew *P. curvatum*. One day after the operation, the patient became febrile and hypotensive and developed acute congestive heart failure and disseminated intravascular coagulation and then died. Postmortem examination revealed cardiomegaly, marked left ventricular dilatation, splenomegaly, and a large splenic abscess. A large "vegetation" had overgrown the prosthetic mitral valve and had invaded through the left ventricular wall.

The second case concerned a 66-year-old man who had undergone long-term hemodialysis after an AV graft. The patient presented a soft systolic murmur over the cardiac apex. Three blood cultures and a graft tissue specimen culture grew *P. curvatum*. TEE showed a small, calcified vegetation on the anterior leaflet of the mitral valve that extended into the left atrium, with mild mitral regurgitation. The third case involved a 70-year-old woman undergoing long-term hemodialysis. The patient had cerebrovascular accident, type II diabetes mellitus, and peripheral vascular disease. A *Phialemonium* species grew on culture of a graft tissue specimen. The fourth patient involved a 77-year-old man with type II diabetes mellitus and chronic renal failure requiring hemodialysis. Ultrasonography of the AV graft revealed thrombosis; histopathologic examination of resected graft material showed inflammatory cells and numerous fungal elements, and culture of the AV graft and blood grew *Phialemonium* species [11].

P. obovatum has been linked to cases of endocarditis, fungemia, and osteomyelitis [17–19]. A case of native valve endocarditis caused by *P. obovatum* in a premature neonate was reported [20]. A 4.5-month-old infant with fatal *P. obovatum* infection of thermal burn wounds, viable tissue, and blood vessels is reported, with dissemination to the spleen. *P. obovatum* osteomyelitis has been observed in a previously healthy 41-year-old man following discography. A case of *P. curvatum* fungemia in a patient with chronic myelogenous leukemia was reported. The patient developed graft-versus-host disease of the skin and liver with fever and severe diarrhea, 6 months after peripheral bone marrow transplantation. Biopsies of the stomach, colon, and rectum showed granulomatous inflammation with marked crypt distortion, simulating Crohn disease. A lung wedge biopsy of the lesion revealed septate branching hyphae (4–5 µm in diameter) with terminal globular structures (10 µm in diameter). Blood cultures grew a pure isolate of *P. obovatum* [21].

The in vitro susceptibility of *Phialemonium* indicated a more consistent level of activity for voriconazole and posaconazole. Treatment with the new triazoles is associated with improved survival [8]. Caspofungin, combined with amphotericin B deoxycholate may have an additive effect.

57.1.3 Diagnosis

Although *Phialemonium* is considered as a dematiaceous fungus, direct microscopic examination of the material obtained from its lesions often shows hyaline septate hyphae and its culture also appears pale. Since other phaeohyphomycotic agents such as *A. alternata*, *Bipolaris spicifera*, *Exophiala jeanselmei*, or *E. spinifera* also appear hyaline, they may complicate the diagnosis of *Phialemonium* infection. The differentiation of *Phialemonium* from other phaeohyphomycotic agents relies on the detection of the innate hyphal pigmentation in *Phialemonium* with the Fontana-Masson silver stain that binds specifically to melanin.

Microscopically, the distinctive features of *P. curvatum* isolates include the presence of setose and cupulate sporodochium, conidia formation from branched conidiophores, conidium production on typical tapering adelophialide lacking a basal septum, and hyphae with lateral and intercalary phialidic conidiogenous cells, with typical clusters of conidia aggregating in slimy heads [9]. By contrast, *Phialemonium obovatum* isolates show obovate straight conidia (3.5–5 by 1–2 μm) borne on adelophialides (reduced phialides lacking a basal septum); and *P. obovatum* colonies on potato flakes agar (at 25°C) demonstrate characteristic diffusible green pigment on reverse [20].

While in vitro culture enables more accurate morphological characterization of *Phialemonium* spp., the process is time-consuming and demands considerable technical expertise. With the recent development and application of nucleic acid amplification and sequencing techniques, the precise and speedy identification of *Phialemonium* spp. has become feasible. For example, PCR amplification of the internal transcribed spacer (ITS) regions followed by nucleotide sequencing contributed to the identification of *P. curvatum* from patients [22].

57.2 METHODS

57.2.1 Sample Preparation

Clinical specimens suspected of dematiaceous fungi are treated with 10% KOH or stained with calcofluor white and observed microscopically for septate hyphal elements. Fungal isolation is performed on sabourauol's dextrose agar (SDA) with chloramphenicol or potato dextrose agar (PDA) at 25°C and 35°C. Suspected *Phialemonium* isolates are subcultured on PDA at 25°C. Preliminary identification is based on macroscopic and microscopic features of the colonies.

After growth for 1–7 days. From day 1 to approx 4 is too early to take mycelium, on PDA slants, approximately 1 cm^2 of mycelium is collected by scraping the slant with a sterile stick in 1 mL of sterile H_2O. The material is transferred to a 2 mL screw-cap tube, and the tube is centrifuged for 1 min at 6000 × *g*. If the mycelium does not pellet, the material is contained with a pediatric blood serum filter. After removal of supernatant, the material is resuspended in 200 μL of IDI sample buffer and transferred to the lysis tube containing glass beads. Lysis tube is vortexed for 5 min, and the tube is placed in a boiling water bath for 15 min. Following centrifugation for 5 min at 16,000 × *g*, the supernatant is stored at −20°C until PCR is undertaken.

Alternatively, about 1 cm^2 of fungal material is transferred to a 2 mL Eppendorf tube containing a 2:1 (wt/wt) mixture of silica gel and Celite (silica gel H, Merck 7736/Kieselguhr Celite 545; Machery) and 300 μL of TES buffer [2 g Tris (hydroxymethyl)-aminomethane, 0.38 g Na-ethylenediaminetetraacetic acid (EDTA), and 2 g sodium dodecyl sulfate in 80 mL of ultrapure water (pH 8)]. The fungal material is ground with a micropestle for 1–2 min. The volume is adjusted by adding 200 μL of TES buffer. After vigorous shaking and the addition of 10 μL of a 10 mg/mL concentration of proteinase K to the tube, the mixture is incubated at 65°C for 10 min. With the addition of 140 μL of 5 M NaCl solution, the mixture is combined with 1/10 volume (~65 μL) of CTAB (cetyltrimethylammonium bromide) buffer 10%, followed by incubation for another 30 min at 65°C. One volume (~700 μL) of chloroform-isoamyl alcohol (vol/vol = 24/L) is added and mixed by inversion. After incubation for 30 min at 0°C (on ice water) and centrifugation at 14,000 rpm at 4°C for 10 min, the top layer is transferred to a clean Eppendorf tube. The sample is added with 225 μL of 5 M NH_4-acetate and incubated for at least 30 min (on ice water) and centrifuged. The supernatant is transferred to a clean sterile Eppendorf tube and mixed with a 0.55 volume (~510 μL) of ice-cold isopropanol. After centrifugation for 7 min at 14,000 rpm and 4°C (or room temperature), the supernatant is decanted. The pellet is washed with 70% ice-cold ethanol two times and dried by using a vacuum dryer. The powder is resuspended in 48.5 μL of Tris-EDTA buffer with 1.5 μL of 10 mg of RNase/mL, incubated at 37°C for 15–30 min, and stored at −20°C until use [23].

Alternatively, *Phialemonium* isolates are grown on liquid complete yeast medium, and DNA may be extracted using the FastDNA Kit (Bio 101).

57.2.2 Detection Procedures

Proia et al. [11] utilized universal primers V9G (5′-TTACGTCCCTGCCCTTTGTA-3′) and LR5 (5′-ATCCTGAGGGAAACTTC-3′) to amplify the ITS region of nuclear ribosomal DNA followed by sequencing, leading to the identification of *Phialemonium* isolates.

Procedure

1. PCR mixture (50 μL) is composed of 25 pmol of each primer, 200 μmol of each dNTP, 1 U of Taq polymerase (Super Taq; HT Biotechnology), 1× standard PCR buffer, and 1 μL of genomic DNA extract.
2. Amplification is carried out with an initial 94°C for 2 min; 35 cycles at 94°C for 35 s, 55°C for 50 s, and 72°C for 2 min; and a final 72°C for 6 min.
3. After checking a portion of the PCR products on agarose gel, the remaining PCR products are purified with the GFX PCR DNA and Gel Band Purification Kit (Amersham Pharmacia).
4. The PCR products are sequenced with the primers ITS1 (5′-TCCGTAGGTGAACCTGCGG-3′) and ITS4 (5′-TCCTCCGCTTATTGATATGC-3′) [White, et al. 1990] by means of the BigDye terminator cycle sequencing kit (Applied Biosystems). Sequences are analyzed on an ABI Prism 3700 instrument (Applied Biosystems).
5. The resulting sequences are aligned against published sequences using Clustal X, version 1.81. Phylogenetic relationships of the taxa are estimated from the aligned sequences by the maximum parsimony criterion, as implemented in PAUP, version 4.0b10. Heuristic searches are performed using parsimony informative, unordered, and equally weighted characters; branch robustness is tested by

means of 1000 search replications, each on bootstrapped data sets. Starting tree(s) are obtained via stepwise, random sequence addition repeated 100 times (repeated 10 times in bootstrap analyses). A maximum number of 9000 trees (1000 trees in bootstrap analyses) are allowed [9].

57.3 CONCLUSIONS

The genus *Phialemonium* is a dematiaceous fungus (with melanin in the cell walls and septa in the hyphae) that demonstrates intermediate morphological features between the hyaline or nonpigmented genus *Acremonium* and pigmented genus *Phialophora*. Because *Phialemonium* hyphal pigmentation is only detectable in tissue with melanin-specific histological stain such as the Fontana-Masson silver stain, it may be also confused with other phaeohyphomycotic agents including *Alternaria alternata*, *Bipolaris spicifera*, *Exophiala jeanselmei*, and *E. spinifera*, which produce hyaline hyphae.

Along with an increasing occurrence of immunosuppression in human populations, invasive infectious diseases due to emerging pathogens such as *Phialemonium* species are reported with alarming frequency. Indeed, both species within the genus *Phialemonium*, *P. obovatum*, and *P. curvatum*, have been shown to cause endocarditis, fungemia, osteomyelitis, keratitis, and endophthalmitis, often with fatal outcome. Considering that many fungal organisms demonstrate varied sensitivity to antifungal drugs, it is necessary to correctly identify them to species level. As laboratory characterization of *Phialemonium* species based on their macroscopic and microscopic features is lengthy and technically demanding, molecular techniques such as PCR and sequencing have become an indispensible tool for their identification and detection.

REFERENCES

1. The UniProt Consortium. Available at http://www.uniprot.org/, accessed on August 1, 2010.
2. Gams W, McGinnis MR. *Phialemonium*, a new anamorph genus intermediate between *Phialophora* and *Acremonium*. *Mycologia* 1983;75:977–987.
3. de Hoog GS et al. (eds.), *Atlas of Clinical Fungi*, 2nd edn. Utrecht, the Netherlands: Centraalbureau voor Schimmelcultures, 2000.
4. Ajello L. Hyalohyphomycosis and phaeohyphomycosis: Two global disease entities of public health importance. *Eur J Epidemiol*. 1986;2:243–251.
5. King D et al. A phaeohyphomycotic cyst and peritonitis caused by *Phialemonium* species and a reevaluation of its taxonomy. *J Clin Microbiol*. 1993;31:1804–1810.
6. Guarro J et al. *Phialemonium* fungemia: Two documented nosocomial cases. *J Clin Microbiol*. 1999;37:2493–2497.
7. Huang J et al. Flow impeding fungal thrombus in the ascending aorta. *Ann Thorac Surg*. 2008;86(4):1373–1375.
8. Rivero M et al., Infections due to *Phialemonium* species: Case report and review. *Med Mycol*. 2009;47(7):766–774.
9. Strahilevitz J et al. An outbreak of *Phialemonium* infective endocarditis linked to intracavernous penile injections for the treatment of impotence. *Clin Infect Dis*. 2005;40:781–786.
10. Schonheyder HC et al. Late bioprosthetic valve endocarditis caused by *Phialemonium* aff. *curvatum* and *Streptococcus sanguis*: A case report. *J Med Vet Mycol*. 1996;34:209–214.
11. Proia LA et al. *Phialemonium*: An emerging mold pathogen that caused 4 cases of hemodialysis—Associated endovascular infection. *Clin Infect Dis*. 2004;39:373–379.
12. Zayit-Soudry S et al. Endogenous *Phialemonium curvatum* endophthalmitis. *Am J Ophthalmol*. 2005;140(4):755–757.
13. Dan M et al. *Phialemonium curvatum* arthritis of the knee following intra-articular injection of a corticosteroid. *Med Mycol*. 2006;44(6):571–574.
14. Weinberger M et al. Isolated endogenous endophthalmitis due to a sporodochial-forming *Phialemonium curvatum* acquired through intracavernous autoinjections. *Med Mycol*. 2006 May;44(3):253–259.
15. Rao CY et al. Contaminated product water as the source of *Phialemonium curvatum* bloodstream infection among patients undergoing hemodialysis. *Infect Control Hosp Epidemiol*. 2009;30(9):840–847.
16. Heins-Vaccari EM et al. *Phialemonium curvatum* infection after bone marrow transplantation. *Rev Inst Med Trop Sao Paulo*. 2001;43(3):163–166.
17. McGinnis MR, Gams W, Goodwin MN. *Phialemonium obovatum* in a burned child. *J Med Vet Mycol*. 1986;24:51–55.
18. Magnon KC, Jalbert M, Padhye AA. Osteolytic phaeohyphomycosis caused by *Phialemonium obovatum*. *Arch Pathol Lab Med*. 1993;117:841–843.
19. Osherov A et al. *Phialemonium curvatum* prosthetic valve endocarditis with an unusual echocardiographic presentation. *Echocardiography*. 2006;23:6, 503–505.
20. Gavin PJ et al. Fatal endocarditis caused by the dematiaceous fungus *Phialemonium obovatum*: Case report and review of the literature. *J Clin Microbiol*. 2002;40:2207–2212.
21. Scott RS, Sutton DA, Jagirdar J. Lung infection due to opportunistic fungus, *Phialemonium obovatum*, in a bone marrow transplant recipient: An emerging infection with fungemia and Crohn disease-like involvement of the gastrointestinal tract. *Ann Diagn Pathol*. 2005;9(4):227–230.
22. Clark T et al. Outbreak of bloodstream infection with the mold *Phialemonium* among patients receiving dialysis at a hemodialysis unit. *Infect Control Hosp Epidemiol*. 2006;27:1164–1170.
23. Zeng JS et al. Spectrum of clinically relevant *Exophiala* species in the United States. *J Clin Microbiol*. 2007;45(11):3713–3720.
24. White TJ et al. Amplification and direct sequencing of fungal ribosomal RNA genes for phylogenetics, in *PCR Protocols: A guide to methods and Applications*, Innis, M.A., Gelfand, D.H., Sninsky, J. and White, T.J., Eds., Academic Press Inc., San Diego, CA, 1990;315–322.

58 *Pseudallescheria* and *Scedosporium*

Ana Alastruey-Izquierdo, Maria Victoria Castelli, Leticia Bernal-Martinez, and Juan Luis Rodríguez Tudela

CONTENTS

58.1 INTRODUCTION

Scedosporium is a ubiquitous filamentous fungus with a worldwide distribution. Among this genus, there are two medically important species: *Scedosporium apiospermum* (the anamorph of *Pseudallescheria boydii*) and *Scedosporium prolificans*. Scedosporiosis is the term used to define the clinical diseases caused by these species. The diseases caused by these fungi range from localized to disseminated infections in immunocompetent and immunocompromised patients. The diagnosis of Scedosporiosis is difficult since clinical features and histopathology are similar to those produced by other common fungi [1,2]. The treatment of Scedosporiosis is also difficult because of the multiresistant nature of the genus.

58.1.1 Taxonomy, Biology, and Morphology

58.1.1.1 Taxonomy

P. boydii (*Ascomycetes*), the sexual state of *S. apiospermum*, was first discovered in 1889 as a cause of a human otitis [3], while the anamorph state, called at that time *Monosporium apiospermum*, was discovered in 1911 from a case of human mycetoma [4]. In 1919, Castellani and Chalmers [5] renamed the genus *Monosporium* as *Scedosporium*.

P. boydii has been described for many years as the sexual state of *S. apiospermum*. Recent molecular studies have demonstrated that *P. boydii*/ *S. apiospermum* is a complex of species. *Scedosporium boydii* is the anamorph of *P. boydii*, while *S. apiospermum* has no teleomorph described yet. In addition, some new species have been described by means of sequencing part of the β-tubulin gene. The species currently accepted in the *P. boydii*–*S. apiospermum* species complex are *S. apiospermum*, *P. boydii*, *Scedosporium aurantiacum*, *Pseudallescheria minutispora*, *Pseudallescheria angusta*, *Pseudallescheria ellipsoidea*, *Pseudallescheria fusoidea*, and *Scedosporium dehoogii*. However, the latest molecular studies suggested that this complex could have more cryptic species [6,7].

S. prolificans was first isolated in 1974 from a greenhouse soil under the name *Lomentospora prolificans* [8]. The first case, described in 1984, was a disseminated infection in an immunocompromised host [9]. The mold causing the infection was named as *Scedosporium inflatum*. The identity of these two taxa was recognized in 1991 by Gueho and de Hoog [10] on molecular grounds, being renamed as *Scedosporium prolificans*. *S. prolificans* has no known sexual state.

58.1.1.2 Biology

S. apiospermum is mainly found in temperate climates and less frequently in the tropics. Although its natural niche is unknown, it is always associated with environments associated to human activity. It is frequently found in manure-enriched or polluted environments as garden soil, agricultural fields, sewer, or ditch mud and pollutes pond bottoms [1]. Because of its ability to assimilate natural gas and aromatic compounds, it has been suggested for use in bioremediation [11]. *S. prolificans* is frequently found in warm, indoor soil, such as in potted plants as well as in glasshouses [8].

58.1.1.3 Morphology

58.1.1.3.1 S. apiospermum/P. boydii

Macroscopy: Fast growing colonies, cottony to lanose, greyish-white becoming pale brown with a greyish black reverse. The optimal temperature is 25°C although it can grow up to 42°C.

Microscopy: Two asexual states are described for this species. The most prevalent is *Scedosporium* while the synanamorph *Graphium* is not frequently found in culture. The cylindrical conidiogenous annellidic cells are produced terminally or laterally from the hyaline hyphae (Figure 58.1a). Conidia are subhyaline to brown, smooth walled, and subspherical to elongate with a truncated base and measure 6–12 × 3.5–6.0 μm [12]. The *Graphium* synanamorph may be produced in a later stage in some colonies. It is characterized by erect synnemata (Figure 58.1b) with terminally conidiogenous cells (smaller than the ones of the *Scedosporium* synanamorph) producing broadly clavate conidia less pigmented than the *Scedosporium* ones.

The sexual state, *Pseudallescheria*, is favored by nutritionally poor media such as cornmeal agar. It is homothallic and produces yellow brown to black, closed fruit bodies (cleistothecia) spherical and 140–200 μm in size. The cleistothecia contain ovate to subglobose (12–18 × 8–13 μm in size) asci with eight lemon-shaped, smooth-walled, pale yellow to golden brown ascospores and 6–7 × 4.0–4.5 μm in size [1,12,13].

58.1.1.3.2 S. prolificans

Macroscopy: Colonies grow moderately rapidly at 25°C on Sabouraud agar. They are expanding and flat and with a moist surface of olive grey to black color. Unlike *S. apiospermum*, it is not able to grow in presence of cycloheximide.
Microscopy: The conidiophores are short, characterized by having swollen flask-shaped bases and frequently showing a long annellated zone (Figure 58.1c). Conidia usually form aggregates at the apex; they are ovoid to pyriform, smooth walled, olive to brown, 2–5 × 3–13 μm in size and having a truncated base. Some isolates may produce round, thick-walled solitary conidia arising directly from the hyphae [1,12,13]. It has no sexual state described.

58.1.2 Clinical Features and Pathogenesis

In 1922, the ascomycete *Allescheria boydii* was isolated from a mycetoma patient in the United States. Since then, this mold has been involved in chronic subcutaneous infections before 1980 and systemic or disseminated infections in immunocompromised patients in the following years. In 1982, Fisher et al. [14] described the near-drowning syndrome caused by *S. apiospermum*, and few years later Berenguer et al. [15] reviewed its ability to cause brain infections. Although it had already been described as a pulmonary colonizer in 1955 [16], it was not until 2000 when its importance as a colonizer in cystic fibrosis patients was addressed [17].

Scedosporium apiospermum and *Scedosporium prolificans* are fungal opportunistic pathogens. They may cause asymptomatic colonization or localized or disseminated infection following trauma, surgery, and immunosuppression [18]. As stated above, infections by *S. prolificans* have been more recently described than those by *S. apiospermum*. Initially *S. prolificans* was found to cause bone and soft tissue infections in immunocompetent individuals. However, the importance of *S. prolificans* has increased in the last years as it can cause disseminated infections, especially in neutropenic patients [19,20].

Scedosporiosis represents a broad spectrum of clinical diseases caused by these two species. Three clinical syndromes can be distinguished [1]:

1. *Asymptomatic or symptomatic colonization of cavities*: These species can be colonizers of the lungs of patients with pulmonary disorders, causing pulmonary fungal balls in susceptible patients suffering from sarcoidosis or tuberculosis, allergic bronchopulmonary pneumonia and pulmonary colonization in cystic fibrosis patients. *S. apiospermum* is among the most common filamentous fungi colonizing the lungs of cystic fibrosis patients. This occurrence of *S. apiospermum* species complex in the lungs of

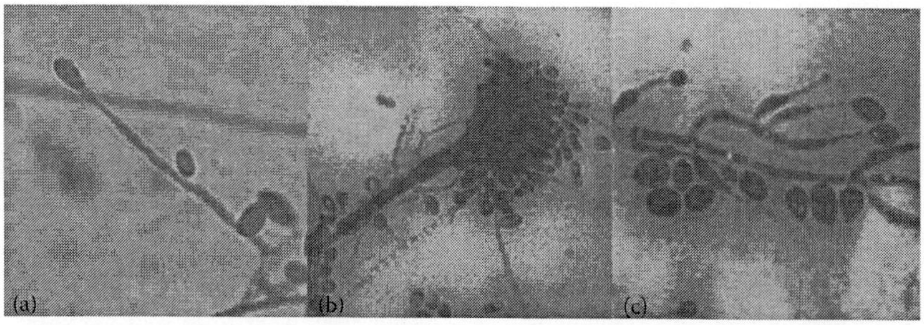

FIGURE 58.1 Microscopy of *Scedosporium* spp. (a) *Scedosporium apiospermum*, (b) *Graphium* synanamorph, (c) *Scedosporium prolificans*.

these patients is remarkable since this species is infrequently isolated from air. Regarding *S. prolificans*, regional differences have been found. Thus, this mold is, by far, more prevalent in Spain and Australia although it has been described in many other countries as the United States and Belgium [19,21]. Sinuses, nasal septum, and external auditory conduct may also be colonized by these species constituting the portal of entry for developing sinusitis or external otitis [20].

2. *Localized infections*: These infections are usually developed after traumatic inoculation of fungal elements in healthy patients. Skin and soft tissue infections, mycetoma; arthritis, osteomyelitis; lymphocutaneous syndrome as well as eye infection; and endophthalmitis have been described [22,23].
3. *Systemic invasive disease*: Neutropenic patients, HIV patients, immunosuppressed recipients of solid organ, or stem cell transplant recipients are susceptible to *Scedosporium* although these mycoses may be occasionally observed in immunocompetent patients. *S. apiospermum* has caused infections in victims of near drowning and evacuees of natural disasters, such as tsunamis and hurricanes [24].

Most reports of *S. prolificans* infections have occurred in Australia, Belgium, Spain, and the United States [19,21,23,25]. Although the causes for this highly localized prevalence are not known yet, climatic conditions could be related [26].

Invasive *Scedosporium* infections are characterized by high mortality and poor response to antifungal agents. The majority of the available antifungal drugs show low in vitro activity against *S. apiospermum* and only some new triazoles such as voriconazole and posaconazole seem to be active [1]. On the other hand, *S. prolificans* is a multi-resistant fungus, tolerating virtually all systemically active antifungal agents including the new triazoles and echinocandins [1,27,28].

58.1.3 Diagnosis

Because of the usually fatal outcome of these infections, particularly in the immunocompromised patients, a correct and early diagnosis and identification of these fungi are necessary [29].

Regarding the European Organisation for Research and Treatment of Cancer (EORTC) criteria, diagnosis of *Scedosporium* infection should be based on cytology and histopathology evidence in samples from normally sterile body sites and isolation of the mold from cultures.

There are different diagnostic methods described for these species. However, the importance of the different available diagnostic techniques may be determined by the type of infection [22].

58.1.3.1 Conventional Techniques

Culture and identification of the fungus is necessary to perform a correct diagnosis, but other morphologically similar species can make the classification of these fungi difficult [30]. Besides, conventional techniques are time consuming, the delay in treatment being the major disadvantage for this mycosis.

58.1.3.1.1 Direct Microscopy

The most common procedure employed to visualize fungal structures in clinical samples is to macerate the sample in 20% KOH followed by fluorescence microscopy using Blankophor or Uvitek 2B. These stains are only useful to detect fungal structures but they are not able to discriminate between other fungi as *Aspergillus*, *Fusarium*, etc. Direct exam can be useful for specific clinical pictures as mycetoma because grains of different etiologic agents have specific characteristics that can help to identify the mold to genus level. For instances, grains of *Scedosporium* are white and large (1–2 mm), with a lobed surface, surrounded by a significant eosinophilic zone, which is not present in other grains from other agents of white mycetoma [22].

58.1.3.1.2 Histopathology

In tissue sections, these species appear as septated and branched hyphae and can be easily misidentified either as *Aspergillus*, *Fusarium*, or black fungi. *Scedosporium* presents an irregular branching at acute angles and is dichotomous. *Scedosporium* spp. might also have terminal or intercalary chlamydospores that can be confused with yeast.

Kaufman et al. [31] reported the use of a polyclonal fluorescent antibody that successfully identifies *S. apiospermum* in tissue sections. In addition, a technique based on hybridization in situ using DNA probes has been described to improve these methods [32,33].

58.1.3.1.3 Culture

Samples from respiratory tract, biopsies, or grains can be cultured. In cases of nonmycetoma infections, samples from the respiratory tract and biopsy specimens from other sites should be carefully handled to avoid contamination with other molds. These species grow well on routine fungal media such as Sabouraud glucose agar or malt extract agar (MEA). The incubation temperature is from 25°C to 35°C, some strains being able to grow at 42°C. The growth response of *S. apiospermum* and *S. prolificans* is a useful feature for distinction between the two species as *S. prolificans* is not able to growth in conventional media containing cycloheximide [34]. In addition, *S. prolificans* is frequently recovered from blood cultures [23].

58.1.3.1.4 Susceptibility to Antifungals

Scedosporium prolificans is a multiresistant mold [28]. However, some antifungals are active against *S. apiospermum* species complex [28,35]. Voriconazole is the most active antifungal against this species complex, whereas amphotericin B and itraconazole showed a limited activity [28,35]. Posaconazole has also exhibited a good activity against *S. apiospermum* complex (geometric mean: 1.17 mg/L;

unpublished observations). No clear differences regarding antifungal activity have been detected among the different species of *S. apiospermum* complex so far [35]. Table 58.1 displays the data of susceptibility in vitro of *Scedosporium* clinical isolates. Table 58.1 includes data of 94 clinical strains tested in the Spanish Mycology Reference Laboratory following the EUCAST (European Committee on Antimicrobial Susceptibility Testing) reference procedure for susceptibility testing of filamentous fungi.

58.1.3.1.5 Serological Methods

These methods like immunodiffusion tests are not commercially available yet [1,36]. An assay based on the detection of a peptidorhamnomannan antigen for *S. apiospermum* is available [37]; however, cross reactivity with other fungal species has been reported [38].

Recently, Thornton et al. [24] have developed an immunofluorescence and double-antibody sandwich enzyme-linked immunosorbent assays (DAS-ELISAs) to differentiate *S. apiospermum* from other infectious fungi and to track the pathogen in environmental samples. The immunoassay uses an immunoglobulin M and IgG1 kappa-light chain monoclonal antibodies (MAbs) specific to *S. apiospermum*, which bind to an immunodominant carbohydrate epitope on an extracellular 120kDa antigen present in the spore and hyphal cell walls of *S. apiospermum* and its sexual state, *P. boydii*. The MAbs are specific and do not react with *S. prolificans*, *Scedosporium dehoogii*, or other clinically relevant fungi as *A. fumigatus*, *Candida albicans*, *Cryptococcus neoformans*, *Fusarium solani*, *and Rhizopus oryzae*. The specificity of this assay was confirmed by sequencing of the internally transcribed spacer 1 rRNA-encoding region of environmental isolates. Unfortunately, there is not data regarding human clinical samples

58.1.3.2 Molecular Techniques

An increasing number of molecular techniques for the diagnosis of fungal infections have been developed in the last few years due to the growing prevalence of mycoses and the length of time required for diagnosis when classical microbiological methods are used.

With the development of the polymerase chain reaction (PCR), it is possible to multiply a given DNA segment quickly and identify the fungus, both from cultures and clinical samples. As a consequence, an early clinical diagnosis can be established.

58.1.3.2.1 Conventional PCR Assays

These assays have been developed to detect DNA from tissues or cultured strains and are targeted toward the internal transcribed spacer (ITS) region of the ribosomal DNA gene cluster or to protein-coding genes such as β-tubulin or calmodulin. Once amplification of the target region has been performed, PCR-amplified segments must be identified. Several methods have been described:

Sequencing of the amplified segment, with a posterior analysis of the sequence in a sequence database [6,7,35,39].

Use of probes [32,33,40] (1) Fluorescent dyes like SYBR green using a DNA intercalating dye that binds nonspecifically to any double-stranded DNA generated during the PCR reaction. This method provides sensitive detection of DNA but is not specific. Some instruments can perform a melting curve analysis to determine the melting temperature (T_m), which depends on the percentage of G+C content and length of the amplicon [41]. (2) Hybridization probes with sequence homology to target DNA, which allows a specific detection.

58.1.3.2.2 Microarrays

A DNA microarray-based assay for the detection and identification of 14 fungal pathogens in clinical samples including *S. apiospermum* has been developed by Spiess et al. [42]. The assay combines multiplex PCR and consecutive DNA microarray hybridization. PCR primers and probes were designed from unique sequences of the 18S, 5.8S, and ITS1 regions of the fungal rDNA genes. The assay has been tested in blood, bronchoalveolar lavage, and tissue samples from neutropenic patients with proven, probable or possible invasive fungal infection (IFI) or without IFI.

TABLE 58.1
Summary of Susceptibility Results In Vitro of Antifungal Agents Tested against *Scedosporium* Species[a]

Species	No.	AMB	ITC	VRC	POS	TRB	CPF	MCF	ANF
S. apiospermum	25	4.80	4.15	0.91	1.45	22.11	0.25	0.17	1.19
P. boydii	13	6.44	1.48	0.62	0.74	22.63	6.35	0.16	1.01
S. aurantiacum	5	10.6	9.19	1.15	1.74	24.25	0.69	0.69	0.02
S. dehoogii	3	16	2	1	1.41	32	11.31	0.03	1
P. ellipsoidea	6	6.96	1.14	0.5	0.75	21.11	0.71	0.02	0.02
S. prolificans	42	13.99	9.84	13.01	11.81	20.40	6.56	2.69	4

Notes: AMB, amphotericin B; ITC, itraconazole; VRC, voriconazole; POS, posaconazole; TRB, terbinafine; CPF, caspofungin; MCF, micafungin; ANF, anidulafungin.

[a] Minimal inhibitory concentration results of antifungals tested are displayed as geometric means in mg/L.

58.1.3.2.3 Real-Time PCR

Recently, two real-time PCR (RT-PCR)-based assays using Molecular Beacon probes targeting the ITS region of rDNA for the specific and quantitative detection of *S. prolificans* and *S. apiospermum* have been described. DNA from different samples as serum, blood, and lung tissues can be analyzed without post-amplification manipulations steps [29]. The two assays were specific (100%) and had a good sensitivity and reproducibility when performed in *Scedosporium* animal model of infection. In addition, these assays allow a rapid identification of cultured strains suspected to be *Scedosporium*.

58.2 METHODS

58.2.1 Sample Preparation

Fungal DNA extraction is the first step in the molecular diagnosis. There are several commercial and in-house methods to perform this step. The extraction method described by Holden et al. is described below:

DNA extraction from culture [43]: Subculture the isolated fungus in MEA (2% malt extract) or potato dextrose agar (PDA) until visible growth at 30°C ± 2°C. Add distilled water to the subculture and scrape with a swab. Inoculate plastic Petri dish containing 3 mL GYEP (2% glucose, 0.3% yeast extract, 1% peptone) with 50–100 µL of the spore suspension obtained. Incubate until growth at 30°C ± 2°C. Harvest mycelium with a plastic tip and put on a Whatman paper. It is important to dry mycelium completely. Transfer dry mycelium to a 50 mL propylene tube with screw cap and add six glass beads (4 mm diameter). Introduce the tube into liquid N_2 for 30 s and vortex for 30 s. Repeat operation until all mycelial powder is obtained. Add 0.8 mL of extraction buffer (0.2 M Tris–HCl, 0.5 M NaCl, 10mM EDTA, SDS 1%) and 0.8 mL of phenol: chloroform: isoamilic alcohol (25:24:1) and shake gently. Transfer to an Eppendorf and centrifuge at 12,000 rpm for 15 min at 4°C–8°C. Transfer supernatant to a new tube and add an equal volume of phenol: chloroform: isoamilic alcohol (25:24:1) and vortex for 30 s. Centrifuge for 15 min at 12,000 rpm at 4°C–8°C. Remove supernatant and extract with an equal volume of chloroform: isoamilic alcohol (24:1). Vortex for 30 s and centrifuge at 12000 rpm for 5 min at 4°C–8°C. Take supernatant and precipitate DNA with isopropanol. Centrifuge for 15 min at 12,000 rpm at room temperature. Aspirate supernatant and wash the pellet with 70% ethanol. Centrifuge at 12000 rpm for 10 min at 4°C–8°C. Aspirate supernatant and resuspend the pellet in 25–50 µL RNAase (1mg/mL). Incubate at 37°C for 1 h.

DNA extraction from clinical samples: Different DNA extraction kits can be employed depending on the clinical sample. Examples of commercial kits available to extract DNA are Tissue: Wizard (Promega), QiAmpTissue (Qiagen), etc.; Blood and other fluids (serum, bone marrow aspirate, bronchoalveolar lavage, etc.): "NucleoSpinBlood" (Macherey-Nagel), QiAmp (Qiagen).

58.2.2 Detection Procedures

58.2.2.1 PCR Amplification of ITS1-ITS2 Regions

There are several universal primers to amplify the ITS regions:

Primer [44]	ITS-I-5.8S-ITS-II Sequence (5′–3′)	Approximate Size (bp)
ITS 1	TCCGTAGGTGAACCTGCGG	600–630
ITS 4	TCCTCCGCTTATTGATATGC	
ITS 3	GCATCGATGAAGAACGCAGC	ITS 3+4=330
ITS 2	GCTGCGTTCTTCATCGATGC	ITS 1+2=290
ITS 5	GGAAGTAAAAGTCGTAACAAGG	ITS 5+4=645

Primers ITS1 (5′-TCCGTAGGTGAACCTGCGG-3′) and ITS4 (5′-TCCTCCGCTTATTGATATGC-3′) can be used to amplify the complete region ITS1-5.8S-ITS2 of these fungi.

The reaction mixtures contain 0.5 µM of each primer, 0.2 mM of each deoxynucleoside triphosphate, 1.5 mM $MgCl_2$, 2.5 U *Taq* DNA polymerase and 25 ng of DNA in a final volume of 50 µL of PCR buffer.

Cycling conditions include one initial cycle at 94°C for 2 min, followed by 35 cycles of 30 s at 94°C, 45 s at 56°C, and 2 min at 72°C, with one final cycle of 5 min at 72°C. The reaction products must be analyzed in a 0.8% agarose gel.

The above conditions are examples, and each PCR should be optimized in individual laboratory.

58.2.2.2 β-Tubulin PCR

Two regions of the β-tubulin gene have been amplified to identify these species:

Primer	β-Tubulin Sequence 5′–3′
BT2-F [6]	GGYAACCARATHGGTGCYGCY
BT2-R [6]	ACCCTCRGTGTAGTGACCCTTGGC
TUB-F [45]	CTGTCCAACCCCTCTTACGGCGACCTGAAC
TUB-R [45]	ACCCTCACCAGTATACCAATGCAAGAAAGC

The reaction mixtures should be performed with 1 µM of each primer, 50 µM of each deoxynucleoside triphosphate, 2 mM $MgCl_2$, 2.5 U *Taq* DNA polymerase and 25 ng of DNA in a final volume of 50 µL of PCR buffer.

Cycling conditions include one initial cycle at 94°C for 5 min, followed by 35 cycles of 30 s at 95°C, 1 min at 55°C (TUB) or 60°C (BT2), and 1 min at 72°C, with one final cycle of 7 min at 72°C.

The identity of *Scedosporium* can be verified through sequencing analysis of the amplified segment. The PCR product is first purified with one of the commercial kits available such as Chroma SPIN+TE-200; Chroma SPIN+TE-400; Chroma SPIN+TE-1000; UltraClean™ (PCR Clean-up™ Kit, Mo Bio Laboratories Inc). For sequencing reactions, use commercial kits like BigDye Terminator cycle sequencing ready reaction: Applied Biosystems. Use 1 µM of the primers and 3 μL of the purified PCR product in a final volume of 10 μL.

58.2.2.3 Real-Time PCR

Two RT-PCR assays for the diagnosis of these species were developed and validated for the classification of clinical strains and for the detection of DNA in clinical samples using a murine model of invasive infection [29]. PCR reactions were performed in the Chromo 4 System (MJ Research, Biorad). The kit 2x SensiMix DNA (Quantace, Ecogen) was used as described by the manufacturer. The PCR reaction (20 μL) contained 0.5 μM of each of the primers, 4.5 mM $MgCl_2$, and 0.4 μM of the probe labeled with 5′ FAM and 3′ BHQ1 (Sigma Genosys). Each reaction mixture contained 18 μL of the master mix and a 2 μL aliquot of DNA from the extracted sample (see above). The cycling conditions included a first step for preincubation (activation of the enzyme) and denaturation of the template DNA at 95°C for 10 min. Next steps included an amplification program of 40 cycles as follows: denaturation, 95°C, 15 s; annealing, 54°C for *S. apiospermum* or 52°C for *S. prolificans*, 15 s; and extension, 72°C, 5 s. Quantification standards were run in conjunction with each set of samples. The amplicons generated were 132 bp for *S. apiospermum* and 176 bp for *S. prolificans*. PCR products were subjected to electrophoresis in 2% agarose gels following the protocols of Sambrook [46] to confirm the PCR results. Amplified fragments were sequenced (ABI Prism 377 DNA sequencer, Applied Biosystem, Madrid, Spain) and the obtained sequences were compared to *S. prolificans* and *S. apiospermum* ITS1 sequence database available in the laboratory. Each PCR run contained both negative and positive controls consisting in water and different concentrations of genomic DNA from *S. prolificans* (CNM-CM1627) and *S. apiospermum* (CNM-CM3169), respectively.

58.3 CONCLUSIONS AND FUTURE PERSPECTIVES

Scedosporium species are the fourth most common cause of mold infections after *Aspergillus*, Zygomycetes, and *Fusarium*. However, as there are geographical differences in mold distribution, it is recommended to study the local epidemiology in order to know which are the most prevalent species in each region are. For instances, *S. prolificans* infections have been described more often in Spain, Australia, the United States, and Belgium. The increase in mold infections has a multifactor cause. On one hand, the number of patients at risk has increased, especially those who are immunosuppressed by hematological malignancies or are transplanted because of their underlying disease, but on the other hand, environmental factors and antifungal pressure can contribute to the emergence of new mold species. The spectrum of the disease caused by *Scedosporium* species complex is broad ranging from colonization in selected populations to localized or disseminated infections especially in immunocompromised patients. The availability of several antifungals with different spectrum of activity has undertaken to identify fungi to the species level. However, the classical identification by means of observation of morphology is a cumbersome and slow process. This situation has favored the development of new methodologies based on sequencing of specific targets. Sequencing is now considered the gold standard to identify fungi to species level and its use has allowed the identification of new cryptic species indistinguishable by classical morphology examination. In the case of *Scedosporium apiospermum* species complex, the clinical importance of these new species still needs to be addressed. Most urgent need is the design and validation of new diagnostic methodologies that able to detect the pathogen as quick as possible in human clinical samples. To accomplish this goal, two RT-PCR-based assays using molecular beacon probes targeting the ITS region of rDNA for the specific and quantitative detection of *S. prolificans* and *S. apiospermum* have been developed. DNA from different samples as serum, blood, and lung tissues can be analyzed without post-amplification manipulation steps [29]. The two assays were specific (100%) and had a good sensitivity and reproducibility when performed in *Scedosporium* animal model of infection. In addition, these assays allow a rapid identification of cultured strains suspected to be *Scedosporium*. Clinical studies must be performed in order to know the diagnostic performance of this RT-PCR methodology. In any case, molecular diagnostic techniques will be the gold standard for diagnosis of IFIs in the next future.

REFERENCES

1. Guarro J et al. *Scedosporium apiospermum*: Changing clinical spectrum of a therapy-refractory opportunist. *Med Mycol*. 44(4), 295, 2006.
2. Lopez FA et al. *Scedosporium apiospermum* (*Pseudallescheria boydii*) infection in a heart transplant recipient: A case of mistaken identity. *J Heart Lung Transplant*. 17(3), 321, 1998.
3. Siebenmann F. Die Schimmelmykosen des menschlichen Ohres [*The Mycoses of the Human Ear*]. Wiesbaden, Germany: Bergmann, 1899.
4. Saccardo PA. Notae mycologicae [Mycological notes]. *Ann Mycol*. 9, 249, 1911.
5. Castellani A, Chalmers AJ. *Manual of Tropical Medicine*. New York: Williams Wood and Co, 1919, pp 2110–2150.
6. Gilgado F et al. Molecular phylogeny of the *Pseudallescheria boydii* species complex: Proposal of two new species. *J Clin Microbiol*. 43(10), 4930, 2005.
7. Gilgado F et al. Molecular and phenotypic data supporting distinct species statuses for *Scedosporium apiospermum* and *Pseudallescheria boydii* and the proposed new species *Scedosporium dehoogii*. *J Clin Microbiol*. 46(2), 766, 2008.
8. Hennebert GL, Dessai BG. *Lomentospora prolificans*, a new hyphomycete from greenhouse soil. *Mycotaxon*. 1(45), 50, 1974.
9. Malloch D, Salkin IF. A new species of *Scedosporium* associated with osteomyelitis in humans. *Mycotaxon*. 21, 247, 1984.
10. Gueho E, de Hoog GS. Taxonomy of the medical species of *Pseudallescheria* and *Scedosporium*. *J Mycol Med*. 118, 3, 1991.
11. April TM et al. Degradation of hydrocarbons in crude oil by the ascomycete *Pseudallescheria boydii* (*Microascaceae*). *Can J Microbiol*. 44(3), 270, 1998.

12. de Hoog G et al. Hyphomycetes. In: G.S. de Hoog, J. Guarro (eds.), *Atlas of Clinical Fungi*. Reus, España/Utrecht, The Netherlands: Universitat Rovira i Virgili and Centraalbureau voor Schimmelcultures, 1995, pp. 380–1007.
13. Kwon-Chung KJ, Bennett JE. Pseudallescheriasis and *Scedosporium* Infection. In: C. Cann, S. Hunsberger (eds.), *Medical Mycology*. Philadelphia, PA: Lea and Febiger, 1992, pp. 678–694.
14. Fisher JF et al. Near-drowning complicated by brain abscess due to *Petriellidium boydii*. *Arch Neurol*. 39(8), 511, 1982.
15. Berenguer J et al. Central nervous system infection caused by *Pseudallescheria boydii*: Case report and review. *Rev Infect Dis*. 11(6), 890, 1989.
16. Creitz J, Harris HW. Isolation of *Allescheria boydii* from sputum. *Am Rev Tuberc*. 71(1), 126, 1955.
17. Cimon B et al. Clinical significance of *Scedosporium apiospermum* in patients with cystic fibrosis. *Eur J Clin Microbiol Infect Dis*. 19(1), 53, 2000.
18. Rippon JW. Pseudallescheriasis. In: *Medical Mycology: The Pathogenic Fungi and the Pathogenic Actinomycetes*. Philadelphia, PA: W. B. Saunders, 1988, pp. 651–680.
19. Berenguer J et al. Deep infections caused by *Scedosporium prolificans*. A report on 16 cases in Spain and a review of the literature. *Scedosporium Prolificans* Spanish Study Group. *Medicine (Baltimore)*. 76(4), 256, 1997.
20. deBatlle J. et al. Disseminated infection caused by *Scedosporium prolificans* in a patient with acute multilineal leukemia. *J Clin Microbiol*. 38(4), 1694–1695, 2000.
21. Cooley L et al. Infection with *Scedosporium apiospermum* and *S. prolificans*, Australia. *Emerg Infect Dis*. 13(8), 1170, 2007.
22. Cortez KJ et al. Infections caused by *Scedosporium* spp. *Clin Microbiol Rev*. 21(1), 157, 2008.
23. Rodriguez-Tudela JL et al. Epidemiology and outcome of *Scedosporium prolificans* infection, a review of 162 cases. *Med Mycol*. 47(4), 359, 2009.
24. Thornton CR. Tracking the emerging human pathogen *Pseudallescheria boydii* by using highly specific monoclonal antibodies. *Clin Vac. Immunol*. 16(5), 756, 2009.
25. Maertens J et al. Disseminated infection by *Scedosporium prolificans*: An emerging fatality among haematology patients. Case report and review. *Ann Hematol*. 79(6), 340, 2000.
26. Ponton J et al. Emerging pathogens. *Med Mycol*. 38(Suppl 1), 225, 2000.
27. Cuenca-Estrella M et al. Head-to-head comparison of the activities of currently available antifungal agents against 3,378 Spanish clinical isolates of yeasts and filamentous fungi. *Antimicrob Agents Chemother*. 50(3), 917, 2006.
28. Cuenca-Estrella M et al. Comparative in-vitro activity of voriconazole (UK-109,496) and six other antifungal agents against clinical isolates of *Scedosporium prolificans* and *Scedosporium apiospermum*. *J Antimicrob Chemother*. 43(1), 149, 1999.
29. Castelli MV et al. Development and validation of a quantitative PCR assay for diagnosis of scedosporiosis. *J Clin Microbiol*. 46(10), 3412, 2008.
30. Guarro J, Gene J. *Acrophialophora fusispora* misidentified as *Scedosporium prolificans*. *J Clin Microbiol*. 40(9), 3544, 2002.
31. Kaufman L et al. Immunohistologic identification of *Aspergillus* spp. and other hyaline fungi by using polyclonal fluorescent antibodies. *J Clin Microbiol*. 35(9), 2206, 1997.
32. Hagari Y et al. Cutaneous infection showing sporotrichoid spread caused by *Pseudallescheria boydii* (*Scedosporium apiospermum*): Successful detection of fungal DNA in formalin-fixed, paraffin-embedded sections by seminested PCR. *Arch Dermatol*. 138(2), 271, 2002.
33. Hayden RT et al. In situ hybridization for the differentiation of *Aspergillus*, *Fusarium*, and *Pseudallescheria* species in tissue section. *Diagn Mol Pathol*. 12(1),21, 2003.
34. Dixon DM, Polak-Wyss A. The medically important dematiaceous fungi and their identification. *Mycoses*. 34(1–2), 1, 1991.
35. Alastruey-Izquierdo A et al. Prevalence and susceptibility testing of new species of *Pseudallescheria* and *Scedosporium* in a collection of clinical mold isolates. *Antimicrob Agents Chemother*. 51(2), 748, 2007.
36. Tang CM, Cohen J, Holden DW. An *Aspergillus fumigatus* alkaline protease mutant constructed by gene disruption is deficient in extracellular elastase activity. *Mol Microbiol*. 6(12), 1663, 1992.
37. Pinto MR et al. A peptidorhamnomannan from the mycelium of *Pseudallescheria boydii* is a potential diagnostic antigen of this emerging human pathogen. *Microbiology*. 147(Pt 6), 1499, 2001.
38. Jabado N et al. Invasive pulmonary infection due to *Scedosporium apiospermum* in two children with chronic granulomatous disease. *Clin Infect Dis*. 27(6), 1437, 1998.
39. Bagyalakshmi R et al. Newer emerging pathogens of ocular non-sporulating molds (NSM) identified by polymerase chain reaction (PCR)-based DNA sequencing technique targeting internal transcribed spacer (ITS) region. *Curr Eye Res*. 33(2), 139, 2008.
40. Wedde M et al. PCR-based identification of clinically relevant *Pseudallescheria/Scedosporium* strains. *Med Mycol*. 36(2), 61, 1998.
41. Espy MJ et al. Real-time PCR in clinical microbiology: Applications for routine laboratory testing. *Clin Microbiol Rev*. 19(1), 165, 2006.
42. Spiess B et al. DNA microarray-based detection and identification of fungal pathogens in clinical samples from neutropenic patients. *J Clin Microbiol*. 45(11), 3743, 2007.
43. Holden DW. DNA mini prep method for *Aspergillus fumigatus* (and other filamentous fungi). In: B. Maresca, G.S. Kobayashi (eds.), *Molecular Biology of Pathogenic Fungi, a Laboratory Manual*. New York: Telos Press, 1994, pp. 3–4.
44. White TJ. Amplification and direct sequencing of fungal ribosomal RNA genes for phylogenetics. In: M. Innis (ed.), *PCR Protocols: A Guide to Methods and Applications*. San Diego, CA: Academic Press, 1990, pp. 315–322.
45. Cruse M et al. Cryptic species in *Stachybotrys chartarum*. *Mycologia*. 94(5), 814, 2002.
46. Sambrook KJ, Fritsch EF, Maniatis T. Molecular cloning second edition. New York: Cold spring harbour laboratory Press, 1989.

59 *Sarcopodium*

Dongyou Liu and R.R.M. Paterson

CONTENTS

59.1 INTRODUCTION

59.1.1 Classificaton and Morphology

The genus *Sarcopodium* (obsolete synonyms: *Tricholeconium*, *Periolopsis*, *Actinostilbe*, *Kutilakesa*, and *Kutilakesopsis*) covers a group of plant-infecting fungi in the family Nectriaceae, order Hypocreales, class Sordariomycetes, subphylum Pezizomycotina, phylum Ascomycota, kingdom Fungi. Among the 12 recognized species in the genus, *Sarcopodium araliae*, *Sarcopodium circinatum* (the type species of the genus), *Sarcopodium oculorum*, and *Sarcopodium tortuosum* are relatively common plant pathogens; and *S. oculorum*, named after the infection site, has been shown to cause human infection [1–3].

On potato dextrose agar (PDA) at 25°C, *S. oculorum* colonies reach a diameter of 39–40 mm after 14 days. Initially, colonies are flat, mucoid, and cream but later become radially folded, brownish gray, and granulose due to the production of conidiomata. The colonies are grayish white toward the periphery with sparse aerial mycelium; the reverse is colorless to brown. On oatmeal agar (OA) and potato carrot agar (PCA) at 25°C, *S. oculorum* colonies grow more rapidly than on PDA, achieving a diameter of up to 48 and 45 mm, respectively, after 14 days. Colonies are granulose at the center and smooth toward the periphery, with whitish, soft cottony aerial mycelium and brownish gray submerged hyphae; the reverse is brownish gray and paler toward the periphery. On PDA at 37°C, *S. oculorum* colonies measure 14–15 mm in diameter after 20 days. Colonies are elevated and cerebriform, with abundant sporulation, but no conidiomata develops. *S. oculorum* does not grow at 40°C [3].

S. oculorum typically produces two types of conidia: one is derived from sporodochial conidiomata (cushion-like structures bearing numerous compact, short conidiophores, which produce the conidial mass) and the other is derived from undifferentiated hyphae. Sporodochia are superficial, solitary, gregarious or confluent, sessile, applanate to cupulate, or pulvinate, subhyaline to dark brown, setose, and up to 400 μm in diameter. Sterile hyphae (setae) (65 × 1.5–2.5 μm) form a frill at the margin of the sporodochium but are also interspersed with the conidiophores. They are single or form small fascicles of 3–5 hyphae and are erect, unbranched or slightly branched toward the base, straight or flexuose, septate, subhyaline to dark brown, smooth walled, thin to slightly thick walled, and cylindrical, with obtuse apices. The conidiophores (up to 35 μm long) are well differentiated, straight or flexuose, subhyaline to pale brown, and smooth walled. They are irregularly branched, and each branch usually bears a single terminal group of slightly appressed conidiogenous cells. The conidiogenous cells (8–13 μm × 1–1.8 μm) are enteroblastic, monophialidic, terminal or lateral, hyaline to subhyaline, smooth walled, subcylindrical, and rarely intercalary with a cylindrical and lateral projection. These intercalary conidiogenous cells are predominantly found on undifferentiated hyphae. The conidia are aggregated in cream slimy masses, which remain attached to the upper part of the conidioma and cover the entire surface. The individual conidia (1.2–3 μm × 0.8–1.5 μm) are subhyaline, aseptate, smooth and thin walled, ellipsoid, and navicular or slightly allantoid. They reach up to 5 μm and are cylindrical or allantoid when they emerge from undifferentiated hyphae [3].

Although *S. oculorum* and *S. circinatum* have sessile sporodochia, unbranched setae, and ellipsoid or cylindrical conidia, *S. circinatum* produces flexuous or circinate and verrucose setae up to 6 μm wide and larger (7–10 μm × 2 μm) conidia that are never navicular or allantoid. *S. oculorum* is differentiated from *S. tortuosum* by the latter's orange slimy conidial masses and branched setae. *S. oculorum* also resembles *Myrothecium* spp. and *Stephanonectria keithii* in the order Hypocreales (Ascomycota). However, the sporodochial sterile hyphae in *Myrothecium* are hyaline to subhyaline and confined to the edge of the sporodochium, and in *S. keithii* they are absent. Moreover, the conidial masses are green in *Myrothecium* and brown in *S. keithii* [3].

59.1.2 Clinical Features

Sarcopodium spp. are commonly present in plants, dead herbaceous stems, and dead wood in many parts of the world, but not in soil, air, or animals.

The only reported clinical case of *Sarcopodium oculorum* infection in humans involved a 12-year-old Brazilian boy with a vernal conjunctivitis [3]. The patient presented with a pain in his right eye and felt the presence of a foreign body. After a 5 month treatment for keratoconjunctivitis with specific antiallergic drugs, topical dexamethasone and prednisone, the patient developed a corneal ulcer as a consequence of the allergic process, showing redness of the eye with inflammatory infiltrate and a suspicion of infection. Direct examination of Gram-stained mounts of the deep corneal scrapings revealed the presence of numerous septate and branched hyphae and a few ellipsoidal conidia. Culture of the corneal scrapings on Sabouraud dextrose agar at 25°C, 30°C, and 37°C for 4–5 days, yielded several colonies of a single, darkly pigmented fungus, which was verified as *Sarcopodium oculorum* on the basis of its distinct morphological, biochemical, and molecular features. The infection was resolved with natamycin and ketoconazole [3].

59.1.3 Diagnosis

The microscopic characteristics of *Sarcopodium oculorum* are examined under a light microscope upon preparation of wet mounts with lactic acid. Photomicrographs may also be obtained by scanning electron microscopy. To differentiate from other *Sarcopodium* species, a conidial suspension of *Sarcopodium oculorum* may be inoculated onto sterilized plant material since many plant pathogenic fungi show different morphological features on natural substrates compared to laboratory culture media.

Sarcopodium is a hyphomycetous genus and is characterized by the presence of sporodochia with sterile, smooth or ornamented, often coiled pale brown setae arising from branched conidiophores and producing slimy conidia from phialides. The conidia are hyaline, aseptate, and fusiform to ellipsoid or cylindrical. The presence of numerous, dark sporodochia covered by a wet mass of cylindrical, hyaline conidia may be confused with the typical conidiomata (pycnidia) of *Phoma* or other similar coelomycetous fungi under a stereomicroscope. However, these structures are easily differentiated at high magnification. The pycnidia are closed, spherical or obpyriform structures that are open only at the apical part by an ostiole. Additionally, the pycnidia possess a pseudoparenchymatous wall that is absent in the sporodochia. Conidiophores are formed inside the pycnidia and line the internal cavity. Other coelomycetous fungi such as *Colletotrichum*, develop acervular conidiomata (cup-shaped fruiting bodies) on natural substrates and turn sporodochial in vitro.

Given that culture-based methods for fungal diagnosis require up to 3 weeks of incubation, PCR amplification and sequencing analysis of rRNA intergenic spacer (ITS) regions provide a rapid and accurate approach for *S. oculorum* identification [4–6].

59.2 METHODS

59.2.1 Sample Preparation

Sarcopodium specimens are grown on inhibitory mold agar, modified Sabouraud agars, or potato dextrose agar (PDA). Oatmeal agar (OA) (30 g of oat flakes, 1 g of $MgSO_4 \cdot 7H_2O$, 1.5 g of KH_2PO_4, 15 g of agar, 1000 mL of tap water) and potato carrot agar (PCA) (20 g of potatoes, 20 g of carrot, 18 g of agar, 1000 mL of tap water) may also be used.

After growth for 4–7 days on PDA slants, lysates are prepared from approximately 1 cm^2 of mycelia with IDI lysis kits (GeneOhm Sciences). Briefly, in a biological safety cabinet, mycelia are collected by scraping the slant with a sterile stick in 1 mL of sterile, molecular-grade H_2O. The material is transferred to a 2 mL screw-cap tube. The tubes are centrifuged for 1 min at 6000 × *g*. If the mycelium does not pellet, the material is contained with a pediatric blood serum filter (Porex Corp., Fairburn, GA). After removal of supernatant, the material is resuspended in 200 μL of IDI sample buffer and transferred to the lysis tube, which contains glass beads. Lysis tubes are vortexed on the highest setting for 5 min and placed in a boiling water bath for 15 min and centrifuged for 5 min at 16,000 × *g*. The supernatant is stored at −20°C until amplification [5].

59.2.2 Detection Procedures

Pounder et al. [5] described a real-time PCR with SYBR green DNA-binding dye and amplicon melting temperature analysis for fungal detection using pan-fungal primers ITS1 forward (5′-TCCGTAGGTGAACCTGCGG-3′) and ITS4 reverse (5′-TCCTCCGCTTATTGATATGC-3′) [7]. The identity of the fungi is verified by subsequent sequence analysis.

Procedure

1. PCR mixture is composed of 1× Lightcycler FastStart DNA Master Hybridization Probes mixture (Roche Applied Science) (containing deoxynucleoside triphosphates, FastStart *Taq* DNA polymerase, and 1 mM $MgCl_2$, additional $MgCl_2$ is added to a final concentration of 4.6 mM), 0.4 μM each of ITS1 forward and ITS4 reverse primers, 1× SYBR green (Molecular Probes), and 3 μL template DNA.

2. Thermal cycling parameters include 95°C for 10 min; 50 cycles of 95°C for 5 s, 60°C for 20 s, and 76°C for 30 s; and a final extension at 72°C for 2 min.
3. The quality of the amplicon is determined using the derivative of the melt analysis curve (55°C–99°C, 45 s hold at 55°C, 5 s/°C) using the RotorGene 3000 (Corbett Robotics, Inc).
4. The amplified product is purified for bidirectional sequencing using ExoSAP-IT (USB Corp). Five microliters of Big Dye Terminator Ready Reaction Mix v. 1.1 (Applied Biosystems) is added to 4 μL of each primer (0.8 pmol/μL) and 3 μL of purified PCR product. Cycle sequencing is performed with a 9700 thermal cycler (ABI), using 25 cycles of 96°C for 10 s, 50°C for 5 s, and 60°C for 4 min. Sequencing reaction products are passed through a Sephadex G-50 fine column to remove unincorporated dye terminators. Purified sequencing reaction products are run on an ABI Prism 3100 Genetic Analyzer with a 50 cm capillary array.
5. Sequences are analyzed with the SmartGene Integrated Database Network software version 3.2.3 vr. SmartGene is a web-based software and database system with reference sequences derived from the National Center for Biological Information (NCBI) GenBank repository.

Note. If real time PCR instrument is unavailable, standard PCR may be performed with primers ITS1 and ITS4, and the resulting amplicon is sequenced with the same primers. Sequence-based identifications are defined by percent identity: species, ≥99%; genus, 93%–99%; and inconclusive, ≤93%. For strains producing discrepant identification between the methods based on phenotypic characteristics and ITS sequence analysis, the D1-D2 region of the large-subunit rRNA gene is amplified with primers NL1 (5′-GCATATCAATAAGCGGAGGAAAAG-3′) and NL4 (5′-GGTCCGTGTTTCAAGACGG-3′) and sequenced for species clarification (Leaw et al. 2006).

59.3 CONCLUSIONS

The genus *Sarcopodium* consists of 12 recognized fungal species that are commonly found in plants, dead herbaceous stems, and dead wood worldwide. Only one case of *Sarcopodium oculorum* vernal conjunctivitis has been reported recently [3]. Because many fungal species such as members of the *Phoma* and *Colletotrichum* genera may also cause keratomycosis, there is a need to correctly identify them for specific treatments. The fact that some of these fungi sometimes fail to sporulate in culture often delays the results based on morphology and biochemistry. Use of PCR and sequencing offers a rapid and precise means of determining the identity of *Sarcopodium* and the morphologically similar fungi.

REFERENCES

1. Sutton, B. C. 1981. *Sarcopodium* and its synonyms. *Trans Br Mycol Soc*. 76:97–102.
2. Watanabe, T. 1993. *Sarcopodium araliae* sp. nov. on root of *Aralia elata* from Japan. *Mycologia*. 85:520–526.
3. Guarro, J. et al., Corneal ulcer caused by the new fungal species *Sarcopodium oculorum*. *J Clin Microbiol*. 2002;40(8):3071–3075.
4. Leaw, S. N. et al., Identification of medically important yeast species by sequence analysis of the internal transcribed spacer regions. *J Clin Microbiol*. 2006;44(3):693–699.
5. Pounder, J. I. et al. Discovering potential pathogens among fungi identified as nonsporulating molds. *J Clin Microbiol*. 2007;45(2):568–571.
6. Bagyalakshmi, R. et al. Newer emerging pathogens of ocular non-sporulating molds (NSM) identified by polymerase chain reaction (PCR)-based DNA sequencing technique targeting internal transcribed spacer (ITS) region. *Curr Eye Res*. 2008;33(2):139–147.
7. White, T. J. et al. 1990. Amplification and direct sequencing of fungal ribosomal RNA genes for phylogenetics. In M. A. Innis, D. H. Gelfand, J. J. Snisky, and T. J. White (eds.), *PCR Protocols: A Guide to Methods and Applications*. Academic Press, San Diego, CA, pp. 315–322.

60 *Sporothrix* and Sporotrichosis

Conchita Toriello, María del Rocío Reyes-Montes, Armando Pérez-Torres, and Amelia Pérez-Mejía

CONTENTS

60.1 INTRODUCTION

Fungal species of the genus *Sporothrix* are ubiquitous fungi that include common soil saprobes on wood, peat moss, and organic litter, and also human, insect, and plant parasites. *Sporothrix* is one of the anamorph genera associated with the ascomycete genus *Ophiostoma* (former *Ceratocystis*), with species that are symbionts of bark beetles[1,2] and also causal agents of sap stain in freshly cut wood.[3] Ribosomal DNA sequences have confirmed the phylogenetic relation of *Sporothrix* to this genus of the *Ophiostomatales*.[4–6] These two genera are largely distributed in nature and found in identical biotopes, showing similar morphologies and physiological traits. The conidia of some *Ophiostoma* species are indistinguishable morphologically from those of *Sporothrix schenckii*, and they synthesize rhamnomannans, which also are the main surface antigens of *S. schenckii*[7–9] and which are absent in most pathogenic fungi. Mariat[10] proposed that *Ceratocystis stenoceras* (now *O. stenoceras*) is the perfect stage of *S. schenckii*; however, although a clear relation exists between *Sporothrix* and *Ophiostoma* as mentioned above,[4–6] analyses of restriction profiles of mDNA[11], macro-restriction patterns with two enzymes,[12] and DNA sequences of internal transcribed spacer (ITS) regions of the ribosomal RNA operon[6] confirm the separation of both fungi.

60.1.1 CLASSIFICATION

As no teleomorph of the genus *Sporothrix* has been found, *S. schenckii* was classified in the Deuteromycota, Hyphomycetes. However, following sequencing analyses of six gene regions: 18S rRNA, 28S rRNA, 5.8 rRNA, elongation factor 1-α (*EF1α*), and two RNA polymerase II subunits (*RPB1* and *RPB2*), a recent milestone in fungal molecular systematics is the estimated phylogeny of the Kingdom Fungi proposed by James et al.,[13] leading to a higher-level classification for all groups of Fungi.[14] All previous work mentioned above, recognizing the close association of *Ophiostoma/Sporothrix*, is now demonstrated by the phylogenetic relationship among 106 taxa from 12 orders in the Sordariomycetes based on four nuclear loci (nSSU and nLSU rDNA, *TEF* and *RPB2*).[15] Therefore, today *Sporothrix schenckii* Hektoen & C.F. Perkins[16] is classified as an anamorphic *Ophiostoma* belonging to Ophiostomataceae, Ophiostomatales, Sordariomycetidae, Sordariomycetes, Ascomycota, and Fungi.[17] Other molecular investigations [6,12,15,18,19] showed a high genetic variability in *S. schenckii*, pointing out diverse phylogenetic lineages and proposing *S. schenckii* as a complex of numerous cryptic species. In recent work,[20] three new species of *Sporothrix* of medical importance were described. In all, 61 names of *Sporothrix* taxa are listed in the Index

TABLE 60.1
Species of the *Sporothrix schenckii* Complex

Species	Morphologically Identified as	Origin	References
S. albicans	*S. schenckii*	Environmental	Marimon et al.[20]
S. brasiliensis	*S. schenckii*	Clinical	Marimon et al.[20]
S. globosa	*S. schenckii*	Clinical	Marimon et al.[20]
S. mexicana	*S. schenckii*	Environmental	Marimon et al.[20]
S. schenckii	*S. schenckii*	Clinical	Hektoen and Perkins[16]
S. luriei	*S. schenckii* var. *luriei*	Clinical	Marimon et al.[21]

fungorum by CABI Bioscience,[17] including the new species of clinical relevance (Table 60.1). In the following sections, the names of *Sporothrix* spp. mentioned in the original publications are used.

60.1.2 Epidemiology

S. schenckii has been largely known as the causative agent of sporotrichosis,[22–24] a subcutaneous mycosis of human and other mammals, acquired generally by a traumatic contact with live or decayed vegetation, or any traumatism with a fungus-contaminated object or organic material, rarely with insect bites,[25] and recently, even with cat contact or scratches.[26] In rare occasions, it might be acquired by spore inhalation. This subcutaneous mycosis may be acute, well delimitated, or chronic; but in recent times, the disease may develop as a very serious disseminated disease to osteoarticular structures and viscera in immunosuppressed individuals, particularly with AIDS.[27] This mycosis has been known since the beginning of the twentieth century and mentioned for the first time by Schenck in the United States in 1898. In 1900, Hektoen and Perkins[16] described exhaustively the disease. Nowadays, sporotrichosis is well known throughout the world; it is a mycosis rare in Europe but frequent in America, in tropical and subtropical climates, in some Asian countries, and in Australia. The disease has been considered as occupational due to the fact that infected patients are generally workers closely related to vegetation exposure where the causal fungus has its ecological niche.[28–31] One of the most famous epidemic of sporotrichosis occurred in 3000 gold miners from South Africa infected with timbers on which the fungus was growing.[22] Nowadays, the largest epidemic of this mycosis due to zoonotic transmission has been described in Rio de Janeiro, Brazil.[26,32–35] Epidemiologic relevance of results of feline sporotrichosis in Brazil shows that the isolation of the fungus from nails and oral cavity of cats indicates that transmission can occur through a scratch or bite, while isolation from the nasal fossae and cutaneous lesions postulates the possibility of infection through secretions.[33,35,36]

60.1.3 Morphology and Biology

Sporothrix spp. are dimorphic fungi that exhibit a mycelial morphotype in nature and a yeast morphotype when penetrating the host. The mycelial morphotype shows hyaline, regularly septated, thin hyphae, 1–2 μm in diameter, and oval, pyriform, or elongated conidia, 1.5–3 to 3–6 μm, which are born singly or in groups produced sympodially on denticulate conidiogenous cells (sympodial conidia). Another type of conidia produced is thick walled, dark brown, and usually borne individually on short denticles along the sides of hyphae.[37] These conidia regarded as sessile conidia show different shapes among *Sporothrix* species of clinical relevance.[20] The yeast morphotype, easily obtained on different culture media and at 28°C and 37°C,[38] shows yeast cells that are fusiform and ovoid of 2.5–5 to 3.5–6.5 μm, with single, double, or multiple budding. Yeast forms originate from the sides and tips of the hyphae. When rarely observed in tissue, yeast cells form "asteroid bodies" described as a central yeast cell surrounded by an extracellular eosinophilic material forming spicules described 100 years ago.[39] Yeast cells are also observed in the host tissues as elongated yeasts of diverse forms, generally described as "cigar bodies." *S. schenckii* requires thiamine, with the pyrimidine moiety being the effective structure.[40] This fungus is able to grow over a wide pH range, mycelial morphotype from 3.0 to 12.5 and the yeast morphotype from 2.4 to 9.5. Urease activity was only observed in the mycelial morphotype.[41] The key phenotypic features for recognizing five species of *Sporothrix* spp. of clinical relevance are the morphology of the sessile pigmented conidia, growth at 30°C, 35°C, and 37°C, and the assimilation of sucrose, raffinose, and ribitol.[20]

The cell wall composition and polysaccharides of *S. schenckii* have been widely reviewed previously[9,42–44] showing mannose, glucose, and rhamnose constituents, this last sugar very rare in other human pathogens. Rhamnomannans with monorhamnosyl side chains predominate in conidia and yeast cells, whereas dirhamnosyl side chains in hyphae.[9] Peptido-rhamnomannans isolated from different culture media and with diverse procedures are used in different immunological diagnostic tests and epidemiological studies. In particular, sporotrichin, a culture filtrate-derived peptide polysaccharide,[45,46] is especially efficient at detecting delayed-type hypersensitivity and widely used in epidemiological research.

Melanin production of *S. schenckii* has been described on conidial[47] and in yeast cells.[48] The fungus ability of producing melanin over a broad pH range is advantageous for its survival on environmental and pathogenic conditions.[49] *S. schenckii* produces melanin via the 1,8-dihydroxynaphthalene (DHN) pentaketide pathway on conidia, but not on hyphae.[47] However, yeast cells can also produce melanin in vitro and during mammalian infection.[48] In recent work, Almeida-Paes et al.[49] showed melanin particles derived from hyphae grown in L-DOPA medium, providing evidence in support of L-DOPA melanin synthesis distinct from the DHN processes in this fungus. This melanin production is variable among different strains of the fungus.[49] Melanized *S. schenckii* strains appear to cause infection more easily than strains that produce low amounts of this pigment.[50]

60.1.4 Clinical Features and Pathogenesis

S. schenckii is a pathogenic fungus because it causes host damage as a result from either direct fungal action or the host immune response. However, defining microbial pathogenesis in the context of host response or host damage permits the inclusion of many variables which affect the host–pathogen relationship. Thus, virulence, as a relative property of the pathogen, is modulated by host susceptibility and resistance.[51]

Sporotrichosis, characterized by a wide range of cutaneous and systemic clinical manifestations, is caused by *S. schenckii*, a fungus with an apparent regular pattern of virulence, which draws attention to the immunological mechanisms of the host on the pathogenesis of the mycosis.[52] Localized cutaneous and subcutaneous is the most common form of sporotrichosis and is characterized by ulcerative lesions, associated regional lymphangitis, and lymphadenopathy. A distinctive clinical finding is the presence of hard, spherical nodules along lymphatic vessels. Extracutaneous sporotrichosis can involve the lung, joints, bones, and other organs. The systemic form of sporotrichosis is usually associated to immunocompromised patients.[9] Interestingly, in these cases, primary skin lesions are scattered or often absent. It has been proposed that clinical differences could be related to different virulence of *S. schenckii* strains. However, virulence of strains isolated from nature is similar to the virulence of strains isolated from humans.[53]

S. schenckii infection is acquired by a cutaneous microtraumatism and, probably also, by inhalation of the fungus. Independently of the entrance route, the fungal cell wall is the first structure that interacts with cells and tissues of the host. The composition of *S. schenckii* cell walls has been extensively studied,[9,44] but remains poorly understood with regard to its role in infection, colonization, invasion, latency, and disease outcome.

The cell wall of *S. schenckii* contains biochemical components that could act as virulence factors that can play a central role in pathogen–host interaction eliciting a miscellaneous of pathogenic mechanisms. Thus, virulent strains of *S. schenckii* form conidia with thick cell walls with irregular electrondense granules representing pigment particles on the outermost layer.[54] This pigment has been characterized as melanin using monoclonal antibodies against melanin to detect it in conidia in vitro and in yeast cells in vitro and in vivo.[48] Melanin is a virulence factor due to its presence on the cell wall during in vitro culture and during infection, and may have a protective role for the fungus since melanin has a scavenger property of free radicals.[47,48] Moreover, this pathogenic attribute of *S. schenckii* could be related to impairment of oxidative mechanisms of phagocytic cells, mainly neutrophils and macrophages, as was observed in mongolian gerbils with experimental sporotrichosis.[55]

An additional attribute for the pathogenic success of *S. schenckii* through the ability to resist damage from reactive oxygen species (ROS) elicited by host effector cells could be the expression of catalases. An increased expression of catalase, at mRNA level, has been demonstrated during the conversion to the yeast form of *S. schenckii* in vitro. Moreover, it has been speculated that catalase may facilitate the phase transition. Since transcription of catalase mRNA is rapidly upregulated after hydrogen peroxide challenge, it would be relevant in vivo to facilitate the intracellular survival of *S. schenckii* by providing a nontoxic microenvironment in the polymorphonuclear leukocyte and macrophage phagosomes.[56] Neuraminidase may be another pathogenic factor in sporotrichosis because this enzyme can metabolize a number of host substrates as glycoproteins, lipoproteins, immunoglobulins, and fibrinogen.[57]

Recent works have been addressed to investigate the potential pathogenic attributes of some constitutive molecules located at the cell wall of *S. schenckii*. For example, a glycoprotein of 70 kDa (Gp70) from the cell wall of this fungus is involved in fungal adherence to dermal extracellular matrix, a process significantly reduced when yeast cells are preincubated with anti-Gp70 serum in a concentration-dependent manner.[58] Since adhesion to cell surfaces or extracellular matrix molecules is the first step to infection and dissemination of pathogens in the host, it is relevant that yeast cells and conidia of *S. schenckii* can bind to type II collagen, fibronectin, and laminin; additionally, yeast cells can also bind to fibrininogen whereas conidia cannot bind to either fibrinogen or thrombospondin.[59,60] Yeast cells of *S. schenckii* are able to adhere and invade an endothelial cells monolayer, a process modulated by IL-1β, TGFβ, and divalent cations,[61] which could play a role in the dissemination of the fungus from the primary site of cutaneous infection to other organs.

Cell walls of fungi contain multiple potential pathogenic attributes but are also highly immunogenic, and antibodies to its molecules are produced during fungal infection. Thus, fungal cell walls have a dual status: like a Trojan horse and like an Achilles heel. This analogy, once again, points out the relevance of host response in the outcome of fungus–host interaction. An example of the above-mentioned is recent evidence of antibody-mediated passive protection in a model of sporotrichosis obtained with a monoclonal antibody specific to a 70 kDa glycoprotein of *S. schenckii* cell wall, injected before, during, and after experimental infection. The result of passive protection was evaluated as a significant reduction in the number of colony-forming units (CFUs) in the organs of mice.[62] Interestingly, a 70 kDa glycoprotein of *S. schenckii* is now recognized as an adhesin for fibronectin, due to the fact that the interaction or binding capacity of several strains of the fungus to fibronectin correlates with virulence, determined by mortality rate, fungal burden, and histopathology.[63]

From the histopathological point of view, lesions of sporotrichosis are nonspecific. Severe infiltration of inflammatory cells, consisting of abundant neutrophils, macrophages, and few plasma cells and lymphocytes, is frequently described in experimental and natural sporotrichosis. Typically, the fungus is often difficult to detect. However, in acute and subacute experimental infection, it is easy to observe numerous fungal cells that show morphological diversity but yeast cells predominate, many inside of phagocytic cells at the inoculated site and in regional lymph nodes (Figure 60.1). During a long time, the so-called asteroid bodies have been considered a hallmark of sporotrichosis, although they can occur in other systemic mycoses.[22] The ultrastructure of these structures reveals the presence of a central fungal cell, surrounded by peripheral rays of an electrondense granular layered material that resemble remnants of destroyed host cells and adherent cell membrane fragments at the outermost cell layer.[64] Using hematoxylin–eosin stain, asteroid bodies, 15–35 μm in diameter, are observed within abscesses and consist of a central yeast cell surrounded by an extracellular eosinophilic material forming spicules. Since asteroid bodies occur in lesion with extensive necrosis, it has been postulated that eosinophilic material results in part from precipitations of antigen–antibody complexes.[65,66] The presence of asteroid bodies appeared to be related to the type of lesion in view of the fact that it was observed in 17 of 32 primary lesions whereas in 22 of 31 secondary lesions.[65,67] It has also been observed in cytologic smears obtained by fine-needle aspiration biopsies.[68] Other in vivo morphology of yeast cells is the termed cigar-shaped form, generally observed in large and suppurative lesions[69] and in experimental infections. Mycelial filaments are not usually observed in vivo; however, its presence in tissue sections in human and murine sporotrichosis has been reported.[70–72]

Histopathology of granulomatous skin lesions of sporotrichosis indicates a link between innate and adaptive immune responses. Although neutrophils and macrophages are the main infiltrating cells, CD4-positive T cells and dendritic cells (DCs) have also been identified in those lesions. These IFN-gamma-producing T cells would eventually activate local macrophages to kill intracellular *S. schenckii*.[73–75] However, activation of CD4-positive cells requires the participation of mature or activated DCs (after uptake and the processing of fungal antigens), the most potent antigen-presenting cells that play a key role in cell-mediated immunity, at least. In fact, activated DCs have been observed in the granuloma of sporotrichosis.[76] Recognition of fungal pathogens by pattern-recognition receptors of DCs and macrophages[77] is a necessary prerequisite to elicit several signaling pathways associated to the activation of the innate immune system, releasing proinflammatory cytokines and modulating adaptive or acquired immunity.[78]

A recent study indicates that *S. schenckii* strains of the cutaneous clinical form, but not those of visceral origin, induce JNK, ERK, and p38 MAPK activation (intracellular signaling pathways), and IL-6 and TNF-α release in DCs derived from monocytes.[79] Since release of IL-10 was not detected, it has been speculated that toll-like receptor-4 (TLR-4) but not TLR-2, expressed by DCs might be the receptor to recognize *S. schenckii* strains of cutaneous origin and induce strong Th1 immune response. Interestingly, *S. schenckii* strains of visceral origin induce Th2 response according to the significantly higher IL-4 production along with the inability to induce strong Th1 response.[79] Moreover, DCs cocultured with *S. schenckii* strains of cutaneous origin

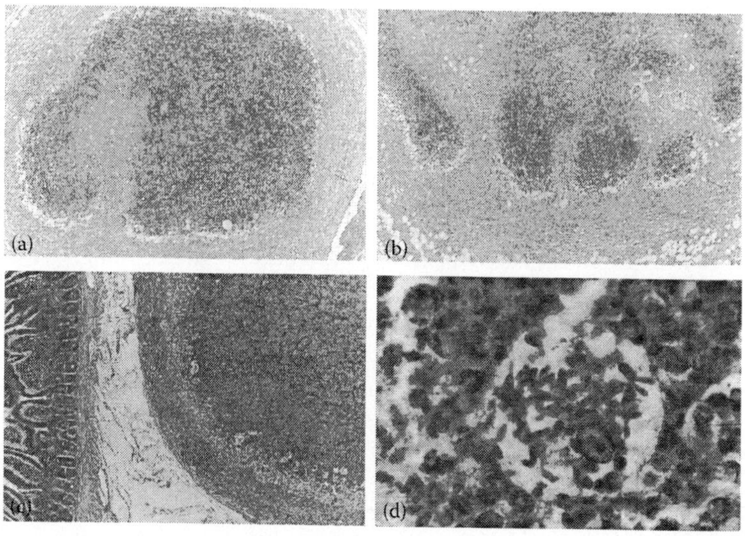

FIGURE 60.1 Subacute experimental infection by *Sporothrix schenckii* in mice. Lesions consist of abundant neutrophils, macrophages, and some lymphoid cells. In experimental lesions, it is easy to observe fungal cells, mainly yeast cells, engulfed by phagocytic cells located at inoculated site and in regional lymph nodes. Note the presence of great inflammatory infiltrate forming a granuloma located near to tunica albuginea covering testis (a, b), and in mesenteric lymph node (c) after intraperitoneal inoculation of *S. schenckii*. Observe many foamy macrophages with abundant fungal cells inside, distributed at the periphery of the granuloma (d). (a–c ×40; d: ×1000. PAS stain).

(yeast and conidia) exhibited a clearer morphological evidence of activation with long and wider dendrites than when they were cocultured with strains of visceral origin. However, internalization of fungal cells of strains of cutaneous and visceral origin by DCs appeared comparable, regardless of the presenting forms (yeast or conidia).[79]

60.1.5 Molecular Epidemiology

Molecular taxonomy is becoming one of the main methods for identifying fungal species. The introduction of these techniques is originating new and continuous taxonomic classifications to help identify species that could only be classified by gender for the last few years.

In order to identify *S. schenckii*, traditionally researchers used markers that could be phenotypically observed in the organisms. The phenotypical techniques are those that detect characteristics expressed by the microorganism. These markers have been useful for differentiating groups within a species, but of insufficient resolution for distinguishing individuals in a population. However, genotypical techniques are more sensitive and specific since they involve the direct DNA analysis of chromosomal or extra-chromosomal genetic elements. These markers are a good system for examining, among other things, the diversity of levels between and within species, population structures, and also allowing the study of the origin and evolution of isolates. Currently, phenotypic and genotypic markers are used for studying pathogenic fungi, that is, the recent research that pointed out three new species of *Sporothrix* spp.[20]

Classical phenotypic morphological, thermal tolerant, and biochemical markers are used for characterizing *S. schenckii*. In the molecular typification of *Sporothrix*, the majority of authors use identification methods based on genomic digestion with restriction enzymes such as restriction fragment length polymorphisms (RFLPs), in addition to other techniques as amplified fragment length polymorphisms (AFLPs) and randomly amplified polymorphic DNA (RAPD). On the other hand, polymerase chain reaction (PCR) techniques are more frequently used by amplifying specific genomic zones, usually ribosomic DNA (rDNA) or mitochondrial DNA (mtDNA). Lastly, the amplification of small rDNA fragments is being used successfully (ITS) for identifying certain fungal species.

The first studies that were conducted for typifying *S. schenckii* strains were based on its phenotypical characteristics.[22,28,54,80–82] Several investigators[9,20,41] found morphological and physiological differences between clinical isolates and those from other sources. However, these characteristics may be influenced by several variables, for which genotypical methods, using molecular biology techniques allowing for more sensitive studies, complement phenotypical methods.

To date, a number of molecular epidemiological studies have been carried out with *S. schenckii* (Table 60.2). One of the first was conducted by Suzuki et al.[11] who studied *S. schenckii* isolates and related fungi such as *Ceratocystis* spp. (now *Ophiostoma*) and *Graphium* sp. by mtDNA RFLP. Based on the RFLP patterns, 11 types were reported corresponding to *S. schenckii*, four types to *C. stenoceras* (=*O. stenoceras*), four types to *S. inflata*, and seven types to *C. minor* (=*O. minor*). Those isolates related to *S. schenckii* due to their phenotypical characteristics had different polymorphic patterns allowing discrimination from this species. Takeda et al.[83] divided 258 clinical isolates from four different places in Japan into 10 types—phylogenetically related in two groups, group A and group B. Afterward, Cooper et al.,[50] using the study done by Dixon et al.[28] as a basis and considering the phenotypical aspects from *S. schenckii* clinical isolates and those from nature from a sporotrichosis epidemic in the United States in 1988, also used RFLP from mtDNA and total DNA methods. These investigators selected representative isolates from each of the phenotypical groups (A–H) established by Dixon et al.[28] and formed eight groups according to the polymorphic patterns revealed. An important aspect of this study was the determination of the similarity between the species isolated from patients with a nature group (group A) and differences with other isolates from nature (B–H). This study suggests that the individuals within group A were possibly the etiological agents (*S. schenckii*) responsible for the epidemic. In addition, these results concur with those obtained by Dixon et al.[28] who found that the microorganisms obtained from clinical cases were phenotypically identical to those of group I from nature. On the other hand, the study conducted by Cooper et al.[50] using RFLP showed greater discretion between strains than the one carried out by Dixon et al.[28] who based their studies on phenotypical characteristics. In another epidemiological study using RFLP, Hajjeh et al.[84] analyzed clinical and environmental isolates from a sporotrichosis outbreak in workers from three plant nurseries in Florida, and different from that found by Dixon et al.[28] and Cooper et al.[50] where the clinical isolates had very similar patterns amongst themselves and different ones from the environmental isolates. These observations, as well as those on the phenotypical and genotypical characterization of clinical and environmental isolates where the strains isolated from clinical cases differ from those isolated in nature, suggest diverse fungal subtypes or varieties.

Ishizaki et al.[85–88] have conducted several studies using mtDNA-RFLP for assessing the variability between isolates from different geographical origins (America, Asia, and Oceania). Their results confirmed the identification of the clinical isolates studied such as *S. schenckii* previously classified by its morphological characteristics. Similarly, they reported that the 24 types of *S. schenckii* by mtDNA-RFLP formed two phylogenetic groups. Group A was composed by types 1–3, 11, 14–19, 22, and 23, while group B by types 4–10, 12, 13, 20, 21, and 24. It was also reported that group A strains were predominantly from South America and Africa, while group B strains were from Australia and Asia. Complementing this study, Kawasaki et al.[98] studied isolates from Brazil, Mexico, and Spain, using the same molecular method, and found that all of the isolates from Spain were in group B (previously established)—different

TABLE 60.2
Different Genotypes within *Sporothrix* spp. of Clinical Relevance and *Ophiostoma*-Related Species

Molecular Typing	Genotypes	References
RFLP of mtDNA	*S. schenckii* isolates from all the world as: 24 mtDNA types with two groups A = South Africa, North, Central, and South America B = Japan, Australia, China, Korea	Suzuki et al.,[11] Takeda et al.,[83] Ishizaki et al.,[85–88] Lin et al.,[89] Mora-Cabrera et al.,[90] and Arenas et al.[91]
DNA ITS1-RFLP	204 *S. schenckii* isolates into four types correlated with geographical origins	Watanabe et al.[19]
RFLP (*Hae*III, *Msp*I), whole cell DNA	DNA patterns of clinical isolates identical to group I environmental isolates Groups II–VI complex of related fungi: *Sporothrix–Ophiostoma*	Cooper et al.[50]
RFLP hybridization of *Apa*I-digest genomic DNA with a probe (rRNA)	31 *S. schenckii* isolates (China), 15 individual patterns (DNA types A-O) recognized Types A–C = 51.61%	Zhang et al.[92]
18S rDNA sequencing	*S. schenckii* related phylogenetically to *Ophiostoma*	Berbee and Taylor[4]
RAPD-PCR	44 *S. schenckii* isolates of Mexico, Guatemala, and Colombia: grouping related to geographical origin. Separation of clinical and environmental isolates of Mexico	Mesa-Arango et al.[93]
DNA fingerprinting M13	88 *S. schenckii* isolates of Brazil (59 cat epidemic) and Spain	Gutierrez Galhardo et al.[94]
rDNA ITS sequencing	Nine subtypes not associated to clinical forms	
DNA sequence of ITS including 5.8 rRNA	Clear separation of *S. schenckii*, *O. stenoceras*, and *O. nigrocarpum* species Phylogenetically related *S. schenckii* appears as more than one species Two groups: clinical and environmental	De Beer et al.[6]
AFLP (EcoRI-AA, MseI-A)	32 *S. schenckii* isolates from Peru Two populations (A, B) not related to substrate, pathology, or geography	Neyra et al.[95]
Pulsed field gel electrophoresis (PFGE)	Eight strains *S. schenckii* from Japan, three types—Six to eight chromosomes	Tateishi et al.[96]
CHEF	Same PFGE pattern in *S. schenckii*-dominant strain isolated from clinical cases and environmental hay for 15 years in Western Australia	O'Reilly and Altman[12]
Sequence analysis of protein coding loci for chitin synthase, β-tubulin, calmodulin	60 *S. schenckii* isolates Six distinct putative phylogenetic species among complex	Marimon et al.[18]
Calmodulin gene sequences	127 *S.schenckii* isolates (USA, Europe, Asia) Pheno- and genetically separation. Three new species: *S. brasiliensis*, *S. globosa*, and *S. mexicana* *S. albicans*, *S. inflata*, and *S. schenckii* var. *luriei*, different from *S. schenckii*, no longer considered as single species	Marimon et al.[20]
Chitin synthase, β-tubulin, calmodulin sequences	Strain CBS 937.72 (=ATCC 18616), clinical isolate from South Africa New species: *S. luriei*	Marimon et al.[21]
Partial sequences of calmodulin gene	*S. globosa* in Central and South America	Madrid et al.[97y]

from those from North and South America that were mainly in the group A, coincident with the results by Ishizaki et al.[85–87]. Mora-Cabrera et al.[90] analyzed isolates from Mexico, Guatemala, and Colombia using mtDNA-RFLP and reported six new types (type 25–30). Watanabe et al.[19] used DNA ITS1-RFLP to study the phylogeny among the existing *S. schenckii* types. They found four types: rDNA types I–III corresponding to the mtDNA from group A (types 1, 2, 11, 14–19, 22, and 23) and type IV corresponding to the mtDNA group B strains (types 4–10, 12, 13, 20, and 21) of Ishizaki et al.[85–87] The authors mention that ITS-RFLP is a method with a greater discriminatory power than mtDNA-RFLP. In a further study, Ishizaki et al.[88] typified other isolates from India, Thailand, Mexico, Guatemala, Colombia, and Brazil using mtDNA-RFLP and ITS-RFLP, finding two new types (31 and 32). Type 30 previously reported by Mora-Cabrera et al.[90] was confirmed as type 3. Of the 48 isolates from Mexico, 41 were included in group A, where type 2 (13 isolates), type 3 (10 isolates), and type 28 (7 isolates) dominated. All of the isolates from India and Thailand were included in group B. The 52 isolates from group A and 24 from group B corresponded to type I and type IV rDNA, respectively, reported by Watanabe et al.[19]

With another molecular typing method, Tateishi et al.[96] applied karyotype analysis to eight *S. schenckii* isolates and divided them into three types. Sugita[99] used single-strand conformation polymorphism (SSCP) to study 20 *S. schenckii* isolates previously classified by mtDNA-RFLP. These authors used oligonucleotides designed from the region of a putative gene that codes for a membrane transporting protein. The isolates were divided into three types corresponding to those previously obtained in karyotype analysis.[96]

Reis et al.[100] carried out a study using RAPD-PCR for identifying the genetic relationship among *S. schenckii* strains from patients and cats involved in an outbreak of sporotrichosis in Brazil. Their results showed great genetic similarity among the 15 cases studied. The DNA from a strain isolated from a patient showed a profile identical to the DNA isolated from a cat. The polymorphic patterns obtained from the isolates from three other patients were very similar to patterns obtained from the isolates of cats studied, suggesting a common source of infection and thought of as the mode of infection. Lee et al.[101] analyzed 10 *S. schenckii* clinical isolates, in addition to other fungi, using RAPD-PCR. The results showed that each random primer amplified the characteristic band patterns in DNA of 8 out of 10 isolates, but different in two isolates, suggesting a different *Sporothrix* anamorph from Korea. Mesa-Arango et al.[93] analyzed the phenotypical characteristics (conidia size and thermal tolerance) and genotypes using RAPD of isolates from Mexico, Guatemala, and Colombia. The isolates from Colombia showed a greater size of conidia and were thermal sensitive in relation to the isolates from Mexico and Guatemala. Using RAPD, a great variability was observed among the *S. schenckii* isolates, forming four groups according to their geographical origin. No relationship was found among the polymorphic patterns and the clinical forms of the disease, although isolates from Mexico were separated into two groups: clinical and environmental.

Neyra et al.[95] used AFLP for analyzing the genetic diversity of *S. schenckii* isolates of Peru and reference strains from other countries. The isolates from Peru were divided into two separate homogeneous groups. A high level of variability was seen among the two groups without finding a correlation to the clinical form of sporotrichosis.

Recently, based on the analysis of the ITS rDNA region, it has been suggested that more than one species may exist in *S. schenckii*.[6] With the purpose of clarifying if the variability of this fungus is due to species divergence or intraspecific diversity, Marimon et al.[18] determined the sequence analysis of three protein coding loci [chitin synthase, β-tubulin, and calmodulin (CAL)] of 60 *S. schenckii* isolates from different geographical locations. The isolates formed three main clades, one grouping all isolates from Europe, another with only isolates from Brazil, and the third with isolates from other South American countries and Africa. The authors proposed the existence of at least six phylogenetic species from different regions, confirming *S. schenckii* as a complex of species. Later on, the same authors[20] studied phenotypically and genotypically (nuclear CAL gene) a total of 127 isolates, differentiating *S. albicans*, *S. inflata*, and *S. schenckii* var. *luriei* from *S. schenckii*, and proposing three new species: *Sporothrix brasiliensis*, *S. globosa*, and *S. mexicana* of clinical relevance (Table 60.2). They suggest the CAL gene as a good marker, and sessile pigmented conidia, growth at 30°C, 35°C, and 37°C, and the assimilation of sucrose, raffinose, and ribitol as phenotypic features, for the recognition of these species. A later communication[21] proposed *S. schenckii* var. *luriei* as the species *S. luriei*, based on its phenotypic characteristics and a multilocus sequence analysis, to be different to other species of the *S. schenckii* complex.[20] Furthermore, Madrid et al.[97] studied another 32 clinical and environmental isolates identified morphologically as *S. schenckii*, from Mexico, Guatemala, and Colombia, through partial gene sequences of CAL. The study revealed, apart from *S. schenckii sensu stricto*, *S. globosa* in Mexico, Central, and South America for the first time.

In regard to the sensitivity of *Sporothrix* to different antimycotics, Gutierrez Galhardo et al.[94] investigated the phenotypes (sensitivity to antimycotics) and genotypes (DNA fingerprinting method with primer M13, sequencing of the ITS-rDNA region) of the isolates recovered from different clinical forms during an outbreak transmitted by cats. The minimum inhibitory concentration (MIC) values to amphotericin B, itraconazole, posaconazole, ravuconazole, and terbinafine were low for the isolates, and no differences were found between the different clinical forms of the disease. The phenotypical techniques revealed that the isolates were associated among themselves, and the fingerprinting analysis showed that Rio de Janeiro epidemic strains were genetically related. Although nine subtypes were found, they were not associated with specific clinical forms. Similar results were obtained with ITS sequence analysis, suggesting that the cases related to the outbreak had a common origin. In another antifungal susceptibility study of five species of *Sporothrix*,[102] significant differences were found among these species, with *S. brasiliensis* showing the best response to antifungals, and *S. mexicana* the worst response.

60.1.6 Laboratory Diagnosis of Sporotrichosis

The definitive diagnosis of sporotrichosis relies on isolating the organism from the site of infection in cutaneous sporotrichosis. *S. schenckii* is relatively easy to grow when compared with other systemic fungi, such as *Histoplasma capsulatum*. Within days to weeks, hyphal growth occurs. The morphological microscopic characteristic of the fungus allows presumptive identification of the organism as *Sporothrix* spp., but conversion to the yeast form should be attempted to firmly establish the diagnosis of sporotrichosis.[36] Although some cases are benign, severe disease or unusual presentations can occur in immunocompromised individuals such as human immunodeficiency virus-infected patients and patients with chronic granulomatous disease.[26,27,103] Cutaneous sporotrichosis lesions may be mistaken with other infections as tuberculosis, leishmaniasis, paracoccidioidomycosis, gummatous syphilis, and chromoblastomycosis.[22,104,105] In such cases, the delay in the diagnosis of sporotrichosis can lead to fatality.[106] These handicaps require more efficient methods to detect accurately the pathogen in infected individuals, and recently molecular tests have been introduced to reveal small amounts of genetic material of the fungus in clinical samples.

60.1.6.1 Conventional Techniques

Biological material can be swabbed or aspirated from cutaneous lesions, or biopsy material can be obtained from the tissue lesion. In general, other samples such as sputum, synovial fluid, cerebrospinal fluid (CSF), and, rarely, blood have been reported to yield *S. schenckii* when cultured. The direct examination of these samples is usually negative and when positive, asteroid bodies are observed; the percentage of positive samples increase significantly when the initial pus is discarded and samples are taken more deeply and examined thoroughly.[107] Generally, Sabouraud dextrose agar and/or Sabouraud dextrose agar with antibiotics (the fungus is resistant to cicloheximide) are used for primary isolation, and the culture is kept at 26°C–28°C. In general, a high percentage of cultures become positive and this is one of the most reliable methods for diagnosis. Growth occurs within 3–5 days and the colony morphology is variable. At first, small dirty whitish, finely radiated humid-like colonies are observed. Afterward, the colony becomes wrinkled and membranous, developing areas that become dark brown and black eventually. Coloration depends on strains and culture media, some colonies remaining white and glabrous. Microscopically, the mycelia morphotype shows the characteristic sympodial and sessile conidia described elsewhere, depending on each species.

Conversion of the mycelial to yeast form of *S. schenckii* is important for its identification and is easily performed in different culture medium at 28°C or 37°C. To induce the transformation, conidia and mycelia are grown on brain heart infusion agar at 37°C for 4–6 days. Other culture media are yeast peptone glucose agar and Rodríguez del Valle medium[38], both with a pH of 7.2–7.4 at 28°C for 48–72h. Microscopically, yeast are fusiform and ovoid cells of 2.5–5 to 3.5–6.5 μm, with single or multiple budding.

60.1.6.1.1 Histopathology

Histopathology is another inexpensive and useful procedure for this disease. When the biopsy sample is from a nodule, the infiltrate is seen as a central and chronic suppurative zone, with microabscesses containing polymorphonuclear cells surrounded by lympho-histiocytic infiltrates. This zone may contain asteroid bodies. The second zone is where an asteroid body is seen as a gemming cell with a radiating halo of eosinophils (Hoeppli-Splendore phenomenon). Around the chronic suppurative zone is the medial or tuberculoid zone constituted by epitheliod cells, some multinucleated Langerhans giant cells and the third zone, or syphilloid area, is comprised of plasmocytes, lymphocytes, fibroblasts, and vascular neoformations. This arrangement is not always found since the cellular elements are often mixed.[108] It is also essential that multiple serial sections stained with periodic acid-Schiff (PAS) reagent be meticulously studied for the evidence of spores and hyphae of *S. schenckii*.[109]

60.1.6.1.2 Immunological Techniques

The immunological diagnosis that includes serological and skin tests may also be helpful for sporotrichosis diagnosis. Agglutination and immunodiffusion tests are recommended for the diagnosis and prognosis of sporotrichosis. The agglutinin and precipitin titers disappear with the resolution of the disease. Other serological tests include indirect immunofluorescence and enzyme-linked immunosorbent assay (ELISA).

The techniques developed to detect antibody response in patients with mycoses have traditionally used crude extracts with a large number of fungal fractions. While being relatively easy to obtain, these extracts are not free of problems. The standardization of different batches is difficult, and the use of crude extracts may facilitate the cross-reactivity between antibodies from patients with different fungal and bacterial infections. However, in the last two decades, there has been an approach to develop more defined antigens. Several antigens have been obtained of the yeast phase of *S. schenckii*, essentially carbohydrates and proteins. One of the most outstanding and biological active antigen is the peptide-rhamnomannan fraction expressed in the cell wall glycopeptide components.[42] Using Concanavalin A (ConA), three main glycoproteins of 84, 70, and 58 kDa were identified in *Sporothrix*.[110] Mendoza et al.[111] reported exoantigens in filtrate cultures of the mycelial morphotype of the fungus, especially those of 90 and 50 kDa, which appear to be species specific. They showed that the antigenic composition of *S. schenckii* exoantigens of the mycelial morphotype did not cross-react with a single serum specimen of coccidioidomycosis, histoplasmosis, and paracoccidioidomycosis.[111] *S. schenckii* exoantigens of the mycelial morphotype have also been used in immunodiffusion and immunoelectrophoresis assays in which no cross-reactions were observed when using chromoblastomycosis and leishmaniasis sera.[112] In other studies, an ELISA test was developed with a ConA-binding fraction (*Ss*CBF) of the peptido-rhamnomannan fraction of *S. schenckii* that showed 90% sensitivity and 86% overall efficacy when tested against sera from patients with lymphocutaneous, fixed cutaneous, disseminated cutaneous, or multiple and extracutaneous forms of sporotrichosis and serum samples from healthy controls and subjects with other diseases.[113] Recently, Almeida-Paes et al.[114] developed an assay (ELISA) for specific antibody detection in serum specimens of patients with sporotrichosis. The assay used mycelial morphotype *S. schenckii* exoantigens and tested against sera from patients with different clinical forms of sporotrichosis. Potential cross-reactions were analyzed with heterologous sera, as well as with sera from healthy controls. They found a sensitivity of 97% and a specificity of 89% in this assay. These results suggest ELISA as a sensitive diagnostic tool for the serodiagnosis of sporotrichosis and can be used together with conventional methods of diagnosis, particularly in those cases where cross-reactions or false-positive results are experienced with serodiagnosis. Additionally, Almeida-Paes et al.[115] demonstrated that antibodies produced during *S. schenckii* infection are diverse, and that an exoantigen ELISA (mycelial morphotype exoantigen of *S. schenckii*) for the detection of combinations of IgA, IgG, and IgM antibodies is a highly sensitive and specific diagnostic procedure for sporotrichosis.

Intradermal skin tests with mycelial-derived sporotrichin have been widely used for the immunological detection of sporotrichosis since its first report by González-Ochoa.[45] It is useful in epidemiological studies and as an auxiliary method in atypical forms of the disease. An intradermal injection (0.1 mL) is applied, and after 24–48 h, a 5 mm or more induration reaction is considered positive. Test positivity indicates an immunocontact with the specific *S. schenckii* fungus and is generally positive in 95% of the cases with fungal exposure.

60.1.6.2 Molecular Techniques

The definite diagnosis of sporotrichosis is achieved through the isolation of *S schenckii* in clinical samples and in some cases complemented with immunodiagnostic procedures. However, both isolation and immunodiagnosis pose limitations, such as cultures of the biopsy specimens frequently yielding negative results and the immune cross-reaction with other fungal species causing similar nosologies. Actually, among diverse methods, the PCR is a viable option. PCR is the most commonly used molecular test for diagnosis. It amplifies fungal DNA from a small amount of the pathogen from clinical samples and constitutes a fast and sensitive procedure for the diagnosis of sporotrichosis. PCR used for the detection and identification of *S. schenckii* can be done in a single step or in two sets (nested PCR).

Among different .studies using PCR for *S. schenckii* diagnosis, we will mention the most recent ones. Hu et al.[116] designed a nested PCR assay for the detection of *S. schenckii* with the use of an 18S rRNA gene sequence as a blank. Different isolates were used for specificity and sensitivity, including five clinical isolates, *S. schenckii* ATCC 10213 strain, 10 clinical isolates of common fungi, three strains of *Mycobacterium* spp. and *Staphylococcus aureus*, and normal human skin tissue samples. The amplification of the fragment alone was done in six *S. schenckii* isolates during the first PCR and in the nested version. The sequences obtained were 100% identical to *S. schenckii* 18S rRNA gene sequence. The nested PCR showed a sensitivity of 40 fg of *S. schenckii* DNA detecting 1 CFU of *S. schenckii* in tissue samples. Validation of the nested PCR assay in clinical isolates and infected experimental mice showed a high sensitivity and specificity, suggesting this assay as a quick diagnosis with sufficient precision to be clinically useful for patients with sporotrichosis.

In another study, Sandhu et al.[117] described a probe based on the 28S rRNA gene for detecting any pathogenic or saprobic fungi during the first PCR, while in the second PCR, oligonucleotides specific for *S. schenckii* were used for revealing the presence of this DNA–DNA hybridization in clinical samples. The probe showed a high level of specificity when testing it with related pathogenic fungi and clinical samples. In order to obtain a rapid detection system, Kano et al.[118] designed a pair of specific oligonucleotides from the chitin synthase 1 gene for detecting *S. schenckii* in tissue samples. The S1, R1 and S2, R2 oligonucleotides did not amplify the DNA from other fungal or bacterial pathogens, nor human or feline cells. Using PCR with the S1, R1 and S2, R2 oligonucleotides, they detected 100 and 10 pg of genomic DNA of *S. schenckii*. Results point out this system as another useful molecular tool for the diagnosis of human and animal sporotrichosis.

Using the previously mentioned report of the nested PCR assay of Hu et al.[116] for *S. schenckii* detection, Xu et al.[119] assayed those probes in 38 *S. schenckii* DNAs (including the 24 types of mtDNA described previously) from mice tissue infected by each fungal isolate and skin biopsies of patients with sporotrichosis. In addition, two strains of *Ceratocystis minor* and DNA from isolates from other 10 species of human pathogenic fungi were used. Amplification was observed in the 38 *S. schenckii* isolates, in infected mice tissue and in human clinical skin biopsies, but not in *C. minor* DNA nor in other fungal species. The detection limit for *S. schenckii* DNA was 50 fg. The authors showed that nested PCR can identify *S. schenckii* from all the described types of mtDNA and in isolates from different parts of the world. The nested PCR assay with the18S rRNA gene sequence of Hu et al.[116] seems to be sensitive and specific and is another useful tool for a quick diagnosis of sporotrichosis.

60.2 METHODS

Isolates of *Sporothrix* spp. are stored on potato dextrose agar at 4°C–7°C, in slant cultures submerged in mineral oil at room temperature and at 4°C, and conidial suspensions in 10% glycerol at 4°C liquid nitrogen. Isolates grow well after 10 years storage in any of the methods. When needed, isolates are grown on potato dextrose agar plates at 28°C for 5–7 days. Two DNA extraction methods are given to suit different laboratory facilities.

60.2.1 Sample Preparation

Growth of S. schenckii. Transfer a single colony of mycelial-phase *S. schenckii* to Sabouraud's dextrose broth with 1% yeast extract (SDY) and incubate at 28°C on a gyratory shaker at 120 rpm for 7 days to achieve sufficient cell mass for DNA extraction. Harvest mycelial cells by centrifugation and wash three times in sterile Milli-Q water.

First DNA extraction method. Grind approximately 25 mg of washed mycelium with a mortar and pestle in liquid nitrogen. Suspend in 500 µL of sterile lysis buffer (50 mM Tris–HCl, 150 mM NaCl, 100 mM EDTA) with 50 µL of 10% sodium dodecyl sulfate (SDS). Gently shake the tubes and centrifuge at 18,300 × *g* for 5 min. Treat the supernatant with 5 µL of 10 mg of RNase (Qiagen Inc., Valencia, CA) per ml at 37°C for 1 h and then subject to phenol–chloroform–isoamyl alcohol (25:24:1) extraction by adding an equal volume of this mixture. Precipitate nucleic acids by adding 10 µL of 3 M NaOAc and two volumes of absolute ethanol. Wash the pellet with 70% ethanol, dry, and resuspend in water. Quantify DNA fluorometrically and check against standard DNA concentrations by agarose gel electrophoresis.

Second DNA extraction method. In this method, instead of mycelia breakage by liquid nitrogen, a FastPrep®-24 (Qbiogene, Inc, CA) equipment is used. Place approximately 25 mg of mycelia into 2 mL microtubes containing 0.2 g of glass beads (400–455 µm in diameter, Sigma Chemical Co.), washed in hydrochloric acid and 500 µL of sterile lysis buffer (50 mM Tris–HCl, 100 mM EDTA, 3% SDS, and 1% mercaptoethanol). Stir microtubes in the FastPrep®-24 (Qbiogene) for two periods of 40 s at 6 m/s speed with a 2 min interval in ice between stirrings. Gently shake the microtubes and then centrifuge at 18,300 × *g* for 5 min. Treat the supernatant with 20 μL of 10 mg of RNase (Qiagen) per ml at 37°C for 1 h and then subject to phenol–chloroform–isoamyl alcohol (25:24:1) extraction by adding an equal volume of this mixture. Precipitate nucleic acids by adding 10 μL of 3 M sodium acetate and two volumes of absolute ethanol. Wash the pellet with 70% ethanol, dry, and resuspend in water. Quantify DNA fluorometrically and check against standard DNA concentrations by agarose gel electrophoresis.

60.2.2 Detection Procedures

60.2.2.1 RAPD-PCR Assay[93]

Perform a 25 μL reaction mixture that contains 10 ng of the required DNA sample; 2.5 mM $MgCl_2$; 200 μM each dATP, dTTP, dCTP, and dGTP (Applied Biosystems Inc., Foster City, CA); 100 pmol of each primer [primers OPBG-01, OPBG-14, and OPBG-19 (kit B; Operon Technologies Ltd.)]; sterile, distilled MilliQ water (Millipore) to make up the final volume; and 1 U of Taq DNA polymerase (Applied Biosystems). Controls, which contain all of the components listed above except DNA, are also set up and run with each set of reactions. Optimal amplification conditions for *S. schenckii* RAPD analysis are 45 cycles of 1 min at 94°C (denaturation), 1 min at 35°C (annealing), and 1 min at 72°C (extension). Amplifications are carried out in a thermocycler (Perkin-Elmer Cetus, Norwalk, CT). A final 7 min cycle at 72°C ensures full extension of all amplified products, and samples are maintained at 4°C. The polymorphic amplified patterns revealed by RAPD analysis-PCR assays are viewed, and digital images of ethidium bromide (0.5 μg/mL)-stained agarose gels (2%) are captured with a documentation system (GeneCam; Syngene, Cambridge, MA) and printed on a digital graphic printer (Sony Electronics Inc., Park Ridge, NJ). RAPD analysis-PCR assays are repeated independently three times with each primer with all isolates studied, with consistent gel patterns detected from one assay to another.

60.2.2.2 Identification of *Sporothrix* Species by Sequencing

The CAL gene sequence is the most phylogenetically informative locus found for the recognition of these species.[20] Marimon et al.[20] developed a PCR assay targeting this gene. CAL gene is amplified by PCR: 20–60 ng of genomic DNA template in 25 μL reaction. Reaction mixtures consist of 0.5–1 mM of each primer (degenerated primers CL1 and CL2A[20]), 1.0 U of AmpliTaq DNA polymerase (Perkin-Elmer), 10 mM Tris–HCl (pH 8.3), 1.5 mM $MgCl_2$, 50 mM KCl, and 0.2 mM deoxynucleotide triphosphates. The amplification program includes an initial denaturation cycle at 94°C for 5 min, 35 cycles at 95°C for 30 s, 60°C for 1 min, and 72°C for 1 min, and 1 cycle at 72°C for 7 min.

Automated sequencing is done with an ABI dye terminator cycle sequencing ready reaction kit and PCR primers in accordance with the recommendations of the manufacturer (Applied Biosystems). Sequences are generated from both strands and edited and initially aligned with the Sequence Navigator (v1.01; Applied Biosystems) software package, and the alignments optimized visually.

60.2.2.3 *S. schenckii* Detection by Nested PCR

A nested PCR is carried out as described by Hu et al.[116] DNA samples, from isolates and positive *S. schenckii*, are processed as follows: Two sets of primers are used. The design of oligonucleotides used in this technique was based on the comparison of the sequence of 18S rRNA gene of *S. schenckii* (accession no. M85053) and those of other fungi in the GenBank database (National Center for Biotechnology Information, National Library of Medicine, Bethesda, MD). The outer primer set SS1, 5′-CTC GTT CGG CAC CTT ACA CG-3′, and SS2, 5′-CGC TGC CAA AGC AAC GCG GG-3′, delimit a 305-nucleotide sequence of the gene. The inner primers SS3, 5′-ACT CAC CAG GTC CAG ACA CGA TG-3′, and SS4, 5′-CGC GGG CTA TTT AGC AGG TTA AG-3′, delimit a specific 152-nucleotide sequence. DNA amplification is performed on a Perkin-Elmer Cetus DNA thermal cycler and the first PCR performed in a 50 μL reaction mixture containing 200 μM of dNTPs (Applied Biosystems), 1.5 mM $MgCl_2$, 0.4 μM of each outer primer, 1.5 U *Taq* DNA polymerase (Applied Biosystems), and 10 ng of DNA. Cycling conditions are as follows: one cycle at 95°C for 5 min; 40 cycles at 95°C for 1 min, 68°C for 1 min, and 72°C for 1 min; and 1 cycle at 72°C for 10 min. The reaction mixture of the nested PCR is identical, except that 3 μL of the first reaction product and the inner primer pair SS3 and SS4 are used.

Amplification products are electrophoresed through 2% agarose in Tris-borate-EDTA 0.5X buffer. Electrophoresis is conducted at 90 V for 60 min. The 123 bp DNA Ladder (Invitrogen, Carlsbad, CA) is used as molecular markers. The bands are visualized with a UV transilluminator after ethidium bromide (0.5 μg/mL) staining. They are captured with a documentation system (GeneCam; Syngene, Cambridge, MA) and printed with a thermal printer (Sony 650).

60.3 CONCLUSIONS AND FUTURE PERSPECTIVES

Despite the large impact of molecular techniques in the field of clinical mycology, they do not eliminate the routine use of classic techniques, and as recommended in clinical practice management, both should be performed simultaneously. Importantly, molecular methods have not been fully validated by comparison with those used in routine identification of fungi and therefore are not employed in clinical laboratories as a reliable tool for the diagnosis of disseminated sporotrichosis. These circumstances demand an urgent validation for routine use in diagnosis and in different types of research. Although the advantages of molecular methods are unquestionable, its failures should be considered timely for purposes of future corrections and/or considerations to improve implementation. Interestingly, physicians need a better understanding of fungal infection to notify the diseases and take into account the epidemiological evidence associated to the mycosis, but above all they should be familiar with the use and limitations of each of the available diagnostic tests for fungal diseases particularly for sporotrichosis. It is evident that much work is still needed in order to identify clearly species of the *S. schenckii* complex of clinical relevance especially in regard to their different antifungal responses.

Sporothrix species can now be added to a list of emerging pathogens, according to their capacity to cause disease in otherwise healthy hosts as well as in immunocompromised individuals. Several contributing factors influence the pathogenesis of infectious diseases by performing effects on both host and the fungi, and the resulting host–microbe interaction. These factors include demographic characteristics and technology with an increase in the number of susceptible hosts, microorganism adaptations, land use and global travel, and the breakdown of public health measures. Microbe pathogenesis is multifactorial and complex. The mechanisms of fungal pathogenesis are much less-well understood than are those of bacterial or parasite pathogens. Moreover, many fungi are not "professional pathogens" and thus do not stimulate their study. The emerging pathogenic status of diverse *Sporothrix* species could be a reason to reverse this view.

REFERENCES

1. Six, D.L., Bark beetle-fungus symbioses, in *Insect Symbiosis*, p. 97, Bourtzis, K. and Miller, T.A., Eds., CRC Press, New York, 2003.
2. Kirisits, T., Fungal associates of European Bark beetles with special emphasis on the ophiostomatoid fungi, in *Bark and Wood Boring Insects in Living Trees in Europe, a Synthesis, p.* 181, Lieutier, F. et al., Eds., Kluwer Academic Press, Dordrecht, the Netherlands, 2004.
3. Seifert, K.A., Sapstain of commercial lumber by species of *Ophiostoma and Ceratocystis*, in *Ceratocystis and Ophiostoma: Taxonomy, Ecology and Pathogenicity*, p. 141, Wingfield, M. J., Seifert, K. A., and Webber, J. F., Eds., APS Press, Mendota Heights MN, 1993.
4. Berbee, M.L. and Taylor, J.W., 18S Ribosomal RNA gene sequence characters place the human pathogen *Sporothrix schenckii* in the genus *Ophiostoma*, *Exp. Mycol.*, 16, 87, 1992.
5. Summerbell, R.C. et al., Medically important *Sporothrix* species and related *Ophiostomatoid* fungi, in *Ceratocystis and Ophiostoma: Taxonomy, Ecology and Pathogenicity*, p. 185, Wingfield, M.J., Seifert, K.A., and Webber, J.F., Eds., APS Press, Mendota Heights, MN, 1993.
6. de Beer, Z.W., Phylogeny of the *Ophiostoma stenoceras-Sporothrix schenckii* complex, *Mycologia*, 95, 434, 2003.
7. Spencer, J.F.T. and Gorin, P.A.J., Systematics of the genera *Certocystis* and *Graphium*. Proton magnetic resonance spectra of the mannose-containing polysaccharides as an aid in classification, *Mycologia*, 63, 387, 1971.
8. Toriello, C. and Mariat, F., Etude comparée des polyosides des champignons *Ceratocystis stenoceras* et *Sporothrix schenckii*. Composition chimique et analyse immunologique, *Ann. Microbiol.*, 125A, 287, 1974.
9. Travassos, L.R. and Lloyd, K.O., *Sporothrix schenckii* and related species of *Ceratocystis*, *Microbiol. Rev.*, 44, 683, 1980.
10. Mariat, F., Adaptation de *Ceratocystis* a la vie parasitaire chez l'animal. Etude de l'aquisition d'un pouvoir pathogéne comparable a celui de *Sporothrix schenckii*, *Sabouraudia*, 9, 191, 1971.
11. Suzuki, K., Kawasaki, M., and Ishizaki, H., Analysis of restriction profiles of mitochondrial DNA from *Sporothrix schenckii* and related fungi, *Mycopathologia*, 103, 147, 1988.
12. O'Reilly, L.C. and Altman, S.A., Macrorestriction analysis of clinical and environmental isolates of *Sporothrix schenckii*, *J. Clin. Microbiol.*, 44, 7, 2006.
13. James, J.T. et al., Reconstructing the early evolution of fungi using a six-gene phylogeny, *Nature*, 44, 19, 2006.
14. Hibbett, S.D., A higher-level phylogenetic classification of the Fungi, *Mycol. Res.*, 111, 509, 2007.
15. Zhang, N. et al., An overview of the systematics of the Sordariomycetes based on a four-gene phylogeny, *Mycologia*, 98, 1076, 2006.
16. Hektoen, L. and Perkins, C.F., Refractory subcutaneous abscesses caused by *Sporothrix schenckii*, *J. Exp. Med.*, 5, 77, 1900.
17. CABI Bioscience, CBS, 2009, Index Fungorum. Available: http://www.indexfungorum.org, accessed October 12, 2009.
18. Marimon, R. et al., Molecular phylogeny of *Sporothrix schenckii*, *J. Clin. Microbiol.*, 44, 3251, 2006.
19. Watanabe, S. et al., RFLP analysis of the internal transcribed spacer regions of *Sporothrix schenckii*, *Jpn. J. Med. Mycol.*, 45, 165, 2004.
20. Marimon, R. et al., *Sporothrix brasiliensis*, *S. globosa*, and *S. mexicana*. Three new *Sporothrix* species of clinical interest, *J. Clin. Microbiol.*, 45, 3198, 2007.
21. Marimon, R. et al., *Sporothrix luriei*: A rare fungus from clinical origin, *Med. Mycol.*, 46, 621, 2008.
22. Rippon, J.W., Sporotrichosis, in *Medical Mycology*, 3rd edn., The W.B. Saunders Co, Philadelphia, PA, 1990, Chap. 11.
23. Kwon-Chung, K.J. and Bennett, J.E., *Medical Mycology*, Lea & Febiger, Philadelphia, PA, 1992, Chap. 26.
24. Emmons, C.W. et al., *Medical Mycology*, 3rd edn., Lea & Febiger, Philadelphia, PA, 1977, Chap. 25.
25. Mariat, F., The epidemiology of sporotrichosis, in *Systemic Mycoses*, p. 144, Wolstenholme, G. E. W., Ed., A. Churchill, London, U.K., 1968.
26. Barros, M.B.L. et al., Cat-transmitted sporotrichosis epidemic in Rio de Janeiro, Brazil: Description of a series of cases, *Clin. Infect. Dis.*, 38, 529, 2004.
27. Durden, F.M. and Elewski, B., Fungal infections in HIV patients, *Semin. Cutan. Med. Surg.*, 16, 200, 1997.
28. Dixon, D.M. et al., Isolation and characterization of *Sporothrix schenckii* from clinical and environmental sources associated with the largest U.S. epidemic of sporotrichosis, *J. Clin. Microbiol.*, 29, 1106, 1991.
29. Gonzalez-Ochoa, A., Contribuciones recientes al conocimiento de la esporotricosis, *Gac. Med. Mex.*, 95, 463, 1965.
30. Bonifaz, A., *Micología Médica Básica*, p. 179, 3rd edn., McGraw-Hill, México, 2010.
31. Arenas, R., *Micología Médica Ilustrada*, p. 149, 3rd edn., McGraw-Hill, México, 2008.

32. Barros, M.B. et al., Sporotrichosis: An emergent zoonosis in Rio de Janeiro, *Mem. Inst. Oswaldo Cruz*, 96, 777, 2001.
33. Schubach, T.M.P. et al., Evaluation of an epidemic of sporotrichosis in cats: 347 cases (1998–2001), *J. Am. Vet. Med. Assoc.*, 224, 1623, 2004.
34. Schubach, T.M. et al., Canine sporotrichosis in Rio de Janeiro, Brazil, clinical presentation, laboratory diagnosis and therapeutic response in 44 cases (1998–2003), *Med. Mycol*, 44, 87, 2006.
35. Schubach, A., Barros, M.B.D.L., and Wanke, B., Epidemic sporotrichosis, *Curr. Opin. Infect. Dis*, 21, 129, 2008.
36. Kauffman, C.A., Sporotrichosis, *Clin. Infect. Dis.*, 29, 231, 1999.
37. Nicot, J. and Mariat, F., Caracteres morphologiques et position systématique de *Sporothrix schenckii* agent de la sporotrichose humaine, *Mycopathol. Mycol. Appl.*, 49, 53, 1973.
38. Rodríguez del Valle, N., Rosario, M., and Torres-Blasini, G., Effects of pH, temperature, aeration and carbon source on the development of the mycelia or yeast forms of *Sporothrix schenckii* from conidia, *Mycopathologia*, 82, 83, 1983.
39. Splendore, A., Sobre a cultura de uma nova especie de cogumello patogénica do home: *Sporotrichum splendore* ou "*Sporotrichum asteroide*" n. sp, *Rev. Soc. Sc. Sao Paulo*, 3, 62, 1908.
40. Drouhet, E. and Mariat, F., Etude des facteurs déterminant la dévelopment de la phase levure de *Sporotrichum schenckii*, *Ann. Inst. Pasteur Paris*, 83, 506, 1952.
41. Ghosh, A. et al., Physiological characters of *Sporothrix schenckii* isolates, *Mycoses*, 45, 449, 2002.
42. Lloyd, K.O. and Bitoon, M.A., Isolation and purification of a peptido-rhamnomannan from the yeast form of *Sporothrix schenckii*. Structural and immunochemical studies, *J. Immunol.*, 107, 663, 1971.
43. Mendonca, L. et al., Polymorphism of *Sporothrix schenckii* surface polysaccharides as a function of morphological differentiation, *Biochemistry*, 15, 2423, 1976.
44. Lopes-Bezerra, L.M., Schubach, A., and Costa, O.R., *Sporothrix schenckii* and Sporotrichosis, *Ann. Acad. Bras. Cienc.*, 78, 293, 2006.
45. Gonzalez-Ochoa, A. and Figueroa, E.S., Polisacáridos del *Sporotrichum schenckii*, Datos immunológicos. Intradermoreacción en el diagnóstico de la esporotricosis, *Rev. Inst. Salubr. Enferm. Trop. Mexico City*, 8, 143, 1947.
46. Arenas, G. and Toriello, C., Actividad inmunológica de antígenos miceliales-levaduriformes de diferentes fases de crecimiento de *Sporothrix schenckii*, *Rev. Mex. Micol.*, 2, 131, 1986.
47. Romero-Martínez, R. et al., Biosynthesis and function of melanin in *Sporothrix schenckii*, *Infect. Immun.*, 68, 3696, 2000.
48. Morris-Jones, R. et al., Synthesis of melanin-like pigments by *Sporothrix schenckii* in vitro and during mammalian infection, *Infect. Immun.*, 71, 4026, 2003.
49. Almeida-Paes, R. et al., Growth conditions influence melanization of Brazilian clinical *Sporothrix schenckii* isolates, *Microbes Infect.*, 11, 554, 2009.
50. Cooper, C.R., Dixon, D.M., and Salkin, I.F., Laboratory-acquired sporotrichosis, *J. Med. Vet. Mycol.*, 30, 169, 1992.
51. Casadevall, A. and Pirofski, L.A., Host-pathogen interactions: Redefining the basic concepts of virulence and pathogenicity, *Infect. Immun.*, 67, 3703, 1999.
52. Meléndez Ruíz, C.A. et al., Perfil inmunológico del paciente con esporotricosis linfocutánea, *Mycopathologia*, 83, 169, 1983.
53. Mackinnon, J.E., Ecology and epidemiology of sporotrichosis, in *Proceedings of the International Symposium on Mycoses*, p. 169, Pan American Health Organization, Washington, DC, 1970.
54. Mariat, F. et al., Premieres observations sur les macrospores pigmentées de *Sporothrix schenckii*, *C. R. Acad. Sci. Paris*, 286, 1429, 1978.
55. Ramos-Zepeda, R. and Gonzalez-Mendoza, A., Metabolic activity of phagocytes in experimental sporotrichosis, *Mycopathologia*, 93, 109, 1986.
56. Wang, X.L. et al., Differential mRNA expression of *Sporothrix schenckii* catalase gene in two growth phases and other factors, *Chin. Med. J.*, 121, 2100, 2008.
57. Muller, H.E., Uber das Vorkommen von Neuraminidase bei *Sporothrix schenckii* and *Ceratocystis stenoceras* und ihre Bedeutung fur die Okologie und den Pathomechanismus dieser Pilze, Zentralbl Bakteriol, Parasitenkd Infektionskr, *Hyg. Abt.*, 232, 365, 1975.
58. Ruiz-Baca, E. et al., Isolation and some properties of a glycoprotein of 70 kDa (Gp70) from the cell wall of *Sporothrix schenckii* involved in fungal adherence to dermal extracellular matrix, *Med. Mycol.*, 27, 185, 2009.
59. Lima, O.C. et al., Adhesion of the human pathogen *Sporothrix schenckii* to several extracellular matrix proteins, *Braz. J. Med. Biol. Res.*, 32, 651, 1999.
60. Lima, O.C. et al., Immunofluorescence and flow cytometry analysis of fibronectin and laminin binding to *Sporothrix schenckii* yeast cells and conidia, *Microb. Pathog.*, 37, 131, 2004.
61. Figueiredo, C. et al., The in vitro interaction of *Sporothrix schenckii* with human endothelial cells is modulated by cytokines and involves endothelial surface molecules, *Microb. Pathog.*, 36, 177, 2004.
62. Nascimento, R.C. et al., Passive immunization with monoclonal antibody against a 70 kDa putative adhesin of *Sporothrix schenckii* induces protection in murine sporotrichosis, *Eur. J. Immunol.*, 38, 3080, 2008.
63. Teixeira, C.P.A. et al., Cell surface expression of adhesions for fibronectin correlates with virulence in *Sporothrix schenckii*, *Microbiology*, 155, 3730, 2009.
64. Hiruma, M., Kawada, A., and Ishibashi, A., Ultrastructure of asteroid bodies in sporotrichosis, *Mycoses*, 34, 103, 1991.
65. Lurie, H.I. and Still, W.J.S., The "capsule" of *Sporotrichum schenckii* and the evolution of the asteroid body. A light and electron microscopic study, *Sabouraudia*, 7, 64, 1969.
66. Smith, A.G. et al., Chronic pulmonary sporotrichosis: Report of a case, including morphologic and mycologic studies, *Am. J. Clin. Pathol.*, 54, 401, 1970.
67. Lurie, H.I. and Still, W.J.S., Histopathology of sporotrichosis, *Arch. Pathol.*, 75, 421, 1963.
68. Zaharopoulos, P., Fine-needle aspiration cytologic diagnosis of lymphocutaneous sporotrichosis: A case report, *Diagn. Cytopathol.*, 20, 74, 1999.
69. Gerhard, R. et al., Fine-needle aspiration biopsy of disseminated sporotrichosis: A case report, *Diagn. Cytopathol.*, 36, 174, 2008.
70. Okudaira, M., Tsubura, E., and Schwartz, J., A histopathological study of experimental murine sporotrichosis, *Mycopathologia*, 14, 284, 1961.
71. Maberry, J.D., Mullins, J.F., and Stone, O.J., Sporotrichosis with demonstration of hyphae in human tissue, *Arch. Dermatol.*, 93, 1, 65, 1966.
72. Lii, S.L. and Shigemi, F., Demonstration of hyphae in human tissue of sporotrichosis, with statistics of cases reported from Tokushima, Tokushima, *J. Exp. Med.*, 20, 69, 1973.

73. Fujimura, T. et al., Local expression of migration inhibitory factor and Th1 type cytokine mRNA in sporotrichosis lesions, *Acta Dermatol. Venereol. (Stockh)*, 76, 321, 1996.
74. Tachibana, T., Matsuyama, T., and Mitsuyama, M., Involvement of CD4+ T cells and macrophages in acquired protection against infection with *Sporothrix schenckii* in mice, *Med. Mycol.*, 37, 397, 1999.
75. Koga, T., Duan, H., and Furue, M., Immunohistochemical detection of interferon-gamma-producing cells in granuloma formation of sporotrichosis, *Med. Mycol.*, 40, 111, 2002.
76. Koga, T. et al., Immunohistochemical localization of activated and mature CD83+ dendritic cells in granulomas of sporotrichosis, *Eur. J. Dermatol.*, 11, 527, 2001.
77. Buentke, E. and Scheynius, A., Dendritic cells and fungi, *Acta Pathol. Microbiol. Immunol. Scand.*, 111, 789, 2003.
78. Netea, M.G. et al., Recognition of fungal pathogens by Toll-like receptors, *Eur. J. Clin. Microbiol. Infect. Dis.*, 23, 672, 2004.
79. Uenotsuchi, T. et al., Differential induction of Th1-prone immunity by human dendritic cells activated with *Sporothrix schenckii* of cutaneous and visceral origins to determine their different virulence, *Int. Immunol.*, 18, 1637, 2006.
80. Howard, D.H. and Orr, G.F., Comparison of strains of *Sporotrichum schenckii* isolated from nature, *J. Bacteriol.*, 85, 816, 1963.
81. Vélez, H. et al., Polimorfismo fenotípico de cepas autóctonas de *Sporothrix schenckii*, *IATREIA*, 9, 71, 1996.
82. Pérez, A. et al., Diferencias en la virulencia de aislamientos de *Sporothrix schenckii*, in *Memorias VI Congreso Nacional de Micología, IX Jornadas Científicas, Sociedad Mexicana de Micología, El Colegio de la Frontera Sur*, Tapachula, Chiapas, México, 1997, p. 176.
83. Takeda, Y., Kawasakii, M., and Ishizaki, H., Phylogeny and molecular epidemiology of *Sporothrix schenckii* in Japan, *Mycopathologia*, 116, 9, 1991.
84. Hajjeh, R. et al., Outbreak of sporotrichosis among tree nursery workers, *J. Infect. Dis.*, 176, 499, 1997.
85. Ishizaki, H. et al., Mitochondrial DNA analysis of *Sporothrix schenckii* in Costa Rica, *J. Med. Vet. Mycol.*, 34, 71, 1996.
86. Ishizaki, H. et al., Mitochondrial DNA analysis of *Sporothrix schenckii* in North and South America, *Mycopathologia*, 142, 115, 1998.
87. Ishizaki, H. and Kawasaki, M., Molecular epidemiology of *Sporothrix schenckii*, *Jpn. J. Med. Mycol.*, 41, 245, 2000.
88. Ishizaki, H. et al., Mitochondrial DNA analysis of *Sporothrix schenckii* in India, Thailand, Brazil, Colombia, Guatemala and Mexico, *Jpn. J. Med. Mycol.*, 50, 19, 2009.
89. Lin, J. et al, Mitochondrial DNA analysis of *Sporothrix schenckii* clinical isolates from China, *Mycopathology*, 148, 69, 1999.
90. Mora-Cabrera, M. et al., Analysis of restriction profiles of mitochondrial DNA from *Sporothrix schenckii*, *Med. Mycol.*, 39, 439, 2001.
91. Arenas, R., Miller, D., and Campos-Macias, P., Epidemiological data and molecular characterization (mtDNA) of *Sporothrix schenckii* in 13 cases from Mexico, *Int. J. Dermatol.*, 46, 177, 2007.
92. Zhang, Z. et al., Genotyping of *Sporothrix schenckii* by analysis of ribosomal DNA regions, *Mycoses*, 49, 305, 2006.
93. Mesa-Arango, A.C. et al., Phenotyping and genotyping of *Sporothrix schenckii* isolates according to geographic origin and clinical forms of sporotrichosis, *J. Clin. Microbiol.*, 40, 3004, 2002.
94. Gutiérrez Galhardo, M.C. et al., Molecular epidemiology and antifungal susceptibility patterns of *Sporothrix schenckii* isolates from a cat-transmitted epidemic of sporotrichosis in Rio de Janeiro, Brazil, *Med. Mycol.*, 46, 141, 2008.
95. Neyra, E.P. et al., Epidemiology of human sporotrichosis investigated by amplified fragment length polymorphism, *J. Clin. Microbiol.*, 43, 1348, 2005.
96. Tateishi, T. et al., Karyotyping by PFGE of clinical isolates of *Sporothrix schenckii*, *FEMS Immunol. Med. Microbiol.*, 13, 147, 1996.
97. Madrid, H. et al., *Sporothrix globosa*, a pathogenic fungus with widespread geographical distribution, *Rev. Iberoam. Micol.*, 26, 218, 2009.
98. Kawasaki, M. et al., Molecular epidemiology of *Sporothrix schenckii* in Mexico, Brazil and Spain, in *Proceedings of the Fourth Congress of the International Society for Human and Animal Mycology*, p. 120, Buenos Aires, Argentina, 2000.
99. Sugita, Y., Molecular analysis of DNA polymorphism of *Sporothrix schenckii*, *Jpn. J. Med. Mycol.*, 41, 11, 2000.
100. Reis, R.S. et al., Molecular typing of *Sporothrix schenckii* strains isolated from clinical specimens in Rio de Jainero, Brazil, in *Proceedings of the Fourth Congress of the International Society for Human and Animal Mycology*, p. 132, Buenos Aires, Argentina, 2000.
101. Lee, J.B., Lee, S.C., and Won, H.Y., Analysis of random amplified polymorphic DNA (RAPD) for *Sporothrix schenckii* and related fungi, *Korean J. Med. Mycol.*, 5, 113, 2000.
102. Marimon, R. et al., In vitro antifungal susceptibilities of five species of *Sporothrix*, *Antimicrob. Agents Chemother.* 52, 732, 2008.
103. Kajiwara, H. et al., Impaired host defense against *Sporothrix schenckii* in mice with chronic granulomatous disease, *Infect. Immun.*, 72, 5073, 2004.
104. Sharma, S. et al., Cutaneous tuberculosis mimicking sporotrichosis, *Indian J. Pediatr.*, 72, 86, 2005.
105. Tobin, E.H. and Jih, W.W., Sporotrichoid lymphocutaneous infections: Etiology, diagnosis, and therapy, *Am. Fam. Physician*, 63, 326, 2001.
106. Al-Tawfiq, J.A. and Wools, K.K., Disseminated sporotrichosis and *Sporothrix schenckii* fungemia as the initial presentation of immunodeficiency virus infection, *Clin. Infect. Dis.*, 26, 1403, 1983.
107. Gezuele, E. and Da Rosa, D., Relevancia del cuerpo asteroide esporotricósico en el diagnóstico rápido de la esporotricosis, *Rev. Iberoam. Micol.*, 22, 147, 2005.
108. Padilla Desgarennes, M.C., Siodia Zambrano, S.P., and Novales Santa Coloma J., Esporotricosis con involución espontánea, *Dermatol. Rev. Mex.*, 51, 14, 2007.
109. Suchitha, A. et al., Sporotrichosis in Mysore: A case report to emphasize the role of histopathology, *Indian J. Pathol. Microbiol.*, 51, 45, 2008.
110. Lima, O.C. and Bezerra, L.M., Identification of a concanavalin A-binding antigen of the cell surface of *Sporothrix schenckii*, *J. Med. Vet. Mycol.*, 35, 167, 1997.
111. Mendoza, M. et al., Production of culture filtrates of *Sporothrix schenckii* in diverse culture media, *Med. Mycol.*, 40, 447, 2002.
112. Albornoz, M.B., Villanueva, E., and Torres, E.D., Application of immunoprecipitation techniques to the diagnosis of cutaneous and extracutaneous forms of sporotrichosis, *Mycopathologia*, 85, 177, 1984.
113. Bernardes-Engermann, A.R. et al., Improved-enzyme-linked immunosorbent assay using the Ss-CBF antigen for the serodiagnosis of several clinical forms of sporotrichosis, *Med. Mycol.*, 43, 487, 2005.

114. Almeida-Paes, R. et al., Use of Mycelial-Phase *Sporothrix schenckii* exoantigens in an enzyme-linked immunosorbent assay for diagnosis of sporotrichosis by antibody detection, *Clin. Vaccine Immunol.*, 14, 244, 2007.
115. Almeida-Paes, R. et al., Immunoglobulins G, M, and A against *Sporothrix schenckii* exoantigens in patients with sporotrichosis before and during treatment with itraconazole, *Clin. Vaccine Immunol.*, 14, 1149, 2007.
116. Hu, S. et al., Detection of *Sporothrix schenckii* in clinical samples by a nested PCR assay, *J. Clin. Microbiol.*, 41, 1414, 2003.
117. Sandhu, G.S. et al., Molecular probes for diagnosis of fungal infections, *J. Clin. Microbiol.*, 33, 2913, 2005.
118. Kano, R. et al., Identification of *Sporothrix schenckii* based on sequences of the chitin synthase 1 gene, *Mycoses*, 44, 261, 2001.
119. Xu, T.H. et al., Identification of *Sporothix schenckii* of various mtDNA types by Nested PCR assay, *Med. Mycol.*, 48, 161, 2010.

61 *Stachybotrys*

Dongyou Liu and R.R.M. Paterson

CONTENTS

61.1 INTRODUCTION

61.1.1 Classification, Morphology, and Biology

The genus *Stachybotrys* is an asexually reproducing, dematiaceous fungus belonging to the mitosporic Hypocreales group, order Hypocreales, class Sordariomycetes, subphylum Pezizomycotina, phylum Ascomycota, and kingdom Fungi. The mitosporic Hypocreales group encompasses the genera of *Acremonium*, *Acrostalagmus*, *Cephalosporium*, *Chaetopsina*, *Cylindrocladiella*, *Escovopsis*, *Fusarium*, *Gliocladiopsis*, *Gliocladium*, *Hobsonia*, *Illosporium*, *Myrothecium*, *Parasarcopodium*, *Polycephalomyces*, *Rotiferophthora*, *Sesquicillium*, *Solheimia*, *Stachybotrys*, *Stilbella*, *Trichothecium*, *Tubercularia*, *Ustilaginoideae*, *Verticillium*, and *Xenocylindrocladium* [1].

In turn, the genus *Stachybotrys* is divided into 15 recognized species: *Stachybotrys bisbyi*, *Stachybotrys chartarum*, *Stachybotrys chlorohalonata*, *Stachybotrys cylindrospora*, *Stachybotrys dichroa*, *Stachybotrys echinata*, *Stachybotrys elegans*, *Stachybotrys kampalensis*, *Stachybotrys longispora*, *Stachybotrys microspora*, *Stachybotrys nephrospora*, *Stachybotrys oenanthes*, *Stachybotrys parvispora*, *Stachybotrys subsimplex*, and *Stachybotrys theobromae*, in addition to 13 unassigned species. Two species, *S. chartarum* (obsolete synonyms: *S. alternans* and *S. atra*) and *S. echinata* (obsolete synonym: *Memnionella echinata*), are implicated in human diseases [2,3].

Stachybotrys spp. grow rapidly and mature in approximately 4 days. Colonies are "cottony," white initially and turning black with age. Septate hyphae are hyaline first and become darkly pigmented later. Conidiophores are simple or branched, bearing phialides at apices. Phialides are hyaline or pigmented and cylindrical in shape, with swollen upper portions, forming clusters of 3–6. Conidia ($4.5 \times 9\,\mu m$) are oval, hyaline or pigmented, one-celled, and in clusters.

First identified from a mold growing on domestic wallpaper in Prague in 1837, *Stachybotrys* is a member of the Fungi Imperfecti known as "black mold" or "toxic black mold." The fungus is an inhabitant of soil and strata rich in cellulose (e.g., hay, straw, grain, hemp, plant debris, dead roots, wood pulp, cotton, fabrics, paper, book bindery glue, plant fiber-processing plants, etc.). It has been isolated from contaminated grains, tobacco, insulator foams, indoor air, and water-damaged buildings. The fungus tolerates temperatures up to >60°C and survives over-winter. *Stachybotrys* spores stay viable for years to decades, and conidia retain viability after passage through the gastrointestinal tract. However, the organism is killed by composting of manure and by disinfectants [4].

Stachybotrys chartarum is a known producer of trichothecene mycotoxins and stachylysin (a hemolysin). The best characterized trichothecenes include satratoxins F, G, and H, roriden E, verrucarin J, and trichoverrols A and B, which share chemical formula of $C_{25}H_{34}O_6$ or $C_{26}H_{38}O_6$ and are tricyclic sesquiterpenes with a 12,13-epoxy-trichothec-9-ene ring. *Stachybotrys* species also produce spirolactams and spirolactones (related to anticomplement components), phenylspirodrimanes (inhibitor of complement activation), cyclosporins, and endothelin receptor antagonists [5–7].

Trichothecenes modulate inflammatory reactions and alter alveolar surfactant phospholipid concentrations, besides being potent inhibitors of protein synthesis (e.g., scirpentriol, 15-acetoxyscirpendiol, diacetoxyscirpenol [DAS or anguidine], verucarin A, and T-2 toxin) and elongation or termination (e.g., trichodermin, trichodermol, crotocol, trichothecolone, trichothecin, and verrucarol). Trichothecenes are susceptible to destruction by alkali although resistant to sunlight, UV light, x-rays, heat (up to 120°C), and acids. Some of the mycotoxins have been isolated from dust (e.g., satratoxins, trichoverrols, verrucarol, verrucarins, and trichoverrins) and grain (T-2 toxin, nivalenol, and derivatives of others). Humans may develop toxin-related disease by ingestion of food products contaminated with the fungus toxins, exposure to mycotoxins in building (sick-building syndrome), and/or inhalation of propogules while undertaking experiments with the fungus [8].

61.1.2 Clinical Features and Pathogenesis

61.1.2.1 Clinical Features

Stachybotrys may induce disease by (1) infection, (2) generation of a deleterious immune response, and (3) toxic-irritant effects from the metabolites. Together with other molds (e.g., *Aspergillus*, *Penicillium*, *Alternaria*, and *Cladosporium*), *Stachybotrys* may play a role in the development of sick-building syndrome [9–25].

Stachybotrys was first noted as a pathogen of horses in Ukraine in the early 1930s. After the ingestion of hay contaminated with *Stachybotrys*, horses developed lip edema, stomatitis, oral necrosis, rhinitis, conjunctivitis, coagulopathy, hemorrhage, and neurologic disorders (irritability, gait disturbance, and blindness). So-called superinfections occurred and deaths were observed. The disease syndrome is referred to as stachybotrytoxicosis. In a rare "atypical" or "shocking" form, the disease was primarily neurological and highly fatal, with areflexia (loss of sensorimotor reflexes), hyperesthesia (hypersensitivity to pain), hyperirritability, blindness, and stupor. Cattle were also affected to a lesser extent, and younger animals fared better than older one. Similar diseases have been reported in other parts of the world. A case of sheep disease was described in the 1990s after animals consumed heavily contaminated grain cubes in South Africa. The affected animals displayed fever, listlessness, oral lesions, pancytopenia, hemorrhage, opportunistic infections, and a significant mortality rate [4,26–28].

In areas of enzootic equine stachybotrytoxicosis, fodder handlers and others with close contact with musty straw (e.g., using straw for fuel or bedding) also developed a dermatologic and respiratory syndrome. Dermatologic symptoms were dermatitis on the scrotum, medial thighs, axilla, the hands, and other areas, which progressed from hyperemia to crusting exudates to necrosis, with subsequent resolution. Some patients showed erosions on the oral and gingival mucosa. Respiratory symptoms included catarrhal angina, bloody rhinitis, cough, throat pain, chest tightness, and occasional fever. Some patients had transient leukocytopenia (reduced white blood cell count) and hemorrhage (bleeding). *S. chartarum* (*S. alternans*) isolated from straw produced areal fructifications in experimental rabbits by a dermal toxicity test and similar local and systemic responses on the skin of volunteers [4].

Since then, other clinical manifestations have been associated with exposure to *Stachybotrys chartarum* mycotoxins and spores, ranging from (1) chronic fatigue or headaches, (2) fever, (3) irritation to the eyes, mucous membranes of the mouth, nose, and throat, (4) sneezing, (5) rashes, (6) chronic coughing, (7) nausea, (8) memory loss, (9) vomiting, (10) bleeding in the lungs and nose, (11) hypersensitivity pneumonitis (HP), (12) allergic rhinitis, and (13) asthma exacerbations. Occupants of mold-contaminated, water-damaged buildings often develop symptoms in the central nervous system and the immune system as well as pulmonary diseases, allergy, and inflammatory reactions. Occupational stachybotrytoxicosis acquired by inhalation showed chest and upper airway symptoms, fever, leucopenia, and dermatitis, which started within 2–3 days of exposure and lasted for 3 weeks [18]. In a recent investigation of 32 patients with symptoms attributed to mold exposures at work, 25 (88%) patients had *S. chartarum* as well as *Aspergillus* and *Penicillium*. 79%, 70%, and 64% of the 32 patients presented with cough, shortness of breath, and chest tightness, respectively, which persisted more than 6 weeks in 91%, suggestive of sick-building syndrome. Thirty one percent of these patients had positive skin test to fungal extracts, suggesting IgE-mediated or other nonimmune mechanisms could be the cause of their symptoms [25,29].

61.1.2.2 Pathogenesis

Contemporary construction methods that use cellulose substrates (e.g., fiber board) favor the growth of cellulolytic fungi such as *S. chartarum*. The fungus is shown to produce trichothecene mycotoxins as mentioned above, although several other fungi (e.g., *Fusarium*, *Myrothecium verrucaria*, *Myrothecium roridum*, *Trichothecium*, *Trichoderma*, *Cephalosporium*, *Verticimonosporum*, and *Cylindrocarpon*) also synthesize these compounds. *S. chartarum* macrocyclic trichothecene mycotoxins (MTM) are dissociated readily from the surface of the organism and are consequently spread in damp buildings. *S. chartarum* MTM remain toxic over extended periods of time, and individuals with exposure to the fungus contain MTM in their sera [24]. High indoor exposures to trichothecene mycotoxins are associated with infrequent ventilation or vacuuming of the building, pets, visible mold, and old carpets [18].

Trichothecenes are potent translational inhibitors and stress kinase activators. Experimental exposure of trichothecene mycotoxins in mice led to severe intra-alveolar, bronchiolar, and interstitial inflammation. In experimental Wistar rats, decreased viability of alveolar macrophages and increased activity of the lysosomal enzyme cathepsin D in bronchoalveolar lavage cells after *S. chartarum* exometabolite exposure were noticeable [30]. *S. chartarum* metabolites suppress red blood cells (RBC); decrease the total RBC count, hemoglobin, and hematocrit; and increase total bronchoalveolar lavage fluid cell count (indicating inflammation, lower alveolar macrophage counts, and increased granulocyte count related to the BALF cells) in these animals [31].

Satratoxin-positive *S. chartarum* activates inflammasome-associated caspase-1, which is needed for proteolytic processing of IL-1beta and IL-18, in human macrophages. In addition, purified trichothecene mycotoxins, roridin A, verrucarin A, and T-2 toxin activate caspase-1 and strongly enhance LPS-dependent secretion of IL-1beta and IL-18. Satratoxin-positive *S. chartarum* and the trichothecenes it produces also trigger the activation of caspase-3, which is an effector caspase of apoptosis. Thus, human macrophages sense trichothecene mycotoxins as a danger signal, which activates caspase-1 and further enables the secretion of IL-1beta and IL-18 from the LPS-primed cells [32].

Several enzymes from *S. chartarum* spores demonstrate proteolytic activity and are able to hydrolyze gelatin and collagen I and IV [33]. *S. chartarum* spore extracts induce high levels of IL-6, IL-8, and TNF-alpha in human tracheal epithelial cells. This stimulation of cytokine production is abolished by a serine protease inhibitor Pefabloc. Thus, proteinases from *S. chartarum* spores significantly contribute to lung inflammation and injury [34]. *S. chartarum* spores are capable of inducing both apoptosis and necrosis in THP-1 cells, leading to cell death within 3–6h. More specifically, *S. chartarum* spores increase the formation of reactive oxygen species (ROS) and oxidative DNA damage. *S. chartarum* trichothecenes T-2 toxin and satratoxin G are mainly responsible for apoptosis [35].

61.1.3 Diagnosis

S. chartarum is a cellulose-decaying fungus that grows well at room temperature and with humidity above 93%. Isolation of *S. chatarum* requires special media with high concentration of cellulose and low concentration of sugar and nitrogen to compete with *Penicillium* and *Aspergillus*. Growing *S. chatarum* isolates on rice or potato dextrose agar results in higher proteolytic activity of the spores than those grown on drywall.

Thin-layer chromatography, high-performance liquid chromatography (HPLC), tandem mass spectrometry (MSMS), gas chromatography (GC), GC combined with mass spectrometry (GC-MS, GC-MSMS), and enzyme-linked immunosorbent assay (ELISA) have been applied to detect *S. chartarum* mycotoxins in mold-affected materials inside buildings, in carpet dust from water-damaged buildings, and in animal tissues [36–41]. Finally, various reports have demonstrated the potential usefulness of molecular biological techniques for detection of molds in air and clinical specimens [42–45].

61.2 METHODS

61.2.1 Sample Preparation

61.2.1.1 Trichothecene Collection and Analysis

Airborne trichothecene mycotoxins are collected using a SpinCon PAS 450-10 bioaerosol sampler (Sceptor Industries) and an Andersen GPS-1 polyurethane foam (PUF) high-volume air sampler (Thermo Electron Corporation). Entrained solids are concentrated in a phosphate-buffered saline (PBS) solution (pH 7.4) to a final volume of 10mL [46].

Following the collection, SpinCon samples (all 10mL) are filtered using Fisher 13mm-diameter nylon syringe filters with a 0.45 μm pore size (Fisher Scientific). The filtered fluid is transferred asceptically to 15mL polypropylene conical centrifuge tubes, frozen at −80°C, and lyophilized using a VirTis Freezmobile (SP Industries). The dried samples are individually resuspended in 1mL of pyrogen-free water (25°C) for immediate testing.

Filters obtained from the Andersen PUF sampler are transferred individually to 50mL polypropylene centrifuge tubes on-site. The filters are suspended in 40mL of PBS, vortexed vigorously for 60s, removed from the tubes using sterile forceps, and then discarded. The PBS extracts are filtered into new 50mL tubes. These are frozen at −80°C, lyophilized, and resuspended in 1mL pyrogen-free water for immediate testing.

Samples are analyzed for macrocyclic trichothecenes using a QuantiTox kit for trichothecenes (EnviroLogix). This competitive ELISA kit incorporates trichothecene-specific antibodies immobilized in polystyrene microtiter wells and is highly specific for MTM of *S. chartarum*. To ensure that the ELISA runs correctly, the macrocyclic trichothecene roridin A is used at a concentration of 50ng/mL in PBS as a positive control for each set of tests. PBS alone is used as a negative control [39].

Conidia and other airborne particulates are collected on glass microscope slides that have been coated with a thin layer of petroleum grease using a foam makeup applicator. For testing purposes, conidia are collected from plates that have reached confluence (approximately 7–14 days), using sterile cotton swabs. To collect the conidia, swabs are gently rolled over the surface of the fungal growth. The cotton tips of the swabs are placed in 1mL of sterile room-temperature (25°C) PBS in 1.5mL microcentrifuge tubes and vortexed for approximately 1min to remove conidia. The conidia are then counted using a hemacytometer and identified to the genus level by a trained technician using an Olympus BH2-RFCA optical light microscope. For ELISA testing, the fungal spore suspensions are centrifuged at 14,500rpm for 1min to pellet the conidia. Care is taken not to disturb the conidium pellet, and only the top 80% of the supernatant is used for ELISA testing. Each sample is run in duplicate wells on two separate occasions [39].

61.2.2 Detection Procedures

Pounder et al. [45] described a real-time PCR with SYBR green DNA-binding dye and amplicon-melting temperature analysis for fungal detection using pan-fungal primers ITS1 forward (5′-TCCGTAGGTGAACCTGCGG-3′) and ITS4 reverse (5′-TCCTCCGCTTATTGATATGC-3′). The identity of the fungi is verified by subsequent sequence analysis.

Procedure

1. The PCR mixture is composed of 1× Lightcycler FastStart DNA Master Hybridization Probes mixture (Roche Applied Science) containing deoxynucleoside triphosphates, FastStart *Taq* DNA polymerase, and 1 mM $MgCl_2$ (additional $MgCl_2$ is added to a final concentration of 4.6 mM), 0.4 µM each of ITS1 forward and ITS4 reverse primers, 1× SYBR green (Molecular Probes), and 3 µL template DNA.
2. Thermal cycling parameters are 95°C for 10 min; 50 cycles of 95°C for 5 s, 60°C for 20 s, and 76°C for 30 s; and a final extension at 72°C for 2 min.
3. The quality of the amplicon is determined using the derivative of the melt analysis curve (55°C–99°C, 45-s hold at 55°C, 5 s/°C) using the RotorGene 3000 (Corbett Robotics, Inc).
4. The amplified product is purified for bidirectional sequencing using ExoSAP-IT (USB Corp). Five microliters of Big Dye Terminator Ready Reaction Mix v. 1.1 (Applied Biosystems) is added to 4 µL of each primer (0.8 pmol/µL) and 3 µL of purified PCR product. Cycle sequencing is performed with a 9700 thermal cycler (ABI), using 25 cycles of 96°C for 10 s, 50°C for 5 s, and 60°C for 4 min. Sequencing reaction products are passed through a Sephadex G-50 fine column to remove unincorporated dye terminators. Purified sequencing reaction products are run on an ABI Prism 3100 Genetic Analyzer with a 50 cm capillary array.
5. Sequences are analyzed with the SmartGene Integrated Database Network software version 3.2.3 vr. SmartGene is a web-based software and database system with reference sequences derived from the National Center for Biological Information (NCBI) GenBank repository.

Note. If real time PCR instrument is not available, standard PCR may be performed with primers ITS1 and ITS4, and the resulting amplicon is sequenced using the same primers. Sequence-based identifications are defined by percent identity: species, ≥99%; genus, 93%–99%; and inconclusive, ≤93%. For strains producing discrepant identification between the methods based on phenotypic characteristics and ITS sequence analysis, the D1–D2 region of the large-subunit rRNA gene is amplified with primers NL1 (5′-GCATATCAATAAGCGGAGGAAAAG-3′) and NL4 (5′-GGTCCGTGTTTCAAGACGG-3′) and sequenced for species clarification [47].

61.3 CONCLUSIONS

The genus *Stachybotrys* consists of large number of cellulose-decaying fungal species that are distributed in soil, plants, air, and water-damaged buildings [48]. These fungi produce several mycotoxins that are potent translational inhibitors and stress kinase activators, leading to diseases in humans. Along with biochemical procedures, molecular techniques have been developed to detect the fungus and its mycotoxins [49].

REFERENCES

1. The UniProt Consortium. Available at http://www.uniprot.org/, accessed on August 1, 2010.
2. Haugland, R. A., S. J. Vesper, and S. M. Harmon, 2001. Phylogenetic relationships of *Memnoniella* and *Stachybotrys* species and evaluation of morphological features for *Memnoniella* species identification. *Mycologia.* 93(1):54–65.
3. Li, D.-W. et al., 2003. A new species of *Memnoniella. Mycotaxon.* 135:253–257.
4. Kuhn, D. M. and M. A. Ghannoum, 2003. Indoor mold, toxigenic fungi, and *Stachybotrys chartarum*: Infectious disease perspective. *Clin Microbiol Rev.* 16:144–172.
5. Sorenson, W. G. et al., 1987. Trichothecene mycotoxins in aerosolized conidia of *Stachybotrys atra. Appl Environ Microbiol.* 53:1370–1375.
6. Hinkley, S. et al., 2000. Atranones A-G, from the toxigenic mold *Stachybotrys chartarum. Phytochemistry.* 55:663–673.
7. Bennett, J. W. and M. Klich, 2003. Mycotoxins. *Clin Microbiol Rev.* 16:497–516.
8. Fung, F., R. Clark, and S. Williams, 1998. *Stachybotrys*, a mycotoxin-producing fungus of increasing toxicologic importance. *J Toxicol Clin Toxicol.* 36:629–631.
9. Johanning, E. et al., 1996. Health and immunology study following exposure to toxigenic fungi (*Stachybotrys chartarum*) in a water damaged office environment. *Int Arch Occup Environ Health.* 68:207–218.
10. Etzel, R. A. et al., 1998. Acute pulmonary hemorrhage in infants associated with exposure to *Stachybotrys atra* and other fungi. *Arch Pediatr Adolesc Med.* 152:757–762.
11. Hodgson, M. J. et al., 1998. Building-associated pulmonary disease from exposure to *Stachybotrys chartarum* and *Aspergillus versicolor. J Occup Environ Med.* 40:241–249.
12. Jarvis, B. B. et al., 1998. Study of toxin production by isolates of *Stachybotrys chartarum* and *Memnoniella echinata* isolated during a study of pulmonary hemosiderosis in infants. *Appl Environ Microbiol.* 64:3620–3625.
13. Elidemir, O. et al., 1999. Isolation of *Stachybotrys* from the lung of a child with pulmonary hemosiderosis. *Pediatrics.* 104:964–966.
14. Flappan, S. M. et al., 1999. Infant pulmonary hemorrhage in a suburban home with water damage and mold (*Stachybotrys atra*). *Environ Health Perspect.* 107:927–930.
15. Mahmoudi, M. and M. E. Gershwin, 2000. Sick building syndrome III. *Stachybotrys chartarum. J Asthma.* 37:191–198.
16. Pitt, J. I. et al., 2000. Mycotoxins and toxigenic fungi. *Med Mycol.* 38:41–46.
17. Barnes, C. et al., 2002. IgE-reactive proteins from *Stachybotrys chartarum. Ann Allergy Asthma Immunol.* 89:29–33.
18. Chapman, J.A., 2003. *Stachybotrys chartarum* (chartarum = atra = alternans) and other problems caused by allergenic fungi. *Allergy Asthma Proc.* 24(1):1–7.
19. Miller, J. D., T. G. Rand, and B. B. Jarvis, 2003. *Stachybotrys chartarum*: Cause of human disease or media darling? *Med Mycol.* 41:271–291.
20. Revankar, S. G. 2003. Clinical implications of mycotoxins and *Stachybotrys. Am J Med Sci.* 325:262–274.
21. Hossain, M. A., M. S. Ahmed, and M. A. Ghannoum, 2004. Attributes of *Stachybotrys chartarum* and its association with human disease. *J Allergy Clin Immunol.* 113(2):200–208.

22. Solomon, G. M. et al., 2006. Airborne mold and endotoxin concentrations in New Orleans, Louisiana, after flooding, October through November 2005. *Environ Health Perspect.* 114:1381–1386.
23. Meggs, W. J., 2009. Epidemics of mold poisoning past and present. *Toxicol Ind Health.* 25:571–576.
24. Straus, D. C., 2009. Molds, mycotoxins, and sick building syndrome. *Toxicol Ind Health.* 25:617–635.
25. Al-Ahmad, M. et al., 2010. Symptoms after mould exposure including *Stachybotrys chartarum*, and comparison with darkroom disease. *Allergy.* 65(2):245–255.
26. Nikulin, M. et al., 1996. Experimental lung mycotoxicosis in mice induced by *Stachybotrys atra. Int J Exp Pathol.* 77:213–218.
27. Wilkins, C. K. et al., 1998. Respiratory effects in mice exposed to airborne emissions from *Stachybotrys chartarum* and implications for risk assessment. *Pharm Toxicol.* 83:112–119.
28. Rand, T. G. et al., 2002. Histological, immunohistochemical and morphometric changes in lung tissue in juvenile mice experimentally exposed to *Stachybotrys chartarum* spores. *Mycopathologica.* 156:119–131.
29. Edmondson, D.A. et al., 2009. Immune response among patients exposed to molds. *Int J Mol Sci.* 10:5471–5484.
30. Piecková, E. et al., 2006. Pulmonary cytotoxicity of secondary metabolites of *Stachybotrys chartarum* (Ehrenb.) Hughes. *Ann Agric Environ Med.* 13(2):259–262.
31. Piecková, E. et al., 2009. Inflammatory and haematotoxic potential of indoor *Stachybotrys chartarum* (Ehrenb.) Hughes metabolites. *Arh Hig Rada Toksikol.* 60(4):401–409.
32. Kankkunen, P. et al., 2009. Trichothecene mycotoxins activate inflammatory response in human macrophages. *J Immunol.* 182(10):6418–6425.
33. Mason, C. D. et al., 2001. Effects of *Stachybotrys chartarum* on surfactant convertase activity in juvenile mice. *Toxicol Appl Pharm.* 172:21–28.
34. Yike, I., T. Rand, and D. C. Dearborn, 2007. The role of fungal proteinases in pathophysiology of *Stachybotrys chartarum. Mycopathologia.* 164(4):171–181.
35. Rakkestad, K. E. et al., 2010. DNA damage and DNA damage responses in THP-1 monocytes after exposure to spores of either *Stachybotrys chartarum* or *Aspergillus versicolor* or to T-2 toxin. *Toxicol Sci.* 115(1):140–155.
36. Hinkley, S. F. and B. B. Jarvis, 2001. Chromatographic method for *Stachybotrys* toxins. *Methods Mol Biol.* 157:173–194.
37. Chung, Y.-J. et al., 2003. Immunochemical assay for satratoxin G and other macrocyclic trichothecenes associated with indoor air contamination by *Stachybotrys chartarum. Toxicol Mech Methods.* 13:247–252.
38. Brasel, T. L. et al., 2004. Detection of airborne *Stachybotrys chartarum* macrocyclic trichothecene mycotoxins on particulates smaller than conidia. *Appl Environ Microbiol.* 71:114–122.
39. Brasel, T.L. et al., 2005. Detection of airborne *Stachybotrys chartarum* macrocyclic trichothecene mycotoxins in the indoor environment. *Appl Environ Microbiol.* 71(11):7376–7388.
40. Gregory, L. et al., 2004. Localization of satratoxin-G in *Stachybotrys chartarum* spores and spore-impacted mouse lung using immunocytochemistry. *Toxicol Pathol.* 32:26–34.
41. Bloom, E. et al., 2007. Mass spectrometry-based strategy for direct detection and quantification of some mycotoxins produced by *Stachybotrys* and *Aspergillus* spp. in indoor environments. *Appl Environ Microbiol.* 73:4211–4217.
42. Vesper, S. J. et al., 1999. Hemolysis, toxicity, and randomly amplified polymorphic DNA analysis of *Stachybotrys chartarum* strains. *Appl Environ Microbiol.* 65:3175–3181.
43. Vesper, S. J. et al., 2000. Quantification of siderophore and hemolysin from *Stachybotrys chartarum* strains, including a strain isolated from the lung of a child with pulmonary hemorrhage and hemosiderosis. *Appl Environ Microbiol.* 66:2678–2681.
44. Pietarinen, V. M. et al., 2008. Quantitative PCR analysis of fungi and bacteria in building materials and comparison to culture-based analysis. *J Environ Monit.* 10:655–663.
45. Pounder, J. I. et al., 2007. Discovering potential pathogens among fungi identified as nonsporulating molds. *J Clin Microbiol.* 45(2):568–571.
46. Pasanen, A.-L. et al., 1993. Laboratory experiments on membrane filter sampling of airborne mycotoxins produced by *Stachybotrys atra* corda. *Atmos Environ.* 27A:9–13.
47. Leaw, S. N. et al., 2006. Identification of medically important yeast species by sequence analysis of the internal transcribed spacer regions. *J Clin Microbiol.* 44:693–699.
48. Pestka, J. J. et al., 2008. *Stachybotrys chartarum*, trichothecene mycotoxins, and damp building-related illness: New insights into a public health enigma. *Toxicol Sci.* 104(1):4–26.
49. Vesper, S. J. et al., 2001. Initial characterization of the hemolysin stachylysin from *Stachybotrys chartarum. Infect Immun.* 69:912–916.

62 *Trichoderma*

László Kredics, Lóránt Hatvani, László Manczinger, Csaba Vágvölgyi, and Zsuzsanna Antal

CONTENTS

62.1 INTRODUCTION

62.1.1 Classification, Morphology, and Biology

Trichoderma species are asexual, saprophytic filamentous fungi with teleomorphs belonging to the genus *Hypocrea* (*Ascomycota*, *Pyrenomycetes*, *Hypocreales*, *Hypocreaceae*). Comprehensive historical overviews about the developments in the taxonomy of the genus *Trichoderma* were presented by Druzhinina et al.[1,2] The genus was introduced by Persoon in 1794 for a green fungus growing on fallen branches and other substrates.[3] Bisby[4] suggested that the genus *Trichoderma* consisted of a single species, *T. viride*. Until 1969, most species were referred to as either *T. viride* or *T. koningii*, depending upon whether they produced globose or oval conidia, respectively. In 1969, Rifai[5] revised the taxonomy of the genus *Trichoderma* and described nine "species aggregates" based on the types of branching systems of the conidiophores, the character of phialospores, the phialide disposition, and conidial morphology. In the 1980s and early 1990s, Bissett[6–9] dissected several of Rifai's aggregate species into several defined taxa and proposed an infrageneric classification of *Trichoderma*, which divides the genus into sections *Trichoderma*, *Pachybasium*, *Saturnisporum*, *Longibrachiatum*, and *Hypocreanum*. The introduction of sequence-based molecular methods[10] and multigene phylogenic approaches[11] as well as the application of the genealogical concordance phylogenetic species recognition (GCPSR) concept[12] enabled the revision of the morphology-based taxonomy in *Trichoderma* and the identification of new species. The number of described and confirmed, phylogenetically different *Trichoderma* species has already reached 100.[2]

Macroscopically, *Trichoderma* species can be generally characterized by rapidly growing, initially smooth and translucent or watery white colonies, later becoming floccose, with distinct tufts of conidiophores that frequently form concentric ringlike zones, gradually changing their color from whitish green to light olive green at maturity. Microscopically, conidia occur in balls at the apexes of flask-shaped phialides that arise at angles from the conidiophores. The identification of *Trichoderma* strains at the species level by morphological examinations is problematic. In order to aid the fast and reliable identification of isolates, the online resources *TrichO*key, a DNA-barcode system based on defined nucleotide sequence differences in the ITS1 and ITS2 region,[13,14] and *Tricho*BLAST, a multiloci database of phylogenetic markers for *Trichoderma* and *Hypocrea* powered by sequence diagnosis and similarity search tools[15] have been developed.

Members of the genus *Trichoderma* take part in the decomposition of plant residues in the soil. Certain species, including *T. reesei* are excellent producers of cellulolytic enzymes and can therefore be applied for cellulose degradation in the biotechnological industry.[16] Other *Trichoderma* species (e.g., *T. harzianum*, *T. virens*, *T. atroviride*, *T. asperellum*) are well known as potential biocontrol agents with the ability of antagonizing a series of plant pathogenic fungi.[17] Proposed mechanisms of antagonism include mycoparasitism by the action of cell-wall degrading enzymes, antibiosis by the production of antibiotics, and competition for space and nutrients through rhizosphere competence.[18–21] Facilitation of seed germination and growth of the plants via releasing important minerals and trace elements from soil and induction of the defense responses in plants are further beneficial features of these species. *Trichoderma* spp. generally occur in soils, but certain *Trichoderma* species are also known from special habitats. During the last decades, representatives of the genus have been reported to be harmful as the causative agents of green mold, a disorder that results in substantial losses in the production of cultivated mushrooms including champignon (*Agaricus bisporus*) and oyster mushroom (*Pleurotus ostreatus*).[22–24] Besides mushroom pathogenic *Trichoderma* species, clinically relevant members of the genus including *T. longibrachiatum* and *T. harzianum* were also isolated from compost and substrate samples used for mushroom cultivation.[23] *Trichoderma* species have also been reported as pathogens of reptiles[25,26] and, in increasing number of cases, as etiologic agents in human infections.[27,28] Members of this genus were previously considered to be contaminants or harmless colonizers when isolated from human clinical specimens, but nowadays they are recognized as emerging opportunistic pathogens of humans, especially of immunocompromised hosts.

62.1.2 Epidemiology and Pathogenesis

The number of clinical cases caused by *Trichoderma* species is increasing from year to year, which can be attributed to the increasing number of immunocompromised patients as well as the increasing attention of microbiologists to these fungi that were previously considered to be nonpathogenic. Altogether, 59 cases of isolation of *Trichoderma* strains from clinical samples could be recovered from the literature and culture collection data. Case details available for 36 cases of *Trichoderma* infections[29–62] are summarized in Table 62.1. The first report about a *Trichoderma* infection described the inadvertent infusion of *T. viride* to an immunocompetent host with contaminated intravenous fluid,[58] the patient remained well after 24 h of amphotericin B treatment. Since this initial report, clinical isolation of *Trichoderma* species has been reported from several countries including France,[30,32,40,42,46,48,53,59–61,63] Spain,[34,44,51,64] United Kingdom,[58] Netherlands,[33] Belgium,[41] Italy,[36] Slovenia,[35] Austria,[47,65,66] Hungary,[67] Lithuania,[52] Turkey,[37,54,68] Canada,[57] United States,[29,43,45,49,56] Brasil,[69] Chile,[38] India,[70] Korea,[39] Japan,[50] and Mali.[46] The prognosis for *Trichoderma* infection is usually poor: the outcome of the infection was fatal in 17 out of the 36 cases (47.22%). Case studies reported the involvement of eight species from the genus *Trichoderma* (*T. longibrachiatum*,[33,38–40,43,45,46,48–50,53,57,59,60] *T.* citrinoviride,[52] *T. pseudokoningii*[36,42] and *T. reesei*[61] from section *Longibrachiatum*, *T. harzianum*[32,44,54] from section *Pachybasium* B, *T. koningii*,[30,34] *T. atroviride*[40] and *T. viride*[29,31,41,51,55] from section *Trichoderma*) and a yet undescribed *Hypocreaceae* sp. close to the genus *Hypocrea*/*Trichoderma*[56] in human infections.

For the successful prevention of *Trichoderma* infections, the understanding of the potential sources of contamination and possible routes of infection is needed. Airborne *Trichoderma* conidia were suggested to be possibly related to allergic diseases,[71] and ambient air is also supposed to be the likely source of infection for some cases of opportunistic *Trichoderma* infections (e.g., sinusitis, pneumonia, abscesses following surgery). Based on a review by Madsen et al.[72] *Trichoderma* species could be isolated from air and/or dust samples of human environments including hospital wards, homes, air conditioners, farms, sawmills, herb processing plants, carpentries, flats, roofs of houses, stationary cars, wood chip terminals, buildings heated by wood chips, schools, combine harvesters, and storey buildings. *Trichoderma* species were reported among the most frequently occurring filamentous fungi in certain studies, e.g., in the fungal microbiota of air conditioners in Brazilian intensive care units.[73]

A series of studies suggest that water-related sites represent potential sources of *Trichoderma* infections. Hageskal et al.[74] investigated the diversity of mold species in Norwegian drinking water and found the occurrence of *Trichoderma* species in all samples from all of the sampling points in surface water-derived water systems, the predominant fungi being isolated (72 of 273 isolates, 26.4%). After *A. fumigatus*, *Trichoderma* species were found to be the second most prevalent filamentous fungi cultured from samples from outlets, waste pipes, and shower heads in a bone marrow transplantation ward.[75] In another study, Varo et al.[76] collected 50 samples from seven water distribution points of the water system of a hemodialysis unit, and isolated a total of 116 filamentous fungi, with *Trichoderma* isolates being the most frequent (47 isolates, 40.5%). In a study by Enríquez-Matas et al.,[77] a fungal strain identified as *T. viride* was isolated in cultures of water samples from an ultrasonic humidifier installed in the home of a patient diagnosed with hypersensitivity pneumonitis. Precipitating immunoglobulin G antibodies to *T. viride* were detected in the patient's serum by enzyme-linked immunosorbent assay, suggesting that the hypersensitivity pneumonitis might have been due to *Trichoderma* from the humidifier. *Trichoderma* species are abundant in indoor environments of water-damaged buildings, where they are frequently among the dominating microfungi.[78–81] Besides *T. atroviride*, *T. viride* and *T. hamatum*, the species detected in such environments include the clinically relevant species *T. longibrachiatum*, *T. citrinoviride*, and *T. harzianum*.[79,80] The release of spores from wet wallpapered gypsum boards by nine species of typical indoor fungi including

TABLE 62.1
Case Reports about *Trichoderma* Infections in the Literature

Age/Sex	Clinical Diagnosis	Source	Etiology	Therapy	Outcome	Reference
47/M	CAPD peritonitis	Peritoneal fluid, autopsy	*T. viride*	AMB	Died	[29]
63/F	CAPD peritonitis	Peritoneal fluid	*T. koningii*	Catheter removal, MCZ	Cured	[30]
44/M	CAPD peritonitis	Peritoneal fluid	*T. viride*	AMB	Died	[31]
82/M	CAPD peritonitis	Peritoneal fluid	*T. harzianum*	KTZ, 5FC	Died	[32]
48/M	CAPD peritonitis	Peritoneal fluid, autopsy	*T. longibrachiatum*	AMB	Died	[33]
41/M	CAPD peritonitis	Peritoneal fluid	*T. koningii*	FLZ, 5FC, AMB	Died	[34]
60/M	CAPD peritonitis	Peritoneal fluid	*Trichoderma* sp.	Catheter removal, KTZ	Cured	[35]
33/M	APD peritonitis	Peritoneal fluid	*T. pseudokoningii*	Catheter removal	Cured	[36]
40/M	CAPD peritonitis	Peritoneal fluid	*Trichoderma* sp.	Catheter removal, ITZ, AMB	Died	[37]
13/F	CAPD peritonitis	Peritoneal fluid	*T. longibrachiatum*	FLZ, AMB, catheter removal	Died	[38]
67/M	CAPD peritonitis, intraabdominal abscess	Peritoneal fluid, aspirated fluid	*T. longibrachiatum*	AMB, Catheter removal	Cured	[39]
70/M	Cholecystitis and peritonitis	ND	*T. longibrachiatum*	VRZ, CSP	Died	[40]
44/F	TX/abdominal dissemination	Peritoneal fluid, haematoma	*T. viride*	Surgery, AMB, FLZ	Died	[41]
45/F	TX/lung and skin dissemination	Bronchoalveolar lavage, skin biopsy, lung, brain, heart, stomach	*T. pseudokoningii*	FLZ, AMB, ABLC, 5FC	Died	[42]
29/F	TX/acute invasive sinusitis	Sinus debridement	*T. longibrachiatum*	Surgery, AMB, ITZ	Cured	[43]
68/M	TX/disseminated infection	Brain and lung abscesses, autopsy	*T. harzianum*	—	Died	[44]
29/M	TX/disseminated infection	Lung, liver, intestinal wall, stool, autopsy	*T. longibrachiatum*	AMB, ITZ, ABLC	Died	[45]
63/F	TX/invasive infection	Subcapsular hepatic collection	*T. longibrachiatum*	Surgical debridement	Cured	[46]
11/M	TX/invasive infection	Bronchoalveolar lavage, pleural drains	*T. longibrachiatum*	ABLC	Died	[46]
62/M	TX/disseminated infection	Liver, gastrointestinal tract, lung, heart, kidneys and skin	*Trichoderma* sp. and *Absidia corymbifera*	VRZ	Died	[47]
49/M	TX/liver infection	Liver biopsy	*T. atroviride*	FLZ	Died	[40]
17/F	HM/brain abscess	Brain biopsy, cerebral pus	*T. longibrachiatum*	Surgery, AMB, 5FC, KTZ, ITZ	Cured	[48]
11/M	HM/skin infection	Skin biopsy	*T. longibrachiatum*	AMB, ABLC	Cured	[49]
66/F	HM/necrotizing stomatitis	Ulcerative mucogingiva	*T. longibrachiatum*	AMB, ITZ	Died	[50]
54/F	HM/pulmonary infection	Fine-needle aspiration from pulmonary consolidation	*T. viride*	ABLC, VRZ, CSP	Cured	[51]
49/F	HM/pneumonia	Bronchoalveolar lavage	*T. citrinoviride*	AMB	Cured	[52]
16/M	HM/invasive pulmonary infection	Sputum, bronchoaspiration, bronchoalveolar lavage	*T. longibrachiatum*	VRZ, CSP	Cured	[53]
9/M	HM/invasive fungal infection	Serum, sputum, skin lesions	*T. harzianum*	ABLC	Died	[54]
46/M	Pulmonary mycetoma	Sputum, lung biopsy	*T. viride*	Surgery	Cured	[55]
19/F	Pulmonary fibrosis	Lung biopsy	*Hypocreaceae* sp.	—	Cured	[56]
52/F	Allergic fungal sinusitis	Bilateral endoscopic antral lavage	*T. longibrachiatum*	Sinus lavage, ITZ	Cured	[57]
26/F	Fungemia by contaminated saline	Blood	*Trichoderma* sp.	AMB	Cured	[58]
58/M	Fungemia	Blood	*T. longibrachiatum*	AMB, VRZ, catheter removal	Cured	[59]
12/M	Otitis externa	Ear discharge	*T. longibrachiatum*	Nystatin, polymyxin B	Cured	[60]
61/M	Infection of cerebrospinal fluid	Cerebrospinal fluid, shunt device	*T. reesei*	ABLC, VRZ, CSP	Cured	[61]
66/M	Endocarditis	Aortic conduit	*Trichoderma* sp.	Surgery, antifungal drugs	Cured	[62]

Notes: M, male; F, female; APD, automated peritoneal dialysis; CAPD, chronic ambulatory peritoneal dialysis; 5FC, 5-fluorocytosine; AMB, amphotericin B; ABLC, amphotericin B lipid complex; FLZ, fluconazole; ITZ, itraconazole; KTZ, ketoconazole; MCZ, miconazole; VRZ, voriconazole; CSP, caspofungin; TX, transplant; HM, hematological malignancy; ND, no data available.

T. harzianum has been measured under controlled conditions by Kildesø et al.[82] For *T. harzianum*, it appeared that many particles were released in agglomerates of conidia due to the mucous layer covering the surface of the spores. Released fragments, smaller than the conidia, were also found in the case of *T. harzianum*. The authors suggested that these finer particles could be responsible for some of the adverse health effects seen in moldy buildings, especially, as the small particles are much easier transported from the building envelope to the breathing zone.

Trichoderma species have also been isolated from food,[83,84] which may explain positive stool culture results.[45,67] The catheter may be a portal of entry for invasive[42,49] and peritoneal dialysis-related *Trichoderma* infections.[29–39] Contaminated infusion products can also lead to disseminated infections when infused intravenously.[58]

Antal et al.[85] studied the potential virulence factors of 9 saprophytic and 12 clinical *T. longibrachiatum* strains—including strains from previously published case reports[34,42,43,45,49]—in order to compare their capacity to cause infection in humans. All of the examined strains were able to grow at temperatures ranging from 10°C to 40°C with an optimum at 30°C, and at pH values ranging from 2.0 to 9.0 at 25°C with an optimum at pH 4. Besides *T. longibrachiatum*, other species belonging to the *Longibrachiatum* section of the genus (including *H. orientalis*) are also capable of growing at 37°C. Because growth at elevated temperatures is one of the virulence factors of fungi, it is not surprising that most of the strains involved in *Trichoderma* infections belong to the *Longibrachiatum* section of the genus. All of the clinical strains were able to grow at physiological pH, which is another prerequisite of growth in human hosts. Carbon and nitrogen source utilization experiments revealed that all of the strains were able to utilize a series of basic amino acids both as sole carbon and nitrogen sources. The growth of different bacteria was inhibited by metabolites of the strains. Compounds produced by three clinical isolates also reduced the motility of boar spermatozoa, indicating their toxicity to mammalian cells. Exposure to the fungal extracts of these clinical isolates resulted in quenching of the yellow fluorescence of the midpiece of boar spermatozoa stained with JC-1/propidium iodide, suggesting that the dissipation of the mitochondrial membrane potential may be involved in the toxic effect. There were no significant differences in the examined features between *T. longibrachiatum* strains derived from clinical or soil samples.[85]

Evidences are available for the possible involvement of proteolytic enzymes in aspergillosis, coccidioidomycosis, and sporotrichosis[86]; therefore, it can be supposed that extracellular proteolytic enzymes may be involved in the pathogenicity of *Trichoderma* strains as opportunistic pathogens of humans. Kredics et al.[87] screened supernatants from induced liquid cultures of six clinical *T. longibrachiatum* isolates—including strains from previously published case reports[43,45,49]—for proteolytic enzyme activities with 11 different chromogenic *p*-nitroanilide substrates. The production of trypsin-like, chymotrypsin-like, and chymoelastase-like protease activities proved to be common among the examined strains. Separation of trypsin- and chymotrypsin-like activities by Sephadex G-100 column chromatography revealed that both systems are complex consisting of several isoenzymes. The pH-dependence of these two protease systems was also studied and the results suggested that the different isoenzymes have different optimal pH values.

Schmoll et al.[88] applied the method of rapid subtraction hybridization (RaSH) for the detection of *T. longibrachiatum* genes differentially expressed between a clinical *T. longibrachiatum* isolate grown in confrontation with bronchial epithelial cells, and the same strain grown in bronchial epithelial growth medium alone. Among the genes found to be upregulated under conditions of simulated infection there was a cyanoviridin-N-homologue for homologues of which an antiviral effect has been shown, a putative extracellular DNase for which bacterial homologues with virulence promoting function are known, a member of the protein family of mannosyltransferases, which are required for virulence in *Candida albicans* as well as genes putatively involved in ATP/ADP translocation and ferric iron uptake.

62.1.3 Clinical Features

62.1.3.1 Trichoderma Infections in Patients Undergoing Peritoneal Dialysis

Twelve cases of *Trichoderma* peritonitis have been reported in details in the case of patients requiring peritoneal dialysis due to end-stage renal failure. The reported species include *T. longibrachiatum*,[33,38–40] *T. pseudokoningii*,[36] *T. harzianum*,[32] *T.* koningii,[30,34] *T. viride*[29,31] as well as unidentified *Trichoderma* species.[35,37] A previous history of bacterial peritonitis was frequent. Reported symptoms included abdominal pain and fever. The fungi were isolated from the peritoneal fluid in all of these cases. Therapeutical interventions involved an early catheter removal[30,35–39] as well as the administration of antifungals (amphotericin B, ketoconazole, fluconazole, miconazole, itraconazole, voriconazole, caspofungin, 5-fluorocytosin). Only four of the patients survived.[30,35,36,39] Further two peritonitis-related clinical isolates were deposited at the University of Alberta Microfungus Collection and Herbarium (UAMH): *T. longibrachiatum* UAMH 9515 from the peritoneal effluent of a female and *Hypocrea orientalis* UAMH 9573 from a peritoneal catheter tip (deposited as *T. citrinoviride*).

62.1.3.2 Trichoderma Infections in Organ Transplant Recipients

Immunocompromised solid-organ transplant recipients represent another important risk population of *Trichoderma* infection. An infection of a perihepatic haematoma reported to be due to *T. viride* occurred in a liver transplant recipient.[41] The causal agent persisted in the immunocompromised host despite amphotericin B treatment and surgical removal of the infected haematoma; however, the patient died of unrelated complications. A fatal infection caused by a strain identified as *T. pseudokoningii* was described after allogeneic bone marrow transplantation in a patient with acute erythroleukemia.[42] The fungus was obtained from bronchoalveolar lavage and skin biopsy specimens at the site of insertion of an intravenous catheter. Autopsy revealed a disseminated infection with hyphae present in the lungs, pretracheal abscesses, stomach, heart, and brain. The acute invasive sinusitis caused by *T. longibrachiatum* in a patient 6 months after small bowel and liver transplantation could be successfully treated with endoscopic sinus operations for debridement, amphotericin B irrigation of the maxillary sinus, and oral itraconazole treatment for a total of 6 months.[43] A disseminated *T. harzianum* infection was diagnosed in the postmortem examination of a renal transplant recipient, the fungus was recovered from lung tissue microabscesses and mycotic brain lesions.[44] The gastrointestinal tract (GIT) was suggested as the possible portal of entry in the case of a fatal, disseminated *T. longibrachiatum* infection in an allogenic bone marrow transplant patient with acute lymphoblastic leukemia, as the causal agent could be recovered from stool surveillance cultures and a perirectal ulcer biopsy specimen.[45] Chouaki et al.[46] reported further two cases of invasive *T. longibrachiatum* infections in solid-organ transplant recipients. In the case of a patient who received liver transplant due to hepatitis C virus-induced cirrhosis, a sample from a subcapsular collection related to suture threads and biopsy specimens of perilesional tissues yielded pure cultures of *T. longibrachiatum*. Concomitant surgical debridement and local administration of povidone iodine resulted in total recovery. The other case of *T. longibrachiatum* infection was fatal in a cystic fibrosis patient who underwent pulmonary transplantation due to terminal respiratory failure. The fungus could be isolated from samples obtained from the transcutaneous tracheal puncture and bronchoalveolar lavage fluid. Stelzmueller et al.[47] diagnosed postmortem a disseminated double-infection with a *Trichoderma* sp. and *Absidia corymbifera* involving the liver, GIT, lung, heart, kidneys, and skin in a liver transplant recipient who died from progressive graft-versus-host disease, sepsis, and multiorgan failure. Ranque et al.[40] reported the isolation of *Trichoderma atroviride* from a liver biopsy specimen of a liver transplant recipient. Further clinical *Trichoderma* isolates from transplant recipients were deposited at the Canadian Collection of Fungal Cultures (CCFC): *T. pseudokoningii* CCFC 007753 and CCFC 007754 isolated from a bone marrow transplant recipient and from a liver and bowel transplant recipient, respectively; at the National Reference Center of Mycoses and Antifungal Drugs at the Institute Pasteur (IP): IP-94 0958 isolated together with *Aspergillus fumigatus* from a patient who had a lung transplantation and IP-93 1282 isolated from a bronchoalveolar washing in a 2 month bone marrow recipient[63]; and at the Centro Nacional de Microbiologia, Immunologia y Virologia Sanitaria (CNM-CM): *T. longibrachiatum* CNM-CM 1798 from the blood culture of a patient with liver transplant.

62.1.3.3 Trichoderma Infections in Patients with Hematologic Malignancies

Hematologic malignancies are also among the diseases predisposing to *Trichoderma* infections. A brain abscess due to *T. longibrachiatum* corresponding to a contiguous infection from an ethmoidal invasive sinusitis in a leukemic patient with prolonged neutropenia was successfully treated with extremely prolonged antifungal therapy including amphotericin B, 5-fluorocitozin, ketoconazole, and itraconazole as well as with the neurosurgical resection of the abscess.[48] A pediatric patient with severe aplastic anemia and neutropenia developed an invasive *T. longibrachiatum* skin infection, which was successfully treated with intravenous amphotericin B therapy.[49] A case of *T. longibrachiatum* stomatitis was reported by Myoken et al.[50] in a neutropenic patient with malignant lymphoma. The infection rapidly disseminated from the oral mucosa to the lungs with a fatal outcome despite the intensive antifungal therapy with amphotericin B and itraconazole. De Miguel et al.[51] reported a pulmonary *T. viride* infection in an adult patient with acute myeloid leukemia. The fungus was isolated from pulmonary aspirate. The patient was treated unsuccessfully with liposomal amphotericin B, which was then replaced with voriconazole and caspofungin, resulting in successful therapy. In another case of pneumonia in an acute myeloid leukemia patient, *T. citrinoviride* was isolated from the bronchoalveolar lavage.[52] The infection was successfully cured with amphotericin B therapy. In a patient with B cell acute lymphoblastic leukemia, Alanio et al.[53] reported an invasive lung infection by *Trichoderma*. The cultures of sputum, bronchoaspiration, and bronchoalveolar lavage fluid samples yielded *T. longibrachiatum*. The infection was successfully treated with a combination of voriconazole and caspofungin. A fatal *T. harzianum* infection was reported in another patient with acute lymphoblastic leukemia.[54] The causal agent could be recovered from samples of blood serum, skin lesions, sputum, and throat of the patient. Further clinical isolates from patients with hematologic malignancies include *T. longibrachiatum* IP-92 0647 from a patient with acute leukemia,[63] *T. citrinoviride* IP-95 1151 from blood cultures of a patient in aplasia associated with lymphoma,[63] and two *H. orientalis* isolates: one from the blood culture of a 3-year-old child with acute lymphoblastic leukemia and another one from the stool of a 15-year-old child with non-Hodgkin lymphoma.[67] These two *H. orientalis* isolates derived from patients of the same pediatrics clinic, thus an epidemiological connection could not be ruled out.

62.1.3.4 Trichoderma Infections of the Respiratory Tract and Allergic Reactions

T. viride was reported from sputum and lung biopsy specimens of a patient with pulmonary mycetoma, which was cured by surgical resection.[55] Druzhinina et al.[56] reported the isolation of an unknown *Hypocreaceae* species close to the genus *Hypocrea/Trichoderma* from the lung tissue of a patient with nonfatal pulmonary fibrosis, who improved dramatically after receiving bronchodilators. The clinical and mycological investigations did not determine whether this fungus did indeed cause the infection or contribute to pulmonary fibrosis. Based on culture collection data, the isolation of *T. longibrachiatum* is known from the sputum of a tuberculosis patient (CNM-CM 2277) and from a lung of a man who died (*T. longibrachiatum* CBS 446.95), while *T. citrinoviride* was isolated from a patient suffering from chronic bronchitis with fever and respiratory infection (CNM-CM 1792, deposited as *T. viride*, reidentified as *T. citrinoviride* by Kredics et al., unpublished data).

Tang et al.[57] described a case of allergic fungal sinusitis associated with *T. longibrachiatum* in a patient with a history of atopy and asthma, the case was successfully managed with a combination of sinus lavage, oral corticosteroids, itraconazole, and allergen immunotherapy. Further, *T. longibrachiatum* strains were isolated from infected maxillar sinus (strain IP-94 1510[63]) and from the sinus lavage sample of a rhinosinusitis patient.[67] Two strains of *T. viride*[65] and a strain of *T. inhamatum*[66] were reported from the nasal mucus of chronic rhinosinusitis (CRS) patients; however, a great number of other fungi were found to be associated with CRS and it is not known which of them were responsible for the eosinophilic reaction.

Trichoderma species may also be present as part of the human microbiota: strains were isolated from sputum (*T. koningii* UAMH 473 and UAMH 475), from sinus ethmoidalis (*T. pseudokoningii* CBS 500.94), and from the nasal mucus (*T. viride*) of healthy patients.[66]

Some *Trichoderma* species have been found to be allergenic.[89] Vesper et al.[90] have found that *T. viride* had significantly higher concentrations in the water-damaged homes of asthmatic children compared with control homes. The possible involvement of *Trichoderma* species in eosinophil activation has also been demonstrated: precipitating antibodies against *T. viride* were detected in the sera of patients with acute eosinophilic pneumonia.[91,92] Volatile organic compounds (VOC) from a *Trichoderma* isolate identified as *T. viride* were found to potentiate IgE-related histamine release from human bronchoalveolar cells.[93] The authors suggested that VOC of *Trichoderma* might cause adverse health implications via mediator release, inflammation, and neural stimulation. Besides the visible growth of fungi, growth hidden behind walls and ceilings might also represent a risk, as VOC penetrate the polyethylene vapor barrier. *T. longibrachiatum* has been reported to be one of the most prevalent molds in the manufacturing process of stoppers,[94] although no evidence of allergic sensitization by *T. longibrachiatum* in patients with asthma in the cork industry could be demonstrated.[95] Jaakkola et al.[96] conducted a population-based incident case–control study to assess the risk of asthma in relation to specific IgG antibodies to eight dampness-related microbes (*Aspergillus fumigatus*, *A. versicolor*, *Cladosporium cladosporioides*, *Fusarium oxysporum*, *Sporobolomyces salmonicolor*, *Stachybotrys chartarum*, *Streptomyces albus* and *T. citrinoviride*). An increased risk of developing asthma in adulthood was significantly related to IgG antibodies to *T. citrinoviride* but not to the other molds, suggesting that *T. citrinoviride* may play a role in the etiology of adult-onset asthma or serve as an indicator of other causal factors. In another study, doctor-diagnosed asthma among school children was also strongly associated with high IgG levels to *T. citrinoviride*.[97] In a lung epithelial cell assay, *T. harzianum* was shown to be an inflammogen.[98] On the other hand, short-term human exposure to *T. harzianum* did not result in more clinical effects than placebo exposure.[99]

62.1.3.5 Occurrence of Trichoderma Associated with HIV Infection

The clinical occurrence of *Trichoderma* may also be associated with human immunodeficiency virus (HIV) infection, as suggested by the isolation of *Trichoderma* sp. from the cerebrospinal fluid of an AIDS patient,[69] the isolation of *T. longibrachiatum* from a HIV-positive patient (deposited in the American Type Culture Collection as strain ATCC 208859[100]), and *Trichoderma* fungaemia in a patient with pulmonary cancer, chemotherapy-induced neutropenia, and HIV infection.[59]

62.1.3.6 Miscellaneous Cases of *Trichoderma* Infections

The involvement of *Trichoderma* species is also known from two cases of mycotic keratitis,[46,70] a case of onychomycosis,[68] a case of otitis externa,[60] a case of infection of cerebrospinal fluid shunt device,[61] and a case of endocarditis.[62] Further known clinical *Trichoderma* isolates include *T. longibrachiatum* strain IP-96 0086 from a hematic wound of an otherwise healthy patient,[63] strain IP-93 1792 obtained from a cerebrospinal derivative catheter,[63] IP-97 0711 from the liquid of a chylothorax,[63] and *T. longibrachiatum* CNM-CM 2171 from the foot skin of a premature infant with subcutaneous lesions.[85]

62.1.3.7 Antifungal Susceptibilities of Clinical *Trichoderma* Isolates

Antifungal susceptibility testing of fungi involved in opportunistic infections is relevant for the choice of adequate therapy, as the susceptibility of the organisms to antifungal agents is an important factor affecting the outcome of the infection. Data available about the susceptibilities of clinical *Trichoderma* isolates are summarized in Table 62.2. Different methods were applied for the determination of the minimal inhibitory concentration (MIC) values of the antifungal agents in the available studies.[101–109] Table 62.2 reflects that most of the clinical *Trichoderma* isolates are resistant to fluconazole, 5-fluorocytosine, and amphotericin B, and susceptible or intermediate to itraconazole, ketoconazole, miconazole, and voriconazole; however, high MIC-levels of itraconazole[44,50,60,85] and ketoconazole[44,52,56] and voriconazole[40] were also reported in certain cases. Based on the available susceptibility data, voriconazole might be an important drug in the treatment of *Trichoderma* infections.[50,51,53,54,56,59,111–115]

TABLE 62.2
Antifungal Susceptibilities of Clinical *Trichoderma* Isolates (Minimum Inhibitory Concentrations in mg/L)

Isolate	Method	AMB	5FC	FLZ	ITZ	KTZ	MCZ	PSZ	VRZ	RVZ	TRB	CSP	AND	AP	CHG	Reference
T. longibrachiatum	Broth microdilution[104]	2		64					0.5					0.5	2	[113]
UAMH 7955	Disk method								*7 mm							[111]
	Etest[102]	1		>256	32	0.125										[85]
	Macrobroth dilution[101]	1.16	>322.75	80	0.3											[43]
T. longibrachiatum	Broth micodilution[103]	2.0	>256	16	1.0											[45]
UAMH 7956	Broth microdilution[104]	2		64					0.5					0.125	2	[113]
	Disk method								*9 mm							[111]
	Etest[102]	2		64	16	0.25										[85]
T. longibrachiatum	Broth microdilution[104]	2		64					0.5					0.125	2	[113]
UAMH 9515	Disk method								*9 mm							[111]
	Etest[102]	2	>256		16	0.5										[85]
T. longibrachiatum	Both macrodilution	2	>64	>64	2											[49]
ATCC 201044	Broth microdilution[104]	2		128					1					0.25	4	[113]
	Etest[102]	2		64	0.5	0.008										[85]
T. longibrachiatum	Broth microdilution[104]	2		128					0.5					0.5	2	[113]
ATCC 208859	Etest[102]	2		>256	8	0.25										[85]
T. longibrachiatum	Broth dilution	2.5	50	12.5	1.25											[48]
IP-93 1192	Disk method					*34 mm										[48]
T. longibrachiatum	Broth dilution	5														[33]
T. longibrachiatum	EUCAST[107]	0.5	>64	>64	>8				1			0.5 (MEC)				[53]
T. longibrachiatum	EUCAST[107]	1		>64	2			4	0.5			0.5				[59]
T. longibrachiatum	Broth microdilution	0.5	>32	>64	>32				0.5							[50]
T. longibrachiatum	Broth microdilution[104]	2		128					1					0.25	8	[113]
IP 2110.92	Broth microdilution[109]	0.09	>100	25	0.18		0.09									[42]

(continued)

TABLE 62.2 (continued)
Antifungal Susceptibilities of Clinical *Trichoderma* Isolates (Minimum Inhibitory Concentrations in mg/L)

Isolate	Method	AMB	5FC	FLZ	ITZ	KTZ	MCZ	PSZ	VRZ	RVZ	TRB	CSP	AND	AP	CHG	Reference
(originally identified as *T. pseudokoningii*)[42]	Etest[102]	2		>256	16	1										[85]
T. longibrachiatum	Broth microdilution[104]	2		64					0.5					0.125	2	[113]
CBS 446.95	Etest[102]	2		>256	2	0.5										[85]
T. longibrachiatum	Broth microdilution[104]	2		128					0.5					0.125	2	[113]
CNM-CM 382	Broth microdilution[108]	4.0	16.0	128.0		1.0	≤0.25									[34]
(originally identified as *T. koningii*)[34]	Disk method								*8 mm							[111]
	Etest[102]	8		>256	2	0.25										[85]
T. longibrachiatum	Broth microdilution[104]	2		64					1					0.25	2	[113]
CNM-CM 1798	Etest[102]	0.016		>256	1	0.125										[85]
T. longibrachiatum	Broth microdilution[104]	2		128					1					0.06	2	[113]
CNM-CM 2171	Etest[102]	2		>256	16	0.5										[85]
T. longibrachiatum	Broth microdilution[104]	2		256					1					0.06	1	[113]
CNM-CM 2277	Etest[102]	2		>256	32	0.25										[85]
T. longibrachiatum (5 isolates)	Broth microdilution[103]	0.5–2			>8				2							[114]
	Etest[102]	1.0–4			>8											[110]
T. longibrachiatum (3 isolates)	Broth dilution[104]	0.5–2			>8			>8	2	>8						[115]

T. longibrachiatum (2 isolates)	NCCLS M-38P[103]	1–2			16				2						[46]
H. orientalis	Broth microdilution[104]	2		128					0.5			0.25		1	[113]
UAMH 9573	Disk method								*9 mm						[111]
	Etest[102]	2		64	32	0.25									[85]
T. citrinoviride	NR	>32			>32	>32			4		0.125				[52]
T. reesei	Etest[102]	1.5			>32				1						[61]
T. harzianum	Broth microdilution[104]	2		1024					1				0.25	8	[113]
CBS 102174	Microdilution[106]	2	256	128	32	8	8								[44]
T. harzianum	Broth macrodilution[104]	>16		0.125	<0.03	<0.03	<0.03	<0.03	<0.03	<0.03					[54]
T. atroviride	EUCAST[107]	1	>64	>64	>8				8		0.5				[40]
T. viride	Broth dilution	0.78	100			1.56									[29]
T. viride	Broth microdilution[104]	0.25			8				2						[51]
T. viride	Broth dilution	3.1	>50	25	1.6	0.8									[41]
Trichoderma spp. (4 isolates)	Broth microdilution[104]									0.06–2					[64]
Trichoderma sp. (1 isolate)	Broth microdilution[105]	1	>128		2			1	0.25		0.25	0.06			[112,116,117]
Trichoderma sp. (1 isolate)	Broth microdilution[104]											32			[118]
												0.125 (MEC)			
Hypocreaceae sp.	Etest[102]	0.125	>32	>256	16	>32		>32	1		>32				[56]

Notes: *, diameter of inhibition zone; NR, not reported; MEC, minimum effective concentration; Antifungal agents; 5FC, 5-fluorocytosine; AMB, amphotericin B; FLZ, fluconazole; ICZ, itraconazole; KTZ, ketoconazole; MCZ, miconazole; VRZ, voriconazole; PSZ, posaconazole; RVZ, ravuconazole; TRB, terbinafine; CSP, caspofungin; AND, anidulafungin; CHG, chlorhexidine digluconate; AP, Akacid plus®; ATCC, American Type Culture Collection; CBS, Centraalbureau voor Schimmelcultures; CNM-CM, Centro Nacional de Microbiologia, Immunologia y Virologia Sanitaria; IFM, Medical Mycology Research Center, Chiba University, Japan; NRCMA, French National Reference Centre for Mycoses and Antifungals; IP, Institut Pasteur; UAMH, University of Alberta Microfungus Collection and Herbarium.

In the study of Marco et al.,[116] posaconazole displayed good in vitro activity against a *Trichoderma* sp. isolate. Pfaller et al.[117] investigated the in vitro activity of two echinocandin derivatives, anidulafungin and caspofungin as well as itraconazole, amphotericin B, and 5-flucytosine against 51 clinical isolates of filamentous fungi, including a *Trichoderma* strain. Anidulafungin was fourfold more active than caspofungin against *Trichoderma*. Kahn et al.[118] have demonstrated that caspofungin inhibits β-D-1,3 glucan synthesis and reduces in vitro growth of a clinical *Trichoderma* isolate. Terbinafine was also shown to have a very strong activity in vitro against clinical isolates of several species of filamentous fungi, including *Trichoderma* spp.[64]

The MICs of fluconazole and amphotericin B obtained by the Etest and the agar dilution method for seven clinical *Trichoderma* strains were compared by Dóczi et al.[119] The MICs of fluconazole obtained with the two methods were in agreement, while higher MICs were obtained for amphotericin B with the agar dilution method than with the Etest in the case of six clinical isolates; however, most of these differences were within the ± 2 two-step dilutions.

Antal et al.[85] used the Etest method modified for molds[102] to reveal a wide-scale dataset about the antifungal susceptibilities of *T. longibrachiatum*, including 12 and 9 isolates from clinical and environmental samples, respectively. Data for the clinical isolates from this study are included in Table 62.2. The MIC values of the tested antifungal drugs were found to be 0.016–8 μg/mL for amphotericin B, 64–256 μg/mL for fluconazole, 0.5–32 μg/mL for itraconazole, and 0.008–1 μg/mL for ketoconazole. All of the examined *T. longibrachiatum* strains proved to be susceptible to ketoconazole, resistant to fluconazole and with one exception, also resistant to itraconazole. MIC values for amphotericin B were more variable among the strains: four of the soil isolates were susceptible (MIC value < 1 μg/mL) and five of them were resistant (MIC value > 2 μg/mL), while one susceptible and 11 resistant strains could be found among the clinical isolates. The same set of clinical isolates was tested by the broth microdilution method[103] for their antifungal susceptibilities to conventional antimycotics (amphotericin B, fluconazole, voriconazole) and cationic antimicrobials (chlorhexidine digluconate and Akacid plus®) by Kratzer et al.[113] The data from this study for amphotericin B and fluconazole (Table 62.2) are in good agreement with the Etest results of Antal et al.[85] The high in vitro activity of voriconazole against clinical and environmental *Trichoderma* isolates suggests that it can be an appropriate choice in the treatment of invasive *Trichoderma* infections. Evaluation of synergism of double antifungal combinations revealed no interaction between convential antimycotics, some degree of synergism was detected between azoles and cationic antimicrobials, while the interaction of the two cationic antimicrobials was synergistic for each tested *Trichoderma* isolate.[113] The authors concluded that an earlier diagnosis could enable the clinical application of cationic antimicrobials to support the treatment of localized *Trichoderma* infections in the future.

The low susceptibility level of some clinical *Trichoderma* isolates to antifungal drugs may cause difficulties in the therapy of infected patients. In cases of suspected or confirmed *Trichoderma* infection, amphotericin B in combination with itraconazole or ketoconazole,[49] or the combination of caspofungin and voriconazole[51,53] were suggested as adequate choices of treatment, the duration of which should be individualized to each case according to the type and extent of the infection and underlying predisposing conditions.

62.1.4 Diagnosis

62.1.4.1 Conventional Techniques

The difficulties of definitive diagnosis of *Trichoderma* infections were discussed in details by Chouaki et al.[46] Despite the documented dissemination, culture of blood samples does not appear to be a valuable diagnostic tool, as the common isolation rate of fungi in hemocultures is low.[120] Therefore, the diagnosis mostly relies on the microscopic visualization of fungal elements (fine, septate hyaline hyphae, eventually chlamydospores) in tissue samples obtained from accessible sites, along with the isolation of the fungus on culture. The diagnosis can be reinforced by the repeated isolation of the same fungus from the same—or in the case of disseminated infections from different—sites. Differential diagnosis of *Trichoderma* infections with aspergillosis and other hyalohyphomycoses is difficult because of the morphologically similar hyphae. The highest level of certainty in diagnosing an invasive mycosis requires the establishment of the presence of fungi in tissue by biopsy or a needle aspirate; however, such invasive techniques are not always possible due to the patients' conditions.[54] Therefore, the definition of an invasive fungal infection may have to rely on a combination of less-specific clinical, laboratory and radiological data.[121] Methods for identifying molds including *Trichoderma* from histological specimens by the development of fluorescent antibody conjugates and the use of molecular techniques may also provide definitive diagnoses.[44]

Altogether, eight species from the genus *Trichoderma* (*T. longibrachiatum*, *T. citrinoviride*, *T. pseudokoningii*, *T. reesei*, *T. harzianum*, *T. koningii*, *T. atroviride*, *T. viride*) have been reported from clinical cases (Table 62.1). The question rises whether all of these species are really capable of causing infections in humans as several clinical *Trichoderma* isolates were identified by the use of culture-based morphological characters only.[34,38,42,44,51,52] The identification of *Trichoderma* strains is very difficult if only morphological characters are considered and needs special expertise. Summerbell[122] provided the key features for morphological distinction of clinically relevant *Trichoderma* species and suggested to follow the key of Gams and Bissett[123] for the purposes of morphology-based identification.

Four members of *Trichoderma* section *Longibrachiatum*: *T. longibrachiatum*, *T. citrinoviride*, *T. pseudokoningii* and *T. reesei* have been reported from clinical cases, with *T. longibrachiatum* being the most frequently occurring ethiologic agent in the entire genus.[3,43,45,48–50,53,57,59,60] Tang et al.[57]

identified strain UAMH 10147, the causal agent of a case of allergic fungal sinusitis as *T. longibrachiatum* based on features agreeing with criteria described by Samuels et al.[100]: a radius of 60 mm on potato dextrose agar (PDA) after 65 h at 35°C (radius of 55 mm at 30°C); a yellow diffusing pigment absent at 35°C but present at 30°C; mostly solitary, cylindrical phialides tapered at the neck; smooth and ellipsoidal conidia measuring 3.5–5 × 2.5–3 μm; the common presence of chlamydospores appearing either terminal or oval to globose or intercalary and cylindrical to barrel shaped, measuring 6–10 μm × 3–5 μm. This original diagnosis was confirmed by molecular techniques, similarly to other cases of *T. longibrachiatum* infections.[45,48,50,53,57,59,60] Among the two known clinical cases of *T. citrinoviride*, the study of Kuhls et al.[63] mentioned a *T. citrinoviride* isolate from blood cultures, which has been identified by molecular techniques, while Kviliute et al.[52] have not provided any details of the identification in their case report. The occurrence of *T. pseudokoningii* has been reported from two clinical cases.[36,42] Rota et al.[36] provided no details about the identification of their peritonitis isolate, while Gautheret et al.[42] used an immunofluorescence assay besides morphology for the identification of a *Trichoderma* isolate from a fatal infection in a bone marrow transplant recipient as *T. pseudokoningii*, the patient's serum reacted strongly against *T. pseudokoningii* cultures fixed on slides with methanol. This diagnosis could not be confirmed later by molecular techniques: the isolate turned out to be *T. longibrachiatum*.[63,124,125] Studies on section *Longibrachiatum*[126,127] suggested that the species *T. pseudokoningii* may be limited in distribution to Australia and New Zealand. Later, Samuels et al.[100] have found some *T. pseudokoningii* isolates also in North America and Sri Lanka and have suggested that the species is widely distributed, but probably not as common as indicated by reports in the literature.

From *Trichoderma* section *Pachybasium B*, only *T. harzianum* has been shown to be involved in a few cases of human infections.[32,44,54] The fungal isolate from brain and lung abscesses of a renal transplant recipient was identified as *T. harzianum* by Guarro et al.[44] upon its morphology on PDA (very quick growth, dense, and cottony colonies with a yellowish green area at the center and white toward the periphery, the whole surface rapidly turning granular with dark green masses because of the abundant production of conidia in tufts, colorless colony reverse, production of a yellowish exudate) and oatmeal agar (very quick growth, cottony, and whitish green colonies becoming olive green in tufted conidial areas, colorless reverse, no exudate), as well as microscopical characters (pyramidally branched conidiophores with short branches near the apex, phialides usually in groups of 2–5, ampulliform or lageniform and markedly constricted at the base, generally 4–7 × 3–3.5 μm, usually longer and slender near the apex of the conidiophore, subglobose or short obovoid conidia of 2.5–3 × 2–2.5 μm, subhyaline to pale green and smooth walled, presence of subglobose or ellipsoidal, hyaline, smooth-walled chlamydospores up to 12 μm in diameter in old cultures). This careful morphological diagnosis was later confirmed by molecular techniques.[124,125] Kantarcioğlu et al.[54] identified a clinical fungal isolate as *Trichoderma* spp. phenotypically by comparing its morphological characteristics with the standard descriptions given by Summerbell,[122] and performed molecular identification, which revealed the identity of the isolate as *T. harzianum*.

From section *Trichoderma*, data were published about the clinical relevance of three species: *T. koningii*, *T. atroviride*, and *T. viride*. Two clinical isolates from peritonitis were identified as *T. koningii*.[30,34] Ragnaud et al.[30] provided a morphological description that allows the identification of the isolate at the genus level only. Campos-Herrero et al.[34] identified their isolate based on its rapid growth, morphology on Sabouraud dextrose agar, a grass-green pigmentation and microscopic characteristics (branching, septate, and hyaline, clusters of single-celled conidia at the end of conidiophores), and provided a microphotograph. Based on the three long terminal conidiophore branches with phialides borne singly or in offset pairs and the ellipsoidal–oblong conidia seen on this microphotograph, Summerbell[122] recognized the isolate as a member of section *Longibrachiatum*. This was later confirmed by molecular techniques: the isolates turned out to be *T. longibrachiatum*.[63,124,125] A neotypification of *T. koningii* has been carried out by Lieckfeldt et al.[128] who characterized this species with a maximum growth temperature of approximately 33°C, which does not meet the criteria of an opportunistic human pathogen. The authors of the two papers reporting the occurrence of *T. koningii* in peritonitis cases have likely applied the outdated *T. koningii* concept, which included all species with ellipsoidal conidia. *T. atroviride* is known from a single clinical case only.[40] Based on the morphological characterization, the isolate had conidiophores with branching pattern at right angles; flask-shaped, three or four verticillate phialides (8–10 μm × 2 μm) often curved in whorls of two, dark green, subglobose, short ellipsoidal, smooth-walled conidia (3.5 μm × 4 μm). The identity of this isolate has been confirmed by molecular techniques. Five studies reported the isolation of *T. viride* from clinical cases.[29,31,41,51,55] Summerbell[122] evaluated the morphological description from three of these case reports[29,41,55] and found that they do not support the identification of the isolates at the species level as *T. viride*. De Miguel et al.[51] reported *T. viride* as the ethiologic agent of a pulmonary infection on the basis of its macroscopic morphology on Sabouraud chloramphenicol agar and microscopic characteristics based on Rifai[5] (the types of branching systems of the conidiophores, the phialide disposition, and the character of the phialospores). To the best of our knowledge, this diagnosis has not been tested by molecular techniques. Identification of clinical *Trichoderma* isolates as *T. viride* may be due to the fact that the name *T. viride* was applied for a long period to a phylogenetically diverse group of *Trichoderma* strains with rounded, roughened conidia. Lieckfeldt et al.[129] redefined this species based on isolates derived from the *Hypocrea rufa* teleomorph and characterized it with a maximum growth temperature below 35°C.

The above studies reflect that identification relying solely on conventional methods may sometimes lead to erroneous species identification; therefore, the application of biochemical and molecular techniques is suggested to confirm the species-level diagnosis of clinical *Trichoderma* isolates.

Szekeres et al.[125] performed studies on clinical *Trichoderma* isolates applying a biochemical method, isoenzyme analysis based on cellulose-acetate electrophoresis (involving glucose-6-phosphate dehydrogenase, glucose-6-phosphate isomerase, 6-phosphogluconate dehydrogenase, peptidases A, B and D, and phosphoglucomutase enzymes) according to Hebert and Beaton[130] for the identification. The method was suggested as a cheap and useful alternative of molecular techniques for the quick identification of clinical *T. longibrachiatum* isolates.

62.1.4.2 Molecular Techniques

The appropriate identification of the fungal etiologic agents involved in opportunistic infections is crucial for the adequate choice of specific therapeutic interventions. Certain molecular techniques have the potential to reveal an exact diagnosis, thus helping to overcome the difficulties of morphology-based identification. In a number of case reports, the identity of the clinical *Trichoderma* isolates was confirmed by molecular techniques, like DNA-fingerprinting[43,45,49] or the sequencing of ribosomal DNA internal transcribed spacer (ITS) region ITS1–5.8S–ITS2.[40,45,46,48,50,53,54,57,59,60,63,67] Kuhls et al.[63] used PCR-fingerprinting to identify six clinical *Trichoderma* isolates—including two strains from previously reported cases[42,48]—as *T. longibrachiatum* (five isolates) and *T. citrinoviride* (one isolate). Sequencing experiments of the ITS-region confirmed the identifications made on the basis of PCR-fingerprinting data. The results of this study have revealed that the "*T. pseudokoningii*" strain from the study of Gautheret et al.[42] needs to be reidentified as *T. longibrachiatum*.

Besides isoenzyme analysis, Kredics et al.[124] and Szekeres et al.[125] used ITS sequence analysis to reidentify further clinical *Trichoderma* isolates as *T. longibrachiatum*, including strain CNM-CM 382 from the case report of Campos-Herrero et al.,[34] which was originally identified as "*T. koningii*" based on morphological characters. The genetic diversity of *T. longibrachiatum* as an emerging fungal pathogen was also examined by the analysis of restriction fragment length polymorphisms (RFLPs) of the mitochondrial DNA.[131] The examined 17 isolates (9 strains from clinical sources including strains from previously published case reports,[34,42,43,45,49] and 8 strains from environmental samples) exhibited 7 and 10 different mitochondrial DNA profiles with the restriction enzymes *Bsu*RI and *Hin*6I, respectively. The method had a discriminatory power higher than that of ITS sequence analysis and was therefore suggested as a suitable tool for epidemiological investigations. In these studies, clinical and environmental isolates did not form separate clusters on the resulting phylogenetic trees and dendrogram, suggesting that not a specific population of *T. longibrachiatum* is responsible for opportunistic infections in humans, but every environmental isolate has the capacity to cause infection.

Druzhinina et al.[132] performed a multilocus phylogenetic analysis based on the ITS region of the ribosomal RNA gene cluster and fragments of the translation elongation factor 1α (*tef1*), calmodulin (*cal1*), and endochitinase (*chit18-5*) genes with 15 clinical *Trichoderma* isolates (including strains from previously published case reports[34,42,43,45,49]). The study also included 36 environmental isolates. In this study, 12 of the clinical isolates proved to be *T. longibrachiatum* while three strains were identified as *Hypocrea orientalis*, which was previously proposed as the teleomorph of *T. longibrachiatum*.[100] However, the concordance of gene genealogies recognized *T. longibrachiatum* and *H. orientalis* to be closely related, but different phylogenetic species. The phylogenetic analysis has supported the conclusion that *H. orientalis* is sexually recombining whereas strict clonality prevails in *T. longibrachiatum*.[132] This study broadened the phylogenetic span of human opportunists in the genus *Trichoderma/Hypocrea* by identifying the sexually recombining teleomorph species *H. orientalis* as a potential human opportunist, and confirmed that the clinical isolates do not belong to a single population of *T. longibrachiatum*, but presumably every isolate of *T. longibrachiatum* or *H. orientalis* is capable of causing opportunistic infections in humans.

H. orientalis may not be the only sexually reproducing species of clinical importance within the genus *Trichoderma/Hypocrea*: when a multigenic approach was used to identify a clinical fungal isolate with *Hypocrea*-like teleomorph morphology,[56] phylogenetic analysis of molecular markers (ITS1 and ITS2 of the rRNA gene cluster, a large exon and a short intron of *tef1*, and a fragment of the RNA polymerase subunit B-encoding gene *rpb2*) showed that it belongs to an unknown fungal species that occupies a unique taxonomic position close to *Trichoderma/Hypocrea*.

Altogether, from the 59 clinical *Trichoderma* isolates known from the literature, a sequence-based molecular identification has been performed for 33 isolates. The identities of 12 and 3 isolates were confirmed as *T. longibrachiatum* and *H. orientalis*, respectively by multigenic approaches.[132] Further, 12 isolates were identified as *T. longibrachiatum* by ITS sequence analysis[46,48,50,53,57,59,60,63]; however, as this method alone is not able to differentiate between *T. longibrachiatum* and *H. orientalis*,[13] it cannot be excluded that some of these isolates belong to *H. orientalis*. Therefore, if the causal agent is supposed to belong to one of these two species, *tef1* sequence analysis is suggested to be performed in addition to ITS in order to reveal an exact molecular identification at the species level. Further isolates have been confirmed as *T. citrinoviride* (2 strains)[63] (Kredics et al., unpublished data), *T. harzianum* (2 strains),[54,124] *T. atroviride* (1 strain),[40] and an unknown *Hypocreaceae* sp. close to the genus *Hypocrea/Trichoderma* (1 strain).[56] These data clearly indicate that—with the exception of a few cases—the clinical isolates within the genus *Trichoderma* are restricted almost exclusively to section *Longibrachiatum* of the genus, suggesting that certain reports about the clinical occurrence of other *Trichoderma* species (e.g., *T. viride*, *T. koningii*) may require revision.

62.2 METHODS

62.2.1 Sample Preparation

For DNA-based molecular techniques, isolation of DNA from *Trichoderma* strains can be performed by any protocol that was successfully applied for filamentous fungi. Kuhls et al.[63] used the method of Gruber et al.[133] for DNA isolation. In the study of Antal et al.[131] applying mtDNA RFLP, DNA samples were isolated from lyophilized mycelia by the method of Leach et al.[134] Recent studies applied the GeneElute™ Plant Genomic DNA Miniprep Kit (Sigma–Aldrich) or the Qiagen DNeasy Plant Mini Kit[132] according to the manufacturers' instructions.

62.2.2 Detection Procedures

For mtDNA RFLP, the fast typing method of Varga et al.[135] was used by Antal et al.[131] Total DNA samples were digested with *Bsu*RI (GG/CC) or *Hin*6I (G/CGC) restriction enzymes. PCR-fingerprinting was carried out by Kuhls et al.[63] with the oligonucleotides $(GACA)_4$, $(GTG)_5$ and the phage M13 core sequence 5′-GAGGGTGGCGGTTCT-3′ as primers, according to a previous study of the authors.[136]

Sequence-based molecular techniques including the primers and reaction conditions applied for the identification of clinical *Trichoderma* isolates are summarized in Table 62.3.[137–142]

For amplification of the ITS region, most of the studies applied the combination of SR6R and LR1[46,50,60,63,132] as primers, while other studies applied ITS1 and ITS4[53,67,124] or ITS5 and ITS4.[54] Optimized protocols for amplification of *Trichoderma* phylogenetic markers including ITS, *tef1*, *chi18-5*, and *cal1* for molecular identification are available at the web page of the International Subcommission on *Trichoderma* and *Hypocrea* taxonomy (http://www.isth.info). Amplified PCR products can be purified by the GenElute™ Minus EtBr Spin Columns (Sigma–Aldrich) or by the Qiagen Gel Extraction Kit (Qiagen), and sequenced using the dye deoxy terminator chemistry on a DNA sequencer, e.g., ABI 373 or 377 (Applied Biosystems). Sequencing of fragments can be carried out with the aid of an appropriate service. For sequence analysis, results from BLASTN similarity

TABLE 62.3
Sequence-Based Molecular Techniques for the Identification of Clinical *Trichoderma* Isolates at the Species Level

Locus	Primers	PCR Conditions	Suggested Tool for Analysis	Diagnostic Value
ITS	SR6R: 5′-AAGTAGAAGTCGTAACAAGG-3′[137] LR1: 5′-GGTTGGTTTCTTTTCCT-3′[137]	Initial denaturation: 1 min at 94°C 1 min at 94°C 1 min at 50°C } 30 cycles 1.5 min at 74°C Final extension: 7 min at 74°C[11]	*Trich*Okey 2.0[13,14]	Identification of clinical *Trichoderma* isolates at the species level (not able to differentiate between *T. longibrachiatum* and *H. orientalis*)
	ITS1: 5′-TCCGTAGGTGAACCTGCGG-3″[137] ITS4: 5′-TCCTCCGCTTATTGATA TGC-3′[137]	Initial denaturation: 5 min at 95°C 1.5 min at 94°C 2 min at 55°C } 35 cycles 3 min at 72°C Final extension: 5 min at 72°C[138]	*Trich*Okey 2.0[13,14]	Identification of clinical *Trichoderma* isolates at the species level (not able to differentiate between *T. longibrachiatum* and *H. orientalis*)
	ITS5: 5′-GGAAGTAAAAGTCGTAACAAGG-3′[137] ITS4: 5′-TCCTCCGCTTATTGATATGC-3′[137]	Initial denaturation: 4 min at 94°C 2 min at 94°C 2 min at 55°C } 35 cycles 2 min at 72°C Final extension: 10 min at 72°C[139]	*Trich*Okey 2.0[13,14]	Identification of clinical *Trichoderma* isolates at the species level (not able to differentiate between *T. longibrachiatum* and *H. orientalis*)
tef1	EF1-728F: 5′-CATCGAGAAGTTCGAGAAGG-3′[140] TEF1-LLErev: 5′ AACTTGCAGGCAATGTGG-3′[141]	Initial denaturation: 1 min at 94°C 1 min at 94°C 1 min at 59°C } 30 cycles 50 s at 74°C Final extension: 7 min at 74°C[23]	*Tricho*BLAST[15]	Differentiation between *T. longibrachiatum* and *H. orientalis*
chi18-5	Chit42-1a: 5′-GCTYTCCATCGGTGGCTGGAC-3′[11] Chit42-2a: 5′-GGAGTTGGGGTAGCTCAGC-3[11]	Initial denaturation: 1 min at 94°C 1 min at 94°C 1 min at 62°C } 30 cycles 1 min at 74°C Final extension: 7 min at 74°C[11]	*Tricho*BLAST[15]	Differentiation between *T. longibrachiatum* and *H. orientalis*
cal1	CAL-228F: 5′-GAGTTCAAGGAGGCCTTCTCCC-3′[140] CAL-737R: 5′-CATCTTTCTGGCCATCATGG-3′[140]	Initial denaturation: 3 min at 94°C 30 s at 94°C 30 s at 50°C } 30 cycles 60 s at 72°C Final extension: 10 min at 72°C[142]	NCBI BLAST[143]	Differentiation between *T. longibrachiatum* and *H. orientalis*

searches[143] performed at the Web site of the National Center for Biotechnology Information (http://www.ncbi.nlm.nih.gov/BLAST) should be handled with care, as the GenBank database contains many sequences of *Trichoderma* isolates, which had been misidentified and thus occur under a false name.[1] *TrichO*key[13,14] is therefore suggested for the sequence analysis of ITS-fragments, *Tricho*BLAST[15] is suitable for the analysis of *tef1* and *chi18-5* sequences, while NCBI BLAST[143] can be applied for the sequence analysis of *cal1* sequences.

62.3 CONCLUSIONS AND FUTURE PERSPECTIVES

In certain risk populations, *Trichoderma* strains are able to cause diseases varying from localized infections to fatal disseminated diseases. There is a clearly emerging pattern of *Trichoderma* infections: most of the cases are reported from patients undergoing peritoneal dialysis, from transplant recipients and from patients with hematologic malignancies. Further risk factors include HIV infection and the receiving of antibiotics, corticosteroids, and chemotherapy.

T. longibrachiatum is the most frequently occurring etiologic agent within the genus; therefore, the application of this species in biotechnology and agriculture should be carefully monitored or rather abandoned, as suggested by Druzhinina et al.[132] Possible sources of infection include water-related sites, air, foods, and catheters. Antifungal susceptibility data available in the literature provide useful information for the planning of the therapy in cases of suspected or confirmed *Trichoderma* infections.

The ecophysiological and enzymological investigation of clinical *Trichoderma* isolates may result in better understanding of their virulence factors. Growth at elevated temperatures, and neutral pH, the ability to utilize amino acids as carbon and nitrogen sources, as well as the production of extracellular proteases are among the potential virulence factors of *Trichoderma* strains involved in opportunistic infections. Sequencing the complete genome of *T. longibrachiatum*—the most frequently occurring opportunistic human pathogen within the genus—would reveal the opportunity to identify further possible virulence factors by comparing its genome sequence with that of *Trichoderma* species, which are not known from clinical cases (e.g., *T. virens*).

The identification of *Trichoderma* species is problematic if only morphological characters are considered; therefore, the identity of clinical isolates is suggested to be confirmed by the sequence analysis of the ITS region and a segment of the *tef1* gene. Molecular methods may also reveal diagnostic tools for rapid species identification in the clinical practice.

REFERENCES

1. Druzhinina, I.S. and Kubicek, C.P., Species concepts and biodiversity in *Trichoderma* and *Hypocrea*: From aggregate species to species clusters? *J. Zhejiang Univ. Sci.*, 6B, 100, 2005.
2. Druzhinina, I.S. et al., The first 100 *Trichoderma* species characterized by molecular data, *Mycoscience*, 47, 55, 2006.
3. Persoon, C.H., Disposito methodica fungorum in classes, ordines, familias et genera. In: Römer, J.J. (ed.), *Neues Magazin für Botanik*, pp. 63–128. Ziegler und Söhne, Zürich, 1794.
4. Bisby, G.R., *Trichoderma viride* Pers. ex Fries, and notes on *Hypocrea*., *Trans. Br. Mycol. Soc.*, 23, 149, 1939.
5. Rifai, M.A., A revision of the genus *Trichoderma*, *Mycol. Pap.*, 116, 1, 1969.
6. Bissett, J., A revision of the genus *Trichoderma*. I. Sect. *Longibrachiatum* sect. Nov, *Can. J. Bot.*, 62, 924, 1984.
7. Bissett, J., A revision of the genus *Trichoderma*. II. Infrageneric classification, *Can. J. Bot.*, 69, 2357, 1991.
8. Bissett, J., A revision of the genus *Trichoderma*. III. Sect. *Pachybasium*, *Can. J. Bot.*, 69, 2373, 1991.
9. Bissett, J., A revision of the genus *Trichoderma*. IV. Additional notes on section *Longibrachiatum*, *Can. J. Bot.*, 69, 2418, 1991.
10. Kindermann, J. et al., Phylogeny of the genus *Trichoderma* based on sequence analysis of the internal transcribed spacer region 1 of the rDNA clade, *Fungal Genet. Biol.*, 24, 298, 1998.
11. Kullnig-Gradinger, C.M. et al., Phylogeny and evolution of the fungal genus *Trichoderma*: A multigene approach, *Mycol. Res.*, 106, 757, 2002.
12. Taylor, J.W. et al., Phylogenetic species recognition and species concepts in fungi, *Fungal Genet. Biol.*, 31, 21, 2000.
13. Druzhinina, I. et al., An oligonucleotide barcode for species identification in *Trichoderma* and *Hypocrea*, *Fungal Genet. Biol.*, 42, 813, 2005.
14. Druzhinina, I.S. and Kopchinskiy, A.G., *TrichO*KEY v. 2—A DNA oligonucleotide BarCode program for the identification of multiple sequences of *Hypocrea* and *Trichoderma*. In: Meyer, W. and Pearce, C. (eds.), *Proceedings of the 8th International Mycological Congress, Cairns, Australia*, pp. 53–59. Medimond-Monduzzi, Bologna, Italy, 2006.
15. Kopchinskíy, A. et al., *Tricho*BLAST: A multiloci database of phylogenetic markers for *Trichoderma* and *Hypocrea* powered by sequence diagnosis and similarity search tools, *Mycol. Res.*, 109, 658, 2005.
16. Kubicek, C.P. and Penttilä, M.E., Regulation of production of plant polysaccharide degrading enzymes by *Trichoderma*. In: Harman, G.E. and Kubicek, C.P. (eds.), *Trichoderma and Gliocladium, Volume 2: Enzymes, Biological Control and Commercial Applications*, pp. 49–71. Taylor & Francis Ltd., London, U.K., 1998.
17. Papavizas, G.C., *Trichoderma* and *Gliocladium*: Biology, ecology, and potential for biocontrol, *Annu. Rev. Phytopathol.*, 23, 23, 1985.
18. Herrera-Estrella, A. and Chet, I., Biocontrol of bacteria and phytopathogenic fungi. In: Altman, A. (ed.), *Agricultural Biotechnology*, pp. 263–282. Marcel Dekker Inc. New York, 1998.
19. Howell, C.R., Mechanisms employed by *Trichoderma* species in the biological control of plant diseases: The history and evolution of current concepts, *Plant Dis.*, 87, 4, 2003.
20. Benítez, T. et al., Biocontrol mechanisms of *Trichoderma* strains. *Int. Microbiol.*, 7, 249, 2004.
21. Szekeres, A. et al., Peptaibols and related peptaibiotics of *Trichoderma*. A review, *Acta Microbiol. Immunol. Hung.*, 52, 137, 2005.
22. Samuels, G.J. et al., *Trichoderma* species associated with the green mold epidemic of commercially grown *Agaricus bisporus*, *Mycologia*, 94, 146, 2002.
23. Hatvani, L. et al., Green mould diseases of *Agaricus* and *Pleurotus* are caused by related but phylogenetically different *Trichoderma* species, *Phytopathology*, 97, 532, 2007.

24. Komoń-Zelazowska, M. et al., Genetically closely related but phenotypically divergent *Trichoderma* species cause world-wide green mould disease in oyster mushroom farms, *Appl. Environ. Microbiol.*, 73, 7415, 2007.
25. Foreyt, W.J. and Leathers, C.W., *Trichoderma* sp. infection in the alligator (*Alligator mississippiensis*), *J. Herpetol.*, 19, 530, 1985.
26. Jacobson, E.R., Necrotizing mycotic dermatitis in snakes: Clinical and pathologic features, *J. Am. Vet. Med. Assoc.*, 177, 838, 1980.
27. Kredics, L. et al., Clinical importance of the genus *Trichoderma*, *Acta Microbiol. Immunol. Hung.*, 50, 105, 2003.
28. Nagy, E. et al., Molecular diagnosis, epidemiology and taxonomy of emerging medically important filamentous fungi, *Rev. Med. Microbiol.*, 15, 153, 2004.
29. Loeppky, C.B. et al., *Trichoderma viride* peritonitis, *South Med. J.*, 76, 798, 1983.
30. Ragnaud, J.M. et al., Infection peritoneale a *Trichoderma koningii* sur dialyse peritoneale continue ambulatoire, *Med. Maladies Infect.*, 7, 402, 1984.
31. Warnock, D.W. and Johnson, E.M., Clinical manifestation and management of hyalohyphomycosis, phaehyphomycosis and other uncommon forms of fungal infection in the compromised patient. In: Warnock, D.W. and Richardson, M.D. (eds.), *Fungal Infection in the Compromised Patient,* pp. 248–310. John Wiley & Sons Ltd., Chichester, U.K. 1991.
32. Guiserix, J. et al., *Trichoderma harzianum* peritonitis in peritoneal dialysis, *Nephron*, 74, 473, 1996.
33. Tanis, B.C. et al., Fatal fungal peritonitis by *Trichoderma longibrachiatum* complicating peritoneal dialysis, *Nephrol. Dial. Transplant.*, 10, 114, 1995.
34. Campos-Herrero, M.I. et al., *Trichoderma koningii* peritonitis in a patient undergoing peritoneal dialysis, *Clin. Microbiol. Newslett.*, 18, 150, 1996.
35. Bren, A., Fungal peritonitis in patients on continuous ambulatory peritoneal dialysis, *Eur. J. Clin. Microbiol. Infect. Dis.*, 17, 839, 1998.
36. Rota, S. et al., *Trichoderma pseudokoningii* peritonitis in automated peritoneal dialysis patient successfully treated by early catheter removal, *Perit. Dial. Int.*, 20, 91, 2000.
37. Esel, D. et al., Fatal peritonitis due to *Trichoderma* sp. in a patient undergoing continuous ambulatory peritoneal dialysis, *Mycoses*, 46, 71, 2003.
38. Aroca, T.S. et al., Case report: *Trichoderma longibrachiatum* infections in a pediatric patient with peritoneal dialysis, *Bol. Micol.*, 19, 13, 2004.
39. Lee, H.J. et al., Peritonitis and intra-abdominal abscess by *Trichoderma longibrachiatum* in a patient undergoing continuous ambulatory peritoneal dialysis (CAPD), *Korean J. Nephrol.*, 26, 254, 2007.
40. Ranque, S. et al., Isolation of *Trichoderma atroviride* from a liver transplant, *J. Mycol. Med.*, 18, 234, 2008.
41. Jacobs, F. et al., *Trichoderma viride* infection in a liver transplant recipient, *Mycoses*, 35, 301, 1992.
42. Gautheret, A. et al., *Trichoderma pseudokoningii* as a cause of fatal infection in a bone marrow transplant recipient, *Clin. Infect. Dis.*, 20, 1063, 1995.
43. Furukawa, H. et al., Acute invasive sinusitis due to *Trichoderma longibrachiatum* in a liver and small bowel transplant recipient, *Clin. Infect. Dis.*, 26, 487, 1998.
44. Guarro, J. et al., Fatal case of *Trichoderma harzianum* infection in a renal transplant recipient, *J. Clin. Microbiol.*, 37, 3751, 1999.
45. Richter, S. et al., Fatal disseminated *Trichoderma longibrachiatum* infection in an adult bone marrow transplant patient: Species identification and review of the literature, *J. Clin. Microbiol.*, 37, 1154, 1999.
46. Chouaki, T. et al., Invasive infections due to *Trichoderma* species: Report of 2 cases, findings of *in vitro* susceptibility testing and review of the literature, *Clin. Infect. Dis.*, 35, 1360, 2002.
47. Stelzmueller, I. et al., Zygomycosis and other rare filamentous fungal infections in solid organ transplant recipients, *Transpl. Int.*, 21, 534, 2008.
48. Seguin, P. et al., Successful treatment of a brain abscess due to *Trichoderma longibrachiatum* after surgical resection, *Eur. J. Clin. Microbiol. Infect. Dis.*, 14, 445, 1995.
49. Munoz, F.M. et al., *Trichoderma longibrachiatum* infection in a pediatric patient with aplastic anemia, *J. Clin. Microbiol.*, 35, 499, 1997.
50. Myoken, Y. et al., Fatal necrotizing stomatitis due to *Trichoderma longibrachiatum* in a neutropenic patient with malignant lymphoma: A case report, *Int. J. Oral. Maxillofac. Surg.*, 31, 688, 2002.
51. De Miguel, D. et al., Nonfatal pulmonary *Trichoderma viride* infection in an adult patient with acute myeloid leukemia. Report of one case and review of the literature, *Diagn. Microbiol. Infect. Dis.*, 53, 33, 2005.
52. Kviliute, R. et al., Nonfatal *Trichoderma citrinoviride* pneumonia in an acute myeloid leukemia patient, *Ann. Hematol.*, 87, 501, 2008.
53. Alanio, A. et al., Invasive pulmonary infection due to *Trichoderma longibrachiatum* mimicking invasive Aspergillosis in a neutropenic patient successfully treated with voriconazole combined with caspofungin, *Clin. Infect. Dis.*, 46, e116, 2008.
54. Kantarcioğlu, A.S. et al., Fatal *Trichoderma harzianum* infection in a leukemic pediatric patient, *Med. Mycol.*, 47, 207, 2009.
55. Escudero, C.M.R., Pino Corral, E., and Munoz, R., Micoma pulmonar causado por *Trichoderma viride*, *Actas Dermosifiliogr.*, 67, 673, 1976.
56. Druzhinina, I.S. et al., An unknown *Hypocreaceae* species isolated from human lung tissue of a patient with non-fatal pulmonary fibrosis, *Clin. Microbiol. Newslett.*, 29, 180, 2007.
57. Tang, P. et al., Allergic fungal sinusitis associated with *Trichoderma longibrachiatum*, *J. Clin. Microbiol.*, 41, 5333, 2003.
58. Robertson, M.H., Fungi in fluids—A hazard of intravenous therapy, *J. Med. Microbiol.*, 3, 99, 1970.
59. Lagrange-Xélot, M. et al., *Trichoderma* fungaemia in a neutropenic patient with pulmonary cancer and human immunodeficiency virus infection, *Clin. Microbiol. Infect.*, 14, 1190, 2008.
60. Hennequin, C. et al., Otitis externa due to *Trichoderma longibrachiatum*, *Eur. J. Clin. Microbiol. Infect. Dis.*, 19, 641, 2000.
61. Piens, M.A. et al., *Trichoderma* infection of cerebrospinal fluid shunt device in a non immunocompromised patient, *J. Mycol. Med.*, 14, 49, 2004.
62. Bustamante-Labarta, M.H. et al., Unsuspected embolic fungal endocarditis of an aortic conduit diagnosed by transesophageal echocardiography, *J. Am. Soc. Echocardiogr.*, 13, 953, 2000.
63. Kuhls, K. et al., Molecular reidentification of human pathogenic *Trichoderma* isolates as *Trichoderma longibrachiatum* and *Trichoderma citrinoviride*, *Med. Mycol.*, 37, 25, 1999.

64. Garcia-Effron, G. et al., *In vitro* activity of terbinafine against medically important non-dermatophyte species of filamentous fungi, *J. Antimicrob. Chemother.*, 53, 1086, 2004.
65. Braun, H. et al., Incidence and detection of fungi and eosinophilic granulocytes in chronic rhinosinusitis, *Laryngo-Rhino-Otol.*, 82, 330, 2003.
66. Buzina, W. et al., Biodiversity of fungi in nasal mucus. In: *IUMS—The World of Microbes, Xth International Congress of Mycology, Abstracts*, p 79. EDK Medical and Scientific International Publisher, Paris, U.K., 2002.
67. Kredics, L. et al., Emergence of the filamentous fungal opportunist *Trichoderma longibrachiatum* in Hungary, *Acta Microbiol. Immunol. Hung.*, 53, 305, 2006.
68. Hilmioğlu-Polat, S. et al., Non-dermatophytic molds as agents of onychomycosis in Izmir, Turkey—A prospective study, *Mycopathologia*, 160, 125, 2005.
69. Amato, J.G. et al., An AIDS patient with *Trichoderma*, *Rev. Soc. Brasil. Med. Tropic.*, 35, 269, 2002.
70. Venugopal, P.L. et al., Mycotic keratitis in Madras, *Indian J. Pathol. Microbiol.*, 32, 190, 1989.
71. Colakoglu, G., Airborne fungal spores at the Belgrad forest near the city of Istanbul (Turkey) in the year 2001 and their relation to allergic diseases, *J. Bas. Microbiol.*, 43, 376, 2003.
72. Madsen, A.M. et al., Human exposure to airborne fungi from genera used as biocontrol agents in plant production, *Ann. Agric. Environ. Med.*, 14, 5, 2007.
73. Mobin, M. and Amparo Salmito, M., Fungus microbiota in air conditioners in intensive care units in Teresina, Piauí, *Rev. Soc. Bras. Med. Trop.*, 39, 556, 2006.
74. Hageskal, G. et al., Diversity and significance of mold species in Norwegian drinking water, *Appl. Environ. Microbiol.*, 72, 7586, 2006.
75. Warris, A. et al., Water as a source of filamentous fungi in a childhood bone marrow transplantation unit. In: *Program and Abstracts of the 39th Interscience Conference on Antimicrobial Agents and Chemotherapy (San Francisco)*, p. 578. American Society for Microbiology, Washington, DC, 1999.
76. Varo, S.D. et al., Isolation of filamentous fungi from water used in a hemodialysis unit, *Rev. Soc. Bras. Med. Trop.*, 40, 326, 2007.
77. Enríquez-Matas, A. et al., Hypersensitivity pneumonitis caused by *Trichoderma viride*, *Arch. Bronconeumol.*, 45, 304, 2009.
78. Gravesen, S., Identification and quantitation of indoor airborne microfungi during 12 months from 44 danish homes, *Acta Allergol.*, 27, 337, 1972.
79. Lübeck, M. et al., Identification of *Trichoderma* strains from building materials by ITS1 ribotyping, UP-PCR fingerprinting and UP-PCR cross hybridization, *FEMS Microbiol. Lett.*, 185, 129, 2000.
80. Thrane, U. et al., Identification of *Trichoderma* strains by image analysis of HPLC chromatograms, *FEMS Microbiol. Lett.*, 203, 249, 2001.
81. Ebbehøj, N.E. et al., Building-related symptoms and molds: A two-step intervention study, *Indoor Air*, 12, 273, 2002.
82. Kildesø, J. et al., Determination of fungal spore release from wet building materials, *Indoor Air*, 13, 148, 2003.
83. Bullerman, L.B., Biology and healths aspects of molds in foods and the environment, *J. Korean Soc. Food Nutr.*, 22, 359, 1993
84. Bouakline, A. et al., Fungal contamination of food in hematology units, *J. Clin. Microbiol.*, 38, 4272, 2000.
85. Antal, Z. et al., Comparative study of potential virulence factors in human pathogenic and saprophytic *Trichoderma longibrachiatum* strains, *Acta Microbiol. Immunol. Hung.*, 52, 341, 2005.
86. Hogan, L.H. et al., Virulence factors of medically important fungi, *Clin. Microbiol. Rev.*, 9, 469, 1996.
87. Kredics, L. et al., Production of extracellular proteases by human pathogenic *Trichoderma longibrachiatum* strains, *Acta Microbiol. Immunol. Hung.*, 51, 283, 2004.
88. Schmoll, M. et al., Gene regulation in the emerging fungal pathogen *Trichoderma longibrachiatum* during its growth in the presence of bronchial epithelial cells. In: Druzhinina, I.S., Kopchinskiy, A.G., and Kubicek, C.P. (eds.), *Proceedings of the 8th European Conference on Fungal Genetics, Book of Abstracts*, p. 394. Vienna University of Technology, Vienna, Austria, 2006.
89. Burge, H.A., Fungus allergens, *Clin. Rev. Allergy*, 3, 319, 1985.
90. Vesper, S.J. et al., Specific molds associated with asthma in water-damaged homes, *J. Occup. Environ. Med.*, 48, 852, 2006.
91. Imokawa, S. et al., Two cases of acute eosinophilic pneumonia with precipitating antibody against *Trichosporon cutaneum* and *Trichoderma viride*, *Nihon Kyobu Shikkan Gakkai Zasshi*, 31, 352, 1993.
92. Mouri, M. et al., Case report and review of literature on seasonal distribution and pathogenesis of acute eosinophilic pneumonia in Japan, *Nihon Kyobu Shikkan Gakkai Zasshi*, 31, 1568, 1993.
93. Larsen, F.O. et al., The indoor microfungus *Trichoderma viride* potentiates histamine release from human bronchoalveolar cells, *APMIS*, 104, 673, 1996.
94. Danesh, P. et al., Mycobiota in Portuguese normal and green cork throughout the manufacturing process of stoppers, *J. Appl. Microbiol.*, 82, 689, 1997.
95. Winck, J.C. et al., Cork workers' occupational asthma: Lack of association with allergic sensitisation to fungi of the work environment, *Int. Arch. Occup. Environ. Health*, 77, 296, 2004.
96. Jaakkola, M.S. et al., Immunoglobulin G antibodies against indoor dampness-related microbes and adult-onset asthma: A population-based incident case–control study, *Clin. Exp. Immunol.*, 129, 107, 2002.
97. Hyvärinen, A. et al., Comparison of the indoor air quality in mould damaged and reference buildings in a subarctic climate, *Cent. Eur. J. Publ. Health*, 9, 133, 2001.
98. Allermann, L. et al., Inflammatory potency of dust from the indoor environment and correlation to content of NAGase and fungi, *Toxicol. In Vitro*, 20, 1522, 2006.
99. Meyer, H.W. et al., Double blind placebo controlled exposure to molds: Exposure system and clinical results, *Indoor Air*, 15(S10), 73, 2005.
100. Samuels, G.J. et al., The *Hypocrea schweinitzii* complex and *Trichoderma* sect. *Longibrachiatum*, *Stud. Mycol.*, 41, 1, 1998.
101. Rinaldi, M.G. and Howell, A.W., Antifungal antimicrobics: Laboratory evaluation. In: Wentworth, B. (ed.), *Diagnostic Procedures for Mycotic and Parasitic Infections*, 7th edn., pp. 325–356. American Public Health Association, Washington, DC, 1988.
102. Etest technical guide 10: Antifungal susceptibility testing of moulds. AB Biodisk, Solna, Sweden, 1999.
103. National Committee for Clinical Laboratory Standards, Reference method for broth dilution antifungal susceptibility testing of conidium forming filamentous fungi: Proposed standard M38-P. National Committee for Clinical Laboratory Standards, Wayne, PA, 1998.

104. National Committee for Clinical Laboratory Standards, Reference method for broth dilution antifungal susceptibility testing of conidium forming filamentous fungi: Approved standard M38-A. National Committee for Clinical Laboratory Standards, Wayne, PA, 2002.
105. National Committee for Clinical Laboratory Standards, Reference method for broth dilution antifungal susceptibility testing of yeast: Approved standard M27-A. National Committee for Clinical Laboratory Standards, Wayne, PA, 1997.
106. Pujol, I. et al., Comparison study of broth macrodilution and microdilution antifungal susceptibility tests for the filamentous fungi, *Antimicrob. Agents Chemother.*, 40, 2106, 1996.
107. Cuenca-Estrella, M. et al., Multicentre determination of quality control strains and quality control ranges for antifungal susceptibility testing of yeasts and filamentous fungi using the methods of the Antifungal Susceptibility Testing Subcommittee of the European Committee on Antimicrobial Susceptibility Testing (AFST-EUCAST), *Clin. Microbiol. Infect.*, 13, 1018, 2007.
108. Shadomy, S. and Pfaller, M., Laboratory studies with antifungal agents: Susceptibility tests and quantitation in body fluids. In: Balows, A. et al. (eds.), *Manual of Clinical Microbiology*, 5th edn., pp. 1173–1183. American Society for Microbiology, Washington, DC, 1991.
109. Drouhet, E. et al., Disc agar diffusion and microplate automatized technics for *in vitro* evaluation of antifungal agents on yeasts and sporulated pathogenic fungi. In: Iwata, K. and Vanden Bossche, H. (eds.), *In Vitro and In Vivo Evaluation of Antifungal Agents*, pp. 31–49. Elsevier Science Press, Paris, France, 1986.
110. Espinel-Ingroff, A., Comparison of the Etest with the NCCLS M38-P method for antifungal susceptibility testing of common and emerging pathogenic filamentous fungi, *J. Clin. Microbiol.*, 39, 1360, 2001.
111. Antal, Z. et al., The physiological features of opportunistic *Trichoderma* strains, *Acta Microbiol. Immunol. Hung.*, 49, 393, 2002.
112. Marco, F. et al., Antifungal activity of a new triazole, voriconazole (UK-109,496), compared with three other antifungal agents tested against clinical isolates of filamentous fungi, *Med. Mycol.*, 36, 433, 1998.
113. Kratzer, C. et al., *In vitro* activity and synergism of amphotericin B, azoles and cationic antimicrobials against the emerging pathogen *Trichoderma* spp., *J. Antimicrob. Chemother.*, 58, 1058, 2006.
114. Espinel-Ingroff, A., *In vitro* fungicidal activities of voriconazole, itraconazole, and amphotericin B against opportunistic moniliaceous and dematiaceous fungi, *J. Clin. Microbiol.*, 39, 954, 2001.
115. Espinel-Ingroff, A. et al., Optimal testing conditions for determining MICs and minimum fungicidal concentrations of new and established antifungal agents for uncommon molds: NCCLS collaborative study, *J. Clin. Microbiol.*, 40, 3776, 2002.
116. Marco, F. et al., *In vitro* activity of a new triazole antifungal agent, Sch 56592, against clinical isolates of filamentous fungi, *Mycopathologia*, 141, 73, 1998.
117. Pfaller, M. et al., *In vitro* activity of two echinocandin derivatives, LY303366 and MK-0991 (L-743,792), against clinical isolates of *Aspergillus, Fusarium, Rhizopus,* and other filamentous fungi, *Diagn. Microbiol. Infect. Dis.*, 30, 251, 1998.
118. Kahn, J.N. et al., Caspofungin susceptibility in *Aspergillus* and non-*Aspergillus* molds: Inhibition of glucan synthase and reduction of β-D-1,3 glucan levels in culture, *Antimicrob. Agents Chemother.*, 50, 2214, 2006.
119. Dóczi, I. et al., Etest for assessing the susceptibility of filamentous fungi, *Acta Microbiol. Immunol. Hung.*, 51, 271, 2004.
120. Lionakis, M.S. et al., The significance of blood cultures positive for emerging saprophytic moulds in cancer patients, *Clin. Microbiol. Infect.*, 10, 922, 2004.
121. Ascioglu, S. et al., Defining opportunistic invasive fungal infections in immunocompromised patients with cancer and hematopoietic stem cell transplants: An international consensus, *Clin. Infect. Dis.*, 34, 7, 2002.
122. Summerbell, R.C., *Ascomycetes: Aspergillus, Fusarium, Sporothrix, Piedra*, and relatives. In: Howard, D.H. (ed.), *Pathogenic Fungi in Humans and Animals*, pp. 437–449. Marcel Dekker Inc., New York, 2003.
123. Gams, W. and Bissett, J., Morphology and identification of *Trichoderma*. In: Kubicek, C.P. and Harman, G.E. (eds.), *Trichoderma and Gliocladium. Volume 1. Basic Biology, Taxonomy and Genetics*, pp. 3–34. Taylor & Francis, London, U.K., 1998.
124. Kredics, L. et al., Genetic variability and antifungal susceptibilities of clinical *Trichoderma* isolates. In: *Trends in Medical Mycology*, pp. 133–136. Monduzzi Editore, Bologna, Italy, 2003.
125. Szekeres, A. et al., Rapid identification of clinical *Trichoderma longibrachiatum* isolates by cellulose-acetate electrophoresis mediated isoenzyme analysis, *Clin. Microbiol. Infect.*, 12, 369, 2006.
126. Turner, D. et al., Biogeography and phenotype variation in *Trichoderma* sect. *Longibrachiatum* and associated *Hypocrea* species, *Mycol. Res.*, 101, 449, 1997.
127. Kuhls, K. et al., Revision of *Trichoderma* section *Longibrachiatum* including related teleomorphs based on analysis of ribosomal DNA internal transcribed spacer sequences, *Mycologia*, 89, 442, 1995.
128. Lieckfeldt, E. et al., Neotypification of *Trichoderma koningii* and its *Hypocrea koningii* teleomorph, *Can. J. Bot.*, 76, 1519, 1998.
129. Lieckfeldt, E. et al., A morphological and molecular perspective of *Trichoderma viride*: Is it one or two species? *Appl. Environ. Microbiol.*, 65, 2418, 1999.
130. Hebert, P.D.N. and Beaton, M.J., *Methodologies for Allozyme Analysis Using Cellulose Acetate Electrophoresis. A Practical Handbook*. Helena Laboratories, Beaumont, TX, 1993.
131. Antal, Z. et al., Intraspecific mitochondrial DNA polymorphism within the emerging filamentous fungal pathogen *Trichoderma longibrachiatum*, *J. Med. Microbiol.*, 55, 31, 2006.
132. Druzhinina, I.S. et al., Alternative reproductive strategies of *Hypocrea orientalis* and genetically close but clonal *Trichoderma longibrachiatum*, both capable to cause invasive mycoses of humans, *Microbiol-SGM*, 154, 3447, 2008.
133. Gruber, F. et al., The development of a heterologous transformation system for the cellulolytic fungus *Trichoderma reesei* based on a *pyrG*-negative mutant strain, *Curr. Genet.*, 18, 71, 1990.
134. Leach, J. et al., Rapid miniprep of DNA from filamentous fungi, *Fungal Genet. Newslett.*, 33, 32, 1986.
135. Varga, J. et al., Restriction fragment length polymorphisms in the mitochondrial DNAs of the *Aspergillus niger* aggregate, *Mycol. Res.*, 97, 1207, 1993.

136. Kuhls, K. et al., PCR-fingerprinting used for comparison of ex type strains of *Trichoderma* species deposited in different culture collections, *Microbiol. Res.*, 150, 363, 1995.
137. White, T.J. et al., Amplification and direct sequencing of fungal ribosomal RNA genes for phylogenetics. In: Innis, M.S. and Gelfand, D.H. (eds.), *PCR Protocols*: A Guide to Methods and Applications. pp. 315–322. Academic Press, New York, 1990.
138. Hermosa, M.R. et al., Molecular characterization and identification of biocontrol isolates of *Trichoderma* spp., *Appl. Environ. Microbiol.*, 66, 1890, 2000.
139. Tamura, M. et al., Molecular characterization of new clinical isolates of *Candida albicans* and *C. dubliniensis* in Japan: Analysis reveals a new genotype of *C. albicans* with Group I intron, *J. Clin. Microbiol.*, 39, 4309, 2001.
140. Carbone, I. and Kohn, L.M., A method for designing primer sets for speciation studies in filamentous ascomycetes, *Mycologia*, 91, 553, 1999.
141. Jaklitsch, W.M. et al., *Hypocrea voglmayrii* sp. nov. from the Austrian Alps represents a new phylogenetic clade in *Hypocrea/Trichoderma*, *Mycologia*, 97, 1391, 2005.
142. Samuels, G.J. et al., *Trichoderma theobromicola* and *T. paucisporum*: Two new species isolated from cacao in South America, *Mycol. Res.*, 110, 381, 2006.
143. Altschul, S.F. et al., Basic local alignment search tool, *J. Mol. Biol.*, 215, 403, 1990.

63 *Verticillium*

Malena P. Pantou and Milton A. Typas

CONTENTS

63.1 INTRODUCTION

63.1.1 Classification and Morphology

63.1.1.1 Classification

Fungi of the genus *Verticillium* are soilborne, mainly plant pathogens that attack plants through the roots and are responsible for *Verticillium* wilt diseases. The name of the genus is derived from the so-called verticillate disposition of the conidiophore branches and phialides, i.e., the branching of conidiophores in whorls at several levels (Figure 63.1e and f). The mycelium of *Verticillium* spp. is hyaline, simple or branched, and septate. The conidia are ovoid to elongate and are produced on long phialides positioned in a whorl or spiral-like shape around the conidiophores (Figure 63.1a and b). With the exception of hyphal tip cells that are multinucleate, older hyphal cell compartments are mostly uninucleate. All hyphal cells and conidia of *Verticillium* spp. are almost exclusively haploid.

Based on these general morphological features, the genus *Verticillium* for years became the depository of some 200 described species, and, therefore, its taxonomy needed urgently major revision, a task undertaken over the past years mainly by Gams and coworkers, amidst the contribution of some other groups.[1–9] The revision resulted initially in retaining only phytopathogens and saprophytes in *Verticillium sensu stricto*, viz. the "*Nigrescentia*" (*Phyllachorales*) and removed all members of *Verticillium* sect. *Prostrata* into *Hypocreales*. Thus, the genera *Lecanicillium* and *Simplicillium* were introduced to accommodate all entomogenous and fungicolous verticillium-like anamorphic fungi, *Haptocillium* for nematophagous species with adhesive conidia, and *Pochonia* comprising species that form dictyochlamydospores and parasitize nematode cysts and eggs. The saprophyte *V. tenerum* (syn. *V. luteo-album*) was also reclassified as *Acrostalagmus luteo-albus*.[10] However, in spite of this first generous reclassification, the genus of *Verticillium* was still a polyphyletic entity that required resolution into more natural units. Thus, molecular phylogenetic data based on mitochondrial and nuclear ribosomal sequences placed *V. nigrescens*, *V. theobromae* and *Acrostalagmus luteo-albus* close to *Phyllachorales* but clearly differentiated them from the group of all other phytopathogenic species of *Verticillium*.[9] Consequently, *V. nigrescens* and *V. theobromae* were placed under the new generic names *Gibellulopsis* and *Musicillium*, respectively.[11] Finally, the placement of *V. fungicola*, an important mushroom parasite, was also reconsidered according to molecular phylogenetic evidence that showed the affinity of the latter to the genus of *Lecanicillium*.[9,12,13] Therefore, in current taxonomy, only five closely related species, namely *V. dahliae*, *V. longisporum*, *V. albo-atrum*, *V. tricorpus* and *V. nubilum* remain as members of *Verticillium* s. str.

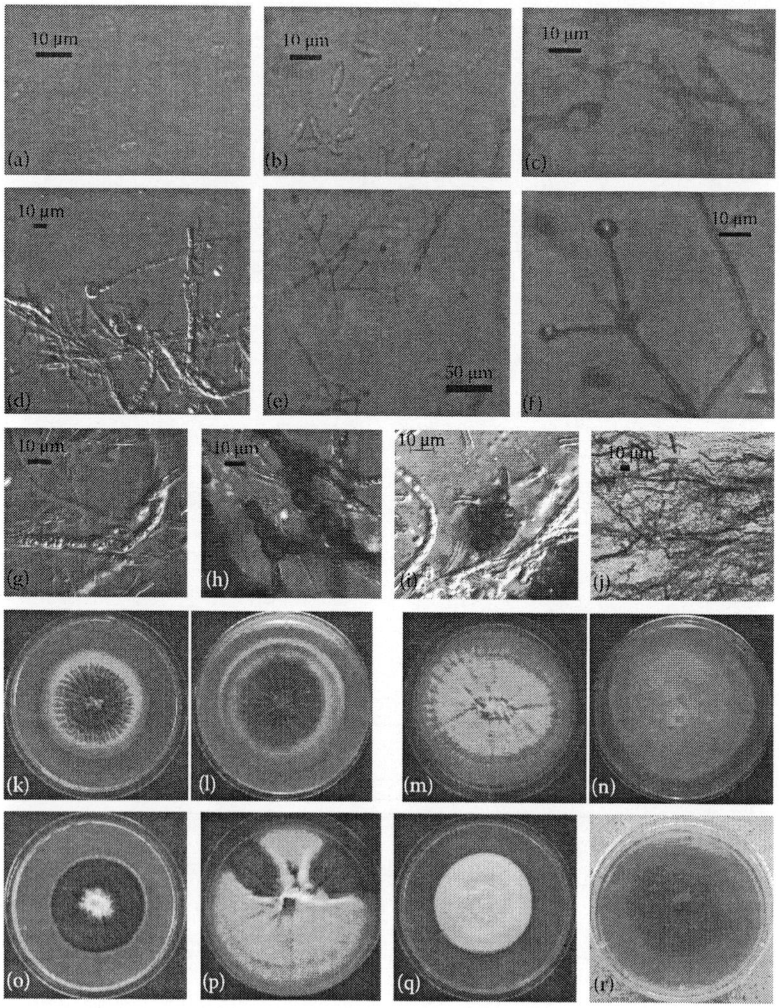

FIGURE 63.1 Microscopic and macroscopic imaging of *Verticillium* species. (a) Conidia of *V. dahliae* and (b) *V. longisporum*. (c) Young developing chlamydosporia of *V. tricorpus*. (d) A *V. dahliae* germling; the germinated conidium appears swollen due to osmotic entrance of water beneath the cell wall, while the newly formed septae are prominent along the young hypha. (e, f) Typical verticillately branched conidiophores; the abundant hyaline conidiophores are more or less erect, with a varying number of phialides arising at each node and a darkened or hyaline base, depending on the species. (g, h, and i) Three successive developmental stages of microsclerotia, the characteristic multicellular resting structures of *V. dahliae*. They are typically dark brown to blackish, torulose, or botryoidal, consisting of swollen almost globular cells arising in the centre of the culture after 6–10 days of growth. (j) *V. albo-atrum* characteristic dark resting mycelium, produced by this species as thickened and darkened hyphae instead of microsclerotia. (k, l, m, and n) Top and bottom plate views of 3-week-old darkly pigmented and hyaline *V. dahliae* cultures respectively. (o and p) Intensity and patterns of pigmentation in *Verticillium* cultures. (q) Fluffy mycelium, *V. lecanii*. (r) Sparse aerial mycelium, *V. tenerum* (now *Acrostalagmus luteo-albus*). Photos were taken with the use of a Zeiss Axio Imager. A1Microscope, with in-build camera, using differential interference contrast. (Courtesy of I.A. Papaioannou.)

63.1.1.2 Morphology

V. dahliae and *V. longisporum* form microsclerotia, dense aggregates of darkly pigmented, and thick-walled hyphal cells. Microsclerotia develop by swelling of the hyphae and septation of the cells that enlarge and produce lateral cells (Figure 63.1 g through i). Subsequently, melanin is deposited in the cell walls and intercellular space.[14] *V. albo-atrum* is differentiated from the above species by the absence of microsclerotia and the presence of dark resting mycelium (Figure 63.1j). *V. longisporum* is considered by some scientists as a variety of *V. dahliae* and by several others as a new species.[15] It has been suggested that it may be a hybrid that originated from *V. dahliae* and *V. albo-atrum* parental species, following the fusion of two different haploid nuclei to form a heterozygous diploid nucleus. However, molecular phylogenetic data support its divergence from the parental species, while morphological features such as the increased length of conidia (Figure 63.1b), the shape of microsclerotia, and the DNA content per nuclei further differentiate the species.[9,15,16] The phytopathogen *V. nubilum* forms only chlamydospores, while *V. tricorpus* forms all three types of structures, microsclerotia, resting mycelium, and chlamydospores, as well as a copious yellow orange pigment when grown on plates (Figure 63.1c). The presence of dark hyphae connected to microsclerotia along with the irregular shape of the latter is characteristic of *V. tricorpus* and discriminatory for the species.[17]

From the former members of the genus, *V. nigrescens* produces brown, intercalary or terminal chlamydospores and presents rather short, hyaline conidiophores while *V. theobromae* has long, brown conidiophores and no chlamydospores, but chains of moniloid cells that gradually darken (Table 63.1).[11] *V. theobromae* mainly induces cigar-end rots of bananas and is widespread only in the tropics, and *V. luteo-album* (now *Acrostalagmus luteo-albus*) can be easily distinguished from the species of the sect. *Nigrescentia* by its brick-red colony color. *Verticillium* sect. *Prostrata* was introduced for species producing prostrate conidiophores that are often poorly differentiated from a fine, white, or yellowish mycelium. A number of species produce dictyochlamydospores, but production varies amongst and within species and across a range of culture conditions. Conidia often aggregate on the tips of phialides in heads, but in a few species they adhere in chains. Variation in morphological traits is accompanied by a considerably diverse host range for the species of the section and by the fact that some species are known to produce differentiated erect conidiophores (e.g., *V. suchlasporum*, now *Pochonia chlamydosporia*). It is therefore evident that as morphological boundaries are not always clear, molecular traits often need to be considered for a final taxonomic assignment. *V. lecanii*, now placed under the generic name *Lecanicillium*, is an important entomopathogen of the sect. *Prostrata* that is characterized by the presence of white fluffy mycelium (Figure 63.1q) and the absence of resting structures and can be diversified from its close relative fungicolous species *V. psalliotae* based on the formation of cylindrical versus falcate conidia.[1]

63.1.2 Biology, Epidemiology, and Genomics

63.1.2.1 Biology

The formation of resting structures enables *Verticillium* species to persist in the environment. Melanin appears to contribute to long-term survival of resting structures since it protects against parasitism, dehydration, toxic chemicals, and presumably against UV radiation.[18] The life cycle of all *Verticillium* species is very similar and can be divided into a dormant, a parasitic, and a saprophytic phase. In the dormant phase, germination of fungal resting structures present in the soil is inhibited through mycostasis, as conidia and resting structures have no innate or constitutive dormancy. This stage can be overcome by either increasing the available nutrient or decreasing (often diluting) the inhibitory agent. Hyphae that grow out of the germinating resting structures can traverse a limited distance, possibly directed by nutrient gradients, to reach potential host plants. After crossing the endodermis of the root, the fungus enters the vascular tissues where it can form conidia, a process referred to as budding. The initial sporulation in the root is thought to account for the colonization of stem vessels, which results in rapid accumulation of fungal biomass a few days later. During tissue necrosis or plant senescence, the fungus enters the saprophytic stage. Apart from the vascular tissues, shoots and roots of the plant now also become colonized. At this stage, *V. albo-atrum* can produce its conidia on infected plant tissues that may be dispersed by air currents and start another cycle of disease.[19] In *V. dahliae* infection, large amounts of microsclerotia are produced, and with the decomposition of plant materials they are released in the soil where they can survive for 10–15 years. The survival potential of the melanized, dark, resting mycelium produced by *V. albo-atrum* is shorter. According to this life cycle, *Verticillium* diseases generally spread through the use of contaminated equipment, the transfer of contaminated soil, irrigation, and the use of infected seed or plant materials such as rootstocks, bulbs, and tubers. *V. albo-atrum*, which produces conidia on infected tissues, also spreads through air currents. These spores cannot be detached by currents of dry air, but are readily removed by a fine spray from an atomizer. It is probable that in nature wind-blown mist is effective in detaching wettable spores from the conidiophores of *Verticillium* species.

The entomopathogenic fungi of the former *V. lecanii* group are of particular interest since they display a completely different life cycle. They infect insects with their conidia, which adhere on the insect and germinate on its surface to give germ tubes, from which hyphae penetrate into the tissues mechanically and enzymatically using an array of enzymes like chitinases, lipases, and proteinases. The fungus is dimorphic, and inside the host's hemolymph (blood) it grows as single yeast-like cells rather than in the form of mycelium. This switch in growth form may aid dispersion and colonization of the hemocoel, optimize nutrient acquisition through the increase of the surface area, and disparate the efforts of the host cellular immune system.[20] Later, as the host's tissues are killed, the fungus reverts to the hyphal form and grows throughout the softer portions of the cuticle, covering the insect with a layer of millions of spores that are dispersed through the air currents, while the insect itself acts as a successful carrier spreading conidia at other members of its order. This is a critical step, as in the absence of resting structures the fungus persistence in soil is rather limited. Nevertheless, it was shown that when the fungus is used as a biological control agent, it may persist in soil for long periods of time and can infect new insects.[21] In the case of *V. chlamydosporium*, a fungus that parasitizes nematode cysts and produces dictyochlamydospores, it has been proven that it can proliferate in soil and survive in considerable numbers for at least 3 months.[22] Therefore, when used as a biological control agent, pre-cropping applications of the fungus is suggested as it is estimated that the survival is long enough to kill nematode eggs and females that develop on roots of spring-sown crops.

63.1.2.2 Epidemiology

Airborne conidia of *Verticillium* have been detected at several sites above cultivated fields at various sites, in the atmosphere of a London garden, in house dust in Edinburgh, and even in sandstorm dust from Saudi Arabia.[19,23,24] The isolates were derived mainly from 7 and >7 µm particles, but

TABLE 63.1
Morphological Description of *Verticillium* Plant Pathogenic Species

	V. dahliae	*V. albo-atrum*	*V. longisporum*	*V. tricorpus*	*V. nubilum*	*V. nigrescens*	*V. theobromae*
Mycelium[a]							
Color	Black	Grayish black	Black	Golden black	Brownish black	Brownish black	Grayish brown
Hyaline sectors	+	+	+	+	+	–	–
Resting mycelium (μm)	–	3–7	–	3.5–7	–	–	2–3.5
Microsclerotia (μm)	32–135 × 15–45	–	37–150 × 15–45	25–85 × 20–50	–	–	–
Chlamydospores (μm)	–	–	–	7.5–11	8.5–17	5.5–8	–
Conidia							
Shape and color	Hyaline, ellipsoidal to irregular, subcylindrical, single-celled, rarely one-septate	Hyaline, ellipsoidal to irregular, subcylindrical, single-celled, rarely one-septate	Hyaline, ellipsoidal to irregular, subcylindrical, single-celled, rarely one-septate	Hyaline, ellipsoidal to irregular, subcylindrical, single-celled	Hyaline, ellipsoidal to irregular, subcylindrical, single-celled, occasionally one-septate	Hyaline, ellipsoidal to irregular, subcylindrical, single-celled, occasionally one-septate	Hyaline, ellipsoidal to irregular, subcylindrical, single-celled
Size (μm)	4–6 × 1.8–2	3.5–10.5 × 2–4	7.2–9 × 2–3	3.5–10 × 1.5–3.5	4–10 × 2.5–3.5	4–8.5 × 1.5–2.5	3–8 × 1.5–3

[a] Description of cultures growing on PDA at 20°C for 2–3 weeks.

they could be detected even on particles of 2–4 μm.[25] The exposure at a waste composting facility was measured to be 49 cfu/m^3, which is not very elevated, while the average exposure at domestic sites in the United States was estimated to be 314 cfu/m^3, which is considered medium to high if it represents only one species.[26,27] In a similar study, where accurate identification to the species level was performed, *V. lecanii* was detected at 67% of the air samples taken during harvesting and 35% of the farm workers given skin tests produced positive reactions against the fungus.[28] *V. lecanii*, now *Lecanicillium* group, is an entomopathogenic fungus that is applied in greenhouse crops for the biological control of aphids and other pests. A wide variety of airborne particles including organic vapors of the outdoor environment, farms, industries, and homes are implicated in rhinitis, asthma, allergies, and a host of other upper respiratory tract infections. Respiratory symptoms and lung function impairment are probably the most widely studied among organic dust-associated health effects. Fungi are well-known sources of allergens and also are sources of β-glucan, which causes nonallergic respiratory symptoms.[29] It should be noted that thorough experiments conducted to examine the potential risk of workers exposed to fungal biological control agents used for crop spraying have shown that the doses needed to be inhaled in order to cause acute or chronic toxicity, or allergenicity, ought to be several tens of thousand times higher than those applied in the field (http://www.rebeca-net.de; http://www.rafbca.com).

Irrespective of older or more recent taxonomy, all *Verticillium* species are mitosporic (i.e., absence of sexual reproduction and a known teleomorph), and the only proven way by which they can exchange genetic material is through the parasexual cycle.[30] The cycle requires as its first step the fusion of genetically different hyphae (hyphal anastomosis) and the formation of a heterokaryon. On rare occasions, nuclear fusion between haploid nuclei with different genotypes may occur and heterozygous diploid nuclei are formed. The diploid nuclei multiply by mitosis in the heterokaryon and mitotic crossing-over can take place, obviously at reduced frequencies (approximately 1/500–1/100) compared with meiotic recombination.[31] Diploid nuclei are very unstable and are finally reduced to the haploid condition, at a relatively constant frequency of 10^{-3} per nuclear division. During the progressive loss of chromosomes (a process called "haploidization"), they may go through different stages of aneuploidy and the resulting haploids may be of different genotypes, depending on the number of genes involved in heterozygocity.[32,33] However, despite the proven great value of the parasexual cycle in the genetic analysis of *Verticillium*, its true significance in nature is questionable due to its rarity and the widespread occurrence of vegetative incompatibility, a phenomenon that seems to prevent heterokaryosis between genetically loosely related strains. In laboratory conditions, the parasexual cycle is imposed by the application of appropriate selective conditions. Similarly, in nature, a suitable substratum or a host may support the vegetative growth of a fungus for a considerable period. Ultimately, however, conditions at any site will become unsuitable for further growth. The fungus must then reach other sites suitable for growth, or survive in a dormant state until favorable conditions return. Spores have two major roles in the life of the fungus, dispersal to a new site and survival until favorable conditions return. Many fungi produce more than one kind of spore. It is thus possible, and common, for specialization to occur, with a single species producing some spore types that are efficiently dispersed and others that have a high capacity for dormant survival. Sporulation in fungi results in a complexity and diversity of form that contrasts with the relative simplicity and uniformity of structure found among vegetative cells. It is for this reason that taxonomists, concerned with the classification of fungi, have devoted much attention to spores and sporophores.

63.1.2.3 Genomics

Due to the economic and biotechnological interest of certain species of the genus, most of the presently available genetic and molecular information concerns two phytopathogens, namely *V. dahliae* and *V. albo-atrum*, and the entomopathogen *V. lecanii*. The two phytopathogenic species have been reported to cause *Verticillium* wilt to nearly 80 plant genera, including more than 410 plant species, and serious outbreaks of the disease with almost complete yield losses in crops like cauliflower, cotton, and sunflower.[34] On the other hand, members of the *V. lecanii* group have been extensively exploited as bio-control agents against a number of insects during the past years and are currently available as commercial products.[35] Although the taxonomic and population studies on these fungi are constantly increasing, with the advent of molecular genetics the need for a better understanding of the genetic properties of these fungi became more apparent. A first step toward this target was made with the study of the chromosome numbers of various *Verticillium* species using pulsed field gel electrophoresis to be followed by the analysis of the complete mitochondrial genomes (mtDNAs) of *V. lecanii* (syn. *Lecanicillium muscarium*), *V. dahliae*, and *V. albo-atrum*, obviously due to the pivotal role of mitochondria in producing energy in cells, and the relatively small size, high copy number, and fast evolutionary rates of mtDNAs.[36] Several differences that could facilitate the genetic fingerprinting and discrimination of *Verticillium* species and/or isolates were detected, and these enhanced the notion that the "phytopathogenic" branch of *Verticillium* is only distantly related to the entomopathogenic members of the genus.[37,38] Finally, in February 2008, the Fungal Genome Initiative and the Broad Institute released the first sequence assembly for a *V. dahliae* strain (VdLs.17) genome and only 5 months later, in July 2008, the first annotation of this genome along with the annotation of the genome of *V. albo-atrum* strain VaMs.102 was publically available (http://www.broadinstitute.org/annotation/genome/verticillium_dahliae/MultiHome.html). The two species were selected to have distinct phenotypic variation and very different host range so that the comparative analysis of their genomic variation would help expose the molecular mechanisms that underlie pathogenicity, differentiation, and host-adapted virulence. Thus, it was shown that the two genomes have fairly similar %GC content (*V. dahliae*: 55.85%, *V. albo-atrum*: 56.06%), they exhibit extended synteny in their chromosomes, and their total genome sizes differ by 1 Mb (*V. dahliae*: 33.83 Mb, *V. albo-atrum*: 32.83 Mb) and 314 genes (*V. dahliae*: 10,535, *V. albo-atrum*: 10,221). The optimal *V. dahliae*'s genome map that was generated with the use of restriction enzymes and light microscopy was correlated to the *in silico* restriction maps of the genome assembly to produce eight linkage groups. However, since the number of chromosomes in fungi may vary among isolates of the same species, and previous studies have indicated extensive chromosome number and/or genome size variability in strains of *V. dahliae* but not *V. albo-atrum*, the possibility that other strains of *V. dahliae* may display different number of chromosomes or even possess dispensable chromosomes like those reported in other phytopathogenic fungi cannot be excluded.[16,39,40] Considering the ease with which forced heterozygous diploids and mitotic recombinants of *V. dahliae* and *V. albo-atrum* were produced in the laboratory, it is very likely that the genome homology observed in *V. dahliae*, *V. albo-atrum*, and *V. longisporum* can facilitate the formation of hybrids in nature.[33]

63.1.3 Clinical Features and Pathogenesis

Species of the genus *Verticillium* rarely cause infections in humans, but fungal infections have been increasing in incidence over the past decades due to an expanding number of immunosuppressed patients such as those with hematologic malignancies and transplants along with the prolonged use of broad-spectrum antibiotics. Treatment for *Verticillium* infection is not standardized because of the rarity of human infections and the lack of clinical therapeutic trials. Very few data are available on in vitro susceptibility tests. Based on the limited number of isolates, amphotericin B, ketoconazole, itraconazole, and voriconazole appear active in vitro against *Verticillium* spp., with ketoconazole and voriconazole being the most prominent solutions.[41] Among few case reports analyzed below, *Verticillium* infection has been successfully treated by systemic fluconazole, intravenous fluconazole and subsequent oral flucytosine, topical amphotericin B and systemic fluconazole, intravenous amphotericin B, caspofungin combined with oral voriconazole, a therapy that was discontinued twice and replaced by liposomal amphotericin B and itraconazole accordingly, and intravitreal and oral voriconazole for surgical management of chronic postoperative fungal endophthalmitis.[42–48] However, the optimal antifungal agent remains unresolved, and successful clinical outcome requires early diagnosis.

One of the early reports of *Verticillium* infections was that of a 33-year-old farmer on continuous ambulatory peritoneal dialysis (CAPD) for end-stage renal failure due to chronic pyelonephritis.[43] The patient had been surgically treated in neonatal age for intestinal and renal malformations, and 2 weeks before the identification of the fungal infection he had been treated with intraperitoneal broad-spectrum antibiotic therapy for peritonitis caused by mixed enteric flora. The fungus was isolated from the peritoneal fluid culture and from the tip of catheter and was identified based on macroscopic and microscopic features. The macroscopic examination of the plate cultures on Sabouraud's dextrose agar (SDA) showed fast-growing white mycelial colonies that acquired a yellowish-red color after a while. Microscopic observations revealed branched, septate and erect conidiophores, and verticillate phialides that supported ball-shaped single-celled conidia (gloiosporae). Antimycotic therapy with fluconazole and flucytosine was significantly effective only after the removal of the catheter and the discontinuance of the CAPD.

A later report of *Verticillium* infection concerned a 40-year-old male juvenile diabetic with a recent history of a nephrectomy for renal cell carcinoma followed by radiotherapy and chemotherapy, factors that made him a highly susceptible candidate for mycotic infections.[42] Chemotherapy along with interferon α-2a was administered to the patient for a suspected pulmonary metastasis at the time that he developed a diffuse subcutaneous nodule at his left arm. Fine-needle aspiration of the nodule and routine staining of the smears with May-Grünwald-Giemsa and Papanikolaou stains scarcely revealed a few septate hyphae in the midst of an acute inflammatory exudate. It was the inoculation of the pus on culture plates that showed the growth of white cottony colonies, while microscopic observation of a slide culture revealed elongated conidiophores arranged in whorls, septate hyphae, and single-celled conidia. Subsequently the fungus was identified as a *Verticillium* spp., and fluconazole was given orally. The infection must have progressed through a traumatic skin laceration, probably through the injection site of insulin, and the authors conclude that, had it not diagnosed early enough, the lesion could have evolved to disseminated disease.

Similarly, a 6-year-old child, diagnosed as a case of acute lymphoblastic leukaemia-L1, received chemotherapy and during the second session of induction developed high-grade fever.[45] The child was initially treated with broad-spectrum antibiotics but did not respond to the treatment. Blood cultures on the 13th and 14th day of induction from two different venepuncture sites revealed the growth of a mold after 48 h of aerobic incubation from routine blood culture bottles. The mold was morphologically identified to belong to the genus of *Verticillium*, and since the fungus was also isolated from the central line tip, the patient must have contracted the infection through colonized central line. Antifungal treatment in the form of amphotericin B was added to the treatment regimen, and the patient became afebrile after 48 h.

A case, where *Verticillium* was initially misidentified, was that of a 50-year-old male patient with an infiltrate in his eye and no history of trauma.[44] Clinical features suggested herpetic keratitis and was treated with antiviral therapy, but his condition deteriorated. Subsequent microbiologic investigation of corneal scraping revealed the presence of septate hyphae that where morphologically identified to belong to *Verticillium* spp., and appropriate therapy with amphotericin B and fluconazole was administered. In a case of postoperative fungal endophthalmitis where *Verticillium* was recognized as the causative microorganism, the injection of vancomycin, ceftazidime, and amphotericin B did not lead to the regression of inflammatory reaction, possibly due to fungal resistance to amphotericin B.[47] The treatment was subsequently changed to oral and intravitreal voriconazole that was combined with vitrectomy with IOL removal and en bloc capsulectomy which resulted in the resolution of the infection. The accurate identification of the etiological agent is crucial as it designates the appropriate course of treatment. In a report that referred to cases of onychomycoses, the most prevalent nail disease, the implementation of molecular techniques resolved the problem of fungal identification.[49] PCR, sequencing, and restriction fragment length polymorphism (RFLP) of a part of the 28S nuclear rDNA region were carried out in order to reach a final conclusion. According to the report, several cases of *Penicillium* and *Alternaria* spp., which were identified based on morphological features of the isolated fungi, proved later to represent contaminants as molecular methods revealed the presence of other filamentous fungi. Similarly, an isolate originally described as *Fusarium* was found to belong to the genus of *Verticillium*. While in the report the identity of the isolate is not specified to the species level, the sequence, based on which the classification was made, corresponds to *Lecanicillium muscarium* (former *V. lecanii*).

Molecular techniques were also used to identify the infectious agent at a different occasion. A 40-year-old female patient with promyelotic leukemia presented multiple hypoechoic hepatosplenic nodules detected by abdominal sonography, 57 days after the third course of consolidation chemotherapy.[46] Initially, she was treated with oral fluconazole for presumed hepatosplenic candidiasis, as the most common cause of hepatosplenic abscesses in patients with leukemia treated with intensive chemotherapy is *Candida* spp.[50] But fever and right upper quadrant pain developed 20 days later and consequently the therapy was replaced by conventional amphotericin B. Abdominal computed tomography (CT) demonstrated multiple hypodense lesions in liver and spleen that progressed in size despite the applied therapy. Histopathologic examination of a liver biopsy demonstrated granulomatous inflammation, and few broad septate hyphae were detected after Gomori's methenamine silver staining. However, bacterial and fungal cultures of liver tissue were negative probably due to the previous prolonged exposure of the patient to antifungal reagents. Fungal identification was achieved after DNA extraction from the paraffin-embedded specimen from the liver biopsy, PCR amplification, and sequencing of the ITS1, ITS2, and D1–D2 nuclear ribosomal regions. The acquired sequence for D1–D2 region displayed high degree of identity (98.8%) against an entomopathogenic *Verticillium* strain that is no longer a member the genus. This discrepancy was furthermore reflected in the low percentages of sequence identity that the ITS1 (33.8%) and ITS2 (49.5%) regions displayed when compared to the corresponding regions of *V. dahliae*. Therefore, although initially recognized as a case report of *Verticillium* spp. infection, the taxonomic reform of the genus along with the molecular data acquired from the nuclear ribosomal regions of the fungus points to *Lecanicillium*, the entompathogenic ex member of the genus.

63.1.4 Diagnosis

63.1.4.1 Conventional Techniques

The diagnosis of *Verticillium* is mainly based on the microscopic observation of morphological features. Clinical material can be skin and nail scrapings, urine, sputum, and bronchial washings, cerebrospinal fluid, pleural fluid and blood, tissue biopsies from various visceral organs, and indwelling catheter tips. In general, direct microscopy and culturing are performed on all specimens. Microscopy provides vital information, often an immediate presumptive diagnosis, which is of particular importance in the immunosuppressed patient. Microscopy depending on the sample usually consists of either (a) wet mounts in 10% KOH with Parker ink, or India ink, (b) smears for Gram, Giemsa, and periodic acid-Schiff (PAS) staining, and (c) histopathology of tissue sections using hematoxylin and eosin, Gomori methenamine silver (GMS), and PAS. Special care should be placed on the selection of appropriate stain as Das et al.[42] describe that smears of fine-needle aspirate from the fungal nodule that were routinely stained with May–Grünwald–Giemsa (MGG) were able to demonstrate very few hyphae in the midst of an acute inflammatory exudate. These hyphae could have easily been missed, and the true diffusion of the fungal hyphae was revealed only after GMS, PAS, and Gram staining. However, even though histopathology offers the most rapid confirmatory diagnosis, it is not always possible to obtain suitable biopsy material or to make a specific identification of the fungus.

This is especially true for hyalohyphomycosis that *Verticillium* causes, since the presence of hyaline mycelium can be attributed to a wide variety of causative agents like *Acremonium*, *Beauveria*, *Fusarium*, *Geotrichum*, *Paecilomyces*, *Penicillium*, and *Scopulariopsis*. Therefore, culture preparation of the specimen is a prerequisite for *Verticillium* identification. The fungus can grow on a wide variety of culturing media such as potato dextrose agar (PDA), SDA, and Czapek Dox agar. The addition of inhibitory substances that would eliminate contamination from bacterial growth (e.g., chloramphenicol and gentamycin) can delay the growth of the fungus, while the presence of cycloheximide inhibits it. Generally, it is important to use media with and without inhibitory agents, and specimens from normally sterile sites should be preferably inoculated to media without inhibitory substances. Although culture characteristics such as surface texture, topography, and pigmentation are variable (Figure 63.1k through r) and are therefore the least reliable criteria for identification, they can be proven useful. Colonies of *Verticillium* are fast to moderate growing, white to pale yellow in color at first, becoming pinkish brown, red, green or yellow with a colorless, yellow, black, or reddish brown reverse. Some of the *Verticillium* species can be identified by the production of resting structures (after 1–2 weeks of incubation), macroscopically visible as a darkening of the cultures (Figure 63.1 k–l, o–p). Examination of darkened cultures under a low power microscope, either *in situ* or squashed onto a slide, should reveal the nature of the resting structures (e.g., microsclerotia and dark resting mycelium, Figure 63.1i–j). For the rest of the species that do not form resting structures, the observation of conidiophores and conidia is indicative of their taxonomic status (Table 63.1). Conidiophores are usually well differentiated, verticillately branched over most of their length, bearing whorls of slender awl-shaped divergent phialides. Conidia are hyaline or brightly colored, mostly one-celled, and are usually borne in slimy heads (glioconidia, Figure 63.1a–b). An online aid for the identification to the genus level of *Verticillium* against medically important conidial molds is currently available at: http://www.mycology.adelaide.edu.au/Fungal_Jungle/Identification_Keys.html.

63.1.4.2 Molecular Techniques

The use of morphologic criteria such as colony pigmentation during growth on specific media and microscopic features has been useful for distinguishing *Verticillium* against other similar genera. However, using morphologic criteria as the sole source for fungus identification does not always result in safe conclusions, as the phenotypic traits of the genus are quite uniform and confusing for the untrained eye, and colony pigmentation can vary significantly depending on culture conditions (Figure 63.1 k through r). Hence, there is considerable interest in developing molecular markers/approaches that would easily and accurately identify species and strains of this pathogenic fungus. Although molecular markers are definitive and more stable than phenotypic observations, it should be noted that there is no ideal molecular marker for every organism or every purpose. Thus, some markers may be good for the discrimination of species or higher taxonomic groups, whereas others are better for distinguishing individual isolates or strains. Nevertheless, the irreversible character of molecular markers that makes them such a valuable tool for fungal species identification underlines their validity for their application in numerous pathogenic conditions.

In this direction, most of the studies on molecular markers of *Verticillium* regard the genus in the context of its phytopathogenic and/or entomopathogenic potential and are focusing on the discrimination of representative species from other microorganisms that share the same ecological habitats and have similar taxonomic characteristics. From several molecular methods tested for species and/or isolate identification within the genus, RFLP and random amplified polymorphic DNA (RAPD) analyses are perhaps the most popular probably due to the easiness to perform them in the laboratory. RFLPs have been used to reveal inter- and intra-species relationships, indicating that molecular subgrouping of isolates has a loose association with hosts and geographic origins.[51,52] In addition, PCR-RFLP using primers NMS1/NMS2 that amplify part of the mitochondrial small rRNA gene were able to differentiate *Verticillium* spp. from all other ascomycetes tested on the basis of amplicon size and DNA hybridization patterns when the amplicon was used as probe.[53] Similarly, RAPD analysis of *Verticillium* isolates provided clearly discriminatory banding patters at the species and subspecies level that could occasionally be associated with geographic origin and ability to form heterokaryons—vegetative compatibility group, VCG.[54,55] Although RAPDs are technically fast, simple, and very sensitive, as they can detect even a single nucleotide substitution in the priming region of the used primers by permitting or preventing the production of a characteristic band, they also have some disadvantages. The major drawback is reproducibility as RAPDs are sensitive to the vagaries of the testing procedure. Additionally, they are very susceptible to contamination by nontarget DNA and can only be performed reliably on DNA from axenic cultures. Like RAPD markers, amplified fragment length polymorphism (AFLP) markers are amplified with the use of arbitrary sequences, but with greater reproducibility and fidelity. AFLP analysis comprising 349 loci showed that the closest relative to *V. longisporum* was *V. dahliae*.[56] Fluorescent AFLP analysis also demonstrated that isolates of *V. dahliae* within a VCG subgroup are molecular similar, suggesting that the species is a highly structured, clonal pathogen.[57]

RAPD and AFLP analyses due to their high discriminatory potential have been mainly focusing on revealing molecular differences at the subspecies level. An effort to develop a rapid molecular method for the detection and identification of species that caused vascular wilt to tomato resulted in the design of a DNA array.[58] Based on the reverse dot blot technology, immobilized oligonucleotide detectors hybridized with a labeled PCR product corresponding to the fungal ITS ribosomal regions were detected in a sample. The hybridization pattern could reveal the presence of colonizing pathogens and subsequently could identify them as members of the species *V. albo-atrum*, *V. dahliae*, and *Fusarium oxysporum*. The advantage of this technique was that it allowed the simultaneous detection of multiple microorganisms, a trait very useful when handling potentially complex samples. The extension of this preliminary effort will be the design of an array that would allow the identification of all fungal species present in a multiplex sample, a task presently undertaken for all lower eukaryotes by Dr. G. Anderson (Berkeley University, California) who has already developed the PhyloChip based on 16S sequences, capable of detecting 9000 different operational taxonomic units of prokaryotes (i.e., taxa, genera, families, orders) in one experiment.

The most successful tools applied so far for species identification in *Verticillium* are the nuclear rRNA gene complex repeat and the mitochondrial DNA (mtDNA). Both molecules are multicopied and seem to have a complementary role in the reconstruction of fungal phylogeny. The rRNA gene complex repeat (*V. dahliae* sequence total length 7216 bp; AF 104926) is undoubtedly the most popular because it contains both highly conserved and universally present coding genes (18S or SSU, 5.8S, and 28S or LSU), as well as variable regions like the internal transcribed spacers ITS1 and ITS2, and the non-transcribed intergenic spacer (IGS) region, all of which can be used for phylogenetic comparisons at many taxonomic levels. Since the 28S rRNA gene was shown in many other fungi to frequently harbor intron sequences, both its 5′and 3′ends were used by several research groups as targets for species identification in the genus but with moderate success.[59] On the contrary, a unique 839-bp group-I intron was located in the 18S rRNA gene of all *V. longisporum* strains examined and was found inserted after a highly conserved insertion position that is common to most ascomycetes. Since this intron was absent from the corresponding 18S regions of all the other phytopathogenic and entomopathogenic *Verticillium* species that were also tested in this study, it can be considered as a specific identity marker for *V. longisporum*.[60]

Unlike the conserved 18S and 28S rRNA genes, IGS region has been proven highly variable for *Verticillium* and extremely informative since it permitted isolate grouping at the subspecies level.[59,61] However, due to its large size, the interference of its secondary structures in obtaining good PCR products, and the rarity of corresponding sequences in data banks, it is less used than the other parts of the rRNA repeat. Thus, the most popular locus for DNA-based mycological studies at the subgeneric level, and hence for species identification, remains the ITS region. This roughly 550 bp segment combines the advantage of resolution at various scales (ITS1: rapidly evolving, 5.8S: very conserved, ITS2: moderately rapid to rapid), with the ease of amplification of a multicopy region into a readily obtainable product. The large number of fungi for which ITS sequences have been generated in data banks further increase the usefulness of this region for purposes of comparison. Using the basic local alignment search tool (BLAST), it is now easy to compare sequences of any unidentified fungus with those deposited previously in GenBank and predict or identify its species. However, it is always important to bear in mind the fact that some of ITS sequences in GenBank have been deposited without a strict quality control as far as species identification is concerned, and consequently there are ITS sequences that have been deposited under erroneous taxonomic names.

The usefulness of ITS for the identification of *Verticillium* species has been demonstrated in several population studies with the fungus.[2,8–9] However, several parameters must be taken into consideration before using it as a putative fungal barcode. First of all, a large number of co-specific specimens from as many populations and geographical regions as possible, as well as the erection of a tailored, closed-submission database for the purpose, will be required.[62] Secondly, the intra-specific and often intra-isolate variations that have been recorded in several fungal groups indicate that heterogeneity among rRNA repeats because of its multicopy nature creates complications for direct sequencing of ITS-PCR products and makes it difficult to accurately define taxonomic groups at the species level.[63] Finally, it should be stressed that no single gene can be used as the "gold standard" of fungal identification since a single gene does not always faithfully represent the history of the entire genome containing it, and comparisons may mislead and give wrong conclusions about the relationship of the fungus with the members of the same species or even genus. In the case of *Verticillium*, these precautions appear to be very important since it has been demonstrated that morphological classification alone or in combination with sequence data from one gene led to the misclassification of isolates of *V. albo-atrum* strains, which later, with additional molecular data, proved to be closely related to *V. fungicola* and *V. psalliotae*, species that are placed currently under different generic names.[64]

The other molecule that can be used either as an alternative or a complementary to rRNA gene complexes for the reconstruction of fungal phylogeny is mtDNA. mtDNA evolves 5–10 times faster than the nuclear DNA, has a high AT content, relative small size, and a rather conserved gene content compared to a considerably flexible gene order. Studies on *Verticillium* have shown that several mitochondrial genes can be used for the identification of different species with an emphasis given to the NADH dehydrogenase subunit 1 gene (*nad*1), the cytochrome oxidase subunit III gene (*cox*3), and the cytochrome b (*cob*) gene.[8,9] However, none of these groups have evaluated so far the discriminatory value of the mitochondrial cytochrome C oxidase subunit I (*cox*1), a gene that consists of a newly developed molecular marker for fungal taxonomy and identification.[65] The complete mitochondrial genome of *V. dahliae* revealed a gene order that seems to be common amongst the phytopathogens of the genus but is not shared by the clade of *Prostrata*, e.g., *V. lecannii* (syn. *Lecanicillium muscarium*).[37,38] In this context, the amplification of the intergenic region *nad*3-*nad*1 was characteristic of the *Verticillium* phytopathogens only, while amlpicon size alone could differentiate *V. nigrescens* isolates from all other species.[37,38] A second PCR product corresponding to the *nad*1-*nad*4 intergenic region, which was present in all *Verticillium* species (*Prostrata* and *Nigrescentia*), was capable to resolve all discrepancies and identify strains at the species level. Finally, a specific for *V. longisporum* primer, VlonspF combined with Vdnad4R could clearly and unambiguously discriminate *V. longisporum* strains from *V. dahliae* and *V. albo-atrum* (Table 63.2).[38]

Apart from the efforts for molecular identification and differentiation of *Verticillium* species, some work has also been done on the molecular quantification of the presence of these fungi on different plants. This was achieved with the use of real-time PCR technique (RT-PCR or qPCR), a method by which the quantity of *V. dahliae* and *V. albo-atrum* DNA was measured in the stems and roots of resistant and susceptible olive tree genotypes, potato cultivars, and alfalfa plants in order to monitor pathogen colonization and estimate the resistance and tolerance to *Verticillium* wilt.[66,67] Although most of the current studies are focusing on the quantification of the fungus in its natural inhabitant, which is in plant specimens for the phytopathogens, it is assumed that only minor adjustments will be necessary in order to achieve similar results in routine clinical specimens.

TABLE 63.2
Proposed Primer Pairs for the Identification of *Verticillium* Species

Primer Pair	Region	Sequences 5′–3′	Annealing Temperature (°C)	Amplicon Size (Species)	References
18S-ITS1/28S-ITS2	ITS1-5.8S-ITS2	GTCCCTGCCCTTTGTA/ CCTGGTGGTTTCTTTTCC	50	710 bp (*V. dahliae*) 750 bp (*V. lecanii*)	[9]
RPB1-Af/RPB1-Cr	*rpb*1	GARTGYCCDGGDCAYTTYGG/ CCNGCDATNTCRTTRTCCATRTA	45	755 bp (*V. dahliae*)	http://faculty.washington.edu/benhall
NMS1/NMS2	*rns*	CAGCAGTGAGGAATATTGGTCAATG/ GCGGATCATCGAATTAAATAACAT	45	458 bp (*Verticillium* spp.)	[53]
18SVDF/18SVDR	18S	GCGAAACTGCGAATGGCT/ GTAATGATCCCTCCGCTG	60	1.65 kb (*V. dahliae*) 2.5 kb (*V. longisporum*)	[60]
VDAL6/VDAL7	IGS	GCTGTGCGTTGCAAGGCA/ GAGCCATTCGCAGTTTCG	63	1.6 kb (*V. dahliae*)	[59]
cox3A/cox3B	*cox*3	TGATTTAGAGATSTAATATCAGAAG/ CCGTGGAAACCTGTSCCAAAATA	48	363 bp (*Verticillium* spp.)	[9]
nad1A/nad1B	*nad*1	ATGGCSAGTATGCAAAGAAGA/ GCATGTTCTGTCATAAASCCACTAAC	50	479 bp (*Verticillium* spp.)	[37]
VlonspF/Vdnad4R	*nad*1-*nad*4 intergenic	AACAATTAATTTATAATAAAGTTA/ TAAAATTTTAAATTTTCTTTTATTG	48	500 bp (*V. longisporum* only)	[38]
$SVert_{1513}$for/ $SVert_{1513}$rev		GGTGACGCAGCAGACGAGCGGGACCGTG/ GCACTTATAGTTGTTACCCCTGCGTCACC	72	1513 bp (*V. albo-atrum* only)	[67]
DB19/DB22		CGGTGACATAATACTGAGAG/ GACGATGCGGATTGAACGAA	54	539 or 523 bp (*V. dahliae* only)	[57]
Nad3a/atp9R	*nad3-atp*9 intergenic	ATTTGAATGTGGTTTTCAT/ GAGAATAATTGGTTTTTTAATG	45	450 bp (*L. muscarium*)[a] 600 bp (*L. lomgisporum*)[a]	[69]

[a] Former *V. lecanii* species.

63.2 METHODS

63.2.1 Sample Preparation

The choice of the clinical specimen to be analyzed and diagnose an invasive infection is a key question in medical microbiology. For diagnosis of an invasive fungal infection, the ideal material must be easily assessable and free of fungal DNA in a healthy individual so that the distinction between invasive infection and colonization can be made easily. Moreover, the clinical material should be taken from a body site affected by the infection. Blood should be collected using the appropriate anticoagulants as heparin is a potential inhibitor for the DNA polymerase, and tissue samples treated with formalin should be washed thoroughly with sterile saline as formalin is also an inhibitor of DNA polymerase. In general, DNA extraction should aim to reduce the amount of the extracted human DNA and at the same time to minimize the losses of the target fungal DNA, should also eliminate potential DNA polymerase inhibitors and avoid contamination with airborne spores.

For filamentous fungi and in consequence for *Verticillium*, an extra step for cell wall disruption is necessary. Since the cell wall of *Verticillium* is somewhat resistant to enzymatic treatment approaches (e.g., zymolyase), mechanical disruption is recommended. In a comparison of six DNA extraction methods for the recovery of fungal DNA from bronchoalveolar lavage (BAL), those employing mechanical agitation/disruption of cell walls with beads produced the highest yields with *Aspergillus* hyphae, a model filamentous fungal pathogen.[68] At present, there are no publications to evaluate *Verticillium* DNA extraction from different clinical specimens with different methods, but the authors presume that all methods suitable for *Fusarium* are also applicable to *Verticillium*, since the cell wall and life cycle of the two fungi are rather similar. In the two reports where molecular evaluation of the presence of *Verticillium* was carried out, DNA extraction was performed in one occasion from paraffin-embedded tissue specimen without stating the protocol that was followed and in the other occasion from nail scrapings and fragment after nail dissolvement and with the use of a commercially available kit.[46,49]

The precautions and preconditions, which are encountered due to the complexity and variety of clinical specimens that are directly processed, can be overcome when fungal DNA isolation is performed from starting material grown on plate/liquid cultures used for the identification of the fungus. Even though this method is time consuming since it requires an extra step for the growth of the fungus on artificial medium, it is less prone to produce false-negative results due to extraction difficulties. According to the proposed protocol, 100 mg of fungal scrapings is collected from the culture of *Verticillium* on an agar plate and placed in an Eppendorf tube. After an overnight incubation at −20°C, 500 µL of the extraction buffer (200 mM Tris–HCl pH 8.5, 250 mM NaCl, 25 mM EDTA, 0.5% SDS) is added to the freeze mycelium and lysis is achieved through the use of a micropestle. A centrifugation step of 30 min at maximum speed (14,000 rpm) is carried out in order to discard the cellular debris. The supernatant is quickly transferred to a new Eppendorf tube containing 20 μL of RNAse A (20 mg/mL), and the mixture is incubated at 37°C for 30 min. The solution is extracted at least twice with one volume of phenol:chloroform:isoamyl alcohol (25:24:1) and centrifuged for 10 min at maximum speed. DNA is precipitated with the addition to the aqueous phase of 0.54 volumes (250 μL) of isopropanol and the incubation of the mixture at −20°C for at least 1 h. Following a centrifugation step of 15 min at maximum speed, the pellet is rinsed with 70% ethanol and centrifuged once again at maximum speed for 15 min. The supernatant is discarded with extra care to remove all residual ethanol, without disturbing the pellet. The pellet is air dried and resuspended in 100 μL TE-buffer (10 mM Tris–HCl pH 8.0, 1 mM EDTA pH 8.0) or sterile distilled water. DNA quality can be checked visually after electrophoresis on a 0.7% agarose gel in 1× TAE buffer (Trisacetate-EDTA), while DNA quantity can be estimated with a spectrophotometer by measuring the optical density at 260 nm.

63.2.2 Detection Procedures

For the detection of *Verticillium*, PCR amplification and sequencing of at least two independent gene regions is proposed. The first should be the traditionally examined ITS1-5.8S-ITS2 region which will provide a generic sequence that can be compared with all corresponding sequences in databases and place the microorganism under inspection within, and rarely close to, the genus *Verticillium*. As a rule, the nucleotide sequence obtained should group *V. dahliae*, *V. albo-atrum*, and *V. longisporum* together and clearly separate them from all the other *sensu stricto Verticillium* and former *V. lecanii* members. As a second molecule, and depending on the accuracy that a scientist wants to achieve, any of the primers proposed in Table 63.2 can give additional information for genus, species, or even subspecies identification. For example, the nuclear ribosomal 18S subunit and the mitochondrial small ribosomal *rns* genes due to their relatively conserved character facilitate the recognition of taxa at the level of genus. Similarly, the nuclear *rpb*1 and the mitochondrial *nad*1 and *cox*3 genes, alone or in combination with the above gene sequences, are capable of distinguishing between closely related species. On the other hand, the IGS and the mitochondrial intergenic regions that are known to display an increased heterogeneity, provide sequences that enable us to identify strains/isolates of a particular species. In addition, when searching specifically for one of the species *V. dahliae*, *V. albo-atrum*, and *V. longisporum*, the use of the species-specific primers described in Table 63.2 permits the immediate screening of the sample for their presence through a single PCR reaction and without the necessity of the extra sequencing step. These propositions do not disqualify the use of different genes, such as the β-tubulin gene, the translation elongation factor *tef*1, and the ribosomal polymerase B2 *rpb*2 for molecular markers, and as data on different genomic loci accumulate, the regions that are used as fungal barcodes maybe altered.

For the amplification of the ITS1-5.8S-ITS2 region (~450 bp), the suggested primers (Table 63.2) can be used and the produced amplicons will comprise, except from the entire ITS1-5.8S-ITS2, partial sequences of the flanking 18S and 28S regions as well. Amplification reaction should be performed in a total volume of 50 μL, so that adequate quantity is produced for subsequent sequencing reactions. When tested in our laboratory, each reaction included 5 μL reaction buffer (10 mM Tris–HCl pH 9.0 at 25°C, 1.5 mM $MgCl_2$, 50 mM KCl, 0.1% Triton X-100), 100 μM of each dNTP, 40 pmol of each primer, 2 U of Thermophilic Taq DNA polymerase (5.0 U/μL; Promega), 50 ng of template DNA, and sterile ultra pure water. Amplification is carried out in a PTC-200 Gradient Peltier Thermal Cycler (MJ Research) programmed as follows: initial denaturation at 94°C for 3 min; 35 cycles of denaturation at 94°C for 1 min; annealing at 50°C for 1 min; extension at 74°C for 2 min; and final extension at 74°C for 5 min. PCR products are analyzed on a 0.7% agarose gel in 1× TAE buffer. A 1 kb DNA ladder (Invitrogen, Paisley) is included as DNA size marker, and clear single bands corresponding to an approximate size of 700 bp are purified as mentioned below for the subsequent sequencing reactions. For the use of all other primer pairs, the annealing temperature should be adjusted accordingly.

Sequencing reactions can be carried out in a variety of available automated platforms according to the manufacturer's instructions, but all PCR products prior to sequencing should be purified from unbound dNTPs. This purification step can be performed with various commercially available spin columns (e.g., QIAquick PCR Purification Kit, Qiagen, Nucleospin Extract, Macherey-Nagel). Sequencing of the PCR products must be carried out in both directions, and DNA sequences from these reactions must be aligned in order to remove the regions corresponding to the primers used at the amplification step and clarify all discrepancies between them. The consensus sequence can subsequently be tested for DNA similarity with BLAST against other available fungal sequences deposited in GenBank. In the case of ITS or *rpb*1, other databases especially dedicated to the purpose of fungal species identification can be used for sequence comparison like the one publically offered by the Assembling the Fungal Tree of Life (AFTOL) organization. According to the hits recorded from this search, a positive identification of the fungus can be achieved, sometimes to the species level.

63.3 CONCLUSIONS AND FUTURE PERSPECTIVES

The ability of fungi to cause human disease appears to be an accidental phenomenon. With the exception of a few dermatophytes, pathogenicity among the fungi is not necessary for the maintenance or dissemination of the species. In general, the development of human mycoses is related primarily to the immunological status of the host and environmental exposure, rather than to the infecting organism. *Verticillium* is an "opportunistic" fungus which causes infections almost exclusively in debilitated patients whose normal defense mechanisms are impaired. Diagnosing infections remains a problem in the management of fungal diseases, particularly in the immunocompromised host. Signs and symptoms are not specific, colonization is difficult to distinguish from invasive disease, blood cultures are commonly negative, and patients are often unable to undergo invasive diagnostic procedures. In order to improve the survival for these infections, an early diagnosis is required.

Conventional microbiological, histological, and radiological techniques remain the cornerstone of diagnosis but are insensitive and have a limited impact on clinical decision making. There is an urgent need for new and efficient diagnostic methods. These tests should be fast and highly sensitive. In addition, the recognition of the causal agent should be very precise. DNA- and RNA-based methods hold promise for improved sensitivity and specificity. The amplification of gene sequences unique to fungi is conceptually appealing, offering the potential for rapid and sensitive diagnosis of invasive fungal infections. In general, assays targeting multicopy genes have better detection limits than those targeting single copy genes. However, the selection of a suitable fungal barcode locus is rather difficult, and several mycological networks and barcode-like resources have been focusing on this issue. An international multilaboratory consortium, AFTOL, dedicated to produce a complete phylogeny of the Kingdom of Fungi has so far selected six different genetic markers: the large and small ribosomal subunits (LSU or 28S, SSU or 18S), ribosomal polymerase B1 and B2 (*rpb*1, *rpb*2), translation elongation factor 1-α (*tef*1), and the small subunit of the mitochondrial ribosomal operon (mtSSU or *rns*) for evaluation. Its purpose is served through the de novo generation of in-house sequences from authoritatively identified vouchers. Although no molecular marker seems able to function as "passe-partout" for all fungal species, AFTOL sequenced ITS to provide a reference framework that could potentially be used and expanded to a complete identification database by subsequent studies, effectively promoting the use of this marker as the standard for DNA barcoding. But the selection of the suitable molecular markers is only the first step for the use of molecular techniques in a clinical context. These techniques hold great promise in the near future; however, it should be kept in mind that they are not yet standardized or readily available in most clinical laboratories, and, moreover, large clinical trials to determine the sensitivity and specificity of such molecular tests are absolutely essential.

REFERENCES

1. Gams, W. and Zare, R., A revision of *Verticillium* sect. *Prostrata*. III. Generic classification, *Nova Hedwigia*, 72, 329, 2001.
2. Zare, R., Gams, W., and Culham, A., A revision of *Verticillium* sect. *Prostrata*. I. Phylogenetic studies using ITS sequences, *Nova Hedwigia*, 71, 465, 2000.
3. Zare, R. and Gams, W., A revision of *Verticillium* section *Prostrata*. IV. The genera *Lecanicillium* and *Simplicillium* gen. nov., *Nova Hedwigia*, 73, 1, 2001.
4. Zare, R. and Gams, W., A revision of *Verticillium* section *Prostrata*. VI. The genus *Haptocillium*, *Nova Hedwigia*, 73, 271, 2001.
5. Zare, R., Gams, W., and Evans, H.C., A revision of *Verticillium* section *Prostrata*. V. The genus *Pochonia*, with notes on Rotiferophthora, *Nova Hedwigia*, 73, 51, 2001.
6. Sung, G. H. et al., A revision of *Verticillium* sect. *Prostrata*. II. Phylogenetic analyses of SSU and LSU nuclear rDNA sequences from anamorphs and teleomorphs of the Clavicipitaceae, *Nova Hedwigia*, 72, 311, 2001.
7. Zare, R. and Gams, W., A monograph of *Verticillium* section *Prostrata*, *Rostaniha*, 3(Suppl.), 1, 2004.
8. Fahleson, J., Hu, Q., and Dixelius, C., Phylogenetic analysis of *Verticillium* species based on nuclear and mitochondrial sequences, *Arch. Microbiol.*, 181, 435, 2004.
9. Pantou, M.P. et al., Molecular and immunochemical phylogeny of *Verticillium* species, *Mycol. Res.*, 109, 889, 2005.
10. Zare, R., Gams, W., and Schroers, H.J., The type species of *Verticillium* is not congeneric with the plant-pathogenic species placed in *Verticillium* and it is not the anamorph of '*Nectria*' *inventa*, *Mycol. Res.*, 108, 576, 2004.
11. Zare, R. et al., *Gibellulopsis*, a suitable genus for *Verticillium nigrescens*, and *Musicillium*, a new genus for *V. theobromae*, *Nova Hedwigia*, 85, 463, 2007.
12. Yokoyama, E., Yamagishi, K., and Hara, A., Development of a PCR-based mating-type assay for *Clavicipitaceae*, *FEMS Microbiol. Lett.*, 237, 205, 2004.
13. Zare, R. and Gams, W., A revision of the *Verticillium* fungicola species complex and its affinity with the genus *Lecanicillium*, *Mycol. Res.*, 112, 811, 2008.
14. Perry, J.W. and Evert, R.F., Structure of microsclerotia of *Verticillium dahliae* in roots of 'Russett Burbank' potatoes, *Can. J. Bot.*, 62, 396, 1982.
15. Karapapa, V.K., Bainbridge, B.W., and Heale, J.B., Morphological and molecular characterization of *Verticillium longisporum* comb. nov., pathogenic to oilseed rape, *Mycol. Res.*, 101, 1281, 1997.
16. Collins, A. et al., Isolates of *Verticillium dahliae* pathogenic to crucifers are of at least three distinct molecular types, *Phytopathology*, 93, 364, 2003.
17. Goud, J.-K.C., Termorshuizen, A.J., and Gams, W., Morphology of *Verticillium dahliae* and *V. tricorpus* on semi-selective media used for the detection of *V. dahliae* in soil, *Mycol. Res.*, 107, 822, 2003.
18. Bell, A.A. and Wheeler, M.H., Biosynthesis and function of fungal melanins, *Annu. Rev. Phytopathol.*, 24, 411, 1986.
19. Jiménez-Díaz, R.M. and Millar, R.L., Sporulation on infected tissues and presence of airborne *Verticillium albo-atrum* in alfalfa fields in New York, *Plant Pathol.*, 37, 64, 1988.
20. Clarkson, J.M. and Charnley, A.K., New insights into the mechanisms of fungal pathogenesis in insects, *Trends Microbiol.*, 4, 197, 1996.
21. Lisansky, S., Robinson, A., and Coombs, J., *Green Growers' Guide*, CPL Scientific Press, Newbury, U.K., 1991.
22. Kerry, B.R. et al., Growth and survival of *Verticillium chlamydosporium* goddard, a parasite of nematodes, in soil, *Biocontrol. Sci. Technol.*, 3, 355, 1993.
23. Davies, R.R. and Isaak, I., Dissemination of *Verticillium albo-atrum* through the atmosphere, *Nature (London)*, 181, 649, 1958.

24. Kwaasi, A.A.A. et al., Aeroallergens and viable microbes in sandstorm dust. Potential triggers of allergic and nonallergic respiratory ailments, *Allergy*, 53, 255, 1998.
25. Lindemann, J., Arny, D.C., and Delwiche, P.A., Detection of *Verticillium albo-atrum* in the air over infected alfalfa fields in Wisconsin, *Phytopathology*, 72, 1382, 1982.
26. Hryhorczuk, D. et al., Bioaerosol emissions from a suburban yard waste composting facility, *Ann. Agric. Environ. Med.*, 8, 177, 2001.
27. Solomon, W.R., A volumetric study of winter fungus prevalence in the air of midwestern homes, *J. Allergy Clin. Immun.*, 57, 46, 1976.
28. Lacey, J., Exposure of farm workers to fungi and Actinomycetes while harvesting cereal crops and handling stored grain, *Eur. J. Respir. Dis.*, 154, 37, 1987.
29. Douwes, J. et al., Bioaerosols health effects and exposure assessment: Progress and prospects, *Ann. Occup. Hyg.*, 47, 187, 2003.
30. Hastie, A.C., The parasexual cycle in *Verticillium albo-atrum*, *Genet. Res.*, 5, 305, 1964.
31. Hastie, A.C., Mitotic recombination in conidiophores of *Verticillium albo-atrum*, *Nature*, 214, 249, 1967.
32. Typas, M.A. and Heale, J.B., Heterozygous diploid analyses via the parasexual cycle and a cytoplasmic pattern of inheritance in *Verticillium* species, *Genet. Res.*, 31, 131, 1978.
33. Typas, M.A., Heterokaryon incompatibility and interspecific hybridization between *Verticillium albo-atrum* and *Verticillium dahliae* following protoplast fusion and microinjection, *J. Gen. Microbiol.*, 129, 3043, 1983.
34. Koike, S.T. et al., *Verticillium* wilts of cauliflower in California, *Plant Dis.*, 78, 116, 1994.
35. Butt, T.M., Jackson, C.W., and Magan, N., Introduction-fungal biological control agents: Progress, problems and potential, in *Fungi as Biocontrol Agents*, pp. 1–8, Butt, T.M., Jackson, C.W. and Magan, N. (Eds.), CABI Publishing, Walingford, CT, 2001.
36. Pantou, M.P. and Typas, M.A., Electrophoretic karyotype and gene mapping of the vascular wilt fungus *Verticillium dahliae*, *FEMS Microbiol. Lett.*, 254, 213, 2005.
37. Kouvelis, V.N., Ghikas, D.V., and Typas, M.A., The analysis of the complete mitochondrial genome of the entomopathogenic fungus *Lecanicillium muscarium* (syn. *Verticillium lecanii*) suggests a minimum common gene organization in mtDNAs of Sordariomycetes: Phylogenetic implications, *Fungal Genet. Biol.*, 41, 930, 2004.
38. Pantou, M.P., Kouvelis, V.N., and Typas, M.A., The complete mitochondrial genome of the vascular wilt fungus *Verticillium dahliae*: A novel gene order for *Verticillium* and a diagnostic tool for species identification, *Curr. Genet.*, 50, 125, 2006.
39. Zolan, M.E., Chromosome-length polymorphism in fungi, *Microbiol. Rev.*, 59, 686, 1995.
40. Hatta, R. et al., A conditionally dispensable chromosome controls host-specific pathogenicity in the fungal plant pathogen *Alternaria alternata*, *Genetics*, 161, 59, 2002.
41. Wildfeuer, A. et al., In vitro evaluation of voriconazole against clinical isolates of yeasts, moulds and dermatophytes in comparison with itraconazole, ketoconazole, amphotericin B and griseofulvin, *Mycoses*, 41, 309, 1998.
42. Das, D.K. et al., Fine needle aspiration cytology diagnosis of a fungal lesion of the *Verticillium* species. A case report, *Acta Cytol.*, 41, 577, 1997.
43. Amici, G. et al., *Verticillium* peritonitis in a patient on peritoneal dialysis, *Am. J. Nephrol.*, 14, 216, 1994.
44. Shin, J.Y., Kim, H.M., and Hong, J.W., Keratitis caused by *Verticillium* species, *Cornea*, 21, 240, 2002.
45. Bashir, G. et al., Fungemia caused by *Verticillium* species in an immunocompromised child, *Indian J. Med. Microbiol.*, 24, 65, 2006.
46. Wu, C.-J. et al., *Verticillium* infection—A rare cause of hepatosplenic abscesses, *Diagn. Microbial. Infect. Dis.*, 62, 427, 2008.
47. Nehemy, M.B. et al., Chronic endophthalmitis due to *Verticillium* species after cataract surgery treated (or managed) with pars plana vitrectomy and oral and intravitreal voriconazole, *Retina*, 26, 225, 2006.
48. Vasconcelos-Santos, D.V. and Nehemy, M.B., Use of voriconazole in the surgical management of chronic postoperative fungal endophthalmitis, *Ophthalmic Surg. Laser Imaging*, 40, 425, 2009.
49. Monod, M. et al., Fast and reliable PCR/sequencing/RFLP assay for identification of fungi in onychomycoses, *J. Med. Microbiol.*, 55, 1211, 2006.
50. Anttila, V.J. et al., Hepatosplenic candidiasis in patients with acute leukemia: Incidence and prognostic implications, *Clin. Infect. Dis.*, 24, 375, 1997.
51. Carder, J.H. and Barbara, D.J., Molecular variation and restriction length polymorphisms (RFLPs) within and between six species of *Verticillium*, *Mycol. Res.*, 95, 935, 1991.
52. Okoli, C.A.N., Carder, J.H., and Barbara, D.J., Restriction fragment length polymorphisms (RFLPs) and the relationships of some host-adapted isolates of *Verticillium dahliae*, *Plant Pathol.*, 43, 33, 1994.
53. Li, K.-N., Rouse, D.I., and German, T.L., PCR primers that allow intergeneric differentiation of Ascomycetes and their application to *Verticillium* spp., *Appl. Environ. Microbiol.*, 60, 4324, 1994.
54. Mitina, G.V., Mikhailova, L.A., and Yli-Mattila, T., RAPD-PCR, UP-PCR and rDNA sequence analyses of the entomopathogenic fungus *Verticillium lecanii* and its pathogenicity towards insects and phytopathogenic fungi, *Arch. Phytopathol. Pfl.*, 41, 113, 2008.
55. Zeise, K. and von Tiedemann, A., Application of RAPD-PCR for virulence type analysis within *Verticillium dahliae* and *V. longisporum*, *J. Phytopathol.*, 150, 557, 2002.
56. Fahleson, J. et al., Estimation of genetic variation among *Verticillium* isolates using AFLP analysis, *Eur. J. Plant Pathol.*, 109, 361, 2003.
57. Collado-Romero, M. et al., Phylogenetic analysis of *Verticillium dahliae* vegetative compatibility groups, *Phytopathology*, 98, 1019, 2008.
58. Lievens, B. et al., Design and development of a DNA array for rapid detection and identification of multiple tomato vascular wilt pathogens, *FEMS Microbiol. Lett.*, 223, 113, 2003.
59. Pramateftaki, P.V., Antoniou, P.P., and Typas, M.A., The complete DNA sequence of the nuclear ribosomal RNA gene complex of *Verticillium dahliae*: Intraspecific heterogeneity within the intergenic spacer region, *Fungal Genet. Biol.*, 29, 19, 2000.
60. Karapapa, V.K. and Typas, M.A., Molecular characterization of the host-adapted pathogen *Verticillium longisporum* on the basis of a group-I intron found in the nuclear SSU-rRNA gene, *Curr. Microbiol.*, 42, 217, 2001.
61. Qin, Q.-M. et al., Phylogenetic analyses of phytopathogenic isolates of *Verticillium* spp., *Phytopathology*, 96, 582, 2006.
62. Nilsson, R.H. et al., Intraspecific ITS variability in the kingdom of fungi as expressed in the international sequence databases and its implications for molecular species identification, *Evol. Bioinform.*, 4, 193, 2008.

63. Seifert, K.A., Progress towards DNA barcoding of fungi, *Mol. Ecol. Resour.*, 9, 83, 2009.
64. Carder, J.H. and Barbara, D.J., Taxonomic status of putative *Verticillium albo-atrum* isolates, *FEMS Microbiol. Lett.*, 170, 211, 1999.
65. Seifert, K.A. et al., Prospects for fungus identification using CO1 DNA barcodes, with *Penicillium* as a test case, *Proc. Natl. Acad. Sci. U.S.A.*, 104, 3901, 2007.
66. Gayoso, C. et al., Assessment of real-time PCR as a method for determining the presence of *Verticillium dahliae* in different Solanaceae cultivars, *Eur. J. Plant Pathol.*, 118, 199, 2007.
67. Larsen, R.C. et al., Development of a real-time polymerase chain reaction assay for quantifying *Verticillium albo-atrum* DNA in resistant and susceptible alfalfa, *Phytopathology*, 97, 1519, 2007.
68. Fredricks, D.N., Smith, C., and Meier, A., Comparison of six DNA extraction methods for recovery of fungal DNA as assessed by quantitative PCR, *J. Clin. Microbiol.*, 43, 5122, 2005.
69. Kouvelis, V.N., Sialakouma, A., and Typas, M.A., Mitochondrial gene sequences alone or combined with ITS region sequences provide firm molecular criteria for the classification of *Lecanicillium* species, *Mycol. Res.*, 112, 829, 2008.

Saccharomycotina and Taphrinomycotina

64 *Candida*

P. Lewis White, Michael D. Perry, and Rosemary A. Barnes

CONTENTS

64.1 INTRODUCTION

Candida remains the most common cause of fungal disease worldwide with infections ranging from superficial candidosis of mucosal membranes requiring "over-the-counter" pharmaceutical intervention to life-threatening invasive disease necessitating extensive medical, even surgical treatment. An improved awareness of invasive disease and an increasing susceptible population are countered by the availability of effective prophylaxis and limited techniques for accurate diagnosis. The incidence of disease is, at best, described as steady but varies geographically and is influenced by general hospital practice. Indeed, comparison of incidence between countries is difficult. For countries without surveillance programs or the capacity to diagnose the disease using conventional techniques, the incidence may be greatly underestimated.

Early diagnosis of invasive disease can improve mortality rates,[1] but conventional microbiology techniques can be slow, lacking both sensitivity and specificity. Many cases are treated empirically or treated preemptively on the basis of the presence of risk factors. Unfortunately, these risk factors are common amongst critically ill patients within the intensive care setting. Unless additional clinical interpretation guidelines are applied, both empirical and preemptive strategies may result in unnecessary prescribing of antifungal therapy at great expense and drug-related toxicity. Crude mortality rates for invasive disease range from 20% to 75% depending on patient population, type of disease, and the species involved.[2] Attributable mortality has not been accurately assessed but is estimated at 10%–49% for invasive disease. The cost to health care administrators is substantial, and in the United States it has been estimated at $1.7 billion per annum.[3]

Molecular diagnostics have the potential to overcome the limitations of conventional practice. Many of the published protocols provide excellent analytical performance with some clinical validity but suffer from a lack of standardization with limited evaluation of clinical utility. Most current *Candida* specific assays have excellent sensitivity and negative predictive value, making them a useful screening tool but diagnostic utility is affected by the prevalence of disease and the impact of antifungal prophylaxis. Assays solely targeting *Candida* cannot exclude diseases caused by other yeasts or molds, and a diagnostic role reliant on assay specificity to confirm candidal disease may be preferred. The use of molecular diagnostics designed to test for invasive disease could be applied to superficial disease, although it is hard to justify this in health economic terms and it cannot distinguish between disease and colonization.

The chapter will cover the topic of *Candida* and candidosis focusing mainly on invasive disease although many of the aspects will apply to superficial disease.

64.1.1 Taxonomy, Morphology, Biology, and Genomics

64.1.1.1 Taxonomy

The genus *Candida* comprises approximately 200 species although only approximately 10% of species have been associated with human disease.[2] *Candida* species are eukaryotic organisms of the Kingdom Fungi belonging to the phylum *Ascomycota*, subphylum *Saccharomycotina*, class *Saccharomycetes*, and order *Saccharomycetales* and are members of the mitosporic *Saccharomycetales* family. They are opportunistic human pathogens and commensals of mucosal membranes and skin with most infections arising from an endogenous source.[4] Changes within the host allow the organism to cause opportunistic infection although the yeast is not entirely passive during the infective process and possesses various virulence and immunomodulatory factors.[5]

64.1.1.2 Morphology, Biology, and Genomics

The unicellular organism survives by obtaining nutrients from both living and dead organic matter. Most *Candida* species are described as dimorphic, but are better described pleiomorphic, producing true budding round, oval, ellipsoid or elongate yeast cells (blastoconidia), elongated, connected yet separated pseudohyphae, true hyphal structures, chlamydospores, and opaque mating cells. Cellular structure depends on environmental conditions and transition between morphological states is rapid in response to environmental pressure. The ability to change between hyphal and yeast forms may play an important role in virulence through tissue invasion, colonization/dissemination, and biofilm formation. One species *C. glabrata* rarely produces pseudohyphae or hyphal structures and was previously classified in the genus *Torulopsis*. It is still debated whether it should be classified in the genus *Candida*.

Colonial appearance is characteristically creamy/white and noted by the absence of pigmentation. Colonies are generally smooth and wet looking although under certain growth conditions, a fringed border representing pseudohyphae or hyphal growth can be seen. Microscopically, *C. albicans* blastoconidia are typically $2–7 \times 3–8\,\mu m$ in size but dimensions vary between species. *C. glabrata* produces smaller balstoconidia ($3.4 \times 2.0\,\mu m$). *C. kefyr*, *C. krusei*, and *C. lipolytica* may produce elongated blastoconidia up to $15\,\mu m$ in length.[6] Table 64.1 summarizes the phenotypic characteristics of eight common species of pathogenic *Candida*.

C. albicans possesses the ability to reversibly switch colony surface appearance from creamy/white to opaque. The "white cells" form the distinctive smooth creamy domed colonies consisting of budding yeasts whereas the "opaque cells" have pitted cell walls resulting in flatter colonies of grey appearance consisting of rough elongate cells possibly with mating projections.[7]

The "opaque cells" are the mating-competent form of *C. albicans*.[8] A stimulus is not always necessary for *C. albicans* cells to switch but pH, oxygen concentration, low intensity treatment with UV, mating pheromones, and a limitation of zinc in growth medium affect the frequency of "phenotypic switching," which has been proposed to enhance pathogenesis of the organism.[7–10] Despite the ability to mate sexually candidal reproduction is almost entirely asexual by budding, as most of the cells are not homozygous at the "cell type controlling" MTL locus as the region is highly polymorphic.[7] Budding may be multilateral but growth of yeast cells is usually polarized and directed by polarisomes. The formation of chains of connected but separated buds is not uncommon.[7]

Candida species prefer aerobic growth conditions and although they will grow in the presence of elevated CO_2, strict anaerobic conditions may inhibit growth.[11] Growth ranges are typically between 20°C and 37°C and the pH range 2.5–7.5. Growth at 37°C is a prerequisite for pathogenic strains and, although it appears that 37°C is the optimal growth temperature, many of the more pathogenic species can grow at 40°C (Table 64.1).[6,12] Growth rates vary depending on conditions with optimal doubling times of less than 1 h for *C. albicans*.[11] Growth below 33°C leads to individual yeast cell formation via polar budding, whereas growth at higher temperatures in the presence of inducers and nutrients at neutral pH usually confers germ-tube formation leading to hyphal structures.[11]

Another interesting concept of candidal growth is the formation of biofilms, often associated with the use of medical devices such as venous catheters or mechanical ventilators. Through the upregulation of adherence factors, the yeast adheres to the surface of the device or other cells and forms a basal layer from which hyphal forms propagate to form an upper layer that secretes an extracellular matrix.[13] Biofilms are particularly resistant to antifungal therapy postulated to be a result of limited drug penetration, although it has been shown that drug diffusion through candidal biofilms is not limited and over production of drug efflux pumps may have an important role, particularly in early biofilm production.[13] It is likely that biofilm resistance is a composite response reliant on more than efflux pumps and removal of the infected implanted device is recommended.[13]

When exposed to unfavorable environmental conditions, *C. ablicans* forms chlamydospores. These large, thick-walled cells are high in lipid and carbohydrate content, are three- to fourfold the size of typical blastospores and form at the end of branched filamentous suspensor cells when oxygen, light, temperature, and nutrients are limited.[7,8] It is proposed that these stable structures are long-term resting cells although extended viability compared to active states has not been demonstrated.[7] Chlamydospores are not thought to have a role in pathogenicity.[8]

As the candidal cell wall is in primary contact with the surrounding environment, it is critical to cell survival irrespective of pathogenicity or commensalism. Initially, the candidal cell wall was perceived to have a purely inert role providing structural rigidity and cellular protection.[5] This simplistic view has been replaced with the widely accepted role of a dynamic structure responding to changing environmental pressures and serving dual function; defensively as an anatomical barrier, and by incorporating molecules (mainly proteins) essential for interaction with the environment including virulence factors, it has the potential to act in a pathogenic manner. It is comprised mainly of carbohydrate (80%–90%), with limited protein and lipid, forming a complex multilayered barrier of mannoprotein (30%–40%), β-glucan (50%–60%) and chitin (0.6%–3%).[5,14]

TABLE 64.1
Morphological and Physiological Characteristics of the Most Common *Candida* Species

	Species							
Attribute	***C. albicans***	***C. glabrata***	***C. tropicalis***	***C. parapsilosis***[a]	***C. kefyr***	***C. dubliniensis***	***C. krusei***	***C. guilliermondii***
Teleomorph	Not known	Not known	Not known	Not known	*Kluyveromyces marxianus*	Not known	Issatchenkia orientalis	*Pichia guilliermondii*
Colony color	White/ cream	White/ cream	White/cream	White/cream	White/cream	White/cream	White/ cream/gray	White/cream
Colony appearance	Smooth/ wet[b]	Smooth/ wet	Smooth/wet (dull/ rough)	Smooth/wet or dull/rough	Smooth/ dull	Smooth/wet	Smooth/flat/ dull	Smooth/wet[c]
Blastoconidia shape	Sub/ spherical	Round to ellipsoid	Sub/spherical	Spherical to ovoid	Elongated or ovoid	Sub/ spherical	Elongated or ovoid	Sub/spherical
Blastoconidia size (µm)	2–7×3–8	3.4×2.0	3–5.5×4–9	2–3.5×3–4.5	3–6.5×5.5–16	2–7×3–8	2–5.5×4–15	2–4×3–6.5
Growth at 37°C	+	+	+	+	+	+	+	Variable
Germ-tube	+	–	–	–	–	+	–	–
Fermentation								
Glucose	+	+	+	+	+	+	+	+
Galactose	Variable	–	+	Variable	+/slow	Variable	–	Variable
Sucrose	– (slow)	–	Variable	–/slow	+	–	–	+
Maltose	+	–	+	–/slow	–	+	–	–
Lactose	–	–	–	–	Variable	–	–	–
Trehalose	Variable	Variable	+/slow	–/slow	–	Variable	–	+
Assimilation								
Glucose	+	+	+	+	+	+	+	+
Galactose	+	–	+	+	Slow	+	–	+
L-Sorbose	Variable	–	Variable	+/slow	Variable	–	–	Variable
Sucrose	Variable	–	Variable	+	+	+	–	+
Maltose	+	–	+	+	–	+	–	+
Cellobiose	–	–	+/slow	–	Variable	–	–	Variable
Trehalose	+ (slow)	–	+	+	–/weak	+	–	+
Lactose	–	–	–	–	Variable	–	–	–
Melibiose	–	–	–	–	–	Variable	–	Variable
Raffinose	–	–	–	–	+	–	–	+
Melezitose	Variable	–	Variable	+	–	+	–	Variable
Soluble starch	+	–	+	–	–	+	–	–
D-Xylose	+	–	+	+	Slow	Variable	–	+
L-Arabinose	Variable	–	–	+	Variable	–	–	Variable
D-Arabinose	Variable	–	–	–	–	–	–	Variable
D-Ribose	– (slow)	–	– (slow)	Variable	Variable	–	–	+
L-Rhamnose	–	–	–	–	–	–	–	Variable
D-Glucosamine	Variable	–	Variable	Variable	–	–	+	+
NAD-glucosamine	+	–	+	+	–	+	+	+
Glycerol	Variable	+/slow	Variable	+	Slow	+/slow	+	+
Erythritol	–	–	–	–	–	–	–	–
Ribitol	Variable	–	+/slow	+/slow	Slow	+/slow	–	+
Galactitol	–	–	–	–	–	–	–	Variable
D-Mannitol	+	–	+	+	Variable	+	–	Variable
D-Glucitol	– (slow)	–	+	+	Variable	+	–	Variable
αMD-glucoside	Variable	–	Variable	+	–	+/slow	–	Variable
D-Gluconate	– (slow)	–	Variable	+/slow	–	–/slow	–	Variable
DL-Lactate	+	–	Variable	–	+	+	+	Variable
Myo-inositol	+	–	–	–	–	–	–	–

(continued)

TABLE 64.1 (continued)
Morphological and Physiological Characteristics of the Most Common *Candida* Species

	Species							
Attribute	***C. albicans***	***C. glabrata***	***C. tropicalis***	***C. parapsilosis*[a]**	***C. kefyr***	***C. dubliniensis***	***C. krusei***	***C. guililermondii***
2KD-gluconate	+	Variable	+	+	–	+	–	+
D-Glucuronate	–	–	–	–	–	–	–	–
Nitrate	–	–	–	–	–	–	–	–
Urease	–	–	–	–	–	–	–	–
0.1% Cyclohexamide	+	–	+	–	+	+	Variable	Variable

Source: Data summarized from Ellis, D. et al., *Description of Medical Fungi*, 2nd edn., Nexus Print Solutions, Adelaide, Australia, 2007.

Key: (+), Positive; (–), Negative; Variable: Result is strain dependent; Slow: May be slow in developing a positive result; +/slow: Positive, although some strains may be slow in developing a positive result; + (slow): Positive, although occasional strains may be slow in developing a positive result; –/slow: Negative, although some strains may be slow in developing a positive result; – (slow): Negative, although occasional strains may be slow in developing a positive result; –/weak: Negative, although some strains may develop a weak positive result.

[a] *Candida parapsilosis* complex contains three species (*C. parapsilosis*, *C. orthopsilosis*, and *C. metapsilosis*) that are phenotypically indistinguishable.

[b] Note the possibility of the opaque mating cells that may be dull and rough and the possibility of a rough margin due to pseudohyphal or hyphal growth.

[c] Occasionally dull and rough.

β-Glucan/chitin are essential structural polysaccharides and form the rigid framework to which mannoproteins are linked, providing the cell surface characteristics.[14] Overall, cell-wall composition is similar regardless of the candidal cell structure (yeast, pseudohyphae, or hyphal), although the amount of chitin in hyphal structures is threefold that of its yeast counterpart.[5]

The complete genome sequence of *C. albicans* was published in 2004.[15] *C. albicans* possesses a diploid genome spanning 14,851 kb comprising eight pairs of chromosomes (R, 1–7), of which chromosomes R and 1 are the largest (>3000 kb) and chromosomes 6 and 7 are the smallest (≅1000 kb).[15] Significant natural heterozygosity is demonstrated between strains such that no two homologous chromosomes are identical in DNA sequence.[15] Heterozygosity is not distributed evenly throughout the genome and is most evident on chromosomes 5 and 6.[15] There is no known haploid state and attempts to confirm meiosis have yet to be successful, a major hurdle to completing the sexual cycle.[15] It has been proposed that mating can occur by combining two diploid genomes to produce tetraploid organisms that randomly lose chromosomes to become diploid with indiscriminate recombination of the eight chromosomes.[16] The genome contains more than 6000 open reading frames (ORF) possibly coding for large proteins (>100 amino acids).[15] Comparison of the *C. albicans* genome with that of *Saccharomyces cerevisiae*, *Schizosaccharomyces pombe*, and the human genome revealed that almost 50% of the ORF found matches in all three genomes and approximately 15% had further matches in both the yeast genomes.[15] Genes specific to *C. albicans* could explain its heightened pathogenicity compared to *S. cerevisiae*.[15,17–19] In addition to large gene families related to pathogenesis including agglutinin like sequence genes (ALS), iron transport, secreted aspartyl proteinase, and secreted lipase genes, there appear to be genes permitting increased oxidative metabolism/respiration and genes responsible for morphological restructuring, environmental monitoring, and response.[15]

Five hundred and sixty seven genes have been deemed essential with an estimated 1400 genes necessary for growth.[18] The essential rDNA genes have shown varying copy numbers between both strains and species with some strains of *C. albicans* possessing over 200 copies, making them ideal targets for molecular detection.[20] Comparison of the genomes of *C. albicans* and its closest relative *C. dubliniensis* revealed 168 species-specific genes possibly indicative of recent evolutionary divergence and provides a useful model for investigating the increased virulence of *C. albicans*.[21]

64.1.2 Disease Incidence, Mortality, and Epidemiology

The true burden of candidal disease is not possible to quantify as surveillance of superficial diseases such as genital candidosis (thrush) is not performed routinely. However, the serious nature of invasive disease has led to the development of fungal registries in developed countries although these are often incomplete and confined to documentation of candidemia only. Pfaller and Diekema in their extensive review of *Candida* epidemiology predict nosocomial infection rates of between 2.5% and 10% resulting in 7,000–28,000 cases of candidemia and 2,800–11,200 deaths in the United States per annum.[2] In the United states, candidemias are the fourth leading cause of nosocomial bloodstream infection with a median incidence rate for all types of invasive candidal disease of 23.5 cases per 100,000 population over the years 1996–2003.[2] Outside of the United States, incidence rates are much lower and probably reflect differences in patient

TABLE 64.2
Incidence of Invasive Candidal Disease

Country [Reference]	Range	Incidence Rate/10^5 Population	Patient Group	No. Sites	Most Common Agent
Norway [91]	1991–2003	1.9–3.3	All	22	*C. albicans* (70%)
United States[a] [23]	1992–1993	8	All	87	*C. albicans* (53%)
United States[b] [22]	1998–2000	10	All	47	*C. albicans* (45%)
United States[c] [92]	1998–2001	6.0	All	17	*C. albicans* (57%)
United States[d] [2]	1996–2003	22–29	All	—	*C. albicans*
Iceland [93]	1980–1999	1.4–4.9	All	16	*C. albicans* (64.6%)
Iceland [94]	1991–2006	3.7–5.8	All	16	*C. albicans* (54.2%–70.6%)
Finland [95]	1995–1999	1.9	All	—	*C. albicans* (70%)
Canada [96]	1999–2004	2.8	All	4	*C. albicans* (51.2%)
Spain[e] [29]	1997–1999	3.5	All	19	*C. albicans* (Approx. 44%)
Spain [97]	2002–2003	4.3	All	14	*C. albicans* (51%)
Denmark [98]	2003–2004	11.0	All	34	*C. albicans* (63%)
Denmark [99]	2004–2006	10.4	All	41	*C. albicans* (53.8%–66.1%)
U.K.[f] [100]	2002–2005	2.3–3.1	All	—	*C. albicans* (51.2%–55.1%)
Scotland [101]	2005–2006	4.8	All	—	*C. albicans* (50%)
England [102]	2006	3.3	All	—	*C. albicans* (53%)

[a] San-Francisco and Atlanta.
[b] Connecticut State and Baltimore.
[c] Iowa State.
[d] The U.S. National Hospital Discharge Survey results.
[e] ECMM Survey.
[f] Excludes Scotland.

demographics and medical practice, particularly those within the intensive care setting (Table 64.2). In recent years, the global incidence of disease is best described as steady but varies with geographical location, patient age, ethnicity, hospital practice, and underlying disorder. The incidence is greatest at the extremes of the age spectrum with the very young (<1 year old) and elderly (>65 years old) more prone to disease.[2] An association with ethnicity has also been established in two U.S. studies where the incidence of disease was greater in all age groups in the black compared to the white population.[22,23] Links between incidence and the ICU are well established and relate to the use of clinical procedures and the disorders encountered within this setting. A range of medical conditions and surgical procedures places patients at varying degrees of risk of invasive candidal disease and many will require additional medical interventions such as venous catheters, mechanical ventilation, total parenteral nutrition, prolonged antibacterial or steroid use, all of which are additional independent risk factors. Patients undergoing major abdominal surgery and solid-organ transplantation or suffering from pancreatitis, premature birth, HIV infection, hematological malignancy, bone marrow transplantation, GI tract rupture, or solid tumors and burns may all suffer from invasive candidal disease. The geographical incidence of invasive candidal infections is described in Table 64.2.

Globally invasive candidal disease remains the most common cause of invasive fungal disease (IFD) but in specific patient groups, i.e., hematology, invasive candidal disease has been superseded by invasive mold infections and the distribution of species causing invasive candidal disease is changing.[2,24] Generally, five species of *Candida* (*C. albicans, C. glabrata, C. krusei, C. parapsilosis*, and *C. tropicalis*) cause more than 90% of all candidemias (Table 64.3). *Candida albicans* remains the most common cause of both superficial and invasive candidal disease although global data from the ARTEMIS DISK Surveillance program showed a 10% reduction in the incidence of *C. albicans* disease between 1997 and 2003, coinciding with increases mainly in the incidence of *C. tropicalis* and *C. parapsilosis*.[2] In a recent study of data collated from 2019 patients from 23 U.S. centers involved in the prospective antifungal therapy (PATH) alliance *C. albicans* remained the most common single cause of invasive candidal disease (45.6%) but was outweighed by the overall incidence of non-*albicans* disease (54.4%).[25] Generally, *C. glabrata* is the second most common species although in neonatal units and outbreak situations, *C. parapsilosis* is more prevalent, associated with skin, in particular hand colonization of health care workers and ability to form biofilms on catheters. *C. glabrata* is particularly prevalent in North America and to a lesser extent in Europe, whereas in the Middle East and South America, *C. tropicalis* and *C. parapsilosis* are more frequently encountered (Table 64.3). The impact of fluconazole prophylaxis is demonstrated in a single-center study evaluating candidemia over a 10-year period where *C. albicans* (13.5%) was demoted to the fourth most common cause of invasive candidal disease,

TABLE 64.3
The Global Epidemiology of Invasive Candidal Disease

Species (%)	Global	Global	North America	North America	Europe	England and Wales	England	Scotland	France	Spain	Denmark	Denmark	Asia/ Pacific	Australia	Latin America	Brazil	Saudi Arabia
C. albicans	73.3	62.3	51	45.6	56	60.1	53	52.0	57.0	51	63	59.8	56	62	50	40.9	53
C. glabrata	11.0	12.0	22	26.0	14	9.4	18	22.7	16.7	9	20	20.5	10	17	7	4.9	7
C. tropicalis	4.6	7.5	7	8.1	7	3.8	3	2.0	4.9	10	4	4.6	14	9	20	20.9	19
C parapsilosis	4.2	7.3	14	15.7	13	10.6	11	11.7	7.5	23	4	4.0	16	12	16	20.5	16
C krusei	1.7	2.7	2	2.5	2	1.5	1	1.0	5.2	4	3	4.1	2	13	2	1.1	2
C. guilliermondii	0.5	0.8	<1	0.3	1	0.6	1	3.3	—	—	—	—	<1	—	4	2.4	1
C. lusitaniae	0.5	0.6	<1	0.8	1	0.3	1	2.0	—	—	—	—	<1	—	<1	—	1
C. pelliculosa	—		—	—	—	0.02	—	—	—	—	—	—	—	—	—	6.2	—
C. dubliniensis	—	—	—	0.3	—	—	—	3.0	—	—	—	2.6	—	—	—	—	—
C. pseudotropicalis	—		—	—	—	—	—	—	—	—	—	—	—	—	—		
Other	—		>2	—	—	0.68	—	2.3	—	3	3	1.9	—	—	—	2.6	
Candida spp.	3.9	4.9	—	0.5	—	13	11	—	3.6	—	4	2.6	—	4	—		
Year	1997/1998	2003	2001/2004	2004/2008	1997–1999	1990/1999	2006	2005/2006	2005/2006	2002/2003	2003/2004	2004/6	2001/4	2002	2001/4	2003/4	1996–2004
Reference	[2]	[2]	[103]	[25]	[29]	[104]	[102]	[101]	[105]	[97]	[98]	[99]	[103]	[27]	[103]	[106]	[107]

behind *C. glabrata* (30.6%), *C. krusei* (24.2%), and *C. parapsilosis* (13.9%) in hematology patients when compared to the previous 5 years.[26] The prophylactic use of fluconazole was determined as a risk factor for *C. glabrata* and *C. krusei* candidemia.[26,27] This supports the results of the PATH alliance where previous azole use was associated with increased incidence of *C. krusei* candidemia.[25] However, this observation is not universal and long-term azole prophylaxis in one surgical ICU did not result in an increased incidence of azole-resistant species.[28]

Changing epidemiology may impact mortality rates and in both the PATH alliance and the European Confederation of Medical Mycology (ECMM) survey, crude mortality rates were greater for disease caused by *C. krusei* (52.9%–55.3%) and lower for disease caused by *C. parapsilosis* (23.7%–25.9%).[25,29] The mean crude mortality for all species and that for *C. albicans* in the PATH alliance and ECMM survey was 35.2%–38.5% and 35.9%–37.9%, respectively.[25,29] It was also noted that *C. krusei* disease was associated with patients with neutropenia and corticosteroid use.[25] Rather than attributing this increased mortality entirely on a putative enhanced virulence of *C. krusei*, the patient comorbidities and potential delay in the initiation of appropriate treatment are more likely to be implicated.

64.1.3 Clinical Features and Pathogenesis

Candida species can cause both superficial and invasive disease and for most cases, the source of the infection is endogenous. Superficial disease can be oropharyngeal, genital, cutaneous, and paronychial. In certain patients with congenital immunological or endocrinological disorders, recurrent disease occurs and results in chronic mucocutaneous candidosis. Oral candidosis has several major clinical presentations: pseudomembranous candidosis where raised white removable lesions/plaques can develop in debilitated patients or those with HIV disease; erythematous candidosis presents as a flat red lesion on the palate and tongue resulting in tenderness and swelling in patients on antibacterial agents or long-term corticosteroid use; and hyperplastic candidosis produces asymptomatic fixed lesions that occur on the inside of the cheeks of smokers or patients with oral trauma or dental neglect and the lesions are dysplastic and considered premalignant. Genital candidosis usually presents with itching, redness, and burning with occasional discharge, fissuring, and the presence of plaques.

Invasive disease can cause a variety of clinical manifestations with translocation from the GI tract or skin resulting in candidemia and hematogenous dissemination to a variety of organs in the critically ill population. Symptoms are nonspecific including a refractory fever and macronodular rash. Symptoms representative of specific manifestations (e.g., peritonitis, endocarditis, meningitis) are no different to bacterial disease. For a concise but informative review of all the clinical presentations of both superficial and invasive candidal diseases, see the book *Fungal Infection; Diagnosis and Management*.[30]

The pathogenicity of *Candida* varies between species. *C. albicans* is the most potent followed by *C. tropicalis*.[31] Sixty-four gene families have been positively identified in highly pathogenic *Candida* species and are associated with the cell wall and cell morphology, including hyphal and biofilm functions.[32] To become a human pathogen, an organism must be able to grow at 37°C and the ability to grow at higher temperatures (40°C) may contribute to virulence, particularly as invasive candidal infections result in fever.[12] *C. albicans*, *C. dubliniensis*, *C. famata*, *C. glabrata*, *C. inconspicua*, *C. kefyr*, *C. krusei*, *C. lusitaniae*, and *C. tropicalis* all have the capacity to grow at 40°C.[6]

Following infection, it is paramount that the interaction with the host is maintained and this is achieved through the production of adhesins (Alslp, Hwplp, Intlp, and Mntlp).[16] Adherence to human buccal cells appears to be greater for *C. albicans* than *C. parapsilosis*.[33] *C. albicans* possesses eight *ALS* family genes coding for cell well glycoprpoteins (adhesins) that are adhesive and/or promote flocculation and filamentation and the ability to adhere to different cell types, extracellular matrices, or foreign bodies such as endotracheal tubes or catheters.[31,34] Adhesins are situated at the cell surface as is the case for any hydrolytic enzymes or molecules used to avoid host defences. Proteolytic enzymes secreted by the fungi can hydrolyze host membranes to not only enhance tissue damage and the ease of invasion but to improve the binding efficiency between pathogen and host.[16,31] Maintaining an optimally functioning cell wall is essential and gene deletions affecting this have resulted in attenuated virulence in vivo.[35]

Secreted aspartyl proteinases (SAPs) are well established as *C. albicans* virulence factors and are only found in the two most pathogenic species *C. albicans* and *C. tropicalis*.[31] Produced by a family of 10 genes, SAPs function by destroying tissue and host defenses or by digesting proteins for nutrients. They have an important role in both colonization and infection, and it is proposed that they are required for invasion.[36] Recent research indicates that *C. albicans* metabolites (possibly SAP and ALS) cause actin rearrangement in macrophages significantly reducing their ability to phagocytose both *C. albicans* and *C. glabrata*.[37]

Within host tissue, *Candida* will take on a mix of morphologies. The formation of germ-tubes in human serum at body temperature indicates that this may be the preferred morphology for *C. albicans* within human tissue, although varying forms are found *in vivo*, demonstrating the effect of environmental pressures and the capacity the organism has to adapt and invade different niches.[38] During infection, hyphal structures are more invasive than yeasts and can penetrate deeper into endothelial tissue, expanding the area of tissue damage and allowing increased scope for yeast proliferation.[38] However, it cannot be definitively proven that one form of candidal cell is more virulent than the other.[31]

Candida cell surface molecules can trigger innate host responses via the toll-like receptors and other recognition molecules.[39] Certain mannoproteins can trigger a Th1 cytokine response, a reaction to acute candidemia resulting in sepsis. Conversely, exposure to other mannoproteins, dectin, and glucans may result in a Th2 cytokine response, a reaction to persistent fungal infection. The application of this Th1/Th2 paradigm to human infection is far from clear and recent recognition of the role of regulatory T cells (T-reg) and Th17 responses further complicates the picture. Recent evidence suggests *Candida* can modulate and evade host immunity and use immunomodulatory molecules that enable them to promote protective tolerance leading to colonization and persistence.[40]

64.1.4 Diagnosis

64.1.4.1 Conventional Techniques

The conventional diagnosis of candidal disease is based on the combination of clinical features with microbiological, radiological, and histopathological evidence. The ability to culture the organism can be important but in anatomical sites where *Candida* is commensal, it is beneficial to show evidence of an invasive process. Microscopy presents typical yeast cells, pseudo or true hyphae remembering that only *C. albicans* produces true hyphae and that *C. glabrata* only produces yeast cells in tissue.[30] The isolation of *Candida* from sterile sites, including deep fluids/tissue biopsies (excluding GI tract) and blood is significant and lytic blood culture methods should be performed in all suspected cases of invasive disease and will provide detection rates between 50% and 80%, depending on patient population and underlying candidal disease.[30] Culture from oral, genital, GI and respiratory tract must be balanced against the possibility of it arising as a result of colonization.

If *Candida* is successfully isolated from a patient, then all isolates from clinically significant sites should be identified to a species level to determine antifungal susceptibility.[41] Classically, *Candida* species are indentified on the basis of their assimilation and fermentation patterns, as described in Table 64.1 and many commercial kits are available for this purpose (API 20C and API *Candida*, BioMerieux; AUXACOLOR, Biorad). Both the AUXACOLOR and API Candida systems correctly identified >90% of clinical yeast isolates when comparatively evaluated.[42] Alternatively, the most common species of *Candida* can be identified using chromogenic agar where usually glucosaminide derivatives are metabolized by *Candida* to produce colonial colors that vary according to species or group of species.[43]

As *C. albicans* remains the most common cause of both superficial and invasive disease, the identification of this species takes precedence in most clinical laboratories and tests have been developed to distinguish between *C. albicans* and most other species. The phenomenon of "phenotypic switching" has been used as a laboratory technique to differentiate *C. albicans* from non-albicans species. When grown on heated blood agar in an atmosphere of 5% CO_2, *C. albicans* colonies grow with an irregular margin that distinguishes them from other *Candida* species.[44] In addition, the formation of germ tubes within 3 h when incubated in serum at 37°C is a commonly used laboratory test for determining the identity of *C. albicans*, although *C. dubliniensis* also possesses this ability.[45] Another useful test to distinguish *C. albicans* and *C. dubliniensis* from other *Candida* species is the production of large, thick-walled chlamydospores when grown under a cover slip (oxygen limited) on corn-meal agar. It has been noted that *C. albicans* strains usually produce only one terminal chlamydospore whereas *C. dubliniensis* produces two to three, generally on shorter pseudohyphae.[8] However, this is not truly definitive and relies on the discretion of the mycologist, a more accurate test being the production of chlamydospores by *C. dubliniensis* and not *C. albicans* when grown on Staib agar.[7]

An alternative approach not reliant on culture is the serological detection of candidal cell wall antigens and/or anti-*Candida* antibodies. The main targets are mannan and β-D-Glucan, with mannan being specific for *Candida* species and β-D-glucan being found in a range of different fungi. For improved sensitivity, the mannan antigen test can be combined with anti-mannan antibody test and a sensitivity/specificity of 80%/93% has been achieved.[46] The usefulness of the antibody test in certain immunosuppressed populations incapable of raising a sufficient immune response must be questioned. All the above tests benefit from quality control achieved by the availability of standardized commercial kits and as such are included in disease-defining criteria despite variable clinical performance. Both latex agglutination (LA) and ELISA kits are available although the former has a reduced sensitivity (25% LA vs. 75% ELISA).[47] Detection of mannanaemia and anti-mannan antibodies has shown varied performance between candidal species ranging from 100% sensitivity for *C. albicans* to 40% for *C. parapsilosis*.[48]

In a recent comparison of nonculture diagnostics in patients with candidemia, the following sensitivities were generated 88%, 47%, 41%, 47%, and 75% for PCR, β-D-glucan, mannan ELISA, anti-mannan antibody detection, and combined mannan-antibody, respectively.[49] For a more expansive review of fungal cell wall immunodiagnostics, see the recent article by Kedzierska et al.[50]

The difficulty in definitively diagnosing invasive candidal disease results in many patients receiving unnecessary empirical antifungal therapy. Without a "gold-standard" diagnostic test, diagnosis is often based on a combined strategy of risk factors, microbiology, and clinical symptoms. Varying disease definitions have been described in the literature and makes comparison between studies difficult. To overcome these problems, consensus definitions for IFD using host, clinical, and microbiological criteria have been agreed.[51] Whilst these encompass *Candida* and certain molds, they were designed for clinical and epidemiological research purposes and were not intended for routine clinical practice. Initially, covering patients with cancer and hematological malignancy, the revised definitions were extended to patients with hereditary immunodeficiencies, connective tissue disorders, and solid-organ transplant patients. However, critically ill patients on the ICU and surgical patients are not covered by these criteria, which concentrate on candidemia, histologically proven disease, and chronic disseminated candiodosis. As a consequence, the majority of the invasive candidal burden will not fall within the remit of these consensus definitions.

Several published studies have investigated the predictive value of risk factors for diagnosis of invasive fungal infection in non-neutropenic, critically ill adult patients. *Candida* colonization is an independent risk factor for infection and precedes invasive disease in most cases. More than 50% of ICU patients will become colonized during hospitalization and distinguishing infection from colonization may be difficult. Risk of infection increases with the number of sites colonized and is dependent on the colonizing species.[52] Further studies in surgical ICU patients have confirmed this trend and recommend the use of a *Candida* colonization index to assess the degree of colonization. A ratio of ≥0.5 calculated from the number of noncontiguous sites colonized with the same strain over the number of sites sampled, has been shown to have a good positive predictive value.[53] Some studies have attempted to develop or evaluate a risk score for invasive fungal infection or an algorithm for use of antifungal drugs (either prophylactically or preemptively) in non-neutropenic, critically ill adult patients.[54,55] Many of these scoring systems have been developed and evaluated within the same data set. Given the geographical variation in incidence, scoring systems will identify different proportions of patients. Sensitivity will vary accordingly.

64.1.4.2 Molecular Techniques

With the increasing population of patients at risk and the emergence of resistant species causing these infections, the need for rapid and accurate diagnostics is imperative. Molecular techniques have the potential to offer optimum accuracy of detection of these pathogens with as yet unparalleled speed, although the economic necessity of batch testing may compromise diagnostic efficiency.

Publications on molecular detection and identification of *Candida* species are many and varied as highlighted in Table 64.4. Clinical sensitivity and specificity are difficult to compare between studies of varying design and sensitivity will be influenced by the specimen type and the consequent DNA extraction protocol. Furthermore, the evaluation of theoretically more sensitive molecular methods by comparison to "reference" conventional techniques is somewhat of a misnomer. Nonetheless, three publications highlighted in Table 64.4 reported clinical sensitivity and specificity results for *Candida* specific PCR and sensitivity ranged from 90.9% to 100% with specificity 97% or greater.[47,56,57] Between these three studies, both hematology and critical-care patients have been studied and both serum and whole blood have been successfully tested confirming the presence of both a free DNA and cell-associated DNA source and a potential clinical benefit in varying patient populations.

Limit and range of detection differ widely although most tests reportedly detect clinically relevant levels of *Candida* (10^0 to 10^1 cfu/mL). Despite this, some authors have reported discordant results compared to traditional culture techniques.[58,59] This may be due to the uneven distribution of pathogens within clinical specimens (especially tissue biopsies), the transient nature of candidemias, and/or the lack of consensus regarding optimal specimen type and subsequent extraction procedures and platforms.

Choices of extraction method are influenced by specimen type, which is in turn dictated by the pathology of the disease. The vast majority of papers have described amplification procedures utilizing primer sequences that target conserved multi-copy genes to heighten assay sensitivity and specificity. Frequently, these primers flank less-conserved regions of the genome to allow for species differentiation by various detection methods. Assays have been developed for direct detection from clinical specimens and for identification of cultured isolates. The majority of publications focus on samples for the diagnosis and/or preemptive monitoring of patients at risk of invasive candidiasis (IC). Expedited identification from clinical specimens and cultured isolates allow for more timely and appropriate treatment of infected patients, a reduction in the cost and morbidity associated with inappropriate treatment of those without disease.

64.1.4.2.1 *Specimen Type*

The most useful specimen for diagnosing candidemia should be blood. As already discussed, blood culture is not 100% successful although its limited positivity confirms the presence of viable organism within the peripheral circulation. Possible routes of entry of the organism into the blood stream are highlighted in Figure 64.1 with the three main sources of *Candida* being endogenous, hospital acquired, or via a traumatic disruption to the anatomical barriers. The most pertinent blood fraction is open to debate and may differ according to disease presentation. Most investigations use whole blood although serum and plasma testing have been shown to be a suitable alternative (Table 64.4). Testing whole blood requires large volumes for optimal detection of the low fungal burden but also increases the opportunity for the presence of inhibitors. The consequent laborious extraction procedures and the protocol dependent inefficiency to extract "free DNA" are further weaknesses. Smaller volumes and simpler extraction procedures make serum and plasma testing attractive. However, loss of cell-associated fragments/organism and the presence of DNAses within blood may compromise utility. In addition, the interpretation of DNAemia is still poorly understood.

Other clinical specimens utilized include various respiratory secretions,[60,61] fresh and paraffin-embedded tissue specimens and biopsies,[62,63] ocular samples,[64] urine,[65] and nail clippings.[66] The significance of detection of *Candida* in non-sterile anatomical sites, particularly respiratory, genital, and GI tract must be balanced against the detection of commensal organisms and it is difficult to assign significance without other evidence of an invasive process, e.g., histopathologically. The majority of publications report pan-fungal assays that include *Candida* species within their detection capability although with decreased sensitivity compared to the assays utilizing blood specimens (Table 64.4).

TABLE 64.4
Molecular Techniques Used to Detect *Candida* Species from Clinical Specimens

Predominant Sample Type	Limit of Detection	*Candida* Species Detected	Target Region	Detection Method	Reference
Serum	0.22–2.3 copies/mL	Calb, Cdub, Cglab, Cpara, Ckrus, Ctrop	18S rRNA and ITS	Real-time PCR—hydrolysis	[57]
Serum	1 organism/mL	Clab, Cglab, Cpara, Ctrop	ITS	Semi-nested PCR and AGE	[70]
Whole blood	1 cfu/mL	Calb, Cdub, Cglab, Ckef, Ckrus, Cpara, Ctrop	18S rRNA	PCR and Southern blot	[71]
Whole blood	1–5 cfu/mL	Calb, Cdub, Cglab, Ckef, Ckrus, Cpara, Ctrop	18S rRNA	Real-time PCR—biprobe	[47]
Whole blood	1–5 cfu/mL	Calb	18S rRNA	Real-time PCR—hybridization	[88]
Whole blood and plasma	1–10 cfu/mL	Calb, Cglab, Ckrus, Clus, Ctrop	18S rRNA	NASBA-enzymatic, bead-based, and membrane-based detection	[82]
Whole blood	1–100 cells/mL	Calb, Cglab	ITS region	PCR-DEIA	[75]
Whole blood	2 cfu/mL	Calb, Cdub, Cglab, Cguil, Ckef, Ckrus, Clus, Cnorv, Cpara, Ctrop	18S rRNA	Real-time PCR—hybridization	[108]
Whole blood	2 cfu/mL	Calb, Ckrus, Cpara, Ctrop	ITS region	PCR-ELISA	[74]
Whole blood	3 cfu/mL	Calb, Cdub, Cglab, Cguil, Ckef, Ckrus, Clus, Cnorv, Cpara, Ctrop	18S rRNA	Real-time PCR—biprobe	[59]
Whole blood	4 cfu/mL	Pan-fungal	18S rRNA	PCR and Southern blot	[72]
Whole blood	4–5 cfu/mL	Pan-fungal	18S rRNA	PCR and AGE	[69]
Whole blood	5 cfu/mL	Calb	18SrRNA	PCR-ELISA	[73]
Whole blood	5 cfu/mL	Calb, pan-fungal	ITS region	Real-time PCR—hydrolysis	[56]
Whole blood	10 cfu/mL	Pan-fungal	18S rRNA	Real-time PCR—hybridization	[109]
Whole blood/biopsy	<10 organisms/reaction	Calb, Cdub, Cglab, Cguil, Ckef, Ckrus, Clus, Cpara, Ctrop	28S rRNA	RQ-PCR	[110]
Whole blood and plasma	50 copies	Calb, Cdub, Cfam, Cglab, Cguil, Ckrus, Cpara, Ctrop	RNase P RNA	Real-time PCR—hydrolysis	[111]
Whole blood	200 fg DNA	Calb, Cglab, Cguil, Ckef, Ckrus, Cpara, Ctrop	P450 LDMG	PCR-REA	[112]
Respiratory Samples	5–10 cfu/mL	Calb, Cdub, Cglab, Ckrus, Cpara, Ctrop	ITS region	Real-time PCR—biprobe	[60]
Respiratory Samples	25 cfu/mL	Pan-fungal	28S rRNA	Real-time PCR—hydrolysis	[61]
Tissues	2–20 cfu	Pan-fungal	ITS region	PCR-MLH	[62]
Tissues	—	Pan-fungal	ITS region	PCR-sequence	[63]
Nail clippings	100–1000 plasmid copies	Calb, Cdub, Cglab, Cguil, Ckrus, Cpara, Ctrop	28S rRNA	Nested PCR and AGE	[66]
Ocular samples	1 fg DNA	Pan-fungal	ITS region	Semi-nested PCR and AGE	[64]
Urine	1–50 cfu/mL	Calb	CaHSP70	PCR and dot-blot	[65]
Blood culture	1 cell/2 μL sample	Calb, Cglab, Ckrus, Cpara, Ctrop	ITS 2 region	PCR and EPF	[113]
Blood culture media	2 cfu/mL	Calb, Cdub, Cglab, Cguil, Ckrus, Clus, Cpara, Ctrop	ITS region	Multiplex PCR and AGE	[68]
Blood culture media	10 cfu/mL	Calb, Cdub, Cglab, Cguil, Ckrus, Cpara, Ctrop	ITS region	MT-PCR	[58]
Blood culture	10 cfu/mL	Panfungal and Calb, Cdub, Cfam, Cglab, Cguil, Cinc, Ckef, Ckrus, Clipo, Clus, Cmem, Cnorv, Cpara, Cparar, Cpel, Csak, Ctrop, Cuti, Cval, Czey	18S, ITS and 28S rRNA regions	Semi-nested PCR-xMAP	[79]
Blood culture	100 cfu/mL	Calb, Cglab, Ckrus, Cpara, Ctrop	18S rRNA and ITS region	Nested and real-time PCR—hydrolysis	[114]

TABLE 64.4 (continued)
Molecular Techniques Used to Detect *Candida* Species from Clinical Specimens

Predominant Sample Type	Limit of Detection	*Candida* Species Detected	Target Region	Detection Method	Reference
Fungal isolates	—	Calb	Numerous regions (77 different primer pairs employed)	LSplex PCR and microarray	[78]
Fungal isolates	—	Yeasts including: Calb, Cdub, Ckrus, Clamb, Cpara, Ctrop	ITS region	PCR and ACE	[76]
Fungal isolates	—	Pan-fungal	18S rRNA	fTEFAP	[80]
Fungal isolates	—	Calb, Cglab, Ckef, Ckrus, Cpara, Ctrop	ITS region	Real-time PCR—hydrolysis	[89]

PCR, conventional PCR; CaHSP70, *C. albicans* heat shock protein 70; AGE, agarose gel electrophoresis; EPF, end-point fluorescence; ACE, automated capillary electrophoresis; fTEFAP, fungal tag-encoded FLX amplicon pyrosequencing; LSplex, large-scale multiplex; MT-PCR, multiplex-tandem PCR; MLH, multiplex liquid hybridization; xMAP, Luminex xMAP technology; NASBA, nucleic acid sequence-based amplification; EIA, enzyme immunosorbent assay; DEIA, DNA-enzyme immunosorbent assay; RQ-PCR, real-time quantitative PCR; REA, restriction enzyme analysis; P450 LDMG, cytochrome P-450 lanosterol-14α-demethylase gene.

Abbreviations: Calb, *C. albicans*; Cdub, *C. dubliniensis*; Cfam, *C. famata*; Cglab, *C. glabrata*; Cguil, *C. guilliermondii*; Cinc, *C. inconspicua*; Ckef, *C. kefyr*; Ckrus, *C. krusei*; Clamb, *C. lambica*; Clipo, *C. lipolytica*; Clus, *C. lusitaniae*; Cmem, *C. membranifaciens*; Cnorv, *C. norvegenesis*; Cpara, *C. parapsilosis*; Cparar, *C. pararugosa*; Cpel, *C. pelliculosa*; Csak, *C. sake*; Ctrop, *C. tropicalis*; Cuti, *C. utilis.*; Cval, *C. valida*; Czey, *C. zeylanoides*.

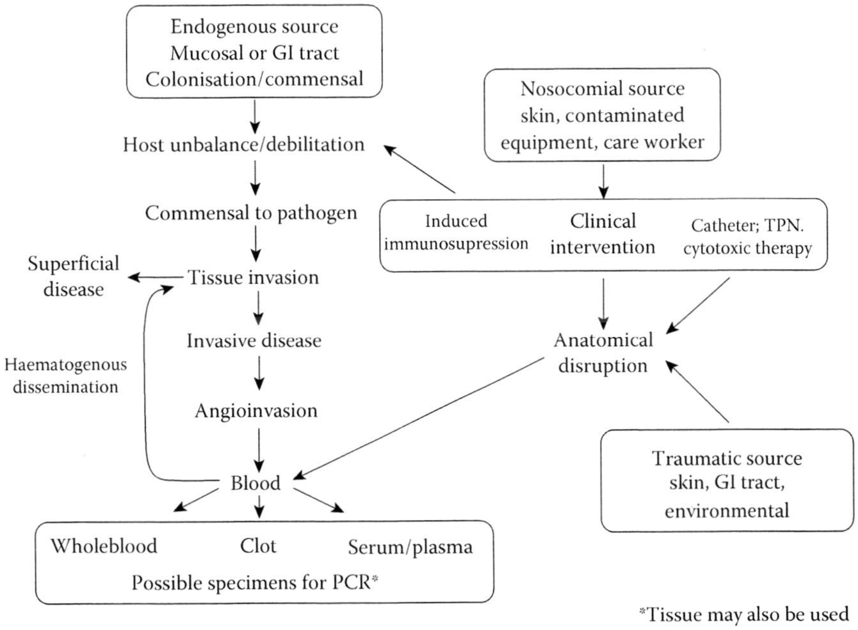

FIGURE 64.1 Candidal disease, sources, disease progression, and samples for molecular testing.

The identification of *Candida* species from culture using molecular techniques may be useful, particularly for peculiar species that are difficult to identify using classical techniques. The application of pyrosequencing looks particularly promising.[67] These assays utilize straightforward extraction procedures and provide adequate performance when testing cultures. Improved turnaround time in relation to conventional identification is reported, as well as accurate identification of polyfungal infections.[68] The process is limited by its reliance on suboptimal conventional culture detection. Direct molecular testing of blood cultures will be reliant on the organism multiplying within the blood culture bottle and if a specimen with insufficient inoculum or nonviable organisms is introduced into a blood culture system, the concentration of target will be diluted 10-fold to below the detection limit of PCR testing. In this instance, it would have been easier to test the original blood directly.

64.1.4.2.2 Detection Technologies

A wide range of real-time (RT) and end-point detection assays have been employed for the diagnosis of *Candida* disease and are represented in Table 64.4. Conventional PCR followed by end-point detection techniques such as agarose gel electrophoresis (AGE),[64,68–70] blotting,[65,71,72] enzyme immunoassay (EIA),[73–75] multiplex liquid hybridization (MLH),[62] and automated capillary electrophoresis (ACE)[76] have demonstrated suitable analytical sensitivity for clinical diagnostics. However, these techniques are often time-consuming, with amplification time ranging from 2 to 3 h and time for detection of at least 1 h. They also require facilities, equipment, and expertise with limited availability in most diagnostic facilities. Insufficient specificity and sensitivity for species-level identification may also be a drawback with the most basic AGE analysis and additional post-amplification processing required. All of these assays require the opening of reactions vessels subsequent to amplification. This increases the risk of false-positive results through amplicon contamination, especially for nested protocols, which may be incorporated to improve sensitivity.

RT technologies have become popular, partially due to increased rapidity, sensitivity, and specificity in comparison to many end-point detection systems. These protocols often require no more than 1 h to perform the same number of cycles of amplification that would take 2–3 h on a conventional thermal block. The incorporation of specific detection probes within the amplification mix also allows for greater specificity than most conventional PCR detection systems and allows visualization of results as amplification occurs. The compatibility of RT platforms with diagnostic facilities, especially in terms of their foot-print and closed-tube nature, has further enhanced their popularity by reducing and controlling the contamination risk. Batch testing remains a financial imperative in clinical molecular diagnostics although the development of platforms that circumvent batch testing requirements may in the future facilitate on-demand even bedside testing.

RT detection is limited by the number of channels, usually no more than four, available to detect reporter dyes and by the nature of the reaction kinetics employed. When only genus-specific detection is required, RT detection is relatively straightforward. Where a species-specific assay is sought to differentiate the main pathogenic species or identify potentially resistant species (*C. glabrata* or *C. krusei*), it becomes necessary to devise a more complex detection grouping in order to leave capacity for an internal control reporter. This is essential if false-negative results are to be avoided. If a RT pan-fungal approach is desired in addition to *Candida* detection, the similarity of conserved regions to the human genome and differences in less-conserved regions amongst clinically significant yeasts and molds makes differentiation challenging using a single RT assay.

A vast array of RT platforms have been used for *Candida* detection, including bi-probes, hydrolysis (TaqMan) probes, and hybridization (FRET) probes (Table 64.4). The choice of probe can be limited by platform and the choice of platform is most often dictated by existing technologies within laboratories although cost, familiarity with, and diversity of application of new equipment and user preference are also important. Plate-based platforms are becoming popular as they allow for larger batches to be processed at one time. Fast technologies are also increasingly common within molecular laboratories although adapting assays for fast kinetics is not always straightforward.

Nested/semi-nested assays may be required in order to achieve clinically significant sensitivity, thereby increasing the turnaround times and cost. Indeed, the experience of the authors, supported by many of the papers published for RT molecular diagnosis of fungal disease, shows that to achieve the required analytical sensitivity for clinical diagnostics, more than 35 cycles of amplification is frequently required. For most cases, this only leaves a small window of opportunity for accurate detection as beyond 40 cycles, many RT assays give spurious non-reproducible results due to a combination of Poisson's law and probe degradation/nonspecific interactions between the oligonucleotides. Nested or semi-nested systems have been used to overcome this problem, but the necessity to handle PCR products leads to contamination problems and more frequent false positivity, compromising the specificity/positive predictive value of a diagnostic assay when compared to RT systems. For any PCR test, it is important to interpret results correctly within the clinical context, placing greater significance on multiple positive results than a single non-reproducible PCR positive result.

Multiplex PCR reactions where two or more primer pairs are used to increase the detection range while maintaining analytical specificity are technically demanding, but allow for the detection of a larger number of targets within a single test. Multiplex PCRs have been developed in tandem with both RT and conventional detection techniques as well as more recently with microarray analysis,[77,78] Luminex Technology,[79] and rapid sequencing techniques.[67,80] A reduction in assay sensitivity is often seen when reactions are multiplexed, leading some researchers to develop nested/semi-nested procedures despite the potential drawbacks already discussed.[79] Currently, no publications are available that demonstrate the application of microarray analysis or rapid sequence detection directly to clinical specimens, probably a result of a reduced analytical threshold inherent with these protocols as they now exist.

Candidal RNA has also been targeted via NASBA and reverse transcriptase-PCR (RT-PCR).[81,82] Both RT-PCR and NASBA are end point analysis system suffering similar limitations to conventional PCR. To our knowledge, no RT-NASBA methods, utilizing beacon technology, have been described. However, reverse transcriptase PCR has been combined with RT detection but only after the synthesis of cDNA.[81,83] It is possible to perform the reverse transcription and RT-PCR in a single tube and this should be performed to avoid the problems mentioned.

There are a host of assays suitable for the accurate and rapid detection and/or identification of *Candida* species from clinical specimens and cultured isolates giving interested health professionals from varied facilities the opportunity to validate molecular assays for their particular needs (Table 64.4). The choice of assay should depend on the range of detection required (*Candida*/resistance species/panfungal) dictated by the patient group of concern. The optimum specimen is still debated. Users should ensure that extraction procedures are fit for purpose for the chosen sample type (Figure 64.1).

64.2 METHODS

64.2.1 Sample Preparation

64.2.1.1 DNA Extraction from *Candida* Cultures

1. Harvest 1 mL of overnight liquid culture by centrifugation at 5000×*g* for 5 min, remove supernatant, and resuspend in lyticase buffer. Alternatively, simply scrape several colonies from an agar plate and resuspend in lyticase buffer.
2. Extract DNA from *Candida* isolates using recombinant lyticase (Sigma-Aldrich) and the QiaAmp tissue kit (Qiagen) as described previously.[84]

   For a cheaper alternative to recombinant lyticase, simply vortex the harvested *Candida* culture with Roche MagNA Lyser Green Beads for 30 s. Pulse centrifuge and wash with 200 µL of molecular grade water. Transfer washings to the Qiamp tissue kit and continue as directed by the manufacturer's protocol.

   Note: (1) If enhanced sensitivity is required when precipitating DNA with ethanol or isopropanol, incubate the solution on ice for 30 min to increase the yield. Secondly, after the elution of the DNA, YM-100 microconcentrators (Millipore) can be used to increase the DNA concentration, although this increases the opportunity for contamination. (2) All reagents are filter-sterilized through 0.2 µm filters before use. (3) After the fungal cell lysis, it is possible to replace QiaAmp tissue kit with Roche High Pure Template DNA spin columns or even use automated extractors such as the Qiagen EZ1 (EZ1 tissue kit) or Roche MagNA Pure LC (Total NA kit).
3. For a quick and simple candidal DNA extraction, simply harvest cells as above, remove supernatant, and resuspend pellet in 60 µL of Microlysis plus buffer (Microzone) and incubate at 65°C for 15–30 min, then 95°C for 5 min. The extract is now ready for PCR amplification.

Note: The use of bead-beating after cell harvesting followed by Microlysis plus extraction may improve DNA yield.

64.2.1.2 DNA Extraction from Clinical Specimens

64.2.1.2.1 Whole Blood

In principle, when using whole blood, both free DNA and cell-associated DNA can be targeted, although in practice, free DNA is lost during the processing procedure. Cell-associated DNA sources include both the viable organism as would be the case in a candidemia and organism rendered nonviable by the action of the host immune system. This nonviable target should provide molecular testing with an improved sensitivity compared to culture. Optimal volumes for testing are 2–4 mL and EDTA anticoagulated should only be used, as heparin can inhibit PCR processes and sodium citrate has been associated with fungal contamination, albeit *Aspergillus*. The DNA extraction process is laborious and involves red cell lysis, white cell lysis, fungal lysis, followed by DNA purification and precipitation and is described as follows:

1. Add 10 mL of red cell lysis buffer to 3 mL blood in a sterile 15 mL tube (If processing fresh non-frozen blood shake at room temperature for 5 min).
2. Centrifuge at 3000×*g* for 10 min in a sealed rotor centrifuge. Decant the supernatant, leaving the pellet and/or the remaining 0.5 mL of fluid, whichever is the greater volume.
3. Repeat steps 1 and 2, then proceed to step 4.
4. Resuspend the pellet in 1 mL white cell lysis buffer.
5. Incubate tubes at 65°C for 30–45 min.
6. Centrifuge at 7500×*g* in a sealed centrifuge for 10 min.
7. Discard the supernatant being careful not to disturb the pellet (that may not be visible). Add 30 µL of the Roche MagNA lyser Green Beads and vortex vigorously for 30 s.
8. Pulse centrifuge, wash beads with molecular grade water, transfer washings to QIAamp tissue kit and follow manufacturer's protocol.

Red Cell Lysis buffer (RCLB): 10 mM Tris pH 7.6, 5 mM $MgCl_2$, 10 mM NaCl. Sterilize by autoclaving and store at room temperature.

White Cell Lysis buffer (WCLB): 10 mM Tris, 50 mM NaCl, 0.2% SDS, 10 mM EDTA. Make up and filter sterilize and store at room temperature. Make up proteinase K at 20 mg/mL and freeze aliquots of 100 µL at −80°C. On the day of extraction, thaw aliquots and add contents to 10 mL of rest of buffer to give 200 µg/mL and use immediately.

Note: An interesting concept permitting the targeting of both the free DNA and cell-associated DNA from a whole blood sample is to separate the plasma and blood cell fractions by centrifugation, then remove the plasma and save. Process the cellular material as described above (steps 1–8), then re-introduce the plasma when the commercial DNA purification and precipitation is performed.[85]

64.2.1.2.2 Serum or Plasma

The use of serum or plasma only permits the detection of free circulating DNA as any cell-associated DNA is lost during sample processing. When the specimen is centrifuged to remove the human blood cells, candidal blastospores, phagocytosed, or viable will also be sedimented whereas the free DNA will be retained in the plasma or serum fraction. Free DNA may be released from the fungi by the action of the host immune response, autolysis due to nutrient limitation, cellular damage arising from the interstitial pressures experienced during tissue invasion, or affect of antifungal therapies, most of which target the fungal cell wall or membrane. The larger the sample volume, the increased opportunity for detection of DNAemia and typically 1 mL volumes are used, but volumes are often limited by the system used and many automated systems only permit the use of volumes up to 0.5 mL. DNA within samples can be concentrated by using YM-100 microconcentrators (Millipore). DNA extraction from serum or plasma can be performed using most commercial manual or automated kits, but detection thresholds should be determined and contamination excluded prior to routine diagnostic use.

Note: It is not necessary to perform bead-beating or lyticase digestion when using serum.

64.2.1.2.3 Tissue

The essential step when testing tissue specimens is the efficient digestion of the tissue to allow access to the invading fungi. Both paraffin/formalin embedded and fresh tissue have been used and from a technical perspective, it is easier to use fresh tissue as this does not require additional processing to remove the fixative. Conversely, if a specimen has been shown to contain fungal elements by histopathological investigation, then testing the same specimen excludes the possibility of false negativity arising from testing a different biopsy or part of the biopsy that does not contain the invasive agent. Tissue should be digested with a combination of SDS/proteinase K and may require overnight incubation. After complete tissue lysis, the fungal cell is targeted by bead-beating or lyticase before completing the process using one of the previously mentioned commercial methods.

64.2.2 Detection Procedures

Providing the PCR assay meets the detection requirements of the user any PCR amplification system can be used to test any clinical specimen or culture providing the DNA extraction is suited to purpose and is capable of processing the specimen type.

Below a detailed quantitative RT-PCR system is described. It is based on a qualitative bi-probe assay (Sybr Green with Cy-5 probe) designed to detect, but not differentiate the main pathogenic species of *Candida* (i.e., *C. albicans*, *C. dubliniensis*, *C. glabrata*, *C. kefyr*, *C. krusei*, *C. parapsilosis*, and *C. tropicalis*) using the Idaho Technologies Light-Cycler system.[86] Due to sequence variations, the original assay was found to have a variable detection limit when testing certain species, particularly reduced for *C. glabrata*.[87] Furthermore, the development of new RT technologies lead to the modification of the assay for use on more widely used improved platforms (e.g., Roche Light-Cycler, Applied BioSystems TaqMan, or Corbett RotorGene) and was multiplexed to allow the differentiation of *C. glabrata* and *C. krusei*. This assay uses the same pan-fungal primers (L18F, 5′-CTCGTAGTTGAACCTTGG-3′; L18R, 5′-GCCTGCTTTGAACACTCT-3′) that target two conserved regions encompassing a variable region within the 18S rRNA gene, resulting in the production of a 140 bp amplicon that corresponds to the nucleotide positions of 620–760 in the 18S rRNA gene of *C. albicans*. The amplified product is then detected by hydrolysis probes (Generic *Candida* probe, 5′-FAM ATC TTT TTG ATG CGT ACT GGA CCC TG—BHQ1-3′, *C. glabrata* probe, 5′-JOE GGC TAA CCC CAA GTC CTT GTG GCT T—BHQ1-3′ and *C. krusei* 5′-ROX TAC CTA TGG TAA GCA CTG TTG CGG C—BHQ2-3′) that bind to a *Candida*-specific sequence within the PCR amplicon.

Procedure

1. Prepare PCR mixture (20 μL) consisting of 2.0 μL Light-Cycler FastStart DNA Master hybprobe master-mix (Roche) [containing dNTPs, Faststart *Taq* DNA polymerase, 10 mM $MgCl_2$], 2.0 μL each of L18F and L18R primers (final concentration 750 nM), 2.0 μL *Candida* probe (final concentration 400 nM), 2.4 μL of 25 mM $MgCl_2$, up to 9.6 μL template DNA, and molecular grade water as required.
   Note: The probe can be monoplex or multiplex.
2. Conduct PCR amplification on a Roche Light-Cycler, Corbett Rotorgene, or Applied Biosystems TaqMan as follows: 1 cycle of 95°C for 10 min; 50 cycles of 95°C for 15 s, 62°C for 30 s with a single acquisition to the required channels (FAM-generic, JOE-*glabrata* and ROX-*krusei*), 1 cycle 40°C for 30 s.

Note: The generic assay has been optimized for Roche Light-Cycler and may require further optimization for other platforms. The multiplex assay has only been performed on the Corbett Rotorgene and reaction volumes can be adjusted to maintain the same concentrations but allow a 15 μL DNA input in a 50 μL reaction volume. Different fluorophores are required for multiplex use on the Roche Light-Cycler. The *Candida* assay specifically recognizes *C. albicans*, *C. kefyr*, *C. krusei*, *C. dubliniensis*, *C. tropicalis*, *C. parapsilosis*, and *C. glabrata* but not *C. guilliermondii*, *C. lipolytica*, *C. famata*, *A. fumigatus*, *A. niger*, *S. cerevisiae*, or *Cryptococcus neoformans*. Although all the species above produce an amplicon with primers L18F and L18R, only *C. albicans*, *C. kefyr*, *C. krusei*, *C. dubliniensis*, *C. tropicalis*, *C. parapsilosis*, and *C. glabrata* produce a probe signal. The assay has a sensitivity of 5 *C. albicans* cfu per mL of spiked whole blood.[86]

For a hybridization (FRET) probe detection system, see the manuscript of Loeffler et al.[88] For a hydrolysis (TaqMan) probe detection system capable of further differentiation of *Candida* species, see the manuscripts of McMullan et al. and Guiver et al.[57,89]

64.3 FUTURE PERSPECTIVES AND CONCLUSIONS

The advantages of molecular diagnostics are well known. One of the main disadvantages of molecular detection of any specific pathogen has been that molecular assays are not catch-all tests. For diseases such as IFD, with a low prevalence in a population, the ability to exclude multiple pathogens in a single reaction is beneficial. The increasing ability to efficiently perform multiplex amplification and detection or rapidly sequence even large sequences of any given genome hold the potential to provide more comprehensive pan-genomic/syndromic approaches to molecular diagnosis in the future. This is demonstrated by recent publications on Luminex technology,[79] MT-PCR,[58] Microarray,[77,78] and Pyrosequencer[80] based detection and/or identification of fungal pathogens, including *Candida* species.

MT-PCR is a novel nested-style multiplex method that has the potential to detect up to 72 targets simultaneously and has been shown to have a detection limit of 10 cfu/mL.[58] A small number of cycles of a first round multiplex PCR enhances analytical sensitivity by avoiding competition between amplicons and is followed by subsequent analysis in multiple monplex PCRs using nested specific primers to improve target accuracy.[90] Luminex xMAP technology utilizes multiplex PCR with flow cytometry detection of biotinylated target DNA on oligonucleotide labeled capture probe microbeads. It has the capability of detecting up to 100 targets in a single reaction tube, although specificity remains an issue and nested amplification may be required to provide clinically significant limits of detection of *Candida* species.[79] Both methods have been shown to effectively diagnose candidal infection from a variety of clinical specimens.

Microarray-based technology gives the opportunity of amplifying hundreds of different gene segments in one test via multiplex PCR followed by DNA chip hybridization. Proof of concept microarray-based detection and identification of fungal pathogens have been published although specificity and sensitivity issues persist as well as the expense of these assays being somewhat prohibitive.[77] Their ability to not only detect and differentiate *Candida* species but also determine their likely resistance profiles, together with improving technologies that enhance analytical sensitivity 100- to 1000-fold make them a continuing area of interest for researchers and a possibility for future diagnostic tests.[78] Equally, rapid sequence-based methods although currently not sufficiently sensitive to allow detection and identification from clinical samples, show the ability to accurately and rapidly identify *Candida* isolates.[67,80]

The molecular diagnosis of invasive candidal disease has been shown to be clinically useful but, as with most other diagnostic fungal PCR systems, lacks standardization and multicenter evaluation to determine a true clinical utility. Despite it being the most common cause of IFD, the use of effective antifungal, particularly anti-candidal prophylaxis will reduce the incidence of disease and in certain populations, candidal disease has been superseded by invasive mold disease with higher mortalities. Consequently, the focus of attention will and has changed with many publications describing the use of PCR to diagnose invasive aspergillosis. Nevertheless, as both diseases have a relatively low incidence, molecular testing regimes are better suited to a screening service to exclude disease and this should be coupled with the ability to detect many different pathogens (10–20) within a single test. These are beyond the current capability of RT detection systems and the technology described above will certainly have a future, possibly syndromic, role. It is only a matter of time before ways of increasing sensitivity of detection are found and these technologies can also be applied in diagnosing candidal, indeed broad-range IFD directly from clinical specimens.

REFERENCES

1. Morrell, M., Fraser, V.J., and Kollef, M.J., Delaying empiric treatment of *Candida* bloodstream infection until positive blood culture results are obtained: A potential risk factor for mortality. *Antimicrob. Agents. Chemother.*, 49, 3640, 2005.
2. Pfaller, M.A. and Diekema, D.J., Epidemiology of invasive candidiasis: A persistent public health problem. *Clin. Microbiol. Rev.*, 20, 133, 2007.
3. Wilson, L.S. et al., The direct cost and incidence of systematic fungal infections. *Value Health.*, 5, 26, 2002.
4. Odds, F.C., *Candida and Candidosis—A Review and Bibliography*, 2nd edn., Bailliere Tindall Publishers: London, U.K., 1988.
5. LaJean Chaffin, W. et al., Cell wall and secreted proteins of *Candida albicans*: Identification, function and expression. *Microbiol. Mol. Biol. Rev.*, 62, 130, 1998.
6. Ellis, D. et al., *Description of Medical Fungi*, 2nd edn., Nexus Print Solutions, Adelaide, Australia, 2007.
7. Whiteway, M. and Bachewich, C., Morphogenesis in *Candida albicans*. *Annu. Rev. Microbiol.*, 61, 529, 2007.
8. Staib, P. and Morschhauser, J., Chlamydospore formation in *Candida albicans* and *Candida dubliniensis*—An enigmatic developmental programme. *Mycoses*, 50, 1, 2006.
9. White, P.L. et al., *Candida*, in *Molecular Detection of Foodborne Pathogens*, Chap. 39, p. 549, Liu, D. (Ed.), CRC Press, New York, 2009.
10. Perez-Martin, J., Uria, J.A., and Johnson, A.D., Phenotypic switching in *Candida albicans* is controlled by *SIR2* gene. *EMBO J.*, 18, 2580, 1999.
11. Rinaldi, M.G., Biology and Pathogenicity of *Candida* species, in *Candidiasis, Pathogenesis, Diagnosis and Treatment*, Chap. 1, 2nd edn., Bodey, G.P. (Ed.), Raven Press, New York, 1993.
12. MacCullum, D.M. et al., Property differences among the four major *Candida albicans* strain clades. *Eukaryot. Cell.*, 8, 373, 2009.
13. Blankenship, J.R. and Mitchell, A.P., How to build a biofilm: A fungal perspective. *Curr. Opin. Microbiol.*, 9, 588, 2006.
14. Kapteyn, J.C. et al., The cell wall architecture of *Candida albicans* wild type cells and cell wall defective mutants. *Mol. Microbiol.*, 35, 601, 2000.
15. Jones, T. et al., The diploid genome sequence of *Candida albicans*. *Proc. Natl. Acad. Sci. U.S.A.*, 101, 7329, 2004.
16. Kabir, M.A. and Hussain, M.A., Human fungal pathogen *Candida albicans* in the postgenomic era: An overview. *Expert Rev. Anti. Infect. Ther.*, 7, 121, 2009.
17. Berman, J. and Sudberry, P.E., *Candida albicans*: A molecular revolution built on lessons from budding yeasts. *Nat. Rev. Genet.*, 3, 918, 2002.
18. Roemer, T. et al., Large scale essential gene identification in *Candida albicans* and application to antifungal drug discovery. *Mol. Microbiol.*, 50, 167, 2003.
19. Herrero, E., de la Torre, M.A., and Valentin, E., Comparative genomics of yeast species: New insights into their biology. *Int. Microbiol.*, 6, 183, 2003.

20. Rustchenko, E.P., Curran, T.M., and Sherman, F., Variations in the number of ribosomal DNA units in morphological mutants and normal strains of *C. albicans* and *S. Cerevisiae*. *J. Bacteriol.*, 175, 7189, 1993.
21. Jackson, A.P. et al., Comparative genomics of the fungal pathogens *Candida dubliniensis* and *C. albicans*. *Genome Res.*, 19, 2231, Epub ahead of print, 2009.
22. Hajjeh, R.A. et al., Incidence of bloodstream infections due to Candida species and in vitro susceptibilities of isolates collected from 1998 to 2000 in a population-based active surveillance program. *J. Clin Microbiol.*, 42, 1519, 2004.
23. Kao, A.S. et al., The epidemiology of candidemia in two United States cities: Results of a population-based active surveillance. *Clin. Infect. Dis.*, 29, 1164, 1999.
24. Lass-Florl, C., The changing face of epidemiology of invasive fungal disease in Europe. *Mycoses*, 52, 197, 2009.
25. Horn, D.L. et al., Epidemiology and outcomes of candidaemia in 2019 patients: Data from the Prospective Antifungal Therapy Alliance Registry. *Clin. Infect. Dis.*, 48, 1695, 2009.
26. Hachem, R. et al., The Changing epidemiology of invasive candidiasis. *Candida glabrata* and *Candida krusei* as the leading causes of candidaemia in Hematologic Malignancy. *Cancer*, 112, 2493, 2008.
27. Hope, W., Morton, A., and Eisen, D.P., Increase in prevalence of nosocimial non-*Candida albicans* candidaemia and the association of *Candida krusei* with fluconazole use. *J. Hosp. Infect.*, 50, 56, 2002.
28. Magill, S.S. et al., The epidemiology of *Candida* colonisation in a surgical intensive care unit where fluconazole prophylaxis is utilised. *Ann. Surg.*, 249, 657, 2009.
29. Tortorano, A.M. et al., Candidaemia in Europe: Epidemiology and resistance. *Int. J. Antimicrob. Agents*, 27, 359, 2006.
30. Richardson, M.D. and Warnock, D.W., *Fungal Infections: Diagnosis and Management*, 3rd edn., Blackwell Science Ltd. Oxford, London, U.K., 2003.
31. Odds, F.C., Gow, N.A.R., and Brown, A.J.P., Toward a molecular understanding of *Candida albicans* virulence, in *Molecular Principles of Fungal Pathogenesis*, Chap. 2, Heitman, J., Filler, S.G., Edwards, J.E., and Mitchell, A.P. (Eds.), ASM Press, Washington, DC, 2006.
32. Butler, G. et al., Evolution of pathogenicity and sexual reproduction in eight *Candida* genomes. *Nature*, 459, 657, 2009.
33. Lima Neto, R.G. et al., Adherence of *Candila albicans Candida parapsilosis* to epithelial cells correlates with fungal cell surface carbohydrates. *Mycoses*, 54, 23, 2011.
34. Castano, I., Des Las Penas, A., and Cormack, B.P., Molecular basis of fungal adherence to endothelial and epithelial cells, in *Molecular Principles of Fungal Pathogenesis*, Chap. 13, Heitman, J., Filler, S.G., Edwards, J.E., and Mitchell, A.P. (Eds.), ASM Press, Washington, DC, 2006.
35. Hong Nguyen, M., Cheng, S., and Clancy, C.J., Assessment of *Candida albicans* genes expressed during infection as a tool to understand pathogenesis. *Med. Mycol.*, 42, 293, 2004.
36. Lerman, U. and Morshhauser, J., Secreted aspartic proteases are not required for invasion of reconstituted human epithelia by *Candida albicans. Microbiology*, 154, 3281, 2008.
37. Schindler, B. and Segal, E., *Candida albicans* metabolite affects the cytoskeleton and phagocytic activity of murine macrophages. *Med. Mycol.*, 46, 251, 2008.
38. Sundstrom, P., Candida albicans hyphal formation and virulence, in *Molecular Principles of Fungal Pathogenesis*, Chap. 4, Heitman, J., Filler, S.G., Edwards, J.E., and Mitchell, A.P. (Eds.), ASM Press, Washington, DC, 2006.
39. Netea, M.G. et al., Immune sensing of *Candida albicans* requires cooperative recognition of mannans and glucans by lectin and toll-like receptors. *J. Clin. Invest.*, 116, 1642, 2006.
40. Romani, L. and Puccetti, P., Controlling pathogenic inflammation to fungi. *Expert Rev. Anti-Infect. Ther.*, 5, 1007, 2007.
41. Denning, D.W., Kibbler, C.C., and Barnes, R.A., British Society for Medical Mycology proposed standards of care for patients with invasive fungal infections. *Lancet Infect. Dis.*, 3, 230, 2003.
42. Campbell, C.K. et al., Comparison of the API *Candida* system with the AUXACOLOR system for identification of common yeast pathogens. *J. Clin. Microbiol.*, 37, 821, 1999.
43. Cooke, V.M. et al., New chromogenic agar medium for the identification of *Candida* spp. *Appl. Environ. Microbiol.*, 68, 3622, 2002.
44. Barnes, R.A. and Vale, L., Spiking as a rapid method for differentiation of *Candida albicans* from other yeasts species. *J. Hosp. Infect.*, 60, 78, 2005.
45. Rimek, D., Fehse, B., and Gopel, P., Evaluation of Mueller Hinton Agar as a simple medium for the germ tube production of *Candida albicans* and *Candida dubliniensis*. *Mycoses*, 51, 205, 2007.
46. Sendid, B. et al., New Immunoassays for sensitive detection of circulating *Candida albicans* mannan and antimannan antibodies: Useful combined test for diagnosis of systemic candidiasis. *J. Clin. Microbiol.*, 37, 1510, 1999.
47. White, P.L., Archer, A.E., and Barnes, R.A., Comparison of non-culture based methods for detection of systemic fungal infections, with an emphasis on invasive *Candida* infections. *J. Clin. Microbiol.*, 43, 2181, 2005.
48. Sendid, B. et al., Combined detection of mannanaemia and anti-mannan antibodies as a strategy for the diagnosis of systemic infection caused by pathogenic *Candida* species. *J. Med. Microbiol.*, 51, 433, 2002.
49. Alam, F.F., Mustafa, A.S., and Khan, Z.U., Comparative evaluation of (1,3)-β-D-glucan, mannan and anti-mannan antibodies and *Candida* species specific snPCR in patients with candidemia. *BMC Infect. Dis.*, 7, 103, 2007.
50. Kedzierska, A. et al., Current status of fungal cell wall components in the immunodiagnostics of invasive fungal infections in humans: Galactomannan, mannan and (1–3)-β-D-Glucan. *Eur. J. Clin. Microbiol. Infect. Dis.*, 26, 755, 2007.
51. De Pauw, B. et al., Revised definitions of invasive fungal disease from the European Organization for Research and Treatment of Cancer/Invasive Fungal Infections Cooperative Group and the National Institute of Allergy and Infectious Diseases Mycoses Study Group (EORTC/MSG) Consensus Group. *Clin. Infect. Dis.*, 46, 1813, 2008.
52. Voss, A. et al., Investigation of the sequence of colonization and candidemia in nonneutropenic patients. *J. Clin. Microbiol.*, 32, 975, 1994.
53. Pittet, D. et al., *Candida* colonization and subsequent infections in critically ill surgical patients. *Ann. Surg.*, 220, 751, 1994.
54. Leon, C. et al., A bedside scoring system ("*Candida* score") for early antifungal treatment in nonneutropenic critically ill patients with *Candida* colonization. *Crit. Care Med.*, 34, 730, 2006.
55. Faiz, S. et al., Risk-based fluconazole prophylaxis of *Candida* bloodstream infection in a medical intensive care unit. *Eur. J. Clin. Microbiol. Infect. Dis.*, 28, 689, 2009.
56. Maaroufi, Y. et al., Rapid detection of *Candida albicans* in clinical blood samples by using a TaqMan-based PCR assay. *J. Clin. Microbiol.*, 41, 3293, 2003.

57. McMullan, R. et al., A prospective clinical trial of a real-time polymerase chain reaction assay for the diagnosis of candidemia in nonneutropenic, critically ill adults. *Clin. Infect. Dis.*, 46, 890, 2008.
58. Lau, A. et al., Multiplex tandem PCR: A novel platform for rapid detection and identification of fungal pathogens from blood culture specimens. *J. Clin. Microbiol.*, 46, 3021, 2008.
59. Wellinghausen, N. et al., Rapid diagnosis of candidaemia by real-time PCR detection of *Candida* DNA in blood samples. *J. Med. Microbiol.*, 58, 1106, 2009.
60. Schabereiter-Gurtner, C. et al., Development of novel real-time PCR assays for detection and differentiation of eleven medically important *Aspergillus* and *Candida* species in clinical specimens. *J. Clin. Microbiol.*, 45, 906, 2007.
61. Vollmer, T. et al., Evaluation of novel broad-range real-time PCR assay for rapid detection of human pathogenic fungi in various clinical specimens. *J. Clin. Microbiol.*, 46, 1919, 2008.
62. Hendolin, P.H. et al., Panfungal PCR and multiplex liquid hybridization for detection of fungi in tissue specimens. *J. Clin. Microbiol.*, 38, 4186, 2000.
63. Lau, A. et al., Development and clinical application of a panfungal PCR assay to detect and identify fungal DNA in tissue specimens. *J. Clin. Microbiol.*, 45, 380, 2007.
64. Bagyalakshmi, R., Madhavan, H.N., and Therese, K.L., Development and application of multiplex polymerase chain reaction for the etiological diagnosis of infectious endophthalmitis. *J. Postgrad. Med.*, 52, 179, 2006.
65. Arancia, S. et al., Construction and use of PCR primers from a 70 kDa heat shock protein gene for identification of *Candida albicans*. *Mol. Cell Probes*, 11, 329, 1997.
66. Ebihara, M. et al., Molecular detection of dermatophytes and nondermatophytes in onychomycosis by nested polymerase chain reaction based on 28S ribosomal RNA gene sequences. *Br. J. Dermatol.*, 161, 1038, 2009.
67. Borman, A.M. et al., Pyrosequencing analysis of 20 nucleotides of internal transcribed spacer 2 discriminates *Candida parapsilosis*, *Candida metapsilosis*, and *Candida orthopsilosis*. *J. Clin. Microbiol.*, 47, 2307, 2009.
68. Carvalho, A. et al., Multiplex PCR identification of eight clinically relevant *Candida* species. *Med. Mycol.*, 45, 619, 2007.
69. Sakai, T. et al., Rapid, sensitive and simple detection of *Candida* deep mycosis by amplification of 18S ribosomal RNA gene; comparison with assay of serum beta-D-glucan level in clinical samples. *Tohoku J. Exp. Med.*, 190, 119, 2000.
70. Ahmad, S. et al., Seminested PCR for diagnosis of candidemia: Comparison with culture, antigen detection, and biochemical methods for species identification. *J. Clin. Microbiol.*, 40, 2483, 2002.
71. Einsele, H. et al., Detection and identification of fungal pathogens in blood by using molecular probes. *J. Clin. Microbiol.*, 35, 1353, 1997.
72. Van Burik, J. et al., Panfungal PCR assay for detection of fungal infection in human blood specimens. *J. Clin. Microbiol.*, 36, 1169, 1998.
73. Löffler, J. et al., Detection of PCR-amplified fungal DNA by using a PCR-ELISA system. *Med. Mycol.*, 36, 275, 1998.
74. Badiee, P. et al., Early detection of systemic candidiasis in the whole blood of patients with hematologic malignancies. *Jpn. J. Infect. Dis.*, 62, 1, 2009.
75. Skovbjerg, S. et al., Optimization of the detection of microbes in blood from immunocompromised patients with haematological malignancies. *Clin. Microbiol. Infect.*, 15, 680, 2009.
76. Chen, Y.C. et al., Identification of medically important yeasts using PCR-based detection of DNA sequence polymorphisms in the internal transcribed spacer 2 region of the rRNA genes. *J. Clin. Microbiol.*, 38, 2302, 2000.
77. Spiess, B. et al., DNA microarray-based detection and identification of fungal pathogens in clinical samples from neutropenic patients. *J. Clin. Microbiol.*, 45, 3743, 2007.
78. Palka-Santini, M. et al., Large scale multiplex PCR improves pathogen detection by DNA microarrays. *BMC Microbiol.*, 3, 1, 2009.
79. Landlinger, C. et al., Species-specific identification of a wide range of clinically relevant fungal pathogens by use of Luminex xMAP technology. *J. Clin. Microbiol.*, 47, 1063, 2009.
80. Leake, J.L. et al., Identification of yeast in chronic wounds using new pathogen-detection technologies. *J. Wound Care*, 18, 103, 2009.
81. Bleve, G. et al., Development of reverse transcription (RT)-PCR and real-time RT-PCR assays for rapid detection and quantification of viable yeasts and molds contaminating yogurts and pasteurized food products. *Appl. Environ. Microbiol.*, 69, 4116, 2003.
82. Widjojoatmodjo, M.N. et al., Nucleic acid sequence-based amplification (NASBA) detection of medically important *Candida* species. *J. Microbiol. Methods*, 38, 81, 1999.
83. Hierro, N. et al., Real-time quantitative PCR (QPCR) and reverse transcription-QPCR for detection and enumeration of total yeasts in wine. *Appl. Environ. Microbiol.*, 72, 7148, 2006.
84. Löffler, J. et al., Comparison of different methods for extraction of DNA of fungal pathogens from cultures and blood. *J. Clin. Microbiol.*, 35, 3311, 1997.
85. Morrissey, C.O. et al., Use of PCR on the combination of serum and whole blood specimens for the earlier diagnosis of invasive aspergillosis in haematology patients. Abstract M1721. ICAAC Washington, DC, 2008.
86. White, P.L., Shetty, A., and Barnes, R.A., Detection of seven *Candida* species using the Light-Cycler system. *J. Med. Microbiol.*, 52, 229, 2003.
87. White, P.L. *et al.*, Detection of *Candida* in concentrated oral rinse cultures by real-time PCR. *J. Clin. Microbiol.*, 42, 2101, 2004.
88. Loeffler, J. et al., Quantification of fungal DNA by using fluorescence resonance energy transfer and the light cycler system. *J. Clin. Microbiol.*, 38, 586, 2000.
89. Guiver, M., Levi, K., and Oppenheim, B.A., Rapid identification of Candida species by TaqMan PCR. *J. Clin. Pathol.*, 54, 362, 2001.
90. Stanley, K.K. and Szewczuk, E., Multiplexed tandem PCR: Gene profiling from small amounts of RNA using SYBR Green detection. *Nucleic Acids Res.*, 33, e180, 2005.
91. Sandven, P. et al., Candidaemia in Norway, 1991 to 2003: Results from a nationwide study. *J. Clin. Microbiol.*, 44, 1977, 2006.
92. Diekema, D.J. et al., Epidemiology of candidaemia: 3 year results from the emerging infections and the epidemiology of Iowa organisms study. *J. Clin. Microbiol.*, 40, 1298, 2002.
93. Asmundsdottir, L.R., Erlendsdottir, H., and Gottfredsson, M., Increasing incidence of candidaemia: Results from 20 year study in Iceland. *J. Clin. Microbiol.*, 40, 3489, 2002.
94. Asmundsdottir, L.R. et al., Molecular epidemiology of candidaemia: Evidence of clusters of smoldering nosocomial infections. *Clin. Infect. Dis.*, 47, e17, 2008.
95. Poikonen, E. et al., Candidaemia in Finland, 1995–1999. *Emerg. Infect. Dis.*, 9, 985, 2003.

96. Laupland, K.B. et al., Invasive *Candida* species infections: A 5 year population based assessment. *J. Antimicrob. Chemother.*, 56, 532, 2005.
97. Almirante, B. et al., Epidemiology and predictors of mortality in cases of *Candida* bloodstream infection: Results from population based surveillance, Barcelona, Spain, from 2002 to 2003. *J. Clin. Microbiol.*, 43, 1829, 2005.
98. Arendrup, M.C. et al., Seminational surveillance of fungemia in Denmark: Notably high rates of fungemia and number of isolates with reduced azole susceptibility. *J. Clin. Microbiol.*, 43, 4434, 2005.
99. Arendrup, M.C. et al., Seminational surveillance of fungemia in Denmark 2004–2006: Increasing incidence of fungaemia and number of isolates with reduced azole susceptibility. *Clin. Microbiol. Infect.*, 14, 487, 2008.
100. HCAI, *CDR Weekley*, 16, 42, 2006.
101. Odds, F.C. et al., One year prospective study of *Candida* bloodstream infections in Scotland. *J. Med. Microbiol.*, 56, 1066, 2007.
102. Surveillance of Healthcare associated infections report. Health Protection Agency, U.K., 2007.
103. Pfaller, M.A. et al., In vitro susceptibilities of *Candida* spp. to caspofungin: Four years of global surveillance. *J. Clin. Microbiol.*, 44, 760, 2006.
104. Lamagni, T.L. et al., Emerging trends in the epidemiology of invasive mycoses in England and Wales. *Epidemiol. Infect.*, 126, 397, 2001.
105. Leroy, O. et al., Epidemiology, management and risk factors for death of invasive *Candida* infections in critical care: A multicentre, prospective, observational study in France (2005–2006). *Crit. Care Med.*, 37, 1612, 2009.
106. Colombo, A.L. et al., Epidemiology of candidaemia in Brazil: A Nationwide sentinel surveillance of candidaemia in eleven medical centres. *J. Clin. Microbiol.*, 44, 2816, 2006.
107. Al-Tawfiq, J.A., Distribution and epidemiology of *Candida* species causing fungemia at a Saudi Arabian hospital, 1996–2004. *Int. J. Infect. Dis.*, 11, 239, 2007.
108. Klingspor, L. and Jalal, S., Molecular detection and identification of *Candida* and *Aspergillus* spp. from clinical samples using real-time PCR. *Clin. Microbiol. Infect.*, 12, 745, 2006.
109. Jordanides, N.E. et al., A prospective study of real-time panfungal PCR for the early diagnosis of invasive fungal infection in haemato-oncology patients. *Bone Marrow Transplant.*, 35, 389, 2005.
110. Basková, L. et al., The Pan-AC assay: A single-reaction real-time PCR test for quantitative detection of a broad range of *Aspergillus* and *Candida* species. *J. Med. Microbiol.*, 56, 1167, 2007.
111. Innings, A. et al., Multiplex real-time PCR targeting the RNase P RNA gene for detection and identification of *Candida* species in blood. *J. Clin. Microbiol.*, 45, 874, 2007.
112. Morace, G. et al., PCR-restriction enzyme analysis for detection of Candida DNA in blood from febrile patients with hematological malignancies. *J. Clin. Microbiol.*, 37, 1871, 1999.
113. Shin, J.H. et al., Rapid identification of up to three *Candida* species in a single reaction tube by a 5′ exonuclease assay using fluorescent DNA probes. *J. Clin. Microbiol.*, 37, 165, 1999.
114. Khlif, M. et al., Evaluation of nested and real-time PCR assays in the diagnosis of candidaemia. *Clin. Microbiol. Infect.*, 15, 656, 2009.

65 *Debaryomyces*

María J. Andrade, Mar Rodríguez, Elena Bermúdez, Félix Núñez, Miguel A. Asensio, and Juan J. Córdoba

CONTENTS

65.1 INTRODUCTION

Species of the ascomycetous genus *Debaryomyces* are among the most common yeasts isolated from many natural habitats such as air, soil, pollen, tree exudates, plants, fruits, insects, feces, gut of vertebrates, and seawater.[1] This genus is generally recognized by the production of persistent asci by mother cell-bud conjugation and the formation of warty ascospores, usually one or two per ascus.[2] *Debaryomyces hansenii* is the most important species of the genus and is characterized by its cryo- and halo-tolerance.[3,4] This species and its anamorph *Candida famata* have been isolated from a large diversity of natural sources including fruit, air, water, and soil but most frequently from a wide variety of foods, mainly those with low water activity.[5–8]

D. hansenii is considered normally as a non-pathogenic yeast[9] and has rarely been isolated from humans. However, some clinical cases have been reported with *D. hansenii* or *C. famata*, generally in patients with immunological disorders, including bone infection,[10] alveolitis,[11] septicaemia,[12] retinopathy,[13] ocular endophthalmitis,[14] or central nervous system infection.[15]

Classification, pathogenic relevance, and diagnosis of *Debaryomyces* are discussed in this chapter.

65.1.1 Classification and Morphology

Genus *Debaryomyces* was established by Klocker[16] with the single species of *D. globosus* that was classified later in the genus *Saccharomyces*.[17] Currently 19 species are included: *D. carsonii*, *D. castellii*, *D. coudertii*, *D. etchelsii*, *D. hansenii*, *D. maramus*, *D. melissophilus*, *D. nepalensis*, *D. occidentalis*, *D. polymorphus*, *D. pseudopolymorphus*, *D. robertsiae*, *D. udenii*, *D. vanrijiae*, *D. yamadae*,[18] *D. prosopidis*,[19] *D. mycophilus*,[20] *D. singareniensis*,[21] and *D. vietnamensis*.[22]

The yeasts of this genus show spherical cells, and pseudomycelium is absent, primitive, or occasionally well developed. All species are perfect and haploid and have a vegetative reproduction by multilateral budding.[17] The sexual reproduction proceeds via heterogamous conjugation of two cells of different form or size, generally mother and bud, although the isogamous conjugation also occurs.[17] The conjugation commonly leads to a diplophase followed by meiosis and ascospore formation.[23] One to two spherical, globular, ovoidal or lenticular smoothy, or warty spores are usually formed per ascus, but in some species up to four spores could be present.[17] *Debaryomyces* species are distinguished from other ascomycetous yeast genera by the special internal ultrastructure of the spore's wall, consisting of a dark outer layer and two light inner layers separated by a dark line.[17]

The nuclear base composition of the *Debaryomyces* species is 37 mol% G+C or higher.[24] The karyotype analysis reveals a high degree of polymorphism.[25]

The most common species of this genus, *D. hansenii*, grows as spherical to short, oval cells, single, in pairs or in short chains. As well as other species of the genus, the pseudomycelium is absent, primitive, or exceptionally well developed.[17] Nevertheless, a yeast-to-mycelium dimorphism has been described in continuous fermentation, with the yeast morphology being reversibly affected by the dissolved O_2: low aeration leads to hyphae formation and the increase in O_2 concentration resulted in recuperation of the oval shape.[26] After conjugation, usually one or seldom two spherical spores with a warty wall are formed per ascus.[17] The wartiness is not always distinct under light microscope, and in scanning micrographs it appears as small blunt protuberances or as small ridges.[27] The presence of spores gives the culture a brown color.

D. hansenii is a highly heterogeneous species, with remarkable phenotypic differences between strains, such as variations in their ability to metabolize various carbon sources, the expression of different lipase and protease activities, and their diverse optimal growth conditions.[28,29] Therefore, a number of synonymous species and varieties was recorded due to the variability in morphological, physiological, and biochemical criteria commonly employed in yeast classification.[17] This heterogeneity in taxonomic characteristics has resulted in the description of varieties linked to *D. hansenii*. According to the current taxonomy, two varieties of *D. hansenii* are distinguished, *D. hansenii* var. *fabryi* and *D. hansenii* var. *hansenii*, with different properties, which may be discriminated by their maximum growth temperatures, the sequence divergences of their 26S rRNA genes, and differences in the electrophoretic mobility of their glucose-6-phosphate dehydrogenase.[30–32] However, according to genomic studies, it has been proposed that the varieties *hansenii* and *fabryi* should be considered as two genetically distinct species.[33] In this sense, a polyphasic approach study re-examining 65 strains of *D. hansenii*, including isolates from varieties *fabryi* and *hansenii*, concluded that three distinct species should be considered within the *D. hansenii* group: *D. hansenii*, *D. fabryi*, and *D. subglobosus*.[2] These three taxa may be differentiated by polymerase chain reaction (PCR) fingerprints, DNA reassociations, and phenotypic characterization, mainly in maximum growth temperatures. Additionally, the divergence in intron sequences between the *D. hansenii* varieties (*D. hansenii* var. *hansenii*, *D. hansenii* var. *fabryi*, *C. famata* var. *flareri*, and *D. tyrocola*) unambiguously distinguished the groups as four separate taxa: *D. hansenii*, *D. tyrocola*, *D. fabryi*, and the new taxon *D. macquariensis*.[34] In addition, this gene genealogy-based approach revealed genetic exchanges between these populations, indicating an unexpected genetic diversity within this part of the genus *Debaryomyces*.[34]

D. hansenii and its anamorph *C. famata* are phylogenetically related to *Candida albicans*.[35] In addition, *D. hansenii* and *Candida guilliermondii* belong to a monophyletic clade containing organisms that translate CTG as serine instead of leucine, suggesting they are sister taxa.[36]

65.1.2 Biology, Pathogenesis, and Clinical Features

Debaryomyces species are osmotolerant and can grow in media containing up to 4 M NaCl.[37] Species of this genus are characterized physiologically by their inability to assimilate nitrate, as well as their weak or nonexistent fermentation capacities,[17] and chemotaxonomically by their expression of coenzyme Q-9.[38] The type species of this genus, *D. hansenii* and its anamorph *C. famata*, is a cryotolerant, marine yeast, which can grow with salinity levels up to 24%.[39] These and others characteristics that are used for testing strains of *D. hansenii* are summarized in Table 65.1.

D. hansenii is able to grow at 10% NaCl or 5% glucose, and these characteristics are used to discriminate *D. hansenii* from other ascomycetous yeasts. On the other hand, *D. hansenii* is one of the lipid-accumulating, "oleaginous" yeasts and can accumulate lipids to concentrations up to 70% of their dry biomass,[40] and their metabolism is clearly dominated by pathways that contribute to lipid metabolism. *D. hansenii* can be cultivated in media with up to 25% NaCl or 18% glycerol.[41] It is isolated from environments with high salt concentrations, such as seawater or several types of food. In fact, moderate NaCl concentrations improve the growth of *D. hansenii* cells. The positive effect of NaCl on *D. hansenii* growth is even more evident in the presence of several stress conditions, such as high temperature[42] and low or high pH levels in the media. *D. hansenii* is reported to grow optimally at 20°C–25°C, which might be a consequence of its natural occurrence in habitats such as seawater. *D. hansenii* can grow at 5°C and even below 0°C.[43] At 10°C, this yeast is able to grow at pH 4.0–6.0 when a_w is up to 0.99. *D. hansenii* can grow at a_w values as low as 0.65.[44,45] Although at 37°C selective growth of *D. subglobosus* belonging to the complex *D. hansenii*[2] has been reported, it is difficult to use this temperature as the only criterion to differentiate *Debaryomyces* and specially those of complex *D. hansenii*.

Debaryomyces spp. have been rarely isolated from humans, but they are considered opportunistic pathogens. There are diverse reports about sanitary problems caused especially for *C. famata*, though some few ones refer to *D. hansenii*. Wong et al.[10] reported some infections caused by *Debaryomyces* species. *Torulopsis candida* (*C. famata*) was isolated from a patient with chronic skin lesions on the hands and feet. *D. hansenii* was also found in one case of bone infection that a 23-year-old woman suffered during 4 years. Several clinical samples were identified as *D. hansenii* (and its anamorph *C. famata*) in superficial infections.[46] *D. hansenii* was also responsible for a persistent candidemia observed in a patient heavily treated with various fungicides.[12] A 65-year-old female was diagnosed with extrinsic allergic alveolitis resulting from exposure to inhaled organic dusts, with *D. hansenii* being the dominant species in indoor air

TABLE 65.1
Characteristics of Complex Group *Debaryomyces hansenii*

Characteristics	Possible Results for Complex Group *D. hansenii*
Fermentation of:	
• Galactose, glucose, maltose, melibiose, raffinose	v
• Lactose	–
Assimilation of substrates:	
• L-Arabinose, cellobiose, ethanol, galactose, D-glucitol, D-gluconate, 2-keto-D-gluconate, glucose, α-methyl-D-glucoside, glycerol, lactose, maltose, D-mannitol, raffinose, ribitol, D-ribose, salicin, soluble starch, succinic acid, sucrose, trehalose, D-xylose	+
• D-Arabinose, citric acid, erythritol, galacitol, 5-keto-D-gluconate, D-glucosamine, *N*-acetyl-D-glucosamine, inulin, DL-lactate, hexadecan, melezitose, melibiose, L-rhamnose, L-sorbose	v
• Inositol, methanol, saccharate	–
Splitting of arbutin	+
Assimilation of nitrate	–
Assimilation of nitrite	v
Assimilation of creatine	+
Growth in vitamin-free medium	–
Growth on 50% (w/w) glucose-yeast extract agar	+
Growth at 37°C	v
G+C (mol%)	38.2–38.6
Shape of the ascospores	Spherical
Wall of the spores	Warty
Number of spores per ascus	1, 2
Formation of pseudomycelium	Absent or very primitive

Sources: Adapted from Kreger-Van Rij, N.J.W., *The Yeast, a Taxonomic Study*, 3rd edn., Elsevier Science Publishers, Amsterdam, the Netherlands, 1984; Nakase, T. et al., *The Yeasts, a Taxonomic Study*, 4th edn., Kurtzman, C.P. and Fell, J.W., Eds., Elsevier Science Publishers, Amsterdam, the Netherlands, 1998; Del Bove, M. et al., *Food Microbiol.*, 26, 453, 2009; Córdoba, J.J. et al., *Molecular Detection of Foodborne Pathogens*, Liu, D. Ed., CRC Press, Taylor & Francis Group, Boca Raton, FL, 2009.
+, positive reaction; –, negative reaction; v, variable reaction.

samplings.[11] A 66-year-old man stem cell transplant recipient had no response to therapy with amphotericin B and voriconazole and died after a septic shock. Positive blood cultures have demonstrated persistent candidemia due to *D. hansenii* (*C. famata*).[12]

Different species of *Candida* are part of the natural microbiota and, thus, are regarded as commensal organisms in humans. *C. famata* was thought to be non-pathogenic for humans. However, this yeast was isolated in combination with other *Candida* spp. from a relevant number of clinical cases, including ocular endophthalmitis,[14] acute zonal occult outer retinopathy,[13,47] and central nervous system infection.[15] In addition, *C. famata* is not rarely implicated in human fungemia.[48–51] The occurrence of *C. famata* in patients with problems of the oral cavity is also documented.[52] New treatments of hospitalized patients seem to have favored the emergence of *C. famata* as a pathogen or as responsible of nosocomial infections, but they may be underreported.[53] In this sense, *C. famata* has been associated with catheter-related bloodstream infection[54] and rarely with other infections,[55–57] generally in immunocompromised patients. *C. famata* was implicated in a complication after a coronary artery bypass and in a patient with continuous ambulatory peritoneal dialysis.[58,59] Thus, medical importance of *D. hansenii* and its anamorph *C. famata* may rely on the susceptibility of immunocompromised patients and on its resistance to the treatments applied to different pathologies.

65.1.3 Diagnosis

Since some *Debaryomyces* spp. can act as opportunistic pathogens, accurate and sensitive methods for detection in clinical and environmental samples are needed. For diagnosis purposes, a complete characterization of yeasts at species and strain level is required for most clinical cases. Different physiological and morphological methods have been traditionally used in taxonomic differentiation of *Debaryomyces*.[1,38,60,61] However, several studies have shown that traditional identification methods based on phenotypic properties of yeasts (morphological, biochemical, and physiological tests) are laborious, lack discriminatory power, and misidentification occurs frequently.[62] In addition, identification on the basis of morphological properties generates the so-called double binomial nomenclature, with one name for the vegetative state (anamorph) and another for the sexual state (teleomorph). This is the case of *C. famata*/*D. hansenii* and more recently *C. flareri*/*D. fabryi*.[34] Furthermore, morphological methods generally produce ambiguities and inaccuracies in the results, because the morphological and physiological characteristics are strongly influenced by growing conditions.

Progress in the molecular biology in the last decade has opened up possibilities of characterizing yeasts at the genomic level. The sequencing of the genes coding for 18S and 26S ribosomal RNA (rRNA), as well as internal transcribed spacer (ITS) and intergenic spacer (IGS) region, has brought about many changes in the identification and classification of yeasts at species and strain level.[22,63] In addition, techniques based on random amplified polymorphic DNA (RAPD-PCR) and restriction fragment length polymorphism (RFLP) have been already recognized as reliable tools for the rapid identification of yeasts.[64] We here review the main physiological and morphological methods, as well as the molecular techniques.

65.1.3.1 Physiological and Morphological Analysis

Conventional identification of *Debaryomyces* spp. is based on physiological, morphological, and biochemical characteristics. Morphological and sexual characteristics are not useful for identifying *Debaryomyces* at the genus and species level, because ambiguous identification may be obtained due to strain variability. The physiological tests used for identifying purposes are associated with the utilization of carbon and nitrogen sources, growth factor requirements, growth at elevated temperatures and on media of high sugar or sodium chloride content, formation of typical characteristic metabolites, and susceptibility to antibiotics. The utilization of these tests for *Debaryomyces* characterization requires considerable experience and skill for evaluating the specific tests. Furthermore, great difficulties for the differentiation at the species level in this genus could be found, since many biochemical and physiological tests show the same results for different species. In this sense, the classical growth temperature test at 37°C to differentiate *D. hansenii* of *D. fabryi* used alone confuses several strains of these species.[22]

The main biochemical tests are electrophoresis of proteins, coenzyme Q analysis, allozyme analysis, and ultrastructure and chemical composition (polysaccharides, fatty acids) of the cell wall.[17] Missoni et al.[65] found the evaluation of cell fatty acids by gas chromatography very useful in the routine diagnosis and epidemiological monitoring of the infection due to *Candida* spp. (including *C. famata*). These techniques are not always stable or reproducible because they depend on the physiological status of yeast strains. For example, fermentation of sugars is not very accurate because the slow release of CO_2 is not so immediate to be trapped in a Durham tube.

Many commercial methods based on the above morphological and physiological characteristics have been developed for the identification of yeasts, although most of them do not cover species of genus *Debaryomyces*, including Yeast Identification System API 20C (BioMérieux), the Uni-Yeast-Yek system (Remel), the Minitek system (BBL),[60,66,67] Yeast Identification Panel (Baxter-MicroScan), and MicroScan Rapid Yeast Identification.[68,69] API ID32C system allows the identification of three species of the genus *Debaryomyces*: *D. hansenii*, *D. marama*, and *D. polymorphus*.[70] Similarly, the Vitek Yeast Biochemical card (BioMérieux) allows the identification of *D. hansenii* strains.[71,72]

Additionally, several methods have been developed for automatic identification of yeasts on the basis of biochemical tests, such as the highlighting system Vitek 2® (BioMérieux) with colorimetric and fluorimetric VITEK 2 yeast cards[73,74] and Biolog YT Microplate® (AES Laboratories).

Furthermore, selective and differential chromogenic solid media have been developed for the detection of *Debaryomyces* spp., although they have been used only with food samples.[75,76]

65.1.3.2 Molecular Analysis

Molecular diagnostic methods possess several advantages over conventional methods based on phenotypic characteristics, such as higher reliability and sensitivity. They are also potentially more accurate, reproducible, and rapid.[71] These characteristics could serve as an important prerequisite for the timely onset of antifungal therapy. In addition, molecular analysis is not influenced by environmental conditions of yeasts cells, because the nucleotide sequence of the DNA does not change during growth.

Nowadays, different molecular methods are available for both qualitative and quantitative yeasts detection, as well as for species identification, although most of them focus on detection of *Candida* spp.

Molecular identification can be carried out by nucleic acid hybridization or amplification. Methods of nucleic acid hybridization use a DNA probe that has a sequence complementary to the target sequence. 18S rRNA-targeted oligonucleotide probes have been designed for rapid and reliable identification of yeasts, such as the genus *Debaryomyces* and the species *D. hansenii*.[77,78] However, such probes do not allow a full identification of all different species of *Debaryomyces*.

Other molecular methods are based on nucleic acid amplification. The most popular method of amplification is the PCR technique. Several conventional PCR procedures have been developed for the identification of *D. hansenii*.[22,79] Different variations of this technique (PCR-RFLP, RAPD, real-time PCR, or quantitative PCR [Q-PCR], nucleic acid sequence-based amplification [NASBA], etc.) are also used to differentiate species of the genus *Debaryomyces*. PCR methods based on four housekeeping genes (*ACT1*, *TUB2*, *RPL31*, and *RPL33*) have been recently employed to the delineation of species among strains assigned to *D. hansenii*.[34]

One of the problems associated with using PCR for microbial investigation is the possibility of detecting naked DNA derived from dead and degrading yeast cells instead of live yeasts, an outcome that may be unintended. An alternative to prevent naked DNA detection by PCR consists of using NASBA system that selectively amplifies RNA. This method has been used with yeasts of genus *Candida*.[80] However, this is not a big problem for clinical specimens, since the presence of dead cells may be of clinical significance.

The RAPD-PCR is a variation of the PCR useful for the characterization of yeasts based on amplification of genomic DNA with a single short primer. Due to low-temperature hybridization, primer joins unspecific sites throughout the genome, allowing amplification of DNA fragments of different length. The use of RAPD-PCR permits the generation of fingerprints which are specific for species and even strains. The RAPD-PCR technology has been used for the correct identification of several species of *Debaryomyces* with different primers, mainly derived from micro- and mini-satellite DNA.[62,79,81–85] These micro- and mini-satellite primers amplify the genomic DNA of these highly polymorphic regions, thus providing resolution among yeast strains. Using RAPD-PCR, Prillinger et al.[33] concluded that *D. hansenii* and *D. fabryi* are distinct entities and reinstated the varieties at species level.

One of the most promising PCR techniques in the detection of microorganisms is Q-PCR. Several Q-PCR assays have been developed for detecting and enumerating *C. famata* in fungal infections.[47,86]

Other molecular methods are based on the determination of the RFLP. This procedure allows the differentiation of organisms by analyzing patterns of rupture that are generated in a specific site of the genome when it is cut by restriction enzymes. Then, gel electrophoresis displays a pattern of polymorphic bands corresponding to the fragments of different sizes that are generated on the cut of each endonuclease. This fragment length polymorphism appears because organisms from different species and even strains differ in the distance of cleavage sites for each restriction enzyme. The similarity of the generated patterns allows for correlations between species and strains.

The PCR-RFLP is a useful method for identification of some yeast genus, using the restriction analysis of different regions of rRNA/DNA genes (ITS-5.8S rDNA, 18S rDNA, etc.). This technique includes two steps: amplification of the rDNA region and digestion of the PCR product. Therefore, this method requires a longer time frame than current PCR methods that achieve identification and typing without further treatment of the PCR products. The results of PCR-RFLP depend on the region of genome amplified and the enzymes used. For example, PCR-RFLP of ITS1-5.8S rDNA-ITS2 and 18S rDNA regions has been reported as a good method for the differentiation of *Candida* species,[87–89] but it does not allow to distinguish *Debaryomyces* spp.[81] However, the PCR-RFLP of the IGS of rDNA is proposed as a valuable technique for discriminating species of the genera *Debaryomyces* and *Candida*.[22,88,90]

Another technique relied on RFLP is the mitochondrial DNA (mtDNA) restriction analysis. Among all the molecular techniques described in literature, mtDNA restriction analysis appears as one of the most suitable methods to differentiate yeast strains. Querol et al.[91] developed a new mtDNA restriction analysis method based on the extraction of total yeast DNA and the use of GC-rich restriction endonucleases that recognize a high number of sites in the yeast nuclear DNA, but few sites in the mtDNA. This technique has successfully been used to characterize strains of genus *Debaryomyces*[81,83,84,92,93] and clinical isolates of *Candida*.[94]

Other method used for the characterization of *Debaryomyces* is pulsed-field gel electrophoresis (PFGE). With this technique, a restriction enzyme digests the complete genome, and large DNA molecules are resolved by continuous reorientation of the electric field during gel electrophoresis, determining chromosome length polymorphism (CLP). This pattern is specific for yeast species, due to genetic and evolutionary phenomena that take place in the chromosomes (insertions, deletions, and translocations). This method has been reported as a useful tool in the differentiation of species and strain level of *Debaryomyces*.[29]

Overall, the most reliable, simple, and fast methods to differentiate *Debaryomyces* spp. at strain level rely first on RFLP analysis of mtDNA and then on RAPD-PCR.[83]

65.2 METHODS

In this chapter, two methods for the differentiation of *Debaryomyces* spp. are proposed: single RFLP analysis of the IGS rDNA region and a combined method that includes mtDNA restriction analysis and then RAPD-PCR.[95] For these methods, total yeast DNA from presumptive clinical yeasts isolates should be first isolated.

65.2.1 Reagents and Equipment

Reagent and equipment for RFLP of IGS and mtDNA and RAPD-PCR analyses are listed in Table 65.2.

65.2.2 Sample Preparation

65.2.2.1 Yeast Isolates

Clinical specimens collected from patients are cultured on Sabouraud dextrose agar or yeast peptone glucose (YPG) agar (1% w/v yeast extract; 2% w/v peptone; 2% w/v glucose; 1.7% agar) aerobically at 27°C for 48–72 h. The obtained isolates can be subcultured in cryotubes containing Sabouraud dextrose broth or YPG broth (1% w/v yeast extract; 2% w/v peptone; 2% w/v glucose) and stored with 40% (v/v) glycerol at −80°C.

65.2.2.2 DNA Extraction

Genomic DNA is extracted from yeasts grown in YPG broth (Figure 65.1). Yeast cells, taken from a single colony, are cultured in 50 mL conical tubes containing 10 mL of YPG broth at 25°C for 48 h to reach an appropriate cell density and placed inclined in an orbital shaker under constant agitation at 200 rpm to obtain large amounts of mtDNA. Yeast

TABLE 65.2
Culture Media, Reagents, and Equipment Required for Yeast DNA Extraction, RFLP Analysis of IGS and Mitochondrial DNA, and RAPD-PCR

	Culture Media and Reagents	
Sabouraud dextrose agar	Ethylenediaminetetraacetic acid (EDTA) disodium salt, dihydrate	RNase
Sabouraud dextrose broth	Lyticase	Sterile deionized water
Bacteriological agar	Tris-(hydroxymethyl)-aminomethane	PCR reagents (see Table 65.3)
Yeast extract	Hydrochloric acid (HCl) 35%	Restriction enzymes and 10× buffer (see Table 65.3)
D-Glucose anhydrous, extra pure	Sodium dodecyl sulfate (SDS), molecular biology grade	DNA molecular markers
Bacteriological peptone	Potassium acetate, extra pure	Bromophenol blue, indicator
Glycerol	Isopropanol	Ethidium bromide
D-Sorbitol extra pure	Ethanol absolute	
	Equipment	
Laminar flow cabinet	Water bath	Microwave oven
Stomacher lab blender	Centrifuge for 50 mL conical tubes	UV transilluminator
Incubator	Centrifuge for 2 and 0.5 mL microtubes	Thermal cycler and PCR tubes
Orbital shaker	1 mL, 200, 50, 5, 1, and 0.5 µL pipettes	Horizontal electrophoresis unit with the appropriate gel casting tray and combs
Vortex	5 mL cryotubes	
Freezer	Spectrophotometer	

cells are harvested from pure cultures by centrifugation at 4000 rpm for 5 min; resuspended in 500 µL of 1 M sorbitol, 0.1 M ethylenediaminetetraacetic acid (EDTA) pH 7.5; and transferred to 2 mL sterile microcentrifuge tubes. At this stage, 40 µL of a lyticase solution (20 mg/mL) is added to digest yeast cell wall and obtain spheroplasts. After 45 min incubation at 37°C in a water bath with occasional shaking, the suspension is centrifuged at 13,000 rpm for 1 min and the supernatant is discarded. The pellet is resuspended in 500 µL of a solution containing 50 mM Tris–HCl, 20 mM EDTA pH 7.4 to release cellular DNA from spheroplasts. Next, 50 µL of 10% (w/v) sodium dodecyl sulfate (SDS) is added, and the mixture is incubated at 65°C for 10 min in a water bath. To remove proteins, 200 µL of 5 M potassium acetate is added, and the mixture is shaken and stored on ice for 15 min. Next, the mixture is centrifuged at 13,000 rpm for 5 min and 500 µL of the resulting aqueous phase (top layer) is transferred to a new sterile microtube together with an equal volume of ice-cold isopropanol and left 5 min at room temperature. After the precipitated nucleic acids are centrifuged at 13,000 rpm for 10 min, the supernatant is decanted and the pellet is further washed with 1 mL of ice-cold 70% (v/v) ethanol and centrifuged at 13,000 rpm for 1 min. Ethanol is aspirated with a pipette and the pellet is dried at 37°C for 45 min. Dried DNA is suspended in 50 µL of TE buffer (10 mM Tris–HCl, 1 mM EDTA pH 8.0) and 1 µL of RNase (20 mg/mL) is added to digest the contaminating RNA. Then, the solution is incubated for 30 min at 37°C in a water bath and immediately placed at −20°C.

To be used for PCR, concentration and purity of the extracted DNA have to be determined by the optical density (OD) at 260 nm and the ratio OD 260/280 nm, respectively. Pure DNA should have a ratio value between 1.8 and 2.0.

65.2.3 Detection Procedures

65.2.3.1 RFLP Analysis of the IGS rDNA Region

The IGS region of the rDNA, which is located between 26S and 18S genes, is amplified by PCR using the primers CNL12 (5′ CTGAACGCCTCTAAGTCAG 3′) and CNS1 (5′ GAGACAAGCATATGACTACTG 3′).[96] Amplification reaction mixture is prepared on ice in volumes of 50 µL containing the reagents and concentrations summarized in Table 65.3. The thermal cycling parameters are summarized in Table 65.3. After the program ends, amplification products are cooled in the thermal cycler at 4°C until removed, and then kept at −20°C until required. A negative control, which consists of an equal volume of water replacing the DNA template, can be included. PCR-amplified DNA fragments are analyzed in 1% (w/v) agarose gels at 80 V.[90,97]

Aliquots of 10 µL of PCR products are individually digested with the restriction endonucleases *Hap*II, *Hha*I, and *Mbo*I at 37°C overnight[90,97] (Table 65.3). Restriction fragments are separated by electrophoresis on 2.5% (w/v) agarose gels at 60 V[98] (Figure 65.2).

A DNA molecular marker can be incorporated into the gel to estimate the size of both amplification and restriction products. Gels are stained with an ethidium bromide solution (0.5 µg/mL) and visualized under UV light. The size of the amplified IGS rDNA region of *D. hansenii* is of about 2800 bp.[90] Restriction patterns of tested clinical yeasts isolates are compared with those of reference strains for identification.

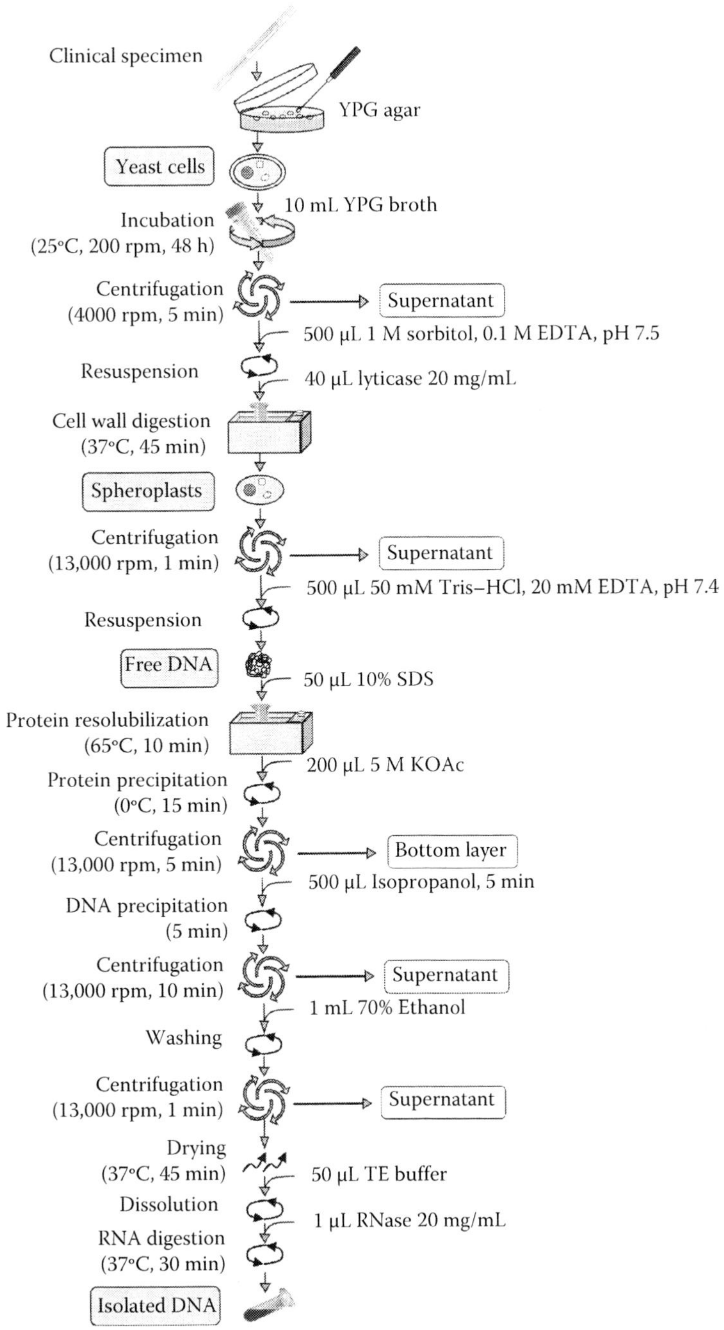

FIGURE 65.1 Diagram of DNA extraction from *Debaryomyces* spp.

65.2.3.2 RFLP Analysis of Mitochondrial DNA

mtDNA restriction analysis involves the digestion of total DNA with restriction enzymes selective for GC-rich sequences, such as *Hae*III, which results in an over-digestion of the nuclear DNA to render specific bands from mtDNA. Thus, this methodology allows detecting mtDNA RFLPs without first separating the mtDNA from nuclear DNA.

Digestion mixture is performed on ice in a volume of 15 μL, containing the reagents and amounts summarized in Table 65.3. After the incubation overnight at 37°C in a water bath, restriction fragments are electrophoretically separated on horizontal 0.8% (w/v) agarose gels at 100 V (Figure 65.3). A DNA molecular marker can be incorporated into the gel as a standard for the calculation of the restriction fragment size. Gels are stained and visualized as described in Section 65.2.3.1. Restriction profiles of tested clinical yeasts isolates are compared with those of reference strains for characterization.

To reduce DNA digestion time, a microwave oven can be used.[99,100] The digestion mixture is placed into a water bath and heated three times at maximum level of the microwave (1250 W) for 20 s each, giving a spin between each time.

TABLE 65.3
PCR and RFLP Reagents and PCR Conditions for Identification and Characterization of *Debaryomyces* spp.

IGS PCR[a]

Reagents (Concentration)	Volume (μL)
Mg^{2+}-free reaction buffer (10 mM Tris–HCl pH 8.8, 50 mM KCl, 0.1% Triton X-100)	5
$MgCl_2$ (50 mM)	3
PCR nucleotide mix (10 mM)	1
CNL12 (100 ng/μL)	1
CNS1 (100 ng/μL)	1
DNA (10 ng/μL)	10
Taq DNA polymerase (2 U/μL)	0.5
Sterile deionized water	28.5

PCR stages[a]	PCR program[a]	
Initial denaturation	94°C 85 s	
Denaturation	95°C, 35 s	35 cycles
Annealing	58°C, 55 s	
Extension	72°C, 2 min	
Final extension	72°C, 10 min	

IGS RFLP[a]

Reagents	Volume (μL)
*Hha*I, *Mbo*I, or *Hap*II (10 U/μL)	0.1
10× buffer	4
IGS-PCR	10
Sterile deionized water	25.9

mtDNA RFLP[b]

Reagents	Volume (μL)
*Hae*III (20 U/μL)	1
10× buffer	1.5
DNA	5
Sterile deionized water	7.5

RAPD-PCR[b]

Reagents (Concentration)	Volume (μL)
Mg^{2+}-free reaction buffer (10 mM Tris–HCl pH 8.8, 50 mM KCl, 0.1% Triton X-100)	5
$MgCl_2$ (50 mM)	3
PCR nucleotide mix (10 mM)	1
$(GACA)_4$ (100 ng/μL)	2
DNA (10 ng/μL)	10
Taq DNA polymerase (2 U/μL)	0.5
Sterile deionized water	28.5

PCR stages	PCR program	
Denaturation	94°C, 1 min	30 cycles
Annealing	36°C, 1 min	
Extension	55°C, 5 min	
Final extension	55°C, 5 min	

[a] Adapted from Quirós, M., *Debaryomyces hansenii*: Estudio de su capacidad para deteriorar alimentos y desarrollo de métodos para su detección rápida, Doctoral thesis, UCM, Madrid, Spain, 2005; Romero, P. et al., *FEMS Yeast Res.*, 5, 455, 2005; Quirós, M. et al., *Anton. Leeuwen.*, 90, 211, 2006.

[b] Adapted from Andrade, M.J. et al., *Int. J. Food Microbiol.*, 107, 48, 2006; Córdoba, J.J. et al., *Molecular Detection of Foodborne Pathogens*, Liu, D. Ed., CRC Press, Taylor & Francis Group, Boca Raton, FL, 2009.

65.2.3.3 RAPD-PCR

RAPD-PCR is based on the amplification of genomic DNA using a single primer with an arbitrary nucleotide sequence. Consequently, this method does not depend on prior knowledge of the DNA template sequence from the assayed yeasts. RAPD-PCR with primers derived from simple repetitive sequences, such as $(GACA)_4$, has been applied for genetic characterization of *D. hansenii*.[83,84]

The amplification reaction mixture must be prepared on ice, using the reagents and concentrations summarized in Table 65.3 for a final volume of 50 μL. The reaction is carried out in a thermal cycler under the amplification conditions summarized in Table 65.3.

RAPD-PCR products are separated by 1% (w/v) agarose gel electrophoresis at 100 V (Figure 65.3). A molecular weight DNA standard can be incorporated into the gel to estimate the size of the amplification products. Gels are stained and visualized as described in Section 65.2.3.1. Patterns of tested clinical yeast isolates are compared with those of reference strains for characterization.

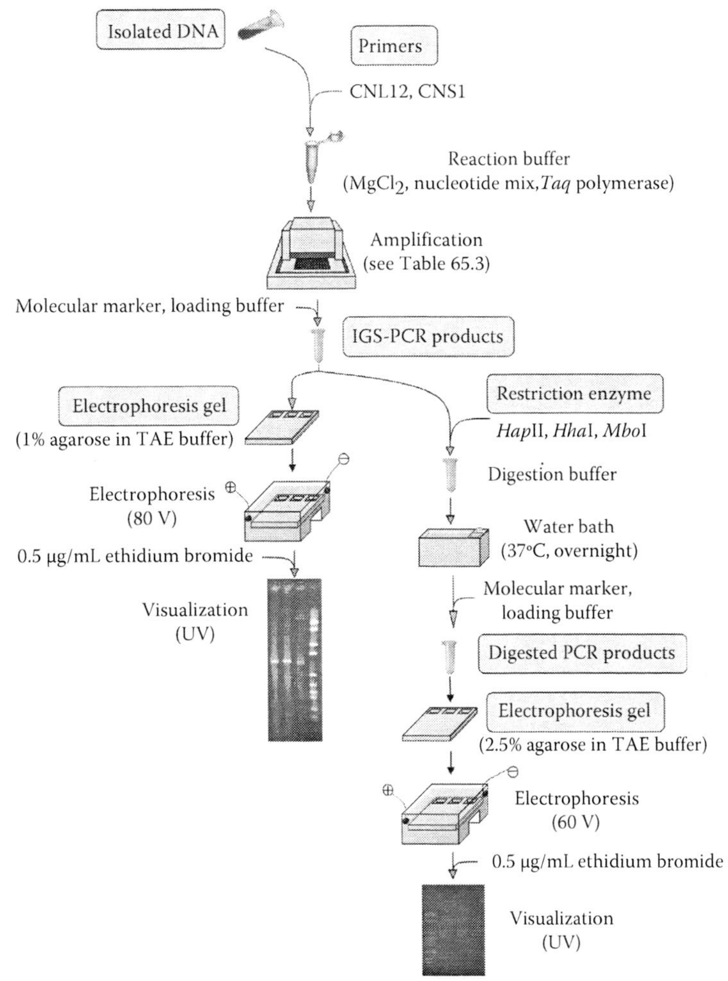

FIGURE 65.2 Diagram of IGS-RFLP for analysis of *Debaryomyces* spp.

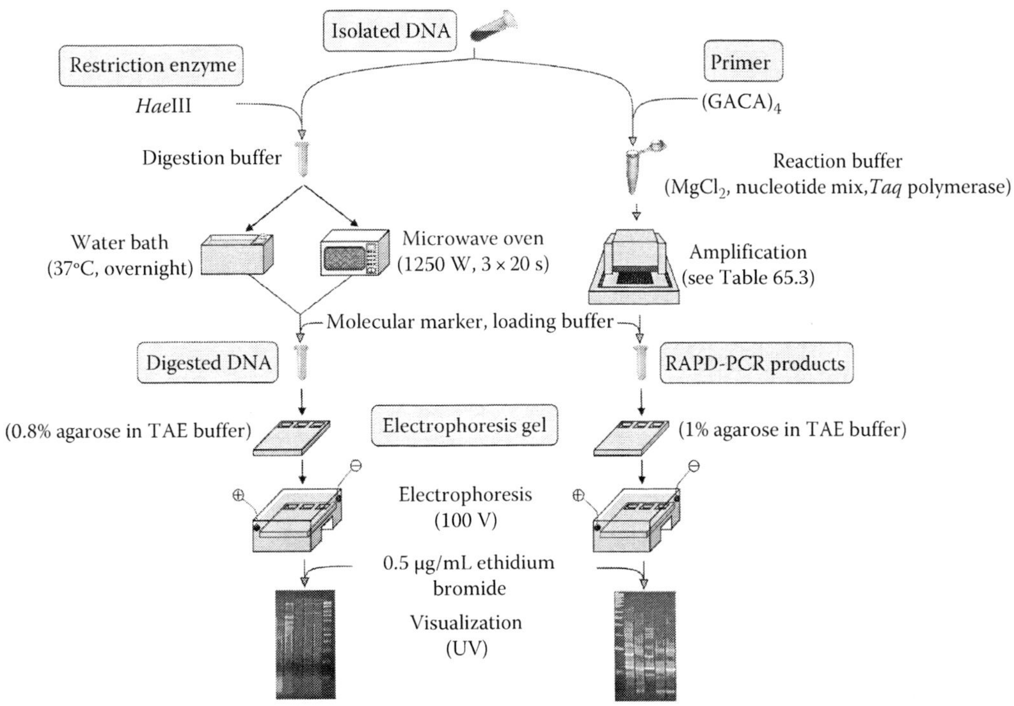

FIGURE 65.3 Diagram of mitochondrial DNA restriction and RAPD-PCR analyses for *Debaryomyces* spp.

65.3 CONCLUSIONS

Yeasts of genus *Debaryomyces* are among the most common yeasts isolated from many natural habitats. These yeasts are considered normally as non-pathogenic. However, some species of this genus have been occasionally isolated from human diseases such as bone infection, retinopathy, allergic alveolitis, or septicaemia in immunocompromised patients.

Nineteen different species have been included in the genus *Debaryomyces*. Most of them are osmotolerant and can grow in media containing up to 4M NaCl. Species of this genus are characterized physiologically by their inability to assimilate nitrate, as well as their weak or nonexistent fermentation capacities.

Since some *Debaryomyces* spp. can act as opportunistic pathogens, it is necessary that accurate and sensitive methods are used to detect *Debaryomyces* in clinical and environmental samples. For diagnosis purposes in clinical specimens, a complete characterization of the yeast at species and strain level is required. Traditional identification methods, based on phenotypic properties of yeasts, are laborious, lack discriminatory power, and misidentification occurs frequently. Progress in the molecular biology has opened up possibilities of characterizing yeasts at the genomic level. Several techniques that include single analysis of RFLP analysis of the IGS rDNA region or a combined method that includes first mtDNA restriction analysis and then RAPD-PCR have been proposed to characterize *Debaryomyces* spp. These methods could be appropriated to investigate *D. hasenii* in clinical specimens, since they are sensitive, simple, and low time-consuming. Furthermore, combined mtDNA restriction and RAPD-PCR analyses allow an accurate differentiation of *Debaryomyces* spp. at strain level.

ACKNOWLEDGMENTS

The authors are grateful to the Spanish Ministry of Education and Science (grants AGL2004-03291, AGL2007-64639, and Consolider-Ingenio 2010: Carnisenusa CSD2007-00016) for the financial support to study different aspects of molecular fungal detection.

REFERENCES

1. Barnett, J.A., Payne, R.W., and Yarrow, D. *Yeasts: Characteristics and Identification*, 3rd edn. Cambridge University Press, Cambridge, U.K., 2000.
2. Groenewald, M. et al. Polyphasic re-examination of *Debaryomyces hansenii* strains and reinstatement of *D. hansenii, D. fabryi* and *D. subglobosus*. *Persoonia*, 21, 17, 2008.
3. Prista, C. et al. Physiological basis for the high salt tolerance of *Debaryomyces hansenii*. *Appl. Environ. Microbiol.*, 63, 4005, 1997.
4. Prista, C. et al. Mechanisms underlying the halotolerant way of *Debaryomyces hansenii*. *FEMS Yeast Res.*, 5, 693, 2005.
5. Gardini, F. et al. A survey of yeasts in traditional sausages of southern Italy. *FEMS Yeast Res.*, 1, 161, 2001.
6. Cocolin, L. et al. Study of the ecology of fresh sausages and characterization of populations of lactic acid bacteria by molecular methods. *Appl. Environ. Microbiol.*, 70, 1883, 2004.
7. Fadda, M.E. et al. Occurrence and characterization of yeasts isolated from artisanal Fiore Sardo cheese. *Int. J. Food Microbiol.*, 95, 51, 2004.
8. Simoncini, N. et al. Dynamics and characterization of yeasts during ripening of typical Italian dry-cured ham. *Food Microbiol.*, 24, 577, 2007.
9. Warren, N.G. and Hazen, K.C. *Candida*, *Cryptococcus*, and other yeasts of medical importance. In: *Manual of Clinical Microbiology*, 7th edn., Murray, P.R., Ed. ASM Press, Washington, DC, 1998.
10. Wong, B. et al. Bone infection caused by *Debaryomyces hansenii* in a normal host: A case report. *J. Clin. Microbiol.*, 16, 545, 1982.
11. Yamamoto, Y. et al. Extrinsic allergic alveolitis induced by the yeast *Debaryomyces hansenii*. *Eur. Respir. J.*, 20, 1351, 2002.
12. Wagner, D. et al. Breakthrough invasive infection due to *Debaryomyces hansenii* (teleomorph *Candida famata*) and *Scopulariopsis brevicaulis* in a stem cell transplant patient receiving liposomal amphotericin B and caspofungin for suspected Aspergillosis. *Infection*, 33, 397, 2005.
13. Carrasco, L. et al. Isolation of *Candida famata* from a patient with acute zonal occult outer retinopathy. *J. Clin. Microbiol.*, 43, 635, 2005.
14. Rao, N.A., Nerenberg, A.V., and Forster, D.J. *Torulopsis candida* (*Candida famata*) endophthalmitis simulating *Propionibacterium* acnes syndrome. *Arch. Ophthalmol.*, 109, 1718, 1991.
15. Prinsloo, B., Weldhagen, G.F., and Blaine, R.W. *Candida famata* central nervous system infection. *S. Afr. Med. J.*, 93, 601, 2003.
16. Klocker, A. Deux nouveaux genres de la famille des Saccharomycetes, *CR. Trav. Lab. Carlsberg*, 7, 273, 1909.
17. Kreger-Van Rij, N.J.W. *The Yeast, a Taxonomic Study*, 3rd edn. Elsevier Science Publishers, Amsterdam, the Netherlands, 1984.
18. Nakase, T. et al. *Debaryomyces* Lodder and Kreger-van Rij nom. cons. In: *The Yeasts, a Taxonomic Study*, 4th edn., Kurtzman, C.P. and Fell, J.W., Eds. Elsevier Science Publishers, Amsterdam, the Netherlands, 1998.
19. Phaff, H.J., Vaughan-Martini, A., and Starmer, W.T. *Debaryomyces prosopidis* sp. nov., a yeast from exudates of mesquite trees. *Int. J. Syst. Bacteriol.*, 48, 1419, 1998.
20. Thanh, V.N., Van Dyk, M.S., and Wingfield, M.J. *Debaryomyces mycophilus* sp. nov., a siderophore-dependent yeast isolated from woodlice. *FEMS Yeast Res.*, 2, 415, 2002.
21. Saluja, P. and Prasad, G.S. *Debaryomyces singareniensis* sp. nov., a novel yeast species isolated from a coal mine soil in India. *FEMS Yeast Res.*, 7, 482, 2007.
22. Nguyen, H.V. et al. Differentiation of *Debaryomyces hansenii* and *Candida famata* by rRNAgene intergenic spacer fingerprinting and reassessment of phylogenetic relationships among *D. hansenii, C. famata, D. fabryi, C. flareri* (=*D. subglobosus*) and *D. prosopidis*: Description of *D. vietnamensis* sp. nov. closely related to *D. nepalensis*. *FEMS Yeast Res.*, 9, 641, 2009.
23. Forrest, S.I., Robinow, C.F., and Lachance, M.A. Nuclear behaviour accompanying ascus formation in *Debaryomyces polymorphus*. *Can. J. Microbiol.*, 33, 967, 1987.

24. Price, C.W., Fuson, G.B., and Phaff, H.J. Genome comparison in yeast systematics: Delimitation of species within the Genera *Schwanniomyces*, *Saccharomyces*, *Debaryomyces* and *Pichia*. *Microbiol. Rev.*, 42, 161, 1978.
25. Corredor, M. et al. Chromosomal polymorphism in the yeast species *Debaryomyces hansenii*. *Ant. Leeuwenhoek*, 83, 215, 2003.
26. Cruz, J.M. et al. Dimorphic behaviour of *Debaryomyces hansenii* grown on barley bran acid hydrolyzates. *Biotechnol. Lett.*, 22, 605, 2000.
27. Kurtzman, C.P., Smiley, M.J., and Baker, F.L. Scanning electron microscopy of ascospores of *Debaryomyces* and *Saccharomyces*. *Mycopathol. Mycol. Appl.*, 55, 29, 1975.
28. Núñez, F. et al. Yeast population during ripening of dry-cured Iberian ham. *Int. J. Food Microbiol.*, 29, 271, 1996.
29. Petersen, K.M. and Jespersen, L. Genetic diversity of the species *Debaryomyces hansenii* and the use of chromosome polymorphism for typing of strains isolated from surface ripened cheeses. *J. Appl. Microbiol.*, 97, 205, 2004.
30. Nakase, T. and Suzuki, M. Taxonomic studies on *Debaryomyces hansenii* (Zopf) Lodder et Kreger-Van Rij and related species. II. Practical discrimination and nomenclature. *J. Gen. Appl. Microbiol.*, 31, 71, 1985.
31. Kurtzman, C.P. and Robnett, C.J. Identification of clinically important ascomycetous yeasts based on nucleotide divergence in the 5′ end of the large-subunit (26S) ribosomal DNA gene. *J. Clin. Microbiol.*, 35, 1216, 1997.
32. Kurtzman, C.P. and Robnett, C.J. Identification and phylogeny of ascomycetous yeasts from analysis of nuclear large subunit (26S) ribosomal DNA partial sequences. *Anton. Leeuwen.*, 73, 331, 1998.
33. Prillinger, H. et al. Phenotypic and genotypic identification of yeasts from cheese. *Anton. Leeuwen.*, 75, 267, 1999.
34. Jacques, N., Mallet, S., and Casaregola, S. Delimitation of the species of the *Debaryomyces hansenii* complex by intron sequence analysis. *Int. J. Syst. Evol. Microbiol.*, 59, 1242, 2009.
35. Fabre, E. et al. Comparative genomics in hemiascomycete yeasts: Evolution of sex, silencing, and subtelomeres. *Mol. Biol. Evol.*, 22, 856, 2005.
36. Fitzpatrick, D.A. et al. A fungal phylogeny based on 42 complete genomes derived from supertree and combined gene analysis. *BMC Evol. Biol.*, 6, 99, 2006.
37. Onishi, H. Osmophilic yeasts. *Adv. Food Res.*, 12, 53, 1963.
38. Yamada, Y. Identification of coenzyme Q (ubiquinone) homologs. In: *The Yeasts, a Taxonomic Study*, 4th edn., Kurtzman, C.P. and Fell, J.W., Eds. Elsevier Science Publishers, Amsterdam, the Netherlands, 1998.
39. Lépingle, A. et al. Genomic exploration of the Hemiascomycetous yeasts: 14. *Debaryomyces hansenii* var. *hansenii*. *FEBS Lett.*, 487, 82, 2000.
40. Ratledge, C. and Tan, K.-H. Oils and fats: Production, degradation and utilization by yeasts. In: *Yeast Biotechnology and Biocatalysis*, Verachtert, H.J. and De Mot, R., Eds. Marcel Dekker, New York, 1990.
41. Butinar, L. et al. Yeast diversity in hypersaline habitats. *FEMS Microbiol. Lett.* 244, 229, 2005.
42. Papauskova, K. and Sychrova, H. The co-action of osmotic and high temperature stresses results in a growth improvement of *Debaryomyces hansenii* cells. *Int. J. Food Microbiol.*, 118, 1, 2007.
43. Davenport, R.R. *Biology and Activities of Yeasts*, Skinner, F.A., Passmore, S.M., and Davenport, R.R., Eds. Academic Press, London, U.K., 1980.
44. Van den Tempel, T. and Jacobsen, M. The technological characteristics of *Debaryomyces hansenii* and *Yarrowia lipolytica* and their potential as starter cultures for production of Danablu. *Int. Dairy J.*, 10, 263, 2000.
45. Breuer, U. and Harms, H. *Debaryomyces hansenii*—An extremophilic yeast with biotechnological potential. *Yeast*, 23, 415, 2006.
46. Nishikawa, A. et al. Taxonomic position of clinical isolates of *Candida famata*. *J. Med. Vet. Mycol.*, 34, 411, 1996.
47. Pisa, D. et al. Fungal infection in patients with serpiginous choroiditis or acute zonal occult outer retinopathy. *J. Clin. Microbiol.*, 46, 130, 2008.
48. Kawakami, S. et al. Survey of fungemia cases during the past seventeen years at Teikyo University Hospital. *J. Jpn. Assoc. Infect. Dis.*, 72, 105, 1998.
49. Viudes, A. et al. Candidemia at a tertiary-care hospital: Epidemiology, treatment, clinical outcome and risk factors for death. *Eur. J. Clin. Microbiol. Infect. Dis.*, 21, 767, 2002.
50. Al-Hedaithy, S.S.A. The yeast species causing fungemia at a university hospital in Riyadh, Saudi Arabia, during a 10-year period. *Mycoses*, 46, 293, 2003.
51. Pfaller, M.A. and Diekema, D.J. Epidemiology of invasive candidiasis: A persistent public health problem. *Clin. Microbiol. Rev.*, 20, 133, 2007.
52. Kurnatowska, A.J. The occurrence of fungi in ontocenosis of the oral cavity and mucosal changes. *Mikologia Lekarska*, 10, 295, 2003.
53. Findik, D. and Tuncer, U. Nosocomial fungal infections in a teaching hospital in turkey: Identification of the pathogens and their antifungal susceptibility patterns. *Turk. J. Med. Sci.*, 32, 35, 2002.
54. Saint Germain, G. and Laverdiere, M. *Torulopsis candida*, a new opportunistic pathogen. *J. Clin. Microbiol.*, 24, 884, 1986.
55. Quindos, G. et al. Fatal *Candida famata* peritonitis in a patient undergoing continuous ambulatory peritoneal dialysis who was treated with fluconazole. *Clin. Infect. Dis.*, 18, 658, 1994.
56. Carrega, G. et al. *Candida famata* fungemia in a surgical patient successfully treated with fluconazole. *Eur. J. Clin. Microbiol. Infect. Dis.*, 16, 698, 1997.
57. Migliorati, C.A., Birman, E.G., and Cury, A.E. Oropharyngeal candidiasis in HIV-infected patients under treatment with protease inhibitors. *Oral Surg. Oral Med. Oral Pathol. Oral Radiol. Endod.*, 98, 301, 2004.
58. Ahmed, I.M. et al. A fatal fungus. *Ann. Thorac. Surg.*, 80, 723, 2005.
59. Gupta, A. et al. Fatal *Candida famata* peritonitis complicating sclerosing peritonitis in a peritoneal dialysis patient. *Nephrol. Dial. Transplant.*, 21, 2037, 2006.
60. El-Zaatari, M. et al. Evaluation of the updated Vitek yeast identification database. *J. Clin. Microbiol.*, 28, 1938, 1990.
61. Török, T. and King, A.D. Comparative study on the identification of food-borne yeasts. *Appl. Environ. Microbiol.*, 57, 1207, 1991.
62. Lopandic, K. et al. Identification of yeasts associated with milk products using traditional and molecular techniques. *Food Microbiol.*, 23, 341, 2006.
63. Fell, J.W. et al. Biodiversity and systematics of basidiomycetous yeasts as determined by large-subunit rDNA D1/D2 domain sequence analysis. *Int. J. Syst. Evol. Microbiol.*, 50, 1351, 2000.

64. Vasdinyei, R. and Deák, T. Characterization of yeast isolates originating from Hungarian dairy products using traditional and molecular identification techniques. *Int. J. Food Microbiol.*, 86, 123, 2003.
65. Missoni, E.M. et al. Differentiation between *Candida* species isolated from diabetic foot by fatty acid methyl ester analysis using gas chromatography. *J. Chromatogr. B*, 822, 118, 2005.
66. Buesching, W.J., Kurek, K., and Roberts, G.D. Evaluation of the modified API 20C system for identification of clinically important yeasts. *J. Clin. Microbiol.*, 9, 565, 1979.
67. Salkin, I.F. et al. Evaluation of YeastIdent and Uni-Yeast-Tek yeast identification systems. *J. Clin. Microbiol.*, 25, 624, 1987.
68. Land, G.A. et al. Evaluation of the Baxter-MicroScan 4-hour enzyme based yeast identification system. *J. Clin. Microbiol.*, 29, 718, 1991.
69. St.-Germain, G. and Beauchesne, D. Evaluation of the MicroScan rapid yeast identification panel. *J. Clin. Microbiol.*, 29, 2296, 1991.
70. Rohm, H., Lechmer, F., and Lehner, M. Evaluation of the API ATB 32C system for the rapid identification of foodborne yeasts. *Int. J. Food Microbiol.*, 11, 215, 1990.
71. Latouche, G.N. et al. Comparison of use of phenotypic and genotypic characteristics for identification of species of the anamorph genus *Candida* and related teleomorph yeast species. *J. Clin. Microbiol.*, 35, 3171, 1997.
72. Álvarez-Martín, P. et al. Phenotypic and molecular identification of yeast species associated with Spanish blue-veined Cabrales cheese. *Int. Dairy J.*, 17, 961, 2007.
73. Loïez, C. et al. Evaluation of VITEK 2 colorimetric card for identification of yeast. *Diag. Microbiol. Inf. Dis.*, 56, 455, 2006.
74. Hata, D.J. et al. Multicenter evaluation of the new VITEK 2 advanced colorimetric yeast identification card. *J. Clin. Microbiol.*, 45, 1087, 2007.
75. De Silóniz, M.I., Valderrama, M.J., and Peinado, J.M., A chromogenic medium for the detection of yeasts with beta-galactosidase and beta-glucuronidase activities from intermediate moisture foods. *J. Food Prot.*, 68, 808, 2000.
76. Quirós, M. et al. A beta-glucuronidase-based agar medium for the differential detection of the yeast *Debaryomyces hansenii* from foods. *J. Food Prot.*, 68, 808, 2005.
77. Kosse, D. et al. Identification of yoghurt-spoiling yeasts with 18S rRNA-targeted oligonucleotide probes. *Syst. Appl. Microbiol.*, 20, 468, 1997.
78. Davila, A.M. et al. Use of specific DNA probes for the rapid characterization of yeasts isolated from complex biotypes. *Gen. Sel. Evol.*, 33, 353, 2001.
79. Del Bove, M. et al. Comparison of molecular and metabolomic methods as characterization tools of *Debaryomyces hansenii* cheese isolates. *Food Microbiol.*, 26, 453, 2009.
80. Borst, A. et al. Detection of *Candida* spp in blood cultures using nucleic acid sequence-based amplification (NASBA). *Diag. Microbiol. Inf. Dis.*, 39, 155, 2001.
81. Andrade, M.J. et al. DNA typing methods for differentiation of yeasts related to dry-cured meat products. *Int. J. Food Microbiol.*, 107, 48, 2006.
82. Senses-Ergul, S. et al. Characterization of some yeasts isolated from foods by traditional and molecular tests. *Int. J. Food Microbiol.*, 108, 120, 2006.
83. Andrade, M.J. et al. Efficiency of mitochondrial DNA restriction analysis and RAPD-PCR to characterize yeasts growing on dry-cured Iberian ham at the different geographic areas of ripening. *Meat Sci.*, 84, 377, 2010.
84. Andrade, M.J. et al. Differentiation of yeasts growing on dry-cured Iberian ham by mitochondrial DNA restriction analysis, RAPD-PCR and their volatile compounds production. *Food Microbiol.*, 26, 578, 2009.
85. Capece, A. and Romano, P. "Pecorino di Filiano" cheese as a selective habitat for the yeast species, *Debaryomyces hansenii*. *Int. J. Food Microbiol.*, 132, 180, 2009.
86. Pisa, D. et al. Evolution of antibody response and fungal antigens in the serum of a patient infected with *Candida famata*. *J. Med. Microbiol.*, 56, 571, 2007.
87. Ayatollahi, S.A. et al. Rapid molecular diagnosis for *Candida* species using PCR-RFLP. *Biotechnology*, 6, 583, 2007.
88. Asadzadeh, M. et al. Rapid molecular differentiation and genotypic heterogeneity among *Candida parapsilosis* and *Candida orthopsilosis* strains isolated from clinical specimens in Kuwait. *J. Med. Microbiol.*, 58, 745, 2009.
89. Tofalo, R. et al. Molecular identification and osmotolerant profile of wine yeasts that ferment a high sugar grape must. *Int. J. Food Microbiol.*, 130, 179, 2009.
90. Quirós, M. et al. PCR-RFLP analysis of the IGS region of rDNA: A useful tool for the practical discrimination between species of the genus *Debaryomyces*. *Ant. Leeuwenhoek*, 90, 211, 2006.
91. Querol, A., Barrio, E., and Ramón, D. A comparative study of different methods of yeast strain characterization. *Syst. Appl. Microbiol.*, 15, 439, 1992.
92. Romano, A. et al. Use a RAPD and mitochondrial DNA RFLP for typing of *Candida zeylanoides* and *Debaryomyces hansenii* yeast strains isolated from cheese. *Syst. Appl. Microbiol.*, 19, 255, 1996.
93. Martorell, P., Fernández-Espinar, M.T., and Querol, A. Molecular monitoring of spoilage yeasts during the production of candied fruit nougats to determined food contamination sources. *Int. J. Food Microbiol.*, 101, 293, 2005.
94. Farkas, Z. et al. Genetic variability of *Candida albicans* isolates in a university hospital in Hungary. *Mycoses*, 52, 318, 2009.
95. Córdoba, J.J. et al. *Debaryomyces*. In: *Molecular Detection of Foodborne Pathogens*, Liu, D. Ed. CRC Press, Taylor & Francis Group, Boca Raton, FL, 2009.
96. Appel, D.J. and Gordon, T.R. Intraspecific variation within populations of *Fusarium oxysporum* based on RFLP analysis of the intergenic spacer region of the rDNA. *Exp. Mycol.*, 19, 120, 1995.
97. Romero, P. et al. Differential detection of *Debaryomyces hansenii* isolated from intermediate-moisture foods by PCR-RFLP of the IGS region of rDNA. *FEMS Yeast Res.*, 5, 455, 2005.
98. Quirós, M. *Debaryomyces hansenii*: Estudio de su capacidad para deteriorar alimentos y desarrollo de métodos para su detección rápida. Doctoral thesis, Universidad Complutense, Madrid, Spain, 2005.
99. López, V. et al. A simplified procedure to analyse mitochondrial DNA from industrial yeasts. *Int. J. Food Microbiol.*, 68, 75, 2001.
100. Casado, E.M. et al. Rapid characterization of yeasts isolated from Iberian ham by restriction analysis mitochondrial DNA. In: *XX National Congress of Microbiology*, Cáceres, Spain, 2005.

66 *Geotrichum*

Silvia D'Arezzo, Paolo Visca, and Corrado Girmenia

CONTENTS

66.1 INTRODUCTION

66.1.1 Classification and Morphology

Geotrichum is a filamentous yeast-like fungus found worldwide in soil, water, air, and sewage, as well as in plants, cereals, and dairy products [1–3]. It is also found in normal human flora and is isolated from sputum and feces. The genus *Geotrichum* includes 18 species, some of which can be pathogenic to humans [4–7]. The classification of arthroconidial yeast-like strains has long been problematic, and a crucial issue of fungal taxonomists has been the differentiation between the genus *Trichosporon* and the genus *Geotrichum* [1,2]. Various studies demonstrated that the type species of *Trichosporon* was of basidiomycetous affinity, and, on the basis of differences in carbohydrate composition of the cell walls, a separation of the basidiomycetous and ascomycetous arthroconidial species as *Trichosporon* and *Geotrichum* species, respectively, was proposed [8–11].

In clinical laboratory, *Geotrichum capitatum* and *Geotrichum candidum*, the two most frequently encountered *Geotrichum* species in humans, and *Trichosporon* spp. are generally indistinguishable from one another on the basis of colony morphology alone. *G. candidum* produces only arthroconidia; both *G. capitatum* and *Trichosporon* spp. produce arthroconidia as well as blastoconidia. Only *G. capitatum* can also produce anelloconidia, which unfortunately may be misidentified as arthroconidia or blastoconidia. Therefore, these species are usually differentiated by their carbohydrate assimilation patterns and other biochemical properties.

De Hoog et al. [3] considered in a taxonomic revision all the arthroconidial ascomycetous yeast-like genera together combining studies of morphology, physiology, molar guanine + cytosine percentage (mol% G + C), and DNA–DNA homology. In particular, the revision considered *Geotrichum* and its teleomorphs *Galactomyces* and *Dipodascus*, and four species were recognized in *Geotrichum*, two in *Galactomyces*, and 13 in *Dipodascus*. More recent phylogenetic studies of hyphal ascomycetous yeasts based on the determinations of the sequence divergence of the large-subunit (25S–28S) ribosomal RNA (rRNA) gene concluded that *Galactomyces*, *Geotrichum*, and *Dipodascus* are related to the ascomycetous yeasts and should be placed in the order Saccharomycetales [12–16].

Based on more recent taxonomic revisions [4–7], the genus *Geotrichum* is composed of 18 species. Of these, six species including *G. candidum* have a teleomorph in *Galactomyces*, six species including *G. capitatum* have a teleomorph in *Dipodascus*, and six species have an unknown teleomorph.

Hereafter, we will focus on the two species *G. candidum* and *G. capitatum*, because of their involvement in superficial or deep infections in humans. A schematic taxonomic classification of these two pathogens is shown in Figure 66.1.

G. capitatum was originally known as *Trichosporon capitatum* and classified among the basidiomycetes. Later,

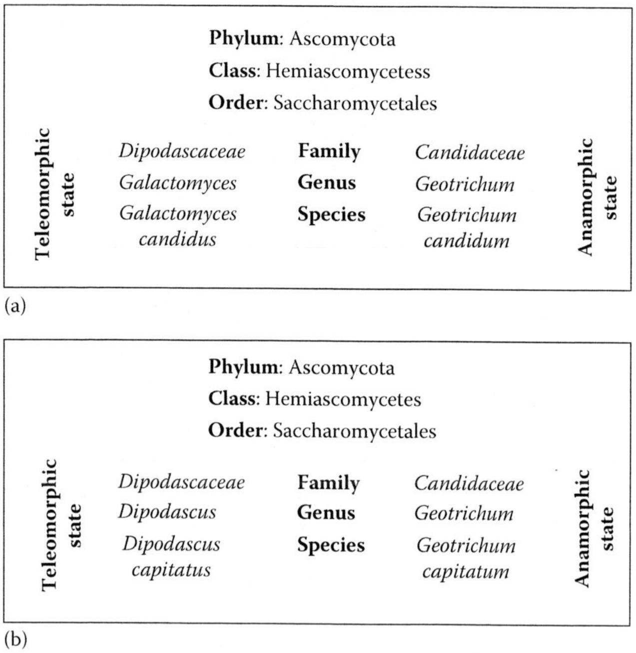

FIGURE 66.1 Present nomenclature of *Galactomyces candidus/Geotrichum candidum* (a) and *Dipodascus capitatus/Geotrichum capitatum* (b).

however, in light of its cell wall structure and septal pores and its tendency to produce numerous arthroconidia and few blastoconidia, it was considered to be an ascomycete. The appropriateness of this classification was further supported by the discovery of its sexual form (teleomorph), *Dipodascus capitatus* [3,17,18], and it was thus assigned to the genus *Geotrichum*. The subsequent discovery of its ability to produce annelloconidia, as well as arthroconidia and blastoconidia, led Salkin to reclassify *G. capitatum* as the single species of a new genus: *Blastoschizomyces capitatus* [19]. However, from a taxonomic point of view, the designation *G. capitatum* is still regarded as the correct anamorphic name [20,21].

66.1.2 Epidemiology

Members of the arthroconidial yeast genera *Dipodascus* and *Galactomyces* and their *Geotrichum* anamorphs are isolated mainly from plant materials and to a lesser extent from human and other animal specimens.

G. candidum is a ubiquitous fungus found in a wide range of habitats such as plant tissues, silage, soil, milk, air, and water. It is also a component of the natural flora of the digestive tract in humans and many other mammals [22,23]. It is found naturally in cheese and fermented milk, but in France it has also been added as a ripening agent for at least 30 years. It is reported by the International Dairy Federation and European Food and Feed Culture Association as a microorganism with a documented history of use in dairy products [24]. Sources of exposure are ingestion, inhalation, and contact [25]. Cheese consumption proved to be a major exposure source. Usually, the strains survive up to the stools but do not colonize the digestive tract. Although many workers are exposed to large quantities of *G. candidum* spores (e.g., cheesemakers and workers in starters factories), no professional disease implicating *G. candidum* has been reported to date. The pathogenicity of *G. candidum* in humans has not been defined [25]. Infections are very rare and may be superficial and deep invasive. Intestinal colonization by *G. candidum* may occur when mucosal immunity is disturbed and may be the source of dissemination in cases of severe immunodeficiency. The cases of *G. candidum* illness reported in literature may be overestimated because of the possibility of misidentification with other pathogens such as *G. capitatum* or *Trichosporon* spp., particularly in older publications before the recent taxonomic classification of arthroconidial fungi, or the difficult etiological attribution of some mixed infections in which multiple bacteria and fungi are involved.

G. capitatum has been isolated from environmental sources such as woods and poultry feces and is widely distributed in nature as a soil saprobe. More frequently than *G. candidum*, *G. capitatum* is part of the normal flora of human skin and may be isolated from sputum and the digestive tract of healthy people. As with other opportunistic yeasts, the number of infections due to this species has increased during the past two decades, as a consequence of the rise of hosts presenting factors that predispose them to fungal infections such as cytotoxic chemotherapy, neutropenia, broad-spectrum antibiotic treatment, steroids, and invasive catheterization [26–28].

A literature review published in 2005 [28] demonstrated that patients with hematological malignancies are by far the most common victims of *G. capitatum* infections. Patients of this type accounted for 92% of reported invasive *G. capitatum* infections, and acute leukemia, acute myeloid leukemia in particular, was the underlying disease most frequently associated with all these infections. Most of the infected patients had been treated with conventional cytotoxic chemotherapy, and very few had received blood stem cell transplants. The infections usually occurred during a period of profound neutropenia (neutrophil count, less than $100/mm^3$).

Although acute leukemia is the major underlying condition in *G. capitatum* infections, its incidence seems to be low even in this group. On the basis of a retrospective study of Italian cases, the incidence rates for *G. capitatum* infections among patients with acute leukemia were only five cases per 1000 adult patients with acute leukemia (or 0.5%). The geographic distribution of *G. capitatum* infections is by no means homogeneous, and there is a significantly higher frequency of *G. capitatum* infections in Europe (87% of reported cases) than in the United States (5%) [28]. Furthermore, 87% of the European cases occurred in Italy, Spain, and France. Similar pictures emerge from the data collected during the ARTEMIS DISK Surveillance Study, a prospective study conducted in more than 30 countries to identify global trends in the susceptibility of yeast pathogens to fluconazole and voriconazole [29]. Isolates collected between June 1997 and December 2002 included 40 *G. capitatum* strains of whom 32 (80%) came from Europe, and over half of the European isolates were recovered in Italy [28,29]. This finding seems to suggest that climatic factors might play a selective role in the epidemiology of *G. capitatum* infections.

66.1.3 Clinical Features, Pathogenesis, and Therapy

The clinical features of geotrichosis frequently resemble those of superficial and invasive candidiasis, although there are some significant differences. Both *G. candidum* and *G. capitatum* are recognized as a cause of superficial, mucosal and skin infections not only in immunocompromised patients but also in healthy subjects, whereas deep invasive infections have been reported almost exclusively for *G. capitatum* in patients with underlying hematological diseases.

Invasive infections by *G. candidum* are mainly pulmonary or bronchopulmonary but also cutaneous and disseminated. The first cases of bloodstream and invasive disseminated infection by this pathogen have been reported in 1952 and 1965, respectively [30–32]. Invasive infections may be localized to the skin and soft tissues, for example, when a traumatism enables fungal inoculation [33], but disseminated geotrichosis has been rarely reported in some patients immunosuppressed by HIV infection or chemotherapy [34–42].

The first case of *G. capitatum* infection was reported in 1965 [43]. The fungus is isolated from blood in over 70% of invasive infections, and approximately two-thirds of the reported cases of fungemia are associated with clinically or microbiologically documented invasive tissue localization; in a very small number of cases, a central venous catheter is the portal of entry of the infection [28]. These clinical features are in contrast with those observed in cancer patients with candidemia, as shown in a large prospective European study on candidemia in cancer patients in which only 10% of patients had a clinically or microbiologically histologically documented organ involvement and a correlation of fungemia with a central venous catheter was demonstrated in 31% of cases [44]. The EORTC/IFICG and NIAID/MSG definitions provide no specific indications on the significance of *G. capitatum* recovery from respiratory-tract specimens [45]. Since the microorganism is the potential component of the normal microbial flora of the human digestive and respiratory tracts, it is difficult to distinguish between colonization and infection. However, several experiences seem to support the view that the clinical significance of isolation of *G. capitatum* from sputum in neutropenic patients seems to be comparable with that of molds and *Cryptococcus neoformans*, and for this reason, the recovery of *G. capitatum* from respiratory tract specimens of patients with clinically documented pneumonia may be indicative of probable pulmonary geotrichosis [28].

The outcome of *G. capitatum* invasive infections appears to be worse than that associated with invasive *Candida* infections, with a mortality rate of about 60% of cases [28].

The optimal therapy for geotrichosis has yet to be identified. Most of data are available for *G. capitatum*, whereas the paucity of clinical and microbiological data reported to date does not allow to define specific indications for *G. candidum*. However, it could be hypothesized that there is a similar antifungal susceptibility profile for both *Geotrichum* species. Conventional amphotericin B, alone or associated with other antifungal agents, has been the drug most frequently employed in first-line therapy of *G. capitatum* infections. The low number of cases treated with alternative antifungal regimens does not allow any comparative evaluation of the efficacy of these strategies. Several investigators have proposed dual-drug therapy with amphotericin B and flucytosine as a valid option, but there is no evidence in the literature that this (or any other) drug combination was more effective than single-drug regimens. As for the newer antifungal drugs, it is currently impossible to make any predictions of their clinical efficacy. The literature contains reports of only a few cases in which voriconazole or caspofungin was used as initial or salvage antifungal therapy [27,46,47]. However, in vitro susceptibility findings can be a useful guide in selecting an antifungal regimen for geotrichosis. Amphotericin B and voriconazole are very active drugs against this yeast whereas a variable susceptibility to flucytosine, fluconazole, and itraconazole has been observed [27,48]. Importantly, echinocandins are intrinsically not active against *Geotrichum* species [28].

66.1.4 Diagnosis

66.1.4.1 Conventional Techniques

Early diagnosis of deep-seated *Geotrichum* species infection is difficult to establish because clinical symptoms are not specific and frequently resemble those of invasive candidosis. Blood cultures are the "gold standard" for diagnosis of invasive *G. capitatum* infections, although they usually take several days to show detectable growth of yeast cells and remain negative in about 30% of cases. As anticipated, the isolation of *Geotrichum* species from respiratory tract is easy to obtain but of ambiguous significance considering the possibility of a simple colonization.

Noncultural methods could be useful for the improvement of laboratory diagnosis, but clinical data on *G. capitatum* antigens or DNA detection are not available to date. Consequently, the microbiological diagnosis of geotrichosis continues to be represented by conventional cultural methods of blood and, less frequently, of other specimens.

Recently, false-positive reactivity in the Platelia *Aspergillus* assay, a sandwich enzyme-linked immunosorbent assay (ELISA) assay for the detection of the *Aspergillus* galactomannan antigen, has been observed in a series of hematologic patients with invasive *G. capitatum* infection [49,50]. In vitro studies showed that *G. capitatum* exoantigens may be cross-reactive with the *Aspergillus* galactomannan assay. This cross-reactivity may be related to the presence of galactomannans demonstrated in the cell wall of *Geotrichum* spp. and their teleomorphs *Dipodascus* and *Galactomyces* [51]. This unexpected reactivity in the Platelia

Aspergillus test could be considered a useful diagnostic and prognostic tool in the management of *Geotrichum* infections. A positive galactomannan assay result for a patient with an infective clinical picture suggestive of *Candida* infection could indicate the possibility of a *G. capitatum* infection, considering also that galactomannan serum detection may precede by several days the documentation of the infection by conventional cultures.

66.1.4.2 Molecular Techniques

Early detection of infectious fungi has a significant impact on the clinical management of many fungal diseases. From a clinical perspective, early and accurate diagnosis of pathogenic fungi is crucial to contain the infection and minimize empirical antifungal therapy, which is the primary cause of the emergence of antifungal resistance [52]. All these issues apply to *Geotrichum* infections.

Molecular methods have solved many problems of mycological diagnosis by providing species-level identification of fungi through sequence-based analysis of suitable taxonomic markers, yielding a more accurate identification and shortening of the time required for microbiological confirmation of life-threatening fungal infections. Moreover, DNA-based techniques have become essential tools to trace the epidemiology of many fungal diseases. In general, nucleic acid-based methods are (i) less subjective than microscopy- or culture-based methods, (ii) unaffected by the growth conditions of the fungus, (iii) capable of discriminating between phenotypically undistinguishable species, and (iv) more rapid than traditional methods. Because of their high specificity and sensitivity, molecular assays have become increasingly adopted by clinical laboratories to complement the information gained from conventional methods.

In the large majority of invasive infections, *Geotrichum* spp. is isolated from blood. However, blood cultures usually take several days to show detectable growth, and in the remaining cases of deep-seated infection without fungemia, the diagnostic approach may be very difficult. Moreover, real-time monitoring of invasive infection by *Geotrichum* spp. is hampered by the lack of noncultural markers [28,49,50].

66.1.4.2.1 Detection and Identification

On this premise, the application of polymerase chain reaction (PCR) technology to molecular diagnostics holds great promise for the early identification of *Geotrichum* spp. For rapid detection of *Geotrichum* spp. in clinical samples, target gene sequences and practicable primers have been combined with effective DNA extraction methods (Table 66.1). PCR assays may be applied to different clinical samples, including respiratory specimens, blood fractions (serum, plasma, and whole blood), urine, and other sterile body fluids [53–55].

DNA targets used in molecular diagnostic tests for *Geotrichum* spp. include the group of genes encoding the nuclear rRNAs (Figure 66.2a) that are present in multiple copies and repeated in tandem. In yeasts, the 5.8S, 18S, and 25S–28S rRNA genes (or ribosomal DNAs [rDNAs]) are transcribed as a 35S–40S precursor encompassing both internal and external transcribed spacers (ITS and ETS, respectively), which are spliced out of the transcript [56]. The conserved regions alternate with divergent domains (D1 and D2) and highly variable regions (ITS). Between each cluster, there is an intergenic spacer (IGS), while a 5S rRNA gene has a variable position and transcription direction, depending on the fungal group (Figure 66.2a). The ribosomal conserved sequences can be used to screen for yeast presence, while the variable ITS and D1/D2 sequences provide species-level identification [55,57,58].

Many in-house PCR assays have been developed for the identification of *Geotrichum* spp. PCR with fungal primers designed on the conserved sequences of the 5.8S and 25S–28S rDNA results in amplification of the species-specific ITS2 regions, which are variable in amplicon length [59]. The variability in length of the ITS2 region is utilized to make specific diagnosis of pathogenic fungal isolates from blood and tissues, including *G. candidum*. Fluorescently labeled amplicons can be rapidly and accurately sized

TABLE 66.1
DNA-Based Assays for Identification of *Geotrichum* spp.

DNA Target	Primer	Sequence (5′–3′)	Detection Method	Reference
18S rRNA and ITS1	NS1	GTAGTCATATGCTGTCTC	PCR + restriction analysis	Senses-Ergul et al. [60]
	ITS2	GCTGCGTTCTTCATCGATGC		
5.8S and 28S rRNA	ITS41	TCCTCCGCTTATTGATAGC	PCR + fluorescent capillary electrophoresis	Turenne et al. [59]
	ITS86	GTGAATCATCGAATCTTTGAAC		
ITS1 and ITS2	ITS1	TCCGTAGGTGAACCTGCGG	PCR + sequencing	White et al. [74]
	ITS4	TCCTCCGCTTATTGATATGC		
ITS1 and ITS2	ITS4	TCCTCCGCTTATTGATATGC	PCR + restriction analysis	Florez et al. [61]
	ITS5	GGAAGTAAAAGTCGTAACAAGG		
D1/D2	NL1	GCATATCAATAAGCGGAGGAAAAG	PCR + sequencing	Ersoz et al. [53]
	NL4	GGTCCGTGTTTCAAGACGG		
D2	U1	GTGAAATTGTTGAAAGGGAA	PCR (±sequencing)	Putignani et al. [55]
	U2	GACTCCTTGGTCCGTGTT		

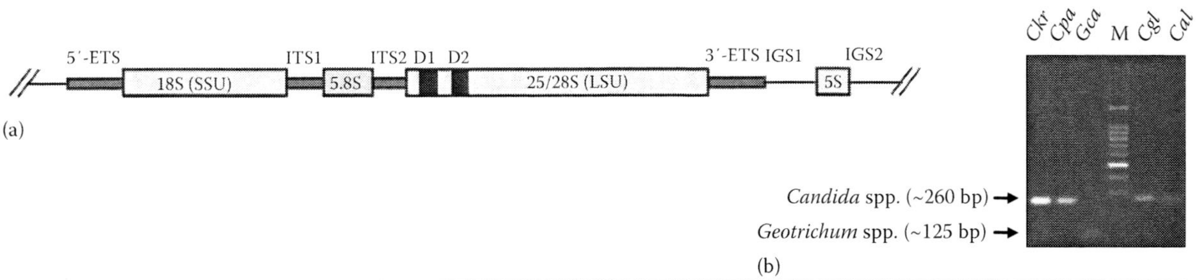


```
Cal  1- GTGAAATTGTTGAAAGGGAA GGGCTTGAGATCAGACTTGGTATTTTGCATGTTGC --  TCTCTCGGGGGC----   GGCCGCTGCGGTTTACCGGGCC
Cpa  1- GTGAAATTGTTGAAAGGGAA GGGCTTGAGATCAGACTTGGTATTTTGTATGTTAC --  TCTCTCGGGGGT----   GGCCTCTACAGTTTACCGGGCC
Cgl  1- GTGAAATTGTTGAAAGGGAA GGGCATTTGATCAGACATGGTGTTTTGCGCCCCTTG - CCTCTCGTGGGCTTGGGACTCTCGCAGCTCACTGGGCC
Ckr  1- GTGAAATTGTTGAAAGGGAA GGGTATTGCGCCCGACATGG- GGATTGCGCACCGCTGCCTCTCGTGGGC--  GGCGCTCTGGGCTTTCCCTGGGCC
Gca  1- GTGAAATTGTTGGAGGGGAAGG---------------------------------------------         ---------------
Gcl  1- GTGAAATTGTTGGAGGGGAAGG-------------------------------------------------          -------------
        *********** ********* *   **** *** ***    ****** ***   *    *  * *****

Cal  AGCATCGGTTTGGAGCGGCAGGATAATGG- CGGAGGAATGTGGCACGGCTTCT----   GCTGTGTGTTATAGCCTCTGAC - GATACTGCCAGCCTAGA
Cpa  AGCATCAGTTTG- AGCGGTAGGATAAGTG- CAAAGAAATGTGGCACTGCTTCG----   GTAGTGTGTTATAGTCTTTGTC - GATACTGCCAGCTTAGA
Cgl  AGCATCGGTTTT - GGCGGCCGGAAAAAAC- CTAGGGAATGTGGCTCTGCGCCTCGGTGTAGAGTGTTATAGCC CTGGGG- AATACGGCCAGTCGGGA
Ckr  AGCATCGGTTCT - TGCTGCAGGAGAAGGGGTTCTGGAACGTGGCTCTTC---------     GGAGTGTTATAGCCAGGGCCAGATGCTGCGTGCGGGGA
Gca  ----------------------------------------------------------------------                       CGATGGTAGGAA
Gcl  ------------------       ------------------------------------------------------               CGATGGTAGGAA
     ****** ***  ** *  *** **   *** ***** * *     * *****                  ** **  *  *  *    *  * *  *  *

Cal  CCGAGGACTGCGGT---  TTTTACCTAGGATGTTGGCA TAATGATCTTAAGTCGCCCGTCTT --------     ----  GAAACACGGACCAAGGAGTC- 258
Cpa  CTGAGGACTGCGGC---  TTCGGCCTAGGATGTTGGCATAATGATCTTAAGTCGCCCGTCTT -----------        GAAACACGGACCAAGGAGTC- 257
Cgl  CCGAGGACTGCGATACTTGTTATCTAGGATGCTGGCATAATGGTTATATGCCGCCCGTCTT -----------        GAAACACGGACCAAGGAGTC- 269
Ckr  CCGAGGACTGCGGCCGTGTAGGTCACGGATGCTGGCAGAACGGCGCAACACCGCCCGTCTT-----------         GAAACATGGACCAAGGAGTC- 259
Gca  TAAGAGGCTGCGGT--  TTGAAATAATTGTTTTTCGGGCCACGGTCTCCTGA - GCCTGCTTTCGCACCC GTCTTGAAACACGGACCAAGGAGTC- 124
Gcl  TAAGAGGCTGCGGT--  TTGAAATAAATGTTTTTCGGGCCACGGTCTCCTGA -GCCTGCTTTCGCACNNNNNNNGAAACACGGACCAAGGAGTC- 124
       * *****           ** * *  * *    *** *  ** **                           ******     ************
(c)
```

FIGURE 66.2 (a) Physical organization of the fungal rRNA gene cluster. All cluster components are included in the representation. SSU, small subunit (18S rRNA gene); LSU, large subunit (25S–28S rRNA gene); D1 and D2 regions are the highly divergent regions of the LSU rRNA gene, shown as black boxes. ITS1, ITS2, and ETS1, ETS2 represent intergenic and extragenic spacers, as shown as dark gray boxes, respectively. Intergenic spacers (IGS1 and IGS2), between each cluster are nontranscribed regions which separate rRNA clusters from one another along the chromosome. (b) Agarose gel electropherogram of PCR amplicons generated with primers defining the D2 region of the 25S–28S rRNA gene. Differences between *Candida* and *Geotrichum* spp. are shown by the different size of the amplicon (ca. 260 and 125 bp, respectively). (c) Alignment of the D2 region of the 25S–28S rRNA genes of *C. albicans*, *C. parapsilosis*, *C. glabrata*, *C. krusei*, *G. capitatum*, and *G. clavatum*. The 5′- and 3′-DNA regions matching the forward and reverse primer sequences are in bold. The boxed region denotes a deletion on the D2 region of *Geotrichum* spp. 25S–28S rRNA genes leading to smaller sized (124 bp) amplicon. Conserved positions are marked with an asterisk below the alignment. *Abbreviations:* Cal, *Candida albicans*; Cgl, *Candida glabrata*; Ckr, *Candida krusei*; Cpa, *Candida parapsilosis*; Gca, *G. capitatum*; Gcl, *G. clavatum*.

with an automated capillary electrophoresis system [59] (Table 66.1). PCR of the ITS1 and ITS2 regions and direct sequence analysis of the amplicons, coupled with classical detection methods, proved effective for identification of *G. capitatum* as the etiologic agent of superficial geotrichosis in an old patient [54]. During a mini epidemic of *G. capitatum* involving three cases, PCR and sequencing analysis of the D1/D2 region was performed to ensure identification of this organism using the primer pair NL1 and NL4 (Table 66.1). In the same study, randomly amplified polymorphic DNA (RAPD) analysis was applied to allow discrimination among isolates [53]. More recently, the hypervariable D2 region was found to contain significant information for confident identification of *Geotrichum* isolates at the species level [55]. *G. capitatum* can be easily differentiated from other yeast species, most prominently *Candida* spp., by simple visual inspection of the D2 amplicon size following agarose gel electrophoresis, and identification can be confirmed by amplicon sequencing (Table 66.1). The D2 amplicon size of pathogenic *Geotrichum* spp. is shorter than that obtained for other pathogenic yeast isolates (*ca.* 120 bp vs. a mean size of *ca.* 260 bp; Figures 66.2b and c) [55]. To the best of our knowledge, there are no commercial systems available capable of identifying *G. capitatus*, including those based on rRNA gene sequence analysis (MicroSeq ID Fungal gene library, Applied Biosystems, version 2.0). Since *G. capitatum* reference sequence has not yet been included in the commercial MicroSeq ID Fungal gene library, D2 sequencing and interrogation of free-access databases (e.g., the GenBank at: http://www.ncbi.nlm.nih.gov/) is advisable for identification of this species [55].

Another common approach for *Geotrichum* species differentiation relies upon differences in restriction fragment length polymorphism (RFLP) profiles obtained with different restriction enzymes. Restriction enzyme digestion of the ITS1-18S rRNA amplicons [60] and ITS1–5.8S–ITS2 products [61] has successfully been applied to identification of *G. candidum* (Table 66.1).

New advanced molecular approaches based on matrix-assisted laser description ionization time of flight-mass spectrometry (MALDI-TOF) are currently under setting for the rapid detection and identification of fungi in clinical microbiology laboratories. A recent study demonstrates that MALDI-TOF can be used as a fast technique for reliable identification of *Geotrichum* spp. [62]. With state-of-the art instruments and user-friendly software, mass spectrometry specialists are not needed and laboratory technicians are able to run the system after a few days of training. Investment and maintenance of the instrument are balanced by low consumable costs [62]. MALDI-TOF appears to be a very promising technique, which could open a new horizon for the fast and reliable diagnosis of fungal infections.

It is foreseeable that improvement in DNA sequencing technology together with the development of free user-friendly sequence analysis software and validated yeast databases will facilitate transfer of genotypic identification to the clinical laboratory routine. Thus, molecular-based *Geotrichum* spp. identification is ready to be used not only by specialized research facilities but also in the clinical diagnostic laboratory.

66.1.4.2.2 Typing

Among the many potential applications for strain typing are outbreak analysis and environmental monitoring, patient monitoring and treatment follow-up, local and global epidemiology, database construction, strain identification, and many more. Typing of isolates helps to define prevention measures and improved guidelines for patient management due to a greater understanding of the potential sources of infections.

It has been suggested that a number of *Geotrichum* spp. cases have a nosocomial origin, emanating from a common source within the hospital environment [17]. Typing of *G. capitatum* isolates by molecular methods made it possible to demonstrate [17,63] or exclude [64] clonal correlation among clinical isolates in the nosocomial setting, supporting the value of molecular typing in investigation of epidemiological correlates of *G. capitatum.*

With the development of molecular typing, it has become possible to demonstrate inter- or intraspecific DNA polymorphism in fungi despite a lack of information about the target sequences. One PCR-based method independent of whole genome knowledge is RAPD, which has long been applied to prokaryotic genomes and successfully extended to eukaryotic microorganisms, including *G. candidum* [17,25,53,61,65]. However, RAPD analysis exhibits a low level of reproducibility because of the low-stringency conditions used in the PCR, and these conditions lead to mismatched pairings. Therefore, the patterns may be complex and hardly comparable between laboratories [66,67].

RFLP analysis of whole-cell DNA has also been applied to differentiate clinical isolates of *G. capitatum* [63,64].

More recently, using the random amplified microsatellites technique (RAM or inter-single sequence repeats, [ISSR]), the specific oligonucleotides of single sequence repeats (microsatellites) have been used to amplify the regions located between two microsatellites in a large number of eukaryotic microorganisms [68]. Used in combination, RAPD-PCR and RAM-PCR techniques permit *G. candidum* strains to be unambiguous differentiated by using three RAPD probes and one RAM probe [66,67]. Different from RAPD-PCR, the RAM-PCR involves large primers and an annealing temperature at 50°C, leading to higher hybridization specificity [66,67]. The RAM-PCR method therefore seems to be suitable for the purpose of inter-laboratory analysis. Nevertheless, obtaining reproducible fingerprints requires standardized experimental conditions, such as using the same thermal cycler, cycling conditions, and concentrations of reagents (template DNA, PCR buffer, dNTPs, and *Taq* DNA polymerase).

To determine the size of the chromosomal DNA molecules from 13 strains of *G. candidum* differing in habitat and morphotype, pulsed field gel electrophoresis (PFGE) has been used without previous macrorestriction analysis [66,67]. The strains investigated had five to eight chromosomes, 0.6–4.5 Mb in size. PFGE profiles, defined as electrophoretic karyotypes, showed a high degree of polymorphism, indicating considerable variability between strains. Notably, genome size and the presence of large chromosomes appeared correlate with the morphotype [66,67]. *G. candidum* strains with a mold-like or intermediate morphotype tended to have larger genomes than strains with a yeast-like morphotype.

66.2 METHODS

66.2.1 Sample Preparation

Specimen handling and preparation have a significant impact on the performance of molecular diagnostic tests for fungal detection. The sample preparation method should release intracellular DNA from the fungal cell wall, concentrate DNA targets that may be present in very small amounts, and eliminate contaminants, potential inhibitors, and other extraneous materials without degrading the DNA. The availability of an easy-to-perform DNA extraction procedure, providing pure DNA devoid of PCR inhibitors, would be ideal for any PCR-based diagnostic test. There are a multitude of nucleic acid extraction techniques [69]. The preferable method represents a compromise between efficiency, purification yields, and transferability to the laboratory routine. DNA may be extracted using in-house methods, commercial kits, and automated commercial techniques. Mechanical destruction with glass beads and freeze-thaw steps with liquid nitrogen or a heat-alkali treatment have successfully been applied [69,70]. DNA extraction following enzymatic digestion of the fungal wall is another effective method. However, the use of toxic chemicals such as phenol–chloroform mixtures further limits the use of in-house methods in the clinical routine [71]. The use of commercial kits (e.g., QIAmp Tissue, Qiagen; GeneReleaser, BioVentures; Puregene D 6000, Gentra; Dynabeads DNA DIRECT, Dynal; and DNAzol, Molecular Research Center) shortens the extraction procedure, but the efficiency of extraction of fungal DNA can vary considerably between commercial kits [69,70]. Automated commercial techniques (e.g., MagNA Pure LC; Roche Diagnostics) are better suited for routine clinical laboratories. High-speed cell disruption (HSCD) incorporating chaotropic reagents and lysing matrices provides rapid disruption of cells and high yields of DNA [72].

A simple thermolytic procedure for *G. capitatum* DNA extraction has been used as routine protocol in the clinical laboratory for direct PCR amplification of the D2 region [55]. Sample contamination should also be considered as major problem in molecular diagnosis because of the extremely high sensitivity of all nucleic acid amplification techniques. Fungal spores, such as conidia from *Aspergillus* spp. and other molds, might be present in the air. Thus, airborne spore inoculation during the DNA extraction process could potentially lead to false-positive results, especially in pan-fungal assays [73]. However, the risk of contamination is not higher in fungal PCR assays than in other diagnostic PCRs if general precautions are taken. In order to control naturally arising DNA from airborne sources (e.g., fungal spores), negative controls should be included during each DNA extraction procedure. Negative controls consist of sterile water or blood from healthy individuals and should be subjected to all preparation steps in parallel with the extracted samples [72].

66.2.2 Detection Procedures

66.2.2.1 Identification of *Geotrichum* by PCR Amplification of the D2 Region

Procedure

The following procedure is adapted from Ref. [55].

1. A single fungal colony >1 mm in diameter is picked with a micropipette tip, suspended in 20 µL of sterile doubly distilled water, and thermolyzed for 5 min at 95°C.
2. Prepare a PCR mixture (50 µL) containing 2 µL of thermolysate, 750 mM Tris–HCl pH 8.8, 200 mM $(NH_4)_2SO_2$, 0.1% Tween 20 (v/v), 1.5 mM $MgCl_2$, 0.2 mM of each dNTP, 25 pmol primer U1 (5′-GTGA AATTGTTGAAAGGGAA-3′) and 25 pmol primer U2 (5′-GACTCCTTGGTCCGTGTT-3′), and 0.7 unit of *Taq* DNA polymerase (Red Hot DNA polymerase, Abgene, Epsom, United Kingdom) [55].
3. Perform the PCR amplification in a thermocycler using a cycling program consisting of a denaturation step at 95°C for 3 min; 35 cycles of 94°C for 45 se 50°C for 45 s, and 72°C for 1 min; and a final extension step at 72°C for 7 min [55].
4. Separate the PCR products on a 2% agarose gel and stain with ethidium bromide (0.5 µg/mL). The presence of a single product of expected size is interpreted as evidence of successful amplification.

Note. PCR analysis of the hypervariable D2 region of the fungal 25S–28S rDNA provides clear differentiation between *Geotrichum* and *Candida* based on the D2 amplicon size, and identification can be confirmed by amplicon sequencing. The D2 amplicon size of pathogenic *Geotrichum* spp. is shorter than that obtained for other pathogenic yeast isolates (ca. 120 bp vs. a mean size of ca. 260 bp [Figures 66.2b and c]) [55].

66.2.2.2 RFLP Analysis

Procedure

1. Prepare a PCR mixture for the amplification of ITS1 or ITS1 and ITS2 regions and perform the PCR amplification in a thermocycler using primers and PCR conditions described by Senses-Ergul et al. [60] and Florez et al. [61].
2. Prepare an RFLP mixture containing purified PCR amplicons, 1× corresponding restriction buffer (dependent on the enzyme used and the manufacturer) and restriction enzymes *Hae*III and *Msp*I [60], or *Alu*I, *Hae*III, *Hha*I, *Hinf*I, *Rsa*I, and *Sau*3AI [61]. Incubate overnight at the temperature indicated by the manufacturer.
3. Seperate the fragments on 3% agarose gels, achieve visualization by staining with ethidium bromide (0.5 µg/mL), and photograph the results under UV light [60].

66.2.2.3 Automated Fluorescent Capillary Electrophoresis

To detect fluorescently tagged amplicons by an automated capillary electrophoresis sequencer, a capillary is filled with a replaceable liquid polymer for high resolution under denaturing conditions. The sample containing fluorescent PCR product is injected into the capillary and is subject to constant voltage and temperature (60°C). The fluorescently tagged fragments of DNA resulting from amplification by fluorescent primers are detected by a laser beam and automatically analyzed by the instrument [58]. Up to hundreds of samples can be loaded depending on the instrument, and each run is completed within 30 min. Results can be viewed as soon as each sample has been processed. The instrument detects up to four various fluorescent "colors," allowing for the inclusion of an internal standard with each run. Accurate fragment size analysis is based on the electrophoretic mobility of the sample relative to the internal standard [58].

66.2.2.4 DNA Sequence Analysis

A variety of automated sequencers are available on the market, but their use requires specialized personnel with direct and indirect costs often unaffordable by clinical laboratories. However, many companies offer sequencing services at reasonable price and provide fast (internet-based) reporting of sequencing data.

Sequence-based identification is achieved by a simple process involving free internet-based tools for sequence quality assessment and database interrogation at GenBank (http://www.ncbi.nlm.nih.gov/) or EMBL (http://www.ebi.ac.uk/embl/), using BLASTN (http://www.ncbi.nlm.nih.gov/BLAST/) and FASTA homology search algorithms (http://www.ebi.ac.uk/fasta33/nucleotide.html).

66.3 CONCLUSIONS AND FUTURE PERSPECTIVES

For many years, the methods used to identify yeasts have essentially been based on morphological criteria, physiological characteristics, or a combination of both. Phenotype-based commercial identification micromethods, which were developed for yeasts of clinical importance, should however be used with caution when identifying the less common yeasts. Cultural characteristics and fermentative and assimilative patterns are somehow similar in *Candida krusei* and *G. capitatus*, and this can result in erroneous identification by phenotypic methods [55].

Current literature on DNA-based methods for identification of *Geotrichum* spp. highlights the utility of such a diagnostic approach in the clinical practice. As far as *G. capitatum* is concerned, a simple PCR analysis of the D2 region of the fungal 25S–28S rDNA provides clear differentiation between *Geotrichum* and *Candida* based on the D2 amplicon size, and further sequencing allows species-level identification with reasonable confidence (Figure 66.2) [55]. Simplification and/or standardization of methods for DNA extraction and detection are likely to facilitate the transfer of molecular technology to routine clinical laboratories. Automated DNA extraction and the development of controlled sequence databases are likely to improve reliability of the current DNA-based identification assay. In this scenario, it is expected that ongoing progress in molecular biology will continue to have an overall positive impact on medical mycology.

There are many issues, however, that need to be addressed in the future. First, the direct identification of fungal DNA extracted from pathological samples remains problematic, and most of molecular identification methods require preliminary isolation fungi as pure cultures, with obvious delay in diagnosis. Second, some technical problems—primarily the risk of contamination—still need to be solved. Third, the percentage of sequence homology required to define fungal species or genera should be clearly established by the scientific community for the main phylogenetic markers, for example, the rRNA genes. Fourth, interrogation of public databases of rRNA genes can provide dubious identities because of the presence of uncontrolled entries. If fungal cultures are unavailable, uncertain or spurious DNA-based identifications cannot be verified through appropriate phenotypic testing, and the molecular epidemiology of infectious fungi will be difficult to trace. Fifth, the reliability of the current DNA-based identification assays ought to be assessed in the coming years with well-designed prospective studies and regular quality controls. Rigorous testing of molecular assays for the diagnosis of fungal disease through multicenter proficiency studies and external validation from certifying institutions will be unavoidable steps. Lastly, the clinical meaning of a positive PCR test should be interpreted in the light of our knowledge on the clearance kinetics of fungal DNA in the infected patient; this would help in distinguishing between colonized individuals and infected patients.

REFERENCES

1. Lodder, J. and Kreger-van Rij, N.J.W., *The Yeasts: A Taxonomic Study*. North-Holland, Amsterdam, the Netherlands, 1967.
2. Do Carmo-Sousa, L., The genus *Trichosporon* Behrend. In: *The Yeasts: A Taxonomic Study* (Lodder, J., Ed.), pp. 1309–1352. North-Holland, Amsterdam, the Netherlands, 1970.
3. De Hoog, G.S., Smith, M.Th., and Gueho, E., A revision of the genus *Geotrichum* and its teleomorphs. *Stud. Mycol.*, 29, 1, 1986.
4. De Hoog, G. and Smith, M., Ribosomal gene phylogeny and species delimitation in *Geotrichum* and its teleomorphs. *Stud. Mycol.*, 50, 489, 2004.
5. Pimenta, R.S. et al., *Geotrichum silvicola* sp. nov., a novel asexual arthroconidial yeast species related to the genus *Galactomyces*. *Int. J. Syst. Evol. Microbiol.*, 55, 497, 2005.
6. Wuczkowski, M., Bond, C., and Prillinger, H., *Geotrichum vulgare* sp. nov., a novel asexual arthroconidial yeast. *Int. J. Syst. Evol. Microbiol.*, 56, 301, 2006.
7. Suh, S.O. and Blackwell, M., Three new asexual arthroconidial yeasts, *Geotrichum carabidarum* sp. nov., *Geotrichum histeridarum* sp. nov., and *Geotrichum cucujoidarum* sp. nov., isolated from the gut of insects. *Mycol. Res.*, 110, 220, 2006.
8. Aschner, M. and Cury, A., Starch production in the genus *Trichosporon*. *J. Bacteriol.*, 62, 350, 1051.
9. Kreger-van Rij, N.J.W. and Veenhuis, M., Septal pores in *Trichosporon cutaneum*. *Sabouraudia*, 9, 36, 1971.
10. van der Walt, J.P. and Hopsu-Havu, V.K., A colour reaction for the differentiation of ascomycetous and basidiomycetous yeasts. *Anton. Leeuwen.*, 42, 157, 1976.
11. Weijman, A.C.M., Carbohydrate composition and taxonomy of *Geotrichum*, *Trichosporon* and allied genera. *Anton. Leeuwen.*, 45, 119, 1979.
12. Kurtzman, C.P. and Robnett, C.J., Identification and phylogeny of ascomycetous yeasts from analysis of nuclear large subunit (26S) ribosomal DNA partial sequences. *Anton. Leeuwen.*, 73, 331, 1998.
13. Kurtzman, C.P. and Robnett, C.J., Molecular relationships among hyphal ascomycetous yeasts and yeastlike taxa. *Can. J. Bot.*, 73, S824, 1995.
14. Phaff, H.J. et al., *Dipodascus starmeri* sp. nov., a new species of yeast occurring in cactus necroses. *Int. J. Syst. Bacteriol.*, 47, 307, 1997.
15. Ueda-Nishimura, K. and Mikita, K., Two distinct 18S rRNA secondary structures in *Dipodascus* (Hemiascomycetes). *Microbiology*, 146, 1045, 2000.
16. Ueda-Nishimura, K. and Mikita, K., Reexamination of *Dipodascus* and *Geotrichum* strains by DNA-DNA hybridization. *IFO Res. Commun.*, 20, 92, 2001.
17. Gadea, I. et al., Genotyping and antifungal susceptibility profile of *Dipodascus capitatus* isolates causing disseminated infection in seven hematological patients of a tertiary hospital. *J. Clin. Microbiol.*, 42, 1832, 2004.
18. Smith, M.T. and Poot, G.A., *Dipodascus capitatus*, *Dipodascus spicifer* and *Geotrichum clavatum*: Genomic characterization. *Anton. Leeuwen.*, 74, 229, 1998.

19. Salkin, I.F. et al., *Blastoschizomyces capitatus*, a new combination. *Mycotaxon*, 22, 375, 1985.
20. Gueho, E. et al., *Trichosporon* on humans: A practical account. *Mycoses*, 37, 3, 1994.
21. Schiemann, R. et al., *Geotrichum capitatum* septicaemia in neutropenic patients: Case report and review of the literature. *Mycoses*, 41, 113, 1998.
22. O'Brien, M. et al., Fungi isolated from contaminated baled grass silage on farms in the Irish Midlands. *FEMS Microbiol. Lett.*, 247, 131, 2005.
23. Samson, R. et al., Chapter 1: Identification of the common food-borne fungi. In: *Introduction to Food-Borne Fungi* (Samson, R., Hoekstra, E., Frisvad, J., and Filtenborg, O., Eds.), 200pp. Centraalbureau voor Schimmelcultures, AG Barn, the Netherlands, 1996.
24. Mogensen, G. et al., Health benefits and safety evaluation of certain food components. *Bull. Int. Dairy Fed.*, 377, 4, 2002.
25. Pottier, I. et al., Safety assessment of dairy microorganisms: *Geotrichum candidum*. *Int. J. Food Microbiol.*, 126, 327, 2008.
26. Martino, P. et al., *Blastoschizomyces capitatus*: An emerging cause of invasive fungal disease in leukemia patients. *Rev. Infect. Dis.*, 12, 570, 1990.
27. Martino, R. et al., *Blastoschizomyces capitatus* infection in patients with leukemia: Report of 26 cases. *Clin. Infect. Dis.*, 38, 335, 2004.
28. Girmenia, C. et al., Invasive infections caused by *Trichosporon* species and *Geotrichum capitatum* in patients with hematological malignancies: A retrospective multicenter study from Italy and review of the literature. *J. Clin. Microbiol.*, 43, 1818, 2005.
29. Hazen, K.C. et al., Comparison of the susceptibilities of *Candida* spp. to fluconazole and voriconazole in a 4-year global evaluation using disk diffusion. *J. Clin. Microbiol.*, 41, 5623, 2003.
30. Bendove, R.A. and Ashe, B.I., *Geotrichum* septicemia. *Arch. Intern. Med.*, 89, 107, 1952.
31. Kalish, S.R., Beene, M.L., and Mattman, L., *Geotrichum* in bloodstream of an infant. *J. Am. Med. Assoc.*, 148, 1207, 1952.
32. Chang, W.W.L. and Buerger, L., Disseminated geotrichosis: Case report. *Arch. Intern. Med.*, 113, 356, 1964.
33. Hrdy, D., Nassar, N., and Rinaldi, M., Traumatic joint infection due to *Geotrichum candidum*. *Clin. Infect. Dis.*, 20, 468,1995.
34. Sfakianakis, A. et al., Invasive cutaneous infection with *Geotrichum candidum*: Sequential treatment with amphotericin B and voriconazole. *Med. Mycol.*, 45, 81, 2007.
35. André, N. et al., *Geotrichum candidum* septicemia in a child with hepatoblastoma. *Pediatr. Infect. Dis. J.*, 23, 86, 2004.
36. Depagne, C., Louerat, C., and Nesme, P., Herpes simplex and *Geotrichum candidum* pneumonia in a patient with moderate renal failure. *Rev. Pneumol. Clin.*, 59, 297, 2003.
37. Kasantikul, V. and Chamsuwan, A., Brain abscesses due to *Geotrichum candidum*. *Southeast Asian J. Trop. Med. Public Health*, 26, 805, 1995.
38. Ng, K.P. et al., Disseminated *Geotrichum* infection. *Med. J. Malaysia*, 49, 424, 1994.
39. Heinic, G.S. et al., Oral *Geotrichum candidum* infection associated with HIV infection: A case report. *Oral Surg. Oral Med. Oral Pathol.*, 73, 726, 1992.
40. Buchta, V. and Otcenásek, M., *Geotrichum candidum*—An opportunistic agent of mycotic diseases. *Mycoses*, 31, 363, 1988.
41. Kassamali, H. et al., Disseminated *Geotrichum candidum* infection. *J. Clin. Microbiol.*, 25, 1782, 1987.
42. Jagirdar, J., Geller, S.A., and Bottone, E.J., *Geotrichum candidum* as a tissue invasive human pathogen. *Hum. Pathol.*, 12, 668, 1981.
43. Gemeinhardt, H., Lungenpathogenitat von *Trichosporon capitatum* beim menschen. *Zentbl. Bakteriol.* A, 196, 121, 1965.
44. Viscoli, C., et al. Candidemia in cancer patients. A prospective, multicenter surveillance study by the Invasive Fungal Infection Group (IFIG) of the European Organization for Research and Treatment of Cancer (EORTC). *Clin. Infect. Dis.*, 28, 1071, 1999.
45. De Pauw, B. et al., Revised definitions of invasive fungal disease from the European Organization for Research and Treatment of Cancer/Invasive Fungal Infections Cooperative Group and the National Institute of Allergy and Infectious Diseases Mycoses Study Group (EORTC/MSG) Consensus Group. *Clin. Infect. Dis.*, 46, 1813, 2008.
46. Goodman, D. et al. Breakthrough trichosporonosis in a bone marrow transplant recipient receiving caspofungin acetate. *Clin. Infect. Dis.*, 35, e35, 2002.
47. Perez-Sanchez, I. et al., *Blastoschizomyces capitatus* infection in acute leukemia patients. *Leuk. Lymphoma*, 39, 209, 2000.
48. Girmenia, C. et al., In vitro susceptibility testing of *Geotrichum capitatum*: Comparison of the E-test, disk diffusion, and Sensititre colorimetric methods with the NCCLS M27-A2 broth microdilution reference method. *Antimicrob. Agents Chemother.*, 47, 3985, 2003.
49. Giacchino, M. et al., *Aspergillus galactomannan* enzyme-linked immunosorbent assay cross-reactivity caused by invasive *Geotrichum capitatum*. *J. Clin. Microbiol.*, 44, 3432, 2006.
50. Bonini, A. et al., Galactomannan detection in *Geotrichum capitatum* invasive infections: Report of 2 new cases and review of diagnostic options. *Diagn. Microbiol. Infect. Dis.*, 62, 450, 2008.
51. Ahrazem, O. et al., Fungal cell-wall galactomannans isolated from *Geotrichum* spp. and their teleomorphs, *Dipodascus* and *Galactomyces*. *Carbohydr. Res.*, 337, 2347, 2002.
52. Yeo, S.F. and Wong, B., Current status of nonculture methods for diagnosis of invasive fungal infections. *Clin. Microbiol. Rev.*, 15, 465, 2002.
53. Ersoz, G. et al., An outbreak of *Dipodascus capitatus* infection in the ICU: Three case reports and review of the literature. *Jpn. J. Infect. Dis.*, 57, 248, 2004.
54. Hattori, H. et al., A case of oral geotrichosis caused by *Geotrichum capitatum* in an old patient. *Jpn. J. Infect. Dis.*, 60, 300, 2007.
55. Putignani, L. et al., Identification of clinically relevant yeast species by DNA sequence analysis of the D2 variable region of the 25–28S rRNA gene. *Mycoses*, 51, 209, 2008.
56. Guarro, J., Gené, J., and Stchigel, A.M., Developments in fungal taxonomy. *Clin. Microbiol. Rev.*, 12, 454, 1999.
57. Hinrikson, H.P. et al., Assessment of ribosomal large-subunit D1–D2, internal transcribed spacer 1, and internal transcribed spacer 2 regions as targets for molecular identification of medically important *Aspergillus* species. *J. Clin. Microbiol.*, 43, 2092, 2005.
58. Linton, C.J. et al., Molecular identification of unusual pathogenic yeast isolates by large ribosomal subunit gene sequencing: 2 years of experience at the United kingdom mycology reference laboratory. *J. Clin. Microbiol.*, 45, 1152, 2007.

59. Turenne, C.Y. et al., Rapid identification of fungi by using the ITS2 genetic region and an automated fluorescent capillary electrophoresis system. *J. Clin. Microbiol.*, 37, 1846, 1999.
60. Senses-Ergul, S. et al., Characterization of some yeasts isolated from foods by traditional and molecular tests. *Int. J. Food Microbiol.*, 15, 108, 2006.
61. Florez, A.B. et al., Morphotypic and molecular identification of filamentous fungi from Spanish blue-veined cabrales cheese, and typing of *Penicillium roqueforti* and *Geotrichum candidum* isolates. *Int. Dairy J.*, 17, 350, 2007.
62. Marklein, G. et al., Matrix-assisted laser desorption ionization-time of flight mass spectrometry for fast and reliable identification of clinical yeast isolates. *J. Clin. Microbiol.*, 47, 2912, 2009.
63. D'Antonio, D. et al., Emergence of fluconazole-resistant strains of *Blastoschizomyces capitatus* causing nosocomial infections in cancer patients. *J. Clin. Microbiol.*, 34, 753, 1996.
64. D'Antonio, D. et al., Onychomycosis caused by *Blastoschizomyces capitatus*. *J. Clin. Microbiol.*, 37, 2927, 1999.
65. Marcellino, N. et al., Diversity of *Geotrichum candidum* strains isolated from traditional cheesemaking fabrications in France. *Appl. Environ. Microbiol.*, 67, 4752, 2001.
66. Gente, S. et al., Genetic diversity among *Geotrichum candidum* strains from various substrates studied using RAM and RAPD-PCR. *J. Appl. Microbiol.*, 92, 491, 2002.
67. Gente, S. et al., Intra-species chromosome-length polymorphism in *Geotrichum candidum* revealed by pulsed field gel electrophoresis. *Int. J. Food Microbiol.*, 76, 127, 2002.
68. Bruford, M.W. and Wayne, R.K., Microsatellites and their application to population genetic studies. *Curr. Opin. Genet. Dev.*, 3, 939, 1993.
69. Löffler, J. et al., Comparison of different methods for extraction of DNA of fungal pathogens from cultures and blood. *J. Clin. Microbiol.*, 35, 3311, 1997.
70. Griffin, D.W. et al., A rapid and efficient assay for extracting DNA from fungi. *Lett. Appl. Microbiol.*, 34, 210, 2002.
71. Griffiths, L.J. et al., Comparison of DNA extraction methods for *Aspergillus fumigatus* using real-time PCR. *J. Med. Microbiol.*, 55, 1187, 2006.
72. Müller, F.M. et al., Rapid extraction of genomic DNA from medically important yeasts and filamentous fungi by high-speed cell disruption. *J. Clin. Microbiol.*, 36, 1625, 1998.
73. Löffler, J. et al., Contaminations occurring in fungal PCR assays. *J. Clin. Microbiol.*, 37, 1200, 1999.
74. White, T.J. et al., Amplification and direct sequencing of fungal ribosomal RNA genes for phylogenetics. In: *PCR Protocols: A Guide to Methods and Applications* (Innis, M.A., Gelfand, D.H., Sninsky, J.J. and White, T.J., Eds.). Academic Press, San Diego, CA, 1990.

67 *Kluyveromyces*

Dongyou Liu

CONTENTS

67.1 INTRODUCTION

67.1.1 Classification and Morphology

The genus *Kluyveromyces* is an ascomycetous yeast belonging to the family Saccharomycetaceae, class Saccharomycetes, subphylum *Saccharomycotina*, phylum Ascomycota, and kingdom Fungi. The family Saccharomycetaceae consists of 28 genera: *Ascobotryozyma*, *Citeromyces*, *Debaryomyces*, *Dekkera*, *Eremothecium*, *Issatchenkia*, *Kazachstania*, *Kluyveromyces*, *Kodamaea*, *Komagataella*, *Kregervanrija*, *Kuraishia*, *Lachancea*, *Lindnera*, *Lodderomyces*, *Nakaseomyces*, *Nakazawaea*, *Naumovia*, *Pachysolen*, *Pichia*, *Saccharomyces*, *Saturnispora*, *Tetrapisispora*, *Torulaspora*, *Vanderwaltozyma*, *Williopsis*, *Zygosaccharomyces*, and *Zygowilliopsis* as well as some unclassified *Saccharomycetaceae* [1].

In turn, the genus *Kluyveromyces* covers 16 recognized species: *Kluyveromyces aestuarii*, *Kluyveromyces blattae*, *Kluyveromyces dobzhanskii*, *Kluyveromyces hubeiensis*, *Kluyveromyces lactis*, *Kluyveromyces lodderae*, *Kluyveromyces marxianus*, *Kluyveromyces nonfermentans*, *Kluyveromyces phaffii*, *Kluyveromyces piceae*, *Kluyveromyces siamensis*, *Kluyveromyces sinensis*, *Kluyveromyces thermotolerans*, *Kluyveromyces waltii*, *Kluyveromyces wickerhamii*, and *Kluyveromyces yarrowii*, in addition to 20 unassigned species [1–4]. Of these, *Kluyveromyces marxianus* (synonyms: *Kluyveromyces cicerisporus* and *Kluyveromyces fragilis*; anamorph: *Candida kefyr* or *Candida pseudotropicalis*) has been implicated in burn wounds and blood and vaginal infections in humans [5–7].

Based on the analysis of mitochondrial cytochrome-c oxidase II gene sequences, the genus *Kluyveromyces* appears as a polyphyletic taxon that consists of the four main groups: (1) the *Kluyveromyces phaffii* group encompassing *Kluyveromyces blattae*, *Kluyveromyces phaffii*, and *Kluyveromyces yarrowii*; (2) the *Kluyveromyces marxianus* group consisting of *Kluyveromyces aestuarii*, *Kluyveromyces dobzhanskii*, *Kluyveromyces lactis*, *K. marxianus*, and *Kluyveromyces wickerhamii*; (3) the *Kluyveromyces thermotolerans* group including *Kluyveromyces thermotolerans*, *Kluyveromyces waltii*, and *Saccharomyces kluyveri*; and (4) the *Saccharomyces cerevisiae* group containing the remaining *Kluyveromyces* species and the reference *Saccharomyces* species (sensu lato and sensu stricto) and *Candida glabrata* [8].

Morphologically, *Kluyveromyces marxianus* (anamorph: *Candida kefyr*) may be similar to other *Candida* spp. On Sabouraud dextrose agar at 25°C, *Kluyveromyces marxianus* colonies are cream to brown, glossy, and smooth. On cornmeal agar at 25°C for 72 h, the fungus forms highly branched pseudohyphae. Under the microscope, *K. marxianus* shows evanescent asci containing 2–4 cresent-shaped spores [9].

67.1.2 Clinical Features

Kluyveromyces marxianus is used commercially to produce the lactase enzyme. In addition, it is also produced as a nutritional yeast and bonding agent for fodder and pet food, and as a source of ribonucleic acid in pharmaceuticals.

The first documented case of *Kluyveromyces marxianus* (*Kluyveromyces fragilis*) involved an immunosuppressed cardiac transplant patient with pulmonary manifestations. The *K. marxianus* isolates were sensitive to 5-fluorocytosine and miconazole as well as amphotericin B. The patient responded to amphotericin B treatment and recovered without sequelae [20]. *K. marxianus* has also been found to cause vaginitis [5,9,10]. A *K. marxianus* strain was isolated from a 33 yo woman with clear signs of vaginitis, leucorrea, pruritis and vulvovaginal erythema [9]. In addition, seven *K. marxianus* (*C. kefyr*) isolates were identified from women showing thick curdy vaginal discharge, vulvovaginal itching and discomfort [10].

67.1.3 Diagnosis

Pathogenic yeasts are responsible for bulk of human clinical cases of invasive mycosis that often produces indistinguishable symptoms and results in significant morbidity and mortality, especially in high-risk/immunocompromised individuals. Despite their diversity and taxonomical complexity, yeasts are renowned for their apparent lack of characteristic morphological features, contributing to the difficulty in the identification and diagnosis of disease-causing yeasts including *Kluyveromyces* on the basis of macroscopic, microscopic, and biochemical examination [11].

Molecular techniques such as PCR amplification and sequencing analysis offer a rapid, sensitive, and specific alternative to conventional methods for yeast speciation. By exploiting the nucleotide differences at the small-subunit (18S) rRNA gene, the D1/D2 domains of the large-subunit (28S) rRNA gene, and the internal transcribed spacer (ITS) regions, accurate determination of yeasts of interest can be achieved within a day instead of weeks that is required by biochemical testing [12–16].

67.2 METHODS

67.2.1 Sample Preparation

DNA can be extracted from yeast culture by the glass bead lysis method. A loopful of culture from malt yeast extract agar is transferred to 200 µL lysis buffer (2% Triton X-100, 1% SDS, 10 mM NaCl, 10 mM Tris/HCl pH 8.0, and 1 mM EDTA) in a 1.5 mL tube, and 200 µL phenol/chloroform/isoamyl alcohol (25:24:1) is added. Cells are lysed by vortexing for 4 min in the presence of 20 mg glass beads. The aqueous layer is treated once more with phenol/chloroform/isoamyl alcohol, and the DNA is precipitated with chilled 2-propanol. The pellet is washed with 70% ethanol and finally resuspended in 50 µL TE buffer (10 mM Tris/HCl pH 8.0 and 1 mM EDTA) [17].

Alternatively, DNA may be prepared from yeast cultures after 2 days of incubation on Sabouraud agar using Whatman FTA filter paper technology. Briefly, yeast aqueous suspensions (200 µL) are applied directly to Whatman FTA microcards, which are then placed, open, in a Panasonic microwave (800 W) while still damp and subjected to two cycles of 30 s on full power, with a pause of at least 30 s between each cycle. Punches (2 mm in diameter) are then removed from dried FTA filters by using a Harris micropunch and washed twice for 1 min each with 100 µL of Whatman FTA wash reagent, followed by two washes for 1 min each with 100 µL of TE buffer (10 mM Tris–HCl pH 7.5 and 0.1 mM EDTA). Washed filters are then dried for 5 min at 55°C on a dry heat block, and PCR mixes are added directly to the washed and dried FTA filter punches [16,18].

67.2.2 Detection Procedures

67.2.2.1 Real-Time PCR Amplification and Sequencing Analysis of ITS Regions

Pounder et al. [19] described a real-time PCR with SYBR green DNA-binding dye and amplicon-melting temperature analysis for fungal detection using pan-fungal primers ITS1 forward (5′-TCCGTAGGTGAACCTGCGG-3′) and ITS4 reverse (5′-TCCTCCGCTTATTGATATGC-3′). The identity of the fungi is verified by subsequent sequencing analysis.

Procedure

1. PCR mixture is composed of 1×Lightcycler FastStart DNA Master Hybridization Probes mixture (Roche Applied Science) containing deoxynucleoside triphosphates, FastStart *Taq* DNA polymerase, and 1 mM $MgCl_2$ (additional $MgCl_2$ is added to a final concentration of 4.6 mM), 0.4 µM each of ITS1 forward and ITS4 reverse primers, 1× SYBR green (Molecular Probes), and 3 µL template DNA.
2. Thermal cycling parameters include 95°C for 10 min; 50 cycles of 95°C for 5 s, 60°C for 20 s, and 76°C for 30 s; and a final extension at 72°C for 2 min.
3. The quality of the amplicon is determined using the derivative of the melt analysis curve (55°C–99°C, 45 s hold at 55°C, and 5 s/°C) using the RotorGene 3000 (Corbett Robotics, Inc).
4. The amplified product is purified for bidirectional sequencing using ExoSAP-IT (USB Corp). Five microliters of Big Dye Terminator Ready Reaction Mix v. 1.1 (Applied Biosystems) is added to 4 µL of each primer (0.8 pmol/µL) and 3 µL of purified PCR product. Cycle sequencing is performed with a 9700 thermal cycler (ABI), using 25 cycles of 96°C for 10 s, 50°C for 5 s, and 60°C for 4 min. Sequencing reaction products are passed through a Sephadex G-50 fine column to remove unincorporated dye terminators. Purified sequencing reaction products are run on an ABI Prism 3100 Genetic Analyzer with a 50 cm capillary array.
5. Sequences are analyzed with the SmartGene Integrated Database Network software version 3.2.3 vr. SmartGene is a web-based software and database system with reference sequences derived from the National Center for Biological Information (NCBI) GenBank repository.

Note. Sequence-based identifications are defined by percent identity: species, ≥99%; genus, 93%–99%; and inconclusive, ≤93%.

67.2.2.2 PCR and Sequencing Analysis of ITS1 and ITS2

Leaw et al. [15] utilized fungus-specific universal primers ITS1 (5′-TCCGTAGGTGAACCTGCGG-3′)

and ITS2 (5′-GCATCGATGAAGAACGCAGC-3′) to amplify the ITS1 region and universal primers ITS3 (5′-GCATCGATGAAGAACGCAGC-3′) and ITS4 (5′-GCATATCAATAAGCGGAGGA-3′) to amplify the ITS2 region. Subsequent sequencing analysis of the resulting amplicons facilitated identification of a range of fungal pathogens.

Procedure

1. PCR mixture (50 μL) is made up of 10 mM Tris–HCl pH 8.3, 50 mM KCl, 1.5 mM $MgCl_2$, 0.2 mM each of deoxynucleoside triphosphates, 1.2 U of *Taq* DNA polymerase, 0.4 μM (each) of the ITS1 region primers (ITS1/ITS2) or the ITS2 region primers (ITS3/ITS4), and 2 μL (1–5 ng) of DNA template. A negative control is included in each run by replacing the template DNA with sterile water in the PCR mixture.
2. Amplification is conducted with an initial 94°C for 3 min; 30 cycles of 94°C, 60°C, and 72°C for 1 min each; and a final 72°C for 3 min.
3. After checking the PCR products on 1.5% agarose gel, the amplicons are purified using a commercial PCR cleanup kit. The DNA fragments are sequenced using an ABI Prism 377 automated DNA sequencer (Applied Biosystems) with a BigDye Terminator cycle sequencing kit (version 3.1; Applied Biosystems). All amplicons are sequenced on both strands using primers ITS1 and ITS2 for the ITS1 region and primers ITS3 and ITS4 for the ITS2 region.
4. After sequencing, portions of the 18S, 5.8S, and 26S rRNA gene sequences of the PCR products are removed to obtain the exact ITS1 and ITS2 sequences.
5. For all yeasts, the sequences of the 3′ ends of the 18S and 5.8S rRNA genes are GCGGAAGGA TCATTA and GTTTGAGCGTCATTT, respectively, and the sequences of the 5′ ends of the 5.8S and 26S rRNA genes are AAACTTTCAACAA and GACCTCAAATCAG, respectively.
6. The ITS1 or ITS2 sequence is compared with those available at the GenBank databases using the BLAST sequence analysis tool (http://www.ncbi.nlm.nih.gov/BLAST/) (blastn) with default settings except that sequences are not filtered for low complexity. Species identification is determined from the lowest expected value of the BLAST output.

Note. For strains producing discrepant identification between the methods based on phenotypic characteristics and ITS sequence analysis, the D1–D2 region of the large-subunit rRNA gene is amplified with primers NL1 (5′-GCATATCAATAAGCGGAGGAAAAG-3′) and NL4 (5′-GGTCCGTGTTTCAAGACGG-3′) and sequenced for species clarification.

67.3 CONCLUSION

The genus *Kluyveromyces* is a member of the ascomycetous yeast family Saccharomycetaceae that comprises at least 28 genera. Among the 16 recognized species and 20 unassigned species in the genus *Kluyveromyces*, *Kluyveromyces marxianus* (synonyms: *Kluyveromyces cicerisporus* and *Kluyveromyces fragilis*; anamorph: *Candida kefyr* or *Candida pseudotropicalis*) has been occasionally involved in human burn wound and blood and vaginal infections [20]. Given their close morphological and biochemical similarity to other *Candida* spp., identification of *Kluyveromyces* spp. using conventional techniques is difficult, if not impossible. Application of molecular techniques such as PCR amplification and sequencing analysis is essential for correct and efficient determination of *Kluyveromyces* from other related yeasts.

REFERENCES

1. The UniProt Consortium. Available at http://www.uniprot.org/, accessed on August 1, 2010.
2. Nagahama T et al. *Kluyveromyces nonfermentans* sp. nov., a new yeast species isolated from the deep sea. *Int J Syst Bacteriol*. 1999;49:1899–1905.
3. Lachance MA. Current status of *Kluyveromyces systematics*. *FEMS Yeast Res*. 2007;7:642–645.
4. Am-In S, Yongmanitchai W, Limtong S. *Kluyveromyces siamensis* sp. nov., an ascomycetous yeast isolated from water in a mangrove forest in Ranong Province, Thailand. *FEMS Yeast Res*. 2008;8(5):823–828.
5. Corpus K et al. *Candida kefyr*, an uncommon but emerging fungal pathogen: Report of two cases. *Pharmacotherapy* 2004;24:1084–1088.
6. Gupta N et al. Epidemiology and molecular typing of *Candida* isolates from burn patients. *Mycopathologia* 2004;158:397–405.
7. Miranda TT et al. Diversity and frequency of yeasts from the dorsum of the tongue and necrotic root canals associated with primary apical periodontitis. *Int Endod J*. 2009;42(9):839–844.
8. Belloch C et al. Phylogeny of the genus *Kluyveromyces* inferred from the mitochondrial cytochrome-c oxidase II gene. *Int J Syst Evol Microbiol*. 2000;50:405–416.
9. García-Martos P et al. Sexual forms of yeasts in clinical samples. *Mycopathologia* 1996;136(2):67–70.
10. Erdem H et al. Identification of yeasts in public hospital primary care patients with or without clinical vaginitis. *Aust N Z J Obstet Gynaecol*. 2003;43:312–316.
11. Hospenthal DR et al. Presumptive identification of *Candida* species other than *C. albicans*, *C. krusei*, and *C. tropicalis* with the chromogenic medium CHROMagar *Candida*. *Ann Clin Microbiol Antimicrob*. 2006;5:1.
12. Rimek D et al. Identification of contaminating fungal DNA sequences in Zymolyase. *J Clin Microbiol*. 1999;37(3):830–831.
13. Chen Y-C et al. Identification of medically important yeasts using PCR-based detection of DNA sequence polymorphisms in the internal transcribed spacer region 2 of the rRNA genes. *J Clin Microbiol*. 2000;38:2302–2310.
14. Chen Y-C et al. Polymorphic internal transcribed spacer region 1 DNA sequences identify medically important yeasts. *J Clin Microbiol*. 2001;39:4042–4051.

15. Leaw SN et al. Identification of medically important yeast species by sequence analysis of the internal transcribed spacer regions. *J Clin Microbiol.* 2006;44(3):693–699.
16. Linton CJ et al. Molecular identification of unusual pathogenic yeast isolates by large-ribosomal-subunit gene sequencing: 2 years of experience at the United kingdom mycology reference laboratory. *J Clin Microbiol.* 2007;45(4):1152–1158.
17. Bhardwaj S et al. PCR-based identification and strain typing of *Pichia anomala* using the ribosomal intergenic spacer region IGS1. *J Med Microbiol.* 2007;56:185–189.
18. Borman AM et al. An improved protocol for the preparation of total genomic DNA from isolates of yeast and mould using Whatman FTA filter papers. *Mycopathologia* 2010;169:445–449.
19. Pounder JI et al. Discovering potential pathogens among fungi identified as nonsporulating molds. *J Clin Microbiol.* 2007;45(2):568–571.
20. Lutwick LI, Phaff HJ, Stevens DA. *Kluyveromyces fragilis* as an opportunistic fungal pathogen in man. *Sabouraudia* 1980;18(1):69–73.

68 *Pichia* and *Kodamaea*

Dongyou Liu

CONTENTS

68.1 INTRODUCTION

68.1.1 Classification and Morphology

The genera *Pichia* and *Kodamaea* are ascomycetous yeasts belonging to the family Saccharomycetaceae, class Saccharomycetes, subphylum Saccharomycotina, phylum Ascomycota, kingdom Fungi. The family Saccharomycetaceae covers 28 genera: *Ascobotryozyma*, *Citeromyces*, *Debaryomyces*, *Dekkera*, *Eremothecium*, *Issatchenkia*, *Kazachstania*, *Kluyveromyces*, *Kodamaea*, *Komagataella*, *Kregervanrija*, *Kuraishia*, *Lachancea*, *Lindnera*, *Lodderomyces*, *Nakaseomyces*, *Nakazawaea*, *Naumovia*, *Pachysolen*, *Pichia*, *Saccharomyces*, *Saturnispora*, *Tetrapisispora*, *Torulaspora*, *Vanderwaltozyma*, *Williopsis*, *Zygosaccharomyces*, and *Zygowilliopsis*, in addition to some unclassified Saccharomycetaceae [1].

The genus *Pichia* (obsolete synonyms: *Hansenula*, *Hyphopichia*, and *Yamadazyma*) is an ascospore-producing teleomorph taxon consisting of >100 recognized species and more than 150 unassigned species. The anamorphs of the genus *Pichia* are found in the genus *Candida*, and both *Pichia* and *Candida* form ascospores and share considerable sequence similarity in the 28S rRNA gene. Among the large number of species within the *Pichia* genus, *Pichia anomala* (obsolete synonym: *Hansenula anomala*; anamorph: *Candida pelliculosa*), *Pichia fabianii*, *Pichia farinosa* (synonym: *Pichia miso*; anamorph: *Candida cacaoi*), *Pichia guilliermondii* (anamorph: *Candida guilliermondii* var. *guilliermondii*), and *Pichia norvegensis* (anamorph: *Candida norvegensis*) are of medical importance.

The genus *Kodamaea*, also classified in the family Saccharomycetaceae, contains five recognized species: *Kodamaea anthophila*, *Kodamaea kakaduensis*, *Kodamaea laetipori*, *Kodamaea nitidulidarum*, and *Kodamaea ohmeri* and five unassigned species (i.e., *Kodamaea* sp. BG-3, *Kodamaea* sp. I11, *Kodamaea* sp. NRRL Y27634, *Kodamaea* sp. YS 85, *Kodamaea* sp. YS9). Of these, *Kodamaea ohmeri* (obsolete synonyms: *Pichia ohmeri* and *Yamadazyma ohmeri*; anamorph: *Candida guilliermondii* var. *membranaefaciens*), a yeast commonly used in food industry for fermentation in pickles, rinds, and fruits, is implicated in human infections [2,3].

Morphologically, *Pichia* and *Kodamaea* spp. resemble *Candida* spp. Being ascomycetous yeasts, *Pichia* and *Kodamaea* often produce asci each containing one to four hat-shaped or globose ascospores on acetate ascospore agar, Gorodkowa medium, V-8 medium, or 2% glucose malt agar. The ascospores of *Pichia* and *Kodamaea* are stained red with Kinyoun stain, while the cells other than ascospores take the counterstain [4].

P. anomala is identified on the basis of its inability to produce germ tubes and urease, a standard sugar assimilation pattern, and the production of one to four hat-shaped ascospores. *Kodamaea* (*Pichia*) *ohmeri* colonies show a color change from pink to blue on CHROMagar *Candida* medium on which few other yeast species share such color change [5,6].

68.1.2 Clinical Features and Pathogenesis

Members of the *Pichia* and *Kodamaea* genera are free-living ascosporogenous yeasts that constitute part of the normal or transient flora of the human throat and alimentary tract [7]. Generally considered as contaminants, some *Pichia* and *Kodamaea* spp. are now recognized as clinically significant opportunistic pathogens in immunocompromised patients or individuals with other predisposing factors (e.g., prematurity, low birth weight, long duration of hospital stay, and existence of prosthetic valve). The clinical diseases caused by *Pichia* and *Kodamaea* spp. encompass urinary tract infection, peritonitis, enteritis, prosthetic valve endocarditis, nosocomial fungemia, disseminated infection, etc. [8,9].

68.1.2.1 *Pichia anomala*

Pichia anomala (formerly *Hansenula anomala*) has been identified as an important opportunistic fungal pathogen causing nosocomial fungemia, interstitial lung disease, endocarditis, enteritis, and keratitis in children and adults [10–13].

Chakrabarti et al. [12] documented a *Pichia anomala*-related outbreak of nosocomial fungemia involving 379 neonates and children (4.2% admissions) in the hospital pediatric wards over a period of 23 months (April 1996 to February 1998). Based on a culture prevalence survey, 14 (28%) out of the 50 consecutive premature neonates were colonized with *P. anomala* at the umbilicus, mouth, rectum, and groin sites on the date of delivery and 10 of these colonized neonates later developed *P. anomala* fungemia. Neonates with documented sepsis responded well to treatment with fluconazole or amphotericin B, or itraconazole. Significant risk factors associated with *P. anomala* fungemia in premature neonates included lower gestational age, a very low birth weight (<1500 g), and a longer duration of hospital stay. Upon analysis by multilocus enzyme electrophoresis, *P. anomala* outbreak isolates (from both patients and health-care workers' hands) appeared identical, suggesting that carriage on the hands of health-care personnel was likely to be responsible for dissemination of the fungus. The outbreak was only controlled after institution of a health education campaign to improve hand-washing practices and after introduction of a nystatin-fluconazole prophylaxis to all premature neonates and high-risk infants.

In a separate study, Bakir et al. [14] reported nosocomial port catheter infection due to *P. anomala* in three children receiving cancer chemotherapy, a preterm infant with bloodstream infection, and an infant with severe combined immunodeficiency. All *P. anomala* isolates were susceptible to amphotericin B and fluconazole, and patients responded well to amphotericin B treatment. The common clinical feature in these patients was the presence of prior antimicrobial therapy.

Pasqualotto et al. [15] also reported an outbreak of *Pichia anomala* fungemia occurring in a pediatric intensive care unit (ICU). During the study period (from October 2002 to January 2004), 1046 children were admitted to the pediatric ICU, and 17 (1.6%) of them developed *P. anomala* fungemia. The median age of infected children was 1.1 years, and the overall mortality rate was 41.2% (7/17). On multivariate analysis, the presence of a central venous catheter was identified as a significant factor for this outbreak. Molecular studies showed that the outbreak was caused by a single strain, which was absent on health-care workers' hands or in the environment.

Kalenic et al. [16] described an outbreak of *Pichia (Hansenula) anomala* infections in eight adult patients treated at a surgical ICU during a 5 month period. Among the risk factors analyzed for this outbreak (i.e., blood alkalosis, reduced urea, duration of hospitalization, bacteremia and colonization with *Pseudomonas aeruginosa*, and an APACHE II score >17 during bacteremia or fungemia), only the duration of blood alkalosis was significant in case patients.

Recently, Park et al. [17] reported a case of *Pichia anomala*-associated keratitis of the eye in a 50-year-old woman with systemic lupus erythematous and Stevens-Johnson syndrome. The patient presented with ocular pain, and culture of corneal scrapings yielded *P. anomala*. Following topical amphotericin B, natamycin, and systemic inidazole treatments, the patient showed a full recovery.

In vitro susceptibility study of 58 isolates of *Pichia anomala* to five antifungal drugs using two broth microdilution methods clinical and laboratory standards institute (CLSI) and European committer on antimicrobial susceptibility testing (EUCAST) showed that *Pichia anomala* is generally sensitive to fluconazole, voriconazole, amphotericin B, and caspofungin, but less susceptibile to itraconazole [18].

68.1.2.2 *Pichia fabianii*

Bhally et al. [19] documented *Pichia fabianii* as the cause of an infection in a 5-week-old female twin delivered at 25 and 37 weeks. The patient presented with respiratory distress syndrome and necrotizing enterocolitis, and was febrile, thrombocytopenic, and requiring minimal ventilatory support. Blood cultures yielded a yeast, which was identified as *P. fabianii* by sequencing a subunit of D2 of the large subunit (LSU) rRNA gene. The organism was susceptible in vitro to amphotericin B, fluconazole, and 5-fluorocytosine, and the patient responded to amphotericin B and removal of vascular catheter. Hamal et al. [20] also isolated *Pichia fabianii* from the blood of a patient with aortic valve endocarditis. The isolates were initially identified biochemically as *Candida pelliculosa* but confirmed subsequently as *P. fabianii* by direct sequencing of the ITS2 region of rRNA gene.

68.1.2.3 *Pichia farinosa*

Pichia farinosa (synonym: *Pichia miso*) is a halotolerant yeast with widespread distribution and has been isolated from various sources such as miso soup and giraffe dung. The fungus is characterized by the production of a salt-mediated killer toxin that kills yeasts of several genera, including *Saccharomyces cerevisiae* [21]. Adler et al. [22] reported a case of *Pichia farinosa* bloodstream infection in a 13-year-old boy with anaplastic large-cell lymphoma, who had a Broviac catheter inserted into the right jugular vein. The patient presented with fever (40°C), vomiting, and local redness and swelling along the catheter tunnel. Cultures of blood specimens on Sabouraud dextrose agar grew yeast that was identified as *Candida boidinii* by using the API 20 C AUX identification system, which lacks the species *P. farinosa* in its database (bioMérieux), and as *Pichia farinosa* by using the ID 32 C identification system (bioMérieux). Sequencing analyses of PCR-amplified fragments covering internal transcribed spacer 1 (ITS1), the 5.8S rRNA gene, and ITS2 as well as the ribosomal LSU D1/D2 variable regions revealed the identity of the fungus as *P. farinosa*. The isolate was sensitive to amphotericin B and caspofungin and less sensitive to fluconazole. After initiation with amphotericin B treatment and removal of the Broviac catheter, the patient became afebrile after 2 days, and neutropenia was resolved. The patient was discharged with a course of oral fluconazole for 2 weeks and no evidence of recurrence or complications was noted in subsequent checkup.

68.1.2.4 *Kodamaea ohmeri*

Kodamaea (*Pichia*) *ohmeri* is an ascosporogenous yeast commonly used in food products. The fungus has been implicated in fungemia, endocarditis, fungiurea, peritonitis, and wound infection in immunocompromised individuals and the elderly [6,23–35].

Shin et al. [30] reported a case of fungemia and phlebitis due to *Pichia ohmeri* in a 59-year-old hospitalized patient, who had previously received intensive antimicrobial therapy for a ventriculoperitoneal shunt infection and subsequent nosocomial pneumonia. The patient developed intermittent fever and suffered from tender swelling of the right leg due to peripheral phlebitis at the site of insertion of a peripheral venous catheter (which had already been removed at the onset of fever). Cultures of samples from the insertion site as well as blood yielded *P. ohmeri*. Following a 14 day amphotericin B therapy, the fungemia and phlebitis were cleared.

Han et al. [31] also described two cases of *Kodamaea ohmeri*-associated infections in patients with preexisting conditions. The first one involved a 14-year-old boy who was neutropenic following chemotherapy for leukemia. The infection was cured by catheter removal and the use of fluconazole. The second case concerned a 74-year-old man who had undergone surgeries for a subcutaneous tumor. The patient developed polymicrobic cellulitis and succumbed to the complications despite surgical debridement and antibiotic therapy. *Kodamaea ohmeri* isolates were identified correctly by biochemical tests and were susceptible to amphotericin B and dose dependently susceptible to itraconazole and fluconazole.

Yang et al. [35] reported a case of fungemia due to *K. ohmeri* in a 71-year-old man with cellulitis. The patient presented with leg edema, fever, and change of consciousness. Blood culture yielded a yeast colony, which was identified as *K. ohmeri* by Vitek YBC card, API 20C, sequencing of the 18S rRNA gene, the D1/D2 domains of the 26S rRNA gene and the ITS regions. The isolate was resistant to fluconazole but sensitive to amphotericin B deoxylate. After treatment with amphotericin B deoxylate, the fever and cellulitis inflammation subsided and the patient was discharged in a stable condition.

Taj-Aldeen et al. [36] reported the first case of *Kodamaea* (*Pichia*) *ohmeri* fungemia in a premature neonate with low birth weight (680 g). During the first week, the baby developed clinical necrotizing enterocoli, and a blood culture taken on the 13th day of life showed a yeast-like organism using the Bactec automated culturing system (BD Diagnostic Systems) and pediatric bottle. The organism failed to produce ascospores on Sabouraud dextrose agar and malt extract agar +2.5% glucose even with a prolonged incubation period (14 days), and was tentatively identified as *K. ohmeri* with Vitek II and API ID 32C (bioMérieux). Sequence analysis of the 5.8S rRNA gene confirmed its identity as *K. ohmeri*. The organism was susceptible to amphotericin B and flucytosine, but dose dependently susceptible to itraconazole and fluconazole. The isolate failed to produce ascospores on Sabouraud dextrose agar and malt extract agar + 2.5% glucose even with a prolonged incubation period (14 days). The patient responded to amphotericin B/liposomal amphotericin B and fluconazole treatment.

De Barros et al. [37] also described a case of fungemia caused by *Kodamaea ohmeri* in a 3-year-old female patient, who had previously received antimicrobial therapy due to a peritoneal infection and nosocomial pneumonia, and had a central venous catheter implanted. *K. ohmeri* was isolated from blood and the tip of the catheter 48 h after its implantation. The yeast was identified by standard microbiological methods and sequence analysis of the D1/D2 domains and the ITS1 + 2 spacer regions of the ribosomal DNA. The yeast was susceptible to amphotericin B, and liposomal AmB was used successfully to resolve the infection.

68.1.3 Laboratory Diagnosis

Due to the increasing involvement of *Pichia* spp. in human infections, there is a need to develop rapid methods for the identification of these organisms at the species and subspecies level. The availability of strain typing information is also critical for epidemiological tracking of the outbreaks, detection of cross-transmission of nosocomial pathogens, and determination of the source and virulence of infection strains.

Phenotypic techniques are useful for the identification of *Kodamaea* (*Pichia*) *ohmeri*, including the formation of characteristic pink-blue colonies on CHROMagar media [5,6] and demonstration of distinct biochemical profiles on commercial yeast identification systems (e.g., API or Vitek) [38,39]. However, these methods failed to give conclusive results for *Pichia anomala* (formerly *Hansenula anomala*), *Pichia fabianii*, and *Pichia farinosa* on a number of occasions [20,22]. PCR amplification and sequencing analysis of the 18S rRNA gene, the D1/D2 domains of the 28S rRNA gene, and the ITS regions have proven crucial for their correct determination [3,40–46].

Pulsed-field gel electrophoresis (PFGE) karyotyping and restriction endonuclease analysis of genomic NotI-digested DNA (REAG-N) are also valuable for the identification of *Pichia* and *Kodamaea* stains [6]. In addition, while the ITS harbors limited sequence variation to be useful for *Pichia anomala* typing, the intergenic spacer 1 (IGS1) region (which varies from 1213 to 1231 bp in length) of the rRNA is interspersed with repeats and has more variation than the ITS regions, providing a more discriminatory tool in the typing of *P. anomala* strains [47–49]. By using a pair of specific primers corresponding to the 18S rRNA gene sequence, *Pichia guilliermondii* (anamorph: *Candia guilliermondii*) can be reliably differentiated from other members of the *Candia famata* complex [8].

68.2 METHODS

68.2.1 Sample Preparation

Clinical samples (e.g., tracheal secretion, skin scraping, cervical swab, urine, blood) are cultivated on Sabouraud agar (SGC 2; bioMérieux) for 5 days at 30°C. Yeast isolates are identified using traditional morphology and the ID 32C and VITEK 2 ID-YST biochemical identification methods (bioMérieux).

DNA from clinical samples can be prepared with a QIAamp DNA Blood Kit (Qiagen). Viscous tracheal secretion samples are diluted with an equal volume of sputasol (Oxoid) and liquefied at 37°C for 1 h. Thereafter, 400 μL of the sample is transferred to a tube containing lysing matrix E and mechanically disrupted in a RiboLyser. Subsequently, 200 μL of the sample is incubated with 200 μL buffer AL at 56°C for 10 min prior to DNA extraction. Valvular tissue (20 mg) is treated with 180 μL buffer ATL and 20 μL proteinase K (30 U; Sigma-Aldrich) at 56°C until the tissue is completely lysed, followed by incubation with 200 μL buffer AL at 70°C for 10 min prior to DNA extraction. Skin and nail scraping samples are suspended in 200 μL sterile distilled water and transferred into a tube with lysing matrix E. After mechanical disruption, 180 μL of buffer ATL and 20 μL of proteinase K are added. Samples are incubated at 56°C overnight, followed by incubation with 200 μL of buffer AL at 70°C for 10 min. The remaining steps are undertaken according to the manufacturer's instructions. DNA is eluted from the QIAamp column with 50 μL sterile distilled water [50].

Cervical swabs are rinsed in lysis buffer (Digene Hybrid Capture 2 CT/GC DNA kit; Digene). Total DNA is extracted from 200 μL cervical swab lysis solution, 200 μL ethylenediaminetetraacetic acid (EDTA)-anticoagulated blood, 400 μL serum, and 400 μL urine using the NucliSens easyMAG (bioMérieux) automated DNA extraction system. DNA is eluted in 55 μL easyMAG elution buffer [50].

DNA can be purified from yeast culture by the glass bead lysis method. A loopful of culture from malt yeast extract agar is transferred to 200 μL lysis buffer (2% Triton X-100, 1% SDS, 10 mM NaCl, 10 mM Tris/HCl pH 8.0, 1 mM EDTA) in a 1.5 mL tube and 200 μL phenol/chloroform/isoamyl alcohol (25:24:1) is added. Cells are lysed by vortexing for 4 min in the presence of 20 mg glass beads. The aqueous layer is treated once more with phenol/chloroform/isoamyl alcohol and the DNA is precipitated with chilled 2-propanol. The pellet is washed with 70% ethanol and finally resuspended in 50 μL TE buffer (10 mM Tris/HCl pH 8.0, 1 mM EDTA) [49].

68.2.2 Detection Procedures

68.2.2.1 PCR Identification of *Pichia anomala*

Bhardwaj et al. [49] designed several primers (IGF1, IGF2, IGR1, IGR2, and 5SR1) from the conserved region of rRNA gene intergenic spacer region 1 (IGS1) for specific identification of *Pichia anomala* (Table 68.1).

Procedure

1. PCR mixture (50 μL) is made up of 50 ng genomic DNA, 25 pmol each of primer pair (IGF1/IGR1, IGF1/IGR2, IGF2/IGR1, IGF1/5SR1, or IGF2/5SR1 in separate tubes), 200 μM each dNTP, 2.5 mM $MgCl_2$, and 2 U *Taq* polymerase.
2. The mixture is subjected to an initial denaturation at 94°C for 5 min, followed by 30 cycles of 94°C for 1 min, 55°C for 1 min, and 72°C for 1 min, with a final extension at 72°C for 10 min.
3. The resulting amplicons are separated by electrophoresis in a 1% agarose gel, stained with ethidium bromide (0.5 μg/mL), and visualized under UV light.

Note. The expected products from different combinations of primers (IGF1/IGR1, IGF1/IGR2, IGF2/IGR1, IGF1/5SR1, and IGF2/5SR1) are of 1.1, 0.6, 0.7, 1.25, and 0.75 kb, respectively. They are highly specific for *Pichia anomala*.

68.2.2.2 PCR Identification of *Pichia guilliermondii* (*Candida guilliermondii*)

Yamamura et al. [8] developed a specific PCR assay for *Pichia guilliermondii* (*Candida guilliermondii*) using primers CUGF (5′-CGGGGAGGTAGTGACAATAC-3′) and CUGR (5′-CAAACACCACAAGGGCGAAT-3′) derived from 18S rRNA gene. Use of these primers facilitates amplification

TABLE 68.1
PCR Primers for Species-Specific Identification of Pichia anomala

Primer	Sequence (5′→3′)
IGF1	AGTATACTGGCTAACAGAAGTTGGCTA
IGF2	GGAATCGTGACCAAAAAATGGGAAAT
IGR1	ATAATTTGGCCCAAATCTATTCTCCCA
IGR2	CTAACGGGGGGAGGATCCAGTATAAAA
5SR1	CACCGTTTCCGTTCCGATC

of a ca. 250 bp product from DNA of *P. guilliermondii* (*C. guilliermondii*), not that of related taxa including *Pichia anomala, Pichia ohmeri, Candida albicans, Candida parapsilosis, Candida tropicalis, Candida krusei, Candida glabrata, Candida kefyr, Candida utilis, Candida dubliniensis,* and *Candida auris.*

Procedure

1. PCR mixture (100 µL) is made up of 100 µM dNTPs, 1× PCR buffer (GE Healthcare), 2.5 U of DNA Taq polymerase (GE Healthcare), 30 pmol each of primers CUGF and CUGR, and 3–20 ng DNA.
2. Amplification is performed with an initial 94°C for 5 min, 25 cycles of 95°C for 1 min, 60°C for 15 s, and 72°C for 15 s, followed by a final 72°C for 10 min.
3. PCR products are separated by electrophoresis on a 1.2% agarose gel, stained with ethidium bromide, and visualized under ultraviolet light.

Note. A ca. 250 bp product is obtained only from DNA of *P. guilliermondii* (*C. guilliermondii*), but not that of other fungal taxa.

68.2.2.3 Sequencing Analysis of ITS Regions

Pounder et al. [51] described a real-time PCR with SYBR green DNA-binding dye and amplicon melting temperature analysis for fungal detection using pan-fungal primers ITS1 forward (5′-TCCGTAGGTGAACCTGCGG-3′) and ITS4 reverse (5′-TCCTCCGCTTATTGATATGC-3′). The identity of the fungi is verified by subsequent sequencing analysis.

Procedure

1. PCR mixture is composed of 1× Lightcycler FastStart DNA Master Hybridization Probes mixture (Roche Applied Science) (containing deoxynucleoside triphosphates, FastStart *Taq* DNA polymerase, and 1 mM $MgCl_2$ (additional $MgCl_2$ is added to a final concentration of 4.6 mM), 0.4 µM each of ITS1 forward and ITS4 reverse primers, 1× SYBR green (Molecular Probes), and 3 µL template DNA.
2. Thermal cycling parameters include 95°C for 10 min; 50 cycles at 95°C for 5 s, 60°C for 20 s, and 76°C for 30 s; and a final extension at 72°C for 2 min.
3. The quality of the amplicon is determined by using the derivative of the melt analysis curve (55°C–99°C, 45 s hold at 55°C, 5 s/°C) using the RotorGene 3000 (Corbett Robotics, Inc.).
4. The amplified product is purified for bidirectional sequencing using ExoSAP-IT (USB Corp). Five microliters of Big Dye Terminator Ready Reaction Mix v. 1.1 (Applied Biosystems) is added to 4 µL of each primer (0.8 pmol/µL) and 3 µL of purified PCR product. Cycle sequencing is performed with a 9700 thermal cycler (ABI), using 25 cycles of 96°C for 10 s, 50°C for 5 s, and 60°C for 4 min. Sequencing reaction products are passed through a Sephadex G-50 fine column to remove unincorporated dye terminators. Purified sequencing reaction products are run on an ABI Prism 3100 Genetic Analyzer with a 50 cm capillary array.
5. Sequences are analyzed with the SmartGene Integrated Database Network software version 3.2.3 vr. SmartGene is a web-based software and database system with reference sequences derived from the National Center for Biological Information (NCBI) GenBank repository.

Note. Sequence-based identifications are defined by percent identity: species, ≥99%; genus, 93%–99%; and inconclusive, ≤93%.

68.2.2.4 Sequencing Analysis of ITS1 and ITS2

Leaw et al. [45] utilized fungus-specific universal primers ITS1 (5′-TCCGTAGGTGAACCTGCGG-3′) and ITS2 (5′-GCATCGATGAAGAACGCAGC-3′) to amplify the ITS1 region, and universal primers ITS3 (5′-GCATCGATGAAGAACGCAGC-3′) and ITS4 (5′-GCATATCAATAAGCGGAGGA-3′) to amplify the ITS2 region. Subsequent sequencing analysis of the resulting amplicons facilitated identification of a range of fungal pathogens.

Procedure

1. PCR mixture (50 µL) is made up of 10 mM Tris–HCl pH 8.3, 50 mM KCl, 1.5 mM $MgCl_2$, 0.2 mM each of deoxynucleoside triphosphates, 1.2 U of *Taq* DNA polymerase, 0.4 µM (each) of the ITS1 region primers (ITS1/ITS2) or the ITS2 region primers (ITS3/ITS4), 2 µL (1–5 ng) of DNA template. A negative control is included in each run by replacing the template DNA with sterile water in the PCR mixture.
2. Amplification is conducted with an initial 94°C for 3 min; 30 cycles at 94°C, 60°C, and 72°C for 1 min each; and a final 72°C for 3 min.
3. After checking the PCR products on 1.5% agarose gel, the amplicons are purified using a commercial PCR cleanup kit. The DNA fragments are sequenced using an ABI Prism 377 automated DNA sequencer (Applied Biosystems) with a BigDye Terminator cycle sequencing kit (version 3.1; Applied Biosystems). All amplicons are sequenced on both strands using primers ITS1 and ITS2 for the ITS1 region and primers ITS3 and ITS4 for the ITS2 region.
4. After sequencing, portions of the 18S, 5.8S, and 26S rRNA gene sequences of the PCR products are removed to obtain the exact ITS1 and ITS2 sequences.
5. For all yeasts, the sequences of the 3′ ends of the 18S and 5.8S rRNA genes are GCGGAAGGA TCATTA and GTTTGAGCGTCATTT, respectively, and the sequences of the 5′ ends of the 5.8S

and 26S rRNA genes are AAACTTTCAACAA and GACCTCAAATCAG, respectively.

6. The ITS1 or ITS2 sequence is compared with those available at the GenBank databases using the BLAST sequence analysis tool (http://www.ncbi.nlm.nih.gov//BLAST//) (blastn) with default settings except that sequences are not filtered for low complexity. Species identification is determined from the lowest expected value of the BLAST output.

Note. The ITS2 region, located between the 5.8S and 28S subunits in the RNA gene, provides a more specific target than the ITS1 for determination of *K. ohmeri* [6]. For strains producing discrepant identification between the methods based on phenotypic characteristics and ITS sequence analysis, the D1–D2 region of the LSU RNA gene is amplified with primers NL1 (5′-GCATATCAATAAGCGGAGGAAAAG-3′) and NL4 (5′-GGTCCGTGTTTCAAGACGG-3′) and sequenced for species clarification.

68.3 CONCLUSION

Pichia and *Kodamaea* spp. are free-living ascosporogenous yeasts that form part of the normal or transient flora of the human throat and alimentary tract. Previously regarded as contaminants, several *Pichia* and *Kodamaea* species have emerged as clinically relevant organisms causing urinary tract infection, peritonitis, enteritis, prosthetic valve endocarditis, nosocomial fungemia, and disseminated infection, in immunocompromised patients or individuals with other predisposing factors. Given the close morphological and biochemical similarity among *Pichia*, *Kodamaea*, and *Candida* spp., it is important to determine these organisms to species/subspecies level for effective treatment and epidemiological tracking of the outbreaks. While *Kodamaea* (*Pichia*) *ohmeri* can be reliably identified by phenotypic methods, *Pichia anomala* (formerly *Hansenula anomala*), *Pichia fabianii*, and *Pichia farinosa* have proven difficult to determine on a number of occasions using similar approaches. PCR amplification and sequencing analysis of the 18S rRNA gene, the D1/D2 domains of the 28S rRNA gene, and the ITS regions are instrumental in helping resolve the taxonomical status of these isolates.

REFERENCES

1. The UniProt Consortium. Available at http://www.uniprot.org/, accessed on August 1, 2010.
2. Yamada, Y. et al., 1995. The phylogeny of *Yamadazyma ohmeri* (Etchells et Bell) Billon-Grand based on the partial sequences of 18S and 26S ribosomal RNAs: The proposal of *Kodomaea* gen. nov. (Saccharomycetaceae). *Biosci Biotechnol Biochem* 59, 1172–1174.
3. Kurtzman, C.P. and Robnett, C.J. 1998. Identification and phylogeny of ascomycetous yeasts from analysis of nuclear large subunit (26S) ribosomal DNA partial sequences. *Anton Leeuwen* 73, 331–371.
4. Garcia-Martos, P. et al., 1996. Sexual forms of yeasts in clinical samples. *Mycopathologia* 136, 67.
5. Otag, F. et al., 2005. An outbreak of *Pichia ohmeri* infection in the paediatric intensive care unit: Case reports and review of the literature. *Mycoses* 48, 265–269.
6. Lee, J.S. et al., 2007. *Kodamaea ohmeri* isolates from patients in a university hospital: Identification, antifungal susceptibility, and pulsed-field gel electrophoresis analysis. *J Clin Microbiol* 45(3), 1005–1010.
7. Moses, A. et al., 1991. *Hansenula anomala* infections in children: From asymptomatic colonization to tissue invasion. *Pediatr Infect Dis J* 10, 400–402.
8. Yamamura, M. et al., 2009. Polymerase chain reaction assay for specific identification of *Candida guilliermondii* (*Pichia guilliermondii*). *J Infect Chemother* 15(4), 214–218.
9. Grenouillet, F. et al., 2010. *Pichia fabianii* fungemia in a neonate. *Pediatr Infect Dis J* 29, 191.
10. Thular, L.C. et al., 1997. Fungaemia caused by *Hansenula anomala*—An outbreak in a cancer hospital. *Mycoses* 40, 193–196.
11. Ma, J.S. et al., 2000. Neonatal fungemia caused by *Hansenula anomala*: A case report. *J Microbiol Immunol Infect* 33(4), 267–270.
12. Chakrabarti, A. et al., 2001. Outbreak of *Pichia anomala* infection in the pediatric service of a tertiary-care center in Northern India. *J Clin Microbiol* 39, 1702–1706.
13. Paula, C.R. et al., 2006. Nosocomial infection in newborns by *Pichia anomala* in a Brazilian intensive care unit. *Med Mycol* 44(5), 479–484.
14. Bakir, M. et al., 2003. *Pichia anomala* fungaemia in immunocompromised children. *Mycoses* 47, 231–235.
15. Pasqualotto, A.C. et al., 2005. An outbreak of *Pichia anomala* fungemia in a Brazilian pediatric intensive care unit. *Infect Control Hosp Epidemiol* 26(6), 553–558.
16. Kalenic, S. et al., 2001. *Hansenula anomala* outbreak at a surgical intensive care unit: A search for risk factors. *Eur J Epidemiol* 17, 491–496.
17. Park, K.A. et al., 2008. *Pichia anomala* fungal keratitis. *Cornea* 27(5), 619–620.
18. da Matta, V.L. et al., 2007. Antifungal drug susceptibility profile of *Pichia anomala* isolates from patients presenting with nosocomial fungemia. *Antimicrob Agents Chemother* 51, 1573–1576.
19. Bhally, H.S. et al., 2006. Infection in a neonate caused by *Pichia fabianii*: Importance of molecular identification. *Med Mycol* 44, 185–187.
20. Hamal, P. et al., 2008. A case of endocarditis caused by the yeast *Pichia fabianii* with biofilm production and developed in vitro resistance to azoles in the course of antifungal treatment. *Med Mycol* 46(6), 601–605.
21. Suzuki, C., Ando, Y., and Machida, S. 2001. Interaction of SMKT, a killer toxin produced by *Pichia farinosa*, with the yeast cell membranes. *Yeast* 18, 1471–1478.
22. Adler, A. et al., 2007. *Pichia farinosa* bloodstream infection in a lymphoma patient. *J Clin Microbiol* 45(10), 3456–3458.
23. Bergman, M.M., Gagnon, D., and Doern, G.V. 1998. *Pichia ohmeri* fungemia. *Diagn Microbiol Infect Dis* 30, 229–231.
24. Choy, B.Y. et al., 2000. *Pichia ohmeri* peritonitis in a patient with CAPD: Response to treatment with amphotericin (Letter). *Perit Dial Int* 20, 91.
25. Matute, A.J. et al., 2000. A case of disseminated infection with *Pichia ohmeri*. *Eur J Clin Microbiol Infect Dis* 19, 971–973.
26. Hitomi, S. et al., 2002. A case of central-venous-catheter-associated infection caused by *Pichia ohmeri*. *J Hosp Infect* 51, 75–77.

27. João, I. et al., 2002. Native valve endocarditis due to *Pichia ohmeri*. *Heart Vessels* 16, 260–263.
28. Puerto, J.L. et al., 2002. First report of urinary tract infection due to *Pichia ohmeri*. *Eur J Clin Microbiol Infect Dis* 21, 630–631.
29. Reina, J.P. et al., 2002. *Pichia ohmeri* prosthetic valve endocarditis and review of the literature. *Scand J Infect Dis* 34, 140–141.
30. Shin, D.H. et al., 2003. *Pichia ohmeri* fungemia associated with phlebitis: Successful treatment with amphotericin B. *J Infect Chemother* 9, 88–89.
31. Han, X.Y., Tarrand, J.J., and Escudero, E. 2004. Infections by the yeast *Kodomaea* (*Pichia*) *ohmeri*: Two cases and literature review. *Eur J Clin Microbiol Infect* 23, 127–130.
32. Ostronoff, F. et al., 2006. *Pichia ohmeri* fungemia in a hematologic patient: An emerging human pathogen. *Leuk Lymphoma* 47, 1949–1951.
33. Mahfouz, R.A. et al., 2008. *Kodamaea* (*Pichia*) *ohmeri* fungaemia complicating acute myeloid leukaemia in a patient with haemochromatosis. *Pathology* 40(1), 99–101.
34. Poojary, A. and Sapre, G., 2009. *Kodamaea ohmeri* infection in a neonate. *Indian Pediatr* 46, 629, 2009.
35. Yang, B.H. et al., 2009. Fluconazole-resistant *Kodamaea ohmeri* fungemia associated with cellulitis: Case report and review of the literature. *Int J Infect Dis* 13(6), e493–e497.
36. Taj-Aldeen, S.J., Doiphode, S.H., and Han, X.Y., 2006. *Kodamaea* (*Pichia*) *ohmeri* fungaemia in a premature neonate. *J Med Microbiol* 55, 237–239.
37. De Barros, J.D. et al., 2009. *Kodamaea* (*Pichia*) *ohmeri* fungemia in a pediatric patient admitted in a public hospital. *Med Mycol* 47(7), 775–779.
38. Latouche, G.N. et al., 1997. Comparison of use of phenotypic and genotypic characteristics for identification of species of the anamorph genus *Candida* and related teleomorph yeast species. *J Clin Microbiol* 35, 3171–3180.
39. Buchaille, L. et al., 1998. Evaluation of six commercial systems for identification of medically important yeasts. *Eur J Clin Microbiol Infect Dis* 17, 479–488.
40. Lott, T.J., Kuykendall, R.J., and Reiss, E. 1993. Nucleotide sequence analysis of the 5.8S rDNA and adjacent ITS2 region of *Candida albicans* and related species. *Yeast* 9, 1199–1206.
41. Lott, T.J. et al., 1998. Sequence analysis of the internal transcribed spacer 2 (ITS2) from yeast species within the genus *Candida*. *Curr Microbiol* 36, 63–69.
42. Kurtzman, C.P. and Robnett, C.J. 1997. Identification of clinically important ascomycetous yeasts based on nucleotide divergence in the 5′ end of the large-subunit (26S) ribosomal DNA gene. *J Clin Microbiol* 35, 1216–1223.
43. Chen, Y.C. et al., 2000. Identification of medically important yeasts using PCR-based detection of DNA sequence polymorphisms in the internal transcribed spacer 2 region of the rRNA genes. *J Clin Microbiol* 38(6), 2302–2310.
44. Naumov, G.I., Naumova, E.S., and Schnürer, J. 2001. Genetic characterization of the non-conventional yeast *Hansenula anomala*. *Res Microbiol* 152, 551–562.
45. Leaw, S.N. et al., 2006. Identification of medically important yeast species by sequence analysis of the internal transcribed spacer regions. *J Clin Microbiol* 44(3), 693–699.
46. Passov, V. et al., 2006. Biotechnology, physiology and genetics of the yeast *Pichia anomala*. *FEMS Yeast Res* 6, 3–13.
47. Caggia, C. et al., 2001. Identification of *Pichia anomala* isolated from yogurt by RFLP of the ITS region. *Int J Food Microbiol* 71, 71–73.
48. Sutar, R. et al., 2004. Comparison of ITS and IGS1 regions for strain typing of clinical and non-clinical isolates of *Pichia anomala*. *J Med Microbiol* 53, 119–123.
49. Bhardwaj, S. et al., 2007. PCR-based identification and strain typing of *Pichia anomala* using the ribosomal intergenic spacer region IGS1. *J Med Microbiol* 56, 185–189.
50. Vollmer, T. et al., 2008. Evaluation of novel broad-range real-time PCR assay for rapid detection of human pathogenic fungi in various clinical specimens. *J Clin Microbiol* 46, 1919–1912.
51. Pounder, J.I. et al., 2007. Discovering potential pathogens among fungi identified as nonsporulating molds. *J Clin Microbiol* 45, 568–571.

69 Pneumocystis

Steve M. Taylor and Steven R. Meshnick

CONTENTS

69.1 INTRODUCTION

The fungus now known as *Pneumocystis jirovecii* that infects humans entered modern medicine as a trypanosomal infection of guinea pigs reported by Carlos Chagas in 1909.[1] The eminent parasitologist's misclassification was soon corrected, and it was rechristened in deference both to its distinct parasitic morphology and to early investigator Dr. Antonio Carini as *Pneumocystis carinii*.[2] It retained this name through the balance of the twentieth century but assumed varied medical guises, first as a rare cause of sporadic respiratory infection, then as a significant cause of pneumonia (*Pneumocystis carinii*-pneumonia [PcP]) in undernourished children in Europe in the wake of World War II.[3] Subsequently, severely immunosuppressed children and adults receiving cytotoxic chemotherapy became the main victims of this persistent yet uncommon pathogen.[4] In 1981, clusters of PcP in injection-drug users and men who have sex with men in the United States presaged the recognition of the acquired immunodeficiency syndrome (AIDS),[5] and in the early years of the AIDS epidemic PcP was thrust center stage as a common and devastating AIDS-defining opportunistic infection.[6] With aggressive chemoprophylaxis in high-risk patients and the effective treatment of human immunodeficiency virus (HIV) infection in the developed world, the overall incidence of PcP has declined significantly, though it remains an important cause of opportunistic pneumonia in patients with immunodeficiency due to therapeutic immunosuppression, congenital disorders, or inadequately treated or undiagnosed HIV infection.[7]

69.1.1 Classification, Morphology, and Biology

Since its initial recognition as a human pathogen, *P. jirovecii* has changed both names and kingdoms. Chagas' classification of *Pneumocystis* as a trypanosome in 1909 quickly yielded to revision by Delanoe and Delanoe in 1912, but it remained a protozoan through much of the twentieth century. Several factors contributed to this misclassification, including the inability to culture the organism in vitro; therapeutic success with antiprotozoal, but not antifungal, antimicrobials; and limited interest in the organism due to its clinical rarity.

In the 1980s, its newfound clinical importance and advances in DNA technology fueled interest, and molecular phylogenetic analyses definitively demonstrated that *P. jirovecii* is a fungus, though an atypical one.[8] It lacks ergosterol in its cell wall—possessing instead an unusual abundance of cholesterol—rendering it impervious to the polyene antifungal agents that target ergosterol synthesis, and the cell wall lacks the rigidity and impermeability of most yeasts. Additionally, sequencing of *Pneumocystis* isolates from different vertebrate species demonstrated that *Pneumocystis* strains are species-specific, and it is unclear that either cross-infections or genetic commingling occur between strains.[9]

Genotyping these host species-specific isolates confirmed significant sequence divergence of key loci, complicating *Pneumocystis* nomenclature with a proliferation of host species-specific designations. In 2002, human-derived *Pneumocystis* was officially renamed *P. jirovecii* (pronounced "yee-row-vet-zee"), acknowledging both its genetically distinct heritage and the contributions of Czech parasitologist Otto Jirovec, who first described the fungus in human disease.[10] *P. carinii* was retained to describe one of two strains that infect rats, though, to avoid confusion, the abbreviation PcP continues to refer to pneumonia caused by *P. jirovecii* (<u>P</u>neumo<u>c</u>ystis <u>p</u>neumonia).

The *P. jirovecii* life cycle is incompletely understood, as attempts to reproducibly cultivate the organism in vitro have failed.[11] Because of this and the difficulty of generalizing observations from models of infected animals, insight into its biology and life cycle has been intuited from the microscopic examination of human lung specimens and by analogy to life cycles of other ascomyceteous fungi.[12] Both major forms of the organism—trophic and cystic—comprise the life cycle, which is propagated by both asexual and sexual reproduction. Trophic forms exist in clusters with avidity for Giemsa staining and can reproduce by binary fission. Alternatively, these haploid trophic forms can conjugate to form diploid cyst forms, whose thick cell walls are readily identified by methenamine silver and toluidine blue-O stains. The clinical pathological consequence is that in human lung infections, the haploid trophic forms typically predominate and comprise up to 90% of the microscopically visible life forms.

69.1.2 Clinical Features and Pathogenesis

P. jirovecii causes pneumonia with atypical clinical and radiographic features.[13] Classic symptoms in adults include fever, shortness of breath, and a nonproductive cough, though these are neither specific nor sensitive for the pathogen. In patients infected with HIV, the onset of symptoms is subacute or insidious, whereas in patients with immunosuppression due to cancer, transplantation, or cytotoxic chemotherapy, the clinical course is often more rapid.[14] In children, malnourishment predisposes to infection that begins with poor feeding and gradually progresses to respiratory distress and cyanosis. An abnormal lung exam is frequent, but hypoxemia is almost uniformly present. Radiographically, pneumonia due to *P. jirovecii* is notoriously varied and nonspecific. Its appearance ranges from classic lobar pneumonia to diffuse interstitial infiltrates to, in early disease, normal radiography, and nearly every radiographic phenomenon has been reported (see Figure 69.1).

The overwhelming risk factor for disease is immunocompromise, usually due to either advanced HIV infection or the administration of cytotoxic chemotherapy or high-dose corticosteroids; in particular, an association between the tapering phase of corticosteroids and the onset of infection has been described. Since the widespread application of HIV therapy and thus the prevention of severe immunocompromise, the incidence of PcP has declined significantly among HIV-positive patients. Among HIV-negative patients, the incidence varies according to type and duration of immunosuppression.[15]

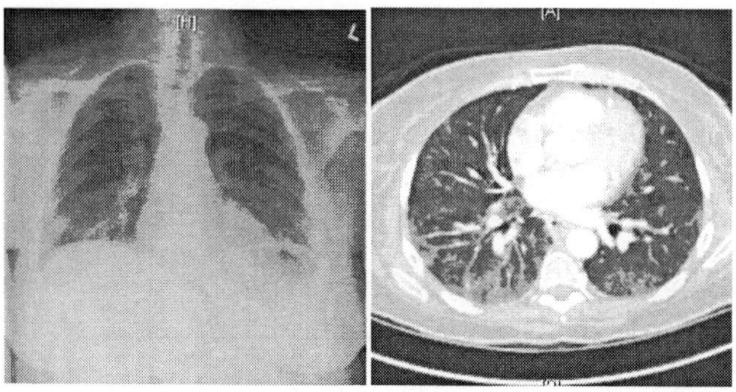

FIGURE 69.1 Radiographic appearance of *P. jirovecii*-pneumonia. Left: a standard anterior/posterior chest x-ray shows bilateral perihilar reticular opacities. Right: computer tomography of the chest can show multilobar ground-glass opacities.

Mortality from PcP remains substantial, and has not significantly improved among HIV-infected patients despite the advent of antiretroviral therapy in the mid-1990s.[16,17] Recent reviews of HIV-infected patients in San Francisco[16] and in London[17] document mortality rates of approximately 10% in the 2000s, with death associated with advanced age, medical comorbidity, and subsequent episodes of PcP. Among HIV-negative patients, mortality is nearly 40%, in accordance with its more rapid and fulminant presentation (see above).[18]

Extrapulmonary infection with *Pneumocystis* has been reported, but its clinical importance is unknown.[19] Autopsy series of HIV-infected patients have identified *P. jirovecii* histologically and molecularly in the reticuloendothelial, gastrointestinal, renal, and hematopoetic systems; these lesions are typically associated with patients receiving either inhaled or no prophylaxis.[20]

Pneumocystis enters the body via inhalation and infects alveoli.[12] Organisms have been identified in a wide variety of environmental reservoirs, though it is unclear whether new infections are acquired from these sources or from human-to-human spread. Serologic epidemiology suggests widespread exposure to the organism in humans, usually within the first several years of life, consistent with the peak incidence of clinical infection in HIV-infected infants.[21] The natural history of infection after acquisition is unclear. Colonization with *P. jirovecii*, defined as the presence of organisms in the absence of clinical infection, is common in adults with HIV or non-HIV-related immune suppression and chronic lung disease, although notably minimal in health, immunocompetent adults.[22] The implications of colonization studies are twofold: minimal colonization among health adults suggests that lifelong latent infection is unlikely in most adults, and colonized individuals may serve as a reservoir for transmission to susceptible individuals. Clinical infection may result from either activation of these latent infections due to immunosuppression or from the acquisition of new organisms, though evidence from molecular genotyping studies suggests the latter.[10]

The development of clinical disease following infection is typically contingent upon an attenuation of host immunity. Impairment of either humoral or cellular immunity has been associated with development of pneumocystosis, though the mechanisms of this are complex and poorly understood. In normal lungs, alveolar macrophages primarily affect clearance of organisms after infection, and this activity is impaired in patients with AIDS or cancer.[12] Additionally, macrophage activation by *P. jirovecii* cell-surface antigen β-glucan stimulates release of cytokines including tumor necrosis factor and interleukin-8, which serve to augment organism clearance by recruiting additional inflammatory cells. This downstream macrophage activity is highly dependent upon the presence of upstream T cells that are primarily of the CD4+ lineage, as evidenced both by clearance of infection in T-cell deficient, macrophage-replete mice reconstituted with CD4+ cells and, more simply, by the close association between infection risk and CD4+ cell level in patients with AIDS.

Pathologically, progressive diminution of host immunity allows proliferation of organisms resident in alveoli.[13] The histologic picture is typically one of a foamy, intra-alveolar eosinophilic infiltrate; more advanced cases often progress to hyaline membranization with interstitial fibrosis, which is pathologically similar to changes that characterize the adult respiratory distress syndrome (ARDS). These pathologic changes in the lung result not only from direct effects of the organism but also the proinflammatory response, a paradoxical phenomenon given that clinical infection virtually requires advanced immune suppression. Nevertheless, experimental evidence in rat models of pneumonia and clinical experience in AIDS patients with PcP both suggest that exuberant or dysregulated natural and adaptive immunity in the lung contribute to pathophysiology.[20]

69.1.3 Diagnosis

The diagnosis of PcP is supported by clinical, radiographic, and serum chemistry data, but the gold standard test remains the staining and microscopic examination of respiratory specimens.[23] As above, the clinical and radiographic findings are nonspecific and substantially overlap with those of lower respiratory tract infections due to other pathogens. Serum testing can support the diagnosis, as an elevated serum lactate dehydrogenase (LDH), though nonspecific, is frequently considered supportive of the diagnosis. Additionally, serum levels of *S*-adenosyl-methionine (AdoMet), an enzyme essential for fungal metabolism, may offer a means to diagnose PcP nonmicrobiologically,[24] though this is controversial.[25] More recently, serum (1→3) β-D-glucan, a major component of the cyst wall of *P. jirovecii*, has been advocated as a sensitive serum marker of PcP, though it has not yet been widely employed in clinical practice.[26]

69.1.3.1 Conventional Techniques

Securing the diagnosis traditionally requires the demonstration of organisms in respiratory specimens.[20] A variety of specimen types has been investigated, and the diagnostic yield is dependent on the quality of specimen and its route of collection. Historically, open lung biopsy was necessary to obtain an appropriate specimen for analysis, but this has been supplanted by fiber-optic bronchoscopy and less-invasive collection techniques. Sputa induced by the inhalation of saline mist can be examined as an initial diagnostic step, but the highest diagnostic yield is provided by fluid and tissue obtained during bronchoscopy. The examination of biopsies or fluid gathered at bronchoscopy reputedly have a sensitivity over 90% for the diagnosis, though the need for invasive specimen collection limits their clinical utility.[27] Because of this, induced sputa can be examined as the first step in the diagnostic evaluation; its sensitivity is widely variable, though at some experienced centers approaches 80%. In contrast, neither oropharyngeal wash nor expectorated sputum reliably demonstrates organisms detectable by staining in patients with PcP.[20]

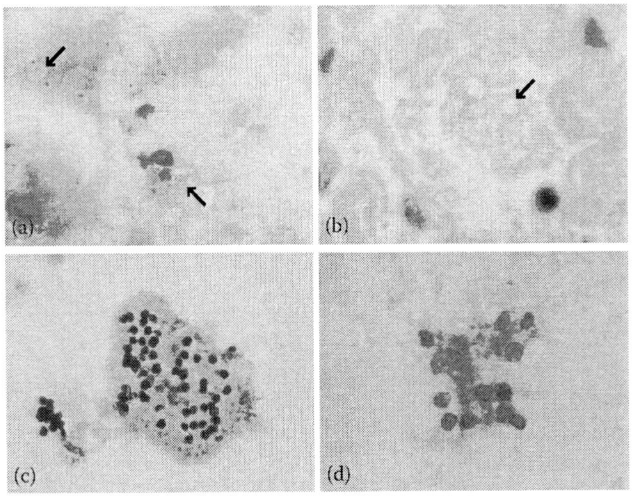

FIGURE 69.2 *P. jirovecii* detected in respiratory specimens by colorimetric staining. (a) Wright-Giemsa preparations do not stain the cyst walls but do stain the intracystic bodies (10×), (b) hematoxylin and eosin staining typically reveals a foamy, intra-alveolar infiltrate, with trophic forms failing to stain (40×), (c) the Gomori methenamine silver stain readily stains cysts (40×), and (d) toluidine blue-O stains cysts (40×).

Organisms are detected microscopically by either colorimetric staining or immunofluorescence (see Figure 69.2). Staining is more widely employed, and a variety of stains reliably identify the different life forms of *P. jirovecii*. The cyst-stage-specific stains toluidine blue-O and methenamine silver will stain only fungi and thus fail to identify the more-abundant trophic forms. Giemsa and Diff-Quik stain all stages of *P. jirovecii* and thus address this shortcoming, but will also stain yeast and other extracellular debris and thus complicate examination. Some laboratories employ a staining algorithm incorporating a rapid screening examination followed by a more-intensive preparation for definitive diagnosis. In tissue specimens, methenamine silver and toluidine blue-O commonly diagnose cysts, and standard hematoxylin–eosin often reveals a foamy, intra-alveolar eosinophilic infiltrate that is characteristic of PcP.

As an alternative to colorimetric straining, immunofluorescent staining also allows for rapid microscopic detection, though it requires specialized equipment. On induced sputum, it may be more sensitive and specific than colorimetric staining,[28] but these advantages are largely abrogated when bronchoalveolar lavage (BAL) fluid is examined.[29]

69.1.3.2 Molecular Techniques

Because of the inability to reliably cultivate the organism for definitive microbiologic diagnosis, molecular diagnostic assays have generated significant interest for the detection of *Pneumocystis*. Since the first description of diagnosis by DNA amplification in 1990,[30] subsequent reports have promulgated protocols with a wide variety of gene targets, detection chemistries, specimen inputs, and operating characteristics. Despite this variety, a common finding of most assays when compared with conventional methods is a high sensitivity for organisms but a diminished specificity that likely reflects the detection of organisms in a carrier, rather than infecting, state.

The most widely investigated protocols target sequences of the rRNA of *P. jirovecii*, including the mitochondrial large subunit (mtLSUrRNA),[30] the internal transcribed spacers (ITS),[31] or the 18S subunit.[32] In order to shorten testing turnaround time and minimize the risk of contamination by postamplification processing, assay chemistry has evolved from nested PCR with Southern blotting or agarose gel electrophoresis to real-time amplification with detection by fluorescence resonance energy transfer (FRET) or TaqMan probes or by SYBR green fluorescence.[33–38] These real-time PCR technologies have been employed by protocols to generate a quantitative estimate of organism burden and thus potentially distinguish between asymptomatic *P. jirovecii* carriage and clinically significant infection. Additionally, reverse transcriptase (RT)-PCR protocols have been described that detect *P. jirovecii* mRNA in an effort to identify only viable organisms from clinical or environmental samples.[39,40] For amplicon detection, protocols employing enzyme immunoassays[41] or solution hybridization enzyme-linked assays[42] have generally been supplanted by protocols exploiting real-time technology for detection.

Notably, one report describes the development and application to clinical samples of a loop-mediated isothermal amplification (LAMP) protocol to detect *P. jirovecii*,[43] which may reduce the cost of testing compared with real-time technologies.

69.1.3.2.1 Operating Characteristics of PCR Assays

Compared with conventional diagnosis of PcP by staining or clinical factors, many molecular assays for *P. jirovecii* performed on respiratory specimens have a sensitivity that approaches 100%. When performed on induced sputum or BAL, sensitivity of 100% has been reported for assays targeting the ITS,[44] 18S rRNA,[32] 5S rRNA,[45] thymidylate synthase,[46] and major surface glycoprotein (MSG).[47,48] When first described, the assay targeting the mtLSUrRNA reported sensitivity of 87%–91%,[30,49] but subsequent modifications incorporating a nested[50,51] or touchdown[52] amplification have improved sensitivity to 100%. A notable exception to the rule of the high sensitivity of molecular assays is that for *P. jirovecii* dihydrofolate reductase (dhfr) promulgated by Schluger et al.[53] due to its reported sensitivity of 17% in respiratory specimens[54] and the possibility that it may amplify a nonspecific target[55]; the clinical sensitivity of a subsequent nested assay targeting dhfr has not been rigorously evaluated.[56]

The high sensitivity of these assays complicates definitive assessment of their specificity because they frequently detect *P. jirovecii* in the absence of compelling clinical evidence of PcP. This presence of *P. jirovecii* in the respiratory tract in the absence of clinical pneumonia has been termed "carriage," "colonization," and "subclinical infection," and molecular assays have generated evidence of significant carriage of *P. jirovecii* in children, HIV-infected adults, non-HIV-infected immunosuppressed adults, health-care workers, and adults with chronic lung diseases (reviewed by Morris et al.[22]). Because the clinical significance of this carrier state is unclear, the detection of *P. jirovecii* by molecular assays in asymptomatic individuals complicates their use in diagnosis of clinical disease.

Nevertheless, the specificity of these assays has been evaluated both microbiologically and clinically. Most assays have tested negative on panels of non-*P. jirovecii* pulmonary pathogens either in pure culture or in clinical respiratory specimens,. indicating 100% microbiologic specificity.[45,46,48,49,57] However, a subsequent comparison of assays found cross-reactivity with *S. cervisiae*, *C. albicans*, or *C. neoformans* for assays targeting the 18S rRNA and thymidylate synthase genes.[49] Because clinical specificity is bedeviled by the detection of *P. jirovecii* DNA in clinically noninfected patients, specificities have been reported between 86% and 100%, depending on the assay.[30,32,44,50,52] Specificity may be improved by quantifying organism burden, and, to this end, several assays have been adapted to exploit real-time detection chemistry. Assays targeting MSG have been modified to include amplicon detection with FRET probes, and though the application of a post hoc cutoff improved the specificity to detect clinical disease to nearly 100% in respiratory specimens[58] and oral washes,[59] a predictable consequence was a slight decrease in clinical sensitivity. In contrast, improvements in specificity without significant diminution of sensitivity in real-time assays have been achieved with TaqMan assays targeting both HSP70[60] (sensitivity 95% and specificity 96%) and dihydroopteroate synthase (dhps)[35] (sensitivity 94%, specificity 96%).

69.1.3.2.2 Comparisons between Assays

Despite the promulgation and promotion of diverse protocols, few investigations have assessed the agreement between assays. Fischer et al.[57] compared the sensitivities of the mtLSUrRNA and MSG assays on serial dilutions of clinical samples and found the MSG assay was 10-fold more sensitive than the mtLSUrRNA. Flori et al.[61] found nearly identical diagnostic accuracy between assays targeting the mtLSUrRNA and MSG, though the incorporation of a quantitative cutoff in the MSG real-time protocol allowed for an improvement in specificity for clinical disease. Linssen et al.[62] found high agreement between different microbiology laboratories performing real-time and conventional PCR assays.

Two studies have expressly compared the diagnostic accuracy of panels of published protocols. One study compared the operating characteristics of six common *P. jirovecii* molecular assays published by 1995 on stain-positive and -negative BAL specimens.[49] The sensitivity of assays targeting the ITS and the 18S rRNA was 100%, but assays were less sensitive when targeting the mtLSUrRNA (87%), thymidylate synthase (60%), 5S rRNA (33%), and dhfr (23%).

Differences in sensitivity were attributed to differences in both gene copy number and amplification conditions, as the ITS and 18S assays were the only nested assays. All assays returned negative results on the subset of BALs with both negative stains for *P. jirovecii* and positive results for other pulmonary pathogens, indicating microbiologic specificities of 100%.

A similar comparison was repeated in 2007,[63] comparing protocols targeting ITS,[49] dhps (nested),[64] dhps (single),[65] dhfr,[56] MSG (heminested),[47] mtLSUrRNA (nested),[66] 18S rRNA,[67] and 5S rRNA (real-time).[48] The greatest sensitivities, when compared with methenamine silver staining, were achieved by the mtLSUrRNA nested assay (100%), the real-time assay targeting 5S rRNA (88%), and the heminested MSG assay (71%). The study did not report a comprehensive comparison of the assays' clinical or microbiological specificities.

69.1.3.2.3 Nonpulmonary Specimens

When applied to less-invasive specimens such as blood and oral washes, data are mixed on the diagnostic utility of PCR-based assays for *P. jirovecii*. In patients with PcP, most assays have failed to detect *P. jirovecii* DNA in blood,[32,68,69] or returned positive results in PcP-negative control patients.[69] In contrast, one report found that detection in blood of the *P. jirovecii* rRNA ITS was highly sensitive and specific for PcP,[54] though this has not been repeated in other cohorts.

Oral wash specimens have been investigated as a noninvasive method of diagnosis, and several reports have documented sensitivities as high as 88%–91%,[52,57,59] while other reports found the assays much less sensitive.[70] The examination of multiple oral washes increases the sensitivity but decreases specificity, though this diminution of specificity may be mitigated by the use of an RT-PCR assay.[40]

69.1.3.2.4 Non-HIV-Infected Patients

Several of the *P. jirovecii* molecular assays are less robust in non-HIV-infected patients.[71] Two reports compare the utility in different patient groups of nested PCR protocols targeting the *P. jirovecii* mtLSUrRNA in the respiratory specimens of patients with pneumonia. The assay accurately predicted clinically proven PcP in HIV-infected patients but was poorly predictive in leukemia and lymphoma patients.[72] Similarly, the positive predictive value of an mtLSUrRNA nested assay was >97% in HIV patients but <50% in transplant patients, patients with malignancies, or non-HIV-infected immunosuppressed patients.[73] Other reports in non-HIV infected immunosuppressed patients describe "false-positive" (i.e., PCR-positive patients without other compelling clinical evidence of PcP) rates as high as 82%[50] and quantitative trends toward lower organism burdens (compared with HIV-infected patients).[58] It is unclear if these findings in non-HIV-infected patients reflect asymptomatic carriage, incipient clinical infection, or a biologic difference in virulence between the lungs of HIV- and non-HIV-infected patients. Nevertheless, greater assay specificity for clinical disease is a priority in immunocompromised non-HIV-infected patients, and, to this end, real-time assays have been advocated as offering greater diagnostic utility in this population.[74]

69.2 METHODS

69.2.1 Sample Preparation

Genomic DNA (gDNA) extraction from respiratory specimens (oral washes, induced sputa, or BAL fluid) may require preextraction preparation. Specimens may be frozen prior to extraction. Mucous-containing BAL or sputum samples may require the addition of an equal volume of 0.9% NaCl to resuspend and subsequent pelleting by centrifugation,[61] or the addition of lysis buffers,[72] dithiothreitol,[46,51] or *N*-acetyl-L-cysteine,[36] though these latter additions may necessitate amplification controls to ensure the absence of introduced PCR inhibitors. Oral wash specimens are obtained by vigorous rinsing of the oral cavity for 10–30 s with 50 mL of sterile saline,[48] centrifugation at 2800 × *g* for 10 min, and resuspension of the pellet in 1 mL of supernatant. Preparation of oral wash, sputum, or BAL fluid for gDNA extraction can be accomplished by centrifuging 2 mL of specimen at 3000 × *g* for 15 min, digestion of the pellet with proteinase K at 60°C for 2 h, and boiling for 10 min to deactivate the enzyme.[52] However, BAL fluid is frequently input directly into commercial extraction protocols without processing.[60]

Paraffin-embedded lung tissue samples may also be used as a source of gDNA for amplification, after treatment of sections with xylene and ethanol and digestion with proteinase K.[75] Additionally, stained or unstained smears of BAL fluid or lung biopsies have successfully yielded DNA for amplification.[76] After soaking slides in methanol or in ethanol and water to remove Giemsa or silver stains, slides should be scraped with a sterile scalpel into sterile water, with the scraped contents then transferred to a glass-fiber filter disk and processed for DNA extraction after proteinase K digestion. Despite this, stained or unstained slides generally yield less gDNA than BAL that has been frozen or stored in ethanol.[35]

For all specimens, phenol–chloroform, Chelex, or commercial extraction kits are appropriate extraction alternatives. Commercial extraction kits employed include the Wizard Genomic DNA Purification Kit (Promega), MagnaPure LC Isolation Kit (Applied Biosystems), QiaAMP DNA Mini Kit (Qiagen),[62] Gene Clean II (BIO 101),[72] QIAamp Tissue Kit (Qiagen),[73] DNeasy Tissue Kit (Qiagen),[51] and NucliSENS (bioMerieux).[43]

Few studies compare the success of extraction protocols. Huggett et al.[60] employed a real-time PCR assay targeting HSP70 on gDNA extracted from BAL fluid by the QIAamp UltraSens Virus Kit or the DNeasy Tissue Kit, with the QIAamp system isolating greater amounts of DNA as evidenced by increased gene copy number when assayed. Huang et al.[47] report a comparison of DNA Extraction Reagent (Perkin Elmer) and phenol–chloroform extraction from oral washes, sputa, and BAL fluid, and found no difference in detection rates when assaying with two protocols targeting MSG and mtLSUrRNA.

Though their diagnostic utility is largely unproven, serum specimens may be processed along much the same lines, with the addition of proteinase K to serum (4:1), incubation at 56°C for 2 h, boiling for 10 min, and proceeding with phenol–chloroform extraction on an aliquot of the remainder.[54]

69.2.2 Detection Procedures

Protocols employing PCR, RT-PCR, and real-time PCR for detection target a variety of genes and incorporate many modifications; see Table 69.1 for specifics of the most current and widespread protocols. The original protocol targeting the mtLSUrRNA[30] has been updated for improved sensitivity by either adding a nested amplification or maintaining a single round of amplification with a touchdown protocol. Both Wakefield[66] and Weig et al.[72] amplify mtLSUrRNA in two rounds of amplification and electrophoresce products on an agarose gel, with positive samples demonstrating the successful amplification of a 260 or 193 bp product, respectively. Helweg-Larsen et al.,[52] however, maintain the original single round of amplification of mtLSUrRNA but incorporate a touchdown step in the cycling conditions, with positive amplification indicated by the presence of a 346 bp band after agarose gel electrophoresis.

Assays targeting the ITS have been employed not only for detection but for typing *P. jirovecii* strains. From the original protocol,[31] the second round of amplification has been altered[77–79] to yield products suitable for molecular diversity and phylogenetic analysis, though the validity of this practice has been questioned.[80] Similarly, assays targeting dhfr and dhps have variable sensitivity[35,49] and have generally been employed for genotyping rather than detection. Mutations in both dhfr[64] and dhps[56] have been detected in isolates from patients after exposure to antifolate medications, and have thus been of interest as possible mediators of *P. jirovecii* drug resistance.[81]

The assay developed by Huang et al.[47] targeting the MSG has been infrequently employed as a conventional PCR assay but frequently reported upon after adaptation for real-time PCR with a touchdown amplification and real-time FRET probes.[48,57–59,61,62] SYBR green detection has been employed by two assays that target the mtLSUrRNA[82] and the 5S rRNA, but the potential for nonspecific amplification of targets of other pathogens likely limits their application to clinical samples. Of late, TaqMan probes, or 5′ nuclease probes, have been employed by assays targeting *P. jirovecii* β-tubulin,[34] dhps,[35] dhfr and 5.8S rDNA,[38] and *Pj*HSP70.[60] Such assays, like the other FRET assays, have allowed for estimates of organism burden and assay-specific cutoff values to distinguish *P. jirovecii* carriage from clinical disease.[35,48,60]

69.3 CONCLUSIONS AND FUTURE PERSPECTIVES

The advent of antiretroviral therapy has dramatically reduced the incidence of PcP, but it remains a significant infectious threat to patients with drug-resistant HIV, those with poor adherence and access to antiretroviral therapy or PcP prophylaxis, and those in receipt of organ transplantation or other immunosuppressive drugs.[7] Additionally, though its incidence and contribution to opportunistic infection in patients with AIDS in developing countries remains unclear, these patients may represent a large at-risk population. For these reasons, the importance of *P. jirovecii* abides.

At this time, numerous well-characterized assays achieve accurate detection of *P. jirovecii* in clinical samples, though there are none that are commercially available, and protocol specifics vary widely. Sample preparation and DNA extraction have been achieved by a wide variety of commercial and conventional protocols, and assays have targeted more than 10 different genes and employed conventional PCR, real-time PCR with fluorescent probes and SYBR green, RT-PCR, and LAMP.

For the clinical microbiologist, what role should molecular assays play in the diagnosis of *P. jirovecii* infection? In AIDS patients with PcP, the added benefit of molecular detection of *P. jirovecii* remains unclear because conventional staining is widely available and accurate. In immunosuppressed patients without AIDS, the high reported rates of false-positive PCR results can contribute to diagnostic confusion. Recent incorporation of real-time technologies may add luster to the clinical application of these assays by offering timelier testing in AIDS patients and allowing quantitation of organism burden to differentiate asymptomatic carriage from clinical infection. Additionally, these assays may provide a basis for the inclusion of *P. jirovecii* as a target in multiplex respiratory pathogen detection panels designed for immunocompromised hosts.

Compared with conventional methods, molecular assays have several unique advantages: they more reliably detect *P. jirovecii* carriage in noninvasive specimens collected for epidemiologic investigations; they can generate data to allow strain typing for molecular phylogeny studies; and they allow genotyping to detect described or prospective molecular correlates of clinical drug resistance. These tools exploit the benefits of molecular technologies to expand our understanding of *P. jirovecii*'s basic biology, clinical pathogenesis, and response to antimicrobial therapy.

Though molecular tools will refine our understanding of *P. jirovecii* in HIV-infected and immunosuppressed patients in the developed world, perhaps their most promising application is in better characterizing the biology and burden of *P. jirovecii* in the developing world. With over 20 million people living with HIV in sub-Saharan Africa (compared with 1.4 million in North America), careful investigation is needed to characterize both their individual risk of infection and the epidemiology and spread of strains within populations. Properly adapted and applied, molecular assays can both inform our biological understanding of *P. jirovecii* as well as optimize the clinical care of patients both infected and at risk for infection with *P. jirovecii*.

TABLE 69.1
Molecular Protocols for *P. jirovecii* Detection Employing PCR, RT-PCR, and Real-Time PCR[a]

Reference	Target (Approximate bp Length)	Primers (5′–3′)	Reaction Mixture	Amplification Conditions (after Denaturation/Activation)	Detection Output
PCR					
Kitada et al., 1991[45]	5S rRNA (120 bp)	5S sense: AGTTACGGCCATACCTCAGA 5S antisense: AAAGCTACAGCACGTCGTAT	50 pmol each primer 1.5 mM $MgCl_2$ Template/reaction vol: 10/100 µL	(94°C for 1.5 min, 55°C for 2.5 min, 72°C for 1.9 min) × 25 or 40	Polyacrylamide or agarose gel electrophoresis
Schluger et al., 1991[53]	dhfr (NR)	Sense: CTGCAAAATCCTTGGATCAT Antisense: CTTTAGTACCAACCCAAGAT	1 µM each primer 1.5 mM $MgCl_2$ Template/reaction vol: 5/95 µL	(92°C for 1 min, 55°C for 1 min, 72°C for 1 min) × 45	Southern blotting: probe GATAGAATTATGGCTACAATA
Lipschik et al., 1992[32]	18S rRNA Round 1 (378 bp) Round 2 (312 bp)	340a: CCAGATTAGCTTTTGCTGATCGCGGG 708: ACTTTCCAGTAATAGGCTTATCG JK1: TGTTGGCATGAAGCCAATGGAA JK2: CAATAACCCATCACCAGTCCGAAG	50 pmol each primer Template/reaction vol: 5/50 µL	(94°C for 1 min, 50°C for 1 min, 72°C for 2 min) × 36, 72°C × 10 min	Southern blotting; phosphorus-32 labeled probe 372a: AAGGAAAATGAACTTGCTGGCTCT
Olsson et al., 1993[46]	Thymidylate synthase (403 bp)	PC1: ATTTATGGGTTTCAATGG PC2: GTTCCCTTTAATATTGC	0.4 µM each primer 3 mM $MgCl_2$ Template/reaction vol: 2/50 µL	(94°C for 30 s, 50°C for 25 s, 70°C for 1 min) × 40, 70°C for 5 min	Agarose gel electrophoresis
Lu et al., 1994[31]	ITS Round 1 (693 bp) Round 2 (550 bp)	1724F: AAGTTGATCAAATTTGGTC ITS2R: CTCGGACGAGGATCCTCGCC ITS1F: CGTAGGTGAACCTGCGGAAGGATC ITS2R1: GTTCAGCGGGTGATCCTGCCTG	0.2 µM each primer Template/reaction vol: NR/100 µL	(94°C for 1 min, 47°C for 1 min, 72°C for 2 min) × 35, 72°C for 10 min Then, (94°C for 1 min, 58°C for 1 min, 72°C for 2 min) × 35, 72°C × 5 min	Agarose gel electrophoresis, cloning, sequencing
Wakefield, 1996[66]	mtLSUrRNA Round 1 (340 bp) Round 2 (260 bp)	pAZ102E: GATGGCTGTTTCCAAGCCCA pAZ102H: GTGTACGTTGCAAAGTACTC pAZ102X: GTGAAATACAAATCGGACTAGG pAZ102Y: TCACTTAATATTAATTGGGGAGC	1 µM each primer 3 mM $MgCl_2$ Template/reaction vol: NR/NR	(94°C for 1.5 min, 55°C for 1.5 min, 72°C for 2 min) × 40 Then, (94°C for 1.5 min, 55°C for 1.5 min, 72°C for 2 min) × 10, (94°C for 1.5 min, 63°C for 1.5 min, 72°C for 2 min) × 30	Cloning and sequencing
Lane et al., 1997[64]	Dhps Round 1: Round 2: 335 bp	F1: CCTGGTATTAAACCAGTTTTGCC B45: CAATTTAATAAATTTCTTTCCAAATAGCATC AHUM: GCGCCTACACATATTATGGCCATTTTAAATC BN: GGAACTTTCAACTTGGCAACCAC	0.4 µM each primer Template/reaction vol: 5/100 µL	(92°C for 30 s, 52°C for 30 s, 72°C for 60 s) × 35, 72°C for 5 min Then, (92°C for 30 s, 55°C for 30 s, 72°C for 60 s) × 35, 72°C for 5 min	Agarose gel electrophoresis

(continued)

TABLE 69.1 (continued)
Molecular Protocols for *P. jirovecii* Detection Employing PCR, RT-PCR, and Real-Time PCR[a]

Reference	Target (Approximate bp Length)	Primers (5′–3′)	Reaction Mixture	Amplification Conditions (after Denaturation/ Activation)	Detection Output
Weig et al., 1997[72]	mtLSUrRNA Round 1 (340 bp) Round 2 (193 bp)	pAZ102E: GATGGCTGTTTCCAAGCCCA pAZ102H: GTGTACGTTGCAAAGTACTC pLE1: TCGGACTAGGATATAGCTGG pLE2: CCCTTTCGACTATCTACC	25 pmol each primer 3 μL of 25 mM $MgCl_2$ Template/reaction vol: 10/50 μL (first round) 1/50 μL (second round)	(94°C for 1 min, 55°C for 1 min, 72°C for 90 s) × 35, 72°C for 1 min Then, (94°C for 1 min, 58°C for 1 min, 72°C for 90 s) × 30, 72°C for 10 min	Agarose gel electrophoresis
Helweg-Larsen et al., 1998[52]	mtLSUrRNA (346 bp)	pAZ102E: GATGGCTGTTTCCAAGCCCA pAZ102H: GTGTACGTTGCAAAGTACTC	0.4 mM each primer Template/reaction vol: NR/100 μL	(94°C for 15 s, 72 → 62°C for 30 s (decrement of 1°C cycles 1–10), 72°C for 15 s) × 50, 72°C for 5 min	Agarose gel electrophoresis
Huang et al., 1999[47]	MSG (249 bp)	JKK14/15: GAATGCAAATCYTTACAGACAACA JKK17: AAATCATGAACGAAATAACCATTGC	1 μM each primer 3 mM $MgCl_2$ Template/reaction vol: NR/50 μL	(95°C for 1 min, 65°C for 1 min, 72°C for 1 min) × 43, 72°C for 10 min	Southern blotting, Probe JKK16: TGCAAACCAACCAAGTGTACGACAGG
RT-PCR					
Maher et al., 2001[39]	Phsb1 (425 bp)	Phsb1-161C: TGTTAAAAAAGACATGAAAATG Phsb1-566nc: CAGCAGTGGCTTTAACTGAA	Reverse transcription: Phsb1-566nc at 4.3 μM	(94°C for 1 min, 50°C for 1 min, 72°C for 1 min) × 40	Agarose gel electrophoresis
Real-time PCR					
Kaiser et al., 2001[82]	mtLSUrRNA	pAZ102E: GATGGCTGTTTCCAAGCCCA pAZ102H: GTGTACGTTGCAAAGTACTC	0.6 mM each primer 5 mM $MgCl_2$ Template/reaction vol: 2/20 μL	(95°C for 5 s, 56°C for 10 s, 72°C for 20 s) × 45, then dissociation step	SYBR green
Palladino et al., 2001[37]	5S rRNA	5S sense: AGTTACGGCCATACCTCAGA 5S antisense: AAAGCTACAGCACGTCGTAT	0.5 μM each primer	(95°C for 15 s, 55°C for 10 s, 72°C for 10 s) × 38, then dissociation step	SYBR green
Larsen et al., 2002[48]	MSG	JKK14/15: GAATGCAAATCYTTACAGACAACA JKK17: AAATCATGAACGAAATAACCATTGC	1 μM each primer 5 mM $MgCl_2$ Template/reaction vol: 5/20 μL 0.2 μM FRET probe	(95°C for 5 s, 65 → 50°C × 10 s [decrement of 1°C for cycles 2–6, then 2°C for steps 7–11], 72°C for 20 s) × 46	FRET probes: PCMSGFRET1U: CAAAAATAACAYTSACATCAACRAGGCG-fluorescein PCMSGFRET1D: Red 640-TGCAAACCAACCAAGTGTACGACAGG PCMIM1U: GATATCGTCCATTCCGACAGCATC-fluorescein PCMIM1D: Red 705-CCAGTCACTATGGCGTGCTGCTAG

Brancart et al., 2005[34]	β-tubulin	1186F: GATCCGAGACATGGTCGCTATT 1257R: TTCAACCTCCTTCATGGAAACAG	0.3 μM each primer 0.66 μM probe Template/reaction vol: 5/25 μL	(95°C for 15 s, 60°C for 1 min) × 45	TaqMan probe: 1212T: TGTTGCAGCGATTTTCCGCGGTA
Alvarez-Martinez et al., 2006[35]	dhps	DHPS_F: GCTTGGTCCAAGTCGCAAAA DHPS_R: CAGCAGTGCCCCAAATCC	0.9 mM each primer 0.25 μM probe Template/reaction vol: 2/25 μL	(95°C for 15 s, 60°C for 60 s) × 45	TAMRA probe: VIC-ATTTACAGGGTGTCTTACA GGTGATGTTATGCCAA
Arcenas et al., 2006[36]	Cdc2	Fwd: AGGTAGGAGAAGGTAAGAAA Rev: GCTGTGCTTGGAACCC	0.5 μM each primer 0.4 μM Red-640 probe 0.2 μM fluorescein probe Template/reaction vol: 5/20 μL	(95°C for 10 s, 55°C for 15 s, 72°C for 15 s) × 45	FRET probes: GATCTTGAAAATGGCACAATAGTAG-fluorescein Red-640-TTAAAAAAATCCGGCTAGAAGCAGAAG
Bandt and Monecke, 2007[38]	5.8S rRNA	5.8S_F: TGGCTCTCGCGTCGATG 5.8S_R: TTCGATGATTCACTAAATTCTGCAA	10 pmol each primer 4 pmol probe Template/reaction vol: 5/20 μL	(95°C for 15 s, 61°C for 5 s, 72°C for 5 s) × 55	TAMRA probe: 5.8S_PR: FAM-AGAACGTGGCAAAATGCGATAAGTAGTGTGA
Bandt and Monecke, 2007[38]	dhfr	Dhfr2.F: GGCTGATCAAAGAAGCATGGATA Dhfr2_R: CGGCATAGACATATTCGATACTTGTT	10 pmol each primer 4 pmol probe Template/reaction vol: 5/20 μL	(95°C for 15 s, 61°C for 5 s, 72°C for 5 s) × 55	TAMRA probe: Dhfr2_PR: FAM-TGCGTGAAACAGATACATGGAGCTCTACCC
Huggett et al., 2008[60]	*Pj*HSP70a	Fwd: CGTCTTGTAAACCACTTCATTGC Rev: AGTCCGTTTAGCACGCTCAC	0.3 μM each primer 0.075 μM probe Template/reaction vol: NR/12.5 μL	(95°C for 10 s, 72°C for 20 s, 72°C for 30 s) × 45	BHQ1 probe: FAM-AAGAAAGATCTTTCAGGG

[a] NR, not reported; mtLSUrRNA, large subunit of mitochondrial rRNA; ITS, internal transcribed spacer; MSG, major surface glycoprotein; dhfr, dihydrofolate reductase gene; dhps, dihydroopteroate synthase gene; *Pj*HSP70, *P. jirovecii* heat-shock protein 70 gene; FRET, fluorescence resonance energy transfer; FAM and VIC, fluorescent dyes; consultation of the referenced original manuscript is recommended before application of any assays.

ACKNOWLEDGMENTS

We thank Robert Miller, Joe Kovacs, Peter Walzer, C. Ben Beard, and Laurence Huang for their helpful reviews of a draft of this chapter. We thank Melissa Miller and Nacy Henshaw for the images of *P. jirovecii* in stained clinical specimens.

REFERENCES

1. Chagas, C., Nova tripanozomiaze humana. *Mem Inst Oswaldo Cruz* 1, 159–218 (1909).
2. Delanoe, P. and Delanoe, M., Sur les rapports des kystes de Carini du poumon des rats avec le Trypanosoma lewisi. *C R Acad Sci Gen* 155, 658–660 (1912).
3. Van der Meer, G. and Brug, S.L., Infection a *Pneumocystis* chez l'homme et chez les animaux. *Ann Soc Belg Med Trop* 22, 301–307 (1942).
4. Walzer, P.D. et al., *Pneumocystis carinii* pneumonia in the United States: Epidemiologic, diagnostic, and clinical features. *Ann Intern Med* 80, 83–93 (1974).
5. *Pneumocystis* pneumonia—Los Angeles. *MMWR*, 30, 250–252 (1981).
6. Gottlieb, M.S. et al., *Pneumocystis carinii* pneumonia and mucosal candidiasis in previously healthy homosexual men: Evidence of a new acquired cellular immunodeficiency. *N Engl J Med* 305, 1425–1431 (1981).
7. Morris, A. et al., Current epidemiology of *Pneumocystis* pneumonia. *Emerg Infect Dis* 10, 1713–1720 (2004).
8. Edman, J.C. et al., Ribosomal RNA sequence shows *Pneumocystis carinii* to be a member of the fungi. *Nature* 334, 519–522 (1988).
9. Stringer, J.R., *Pneumocystis carinii*: What is it, exactly? *Clin Microbiol Rev* 9, 489–498 (1996).
10. Stringer, J. et al., A new name (*Pneumocystis jiroveci*) for *Pneumocystis* from humans. *Emerg Infect Dis* 8, 891–896 (2002).
11. Huang, L. et al., An official ATS workshop summary: Recent advances and future directions in *Pneumocystis* pneumonia (PCP). *Proc Am Thorac Soc* 3, 655–664 (2006).
12. Thomas, C.F. Jr. and Limper, A.H., Current insights into the biology and pathogenesis of Pneumocystis pneumonia. *Nat Rev Microbiol* 5, 298–308 (2007).
13. Thomas, C.F. and Limper, A.H., *Pneumocystis* pneumonia. *N Engl J Med* 350, 2487–2498 (2004).
14. Kovacs, J.A. et al., *Pneumocystis carinii* pneumonia: A comparison between patients with the acquired immunodeficiency syndrome and patients with other immunodeficiencies. *Ann Intern Med* 100, 663–671 (1984).
15. Sepkowitz, K.A., Opportunistic infections in patients with and patients without acquired immunodeficiency syndrome. *Clin Infect Dis* 34, 1098–1107 (2002).
16. Fei, M.W. et al., Predicting mortality from HIV-associated *Pneumocystis* pneumonia at illness presentation: An observational cohort study. *Thorax* 64, 1070–1076 (2009).
17. Walzer, P.D. et al., Early predictors of mortality from *Pneumocystis jirovecii* pneumonia in HIV-infected patients: 1985–2006. *Clin Infect Dis* 46, 625–633 (2008).
18. Mansharamani, N.G. et al., Management and outcome patterns for adult *Pneumocystis carinii* pneumonia, 1985 to 1995: Comparison of HIV-associated cases to other immunocompromised states. *Chest* 118, 704–711 (2000).
19. Ng, V.L., Yajko, D.M., and Hadley, W.K., Extrapulmonary pneumocystosis. *Clin Microbiol Rev* 10, 401–418 (1997).
20. Mandell, G., Bennett, J., and Dolin, R., *Principles and Practice of Infectious Diseases*, Churchill Livingstone, New York, 2004.
21. Vargas, S.L. et al., Search for primary infection by *Pneumocystis carinii* in a cohort of normal, healthy infants. *Clin Infect Dis* 32, 855–861 (2001).
22. Morris, A., et al., Epidemiology and clinical significance of *Pneumocystis* colonization. *J Infect Dis* 197, 10–17 (2008).
23. Kovacs, J.A. and Masur, H., Evolving health effects of *Pneumocystis*: One hundred years of progress in diagnosis and treatment. *JAMA* 301, 2578–2585 (2009).
24. Skelly, M.J., Holzman, R.S., and Merali, S., S-adenosylmethionine levels in the diagnosis of *Pneumocystis carinii* pneumonia in patients with HIV infection. *Clin Infect Dis* 46, 467–471 (2008).
25. Kutty, G. et al., *Pneumocystis* encodes a functional S-adenosylmethionine synthetase gene. *Eukaryot Cell* 7, 258–267 (2008).
26. Watanabe, T. et al., Serum (1→3) beta-D-glucan as a noninvasive adjunct marker for the diagnosis of *Pneumocystis* pneumonia in patients with AIDS. *Clin Infect Dis* 49, 1128–1131 (2009).
27. Shelhamer, J.H. et al., The laboratory evaluation of opportunistic pulmonary infections. *Ann Intern Med* 124, 585–599 (1996).
28. Kovacs, J.A. et al., Diagnosis of *Pneumocystis carinii* pneumonia: Improved detection in sputum with use of monoclonal antibodies. *N Engl J Med* 318, 589–593 (1988).
29. Armbruster, C., Pokieser, L., and Hassl, A., Diagnosis of *Pneumocystis carinii* pneumonia by bronchoalveolar lavage in AIDS patients. Comparison of Diff-Quik, fungifluor stain, direct immunofluorescence test and polymerase chain reaction. *Acta Cytol* 39, 1089–1093 (1995).
30. Wakefield, A.E. et al., Detection of *Pneumocystis carinii* with DNA amplification. *Lancet* 336, 451–453 (1990).
31. Lu, J.J. et al., Typing of *Pneumocystis carinii* strains that infect humans based on nucleotide sequence variations of internal transcribed spacers of rRNA genes. *J Clin Microbiol* 32, 2904–2912 (1994).
32. Lipschik, G.Y. et al., Improved diagnosis of *Pneumocystis carinii* infection by polymerase chain reaction on induced sputum and blood. *Lancet* 340, 203–206 (1992).
33. Larsen, H.H. et al., Development of a rapid real-time PCR assay for quantitation of *Pneumocystis carinii* f. sp. *carinii*. *J Clin Microbiol* 40, 2989–2993 (2002).
34. Brancart, F.O. et al., Quantitative TaqMan PCR for detection of *Pneumocystis jiroveci*. *J Microbiol Meth* 61, 381–387 (2005).
35. Alvarez-Martínez, M. et al., Sensitivity and specificity of nested and real-time PCR for the detection of *Pneumocystis jiroveci* in clinical specimens. *Diagn Microbiol Infect Dis* 56, 153–160 (2006).
36. Arcenas, R. et al., A real-time polymerase chain reaction assay for detection of *Pneumocystis* from bronchoalveolar lavage fluid. *Diagn Microbiol Infect Dis* 54, 169–175 (2006).
37. Palladino, S. et al., Use of real-time PCR and the LightCycler system for the rapid detection of *Pneumocystis carinii* in respiratory specimens. *Diagn Microbiol Infect Dis* 39, 233–236 (2001).
38. Bandt, D. and Monecke, S., Development and evaluation of a real-time PCR assay for detection of *Pneumocystis jiroveci*. *Transplant Infect Dis* 9, 196–202 (2007).
39. Maher, N.H. et al., Development and characterization of a molecular viability assay for *Pneumocystis carinii* f sp *hominis*. *J Infect Dis* 183, 1825–1827 (2001).

40. de Oliveira, A. et al., Performance of a molecular viability assay for the diagnosis of *Pneumocystis* pneumonia in HIV-infected patients. *Diagn Microbiol Infect Dis* 57, 169–176 (2007).
41. Cartwright, C.P., Nelson, N.A., and Gill, V.J., Development and evaluation of a rapid and simple procedure for detection of *Pneumocystis carinii* by PCR. *J Clin Microbiol* 32, 1634–1638 (1994).
42. Ortona, E. et al., Detection of *Pneumocystis carinii* in respiratory specimens by PCR-solution hybridization enzyme-linked immunoassay. *J Clin Microbiol* 35, 1589–1591 (1997).
43. Uemura, N. et al. Development of a loop-mediated isothermal amplification method for diagnosing *Pneumocystis* pneumonia. *J Med Microbiol* 57, 50–57 (2008).
44. Torres, J. et al., Diagnosis of *Pneumocystis carinii* pneumonia in human immunodeficiency virus-infected patients with polymerase chain reaction: A blinded comparison to standard methods. *Clin Infect Dis* 30, 141–145 (2000).
45. Kitada, K. et al., Detection of *Pneumocystis carinii* sequences by polymerase chain reaction: Animal models and clinical application to noninvasive specimens. *J Clin Microbiol* 29, 1985–1990 (1991).
46. Olsson, M. et al., Detection of *Pneumocystis carinii* DNA in sputum and bronchoalveolar lavage samples by polymerase chain reaction. *J Clin Microbiol* 31, 221–226 (1993).
47. Huang, S.N. et al., Development of a PCR assay for diagnosis of *Pneumocystis carinii* pneumonia based on amplification of the multicopy major surface glycoprotein gene family. *Diagn Microbiol Infect Dis* 35, 27–32 (1999).
48. Larsen, H.H. et al., Development and evaluation of a quantitative, touch-down, real-time PCR assay for diagnosing *Pneumocystis carinii* pneumonia. *J Clin Microbiol* 40, 490–494 (2002).
49. Lu, J.J. et al., Comparison of six different PCR methods for detection of *Pneumocystis carinii. J Clin Microbiol* 33, 2785–2788 (1995).
50. Weig, M. et al., Correlation of *Pneumocystis carinii* PCR with clinical diagnosis in immunocompromised patients. *Lancet* 347, 1266 (1996).
51. Gupta, R. et al., Diagnostic significance of nested polymerase chain reaction for sensitive detection of *Pneumocystis jirovecii* in respiratory clinical specimens. *Diagn Microbiol Infect Dis* 64, 381–388 (2009).
52. Helweg-Larsen, J. et al., Diagnostic use of PCR for detection of *Pneumocystis carinii* in oral wash samples. *J Clin Microbiol* 36, 2068–2072 (1998).
53. Schluger, N. et al., Application of DNA amplification to pneumocystosis: Presence of serum *Pneumocystis carinii* DNA during human and experimentally induced *Pneumocystis carinii* pneumonia. *J Exp Med* 176, 1327–1333 (1992).
54. Atzori, C. et al., Diagnosis of *Pneumocystis carinii* pneumonia in AIDS patients by using polymerase chain reactions on serum specimens. *J Infect Dis* 172, 1623–1626 (1995).
55. Ortona, E. et al., Non specific PCR products using rat-derived *Pneumocystis carinii* dihydrofolate reductase gene-specific primers in DNA amplification of human respiratory samples. *Mol Cell Probes* 10, 187–190 (1996).
56. Ma, L. et al., *Pneumocystis carinii* dihydropteroate synthase but not dihydrofolate reductase gene mutations correlate with prior trimethoprim-sulfamethoxazole or dapsone use. *J Infect Dis* 180, 1969–1978 (1999).
57. Fischer, S. et al., The use of oral washes to diagnose *Pneumocystis carinii* pneumonia: A blinded prospective study using a polymerase chain reaction-based detection system. *J Infect Dis* 184, 1485–1488 (2001).
58. Fillaux, J. et al., Accuracy of a routine real-time PCR assay for the diagnosis of *Pneumocystis jirovecii* pneumonia. *J Microbiol Meth* 75, 258–261 (2008).
59. Larsen, H.H. et al., A prospective, blinded study of quantitative touch-down polymerase chain reaction using oral-wash samples for diagnosis of *Pneumocystis* pneumonia in HIV-infected patients. *J Infect Dis* 189, 1679–1683 (2004).
60. Huggett, J.F. et al., Development and evaluation of a real-time PCR assay for detection of *Pneumocystis jirovecii* DNA in bronchoalveolar lavage fluid of HIV-infected patients. *Thorax* 63, 154–159 (2008).
61. Flori, P. et al., Comparison between real-time PCR, conventional PCR and different staining techniques for diagnosing *Pneumocystis jiroveci* pneumonia from bronchoalveolar lavage specimens. *J Clin Microbiol* 53, 603–607 (2004).
62. Linssen, C. et al., Inter-laboratory comparison of three different real-time PCR assays for the detection of *Pneumocystis jiroveci* in bronchoalveolar lavage fluid samples. *J Med Microbiol* 55, 1229–1235 (2006).
63. Robberts, F.J., Liebowitz, L.D., and Chalkley, L.J., Polymerase chain reaction detection of *Pneumocystis jiroveci*: Evaluation of 9 assays. *Diagn Microbiol Infect Dis* 58, 385–392 (2007).
64. Lane, B.R. et al., Dihydropteroate synthase polymorphisms in *Pneumocystis carinii. J Infect Dis* 175, 482–485 (1997).
65. Takahashi, T. et al., Relationship between mutations in dihydropteroate synthase of *Pneumocystis carinii* f. sp. *hominis* isolates in Japan and resistance to sulfonamide therapy. *J Clin Microbiol* 38, 3161–3164 (2000).
66. Wakefield, A.E., DNA sequences identical to *Pneumocystis carinii* f. sp. *carinii* and *Pneumocystis carinii* f. sp. *hominis* in samples of air spora. *J Clin Microbiol* 34, 1754–1759 (1996).
67. Mathis, A. et al., Simplified sample processing combined with a sensitive one-tube nested PCR assay for detection of *Pneumocystis carinii* in respiratory specimens. *J Clin Microbiol* 35, 1691–1695 (1997).
68. Tamburrini, E. et al., Detection of *Pneumocystis carinii* DNA in blood by PCR is not of value for diagnosis of *P. carinii* pneumonia. *J Clin Microbiol* 34, 1586–1588 (1996).
69. Rabodonirina, M. et al., Detection of *Pneumocystis carinii* DNA in blood specimens from human immunodeficiency virus-infected patients by nested PCR. *J Clin Microbiol* 37, 127–131 (1999).
70. Nyamande, K. et al., Low sensitivity of a nested polymerase chain reaction in oropharyngeal washings for the diagnosis of pneumocystis pneumonia in HIV-infected patients. *Chest* 128, 167–171 (2005).
71. Azoulay, E. et al., Polymerase chain reaction for diagnosing *Pneumocystis* pneumonia in non-HIV immunocompromised patients with pulmonary infiltrates. *Chest* 135, 655–661 (2009).
72. Weig, M. et al., Usefulness of PCR for diagnosis of *Pneumocystis carinii* pneumonia in different patient groups. *J Clin Microbiol* 35, 1445–1449 (1997).
73. Sing, A. et al., Evaluation of diagnostic value and epidemiological implications of PCR for *Pneumocystis carinii* in different immunosuppressed and immunocompetent patient groups. *J Clin Microbiol* 38, 1461–1467 (2000).
74. Fujisawa, T. et al., Real-time PCR is more specific than conventional PCR for induced sputum diagnosis of *Pneumocystis* pneumonia in immunocompromised patients without HIV infection. *Respirology* 14, 203–209 (2009).
75. De la Horra, C. et al., Comparison of single and touchdown PCR protocols for detecting *Pneumocystis jirovecii* DNA in paraffin-embedded lung tissue samples. *J Eukaryot Microbiol* 53, S98 (2006).

76. Lee, C.H. et al., Amplification of *Pneumocystis carinii* DNA on specimens scraped from slides. *Diagn Microbiol Infect Dis* 18, 197–199 (1994).
77. Lee, C.H. et al., Update on *Pneumocystis carinii* f. sp. *hominis* typing based on nucleotide sequence variations in internal transcribed spacer regions of rRNA genes. *J Clin Microbiol* 36, 734–741 (1998).
78. Robberts, F.J., Liebowitz, L.D., and Chalkley, L.J., Genotyping and coalescent phylogenetic analysis of *Pneumocystis jiroveci* from South Africa. *J Clin Microbiol* 42, 1505–1510 (2004).
79. Nimri, L.F. et al., Genetic diversity of *Pneumocystis carinii* f. sp. *hominis* based on variations in nucleotide sequences of internal transcribed spacers of rRNA genes. *J Clin Microbiol* 40, 1146–1151 (2002).
80. Beser, J., Hagblom, P., and Fernandez, V., Frequent in vitro recombination in internal transcribed spacers 1 and 2 during genotyping of *Pneumocystis jirovecii. J Clin Microbiol* 45, 881–886 (2007).
81. Huang, L. et al., Dihydropteroate synthase gene mutations in *Pneumocystis* and sulfa resistance. *Emerg Infect Dis* 10, 1721–1728 (2004).
82. Kaiser, K., Rabodonirina, M., and Picot, S., Real time quantitative PCR and RT-PCR for analysis of *Pneumocystis carinii* hominis. *J Microbiol Methods* 45, 113–118 (2001).

70 *Saccharomyces*

Franca Rossi and Sandra Torriani

CONTENTS

70.1 INTRODUCTION

70.1.1 CLASSIFICATION, MORPHOLOGY, AND BIOLOGY

70.1.1.1 Classification

Saccharomyces cerevisiae belongs to the fungal phylum Ascomycota, lineage Hemiascomycetes, subphylum Saccharomycotina, order Saccharomycetales.[1,2] The latter is composed of three clades among which the *Saccharomyces* complex, constituted by the genera *Saccharomyces* and *Kluyveromyces*, is the most densely sequenced. The phylogeny of this group was defined by multi-gene sequence comparison[1] and by comparative hybridization of genomic DNA on *S. cerevisiae* cDNA microarrays.[3] *S. cerevisiae* and the sibling species *S. bayanus*, *S. cariocanus*, *S. kudriavzevii*, *S. mikatae*, *S. paradoxus*, and *S. pastorianus*, which is considered a hybrid of *S. cerevisiae* and *S. bayanus*, constitute the *Saccharomyces* sensu stricto clade. Adaptation of the *Saccharomyces* sensu stricto species to low-oxygen environments was driven by the duplication of genes involved in the "fermentative lifestyle," alcohol dehydrogenase (ADH) and pyruvate decarboxylase (PDC) encoding alcohol dehydrogenase and pyruvate decarboxylase, respectively, following a whole genome duplication event that occurred in a common ancestor of the subgroup "*Saccharomyces* complex."[4]

S. cerevisiae is used in the food industry worldwide as a principal agent of fermentation in naturally leavened bakery products, alcoholic beverages like wines, beers, fermented juices, and distillates. Strains of *S. cerevisiae*, thanks to their high level of resistance to ethanol, anaerobic conditions, low pH, and osmotic stress, are essentially the only organisms remaining alive at the end of wine fermentation. This yeast can be found in nature on various plants and in the soil. In studies aimed at defining the source from which it contaminates grapes in the vineyards and the winery environment, it has been isolated from the bodies of insects, most frequently honey bees, but also wasps and *Drosophila*. These are supposed to be the carriers that transfer the yeast to the interior of damaged fruits. In a phylogenetic study based on diversity at 5 loci in a sequence of 7 kb, 184 polymorphic sites were identified; nucleotide variation in 81 isolates originating from natural and artificial fermentations, tree exudates, and immunocompromised patients from Europe, Africa, Southeast Asia, and North America supported the hypothesis that "domesticated" strains are derived from natural populations. The phylogenetic tree suggested that lineages at the root were derived from tree exudates in North America and Africa. The comparison of oak-associated and vineyard-associated yeasts suggested that both habitat and geographical distance have a role in determining genetic diversity.[5]

S. cerevisiae has been isolated from humans and shown to be present in the digestive tract, vagina, skin, and oropharynx of healthy hosts; however, it has not yet been clearly demonstrated whether this is a true commensal species or its colonization is transient and related to food consumption.

Due to high concentration of the B vitamins, minerals, and cofactors, this yeast is used as a nutritional supplement. The *S. cerevisiae* variant "*S. boulardii*" (*S. cerevisiae* Hansen CBS 5926[6,7] patent strain held in the American Type Culture Collection [ATCC]), first isolated from litchi fruit in Indochina and once considered a separate species, is used worldwide as a biotherapeutic agent. *S. boulardii* has been used in Europe as a probiotic preparation for oral administration under different commercial names (Ultra-Levure, Precosa®, Codex, etc.). *S. boulardii* is also used as a feed additive, an example is Levucell SB®, and as a food supplement. This yeast is highly effective for the treatment of several types of acute diarrhea and colitis by exerting multiple beneficial activities like the inactivation of enteropathogenic bacterial toxins by proteolytic enzymes, trophic effects on the small intestine mucosa mediated by the release of polyamines in the endoluminal compartment, improvement of the absorption of glucose by secretion of high amounts of sucrases, enhancement of antibody response specific for bacterial pathogens, and attenuation of inflammatory symptoms by different mechanisms.[8,9]

Clinical strains and commercial strains used in the food industry could not be discriminated sharply by genetic, phenotypic, or metabolic profiling because of the existence of food-associated strains that share high similarity with representatives of the other group and the high diversity within the two groups.[10–12]

The entire genomes of three *S. cerevisiae* strains, i.e., those of the laboratory strain S288c originating from a rotting fig,[13] of the clinical strain YJM789,[14] and of the wine strain RM11-1a (http://www.yeastgenome.org), are available to date.

70.1.1.2 Morphology

Cells of *S. cerevisiae* are spheroidal, ovoidal, or elongate. Pseudohyphae may be formed but true hyphae are not produced. This yeast can stably propagate by budding as haploid or diploid cells. Vegetative cell division characteristically occurs by axial or bipolar budding in which a daughter cell is initiated as an outgrowth from the mother cell. After cell separation a scar is left in the budding region on the mother cell surface. Haploid cells mate in response to pheromones. Diploid cells undergo meiosis and sporulation in condition of complete starvation, i.e., absence of fermentable carbon and nitrogen sources. Asci are unconjugated, persistent, and contain one–four spherical to ovoidal ascospores.

The variety of morphological phenotypes is a consequence of adaptive responses to environmental signals. While it is mainly known as a unicellular species with ellipsoid cells that divide by multilateral budding and form smooth colonies on solid medium, many *S. cerevisiae* strains are dimorphic and can form complex colonies or mats constituted by multicellular pseudohyphae by modifying their budding pattern from multi- to unipolar. This process, observed both in haploid and diploid cells during nitrogen starvation and in presence of carbon sources other than glucose,[15] is known as "pseudofilamentation" or pseudohyphal growth (PHG).

Other phenotypes include the formation of biofilms[16] and complex fluffy-structured colonies with cells connected by extracellular glycosylated matrix material.[17] When in suspension, some industrial strains can form specialized biofilms called flors at the interface liquid–air, which are necessary for particular wine-making processes.[18] The ability to flocculate is also a strain-dependent character.[14,19]

For the different morphologies observed within the *S. cerevisiae* species, like PHG, substrate invasion, flor formation, flocculation, and biofilms on plastic surfaces, the adhesin Flo11p plays a key role. This is a 196 kDa mannoprotein that functions as target for adhesion on the yeast cell wall.[18] It is also implicated in cell adherence to surfaces by conferring hydrophobicity.[16]

Pseudohyphal filaments consist of elongated undetached cells that result in a colony morphology presenting a central body from which numerous branched spokes depart. The elongated cells form filaments in the agar beneath and, in some strains, beyond the colony perimeter. McCusker et al.[20] observed that many colonies of clinical and nonclinical strains grown on casein formed pseudohyphae (Psh+ phenotype); however, pseudohyphal formation was seen more frequently and tended to be more extensive (longer pseudohyphal chains and more pseudohyphal chains emanating from a colony) for clinical isolates. A similar observation was reported in a study carried out with 118 clinical and industrial *S. cerevisiae* strains. The exceptions among the industrial strains were *S. boulardii*, two baker's yeasts, and one wine strain, Fermivin Crio.[21] A screening of 1026 strains isolated from Italian vineyards showed that roughly half were able to show invasive filamentous growth when starved for nitrogen. Also, 2.5% of the strains had a filigreed ("rough") or fluffy colony phenotype. Furthermore, several strains also showed differences in the treatments that induce modification of growth morphology, indicating the presence of genetic variation in response to stimuli that can produce filamentous growth.[17]

70.1.1.3 Biology

Life cycle, cell structure assembly, metabolic pathways, gene function, and expression regulation of *S. cerevisiae* are the most well known among eukaryotes since this organism constituted for decades a model system for its relatively small and compact genome, comprising 16 chromosomes of sizes 200–2200 kb in the haploid stage, and ease of genetic transformation and manipulation. Knowledge about the physiology of this yeast has increased after whole genome sequencing and annotation for *S. cerevisiae* S288c. In this strain, about 6000 genes are present of which 3.8% have introns.[13] Chromosomes contain movable retrotransposons, subdivided in the types Ty1–Ty5. These are 5.3–5.7 kb in size and are constituted by retroviral DNA flanked by ~330 bp terminal repeats long terminal repeats (LTRs) or δ sequences. The Ty transposable elements are involved in insertion/deletion, or "indel," events. The clinical strain sequenced, *S. cerevisiae* YJM789, is the haploid isoform of YJM145, a yeast isolated from the lung of an AIDS patient with pneumonia, which causes death in complement-deficient mice[19] and differs in phenotype from S288c in being flocculant, displaying colony morphology switching, and growing at high temperatures.[20] Its genome sequence is studied for correlating genetic traits to the ability of causing infections. Several open reading frames (ORFs) are found to be unique to YJM789, some of which might have been acquired through horizontal transfer. From genome comparisons between *S. cerevisiae* S288c and this strain, it emerged that single nucleotide polymorphism (SNPs) are the primary cause of phenotypic variability. The remaining variability is generated by insertion/deletions and consequent difference in gene copy number, with the subtelomeric regions rich in redundant sequences as preferential locations for nonhomologous recombination.[14] Moreover, several genes encoding cell surface proteins (such as *TIR1*, *HSP150*, *FIT1*, *AGA1*, *MNN4*, and *FLO10*), bearing intragenic tandem repeats, have different frame repeat numbers between S288c and YJM789. Their variation in tandem repeats may generate functional cell surface variability, that contributes to adaptation to the environment and possible host immune evasion.[22] While strain S288c has 50 Ty elements of all classes, YJM789 genome contains only 17 Ty elements, all of which are Ty1, Ty2, and Ty5.[14]

Other nucleic acid types are mitochondrial DNA (mtDNA), which encodes components of the mitochondrial translational machinery, approximately 15% of the mitochondrial proteins and the 2 μm autoreplicative circle plasmids. Almost all *S. cerevisiae* strains contain three families of viral dsRNA that confer the capacity to produce a killer toxin active against other yeasts and are transmitted when packaged in viral particles. *Saccharomyces cerevisiae* also contains a 20S circular single-stranded RNA that acts as an independent replicon and is inherited as a non-Mendelian genetic element.

Due to the gene asset permitting efficient growth in anaerobic conditions, *S. cerevisiae* has the ability to consume glucose preferentially via fermentation rather than respiration, to grow in total absence of oxygen and to be induced to form respiratory-deficient mitochondrial mutants ("petites").[23] Fermentation is such a rapid process in *S. cerevisiae* that optimization of glycolysis allows synthesizing a similar number of ATP molecules per second as in the aerobic metabolism.[24] *S. cerevisiae* tolerates high levels of ethanol, which may be utilized as a source of energy once that glucose is depleted.[25]

The life cycle of *S. cerevisiae* alternates between diplophase and haplophase. Clonal reproduction is predominant. In heterothallic strains, haploid cells are of two mating types, **a** and α. Mating of **a** and α cells results in **a**/α diploids that undergo meiosis upon starvation. The four haploid spores resulting from meiosis are contained within the wall of the mother cell (the ascus). The spores of about 70% of natural isolates are able to switch mating type; these are referred to as homothallic strains. The wild-type allele of the *HO* gene for homothallism encodes a site-specific endonuclease that promotes mating type switching from **a** to α or vice versa. *HO* is transcribed only in the mother cell, so that when a homothallic spore divides, the mother cell switches mating type and immediately fuses with the newly produced daughter cell.[26] A positive correlation exists between growth rate and the transcription of genes of both cytosolic and mitochondrial translation machinery. Genes related to functions associated with oxidative energy metabolism, especially those involving peroxisomes, i.e., fatty acid metabolism, oxygen metabolism, and autophagy, are negatively correlated with growth rate. The hypothesis is that it serves to protect the cell from the longer exposure to oxidative metabolism necessary to produce enough energy at lower growth rates.[27] In *S. cerevisiae*, PHG is regulated by at least two signaling pathways: (1) the nutrient-sensing cyclic AMP–protein kinase A (PKA) pathway and (2) the mitogen-activated protein kinase (MAPK) pathway. Recently, based on DNA-microarray analysis, a model of yeast PHG regulation by autophagy was also derived that illustrates how this recycling process plays a critical role in response of PHG to nitrogen starvation in a gradual manner by mitigating the effects of low nitrogen availability.[28] Pseudohyphal differentiation in *S. cerevisiae* is regulated by the cell surface G protein-coupled receptor (GPCR) *GPR1* that receives extracellular signal from glucose by the Ras signaling system and by the *MEP2* encoded ammonium permease. Ras1/2, partially redundant small GTP-binding proteins, activates the MAPK pathway to cause pseudohyphal differentiation.[29] The gene *MEP2*, encoding an ammonium transporter, has been implicated in transducing the signal in response to low ammonium levels and is subject to nitrogen catabolite repression (NCR). A strain defective in *MEP2* is unable to form pseudohyphae. Recent studies suggest that *MEP2* plays a role in pseudohyphal differentiation through the reuptake of secreted ammonium under nitrogen starvation conditions.[30]

The regulation of the gene *FLO11*, encoding Flo11p, appears to be critically dependent on the relative levels of glucose and ammonium.[29] One known example of genetic polymorphism accounting for a different capability of PHG is the *FLO8* gene, a transcription factor required for pseudohyphal formation, which in strain S288c contains an inactivating amber mutation, yet *S. cerevisiae* YJM145 forms abundant pseudohyphae[20] and its *FLO8* ORF has no amber mutation. Another example of inactivating polymorphism with a consequence on morphology found in the S288c involves the *AMN1* gene. The same gene in *S. cerevisiae* YJM789 contains an SNP as that confers a more clumpy phenotype to this strain.[14]

70.1.2 Clinical Features, Pathogenesis, and Epidemiology

70.1.2.1 Clinical Features

Though it is commonly perceived as a harmless microorganism given its thousand-year association with fermented food and beverage production, numerous case reports of this yeast exist in the medical literature. Local infections caused by *S. cerevisiae* may occur as a consequence of mechanical factors in mildly immunocompromised or immunocompetent persons.[31,32] *Saccharomyces cerevisiae* is responsible of vaginitis in a minority of cases with yeast etiology.[33] Medical literature reports described cases of vaginal infections directly caused by *S. cerevisiae* strains used in domestic or professional bread-making.[34]

Serious gastrointestinal or systemic infections have been observed in immunosuppressed patients. The portal of entry for invasive infections is mainly digestive: a case described in an immunocompetent patient was due to the ingestion of health food containing viable yeasts. Symptoms were recurrent fever, malaise with nausea, and night sweats. Yeasts were isolated from bone marrow and urine specimens. The patient recovered when he stopped ingesting health food containing yeasts.[35] In some instances, the *S. cerevisiae* strains responsible for infection were winery-resident strains or bakery strains.[31,34]

S. cerevisiae is currently considered an emerging pathogen in part as an effect of the refinement of yeast identification methods, but also for new patient predisposing factors as a consequence of the increase of the population immunocompromised for underlying chronic or debilitating diseases, increased use of immunosuppressive drugs or broad-spectrum antibiotics, as well as parenteral nutrition, use of intravascular catheters, and, finally, the increasing numbers of HIV-infected individuals. Infections with lethal outcome were observed in patients with critical conditions such as AIDS, neoplastic and metabolic diseases, intestinal surgery.

Invasive *Saccharomyces* infections showed a significantly increased incidence since the 1990s[35] and a recent 15 month survey of fungal bloodstream infections in 54 Belgian hospitals reported a threefold increase in the frequency of *S. cerevisiae*.[36] Of 21 isolates, 19 gave the profile of *S. boulardii*. In one patient with multiple yeast infections, *S. cerevisiae* was associated with *C. albicans*. Concomitant isolation of *S. cerevisiae* and a second microorganism, mainly enterobacteria, was reported in some cases of fungemia. In single organ infections, the main sites were the lungs and the heart valves.[35] From these evidences, it is clear that *S. cerevisiae* should be considered as an opportunistic pathogen of low virulence rather than a nonpathogenic yeast.

70.1.2.2 Pathogenesis

Human infections caused by *S. cerevisiae* occurred with a broad range of clinical manifestations, from fungemia, pneumonia, and deeply invasive dissemination to localized infections in specific organs or body districts. Virulent *S. cerevisiae* possess the capability to colonize and persist in various normally sterile body sites (e.g., the lung and blood) and, consequently, is able to survive and disseminate in vivo. However, it has been observed that clinical and nonclinical strains are poorly capable of adhering to intestinal cell monolayers.[37] Invasive *Saccharomyces* infection is clinically indistinguishable from invasive candidiasis, notably because chorioretinitis (fluffy yellow exudates) and esophagitis (yellow-white plaques on an erythematous background) can be present in both conditions, and is even indistinguishable from infections due to other microorganisms that cause endocarditis.[35] Fever is not always present. Histological documentation of deep-site involvement demonstrated the association of *S. cerevisiae* with necrosis and granulomatous reaction.[38]

In two studies of in vivo pathogenesis in a mouse model, different isolates of *S. cerevisiae* displayed a continuum of virulence in mice, defined as the capacity for the organism to proliferate in the brain and/or persist for an extended period of time.[39] After intravenous inoculation, most clinical strains tested were able to proliferate principally in the brain, but also in kidneys, spleen, liver, and lungs, whereas nonclinical isolates were readily cleared. Histological inspection of brains infected with a virulent clinical isolate and an avirulent strain showed that the former was more able to invade the parenchyma while the latter was mostly localized in the capillaries.[19]

The ability of *S. cerevisiae* strains to cause infection depends on the possession of fungal properties frequently associated with pathogenesis, i.e., the ability to grow at high temperatures (38°C–42°C), the ability to invade host cells, morphological versatility, and the ability to produce and secrete degradative enzymes (proteinases and phospholipases), which could play a role in blocking capillaries, a fungal characteristic associated with animal mortality.[40]

The ability to grow at 37°C showed no significant association with virulence, instead growth at 39°C was significantly associated with virulence. The capacity to grow at 42°C is tightly associated with virulence[39] and was found almost exclusively in clinical strains: In a screening of 118 *S. cerevisiae* strains, the ability to grow at 42°C was found mainly in clinical isolates and in very few industrial strains, namely, *S. boulardii*, the wine strain Uvaferm 71B, and two baker's yeasts.[21] The high-temperature growth phenotype of YJM789, in particular, has been dissected to an SNP resolution for several local regions of divergence.[14] A gene implicated in the capacity to grow at 42°C is *SSD1*, which encodes a protein able to affect various cellular processes including maintenance of cell integrity, and seems to play an important role in the virulence of *Saccharomyces* organisms. The encoded protein, Ssd1, regulates the biological activity of Hsp104. This chaperone mediates protein disaggregation after heat stress, a process that permits the recovery of cell viability responding to signals from cellular integrity pathways. The possible involvement of this protein in cell properties bound to pathogenesis is indicated by evidence that deletion of its genetic determinant *SSD1* led to a significant increase in virulence for both clinical and plant isolates, as revealed in a mouse model of invasive infection, likely for the alteration in composition and cell wall architecture that overstimulates the pro-inflammatory response.[41]

Regarding the production of extracellular lytic enzymes possibly implied in virulence, it was found that generally the potentially virulent strains produce higher amounts of phospholipases compared to industrial strains.[21]

A microbial virulence-related trait that promotes host colonization and invasion is the ability to survive oxidative stress, exerted by reactive oxygen species (ROS), a component of mammalian host defense. Among 103 *S. cerevisiae* strains, isolated from clinical and other sources and tested for survival oxidative stress exerted by tert-butyl hydroperoxide, a stable organic analog of H_2O_2, clinical isolates, some soil isolates, and strains from insect guts exhibited the highest mean resistance to ROS expressed as survival rates.[42] The genetic diversity among the clinical isolates determined using five genetic loci suggested multiple origins of ROS resistance. In a pathogenesis assay with *Caenorhabditis elegans*, an established model host, *S. cerevisiae* infected the nematodes causing disease and death.[43] Data indicated that the host produced reactive oxygen species (ROS) in response to fungal infection. Yeast deletion mutants of genes *sod*1 and *yap*1, could not withstand ROS and failed to cause disease, thus, indicating that these genes can be considered virulence markers.

S. cerevisiae can give rise to the initial steps of biofilm formation by adhering on plastic materials.[16] This can explain why use of intravascular or other types of catheters was among the causes of increase of *S. cerevisiae* infections in nosocomial environments. In support of the possible implication of biofilm formation in infection transmission via catheters is the prompt resolution of symptoms when these were removed[44] and the persistence of the infection when these were left in place.[45,46]

70.1.2.3 Epidemiology

For the establishment of the epidemiological trend, virulence levels, and mechanisms, 92 documented cases of proven invasive *S. cerevisiae* infection satisfying the definitions of the Invasive Fungal Infection Cooperative Group of the European Organization for Research and Treatment of Cancer-Mycosis Study Group of the National Institute of Allergy and Infectious Diseases were taken into account. These definitions imply fungemia, isolation from normally sterile fluids (e.g., pleural and synovial fluids), and deep-site infections either proven by histological examination for sites susceptible to colonization (lungs, peritoneum, esophagus). Fifteen cases were diagnosed before 1990 and 76 after 1990.[35] All patients had at least one condition facilitating the development of invasive fungal infection; the most frequent were an intravenous catheter and a previous receipt of antibiotic therapy. *S. boulardii* was considered to be the etiological agent in 37 cases (40.2%). Molecular identification was performed in 23 cases, mainly by means of restriction enzyme analysis of mtDNA. Among the 37 patients with presumptive *S. boulardii* infection, five did not take this probiotic at the time of diagnosis. However, the similarity of the genotypic profile between *S. boulardii* isolates from patients and from probiotic preparations strongly suggested nosocomial acquisition, with catheters being a likely portal of entry because of possible contamination through hand transmission. This explanation is given for cases of *S. cerevisiae* infection in hematological patients because they are supplied exclusively with sterile food, and isolates from patients concurrently hospitalized in the same unit were identical. Hospital-acquired transmission of this yeast and possible systemic fungal infection is an important cause of death of these patients since they receive cytotoxic therapy and preventive antifungal treatment that can select *Saccharomyces* strains resistant to antifungal drugs like amphotericin B and fluconazole. The ability of the yeasts to persistently colonize multiple body sites and contamination via hands of health care providers can explain why in a study regarding colonization of hematological patients *S. cerevisiae* increased markedly during the follow-up; it was detected 1–30 weeks after initiation of the cytotoxic chemotherapy in different patients and only in one case the patient was already colonized at admission.[47] In a similar investigation, contamination from food was considered possible since *Saccharomyces* strains were absent in the hospital environment and on the hands or swabs from the personnel.[48] *Saccharomyces* organisms were most frequently isolated from blood, either exclusively or with involvement of other organ(s). The rates of favorable outcome among immunocompromised and immunocompetent patients did not differ significantly, indicating a low virulence of this species.[35]

Epidemiologic and clinical differences between *S. cerevisiae* and *S. boulardii* invasive infections were revealed. Systemic infections like fungemia and septicemia caused by *S. boulardii* resulted from oral administration both in immunocompromised and immunocompetent patients, thus indicating the ability of this organism to cross the intestinal mucosa.[49,50] *S. boulardii* was exclusively isolated from blood. Predisposition factors for *S. boulardii* infections included intensive care unit hospitalization, presence of an indwelling catheter, and intestinal disease. The lack of organ involvement and the better prognosis could be due to the relatively low virulence of *S. boulardii*[39] or the lower frequency of immunocompromise in the group of patients with *S. boulardii* infection. Though the level of virulence of *S. boulardii* is modest, it has been reported to vary among different batches and is therefore not predictable.[6] By applying Affimetrix oligonucleotide chips designed on the whole *S. cerevisiae* sequenced genome, a distinctive genetic feature of *S. boulardii* with implications for pathogenicity was found, i.e., a higher copy number for genes of the cAMP pathway involved in PHG.[7] Other overrepresented genes are involved in protein synthesis and stress response. Epidemiological data available led the EFSA (European Food Safety Authority) to recommend that *S. boulardii* should certainly be contraindicated for patients of fragile health, as well as for patients with a central venous catheter in place.[51]

70.1.3 Laboratory Diagnosis

70.1.3.1 Conventional Techniques

When not required for epidemiological studies, the identification of the yeasts in daily practice is not carried out in clinical microbiology laboratories; the only routine procedure can be microscopic observation of the organisms from positive cultures. Specimen categories include blood and other body fluids, feces, and swabs from different sites (throat, vagina, or external body sites). Cases of unfavorable outcome require culturing of internal organs like lungs, liver, and kidneys in order to individuate the site of infection and the cause of death. Cultures of catheter tips are necessary, when contamination is suspected to derive from such devices, to make the right decision about their removal. In some outbreaks, the hospital environment was also sampled in order to individuate the source of infective *Saccharomyces*. This was done by directly touching Petri plates with finger tips of the involved personnel and sweep plating surfaces and clothes.[48] For identification of the species, the first step is direct seeding of samples onto selective media such as Sabouraud dextrose agar (SDA) supplemented with chloramphenicol and incubation for 24–48 h. Isolated colonies are inoculated into miniaturized identification trays available from different manufacturers. The most frequently used are API 32C and API 20C AUX test strips or Vitek 2 ID-Yst cards (bioMèrieux, Marcy l'Etoile, France). Results are acquired in 48 h and identification is based on carbohydrate fermentation profiles. Some systems concomitantly provide data on susceptibility to antifungal drugs. Another conventional detection system is a differential plating medium called chromogenic Candida agar (CHROMagar Company, Paris, France) that has been proven to allow the growth and distinction of many yeast species, including *S. cerevisiae*.[52]

70.1.3.2 Molecular Techniques

The choice of molecular techniques for identifying yeasts involved in infections offers advantages like faster diagnosis and unequivocal identification. Moreover, after identification, the same nucleic acid samples extracted from pure cultures can be used for strain typing by methods enabling the definition of the infection source. Three species-specific PCR tests are described in literature for the culture-independent detection of *S. cerevisiae*: two have been applied to wine and one to feces.[53–55] All these tests were designed for quantification by real-time PCR, however, the specific primer sets can be used for identification by conventional PCR.

Particularly noteworthy for the diagnosis of fungal infections is the APEX (arrayed primer extension) microarray support projected for one-step detection of common pathogenic fungi from clinical specimens based on their ITS1 and ITS2 sequences, in which a specific probe for *S. cerevisiae* was included.[56] Authors stated that identification by this method is unambiguous, sensitive, and reproducible.

When DNA from pure cultures is available, a rapid method for the identification of *S. cerevisiae* as well as of other yeast species is PCR amplification of the ITS1-5.8S-ITS2 rDNA internal transcribed spacer region, followed by digestion with frequent cut restriction endonucleases *Cfo*I, *Hae*III, and *Hinf*I,[57] and by separation on a 3% agarose gel. Sequencing of the domains D1 and D2 located at the 5′ end of gene 26S is another reliable identification method.[1]

Different genetic fingerprinting methods have been used to compare clinical isolates and strains from other sources. Initially, karyotyping by pulsed field gel electrophoresis (PFGE), restriction length polymorphisms of whole genome or mtDNA were applied for strain discrimination.[58] A study regarding DNA typing of 60 clinical and nonclinical isolates of *S. cerevisiae* reported that *Eco*RI DNA digestion profiles of clinical isolates were very heterogeneous, exhibiting little clonality. However, it was possible to subgroup these isolates into two clusters, A and B, on the basis of the presence or absence of a 3.0 kb band on agarose gel electrophoresis of *Eco*RI-digested DNA and a statistical association of virulence with the group A DNA type was found.[58]

Different *S. cerevisiae*-specific PCR-based genetic profiling techniques were developed and applied to the distinction of clinical strains. Among these techniques, amplification of microsatellites from different loci was found to be highly reproducible, able to discriminate single strains according to the guidelines proposed by the European Study Group on Epidemiological Markers.[35] Microsatellites are short sequence repeats (SSR), usually less than 10 nt and highly polymorphic in length, found in many ORFs and in noncoding regulatory regions. All the *S. cerevisiae* strains, including *S. boulardii*, gave 100% positive PCR results with primers flanking all microsatellites tested by Hennequin et al.[59] Limitedly to the loci considered, the method was proven to be specific for *S. cerevisiae* since isolates belonging to the closely related species *S. pastorianus*, *S. paradoxus*, and *S. bayanus* failed to be reproducibly amplified.[59] Microsatellite amplification profiles were exploited for strain clustering and different loci were used to this aim.[12,59,60] Amplification products of sizes in the range 110–370 bp are obtained that are separated on polyacrylamide or agarose gels depending on size and difference in length.[59,60]

In the first study on clinical strains, it was found that some of them had microsatellite repeat profiles similar to bakery strains. Moreover, this method provided a distinction of *S. boulardii* based on the size of the amplification product obtained for the microsatellite trinucleotide repeat sequence (CAG)[9,59,61] concomitantly confirming its affiliation to the *S. cerevisiae* species.

Before the assessment of this technique, multiple enzymatic restrictions of mtDNA[10] were required for a definitive identification since the lack of univocal phenotypic traits did not permit to identify *S. boulardii* in a routine manner. Another widely used PCR fingerprinting procedure for *S. cerevisiae* is determination of inter-δ sequence amplification profile.[10] The amplification products are separated by electrophoresis on agarose gel. Also with this method, de Llanos et al.[10] found that a *S. cerevisiae* isolate from an infection case was highly similar to two commercial bread-making strains. The two molecular types were also found similar to several clinical isolates from feces, pharynx, and vagina previously characterized.

Finally, another PCR-based fingerprinting method proposed is based on amplification of minisatellite repeats, very commonly present inside the ORFs of cell wall protein genes in *S. cerevisiae* and other yeasts.[62] These repeats are recombinogenic and can be a source of genetic variability among strains. This typing method was developed to exploit minisatellite variability for the distinction of different strains isolated from wine. The minisatellite regions selected in the sequences available for the reference strain *S. cerevisiae* S288c had to satisfy the following criteria in order to allow amplification from different strains and detection on agarose gel: at least 80% identity between repeats and at least 20 bp repeat length, respectively. Moreover, the test was restricted to nonenzymatic proteins supposed to be more variable since their sequence variation would not cause a lethal loss of function. The genes chosen were *AGA1* coding for the anchoring subunit of α-agglutinin; *DAN4*, member of the PAU proteins of unknown function; *HSP150*, a cell wall–bound heat shock protein; and the major cell wall glycoprotein *SED1*. Amplification products ranged in size between 1000 and 1600 bp and were separated on agarose gel. In Table 70.1 all the mostly adopted methods to type *S. cerevisiae* are listed and briefly described.

70.2 METHODS

70.2.1 Sample Preparation

70.2.1.1 Sample Handling

In clinical laboratories, sample types for the isolation and identification of yeasts are blood and other body fluids or lavages, swabs from different sites, and tissue specimens.

Currently, blood samples are first inoculated in BACTEC™ Plus Aerobic/F and Anaerobic/F bottles and then are monitored using the BACTEC 9240 system (Becton Dickinson) or similar systems. Yeast-positive cultures identified by Gram's stain are streaked on isolation media or can be used for nucleic acid extraction and identification by species-specific molecular methods. Direct streaking onto Petri dishes is a common practice also for the analysis of swabs, lavages, and tissue specimens. Frequently used media for yeast isolation are SDA agar (casein hydrolyzate 5 g/L, tissue hydrolyzate 5 g/L, dextrose 40 g/L, agar 15 g/L, pH 5.6) plus 4 g/L chloramphenicol; YEPD (yeast extract 10 g/L, peptone 20 g/L, glucose 20 g/L, agar 15 g/L); and Wallerstein Laboratory (WL) nutrient agar (Oxoid Ltd.) both supplemented with 0.1 g/L chloramphenicol or 15 µg/L biphenyl to inhibit bacteria and molds, respectively. Solid samples like feces or dough are homogenized in ninefold volume of sterile tryptone solution (casein tryptone 1 g/L, NaCl 9 g/L) in Stomacher bags. If not analyzed soon all sample types must be stored at −80°C in sterile plastic bags or tubes before inoculation or direct DNA extraction.

All the identification methods described in the following part of this section are based on PCR amplification. Both direct detection in a clinical or food sample and identification of an organism in pure culture will be reported. Sample types considered are pure cultures and feces. The latter was chosen because it is the only type of clinical sample in which a direct detection method specific for *S. cerevisiae* has been experimented to date.

70.2.1.2 Extraction of DNA from Colonies or Pure Cultures

For preparing pure cultures of the isolated clinical *S. cerevisiae* strains, single well-isolated colonies that look round, convex, cream colored, and smooth or slightly wrinkled are picked from plates and transferred directly in tubes with lysis solution or in broth for subculturing. In this second case, the colony is allowed to develop for 48 h at 30°C. Part of the culture will be used to prepare a frozen stock of the isolate by transferring 1 mL of it in a sterile cryovial, adding ca. 200 µL of sterile glycerol, mixing, and storing at −80°C. DNA suitable for typing by a PCR-based method from a pure culture of *S. cerevisiae* can be derived from the protocol of Hennequin et al.[59] applied prior to microsatellite amplification. The procedure includes the following steps:

1. Harvest yeast cells from 3 mL of culture by centrifugation at 10,000 × *g* for 5 min.
2. Remove the supernatant and wash the cell pellet with 1 mL of TE buffer; centrifuge the cells at 10,000 × *g* for 5 min.
3. Resuspend the cell pellet in 200 µL of lysis buffer; incubate for 1 h at 37°C.
4. Add 200 µL of lysis solution; incubate at 65°C for 30 min; keep on ice for 5 min.
5. Add 100 µL of 5 M potassium acetate; keep at 4°C for 45 min; centrifuge for 5 min at 10,000 × *g*.
6. Transfer the supernatant in a clean tube; precipitate DNA by addition of 30 µL of ammonium acetate solution plus three volumes of 100% ethanol; centrifuge for 10 min at 10,000 × *g*.
7. Remove the supernatant and add 1 mL of 70% ethanol; centrifuge for 2 min at 10,000 × *g*.
8. Carefully remove the ethanol by gentle aspiration; dry in open tubes under a fume cabinet until all residual ethanol evaporates.
9. Resuspend in 50 µL of TE buffer; store at 4°C for up to 6 months.

70.2.1.3 Extraction of Total DNA from Human Fecal Samples

Application of direct detection of *S. cerevisiae* in clinical samples by PCR would expedite the diagnosis of infections due to this yeast. In a study aimed at estimating the efficiency of different methods for extracting a high amount of good quality DNA from fungal species, it was found that best results are obtained by combining enzymatic treatment with mechanical disruption by mortar and pestle grinding of material frozen in liquid nitrogen.[63] This method was applied later and allowed high sensitivity in the PCR-based detection of *S. cerevisiae* and *Saccharomyces* sensu stricto DNA from feces.[54] In the reported experiments, fecal samples were frozen at −80°C

TABLE 70.1
Molecular Methods Developed for the Discrimination of *S. cerevisiae* Isolates at the Species or Strain Level

Method	Procedural Details	Discrimination Level	Disadvantages
Karyotyping	Purification of total DNA by extraction from cells immobilized in agarose plugs Electrophoretic separation in a minimum of 23 h by PFGE Profile analysis	Single strain	Time-consuming Delicate and expensive equipment
*Eco*RI RFLP	Total DNA extraction from pure cultures Digestion for 90 min with *Eco*RI endonuclease 21 h run on 1% agarose gel Profile analysis	Strain clusters	High amount of unfragmented DNA needed Time-consuming No distinction of the restriction profile group B from *S. paradoxus*
mtDNA restriction	Total DNA extraction from pure cultures Digestion overnight with *Hinf*I or *Rsa*I endonucleases Run on 1% agarose gel Profile analysis	Single strain	Relatively high amount of unfragmented DNA needed for digestion (~0.5 μg) Time-consuming
ITS1-5.8S-ITS2 region amplification and restriction	DNA extraction from pure cultures PCR with primers flanking the ITS1-5.8S-ITS2 region Restriction with endonucleases *Cfo*I, *Hae*III, and *Hinf*I Separation on 3% agarose gel	Species	None
Amplification of δ sequences	DNA extraction from pure cultures PCR with primers targeted to transposon δ sites Separation on 1.5% agarose gel Profile comparison	Single strain	Unstable patterns with some primer sets Not all strains produce amplification products
Amplification of microsatellite sequences	DNA extraction or cell disruption by freeze–thaw treatment from pure cultures PCR with primers flanking microsatellite sequences of different genetic loci Separation on polyacrylamide gels containing formamide or on agarose gel Profile comparison	Single strain	Need of separation on polyacrylamide gels in a sequencing apparatus in the original method or labeling with different fluorochromes due to the range of dimension of the amplification products obtained for some genetic loci
Amplification of the minisatellites in cell wall protein genes	DNA extraction or cell disruption by heat treatment from pure cultures PCR amplification with a primer set targeted on the sequences flanking the minisatellite regions of cell wall protein genes Separation on agarose gel Profile comparison	Single strain	Need to combine the profiles obtained for amplicons from different loci with four different primer pairs
Real-time PCR targeted on rRNA genes and ITS or on *S. cerevisiae*-specific random amplification products	DNA extraction by mechanical cell disruption Amplification in a normal PCR reaction followed by conventional electrophoresis or in a real-time PCR apparatus for quantification using SYBR Green as a fluorescent dye	Culture-independent detection and identification of *S. cerevisiae*	To date experimented only on feces and wine

in sterile plastic bags before DNA extraction. The phases of the method are as follows:

1. Thaw the fecal sample; transfer about 1 g of it in a laboratory mortar previously well cleaned and sterilized by keeping overnight at 200°C in a stove.
2. Add 500 μL of extraction buffer; immediately freeze by liquid nitrogen insufflations.
3. Grind for 2 min; add 500 μL of phenol–chloroform–isoamyl alcohol mixture and rotary mix at room temperature for 10 min.
4. Transfer the sample in a 2 mL centrifuge tube; centrifuge at 10,000 × *g* for 10 min at 4°C.
5. Carefully transfer the supernatant in a new tube.
6. Add 500 μL isopropanol and 150 μL of 3 M sodium acetate solution pH 4.8; mix for 5 min and keep for additional 5 min at room temperature; centrifuge at 10,000 × *g* for 10 min at room temperature.
7. Carefully discard the supernatant and add 500 μL of 70% ethanol to the pellet; centrifuge for 2 min at 10,000 × *g*.

8. Remove as much ethanol as possible by gentle aspiration; dry in the open tubes under a fume cabinet until all residual ethanol evaporates.
9. Resuspend in 100 µL of TE buffer containing DNase free RNase A; store at 4°C allowing RNA digestion overnight.

70.2.2 Detection Procedures

Detection from pure cultures can be carried out either by the ITS1-5.8S-ITS2 rDNA internal transcribed spacer region amplification/RFLP (restriction fragment length polymorphism) method recommended by Esteve-Zarzoso et al.[57] for the identification of several yeast species or by *S. cerevisiae*-specific PCR tests.[53–55] In addition, the identification of *S. boulardii* can be carried out by specific amplification of the $(CAG)_9$ microsatellite repeat at locus YLR177w.[59,61] The former method is worth being carried out because when a yeast contaminant does not belong to the species *S. cerevisiae* it anyhow allows the true specie to be determined. The *S. cerevisiae*-specific PCR tests are useful in culture-independent detection, too. The PCR test specific for *S. boulardii*, when applied on pure cultures, can elucidate the epidemiology of infection cases caused by this particular strain.

PCR/RFLP of the ITS1-5.8S-ITS2 rDNA internal transcribed spacer region comprises amplification with oligonucleotides ITS1 (5′-TCCGTAGGTGAACCTGCGG-3′) and ITS4 (5′-TCCTCCGCTTATTGATATGC-3′); digestion with frequent cut restriction endonucleases *Cfo*I, *Hae*III, and *Hinf*I; and separation on a 3% agarose gel. Species identification is achieved by comparison with molecular weight tables reporting the profiles obtained for each yeast species. For this purpose, primers delimiting a region of the rRNA coding region comprising the 5.8S rRNA gene and the two internal transcribed spacers ITS1 and ITS2 are used. *S. cerevisiae* shows an 880 bp amplicon that gives bands of 385, 365, and 130 bp with *Cfo*I; 320, 230, 180, and 150 bp with *Hae*III; and 365 bp and 155 bp with *Hinf*I.

1. Prepare PCR mixture containing 0.5 µM of each primer, 10 µM deoxynucleotides, 1.5 mM $MgCl_2$ and 1× PCR buffer, sterile double-distilled water up to the desired volume and 1 µL of DNA sample in a 30 µL reaction.
2. Run a cycling program consisting of initial denaturation at 95°C for 5 min; 35 cycles of denaturing at 94°C for 1 min, annealing at 55 ± 5°C for 2 min, and extension at 72°C for 2 min; and a final extension at 72°C for 10 min.
3. Digest the PCR products (10 µL or approximately 0.5–1.0 µg) without further purification with 10 U (1 µL) of the restriction endonucleases *Cfo*I, *Hae*III, and *Hinf*I at 37°C for 15 min.
4. Separate the PCR products and their restriction fragments on 1.4% and 3% agarose gels, respectively, in 1× TAE buffer.
5. Stain the gels in 0.5 µg/mL ethidium bromide aqueous solution.
6. Estimate the sizes of PCR amplicons by comparison against a DNA length standard (e.g., 100 bp ladder, Invitrogen). Examples of profiles are reported in Figure 70.1.

For specific PCR detection, the method developed by Chang et al.[54] is represented here. The oligonucleotides SCDF (5′-AGGAGTGCGGTTCTTTG-3′) and SCDR (5′-TACTTACCGAGGCAAGCTACA-3′) targeting the D1/D2 region of 26S rRNA gene and amplifying a 310 bp DNA fragment from *S. cerevisiae* are used. Amplification is done using DyNAmo™ HS SYBR® qPCR kit (FINNZYMES) master mix, 10 pmol of each primer, and 1 µL template. The cycling program comprises initial denaturation at 95°C for 15 min; 45 cycles of denaturation at 94°C for 20 s, annealing at 58°C for 30 s, and extension at 72°C for 45 s; and a final extension at 72°C for 5 min. Amplification is performed in an OPTICON™2 DNA Engine, MJ Research.

Finally, the *S. boulardii*-specific microsatellite amplification test with primers Boul3 (5′-CTTAAACAAC AGCTCCCAAA-3′) and Boul4 (5′-ATTTCTGATGC GCTGATTCAT-3′)[61] generates a PCR product of 130 bp. The reaction mixture is composed of 1.5 mM $MgCl_2$, 200 µM of each dNTP, 0.2 µM of each primer, and 1 U of Taq DNA polymerase. The cycling program included initial denaturation at 94°C for 4 min; 28 cycles of 94°C for 30 s, 56°C for 45 s, 72°C for 30 s; and a final extension at 72°C for 10 min.[61] PCR fragments are denatured (94°C for 4 min), separated

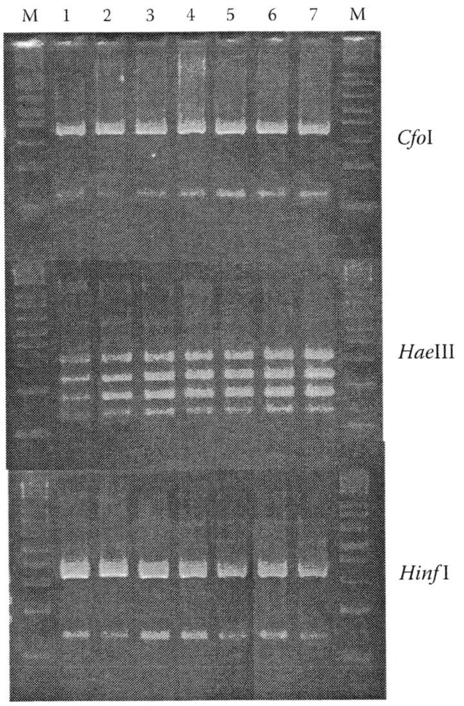

FIGURE 70.1 ITS1-5.8S-ITS2 PCR/RFLP profiles of seven *S. cerevisiae* isolates from grapes (numbers 1–7). M; 1 kb Plus DNA Ladder (Invitrogen).

on 7.5 M urea-12.5% polyacrylamide gels in 0.5× TBE buffer (5.4 g/L Tris base, 2.75 g/L boric acid, 2.92 g/L EDTA pH 8.1), and then stained with ethidium bromide solution as above. Length polymorphisms are evaluated by comparison with the migration of a reference strain, with a slightly different size of the amplicon, e.g., 133 bp from *S. cerevisiae* S288c.[59]

70.3 CONCLUSIONS AND FUTURE PERSPECTIVES

The introduction of molecular diagnostics, with the consequent increase in rapidity, specificity, and ease of execution, has not only expedited the identification and diagnosis of *S. cerevisiae* infections, but also revealed the true dimension of the risk related to the diffusion and hazardousness of this opportunistic pathogen. Reviewing the infection cases attributable to *S. cerevisiae* through 2005,[35] it is clear that the improvement of diagnostic procedures has played a key role in tracing a veritable outline of the incidence trend.

PCR-based species-specific identification is now widely adopted in research and clinical laboratories to supplant the current procedures based on rather equivocal physiological identification keys. As the methods proposed for *S. cerevisiae* detection/identification have been designed for culture-independent quantitative analysis with the double advantage of giving a precise indication of the presence of this pathogen and of its load in clinical samples, they facilitate a judgment on whether or not it is the main infective agent in each case. Epidemiological data suggest the opportunity to comprise the detection of this yeast in routine tests for microbial pathogens, at least when infection of yeast etiology is suspected, so avoiding retarded treatment with fungicidal drugs. Also it should be important to validate the APEX oligonucleotide microarray support developed by Campa et al.[56] for routine clinical practice since it could represent a powerful tool for the simultaneous diagnosis of *S. cerevisiae* and many other fungal infections thus obviating to the execution of separate tests. Use of this analysis tool, to our knowledge only on pure cultures at this time, could be extended to direct detection from clinical specimens.

Many investigations have dealt with testing the possibility of distinguishing clinical *Saccharomyces* organisms by molecular fingerprinting; however, this scope was reached only partially since clinical isolates showed genetic profiles similar to bakery and wine starter strains.[6,59] Moreover, the little clonality exhibited by clinical isolates[10–12,58] seems to indicate that the assessment of molecular markers for potential pathogens selected by such techniques is a heavy task. A good alternative, also to prevent the use of potentially pathogenic strains as food and fermented beverage starters, could be checking few genes shown to be involved in virulence in terms of presence, copy number, SNP and "indel" mutations, and expression levels in particular conditions. Some of these like the *SSD1* gene,[41] the ammonia transporter *MEP2* gene that show higher expression during pseudofilamentation,[29] the gene *PLB1* encoding a phospholipase,[64] and genes *SOD*1 and *YAP*1[43] are possible candidates. Research in the direction of finding suitable target genes can be validly aided by the development of DNA microarray supports constructed on the basis of available *S. cerevisiae* genome sequences. Other targets to be investigated with respect to their contribution to virulence and to the capacity to evade host immune defenses are genes encoding cell surface proteins. Since this property can depend on the number of minisatellite repeats found in their ORFs,[22] an amplification strategy proposed by Marinangeli et al.[62] could allow linking the profiles found for some of those genes to virulence. At present, this kind of molecular method has not yet been developed. The current testing for the safety of food starter strains still relies on phenotypic evaluation of characters strongly associated to pathogenicity like growth capacity at 42°C, PHG, and phospholipase activity.[21]

Given that little or no information is available to the public about the hazardousness of the bakery yeast, correct instructions, especially directed to food and beverage operators, could prevent the onset of many infectious cases and favor the development of good hygienic practices that minimize the environmental diffusion of the yeast and its delivery via food to infection-predisposed individuals.

REFERENCES

1. Kurtzman, C.P. and Robnett, C.J., Phylogenetic relationships among yeasts of the "*Saccharomyces* complex" determined from multigene sequence analyses, *FEMS Yeast Res.*, 3, 417, 2003.
2. Scannell, D.R. et al., Yeast genome evolution-the origin of the species. Review. *Yeast*, 24, 929, 2007.
3. Edwards-Ingram, L.C. et al., Comparative genomic hybridization provides new insights into the molecular taxonomy of the *Saccharomyces* sensu stricto complex, *Genome Res.*, 14, 1043, 2004.
4. Thomson, J.M. et al., Resurrecting ancestral alcohol dehydrogenases from yeast, *Nat. Genet.*, 37, 630, 2005.
5. Fay, J.C. and Benavides, J.A., Evidence for domesticated and wild populations of *Saccharomyces cerevisiae*, *PLoS Genet.*, 1, e5., 2005.
6. McCullough, M.J. et al., Species identification and virulence attributes of *Saccharomyces boulardii* (nom. inval.), *J. Clin. Microbiol.*, 36, 2613, 1998.
7. Edwards-Ingram, L. et al., Genotypic and physiological characterization of *Saccharomyces boulardii*, the probiotic strain of *Saccharomyces cerevisiae*, *Appl. Environ. Microbiol.*, 73, 2458, 2007.
8. Buts, J.P. and De Keyser, N., Effects of *Saccharomyces boulardii* on intestinal mucosa, *Dig. Dis. Sci.*, 51, 1485, 2006.
9. Czerucka, D., Piche, T., and Rampal, P., Review article: yeast as probiotics—*Saccharomyces boulardii*, *Aliment. Pharmacol. Ther.*, 26, 767, 2007.
10. de Llanos, R. et al., Food and probiotic strains from the *Saccharomyces cerevisiae* species as a possible origin of human systemic infections, *Int. J. Food Microbiol.*, 110, 286, 2006.

11. MacKenzie, D.A. et al., Relatedness of medically important strains of *Saccharomyces cerevisiae* as revealed by phylogenetics and metabolomics, *Yeast*, 25, 501, 2008.
12. Muller, L.A. and McCusker, J.H., Microsatellite analysis of genetic diversity among clinical and nonclinical *Saccharomyces cerevisiae* isolates suggests heterozygote advantage in clinical environments, *Mol. Ecol.*, 18, 2779, 2009.
13. Dujon, B., The yeast genome project: What did we learn?, *Trends Genet.*, 12, 263, 1996.
14. Wei, W. et al., Genome sequencing and comparative analysis of *Saccharomyces cerevisiae* strain YJM789, *PNAS*, 104, 12825, 2007.
15. Palecek, S.P. et al., Depression of *Saccharomyces cerevisiae* invasive growth on non-glucose carbon sources requires the Snf1 kinase, *Mol. Microbiol.*, 45, 453, 2002.
16. Reynolds, T.B. and Fink, G.R., Bakers' yeast, a model for fungal biofilm formation, *Science*, 291, 878, 2001.
17. Casalone, E. et al., Characterization of *Saccharomyces cerevisiae* natural populations for pseudohyphal growth and colony morphology, *Res. Microbiol.*, 156, 191, 2005.
18. Douglas, L.M. et al., Expression and characterization of the flocculin Flo11/Muc1, a *Saccharomyces cerevisiae* mannoprotein with homotypic properties of adhesion, *Eukaryot. Cell*, 6, 2214, 2007.
19. Byron, J.K. et al., Pathogenicity of *Saccharomyces cerevisiae* in complement factor five-deficient mice, *Infect. Immun.*, 63, 478, 1995.
20. McCusker, J.H. et al., *Saccharomyces cerevisiae* virulence phenotype as determined with CD-1 mice is associated with the ability to grow at 42°C and form pseudohyphae, *Infect. Immun.*, 62, 5447, 1994.
21. de Llanos, R., Fernandez-Espinar, M.T., and Querol, A., A comparison of clinical and food *Saccharomyces cerevisiae* isolates on the basis of potential virulence factors. *Anton Leeuwen*, 90, 221, 2006.
22. Verstrepen, K.J. et al., Intragenic tandem repeats generate functional variability. *Nat. Genet.*, 37, 986, 2005.
23. Snoek, I.S. and Steensma, H.Y., Factors involved in anaerobic growth of *Saccharomyces cerevisiae*, *Yeast*, 24, 1, 2007.
24. Pfeiffer, T. et al., Cooperation and competition in the evolution of ATP-producing pathways, *Science*, 292, 504, 2001.
25. Pretorius, I.S., Tailoring wine yeast for the new millennium: novel approaches to the ancient art of winemaking, *Yeast*, 16, 675, 2000.
26. Klar, A.J., The mother-daughter mating type switching asymmetry of budding yeast is not conferred by the segregation of parental *HO* gene DNA strands, *Genes Dev.*, 1, 1059, 1987.
27. Brauer, M.J. et al., Coordination of growth rate, cell cycle, stress response, and metabolic activity in yeast, *Mol. Biol. Cell*, 19, 352, 2008.
28. Ma, J. et al., An interrelationship between autophagy and filamentous growth in budding yeast, *Genetics*, 177, 205, 2007.
29. Iyer, R.S. et al., Pseudohyphal differentiation defect due to mutations in GPCR and ammonium signaling is suppressed by low glucose concentration: a possible integrated role for carbon and nitrogen limitation, *Curr. Genet.*, 54, 71, 2008.
30. Boeckstaens, M. et al., The yeast ammonium transport protein Mep2 and its positive regulator, the Npr1 kinase, play an important role in normal and pseudohyphal growth on various nitrogen media through retrieval of excreted ammonium, *Mol. Microbiol.*, 64, 534, 2007.
31. Almanza, L. et al., Quatre heures pour un record, ou une cellulite vraiment foudroyante: *Saccharomyces cerevisiae* peut-elle en être la cause?, *Ann. Fr. Anesfh. Rianim.*, 1, 130, 1998.
32. Pérez González, E., Peritonitis caused by *Saccharomyces cerevisiae* in an ambulatory peritoneal dialysis patient, *Nefrología*, 29, 371, 2009.
33. Savini, V. et al., Two cases of vaginitis caused by itraconazole-resistant *Saccharomyces cerevisiae* and a review of recently published studies, *Mycopathologia*, 166, 47, 2008.
34. McCullough, M.J. et al., Epidemiological investigation of vaginal *Saccharomyces cerevisiae* isolates by a genotypic method, *J. Clin. Microbiol.*, 36, 557, 1998.
35. Enache-Angoulvant, A. and Hennequin, C., Invasive *Saccharomyces* infection: A comprehensive review. *Clin. Infect. Dis.*, 41, 1559, 2005.
36. Swinne, D. et al., Bloodstream yeast infections: A 15-month survey, *Epidemiol. Infect.*, 137, 1037. 2009.
37. Klingberg, T.D. et al., Comparison of *Saccharomyces cerevisiae* strains of clinical and nonclinical origin by molecular typing and determination of putative virulence traits, *FEMS Yeast Res.*, 8, 631, 2008.
38. Cairoli, R. et al., *Saccharomyces cerevisiae* fungemia with granulomas in the bone marrow in a patient undergoing BMT, *Bone Marrow Transpl.*, 15, 785, 1995.
39. Clemons, K.V. et al., Comparative pathogenesis of clinical and nonclinical isolates of *Saccharomyces cerevisiae*, *J. Infect. Dis.*, 169, 859, 1994.
40. van Burik, J.H. and Magee, P., Aspects of fungal pathogenesis in humans, *Annu. Rev. Microbiol.*, 55, 743, 2001.
41. Mir, S.S. et al., Ssd1 is required for thermotolerance and Hsp104-mediated protein disaggregation in *Saccharomyces cerevisiae*, *Mol. Cell Biol.*, 29, 187, 2009.
42. Diezmann. S. and Dietrich, F.S., *Saccharomyces cerevisiae*: population divergence and resistance to oxidative stress in clinical, domesticated and wild isolates, *PLoS One*, 4, 5317, 2009.
43. Jain, C. et al., A pathogenesis assay using *Saccharomyces cerevisiae* and *Caenorhabditis elegans* reveals novel roles for yeast AP-1, Yap1, and host dual oxidase BLI-3 in fungal pathogenesis, *Eukaryot. Cell.*, 8, 1218, 2009.
44. Belet, N. et al., Catheter-related fungemia caused by *Saccharomyces cerevisiae* in a newborn, *Pediatr. Infect. Dis. J.*, 24, 1125, 2005.
45. Smith, D. et al., Fatal *Saccharomyces cerevisiae* aortic graft infection, *J. Clin. Microbiol.*, 2691, 2691, 2002.
46. Graf, C. and Gavazzi, G., *Saccharomyces cerevisiae* fungemia in an immunocompromised patient not treated with *Saccharomyces boulardii* preparation, *J. Infect.*, 54, 310, 2007.
47. Salonen, J.H. et al., Fungal colonization of haematological patients receiving cytotoxic chemotherapy: Emergence of azole-resistant *Saccharomyces cerevisiae*, *J. Hosp. Infect.*, 45, 293, 2000.
48. Olver, W.J. et al., Nosocomial transmission of *Saccharomyces cerevisiae* in bone marrow transplant patients, *J. Hosp. Infect.*, 52, 268, 2002.
49. Cesaro, S. et al., *Saccharomyces cerevisiae* fungemia in a neutropenic patient treated with *Saccharomyces boulardii*, *Support Care Cancer*, 8, 504, 2000.
50. Cassone, M. et al., Outbreak of *Saccharomyces cerevisiae* subtype *boulardii* fungemia in patients neighboring those treated with a probiotic preparation of the organism, *J. Clin. Microbiol.*, 41, 5340, 2003.

51. Opinion of the Scientific Committee on a request from EFSA on the introduction of a Qualified Presumption of Safety (QPS) approach for assessment of selected microorganisms referred to EFSA. *EFSA J.*, 587, 1, 2007.
52. Ghelardi, E. et al., Efficacy of Chromogenic Candida Agar for isolation and presumptive identification of pathogenic yeast species, *Clin. Microbiol. Infect.*, 14, 141, 2008.
53. Martorell, P. et al., Rapid identification and enumeration of *Saccharomyces cerevisiae* cells in wine by real-time PCR, *Appl. Environ. Microbiol.*, 71, 6823, 2005.
54. Chang, H.W. et al., Quantitative real time PCR assays for the enumeration of *Saccharomyces cerevisiae* and the *Saccharomyces* sensu stricto complex in human feces, *J. Microbiol. Methods*, 71, 191, 2007.
55. Salinas, F. et al., Taqman real-time PCR for the detection and enumeration of *Saccharomyces cerevisiae* in wine, *Food Microbiol.*, 26, 328, 2009.
56. Campa, D. et al., A DNA microarray based on arrayed-primer extension (APEX) technique for identification of pathogenic fungi responsible for invasive and superficial mycoses, *J. Clin. Microbiol.*, 46, 909, 2008.
57. Esteve-Zarzoso, B. et al., Identification of yeasts by RFLP analysis of the 5.8S rRNA gene and the two ribosomal internal transcribed spacers, *Int. J. Syst. Bacteriol.*, 49, 329, 1999.
58. Clemons, K.V. et al., Application of DNA typing methods and genetic analysis to epidemiology and taxonomy of *Saccharomyces*, *J. Clin. Microbiol.*, 35, 1822, 1997.
59. Hennequin, C. et al., Microsatellite typing as a new tool for identification of *Saccharomyces cerevisiae* strains, *J. Clin. Microbiol.*, 39, 551, 2001.
60. Vaudano, E. and Moruno-Garcia, E., Discrimination of *Saccharomyces cerevisiae* wine strains using microsatellite multiplex PCR and band pattern analysis, *Food Microbiol.*, 25, 56, 2008.
61. Posteraro, B. et al., Molecular tools for differentiating probiotic and clinical strains of *Saccharomyces cerevisiae*, *Int. J. Food Microbiol.*, 103, 295, 2005.
62. Marinangeli, P. et al., Minisatellites in *Saccharomyces cerevisiae* genes encoding cell wall proteins: a new way towards wine strains characterisation, *FEMS Yeast Res.*, 4, 427, 2004.
63. Karakousis, A. et al., An assessment of the efficiency of fungal DNA extraction methods for maximizing the detection of medically important fungi using PCR, *J. Microbiol. Methods*, 65, 38, 2006.
64. Lee, K.S. et al., The *Saccharomyces cerevisiae PLB1* gene encodes a protein required for lysophospholipase and phospholipase B activity, *J. Biol. Chem.*, 269, 19725, 1994.

Part II

Bastidiomycota

71 *Coprinopsis* and *Hormographiella*

Dongyou Liu and R.R.M. Paterson

CONTENTS

71.1 INTRODUCTION

71.1.1 Classification and Morphology

The genus *Coprinopsis* is a basidiomycete belonging to the family Psathyrellaceae, order Agaricales, class Homobasidiomycetes, phylum Basidiomycota, kingdom Fungi. The family Psathyrellaceae currently consists of five genera: *Coprinellus*, *Coprinopsis*, *Parasola*, *Psathyrella*, and Psathyrellaceae incertae sedis [1]. In turn, the genus *Coprinopsis* covers 36 recognized species and 11 unassigned species. Of these, *Coprinopsis cinerea* (inky cap fungus) (obsolete synonym: *Coprinus cinereus*; anamorph: *Hormographiella aspergillata*) is a mushroom species, whose involvement in human disease process is through its anamorph *H. aspergillata*.

The anamorphic genus *Hormographiella* (class Ascomycetes, phylum Ascomycota) is composed of four mold species: *H. aspergillata*, *H. candelabra*, *H. candelabrata*, and *H. verticillata* [2]. The anamorph of *Coprinopsis cinerea* (formerly *Coprinus cinereus*), *H. aspergillata* is associated occasionally with diseases in humans and animals [3].

Hormographiella aspergillata shows moderate growth on mycological media (e.g., Sabouraud glucose agar, potato dextrose agar, and chocolate agar). Colonies measure 3 cm in 10 days at 30°C and are dense, velvety with cottony tufts and irregular margin, white to slightly cream, and pale on reverse. Hyphae are hyaline and septate; conidiophores are simple and well-differentiated. Arising from the end of conidiophores either in clusters or in irregular groups, arthroconidia are catenate, hyaline, smooth-walled, single-celled, and rectangular or cylindrical with truncate ends, but with terminal cells rounded at the tip [4,5].

71.1.2 Clinical Features and Pathogenesis

Coprinopsis cinerea is observed growing on lawns. *Hormographiella aspergillata* has been isolated from human lung tissue, bronchoalveolar lavage fluid, brain tissue, cerebrospinal fluid, and eye and skin scrapings, and was responsible for pneumonia, lung abscess, endocarditis, and keratomycosis in immunocompetent and immunosuppressed individuals [6–9]. The fungus is responsible for keratomycosis in dogs [5].

Nenoff et al. [8] described a fatal case of pulmonary infection with *Hormographiella aspergillata* in a 40-year-old woman, who underwent autologous bone marrow transplantation for acute lymphoblastic leukemia. A *Coprinopsis* (*Coprinus*) sp. was isolated repeatedly from bronchial secretions and bronchoalveolar lavage. Histopathological examination of the lung tissue demonstrated septate hyphae characteristic of *Aspergillus* and *Coprinopsis* sp.

Verweij et al. [9] documented another fatal case of pulmonary infection due to *H. aspergillata* involving a 24-year-old man receiving intensive cytotoxic treatment for acute lymphoblastic leukemia. The patient became neutropenic with fever and headache and complained of right-sided pleuritic chest pain. Chest radiographs showed focal pulmonary infiltrates in the right lower lobe, progressing to the left upper and lower lobes of the lung. The patient was treated with amphotericin B and then itraconazole. However, the patient died from respiratory failure. A necrotizing bronchopneumonia was noted at autopsy. Periodic acid-Schiff and Gomori methenamine silver stains revealed a large mass of hyphae in the cavity of the lung abscess. The hyphae, measuring between 1.9 and 4.6 μm in diameter, were septate, showing irregularly shaped swellings and branching at acute angles or dichotomously. Cultures of the left and right lungs grown for 2 days yielded white to cream, cottony and dense fungal colonies, while cultures were not obtained from other organs. Microscopic examination of the fungus showed hyaline, septate hyphae, macronematous conidiophores with no clamp connections, and straight or curved, thin-walled conidia accumulating around the conidiophore. As the isolate displayed undiagnostic arthroconidial shape and failed to sporulate, its identity was confirmed as *H. aspergillata* (*C. cinereus*) on the basis of small subunit (SSU) and internal transcribed spacer (ITS) restriction fragment length polymorphism (RFLP) patterns.

There was also a report of a lung abscess due to *H. aspergillata* (anamorph. *Coprinus cinereus*) in a patient with non-Hodgkin's lymphoma [10]. In addition, Lagrou et al. [11] described the involvement of *Hormographiella aspergillata* in lung tissues of a 34-year-old man, who underwent an uncomplicated allogeneic peripheral blood stem cell transplantation from a human leukocyte antigen (HLA)-identical sibling for acute myeloid leukemia. The patient presented with serious capillary leak syndrome, hepatic and renal failure, progressive encephalopathy, neutropenic fever, bilateral pulmonary infiltrates, dyspnea, and cough. Microscopic examination of bronchoalveolar fluid (BAL) samples with Grocott staining showed *Aspergillus*-like molds, and cultures of the BAL samples yielded fast-growing, white to slightly cream-colored dense fungal colonies. Clusters of rectangular, cylindrical arthroconidia with truncate ends were observed but with terminal cells rounded at the tip produced from conidiophores. The patient died from respiratory failure and refractory septic shock. At autopsy, multiple intraparenchymal and subpleural hemorrhagic nodules were found in both lungs. Hematoxylin and eosin staining of the lung tissues confirmed the presence of septated hyphae. The fungus was identified as *Coprinus cinereus* by amplification and sequencing of the ITS 2 region. It is interesting that the teleomorph was identified here rather than the anamorph, and further verification may be required.

Conen et al. [12] also documented three cases of invasive pulmonary infections with *Hormographiella aspergillata* in patients undergoing treatment for acute leukemia. The fungus was detected in BAL, and despite systemic antifungal therapy with voriconazole and amphotericin B, all three patients died. Greer et al. [13] presented a case of *H. aspergillata*-related endocarditis involving a 77-year-old female with a bioprosthesis for symptomatic mitral valve stenosis. The patient developed an unusual clot-like mass on the atrial side, due to extensive colonization by a thick filamentous fungus with true hyphae, pseudohyphae, and yeast forms. The fungus was identified as *Hormographiella aspergillata*, which is perhaps surprising as yeast forms are not associated with the fungus in question.

Abuali et al. [4] described the isolation of *Hormographiella aspergillata* from a skin lesion of a 14-year-old female patient with acute myelogenous leukemia. The patient underwent an allogeneic HLA-matched bone marrow transplant and later developed odynophagia and persistent febrile neutropenia. A computed tomography (CT) scan of the lungs revealed pulmonary nodules. The patient had white plaques involving the soft palate and pharynx. A smear from a throat culture showed hyphal elements and conidiophores, and a biopsy of the palate lesion revealed submucosa and mucosa infiltrated with hyphal forms with sparse septation, rare branching, and chlamydoconidia, which was confirmed by ITS 2 sequencing as *Rhizomucor variabilis*. Following treatment with posaconazole and caspofungin, the palate lesion had decreased in size, with improved symptoms. However, the patient had persistent severe refractory pancytopenia and developed an altered mental status and a generalized seizure. A CT scan of the brain showed multiple hypodense lesions of the cerebral hemispheres and cerebellum and a repeat CT scan of the lungs uncovered cavitating lesions in the right upper lobe/right middle lobe. A modified therapy consisting of liposomal amphotericin (Ambisome) and caspofungin was initiated. The patient developed high fever and new small erythematous skin papule on the right knee and the left arm. Cultures of the skin biopsy yielded a white mold, which was identified as *Hormographiella aspergillata*. The patient died from respiratory failure 2 weeks after the appearance of the initial skin lesion.

Hormographiella spp. are considered sensitive to some azoles such as miconazole but resistant to fluconazole or itraconazole; they are variably sensitive to amphotericin B and are resistant to flucytosine [3].

71.1.3 Diagnosis

Infections caused by *Hormographiella aspergillata* are reported increasingly, especially in patients with acute leukemia or other underlying diseases. It is important to identify *Hormographiella* from other molds as this fungus displays varied sensitivity to antifungal drugs.

Filamentous basidiomycetes are notably difficult to identify because they do not always show key diagnostic features and tend to only form undifferentiated arthroconidia. For example, some *Hormographiella* isolates fail to sporulate, and others do not display diagnostic clamp connections or arthroconidial shape. Other common fungi forming arthroconidia include *Geotrichum candidum*, *Trichosporon* species, *Geomyces pannorum*, *Malbranchea* species, *Arthrographis kalrae*, and *Scytalidium* species. *Arthrographis*, *Geomyces*, and *Hormographiella* also produce conidiophores. However, the *Hormographiella* conidiophore is clearly broader than the conidiogenous hyphae, which is sometimes useful [2].

Definitive determinations of *Hormographiella* are only possible with molecular techniques using RFLP patterns of PCR-amplified ITSs (~600 bases) and SSU rRNA (~1700 bases). Alternatively, sequencing analysis of ITS 2 and the D1/D2 variable domains (~600 bases) at the 5′ end of the large subunit rRNA gene (D1/D2) is required [14–16].

71.2 METHODS

71.2.1 Sample Preparation

Clinical specimens (e.g., BAL fluid) are first examined by microscopy for mycotic elements with KOH and other stains. Tissue samples (e.g., skin, lung, liver, spleen, and kidney tissues) are fixed in 10% formalin, embedded in paraffin, sectioned at 5 μm, and stained with periodic acid-Schiff Gomori methenamine silver with eosin counterstain or hematoxylin and eosin.

Portions of the samples are inoculated on mycological agars (e.g., Sabouraud dextrose agar, Mycosel, inhibitory mold agar, chocolate agar) and incubated at 30°C or 37°C in ambient air. Tissue samples are homogenized (Stomacher 80 Lab-Blender) and streaked onto mycologic agars. Production of conidia may be stimulated by subcultivation on Takashio medium (Sabouraud agar with 0.2% glucose). The isolate may be transferred to a potato dextrose agar plate and overlaid with sterile coverslips and incubated. The mold is examined by placing the coverslip onto a glass slide with a lactophenol cotton blue stain.

For DNA extraction, approximately 1 cm^2 of fungal material is transferred to a 2 mL Eppendorf tube containing a 2:1 (wt/wt) mixture of silica gel and Celite (silica gel H, Merck 7736/Kieselguhr Celite 545; Machery) and 300 μL of TES buffer [2 g Tris (hydroxymethyl)-aminomethane, 0.38 g Na-EDTA, and 2 g sodium dodecyl sulfate in 80 mL of ultrapure water (pH 8)]. The fungal material is ground with a micropestle for 1–2 min. The volume is adjusted by adding 200 μL of TES buffer. After vigorous shaking and the addition of 10 μL of a 10 mg/mL concentration of proteinase K to the tube, the mixture is incubated at 65°C for 10 min. With the addition of 140 μL of 5 M NaCl solution, the mixture is combined with 1/10 volume (~65 μL) of cetyltrimethylammonium bromide (CTAB) buffer 10%, followed by incubation for 30 min at 65°C. One volume (~700 μL) of chloroform–isoamyl alcohol (vol/vol = 24/L) is added and mixed by inversion. After incubation for 30 min at 0°C (in iced water) and centrifugation at 14,000 rpm at 4°C for 10 min, the top layer is transferred to a clean Eppendorf tube. To the sample is added 225 μL of 5 M NH_4-acetate and it is incubated for at least 30 min (in iced water) and centrifuged. The supernatant is transferred to a clean sterile Eppendorf tube and mixed with a 0.55 volume (~510 μL) of ice-cold isopropanol. After centrifugation for 7 min at 14,000 rpm at 4°C (or room temperature), the supernatant is decanted. The pellet is washed twice with ice-cold 70% ethanol and dried by using a vacuum dryer. The powder is resuspended in 48.5 μL of Tris-EDTA buffer with 1.5 μL of 10 mg of RNase/mL, incubated at 37°C for 15–30 min, and stored at −20°C until use [17].

71.2.2 Detection Procedures

71.2.2.1 Sequencing Analysis of the ITS 2 Region

Rampazzo et al. [5] utilized universal primers ITS 3 (5′-GCATCGATGAAGAACGCAGC-3′) and ITS 4 (5′-TCCTCCGCTTATTGATATGC-3′) to amplify a 350 bp fragment of the ITS 2 region, interspacing the 5.8S rRNA and 28S rRNA of fungal ribosomal RNA genes for identification of *Hormographiella*.

Procedure

1. The PCR mixture (30 μL) is composed of 12 pmol of each primer, 0.25 mM each of dATP, dCTP, dGTP, and dTTP, 5 mM $MgCl_2$, 1× reaction buffer, 2.5 U Taq DNA polymerase, and 3 μL paraffin extract.
2. Amplification is performed with an initial 94°C for 3 min; 35 cycles of 94°C, 55°C, and 72°C for 30 s each; and a final 72°C for 7 min.
3. PCR products are analyzed on a 1.5% agarose gel stained with ethidium bromide. Fragments are purified with the high pure PCR purification kit (Roche Diagnostics) and 50 ng of purified PCR product is sequenced with the BigDye v.3.1 terminator sequencing kit (Applied Biosystems) using the same primers used for PCR. Reactions are run on an ABI 3130xl (Applied Biosystems) and sequences are edited using Sequencher 4.6 (GeneCodes, MI). Sequences proofread in both directions are compared against GenBank using BLAST.

Note. The primer combination ITS 3/ITS 4, covering the ITS 2 region, amplified a product of about 350 bp from the clinical sample. Sequence analysis of the fragment gave 343 bp of sequence information after trimming of the primer sequences. Comparison against GenBank showed a 99.7% identity (one mismatch over the 343 bp) to the corresponding sequence of *C. cinerea*, the anamorph of which is *H. aspergillata* (see also Lagrou et al. [11] above). Interestingly, other primer combinations covering ITS 1 or parts of 18S rRNA gene did not result in any amplification products in this case.

71.2.2.2 PCR-RFLP Analysis of SSU and ITS 2 Regions

Verweij et al. [9] described the use of primers NS1 and NS24 (for amplification of SSU) and primers ITS 1 and ITS 4 (for amplification of ITS 1 and 2, including 5.8S ribosomal RNA [rRNA]) for identification of *Coprinopsis cinerea* and other molds. Amplicons are digested with the restriction enzymes *Hin*fI, *Hae*III, *Rsa*I, and *Dde*I and electrophoresed on 1.4% agarose gels. Patterns are compared by using the Image Master (Pharmacia) software package for evaluation of molecular weights. The fungus had different patterns to the other fungi analyzed.

71.3 CONCLUSIONS

Hormographiella aspergillata is the anamorph of *Coprinopsis cinereus* (*Coprinus cinereus*), which is a basidiomycete with a typical mushroom form normally occurring in compost and sewage [2]. *H. aspergillata* has been associated with human infections such as pneumonia, endocarditism, and endophthalmitis. Due to the difficulty of identification on the basis of macroscopic and microscopic features, filamentous basidiomycetes such as *Hormographiella* species are poorly recognized in the clinical laboratory. Application of PCR and sequencing analysis of the ITS2 and other gene regions permits rapid and accurate identification of *H. aspergillata*, shortening the time needed for its diagnosis from as much as 4–6 weeks to a day, and hence facilitating early antifungal therapy.

REFERENCES

1. The UniProt Consortium. Available at http://www.uniprot.org/, accessed on August 1, 2010.
2. Guarro, J. et al. (1992). *Hormographiella*, a new genus of hyphomycetes from clinical sources. *Mycotaxon*, 45, 179–190.
3. Gené, J. et al. (1996). Molecular characterization, relatedness and antifungal susceptibility of the basidiomycetous *Hormographiella* species and *Coprinus cinereus* from clinical and environmental sources. *Anton Leeuwen*, 70(1), 49–57.
4. Abuali, M. M. et al. (2009). *Rhizomucor variabilis* var. *regularior* and *Hormographiella aspergillata* infections in a leukemic bone marrow transplant recipient with refractory neutropenia. *J. Clin. Microbiol.*, 47(12), 4176–4179. Erratum in: *J. Clin. Microbiol.*, 2010;48:1018.
5. Rampazzo, A. et al. (2009). *Hormographiella aspergillata* keratomycosis in a dog. *Vet. Ophthalmol.*, 12, 43–47.
6. Speller, D. E. and MacIver, A. G. (1971). Endocarditis caused by a *Coprinus* species: A fungus of the toadstool group. *J. Med. Microbiol.*, 4, 370–374.
7. Bartz-Schmidt, K. U. et al. (1996). Chronic basidiomycetous endophthalmitis after extracapsular cataract extraction and intracapsular lens implantation. *Graefe's Arch. Clin. Exp. Ophthalmol.*, 234, 591–593.
8. Nenoff, P. et al. (1997). Rare fatal simultaneous mould infection of the lung caused by *Aspergillus flavus* and the basidiomycete *Coprinus* sp. in a leukemic patient. *J. Med. Vet. Mycol.*, 35, 65–69.
9. Verweij, P. E. et al. (1997). Fatal pulmonary infection caused by the basidiomycete *Hormographiella aspergillata*. *J. Clin. Microbiol.*, 35, 2675–2678.
10. Surmont, I. et al. (2002). A pulmonary infection caused by *Coprinus cinereus* (*Hormographiella aspergillata*) diagnosed after a neutropenic episode. *Med. Mycol.*, 40, 217–219.
11. Lagrou, K. et al. (2005). Fatal pulmonary infection in a leukaemic patient caused by *Hormographiella aspergillata*. *J. Med. Microbiol.*, 54, 685–688.
12. Conen, A. et al. (2010). *Hormographiella aspergillata*: An emerging mould in acute leukaemia patients? *Clin. Microbiol. Infect.*, Published Online: May 18, 2010.
13. Greer, E.L. et al. (2008). Truffle's revenge: A pig-eating fungus. *Cardiovasc. Pathol.*, 17(5), 342–343.
14. De Baere, T. et al. (2002). Identification of cultured isolates of clinically important yeast species using fluorescent fragment length analysis of the amplified internally transcribed rRNA spacer 2 region (ITS2). *BMC Microbiol.*, 2, 21.
15. Scorzetti G. et al. (2002). Systematics of basidiomycetous yeasts: A comparison of large subunit D1/D2 and internal transcribed spacer rDNA regions. *FEMS Yeast Res.*, 2(4), 495–517.
16. Romanelli, A. M. et al. (2010). Sequence-based identification of filamentous basidiomycetous fungi from clinical specimens: A cautionary note. *J. Clin. Microbiol.*, 48(3), 741–752.
17. Zeng, J.S. et al. (2007). Spectrum of clinically relevant *Exophiala* species in the United States. *J. Clin. Microbiol.*, 45(11), 3713–3720.

72 *Cryptococcus*

Massimo Cogliati, Anna Maria Tortorano, and Maria Anna Viviani

CONTENTS

72.1 INTRODUCTION

72.1.1 Classification, Morphology, and Biology

The yeasts of the genus *Cryptococcus* are spheroidal, ovoidal, or elongate in shape, and reproduce mainly by multilateral budding. Key characteristics of the genus are the ability to assimilate inositol as sole carbon source, produce urease, react with diazonium blue B, and the lack of fermentative ability. Most of the isolates have a glucoronoxylomannan (GXM) capsule that differs in size according to the strain or to environmental factors [1].

At least 40 *Cryptococcus* species have been found in a wide variety of environmental locations. Among clinical isolates, *C. neoformans* and *C. gattii* are the most frequent causes of disease, although other species such as *C. albidus*, *C. laurentii*, *C. uniguttulatus*, *C. humiculus*, *C. curvatus*, and *C. luteolus* have been reported to cause infection occasionally [1].

Recognized as separate species, *C. neoformans* and *C. gattii* belong to two distinct monophyletic lineages and exhibit morphological differences during sexual reproduction.

More than 35 years ago, Kwon-Chung [2,3] described the teleomorph, *Filobasidiella*, belonging to the order *Tremellales* and phylum *Basidiomycota*. Conjugation between mating type (MAT) α and **a** cells produces dikaryotic hyphae with clamp connections. A basidium is formed at the tip of the hypha. After karyogamy and meiosis, four meiotic nuclei are formed that produce, by mitosis, four chains of uninucleated basidiospores released from the basidium. MAT α and, more rarely, MAT **a** cells alone are also able to produce basidiospores by monokaryotic fruiting [4]. Furthermore, same-sex reproduction may occur between two different α strains [5].

F. neoformans, *C. neoformans* teleomorph, produces spheroidal, oblong, finely roughened basidiospores, while *F. bacillispora*, *C. gattii* teleomorph, produces smooth bacillary basidiospores [2,3]. In addition, these two species differ in the ability to hydrolyze glycine and in the resistance to canavanine. In presence of *C. gattii*, the canavanine-glycine-bromothymol (CGB) agar turns blue within 2–5 days, while no reaction occurs in presence of *C. neoformans*. *C. neoformans* has an EDTA-resistant urease activity and does not assimilate D-proline in comparison with *C. gattii* [6].

The major environmental source of *C. neoformans* is avian, mainly pigeon, droppings due to the ability of this species to utilize creatinine as a source of nitrogen by a creatinine deiminase that is repressed by the overproduction of ammonia. On the contrary, the absence of the enzyme repression in *C. gattii* causes overproduction of ammonia, strong alkalinization of the substrate, and inhibition of the yeast growth in avian droppings [6]. Recently, pigeon guano has been shown to support the growth of both species, but *C. gattii* cannot sexually reproduce. Therefore, pigeon guano does not represent a niche for this species [7].

C. neoformans and *C. gattii* have been isolated from soil and trees, mainly from decaying wood in the hollows. The colonization of decaying wood and tree hollows by both *C. neoformans* and *C. gattii* is enhanced by the production of a laccase enzyme with phenoloxidase activity, which degrades lignin [6].

On the basis of the specific structure of GXM, the main capsule component, *C. neoformans*, is classified into three serotypes (A, D, and AD), and *C. gattii* in two serotypes (B and C). Serotype A corresponds to *C. neoformans* var. *grubii* and serotype D to *C. neoformans* var. *neofomans* [6,8]. GXM structure alterations may occur at low rate determining the colony phenotypic switching phenomenon [9]. *Cryptococcus* colony may switch from mucoid to smooth to wrinkled phenotype or vice versa. Mucoid and wrinkled phenotypes are reported to be more virulent than the smooth ones in mouse models [10].

C. neoformans is ubiquitous, whereas *C. gattii*, isolated firstly from *Eucalyptus camaldulensis* and from several tree species in Australia and in other tropical and subtropical regions, has recently been isolated from different native tree species on Vancouver Island, British Columbia, Canada, and from the surrounding soil and air [11]. The specific route of the recent introduction of *C. gattii* to the Pacific Northwest has not yet been established [11].

Cryptococcus MAT distribution is a further enigma that still remains to be solved. MAT α population represents the majority of the isolates, whereas MAT **a** cells continue to survive in specific niches. In addition, the MAT **a** locus is conserved in hybrid diploid strains, which act as reservoir of this allele. A possible explanation for this distribution bias is the ability of α cells to reproduce by both monokaryotic fruiting and same-sex mating [5]. Genotyping of several *C. neoformans* and *C gattii* isolates by PCR fingerprinting [12,13], amplified fragment length polymorphism (AFLP) [14], and multilocus sequence typing (MLST) [15–17] showed that the prevalent genotype in the world is VNI (var. *grubii*) and that VNIV genotype (var. *neoformans*) is present not only in Europe [18] but also in North [19] and South Americas [12], India [20], and Japan [21]. Interestingly, in these regions, a consistent number of VNIII strains were identified that were shown to be diploid hybrids originated from an intervarietal mating between VNI and VNIV strains [22,23]. Intracluster VNI diploids were also observed among Botswana and North Carolina isolates [24]. Four molecular types (VGI, VGII, VGIII, and VGIV), not corresponding to serotypes B or C, were also identified in *C. gattii* species and a specific *C. gattii* genotype, VGIIa, was shown to be responsible of the outbreak in Canada [15].

72.1.2 Clinical Features and Pathogenesis

Several factors contribute to the virulence of *C. neoformans* and *C. gattii* [25]. The ability to grow at 37°C is essential for survival in the human host. The polysaccharide capsule, composed of GXM (more than 90%) and galactoxylomannan, protects the yeast cell from phagocytosis and interferes with host immunity. As a facultative intracellular pathogen, the yeast releases the capsular polysaccharide intracellularly, and the accumulation of the polysaccharide in the cytoplasmic vesicles results in macrophage dysfunction and lysis. In addition, the polysaccharide represses the migration of neutrophils, interferes with cytokine secretion, inhibits T-cell proliferation, induces macrophage apoptosis, and delays maturation and activation of human dendritic cells [25]. Another important virulence factor is the melanin produced from polymerization of phenolic compounds, such as cathecolamins, by the fungal laccase. The melanin is deposited in the cell wall conferring the brown color to the yeast cells. Melanization protects the yeast from oxidative damage by scavenging host-produced antioxidants [26] and reduces the fungal susceptibility to amphotericin B and caspofungin [27]. Melanin is essential for extrapulmonary dissemination as it allows survival of the yeasts inside the alveolar macrophages and their transport into the lymph nodes and then into the bloodstream. The existence in the brain of melanin precursors (L-dopamine and epinephrine) has been hypothesized to explain the neurotropism of cryptococci. Other hypotheses about the fungus neurotropism concern the presence in the brain of specific receptors that attract the yeast or the yeast protection in the central nervous system (CNS) from the host immune responses [8]. The production of capsule and melanin was shown to be linked to the expression of several genes contained in the MAT locus. MAT **α** cells were shown in animal model to be more virulent than a cells [28,29] and to be more efficient than **a** cells in crossing the blood–brain barrier during coinfection [30]. In addition, the Vancouver hypervirulent clone of *C. gattii* was proven to descend from two α MAT parents [15]. Other enzymes, namely, protease, phospholipase, and urease, are also considered virulence factors.

Cryptococcal infection is acquired by inhalation from an environmental source of the infectious propagules (basidiospores or poorly or no capsulated dried yeast cells of less than 3 μm in diameter) that can reach the alveoli where they rehydrate and acquire the polysaccharide capsule [6,8]. The pulmonary infection is typically asymptomatic. The yeasts can be either cleaned by macrophages or remain as a dormant latent form within a granuloma. When host immunity is compromised, the dormant yeasts can reactivate, proliferate inside the macrophages, and be released into the extracellular environment. The released yeasts can enter other phagocytes and dissemination occurs.

The host immune response to cryptococcal infection is the result of a complex interplay between cellular and humoral immunity. Innate immunity acts early and is activated by specific recognition of molecular structures of cryptococcal cells and of secreted products. In the lung, cryptococci first interact with alveolar macrophages and dendritic cells that participate in fungal recognition, phagocytosis, antigen presentation, and activation of the host response. Humoral response with production of a specific set of antibodies can play an important role in the elimination or control of cryptococci in the body [31,32]. However, cellular-mediated immunity remains the critical component for protection against cryptococci [33]. For instance, effector cells against the yeast include CD4+ and CD8+ lymphocytes, NK cells, and activated professional phagocytes that produce granulomatous inflammation. The granuloma formation is the result of TH1-response mediated by IL12. A shift toward a TH2-response probably antagonizes the clearance of the yeast from host tissue. The TH17 lymphocytes have recently been recognized as cause of the exaggerated immune response in AIDS patients with active cryptococcosis treated with highly active antiretroviral therapy (HAART).

Impairment of the host's defences against cryptococci can lead to dissemination of the infection, most likely by migration of macrophages with ingested fungal cells from the lung to the draining lymph nodes and via the bloodstream to cause systemic infection involving many body sites and organs, especially the brain. Headache, fever, lethargy, nausea and vomiting, memory loss, personality changes, stupor, and coma are the signs and symptoms of meningoencephalitis observed in the majority of the patients. In patients with AIDS, symptoms usually appear late in the course of the disease when the fungal burden in the brain is high. In contrast, in non-AIDS patients, the onset is insidious and symptoms may last months and years before diagnosis is made. At lumbar puncture, the opening pressure may be markedly elevated, especially in patients with AIDS due to the high fungal burden [34]. Cerebrospinal fluid (CSF) cell counts are low (≤20/mmc) in AIDS-associated cryptococcosis and up to 150–200/mmc in non-AIDS patients. CSF glucose level is low (<40 mg/dL) and protein level is elevated (>45 mg/dL) mainly in non-AIDS patients [35].

Cryptococcal infections caused by var. *grubii* and var. *neoformans* occur worldwide, mainly in the immunocompromised host. The medical importance of cryptococcosis increased dramatically with the AIDS epidemic. Patients with HIV infection are at risk of cryptococcosis late in the course of the viral infection when the CD4 lymphocytes counts are less than 100 cells/mm^3. Despite the widespread use of HAART, the global burden of HIV-associated cryptococcosis approximates 1 million cases annually. The morbidity and mortality of this fungal disease are still extremely high in developing countries where access to HAART and to healthcare is limited [36]. In medically developed countries, most of the cases occur in patients with newly diagnosed HIV infection and in HIV-negative patients receiving high-dose corticosteroids, or monoclonal antibodies (e.g., alemtuzumab and infliximab) or other immunosuppressive agents for their underlying conditions.

Unlike *C. neoformans*, *C. gattii* causes disease also in otherwise healthy immunocompetent individuals. In these subjects, the most common clinical manifestations are cryptococcoma of the lung and mass lesions in the brain with long-term neurological sequelae [37].

Infection due to *C. gattii* has most often been reported in tropical and subtropical regions; however, since 1999, cases have been identified in domestic and wild animal population and in residents in, and travelers to Vancouver Island and lower mainland British Columbia, Canada. The annual incidence rate (5.8 and 25.1 cases per million on mainland British Columbia and on Vancouver Island, respectively) is the highest worldwide [38]. In recent years, several human and animal cases of *C. gattii* disease have been recognized in Pacific Northwest regions of the United States [38].

72.1.3 Diagnosis

72.1.3.1 Conventional Techniques

The conventional diagnosis of cryptococcosis is based on the microscopic demonstration of the fungus, its growth in culture, and on the capsular antigen detection [6,39].

The India ink examination of the CSF is a rapid test for the diagnosis of cryptococcal meningitis. The yeast cells are visualized within a clear halo (the capsule) against the dark background of the India ink and they can be easily differentiated from leukocytes or fat droplets by the presence of some budding cells. The sensitivity of the test depends on the fungal burden and the volume of the examined sample, and it may be improved by examining the pellet from the centrifuged CSF.

Cryptococci are difficult to observe in routine hematoxylin–eosin histological preparations of bioptic samples. By Gomori methenamine silver and periodic acid Schiff stains, fungal cells are recognized by their oval shape and narrow-based budding. However, the Mayer's mucicarmine stain is needed to reveal the polysaccharide capsule. The presence of the capsule allows the differentiation of cryptococcal cells from the yeast forms of the dimorphic fungi *Blastomyces dermatitidis* and *Histoplasma capsulatum*.

Cryptococci are not fastidious organisms and can be cultured from biological samples on standard fungal media in aerobic conditions. Antibiotics, such as chloramphenicol or gentamycin, should be added to the media to reduce bacterial contamination. On the contrary, cycloheximide should not be added as it inhibits most of the isolates. Cultures are incubated at 30°C–35°C, avoiding temperature ≥37°C. Yeast colonies usually develop in 2–5 days. Nevertheless, incubation up to 3–4 weeks is recommended, mainly if the patient is receiving or has received an antifungal treatment. Cryptococci grow on agar media as white to cream-colored colonies that may turn to brown after prolonged incubation. The mucoid appearance of the colony is related to the size of the capsule that varies with strain and medium composition.

Cryptococcal antigen assays that measure in the body fluids the polysaccharide released from the yeast have proven to be a valuable diagnostic test, more sensitive than India ink examination and culture. The slide agglutination using latex particles (LA) coated with polyclonal antibodies or with anti-GXM monoclonal antibodies is widely used. The specificity and sensitivity of the test performed on serum samples have been improved by pretreatment of the sample with pronase. A titer of at least 1:4 is suggestive of cryptococcal infection and a titre ≥1:8 is indicative of active infection. Antigen titer is much higher in AIDS patients than in HIV-negative individuals. The CSF antigen titer has been shown to be a prognostic factor as high CSF antigen titers correlate with failing host defence and poor outcome [34].

The apparent subjectivity of reading and grading the LA reaction is overcome by the more recently developed enzyme immunoassay (EIA) that measures the major component of the capsular polysaccharide, the GXM. However, the format of the kit, designed for the simultaneous testing of multiple samples, makes it inappropriate for its use in emergency.

Polysaccharide antigen screening in serum of HIV-positive patients with CD4 cell count <100 cells/μL living in areas with a high incidence of cryptococcal disease should be performed before the initiation of antiretroviral therapy in order to diagnose and treat subclinical infections before it is unmasked by the immune reconstitution [40].

A positive serum or CSF antigen titer, even in absence of the isolation in culture or microscopic evidence of the fungus, should be taken into serious consideration.

The measurement of anticryptococcal antibodies is not of consistent clinical benefit for the diagnosis of active infection, while the antibody detection may have a prognostic value in AIDS patients during recovery from active infection, when the antigen titer declines.

Even if European Organization for Research and Treatment of Cancer/Mycoses Study Group (EORTC/MSG) definitions state that a positive result of CSF India ink or of serum or CSF LA is sufficient to establish a diagnosis of proven cryptococcosis [41], an extended diagnostic workup is strongly recommended including culture of blood, CSF, urine, bronchoalveolar lavage in case of lung involvement, and biopsy sample of skin lesions.

72.1.3.2 Molecular Techniques

Molecular techniques offer an excellent alternative to conventional methods for an early diagnosis of most important fungal diseases as they are rapid, can detect low fungal load, and can be used for small size samples. However, molecular methods for the diagnosis of cryptococcosis have poorly been developed because of the limited demand of these tests due to the high yield of polysaccharide antigen detection tests.

Prariyachatigul et al. [42] first applied PCR in the detection and identification of *C. neoformans* in 27 CSF samples, 20 of which from patients having or suspected to have cryptococcal meningitis. Sensitivity of PCR for direct detection of *C. neoformans* in CSF was high compared to the CSF culture (17 positive out of 20 proven or highly suspected cases vs. 11/20), whereas it was low compared with LA (20/20) and India ink examinations (19/20). The authors suggested that these results were due to the ability of LA and India ink to detect cells or antigens, no matter whether the yeasts were dead or alive, while the reduced sensitivity of PCR was attributed to the rapid degradation of DNA.

Rappelli et al. [43] developed a nested-PCR assay to detect *C. neoformans*-specific internal transcribed sequence 1 (ITS1) sequence in CSF. They obtained positive reactions with all 21 clinical samples from patients previously diagnosed as having cryptococcal meningitis by conventional techniques, and negatives reactions with all 19 negative controls.

Paschoal et al. [44] examined a total of 72 CSF samples, obtained from patients with and without AIDS, by specific PCR of ITS1–5.8S–ITS2 regions. The results demonstrated that PCR test had the highest sensitivity rate, superior to culture (85.7%) and India ink test (76.8%).

Recently, a nested PCR of ITS regions was compared with conventional diagnostic techniques by Saha et al. [45] who investigated 359 samples from 52 patients with cryptococcosis and 30 negative controls. The results indicated that the sensitivity and specificity of PCR, EIA, and LA were comparable using urine, CSF, and serum for the diagnosis of cryptococcosis, both before and after the initiation of treatment.

A reverse cross-blot hybridization assay, based on *ERG11* sequence, was applied by Posteraro et al. [46] to detect important medical yeasts including *C. neoformans* in clinical samples. The results agreed with those of culture and phenotyping for the two CSF positive samples from patients with cryptococcosis. The authors concluded that the PCR assay is a rapid method (7 h) compared to conventional identification (4 days).

The usefulness of molecular methods in the diagnosis of cryptococcosis was in particular demonstrated by Abbott et al. [47], who employed in situ hybridization (ISH) utilizing oligonucleotide probes directed against fungal ribosomal RNA to detect *C. neoformans* and other dimorphic fungi in paraffin-embedded tissue sections. The ISH confirmed the diagnosis of the five cases, two of which were culture-proven cryptococcosis. ISH should be limited to cases with a positive histological diagnosis for yeast or dimorphic fungi. The method offers a valuable tool in the rapid identification of fungal infections, especially in circumstances when tissue cultures are an afterthought.

Real-time PCR for *C. neoformans* identification in clinical samples was applied in two different studies. Lau et al. [48] set up a multiplex PCR, with amplicons being detected via analysis of the melting curve in a real-time thermocycler. The method was able to detect and identify simultaneously 12 different fungal pathogens including *C. gattii* and *C. neoformans*. Seventy blood specimens previously identified by conventional methods were analyzed. Molecular identification was 100% in agreement with culture-based identification including three cryptococcal cases.

Veron et al. [49] reported a TaqMan® real-time PCR protocol for diagnosing cryptococcosis. DNA extracted from 101 various biological samples were investigated and the seven samples culture-positive for *C. neoformans* or *C. gattii* also tested positive by real-time PCR. No false negatives were found among the 94 clinical samples that were *Cryptococcus* culture-negative.

Although molecular biology is still scarcely applied in the routine diagnosis of cryptococcosis due to the lack of standardized, easy to use procedures in clinical laboratory, it has greatly contributed to the progress of epidemiological studies on cryptococcosis where important progresses toward standardization of the genotyping method have been made.

Molecular techniques as PCR fingerprinting [12,13], AFLP [14], and MLST [15–17] have been employed to genotype a large number of clinical and environmental *Cryptococcus* strains isolated from different part of the world.

Recently, Meyer et al. [50] reached a consensus in the establishment of a standard nomenclature for the main genotyping clusters: VNI and VNII (var. *grubii*), VIII (AD-hybrids), VNIV (var. *neoformans*), and VGI, VGII,

VGIII, and VGIV (var. *gattii*). The authors also described a reference MLST protocol that includes seven housekeeping genes (*URA5*, *PLB1*, *LAC1*, *SOD1*, *CAP59*, *IGS1*, and *GPD1*). At present, gene sequences obtained with the above protocol can be recorded in a dynamic database (Multi Locus Sequence Typing) at http://www.mlst.net.

72.2 METHODS

72.2.1 Sample Preparation

1. DNA extraction from serum, CSF, and urine (Saha et al., 2009) [45]

   Procedure

   1. Boil the sample for 5 min.
   2. Centrifuge at 1000×*g* for 15 min.
   3. Discard supernatant to obtain 200 µL pellet.
   4. Add 500 µL of guanidine thiocyanate (6M), 500 µL of phenol saturated in Tris(hydroxymethyl) aminomethane (Tris), and mix.
   5. Boil for 15 min.
   6. Centrifuge at 1000×*g* for 15 min.
   7. Discard supernatant to obtain 200 µL pellet.
   8. Add 180 µL of ATL buffer (Qiagen), 20 µL of proteinase K (20 mg/mL), and mix.
   9. Incubate at 56°C for 30 min.
   10. Briefly spin to remove drops from the inside of the microcentrifuge tube lid.
   11. Add 200 µL ethanol (96%–100%) to the sample and mix again by pulse-vortexing for 15 s. After mixing, briefly spin to remove drops from the inside of the microcentrifuge tube lid.
   12. Carefully apply the mixture to the QIAamp spin column (Qiagen), in a 2 mL collection tube, and centrifuge at 6000×*g* for 1 min. Place the spin column in a clean 2 mL collection tube and discard the tube containing the filtrate.
   13. Add 500 µL AW1 buffer (Qiagen) and centrifuge at 6000×*g* for 1 min. Place the spin column in a clean 2 mL collection tube and discard the collection tube containing the filtrate.
   14. Add 500 µL AW2 buffer (Qiagen) and centrifuge at full speed for 3 min.
   15. Place the spin column in a clean 1.5 mL microcentrifuge tube and discard the collection tube containing the filtrate.
   16. Add 50 µL of warm AE buffer (Qiagen).
   17. Incubate at room temperature for 5 min and then centrifuge at 6000×*g* for 1 min.
   18. Store the elution at −20°C.

2. DNA extraction from tissue samples (Hendolin et al., 2000) [51]

   Procedure

   1. Cut tissue samples in 5–25 mg pieces, place them in a 1.5 mL microcentrifuge tube, and add 180 µL of ATL buffer (Qiagen).
   2. Add 20 µL of proteinase K (10 mg/mL) mix by vortexing, and incubate at 56°C until the tissue is completely lysed. To disperse the sample, vortex the microcentrifuge tube occasionally during incubation or place it in a shaking water bath or on a rocking platform.
   3. Briefly spin to remove drops from the inside of the microcentrifuge tube lid.
   4. Add 25 µL of 250 mM NaOH, 25 µL sodium dodecyl sulphate (SDS) 5%, and incubate at 95°C for 10 min.
   5. Neutralize with HCl.
   6. Add 200 µL AL buffer (Qiagen) to the sample, mix by pulse-vortexing for 15 s, and incubate at 70°C for 10 min. Briefly spin to remove drops from the inside of the microcentrifuge tube lid.
   7. Add 200 µL ethanol (96%–100%) to the sample, and mix by pulse-vortexing for 15 s. After mixing, briefly spin to remove drops from the inside of the microcentrifuge tube lid.
   8. Carefully apply the mixture to the QIAamp spin column (Qiagen) in a 2 mL collection tube and centrifuge at 6000×*g* for 1 min. Place the spin column in a clean 2 mL collection tube and discard the tube containing the filtrate.
   9. Add 500 µL AW1 buffer (Qiagen) and centrifuge at 6000×*g* for 1 min. Place the spin column in a clean 2 mL collection tube and discard the collection tube containing the filtrate.
   10. Add 500 µL AW2 buffer (Qiagen) and centrifuge at full speed for 3 min.
   11. Place the spin column in a clean 1.5 mL microcentrifuge tube and discard the collection tube containing the filtrate.
   12. Add 200 µL AE buffer (Qiagen) or DNA-free distilled water.
   13. Incubate at room temperature for 5 min and then centrifuge at 6000×*g* for 1 min.
   14. Store the elution at −20°C.

3. DNA extraction from cultures (Viviani et al., 1997) [13]

   Procedure

   1. Culture *Cryptococcus* isolate on two Sabouraud dextrose agar slants for 48 h at 35°C.
   2. Gently wash with 2 mL of sterile distilled water the agar surface of the slants.
   3. Transfer the yeast suspension in a 2 mL microcentrifuge tube.
   4. Count the cell concentration using a Bürker hemocytometer and dilute the suspension to 10^8 yeasts/mL.
   5. Centrifuge 2 mL of the 10^8/mL yeast suspension at 1300×*g* for 5 min.
   6. Discard supernatant and suspend the pellet in 1 mL of protoplasting solution: 20 mg lysing enzymes [Sigma-Aldrich] in 1 mL sodium

citrate–sorbitol [SCS] (2 mL of 1 M sodium citrate [pH 5.8] and 98 mL of 1 M sorbitol).

7. Seal the microcentrifuge tube with parafilm and incubate in a water bath at 37°C for 1 h.
8. Mix 10 μL of 2% SDS (in DNAase-free distilled water) and 10 μL of cell suspension on a slide and check protoplast formation at phase-contrast light microscopy. If protoplasts do not reach about the 90% of the cells, incubate the suspension at 37°C for a longer time.
9. Centrifuge at 1300 × *g* for 5 min and discard the supernatant.
10. Suspend the pellet in 600 μL of lysing buffer (8% dodecyltrimethyl-ammonium bromide [DTAB, Sigma-Aldrich], 1.5 M NaCl, 100 mM Tris [pH 8.6], 50 mM EDTA [Sigma-Aldrich]).
11. Incubate in a water bath at 68°C for 30 min.
12. Add 900 μL of cold chloroform and mix by inverting the tube.
13. Centrifuge at 11,600 × *g* for 10 min.
14. Transfer 350 μL of the upper aqueous phase in a 1.5 mL microcentrifuge tube.
15. Add 100 μL of CTAB (5% hexadecyltrimethyl-ammonium bromide [CTAB, Sigma-Aldrich] in 0.4 M NaCl) and gently mix.
16. Add 700 μL of DNase-free distilled water and mix by inverting the tube.
17. Centrifuge at 8000 × *g* for 2 min and discard the supernatant.
18. Add 150 μL of 1.2 M NaCl and gently mix till the pellet dissolve.
19. Add 500 μL of ice-cold 99%–100% ethanol and mix by inverting the tube.
20. Centrifuge at 8000 × *g* for 10 min and discard the supernatant.
21. Add 500 μL of 70% ethanol and mix by inverting the tube.
22. Centrifuge at 8000 × *g* for 10 min and discard the supernatant.
23. Spin dry.
24. Add 30–50 μL of DNA-free distilled water and gently mix till the pellet dissolve.
25. To store the DNA for long time, add 3–5 μL of Tris EDTA 10× solution (10 mM Tris PH 7.4, 1 mM EDTA)
26. Store at −20°C.

72.2.2 Detection Procedures

72.2.2.1 PCR Identification

1. Real-time PCR identification (Veron et al., 2009) [49]

Procedure

1. Amplification mixture in a final 25 μL volume:
    - 1× TaqMan Gene Expression Master Mix (Applied Biosystems)
    - 0.9 μM of each primer, CneoFrw (5′-GCCGCGACCTGCAAAG-3′) and CneoRev (5′-GGTAATCACCTTCCCACTAACACAT-3′)
    - 0.25 μM of CneoProbe (5′-FAM-ACGTCGGCTCGCC-MGB-3′) (MGB: minor groove binder, Applied Biosystems)
    - 3 μL TaqMan Exogenus Internal Positive Control (IPC, Applied Biosystems)
    - 5 μL of DNA template
2. Run real-time PCR on an ABI Prism 7300 (Applied Biosystems)

2. Nested-PCR identification (Bialek et al., 2002) [52]

Procedure

1. First-round PCR amplification mixture in a final 50 μL volume:
    - 10 mM Tris–HCl (pH 8.3), 50 mM KCl, and 2.5 mM $MgCl_2$ (10× buffer II plus $MgCl_2$ solution, Applied Biosystems)
    - 100 μM of each deoxynucleoside triphosphate
    - 1 μM of each primer, Fungus I (5′-GTTAAAAAGCTCGTAGTTG-3′) and Fungus II (5′-TCCCTAGTCGGCATAGTTTA-3′)
    - 1.5 U of AmpliTaq DNA polymerase (Applied Biosystems)
    - 10 μL of template DNA
2. Thermal cycling conditions:
    - Initial denaturation at 94°C for 5 min
    - 35 cycles at 94°C for 30 s, 50°C for 30 s, and 72°C for 1 min
    - Final extension at 72°C for 5 min
3. Nested-PCR amplification mixture in a final 50 μL volume:
    - 10 mM Tris–HCl (pH 8.3), 50 mM KCl, and 2.5 mM $MgCl_2$ (10× buffer II plus $MgCl_2$ solution, Applied Biosystems)
    - 50 μM of each deoxynucleoside triphosphate
    - 1 μM of each primer, Cryp I (5′-TCC TCA CGGAGTGCACTGTCTTG-3′) and Cryp II (5′-CAGTTGTTGGTCTTCCGTCAATCTA-3′)
    - 1.5 U of AmpliTaq DNA polymerase (Applied Biosystems)
    - 1 μL of the first-round amplification reaction mixture
4. Thermal cycling conditions (two-step PCR):
    - Initial denaturation at 94°C for 5 min
    - 30 cycles at 94°C for 30 s and 72°C for 1 min
    - Final extension at 72°C for 5 min

3. 18S rDNA fragment amplification and hybridization (Prariyachatigul et al., 1996) [42]

Procedure

1. PCR amplification mixture in a final 50 μL volume:
    - 10 mM Tris–HCl (pH 8.3), 50 mM KCl (10× buffer II solution, Applied Biosystems)
    - 2.5 mM $MgCl_2$

- 200 μM of each deoxynucleoside triphosphate
- 25 pmol of each primer, CPL1 (5′-GGAGGTAG TGACAATAAATA-3′) and CPR4 (5′-TGCTAATGTATTCGGGCGATT-3′)
- 1 U of AmpliTaq DNA polymerase (Applied Biosystems)
- 10 μL of genomic DNA

2. Thermal cycling conditions:
   - Initial denaturation at 94°C for 5 min
   - 50 cycles at 94°C for 45 s, 62°C for 45 s, and 72°C for 45 s
   - Final extension at 72°C for 5 min
3. Hybridization
   - Transfer DNA from the gel to a nylon membrane Hybond-N+ (Amersham Biosciences) by capillary transfer
   - Hybridize the membrane with the species-specific probe INSR4 (5′-AGGCGGTCC TCCTCACGGAGTGC-3′) labeled with a colorimetric detection system (Dig Oligonucleotide 3′-end labeling and detection system, Boehringer, Mannheim)
   - Wash the membrane according to manufacturer's instructions and develop it for color

4. Internal transcribed sequence 1 (ITS1 rDNA fragment amplification (White et al., 1990, modified) [53]

   Procedure
   1. Amplification mixture in a final 50 μL volume:
      - 10 mM Tris–HCl (pH 8.3), 50 mM KCl (10× buffer solution, Applied Biosystems)
      - 3 mM $MgCl_2$
      - 200 μM of each deoxynucleoside triphosphate
      - 50 pmol of each primer, CN4 (5′-ATCACCT TCCCACTAACACATT-3′) and CN5 (5′-GAAGGGCATGCCTGTTTGAGAG-3′)
      - 2.5 U of AmpliTaq DNA polymerase (Applied Biosystems)
      - 100 ng of genomic DNA
   2. Thermal cycling conditions:
      - Initial denaturation at 94°C for 3 min
      - 25 cycles at 94°C for 30 s, 62°C for 30 s, and 72°C for 1 min
      - Final extension at 72°C for 10 min

72.2.2.2 Molecular Typing

1. PCR fingerprinting (Meyer et al., 2003) [12]

   Procedure
   1. Amplification mixture in a final 50 μL volume:
      - 10 mM Tris–HCl (pH 8.3), 50 mM KCl, and 1.5 mM $MgCl_2$ (10× buffer plus $MgCl_2$ solution, Applied Biosystems)
      - 3 mM magnesium acetate
      - 200 μM of each deoxynucleoside triphosphate
      - 30 ng of M13 (5′-GAGGGTGGCGGTTCT-3′) primer
      - 2.5 U of AmpliTaq DNA polymerase (Applied Biosystems)
      - 25 ng of genomic DNA
   2. Thermal cycling conditions:
      - 35 cycles at 94°C for 20 s, 50°C for 1 min, and 72°C for 20 s
      - Final extension at 72°C for 6 min
   3. Reference strains:
      ATCC MYA-4564 (WM148, serotype A, VNI), ATCC MYA-4565 (WM626, serotype A, VNII), ATCC MYA-4566 (WM628, serotype AD, VNIII), ATCC MYA-4567 (WM629, serotype D, VNIV), ATCC MYA-4560 (WM179, serotype B, VGI), ATCC MYA-4561 (WM178, serotype B, VGII), ATCC MYA-4562 (WM161, serotype B, VGIII), and ATCC MYA-4563 (WM779, serotype C, VGIV).

2. Multilocus sequence typing (Meyer et al., 2009) [50]

   Procedure
   1. Amplification mixture in a final 50 μL volume:
      - 10 mM Tris–HCl (pH 8.3), 50 mM KCl (10× buffer solution, Applied Biosystems)
      - 3 mM $MgCl_2$
      - 200 μM of each deoxynucleoside triphosphate
      - 20 pmol of forward and reverse primers
      - 2.5 U of AmpliTaq DNA polymerase (Applied Biosystems)
      - 100 ng of genomic DNA
   2. Primer sequences and thermal cycling conditions are reported in Table 72.1.

72.3 CONCLUSION

Molecular diagnosis of cryptococcal disease is far from being a standardized procedure in clinical laboratory where conventional techniques continue to represent the gold standard. Antigen detection, despite the rare occurrence of false negative (due to poorly or noncapsulated yeasts) or false positive results (due to cross reactions with *Trichosporon* infection), remains a valid diagnostic tool, especially in emergency as it provides diagnosis in less than 1 h.

Molecular techniques could provide some advantages compared to conventional methods as DNA detection has high target specificity and may be able to couple detection and identification simultaneously. This specificity is reached also by culture method that, however, requires longer time for fungal growth and identification.

Molecular techniques might be particularly useful for the analysis of biological samples from nonsterile sites, where high contamination by *Candida* may mask the growth of cryptococci in culture medium. Among molecular-biology techniques, ISH has proven to be a valuable diagnostic adjunct to positive histology as it provides a rapid identification of the fungus in the tissue when culture is an afterthought.

A technical problem of molecular methods is represented by difficulties in the DNA extraction due to *Cryptococcus*

TABLE 72.1
Multilocus Sequence Typing Scheme

Gene Locus and Primer Sequence (5′-3′) (Reference)	PCR Cycling Conditions	Start Sequence	End Sequence	Sequence Length (bp)
rRNA intergenic spacer 1 (IGS1) IGS1F ATCCTTTGCAGACGACTTGA IGS1R GTGATCAGTGCATTGCATGA [16]	94°C 3 min—35 cycles: 94°C 30 s, 60°C 30 s, 72°C 1 min—72°C 10 min	TAAGCCCTTGTTAA	AGATTTATTG	723
Glyceraldehyde-3-phosphate dehydrogenase (GPD1) GPD1F CCACCGAACCCTTCTAGGATA GPD1R CTTCTTGGCACCTCCCTTGAG [15]	94°C 3 min—35 cycles: 94°C 45 s, 63°C 1 min, 72°C 2 min—72°C 10 min	GGTTTCGGTACGG	GACCCTGCCAA	543
Laccase (LAC1) LAC1F AACATGTTCCCTGGGCCTGTG LAC1R ATGAGAATTGAATCGCCTTGT [15]	94°C 3 min—30 cycles: 94°C 30 s, 58°C 30 s, 72°C 1 min—72°C 10 min	GTAAGTATCAGCT	CAAGCTAAACA	469
Phospholipase B (PLB1) PLB1F CTTCAGGCGGAGAGAGGTTT PLB1R GATTTGGCGTTGGTTTCAGT [16]	94°C 3 min, 30 cycles: 94°C 45 s, 61°C 45 s, 72°C 1 min—72°C 5 min	TGTTACTTGGATT	CTGGAACATCG	532
Orotidine monophosphate pyrophosphorylase (URA5) URA5F ATGTCCTCCCAAGCCCTCGAC URA5R TTAAGACCTCTGAACACCGTACTC [12]	94°C 3 min—35 cycles: 94°C 45 s, 63°C 1 min, 72°C 2 min—72°C 5 min	TTTTCGGCAACTCT	TGGAAAGCTC	601
Capsule-associated protein (CAP59) CAP59F CTCTACGTCGAGCAAGTCAAG CAP59R TCCGCTGCACAAGTGATACCC [15]	94°C 3 min—35 cycles: 94°C 30 s, 56°C 30 s, 72°C 1 min—72°C 5 min	ACGGTACGCGCC	GAGACAGAATG	559
Cu, Zn superoxyde dismutase (SOD1) SOD1CNF AAGCCTCTCATCCATATCTT SOD1CNR TTCAACCACGAATATGTA SOD1CGF GATCCTCACGCCATTACG SOD1CGR GAATGATGCGCTTAGTTGGA [54]	94°C 3 min—35 cycles: 94°C 30 s, 52°C 30 s, 72°C 1 min 30 s—72°C 5 min	CCACGTGCTCGCA	CCTGTCAATGC	700

intracellular growth or to the presence of a large polysaccharide capsule. So far, molecular techniques have yet to supersede conventional methods for diagnosis of *Cryptococcus* infections. On the contrary, molecular techniques have been applied successfully for advances in the knowledge of *Cryptococcus* ecology, epidemiology, and pathogenesis.

REFERENCES

1. Barnett, J., Payne, R., and Yarrow, D., *Yeasts: Characteristics and Identification*, 2nd edn. Cambridge, MA: Cambridge University Press; 1990.
2. Kwon-Chung, K.J., Morphogenesis of *Filobasidiella neoformans*, sexual state of *Cryptococcus neoformans*. *Mycologia*, 68, 821, 1976.
3. Kwon-Chung, K.J., A new species of *Filobasidiella*, the sexual state of *Cryptococcus neoformans* B and C serotypes. *Mycologia*, 68, 943, 1976.
4. Wickes, B.L. et al., Dimorphism and haploid fruiting in *Cryptococcus neoformans*: Association with the alpha-mating type. *Proc. Natl. Acad. Sci. U.S.A.*, 93, 7327, 1996.
5. Lin, X., Hull, C.M., and Heitman, J., Sexual reproduction between partners of the same mating type in *Cryptococcus neoformans*. *Nature*, 434, 1017, 2005.
6. Casadevall, A. and Perfect, J., *Cryptococcus neoformans*. Washington, DC: American Society for Microbiology; 1990.
7. Nielsen, K., De Obaldia, A.L., and Heitman, J., *Cryptococcus neoformans* mates on pigeon guano: Implications for the realized ecological niche and globalization. *Eukar. Cell*, 6, 949, 2007.
8. Lin, X. and Heitman, J., The biology of the *Cryptococcus neoformans* species complex. *Ann. Rev. Microbiol.*, 60, 69, 2006.
9. Goldman, D.L. et al., Phenotypic switching in the human pathogenic fungus *Cryptococcus neoformans* is associated with changes in virulence and pulmonary inflammatory response in rodents. *Proc. Natl. Acad. Sci. U.S.A.*, 95, 14967, 1998.
10. Jain, N. and Fries, B.C., Phenotypic switching of *Cryptococcus neoformans* and *Cryptococcus gattii*. *Mycopathologia*, 166, 181, 2008.
11. Kidd, S.E. et al., Characterization of environmental sources of the human and animal pathogen *Cryptococcus gattii* in British Columbia, Canada, and the Pacific northwest of the United States. *Appl. Environ. Microbiol.*, 73, 1433, 2007.
12. Meyer, W. et al., IberoAmerican Cryptococcal S. Molecular typing of IberoAmerican *Cryptococcus neoformans* isolates. *Emerg. Infect. Dis.*, 9, 189, 2003.
13. Viviani, M.A. et al., Identification by polymerase chain reaction fingerprinting of *Cryptococcus neoformans* serotype AD. *J. Med. Vet. Mycol.*, 35, 355, 1997.
14. Boekhout, T. et al., Hybrid genotypes in the pathogenic yeast *Cryptococcus neoformans*. *Microbiology*, 147, 891, 2001.
15. Fraser, J.A. et al., Same-sex mating and the origin of the Vancouver Island *Cryptococcus gattii* outbreak. *Nature*, 437, 1360, 2005.
16. Litvintseva, A.P. et al., Multilocus sequence typing reveals three genetic subpopulations of *Cryptococcus neoformans* var. *grubii* (Serotype A), including a unique population in Botswana. *Genetics*, 172, 2223, 2006.

17. Bovers, M. et al., Six monophyletic lineages identified within *Cryptococcus neoformans* and *Cryptococcus gattii* by multi-locus sequence typing. *Fungal Genet. Biol.*, 45, 400, 2008.
18. Viviani, M.A. et al., Molecular analysis of 311 *Cryptococcus neoformans* isolates from a 30-month ECMM survey of cryptococcosis in Europe. *FEMS Yeast Res.*, 6, 614, 2006.
19. Yan, Z., Li, X.G., and Xu, J.P., Geographic distribution of mating type alleles of *Cryptococcus neoformans* in four areas of the United States. *J. Clin. Microbiol.*, 40, 965, 2002.
20. Jain, N. et al., Molecular epidemiology of clinical *Cryptococcus neoformans* strains from India. *J. Clin. Microbiol.*, 43, 5733, 2005.
21. Hironaga, M. et al., Mating types and serotypes of *Cryptococcus neoformans* isolated in Japan. *Sabouraudia-J. Med. Vet. Mycol.*, 21, 73, 1983.
22. Lengeler, K.B., Cox, G.M., and Heitman, J., Serotype AD strains of *Cryptococcus neoformans* are diploid or aneuploid and are heterozygous at the mating-type locus. *Infect. Immun.*, 69, 115, 2001.
23. Cogliati, M. et al., Origin of *Cryptococcus neoformans* var. *neoformans* diploid strains. *J. Clin. Microbiol.*, 39, 3889, 2001.
24. Lin, X. et al., Diploids in the *Cryptococcus neoformans* serotype A population homozygous for the alpha mating type originate via unisexual mating. *PLOS Pathog.*, 5, 1, 2009.
25. Ma, H.S. and May, R.C., Virulence in *Cryptococcus* Species. *Adv. Appl. Mircobiol.*, 67, 131, 2009.
26. Williamson, P.R., Laccase and melanin in the pathogenesis of *Cryptococcus neoformans. Front. Biosci.*, 2, 99, 1997.
27. van Duin, D., Casadevall, A., and Nosanchuk, J.D., Melanization of *Cryptococcus neoformans* and *Histoplasma capsulatum* reduces their susceptibilities to amphotericin B and caspofungin. *Antimicrob. Agents Chemother.*, 46, 3394, 2002.
28. Kwon-Chung, K.J., Edman, J.C., and Wickes, B.L., Genetic association of mating types and virulence in *Cryptococcus neoformans. Infect. Immun.*, 60, 602, 1992.
29. Barchiesi, F.et al., Comparative analysis of pathogenicity of *Cryptococcus neoformans* serotypes A, D and AD in murine cryptococcosis. *J. Infect.*, 51, 10, 2005.
30. Nielsen, K. et al., Sexual cycle of *Cryptococcus neoformans* var. *grubii* and virulence of congenic a and alpha isolates. *Infect. Immun.*, 71, 4831, 2003.
31. Casadevall, A., Antibody immunity and invasive fungal infections. *Infect. Immun.*, 63, 4211, 1995.
32. Mukherjee, J., Scharff, M.D., and, Casadevall, A., Protective murine monoclonal antibodies to *Cryptococcus neoformans. Infect. Immun.*, 60, 4534, 1992.
33. Syme, R.M. et al., Primary dendritic cells phagocytose *Cryptococcus neoformans* via mannose receptors and Fc gamma receptor II for presentation to T lymphocytes. *Infect. Immun.*, 70, 5972, 2002.
34. Graybill, J.R. et al., Diagnosis and management of increased intracranial pressure in patients with AIDS and cryptococcal meningitis. *Clin. Infect. Dis.*, 30, 47, 2000.
35. Patterson, T.F. and Andriole, V.T., Current concepts in cryptococcosis. *Eur. J. Clin. Microbiol. Infect. Dis.*, 8, 457, 1989.
36. Park, B.J. et al., Estimation of the current global burden of cryptococcal meningitis among persons living with HIV/AIDS. *Aids*, 23, 525, 2009.
37. Speed, B. and Dunt, D., Clinical and host differences between infections with the 2 varieties of *Cryptococcus neoformans. Clin. Infect. Dis.*, 21, 28, 1995.
38. Datta, K. et al., Spread of *Cryptococcus gattii* into Pacific Northwest Region of the United States. *Emerg. Infect. Dis.*, 15, 1185, 2009.
39. Kwon-Chung, K. and Bennett, J., *Medical Mycology*. Philadelphia, PA: Lea & Febiger; 1992.
40. Jarvis, J.N. et al., Screening for cryptococcal antigenemia in patients accessing an antiretroviral treatment program in South Africa. *Clin. Infect. Dis.*, 48, 856, 2009.
41. De Pauw, B. et al., Revised definitions of invasive fungal disease from the European Organization for Research and Treatment of Cancer/Invasive Fungal Infections Cooperative Group and the National Institute of Allergy and Infectious Diseases Mycoses Study Group (EORTC/MSG) Consensus Group. *Clin. Infect. Dis.*, 46, 1813, 2008.
42. Prariyachatigul, C. et al., Assessment of a PCR technique for the detection and identification of *Cryptococcus neoformans. J. Med. Vet. Mycol.*, 34, 251, 1996.
43. Rappelli, P. et al., Development of a nested PCR for detection of *Cryptococcus neoformans* in cerebrospinal fluid. *J. Clin. Microbiol.*, 36, 3438, 1998.
44. Paschoal, R.C. et al., Neurocryptococcosis: Diagnosis by PCR method. *Rev. Inst. Med. Trop. Sao Paulo*, 46, 203, 2004.
45. Saha, D.C. et al., Detection of *Cryptococcus* by conventional, serological and molecular methods. *J. Med. Microbiol.*, 58, 1098, 2009.
46. Posteraro, B. et al., Reverse cross blot hybridization assay for rapid detection of PCR-amplified DNA from *Candida* species, *Cryptococcus neoformans*, and *Saccharomyces cerevisiae* in clinical samples. *J. Clin. Microbiol.*, 38, 1609, 2000.
47. Abbott, J.J., Hamacher, K.L., and Ahmed, I., In situ hybridization in cutaneous deep fungal infections: A valuable diagnostic adjunct to fungal morphology and tissue cultures. *J. Cutan. Pathol.*, 33, 426, 2006.
48. Lau, A. et al., Multiplex tandem PCR: A novel platform for rapid detection and identification of fungal pathogens from blood culture specimens. *J. Clin. Microbiol.*, 46, 3021, 2008.
49. Veron, V. et al., Real-time polymerase chain reaction detection of *Cryptococcus neoformans* and *Cryptococcus gattii* in human samples. *Diagn. Microbiol. Infect. Dis.*, 65, 69, 2009.
50. Meyer, W. et al., Consensus multi-locus sequence typing scheme for *Cryptococcus neoformans* and *Cryptococcus gattii. Med. Mycol.*, 47, 561, 2009.
51. Hendolin, P.H. et al., Panfungal PCR and multiplex liquid hybridization for detection of fungi in tissue specimens. *J. Clin. Microbiol.*, 38, 4186, 2000.
52. Bialek, R. et al., Detection of *Cryptococcus neoformans* DNA in tissue samples by nested and real-time PCR assays. *Clin. Diagn. Lab. Immunol.*, 9, 461, 2002.
53. White, T. et al., Amplification and direct sequencing of fungal ribosomal RNA genes for phylogenetics. In: *PCR Protocols: A Guide to Methods and Applications*. Innis, M., Gelfand, D., Sninsky, J., and White, T. (Eds.), San Diego, CA: Academic Press, Inc.; 1990.
54. D'Souza, C.A. et al., Investigation of the basis of virulence in serotype A strains of *Cryptococcus neoformans* from apparently immunocompetent individuals. *Curr. Genet.*, 46, 92, 2004.

73 *Malassezia*

Takashi Sugita, Mami Tajima, Hisae Tsubuku, Mayumi Miyamoto, Enshi Zhang, Ryoji Tsuboi, Masako Takashima, Yoshio Ishibashi, and Akemi Nishikawa

CONTENTS

73.1 INTRODUCTION

The genus *Malassezia* colonizes only on the skin of human or nonhuman animals and does not exist in the environment [1,2]. With the exception of one species, *Malassezia pachydermatis*, these microorganisms require lipids for growth. The fatty acid biosynthesis pathway of *Malassezia* was elucidated when the entire genome was determined in 2007 [3]. The fatty acid biosynthesis pathway of *Malassezia* differs from that of typical fungi and is actually similar to that of animals. Consequently, *Malassezia* prefer sebum-rich areas, such as the scalp and face. Although *Malassezia* is a major component of the cutaneous fungal microbiota of humans, these microorganisms are associated with several skin diseases. *Malassezia* are responsible for the development or exacerbation of pityriasis versicolor, seborrheic dermatitis, folliculitis, and atopic dermatitis [4–6]. Relationships between *Malassezia* and psoriasis or acne vulgaris have also been suggested.

73.1.1 Taxonomy

The taxonomy of anamorphic basidiomycetous yeast in the genus *Malassezia* dates to the nineteenth century [7]. In 1853, Robin isolated a culture from a patient with pityriasis versicolor and named the species *Microsporum furfur*. Subsequently, the new genus *Malassezia* was established for the microorganism in 1889. Independently, Sabouraud established the new genus *Pityrosporum* in 1904. For approximately 100 years, many taxonomists debated whether *Malassezia* and *Pityrosporum* were synonymous, and now *Pityrosporum* is treated as a synonym of *Malassezia*. Nevertheless, the name *Pityrosporum* is often used in papers and books, although this use is not in accordance with fungal nomenclature rules. The greatest influence on the taxonomy of the genus *Malassezia* was the discovery of the heterogenicity of *M. furfur*. This microorganism was once considered to be the common causative agent of several skin diseases; however, in 1996, it was determined that *M. furfur* consisted of five distinct species [1]. The reclassification of *M. furfur* has contributed to the elucidation of the causative agents of *Malassezia*-skin related diseases (see Section 73.1.2). Of the known *Malassezia* species, four have affinity for animals (*M. caprae*, *M. equina*, *M. nana*, and *M. pachydermatis*), whereas the remaining nine species colonize human skin (*M. dermatis*, *M. furfur*, *M. globosa*, *M. japonica*, *M. obtusa*, *M. restricta*, *M. slooffiae*, *M. sympodialis*, and *M. yamatoensis*).

73.1.2 Epidemiology

Although the clinical conditions pityriasis versicolor, seborrheic dermatitis, atopic dermatitis, and psoriasis are quite different, the cutaneous *Malassezia* microbiota in patients with

TABLE 73.1
Detection of *Malassezia* DNA in Patients with *Malassezia*-Related Skin Diseases and Healthy Subjects

Subject	Number of Patients	Detection of *Malassezia* DNA (%)								
		M. globosa	*M. restricta*	*M. slooffiae*	*M. dermatis*	*M. sympodialis*	*M. japonica*	*M. yamatoensis*	*M. obtusa*	*M. furfur*
Seborrhoeic dermatitis [9]	31	100	100	39	39	36	13	10	10	7
Pityriasis versicolor (lesional region) [10]	49	100	100	8	25	35	6	4	4	10]
Atopic dermatitis (lesional region) [9]	36	100	100	31	31	58	33	14	28	33
Healthy subject [9]	30	100	100	17	30	37	10	7	10	27

Note: *M. globosa* and *M. restricta* DNA were detected by a real-time PCR (unpublished data).

these skin diseases is almost identical [8–12]. Pityriasis versicolor develops on the trunk and upper arms of people in their 20s and 30s, occurring more often during the summer than the winter, whereas seborrheic dermatitis develops in areas of the body that are rich in sebaceous glands, such as the face, scalp, and upper trunk. Atopic dermatitis is a chronic disease involving cycles of remission and deterioration and is characterized by hypersensitivity to dry skin. Psoriasis is a multifactorial skin disease with a strong genetic component, including several genes encoding proteins involved in epidermal differentiation and immune, inflammatory, and pathogen responses. Of the nine human-associated *Malassezia* species, both *M. globosa* and *M. restricta* are commonly detected from almost all patients, whereas the other species are found in fewer than 60% of cases (Table 73.1). Quantitative analysis gives additional interesting results. Although the *Malassezia* microbiota is similar regardless of skin disease, the ratio of *M. globosa* to *M. restricta* differs according to disease (Figure 73.1). In seborrheic dermatitis, *M. restricta* predominates over *M. globosa*, whereas *M. globosa* is the predominant species in pityriasis versicolor. The colonization level of *M. restricta* is twice that of *M. globosa* in atopic dermatitis and psoriasis. Therefore, the *Malassezia* microbiota is identical qualitatively but significantly different quantitatively in each skin disease.

Malassezia colonizes the sebaceous areas of the skin. Therefore, given the dependency of sebaceous gland activity on age and gender, the *Malassezia* microbiota of healthy individuals should be assessed at every age and in both genders. The latest large-scale study examined 770 healthy Japanese aged 0–82 years [13]. In males, the total amount of *Malassezia* DNA remained essentially constant between 0 and 9 years of age and then increased markedly up to 16–18 years of age (Figure 73.2). In females, the amount of *Malassezia* DNA increased until 10–12 years of age, decreased until 19–22 years of age, and then increased again until 30–39 years of age, before finally decreasing gradually with age beyond 39 years. *M. globosa* and *M. restricta* accounted for more than 70% of the total *Malassezia* DNA in both males and females at all ages.

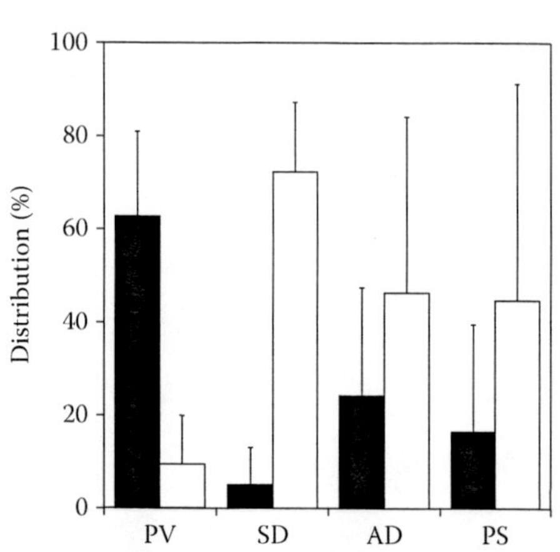

FIGURE 73.1 Proportions of *M. globosa* and *M. restricta* in patients with *Malassezia*-related skin diseases. Closed squares, *M. globosa*; open squares, *M. restricta*; PV, pityriasis versicolor; SD, seborrheic dermatitis; AD, atopic dermatitis; PS, psoriasis.

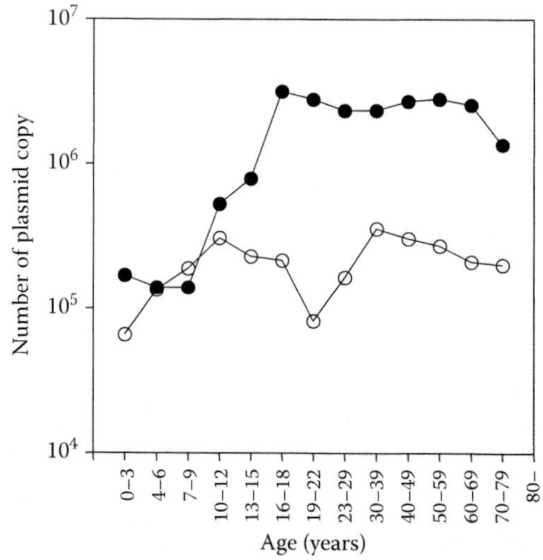

FIGURE 73.2 Changes in the amounts of *Malassezia* DNA with age. Closed circles, males; open circles, females.

73.1.3 Immunological Aspects

Specific IgE antibody against several cutaneous microorganisms, including *Malassezia*, is produced in patient's serum as patients with atopic dermatitis lack ceramide barrier function. No anti-*Malassezia* IgE antibody is detected in serum from healthy subjects. Numerous *Malassezia* allergens (Mala f 2–4, Mala s 1, and Mala s 5–13) have been identified and characterized [14–16]. Recently, MGp42, a new major *M. globosa* allergen, was identified and may be a cleavage product of intact HSP70 [17]. The precise mechanism by which *Malassezia* colonization induces IgE antibody production and the inflammatory cascades that lead to atopic dermatitis remain unclear. The production of IgE antibodies has been implicated in the Th2-type immune response [18–20]. Keratinocytes produce a range of proinflammatory and immune cytokines in response to microorganisms and skin damage [21,22]. A recent study demonstrated that keratinocytes play a critical role in the pathogenesis of atopic dermatitis by secreting several Th2-type cytokines [23]. Cytokine secretion profiles using antibody array analysis have revealed that *M. globosa* and *M. restricta* induce the secretion of distinct Th2-type cytokines from human keratinocytes: *M. globosa* induces IL-5, IL-10, and IL-13 secretion, and *M. restricta* induces IL-4 secretion. These observations suggest a possible relationship between *Malassezia* colonization and the increased IgE production in atopic dermatitis. Another important connection between *Malassezia* colonization and atopic dermatitis is the increased secretion of granulocyte-macrophage colony-stimulating factor (GM-CSF) and cutaneous T-cell-attracting chemokine (CTACK) from keratinocytes [23]. *M. globosa* is capable of inducing GM-CSF secretion from keratinocytes. GM-CSF contributes primarily to the maintenance of the chronic inflammatory process in atopic dermatitis by enhancing the antigen-presenting capacity of Langerhans cells (LCs) and dendritic cells (DCs) [24]. *M. restricta* induces the secretion of CTACK from keratinocytes. CTACK selectively attracts cutaneous lymphocyte antigen-positive memory T cells to the inflamed sites [25]. A recent study revealed that CTACK is upregulated in atopic dermatitis patients [26]. Based on these findings, the following is a possible mechanism by which *Malassezia* species induces an IgE-immune response in patients with atopic dermatitis: (1) skin barrier dysfunction facilitates skin penetration by colonizing *Malassezia*, allowing the interaction of *Malassezia* with epidermal cells including LCs, DCs, and keratinocytes; and (2) local antigen-presenting cells (APCs) such as LCs and DCs subsequently process and present *Malassezia* antigens to induce immune responses. This could be augmented by keratinocyte-derived GM-CSF. *Malassezia*-stimulated keratinocytes produce Th2 cytokines, including IL-4 and IL-13, which in turn stimulate B cells to undergo IgE class switching and the production of *Malassezia*-specific IgE. In addition, keratinocyte-derived IL-5 may attract and activate eosinophils locally in atopic dermatitis lesions.

73.1.4 Diagnosis

73.1.4.1 Conventional Techniques

In the clinical laboratory, conventional identification methods for yeast identification are based on biochemical characteristics. Two commercial kits, API 20C Aux™ and ID 32C API™ (bioMérieux), are the most popular for yeast identification. These kits can identify 47 and 69 yeast species, respectively, within 48 h. Unfortunately, as *Malassezia* requires a lipid for growth, these kits cannot be applied for *Malassezia* identification. An identification method combining (i) growth on Sabouraud dextrose agar (SDA); (ii) the growth temperature on modified Dixon agar; (iii) the utilization of Tween 20, 40, 60, and 80; (iv) the hydrolysis of esculin; (v) the catalase reaction; and (vi) the utilization of polyethoxylated castor oil has been proposed (Table 73.2) [27]. This method has been useful for discovering new *Malassezia* species. In addition, the accuracy of identification is improved significantly by using CHROMagar™ Malassezia medium (Kanto Chemical, Tokyo, Japan) [28]. *M. sympodialis*, *M. globosa*, *M. dermatis*, and *M. pachydermatis* produce a precipitate after incubation at 32°C for 4 days on CHROMagar Malassezia medium. Pale purple and pink colors are developed. Of 370 *Malassezia* strains belonging to nine species, 363 strains (98.1%) were identified accurately using the characteristics listed in Table 73.2.

73.1.4.2 Molecular Techniques

As yeast species have few distinct morphological characteristics, molecular techniques are a powerful tool for species identification. The most broadly utilized identification method is rRNA gene sequence analysis. Figure 73.3 shows the rRNA gene locus in *M. globosa*; it is approximately 8000 bp long. Within this, the internal transcribed spacer (ITS) region and D1/D2 26S rRNA gene have been used widely for fungal identification. IGS analysis is also preferable for strain typing or for use as an epidemiological tool, but is limited to *Malassezia* [29–31], *Trichosporon* [32–36] (see Chapter 77 on the genus *Trichosporon*), and *Cryptococcus* [37,38].

73.1.4.2.1 Culture-Dependent Methods

The PCR primers used to amplify the ITS region, including the 5.8S rRNA gene, D1/D2 26S rRNA gene, and IGS 1 region, are listed in Table 73.3. Primers ITS1, ITS4, NL1, and NL4 amplify the ITS region and D1/D2 rRNA gene of all fungi, and DNA sequencing is performed with the same primers. Primers 26BF and GBR are used to amplify the IGS1 region of *Malassezia* species, but cannot be applied to *M. sympodialis* and *M. dermatis* [29]. Identification criteria (% DNA similarity) should be considered when a strain is identified using rRNA gene sequencing. Generally, strains with greater than 99% DNA sequence similarity in both the ITS region and D1/D2 26S rRNA gene are considered to be conspecifics (see Chapter 77 on the genus *Trichosporon*). The length of the D1/D2 26S rRNA gene is identical across all species, and the DNA sequence similarity within a species exceeds 99% for all *Malassezia* species (Table 73.4).

TABLE 73.2
Typical Phenotypic Features in Nine Human-Associated *Malassezia* Species

		Growth on mDixon at					Utilization of						
Species	Growth on SDA	32°C	37°C	40°C	Color on CHROM	Precipitate Production on CHROM	10% Tween 20	0.5% Tween 40	0.5% Tween 60	0.1% Tween 80	Esculin	Polyethoxylated Castor Oil	Catalase Reaction
M. pachydermatis	+	+	+	+	Pale pink	+	+	+	+	+	±	±	+
M. sympodialis	–	+	+	+	Pale pink	+	–	+	+	+	+	–	+
M. globosa	–	+	± or –	–	Purple	+	–	–	–	–	–	–	+
M. dermatis	–	+	+	+	Pale pink	+	+	+	+	+	–	–	+
M. furfur	–	+	+	+	Pale pink	–	+	+	+	+	–	+	+
M. slooffiae	–	+	+	+	Pale pink	–	± or +	+	+	–	–	–	+
M. obtusa	–	+	± or +	–	Pink	–	–	–	–	–	+	–	+
M. restricta	–	+	+	–	Pink	–	–	–	–	–	–	–	–
M. japonica	–	+	+	–	Pink	–	–	±	+	–	+	–	+

Source: Referred from a paper of Kaneko, T. et al. *J. Clin. Microbiol.*, 45, 3737, 2007.
Notes: +, positive; –, negative.

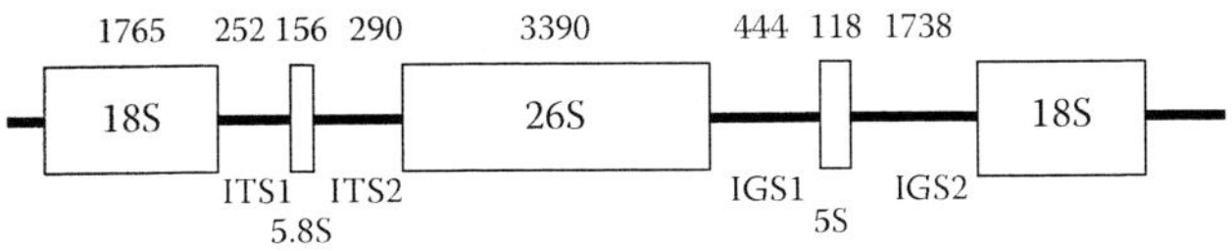

FIGURE 73.3 Schematic representation of the rRNA gene locus in *Malassezia*. ITS, internal transcribed spacer; IGS, intergenic spacer region. The length of a subunit or region is from *M. globosa* CBS 7966.

By contrast, the length of the ITS region differs in each species, ranging from 161 to 266 bp long, and both *M. globosa* and *M. restricta* show intraspecies sequence diversity. The length of the *M. globosa* and *M. restricta* ITS1 region ranges from 237 to 266 and from 210 to 261 bp, respectively. The former species shows 88% sequence within-species similarity, and the latter shows 73% similarity. Of the 13 *Malassezia* species, *M. caprae*, *M. dermatis*, *M. equina*, *M. japonica*, *M. obtusa*, *M. slooffiae*, and *M. yamatoensis* do not show intraspecies ITS diversity. Many *Malassezia* species can be identified using only the D1/D2 rRNA gene sequence, except for *M. sympodialis* and the phylogenetically closely related species *M. caprae*, *M. dermatis*, and *M. equina*. *M. caprae* and *M. equina* were separated from *M. sympodialis sensu lato* based on phylogenetic analysis, but they also differ phenotypically. The D1/D2 rRNA gene sequences of these four species have greater than 98%–99% sequence similarity and they share 82%–93% and 88%–95% sequence similarity in the ITS1 and ITS2 regions, respectively (Table 73.5). DNA sequencing of both the ITS regions and the D1/D2 rRNA gene sequence is recommended when isolates may belong to *M. sympodialis* and related species. *Malassezia* species can be identified using IGS sequences, but analysis of the region is preferred when used as an epidemiological tool.

In addition to rRNA gene sequencing, other molecular identification methods have been introduced, including random amplification of polymorphic DNA (RAPD) [39,40], amplified fragment length polymorphism (AFLP) [40], restriction fragment length polymorphism (RFLP) analysis [41,42], terminal fragment length polymorphism (*t*FLP) [43], denaturing gradient gel electrophoresis (DGGE) [40], and single strand conformation polymorphism (SSCP) analysis [44].

73.1.4.2.2 Culture-Independent Methods

One member of the genus *Malassezia*, *M. restricta*, is often absent from cultured samples because its growth on plates is slower than that of related species. All of the *Malassezia* species in a sample must be recovered from the medium to obtain an accurate quantitative analysis. Unfortunately, no optimal growth medium for all *Malassezia* species is available. Culture-dependent methods can yield significant numbers of viable cells, but have limited ability to produce accurate analytical results. Therefore, methods that do not require an incubation period (i.e., molecular-based, culture-independent methods) have been developed to analyze the distribution of the *Malassezia* microbiota. Each species is detected using nested PCR or real-time PCR [8–13,45–48].

73.2 METHOD

73.2.1 Sample Preparation

In vitro culture: As mentioned above, *Malassezia* species, with the exception of *M. pachydermatis*, require a lipid for growth. Therefore, lipids such as olive oil, glycerin, or Tween are added to the medium. Dixon media or Leeming and Notman agar (LNA) are widely used to culture *Malassezia*. A small amount of pure culture (approximately one loop) is prepared.

DNA extraction from scale samples: Sample scale from a subject is obtained using an appropriate method, such as swabbing with a cotton applicator, scraping with a scalpel, stripping with tape, or washing with appropriate buffer. Stripping with tape may be best because this method is simple and does not stress the subject [8]. A transparent dressing (e.g., OpSite [3 × 7 cm; Smith and Nephew Medical, Hull, United Kingdom]) is applied to the subject's skin. Then, the OpSite dressing is removed and placed in 1.5 mL of lysing solution (100 mM Tris–HCl pH 8.0, 30 mM EDTA pH 8.0, and 0.5% sodium dodecyl sulfate) and incubated for 15 min at 100°C. The OpSite dressing is removed from the tube, and the suspension extracted with phenol–chloroform–isoamyl alcohol (25:24:1, vol/vol/vol). Subsequently, the samples are extracted with chloroform–isoamyl alcohol (24:1, vol/vol), and the DNA is precipitated with 2-propanol using Ethatimate (Nippon Gene, Toyama, Japan) as a precipitation activator. The DNA pellet is resuspended in 50 µL of 1× TE (10 mM Tris–HCl pH 8.0, 1 mM EDTA pH 8.0). An unused OpSite dressing is used as a negative control.

73.2.2 Detection Procedures

73.2.2.1 Identification by PCR and Sequencing

DNA sequence analysis of the ITS region or D1/D2 26S rRNA in the rRNA gene is the most reliable and accurate identification method. These regions are amplified by PCR with primers ITS1 and NL4, as shown in Table 73.3. PCR is performed with an initial 3 min denaturation at 94°C, followed by 30 cycles of 30 s at 94°C, 30 s at 57°C, and 1.5 min at 72°C, and a final 10 min extension at 72°C. The PCR products are sequenced with the primers ITS1 and ITS4 or NL1 and NL4, and the resulting DNA sequence is subjected to a blast search (http://www.ncbi.nlm.nih.gov/blast/Blast.cgi). A species is defined as organisms sharing >99% identity in their D1/D2 26S rRNA gene. For the ITS region, the sequence similarity shown in Table 73.2 is employed to define each species.

73.2.2.2 Nested PCR Detection

Nested PCR conducted using the two primer sets shown in Table 73.3 can be applied for detection of the nine human-associated *Malassezia* species [10]. Extracted DNA (3 µL) from each sample is added to 37 µL of PCR master mixture, which consists of 4 µL of 10× PCR buffer (100 mM Tris–HCl pH 8.3, 500 mM KCl, 15 mM $MgCl_2$), 3.2 µL of

TABLE 73.3
The PCR Primers Used to Detect *Malassezia* DNA

Region or subunit	Primer	Sequence (5′–3′)
Primers for identifying *Malassezia* species		
ITS region	ITS1 (forward)	TCCGTAGGTGAACCTGCGG
	ITS4 (reverse)	TCCTCCGCTTATTGATATGC
IGS region	26BF (forward)	GCTTTCGAGTGCATACCACAC
	GBR (reverse)	GGAAATAGGATGAGAGAAAC
D1/D2 26S rRNA gene	NL1 (forward)	GCATATCAATAAGCGGAGGAAAAG
	NL4 (reverse)	GGTCCGTGTTTCAAGACGG
Target species	**Primer**	**Sequence (5′–3′)**
Primers for detecting DNA of each *Malassezia* species by nested PCR		
First PCR for six *Malassezia* species[a]	Forward	ATCCTTTGCAGACGACTTGA
	Reverse	TGCTTAACTTCGCAGATCGG
First PCR for three *Malassezia* species[b]	Forward	ACCTGCAGAAGGATCATTAGTGA
	Reverse	TCCTCCGCTTATTGATATG
Second PCR for each *Malassezia* species		
M. dermatis	Forward	CGCACCTTGCGCTCCATGGT
	Reverse	AGCCTGGTTTCCCAGGCAGCGG
M. furfur	Forward	TGTGTACCATAGGCACCCAC
	Reverse	CACGGTGATAAAGGGATGCA
M. globosa	Forward	TCGAGTGCATACCACACTCGAG
	Reverse	TACGGTGCTTTCACGGTTCT
M. japonica	Forward	CGTATGTGGATCTTATCCTAT
	Reverse	TGACTAGTGTCGTAGGCACGGTA
M. obtusa	Forward	CATGGTCCATCTCCCACACA
	Reverse	AGAGAGTGCGTGGCGCATGGT
M. restricta	Forward	CGACCTAGTCGACTACATCCTACT
	Reverse	TTCGGAGATACAAGCCTCCAT
M. slooffiae	Forward	ACGCACGCTAACACAACGTG
	Reverse	TGTGCGATTCGAAGCGCACA
M. sympodialis	Forward	CGCACCTTGCGCTCCATGGT
	Reverse	GGTACAATCCCCAGGCAGCAA
M. yamatoensis	Forward	CGATCAAACTTCTCTGTGTCCAG
	Reverse	TGTGTGGGAGGTAGAAGAGGCA
Target species	**Primer and probe**	**Sequence (5′–3′)**
Primers and TaqMan probe for detecting *Malassezia* DNA by real-time PCR		
All *Malassezia* species	Mala-F (forward)	CTAAATATCGGGGAGAGACCGA
	Mala-R (reverse)	GTACTTTTAACTCTCTTTCCAAAGTGCTT
	Mala-MGB (probe)	FAM-TTCATCTTTCCCTCACGGTAC-MGB
M. globosa	M. glob-F (forward)	GGCCAAGCGCGCTCT
	M. glob-R (reverse)	CCACAACCAAATGCTCTCCTACAG
	M. glob-MGB (probe)	FAM-ATCATCAGGCATAGCATG-MGB
M. restricta	M. rest-F (forward)	GGCGGCCAAGCAGTGTTT
	M. rest-R (reverse)	AACCAAACATTCCTCCTTTAGGTGA
	M. rest-MGB (probe)	FAM-TTCTCCTGGCATGGCAT-MGB

[a] *M. furfur, M. globosa, M. japonica, M. obtusa, M. restricta, M. yamatoensis.*
[b] *M. dermatis, M. sympodialis, M. slooffiae.*

TABLE 73.4
Summary of the ITS1, ITS2, and D1/D2 LSU rRNA Gene Sequences

	ITS1		ITS2		D1/D2 26S rRNA gene
Species	**Length (bp)**	**Similarity (%)**	**Length (bp)**	**Similarity (%)**	**Similarity (%)**
Malassezia caprae	164	100	231	100	100
Malassezia dermatis	161	100	233	100	100
Malassezia equina	162–164	>99	232	100	100
Malassezia furfur	209–210	>95	370	100	>99
Malassezia globosa	237–266	>88	278–290	>91	>99
Malassezia japonica	208	100	337	100	100
Malassezia nana	182–186	>97	241–242	>98	>99
Malassezia obtusa	213	100	367	100	100
Malassezia pachydermatis	186–187	>94	328–344	>94	>99
Malassezia restricta	210–261	>73	276	>99	>99
Malassezia sloffiae	195–196	>99	313–316	>99	100
Malassezia sympodialis	162–163	100	234–235	100	100
Malassezia yamatoensis	176	100	298	100	100

TABLE 73.5
DNA Sequence Similarity of the Phylogenetically Close Species *M. sympodialis*, *M. caprae*, *M. dermatis*, and *M. equina*

Subunit or Spacer Region	Species	*M. caprae*	*M. dermatis*	*M. equina*
D/D2 26S rRNA gene	*M. dermatis*	98		
	M. equina	99	99	
	M. sympodialis	99	99	99
ITS1	*M. dermatis*	93		
	M. equina	82	85	
	M. sympodialis	88	90	85
ITS2	*M. dermatis*	95		
	M. equina	91	91	
	M. sympodialis	91	91	88

200 μM deoxynucleoside triphosphates (an equimolar mixture of dATP, dCTP, dGTP, and dTTP), 15 pmol of each primer, and 0.5 U of Ex *Taq* DNA polymerase (Takara). In the nested PCR step, 1 μL of the first amplification product is added to a new reaction mixture with the same composition as the first.

73.2.2.3 Real-Time PCR Quantitation

A real-time PCR assay can also quantify *Malassezia* DNA from scale samples without the need to culture isolates. Two methods have been introduced. One detects all *Malassezia* species, especially the two major components, *M. globosa* and *M. restricta*, with TaqMan probes [43]. First 25 μL of TaqMan Universal PCR master mix (Applied Biosystems), 200 nM each of the forward and reverse primers, 250 nM TaqMan probe (see Table 73.2), 3 μL of DNA, and Milli-Q water in a total volume of 50 μL are placed in each well of a 96-well plate. Amplification and detection are performed with the cycle parameters 50°C for 2 min, 95°C for 10 min, and 40 cycles of 95°C for 15 s, and 60°C for 1 min. A standard curve is plotted using the cycle threshold values obtained by amplifying successive 10-fold dilutions of a known concentration of plasmid DNA (from 10^1 to 10^9 copies). The assay can detect 10^1 copies. The second method detected the nine human-associated *Malassezia* species using SYBR GREEN [48]

73.3 CONCLUSION

Malassezia furfur has long been considered as the major causative agent of all *Malassezia*-related skin diseases. The species heterogenicity of this microorganism is elucidated, and analysis of the *Malassezia* microbiota has progressed rapidly. Molecular-based approaches have contributed greatly to this research. The culture of *Malassezia* is difficult compared to that of other pathogenic fungi, making direct *Malassezia* DNA detection a valuable tool. Molecular-based methods for detecting and identifying *Malassezia* will continue to be mainstream.

REFERENCES

1. Guého E, Midgley G, Guillot J. The genus *Malassezia* with description of four new species. *Antonie Van Leeuwenhock* 69, 337–355, 1996.
2. Guillot J, Hadina S, Guého E. The genus *Malassezia*: Old facts and new concepts. *Parassitologia* 50, 77–79, 2008.
3. Xu J et al. Dandruff-associated *Malassezia* genomes reveal convergent and divergent virulence traits shared with plant and human fungal pathogens. *Proc. Natl. Acad. Sci. U.S.A.* 104, 18730–18735, 2007.
4. Ashbee HR. Update on the genus *Malassezia. Med. Mycol.* 45, 287–303, 2007.
5. Faergemann J. Atopic dermatitis and fungi. *Clin. Microbiol. Rev.* 15, 545–563, 2002.
6. Batra R et al. *Malassezia* Baillon, emerging clinical yeasts. *FEMS Yeast Res.* 5, 1101–1113, 2005.
7. Takashima M. Importance of nomenclature for medically important fungi. *Nippon Ishinkin Gakkai Zasshi* 50, 263–267, 2009.
8. Sugita T et al. Molecular analysis of *Malassezia* microflora on the skin of atopic dermatitis patients and healthy subjects. *J. Clin. Microbiol.* 39, 3486–3490, 2001.
9. Tajima M et al. Molecular analysis of *Malassezia* microflora in seborrheic dermatitis patients: Comparison with other diseases and healthy subjects. *J. Invest. Dermatol.* 128, 345–351, 2008.
10. Morishita N, Sei Y, Sugita T. Molecular analysis of *Malassezia* microflora from patients with pityriasis versicolor. *Mycopathologia* 161, 61–65, 2006.
11. Amaya M et al. Molecular analysis of *Malassezia* microflora in the lesional skin of psoriasis patients. *J. Dermatol.* 34, 619–624, 2007.
12. Takahata Y et al. Quantitative analysis of *Malassezia* in the scale of patients with psoriasis using a real-time polymerase chain reaction assay. *Br. J. Dermatol.* 157, 670–673, 2007.
13. Sugita T et al. Quantitative analysis of the cutaneous Malassezia microbiota in 770 healthy Japanese by age and gender using a real-time PCR assay. *Med. Mycol.* 48, 229–233, 2010.
14. Lindborg M et al. Selective cloning of allergens from the skin colonizing yeast *Malassezia furfur* by phage surface display technology. *J. Invest. Dermatol.* 113, 156–161, 1999.
15. Schmidt M et al. The complete cDNA sequence and expression of the first major allergenic protein of *Malassezia furfur*, Mal f 1. *Eur. J. Biochem.* 246, 181–185, 1997.
16. Rasool O et al. Cloning, characterization and expression of complete coding sequences of three IgE binding *Malassezia furfur* allergens, Mal f 7, Mal f 8 and Mal f 9. *Eur. J. Biochem.* 267, 4355–4361, 2000.
17. Ishibashi Y et al. Identification of the major allergen of *Malassezia globosa* relevant for atopic dermatitis. *J. Dermatol. Sci.* 55, 185–192, 2009.
18. Hamid Q, Boguniewicz M, Leung DY. Differential in situ cytokine gene expression in acute versus chronic atopic dermatitis. *J. Clin. Invest.* 94, 870–876, 1994.
19. Leung DY. Atopic dermatitis. New insights and opportunities for therapeutic intervention. *J. Allergy Clin. Immunol.* 105, 860–876, 2000.
20. Chen L et al. Early up-regulation of Th2 cytokines and late surge of Th1 cytokines in an atopic dermatitis model. *Clin. Exp. Immunol.* 138, 375–387, 2004.
21. Grone A. Keratinocytes and cytokines. *Vet. Immunol. Immunopathol.* 88, 1–12, 2002.
22. Watanabe S et al. The effects of *Malassezia* yeasts on cytokine production by human keratinocytes. *J. Invest. Dermatol.* 116, 769–773, 2001.
23. Ishibashi Y, Sugita T, Nishikawa A. Cytokine secretion profile of human keratinocytes exposed to *Malassezia* yeasts. *FEMS Immunol. Med. Microbiol.* 48, 400–409, 2006.
24. Witmer-Pack MD et al. Granulocyte-macrophage colony-stimulating factor is essential for the viability and function of cultured murine epidermal Langerhans cells. *J. Exp. Med.* 166, 1484–1498, 1987.
25. Morales J et al. CTACK, a skin-associated chemokine that preferentially attracts skin-homing memory T cells. *Proc. Natl. Acad. Sci. U.S.A.* 96, 14470–14475, 1999.
26. Kakinuma T et al. Increased serum cutaneous T cell-attracting chemokine (CCL27) levels in patients, with atopic dermatitis and psoriasis vulgaris. *J. Allergy Clin. Immunol.* 111, 592–597, 2003.
27. Kaneko T et al. Revised culture-based system for identification of *Malassezia* species. *J. Clin. Microbiol.* 45, 3737–3742, 2007.
28. Kaneko T et al. Tween 40-based precipitate production observed on modified chromogenic agar and development of biological identification kit for *Malassezia* species. *Med. Mycol.* 44, 227–231, 2006.
29. Sugita T et al. Sequence diversity of the intergenic spacer region of the rRNA gene of *Malassezia globosa* colonizing the skin of patients with atopic dermatitis and healthy individuals. *J. Clin. Microbiol.* 41, 3022–3027, 2003.
30. Sugita T et al. Genotype analysis of *Malassezia restricta* as the major cutaneous flora in patients with atopic dermatitis and healthy subjects. *Microbiol. Immunol.* 48, 755–759, 2004.
31. Sugita T et al. DNA sequence diversity of intergenic spacer I region in the non-lipid-dependent species *Malassezia pachydermatis* isolated from animals. *Med. Mycol.* 43, 21–26, 2005.
32. Sugita T et al. Sequence analysis of the ribosomal DNA intergenic spacer 1 regions of *Trichosporon* species. *J. Clin. Microbiol.* 40, 1826–1830, 2002.
33. Sugita T, Ikeda R, Nishikawa A. Analysis of *Trichosporon* isolates obtained from the houses of patients with summer-type hypersensitivity pneumonitis. *J. Clin. Microbiol.* 42, 5467–5471, 2004.
34. Rodriguez-Tudela JL et al. Genotype distribution of clinical isolates of *Trichosporon asahii* based on sequencing of intergenic spacer 1. *Diagn. Microbiol. Infect. Dis.* 58, 435–440, 2007.
35. Kalkanci A et al. Molecular identification, genotyping, and drug susceptibility of the basidiomycetous yeast pathogen *Trichosporon* isolated from Turkish patients. *Med. Mycol.* 48, 141–146, 2010.
36. Mekha N et al. Genotyping and antifungal drug susceptibility of the pathogenic yeast *Trichosporon asahii* isolated from Thai patients. *Mycopathologia* 169, 67–70, 2010.
37. Diaz MR et al. Molecular sequence analyses of the intergenic spacer (IGS) associated with rDNA of the two varieties of the pathogenic yeast, *Cryptococcus neoformans. Syst. Appl. Microbiol.* 23, 535–545, 2000.
38. Sugita T, Ikeda R, Shinoda T. Diversity among strains of *Cryptococcus neoformans* var. *gattii* as revealed by a sequence analysis of multiple genes and a chemotype analysis of capsular polysaccharide. *Microbiol. Immunol.* 45, 757–768, 2001.

39. Boekhout T, Kamp M, Guého E. Molecular typing of *Malassezia* species with PFGE and RAPD. *Med. Mycol.* 36, 365–372, 1998.
40. Theelen B et al. Identification and typing of *Malassezia* yeasts using amplified fragment length polymorphism (AFLP), random amplified polymorphic DNA (RAPD) and denaturing gradient gel electrophoresis (DGGE). *FEMS Yeast Res.* 1, 79–86, 2001.
41. Gaitanis G, Robert V, Velegraki A. Verifiable single nucleotide polymorphisms of the internal transcribed spacer 2 region for the identification of 11 *Malassezia* species. *J. Dermatol. Sci.* 43, 214–217, 2006.
42. Mirhendi H et al. A simple PCR-RFLP method for identification and differentiation of 11 *Malassezia* species. *J. Microbiol. Methods* 61, 281–284, 2005.
43. Gemmer CM et al. Fast, noninvasive method for molecular detection and differentiation of *Malassezia* yeast species on human skin and application of the method to dandruff microbiology. *J. Clin. Microbiol.* 40, 3350–3357, 2002.
44. Gaitanis G et al. Distribution of *Malassezia* species in pityriasis versicolor and seborrhoeic dermatitis in Greece. Typing of the major pityriasis versicolor isolate *M. globosa*. *Br. J. Dermatol.* 154, 854–859, 2006.
45. Sugita T et al. Quantitative analysis of cutaneous *Malassezia* in atopic dermatitis patients using real-time PCR. *Microbiol. Immunol.* 50, 549–552, 2006.
46. Takahata Y et al. Cutaneous *Malassezia* flora in atopic dermatitis differs between adults and children. *Br. J. Dermatol.* 157, 1178–1182, 2007.
47. Paulino LC, Tseng CH, Blaser MJ. Analysis of *Malassezia* microbiota in healthy superficial human skin and in psoriatic lesions by multiplex real-time PCR. *FEMS Yeast Res.* 8, 460–471, 2008.
48. Akaza N et al. *Malassezia* folliculitis is caused by cutaneous resident *Malassezia* species. *Med. Mycol.* 47, 618–624, 2009.

74 *Rhodotorula*

Diego Libkind

CONTENTS

74.1 INTRODUCTION

When dealing with the isolation of fungi from clinical samples, in a few cases, yeast colonies with a characteristic salmon-pink to coral-red color appear. A classic procedure of many medical mycologists is to assign such isolates to the genus *Rhodotorula* on the basis of this sole phenotypic attribute. Today, thanks to the improvement of molecular biology techniques (especially DNA sequencing) and molecular phylogenetics, it is well known that yeasts producing pigmented colonies (also called pigmented or red yeasts) may belong to several basidiomycetous genera other than *Rhodotorula*. Examples of these genera are *Rhodosporidium* (teleomorphic stage of many *Rhodotorula* species), *Sporobolomyces*, *Sporidiobolus*, *Dioszegia*, *Cystofilobasidium*, *Xanthophyllomyces*, and *Cryptococcus*. The distinctive pink-to-red coloration of their colonies is the result of the intracellular production and accumulation of carotenoid compounds. The composition of carotenoid pigments may vary qualitatively in red yeasts depending on the species. The most important carotenoids found in yeasts are torularhodin, torulene, and β-carotene.[1,2]

However, among yeast with current medical relevance, the pigmented species known to be most frequently involved in human mycoses in fact belong to *Rhodotorula*. There are no records of clinical isolates for almost any other pigmented genera. The only exception is *Sporobolomyces* (and its sexual stage: *Sporidiobolus*), which has been found, in a few occasions, involved in human fungemia or has been isolated from clinical specimens.[3–8]

74.1.1 Taxonomy

The genus *Rhodotorula* is polyphyletic and has 47 different species, constituting a very heterogeneous taxon. Most of *Rhodotorula* species belong to the subphylum Pucciniomycotina, either to the class Microbotryomycetes (29 species) or to the class Cystobasidiomycetes (15 species). The remaining three species are classified in the subphylum Ustilaginomycotina and do not produce carotenoid pigments. *R. glutinis* is the type species of the genus *Rhodotorula* and belongs to the order Sporidiobolales (Class Microbotryomycetes), together with the abundant yeast *R. mucilaginosa* (previously known as *R. rubra*). A third clinically important *Rhodotorula* species is *R. minuta*, which, unlike *R. glutinis* and *R. mucilaginosa*, is ascribed to the order Cystobasidiales (Class Cystobasidiomycetes). The case of *R. glutinis* is peculiar. Until recently, it was regarded as a ubiquitous yeast, and at least 370 collection strains were considered to belong to this species.[9] The classical circumscription of *R. glutinis* was based mostly on a few physiological and cultural properties. However, several studies have reported the physiological and genetic heterogeneity of

R. glutinis.[10–14] More recently, Sampaio et al.[15] and Gadanho and Sampaio[16] reevaluated the circumscription of this species using a large set of isolates that had been previously identified by phenotypic criteria as *R. glutinis* and found that most of the isolates did not actually belong to this species. The most common were anamorph strains of *Rhodosporidium babjevae*, but other species of *Rhodotorula* (*R. dairenensis*, *R. graminis*, and *Rhodotorula* spp.) and *Rhodosporidium* (*Rh. diobovatum*, *Rh. sphaerocarpum*) had also been misidentified as *R. glutinis*. It has been also shown that from a total of 45 isolates, only 4 had been properly assigned to *R. glutinis*.[16] The species with the closest phenetic and phylogenetic proximity to *R. glutinis* are *R. graminis* and *Rh. babjevae*. Therefore, it is likely that many of the isolates of *R. glutinis* from humans, which have been reported, may actually have been confused with other yeast species. For the case of medically important yeasts, a reliable identification is needed for the selection of the appropriate antifungal.

From the above, it becomes clear that the accurate identification of *R. glutinis* and closely related species is only possible by the use of molecular techniques. In order to avoid confusions, hereinafter the term *R. glutinis* complex will be used to refer to *R. glutinis* and closely related *Rhodotorula* and *Rhodosporidium* species often confused with *R. glutinis*. The *R. glutinis* complex is hence a group of species, and its characteristics and properties correspond to that of the sum of species included in it.

Figure 74.1 provides a schematic representation of the phylogenetic placement of *Rhodotorula* and *Rhodosporidium* species within the phylum Basidiomycota, as well as the main phenotypic characteristics of the species discussed in this chapter: *R. mucilaginosa*, *R. glutinis* complex, and *R. minuta*.

74.1.2 Environmental Distribution

Rhodotorula species can be commonly isolated from natural environments (e.g., soil, water, phylloplane, etc.). A notable case is that of *R. mucilaginosa*, which is one possibly the most ubiquitous basidiomycetous yeast species. Besides a probable association with humans, *Rhodotorula mucilaginosa* is found in a wide range of natural habitats including living or decomposing plant constituents and soil,[17–21] and various aquatic environments such as freshwater lakes,[22] estuaries,[23] coastal waters,[24] and open ocean and deep-sea environments.[25,26] Moreover, *R. mucilaginosa* is also present in extreme environments like hyper-acidic waters,[27,28] uranium leachate,[29] and cold environments[30–32] among others.

The *R. glutinis* complex is also worldwide in distribution and has been isolated from a variety of substrates. Known sources include air, freshwaters, marine waters, terrestrial environments (soil, phylloplane, flowers, etc.), food and beverages, animals, and humans.[9]

In comparison to *R. mucilaginosa* and *R. glutinis* complex, *R. minuta* has been less frequently isolated in natural environments. This species was detected in air, seawater (including deep-sea environments),[25] and freshwater.[22]

Concerning the presence of these *Rhodotorula* species in food, *R. mucilaginosa* is the species with the largest frequency of appearance and the widest range of food and beverages from which it has been detected. Several studies reported the isolation of *R. mucilaginosa* from peanuts, apple

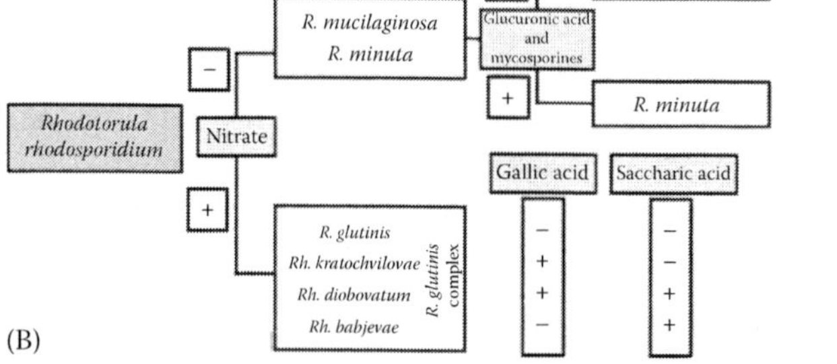

FIGURE 74.1 Schematic representation of the phylogenetic placement of *Rhodotorula* (*R.*, asexual) and *Rhodosporidium* (*Rh.*, sexual) within the phylum Basidiomycota (A) and the main phenotypic characteristics of the species discussed in this chapter (B)—*R. glutinis* complex, *R. minuta*, and *R. mucilaginosa*. Most species in *Cryptococcus* form white colonies (open circle) and a minority of species form pink colonies (gray circle). Colony colors are similar in *Rhodotorula*, but their proportions are symmetrical. The members of the Ustilaginomycotina do not form pink colonies. The ability to utilize inositol as sole carbon source and to produce amyloid compounds (PAC) allows the discrimination of the yeasts belonging to the three main lineages of the Basidiomycota. The most relevant *Rhodotorula* and *Rhodosporidium* species can be distinguished based on the ability to utilize nitrate as sole nitrogen source, the capacity to utilize glucuronic, gallic, and saccharic acids, and on the production of mycosporines. (From Libkind, D. et al., *Syst. Appl. Microbiol.*, 28, 749, 2005.)

cider,[33] cherries,[34] fresh fruits,[35–37] fruit juice,[38] cheese,[38–40] sausages,[41] edible mollusks,[42] and crustaceans,[43] among others. The remaining two *Rhodotorula* species are less frequent, and normally when they are present, *R. mucilaginosa* is also present. However, a few exceptions are known; as an example Callon et al.[44] reported the isolation of *R. glutinis* complex and *R. minuta* as the only representatives of pigmented yeasts from goat milk.

Rhodotorula species can be found in association with humans, having been isolated from various types of human specimens like feces, skin, and sputum, forming part of the normal flora.[45] These yeasts are generally considered commensal saprobes,[46–49] occurring most frequently in the environment than in human tissues. It is important to know that *Rhodotorula* spp. have been found to be one of the most frequent isolated microorganism from nurses and nonnursing hospital personnel hands.[50–51] Moreover, many patients studied on the very day of hospital admission have been shown to be carriers of such yeasts.[52] *Rhodotorula* yeasts possess a strong affinity for plastic, having been recovered from various medical equipments like hemodialysis machines and fiber-optic bronchoscopes,[53,54] and several environmental sources including shower curtains, bathtub grouts, and toothbrushes.[55] A direct consequence of such broad exposure to these yeasts is that, in humans with depressed immunological systems, *Rhodotorula* spp. can occasionally behave as opportunistic pathogens causing a variety of systemic infections. As a result, patients may suffer increased morbidity and mortality due to colonization, allergy, and invasive infection from such yeasts.

74.1.3 Clinical Features and Pathogenesis

Rhodotorula species that are regarded as emerging yeast pathogens are *R. mucilaginosa* (synonym *R. rubra*),[56,57] which seems to be frequently associated with human infections, and the less frequent *R. glutinis* complex and *R. minuta*.[58] Such species have been implicated in a broad spectrum of fungemia and other malignancies mostly in immunocompromised hosts; however, a few cases of Rhodotorulosis in inmunocompetent patients have been reported.[59–61]

To the best of our knowledge, one of the first cases of fungemia caused by *Rhodotorula* spp. was reported in 1960 for a patient with endocarditis.[62] A dramatic increase in the number of cases of Rhodotorulosis was observed in the following years. Factors predisposing such opportunistic infections include underlying immunosuppression, the use of broad-spectrum antibiotics and/or narrow-spectrum antifungals, and the presence of foreign devices such as central venous catheters.[55,63] However, the use of indwelling devices is by far the major risk factor for infection with these opportunistic yeasts.[64–66] *Rhodotorula* spp. have been incriminated in cases of ocular infections,[67–71] meningitis,[72–77] peritionitis,[78,79] endocarditis,[80] ventriculitis,[81] cancer-associated infections,[59,57,82] infections related to central venous catheters and other indwelling devices,[47,49,53,57,66,74,83–86] and various others types of infections.[87–97] Tuon and Costa (2008)[98] reviewed 128 literature cases (up to 2006) of *Rhodotorula*-associated infections and found that 79% of the cases were fungemia, 7% eye infections, and 5% peritonitis related with continuous ambulatory peritoneal dialysis. Furthermore, in 87% of the cases, the patient was immunosuppressed or had cancer.

R. mucilaginosa is the species responsible for the majority of these medical cases (~70%) followed by *R. glutinis* complex (~10%) and *R. minuta* (~6%). Reports in which *Rhodotorula* isolates could not be properly identified to the species level correspond to the remaining percentage of the cases.

It should be noted that in many of these cases, sufficient evidence supporting the involvement of the yeast isolate in the pathogenic process is lacking. In some cases, the basis for considering *Rhodotorula* yeasts as pathogenic is based on the resolution of fever accompanied by the disappearance of the yeast from blood cultures without evidence of clearance of an infectious focus. Such evidence is suggestive that the blood isolate was the etiologic agent of fever, but it is not definitive.[56]

With respect to other pathogenic yeasts, like those of the genus *Candida*, the incidence of *Rhodotorula* infections is quite low. Early epidemiological studies in the United States and Europe showed that the incidence of *Rhodotorula* infections was between 0.5% and 2.3% of total fungemia cases.[63,99,100] Moreover, Lunardi et al.[58] reported that in a 3 year surveillance of the occurrence of Rhodotorulosis in a Brazilian tertiary care hospital, the overall incidence of such infections was 0.056 episodes per 1000 hospital admissions. In a more recent work, Pfaller et al. (2009)[101] reported the incidence of non-*Candida* yeast clinical isolates from all over the world in a period of ca. 10 years and showed that *Rhodotorula* represented only 4.1%. Of a total of 462 *Rhodotorula* isolates studied, *R. mucilaginosa* was the most frequent, with 64 isolates, followed by *R. glutinis* complex (37) and *R. minuta* (1). The remaining strains (76%) could not be identified to the species level using conventional identifications methods. The incidence of yeast-related mycoses in a Portuguese oncology hospital during a 6-year period was recently reported, and *R. mucilaginosa* was the only pigmented yeast detected representing just 1.9% of the cases.[61]

It should be noted that the infections caused by unusual yeasts like *Rhodotorula* spp. could be under-reported because of the difficulties in accurate diagnosis and a tendency of attributing isolates to specimen contamination.[102]

Nowadays, *Rhodotorula* spp. are still rare in the diagnostic mycological laboratory.[103] However, an increase in the frequency of invasive mycoses due to these opportunistic can be anticipated, considering the ever-increasing application of newer technologies and therapies in the medical practice, which result in longer and more severe immunosuppression of nosocomial patients. Furthermore, given the *Rhodotorula* spp. intrinsic resistance to the widely used triazole and echinocandin antifungal agents, patients receiving such drugs, which actually select for *Rhodotorula* spp.,[59] will be more susceptible to the development of *Rhodotorula* fungemia.[68,104]

Even though *Rhodotorula* species appear to be less virulent than more common yeast pathogens like *Candida* and *Cryptococcus neoformans*, for example, fatal cases in which *Rhodotorula* spp. was suspected to be the etiological agent have indeed been reported. In the literature, the overall mortality rate of *Rhodotorula* fungemia of reported cases is around 15%,[58,85,98,105,106] although a few authors consider it higher.[64,107–109] *Rhodotorula*-related sepsis syndrome and other life-threatening complications have also been reported.[47,65,110]

The generally low pathogenicity of *Rhodotorula* spp. is probably related to its reduced ability to grow at 37°C, an attribute typically enhancing the virulence of pathogenic strains.[111] Several clinical and environmental isolates of *R. glutinis* complex have been reported to grow at 37°C[112]; however, it should be noted that the species *Rh. toruloides*, often misidentified as *R. glutinis*, and hence included in the *R. glutinis* complex, can grow at 37°C.[9] Other explanations for the low pathogenicity of *Rhodotorula* probably derive from the scarce ability of these yeasts to secrete degradative enzymes such as proteinase[113] and phospholipases,[114] which may be associated with pathogenic processes caused by opportunistic yeasts.

Rhodotorula spp. infections have been successfully treated with catheter removal alone (for indwelling device-associated cases), antifungal therapy, and with a combination of these two approaches.[58,65,115,116] Antifungal treatment of *Rhodotorula* infections typically involve the use of amphotericin B, given that *Rhodotorula* spp. was shown to be highly susceptible to such compound in vitro[71,115] and that it has been proven to be effective for clearing various types of *Rhodotorula* fungeamia.[65,81,117–119] Liposomal amphotericin B should be also considered because it is less nephrotoxic than conventional amphotericin B deoxycholate while maintaining antifungal activity against *Rhodotorula* spp.[120] More than a few cases have been reported in which treatment with this antifungal worked successfully for curing various types of *Rhodotorula* infections.[53,90,104,121] Alternative antifungals for the treatment of Rhodotorulosis include Flucytosine[73,118] showing excellent in vitro activity[84,122] and the promising new triazoles: Ravuconazole[115–116] and Isavuconazole.[123] Of worth noting is the evidence provided by the work of Diekema et al.[115] who tested the antifungic susceptibility of 64 clinical *Rhodotorula* isolates including *R. mucilaginosa* (24 isolates), *R. glutinis* complex (29 isolates), *R. minuta* (5 isolates) as well as other non-identified *Rhodotorula* species (6 isolates), finding no significant differences among them in the activities of any of the tested agents. These results suggest that the frequent infection-causing *Rhodotorula* species behave similarly toward most antifungal treatments. Interestingly, Silici[124] showed that six different honey bee propolis were highly effective in inhibiting a strain o *Rhodotorula* spp. displaying very low minimum inhibitory concentration (MIC) values (≤0.01 μg/mL). Such results are promising and should be corroborated with a larger set of isolates in order to evaluate propolis as an alternative antifungal agent for *Rhodotorula* mycoses.

Narrow-spectrum azoles (e.g., fluconazole) and echinocandins (e.g., caspofungin) type antifungics should not be considered appropriate for treatment of such cases due to *Rhodotorula* spp. high resistance to these kind of drugs.[8,84,115,116,125]

74.1.4 Diagnosis

74.1.4.1 Conventional Identification

Conventional identification of yeast species and strains is based on morphological, physiological, and biochemical characteristics. In order to reliably identify yeast to the species level, it is necessary to conduct 50–100 phenotypic tests. This procedure is laborious, time-consuming (it usually takes 3 weeks or more), and only an experienced person can obtain proper interpretation of the test results.[126] Furthermore, sometimes ambiguous identifications may be obtained due to intra-specific physiological variability of the strains.

For the identification of mycoses-causing yeasts, as is the case of some *Rhodotorula* species, a reliable identification is required for epidemiological investigations and for the selection of an appropriate antifungal treatment.

Conventional identification of yeast isolates in the clinical laboratory is based on one of several commercially available biochemical and enzymatic panels. However, some disadvantages of these identification kits include the existence of limited databases[122,123] and frequent misidentification.[85,127–136] Because of the ever-expanding spectrum of fungal pathogens (including yeasts), the field of medical mycology has become an extremely challenging study of infections caused by opportunistic fungi. Given that the identification of such a wide and taxonomically diverse array of fungal pathogens by conventional methods is often difficult and sometimes inconclusive,[137] molecular biology techniques are being incorporated in clinical laboratories to provide more accurate and prompt diagnosis.

74.1.4.2 Molecular Identification

Molecular diagnostic methods possess several advantages over conventional methods based on phenotypic characteristics, such as higher reliability, reproducibility, and throughput. Due to their high specificity and sensitivity, these molecular methods should be used routinely in the clinical laboratories as a complement of the information provided by conventional methods and particularly, to help in the diagnosis of dubious cases.

A variety of molecular approaches for genotyping and identification of yeast isolates have been described. These methods include species-specific DNA primers,[131,138–140] multiplex PCR,[141,142] real-time PCR assays,[142,143] PCR fingerprinting (i.e., random amplification of polymorphic DNA [RAPD], micro/minisatellite primed-PCR [MSP-PCR]),[15,16,22,24,144] PCR-enzyme immunoassays,[145] single-strand conformation polymorphism (SSCP) analysis of ribosomal DNA (rDNA),[146] restriction fragment length polymorphism (RFLP) analysis of 5.8S-internal transcribed spacer (ITS) rDNA,[147–149] electrophoretic karyotyping by pulsed-field gel electrophoresis,[150–152] denaturing/temperature gradient gel electrophoresis (DGGE/TGGE),[153–158] amplified fragment length polymorphism,[155,159] flow cytometry or Luminex®,[160–163] DNA microarray,[163–165] and DNA sequencing.[166,167] However, not all these techniques can be readily applied in the average clinical laboratory due to their complexity, high costs, and/or limitations in the array of yeast species that can detect. Several of such methods have been only developed for usual pathogenic yeasts (*Candida* spp., *Malassezia* spp., *Cryptococcus neoformans*) and are not yet applicable for less common opportunistic yeasts like *Rhodotorula* spp. For a list of available molecular techniques for yeast identification, as well as their advantages and disadvantages, see Table 74.1.

PCR-based techniques in general provide the required simplicity, specificity, and sensitivity to identify most fungal species in a short time period. Most of these methods do not directly provide identification, unless results are compared with databases or tests are run together with reference strains (i.e., PCR fingerprinting, multiplex PCR, ITS sequence length polymorphism, and SSCP-ITS). A PCR fingerprinting method that proved to be very useful for the rapid and simple genotyping of *Rhodotorula* isolates is the MSP-PCR technique.[16,22,144] It consists in the PCR amplification of genomic DNA using synthetic oligonucleotides that detect micro/minisatellite DNA. Microsatellite DNA consists of sequence repeated motifs of ca. 2–10 bp arranged in tandem at various loci, while minisatellite DNA is made up of motifs of 15–30 pb arranged in tandem. Such regions are highly polymorphic and thus provide sufficient resolution for the discrimination among yeast species. However, the discrimination power of each MSP-PCR primer differs, and it is related to the phylogenetic position of the taxa under study. After PCR reaction, MSP-PCR DNA banding patterns are obtained by agarose gels electrophoresis. These are species specific, allowing the unambiguous identification of most yeast species tested so far by comparison with profile databases or reference strains. Often, small variations in minor bands may provide differentiation up to strain level, which is useful for epidemiological studies.[144,168] Several studies have successfully employed the MSP-PCR method for rapid genotyping of large sets of collection strains and/or environmental isolates which included *Rhodotorula* species.[15,16,22,28,32] In such cases, the species *R. mucilaginosa* and *R. minuta* were easily identified by direct comparison with reference strains. Libkind et al.[144] studied almost 100 environmental isolates of *R. mucilaginosa* and found that more than 90% of these shared similar MSP-PCR profiles as the type strain using either M13 or $(GTG)_5$ primers, thus allowing prompt identification. On the other hand, studies within the *R. glutinis* complex have demonstrated that this closely related group of species shares similar MSP-PCR banding patterns with primers like M13. The discrimination among

TABLE 74.1
Molecular Detection Techniques for Yeasts with Special Reference to *Rhodotorula* spp. and Their Characteristics

Method	Characteristics	References
PFGE	Laborious, time consuming, lack of databases, needs reference strains, requires specific equipment	[150–152]
PCR fingerprinting	Fast, simple, needs reference strains, suitable primers may vary with yeast group	
MSP-PCR	High repetitiveness, high discriminatory power	[16,144]
RAPD	Low repetitiveness	[155,184]
Species-specific primers/multiplex PCR	Fast, simple, needs specific primers[b]	[131,138–142,185]
Sequence length polymorphism (ITS1 or ITS2)	Fast, simple, limited databases	[129,130]
SSCP-ITS	Fast, simple, needs reference strains, limited to known species	[144,186]
RFLP-ITS	Simple, specificity depends on yeast group, limited database	[147—157]
PCR–T/DGGE	Complicated, needs experience, low repetitiveness, useful for culture-independent detection	[153–158]
D1/D2 rDNA sequence	Highly conserved, more complete database, provides accurate identifications	[166,171]
ITS rDNA sequence	More variable, less complete database. Useful for analyses of closely related species	[133,167,172]
Other gene sequences (IGS, 18S, cytochrome *b*)	Limited databases	[187–189,192]
Real-time PCR	Sensitive, fast, needs specific probes,[b] and equipment	[142,143,182]
FISH[a]	Simple, fast, needs specific probes.[b] Useful only if cells are in high numbers and metabolically active	[183,190]
Pyrosequencing®	Low cost, limited to short DNA fragments	[191]
DNA microarray	High-throughput, needs specific DNA probes,[b] specific equipment	[144,155–157]
Flow cytometry, Luminex®[a]	High-throughput, needs specific DNA probes,[b] specific equipment	[160–162]

Notes: PFGE, pulsed-field gel electrophoresis; MSP-PCR, micro/minisatellite primed-PCR; RAPD, random amplification of polymorphic DNA; SSCP, single-strand conformational polymorphism; RFLP, restriction fragment length polymorphism; T/DGGE, temperature/denaturing gradient gel electrophoresis; FISH, fluorescent in situ hybridization.

[a] Techniques that to the best of our knowledge have not been yet tested or developed for *Rhodotorula* spp. identification.

[b] Implies that such technique is limited to known species.

these species can be obtained with the alternative MSP-PCR primer $(GAC)_5$.[16]

To the best of our knowledge, the MSP-PCR technique has not been yet tested for clinical isolates of *Rhodotorula*, however; Latouche et al.[168] applied this technique for *Candida* clinical strains and found it fast, reproducible, and more cost-effective than available biochemical approaches.

An alternative PCR method that can be used to directly identify *Rhodotorula* spp. (as well as others fungal pathogens) involves the use of species-specific primers.[138,139] The technique employs three primers in a PCR amplification of a ~600 bp DNA segment, consisting of two external universal (upstream and downstream) primers and one internal species-specific primer. Species identification requires the formation of species-specific rDNA nucleotide amplicon which is significantly smaller than the nontarget segment. Multicopy genes such as those of ribosomal origin (either the D1/D2 region of the large rDNA subunit or the noncoding ITS regions of the rRNA-encoding gene) are preferred by many researchers to achieve well-defined results.[138,139] However, because they may generate false-positive results, other researchers have looked at single-copy genes of high specificity.[169] Specific rDNA signature nucleotides or sequences of the target yeast species are used for the development of the specific oligonucleotide primers. A worth noting disadvantage of this detection technique is that primers are designed based on sequences available in the GenBank database, which may not represent the whole nucleotide variability for each species. This is particularly important for *Rhodotorula* spp. given that *R. mucilaginosa*[144,152] and *R. glutinis* complex[16] are genetically heterogeneous species. Thus, it can be anticipated that false-negative results could be obtained using species-specific primers for the molecular diagnosis of *Rhodotorula* species. Furthermore, to the best of our knowledge, currently there is only one available specific primer for *Rhodotorula* spp., which was designed to detect *R. mucilaginosa*.[139]

Other techniques have been developed for rapid yeast genotyping that also involve PCR (Table 74.1), one of which is the RFLP of the 5.8S-ITS rDNA sequence (RFLP-ITS). RFLP-ITS consists of a PCR amplification of the ITS fragment and subsequent digestion with three restriction enzymes.[148] The length of the ITS fragment together with the lengths of the fragments resulting from the enzymatic restriction are used for species identification employing existing RFLP-ITS databases (i.e., http://www.yeast-id.com). However, it may be impossible to differentiate very closely related species using this method, given that sequence differences occurring outside the restriction sites cannot be detected by RFLP-ITS analysis, and thus, fragments with identical sizes do not necessarily have identical sequences.[170] Furthermore, RFLP-ITS database is still quite limited for basidiomycetous yeast species, particularly for *Rhodotorula* spp.

A major drawback of PCR-based techniques, in general, is the fact that their diagnostic capabilities are limited to the respective existing databases, which commonly only include species considered pathogenic. Thus, new species becoming emergent pathogens would not be detected with these techniques until they are incorporated into each database.

Currently, the molecular technique most commonly used for accurate identification of yeasts is gene sequencing, because it offers a rapid and robust method for recognizing species and resolution is not limited to closely related taxa. Yeast species identification based on DNA sequencing have emphasized either in coding (D1/D2 variable domains of the large subunit rDNA) or in noncoding (ITS) regions of the rDNA. As a result, largely complete databases of D1/D2[168] and ITS[167] sequences are available for molecular classification and identification of yeasts, including all known *Rhodotorula* species.

The value of the D1/D2 region of the 26S rDNA for identifying clinically relevant yeast isolates by DNA sequencing has been clearly demonstrated.[171] Sequence analysis of this region provides enough resolution for the correct discrimination of *R. mucilaginosa* and *R. minuta*, given that both species show low genetic heterogeneity in such region.[144] Contrarily, for *R. glutinis*, a single base mismatch differentiates it from the closely related species *R. graminis* and *R. babjevae*.[16] Thus, D1/D2 sequence analysis may be useful for the identification of *R. glutinis* complex, but does not provide enough resolution for the discrimination of the various related species included in this group. As an example, Leaw et al.[172] were not able to differentiate *R. glutinis* from *Rh. babjevae* and *R. graminis* using D1/D2 sequence analysis.

A higher discriminatory level may be achieved when using the more variable ITS region in rDNA sequencing studies. This sequence is equally suitable than the D1/D2 region for accurate identification of both *R. minuta* and *R. mucilaginosa*. Furthermore, it has been demonstrated that even sequencing of either the ITS1 region alone or the ITS2 region alone may be enough for the correct identification of these two yeast species.[172] For the particular case of *R. mucilaginosa*, strain variability can also be detected with ITS sequencing,[144] which is interesting for epidemiological investigations. Again, difficulties may appear when trying to discriminate species of the *R. glutinis* complex by sequencing the ITS region. Gadanho and Sampaio,[16] using the complete ITS region, determined that *R. glutinis* and *R. graminis* differed in one mismatch and three indels, whereas three mismatches differentiated *R. glutinis* from *R. babjevae*. Studies testing the usefulness of ITS1 or ITS2 sequences alone for yeast identification found that a strain suspected to be *R. glutinis* (strain BCRC 20576) could not be differentiated from *Rh. babjevae* using such regions.[172] However, in such case, the criterion used to assign the strain BCRC 20576 to *R. glutinis* is not provided and thus is probable that this strain did never actually belong to *R. glutinis*.

A key factor determining the correct identification of the sequenced yeast isolate using the BLAST tool of the GenBank is the sequences used for comparison. Unfortunately, today the GenBank database includes many rDNA sequences that have been wrongly submitted or have been submitted with inappropriate species designations. Hence, inexperienced clinical mycologists might be leaded to incorrect identifications. When using rDNA sequencing for yeast identification, one should be aware of this fact and select for comparison purposes only trustworthy sequence submissions, preferably those corresponding to type strains. Additional hints for a correct use of GenBank database while identifying *Rhodotorula* yeasts are given in Section 74.2.2.4.

74.2 METHODS

There are several molecular methods that allow the accurate differentiation and detection of *Rhodotorula* spp. species. In this section, appropriate molecular techniques for identification and genotyping of *Rhodotorula* isolates from clinical samples are recommended.

74.2.1 Sample Preparation

74.2.1.1 Sample Collection and Culture

The sample collection and analysis procedures vary for each material drawn. With the exception of blood or swab samples which are used to inoculate appropriate liquid medium for yeast growth, other specimens are aseptically collected, homogenized, diluted, and submitted to plating procedures. In the clinical laboratory, the isolation of yeasts is commonly performed on Sabouraud dextrose agar (10 g/L mycological peptone, 40 g/L dextrose, pH 5.6) containing 100 mg/L chloramphenicol, though *Rhodotorula* species may be more effectively grown in YM agar (3 g/L yeast extract, 3 g/L malt extract, 5 g/L mycological peptone, 10 g/L dextrose, pH 5). It should be pointed out that the Pagano Levin medium was not suitable for the detection of *R. mucilaginosa* in various human specimens.[173]

The characterization of the yeast isolate by genetic methods should be performed only after the possibility that the strain under study is a saprophytic contaminant has been discarded.

74.2.1.2 DNA Extraction

Firstly, for DNA extraction, pure cultures of the *Rhodotorula* isolate must be obtained. For this, single colonies should be picked from the isolation plates and streaked on the surface of yeast-specific agar plates. After growth, a well isolated colony is transferred to fresh culture medium either in tubes or in plates. At the moment of DNA extraction, the yeast is inoculated in the surface of fresh agar medium plates and incubated for 48–72 h at room temperature (20°C–25°C). A rapid and simple DNA extraction protocol suitable for any of the above-mentioned PCR-based methods has been thoroughly described by Sampaio et al.[15] Here we provide a shorter and equally effective method for extraction of genomic DNA from a pure culture of *Rhodotorula* yeasts.

Procedure

1. Grow the *Rhodotorula* spp. isolate for 24–48 h in regular yeast agar plates (i.e., YM) at room temperature (20°C–25°C).
2. Transfer two full loopfulls of the biomass into a 1.5 tubes containing 500 μL of lyzing buffer and the equivalent to 200 μL of 425–600 μm glass beads.
3. Vortex vigorously for 3 min, incubate at 65°C for 1 h, and repeat vortexing (3 min).
4. Centrifuge suspensions for 15 min at 10,000 rpm and preferably at low temperatures (5°C).
5. Collect supernatant and transfer to new tube; store at −20°C for up to 10–12 months.
6. For PCR studies, dilute 1:600 in double-distilled water in order to obtain a DNA solution ranging from 5 to 50 ng/mL; add 5 μL to the PCR mix and store at −20°C for up to 4–6 months.

Alternatively, several commercial kits for yeast genomic DNA extraction are available.

74.2.2 Identification Procedures

74.2.2.1 Micro/Minisatellite Primed PCR

The MSP-PCR typing method is a valuable tool for mycological laboratories due to its rapidity, high output, and high strain discrimination potential. The following are the suggested MSP-PCR primers for each medically relevant *Rhodotorula* species:

R. mucilaginosa: the core sequence of the phage M13 (5′-GAG GGT GGC GGT TCT-3′) or the synthetic oligonucleotide $(GTG)_5$ provide enough species level resolution

R. minuta: same as *R. mucilaginosa*

R. glutinis complex: M13 primer allows the detection of *R. glutinis* species complex, though discrimination of species within this group or strains may be achieved with the $(GAC)_5$ primer

Procedure

1. Prepare a PCR mixture containing 0.8 μM each primer, 2 mM of each deoxynucleotide, 3.5 mM $MgCl_2$, 1 U of *Taq* DNA polymerase and 1× PCR buffer, sterile double-distilled water up to the desired volume, and 5 μL of the DNA sample (1:600 dilution) for a 25 μL reaction.
2. The cycling program consists of initial denaturation at 95°C for 5 min; 40 cycles of denaturing at 94°C for 45 s, annealing at 50°C–55°C for 1 min and extension at 72°C for 1 min; and a final extension at 72°C for 6 min.
3. Run gel electrophoresis on 1.4% agarose gel, 0.5× TBE buffer; at 90 V for 3.5 h. Include a molecular size marker on each gel, a mixture of DNA cleaved with *Hin*dIII and ΦX174 DNA cleaved with *Hae*III is recommended.
4. Stain the gel in 0.5 μg/mL ethidium bromide aqueous solution.
5. Compare obtained MSP-PCR profiles with database or reference strains.

74.2.2.2 Species-Specific PCR

Species-specific oligonucleotide primers may be designed for PCR identification of known *Rhodotorula* species or group of species. The procedure uses standard PCR components and protocols (see Section 74.2.3.1) including DNA from the test species and three primers: two universal external limiting primers (either universal D1/D2 region or ITS primers) and a species-specific internal primer. Species identification requires the amplification of a species-specific rDNA nucleotide segment that is significantly smaller than the non-target segment. Primers to be employed for the detection of *R. mucilaginosa* with species-specific PCR primers[139] are

External universal primers: NL1 (forward): 5′-GCA TAT CAA TAA GCG GAG GAA AAG-3′; and NL4 (reverse): 5′-GGT CCG TGT TTC AAG ACG G-3′;

Internal specific primer (forward) of the D1/D2 region: 5′-TCA GAC TTG CTT GCC GAG CAA TCG-3′.[139]

74.2.2.3 Restriction of the Amplified 5.8S-ITS Region of the Ribosomal rRNA Operon

A frequently used method for rapid identification of yeasts consists of PCR amplification of the ITS1–5.8S–ITS2 rDNA ITS region, followed by digestion with restriction endonucleases and by separation by agarose gel electrophoresis.[35,147,148] Studies using the RFLP-ITS method for the identification of *Rhodotorula* yeasts are scarce, comprising only a few reports of wine [174,175] and orange-juice spoilage yeasts.[35]

The RFLP-ITS profiles to date known corresponding to *Rhodotorula* spp. are the following: *R. mucilaginosa* strains typically show bands of the following molecular weights: ITS amplicon: ~600 bp, *Cfo*I: 320+240+80, *Hae*III: 425+215, and *Hinf*I: 340+225+75,[174,175] whereas *R. minuta* has ITS amplicons of 660 bp and *Cfo*I: 300+300, *Hae*III: 400+215, *Hinf*I: 345+215.[174] *R. glutinis* complex, on the other hand, generates ITS amplicons of approximately 640–660 bp and restriction fragments of variable sizes with all three enzymes (see the database http://www.yeast-id.com). Unfortunately, some strains of *R. glutinis* complex share similar RFLP-ITS profiles with *R. mucilaginosa*. The only differences observed are seen in the fragments produced with *Hae*III, which generally are 425+215 bp for *R. mucilaginosa* and 430+210 for *R. glutinis* complex.[35,148] The protocol for the RFLP-ITS technique is described in the following. The amplification/digestion procedure is applicable also directly from fresh yeast colonies for rapid identification (see following text).

Procedure

1. Prepare PCR mixture containing 0.5 μM each primer, 10 μM deoxynucleotides, 1.5 mM $MgCl_2$ and 1× PCR buffer, sterile double-distilled water up to the desired volume, and 5 μL of the DNA sample (conveniently diluted) for a 25 μL reaction. For the alternative method of direct amplification from a single colony, the cell material (picked with sterile tip or needle) is added to a reaction mixture of 25 μL (though mix components are at a concentration corresponding to 50 μL of final volume). After heating the mixture containing the cell material for 15 min at 95°C, 1 U of *Taq* DNA polymerase is added in 25 μL double-distilled water.
2. The cycling program consists of initial denaturation at 95°C for 5 min; 40 cycles of denaturing at 94°C for 1 min, annealing at 55°C for 2 min, and extension at 72°C for 2 min; and a final extension at 72°C for 10 min.
3. Digest the PCR products (15 μL or approximately 0.5–1.0 μg) without further purification with 1 U/μL of the restriction endonucleases *Cfo*I, *Hae*III, and *Hinf*I at 37°C overnight.
4. Separate the PCR products and their restriction fragments on 1.4% and 3% agarose gels, respectively, in 0.5× TBE buffer.
5. Stain the gels in 0.5 μg/mL ethidium bromide aqueous solution.
6. Estimate the sizes of PCR amplicons by comparison against a DNA length standard (e.g., 1000–100 bp ladder).

74.2.2.4 rDNA Sequencing Analysis

The rRNA genes are the preferred target for DNA sequencing. The main reasons are the availability of conserved and highly variable regions within the rRNA operon and its presence in multiple copies in fungal cells allowing high sensitivity, as well as accurate molecular identification for most yeasts.[171,172]

The 5′-end of the 26S rDNA D1/D2 domain (D1/D2) and the 5.8S-ITS region of the ribosomal rRNA gene (ITS) are the most frequently sequenced regions for yeast identification. Hence, large databases are available for non-taxonomist to quickly and accurately identify most known yeast species, as well as recognize new species, by sequencing approximately 600 nucleotides and doing a BLAST search in GenBank.

In general, strains differing in no more than three nucleotide differences (0%–0.5%) are considered conspecific, while strains showing six or more noncontiguous substitutions (1%) are separate species.[176] Strains with intermediate nucleotide substitutions are also likely to be separate species, though further molecular and conventional characterization is needed for reassurance. On the other hand, the ITS regions ITS1 and ITS2, which are separated by the 5.8S gene of rDNA, are also highly substituted and often used for species identification, but for many species, ITS sequences provide no greater resolution than that obtained from 26S domains D1/D2.[177,178]

List of universal primers for rDNA sequencing:

- D1/D2 region: NL1 (forward): 5′-GCA TAT CAA TAA GCG GAG GAA AAG-3′ and NL4 (reverse): 5′-GGT CCG TGT TTC AAG ACG G-3′ primers.
- ITS region: ITS1 (forward): (5′-TCC GTA GGT GAA CCT GCG G-3′) and ITS4 (reverse) (5′-TCC TCC GCT TAT TGA TAT GC-3′) primers.

- ITS1 region: ITS1 (forward): (5′-TCC GTA GGT GAA CCT GCG G-3′) and ITS2 (reverse) (5′-GCT GCG TTC TTC ATC GAT GC-3′) primers.
- ITS2 region: ITS3 (forward) (5′-GCA TCG ATG AAG AAC GCA GC-3′) and ITS4 (reverse) (5′-TCC TCC GCT TAT TGA TAT GC-3′) primers.

Procedure

1. Prepare PCR mixture containing 0.5 µM each primer (see following text), 10 µM deoxynucleotides, 1.5 mM $MgCl_2$ and 1× PCR buffer, sterile double-distilled water up to the desired volume, and 5 µL of the DNA sample (diluted, see above DNA extraction protocol) for a 25 µL reaction. For the alternative method of direct amplification from a single colony, the cell material (picked with sterile tip or needle) is added to a reaction mixture of 25 µL (though mix components are at a concentration corresponding to 50 µL of final volume). After heating of the mixture containing the cell material for 15 min at 95°C, 1 U of *Taq* DNA polymerase is added in 25 µL double-distilled water.
2. The cycling program consists of initial denaturation at 95°C for 5 min; 40 cycles of denaturing at 94°C for 1 min, annealing at 55°C for 2 min, and extension at 72°C for 2 min; and a final extension at 72°C for 10 min.
3. The presence of PCR products must be confirmed by 1% gel electrophoresis.
4. Purification of the PCR product should be carried out using DNA purification kit following manufacture's indications.
5. Sequence PCR product using one or both of the primers either by sending samples to a DNA Sequencing Service or, if applicable, using your own Sequencer equipment following the manufacture's protocol.
6. Visually correct sequences for nucleotide ambiguities or errors and compare to GenBank nucleotide database by using the BLASTN program (NCBI, Bethesda, MD).[179]

Type strains of medically relevant *Rhodotorula* species and their GenBank accession numbers for identification purposes are shown in Table 74.2.

74.3 CONCLUSIONS AND FUTURE PERSPECTIVES

In spite of the large contact that humans have with *Rhodotorula* yeasts as a consequence of the broad distribution in natural and artificial environments of these microorganisms, a relatively low incidence of Rhodotorulosis cases have been reported. This suggests that *Rhodotorula* infections are opportunistic and that the pathogenicity of these yeasts is limited by factors that are currently unknown and

TABLE 74.2
The Type or Reference Strains for Medically Relevant *Rhodotorula* and Other Pigmented Yeast Species

Species	Type Strain	D1/D2[e]	ITS[e]
R. mucilaginosa	CBS 316[T]	AF070432	AF444541
R. glutinis	CBS 20[T]	AF070430	AF444539, AF387775
R. dairenensis	CBS 4406[T]	AF070429	AF444501
Rh. sphaerocarpum	CBS 5939[T]	AF070432	AF444499
Rh. toruloides	CBS 14[a]	AF207884	AB049028
Rh. kratochvilovae	CBS 7436[T]	AF071436	AF444520
Rh. diobovatum	CBS 6085[T]	AF070421	AF387782, AF444502
Rh. babjevae	CBS 7808[T]	AF070420	AF444542
R. graminis	CBS 2826[T]	AF070431	AF444505
R. minuta	CBS 319[T]	AF189945	AF190011
Other			
Sporidiobolus johnsonii[b]	CBS 5470[T]	AF070435	EF592109
Sporidiobolus salmonicolor[c]	CBS 490[T]	AF070439	AF387784
Sporidiobolus metaroseus[d]	CBS 7687[T]	EU003461	EU003482
Sporidiobolus pararoseus	CBS 491[T]	AF189977	AY015429

Note: *R.*: *Rhodotorula*, *Rh.*: *Rhodosporidium*.
[a] Not the type strain but a widely used reference strain.
[b] Formerly *Sporobolomyces holsaticus*.
[c] Formerly *Sporobolomyces salmonicolor*.
[d] Formerly *Sporobolomyces roseus*.
[e] Genbank accession number.

need investigation. Nevertheless, it is clear that *Rhodotorula* yeasts, in particular the species referred to in this chapter, must be considered medically relevant given that inmunocompromised patients are susceptible of developing fungeamia when exposed to such yeasts.

Future investigations must also focus on the factors that favor infection by *Rhodotorula* yeasts, and on the search for phenotypic and genotypic characteristics that may help discriminate potential pathogenic strains from those that are not. In this task, molecular genotyping tools will play a significant role by allowing the rapid screening of clinical and environmental isolates for the detection of molecular traits proper of more virulent strains. The increasing importance of *Rhodotorula* yeasts as infective agents makes of the possibility to distinguish between potentially pathogenic and harmless strains on the basis of genetic traits a pressing need.

Molecular techniques for the detection, identification, and typing of potentially pathogenic strains of *Rhodotorula* must provide rapid and accurate identification and discrimination from taxonomically related species. For the case of *Rhodotorula*, a requisite for such molecular technique is that it must grant precise species differentiation within the *R. glutinis* complex. This will be the only way to determine which of these species can actually cause disease and in the hypothetical case that all did, which has the higher incidence of infections. Such epidemiological investigations are needed and are important for a correct circumscription of the *Rhodotorula* species that can behave as pathogens, thus allowing the assessment of their antifungal susceptibilities.

The implementation of molecular biology techniques in the clinical laboratory for the routine detection of *Rhodotorula* spp. is vital for a prompt and precise diagnosis, and the subsequent application of the appropriate antifungal treatment. Current conventional methodologies based on phenotypic criteria do not provide the required specificity, often leading to inconclusive identifications or even to incorrect ones. The techniques proposed in the present chapter have been used extensively in the identification and typing of *Rhodotorula* strains from different sources including clinical and may be perfectly implemented for identification and/or epidemiological purposes in the clinical laboratory. As discussed above, rDNA sequence analysis represents a powerful molecular tool for the rapid and accurate diagnosis of *Rhodotorula* spp. as well as of many other infections caused by fungi, and is probably the best choice for routine clinical practice. However, each clinical laboratory will determine which molecular technique can be used on the basis of its workflow, workload, and available financial resources.

An additional advantage of molecular techniques over conventional ones is that the direct detection of pathogens from clinical specimens becomes possible. Such technology has been already investigated for several pathogenic yeasts, though mostly of the genera *Candida*[142,180–182] and *Cryptococus* [160,183] from various types of specimens. The molecular techniques that have been to date evaluated for culture-independent diagnosis of yeast infections were based on real-time PCR,[142,180,182] DNA microarrays,[136,163,181] fluorescent in situ hybridization (FISH),[183] and Luminex®[160] technologies. Recently, Hsiue et al.[136] tested a new oligonucleotide array system targeting the ITS1 or ITS2 regions[163] to directly detect fungal pathogens from positive blood cultures. This array system includes probes for *R. mucilaginosa*, *R. glutinis*, and *R. minuta* and was able to detect efficiently *R. mucilaginosa* from blood cultures.[136] It remains to be determined if this technique can be used directly from human specimens.

However, to the best of our knowledge, the rest of the techniques mentioned above have not been yet tested for the detection of *Rhodotorula* spp. directly from human specimens. Thus, future investigations and developments in this area should take account of *Rhodotorula* species due to their increasing importance as opportunistic mycoses-producing yeasts.

REFERENCES

1. Libkind, D. and van Broock, M.R., Biomass and carotenoid pigments production by Patagonian native yeasts, *World J. Microbiol. Biotechnol.*, 22, 687, 2006.
2. Buzzini, P. et al., Carotenoid profiles of yeasts belonging to the genera *Rhodotorula, Rhodosporidium, Sporobolomyces* and *Sporidiobolus*, *Can. J. Microbiol.*, 53, 1024, 2007.
3. Bergman, A.G. and Kauffman, C.A., Dermatitis due to *Sporobolomyces* infection, *Arch. Dermatol.*, 120, 1059, 1984.
4. Morris, J.T., Beckius, M., and McAllister, C.K., *Sporobolomyces* infection in an AIDS patient, *J. Infect. Dis.*, 164, 623, 1991.
5. Plazas, J. et al., *Sporobolomyces salmonicolor* lymphadenitis in an AIDS patient: Pathogen or passenger? *AIDS*, 8, 387, 1994.
6. Morrow, J.D., Prosthetic cranioplasty infection due to *Sporobolomyces*, *J. Tenn. Med. Assoc.*, 87, 466, 1994.
7. Sharma, V., Shankar, J., and Kotamarthi, V., Endogeneous endophthalmitis caused by *Sporobolomyces salmonicolor*, *Eye*, 20, 945, 2006.
8. Serena, C. et al., In vitro antifungal susceptibilities of uncommon basidiomycetous yeasts, *Antimicrob. Agents Chemother.*, 48, 2724, 2004.
9. Fell, J.W. and Statzell-Tallman, A., *Rhodosporidium* Banno, in *The Yeasts: A Taxonomic Study*, Kurtzman, C.P. and Fell, J.W. (Eds.), p. 678, Elsevier, Amsterdam, the Netherlands, 1998.
10. Yamazaki, M. and Komagata, K., Taxonomic significance of electrophoretic comparison of enzymes in the genera *Rhodotorula* and *Rhodosporidium*, *Int. J. Syst. Bacteriol.*, 31, 361, 1981.
11. Hamamoto, M., Sugiyama, J., and Komagata, K., DNA base composition of strains in the genera *Rhodosporidium, Cystofilobasidium*, and *Rhodotorula* determined by reversed-phase high performance liquid chromatography, *J. Gen. Appl. Microbiol.*, 32, 215, 1986.
12. Hamamoto, M. et al., Numerical taxonomy based on the electrophoretic mobility of enzymes in the genera *Rhodosporidium, Cystofilobasidium* and *Rhodotorula*, *J. Gen. Appl. Microbiol.*, 32, 89, 1986.
13. Hamamoto, M., Sugiyama, J., and Komagata, K., DNA–DNA reassociation studies of strains in the genera *Rhodosporidium* and *Rhodotorula*, *J. Gen. Appl. Microbiol.*, 33, 57, 1987.

14 Vancanneyt, M. et al., A taxonomic study of the basidiomycetous yeast genera *Rhodosporidium* Banno and *Rhodotorula* Harrison based on whole-cell protein patterns, DNA base compositions and coenzyme Q types, *J. Gen. Appl. Microbiol.*, 38, 363, 1992.
15 Sampaio, J.P. et al., Polyphasic taxonomy of the basidiomycetous yeast genus *Rhodosporidium*: *Rhodosporidium kratochvilovae* and related anamorphic species, *Int. J. Syst. Evol. Microbiol.*, 51, 687, 2001.
16 Gadanho, M. and Sampaio, J.P., Polyphasic taxonomy of the basidiomycetous yeast genus *Rhodotorula*: *R. glutinis sensu stricto* and *R. dairenensis* comb. nov., *FEMS Yeast Res.*, 2, 47, 2002.
17 Libkind, D. et al., Mycosporines in carotenogenic yeasts, *Syst. Appl. Microbiol.*, 28, 749, 2005.
18 Sampaio, J.P., Utilization of low molecular weight lignin-related aromatic compounds for the selective isolation of yeasts: *Rhodotorula vanillica*, a new basidiomycetous yeast species, *Syst. Appl. Microbiol.*, 17, 613, 1994.
19 Slávikavá, E. and Vadkertiová, R., The occurrence of yeasts in the forest soils, *J. Basic Microbiol.*, 40, 207, 2000.
20 Polyakova, A.V., Chernov, I.Y., and Panikov, N.W.S., Yeast diversity in hydromorphic soils with reference to a grass-sphagnum wetland in western Siberia and hummocky Tundra Region at Cape Barrow (Alaska), *Mikoobiologia*, 70, 714, 2001.
21 Fonseca, A. and Inácio, J., Phylloplane yeasts, in *Biodiversity and Ecophysiology of Yeasts*, Rosa, C.A. and Gabor, P. (Eds.), p. 263, Springer, Berlin, Germany, 2006.
22 Libkind, D. et al., Molecular characterization of carotenogenic yeasts from aquatic environments in Patagonia, Argentina, *Anton. Leeuwen.*, 84, 313, 2003.
23 de Almeida, J.M., Yeast community survey in the Tagus estuary, *FEMS Microbiol. Ecol.*, 53, 295, 2005.
24 Gadanho, M., Almeida, J.M., and Sampaio, J.P., Assessment of yeast diversity in a marine environment in the south of Portugal by microsatellite-primed PCR, *Anton. Leeuwen.*, 84, 217, 2003.
25 Nagahama, T., Distribution and identification of red yeasts in deep-sea environments around the northwest Pacific Ocean, *Anton. Leeuwen.*, 80, 101, 2001.
26 Gadanho, M. and Sampaio, J.P., Occurrence and diversity of yeasts in the mid-Atlantic ridge hydrothermal fields near the Azores Archipelago, *Microb. Ecol.*, 50, 408, 2005.
27 Gadanho, M., Libkind, D., and Sampaio, J.P., Yeast diversity in the extreme acidic environments of the Iberian Pyrite Belt, *Microb. Ecol.*, 52, 552, 2006.
28 Russo, G. et al., Yeast diversity at the Volcanic acidic environment of the Lake Caviahue and Rio Agrio (Patagonia, Argentina), *FEMS Microbiol. Ecol.*, 65, 415, 2008.
29 de Silóniz, M.I. et al., Environmental adaptation factors of two yeasts isolated from the leachate of a uranium mineral heap, *FEMS Microbiol. Lett.*, 210, 233, 2002.
30 Butinar, L., Spencer-Martins, I., and Gunde-Cimerman, N., Yeasts in high Arctic glaciers: The discovery of a new habitat for eukaryotic microorganisms, *Anton. Leeuwen.*, 91, 277, 2007.
31 Starmer, W.T. et al., Yeasts in the genus *Rhodotorula* recovered from the Greenland ice sheet, in *Life in Ancient Ice*, Rogers, S.O. and Castello, J.D. (Eds.), p. 181, Princeton University Press, Princeton, NJ, 2005.
32 de García, V. et al., Biodiversity of cold-adapted yeasts from glacial meltwater rivers in Patagonia, Argentina, *FEMS Microbiol. Ecol.*, 59, 331, 2007.
33 Tournas, V.H., Heeres, J., and Burgess, L., Moulds and yeasts in fruit salads and fruit juices, *Food Microbiol.*, 23, 684, 2006.
34 Venturini, M.E., Oria, R., and Blanco, D., Microflora of two varieties of sweet cherries: Burlat and Sweetheart, *Food Microbiol.*, 19, 15, 2002.
35 Las Heras-Vazquez, F.J. et al., Identification of yeast species from orange fruit and juice by RFLP and sequence analysis of the 5.8S rRNA gene and the two internal transcribed spacers, *FEMS Yeast Res.*, 3, 3, 2003.
36 Deak, T. and Beuchat, L., *Handbook of Food Spoilage Yeasts*, CRC, Boca Raton, FL, 1996.
37 Fleet, G.H., Yeasts in fruit and fruit products, in *Yeasts in Food: Beneficial and Detrimental Aspects*, Boekhout, T. and Robert, V. (Eds.), p. 267, Behr, Hamburg, 2003.
38 Senses-Ergul, S. et al., Characterization of some yeasts isolated from foods by traditional and molecular tests, *Int. J. Food Microbiol.*, 108, 120, 2006.
39 Welthagen, J.J. and Viljoen, B.C., The isolation and identification of yeasts obtained during the manufacture and ripening of Cheddar cheese, *Food Microbiol.*, 16, 63, 1999.
40 Viljoen, B.C. and Greyling, T., Yeasts associated with Cheddar and Gouda making, *Int. J. Food Microbiol.*, 28, 79, 1995.
41 Gardini, F. et al., A survey of yeasts in traditional sausages of southern Italy, *FEMS Yeast Res.*, 1, 161, 2001.
42 Kajikazawa, T. et al., Detection of pathogenic yeasts from processed fresh edible sea urchins sold in a fish market, *Nipp. Ishin. Gak. Zas.*, 48, 169, 2007.
43 Eklund, M.W. et al., Characteristics of yeasts isolated from Pacific crab meat, *Appl. Microbiol.*, 13, 985, 1965.
44 Callon, C. et al., Stability of microbial communities in goat milk during a lactation year: Molecular approaches, *Syst. Appl. Microbiol.*, 30, 547, 2007.
45 Mackenzie, D.W.R., Yeasts from human sources, *Med. Mycol.*, 1, 8, 1962.
46 Alliot, C. et al., Opportunistic infection with *Rhodotorula* in cancer patients treated by chemotherapy: Two case reports, *Clin. Oncol.*, 12, 115, 2000.
47 Braun, D.K. and Kaufmann, C.A., *Rhodotorula* fungemia: A lifethreatening complication of indwelling central venous catheters, *Mycoses*, 35, 305, 1999.
48 Galan-Sanchez, F. et al., Microbiological characteristics and susceptibility patterns of strains of *Rhodotorula* isolated from clinical specimens, *Mycopathologia*, 145, 109, 1999.
49 Hsueh, P.R. et al., Catheter-related sepsis due to *Rhodotorula glutinis*, *J. Clin. Microbiol.*, 41, 857, 2003.
50 Strausbaugh, L.J. et al., High frequency of yeast carriage on hands of hospital personnel, *J. Clin. Microbiol.*, 32, 2299, 1994.
51 Strausbaugh, L.J. et al., Comparison of three methods for recovery of yeasts from hands of health-care workers, *J. Clin. Microbiol.*, 34, 471, 1996.
52 Rose, H.D. and Kurup, V.P., Colonization of hospitalized patients with yeast-like organisms, *Med. Mycol.*, 15, 251, 1977.
53 Lo Re, V., Fishman, N.O., and Nachamkin, I., Recurrent catheter-related *Rhodotorula rubra* infection, *Clin. Microbiol. Infect.*, 9, 897, 2003.
54 Hagan, M.E. et al., A pseudoepidemic of *Rhodotorula rubra*: A marker for microbial contamination of the bronchoscope, *Infect. Control Hosp. Epidemiol.*, 16, 727, 1995.
55 Pfaller, M.A. and Diekema, D.J., Rare and emerging opportunistic fungal pathogens: Concern for resistance beyond *Candida albicans* and *Aspergillus fumigatus*, *J. Clin. Microbiol.*, 42, 4419, 2004.

56 Hazen, K.C., New and emerging yeast pathogens, *Clin. Microbiol. Rev.*, 8, 462, 1995.
57 Samonis, G. et al., Transient fungemia due to *Rhodotorula rubra* in a cancer patient: Case report and review of the literature, *Infection*, 29, 173, 2001.
58 Lunardi, L.W. et al., Epidemiology and outcome of *Rhodotorula* fungemia in a tertiary care hospital, *Clin. Infect. Dis.*, 43, 60, 2006.
59 Maeder, M., et al., Aortic homograft endocarditis caused by *Rhodotorula mucilaginosa*, *Infection*, 31, 181, 2003.
60 Thanos, L. et al., Multifocal skeletal tuberculosis with *Rhodotorula minuta* co-infection, *Scand. J. Infect. Dis.*, 38, 309, 2006.
61 Sabino, R. et al., Epidemiology of candidemia in oncology patients: A 6-year survey in a Portuguese central hospital, *Med. Mycol.*, 5, 1–10, DOI: 10.1080/13693780903161216, 2009.
62 Louria, D.B. et al., Fungemia caused by 'nonpathogenic' yeasts, *Arch. Intern. Med.*, 119, 247, 1967.
63 Anaissie, E. and Bodey, G.P., Nosocomial fungal infections: Old problems and new challenges, *Infect. Dis. Clin. North Am.*, 3, 867, 1989.
64 Kiehn, T.E. and Armstrong, D., Changes in the spectrum of organisms causing bacteremia and fungemia in immunocompromised patients due to venous access devices, *Eur. J. Clin. Microbiol. Infect. Dis.*, 9, 869, 1990.
65 Kiehn, T.E. et al., Sepsis due to *Rhodotorula* related to use of indwelling central venous catheters, *Clin. Infect. Dis.*, 14, 841, 1992.
66 Tuon, F.F., Duboc de Almeida, G.M., and Costa, S.F., Central venous catheter-associated fungemia due to *Rhodotorula* spp.—A systematic review, *Med. Mycol.*, 45, 441, 2007.
67 Romano, A. et al., Yeasts in external ocular inflammations, *Ophthalmology*, 170, 13, 1975.
68 Guerra, R. et al., *Rhodotorula glutinis* keratidis, *Int. Ophtalmol.*, 16, 187, 1992.
69 Panda, A. et al., *Rhodotorula* sp. infection in corneal interface following lamellar keratoplasty—A case report, *Acta Ophthalmol. Scand.*, 77, 227, 1999.
70 Gregory, J.K. and Haller, J.A., Chronic postoperative *Rhodotorula* endophthalmitis, *Arch. Ophthalmol.*, 110, 1686, 1992.
71 Muralidhar, S. and Sulthana, C.M., *Rhodotorula* causing chronic Dacryocystitis: A case report, *Ind. J. Ophthalmol.*, 43, 196, 1995.
72 Pore, R.S. and Chen, J., Meningitis caused by *Rhodotorula, Sabouraudia*, 14, 331, 1976.
73 Gyaurgieva, O.H., Bogomolova, T.S., and Gorshkova, G.I., Meningitis caused by *Rhodotorula rubra* in an HIV-infected patient, *J. Med. Vet. Mycol.*, 34, 357, 1996.
74 Lanzafame, M. et al., *Rhodotorula glutinis*-related meningitis, *J. Clin. Microbiol.*, 39, 410, 2001.
75 Flemming, R.V., Walsh, T.J., and Anaissie, E.J., Emerging and less common fungal pathogens, *Infect. Dis. Clin. N. Am.*, 16, 915, 2002.
76 Thakur, K. et al., Meningitis caused by *Rhodotorula rubra* in an human immunodeficiency virus infected patient, *Indian J. Med. Microbiol.*, 25, 166, 2007.
77 Baradkar, V.P. and Kumar, S., Meningitis caused by *Rhodotorula mucilaginosa* in human immunodeficiency virus seropositive patient, *Ann. Indian Acad. Neurol.*, 11, 245, 2008.
78 Eisenberg, E.S. et al., *Rhodotorula rubra* peritonitis in patients undergoing continuous ambulatory peritoneal dialysis, *Am. J. Med.* 75, 349, 1983.
79 Ünal, A. et al., CAPD-related peritonitis caused by *Rhodotorula mucilaginosa*, *Perit. Dial. Int.*, 29, 581, 2009.
80 Naveh, Y. et al., Endocarditis caused by *Rhodotorula* successfully treated with 5-fluorocytosine, *Br. Heart J.*, 37, 101, 1975.
81 Donald, F.E. and Sharp, J.F., *Rhodotorula rubra* ventriculitis, *J. Infect.*, 16, 187, 1988.
82 Anaissie, E. et al., New spectrum of fungal infections in patients with cancer, *Rev. Infect. Dis.*, 11, 369, 1989.
83 Kiraz, N., Gulbas, Z., and Akgun, Y., Case report: *Rhodotorula mucilaginosa* fungaemia due to use of indwelling venous catheters, *Mycoses*, 43, 209, 2000.
84 Zaas, A.K. et al., Risk of fungemia due to *Rhodotorula* and antifungal susceptibility testing of *Rhotodorula* isolates, *J. Clin. Microbiol.*, 41, 5233, 2003.
85 Duboc De Almeida, G.M. et al., *Rhodotorula* spp. isolated from blood cultures: Clinical and microbiological aspects, *Med. Mycol.*, 12, 1, 2008.
86 Neofytos, D., Horn, D., and De Simone, J.A. Jr., *Rhodotorula mucilaginosa* catheter-related fungemia in a patient with sickle cell disease: Case presentation and literature review, *South Med. J.*, 100, 198, 2007.
87 Pien, F.D. et al., *Rhodotorula* septicemia: Two cases and a review of the literature, *Mayo Clin. Proc.*, 55, 258, 1980.
88 Savini, V. et al., Femoral prosthesis infection by *Rhodotorula mucilaginosa*, *J. Clin. Microbiol.*, 46, 3544, 2008.
89 Cutrona, A.F. et al., *Rhodotorula minuta*: An unusual fungal infection in hip joint prosthesis, *Am. J. Orthop.*, 31, 137, 2002.
90 Lui, A.Y. et al., Amphotericin B lipid complex therapy in an AIDS patient with *Rhodotorula rubra* fungemia, *Clin. Infect. Dis.*, 27, 892, 1998.
91 Papadogeorgakis, H. et al., *Rhodotorula rubra* fungaemia in an immunocompromised patient, *J. Eur. Acad. Dermatol. Venerol.*, 12, 169, 1999.
92 Polo, R., Disseminated *Rhodotorula rubra* infection in an HIV-infected drug abuser, *Eur. J. Med.*, 2, 252, 1993.
93 Rusthoven, J.J., Feld, R., and Tuffnell, P.G., Systemic infection by *Rhodotorula* spp. in the immunocompromised host, *J. Infect.*, 8, 241, 1984.
94 Kofteridis, D. et al., *Rhodotorula glutinis* fungemia successfully treated with fluconazole: report of two cases, *Int. J. Infect. Dis.*, 11, 179, 2006.
95 Riedel, D.J., Johnson, J.K., and Forrest, G.N., *Rhodotorula glutinis* fungemia in a liver kidney transplant patient, *Transpl. Infect. Dis.*, 10, 197, 2008.
96 da Cunha, M.M.L. et al., Identification, antifungal susceptibility and scanning electron microscopy of a keratinolytic strain of *Rhodotorula mucilaginosa*: A primary causative agent of onychomycosis, *FEMS Immunol. Med. Microbiol.*, 55, 396, 2009.
97 Fung, H.B. et al., *Rhodotorula mucilaginosa* lymphadenitis in an HIV-infected patient, *Int. J. Infect. Dis.*, 13, 27, 2009.
98 Tuon, F.F. and Costa, S.F., *Rhodotorula* infection. A systematic review of 128 cases from literature, *Rev. Iberoam. Micol.*, 25, 135, 2008.
99 Krcmery, V. et al., Aetiology, antifungal susceptibility, risk factors and outcome in 201 fungaemic children: Data from a 12-year prospective national study from Slovakia, *J. Med. Microbiol.*, 51, 110, 2002.
100. Nucci, M. et al., Fungal infections in neutropenic patients: A 8-year prospective study, *Rev. Inst. Med. Trop. Sao Paulo*, 37, 397, 1995.

101. Pfaller, M.A. et al., Results from the ARTEMIS DISK global antifungal surveillance study, 1997 to 2005: An 8.5-year analysis of susceptibilities of *Candida* species and other yeast species to fluconazole and voriconazole determined by CLSI standardized disk diffusion testing, *J. Clin. Microbiol.*, 47, 117, 2009.
102. Goyal, R. et al., *Rhodotorula mucilaginosa* as a cause of persistent femoral nonunion, *J. Postgrad. Med.*, 54, 25, 2008.
103. Walsh, T.J. et al., Infections due to emerging and uncommon medically important fungal pathogens, *Clin. Microbiol. Infect.*, 10, 48, 2004.
104. Petrocheilou-Paschou, et al., *Rhodotorula* septicemia: Case report and minireview, *Clin. Microbiol. Infect. Dis.*, 7, 88, 2001.
105. Ahmed, A. et al., A fatal case of *Rhodotorula* meningitis in AIDS, *Med. Health R I*, 81, 22, 1998.
106. Krcmery, V., Krupova, I., and Denning, D.W., Invasive yeast infections other than *Candida* spp. in acute leukaemia, *J. Hosp. Infect.*, 41, 181, 1999.
107. Samonis, G. and Bafaloukos, D., Fungal infections in cancer patients: An escalating problem, *In Vivo*, 6, 183, 1992.
108. Vartivarian, S.E., Anassie, E.J., and Bodey, G.P., Emerging fungal pathogens in immunocompromising patients: Classification, diagnosis, and management, *Clin. Infect. Dis.*, 17, 487, 1993.
109. Georgiev, V.S., *Opportunistic Infections: Treatment and Prophylaxis*. Humana Press, Totowa, NJ, 2003.
110. Leeber, D.A. and Scher, I., *Rhodotorula* fungemia presenting as 'endotoxic' shock, *Arch. Intern. Med.*, 123, 78, 1969.
111. Wheeler, R.T. et al., A *Saccharomyces cerevisiae* mutant with increased virulence, *Proc. Natl. Acad. Sci. U.S.A.*, 100, 2766, 2003.
112. Garcia-Martos, P. et al., Asimilación de carbohidratos por cepas de *Rhodotorula glutinis* de origen clínico y ambiental, *Rev. Iberoam. Micol.*, 21, 90, 2004.
113. Brizzio, S. et al., Extracellular enzymatic activities of basidiomycetous yeasts isolated from glacial and subglacial waters of northwest Patagonia (Argentina), *Can. J. Microbiol.*, 53, 519, 2007.
114. Cafarchia, C. et al., Phospholipase activity of yeasts from wild birds and possible implications for human disease, *Med. Mycol.*, 46, 429, 2008.
115. Diekema, D.J. et al., Activities of available and investigational antifungal agents against *Rhodotorula* species, *J. Clin. Microbiol.*, 43, 476, 2005.
116. Gomez-Lopez, A. et al., Pharmacotherapy of yeast infections, *Expert Opin. Pharmacother.*, 9, 2801, 2008.
117. Goldani, L.Z., Craven, D.E., and Sugar, A.M., Central venous catheter infection with *Rhodotorula minuta* in a patient with AIDS taking suppressive doses of fluconazole, *J. Med. Vet. Mycol.*, 33, 267, 1995.
118. Marinova, I. et al., *Rhodotorula* spp. fungemia in a immunocompromised boy after neurosurgery successfully treated with miconazole and 5-flucytosine: Case report and review of the literature, *Chemother.*, 40, 287, 1994.
119. Fanci, R. et al., Treatment of fungemia due to unusual pathogens in neutropenic patients: Report of two cases, *J. Chemother.*, 9, 427, 1997.
120. Lanternier, F. and Lortholary, O., Liposomal amphotericin B: What is its role in 2008? *Clin. Microbiol. Infect.*, 14, 71, 2008.
121. Navarro, J.T., Lauzurica, R., and Gimenez, M., *Rhodotorula rubra* infection in a kidney transplant patient with pancytopenia, *Haematologia*, 86, 111, 2001.
122. Preney, L. et al., Experimental evaluation of antifungal and antiseptic agents against *Rhodotorula* spp., *Mycoses*, 46, 492, 2003.
123. Thompson, G.R. et al., In vitro activity of isavuconazole against *Trichosporon*, *Rhodotorula*, *Geotrichum*, *Saccharomyces* and *Pichia* species, *J. Antimicrob. Chemoth.*, 64, 79, 2009.
124. Silici, S., Antifungal activities of propolis collected by different races of honeybees against yeasts isolated from patients with superficial mycoses, *J. Pharmacol. Sci.*, 99, 39, 2005.
125. Giusiano, G. et al., Fluconazole and itraconazole resistance of yeasts isolated from the bloodstream and catheters of hospitalized pediatric patients, *Chemotherapy*, 52, 254, 2006.
126. Fleet, G., Spoilage yeasts, *Crit. Rev. Biotechnol.*, 12, 1, 1992.
127. Dooley, D.P., Beckius, M.L., and Jeffrey, B.S., Misidentification of clinical yeast isolates by using the updated Vitek Yeast Biochemical Card, *J. Clin. Microbiol.*, 32, 2889, 1994.
128. Ramani, R. et al., Efficacy of API 20C and ID 32C systems for identification of common and rare clinical yeast isolates, *J. Clin. Microbiol.*, 36, 3396, 1998.
129. Chen, Y.C. et al., Identification of medically important yeasts using PCR-based detection of DNA sequence polymorphisms in the internal transcribed spacer 2 region of the rRNA genes, *J. Clin. Microbiol.*, 38, 2302, 2000.
130. Chen, Y.C. et al., Polymorphic internal transcribed spacer region 1 DNA sequences identify medically important yeasts, *J. Clin. Microbiol.*, 39, 4042, 2001.
131. Coignard, C. et al., Resolution of discrepant results for *Candida* species identification by using DNA probes, *J. Clin. Microbiol.*, 42, 858, 2004.
132. Li, Y.L. et al., Rapid identification of yeasts commonly found in positive blood cultures by amplification of the internal transcribed spacer regions 1 and 2, *Eur. J. Clin. Microbiol. Infect. Dis.*, 22, 693, 2003.
133. Ciardo, D.E. et al., Internal transcribed spacer sequencing versus biochemical profiling for identification of medically important yeasts, *J. Clin. Microbiol.*, 44, 77, 2006.
134. Massonet, C. et al., Comparison of VITEK 2 with ITS2-fragment length polymorphism analysis for identification of yeast species, *J. Clin. Microbiol.*, 42, 2209, 2004.
135. Szabó, Z. et al., Evaluation of the new Micronaut-*Candida* system compared to the API ID32C method for yeast identification, *J. Clin. Microbiol.*, 46, 1824, 2008.
136. Hsiue, H.-C. et al., Rapid identification of fungal pathogens in positive blood cultures using oligonucleotide array hybridization, *Clin. Microbiol. Infect.*, 16, 493–500, DOI: 10.1111/j.1469–0691.2009.02828.x, 2009.
137. Reiss, E. et al., Molecular diagnosis and epidemiology of fungal infections, *Med. Mycol.*, 36, 249, 1998.
138. Fell, J.W., Rapid identification of yeast species using three primers in a polymerase chain reaction, *Mol. Mar. Biol. Biotech.*, 2, 174, 1993.
139. Fell, J.W., rDNA targeted oligonucleotide primers for the identification of pathogenic yeasts in a polymerase chain reaction, *J. Ind. Microbiol.*, 14, 475, 1995.
140. Elie, C.M. et al., Rapid identification of *Candida* species with species-specific DNA probes, *J. Clin. Microbiol.*, 36, 3260, 1998.
141. Luo, G. and Mitchell, T.G., Rapid identification of pathogenic fungi directly from cultures by using multiplex PCR, *J. Clin. Microbiol.*, 40, 2860, 2002.
142. Innings, A. et al., Multiplex real-time PCR targeting the RNase P RNA gene for detection and identification of *Candida* species in blood, *J. Clin. Microbiol.*, 45, 874, 2007.

143. Schabereiter-Gurtner, C. et al., Development of novel real-time PCR assays for detection and differentiation of eleven medically important *Aspergillus* and *Candida* species in clinical specimens, *J. Clin. Microbiol.*, 45, 906, 2007.
144. Libkind, D. et al., Studies on the heterogeneity of the carotenogenic yeast *Rhodotorula mucilaginosa* from Patagonia, Argentina, *J. Basic Microbiol.*, 48, 93, 2008.
145. Loeffler, J. et al., Detection of PCR-amplified fungal DNA by using a PCR-ELISA system, *Med. Mycol.*, 36, 275, 1998.
146. Wang, Q.-M. et al., Rapid differentiation of phenotypically similar yeast species by single-strand conformation polymorphism analysis of ribosomal DNA, *Appl. Environ. Microbiol.*, 74, 2604, 2008.
147. McCullough, M.J. et al., Intergenic transcribed spacer PCR ribotyping for differentiation of *Saccharomyces* species and interspecific hybrids, *J. Clin. Microbiol.*, 36, 1035, 1998.
148. Esteve-Zarzoso, B. et al., Identification of yeasts by RFLP analysis of the 5.8S rRNA gene and the two ribosomal internal transcribed spacers, *Int. J. Syst. Bacteriol.*, 49, 329, 1999.
149. de Llanos Frutos, R., Fernandez-Espinar, M.T., and Querol., A., Identification of species of the genus *Candida* by analysis of the 5.8S rRNA gene and the two ribosomal internal transcribed spacers, *Anton. Leeuwen.*, 85, 175, 2004.
150. Bai, F.-Y., Liang, H.-Y., and Jia, J.-H., Taxonomic relationships among the taxa in the *Candida guilliermondii* complex, as revealed by comparative electrophoretic karyotyping, *Int. J. Syst. Evol. Microbiol.*, 50, 417, 2000.
151. Boekhout, T. et al., The use of karyotyping in the systematics of yeasts, *Anton. Leeuwen.*, 63, 157, 1993.
152. Saracli, M.A. et al., Genotyping of clinical *Rhodotorula mucilaginosa* isolates by pulsed field gel electrophoresis, *Mycoses*, 46, 487, 2003.
153. Gadanho, M. and Sampaio, J.P., Application of temperature gradient gel electrophoresis to the study of yeast diversity in the estuary of the Tagus river, Portugal, *FEMS Yeast Res.*, 5, 253, 2004.
154. Gadanho, M. and Sampaio, J.P., Microeukaryotic diversity in the extreme environments of the Iberian Pyrite Belt: A comparison between universal and fungi-specific primer sets, temperature gradient gel electrophoresis and cloning, *FEMS Microbiol. Ecol.*, 57, 139, 2006.
155. Theelen, B. et al., Identification and typing of *Malassezia* yeasts using amplified fragment length polymorphism (AFLP), random amplified polymorphic DNA (RAPD) and denaturing gradient gel electrophoresis (DGGE), *FEMS Yeast Res.*, 1, 79, 2001.
156. Manzano, M. et al., PCR-DGGE differentiation of strains of *Saccharomyces* sensu stricto, *Anton. Leeuwen.*, 85, 23, 2004.
157. Aquilanti, L. et al., The microbial ecology of a typical Italian salami during its natural fermentation, *Int. J. Food Microbiol.*, 120, 136, 2007.
158. Prakitchaiwattana, C.J., Fleet, G.H., and Heard, G.M., Application and evaluation of denaturing gradient gel electrophoresis to analyse the yeast ecology of wine grapes, *FEMS Yeast Res.*, 4, 865, 2004.
159. Gupta, A.K. et al., Identification and typing of *Malassezia* species by amplified fragment length polymorphism and sequence analyses of the internal transcribed spacer and large-subunit regions of ribosomal DNA, *J. Clin. Microbiol.*, 42, 4253, 2004.
160. Bovers, M. et al., Identification of genotypically diverse *Cryptococcus neoformans* and *Cryptococcus gattii* isolates by Luminex xMAP technology, *J. Clin. Microbiol.*, 45, 1874, 2007.
161. Diaz, M.R. et al., Microcoding and flow cytometry as a high-throughput fungal identification system for *Malassezia* species, *J. Med. Microbiol.*, 55, 1197, 2006.
162. Page, B.T. et al., Rapid identification of ascomycetous yeasts from clinical specimens by a molecular method based on flow cytometry and comparison with identifications from phenotypic assays, *J. Clin. Microbiol.*, 44, 3167, 2006.
163. Leaw, S.N. et al., Identification of medically important *Candida* and non-*Candida* yeast species by an oligonucleotide array, *J. Clin. Microbiol.*, 45, 2220, 2007.
164. Leinberger, D.M. et al., Development of a DNA microarray for detection and identification of fungal pathogens involved in invasive mycoses, *J. Clin. Microbiol.*, 43, 4943, 2005.
165. Campa, D. et al., A DNA microarray based on Arrayed-Primer Extension (APEX) technique for identification of pathogenic fungi responsible for invasive and superficial mycoses, *J. Clin. Microbiol.*, 46, 909, 2008.
166. Fell, J.W. et al., Biodiversity and systematics of basidiomycetous yeasts as determined by large-subunit rDNA D1/D2 domain sequence analysis, *Int. J. Syst. Evol. Microbiol.*, 50, 1351, 2000.
167. Scorzetti, G. et al., Systematics of basidiomycetous yeasts: A comparison of large subunit D1/D2 and internal transcribed spacer rDNA regions, *FEMS Yeast Res.*, 2, 495, 2002.
168. Latouche, G.N. et al., Comparison of use of phenotypic and genotypic characteristics for identification of species of the anamorph genus *Candida* and related teleomorph yeast species, *J. Clin. Microbiol.*, 35, 3171, 1997.
169. San-Blas, G. and Niño-Vega, G., New approaches in the diagnosis of medically relevant mycoses: Fungal identification by molecular techniques, in *Pathogenic Fungi: Insights in Molecular Biology*, San-Blas, G. and Calderone, R.A. (Eds.), Chap. 8, Caister Academic Press, Wymondham, U.K., 2008.
170. Villa-Carvajal, M., Querol, A., and Belloch, C., Identification of species in the genus *Pichia* by restriction of the internal transcribed spacers (ITS1 and ITS2) and the 5.8S ribosomal DNA gene, *Anton. Leeuwen.*, 90, 171, 2006.
171. Linton, C.J., Molecular identification of unusual pathogenic yeast isolates by large ribosomal subunit gene sequencing: 2 years of experience at the United Kingdom mycology reference laboratory, *J. Clin. Microbiol.*, 45, 1152, 2007.
172. Leaw, S.N. et al., Identification of medically important yeast species by sequence analysis of the internal transcribed spacer regions, *J. Clin. Microbiol.*, 44, 693, 2006.
173. Silva, J.O. et al., Performance of selective and differential media in the primary isolation of yeasts from different biological samples, *Mycopathology*, 157, 29, 2004.
174. Guillamón, J.M. et al., Rapid identification of wine yeast species based on RFLP analysis of the ribosomal internal transcribed spacer (ITS) region, *Arch. Microbiol.*, 169, 387, 1998.
175. Sabate, J. et al., Isolation and identification of yeasts associated with vineyard and winery by RFLP analysis of ribosomal genes and mitochondrial DNA, *Microbiol Res.*, 157, 267, 2002.
176. Kurtzman, C. and Fell, J.W., Yeast systematics and phylogeny—Implications of molecular identification methods for studies in ecology, in *Biodiversity and Ecophysiology of Yeasts*, Rosa, C.A. and Gabor, P. (Eds.), Chap. 2, Springer, Berlin, Germany, 2006.
177. James, S.A., Collins, M.D., and Roberts, I.N., Use of an rRNA internal transcribed spacer region to distinguish phylogenetically closely related species of the genera *Zygosaccharomyces* and *Torulaspora*, *Int. J. Syst. Bacteriol.*, 46, 189, 1996.

178. Kurtzman, C.P. and Robnett, C.J., Phylogenetic relationships among yeasts of the '*Saccharomyces* complex' determined from multigene sequence analyses, *FEMS Yeast Res.*, 3, 417, 2003.
179. Altschul, S.F. et al., Gapped BLAST and PSI-BLAST: A new generation of protein database search programs, *Nucleic Acids Res.*, 25, 3389, 1997.
180. Salvarangan, R. et al., Rapid identification of commonly encountered *Candida* species directly from blood culture bottles, *J. Clin. Microbiol.*, 41, 5660, 2003.
181. Spiess, B. et al., DNA microarray-based detection and identification of fungal pathogens in clinical samples from neutropenic patients, *J. Clin. Microbiol.*, 45, 3743, 2007.
182. Vollmer, T. et al., Evaluation of novel broad-range real-time PCR assay for rapid detection of human pathogenic fungi in various clinical specimens, *J. Clin. Microbiol.*, 46, 1919, 2008.
183. Wilson, D.A. et al., Multicenter evaluation of a *Candida albicans* peptide nucleic acid fluorescent in situ hybridization probe for characterization of yeast isolates from blood cultures, *J. Clin. Microbiol.*, 43, 2909, 2005.
184. Herzberg, M., Fischer, R., and Titze, A., Conflicting results obtained by RAPD-PCR and large-subunit rDNA sequences in determining and comparing yeast strains isolated from flowers: A comparison of two methods, *Int. J. Syst. Evol. Microbiol.*, 52, 1423, 2002.
185. Vermitsky, J.-P. et al., Survey of vaginal-flora *Candida* species isolates from women of different age groups by use of species-specific PCR detection, *J. Clin. Microbiol.*, 46, 1501, 2008.
186. Walsh, T.J. et al., PCR and single-strand conformational polymorphism for recognition of medically important opportunistic fungi, *J. Clin. Microbiol.*, 33, 3216, 1995.
187. Sugita, T. et al., Sequence analysis of the ribosomal DNA intergenic spacer 1 regions of *Trichosporon* species, *J. Clin. Microbiol.*, 40, 1826, 2002.
188. Sugita, T. et al., Sequence diversity of the intergenic spacer region of the rRNA gene of *Malassezia globosa* colonizing the skin of patients with atopic dermatitis and healthy individuals, *J. Clin. Microbiol.*, 41, 3022, 2003.
189. Biswas, S.K. et al., Molecular phylogenetics of the genus *Rhodotorula* and related basidiomycetous yeasts inferred from the mitochondrial cytochrome b gene, *Int. J. Syst. Evol. Microbiol.*, 51, 1191, 2001.
190. Trnovsky, J. et al., Rapid and accurate identification of *Candida albicans* isolates by use of PNA $FISH_{Flow}$, *J. Clin. Microbiol.*, 46, 1537, 2008.
191. Montero, C.I. et al., Evaluation of Pyrosequencing® technology for the identification of clinically relevant non-dematiaceous yeasts and related species, *Eur. J. Clin. Microbiol. Infect. Dis.*, 27, 821, 2008.
192. Diaz, M.R. et al., Molecular sequence analyses of the intergenic spacer (IGS) associated with rDNA of the two varieties of the pathogenic yeast, *Cryptococcus neoformans*, *Syst. Appl. Microbiol.*, 23, 535, 2000.

75 *Schizophyllum*

Sophie Cassaing, Marie-Denise Linas, and Antoine Berry

CONTENTS

75.1 INTRODUCTION

Schizophyllum is a wood-decaying fungus and *S. commune* is the single species up to now involved in human diseases. *S. commune* is found on every continent except Antarctica where the wood substrate is lacking. Although it is encountered throughout the year, it is more abundant during the cold raining season in tropical countries.[5]

S. commune was discovered in 1930 by Hans Kniep. The main role of *S. commune* is to recycle carbon from rotten wood, but it has also been described as a fruit orchards pathogen or protruding through the plastic film on baled grass silage.[1] *S. commune* is considered edible in some countries such as Malaysia, China,[2] Thailand,[3,4] Nigeria,[5] Congo, Cameroon,[6] Peru, and Southwest Mexico.[7] *S. commune* is rich in fibers, proteins, and vitamins such as thiamine and riboflavin. In addition, some compounds of this mushroom have been shown to possess an immunomodulatory activity and potent macrophages/T-cell-mediated antitumor effect.[4,8–10]

75.1.1 Classification, Morphology, and Biology

S. commune is a filamentous heterothallic basidiomycete from the Agaricales order and the Schizophyllaceae family (http://www.cbs.knaw.nl). The life cycle of *S. commune* is made up of two phases: the unmated homokaryon and the mated dikaryon. Sexual reproduction involves the formation of mycelium special structures named clamp connections: upon cell division, one of the nuclei migrates and divides into a protrusion (clamp) from the apical cell. The other nucleus migrates and divides synchronously near the clamp. Septa are formed across the mitotic spindle, one at the basis of the clamp and one between apical and subapical cells. Thus, one nucleus remains entrapped into the clamp. Then the clamp and the subapical cell fuse and the clamp nucleus is released toward the subapical cell, restoring the dikaryotic state of the subapical cell.[11,12] As a characteristic of basidiomycete fungi, the dikaryotic phase can be prolonged over several years or even indefinitely.

Only the dikaryon is competent to form fruit body and to sporulate. Fruit body appears as a split-gill fungus, macroscopically visible on dying trees, and occurs single or in small colonies. It is 3–7 cm wide with a fan-shaped cap. The hymenium underneath the fruit body is formed of gills where numerous basidia are arranged in palisade, each bearing four basidiospores on erect sterigmata that are released into the environment.[13] Fruiting bodies are also macroscopically visible in culture when the mycelium is dikaryotic.

Mating, fertilizing, fruiting, and meiosis are determined in *Schizophyllum* by two unlinked genetic compatibility factors named A and B, each one consisting of two closely linked loci α and β. Aα is estimated to have 9 different specificities since $A\beta$ is estimated to have 32 specificities. $B\alpha$ and $B\beta$ are both estimated to have nine specificities. As a consequence, more than 28,000 mating types are encountered in the sexual progeny.[11,14–16]

Both *A* and *B* gene functions are required for dikaryotic growth. Sexual reproduction occurs only if the two haploid hyphae are compatible. The condition for a fully compatibility is a difference in specificity at $A\alpha$ or $A\beta$ and a difference at $B\alpha$ or $B\beta$.[11,15,17–20] The B mating-type genes encode G-protein-coupled receptors and their ligands that are small lipopeptide pheromones. These pheromone–receptor complexes regulate nuclear exchange and migration.[17,19,21–25] The A mating-type genes encode DNA-binding regulatory proteins involved in controlling synchronous nuclear division with clamp connection development.[26]

75.1.2 Clinical Features and Pathogenesis

Despite its large distribution and the common presence of environmental basidiospores liable to be inhaled through the nasal route, *S. commune* is rarely reported as a cause of infection in humans. Firstly regarded as a nonpathogenic fungus, it is now emerging as a cause of mycosis. It is responsible for disease in both immunocompetent and immunosuppressed people such as patients with diabetes or in patients with human immunodeficiency virus infection. Infections include local or disseminated disease and many clinical presentations can be confused with *Aspergillus* infections. In 1950, Kligman reported *S. commune* as a possible agent of onychomycosis,[27] and in 1994, Kamei was the first to identify the fungus in the human respiratory tract.[28] More recently, although the number of reports of *S. commune* mycosis is increasing, the review of the English-language medical literature reports no more than 50 cases (Table 75.1). Diseases mostly consist of lung disorders and infections of the sinuses. Pulmonary diseases (48%) include allergic bronchopulmonary mycosis, bronchial mucoid impaction, a case of chronic eosinophilic pneumonia, a case of fungus ball, and a case of bilateral nodules. Chronic or allergic sinusitis represents 43% of reports with a clear predominance in women (81%). *S. commune* sinusitis concerns healthy patients as well as patients with diabetes or HIV infection. When infection occurred in AIDS individuals, the CD4 lymphocyte count was 150/mm^3 and 1/mm^3.[29,30] In most of the reported cases, sinusitis caused by *S. commune* is unilateral (94%). Maxillary sinus is more often affected (38%), but all sinuses are concerned (Table 75.2). Several sinuses are involved in 31% of cases. The diagnosis remains difficult because the clinical presentation and histopathological findings can be suggestive of aspergillar sinusitis.

75.1.3 Laboratory Diagnosis

75.1.3.1 Conventional Techniques

75.1.3.1.1 Microscopic Examination of Biological Samples and Histopathology

Mycological examination of biological specimen (e.g., nasal flushing, sinus content, bronchoalveolar fluid (BAF), and tissue biopsy) consists of the search for hyaline, septae hyphae, varying in width from 1.5 to 5 μm. Hyphae can branch at acute or right angles but without dichotomous fork.[16,31,32] These fungal elements are visualized after KOH preparation or after hematoxylin–eosin, Grocott–Gomori stain, or lactophenol Cotton blue stain. Histologically, mucinous material with eosinophils and Charcot-Leyden crystals may be visualized in pulmonary or sinusal specimens when an allergic process is involved.

75.1.3.1.2 Culture

The fungus is easy to culture on most of standard media used in clinical laboratories, such as Sabouraud dextrose (with or without gentamicin/chloramphenicol) and potato dextrose agars incubated between 25°C and 37°C, even until 43°C.[5,16,31,32] The colony reaches a diameter of several centimeters in a week. The optimal temperature is not well specified: the rate of growth is described slightly faster at 43°C by Sigler et al.[31] when the optimum growth temperature for Nigerian strains is 25°C at pH 5.5.[5] Media such as potato dextrose, 2% malt extract, and phytone-yeast agars can be used for subcultures. *S. commune* is susceptible to cycloheximide at a concentration of 0.4 mg/mL.[16]

75.1.3.1.2.1 Macroscopic Colony Aspect The mycelium presents as a woolly, rather dense, white mold. Colonies have a rapid growth rate up to 50–60 mm after 7 days. Zones of condensation of the mycelium as masses of hyphae can appear in 1- or 2-week-old cultures. The reverse is pale cream-colored. Culture plates or tubes often release a characteristic strong disagreeable odor. Under conditions of light (alternative dark and light),[33] and if the genotype is formed from fully compatible homokaryons, dikaryotic mycelia can produce characteristic fan-shaped or shell-like basidiocarps (Figure 75.1). These fruiting bodies are provided with longitudinally split gills and are visible to the naked eye after a culture of 1 week to several months. They are up to 2 cm in height and in diameter. The formation of fruiting bodies is easier at 25°C compared to 37°C. Sometimes, tube-like structures of a few millimeter in height can appear with a central aperture. These elements likely correspond to aborted fruiting bodies (infertile) and are the result of two homokaryons partially compatible, differing only at the A locus or only at the B locus.[16,17,31,33–36]

75.1.3.1.2.2 Microscopy Characteristics As observed in biological samples, hyphae observed in culture are 1.5–3 μm wide. They can present with discreet rounded swellings (crenate hyphae). Dikaryotic isolates are easier to identify because they present with clamp connections that are involved in cell division and constitute an essential marker for the recognition of a basidiomycete. The presence of spicules (also named tubercles or pegs) is also highly representative of *S. commune*, but monokaryotic isolates may fail to produce these structures.[16,31,35] If the mycelium is formed from incompatible or partially compatible homokaryons, if it fails to produce clamps, or if it is overlooked, the identification becomes more difficult and these isolates are inclined to be labeled "Mycelia sterilia."[35,37] Moreover, clamps and spicules can be lost upon subcultures.[35] In contrast, some isolates present clamps only in subcultures and after several months.[16]

TABLE 75.1
***S. commune* Human Infections Reported in the English Literature between 1950 and 2009**

Biological Sample	Clinical Feature	Age	Sex	Underlying Disease	Reference
Toe nails	Onychomycosis	33	M	—	[27]
CSF	Atypical meningitis	24	M	—	[42]
Palate	Palate ulceration	4 months	M	Dehydration	[43]
Sinus content	Sinusitis	30	F	None	[44]
Sinus content	Sinusitis	75	F	Obesity, diabetes	[44]
Sinus content	Sinusitis	35	F	—	[45]
Sinus content	Sinusitis	62	F	HIV, diabetes	[29]
Sinus content	Sinusitis	42	M	HIV	[30]
BAL fluid	Allergic bronchopulmonary mycosis	57	F	Healthy	[28]
Lingular brushing Lobectomy specimen	Lung fungus ball	53	F	Tuberculosis Diabetes	[31]
Sinus content	Sinusitis	35	M	None	[46]
Brain biopsy Lung biopsy	Brain abscess after pulmonary infection	58	M	Hypertension Coronary artery disease	[32]
Bronchial plugs Sputum	Bronchial mucoid impaction	67	F	Previous tuberculosis	[47]
Bronchial plugs	Bronchial mucoid impaction	72	F	—	[48 cited by 3]
—	Allergic bronchopulmonary mycosis	53	F	—	[49 cited by 3]
—	Allergic bronchopulmonary mycosis	44	F	—	[50 cited by 3]
Nasal discharge Sinus content	Sinusitis	36	F	None	[35]
Sinus content	Sinusitis	62	F	Diabetes	[16]
Bronchial plugs	Bronchial mucoid impaction	51	F	Bronchiectasis	[51]
Lobectomy specimen	Pulmonary mucous consolidative lesion	74	M	Tuberculosis	[39]
N/A	Bronchial mucoid impaction	51	F	—	[52 cited by 3]
Nostrils flushing	Sinusitis	60	F	—	[37]
Nostrils flushing	Sinusitis	51	F	—	[37]
Sinus content	Sinusitis	63	F	—	[37]
Sinus content	Sinusitis	63	F	—	[37]
N/A	Allergic bronchopulmonary mycosis	79	F	—	[53 cited by 3]
BAL fluid	Chronic eosinophilic pneumonia	74	F	Diabetes	[54]
Sputum	Allergic bronchopulmonary mycosis	27	F	Bronchial asthma	[55 cited by 3]
Sputum	Allergic bronchopulmonary mycosis	33	F	Bronchial asthma	[55 cited by 3]
Sinus content	Sinusitis	47	M	Diabetes Renal failure	[56]
Sinus content	Sinusitis	35	F	None	[57]
Cyst content	Bronchogenous cyst	56	M	None	[58]
Bronchial plugs Sputum	Bronchial mucoid impaction	54	F	None	[3]
Sinus content	Sinusitis	55	F	None	[59]
Bronchial aspirate	Bronchopneumonia	59	M	Hodgkin's disease	[36]
Lung biopsy	Bilateral pulmonary nodules	56	F	Cardiac transplantation	[60]
Sinus content	Sinusitis	57	F	Asthma, myasthenia, rheumatoid arthritis Corticosteroids	[40]

TABLE 75.2
Sinusitis Caused by *Schizophyllum commune*

Localization	Treatment	Outcome	Reference
Maxillary	Surgery	—	[44]
Maxillary	Surgery	Sinus pain for a year	[44]
Maxillary	Surgery	Favorable	[45]
Maxillary	Surgery	Probable cure	[29]
	Antifungal treatment	Death due to lymphoma	
Maxillary	Drainage	Possible recovery	[30]
	Antifungal treatment	Death due to meningoencephalitis	
Maxillary, ethmoidal, and frontal	Surgery	—	[46]
	Corticosteroids		
Maxillary	Surgery	Favorable	[35]
	Antifungal treatment		
Maxillary and sphenoidal	Surgery	Favorable	[16]
Maxillary and ethmoidal	Surgery and antifungal treatment	Favorable	[37]
Ethmoidal and sphenoidal	Antifungal and antihitaminal treatment	Favorable	[37]
Ethmoidal and sphenoidal	Surgery	Favorable	[37]
Ethmoidal and sphenoidal	Surgery	—	[37]
Sphenoidal	Surgery and antifungal treatment	Favorable	[56]
Ethmoidal	Surgery	Favorable	[57]
Ethmoidal and sphenoidal	Surgery	—	[59]
Bilateral maxillary, ethmoidal, and sphenoidal	Surgery and antifungal treatment	Favorable	[40]

FIGURE 75.1 *S. commune* on malt extract agar after 7 days at 25°C showing fan-shaped fruiting bodies.

Unfortunately, more than half of *S. commune* clinical isolates are monokaryotic and then, without clamp connections. Thus, the identification has to be suspected by the woolly extensive culture, the unpleasant odor, and the presence of spicules.

75.1.3.1.3 Additional Studies

75.1.3.1.3.1 Tolerance to Benomyl Tolerance to benomyl is tested by assessing growth rates on Sabouraud dextrose agar with and without benomyl at concentrations of 2 and 10 μg/mL.[38]

75.1.3.1.3.2 Dikaryotization Assay To confirm the aptitude to dikaryotization of an isolate, a plug of a supposed monokaryotic *S. commune* strain is placed onto a potato dextrose agar plate beside a plug of a known single-basidiospore isolate (e.g., UAMH 7793, 7694, and 7695). A microscopic examination of the mycelium from the contact zone is performed after 7–10 days of coculture to display the presence of clamp connections demonstrating a process of dikaryotization. Compatibility for dikaryotization between mycelia derived from different fruiting bodies reaches almost 100%.[16,31,39]

75.1.3.1.3.3 Serological Tests Some authors report the detection of serum antibodies (IgG, IgM, and IgA) to *S. commune* cytosol or extracellular culture antigen to confirm a sensitization to the fungus using the method described by Kamei et al.[28] In case of an allergic pathology, high levels of serum IgE are also observed.[40]

75.1.3.2 Molecular Techniques

Excepted when isolates consist in typical dikaryotic mycelium, conventional culture methods often lack of characteristic clues to identify *S. commune* with certitude. Moreover, cultures need sometimes to be prolonged for several weeks to visualize clamp connections or fruiting bodies and dikaryotization assays need the possession of monokaryotic strains derived from a single basidiospore. So certitude diagnosis often calls to molecular methods. The sequence enclosed between the highly conserved small (18S) and large (28S) ribosomal subunit genes, and particularly the one and two internal transcribed spacer (ITS) regions is a valuable target for identification of *S. commune*, using analysis of the restriction fragment length polymorphism (RFLP) or nucleotide sequencing.[13,37,41] Although

Buzina et al.[37] showed a reliable identification of *S. commune* using *Eco*RI and *Ava*I for RFLP, sequencing, if it is possible, appears the most accurate method to identify this species. Using ITS2 and ITS3 as sequencing primers, a similarity of 99.4%–100% (of 660 bp) has been shown between the nucleotide sequences of 10 strains obtained either from clinical samples or from fungal collections.[37] If morphological criteria are sparse, sequencing presents the advantage to allow identification of most fungal species using the same set of universal fungal primers (Tables 75.3 and 75.4). In contrast, if there is a strong suspicion of *S. commune* based on cultural criteria, a *S. commune*-focused PCR using the specific set of primers *scom1* and *scom2*[37] (Table 75.5) is a faster and cheaper method to confirm identification.

75.2 METHODS

75.2.1 Sample Preparation

75.2.1.1 Pretreatment

Some specimens such as sputum or BAF if viscous need a first treatment at room temperature with a mucolytic agent (v/v) as Digest-Eur® (Eurobio). After vortex, wait until complete fluidization. Then centrifuge 5 mL of fluidized sputum, BAF, or other biological fluid at 900 × *g* for 20 min and preserve a 500 μL pellet.

75.2.1.2 DNA Extraction

All clinical samples (surgical specimens such as sinus or tissue biopsies, biological fluids) and cultures need to be disrupted

TABLE 75.3
Sets of Panfungal Primers Previously Described for the Identification of *S. commune* by Nucleotide Sequencing (English Literature)

Initial PCR Primers	Sequencing Primers	Sequencing Target	Final Product Length (bp)	Reference
ITS5	ITS2	Partial 18S rDNA–partial 28S	660	[37]
ITS4	ITS3	rDNA		[61]
NS1	V9G	V9 variable domain of the 18S	450	[41]
ITS4	5.8G	rDNA–central region of the 5.8S rDNA		[61]
ITS1	ITS1	Partial 18S rDNA–partial 28S	590–630	[61,62]
ITS4	ITS4	rDNA		[60]
Unnamed 1	Unnamed 1	18S rDNA	500	[63]
Unnamed 2	Unnamed 2			

TABLE 75.4
Sequences of Panfungal Primers for the Identification of *S. commune*

Panfungal Primers	Sequences (5′–3′)	Reference
ITS1	TCCGTAGGTGAACCTGCGG	[61]
ITS2	GCTGCGTTCTTCATCGATGC	[61]
ITS3	GCATCGATGAAGAACGCAGC	[61]
ITS4	TCCTCCGCTTATTGATATGC	[61]
ITS5	GGAAGTAAAAGTCGTAACAAGG	[61]
NS1	GTAGTCATATGCTTGTCTC	[41]
V9G	TTACGTCCCTGCCCTTTGTA	[41]
5.8G	AATGTGCGTTCAAAGATTCG	[41]
Unnamed 1	ATTGGAGGGCAAGTCTGGTG	[63]
Unnamed 2	CCGATCCCTAGTCGGCATAG	[63]
MITS2A	GATGAAGAACGCAGCGAAAT	Cassaing, S. et al., Unpublished data
MITS2B	ATGCTTAAGTTCAGCGGGTA	Cassaing, S. et al., Unpublished data

TABLE 75.5
Sequences of *S. commune* Specific Primers

Target	PCR Primers (5′–3′)	Product Length (bp)	Reference
Part of ITS region flanking the 5.8SrDNA	*scom*1 GTTGACTACGTCTACCTCAC *scom*2r GTTAGGCTCCAGCAGACCTC	305	[37]

before DNA extraction. Several physical (e.g., liquid nitrogen, solid beads) or enzymatic (e.g., lysozyme 10 mg/mL in 10 mM Tris–HCl, pH 8.0, or lyticase 0.5 mg/mL) lysis methods are used. The combination of solid beads contained in a screw cap microtube and lysis buffer allows both the splitting and homogenization of tissues and the break up of hyphal structures without the projection of microparticles. Commercial kits are available, including tubes filled with beads alone or associated with a lysis buffer (e.g., IDI lysis kit, GeneOhm Sciences; MagNA Lyser Green Beads®, Roche Applied Science).

1. In a biological safety cabinet, place the sample (a fragment of biopsy, 500 µL of pelleted fluidized BAF or sputum or 0.5 cm^2 of mycelium) in a microtube filled with ceramic beads (MagNA Lyser Green Beads, Roche Applied Science) and 500 µL of BLB (MagNA Pure Bacteria Lysis Buffer, Roche Applied Science).
2. Insert the tubes into the rotor of the MagNa Lyser instrument (Roche Applied Science) and secure it with the retention plate.
3. Cell disruption (fast oscillating) is performed twice for 35 s at 7000 rpm.
4. After a brief centrifugation to pellet the cell debris, place the tube for 20 min at room temperature prior to DNA extraction.

Proceed to the DNA extraction using, for example, the high pure PCR Template Preparation Kit® (Roche Applied Science):

1. Transfer 200 µL of supernatant for DNA extraction into a microtube.
2. Extract DNA following the manufacturer's instructions.
3. Store DNA at +4°C until amplification and sequencing or at −20°C if analysis is deferred.

75.2.2 Detection Procedures

75.2.2.1 *S. commune*-Specific PCR

Buzina et al. published specific *S. commune* primers[37] (see Table 75.5) suitable for a targeted diagnosis, for example, in the case of a previous clinical history with *S. commune*. In others cases, this fungus is not frequent enough to use specific PCR as a first diagnostic test.

75.2.2.2 Tm Determination after Real-Time Panfungal PCR

If *S. commune* is suspected with a woolly white mycelium, a panfungal real-time PCR with the analysis of the PCR product, Tm (melting temperature) is a good and rapid complementary test for the identification of this mold. The protocol below details a PCR targeting the ITS2 region.

Procedure

1. Prepare a panfungal Sybrgreen PCR mixture (10 µL) with 1 µL of LightCycler® FastStart DNA Master SYBR Green 10× (Roche Applied Science), 0.4 µM of each panfungal primers MITS2A and MITS2B (Table 75.4), additional $MgCl_2$ to reach a final concentration of 3.5 mM, 0.25 U of UNG (Fermentas Life Sciences), and 2.5 µL of DNA.
2. Thermocycling and detection are performed in glass capillaries on a LightCycler instrument (Roche Applied Science). An initial preheating step of 8 min at 95°C is used to activate the DNA polymerase and inactivate UNG. Then, a touchdown procedure follows, consisting of a first amplification of 5 cycles (15 s at 95°C, 10 s at 70°C, and 19 s at 72°C), then a second amplification of 11 cycles (15 s at 95°C, 10 s at temperatures decreasing from 70°C to 60°C with 1°C decremental steps, and 19 s at 72°C, and, finally, a third amplification of 22 cycles (15 s at 95°C, 10 s at 60°C, and 19 s at 72°C). A total of 38 cycles is performed.
3. Cycling is followed by a melting curve analysis. The expected Tm of the PCR product for *S. commune* is 86°C ± 0.3.

75.2.2.3 Sequencing

If the PCR is performed on a clinical specimen DNA or a nonsporulating mold without any morphological orientation, a panfungal PCR using universal fungal primers followed by nucleotide sequencing will be more informative. The protocol below details the procedure targeting the ITS1 region.

Initial conventional PCR

1. Prepare a mixture (final volume 50 µL) for conventional PCR with 0.5 µL Taq polymerase (Ampli Taq Gold® DNA polymerase, Applied Biosystems), 0.4 µM of each primer (ITS1 and ITS2), 0.2 mM of each dNTPs (MP Biomedicals), 2 mM $MgCl_2$ (Applied Biosystems), and 10 µL of DNA.
2. Run the cycling program on a thermocycler (e.g., Veriti, Applied Systems): 1 cycle at 96°C for 10 min, then 35 cycles with 96°C for 1 min, 58°C for 45 s, 72°C for 1 min, and then 72°C for 10 min.
3. Perform an electrophoresis of the PCR product together with a DNA molecular weight (e.g., pUC Mix Marker 8, Fermentas) using a 2% agarose gel containing 0.5 µg/mL ethidium bromide.
4. Visualize the PCR product under UV light box. A 228 bp band is expected.

PCR sequencing

1. Perform two PCR sequencing reactions (10 µL) containing 1 µL of the previous PCR product, 2 µL of primer ITS1 or ITS2 (1 pmol/µL), 2 µL of Big Dye Terminator v1.1 (Applied Biosystems), and 5 µL of water.
2. Cycle sequencing is performed as follows: a first step at 96°C for 1 min and then 25 cycles as follows: 96°C for 10 s, 50°C for 5 s, and 60°C for 75 s.

3. PCR products are passed through a Performa® DTR gel filtration cartridge (Edge Bio) to remove unincorporated dye terminators, excess of dNTPs, and salts.

Sequencing

1. Purified sequencing products are read on an ABI Prism 3100 Avant Genetic Analyzer® (Applied Biosystems) according to the manufacturer's instructions.
2. The BLASTn algorithm (National Center for Biotechnology Information) is used for the analysis of sequences and species identification (e.g., http://www.ncbi.nlm.nih.gov/BLAST).

75.3 CONCLUSION

S. commune is described as a growing emergent pathogen, which remains too much unknown. The clinical reports review suggests that this fungus has now to be included in the differential diagnosis of *Aspergillus* in pulmonary and sinuses diseases. Because of the frequent poorness of the primary culture and the difficulties of the identification based on morphological features, the laboratory diagnosis is still laborious and frequency of *S. commune* infections is likely underestimated. There is no doubt that molecular techniques including nucleotide sequencing are become essential for the identification of this filamentous basidiomycete.

REFERENCES

1 O'Brien, M. et al., Fungi isolated from contaminated baled grass silage on farms in the Irish Midlands, *FEMS Microbiol. Lett.*, 247, 131, 2005.

2 Han, C.H. et al., A novel homodimeric lactose-binding lectin from the edible split gill medicinal mushroom *Schizophyllum commune*, *Biochem. Biophys. Res. Commun.*, 336, 252, 2005.

3 Ishiguro, T. et al., Pulmonary *Schizophyllum commune* infection developing mucoid impaction of the bronchi, *Yale J. Biol. Med.*, 80, 105, 2007.

4 Chumkhunthod, P. et al., Purification and characterization of an *N*-acetyl-D-galactosamine-specific lectin from the edible mushroom *Schizophyllum commune*, *Biochim. Biophys. Acta*, 1760, 326, 2006.

5 Adejoye, B.C. et al., Physicochemical studies on *Schizophyllum commune* (Fries) a Nigerian edible fungus, *World Appl. Sci. J.*, 2, 73, 2007.

6 Dijk, H., Onguene, N.A., and Kuyper, T.W., Knowledge and utilization of edible mushrooms by local populations of the rain forest of South Cameroon, *AMBIO*, 32, 19, 2003.

7 Ruan-Soto, F., Garibay-Orijel, R., and Cifuentes, J., Process and dynamics of traditional selling wild edible mushrooms in tropical Mexico, *J. Ethnobiol. Ethnomed.*, 2, 3, 2006.

8 Wasser, S.P. and Weis, A.L., Therapeutic effects of substances occurring in higher Basidiomycetes mushrooms: A modern perspective, *Crit. Rev. Immunol.*, 19, 65, 1999.

9 Borchers, A.T. et al., The immunobiology of mushrooms, *Exp. Biol. Med. (Maywood)*, 233, 259, 2008.

10 Miyazaki, K. et al., Activated (HLA-DR+) T-lymphocyte subsets in cervical carcinoma and effects of radiotherapy and immunotherapy with sizofiran on cell-mediated immunity and survival, *Gynecol. Oncol.*, 56, 412, 1995.

11 Casselton, L.A. and Olesnicky, N.S., Molecular genetics of mating recognition in basidiomycete fungi, *Microbiol. Mol. Biol. Rev.*, 62, 55, 1998.

12 Schubert, D. et al., Ras GTPase-activating protein gap1 of the homobasidiomycete *Schizophyllum commune* regulates hyphal growth orientation and sexual development, *Eukaryot. Cell*, 5, 683, 2006.

13 de Hoog, G.S., Filamentous basidiomycetes, *Atlas of Clinical Fungi*, 2nd edn., p. 242, Universitat Rovira i Virgili, Reus, 2000.

14 Raper, J.R., Boyd, D.H., and Raper, C.A., Primary and secondary mutations at the incompatibility loci in *Schizophyllum*, *Proc. Natl. Acad. Sci. U.S.A.*, 53, 1324, 1965.

15 Fowler, T.J. et al., Crossing the boundary between the Balpha and Bbeta mating-type loci in *Schizophyllum commune*, *Fungal Genet. Biol.*, 41, 89, 2004.

16 Sigler, L. et al., Maxillary sinusitis caused by medusoid form of *Schizophyllum commune*, *J. Clin. Microbiol.*, 37, 3395, 1999.

17 Vaillancourt, L.J. et al., Multiple genes encoding pheromones and a pheromone receptor define the B beta 1 mating-type specificity in *Schizophyllum commune*, *Genetics*, 146, 541, 1997.

18 Raper, C.A. and Raper, J.R., Mutations modifying sexual morphogenesis in *Schizophyllum*, *Genetics*, 54, 1151, 1966.

19 Wendland, J. et al., The mating-type locus B alpha 1 of *Schizophyllum commune* contains a pheromone receptor gene and putative pheromone genes, *EMBO J.*, 14, 5271, 1995.

20 Wang, C.S. and Raper, J.R., Isozyme patterns and sexual morphogenesis in *Schizophyllum*, *Proc. Natl. Acad. Sci. U.S.A.*, 66, 882, 1970.

21 Gola, S. and Kothe, E., The little difference: in vivo analysis of pheromone discrimination in *Schizophyllum commune*, *Curr. Genet.*, 42, 276, 2003.

22 Fowler, T.J. et al., Multiple sex pheromones and receptors of a mushroom-producing fungus elicit mating in yeast, *Mol. Biol. Cell*, 10, 2559, 1999.

23 Gola, S., Hegner, J., and Kothe, E., Chimeric pheromone receptors in the basidiomycete *Schizophyllum commune*, *Fungal Genet. Biol.*, 30, 191, 2000.

24 Kothe, E., Gola, S., and Wendland, J., Evolution of multispecific mating-type alleles for pheromone perception in the homobasidiomycete fungi, *Curr. Genet.*, 42, 268, 2003.

25 Kothe, E., Mating types and pheromone recognition in the homobasidiomycete Schizophyllum commune, *Fungal Genet. Biol.*, 27, 146, 1999.

26 Yue, C. et al., The specificity determinant of the Y mating-type proteins of *Schizophyllum commune* is also essential for Y-Z protein binding, *Genetics*, 145, 253, 1997.

27 Kligman, A.M., A basidiomycete probably causing onychomycosis, *J. Invest. Dermatol.*, 14, 67, 1950.

28 Kamei, K. et al., Allergic bronchopulmonary mycosis caused by the basidiomycetous fungus *Schizophyllum commune*, *Clin. Infect. Dis.*, 18, 305, 1994.

29 Rosenthal, J. et al., Chronic maxillary sinusitis associated with the mushroom *Schizophyllum commune* in a patient with AIDS, *Clin. Infect. Dis.*, 14, 46, 1992.

30 Marlier, S. et al., Chronic sinusitis caused by *Schizophyllum commune* in AIDS, *Presse Med.*, 22, 1107, 1993.

31 Sigler, L. et al., Diagnostic difficulties caused by a nonclamped *Schizophyllum commune* isolate in a case of fungus ball of the lung, *J. Clin. Microbiol.*, 33, 1979, 1995.

32 Rihs, J.D., Padhye, A.A., and Good, C.B., Brain abscess caused by *Schizophyllum commune*: An emerging basidiomycete pathogen, *J. Clin. Microbiol.*, 34, 1628, 1996.

33 Perkins, J.H. and Raper, J.R., Morphogenesis in *Schizophyllum commune*. 3. A mutation that blocks initiation of fruiting, *Mol. Gen. Genet.*, 106, 151, 1970.

34 Raper, J.R. and Krongelb, G.S., Genetic and environmental aspects of fruiting in *Schizophyllum commune*, *Mycologia*, 50, 707, 1958.

35 Sigler, L. et al., Maxillary sinusitis caused by *Schizophyllum commune* and experience with treatment, *J. Med. Vet. Mycol.*, 35, 365, 1997.

36 Tullio, V. et al., *Schizophyllum commune*: An unusual of agent bronchopneumonia in an immunocompromised patient, *Med. Mycol.*, 46, 735, 2008.

37 Buzina, W. et al., Development of molecular methods for identification of *Schizophyllum commune* from clinical samples, *J. Clin. Microbiol.*, 39, 2391, 2001.

38 Summerbell, R.C., The benomyl test as a fundamental diagnostic method for medical mycology, *J. Clin. Microbiol.*, 31, 572, 1993.

39 Iizasa, T. et al., Colonization with *Schizophyllum commune* of localized honeycomb lung with mucus, *Respiration*, 68, 201, 2001.

40 Ahmed, M.K. et al., Bilateral allergic fungal rhinosinusitis caused by *Schizophillum commune* and *Aspergillus niger*. A case report, *Rhinology*, 47, 217, 2009.

41 de Hoog, G.S. and Gerrits van den Ende, A.H., Molecular diagnostics of clinical strains of filamentous Basidiomycetes, *Mycoses*, 41, 183, 1998.

42 Batista, A.C., Maia, J.A., and Singer, R., Basidioneuromycosis on man, *An. Soc. Biol.*, 13, 52, 1995.

43 Restrepo, A. et al., Ulceration of the palate caused by a basidiomycete *Schizophyllum commune*, *Sabouraudia*, 11, 201, 1973.

44 Kern, M.E. and Uecker, F.A., Maxillary sinus infection caused by the homobasidiomycetous fungus *Schizophyllum commune*, *J. Clin. Microbiol.*, 23, 1001, 1986.

45 Catalano, P. et al., Basidiomycetous (mushroom) infection of the maxillary sinus, *Otolaryngol. Head Neck Surg.*, 102, 183, 1990.

46 Clark, S. et al., *Schizophyllum commune*: An unusual isolate from a patient with allergic fungal sinusitis, *J. Infect.*, 32, 147, 1996.

47 Amitani, R. et al., Bronchial mucoid impaction due to the monokaryotic mycelium of *Schizophyllum commune*, *Clin. Infect. Dis.*, 22, 146, 1996.

48 Tomita, K. et al., Allergic bronchopulmonary mycosis caused by *Schizophyllum commune*, *Nihon Kyobu Shikkan Gakkai Zasshi*, 34, 804, 1996.

49 Ikushima, S., Case of allergic bronchopulmonary mycosis caused by *Schizophyllum commune*, *Jpn. J. Antibiot.*, 50, 47, 1997.

50 Yamashina, S., Case of allergic bronchopulmonary mycosis caused by *Schizophyllum commune*, *Jpn. J. Antibiot.*, 50, 51, 1997.

51 Miyazaki, Y. et al., Mucoid impaction caused by monokaryotic mycelium of *Schizophyllum commune* in association with bronchiectasis, *Intern. Med.*, 39, 160, 2000.

52 Itou, Y. et al., A case of mucoid impaction of bronchi (MIB) due to Schizophyllum commune, *Nihon Kokyuki Gakkai Zasshi*, 39, 266, 2001.

53 Yamasaki, A. et al., A case of allergic bronchopulmonary mycosis caused by *Schizophyllum commune*, *Arerugi*, 51, 439, 2002.

54 Kawayama, T. et al., Chronic eosinophilic pneumonia associated with *Schizophyllum commune*, *Respirology*, 8, 529, 2003.

55 Kawano, T. et al., Two cases of allergic bronchopulmonary mycosis caused by *Schizophyllum commune* in young asthmatic patients, *Nihon Kokyuki Gakkai Zasshi*, 41, 233, 2003.

56 Roh, M.L. et al., Sphenocavernous syndrome associated with *Schizophyllum commune* infection of the sphenoid sinus, *Ophthal. Plast. Reconstr. Surg.*, 21, 71, 2005.

57 Baron, O. et al., Nucleotide sequencing for diagnosis of sinusal infection by *Schizophyllum commune*, an uncommon pathogenic fungus, *J. Clin. Microbiol.*, 44, 3042, 2006.

58 Bulajic, N. et al., *Schizophyllum commune* associated with bronchogenous cyst, *Mycoses*, 49, 343, 2006.

59 Taguchi, K. et al., Allergic fungal sinusitis caused by *Bipolaris spicifera* and *Schizophyllum commune*, *Med. Mycol.*, 45, 559, 2007.

60 Roan, J.N. et al., Pulmonary nodules caused by *Schizophyllum commune* after cardiac transplantation, *J. Infect.*, 58, 164, 2009.

61 White, T.J. et al., Amplification and direct sequencing of fungal ribosomal genes for phylogenetics, in *PCR Protocols: A Guide to Methods and Applications*, Innis, M.A. et al., eds., 2nd edn., p. 315, Academic Press Inc., San Diego, CA, 1990.

62 Pounder, J.I. et al., Discovering potential pathogens among fungi identified as nonsporulating molds, *J. Clin. Microbiol.*, 45, 568, 2007.

63 Loeffler, J. et al., Comparison between plasma and whole blood specimens for detection of Aspergillus DNA by PCR, *J. Clin. Microbiol.*, 38, 3830, 2000.

76 *Sporobolomyces*

Dongyou Liu

CONTENTS

76.1 INTRODUCTION

76.1.1 CLASSIFICATION AND MORPHOLOGY

The genus *Sporobolomyces* is a basidiomycetous yeast classified in the mitosporic Sporidiobolales group, order Sporidiobolales, class Microbotryomycetes, phylum Basidiomycota, kingdom Fungi. The mitosporic Sporidiobolales group comprises two genera: *Rhodotorula* and *Sporobolomyces*; and the genus *Sporobolomyces* currently contains >20 recognized species as well as >30 unassigned species, of which *Sporobolomyces roseus* and *Sporobolomyces salmonicolor* are involved in human infections [1–14]. The teleomorphs of the genus *Sporobolomyces* are found in the genus *Sporidiobolus* [15].

Sporobolomyces is a ballistoconidium-forming yeast that is present in environments (e.g., soil, air, water), plants, animals, and humans [2]. *Sporobolomyces* species (e.g., *S. salmonicolor*, *S. roseus*, and *S. holsaticus*) may cause dermatitis, cerebral infection, lymphadenitis, fungemia, and allergic alveolitis, particularly in immunosuppressed patients.

Sporobolomyces colonies grow rapidly and mature in 5 days. The optimal growth temperature is 25°C–30°C, with some isolates growing poorly at 35°C–37°C. Colonies are smooth, often wrinkled, and glistening with bright red to orange color, resembling *Rhodotorula* spp.

Sporobolomyces produces oval to elongated yeast-like cells, pseudohyphae, true hyphae, and one-celled, usually reniform (kidney-shaped) ballistoconidia (2–12 × 3–35 μm). Blastoconidia are forcibly discharged from denticles located on ovoid to elongate vegetative cells and function to produce satellite colonies. The shot-off (or discharge) of *Sporobolomyces* ballistoconidia can be observed by inoculating one plate, taping it on top of an uninoculated plate, face to face, and incubating for an extended period at 25°C. Colonies on the initially uninoculated plate form a mirror image of the colonies that are on the initially inoculated plate.

76.1.2 CLINICAL FEATURES AND PATHOGENESIS

Sporobolomyces spp. are phylloplane yeasts that have widespread distribution in environments and occur in animals and humans. Several *Sporobolomyces* spp. (e.g., *S. salmonicolor*, *S. roseus*, and *S. holsaticus*) have been implicated in human infections, inducing dermatitis, cerebral infection, lymphadenitis, fungemia, and allergic alveolitis, notable in immunosuppressed individuals, as a consequence of advanced medical intervention [16–23].

A case of *Sporobolomyces*-related dermatitis involved a 30-year-old woman with a pruritic skin lesion. A yeast-like organism, *Sporobolomyces holsaticus*, was isolated from the lesion [18]. Additionally, *Sporobolomyces* species was responsible for extrinsic allergic alveolitis in a 28-year-old horseback rider with exposure to a horse barn. The patient presented with fever 4 h after exposure and precipitins positive against *Sporobolomyces* but negative against other fungi and horses. *Sporobolomyces* was isolated from straw in the barn and cessation of exposure to this barn (but continued exposure to horses) had contributed to improvement in clinical condition [17]. Similarly, *Sporobolomyces salmonicolor* was shown to be a likely cause of allergic disorders in 14 employees working in a hospital building with severe, repeated, and enduring water and mold damage. The patients presented cough, asthma, alveolitis, and rhinitis [22].

The common antifungal drugs amphotericin B (AMB) and fluconazole (FLC) may be useful for treatment of *Sporobolomyces* infections [24,25].

76.1.3 Laboratory Diagnosis

Observation of characteristic macroscopic and microscopic features in clinical samples and cultured isolates has formed the basis of conventional laboratory identification of *Sporobolomyces* and other yeasts. To aid the microscopic detection of fungal elements in clinical specimens and tissue biopsy, several stains (e.g., KOH, lactophenol, calcofluor white, PAS, Gomori methenamine silver) may be utilized. Further characterization of Sporobolomyces is conducted with cultured isolates using a range of biochemical tests such as carbon and nitrogen compound assimilation patterns and vitamin requirement tests [8]. Application of commercial yeast identification systems (e.g., API and VITEK) facilitates easy biochemical differentiation of yeasts.

Considering that detection of characteristic morphological features requires specialized skills, in vitro development of reproductive structures of yeasts is time-consuming, and biochemical tests seldom provide species-specific determination, molecular techniques have been increasingly employed for improved detection and identification of yeasts including *Sporobolomyces*. By targeting the conserved 18S, 5.8S, and 28S nuclear rRNA genes, the phylogenetic relationships among yeasts can be accurately assessed [26,27]. Furthermore, PCR amplification and sequencing analysis of the more rapidly evolved internal transcribed spacer 1 and 2 regions (ITS1 and ITS2, respectively) allow rapid species differentiation within a genus [28–31].

76.2 METHODS

76.2.1 Sample Preparation

Clinical samples are first examined by microscopy with the use of fungal stains; portions of the samples are inoculated on Sabouraud dextrose agar (SDA), potato dextrose agar (PDA), and other culture media and incubated at 30°C and 35°C for a few days [25].

For DNA extraction, 2 day old yeast colonies from SDA plate are resuspended in 100 μL of lysis buffer (100 mM Tris, 30 mM EDTA, 0.5% [wt/vol] sodium dodecyl sulfate, pH 7.5), vortexed briefly and incubated at 100°C for 15 min. Following addition of 100 μL of 2.5 M potassium acetate, the suspension is incubated on ice for 1 h and centrifuged at 14,000 × *g* for 5 min. The supernatant is transferred to a new tube, an equal volume of isopropanol is added, and the tube is centrifuged for 5 min. After decanting the supernatant, 500 μL of 100% ethanol is added, and the tube is centrifuged for 20 min. The supernatant is decanted, and the extracted DNA is air dried and resuspended in 100 μL of sterile water and stored at −20°C [28].

Alternatively, yeast aqueous suspensions (200 μL) are applied directly to Whatman FTA microcards, which are then placed, open, in a Panasonic microwave (800 W) while still damp and subjected to two cycles of 30 s on full power, with a pause of at least 30 s between each cycle. Punches (2 mm in diameter) are then removed from dried FTA filters by using a Harris micropunch and washed twice for 1 min each with 100 μL of Whatman FTA wash reagent, followed by two washes for 1 min each with 100 μL of TE buffer (10 mM Tris–HCl pH 7.5, 0.1 mM EDTA). Washed filters are then dried for 5 min at 55°C on a dry heat block, and PCR mixes are added directly to the washed and dried FTA filter punches [30].

76.2.2 Detection Procedures

Bai et al. [8] utilized primers ITS1 (5′-GTCGTAACAA GGTTTCCGTAGGTG-3′) and NL4 (5′-GGTCCGTGTTT CAAGACGG-3′) to amplify a DNA fragment covering the ITS region and 25S rRNA D1/D2 domain for identification of *Sporobolomyces* and other yeasts.

Procedure

1. PCR mixture (100 μL) is made up of 200 μM concentrations of each deoxynucleoside triphosphate, 250 nM concentrations of the appropriate primers, 2 U of HotStarTaq polymerase (QIAGEN), and either 5 μL of extracted genomic DNA or a single filter punch.
2. Amplification is conducted with initial enzyme activation at 94°C for 15 min, followed by 40 cycles at 94°C for 30 s, 52°C for 1 min, and 72°C for 2 min.
3. After checking a fraction of total amplification products in 1.2% agarose gels run for 45 min at 120 V in Tris-borate buffer, the remainder of the PCR products are adjusted to final concentrations of 10% (wt/vol) PEG 8,000 and 10 mM $MgCl_2$ and then centrifuged for 10 min at 12,000 rpm in a bench-top centrifuge. The resulting DNA pellets are washed in 75% ethanol, air dried, resuspended in sterile water.
4. The nucleotide sequences of the PCR products are determined on an ABI PRISM 377 DNA sequencer using the ABI BigDye cycle sequencing kit with the forward primers ITS1 and NL1 (5′-GCATATCAATAAGCGGAGGAAAAG-3′) and the reverse primers ITS4 (5′-TCCTCCGCTTA TTGATATGC-3′) and NL4. Electrophoresis was carried out.
5. The sequences of the ITS regions or 26S rDNA D1/D2 domains of the testing strains and the reference sequences from GenBank are aligned with the program clustal x and adjusted manually. Phylogenetic trees are constructed from the evolutionary distance data calculated from Kimura's two-parameter model using the neighbor-joining method. Bootstrap analyses are performed on 1000 random resamplings.

76.3 CONCLUSION

Sporobolomyces spp. are ballistoconidium-forming yeasts with widespread distribution. Several *Sporobolomyces* spp. (e.g., *S. salmonicolor*, *S. roseus*, and *S. holsaticus*) have been implicated in human diseases such as dermatitis, allergic alveolitis, and disseminated infections in both immunocompetent and immunosuppressed individuals [32]. While it is possible to identify *Sporobolomyces* and other yeasts through macroscopic and microscopic examination as well as biochemical testing, the whole process is lengthy and at times variable. By targeting the conserved small subunit (SSU) and large subunit (LSU) rRNA genes as well as the rapidly evolving internal transcribed spacer (ITS) regions, molecular techniques provide a rapid, sensitive, and specific alternative for phylogenetic analysis and species-specific identification of *Sporobolomyces* and other yeasts. Future identification and application of species-specific primers and probes in real-time PCR platform will facilitate direct detection and quantitation of *Sporobolomyces* in clinical specimens.

REFERENCES

1. Misra VC, Randhawa HS. *Sporobolomyces salmonicolor* var. *fischerii*, a new yeast. *Arch. Microbiol.* 1976;108(1):141–143.
2. Nakase T, Suzuki M. *Sporobolomyces inositophilus*, a new species of ballistosporous yeast isolated from a dead leaf of *Sasa* sp. in Japan. *Anton Leeuwen* 1987;53(4):245–251.
3. Nakase T, Suzuki M. *Sporobolomyces yuccicola*, a new species of ballistosporous yeast equipped with ubiquinone-9. *Anton Leeuwen* 1988;54(1):47–55.
4. Nakase T et al. The expanding realm of ballistosporous yeasts. *Anton Leeuwen* 1993;63(2):191–200.
5. Nakase T et al. *Sporobolomyces magnisporus* sp. nov., a new yeast species in the *Erythrobasidium* cluster isolated from plants in Taiwan. *J. Gen. Appl. Microbiol.* 2003;49(6):337–344.
6. Nakase T et al. *Sporobolomyces diospyroris* sp. nov., *Sporobolomyces lophatheri* sp. nov. and *Sporobolomyces pyrrosiae* sp. nov., three new species of ballistoconidium-forming yeasts in the *Agaricostilbum* lineage isolated from plants in Taiwan. *J. Gen. Appl. Microbiol.* 2005;51(5):277–286.
7. Sláviková E, Grabinska-Loniewska A. *Sporobolomyces lactosus*, a new species of ballistosporous yeast equipped with ubiquinone-10. *Anton Leeuwen* 1992;61(3):245–248.
8. Bai FY et al. Reclassification of the *Sporobolomyces roseus* and *Sporidiobolus pararoseus* complexes, with the description of *Sporobolomyces phaffii* sp. nov. *Int. J. Syst. Evol. Microbiol.* 2002;52:2309–2314.
9. Fell JW et al. Recognition of the basidiomycetous yeast *Sporobolomyces ruberrimus* sp. nov. as a distinct species based on molecular and morphological analyses. *FEMS Yeast Res.* 2002;1:265–270.
10 Hamamoto M, Thanh VN, Nakase T. *Bannoa hahajimensis* gen. nov., sp. nov., and three related anamorphs, *Sporobolomyces bischofiae* sp. nov., *Sporobolomyces ogasawarensis* sp. nov. and *Sporobolomyces syzygii* sp. nov., yeasts isolated from plants in Japan. *Int. J. Syst. Evol. Microbiol.* 2002;52:1023–1032.
11 Zhao JH et al. *Sporobolomyces bannaensis*, a novel ballistoconidium-forming yeast species in the *Sporidiobolus* lineage. *Int. J. Syst. Evol. Microbiol.* 2003;53:2091–2093.
12 Wang QM, Bai FY. Four new yeast species of the genus *Sporobolomyces* from plant leaves. *FEMS Yeast Res.* 2004;4(6):579–586.
13 Libkind D et al. *Sporidiobolus longiusculus* sp. nov. and *Sporobolomyces patagonicus* sp. nov., novel yeasts of the Sporidiobolales isolated from aquatic environments in Patagonia, Argentina. *Int. J. Syst. Evol. Microbiol.* 2005;55:503–509.
14 Satoh K, Makimura K. *Sporobolomyces koalae* sp. nov., a basidiomycetous yeast isolated from nasal smears of Queensland koalas kept in a Japanese zoological park. *Int. J. Syst. Evol. Microbiol.* 2008;58:2983–2986.
15 Valério E, Gadanho M, Sampaio JP. Reappraisal of the *Sporobolomyces roseus* species complex and description of *Sporidiobolus metaroseus* sp. nov. *Int. J. Syst. Evol. Microbiol.* 2008;58:736–741.
16 Moore JE et al. Edible dates (*Phoenix dactylifera*), a potential source of *Cladosporium cladosporioides* and *Sporobolomyces roseus*: Implications for public health. *Mycopathologia* 2002;154(1):25–28.
17 Cockcroft DW et al. *Sporobolomyces*: A possible cause of extrinsic allergic alveolitis. *J. Allergy Clin. Immunol.* 1983;72(3):305–309.
18 Bergman AG, Kauffman CA. Dermatitis due to *Sporobolomyces* infection. *Arch. Dermatol.* 1984;120(8):1059–1060.
19 Bross JE et al. Pseudomeningitis caused by *Sporobolomyces salmonicolor. Am. J. Infect. Control.* 1986;14(5):220–223.
20 Morris JT, Beckius M, McAllister CK. *Sporobolomyces* infection in an AIDS patient. *J. Infect. Dis.* 1991;164:623–624.
21 Plazas J et al. *Sporobolomyces salmonicolor* lymphadenitis in an AIDS patient. Pathogen or passenger? *AIDS* 1994;8(3):387–388.
22 Seuri M et al. An outbreak of respiratory diseases among workers at a water-damaged building—A case report. *Indoor Air* 2000;10(3):138–145.
23 Sharma V, Shankar J, Kotamarthi V. Endogeneous endophthalmitis caused by *Sporobolomyces salmonicolor. Eye (Lond.)* 2006;20(8):945–946.
24 Serena C et al. In vitro antifungal susceptibilities of uncommon basidiomycetous yeasts. *Antimicrob. Agents Chemother.* 2004;48(7):2724–2726.
25 Serena C et al. In vitro interaction of micafungin with conventional and new antifungals against clinical isolates of *Trichosporon*, *Sporobolomyces* and *Rhodotorula. J. Antimicrob. Chemother.* 2005;55(6):1020.
26 Suh SO, Nakase T. Phylogenetic analysis of the ballistosporous anamorphic genera *Udeniomyces* and *Bullera*, and related basidiomycetous yeasts, based on 18S rDNA sequence. *Microbiology* 1995;141:901–906.
27 Wang QM et al. Rapid differentiation of phenotypically similar yeast species by single-strand conformation polymorphism analysis of ribosomal DNA. *Appl. Environ. Microbiol.* 2008;74(9):2604–2611.
28 Chen Y-C et al. Identification of medically important yeasts using PCR-based detection of DNA sequence polymorphisms in the internal transcribed spacer region 2 of the rRNA genes. *J. Clin. Microbiol.* 2000;38:2302–2310.
20 Chen Y-C et al. Polymorphic internal transcribed spacer region 1 DNA sequences identify medically important yeasts. *J. Clin. Microbiol.* 2001;39:4042–4051.
30 Linton CJ et al. Molecular identification of unusual pathogenic yeast isolates by large ribosomal subunit gene sequencing: 2 years of experience at the United Kingdom mycology reference laboratory. *J. Clin. Microbiol.* 2007;45(4):1152–1158.
31 Pounder JI et al. Discovering potential pathogens among fungi identified as nonsporulating molds. *J. Clin. Microbiol.* 2007;45:568–571.
32 Morrow JD. Prosthetic cranioplasty infection due to *Sporobolomyces. J. Tenn. Med. Assoc.* 1994;87(11):466–467.

77 Trichosporon

Takashi Sugita, Reiko Ikeda, Akemi Nishikawa, Masako Takashima, Nanthawan Mekha, Natteewan Poonwan, Ayse Kalkanci, and Semra Kustimur

CONTENTS

77.1 INTRODUCTION

77.1.1 Taxonomy, Morphology, and Epidemiology

Of the basidiomycetous yeasts, *Trichosporon* is the second largest genus worldwide, following the genus *Cryptococcus*. One-third of *Trichosporon* species are associated with infection or allergy. In recent years, the number of patients with breakthrough *Trichosporon* infections after receiving candin-derived antifungal agents has increased. The prognosis in patients with disseminated trichosporonosis is poor and mortality is high. Thus, early, rapid detection of the pathogen is important in the diagnosis and treatment of trichosporonosis. This chapter describes a molecular detection method for *Trichosporon*.

77.1.1.1 Taxonomy

Taxonomically, the basidiomycetous yeast in the genus *Trichosporon* belongs to the order Trichosporonales, class Tremellomycetes, and subphylum Agaricomycotina. Of the basidiomycetous yeasts, the genus *Trichosporon* with 39 species is the second largest genus worldwide, following the genus *Cryptococcus*. There are also many candidate species, and *Trichosporon* will likely expand to 50 species within a few years. The genus is further divided into five clades: Cutaneum, Ovoides, Brassicae, Gracile, and Porosum. These clades correlate well with some other taxonomic characteristics. Species of this genus possess CoQ-9 or CoQ-10 as major ubiquinones; species belonging to the Cutaneum clade have CoQ-10, whereas species belonging to the other clades have CoQ-9 [1].

77.1.1.2 Morphology

Morphologically, *Trichosporon* species are characterized by the production of arthroconidia. Abundant arthroconidia are observed when the microorganism is grown in corn meal agar. The colonies are cream colored, moist or dry, and with or without a white farinose covering. True hyphae are present and lateral, clavate, or globose blastoconidia may also be present. Hyphal septa have, as far as is known, dolipores, which may be with or without vesicular or tubular parenthesomes. With India ink (*bo-ku-ju* in Japanese) staining, a thin capsule can also be observed surrounding the cells [2].

77.1.1.3 Epidemiology

These microorganisms cause superficial and deep-seated infections (Table 77.1). Of the genus *Trichosporon*, approximately one-third of the species have been isolated from clinical specimens. Historically, *Trichosporon* species have long been known to be the causative agents of white piedra. Two species are most often associated with this infection: *T. ovoides* is responsible for white piedra of the head, while *T. inkin* is limited to the genital area [2]. Other species are often isolated from patients with superficial infections, such as *T. asahii*, *T. asteroids*, and *T. cutaneum* [2–4]. However,

TABLE 77.1
***Trichosporon* Species Isolated from Clinical Specimens**

Infection	Deep-seated infection	*T. asahii*, *T. asteroides*, *T. debeurmannianum*, *T. inkin*, *T. japonicum*, *T. loubieri*, *T. mucoides*
	Superficial infection	*T. ovoides* (capital white piedra), *T. inkin* (white piedra of the genital area), *T. asahii*, *T. asteroides*, *T. cutaneum*
Allergy	Summer-type hypersensitivity pneumonitis	*T. dermatis*, *T. asahii*, *T. asteroides*, *T. coremiiforme*, *T. faecale*, *T. japonicum*, *T. ovoides*, *T. domesticum*, *T. montevideense*

Source: Sugita, T., *The Yeasts, A Taxonomic Study*, 5th edn., Boekhout, T., Fell. J.W., Kurtzman. C.P. (eds.), Elsevier, Amsterdam, the Netherlands, 2011, pp. 2015–2061.

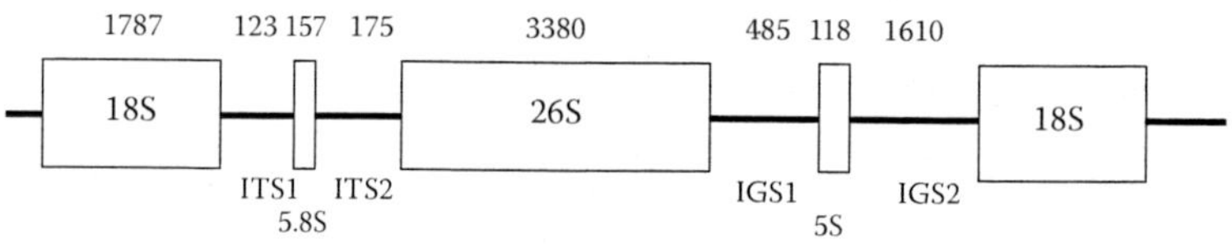

FIGURE 77.1 Schematic representation of the rRNA locus in *Trichosporon*. ITS, internal transcribed spacer; IGS, intergenic spacer region.

white piedra is limited to Africa at present and other superficial infections due to *Trichosporon* species are rare.

Deep-seated trichosporonosis is an important fungal infection clinically. The first case was reported in a brain abscess patient in 1970 in the United Kingdom [5]. Since then, the number of cases has increased, primarily in immunocompromised patients. For about 5 years, deep-seated trichosporonosis has been recognized as a breakthrough infection after treatment with candin-derived antifungal agents, such as micafungin [6–9]. Although several species have been isolated from clinical specimens, *T. asahii* is the causative agent in the vast majority of cases [2,3]. This microorganism has nine genotypes in the intergenic spacer region (IGS), which is located between the 26S and 5S rRNA genes and the population of genotypes differs in different countries [10–12] (Figures 77.1 and 77.2). The genotype distribution of isolates in Japan and Turkey is similar: approximately 80% are genotype 1. The major genotypes of isolates from the United States are types 3 and 5, while the major Thai isolates are types 1 and 3. There is a geographic substructure among *T. asahii* clinical isolates. *T. asteroids* [13], *T. debeurmannianum* [14], *T. inkin* [15,16], *T. japonicum* [17], *T. loubieri* [18,19], and *T. mucoides* [20] are also occasionally isolated from clinical specimens.

Trichosporon species not only cause infections but also allergies. The microorganism triggers summer-type hypersensitivity pneumonitis (SHP) [21], which follows the development of type 3 or 4 allergy with the repeated inhalation of *Trichosporon* arthroconidia, which often contaminate home environments during the summer. In Japan, the summer is hot, humid, and rainy. Such conditions favor the growth of *Trichosporon* species. Although several *Trichosporon* species may be distributed in the homes of patients with SHP, *T. dermatis* (41%) and *T. asahii* (51%) are the major causes [22]. Interestingly, the genotype distribution of *T. asahii* in patients with infection and SHP is significantly different. The major genotype obtained from infectious specimens is type 1, while that from the homes of SHP patients is type 3. Random amplification of polymorphic DNA analysis also suggests that *T. asahii* strains obtained from clinical specimens are genetically different from those obtained from the homes of SHP patients [23].

77.1.2 Clinical Features and Pathogenesis

As mentioned above, *Trichosporon* species cause superficial infection, although the incidence of this is low. Deep-seated trichosporonosis can be life threatening, with high mortality. Populations at risk are people with human immunodeficiency virus (HIV) infection, bone marrow and organ

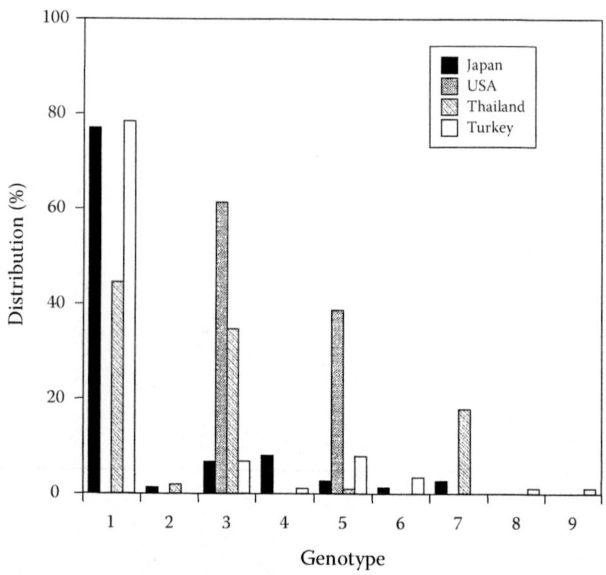

FIGURE 77.2 Genotype distribution of *Trichosporon asahii* clinical isolates from Japan, the United States, Thailand, and Turkey.

transplant recipients, and patients with profound neutropenia [24–26]. The most recent epidemiological survey was conducted by a Japanese group who retrospectively analyzed 33 cases of *Trichosporon* fungemia in patients with hematologic malignancies, treated between 1992 and 2007 [27]. *Trichosporon* fungemia occurred as a breakthrough infection during antifungal therapy in 30 patients (91%), 18 of whom were receiving micafungin. Of the 33 cases, 25 (76%) died of the infection. This morality rate is similar to other reports [28,29]. The resolution of infection was associated with neutrophil recovery, the absence of hyperglycemia, and therapy including azoles. Survival was significantly longer in patients receiving antifungal therapy containing azoles than in those who did not receive azoles.

Trichosporon species also cause SHP with the repeated inhalation of their spores [21]. Patients develop cough, chills, headache, fever, and difficulty breathing as acute symptoms. These symptoms disappear on eliminating the causative antigen, *Trichosporon*, from colonizing places. In chronic SHP, the patient's lungs become fibrotic. As anti-*Trichosporon*-specific IgG antibody is produced in the patient's serum, detection of this antibody can be used to diagnose SHP.

77.1.3 Diagnosis

77.1.3.1 Conventional Techniques

The conventional identification method for yeast, including *Trichosporon* species, is based on biochemical characteristics. Two commercial kits, API 20C Aux™ and ID 32C API™ (bioMérieux), are the most popular for yeast identification. These kits can identify 47 and 69 yeast species, respectively, within 48 h. However, only three species (*T. inkin*, *T. asahii*, *T. mucoides*) in the genus *Trichosporon* can be identified using these kits. Morphologically, *Trichosporon* is characterized by the production of arthroconidia, which is a key characteristic at the genus level.

77.1.3.2 Molecular Techniques

77.1.3.2.1 Culture-Dependent Methods

As yeast species have few unique morphological characteristics, molecular techniques are a powerful tool for species identification. The most broadly used identification method is rRNA gene sequence analysis. Figure 77.1 shows the rRNA locus in *Trichosporon*, which is approximately 7800 bp long. Of this, the internal transcribed spacer (ITS) region and D1/D2 26S rRNA gene have been widely used for fungal identification. The most important part of rRNA gene sequence analysis is to establish identification criteria. When there is less than 1% difference in the ITS region or D1/D2 26S rRNA gene between two strains, they are conspecifics [30,31], although some *Trichosporon* species do not conform to this standard. No difference in the ITS region is observed between *T. laibachii* and *T. multisporum*, although these species have seven differences (1.1%) in their D1/D2 26S rRNA gene. *T. montevideense* and *T. domesticum* also share identical nucleotide sequences in the ITS region but differ at two positions in the D1/D2 26S rRNA gene. The nucleotide differences in the ITS region and D1/D2 26S rRNA gene among *T. asahii*, *T. coremiiforme*, and *T. faecale* are only one or two base pairs. The DNA sequence of the IGS1 region is more diverse, compared with ITS or D1/D2 26S rRNA gene, indicating that IGS1 analysis can readily and reliably differentiate between such closely related species [12]. The 99% similarity in ITS sequences observed between two species corresponds to approximately 55%–95% IGS1 sequence similarity. The 98% ITS sequence similarity in another pairwise comparison corresponds to approximately 45%–55% IGS1 sequence similarity (Figure 77.3). Approximately 20% difference has been observed in the IGS1 region among *T. asahii*, *T. coremiiforme*, and *T. faecale* and a 7% difference between *T. domesticum* and *T. montevideense*. Using the primers shown in Table 77.2, from 195 to 719 bp of the IGS1 region can be amplified.

Thus, IGS1 sequencing analysis is a reliable method for identifying *Trichosporon* species. A high-throughput

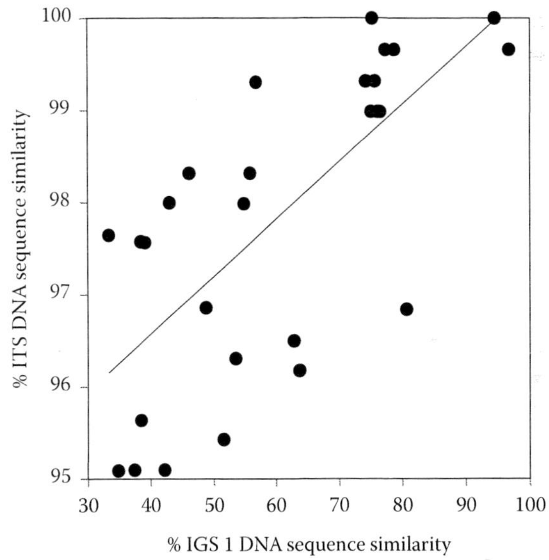

FIGURE 77.3 Sequence similarities between the IGS1 and ITS regions. Data are derived from *Trichosporon* DNA sequences.

TABLE 77.2
PCR Primers and Probes Used to Detect *Trichosporon* Species

Primer and Probe		Sequence (5′–3′)
Trichosporon species	26SF, forward	ATCCTTTGCAGACGACTTGA
	TR5R, reverse	AGCTTGACTTCGCAGATCGG
Trichosporon asahii	Forward	AGTATAAGGGATCAAAAATGAGAGCCA
	Reverse	CCTCTGAGGCCTTGCTCCTG
	TaqMan	TGATGGCCTTGGTTGAGA

suspension array system of uniplex and multiplex D1/D2 26S rRNA, ITS, and IGS probes is a rapid, accurate method to identify species of *Trichosporon* [32].

77.1.3.2.2 Culture-Independent Methods

The *Trichosporon* species of the greatest clinical importance is *T. asahii* because infection due to *T. asahii* can be life threatening and its prognosis is poor. Thus, the early detection and accurate identification of the causative agent are necessary to provide appropriate treatment. As a first-generation method, a direct detection method was developed for *T. asahii* DNA from blood [33–37]. This polymerase chain reaction (PCR) assay could detect *T. asahii* DNA from the polysaccharide antigen-negative samples, suggesting that PCR had greater sensitivity than antigen detection methods (Table 77.3) [34]. Quantification of the pathogen in clinical samples is also of great clinical significance. Three sets of oligonucleotide primers and TaqMan probe pairs were designed for detecting the major causative agent of trichosporonosis, *T. asahii* [38]. The sequences were designed from the IGS1 region of the rRNA gene (Table 77.2). A real-time PCR assay can detect as few as 10 copies of *T. asahii* DNA. For 21 blood samples, between 1.8×10^3 and 2.1×10^4 copies of *T. asahii* DNA were detected. Of these, 16 were positive for polysaccharide antigen using the latex agglutination test with titers of 1:4–1:64. For the five samples that were negative by the latex agglutination test, 1.8×10^3 to 4.9×10^3 copies of *T. asahii* DNA were detected. There was a good correlation between the number of DNA copies and the titer of polysaccharide antigen ($r^2 = 0.910$). Real-time PCR can be used to diagnose deep-seated trichosporonosis due to *T. asahii*. A practical example is described in Section 77.2.2.

77.2 METHODS

77.2.1 Sample Preparation

DNA is prepared from a small amount of a pure culture (approximately one loop) using any DNA extraction method.

77.2.2 Detection Preparation

77.2.2.1 Identification by PCR and Sequencing

DNA sequence analysis of the IGS1 region in the rRNA gene is the most reliable and accurate identification method. The IGS1 region is amplified by PCR with primers 26SF and TR5R (see Table 77.2). PCR is performed with an initial 3 min denaturation at 94°C, followed by 30 cycles of 30 s at 94°C, 30 s at 57°C, and 1 min at 72°C, and a final 10 min extension at 72°C. The PCR products are sequenced with primers 26SF and TR5R. The IGS1 sequences range in length from 195 to 719 bp, and the resulting DNA sequence is searched using BLAST (http://www.ncbi.nlm.nih.gov/blast/Blast.cgi). A species is defined as a group of organisms sharing >95% identity in their IGS1 region.

77.2.2 2 Real-Time PCR Detection

First, 100 μL of serum is mixed with an equal volume of lysis buffer (100 mM KCl, 20 mM Tris–HCl pH 8.3, 5 mM $MgCl_2$, 0.2 mg/mL gelatin, and 0.9% Tween 20 solution) and proteinase K is added to the mixture. The mixture is incubated for 60 min at 55°C and then for 10 min at 95°C to inactivate the proteinase K. The sample is extracted twice with phenol–chloroform–isoamyl alcohol and precipitated with ethanol. Ethatimate™ (Nippon Gene, Toyama, Japan) is used as a precipitation activator. For real-time PCR [38], 25 μL of a TaqMan Universal PCR master mix (Applied Biosystems), 200 nM each of the forward and reverse primers, 250 nM TaqMan probe (see Table 77.2), 3 μL of each DNA sample, and Milli-Q water in a total volume of 50 μL are placed in each well of a 96 well plate. Amplification and detection are performed with the cycle parameters 50°C for 2 min, 95°C for 10 min, and 40 cycles of 95°C for 15 s, and 60°C for 1 min. A standard curve is plotted using the cycle threshold values obtained by amplifying successive 10-fold dilutions of a known concentration of plasmid DNA (from 10^1 to 10^9 copies). A practical example is shown in Table 77.4 [39].

A 16-year-old boy received an allogeneic bone marrow transplant from a human leukocyte antigen (HLA)-identical unrelated donor for T cell acute lymphoblastic leukemia. On

TABLE 77.3
Blood Culture, Nested PCR, and Polysaccharide Detection Using Sera from Patients with Deep-Seated Trichosporonosis

Patient No.	Sex/Age (Year)[a]	Underlying Disease[b]	Blood Culture[c]	PCR Products[c]	Titer of PS Antigen
1	M/64	AML	+	+	1:4096
2	M/47	ML	+	+	1:32
3	F/15	AML	–	–	0
4	F/55	AML	–	–	0
5	M/29	MI	ND	+	0
6	M/63	MM	ND	+	0

[a] M, male; F, female.
[b] AML, acute myelogenic leukemia; MI, multiple injuries, ML, malignant lymphoma; MM, multiple myeloma.
[c] +, positive; –, negative; ND, not done.

TABLE 77.4
Laboratory Features of Trichosporonosis in the Patient

Day after transplant	59	61	62	68	72	75	76	78	81	83	86	90	104
Blood culture			−					+	−			−	
Urine culture		+		+		+			−				
Beta-D-glucan (pg/dL)			ND				95	60		22		ND	
Serum glucuronoxylomannan			−		+		+			+		−	
PCR (number of plasmid copy)	ND		63		480		642				190	52	ND

Note: ND, not detected.

day 61, *Trichosporon* species was cultured from a urine sample, but not from a blood sample. The patient was treated with micafungin. On day 76, beta-D-glucan in serum was positive and on day 78, *Trichosporon* was cultured from a blood sample. On day 80, the antifungal therapy was switched from micafungin to voriconazole. From day 59 to day 104, the serum *T. asahii* DNA concentration was assessed serially using real-time PCR. Fungal DNA was detected in serum 16 and 14 days before the respective days the positive blood culture and positive blood beta-D-glucan were obtained. The concentration of *T. asahii* DNA correlated well with the patient's clinical course.

77.3 CONCLUSION

For pathogenic *Trichosporon* species, it is clinically important to be able to detect and identify the causative agent of deep-seated trichosporonosis. After the introduction of candin antifungal agents, the number of patients with breakthrough trichosporonosis infections has increased. Early detection of the causative agent is needed for a diagnosis. Several molecular-based detection methods have been developed, including a quantitative assay. A retrospective study of these methods has been conducted using clinical specimens. A prospective study using molecular-based methods is expected.

REFERENCES

1. Sugita T. The genus *Trichosporon*. In *The Yeasts, A Taxonomic Study*, 5th edn., Boekhout T, Fell JW, Kurtzman CP (eds.). Elsevier, Amsterdam, the Netherlands, 2011, pp. 2015–2061.
2. Guého E et al. *Trichosporon* on humans: A practical account. *Mycoses* 37, 3–10, 1994.
3. Sugita T et al. Taxonomic position of deep-seated, mucosa-associated, and superficial isolates of *Trichosporon cutaneum* from trichosporonosis patients. *J. Clin. Microbiol.* 33, 1368–1370, 1995.
4. Sugita T et al. Taxonomic studies on clinical isolates from superficial trichosporonosis patients by DNA relatedness. *Jpn. J. Med. Mycol.* 37, 107–110, 1996.
5. Watson KC, Kallichurum S. Brain abscess due to *Trichosporon cutaneum*. *J. Med. Microbiol.* 3, 191–193, 1970.
6. Goodman D et al. Breakthrough trichosporonosis in a bone marrow transplant recipient receiving caspofungin acetate. *Clin. Infect. Dis.* 35, E35–E36, 2002.
7. Matsue K et al. Breakthrough trichosporonosis in patients with hematologic malignancies receiving micafungin. *Clin. Infect. Dis.* 42, 753–757, 2006.
8. Asada N et al. Successful treatment of breakthrough *Trichosporon asahii* fungemia with voriconazole in a patient with acute myeloid leukemia. *Clin. Infect. Dis.* 43, e39–e41, 2006.
9. Bayramoglu G et al. Breakthrough *Trichosporon asahii* fungemia in neutropenic patient with acute leukemia while receiving caspofungin. *Infection* 36, 68–70, 2008.
10. Mekha N et al. Genotyping and antifungal drug susceptibility of the pathogenic yeast *Trichosporon asahii* isolated from Thai patients. *Mycopathologia* 169, 67–70, 2010.
11. Kalkanci A et al. Molecular identification, genotyping, and drug susceptibility of the basidiomycetous yeast pathogen *Trichosporon* isolated from Turkish patients. *Med. Mycol.* 48, 141–146, 2010.
12. Sugita T et al. Sequence analysis of the ribosomal DNA intergenic spacer 1 regions of *Trichosporon* species. *J. Clin. Microbiol.* 40, 1826–1830, 2002.
13. Kustimur S et al. Nosocomial fungemia due to *Trichosporon asteroides*: Firstly described bloodstream infection. *Diagn. Microbiol. Infect. Dis.* 43, 167–170, 2002.
14. Sugita T et al. Two new yeasts, *Trichosporon debeurmannianum* sp. nov. and *Trichosporon dermatis* sp. nov., transferred from the *Cryptococcus humicola* complex. *Int. J. Syst. Evol. Microbiol.* 51, 1221–1228, 2001.
15. Koyanagi T et al. Autopsy case of disseminated *Trichosporon inkin* infection identified with molecular biological and biochemical methods. *Pathol. Int.* 56, 738–743, 2006.
16. David C et al. Disseminated *Trichosporon inkin* and *Histoplasma capsulatum* in a patient with newly diagnosed AIDS. *J. Am. Acad. Dermatol.* 59, S13–S15, 2008.
17. Ağirbasli H et al. Two possible cases of *Trichosporon* infections in bone-marrow-transplanted children: The first case of *T. japonicum* isolated from clinical specimens. *Jpn. J. Infect. Dis.* 61, 130–132, 2008.
18. Marty FM, Barouch DH, Coakley EP, Baden LR. Disseminated trichosporonosis caused by *Trichosporon loubieri*. *J. Clin. Microbiol.* 41, 5317–5320, 2003.
19. Padhye AA et al. *Trichosporon loubieri* infection in a patient with adult polycystic kidney disease. *J. Clin. Microbiol.* 41, 479–482, 2003.
20. Lacasse A, Cleveland KO. *Trichosporon mucoides* fungemia in a liver transplant recipient: Case report and review. *Transpl. Infect. Dis.* 11, 155–159, 2009.
21. Ando M et al. Summer-type hypersensitivity pneumonitis. *Intern. Med.* 34, 707–712, 1995.
22. Sugita T, Ikeda R, Nishikawa A. Analysis of *Trichosporon* isolates obtained from the houses of patients with summer-type

hypersensitivity pneumonitis. *J. Clin. Microbiol.* 42, 5467–5471, 2004.
23. Sugita T et al. Genetic diversity and biochemical characteristics of *Trichosporon asahii* isolated from clinical specimens, houses of patients with summer-type-hypersensitivity pneumonitis, and environmental materials. *J. Clin. Microbiol.* 39, 2405–2411, 2001.
24. Gross JW, Kan VL. *Trichosporon asahii* infection in an advanced AIDS patient and literature review. *AIDS* 22, 793–795, 2008.
25. Tokimatsu I, Kadota J. Emerging deep-seated fungal infection, trichosporonosis. *Kansenshogaku Zasshi* 80, 196–202, 2006.
26. Walsh TJ. Trichosporonosis. *Infect. Dis. Clin. North Am.* 3, 43–52, 1989.
27. Suzuki K et al. Fatal *Trichosporon* fungemia in patients with hematologic malignancies. *Eur. J. Haematol.* 84, 441–447, 2010.
28. Girmenia C et al. GIMEMA Infection Program. Invasive infections caused by *Trichosporon* species and *Geotrichum capitatum* in patients with hematological malignancies: A retrospective multicenter study from Italy and review of the literature. *J. Clin. Microbiol.* 43, 1818–1828, 2005.
29. Ruan SY, Chien JY, Hsueh PR. Invasive trichosporonosis caused by *Trichosporon asahii* and other unusual *Trichosporon* species at a medical center in Taiwan. *Clin. Infect. Dis.* 49, e11–e17, 2009.
30. Sugita T et al. Identification of medically relevant *Trichosporon* species based on sequences of internal transcribed spacer regions and construction of a database for *Trichosporon* identification. *J. Clin. Microbiol.* 37, 1985–1893, 1999.
31. Peterson SW, Kurtzman CP. Ribosomal RNA sequence divergence among sibling species of yeasts. *Syst. Appl. Microbiol.* 14, 124–129, 1991.
32. Diaz MR, Fell JW. High-throughput detection of pathogenic yeasts of the genus *Trichosporon. J. Clin. Microbiol.* 42, 3696–3706, 2004.
33. Nagai H et al. PCR detection of DNA specific for *Trichosporon* species in serum of patients with disseminated trichosporonosis. *J. Clin. Microbiol.* 37, 694–699, 1999.
34. Sugita T et al. A nested PCR assay to detect DNA in sera for the diagnosis of deep-seated trichosporonosis. *Microbiol. Immunol.* 45, 143–148, 2001.
35. Nakajima M, Sugita T, Mikami Y. Granuloma associated with *Trichosporon asahii* infection in the lung: Unusual pathological findings and PCR detection of *Trichosporon* DNA. *Med. Mycol.* 45, 641–644, 2007.
36. Sano M et al. Supplemental utility of nested PCR for the pathological diagnosis of disseminated trichosporonosis. *Virchows Arch.* 451, 929–935, 2007.
37. Hosoki K et al. Early detection of breakthrough trichosporonosis by serum PCR in a cord blood transplant recipient being prophylactically treated with voriconazole. *J. Pediatr. Hematol. Oncol.* 30, 917–919, 2008.
38. Mekha N et al. Real-time PCR assay to detect DNA in sera for the diagnosis of deep-seated trichosporonosis. *Microbiol. Immunol.* 51, 633–635, 2007.
39. Tsuji Y et al. Quantitative PCR assay used to monitor serum *Trichosporon asahii* DNA concentrations in disseminated trichosporonosis. *Pediatr. Infect. Dis. J.* 27, 1035–1037, 2008.

78 *Ustilago* and *Pseudozyma*

Dongyou Liu

CONTENTS

78.1 INTRODUCTION

78.1.1 CLASSIFICATION, MORPHOLOGY, AND BIOLOGY

The genera *Ustilago* and *Pseudozyma* are basidiomycetous yeasts belonging to the family Ustilaginaceae, order Ustilaginales, class Ustilaginomycetes, phylum Basidiomycota. The family Ustilaginaceae covers 26 genera: *Anomalomyces, Cintractia, Dermatosorus, Farysia, Franzpetrakia, Gymnocintractia, Heterotolyposporium, Leucocintractia, Macalpinomyces, Melanopsichium*, mitosporic *Ustilaginaceae, Moesziomyces, Moreaua, Mycosyrinx, Parvulago, Pericladium, Portalia, Restiosporium, Schizonella, Sporisorium, Stegocintractia, Tolyposporium, Tranzscheliella, Trichocintractia, Ustilago*, and *Websdanea* as well as some unclassified Ustilaginaceae [1–4].

The genus *Ustilago* consists of 42 recognized species as well as 11 unassigned species, which are known as smut fungi with sophisticated life cycles including sexual and asexual (yeast) morphologies and pathogenic to corn [1,5]. The best known member of this genus is *Ustilago maydis* (obsolete synonym: *U. zeae*), which is smut fungus causing diseases in maize, and is occasionally involved in human infections.

Ustilago maydis, a Basidiomycete fungus infecting maize and teosinte, demonstrates two basic morphologies: a yeast-like form and a filamentous form. The yeast-like form is haploid, unicellular, and elongated with tapered ends (cigar shaped), divides by budding (once per cell cycle at one of the cell poles), with the bud growing by tip extension, and is nonpathogenic. The filamentous form is dikaryotic, divides at the apical cell, grows by tip extension, and is pathogenic.

A fungal-specific phase-contrast opaque body, the Spitzenkörper (apical body), located at or just below the tip of the apical cell, is thought to drive hyphal growth. The Spitzenkörper consists of a heterogeneous population of vesicles surrounding a core that contains polysomes, microtubules, and actin, and serves as a supply center for the distribution of vesicles containing materials necessary for tip extension.

U. maydis is capable of switching from one form to the other (the dimorphic switch). The dimorphic switch in the *U. maydis* life cycle is controlled by two mating type loci, *a* and *b*. The *a* locus has two alleles (*a1* and *a2*), codes for components of a signal transduction pathway (pheromone precursor and receptor genes), and governs cell fusion of haploid cells and filamentous growth of the dikaryon in vitro but not *in planta*. The *b* locus has 25 naturally occurring alleles (*b1*...*b25*), codes for a combinatorial homeodomain protein, and is the major determinant of filamentous growth, in vitro and *in planta*, and of pathogenicity. During interaction with its plant host, *U. maydis* shows an increased repertoire of morphologies, suggesting that plant signals play an important role in generation of additional morphologies [6,7].

On fungal culture media, *Ustilago* grows slowly and matures in 20 days. Colonies are moist, compact, and yeast-like, later becoming wrinkled and membrane-like with profuse budding in the medium. The color is cream to yellow initially and turns to tan to dark brown with age. Reverse is pale. Blastoconidia are spindle shaped, elongated, and irregular, resembling arthroconidia. Pseudohyphae consist of spindle-shaped cells. Short hyphae with clamp connections may appear when old.

Ustilago maydis harbors a genome of approximately 20 Mb, which contains the machinery for recognition and interpretation of the budding yeast axial and bipolar landmarks as well as for cell polarity establishment, exocytosis, actin and microtubule organization, microtubule plus end–associated proteins, kinesins, and myosins [8].

The genus *Pseudozyma*, the only member of the mitosporic Ustilaginaceae group, comprises 15 recognized species: *Candida tsukubaensis*, *Pseudozyma antarctica*, *Pseudozyma aphidis*, *Pseudozyma flocculosa*, *Pseudozyma fusiformata*, *Pseudozyma graminicola*, *Pseudozyma hubeiensis*, *Pseudozyma jujuensis*, *Pseudozyma parantarctica*, *Pseudozyma prolifica*, *Pseudozyma pruni*, *Pseudozyma rugulosa*, *Pseudozyma shanxiensis*, *Pseudozyma siamensis*, and *Pseudozyma thailandica*, in addition to 33 unassigned species [1]. *Pseudozyma* species are heterobasidiomycetous yeasts related to the smut fungi in the genus *Ustilago* [9]. While many *Pseudozyma* spp. are usually isolated from corns [6], several (*P. Antarctica*, *P. parantarctica*, and *P. thailandica*) have been identified from the blood of human patients [10], and *P. aphidis* is recently implicated in a pediatric case of central venous catheter (CVC) infection [11].

After 4 days of incubation on Sabouraud dextrose agar at 37°C in ambient air, *Pseudozyma* colonies are tan-yellow and wrinkled; blastoconidia are fusiform, spindle shaped, and elongated.

78.1.2 Clinical Features and Pathogenesis

Ustilago and *Pseudozyma* are widely distributed in soil and are pathogens of seeds and flowers of cereals, wheat, corn, and grasses [26]. Occasionally, these yeasts are associated with human diseases such as skin and CVC infections [10–13]. Since *Ustilago* and *Pseudozyma* are associated with corn, these organisms may enter the host through diet.

In a *Ustilago*-related skin infection, the patient showed a chronic skin rash with scaly erythematous plaques over the nasal alae and philtrum [13]. Airborne *Ustilago* spores have also been implicated in hypersensitivity pneumonitis, asthma, and allergic rhinitis [14,15].

The first documented case of CVC infection due to *Pseudozyma* species involved a 7-year-old child, who was born with gastroschisis and ileocolonic atresia resulting in short bowel syndrome. As a result, the patient relied on a long-term indwelling central line for parenteral nutrition. On admission, the patient presented with intermittent fevers (up to 39.7°C) accompanied by chills, malaise, and fatigue. Culture of the blood specimen on chocolate and Sabouraud dextrose agars yielded mature, moist yeast-like, tan-yellow, and wrinkled colonies after 4 days of incubation at 37°C in ambient air. Microscopic examination of the isolate on wet mount revealed fusiform spindle-shaped blastoconidia that were elongate and slightly irregular, suggestive of a yeast belonging to the order *Ustilaginales*. Sequencing analysis using fungal-specific primers ITS5 and ITS4 targeting the intervening transcribed spacer region of the rRNA confirmed its *P. aphidis* identity. Prompt removal of the CVC in conjunction with oral itraconazole therapy resulted in a successful outcome [11].

U. maydis interacts with its plant host through a reciprocal process of signal exchange, leading to alterations in cell morphology of both host and fungus. The fungus induces tumors on host plants and forms masses of diploid teliospores, which germinate to form haploid meiotic strains that can propagate in culture as yeast-like cells. Haploid strains of opposite mating type fuse and form a filamentous, dikaryotic cell type that invades plant tissue to reinitiate infection. Several invasion-associated genes have been identified in *U. maydis*. An ortholog of YAP1 (for Yeast AP-1-like) functions as a redox sensor and regulates the oxidative stress response in *U. maydis*. The fungus utilizes its Yap1-controlled detoxification system for coping with early plant defense responses [16,17]. In addition, extracellular effector proteins Hum3 and Rsp1 are pathogenicity proteins that share an essential function in early stages of the infection [18].

Ustilago and *Pseudozyma* are susceptible to amphotericin B, ketoconazole, itraconazole, simeconazole, which may be used for effective treatment of these fungi [10,13].

78.1.3 Laboratory Diagnosis

Traditionally, laboratory identification of *Ustilago* and *Pseudozyma* yeasts are based on assessment of their macroscopic and microscopic features. Several stains (e.g., KOH, lactophenol, calcofluor white, periodic acid-Schiff [PAS], Gomori methenamine silver [GMS]) may be employed. Standard fungal media are useful for isolation of these organisms for further characterization. More recently, molecular techniques are increasingly relied upon for identification, detection, and phylogenetic analysis of these fungi [19].

Through examination of the cytochrome b gene sequences, it was shown that basidiomycetous yeasts can be separated into two main clusters: cluster 1 includes Tremellales, Filobasidiales, and their anamorphs (which possess xylose in the cell wall and have dolipore septa) and cluster 2 includes Ustilaginales, Sporidiales, and their anamorphs (which possess no xylose in the cell wall and have a simple pore septum). Thus, use of cytochrome b sequences enables both species identification and the study of phylogenetic relationships [20]. Similarly, DNA sequencing analysis of the internal transcribed spacer (ITS) region of the nuclear rRNA genes revealed the monophyly of a bipartite genus *Sporisorium* and the monophyly of a core *Ustilago* clade and indicated a putative connection between *Ustilago maydis* and *Sporisorium* [21]. Furthermore, combined analyses of ITS and LSU rRNA sequences resolved three major groups of almost identical composition: *Sporisorium*, *Ustilago*, and a basal assemblage of both *Ustilago* and *Sporisorium* species [21].

By using oligonucleotides corresponding to a specific region downstream of the homeodomain of the bE genes, a 500 bp product was amplified from *U. maydis* DNA only. Use of this method permitted rapid and precise detection and identification of *U. maydis* organism from infected tissues [22].

78.2 METHODS

78.2.1 Sample Preparation

Clinical specimens are examined under microscope for mycotic elements. Portions of the samples are inoculated on inhibitory mold agar, modified Sabouraud agars, or potato dextrose agar and incubated at various temperatures for a few days.

After growth for 1–7 days on potato dextrose agar slants, lysates are prepared from approximately 1 cm^2 of mycelia with IDI lysis kits (GeneOhm Sciences, San Diego, CA). Briefly, in a biological safety cabinet, mycelia are collected by scraping the slant with a sterile stick in 1 mL of sterile, molecular-grade H_2O. The material is transferred to a 2 mL screw-cap tube. The tubes are centrifuged for 1 min at 6000 × *g*. If the mycelia do not pellet, the material is contained with a pediatric blood serum filter (Porex Corp.). After removal of supernatant, the material is resuspended in 200 μL of IDI sample buffer and transferred to the lysis tube, which contains glass beads. Lysis tubes are vortexed on the highest setting for 5 min. The tubes are placed in a boiling water bath for 15 min. Tubes are centrifuged for 5 min at 16,000 × *g*. The supernatant is stored at −20°C until amplification [23].

Alternatively, 1 cm^2 of fungal material is transferred to a 2 mL Eppendorf tube containing a 2:1 (wt/wt) mixture of silica gel and Celite (silica gel H, Merck 7736/Kieselguhr Celite 545; Machery) and 300 μL of TES buffer (2 g Tris [hydroxymethyl]-aminomethane, 0.38 g Na-ethylenediaminetetraacetic acid (EDTA), and 2 g sodium dodecyl sulfate in 80 mL of ultrapure water [pH 8]). The fungal material is ground with a micropestle for 1–2 min. The volume is adjusted by adding 200 μL of TES buffer. After vigorous shaking and the addition of 10 μL of a 10 mg/mL concentration of proteinase K to the tube, the mixture is incubated at 65°C for 10 min. With the addition of 140 μL of 5 M NaCl solution, the mixture is combined with 1/10 volume (~65 μL) of cetyltrimethylammonium bromide (CTAB) buffer 10%, followed by incubation for another 30 min at 65°C. One volume (~700 μL) of chloroform-isoamyl alcohol (vol/vol = 24/1) is added and mixed by inversion. After incubation for 30 min at 0°C (on ice water) and centrifugation at 14,000 rpm at 4°C for 10 min, the top layer is transferred to a clean Eppendorf tube. The sample is added with 225 μL of 5 M NH_4-acetate and incubated for at least 30 min (on ice water) and centrifuged. The supernatant is transferred to a clean sterile Eppendorf tube and mixed with a 0.55 volume (~510 μL) of ice-cold isopropanol. After centrifugation for 7 min at 14,000 rpm and 4°C (or room temperature), the supernatant is decanted. The pellet is washed with ice-cold 70% ethanol twice and dried by using a vacuum dryer. The powder is resuspended in 48.5 μL of Tris-EDTA buffer with 1.5 μL of 10 mg of RNase/mL, incubated at 37°C for 15–30 min, and stored at −20°C until use [24].

78.2.2 Detection Procedures

78.2.2.1 PCR Identification of *Ustilago maydis*

Martínez-Espinoza et al. [22] designed a polymerase chain reaction (PCR) assay for specific identification of *Ustilago maydis* using primers 1369 (5′-CTCGAGGTTCATCAGCTCA-3′) and 1370 (5′-GCTGAGTTCTGGAGTCG-3′). These primers target a conserved region downstream of the homeodomain of the *bE* genes from *U. Maydis*, with primer 1369 being located at nt 645 of *bE1* and *bE2*, and nt 489 of *bE5*; with primer 1370 being located at nt 1147 of *bE1* and *bE2*, and nt 992 of *bE5*.

Interestingly, the life cycle of *U. maydis* is regulated by the mating-type loci *a* and *b*. The *a* locus (consisting of two idiomorphs, a_1 and a_2) is required for cell-to-cell recognition during the mating process and for the maintenance of filamentous growth. Each idiomorph encodes a pheromone and a receptor for the pheromone synthesized by the compatible partner strain. The *b* locus (consisting of at least 25 alleles at each of two genes, *bE* and *bW*) regulates the steps in sexual development that occur after fusion of haploid cells. Different *a* and *b* alleles in mating partners are necessary to trigger mating, filamentous growth, and tumor induction [22].

Procedure

1. PCR mixture (50 μL) is made up of 100 μM dNTPs, 2 mM $MgCl_2$, 1× PCR buffer (Gibco), 2.5 U of DNA Taq polymerase (Gibco), extracted fungal DNA, and primers 1369 and 1370.
2. Amplification is conducted with 30 cycles of denaturation at 95°C for 1 min, annealing at 55°C for 1 min, and extension at 72°C for 1 min, followed by a final extension for 10 min at 72°C.
3. PCR products (5 μL) are separated by electrophoresis on a 1.5% agarose gel, stained with ethidium bromide, and observed under ultraviolet light.

Note. The PCR assay using primers 1369 and 1370 generates a 502-bp product from DNA of *U. maydis*, but not from DNA of other fungi unrelated to *U. maydis*.

78.2.2.2 Pan-Fungal Detection

Pounder et al. [23] described a real-time PCR with SYBR green DNA-binding dye and amplicon melting temperature analysis for fungal detection using pan-fungal primers ITS1 forward (5′-TCCGTAGGTGAACCTGCGG-3′) and ITS4 reverse (5′-TCCTCCGCTTATTGATATGC-3′), which covers ITS1-5.8S-ITS2 rRNA gene cluster. The identity of the fungi is verified by subsequent sequencing analysis.

Procedure

1. PCR mixture is composed of 1× Lightcycler FastStart DNA Master Hybridization Probes mixture (Roche Applied Science) (containing deoxynucleoside triphosphates, FastStart *Taq* DNA polymerase, and 1 mM $MgCl_2$, additional $MgCl_2$ is added to a final concentration of 4.6 mM), 0.4 μM each of ITS1 forward and ITS4 reverse primers, 1× SYBR green (Molecular Probes), and 3 μL template DNA.

2. Thermal cycling parameters include 95°C for 10 min; 50 cycles at 95°C for 5 s, 60°C for 20 s, and 76°C for 30 s; and a final extension at 72°C for 2 min.
3. The quality of the amplicon is determined using the derivative of the melt analysis curve (55°C–99°C, 45 s hold at 55°C, 5 s/°C) using the RotorGene 3000 (Corbett Robotics, Inc.).
4. The amplified product is purified for bidirectional sequencing using ExoSAP-IT (USB Corp.). Five microliters of Big Dye Terminator Ready Reaction Mix v. 1.1 (Applied Biosystems) is added to 4 μL of each primer (0.8 pmol/μL) and 3 μL of purified PCR product. Cycle sequencing is performed with a 9700 thermal cycler (ABI), using 25 cycles at 96°C for 10 s, 50°C for 5 s, and 60°C for 4 min. Sequencing reaction products are passed through a Sephadex G-50 fine column to remove unincorporated dye terminators. Purified sequencing reaction products are run on an ABI Prism 3100 Genetic Analyzer with a 50 cm capillary array.
5. Sequences are analyzed with the SmartGene Integrated Database Network software version 3.2.3 vr. SmartGene is a web-based software and database system with reference sequences derived from the National Center for Biological Information (NCBI) GenBank repository.

Note. If real time PCR instrument is unavailable, standard PCR may be performed with primers ITS1 and ITS4, and the resulting amplicon is sequenced with the same primers. Sequence-based identifications are defined by percent identity: species, ≥99%; genus, 93%–99%; and inconclusive, ≤93%. For strains producing discrepant identification between the methods based on phenotypic characteristics and ITS sequence analysis, the D1–D2 region of the large-subunit RNA gene is amplified with primers NL1 (5′-GCATATCAATAAGCGGAGGAAAAG-3′) and NL4 (5′-GGTCCGTGTTTCAAGACGG-3′) and sequenced for species clarification [25].

78.3 CONCLUSION

Ustilago and *Pseudozyma* are related basidiomycetous yeasts belonging to the family Ustilaginaceae. Besides causing diseases in plants, they may be occasionally involved in human illnesses. Due to their widespread presence in environment and corns, they may gain entry into human host through food intake [26]. Patients with underlying conditions such as short gut syndrome seem vulnerable to the infection by *Ustilago* and *Pseudozyma* spp. Assessment of the macroscopic and microscopic features of these organisms allows their determination; however, it suffers from the drawbacks of being time-consuming and laborious. Use of molecular techniques facilitates rapid, sensitive and specific identification, detection, and phylogenetic analysis of *Ustilago* and *Pseudozyma* spp.

REFERENCES

1. The UniProt Consortium. Available at http://www.uniprot.org/, accessed on August 1, 2010.
2. Begerow, D. and Bauer, R. (2000). Phylogenetic placements of ustilaginomycetous anamorphs as deduced from nuclear LSU rDNA sequences. *Mycol. Res.* 104, 53–60.
3. Begerow, D., Stoll, M., and Bauer, R. (2006). A phylogenetic hypothesis of Ustilaginomycotina based on multiple gene analyses and morphological data. *Mycologia* 98(6), 906–916.
4. Fell, J.W. et al. (2000). Biodiversity and systematics of basidiomycetous yeasts as determined by large-subunit rDNA D1/D2 domain sequence analysis. *Int. J. Syst. Evol. Microbiol.* 50, 1351.
5. Stoll, M., Begerow, D., and Oberwinkler, F. (2005). Molecular phylogeny of *Ustilago, Sporisorium*, and related taxa based on combined analyses of rDNA sequences. *Mycol. Res.* 109, 342–356.
6. Martinez-Espinoza, A.D., Garcia-Pedrajas, M.D., and Gold, S.E. (2002). The Ustilaginales as plant pests and model systems. *Fungal Genet. Biol.* 35, 1–20.
7. Perez-Martin, J. and Castillo-Lluva, S. (2006). Pathocycles: *Ustilago maydis* as a model to study the relationship between cell cycle and virulence in pathogenic fungi. *Mol. Genet. Genom.* 276, 211–220.
8. Kamper, J. et al. (2006). Insights from the genome of the biotrophic fungal plant pathogen *Ustilago maydis. Nature* 444, 97–101.
9. Boekhout, T. (1995). *Pseudozyma* Bandoni emend. Boekhout, a genus for yeast-like anamorphs of ustilaginales. *J. Gen. Appl. Microbiol.* 41, 359–366.
10. Sugita, T. et al. (2003). The first isolation of ustilaginomycetous anamorphic yeasts, *Pseudozyma* species, from patients' blood and a description of two new species: *P. parantarctica* and *P. thailandica. Microbiol. Immunol.* 47, 183–190.
11. Lin, S.S. et al. (2008). Central venous catheter infection associated with *Pseudozyma aphidis* in a child with short gut syndrome. *J. M ed. Microbiol.* 57, 516–518.
12. Patel, R. et al. (1995). Central venous catheter infection due to *Ustilago* species. *Clin. Infect. Dis.* 21, 1043–1044.
13. Teo, L.H. and Tay, Y.K. (2006). *Ustilago* species infection in humans. *Br. J. Dermatol.* 155, 1096–1097.
14. Santilli, J., Rockwell, W.J., and Collins, R.P. (1985). The significance of the spores of the basidiomycetes (mushrooms and their allies) in bronchial asthma and allergic rhinitis. *Ann. Allergy* 55, 469–471.
15. Yoshida, K. et al. (1996). Hypersensitivity pneumonitis induced by a smut fungus *Ustilago esculenta. Thorax* 51, 650–651.
16. Molina, L. and Kahmann, R. (2007). An *Ustilago maydis* gene involved in H_2O_2 detoxification is required for virulence. *Plant Cell* 19(7), 2293–2309.
17. Howlett, B.J., Idnurm, A., and Heitman, J. (2007). Fungal pathogenesis: Gene clusters unveiled as secrets within the *Ustilago maydis* code. *Curr. Biol.* 17(3), R87–R90.
18. Müller, O., Schreier, P.H., and Uhrig, J.F. (2008). Identification and characterization of secreted and pathogenesis-related proteins in *Ustilago maydis. Mol. Genet. Genom.* 279, 27–39.
19. Menzies, J.G. et al. (2003). Use of inter-simple sequence repeats and amplified fragment length polymorphisms to analyze genetic relationships among small grain-infecting species of *Ustilago. Phytopathology* 93, 167.

20. Biswas, S.K. et al. (2001). Molecular phylogenetics of the genus *Rhodotorula* and related basidiomycetous yeasts inferred from the mitochondrial cytochrome *b* gene. *Int. J. Syst. Evol. Microbiol.* 51, 1191–1199.
21. Stoll, M. et al. (2003). Molecular phylogeny of *Ustilago* and *Sporisorium* species (Basidiomycota, Ustilaginales) based on internal transcribed spacer (ITS) sequences. *Can. J. Bot.* 81(9), 976–984.
22. Martínez-Espinoza, A.D. et al. (2003). Use of PCR to detect infection of differentially susceptible maize cultivars using *Ustilago maydis* strains of variable virulence. *Int. Microbiol.* 6(2), 117–120.
23. Pounder, J.I. et al. (2007). Discovering potential pathogens among fungi identified as nonsporulating molds. *J. Clin. Microbiol.* 45, 568–571.
24. Zeng, J.S. et al. (2007). Spectrum of clinically relevant *Exophiala* species in the United States. *J. Clin. Microbiol.* 45(11), 3713–3720.
25. Leaw, S.N. et al. (2006). Identification of medically important yeast species by sequence analysis of the internal transcribed spacer regions. *J. Clin. Microbiol.* 44(3), 693–699.
26. Valverde, M.E. et al. (1995). Huitlacoche (*Ustilago maydis*) as a food source—biology, composition, and production. *Crit. Rev. Food Sci. Nutr.* 35, 191–229.

79 *Wallemia*

Dongyou Liu

CONTENTS

79.1 INTRODUCTION

79.1.1 Classification, Morphology, and Biology

The genus *Wallemia* is the only member of the family Wallemiales incertae sedis, order Wallemiales, class Wallemiomycetes, phylum Basidiomycota, kingdom Fungi. As a mold lacking a known sexual state, the genus *Wallemia* is often considered as a member of the Fungi Imperfecti. Currently, the genus *Wallemia* (obsolete synonym: *Hemispora*) consists of three xerophilic species *Wallemia ichthyophaga*, *Wallemia muriae* (obsolete synonym: *Torula epizoa* var. *muriae*), and *Wallemia seb*i (obsolete synonyms: *Hemispora stellata*, *Sporendonema sebi*, and *Wallemia ichthyophaga*) and two unassigned species (*Wallemia* sp. F53 and *Wallemia* sp. HLS205) [1]. These unassigned species are phenotypically distinguished by the size range of conidia and by the degree of their xerophily as well as genetically by sequence analysis of the rRNA internal transcribed spacer (ITS) regions [2]. Within the genus, *Wallemia sebi* is the only species that has been implicated in cutaneous or subcutaneous infections as well as allergies in humans. Collectively, the infections due to *Wallemia sebi* is referred to as "hemisporiosis" after its now obsolete synonym *Hemispora stellata* [3].

Wallemia sebi grows slowly on malt extract agar and other common fungal culture media and is often obscured by the fast-growing fungi. Use of selective media for xerophilic fungi enhances the growth and recovery of *W. sebi* from environmental and clinical specimens. *Wallemia sebi* colonies reach a diameter of 2–4 mm after 2 weeks at 25°C and are elevated, irregular in shape, and orange–brown to blackish brown in color. Smooth and subhyaline conidiophores are slender, cylindrical, and usually unbranched, and show a swollen upper part from which a cylindrical, verrucose fertile hypha arises and fragments to form conidia. Conidia are catenate, cubical, 1.5–2.5 μm in diameter, becoming spherical or subspherical, finely verruculose, subhyaline, and brown in mass. *W. sebi* conidiogenesis is characterized by the basauxic development of fertile hyphae, basipetal segregation of conidial units, and disarticulation of conidial units into mostly four arthrospore-like conidia [4].

Analyses of the nuclear small subunit ribosomal RNA (SSU rRNA) place *Wallemia* into a clade together with Ustilaginomycetes and Hymenomycetes (Basidiomycota). Within this clade, *Wallemia* occupies an isolated position distantly related to the Filobasidiales (reflecting their similar parenthesome ultrastructures). *Wallemia* demonstrates a rare xerotolerance among the Basidiomycota. Of the three recognized *Wallemia* species (*W. ichthyophaga*, *W. sebi*, and *W. muriae*), *W. sebi* has the smallest conidia; *W. ichthyophaga* differs from *W. sebi* and *W. muriae* in numerous nucleotides of the SSU and ITS rRNA. This high variation within *Wallemia* suggests the existence of at least two cryptic genera not distinguishable by morphological characters [2].

Wallemia sebi is a mold with a ubiquitous distribution and has been isolated from agricultural and indoor environments (e.g., soil, hay, hypersaline water of man-made salterns and textiles) and foods (e.g., jams, dates, fruits, bread, cakes, salted beans, maize flour, crystalline sugar, fish, bacon, dairy products, and salt). *Wallemia* is known to cause food spoilage [5]. Given that *W. sebi* conidia are rough-surfaced spheres of 1.5–2.5 μm in diameter, they can reach the respiratory bronchioles when inhaled [6].

79.1.2 Clinical Features and Pathogenesis

Being a mold with common presence in all environments, *Wallemia sebi* has been shown as a causative agent for human cutaneous or subcutaneous infections. In addition, airborne *W. sebi* spores have been implicated in human allergies, particularly bronchial asthma [6,7]. Some asthmatic individuals often show immediate type hypersensitivity to *W. sebi* [8,9]. The organism has been identified as playing a role in farmer's lung disease (FLD) (together with *Eurotium amstelodami* and *Absidia corymbifera*), with symptoms of dyspnea, cough, tiredness, headaches, occasional fever/night sweats, and general feeling of sickness [4,10–13]. FLD tends to occur in cattle-rearing areas, during the cold and rainy during the indoor feeding season (winter) [13].

In a case of *Wallemia sebi*-related subcutaneous phaeohyphomycosis, a 43-year-old woman in northern India presented with a nonhealing ulcer, which had a ragged margin and was covered with slough, on the dorsum of left foot. Having started 8 months earlier as an itchy papule that gradually progressed to its present size (8 × 6 cm), the lesion was erythematous at its base, mildly warm (to touch), and minimally tender. Histopathological examination of lesion biopsy using Grocott stain revealed septate hyphae. A portion of the biopsy sample (minced into tiny pieces) was cultured on Sabouraud dextrose agar (SDA; Difco) containing chloramphenicol (0.05 mg/mL) and SDA with chloramphenicol and cycloheximide (0.5 mg/mL). This led to the isolation of *Wallemia sebi* [14].

Wallemia sebi produces a toxic metabolite, walleminol A, which is a tricyclic dihydroxy compound, with structural features characteristic of a sesquiterpene possessing an isolated double bond. The minimum inhibitory dose of walleminol A in the bioassays is approximately 50 mg/mL, which is comparable with a number of mycotoxins such as citrinin and penicillic acid [15].

79.1.3 Diagnosis

Wallemia sebi is a dematiaceous anamorphic fungus that is a causative agent of subcutaneous phaeohyphomycosis and FLD. Phaeohyphomycosis is an infectious disease of skin, subcutaneous tissues, and internal organs due to dematiaceous (melanized) fungi that produce pigmented hyphae and/or yeast-like cells in culture and frequently in the infected tissue. Besides *Wallemia sebi*, other dematiaceous fungi (e.g., *Alternaria*, *Bipolaris*, *Curvularia*, *Cladophialophora*, *Cladosporium*, *Exophiala*, *Exserohilum*, *Phaeoacremonium*, or *Phialophora*) are also capable of causing phaeohyphomycosis. Thus, it is important to differentiate *Wallemia sebi* from other dematiaceous fungal species in order to implement appropriate treatment strategies.

Conventional methods for the identification of *W. sebi* are dependent on observation of mycotic elements in clinical samples and cultured isolates with the help of a variety of stains such as KOH, lactophenol, calcofluor white, periodic acid-Schiff (PAS), Gomori methenamine silver (GMS), etc. Since the macroscopic and microscopic features of *W. sebi* demonstrate close resemblances to those of dematiaceous fungi, it is technically challenging to make correct calls on the basis of phenotypic criteria alone. This is further exacerbated by the fact that *W. sebi* tends to display a slower growth than other molds and may be discarded as contaminants without detailed investigation.

Molecular techniques are independent of the morphological and biochemical characteristics of *W. sebi*, and offer a promising alternative to the conventional detection methods, with an added benefit of detecting the presence of microorganisms in a sample regardless of their culturability. PCR amplification and sequencing analysis of the 18S and 28S rRNA genes and their internal transcribed spacer (ITS) regions enable rapid and precise determination of *W. sebi* and other fungi [16–18]. The recent development of *W. sebi*-specific PCR targeting 18S rRNA gene sequence allows sensitive amplification of *W. sebi* from composite environmental samples [19]. Application of this PCR revealed a high concentration of *W. sebi* spores in aerosol and surface samples from farms handling hay and grain and from cow barns, highlighting the potential role of this fungus in FLD [19].

79.2 METHODS

79.2.1 Sample Preparation

Smears of scrapings from the lesion are stained with fungal stains and examined under microscope for mycotic elements. Air samples are collected by 25 mm diameter polycarbonate filters with a pore size of 0.4 μm (Isopore; Millipore), which is mounted in a 25 mm cassette IOM sampler (SKC Inc.). Air is drawn through the filter with an Aircheck Sampler model 224-PCXR7 (SKC Inc.). After sampling, 1.5 mL of suspension buffer (50 mM Tris–HCl pH 7.5, 50 mM ethylenediaminetetraacetic acid [EDTA], 2% sodium dodecyl sulfate, 1% Triton-100) is added to each sampling cassette. The cassettes are shaken on a shaker for 10 min to suspend the particles. From each suspension, a 500 μL aliquot is serially diluted in 0.05% Tween 80. Colony counting is performed by spreading 100 μL of each dilution onto DG18 agar plates in duplicate and incubating at room temperature (22°C) for 14 days. Another 750 μL of the particle suspension is used for DNA extraction with a Qiagen kit; and 1:100 dilution of the eluate is used for PCR testing [16,19].

Alternatively, 1 cm^2 of fungal material is transferred to a 2 mL Eppendorf tube containing a 2:1 (wt/wt) mixture of silica gel and Celite (silica gel H, Merck 7736/Kieselguhr Celite 545; Machery) and 300 μL of TES buffer (2 g Tris [hydroxymethyl]-aminomethane, 0.38 g Na-EDTA, and 2 g sodium dodecyl sulfate in 80 mL of ultrapure water [pH 8]). The fungal material is ground with a micropestle for 1–2 min. The volume is adjusted by adding 200 μL of TES buffer. After vigorous shaking and the addition of 10 μL of a 10 mg/mL concentration of proteinase K to the tube, the mixture is incubated at 65°C for 10 min. With the addition of 140 μL of 5 M NaCl solution, the mixture is combined with 1/10 volume (~65 μL) of cetyltrimethylammonium bromide (CTAB) buffer 10%, followed by incubation for another 30 min at 65°C. One volume (~700 μL) of chloroform–isoamyl alcohol (vol/vol = 24/L) is added and mixed by inversion. After incubation for 30 min at 0°C (on ice water) and centrifugation at 14,000 rpm at 4°C for 10 min, the top layer is transferred to a clean Eppendorf tube. The sample is added with 225 μL of 5 M NH_4-acetate and incubated for at least 30 min (on ice water) and centrifuged. The supernatant is transferred to a clean sterile Eppendorf tube and mixed with a 0.55 volume (~510 μL) of ice-cold isopropanol. After centrifugation for 7 min at 14,000 rpm and 4°C (or room temperature), the supernatant is decanted. The pellet is washed with ice-cold 70% ethanol twice and dried by using a vacuum dryer. The powder is resuspended in 48.5 μL of Tris-EDTA buffer with 1.5 μL of 10 mg of RNase/mL, incubated at 37°C for 15–30 min, and stored at −20°C until use [20].

79.2.2 Detection Procedures

79.2.2.1 PCR for *Wallemia sebi*

Zeng et al. [19] designed two pairs of primers from the nuclear 18S rRNA sequence (GenBank accession number AF548107) for identification of *Wallemia sebi*. These primers displayed a mismatch of >25% to any other fungal sequences in the GenBank database. While primer pair Wall-SYB4/6 amplifies a 371 bp fragment, primer pair Wall-SYB7/8 generates a 328 bp fragment from *W. sebi* DNA (Table 79.1).

Procedure

For conventional PCR detection, PCR mixture (25 μL) is composed of 10 pmol of each primer (Wall-SYB4/6 or Wall-SYB7/8), 0.75 U of *Taq* DNA polymerase (Invitrogen Life Technologies), 200 μM of each deoxynucleoside triphosphate, 1.5 mM $MgCl_2$, and 1–5 ng of template DNA. Amplification is conducted with an initial 94°C for 3 min; 30 cycles at 94°C for 30 s, 54°C for 30 s, and 72°C for 30 s; and a final 72°C for 3 min. PCR products (3 μL) are analyzed by electrophoresis on 1.4% agarose gels in 1× Tris-acetate-EDTA buffer. A 1 kb Plus ladder (Invitrogen Life Technologies) is used as a DNA size standard. The gels are stained with ethidium bromide and visualized under UV light.

For real-time PCR detection, PCR mixture (25 μL) is made up of 10 pmol of each primer (Wall-SYB4/6 or Wall-SYB7/8), 12.5 μL of iQ SYBR Green Supermix with an iCycler iQ real-time PCR detection system (Bio-Rad), and 1–5 ng of template DNA. Amplification is performed with an initial 95°C for 3 min followed by 40 cycles at 95°C for 10 s and 60°C for 60 s (for Wall-SYB4/6) or 30 s (for Wall-SYB7/8). A melt-curve analysis is conducted immediately after amplification at 95°C for 60 s, cooling to 60°C for 60 s, and a slow rise in temperature to 95°C at a rate of 0.5°C/10 s with continuous acquisition of fluorescence decline. Melt-curve analysis is used to observe melting characteristics of the amplicon and to determine the presence of the specific product.

Note. Without other background DNA, 4.2×10^{-4} ng of *W. sebi* genomic DNA could be detected with either of the primer sets Wall-SYB4/6 or Wall-SYB7/8 in conventional PCR. When *W. sebi* DNA was mixed with DNA from 10 other fungi, 2.1×10^{-4} ng of *W. sebi* DNA could be detected against a background of 11.2 ng of unrelated fungal DNA. Assuming one fungal genome is ca. 4.0×10^{-5} ng of DNA, the conventional PCR could potentially detect 5–10 fungal spores in a reaction. In the real-time PCR, all isolates of *W. sebi* gave strong positive fluorescent signals after 12–20 cycles, while the other 36 fungal DNAs produced very faint signals only after 36 cycles that were identified by melt-curve analysis as primer dimers [19].

79.2.2.2 Pan-Fungal Real-Time PCR and Sequencing Analysis

Pounder et al. [17] described a real-time PCR with SYBR green DNA-binding dye and amplicon melting temperature analysis for fungal detection using pan-fungal primers ITS1 forward (5′-TCCGTAGGTGAACCTGCGG-3′) and ITS4 reverse (5′-TCCTCCGCTTATTGATATGC-3′). The identity of the fungi is verified by subsequent sequencing analysis.

Procedure

1. PCR mixture is composed of 1× Lightcycler FastStart DNA Master Hybridization Probes mixture (Roche Applied Science) (containing deoxynucleoside triphosphates, FastStart *Taq* DNA polymerase, and 1 mM $MgCl_2$, additional $MgCl_2$ is added to a final concentration of 4.6 mM), 0.4 μM each of ITS1 forward and ITS4 reverse primers, 1× SYBR green (Molecular Probes), and 3 μL template DNA.
2. Thermal cycling parameters include 95°C for 10 min; 50 cycles at 95°C for 5 s, 60°C for 20 s, and 76°C for 30 s; and a final extension at 72°C for 2 min.
3. The quality of the amplicon is determined using the derivative of the melt analysis curve (55°C–99°C, 45 s hold at 55°C, 5 s/°C) using the RotorGene 3000 (Corbett Robotics, Inc.).
4. The amplified product is purified for bidirectional sequencing using ExoSAP-IT (USB Corp.). Five microliters of Big Dye Terminator Ready Reaction

TABLE 79.1
***W. sebi*-Specific Primers**

Primer	Sequence (5′–3′)	Position	Amplicon (bp)
Wall-SYB4	GTAGTGAACTATATTGAAGAA	621–991	371
Wall-SYB6	ATGAGTCAATAATATAACGTC		
Wall-SYB7	GATTGGATGACGTTATATTAT	963–1290	328
Wall-SYB8	ACAACAAAATGTCGTACCG		

Mix v. 1.1 (Applied Biosystems) is added to 4 μL of each primer (0.8 pmol/μL) and 3 μL of purified PCR product. Cycle sequencing is performed with a 9700 thermal cycler (ABI), using 25 cycles at 96°C for 10 s, 50°C for 5 s, and 60°C for 4 min. Sequencing reaction products are passed through a Sephadex G-50 fine column to remove unincorporated dye terminators. Purified sequencing reaction products are run on an ABI Prism 3100 Genetic Analyzer with a 50 cm capillary array.

5. Sequences are analyzed with the SmartGene Integrated Database Network software version 3.2.3 vr. SmartGene is a web-based software and database system with reference sequences derived from the National Center for Biological Information (NCBI) GenBank repository.

Note. If real time PCR instrument is available, standard PCR may be performed with primers ITS1 and ITS4, and the resulting amplicon is sequenced with the same primers. Sequence-based identifications are defined by percent identity: species, ≥99%; genus, 93%–99%; and inconclusive, ≤93%.

79.3 CONCLUSION

Wallemia sebi is a dematiaceous anamorphic fungus that occasionally causes subcutaneous phaeohyphomycosis and FLD in humans. Due to its presence in the environment and its relatively slow growth in standard fungal culture media in comparison with other molds, *Wallemia sebi* may be deemed as contaminants and discarded without further analysis, leading to a possible underestimation of this fungus in human disease process. With the development and application of rapid, sensitive, and specific PCR assays, it is envisaged that a true picture of *W. sebi*'s involvement in human infections will emerge in future, contributing to the improved control and prevention of human diseases due to this fungus.

REFERENCES

1. The UniProt Consortium. Available at http://www.uniprot.org/, accessed on August 1, 2010.
2. Zalar, P. et al. (2005). Taxonomy and phylogeny of the osmophilic genus *Wallemia* (*Wallemiomycetes* and *Wallemiales*, cl. et ord. nov.). *Anton Leeuwen* 87:311.
3. Moore, R.T. (1986). A note on *Wallemia sebi. Anton Leeuwen* 52:183–187.
4. Reboux, G. et al. (2001). Role of molds in farmer's lung disease in Eastern France. *Am. J. Respir. Crit. Care Med.* 163:1534–1539.
5. Vindelov, J. and N. Arneberg. (2001). Interactions between *Zygosaccharomyces mellis* and *Wallemia sebi* in diluted molasses. *Int. J. Food Microbiol.* 63:73–79.
6. Hanhela, R., K. Louhelainen, and A.-L. Pasanen. (1995). Prevalence of microfungi in Finnish cow barns and some aspects of the occurrence of *Wallemia sebi* and *Fusaria. Scand. J. Work Environ. Health* 21:223–228.
7. Lappalainen, S. et al. (1998). Serum IgG antibodies against *Wallemia sebi* and *Fusarium* species in Finnish farmers. *Ann. Allergy Asthma Immunol.* 81:585–592.
8. Sakamoto, T. et al. (1989a). Allergenic and antigenic activities of the osmophilic fungus *Wallemia sebi* asthmatic patients. *Arerugi* 38:352–359.
9. Sakamoto, T. et al. (1989b). Studies on the osmophilic fungus *Wallemia sebi* as an allergen evaluated by skin prick test and radioallergosorbent test. *Int. Arch. Allergy Appl. Immunol.* 90:368–372.
10. Roussel, S. et al. (2004). Microbiological evolution of hay and relapse in patients with farmer's lung. *Occup. Environ. Med.* 61(1):e3.
11. Roussel, S. et al. (2005a). Evaluation of salting as a hay preservative against farmer's lung disease agents. *Ann. Agric. Environ. Med.* 12:217–221.
12. Roussel, S. et al. (2005b). Farmer's lung disease and microbiological composition of hay: A case-control study. *Mycopathologia* 160(4):273–279.
13. Gbaguidi-Haore, H. et al. (2009). Multilevel analysis of the impact of environmental factors and agricultural practices on the concentration in hay of microorganisms responsible for farmer's lung disease. *Ann. Agric. Environ. Med.* 16:219–225.
14. Guarro, J. et al. (2008). Subcutaneous phaeohyphomycosis caused by *Wallemia sebi* in an immunocompetent host. *J. Clin. Microbiol.* 46(3):1129–1131.
15. Wood, G.M. et al. (1990). Studies on a toxic metabolite from the mould *Wallemia. Food Addit. Contam.* 7:69–77.
16. Wu, Z. et al. (2003). 18S rRNA gene variation among common airborne fungi, and development of specific oligonucleotide probes for the detection of fungal isolates. *Appl. Environ. Microbiol.* 69:5389–5397.
17. Pounder, J.I. et al. (2007). Discovering potential pathogens among fungi identified as nonsporulating molds. *J. Clin. Microbiol.* 45(2):568–571.
18. Bagyalakshmi, R. et al. (2008). Newer emerging pathogens of ocular non-sporulating molds (NSM) identified by polymerase chain reaction (PCR)-based DNA sequencing technique targeting internal transcribed spacer (ITS) region. *Curr. Eye Res.* 33(2):139–147.
19. Zeng, Q.-Y. et al. (2004). Detection and quantification of *Wallemia sebi* in aerosols by real-time PCR, conventional PCR, and cultivation. *Appl. Environ. Microbiol.* 70:7295–7302.
20. Zeng, J.S. et al. (2007). Spectrum of clinically relevant *Exophiala* species in the United States. *J. Clin. Microbiol.* 45(11):3713–3720.

Part III

Entomohpthoromycotina and Mucoromyotina

80 *Apophysomyces*

Arunaloke Chakrabarti, Shiv Sekhar Chatterjee, and Varghese K. George

CONTENTS

80.1 INTRODUCTION

Apophysomyces elegans, an emerging human pathogen, is an elegant zygomycotic fungus with an impressive bell- or funnel-shaped distinct apophyses.

Misra et al., in 1979, isolated this fungus from soil specimens from a mango orchard in northern India.[1] The first isolation from human was reported by Ellis and Ajello in 1982. They isolated this fungus from the bronchial washings of a patient from the United States.[2] Two more cases were reported around the same time when *A. elegans* was isolated from traumatic lesions of limbs.[3,4] However, the first comprehensive report of such infection is attributed to the study by Wieden et al.[5] They reported *A. elegans* cutaneous infection on the lower limb of a diabetic patient. The patient survived after a below-knee amputation followed by amphotericin B therapy.[5]

80.1.1 Classification and Morphology

Taxonomical position[6]

Domain Eukaryota
Kingdom Fungi
Phylum Zygomycota
Subphylum Zygomycotina
Class *Zygomycetes*
Order *Mucorales*
Family *Mucoraceae*

The colonies of *A. elegans* on Sabouraud dextrose agar (SDA) are fast growing, covering the agar plate over 7 days at 25°C.[5–12] The colonies are floccose with erect mycelial growth, white when young, turning cream white to yellow on aging. However, the growth is non-pigmented on reverse. Colonies often fail to sporulate on routine SDA medium or on deficient media like corn meal agar and potato dextrose agar (PDA). The fungus grows abundantly without sporulation at 25°C, 37°C, and 42°C (identification point), though occasionally[9] it may sporulate at the higher temperatures. It can grow up to 45°C.

The sporulation is best induced by growth on 1% water agar[2] or by the modified method of Padhye and Ajello, 1988.[7] According to the Ellis and Ajello recommendation, fungus is grown on corn meal-glucose-sucrose-yeast extract agar at 30°C for 7 days, and then 3 mm^3 agar blocks permeated with hyphae are cut aseptically and placed on the surface of sterilized 1% water agar in a petri dish. This plate is incubated at 30°C for 5–7 days to initiate sporulation. According to the modified method of Padhye and Ajello, agar blocks permeated with hyphal growth of the isolate on SDA are cut aseptically and transferred on to a Petri plate containing sterile distilled water. Three drops of 10% yeast extract solution are added to the water and incubated at 37°C for 15 days. Most isolates sporulate within 10–12 days.

On lactophenol cotton blue mount, coenocytic, broad, sparsely septate hyphae typical of *Zygomycetes* is found. Rhizoids are thin walled, subhyaline, and predominantly unbranched. The sporangiophores arise at right angles to the

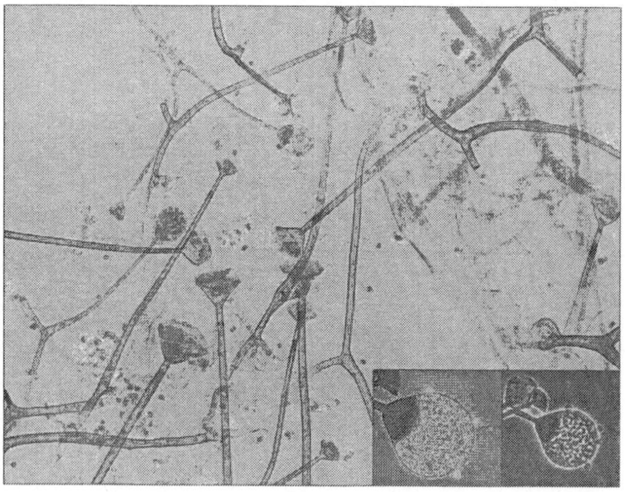

FIGURE 80.1 Thick-walled, darkly pigmented sporangiophore with champagne glass-shaped apophysis and sporangiospores ×200 (insert X400).

aerial hyphae and often have a septate basal segment resembling the "foot cell" seen in *Aspergillus* spp. The sporangiophores are unbranched, straight, or curved, slightly tapering toward the apex, nearly 200 µm long (100–200 µm in young cultures, and 200–300 µm in mature cultures), 3–5 µm in width near the apophyses, hyaline when young and turning into sepia or brown in older cultures. A conspicuous pigmented subapical thickening (10–16 µm below the apophyses), which constricts the lumen of the sporangiophore, is a distinctive feature. The sporangia are multi-spored, small (20–50 µm in diameter), typically pyriform in shape, hyaline at first, sepia colored when mature, columellate, and strongly apophysate. The columella are hemispherical, and the apophyses distinctly funnel or bell shaped giving the sporangia a characteristic "cocktail glass" look. The sporangiospores are smooth-walled, mostly oblong, occasionally subglobose (3–4 × 5–6 µm), and subhyaline to sepia *en masse*. *A. elegans* can be easily distinguished from other members of the family *Mucoraceae*, especially the morphologically similar *Absidia corymbifera*, by growth at 42°C, bell-shaped apophyses, hemispherical columella, a distinctive foot cell and the pigmented subapical thickening in the sporangiophore (Figure 80.1).

Like other fungi, *Apophysomyces elegans* can be preserved for long periods in soil, on oil-covered slants, at −20°C to −70°C, in liquid nitrogen and by lyophillization.[13] This fungus can even be preserved most conveniently on PDA slants at −70°C for a prolonged period of time.

80.1.2 Ecology and Epidemiology

A. elegans is found abundantly in tropics and subtropical areas. Following the first isolation in India from soil samples obtained from a mango orchard,[1] it was isolated from soil and dust of air filters in north Australia.[8,14] *A. elegans* may produce infection by traumatic inoculation of the fungus in soil-contaminated wounds, particularly after accidents and burns. Few infections are reported as a result of traumatic inoculation of cactus spines, other plant materials, and insect parts, which have presumably been contaminated with soil.[15–17] Interestingly, this agent has also caused fatal infections in marine animals.[18]

Disease due to *A. elegans* has been considered rare, but not so rare in tropical and subtropical climates.[5,19,20] Cases have been documented from southern United States, India, North and Western Australia, Mexico, Caribbean islands, Columbia, and Venezuela. Of nearly 100 cases published in the literature, a major proportion (~60%) of cases was reported from India (Table 80.1).[9,12,19–32] Next in frequency was the United States (~20% cases). Like other zygomycosis, *A. elegans* infection has been documented predominantly in males, though the reason is not known.

Zygomycosis due to *A. elegans* is often a disease of the immunocompetent host.[15,19,33,34] Trauma is the single most common (72%) factor associated with cutaneous *A. elegans* infections[5,19] and can range from trivial ones to severe road traffic accidents with multiple fractures. However, contact with soil or plant material is observed in many cases of traumatic *A. elegans* infections. Of note, two tsunami victims with extensive polytrauma have been reported with severe and multifocal *A. elegans* cutaneous infections.[35,36] Cutaneous zygomycosis due to *A. elegans* acquired in the hospital is often associated with burns, intramuscular injections,[19,23] and surgical wounds.[21] Diabetes mellitus,[9,19,33,37] corticosteroid therapy, hematological malignancy, myelofibrosis,[31] renal transplantation,[38] and chronic alcoholism[19] are other predisposing factors noted for the development of *A. elegans* infection. Extensive burn wounds (more than 25% of body surface area) are especially associated with zygomycotic infection. Contact with soil during the act of extinguishing the flames is possibly the reason for acquiring *A. elegans* infection in few patients. In one such case *A. elegans* has been isolated from the implicated soil.[8] Rhino-orbito-cerebral (ROC) zygomycosis due to *A. elegans* has been recorded in farm laborers, and in persons with a history of intimate contact with soil.[39,40] Pregnancy has been associated on two occasions.[21,41] Due to the rarity of the *A. elegans* infection and the lack of case control studies, the risk factors are only empirically understood.

80.1.3 Pathogenesis and Clinical Features

In the more common scenario of traumatic *A. elegans* zygomycosis, route of entry is through the skin.[5,8,42,43] The traumatic episode can range from an innocuous insect and spider bite/sting,[15,33] skin patch test,[17] intramuscular injections,[19] burns, blunt trauma,[19] surgical wounds,[12] and road traffic accidents.[20] Contact with soil and vegetation containing the fungus greatly enhances the chances of *A. elegans* infection.[8,16,21,33,43] Even inhalation of soil particles contaminated with *A. elegans* has been proposed to be the mode of entry of the organism in a case of ROC *A. elegans* infection.[39] After gaining entry into the body, *A. elegans* may invade blood vessels and disseminates to other organs like other zygomycotic infection.[33] Contiguous spread of *A. elegans*

TABLE 80.1
***Apophysomyces elegans* Infections Reported in Human**

Reference	Number of Patients	Clinical Type	Location	Risk Factors	Treatment	Outcome
Ellis and Ajello[2]	1	Pulmonary (bronchial washings)	USA	Information not available	Information not available	Information not available
Winn et al.[3]	2	Cutaneous	Texas, USA	Trauma	Amp-B, extensive debridement	Survived—1, Died—1
Winn et al.[4]	1	Cutaneous	Texas, USA	Trauma	Amp-B, extensive debridement	Survived
Wieden et al.[5]	1	Cutaneous	Arizona, USA	Diabetes mellitus, vascular disease, trauma	Amp-B, amputation	Survived
Lawrence et al.[46]	1	Renal	Texas, USA	None	Amp-B, nephrectomy	Survived
Newton et al.[59]	1	Cutaneous	Mississippi Gulf Coast, USA	Trauma	Amp-B, hyperbaric oxygen, extensive debridement	Recovered
Cooter et al.[8]	1	Cutaneous	Northern Territory, Australia	Burn, soil contact	Amp-B, amputation	Survived
Huffnagle et al.[61]	1	Cutaneous	Texas, USA	Burn, soil and vegetation contact	Amp-B, multiple amputations	Survived
McGinnis et al.[62]	1	Cutaneous (*necrotizing fasciitis*) later spread to lungs, kidney, aortic wall, cecum, appendix, vertebrae	—	Trauma	Surgical debridement	Died
Weinberg et al.[33]	1	Cutaneous	Florida, USA	Bite/sting	Amp-B, resection	Survived
Lakshmi et al.[12]	1	Cutaneous	Andhra Pradesh, India	Surgery (trauma)	Amp-B, extensive debridement	Died
Okhuysen et al.[42]	1	Cutaneous with contiguous spread to kidney	Texas, USA	Trauma	Amp-B, multiple debridement, nephrostomy	Survived
Eaton et al.[11]	1	Cutaneous	Florida, USA	Trauma due to cut end of a tree branch	Amp-B, extensive debridement	Survived
Meis et al.[10]	1	Cutaneous	Aruba, Netherland Antilles, Caribbean Islands	None	Amp-B, amputation	Survived
Radner et al.[34]	1	ROC	Guadalajara, Mexico (treated in California, USA)	Trauma (fall into a ditch)	Amp-B, orbital exenteration, maxillectomy, multiple debridement	Survived
Naguib et al.[38]	1	Cutaneous	Oklahoma, USA	Trauma, renal transplant, immunosuppressive agents	Amp-B lipid complex, debridement	Survived
Chugh et al.[32]	1	Renal	Chandigarh, India	None	Amp-B, bilateral nephrectomy	Death
Chakrabarti et al.[23]	1	Cutaneous	Chandigarh, India	Trauma (intra-muscular injection)	Surgical debridement	Recovered
Chakrabarti et al.[31]	1	Craniofacial (ROC)	Chandigarh, India	Myelofibrosis, hyperglycemia, thrombocytopenia	Amp-B, orbital exenteration, extensive debridement	Died
Mathews et al.[21]	1	Cutaneous	Vellore, South India	Pregnancy, surgery (trauma)	Amp-B, extensive debridement	Survived
Sedralis et al.[14]	1	ROC	Australia	None	Amp-B, orbital exenteration, multiple (7) debridement, hyperbaric oxygen	Survived
Cáceres et al.[43]	1	Cutaneous	Caracas, Venezuela	Trauma	Amp-B, itraconazole, debridement	Survived

(continued)

TABLE 80.1 (continued)
***Apophysomyces elegans* Infections Reported in Human**

Reference	Number of Patients	Clinical Type	Location	Risk Factors	Treatment	Outcome
Burrell et al.[16]	1	Cutaneous with contiguous spread to kidneys, spleen, diaphragm, colon, and retroperitoneal tissues	Arizona, USA	Trauma (cactus spine)	Lip Amp-B, multiple (7) debridement	Survived
Brown et al.[40]	1	ROC	Arizona, USA	Contact with soil (chiseling of soil), steroids	Amp-B, bilateral debridement of maxillary sinuses	Survived
Kimura et al.[60]	2	Patient 1—cutaneous with widespread involvement of viscera (kidney, spleen, tail of pancreas, left colon, jejunum)	Osaka-Sayama, Japan	Trauma	Amp-B, itraconazole, extensive debridement	Died
		Patient 2—cutaneous (leg wound)		Alcoholic cirrhosis, trauma, contact with soil	Amp-B changed to lip Amp-B, multiple debridement, hyperbaric oxygen	Survived
Fairley et al.[63]	1	ROC	Brisbane, Australia	Water-jet injury while cleaning air-conditioner	Lip Amp-B, local Amp-B, orbital exenteration, multiple debridement (2)	Survived
Garcia-Covarrubias et al.[58]	1	ROC	Columbia, USA	Polytrauma with facial fractures	Lip Amp-B, orbital exenteration, multiple debridement, hyperbaric oxygen, G-CSF	Survived
Chakrabarti et al.[22]	3	ROC	Chandigarh, India	Information not available	Information not available	Information not available
	4	Cutaneous				
	1	Renal				
Page et al.[64]	2	Cutaneous	Western Australia	Case-1—diabetes, obesity	Amp-B, wound debridement, below knee-amputation	Death
				Case-2—none	Lip Amp-B, followed by Amp-B, multiple debridement	Survived
Blair et al.[66]	1	Cutaneous	Arizona, USA	Trauma with plant material (patch test)	Lip Amp-B, debridement	Survived
Wang et al.[65]	1	Cutaneous (with disseminated spread)	Texas, USA	Alcoholic cirrhosis, trauma, renal failure	Amp-B, extensive debridement	Died
Lesueur et al.[17]	1	Cutaneous	Arizona, USA	Contact with snapdragon plant (*Antirrhinum majus*) (patch test), diabetes mellitus	Lip Amp-B, multiple debridement	Survived
Chakrabarti et al.[19]	8	ROC-1	Chandigarh, India	None	Amp-B, orbital exenteration, extensive debridement	Survived
		Cutaneous—3		Intramuscular infections-2, unknown-1	Local debridement, amp-B-1, Itraconazole-1, No antifungal-1	Survived-2, Left against medical advice-1
		Renal—3		Alcoholism—1	Amp-B (2 patients), Surgical drainage (2 patients), local debridement (1 patient)	Survived-2, Died-1

		Cutaneous extending to kidney and spleen—1		Trauma	Amp-B, local debridement	Died
Carter and Ulusarac[41]	1	Cutaneous	Alabama, USA	Trauma, pregnancy	Amp-B, multiple debridement, amputation	Died
Kordy et al.[45]	1	Cutaneous	Riyadh, Saudi Arabia	Trauma	Lip Amp-B, multiple debridement	Survived
Ruiz et al.[67]	1	Cutaneous (necrotizing fasciitis)	Medellin, Colombia	Trauma	Information not available	Died
Andresen et al.[36]	1	Multifocal cutaneous	Sri Lanka (reported from Australia)	Trauma (tsunami), soil contact (found in a paddy field)	Lip Amp-B, debridement, hyperbaric oxygen	Survived
Pitisuttithum et al.[47]	1	Brain abscess (right parietal, mixed infection with a basidiomycetes)	Information not available	Severely immunocompromised (not stated precisely)	Lip Amp-B changed to posaconazole	Died
Sridhara et al.[30]	3	ROC	Chandigarh, India	None	Amp-B, orbital exenteration, extensive debridement	All survived
Schütz et al.[39]	1	ROC	Kuwait	Farm laborer (probable contact with soil), steroid	Amp-B, local Amp-B, orbital exenteration, extensive debridement (delayed)	Died
Liang et al.[37]	1	ROC	Minnesota, USA	Diabetes mellitus, corticosteroid therapy	Lip Amp-B, orbital exenteration, maxillectomy, sphenoidectomy, multiple debridement	Survived
Jain et al.[29]	5	Cutaneous (necrotising fascitis)	Chandigarh, India	Trauma-2 cases, burn-1 case, none-2 cases	Information not available	Information not available
Suryanarayan Rao et al.[27]	5	ROC, 1 patient also had a para-pharyngeal abscess	Chandigarh, India	None	Amp-B, orbital exenteration, debridement	Survived-4, Died-1
Chakrabarti et al.[28]	1	ROC	Chandigarh, India	Information not available	Information not available	Information not available
	12	Cutaneous				
	3	Renal				
	1	Disseminated				
Kindo et al.[20]	1	Cutaneous	Tamil Nadu, India	Trauma	Amp-B, extensive debridement	Died
Goyal et al.[26]	1	Cutaneous (cervico-facial) with salivary gland involvement	Lucknow, India	Diabetes mellitus, trauma	Lip Amp-B, debridement	Died
Snell and Tavakoli[35]	1	Cutaneous	Thailand	Trauma (tsunami survivor)	Lip Amp-B, hyperbaric O_2, debridement, split-skin graft	Survived
Ferguson et al.[57]	1	ROC	Georgia, USA	Diabetes mellitus, trauma	Lip Amp-B changed to posaconazole, multiple debridement (2)	Survived
Reddy et al.[9]	1	Cutaneous	Andhra Pradesh, India	Diabetes mellitus, trauma	Amp-B, itraconazole, extensive debridement	Survived
Thomas et al.[24]	1	Renal	Tamil Nadu, India	None (except hypertension)	Amp-B, nephrectomy	Survived
Devi et al.[25]	1	Cutaneous	Pondicherry, India	Trauma (tractor accident)	Amp-B, multiple debridement (2)	Survived
Saravia-Flores et al.[15]	1	Cutaneous	Guatemala	Trauma (spider bite)	Amp-B, multiple debridement, amputation leg	Died

Note: ROC, Rhino-orbito cerebral; Lip Amp-B, Liposomal Amphotericin B; Amp-B, Amphotericin B.

through tissue layers has also been documented on multiple occasions.[33] Rapid spread of lesions is a reason provided for repeated debridement and amputations in cases of *A. elegans* necrotizing fasciitis.[33]

The overwhelming association of cutaneous zygomycosis and *A. elegans* has been hypothesized on the basis of the thermophillic nature of the fungus. Rapid growth at higher temperature has been suggested as a reason behind successful proliferation of *A. elegans* from cutaneous to deep tissues in a febrile host. However, proof to such hypothesis is lacking.

Metabolic acidosis and local ischemia contribute to the deterioration of infectious process. Along with *A. elegans*, *Staphylococcus* spp., *Escherichia coli*, *Pseudomonas aeruginosa*, *Enterobacter* spp., and *Enterococcus* spp. are sometimes isolated from the culture of the necrosed tissue.[25,33,42] Around one-third of patients with *A. elegans* infection succumb to illness, though the patient may be immunocompetent.[39] Poor general condition of those patients due to burns, trauma, surgical procedure, and general anesthesia possibly contributes to such high mortality.[39] No specific virulence factor of this fungus has been identified till date. Experimental animal studies indicate that even a 250–2300-fold lower size inoculum of *A. elegans* can match the mortality due to *R. microsporus* and *Absidia corymbifera* infections.[44]

Cutaneous zygomycosis: *A. elegans* commonly involves cutaneous and subcutaneous tissue. Of all reported cases with *A. elegans* infection, nearly 60% had cutaneous involvement (Table 80.1). Cutaneous zygomycosis may be divided into primary cutaneous zygomycosis and those secondarily involved in disseminated zygomycosis. In primary cutaneous zygomycosis, the route of infection is trans-cutaneous (infections at sites of injury—intravenous access, traumatic inoculation, adhesive tapes, occlusive dressing, burn wound infections, and postoperative surgical wound infections); while in secondary cutaneous zygomycosis, infection spreads to skin either hematogenously (in cases of disseminated zygomycosis), or by contiguous spread from underlying diseases. The secondary cutaneous zygomycosis due to *A. elegans* is not clearly established. Primary cutaneous zygomycosis is by far the most common manifestation of *A. elegans* disease. Traumatic implantation of the agent is often the initiator of the disease.[19,43] Iatrogenic introduction of the fungus has been documented following surgery[12]. An erythematous lesion with raised, darkened, tender eschar at the center usually appears within a few days of inoculation.[33,43] Extensive damage of soft tissue, local ischemia, and metabolic acidosis characterize such infections.[39,43] Sloughing, purulent discharge, and indurated margins are observed around wounds, which are grossly infected with multiple microorganisms. Later white cheesy material may come out from pockets of such lesions following antibiotic therapy.[25] Obvious mold-like growth over the lesion is an interesting finding.[43] Leucocytosis with neutrophilia is commonly detected on peripheral blood smear examination.[9,20,45] Cutaneous lesions often extend to produce extensive necrotizing fascitis[5] and osteomyelitis.[10,11] Contiguous spread may also involve kidneys, spleen, diaphragm, and other visceral organs. Initial diagnosis in most of those cases was bacterial infection[45] and was treated empirically with parenteral antibiotics. Fungal etiology was suspected after failure in response to antibiotic therapy and repeated debridement. Direct microscopy and culture of the debrided tissue often help in diagnosis. Cutaneous zygomycosis due to *A. elegans* may lead to fatality because of delay in diagnosis and management.[39]

ROC zygomycosis: ROC zygomycosis was involved in nearly 25% patients with *A. elegans* infection (Table 80.1). ROC zygomycosis due to *A. elegans*, unlike that due to other agents like *Rhizopus* spp., is a disease occurring mainly in immunocompetent individuals.[27,34,37] However, myelofibrosis, systemic corticosteroid therapy, and diabetes mellitus have been documented in few patients.[19,37] Majority of reported cases are adult males. ROC zygomycosis due to *A. elegans* may manifest in two situations: (i) patients have traumatic inoculation of *A. elegans* in the orbit and face, which leads to orbital cellulitis and further spread into the contiguous areas of the brain and sinuses[39]; and (ii) *A. elegans* enters through nasal or sinus mucosa and spreads into the orbit and brain. In a farm laborer, inhalation of soil particles contaminated with *A. elegans* was suspected as the route of inoculation.[39] Patients typically present with facial pain, and swelling, nasal congestion, rhinitis, frontal headache, redness and swelling of eyes, periorbital pain, chemosis, diplopia, impaired ocular motility, decreased vision, ptosis, facial drooping, numbness, and fever.[37,39] Tenderness, erythema, necrosis, periorbital echymoses, periorbital edema, and proptosis have been noted during physical examination.[37] Presenting signs and symptoms of the disease are thus similar to the classic features of ROC fungal infection due to other *Zygomycetes*.

Renal zygomycosis: *A. elegans* renal infections may present as isolated renal zygomycosis and have been documented in 12% cases (Table 80.1). Lawrence et al. described the first case of isolated renal *A. elegans* infection in 1986[46]. The infection may spread contiguously from skin and subcutaneous tissue.[16,19,42] In a patient with cactus spine injury extensive lesions developed involving the retroperitoneum, spleen, left diaphragm, colon and kidney.[16] Another case of *A. elegans* renal infection (of the contralateral kidney) has been described following extensive traumatic cutaneous zygomycosis of the right flank and retroperitoneal tissue.[19] Most patients with isolated renal disease due to *A. elegans* have been documented from different centers in India.[19,24,32] The typical patient is a young adult male, previously healthy, with no known risk factors.[19,24] However, chronic alcoholism and systemic corticosteroid therapy as predisposing factors were found in few patients.[19] Clinical features of isolated renal zygomycosis due to *A. elegans* include flank pain (unilateral or bilateral), weight loss, abdominal distension, pedal edema, uremia, dysuria, pyuria, passage of whitish flakes per urine, and fever.[19,24]

Pulmonary zygomycosis: No case of pulmonary zygomycosis due to *A. elegans* had been reported except for the isolation of the fungus from bronchial washings of a patient with respiratory tract infection.[2]

Brain abscess: Mixed infection due to *A. elegans* and a *Basidiomycetes* spp. has been implicated to cause fatal brain abscess in a severely immunocompromised adult male patient.[47]

Other infection sites: Goyal et al. reported a patient who developed otogenic cervico-facial zygomycosis with salivary gland abscess due to *A. elegans*.[26]

80.1.4 Diagnosis

80.1.4.1 Conventional Techniques

Early clinical suspicion of zygomycotic infection is most important for management. Otherwise, the disease may disseminate to many organs resulting in mortality. Tissue specimens are the best samples and superficial swabs are not recommended. Direct examination of potassium hydroxide wet mount of the tissue, and histopathological examination help in early diagnosis. A black necrotic eschar on nasopharynx/palate/skin may be the clue to diagnosis. Material from the eschar and underlying tissue should be obtained. Biopsy or curettage samples are transported quickly to the laboratory. For renal and pulmonary lesions, ultrasound and computed tomography-guided aspiration from the site of lesion may be useful. Both histopathology and isolation of fungi should be attempted from the sample. Calcofluor white stain may help in detection of broad, sparsely septate hyphae on fluorescent microscope. Various special fungal stains like periodic acid schiff stain (PAS) and gomori's methanamine silver stain (GMS) may be performed on histopathology sections. These allow better identification of the wide (3–25 µm wide), right angled branching, sparsely septate, ribbon like hyphae in tissue. Frequently these hyphae have focal bulbous dilatation and non-dichotomous, irregular branching. Generally, surrounding reaction is often acute inflammatory or necrotic. The fungi are stained black with GMS and red with PAS stain. Zygomycetes may not take up special fungal stain (PAS, GMS) as deeply as other filamentous fungi, but may be detected in tissue sections stained with hematoxylin and eosin stain (H&E). Hyphae invading tissue and blood vessels with or without thrombosis are often observed. Differential diagnosis of *A. elegans* on histopathological examination include other members of the class Zygomycetes, namely *Rhizopus*, *Absidia*, *Mucor*, *Cuninghamella*, *Conidiobolus* and *Basidiobolus* spp. Hence fungal culture is recommended in all cases. The fungus grows within 4–6 days when adequate samples are inoculated on SDA. Importantly, tissue samples should be minced and not crushed before inoculation. The necrotic material may be gently teased apart with scissors in a sterile Petri plate. Upon primary isolation, the fungus may not sporulate; hence special techniques like water agar may be required to induce sporulation.

Serologic methods had been attempted for diagnosis of zygomycosis including immunodiffusion and indirect ELISA.[48,49] Most of these serological studies had been done using different antigen preparations of the more common zygomycetes namely *Rhizopus arrhizus*, *Absidia corymbifera*, and *Rhizomucor pusillus*. No such test has been documented in *A. elegans* infection.

As the fungus fails to sporulate on routine media, exoantigen testing may provide a simpler and more rapid method of identification. A simple immunodiffusion test has been attempted using exo-antigens of *A. elegans*.[50] Three specific *A. elegans* precipitins were produced when *A. elegans* antisera was absorbed with *S. vasiformis*. *A. elegans* antigen had little cross-reaction with those of *S. vasiformis* and *A. corymbifera*.

80.1.4.2 Molecular Techniques

Identification. Molecular techniques have also been employed for the identification of *A. elegans* using database of 18S and 28S ribosomal DNA sequences of 42 isolates of zygomycetes including *A. elegans*.[51] By using the aligned 28S rDNA sequences, 13 taxon-specific PCR primer pairs were designed, which correctly identified those 42 isolates. However, these primers should be evaluated on a larger number of zygomycetes before its recommendation for routine use. Chakrabarti et al.[19] analyzed the internal transcribed spacer (ITS) region of eight *A. elegans* isolates. PCR-RFLP of the ITS regions of *A. elegans* ribosomal DNA using restriction enzymes *Mbo*I, *Msp*I, or *Hinf*I distinguished this fungus from other common zygomycetes like *Rhizopus oryzae*, *Saksenaea vasiformis*, *Rhizomucor pusillus*, *Mucor cercinelloides*, and *Basidiobolus ranarum*.

Real time PCR assay was attempted for the detection of zygomycetes both in culture and tissue specimens.[52] In that assay a 167 bp conserved region (found in *Absidia*, *Apophysomyces*, *Cunninghamella*, *Mucor*, *Rhizopus*, and *Saksenaea*) of the multicopy *cytochrome b* gene was targeted. The identification at the Genus level could be achieved in approximately 4 h. The assay did not show any cross reaction with other fungi like *Candida*, *Aspergillus*, *Fusarium* or even bacteria. Future studies in this area may help in development of molecular diagnosis of *A. elegans* infection.

Typing. Chakrabarti et al. attempted to type 12 *A. elegans* strains using microsatellite primers. The method was not discriminatory, as 10 strains isolated at different times, had the same pattern, though 2 strains from the United States had a different pattern.[19]

80.1.5 Treatment

The best management of *A. elegans* infections is presumed to be aggressive surgical debridement combined with antifungal therapy and control of the predisposing factor where possible.[53–55] Surgery is considered necessary due to massive amount of tissue necrosis that may prevent entry of antifungal agents to the site of infection. Additionally surgery is supposed to minimize the fungal load in the tissue.[53,55,56] Extensive and multiple debridements may be necessary in invasive cutaneous *A. elegans* lesions.[41] Orbital exenteration and partial facial resection are often necessary in *A. elegans* ROC disease.[37,57] Amphotericin B (conventional or liposomal) is chosen as the first line of therapy.[44,53,55] In recent years posaconazole as a substitute for amphotericin B is gaining popularity especially as salvage therapy.[47,57] However, in-vitro susceptibility testing against multiple *A. elegans* isolates has not been performed. Control of predisposing factor or boosting the immunity with the use of cytokines may help to restore immune function.[53] Hyperbaric oxygen therapy has been used successfully in a few cases of cutaneous *A. elegans* infections; though, true value of this approach remains uncertain.[21,35,36,58–60]

80.2 METHODS

80.2.1 Sample Preparation

Whole cell DNA from each isolate is extracted as follows: the isolate is grown on SDA plates for 5–7 days. Once luxuriant growth is obtained, subcultures are made onto Sabouraud dextrose broth and incubated at 37°C in a rotating shaker-incubator at 120 rpm for 3–5 days. The mycelial mat is recovered by filtration and washed with normal saline. About 0.2–0.3 g of the mycelial mat is placed in a clean mortar. Liquid nitrogen is added to the mat and ground quickly with a pestle till a fine powder is obtained. The resultant powder is then transferred to a clean sterile 1.5 mL microcentrifuge tube to which 600 µL of lysis buffer (100 mM Tris HCl pH 8.0, 50 mM EDTA, 3% SDS) is added and vortexed briefly. Twenty microliters proteinase K at a concentration of 20 mg/mL is then added and incubated at 56°C for 1 h. Finally, DNA is extracted using 600 µL phenol:chloroform (25:24). After centrifugation (10,000 rpm for 5 min), equal volume of chloroform:isoamylalcohol (24:1) is added to the aqueous phase and mixed thoroughly. The mixture is centrifuged at 10,000 rpm for 5 min and the aqueous phase transferred to a microcentrifuge tube and DNA is precipitated with 1 mL isopropanol and 10 µL 3 M sodium acetate. The pellet is then washed with 70% alcohol, treated with 100 µL of Tris-EDTA-RNase (10 mM Tris HCl pH 7.5, 1 mM EDTA, 50 µg of RNase per mL) at 37°C for 2 h, re-precipitated with absolute alcohol, and finally dissolved in 100 µL of TE (10 mM Tris HCl pH 7.5, 1 mM EDTA) and stored at −20°C for further use.

80.2.2 Detection Procedures

80.2.2.1 PCR Identification

After confirming the purity of the DNA obtained by gel electrophoresis, the desired genes are amplified by PCR.

Primers. At first the ITS region is amplified in order to carry out any further analysis of the DNA. The ITS region has been used as an important tool for phylogenetic and epidemiological studies because of its moderate rate of evolution. The ITS primers are designed from the conserved regions of the 18S and 28S rRNA genes, respectively. The ITS regions are amplified with two primers namely the ITS1 (forward) and the ITS4 (reverse): pITS1F (5′-TCCGTAGGTGAACCTGCGC-3′) and pITS4R (5′-TCCTCCGCTTATTGATATGC-3′).

PCR setup. The total volume per reaction mixture is usually 10, 25 or 50 µL. All the reaction components can be mixed together in a 0.5 mL PCR tube in any sequence except for DNA polymerase, which should be added at the end. It is recommended to mix all the components just before PCR cycling. Positive control (DNA from known *A. elegans* isolate) and negative control (sterile double-distilled water [DDW]) must always be tested in each run.

Each PCR mixture (50 µL) is prepared by adding the following components:

5× PCR buffer with $MgCl_2$	10 µL
Primers forward and reverse	10–30 pmol of each primer
dNTPs	10 mM
DNA polymerase	1.5 U (0.5 µL of 3 U/µL)
Template DNA	~500 ng
DDW up to	50 µL

PCR cycling. A common PCR program starts with an initial denaturation step at 95°C for 2–5 min followed by 30 cycles of denaturation at 94°C for 1 min, primer annealing at 45°C (decided by primer Tm) for 1 min and primer extension at 72°C for 1 min followed by a final extension step at 72°C for 5–10 min.

Verifying PCR amplification. About 2–5 µL of PCR product is run on a 1%–2% agarose gel to check for the presence of fungal specific DNA in the sample. Molecular weight markers, positive and negative control samples are also run together. The gel is visualized by staining with ethidium bromide. A successful PCR amplification should display a specific band with the expected size without nonspecific bands and smear.

Some important facts. (i) The quality and concentration of DNA templates can directly affect outcome of PCR; (ii) Proper concentration (usually 0.1–1 µM) and appropriate primers are very critical for successful PCR. The optimal primer size is usually between 18–20 bases. Both forward and reverse primers should have melting temperature within limits of 2°C–5°C of each other. Avoid complementary sequences between primers to reduce primer dimer formation. Primers with Tm higher than 50°C will generally give more specific results; and (iii) Magnesium concentration is critical to the success rate of PCR amplification. It may affect DNA polymerase activity. Excess magnesium results in accumulation of nonspecific products. Desirable magnesium concentration is at 0.5–2 mM.

80.2.2.2 Restriction Fragment Length Polymorphism

The technique may be used to detect macroevolutionary changes in genomes. After PCR, the ITS amplified products are subjected to digestion using different restriction enzymes. For *A. elegans*, restriction enzymes like *Mbo*I, *Msp*I, *Hinf*I, *Bam*HI, and *Bgl*II have been used in our laboratory. The restriction mixture (30 µL), which contains 3 µL of 10× buffer, 1 µL (10–12 U) of restriction enzymes, 10 µL of amplified product and 16 µL of DDW is incubated at 37°C for 3 h. After digestion, 15 µL of each digested product is electrophoresed in a 1.5%–2.5% agarose gel for 30 min to 2 h depending on product size. The bands are visualized by ethidium bromide staining for 1 h in the dark and destaining in 1× TAE (Tris-acetate-EDTA) buffer for 1 h.

80.2.2.3 Microsatellite Fingerprinting

High discriminatory value and reproducibility of microsatellite fingerprinting technique makes it a much better typing method than RFLP. For *A. elegans*, a single primer, pMS1

([GTG]$_5$ or pMS2 [GAC]$_5$) may be used. All components of PCR mixture are the same as mentioned previously for ITS amplification. The cycling parameters used for microsatellite fingerprinting are; initial denaturation for 5 min at 94°C, followed by 40 cycles of 1 min at 94°C, 2 min at 45°C (for pMS1) or 55°C (for pMS2), and 3 min at 72°C, final extension for 10 min at 72°C and cooling at 4°C. The amplified product is then separated by gel electrophoresis and analyzed.

80.3 CONCLUSION

Apophysomyces elegans, an emerging pathogen belonging to class *Zygomycetes* and the order *Mucorales*, is a relatively newer agent in this order, being isolated from soil, dust, and vegetation matter in tropical and subtropical countries. By far, the agent most commonly causes cutaneous zygomycosis in apparently healthy hosts and traumatic implantation into skin following contact with soil and vegetation containing the fungus is considered the route of entry. However, in recent years rhino-cerebral and renal zygomycosis due to *A. elegans* are increasingly reported in immunocompetent or immunosuppressed individuals. The mode of entry in those infections is not clearly understood. Majority of the cases with *A. elegans* infection have been reported from India, southern United States, and Australia. Diagnosis of the disease is established by histopathology, direct microscopy, and culture techniques; and the management protocol includes aggressive surgical debridement combined with systemic antifungal therapy. Although excellent mycelial growth is seen on standard culture media, *A. elegans*, unlike other zygomycetes, does not readily produce asexual spores. A special nutrient-deficient growth medium, a high temperature and prolonged incubation can be used to induce *A. elegans* isolates to sporulate. Given this problem of delay in identification, DNA-based molecular typing techniques show enormous potential for rapid and accurate identification of the fungus. Molecular techniques are also developed for rapid detection in tissue samples and typing of the fungus.

REFERENCES

1. Misra, P.C., Srivastava, K.J., and Lata, K., *Apophysomyces*, a new genus of the Mucorales, *Mycotaxon*, 8, 377, 1979.
2. Ellis, J.J. and Ajello, L., An unusual source for *Apophysomyces elegans* and a method for stimulating sporulation of *Saksenia vasiformis*, *Mycologia*, 74, 144, 1982.
3. Winn, R.E., Ramsey, P.D., and Adams, E.D. Jr., Traumatic mucormycosis secondary to *Apophysomyces elegans*, a new genus of Mucorales [abstract], *Clin. Res.*, 30, 382A, 1982.
4. Winn, R.E. et al., Traumatic zygomycosis due to *Apophysomyces elegans*, a new genus of the Mucorales [abstract no F-29], in: *Proceedings of the 84th Annual Meeting of the American Society for Microbiology* (*St. Louis*), Washington, DC, Am. Soc. Microbiol., p. 297, 1984.
5. Wieden, M.A. et al., Zygomycosis caused by *Apophysomyces elegans*, *J. Clin. Microbiol.*, 22, 522, 1985.
6. Ellis, D.H., Systemic zygomycosis, in *Topley and Wilson's Microbes and Microbial Products*, 10th edn., Volume 6: Medical Mycology, Merz, W.G., Hay, R.I. (Eds.), ASM Publisher, Washington DC, pp. 662–663.
7. Padhye, A.A. and Ajello, L., Simple method of inducing sporulation by *Apophysomyces elegans* and *Saksenaea vasiformis*, *J. Clin. Microbiol.*, 26, 1861, 1988.
8. Cooter, R.D. et al., Burn wound zygomycosis caused by *Apophysomyces elegans*, *J. Clin. Microbiol.*, 28, 2151, 1990.
9. Reddy, I.S. et al., Primary cutaneous mucormycosis (zygomycosis) caused by *Apophysomyces elegans*, *Indian J. Dermatol. Venereol. Leprol.*, 74, 367, 2008.
10. Meis, J.F. et al., Severe osteomyelitis due to the zygomycete *Apophysomyces elegans*, *J. Clin. Microbiol.*, 32, 3078, 1994.
11. Eaton, M.E. et al., Osteomyelitis of the sternum caused by *Apophysomyces elegans*. *J. Clin. Microbiol.*, 32, 2827, 1994.
12. Lakshmi, V. et al., Zygomycotic necrotizing fasciitis caused by *Apophysomyces elegans*, *J. Clin. Microbiol.*, 31, 1368, 1993.
13. Pasarell, L. and McGinnis, M.R., Viability of fungal cultures maintained at −70°C, *J. Clin. Microbiol.*, 30, 1000, 1992.
14. Sedralis, T., Krishnan, S., and Holland J., Martini glass mucormycosis: *Apophysomyces elegans* infection in an immunocompetent host, *Aust. J. Otolaryngol.*, 2, 600, 1997.
15. Saravia-Flores, M., Guaran, D.M., and Argueta, V., Invasive cutaneous infection caused by *Apophysomyces elegans* associated with a spider bite, *Mycoses*, 53, 259, 2010.
16. Burrell, S.R. et al., *Apophysomyces elegans* infection associated with cactus spine injury in an immunocompetent pediatric patient, *Pediatr. Infect. Dis. J.*, 17, 663, 1998.
17. Lesueur, B.W., Warschaw, K., and Fredrikson, L., Necrotizing cellulitis caused by *Apophysomyces elegans* at a patch test site, *Am. J. Contact Dermat.*, 13, 140, 2002.
18. Robeck, T.R. and Dalton, L.M., *Saksenaea vasiformis* and *Apophysomyces elegans* zygomycotic infections in bottlenose dolphins (*Tursiops truncatus*), a killer whale (*Orcinus orca*), and pacific white-sided dolphins (*Lagenorhynchus obliquidens*), *J. Zoo Wildlife Med.*, 33, 356, 2002.
19. Chakrabarti, A. et al., *Apophysomyces elegans*: An emerging zygomycete in India, *J. Clin. Microbiol.*, 41, 783, 2003.
20. Kindo, A.J. et al., Fatal cellulitis caused by *Apophysomyces elegans*. *Indian J. Med. Microbiol.*, 25, 285, 2007.
21. Mathews, M.S., Raman, A., and Nair, A., Nosocomial zygomycotic post-surgical necrotizing fasciitis in a healthy adult caused by *Apophysomyces elegans* in south India, *J. Med. Vet. Mycol.*, 35, 61, 1997.
22. Chakrabarti, A. et al., Ten years' experience in zygomycosis at a tertiary care centre in India. *J. Infect.*, 42, 261, 2001.
23. Chakrabarti, A. et al., Primary cutaneous zygomycosis due to *Saksenaea vasiformis* and *Apophysomyces elegans*, *Clin. Infect. Dis.*, 24, 580, 1997.
24. Thomas, A.J. et al., *Apophysomyces elegans*—Renal mucormycosis in a healthy host: A case report from south India, *Indian J. Med. Microbiol.*, 26, 269, 2008.
25. Devi, S.C. et al., Favorable outcome of amphotericin B treatment of zygomycotic necrotizing fascitis caused by *Apophysomyces elegans*, *Int. J. Dermatol.*, 47, 407, 2008.
26. Goyal, A. et al., *Apophysomyces elegans* causing acute otogenic cervicofacial zygomycosis involving salivary glands, *Med. Mycol.*, 45, 457, 2007.
27. Suryanarayan Rao S. et al., Sinoorbital mucormycosis due to *Apophysomyces elegans* in immunocompetent individuals—An increasing trend, *Am. J. Otolaryngol.*, 27, 366, 2006.
28. Chakrabarti, A. et al., The rising trend of invasive zygomycosis in patients with uncontrolled diabetes mellitus, *Med. Mycol.*, 44, 335, 2006.

29. Jain, D. et al., Zygomycotic necrotizing fasciitis in immunocompetent patients: A series of 18 cases, *Mod. Pathol.*, 19, 1221, 2006.
30. Sridhara, S.R. et al., Mucormycosis in immunocompetent individuals: An increasing trend, *J. Otolaryngol.*, 34, 402, 2005.
31. Chakrabarti, A. et al., Craniofacial zygomycosis caused by *Apophysomyces elegans*. *Mycoses*, 40, 419, 1997.
32. Chugh, K.S. et al., Renal zygomycosis in otherwise healthy hosts, *J. Mycol. Med.*, 6, 22, 1996.
33. Weinberg, W.G. et al., Invasive infection due to *Apophysomyces elegans* in immunocompetent hosts, *Clin. Infect. Dis.*, 17, 881, 1993.
34. Radner, A.B., Witt, M.D., and Edwards, J.E. Jr., Acute invasive rhinocerebral zygomycosis in an otherwise healthy patient: Case report and review, *Clin. Infect. Dis.*, 20, 163, 1995.
35. Snell, B.J. and Tavakoli, K., Necrotizing fasciitis caused by *Apophysomyces elegans* complicating soft-tissue and pelvic injuries in a tsunami survivor from Thailand, *Plast. Reconstr. Surg.*, 119, 448, 2007.
36. Andresen, D. et al., Multifocal cutaneous mucormycosis complicating polymicrobial wound infections in a tsunami survivor from Sri Lanka, *Lancet*, 365, 876, 2005.
37. Liang, K.P. et al., Rhino-orbitocerebral mucormycosis caused by *Apophysomyces elegans*, *J. Clin. Microbiol.*, 44, 892, 2006.
38. Naguib, M.T. et al., *Apophysomyces elegans* infection in a renal transplant recipient, *Am. J. Kidney Dis.*, 26, 381, 1995.
39. Schütz, P. et al., Fatal rhino-orbito-cerebral zygomycosis caused by *Apophysomyces elegans* in a healthy patient, *J. Oral Maxillofac. Surg.*, 64, 1795, 2006.
40. Brown, S.R., Shah, I.A., and Grinstead, M. Rhinocerebral mucormycosis caused by *Apophysomyces elegans*. *Am. J. Rhinol.*, 12, 289, 1998.
41. Carter, J.E. and Ulusarac, O., Widespread cutaneous involvement by invasive *Apophysomyces elegans* in a gravid patient following trauma, *Cutis*, 72, 221, 2003.
42. Okhuysen, P.C. et al., Successful treatment of extensive posttraumatic soft-tissue and renal infections due to *Apophysomyces elegans*, *Clin. Infect. Dis.*, 19, 329, 1994.
43. Cáceres, A.M. et al., *Apophysomyces elegans* limb infection with a favorable outcome: Case report and review, *Clin. Infect. Dis.*, 25, 331, 1997.
44. Dannaoui, E. et al., Efficacy of antifungal therapy in a nonneutropenic murine model of zygomycosis, *Antimicrob. Agents Chemother.*, 46, 1953, 2002.
45. Kordy, F.N. et al., Successful treatment of a child with posttraumatic necrotizing fasciitis caused by *Apophysomyces elegans*: Case report and review of literature, *Pediatr. Infect. Dis. J.*, 23, 877, 2004.
46. Lawrence, R.M. et al., Systemic zygomycosis caused by *Apophysomyces elegans*, *J. Med. Vet. Mycol.*, 24, 57, 1986.
47. Pitisuttithum, P. et al., Activity of posaconazole in the treatment of central nervous system fungal infections, *J. Antimicrob. Chemother.*, 56, 745, 2005.
48. Jones, K.W. and Kaufman, L., Development and evaluation of an immunodiffusion test for diagnosis of systemic zygomycosis: Preliminary report, *J. Clin. Microbiol.*, 7, 97, 1978.
49. Kaufman, L., Turner, L.F., and McLauglin, D.W., Indirect enzyme-linked immunosorbent assay for zygomycosis, *J. Clin. Microbiol.*, 27, 1979, 1989.
50. Lombardi, G. et al., Exoantigen tests for the rapid and specific identification of *Apophysomyces elegans* and *Saksenaea vasiformis*, *J. Med. Vet. Mycol.*, 27, 113, 1989.
51. Voigt, K., Cigelnik, E., and O'donnell, K., Phylogeny and PCR identification of clinically important zygomycetes based on nuclear ribosomal-DNA sequence data, *J. Clin. Microbiol.*, 37, 3957, 1999.
52. Hata, D.J. et al., Real-time PCR method for detection of zygomycetes, *J. Clin. Microbiol.*, 46, 2353, 2008.
53. Ribes, J.A., Vanover-Sams, C.L., and Baker, D.J., Zygomycetes in human disease, *Clin. Microbiol. Rev.*, 13, 236, 2000.
54. Roden, M.M. et al., Epidemiology and outcome of zygomycosis: A review of 929 reported cases, *Clin. Infect. Dis.*, 41, 634, 2005.
55. Spellberg, B., Edwards, J. Jr., and Ibrahim, A., Novel perspectives on mucormycosis: Pathophysiology, presentation and management, *Clin. Microbiol. Rev.*, 18, 556, 2005.
56. Ibrahim, A.S. et al., *Rhizopus oryzae* adheres to, is phagocytosed by, and damages endothelial cells in vitro, *Infect. Immun.*, 73, 778, 2005.
57. Ferguson, T.D. et al., Posaconazole treatment for *Apophysomyces elegans* rhino-orbital zygomycosis following trauma for a male with well-controlled diabetes, *J. Clin. Microbiol.*, 45, 1648, 2007.
58. Garcia-Covarrubias, L. et al., Rhino-orbitocerebral mucormycosis attributable to *Apophysomyces elegans* in an immunocompetent individual: Case report and review of the literature, *J. Trauma*, 50, 353, 2001.
59. Newton, W.D., Cramer, F.S., and Norwood, S.H., Necrotizing fasciitis from invasive phycomycetes, *Crit. Care Med.*, 15, 331, 1987.
60. Kimura, M., Smith, M.B., and McGinnis, M.R., Zygomycosis due to *Apophysomyces elegans*: Report of 2 cases and review of the literature, *Arch. Pathol. Lab. Med.*, 123, 386, 1999.
61. Huffnagle, K.E. et al., *Apophysomyces elegans* as an agent of zygomycosis in a patient following trauma, *J. Med. Vet. Mycol.*, 30, 83, 1992.
62. McGinnis, M.R. et al., Necrotizing fasciitis caused by *Apophysomyces elegans*, *J. Mycol. Med.*, 3, 175, 1993.
63. Fairley, C. et al., Survival after rhino-orbital-cerebral mucormycosis in an immunocompetent patient, *Ophthalmology*, 107, 555, 2000.
64. Page, R., Gardam D.J., and Heath C.H., Severe cutaneous mucormycosis (Zygomycosis) due to *Apophysomyces elegans*, *ANZ J. Surg.*, 71, 184, 2001.
65. Wang, J. et al., Systemic *Apophysomyces elegans* after trauma: Case report and literature review, *Surg. Infect.* (Larchmt.), 3, 283, 2002.
66. Blair, J.E. et al., Locally invasive cutaneous *Apophysomyces elegans* infection acquired from snapdragon patch test, *Mayo Clin. Proc.*, 77, 717, 2002.
67. Ruiz, C.E. et al., Necrotizing fasciitis in an immunocompetent patient caused by *Apophysomyces elegans*, *Biomedica*, 24, 239, 2004.

81 *Cokeromyces*

Dongyou Liu

CONTENTS

81.1 INTRODUCTION

According to the most recent classification system, the fungi kingdom is divided into seven phyla: Ascomycota (Dikarya), Basidiomycota (Dikarya), Blastocladiomycota, Chytridiomycota, Glomeromycota (containing part of the former phylum Zygomycota), Microsporidia, and Neocallimastigomycota as well as Fungi incertae sedis (containing the remainder of the former phylum Zygomycota) [1].

Fungi incertae sedis, covering remaining molds of the displaced phylum Zygomycota (commonly known as "sugar" and "pin" molds), is separated into four subphyla: Entomophthoromycotina, Kickxellomycotina, Mucoromycotina, and Zoopagomycotina, with most opportunistic human and animal pathogens belonging to Entomophthoromycotina and Mucoromycotina [1–3].

While the subphylum Entomophthoromycotina consists of a single-order Entomophthorales, which is separated into four families: Ancylistaceae, Basidiobolaceae, Entomophthoraceae, and Neozygitaceae; the subphylum Mucoromycotina contains three orders: Endogonales, Mortierellales, and Mucorales, in addition to some unclassified Mucoromycotina [1].

81.1.1 Classification and Morphology

Previously assigned to the family *Choanephoraceae*, the genus *Cokeromyces* is classified now in the family Thamnidiaceae, order Mucorales, subphylum Mucoromycotina. The family Thamnidiaceae is made up of 12 genera: *Backusella*, *Cokeromyces*, *Dicranophora*, *Ellisomyces*, *Fennellomyces*, *Helicostylum*, *Kirkomyces*, *Phascolomyces*, *Pirella*, *Thamnidium*, *Thamnostylum*, and *Zychaea* [4]. Currently, there are one recognized species *Cokeromyces recurvatus* and an unassigned species *Cokeromyces* sp. SW078 within the genus *Cokeromyces*. A former species *Cokeromyces poitrasii* is now known as *Benjaminiella poitrasii*.

Cokeromyces recurvatus grows slowly at 30°C and more rapidly at 37°C. Colonies measure 15–21 mm in diameter at 5 days on potato flakes and Sabouraud dextrose agars at 25°C and are yeast-like, pasty in texture, containing large, spherical, thick-walled cells (of 15–30 μm in size) bearing one to multiple buds. On potato dextrose agar (PDA) slide cultures at 30°C for 10 days, the fungus is mold-like, producing hyaline, smooth-walled, unbranched, broad sporangiophores (usually 300–500 μm long), each terminating in a spherical vesicle (13–31 μm in diameter). Elongate, recurved stalks or pedicels (each bearing globose sporangiola of 10–12 μm in diameter) arise from a vesicle at the apex of the sporangiophore. Each sporangiolum contains 12–20 oval, smooth-walled sporangiospores. Being homothallic, *C. recurvatus* requires only one mating type for sexual reproduction and zygospores are generated within a single isolate. Spherical, rough-walled, dark brown–black zygospores (33.5–54.5 μm in diameter) borne on opposed suspensors from the hyphae are visible at higher temperatures (up to 37°C) under anaerobic candle jar incubation [5–9].

Being a dimorphic fungus, *C. recurvatus* grows as a felt-like (mold-like) mat on animal excreta in nature and appears as large yeast-like, budding cells in vaginal, urine, and stool specimens, resembling morphologically to the yeast phase of *Paracoccidioides brasiliensis* as well as a spherule of *Coccidioides immitis* [10].

As a sporangiola-forming species of zygomycete, *C. recurvatus* is commonly present in soil contaminated with rodent, lagomorph, or lizard dung in parts of Mexico and such U.S. states as California, Arizona, Illinois, Michigan, Florida, and Texas. The fungus has been identified from the colon and genitourinary tract of healthy individuals as well as patients in the United States only. It is possible that human clinical cases of *C. recurvatus* infection may arise from previous colonization of the involved sites (stomach, genitourinary tract, or colon).

81.1.2 Clinical Features and Pathogenesis

Cokeromyces recurvatus, a sporangiola-forming dimorphic fungus, is a rare cause of urogenital and gastrointestinal infections, cystitis, peritonitis, and myositis in humans, both immunocompetent and immunosuppressed (with underlying ulcer, diabetes mellitus, myeloma, and cancer as well as chronic alcoholism) [6,11].

C. recurvatus was implicated in a case of hemorrhagic cystitis involving a 72-year-old male. Large yeast-like cells with one to multiple buds, similar in appearance to *P. brasiliensis*, were detected in the urine specimens. However, the fungus was absent in the bladder wall. The patient responded to amphotericin B therapy [12]. *C. recurvatus* was also responsible for causing severe watery diarrhea in a patient under post-transplantation immunosuppression therapy. *Cokeromyces* was detected in the stool and intestinal biopsy specimens. The patient recovered from the infection fully with high-dose oral nystatin [13]. Similarly, a 66-year-old man presented with rash, nausea, and vomiting after undergoing bone marrow transplantation for chronic myelogenous leukemia. The patient succumbed to the disease shortly after hospital admission. Culture of sputum collected on the day of death showed heavy growth of *C. recurvatus* 6 days later. Microscopic examination of the lungs uncovered numerous thick-walled, non-budding spherules of 40–80 μm in size [10].

In a separate report, a 64-year-old man with a 10-year history of peptic ulcer disease and alcohol abuse presented with severe abdominal pain, pulmonary edema, and bilateral pleural effusions. Culture of peritoneal and pleural fluid specimens on chocolate and blood agars at 35°C grew molds. A tease mount of the fungus with one drop of phenol revealed sporangia with nonseptate hyphae (typical of zygomycetes). A scotch tape preparation with lactophenol aniline blue showed no rhizoids. Colonies on potato flakes and Sabouraud dextrose agar subcultures were tan, thin, and radially wrinkled, with gray central areas due to the production of sporangiophores, and turning brown later with the production of zygospores. Microscopically, the isolate showed distinct sporangiophores forming terminal, enlarged vesicles, from which sporangiolar stalks (pedicels) arose, elongated and recurved toward the vesicle, and subsequently produced smooth-walled, ovoid to elliptical sporangiospores. Sexual zygospores, which developed between opposing suspensory cells on the same hyphae, were spherical, rough edged, and dark brownish black in color. Large, spherical yeast-like cells (up to 20 mm in diameter) with sparse, but distinct, budding developed at 35°C on brain heart infusion agar containing 5% sheep blood, and colonies appeared pasty in texture, gray, and slightly wrinkled and measured 3–5 mm in diameter after 2 days of incubation. Examination of biopsy specimen using Gomori's methenamine silver and periodic acid-Schiff (PAS) stains showed no evidence of tissue infection. The patient went into cardiopulmonary arrest and died on day 32 of hospitalization [14].

Significantly, in most of the clinical cases reported, no obvious tissue invasions by *C. recurvatus* have been noted. It is therefore possible that disease caused by this fungus is mediated by one or more of its extracellular mycotoxins. Underlying conditions (e.g., pregnancy, diabetes mellitus, alcoholism, and immunosuppression) often increase susceptibility to *C. recurvatus* infection.

C. recurvatus infection may be treated with a 10 day course of nystatin or a 2-week course of terconazole [13,15]. Its response to amphotericin B is variable.

81.1.3 Laboratory Diagnosis

Because *Cokeromyces recurvatus* is commonly present in the environment, and clinical manifestations of *C. recurvatus* infection and other zygomycosis are indistinguishable, a definitive diagnosis of infection due to *C. recurvatus* requires a combination of mycological, histopathological, and clinical evidence. Although culture isolation of the fungus culture is useful for confirmation of *C. recurvatus* infection in patients with diabetes or neutropenia, direct observation of the organism in clinical specimens gives additional weight to a positive culture result. Repeated positive fungal cultures and a response to specific antifungal therapy in the absence of concurrent conditions also provide support that the organism is the causative agent of the disease.

C. recurvatus is a dimorphic zygomycete. In clinical specimens stained with calcofluor white or lactophenol cotton blue, *C. recurvatus* appears as a large (30–90 μm), thick-walled yeast often surrounded by a crown of smaller yeast buds (so called "mariner's wheel"), which is morphologically similar to the yeast phase of *Paracoccidioides brasiliensis* and a spherule of *Coccidioides immitis*. However, *P. brasiliensis* and *C. immitis* often show early evidence of internal cleavage or contain endospores [15]. While hematoxylin and eosin (H&E) can be used to observe *C. recurvatus* hyphae in tissue sections, the Grocott technique gives high contrast with minimal background.

In standard fungal culture media, *C. recurvatus* may be both yeast-like and mold-like. The yeast-like colonies (of 3–5 mm in size) grow on brain heart infusion agar with 5% sheep blood at 37°C in 2 days and are pasty, gray, and slightly wrinkled, with large, spherical, thick-walled cells bearing multiple buds. The mold-like colonies (of 15–30 mm in size) appear in 3–5 days at room temperature, containing branching, broad, generally nonseptate hyphae (5–15 μm in diameter) and thick-walled, dark brown–black zygospores borne on suspensors. After 10 days of incubation at 30°C, hyaline, smooth-walled, broad sporangiophores with terminating vesicles, elongated, recurved stalks or pedicels, globose sporangiola and sporangiospores as well as zygospores are visible [6].

Use of molecular techniques such as PCR and sequencing allows early and accurate diagnosis of *C. recurvatus* infection in comparison with conventional methods. The most common gene targets for fungal identification comprise 18S and 28S ribosomal RNA (rRNA) internal transcribed spacer (ITS) as well as other conserved gene regions [16–19]. By using the aligned 28S rRNA sequences, *C. recurvatus*-specific PCR primers were designed, facilitating its rapid and precise determination [20].

81.2 METHODS

81.2.1 Sample Preparation

For direct microscopic examination, portions of clinical specimens are stained with KOH, calcofluor white, or lactophenol blue; and tissue sections are stained with PAS and other fungal stains. Standard fungal media containing antibiotics (e.g., malt agar, Sabouraud dextrose agar, Sabouraud glucose agar, and PDA) are inoculated with portions of the samples and incubated at 25°C and 30°C for a few days. Both yeast-like and mold-like colonies are examined microscopically by wet mount with lactophenol aniline blue stain for characteristic features of *C. recurvatus* [6].

Mycelium for DNA isolation is grown in YM broth (0.3% yeast extract, 0.3% malt extract, 0.5% peptone, 2% dextrose; Difco) at room temperature for 2–5 days. DNA is isolated from lyophilized mycelium according to the CTAB (hexacetyltrimethylammonium bromide; Sigma) miniprep protocol. Namely, approximately 50 mg of pulverized mycelium is resuspended in 700 μL of CTAB extraction buffer (100 mM Tris–Cl pH 8.4, 1.4 M NaCl, 25 mM EDTA, 2% CTAB) and vortexed for 10 s. Following extraction, an equal volume of chloroform is added to each tube, vortexed for 5 s, and then spun for 10 min at 12,300 × *g* in a microcentrifuge. A 500 μL portion of the upper phase is removed to a new 1.5 mL tube, and DNA is precipitated by the addition of an equal volume of −20°C isopropanol. After centrifugation at 12,300 × *g* in a microcentrifuge for 1 min, the supernatant is discarded and the pellet is washed with 70% ethanol and resuspended in 200 μL of TE buffer (10 mM Tris–Cl pH 8.0, 1 mM EDTA pH 8.0). For PCR amplifications, 8 μL of the genomic DNA stock is diluted in 1 mL of deionized water and stored at −20°C until use. For PCR experiments, 25 μL of the diluted genomic DNA is added to an equal volume of a 2 × PCR master mix [20].

81.2.2 Detection Procedures

81.2.2.1 PCR for *Cokeromyces recurvatus*

Voigt et al. [20] designed a pair of *Cokeromyces recurvatus* primers from its 28S rRNA gene for rapid identification of this fungus. The primers Cr1 (5′-GTGAGAATCCC GTGAATTCAC-3′) and Cr2 (5′-CAAAGCACTCAACTAT TTCGC-3′) amplify a specific fragment of 434 bp from *C. recurvatus* DNA only.

Procedure

1. PCR mixture (50 μL) is made up of 0.225 mM of each deoxynucleotide, 25 pmol of each primer, 50 mM KCl, 10 mM Tris–Cl (pH 8.4), 2.5 mM $MgCl_2$, 0.1 mg of gelatin/mL, and 1.25 U of AmpliTaq polymerase (Perkin-Elmer), and approximately 10–20 ng of genomic DNA.
2. Amplification is performed with an initial 94°C for 2 min; 40 cycles at 94°C for 30 s, 60°C for 30 s, and 72°C for 90 s; a final 72°C for 10 min.
3. PCR product (434 bp) is fractionated in agarose gel, stained with ethidium bromide, and visualized under UV light.

81.2.2.2 Sequence Analysis of ITS Regions

Fungus-specific universal primers ITS1 (5′-TCCGTAGGTG AACCTGCGG-3′) and ITS4 (5′-GCATATCAATAAGCGG AGGA-3′) [21] can be used to amplify the ITS1 and ITS2 regions. Subsequent sequencing analysis of the resulting amplicon facilitates identification of a range of fungal pathogens including *Cokeromyces recurvatus* [19].

Procedure

1. PCR mixture (50 μL) is made up of 10 mM Tris–HCl pH 8.3, 50 mM KCl, 1.5 mM $MgCl_2$, 0.2 mM of each deoxynucleoside triphosphate, 1.2 U of *Taq* DNA polymerase, 0.4 μM of each primer (ITS1/ITS4), 2 μL (1–5 ng) of DNA template. A negative control is included in each run by replacing the template DNA with sterile water in the PCR mixture.
2. Amplification is conducted with an initial 94°C for 3 min; 30 cycles at 94°C, 60°C, and 72°C for 1 min each; and a final 72°C for 3 min.
3. After checking the PCR products on 1.5% agarose gel, the amplicons are purified using a commercial PCR cleanup kit. The DNA fragments are sequenced using an ABI Prism 377 automated DNA sequencer (Applied Biosystems) with a BigDye Terminator cycle sequencing kit (version 3.1; Applied Biosystems). All amplicons are sequenced on both strands using primers ITS1 and ITS4.
4. The resulting sequence is compared with those available at the GenBank databases using the BLAST sequence analysis tool (http://www.ncbi.nlm.nih.gov//BLAST//) (blastn) with default settings except that sequences are not filtered for low complexity. Species identification is determined from the lowest expected value of the BLAST output.

Note. For strains producing discrepant identification between the methods based on phenotypic characteristics and ITS sequence analysis, the D1–D2 region of the large-subunit RNA gene is amplified with primers NL1 (5′-GCATATCAATAAGCGGAGGAAAAG-3′) and NL4

(5′-GGTCCGTGTTTCAAGACGG-3′) and sequenced for species clarification.

81.3 CONCLUSION

Cokeromyces recurvatus is a ubiquitous fungus that is occasionally associated with human diseases, ranging from urogenital and gastrointestinal infections, cystitis to peritonitis, especially in patients with underlying illnesses such as ulcer, diabetes mellitus, myeloma, and cancer. As the fungus does not normally invade host tissues, its pathogenicity may be due to the effects of extracellular mycotoxins it produces.

As a dimorphic zygomycete, *C. recurvatus* often appears in clinical samples as large yeast-like, thick-walled cells with multiple buds ("mariner's wheel"), which may be confused with the yeast phase of *Paracoccidioides brasiliensis* and a spherule of *Coccidioides immitis* morphologically. In culture, *C. recurvatus* may be both yeast-like and mold-like. Microscopic identification of this fungus is aided by its formation of hyaline, smooth-walled, broad sporangiophores with terminating vesicles, elongated, recurved stalks or pedicels, globose sporangiola and sporangiospores as well as zygospores. Recent application of PCR and sequencing analysis permits improved diagnosis of *C. recurvatus* infection.

REFERENCES

1. Hibbett DS et al., A higher-level phylogenetic classification of the fungi. *Mycol Res*. 2007;111:509–547.
2. Elgart ML. Zygomycosis. *Dermatol Clin*. 1996;14:141–146.
3. Ribes JA, Vanover-Sams CL, Baker DJ. Zygomycetes in human disease. *Clin Microbiol Rev*. 2000;13:236–301.
4. The UniProt Consortium. Available at http://www.uniprot.org/, accessed on August 1, 2010.
5. Kwon-Chung KJ, Bennett JE. *Medical Mycology*. Philadelphia, PA: Lea & Febiger; 1992.
6. Kemna ME et al., *Cokeromyces recurvatus*, a mucoraceous zygomycete rarely isolated in clinical laboratories. *J Clin Microbiol*. 1994;32(3):843–845.
7. Weitzman I et al., Zygospores: The last word in identification of rare or atypical zygomycetes isolated from clinical specimens. *J Clin Microbiol*. 1995;33:781–783.
8. de Hoog GS et al., *Atlas of Clinical Fungi*. Baarn, the Netherlands: Centraalbureau voor Schimmelcultures; 2000.
9. Nielsen C et al., Isolation of *Cokeromyces recurvatus*, initially misidentified as *Coccidioides immitis*, from peritoneal fluid in a cat with jejunal perforation. *J Vet Diagn Invest*. 2005;17:372–378.
10. Ryan LJ et al., Fatal *Cokeromyces recurvatus* Pneumonia: Report of a case highlighting the potential for histopathologic misdiagnosis as *Coccidioides*. *Int J Surg Pathol*. 2009.
11. Ramani R et al., *Cokeromyces recurvatus* as a human pathogenic fungus: Case report and critical review of the published literature. *Pediatr Infect Dis J*. 2000;19:155–158.
12. Axelrod P et al., Chronic cystitis due to *Cokeromyces recurvatus*: A case report. *J Infect Dis*. 1987;155(5):1062–1064.
13. Alvarez OA et al., Severe diarrhea due to *Cokeromyces recurvatus* in a bone marrow transplant recipient. *Am J Gastroenterol*. 1995;90(8):1350–1351.
14. Munipalli B, Rinaldi MG, Greenberg SB. *Cokeromyces recurvatus* isolated from pleural and peritoneal fluid: Case report. *J Clin Microbiol*. 1996;34(10):2601–2603.
15. Tsai TW et al., *Cokeromyces recurvatus* infection in a bone marrow transplant recipient. *Bone Marrow Transplant*. 1997;19(3):301–302.
16. Makimura K, Murayama SY, Yamaguchi H. Detection of a wide range of medically important fungi by the polymerase chain reaction. *J Med Microbiol*. 1994;40:358–364.
17. Haynes KA, et al., Rapid detection and identification of pathogenic fungi by polymerase chain reaction amplification of large subunit ribosomal DNA. *J Med Vet Mycol*. 1995;33:319–325.
18. Sandhu GS et al., Molecular probes for diagnosis of fungal infections. *J Clin Microbiol*. 1995;33:2913–2919.
19. Leaw SN et al., Identification of medically important yeast species by sequence analysis of the internal transcribed spacer regions. *J Clin Microbiol*. 2006;44:693–699.
20. Voigt K, Cigelnik E, O'donnell K. Phylogeny and PCR identification of clinically important Zygomycetes based on nuclear ribosomal-DNA sequence data. *J Clin Microbiol*. 1999;37(12):3957–3964.
21. White TJ et al., Amplification and direct sequencing of fungal ribosomal RNA genes for phylogenetics. In: Innis MA et al., Eds., *PCR Protocols: A Guide to Methods and Applications*. San Diego, CA: Academic Press, Inc.; 1990, pp. 315–322.

82 *Cunninghamella*

Nancy L. Wengenack and D. Jane Hata

CONTENTS

82.1 INTRODUCTION

82.1.1 Classification and Morphology

82.1.1.1 Classification

Cunninghamella species are ubiquitous environmental fungi found in soil and plant materials worldwide with seven species described.[1] *C. bertholletiae* has been the primary species reported as a human pathogen, although *C. echinulata* has also been reported in association with diseases. Only rarely has *C. elegans* been reported as an agent in human infection.[2] In 1958, the first description of a disseminated *Cunninghamella* infection was described in 1958.[3]

The phylum Zygomycota consists of two classes: the Trichomycetes and the Zygomycetes. Members of the Trichomycetes are obligate symbiotes of arthropods. Zygomycetes are divided into two orders: Mucorales and Entomophthorales (Ancylistaceae, Basidiobolaceae), which contain the human pathogens *Basidiobolus* and *Conidiobolus*. Production of actively expelled asexual sporangioles is a unique characteristic and distinguishes Entomophthorales from the Mucorales.

Cunninghamella was first described by Blakeslee in 1904. As a member of the phylum Zygomycota, class Zygomycetes in the order Mucorales, these eukaryotic molds are characterized by the formation of aseptate or pauciseptate, wide hyaline hyphae (coenocytic hyphae) with branching from 45° to 90°. Wide, ribbon-like hyphae are a key feature that allows microscopic distinction of the Zygomycetes from other filamentous pathogenic fungi, such as *Aspergillus* spp. Members of the Mucorales associated with human diseases include *Rhizopus*, *Mucor*, *Absidia*, *Syncephalastrum*, *Rhizomucor*, *Apophysomyces*, *Saksenaea*, *Cokeromyces*, and *Cunninghamella*. For additional details on taxonomy, the reader is referred to the excellent review by Ribes et al.[3]

82.1.1.2 Morphology

Characteristic of the zygomycetes are zygospores or sexual reproductive structures that form as a result of contact between two rapidly growing hyphal branches known as zygophores. Upon contact, and subsequent formation of a structure known as a gametangium, nuclear fusion and mixing of cytoplasm occur, with formation of the zygosporangium. These sexual structures are rarely seen in the human host and are generally not used as a basis of laboratory identification.

As a result of asexual reproduction, sporangiospores are produced in a sack-like structure known as a sporangium. Sporangiospores have been implicated as agents of transmission due to their small size and ability to remain airborne for long periods of time. Unfortunately, this characteristic may also contribute to contamination within the laboratory and potential confusion in clinical diagnosis.[3]

Although most zygomycetes may typically produce hundreds of sporangiospores within a sporangium, *Cunninghamella* is unique, as only a single spore is contained within a vesicle-like sporangiola. Single sporangiola may appear hyaline or brownish when observed in a mass. The sporangiola are supported by short stalks known

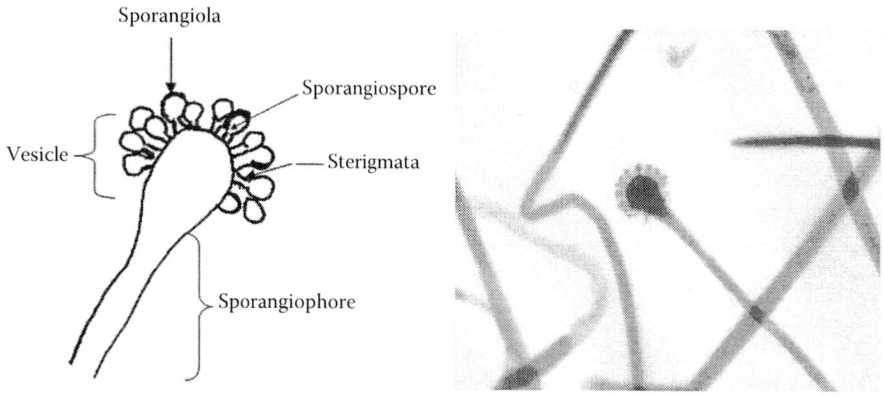

FIGURE 82.1 Microscopic morphology of *Cunninghamella* spp.

as sterigmata, which are mounted on swollen vesicles (15–59 μm), and may be subglobose, pyriform, or ellipsoidal. Vesicles are supported by a structure known as a sporangiophore. In *Cunninghamella*, sporangiophores are up to 20 μm wide with single or verticillate branches supporting the subglobose or pyriform vesicles. Lateral branches of the sporangiophores may also produce vesicles. These vesicles may also be ellipsoidal (3–30 μm). Unlike *Rhizopus* or *Rhizomucor*, rhizoids are primitive and rarely noted extending from the sporangiophores (Figure 82.1).

Macroscopic colony morphology of *Cunninghamella* sp. is similar to other zygomycetes. Colonies grow optimally at 25°C–30°C; the maximal temperature tolerated is 50°C. Colonies are white and cottony, becoming dark gray with time. Colony reverse may be buff or colorless. *Cunninghamella* grows extremely rapidly and may completely fill a culture plate within 2–3 days. Due to the propensity of these fungi to lift the lid off of culture dishes, sealing of culture plates is recommended.[4] Although some species of zygomycetes have been reported to exhibit thermal dimorphism, this has not been reported among *Cunninghamella* sp.[3] Growth is most robust on mycological culture media such as Sabouraud dextrose agar (SDA) or inhibitory mold agar (IMA). Growth of *Cunninghamella* sp. and other zygomycetes are inhibited on media containing cycloheximide (i.e., Mycosel).

Recovery of zygomycetes from tissues can be problematic. In some studies, recovery of a zygomycete by culture was only achieved in 50% of cases.[5] The fragile reproductive structures of the mold are easily disrupted by grinding of tissues or other aggressive processing procedures prior to inoculation of culture media. Use of more gentle methods, such as mincing of tissue samples will result in improved fungal recovery.[3]

As with other zygomycetes, *Cunninghamella* spp. exhibit hyaline, broad ribbon-like pauciseptate hyphae. In histological preparations, hyphae are prone to twisting due to their thinner cell walls. The appearance of folding may be misinterpreted as hyphal septations. Hyphal branching of 45°–90° is a helpful characteristic in differentiation of zygomycetes from other acutely branching fungi such as *Aspergillus* sp. Visualization in tissue is highly dependent upon appropriate staining techniques. Although traditionally used, histological stains such as periodic acid-Schiff (PAS), Gomori methenamine silver (GMS), or hematoxylin–eosin (H&E) may stain zygomycetes inconsistently or weakly. This may be due in part to the thinner hyphal walls characteristic of this genus, as compared to other fungi.[6] Chlamydospores may be H&E or PAS positive, but stain poorly with GMS. However, sporulation in tissue is rarely reported.[4]

Rapid invasion of hyphal elements into vascular tissues (angioinvasion) is a characteristic of zygomycetes. Involved soft tissues are swollen and erythematous, becoming black with disease progression. Extensive thrombosis and necrosis often occur, resulting in tissue infarction. Vascular stenosis may occur due to thromboembolic fungal dissemination.[7]

Perineural invasion is commonly encountered, resulting in rhinocerebral zygomycosis. It is theorized that secondary to vascular invasion, fungi may extend into the optical orbit, meninges, and brain. Invasion via neural tissue may also occur. The mechanisms by which vascular or neural tropism occur are not currently known, although iron uptake or physical characteristics of tissues may play a role.[6]

Pulmonary manifestations can range from solitary nodular lesions on x-ray to lobar involvement, with progression to wedge infarcts. Radiographic findings in zygomycosis are nondescript and may result in initial misdiagnoses of bacterial pneumonia. Darrisaw et al. reported improved survival in *Cunninghamella* pulmonary infections associated with wedge resection or lobectomy in conjunction with amphotericin B therapy.[8]

82.1.2 Epidemiology

As with other members of zygomycetes, *Cunninghamella* spp. are commonly isolated from a variety of environmental sources. These origins may vary from soils, to dead wood, to edible fruits and vegetables. *C. bertholletiae* is frequently isolated from Brazil nuts. Environmental isolation of *Cunninghamella* sp. is noted predominantly in temperate climates.

Interestingly, *C. elegans* has been reported to have a number of biotechnological applications, ranging from chitin production[9] to neutralization of organic pesticides[10] and aromatic hydrocarbons[11] and as a microbial model of cytochrome P-450 hepatic metabolism.[12]

Disease transmission by zygomycetes has been reported to occur via inhalation of spores, through percutaneous exposure (needlesticks, catheter insertion), or via ingestion. Acquisition of infection with *Cunninghamella* sp. may occur through ingestion of sporangioles carried on food. Transmission may also occur via percutaneous implantation of sporangioles secondary to trauma. Motohashi et al. reported a localized cutaneous infection with *C. bertholletiae* in a patient with chronic myelogenous leukemia. In this case, the lesion was associated with application of an elasticized tape dressing after thoracentesis for a pleural effusion.[13]

Most cases of infection due to *Cunninghamella* sp. are believed to be acquired via inhalation, resulting in sinus or pulmonary disease. Several case studies report the predominance of pulmonary infections due to *Cunninghamella* sp.[8,14] The small size of sporangiospores (approximately 6 μm) allow for ease of environmental dissemination and persistence in currents of air.

In one of the largest reviews of epidemiology and outcomes in zygomycosis, Roden et al. report infection specifically with *Cunninghamella* sp. as a significant risk factor for mortality among patients with a confirmed diagnosis of zygomycosis.[5] In this study of 465 patients, infection with *C. bertholletiae* was associated with the greatest mortality rate (76%), as compared to other zygomycete genera. Mortality due to *C. bertholletiae* was second only to *Rhizopus stolonifer* among patients studied. *Cunninghamella* sp. was also considered an independent risk factor for development of pulmonary zygomycosis (odds ratio = 7.75).[5]

Proven infections with *Cunninghamella* sp., and indeed with any of the zygomycetes, are highly dependent upon the immunological status of the host. Hematological malignancy is most commonly reported as an underlying disease associated with *Cunninghamella* sp. infections.[8] Other associated conditions may include bone marrow transplant (BMT), diabetes mellitus, graft vs. host disease (GVHD), and malnutrition.[15] Generalized disseminated zygomycosis was noted most frequently in patients undergoing deferoxamine therapy.[5] Prabhu and Patel reported only a rare association of zygomycete infection in patients with solid tumors.[16]

A review by Roden et al. reported a clear association of zygomycosis due to *Cunninghamella* sp. with the male sex. However, it was unclear whether this association was due to physiological factors such as male hormones or environmental factors (i.e., trauma). Estrogen has been shown to have protective effects in paracoccidioidomycosis; however, the role of estrogen in zygomycosis has not been investigated.[5]

Several studies have suggested increasing overall incidence of zygomycosis in specific transplant populations. Marr et al. reported that among allogenic hematopoietic stem cell transplant (HSCT) patients, severe GVHD and underlying myelodysplastic syndrome were significant factors in the development of zygomycete infection. In addition, infections tended to manifest themselves >90 days after transplantation. In 1993, the practice of routine antifungal prophylaxis with fluconazole for 75 days after HSCT was initiated. In a study of non-*Aspergillus* mold infections, it was noted that during the period 1995–1999 invasive infections with zygomycete species (including *Cunninghamella*) nearly doubled as compared to the previous 5-year period.[17]

In a more recent study also at a large cancer center, Kontoyiannis et al. also noted increased overall incidence of zygomycosis among BMT recipients during a 2002–2004 study period. This was higher overall (13/834 patients; 1.55%) as compared to 1984–1998 (10/4020 patients; 0.25%). The incidence of invasive aspergillosis (IA) had decreased during the 2002–2004 study period. A multivariate analysis comparing patients with zygomycosis to patients with IA determined voriconazole prophylaxis to be a significant risk factor in the development of zygomycosis vs. IA. It is of note that zygomycetes are inherently resistant to most expanded-spectrum azole drugs, including voriconazole. These authors raised the concern that prophylaxis with agents that lack antifungal activities against non-*Aspergillus* molds (i.e., zygomycetes), could lead to increased incidence of breakthrough infections in highly susceptible transplant populations.[15]

82.1.3 Pathogenesis

Zygomycosis refers to invasive infection caused by any of the zygomycetes known to be associated with human diseases: *Rhizopus*, *Mucor*, *Absidia*, *Rhizomucor*, *Saksenaea*, *Apophysomyces*, *Cokeromyces*, and *Cunninghamella*. As previously discussed, the immune status of the host plays a crucial role in pathogenesis and progression of disease due to *Cunninghamella* sp.

Inflammatory responses in tissue are variable but may commonly consist entirely of neutrophils. In a study by Frater et al., mixtures of granulomas and neutrophils were noted. Primary granulomatous reactions have been described, or, in some cases, no tissue inflammatory reactions at all have been reported.[6] Patients who develop functional deficiencies or numerical deficiencies in monocytes, macrophages, or neutrophils are at increased risk of development of zygomycosis. Recovery of neutrophil numbers may play a significant role in reduction of or recovery from infections due to *Cunninghamella* sp.[7,16] Neutrophils are primarily protective against fungal hyphae, whereas macrophages may engulf and destroy sporangiospores through release of oxidative by-products and proteases.[18] Phagocytosis by macrophages results in oxidative destruction of spores, thus preventing initiation of hyphal growth. When macrophage function is impaired, clearance of spores is diminished.[19] It is believed that dysfunction in T cells does not appear to play a major role in immune defense against zygomycete infection as demonstrated by the lack of reported disease in individuals infected with HIV.[18]

Systemic physiological states, such as increased iron concentrations secondary to chelation therapy, administration of deferoxamine, diabetes, or metabolic acidosis are risk factors for development of zygomycosis.[18] The chemotactic and phagocytic characteristics of neutrophils are inhibited by decreased serum pH.[19] Steroid therapy or other immunosuppressive therapies play a contributory role in the development of disease either by directly suppressing production of inflammatory mediators, inhibition of inflammatory cell migration, or induction of a diabetic state in the host.[3]

82.1.4 Clinical Features

In addition to pulmonary diseases, *Cunninghamella* spp. have been isolated from infections in sites such as the gastrointestinal tract, lymph nodes, and brain. Cutaneous infections may be localized secondary to inoculation or may result in disseminated disease. Cutaneous disease is the most common presentation of zygomycosis in individuals with no underlying medical conditions.[5] Abscess formation, production of pus, and tissue swelling are noted in local infections. Lesions may become necrotic and result in large ulcers that are foci of secondary bacterial infection. Subcutaneous tissues, fat, fascial layers, and muscle may be invaded by fungal hyphae. Necrotizing fasciitis with resultant high mortality rates may occur. Surgical excision of infected sites may be curative but may result in substantial disfigurement.[3] Cutaneous disease may be underestimated as a source of hematogenous dissemination, leading to a systemic syndrome. In their case series, Roden et al. reported that 20% of 176 patients developed dissemination of zygomycete infection from a cutaneous site to another organ. This resulted in a mortality rate of 94% within this subgroup.[5]

Development of disseminated disease may manifest itself as multiple, rapidly developing skin lesions. Disseminated disease is most commonly associated with hematogenous spread from the lung. In a case report from Darrisaw et al., a 46-year-old female with acute GVHD secondary to unrelated allogenic BMT for acute myelogenous leukemia developed numerous erythematous skin lesions with necrotic centers 3 days after development of bilateral pulmonary infiltrates. Skin biopsies indicated fungal hyphae with dichotomous branching and occasional septation. Autopsy findings included fungal pneumonia with hyphal masses in addition to disseminated fungal dermatitis with *Cunninghamella* sp.[8]

Spread of infections typically occurs via the hematogenous route. A review by Darrisaw and colleagues of 26 cases of invasive *Cunninghamella* sp. infections reported pulmonary infection as a primary site in 19/26 (73%) cases. Mortality in this case group was 77%. Dissemination to other sites such as the brain, sinuses, cervical lymph nodes, and joint spaces were reported. Both localized skin lesions and disseminated cutaneous infections were described in this case series.[8] An additional case series from Garey et al. reported a mortality rate of 81%, with lung as the most common site of involvement. A troubling observation was noted in 10 cases with an increase in reported amphotericin B MIC values. In seven of these cases, *Cunninghamella* MICs ranged from 12.5 to 100 μg/mL. Mortality in this group was 100%.[20]

Human infection with *Cunninghamella* spp., and indeed any zygomycete, constitutes a medical emergency and should not be discounted. Some studies have indicated increased pathogenicity of *Cunninghamella* as compared to other members of the zygomycetes. Treatment generally consists of amphotericin B plus surgical debridement when possible. Early diagnosis of infection is essential to improve patient outcome.

82.1.5 Virulence Factors

Iron appears to play a special role contributing to increased risk of zygomycosis. In a recent review by Ibrahim et al., it was noted that high serum iron levels predispose individuals to zygomycete infection but not to infections with *Candida* or *Aspergillus*. Iron is a requirement for enhanced growth and virulence by many fungal pathogens.[21] In acid pH conditions (i.e., metabolic acidosis), the transferrin system functions at a lower efficiency, thereby allowing additional unbound iron to circulate in the serum.[19] Physiologically, iron must be in the ferrous (Fe^{2+}) ionization state in order to be accumulated intercellularly. However, excess intercellular iron must be stored in order to prevent iron-catalyzed production of free radicals, which cause oxidative damage to cellular substrates. Zygomycetes are unique in their ability to store iron as part of protein complexes known as ferritins. Mycoferritin, bacterioferritin, and zygoferritins have all been identified in zygomycetes; zygoferritin is unique to this genera.

Three mechanisms of enhanced iron uptake have been described in zygomycetes. Siderophores chelate ferric (Fe^{3+}) iron from the environment, reduce it to the ferrous (Fe^{2+}) form, and transfer it into the fungal cell via a siderophore permease. Zygomycetes secrete rhizoferrin, a siderophore that supplies iron to the cell via an energy-dependent receptor-mediated process. *Rhizopus* has been shown to also utilize siderophores secreted by other organisms (xenosiderophores) to enhance fungal iron uptake. The exogenous chelating agent, deferoxamine, can be utilized as a siderophore by *Rhizopus* and other zygomycetes, thus supplying supplemental iron to spur fungal growth. In patients with chronic renal failure, deferoxamine treatment has proven to be a significant factor in development of zygomycete infections. However, deferasirox, an orally bioavailable iron chelator, has been shown to chelate iron from *R. oryzae* and is cidal against zygomycetes in vitro. Use of this agent may have substantial contribution to survival in patients who must undergo iron chelation therapy.

Reductive iron uptake involves an iron reductase, which reduces Fe^{3+} to Fe^{2+}, combined with an iron permease, which brings Fe^{2+} into the fungal cell. A high-affinity iron permease in *R. oryzae* has been cloned and found to have significant DNA sequence homology to high-affinity Fe^{2+} permeases noted in *C. albicans* and *S. cerevisiae*. In addition, disruption of Fe^{2+} permease expression reduced *R. oryzae* virulence in a mouse model of ketoacidiosis.

The third mechanism for iron uptake involves hemin as a source of iron. In addition to receptors for hemin binding, an oxygenase is required for iron assimilation into fungal cells. Sequencing of the *Rhizopus* genome has described two homologues of a putative *C. albicans* heme oxygenase gene. These homologues may provide a mechanistic example by which *Rhizopus* and other zygomycetes may obtain iron from host hemoglobin. It may also provide some insight into the angioinvasive tendencies of zygomycetes.[21]

Mechanisms of antifungal resistance may also be considered factors contributing to increased virulence of zygomycetes. Antifungals of the azole class are generally ineffective against zygomycetes. Aside from posaconazole, azole MICs (itraconazole, voriconazole) can range from 0.03 to 64 μg/mL.[22] Azoles inhibit lanosterol 14-α-demethylase, resulting in reduced synthesis of ergosterol required for development and maintenance of fungal cell membranes. Although genus-specific data are limited, zygomycete resistance to azoles are most likely due to decreased membrane permeability to drugs, upregulation of membrane-bound efflux transporter pumps, interference of azole activity in inhibition of lanosterol 14-α-demethylase, or overexpression of this particular enzyme target.[23]

Echinocandin agents have no effect against zygomycetes. This class of antifungals inhibit β-1,3-D-glucan synthetase, responsible for synthesis of β-1,3 glucan, an essential component of the fungal cell wall. Mutation in *Fks1*, a gene that encodes the major subunit of glucan synthetase, is believed to be the principal resistance mechanism to echinocandins, as demonstrated in *Aspergillus* and *Candida*. As glucan synthetase activity in *Cryptococcus* and other molds is strongly inhibited by echinocandins, it is postulated that an alternative glucan synthesis mechanism, independent of *Fks1* and unaffected by echinocandin agents, may be involved in the resistance of zygomycetes to this antifungal class. In addition, reduced cell permeability for echinocandins may also play a role in resistance. Unfortunately, specifics of these mechanisms have not been elucidated for the zygomycetes.[24]

Aside from these general discussions of iron uptake and antifungal resistance as applied to zygomycetes as a whole, little information is available concerning specific virulence factors in *Cunninghamella* sp. When compared to other zygomycetes, *Cunninghamella* does exhibit thermotolerant growth properties, which may contribute to invasive cerebral and pulmonary infection.[3] An early study from Weitzman and Crist demonstrated robust growth of *C. bertholletiae* at temperatures up to 45°C, whereas *C. elegans* exhibited growth only at 25°C–37°C.[1]

82.2 METHODS

82.2.1 CONVENTIONAL TECHNIQUES FOR DIAGNOSIS OF *CUNNINGHAMELLA* INFECTION

Cunninghamella spp. are traditionally identified from human specimens using a combination of macroscopic and microscopic techniques. As mentioned earlier, colonies grow optimally at 25°C–30°C and are initially fluffy and white, turning gray over time. The reverse side of the colony is typically white to buff colored. *Cunninghamella* grows extremely rapidly and may completely fill a culture plate within 2–4 days. Growth is inhibited by cycloheximide. The microscopic morphology is usually sufficient to allow identification and demonstrates broad, pauciseptate hyphae containing long, branched sporangiophores that terminate in swollen vesicles, which are 30–65 μm in diameter. The vesicles are covered with sterigmata that each terminates in a single round to oval-shaped sporangiola containing a single spore.[25] The sporangiola often is covered with small hair-like crystals. There are certain instances in which identification using conventional techniques such as microscopic morphology is not possible or is not optimum for patient care. For example, it is not infrequent that surgical samples are obtained for pathologic stain and examination but a microbiologic culture is not considered. Although, histopathology may indicate the presence of a zygomycete, often identification to the genus and/or species level is not possible from these specimens. In addition, there are instances when direct identification from patient specimens without the need to wait for cultures to grow is desired because, in spite of their rapid growth characteristics, zygomycete culture still often requires several days. Therefore, molecular methods of identification, such as those discussed in the next section, may speed the turnaround time for identification of these organisms and have the potential to positively impact patient care.

82.2.2 MOLECULAR DIAGNOSIS OF *CUNNINGHAMELLA* INFECTION

Molecular methods for the detection of *Cunninghamella* spp. have not yet been widely reported in the literature but, in general, two molecular approaches have been used to date. The first approach uses DNA sequence analysis to rapidly identify *Cunninghamella* after it is grown in pure culture from patient specimens. The second approach uses PCR to directly identify *Cunninghamella* from patient specimens. The PCR approaches utilize traditional PCR, nested PCR, or real-time PCR methods, each of which differ in their turnaround time and potential for false-positive results due to environmental contaminants. Both DNA sequencing from culture and PCR from specimen are described in Sections 82.2.3 and 82.2.4 in detail.

82.2.3 SEQUENCE-BASED IDENTIFICATION OF *CUNNINGHAMELLA* SPP.

Alastruey-Izquierdo et al. used sequencing of the internal transcribed spacer (ITS) region to identify 77 clinical strains of *Mucorales* species.[26] The ITS region of the rRNA gene has been demonstrated to be useful for the differentiation of *Mucorales* genera including *Cunninghamella* spp. from other fungal genera and from other members of the *Mucorales*. PCR amplification of the ITS target was done using primers ITS1 and ITS4 (Table 82.1) in a GeneAmp 9700 PCR system (Applied Biosystems). For the sequencing reaction, 2 μL sample of ReadyReaction mixture from the BigDye terminator cycle sequencing kit (Applied Biosystems) was used along with 1 μM of primers (ITS1 and ITS4) and 3 μL of the PCR reaction product. The resulting nucleotide sequence was assembled and edited using the SeqManII and EditSeq software packages

TABLE 82.1
Primer and Probe Sequences for Amplification and Detection of *Cunninghamella*

Primer/Probe	Sequence	Reference
U1	5′-GTG AAA TTG TTG AAA GGG AA-3′	Kobayashi[33]
U2	5′-GAC TCC TTG GTC CGT GTT-3′	
Cu1	5′-GGA TTG TAA ACT AAA GTT TTC-3′	
Cu2	5′-AAA TTC TCT AAT TAT TCC CTC-3′	
ITS1	5′-TCC GTA GGT GAA CCT GCG G-3′	Alastruey-Izquierdo[26]
ITS4	5′-TCC TCC GCT TAT TGA TAT GC-3′	
ZM1	5′-ATT ACC ATG AGC AAA TCA GA-3′	Bialek[34]
ZM2	5′-TCC GTC AAT TCC TTT AAG TTT C-3′	
ZM3	5′-CAA TCC AAG AAT TTC ACC TCT AG-3′	
ITS4	5′-TCC TCC GCT TAT TGA TAT GC-3′	Alvarez[28]
ITS5	5′-GGA AGT AAA AGT CGT AAC AAG G-3′	
CU(+) sense	5′-TAG TCA GCC AGG TAA ATA AGT-3′	Kasai[37]
CU(−) sense	5′-TCG TCA ATA TTT AGC TTT AGG-3′	
CU FITC	5′-GCT TGG AAA CGA AGA GTC AGG TTG-3′	
CU RD 640	5′-TGG GAA TGC AGC CTA AAA TGG GAG TGA-3′	
Cu1	5′-GGA TTG TAA ACT AAA GTT TTC-3′	Voigt[32]
Cu2	5′-AAA TTC TCT AAT TAT TCC CTC-3′	

(Lasergene, DNAstar, Inc., Madison, WI).[27] Sequencing of the ITS region of 77 clinical strains of *Mucorales* species allowed the authors to identify all 77 strains including 2 isolates of *Cunninghamella* spp.

Alvarez et al. examined the spectrum of zygomycete species identified at a mycology reference laboratory from clinically significant specimens in the United States using sequencing of the ITS region of the fungal ribosomal DNA.[28] The fungal DNA was extracted and purified using the FastDNA kit (Bio101), which was slightly modified by performing a homogenization step three times using a FastPrep F120 instrument (ThermoSavant, Holbrook, NY). The ITS region of the rDNA was amplified using primers ITS4 and ITS5 (Table 82.1)[29,30] following a protocol described by Gilgado et al.[31] The length of the amplicons were in the range of 500–800 bp with *Cunninghamella bertholletiae* at 602–631 bp and *Cunninghamella echinulata* at 770–810 bp. The PCR mixture included 10 mM Tris–HCl pH 8.3, 50 mM KCl, 2.5 mM $MgCl_2$, 100 µM of each dNTP, 1 µM of each primer, and 1.5 U of AmpliTaq DNA polymerase. The amplification program consisted of 35 cycles with denaturation at 95°C for 30 s, annealing for 1 min at 52°C, and extension for 1 min at 72°C. The PCR products were sequenced using the dideoxy chain termination method on an ABI Prism 310 instrument. Sequences were compared to those deposited in the GenBank/EMBL database with an identity level of ≥98% required for identification to the species level. Identity levels below 98% were reported only to the genus level. Of 190 isolates morphologically identified as zygomycetes, the molecular analysis indicated that *Rhizopus oryzae* represented almost half of the isolates (44.7%) followed by *Rhizopus microsporus* (22.1%), *Mucor circinelloides* (9.5%), *Myocladus corymbifera* (5.3%), *Rhizopus pusillus* (3.7%), *Cunninghamella bertholletiae* (3.2%), *Mucor indicus* (2.6%), *Cunninghamella echinulata* (1%), and *Apophysomyces elegans* (0.5%). The authors also noted a high genetic variability in the 5.8S rRNA gene of *Cunninghamella* spp. and *Apophysomyces* spp., leading them to conclude that the identification of a *Cunninghamella* ITS region amplicon distinct from either *C. bertholletiae* or *C. echinulata* using phylogenetic analysis supports the idea that there may be other species of *Cunninghamella* that have yet to be recognized.

Voigt et al. sequenced a portion of the 18S small ribosomal DNA subunit and the D1/D2 domains of the 28S large ribosomal DNA subunit in order to construct a sequence database for the clinically significant zygomycetes including *Cunninghamella*.[32] The 28S sequences were aligned and 13 separate primer pairs were constructed from the consensus sequence to allow for PCR amplification in order to rapidly identify causative agents of mucormycosis without cross-reacting with related genera. The *Cunninghamella*-specific primer pair sequences are listed in Table 82.1.

82.2.4 PCR Methods for Identification of *Cunninghamella* spp.

Kobayashi et al. used a pan-fungal PCR reaction followed by direct DNA sequencing of the PCR reaction product to identify *C. bertholletiae* as the cause of a fatal case of pulmonary mucormycosis in a host with acute lymphocytic leukemia.[33] Serum samples (200 µL) were extracted using a High-Pure viral nucleic acid kit (Roche Diagnostics). Sputum was extracted using traditional phenol-chloroform technique and formalin-fixed, paraffin-embedded lung tissue was extracted using DEXPAT (Takara Shuzo, Tokyo, Japan). PCR amplification was done using pan-fungal (U1 and U2) or *Cunninghamella*-specific primers (Cu1 and Cu2) (Table 82.1).

The PCR reaction mixture consisted of 5 μL of 10× PCR buffer, 2.5 mmol $MgCl_2$, 0.25 mmol of the four dNTPs, 100 pmol of each primer, and 2.5 U of Taq polymerase in a total volume of 50 μL. Thermal cycling conditions were 5 min at 95°C followed by 50 cycles of 30 s at 95°C, 1 min at 50°C, and 2 min at 72°C. A 5 min 72°C hold cycle followed. PCR products were separated on an agarose gel to obtain either a 371 bp product using the U1/U2 primer set or a 571 bp product using the CU1/CU2 primer set. Direct sequencing of the PCR product was done using the BigDye terminator cycle sequencing kit (Applied Biosystems) using the Cu1 primer. A BLAST search of the sequence obtained using the *Cunninghamella*-specific primer product matched the *C. bertholletiae* 28S rRNA sequence exactly. Sequence analysis using the pan-fungal primer product demonstrated a single mismatch in the U1 primer region.

Bialek et al. described a seminested PCR assay that targeted the 18S rDNA of zygomycetes including *C. bertholletiae* to the exclusion of *Aspergillus* species.[34] Paraffin-embedded tissue from sinus, lung, CNS, liver, and skin biopsies were screened. The paraffin wax was removed by adding 1 mL of xylene to an Eppendorf tube containing two 5 μm sections of the embedded tissue. Following incubation, the extract was washed with xylene and ethanol and DNA was extracted using a QIAamp tissue kit (Qiagen), which was modified by the authors through the addition of three cycles of freezing in liquid nitrogen for 1 min followed by boiling for 5 min to disrupt the fungal cell wall. Two outer primers (ZM1 and ZM2, Table 82.1) were used for the first PCR reaction and produced a product corresponding to nucleotide positions 711–1117 of *Rhizopus arrhizus* (AF113440). A second nested PCR reaction utilized primers ZM1 and ZM3, which produced a product of about 176 bp depending on the species amplified.

The PCR reaction mixture contained 10 μL of extracted DNA in a total volume of 50 μL with 10 mM Tris/HCl pH 8.3, 50 mM KCl, 2.5 mM $MgCl_2$, 1 μM of each outer primer, 1.5 U of AmpliTaq polymerase (Roche), and 100 μM of each dNTP (Promega). For the nested reaction, the same mixture was used with 1 μL of the first PCR reaction replacing the extracted DNA, 50 μM of each dNTP, and 1 μM of each inner primer. PCR conditions were 94°C for 5 min, 94°C for 30 s over 35 cycles, 50°C for 30 s, 72°C for 1 min followed by 72°C for 5 min. A human β globin gene (nucleotides 70400–70667) was used as a control of amplification inhibition. Globin primers were G1 (5′-GAA GAG CCA AGG ACA GGT AC-3′) and G2 (5′-CAA CTT CAT CCA CGT TCA CC-3′), and conditions were identical to those used above with the exception of 5 μL of extracted DNA. Three millimolar $MgCl_2$ a measured quantity was used and the extension time was lowered to 45 s. Primary PCR products were analyzed using electrophoresis on a 1.8% agarose gel and ethidium bromide staining for detection. Cloned plasmid DNA was used as a positive control. Nested PCR products were sequenced and sequences were compared to BLAST search results from the GenBank databases. For the zygomycete targets, a 100% homology to the database sequence was required to be reported.

Of 23 cases of zygomycosis diagnosed by histopathology, PCR resulted in genus- or species-specific identification in 14 cases. The authors concluded that the availability of a definitive diagnosis of zygomycosis to the genus or species level can help to optimize antifungal therapy. Organisms identified included *Rhizopus* spp., *Rhizomucor* spp., *Mucor* sp., and *Myocladus* (formerly *Absidia*) *corymbifera* but *Cunninghamella*-specific DNA was not detected in the study specimens.

Rickerts et al. used the assay described above[34] and compared the ability of histopathologic analysis, culture, and PCR to detect invasive mold infections in tissue biopsy specimens.[35] Mold hyphae, including zygomycetes ($n = 6$), were detected in 27/56 histological specimens (48%) while culture detected 17/27 specimens (63%), and a seminested PCR strategy detected mold in 26/27 (96%) specimens. Identification to the genus and/or species level by PCR correlated with culture identification in 16/18 culture-positive cases, leading the authors to conclude that PCR is superior to culture in patients with invasive mold infections.

We recently developed a real-time PCR method for the detection of zygomycetes including *Cunninghamella* species (*C. bertholletiae*, *C. blakesleeana*, *C. echinulata*, *C. elegans*).[36] The assay utilized primers and fluoresence resonance energy transfer (FRET) probes targeting a 167 bp region of the multicopy cytochrome *b* gene. The clinical sensitivity and specificity of the assay from culture isolates was determined to be 100% and 92%, respectively. The assay detects the genera *Absidia*, *Apophysomyces*, *Cunninghamella*, *Mucor*, *Rhizopus*, and *Saksenaea* through the use of melt curve analysis, although the melt curves are not sufficiently resolved to allow identification to the genus level. The assay is also able to detect zygomycetes from fresh tissue as well as from formalin-fixed, paraffin-embedded tissue. The closed format of real-time PCR significantly reduces the potential for contamination from environmental zygomycetes.

Kasai et al. described a quantitative real-time PCR assay for the detection of *Cunninghamella* species.[37] The assay targeted the 28S rRNA gene and used consensus sequence alignment of *Cunninghamella* species from GenBank. The mastermix for PCR contained 0.5 μM primers, 1× PCR buffer, 4 mM $MgCl_2$, 0.025% bovine serum albumin, 0.025 U/mL Platinum Taq DNA polymerase, 0.2 mM dNTPs, and 0.2 μM FRET probes labeled with fluorescein isothiocyanate (FITC) and RD640 (Table 82.1). Uracil-*N*-glycosylase (UNG) was also incorporated to prevent amplicon carryover. The amplification conditions consisted of denaturation at 95°C for 0 s (slope, 20°C/s), annealing at 57°C for 5 s (slope, 10°C/s), and extension at 72°C for 15 s for 50 cycles. A melt curve was generated at the end of the run to assure product specificity; the conditions were 90°C for 0 s (slope, 20°C/s), 40°C for 30 s (slope, 2°C/s) and 75°C for 0 s (slope, 0.2°C/s). Furthermore, the assay was able to detect 10–100 copies of target and had high specificity. Further, the assay was able to distinguish between *C. bertholletiae* and *C. echinulata* using melt curve analysis. Although not tested directly, *C. elegans* may also be detected by virtue of its sequence homology to the primer and probe sequences used for this assay. The clinical sensitivity of the assay from an experimental model of pulmonary zygomycosis in both BAL fluid and lung tissue samples was excellent at 100% for both specimen sources.

82.3 CONCLUSIONS AND FUTURE PERSPECTIVES

Zygomycosis appears to be increasing in frequency particularly in the immunocompromised host possibly due to prophylaxis regimens using antifungal agents that have gaps in their spectrum of coverage for these organisms. The recent study by Alvarez et al. indicates that *Cunninghamella* species account for 3%–4% of all zygomycetes identified from clinically significant sources at one reference laboratory in the United States. Given that *Cunninghamella* spp. appear to be more pathogenic than other zygomycete genera, the need for rapid, accurate diagnostics of *Cunninghamella* is important in order to provide the best possible chance for therapeutic success in these extremely challenging clinical situations. Traditional methods of identification such as morphology can be challenging, slow and requires trained, experienced mycologists who are in short supply in many laboratories; therefore, molecular diagnostics provide an important opportunity to accurately and rapidly differentiate *Cunninghamella* spp. from other zygomycetes. To date, there has not been a significant emphasis in the clinical diagnostic laboratory on the development of molecular methods for zygomycetes including *Cunninghamella*. No doubt this is secondary to the relative rarity of these infections. Nucleic acid sequencing has proven to be a useful tool once the organism has been grown in culture but for optimal patient care, the ideal scenario is to directly identify *Cunninghamella* spp. from patient specimens without waiting for culture growth. A few PCR methods described in this chapter have been useful for direct detection of *Cunninghamella* spp. although many were designed to detect the more commonly encountered genera of Zygomycetes such as *Rhizopus* or *Mucor* rather than focusing specifically on *Cunninghamella*. The laboratory is an essential partner in patient care and management, and as such the development of molecular methods able to rapidly and accurately detect these organisms will be critical to continue success in the battle against the high mortality associated with *Cunninghamella* infection.

REFERENCES

1. Weitzman, I. and Crist, M.Y., Studies with clinical isolates of *Cunninghamella*. I. Mating behavior, *Mycologia*, 71, 1024, 1979.
2. Jayasuriya, N.S. et al., An unusual presentation of rhinofacial zygomycosis due to Cunninghamella sp. in an immunocompetent patient: A case report and literature review, *Oral Dis.*, 12, 67, 2006.
3. Ribes, J.A., Vanover-Sams, C.L., and Baker, D.J., Zygomycetes in human disease, *Clin. Microbiol. Rev.*, 13, 236, 2000.
4. Richardson, M.D. and Koukila-Kahkola, P., *Rhizopus, Rhizomucor, Absidia*, and other agents of systemic and subcutaneous Zygomyceses, in *Manual of Clinical Microbiology*, p. 1839, Murray, P.R., Baron, E.J., Jorgensen, J.H., Landry, M.L., and Pfaller, M.A. (Eds.), ASM Press, Washington, DC, 2007.
5. Roden, M.M. et al., Epidemiology and outcome of zygomycosis: A review of 929 reported cases, *Clin. Infect. Dis.*, 41, 634, 2005.
6. Frater, J.L., Hall, G.S., and Procop, G.W., Histologic features of zygomycosis: Emphasis on perineural invasion and fungal morphology, *Arch. Pathol. Lab. Med.*, 125, 375, 2001.
7. Ortin, X. et al., *Cunninghamella bertholletiae* infection (mucormycosis) in a patient with acute T-cell lymphoblastic leukemia, *Leuk. Lymphoma*, 45, 617, 2004.
8. Darrisaw, L. et al., *Cunninghamella* infection post bone marrow transplant: Case report and review of the literature, *Bone Marrow Transplant.*, 25, 1213, 2000.
9. Andrade, V.S. et al., A factorial design analysis of chitin production by *Cunninghamella elegans*, *Can. J. Microbiol.*, 46, 1042, 2000.
10. Pothuluri, J.V. et al., Metabolism of metolachlor by the fungus *Cunninghamella elegans*, *Arch. Environ. Contam. Toxicol.*, 32, 117, 1997.
11. Lisowska, K. and Dlugonski, J., Concurrent corticosteroid and phenanthrene transformation by filamentous fungus *Cunninghamella elegans*, *J. Steroid Biochem. Mol. Biol.*, 85, 63, 2003.
12. Parshikov, I.A. et al., Transformation of artemisinin by *Cunninghamella elegans*, *Appl. Microbiol. Biotechnol.*, 64, 782, 2004.
13. Motohashi, K. et al., Cutaneous zygomycosis caused by *Cunninghamella bertholletiae* in a patient with chronic myelogenous leukemia in blast crisis, *Am. J. Hematol.*, 84, 447, 2009.
14. Rickerts, V. et al., Cluster of pulmonary infections caused by *Cunninghamella bertholletiae* in immunocompromised patients, *Clin. Infect. Dis.*, 31, 910, 2000.
15. Kontoyiannis, D.P. et al., Zygomycosis in a tertiary-care cancer center in the era of *Aspergillus*-active antifungal therapy: A case-control observational study of 27 recent cases, *J. Infect. Dis.*, 191, 1350, 2005.
16. Prabhu, R.M. and Patel, R., Mucormycosis and entomophthoramycosis: A review of the clinical manifestations, diagnosis and treatment, *Clin. Microbiol. Infect.*, 10, S31, 2004.
17. Marr, K.A. et al., Epidemiology and outcome of mould infections in hematopoietic stem cell transplant recipients, *Clin. Infect. Dis.*, 34, 909, 2002.
18. Walsh, T.J. et al., Infections due to emerging and uncommon medically important fungal pathogens, *Clin. Microbiol. Infect.*, 10, S48, 2004.
19. Gonzalez, C.E., Rinaldi, M.G., and Sugar, A.M., Zygomycosis, *Infect. Dis. Clin. North Am.*, 16, 895, 2002.
20. Garey, K.W. et al., *Cunninghamella bertholletiae* infection in a bone marrow transplant patient: Amphotericin lung penetration, MIC determinations, and review of the literature, *Pharmacotherapy*, 21, 855, 2001.
21. Ibrahim, A.S., Spellberg, B., and Edwards, J. Jr., Iron acquisition: A novel perspective on mucormycosis pathogenesis and treatment, *Curr. Opin. Infect. Dis.*, 21, 620, 2008.
22. Kanafani, Z.A. and Perfect, J.R., Antimicrobial resistance: Resistance to antifungal agents: Mechanisms and clinical impact, *Clin. Infect. Dis.*, 46, 120, 2008.

23. Canuto, M.M. and Rodero, F.G., Antifungal drug resistance to azoles and polyenes, *Lancet Infect. Dis.*, 2, 550, 2002.
24. Perlin, D.S., Resistance to echinocandin-class antifungal drugs, *Drug Resist. Updat.*, 10, 121, 2007.
25. Larone, D.H., *Medically Important Fungi: A Guide to Identification*, ASM Press, Washington, DC, 2002.
26. Alastruey-Izquierdo, A. et al., Activity of posaconazole and other antifungal agents against Mucorales strains identified by sequencing of internal transcribed spacers, *Antimicrob. Agents Chemother.*, 53, 1686, 2009.
27. Alastruey-Izquierdo, A. et al., Prevalence and susceptibility testing of new species of *Pseudallescheria* and *Scedosporium* in a collection of clinical mold isolates, *Antimicrob. Agents Chemother.*, 51, 748, 2007.
28. Alvarez, E. et al., Spectrum of zygomycete species identified in clinically significant specimens in the United States, *J. Clin. Microbiol.*, 47, 1650, 2009.
29. Mitchell, T.G. et al., Unique oligonucleotide primers in PCR for identification of *Cryptococcus neoformans*, *J. Clin. Microbiol.*, 32, 253, 1994.
30. Viebahn, M. et al., Assessment of differences in ascomycete communities in the rhizosphere of field-grown wheat and potato, *FEMS Microbiol. Ecol.*, 53, 245, 2005.
31. Gilgado, F. et al., Molecular phylogeny of the *Pseudallescheria boydii* species complex: Proposal of two new species, *J. Clin. Microbiol.*, 43, 4930, 2005.
32. Voigt, K., Cigelnik, E., and O'Donnell, K., Phylogeny and PCR identification of clinically important Zygomycetes based on nuclear ribosomal-DNA sequence data, *J. Clin. Microbiol.*, 37, 3957, 1999.
33. Kobayashi, M. et al., Molecular polymerase chain reaction diagnosis of pulmonary mucormycosis caused by *Cunninghamella bertholletiae*, *Respirology*, 9, 397, 2004.
34. Bialek, R. et al., PCR based identification and discrimination of agents of mucormycosis and aspergillosis in paraffin wax embedded tissue, *J. Clin. Pathol.*, 58, 1180, 2005.
35. Rickerts, V. et al., Comparison of histopathological analysis, culture, and polymerase chain reaction assays to detect invasive mold infections from biopsy specimens, *Clin. Infect. Dis.*, 44, 1078, 2007.
36. Hata, D.J. et al., Real-time PCR method for detection of zygomycetes, *J. Clin. Microbiol.*, 46, 2353, 2008.
37. Kasai, M. et al., Detection of a molecular biomarker for zygomycetes by quantitative PCR assays of plasma, bronchoalveolar lavage, and lung tissue in a rabbit model of experimental pulmonary zygomycosis, *J. Clin. Microbiol.*, 46, 3690, 2008.

83 Entomophthorales

Johannes E. Rothhardt, Volker U. Schwartze, and Kerstin Voigt

CONTENTS

83.1 INTRODUCTION

83.1.1 CLASSIFICATION

The intraordinal and interordinal phylogenetic relationships of the order Entomophthorales are still unclear. The Entomophthorales includes, primarily, insect-killing fungi, although a few exhibit a certain pathogenic potential to infect animals and humans. The Entomophthoralean group traditionally belongs to the phylum Zygomycota, but recent phylogenetic reevaluations show that the Zygomycota are polyphyletic.[1] The phylogenetic position of the Entomophthorales and the degree of phylogenetic relatedness to its putative connatural fungal groups are still unresolved. Recently, the Zygomycota have been divided into several subphyla and the Entomophthorales are included as a separate subphylum, the Entomopthoromycotina.[2] In addition, the traditional family structure within the Entomophthorales has changed several times based on a multitude of phylogenetic analyses and algorithms applied to different loci. In these analyses, the phylogenetic position of *Basidiobolus* is pivotal in resolving a natural classification. Multigene genealogies place *Basidiobolus* either with the Chytridiomycota[3] or the Entomophthorales.[1] In contrast to the ambiguous phylogenetic position of *Basidiobolus*, *Conidiobolus*, and other Entomophthoralean genera consistently group within the Entomophthorales.

The classification of the Entomophthorales using morphological and ecological characteristics has changed several times after the first description by Winter[4] and is still under controversial discussion. The molecular phylogenetic analysis using, for instance, exonic nucleotide sequences from the gene encoding the nuclear RNA polymerase subunit 1 (RPB1) separated the Basidiobolaceae from the Entomophthorales.[5] A recently published comprehensive phylogenetic analysis confirms the uncertain position of *Basidiobolus* within the fungi.[2]

Using 18S rRNA sequences for phylogenetic analysis produces results placing *Basidiobolus* inside a cluster of

Chytridiomycota.[3] However, a comprehensive study[1] using six genes of taxa representing almost all fungal or fungus-like group places *Basidiobolus* with the chytrid *Olpidium brassicae* as a sistergroup to *Entomophthora* and *Conidiobolus* constituting the Entomophthorales. Summarizing using molecular, morphological, and ecological data, the phylogeny of Entomophthorales excluding *Basidiobolus* seems to be clear. *Basidiobolus* may probably represent a phylogenetic link between the chytrids and Entomophthorales, but this position is still unclear and requires further research based on more phylogenetic marker genes. The ambiguity between phylogenetic position and morphology-based taxonomy of *Basidiobolus* among the alliance of Entomophthoralean fungi requires further discussion in the light of the genealogical concordance phylogenetic species recognition (GCPSR) concept, which is based on the multigene genealogy-dependent definition of phylogenetic species.[6]

In accordance to the classification adopted in the 10th edition of the *Dictionary of the Fungi*, the Entomophthorales consist of five families: Ancylistaceae (5 genera, e.g., *Conidiobolus*), Completoriaceae (1 genus *Completoria*), Entomophthoraceae (17 genera, e.g., *Entomophthora* and *Entomophaga*), Meristacraceae (6 genera), Neozygitaceae (3 genera), and Incertae sedis (1 genus).[7] The Basidiobolaceae, which are part of the Incertae sedis group among the phylum Zygomycota, comprises solely two genera, *Basidiobolus* and *Amphoromorpha*.[7]

83.1.2 Morphology

Although the Entomophthorales do not probably belong to the abolished polyphyletic phylum Zygomycota, they still share many morphological and ecological characteristics with the members traditionally assigned to this taxon. The medically important species *Basidiobolus ranarum*, *Conidiobolus coronatus*, and perhaps *Conidiobolus incongruus* possess mainly nonseptate mycelium and they produce large and very hydrophobic single-celled spores (10–40 μm), which are forcibly discharged. Zygospores, so far produced, develop after fusion of a pair of two gametangia (a process called gametangiogamy) forming the suspensors after the initialization of genesis and maturation of the zygospore. Asymmetric zygospore formation on a single suspensor occurs frequently and is the characteristic for the Entomophthorales.[8]

83.1.2.1 *Conidiobolus coronatus*

The unbranched sporophores of *C. coronatus* are positively phototropic and discharge spores that are able to germinate immediately (Figure 83.1C–F). The typical papilla, the former connection between sporophore and spore, is a 5–10 μm projection on the spore and it is not involved in spore germination. The emerging germ tube is able to develop into a mycelium comprising hyphae up to 6–25 μm in diameter and nonparallel, pleomorphic walls (Figure 83.1A and B).[9] On the other hand, the germ tube is also capable to form a secondary dumbbell-shaped spore. It is described in few studies that the primary spores form hair-like structures or multiple small spherical structures resembling a corona.[10,11] Another type of spore, a chlamydospore, serves as a resting spore. This is an asexual spore and is formed by modification of the vegetative hyphae, which become thick walled.

Observations have shown that the fungal mycelium grows fast and flat on most standard media (e.g., Sabouraud [SABG] agar) at 28°C, different strains of *C. coronatus* growing at different rates *in vitro*. The time taken to overgrow a Petri dish differs from 2 to 4 days depending on the strains tested. The cause of the rapid growth is the early and multitudinous spore production that results in the ejection of ballistospores extending the area of the growing mycelium with satellite colonies. Macroscopically, the mass of these large spores gives the colony a white-cream, rough, and powdery appearance. In addition to these morphological features, the mycelium desiccates the agar to produce a radially folded colony. This is a unique and reliable macromorphological diagnostic character.

83.1.2.2 *Basidiobolus ranarum*

Basidiobolus ranarum shows morphological and ecological affinities to *Conidiobolus* spp., and it also exhibits some characteristics resembling features of *C. coronatus*. The discharged ballistospores of *B. ranarum* germinate immediately under suitable conditions like those of *C. coronatus*. Under inappropriate growth conditions, the spores can form a new sporophore directly with the new spore reaching maturity within 2–3 h.[8] The germinating spores form hyphae of 8–20 μm width (similar to the hyphae of *C. coronatus*) with somewhat randomly produced septa providing the mycelial segmentation typical for the Entomophthorales (Figure 83.1G and H).

During the process of spore formation, which is less prolific and not as rapid as in *C. coronatus*, the hyphae become increasingly septate. The short sporophores grow with positive phototropism and enlarge to form single-celled sporangia. The sporangium and the upper part of the sporophores are forcibly discharged for up to 1–2 cm. It is a characteristic of *B. ranarum* that the apexes of these sporophores break after the spores were actively released. Subsequently, the sporangia (ballistospores) form new satellite colonies on the agar. Due to this multiplication of growth initiation points, the expansion of the colony increases faster than the normal growth of the fungal hyphae.[11]

Zygospores are formed after conjugative hyphal fusion by homothallic gametangiogamy. They are produced in high quantity and possess a thick wall with one beak-like tip (Figure 83.1J and K). This tip is a characteristic remnant of the gametangiogamy process. It is conceivable that the zygospores serve as dormant spores ensuring survival during conditions unsuitable for growth. However, the germination of zygospores is not unambiguously and reproducibly documented.

The colony morphology of *B. ranarum* cannot be easily distinguished from that of *C. coronatus*. The hyphal system of *B. ranarum* might be more compact than that of

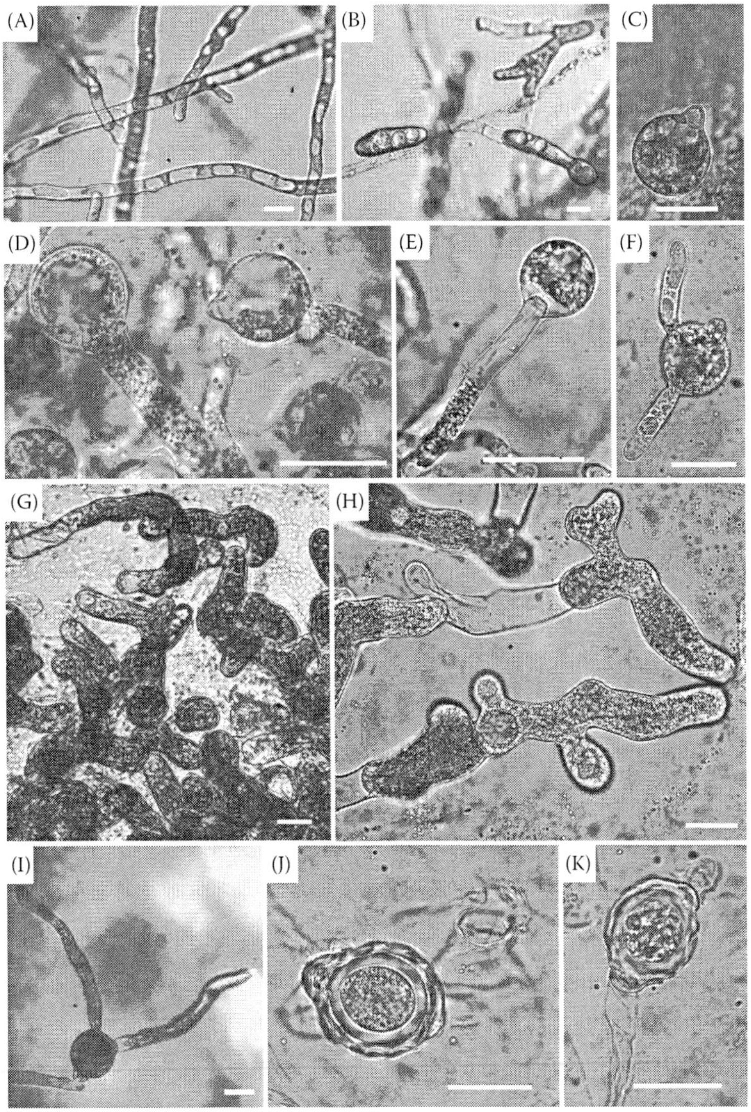

FIGURE 83.1 Micromorphological observations on Conidiobolus coronatus (A–F) and Basidiobolus ranarum (G–K). A: hyphae, B: chlamydospore-like resting spores, C–F: unispored sporangiole (mitospore), C and D: mitospores with typical tips, F: bipolar spore germination, G and H: hyphae with swollen and pleomorphic hyphal tips typical for Basidiobolus, I: germinating mitospore bearing a beak-like tip, J and K: zygospores. The scale bars indicate 10 μm each.

C. coronatus. The color is whitish cream and the surface is rough with the typical powdery appearance. The agar medium also becomes folded and desiccated as in the case of *Conidiobolus*.

In comparison, both Entomophthoralean fungi are able to sporulate rapidly after ballistospore germination, while the initial mycelium is just starting to grow. In addition, the medium becomes desiccated, a process which is associated with a folded colony morphology. This feature is apparently more pronounced in *Basidiobolus ranarum*.

The beak-like tip of the zygospores and sporophores with a swollen area below the ballistospore is typical for *B. ranarum*.

Even though zygospores are unknown in *C. coronatus*, its asexual chlamydospores may serve as analogues. In addition, *C. coronatus* develops spores with typical papillae giving the spores a zygospore-like appearance.[11]

83.1.3 Biology and Epidemiology

The Entomophthorales are commonly distributed and ubiquitous. All genera are more or less able to infect and kill arthropods including a very broad spectrum of taxa such as Tardigrada,[12] Coleoptera,[13] and Diptera.[14] Species of the genera *Basidiobolus* and *Conidiobolus* can also infect humans and other mammals such as horses[15] and sheep.[16,17]

Species of *Conidiobolus*, with *C. coronatus* as the most common one, can be isolated from soil, decaying vegetation, the feces of amphibians, from reptiles, insectivorous bats, and other animals and also from dead insects overgrown by various other fungi.[18,19] In addition to humans, horses[20] and sheep[21] infections with *Conidiobolus* spp. in a dolphin[22] and a chimpanzee[23] are also reported. The fungus can live as a saprobe, commensal, or parasite in very different environmental conditions and has a worldwide distribution. However, high humidity (95%–100%) and high temperature (25°C–37°C) appear to be important for its satisfactory propagation and the germination of the spores.[18,24]

Almost all described cases of conidiobolomycosis caused by *Conidiobolus coronatus* are located in tropical and subtropical area of Africa, Asia, Southeast Asia, the Middle East, Central, and South America.[25] Only a few cases are reported from Europe and North America and for some of these scattered instances, the reports relate to immigrants from the main regions of prevalence.[26] Only few cases are infections caused by *C. incongruus* known,[27] but these were comparatively invasive.[18]

Basidiobolus ranarum is as prevalent as species of *Conidiobolus*. It is common worldwide, but usually causes infections in subtropical and tropical regions. The fungus was first isolated in 1886 from a frog[28] and can be isolated from the same habitats as species of *Conidiobolus*: soil, decaying plants, insects, and the gastrointestinal tract or the feces of amphibians and reptiles. Moreover, *B. ranarum* can also be detected in substances of mammalian origin with case reports related to bats,[29] horses,[15] dogs,[30] and also humans.

Both pathogens induce the rare entomophthoromycosis in mostly immunocompetent human hosts. In addition, conidiobolomycosis and basidiobolomycosis are usually subcutaneous infections and every genus has its own predilection site.

Conidiobolus coronatus mainly afflicts adults and causes facial infections with a chronic sinusitis.[10,31] Infections by *C. incongruus* are very rarely described, but the few reported cases have almost always resulted in a fatal outcome, independent of the immune status of the infected host. The infections caused are usually more invasive and have a higher dissemination rate than infections caused by *C. coronatus*.[27]

Basidiobolus ranarum causes chronic subcutaneous and peripheral infections of the trunk, limbs, buttocks, and arms.[31] Only a few cases (15–20) of fatal gastrointestinal basidiobolomycosis are reported.[32,33] Basidiobolomycosis of male and female patients of every age are known, although male children seem to be more frequently afflicted.[9,18]

Conidiobolus coronatus and *Basidiobolus ranarum* are able to disseminate in the human host and cause extremely dangerous infections. Conidiobolomycoses seem to be less abundant than basidiobolomycoses. All cases are very infrequently reported. The dependency between infection and the immune status of the host is still unclear. Disseminated entomophthoromycoses occur in patients with and without immunodeficiency but without a tendency of prevalence for a special immune status.[18,34,35]

It is likely that there are specific preconditions for the rare but occasionally occurring infections caused by *Conidiobolus coronatus*, *C. incongruus*, and *Basidiobolus ranarum*. Even though the fungi are ubiquitous, only 167 cases of conidiobolomycosis[10] are reported worldwide and a similar number can be assumed for basidiobolomycoses.[32,36,37] A possible explanation might be the climatic influence of tropical and subtropical regions on the etiology and manifestation of the disease. Warm and humid climate appears to be important for a successful infection and establishment of entomophthoromcoses. Furthermore, a number of gastrointestinal basidiobolomycoses in Arizona have recently emerged even if the number of patients (at seven cases) remains still small.[32] Possible reasons might be (i) an increased exposure to amphibians and reptiles,[37] (ii) camping near a water body, or (iii) accompanying diseases like diabetes mellitus, peptic ulcer disease, or pica.[32] Treatment with ranitidine might be a risk factor as its major effect is a reduced stomach acidity. This allows the fungi to pass the stomach more easily and results in more frequent infections of the gastrointestinal tract.[32]

A more or less intensive thermotolerance (28°C–37°C) and the ability to use a wide spectrum of carbon sources (these features can differ strongly among strains) are possible factors contributing to an essential pathogenicity of these fungi (own unpublished data).

83.1.4 Clinical Features and Pathogenesis

The infections caused by Entomophthoralean fungi are referred to by a number of different terms in the literature. Usually these cases are called zygomycosis. The detailed names to characterize the etiologic agent of the mycosis are entomophthoromycosis (to differentiate them from mucormycosis), basidiobolomycosis (caused by *Basidiobolus ranarum*), and conidiobolomycosis (caused by *Conidiobolus coronatus* and *C. incongruus*).

The etiologic agents of these fungal infections are *B. ranarum*, *C. coronatus*, and, in very rare cases, *C. incongruus*. The two main Entomophthoralean pathogens differ in pathogenesis and clinical features, but still have many similarities. Usually both fungi cause chronically subcutaneous infection of the human tissue of immunocompetent patients, whereas the predilection site differs. *B. ranarum* normally afflicts the skin between feet and shoulder: trunk, limbs, buttocks, and arms.[31] In addition, few (10–15) gastrointestinal cases of basidiobolomycoses are known and well documented.[32,33,36,38] In contrast, *C. coronatus* causes mainly subcutaneous infections of the sinusitis and facial skin.[10,26] Furthermore, both fungi can rarely cause disseminated infections, which may prove to be fatal.

83.1.4.1 Characteristic Features of Entomophthoromycoses

Patients infected with *C. coronatus* display in most cases a swollen nose or a swelling of the upper lip or the entire face.[9,10,26,39] The infection is almost always present in the paranasal sinus and extends to nearly all contiguous tissues.[18] The swollen area is firm and feels rubbery. As different tissue can be infected, the symptoms vary from case to case. The infected skin may ache or be pain-free and subcutaneous nodules may be present or absent.[10,26]

The clinical picture of subcutaneous infections caused by *B. ranarum* is usually as follows. The infection becomes apparent as a solid, hardened subcutaneous painless nodule on arms, the trunk, buttocks, and limbs. It enlarges gradually toward the periphery as a clearly defined flat mass attached to the overlying skin. The cutaneous area is intact and might be bluish or purple at the region with active growth. A functional impairment is not present as long as the joints are not blocked by the swollen tissue, which may reach enormous proportions.[36]

Normally, cross sections of samples show broad, thin-walled hyphae with surrounding eosinophilic granulocytes, called Splendore-Hoeppli phenomenon. However, the presence or absence of this feature is a sufficient criterium for the medical outcome of an entomophthoromycosis. Hence, three techniques combined—histopathology, molecular detection, and culture of the fungus—are the most reliable approach for the verification of an entomophthoromycosis.

83.1.4.2 Unusual Cases of Entomophthoromycoses

One of the first well-documented cases of disseminating conidiobolomycosis was reported by Eckert et al.[40] The authors describe a deep infection with *Conidiobolus* in a 15-month-old boy, who showed a brassy and hoarse cough, weight loss, temperature, and irritability. Investigations led to a thoracotomy, which revealed a hard, posterior mediastinal mass invading the left main stem bronchus and pericardium. Histology showed a Splendore-Hoeppli phenomenon, which is characteristic for entomophthoromycosis. The culture of a biopsy sample led to the identification of a fungus of the genus *Conidiobolus*. A determination to the species level was not carried out. Kamalam and Thambiah report a case of lymph node invasion by *C. coronatus* in a 40-year-old woman from India with a subcutaneous infection of facial tissue with this kind of etiologic agent.[41] Another case is described by Walker et al.[35] A renal transplant patient infected with cytomegalovirus and the fungus *Histoplasma capsulatum*, and additionally with *C. coronatus*. This caused a disseminated infection and blood vessel invasion, which suggests that *C. coronatus* can cause disseminated infections in immunocompetent and immunocompromised humans as well.

Case reports for *C. incongruus* infections are very rare, but almost all cases are invasive and fatal, independent of the immune status of the host. Busapakum et al. describe a case of a 20-year-old immunocompetent woman from Thailand with a disseminated infection of *C. incongruus*.[27] She suffered from a subcutaneous mass in her left breast, low-grade fever, weight loss, and cough with hemoptysis. Biopsy of the skin and subcutaneous tissue showed the Splendore-Hoeppli phenomenon around fungal hyphae. Six weeks after admission, the patient died despite antifungal therapy. The autopsy examination showed fungal filaments with Splendore-Hoeppli phenomena in lymph nodes, mediastinum, esophagus, lung, liver, and jejunum. To identify the fungal pathogen, the biopsy samples of skin and subcutaneous tissue were cultured and the characteristic morphology of the culture resulted in a diagnosis of *C. incongruus* as the etiologic agent.

Another fatal case of disseminated infection is reported by Walsh et al.[42] In this case, an immunocompromised 32-year-old woman from the United States was infected with disseminating *C. incongruus*. The patient, who was granulocytopenic due to lymphocytic lymphoma, showed a rapidly progressive pneumonia and fatal pericardial tamponade. The fungus, which was resistant to amphotericin B and flucytosine *in vitro*, was determined using the morphology of the culture.

Cases of disseminated basidiobolomycosis are rarely reported, whereas gastrointestinal infections are more frequently described. A case of dissemination of a basidiobolomycosis is reported by Bigliazzi et al.[37] A healthy 40-year-old immunocompetent woman died after 4 months of indeterminate medical conditions. The patient showed low-grade fever, increasing eosinophilia, and increasing immunoglobulin E. Furthermore, an airflow obstruction was diagnosed, an oxygen saturation of 98% while on room air was noted, and chest radiographs revealed a right upper lobe infiltrate. Biopsies from bronchus and middle right lobe showed only eosinophil infiltration and necrosis without organisms. However, the histology of biopsied subcutaneous nodules showed eosinophilic pannicultis with hyphal elements. The next day the patient died in the emergency room due to cardiorespiratory arrest after sudden acute dyspnea and hypotension. The autopsy report identified fungal hyphae with Splendore-Hoeppli phenomenon in the brain, kidneys, lungs, pancreas, spleen, and stomach. Macroscopic examination indicated necrotic white-yellow opacities (3–4 mm diameter) in the pericardium, parietal pleura of the right hemithorax, visceral surface of the right lung, liver, spleen, kidneys, adrenal glands, pancreas, uterus, and peritoneum.[37]

Another case of fatal basidiobolomycosis is reported from the Netherlands by Van den Berk et al.[34] A 61-year-old man suffering from abdominal pain and constipation for several months died of a septic shock. First diagnosis by colonoscopy revealed a large obstructing tumor in the descending colon and histology showed an inflammatory reaction. However, organisms were not implicated at this stage. A few weeks later, abdominal discomfort in the right upper colon developed. Six weeks after colonoscopy a CT scan showed a large mass of 6 cm diameter in the central position in the right liver lobe. Treatment against an amebic liver abscess with metronidazole was unsuccessful. After developing an eosinophilia, a liver biopsy was performed. Furthermore, a second CT scan showed that the size of the hepatic mass had increased to 9 cm diameter. A review of the samples showed fungal hyphae and the Splendore-Hoeppli phenomenon. The patient developed a septic shock, while *Escherichia coli* and *Clostridium perfringens* were isolated from blood. A therapy with antibiotics and amphotericin B was unsuccessful and the patient died of multiple-organ failure a few days later. After autopsy, *B. ranarum* was cultured from liver, gallbladder, and sigmoid colon, each of which was severely afflicted.

These cases show clearly that rapid molecular detection of the etiologic agent is extremely important, especially when the organism cannot be identified by pathohistology and culture.

Cases of gastrointestinal basidiobolomycosis are more frequently reported than disseminated infection with

B. ranarum, but they are still rare and usually nonfatal. In 2001, Lyon et al. summarized seven cases of gastrointestinal basidiobolomycosis, which occurred in Arizona from April 1994 to May 1999.[32] The patients' most common symptom was abdominal pain, but discharged mucus was also presented and also Leukocytosis and eosinophilia in all cases. After excision of the inflammatory mass, these features disappeared. The characteristic Splendore-Hoeppli phenomenon was not uniformly noted, although eosinophilic inflammation was prominent in all cases. The identification of the etiologic agent was always done by histology and in some cases an additional culture was performed. Remarkably, four of the seven cases of basidiobolomycosis were preoperatively diagnosed as cancer, two as inflammatory bowel disease, and one as diverticulitis.[32] Another case of gastrointestinal basidiobolomycosis was described by Khan et al. from Kuwait in 2001.[33] A 41-year-old male patient, who was otherwise healthy, was taken to a hospital after 20 days of abdominal pain. He was febrile and had a tender and distended abdomen with a palpable, nodular mass. Ultrasound confirmed the presence of a thick mass in the right iliac fossa. His prostate was enlarged and his colon showed fecal impaction. The presumptive diagnosis of intestinal tuberculosis turned out to be negative.

After the patient had developed difficulty in passing urine, a culture of his urine sample was made. Some days later, *B. ranarum* was identified by colony morphology and microscopic examination.[33]

These cases demonstrate clearly that nonspecific and diverse symptoms of gastrointestinal basidiobolomycosis may lead to wrong diagnoses followed by medication detrimental to the well-being of the patients and fatal in worst-case scenarios.

The treatment of entomophthoromycosis is highly variably implemented. Amphotericin B, terbinafin, potassium iodide, fluconazole, itraconazol, and other antimycotic agents are administered in various doses, combinations, and chronologies.[9,10,18,26,39] Although the *in vitro* susceptibility testing can show resistant isolates, the cure with these antifungals may still be achievable.[26] An additional surgical resection of the infected tissue is described in several case reports,[33,43] especially in cases of gastrointestinal basidiobolomycosis.[32,34,44,45] The need for surgery has to be carefully assessed in each case taking into account the potential for spreading the infection.

In summary, a fatal outcome of entomophthoromycosis is rare and the cure is highly dependent on a positive detection of the fungus in biopsy material, using molecular detection and culture methods.

83.1.5 Diagnosis

83.1.5.1 Conventional Techniques

Infections with *C. coronatus* are mostly subcutaneous and restricted to the facial region[10,31] whereas those of *B. ranarum* cause chronic subcutaneous and peripheral infections of the trunk, limbs, buttocks, and arms.[31] In addition, there are a few described cases of gastrointestinal infections. Both pathogens have to be isolated from the host by biopsy of the infected soft tissue rather than from the surface of the skin.

Axenic cultivation of *Basidiobolus* and *Conidiobolus* on commercially available cultivation media (e.g., SABG or cornmeal) is easy and neither time- nor cost-intensive. In contrast to most of the other Entomophthoralean genera, which include obligate parasites which rarely grow in pure cultures, species of *Basidiobolus* and *Conidiobolus* are saprobes. Macroscopic (colony morphology) and microscopic (spore morphology) observations are easy to perform and applicable to the unequivocal identification of both these organisms. The formation of asexual, aplanetic, mitospores (also known as capilliconidia, conidia, sporangia, or sporangiola) can be observed on SABG agar.[10] In addition to the asexual aplanospores, zygospores can be found for *B. ranarum* growing on SABG.[38] Since the mitospores of *C. coronatus* are large (around 25 μm diameter) and actively discharged, they can be found on the lid of the Petri dish as a fine white powdery deposit.

The observation of fungal hyphae in tissue samples is more difficult, although many staining methods are available to visualize them. In addition to staining techniques used for light microscopy, which are mostly time-consuming and labor-intensive, fluorescent dyes for fluorescence microscopy provide fast and reliable alternatives to conventional staining methods. Staining with optical brighteners like Calcofluor (Bayer, Leverkusen, Germany) or Blancophor (Sigma, Taufenkirchen, Germany) is easy, rapid, and inexpensive.[46] Optical brighteners form strong, noncovalent bonds to glycosides in the fungal cell wall.[46,47] Since such glycosides are ubiquitous in fungal cell walls (e.g. chitin) and absent in human tissue, this method results a high contrast between the fungal tissue and the host tissue, including immune cells.

A common feature of infections by Entomophthoralean fungi is the Splendore-Hoeppli phenomenon.[10,37,48,49] It is characterized by intensively eosinophilic material surrounding the pathogen.[50,51] To visualize this, different staining techniques can be used. Staining with methenamine-silver nitrate or periodic acid-Schiff (PAS) distinguishes hyphae of fungi well. Staining with hematoxylin and eosin for histology is also possible.

Since biopsy material does not show typical structures in all instances, isolation from infected tissue is not always successful; alternative methods using samples of serum have been developed.

An immunodiffusion test for serodiagnosing has been developed by Kaufman et al.[52] Antigens of the pathogen are employed to observe interactions with serum of the patient.[52] This method is inexpensive and takes just 24 h.

83.1.5.2 Molecular Techniques

Molecular techniques are fast and potentially exact techniques for the identification of fungal pathogens. For the two best-known Entomophthoralean pathogens, *Conidiobolus coronatus* and *Basidiobolus ranarum*, specific primers are available.[53] Both primers are based on specific regions of the 28S ribosomal DNA and in a polymerase chain reaction (PCR); using these primers, there is only amplification of fragments if DNA of the target organism is present. By employing this method it is possible to use samples from an infected region to amplify DNA even though there might be DNA from the patient or other microorganisms. The result can be easily visualized by gel electrophoresis in which a band of expected size can be found in the case of infection with the pathogen. Thus, diagnosis without the need for cultivation of the pathogen is possible. The main disadvantage of the use of taxon-specific primers is the limitation on the range of detectable pathogens by the availability of specific primers. In addition, it may be the case that not all pathogenic species of the Entomophthorales are described. For the genus *Basidiobolus*, there is another primer available that is capable of detecting every described species based on a specific sequence of the internal transcribed spacer (ITS) region of the genome (own unpublished data). It has not been possible, thus far, to design an equivalent primer for the genus *Conidiobolus*. This is mainly due to the lack of ITS sequences in public databases. To increase the concentration of DNA in samples taken from biopsy material, nested PCR can be performed. Since the primers for both *C. coronatus* and *B. ranarum* are for the 28S ribosomal DNA, the first amplification step is performed with universal primers for this region. The amplicon can then be amplified with the specific primer in a second PCR.

If cultures of the pathogen are available, sequence analyses are possible using universal primers to amplify the DNA for specific regions such as the 18S or 28S ribosomal DNA and the ITS followed by sequencing. Many commercial organizations offer these services.

83.2 METHODS

83.2.1 Sample Preparation

83.2.1.1 Sample Recovery

Material for the detection of *C. coronatus* can be taken from the nose cavity or other infected area under a local anesthetic.[10] Biopsy material infected with *B. ranarum* should be recovered directly from the infected region under general anesthesia.

Kaufman et al. designed an immunodiffusion assay for the diagnosis of infections with *C. coronatus* and *B. ranarum*.[52] For this assay, serum from the whole blood of the patient has to be generated in order to raise antibodies against the fungal pathogens.

83.2.1.2 Sample Preparation for Light Microscopy and Fluorescent Microscopy

Samples for microscopy can be fixed by neutral buffered formalin or 10% formaldehyde. Alternatively, fixation according to Beckstead can be performed in order to obtain better preservation of the antigenic surface proteins and morphological integrity of the fungi.[54] In this method, fresh samples of approximately 3 mm^3 are incubated for 8–12 h in zinc fixative containing 0.5% zinc acetate, 0.5% zinc chloride, and 0.05% calcium acetate in Tris base buffer (pH 7–7.4) at room temperature. After fixation, samples can be embedded in paraffin wax. For cultivation analysis, samples should not be pretreated with formalin or formaldehyde to retain maximum viability of the fungal material. If necessary, samples can be stored in 20% glycerol at −20°C. Cultivation of species from both genera should be performed on SABG agar or cornmeal agar (CMA). SABG contains 10 g/L peptone, 100 mM glucose, 4 mM $MgSO_4 \times 7\ H_2O$, and 7 mM KH_2PO_4. CMA contains 20 g/L cornmeal, 20 g/L peptone, and 20 g/L glucose. For solid medium, 15 g/L agar should be added. Optimal temperature for cultivation is 28°C–34°C. For microscopy of cultured mycelium, hyphae are scraped off and transferred to a slide. To maintain the intactness of the micromorphological structures, a few drops of 10%–35% potassium hydroxide (KOH) can be added to the slide. Further, optical brighteners like Calcofluor white (disodium salt of 4,4′-bis-[Canilino-bis-diethylamino-*S*-triazin-2-ylaminol-2,2′-stilbene-disulfonic acid; Bayer, Leverkusen, Germany) or Blancophor (Sigma, Taufenkirchen, Germany) can be added to enhance the contrast. A similar rapid method represents the fluorescence microscopy using Calcofluor white and Blancophor.[46] Thin sections of the material are transferred to a slide and are covered with a 0.1% (w/v) Calcofluor white solution. For staining with Blancophor, paraffin has to be previously removed from the sample. The paraffin-free sample is incubated for 3–30 min in 0.08 mg/mL in Tris buffer followed by two washes for 1 min each with Tris buffer and embedding in mounting medium.[46] Samples should be kept in the dark since Calcofluor and Blancophor are light-sensitive and not stable at daylight.

A more complex staining procedure for the detection of fungi in host tissue is a combined staining with methenamine-silver nitrate and hematoxylin and eosin, according to Huppert et al.[55] This method applies a combination of techniques, which are also parts of other protocols, such as Grocott's staining or Gomori methenamine silver (GMS) stain. For the staining, several solutions have to be prepared. The sample has to be deparaffinized and hydrated in water. After incubation for 1 h in 4% chromic acid and a subsequent washing step for 1 min with water, the slice is incubated for 1 min in 1% sodium bisulfite and then rinsed for 5–10 min in water. After this, the samples are washed four times with distilled water. The section is then incubated at 58°C–60°C in methenamine-silver nitrate containing 5% silver nitrate in 20 volumes hexamythelentetramine diluted in equal volume of distilled water and 8% (v/v) 5% sodium borate. The staining reaction is stopped by rinsing six times in distilled water until the microscopical sections look golden brown (minimum 1 h). The sample is incubated for 4 min in 0.1% gold chloride followed by rinsing with distilled water and incubation for 4 min in 2% sodium thiosulfate. The sections are washed under running water for 4–5 min and then incubated for 30 min in hematoxylin solution and 2% glacial acetic acid followed by washing for 30 s in running water. For differentiation, it is now quickly dipped in acid alcohol containing

70% ethanol and 1% hydrochloric acid followed by rinsing in water. The sample is finally treated with ammonia water containing 0.1% ammonium hydroxide until it is blue followed by washing under running tap water for 5–10 min. If the differentiation was successful, the fungi can be observed in black and nuclei in blue to blue-black in comparison with a colorless background. If the differentiation was not successful, the steps starting from the incubation with acid alcohol have to be repeated. For counterstaining, the section is incubated for 30–60 s in eosin–phloxin solution containing 11% (v/v) eosin stock (1% eosin Y in 95% ethanol), 1% phloxin stock (1% phloxin), and 0.5% glacial acetic acid in 95% ethanol. Finally, the slice is dehydrated, cleared, and mounted.

Alternatively, the detection can be performed via hematoxylin and eosin staining in combination with PAS staining; the PAS procedure was described by McManus[56,57] and Kligman and Mescon.[58] For this method, the fixed samples are first dehydrated and embedded in paraffin and sectioned. Sections are deparaffinized and dehydrated with absolute alcohol, washed in distilled water, and then incubated for 5 min in 1% periodic acid following washing under running tap water for 15 min. The sections are then stained with Schiff reagent for 10–15 min. Schiff regent is prepared as follows: 0.5 g of basic fuchsine is dissolved in 100 mL boiling distilled water. After cooling to 50°C, the solution is filtered and 10 mL of 1 N hydrochloric acid and 0.5 g anhydrous potassium metabisulfite are added to the filtrate. The sections are incubated twice in 0.5% potassium metabisulfite in 0.05 N hydrochloric acid and twice in 5% thionyl chloride for 5 min each, and then washed for 10 min under running water. Counterstaining is performed with light green. Finally, the sections are dehydrated, cleared, and mounted.

83.2.1.3 Preparation of Genomic DNA for Molecular Detection

Genomic DNA can be purified by the method according to Cenis.[59] Fungal cultures for DNA preparation should be cultivated in liquid SABG medium. Filtered mycelium is ground in a pestle and mortar under constant supply of liquid nitrogen to fine powder, from which 50–100 mg are transferred to a reaction tube with 300 µL lysis buffer containing 200 mM Tris–HCl, 250 mM sodium chloride (SDS), 25 mM ethylenediaminetetraacetic acid (EDTA), and 0.5% SDS in accordance to Raeder and Broda.[60] The sample is vigorously stirred for 3 min and then incubated for 10 min at 65°C for an increased cell lysis. After an additional 3 min stirring, 150 µL 3 M sodium acetate at pH 5.2 is added, and the mix is incubated for 10 min at −20°C. After centrifugation at 13,000 rpm for 15 min, the supernatant is transferred into a new reaction tube and an equal volume of isopropanol is added. The mix is incubated at room temperature for 15 min and then centrifuged at 13,000 rpm for 15 min. The pellet is washed twice with 70% isopropanol, dried, and resuspended in 10 mM Tris–HCl (pH 8) or TE buffer, containing 10 mM Tris–HCl (pH 8) and 100 mM EDTA (pH 8). The sample can be stored at −20°C until required for PCR-based analysis. This is a rapid and inexpensive procedure. However, also other methods, for example, the hexadecyl-trimethylammonium bromide (CTAB) method[61] or kits for DNA purification commercially available (e.g., Quiagen plant tissue kit, analytic Jena innuPrep Plant DNA Kit, Invitek InviMag Plant DNA Mini Kit, Macherey & Nagel NucleoSpin Plant L) can be used.

83.2.2 Detection Procedures

83.2.2.1 Morphological Detection in Pure Cultures

Cultivation of *C. coronatus* on SABG agar produces flat fungal colonies where almost no aerial hyphae can be observed. Large spores of about 10–40 µm diameter are formed and forcibly discharged, resulting in multiple satellite colonies. Macroscopically, the formation of these spores in large quantities results in a white-cream colony with a rough and powdery texture. In addition, spores can be found adhering to the lid of the Petri dish. The broad pleomorphic, essentially nonseptate, hyphae with refractive inclusions and large and typically papillate spores formed on the tip of unbranched sporophores can be easily observed under a simple light microscope using 80–400× magnification (Figure 83.1A–F).

Cultivation of *B. ranarum* on SABG agar leads also to more or less identical colonies, which are thus not easy to distinguish from *C. coronatus*. In contrast to *C. coronatus*, cultures of *B. ranarum* usually dry out and fold the agar surface. Light microscopy at 80–400× magnification can be used to observe the broad, essentially nonseptate, although rarely septate, hyphae which form short sporophores, which bulge to form solitary spores (Figure 83.1G–K).

83.2.2.2 Morphological Detection in Human Tissue

For detection of hyphae in human tissues, the staining procedures described in Section 83.2.1.2 should be used. Fluorescence microscopy with optical brighteners, performed at 365–395 nm, enables the hyphal cell wall to be detectable as a blue region, although spores, which are not formed during growth in host tissue, cannot be detected. Thus, it is not possible to reliably distinguish between *C. coronatus* and *B. ranarum*, but it helps to discriminate entomophthoromycosis from other zygomycosis. Staining by the method according to Huppert et al. visualizes fungal hyphae and cells of the host immune system. Fungal hyphae are stained black while nuclei appear dark blue.[55] With this method, it is also possible to detect the asteroid Splendore-Hoeppli reaction, which is a typical feature of entomophthoromycoses.[10,37,48,49] The fungal hyphae are surrounded by intensely eosinophilic material. This phenomenon is also detectable by PAS and hematoxylin–eosin staining.[37]

83.2.2.3 Immunological Detection

To use the immunodiffusion test according to Kaufman et al.,[52] specific antigens of *C. coronatus* and *B. ranarum* have to be generated. For that, *C. coronatus* and *B. ranarum* are cultivated in brain heart infusion broth at 37°C under constant shaking with 150 rpm. The cultures are inactivated with 0.02% merthiolate and filtered through Whatman no. 1 paper

(Whatman, Inc., Clifton, NJ). Afterwards, an additional filtering step through membrane with a pore size of 0.45 μm is performed and the filtrate is concentrated via ultracentrifugation. Petri dishes with 0.25% phenolized–1% purified agar are used for the immunodiffusion assay. Holes, 4 mm diameter, are punched in pairs with 4 mm distances from each other on the agar, so that one pair of holes corresponds to a single antigen–antibody reaction and is more distantly separated from the other pairs. Sera and antigens are applied to one of each of a pair of wells followed by incubation for 24 h in a humid chamber at room temperature. Positive antigen–antibody reactions can be observed as white bands in the agar between each pair of wells. The white bands result from coagulation of the antigen with its respective antibody in the serum following a compatible reaction.

83.2.2.4 Molecular Detection Based on Polymerase Chain Reaction

All PCR amplification mixtures should contain 1 μL genomic DNA (around 50–100 ng), 0.2 mM of each deoxynucleotide (from a 2 mM stock solution mixture), 10 pmol of each primer (from 10 pmol/μL stock solutions), 1.5–3.0 mM magnesium chloride, 1–2 units of *Taq* polymerase (e.g., Dream Taq from MBI Fermentas or InnuTaq from Analytik Jena), and a buffer system according to the manufacturer's recommendations. The temperature profile of the PCR includes an initial denaturation step of 5 min at 95°C, 30 cycles of 0.5 min at 95°C to denature the template DNA followed by 30 cycles of 1 min at 52°C–60°C for primer annealing, 1 min at 72°C for primer extension, 0.5 min at 95°C for denaturation of the novel synthesized DNA double strands, and a final primer extension step of 10 min at 72°C to ensure the completion of double-stranded amplicons. To visualize the PCR fragments after finalization of the PCR temperature profile, the PCR mix is loaded with one-fourth volume of loading buffer (0.2 g/L bromphenol blue, 0.2 g/L xylene cyanol, 10 mM EDTA, 30% glycerol) on a 1% agarose gel (Seakem LE, Serva) in 1× TAE buffer (40 mM Tris [pH 7.8], 20 mM sodium acetate, 1 mM EDTA) and electrophoretically separated in a electric field of 5–10 V/cm field strength. The fragments are visualized and photodocumented on a UV transilluminator (wavelength of 321 nm) after staining of the agarose gel in 50 μg/mL ethidiumbromide solution for 10 min. PCR fragments appear as orange fluorescent bands by intercalating the ethidium bromide molecules in the DNA double helix.

83.2.2.4.1 Taxon-Specific Amplification of Conidiobolus *and* Basidiobolus

The molecular detection can reliably be performed using *in vitro* amplification of species-specific DNA fragments by PCR using oligonucleotide primers inhabiting different taxonomic specificities (Table 83.1). For the PCR-mediated detection of *C. coronatus*, an annealing temperature of 55°C and the taxon-specific primers Cc1 and Cc2 targeting the D1/D2 domain of the nuclear large subunit (LSU, 28S) ribosomal DNA are recommended. In assays positive for the pathogen, a fragment of 419 bp in length can be detected in the lane respective to the sample. For the species-specific PCR detection of *Basidiobolus ranarum*, the taxon-specific primers Ba1 and Ba2 are used. The primer pair Ba1/2 detects *B. ranarum* and *B. haptosporus*. The PCR temperature profile is much the same like for Cc1/2-primed PCR amplifications with the exception of the application of an increased annealing temperature (60°C) to prevent false positive fragments in consequence of unspecific primer annealing. After separation of the fragments on 1% agarose gel, a fragment of 651 bp can be detected in PCR assays positives for both species of *Basidiobolus*. Considering a putative pathogenic potential of *Basidiobolus* species other than *B. ranarum* and *B. haptosporus*, an additional taxon-specific primer pair, Bs1 and Bs2, was developed that is specific for the ITS1 and ITS2 region including 5.8S of the nuclear rDNA cluster from the species of the entire genus *Basidiobolus*. For the application of that primer pair, the annealing temperature as outlined for the PCR with Cc1 and Cc2 should be used.

TABLE 83.1
Oligonucleotide Primers for PCR Detection of Entomophthorales

Primer	Specificity	Target	Primer Sequence (5′–3′)	Amplicon (bp)	Reference
Bs1	*Basidiobolus* spec.	ITS	ACTGTTRAMGTATGCTTTGGTAG	465	Authors's own unpublished data
Bs2	*Basidiobolus* spec.	ITS	CTTGCGACGCCTCCAACTAG		Authors's own unpublished data
Cc1	*Conidiobolus coronatus*	28S rDNA	TCTCTTAACTTGCTTCTATGCC	419	Voigt et al. [53]
Cc2	*Conidiobolus coronatus*	28S rDNA	CTTTAATTAAGCTAATCAACATG		Voigt et al. [53]
Ba1	*Basidiobolus ranarum*	28S rDNA	AAAATCTGTAAGGTTCAACCTTG	651	Voigt et al. [53]
Ba2	*Basidiobolus ranarum*	28S rDNA	TGCAGGAGAAGTACATCCGC		Voigt et al. [53]
NS1	Pan-eukaryotic	18S rDNA	GTAGTCATATGCTTGTCTC	~1140	White et al. [63]
NS41	Pan-eukaryotic	18S rDNA	CCCGTGTTGAGTCAAATTA		O'Donnell et al. [64]
NL1	Pan-eukaryotic	28S rDNA	GCATATCAATAAGCGGAGGAAAAG	~680	O'Donnell et al. [70]
NL4	Pan-eukaryotic	28S rDNA	GGTCCGTGTTTCAAGACGG		O'Donnell et al. [70]
ITS1	Pan-eukaryotic		TCCGTAGGTGAACCTGCGG		White et al. [63]
ITS4	Pan-eukaryotic		TCCTCCGCTTATTGATATGC		White et al. [63]

83.2.2.4.2 Universal Amplification of Conidiobolus *and* Basidiobolus *from Axenic Cultures*

As taxon-specific PCR approaches often do not detect all potential pathogenic species, it is advisable to use universal primers for the PCR-driven molecular identification (Table 83.1). A prerequisite for that procedure are pure cultures of the isolated fungi with pathogenic potential in axenic cultures in order to eliminate cross-reactions with the host or other fungal contaminants. The primer pairs ITS1/4, NS1/41, and NL1/4 according to the ITS, the nuclear small (SSU, 18S), and large subunit (LSU, 28S) ribosomal DNA, respectively, universally hybridize to the ribosomal DNA cluster of all eukaryonts. Twenty sequences of the ITS from reference species for *Basidiobolus* and 3 ITS sequences of *Conidiobolus* are available from the International Nucleotide Sequence Database Collaboration, whereas just 6 (4/2) entries exist for the 18S/28S rDNA of *Basidiobolus* but 16 (11/5) for the SSU/LSU *Conidiobolus*, respectively (as of September 28, 2009). Those nucleotides sequences can easily be used as reference sequences in BLAST searches, targeting regions of the nuclear ribosomal DNA as molecular barcodes as outlined by Hoffmann et al.[62] The oligonucleotides ITS1, ITS4, and NS1 originate from White et al.[63] To overcome the necessity of primer walking using internal sequencing primers for sequencing completion, the reverse primer NS41 should be used.[64] Amplification with primer pair NS1/41 results an amplicon of approximately 1.3 kb in size, which can be directly sequenced through using NS 1 and NS41 as sequencing primers. The temperature profile for the amplification of the SSU and the LSU is very the same as used for the taxon-specific primers besides that the annealing temperature should be 52°C in order to reach a universal amplification of the pure fungal DNA.

83.2.2.4.3 Taxon-Specific, Cultivation-Independent Amplification of Species of Conidiobolus *and* Basidiobolus *in Clinical Samples*

Crude extracts from biopsy material can be used for total DNA extraction. In order to gain an increased specificity and a decreased cross reaction between the fungal and the human background DNA, it becomes advantageous to conduct nested PCR approaches. The nested PCR procedure applies a two-step PCR strategy using the universal amplification of eukaryotic or fungal DNA in a first and the taxon-specific amplification of DNA from species of *Basidiobolus* and *Conidiobolus* in a second step. The first step aims at the increased accumulation of target DNA, which will be used as template in the second PCR. Using the ITS region as molecular barcode marker, the primer pairs ITS1/4 can be used for the enrichment of the ITS region in the first step following subsequent amplification with primer pairs Bs1/2, Bs1/ITS4, or ITS1/Bs2 in the second step. Using the D1/D2 domain of the 28S rDNA (SSU) as molecular barcode marker, the primer pairs NL1/4 may be used for the enrichment of the 5′ region of the SSU rDNA in a first step following subsequent amplification with primer pairs Cc1/Cc2, Ba1/Ba2 or combinations with universal oligonucleotide primers (Cc1/NL4, NL1/Cc2, Ba1/NL4, or NL1/Ba2) in the second step. The detection of positive PCR signals is the same as described in more detail in Section 83.2.2.4.1.

83.3 CONCLUSIONS AND FUTURE PERSPECTIVES

Entomopthoralean fungi are distributed worldwide and can be found in soil, decaying vegetation, in or on different animals including insects and mammals, and also in vertebrate feces. The systematic relationships on an inter- and intraordinal level are still unclear. Infections with Entomophthoralean fungi in human are uncommon in Europe and North America but widespread in Africa, Asia, Southeast Asia, the Middle East, Central, and South America.[25] The infections are subcutaneous, prolonged but usually painless. Almost no fundamental research has been done. Most articles are based on case reports. Since the immune status of the host seems to be less important for the infection, the causes, circumstances, and preconditions for an infection remain unclear. This situation is aggravated by the fact that the pathogenic potential of many species is not known. Since most Entomophthoralean fungi are pathogenic to insects, the correlation between pathothenicity to insects and human is yet to be determined. This necessitates an exhaustive screening with imaging techniques appropriate for the detection of infected areas of the bowels for local and deep Entomophthoralean infections in both humans and insects.

Moreover, there are plans for the use of Entomophthoralean fungi as biocontrol agent including species of *Conidiobolus*.[14,65] The knowledge relating to the pathogenic potential of these fungi should help to avoid the use of potentially harmful fungi in such applications. Diagnostic tools for the detection of entomophthoromycosis are not well developed. This is mostly caused by the rare occurrence of such infections. The sensitivity of the immunodiffusion test for zygomycosis designed by Kaufman et al.[52] is around 66%.[66] However, the sensitivity could be increased by the use of indirect enzyme-linked immunosorbent assay (ELISA) by which a sensitivity of 81% was achieved.[66] This assay has not been demonstrated for Entomophthoralean fungi but only for the mucoralean *Rhizopus arrhizus* and *Rhizomucor pusillus*. It is likely that the investigation of this technique would improve the detection of entomophthoromycosis. Although PCR-based methods are used to the advantage for the detection of fungal infections, there are some major disadvantages. Real-time PCR is a standard technique in diagnostics and it has been reported that the sensitivity and specificity is increased compared to classical PCR.[67] This technique facilitates the rapid and highly sensitive detection of fungal pathogens without the need of cultivation, but no real-time PCR assay has currently been established for the specific demands of Entomophthoralean fungi. The detection of an infection with *C. coronatus* via real-time fluorescence PCR using a broad-range panfungal primer was shown to be sufficient.[67] In combination with the specific primers for *C. coronatus* and *B. ranarum* described above, this technique could be a big advance toward the reliable and rapid PCR-based and cultivation-independent detection of entomophthoromycoses. For the future, a molecular detection using DNA chips with specific nucleotide probes specific for the pathogen would be preferable. Comparable procedures and kits for other pathogens are available.[68] Since zygomycoses have become more common since the mid-1990s, the reliable diagnostics of such infections become more and more important.[69] Moreover, climate changes might affect the distribution of entomophthoromycoses. Infections with Entomophthoralean fungi seem to be restricted to tropical or subtropical regions and global warming could thus lead to an increasing number of infections in, particularly, northern temperate regions, concurrently not significantly affected by cases of entomophoromycoses. Since the mode of infection is still unclear, it is not possible to rule out the influence of a putative insect vector, which could spread to northern temperate regions as a result of major climatic changes.

ACKNOWLEDGMENTS

We wish to express their gratitude to Paul M. Kirk (CAB International Egham, Surrey, U.K.) for his expert advice and critical revision of the manuscript. We would like to thank Rolf Beutel (Institute of Systematic Zoology and Evolutionary Biology, University of Jena, Germany) for sharing his zoological and cladistic expertise. We are grateful to our colleagues at the Centraalbureau voor Schimmelcultures in Utrecht (The Netherlands) for sharing strains of *Conidiobolus*. We would also like to thank Claudia Kesselboth for her excellent assistance in light microscopical observations and for assembling Figure 83.1. JR and VS equally contributed to this chapter.

REFERENCES

1. James, T.Y. et al., Reconstructing the early evolution of fungi using a six-gene phylogeny, *Nature*, 443, 818, 2006.
2. Hibbett, D.S. et al., A higher-level phylogenetic classification of the fungi, *Mycol. Res.*, 111, 509, 2007.
3. Nagahama, T. et al., Phylogenetic divergence of the Entomophthoralean fungi: Evidence from nuclear 18S ribosomal RNA gene sequences, *Mycologia*, 87, 203, 1995.
4. Winter, G., *Rabenhorst's Kryptogamen-Flora, Pilze-Schizomyceten, Saccharomyceten und Basidiomyceten*, 2nd edn., Eduard Kummer Verlag, Leipzig, 1880, p. 74.
5. Tanabe, Y. et al., Molecular phylogeny of Zygomycota based on EF-1α and RPB1 sequences: Limitations and utility of alternative markers to rDNA, *Mol. Phylogenet. Evol.*, 30, 438, 2004.
6. Taylor, J.W. et al., Phylogenetic species recognition and species concepts in fungi, *Fungal Genet. Biol.*, 31, 21, 2000.
7. Kirk, P.M. et al., *Ainsworth & Bisby's Dictionary of the Fungi*, Vol. 10, CAB International, Wallingford, U.K., 2008.
8. Thielke, C. and Institut für den wissenschaftlichen Film: Entwicklung von *Basidiobolus ranarum* (Entomophthoraceae), Film C 1546 des IWF, Göttingen 1984, Publikation von Thielke C, Publ. Wiss. Film., Sekt. Biol., Ser. 18, Nr. 15/C 1546 (1986), 16pp.
9. Pfaller, M.A. and Diekema, D.J., Unusual fungal and pseudofungal infections of humans, *J. Clin. Microbiol*, 43, 1495, 2005.
10. Yang, X. et al., Rhinofacial conidiobolomycosis caused by *Conidiobolus coronatus* in a Chinese rice farmer, *Mycoses: Letter to the Editor*, April 2009.
11. Larone, D.H., *Medically Important Fungi: A Guide to Identification*, Vol. 4, ASM Press, Washington, DC, 2002.
12. Drechsler, C., An Entomophthoraceous tardigrade parasite producing small conidia on propulsive cells in spicate heads, *B. Torrey. Bot. Club*, 78, 183, 1951.
13. Wheeler, A.G., "Violent deaths" of soldier beetles (Coleoptera: Cantharidae) revisited: New records of the fungal pathogen *Eryniopsis lampyridarum* (Zygomycetes: Entomophthoraceae), *Coleopterists Bull.*, 43, 233, 1988.
14. Scholte, E.J. et al., Entomopathogenic fungi for mosquito control: A review, *J. Insect Sci.*, 4, 19, 2004.
15. Miller, R. and Pott, B., Phycomycosis of the horses caused by *Basidiobolus haptosporus*, *Aust. Vet. J.*, 56, 224, 1980.
16. Sutton, D.A., Fothergill, A.W., and Rinaldi, M.G., *Guide to Clinically Significant Fungi*, The Williams and Wilkins Co., Baltimore, MD, 1998.
17. Silva, S.M.M.S. et al., Conidiobolomycosis in sheep in Brazil, *Vet. Pathol.*, 44, 314, 2007.
18. Ribes, J.A., Vanover-Sams, C.L., and Baker, D.J., Zygomycetes in human disease, *Clin. Microbiol. Rev.*, 13, 236, 2000.
19. Beneke, E.S. and Rogers, A.L., *Medical Mycology and Human Mycoses*, Star Publishing Co., Belmont, CA, 1996, pp. 182–206.
20. Mendoza, L. and Alfaro, A.A., Equine subcutaneous zygomycosis in Costa Rica, *Mykosen*, 28, 545, 1985.
21. Morris, R.J., Müller, C.B., and Godfray, H.C.J., Field experiments testing for apparent competition between primary parasitoids mediated by secondary parasitoids, *J. Anim. Ecol.*, 70, 301, 2001.
22. Rippon, J.W., *Medical Mycology: The Pathogenic Fungi and Actinomycetes*, The W.B. Saunders Co., Philadelphia, PA, 1987, pp. 690–699.
23. Roy, A.D. and Cameroon, H.M., Rhinophycomycosis Entomophthorae occurring in a chimpanzee in the wild in East Africa, *Am. J. Trop. Med. Hyg.*, 21, 234, 1972.
24. Yendol, W.G., Factors affecting the germination of *Entomophthora coronatus*, *J. Invertebr. Pathol.*, 10, 111, 1968.
25. Dismukes, W.E., Pappas, P.G., and Sobel, J.D., *Clinical Mycology*, Oxford University Press, New York, 2003.
26. Fischer, N. et al., Rhinofacial *Conidiobolus coronatus* infection presenting with nasal enlargement, *Infection*, 36, 594, 2008.
27. Busapakum, R. et al., Disseminated infection with *Conidiobolus incongruus*, *Sabouraudia*, 21, 323, 1983.
28. Eidam, E., *Basidiobolus*, eine neue Gattung der Entomophthoraceen, *Biol. Pflanz.*, 4, 181, 1886.
29. Chaturvedi, V.P. et al., Prevalence of *Basidiobolus ranarum* Eidam in the intestinal tract of an insectivorous bat *Rhinopoma hardwicki hardwicki*, in Delhi, *Sabouraudia*, 22, 185, 1984.
30. Miller, R.I. and Turnwald, G.H., Disseminated basidiobolomycosis in a dog, *Vet. Pathol.*, 21, 117, 1984.
31. Krishnan, S.G.S. et al., Entomophthoromycosis in India: A 4-year study, *Mycoses*, 41, 55, 1998.
32. Lyon, G.M. et al., Gastrointestinal *basidiobolomycosis* in Arizona: Clinical and epidemiology characteristics and review of the literature, *Clin. Infect. Dis.*, 32, 1448, 2001.

33. Khan, Z.U. et al., *Basidiobolus ranarum* as an etiologic agent of gastrointestinal zygomycosis, *J. Clin. Microbiol.*, 39, 2360, 2001.
34. Van den Berk, G.E. et al., A fatal pseudo-tumour: Disseminated basidiobolomycosis, *BMC Infect. Dis.*, 6, 140, 2006.
35. Walker, S.D. et al., Fatal disseminated *Conidiobolus coronatus* infection in a renal transplant patient, *Am. J. Clin. Pathol.*, 98, 559, 1992.
36. Gugnani, H.C., A review of zygomycosis due to *Basidiobolus ranarum*, *Eur. J. Epidemiol.*, 15, 923, 1999.
37. Bigliazzi, C. et al., Disseminated basidiobolomycosis in an immunocompetent woman, *J. Clin. Microbiol.*, 42, 1367, 2004.
38. Al Jarie, A. et al., Pediatric gastrointestinal basidiobolomycosis, *Pediatr. Infect. Dis. J.*, 22, 1007, 2003.
39. Valle, A.C.F. et al., Entomophthoramycosis by *Conidiobolus coronatus*. Report of a case successfully treated with the combination of itraconazole and fluconazole, *Rev. Inst. Med. Trop. Sao Paulo*, 43, 233, 2001.
40. Eckert, H.L. et al., Deep *Entomophthora* phycomycotic infection reported for the first time in the United States, *Chest*, 61, 392, 1972.
41. Kamalam, A. and Thambiah, A.S., Lymph node invasion by *Conidiobolus coronatus* and its spore formation in vivo, *Sabouraudia*, 16, 175, 1978.
42. Walsh, T.J. et al., Invasive zygomycosis due to *Conidiobolus incongruus*, *Clin. Infect. Dis.*, 19, 423, 1994.
43. Bittencourt, A.L., Marback, R., and Nossa, L.M.B., Mucocutaneous Entomophthoramycosis acquired by conjunctival inoculation of the fungus, *Am. J. Trop. Med. Hyg.*, 75, 936, 2006.
44. Pasha, T.M. et al., Basidiobolomycosis: An unusual fungal infection mimicking inflammatory bowel disease, *Gastroenterology*, 112, 250, 1997.
45. Geramizadeh, B. et al., Gastrointestinal zygomycosis: A report of three cases, *Mycopathologia*, 164, 35, 2007.
46. Rüchel, R. et al., Vereinfachte Pilzdetektion—Screeningmethode mit optischem Aufheller, *Pathologe*, 25, 235, 2004.
47. Maeda, H. and Ishida, N., Specificity of binding of hexopyranosyl polysaccharides with fluorescent brightener, *J. Biochemistry*, 62, 276, 1967.
48. Hussein, M.R., Mucocutaneous Splendore-Hoeppli phenomenon, *J. Cutan. Pathol.*, 35, 979, 2008.
49. Costa, A.R. et al., Rhinofacial zygomycosis caused by *Conidiobolus coronatus*. A case report, *Mycopathologia*, 115, 1, 1991.
50. Splendore, A., Sobre a cultura d'una nova especie de cogumello pathogenico (Sporotrichose de Splendore), *Rev. Soc. Sao Paulo*, 3, 62, 1908.
51. Hoeppli, R., Histological observations in experimental *Schistosomiasis japonica*, *Chin. Med. J.*, 46, 1179, 1932.
52. Kaufman, L., Mendoza, L., and Standard, P.G., Immunodiffusion test for serodiagnosing subcutaneous zygomycosis, *J. Clin. Microbiol.*, 28, 1887, 1990.
53. Voigt, K., Cigelnik, E., and O'Donnell, K., Phylogeny and PCR identification of clinically important zygomycetes based on nuclear ribosomal-DNA sequence data, *J. Clin. Microbiol.*, 37, 3957, 1999.
54. Beckstead, J.H., A simple technique for preservation of fixation-sensitive antigens in paraffin-embedded tissues, *J. Histochem. Cytochem.*, 42, 1127, 1994.
55. Huppert, M., Oliver, D.J., and Sun, S.H., Combined methenamine–silver nitrate and hematoxylin and eosin stain for fungi in tissues, *J. Clin. Microbiol.*, 8, 598, 1978.
56. McManus, J.F.A., The periodic acid routine applied to the kidney, *Am. J. Path.*, 23, 907, 1947.
57. McManus, J.F.A., Histological and histochemical uses of periodic acid, *Stain. Tech.*, 23, 99, 1948.
58. Kligman, A.M. and Mescon, H., The periodic-acid-Schiff stain for the demonstration of fungi in animal tissue, *J. Bacteriol.*, 60, 415, 1950.
59. Cenis, J.L., Rapid extraction of fungal DNA for PCR amplification, *Nucleic Acids Res.*, 20, 2380, 1993.
60. Raeder, U. and Broda, P., Rapid preparation of DNA from filamentous fungi, *Lett. Appl. Microbiol.*, 1, 17, 1985.
61. Bassam, B.J., Caetano-Annoles, G., and Gresshof, P.M., Fast and sensitive silver staining of DNA in polyacrylamide gels, *Anal. Biochem.*, 196, 80, 1991.
62. Hoffmann, K. et al., Diversity, genotypic identification, ultrastructural and phylogenetic characterization of zygomycetes from different ecological habitats and climatic regions: Limitations and utility of nuclear ribosomal DNA barcode markers, in *Current Advances in Molecular Mycology*, Gherbawy, Y., Mach, R.L., Rai, M. (Eds.), Nova Science Publishers, Inc., New York, 2009, p. 263.
63. White, T.J. et al., Amplification and direct sequencing of fungal ribosomal RNA genes for phylogenetics, in *PCR Protocols: A Guide to Methods and Applications*, Innis, M.A., Gelfand, D.H., Sninsky, J.J., White, T.J. (Eds.), Academic Press, San Diego, CA, 1990, p. 315.
64. O'Donnell, K., Cigelnik, E., and Benny, G.L., Phylogenetic relationships among Harpellales and Kickxellales, *Mycologia*, 90, 624, 1998.
65. Sierotzki, H. et al., Biological characteristics of selected *Erynia neoaphidis* isolates, *Mycol. Res.*, 104, 213, 2000.
66. Kaufman, L., Turner, L.F., and McLaughlin, D.W., Indirect enzyme-linked immunosorbent assay for zygomycosis, *J. Clin. Microbiol.*, 27, 1979, 1989.
67. Imhof, A. et al., Rapid detection of pathogenic fungi from clinical specimens using LightCycler real-time fluorescence PCR, *Eur. J. Clin. Microbiol. Infect. Dis.*, 22, 558, 2003.
68. Kriegsmann, J. et al., Molekularpathologische Diagnostik von Gelenkinfektionen, *Orthopäde*, 38, 531, 2009.
69. Brown, J., Zygomycosis: An emerging fungal infection, *Am. J. Health Syst. Pharm.*, 62, 2593, 2005.
70. O'Donnell, K., *Fusarium* and its near relatives, in *The Fungal Holomorph: Mitotic, Meiotic and Pleomorphic Speciation in Fungal Systematics*, Reynolds, D.R. and Taylor, J.W (Eds.), CAB International, Wallingford, U.K., 1993, p. 225.

84 *Lichtheimia* (*Absidia*-Like Fungi)

Kerstin Hoffmann and Kerstin Voigt

CONTENTS

84.1 INTRODUCTION

84.1.1 History of the Classification of *Absidia*-Like Fungi and *Lichtheimia*

Most fungi in the order Mucorales (Mucoromycotina[1]) are saprobic and live as ubiquitous inhabitants of soil. Few of them are able to facultatively parasitize on other fungi or plants. Furthermore, there are few genera and species that occur as opportunistic pathogens and cause mucormycoses in humans and animals.[2–4] The trademark of all mucoralean fungi is the columella, a bulbous and sterile vesicle at the sporangiophore apex, which rejuvenates in a funnel-like manner into the apophysis giving the appearance of pear-shaped (pyriform) sporangia typical for the genus *Absidia* (Figure 84.1A, B, M–R).

The traditional morphology-defined genus *Absidia* Tiegh. (Etym.: *absis*, arcus) is one of few genera within the order Mucorales comprising all habitats and lifestyles mentioned above. But recently, molecular-based phylogenies suggest a trichotomous separation in three distinct genera. This differentiation is supported by morphological and physiological traits.[5,6] Psychrophilic and mesophilic species with optimal temperatures for growth below 30°C and 37°C currently belong to the genera *Lentamyces* K. Hoffm. & K. Voigt 2008 (Etym.: *lentus*, slow) and *Absidia* Tiegh. *sensu stricto*, respectively. Species with an increased optimum temperature around and above 37°C are combined in the genus *Lichtheimia* (Cohn) Vuill. 1903 (Etym.: after Prof. Lichtheim, Bern, Switzerland).

Lentamyces is a genus with saprobic, mycoparasitic, and potentially mycoparasitic species and constitutes *Le. zychae* and its type taxon *Le.* Parricida.[6] Whereas *Le. parricida* is able to parasitize on several other mucoralean fungi (e.g., species of the genera *Thamnidium*, *Zygorhynchus*, *Mucor*, and *Blakeslea*), the mycoparasitic potential on compatible host species for *Le. zychae* is not known.[6,7] Both species exhibit features typical for mycoparasites such as low growth rates and thin mycelia with sucker-like branches in the substrate mycelium (see Figure 2A in Hoffmann and Voigt 2009[6]). Both species are homothallic, forming their sexual reproductive structures, the zygospores, within a single thallus or colony (in contrary to heterothallic species, which cross only between two mycelia of compatible mating types designated as "plus" and "minus" in the literature). The globose zygospores appear dark brown with warty exospore and opposed suspensors lacking appendages. A recently proposed genus for these two species, *Siepmannia* Kwaśna & Nirenberg 2008, comprises in addition to the synonyms *S. parricida* and *S. zychae* also *S. pineti* and *S. lariceti*. However, this study is based mainly on a single analysis of internal transcribed spacer (ITS) sequences (ITS1 and 2, including 5.8S ribosomal DNA [rDNA]) and morphological comparisons.[8,9] Critical reevaluation of the stated morphological features could not assign the two species of *Lentamyces* with absolute certainty to *Siepmannia* as claimed by Kwaśna & Nirenberg.[8] Since living cultures of the type strain from *Siepmannia* are not available, the relationship between *Siepmannia* and *Lentamyces* remains to be investigated.[6,8–10]

Currently, the genus *Absidia sensu stricto* comprises saprobic species displaying the typical *Absidia*-specific characters (Figure 84.1): sporangia arising from stolons (tillers); sporangiophores not opposite the rhizoids; sporangia pyriform, with apophysis and deliquescent wall; and zygospores surrounded with appendages originating from one or both

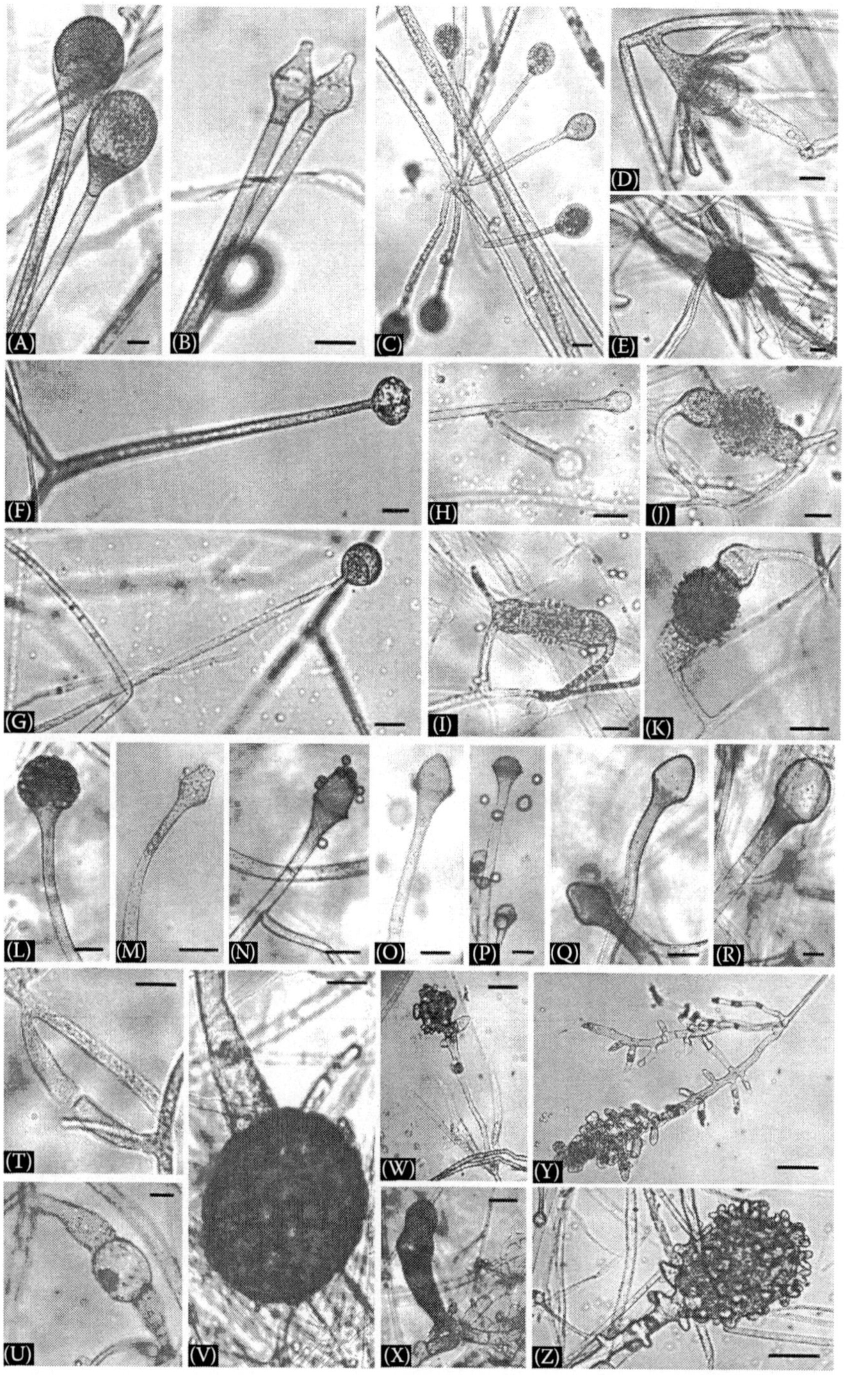

FIGURE 84.1 Micromorphological characteristics of *Absidia sensu lato*. (A–E) *Absidia sensu stricto*. (A) Pyriform sporangia with apophysis; (B) columellae with apical projection; (C) typical whorl of sporangia; (D) young zygospores with appendages originating from a single suspensor (left sided); (E) mature zygospores with appendages; (F–K) *Lentamyces*; (F–H) globose sporangia, not in whorls; (I) young zygospore; (J and K) mature zygospores, warty exospore, and suspensors without appendages; (L–Z) *Lichtheimia*; (L) apophysate sporangium; (M–R) differently shaped columella, predominantly dome-shaped; (T) gametangiogamy, initialization of zygospore formation; (U) young zygospore; (V) mature zygospore with lipid droplets, suspensors without appendages; (W–Z) variously shaped giant cells. Scale bar: (A–V) 20 μm; (W–Z) 50 μm.

suspensors.[11] The sporangia are commonly multispored and possess a columella with an apical projection (Figure 84.1B). Homothallic and heterothallic species exists within the genus. Homothallic species are *A. anomala* and *A. spinosa*. Heterothallic species with plus and minus mating typespecific mycelia are *A. caerulea*, *A. californica*, *A. cuneospora*, *A. cylindrospora*, *A. fusca*, *A. glauca*, *A. heterospora*, *A. macrospora*, *A. pseudocylindrospora*, *A. psychrophilia*, and *A. repens*. *Absidia repens* may probably represent a complex of different cryptic species as accomplished by molecular barcode analyses on different loci.[12] The monophyletic genus *Absidia sensu stricto* is trichotomous divided on its own.[5] The subclades are well supported by sporangiospore morphology and molecular phylogenetic analyses. Globose sporangiospores occur in *A. caerulea*, *A. californica*, *A. glauca*, and *A. macrospora*, whereas oval to cylindrical spores are formed by *A. anomala*, *A. cylindrospora*, *A. pseudocylindrospora*, *A. psychrophilia*, *A. repens*, and *A. spinosa*. Conical shaped spores are present in the intermedia species *A. cuneospora*, which represents an intermediate between the globose and the cylindrical spored groups of *Absidia sensu stricto*.[5–7,13] The latest species described so far for *Absidia* is *A. idahoensis*,[14] probably a misidentification since *A. idahoensis* shows greater similarities in morphology and molecular analyses to *Circinella umbellata* than to any other species of *Absidia* (K. Hoffmann, unpublished data).

Together with the sibling genera *Chlamydoabsidia* Hesselt. & J.J. Ellis 1966, *Cunninghamella* Matr. 1903, *Gongronella* Ribaldi 1952, *Halteromyces* Shipton & Schipper 1975, and *Hesseltinella* H.P. Upadhyay 1970, *Absidia* is summarized in the family Absidiaceae Arx 1982.[5,10,15]

Furthermore, in literature, several species were described as members belonging to *Absidia*, which either lack a proper description or are now considered as lost ones, because such species were once encountered and described. Consequently, with respect to genus and species affiliations, the following species were not considered here: *A. scabra* Cocc. 1899; *A. septata* Tiegh. 1876; *A. reflexa* Tiegh. 1876; *A. aegyptiacum* Satory, Meyer, & Tawfik 1939; *A. capillata* Tiegh. 1876; *A. clavata* B.S. Mehrotra & Nand 1967; *A. fassatiae* Vánová 1971; *A. griseola* H. Nagan. & Hirahara 1970; *A. inflata* J.H. Mirza, S.M. Khan, S. Begum, & Shagufta 1979; *A. narayanai* Subrahamanyam 1990; *A. robusta* Raciborski 1899; *A. tuneta* Renner & Muskat 1958; *A. ushtrina* S.C. Ararwal 1974 (but see Ellis & Hesseltine 1965,[16] 1966[17]).

Species within the genus *Lichtheimia* display features quite different from *Absidia sensu stricto*. Although all species currently grouped here were originally placed within *Absidia*, molecular, morphological, and physiological features justify their separation in their own genus. One main criterion is zygospores with nonappendaged suspensors. The first species described for *Absidia* with such zygospores was *A. verticillata*[18] originally described as *Mycocladus verticillatus* Beauverie 1900. The original culture of Beauverie was lost, but Zycha described a similar fungus, which in return he named *A. verticillata*.[18,19] Within a reexamination, this name was not accepted and changed in *A. zychae* because of morphological differences.[20,21] Nevertheless, the aim to distinguish between morphological and physiological different *Absidia* species resulted in the application of "*Mycocladus*" to species with nonappendaged zygospores and optimum and maximum temperatures for growth higher than 37°C (proposed as subgenus in *Absidia* by Hesseltine & Ellis 1964,[22] accepted by Schipper 1990,[7] and described as separate genus by Mirza et al. 1979[23] and Vánová 1991[24]). Based on these publications, the thermotolerant species of *Absidia* were designated as *Mycocladus* in the family Mycocladaceae (as "Mycocladiaceae"[5]). But unfortunately, the morphological appearance of *M. verticillatus* gave considerable reason to a probable misapplication consisting of at least two different species making *M. verticillatus* an invalid type taxon for a family.[19,25] The oldest available synonyms described for *Absidia* species with naked zygospores were *Pseudoabsidia* (as "*Pseudo-Absidia*"[26]) and *Lichtheimia*.[27] The type taxon of *Pseudoabsidia*, *P. vulgaris*, was originally described as *Absidia dubia*.[28] The correction of the misapplied epithet was done by Sydow,[29] who recombined *P. vulgaris* and *Absidia dubia* to *P. dubia*, which was later on transferred to *Lichtheimia* as *L. dubia*.[27,30] The type of *Lichtheimia* is *L. corymbifera*, described as *Mucor corymbifer*, a pathogenic fungus in rabbits.[27,30,31] *Lichtheimia* is now the type genus of the family Lichtheimiaceae K. Hoffm., G. Walther & K. Voigt.[25] This family includes the following species: *L. corymbifera*, *L. hyalospora*, *L. ornata*, *L. ramosa*, and *L. sphaerocystis*.[32] Due to the difficulty of the synonymous treatment of *L. corymbifera*, *L. ramosa*, and *L. ornata* in the past and their reerection as distinct species,[17,25,32,33] an easy separation of species isolated and described during that time is rather difficult. Therefore, isolates assigned to *L. corymbifera* between the years 1974–2009[32–34] should be handled with caution regarding their current species affiliation. Reliable reference strains (if not type strains) are given under Section 84.1.2 (Table 84.1).

In recent years, one new species of *Lichtheimia* was described—the thermotolerant variety of *A. idahoensis*, originating from soil in the Chinese province Yunnan.[35] *A. idahoensis* var. *thermophila* displayed many features typical for *Lichtheimia* and its membership in this genus was confirmed by molecular analyses, reducing the species to a synonym of *L. ramosa*.[32]

Distinguishing species of *Lichtheimia* requires profound knowledge and experience in recognition and evaluation of observed characteristics. Unfortunately, zygomycetes possess only few characteristics of phylogenetic importance.[36] Moreover, exclusively through a combination of unique morphological features, physiological and biochemical parameters, and molecular or genetic barcodes (nucleotide signature sequences), a precise identification and taxonomic delimitation is possible. Within the zygomycetous order Mucorales, 9 families with 51 genera and 205 species are recognized.[37] Common features of the Mucorales encompass (i) the commonly nonseptated vegetative mycelia (irregularly septae predominantly occur near reproductive structures, injured hyphae, or in older cultures), (ii) the occurrence of rhizoids, and (iii) the reproduction by the formation of sexually produced zygospores or various asexual spores produced in differently shaped sporangia or as chlamydospores or arthrospores. All these structures could vary in color, size, shape, or surface texture.[20,38,39]

But in the course of identification of specimens, one of the most controversially discussed questions is still the definition of the term "species." Although a "species" is the basal unit in systematics, "species recognition" could rely on different criteria, for example, biology, morphology, physiology, ecology, and phylogeny. For fungal systematics, the phylogenetic approach with analyses of several molecular marker

TABLE 84.1
Chronological Summary of *Absidia sensu lato* and Its Generic Synonyms, Including Information about Type Strains, Original Publication and Some Important Taxonomical Synonyms, as well as the Corresponding MycoBank Number

Species, Current Name	Type Strain	MycoBank No.	Reference	Important Taxonomical Synonyms
Absidia sensu stricto				
A. repens	CBS115583 (IT)	MB223578	Van Tieghem[11]	*A. japonica, Mycocladus hyalinus, Tieghemella repens, T. japonica*
A. caerulea	NRRL1315 (NT)	MB351936	Bainier[102]	*A. orchidis, Mucor saccardoi, T. caerulea, T. orchidis*
A. spinosa var. *spinosa*	NRRL2797 (T)	MB224063	Lendner[103]	*T. spinosa*
A. cylindrospora var. *cylindrospora*	NRRL1317 (T)	MB427391	Hagem[104]	*T. cylindrospora, Mycocladus altaini*
A. glauca	NRRL1328 (T)	MB221208	Hagem[104]	*A. sphaerosporangioides, A. septata, T. glauca*
A. heterospora	NRRL2800 (T)	MB252572	Ling-Young[105]	None
A. fusca	NRRL2793 (T)	MB252285	Linnemann[106]	None
A. spinosa var. *azygospora*	NRRL2841 (T)	MB346488	Boedijn[107]	None
A. cuneospora	NRRL2632 (T)	MB292052	Orr and Plunkett[108]	None
A. cylindrospora var. *rhizomorpha*	NRRL2771 (T)	MB348992	Hesseltine and Ellis[63]	None
A. pseudocylindrospora	NRRL2770 (T)	MB325715	Hesseltine and Ellis[63]	*A. brasiliensis*
A. anomala	NRRL1807 (T)	MB325709	Hesseltine and Ellis[22]	None
A. psychrophilia	NRRL3044 (T)	MB325716	Hesseltine and Ellis[22]	None
A. cylindrospora var. *nigra*	NRRL3060 (T)	MB353238	Hesseltine and Ellis[22]	None
A. spinosa var. *biappendiculata*	NRRL3033 (T)	MB348993	Rall and Solheim[109]	None
A. californica	NRRL2968 (T)	MB325710	Ellis and Hesseltine[16]	None
A. macrospora	CBS697.68 (T)	MB325712	Vánová[110]	None
Lichtheimia				
L. corymbifera[a]	NRRL2981 (NT)	MB416447	Vuillemin[27] (originally as *Mucor corymbifer* by Cohn 1884)	*A. corymbifera, A. ginsan, A. gracilis, A. hesseltinei, A. regnieri, L. regnieri, Mucor lichtheimii, M. regnieri, Mycocladus corymbifer*
L. ramosa[a]	NRRL1309 (NT)	MB416448	Vuillemin[27] (originally as *Rhizopus ramosus* by Zopf 1890)	*A. italiana, A. ramosa, L. ramosa, L. italiana, L. italica Mucor ramosus, Mycocladus ramosus, M. lutetiensis, Tieghemella italiana*
L. hyalospora	NRRL2916 (NT)	MB512830	Hoffmann et al.[25] (originally as *Tieghemella hyalospora* by Saito 1906)	*A. cristata, A. hyalospora, A. merdaria, Mycocladus hyalospora* (for *A. blakesleeana, L. blakesleeana, Mycocladus blakesleeanus*, see Alastruey-Izquierdo et al. 2010)
L. ornata[a]	NRRL10293 (IT)	NN	Alastruey-Izquierdo et al.[32] (originally as *Absidia ornata* by Sarbhoy 1965)	
L. sphaerocystis	NN	NN	Alastruey-Izquierdo et al.[32]	None
Lentamyces				
L. parricida	NRRL2409 (T)	MB511980	Hoffmann and Voigt (originally as *Absidia parricida* by Renner & Muskat ex Hesselt. & J.J. Ellis 1964)	*Absidia parasitica, Siepmannia parricida*
L. zychae	NRRL2806 (T)	MB511981	Hoffmann and Voigt (originally as *Absidia zychae* by Hesselt. and Ellis 1966)	*Siepmannia zychae*

Abbreviations: NT, neotype; IT, isotype; T, type; NRRL, Northern Regional Research Laboratories, strain collection of the National Center of Agricultural Utilization Research Peoria, Illinois; CBS, Centraalbureau voor Schimmelcultures, Utrecht, the Netherlands.

[a] …since *L. corymbifera, L. ramosa*, and *L. ornata* were synonymous in recent years, a separation of taxonomical synonyms is rather difficult.

regions (multigene genealogies) proved to be a useful method in species recognition named phylogenetic species recognition (PSR).[40] In this vein, defined species descriptions may be verified and refined by a combination of PSR with biological and morphological aspects.

The classical genus *Absidia* was defined by its morphology (morphospecies), but by PSR, a trichotomous genus separation was recognized, which could be supported by a redefinition of characters, which delimit more precisely the morphological, physiological, and biological features and which are independent on physiological variances.[5–7] With the exception of molecular data provided by the PSR conception, the species delimitation within *Lichtheimia* depends now solely on slight differences in spore morphology and optimum growth temperature treated as a complex trait in the right relation.[5,32]

Although the upper temperature limit is not a fixed value for all isolates of one species, the observed average is nevertheless of diagnostic importance. Whereas *L. corymbifera*, *L. ramosa*, and *L. ornata* show optimal growth (in the mean of growth velocity) at temperatures above 37°C, *L. hyalospora* and *L. sphaerocystis* possess an optimum for growth below that temperature.[5,32] Some isolates of *L. ramosa* are capable of growing still at 53°C, whereas the maximum for *L. corymbifera* is reached at 50°C and at 46°C for *L. ornata*. *L. hyalospora* should not grow above 40°C and *L. sphaerocystis* not above 37 (40)°C.[32] But these values are mean values since few isolates of, for example, *L. hyalospora* (as *L. blakesleeana*) were reported to grow up to 50°C.[5] The reason for the capability of the species *L. corymbifera*, *L. ramosa*, and *L. ornata* and the putative incapability of *L. hyalospora* and *L. sphaerocystis* to colonize endothermic organisms is not yet understood.

84.1.2 Chronological Summary of *Absidia sensu lato* and Its Generic Synonyms

Since the first description of species of *Absidia*, several nomenclatural changes and interchanges of synonymies have occurred That makes it rather difficult to keep track of currently used and accepted species names. In the following, an overview is given for the genus *Asidia sensu lato* including the current state of family affiliations, species names with a reference of their original publication, as well as important synonyms. For each generic name, an entry in MycoBank (http://www.mycobank.org) exists. Mycobank is an online database for the management of the mycological nomenclature including species descriptions and occasionally illustrations.[41,42] MycoBank is associated to the fungal strain collection of the CBS (Centraalbureau voor Schimmelcultures, Utrecht, the Netherlands, http://www.cbs.knaw.nl) and to IndexFungorum (http://www.indexfungorum.org).

To distinguish species with morphological and physiological differences, several genera were proposed over the years, which represent now synonyms of the currently accepted genera of *Absidia sensu lato*, namely, *Absidia sensu stricto*, *Lichtheimia*, and *Lentamyces* (Table 84.1).

Absidia was described by Van Tieghem[11] [MycoBank no. MB20001] and placed in the family Absidiaceae [MB81973] by Von Arx.[15] Although this family was treated as a synonym of the Mucoraceae[37,39] [MB81030], its independent status is more than justified (as substantiated by Hoffmann et al.[5]; Voigt et al.[36]; Hoffmann 2010[10]). *Tieghemella* Berl. & de Toni (1888) [MB20576] and *Proabsidia* Vuill. (1903) [MB20441] are now considered synonyms of *Absidia sensu lato* (discussed in Hesseltine & Ellis 1964[22]) (Table 84.1).

Lichtheimia was described by Vuillemin[27] [MB20308], placed within *Absidia* by Lendner,[43] and is now raised to its own family, the Lichtheimiaceae[25] [MB508680]. Generic synonyms for this genus are *Mycocladus* Beauverie (1900) [MB20354], *Pseudoabsidia* Bainier (1903) [MB20445], and *Protoabsidia* Naumov (1935) [MB20443].[22] Although *L. corymbifera*, *L. ramosa*, and *L. ornata* were originally described as different species, they were treated as synonymous species,[7,34] making it rather difficult separating them again[32,33] (Table 84.1).

Since the genus *Lentamyces* [MB511979] is not closely related to any other mucoralean genus, a separation in a distinct genus is reasonable.[6] Whether the proposed genus *Siepmannia* is a synonymous one, it requires further investigation[6,8–10] (Table 84.1).

84.1.3 Clinical Features and Pathogenesis

Although opportunistic infections associated with members of the Mucorales are rare if compared to other fungal infections like aspergilloses and candidoses, the constant increase in reported cases of mucormycoses indicates their importance in clinical settings.[3,44]

Since opportunistic infections of man and animals require the capability to grow at least at body temperature, thermotolerant fungal species benefit from their physiological possibilities over mesophilic species not capable to grow at elevated temperatures. Whereas all species of *Absidia sensu stricto* grow only below 40°C, all species of *Lichtheimia* could also grow above this limit, which is suppressive for all nonpathogenic *Absidia* spp. Although all species of *Lichtheimia* are thermotolerant growing at temperatures permissive for human pathogenic *Absidia*-like spp., only few turned out to be causative agents of mucormycoses. The first etiological agent of human mucormycoses was described already by Platauf,[45] which is most likely the same pathogenic agent identified in rabbits as *L. corymbifera* (formerly *Absidia corymbifera* and *Mycocladus corymbifer*) 1 year earlier in 1884 by Cohn, who described it as *Mucor corymbifer*.

The mode of transmission of *Lichtheimia* to an efficient causative agent of lichtheimiamycosis does not really differ from that of any other mucoralean human pathogen. The asexually produced sporangiospores of Mucorales are highly airborne and could be easily inhaled leading to pulmonary and rhinocerebral infections[46] (demonstrated in animal models by Singh et al.[47] and Sondhi et al.[48]). Also the spores could dwell on injured skin resulting in (sub)cutaneous mycosis. But ingestion of spores via contaminated food is also possible followed up by gastrointestinal infections. Furthermore, a nosocomial transmission via transplanted organs, bone marrow, or clinical instruments and personnel is also probable.[49–51] Moreover, lichtheimiamycosis caused by *L. corymbifera* was observed as secondary infection following immune suppression in a leukemic patient.[52] Dissemination from the primary spot of infection (metastases), invasion of blood vessels (vasculotropism), as well as the development of allergic symptoms or even an asymptomatic colonization of the human tissue cannot be precluded. Whereas mycoses after traumata with fungal inoculation are rare, most infections occur in diabetic patients (enforced by ketoacidosis), in immunocompromised patients or in patients with neutropenia.[3,53,54] A probable correlation with voriconazole prophylaxis and zygomycoses is also assumed.[55,56] The clinical manifestations of lichtheimiamycoses were reviewed in detail by Ribes et al.[2] and Chayakulkeeree et al.[57]

Since species of *Lichtheimia* are opportunistic fungi, their dominant mode of nutrition is saprobic in their natural habitat soil and decaying plants displaying a worldwide distribution. *Lichtheimia* ranks as the second most causative agent of mucormycoses,[57] causing approximately 5% of the total incidents of human mycoses.[44] However, the mainly morphology-based diagnostics was not supported by a DNA-based detection method in most cases, this number is likely to increase.[55,58]

Like all infectious Mucorales, *Lichtheimia* forms broad ribbon-like hyphae with wide-angle branchings in the infected tissue. Although the prevailing understanding of zygomycetes is the nonseptated coenocytic mycelium, the occurrence of septae in hyphae should never exclude the diagnosis for a mucormycosis.

84.1.4 Diagnosis

Because of the existence of wide-ranged diagnostic tools for specific causative agents of mycoses other than those maintained under the term zygomycoses and mucormycoses (e.g., *Aspergillus*, *Fusarium*), the awareness to unravel the agents behind zygomycoses and mucormycoses is accelerating.[59] Nevertheless, with the exception of aspects on cultivation, morphological, and physiological differences, the diagnosis of lichtheimiamycoses still depends on methods used also for the detection of mucormycotic agents other than *Lichtheimia*.

The diagnosis starts first with the anamnesis elucidating the history of the patients symptomatology, which is easily supplemented by the application of imaging techniques like radiography. The interpretation of the radiograms in combination with the anamnesis requires expertise and experiences in the differentiation of mycotic proliferation, because symptoms developed through fungal infections are well known to be quite similar over a broad range of fungal agents. Therefore, additional steps should be considered for a more convenient identification, because the therapeutic application of antifungal drugs harbor dangerous side effects and display different susceptibilities against fungal species.[60]

In general, the identification of fungal agents down to species level in pure (axenic) culture is always easier than that from infected tissue or liquid cultures. However, the cultivation of the fungal agent on artificial media is not always possible and successful, because it depends on the type and the age of tissue conservation, but should not be neglected. As a rule of thumb, biopsy material recovered from the infected tissue should be as fresh and native as possible in order to retain the viability and gain a successful axenic cultivation of the fungal agent. Freezing of the biopsy material without cryoprotectant additives (e.g., glycerol) and fixation in formaldehyde should be prevented, because these actions dramatically decrease the viability of the fungus. Although a wide range of methods for genomic DNA preparation exist to purify total genomic DNA from pure culture or from infected tissue, the purification of fungal DNA in sufficient quality and quantity for a DNA-based identification is still problematic.

84.1.4.1 Conventional Techniques

The traditional identification of all fungal species predominantly based on morphological criteria with support by physiological (mostly growth temperature) or biochemical (e.g., iodine staining) traits. Since zygomycetes usually do not display many morphological features, the relevance of traditional morphological parameters in species recognition is limited. Although most features are of nonmonophyletic origin and are therefore less relevant in evolutionary phylogeny, the recognition of monophyletic species is possible.[36] Differences are eminent in (i) the mycelium, (ii) its mode of septation (unseptated, unregularly septated, and septated near reproductive structures or injured hyphae), (iii) the morphology of the sexually formed zygospores (with distinguishing characters as displayed in the texture of the epispor as well as the morphology of the appendages and suspensors), and (iv) the morphology of the asexually formed mitospores (sporangiospores, chlamydospores, and arthrospores). The appearance of the asexual reproductive structures is quite diverse, ranging form variously shaped sporangia and sporangiola to merosporangia. Few species produce smaller and few-spored secondary sporangia in older cultures. The sporangiophores may differ in their mode of branching, benting, or in their ability for positive phototropism. Furthermore, size and color of all structures could vary as well as the overall colony appearance.[61]

Apart from all disadvantages of morphological over molecular identification, a big and undeniable advantage of morphology is its universal applicability on all scientific fields of interest. Independently of physiological, ecological, evolutionary, or simply diagnostic aspects, morphological features are always easy to access, but require an experienced eye in terms of the comparability of the observed characters. All morphological features are prone to environmental changes, for example, decreasing/increasing/arresting mycelial growth, occurrence, size, and appearance of reproductive structures by variations in nutrition, temperature, pH, day–night rhythm, humidity, or salinity.

Nevertheless, all clinical important species of *Lichtheimia* grow well in axenic cultures on artificial media. For an unambiguous detection and identification based on morphology combined with growth measurements, we recommend the media used in the respective description keys to avoid ambivalent results.

Besides the constantly updated "Dictionary of the Fungi,"[37] keys and descriptions specially adapted to zygomycetous orders are provided by Zycha et al.,[20] Benjamin,[38] Benny,[62] and Benny et al.[39] Additionally, the homepage of Gerald. L. Benny (Department of Plant Pathology, University of Florida, Gainesville, FL; http://www.zygomycetes.org) offers a highly valuable survey about the taxonomy of zygomycetes and the literature. The system is constantly updated following the actual developments in zygomycete systematics and research.

The genus *Absidia sensu lato* was described in detail by Hesseltine & Ellis/Ellis & Hesseltine[16,17,22,63] and Schipper,[7] who also provided several significant keys.

Alastruey-Izquierdo et al.[32] published a description key to single, accepted species within *Lichtheimia*. The overall colony morphology is represented by rapid and uniform growth at temperatures above 34°C with initially white mycelium, turning gray in age.

Key to the genus *Lichtheimia* (Vuill. 1903; Lichtheimiaceae K. Hoffm., G. Walther, & K. Voigt 2009): this key is originally published in Alastruey-Izquierdo et al.[32]; pictures are presented there. The morphology was assessed on malt extract agar (30 g/L MEA).

1a Sporangia dark brown or dark gray to black; colony diameter after 72 h at 43°C <2 mm; mature sporangiospores rough and/or >6.5 µm in their longest extension **2**

1b Sporangia light brownish gray; colony diameter after 72 h at 43°C >14 mm; mature sporangiospores smooth and <6.5 µm in their longest extension **3**

2a Giant-cells consistently globose, 60–150 µm in diameter ***L. sphaerocystis***

2b Giant-cells (if present) more hypha-like, irregularly swollen, simple to strongly branched, never consistently globose ***L. hyalospora***[1]

2ba Mature sporangiospores small (<5.5 µm), rough, and brownish small-spored variants of ***L. hyalospora***[1]

2bb Mature sporangiospores larger (on the majority >5.5 µm), smooth or rough, hyaline or brownish large-spored variants of ***L. hyalospora***[1]

3a Colony diameter at 43°C after 72 h >40 mm (average growth rate of 1.3 mm/h, growth rate range 0.5–3.2 mm/h); spores ellipsoidal to cylindrical or subglobose to broadly ellipsoidal ***L. ramosa***

3b Colony diameter at 43°C after 72 h <27 mm (average growth rate of 0.4 mm/h, growth rate range 0.1–1.0 mm/h); spores never consistently ellipsoidal to cylindrical **4**

4a Densely branched giant cells, 380–760 (–900) × 320–660 (–770) µm, present in 2-week-old YEA (yeast extract agar) cultures ***L. ornata***

4b Giant-cells absent from 2-week-old YEA cultures ***L. corymbifera***

[1] *L. hyalospora* is now treated as a synonym of *L. blakesleeana*.[7,32] *L. hyalospora* was originally separated from *L. blakesleeana* by the formation of larger and unusual hyaline mitospores. A prospective new separation of both species in varieties or formae cannot be excluded.

In addition to the possibility of a morphological discrimination between *L. corymbifera* and *L. ramosa*, the utilization of the oligosaccharides melecitose and palatinose (isomaltulose) as carbon sources by *L. ramosa*, but not by *L. corymbifera* is possible.[33] Unfortunately, nothing is reported about the assimilation of these sugars by other species of *Lichtheimia*.

Well-defined and well-described reference strains are stored and maintained in several online-accessible culture collections like ATCC (American Type Culture Collection; http://www.lgcpromochem-atcc.com), ARS-NRRL (Agricultural Research Service Culture Collection; http://ncaur.usda.gov), CBS (Centraalbureau voor Schimmelcultures; http://cbs.knaw.nl), DSMZ (German Collection of Microorganisms and Cell Cultures; http://www.dsmz.de), or the JMRC (Jena Microbial Reference Centre; http://www.prz.uni-jena.de).

Many of the reliable reference strains are given in the Section 84.1.2. Type strain material has the important advantage of an unambiguous and functional attribution of a species description and to an individual microorganism—a fact which should always be taken into account in species identification. Information about type strains can be retrieved from MycoBank (http://www.mycobank.org).

If the cultivation of infectious material is impossible or too time-consuming, the existence of zygomycete contaminants in infected tissue can also be proven in liquid samples, scrape or swab tests, as well as biopsy tissue. In combination with appropriate staining techniques, hyphae typical for *Lichtheimia* becomes detectable directly in infected tissue. However, these tests depend on the locality of the sample drawing. If the biopsy needle did not hit directly the infected tissue, the detection might be not always positive. False negative detection may also be achieved in cases with observed septae matching the hyphal morphology of ascomycetes.

84.1.4.2 Molecular Techniques

One and probably the largest disadvantage of the single application of molecular data is its inability to unequivocally identify a fungal organism if used as a single criterion without the combination of morphological observations. The more additional information about the morphology, physiology, or biochemistry or even whole-genome data exist, the more proper and reliable is the identity and species designation of a specimen.

But, with a deliberate choice of the molecular markers used for species identification and a well-established reliable database, which consists of well-defined reference specimens, even a single molecular marker can result in a high approximation of the identity of the fungal agent in question. However, such databases do not enable the identification of new or undescribed species. Therefore, a close collaboration is recommended between clinical laboratories, research institutions, and reference laboratories, which not only enables early detection of novel fungi with pathogenic potential but also facilitates the establishment of a platform for new or improved methods.

One basic and in most diagnostic laboratories well-established method is the polymerase chain reaction (PCR)[64,65] aiming at the amplification of DNA fragments, which localize between two short oligonucleotide primers. One universally applicable molecular marker useful for the identification of most fungi including zygomycetes is the region between the nuclear small subunit (SSU) rDNA (18S rDNA) and the nuclear large subunit rDNA (28S rDNA). The variable regions of the ITS1 and ITS2 include also the 5.8S rDNA, which is highly conserved in fungi. Without evolutionary pressure, the ITS region is highly heterogenous at the above-species level and in zygomycetes at the below-species level.[12,66] Therefore, the ITS is a suitable barcode marker for species discrimination and recognition for all Mucorales including *Lichtheimia*. However, the application of ITS in species recognition is tedious. The ITS is homogenous at the above-species level and not suitable for the species discrimination in *Penicillium*.[67] Although the ITS region possesses few disadvantages arising from its repetitive nature and discontinuities in sequence homogeneity among individual elements at the infraindividual level, it has been widely used in the characterization of type and reference strains. Consequently, a wide range of ITS reference sequences are accessible in public databases. If compared with ITS sequences of unknown species designation, reliable reference sequences play an essential rule in the validity of BLAST searches. The majority of reference sequences target the ITS. Thus, it is not remarkable that ITS markers are discussed as the golden standard for molecular barcodes suitable for the reliable species identification of zygomycetes including *Lichtheimia*[68] (recommended by the Clinical and Laboratory Standards Institute, CLSI 2007[69]). In this chapter, the term molecular barcoding is treated as synonym for the terms molecular identification or molecular detection of fungal taxa, because all procedures have in common the use of specific nucleotide sequences as molecular markers or barcodes in order to gain genetic signatures, which are specific for a certain taxon, for example, genus-, species, or pathotype-specific.

Alternative genetic markers are the adjacent loci of the nuclear SSU and LSU rDNA (18S, 28S) or exonic sequences of single- till few-copy protein-coding genes encoding actin (*act*), translation elongation factor 1 alpha (*tef*),[5,70–72] or the high affinity iron permease (FTR).[73] Apart from nucleotide sequence-based detection, additional methods, which generate DNA polymorphisms, were also proposed including restriction length polymorphism procedures[74] or variations of PCR targeting the multiple species-specific detection of mucormycosis-inducing agents in multiplex PCR[75] and differences in gene expression as revealed by real-time PCR.[76] However, none of those methods are explicitly adapted for the molecular detection of lichtheimiamycosis. Until now, no reliable serological detection method is available for routine use in zygomycete detection.[77]

Several online-accessible databases of the International Nucleotide Sequence Database Collaboration store and maintain sequence information from fungal strains. Open access allows free up- and download of generated sequence material, but requires very high responsibilities on underlying background information about the fungal organisms used for data generation. Some databases are DBBJ (DNA Database of Japan; http://www.ddbj.nig.ac.jp), EMBL (European Molecular Biology Laboratory; http://www.ebi.ac.uk/embl), GenBank at NCBI (http://www.ncbi.nlm.nih.gov), PDB (Protein Data Bank; http://www.rcsb.org/pdb), and UniProt (Universal Protein Resource; http://www.expasy.uniprot.org). Approximately 20% of the sequences currently deposited at GenBank may be not annotated correctly.[78] Thus, BLAST results should be handled with care and verified by the alignment to reference sequences from database platforms specialized on clinically relevant fungi (see also Section 84.3).

84.2 METHODS

84.2.1 Sample Preparation

Direct microscopy of samples like scrapings, smears, blood, or sputum obtained from patients or biopsy material from infected tissue could be used to visualize the characteristically unseptated hyphae typical for *Lichtheimia* and all other Mucorales, though just in material sufficiently charged with fungal material. But fungal hyphae observed here do not allow a differentiation down to species level. Fluorescence staining combined with optical brighteners like Calcofluor or Blancophor[79] facilitates the universal identification of Mucorales.

Cultures could be grown from scrapings or smears obtained from patients or from liquids containing fungal material. However, cultures from blood, urine, sputum, or bronchoalveolar lavage are less successful.[80,81] The utilization of biopsy samples for the inoculation of media should be handled with caution, because a forceful crushing of tissue could easily destroy fungal hyphae causing a false negative result in axenic cultures. Because hyphae of zygomycetes lack regular septation, a reliable and successful inoculation of culture media on Petri dishes is not guarantied.[3,81] Species of *Lichtheimia* as well as other mucoralean fungi grow well on MEA (30 g/L), PDA (potato dextrose agar; DSMZ media no. 129; http://www.dsmz.de), or Sabouraud glucose agar (20 g/L glucose; e.g., Carl Roth, Karlsruhe). These media could easily be supplemented with a cocktail of single or multiple antibiotics (e.g., ampicilline, streptomycine, tetracycline, neomycine, or chloramphenicol in final concentrations of 1.9, 1.2, 0.6, 1.4, and 1.4 mg/mL, respectively) in order to efficiently suppress bacterial growth. Each individual medium will influence the overall appearance of the growing fungal colony and is to be evaluated during the course of identification. Fungal colonies should be cultivated at different temperatures to ensure their optimum (and maximum) growth. The standard procedure for observations on the phenotype of fungal cultures should be applied, where possible.[82]

Microscopic observations on fungal colonies grown in axenic cultures should be done using a binocular dissecting microscope for the assessment of the colony morphology and for micromorphological observations on the reproductive structures (sporangia, chlamydospores, etc.). Under optimum conditions, the overall appearance of colonies of *Lichtheimia* is white at the beginning, turning gray in age. The colony is rapidly and uniformly growing with an uncolored reverse side.

Micromorphological observations were carried out from pure fungal cultures grown on solid agar media. The preparation of the microscopic slides should be done with juvenile and proliferating cultures in order to obtain young, spore-bearing sporangia and to observe the periphery and growth front of the colony. Either a "tape prep" or a "tease prep" preparative procedure facilitates the display of morphological features. The "tape prep" method applies the adhesive side of a tape to fix parts of a fungal colony. The tape is then placed sticky sidedown on a slide for microscopy. The "tease prep" includes careful preparation of parts from the fungal colony in a drop of lactophenol cotton blue, which is covered with a coverslip and used for microscopy. The latter method is more recommended because it ensures a higher biological safety. Another method is the traditional slide culture. Here a sterile coverslip is placed on agar medium prior to fungal inoculation. After inoculation the fungus will partially grow over the coverslip. The coverslip can be removed and subjected to microscopy. All microscopic preparations can be stained with lactophenol cotton blue dye solution to gain a higher contrast during lightmicroscopical observation.[2] All observations have to be compared to the descriptions and keys presented here or elsewhere.[7,17,20,21,32]

For direct histopathological microscopy of biopsy material, the probes should be fixed in neutral buffered or 10% buffered formol saline[83] and embedded in paraffin for the preparation of microscopic sections in a microtome. A fixation method with less toxicity and improved survival of antigens for immunohistochemistry is described by Beckstead[84] using zinc as primary fixative.

Tissue sections could then be stained with hematoxylin and eosin. For verification of fungal infections in the tissue, staining with Grocott's methenamine-silver nitrate, periodic acid-Schiff or similar fluorescence staining methods with optical brighteners are useful.[85–87] Tissue samples, which were not embedded in paraffin, could also be viewed for fungal hyphae after homogenization in a tissue grinder in 0.9% NaCl.[2,79] Even simple phase-contrast microscopy is sufficient for the detection of zygomycetous hyphae.[88]

Samples for cultures could not be treated with formaldehyde. But preservation in 10% glycerine or sterile mineral oil conserves the morphological features and viability of fungi.[89,90]

84.2.2 Detection Procedures

Purification of genomic DNA could either be performed from pure cultures or directly from tissue material. If using tissue-purified fungal DNA, the genomic target has to be fungal specific because the preparation does not discriminate between human and fungal DNA.

DNA extraction can be manually performed using different kits, which are commercially available, for example, the DNeasy Plant Kit series (Qiagen, Venlo, the Netherlands) or the innuPREP Kit series (Analytik Jena, Jena, Germany). The latter one is also available for automatic purification and may also be used for tissue embedded in paraffin. The innuSPEED Kit series (Analytik Jena, Jena, Germany) are also suitable for DNA extraction in combination with a tissue homogenizator or speed mill distributed by the same company. In order to increase yield and quality of the purified fungal DNA from human samples, improved lysis of the fungal hyphae is recommended during grounding the mycelium in liquid nitrogen or bead-mediated maceration of the mycelium in a speed mill prior proceeding with the purification steps. If the DNA purification has to be done from formalin-fixed, paraffin-embedded tissue samples, the potential degradation of DNA during the fixation process has to be considered.

A rapid protocol for fungal DNA based on a modified method was described by Cenis.[91] This method is optimized for pure fungal cultures. The mycelium is frozen and ground to a fine powder in liquid nitrogen. Approximately 100 mg of fungal mycelium is suspended in extraction buffer (200 mM Tris–HCL [pH 8.5]; 250 mM NaCl; 25 mM EDTA [pH 8.0]; 0.5% sodium dodecyl sulfate [SDS])[92] and vigorously mixed for 3 min. After incubation at 65°C for 10 min and repeated agitation, an amount of 0.15 mL 3 M sodium acetate (pH 5.2) is added and tubes are placed for 10 min at −20°C. Separation of the DNA is achieved at 13,000 rpm for 15 min and followed by precipitation of the supernatant with an equal volume of isopropanol. After incubation for 15 min at room temperature, and centrifugation for 15 min at 13,000 rpm, the DNA pellet is washed twice in 70% isopropanol. The air-dried DNA pellet is resuspended in up to 50 μL TE buffer (10 mM Tris–HCl [pH 7.6]; 1 mM EDTA). Storage of the DNA is possible at −25°C.[91] The amount of DNA for successful amplification in a PCR depends on the complexity and number of genomic copies of the desired region in the DNA sample used as template for PCR. Average concentrations from 1 pg for plasmid DNA till 1 μg for mammalian DNA are required per PCR reaction.[93] Alternatively, the DNA extraction may be carried out using CTAB (hexacetyltrimethylammonium bromide; Sigma) as described by Voigt et al.[70] and modified by Schwarz et al.[68]

Amplification of the ITS as target region (ITS1–5.8S rDNA–ITS2) is conducted using the universal primers ITS1 (5′-TCCGTAGGTGAACCTGCGG-3′) and ITS4

(5′-TCCTCCGCTTATTGATATGC-3′).[94] For fungus-specific amplification of the ITS, the fungal universal primers V9D (5′-TTAAGTCCCTGCCCTTTGTA-3′) and LS266 (5′-GCATTCCCAAACAACTCGACTC-3′) are appropriate as well.[95] One PCR assay typically contains 2–6 μg genomic DNA of the fungal pathogen, 10 pmol of each primer, 0–4 mM $MgCl_2$; 5 μL of 2 mM dNTPs (Fermentas), 10× PCR DreamTaq™ Buffer (Fermentas), and 1.25 U of DreamTaq™ Polymerase (Fermentas) in a total volume of 50 μL. Polymerases with 3′–5′ exonucleolytic proofreading activity, for example, *Pfu* polymerase (Fermentas), are recommended in order to obtain amplicons with low error rate. PCR fragments are amplified using a FlexCycler (Analytik Jena, Jena, Germany) and the following temperature profile: an initial denaturation step of 5 min at 95°C following 30 cycles of denaturation for 30 s at 95°C, annealing for 60 s at 55°C, and primer elongation for 60 s at 72°C, with a final primer extension reaction for 4 min at 72°C. PCR amplicons may be purified by adsorption on glass particles (GeneClean II, BIO 101, Vista, CA) based on a protocol of Vogelstein & Gillespie[96] if an intermediate cloning step is required prior sequencing of the fragment. The PCR assay is spiked with a threefold volume of 6 M NaI (Roth, Karlsruhe, Germany) and 5 μL glass milk particles (MP Biomedicals Heidelberg, Germany). After vigorous agitation, the tubes are incubated at 55°C for 5 min under occasional gentle agitation. The glass particles are spun down for 10 s and washed twice, each washing step with 200 μL New Wash (50% Ethanol, 20 mM Tris–HCl [pH 7.6], 1 mM EDTA, 0.1 M NaCl). The DNA is eluted twice from the air-dried glass particles each with 10–20 μL bidistilled water at 65°C for 10 min. The glass particle-free eluate is stored at –25°C until performance of the PCR.

Cycle sequencing is performed using the Big Dye™ fluorescent-labeled terminator DyeDeoxy protocol with AmpliTaq polymerase (Applied Biosystems) and single oligonucleotide PCR primers as sequencings primers. Reaction products are analyzed on an ABI Prism™ 310 automated sequencer applying a one-capillary system (Applied Biosystems). For high-throughput sequence analyses, the application of alternative equipment is recommended, for example, a combination of the SpeedCycler (Analytik Jena AG, Jena, Germany) and the ABI 3730*xl* DNA analyzer applying a 96-capillary block system (Applied Biosystems) together with adapted protocols are adequate methods to reduce pure analytic time to less than 1 h. Sequences are edited and manually manipulated using Chromas Lite version 2.01 (Technelysium, Helensvale, Queensland, Australia).

Sequence analysis for species identification is done by BLAST searches against available reference databases of the network from the International Nucleotide Sequence Database Collaboration, for example, GenBank. Sequence similarities of 100% correlate with a highly supported approximation in species identification. Values below 100% should be treated with care, because the designation to a distinct species is less confirmed. The lack of threshold definitions for sequence similarity makes a molecular barcode-based designation to a species or even the molecular barcode-based identification of a new species rather tedious. Such thresholds are under discussion[12,97] but will never replace taxonomical expertise, experimental experience, and exploration of additional data.

Schwarz et al.[68] proposed a protocol for rapid analysis of zygomycetes directly from infected tissue. Amplification and sequencing of ITS turned out to be a reliable method for species identification in animal models. If this method is also reliable in human, tissue samples need to be tested.

Molecular methods for the generation of sequences of fixed and already embedded tissue samples are applicable with varying success. Prior to PCR, the fixed tissues need to be washed with xylene and 95% alcohol in order to remove the wax. The samples are then equilibrated in buffer (Tris–EDTA, SDS, proteinase K) and incubated at 55°C for 24 h under constant shaking.[76] With a subsequent LightCycler platform-based real-time PCR assay, the detection of *Lichtheimia* was principally possible but resulted in a decreased sensitivity for fixed tissue samples.[76]

84.3 CONCLUSIONS AND FUTURE PERSPECTIVES

Mucormycoses seem to be highly abundant during systemic and rhinocerebral infections. The clinical manifestation of mucormycoses is advanced by airborne transmission in patients with an underlying disease.[44] Recently, an increase of incidents for cutaneous lichtheimiamycoses[98] as well as an increasing number of nosocomial infections in hospitals were reported.[99] The growing awareness of life-threatening zygomycoses and their causative agents makes a taxonomic discrimination of the infectious agents down to the species level indispensable. Mycologists and taxonomists keep constantly going on renaming and recombining fungal species in order to gain an improved systematics, which is concordant with phylogenetic relationships. That mannerism makes it very difficult for the personnel of medical laboratories to keep track of the changes. Nevertheless, its necessity becomes more and more accepted also by nontaxonomists especially clinicians, because the phylogenetic relationships will help to understand the similarities and differences observed not only at the morphological and physiological level but also at the level of the susceptibility to antifungal agents with consequences to treatment and therapy. Therefore, the separation of *Absidia sensu lato* in three distinct genera not only allows the establishment of a monophyly-based natural system but also facilitates the distinction of harmless and noncausative *Absidia*-like genera and their species (*Lentamyces*, *Absidia sensu stricto*) from harmful *Absidia*-like genus (*Lichtheimia*) with species possessing pathogenic potential. Nomenclatural changes did yet not ease the course of identification of the specimens. Although new and fast methods exist to identify fungal organisms, the traditionally morphology-based methods should not be neglected. Just a combined application of both strategies will withdraw disadvantages of the individual methods if applied in single analyses. The concatenation of colony- and micromorphological criteria with growth physiological and molecular markers (genetic signature sequences, carbon source utilization, and other chemotaxonomic profiles) will inevitably result in an increased reliability of the identification and the discrimination between nonpathogenic species from pathogenic species belonging to *Lichtheimia*.

Although the classical phenotypic (morphological) approach of species identification by morphological criteria is still essential for definite new species assignments, it has been proven to be error-prone for daily routine identification because its outcome is subjective and it is costly in terms of time and labor.[55] In most cases, the fungal agents need to be identified directly in the infected tissue and the proof by an axenic culture is often missing. But most fungi, also *Lichtheimia*, develop their typical morphology just in pure cultures, and several modes of molecular detection are less time-consuming and are therefore preferably used over classical morphological identification. However, molecular detection methods are cost-intensive and rely on consistent reference material. The lack of pursuant platforms providing congruent data about clinical symptomatology and fungal type material hamper the discovery of new species. Consequently, molecular data can solely account for an approximation on the fungal identity, whereas morphological traits can support and verify the diagnosis. But in cases of a successful cultivation of the fungus, the incidence of possible air-borne contaminations during cultivation and false-positive signals in molecular detection due to fungal spores inhaled or ingested prior sample recovery has to be eliminated. Both air-borne contaminations and incorporated fungal spores lack clinically relevant manifestations.

Another problem arises from the fact that for many pathogenic fungi the number of available isolates is limited. The deprivation of morphological and molecular information from a broad range of different isolates leads to a deprived and inadequate assessment of the intra- and interspecies variability. Morphological and growth physiological criteria tend to develop intraspecific variations. However, especially sugar assimilation profiles provide easy and cheap methods for the discrimination of *L. corymbifera*, which does not utilize melecitose and palatinose (isomaltulose) as carbon sources, and *L. ramosa*, which does so.[33] Apart from the elucidation of the utilization of these sugars by two species of *Lichtheimia*, unfortunately, nothing has been reported yet about the assimilation of melecitose, palatinose, and other sugars by the remaining three species of *Lichtheimia*. Moreover, the establishment of a robust sugar assimilation-based detection of *Lichtheimia* necessitates expansion downward to the below-species level. That broadening is composed of comparative investigations encompassing at least more than 10 isolates from different case reports or geographical regions. The inclusion of a broader isolate spectrum can be also assigned to the refinement of the diagnostic power of molecular markers. For example, the comparison of isolates from *Absidia repens* and other *Absidia*-like fungi revealed enormous differences in the ITS sequences of isolates from different continents of the northern hemisphere. That variability highly correlates with the geographic distribution.[12] In order to avoid such variances, a wide taxon sampling is required over a broad range of individual isolates recovered from different patients and hospitals in different geographic localities. The more all-embracing that survey is, the more reliable and powerful the molecular diagnostics will be. The more criteria other than molecular markers will be used in combination with the methods of molecular identification, the higher the resolution and dependability in the species discrimination.

After successful evaluations of the intraspecific variability, the long-term storage of the involved fungal specimens and type material in culture collections and the reexamination of the maintained specimens are essential prerequisites for research on the distribution and the epidemiology as well as the evolution of fungal speciation.

Although the number of available methods for molecular detection is relatively low and not fully taxon-discriminative within zygomycetes, their diagnostic value imply a powerful supplement of classical methods traditionally used in species identification of *Absidia*-like fungi including *Lichtheimia*. One major breakthrough toward the achievement of an increased diagnostic power of molecular markers would probably be the establishment of a pan-fungal DNA chip-based assay for all clinical important fungi.[68,100,101]

During the clinical daily routine analysis of *Lichtheimia* and its discrimination from harmless *Absidia*-like fungi or other harmful human pathogens, the decision between a detection-first or a therapy-first action is influenced by the celerity and accuracy of the method of choice. Thus, risks and feasibility of a probable rapid proliferation have to be meticulously calculated if suspicion for a lichtheimiamycosis is entertained. As long as the spectrum for applicable antifungal chemotherapy is limited and not fungus-specific, a profound determination of the exact fungal species of *Lichtheimia* will just play an ancillary role in order to serve the patient's benefits. Consequently, with the development of novel and more species-effective antifungal drugs accompanied with advancement of rapidity and precision of the molecular diagnostic methods, the prognosis of lichtheimiamycosis will markedly be improved.

REFERENCES

1. Hibbett, D.S. et al., A higher-level phylogenetic classification of the Fungi, *Mycol. Res.*, 111, 509, 2007.
2. Ribes, J.A., Vanover-Sams, C.L., and Baker, D.J., Zygomycetes in human disease, *Clin. Microbiol. Rev.*, 13, 236, 2000.
3. Gonzalez, C.E., Rinaldi, M.G., and Sugar, A.M., Zygomycosis, *Infect. Dis. Clin. North Am.*, 16, 895, 2002.
4. Thirion-Delalande, C. et al., Disseminated acute concomitant aspergillosis and mucormycosis in a pony, *J. Vet. Med. A Physiol. Pathol. Clin. Med.*, 52, 121, 2005.

25 C 9781439812402_HB_8.5x11.job

5. Hoffmann, K., Discher, S., and Voigt, K., Revision of the genus *Absidia* (Mucorales, Zygomycetes) based on physiological, phylogenetic, and morphological characters; thermotolerant *Absidia* spp. form a coherent group, Mycocladiaceae fam. nov., *Mycol. Res.*, 111, 1169, 2007.
6. Hoffmann, K. and Voigt, K., *Absidia parricida* plays a dominant role in biotrophic fusion parasitism among mucoralean fungi (Zygomycetes): *Lentamyces*, a new genus for *A. parricida* and *A. zychae*, *Plant Biol.*, 11, 537, 2009.
7. Schipper, M.A.A., Notes on Mucorales—I. Observations on *Absidia*, *Persoonia*, 14, 133, 1990.
8. Kwaśna, H. and Nirenberg, H.I., *Siepmannia*, a new genus in the Mucoraceae, *Mycologia*, 100, 259, 2008.
9. Kwaśna, H. and Nirenberg, H.I., Validation of the genus *Siepmannia* (Mucoraceae) and its four species, *Pol. Bot. J.*, 53, 187, 2008.
10. Hoffmann, K., Identification of the genus *Absidia* (Mucorales, Zygomycetes), a comprehensive taxonomic revision, in *Molecular Identification of Fungi*, p. 439, Gherbawy, Y., Voigt, K. (Eds.), Springer-Verlag, Heidelberg, 2010.
11. Van Tieghem, P., Troisiéme mémoire sur les Mucorinées, *Ann. Sci. Nat. Bot.*, 4, 312, 1876.
12. Hoffmann, K. et al., Diversity, genotypic identification, ultrastructural and phylogenetic characterization of zygomycetes from different ecological habitats and climatic regions: Limitations and utility of nuclear ribosomal DNA barcode markers, in *Current Advances in Molecular Mycology*, p. 263, Gherbawy, Y., Mach, R.L., Rai, M. (Eds.), Nova Science Publishers, Inc., New York, 2009.
13. Kwaśna, H., Ward, E., and Bateman, G.L., Phylogenetic relationships among Zygomycetes from soil based on ITS1/2 rDNA sequences, *Mycol. Res.*, 110, 501, 2006.
14. Hesseltine, C.W., Mahoney, M.K., and Peterson, S.W., A new species of *Absidia* from an alkali bee brood chamber, *Mycologia*, 82, 523, 1990.
15. Von Arx, J.A., On Mucoraceae s. str. and other families of the Mucorales, *Sydowia*, 35, 10, 1982.
16. Ellis, J.J. and Hesseltine, C.W., The genus *Absidia*: Globose-spored species, *Mycologia*, 57, 222, 1965.
17. Ellis, J.J. and Hesseltine, C.W., Species of *Absidia* with ovoid sporangiospores II, *Sabouraudia*, 5, 57, 1966.
18. Zycha, H., Pilze II. *Mucorineae*, Krypt. Fl. Mark Brandenburg 6a, *Leipzig*, Germany, 1935.
19. Beauverie, J., *Mycocladus verticillatus*, *Ann. Univ. Lyon*, 3, 162, 1900.
20. Zycha, H., Siepmann, R., and Linnemann, G., *Mucorales. Eine Beschreibung aller Gattungen und Arten dieser Pilzgruppe*, J Cramer Publishing Co., Lehre, Germany, 1969.
21. Hesseltine, C.W. and Ellis, J.J., Species of *Absidia* with ovoid sporangiospores, I., *Mycologia*, 58, 761, 1966.
22. Hesseltine, C.W. and Ellis, J.J., The genus *Absidia*: *Gongronella* and cylindrical-spored species of *Absidia*, *Mycologia*, 56, 568, 1964.
23. Mirza, J.H. et al., *Mucorales of Pakistan*, University of Agriculture, Faisalabad, Pakistan, 1979.
24. Vánová, M., Nomen novum, nomenclatural changes and taxonomic rearrangements in Mucorales, *Ceská Mykologie*, 45, 25, 1991.
25. Hoffmann, K., Walther, G., and Voigt, K., *Mycocladus* vs *Lichtheimia*: A correction (Lichtheimiaceae fam. nov., Mucorales, Mucoromycotina), *Mycol. Res.*, 113, 277, 2009.
26. Bainier, G., Sur quelques espèces de Mucorinées nouvelles ou peu connues, *Bull. Soc. Mycol. France*, 19, 153, 1903.
27. Vuillemin, P., Le genre *Tieghemella* et la série de Absidées, *Bull. Soc. Mycol. France*, 19, 119, 1903.
28. Bainier, G., *Étude sur les Mucorinées*. Thèse, Paris, France, 1882.
29. Sydow, H., (Referate und kritische Besprechungen), *Ann. Mycol.*, 1, 371, 1903.
30. Vuillemin, P., Le *Lichtheimia ramosa* (*Mucor ramosus* Lindt) champignon pathogène, distinct du *L. corymbifera*, *Arch. Parasitol.*, 8, 562, 1904.
31. Cohn, F.J., *Mucor corymbifer*, In *Über pathogene Mucorineen und die durch sie erzeugten Mykosen des Kaninchens*, Lichtheim, L. (ed.), *Z. Klin. Med.*, 7, 149, 1884.
32. Alastruey-Izquierdo, A. et al., Species recognition and clinical relevance of the zygomycetous genus *Lichtheimia* (syn. *Absidia* pro parte, *Mycocladus*) *J. Clin. Microbiol.*, 48, 2154, 2010.
33. Garcia-Hermoso, D. et al., Molecular and phenotypic evaluation of *Lichtheimia corymbifera* (ex. *Absidia corymbifera*) complex isolates associated with human mucormycosis: Rehabilitation of *L. ramosa*, ex synonym of *L. corymbifera*, *J. Clin. Microbiol.*, 47, 3862, 2009.
34. Nottebrock, H., Scholer, H.J., and Wall, M., Taxonomy and identification of mucormycosis-causing fungi I. synonymity of *Absidia ramosa* with *A. corymbifera*, *Sabouraudia*, 12, 64, 1974.
35. Chen, G.Q. and Zheng, R.Y., A new thermophilic variety of *Absidia idahoensis* from China, *Mycotaxon*, 69, 173, 1998.
36. Voigt, K. et al., Revision of the family structure of the Mucorales (Mucoromycotina, Zygomycetes) based on multigene-genealogies: Phylogenetic analyses suggest a bigeneric Phycomycetaceae with *Spinellus* as sister group to *Phycomyces*, in *Current Advances in Molecular Mycology*, p. 263, Gherbawy, Y., Mach, R.L., Rai, M. (Eds.), Nova Science Publishers, Inc., New York, 2009.
37. Kirk, P.M. et al., *Dictionary of the Fungi*, 10th edn., CABI Publishing, Wallingford, U.K., 2008.
38. Benjamin, R.K., Zygomycetes and their spores, in *The Whole Fungus: The Sexual-Asexual Synthesis*, Kendrick, B. (Ed.), National Museum of Natural Science, National Museums of Canada and the Kananaskis Foundation, Ottawa, Ontario, Canada, 1979.
39. Benny, G.L., Humber, R.A., and Morton, J.B., Zygomycota: Zygomycetes, in *The Mycota. Part A. Systematics and Evolution*, Vol. VII, p. 113, McLaughlin, D.J., McLaughlin, E.G., Lemke, P.A. (Eds.), Springer-Verlag, Berlin, Heidelberg, 2001.
40. Taylor, J.W. et al., Phylogenetic species recognition and species concepts in fungi, *Fungal. Genet. Biol.*, 31, 21, 2000.
41. Crous, P.W. et al., MycoBank: An online initiative to launch mycology into the 21st century, *Stud. Mycol.*, 50, 19, 2004.
42. Robert, V., Stegehuis, G., and Stalpers, J., The MycoBank engine and related databases, http://www.mycobank.org, 2005.
43. Lendner, A., Les Mucorinées de la Suisse, *Mater. Flore Crypt. Suisse*, 3, 1, 1908.
44. Roden, M.M. et al., Epidemiology and outcome of zygomycosis: A review of 929 reported cases, *Clin. Infect. Dis.*, 41, 634, 2005.
45. Platauf, A.P., Mycosis mucorina, *Virchows Arch.*, 102, 543, 1885.
46. Barr, A. et al., Rhinoorbital and pulmonary zygomycosis post pulmonary aspergilloma in a patient with chronic lymphocytic leukaemia, *Acta Biomed.*, 77(suppl 4), 13, 2006.
47. Singh, G. et al., Sequential pathological studies in Asian water buffaloes infected intratracheally with *Absidia corymbifera*, *Rev. Iberoam. Micol.*, 15, 146, 1998.

48. Sondhi, J., Gupta, P.P., and Sood, N., Effect of immunosuppression on the pathogenesis of respiratory zygomycoses in rabbits intranasally infected with *Absidia corymbifera*, *Rev. Iberoam. Micol.*, 16, 107, 1999.
49. Stas, K.J.F. et al., Isolated zygomycosis in a bought living unrelated kidney transplant, *Transplant. Int.*, 9, 600, 1996.
50. Jantunen, E. et al., Case Report. Invasive cutaneous mucormycosis caused by *Absidia corymbifera* after allogenic bone marrow transplantation, *Bone Marrow Transplant.*, 18, 229, 1996.
51. Leong, K.W. et al., Cutaneous mucormycosis due to *Absidia corymbifera* occurring after bone marrow transplantation, *Bone Marrow Transplant.*, 19, 513, 1997.
52. Lopes, J.O. et al., Cutaneous zygomycosis caused by *Absidia corymbifera* in a leukemic patient, *Mycopathologia*, 130, 89, 1995.
53. Lake, F.R., McAleer, R., and Tribe, A.E., Pulmonary mucormycosis without underlying systemic disease, *Med. J. Aust.*, 149, 323, 1988.
54. Marshall, D.H. et al., Post-traumatic corneal mucormycosis caused by *Absidia corymbifera*, *Ophthamology*, 104, 1107, 1997.
55. Kontoyiannis, D.P. et al., Zygomycosis in a tertiary-care cancer center in the era of Aspergillus active antifungal therapy: A case-control observational study of 27 recent cases, *J. Infect. Dis.*, 191(8), 1350, 2005.
56. Trifilio, S. et al., Breakthrough fungal infections after allogeneic hematopoietic stem cell transplantation in patients on prophylactic voriconazole, *Bone Marrow Transplant.*, 40(5), 451, 2007.
57. Chayakulkeeree, M., Ghannoum, M.A., and Perfect, J.R., Zygomycosis: The re-emerging fungal infection, *Eur. J. Clin. Microbiol. Infect. Dis.*, 25, 215, 2006.
58. Balajee, S.A., Sigler, L., and Brandt, M.E., DNA and the classical way: Identification of medically important molds in the 21st century, *Med. Mycol.*, 45, 475, 2007.
59. Boekhout, T. et al., Fungal taxonomy: New developments in medically important fungi, *Curr. Fungal Infect. Reports*, 3, 170, 2009.
60. Singh, J., Rimek, D., and Kappe, R., In vitro susceptibility of 15 strains of Zygomycetes to nine antifungal agents as determined by the NCCLS M38-A microdilution method, *Mycoses*, 48, 246, 2005.
61. O'Donnell, K., *Zygomycetes in Culture*, Palfrey Contributions in Botany no. 2. Department of Botany, University of Georgia, Athens, GA, 1979.
62. Benny, G.L., Zygomycetes, in *Synopsis and Classification of Living organisms*, Vol. 1, Parker, S.P. (Ed.), McGraw-Hill Book Company, Inc., New York, 1982.
63. Hesseltine, C.W. and Ellis, J.J., Notes on Mucorales, especially on *Absidia*, *Mycologia*, 53, 406, 1961.
64. Kleppe, K. et al., Studies on polynucleotides. XCVI. Repair replications of short synthetic DNA's as catalyzed by DNA polymerases, *J. Mol. Biol.*, 56, 341, 1971.
65. Mullis, K.B., The unusual origin of the polymerase chain reaction, *Sci. Am.*, 262, 56, 1990.
66. Gherbawy, Y. et al., Molecular barcoding of microscopic fungi with emphasis on the mucoralean genera Mucro and Rhizopus, in *Molecular Identification of Fungi*, p. 213, Gherbawy, Y., Voigt, K. (Eds.), Springer-Verlag, Heidelberg, 2010.
67. Skouboe, P. et al., Phylogenetic analysis of nucleotide sequences from the ITS region of terverticillate *Penicillium* species, *Mycol. Res.*, 103, 873, 1999.
68. Schwarz, P. et al., Molecular identification of Zygomycetes from culture and experimentally infected tissues, *J. Clin. Microbiol.*, 44, 340, 2006.
69. Clinical and Laboratory Standards Institute. Interpretive criteria for bacteria and fungi identification by DNA target sequencing, Approved Guideline MM-18A. CLSI, 2007.
70. Voigt, K., Cigelnik, E., and O'Donnell, K., Phylogeny and PCR identification of clinically important Zygomycetes based on nuclear ribosomal-DNA sequence data, *J. Clin. Microbiol.*, 37, 3957, 1999.
71. O'Donnell, K. et al., Evolutionary relationships among mucoralean fungi (Zygomycota): Evidence for family polyphyly on a large scale, *Mycologia*, 93, 286, 2001.
72. Voigt, K. and Wöstemeyer, J., Phylogeny and origin of 82 Zygomycetes from all 54 genera of the Mucorales and Mortierellales based on combined analysis of actin and translation elongation factor EF-1a genes, *Gene*, 270, 113, 2001.
73. Nyilasi, I. et al., High affinity iron permease (FTR1) gene sequence-based molecular identification of clinically important Zygomycetes, *Clin. Microbiol. Infect.*, 14, 393, 2008.
74. Machouart, M. et al., Genetic identification of the main opportunistic Mucorales by PCR-restriction fragment length polymorphism, *J. Clin. Microbiol.*, 44, 805, 2006.
75. Nagao, K. et al., Genetic identification and detection of human pathogenic *Rhizopus* species, a major mucormycosis agent, by multiplex PCR based on internal transcribed spacer region of rRNA gene, *J. Dermatol. Sci.*, 39, 23, 2005.
76. Hata, D.J. et al., A real-time PCR method for detection of Zygomycetes, *J. Clin. Microbiol.*, 46, 2353, 2008.
77. Lass-Flörl, C., Zygomycosis: Conventional laboratory diagnosis, *Clin. Microbiol. Infect. Suppl.*, 5, 60, 2009.
78. Bridge, P.D. et al., On the unreliability of published DNA sequences, *New Phytol.*, 160, 43, 2003.
79. Rüchel, R. and Schaffrinski, M., Veratile fluorescent staining of fungi in clinical specimens by using the optical brightener Blankophor, *J. Clin. Microbiol.*, 37, 2694, 1999.
80. Kappe, R. et al., Molecular detection of pathogenic fungi in the presence of human tissue, *J. Med. Microbiol.*, 47, 811, 1998.
81. Kontoyiannis, D.P. et al., Zygomycosis in the 1990s in a tertiary-care cancer center, *Clin. Infect. Dis.*, 30, 851, 2000.
82. de Hoog, G.S., Guarro, J., Gené, J., and Figueras, M.J., *Atlas of Clinical Fungi*, 2nd edn., Centraalbureau voor Schimmelcultures, Utrecht, the Netherlands, 2000.
83. Baker, J.R., The structure and chemical composition of the Golgi element, *Quart. J. Micr. Sci.*, 85, 1, 1944.
84. Beckstead, J.H., A simple technique for preservation of fixation-sensitive antigens in paraffin-embedded tissues, *J. Histochem. Cytochem.*, 42, 1127, 1994.
85. Luna, L.G., *Manual of Histological Staining Methods of the Armed Forces Institute of Pathology*, 3rd edn., McGraw Hill Book Company, New York, 1968.
86. Chandler, F.W., Histopathologic and immunohistologic diagnosis of mycotic diseases, in *Advances in Medical Mycology*, p. 201, Mukherji, K.G., Srivastava, A.K., Singh, K.P., Garg, K.L. (Eds.), Aditya Books, New Delhi, India, 1992.
87. Sehgal, S., Immunohistochemical studies and detection of immune complexes, in *A Handbook of Practical and Clinical Immunology*, p. 346, Talwar, G.P., Gupta, S.K. (Eds.), Vol. I, 2nd edn., CBS Publishers, New Delhi, India, 1992.

88. Roberts, G.D., Detection of fungi in clinical specimens by phase contrast microscopy, *J. Clin. Microbiol.*, 2, 261, 1975.
89. Santos, M.J.S., Trufem, S.F.B., and de Oliveira, P.C., Viability and morphological stability of *Absidia* strains during long-term maintenance under mineral oil, *J. Basic Microbiol.*, 40, 133, 2000.
90. Santos, M.J.S., de Oliveira, P.C., and Trufem, S.F.B., Morphological observations on *Absidia corymbifera* and *Absidia blakesleenana* strains preserved under mineral oil, *Mycoses*, 46, 402, 2002.
91. Cenis, J.L., Rapid DNA extraction of fungal DNA for PCR amplification, *Nucleic Acids Res.*, 20, 2380, 1993.
92. Raeder, U. and Broda, P., Rapid preparation of DNA from filamentous fungi, *Lett. Appl. Microbiol.*, 1, 17, 1985.
93. Sambrook, J. and Russell, D.W., *Molecular Cloning: A Laboratory Manual*. Cold Spring Harbor Laboratory Press, Cold Spring Harbor, New York, 2001.
94. White, T.J. et al., Amplification and direct sequencing of fungal ribosomal RNA genes for phylogenetics, in *PCR Protocols: A Guide to Methods and Applications*, p. 315, Innis, M.A., et al. (Eds.), Academic Press, New York, 1990.
95. Gerrits van den Ende, A.H.G. and de Hoog, G.S., Variability and molecular diagnostics of the neurotropic species *Cladophialophora bantiana*, *Stud. Mycol.*, 43, 151, 1999.
96. Vogelstein, B. and Gillespie, D., Preparative and analytical purification of DNA from agarose, *Proc. Nat. Acad. Sci., USA*, 76, 615, 1979.
97. Balajee, S.A. et al., Sequence-based identification of *Aspergillus*, *Fusarium* and Mucorales species in the clinical mycology laboratory: Where are we and where should we go from here? *J. Clin. Microbiol.*, 47, 877, 2009.
98. Almaslamani, M. et al., An increasing trend of zygomycosis caused by *Mycocladus corymbifer* (formerly *Absidia corymbifera*): Report of two cases and review of primary cutaneous *Mycocladus* infections, *Med. Mycol.*, 31, 1, 2009.
99. Antoniadou, A., Outbreaks of zygomycosis in hospitals, *Clin. Microbiol. Infect.*, 15, 55, 2009.
100. Landlinger, C. et al., Species-specific identification of a wide range of clinically relevant fungal pathogens by use of Luminex xMAP technology, *J. Clin. Microbiol.*, 47, 1063, 2009.
101. Lau, A. et al., Development and clinical application of a pan-fungal PCR assay to detect and identify fungal DNA in tissue specimens, *J. Clin. Microbiol.*, 45, 380, 2007.
102. Bainier, G., Sur l'*Absidia caerulea*, *Bull. Soc. Bot. France*, 36, 184, 1889.
103. Lendner, A., Sur quelques Mucorinées, *Bull. Her. Boiss., Sér. II*, 7, 249, 1907.
104. Hagem, O., Untersuchungen über norwegische Mucorineen, *Math.-Nat. Kl.*, 7, 1, 1908.
105. Ling-Young, M., Étude biologique des phénomènes de la sexualité chez les Mucorinées, *Rev. Gén. Bot.*, 42, 730, 1930.
106. Linnemann, G., Beitrag zu einer Flora der Mucorineae Marburgs, *Flora (Regensburg)*, 130, 176, 1936.
107. Boedijn, K.B., Notes of the Mucorales of Indonesia, *Sydowia*, 12, 321, 1958.
108. Orr, G.F. and Plunkett, O.A., A new species of *Absidia* from Californica, *Mycologia*, 51, 203, 1959.
109. Rall, G. and Solheim, W.G., A variety of *Absidia* isolated from *Comandra pallida*, *Mycologia*, 56, 99, 1964.
110. Vánová, M., Contribution to the taxonomy of the genus *Absidia* (Mucorales) I. *Absidia macrospora* sp. nov., *Ceská Mykol.*, 22, 296, 1968.

85 *Mortierella*

Tamás Papp, Kerstin Hoffmann, Ildikó Nyilasi, Tamás Petkovits, Lysett Wagner, Csaba Vágvölgyi, and Kerstin Voigt

CONTENTS

85.1 INTRODUCTION

Members of the genus *Mortierella* are saprobic zygomycetous fungi belonging to the order Mortierellales. Among them, the thermophilic *Mortierella wolfii* is recognized as a pathogen, being an important casual agent of bovine mycotic abortion, pneumonia, and systemic mycosis (mortierellosis or mortierellomycosis).

85.1.1 Classification and Morphology

The genus *Mortierella* (Mortierellaceae, Mortierellales) is the largest genus of zygomycetous fungi with about 100 species recognized. It was placed under the family Mortierellaceae and originally treated as a family of the traditionally circumscribed Mucorales[1,2] but was later excluded and placed in the newly proposed Mortierellales.[3] The status of this classification was supported by results from several molecular phylogenetic studies based on ribosomal RNA gene (rDNA) sequences[4–7] and sequences of the protein-coding genes for actin (*act*)[8] and the translation elongation factor 1α (*tef*).[5,8] These studies detected a deep dichotomy between the two orders and thus confirmed the new classification.

At present Mortierellales contains one family, Mortierellaceae, comprising six genera—*Aquamortierella*, *Dissophora*, *Gamsiella*, *Lobosporangium*, *Modicella*, and *Mortierella*.[9] Within the genus *Mortierella*, Gams[10] distinguished two subgenera, *Mortierella* subgen. *Mortierella* and *Mortierella* subgen. *Micromucor* (syn. *Umbelopsis*), based on morphological characters. The latter subgenus also contained the *Mortierella isabellina* group described by Zycha et al.[11] Antigenic properties of their extracellular polysaccharides[12] and their sterol and fatty acid composition[13,14] supported this separation. Phylogenetic analysis of 18S and 28S rDNA, *act*, and *tef* sequences demonstrated that *Umbelopsis–Micromucor* represents a basal divergent group of Mucorales.[4,5,8] Later, based on internal transcribed spacer (ITS) rDNA sequences, a new monophyletic family, Umbelopsidaceae, belonging to the Mucorales, was proposed for this group.[15]

The large genus *Mortierella* has been divided into "sections" based on the grouping of presumed related species.[10,11,16] However, sequence comparison of the ITS regions of more than 80 independently obtained isolates suggested a variety of natural groups with monophyletic origin and revealed that the traditional morphology-based classification of Mortierellaceae is highly artificial and further molecular studies are needed to resolve the phylogeny of this taxon.[17]

The genus contains the animal pathogen *M. wolfii*, which differs from the other species of *Mortierella* in its thermophilic nature. This species was placed in the section Mutabilis[11] together with *M. mutabilis* and *M. sterilis*. In a recent phylogeny inferred from sequences of the ITS region, the small and the large subunit rDNA genes and a fragment of the *tef* gene of several strains of *Mortierella*, *Gamsiella*, and *Dissophora* and strains of *M. wolfii* formed a distinct, well-supported clade, which could be clearly distinguished from the other species of *Mortierella* using the determined molecular markers.[18]

Species of *Mortierella* generally produce unusually fine or cottony, coenocytic or irregularly septate mycelium in which anastomosing hyphae may occur. They produce small sporangiospores in one- to many-spored, globose sporangia. Columellae are often absent or rudimentary and never protrude into the sporangium. The lack of pronounced columellae together with non-apophysate sporangia are diagnostic characters of members of the Mortierellales and distinguish them from Mucorales.[19] They may form smooth- or rough-walled chlamydospores and zygospores, if known, are smooth or dimpled with generally apposed and unequal suspensors.[9,20]

M. wolfii form white to grayish white colonies, which, depending on the culturing conditions, often have a broadly zonate (rosette-like) surface appearance (Figure 85.1). The colonies may have a garlic-like odor, characteristic of other species of *Mortierella*. The fungus is thermophilic; its maximum temperature of growth is 48°C. It has 90–250 µm long sporangiophores, which can be simple or branched immediately below the apex. Sporangia are small (15–50 µm in diameter), multispored, and without a columella, and after deliquescence, a conspicuous collar can frequently be observed at the apex of the sporangiophore (Figure 85.1). Sporangiospores are short, cylindrical to reniform, and 6–10 × 3–5 µm in size. Chlamydospores of varying shape but often carrying short appendages (often referred to as stylospores) may be present; zygospores have never been observed.[20,21]

85.1.2 Clinical Features, Pathogenesis, and Epidemiology

Although some early publications reported sporadic cases of primarily cutaneous infection of humans caused by *M. wolfii*, these reports are suspected to be due to possible misidentifications and this species of *Mortierella* is actually not considered as a human pathogen.[22] However, *M. wolfii* is regarded as a true animal pathogen[22] and has been reported as an agent of bovine mycotic abortion, pneumonia, and systemic mycosis in New Zealand,[23,24] Australia,[25,26] Japan,[27] Great Britain,[24] and the United States[28,29] (Table 85.1).

Unlike other types of mycotic abortions, 20% of cows that abort as a consequence of *M. wolfii* infection will subsequently develop fatal mycotic pneumonia and occasionally other systemic mycoses.[29] In addition to the lungs, systemic mycosis may involve damage to the uterus, the liver, and the brain; histology often reveals lesions, necrosis, destruction of the blood vessels, and hemorrhage in the affected organs.[27,29,30] Concerning the pathogenesis of *M. wolfii*, causing bovine abortion and pneumonia, it has been suggested that the fungus first infects the lungs and subsequently reaches the uterus via the blood stream, where it causes endometritis and abortion. After abortion, depending on the degree of the damage to the uterine blood vessels, the fungus may reenter

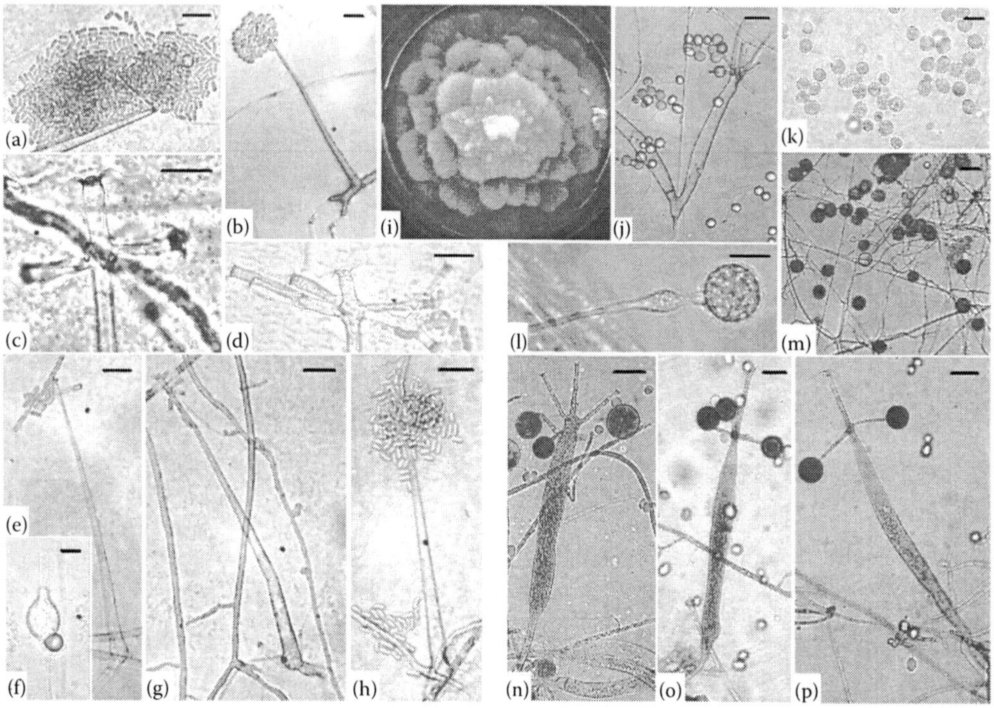

FIGURE 85.1 Lobed growth of *Mortierella* sp. (i) Micromorphology of *Mortierella wolfii* (a–h). Sporangiophores can be simple or branched subterminally, tapering from the base to the tip (b–e, g–h). Sporangia are without a columellae, leaving a conspicuous collar after deliquescence at the apex of the sporangiophore (c–d). Sporangiospores are short and cylindrical to reniform (a). Chlamydospore with short appendages (f). Micromorphology of *Mortierella polycephala* (j–p). The sporangiophores possess a broadened base and taper to the apex; branches occur mostly in the upper part of the sporangiophores; the sporangia are spherical, leaving a collar at the apex of the sporangiophore after dehiscence (j, n–p). Sporangiospores are oval to irregular in shape (k). Stylospores are produced on stalks, verrucose to echinulate (l–m). The lengths of scale bars indicate 10 µm for the panels c, d, f, k, and l and 25 µm for the panels a, b, e, g, h, j, and m–p.

TABLE 85.1
Strains of *M. polycephala* and *M. wolfii* Available from Public Culture Collections Including Information about MycoBank Number, Taxonomically Important Synonyms, and Substrate of Isolation

Strain, Current Name, MycoBank Numbers	Taxonomic Synonyms
Culture collection numbers	Substrate of isolation
M. wolfii Mehrotra & Baijal 1963, [MB334519]	*Actinomortierella wolfii*
CBS209.69	Coal spoil tip soil
CBS611.70, IMI149020, IP1706.87	Cow, lung, dying from mycotic pneumonia
CBS612.70	Decayed hay
CBS614.70	Decayed hay
CBS552.93	Unknown
CBS651.93	Compost for mushrooms
ATCC24296	Unknown
ATCC36820	Stomach contents of an aborted bovine fetus
ATCC64116	Cattle, Australia
ATCC64117	Cattle, Australia
ATCC64118	Cattle, Australia
M. polycephala Coem. 1863, [MB145769]	*M. raphani*, *M. vantieghemi* var. *raphani*, *M. polycephala* var. *raphani*, *M. vantieghemi*, *M. polycephala* var. *canina*, *M. canina*, *M. crystallina*, *M. lemonnieri*
CBS293.34	*Lupinus*
CBS227.35, FSU696	Unknown
CBS456.66, DSMZ1212, FSU759	Dung of wood mouse
CBS649.68, ATHUM2571, IHEM3807, NRRL A-12030	Unknown
CBS327.72, FSU866	Salt-marsh soil under *Spartina townsendii*
CBS328.72, FSU867	Soil

the circulation and cause embolic pneumonia and infections in other organs.[29]

Strains of *M. wolfii* may produce a heat labile (at 60°C) and trypsin-sensitive nephrotoxin, which is active over a wide pH range[31–33] and is able to cause death in experimental models. It has been suggested that this toxin is responsible for most of the symptoms caused by the fungus.[22] However, when only the toxin was administered to rabbits and mice, it damaged primarily the kidneys and its effects proved to be distinct from those associated with *M. wolfii* infections. Thus, the possible role of the toxin in natural infections remained unclear.[33]

Species of *Mortierella* are ubiquitous and are among the most common fungi in the soil.[20] *M. wolfii* has a worldwide distribution and is common in the soil in tropical and subtropical regions of the world. In the temperate zone, it may be found in thermophilic environments, such as overheated, spoiled silage and hay. Infections caused by *M. wolfii* have been reported from almost all regions of the world but with very different frequencies. Previous reports recorded that more than 60% of cases of bovine mycotic abortion in the United States were caused by *Aspergillus fumigatus*, but Zygomycetes accounted for about 20% of the total cases followed by a wide range of opportunistic fungi and yeasts involved in the remaining 20% cases.[28] Concurrently, *M. wolfii* has been identified in 0.5% of all bovine mycotic abortions in whole North America[29] and only occasionally in Japan.[27] Apart from *Aspergillus* spp., *M. wolfii* was found to be the second major fungal pathogen responsible for 15%–20% of this condition in Australia.[25,34] In New Zealand, *M. wolfii* is considered as the major and most common cause of bovine mycotic abortion.[29,35] In Europe, *Mortierella* was frequently reported in Great Britain since 1967[36] and even more recently.[24,37,38]

A mild/warm and wet climate and frequent use of grass silage for feeding are thought to be the main factors resulting in the high incidences in some regions.[22,25,29] It is generally suggested that transmission of the fungus occurs primarily by inhalation of spores originating from moldy silage, but transmission via infected semen or ulcerations in the alimentary tract were also suggested by certain authors.[22,39]

In earlier studies, single cases of animal mycoses were reported as caused by some other species of *Mortierella*, e.g., *M. polycephala*, *M. alpina*, *M. hyalina*, and *M. zychae*. However, the maximum growth temperature of these species is below 36°C, and, thus, it is likely that the identifications of the causal agent associated with these cases were incorrect and were most probably misidentifications of *M. wolfii*.[20,40]

85.1.3 Diagnosis

85.1.3.1 Conventional Techniques

A serological assay was described for detection of *M. wolfii* in cows[41] but no other Zygomycetes were included in this study although similar antigens in other species and genera were reported earlier.[42] Nevertheless, *M. wolfii* appears to possess fewer antigenic similarities when compared to Mucorales,[43] but no serological tests are applied in routine labs.

85.1.3.2 Molecular Techniques

In addition to conventional techniques, molecular-based methods present very promising approaches to identify the fungi involved in human and animal diseases. Although molecular approaches are often regarded as sole identification methods, relying on them in isolation should be avoided. Accepting the advantage of their easy performance because they can be carried out following standardized protocols, these methods are not able to identify all species with unquestionable certainty. Nevertheless, reference databases of molecular markers could provide a resource through which a very close approximation to the identity of the investigated organism could be reached. However, such databases would be required to fulfill several prerequisites:

1. The reference specimens should encompass a variety of isolates of the same species from different locations so as to include possible population and geographical variations.
2. The molecular marker of choice should have a high-level of interspecies variation to differentiate between species and a low-level of intraspecific variation.
3. The amplification of the marker must be easy by the application of universal primers to avoid debarment of key player specimens.
4. The database must encompass more than the target species to ensure the correct species is identified.

Nevertheless, use of such databases could not identify new, undescribed species, enforcing further methods to be applied in collaboration work with scientific research/reference laboratories.

The advantages, disadvantages, difficulties, and requirements of molecular applications are discussed in Chapter 34 and by Balajee et al.[44]

The molecular method of choice is currently the comparative sequence-based identification of fungi.[45] Based on amplification and sequencing of short DNA regions followed by comparison with available reference sequences. Although the first polymerase chain reaction (PCR) application on fungal infections in animals was to identify *M. wolfii*,[29] no standardized recommended method and no reference database exists so far for *Mortierella*. In addition, GenBank currently has only a few relevant sequences available (Table 85.2).

85.2 METHODS

85.2.1 Sample Preparation

Since infections with *M. wolfii* are only confirmed for animals, the methods described here were not commonly applied in diagnosing animal infections and could therefore only be assumed to be applicable in detection of human infections with this fungus. Infections of animals are mainly discovered after death, an option not acceptable in human infections. However, as there is no prior experience of detection of *M. wolfii* in human, the following methods are only an adaptation from commonly known and applied methods for detection of zygomycosis.

In combination with any method applied for detection of fungal infection, a test for infections caused by bacteria and/

TABLE 85.2
Nucleotide Sequences of *M. polycephala* and *M. wolfii* Available from GenBank at http://www.ncbi.nlm.nih.gov

Species	GenBank Acc. Numbers	Gene Region	Strain Collection Numbers	Reference
M. polycephala	AB476414	ITS	NBRC 6335	H. Sekiguchi, A. Masunaka, Y. Hashimoto, and S. Takenaka (unpublished)
	AJ287169	Actin, partial	NRRL22890	Voigt and Wöstemeyer[8]
	X89436	18S rRNA, partial	Unknown	K. O'Donnell (unpublished)
	AF157261	Tef, partial	NRRL22890	Voigt and Wöstemeyer[8]
	AF113464	28S rRNA, partial	NRRL22890	Voigt et al.[4]
M. wolfii	AJ287171	Actin, partial	CBS611.70, NRRL28640	Voigt and Wöstemeyer[8]
	AB154774 // AB154775 // AB154776	28S rRNA, partial	IFM47051, CBS651.93 // IFM52979, CBS611.70, IMI149020 // IFM52980, CBS612.70	H. Miyano, S. Endo, N. Obara, M. Haritani, N. Tanimura, K. Kimura, A. Sano, K. Yokoyama, K. Kamei, M. Miyaji, and K. Nishimura (unpublished)
	AF113465	28S rRNA, partial	NRRL28640	Voigt et al.[4]
	AF113425	18S rRNA, partial	NRRL28640	Voigt et al.[4]

Note: Accession numbers of the sequenced region, strain numbers of the strain collection, and references are also indicated.

Abbreviations: NBRC, NITE Biological Resource Center (NITE, National Institute of Technology and Evaluation), Japan; NRRL, Northern Regional Research Laboratories, Strain Collection of the National Center of Agricultural Utilization Research Peoria, IL; CBS, Centraalbureau voor Schimmelcultures, Utrecht, the Netherlands; IFM, Research Center for Pathogenic Fungi and Microbial Toxicoses, Chiba University, Japan.

or viruses is also recommended so as to assist in determining if the fungal infection is the primary or secondary causative agent of the observed disease.

85.2.1.1 Samples for Direct Microscopic Investigation

If a tissue displays mycotic lesions, samples should be taken under sterile conditions, avoiding secondary inoculation with environmental microorganisms. Most infections with *Mortierella* were described in animals rather than humans and were observed after fetal abortion. Therefore, biopsy samples from the fetus are the main material for examination. The biopsy material should be subjected to direct microscopy or observed after appropriate staining. Using simple phase-contrast microscopy on smears from the tissue or direct tissue samples ground in a homogenizer should reveal the presence of fungal hyphae.[22,46,47] Careful preparation of small tissue samples on a microscopic slide and staining with Parker ink in potassium hydroxide (KOH) can also be used to display fungal structures.

Samples should also be embedded in paraffin after fixation in 10% or neutral buffered formalin saline.[48] The use of zinc as primary fixative improves the survival of antigens and therefore allows for immunohistochemistry.[49] Sections prepared with a microtome are then stained with hematoxylin and eosin (H&E) for general histopathological examination and with periodic acid Schiff reaction (PAS) and Grocott's methenamine silver stain (GMS) for the investigation of fungal hyphae. Other fluorescent stains are also possible like the optical brighteners Calcofluor and Blankophor.[47,50,51] Although these staining techniques are not specific for the genus *Mortierella* or even for other species of Zygomycetes, specific characteristics of this group should be observable and provide a means of excluding, e.g., ascomycetous fungi like *Aspergillus*. Sufficiently infected tissue will show the characteristically nonseptate hyphae of Zygomycetes. Although some septa may be observed typically associated with spore-bearing structures or in old vacuolated hyphae, they are not as regularly distributed as in the hyphae of Ascomycetes. Distinguishing features like asexual or sexual reproductive structures are not very likely to be produced in infected tissue. Therefore, in addition to the direct, first investigation of suspected tissue, the underlying species should be identified by further methods based on morphological and/or molecular data.

85.2.1.2 Samples for Morphological Investigation

Samples for axenic cultures should not be treated with formaldehyde or other harsh fixatives as this action will reduce viability. Like any other terrestrial fungi, *Mortierella* can also be stored via cryopreservation using spores and hyphae in 10%–20% glycerol or in active cultures overlaid with mineral oil like it is done by strain collections for short- and long-term conservation of viability as well as preservation of morphological features.[52,53] Axenic cultures should be grown from any infected material containing spores or hyphae but the tissue samples should not be forcefully crushed because the hyphae of Zygomycetes posses less septa, which are required to maintain the viability of hyphal fragments.[54,55] Species of *Mortierella* grow well on a wide variety of artificial media, e.g., MEA (30 g/L malt extract agar), PDA (potato dextrose agar; DSMZ media no. 129; http://www.dsmz.de), OA (oatmeal agar; DSMZ media no. 425), Sabouraud glucose agar (20 g/L glucose; e.g., Carl Roth, Karlsruhe, Germany), or BHIA (brain heart infusion agar; DSMZ media no. 215, supplemented with 5% sheep blood). Each medium, however, will have an influence on the overall appearance of the fungal colony, and, thus, for confident identification or during examination by inexperienced investigators, media used in the published descriptions and keys are recommended for culturing.[11–56] In addition these media can be supplemented with antibiotics to suppress bacterial growth, applied single or in combinations (e.g., ampicillin, chloramphenicol, neomycin, tetracycline, or streptomycin in final concentrations of 1.9, 1.4, 1.4, 0.6, and 1.2 mg/mL, respectively). Furthermore, incubation should be at various temperatures to evaluate the cultures under optimal and maximal growth conditions. The cultures should also be observed and maintained for 4 weeks to account for slow growing fungi. Where possible, the standard procedure for phenotypic observations of fungal cultures should be applied.[21]

The macromorphological appearance of *M. wolfii* and *M. polycephala* are illustrated in Figure 85.1, where the typical undulate or wavy growth as found in various species of *Mortierella* can be seen. This typical lobed growth is not present in every isolate of a species and may not form on all types of media. Colonies often produce a garlic-like odor.[11]

The members of the order Mortierellales are described as forming coenocytic, anastomosing, and often arachnoid vegetative hyphae and sporangiophores, which are commonly awl-shaped and characterized by a reduced columella, which is either septum-like or slightly convex. The sporangial wall is fugacious and the zygosporangia are hyaline with apposed suspensors.[57]

The sporangiophores of the type species of *Mortierella*, *M. polycephala*, arise single or in tufts. They possess a broadened base and taper to the apex; rhizoid-like structures are sometimes formed at the base. Branches are arranged racemosely and occur mostly in the upper part of the sporangiophores; secondary or tertiary branching is possible. The sporangia are spherical, 37–75 μm, leaving a collar at the apex of the sporangiophore after dehiscence. The sporangiospores are smooth, oval to irregular in shape, and 5.5–13 × 5.5–11 μm in size. Stylospores (sporangiola) are produced on stalks and are single spored and verrucose to echinulate. Chlamydospores occur intercalary or terminally and are typically smaller than sporangiospores. Zygospores are surrounded by a dense hyphal network and the suspensors are slightly unequal. At temperatures above 13°C only stylospores may be encountered, whereas sporangia are produced abundantly at low temperatures[11,21,56] (Figure 85.1).

The sporangiophores of *M. wolfii* often arise from rhizoids formed in the aerial mycelium. The sporangiophores are broad at the base and taper to the apex, mostly branched subterminally. The branches arise cymosely, mostly in whorls, sometimes displaying a helicoid appearance just below the tip. Secondary branches are possible. The sporangia are spherical, 11–68 μm, leaving a prominent reflexed collar at the apex of the sporangiophore after dehiscence; secondary sporangia are smaller. The sporangiospores are smooth, ellipsoidal to reniform, and 2–5.5 × 3.5–13 μm in size. Chlamydospores are infrequently described and are of various shapes with lobed appendages. No stylospores are produced. Zygospores are unknown. Colonies of *M. wolfii* are cottony, white, fine, and often with a garlic-like odor on grass compost media. Good growth at 37°C, maximum at 48°C. Isolates originate from materials characterized by elevated temperatures, e.g., rotten silage, hay, silage casing soil, and coal spoil tips[11,21,56] (Figure 85.1).

Micromorphological observations should be made on fungal colonies bearing slightly immature, spore-containing sporangia. Older sporangia are very likely to loose their spores during slide preparation. With an adhesive tape, fungal structures can be directly fixed from the colony. The tape is then placed sticky side down on a microscopic slide ("tape prep"). For observation of fungal structures at higher resolution, a few parts of the colony are carefully removed and placed in a drop of lactophenol cotton blue and covered with a coverslip ("tease prep"). The high-contrast staining with cotton blue makes observations somewhat easier. Using the traditional slide culture is another option. Here, a sterile coverslip or sterile cellophane is placed on top of the medium prior to inoculation. The fungus eventually grows on the coverslip or cellophane, which can then be removed and subjected to direct microscopy.[22,58] The observed morphological characteristics have to be compared with the original description of the presumed fungus.[11,56,59]

85.2.2 Detection Procedures

The preparation of genomic DNA for any molecular analysis should be performed directly from infected tissue or from pure cultures. The methods currently available for DNA extraction are extensively described in Chapters 23 and 34. Purifying DNA from tissue samples may not be as effective and reliable in terms of diagnostic signal as from pure fungal cultures since host DNA will also be retrieved. Furthermore, the DNA of the host will be in much higher concentration compared to that of the fungal DNA, thus requiring a method with high fungal sensitivity for DNA extraction and subsequent identification of the pathogen. This can be achieved by fungal-specific oligonucleotide primers targeting genomic regions that are unique for fungi. The primers of choice for identification of *Mortierella* are not easy to define. First of all a reliable reference database is needed, which is accessible by the investigator and shared with the scientific community as a whole in order to make the results transparent. Such a database should contain information about as many specimens as possible regarding molecular data and information related to, e.g., isolation, distribution, morphology, physiology, and ecology. Since there are only few sequences available in GenBank (Table 85.2; http://www.ncbi.nlm.nih.gov), such a reference database is far from being available and molecular identification is thus only possible in a rudimentary way. Based on the available sequences of the small ribosomal subunit (SSU), a restriction map for molecular diagnosis is proposed in the Atlas of Clinical Fungi[21] and augmented here (Figure 85.2). Generating new sequences including amplification, sequencing, and sequence analysis from isolated specimens is also described in Chapter 34.

85.3 CONCLUSION AND FUTURE PERSPECTIVES

With a growing awareness to life-threatening zygomycoses, past and future case reports will gain importance. The infectious agents belong to the Mucorales, Entomophthorales, and Mortierellales with most agents in the first order, leading to the term "mucormycosis." This was further supported because in early years most infections were originally identified as species of the genus *Mucor*. However, later research revealed that the most important causal agents belong to the genus *Rhizopus* (reviewed by Ribes et al.[22] and Roden et al.[60]). Therefore, zygomycosis is currently applied with a somewhat overlapping synonymous relationship to mucormycosis. But referring to zygomycosis caused by a variety of different genera in one sentence with mycosis caused by single genera from the Ascomycetes termed, e.g., candidosis or aspergillosis will not help solving the problem of correctly referring to the diversity of disease-causing species. Naming the mycosis after the causal genus will probably help to broaden the insight into their epidemiology. This problem is already addressed like rhizomucormycosis due to infections with *Rhizomucor*[61] or lichtheimiamycosis caused by *Lichtheimia corymbifera* (former *Mycocladus* [ex *Absidia*] *corymbifer*[*a*], Chapter 34).

Determination of the correct species causing the mycosis will also reveal specimens with pathogenicity to humans or animals only.

Although *M. wolfii* and *M. polycephala* are reported as pathogenic to animals, some reasonable doubts may be entertained about the clinical relevance of *M. polycephala* as causative agent of mortierellomycoses. Exclusively, a single case was reported due to the infection with *M. polycephala*,[62] which appears to be undoubtedly a misidentification. In that report, few morphological traits suggest *M. wolfii*, e.g., the production of chlamydospores and the double-walled sporangiospores. However, the fungus was reported to grow well at 37°C, clearly above the accepted maximal growth temperature for *M. polycephala*, which is below 30°C, and thus below body temperature.[20] On the contrary, *M. wolfii* is able to grow well at body temperature, although confirmed infections in humans have not been documented.[63] Currently, *M. wolfii* is thought of as a sole pathogen to animals, but the awareness of its potential to infect humans also should be continued.

Although molecular methods for identification of fungi gain importance, traditional methods should not be neglected. Although there are still fungal infections reported that are

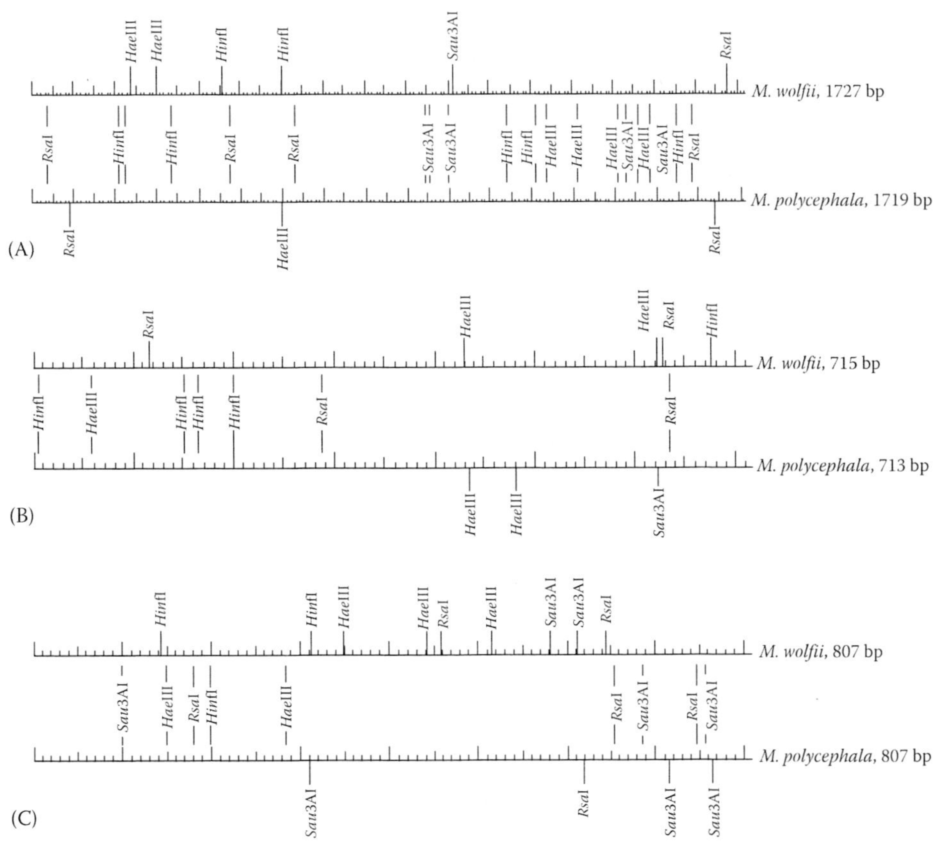

FIGURE 85.2 Restriction maps for molecular diagnosis of *M. wolfii* and *M. polycephala*. Restriction enzymes used for the endonucleolytic digestions are: *Hae*III, *Hin*fI, *Rsa*I, and *Sau*3AI. Restriction sites present in both strains are indicated between both schematic sequences, whereas unique restriction sites are displayed aside the schematic nucleotide sequences. (A) Restriction map based on the sequences NCBI AF113425 (*M. wolfii*) and X89436 (*M. polycephala*) coding for 18S rRNA. (B) Restriction map based on the sequences NCBI AF113465, AB154774, AB154775, AB154776 (*M. wolfii*), and AF113464 (*M. polycephala*) coding for 28S rRNA. (C) Restriction map based on the sequences NCBI AJ287171 (*M. wolfii*) and AJ287169 (*M. polycephala*) coding for actin (cDNA).

identified solely through histopathology, making the same mistake using solely molecular identification methods should be avoided. The diagnosed specimen should always be supported by morphological identification since histopathology or molecular markers are not able to determine the cause of an infection to species level. Molecular-based identification will only be an approximation as long as whole-genome comparisons are not possible. Nevertheless, morphology alone is also not error free, and thus in combination with all other available methods previously unrecognized species could be revealed. Until now the molecular detection of *M. wolfii* and its related species is accomplished in few designated laboratories worldwide. A publicly available database may serve as a platform, which unifies different methods, strategies, and techniques of molecular detection in combination with traditional identification in order to make the expertise more transparent and shape the reliability of the diagnostics toward an increased screening for mortierellomycoses in clinical environments.

ACKNOWLEDGMENTS

We wish to express their gratitude to Paul M. Kirk (CAB International Egham, Surrey, U.K.) for his expert advice and critical revision of the manuscript. We would like to thank our colleagues at the Centraalbureau voor Schimmelcultures in Utrecht (the Netherlands) for sharing strains of *Mortierella*. This project was supported by a bilateral grant between Hungarian Scientific Research Fund (OTKA-NN75255) and German Research Foundation (DFG; project no. Vo 772/9-1). TP and KH contributed equally to this review.

REFERENCES

1. Hesseltine, C.W. and Ellis, J.J., Mucorales, in: *The Fungi, an Advanced Treatise*, vol. IVb. *A Taxonomic Review with Keys: Basidiomycetes and Lower Fungi*, Ainsworth, G.C., Sparrow, F.K., and Sussman, A.S., Eds., Academic Press, New York, pp. 187–217, 1973.
2. Benjamin, R.K., Zygomycetes and their spores, in: *The Whole Fungus, The Sexual-Asexual Synthesis*, vol. 2, Kendrick, B., Ed., National Museums of Canada, Ottawa, Canada, pp. 573–616, 1979.
3. Cavalier-Smith, T., A revised six-kingdom system of life, *Biol. Rev.*, 73, 203, 1998.
4. Voigt, K., Cigelnik, E., and O'Donnell, K., Phylogeny and PCR identification of clinically important Zygomycetes based on nuclear ribosomal-DNA sequence data, *J. Clin. Microbiol.*, 37, 3957, 1999.

5. O'Donnell, K. et al., Evolutionary relationships among mucoralean fungi (Zygomycota): Evidence for family polyphyly on a large scale, *Mycologia*, 93, 286, 2001.
6. Kwaśna, H., Ward, E., and Bateman, G.L., Phylogenetic relationships among Zygomycetes from soil based on ITS1/2 rDNA sequences, *Mycol. Res.*, 110, 501, 2006.
7. White, M.M. et al., Phylogeny of the Zygomycota based on nuclear ribosomal sequence data, *Mycologia*, 98, 872, 2006.
8. Voigt, K. and Wöstemeyer, J., Phylogeny and origin of 82 Zygomycetes from all 54 genera of the Mucorales and Mortierellales based on combined analysis of actin and translation elongation factor EF-1a genes, *Gene*, 270, 113, 2001.
9. Benny, G.L., Zygomycetes, published on the Internet at http://www.zygomycetes.org, 2005, accessed on November 11, 2009.
10. Gams, W., A key to the species of *Mortierella*, *Persoonia*, 9, 381, 1977.
11. Zycha, H., Siepmann, R., and Linnemann, G., *Mucorales. Eine Beschreibung aller Gattungen und Arten dieser Pilzgruppe*, J Cramer, Lehre, 1969.
12. de Ruiter, G.A. et al., Approaches to the classification of the *Mortierella isabellina* group: Antigenic extracellular polysaccharides, *Mycol. Res.*, 97, 690, 1993.
13. Amano, N. et al., Chemotaxonomic significance of fatty acid composition in the genus *Mortierella* (Zygomycetes, Mortierellaceae), *Mycotaxon*, 44, 257, 1992.
14. Weete, J.D. and Gandhi, S.R., Sterols and fatty acids of the Mortierellaceae: Taxonomic implications, *Mycologia*, 91, 642, 1999.
15. Meyer, W. and Gams, W., Delimitation of *Umbelopsis* (Mucorales, Umbelopsidaceae fam. nov.) based on ITS sequence and RFLP data, *Mycol. Res.*, 107, 339, 2003.
16. Domsch, K.H., Gams, W., and Anderson, T.H., *Compendium of Soil Fungi*, vols. 1, 2, Academic Press, New York, 1980.
17. Nyilasi, I. et al., Phylogenetic and physiological characterization of filamentous fungi belonging to the order Mortierellales, *Acta Microbiol. Immunol. Hung.*, 56, 219, 2009.
18. Papp, T. et al., Molecular identification and phylogeny of *Mortierella wolfii*, *Acta Microbiol. Immunol. Hung.*, 56, 224, 2009.
19. Hoffmann, K. et al., Diversity, genotypic identification, ultrastructural and phylogenetic characterization of zygomycetes from different ecological habitats and climatic regions: Limitations and utility of nuclear ribosomal DNA barcode markers, in: *Current Advances in Molecular Mycology*, Gherbawy, Y., Mach, R.L., and Rai, M., Eds., Nova Science Publishers, Inc., New York, pp. 263–312, 2009.
20. Schipper, M.A.A. and Stalpers, J.A., Zygomycetes: The order Mucorales. in: *Pathogenic Fungi in Humans and Animals*, 2nd edn., Mycology Series, vol. 16, Howard, D.H., Ed., Marcel Dekker, New York, 2002, pp. 67–125.
21. de Hoog, D.S. et al., *Atlas of Clinical Fungi*, 2nd edn., Centraalbureau voor Schimmelcultures, the Netherlands, 2000.
22. Ribes, J.A., Vanover-Sams, C.L., and Baker, D.J., Zygomycetes in human disease, *Clin. Microbiol. Rev.*, 13, 236, 2000.
23. Carter, M.E. et al., Fungi isolated from bovine mycotic abortion and pneumonia with special reference to *Mortierella wolfii*, *Res. Vet. Sci.*, 14, 201, 1973.
24. Seviour, R.J., Cooper, A.L., and Skilbeck, N.W., Identification of *Mortierella wolfii*, a causative agent of mycotic abortion in cattle, *Med. Mycol.*, 25, 115, 1987.
25. Connole, M.D., Review of animal mycoses in Australia, *Mycopathology*, 111, 133, 1990.
26. Gabor, L.J., Mycotic pneumonia in a dairy cow caused by *Mortierella wolfii*, *Aust. Vet. J.*, 81, 409, 2003.
27. Komoda, M. et al., An infection of cow with *Mortierella wolfii*, *Mycopathology*, 101, 89, 1988.
28. Knudtson, W.U. and Kirkbride, C.A., Fungi associated with bovine abortion in the northern plains states (USA), *J. Vet. Diagn. Invest.*, 4, 181, 1992.
29. Munday, J.S. et al., Meningoencephalitis in an adult cow due to *Mortierella wolfii*, *J. Vet. Diagn. Invest.*, 18, 619–622, 2006.
30. Uzal, F.A. et al., *Mortierella wolfii* isolated from the liver of a cow in Australia, *Vet. Rec.*, 145, 260, 1999.
31. Davey, G., Smith, J.M., and Kalmakoff, J., Purification and properties of a toxin isolated from *Mortierella wolfii*, *Infect. Immun.*, 8, 882, 1973.
32. Davey, G. and Kalmakoff, J., Evidence that the nephrotoxin from the fungus *Mortierella wolfii* is a protein, *Can. J. Microbiol.*, 20, 1513, 1974.
33. Corbel, M.J. and Edades, S.M., Observations on the experimental pathogenicity and toxigenicity of *Mortierella wolfii* strains of bovine origin, *Br. Vet. J.*, 147, 504, 1991.
34. McCausland, I.P., Slee, K.J., and Hirst, F.S., Mycotic abortion in cattle. *Aust. Vet. J.*, 64, 129, 1987.
35. di Menna, M.E., Carter, M.E., and Cordes, D.O., The identification of *Mortierella wolfii* isolated from cases of abortion and pneumonia in cattle and a search for its infection source, *Res. Vet. Sci.*, 13, 439, 1972.
36. Hugh-Jones, M.E. and Austwick, P.C.K., Epidemiological studies in bovine mycotic abortion. I. The effect of climate on incidence, *Vet. Rec.*, 85, 273, 1967.
37. MacDonald, S.M. and Corbel, M.J., *Mortierella wolfii* infections in cattle in Britain. *Vet. Rec.*, 109, 419, 1981.
38. Johnson, C.T., Lupson, G.R., and Lawrence, K.E., The bovine placentome in bacterial and mycotic abortions, *Vet. Rec.*, 12, 263, 1994.
39. Smith, J.M., An interesting bovine mycotic complex in New Zealand, *N. Z. Vet. J.*, 14, 226, 1966.
40. Streekstra, H., On the safety of *Mortierella alpina* for the production of food ingredients, such as arachidonic acid, *J. Biotechnol.*, 56, 153, 1997.
41. Skilbeck, N.W., Serological diagnosis of bovine abortion due to *Mortierella wolfii*. *Aust. Vet. J.*, 61, 238, 1984.
42. Kaufman, L., Turner, L.F., and McLaughlin, D.W., Indirect enzyme linked immunosorbent assay for zygomycosis, *J. Clin. Microbiol.*, 27, 1979, 1989.
43. Hessian, P. and Smith, J.M.B., Antigenic characterization of some potentially pathogenic mucoraceous fungi, *Sabouraudia*, 20, 209, 1982.
44. Balajee, S.A. et al., Sequence-based identification of *Aspergillus*, *Fusarium* and Mucorales species in the clinical mycology laboratory: Where are we and where should we go from here?, *J. Clin. Microbiol.*, 47, 877, 2009.
45. Summerbell, R.C. et al., Microcoding: The second step in DNA barcoding, *Philos. Trans. R. Soc. Lond. B. Biol. Sci.*, 360,1897, 2005.
46. Roberts, G.D., Detection of fungi in clinical specimens by phase contrast microscopy, *J. Clin. Microbiol.*, 2, 261, 1975.
47. Rüchel, R. and Schaffrinski, M., Versatile fluorescent staining of fungi in clinical specimens by using the optical brightener Blankophor, *J. Clin. Microbiol.*, 37, 2694, 1999.
48. Baker, J.R., The structure and chemical composition of the Golgi element, *Quart. J. Micr. Sci.*, 85, 1, 1944.

49. Beckstead, J.H., A simple technique for preservation of fixation-sensitive antigens in paraffin-embedded tissues, *J. Histochem. Cytochem.*, 42, 1127, 1994.
50. Luna, L.G., *Manual of Histological Staining Methods of the Armed Forces Institute of Pathology*, 3rd edn., McGraw Hill Book Company, New York, 1968.
51. Chandler, F.W., Histopathologic and immunohistologic diagnosis of mycotic diseases, in *Advances in Medical Mycology*, Mukherji, K.G. et al., Eds., Aditya Books, New Delhi, India, pp. 201–210, 1992.
52. Santos, M.J.S., Trufem, S.F.B., and de Oliveira, P.C., Viability and morphological stability of *Absidia* strains during long-term maintenance under mineral oil, *J. Basic Microbiol.*, 40, 133, 2000.
53. Santos, M.J.S., de Oliveira, P.C., and Trufem, S.F.B., Morphological observations on *Absidia corymbifera* and *Absidia blakesleenana* strains preserved under mineral oil, *Mycoses*, 46, 402, 2002.
54. Kontoyiannis, D.P. et al., Zygomycosis in the 1990s in a tertiary-care cancer center, *Clin. Infect. Dis.*, 30, 851, 2000.
55. Gonzalez, C.E., Rinaldi, M.G., and Sugar, A.M., Zygomycosis, *Infect. Dis. Clin. North. Am.*, 16, 895, 2002.
56. Mehrotra, B.S. and Baijal, U., Species of *Mortierella* from India—III, *Mycopath. Mycol. Appl.*, 20, 49, 1962.
57. O'Donnell, K., *Zygomycetes in Culture*, Palfrey Contributions in Botany No. 2, Department of Botany, University of Georgia, Athens, GA, 1979, p. 257.
58. Riddell, R.W., Permanent stained mycological preparations obtained by slide culture, *Mycologia*, 42, 265, 1950.
59. Coemans, E., Quelques hyphomycètes nouveaux.—Ire.—I. *Mortierella polycephala* et II. *Martensella pectinata*, *Bull. l'Acad. Royale Sci., Lett., Beaux-Arts Belgique Sér.* 2, 15, 536, 1863.
60. Roden, M.M. et al., Epidemiology and outcome of zygomycosis: A review of 929 reported cases, *Clin. Infect. Dis.*, 41, 634, 2005.
61. Pandian, J.D. et al., Rhizomycosis infection in the basal ganglia, *Arch. Neurol.*, 64, 134, 2007.
62. Scholz, H.D. and Meyer, L., *Mortierella polycephala* als Erreger einer Lungenmykose beim Rind, *Berl. Münch. Tierärtzl. Wschr.*, 78, 27, 1965.
63. Chayakulkeeree, M., Ghannoum, M.A., and Perfect, J.R., Zygomycosis: The re-emerging fungal infection, *Eur. J. Clin. Microbiol. Infect. Dis.*, 25, 215, 2006.

86 *Mucor*

Peter C. Iwen

CONTENTS

86.1 INTRODUCTION

Zygomycosis (also referred to as mucormycosis) traditionally refers to an infection caused by fungi in the class *Zygomycota* and the order *Mucorales*. Numerous fungi within this order (collectively referred to as the zygomycetes) have been associated with human disease—with *Rhizopus* species the most common known cause and *Mucor* species or *Absidia* species listed as a distant second, dependent on the patient group being affected.[1–5] Although zygomycosis has historically been considered a rare disease, an emergence of this infection in immunocompromised individuals has been noted, as technological advances occur in medicine.[6,7] The number of cases of disease is probably much higher than that reported due to limitations in diagnostic techniques for zygomycosis and the high mortality associated with this disease, where infection may go unrecognized. The problems in diagnosis makes separation of the species difficult, which is evident in the numerous articles available in the literature that review these cases.[1,5–12] This chapter will provide information on what is known about the zygomycetes classified within the genus *Mucor*, with an emphasis on the recent area of fungal molecular diagnostics.

86.1.1 Classification

The classification of species within the genus *Mucor* has remained unresolved due to questions surrounding the previously named phylum *Zygomycota*. Original taxonomic schemes had the zygomycetes placed into this distinctive phylum, which has subsequently been divided into two classes, the *Trichomycetes* and the *Zygomycetes*, with the latter class containing the human pathogens.[11] Recent reports now suggest that the polyphyletic nature of this group supports a new classification.[13–16] Hibbett et al. suggested that the traditional phylum *Zygomycota* consists of four subphyla described as *incertae sedis* to include the *Mucormycotina*, *Kickxellomycotina*, *Zoopagomycotina*, and *Entomophthoromycotina*.[17] Others have also shown that the classification of this group is highly artificial with more work needed to define the taxonomic details.[14,16] The most current taxonomy includes the genus *Mucor* in the subphylum *Mucormycotina* within the order *Mucorales* in the family *Mucoraceae*.[17] Currently, the GenBank database (National Center for Biotechnology Information, Washington, DC) lists the phylum for *Mucor* as "fungi *incertae sedis*" and gives a taxonomic description for this genus as "basal fungal lineages" within the subphylum *Mucormyocotina*.

Over 250 validly named species of *Mucor* with multiple varieties, forma, and subspecies are listed at the Index Fungorum website (http:// www.indexfungorum.org/Names/Names.asp). The number of species within this genus, however, continues to undergo changes as new taxon are added and previously named species are reclassified following molecular sequence studies. Currently, five *Mucor* species are reported to cause human disease.[6,11,18] These include the thermo tolerant species of *M. circinelloides*, *M. indicus*, and *M. ramosissimus*, as well as *M. hiemalis*, which only grows poorly at 37°C. *Mucor racemosus* has also been suggested as a cause of human disease, but this association has been questioned due to the organisms inability to grow at the elevated temperatures.[18]

Although sequencing using genomic targets within the rRNA complex has been shown to be reliable for the species identification of the zygomycetes, this process has also shown that a number of the currently named species within the genus *Mucor* and other closely related genera are highly similar.[2,19–22] This similarity suggests that synonymy among some of these species is present.[23] When the internal transcribed spacer (ITS)1-5.8S-ITS2 sequences of the type strains of the *Mucor* species associated with human infection were evaluated using a BLAST search of the GenBank database, the *M. circinelloides* sequence was shown to be highly similar to those of *Rhizomucor variabilis*, *Amylomyces rouxii*, and *Mucor racemosus* (Table 86.1). This close relationship between *M. circinelloides* and *R. variabilis* was also shown by Alvarez et al. in their evaluation of 18 strains of *M. circinelloides*.[2] They showed these strains to have a sequence similarity of >99% with the type strain of *R. variabilis* var *regularior* (CBS 103.93^T). These authors suggested that it seemed logical to consider these two species to be synonyms of each other. Additionally, Schwarz et al. also showed that

TABLE 86.1

Sequence Analysis Using the Internal Transcribe Spacer Regions of the rDNA Complex to Determine Relatedness within the Various *Mucor* Species Associated with Human Infection[a]

	Alignment Sequence[b]		Significant Alingments with Source Species[c]			Significant Alingments with Related Species[c]		
			GenBank			GenBank		
Mucor Species[d]	GenBank Accession No.	Isolate No. Source[e]	Accession No.	(% Identify)	Sequences[f]	Accession No.	(% Identity)	Species
M. circinelloides	DQ118991	CBS 195.68^T	(99.9)	DQ118988	>20	(98.7)	EF583641	*R. variabilis*
			(99.8)[g]	AY243943		(98.7)	EF583640	
			(99.5)	DQ118987		(98.7)	DQ119007^T	
			(99.5)	AM933549		(98.7)	EF151444	
			(99.5)	AM933548		(98.7)	GQ150159	*A. rouxii*[h]
						(98.8)	EF203695	
						(98.7)	AJ878933	*M. racemosus*
						(98.8)	AJ243940	
						(98.6)	AY243941	
M. hiemalis	DQ118992	CBS 201.65^T	(99.5)	AJ876490	14		None	
			(99.5)	AJ876489				
			(100)	EU484263				
			(100)	EU484253				
			(100)	EU484277				
M. racemosus	AY213659	CBS 260.68^T	(100)[g]	DQ118996	11	(99.0)	EU484245	*M. plumbeus*
			(99.8)	AB369913		(99.0)	EU484261	
			(99.8)	FJ345353		(98.9)	EU484262	
			(99.4)	AY213661		(98.9)	EU484241	
			(100)	AY625074		(98.9)	EU484235	
M. indicus	DQ118994	CBS 226.29^T	(100)	DQ118993	5		None	
			(100)[g]	AB113026				
			(100)	DQ113027				
			(100)	AB113027				
			(100)	EU798706				
M. ramosissimus	DQ118997	CBS 135.65^T	(100)[g]	AY213663	1		None	

Abbreviations: R., Rhizomucor; A., Amylomyces; T, type strain.

[a] Only includes GenBank sequences for analysis that had a complete ITS1-5.8S-ITS2 sequence where the sequence was referenced with a species name.

[b] The alignment sequences were obtained from the type species.

[c] Included only up to five of the sequences with the top bit values (E-scores) following a BLAST search where the percentage identity was >98.5%. The sequences listed are ordered from highest to lowest bit scores.

[d] No attempt was made to correlate results among the various forma or varieties within the *Mucor* species.

[e] Accession code from the culture collection at the Centraalbureau voor Schimmelcultures (CBS), Utrecht, the Netherlands.

[f] Total number of sequences with a 99% identity to the alignment sequence.

[g] Duplicate sequence obtained from the type strain.

[h] Current validated species name reported in the Index Fungorum database (http://www.indexfungorum.org/) is *Rhizopus arrhizus* var. *arrhizus*.

these two species had a high level of similarity when evaluating the spacer sequences.[22] Voigt et al. when utilizing the 18S gene and 28S gene tree topology, also supported the transfer of *R. variabilis* to the genus *Mucor*.[24]

The close genomic sequence similarity between *Amylomyces rouxii* (formerly called *Mucor rouxii*) and *M. circinelloides* has also been reported.[22,25] Although *A. rouxii*, has not been described as a cause of human disease, suggestions are that the taxonomic positioning of this species should be questioned.[25] Previous classifications had this species included in the genus *Mucor*, while current classification within the Index Fungorum database has the valid name for this species as *Rhizopus arrhizus* var. *arrhizus*. Morphological similarities between these two species have also been described.[25]

The high sequence similarity between the type strain of *M. circinelloides* and three GenBank sequences of *Mucor racemosus* was reported to be possibly due to a misidentification of *M. racemosus* strains.[25] In this report, the type strain of *M. racemosus* (CBS 260.68^{T}) only had a low sequence similarity (<96%) to the type strain of *M. circinelloides*, which was also subsequently shown by Alvarez et al. when evaluating multiple sequences of *M. circinelloides*.[2] An alignment of the type strain *M. racemosus* sequence showed good similarity to 11 other sequences of *M. racemosus* in the GenBank database (Table 86.1). Similarly, these species also differ in thermotolerance, where *M. circinelloides* grows well at 37°C, *M. racemosus* is not known to grow at this elevated temperature.[18,26] The alignment of the type strain of *M. racemosus* also showed this species to be closely related to *M. plumbeus*. Kwasna et al. also described this close relationship in their analysis of the ITS1/2 region in their studies, showing phylogenetic relationships within the *Zygomycota*.[27]

The ITS sequences of the three other species associated with human disease did not overlap with other species suggesting that these were divergent enough to validate the position of separate species although a limited number of sequences were available for evaluation (Table 86.1). A report shows *M. hiemalis* and *R. variabilis* var. *variabilis* were phylogenetically closely related; however, the ITS sequences of the two type strains (DQ118992, *M. hiemalis* and DZ119006, *R. variabilis*) appeared to only have a similarity of 88.4%.[2] This difference was also reported by Kwasna et al. in an evaluation of the ITS1/2 regions.[27]

Finally, although many forma and varieties exist among the *Mucor* species, the ability to separate these by sequence analysis is still not possible due to the lack of a taxonomically robust number of isolates with sequence information for comparison. Reports have shown that there is some variability within the ITS regions of *M. circinelloides*, which may support subspecies grouping.[22,25] Schipper described four forma of *M. circinelloides*, based on differences in the shape of the spores and columellae as well as in the thermotolerance (f. *circinelloides*, f. *janssenii*, f. *griseocyanus*, and f. *lusitanicus*).[26] Comparative analysis of multiple gene sequences from a broad representation of well-characterized species within the genus *Mucor* will likely lead to future changes in the classification of this group. Multigene-based phylogeny projects such as the Assembling the Fungal Tree of Life project will most certainly have an impact on future changes in the classification of fungal species including those in the genus *Mucor* (http://aftol.org/).

86.1.2 Pathogenicity and Epidemiology

Mucor species are ubiquitous saprophytes and found in the soil and dust worldwide.[11] Infections by these opportunistic pathogens are usually by inhalation of spores into the respiratory tract (pulmonary disease) or by direct inoculation of spores into a wound (cutaneous or subcutaneous disease).[4,11,28] Instances of gastrointestinal (GI) infection are also known to occur by the ingestion of food contaminated with *Mucor* species[29] as well as disseminated disease resulting from any primary site of infection because of the organisms' ability to rapidly invade blood vessels.[4]

Once the pathogen enters into the human body, studies have shown that the killing and removal of the *Mucor* as well as other *Mucorales* is mediated by both the neutrophils and the macrophages.[30] Conditions that inhibit the formation of neutrophils such as the administration of cytotoxic drugs and those that impair phagocytic function such as diabetes mellitus or steroid therapy are known risk factors that allows for the spores to germinate and form hyphae.[11] Virulence factors for *Mucor* species are not well understood, but growth at an elevated temperature appears to also be a major factor in the organisms' ability to infect human tissue.[11] Although thermotolerance is a characteristic of some *Mucor* species, many of the species will not grow at an elevated temperature, which suggests why infections, for instance those caused by *M. hiemalis*, only appear on cooler areas of the body such as on the skin. Additional studies have also shown that an increase in the availability of iron, as occurs with the administration of iron chelating deferoxamine can serve as a sideophore to provide iron to enhance fungal growth.[4,10] Also, the ability of the pathogen to invade arteries once established in the host allows the disease to rapidly spread, leading to thrombosis and tissue destruction.

86.1.3 Clinical Features

Numerous review articles have been written to describe the clinical features of zygomycosis.[4,9–12] These articles describe the features of disease collectively caused by the zygomycetes following recognition of the pathogen in tissue showing characteristic histological features. These features include hyphae with a variable width (5–20 μm) and a haphazard branching pattern with rare septations, as well as the ability of the fungus to be angioinvasive.[10]

Roden et al., in a review of 929 patients with zygomycosis, showed that sinus involvement was the most common infection site (39% of cases) followed by pulmonary (24%), cutaneous (19%), cerebral (9%), and GI (7%).[4] Of the 465 culture positive cases in this review, only 85 (18%) were positive with a *Mucor* species, with *Rhizopus* species the most common organism group identified (47%). Other reports from various patient groups also showed that *Mucor* species were uncommon causes of disease.[1,5,6,10,31,32] Additionally, even though involvement of the sinus was the most common reported clinical feature of zygomycosis, *Mucor* species have only been rarely detected as causing this infection.[1,3,33] Michael et al. in a review of fungal sinusitis cases from patients at a tertiary care hospital, reported on 72 cases of invasive disease, where culture was obtained.[3] Of these, 33 were culture positive for a zygomycete, 30 of these were positive for *Rhizopus arrhizus*, two were positive for nonsporulating zygomycetes and one was positive for a *Rhizomucor* species and a nonsporulating zygomycete. No *Mucor* species were detected in this study. In a review of 106 solid organ transplant patients with histology proven zygomycosis from a variety of sites, Almyroudis et al. reported that 67 cases were culture positive with 49 (73%) identified as a *Rhizopus* species followed by 9 (13%) identified as a *Mucor* species and the rest identified as other zygomycete species.[1] In this study, sinus involvement was also the most common site of infection (31%) followed by pulmonary (24%), cutaneous (15%), and GI (11.2%). Finally, Neofytos et al., in a review of hematopoietic stem cell transplant recipients with invasive fungal disease, showed that zygomycetes were associated with 18 of the 250 cases of infection with *Mucor/Rhizomucor* described as rare causes of disease in three cases (1.2% of the total).[33] Older literature describing cases of zygomycosis prior to molecular sequencing usually did not list the pathogen because of the difficulty in culturing and the lack of expertise to identify the organism species using micromorphological methods once cultured. Also, reclassification of *Rhizomucor pusillus* from *Mucor*, which was proposed by Schipper and subsequently confirmed by rRNA gene sequencing, has dramatically decreased the number of *Mucor* species associated with infection.[14,26]

This inability to culture a zygomycete from clinical material and the limited capacity to recognize individual species when a culture is available has restricted the ability to describe diseases association with the various *Mucor* species. Alvarez et al., in an evaluation of 190 clinically significant zygomycete isolates submitted for testing to a reference laboratory, showed that *Mucor* species were the second most common genus recognized (14.2%) following *Rhizopus* species (66.8%).[2] Of the 29 *Mucor* species morphologically identified in this study, 18 were confirmed by molecular testing (using the ITS region targets) as *M. circinelloides*, 5 were confirmed as *M. indicus*, and 6 were identified as *Mucor* species only. The latter 6 *Mucor* species, although identified by morphologic methods as *M. racemosus* (three isolates) and *M. ramosissimus* (three isolates), could not be validated to species by sequence analysis.

In the literature, only 21 cases of histology proven invasive infection with positive culture of a confirmed *Mucor* species were described (Table 86.2). A majority of the reported cases involve patients with some underlying condition such as immunosuppression or diabetes mellitus, although eight cases did not have any reported underlying condition. Additionally, most of the infections involved the skin and subcutaneous tissues (10 cases), with a survival rate of these cases at 78.9%, which was not unexpected because of the more localized superficial nature of the infection, although in general the survival rate of zygomycosis in the compromised patient has historically been considered poor.[8,9,11]

Mucor circinelloides was identified as the most common *Mucor* species reported as a cause of human disease in the literature (8 cases).[25,34–40] In five cases, trauma was considered the predisposition to infection with the primary disease including cutaneous and subcutaneous areas (Cases 1–4 and 6). Two of the cases resulted in fungemia (Cases 5 and 8). Diabetes mellitus was the most common underlying disease reported in four of these cases with two of them described as a patient in a "metabolic acidotic" condition. The etiological agents in three of the most recent cases were identified following molecular sequencing of the ITS regions (Cases 6–8).

Mucor indicus was the second most common *Mucor* species identified in the literature associated with invasive infection in seven cases.[29,41–46] The most interesting fact about this species was the association with GI disease (five cases). Deja et al. had suggested that this increased predilection to the GI tract was probably through ingestion, since *M. indicus* is used as a starter for food fermentation of rice and manioc.[42] However, in only one of the cases of GI involvement was the source of infection described (Case 12). This case was suggested due to ingestion of a naturopathic medicine identified as containing *M. indicus*.[45] In two cases, invasion outside of the GI tract to the liver was noted with both patients having an underlying cancer and being treated with immunosuppressive therapies (Cases 12 and 13). The etiological agent in two of the most recent cases was also identified using ITS sequencing.

Three cases each of zygomycosis caused by *M. hiemalis*[47–49] and *M. ramosissimus*[50–52] were also described. Two cases of *M. hiemalis* infection were noted to occur in individuals who did not have an underlying disease (Cases 9 and 11) and both reports described disease progression over an extended time period (5 years for Case 9 and 10 months for Case 11). Although trauma was indicated for two of the cases, the third case did not indicate trauma; although it would have been difficult for the patient to recall this after a 5 year course of the disease. The predilection for a primary cutaneous infection in all these cases may have been due to the organisms' known inability to grow at the high core body temperature.[18]

Mucor ramosissimus was reported in two cases as causes of mucocutaneous and rhinocerebral infections (Cases 19 and 20). More recently, this organism was described as a cause of cutaneous zygomycosis in a patient with aplastic anemia on immunosuppressive therapy.[52] None of the etiological agents in these cases were identified by molecular methods. This lack of molecular analysis for species identification and the fact that the species does not grow well at elevated temperatures make the association of this species to human disease questionable.[18] Additionally, Alvarez et al. showed that *M. ramosissimus*, previously identified by morphological

TABLE 86.2
***Mucor* Species Reported in the Literature as Causes of Zygomycosis[a]**

Case No.	(Species) Predisposition		Underlying Disease	Primary Disease Type	Molecular Target[b]	Outcome	Reference
	(*Mucor circinelloides*)						
1.	34, F	Farmer, trauma?	None	Cutaneous	Not done[c]	Resolved	[40]
2.	23, F	Trauma, neutropenia	AML	Cutaneous	Not done[c]	Resolved	[38]
3.	63, M	Trauma, steroids, metabolic acidosis	DMID, cirrhosis	Cutaneous, subcutaneous	Not done[c]	Death	[36]
4.	62, F	Trauma	MDS	Cutaneous	Not done[c]	Resolved	[34]
5.	48, M	Abdominal surgery TPN[d]	Short gut syndrome	Fungemia	Not done[c]	Resolved	[35]
6.	90, M	Trauma, pancytopenia metabolic acidosis	MDS, DMID	Cutaneous	ITS1, ITS2[e]	Resolved	[25]
7.	40, M	Tooth extraction	DMNID	Rhinosinusitis	ITS1, ITS2[e] 28S rRNA gene	Resolved	[39]
8.	83, F	Intensive care	DMNID	Disseminated[f]	ITS1, ITS2[e]	Death	[37]
	(*Mucor hiemalis*)						
9.	14, F	None	None	Cutaneous	Not reported[g]	Resolved	[48]
10.	78, M	Trauma	DMID	Cutaneous, subcutaneous	Not done[c]	Resolved	[47]
11.	10, F	Trauma	None	Cutaneous	Not reported[g]	Resolved	[49]
	(*Mucor indicus*)						
12.	39, M	GVHD, steroids	MDS	Gastrointestinal[h]	Not done[c]	Resolved	[45]
13.	27, F	HDC	Leukemia	Gastrointestinal[h]	Not done[c]	Death	[46]
14.	66, M	None	None	Gastrointestinal	Not done[c]	Resolved	[43]
15.	82, F	Burn[i]	None	Subcutaneous	Not done[c]	Resolved	[44]
16.	34, M	None	None	Gastrointestinal[j]	Not described[k]	Resolved	[29]
17.	48, M	Acute head injury	None	Gastrointestinal	ITS1, ITS2[e]	Resolved	[42]
18.	6mo, F	LVAD implant	Cardiomyopathy	Endovascular[l]	ITS1, ITS2[e]	Death	[41]
	(*Mucor ramosissimus*)[m]						
19.	39, F	None	None	Mucocutaneous	Not listed[n]	Resolved	[51]
20.	39, F	Metabolic acidosis	DM[o]	Rhinocerebral	Not listed[n]	Resolved	[50]
21.	31, F	Immunotherapy	Aplastic anemia	Cutaneous	Not done[c]	Resolved	[52]

Abbreviations: F, female; M, male; DM, diabetes mellitus; DMID, diabetes mellitus, insulin dependent; DMNID, diabetes mellitus, non-insulin dependent; AML, acute myelogenous leukemia; MDS, myelodysplastic syndrome; TPN, total parenteral nutrition; LVAD, left ventricular assist device; HDC, high dose chemotherapy; GVHD, graft-vs.-host disease.

[a] Only included those cases reported in the literature where the diagnosis was made by culture from a sterile body site with characteristic histopathology in tissue.
[b] When done, a molecular sequence of the listed target site was used to confirm the identity of the *Mucor* species.
[c] Identification of species made on the basis of macro- and micromorphologic features with or without temperature studies.
[d] TPN-induced hyperglycemia.
[e] Ribosomal DNA including the complete internal transcribed spacer (ITS)1-5.8S-ITS2 region.
[f] Multiple positive blood cultures with subsequent dissemination to the skin.
[g] Isolate submitted to a reference laboratory for confirmation. The method of confirmation was not reported.
[h] Dissemination to the liver.
[i] Suggested that hot compresses is association with a cosmetic clay composed of mashed onions suggested as the source.
[j] Fungemia occurred following involvement of the gastrointestinal tract.
[k] Species described as identified by both phenotypic and genotypic methods with no target listed.
[l] Involvement of the ascending aorta.
[m] A case of disseminated septic arthritis was described in abstract form which was unavailable for review[81].
[n] Species identification method was not described.
[o] Newly diagnosed diabetes mellitus with hyperglycemic diabetic coma.

methods, could not be validated to species using a molecular method.[2] In Case 21, the report describes *M. ramosissimus* as being identified using micromorphological methods, temperature growth studies, and biochemical tests. Although deposited in the culture collection of the Centralbureau voor Schmimmelcultures (Baarn, the Netherlands), under the name *Mucor* aff. *ramosissimus* (CBS 144.93), this species was not included in their commercial lists of cultures because of some structures that were described as "atypical." Also, as problematic as including this species as a cause of human disease is that none of the molecular sequences currently deposited in the GenBank database appear to represent any species of *M. ramosissimus* that have been associated with human infection.

Overall, of the 21 cases of zygomycosis reported in the literature, where the *Mucor* species was indicated, only 5 had the etiological agent identified using molecular methods, with all using the ITS1/ITS2 region target for sequence analysis. Due to the difficulty in micromorphological identification of *Mucor* species, it is reasonable to assume that some of the species for the other 16 cases could be questioned. Although morphological methods for the identification of *Mucor* species has been shown to be fairly reliable, misidentifications were possible, especially when identifying *Mucor* species other than *M. circinelloides*.[2] Unfortunately, most of the isolates associated with the reported cases appear unavailable in a reference collection for future verification testing, using a molecular assay.

86.1.4 Diagnosis

The diagnose of zygomycosis has relied mostly on the characteristic finding from a histopathological observation of tissue.[10] Although typical hyphal elements in tissue are used in the diagnosis of a zygomycetous infection, atypical structures may also be present making the diagnosis more difficult. Some *Mucor* species associated with human disease, to include *M. circinelloides* and *M. racemosus*, have been known to produce structures referred to as "spherule-like bodies," both in culture and in tissue.[11,53] These structures have been describe as chlamydospores that resemble those observed in culture[25] and as a dimorphic growth characteristic associated with the fungus.[54] The genes involved in the control of dimorphism in *M. circinelloides* have been further evaluated by Wolff et al.[54]

Culture of the etiological agent is needed to identify the species involved in infection. However, culture has not been reliable from clinical material, and in cases where a positive culture has been successfully obtained, the identification of the species is not easily performed. Thus, the diagnosis of zygomycosis caused by a validated *Mucor* species has only rarely been reported. Additionally, most of the species identification using morphological methods was performed at reference laboratories by individuals who have expertise in fungal identification. Recently however, a shift to using molecular methods on culture material has been shown to be reliable to identify the zygomycete species.[2,19,22,55] The ability to rapidly and reliable identify species from tissue, however, is still not routinely available in the clinical laboratory. The reality of diagnosis for zygomycosis is still dependent on histological evidence of characteristic hyphal forms in tissue. Subsequently, when species identification is made from culture, this result is frequently not available until days to weeks after the specimen had been collected and therefore the result is only useful for epidemiological purposes and not for primary management of the patient.

86.1.4.1 General Growth Characteristics in Culture

Mucor species grow well on most standard fungal culture medium such as Sabouraud dextrose agar. For most of the species associated with human disease, the growth is generally rapid with mycelial elements covering the entire plate within 2–3 days after incubation at 30°C. Unfortunately, the recovery of a zygomycete from tissue has been described as difficult where negative results are reported although histological evidence of a zygomycete was present.[4] One reason for this inability to recover the fungus appears to be partly related to the aggressive processing of the specimen, which may damage the organism.[11] Roden et al. in an evaluation of 929 reported cases of zygomycosis showed that only 465 of these cases had microbiological findings listed.[4] They also showed that there was a clear increase in culture positivity over time with 71% of all cases, since 2000 diagnosed on the basis of culture result. This improvement was suggested due to better training, a greater understanding of specimen processing, improved culture techniques, and increased access to sophisticated reference laboratories.

86.1.4.2 Conventional Diagnostic Methods

The vegetative mycelium of all species in the *Mucorales* is composed of wide, predominately aseptate hyaline hyphae. The general characteristic for the identification within the *Mucor* genus is a rapidly growing colony and the production of sporangiophores containing a multispored sporangia without basal rhizoids.[18] Detailed microscopic characteristics such as the presence of sporangiophores, which are branched or unbranched, the size and shape of the sporangia and sporangiospores, the features of the columellae, and the presence or absence of chlamydospores, are subsequently used to identify *Mucor* species.[18] Additional physiological methods to help in the identification is to define the maximum temperature growth and the organisms' ability to assimilate ethanol.[18] Unfortunately, the identification of *Mucor* species on morphological features alone is difficult and time consuming and requires an assessment by individuals with experience in fungal identification. Alvarez et al., retrospectively analyzed 190 clinical isolates of zygomycetes that had been identified to species by morphological methods only.[2] They used the ITS1-5.8S-ITS2 region as a molecular target and showed an overall correlation of 92.6% between morphological and molecular identification of all the zygomycetes evaluated.[2] In this study, all 18 of the *M. circinelloides* and 5 of 6 *M. indicus* correlated between morphological and molecular identification. However, none of the three morphologically identified

M. racemosus or the three *M. ramosissimus* was confirmed by molecular testing. All of these six isolates, however, were identified to the correct genus by sequence comparison analysis. Although they concluded that ITS region sequencing was a useful tool to identify most of the clinically significant zygomycetes, they also showed that neither phenotypic methods nor sequencing methods alone were able to identify all of the species present. They also concluded that *R. variabilis* var *regularior* should be considered a synonym of *M. circinelloides*, an observation that has been suggested by other research groups.[22,25,56]

The accurate identification of heterothallic zygomycetes, such as *Mucor* species, has also been shown to be enhanced by observation of zygospores in mating studies.[57] This method was successfully employed by Weitzman et al. to identify an isolate of *M. circinelloides*.[57] However, others have shown that mating studies do not often yield a positive result.[25,26] Schipper in a study of fertility among *M. circinelloides* and its formae revealed that isolates from different formae could mate with each other, which thus confused the broad-species concept for *M. circinelloides* but that some isolates within a formae failed to mate.[26] Other researchers showed that mating was not successful even though high similarities in ITS sequences among identical species were present.[25] Nevertheless, the limited number of studies evaluating mating studies does not allow for an accurate assessment of this process, and additional studies are needed to further research this process. Conversely, the difficulty in performing mating studies because of the need to maintain a library of tester strains does make this methodology unrealistic for most clinical laboratories.

86.1.4.3 Molecular Techniques

Recent advances in molecular testing for the identification of fungal pathogens have had a significant impact in the diagnostic mycology laboratory.[55,58,59] A general consensus addressing these changes suggest that approaches to the species identification of many fungi today should include both a combination of morphologic and molecular testing methods.[55] This combination approach is especially relevant for the identification of molds and more specifically for species identification of the zygomycetes. Numerous papers are available to describe the overall review of molecular testing for medically important fungi.[23,55,58–60]

Targets. A variety of genomic targets have been used for the identification of *Mucor* species. The ITS regions, the D1/D2 domains of the 28S rDNA gene, and the V9 region of the 18S rDNA gene, all within the rDNA complex, are the most common sequenced fungal loci for the identification of *Mucor* species.[2,20,22,24,27,61–66] Other molecular targets that have been used with limited success include the cytochrome b gene[67] and the high-affinity iron permease (FTR1) gene.[68] No studies to evaluate specifically the *Mucor* species have been done to compare multiple genomic targets for identification. The nuclear ribosomal ITS region (ITS1, 5.8S rRNA gene, ITS2) located between the nuclear small-(18S rRNA gene) and large-(26–28S rRNA gene) genes has been reported to be a reliable sequence for the identification of most *Mucorales*, including the *Mucor* species.[2,20,22,27,66] This region has also been proposed as the prime fungal barcode for fungal species identification by the International Subcommission on Fungal Barcoding (http://www.allfungi.com/its-barcoe.php).

Methodologies. Numerous molecular methods using both nonsequence- and sequence-based approaches for analysis have been used for the identification of the zygomycetes (Table 86.3). Nonsequence-based approaches for analysis have included conventional PCR with RFLP analysis,[27,64,68] multiplex PCR using a specific DNA probes in a microarray-based assay,[66] real-time PCR using melting curve analysis,[62,67] and seminested PCR using Luminex micro bead hybridization technology.[20]

Restriction fragment length polymorphism (RFLP) analysis of a PCR product has been shown to be reliable to differentiate *Mucor* species using a variety of restriction enzymes. Nyilasi et al. used conventional PCR on 26 zygomycete strains to include four different species of *Mucor* using a *Backusella*/*Mucor*-specific primer set and the FTR1 gene target.[68] Following amplification and digestion with the *Alu*I enzyme, all of the zygomycete species provided a unique fragment pattern allowing for separation of the various species tested. Machouart et al. also used conventional PCR, where 36 *Mucorales* species were evaluated to include 2 species of *Mucor* using a zygomycete-specific primer set that targeted a variable region within the 18S rDNA gene.[64] The amplicons in this case were treated with multiple restriction enzymes. The *Afl*II enzyme was shown to be *Mucor* genus-specific while the *Xmn*I enzyme was shown to be species-specific for *M. circinelloides*, *M. racemosus*, *M. ramosissimus*, and *M. plumbeus*, but did not cut amplicons obtained from *M. hiemalis* and *M. indicus*. Their studies showed the utility of using PCR-RFLP-based assay for a rapid and precise identification of mucorlean strains. They suggested that additional studies were needed to explore this technology as a diagnostic tool on biological samples from patients with zygomycosis. Finally, Kwasna et al. showed that PCR-RFLP analysis of the ITS1/2 rDNA target could be used to discriminate between genera and species of the *Mucorales*.[27] They evaluated 18 species of *Mucorales* including 6 *Mucor* strains representing 4 different species. The nine different restriction enzymes evaluated showed that the best discrimination among species of *Mucorales* could be achieved with the *Alu*I, *Hinf*I, and *Sau*3AI restriction enzymes.

As a means to decrease the time for detection, numerous studies have applied rapid cycle PCR methods as the testing platform. Real-time PCR using the 28S rDNA target[62,63] and the cytochrome b gene target[67] have most recently been done. Kasai et al. used the 28S rDNA gene with zygomycete-specific primers using a quantitative method of PCR followed by melt curve analysis.[62] They evaluated nine different species of zygomycetes to include three species of *Mucor.* In this evaluation, the results showed that melt curve analysis was able to distinguish among the *Rhizopus*, *Mucor*,

TABLE 86.3
Molecular Assays and Genomic Targets Used for the Identification of *Mucor* Species[a]

PCR Technique/Analysis Method	Source	Genomic Target[b]	Specificity	*Mucor* Species(Number Evaluated)	Reference
Conventional/sequencing	Culture	28S rDNA 18S rDNA	Zygomycetes	*M. amphiborum* (1), *M. circinelloides* (1), *M. hiemalis* (1), *M. indicus* (1), *M. mucedo* (1), *M. racemosus* (1), *M. ramosissimus* (1)	[24]
Conventional/sequencing[c]	Culture, clinical	28S rDNA	Universal	*M. circinelloides* (2), *M. hiemalis*(1), *M. racemosus* (1)	[61]
Conventional/sequencing	Culture, clinical	ITS1-5.8S-ITS2	Zygomycetes	*M. circinelloides* (5), *M. hiemalis* (1), *M. indicus* (2), *M. racemosus* (1), *M. ramosissimus* (1)	[22]
Conventional/sequencing	Culture	ITS1-5.8S-ITS2	Universal	*M. circinelloides* (18), *M. indicus* (5) *Mucor* spp.[d] (7)	[2]
Conventional/RFLP[e]	Culture	18S rDNA	Zygomycetes	*M. circinelloides* (1), *M. hiemalis* (1)	[64]
Conventional RFLP[f], sequencing	Culture	ITS1-5.8S-ITS2	Universal	*M. circinelloides* (1), *M. hiemalis* (2), *M. plumbeus* (1), *M. racemosus* (2)	[27]
Conventional/RFLP[g], sequencing	Culture	FTR1	Zygomycetes	*M. circinelloides* (1), *M. rouxii* (1), *M. plumbeus* (1), *M. racemosus* (1)	[68]
Multiplex/DNA probes	Culture, clinical	ITS1	*Mucor*	*M. racemosus* (1)	[66]
Real-time/sequencing	Culture, clinical	28S rDNA	Universal	*M. hiemalis* (1), *M. flavus* (1)	[63]
Real-time/melting curve	Culture, clinical	28S rDNA	Zygomycetes	*M. circinelloides* (1), *M. ramosissimus* (1), *M. indicus* (1)	[62]
Real-time/melting curve	Culture, clinical	Cytochrome b	Zygomycetes	*M. circinelloides* (1), *M. flavus* (1), *M. hiemalis* (1) *M. mucedo* (1), *M. racemosus* (1), *Mucor* spp.[d] (1)	[67]
Seminested Luminex probes[h]	Culture	ITS2	Universal	*M. racemosus* (1), *M. mucedo* (1)	[20]

Abbreviations: M, *Mucor*; ITS, internal transcribed spacer; RFLP, restriction fragment length polymorphism; FTR1, high-affinity iron permease gene.
[a] Only included studies were multiple species or strains of zygomycetes were evaluated.
[b] The 18S, 5.8S, and 28S genes as well as the spacer regions are all included in the rDNA complex of the genome.
[c] Utilized the MicroSeq D2 large-subunit rDNA fungal sequencing kit (Applied Biosystems, Foster City, CA).
[d] Species not specified.
[e] Discriminatory enzymes used were *Alu*I, *Hin*fI, *Sau*3AI, and *Hae*III.
[f] Used the *Xmn*I enzyme (species-specific) and the *Afl*II enzyme (*Mucor*-specific).
[g] Restriction enzyme used was *Alu*I.
[h] Utilized the Luminex microbead hybridization technology (Luminex Corporation, Austin, TX).

and *Rhizomucor* genera, but not within the various species. They suggested that the advantage of this assay was that the three genera within the class *Zygomycete* could be detected in a single result by melt curve analysis without the need for further procedures such as sequencing, which ultimately results in a faster diagnosis time. They also successfully applied this method to tissues from a rabbit model of experimental pulmonary zygomycosis caused by *M. circinelloides*.

Hata et al. applied real-time PCR for the detection of zygomycetes from both culture and tissue specimens (both fresh and fixed).[67] This assay targeted a region of the multicopied cytochrome b gene, where 44 culture isolates of zygomycetes that included 5 different *Mucor* species were evaluated. Although their study showed this assay to be a rapid and accurate method to detect zygomycetes, they also noted that overlapping melting curves were obtained with *Absidia*, *Apophycomyces*, and *Mucor* species, thus preventing the identification to the genus level.

More recent molecular technologies to evaluate *Mucor* species included the application of PCR to microarray analysis and to other hybridization-based probe technologies. Spiess et al. established a sensitive and specific DNA microarray combining multiplex PCR and consecutive DNA chip hybridization to detect fungal genomic DNA in clinical samples.[66] This assay involved primer sets and probes to amplify and detect 14 clinically relevantly fungal species that included both *M. racemosus* and *Rhizopus microsporus*. This proof-of-concept evaluation showed that fungal pathogens such as the zygomycetes could be detected and identified from clinical samples, which ultimately would improve the diagnosis of invasive fungal infections.

Landlinger et al. utilized the Luminex xMAP hybridization technology for species-specific identification of a wide range of clinically relevant fungal pathogens to include the zygomycetes.[20] Their preliminary studies showed that this technology was able to successfully identify numerous fungal pathogens from clinical specimens to include two species of *Mucor*. Further development of this high-throughput assay for a broad range of pathogenic fungi will be relevant in the context for the diagnosis of invasive fungal infections.

Most of the sequence-based analysis methods have been done using conventional PCR with universal primers to target sequences within the 28S rDNA gene[61] or the ITS1-5.8S-ITS2 target area.[2,27] Additionally, Vollmer et al. used a real-time PCR approach also with sequencing as the analysis method.[63] They evaluated 39 fungal species to include two *Mucor* species and showed that this method was useful as a means to rapidly screen for fungal infections in various clinical specimens. Additionally, all of the case reports, where *Mucor* species were identified, used sequencing and alignment analysis of the ITS regions (Table 86.2).[25,37,39,41,42]

The availability of a nucleotide-based sequence database that has the phylogenetic breadth and taxonomic accuracy needed to identify the zygomycetes is desirable. The largest database currently used for sequence-based identification is the publicly available GenBank database. Unfortunately, due to the fact that this is an open source database, reports suggest that up to 20% of the sequence entries listed are associated with erroneously identified species.[55] Currently, over 500 *Mucor* nucleotide base sequences representing 18 species (accessed October 2009) are available in the GenBank database. A majority of these sequences involve three species with *M. circinelloides* having the highest number of sequences available followed by *M. hiemalis* and *M. racemosus*. The partial 18S rDNA, complete ITS1-5.8S-ITS2, and partial 28S rDNA sequence target are the most common areas represented followed by regions of the 18S rDNA gene and 28S rDNA gene targets. Only a small number of sequences are available for other targets such as the act-1 gene for actin, the cytochrome b gene, the beta-tubulin gene, and the translocation elongation factor 1-alpha gene. The need for sequences from a variety of genomic targets in a database that is phylogenetically represented with accurate species identification is clearly needed for the zygomycetes. Although multiple locus sequencing and alignment comparisons has been done using a variety of mucoralean species, additional studies using this approach are also needed to determine the optimal target or targets for *Mucor* species identification. Studies similar to what has been done using multiple targets to evaluate for the identification of the aspergilli[69] and the fusaria[70] should be considered for the zygomycetes.

86.2 METHODS

86.2.1 Sample Preparation

The preparation of specimens for DNA extraction of *Mucor* species is similar to the preparation for any other fungal agent. Briefly, the extraction process, whether processing from culture material or from clinical samples such as blood, bronchial alveolar lavage specimens, and tissue, requires some type of mechanical disruption prior to DNA extraction. Generally, the original culture or tissue material is mechanically disrupted using some type of homogenization along with vortexing of the samples. Other pretreatment processes may include heating at 95°C–100°C and freezing at −70°C or in liquid nitrogen.

Concurrent with or following the mechanical disruption process, a variety of extraction methods have been used. More labor intensive kit-based methods (such as the Qiagen DNeasy plant mini kit or the QIAmp DNA blood kit [Qiagen, Hilden, Germany]) or the automated DNA extraction systems such as the NucliSens easyMAG (bioMerieux) and the MagNA Pure LC system (using either the NA isolate kit I or the DNA Isolation kit III) have all been successfully used for the extraction of fungal DNA. Varieties of enzymes have also been employed in the process and include proteinase K, lyticase, and lysozyme. Recently, a new novel method for fungal DNA extraction was described by Borman et al. and involved the utilization of chemically impregnated cellulose devices (FTA Cards, Whatman, Inc., Florham Park, NJ) for quickly preparing DNA for PCR-based applications.[71] This technique has been shown to be reliable for extraction of genomic DNA from a variety of pure organisms in culture to include *Mucor racemosus*, as well as other zygomycetous molds.

An additional approach is the extraction of DNA from Formalin-fixed, paraffin-embedded tissue. This frequently leads to a DNA product that is not useful for PCR. The major reason suggested for this is due to the extensive cross-linking of tissue proteins after fixation in formaldehyde, which results in nucleic acid fragmentation and thus inhibition of the PCR process.[67] Utilization of a shorter amplicon as the target may somewhat mitigate this effect.

Procedure for DNA extraction from Formalin-fixed tissue.

1. Cut one 50 μm section from the paraffin block using a microtome and place into a 1.5 mL screw-top tube.
2. Prepare a tissue lysate by washing the tissue section with xylene to remove wax followed by washing with 95% alcohol.
3. After air-drying, place the tissue in a mixture of 500 μL of Tris-EDTA buffer, 50 μL of 10% sodium dodecyl sulfate, and 100 μL of proteinase K and incubate for 24 h at 55°C while mixing at 500 rpm.

Note: Lyticase may be added in combination with proteinase K to maximize the efficiency of the fungal DNA extraction process.[72]

86.2.2 Detection Procedures

86.2.2.1 Identification of *Mucor* Species by Conventional PCR and Sequence Analysis

The molecular identification of *Mucor* species has been performed using multiple targets with conventional PCR followed by sequence analysis. These methods have had utilization in not only identification of species, but also in showing the phylogenetic relationship of the zygomycetes. Described below are methods utilizing the universal 28S rDNA gene target[24] and the universal ITS1-5.8S-ITS2 genomic region target.[25]

Procedure for conventional PCR using 28S rDNA target

1. Prepare a PCR mixture (50 μL) containing 10–20 ng of genomic DNA, 0.255 mM each deoxynucleotide, 25 pmol of NL-1 primer (5′-GCA TAT CAA TAA GCG GAG GAA AAG-3′), 25 pmol of NL-2 primer (5′-GGT CCG TGT TTC AAG ACG G-3′), 50 mM KCL, 10 mM Tris–Cl (pH 8.4), 2.5 mM $MgCl_2$, 0.1 mg of gelatin/mL, and 1.25 U of Ampli Tag polymerase (Perkin-Elmer, Foster City, CA).
2. PCR amplification is performed in a Thermocycler using a cycler program consisting of a denaturation step at 94°C for 2 min, 40 cycles of 30 s at 94°C for DNA denaturation, 30 s at 60°C for primer annealing, and 90 s at 72°C for primer extension; a final extension of 10 min at 72°C and a 4°C soak.
3. Amplicons are visualized electrophoretically in 1.5% agarose gel, following staining with ethidium bromide.
4. The PCR products are directly sequenced using both the NL-1 and the NL-2 primers with a resulting sequence analyzed to produce a consensus sequence that is evaluated, following a BLAST search of the GenBank databases.

Procedure for conventional PCR using the ITS1-5.8S-ITS2 rDNA target

1. Prepare a PCR mixture (50 μL) containing 5 μL of DNA (prepared following extraction on the MagNA Pure LC DNA isolate kit II for bacteria and fungi) (Roche, Mannheim, Germany). The PCR master mix consisted of 31.2 μL of UltraPURE distilled water, 5 μL of 10× high-fidelity buffer, 2.5 μL of stock 2 mM deoxynucleoside triphosphate mix, 3 μL of 50 mM $MgSO_4$, 0.75 μL of 20 μM ITS1-M primer (5′-GGA AGT AAA AGT CGT AAC AAG G-3′), 0.75 μL of 20 μM ITS4 primer (5′-TTC TCC GCT TAT TGA TAT GC-3′), 1.5 μL of stock 50% dimethyl sulfoxide, and 0.3 μL of 5 U/μL high-fidelity Platinum Taq DNA polymerase (Invitrogen Corporation, Carlsbad, CA).
2. A total of 35 cycles of amplification are performed after initial denaturation of DNA at 95°C for 4.5 min. Each cycle consisted of a denaturation step at 95°C for 30 s, an annealing step at 50°C for 30 s, and an extension step at 72°C for 1 min with a final extension at 72°C for 3 min following the last cycle.
3. The PCR product is directly sequenced using both the ITS1-M and ITS4 primers.
4. The consensus sequence is submitted to GenBank for a BLAST search of the database for sequence alignment.

Note: Application of this process has successfully been used in phylogenetic studies with the help of the software program MEGA 4.0.[2,73]

86.2.2.2 Identification of *Mucor* Species by Conventional PCR with RFLP Analysis

Multiple applications using restriction enzymes to identify *Mucor* species have been attempted using both the ITS regions and the variable region of the 18S rDNA gene.[27,64] The variety of enzymes used has shown this process to have some utility, although some specificity for *Mucor* species is noted.

Procedure

1. Prepare a PCR mix (50 μL) containing 5 ng of extracted DNA, 2 mM $MgCl_2$, 200 μM of each deoxynucloside triphosphate, 20 pmol of MucL1 primer (5′-TGA TCT ACG TGA CAT ATT CT-3′), 20 pmol of MR1 primer (5′-AGT AGT TTG TCT TCG GKC AA-3′), and 2 U of Fast start Tag polymerase (Roche, Meylan, France).
2. PCR conditions consists of denaturation for 3 min at 94°C, followed by 30 amplification cycles at 94°C for 1 min, 60°C for 1 min, and 72°C for 1 min and one final extension cycle at 72°C for 5 min.
3. For RFLP analysis, the amplicon is digested for 1 h at 37°C in a total volume of 20 μL using 10 μL of the PCR product and 5 U of the selected restriction enzyme. The restriction enzyme *Afl*II (restriction site 5′-CTTAAG-3′) has expected fragment sizes of 750 and 87 base pairs for *Mucor* species while the enzyme *Xmn*I (restriction sites, 5′-GAATAGCTTC-3′ or 5′-AGCTTCGGT-3′) has expected fragment sizes of 613 and 224 base pairs.

86.3 CONCLUSION

The zygomycetes, including *Mucor* species, will continue to emerge as causes of invasive disease as immunosuppressive therapies expand with the use of more advanced medical treatment options.[7] Current management of a patient with zygomycosis starts with recognition of characteristic symptoms in a high-risk patient.[11] These symptoms might include headache, visual disturbance, facial or orbital swelling, or other features such as a pulmonary lesion, necrotic lesion on the skin, or rarely a positive culture from a sterile site. Aggressive management at this point involves radiological imaging along with a biopsy with or without total surgical debridement, followed by empiric antifungal therapy when medically indicated. Subsequently, microscopic evidence of a zygomycete in clinical material from a deep tissue biopsy is evidence enough for the diagnosis of zygomycosis. Unfortunately, due to the aggressive nature of the zygomycetes, positive cultures will not be available to help in the initial management of the patient. Thus, even with the successful isolation and identification of *Mucor* species from clinical material, knowing this result will generally not at the present time change the course of therapy.

Recent studies now show that the in vitro susceptibility for *Mucor* species to antifungal agents has some variation.[74] Although amphotericin B has historically been the agent of choice to treat zygomycosis with a reported 94%–100% susceptibility to *Mucor* in vitro, toxicity associated with this agent with subsequent intolerance make the consideration of other agents for therapy a necessity. Studies now suggest that posaconazole be considered the first choice as alternative therapy to treat zygomycosis.[75–78] However, in vitro studies show that *M. circinelloides*, the most common *Mucor* species recognized as causing zygomycosis, is frequently resistant to this agent.[74] In a study, where *Mucor* species were not differentiated into species, only 70% of the isolates tested were susceptible to posaconazole.[74] Also, most of the other antifungal agents evaluated showed a high level of in vitro resistance to strains of *Mucor*, including high levels of resistant to voriconazole (minimum inhibitory concentration >2).[74] This resistance to voriconazole is especially troublesome since reports now imply that a significant increase in zygomycosis may be due to prior exposure to voriconazole, whether used as prophylaxis or in treatment of other fungal conditions.[22,30,79,80]

Currently, the best therapy to treat zygomycosis caused by *Mucor* species is not known. For optimal management of a patient with a *Mucor*-associated infection, the current need is for the development of a reliable and rapid molecular diagnostic assay to identify the etiological agent directly from clinical material. Limitations to the development of this type of assay, however, are that the disease is still considered rare, which makes clinical studies difficult; the organism is a common environmental contaminate and thus correlation of organism with causing a disease is more complicated; and the need for an invasive procedure in a severely compromised patient makes evaluation of this type of testing under clinical conditions more of a challenge.

In conclusion, the importance of zygomycosis as an emerging infectious disease in compromised patients enhances the need for better methods to diagnose and treat this condition. Preliminary studies have shown great promise in the development of molecular assays for the diagnosis of invasive fungal infections. As these assays undergo further improvement, the accessibility to reliable and rapid tests to diagnose this devastating infection show some promise. On the other hand, more clinical studies are needed to make this technology in the near future a reality.

REFERENCES

1. Almyroudis, N.G. et al. Zygomycosis in solid organ transplant recipients in a tertiary transplant center and review of the literature. *Am J Transplant* **6**, 2365–2374 (2006).
2. Alvarez, E. et al. Spectrum of zygomycete species identified in clinically significant specimens in the United States. *J Clin Microbiol* **47**, 1650–1656 (2009).
3. Michael, R.C., Michael, J.S., Ashbee, R.H., and Mathews, M.S. Mycological profile of fungal sinusitis: An audit of specimens over a 7-year period in a tertiary care hospital in Tamil Nadu. *Indian J Pathol Microbiol* **51**, 493–496 (2008).
4. Roden, M.M. et al. Epidemiology and outcome of zygomycosis: A review of 929 reported cases. *Clin Infect Dis* **41**, 634–653 (2005).
5. Zaoutis, T.E. et al. Zygomycosis in children: A systematic review and analysis of reported cases. *Pediatr Infect Dis J* **26**, 723–727 (2007).
6. Chayakulkeeree, M., Ghannoum, M.A., and Perfect, J.R. Zygomycosis: The re-emerging fungal infection. *Eur J Clin Microbiol Infect Dis* **25**, 215–229 (2006).
7. Greenberg, R.N., Scott, L.J., Vaughn, H.H., and Ribes, J.A. Zygomycosis (mucormycosis): Emerging clinical importance and new treatments. *Curr Opin Infect Dis* **17**, 517–525 (2004).
8. Brown, J. Zygomycosis: An emerging fungal infection. *Am J Health Syst Pharm* **62**, 2593–2596 (2005).
9. Eucker, J., Sezer, O., Graf, B., and Possinger, K. Mucormycoses. *Mycoses* **44**, 253–260 (2001).
10. Freifeld, A.G. and Iwen, P.C. Zygomycosis. *Sem Resp CritCare Med* **25**, 221–231 (2004).
11. Ribes, J.A., Vanover-Sams, C., and Baker, D.J. Zygomycetes in human disease. *Clin Microbiol Rev* **13**, 236–301 (2000).
12. Takakura, S. Zygomycosis. *Nippon Rinsho* **66**, 2356–2361 (2008).
13. James, T.Y. et al. Reconstructing the early evolution of fungi using a six-gene phylogeny. *Nature* **443**, 818–822 (2006).
14. O'Donnell, K., Lutzoni, F.M., Ward, T.J., and Benny, G.L. Evolutionary relationships among mucoralean fungi (Zygomycota): Evidence for family polyphyly on a large scale. *Mycologia* **93**, 286–296 (2001).
15. Tanabe, Y., Watanabe, M.M., and Sugiyama, J. Evolutionary relationships among basal fungi (*Chytridiomycota* and *Zygomycota*): Insights from molecular phylogentics. *J Gen Appl Microbiol* **51**, 267–276 (2005).
16. White, M.M. et al. Phylogeny of the Zygomycota based on nuclear ribosomal sequence data. *Mycologia* **98**, 872–884 (2006).
17. Hibbett, D.S. et al. A higher-level phylogenetic classification of the fungi. *Mycol Res* **111**, 509–547 (2007).
18. de Hoog, G.S., Guarro, J., Gene, J., and Figueras, M.J. *In Atlas of Clinical Fungi*, pp. 84–93, Centraalbureau voor Schimmelcultures, Utrecht, the Netherlands, 2000.
19. Iwen, P.C., Hinrichs, S.H., and Rupp, M.E. Utilization of the internal transcribed spacer regions as molecular targets to detect and identify human fungal pathogens. *Med Mycol* **40**, 87–109 (2002).
20. Landlinger, C. et al. Species-specific identification of a wide range of clinically relevant fungal pathogens by use of Luminex xMAP technology. *J Clin Microbiol* **47**, 1063–1073 (2009).
21. Rakeman, J.L. et al. Multilocus DNA sequence comparisons rapidly identify pathogenic molds. *J Clin Microbiol* **43**, 3324–3333 (2005).
22. Schwarz, P. et al. Molecular identification of zygomycetes from culture and experimentally infected tissues. *J Clin Microbiol* **44**, 340–349 (2006).
23. Balajee, S.A. et al. Sequence-based identification of *Aspergillus*, *Fusarium*, and *Mucorales* species in the clinical mycology laboratory: Where are we and where should we go from here? *J Clin Microbiol* **47**, 877–884 (2009).
24. Voigt, K., Cigelnik, E., and O'Donnell, K. Phylogeny and PCR identification of clinically important zygomycetes based on nuclear ribosomal-DNA sequence data. *J Clin Microbiol* **37**, 3957–3964 (1999).
25. Iwen, P.C., Sigler, L., Noel, R.K., and Freifeld, A.G. *Mucor circinelloides* was identified by molecular methods as a cause of primary cutaneous zygomycosis. *J Clin Microbiol* **45**, 636–640 (2007).

26. Schipper, M.A.A. On *Mucor circinelloides, Mucor racemosus* and related species. *Stud Mycol* **12**, 1–40 (1976).
27. Kwasna, H., Ward, E., and Bateman, G.L. Phylogenetic relationships among zygomycetes from soil based on ITS1/2 rDNA sequences. *Mycol Res* **110**, 501–510 (2006).
28. Mantadakis, E. and Samonis, G. Clinical presentation of zygomycosis. *Clin Microbiol Infect* **15**(Suppl 5), 15–20 (2009).
29. Aboltins, C.A., Pratt, W.A., and Solano, T.R. Fungemia secondary to gastrointestinal *Mucor indicus* infection. *Clin Infect Dis* **42**, 154–155 (2006).
30. Rogers, T.R. Treatment of zygomycosis: Current and new options. *J Antimicrob Chemother* **61**(Suppl 1), i35–i40 (2008).
31. Schalk, E. et al. Zygomycoses in patients with acute leukaemia. *Ann Hematol* **85**, 327–332 (2006).
32. Sridhara, S.R., Paragache, G., Panda, N.K., and Chakrabarti, A. Mucormycosis in immunocompetent individuals: An increasing trend. *J Otolaryngol* **34**, 402–406 (2005).
33. Neofytos, D. et al. Epidemiology and outcome of invasive fungal infection in adult hematopoietic stem cell transplant recipients: Analysis of Multicenter Prospective Antifungal Therapy (PATH) Alliance registry. *Clin Infect Dis* **48**, 265–273 (2009).
34. Chandra, S. and Woodgyer, A. Primary cutaneous zygomycosis due to *Mucor circinelloides*. *Aust J Dermatol* **43**, 39–42 (2002).
35. Chan-Tack, K.M., Nemoy, L.L., and Perencevich, E.N. Central venous catheter-associated fungemia secondary to mucormycosis. *Scand J Infect Dis* **37**, 925–927 (2005).
36. Del Palacio, A. et al. Five cases of zygomycosis (Spanish). *Rev Iberoam Micol* **16**, 50–56 (1999).
37. Dizbay, M. et al. Fungemia and cutaneous zygomycosis due to *Mucor circinelloides* in an intensive care unit patient: Case report and review of literature. *Jpn J Infect Dis* **62**, 146–148 (2009).
38. Fingeroth, J.D., Roth, R.S., Talcott, J.A., and Rinaldi, M.G. Zygomycosis due to *Mucor circinelloides* in a neutropenic patient receiving chemotherapy for acute myelogenous leukemia. *Clin Infect Dis* **19**, 135–137 (1994).
39. Khan, Z.U., Ahmad, S., Brazda, A., and Chandy, R. Mucor circinelloides as a cause of invasive maxillofacial zygomycosis: An emerging dimorphic pathogen with reduced susceptibility to posaconazole. *J Clin Microbiol* **47**, 1244–1248 (2009).
40. Wang, J.J., Satoh, H., Takahashi, H., and Hasegawa, A. A case of cutaneous mucormycosis in Shanghai, China. *Mycoses* **33**, 311–315 (1990).
41. de Repentigny, L., St-Germain, G., Charest, H., Kokta, V., and Vobecky, S. Fatal zygomycosis caused by *Mucor indicus* in a child with an implantable left ventricular assist device. *Pediatr Infect Dis J* **27**, 365–369 (2008).
42. Deja, M. et al. Gastrointestinal zygomycosis caused by *Mucor indicus* in a patient with acute traumatic brain injury. *Med Mycol* **44**, 683–687 (2006).
43. Douvin, D., Lefichoux, Y., and Huguet, C. Gastric phycomycosis. Early anatomo-pathological and mycological diagnosis. Favorable course under medical then surgical treatment (French). *Arch Anat Pathol (Paris)* **23**, 133–138 (1975).
44. Mata-Essayag, S. et al. *Mucor indicus* necrotizing fasciitis. *Int J Dermatol* **40**, 406–408 (2001).
45. Oliver, M.R., Van Voorhis, W.C., Boeckh, M., Mattson, D., and Bowden, R.A. Hepatic mucormycosis in a bone marrow transplant recipient who ingested naturopathic medicine. *Clin Infect Dis* **22**, 521–524 (1996).
46. ter Borg, F., Kuijper, E.J., and van der Lelie, H. Fatal mucormycosis presenting as an appendiceal mass with metastatic spread to the liver during chemotherapy-induced granulocytopenia. *Scand J Infect Dis* **22**, 499–501 (1990).
47. Costa, A.R. et al. Subcutaneous mucormycosis caused by *Mucor hiemalis* Wehmer f. luteus (Linnemann) Schipper 1973. *Mycoses* **33**, 241–246 (1990).
48. de Oliveira-Neto, M.P. et al. Cutaneous mucormycosis in a young, immunocompetent girl. *Med Mycol* **44**, 567–570 (2006).
49. Prevoo, R.L.M.A., Starink, T.M., and de Haan, P. Primary cutaneous mucormycosis in a healthy young girl. Report of a case caused by *Mucor hiemalis* Wehmer. *J Am Acad Dermatol* **24**, 882–885 (1991).
50. Bullock, J.D., Jampol, L.M., and Fezza, A.J. Two cases of orbital phycomycosis with recovery. *Am J Ophthalmol* **78**, 811–815 (1974).
51. Vignale, R., Mackinnon, J.E., Casella de Vilaboa, E., and Burgoa, F. Chronic, destructive, mucocutaneous phycomycosis in man. *Sabouraudia* **3**, 143–147 (1964).
52. Weitzman, I., Della-Latta, P., Housey, G., and Rebatta, G. *Mucor ramosissimus* Samutsevitsch isolated from a thigh lesion. *J Clin Microbiol* **31**, 2523–2525 (1993).
53. Stewart, N.J., Munday, B.L., and Hawkesford, T. Isolation of *Mucor circinelloides* from a case of ulcerative mycosis of platypus (*Ornithorhynchus anatinus*), and a comparison of the response of *Mucor circinelloides* and *Mucor amphibiorum* to different culture temperatures. *Med Mycol* **37**, 201–206 (1999).
54. Wolff, A.M., Appel, K.F., Petersen, J.B., Poulsen, U., and Arnau, J. Identification and analysis of genes involved in the control of dimorphism in *Mucor circinelloides* (syn. *racemosus*). *FEMS Yeast Res* **2**, 203–213 (2002).
55. Balajee, S.A., Sigler, L., and Brandt, M.E. DNA and the classical way: Identification of medically important molds in the 21st century. *Med Mycol* **45**, 475–490 (2007).
56. Voigt, K. and Wostemeyer, J. Phylogeny and origin of 82 zygomycetes from all 54 genera of the *Mucorales* and *Mortierellales* based on combined analysis of actin and translation elongation factor EF-1alpha genes. *Gene* **270**, 113–120 (2001).
57. Weitzman, I., Whittier, S., McKitrich, J.C., and Della-Latta, P. Zygospores: The last word in identification of rare or atypical zygomycetes isolated from clinical specimens. *J Clin Microbiol* **33**, 781–783 (1995).
58. Borman, A.M., Linton, C.J., Miles, S.J., and Johnson, E.M. Molecular identification of pathogenic fungi. *J Antimicrob Chemother* **61**(Suppl 1), i7–i12 (2008).
59. Iwen, P.C. Molecular detection and typing of fungal pathogens. *Clin Lab Med* **23**, 781–799 (2003).
60. Yeo, S.F. and Wong, B. Current status of nonculture methods for diagnosis of invasive fungal infections. *Clin Microbiol Rev* **15**, 465–484 (2002).
61. Hall, L., Wohlfiel, S., and Roberts, G.D. Experience with the MicroSeq D2 large-subunit ribosomal DNA sequencing kit for identification of filamentous fungi encountered in the clinical laboratory. *J Clin Microbiol* **42**, 622–626 (2004).
62. Kasai, M. et al. Detection of a molecular biomarker for zygomycetes by quantitative PCR assays of plasma, bronchoalveolar lavage, and lung tissue in a rabbit model of experimental pulmonary zygomycosis. *J Clin Microbiol* **46**, 3690–3702 (2008).
63. Vollmer, T., Stormer, M., Kleesiek, K., and Dreier, J. Evaluation of novel broad-range real-time PCR assay for rapid detection of human pathogenic fungi in various clinical specimens. *J Clin Microbiol* **46**, 1919–1926 (2008).

64. Machouart, M. et al. Genetic identification of the main opportunistic *Mucorales* by PCR-restriction fragment length polymorphism. *J Clin Microbiol* **44**, 805–810 (2006).
65. Spiess, B. et al. Development of a LightCycler PCR assay for detection and quantification of *Aspergillus fumigatus* DNA in clinical samples from neutropenic patients. *J Clin Microbiol* **41**, 1811–1818 (2003).
66. Spiess, B. et al. DNA microarray-based detection and identification of fungal pathogens in clinical samples from neutropenic patients. *J Clin Microbiol* **45**, 3743–3753 (2007).
67. Hata, D.J., Buckwalter, S.P., Pritt, B.S., Roberts, G.D., and Wengenack, N.L. Real-time PCR method for detection of zygomycetes. *J Clin Microbiol* **46**, 2353–2358 (2008).
68. Nyilasi, I. et al. High-affinity iron permease (FTR1) gene sequence-based molecular identification of clinically important zygomycetes. *Clin Microbiol Infect* **14**, 393–397 (2008).
69. Peterson, S.W. Phylogenetic analysis of *Aspergillus* species using DNA sequences from four loci. *Mycologia* **100**, 205–226 (2008).
70. O'Donnell, K. et al. Phylogenetic diversity and microsphere array-based genotyping of human pathogenic fusaria, including isolates from the multistate contact lens-associated U.S. keratitis outbreaks of 2005 and 2006. *J Clin Microbiol* **45**, 2235–2248 (2007).
71. Borman, A.M., Linton, C.J., Miles, S.J., Campbell, C.K., and Johnson, E.M. Ultra-rapid preparation of total genomic DNA from isolates of yeast and mould using Whatman FTA filter paper technology—A reusable DNA archiving system. *Med Mycol* **44**, 389–398 (2006).
72. Karakousis, A., Tan, L., Ellis, D., Alexiou, H., and Wormald, P.J. An assessment of the efficiency of fungal DNA extraction methods for maximizing the detection of medically important fungi using PCR. *J Microbiol Methods* **65**, 38–48 (2006).
73. Tamura, K., Dudley, J., Nei, M., and Kumar, S. MEGA4: Molecular evolutional genetics analysis (MEGA) software version 4.0. *Mol Biol Evol* **24**, 1596–1599 (2007).
74. Almyroudis, N.G., Sutton, D.A., Fothergill, A.W., Rinaldi, M.G., and Kusen, S. In vitro susceptibilities of 217 clinical isolates of zygomycetes to conventional and new antifungal agents. *Antimicrob Agents Chemother* **51**, 2587–2590 (2007).
75. Greenberg, R.N. et al. Posaconazole as salvage therapy for zygomycosis. *Antimicrob Agents Chemother* **50**, 126–133 (2006).
76. Page, A.V. and Liles, W.C. Posaconazole: A new agent for the prevention and management of severe, refractory, or invasive fungal infections. *Can J Infect Dis Med Microbiol* **19**, 297–305 (2008).
77. Page, R.L. 2nd., Schwiesow, J., and Hilts, A. Posaconazole as salvage therapy in a patient with disseminated zygomycosis: Case report and review of the literature. *Pharmacotherapy* **27**, 290–298 (2007).
78. van Burik, J.A., Hare, R.S., Solomon, H.F., Corrado, M.L., and Kontoyiannis, D.P. Posaconazole is effective as salvage therapy in zygomycosis: A retrospective summary of 91 cases. *Clin Infect Dis* **42**, e61–e65 (2006).
79. Marr, K.A., Carter, R.A., Crippa, F., Wald, A., and Corey, L. Epidemiology and outcome of mould infections in hematopoietic stem cell transplant recipients. *Clin Infect Dis* **34**, 909–917 (2002).
80. Trifilio, S.M. et al. Breakthrough zygomycosis after voriconazole administration among patients with hematologic malignancies who receive hematopoietic stem-cell transplants or intensive chemotherapy. *Bone Marrow Transplant* **39**, 425–429 (2007).
81. Sharma, R., Sekhon, A.S., and Premachandra, B.R. In vitro susceptibility of *Mucor ramosissimus* Samutsevitsch, a rare fungal pathogen, to antimycotics. *Chemotherapy* **41**, 200–203 (1995).

87 *Rhizopus*

Dongyou Liu and Frank W. Austin

CONTENTS

87.1 INTRODUCTION

87.1.1 Classification, Morphology, and Biology

The genus *Rhizopus* is a zygomycetous mold classified under the family Mucoraceae, order Mucorales, subphylum Mucoromycotina, kingdom Fungi. The family Mucoraceae currently consists of 24 genera: *Absidia, Actinomucor, Ambomucor, Amylomyces, Apophysomyces, Chlamydoabsidia, Circinella, Circinomucor, Gongronella, Halteromyces, Hyphomucor, Lentamyces, Mucor, Parasitella, Phycomyces, Protomycocladus, Rhizomucor, Rhizopus, Siepmannia, Spinellus, Sporodiniella, Syzygites, Thermomucor* and *Zygorhynchus* in addition to some unclassified *Mucoraceae* [1]. Within the family *Mucoraceae*, members of the genera *Rhizopus, Mucor, Absidia, Rhizomucor,* and *Apophysomyces* have been implicated in human diseases, with *Rhizopus* being the most common cause of human zygomycosis [2,3].

The genus *Rhizopus* is divided into 10 recognized species, namely, *Rhizopus arrhizus* (obsolete synonyms: *R. batatas, R. delemar, R. formosaensis, R. japonicus, R. javanicus, R. kasanensis, R. maydis, R. nodosus, R. oryzae, R. thermosus, R. tonkinensis, R. tritici, R. trubinii* and *R. usamii*), *Rhizopus azygosporus, Rhizopus caespitosus, Rhizopus circinans, Rhizopus homothallicus, Rhizopus microsporus* (consisting of varieties *microsporus, oligosporus,* and *rhizopodiformis*; obsolete synonyms: *R. oligosporus, R. rhizopodiformis, Mucor rhizopodiformis,* and *R. cohnii*), *Rhizopus niveus, Rhizopus schipperae, Rhizopus sexualis,* and *Rhizopus stolonifer* (synonym *R. nigricans*) plus 17 unassigned species [1].

Due to their close molecular and antigenic relationships, *R. arrhizus* and *R. oryzae* are now considered synonymous. *R. schipperae* resembles *R. microsporus* var. *microsporus*, with similar sporangiophore, sporangium, sporangiospore, and columellar morphology and similar temperature growth dependence. However, *R. schipperae* differs from *R. microsporus* var. *microsporus* by having fastidious sporulation requirements and distinctive organization of sporangiophores in large clusters of up to 10 sporangiophores. Additionally, *R. schipperae* also shows some limited morphologic similarities to *R. caespitosus*, which produces large clusters of sporangiophores. *Rhizopus azygosporus* also appears to be related to the *R. microsporus* group, since mating of *R. microsporus* var. *microsporus* and *R. microsporus* var. *rhizopodiformis* results in the production of azygosporus strains, which have very similar morphology to *R. azygosporus*. However, the species designation of *R. azygosporus* is upheld due to its obligate azygosporic nature.

While *Rhizopus* spp. are cosmopolitan filamentous fungi that are present in soil, decaying fruit and vegetables, animal feces, and old bread, some are pathogenic to plants, humans, and animals, with seven distinct species and varieties of *Rhizopus* being implicated in human disease. The term mucormycosis or the preferred term zygomycosis is used to describe the angioinvasive disease due to the mucorales including *Rhizopus* spp. Indeed, *Rhizopus arrhizus* (=*Rhizopus oryzae*) is the most common cause of human zygomycosis, accounting for 60% of the reported culture positive cases, and nearly 90% of the rhinocerebral forms of infection. Often used in the production of fermented foods and alcoholic beverages in Indonesia, China, and Japan, *R. oryzae* also generates the ergot alkaloid agroclavine with toxicity for humans and animals. Other notable human-infecting *Rhizopus* species/varieties include *Rhizopus microsporus* var. *rhizopodiformis*, *R. microsporus* var. *oligosporus*, *R. microsporus* var. *microsporus*, *R. azygosporus*, *R. schipperae*, and *R. stolonifer* [4–6].

The morphological characteristics of *Rhizopus* spp. include the presence of stolons and pigmented rhizoids, the formation of globose to ovoid, one-celled, hyaline to brown and striate sporangiophores, which appear singly or in clusters from nodes directly above the rhizoids, and apophysate, columellate, multi-spored, generally globose sporangia. Following the release of spores, the apophyses and columella often collapse to form an umbrella-like structure [7].

On agar plates, *Rhizopus* colonies grow very rapidly, and fill the Petri dish surface with a dense cotton-candy growth that matures (produces sporulation) in 4 days. From the front, the colony is white in color and turning gray to yellowish brown, while the reverse is white to pale. Broad thin-walled, hyaline hyphae (6–15 μm in diameter) are coenocytic (aseptate) or sparsely septate. Solitary or clustering sporangiophores are brown in color and unbranched. Rhizoids (root-like hyphae) are situated at the point where the stolons and sporangiophores meet. Round sporangia (40–350 μm in diameter) with flattened bases are at the tip of the sporangiophores. Columellae are hemispherical. Unicellular sporangiospores (4–11 μm in diameter) are round to ovoid in shape, hyaline to brown in color, and smooth or striated in texture (Table 87.1).

The length of rhizoids and sporangiophores, the diameter of sporangia, the shape of columellae, and the size, shape and surface texture of sporangiospores are useful morphological criteria for the differentiation of *Rhizopus* species from each other. The maximum growth temperature may be also employed for discrimination of Mucorales genera. For example, the best growth temperatures for *Mucor, Mortierella, Apophysomyces, Absidia/Rhizopus*, and *Rhizomucor* are <37°C, 40°C, ≥42°C, 45°C, and 54°C, respectively.

Transmission of *Rhizopus* spp. is often by the respiratory, percutaneous, and gastrointestinal or oral routes, with most *Rhizopus* infections involving the rhinocerebral and pulmonary sites, occurring secondary to inhalation of spores into the lungs or sinuses. The use of spore-contaminated adhesive bandages to hold dressings in place over surgical wounds has contributed to percutaneous transmission of *Rhizopus (oryzae) arrhizus* or *R. microsporus* var. *rhizopodiformis*. Insect bite, intramuscular injection, catheter insertion, intravenous drug use, and oropharyngeal examination (using spore-contaminated wooden tongue depressors) may be also responsible for sporadic cases of *Rhizopus* zygomycosis. In addition, ingestion of fermented milk with dried bread products, fermented porridges, and alcoholic drinks derived from corn may promote gastric zygomycosis.

Sequencing analysis of *Rhizopus (oryzae) arrhizus* strain 99–880, isolated from a fatal case of zygomycosis, reveals a highly repetitive 45.3 Mb genome, 20% of which are transposable elements (TEs) containing paralogous gene pairs. A total of 13,895 protein-coding genes are predicted. There is evidence that an ancestral whole-genome duplication (WGD) event may have resulted in the duplication of nearly all subunits of the protein complexes associated with respiratory electron transport chains, the V-ATPase, and the ubiquitin-proteasome systems. Along with recent gene duplications, the WGD has contributed to the expansion of multiple gene families related to cell growth, signal transduction, secreted aspartic protease, and subtilase protein families, which are known fungal virulence factors. It is notable that the duplication of the ergosterol biosynthetic pathway, especially the major azole target, lanosterol 14-α-demethylase (ERG11), may have played a role in altering *R. oryzae* response to different azole drugs (e.g., voriconazole and posaconazole) [8].

87.1.2 Clinical Features and Pathogenesis

87.1.2.1 Clinical Features

Rhizopus spp., especially *R. (oryzae) arrhizus* and *R. microsporus* var. *rhizopodiformis*, are responsible for about 90% of human zygomycosis, which may occur as sinusitis/rhinocerebral, pulmonary, cutaneous/subcutaneous, gastrointestinal, genitourinary, and disseminated diseases, as well as cystitis, external otitis, and allergic disease [9–33].

The hallmarks of human zygomycosis caused by mucorales such as *Rhizopus* spp. are angioinvasion, thrombosis, infarction, and necrosis of involved tissues. An acute inflammatory response is noticeable in non-neutropenic patients, with thick, necrotic fluids aspirating from areas of abscess formation.

The most common sites of zygomycosis are sinuses, rhinocerebral structures, and the lungs (with or without dissemination). Rhinocerebral disease accounts for nearly one-third to one-half of all cases of zygomycosis, as a result of inspiration of fungal spores in the sinuses, especially in the setting of diabetic ketoacidosis [34]. From initial sinus pain, drainage, and soft tissue swelling, the disease may progress rapidly and extend into neighboring tissues, with signs of periorbital

TABLE 87.1
Morphological and Physiological Characteristics of the Clinically Important *Rhizopus* Species

Species	Colony	Sporangium	Columella	Apophysis	Sporangiospore	Rhizoids	Sporangiophore	Zygospores	Growth Temperature
R. arrhizus	Floccose colonies, initially white, turning yellow, pale brown, dark brown, or gray, and collapsing with age	Globose; powdery and gray to black; medium (100–200 µm in diameter)	Globose to oval, 40–75 µm by 60–130 µm	Inconspicuous, often collapsing at sporulation	Oval to ellipsoidal, 6–8 µm by 4.5–6 µm; striated	Abundant; brown, with four to eight branches	Medium (750–2000 µm, usually 1500 µm); singly or in groups from stolon	Heterothallic; reddish brown, round to flattened with conical projections; usually 80–140 µm	Growth at 36°C, but not at >46°C
R. azygosporus	Floccose white mycelim, turns gray to black with age	Globose; small (35–100 µm in diameter)	Oval hyaline collumella	Discrete	Oval, 4–6 µm by 6–7 µm; faintly striated	Simple, pale brown	Short (usually 350–500 µm by 6–14 µm)	Abundant round azygospores, 30–70 µm in diameter	Growth at 28°C–45°C; poor growth at up to 50°C; sporangium and azygospore forming at 25°C–37°C
R. microsporus	Floccose mycelium; variable color	Globose; small (about 100 µm)	Pyriform, ellipsoidal, cylindrical, conical, globose, or subglobose	Discrete	Variable size and shape	Simple, no branching	Short (usually 800–1000 µm)	Heterothallic; small (80–100 µm) red-brown	Variable
R. schipperae	Thin white nonsporulating mycelium	Globose; small (<80 µm in diameter)	Subglobose to conical columella (5.8–6.8 µm by 4.8–5.0 µm)	Discrete	Oval, 4–7 µm; faintly striated	Simple, brown, rarely branched	Short (100–410 µm by 5–15 µm) in clusters of 10	Unknown; chlamydospores (20 µm in diameter) abundant	Growth sporulation at 30°C–35°C; tolerating up to 45°C, but not >48°C
R. stolonifer	floccose white mycelium with black dots	Globose; black; large (up to 275 µm)	Oval to subglobose, 70–120 µm in diameter; pale brown	Distinct	Round to oval, 13 µm; black; prominently striated	Abundant; 300–350 µm long; intensely ramified	Long (up to 2000 µm), pigmented, and unbranched; in groups of 1–3	Heterothallic; black, round zygospores with irregular surface decorations; 150–200 µm wide; equal suspensors	Growth and sporulation at 15°C–30°C, but not >33°C

Source: Adapted from Ribes, J.A. et al., *Clin. Microbiol. Rev.*, 13, 236, 2000.

edema, proptosis, and tearing; pain, blurring, or loss of vision in the infected eye; painful, black, necrotic ulcerations into the hard palate in the mouth; and a bloody nasal discharge due to invasion of the terbinates and the brain. Prolonged exposure to *Rhizopus* spores causes lung damage, with thickened alveolar walls, isolated solitary nodular lesions, cavitary lesions, granulomas, and interstitial infiltrates [35]. This is a significant health problem in the sawmill ("woodcutter's disease") and malt industries.

Cutaneous and/or subcutaneous sites are also involved in *Rhizopus* infections. This is often associated with the use of nonsterile adhesive bandages, which introduce spores either directly into the surgical wound or indirectly into the traumatized skin at the time of bandage removal. Growth of the fungus in a preexisting lesion often generates an acute inflammatory response with pus, abscess formation, tissue swelling, and necrosis. Sloughing-off in necrotic tissue may result in large ulcers. Primary cutaneous disease is highly invasive locally, involving the cutaneous and subcutaneous tissues as well as the fat, muscle, and fascial layers beneath. Necrotizing fasciitis may occur secondary to cutaneous or subcutaneous zygomycosis [36].

Disseminated *Rhizopus* disease can affect any organs in the body, including the skin, central nervous system, liver, spleen, kidney, and gastrointestinal tracts. In fatal cases of disseminated *R. azygosporus*, infections involving a premature infant, angioinvasion, tissue infarction, and granulomatous inflammation with the production of giant cells were noted, and the cause of death was attributable to peritoneal, gastrointestinal, kidney, or liver involvement by the fungus. In gastrointestinal zygomycosis, necrotic gastric or intestinal ulcers may occur whose rupture may result in peritonitis [37]. Other unusual symptoms of zygomycosis due to *Rhizopus* spp. include inguinal abscess, bone, renal, and genitourinary diseases.

87.1.2.2 Pathogenesis

Rhizopus spp. are opportunistic pathogens that may take advantage of breakdown in host's immune system defenses to establish infection and cause disease. Due to their angioinvasive nature and their ability to grow at or above core body temperature, *Rhizopus* spp. spread readily from the sites of entry to other organs. This infection process is aided by the production of a variety of glycosidic, proteolytic, and lipolytic enzymes, proteins, and by-products with pathogenic potential. Several ergot alkaloids (e.g., agroclavin, ergosine, and ergosamine) and mycotoxins (e.g., rhizonin A) produced by *Rhizopus* spp. are toxic to mammalian hosts.

The most frequent predisposing factors for zygomycosis include diabetic ketoacidosis, immunosuppression due to organ transplantation and cancers (e.g., leukemia, lymphoma, other hematologic malignancies), deferoxamine/desferrioxamine treatment for iron or aluminum overload, hyperglycemia, renal failure, extensive burns, trauma, intravenous drug use, heatstroke, contaminated adhesive tapes and wooden tongue depressors, protein calorie malnutrition, diarrhea, typhoid fever, gastric or intestinal ulcers, and amebic colitis. In immunocompetent hosts, zygomycosis with *Rhizopus* often follows a disruption of the mucocutaneous barrier (e.g., underlying trauma or surgical wound) or antibiotic or steroid use. In diabetic hosts, the monocytes/macrophages are dysfunctional and incapable of suppressing the germination of *Rhizopus* spores, and the impaired neutrophil activation further hampers the killing of proliferating hyphal elements.

87.1.2.3 Treatment

Amphotericin B, either alone or together with other interventions (e.g., surgery) provides a clinically useful treatment for zygomycosis caused by the Mucorales [38,39]. Posaconazole is also useful for treating *Rhizopus* infections [40–42]. Without early detection and aggressive treatment with intravenous amphotericin B and surgical debridement, zygomycosis is frequently fatal, with mortality rates approaching 80% in infected transplant recipients [43].

87.1.3 Laboratory Diagnosis

87.1.3.1 Phenotypic Identification

Initial identification of Mucorales including *Rhizopus* spp. often relies on microscopic observation of hyphae and other fungal elements in clinical specimens, with the help of 10% KOH and other fungal specific stains such as calcofluor white stain, Gomori methenamine silver stain (GMS) and periodic acid-Schiff (PAS). Typically, Zygomycetes in cytologic specimens show wide, ribbon-like, hyaline, predominantly aseptate hyphae, which may branch at 45°–90° angle. In tissue sections stained with calcofluor white stain or GMS, zygomycosis induces abscess formation with central tissue necrosis, acute inflammatory exudate, and invasion of peripheral blood vessels (angio-invasion) by hyphal elements. The hyphae may appear crinkled or gnarled in the tissue sections (so-called crinkled cellophane).

In vitro culture and subsequent examination are necessary for determination of zygomycetes to the species level. Colonies of the genus *Rhizopus* grow rapidly on standard mycological agars without cyclohexamide, with coarse and floccose aerial mycelia, stolons, pigmented rhizoids, brown pigmented sporangiophores singly or in groups, multi-spored globose terminal sporangia that are both apophysate and columellate, and one-celled, hyaline to brown striated or grooved sporangiospores (Table 87.1).

Other phenotypic techniques such as indirect immunofluorescence, crossed immunoelectrophoresis and counter-immunoelectrophoresis, and ELISA have been utilized for the diagnosis of disease caused by *Rhizopus* with variable success.

87.1.3.2 Genotypic Identification

Nucleic acid amplification techniques targeting small subunit 18S and large subunit 28S rRNA genes, internal transcribed spacer regions (ITS), cytochrome *b*, actin, high-affinity iron permease 1 (*FTR1*), and translation elongation factor (*EF-1*α) genes have proven valuable for rapid and accurate detection and identification of the etiological agents of zygomycosis

[44–55]. Several species- and group-specific PCR assays have been reported [56–60]. Recent development of real-time PCR further permits rapid, convenient detection, and quantification of zygomycosis-causing agents directly in clinical specimens [50,61,62]. PCR followed by sequencing analysis provides a reliable way to precisely identify a mucoralean species [57]. The Microseq D2 sequencing kit targeting the D2 domain of the large rRNA gene subunit enables the convenient sequencing of clinically important filamentous fungi.

87.2 METHODS

87.2.1 Sample Preparation

87.2.1.1 Fungal Strains and Cultivation

Zygomycetes strains are stored by lyophilization or in liquid nitrogen vapor (−175°C). Strains are subcultured from frozen stocks on Sabouraud-chloramphenicol agar slants for 7 days at 20°C, 28°C, or 35°C, depending on the optimum growth temperature for each species (Table 87.1). From the cultures, spore suspensions are prepared in 0.9% NaCl. Twenty milliliters of RPMI 1640 (Sigma–Aldrich) with L-glutamine but without sodium bicarbonate buffered to pH 7 with 0.165 M morpholinepropanesulfonic acid (Sigma) is inoculated with the spore suspension, and the mycelium is grown for 48–72 h at 30°C with agitation. The mycelium is washed once with 20 mL 0.9% NaCl before use [46]. Alternatively, mycelium for DNA isolation is grown in YM broth (0.3% yeast extract, 0.3% malt extract, 0.5% peptone, 2% dextrose; Difco) at room temperature for 2–5 days. Strains grown on YM agar (2% Difco agar) are examined morphologically to confirm their identity [57].

87.2.1.2 Microscopic Examination

Mycelium is mounted on slides, or stained with KOH, lactophenol cotton-blue, or other fungal stains. The presence of wide, ribbon-like, nonseptate, hyaline hyphae suggests zygomycetes. *Rhizopus* spp. are identified on the basis of microscopy showing the presence of stolons, brown pigmented rhizoids, sporangiophores, and globose sporangia, both apophysate and columellate (Table 87.1).

87.2.1.3 DNA Isolation

Genomic DNA from mycelium may be isolated according to the CTAB (hexacetyltrimethylammonium bromide; Sigma) miniprep protocol. Briefly, approximately 50 mg mycelium is homogenized in a 50-mL tube containing 1 ml CTAB extraction buffer (100 mM Tris–HCl pH 8.4, 1.4 M NaCl, 25 mM EDTA, 2% CTAB), three glass beads (0.5 cm) (Sigma), and 500 mg of 425- to 600-μm-diameter glass beads (Sigma). The suspension is vigorously vortexed and put into liquid nitrogen for 1 min followed by immersion at 37°C for 1 min. After vortexing again for 1 min, 700 μL of the suspension is transferred into a 2-mL microcentrifuge tube. An equal volume of chloroform is added to the mixture, vortexed for 5 s, and spun for 10 min at 14,000 × *g*. The upper phase (500 μL) is transferred to a new 2-mL microcentrifuge tube, and DNA is precipitated by the addition of an equal volume of 2-propanol. DNA is pelleted at 14,000 × *g* for 1 min. After discarding supernatant, the pellet is washed with 100% ethanol and resuspended in 200 μL of TE buffer (10 mM Tris–HCl pH 8.0, 1 mM EDTA). Genomic DNA is stored at −20°C [46].

Alternatively, DNA from cultures or from the clinical samples is extracted using the High Pure PCR template preparation kit (Roche). Briefly, an aliquot of mycelium is suspended in 200 μL of tissue lysis buffer and incubated for 30 min at 37°C in the presence of lysozyme (10 U/μL, final concentration). DNA is then extracted by following the instructions of the manufacturer. Finally, the concentrations are determined with an Ultrospec 2100 spectrophotometer (Amersham). Samples are kept at −20°C until used [45].

87.2.2 Detection Procedures

87.2.2.1 PCR Identification of Individual *Rhizopus* Species

Voigt et al. [57] designed species-specific primers from the 28S rRNA sequences for identification of *Rhizopus azygosporus/Rhizopus microsporus*, *Rhizopus oryzae*, and *Rhizomucor miehei/Rhizomucor pusillus* (Table 87.2). Use of these primers at an annealing temperature of 60°C results in the amplification of PCR products of 469, 413, and 305 bp, respectively.

87.2.2.2 PCR-RFLP Identification of *Rhizopus* spp.

Machouart et al. [45] utilized a pair of primers (RpL1, 5′-TGATCTACGTGACAAATTCT-3′; MR1, 5′-AGTAGTTT GTCTTCGGKCAA-3′) for amplification of a 830 bp fragment from the 18S rRNA gene of *Rhizopus* spp. Subsequent digestion with restriction enzymes BmgBI and AseI permits confirmation of the identity of *Rhizopus* spp. and *Rhizopus microsporus* or *Rhizopus azygosporus*.

Procedure

1. PCR mixture (50 μL) is composed of 2 mM $MgCl_2$, 200 μM of each deoxynucleoside triphosphate, 20 pmol of each primer, 2 U of Fast Start *Taq* polymerase (Roche), and 5 ng of the extracted DNA.
2. Amplification is conducted with an initial 94°C for 3 min, 30 cycles of 94°C for 1 min, 60°C for 1 min, and 72°C for 1 min; a final 72°C for 5 min.
3. Following amplification, 5 μL of amplicons is electrophoresed in 2% agarose gels in the presence of ethidium bromide and visualized under UV light (to confirm the 830 bp product).
4. Then, 10 μL of the specific PCR product is digested with 5 U of restriction enzyme BmgBI or AseI in a volume 20 μL for 1 h at 37°C; the digested samples are analyzed on 1.5% BET-agarose electrophoresis gels (QBiogen).

Note: Digestion of the 830 bp amplicon with enzyme BmgBI results in two fragments of 600 and 230 bp, indicating the

TABLE 87.2
The 28S rRNA Gene-Based Primers for Specific Identification of *Rhizopus* Species

Species	Primer Sequence (5′–3′)	Product (bp)
Rhizopus azygosporus	Rh1 TTTTCCAGGCAAGCCGGACCG	469
Rhizopus microsporus	Rh2 TATTCCCAGCCAACTCGCCAAAT	
Rhizopus oryzae	Ro1 AGCATTTGCCTTTTGTGATACGC	413
	Ro2 ACCGTAGTACCTCAGAAAACC	
Rhizomucor miehei	Rm1 TCTATTGCGATGCATGCTCC	305
Rhizomucor pusillus	Rm2 GGTCTCTTTAGACTCCAAAGC	

presence of *Rhizopus* spp. On the other hand, digestion with enzyme AseI yields a 750 bp fragment (and undetected 80 bp fragment for *Rhizopus microsporus* or *Rhizopus azygosporus*.

87.2.2.3 Nested PCR for Mucorales

Hofman et al. [54] described a modified nested PCR for detection of Mucorales genera, with primers derived from the 28S rRNA sequences. The first round PCR primers are Mucor1 (5′-WTTACC RTG AGC AAA TCA GA-3′) and Mucor2 (5′-CAA TCY AAG AAT TTC ACC TCTAG-3′), and the second round PCR primers are Mucor3 (5′-AGC ATG GAA TAA TRA AAY A-3′) and Mucor4 (5′-AGC ATG GGA TAA CGG AAT A-3′), yielding a final amplicon of 124 bp.

Procedure

1. PCR mixture (50 μL) is made up of 2.5 μL each of primers Mucor1 and Mucor2 (final concentration 3 mM) for the first round [or 2.5 μL each of primers Mucor1 and Mucor2 for the second round], 25 μL Master Mix Gold (Applied Biosystems), and 100–500 ng of DNA.
2. Amplification is performed with an initial 94°C for 10 min; 30 cycles of 94°C, 55°C, and 72°C for 30 s each for both primer sets.
3. Ten microliters of each amplification product is analyzed by electrophoresis in an ethidium bromide-containing 2% agarose gel.

Note: The nested PCR recognizes several zygomycosis-causing genera in the family Mucorales. The species identity of the PCR amplicon can be verified by sequencing.

87.2.2.4 Real Time PCR for Genera *Rhizopus*, *Mucor*, *Absidia*, *Cunninghamella*, *Apophysomyces*, and *Saksenaea*

Hata et al. [50] developed a real time PCR for detection of Mucorales genera *Rhizopus*, *Mucor*, *Absidia*, *Cunninghamella*, *Apophysomyces*, and *Saksenaea*, targeting a highly conserved, multicopy gene, the cytochrome *b* gene. The forward (5′-TAGGAATTACAGCAAAT-3′) and reverse (5′-CCAATGCAAACTCC-3′) primers amplify a specific 167-bp region of the cytochrome *b* gene, and FRET hybridization probes 5′-ACAATTTTCTTATTCTTCTTAGTATTAG-3′ (anchor probe) (3′ fluorescein labeled) and 5′-TTTATTCTTATTCTATGCTCCAAATA-3′ (donor probe) (5′ LCRED-640 labeled) facilitate detection of the amplicon on the LightCycler RT instrument (Roche).

Procedure

1. Real time PCR mixture (15 μL) is composed of 1× LightCycler FastStart DNA master hybridization probe buffer, 3 mM $MgCl_2$, 0.5 μM forward and reverse primers, 0.2 μM fluorescein-labeled probe, 0.4 μM RED 640-labeled probe, and 5 μL of the extracted DNA from clinical specimens or control material.
2. Amplification is carried out with 1 cycle of 95°C for 10 min; 45 cycles of 95°C for 10 s, 48°C for 15 s, and 72°C for 15 s. A melting curve is generated using the following profile: 95°C for 0 s, 59°C for 20 s, and 34°C for 20 s, with a 0.3°C/s transition, and 85°C for 0 s, with a 0.3°C/s transition, using a LightCycler v. 1.2 system with LightCycler v. 3.5 software.

Note: The six genera (*Rhizopus*, *Mucor*, *Absidia*, *Cunninghamella*, *Apophysomyces*, and *Saksenaea*) exhibit a range of melt temperatures from 41°C to 55°C, with three genera (*Absidia*, *Aphophysomyces*, and *Mucor*) having melt temperatures within approximately 1°C of each other. As expected, *Rhizomucor*, *Condiobolus*, and *Syncephalastrum* spp. are not detected by the assay.

87.2.2.5 Real Time PCR for Genera *Rhizopus*, *Mucor*, and *Rhizomucor*

Francesconi et al. [61] described a quantitative real time PCR for detection of Mucorales genera *Rhizopus*, *Mucor*, and *Rhizomucor*. The primers Zygo-F1 (5′-TTCAAAGAGTCAGGTTGTTTGG-3′), and Zygo-R1 (5′-CAGTCTGGCTCCAAACGGTTC-3′) generate a 180 bp amplicon from the 28S rRNA gene sequence, which is detected with hybridization fluorescence resonance energy transfer probes ZYGO FITC (5′-GGCGAGAAACCGATAGCGAAC-fluorescein isothiocyanate-3′) and ZYGO RD 640 (5′-RD640-GTACCGTGAGGGAAAGATGAAAAGAACTTTGAAA-3′) using a LightCycler 2.0 instrument (Roche).

Procedure

1. PCR master mix (15 μL) is composed of 0.025% bovine serum albumin, 3 mM $MgCl_2$, 0.025 U Platinum *Taq* DNA polymerase, 0.2× PCR buffer, 0.2 mM PCR Nucleotide Mix Plus, 0.002 U/μL HK-UNG, 0.25 μM each of Zygo-F1 and Zygo-R1 primers, and 0.1 μM each of ZYGO FITC and ZYGO RD 640 probes. To 15 μL of master mix, 5 μL of extracted specimen is added.
2. Uracil is released by incubating at 37°C for 15 min, and then enzyme is inactivated at 95°C for 3 min. Touchdown PCR cycling includes 95°C for 0 s (20°C/s), followed by annealing in 1°C steps between 68°C and 54°C for 5 s (10°C/s), each with a 72°C extension of 15 s (3°C/s) for each cycle. Touchdown cycling is followed by 35 cycles of 95°C for 0 s (20°C/s), 54°C for 5 s (10°C/s), and 72°C for 15 s (3°C/s).
3. A post-amplification melt analysis is performed by cooling from 96°C to 40°C for 30 s (20°C/s), followed by a gradual increase in temperature (2°C/s) to 75°C for 0 s (0.2°C/s).

Note: Quantitation standards (10-fold serial dilutions of *R. oryzae* genomic DNA ranging from 1×10^3 fg to 1×10^1 fg) are run in conjunction with each set of samples to assess assay sensitivity and linearity and qPCR results. A crossover value of ≤22 cycles is considered positive [62].

87.2.2.6 Sequence Analysis of 18S and 28S rRNA Genes

Voigt et al. [57] utilized primer pairs PNS1–NS41 and NS51–NS8Z or NS5–NS8Z to amplify the 18S rRNA as two overlapping fragments, and sequenced the amplicons with NS2, NS3, NS5, NS7, NS8; PNS1, NS6Z, NS8Z; and NS41 and NS51. In addition, they also utilized primer pair NL1–NL4 to amplify the 5′ end of 28S rRNA spanning domains D1 and D2, and sequenced the amplicon with primers NL1, NL4, ZNL2A, and ZNL3A.

Procedure

1. PCR mixture (50 μL) is made up of 0.225 mM each dNTP, 25 pmol of each primer, 50 mM KCl, 10 mM Tris–Cl pH 8.4, 2.5 mM $MgCl_2$, 0.1 mg of gelatin/mL, 1.25 U of AmpliTaq polymerase (Perkin-Elmer), and 10–20 ng of genomic DNA.
2. PCR amplification is performed with an initial 94°C for 2 min; 40 cycles of 94°C for 30 s, 52°C for 30 s, and 72°C for 90 s; a final 72°C for 10 min.
3. Amplicons are purified with a GeneClean kit (Bio 101), and sequenced with "FS" or "Bigdye" fluorescent-labeled DyeDeoxy protocols (Perkin-Elmer) by using the following temperature profile: 25 cycles of 96°C for 15 s and 55°C for 4 min. After purification via gel chromatography through Sephadex G-50 spin columns (SuperFine; Pharmacia), the sequenced products are run on an Applied Biosystems model 377 automated DNA sequencer.
4. Following initial alignment with CLUSTAL W (version 1.60), sequence alignments are manipulated visually with TSE, a DOS text software program (SemWare; Marietta, GA). Unweighted phylogenetic analyses are performed on the individual and combined data sets by using the heuristic search option in PAUP*4.0b1, with 1000 stepwise random addition sequences. The partition-homogeneity test (PHT) implemented with PAUP is used to evaluate the concordance of the 18S and 28S rRNA data sets, by using 1000 replicates with MAXTREES set to 5000. Uninformative characters are excluded from the PHT. A second incongruence test, the Wilcoxon signed-ranks Templeton test, is implemented with PAUP, by using the most parsimonious tree (MPT) and a 70% majority rule bootstrap tree as constraints in a separate analysis. Clade stability is estimated from 1000 bootstrap replications with PAUP and by decay indices calculated with TreeRot.

87.3 CONCLUSION

Among the human pathogenic genera (*Rhizopus*, *Mucor*, *Absidia*, *Rhizomucor*, and *Apophysomyces*) in the family *Mucoraceae*, *Rhizopus* is noted for its ability to cause zygomycosis in humans, with symptoms ranging from sinusitis/rhinocerebral, pulmonary, cutaneous/subcutaneous, gastrointestinal, genitourinary, and disseminated diseases, as well as cystitis, external otitis, and allergic disease. Of the 10 recognized species in the genus *Rhizopus*, 5 (*R. arrhizus*, *R. azygosporus*, *R. microsporus* [consisting of varieties *microsporus*, *oligosporus* and *rhizopodiformis*], *R. schipperae* and *R. stolonifer*) are implicated as opportunistic human pathogens. Due to their close morphological and biochemical similarity, precise identification of *Rhizopus* spp. through macroscopic and microscopic examinations is difficult and time-consuming. By focusing the 18S and 28S rRNA and other gene sequences, nucleic acid amplification procedures such as PCR and sequencing offer a rapid, sensitive, and specific approach for detection and differentiation of *Rhizopus* spp. Future characterization of additional species-specific gene regions will enhance the utility of molecular techniques for diagnosis of zygomycosis, contributing to the effective treatment and control of this potentially fatal disease.

REFERENCES

1. The UniProt Consortium. Available at http://www.uniprot.org/, accessed on August 1, 2010.
2. Ribes JA, Vanover-Sams CL, and Baker DJ. Zygomycetes in human disease. *Clin Microbiol Rev*. 2000 April;13(2):236–301.
3. Eucker J et al. Mucormycoses. *Mycoses*. 2001;44:253–260.
4. Schipper MA et al. Human infection by *Rhizopus azygosporus* and the occurrence of azygospores in zygomycetes. *J Med Vet Mycol*. 1996;34:199–203 [PubMed].

5. Anstead GM et al. Disseminated zygomycosis due to *Rhizopus schipperae* after heatstroke. *J Clin Microbiol.* 1999;37:2656.
6. Cheng VC et al. Outbreak of intestinal infection due to *Rhizopus microsporus*. *J Clin Microbiol.* 2009;47(9):2834–2843.
7. De Hoog GS, Guarro J, Gené J, and Figueras MJ. *Atlas of Clinical Fungi*, 2nd edn., pp. 62–67, 101–111. Centraalbureau voor Schimmelcultures, Utrecht, the Netherlands, 2000.
8. Ma LJ et al. Genomic analysis of the basal lineage fungus *Rhizopus oryzae* reveals a whole-genome duplication. *PLoS Genet.* 2009;5(7):e1000549.
9. St-Germain G et al. Infection due to *Rhizomucor pusillus*: Report of four cases in patients with leukemia and review. *Clin Infect Dis.* 1993;16:640–645.
10. Hagensee ME et al. Brain abscess following marrow transplantation: Experience at the Fred Hutchinson Cancer Research Center, 1984–1992. *Clin Infect Dis.* 1994;19:402–408.
11. Ferguson BJ. Mucormycosis of the nose and paranasal sinuses. *Otolaryngol Clin N Am.* 2000;33:349–365.
12. Grossi P et al. Prevalence and outcome of invasive fungal infections in 1,963 thoracic organ transplant recipients—A multicenter retrospective study. *Transplantation.* 2000;70:112–116.
13. Nosari A et al. Mucormycosis in hematologic malignancies: An emerging fungal infection. *Haematologica.* 2000;85:1068–1071.
14. Kantarcioğlu AS et al. A *Rhizopus oryzae* strain isolated from resected bone and soft tissue specimens from a sinonasal and palatal mucormycosis case. Report of a case and in vitro experiments of yeastlike cell development. *Med Mycol.* 2006;44(6):515–521.
15. Chawla R et al. A rare case of mucormycosis of median sternotomy wound caused by *Rhizopus arrhizus*. *Indian J Med Microbiol.* 2007 October;25(4):419–421.
16. Kontogiorgi M et al. Fatal post-traumatic zygomycosis in an immunocompetent young patient. *J Med Microbiol.* 2007;56:1243–1245.
17. Mohadjer Y, Smith ME, and Akduman L. Mucormycosis endophthalmitis after cataract surgery. *Ocul Immunol Inflamm.* 2007 March–April;15(2):117–120.
18. Raizman NM et al. Mucormycosis of the upper extremity in a patient with alcoholic encephalopathy. *J Hand Surg Am.* 2007 March;32(3):384–388.
19. Romano C et al. Primary subcutaneous zygomycosis due to *Rhizopus oryzae* in a 71-year-old man with normal immune status. *Mycoses.* 2007 January;50(1):82–84.
20. Uçkay I et al. Invasive zygomycosis in transplant recipients. *Clin Transplant.* 2007;21:577–582.
21. Ustun C et al. Early fatal *Rhizopus* infection on voriconazole prophylaxis following allogeneic stem cell transplantation. *Bone Marrow Transplant.* 2007 June;39(12):807–808.
22. Zaoutis TE et al. Zygomycosis in children: A systematic review and analysis of reported cases. *Pediatr Infect Dis J.* 2007;26(8):723–727.
23. Bonifaz A et al. Palatal zygomycosis: Experience of 21 cases. *Oral Dis.* 2008;14(6):569–574.
24. Clauss H and Samuel R. Simultaneous mold infections in an orthotopic heart transplant recipient. *Transpl Infect Dis.* 2008;10(5):343–345.
25. Johnson KE et al. An atypical case of fatal zygomycosis: Simultaneous cutaneous and laryngeal infection in a patient with a non-neutropenic solid prostatic tumor. *Ear Nose Throat J.* 2008;87(3):152.
26. Martín-Moro JG et al. Rhinoorbitocerebral mucormycosis: A case report and literature review. *Med Oral Patol Oral Cir Bucal.* 2008 December 1;13(12):E792–E795.
27. Page AV et al. Primary cutaneous mucormycosis in a lung transplant recipient: Case report and concise review of the literature. *Transpl Infect Dis.* 2008;10(6):419–425.
28. Popa G, Blag C, and Sasca F. Rhinocerebral mucormycosis in a child with acute lymphoblastic leukemia: A case report. *J Pediatr Hematol Oncol.* 2008;30(2):163–165.
29. Song KR, Wong MS, and Yeung C. Primary cutaneous zygomycosis in an immunodeficient infant: A case report and review of the literature. *Ann Plast Surg.* 2008;60(4):433–436.
30. Antoniadou A. Outbreaks of zygomycosis in hospitals. *Clin Microbiol Infect.* 2009;15(5):55–59.
31. de Mol P and Meis JF. Disseminated *Rhizopus microsporus* infection in a patient on oral corticosteroid treatment: A case report. *Neth J Med.* 2009 January;67(1):25–28.
32. Roilides E et al. Zygomycosis in neonates: An uncommon but life-threatening infection. *Am J Perinatol.* 2009 Sep;26(8):565–573.
33. Wilkins RM, Hahn DB, and Blum R. Bread mold osteomyelitis in the femur. *Orthopedics.* 2009;32(5):362.
34. Garg J et al. Nosocomial cutaneous zygomycosis in a patient with diabetic ketoacidosis. *Int J Infect Dis.* 2009;13(6):e508–e510.
35. Lekakis LJ et al. Fatal rhizopus pneumonia in allogeneic stem cell transplant patients despite posaconazole prophylaxis: Two cases and review of the literature. *Biol Blood Marrow Transplant.* 2009;15(8):991–995.
36. Tilak R et al. Cutaneous zygomycosis: A possible postoperative complication in immunocompetent individuals. *Indian J Dermatol Venereol Leprol.* 2009;75:596–599.
37. Monecke S et al. A case of peritonitis caused by *Rhizopus microsporus*. *Mycoses.* 2006;49(2):139–142.
38. Rodríguez MM et al. Posaconazole combined with amphotericin B, an effective therapy for a murine disseminated infection caused by *Rhizopus oryzae*. *Antimicrob Agents Chemother.* 2008 October;52(10):3786–3788.
39. Yassin MA et al. Rhino-orbital zygomycosis secondary to *Rhizopus oryzae* in a renal transplant recipient successfully treated with liposomal amphotericin B. *Chang Gung Med J.* 2008 July–August;31(4):407–411.
40. Kok J et al. Early use of posaconazole in the successful treatment of rhino-orbital mucormycosis caused by *Rhizopus oryzae*. *J Infect.* 2007 September;55(3):e33–e36.
41. Peel T et al. Posaconazole as first line treatment for disseminated zygomycosis. *Mycoses.* 2008;51(6):542–545.
42. Schlemmer F et al. Breakthrough *Rhizopus* infection on posaconazole prophylaxis following allogeneic stem cell transplantation. *Bone Marrow Transplant.* 2008;42(8):551–552.
43. Vyzantiadis TA et al. Rhino-cerebral zygomycosis resistant to antimycotic treatment: A case report. *Mycoses.* 2009;52(1):87–90.
44. Bialek R et al. PCR based identification and discrimination of agents of mucormycosis and aspergillosis in paraffin wax embedded tissue. *J Clin Pathol.* 2005 November;58(11):1180–1184.
45. Machouart M et al. Genetic identification of the main opportunistic Mucorales by PCR-restriction fragment length polymorphism. *J Clin Microbiol.* 2006;44(3):805–810.

46. Schwarz PS et al. Molecular identification of zygomycetes from culture and experimentally infected tissues. *J Clin Microbiol*. 2006;44:340–349.
47. Landlinger C et al. Identification of fungal species by fragment length analysis of the internally transcribed spacer 2 region. *Eur J Clin Microbiol Infect Dis*. 2009;28(6):613–622.
48. Landlinger C et al. Species-specific identification of a wide range of clinically relevant fungal pathogens by use of Luminex xMAP technology. *J Clin Microbiol*. 2009;47:1063–1073.
49. Spiess B et al. DNA microarray-based detection and identification of fungal pathogens in clinical samples from neutropenic patients. *J Clin Microbiol*. 2007;45(11):3743–3753.
50. Hata DJ et al. Real-time PCR method for detection of zygomycetes. *J Clin Microbiol*. 2008;46(7):2353–2358.
51. Nyilasi I et al. High-affinity iron permease (FTR1) gene sequence-based molecular identification of clinically important Zygomycetes. *Clin Microbiol Infect*. 2008;14(4):393–397.
52. Balajee SA et al. Sequence-based identification of *Aspergillus, fusarium*, and mucorales species in the clinical mycology laboratory: Where are we and where should we go from here? *J Clin Microbiol*. 2009 April;47(4):877–884.
53. Dannaoui E et al. Molecular detection and identification of Zygomycetes species from paraffin-embedded tissues in a murine model of disseminated zygomycosis—A collaborative ESCMID Fungal Infection Study Group (EFISG) evaluation. *J Clin Microbiol*. 2010;48:2043.
54. Hofman V et al. Usefulness of molecular biology performed with formaldehyde-fixed paraffin embedded tissue for the diagnosis of combined pulmonary invasive mucormycosis and aspergillosis in an immunocompromised patient. *Diagn Pathol*. 2010;5:1.
55. Woo PC et al. Internal transcribed spacer region sequence heterogeneity in *Rhizopus microsporus*: Implications for molecular diagnosis in clinical microbiology laboratories. *J Clin Microbiol*. 2010;48(1):208–214.
56. Van Burik J-A et al. Panfungal PCR assay for detection of fungal infection in human blood specimens. *J Clin Microbiol*. 1998;36:1169–1175.
57. Voigt K, Cigelnik E, and O'donnell K. Phylogeny and PCR identification of clinically important Zygomycetes based on nuclear ribosomal-DNA sequence data. *J Clin Microbiol*. 1999;37(12):3957–3964.
58. Larché J et al. Diagnosis of cutaneous mucormycosis due to *Rhizopus microsporus* by an innovative PCR-restriction fragment length polymorphism method. *Clin Infect Dis*. 2005;41:1362–1365.
59. Nagao K et al. Genetic identification and detection of human pathogenic *Rhizopus* species, a major mucormycosis agent, by multiplex PCR based on internal transcribed spacer region of rRNA gene. *J Dermatol Sci*. 2005;39:23–31.
60. Lau A et al. Development and clinical application of a panfungal PCR assay to detect and identify fungal DNA in tissue specimens. *J Clin Microbiol*. 2007;45(2):380–385.
61. Francesconi A et al. Automated and manual methods of DNA extraction for *Aspergillus fumigatus* and *Rhizopus oryzae* analyzed by quantitative real-time PCR. *J Clin Microbiol*. 2008;46(6):1978–1984.
62. Kasai M et al. Detection of a molecular biomarker for zygomycetes by quantitative PCR assays of plasma, bronchoalveolar lavage, and lung tissue in a rabbit model of experimental pulmonary zygomycosis. *J Clin Microbiol*. 2008;46(11):3690–3702.

88 Rhizomucor

Tamás Papp, Ildikó Nyilasi, Miklós Takó, László G. Nagy, and Csaba Vágvölgyi

CONTENTS

88.1 INTRODUCTION

Members of the genus *Rhizomucor* are Zygomycetes belonging in the order Mucorales. These saprophytic fungi are ubiquitous in nature and can be isolated from soil and decaying organic materials, worldwide. *Rhizomucor* spp., and especially *R. pusillus* and *R. variabilis*, are also known as agents of zygomycosis, e.g., frequently fatal mycotic diseases in immunocompromised patients (infections caused by mucoralean fungi are often referred to as mucormycosis too).[1] An expanding body of data is becoming available that demonstrates the opportunistic pathogenic potential of these fungi.

88.1.1 Classification, Morphology, and Epidemiology

The genus *Rhizomucor* (Mucoromycotina, Mucorales) comprises *Mucor*-like fungi inasmuch as they have nonapophysate sporangia and branching sporangiophores, but the formation of rhizoids and stolons clearly distinguishes them from *Mucor* species. *Rhizomucor* differs from *Rhizopus* by the absence of apophysis and the presence of branched sporangiophores, and from *Absidia* by the absence of apophysis and the presence of globose sporangia (Figure 88.1). In contrast with *Rhizopus*, *Rhizomucor* develops the rhizoids internodally, e.g., they arise between and not at the base of the sporangiophores (Figure 88.1).

The genus as accepted by Schipper[2] contains three thermophilic species, *R. miehei*, *R. pusillus*, and *R. tauricus*. As these species were originally classified in the genus *Mucor*, some authors still cite *R. miehei* and *R. pusillus* as *Mucor miehei* and *M. pusillus*, but these designations are obsolete.

R. tauricus had been isolated only once. On the basis of carbon source utilization studies and isoenzyme and ITS-RFLP analyses, Vágvölgyi et al.[3] questioned the validity of this species and suggested that the single original isolate might well be a mutant heterothallic *R. pusillus* strain. Later, two additional thermophilic species, *Rhizomucor nainitalensis*[4] and *Rhizomucor pakistanicus*,[5] were described, but they are now regarded as synonyms of *R. miehei* and *R. pusillus*, respectively.[6]

Two mesophilic species were also described as new members of the *Rhizomucor* genus: *Rhizomucor endophyticus*, an endophyte of wheat,[7] and *Rhizomucor variabilis* (with two varieties, *R. variabilis* var. *regularior* and *R. variabilis* var. *variabilis*), a causative agent of cutaneous zygomycoses in humans.[8–10] Molecular phylogenies inferred from ribosomal RNA (rRNA) gene sequences[11] and from protein coding sequences of the actin and the translation elongation factor 1α (*tef*) genes[12] demonstrated that the thermophilic *R. miehei* and *R. pusillus* are closely related to each other, while the mesophilic *R. variabilis* proved to be phylogenetically related to *Mucor* species, e.g., to *M. hiemalis* and *M. mucedo*. In a recent phylogenetic study based on gene sequences of the rRNA cluster, two strains identified previously as *M. hiemalis* f. *hiemalis* displayed 97% sequence identity with the examined *R. variabilis* isolates, and *R. variabilis* has formed a well-supported sister group with *M. hiemalis* f. *hiemalis*.[13]

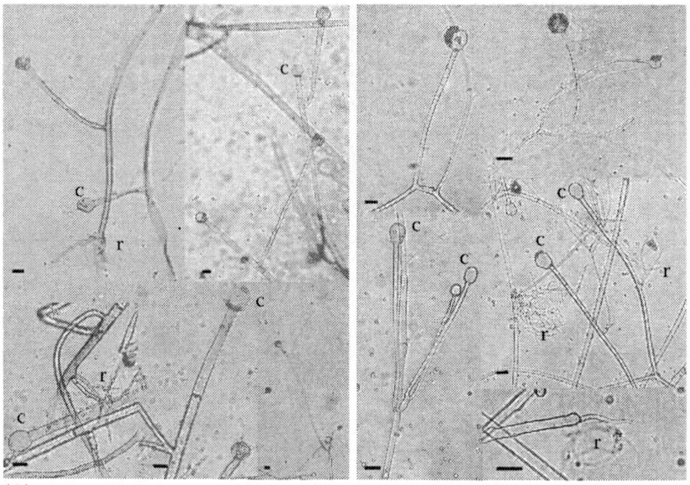

FIGURE 88.1 Light microscopic images of *Rhizomucor pusillus* WRL CN(M)231 (A) and *Rhizomucor miehei* NRRL 5282 (B). Strains were cultured on MEA for 4 days at 36°C. Scale bars indicate 20 μm in each panels. C, columella; r, rhizoid.

TABLE 88.1
Differential Characteristics of *Rhizomucor* Species[a]

Species	*R. miehei*	*R. pusillus*	*R. variabilis*
Max. growth temperature	57°C	55°C	38°C
Colony			
Color	Gray to deep-olive-gray	Gray to hair-brown	Whitish to ochraceous
Height	1 μm	1–2 μm	2 μm
Rhizoids	May be present	May be present	Abundant
Branching of the sporangiophores	Sympodial	Monopodial and sympodial	Simply or once branched
Size of sporangia	50–75 μm in diameter	80–100 μm in diameter	Up to 120 μm in diameter
Columellae	28 × 25 μm, obovoid	45 × 38 μm, obovoid	40 μm in diameter, spherical
Sporangiospores	3–4 μm, subglobose	3–4 μm, subglobose	3–11 μm, subspherical to ellipsoidal
Chlamydospores	Not produced	Not produced	Abundant
Zygospores	Up to 50 μm in diameter	Up to 75 μm in diameter	Unknown
Thiamine dependence of growth	Dependent	Independent	Unknown

Sources: Based on the descriptions of Schipper, M.A.A., *Stud. Mycol.*, 17, 53, 1978; de Hoog, D.S. et al., *Atlas of Clinical Fungi*, 2nd edn., Centraalbureau voor Schimmelcultures, Utrecht, the Netherlands, 2000; Ribes, J.A. et al., *Clin. Microbiol. Rev.*, 13, 236, 2000; Wang, A. et al., *Chin. J. Dermatol.*, 27, 51, 1994 (Chinese); Lu, X. et al., *Clin. Infect. Dis.*, 49, e39, 2009.

Carbon assimilation profiles indicated that *R. variabilis* and *R. variabilis* var. *regularior* are much closer to *Mucor* species than to *Rhizomucor* species.[14] A personal communication from Schipper led de Hoog et al.[15] to postulate whether *R. variabilis* is a degenerate culture of *M. hiemalis*. Although the molecular phylogeny of *R. endophyticus* has not been studied in detail to date, it is considered to be a synonym of *Mucor genevensis*.[16,17] All the results so far suggest that the genus *Rhizomucor* consists of only two thermophilic species, *R. miehei* and *R. pusillus*, while the "mesophilic species" are actually not valid members of this taxon.

Table 88.1 summarizes the main morphological and growth characteristics of *R. miehei*, *R. pusillus*, and *R. variabilis*. *R. miehei* and *R. pusillus* can be easily distinguished from *R. variabilis* on the basis of their maximum growth temperatures, the abundance of rhizoids and the presence or absence of chlamydospores. *R. miehei* and *R. pusillus* are highly similar to each other, but the colonies of *R. miehei* are generally more darkly pigmented than those of *R. pusillus*. Moreover, the latter produces somewhat less branched sporangiophores and smaller sporangia and zygospores than *R. pusillus*. *R. miehei* is homothallic, while *R. pusillus* is mainly heterothallic with rarely occurring homothallic isolates.

R. miehei and *R. pusillus* are cosmopolitan fungi that have been reported from Europe, North America, Asia, and Africa[1,18]; they can readily be isolated from soil and decaying or composting organic matter and wastes.[1] Both species are known as pathogens in various animals, and they are among the most frequently isolated causative agents of bovine mycotic abortion and bovine mastitis.[19] At the same time, *R. pusillus* is a rare human pathogen; it may cause opportunistic infections in patients with a challenged or weakened immune system. Human infection caused by *R. miehei* has not been described[1]; this fungus is rather of substantial biotechnological importance as an excellent producer of proteases and lipases.

R. variabilis causes primary cutaneous zygomycosis in humans; it is able to infect immunocompetent persons. Although our knowledge concerning the distribution of *R. variabilis* is limited, this fungus seems to have the distinct geographical feature that it has been isolated almost exclusively in the central and eastern regions of China, raising the suggestion that *R. variabilis* may be an endemic pathogen.[10]

88.1.2 Clinical Features and Pathogenesis

88.1.2.1 *Rhizomucor pusillus*

Ribes et al.[1] reported that 19 cases of human zygomycosis caused by *R. pusillus* had been described in the literature up to the year 2000. The majority of those cases involved pulmonary and disseminated zygomycosis, but primary cutaneous infection, rhinofacial disease, and mycotic endocarditis had also been described. In the past 10 years, primary pulmonary infections,[20,21] rhinoorbital and rhinocerebral disease,[22–24] and posttraumatic soft tissue and bone infection[23] have been reported. The transmission of *R. pusillus* generally occurs by the inhalation of sporangiospores[1] and disseminated cases most frequently arise from primary pulmonary infections. Several cases have been reported as results of nosocomial infections.[25–28] Most of the human infections have been associated with leukemia and profound neutropenia,[1,24] but uncontrolled diabetes, immunosuppressive therapy, and treatment with steroids have also been implicated as predisposing factors.[1,23,29] For the successful treatment of *R. pusillus* infections, amphotericin B or its lipid formulations, often in combination with surgical intervention, have been used most frequently. The *in vitro* susceptibilities of Zygomycetes were examined by Dannaoui et al.[30] with the involvement of *R. pusillus* and *R. miehei* strains; besides amphotericin B, these *Rhizomucor* spp. proved to be sensitive to itraconazole and posaconazole with a range of MICs of 0.03–0.25 and 0.06–0.25 μg/mL, respectively.

Although no cases of human disease caused by *R. miehei* have yet been noted, information available from animal mycoses and tests in animal models suggest a potentially similar pathogenic behavior as that reported for *R. pusillus*.[1,31]

88.1.2.2 *Rhizomucor variabilis*

All cases of *R. variabilis* infections reported in the literature have involved primary cutaneous zygomycosis.[8–10,32–35] Cutaneous infections caused by other Mucorales species generally have an aggressively invasive feature and primarily occur in immunocompromised patients. In contrast, *R. variabilis* causes superficial or subcutaneous infections, which generally remain localized for several years without dissemination or involvement of the adjacent tissues and the blood vessels.[10] All infections reported so far have affected immunocompetent persons, mainly agricultural workers living in underdeveloped areas. From these facts, together with the strict geographical distribution of the known cases, Lu et al.[10] assumed that environmental exposure after trauma or surgery was the main risk factor for these infections. In the few susceptibility tests performed, *R. variabilis* exhibited sensitivity to amphotericin B, but it proved to be resistant to other tested antifungal agents (such as azoles, terbinafine, flucitozine, and caspofungin), similarly to many other Zygomycetes.[10,35] Accordingly, in the majority of cases, amphotericin B therapy was used to treat the disease.

88.1.3 Diagnosis

88.1.3.1 Conventional Techniques

Although noteworthy advances in the diagnosis of zygomycoses have been achieved in the past decade, many infections are still classified only as zygomycosis or mucormycosis, without species- or at least genus-level identification.[36–38] Histopathological examination is currently regarded as the definitive diagnosis.[37,39,40] Apart from the routine hematoxylin and eosin staining, hyphal elements can be examined by staining with periodic acid-Schiff, Gomori methenamine silver, or Calcoflour White stain.[25,40]

In tissues, *R. pusillus* presents the typical hyaline, ribbon-like coenocytic hyphae characteristic of Mucorales fungi that are often associated with necrosis and angioinvasion.[1] *R. variabilis* appears in histopathological samples, as similar broad, ribbon-like, coenocytic hyphae with right-angled branches.[10,35]

Culturing of isolates from tissue samples is required for the genus- and species-level identification of *Rhizomucor*. However, this is often difficult, because hyphal elements may be scarce in the specimens and they can lose their viability during sample preparation for histopathology. If the pathogen has been successfully isolated, it is important to maintain it for further analysis; strain maintenance methods are well established for Zygomycetes.

Their characteristic morphological features, such as the presence of stolons, rhizoids, and repeatedly branched sporangiophores clearly distinguish *Rhizomucor* from most members of the Mucorales. Nonetheless, misidentifications frequently occur and morphology-based identifications disagree with the results of DNA-based methods in some cases (particularly in certain earlier studies).[41] The detection of zygospore formation in mating studies has been successfully used for the identification of *Rhizomucor* species.[42] However, this method requires a set of testing strains, and it is therefore primarily applicable by reference laboratories.[24]

Based on their thermophily and some morphological characteristics mentioned above (Table 88.1), true *Rhizomucor* species (i.e., *R. pusillus* and *R. miehei*) can be distinguished from *R. variabilis* by conventional methods. At the same time, discrimination between *R. pusillus* and *R. miehei* is often problematic, frequently resulting in misidentifications, or the species names remain to be determined. In earlier studies, determination of the number of nuclei in the sporangiospores[43] or measurement of the diameter of the zygospores (<50 and >50 μm for *R. miehei* and *R. pusillus*, respectively)[2,42] was of some value for species delimitation, but the results were not clear-cut in most cases.

Carbon assimilation tests can also be carried out to differentiate these species. In an earlier study, Vastag et al.[44] established differences between the two species in the utilization of sucrose, glycine, phenylalanine, and β-alanine as sole carbon sources. Recently, Schwarz et al.[14] used the commercialized and standardized kits ID32C and API 50 CH (bioMérieux) to evaluate carbon assimilation profiles as a tool for the identification of clinically important Zygomycetes. These tests not only clearly distinguished *R. variabilis* from *R. miehei* and *R. pusillus*; but also differentiated between the latter species. Thiamine dependence of growth can likewise be a differentiating character: the growth of *R. miehei* is stimulated by thiamine, whereas *R. pusillus* does not require this compound for growth.[28] Lukács et al.[45] described a simple and reliable method for the accurate differentiation of the two species, based on the observation that *R. pusillus* is much more sensitive than *R. miehei* to lovastatin. A limitation of all these methods is that their reliable evaluation requires the maintenance and application of appropriate reference strains.

88.1.3.2 Molecular Techniques

The first attempts to use molecular methods to identify and characterize *Rhizomucor* species involved isoenzyme analysis,[44] random amplified polymorphic DNA (RAPD) PCR,[46,47] and PCR-restriction fragment length polymorphism (PCR-RFLP).[3] RAPD analysis is a rapid method with which to study intra- and interspecies genetic variability without a prior knowledge of DNA sequence data. In this technique, decamer random oligonucleotide primers are used to amplify DNA, generating characteristic amplicon patterns. The method allows the identification of specific amplicons differentiating between *R. pusillus* and *R. miehei*. PCR-RFLP detects specific restriction fragment patterns obtained by the digestion of PCR-amplified fragments and may be a useful identification method for fungi, especially in laboratories that do not have access to instrumentation for the sequencing and analysis of the PCR products.[48] PCR-RFLP of the internal transcribed spacer (ITS) region was found to be appropriately efficient to distinguish *R. miehei* and *R. pusillus* and could be used to clarify the taxonomic position of *R. tauricus*.[3]

The majority of recent molecular identification methods developed for clinically important Zygomycetes are based on the PCR amplification of nuclear ribosomal RNA gene sequences (rDNA). The first comprehensive DNA sequence database for the identification of clinically important Zygomycetes was constructed by Voigt et al.[11] who determined the sequences of the small-subunit (18S) rDNA and domains D1 and D2 of the nuclear large-subunit (LSU) (28S) rDNA from 42 Zygomycetes. They also designed 13 taxon-specific primer pairs on the basis of the 28S rDNA to identify the medically most relevant Zygomycetes. The *Rhizomucor*-specific primer pair described identified *R. pusillus* and *R. miehei* at a genus level, but it was unable to distinguish them from each other. The commercially available molecular identification system, the MicroSeq fungal sequencing kit (Applied Biosystems), also uses the D2 LSU rDNA.[48] In an evaluation of this method, the one *R. miehei* isolate involved could be identified correctly.[49] However, a satisfactory determination of the suitability of the method for the identification of *Rhizomucor* demands the inclusion of further *R. miehei* and *R. pusillus* isolates into the evaluation.

Recent studies have demonstrated that the comparison of ITS sequences is a reliable tool for the precise species-level identification of this fungal group.[48,50] Hsiao et al.[51] developed an oligonucleotide array method that involves hybridization with species- or group-specific probes after PCR amplification of the ITS region, using universal primers to identify 64 clinically important fungal species. Unfortunately, the oligonucleotide probe designed for the delimitation of *R. pusillus* also identified *R. miehei*. Later, Schwarz et al.[52] compared the ITS regions of 54 strains belonging in 16 species to examine the value of ITS sequencing for the molecular identification of medically important Zygomycetes: all three species (*R. pusillus*, *R. miehei*, and *R. variabilis*) were easily distinguished via their ITS sequences. Moreover, in the case of *R. pusillus*, it was demonstrated that DNA extraction, PCR amplification, sequencing, and identification is possible directly from the tissues of experimentally infected mice. Indeed, in clinical practice, the determination and analysis of the ITS region was successfully utilized to identify *R. pusillus* causing a sinus–orbital zygomycosis in a patient with acute myelogenous leukemia,[24] and *R. variabilis* as the agent of primary cutaneous zygomycosis in an immunocompetent host.[35] In both cases, the fungal agent was isolated and cultured prior to the DNA extraction and molecular identification.

Sequences of some protein-coding genes, primarily of the actin and the translation elongation factor-1 alpha (EF-1α) genes, have been determined from numerous Zygomycetes and used to reveal phylogenies of Mucorales.[12,53] *Rhizomucor* species were involved in these studies and the corresponding sequences are available from nucleotide sequence databases (e.g., EMBL and GenBank); however, their usefulness for species identification has not been analyzed in detail. In a recent study, Nyilasi et al.[54] have described an identification method based on PCR amplification and sequencing of the high-affinity iron transporter gene (*FTR1*) for several Zygomycetes (primarily for *Rhizopus* species). Although primer pairs discriminating the thermophilic *Rhizomucor* species from other Zygomycetes were established, motifs of value for the design of primers that clearly resolve *R. miehei* and *R. pusillus*, were not found in the *FTR1* sequences of these two species.

88.2 METHODS

88.2.1 Sample Preparation

DNA extraction and molecular identification directly from clinical specimens are poorly documented for *Rhizomucor* species. Iwen et al.[24] extracted DNA for PCR amplification from sinus tissue according to the procedure originally described by Henry et al.[55] for the identification of *Aspergillus* species. Schwartz et al.[52] demonstrated that DNA extraction and PCR can be performed directly from homogenized animal tissues. However, in most of the published cases, isolation and culturing of the fungal agent have preceded sample preparation for molecular diagnosis.

88.2.1.1 Media and Culturing Conditions

Similarly to other Mucorales fungi, *Rhizomucor* species can be cultured with ease on several commonly used media, such as on malt extract agar or potato dextrose agar or in RPMI 1640, which is often chosen by clinical laboratories. Although Sabouraud agar is among the less recommendable media for the cultivation of Zygomycetes,[31] it has been successfully applied to culture *R. pusillus* and *R. variabilis* for molecular identification.[24,35]

To obtain sporangiospores for inoculum preparation, *malt extract agar* [3,45,46] (MEA; 5 g malt extract, 5 g yeast extract, 5 g glucose, 10 g KH_2PO_4), *yeast extract-glucose medium* (YEG; 5 g yeast extract, 20 g glucose) and *potato dextrose agar*[24] (PDA, PDB; 50% potato filtrate obtained by boiling 300 g diced potato in 500 mL water and filtering, 20 g sucrose) can be used (all media are given for 1 L and contain 15 g agar). Spore suspensions are prepared in 0.9% NaCl.

Genomic DNA can be extracted from mycelia grown either on solid or in liquid media. Mycelia are generally grown for 24–72 h; the appropriate cultivation temperatures are 36°C for *R. pusillus* and *R. miehei* and 24°C for *R. variabilis* and *R. variabilis* var. *regularior*. For DNA extraction, Schwarz et al.[52] applied 50 mg mycelium obtained from cultures in 20 mL RPMI 1640 with L-glutamine and without sodium bicarbonate (buffered to pH 7 with 0.165 M MOPS) incubated for 48–72 h with agitation. In several studies,[3,44,46,54] *R. pusillus* and *R. miehei* were cultured in yeast extract–glucose liquid medium with agitation for 72 h on a rotary shaker at 200 rpm. Zhao et al.[35] cultured *R. variabilis* on Sabouraud glucose agar at 25°C for 7 days before extraction was performed.

88.2.1.2 Purification of Genomic DNA

Several well-established methods of genomic DNA extraction are known for Zygomycetes including *Rhizomucor*, such as the CTAB (hexacetyltrimethylammonium bromide) protocol described by Voigt et al.[11] Briefly, approximately 50 mg lyophilized mycelium is pulverized and resuspended in 700 μL CTAB buffer (100 mM Tris–Cl, pH 8.4; 1.4 M NaCl; 25 mM EDTA; 2% CTAB) and vortexed for 10 s. After that, an equal volume of chloroform is added to the samples, they are vortexed for 5 s, and spun for 10 min at 12,300 × *g* in a microcentrifuge. A 500-μL portion of the upper phase is removed to a new microcentrifuge tube, and DNA is precipitated with an equal volume of 220°C isopropanol. After the DNA is pelleted by centrifugation and gently washed with 70% ethanol, it is resuspended in 200 μL TE buffer (10 mM Tris–Cl, pH 8.0; 1 mM EDTA, pH 8.0) or sterile distilled water.

Another DNA purification method used for Zygomycetes by Nyilasi et al.[54] is as follows. Mycelium (500 mg) is disrupted with pestle and mortar in liquid nitrogen. After 1.25 mL lysis buffer (50 mM Tris–HCl pH 8.0; 20 mM EDTA; 1% *N*-lauroylsarcosine) has been added to the mycelial powder, it is allowed to thaw to room temperature and mixed gently until a homogenous slurry is obtained. This slurry is next incubated for 15 min at 65°C and centrifuged (at 5000 × *g* for 10 min, at 4°C) to pellet the cell debris. The supernatant is transferred to another centrifuge tube and extracted twice with a 25:24:1 mixture of phenol:chloroform:isoamyl alcohol and once with chloroform:isoamyl alcohol (24:1). The DNA is then precipitated from the upper aqueous phase with 2.5 volumes of ethanol at −20°C. After centrifugation (at 10,000 × *g* for 10 min), the resulting pellet is dried under vacuum, and resuspended in 100–200 μL sterile distilled water.

Alternatively, commercial DNA purification kits can be used for rapid sample preparation. Two examples of commercial kits tested for *Rhizomucor* species: the MasterPure™ Yeast DNA Purification Kit (Epicenter), which has been tested for Zygomycetes (in this case, the procedure starts with 20–50 mg mycelium and finally DNA is redissolved in 30–50 mL water), and the DNA QIAamp Mini DNA Extraction Kit (Qiagen).[35]

88.2.2 Detection Procedures

Sequence analysis of the ITS region seems to be the best choice for the species-level identification of *Rhizomucor*. Universal primer pairs designed for amplification of the complete or partial ITS1-5.8S-ITS2 region work well for *Rhizomucor* species.[3,24,35,51,52] Composition of the reaction mixture and the PCR conditions are well documented [52,56,57] and they may vary, depending on the applied primers and the recommendations of the supplier of the polymerase enzyme.

The primers ITS1 (5′-TCCGTAGGTGAACCTGCGG-3′) and ITS4 (5′-TCCTCCGCTTATTGATATGC-3′) described by White et al.[56] have been used in several experiments.[3,22] To amplify the ITS region, the recommended reaction mixture contains 1 U Taq polymerase, 1 × Taq buffer (according to the description of the supplier), 2.5 mM $MgCl_2$, 400 μM of each dNTPs, 0.16 μM of each primer, and 20 ng genomic DNA as template. Amplification conditions involve an initial denaturing step of 5 min at 95°C, followed by 35 cycles at 95°C for 30 s, 48°C for 30 s, and 72°C for 1 min; the final cycle is followed by an extension step at 72°C for 7 min.

Schwarz et al.[52] used the universal fungal primers V9D (5′-TTAAGTCCCTGCCCTTTGTA-3′) and LS266 (5′-GCATTCCCAAACAACTCGACTC-3′).[57] The reaction mixtures contained 5 μL genomic DNA, 2.5 μL of 20 μM concentrations of each primer, 10 μL of each dNTPs (each at 2.5 mM), 10 μL of 25 mM $MgCl_2$, 6.25 U of Ampli*Taq* polymerase (Roche), and 10 μL of 10 × PCR buffer (Roche); the final volume was 100 μL. The cycling parameters for the PCR were initial denaturation for 10 min at 94°C, followed by 30 cycles of denaturation for 30 s at 94°C, annealing for 30 s at 58°C, and elongation for 30 s at 72°C, with a final extension step for

10 min at 72°C. For sequencing, PCR products were purified on P100 gel fine (Bio-Rad). Purified amplification products were sequenced in both directions with the primers V9D and LS266.

The results of sequencing can be compared with nucleotide sequences deposited in international databases (GenBank, EMBL, and DDBJ) by using BLAST[58]; several ITS sequences for *R. miehei*, *R. pusillus*, and *R. variabilis* are available. Schwarz et al.[52] determined the ITS sequences of the type or the neotype strains of these three species, including *R. variabilis* var. *regularior*, which can be used as references during the identification (accession numbers are DQ118995, DQ119005, DQ119006, and DQ119007 for *R. miehei*, *R. pusillus*, *R. variabilis*, and *R. variabilis* var. *regularior*, respectively).[35]

Nyilasi et al.[54] developed a method based on the amplification of a fragment of the *FTR1* gene to detect clinically important Zygomycetes, which is able to distinguish the thermophilic *Rhizomucor* species from other fungi. An advantage of this rapid procedure is that it does not require further sequencing and analysis; identification occurs only via gel electrophoresis of the PCR products through the detection of DNA bands with characteristic sizes. Primers R1 (5′-GGAAACCGATGCYTTGCA-3′) and R2 (5′-CRTCACCRCCTTCTTCGGC-3′) were designed to detect a 432- and a 441-bp amplicon for *R. miehei* and *R. pusillus*, respectively. The reaction mixture contained 1.25 U of Pfu polymerase (Fermentas), 2.5 μL of 10× reaction buffer, 2.5 mM $MgSO_4$, 400 μM each dATP, dCTP, dGTP and dTTP (Fermentas), 2 μM primers, and 20 ng of genomic DNA. The amplification conditions comprised 94°C for 3 min, followed by 35 cycles of 94°C for 30 s, 68°C for 30 s, and 72°C for 1 min, with a final extension at 75°C for 10 min. Correct identification requires the application of reference strains for comparison.

88.3 CONCLUSIONS AND FUTURE PERSPECTIVES

There is an increasing need for improved diagnostic tools available to clinical microbiology to cope with the escalating problems of emerging pathogenic fungi, which are often difficult to culture and/or differentiate. For decades, most of the Zygomycetes infections were identified only as zygomycosis or mucormycosis, without further species or at least genus determination. The main cause of this practice was that histopathological examination of the clinical specimens was able only to prove the presence of the Zygomycete and not to identify the fungus. Morphological examination of the cultured isolates has frequently led to misidentifications, especially at the species level. The possibility to analyze molecular sequence data obtained with PCR techniques has radically changed this situation and improved the reliability of the identifications. Especially in the past 10 years, promising progress has been made in the molecular diagnosis of clinically important Zygomycetes, and the number of sequences available in public databases (e.g., GenBank or EMBL) has increased rapidly. However, it is important to mention that comparison of the determined sequence with the data deposited in public databases may not bend to a correct identification, because of the occasionally false annotation of the deposited sequences. Such problems can be avoided by using sequences of safely identified reference strains.[13]

The main criterion of a marker useful for nucleotide sequence-based identification is that it should be diverse enough to distinguish single species. For *Rhizomucor*, the ribosomal ITS region merits this condition, and in the few cases described to date,[24,35,52] this technique, complemented with histopathology and culturing of the fungus, has been effective in identifying *Rhizomucor* species. However, the development of new PCR protocols involving the use of taxon- or species-specific primer pairs that generate amplicons with sizes characteristic of the different species would make easier and faster the identification process, allowing the detection of the fungus without sequencing of the amplification product. The method described by Nyilasi et al.[54] for the detection of amplification products from the *FTR1* gene can be considered a first step in this direction.

Besides identification of the pathogenic agent, the usage of species- and strain-specific PCR-based methods may reveal important data as concerns the epidemiology of these fungi. Epidemiological investigations can be useful for confirmation of the possible nosocomial spread of *R. pusillus*.[1,25] Such studies may also be of help to prove the hypothesis that *R. variabilis* is an agent of endemic mycoses and to examine the suggestion that environmental exposure after trauma or surgery could be the main risk factor for such disease.[10] These questions can be answered if typing methods with sufficient discriminative capacity are available.

Finally, molecular studies conducted in order to develop identification methods and infer phylogenies may have taxonomic implications. The genus *Rhizomucor* contains three animal and/or human pathogenic species, the thermophilic *R. miehei* and *R. pusillus* and the mesophilic *R. variabilis*. In spite of the morphological similarities, molecular phylogenetic studies have revealed that the genus is polyphyletic and *R. variabilis* is related to *Mucor* species. Although this finding is well supported by molecular, biochemical, physiological, and pathological evidence, taxonomic revision of the genus and the species *R. variabilis* is called for. Broader comparative morphological, physiological, and molecular studies with the involvement of new isolates and the determination of new molecular markers may help to resolve this situation.

ACKNOWLEDGMENT

This work was supported by a grant of the Hungarian Scientific Research Fund (OTKA CK 80188).

REFERENCES

1. Ribes, J.A., Vanover-Sams, C.L., and Baker, D.J., Zygomycetes in human disease, *Clin. Microbiol. Rev.*, 13, 236, 2000.
2. Schipper, M.A.A., On the genera *Rhizomucor* and *Parasitella*, *Stud. Mycol.*, 17, 53, 1978.

3. Vágvölgyi, C. et al., *Rhizomucor tauricus*: A questionable species of the genus, *Mycol. Res.*, 103, 1318, 1999.
4. Joshi, M.C., A new species of *Rhizomucor* from India, *Sydowia*, 35, 100, 1982.
5. Qureshi, M.S.A. and Mirza, J.H., *Rhizomucor pakistanicus* Qureshi & Mirza sp. nov, *Biologia*, 29, 343, 1983.
6. Mouchacca, J., Thermophilic fungi and applied research: A synopsis of name changes and synonymies, *World J. Microbiol. Biotechnol.*, 16, 881, 2000.
7. Zheng, R. and Jiang, H., *Rhizomucor endophyticus* sp. nov., an endophytic Zygomycetes from higher plants, *Mycotaxon*, 56, 455, 1995.
8. Zheng, R. and Chen, G., A non-thermophilic *Rhizomucor* causing human primary cutaneous mucormycosis, *Mycosystema*, 4, 45, 1991.
9. Zheng, R. and Chen, G., Another non-thermophilic *Rhizomucor* causing human primary cutaneous mucormycosis, *Mycosystema*, 6, 1, 1993.
10. Lu, X. et al., Primary cutaneous zygomycosis caused by *Rhizomucor variabilis*: A new endemic zygomycosis? A case report and review of 6 cases reported from China, *Clin. Infect. Dis.*, 49, e39, 2009.
11. Voigt, K., Cigelnik, E., and O'Donnell, K., Phylogeny and PCR identification of clinically important Zygomycetes based on nuclear ribosomal-DNA sequence data, *J. Clin. Microbiol.*, 37, 3957, 1999.
12. Voigt, K. and Wöstemeyer, J., Phylogeny and origin of 82 zygomycetes from all 54 genera of the Mucorales and Mortierellales based on combined analysis of actin and translation elongation factor EF-1 alpha genes, *Gene*, 270, 113, 2001.
13. Hoffmann, K. et al., Diversity, genotypic identification, ultrastructural and phylogenetic characterization of zygomycetes from different ecological habitats and climatic regions: Limitations and utility of nuclear ribosomal DNA barcode markers, in *Current Advances in Molecular Mycology*, Gherbawy, Y., Mach, R.L., and Rai, M., Eds., Nova Science Publishers, New York, 2009.
14. Schwarz, P. et al., Carbon assimilation profiles as a tool for identification of Zygomycetes, *J. Clin. Microbiol.*, 45, 1433, 2007.
15. de Hoog, D.S. et al., *Atlas of Clinical Fungi*, 2nd edn., Centraalbureau voor Schimmelcultures, Utrecht, the Netherlands, 2000.
16. Crous, P.W. et al., MycoBank: An online initiative to launch mycology into the 21st century, *Stud. Mycol.*, 50, 19, 2004.
17. Robert, V., Stegehuis, G., and Stalpers, J., The MycoBank engine and related databases, http://www.mycobank.org, 2005.
18. Domsch, K.H., Gams, W., and Anderson, T., *Compendium of Soil Fungi*, vol. 1, Academic Press, London, U.K., 1980.
19. Scholer, H.J., Muller, E., and Schipper, M.A.A., Mucorales, in *Fungi Pathogenic for Humans and Animals*, Howard, D.H., Ed., pp. 9–59, Marcel Dekker, New York, 1983.
20. Ma, B. et al., Cure of pulmonary *Rhizomucor pusillus* infection in a patient with hairy-cell leukemia: Role of liposomal amphotericin B and GM-CSF, *Leuk. Lymph.* 42, 6, 1393, 2001.
21. Björkholm, M. et al., Liposomal amphotericin B and surgery in the successful treatment of invasive pulmonary mucormycosis in a patient with acute T-lymphoblastic leukemia, *Scand. J. Inf. Dis.*, 33, 316, 2001.
22. Chakrabarti, A. et al., Ten years' experience in zygomycosis at a tertiary care centre in India, *Infection*, 42, 261, 2001.
23. Abbas, A.A.H., Fryer, C.J.H., and Felimban, S.K., Fatal rhinocerebral mucormycosis in a leukaemic child presenting with sudden loss of vision: A case report, *Haema*, 5, 341, 2002.
24. Iwen, P.C. et al., Molecular identification of *Rhizomucor pusillus* as a cause of sinus-orbital zygomycosis in a patient with acute myelogenous leukemia, *J. Clin. Microbiol.*, 43, 5819, 2005.
25. Pasticci, M.B. et al., Nosocomial post-traumatic *Rhizomucor* soft tissue and bone infection treated with liposomal amphotericin B: Case report, *J. Orthopaed. Traumatol.*, 8, 82, 2007.
26. del Palacio-Hernanz, A. et al., Nosocomial infection by *Rhizomucor pusillus* in a clinical haematology unit, *J. Hosp. Infect.*, 4, 45, 1983.
27. Severo, L.C., Job, F., and Mattos, T.C., Systemic zygomycosis: Nosocomial infection by *Rhizomucor pusillus*, *Mycopathologia*, 113, 79, 1991.
28. St. Germaïn, G. et al., Infection due to *Rhizomucor pusillus*: Report of four cases in patients with leukemia and review, *Clin. Infect. Dis.*, 16, 640, 1993.
29. Wickline, C.L., Cornitus, T.G., and Butler, T., Cellulitis caused by *Rhizomucor pusillus* in a diabetic patient receiving continuous insulin infusion pump therapy, *South Med. J.*, 82, 1432, 1989.
30. Dannaoui, E. et al., In vitro susceptibilities of zygomycetes to conventional and new antifungals, *J. Antimicrob. Chem.*, 51, 45, 2003.
31. Schipper, M.A.A. and Stalpers, J.A., Zygomycetes: The Order Mucorales, in: *Pathogenic Fungi in Humans and Animals*, 2nd edn., Howard, D.H., Ed., pp. 67–125, Mycology Series, Vol. 16, Marcel Dekker, New York, 2002.
32. Wang, A. et al., A case report of destructive mucosal and cutaneous mucormycosis caused by *Rhizomucor variabilis*, *Chin. J. Dermatol.* 27, 51, 1994 (Chinese).
33. Li, C., Li, Y., and Hu, Z., A case report of primary cutaneous mucormycosis caused by *Rhizomucor variabilis*, *J. Clin. Dermatol.*, 33, 158, 2004 (Chinese).
34. Fu, M. et al., Facial Mucormycosis due to *Rhizomucor variabilis*: A case report, *Int. J. Dermatol. Venereol.*, 32, 67, 2006 (Chinese).
35. Zhao, Y. et al., Primary cutaneous mucormycosis caused by *Rhizomucor variabilis* in an immunocompetent patient, *Mycopathologia*, 168, 243, 2009.
36. Eucker, J. et al., Mucormycoses, *Mycoses*, 44, 253, 2001.
37. Freifeld, A.G. and Iwen, P.C. Zygomycosis, *Semin. Respir. Crit. Care Med.* 25, 221, 2004.
38. Papp, T. et al., Molecular studies on Zygomycetes fungi causing opportunistic infections, *Rev. Med. Microbiol.*, 19, 47, 2008.
39. Walsh, T.J. et al., Infections due to emerging and uncommon medically important fungal pathogens, *Clin. Microbiol. Infect.*, 10(Suppl. 1), 48, 2004.
40. Chayakulkeeree, M., Ghannoum, M.A., and Perfect, J.R., Zygomycosis: The re-emerging fungal infection, *Eur. J. Clin. Microbiol. Infect. Dis.*, 25, 215, 2006.
41. Kontoyiannis, D.P. et al., Zygomycosis in a tertiary-care cancer center in the era of *Aspergillus* active antifungal therapy: A case-control observational study of 27 recent cases, *J. Infect. Dis.*, 191, 350, 2005.
42. Weitzman, I. et al., Zygospores: The last word in identification of rare or atypical *Zygomycetes* isolated from clinical specimens, *J. Clin. Microbiol.*, 33, 781, 1995.
43. Tansey, M.R., Kamel, S.M., and Shamsai, R., The number of nuclei in sporangiospores of *Rhizomucor* species: Taxonomic and biological significance, *Mycologia*, 76, 1089, 1984.

44. Vastag, M. et al., Differentiation of *Rhizomucor* species by means of carbon source utilization and isoenzyme analysis, *J. Clin. Microbiol.*, 36, 2153, 1998.
45. Lukács, Gy. et al., Differentiation of *Rhizomucor* species on the basis of their different sensitivities to lovastatin, *J. Clin. Microbiol.*, 42, 5400, 2004.
46. Vastag, M. et al., Intraspecific variation in two species of *Rhizomucor* assessed by random amplified polymorphic DNA analysis, *J. Basic Microbiol.*, 40, 269, 2000.
47. Papp, T. et al., Identification of *Rhizomucor* species by means of biochemical and molecular methods, *Clin. Microbiol. Infect.* 10(Suppl. 3), 508, 2004.
48. Balajee, S.A., Sigler, L., and Brandt, M.E., DNA and the classical way: Identification of medically important molds in the 21st century, *Med. Mycol.*, 45, 475, 2007.
49. Hall, L., Wohlfiel, S., and Roberts, G.D., Experience with the MicroSeq D2 Large-Subunit Ribosomal Sequencing Kit for identification of filamentous fungi encountered in the clinical laboratory, *J. Clin. Microbiol.*, 42, 622, 2004.
50. Balajee, S.A. et al., Sequence-based identification of *Aspergillus*, *Fusarium*, and Mucorales species in the clinical mycology laboratory: Where are we and where should we go from here?, *J. Clin. Microbiol.*, 47, 877, 2009.
51. Hsiao, C.R. et al., Identification of medically important molds by an oligonucleotide array, *J. Clin. Microbiol.*, 43, 3760, 2005.
52. Schwarz, P. et al., Molecular identification of Zygomycetes from culture and experimentally infected tissues, *J. Clin. Microbiol.*, 44, 340, 2006.
53. O'Donnell, K. et al., Evolutionary relationships among mucoralean fungi (Zygomycota): Evidence for family polyphyly on a large scale, *Mycologia*, 93, 286, 2001.
54. Nyilasi, I. et al., High-affinity iron permease (FTR1) gene sequence-based molecular identification of clinically important *Zygomycetes*, *Clin. Microbiol. Infect.*, 14, 393, 2008.
55. Henry, T., Iwen, P.C., and Hinrichs S.H., Identification of *Aspergillus* species using internal transcribed spacer regions 1 and 2, *J. Clin. Microbiol.*, 38, 1510, 2000.
56. White, T. et al., Amplification and direct sequencing of fungal ribosomal RNA genes for phylogenetics, in *PCR Protocols. A Guide to Methods and Applications*, Innis, D.H.G.M.A., Sninsky, J.J., and White, T.J., Eds., pp. 315–322, Academic Press, San Diego, CA, 1990.
57. Gerrits van den Ende, A.H.G. and de Hoog, G.S., Variability and molecular diagnostics of the neurotropic species *Cladophialophora bantiana*, *Stud. Mycol.*, 43, 151, 1999.
58. Altschul, S.F. et al., Gapped BLAST and PSIBLAST: A new generation of protein database search programs, *Nucleic Acids Res.*, 25, 3389, 1997.

89 *Saksenaea*

Eric Dannaoui

CONTENTS

89.1 INTRODUCTION

89.1.1 Classification and Morphology

89.1.1.1 Classification

Saksenaea vasiformis has been originally described by Saksena in 1953 [1]. This new genus and species have been erected for an unknown fungus isolated in 1950 and 1952 from forest soil near the city of Sagar, Madhya Pradesh state, located in the central part of India. The genus was named after Dr. R.K. Saksena and the species name refers to the characteristic shape of the sporangium. There is currently only one species in the genus *Saksenaea*. A new family, Saksenaeaceae, was created in 1974 [2] to accommodate this genus because its initial belonging to the Mucoraceae was not supported by its peculiar morphology. More recently, phylogeny of Zygomycetes has been studied by molecular approaches. A recent study based on sequences of the small subunit ribosomal DNA (SSU rDNA) and the variable regions of the large subunit ribosomal DNA (LSU rDNA) has been performed to uncover the phylogenetic relationships among all clinically important Zygomycetes [3]. It was shown that *S. vasiformis* clustered with *Apophysomyces elegans*, a member of the Mucoraceae. The same close relationship between *S. vasiformis* and *A. elegans* was also supported by other phylogenetic studies based on actin and elongation factor (EF-1alpha) sequences [4,5]. Moreover, LSU rDNA analysis showed a 90%–91% identity between sequences of a clinical isolate of *S. vasiformis* and isolates of *A. elegans* [6]. These molecular data showed that *S. vasiformis* and *A. elegans* are closely related although they are currently belonging to different families.

89.1.1.2 Morphology

Although the sporulation is obtained only after culturing the fungus on specific media, the morphological characteristics are very unique, allowing an easy identification [1,7]. A schematic diagram of *S. vasiformis* microscopic morphology is presented in Figure 89.1. Sporangiophores (24–65 μm × 6.5–9.5 μm) are generally single and bear a terminal sporangium. Well-developed rhizoids are present at base of the sporangiosphore. These rhizoids are strongly dichotomously branched and became brown colored at maturity. The sporangium is flask shaped with a spherical venter (22–52 μm × 16–44 μm) and a long neck (54–200 μm × 6.5–11 μm). The collumella is hemispherical. The wall of the sporangium is smooth and hyaline. The distal part of the neck is closed by a gelatinous plugthat eventually dissolved. The sporangiospores are then passively released. The sporangiospores are subcylindrical (2.8–4.2 μm × 1.5–2 μm), hyaline, and smooth.

In only one study [8], morphology of different strains isolated from three geographic areas (India, Ethiopia, and Taiwan) were compared. Very interestingly, some

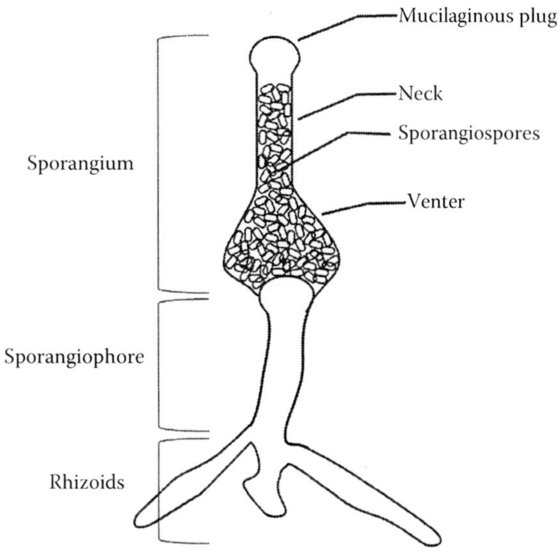

FIGURE 89.1 Schematic diagram of *S. vasiformis*.

morphological differences were noted. In particular, the isolate from India had hyaline and smooth sporangiophores, while the isolate from Ethiopia had brown and distinctly verrucose sporangiophores. These observations suggest that the studied isolates may belong to different varieties or even different species.

89.1.2 Biology, Ecology, and Epidemiology

89.1.2.1 Biology

Concerning the biology of *S. vasiformis*, very few data are available. At least, one detailed study has been performed to evaluate the physiology of the fungus in culture [9]. It has been shown that *S. vasiformis* is able to grow at temperature ranging from 15°C to 39°C, with an optimum at 30°C–35°C. A subsequent study showed that the maximum growth temperature is 44°C [7]. By culturing *S. vasiformis* in different carbon sources, it has been shown that those carbohydrates that supported poor growth (such as arabinose, sorbose, galactose) were able in contrast to support good to excellent sporulation [9]. An interesting study shown that two serotypes of *S. vasiformis* may exist and that the two groups showed different morphologies, serotype 1 isolates having smaller sporangia and shorter neck compared to serotype 2 isolates [10]. These observations further support the hypothesis that *S. vasiformis* may be a species complex.

Other biological characteristics of fungi, which have also clinical implications, are their susceptibility to antifungal drugs. In vitro data on antifungal susceptibility of *S. vasiformis* are very scarce. In selected cases, in vitro susceptibility testing has been attempted but failed due to poor growth of the fungus [11–13]. In one case, the strain has been found to be susceptible to amphotericin B with an minimal inhibitory concentration (MIC) of 0.08 μg/mL [14]. More recently, four isolates have been tested following the CLSI microdilution reference method [15] and showed MICs of 0.015–0.25 μg/mL for posaconazole, 0.015–0.03 μg/mL for itraconazole, 0.125–2 μg/mL for amphotericin B, and higher MICs of 0.5–4 for voriconazole and 1–64 μg/mL for fluconazole [16]. In another report, an isolate from French Guyana, tested with the EUCAST microdilution reference method [17], showed high MIC of >8 μg/mL for voriconazole, caspofungin, and amphotericin B, while itraconazole (MIC of 2 μg/mL) and posaconazole (MIC of 0.5 μg/mL) were more active [18]. Another study also reported the low MIC of 0.01 μg/mL for itraconazole against one isolate [19]. Clearly more data are needed to have a clear picture of the susceptibility of this fungus to the different available antifungal drugs. In particular, it would be very important to know if amphotericin B is less active than posaconazole, which is the case for some Zygomycetes such as *Cunninghamella* spp. and *Apophysomyces elegans* [20,21]. Animal models of *S. vasiformis* infection have not been reported and therefore the in vivo activity of systemic antifungal drugs has not been tested in animals.

89.1.2.2 Ecology and Epidemiology

Overall, the precise geographical distribution of this fungus is very poorly known. This could be explained, at least in part, by the difficulty to obtain the characteristic sporulation necessary for identification. The geographical distribution can be inferred from environment sampling studies as well as from human case reports (Table 89.1).

Although the isolation of *S. vasiformis* from the environment has been rarely reported, the fungus has been found in different parts of the world, suggesting a wide distribution. Several studies reported the presence of *S. vasiformis* in India. After the first isolation in central India by Saksena [1], it has been cultured from rice husk and grain in Puttur Taluk, Karnataka, located in the south-west of the country [22]. More recently, *S. vasiformis* has been cultured at a concentration of 7×10^4 CFU/g in clay loam soil from Annamalai Nagar, Tamil Nadu state, located in southeast of India [23].

Hodges [24] isolated *S. vasiformis* from soil from forest tree nurseries in Georgia, United States, sampled between 1956 and 1961. This was the first recovery of the fungus from the environment in the United States. It has also been isolated from soil, sampled in 1952, in Barro Colorado Island in Panama [25]. Mo [26] found *S. vasiformis* in turtle nest sand on the Nancite beach in Costa Rica sampled in 1987. Goos [27] isolated *S. vasiformis* from soil of banana-producing areas collected between 1958 and 1960 in Cortes province, Honduras. In the Middle East, *S. vasiformis* was isolated in Israel from soil of groundnut fields in 1964 [28]. In Asia, *S. vasiformis* was recovered from soil sample collected in 1986 near the seashore in Su-au County, I-lan prefecture, Taiwan [8]. *S. vasiformis* has also been found in Japan where it has been isolated from pineapple field soil in Okinawa [29,30]. Finally, the fungus was also found in Africa, where it was isolated from intertidal driftwoods near Mitsiua located along the Red Sea coast in Ethiopia [8].

Geographical distribution may also be inferred from clinical cases. Infection by *S. vasiformis* has been reported in

TABLE 89.1
Geographical Distribution of Clinical Cases of *S. vasiformis* Infections and Occurrence of *S. vasiformis* in Soil

Region and Country	Number of Cases: By Region	By Country	Occurrence in Soil
Europe	6		
France		2	
Spain		3	
Greece		1	
North America (United States)	6		
California		1	
Louisiana		1	
Mississippi		1	
Nevada		1	
Texas		1	
NS		1	
Georgia		0	+
South America	3		
Colombia		1	
Ecuador		1	
French Guiana		1	
Panama		0	+
Costa Rica		0	+
Honduras		0	+
Africa	1		
Tunisia		1	
Ethiopia		0	+
Asia	11		
Thailand		1	
India		6	+
Iraq		1	
Israel		2	+
Saudi Arabia		1	
Taiwan		0	+
Japan		0	+
Oceania	13		
Australia		12	
New Zealand		1	
Total		40	

NS: not stated.
+: occurrence in soil has been reported.

most parts of the world (see Section 89.1.3). Of note, most cases have been acquired in warm climate areas.

89.1.3 Clinical Features and Pathogenesis

In a review of 465 published cases of zygomycosis for which a culture was positive, *S. vasiformis* represented 5% of the cases [31]. Nevertheless, the true frequency of the disease is probably underestimated due to the difficulty of identification. At least, 41 cases of human infection by *S. vasiformis* have been reported in the international literature [6,11–14,18,32–68]. Some of these cases have been reported in more than one publication. Conversely, other cases of infection have been mentioned but not reported in details such as the case of post-traumatic wound infection after the 2004 tsunami [69]. Of note, cases have also been reported in animals [70,71].

The first human case has been published in 1976 [33] and subsequently, cases have been reported regularly. Nevertheless, more cases (n = 10) have been reported in the recent years (2008 and 2009). Human cases have been reported worldwide with a clear predominance from tropical and subtropical areas (Table 89.1). In particular, cases originated not only from Australia and New Zealand (n = 13) and Asia (n = 11) mainly India but also Thailand and the Middle East (Iraq, Israel, and Saudi Arabia). Other cases have been reported from North America in the South and West part of the United States (California, Louisiana, Mississippi, Nevada, Texas), and from South America (Colombia, Ecuador, French Guiana). One case originated from North Africa. Interestingly, cases have also been described in Europe (France, Spain, and Greece).

Among the 41 cases, there were 27 males (66%) and 14 females (34%) (sex ratio 1.9). The patient age ranged from 3 months to 84 years (mean age 43.9 years). Considering underlying diseases, most of the patients (90%) were immunocompetent. Only four patients (10%) were considered immunocompromised (one had a bladder cancer, one received previous chemotherapy for a metastatic gastric cancer, one had an acute lymphoblastic leukemia, and one a pre-leukemic syndrome treated with chemotherapy). Five patients had diabetes. Among the 31 patients without any known underlying diseases, most have sustained trauma or skin injury. In five cases, animal (such as spider, scorpion, or bird) bites were considered the portal of entry [6,36,47,63,68]. In one patient, the infection developed after a tattoo [14]. Five cases were of nosocomial origin: infection developed in two cases after surgery [38,55], in two cases at site of injection [52,56], and in one case at site of arterial catheter [53].

Clinical forms were cutaneous, rhinocerebral, disseminated, or localized to the kidney. Clinical presentation was a cutaneous or subcutaneous infection in 77% of the cases, associated with bone involvement in two cases. A rhinocerebral infection was reported in four patients [35,42,45,49] (10%): one patient had hematologic malignancies [45], one patient had ketoacidotic diabetes and received chemotherapy for a solid cancer [42], and two patients were apparently immunocompetent [35,49]. A disseminated infection has been reported in four patients, three of whom were immunocompetent. In all four cases, cutaneous infection was present, either at different sites [46] or associated with infection of kidneys [62,63], lung [60,62], or heart and thyroid [60]. Finally, in one patient, a bilateral kidney infection was diagnosed [32]. Cutaneous lesions were characterized by the presence of tissue swelling, cellulitis, nodule or subcutaneous mass, abscess, or ulcerations. In most cases, lesions were necrotic.

The delay between onset of symptoms and definite diagnosis was very variable, up to 8 years. In only 37% of the cases was the diagnosis made in less than 7 days. Otherwise a delay of 1–3 weeks (23% of the cases) to 3 weeks–3 months (17% of the cases) was necessary. Finally, in 23% of the cases, a delay of more than 3 months elapsed before a positive diagnosis was achieved. This could be probably explained by several factors. First, zygomycosis is a rare infection that is not usually suspected in case of a cutaneous lesion. Second, direct examination may be negative or not done if a fungal infection is not specifically suspected. Third, a non-sporulated Zygomycetes growing in culture, such as *S. vasiformis*, may be interpreted as a contaminant and not as an etiological agent of invasive infection. This lack of suspicion of a fungal origin of the infection is corroborated by the fact that more than 90% of the patients eventually diagnosed with a *S. vasiformis* infection have been previously treated by antibacterial agents.

For most of the patients, treatment consisted of surgery along with administration of antifungal drugs. Among the eight patients who did not receive a medical treatment, six were treated with surgery alone (four of whom survived), and two patients were not treated because they died before diagnosis. Different drugs have been used, including amphotericin B, azoles, and potassium iodide. Nevertheless, among the 32 patients who were treated with antifungal drugs, 31 patients received amphotericin B during the treatment course. Only one patient (the first case of *S. vasiformis* reported from India) did not receive amphotericin B but was treated with potassium iodide (KI) associated with surgery. Conventional amphotericin B was used in 23 cases and a lipid formulation of amphotericin B in nine cases. Posaconazole was used in four patients either concomitantly with or subsequently after liposomal amphotericin B. Most of the patients had debridement of necrotic tissues and/or surgical excision of the lesions. Fourteen patients died (one from other cause) and 26 survived.

89.1.4 Diagnosis

Diagnosis of zygomycosis is notoriously difficult. The conventional approach for diagnosis is based on direct examination and/or histology of infected tissues followed by identification of the fungus in culture. The diagnosis is particularly difficult in case of *S. vasiformis* infections because this fungus do not sporulate on standard mycological media. Recently, molecular approaches have been attempted both for diagnosis directly from tissues and for identification of *S. vasiformis* from pure culture. There is no antigen detection method. Serology (antibody detection in the serum of the patient) has been used in selected cases [49], but there are no commercially available techniques.

89.1.4.1 Conventional Techniques

The samples of choice are tissues of the infected area obtained surgically or by biopsies. Direct examination can be performed on KOH preparation or after staining by Gomori methenamine silver (GMS). Histopathological examination after hematoxilin eosin staining or more specifically after GMS or periodic acid Schiff (PAS) staining should be performed when tissue samples are sending to the histopathological laboratory. The demonstration of broad, pauciseptate, right-angle branching hyphae in tissues is diagnostic of zygomycosis. In all the cases reported in the literature (except one), the direct examination and/or histology were positive.

S. vasiformis grows rapidly as white aerial mycelium on most media. Nevertheless, it has to be noticed that cycloheximide-containing media should be avoided as this substance inhibits the growth of *S. vasiformis* [12]. Sporulation is hardly obtained and needs special procedures and generally extended incubation time. It has been shown that sporulation was better obtained by transferring small agar block permeated with mycelium in sterile distilled water [1] or on solidified 1% water agar [39,72]. Sporulation has also been obtained by culturing the fungus on nutritionally deficient media such as hay infusion agar or Czapek agar [33]. A reliable method of sporulation has been evaluated by Padhye and Ajello [73]. The fungus was cultured on Sabouraud agar and agar blocks containing hyphae were transferred in Petri dishes containing 20 mL of sterile distilled water supplemented with 0.2 mL of 10% yeast extract. By using this technique with 10 different isolates of *S. vasiformis*, it was shown that good sporulation was obtained for all isolates after 10 days of incubation at 37°C [73]. Sporulation was seen only for 6 of the 10 isolates at 30°C, and none of the isolates sporulated at 25°C. Therefore, this technique should be recommended when a non-sporulated Zygomycetes is isolated in culture. Of note, this technique also performed well with *Apophysomyces elegans*, another poor sporulating Zygomycete. In the reported cases of *S. vasiformis* infection, sporulation was mostly achieved by culturing the fungus according to this technique, as well as on Czapek agar (Figure 89.2). It has to be mentioned that identification of the fungus has also been successfully performed by immunological method: antigens prepared from non-sporulating culture were detected by specific rabbit antisera in an immunoprecipitation test [10].

89.1.4.2 Molecular Techniques

Molecular techniques are now used extensively in the medical mycology field. There are two different situations where these techniques could be used; first for identification of a fungus from pure culture, and second for diagnosis of fungal infection and identification of the fungus directly from clinical samples when cultures are negative or when cultures have not been performed.

These techniques have proven useful and accurate for identification of Zygomycetes [74,75] in pure culture. The most common approach is PCR amplification of the fungal DNA with universal primers, followed by sequencing. Identification is then obtained by comparison of the sequences in publicly available databases such as GenBank [76]. The most widely used DNA targets are within the ribosomal DNA region (18S, 28S, and ITS1-5.8S-ITS2). For Zygomycetes, a consensus have been reached to recommend the sequencing of the ITS region as a first-line [74,76,77].

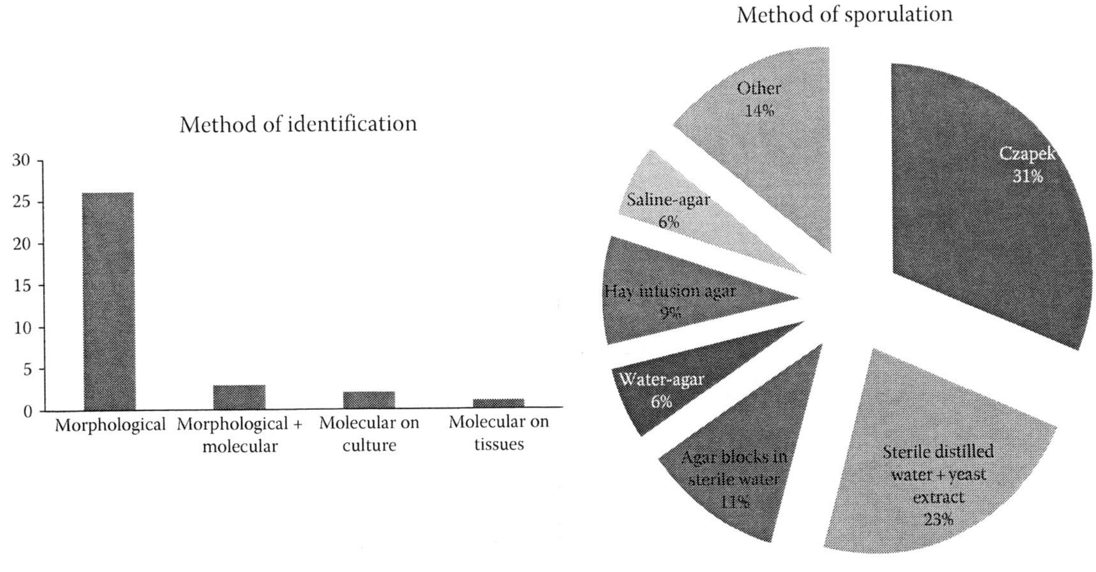

FIGURE 89.2 Method of identification of *S. vasiformis* reported in the literature and media used for inducing sporulation of the fungus in the case of morphological identification.

Not many molecular studies have included *S. vasiformis* isolates as this fungus is a very rare pathogen. Voigt et al. [3] have developed a database of 18S and 28S sequences for all clinically relevant species of Zygomycetes and based on the alignment of the 28S sequences, they were able to design species-specific primers sets for 13 species including *S. vasiformis*. Another PCR approach has been more recently evaluated [78]. In this later study, a real-time (RT) PCR was developed to detect different Zygomycete species including *S. vasiformis*. By using newly designed primers and probes, the authors amplified a 167-bp portion of the cytochrome b gene. This test was able to make an identification of *S. vasiformis* as the melting temperature (41.9°C) was significantly different from those of the other Zygomycete species tested.

As these molecular tools have been developed very recently, they have been used only in a limited number of *S. vasiformis* human infections. For five patients, the fungus was identified by sequencing, in some cases in combination with standard morphological identification. In all cases for which the methods are reported, the DNA target used was the ITS region [11,13,18,48]. In a case from Greece, it was noted that ITS sequence showed 99% identity with other *S. vasiformis* sequences [11]. In another case, acquired in French Guiana, ITS sequence showed 82% of identity [18]. These data are of particular interest and suggest that large variations of ITS sequences exist among strains morphologically identified as *S. vasiformis*. A detailed case report highlighted the power of molecular identification when a culture is not available [6]. A 14-year-old male acquired cutaneous *S. vasiformis* infection following a scorpion sting in North Africa. A skin biopsy grew a non-sporulating Zygomycete but the isolate subsequently lost viability precluding further study of the cultured fungus. Nevertheless, a fragment of the biopsy has been kept frozen and was used for molecular identification directly from tissue. After PCR amplification with universal fungal primers, both the ITS and the 28S region were sequenced. Analysis of the sequences by comparison with available entries in GenBank allowed to identify the fungus as *S. vasiformis* [6].

89.2 METHODS

Due to the low number of studies using molecular methods, there are no standardized techniques available. Nevertheless, technical issues can also been solved by available techniques or protocols used for other Zygomycetes species. Sample preparation as well as detection procedures may be different in case of identification of the fungus from culture or in case of a molecular diagnosis directly from tissues when cultures are negative or have not been performed.

89.2.1 Sample Preparation

89.2.1.1 Culture

Cultures are performed in liquid RPMI 1640 medium containing L-glutamine but without sodium bicarbonate. The medium is buffered at pH 7.0 with 0.165 M morpholinepropanesulfonic acid (MOPS) and filter sterilized. Tissue culture flasks containing 20 mL of RPMI are inoculated with a spore suspension (if sporulation has been obtained) or alternatively with mycelium. Incubation is performed at 35°C for 48–72 h with agitation. The mycelium mat obtained is washed in 20 mL of NaCl 0.9% and ready for extraction.

Genomic DNA extraction procedure may be performed as follows [75]. Approximately 50 mg of mycelium is placed in a 50-mL polypropylene tube containing glass beads and 1 mL of hexacetyltrimethylammonium bromide (CTAB) at 2% in an extraction buffer (NaCl 1.4 M, EDTA 25 mM, Tris–HCl 100 mM). After vortexing, the tube is immersed in liquid nitrogen for 1 min and put in water at 37°C for 1 min. After vortexing, 700 μL of the suspension is mixed with 700 μL of chloroform, vortexed again, and centrifuged at 14,000 × *g* for 10 min. The upper phase (500 µL) is transferred to a new tube and DNA is precipitated by the addition of 500 µL of 2-propanol. After a centrifugation step of 1 min at 14,000 × *g*, the supernatant is discarded, the pellet is washed with 100% ethanol, and resuspended in Tris-EDTA buffer. At this stage, the DNA can be stored frozen at −20°C or −80°C.

89.2.1.2 Tissues

Fresh tissue samples can be homogenized in a tissue grinder in NaCl 0.9% (in a volume depending on the size of the sample). At this stage, an aliquot of the homogenate can be used for direct examination to demonstrate the presence of hyphae in the sample that will be subsequently used for PCR amplification. About 50 µL of the homogenate is placed on a microscopic slide and allowed to dry at room temperature. Afterward, 20 µL of a 0.1% Calcofluor white (an optical brightener that stains fungi) is added and the preparation is observed under a fluorescent microscope. A volume of the homogenate corresponding to approximately 50 mg of tissue is centrifuged at 14,000 × *g* for 2 min, and the pellet is used for extraction as described above.

Formalin-fixed paraffin-embedded (PE) samples can also been used. Nevertheless, it has to be noticed that extraction from PE tissues may be more challenging due to degradation of DNA by formalin during the fixation step. From the histopathology block, 10–30 sections of 10 µm are cut and placed in a 2 mL polypropylene tube [79]. At this stage, examination of a slide after GMS staining allowed to demonstrate the presence of hyphae in the slices used for extraction. Firstly, deparaffinization of the samples is necessary. Two steps of 5 min of incubation with 1500 µL of xylene are performed. After each incubation step, tubes are spun at 20,000 × *g* for 5 min and the supernatant is discarded. After two further steps of 5 min incubation with 1500 µL of 100% ethanol, tubes are placed at 37°C until total evaporation of ethanol. Extraction of DNA can be achieved using QIAamp DNA mini kit (Qiagen) following manufacturer's instruction.

89.2.2 Detection Procedures

Different types of detection procedures may be used such as conventional PCR [18,75] or RT-PCR [78]. Amplification of the ITS region can be performed by conventional PCR as follows. The universal fungal primers LS266 5′-GCATTCCCAAACAACTCGACTC-3′) [10] and V9D (5′-TTAAGTCCCTGCCCTTTGTA-3′) are used to amplify the whole ITS1-5.8S-ITS2 region. After initial denaturizing step of 10 min at 94°C, 30 cycles of denaturation for 30 s at 94°C, annealing for 30 s at 58°C, and elongation for 30 s at 72°C are run with a final extension for 10 min at 72°C. PCR products are then purified and sequenced both strands with the same primers used for amplification. A consensus sequences can then be obtained.

Identification can be achieved by comparison of the sequences with entries in public databases. This is a critical step and it has to be remembered that up to 20% of the entries in public international databases such as GenBank may be erroneously identified to the species level, especially for uncommon species [80]. For this reason, it is recommended to compare sequences to those of reference strains, when available.

89.3 CONCLUSIONS AND FUTURE PERSPECTIVES

S. vasiformis infection is a rarely reported disease. Among Zygomycetes, it has unique morphological characteristics. It also differed from many other Mucorales by its geographical distribution and its propensity to induced cutaneous infections. Because it represents an uncommon human pathogen, this fungus has not been studied extensively and therefore many questions remained unanswered.

In particular, more studies are needed for a better understanding of its ecology and biology. Moreover, diagnosis of *S. vasiformis* infection remains a difficult task and need to be improved. Finally, susceptibility of the fungus and treatment of infection should be further evaluated for a better management of patients.

Recovery of *S. vasiformis* in the environment is very scarce. This could perhaps be related, at least in part, to the difficulty to recognize this fungus in culture as it does not sporulate on standard media. It is then possible that the true occurrence of the fungus is much more frequent. The new techniques that have been developed in the recent years to evaluate the diversity of fungal communities in soils by molecular detection will probably shed some light on the true frequency of *S. vasiformis* in the environment. A more comprehensive knowledge of its geographical distribution will also be possible.

Almost no studies have been performed to assess the biology of *S. vasiformis*. In particular, the pathogenicity, relative to other Mucorales, is not known and should be evaluated in vitro and in animal models of infections. Similarly, the genotypic variability is unknown. Molecular approaches are now available to address these issues and may be applied to strains of diverse geographical origin. This could lead to the recognition of different clades within *S. vasiformis*, or even to the description of new species within the genus *Saksenaea*.

New diagnostic tools both for molecular identification of Zygomycetes in pure culture as well as for molecular detection directly in tissue samples should be developed and their use should be encouraged. Indeed, the relative low numbers of *S. vasiformis* infection reported in the literature may be related to the lack of a positive diagnosis when a nonsporulating Zygomycete is grown from clinical specimens.

Finally, antifungal susceptibility of *S. vasiformis* is poorly known. Reference methods for antifungal susceptibility have been recently developed and these techniques should be used to test a larger panel of *S. vasiformis* isolates. In particular, the activity of amphotericin B and posaconazole should be more precisely determined as the few data currently available suggest that this fungus is significantly less susceptible to amphotericin B than other Zygomycetes. It would be of particular interest to develop animal model of infection to compare the in vivo activity of different antifungal drugs and to assess if the decreased in vitro susceptibility to amphotericin B is clinically relevant or not.

REFERENCES

1. Saksena, S. B. A new genus of the Mucorales. *Mycologia*, 45, 426, 1953.
2. Ellis, J. J. and Hesseltine, C. W. Two new families of mucorales. *Mycologia*, 66, 87, 1974.
3. Voigt, K., Cigelnik, E., and O'Donnell, K. Phylogeny and PCR identification of clinically important Zygomycetes based on nuclear ribosomal-DNA sequence data. *J. Clin. Microbiol.*, 37, 3957, 1999.
4. Voigt, K. and Wostemeyer, J. Phylogeny and origin of 82 zygomycetes from all 54 genera of the Mucorales and Mortierellales based on combined analysis of actin and translation elongation factor EF-1alpha genes. *Gene*, 270, 113, 2001.
5. O'Donnell, K. et al. Evolutionary relationships among mucoralean fungi (Zygomycota): Evidence for family polyphyly on a large scale. *Mycologia*, 93, 286, 2001.
6. Lechevalier, P. et al. Molecular diagnosis of *Saksenaea vasiformis* cutaneous infection after scorpion sting in an immunocompetent adolescent. *J. Clin. Microbiol.*, 46, 3169, 2008.
7. Schipper, M. A. A. and Stalpers, J. A. Zygomycetes. The order Mucorales. In *Pathogenic Fungi in Humans and Animals*, 2nd edn., D. H. Howard, Ed., Marcel Dekker, New York, 67, 2003.
8. Chien, C.-Y., Bhat, D. J., and Kendrick, W. B. Mycological observations on *Saksenaea vasiformis* (Saksenaeaceae, Mucorales). *Trans. Mycol. Soc. Jpn.*, 33, 443, 1992.
9. Baijal, U. A physiological study of *Saksenaea vasiformis* Saksena. *Mycopathol. Mycol. Appl.*, 33, 289, 1967.
10. Lombardi, G. et al. Exoantigen tests for the rapid and specific identification of *Apophysomyces elegans* and *Saksenaea vasiformis*. *J. Med. Vet. Mycol.*, 27, 113, 1989.
11. Giannopoulou, P. et al. Fatal cutaneous zygomycosis from *Saksenaea vasiformis* in a young patient after a car accident. First isolation in Greece. Paper presented at: *19th European Congress of Clinical Microbiology and Infectious Diseases (ECCMID)*, Helsinki, 2009.
12. Pierce, P. F. et al. *Saksenaea vasiformis* osteomyelitis. *J. Clin. Microbiol.*, 25, 933, 1987.
13. Stewardson, A. J. et al. Cutaneous zygomycosis caused by *Saksenaea vasiformis* following water-related wound in a 24-year-old immunocompetent woman. *Mycoses*, 52, 547, 2009.
14. Parker, C., Kaminski, G., and Hill, D. Zygomycosis in a tattoo, caused by *Saksenaea vasiformis*. *Aust. J. Dermatol.*, 27, 107, 1986.
15. Clinical and laboratory Standards Institute (CLSI) 2008. Reference method for broth dilution antifungal susceptibility testing of filamentous fungi: Approved standard. Document M-38A2, Clinical and Laboratory Standards Institute, Wayne, PA.
16. Sun, Q. N. et al. In vitro activities of posaconazole, itraconazole, voriconazole, amphotericin B, and fluconazole against 37 clinical isolates of zygomycetes. *Antimicrob. Agents Chemother.*, 46, 1581, 2002.
17. Subcommittee on Antifungal Susceptibility Testing (AFST) of the ESCMID European Committee for Antimicrobial Susceptibility Testing (EUCAST) et al. EUCAST Technical Note on the method for the determination of broth dilution minimum inhibitory concentrations of antifungal agents for conidia-forming moulds. *Clin. Microbiol. Infect.*, 14, 982, 2008.
18. Blanchet, D. et al. *Saksenaea vasiformis* infection, French Guiana. *Emerg. Infect. Dis.*, 14, 342, 2008.
19. Gomez-Lopez, A. et al. In vitro susceptibility of clinical isolates of Zygomycota to amphotericin B, flucytosine, itraconazole and voriconazole. *J. Antimicrob. Chemother.*, 48, 919, 2001.
20. Dannaoui, E. et al. In vitro susceptibilities of zygomycetes to conventional and new antifungals. *J. Antimicrob. Chemother.*, 51, 45, 2003.
21. Dannaoui, E. et al. In vitro susceptibilities of Zygomycota to polyenes. *J. Antimicrob. Chemother.*, 49, 741, 2002.
22. Pillai, C. M. *Saksenaea vasiformis*, an ecologically important fungus on rice from India. *Curr. Sci.*, 65, 291, 1993.
23. Parthasarathi, K. et al. Diversity of microflora in the gut and casts of tropical composting earthworms reared on different substrates. *J. Environ. Biol.*, 28, 87, 2007.
24. Hodges, C. S. Fungi isolated from southern forest tree nursery soils. *Mycologia*, 54, 221, 1962.
25. Farrow, W. M. Tropical soil fungi. *Mycologia*, 46, 632, 1954.
26. Mo, C. L., Salas, I., and Caballero, M. Are fungi and bacteria responsible for Olive Ridley's egg loss? In *Proceedings of the Tenth Annual Workshop on See Turtle Biology and Conservation*. NOAA Technical Memorandum NMFS-SEFC-278, T. H. Richardson, J. I. Richardson, and M. Donnelly, Eds., p. 249, 1990.
27. Goos, R. D. Further observation on soil fungi in Honduras. *Mycologia*, 55, 142, 1963.
28. Joffe, A. Z. and Borut, S. Y. Soil and kernel mycoflora of groundnut fields in Israel. *Mycologia*, 58, 629, 1966.
29. Watanabe, T. Pictorial atlas of soil and seed fungi. In *Morphologies of Cultured Fungi and Key to Species*, 2nd edn., CRC Press, Boca Raton, FL, 2002.
30. Watanabe, T. Fungi isolated from the rhizosphere soils of wilted pineapple plant in Okinawa. *Trans. Mycol. Soc. Jpn.*, 12, 35, 1971.
31. Roden, M. M. et al. Epidemiology and outcome of zygomycosis: A review of 929 reported cases. *Clin. Infect. Dis.*, 41, 634, 2005.
32. Aarset, H. et al. Acute renal infarction in a woman with slight asthma. *Nephrol. Dial. Transplant.*, 16, 1711, 2001.
33. Ajello, L., Dean, D. F., and Irwin, R. S. The zygomycete *Saksenaea vasiformis* as a pathogen of humans with a critical review of the etiology of zygomycosis. *Mycologia*, 68, 52, 1976.
34. Al-Hedaithy, M. Cutaneous zygomycosis due to *Saksenaea vasiformis*: Case report and literature review. *Ann. Saudi Med.*, 18, 428, 1998.
35. Baradkar, V. P. et al. Fatal rhino-orbito-cerebral infection caused by *Saksenaea vasiformis* in an immunocompetent individual: First case report from India. *Indian J. Med. Microbiol.*, 26, 385, 2008.
36. Bearer, E. A. et al. Cutaneous zygomycosis caused by *Saksenaea vasiformis* in a diabetic patient. *J. Clin. Microbiol.*, 32, 1823, 1994.
37. Cefai, C. et al. Zygomycetic gangrenous cellulitis. *Lancet*, 2, 1337, 1987.
38. Chakrabarti, A. et al. Primary cutaneous zygomycosis due to *Saksenaea vasiformis* and *Apophysomyces elegans*. *Clin. Infect. Dis.*, 24, 580, 1997.

39. Ellis, D. H. and Kaminski, G. W. Laboratory identification of *Saksenaea vasiformis*: A rare cause of zygomycosis in Australia. *Sabouraudia*, 23, 137, 1985.
40. Dean, D. F. et al. Cranial zygomycosis caused by *Saksenaea vasiformis*. Case report. *J. Neurosurg.*, 46, 97, 1977.
41. Campelo, C., de Miguel, S., and del Palacio, A. Zigomicosis rinocerebral causada por *Saksenaea vasiformis* en un paciente diabético. *Enferm. Infect. Microbiol. Clin.*, 23(Suppl. 1), 77, Abstract, 2005.
42. Garcia-Martinez, J. et al. Rhinocerebral zygomycosis caused by *Saksenae vasiformis* in a diabetic patient. *Mycoses*, 51, 549, 2008.
43. Goldschmied-Reouven, A. et al. *Saksenaea vasiformis* infection in a burn wound. *J. Med. Vet. Mycol.*, 27, 427, 1989.
44. Gomez Merino, E. et al. Necrotic lesion in scalp after injury. *Rev. Clin. Esp.*, 203, 451, 2003.
45. Gonis, G. and Starr, M. Fatal rhinoorbital mucormycosis caused by *Saksenaea vasiformis* in an immunocompromised child. *Pediatr. Infect. Dis. J.*, 16, 714, 1997.
46. Hay, R. J. et al. Disseminated zygomycosis (mucormycosis) caused by *Saksenaea vasiformis*. *J. Infect.*, 7, 162, 1983.
47. Holland, J. Emerging zygomycoses of humans: *Saksenaea vasiformis* and *Apophysomyces elegans*. *Curr. Top. Med. Mycol.*, 8, 27, 1997.
48. Kauffmann-Lacroix, C. et al. Mucormycose cutanée à *Saksenaea vasiformis*: Diagnostic mycologique [Abstract]. *J. Mycol. Med.*, 18, 57, 2008.
49. Kaufman, L., Padhye, A. A., and Parker, S. Rhinocerebral zygomycosis caused by *Saksenaea vasiformis*. *J. Med. Vet. Mycol.*, 26, 237, 1988.
50. Koren, G., Polacheck, I., and Kaplan, H. Invasive mucormycosis in a non-immunocompromised patient. *J. Infect.*, 12, 165, 1986.
51. Lye, G. R., Wood, G., and Nimmo, G. Subcutaneous zygomycosis due to *Saksenaea vasiformis*: Rapid isolate identification using a modified sporulation technique. *Pathology*, 28, 364, 1996.
52. Mathews, M. S. et al. Subcutaneous zygomycosis caused by *Saksenaea vasiformis* in India. A case report and review of the literature. *J. Mycol. Med.*, 3, 95, 1993.
53. Oberle, A. D. and Penn, R. L. Nosocomial invasive *Saksenaea vasiformis* infection. *Am. J. Clin. Pathol.*, 80, 885, 1983.
54. Padhye, A. A. et al. First case of subcutaneous zygomycosis caused by *Saksenaea vasiformis* in India. *Diagn. Microbiol. Infect. Dis.*, 9, 69, 1988.
55. Padmaja, I. J., Ramani, T. V., and Kalyani, S. Cutaneous zygomycosis: Necrotising fascitis due to *Saksenaea vasiformis*. *Indian J. Med. Microbiol.*, 24, 58, 2006.
56. Patino, J. F. et al. Mucormycosis: A fatal case by *Saksenaea vasiformis*. *World J. Surg.*, 8, 419, 1984.
57. Pouget-Jasson, C. et al. A necrotic calf lesion. *Ann. Dermatol. Vénéréol.*, 136, 549, 2009.
58. Pritchard, R. C. et al. Subcutaneous zygomycosis due to *Saksenaea vasiformis* in an infant. *Med. J. Aust.*, 145, 630, 1986.
59. Ridley, L. et al. Cardiac mucormycosis revealed on imaging. *AJR Am. J. Roentgenol.*, 174, 1469, 2000.
60. Solano, T. et al. Disseminated mucormycosis due to *Saksenaea vasiformis* in an immunocompetent adult. *Clin. Infect. Dis.*, 30, 942, 2000.
61. Tanphaichitr, V. S. et al. Subcutaneous mucormycosis caused by *Saksenaea vasiformis* in a thalassaemic child: First case report in Thailand. *Mycoses*, 33, 303, 1990.
62. Torell, J., Cooper, B. H., and Helgeson, N. G. Disseminated *Saksenaea vasiformis* infection. *Am. J. Clin. Pathol.*, 76, 116, 1981.
63. Trotter, D. J. et al. Disseminated *Saksenaea vasiformis* in an immunocompetent host. *Med. J. Aust.*, 189, 519, 2008.
64. Upton, A. et al. A patient with cutaneous zygomycosis due to *Saksenaea vasiformis*. *Infect. Dis. Clin. Pract.*, 11, 137, 2002.
65. Vega, W. et al. *Saksenaea vasiformis* infections: Case report and literature review. *Mycopathologia*, 162, 289, 2006.
66. Wierzbicka-Hainaut, E. et al. Mucormycose cutanée granulomateuse à *Saksenaea vasiformis* avec ostéite: Premier cas français. Abstract P122. *Ann. Dermatol. Vénéréol.*, 135(Suppl.), A207, 2008.
67. Wilson, M. et al. *Saksenaea vasiformis* breast abscess related to gardening injury. *Aust. N. Z. J. Med.*, 28, 845, 1998.
68. Wilson, P. A. Zygomycosis due to *Saksenaea vasiformis* caused by a magpie peck. *Med. J. Aust.*, 189, 521, 2008.
69. Edsander-Nord, A. Wound complications from the tsunami disaster: A reminder of indications for delayed closure. *Eur. J. Trauma Emerg. Surg.*, 34, 457, 2008.
70. Hill, B. D. et al. Bovine cranial zygomycosis caused by *Saksenaea vasiformis*. *Aust. Vet. J.*, 69, 173, 1992.
71. Robeck, T. R. and Dalton, L. M. *Saksenaea vasiformis* and *Apophysomyces elegans* zygomycotic infections in bottlenose dolphins (*Tursiops truncatus*), a killer whale (*Orcinus orca*), and pacific white-sided dolphins (*Lagenorhynchus obliquidens*). *J. Zoo. Wildl. Med.*, 33, 356, 2002.
72. Ellis, J. J. and Ajello, L. An unusual source of *Apophysomyces elegans* and a method of stimulating sporulation of *Saksenea vasiformis*. *Mycologia*, 74, 144, 1982.
73. Padhye, A. A. and Ajello, L. Simple method of inducing sporulation by *Apophysomyces elegans* and *Saksenaea vasiformis*. *J. Clin. Microbiol.*, 26, 1861, 1988.
74. Dannaoui, E. Molecular tools for identification of Zygomycetes and the diagnosis of zygomycosis. *Clin. Microbiol. Infect.*, 15(Suppl. 5), 66, 2009.
75. Schwarz, P. et al. Molecular identification of Zygomycetes from culture and experimentally infected tissues. *J. Clin. Microbiol.*, 44, 340, 2006.
76. Balajee, S. A. et al. Sequence-based identification of *Aspergillus, Fusarium*, and mucorales species in the clinical mycology laboratory: Where are we and where should we go from here? *J. Clin. Microbiol.*, 47, 877, 2009.
77. Clinical and laboratory Standards Institute (CLSI). Interpretive criteria for bacteria and fungi identification by DNA target sequencing; Approved guideline MM-18A. 2007.
78. Hata, D. J. et al. Real-time PCR method for detection of zygomycetes. *J. Clin. Microbiol.*, 46, 2353, 2008.
79. Dannaoui, E. et al. Molecular detection and identification of zygomycetes species from paraffin-embedded tissues in a murine model of disseminated zygomyeosis: a collaborative European Society of Clinical Microbiology and Infections Diseases (ESCMID) Fungal Infection Study Group (EFISG) evaluation: *J. Clin. Microbiol.*, 48, 2043, 2010.
80. Nilsson, R. H. et al. Taxonomic reliability of DNA sequences in public sequence databases: A fungal perspective. *PLoS One*, 1, e59, 2006.

90 *Syncephalastrum*

Dongyou Liu

CONTENTS

90.1 INTRODUCTION

90.1.1 Classification, Morphology, and Biology

The genus *Syncephalastrum* is the only member of the family Syncephalastraceae, order Mucorales, subphylum Mucoromyotina, kingdom Fungi. Currently this genus consists of five species/varieties: *Syncephalastrum alma-ataënse*, *Syncephalastrum monosporum* (including *S. monosporum* var. *cristatum* and *S. monosporum* var. *monosporum*), and *Syncephalastrum racemosum* (obsolete synonyms: *S. elegans*, *S. javanicum*, *S. nigricans*, and *S. verruculosum*), in addition to three unassigned species.

Syncephalastrum grows rapidly at up to 40°C, producing either low-growing or tall (0.5–1.5 cm) erect mycelia that fill the Petri dish surface in 7 days. Sporulation occurs on routine medium at room temperature and at temperatures above 37°C. Colonies are wooly-to-cotton candy-like, white-to-light gray initially, turning to olive, dark gray to black with age; on reverse, colonies are pale or yellowish-brown.

Syncephalastrum are broad (4–8 μm in diameter), aseptate, or sparsely septate; septation of the hyphae is mostly seen in old culture. Sporangiophores are erect, stolon-like, often producing adventitious rhizoids, and showing curved sympodial branching (racemose branching), with the main stalk and branches forming terminal, globose-to-ovoid vesicles called ampullae (30–80 μm in diameter). The vesicles are surrounded by rod- or club-shaped structures called merosporangia (finger-shaped, tubular sporangia of 4–6 x 9–60 μm in size), representing an asexual reproductive "sack" in which chains of merosporangiospores (merospores or sporangiospores) develop. Each merosporangium contains a single row of 3–18 sporangiospores (3–7 μm in diameter), which are one-celled, smooth, rough, or verrucose, spherical to cylindrical in shape. Merosporangia and sporangiospores are gray, brown, or black. As a heterothallic fungus, *Syncephalastrum* requires a mating strain to produce black, spherical zygospores (of 50–90 μm in diameter), with conical projections. These round, rough-walled or echinate spores are held by suspensors of nearly equal lengths.

Syncephalastrum is a zygomycetous mold commonly present in soil, animal feces, air, plants, and foodstuffs worldwide [1]. The fungus may be transmitted to humans by airborne or soilborne spores. Often regarded as a contaminant and a colonizer [2–5], *Syncephalastrum* (*S. racemosum*) has been implicated in several cases of human infections.

90.1.2 Clinical Features and Pathogenesis

Syncephalastrum racemosum is a saprophytic organism that can cause opportunistic infections in humans. Unlike other Zygomycetes that are often associated with pulmonary and central nervous system infections, *S. racemosum* tends to induce dermatomycoses and onychomycosis as well as other infections [1].

The first-documented case of *Syncephalastrum* infection began with a skin lesion on the finger of a 50-year-old diabetic man, which later progressed to osteomyelitis. The lesion had several draining sinus tracts with discharge that contained necrotic debris and aseptate ribbon-like hyphae typical of zygomycetes, and tissue sections showed occlusive vasculitis. The patient worked on a tea plantation and might have acquired the fungus from soil contamination via a traumatic event [6].

Another case of *Syncephalastrum* infection involved a 45-year-old man, who presented with a discolored and thickened right great toenail. The patient sustained injury to his toenail during a soccer game 7 months earlier and developed subungual hyperkeratosis, yellowish discoloration, and onycholysis subsequently. KOH mounts of nail clippings revealed wide, ribbon-like hyphae in masses, typical of Zygomycetes, and fruiting bodies similar to *Aspergillus* species. Culture of the nail fragments on Sabouraud's dextrose agar (SDA) slopes at 27°C showed abundant, erected mycelium, 0.5 cm tall in 3 days. The colonies were yellow at first, and the center turned black a few days later. Microscopic examination of the colony demonstrated wide hyphae with sparse septae. Sporangiophores ended with swollen vesicles where radial merosporangiae filled with spores. Typical fruiting bodies and ribbon-like hyphae in the nail plate and the cultured fungus thus confirmed the diagnosis of *Syncephalastrum racemosum* infection. The patient recovered after the surgery and topical nystatin ointment [7].

Similarly, a 30-year-old immunocompetent, nonpregnant woman presented with multiple discharging lesions with no granules over the dorsum of her left foot after a bamboo stick accidentally pricked her foot while she was working in the field 6 months earlier. Skin biopsies from the lesions showed abundant hyaline, aseptate, broad and branching hyphae with Gram stain and the KOH mounts. Culture on SDA with chloramphenicol and gentamicin grew fluffy mold with grayish sporangium within 48 h at 26°C, which appeared pale yellow on reverse. Lactophenol cotton blue (LPCB) mount revealed broad, aseptate, hyaline hyphae bearing erect sporangiophores terminating into globose columellae with the characteristic cylindrical merosporangia with rows of merosporangiospores. The isolate was identified as *Syncephalastrum* species [8].

Syncephalastrum may also be responsible for other types of infection. A 23-year-old male presented with a large abdominal wound (including abdominal wall perforation, splenic lacerations, mesenteric tears, colon and duodenal perforations, pancreatic injury, and an avulsed left renal artery) after falling and being impaled on a steel reinforcing rod. Several surgical procedures were performed, and histopathology of the omental tissue showed characteristic nonseptate hyphae. A gray-brown colony of a fungus was isolated from the retroperitoneal tissue excised at laparotomy on blood and chocolate agar incubated at 35°C for 30 h. Subculture of the fungus on SDA containing chloramphenicol and gentamicin at 35°C for 24 h showed a rapidly growing white cottony fungal colony. Microscopic examination of the fungus stained with LPCB showed sporangiophores branching at right angles and terminating in spherical columella, with cylindrical mesosporangia around the entire circumference, containing rows of sporangiospores. The sporangiospores, merosporangia, and terminal vesicles were smooth and dark brown. The broad, ribbon-like, aseptate hyphae were visible on the slide cultures as well as the histopathology stains. Zygospores were also visible. These morphological features were indicative of *S. racemosum*. Besides the retroperitoneal tissue, *S. racemosum* was identified in abdominal wall swabs, abdominal wall skin, intraabdominal fluid, intraabdominal swabs, and omental tissue. This infection was successfully treated with partial debridement and amphotericin B lipid complex [9].

A case of invasive rhino-orbital mucormycosis due to *Syncephalastrum* infection was reported. The patient was a 45-year-old male with a 12-year history of chronic hepatitis B infection and hepatic cirrhosis presented fever and chills, along with puffiness of face and edema of feet. Being HbsAg positive but nondiabetic, the patient developed nasal pain, swelling of the left eye, along with watering and congestion, and bilateral bleeding from the nose. KOH and hematoxylin and eosin (H&E) staining of nasal biopsy uncovered thick-walled aseptate distorted hyphae, and Gomori's methanamine silver (GMS) staining also showed broad aseptate hyphae with collapsed walls and fruiting head. Culture of the sample material on SDA at 25°C–30°C grew whitish fluffy colonies after 2 days. LPCB preparation of the colony showed sporangiophores branching at right angles and terminating into spherical collumella with cylindrical mesosporangia around entire circumference giving a "daisy head" appearance. The morphology was confirmed by slide subculture on potato dextrose agar (PDA), supporting the diagnosis of mucormycosis due to *Syncephalastrum*. Surgical debridement and intravenous liposomal amphotericin B helped reduce the chemosis and swelling of the eye [10].

S. racemosum produces a variety of mycotoxins *in vitro*, including some that suppress the mitotic rate in plant root cells. This fungus represents an opportunistic pathogen that exploits host's immune system dysfunction and injury. The ability of *Syncephalastrum* to grow at or above 37°C may also contribute to its ability to cause disease in humans.

Therapy for *Syncephalastrum* infection usually involves a combination of early diagnosis, surgical debridement, and systemic amphotericin treatment. Nystatin and pimaricin are also useful.

90.1.3 Laboratory Diagnosis

Conventional laboratory diagnosis of disease due to *Syncephalastrum* is dependent on the demonstration of characteristic zygomycete fungal elements (e.g., aseptate, ribbon-like hyphae and cylindrical merosporangia on a terminal swelling of the sporangiophore) in direct clinical specimens or tissue sections as well as in culture isolation, with the help of KOH and other specific fungal stains [11,12].

Although belonging to different phyla, *Syncephalastrum* (a zygomycete) may be confused with *Aspergillus* (an ascomycete), particularly *A. niger* and *A. flavus*. Both fungi may produce low-growing colonies with black or dark gray/green surface pigmentation and demonstrate similar fruiting bodies that appear as a vesicle surrounded by chains of spores. *Syncephalastrum* differs from *Aspergillus* by the presence of merosporangia and absence of phialides. *Syncephalastrum* hyphae are aseptate, ribbon-like spores (sporangiospores) and are arranged in a single row within mesosporangia [7].

More recently, molecular techniques have been applied for rapid and precise identification of Zygomycetes including *Syncephalastrum* [13]. The gene targets range from the nuclear small-subunit (18S) ribosomal RNA and domains D1 and D2 of the nuclear large-subunit (28S) ribosomal RNA, internal transcribed spacer (ITS) regions, the high-affinity iron permease 1 gene (FTR1) as well as other gene regions [5,14–16]. A combination of PCR amplification, sequencing, or restriction fragment length polymorphism (RFLP) analysis permits accurate determination of *S. racemosum* [17,18].

90.2 METHODS

90.2.1 Sample Preparation

90.2.1.1 Sample Handling and Microscopic Examination

Specimens are stained with 10% KOH, LPCB stain, or periodic acid-Schiff (PAS) stain and examined under microscope for mycotic elements. Portions of samples are grown on inhibitory mold agar, modified Sabouraud agars, or PDA at 25°C and 35°C. For microscopic examination of the fungal isolate, a small portion of a colony is cut out midway between the center and edge of a colony. The fungi are placed in a drop of LPCB on a slide, covered with a coverslip, and observed microscopically. To prepare a slide subculture, a 1.5 cm square of Sabouraud's agar is cut out of an agar plate and placed in the middle of an alcohol-clean slide. The sides of the square are inoculated with the fungus. The slide culture is covered and incubated at 25°C in a moist chamber. After 2 days, the culture is examined and the formation of recognizable elements noted. The coverslip is lifted from the agar, passed through a Bunsen flame, placed on a drop of LPCB on a slide, and examined under microscope [9].

90.2.1.2 DNA Extraction

After growth for 1–7 days on PDA slants, mycelia are collected by scraping the slant with a sterile stick in 1 mL of sterile water. The material is transferred to a 2 mL screw-cap tube. The tubes are centrifuged for 1 min at 6000 × *g*. If the mycelia do not pellet, the material is contained with a pediatric blood serum filter (Porex Corp.). Supernatant is removed, and the material is resuspended in 200 μL of IDI sample buffer and transferred to the lysis tube, which contained glass beads. Lysis tubes are vortexed on the highest setting for 5 min. The tubes are placed in a boiling water bath for 15 min. Tubes are centrifuged for 5 min at 16,000 × *g*. The supernatant is stored at −20°C until amplification [18].

Alternatively, about 1 cm^2 of fungal material is transferred to a 2 mL Eppendorf tube containing a 2:1 (wt/wt) mixture of silica gel and Celite (silica gel H, Merck 7736/Kieselguhr Celite 545; Machery) and 300 μL of TES buffer (2 g Tris [hydroxymethyl]-aminomethane, 0.38 g Na-EDTA, and 2 g sodium dodecyl sulfate in 80 mL of ultrapure water, pH 8). The fungal material is ground with a micropestle for 1–2 min. The volume is adjusted by adding 200 μL of TES buffer. After vigorous shaking and the addition of 10 μL of a 10 mg/mL concentration of proteinase K to the tube, the mixture is incubated at 65°C for 10 min. With the addition of 140 μL of 5 M NaCl solution, the mixture is combined with 1/10 volume (~65 μL) of CTAB (cetyltrimethylammonium bromide) buffer 10%, followed by incubation for another 30 min at 65°C. One volume (~700 μL) of chloroform: isoamyl alcohol (24:1) is added and mixed by inversion. After incubation for 30 min at 0°C (on ice water) and centrifugation at 14,000 rpm at 4°C for 10 min, the top layer is transferred to a clean Eppendorf tube. The sample is added with 225 μL of 5 M NH_4-acetate and incubated for at least 30 min (on ice water) and centrifuged. The supernatant is transferred to a clean sterile Eppendorf tube and mixed with a 0.55 volume (~510 μL) of ice-cold isopropanol. After centrifugation for 7 min at 14,000 rpm and 4°C (or room temperature), the supernatant is decanted. The pellet is washed with ice-cold ethanol 70% two times and dried by using a vacuum dryer. The powder is resuspended in 48.5 μL of Tris-EDTA buffer with 1.5 μL of 10 mg of RNase/mL, incubated at 37°C for 15–30 min, and stored at −20°C until use [19].

90.2.2 Detection Procedures

90.2.2.1 Pan-Fungal Real-Time PCR and Sequencing Analysis

Pounder et al. [18] described a real-time PCR with SYBR green DNA-binding dye and amplicon melting temperature analysis for fungal detection using pan-fungal primers ITS1 forward (5′-TCCGTAGGTGAACCTGCGG-3′) and ITS4 reverse (5′-TCCTCCGCTTATTGATATGC-3′). The identity of the fungi is verified by subsequent sequencing analysis.

Procedure

1. PCR mixture is composed of 1× Lightcycler FastStart DNA Master Hybridization Probes mixture (Roche Applied Science) (containing deoxynucleoside triphosphates, FastStart *Taq* DNA polymerase, and 1 mM $MgCl_2$) (additional $MgCl_2$ is added to a final concentration of 4.6 mM), 0.4 μM each of ITS1 forward and ITS4 reverse primers, 1× SYBR green (Molecular Probes), and 3 μL template DNA.
2. Thermal cycling parameters include 95°C for 10 min; 50 cycles of 95°C for 5 s, 60°C for 20 s, and 76°C for 30 s; and a final extension at 72°C for 2 min.
3. The quality of the amplicon is determined using the derivative of the melt analysis curve (55°C–99°C, 45 s hold at 55°C, 5 s/°C) using the RotorGene 3000 (Corbett Robotics, Inc.).
4. The amplified product is purified for bidirectional sequencing using ExoSAP-IT (USB Corp.). Five microliters of BigDye Terminator Ready Reaction

Mix v. 1.1 (Applied Biosystems) is added to 4 μL of each primer (0.8 pmol/μL) and 3 μL of purified PCR product. Cycle sequencing is performed with a 9700 thermal cycler (ABI), using 25 cycles of 96°C for 10 s, 50°C for 5 s, and 60°C for 4 min. Sequencing reaction products are passed through a Sephadex G-50 fine column to remove unincorporated dye terminators. Purified sequencing reaction products are run on an ABI Prism 3100 Genetic Analyzer with a 50 cm capillary array.

5. Sequences are analyzed with the SmartGene Integrated Database Network software version 3.2.3 vr. SmartGene is a web-based software and database system with reference sequences derived from the National Center for Biological Information (NCBI) GenBank repository.

Note. If real time PCR instrument is unavailable, standard PCR may be performed with primers ITS1 and ITS4, and the resulting amplicon is sequenced with the same primers. Sequence-based identifications are defined by percent identity: species, ≥99%; genus, 93%–99%; and inconclusive, ≤93%.

90.2.2.2 Sequencing Analysis of ITS1 and ITS2

Leaw et al. [20] utilized fungus-specific universal primers ITS1 (5′-TCCGTAGGTGAACCTGCGG-3′) and ITS2 (5′-GCATCGATGAAGAACGCAGC-3′) to amplify the ITS1 region and universal primers ITS3 (5′-GCATCGATGAAGAACGCAGC-3′) and ITS4 (5′-GCATATCAATAAGCGGAGGA-3′) to amplify the ITS2 region. Subsequent sequencing analysis of the resulting amplicons facilitated the identification of a range of fungal pathogens.

Procedure

1. PCR mixture (50 μL) is made up of 10 mM Tris–HCl pH 8.3, 50 mM KCl, 1.5 mM $MgCl_2$, 0.2 mM each of deoxynucleoside triphosphates, 1.2 U of *Taq* DNA polymerase, 0.4 μM (each) of the ITS1 region primers (ITS1/ITS2) or the ITS2 region primers (ITS3/ITS4), and 2 μL (1–5 ng) of DNA template. A negative control is included in each run by replacing the template DNA with sterile water in the PCR mixture.
2. Amplification is conducted with an initial 94°C for 3 min; 30 cycles of 94°C, 60°C, and 72°C for 1 min each; and a final 72°C for 3 min.
3. After checking the PCR products on 1.5% agarose gel, the amplicons are purified using a commercial PCR cleanup kit. The DNA fragments are sequenced using an ABI Prism 377 automated DNA sequencer (Applied Biosystems) with a BigDye Terminator cycle sequencing kit (version 3.1; Applied Biosystems). All amplicons are sequenced on both strands using primers ITS1 and ITS2 for the ITS1 region and primers ITS3 and ITS4 for the ITS2 region.
4. After sequencing, portions of the 18S, 5.8S, and 26S rRNA gene sequences of the PCR products are removed to obtain the exact ITS1 and ITS2 sequences.
5. For all yeasts, the sequences of the 3′ ends of the 18S and 5.8S rRNA genes are GCGGAAGGATCATTA and GTTTGAGCGTCATTT, respectively, and the sequences of the 5′ ends of the 5.8S and 26S rRNA genes are AAACTTTCAACAA and GACCTCAAATCAG, respectively.
6. The ITS1 or ITS2 sequence is compared with those available at the GenBank databases using the BLAST sequence analysis tool (blastn) with default settings except that sequences are not filtered for low complexity. Species identification is determined from the lowest expected value of the BLAST output.

Note. If real time PCR instrument is unavailable, standard PCR may be performed with primers ITS1 and ITS4, and the resulting amplicon is sequenced with the same primers. For strains producing discrepant identification between the methods based on phenotypic characteristics and ITS sequence analysis, the D1–D2 region of the large-subunit RNA gene is amplified with primers NL1 (5′-GCATATCAATAAGCGGAGGAAAAG-3′) and NL4 (5′-GGTCCGTGTTTCAAGACGG-3′) and sequenced for species clarification.

90.3 CONCLUSION

Syncephalastrum is a saprophytic zygomycete with widespread distribution. Although often considered as low virulence organism, *S. racemosum* has been implicated in several clinical cases of human diseases, including skin, nail, abdominal, and rhino-orbital infections. Morphologically, *Syncephalastrum* shows ribbon-like aseptate hyphae, sporangiophores with terminating vehicles surrounded by cylindrical mesosporangia, and merosporangial encase covering a single row of sporangiospores. These features are helpful for its differentiation from *Aspergillus* spp., which may also produce similar fruiting bodies. Due to their specificity and rapidity, molecular techniques such as PCR and sequencing have been utilized for identification and speciation of zygomycetes including *S. racemosum*. This not only facilitates the improved epidemiological investigation of *S. racemosum* infections, but also offers a valuable approach for monitoring of *S. racemosum* resistance during antifungal therapy.

REFERENCES

1. Ribes AJ, Vanover-Sams LC, Baker JB. Zygomycetes in human disease. *Clin. Microbiol. Rev*. 2000; 13: 236–301.
2. Abdel-Hafez AI, el-Sharouny HM. Keratinophilic and saprophytic fungi isolated from students' nails in Egypt. *J. Basic Microbiol*. 1990; 30: 3–11.
3. Foster KW, Ghannoum MA, Elewski BE. Epidemiologic surveillance of cutaneous fungal infection in the United States from 1999 to 2002. *J. Am. Acad. Dermatol*. 2004; 50: 748–752.

4. Rao CY et al. Characterization of airborne molds, endotoxins, and glucans in homes in New Orleans after Hurricanes Katrina and Rita. *Appl. Environ. Microbiol.* 2007; 73(5): 1630–1634.
5. Rao CY et al. Implications of detecting the mold *Syncephalastrum* in clinical specimens of New Orleans residents after Hurricanes Katrina and Rita. *J. Occup. Environ. Med.* 2007; 49: 411.
6. Kamalam A, Thambiah AS. Cutaneous infection by *Syncephalastrum*. *Sabouraudia*, 1980; 18: 19.
7. Pavlović MD, Bulajić N. Great toenail onychomycosis caused by *Syncephalastrum racemosum*. *Dermatol. Online J.* 2006; 12(1): 7.
8. Amatya R, Khanal B, Rijal A. *Syncephalastrum* species producing mycetoma-like lesions. *Indian J. Dermatol. Venereol. Leprol.* 2010; 76(3): 284–286.
9. Schlebusch S, Looke DF. Intraabdominal zygomycosis caused by *Syncephalastrum racemosum* infection successfully treated with partial surgical debridement and high-dose amphotericin B lipid complex. *J. Clin. Microbiol.* 2005; 43(11): 5825–5827.
10. Baradkar VP et al. Sino-orbital infection by *Syncephalastrum racemosum* in chronic hepatorenal disease. *J. Oral Maxillofac. Pathol.* 2008; 12: 45–47.
11. Gupta AK et al. Utility of inoculum counting (Walshe and English criteria) in clinical diagnosis of onychomycosis caused by nondermatophytic filamentous fungi. *J. Clin. Microbiol.* 2001; 39: 2115–2121.
12. Gupta AK et al. Non-dermatophyte onychomycosis. *Dermatol. Clin.* 2003; 21: 257–268.
13. Haynes KA et al. Rapid detection and identification of pathogenic fungi by polymerase chain reaction amplification of large subunit ribosomal DNA. *J. Med. Vet. Mycol.* 1995; 33: 319.
14. Vander Straten RM, Balkis MM, Ghannoum AM. The role of nondermatophyte molds in onychomycosis: Diagnosis and treatment. *Dermatol. Ther.* 2002; 15: 89–98.
15. Jensen AB et al. Relationships of the insect-pathogenic order Entomophthorales (Zygomycota, Fungi) based on phylogenetic analyses of nuclear small subunit ribosomal DNA sequences (SSU rDNA). *Fungal Genet. Biol.* 1998; 24: 325–334.
16. Voigt K, Cigelnik E, O'donnell K. Phylogeny and PCR identification of clinically important Zygomycetes based on nuclear ribosomal-DNA sequence data. *J. Clin. Microbiol.* 1999; 37: 3957.
17. Nyilasi I et al. High-affinity iron permease (FTR1) gene sequence-based molecular identification of clinically important Zygomycetes. *Clin. Microbiol. Infect.* 2008; 14(4): 393–397.
18. Pounder JI et al. Discovering potential pathogens among fungi identified as nonsporulating molds. *J. Clin. Microbiol.* 2007; 45(2): 568–571.
19. Zeng JS et al. Spectrum of clinically relevant *Exophiala* species in the United States. *J. Clin. Microbiol.* 2007; 45(11): 3713–3720.
20. Leaw SN et al. Identification of medically important yeast species by sequence analysis of the internal transcribed spacer regions. *J. Clin. Microbiol.* 2006; 44: 693.

Part IV

Microsporidia

91 *Anncaliia (Brachiola)*

Govinda S. Visvesvara and Lihua Xiao

CONTENTS

91.1 INTRODUCTION

Several species belonging to the genus *Anncaliia* (*Brachiola*) that usually parasitize insects have been found recently to infect humans, causing keratoconjunctivitis, myositis, and dermatitis. These are included under the phylum Microsporidia, which are ancient, spore-forming, mitochondria-lacking, obligate, intracellular, eukaryotic organisms that were previously considered as protozoa. The infective stage is the spore, which is Gram-positive, environmentally resistant, and characterized by the presence of a unique apparatus, the polar tube, which coils around the sporoplasm, the infectious organelle. When the spore is ingested, the polar tube is extruded with great force, penetrating a susceptible host cell and injecting it with the sporoplasm. Because of these unique characteristics, microsporidia were classified in a separate phylum, Microsporidia [1,2]. The ribosomes of microsporidia resemble prokaryotic ribosomes in terms of size but lack a 5.8S subunit [3]. Recent whole genome sequencing of *Encephalitozoon cuniculi* [4] and other microsporidians as well as the phylogenetic analysis of α tubulin and β tubulin genes [5] has placed these organisms along with fungi [6].

The phylum Microsporidia includes more than 1200 species belonging to ~150 genera. However, only a small number of species belonging to eight genera are known to cause human diseases. They are as follows: *Enterocytozoon bieneusi*, *Encephalitozoon* (*Septata*) *intestinalis*, *E. hellem*, *E. cuniculi*, *Anncaliia* (*Brachiola*, *Nosema*) *algerae*, *A.* (*Brachiola*) *vesicularum*, *A.* (*Nosema*) *connori*, *Nosema ocularum*, *Vittaforma corneae* (*previously named Nosema corneum*), *Pleistophora ronneafyi*, *Pleistophora* sp., *Trachipleistophora hominis*, *T. anthropophthera*, and a collective genus *Microsporidium* (to include *M. africanum*, *M. ceylonensis*). In addition to these, a new microsporidium belonging to a new genus and new species (namely, *Tubulinosema acridophagus*) has recently been identified at Centrers for disease control and prevention (CDC) in the tongue muscles of a woman with acute lymphocytic leukemia (ALL) [7].

Although microsporidia have been known since 1930 to infect insects, fish, and laboratory animals such as mice, hamsters, and rabbits, only recently have they been found to infect humans. The first case of human microsporidiosis was diagnosed in 1959 in an immunocompetent, 9-year-old, Japanese boy who developed fever, headache, seizures, and loss of consciousness. Spores of *Encephalitozoon* were demonstrated in the urine and CSF of the boy [8]. Since then, several cases of microsporidiosis have been identified in non-HIV-infected patients [9].

Microsporidiosis, however, has become a burgeoning problem in HIV/AIDS patients with chronic diarrhea since 1985. *Enterocytozoon bieneusi*, first described in 1985 by Desportes et al. in the enterocytes of a Haitian patient with AIDS, is, however, the most commonly identified etiologic agent in AIDS patients with chronic diarrhea [10]. Although *E. bieneusi* primarily causes an intestinal infection leading to malabsorption, it is also known to cause hepatobiliary and pulmonary disease in some patients. Additionally, *E. bieneusi* has also been identified in the fecal samples of children, travelers, immunocompetent patients, transplant recipients, domestic animals, and in water [2,9]. These factors clearly indicate the potential for waterborne transmission as well as zoonotic potential of this parasite.

In addition to *E. bieneusi*, a smaller, but increasing, number of AIDS patients are being diagnosed with ocular and disseminated microsporidiosis including keratoconjunctivitis, sinusitis, tracheobronchitis, nephritis, cystitis, and hepatitis as well as gastrointestinal disease due to the three species included in the genus *Encephalitozoon*, namely, *E. cuniculi, E. hellem, and E. intestinalis* [9].

Although other microsporidia such as *N. ocularum, Vittaforma corneae, Pleistophora* sp., *Pleistophora ronneafyi Trachipleistophora hominis, T. anthropophthera, Microsporidium africanum*, and *M. ceylonensis* are also known to have caused infections in humans, only one or two case reports due to these microsporidia have been described and hence the significance of these agents is largely unknown at present [9]. But several species of *Anncaliia* (*A. algerae, A. vesicularum, and A. connori*) have been known to cause infections in humans [11–15]. In particular, *A. algerae* has been identified recently as an agent in several cases of microsporidiosis of the human eye, muscles, skin, and vocal cords [16–18].

91.1.1 Classification

Anncaliia algerae was originally isolated from *Anopheles stephensi* mosquitoes in 1970 by Vávra and Undeen and was named *Nosema algerae* [19]. It was known to be primarily a mosquito pathogen [19]. But in 1973, Undeen and Maddox [20] infected other insects besides mosquitoes with *N. algerae*. Later it was shown that *N. algerae* can also infect and grow transiently in pig kidney tissue cultures [21]. In 1976, Undeen and Alger infected ears, tail, and foot pads of mice with *N. algerae*, indicating that *N. algerae* has a wide host range [22]. The first human case of *Anncaliia* was described as disseminated nosematosis by Margileth et al. (1973) in an athymic immunologically compromised infant [14]. Sprague [15] described this case as due to a new species, *Nosema connori* (now known as *Anncaliia connori*). Cali et al. [11] described a microsporidium that caused corneal ulceration and named it *N. ocularum*. This may be a species of *Anncaliia*. In 1998, Cali et al. described a microsporidium that caused myositis in an AIDS patient and described it as a new microsporidium and established a new genus and new species, *Brachiola vesicularum*, because it grew at human body temperature (37°C) and possessed unique tubular structures on its surface [13]. It is now renamed *A. vesicularum. A. algerae* (*A. algerae*) has also been isolated from the corneal scraping of an immunocompetent man [12] and also from subcutaneous nodules of a young boy with ALL [16], and from the deep muscles of an immunosuppressed female [17]. Recently, *A. algerae* has been identified in the vocal cord of a man [18].

N. algerae, the original isolate of Vávra and Undeen, has been shown not to cluster with the typical representatives of the genus *Nosema* based on the sequence data of the small subunit (SSU) rRNA gene by Baker et al. [3] and Muller et al. [23] (Figure 91.1). Based on this molecular phylogeny data and other characteristic morphological features such as development in contact with the host cell cytoplasm, diploblastic sporogony, anisofilar polar tubule, and the presence of unique tubular appendages on the surface of the various developmental stages, Lowman et al. [24], transferred *N. algerae* to the newly created genus *Brachiola* and designated it as *B. algerae* Vávra and Undeen [22]. Franzen et al. [25] transferred the genus *Brachiola* and its members to the genus *Anncaliia* described by Issi et al. [26] because of shared characteristics such as development of the organisms in direct contact with the host cytoplasm, diplokaryotic nuclei, disporoblastic sporogony, possession of unique tubular structures on the parasite cell surface as well as the presence of such structures scattered throughout the host cell cytoplasm, and the slightly anisofilar polar tubule. Further, Franzen et al. [25] suggested that weak physiological characteristics such as temperature tolerance are not stable features to be used in systematics because *A.* (*Brachiola, Nosema*) *algerae* can also grow at higher temperatures.

Franzen et al. [27] created a new genus *Tubulinosema* and a new species *T. ratisbonensis* in 2005 to accommodate those microsporidia that parasitize fruit flies (*Drosophila melanogaster*). These microsporidia are diplokaryotic microsporidia that possess small tubes on the surface of the developing stages and develop in full contact with the host cytoplasm and were previously included in the genus *Nosema* [27]. Two other species of *Nosema* (*N. kingi* and *N. acridophagus*), both parasites of insects but which fit the generic description of *Tubulinosema*, were transferred to the new genus and named *T. kingi* comb. nov and *T. acridophagus* comb. nov, respectively. Recently, we have identified a new microsporidium that caused infection of the tongue muscles of a patient [7]. Based on molecular data, it is closely related, if not identical, to *T. acridophagus*, a parasite found in fruit flies [7], suggesting an insect source for this infection or spore-contaminated water. The genera *Tubulinosema* and *Anncaliia* belong to the Tubulinosematidae family of the phylum Microsporidia [27].

91.1.2 Morphology, Biology, and Cultivation

Members of the genus *Anncaliia*, like those of the genera *Nosema* and *Vittaforma*, are monomorphic and are characterized by the presence of a diplokaryon throughout their life cycle. They develop in direct contact with the host cell cytoplasm and are disporoblastic [11–13,16,17,25–34]. The original mosquito isolate of Vávra and Undeen [19] and several isolates from human tissues have been established in continuous cultures at 37°C using several different mammalian cell lines [29,32,34,35]. All of these isolates proliferated in a number of mammalian cell cultures including rabbit kidney (RK 13), monkey kidney (E6), human lung fibroblast (HLF), human foreskin fibroblasts (HFF), and rat muscle cell line (L6E9) monolayers, producing copious numbers of spores. At the light microscope level, no differences were seen in

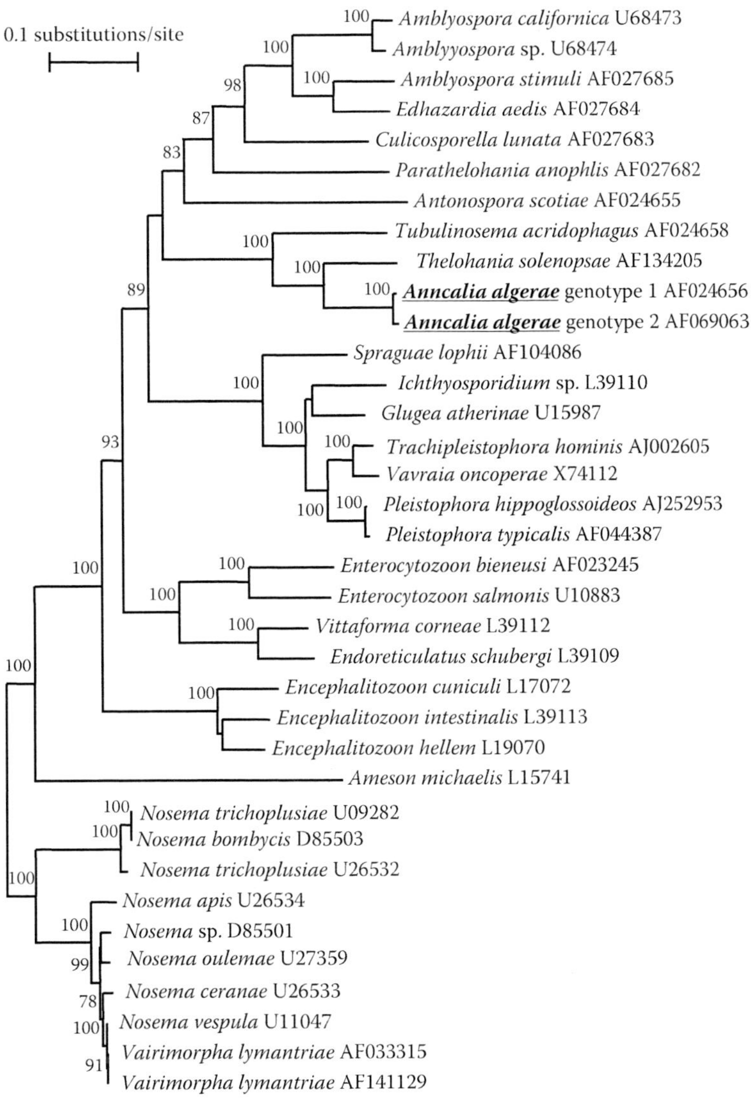

FIGURE 91.1 Genetic relatedness of genotypes 1 and 2 of *Anncaliia algerae* based on neighbor-joining analysis of the sequences of the SSU rRNA gene. Numbers on branches are percent bootstrapping values using 1000 replicates.

the development of the various human and mosquito isolates studied. Visual examination of the cultured isolates revealed that all isolates proliferated initially as a ring around the host cell nucleus (Figure 91.2A) and subsequently filled the entire cell with spores. By the 3 week of culture, the microsporidia had spread to about one-third of the cell culture monolayer and most of the cytoplasms of the host cells were filled with spores usually surrounding the nucleus of each host cell (Figure 91.2A). By about the 2 month of continuous culture, the parasites had adapted well to culture conditions and almost the entire monolayer was filled with spores and developing stages of the parasite, and masses of spores were seen being released from the host cells. A large number of spores were also found lying free in the supernatant medium. The spores appeared to be birefringent when examined with a microscope equipped with phase-contrast optics. Some of the spores had extruded polar filaments (Figure 91.2B and C).

Scanning electron microscope (SEM) images of HLF cells distended with spores and appeared as though the spores were covered with a muslin-like cloth (Figure 91.2D). The spores and the surrounding areas appeared to be covered with tiny particles that gave the disrupted cells a speckled appearance (Figure 91.2E). At higher magnification the spores appeared to be egg shaped and smooth walled (Figure 91.2F).

Transmission electron microscopy (TEM) revealed that all developing stages and spores, whether human or mosquito origin, were found within the host cell cytoplasm without any parasitophorous vacuole (Figure 91.3A and B). Developing stages consisted of meronts or proliferating stages, sporogonial stages, and mature spores. All stages possessed a single diplokaryon. They were also characterized by the presence of abundant ribosomes, either dispersed freely in the cytoplasm of the parasite cell or attached to the endoplasmic reticulum. The meronts appeared to be oval and elongated with the diplokaryotic nucleus occupying almost the entire organism. The diplokaryon was bounded by an envelope that consisted of two unit membranes. The two nuclei of the diplokaryon abutted each other and at the point of contact the

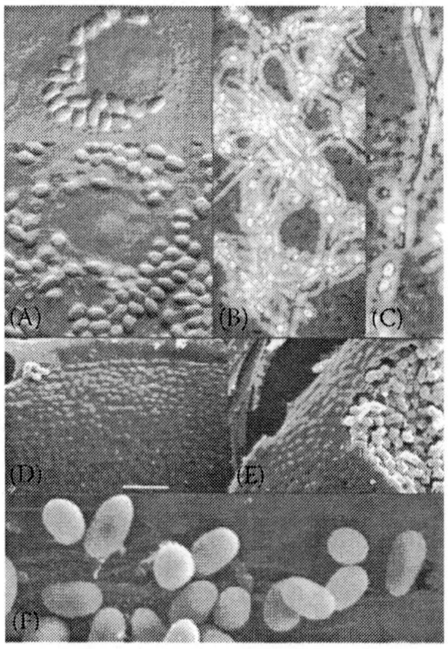

FIGURE 91.2 Light and scanning electron micrographs of mammalian cell cultures heavily infected with *Anncaliia algerae* (CDC:V521). (A–C). Light micrographs of monkey kidney (E6) cell cultures with *A. algerae*. (A). Note the spores surrounding the nucleus of the host cell; (B). Cell culture totally destroyed and covered with a matt of *A. algerae* spores with extruded polar tubules; (C). Two spores showing extruded long polar tubules. D–F. SEM of HLFs infected with *A. algerae* (CDC:V521). (D). HLF cell distended with spores; (E). Portion of the host cell membrane has cracked open and liberated spores; (F). Oval egg-shaped spores.

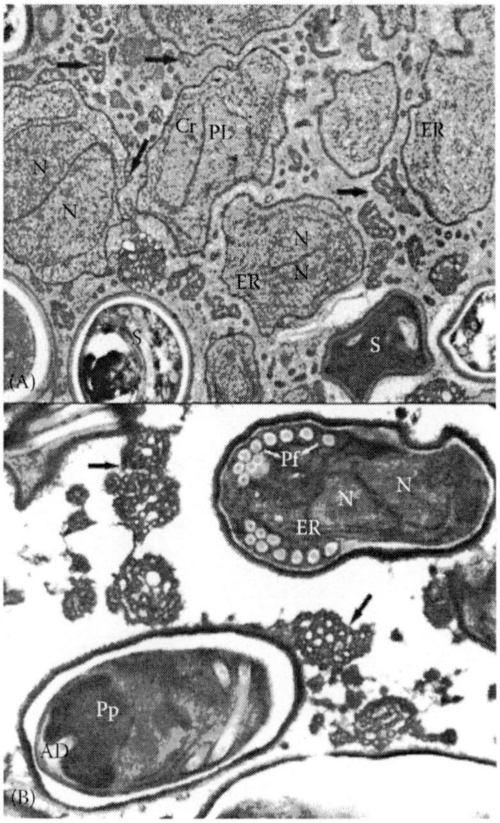

FIGURE 91.3 Transmission electron micrographs of *Anopheles stephensi* mosquitoes infected with *Brachiola algerae*. (A) A host cell filled with spores and developing stages of *A. algerae*. Note the vesiculotubular appendages on the surface of the developing stages as well as free in the host cell cytoplasm denoted by arrows. (B) Two spores with electron-dense exospore, electron-lucent endospore, diplokaryotic nucleus (N), polaroplast (Pp), adhesive disk (AD), and 9–10 cross sections of the polar filament (Pf).

four unit membranes were clearly seen indicating that there was no fusion of the membranes between the apposed nuclei. Multiple dense areas, interpreted as chromatin clusters, were present within the nucleus. The cell surface or the plasmalemma was covered with a thick, electron-dense material but in some cases this surface appeared to contain electron-lucent regions at regular intervals. A characteristic feature was the presence, in all isolates, of the vesiculotubular structures in close association with the membrane of the proliferative stages, indicating that these structures were produced by the parasite cell surface. In some cases, these structures extended into the host cell cytoplasm. As the development proceeded (and with the induction of sporogonic stages), the vesiculotubular structure became very dense and broke off and was found free in the host cell cytoplasm, indicating that these structures dissociated from the parasite stages and accumulated within the host cell cytoplasm as dense aggregates. The sporoblasts had extremely dense cytoplasm with the diplokaryotic nuclei occupying the central area. Abundant ribosomes surrounded the nucleus. Eight to nine cross sections of the polar filament were usually seen within the cytoplasm of a majority of the sporogonial stages. The polaroplast appeared to be highly tubular in the central area and lamellar in the periphery. The spores measured 2.4 μm × 4 μm, usually about 3 μm, and were smooth walled. The spore was bounded by an outer electron-dense exospore and an inner electron-lucent endospore. The number of polar filament coils ranged from 8 to 11, sometimes in two rows; coils present anteriorly were slightly larger than those at the posterior end, indicating that they were anisofilar (Figure 91.3B). The polar filament had a central core surrounded by two layers. The extruded polar filaments seen in many of the spores measured 15–22 μm in length. Although spore contents were dense, in some spores the lamellar polaroplast could be discerned at one end and a vacuole at the opposite end.

The vesiculotubular structure has been described previously by Vávra and Undeen [19], Moura et al. [29] in mosquito-derived *A. algerae* and by Visvesvara et al. [34] and Koudela et al. [28] in mammalian cell culture and severe combined immunodeficiency disease (SCID) mice infected with the human isolate of *A. algerae*, respectively. Further, Cali et al. [13] described these structures in another species of *Anncaliia* (*A. vesicularum*) identified in a muscle biopsy of an HIV-infected patient with fever and progressive muscle weakness. Koudela et al. [28] speculated that these vesiculotubular structures increase the surface area of the parasite cell probably helping the parasite in absorbing nutrients from the host cell, thus facilitating the development of the parasite. Recently, Cali et al. [31,32] have compared the myositis *A. algerae* isolate, newest of the human isolates, with the mosquito isolate by in vitro culture and electron microscopy and came to the conclusion that human and mosquito isolates are morphologically identical. They also state that the developmental pattern and parasite morphology of the myositis isolate is consistent with that of CDC:V404 grown in SCID mice as described by Koudela et al. [28].

Because *A. algerae* is typically a parasite of mosquitoes, it was believed that the body temperature of the mammalian host is a limiting factor for this parasite to jump from an invertebrate host to a mammalian host [13]. However, previous reports indicate that the original mosquito strain of *A. algerae* can grow and proliferate in mammalian cell culture at 37°C [24,29,30]. According to another recent report, human- and mosquito-derived isolates cultured in E6 cells at 30° and 37°C differed in their ability to germinate and infect monkey kidney cells [36]. Further, the mosquito-derived isolate [22,24] was also able to infect mice [37] when spores were injected into the footpad or tail. The parasite grew only at the site of infection. A more recent report indicates that the human-derived isolate (CDC:V404) developed in the liver when SCID mice were infected via the surface of the eye but not when infected subcutaneously, per orally, intraperitoneally, or intramuscularly [28], indicating that although the parasite may be able to cross the temperature barrier it will have to face other barriers such as the immune status of the host and the immune defense mounted by the host.

91.1.3 Epidemiology

Microsporidia spores are small, hardy, and resistant to a number of physical and chemical parameters in the environment and may survive and pass through conventional drinking water treatment plants and hence present a potential threat to human health. For example, spores of *A. algerae* readily infected *Anopheles albimanus* mosquitoes even after exposure to direct sunlight for a period of up to 4h. Additionally, about 24% of *A. algerae* spores, after exposure to 121 μW/cm^2 (0.06 J/cm^2) of artificial ultraviolet radiation for 2 min, were still able to infect Anopheline mosquitoes, indicating that the spores can easily withstand lower levels of UV radiation that occur in sun light [38]. It is also known that the dissolved organic materials in natural bodies of water absorb UV from solar radiation, thus offering more protection for the spores. In another study, very high dosages of UV radiation were required to block germination (or extrusion of the polar tube) of the spore [38]. Moreover, UV radiation of up to 3.8 J/cm^2 was found to stimulate *A. algerae* spores, thus facilitating the germination process [39].

Microsporidia spores, including those from two genera potentially infective to humans (*Pleistophora* sp. and *Nosema* sp.), have been found in ditch water in Florida [40]. In addition, there are no routine methods for the detection of *A. algerae* in water, and we do not know whether the currently available techniques to recover other protozoa also recover microsporidia with equal efficiency given the smaller size of microsporidial spores. Therefore, methods need to be developed (a) for efficient recovery of microsporidia spores in water and (b) to detect viable or infectious spores since only viable/infectious spores represent a threat to public health.

Little is known about epidemiologic features of microsporidiosis caused by *A. algerae* and other *Anncaliia* spp. and risk factors involved in the acquisition of the human infection. Most of the human cases described thus far were seen in immunocompromised persons or persons undergoing chemotherapy. The occurrence of *Anncaliia* spp. in humans is probably more common than believed, as suggested by a serologic study reporting that a significant number of AIDS patients had antibodies against *A. algerae* [9].

91.1.4 Clinical Features and Pathogenesis

As many as seven cases of human infections due to the members of the genus *Anncaliia* have been reported [12–18 and this report]. Of these, five cases were caused by *A. algerae* [12,16–18, and this report], one caused by *A. connori* [14,15], and one caused by *A. vesicularum* [13]. Of the five cases of *A. algerae* infections, two cases were of keratoconjunctivitis, one case of subcutaneous infection, one case of deep muscle infection, and one case of vocal cord infection.

Case 1: A 67-year-old male from Mexico developed stromal keratitis of the right eye leading to decreased vision over a period of 8 months. He was diagnosed as a case of herpes simplex viral keratitis and treated with topical steroids and antiviral medication. On further examination, his visual acuity of the right eye was 20/80 and the right cornea had dense white stromal infiltrates. A corneal scraping was negative for bacterial, fungal, and viral cultures. A corneal biopsy, however, revealed microsporidial spores (Figure 91.4), and he was treated with 400 mg oral albendazole three times daily and topical fumagillin (Fumadil B) every 4h. Since no change was seen even after 6 weeks of therapy, a 9.5 mm lamellar keratoplasty was done and treatment with albendazole and fumagillin was continued for another 6 weeks. Although complete epithelialization of the graft occurred, a recurrence of microsporidial organisms was noticed. A deep penetrating

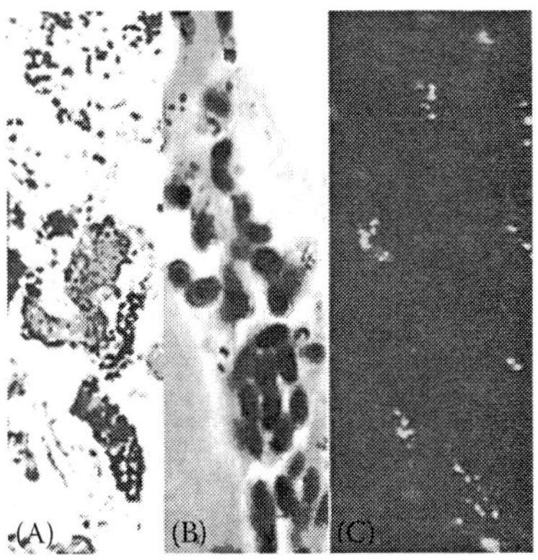

FIGURE 91.4 A patient's muscle tissue sections stained with Gram chromotrope (A) and another patient's corneal scraping stained with Gram chromotrope staining (B) and a different section from the same patient as (A) reacted with anti-*A. algerae* rabbit serum in the immunofluorescent reaction (C).

keratoplasty was performed and treatment with albendazole and fumagillin continued for several months until the graft remained clear with no recurrence of microsporidia. The corneal biopsy was inoculated onto cell cultures and the isolate (CDC:V404) was established and the microsporidia was identified as *Nosema algerae* (now known as *Anncaliia algerae*) [12].

Case 2: An 11-year-old boy was receiving chemotherapy for ALL. He was in remission but was admitted to the hospital because of 6–8 weeks of progressive erythematous skin nodules on the legs, buttocks, and hand that occasionally drained purulent material. He had no other gastrointestinal, genitourinary, or respiratory symptoms. The initial clinical diagnosis was lymphoma cutis vs infection with *Nocardia*. A biopsy of the cutaneous nodule revealed the presence of microsporidial spores. The biopsy was triturated and inoculated into cell cultures; microsporidial organisms that grew (CDC:V422) were identified by immunofluorescence and PCR as *A. algerae*. The patient died despite beginning treatment with albendazole.

Case 3: A 57-year-old female with rheumatoid arthritis and diabetes developed fatigue, generalized muscle and joint pain, weakness, and fever [16,17]. She had a history of treatment with methotrexate and steroids and infliximab. A muscle biopsy taken from the left anterior thigh revealed microsporidial organisms. She was administered 400mg albendazole twice daily along with clindamycin, metronidazole, and atovoquone. The patient died 4 weeks later. The muscle biopsy inoculated into cell cultures yielded a thriving culture of microsporidia (CDC:V521) that produced profuse numbers of microsporidial spores and developmental stages, which were identified as *Anncaliia algerae* by immunofluorescence and PCR.

Case 4: A 69-year-old male, a heavy smoker with chronic lymphocytic leukemia (CLL) of 9 years duration, was admitted to the hospital with a 3 week history of cough, fever, dyspnea, and hoarseness. He also had a history of disseminated nocardiosis and was receiving treatment with fludarabine for his CLL and trimethoprim-sulfamethoxazole for nocardiosis. A laryngoscopy revealed superficial ulceration of the left false vocal cord. A biopsy of the left vocal cord revealed the presence of microsporidia and the patient was treated with albendazole but the patient expired on the 23rd day of hospitalization. The microsporidia was identified as *A. algerae* based on molecular analysis [18].

Case 5: A 79-year-old male had a 1 year history of corneal ulceration of unknown etiology. He received a corneal transplantation but developed recurrent ulceration of the transplanted cornea. Retrospective histopathologic examination of the previous keratoplasty specimens revealed microsporidial organisms. Immunofluorescence identified the microsporidia as *A. algerae*. The patient is being treated with oral albendazole and topical fumagillin (Visvesvara et al. unpublished data).

Case 6: A 31-year-old HIV-positive male developed fever and muscle weakness over a period of 5 months. A biopsy of the left quadriceps revealed granulomatous myositis and microsporidia spores. Electron microscopy of the tissue revealed the presence of microsporidia with characteristic morphology of *Anncaliia*. The microsporidium was originally thought to belong to a new genus and the new species was named *Brachiola vesicularum* [13] but now it has been reclassified as *Anncaliia algerae* [25].

Case 7: An athymic female infant with 4 month duration of diarrhea and malabsorption died with a diagnosis of disseminated nosematosis. Microsporidial spores were seen in cardiac muscles, stomach, intestine, myocardium, and adrenal cortex. Immature spores and sporoblasts with two nuclei were seen and the microsporidium was provisionally identified as *Nosema* [14] but was later identified as *Nosema connori* [15]. Now it is considered as *Anncaliia connori* [25].

Other possible cases of *Anncaliia* include a case of keratoconjunctivitis from Madras, India [41] and a case of keratitis from Bangladesh [42].

91.1.5 Diagnosis

91.1.5.1 Conventional Diagnosis

Currently, the only known and easily performed method of microscopic diagnosis and identification of microsporidia spores in human body fluids including formalin-fixed stool samples is the chromotrope 2R (or modifications thereof) technique [43–45]. This technique differentially stains the microsporidia spores in clinical samples such as stools, urine, sputum, bronchoalveolar lavage, corneal scraping/biopsy, nasal discharge, etc. With the chromotrope 2R technique, which takes about 90–120min to perform, the microsporidia spores appear as pinkish-red ovoid structures possessing a characteristic equatorial belt-like stripe in the middle. Optical brighteners such as calcofluor white or Uvitex 2B also have been used to identify microsporidia in clinical samples since these agents bind to the chitin present in the endospore layer of microsporidian spores [44]. Although this test can be performed in a relatively few minutes with good sensitivity, the specificity is less than desirable. Since these agents also bind with small yeasts that may be present in the samples, a chromotrope stain needs to be performed in order to corroborate the detection results. Neither of these techniques, however, identifies the etiologic agent(s) to the genus and species level, which is a limiting factor especially for the prompt treatment of patients.

Accurate diagnosis of the agent of a particular syndrome of microsporidiosis is, therefore, difficult because of several factors: (a) the spores are small and measure around 1–2.5 μm; (b) routine stains performed on patient specimens in clinics and hospitals are not suitable for the identification of microsporidia; (c) it is neither feasible nor cost effective to perform the tedious, time-consuming electron microscopy on a routine basis; (d) no serologic methods are available for diagnosis; (e) although specific antibody-based immunofluorescence or ELISA tests would be very useful for identifying the species involved in patient specimens, such antibody-based detection methods are not routinely available; (f) PCR assays for molecular diagnosis have been developed, but have limitations since they do not work on formalin-fixed stool samples; and (g) specialized methods such as in vitro culture, while feasible with some microsporidia (*E. intestinalis*), are not possible with others (*E. bieneusi*) and are neither available routinely nor practical because of time constraints. Hence, infections with microsporidial species are usually underdiagnosed or incorrectly diagnosed.

91.1.5.2 Indirect Immunofluorescence Test

Indirect immunofluorescence (IIF) technique has been used to identify spores in microsporidiosis of the corneal [12] and subcutaneous nodule and muscle [16]. Rabbits were immunized by repeated intravenous injections of CDC:V404 isolate and the hyperimmune serum so produced was reacted with the tissue sections containing the *A. algerae* spores. The homologous anti-CDC:V404 serum produced bright apple green fluorescence at a dilution of up to 4096 (Figure 91.4C).

91.1.5.3 Molecular Identification

Nucleic acid amplification techniques such as PCR have shown potential for improving the detection and diagnosis of microsporidiosis. Sequence analysis of large subunit (LSU) and SSU rRNA, internal transcribed spacers (ITS), and other genes has enabled correct identification of microsporidial organisms, leading to the reclarification of *Nosema connori* as *Anncaliia connori* [25].

91.2 METHODS

91.2.1 Sample Preparation

DNA can be extracted from spores of *A. algerae* isolates or infected tissues by alkaline digestion (tissue pellets in 33.3 μL of 1 M KOH and 9.3 μL of 1 M dithiothreitol at 65°C for 15 min, followed by neutralization with 4.3 μL of 25% HCl buffered with 80 μL of 2 M Tris–HCl, pH 8.3), phenol-chloroform extraction, and DNA purification using QIAamp DNA Stool Mini Kit (QIAGEN, Valencia, CA).

91.2.2 Detection Procedures

The entire SSU rRNA gene (>1300 bp) of *A. algerae* isolates can be amplified using primers 5′-CACCAGGTTGA TTCTGCCTGA (MICRO-F) and 5′-CCAACTGAAACCTTG TTACGACTT (1492N4). For specific detection of *Anncaliia* spp., a nested PCR that amplifies a smaller fragment of the SSU rRNA gene can be used. In this method, for the primary PCR, a PCR product of ~703 bp is amplified using primers AL 4278(F1) (5′-TGT GGA TAA CCT TGT TAA GAT AAG-3′) and AL 4280(R1) (5′-TAA GTT TGT CCG TTG CCG GAT-3′). The reaction is carried out in a Perkin Elmer GeneAmp PCR 9700 thermocycler for 35 cycles (94°C for 45 s, 55°C for 45 s, and 72°C for 60 s), with an initial hot start (94°C for 5 min) and a final extension (72°C for 10 min). For the secondary PCR, a fragment of ~606 bp is amplified using 2.5 μL of the primary PCR reaction and primers AL 4279(F2) (5′-GTC GCA CTT TTG TGA AGA AA-3′) and AL 4281(R2) (5′-TGC GCC TCC GCA GAA GCA G-3′). The conditions for the secondary PCR are identical to the primary PCR, except that the annealing temperature is 52°C.

The PCR product is usually sequenced in both directions on an autosequencer such as the ABI3130 Genetic Analyzer (Applied Biosystems, Foster City, CA). The microsporidian species can be diagnosed by aligning sequences obtained with the reference SSU rRNA sequences (AF024656, AF069063, AM422905, AY230191, AY963289, and AY963290 for *A. algerae*) of different microsporidian parasites obtained from GenBank, using the software ClustalX (http://www.clustal.org/). There are two sequence types of the SSU rRNA gene in *A. algerae*, genotypes 1 and 2, which differ from each other by ~1% and have been found in both mosquitoes and humans, sometimes in the same specimen [16]. A neighbor-joining analysis of the sequences can be done using the program MEGA 4 (http://www.megasoftware.net/), which can be especially useful in the identification of *Anncaliia* spp. that have not been sequenced thus far, as the only other *Anncaliia* sp. with full SSU rRNA gene sequence is *A. meligethi* (AY894423).

91.3 CONCLUSIONS AND FUTURE PERSPECTIVES

The microsporidian genus *Brachiola* was recently transferred to the genus *Anncaliia* because of shared characteristics such as development of the organisms in direct contact with the host cytoplasm, diplokaryotic nuclei, disporoblastic sporogony, possession of unique tubular structures on the parasite cell surface and throughout the host cell cytoplasm, the slightly anisofilar polar tubule, and genetic relatedness. Several species included in the genus *Anncaliia*, such as *A. algerae*, *A. vesicularum*, *A. ocularum*, and *A. connori*, have the ability to cause infection in diverse organs of humans and grow well in different mammalian cell cultures. For example, *A. algerae* (syn. *Nosema algerae* Vávra and Undeen, 1970; syn. *Brachiola algerae* Lowman et al. 2000), a parasite of *Anopheles* mosquitoes, has also been isolated from a human cornea, a subcutaneous nodule, deep muscle tissue, and most recently from human vocal cord. All of these human isolates of *A. algerae* are morphologically, serologically, and genetically similar to mosquito-derived isolates including the original isolate of Vávra and Undeen. All of these isolates grew well in mammalian cell cultures at 37°C and produced spores. TEM revealed that all developmental stages including meronts, sporoblasts, and spores were diplokaryotic and developed in direct contact with the host cell cytoplasm, a feature characteristic of the genus *Anncaliia*. Sequencing of the PCR products of the amplified SSU rRNA gene revealed the existence of two distinct genotypes; the original mosquito (Undeen) isolate belonged to genotype 1 and the isolate from cornea and that from the deep muscle biopsy to genotype 2, whereas the isolates from a mosquito and one of the other two human isolates (one from skin abscess) had both genotypes 1 and 2. It is known that spores of mosquito-derived *A. algerae* can not only proliferate in mammalian cell cultures at 37°C but also can infect mice when injected into footpads or deposited on the corneal surface. These observations indicate that *A. algerae* is a potential pathogen in humans, especially those with immunodeficiency.

Studies in the last 15 years have clearly significantly improved our understanding of the biology and taxonomy of *Anncaliia* spp. They have also shown that both *A. algerae* genotypes 1 and 2 can infect mosquitoes and humans, suggesting that mosquitoes can passively transmit microsporidian infections to humans by being swatted while biting. However, only a few human cases of infection with *A. algerae* and other *Anncaliia* spp. have been described. Thus, detailed epidemiologic studies are needed to accurately assess the prevalence of human microsporidiosis caused by *Anncaliia* spp. and to identify risk factors involved in the acquisition of human infections. In particular, the role of water in the transmission of microsporidia such as *A. algerae* and the mechanism involved in the corneal and subcutaneous infections needs to be studied.

ACKNOWLEDGMENTS

We thank Dr. Gordon J. Leitch for generating the electron microscopy pictures. The findings and conclusions in this report are those of the authors and do not necessarily represent the official position of the Centers for Disease Control and Prevention.

REFERENCES

1. Canning, E.U. and Lom, J. *The Microsporidia of Vertebrates*. Academic Press, London, U.K., 1986.
2. Wittner, M. and Weiss, L.M. *The Microsporidia and Microsporidiosis*. ASM Press, Washington, DC, 1999.
3. Baker, M.D. et al. Phylogenetic relationships among *Vairimorpha* and *Nosema* species (Microspora) based on ribosomal RNA sequence data. *J. Invertebr. Pathol.* 64:100–106, 1994.
4. Katinka, M.D. et al. Genome sequence and gene compaction of the eukaryote parasite *Encephalitozoon cuniculi*. *Nature* 414:450–453, 2001.
5. Keeling, P.J. and Slamovits, C.H. Simplicity and complexity of microsporidian genomes. *Eukaryot. Cell* 3:1363–1369, 2004.
6. Gill, E.E. and Fast, N.M. Assessing the microsporidia-fungi relationship: Combined phylogenetic analysis of eight genes. *Gene* 375:103–109, 2006.
7. CDC *Tubulinosema* Bern, C. Personal communication.
8. Matsubayashi, H. et al. A case of *Encephalitozoon*-like body infection in man. *Arch. Pathol.* 67:181–187, 1959.
9. Didier, E.S. and Weiss, L.M. Microsporidiosis: Current status. *Curr. Opin. Infect. Dis.* 19:485–492, 2006.
10. Desportes, I. et al. Occurrence of a new microsporidian: *Enterocytozoon bieneusi* n.g., n. sp., in enterocytes of a human patient with AIDS. *J. Protozool.* 32:250–254, 1985.
11. Cali, A. et al. Corneal microsporidiosis: Characterization and identification. *J. Protozool.* 38:215S–217S, 1991.
12. Visvesvara, G.S. et al. Isolation of *Nosema algerae* from the cornea of an immunocompetent patient. *J. Eukaryot. Microbiol.* 46:10S, 1999.
13. Cali, A. et al. *Brachiola vesicularum*, n.g., n. sp., a new microsporidium associated with AIDS and myosits. *J. Eukaryot. Microbiol.* 45:240–251, 1989.
14. Margileth, A.M. et al. Disseminated nosematosis in an immunologically compromised infant. *Arch. Pathol.* 95:145–150, 1973.
15. Sprague, V. *Nosema connori* n. sp. A microsporidian parasite of man. *Trans. Am. Microsc. Soc.* 93:400–403, 1974.
16. Visvesvara, G.S. et al. Public health importance of *Brachiola algerae* (Microsporidia)—An emerging pathogen of humans. *Folia Parasitol.* 52:83–94, 2005.
17. Coyle, C.M. et al. Fatal myositis due to the microsporidian *Brachiola algerae*, a mosquito pathogen. *N. Engl. J. Med.* 351:42–47, 2004.
18. Cali, A. et al. Human vocal cord infection with the microsporidian *Anncaliia algerae*. *J. Eukaryot. Microbiol.* 57:562–567, 2010.
19. Vávra, J. and Undeen, A.H. *Nosema algerae* n.sp. (Cnidospora, Microsporida) a pathogen in laboratory colony of *Anopheles stephensi* Liston (Diptera, Culicidae). *J. Protozool.* 17:240–249, 1970.
20. Undeen, A.H. and Maddox, J.V. The infection of nonmosquito hosts by injection with spores of the microsporidian *Nosema algerae*. *J. Invertebr. Pathol.* 22:258–265, 1973.
21. Undeen, A.H. Growth of *Nosema algerae* in pig kidney cell cultures. *J. Protozool.* 22:107–110, 1975.
22. Undeen, A.H. and Alger, N.E. *Nosema algerae*: Infection of the white mouse by the mosquito parasite. *Exp. Parasitol.* 40:86–88, 1976.
23. Muller, A. et al. Ribosomal RNA of *Nosema algerae* and phylogenetic relationship to other microsporidia. *Parasitol. Res.* 86:18–23, 2000.
24. Lowman, P.M. et al. The effects of elevated temperatures and various time-temperature combinations on the development of *Brachiola* (*Nosema*) *algerae* n. comb. In mammalian cell culture. *J. Eukaryot. Microbiol.* 47:221–234, 2000.
25. Franzen, C. et al. Transfer of the members of the genus *Brachiola* (Microsporidia) to the genus *Anncaliia* based on ultrastructural and molecular data. *J. Eukaryot. Microbiol.* 53:26–35, 2006.
26. Issi, I.V. et al. The ultrastructure of the microsporidium *Nosema meligethi* and establishment of the new genus *Anncaliia*. *Parazitologia* 27:127–133, 1993.
27. Franzen, C. et al. Morphological and molecular investigation of *Tubulinosema ratisbonensis* gen. nov., sp. nov. (Microsporidia:Tubulinosematidae fam. nov.), a parasite infecting a laboratory colony of *Drosophila melanogaster* (Diptera: Drosophilidae). *J. Eukaryot. Microbiol.* 52:141–152, 2005.
28. Koudela, B. et al. The human isolate of *Brachiola algerae* (Phylum Microspora): Development in SCID mice and description of its fine structure. *Parasitology* 123:153–162, 2001.
29. Moura, H. et al. Characterization of *Nosema algerae* isolates after continuous cultivation in mammalian cells at 37°C. *J. Eukaryot. Microbiol.* 46:14S–16S, 1999.
30. Trammer, T. In vitro replication of *Nosema algerae* (Microsporidia), a parasite of anopheline mosquitoes, in human cells above 36°C. *J. Eukaryot. Microbiol.* 46:464–468, 1999.
31. Cali, A. et al. *Brachiola algerae* spore membrane systems, their activity during extrusion, and a new structural entity, the multilayered interlaced network, associated with the polar tube and sporoplasm. *J. Eukaryot. Microbiol.* 49:164–174, 2002.

32. Cali, A. et al. An analysis of the microsporidian genus *Brachiola*, with comparisons of human and insect isolates of *Brachiola algerae*. *J. Eukaryot. Microbiol.* 51:678–685, 2004.
33. Takvorian, P.M. et al. The early events of *Brachiola algerae* (Microsporidia) infection: Spore germination, sporoplasm structure, and development within host cells. *Folia Parasitol.* 52:118–129, 2005.
34. Visvesvara, G.S. In vitro cultivation of microsporidia of clinical importance. *Clin. Microbiol. Rev.* 15:401, 2002.
35. Belkorchia, A. et al. In vitro propagation of the microsporidian pathogen *Brachiola algerae* and studies of its chromosome and ribosomal DNA organization in the context of the complete genome sequencing project. *Parasitol. Int.* 57:62–71, 2008.
36. Kucerova, Z. et al. Differences between *Brachiola* (*Nosema*) *algerae* isolates of human and insect origin when tested using an in vitro spore germination assay and a cultured cell infection assay. *J. Eukaryot. Microbiol.* 51:339–343, 2004.
37. Trammer, T. Opportunistic properties of *Nosema algerae* (Microspora) a mosquito parasite, in immunocompromised mice. *J. Eukaryot. Microbiol.* 44:258–262, 1997.
38. Kelley, J.F. and Anthony, D.W. Susceptibility of spores of the microsporidian *Nosema algerae* to sunlight and germicidal ultraviolet radiation. *J. Invertebr. Pathol.* 34:164–169, 1979.
39. Avery, S.W. and Undeen, A.H. The isolation of microsporidia and other pathogens from concentrated ditch water. *J. Am. Mosq. Control Assoc.* 3:54–58, 1987.
40. Undeen, A.H. and Vander Meer, R.K. The effect of ultraviolet radiation on the germination of *Nosema algerae* Vávra and Undeen (Microsporida: Nosematidae) spores. *J. Protozool.* 37:194–199, 1990.
41. Fogla, R. et al. Chronic microsporidial stromal keratitis in an immunocompetent, non contact lens wearer. *Ind. J. Ophthalmol.* 53:123–125, 2005.
42. Curry, A. et al. A case of bilateral microsporidial keratitis from Bangladesh—Infection by an insect parasite from the genus *Nosema*. *J. Med. Microbiol.* 56:1250–1252, 2007.
43 Weber, R. et al. Improved light microscopical detection of microsporidia spores in stool and duodenal aspirates. *N. Engl. J. Med.* 326:161–166, 1992.
44. Moura, H. et al. A new and improved "quick-hot Gram chromotrope" technique that differentially stains microsporidian spores in clinical samples including paraffin-embedded tissue sections. *Arch. Pathol. Lab. Med.* 121:888–893, 1997.
45. Didier, E.S. et al. Comparison of three staining methods for detecting microsporidia in fluids. *J. Clin. Microbiol.* 33:3138–3145, 1995.

92 *Encephalitozoon*

Dongyou Liu and Elizabeth S. Didier

CONTENTS

92.1 INTRODUCTION

Microsporidia are evolutionally ancient organisms that display features reminiscent of both prokaryotes and eukaryotes. On the one hand, microsporidia possess a nucleus, mitotic spindle-separated chromosome, an intracytoplasmatic membrane system (cytoskeleton), and polyadenylation on mRNA, typical of eukaryotes. On the other hand, microsporidia have distinct translational apparatus and small genome size but lack recognizable mitochondria, peroxisomes, and a classical stacked Golgi apparatus. In addition, microsporidial ribosomes (70S), ribosomal subunits (30S and 50S), and rRNA (16S and 23S) are characteristic of prokaryotes, with the absence of distinct 5.8S rRNA.[1,2] Interestingly, sequence examination of microsporidian small-subunit (SSU) rRNA suggested that microsporidia may have branched off prokaryotes on the way to become eukaryotes. However, phylogenetic trees constructed on the basis of α- and β-tubulin as well as Hsp70 (a heat-shock protein or chaperonin) gene sequences and those of other genes indicated that the microsporidia are close relatives of fungi and may have evolved degeneratively from higher forms.[3–5]

Not surprisingly, considerable uncertainty has existed in relation to the taxonomic status of microsporidia. For a long period, microsporidia have been classified under the phylum Archezoa and subkingdom Protozoa, which covers primitive amitochondriate eukaryotes such as *Giardia* and *Entamoeba*. These organisms may have acquired mitochondria by endosymbiosis. Following the whole genome sequence analysis of a representative microsporidial organism (i.e., *Encephalitozoon cuniculi*), it became apparent that microsporidia have retained mitochondrion-derived double-membrane organelles, mitosomes, with some mitochondrial functions. Further phylogenetic analyses uncovered that microsporidia are most closely related to the zygomycetes, and their association with fungi is supported by their possession of chitin and trehalose, similar cell cycles, and gene organization of certain genes that are typical of fungi.[6–8] Thus, microsporidia are now regarded as a separate phylum in the Fungi kingdom, consisting of two groups, on the basis of the presence or absence of a membrane surrounding the organisms: the Pansporoblastina (membrane present) and the Apansporoblastina (membrane absent). This may have implications for the discovery of new therapeutic strategies.[9]

As a diverse group of unicellular, obligate intracellular eukaryotes, over 150 microsporidial genera covering >1200 species have been identified.[10,11] Of these, only eight genera have clinical implication, causing human microsporidiosis with intestinal and nonintestinal (i.e., systematic) manifestations, especially in immunocompromised individuals.[12] Members of microsporidial genera *Encephalitozoon* (e.g., *Encephalitozoon cuniculi* and *Encephalitozoon hellem*), *Nosema* (e.g., *Nosema ocularum*), *Vittaforma* (e.g., *Vittaforma corneae*, formerly *Nosema corneum*), *Anncaliia* (e.g., *Anncaliia algerae*, formerly *Brachiola algerae*, *Nosema algerae*), *Pleistophora* (e.g., *Pleistophora ronneafiei*), and *Trachipleistophora* (*Trachipleistophora hominis*, *Trachipleistophora anthropophthera*) are often associated with nonintestinal diseases, whereas *Enterocytozoon* (i.e., *Enterocytozoon bieneusi*) is responsible for common intestinal illnesses in patients with AIDS and *Encephalitozoon* (*Septata*) *intestinalis* causes both intestinal and disseminated infections in persons with AIDS.[9,13,14] Furthermore, *Microsporidium* spp. (i.e., *Microsporidium ceylonensis* and *Microsporidium africanum*) have been reported as agents of ocular infections in immunocompetent persons in a few occasions. In terms of the genome sizes, *Encephalitozoon* is estimated to range from 2.3 to 2.9 Mb; *Nosema*, *Nosema*, *Brachiola*, *Enterocytozoon*, *Pleistophora*, and *Trachipleistophora* may be of 10–15 Mb.[15–17]

While the genus *Encephalitozoon* is discussed in this chapter, *Anncaliia* (*Brachiola*) (Chapter 91); *Enterocytozoon* (Chapter 93); *Nosema*, *Vittaforma*, and *Microsporidium* (Chapter 94); and *Pleistophora* and *Trachipleistophora* (Chapter 95) are covered in separate chapters of this book.

92.1.1 Classification, Morphology, and Life Cycle

The genus *Encephalitozoon* belongs to the family *Unikaryonidae*, suborder Apansporoblastina, phylum Microsporidia, kingdom Fungi. The family *Unikaryonidae* comprises three genera: *Encephalitozoon*, *Microgemma*, *and Orthosomella*, as well as some unclassified *Unikaryonidae*. Currently the genus *Encephalitozoon* consists of four recognized species *Encephalitozoon cuniculi*, *Encephalitozoon hellem*, *Encephalitozoon intestinalis* (formerly *Septata intestinalis*), and *Encephalitozoon lacertae* in addition to four unnamed species *Encephalitozoon* sp., *Encephalitozoon* sp. CF-ENC1, *Encephalitozoon* sp. MB-2008, and *Encephalitozoon* sp. SJ-2008.[18,19]

In general, the microsporidial life cycle undergoes three distinct stages: spore, merogony, and sporogony. Microsporidial spores measure between 1 and 20 μm in length, with those from mammal-infecting species, such as species of *Encephalitozoon*, ranging from 1 to 3 μm in diameter. The spores are covered by a thick wall, consisting of three layers: an electron-dense proteinaceous outer layer called the exospore, an electron-lucent chitinous inner layer called the endospore, and a plasma membrane. Underneath the plasma membrane are the cytoplasm, the nucleus, a posterior vacuole, the polaroplast membranes, and the unique extrusion apparatus. The extrusion apparatus is composed of a coiled polar filament and a characteristic anterior anchoring disk.

After contact with a suitable host cell (e.g., enterocyte), the microsporidial sporoplasm is released from the spore and *Encephalitozoon* species go through a phase of repeated asexual divisions by binary or multiple fissions called merogony to form meronts within membrane-bound parasitophorous vacuoles. Being rounded, irregular, or elongated simple cells with little differentiated cytoplasm, meronts are enclosed by a plasma membrane, larger than mature spores, located at the periphery of the parasitophorous vacuole, and appear to harbor more than one nucleus if cytokinesis is incomplete. Meronts further develop into sporonts having one or two nuclei (if cell division in progress) and a characteristic dense surface coat, which later becomes the exospore layer of the spore wall. Through binary or multiple fissions, sporonts multiply and divide into sporoblasts which then mature into environmentally resistant, infective spores by synthesis of spore organelles. The morphological, biological, and clinical features of *Encephalitozoon* species in comparison to other genera of microsporidia infecting humans are summarized in Table 92.1.

Microsporidial spores are discharged from infected hosts (including humans and animals) to the environment via feces, urine, and/or other body fluids and tissues. Being resistant to environmental conditions and to the usual water treatment procedures, microsporidial spores remain infective for long periods of time, particularly when protected from desiccation. Therefore, person-to-person contact, fecal-oral route, contaminated fingers, inhalation, and waterborne or foodborne contamination represent the most plausible routes of transmission, especially in countries with poor sanitation.[20,21]

To date, three *Encephalitozoon* spp. have been recognized as human pathogens.[22] *Encephalitozoon cuniculi*, first isolated from rabbits in 1922, is distributed in domestic rabbits and a wide range of mammals (including humans) worldwide as well as in carnivores and monkeys from several distinct geographic areas. *Encephalitozoon hellem* was first identified in AIDS patients and has been isolated in avian hosts in the United States and Indonesia in 1991. *Encephalitozoon intestinalis* (formerly *Septata intestinalis*) was also first identified in AIDS patients and has been detected in fecal samples of farm animals in Mexico and in gorillas in Africa in 1993.[23] The other named species in the genus *Encephalitozoon*, *E. lacerate*, is present in reptiles only and appears to be most closely related to *E. cuniculi*.[24,25]

While *E. cuniculi*, *E. hellem*, and *E. intestinalis* produce morphologically indistinguishable spores, intraspecies genetic variations in *E. cuniculi* and *E. hellem* have been noted. Three *E. cuniculi* strains (I, II, and III) were first isolated from rabbits, mice, and dogs, respectively. These strains, however, are known to cross host species (i.e., are not host specific) but may serve as epidemiologic markers. At the DNA level, *E. cuniculi* strain I contains three 5′-GTTT-3′ repeats in the intergenic spacer (ITS) region of the rRNA gene, strain II has two such repeats, and strain III has four such repeats.[26] Further examination at the polar tube protein (PTP) and spore wall protein I (SWP-1) loci confirmed the division of *E. cuniculi* into genotypes I, II, and III.[27] Three *E. hellem* genotypes (1, 2, and 3) are also identified upon analysis of the ITS sequence. *E. hellem* genotype 1 accounts

TABLE 92.1
Morphological, Biological, and Clinical Features of Human-Infecting Microsporidian Genera

Feature	*Anncaliia*	*Nosema*	*Vittaforma*	*Microsporidium*	*Enterocytozoon*	*Encephalitozoon*	*Pleistophora*	*Trachipleistophora*
Spores (μm)	2.5–2.9 × 1.9–2.0	5.0 × 3.0	3.3 × 1.4	3.5–4.5 × 1.5	1.1–1.6 × 0.7–1.0	2.0–2.5 × 1.0–1.5	3.2–3.4 × 2.8	4.0 × 2.4
Coils of polar tubule	7–10	9–12	10	15–16	5–7	5–7	9–12	11
Row of polar tube	1–3	1	1	1	2	1	2	1 or 2
Nucleus[a]	2	2	2	2	1	1	1	1
Presence of vacuole	No	No	No	No	No	Yes	Yes	Yes
Replication within host cell	Binary fission free in the cytoplasm	Binary fission free in the cytoplasm	Binary fission free in the cytoplasm	Binary fission free in the cytoplasm	Multinucleated plasmodium free in the cytoplasm, then individual spores pinch off	Binary fission in parasitophorous vacuole	Binary fission in sporophorous vesicle; with multinucleated plasmodium	Binary fission in sporophorous vesicle; without multinucleated plasmodium
Clinical disease	Keratoconjunctivitis, skin infection, myositis, disseminated infection	Keratitis	Keratitis, urinary tract infection	Keratitis	Diarrhea, acalculous cholecystitis	Keratoconjunctivitis, sinusitis, bronchitis, nephritis, diarrhea, cholangitis, cystitis, prostatitis, urethritis, peritonitis, hepatitis, encephalitis	Myositis	Myositis, keratoconjunctivitis, sinusitis, disseminated infection

Source: Adapted from Franzen, C. and Müller, A., *Microbes Infect.*, 3, 389, 2001.

[a] The nucleus of microsporidia may exist as a monokaryon or diplokaryon. The diplokaryon appears as a double-nucleus structure that functions as a single entity.

for the bulk of *E. hellem* isolates characterized so far and is present mainly in the United States and Europe. *E. hellem* genotypes 2 and 3 (differing from each other only by short insertions) consist of human-derived *E. hellem* isolates from Switzerland and Tanzania.[28,29] The three *E. hellem* genotypes have also been confirmed by polymerase chain reaction (PCR) targeting the PTP gene.[29] In addition, cell culture-propagated *E. intestinalis* has been shown to produce two heterogenic subpopulations that form spores with differing infectivity.[30]

92.1.2 Clinical Features and Pathogenesis

92.1.2.1 Clinical Features

Encephalitozoon intestinalis infects lamina propria macrophages, fibroblasts, and endothelial cells beneath the epithelial surface and disseminates to other organs such as the liver, biliary tract, gallbladder, and kidney. Clinical symptoms of *E. intestinalis* include enteritis with diarrhea, weight loss, malabsorption, cholangitis, and cholecystitis. In biliary infection, sclerosing cholangitis-type lesions induce progressive, irregular obstruction and dilation of the intra- and extra-hepatic bile ducts, causing right upper-quadrant pain and an increasing alkaline phosphatase level. In infection of the gallbladder wall, acalculous cholecystitis eventuates. In intestinal infection, perforation and peritonitis often occur without appropriate treatment. *E. intestinalis* is second only to *Enterocytozoon bieneusi* in its ability to cause human microsporidiosis.[31,32]

Infections with *Encephalitozoon cuniculi* and *Encephalitozoon hellem* may lead to a range of nonintestinal diseases in immunocompromised individuals. These include fatigue, diarrhea, weight loss, keratoconjunctivitis, sinusitis, bronchiolitis, pneumonia, nephritis, urethritis, cystitis, prostatitis, hepatitis, peritonitis, gastroenteritis, cholangitis prostatic abscess, tongue ulcer, cerebral infection, and renal failure. Most patients present with bilateral conjunctival inflammation and also exhibit bilateral punctate epithelial keratopathy, leading to decreased visual acuity and rarely corneal ulcers. More specifically, *E. cuniculi* is responsible for disseminated infections involving nearly all organ systems with clinical manifestations varying from no symptoms to severe disease. *E. hellem* mainly infects the keratoconjunctiva, urinary tract, nasal sinuses, and bronchial system.

92.1.2.2 Pathogenesis

Microsporidia are obligate intracellular pathogens that invade, replicate, and mature within various types of cells and particularly, epithelial cells and monocytes. The infectious particle, the spore, is subsequently released into the lumen of the intestine and discharged with the stool, urine, or mucus, depending on the tissue site of infection. Microsporidial invasion of the enterocytes stimulates the production of cytokines (e.g., interleukin-8) that in turn activate the resident phagocytes and attract new phagocytes into the lamina propria from the blood. The activated leukocytes release prostaglandins, leukotrienes, and platelet-activating factor to induce neurotransmitter-mediated intestinal secretion of chloride and water. In addition, these leukocytes generate histamine, serotonin, and adenosine to inhibit the absorption of epithelial cells. Furthermore, as a consequence of inflammatory response, proteases and oxidants secreted by mast cells may negatively impact epithelial cell growth and produce villus atrophy (shortening or flattening) and crypt hyperplasia. These combined actions contribute to the distortion of villus architecture (e.g., changes in the shape of the cell from columnar to cuboidal to pleomorphic; atrophy of the microvilli; swelling of the mitochondria, Golgi apparatus, and endoplasmic reticulum; and accumulation of lysosomes and lipid vacuoles, sloughing of strips of enterocytes, and denuding the tips of villi), resulting in nutrient malabsorption of vitamin B_{12}, D-xylose, and fat and thus osmotic diarrhea. Inflammation that develops at systemic tissue sites of infection likewise contributes to necrosis and disruption of tissue and organ function (e.g., blurry vision, pneumonia, muscle weakness, etc.).

92.1.3 Laboratory Diagnosis

92.1.3.1 Microscopy

Microscopic examination of spores in clinical samples (e.g., stool specimens, duodenal or bile fluid, urine, conjunctival smears, bronchoalveolar lavage fluid, sputum, nasal discharge, sinus aspirates, or biopsy tissues) offers a valuable approach for the identification of microsporidia to the genus level.[33–35] Frequently, stool specimens from patients suspected of microsporidiosis are assessed by light microscopy with the help of the modified trichrome stain (i.e., concentrated chromotrope 2R).[36] Since microsporidial spores are similar to intestinal bacteria in size, the use of a nonspecific fluorescence method (e.g., Uvitex 2B, Calcofluor white [CFW], and Fungifluor—a formulation of CFW in 10% KOH) helps increase the sensitivity and speed of detection, but this method suffers from being less specific (i.e., also detects fungi and bacterial spores). For visualization of microsporidial organisms in small-bowel biopsy, toluidine blue, methylene blue-azure, basic fuchsin, and Warthin Starry stain (a type of silver stain) can be employed.[37] For reliable identification of microsporidia in paraffin-embedded tissue sections, tissue Gram stains (e.g., Brown-Brenn or Brown-Hopps) may be considered, with spores staining Gram positive. Use of polyclonal or monoclonal antibodies directed against microsporidial spores further enhances the specificity of light or fluorescent microscopic detection of microsporidia. In combination with specific monoclonal or polyclonal antibodies, flow cytometry (FC) allows convenient and specific morphofunctional evaluation and quantification of individual microsporidial microorganisms. Additionally, ultrastructural analysis by transmission electron microscopy enables species-specific differentiation of microsporidial organisms on the basis of the fine-structure features of the spores and proliferative forms, method of division, and nature of the host cell–parasite interface (Table 92.1). It is of importance to note that all stages of the life cycle are observable in tissue, but only spores are detected in body fluids or stool samples.[38]

92.1.3.2 Cell Culture and Animal Models

As obligate intracellular organisms, many microsporidia infecting humans require cell culture systems for growth. Microsporidia species such as *E. cuniculi*, *E. hellem*, and *E. intestinalis* have been successfully cultivated in monkey and rabbit kidney cells (Vero and RK13), human fetal lung fibroblasts (MRC-5), Madin Darley canine kidney (MDCK), etc.[39,40] More recently, it was shown that *Encephalitozoon* species displayed high growth in vero cell line, followed by statens seruminstitut rabbit cornea (SIRC) and HeLa cell lines, while *Vittaforma corneae* grew well in SIRC cells, followed by Vero cells.[41] In vitro cultivation of microsporidia permits improved identification of the organisms, elucidation of their biology, development of immunologic agents, and evaluation of their drug susceptibility.[42] To date, however, long-term growth of *Enterocytozoon bieneusi* in culture has remained unsuccessful.

Animal models offer an important avenue for the assessment of host immune responses, vaccine candidates, therapeutic strategies, and routes of transmission as well as for the production of poly- and monoclonal antibodies for immunological assay development. BALB/c and C57Bl/6 athymic mice as well as SCID mice have been used to establish experimental infection via oral and intraperitoneal inoculation with *E. cuniculi*, *E. hellem*, or *V. corneae*. In addition, rabbits may be infected with *E. cuniculi* spores via oral and rectal routes; simian immunodeficiency virus-infected rhesus macaque monkeys have also been infected with *E. cuniculi*, *E. hellem*, and *V. corneae* introduced by oral or intravenous routes of inoculation.

92.1.3.3 Immunologic Assays

Serological tests such as ELISA and western blot immunodetection employing whole spores, whole spore lysates, or recombinant antigens of the polar tube have been used for specific detection of antibodies to *Encephalitozoon* spp. Indirect immunofluorescent-antibody test (IFAT) using spore antigens of *E. cuniculi* or parasite cell suspensions has been employed for epidemiological investigation of human microsporidial infections as well. Other immunological tests (e.g., carbon immunoassay and counterimmunoelectrophoresis) have been used for the detection of IgG and IgM antibodies to microsporidia such as *E. cuniculi* primarily in laboratory animal studies.[43–47] Serology is considered diagnostic in immunocompetent laboratory animals but has not been proven to be of definitive or reliable diagnostic value in humans.

92.1.3.4 Molecular Techniques

Molecular methods, in particular PCR, have been increasingly utilized for microsporidia detection and identification due to their intrinsic sensitivity and specificity.[48–51] Incorporation of restriction fragment length polymorphism (RFLP) analysis, single-strand conformation polymorphism (SSCP), or sequencing with PCR enables the identification of microsporidia to the subspecies level (strains or genotypes).[52–56] The genes or gene regions that have proven valuable for microsporidial identification, phylogenetic analysis, and susceptibility testing consist of small- and large-subunit rRNA genes (i.e., 16S SSU rRNA and a 23S LSU rRNA), the ITS region, and α- and β-tubulin genes. Other genes or gene regions may be also targeted for microsporidia characterization. These include genes encoding isoleucyl-tRNA synthase, glutamyl-tRNA synthetase, homologues of U2 and U6 spliceosomal RNA, reverse transcriptase, chitin synthase, DNA-directed RNA polymerase II, translation elongation factors EF-1α and EF-2, chaperone protein Hsp70, actin (ACT1), PTP, ribosomal protein, dihydrofolate reductase, serine hydroxymethyltransferase, aminopeptidase, and thymidylate synthase.[1,57]

92.2 METHODS

92.2.1 Sample Preparation

92.2.1.1 Sample Processing

Corneal scrapings are collected by an ophthalmologist using a sterile blade number 15 on a Bard Parker handle, under a slit lamp biomicroscope. The cornea is topically anesthetized (1% proparacaine hydrochloride) before scraping. Multiple scrapings are taken from the lesions and smeared on presterilized slides for staining. The smears are stained with (i) potassium hydroxide + CFW (KOH + CFW; Sigma-Aldrich), (ii) Gram's stain, (iii) Giemsa stain (Diff Quick), and/or (iv) modified Ziehl Neelsen stain (1% H_2SO_4). Scrapings are also collected in 0.5 mL of phosphate-buffered saline (pH 7.2) and stored at −20°C until tested by PCR.[41]

Each stool sample (0.5–1.0 g) is homogenized in 8 mL of distilled water and filtered through a 300 μm-mesh sieve. Following the addition of 3 μL of ether, the water–ether mixture is vortexed for 1 min and then centrifuged at 700 × *g* for 2 min. After decanting the supernatant, the pellet is diluted in 100 μL of water, and a part (20 μL) of the suspension may be used for preparing smears (about 15 mm^2). Other fluids that may be considered for diagnosis (e.g., urine, mucus, cerebral spinal fluid) may require washing with saline and resuspension of the pellets with saline prior to preparing the smears. Smears are fixed in methanol for 2 min and stained with 1% Uvitex 2B in phosphate-buffered saline for 15 min, counterstained with 0.5% Evans blue (Sigma) in phosphate-buffered saline for 1 min, and examined by two independent investigators at ×1000 magnification under a fluorescence microscope (Zeiss) with a 50 W mercury high-pressure lamp, an excitation filter with a transmission rate of 355–425 nm, and a suppression filter of 460 nm.[58]

92.2.1.2 DNA Extraction

For extraction of DNA from tissue biopsy (preferably teased apart or disrupted), body fluids, or cultured parasites, the specimens are washed twice in phosphate-buffered saline before being placed in 300 μL of lysis buffer containing 10 mM Tris–HCl pH 8.0, 100 mM NaCl, 20 mM dithiothreitol, 2 mg of proteinase K per milliliter, and 250 U of lyticase (Sigma). If testing fecal specimens, approximately 1 g of stool is suspended in saline, washed two or three times, filtered through three to four layers of gauze, and resuspended in 1 mL of saline for extraction. After 15 min of incubation, mechanical disruption is performed with 0.5 mm diameter glass beads (Biospec Products) and a mini-bead beater (Biospec Products) at 5000 rpm for 3 min (or vortexed, if a bead-beater is unavailable). The suspension is then incubated at 37°C for 18 h. After addition of 150 µL of 2% sodium dodecyl sulfate and 20 µL of proteinase K (20 mg/mL), the incubation is continued at 50°C for 72 h. DNA is extracted with phenol-chloroform and precipitated with ethanol. Nucleic acid from each sample is resuspended in 50 µL of distilled water and stored at −20°C before use in PCR.[27,29] Alternatively, commercially available DNA extraction kits may be applied.

92.2.2 Detection Procedures

92.2.2.1 PCR for *Encephalitozoon intestinalis*

92.2.2.1.1 PCR Protocol of Coyle et al.

Coyle et al.[59] reported a PCR targeting the SSU ribosomal DNA (SSU-rDNA) for specific identification of *E. intestinalis*. Use of forward primer V1 (5′-CACCAGGTTGATT CTGCCTGAC-3′) and reverse primer SI500 (5′-CTCGCT CCTTTACACTCG-3′) enables the amplification of a 375 bp fragment from *E. intestinalis* only.

Procedure

1. PCR mixture (50 µL) is composed of 1× PCR buffer, 2.5 mM of $MgCl_2$, 20 pmol of each primer, 200 µM concentrations of each dNTP, 1.25 U of Goldstar Taq DNA polymerase (Applied Biosystems), and 10 µL of the DNA.
2. Amplification is conducted with an initial denaturation at 94°C for 5 min; 35 cycles of 94°C for 45 s, 55°C for 30 s, and 72°C for 45 s; and a final extension at 72°C for 10 min.
3. The amplified product is separated by electrophoresis on a 2% agarose gel and visualized after staining with ethidium bromide.

Note. To further verify the identity of *E. intestinalis*, the amplified product may be purified with GenElute PCR Clean-Up kit (Sigma) and sequenced using ABI Prism 377 DNA Sequencer (Applied Biosystems). The resulting sequence is then compared with those available in the GenBank Database with the BlastN program.[60]

92.2.2.1.2 PCR Protocol of Da Silva et al.

Da Silva et al.[61] designed primers SINTF (5′-TATGAGAAGT GAGTTTTTTTC-3′) and SINTR (5′-CCGTCCTCGTTCT CCTGCCCG-3′) from SSU rRNA sequence for specific amplification of a 547 bp product from *E. intestinalis*.

The PCR mixture (25 µL) is composed of 2.5 µL Taq buffer containing 3 mM $MgCl_2$, 1.0 µL dNTPs (0.2 mM), 0.5 µL Taq purple polymerase, 1.0 µL 25 pmol solution of each primer, and 2.0 µL DNA template.

The mixture is subjected to an initial denaturation at 94°C for 5 min; 35 cycles of 94°C for 45 s, 55°C for 30 s, and 72°C for 45 s; and a final extension at 72°C for 10 min.

The resulting PCR products (547 bp) are separated on 1% agarose gel, stained with ethidium bromide and visualized under UV light.

92.2.2.1.3 Real-Time PCR Protocol of Menotti et al.

Menotti et al.[62] described a TaqMan real-time PCR for detection and quantification of *E. intestinalis*. The primer set FEI1 (5′-GCAAGGGAGGAATGGAACAGAACAG-3′) and REI1 (5′-CACGTTCAGAAGCCCATTACACAGC-3′) amplifies a 127 bp fragment between positions 988 and 1114 in the SSU rRNA gene. The TaqMan fluorescent probe (5′-CGGGCGGCACGCGCACTACGATA-3′ at positions 1036–1058) is labeled at the 5′ end with 6-carboxyfluorescein and at the 3′ end with 6-carboxytetramethylrhodamine.

Procedure

1. Real-time quantitative PCR mixture (50 µL) is made up of 5 mM $MgCl_2$, 1× Applied Biosystems *Taq*Man buffer A, 200 µM dATP, 200 µM dCTP, 200 µM dGTP, 400 µM dUTP, 0.2 µM FEI1 primer, 0.2 µM REI1 primer, 0.4 µM *Taq*Man fluorescent probe, 0.5 U of Applied Biosystems AmpErase uracil DNA glycosylase, 1.25 U of Applied Biosystems Ampli*Taq* Gold, and 5 µL of spore-extracted standard DNA or patient sample DNA.
2. Amplification is performed using the following cycling parameters: 2 min at 50°C and 10 min at 95°C, 45 cycles at 95°C for 15 s and 65°C for 1 min.
3. Results are expressed as threshold cycle (Ct) values corresponding to the cycle at which PCR enters the exponential phase, i.e., when the 6-carboxyfluorescein fluorescence exceeds 10 times the standard deviation (SD) of the mean baseline emission for cycles 3–15.

Note. Each set of PCR assays comprises serial dilutions of *E. intestinalis* spores as external standards. Each dilution is tested in triplicate. A standard curve is established between the Ct values and the spore numbers by using the Applied Biosystems sequence detection system software, version 1.6.3. Each clinical specimen, also tested in triplicate, is quantified on the basis of this standard curve to estimate the spore number by interpolation of the Ct value obtained by real-time PCR. Negative controls (distilled water) are included in each set of experiments.

92.2.2.2 PCR for *Encephalitozoon cuniculi*

92.2.2.2.1 PCR Identification of Encephalitozoon cuniculi

Visvesvara et al.[63] described primers ECUNF (5′-ATGAGA AGTGATGTGTGTGCG-3′, positions 344–364) and ECUNR (5′-TGCCATGCACTCACAGGCATC-3′, positions 872–892) from *E. cuniculi* SSU rRNA sequence for specific amplification of a 549 bp fragment. PCR mixture (25 μL) is composed of 2.5 μL Taq buffer containing 3 mM $MgCl_2$, 1.0 μL dNTPs (0.2 mM), 0.5 μL Taq purple polymerase, 1.0 μL 25 pmol solution of each primer, and 2.0 μL DNA template. The mixture is subjected to an initial denaturation at 94°C for 5 min; 35 cycles of 94°C for 45 s, 55°C for 30 s, and 72°C for 45 s; and a final extension at 72°C for 10 min. The resulting PCR products (547 bp) are separated on 1% agarose gel, stained with ethidium bromide, and visualized under UV light.

92.2.2.2.2 PCR-RFLP Genotyping of *Encephalitozoon cuniculi*

Xiao et al.[29] reported a PCR-RFLP technique for genotyping *E. cuniculi* with primers from the PTP gene. The use of primers 5′-GCAGTTCCAGGCTACTAC-3′ (nucleotides 840–857) and 5′-AGGAACTCCGGATGTTCC-3′ (nucleotides 1185–1202) results in the amplification of 363 bp fragment for *E. cuniculi* genotypes I and II and a 285 bp product for *E. cuniculi*. Further differentiation between *E. cuniculi* genotypes I and II is made possible by digestion of the 363 bp product with restriction enzyme *Sau*96I, which cleaves genotype I product at two sites (209, 76, and 78 bp) and genotype II at three sites (131, 78, 76, and 78 bp). The three *E. cuniculi* genotypes identified by this technique correlate well with those by sequencing analysis of the ITS region of *E. cuniculi* rRNA gene.[26]

92.2.2.3 PCR for *Encephalitozoon hellem*

92.2.2.3.1 PCR Identification of *Encephalitozoon hellem*

Visvesvara et al.[63] described primers EHEL-F (5′-TGAGAAG TAAGATGTTTAGCA-3′, positions 358–378) and ECUNR (5′-GTAAAAAGACTCTCACACTCA-3′, positions 884–904) from *E. hellem* SSU rRNA sequence for specific amplification of a 547 bp fragment. PCR mixture (25 μL) is composed of 2.5 μL Taq buffer containing 3 mM $MgCl_2$, 1.0 μL dNTPs (0.2 mM), 0.5 μL Taq purple polymerase, 1.0 μL 25 pmol solution of each primer, and 2.0 μL DNA template. The mixture is subjected to an initial denaturation at 94°C for 5 min; 35 cycles of 94°C for 45 s, 55°C for 30 s, and 72°C for 45 s; and a final extension at 72°C for 10 min. The resulting PCR products (547 bp) are separated on 1% agarose gel, stained with ethidium bromide, and visualized under UV light.

92.2.2.3.2 PCR Genotyping of *Encephalitozoon hellem*

Xiao et al.[29] developed a PCR assay for rapid genotyping analysis of *E. hellem*. By exploiting the length polymorphism among *E. hellem* genotypes in the PTP gene, specific products of 461, 521, 581, and 611 bp are amplified from *E. hellem* genotypes 1A, 1B, 1C, and 2B with primers 5′-CATGCTTGCCAACACAGG-3′ (nucleotides 764–781) and 5′-TGGAGGCATTGCAATAGG-3′ (nucleotides 1207–1224) and detected by agarose gel electrophoresis.

92.2.2.4 PCR for *Encephalitozoon* spp.

Wolk et al.[64] developed pan-*Encephalitozoon* primers and fluorescence hybridization probes to amplify a 268 bp region of the 16S rRNA gene of three *Encephalitozoon* spp. (*E. cuniculi*, *E. hellem*, and *E. intestinalis* and to identify them by using amplicon melting curve analysis using a LightCycler (LC).

The 5′ end of forward primer 5′-GTC CGT TAT GCC CTG AGA T-3′ corresponds to *E. intestinalis* nucleotide 1013 and the 3′ end of the reverse primer 5′-ACA GCA GCC ATG TTA CGA CT-3′ corresponds to nucleotide 1251. The probes are labeled with fluorescein or red 640 reporter dye as follows: probe 1, 5′-GCC CGT CGC TAT CTA AGA TGA CGC A-fluorescein-3′ (the 5′ base corresponds to *E. intestinalis* nucleotide 1178) and probe 2, 5′-red 640-TGG ACG AAG ATT GGA AGG TCT GAG TC-phosphate-3′ (the 5′ base corresponds to nucleotide 1204).

Procedure

PCR mixture (minus target DNA) is composed of 3.5 mM $MgCl_2$, PCR buffer lacking Mg^{2+}, 1 μM each primer, 0.03 U of Platinum *Taq* DNA polymerase (Invitrogen)/μL, 0.2 mM PCR Nucleotide Mix^{Plus} (Roche), 0.2% uracil-DNA-glycosylase (UNG) (1 U/μL; Epicenter), 0.025% bovine serum albumin (Sigma), 0.2 μM probe 1, and 0.4 μM probe 2.

Thermal conditions are amplicon inactivation (37°C, 5 min), DNA denaturation and UNG inactivation (95°C for 3 min), amplification (45 cycles of 95°C, 60°C for 10 s, and 72°C for 16 s at 20°C/s), amplicon melting (55°C–85°C at 0.2°C/s), and cooling (40°C, 30 s). Negative controls consist of unspiked specimens and diluent blanks. Positive results are determined based on mathematical algorithms included with the LC system. Results are determined to be positive when the fluorescence signal exceeds the baseline at 43 cycles.

Note. This LC PCR assay provides a rapid, sensitive, accurate, and convenient approach for microsporidia detection in comparison with traditional staining methods, with potential for easy adaptation for use in a high-throughput clinical laboratory.

92.3 CONCLUSION

The genus *Encephalitozoon* consists of four recognized microsporidial species *E. cuniculi*, *E. hellem*, *E. intestinalis* (formerly *Septata intestinalis*), and *E. lacertae*, the first three of which are opportunistic human pathogens, especially in immunosuppressed individuals. While *E. intestinalis* tends to cause intestinal disease (e.g., diarrhea, weight loss, malabsorption, cholangitis, and cholecystitis) but commonly disseminates, as well, *E. cuniculi* and *E. hellem* are more often responsible for nonintestinal illnesses (e.g., fatigue, diarrhea, weight loss, keratoconjunctivitis, sinusitis, bronchiolitis, pneumonia, nephritis, urethritis, cystitis, prostatitis, hepatitis, peritonitis, gastroenteritis, cholangitis prostatic abscess, tongue ulcer, cerebral infection, and renal failure) associated with systemic infection.

Laboratory identification and diagnosis of *Enterocytozoon* spp. on the basis of morphological and biochemical characteristics are difficult due to their close resemblance to each other as well as to other related microsporidial genera (e.g., *Enterocytozoon*, *Nosema*, *Vittaforma*, *Pleistophora*, and

Trachipleistophora). Application of molecular techniques has not only facilitated rapid and accurate speciation of these organisms, but also enabled clarification of the taxonomic status of microsporidia. As microsporidia have been increasingly recognized as pathogens in patients with or without immunosuppression, use of molecular techniques will play an even more prominent role in speeding up the diagnosis of *Enterocytozoon* infections and contributing to the control and prevention of human microsporidiosis.

REFERENCES

1. Franzen, C. and Muller, A. Molecular techniques for detection, species differentiation, and phylogenetic analysis of microsporidia. *Clin Microbiol Rev* 12, 243–285 (1999).
2. Franzen, C. and Muller, A. Microsporidiosis: Human diseases and diagnosis. *Microbes Infect* 3, 389–400 (2001).
3. Kamaishi, T. et al. Protein phylogeny of translation elongation factor EF-1 alpha suggests microsporidians are extremely ancient eukaryotes. *J Mol Evol* 42, 257–263 (1996).
4. Peyretaillade, E. et al. Microsporidia, amitochondrial protists, possess a 70-kDa heat shock protein gene of mitochondrial evolutionary origin. *Mol Biol Evol* 15, 683–689 (1998).
5. Keeling, P.J. Congruent evidence from alpha-tubulin and beta-tubulin gene phylogenies for a zygomycete origin of microsporidia. *Fungal Genet Biol* 38, 298–309 (2003).
6. Babenko, V.N. and Krylov, D.M. Comparative analysis of complete genomes reveals gene loss, acquisition and acceleration of evolutionary rates in Metazoa, suggests a prevalence of evolution via gene acquisition and indicates that the evolutionary rates in animals tend to be conserved. *Nucleic Acids Res* 32, 5029–5035 (2004).
7. Fedorov, A. and Hartman, H. What does the microsporidian *E. cuniculi* tell us about the origin of the eukaryotic cell? *J Mol Evol* 59, 695–702 (2004).
8. Thomarat, F., Vivares, C.P., and Gouy, M. Phylogenetic analysis of the complete genome sequence of *Encephalitozoon cuniculi* supports the fungal origin of microsporidia and reveals a high frequency of fast-evolving genes. *J Mol Evol* 59, 780–791 (2004).
9. Didier, E.S. and Weiss, L.M. Microsporidiosis: Current status. *Curr Opin Infect Dis* 19, 485–492 (2006).
10. Weiss, L.M. Microsporidia: Emerging pathogenic protists. *Acta Trop* 78, 89–102 (2001).
11. Garcia, L.S. Laboratory identification of the microsporidia. *J Clin Microbiol* 40, 1892–1901 (2002).
12. Orenstein, J.M. Microsporidiosis in the acquired immunodeficiency syndrome. *J Parasitol* 77, 843–864 (1991).
13. Vossbrinck, C.R. and Debrunner-Vossbrinck, B.A. Molecular phylogeny of the Microsporidia: Ecological, ultrastructural and taxonomic considerations. *Folia Parasitol* (*Praha*) 52, 131–142; discussion 130 (2005).
14. Franzen, C. et al. Transfer of the members of the genus *Brachiola* (microsporidia) to the genus *Anncaliia* based on ultrastructural and molecular data. *J Eukaryot Microbiol* 53, 26–35 (2006).
15. Katinka, M.D. et al. Genome sequence and gene compaction of the eukaryote parasite *Encephalitozoon cuniculi*. *Nature* 414, 450–453 (2001).
16. Keeling, P.J. and Fast, N.M. Microsporidia: Biology and evolution of highly reduced intracellular parasites. *Annu Rev Microbiol* 56, 93–116 (2002).
17. Keeling, P.J. and Slamovits, C.H. Simplicity and complexity of microsporidian genomes. *Eukaryot Cell* 3, 1363–1369 (2004).
18. Hartskeerl, R.A. et al. Genetic and immunological characterization of the microsporidian *Septata intestinalis* Cali, Kotler and Orenstein, 1993: Reclassification to *Encephalitozoon intestinalis*. *Parasitology* 110 (Pt 3), 277–285 (1995).
19. Cali, A., Weiss, L.M., and Takvorian, P.M. Microsporidian taxonomy and the status of *Septata intestinalis*. *J Eukaryot Microbiol* 43, 106S–107S (1996).
20. Dowd, S.E., Gerba, C.P., and Pepper, I.L. Confirmation of the human-pathogenic microsporidia *Enterocytozoon bieneusi*, *Encephalitozoon intestinalis*, and *Vittaforma corneae* in water. *Appl Environ Microbiol* 64, 3332–3335 (1998).
21. Dowd, S.E. et al. Confirmed detection of *Cyclospora cayetanesis*, *Encephalitozoon intestinalis* and *Cryptosporidium parvum* in water used for drinking. *J Water Health* 1, 117–123 (2003).
22. Furuya, K. Spore-forming microsporidian *encephalitozoon*: Current understanding of infection and prevention in Japan. *Jpn J Infect Dis* 62, 413–422 (2009).
23. Franzen, C. et al. Genetic evidence for latent *Septata intestinalis* infection in human immunodeficiency virus-infected patients with intestinal microsporidiosis. *J Infect Dis* 173, 1038–1040 (1996).
24. Didier, E.S. Microsporidiosis: An emerging and opportunistic infection in humans and animals. *Acta Trop* 94, 61–76 (2005).
25. Mathis, A., Weber, R., and Deplazes, P. Zoonotic potential of the microsporidia. *Clin Microbiol Rev* 18, 423–445 (2005).
26. Didier, E.S. et al. Identification and characterization of three *Encephalitozoon cuniculi* strains. *Parasitology* 111(Pt 4), 411–421 (1995).
27. Xiao, L. et al. Genotyping *Encephalitozoon cuniculi* by multilocus analyses of genes with repetitive sequences. *J Clin Microbiol* 39, 2248–2253 (2001).
28. Mathis, A. et al. Genetic and phenotypic intraspecific variation in the microsporidian *Encephalitozoon hellem*. *Int J Parasitol* 29, 767–770 (1999).
29. Xiao, L. et al. Genotyping *Encephalitozoon hellem* isolates by analysis of the polar tube protein gene. *J Clin Microbiol* 39, 2191–2196 (2001).
30. Hoffman, R.M. et al. Identification and characterization of two subpopulations of *Encephalitozoon intestinalis*. *Appl Environ Microbiol* 69, 4966–4970 (2003).
31. Kotler, D.P. and Orenstein, J.M. Clinical syndromes associated with microsporidiosis. *Adv Parasitol* 40, 321–349 (1998).
32. Weber, R., Deplazes, P., and Schwartz, D. Diagnosis and clinical aspects of human microsporidiosis. *Contrib Microbiol* 6, 166–192 (2000).
33. Weber, R. et al. Improved light-microscopical detection of microsporidia spores in stool and duodenal aspirates. The Enteric Opportunistic Infections Working Group. *N Engl J Med* 326, 161–166 (1992).
34. Katzwinkel-Wladarsch, S. et al. Comparison of polymerase chain reaction with light microscopy for detection of microsporidia in clinical specimens. *Eur J Clin Microbiol Infect Dis* 16, 7–10 (1997).
35. Li, X. et al. Efficiency for recovering *Encephalitozoon intestinalis* spores from waters by centrifugation and immunofluorescence microscopy. *J Eukaryot Microbiol* 50(Suppl), 579–580 (2003).
36. Chan, K.S. and Koh, T.H. Extraction of microsporidial DNA from modified trichrome-stained clinical slides and subsequent species identification using PCR sequencing. *Parasitology* 135, 701–703 (2008).

37. Didier, E.S. et al. Comparison of three staining methods for detecting microsporidia in fluids. *J Clin Microbiol* 33, 3138–3145 (1995).
38. Barbosa, J., Rodrigues, A.G., and Pina-Vaz, C. Cytometric approach for detection of *Encephalitozoon intestinalis*, an emergent agent. *Clin Vaccine Immunol* 16, 1021–1024 (2009).
39. van Gool, T. et al. *Septata intestinalis* frequently isolated from stool of AIDS patients with a new cultivation method. *Parasitology* 109 (Pt 3), 281–289 (1994).
40. Visvesvara, G.S. et al. In vitro culture and serologic and molecular identification of *Septata intestinalis* isolated from urine of a patient with AIDS. *J Clin Microbiol* 33, 930–936 (1995).
41. Joseph, J. and Sharma, S. In vitro culture of various species of microsporidia causing keratitis: Evaluation of three immortalized cell lines. *Indian J Med Microbiol* 27, 35–39 (2009).
42. De Groote, M.A. et al. Polymerase chain reaction and culture confirmation of disseminated *Encephalitozoon cuniculi* in a patient with AIDS: Successful therapy with albendazole. *J Infect Dis* 171, 1375–1378 (1995).
43. Beckers, P.J. et al. *Encephalocytozoon intestinalis*-specific monoclonal antibodies for laboratory diagnosis of microsporidiosis. *J Clin Microbiol* 34, 282–285 (1996).
44. Deplazes, P. et al. Immunologic and molecular characteristics of *Encephalitozoon*-like microsporidia isolated from humans and rabbits indicate that *Encephalitozoon cuniculi* is a zoonotic parasite. *Clin Infect Dis* 22, 557–559 (1996).
45. Mo, L. and Drancourt, M. Monoclonal antibodies for specific detection of *Encephalitozoon cuniculi*. *Clin Diagn Lab Immunol* 11, 1060–1063 (2004).
46. Jordan, C.N. et al. Direct agglutination test for *Encephalitozoon cuniculi*. *Vet Parasitol* 135, 235–240 (2006).
47. Furuya, K. et al. Recognition profiles of microsporidian *Encephalitozoon cuniculi* polar tube protein 1 with human immunoglobulin M antibodies. *Parasite Immunol* 30, 13–21 (2008).
48. Owen, R.L. Polymerase chain reaction of stool: A powerful tool for specific diagnosis and epidemiologic investigation of enteric microsporidia infections. *AIDS* 11, 817–818 (1997).
49. Hester, J.D. et al. Species-specific detection of three human-pathogenic microsporidial species from the genus *Encephalitozoon* via fluorogenic 5′ nuclease PCR assays. *Mol Cell Probes* 16, 435–444 (2002).
50. Wang, Z., Orlandi, P.A., and Stenger, D.A. Simultaneous detection of four human pathogenic microsporidian species from clinical samples by oligonucleotide microarray. *J Clin Microbiol* 43, 4121–4128 (2005).
51. Verweij, J.J. et al. Multiplex detection of Enterocytozoon bieneusi and *Encephalitozoon* spp. in fecal samples using real-time PCR. *Diagn Microbiol Infect Dis* 57, 163–167 (2007).
52. Fedorko, D.P., Nelson, N.A., and Cartwright, C.P. Identification of microsporidia in stool specimens by using PCR and restriction endonucleases. *J Clin Microbiol* 33, 1739–1741 (1995).
53. Kock, N.P. et al. Species-specific identification of microsporidia in stool and intestinal biopsy specimens by the polymerase chain reaction. *Eur J Clin Microbiol Infect Dis* 16, 369–376 (1997).
54. Liguory, O. et al. Diagnosis of infections caused by *Enterocytozoon bieneusi* and *Encephalitozoon intestinalis* using polymerase chain reaction in stool specimens. *AIDS* 11, 723–726 (1997).
55. Ombrouck, C. et al. Specific PCR assay for direct detection of intestinal microsporidia *Enterocytozoon bieneusi* and *Encephalitozoon intestinalis* in fecal specimens from human immunodeficiency virus-infected patients. *J Clin Microbiol* 35, 652–655 (1997).
56. Raynaud, L. et al. Identification of *Encephalitozoon intestinalis* in travelers with chronic diarrhea by specific PCR amplification. *J Clin Microbiol* 36, 37–40 (1998).
57. Ghosh, K. and Weiss, L.M. Molecular diagnostic tests for microsporidia. *Interdiscip Perspect Infect Dis* 2009, 926521 (2009).
58. Muller, A. et al. A powerful DNA extraction method and PCR for detection of microsporidia in clinical stool specimens. *Clin Diagn Lab Immunol* 6, 243–246 (1999).
59. Coyle, C.M. et al. Prevalence of microsporidiosis due to *Enterocytozoon bieneusi* and *Encephalitozoon* (*Septata*) *intestinalis* among patients with AIDS-related diarrhea: Determination by polymerase chain reaction to the microsporidian small-subunit rRNA gene. *Clin Infect Dis* 23, 1002–1006 (1996).
60. Chabchoub, N. et al. Genetic identification of intestinal microsporidia species in immunocompromised patients in Tunisia. *Am J Trop Med Hyg* 80, 24–27 (2009).
61. Da Silva, A.J. et al. Detection of *Septata intestinalis* (Microsporidia) Cali et al. 1993 using polymerase chain reaction primers targeting the small submit subunit ribosomal RNA coding region. *Mol Diagn* 2, 47–52 (1997).
62. Menotti, J. et al. Development of a real-time polymerase-chain-reaction assay for quantitative detection of *Enterocytozoon bieneusi* DNA in stool specimens from immunocompromised patients with intestinal microsporidiosis. *J Infect Dis* 187, 1469–1474 (2003).
63. Visvesvara, G.S. et al. Polyclonal and monoclonal antibody and PCR-amplified small-subunit rRNA identification of a microsporidian, *Encephalitozoon hellem*, isolated from an AIDS patient with disseminated infection. *J Clin Microbiol* 32, 2760–2768 (1994).
64. Wolk, D.M. et al. Real-time PCR method for detection of *Encephalitozoon intestinalis* from stool specimens. *J Clin Microbiol* 40, 3922–3928 (2002).

93 *Enterocytozoon*

Jaco J. Verweij and Dongyou Liu

CONTENTS

93.1 INTRODUCTION

The term "microsporidia" is commonly used to describe a group of unicellular, small, obligate, intracellular, spore-forming organisms within the phylum Microsporidia, kingdom Fungi. Being true eukaryotes, microsporidia possess a membrane-bound nucleus and an intracytoplasmic membrane system and show chromosome separation on mitotic spindles. On the other hand, by possessing 70S ribosomes (with ribosomal subunits 30S and 50S containing prokaryote-sized 16S and 23S rRNAs [ribosomal RNAs] without separate 5.8S rRNA, which is considered as an important eukaryotic feature), simple Golgi membranes, relatively small genome (about 2–15 Mb), chitin, and trehalose in the spore wall, and by being free of mitochondria or peroxisomes, microsporidia resemble fungi (especially the zygomycetes).[1] Recent phylogenetic analyses of the small-subunit (SSU) rRNA, the largest subunit of RNA polymerase II, TATA-box-binding protein, glutamyl tRNA synthase, α- and β-tubulin and mitochondrial Hsp70 (a heat shock protein or chaperonin) gene sequences in microsporidia provide further evidence that microsporidia and fungi are close relatives, which may have similarly evolved degeneratively from higher forms.[2,3] An extensive phylogenetic analysis using eight genes placed microsporidia within the fungal clade, as a sister to a combined ascomycetes and basidiomycetes clade.[4]

Based on the presence or absence of an outer membrane surrounding the organisms, microsporidia can be separated into the Pansporoblastina (with membrane) and the Apansporoblastina (without membrane). According to the differences in chromosome cycles, microsporidia are divided into the Dihaplophasea (with a diplokaryon containing two adjoined nuclei in a single unit) and the Haplophasea (with unpaired nuclei).[5] Of the more than 150 microsporidial genera, 8 are known to infect humans: *Anncaliia* (formerly *Brachiola* and *Nosema*), *Encephalitozoon*, *Enterocytozoon*, *Microsporidium*, *Nosema*, *Pleistophora*, *Trachipleistophora*, and *Vittaforma*. In this chapter, we focus on *Enterocytozoon* (i.e., *Enterocytozoon bieneusi*), which is responsible for common intestinal illnesses in patients with AIDS.[6]

93.1.1 Classification, Morphology, and Biology/Epidemiology

The genus *Enterocytozoon* belongs to the family Enterocytozoonidae, suborder Apansporoblastina, phylum Microsporidia, kingdom Fungi. The family Enterocytozoonidae includes five genera: *Desmozoon*, *Enterocytozoon*, *Microsporidium*, *Nucleospora*, and *Paranucleospora* as well as some unclassified Enterocytozoonidae (including Enterocytozoonidae gen. sp.). The genus *Enterocytozoon* consists of three recognized species: *Enterocytozoon bieneusi*, *Enterocytozoon hepatopenaei*, and *Enterocytozoon salmonis*; plus a few unassigned species: *Enterocytozoon* sp. IS2005R, *Enterocytozoon* sp. IS2005S, *Enterocytozoon* sp. IS2005T, *Enterocytozoon* sp. IS2005U, *Enterocytozoon* sp. IS2005V, *Enterocytozoon* sp. IS2005W, and *Enterocytozoon* sp. ST-2009a.

As microsporidia are obligate intracellular organisms that are unable to undertake active metabolic processes outside of the host cell, they are dependent on hosts for nutrients and maintenance of their life cycle, which displays three distinct stages: spore, merogony, and sporogony. Usually measuring 1.0–3.0 μm × 1.5–4.0 μm, microsporidial spores possess a thick wall composed of a proteinaceous outer layer called the exospore, a chitinous inner layer called the endospore, and a plasma membrane covering the cytoplasm. Within the cytoplasm (sporoplasm) are located the nucleus, a posterior vacuole, the polaroplast membranes, and the unique extrusion apparatus, which is composed of a coiled polar filament and a characteristic anchoring disc. Microsporidial nucleus may exist as a monokaryon (having a single nucleus, which is found in *Encephalitozoon*, *Enterocytozoon*, *Pleistophora*, and *Trachipleistophora*) or as a diplokaryon (having two adjoined nuclei in a single unit, which is found in *Anncaliia*, *Microsporidium*, *Nosema*, and *Vittaforma*).

Upon ingestion or inhalation by a suitable host, which is often accompanied with a change in pH or osmotic pressure, the polar filament moves out of the anterior end of the spore and penetrates the host cell to inoculate the infective sporoplasm, which then grows to become a meront. Being round, irregular, or elongated simple cells without clearly differentiated cytoplasm, meronts undergo division by binary or multiple fissions called merogony. Multinucleated plasmodial forms may also appear as a result of nuclear division without cell division. In the next stage (i.e., sporogony), meronts turn into sporonts with a characteristic dense surface coat, which later forms the exospore layer of the spore wall. Also replicating by binary or multiple fissions, sporonts divide into ovoid-shaped sporoblasts that subsequently develop into mature spores (which are infective and relatively tolerant of environmental conditions) by synthesis of spore organelles.[3]

There is clear evidence that some *E. bieneusi* genotypes are host specific, whereas others are zoonotic because they have been found to infect both humans and animals. For example, through examination of a 243 bp nucleotide sequence of the internal transcribed spacer (ITS) region of the rRNA gene, over 50 *E. bieneusi* genotypes have been identified to date in farm and companion animals including pigs, cattle, cats, dogs, and goats; in wildlife including beavers, muskrats, otters, foxes, and raccoons; and in laboratory rhesus macaques. On the other hand, of the *E. bieneusi* genotypes that have been shown to infect humans, only some are clearly zoonotic and capable of infecting other vertebrate hosts, while genotypes A, B, and C dominate clinical isolations from humans. Interestingly, genotype C has been found only in stool specimens of HIV-negative transplant recipients.[7–9] Whereas genotype B was found as the predominant genotype in HIV-positive patients in France and the Netherlands, it was not found in HIV-positive patients in Malawi.[7,8] Analyses of the ITS sequences from human- and animal-infecting *E. bieneusi* suggest that *E. bieneusi* genotypes can be clustered according to host species. Indeed, examination of the ITS region led to the identification of 11 *E. bieneusi* genotypes that demonstrate specificity for cattle. However, several other *E. bieneusi* genotypes such as K (or IV/BEB5/Peru2) and D (or PigEBTS9, Peru9, WL8, and CEbC) have been recovered from humans and animals (Figure 93.1).[8,10]

93.1.2 Clinical Features and Pathogenesis

Microsporidia are capable of infecting a wide range of invertebrates and vertebrates from which microsporidial spores are discharged into the environment via the feces, urine, or respiratory secretions. With a chitinous wall, microsporidial spores are relatively resistant to various environmental conditions. For example, drying (dehydration), incubation in distilled water or acid/alkali buffer (e.g., pH 4/pH 9), or freezing and thawing fail to kill all *E. cuniculi*, *E. hellem*, *E. intestinalis* spores.[11] Thus, microsporidial spores remain infectious in the water, and other environments, and are readily transmitted to susceptible hosts via fecal-oral or oral-oral route, inhalation of aerosols, or ingestion of food contaminated with fecal material. Direct human-to-human transmission of intestinal microsporidiosis is also observed in some rare cases involving men who have sex with men.[12]

Enterocytozoon bieneusi is the most common causes of microsporidial infections in humans. In immunologically competent human hosts such as travelers, microsporidia infections such as *E. bieneusi* may pass unnoticed or usually present with mild clinical signs, which may include self-limiting diarrhea.[13] In immunocompromised individuals, microsporidia such as *E. bieneusi* produce persistent diarrhea along with fever, nausea, fatigue, weight loss, wasting disease, and death. Malnutrition in children may lead to persistent diarrhea upon *E. bieneusi* infection.[14]

93.1.3 Diagnosis

As microsporidia are small organisms that tend to cause nonspecific early symptoms, diagnosis of microsporidiosis is dependent on the application of various phenotypic or genotypic laboratory procedures. Whereas conventional techniques (e.g., microscopy, in vitro culture, in vivo assays, and serological tests) have proven useful for the phenotypic characterization of microsporidia, molecular techniques have been established as the preferred methods for improved diagnosis of microsporidiosis in recent years.

93.1.3.1 Conventional Techniques

93.1.3.1.1 Microscopy

Microsporidial spores in biopsy tissues, body fluids (urine, nasal discharge, sinus aspirates, sputum, bronchoalveolar lavage fluid, duodenal or bile juice, and cerebrospinal fluid), or stool samples are often detected under light microscopy with the help of special stains such as chemofluorescent (optical white) agents, staining the chitin wall of the mature spores,[15] or concentrated chromotrope stain (modified trichrome).[16] Occasionally, spores are observed with periodic acid-Schiff (PAS) stain, silver stains (e.g., Warthin-Starry), acid-fast stains, chromotrope-based stains,

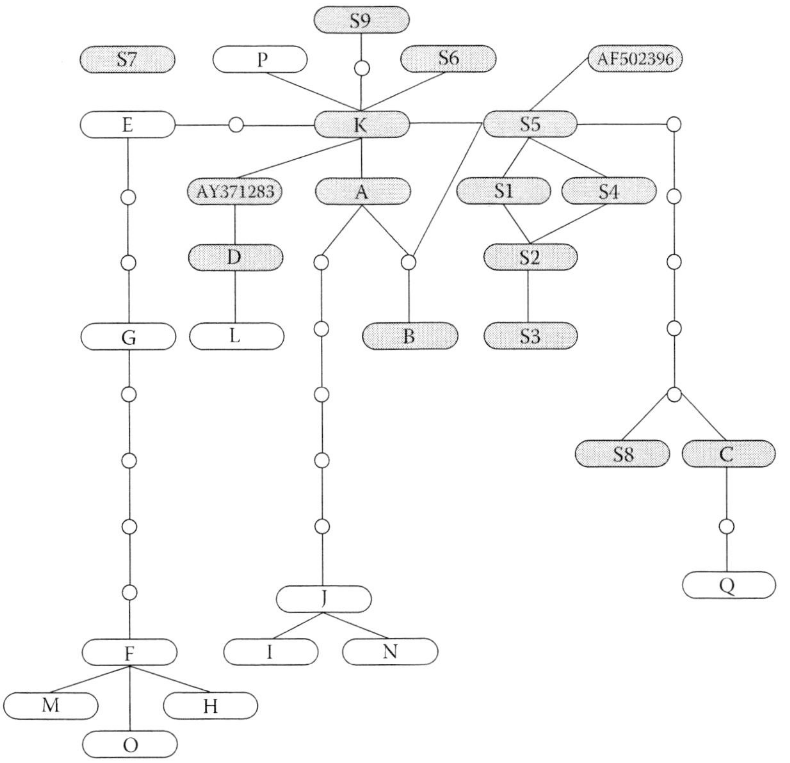

FIGURE 93.1 Map constructed with TCS 1.2.1 software to estimate the gene genealogies of ITS sequences of *Enterocytozoon bieneusi*. The mapped genotypes were derived from isolates in this study[8] (gray boxes) and from the NCBI database (white boxes). Genotypes E, G, F, M, O, and H have been isolated from pigs and genotypes J, I, and N from cattle. Each line between ellipses and/or circles represents a single nucleotide conversion. (From ten Hove, R.J. et al.: Characterization of genotypes of *Enterocytozoon bieneusi* in immunosuppressed and immunocompetent patient groups, *J. Eukaryot. Microbiol.*, 2009, 56, 388, Copyright Wiley-VCH Verlag GmbH & Co. KGaA. Reproduced with permission.)

or chemofluorescent agents. However, identification of microsporidia meront and sporont stages in tissue sections processed with Giemsa, hematoxylin, or eosin stain can be challenging. Given the presence of chitin in the endospore layer, microsporidial spores (showing violet to purple color) are differentiated from lysosomes, mucin droplets, and other debris with a combination of Gram and modified Warthin-Starry stains or tissue Gram stains (Brown-Brenn or Brown-Hopps). For spore detection in ultrathin plastic sections, it is advisable to use methylene blue-azure II-basic fuchsin or toluidine blue stain.[16] Nonetheless, light microscopy does not permit accurate species determination, which is important for selection of appropriate chemotherapy (Figure 93.2).

The unique ultrastructure of the characteristic coiled polar tube and other important structural features of the microsporidial spores can be visualized using electron microscopy (EM). Together with the knowledge of their proliferative forms, method of division, and nature of the host cell–parasite interface, microsporidial spores can be then determined to the genus or even species level. EM has long been regarded as the gold standard for confirmation of microsporidial identity. Nevertheless, identification of microsporidial spores using EM is time-consuming and expensive. Furthermore, it has become increasingly apparent that morphological examination is insufficient for precise determination of microsporidial species or subspecies without detailed

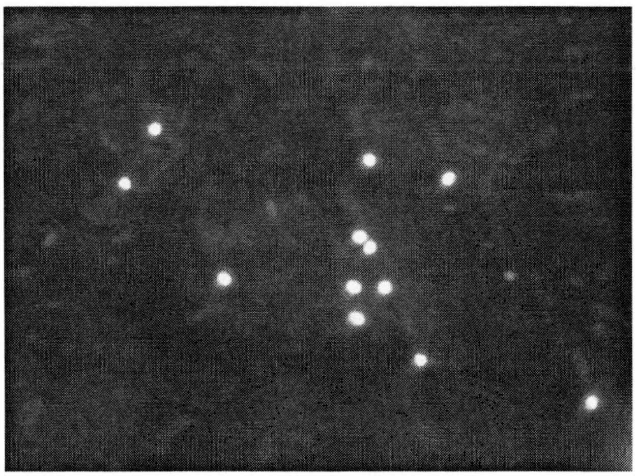

FIGURE 93.2 *Enterocytozoon bieneusi* spores in optical white staining. (Courtesy of Eric A.T. Brienen, LUMC, Leiden.)

assessment of their antigenic and molecular characteristics using other techniques.

93.1.3.1.2 *In Vitro* Culture

Apart from *E. bieneusi*, which is one of the most common human microsporidial pathogens and yet only survives in culture for a short term, most microsporidia from human hosts have the capacity to continuously grow in various established cell lines. The most commonly used cell lines for microsporidial in vitro culture are those derived from monkey and rabbit kidney (Vero and RK-13), human fetal lung fibroblast (MRC-5), and the Madin-Darby canine kidney (MDCK). In addition to providing a confirmatory diagnosis of microsporidial spores, in vitro culture techniques have proven crucial for the elucidation of microsporidial biology, host–parasite interplay, and pathogenesis; for the development and evaluation of serological diagnostic reagents; and for the assessment of the chemotherapy, disinfection, and sterilization regimens that are designed to treat and control microsporidiosis. Despite their time-consuming and laborious nature, in vitro culture techniques continue to play a supporting role in the laboratory confirmation of infections resulting from existing or new microsporidial species, along with microscopic examination, serological and molecular analyses.

93.1.3.1.3 Serological Tests

A large number of serological methods have been developed for specific detection of microsporidia-specific antibodies and antigens. These include carbon immunoassay, immunofluorescent antibody staining, complement fixation, enzyme-linked immunosorbent assay (ELISA), and western immunoblot. The indirect immunofluorescence test and ELISA have been widely used for the diagnosis of microsporidiosis in immunologically competent laboratory animals. In particular, monoclonal antibodies targeting the surface proteins and polar tubes of microsporidial spores have been employed for specific detection of *E. hellem*, *E. cuniculi*, or *E. intestinalis*. However, considering that microsporidiosis frequently occurs in individuals with compromised immune status, who tend to display poor immune response to the invading pathogens, the benefit of using serological methods to diagnose microsporidiosis in this most susceptible population group is severely limited.

93.1.3.1.4 In Vivo Assays

Although, not suitable for patient diagnosis, animal models do offer a valuable approach in the study of microsporidial infections that is not easily obtainable with in vitro culture and other related techniques. For example, examination of route of transmission and host innate and acquired immune responses against microsporidia is impossible without conducting in vivo experiments. Similarly, development and production of laboratory diagnostic reagents (e.g., monoclonal and polyclonal antibodies) are reliant on laboratory animals. Moreover, evaluation of therapeutic drugs and candidate vaccines against microsporidial infections is contingent on first obtaining some critical data concerning their toxicity and efficacy from animal models. Mice (e.g., BALB/c, athymic C57Bl/6, and SCID), rabbits, and rhesus macaque monkeys represent some of the frequently applied animal models for the analysis of microsporidial infections such as *E. cuniculi*, *E. hellem*, *V. corneae*, and *E. bieneusi* via intraperitoneal, oral, or rectal route.

93.1.3.2 Molecular Techniques

In contrast to conventional diagnostic procedures that use the phenotypic characteristics of microsporidia, which can be influenced by changing external conditions, molecular techniques have several distinct advantages. Firstly, as molecular techniques detect differences at the nucleotide level, they are much precise and not affected by variations resulting from altered temperature, osmolarity, pH, and nutrient availability. Secondly, many current molecular detection formats such as PCR and its derivatives involve nucleic acid amplification and they provide an unsurpassed sensitivity. Not surprisingly, molecular techniques have become the methods of choice for microsporidial species identification and strain/genotype determination.[17] Although many genetic components in microsporidia can be targeted for diagnostic purposes, small (16S) and large (23S) subunit rRNA and α- and β-tubulin are favored due to the relative sequence conservation among these genes and the ready accessibility of their nucleotide sequences from various microsporidial species.

93.1.3.2.1 Species-Specific Detection

rRNA molecules are key components of the ribosome that is responsible for protein synthesis in all living organisms. Consisting of two complex folded structures of differing sizes, rRNAs provide a mechanism for decoding messenger RNA (mRNA) into amino acids (at the center of small ribosomal subunit) and to interact with the transfer RNA (tRNA) during translation by providing peptidyltransferase activity (large subunit). Because of its critical role in cellular function and maintenance, rRNAs are not only the most conserved but also the most abundant in all cells. Although being considered as eukaryotes, microsporidia harbor small (16S) and large (23S) subunit rRNA genes that resemble those found in prokaryotes and they also lack a separate 5.8S rRNA that is characteristic of eukaryotes. The small (16S) and large (23S) subunit rRNA genes and their ITS regions have been successfully utilized for the species-specific detection and differentiation of microsporidia infecting humans and animals (e.g., *E. bieneusi*, *E. cuniculi*, *E. hellem*, *E. intestinalis*, *T. hominis*, *V. corneae*, *and V. necatrix*).[17–25]

The tubulin gene family in eukaryotic organisms comprises three distinct and yet conserved subfamilies (i.e., α-, β-, and γ-tubulin) that encode proteins essential for the formation of microtubules, which result from the polymerization of tubulin consisting of a dimer of α- and β-subunits, each about 440 amino acids long. Microtubules constitute the major components of the eukaryotic cytoskeleton and mitotic spindle. The α- and β-tubulins are abundant in eukaryotic cells and their underlying genes are remarkably stable with their nucleotide deletions or insertions being rarely observed. These features make the α- and β-tubulin gene sequences valuable targets for identification and phylogenetic analysis of mitochondrial microsporidia and protozoa. Indeed, a number of reports have described the use of α-tubulin gene for molecular determination of *Nosema locustae* and *Spraguea lophii*[26] and β-tubulin gene for identification of *E. bieneusi*, *E. cuniculi*, *E. hellem*, and *E. intestinalis*.[27–29]

In addition, other genes have also been employed for species-specific detection and identification of microsporidia. These include genes encoding isoleucyl-tRNA synthase, glutamyl-tRNA synthetase, U2 and U6 spliceosomal RNA, reverse transcriptase, chitin synthase, and DNA-directed RNA polymerase II, translation elongation factors EF-1α and EF-2, a chaperone protein Hsp70, actin (ACT1), a polar tube protein, a ribosomal protein, dihydrofolate reductase, serine hydroxymethyltransferase, aminopeptidase, and thymidylate synthase.[17]

93.1.3.2.2 Genotyping

Of the two genera in the family Enterocytozoonidae, the genus *Nucleospora* contains two characterized species: (i) *N. salmonis*, an intranuclear microsporidia of salmonid fish, and (ii) *N. secunda*, a parasite of a warm-water African fish. The genus *Enterocytozoon* includes *E. bieneusi* only, which infects the cytoplasm of enterocytes and other epithelial cells in humans and mammals and is the most common microsporidia species known to cause human disease. Sequence analysis of the ITS region is used for genotyping *E. bieneusi* isolates. In various studies, new names were given for every newly detected sequence variant resulting in partially overlapping nomenclatures, which makes it difficult to compare the results of different studies.[30]

93.1.3.2.3 Susceptibility Testing

Mutations in microsporidial β-tubulin gene have been shown to be linked to susceptibility or resistance to a number of antiparasitic drugs. Being effectors of microtubule polymerization, benzimidazoles (e.g., albendazole) disrupt microsporidial microtubules through β-tubulin binding, and some microsporidia induce mutations in His-6, Phe-167, Glu-198, Phe-200, and Arg-241 to enhance resistance to the derivate benomyl and in Ala-165 to confer resistance to thiabendazole resistance.[31] On the other hand, *E. hellem* and *E. cuniculi* β-tubulins with nonaltered His-6, Phe-167, Glu-198, Phe-200, and Arg-241 are generally susceptible to benzimidazoles.[32] A substitution at Glu (198) in the β-tubulin of *E. bieneusi*, which is one of six amino acids reported to be associated with benzimidazole sensitivity could be a likely explanation for the observed clinical resistance to albendazole.[27]

93.2 METHODS

93.2.1 Sample Preparation

Sample collection: Fresh specimens (e.g., stool, sputum, bronchoalveolar lavage fluid, nasal secretion, or cerebrospinal fluid, conjunctival smears, corneal scrapings, or tissue and body fluids) are preferred for molecular studies. If formalin-fixed fecal samples are used, the formalin has to be removed by washing/centrifuging steps with saline direct or after formalin-ether concentration. However, sensitivity of amplification is known to decrease with time of fixation.

DNA isolation: Both in-house and commercial reagents have been used for preparation of DNA from hookworm eggs. We present a method in the following that is used in our laboratory.[33]

93.2.1.1 DNA Isolation from Feces Using QIAamp Spin Columns

1. Suspend 100 mg of feces in 200 μL of PBS with 2% PVPP (polyvinylpolypyrolidone, Sigma P6755). (note, PVPP does not dissolve, and needs regular stirring to stay like a suspension). Alternatively, suspend 0.5 g of feces in 2 mL of PBS with 2% PVPP and sieve (filter), aliquot 200 μL in a separate eppendorf tube for DNA isolation, and store the rest at −70°C.
2. Freeze sample overnight at −20°C.
3. Heat feces suspension for 10 min in heat block at 100°C.
4. Add an equal volume of tissue lysis buffer (ATL, QIAamp stool kit or QIAamp DNA minikit) containing 20 μL proteinase K (so that is 180 μL of ATL and 20 μL of proteinase K from the kit) solution to the sample (vortex). Incubate 2 h or overnight at 55°C in heatblock (after 1 h vortex briefly).
5. Add 400 μL AL Buffer (with spike) to 400 μL sample, mix thoroughly by vortexing, and incubate at 70°C for 10 min.
6. Centrifuge 30 s full speed and transfer supernatant to an Eppendorf containing 400 μL of ethanol (96%–100%) and mix thoroughly by vortexing (short spin).
7. Place a numbered spin column in a 2 mL collection tube. Carefully apply 600 μL of the mixture from step 5 to the spin column without moistening the rim, close the cap and centrifuge at 10,000 rpm for 1 min.
8. Place spin column in a clean 2 mL collection tube and repeat with the rest of the mixture from step 5.
9. Place the spin column in a clean 2 mL collection tube.
10. Carefully open the spin column and add 500 μL of AW1 buffer. Centrifuge at 10,000 rpm for 1 min.
11. Empty the collection tube and add 500 μL of AW2 buffer. Centrifuge 3 min, 1 min at 10,000 rpm and 2 min at full speed.
12. Place the spin column in a numbered, clean 1.5 mL microfuge tube (with lit).
13. Carefully open the spin column, elute the DNA with 200 μL AE buffer, leave the buffer on the filter for 1 min, and centrifuge at 8000 rpm for 1 min.

93.2.1.2 DNA Isolation from Feces Using Magnapure

1. Suspend 100 mg of feces in 200 μL of PBS with 2% PVPP.
2. Freeze overnight at −20°C.
3. Heat feces suspension for 10 min in heat block at 100°C.
4. Add an equal volume of tissue lysis buffer (ATL) (Qiagen) containing 20 μL proteinase K (so that is 180 μL of ATL and 20 μL of prot K from the kit) solution to the sample (vortex). Incubate 2 h or overnight at 55°C in heat block (after 1 h vortex briefly).
5. Centrifuge 30 s full speed and transfer supernatant to the Magnapure sample-cartridge.
6. Perform isolation using the Magnapure bacterial III kit.
7. Add PhHV (to serve as an internal control) to the Magnapure Lysis buffer.
8. Elute DNA in 100 μL.

93.2.2 Detection Procedures

93.2.2.1 Multiplex Nested PCR

A multiplex nested PCR protocol has been reported for differentiation of *Enterocytozoon bieneusi* from other microsporidia commonly found in humans.[19] In the first round PCR, primers MSP-1 (targeting the 3′ end of the SSU rRNA gene of all microsporidia) is combined with primer MSP-2B (targeting the 5′ region of the LSU rRNA gene of *Enterocytozoon bieneusi*) and primer MSP-2A (targeting the 5′-region of the LSU rRNA gene of other microsporidia). In the second round PCR, primer 3 (targeting the 3′-end of the SSU rRNA gene of all microsporidia) is combined with primer MSP-4B (targeting the 5′-region of the LSU rRNA gene of *Enterocytozoon bieneusi*) and primer MSP-4A (targeting the 5′-region of the LSU rRNA gene of other microsporidia) (Table 93.1).

Procedure

1. Prepare the first round PCR mixture (20 μL) as follows and add 5 μL DNA in the test tube (or distilled water in the negative control tube):

First Round PCR		1 × Mix (μL)	20 × Mix (μL)
Hotstart mastermix		12.5	250
Primer MSP-1	25 μM	0.5	10
Primer MSP-2A	25 μM	0.5	10
Primer MSP-2B	25 μM	0.5	10
BSA	5 mg/mL	0.5	10
H_2O		6.5	130
Total		20.00	400

2. Run PCR amplification in a thermocycler for 1 cycle of 95°C for 15 min; 25 cycles of 95°C for 15 s, 55°C for 30 s, and 72°C for 30 s; and 1 cycle of 72°C for 7 min.

TABLE 93.1
Primers for Multiplex Nested PCR

Oligo Name	Sequence
Primer MSP-1	TGAATGKGTCCCTGT
Primer MSP-2A	TCACTCGCCGCTACT
Primer MSP-2B	GTTCATTCGCACTACT
Primer MSP-3	GGAATTCACACCGCCCGTCRYTAT
Primer MSP-4A	CCAAGCTTATGCTTAAGTCCAGGGAG
Primer MSP-4B	CCAAGCTTATGCTTAAGTYMAARGGGT

3. Separate PCR products (10 μL) on 2% agarose gel containing ethidium bromide and visualize with UV transilluminator.
4. Prepare the second round PCR mixture (20 μL) as follows and add 5 μL of PCR product from the first round PCR in each tube (or distilled water in the negative control tube).

Second Round PCR		1 × Mix (μL)	20 × Mix (μL)
Hotstart mastermix		12.5	250
Primer MSP-3	25 μM	0.5	10
Primer MSP-4A	25 μM	0.5	10
Primer MSP-4B	25 μM	0.5	10
BSA	5 mg/mL	0.5	10
H_2O		6.5	130
Total		20.00	400

5. Perform PCR amplification and detection as above (steps 4 and 5).

Note. After the second round PCR, a 508 bp product is detected from *Enterocytozoon bieneusi* DNA, a289 bp product is detected from *Encephalitozoon intestinalis* DNA and a 305 bp product is detected from *Encephalitozoon cuniculi* DNA. The identity of microsporidial species can be further verified by digestion of the second round PCR products with restriction enzyme *Mn*I, generating fragments of 180, 90, 80, 60, 50, 30, 20 bp for *Enterocytozoon bieneusi*, fragments of 160, 60, 30, 20, 20 bp for *Encephalitozoon intestinalis*, and fragments of 210 and 90 bp for *Encephalitozoon cuniculi*.[19]

93.2.2.2 Real-Time PCR

A multiplex real-time PCR method was developed for the simultaneous detection of *Enterocytozoon bieneusi* and *Encephalitozoon* spp. in stool samples.[25] The *E. bieneusi*-specific real-time PCR amplifies and detects a 103 bp fragment within the ITS sequence of *E. bieneusi*. The *Encephalitozoon* spp. PCR primers and detection probe were designed on the SSU rRNA gene sequence for *E. intestinalis*, amplifying and detecting a 214 bp fragment inside the SSU rRNA gene.

This multiplex PCR included an internal control to detect inhibition of the amplification by fecal constituents in the sample (Table 93.2).

TABLE 93.2
Oligonucleotide Primers and Detection Probes for Real-Time PCR for the Simultaneous Detection of *Enterocytozoon bieneusi, Encephalitozoon spec.*, and Phocin Herpes Virus 1 as an Internal Control

Oligonucleotide Name	Oligonucleotide Sequence	Genbank Accession No. or Literature Reference of Target Sequence
	Target Organism	
Enterocytozoon bieneusi		
EbITS-89F	TGTGTAGGCGTGAGAGTGTATCTG	AF101198
EbITS-191R	CATCCAACCATCACGTACCAATC	AF101198
EbITS-114revT	FAM-5′-CACTGCACCCACATCCCTCACCCTT-3′-Eclipse	AF101198
Encephalitozoon spec.		
MSP1F	CACCAGGTTGATTCTGCCTGAC	U09929
Eint227R	CTAGTTAGGCCATTACCCTAACTACCA	U09929
Eint82Trev	Yakima Yellow-5′-CTATCACTGAGCCGTCC-3′-Eclipse	U09929
Phocin Herpes Virus 1		
PhHV-267s	5′-GGGCGAATCACAGATTGAATC-3′	van Doornum et al (2003)*
PhHV-337as	5′-GCGGTTCCAAACGTACCAA-3′	van Doornum et al (2003)*
PhHV-305tq	Cy5 5′-TTTTTATGTGTCCGCCACCATCTGGATC-3′-BHQ2	van Doornum et al (2003)*

van Doornum, G.J.J. et al., *J. Clin. Microbiol.*, 41, 576, 2003.

Procedure

1. Prepare PCR mixture (25 μL) containing PCR buffer (Hotstart Taq master mix, Qiagen), 5 mM $MgCl_2$, 2 pmol of each *E. bieneusi*-specific primer, 6.25 pmol of each *E. intestinalis*-specific primer, 3.75 pmol of each PhHV-1-specific primer, 2.5 pmol of *E. bieneusi*-specific double-labeled probe, 2.5 pmol of *Encephalitozoon* spp.-specific double-labeled probe, 2.5 pmol of PhHV-1-specific double-labeled probe, and 5 μL of the DNA sample as follows:

			1 × Mix (μL)	100 × Mix (μL)
Hotstar mastermix			12.50	1250
Primer Eb F-EbITS-89F		10 μM	0.20	20,00
Primer Eb R-EbITS-191R		10 μM	0.20	20,00
Probe Eb-EbITS-114revT	FAM	10 μM	0.25	25
Primer MSP-MSP1F		25 μM	0.25	25,00
Primer R Ei-Ein227R		25 μM	0.25	25,00
Probe Ei-Enc82trev	VIC	10 μM	0.25	25
Primer PhHV F-267S		25 μM	0.15	15,00
Primer PhHV R-337 AS		25 μM	0.15	15,00
Probe PhHV-305Tq	CY5	10 μM	0.25	25
$MgCl_2$		25 mM	3.50	350,00
Rox 1:20[a]			0.80	80,00
BSA		5 mg/mL	0.50	50
H_2O			0.75	75,00
Total			20,00	2000,00

[a] Only needed for ABI 7500 can be left out for BioRad Icycler.

2. Perform PCR in I-cycler real-time detection system (BioRad) for 1 cycle of 95°C for 15 min and 50 cycles of 95°C for 15 s, 60°C for 30 s, and 72°C for 30 s. (For amplification in ABI 7500, the cycling program consists of 1 cycle of 95°C for 15 min and 50 cycles of 95°C for 15 s and 60°C for 60 s).
3. Detect fluorescence of FAM, Yakima Yellow, and CY5 at their respective wavelengths during the annealing step of each cycle.

Note. The *Encephalitozoon* spp. primers and probe were designed on the known sequence of *E. intestinalis*. According to NCBI BLAST, however, *E. hellem* and *E. cuniculi* will also be amplified and detected with this primer/probe combination. Although determination to the genus level will be sufficient in clinical practice, species-specific differentiation can be achieved by sequencing of the PCR product.

This PCR was evaluated with species-specific DNA controls ($n = 22$) and a range of well-defined stool samples ($n = 140$) and it demonstrated 100% specificity and sensitivity. Thus, this assay offers the possibility of introducing DNA detection as a feasible technique in the routine diagnosis of intestinal microsporidia infections.

93.2.2.3 Genotyping of *Enterocytozoon bieneusi* on the ITS Region

This PCR amplifies a 296 bp fragment of the ITS-1 region.[8] After purification of the PCR product, both the forward and reverse strands are sequenced with the same primers used for initial PCR:

Oligo Name	Sequence
Eb-80F	GTTGGAGAACCAGCTGAAGGAT
Eb-375R	ATACACCTCTTGATGGCACCCT

Procedure

1. Prepare PCR mixture (20 µL) as follows and add 5 µL DNA in the test tube (or distilled water in the negative control tube):

		1×	20
Hotstart mastermix		12.5	250
Eb-80F	25 µM	0.5	10
Eb-375R	25 µM	0.5	10
BSA	5 mg/mL	0.5	10
H_2O		6	120
Total		20	400

2. Perform PCR amplification for 1 cycle of 95°C for 15 min; 25 cycles of 95°C for 15 s, 55°C for 30 s, and 72°C for 30 s; and 1 cycle of 72°C for 7 min.
3. Separate PCR products (10 µL) on 2% agarose gel containing ethidium bromide and visualize with UV transilluminator.
4. Purify PCR products using, for example, the High Pure PCR Product Purification Kit (Roche Diagnostics) according to the manufacturer's protocol.
5. Measure DNA concentration with, for example, Nanodrop ND-1000 UV-Vis Spectrophotometer (Nanodrop Technologies).
6. Perform sequencing reaction for both the forward and reverse strands with the same primers used for PCR.

Note. In this field of research, every new variation in the ITS sequence has been given a new genotype name and sometimes different names are used for the same variant.

93.3 CONCLUSIONS AND FUTURE PERSPECTIVES

As microsporidial genera and species demonstrate biological and epidemiological variations, which often demand different approaches to prevention and treatment, it is necessary to accurately determine their taxonomic status. The microscopic diagnosis of microsporidia infections using additional staining techniques is labor-intensive; therefore, this will often be limited to samples of severe immune-suppressed patients. Many laboratories will not see as many of these patients due to the success of HAART therapy in HIV infected individuals. As a consequence, gaining expertise is difficult as positive findings will be scarce. Moreover, in recent years intestinal microsporidiosis is increasingly diagnosed in other immunocompromised individuals such as organ transplant recipients, as well as in children, travelers, and elderly.[13,35–43] Nucleic acid-based amplification methods for the detection of microsporidia give useful alternatives for the labor-intensive and expertise-dependent microscopy. Especially real-time PCR, with its reduced risk for contamination and the possibility of multiplex detection of different targets in one tube, offers the possibility of introducing DNA detection as a feasible technique in the routine diagnosis of intestinal microsporidial infections and can also be used as a part of a series of other multiplex assays in the differential diagnosis of diarrhea causing pathogens.

REFERENCES

1. Kent, M.L. et al., A new Microsporidium sp (Microsporidia) from the musculature of the mountain whitefish *Prosopium williamsoni* from British Columbia: Morphology and phylogeny. *J. Parasitol.*, 85, 1114, 1999.
2. Vivares, C.P. et al., Functional and evolutionary analysis of a eukaryotic parasitic genome. *Curr. Opin. Microbiol.*, 5, 499, 2002.
3. Keeling, P.J. and Fast, N.M., MICROSPORIDIA: Biology and evolution of highly reduced intracellular parasites. *Annu. Rev. Microbiol.*, 56, 93, 2002.
4. Gill, E.E. and Fast, N.M., Assessing the microsporidia-fungi relationship: Combined phylogenetic analysis of eight genes. *Gene*, 375, 103, 2006.
5. Sprague, V., Becnel, J.J., and Hazard, E.I., Taxonomy of phylum Microspora. *Crit. Rev. Microbiol.*, 18, 285, 1992.
6. Didier, E.S., Microsporidiosis: An emerging and opportunistic infection in humans and animals. *Acta Trop.*, 94, 61, 2005.
7. Liguory, O. et al., Evidence of different *Enterocytozoon bieneusi* genotypes in patients with and without human immunodeficiency virus infection. *J. Clin. Microbiol.*, 39, 2672, 2001.
8. ten Hove, R.J. et al., Characterization of genotypes of *Enterocytozoon bieneusi* in immunosuppressed and immunocompetent patient groups. *J. Eukaryot. Microbiol.*, 56, 388, 2009.
9. Sing, A. et al., Molecular diagnosis of an *Enterocytozoon bieneusi* human genotype C infection in a moderately immunosuppressed human immunodeficiency virus seronegative liver-transplant recipient with severe chronic diarrhea. *J. Clin. Microbiol.*, 39, 2371, 2001.
10. Sulaiman, I.M. et al., Molecular characterization of *Enterocytozoon bieneusi* in cattle indicates that only some isolates have zoonotic potential. *Parasitol. Res.*, 92, 328, 2004.
11. Franzen, C., How do microsporidia invade cells? *Folia Parasitol. (Praha)*, 52, 36, 2005.
12. Sulaiman, I.M. et al., A molecular biologic study of *Enterocytozoon bieneusi* in HIV-infected patients in Lima, Peru. *J. Eukaryot. Microbiol.*, 50, S591, 2003.
13. Didier, E.S. and Weiss, L.M., Microsporidiosis: Current status. *Curr. Opin. Infect. Dis.*, 19, 485, 2006.
14. Didier, E.S. et al., Epidemiology of microsporidiosis: sources and modes of transmission. *Vet. Parasitol.*, 126, 145, 2004.
15. van Gool, T., Canning, E.U., and Dankert, J., An improved practical and sensitive technique for the detection of microsporidian spores in stool samples. *Trans. R. Soc. Trop. Med. Hyg.*, 88, 189, 1994.
16. Garcia, L.S., Laboratory identification of the microsporidia. *J. Clin. Microbiol.*, 40, 1892, 2002.

17. Franzen, C. and Muller A., Molecular techniques for detection, species differentiation, and phylogenetic analysis of microsporidia. *Clin. Microbiol. Rev.*, 12, 243, 1999.
18. Zhu, X. et al., Small subunit rRNA sequence of *Enterocytozoon bieneusi* and its potential diagnostic role with use of the polymerase chain reaction. *J. Infect. Dis.*, 168, 1570, 1993.
19. Katzwinkel-Wladarsch, S. et al., Direct amplification and species determination of microsporidian DNA from stool specimens. *Trop. Med. Int. Health*, 1, 373, 1996.
20. Wolk, D.M. et al., Real-time PCR method for detection of *Encephalitozoon intestinalis* from stool specimens. *J. Clin. Microbiol.*, 40, 3922, 2002.
21. Menotti, J. et al., Development of a real-time polymerase-chain-reaction assay for quantitative detection of *Enterocytozoon bieneusi* DNA in stool specimens from immunocompromised patients with intestinal microsporidiosis. *J. Infect. Dis.*, 187, 1469, 2003.
22. Menotti, J. et al., Development of a real-time PCR assay for quantitative detection of *Encephalitozoon intestinalis* DNA. *J. Clin. Microbiol.*, 41, 1410, 2003.
23. Franzen, C. and Muller, A., Microsporidiosis: Human diseases and diagnosis. *Microbes Infect.*, 3, 389, 2001.
24. Notermans, D.W. et al., Detection and identification of *Enterocytozoon bieneusi* and *Encephalitozoon* species in stool and urine specimens by PCR and differential hybridization. *J. Clin. Microbiol.*, 43, 610, 2005.
25. Verweij, J.J. et al., Multiplex detection of *Enterocytozoon bieneusi* and *Encephalitozoon* spp. in fecal samples using real-time PCR. *Diagn. Microbiol. Infect. Dis.*, 57, 163, 2007.
26. Mathis, A. et al., Two *Encephalitozoon cuniculi* strains of human origin are infectious to rabbits. *Parasitology*, 114, 29, 1997.
27. Akiyoshi, D.E. et al., Analysis of the beta-tubulin genes from *Enterocytozoon bieneusi* isolates from a human and rhesus macaque. *J. Eukaryot. Microbiol.*, 54, 38, 2007.
28. Mathis, A. et al., Genetic and phenotypic intraspecific variation in the microsporidian *Encephalitozoon hellem*. *Int. J. Parasitol.*, 29, 767, 1999.
29. Mathis, A., Weber, R., and Deplazes, P., Zoonotic potential of the microsporidia. *Clin. Microbiol. Rev.*, 18, 423, 2005.
30. Santin, M. and Fayer, R., *Enterocytozoon bieneusi* genotype nomenclature based on the internal transcribed spacer sequence: A consensus. *J. Eukaryot. Microbiol.*, 56, 34, 2009.
31. Li, J. et al., Tubulin genes from AIDS-associated microsporidia and implications for phylogeny and benzimidazole sensitivity. *Mol. Biochem. Parasitol.*, 78, 289, 1996.
32. Katiyar, S.K. et al., Antiprotozoal activities of benzimidazoles and correlations with beta-tubulin sequence. *Antimicrob. Agents Chemother.*, 38, 2086, 1994.
33. Verweij, J.J. et al., Determining the prevalence of *Oesophagostomum bifurcum* and *Necator americanus* infections using specific PCR amplification of DNA from faecal samples. *Trop. Med. Int. Health*, 6, 726, 2001.

94 *Nosema, Vittaforma,* and *Microsporidium*

Dongyou Liu

CONTENTS

94.1 INTRODUCTION

94.1.1 Classification and Biology

The genus *Nosema* is classified in the family *Nosematidae*, suborder Apansporoblastina, phylum Microsporidia, kingdom Fungi. The family *Nosematidae* consists of into four genera: *Ichthyosporidium*, *Nosema*, *Paranosema*, and *Vittaforma*, together with two *Vittaforma*-like isolates (isolates 5830 and 5843), which may possibly form distinct genera in the family [1].

The genus *Nosema* currently contains the following recognized species: *Nosema antheraeae*, *Nosema apis*, *Nosema bombi*, *Nosema bombycis*, *Nosema carpocapsae*, *Nosema ceranae*, *Nosema chrysorrhoeae*, *Nosema costelytrae*, *Nosema disstriae*, *Nosema distriae*, *Nosema empoascae*, *Nosema epilachnae*, *Nosema fumiferanae*, *Nosema furnacalis*, *Nosema granulosis*, *Nosema heliothidis*, *Nosema kovacevici*, *Nosema necatrix*, *Nosema ocularum Nosema oulemae*, *Nosema plutellae*, *Nosema portugal*, *Nosema pyrausta*, *Nosema serbica*, *Nosema spodopterae*, *Nosema thomsoni*, *Nosema trichoplusiae*, *Nosema tyriae*, *Nosema vespula*, and *Nosema whitei*, in addition to 47 unassigned species [1]. Of these, *Nosema ocularum* is responsible for keratoconjunctivitis in human patients. A former *Nosema* species (*Nosema connori*) causing disseminated infection in humans is now reclassified as *Anncaliia connori* and is discussed in Chapter 91.

The genus *Vittaforma*, also belonging to the family *Nosematidae*, consists of a named species *Vittaforma corneae* (formerly *Nosema corneum*) as well as an unassigned species *Vittaforma* sp. PLDH3. *Vittaforma corneae* has been shown to cause human keratitis and urinary tract infection.

The genus *Microsporidium* is presently classified in the family Enterocytozoonidae suborder Apansporoblastima, phylum Microsporida, Kingtom Fungi, The genus *Microsporidium* comprises 11 recognized species: *Microsporidium africanum*, *Microsporidium arthuri*, *Microsporidium ceylonensis*, *Microsporidium cypselurus*, *Microsporidium epithelialis*, *Microsporidium kosewiense*, *Microsporidium mesnili*, *Microsporidium novacastriense*, *Microsporidium prosopium*, *Microsporidium seriolae*, and *Microsporidium takedai*, in addition to more than 70 unassigned species. Of these, *Microsporidium africanum* and *Microsporidium ceylonensis* were considered as *Nosema* species when first isolated from ocular lesions of immunocompetent persons [2,3]. Subsequent DNA-based investigations suggest their relationship to the members of the family Enterocytozoonidae.

Morphologically, *Nosema* spores are $5.0 \times 3.0\,\mu m$ in size and possess 9–12 coils of polar tubes arranged in a single row and two nuclei [4]. *Vittaforma* spores measure $3.3 \times 1.3\,\mu m$ and have 5–10 coils of polar tubes arranged in a single row and two nuclei. In one light and electron microscopy study on *Vittaforma*-infected corneal tissue, it was noted that *V. corneae* spores are arranged in characteristic linear groups of about 4–8, with each spore containing a diplokaryotic nucleus and a single row of 10 polar tube coils, and posterior coils having a smaller diameter (70 nm) than those located more anteriorly (100 nm) (anisofilar arrangement) [5].

Microsporidium spores are 3.5–4.5 × 1.5 μm in size and contain 15–16 coils of polar tubes arranged in a single row and two nuclei. Electron microscopy examination of corneal tissue infected with *Microsporidium ceylonensis* demonstrated that *M. ceylonensis* spores are uninucleate, with an anisofilar polar tube of 6–10 wide coils and 2–3 narrow coils [6]. Other characteristic morphological, biological, and clinical features of human-infecting microsporidian genera are summarized in Table 92.1.

Nosema, *Vittaforma*, and *Microsporidium* species are obligate intracellular parasites commonly occurring in invertebrates [7]. Similar to other organisms in the phylum Microsporidia, they undergo three distinct phases of development (i.e., spore, merogony, and sporogony) during their life cycle. As causative agents of ocular infections as well as urinary tract infection (as in the case of *Vittaforma*), *Nosema*, *Vittaforma*, and *Microsporidium* spores may enter the host via abrasions or accidental inoculation into eyes, particularly in immunocompetent individuals [8,9]. In one case, *M. ceylonensis* may have been introduced into the eye of an infected patient who was gored in the right eye by a goat a few years earlier. In addition, being highly resistant to external stresses, *Nosema*, *Vittaforma*, and *Microsporidium* spores may gain entry into human hosts via contaminated water (e.g., in rural ponds and swimming pools) [10].

94.1.2 Clinical Features

Nosema ocularum is an occasional cause of human ocular infection [11]. The organism was first isolated from the corneal stroma of a 39-year-old Ohio man. *N. ocularum* spores of 3.0×5.0 μm with 9–12 coils of polar tubule were found in direct contact with the host cell cytoplasm [4,12]. In a recent report, a HIV-negative patient from Bangladesh presenting bilateral keratitis was found to be infected with *Nosema* microsporidia. The patient had bathed in a rural pond 7 days prior to the development of ocular symptoms. The modified Ziehl–Neelson-stained corneal scrape revealed 3–5 μm long, pink, nonbudding, oval organisms on light microscopy. Transmission electron microscopy (TEM) showed groups of spores of 2.5–3.5 μm in length, which were located focally in the corneal tissue, with a lucent endospore and a two-layered exospore, in direct contact with the corneal cell cytoplasm. The spores contained two nuclei as a diplokaryon and their polar tube had six to seven coils arranged in a single row. A few spores had one or two coils of considerably smaller diameter inside the five other coils, indicating the spore maturity [13,14]. Given that *Nosema* species are common parasites of mosquitoes and other hematophagous insects (biting flies), microsporidia inside mosquito larvae developing in the pond in which the patient bathed might have contributed to this case of infection. In addition, the reduced temperature of the human eye and its immune status may favor the establishment of a poikilothermic insect parasite in the cornea of a human host [14,15].

Vittaforma corneae (initially described as *Nosema corneum*) was identified in a stromal biopsy from a non-HIV-infected 45-year-old man from South Carolina, who had traveled to the Caribbean and Central America. The patient presented with central disciform keratitis, recurrent patchy infiltration of the anterior stroma, and iritis [16,17]. The organism showed several taxonomically distinct features after establishment in cell culture and athymic mice and was renamed *Vittaforma corneae* in the newly created genus *Vittaforma* [18,19]. Specifically, the spores measuring 3.7 × 1.0 μm possessed polar tubules with six coils and their nuclei were in diplokaryotic arrangements. The organism appeared individually in the host cell cytoplasm in cell culture and showed tetrasporoblastic sporogony, band-like sporonts, and all developmental stages surrounded by a cisterna of host endoplasmatic reticulum in liver cells of experimentally infected athymic mice as examined by electron microscope [19].

Subsequently, *V. corneae* was detected in the sinunasal aspirate and urine samples of an African patient with AIDS–*E. hellem*, confirming the ability of *V. corneae* to disseminate and survive in deep tissues at least in the immunocompromised host [16–18,20]. Since then, a few other cases of *V. corneae* ocular infections have been reported in United States and Europe [21,22]. In a more recent case, a *V. corneae*-infected 50-year-old male from Cyprus, with negative HIV serology, presented with a right keratitis, showing a yellowish floccular corneal opacity and reduced visual acuity. Light microscopy of biopsy tissues revealed abundant microsporidia within the stroma; electron microscopy confirmed that the spores of 3.3 × 1.4 μm in size contained a diplokaryotic nucleus and a single row of 10 polar tube coils and were linearly arranged in groups of about 4–8 [5]. Additionally, a *Vittaforma corneae*-like organism was described from a significant number of infections in HIV-positive and HIV-negative patients in Portugal [23].

Microsporidium ceylonensis was identified from an ocular infection of a Tamil boy whose eye was gored by a goat [2]. *Microsporidium africanum* was detected in histiocystes of a corneal ulcer of an African woman [3]. No other clinical cases due to *Microsporidium* have since been reported.

94.1.3 Laboratory Diagnosis

Traditional techniques for taxonomic studies and species classification of microsporidia have relied on detection of morphological and biological features such as structure and arrangement of spores, microsporidium–host relationships, and life cycle.

Light microscopic examination of the clinical smears stained with Chromotrope 2R method or its modifications allows easy identification of microsporidia to the genus level, with the thick spore wall showing a bright pinkish red color. Often, a belt-like stripe, which also stains pinkish red, is seen in the middle of the spore. Use of a quick-hot Gram chromotrope technique further speeds up the staining time from 90 min to less than 10 min with the stained spores showing dark violet. Other chemofluorescent agents such as Calcofluor white, Uvitex 2B, and Fungifluor are also useful in the quick identification of microsporidial spores. TEM allows identification of the polar filament and other phylum- and species-specific ultrastructural characters of microsporidia but its routine application in clinical laboratories are hampered by its high expense, time-consuming nature, and relative insensitivity [24].

Similarly, in vitro culture is valuable for diagnosis of human-infecting microsporidia; however, its laboriousness and risk for contamination makes it impractical for microsporidial diagnosis in both clinical specimens and environmental samples [25]. Use of antigen-based detection methods such as immunofluorescence assays (IFA), ELISA, and immunoblot for microsporidial diagnosis has been hindered by the close antigenic and biochemical resemblances that are shared among various microsporidial species [24].

Molecular techniques, in particular PCR, have become increasingly important for the detection and diagnosis of microsporidiosis. Phylogenetic analyses of large subunit (LSU) and small subunit (SSU) rRNA, internal transcribed spacers (ITS), translation elongation factors (EF), heat shock protein (HSP), and β-tubulin genes have proven especially powerful for the clarification of microsporidian taxonomic status [26–30]. For instance, use of universal primer pair V1f (5′-CACCAGGTTGATTCTGCCTGAC-3′) and 1492r (5′-GGTTACCTTGTTACGACTT-3′) facilitates amplification and determination of 1200bp SSU rRNA from previously uncharacterized microsporidial species/strains. Furthermore, application of primer pair 530f (5′-GTGCCAGC(C/A)GCCGCGG-3′) and 580r (5′-GGTCCGTGTTTCAAGACGG-3′) enables direct analysis of 1300–1550bp fragment covering partial SSU rRNA, internal transcribed, and partial LSU rRNA in microsporidia, which, unlike other fungi, do not harbor a 5.8S rRNA between SSU and LSU rRNA [27,31]. A recent study further demonstrated the utility of PCR and sequencing based on SSU for diagnosis of microbial keratitis in which 10 of the 30 patients were shown to be infected with *Vittaforma corneae* [32].

94.2 METHODS

94.2.1 Sample Preparation

94.2.1.1 Light and Electron Microscopy

Microsporidia-infected tissues are fixed for 2h at room temperature by immersion in 3% (v/v) glutaraldehyde in 0.05M sodium cacodylate buffer pH 7.0. After overnight incubation in a refrigerator at 4°C, tissues are washed in a sodium cacodylate buffer rinse, 6 times over 1h, post fixed in 2% (w/v) osmium tetroxide in 0.05M sodium cacodylate buffer, pH 7.0, for 2h, dehydrated in ethanol, and imbedded in Spurr's low-viscosity embedding resin. Sections of 1μm thickness are cut on a microtome. The sections for light microscopy are mounted onto slides, stained with 0.5% (w/v) toluidine blue, and photographed. Sections for electron microscopy are mounted onto 200-mesh Ni grids, stained with 4% (w/v) uranyl acetate and 3% (w/v) lead citrate, and viewed in an H-7000 Hitachi microscope at 75kV [33].

94.2.1.2 DNA Extraction

Corneal scrapings are collected in 0.5mL of phosphate-buffered saline (PBS) (pH 7.2) and stored at −20°C. Samples are centrifuged at 1500×*g* for 3min. The pellet is washed twice using 5mL of PBS buffer and resuspended in 0.5mL of lysis buffer (8M urea, 2% sodium dodecyl sulfate [SDS], 0.15M NaCl, 0.001M ethylenediaminetetraacetic acid [EDTA], 0.1M Tris–HCl). The aqueous lysate is extracted twice with 0.5mL of phenol–chloroform–isoamyl alcohol (25:24:1). The DNA is finally precipitated using 0.1mL volume of 3M NaCl and 2 volumes of ethanol and resuspended in 50μL of double-autoclaved water [13]. Tissue sample is individually ground in liquid nitrogen, and genomic DNA is extracted using the DNAzol DNA purification kit (Invitrogen).

94.2.2 Detection Procedures

94.2.2.1 PCR Identification of *Nosema* spp.

Klee et al. [34] developed a pair of primers SSUrRNA-f1 (5′-CACCAGGTTGATTCTGCCT-3′, positions 1–19) and SSUrRNA-r1c (5′-GTTACCCGTCACTGCCTTG -3′, positions 203–221) from the conserved region of *Nosema* SSU rRNA gene for specific PCR amplification of a 222bp DNA fragment.

Procedure

1. PCR mixture (25μL) contains 5μL of template DNA, 2.5μL of 10× Promega PCR buffer, 2.5mM $MgCl_2$, 200μM of each deoxynucleoside triphosphate (dNTP), 0.625U *Taq* polymerase (Promega), and 0.5μM of each forward and reverse primer. For each PCR experiment, positive (*Nosema* DNA) and negative (sterile water) controls are included.
2. PCR parameters include an initial denaturation at 95°C for 4min; 45 cycles of 95°C for 1min, 48°C for 1min, and 72°C for 1min; and a final extension at 72°C for 4min.
3. Then 10μL of the PCR products is resolved on 1.4% agarose gels (1× TBE) and visualized by ethidium bromide staining.

Note. A 222bp fragment is amplified from all *Nosema* spp. using this pair of primers.

94.2.2.2 PCR Identification of *Vittaforma corneae* Using Pan-Microsporidia Primers

Muller et al. [35] reported a pair of primers V1 (5′-CACCAGGTTGATTCTGCCTGAC-3′) and PMP2 (5′-CCTCTCCGGAACCAAACCCTG-3′) for amplification of a conserved region of the SSU rRNA of *V. corneae*, *E. cuniculi*, *E. hellem*, and *E. intestinalis*. These primers generate amplicons of 268bp for *E. cuniculi*, 270bp for *E. intestinalis*, 279bp for *E. hellem*, and 250bp for *V. corneae*.

Procedure

1. PCR mixture (50 μL) is composed of 25 pmol of each primer, 200 μM concentrations of each dNTP, 10 mM Tris–HCl pH 9.0, 50 mM KCl, 1.5 mM $MgCl_2$, and 2.5 U of *Taq* DNA polymerase.
2. Amplification is conducted with an initial denaturation at 94°C for 3 min; 35 cycles of 94°C for 1 min, 55°C for 1 min, and 72°C for 1 min; and a final extension at 72°C for 10 min.
3. The products are separated on 2% agarose gel, stained with ethidium bromide, and visualized under UV light.

Note. The expected PCR product from *V. corneae* DNA is 250 bp. The identity of the PCR product can be further verified by sequencing analysis. Alternatively, PCR primers specific for *E. cuniculi*, *E. hellem*, and *E. intestinalis* may be used to rule out these organisms [36–38] (see Chapters 92 and 93 of this book).

94.2.2.3 PCR Identification of *Vittaforma corneae*

Ghosh and Weiss [31] listed a pair of primers NCORF1 (5′-TGAGACGTGAAGATGAGTATC-3′) and NCORR1 (5′-TCCCTGCCCACTGTCTCCAAT-3′) that can be utilized for specific identification of *Vittaforma corneae*. The annealing temperature of these primers is 55°C, leading to the amplification of a 375 bp fragment from *Vittaforma corneae* only.

94.2.2.4 Phylogenetic Analysis of Microsporidian SSU rRNA Sequence

Zhu et al. [39] described the use of universal primer pair V1f (5′-CACCAGGTTGATTCTGCCTGAC-3′) and 1492r (5′-GGTTACCTTGTTACGACTT-3′) for phylogenetic analysis of microsporidia. This pair of primers amplifies a 1200 bp fragment from microsporidial SSU rRNA and is valuable for characterization of unknown microsporidial species/strains.

Procedure

1. PCR mixture (50 μL) is made up of 25 pmol of each primer, 200 μM concentrations of each dNTP, 10 mM Tris–HCl pH 9.0, 50 mM KCl, 1.5 mM $MgCl_2$, and 2.5 U of *Taq* DNA polymerase.
2. Amplification is conducted with an initial denaturation at 94°C for 3 min; 35 cycles of 94°C for 1 min, 43°C for 1 min 40 s, and 72°C for 1 min 40 s; and a final extension at 72°C for 10 min.
3. The products (5 μL) are assessed separated on 1% agarose gel, stained with ethidium bromide, and visualized under UV light.
4. The remaining products are purified and sequenced with PCR primers. Use of additional internal primers 530f (5′-GTGCCAGC(C/A)GCCGCGG-3′) and 530r (5′-CCGCGG(T/G)GCTGGCAC-3′) is also helpful for sequencing analysis.
5. The resulting SSU rRNA gene sequences are compared to those in the GenBank to confirm the identity of microsporidial isolates.
6. The sequences are then aligned using the "ClustalX 1.83" program. Phylogenetic trees based on the resultant alignments are, respectively, constructed using the neighbor-joining (NJ) algorithm of Phylip software with Kimura-2 parameter and UPGMA algorithm of MEGA 4.0 software with Maximum Composite Likelihood, and 1000 replications are generated to test the robustness of the tree. The identity and divergence of the sequences are further analyzed with the MegAlign program of DNAStar software.

94.3 CONCLUSION

Nosema, *Vittaforma*, and *Microsporidium* species are closely related microsporidian organisms that are capable of causing keratitis as well as urinary infection (as in case of *Vittaforma*) in immunocompetent and immunocompromised individuals [40]. Being obligate intracellular parasites of invertebrates, these organisms may gain entry into human host via accidental eye injury and contaminated water. Correct determination of microsporidial specie identity is critical for effective treatment. Although light and electron microscopy and in vitro culture techniques are useful for microsporidial identification and diagnosis, their requirement of specialist skills and expensive instrument has effectively limited their application to well-resourced reference laboratories. The recent development of nucleic acid amplification procedures such as PCR has provided a rapid, sensitive, and specific tool for identification and phylogenetic analysis of microsporidial organisms. As a consequence, the species identity of microsporidial organisms can be ascertained precisely from clinical samples in a few hours, contributing to the improved control and prevention of microsporidiosis in human and animal hosts.

REFERENCES

1. The UniProt Consortium. Available at http://www.uniprot.org/, accessed on August 1, 2010.
2. Ashton, N. and Wirasinha, P.A. 1973. Encephalitozoonosis (nosematosis) of the cornea. *Br. J. Ophthalmol.* 57:669–674.
3. Pinnolis, M. et al. 1981. Nosematosis of the cornea. Case report, including electron microscopic studies. *Arch. Ophthalmol.* 99:1044.
4. Cali, A. et al. 1991. Corneal microsporidiosis in a patient with AIDS. *Am. J. Trop. Med. Hyg.* 44:463–468.
5. Rauz, S. et al. 2004. Ultrastructural examination of two cases of stromal microsporidial keratitis. *J. Med. Microbiol.* 53:775–781.
6. Canning, E.U. et al. 1998. Some ultrastructural data on *Microsporidium ceylonensis*, a cause of corneal microsporidiosis. *Parasite* 5:247–254.
7. Didier, E.S. and Weiss, L.M. 2006. Microsporidiosis: Current status. *Curr. Opin. Infect. Dis.* 19(5):485–492.
8. Curry, A. 1999. Human microsporidial infection and possible animal sources. *Curr. Opin. Infect. Dis.* 12:473–480.
9. Didier, E.S. 2005. Microsporidiosis: An emerging and opportunistic infection in humans and animals. *Acta Tropica* 94(1):61–76.
10. Dowd, S.E., Gerba, C.P., and Pepper, I.L. 1998. Confirmation of the human pathogenic microsporidia *Enterocytozoon bieneusi*, *Encephalitozoon intestinalis* and *Vittaforma corneae* in water. *Appl. Environ. Microbiol.* 64:3332–3335.

11. Chan, C.M.L. et al. 2003. Microsporidial keratitis in healthy individuals. *Ophthalmology* 110:1420–1425.
12. Bryan, R.T. et al. 1991. Microsporidia: Opportunistic pathogens in patients with AIDS. *Prog. Clin. Parasitol.* 2:1–26.
13. Joseph, J. et al. 2006. Microsporidial keratitis in India: 16S rRNA gene-based PCR assay for diagnosis and species identification of microsporidia in clinical samples. *Invest. Ophthalmol. Vis. Sci.* 47:4468–4473.
14. Curry, A. et al. 2007. A case of bilateral microsporidial keratitis from Bangladesh—Infection by an insect parasite from the genus *Nosema*. *J. Med. Microbiol.* 56:1250–1252.
15. Hutin, Y.J. et al. 1998. Risk factors for intestinal microsporidiosis in patients with human immunodeficiency virus infection: A case control study. *J. Infect. Dis.* 178:904–907.
16. Davis, R.M. et al. 1990. Corneal microsporidiosis: A case report including ultrastructural observations. *Ophthalmology* 97:953–957.
17. Shadduck, J.A. et al. 1990. Isolation of a microsporidian from a human patient. *J. Infect. Dis.* 162:773–776.
18. Silveira, H. and Canning, E.U. 1995. *Vittaforma corneae* n. comb. for the human microsporidium *Nosema corneum* Shadduck, Meccoli, Davis & Font, 1990, based on its ultrastructure in the liver of experimentally infected athymic mice. *J. Eukaryot. Microbiol.* 42(2):158–165.
19. Silveira, H., Canning, E.U., and Shadduck, J.A. 1993. Experimental infection of athymic mice with the human microsporidian *Nosema corneum*. *Parasitology* 107:489–496.
20. Deplazes, P. et al. 1998. Dual microsporidial infection due to *Vittaforma corneae* and *Encephalitozoon hellem* in a patient with AIDS. *Clin. Infect. Dis.* 27:1521–1524.
21. Font, R.L. et al. 2000. Corneal microsporidiosis: Report of case, including electron microscopic observations. *Ophthalmology* 107:1769–1775.
22. Font, R.L., Su, G.W., and Matoba, A.Y. 2003. Microsporidial stromal keratitis. *Arch. Ophthalmol.* 121:1045–1047.
23. Irshad, M. et al. 2003. Identification of a new microsporidian parasite related to *Vittaforma corneae* in HIV-Positive and HIV-Negative Patients from Portugal. *J. Eukaryot. Microbiol.* 50(6):586–590.
24. Garcia, L.S. 2002. Laboratory identification of the microsporidia. *J. Clin. Microbiol.* 40(6):1892–1901.
25. Visvesvara, G.S. 2002. In vitro cultivation of microsporidia of clinical importance. *Clin. Microbiol. Rev.* 15(3):401–413.
26. Baker, M.D. et al. 1994. Phylogenetic relationships among *Vairimorpha* and *Nosema* species (microspora) based on ribosomal RNA sequence data. *J. Invertebr. Pathol.* 64:100–106.
27. Franzen, C. and Müller, A. 1999. Molecular techniques for detection, species differentiation, and phylogenetic analysis of microsporidia. *Clin. Microbiol. Rev.* 12:243–285.
28. Keeling, P.J. 2003. Congruent evidence from α-tubulin and β-tubulin gene phylogenies for a zygomycete origin of microsporidia. *Fungal Genet. Biol.* 38:298–309.
29. Chan, K.S. and Koh, T.H. 2008. Extraction of microsporidial DNA from modified trichrome-stained clinical slides and subsequent species identification using PCR sequencing. *Parasitology* 135(6):701–703.
30. Franzen, C. and Salzberger, B. 2008. Analysis of the β-tubulin gene from *Vittaforma corneae* suggests Benzimidazole resistance. *Antimicrob. Agents Chemother.* 52:790.
31. Ghosh, K. and Weiss, L.M. 2009. Molecular diagnostic tests for microsporidia. *Interdiscip. Perspect. Infect. Dis.* 2009:926521.
32. Reddy, A.K. et al. 2011. Polymerase chain reaction for the diagnosis and species identification of microsporidia in patients with keratitis. *Clin. Microbiol. Infect.* (in press).
33. Chen, Y.P. et al. 2009. Morphological, molecular, and phylogenetic characterization of *Nosema ceranae*, a microsporidian parasite isolated from the European honey bee, *Apis mellifera*. *J. Eukaryot. Microbiol.* 56(2):142–147.
34. Klee, J., Tay, W.T., and Paxton, R.J. 2006. Specific and sensitive detection of *Nosema bombi* (Microsporidia: Nosematidae) in bumble bees (*Bombus* spp.; Hymenoptera: Apidae) by PCR of partial rRNA gene sequences. *J. Invertebr. Pathol.*, 91:98.
35. Muller, A. et al. 1999. A powerful DNA extraction method and PCR for detection of microsporidia in clinical stool specimens. *Clin. Diagn. Lab. Immunol.* 6:243–246.
36. Visvesvara, G.S. et al. 1994. Polyclonal and monoclonal antibody and PCR-amplified small subunit rRNA identification of a microsporidian, *Encephalitozoon hellem*, isolated from an AIDS patient with disseminated infection. *J. Clin. Microbiol.* 32:2760–2768.
37. de Groote, M.A. et al. 1995. Polymerase chain reaction and culture confirmation of disseminated *Encephalitozoon cuniculi* in a patient with AIDS: Successful therapy with albendazole. *J. Infect. Dis.* 171:1375–1378.
38. Da Silva, A.J. et al. 1997. Detection of *Septata intestinalis* (microsporidia) Cali et al. 1993 using polymerase chain reaction primers targeting the small subunit ribosomal RNA coding region. *Mol. Diagn.* 2:47–52.
39. Zhu, X. et al. 1994. Ribosomal RNA sequences of *Enterocytozoon bieneusi*, *Septata intestinalis* and *Ameson michaelis*: Phylogenetic construction and structural correspondence. *J. Eukaryot. Microbiol.* 41:204–209.
40. Vemuganti, G.K. et al. 2005. Is microsporidial keratitis an emerging cause of stromal keratitis? A case series study. *BMC Ophthalmol.* 5:19–33.

95 *Pleistophora* and *Trachipleistophora*

Dongyou Liu

CONTENTS

95.1 INTRODUCTION

95.1.1 Classification, Morphology, and Biology

The genus *Pleistophora* belongs to the family Pleistophoridae, suborder Pansporablastina, phylum Microsporidia, kingdom Fungi. Currently, the family Pleistophoridae comprises five genera: *Heterosporis*, *Ovavesicula*, *Ovipleistophora*, *Pleistophora*, *Trachipleistophora*, and *Vavraia*. In turn, the genus *Pleistophora* consists of the following recognized species: *Pleistophora ehrenbaumi*, *Pleistophora finisterrensis*, *Pleistophora hippoglossoideos*, *Pleistophora hyphessobryconis*, *Pleistophora mirandellae*, *Pleistophora mulleri*, *Pleistophora ovariae*, *Pleistophora ronneafiei*, and *Pleistophora typicalis*. In addition, it contains a number of unassigned species: *Pleistophora* sp., *Pleistophora* sp. ED-2005, *Pleistophora* sp. LS, *Pleistophora* sp. PA, *Pleistophora* sp. Sd-Nu-IW8201, *Pleistophora* sp. TB, *Pleistophora* sp.1, *Pleistophora* sp.2, and *Pleistophora* sp.3 [1]. As common parasites of fish, *Pleistophora ronneafiei* and other *Pleistophora* spp. have been implicated in clinical cases of human myositis [2].

The genus *Trachipleistophora*, also classified in the family Pleistophoridae, includes three valid species: *Trachipleistophora anthropophthera*, *Trachipleistophora extenrec*, and *Trachipleistophora hominis*, as well as an unnamed species *Trachipleistophora* sp. AH2006a. Based on analyses of the small subunit ribosomal DNA sequences, *Trachipleistophora* spp. are related phylogenetically to species of the genera *Pleistophora* and *Vavraia*, parasites of fish and mosquitoes, respectively [3]. *Trachipleistophora anthropophthera* has been shown to cause encephalitis, myositis, and disseminated infection in patients with AIDS; *Trachipleistophora hominis* is responsible for myositis, keratoconjunctivitis, sinusitis, and rhinitis, also in AIDS patients; *Trachipleistophora extenrec* is isolated from a muscle lesion of an insectivore (the streaked tenrec, *Hemicentetes semispinosus*, Tenrecidae, Mammalia) endemic to Madagascar [4].

Morphologically, *Pleistophora* spores measure 3.2–3.4 × 2.8 μm and possess 9–12 coils of polar tubes arranged in two rows and one nucleus; *Trachipleistophora* spores are of 4.0 × 2.4 μm in size, containing 11 coils of polar tubes arranged in one or two rows and one nucleus. Both *Pleistophora* and *Trachipleistophora* spores develop within a thick parasite-formed coat named the sporophorous vesicle. The genus *Trachipleistophora* differs from the genus *Pleistophora* by forming no multinucleate sporogonial plasmodium at any stage of the life cycle. For morphological, biological, and clinical comparison with other human-infecting microsporidian genera, see Table 92.1.

Although both *Trachipleistophora hominis* and *Trachipleistophora anthropophthera* have a single nucleus, meronts with a thick layer on the outer face of their plasmalemma, and sporogony generating spores inside a thick-walled sporophorous vesicle, *T. anthropophthera* is distinguished from *T. hominis* by its dimorphic feature, with the formation of two types of sporophorous vesicles and spores. Type I is round to oval, polysporous sporophorous vesicle (of 7–10 μm in size), usually with eight spores (3.7 × 2.0 μm), thick endospores, subterminal anchoring disc, and anisofilar polar filaments forming seven thicker and two thinner terminal coils.

This type of sporophorous vesicle has 25–30 nm filaments extending into the host cell cytoplasm. Type II is smaller, bisporous sporophorous vesicle (4–5 × 2.2–2.5 μm in size), with two, nearly round, thin-walled spores (2.2–2.5 × 1.8–2.0 μm) and four to five isofilar coils. The bisporous sporophorous vesicle does not have outside filamentous elements. Both types of sporophorous vesicles can be found within the same cell [5].

The life cycles of *Pleistophora* and *Trachipleistophora* resemble those of other microsporidia, with three distinct stages: spore, merogony, and sporogony. Being resistant to stress, microsporidial spores survive in the environment for a prolonged period of time and are commonly observed in non-tissue specimens. Microsporidial spores possess the coiled polar tube, which pierces the membrane of host cell to allow passage of spore contents (sporoplasm) into the host cell. Once inside the host cell, microsporidian spore undergoes an initial proliferative phase (merogony) followed by a spore-forming phase (sporogony).

Studies conducted in athymic mice demonstrated that *Trachipleistophora hominis* spores produce meronts with two to four nuclei, which divide by binary fission. During sporogony, the surface coat of meronts separates from the plasma membrane to form a sporophorous vesicle, within which meronts undergo repeated binary fissions to become sporoblasts. The number of sporoblasts (and later spores) within the sporophorous vesicles vary from 2 to >32 and the sizes of the sporophorous vesicles range from 5 μm to 14.0 × 11.0 μm in diameter. Spores of 4.0 × 2.4 μm in diameter show a prominent posterior vacuole. *T. hominis* differs from *Pleistophora* spp. by forming no multinucleate sporogonial plasmodia and by having enlarged sporophorous vesicle with non-multilayered wall during sporogony [6,7].

Pleistophora spp. have been known to infect fish muscle before the identification of *P. ronneafiei* from human skeletal muscle [8]. Inside the fish host, *Pleistophora* organisms may either cause serious degradation of the cytoplasm and demise of the cell or elicit host cell hypertrophy, producing a parasite-hypertrophic host cell complex, the xenoma [9]. The first description of *Trachipleistophora* is from a human specimen, which displays developmental differences from *Pleistophora* spp. Human infections with *Pleistophora* and *Trachipleistophora* species may be derived from cold-blooded vertebrates and/or contaminated water [8]. The courses of *Trachipleistophora hominis* infection in mammalian host are uncovered via a severe combined immunodeficient (SCID) mouse model. After eye inoculation with spores, SCID mice show severe infection in the conjunctiva and cornea and lighter infections in the urinary bladder, liver, and spleen within 7–8 weeks. SCID mice inoculated with spores intramuscularly develop heavy infection in muscles around the site of inoculation as well as in several viscera, with a mean survival of 12 weeks. Subcutaneous inoculation of SCID mice with spores results in skin lesions around the inoculation sites, heavy urinary bladder infection, and death in 6 or 7 weeks after infection. Within 5 or 6 weeks of intracerebral inoculation, SCID mice show massive infection in the urinary bladder and liver but none in the brain. Intraperitoneally inoculated SCID mice display heavy infection in the urinary bladder and liver and survive for 13 weeks. Examined after 23 weeks of peroral inoculation, SCID mice are uninfected. These data indicate that *T. hominis* is not tissue specific, and possible routes of infection under natural conditions may be eye contamination or percutaneous inoculation during a blood meal by a hematophagous insect [7].

95.1.2 Clinical Features

Microsporidia are obligate intracellular parasites of both vertebrates and invertebrates that have emerged as important opportunistic pathogens of individuals infected with human immunodeficiency virus type 1 (HIV-1) as well as immunocompetent persons.

Pleistophora spp. are parasites of lower (poikilothermic, cold-blooded) vertebrate hosts such as fish, reptiles, and crustacea [3,8–10]. Human infection with *Pleistophora* may result from consumption of contaminated water or inadequately cooked fish muscle containing the organism. Although the natural hosts of *Trachipleistophora* spp. have yet to be determined, their close phylogenetic relationship with *Pleistophora* as well as some invertebrate-infecting microsporidia such as *Vavraia* species implies that humans may also acquire the infection via exposure to contaminated water [3,11].

The first human infecting *Pleistophora* species was identified from a muscle biopsy of an immunocompromised patient, who presented progressive generalized muscle weakness and contractures, fever, lymphadenopathy, and weight loss [12]. *Pleistophora*-like microsporidian spores were observed in muscle biopsy specimens from the quadriceps and deltoid, and these spores formed multinucleate sporogonial plasmodia, typical of *Pleistophora* spp. Due to its distinct spore size, the organism was designated as a novel *Pleistophora* species: *Pleistophora ronneafiei* [2]. Another case of microsporidian myositis due to *Pleistophora* species concerns a patient with AIDS, who displayed fever, cough, and diffuse myalgias and weakness [13]. A *Pleistophora*-like microsporidium was found in muscle cells in a biopsy specimen from the right quadriceps. Similarly, a HIV-infected patient suffering from myositis presented with fever, myalgia, and progressive weakness. Microsporidian spores were again revealed in a muscle biopsy specimen.

The genus *Trachipleistophora* was established for a microsporidial organism involved in an immobilizing myositis in a patient with AIDS. The organism was demonstrated by light microscopy and electron microscopy in corneal scrapings, skeletal muscle, and nasal discharge. Electron microscopy revealed the presence of spores, meronts with two to four nuclei (dividing by binary fission), sporonts, and sporoblasts. The spores are 4.0 × 2.5 μm in size, with a rugose exospore. Division of sporonts by binary fission creates a sporophorous vesicle, and no multinucleate sporogonial plasmodium is formed at any stage. These differentiate the organism from Pleistophora sp. and enable the establishment of a new genus and species, *Trachipleistophora hominis* [6,14]. A similar case of *Trachipleistophora hominis* infection involving skeletal and cardiac muscle was noted at autopsy of a 47-year-old male with late stage AIDS [15].

Besides myositis, *Trachipleistophora hominis* may be also involved in ocular infection. Light microscopy examination of corneal sample from a 22-year-old man with bilateral, episodic photophobia and blurred vision, associated with vitritis and vasculitis revealed abundant spores within sporophorous vesicles in the corneal stroma. Electron microscopy demonstrated both spores of 4×2.4 μm in size and earlier developmental stages containing a single nucleus (monokaryotic arrangement) within the remains of keratocytes. The polar tube of the spores was arranged as a single row of about 10–13 coils, with the most posterior (2 or 3) coils being smaller (95 nm) than those at the anterior (135 nm) (anisofilar arrangement). These ultrastructural features are consistent with the species *Trachipleistophora hominis* [16].

Trachipleistophora anthropophthera was first found at autopsy in the brain, kidneys, pancreas, thyroid, parathyroid, heart, liver, spleen, lymph nodes, and bone marrow of a patient with AIDS [5]. The organism displays dimorphic sporophorous vesicles and two types of spores simultaneously: polysporous vesicle type I with eight or more oval spores, 3.7–4.0 μm×2.0–2.3 μm and bisporous vesicle type II with two round spores, 1.7–2.2 μm×1.6–2.0 μm in size [17,18]. These characteristic features enable differentiation of *T. anthropophthera* from *T. hominis*. *Trachipleistophora anthropophthera* is also shown to cause keratitis in human patients [19].

95.1.3 Diagnosis

Conventional laboratory diagnosis of microsporidiosis is dependent on the identification of microsporidian spores from feces, urine, secretions, tissue smear, or tissue biopsy using light and electron microscopy. For light microscopic examination of microsporidian spores in body fluids, modified trichrome stains, Giemsa, and optical brightening agents such as calcofluor white can be employed [20]. For histological specimens, Gram stains, periodic acid-Schiff (PAS), silver stains, and Grocotts are useful. Whereas light microscopy allows genus-specific identification, electron microscopy facilitates species-specific determination of microsporidia, especially in tissue biopsy [21].

Other phenotypic techniques for microsporidia identification include in vitro culture, immunoassays and Western blotting. In Western blots, mouse antiserum to *T. hominis* recognized major bands at 39 and 98–150 kDa on *T. hominis* protein profiles. However, this antiserum also showed cross-reaction with high molecular weight antigens of *Vavraia culicis*, *Heterosporis anguillarum*, and three species of *Pleistophora*. Thus, Western blotting may be useful for detection and differentiation of *T. hominis* in muscle biopsy tissue from patients with myositis, where copresence of other microsporidian species is unlikely [22]. Nonetheless, similar to microscopic examination, in vitro culture and immunoassays suffer from the drawbacks of being slow, laborious, or nonspecific [21,23,24]. In addition, serum antibody levels are often variable and undetectable in patients with suppressed immune system.

Due to their extreme rapidity, sensitivity, and specificity, nucleic acid amplification techniques such as PCR have provided a useful alternative for identification and phylogenetic analysis of microsporidia [25,26]. Several conserved gene regions (e.g., small subunit [SSU] and large subunit [LSU] rRNA, internal transcribed spacers [ITS], translation elongation factors [EF], heat shock protein [HSP], and beta-tubulin genes) have often been targeted by molecular procedures for specific detection and taxonomic classification of microsporidian organisms [27–31].

95.2 METHODS

95.2.1 Sample Preparation

95.2.1.1 In Vitro Culture and Microscopy

Corneal biopsy is inoculated into a monolayer culture of fibroblasts derived from newborn mouse brain and incubated in minimum essential medium supplemented with 2% fetal bovine serum at 37°C in an atmosphere of 5% CO_2 [18].

Corneal scrapings from the lesions and microsporidia from cultures are smeared on presterilized slides for staining. The smears may be stained with (i) potassium hydroxide + calcofluor white (KOH + CFW; Sigma-Aldrich), (ii) Gram's stain, (iii) Giemsa stain (Diff Quick), and/or (iv) modified Ziehl Neelsen stain (1% H_2SO_4) and observed under light microscope [32].

95.2.1.2 DNA Extraction

For DNA extraction from spores, 10 mg of spores are placed in a 1.5 mL plastic tube and washed with a STE buffer solution containing 10 mM Tris–HCl pH 8.0 (1×TE), 1 mM ethylenediaminetetraacetic acid (EDTA) pH 8.0, and 10 mM NaCl. The spores are then crushed by vortexing with 50 mg of glass beads (G-8772, Sigma) and 200 μL of STE buffer solution for 30 s. Subsequently, the crushed spores are heated at 95°C for 5 min, and then 150 μL of the supernatant is recovered from which DNA is extracted using the ethanol precipitation method.

For DNA extraction from corneal scrapings, samples are collected in 0.5 mL of phosphate-buffered saline (PBS) pH 7.2 and stored at −20°C. Samples are thawed and centrifuged at 1500×*g* for 3 min. The pellet is washed twice using 5 mL of PBS buffer and resuspended in 0.5 mL of lysis buffer (8 M urea, 2% sodium dodecyl sulfate (SDS), 0.15 M NaCl, 0.001 M EDTA, 0.1 M Tris–HCl). The aqueous lysate is extracted twice with 0.5 mL of phenol–chloroform–isoamyl alcohol (25:24:1). The DNA is finally precipitated using 0.1 mL volume of 3 M NaCl and 2 volumes of ethanol and resuspended in 50 μL of double-autoclaved water [32].

For DNA extraction from tissue biopsy, the specimens are washed twice in PBS before being placed in 300 μL of lysis buffer containing 10 mM Tris–HCl pH 8.0, 100 mM NaCl, 20 mM dithiothreitol, 2 mg of proteinase K per milliliter, and 250 U of lyticase (Sigma). After 15 min of incubation, the sample is vortexed in the presence of 0.5-mm-diameter glass beads at 5000 rpm for 3 min. The suspension is then incubated at 37°C for 18 h. After addition of 150 μL of 2% SDS and 20 μL of proteinase K (20 mg/mL), the incubation is continued at 50°C for 72 h. DNA is extracted with phenol–chloroform and precipitated with ethanol. Nucleic acid from each sample is resuspended in 50 μL of distilled water and stored at −20°C before use in PCR [33]. Alternatively, tissue sample is individually ground in liquid nitrogen, and genomic DNA is extracted using the DNAzol DNA purification kit (Invitrogen) [32].

95.2.2 Detection Procedures

95.2.2.1 PCR Identification of *Pleistophora* sp. Sd-NU

Hatakeyama and Hayasaka [34] designed primers PSDF1 (5′-CACCAGGTTGATTCTGCCTGACG-3′) and PSDR450 (5′-GCTCCGCCTCTCTTTCCGTCTCC-3′) from the SSU rRNA gene of *Pleistophora* sp. Sd-NU (GenBank accession number D85000) for PCR identification. As these primers were only tested with *Pleistophora* sp. Sd-NU, their usefulness for specific detection of other *Pleistophora* spp. needs further evaluation.

PCR mixture (containing 10 mM Tris–HCl pH 8.9, 50 mM KCl, 1.5 mM $MgCl_2$, 0.2 mM deoxynucleoside triphosphate [dNTP], 0.5 μM of the respective primers, and 10 ng of template DNA) is first heated at 98°C for 5 min and subsequently cooled on ice, then 2.5 U of Z-Taq DNA polymerase (Takara)is added. Amplification is proceeded with 35 cycles of 94°C for 20 s, 55°C for 20 s, and extension at 72°C for 30 s. After PCR, a 5 μL aliquot of the PCR product is mixed with 5 μL TE buffer and 2 μL 40% sucrose containing 0.25% bromophenol blue (BPB) and subjected to 2% agarose gel electrophoresis with 200 bp DNA size markers. The gels are stained with ethidium bromide, and the products are visualized under UV light.

95.2.2.2 Sequencing Analysis of Microsporidian SSU rRNA Sequence

Analysis of SSU rRNA offers a useful tool for identification and phylogenetic assessment of microsporidia including *Pleistophora* and *Trachipleistophora*, for which genus- and species-specific primers have not been reported. Typically, universal primers V1f (5′-CACCAGGTTGATT CTGCCTGAC-3′) and 1492r (5′-GGTTACCTTGTTACGAC TT-3′) are employed for amplification of a 1200 bp fragment from microsporidial SSU rRNA. The nucleotide sequence of the amplicon is determined using the PCR primers mentioned above as well as internal primers 530f (5′-GTGCCAGC(C/A)GCCGCGG-3′) and 530r (5′-CCGCGG(T/G)GCTGGCAC-3′). The resulting sequence is compared with those stored in GenBank using BLAST and similar in silico programs to ascertain the identity of the organism of interest [27,35].

Procedure

1. PCR mixture (50 μL) is made up of 25 pmol each of V1f and 1492r each primer, 200 μM each of dNTP, 10 mM Tris–HCl pH 9.0, 50 mM KCl, 1.5 mM $MgCl_2$, and 2.5 U of *Taq* DNA polymerase.
2. The mixture is subjected to an initial 94°C for 3 min; 35 cycles of 94°C for 1 min, 43°C for 1 min 40 s, and 72°C for 1 min 40 s; and a final 72°C for 10 min.
3. The products (5 μL) are assessed separated on 1% agarose gel, stained with ethidium bromide, and visualized under UV light.
4. The remaining products are purified and sequenced with primers V1f and 1492r as well as 530f and 530r.
5. The resulting SSU rRNA gene sequences are compared to those in the GenBank to confirm the identity of microsporidial isolates. The sequences are then aligned using the "ClustalX 1.83" program. Phylogenetic trees based on the resultant alignments are, respectively, constructed using the neighbor-joining (NJ) algorithm of Phylip software with Kimura-2 parameter and UPGMA algorithm of MEGA 4.0 software with maximum composite likelihood, and 1000 replications are generated to test the robustness of the tree. The identity and divergence of the sequences are further analyzed with the MegAlign program of DNAStar software.

95.2.2.3 Sequencing Analysis of Microsporidian RNA Polymerase 2 Gene (Rpb1)

Because the 16S rRNA sequence does not always contain sufficient polymorphic sites for discrimination between closely related species, RNA polymerase 2 gene (Rpb1) may be targeted for improved phylogenetic analysis of microsporidia [36]. The Rpb1 gene can be amplified in three fragments using the primer combinations Pm1F and Pm1R, Pm2F and Pm2R, and Pm3bF and Pm3bR (Table 95.1).

PCR amplification is performed using Invitrogen's recombinant Taq polymerase with an annealing temperature of 50°C and an extension time of 1 min. The amplicons are sequenced with respective PCR primers. The Rpb1 sequences of microsporidia are then aligned with published sequences of Rpb1

TABLE 95.1
Sequence Identity of RNA Polymerase 2 Gene (Rpb1) Primers

Primer	Sequence (5′–3′)	Position
Pm1F	ACA CAT GGA GCT GAC GAG G	1
Pm1R	TTC TCG CGC TGA AGT CCA	891
Pm2F	ACT GAA GGG AAA GGA GGG A	824
Pm2R	AAT TTT ACG GGG CT CAG AA	1739
Pm3bF	AGG CGA TGC ACC TGA TCT T	1469
Pm3bR	ATA CGA ATT CTC GAC GAA GCC	2390

genes from *Pleistophora* and *Trachipleistophora* isolates and from species of the closely related genera *Vavraia*.

Because of the high level of DNA sequence divergence between microsporidia and other fungi, outgroup species are selected from within the phylum Microsporidia such as *Encephalitozoon cuniculi* and *Glugea anomala*. Rpb1 sequences are aligned on the basis of an Rpb1 amino acid sequence template, implemented in the program DAMBE. Phylogenetic analysis is performed on the Rpb1 alignments using maximum likelihood implemented in PAUP_4b10 and Bayesian inference implemented in MrBayes. For each alignment, a maximum likelihood consensus tree is generated by conducting a heuristic search and bootstrapping with 100 replicates [36].

95.3 CONCLUSION

Pleistophora and *Trachipleistophora* spp. are obligate intracellular parasites of both vertebrates and invertebrates that have been shown to cause myositis, keratoconjunctivitis, sinusitis, disseminated infection in individuals infected with human immunodeficiency virus type 1 (HIV-1) as well as immunocompetent persons. Humans may acquire the infection through consumption of inadequately cooked fish muscle containing the organism or exposure to contaminated water (in rural ponds and swimming pools). Due to their close morphological resemblances, correct phenotypic identification of *Pleistophora* and *Trachipleistophora* is only possible by electron microscopy. Following the recent development and application of molecular methods, rapid and precise characterization of microsporidia including *Pleistophora* and *Trachipleistophora* can be achieved on the basis of PCR and sequencing analysis of SSU and LSU rRNA and other conserved genes. This has contributed greatly to the effective treatment and control of microsporidiosis in human and animal hosts.

REFERENCES

1. The UniProt Consortium. Available at http://www.uniprot.org/, accessed on August 1, 2010.
2. Cali A, Takvorian PM. Ultrastructure and development of *Pleistophora ronneafiei* n. sp., a microsporidium (Protista) in the skeletal muscle of an immune-compromised individual. *J. Eukaryot. Microbiol.* 2003;50(2):77–85.
3. Cheney SA, Lafranchi-Tristem NJ, Canning EU. Phylogenetic relationships of Pleistophora-like microsporidia based on small subunit ribosomal DNA sequences and implications for the source of *Trachipleistophora hominis* infections. *J. Eukaryot. Microbiol.* 2000;47:280–287.
4. Horák A et al. *Trachipleistophora extenrec* n. sp. a new microsporidian (fungi: microsporidia) infecting mammals. *J. Eukaryot. Microbiol.* 2006;53(6):464–476.
5. Vavra J et al. Microsporidia of the genus *Trachipleistophora*—Causative agents of human microsporidiosis: Description of *Trachipleistophora anthropophthera* n. sp. (protozoa: microsporidia). *J. Eukaryot. Microbiol.* 1998;45:273–283.
6. Hollister WS et al. Development and ultrastructure of *Trachipleistophora hominis* n.g., n.sp. after in vitro isolation from an AIDS patient and inoculation into athymic mice. *Parasitology* 1996;112(Pt 1):143–154.
7. Koudela B, Vávra J, Canning EU. Experimental infection of severe combined immunodeficient (SCID) mice with the human microsporidian *Trachipleistophora hominis*. *Parasitology* 2004;128:377–384.
8. Cali A, Weiss LM, Takvorian PM. A review of the development of two types of human skeletal muscle infections from microsporidia associated with pathology in invertebrates and cold-blooded vertebrates. *Folia Parasitol.* (*Praha*) 2005;52(1–2):51–61.
9. Lom J, Nilsen F. Fish microsporidia: Fine structural diversity and phylogeny. *Int. J. Parasitol.* 2003;33(2):107–127.
10. Canning EU et al. New diplokaryotic microsporidia (Phylum Microsporidia) from freshwater bryozoans (Bryozoa, Phylactolaemata). *Eur. J. Protistol.* 2002;38:247–265.
11. Weidner E et al. Mosquito (Diptera: Culicidae) host compatibility and vector competency for the human myositic parasite *Trachipleistophora hominis* (Phylum Microspora). *J. Med. Entomol.* 1999;36(4):522–525.
12. Ledford DK et al. Microsporidiosis myositis in a patient with acquired immunodeficiency syndrome. *Ann. Intern. Med.* 1985;102:628–630.
13. Chupp GL et al. Myositis due to *Pleistophora* (Microsporidia) in a patient with AIDS. *Clin. Infect. Dis.* 1993;16(1):15–21.
14. Field AS et al. Myositis associated with a newly described microsporidian, *Trachipleistophora hominis*, in a patient with AIDS. *J. Clin. Microbiol.* 1996;34(11):2803–2811.
15. Curry A et al. *Trachipleistophora hominis* infection in the myocardium and skeletal muscle of a patient with AIDS. *J. Infect.* 2005;51(3):e139–e144.
16. Rauz S et al. Ultrastructural examination of two cases of stromal microsporidial keratitis. *Med. Microbiol.* 2004;53(Pt 8):775–781.
17. Juarez SI et al. A species of microsporidium isolated from an HIV-infected patient and identified as *Trachipleistophora anthropophthera* by transmission electron microscopy. *J. Eukaryot. Microbiol.* 2003;50:S576.
18. Juarez SI et al. In vitro cultivation and electron microscopy characterization of *Trachipleistophora anthropophthera* isolated from the cornea of an AIDS patient. *J. Eukaryot. Microbiol.* 2005;52(3):179–190.
19. Pariyakanok L, Jongwutiwes S. Keratitis caused by *Trachipleistophora anthropopthera*. *J. Infect.* 2005;51(4): 325–328.
20. Didier ES et al. Comparison of three staining methods for detecting microsporidia in fluids. *J. Clin. Microbiol.* 1995;33(12):3138–3145 [PubMed].
21. Garcia LS. Laboratory identification of the microsporidia. *J. Clin. Microbiol.* 2002;40(6):1892–1901.
22. Cheney SA et al. Serological differentiation of microsporidia with special reference to *Trachipleistophora hominis*. *Parasite* 2001;8(2):91–97.
23. Lafranchi-Tristem NJ et al. Growth of *Trachipleistophora hominis* (Microsporidia: Pleistophoridae) in C2,C12 mouse myoblast cells and response to treatment with albendazole. *Folia Parasitol.* (*Praha*) 2001;48(3):192–200.
24. Visvesvara GS. In vitro cultivation of microsporidia of clinical importance. *Clin. Microbiol. Rev.* 2002;15(3):401–413.
25. Fedorko DP, Nelson NA, Cartwright CP. Identification of microsporidia in stool specimens by using PCR and restriction endonucleases. *J. Clin. Microbiol.* 1995;33(7): 1739–1741.
26. Williams BA et al. A mitochondrial remnant in the microsporidian *Trachipleistophora hominis*. *Nature* 2002;418(6900):865–869.

27. Baker MD et al. Small subunit ribosomal DNA phylogeny of various microsporidia with emphasis on AIDS related forms. *J. Eukaryot. Microbiol.* 1995;42(5):564–570.
28. Weiss LM, Vossbrinck CR. Microsporidiosis: Molecular and diagnostic aspects. *Adv. Parasitol.* 1998;40:351–395.
29. Franzen C, Müller A. Microsporidiosis: Human diseases and diagnosis. *Microbes Infect.* 2001;3:389–400.
30. Keeling PJ. Congruent evidence from alpha-tubulin and beta-tubulin gene phylogenies for a zygomycete origin of microsporidia. *Fungal Genet. Biol.* 2003;38:298–309.
31. Ghosh K, Weiss LM. Molecular diagnostic tests for microsporidia. *Interdiscip. Perspect. Infect. Dis.* 2009; 2009:926521.
32. Joseph J et al. Microsporidial keratitis in India: 16S rRNA gene-based PCR assay for diagnosis and species identification of microsporidia in clinical samples. *Invest. Ophthalmol. Vis. Sci.* 2006;47:4468–4473.
33. Xiao L et al. Genotyping *Encephalitozoon cuniculi* by multilocus analyses of genes with repetitive sequences. *J. Clin. Microbiol.* 2001;39:2248–2253.
34. Hatakeyama Y, Hayasaka S. Specific amplification of microsporidian DNA fragments using multiprimer PCR. *JARQ* 2002;36:97.
35. Weiss LM et al. Utility of microsporidian rRNA in diagnosis and phylogeny: A review. *Folia Parasitol.* (*Praha*) 1994;41(2):81–90.
36. Ironside JE, Wilkinson TJ, Rock J. Distribution and Host Range of the Microsporidian *Pleistophora mulleri*. *J. Eukaryot. Microbiol.* 2008;55(4):355–362.

Part V

Oomycota, Chlorophyta, and Mesomycetozoea

96 *Pythium*

Theerapong Krajaejun, Boonmee Sathapatayavongs, and Thomas D. Sullivan

CONTENTS

96.1 INTRODUCTION

Pythiosis is a life-threatening infectious disease caused by the fungus-like organism, *Pythium insidiosum*, a *Pythium* species that was first introduced and validly published by De Cock et al. in 1987.[1,2] In fact, the disease first occurred in horses in the middle of the nineteenth century, and the causative agent was first isolated as an unidentified nonsporulated filamentous fungus in 1901 by two Dutch researchers, de Haan and Hoogkamer.[1,2] In 1974, Austwick and Copland reported that the pathogen isolated from horses should be placed in the genus *Pythium* because it produces biflagellate zoospores, which represent a distinctive morphological phase of members of this genus.[3] *P. insidiosum* has been recognized by other names, including *Hyphomyces destruens*, *Pythium gracile*, and *Pythium destruen*; and the disease pythiosis has also been known under several different names, including hyphomycosis destruens, phycomycosis, swamp cancer, bursattee, kunker, cutaneous habronemiasis, granular dermatitis, espundia, and leeched.[1–4] In addition to horses, pythiosis has been increasingly reported in other animals, including dog, cat, cattle, bear, and sheep.[2] The first patient with human pythiosis was found in New Zealand in 1984, but the case was not reported in the literature until 1997.[5] In 1985, two human pythiosis cases were diagnosed in Thailand.[6] Since then, human pythiosis has been documented from other countries: United States, Haiti, Brazil, Malaysia, and Australia.[7–11]

The hyphal phase of *P. insidiosum* is hard to distinguish from true filamentous fungi, such as *Aspergillus* spp., *Penicillium* spp., and *Zygomycetes*, which cause infections with similar clinical features to pythiosis.[12] This can lead to the misdiagnosis of patients with pythiosis. For this and other reasons, morphological characterization and culture identification of *P. insidiosum* are difficult and time-consuming, and therefore, diagnosis of pythiosis is often delayed. Conventional antifungal agents are not effective for treatment of *P. insidiosum* infection because *Pythium* spp. lack the drug-targeted ergosterol biosynthesis pathway.[13] Because β-glucan is a major cell wall component of fungi and *Pythium* species,[14] Pereira recently tested the in vivo effect of caspofungin (a new antifungal drug that inhibits β-glucan synthase) on *P. insdiosum* infection in a rabbit model of pythiosis.[15] While the drug slowed the growth of lesions, it was unable to eliminate the pathogen in the infected animals.[15] Without a useful chemotherapeutic agent, the main treatment approach for pythiosis is radical surgery, which should be urgently performed to limit the progression of the disease. Many patients die as a result of advanced and aggressive infection. Early and correct diagnosis of pythiosis is crucial for proper treatment and a better prognosis.

Pythiosis should be regarded as an emerging infectious disease. Many health care professionals are not familiar with this disease or its causative agent, *P. insidiosum*. The increasing diagnosis of pythiosis from all over the world indicates that more clinicians and microbiologists are aware of pythiosis and that the disease may not be uncommon. Our limits of knowledge regarding pythiosis and the lack of diagnostic and treatment options bring about a high morbidity and mortality rate for pythiosis patients. In this chapter, up-to-date information about pythiosis and *P. insidiosum* are described in detail, with a focus on diagnostic modalities.

96.1.1 Classification, Morphology, and Biology

P. insidiosum is recognized as a member of the oomycetes,[14,16] which is a unique group of eukaryotic microorganisms with a fungal-like morphology that are physiologically, biochemically, and phylogenetically different from other eukaryotic pathogens, that is, fungi and parasites, and are most closely related to the brown algae.[14] Taxonomically, *P. insidiosum* was previously classified in the family Pythiaceae, order Pythiales, class Oomycetes, and the phylum Pseudofungi in the kingdom Chromista.[2] According to the recently published phylogenetic tree of eukaryotes, *P. insidiosum* belongs to the oomycetes within the Stramenopiles of the supergroup Chromalveolates.[17] Genera within the oomycetes include *Pythium*, *Albugo*, *Peronospora*, *Phytophthora*, *Plasmopara*, *Bremia*, *Aphanomyces*, *Lagenidium*, and *Saprolegnia*.[16] Although the majority of the oomycetes are saprophytes, many of them are pathogenic on a broad range of hosts, including plants and animals.[16] Plant pathogenic oomycetes cause significant economical loss of agricultural products each year, and the animal pathogens among the oomycetes (e.g., *Pythium* spp., *Aphanomyces* spp., *Lagenidium* spp., and *Saprolegnia* spp.) can cause devastating diseases.[16]

P. insidiosum is the only known oomycete capable of infecting both humans and other animals and the resulting disease (pythiosis) can be fatal.[2,16] In nature, *P. insidiosum* inhabits swampy areas and commonly presents in two forms: mycelia and asexual zoospores.[18] *P. insidiosum* has broad (4–10 μm), right-angle branching and sparsely septate hyphae (Figure 96.1), which are indistinguishable from the true filamentous fungi.[1,2] The zoospore, a specific character of *P. insidiosum* and other oomycetes, is produced through zoosporogenesis, which is initiated by the development of a zoosporangium (zoospore sac) at the hyphal tip and results in the formation of 5–60 zoospores within the sac.[18–20] The whole process of zoosporogenesis (generation, maturation, and release of zoospores) takes ~35 min to finish.[18] Zoospores have two motile flagella, which facilitate movement in water for 10–15 min before sequential events of detachment of flagella, encystment, hyphal germination, and colonization of one of several water plants.[18] Unlike other *Pythium* species, *P. insidiosum* occasionally produces oogonia, which are smooth and subglobose.[1,2] During mating, several antheridia, from different hyphae, attach to a single oogonium, and the tip of each antheridium produces a fertilization tube that penetrates the oogonial wall for fertilization.[1,2]

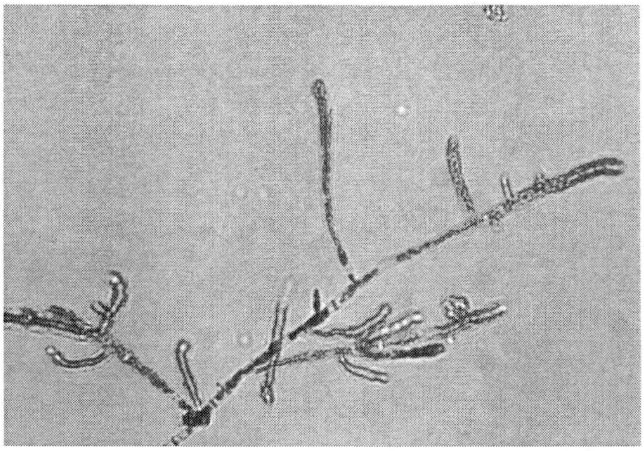

FIGURE 96.1 *P. insidiosum* observed under a light microscope.

Pythiosis has been increasingly reported mostly from tropical and subtropical countries, and also from temperate areas.[2,21] Pythiosis in animals is found worldwide, but the disease in humans has been diagnosed almost exclusively in Thailand.[6,22–31] Patients with human pythiosis in Thailand are found nationwide, indicating that the causative agent, *P. insidiosum*, is widely distributed throughout the country.[6] Because of similar ecology and climate, *P. insdiosum* is expected to inhabit neighboring countries, and unreported pythiosis cases may exist throughout Southeast Asia. A case of human pythiosis reported from Malaysia supports this speculation.[7] A recent study showed that the majority of pythiosis patients (86%) are in the working-age population (20–60 years old), are male (71%), and have agricultural-related careers, including farming, domestic husbandry, and fishing (75%).[6]

The ecological niche of *P. insidiosum* has been described by two groups, Miller[32] and Supabandhu et al.[33] Miller has successfully isolated the organism in water samples from different locations in Queensland, Australia, using a human hair-baiting method.[32] Supabandhu et al. successfully used the same method to isolate the organism in agricultural areas of northern Thailand.[33] These are important findings, which demonstrate *P. insidiosum* in aquatic environments and suggest that patients may acquire *Pythium* infection from exposure to the swampy habitats of the pathogen.

The phylogenetic relationships of *P. insidiosum* isolated from patients and animals living in different geographic areas was first reported by Schurko et al.[34,35] This group genetically classified *P. insidiosum* based on the analysis of rDNA sequences from 17 animal isolates and 6 human isolates from around the world. They demonstrated that there are three phylogenetic groups associated with the geographic origin of the organism: clade-I containing only western hemisphere isolates, clade-II containing American, Asian, and Australian isolates, and clade-III containing two American isolates and two Thai isolates. In a later study, Supabandhu et al. reported a phylogenetic analysis of 4 human isolates and

59 environmental isolates from water collected at different agricultural sites in Thailand.[33] They showed that 3 human isolates and 41 environmental isolates (70% of all isolates) are grouped in clade-II, and the remaining 1 human isolate and 18 environmental isolates (30%) belonged in clade-III. In neither of the two studies were clade-I isolates reported from Thailand.

96.1.2 Clinical Features

96.1.2.1 Human Pythiosis

96.1.2.1.1 Cutaneous/Subcutaneous Pythiosis

Patients present with an infection of cutaneous/subcutaneous tissue, characterized by chronic swelling and a painful subcutaneous, granulomatous infiltrative lump and ulcer on the face, arm, or leg. Histological examination of infected cutaneous/subcutaneous tissue reveals chronic infection with eosinophilia.[6,22] Patients with cutaneous/subcutaneous pythiosis are reported as having a good response to saturated solution of potassium iodide (SSKI) along with surgical debridement.[6] Acute infection can occur, as evidenced in a Thai patient with acute cellulitis in both legs.[31] An isolated case of cutaneous/subcutaneous infection from the United States was successfully treated with a combination of itraconazole and terbinafine.[9]

96.1.2.1.2 Vascular Pythiosis

Arterial insufficiency syndrome of lower extremities is the main feature of vascular pythiosis.[6] The syndrome ranges from intermittent claudication or resting pain to gangrenous ulceration. The infection can occur in one or both legs. Patients delay seeking medical attention on average of 91 days (range, 7–365 days) after the first observation of symptoms.[6] Angiographic findings usually demonstrate occlusion or aneurysm of arteries of the lower extremities or the aorta. On histological examination, *P. insidiosum* resides in the arterial wall or in obstructive thrombi in the arterial lumen. *Pythium* infection is not detected in venous tissue. In advanced cases, a major cause of death is from rupture of an aortic aneurysm.

Conventional antifungal agents and SSKI have had no favorable responses in vascular pythiosis. Radical surgery is the main treatment of choice. The infected tissue is removed by resection of the infected artery or limb amputation, depending on the level and severity of the infection. Most of the treated cases (~80%) of vascular pythiosis underwent amputation. Forty percent of patients with vascular pythiosis die from the infection, while 60% are cured, but left with physical handicaps. Reoccurrence is possible if surgery fails to completely remove the infection.

Recently, an immunotherapeutic vaccine prepared from crude extract of *P. insidiosum* has been used for treatment in humans[6,27,30] and animals[36–39] with pythiosis. The vaccine had favorable responses in 60% of horses, 97% of cattle, and 33% of dogs with pythiosis.[39] In humans, the *P. insidiosum* vaccine was used as the last resort in 12 patients with evidence of inadequate surgery.[6,27,30] Final outcomes of the vaccine immunotherapy were available for nine patients, of which five were cured (56%), while two died (22%), and two had persistent infections (22%).

96.1.2.1.3 Ocular Pythiosis

Patients usually present with corneal ulcer or keratitis.[6,29] Other signs and symptoms include pain, irritation, decreasing visual acuity, eye lid swelling, conjunctival injection, corneal infiltrates, perforated cornea, and hypopyon. Endophthalmitis can occur in advanced cases. No cases of the infection spreading beyond the eye have been reported. Ocular pythiosis patients come to the hospital within a few days or weeks after they first notice symptoms. History of ocular trauma or corneal abrasion is common. Hyphae of *P. insidiosum* in KOH preparation of corneal scraps or discharges can be demonstrated by a simple microscopic examination. Histopathology usually reveals *P. insidiosum* located in the corneal stroma.

Like vascular pythiosis, the main treatment option for ocular pythiosis is radical surgery to get rid of all infected tissue.[6,29] Eye removal by enucleation or evisceration is performed after keratectomy or keratoplasty fails to control the infection. In the minority of cases (20%), the eye is saved after anterior lamella or penetrating keratoplasty, while the majority of patients (~80%) eventually underwent enucleation or evisceration. No fatal outcomes have been reported in ocular pythiosis patients.

96.1.2.1.4 Unusual Form of Pythiosis

Some pythiosis cases with unusual site of infection, such as, the gastrointestinal tract, brain, or rhinosinus, have been reported.[6] It should be noted that gastrointestinal tract infection is a common form in animal pythiosis and may result from drinking *P. insidiosum*-contaminated water.[2] For the infections of brain tissue and rhinosinus, the pathogen may invade directly to these sites through the nasal cavity. Even with medical attention, all these cases have been fatal.

96.1.2.2 Animal Pythiosis

Pythiosis in animals is more commonly reported than the disease found in humans. Various animals are susceptible to an infection caused by *P. insidiosum*.[2] Among them, horses and dogs are the two most frequently affected hosts.[40–54] Pythiosis in some other animals, that is, calve, cat, sheep, camel, tiger, jaguar, and polar bear, has also been reported but is less common.[55–62] Animals with pythiosis present with a broad range of clinical features and with an invariably high rate of morbidity and mortality. In contrast to human pythiosis, for which vascular and ocular infections are the most common, animal pythiosis most commonly features cutaneous/subcutaneous infection of a limb (most common site), chest, abdomen, face, genitalia, tail or gastrointestinal tract infection of the pharynx, esophagus, stomach, or intestines. Other internal organs, that is, lung, bone, liver, pancreas, uterus, and internal lymph nodes can also be affected in invasive or disseminated pythiosis cases.

Animals with cutaneous/subcutaneous pythiosis mainly present with ulcerating lesions.[2] In more advanced cases, ulcers with multiple draining tracts and serosanguineous discharge can be observed. *P. insidiosum*-infected lesions can look like a tumor, with a diameter up to 50 cm (this could be the origin of the name "swamp cancer"). Animals with gastrointestinal pythiosis usually present with dysphagia, vomiting, anorexia, weight loss, diarrhea, abdominal mass, mucosal ulceration, and wall thickening of the stomach and intestines. Histological examination of infected lesions reveals granulomatous tissue with infiltration of eosinophils, neutrophils, macrophages, and giant cells. Splendore-Hoeppli-like phenomenon, including radiating eosinophilic deposits of host cell debris and pathogen antigens surrounding the fungus, is often microscopically observed in infected tissues. At the infected site, an eosinophilic coral-like mass (diameter up to 9 cm) called "kunker," which contains *P. insidiosum* hyphae, is a characteristic of pythiosis found in animals.[2] The kunker has never been described in human tissue infected with *P. insidiosum*. Like human patients, radical surgery is the main treatment option. Incomplete removal of infected tissues leads to recurrent *P. insidiosum* infection. Immunotherapy using *P. insidiosum* antigen extract has had limited efficacy although there are reports of favorable response to this treatment option.[2]

96.1.3 Pathogenesis

In nature, *P. insidiosum* completes its life cycle in aquatic environments.[2,18,32] The life cycle starts with hyphal colonization on plants and produces a zoosporangium with multiple, asexual biflagellate zoospores. Released zoospores swim and adhere to neighboring plants for colonization, and a new loop of this usual life cycle begins. Miller[32] and Mendoza et al.[18] showed that the swimming zoospores of *P. insidiosum* are strongly attracted to plants and animal and human hair, possibly by chemotaxis. The zoospores are observed to accumulate at the follicular end of the hair shafts. The zoospore is thought to be the form that initiates infection and, after conversion to the hyphal form, causes the immune responses and pathology in the host. This belief is based on the successful reproduction of pythiosis in an experimental rabbit model using zoospore inoculation.[63] However, experimental pythiosis can also be reproduced in the rabbit model by either hyphal or kunker inoculation.[63] Upon attachment of zoospore to host tissue, the flagella detach and surface amorphous substances are generated, which may facilitate adhesion to the host.[2,18] Mendoza et al. report that a zoospore tends to adhere to a cut edge of skin rather than to an intact surface.[18] Ravishankar et al. studied hyphal tip exertion forces of *P. insidiosum* and demonstrated that the pathogen is unable to penetrate undamaged cutaneous tissue.[64] They propose that the organism requires secreted proteinases to reduce tissue strength, prior to successful tissue penetration.

Direct contact of host surfaces (i.e., skin, eye) to *P. insidiosum* in its aquatic habitats is the likely route of entry for cutaneous/subcutaneous and ocular infections. Occurrence of arterial infection is thought to be a result of direct extension of *Pythium* infection from the skin.[27] The infection progresses from distal to proximal in an infected artery and causes inflammation, fibrosis, and aneurysm. Many cases of gastrointestinal pythiosis have been reported, and for these, the site of entry is likely direct contact of *P. insidiosum* with the gastric or intestinal wall, as a consequence of drinking water contaminated with the pathogen. In animals, the spread of the pathogen to other internal organs (e.g., liver, spleen, bone) may occur via the vascular system or by lymphatic drainage.[2,43,63] No information on transmission of pythiosis between animals and humans has been reported.

A striking comorbidity exists for human patients with vascular, cutaneous, and the unusual forms of pythiosis: all have an underlying hematological disorder, 85% of which were cases of thalassemia.[6,23,26] Iron overload is a major pathophysiological change in thalassemic patients and could be a factor that enhances host susceptibility to pythiosis, possibly by promoting infectivity of *P. insidiosum* or by compromising host immunity.[6,65–67] However, several pythiosis patients were reported with underlying paroxysmal nocturnal hemoglobinuria, which is associated with iron deficiency,[6] indicating that iron overload alone may not explain susceptibility to *Pythium* infection. Thalassemia and paroxysmal nocturnal hemoglobinuria share a common pathophysiological feature of chronic hemolysis. Identification of roles of iron and hemolysis as susceptibility factors for the host to pythiosis needs further investigation. In contrast to these forms of pythiosis, the majority of patients with ocular pythiosis (84% of cases) report no underlying condition, indicating that healthy individuals are susceptible to ocular infection.[6,29]

96.1.4 Diagnosis

Protocols are available to culture *P. insidiosum* from patient tissue using standard microbiological procedures and conventional media for the culture of fungi. While most clinical laboratories can attempt the culture of *P. insidiosum*, the method is time-consuming and laborious, often fails to isolate the pathogen, and requires expertise to identify *P. insidiosum* in successful cultures. As an alternative method, in-house immunological and molecular assays have been developed to facilitate diagnosis of pythiosis.[7,68–83] These alternative techniques are reliable and more convenient than the standard culture method. In addition, some of the serodiagnostic tests can be used to monitor anti-*P. insidiosum* antibody level to evaluate treatment response.[68,70–73,77] The following section provides a summary of the methods currently used for diagnosis of pythiosis.

96.1.4.1 Conventional Techniques

Conventional techniques for diagnosis of pythiosis include (i) isolation and identification of *P. insidiosum* from clinical specimens by standard culture techniques, (ii) detection of anti-*P. insidiosum* antibodies in patient serum samples, and (iii) direct detection of *P. insidiosum* in infected tissue.

96.1.4.1.1 Culture Identification

Culture identification is a simple microbiological method for the diagnosis of pythiosis by isolation of the causative agent, *P. insidiosum*, from a clinical specimen. *P. insidiosum* can be grown on various types of nutrient agar: Sabouraud dextrose agar, potato dextrose agar, corn meal agar, soil extract agar, Czapek-Dox agar, tissue culture medium, vegetable extract agar, peptone-yeast-glucose agar, 2% water agar, potato flakes agar, and blood agar.[20,84] *P. insidiosum* colonies on agar are flat with no aerial hyphae and white, yellow, or brown in color. De Cock et al. report that the optimal temperature for growth of the pathogen is between 34°C and 36°C, with minimum and maximum temperatures for growth of 10°C and 45°C.[1] Davis et al. report an optimal growth rate at 37°C.[85] Failure to isolate *P. insidiosum* from clinical specimens is associated with low temperature storage of clinical specimens prior to culture, as either refrigeration or freezing can kill the pathogen.[40,41] In our experience, attempts to isolate the pathogen from human specimens transported on ice were usually unsuccessful. Grooters et al. demonstrated that a higher success rate for isolation of *P. insidiosum* was achieved by culturing fresh specimens from affected horses directly on antibiotic-containing medium.[84] Therefore, specimens should be stored at room temperature during transportation to a clinical microbiology laboratory, and should be freshly cultured on antibiotic-containing media at room temperature or at 37°C. *P. insidiosum* grows relatively fast, as indicated by a growth rate of up to 12 mm/day, and macroscopic colony appearance in a day or 2.[20] To confirm the identity of the cultured organism as *P. insidiosum*, it is necessary to induce production of the characteristic zoospores. Chaiprasert et al. and Mendoza and Prendas describe methods for zoospore induction by culturing the pathogen on induction-medium agar containing sterile grass blades.[19,20]

96.1.4.1.2 Serodiagnostic Tests

Serodiagnostic techniques are convenient for diagnosis of pythiosis. The goal of these techniques is to detect anti-*P. insidiosum* antibodies as indirect evidence of *P. insidiosum* infection. Multiple immunological approaches have been employed for the development of diagnostic assays: immunodiffusion (ID),[68–70] enzyme-linked immunosorbent assay (ELISA),[71–73] Western blot,[74,75,78] immunochromatographic test (ICT),[76] and hemagglutination test (HA).[77] Antigen sources for development of these assays are prepared from crude extract of *P. insidiosum* hyphae (called soluble antigen from broken hyphae; SABH) or secretory antigens in culture medium (called culture filtrate antigen; CFA). These techniques vary in their sensitivity and specificity.

ID tests for the diagnosis and treatment monitoring of pythiosis in animals and humans were invented by Mendoza et al., Imwidthaya and Srimuang and Pracharktam et al.[68–70] The ID tests are agar-medium based and rely on the detection of precipitation lines generated by the specific reaction of *P. insidiosum* antigens (CFA) with anti-*P. insidiosum* antibodies (in patient serum). In the standard setup, each of the CFA and patient serum is added to 5 mm diameter wells that are 5 mm apart in agar medium (1% purified agar in Veronal buffer, pH 7.8; 1% phenolized agar; 2% water agar).[68,70,71] Diffusion of the CFA and serum occurs during overnight incubation in a moist chamber, and a precipitation line, observed by naked eye in the following days, indicates a positive result. The absence of the precipitation line is taken as a negative result. Although the test is highly specific, is easy to perform, can test serum from both humans and animals without any modification, and requires cheap and available reagents, the low sensitivity of detection leads to a high rate of false negative results and compromises the efficacy of this test.[71,76,77]

Increased sensitivity for detection of anti-*P. insidiosum* antibody is achieved by ELISA and Western blot techniques developed by several investigators, that is, Mendoza et al., Grooters et al., Krajaejun et al., Vanittanakom et al., and Supabandhu et al.,[71–73,78,79] using SABH or CFA as the antigen source for coating ELISA plates or blotting to a membrane. These techniques have very high sensitivity (up to 100%) and specificity (up to 100%) for diagnosis of pythiosis, and thus overcome the sensitivity problem of the ID test. However, these advantages of ELISA and Western blots assays are offset by their complexities. They are time-consuming and multistep procedures, use expensive reagents and equipment, and require trained personal to perform. These disadvantages limit the use of these assays in clinical diagnostic laboratories.

Recently, an ICT has been developed as a rapid and easy-to-use assay for diagnosis of human pythiosis.[76] In the ICT, CFA is spotted onto a nitrocellulose membrane strip as an antigen source for detecting anti-*P. insidiosum* immunoglobulin G (IgG) in serum samples. From the sample end of the ICT strip, human IgGs in the serum move upward by capillary action and form complexes with the rabbit anti-human IgG-colloidal gold conjugate antibody that is blotted close by. The complexes continue moving upward through the nitrocellulose membrane. If the immune complexes contain human anti-*P. insidiosum* IgG, they bind CFA and develop a purple signal at the detection end of the ICT strip. In contrast, if the immune complexes lack human anti-*P. insidiosum* IgG, they pass through the spotted CFA without developing a signal. Detection sensitivity and specificity of ICT are high. Its turnaround time is less than 30 min.

Another easy-to-use assay format based on HA has been recently developed for diagnosis of pythiosis.[77] Sensitivity and specificity of the HA test are as high as that of ELISA, Western blot, and ICT. CFA is used as an antigen source for coating of sheep red blood cells (SRC). SRCs serve as signal amplification for the specific antibody–antigen reaction. Anti-*P. insidiosum* IgG and IgM in a serum of a pythiosis patient bind CFA on the surface of SRC and form a HA (an aggregation of the red blood cells [RBCs]), which can be simply visualized by the naked eye. In contrast, if there are no anti-*P. insidiosum* antibodies in the serum, no HA is observed. Turnaround time for performing the HA test is ~1 h. An advantage over the other serodiagnostic tests is that HA test is a sensitive test that can be used for detection of anti-*P. insidiosum* antibodies of both human and animal sera, because there is no need for changing the host-specific conjugate antibody.

In addition to their use in diagnosis, the above-mentioned serological tests can be applied for monitoring anti-*P. insidiosum* antibody levels over the course of treatment.[68,70–72,77] A decrease in the antibody level, in association with clinical improvement after surgical treatment, is an indicator for a good prognosis. A drawback of all serodiagnostic tests for pythiosis is that they are likely to fail to detect anti-*P. insidiosum* antibody in patients with ocular pythiosis. It is postulated that this results from the failure of the eye infection to induce an antibody response.[76,77] Therefore, serological tests in patients with an ocular infection should be used with caution because of the danger of a false negative result. On the other hand, if a serum from an ocular patient is positive for anti-*P. insidiosum* antibody, this may indicate the presence of an invasive ocular infection or multiple sites of *Pythium* infection.

As an alternative to antibody detection for the diagnosis of pythiosis, several investigators have developed immunochemical and immunofluorescent assays for direct detection of *P. insidiosum* in preserved infected tissue or in pure cultures.[2,8,82,86] Rabbit polyclonal antibodies raised against *P. insidiosum* crude extract (SABH or CFA) are used to stain paraffin-embedded tissue sections. In the assay developed by Mendoza et al., fluorescein isothiocyanate is conjugated to the rabbit anti-*P. insidiosum* antibody, which allows signal detection with a fluorescent microscope (direct assay),[2,86] while Brown et al.[82] and Triscott et al.[8] use a secondary bovine anti-rabbit IgG antibody, conjugated with peroxidase or streptavidin-biotin-peroxidase complex, to react with the primary anti-*P. insidiosum* antibody and in the presence of chromogenic peroxidase substrates to develop a colored precipitate visible under light microscopy (indirect assay). This histological assay has been used for years to confirm pythiosis in animals; however, its specificity has been brought into question by the appearance of cross reactivity (a false positive result) to other fungi, that is, *Conidiobolus* sp.[80]

96.1.4.2 Molecular Techniques

Molecular techniques are increasingly useful for the diagnosis of infectious diseases including mycoses because they are very specific and sensitive, and all required reagents and equipment are widely available at reasonable cost. For diagnosis of pythiosis, molecular techniques have been developed for the direct detection of *P. insidiosum* from pure culture or infected tissue. The molecular diagnostic assays are especially useful in a laboratory that lacks conventional microbiological and immunological methods or for suspected cases that tested negative for pythiosis by other means. The ribosomal RNA (rRNA) gene repeat (or rDNA region, containing 18S rRNA, ITS1, 5.8S rRNA, ITS2, 28S rRNA, and the intergenic spacer) is often used as a target for the development of molecular detection assays and it has several advantages: (i) it is a multicopy gene that provides abundant target sequences and therefore higher detection sensitivity; (ii) it has highly conserved areas (18S, 5.8S, and 28S rRNA portions), which allow the possibility of obtaining rDNA sequence in unidentified organisms by using primers designed from the rDNA sequence of closely related species; and (iii) it contains variable regions (ITS1, ITS2, and the intergenic spacer), which allow differentiation at the genus and species levels. Advances in molecular assays for pythiosis are summarized here.

96.1.4.2.1 Sequence Analysis

Badenoch et al. were the first group to report the use of DNA sequence analysis for identification and confirmation of *P. insidiosum* infection in a patient with pythiosis.[7] Their protocol used the fungal universal primers NS1 and NS2[87] (Figure 96.2, Table 96.1) to amplify 510 bp PCR product of the 18S rRNA portion from genomic DNA of the culture-isolated organism, which was initially suspected to be a zygomycete. Subsequent sequencing and BLAST searching analyses provided a clue that the causative organism was not a zygomycete, but rather an oomycete. Then, a 900 bp portion of the internal transcribed spacer (ITS), which includes ITS1, the 5.8S gene, and ITS2, was successfully amplified by using primers TW81 and AB28, which were previously used to amplify the ITS of the closely related genus *Phytophthora*[88] (Figure 96.2, Table 96.1). Sequence of the PCR product was obtained by using the *Phytophthora* primers S1, S2, S3c, S4, S5, and S6[88] (Table 96.1), and it was 99.0% identical to the ITS sequence of *P. insidiosum* (strain CBS777.84), indicating that the organism in question was *P. insidiosum*.

Using the same approach, Rivierre et al. identified the first case of pythiosis from Africa,[42] Vanittanakom et al. confirmed the identification of *P. insidiosum* isolated from human patients,[79] and Reis et al. diagnosed pythiosis in a horse.[43] Rivierre et al. amplified the 785 bp PCR product of

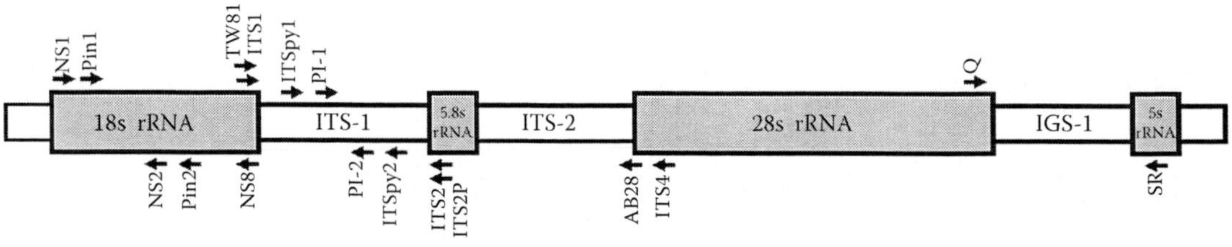

FIGURE 96.2 Map of the rDNA repeat of *P. insidiosum* showing primers (by names as indicated in Table 96.1) and their estimated annealing sites. Arrows represent location and direction of each primer. Sizes of the rDNA repeat regions shown are not proportional to actual sizes. ITS, the internal transcribed spacer; IGS, the rDNA intergenic spacer.

TABLE 96.1
Primers for PCR Amplification of the *P. insidiosum* rDNA Repeat Region

Primer Name	Sequences	Tm (°C)	References
NS1	5′-GTAGTCATATGCTTGTCTC-3′	51.8	[7,87]
NS2	5′-GGCTGCTGGCACCAGACTTGC-3′	68.4	[7,87]
NS8	5′-TCCGCAGGTTCACCTACGGA-3′	64.8	[43,87]
ITS1	5′-TCCGTAGGTGAACCTGCGG-3′	63.5	[42,80,87]
ITS2	5′-GCTGCGTTCTTCATCGATGC-3′	61.4	[80,87]
ITS2P	5′-GCAGCGTTCTTCATCGATGT-3′	60.0	[81]
ITS4	5′-TCCTCCGCTTATTGATATGC-3′	56.2	[42,87]
Pin1	5′-TGGCTCTTCGAGTCGGGCAA-3′	65.1	[79]
Pin2	5′-GTCGGCATAGTTTATGGTTAAGA-3′	56.2	[79]
ITSpy1	5′-CTGCGGAAGGATCATTACC-3′	56.6	[79]
ITSpy2	5′-GTCCTCGGAGTATAGATCAG-3′	56.0	[79]
PI-1	5′-TTCGTCGAAGCGGACTGCT-3′	63.1	[80]
PI-2	5′-GCCGTACAACCCGAGAGTCATA-3′	63.4	[80]
Q	5′-ACGCCTCTAAGTCAGAATC-3′	55.1	[83]
SR	5′-GAAGCCCGGGTGCTGTCTAG-3′	64.3	[83]
TW81	5′-GCGGATCCGTTTCCGTAGGTGAACCTGC-3′	71.1	[7,88]
AB28	5′-GCGGATCCATATGCTTAAGTTCAGCGGGT-3′	68.9	[7,88]
S1	5′-CCGTAGGTGAACCTGCGGAGG-3′	66.4	[7,88]
S2	5′-GCACATCGATGAAGAACGCTG-3′	61.3	[7,88]
S3c	5′-GCGACTTCGGTTAGGACATT-3′	58.4	[7]
S4	5′-CCCGGAAGTGCAATATGCG-3′	60.5	[7,88]
S5	5′-GCCGAAGCCACCCATACCG-3′	62.5	[7,88]
S6	5′-GCTTAAGTTCAGCGGGTAATC-3′	56.8	[7,88]

the ITS portion from genomic DNA, extracted from infected tissue, using primers ITS1 and ITS4[87] (Figure 96.2, Table 96.1). Vanittanakom et al. amplified the 580 bp PCR product of the 18S rRNA portion from genomic DNA, extracted from pure culture, using primers Pin1 and Pin2[79] (Figure 96.2, Table 96.1). Reis et al. sequenced the 18S rRNA gene portion amplified by using the universal fungal primers NS1 and NS8[43,87] (Figure 96.2, Table 96.1). In each case, BLAST searching of the sequences of these PCR products successfully identified the organism as *P. insidiosum*.

96.1.4.2.2 PCR Amplification

Use of PCR amplification, but without subsequent sequence analysis, is an alternative method for identification of *P. insidiosum* from tissue specimens and pure cultures (in the latter case, it is useful for bypassing the need for the difficult-to-achieve induction of zoospores to prove the identity of the isolated organism). A PCR assay for detection and identification of *P. insidiosum* was first described by Grooters and Gee.[80] They used the universal fungal primers, ITS1 and ITS2[87] (Figure 96.2, Table 96.1), for first-round amplification to determine the presence or absence of fungal DNA in the sample. The first-round amplification of *P. insidiosum* DNA produced a product of ~300 bp. The *P. insidiosum*-specific primers, PI-1 and PI-2[80] (Figure 96.2, Table 96.1), were used to generate a 105 bp product in the second-round nested-PCR amplification. The presence of this specific PCR product indicates that genomic DNA of the organism being tested contained *P. insidiosum* DNA. Grooters et al. showed that their nested-PCR assay is very sensitive and specific since the assay correctly detected all *P. insidiosum* isolates tested, and no nested-PCR products are observed for other *Pythium* species or for true fungi with similar morphology to *P. insidiosum*. Znajda et al. showed that, by substitution of primer ITS2 for primer ITS2P (Figure 96.2, Table 96.1), nested PCR can be used to detect *P. insidiosum* directly in frozen and ethanol-fixed infected tissue.[81] Thus, the nested PCR strategy is very useful in the detection of *P. insidiosum* as a means of diagnosing pythiosis, especially in cases in which the culture methods failed.

Vanittanakom et al. later developed a rapid single-round PCR amplification, which was used to identify *P. insidiosum* isolates in Thai patients with pythiosis.[79] Based on the ITS sequence of *P. insidiosum*, they designed two specific primers, ITSpy1 and ITSpy2 (Figure 96.2, Table 96.1), which were used to generate a single 233 bp PCR product in all four *P. insidiosum* isolates tested. Vanittanakom et al. went on to show that, when using this 233 bp ITSpy1–ITSp2 amplicon as a template, the *P. insidiosum*-specific primers for nested PCR (PI-1 and PI-2; Figure 96.2) of Grooters et al. failed to amplify *P. insidiosum*-specific fragment from one of the isolates. This result calls into question the universality of primers PI-1 and PI-2 for amplification of Thai isolates *P. insidiosum*. However, based on the limited number of samples tested, there is no evidence indicating that the primers ITSpy1 and ITSpy2 of Vanittanakom et al. can amplify all *P. insidiosum* isolates from different geographic regions. Therefore, until there is an extensive evaluation of the specific primer sets of Grooters et al. and Vanittanakom et al., regarding detection sensitivity and specificity, tests for PCR amplification using either of them should be interpreted with caution because of the possibility of a false negative result.

Nested PCR has some advantages over a single-round PCR amplification in that it increases detection sensitivity by enriching the DNA target in the first round of amplification, and it increases detection specificity by restriction of annealing sequence targets of the nested primers to only the first-round PCR product, rather than to the whole genome (because of its complexity, the whole genome has the potential for more nonspecific annealing interactions and false positive PCR products). Although the two sets of amplification in nested PCR can make it a more sensitive procedure, the high sensitivity can be a disadvantage because of false positives resulting from cross contamination of *P. insidiosum* sample DNA in the steps of DNA extraction and amplification. Careful procedures should be performed for all PCR steps in order to prevent contamination. For example, use sterile containers and barrier tips for all steps, wear gloves, use a separate set of autopipettes, and use separate laboratory zones for DNA extraction, PCR amplification, and post-PCR gel electrophoresis. Inclusion of positive and negative control samples, along with the unknown sample, would help detect errors in the PCR procedure and false positive or negative results.

96.1.4.2.3 Species-Specific Probe

Schurko et al. have developed a dot-blot hybridization analysis of *P. insidiosum* genomic DNA, using a species-specific probe for the diagnosis of pythiosis.[83] The hybridization analysis employs a chemiluminescence detection system by which the signal from a positive hybridization is recorded on x-ray film. The rDNA intergenic spacer fragment (the sequence between 28S and 5S rRNA) of *P. insidiosum* is amplified from genomic DNA using primers Q and SR (Figure 96.2, Table 96.1) and a 530bp species-specific probe is derived from it by HinfI enzyme digestion. The probe, which does not overlap any of the conserved rRNA regions and provides an intense hybridization signal, is used to hybridize denatured, extracted test-organism genomic DNA, which has been spotted (dot blotted) on a membrane. Schurko et al. demonstrated that their hybridization assay has high sensitivity, as it gives a positive signal for genomic DNAs of *P. insidiosum* isolates from all over the world, while high specificity is revealed by the lack of a hybridization signal for all genomic DNAs of other *Pythium* species or for true fungi. This method is clinically useful for identification of *P. insidiosum* DNA extracted from pure culture, and future applications of the method may make it possible to detect *P. insidiosum* DNA in infected tissue or patient blood. However, this technique is not routinely available for diagnosis of pythiosis because it requires a multistep procedure, expensive reagents, and skilled personnel to perform the assay.

96.2 METHODS

This section describes, in the first part, several established sample preparation methods for extraction of genomic DNA of *P. insidiosum*. This is a very important step for the generation of a high-quality target DNA for the various downstream, conventional molecular diagnostic methods. In the second part, detection procedure guidelines are provided for using the extracted DNA in the identification and diagnosis of pythiosis, using standard molecular detection techniques.

96.2.1 Sample Preparation

P. insidiosum isolated from a clinical sample can be maintained on Sabouraud dextrose agar at 28°C–37°C with subculturing to a new agar plate every 2–4 weeks. Alternatively, the organism can be stored for months, as a few pieces of mycelial-attached agar, submerged in sterile water at room temperature. To prepare hyphae for DNA extraction, *P. insidiosum* from Sabouraud dextrose agar or sterile water storage is subcultured on Sabouraud dextrose agar for several days at 28°C–37°C to allow for active growth. Several small pieces of mycelial-attached agar pieces are excised from the edge of *Pythium* colony and transferred to 100mL of Sabouraud dextrose broth for shaker culture (~150rpm) for 5–10 days at 28°C–37°C. Hyphal growth is harvested by filtration onto sterile filter paper or cloth and washed once with 100mL of phosphate-buffered saline (PBS). Hyphal material can be stored frozen at −80°C for future use. Several methods for DNA extraction of pure culture, frozen hyphae, and tissue are described in detail as follows.

96.2.1.1 Conventional Genomic DNA Extraction

The first method for genomic DNA extraction from pure culture and frozen hyphae of *P. insidiosum* is modified from the method used by Jackson et al.[89] and Pannanusorn et al.[90] To start the DNA extraction, fresh or frozen hyphal cells are placed in a precooled mortar in the presence of liquid nitrogen and are ground to a fine powder with a pestle. Then, the hyphal powder (200mg) is placed in a 1.5mL tube and 600μL of lysis buffer (400mM Tris–HCl [pH 8.0], 60mM ethylenediaminetetraacetic acid [EDTA], 150mM NaCl, 1% sodium dodecyl sulfate [SDS], and 40mg/mL proteinase K) is added and mixed. The sample is incubated at 60°C for 60min with occasional and gentle inversion. Then, 100μL of 5M sodium perchlorate is added and incubated at 60°C for 15min. The sample is stored on ice, and DNA extraction is sequentially performed with 500μL of ice-cold chloroform, with an equal volume of phenol-chloroform-isoamyl alcohol (volume ratio, 25:24:1) and finally with an equal volume of chloroform. Extracted genomic DNA is precipitated with 0.6 volume of isopropanol or 2 volumes of ice-cold absolute ethanol before washing two times with 500μL of 70% ethanol, air-dried for 5–10min, and resuspended in 100–200μL of sterile water. Extracted DNA is stored at 4°C or −20°C.

An alternative method for extraction of genomic DNA from *P. insidiosum* is modified from the fungal DNA extraction method of Hogan and Klein.[91] First, extraction buffer (100 mM Tris-HCl pH 8.0, 100 mM EDTA, 250 mM NaCl, 100 μg/mL proteinase K, and 1% SDS [the latter two components should be added immediately before use]) is used to resuspend the hyphal powder in a 50 mL tube at the ratio of 5 mL of extraction buffer per 1 g hyphal power. The mixture is incubated at 55°C for 2 h before centrifugation at 10,000 × *g* at room temperature for 60 min to remove cell debris. An equal volume of phenol-chloroform-isoamyl alcohol (volume ratio, 25:24:1) is added to the supernatant, shaken hard to mix, allowed to sit for 5 min, and centrifuged at 2000 × *g* for 30 min to separate the upper aqueous and the lower organic phases. Two volumes of absolute ethanol or 0.6 volume of isopropanol are added to the aqueous phase and centrifuged at 2000 × *g* for 5 min to pellet the DNA. The pellet is washed with 70% ethanol and dissolved with 250–500 μL TE buffer (10 mM Tris-HCl pH 7.5, 1 mM EDTA pH 8.0) in a 1.5 mL tube. The sample is treated with RNaseA (50 μg/mL) at 37°C for 30 min, extracted with an equal volume of chloroform–isoamyl alcohol (volume ratio, 24:1), and centrifuged at 14,000 × *g* for 10 min. The resultant supernatant is precipitated with 0.1 volume of 3 M sodium acetate, pH 5.2, and with either 2 volumes of absolute ethanol or 1 volume of isopropanol and centrifuged at 14,000 × *g* for 10 min to pellet the DNA. The pellet is washed twice with 70% ethanol, air-dried for 10–15 min, and dissolved in 100–200 μL TE buffer. The DNA is allowed to sit overnight at 4°C before checking DNA recovery and integrity by absorbance at 260 and 280 nm and by 1% agarose gel electrophoresis.

96.2.1.2 Rapid Genomic DNA Extraction

The methods of Woods et al.[92] and Vanittanakom et al.[79] are modified for a rapid genomic DNA extraction from a pure culture of *P. insidiosum*. A few loops of *P. insidiosum* hyphae is spooled out from broth culture, suspended in a 0.5 mL volume of lysis buffer (250 mM Tris pH 8.0, 1.5% SDS), and boiled for 30 min in a water bath. The boiled-hyphal suspension is vortexed hard for 2 min, and then the DNA is extracted with an equal volume of phenol-chloroform-isoamyl alcohol (volume ratio, 25:24:1) and precipitated from the aqueous phase with 0.1 volume of 3 M sodium acetate (pH 5.2) and 2 volumes of absolute ethanol. The DNA pellet is air-dried, resuspended in 50–75 μL of sterile water, and the DNA concentration is estimated by electrophoresis of a 5 μL sample in 1% agarose gel or with a spectrophotometer (260/280 nm).

96.2.1.3 Genomic DNA Extraction from *P. insidiosum*-Infected Tissue

A method for genomic DNA extraction of *P. insidiosum* in frozen or ethanol-fixed tissues has been reported and successfully used by Znajda et al.[81] In brief, a piece of infected tissue (5 mm^3) is cut and placed in a 1.5 mL tube with 1000 μL digestion buffer (50 mM Tris–HCl [pH 7.2], 50 mM EDTA, 1% SDS, and 1% 2-mercaptoethanol) and 300 μg of proteinase K. The tissue is incubated at 65°C for 2 h before addition of another 300 μg of proteinase K. The tissue is vortexed, incubated at 65°C for 2–4 h, and centrifuged at 10,000 × *g* for 15 min at room temperature to collect the supernatant containing the genomic DNA. The supernatant is further extracted with an equal volume of chloroform-phenol-isoamyl alcohol (volume ratio, 25:24:1) and the sample is centrifuged as above. Genomic DNA in the aqueous phase is then precipitated by adding 1 volume of isopropanol in the presence of 0.3 M sodium acetate pH 5.2. The DNA pellet is washed twice with 70% ethanol, air-dried, resuspended in 100 μL sterile deionized water, and stored at −20°C for future use. DNA is diluted at 1:10 in sterile deionized water before using as a template in PCR.

96.2.2 Detection Procedures

96.2.2.1 Sequence Analysis for Species-Specific Identification

At least four regions of rDNA of *P. insidiosum* can be used for species-specific identification by sequence analysis. These four rDNA regions can be PCR amplified using genomic DNA as template together with the following primers: TW81 and AB28 (900 bp PCR product[7,88]), ITS1 and ITS4 (785 bp PCR product[42,87]), Pin1 and Pin2 (580 bp PCR product[79]), or NS1 and NS8[43,87] (Figure 96.2, Table 96.1).

Procedure

1. Prepare PCR mixture (20 μL) containing 10 ng genomic DNA, 0.1 mM deoxynucleoside triphosphates (dNTPs), 4.0 mM $MgCl_2$, 8 pmole of each primer, and 1 unit of Platinum *Taq* DNA polymerase (Invitrogen, United States) in 1× PCR buffer pH 8.4 (20 mM Tris-HCl, 50 mM KCl).
2. Perform the PCR amplification in a thermocycler using a cycling program consisting of a denaturation step at 95°C for 3 min; 30 cycles of 95°C for 1 min, 51°C (an annealing temperature for the primer ITS1 and ITS4) for 45 s, and 72°C for 1 min 45 s; and a final extension step at 72°C for 10 min. Assess the PCR product for amount and size by 1% agarose gel electrophoresis.
3. Perform direct sequencing reaction of the PCR product using the Big Dye Terminator Cycle Sequencing Kit version 3.1 (Applied Biosystems, United States). Prepare a sequencing mixture (10 μL) containing 4 μL sequencing reaction mix diluted 1:1 in sequencing buffer pH 9.0 (200 mM Tris, 5mM $MgCl_2$), 3.2 pmol primer and 200–500 ng template (PCR product from step 2). The reaction is performed in an ABI 2700 96 well plate thermal cycler with the following cycling temperatures: 10 s denaturation at 96°C, 5 s annealing at 50°C, and 4 min elongation at 60°C for a total of 25 cycles.
4. Precipitate the resulting product by adding 1 μL of 125 mM EDTA, 1 μL of 3 M NaOAc (pH 4.6), and 25 μL of absolute ethanol. Incubate the mixture for

15 min at room temperature before centrifugation at 2000 × *g* for 30 min at room temperature.

5. Wash the pellet with 70 µL 70% cold ethanol and centrifuge at 2000 × *g* for 15 min at room temperature. Add 12 µL HiDi formamide (Applied Biosystems) to each sample and sequence the product with the ABI 3100 Genetic Analyzer with 16 capillary tubes (Applied Biosystems, United States).
6. The sequence obtained is subjected to BLAST searching online, using the National Center for Biotechnology Information Nucleotide Database at the webpage, http://blast.ncbi.nlm.nih.gov/Blast.cgi. Definitive identification of *P. insidiosum* is based on 99%–100% sequence identity of the tested sequence to the previously described *P. insidiosum* 18S rRNA sequence.[7,42,43,79]

Note: Generally, the proper annealing temperature of other sets of primers can be estimated by a 5°C subtraction from the Tm of the lower of the two primers.

96.2.2.2 PCR Detection

The target for PCR amplification of *P. insidiosum* DNA has so far been limited to the region of rDNA repeat. The use of species-specific primers and the presence of a single PCR product of expected size in gel electrophoresis can be definitive for the identification of *P. insidiosum*, without the need for sequence analysis.

96.2.2.2.1 Single-Round PCR Protocol

The specific primer ITSpy1 and ITSpy2 (Figure 96.2, Table 96.1) are designed for a single PCR amplification of a specific product of 233 bp.[79]

Procedure

1. Prepare PCR mixture (50 µL) containing 10 ng genomic DNA template, 0.2 mM dNTPs, 10 pmol of each primer, 2.5 units of *Taq* DNA polymerase (Qiagen GmbH, Hilden, Germany), 20 mM Tris-HCl (pH 8.4), 50 mM KCl, and 1.5 mM $MgCl_2$.
2. Perform the PCR amplification in a thermocycler using a cycling program consisting of a denaturation step at 94°C for 3 min; 35 cycles of 94°C for 45 s, 60°C for 30 s, and 72°C for 30 s; and a final extension step at 72°C for 10 min.
3. Separate the PCR products on a 1% agarose gel and stain with ethidium bromide.

96.2.2.2.2 Nested PCR Protocol

The universal fungal primers ITS1 and ITS2 are used to enrich the DNA target in the first round of PCR amplification, which increases detection sensitivity. The nested primers PI-1 and PI-2 are designed to restrict annealing sequence area to only the first-round PCR product, which increases detection specificity.[80]

Procedure

1. For the first-round PCR amplification, prepare PCR mixture (50 µL) containing 10 ng genomic DNA template, 20 µM dNTPs, 20 pmol of each primer (ITS1 and ITS2), and 2.5 units of *Taq* DNA polymerase (Promega Corporation, Madison, WI), 10 mM Tris–HCl (pH 8.3), 50 mM KCl, and 1.5 mM $MgCl_2$.
2. Perform the first-round amplification in a thermocycler using a cycling program consisting of a denaturation step at 94°C for 3 min; 30 cycles of 94°C for 45 s, 64°C for 30 s, and 72°C for 30 s; and a final extension step at 72°C for 10 min.
3. For the second-round PCR amplification, prepare PCR mixture (50 µL) containing 5 µL of the first-round PCR product diluted at 1:50 in ddH_2O, 20 µM dNTPs, 20 pmol of each primer (PI-1 and PI-2), 2.5 units of *Taq* DNA polymerase (Promega Corporation, Madison, WI), 10 mM Tris–HCl (pH 8.3), 50 mM KCl, and 1.5 mM $MgCl_2$.
4. Perform the second-round amplification in a thermocycler using a cycling program consisting of a denaturation step at 94°C for 3 min; 15 cycles of 94°C for 30 s, 65°C for 30 s, and 72°C for 30 s; and a final extension step at 72°C for 10 min.
5. Separate the PCR products on a 1% agarose gel and stain with ethidium bromide.

96.3 CONCLUSIONS AND FUTURE PERSPECTIVES

Pythiosis can be a fatal infectious disease of humans and animals living in tropical and subtropical areas worldwide and is caused by a fungus-like organism, *P. insidiosum*. The disease has clinical features that mimic other infectious diseases caused by true fungi, such as aspergillosis and zygomycosis. The majority of health care professionals are not familiar with the disease, leading to underdiagnosis and misdiagnosis. The high rate of morbidity and mortality of pythiosis is exacerbated by delayed diagnosis and the lack of effective treatments. The ecological niche of *P. insidiosum* is aquatic or swampy areas, and direct contact of patients to its habitat is likely an initial step for acquiring the *Pythium* infection. Since natural habitats of *P. insidiosum* are widespread and the disease is devastating, increased attention and awareness of pythiosis should be promoted among health care providers.

Early diagnosis leads to a prompt management and a good prognosis of patients with pythiosis. Culture identification of *P. insidiosum* is difficult, laborious, and time-consuming, and it often fails to isolate the causative agent from clinical specimens. Serodiagnostic tests, that is, ID, ELISA, Western blot, ICT, and HA test are available, but use of these tests are limited because of the requirement of skilled personnel and specialized equipment and a technical limitation of detection sensitivity for some types of infection, that is, early infection and ocular infection, which produce low or no antibody response. Direct detection of the pathogen in infected tissue by immunochemical assay should be a solution for the diagnostic problem relating to undetectable anti-*P. insidiosum* antibody. However, false positive reactivity of the immunochemical assay has been described in patients with *Conidiobolosis*, an infectious disease that clinicopathologically mimics pythiosis. Therefore, more still needs to be done, by way of developing new diagnostic techniques that have better sensitivity and specificity for pythiosis.

In recent years, molecular diagnostic assays have become useful techniques for facilitating diagnosis of many infectious diseases. Some molecular techniques, for example, sequence analysis, PCR amplification, and DNA hybridization using species-specific probe, have been developed for detection of *P. insidiosum* in pure culture and frozen ethanol-fixed infected tissues[7,42,43,79–81,83] and could solve diagnostic problems of conventional microbiological and immunological assays mentioned above. However, reliability of these molecular assays may be questioned because only a few *P. insidiosum*-infected and control samples were evaluated during testing for detection efficiency (i.e., sensitivity and specificity). The target gene for detecting *P. insidiosum* DNA has, so far, been limited to only rDNA sequence because genetic information of other target genes is lacking. The specific PCR primers, designed from the rDNA sequence, for detecting *P. insidiosum* isolated from one geographic area may not be able to detect all organisms isolated from other geographic areas.[34,35] Therefore, new assays must be developed in order to account for the apparent genetic diversity that is associated with geographic distribution and more completely cover all strains during diagnostic testing.

It is clear that molecular assays will be the most powerful techniques for diagnosis of pythiosis. Implementation of new molecular technologies, such as real-time PCR and loop-mediated isothermal amplification reaction,[93,94] for the development of better diagnostic tests for pythiosis has potential for rapid and highly reliable tests with a more user-friendly assay format. Two key aspects of establishing an efficient clinical assay, based on molecular techniques, are, first, identifying new gene targets with their corresponding sets of primers to create a new assay and, second, evaluating the new assay for a high level of specificity and sensitivity. It will be necessary to test a large number of pythiosis tissue samples harboring *P. insidiosum* strains from around the world and control samples representing a broad range of mycoses to ensure the reliability of the test. In addition to working with pure cultures and frozen ethanol-fixed infected tissues, diagnostic tests for pythiosis should also be established for paraffin-embedded tissues, as this would allow confirmation of earlier diagnoses and would also allow for retrospective studies of pythiosis using archival specimens.

REFERENCES

1. De Cock, A.W. et al., *Pythium insidiosum* sp. nov., the etiologic agent of pythiosis, *J. Clin. Microbiol.*, 25, 344, 1987.
2. Mendoza, L., Ajello, L., and McGinnis, M.R., Infection caused by the Oomycetous pathogen *Pythium insidiosum*, *J. Mycol. Med.*, 6, 151, 1996.
3. Austwick, P.K. and Copland, J.W., Swamp cancer, *Nature*, 250, 84, 1974.
4. Shipton, W.A., *Pythium destruens* sp. nov., an agent of equine pythiosis, *J. Med. Vet. Mycol.*, 25, 137, 1987.
5. Murdoch, D. and Parr, D., *Pythium insidiosum* keratitis, *Aust. N. Z. J. Ophthalmol.*, 25, 177, 1997.
6. Krajaejun, T. et al., Clinical and epidemiological analyses of human pythiosis in Thailand, *Clin. Infect. Dis.*, 43, 569, 2006.
7. Badenoch, P.R. et al., *Pythium insidiosum* keratitis confirmed by DNA sequence analysis, *Br. J. Ophthalmol.*, 85, 502, 2001.
8. Triscott, J.A., Weedon, D., and Cabana, E., Human subcutaneous pythiosis, *J. Cutan. Pathol.*, 20, 267, 1993.
9. Shenep, J.L. et al., Successful medical therapy for deeply invasive facial infection due to *Pythium insidiosum* in a child, *Clin. Infect. Dis.*, 27, 1388, 1998.
10. Bosco Sde, M. et al., Human pythiosis, Brazil, *Emerg. Infect. Dis.*, 11, 715, 2005.
11. Virgile, R. et al., Human infectious corneal ulcer caused by *Pythium insidiosum*, *Cornea*, 12, 81, 1993.
12. Mendoza, L., Prasla, S.H., and Ajello, L., Orbital pythiosis: A non-fungal disease mimicking orbital mycotic infections, with a retrospective review of the literature, *Mycoses*, 47, 14, 2004.
13. Schlosser, E. and Gottlieb, D., Sterols and the sensitivity of *Pythium* species to filipin, *J. Bacteriol.*, 91, 1080, 1966.
14. Kwon-Chung, K.J., Phylogenetic spectrum of fungi that are pathogenic to humans, *Clin. Infect. Dis.*, 19(Suppl. 1), S1, 1994.
15. Pereira, D.I., Caspofungin *in vitro* and *in vivo* activity against Brazilian *Pythium insidiosum* strains isolated from animals, *J. Antimicrob. Chemother.*, 60, 1168, 2007.
16. Kamoun, S., Molecular genetics of pathogenic oomycetes, *Eukaryot. Cell.*, 2, 191, 2003.
17. Keeling, P.J. et al., The tree of eukaryotes, *Trends. Ecol. Evol.*, 20, 670, 2005.
18. Mendoza, L., Hernandez, F., and Ajello, L., Life cycle of the human and animal oomycete pathogen *Pythium insidiosum*, *J. Clin. Microbiol.*, 31, 2967, 1993.
19. Mendoza, L. and Prendas, J., A method to obtain rapid zoosporogenesis of *Pythium insidiosum*, *Mycopathologia*, 104, 59, 1988.
20. Chaiprasert, A. et al., Induction of zoospore formation in Thai isolates of *Pythium insidiosum*, *Mycoses*, 33, 317, 1990.
21. Kaufman, L., *Penicilliosis marneffei* and pythiosis: Emerging tropical diseases, *Mycopathologia*, 143, 3, 1998.
22. Thianprasit, M., Chaiprasert, A., and Imwidthaya, P., Human pythiosis, *Curr. Top. Med. Mycol.*, 7, 43, 1996.
23. Sathapatayavongs, B. et al., Human pythiosis associated with thalassemia hemoglobinopathy syndrome, *J. Infect. Dis.*, 159, 274, 1989.
24. Tanphaichitra, D., Tropical disease in the immunocompromised host: Melioidosis and pythiosis, *Rev. Infect. Dis.*, 11(Suppl. 7), S1629, 1989.
25. Chetchotisakd, P. et al., Human pythiosis in Srinagarind Hospital: One year's experience, *J. Med. Assoc. Thai.*, 75, 248, 1992.
26. Wanachiwanawin, W. et al., Fatal arteritis due to *Pythium insidiosum* infection in patients with thalassaemia, *Trans. R. Soc. Trop. Med. Hyg.*, 87, 296, 1993.

27. Thitithanyanont, A. et al., Use of an immunotherapeutic vaccine to treat a life-threatening human arteritic infection caused by *Pythium insidiosum*, *Clin. Infect. Dis.*, 27, 1394, 1998.
28. Prasertwitayakij, N. et al., Human pythiosis, a rare cause of arteritis: Case report and literature review, *Semin. Arthritis. Rheum.*, 33, 204, 2003.
29. Krajaejun, T. et al., Ocular pythiosis: Is it under-diagnosed? *Am. J. Ophthalmol.*, 137, 370, 2004.
30. Wanachiwanawin, W. et al., Efficacy of immunotherapy using antigens of *Pythium insidiosum* in the treatment of vascular pythiosis in humans, *Vaccine*, 22, 3613, 2004.
31. Pupaibool, J. et al., Human pythiosis, *Emerg. Infect. Dis.*, 12, 517, 2006.
32. Miller, R.I., Investigations into the biology of three 'phycomycotic' agents pathogenic for horses in Australia, *Mycopathologia*, 81, 23, 1983.
33. Supabandhu, J. et al., Isolation and identification of the human pathogen *Pythium insidiosum* from environmental samples collected in Thai agricultural areas, *Med. Mycol.*, 18, 1, 2007.
34. Schurko, A.M. et al., Evidence for geographic clusters: Molecular genetic differences among strains of *Pythium insidiosum* from Asia, Australia, and the Americas are explored, *Mycologia*, 95, 200, 2003.
35. Schurko, A.M. et al., A molecular phylogeny of *Pythium insidiosum*, *Mycol. Res.*, 107, 537, 2003.
36. Mendoza, L. et al., Evaluation of two vaccines for the treatment of pythiosis insidiosi in horses, *Mycopathologia*, 119, 89, 1992.
37. Mendoza, L., Mandy, W., and Glass, R., An improved *Pythium insidiosum*-vaccine formulation with enhanced immunotherapeutic properties in horses and dogs with pythiosis, *Vaccine*, 21, 2797, 2003.
38. Dixon, D.M. et al., Development of vaccines and their use in the prevention of fungal infections, *Med. Mycol.*, 36(Suppl. 1), 57, 1998.
39. Mendoza, L. and Newton, J.C., Immunology and immunotherapy of the infections caused by *Pythium insidiosum*, *Med. Mycol.*, 43, 477, 2005.
40. Brown, C.C. and Roberts, E.D., Intestinal pythiosis in a horse, *Aust. Vet. J.*, 65, 88, 1988.
41. Patton, C.S. et al., Esophagitis due to *Pythium insidiosum* infection in two dogs, *J. Vet. Intern. Med.*, 10, 139, 1996.
42. Rivierre, C. et al., Pythiosis in Africa, *Emerg. Infect. Dis.*, 11, 479, 2005.
43. Reis, J.L. Jr. et al., Disseminated pythiosis in three horses, *Vet. Microbiol.*, 96, 289, 2003.
44. Mendoza, L. and Alfaro, A.A., Equine pythiosis in Costa Rica: Report of 39 cases, *Mycopathologia*, 94, 123, 1986.
45. Meireles, M.C. et al., Cutaneous pythiosis in horses from Brazil, *Mycoses*, 36,139, 1993.
46. Miller, R.I. and Campbell, R.S., The comparative pathology of equine cutaneous phycomycosis, *Vet. Pathol.*, 21, 325, 1984.
47. Miller, R.I. and Campbell, R.S., Clinical observations on equine phycomycosis, *Aust. Vet. J.*, 58, 221, 1982.
48. Sohn, Y. et al., Enteric pythiosis in a Jindo dog, *Korean J. Vet. Res.*, 36, 447, 1996.
49. Bentinck-Smith, J. et al., Canine pythiosis—Isolation and identification of *Pythium insidiosum*, *J. Vet. Diagn. Invest.*, 1, 295, 1989.
50. Hnilica, K.A., Difficult dermatologic diagnosis. Pythiosis, *J. Am. Vet. Med. Assoc.*, 212, 1192, 1998.
51. Dykstra, M.J. et al., A description of cutaneous-subcutaneous pythiosis in fifteen dogs, *Med. Mycol.*, 37, 427, 1999.
52. Berryessa, N.A. et al., Gastrointestinal pythiosis in 10 dogs from California, *J. Vet. Intern. Med.*, 22, 1065, 2008.
53. Graham, J.P. et al., Ultrasonographic features of canine gastrointestinal pythiosis, *Vet. Radiol. Ultrasound.*, 41, 273, 2000.
54. Helman, R.G. and Oliver, J. 3rd., Pythiosis of the digestive tract in dogs from Oklahoma, *J. Am. Anim. Hosp. Assoc.*, 35, 111, 1999.
55. Rakich, P.M., Grooters, A.M., and Tang, K.N., Gastrointestinal pythiosis in two cats, *J. Vet. Diagn. Invest.*, 17, 262, 2005.
56. Pérez, R.C. et al., Epizootic cutaneous pythiosis in beef calves, *Vet. Microbiol.*, 109, 121, 2005.
57. Santurio, J.M. et al., Cutaneous *Pythiosis insidiosi* in calves from the Pantanal region of Brazil, *Mycopathologia*, 141, 123, 1998.
58. Miller, R.I., Olcott, B.M., and Archer, M., Cutaneous pythiosis in beef calves, *J. Am. Vet. Med. Assoc.*, 186, 984, 1985.
59. Tabosa, I.M. et al., Outbreaks of pythiosis in two flocks of sheep in northeastern Brazil, *Vet. Pathol.*, 41, 412, 2004.
60. Wellehan, J.F. et al., Pythiosis in a dromedary camel (*Camelus dromedarius*), *J. Zoo. Wildl. Med.*, 35, 564, 2004.
61. Buergelt, C., Powe, J., and White, T., Abdominal pythiosis in a Bengal tiger (Panthera tigris tigris), *J. Zoo. Wildl. Med.*, 37, 186, 2006.
62. Camus, A.C., Grooters, A.M., and Aquilar, R.E., Granulomatous pneumonia caused by *Pythium insidiosum* in a central American jaguar, Panthera onca, *J. Vet. Diagn. Invest.*, 16, 567, 2004.
63. Miller, R.I. and Campbell, R.S., Experimental pythiosis in rabbits, *Sabouraudia*, 21, 331, 1983.
64. Ravishankar, J.P. et al., Mechanics of solid tissue invasion by the mammalian pathogen *Pythium insidiosum*, *Fungal. Genet. Biol.*, 34, 167, 2001.
65. Wanachiwanawin, W., Infections in E-beta thalassemia, *J. Pediatr. Hematol. Oncol.*, 22, 581, 2000.
66. Farmakis, D. et al., Pathogenetic aspects of immune deficiency associated with beta-thalassemia, *Med. Sci. Monit.*, 9, 19, 2003.
67. Walker, E.M. Jr. and Walker, S.M., Effects of iron overload on the immune system, *Ann. Clin. Lab. Sci.*, 30, 354, 2000.
68. Pracharktam, R. et al., Immunodiffusion test for diagnosis and monitoring of human *Pythiosis insidiosi*, *J. Clin. Microbiol.*, 29, 2661, 1991.
69. Imwidthaya, P. and Srimuang, S., Immunodiffusion test for diagnosing human pythiosis, *Mycopathologia*, 106, 109, 1989.
70. Mendoza, L., Kaufman, L., and Standard, P.G., Immunodiffusion test for diagnosing and monitoring pythiosis in horses, *J. Clin. Microbiol.*, 23, 813, 1986.
71. Krajaejun, T. et al., Development and evaluation of an in-house enzyme-linked immunosorbent assay for early diagnosis and monitoring of human pythiosis, *Clin. Diagn. Lab. Immunol.*, 9, 378, 2002.
72. Mendoza, L. et al., Serodiagnosis of human and animal pythiosis using an enzyme-linked immunosorbent assay, *Clin. Diagn. Lab. Immunol.*, 4, 715, 1997.
73. Grooters, A.M. et al., Development and evaluation of an enzyme-linked immunosorbent assay for the serodiagnosis of pythiosis in dogs, *J. Vet. Intern. Med.*, 16, 142, 2002.
74. Krajaejun, T. et al., Identification of a novel 74-kiloDalton immunodominant antigen of *Pythium insidiosum* recognized by sera from human patients with pythiosis, *J. Clin. Microbiol.*, 44, 1674, 2006.
75. Mendoza, L., Nicholson, V., and Prescott, J.F., Immunoblot analysis of the humoral immune response to *Pythium insidiosum* in horses with pythiosis, *J. Clin. Microbiol.*, 30, 2980, 1992.

76. Krajaejun, T. et al., Development of an immunochromatographic test for rapid serodiagnosis of human pythiosis, *Clin. Vac. Immunol.*, 16, 506, 2009.
77. Jindayok, T. et al., Hemagglutination test for rapid serodiagnosis of human pythiosis, *Clin. Vaccine. Immunol.*, 16, 1047, 2009.
78. Supabandhu, J. et al., Application of immunoblot assay for rapid diagnosis of human pythiosis, *J. Med. Assoc. Thai.*, 92, 1063, 2009.
79. Vanittanakom, N. et al., Identification of emerging human-pathogenic *Pythium insidiosum* by serological and molecular assay-based methods, *J. Clin. Microbiol.*, 42, 3970, 2004.
80. Grooters, A.M. and Gee, M.K., Development of a nested polymerase chain reaction assay for the detection and identification of *Pythium insidiosum*, *J. Vet. Intern. Med.*, 16, 147, 2002.
81. Znajda, N.R., Grooters, A.M., and Marsella, R., PCR-based detection of *Pythium* and *Lagendium* DNA in frozen and ethanol-fixed animal tissues, *Vet. Dermatol.*, 13, 187, 2002.
82. Brown, C.C. et al., Use of immunohistochemical methods for diagnosis of equine pythiosis, *Am. J. Vet. Res.*, 49, 1866, 1988.
83. Schurko, A.M. et al., Development of a species-specific probe for *Pythium insidiosum* and the diagnosis of pythiosis, *J. Clin. Microbiol.*, 42, 2411, 2004.
84. Grooters, A.M. et al., Evaluation of microbial culture techniques for the isolation of *Pythium insidiosum* from equine tissues, *J. Vet. Diagn. Invest.*, 14, 288, 2002.
85. Davis, D.J. et al., Relationship between temperature optima and secreted protease activities of three *Pythium* species and pathogenicity toward plant and animal hosts, *Mycol. Res.*, 110, 96, 2006.
86. Mendoza, L., Kaufman, L., and Standard, P., Antigenic relationship between the animal and human pathogen *Pythium insidiosum* and nonpathogenic *Pythium* species, *J. Clin. Microbiol.*, 25, 2159, 1987.
87. White, T.J. et al., Amplification and direct sequencing of fungal ribosomal RNA genes for phylogenetics. In: Innis, M., Gelfand, D., Sninsky, J., and White, T. (Eds.), *PCR Protocols A Guide to Methods and Applications*. pp. 315–322, San Diego, CA: Academic Press, 1990.
88. Crawford, A.R. et al., Evolutionary relationships among *Phytophthora* species deduced from rDNA sequence analysis, *Mycol. Res.*, 100, 437, 1996.
89. Jackson, C.J., Barton, R.C., and Evans, E.G., Species identification and strain differentiation of dermatophyte fungi by analysis of ribosomal-DNA intergenic spacer regions, *J. Clin. Microbiol.*, 37, 931, 1999.
90. Pannanusorn, S. et al., Random amplified polymorphic DNA typing and phylogeny of *Pythium insidiosum* clinical isolates in Thailand, *Southeast Asian J. Trop. Med. Public Health*, 38, 383, 2007.
91. Hogan, L.H. and Klein, B.S., Transforming DNA integrates at multiple sites in the dimorphic fungal pathogen *Blastomyces dermatitidis*, *Gene*, 186, 219, 1997.
92. Woods, J.P. et al., Fast DNA isolation from *Histoplasma capsulatum*: Methodology for arbitrary primer polymerase chain reaction-based epidemiological and clinical studies, *J. Clin. Microbiol.*, 31, 463, 1993.
93. Hata, D.J. et al., Real-time PCR method for detection of zygomycetes, *J. Clin. Microbiol.*, 46, 2353, 2008.
94. Nagamine, K. et al., Loop-mediated isothermal amplification reaction using a nondenatured template, *Clin. Chem.*, 47, 1742, 2001.

97 *Prototheca*

Uwe H. Roesler

CONTENTS

97.1 INTRODUCTION

97.1.1 Taxonomy and Morphology

Members of the algae genus *Prototheca* are the only currently known plants that may cause clinical infections both in humans and animals. The disease patterns caused by protothecae are generally referred to as protothecoses. In 1952, the heterotrophic colorless algae *Prototheca* (*P.*) *zopfii* was identified as mastitis pathogen for cows for the first time [1]. Infections were described all over the world and—due to intensive livestock breeding—evolved from an initially sporadic dairy cattle disease into an endemic herd disease. The rare infections of humans are mainly caused by *P. wickerhamii*; in some cases, *P. zopfii* and *P. blaschkeae* were also isolated. The disease mainly affects immunosuppressed persons, e.g., human immunodeficiency virus (HIV) patients. Differentiation is made between cutaneous and subcutaneous protothecoses; inflammations of bursae, fascias, and tendon sheaths; and the more rarely occurring systemic protothecosis.

97.1.1.1 Taxonomy

In 1892, protothecae were first described by [2] who classified them as fungi. He substantiated this classification amongst others with the particular type of reproduction, i.e., endogenous formation of spores. However, protothecae do not have fungi-typical types of reproduction, e.g., mycelium and arthrospores, and, hence, they were reclassified as algae. Even though relationship with green algae was disputed for a long period of time, this assumption is backed by recent molecular biological examinations.

The similarity regarding morphology and type of reproduction convinced the majority of the authors to classify protothecae as offspring of the green algae genus *Chlorella*, which lost its chloroplasts due to mutations and thus is not any more capable of performing autotrophic assimilation [3]. This assumption is further proven by the presence of intermediate forms within the genus *Chlorella*. *Chlorella variegata* and *Chlorella protothecoides* contain chlorophyll; however, it can be observed that they do not produce any more chlorophyll and perform heterotrophic assimilation in case of high availability of nutrients. Molecular biological examinations on SSr DNA level clearly link the genus *Prototheca* to chlorellae.

According to today's applicable taxonomy, protothecae are categorized into five species: *P. zopfii*, *P. wickerhamii*, *P. blaschkeae*, *P. ulmea*, and *P. stagnora*, from which only *P. zopfii*, *P. wickerhamii*, and *P. blaschkeae* are relevant as pathogens [4,5]. *P. zopfii* is subdivided into two genotypes, whereas only *P. zopfii* genotype 2 is pathogenic to humans, cattle, and dogs [6,7].

97.1.1.2 Morphology

Due to their culture morphology and their yeast-like odor, they are often macroscopically confused with *Candida* species on common fungi culture media. Regarding culture morphology, the prototheca colonies have a cream white to beige color, and are smooth, matte-finished and of a soft consistency. As it can also be observed with bacteria and fungi, size and number of contained spores are highly subject to species, isolate, culture medium, and above all development stage.

While colonies of *P. zopfii* present themselves flat with granulated surface on agar plates, colonies of *P. wickerhamii* have a hemispheric shape with a smooth surface.

Microscopically, protothecae are round to oval protozoa with a size of 3–30 μm depending on the development stage [4]. They have a refractive cell wall that mainly consists of cellulose and sporopollenin and shows a considerable stability. *P. zopfii* shows a two-layer cell wall with a thickness of approximately 1000 nm. Subject to the development stage, the protothecae have one-layer to three-layer cell walls.

The different numbers of daughter cells, i.e., sporangiospores, depends on the development stage and species, and these sporangiospores are the most distinctive feature for microscopic identification of the genus *Prototheca* (Figure 97.1). After break of the mother cell, i.e., sporangium, in which the sporangiospores develop, an empty shell will remain and can be well identified microscopically [8].

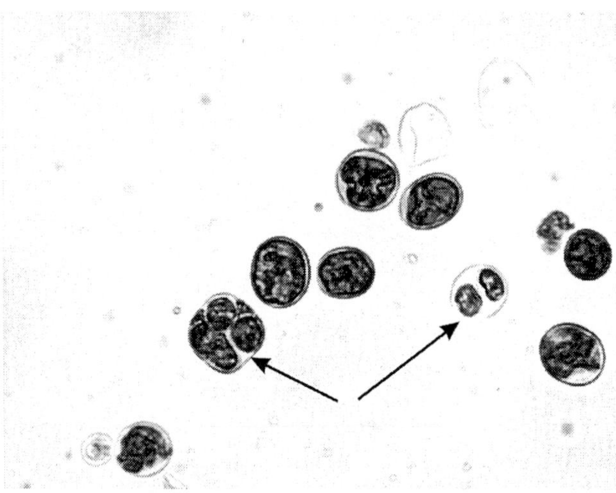

FIGURE 97.1 *Prototheca zopfii* stained with lactophenol cotton blue, 40× magnification. Arrows indicate the spores within the sporangium.

97.1.2 Biology and Epidemiology

97.1.2.1 Biology

Protothecae produce energy by heterotrophic assimilation, i.e., by oxidation of organic compounds. Carbohydrates and lipids are the most important substrates used for production of energy required for any important processes within the cell and alcohols like propanol and glycerol are used as carbon source [9,10].

Nitrogen sources of protothecae are also mainly organic substances. Peptones, lysine, and arginine are very well used, whereas inorganic nitrogen from ammonium salts can only be used limitedly, and inorganic nitrogen from nitrate cannot be used at all.

Protothecae require vitamin B_1 for their growth. Addition of the two main components, i.e., thiazole and pyrimidine, is absolutely necessary, as any members of the genus *Prototheca* are thiamine deficient and thus cannot synthesize this by themselves [11].

In case of poor nutrient supply and unfavorable environmental conditions, protothecae are able to build up permanent spores, so-called hypnospores. Such spores are highly able to store reserve materials like glycogen and fats in the form of oil droplets. Unlike their autotrophic relatives, i.e., chlorellae, the members of the genus *Prototheca* do not have the ability to store starch. This ability was probably lost in connection with the loss of chlorophyll.

Just like other algae, protothecae reproduce asexually by division. Subject to temperature and growth rate, the mother cell, i.e., the sporangium, provides up to 30 endospores or sporangiospores [12]. These are released to the environment by breaking of the cell wall of the mother cell. They grow up into new cells and again reproduce by division. When achieving a third up to a half of the parent's size, spore formation starts again.

Subject to growth conditions, the members of the genus *Prototheca* produce two different spore types. For purposes of outlasting adverse environmental conditions, a few small and thick-walled resting spores or hypnospores with sufficient provision of reserve material are produced. In case of favorable growth conditions, large and thin-walled spores with fewer storage plastids are produced.

The spores of *P. zopfii* and *P. wickerhamii* survive bad environmental conditions by using up their reserve materials. They are better qualified for this than their chlorophyll-containing relatives due to their good provision of such reserve material. Tenacity of *P. zopfii* and *P. wickerhamii* is very high. They can survive in waste water trenches, milk, and also dryness for several months, and they can even be detected in liquid swine and cattle manure after 100 days [13].

Furthermore, the low sensitivity of *P. zopfii* to high temperatures is remarkable. Thus *P. zopfii* is able to survive various pasteurization procedures. A study showed that 62% of 40 mastitis isolates were resistant to a treatment at a temperature of 75°C for 15 s, and a total of 22% of the examined *P. zopfii* isolates were insensitive to all used pasteurization methods [14].

97.1.2.2 Epidemiology

Being ubiquitary saprophytes, the members of the genus *Prototheca* can often be found in places where organic materials decompose. Both genotypes of *P. zopfii* and *P. blaschkeae* can be found in the vicinity (e.g., litter) of cattle and swine as well as in waste water. Only *P. zopfii* genotype 2 can be isolated from cattle mastitis and is thus regarded as a pathogen of such disease. By contrast, *P. wickerhamii* has not yet been proven in the vicinity of cattle and swine, but can be isolated from waste water, e.g., from sewage plants, and it also causes human prototheca infections just like *P. zopfii* genotype 2 and *P. blaschkeae*.

Colonization of the intestines of horse, swine, and cattle with *P. zopfii* is regarded as physiological colonization [15,16]. Protothecae could be isolated from 7% of the examined horse feces samples, 40% of the swine feces samples, and 70% of the cattle feces samples.

97.1.3 Clinical Features and Pathogenesis

97.1.3.1 Protothecoses of Humans

Protothecoses of humans are global sporadically occurring diseases for which *P. wickerhamii* can be regarded as causal agent. The pathogen responsible for prototheca mastitis of cattle, i.e., *P. zopfii* genotype 2, is isolated slightly more rarely; however, it could be isolated several times in connection with severe systemic infections [5,17]. Furthermore, *P. blaschkeae* was repeatedly isolated from human onychomycoses; however, the causal connection between clinical disease and infection with *P. blaschkeae* has not yet been finally clarified. In the meantime, some reports on local and systemic infections of cattle with such pathogens are present [18,19].

Classification of prototheca infections of humans can be made based on their localization, thus resulting in the following three groups [20]: (i) cutaneous and subcutaneous protothecoses; (ii) bursa inflammations, fascia inflammations, and tendon sheath inflammations; and (iii) systemic diseases. However, isolation of *Prototheca* species was also performed as possibly insignificant contaminants of the damaged skin and in stool samples of humans. Despite indications that protothecae—just like with dogs—may cause enteritides even in humans, the etiological significance of protothecae in connection with these symptoms could not yet be fully clarified.

The currently present case descriptions reveal that predispositioning factors play an important role in connection with formation of the human protothecoses [20]. So, e.g., grazes at the elbow should be regarded as a prerequisite for formation of the often described *Bursa subcutanea olecrani* infections. Insufficiency of cellular and humoral immunity is regarded as the most important predispositioning factor for formation of local and also systemic human protothecoses.

It was revealed that iatrogenically caused prototheca infections were promoted by local or systemic glucocorticoid treatment [21]. The highest predispositioning importance is to be allocated to infections with the HIV, which is to be held responsible for accumulation of systemic and local protothecoses with such patients. Metabolic diseases are also mentioned as predispositioning factors, with *diabetes mellitus* playing a decisive role. Furthermore, there are reports on a relation between noninfectious defects of the cellular defense mechanisms and tumors of the hemopoietic system [5,17].

Combined therapies involving dissection of the affected tissue and local and systemic treatment with antifungal agents like amphotericin B or ketoconazole show mostly good results.

97.1.3.2 Prototheca Mastitis of Dairy Cows

Prototheca mastitis of cattle is mostly a chronic local infection of the mammary gland including the udder lymphatic nodes, which is caused by *P. zopfii* genotype 2, whereas *P. blaschkeae* seems to be responsible for a part of the cases of subclinical prototheca mastitis [6,7,22].

It is a multifactorial disease with the factors pasture or exercise management in prototheca-contaminated areas, poor milking hygiene, high lactation, and previous impairment of the udder by further mastitis pathogens playing an important role. Being a saprophyte, the algae can be isolated with a percentage of up to 70% within the stable, in the vicinity of the stable as well as from the feces of other cattle, including cattle without prototheca mastitis.

Infection is galactogenous ascending; injuries of the teats can also be regarded as portal of entry. After entry into the cavity system of the udder, the pathogen starts further distribution. Newly lactating cows are particularly affected as the secreting alveoli provide good reproduction conditions for protothecae.

Furthermore, it is assumed that lymphogenous and hematogenous distribution of the pathogen within the udder also plays a certain role. *P. zopfii* can be regularly isolated from the udder lymphatic nodes of diseased cows. The pathogenicity mechanisms of the protothecae are yet unknown. However, it is assumed that they survive and even reproduce in macrophages and neutrophil granulocytes. Thus, they avoid elimination by the immune system. Furthermore, toxins produced by *P. zopfii* damaging the alveolar epithelium of the udder and thus favoring the distribution of the algae are also suspected to be pathogens.

The ability to survive in macrophages and thus to avoid phagocytosis is also a possible explanation for chronicity of prototheca mastitis. Although the body is able to stop further distribution by isolating the pathogen, it cannot eliminate protothecae.

The prototheca mastitis mostly starts acutely or subacutely and then changes into a chronic stage. This usually results in complete loss of function of the affected udder quarter. *Restitutio ad integrum* with clinically conspicuous animals is an exception.

In the acute stage, the udder is of balloon-like appearance and rough consistency. However, signs of inflammation, e.g., reddening and formation of edema, usually cannot be observed. Milk yield rapidly falls within a short period of time (24h); partially it completely ceases. The somatic cell count in the milk increases; secretion changes are only rarely observed in the beginning. Later, secretion changes appear as slimy-watery consistency of the milk, which then often has white or yellowish coagulations.

The following chronic stage of prototheca mastitis is characterized by atresia of the affected udder quarter. These changes are accompanied by a steady decrease of the milk amount followed by complete loss of function, and an increased susceptibility to other pathogens.

Thus, an increased number of animals with one or several atretic udder quarters should be regarded as a typical sign of a present endemic prototheca mastitis. The related high decrease of milk yield causes high financial losses to the affected farms.

Since their first description, several attempts were made to fight and treat prototheca mastitis with cattle. However, any therapy measures proved to be ineffective or uneconomical until now [18,23,24]. Thus, correct diagnostics of diseased animals is of utmost importance due to the therapy resistance of prototheca mastitis.

Sanitation of affected cattle stocks is performed by means of culling of any infected animals, including cultural and serological diagnostics. Optimization of milking (utilization of iodine-containing teat dipping agents) and animal hygiene is to be regarded as the most important prophylaxis measure for fighting prototheca mastitis on stock level [25].

97.1.3.3 Protothecosis of Dog

After cattle, the dog is the second most described species that may develop a prototheca infection, with the most cases described in the United States. While *P. zopfii* genotype 2 is the only pathogen responsible for the occurrence of infections with cattle, *P. zopfii* genotype 2 and *P. wickerhamii* can be isolated with dogs at equal frequency. There are also reports on simultaneous occurrence of both species within the same case.

Unlike with humans and cattle, dogs usually show a disseminated infection that manifests in various organ systems.

Skin protothecoses with dogs are less frequently observed; chronic hemorrhagic enteritides, often followed by central-nervous disturbances including deafness, blindness, and sudden lameness are more frequent [26–29]. Here, pathomorphological changes in the form of disseminated pyogranulomatous lesions are found in retina, intestines, conjunctivae, kidneys, liver, spleen, heart, lungs, and their tributary lymphnodes.

The almost lacking entry of inflammatory cells into the center of inflammation is noticeable. Also with dogs there are indications that a suppression of the cellular immunity promotes the course of disease [30,31].

Systemic therapies with dogs have been without success until now; treatment with antifungal agents like amphotericin B does not show any effect, so that most of the affected dogs either die or have to be euthanized.

97.1.4 Diagnosis

97.1.4.1 Conventional Techniques

Usually, the presumption diagnosis "prototheca infection" is followed by cultural isolation of pathogens. Most culture media used in fungi diagnostics, e.g., Sabouraud glucose agar are suitable for this.

Protothecae grow on fungi culture media (Sabouraud glucose agar) in round, cream-white to beige colonies that have a soft consistency and a matte-finished surface. Due to morphologic similarity, their growing on fungi culture media, and the yeast-like odor, protothecae are often confused with *Candida* species. Thus, selective culture media are often used for safe isolation of protothecae. The most common selective culture medium is the PIM agar (PIM = *Prototheca* isolation medium) described by Pore [32]. Due to the addition of the substances potassium hydrogen phthalate and 5-fluorocytosin, growth of moulds, yeasts, and bacteria can be avoided and protothecae can be isolated even from highly contaminated sources. However, growing of pure cultures is performed by using the common fungi culture media.

For good growth, the prototheca species require species-dependent temperatures between 20°C and 37°C under aerobic conditions; all pathogenic members are growing at 37°C.

Following cultivation, a microscopic examination of the suspicious colonies is performed. The widely used monochromatic vital stainings with methylene blue or lactophenol cotton blue are most suitable for this. Here, protothecae—in particular their reproductive forms—are represented as round to oval refracting objects.

Furthermore, commercial identification systems can also be used for cultural prototheca differentiation, e.g., API 20C®, BBL Crystal™, and RapID™ Yeast Plus [33].

The often intermittent shedding of the algae complicates safe cultural diagnostics of bovine prototheca mastitis. Thus, several examinations are required for successful decontamination of the affected companies; this increases the already high efforts of cultural prototheca diagnostics [25].

However, serological methods like ELISA for detection of specific IgA and IgG_1 in the milk serum are also available for correct diagnostics of prototheca mastitis of cattle. Such serological diagnostics are particularly suitable for diagnostics of intermittent shedders.

Protothecae can be reliably presented in pathohistological preparation of biopsy sample by means of periodic acid-Schiff (PAS) or Gridley staining. In contrast, hematoxylin–eosin staining is not suitable. Reliable immunohistological diagnostics were also established in the meantime [34,35].

97.1.4.2 Molecular Techniques

At the moment, molecular prototheca diagnostics are limited to pathogen identification from previously isolated cultures or from material for pathohistological examination.

The techniques used most frequently for this are species-specific and genotype-specific polymerase chain reactions (PCRs) or restriction fragment length polymorphisms (RFLPs) based on 18S rDNA. These techniques currently enable identification of the pathogenic etiologic agents *P. wickerhamii*, *P. blaschkeae*, and of both genotypes of *P. zopfii* [4].

Furthermore, analysis of the totality of the ribosomal proteins by means of Matrix-assisted laser desorption/ionization time-of-flight (MALDI-TOF) enables a simple, fast, and very reliable identification of any currently described prototheca species and genotypes [36].

97.2 METHODS

97.2.1 Sample Preparation

97.2.1.1 Sample Preparation for PCR and RFLP

A colony of the grown algae culture is transferred to a 2 mL Eppendorf tube and resuspended in 200 μL aqua bidestilata.

Then, cells are disrupted by means of ultrasound treatment using an ultrasound processor (sonication needle) with maximum amplitude and a pulse of 0.5 s. The Eppendorf tubes are stored on ice. The following DNA extraction is performed using a commercially available DNA extraction kit (e.g., DNeasy Plant Mini Kit™; Qiagen) as per manufacturer's instructions.

Extraction of prototheca DNA from fixed tissue samples is also performed using a suitable commercial kit (e.g., DNeasy Blood & Tissue Kit™; Qiagen) as per manufacturer's instructions [4,6].

97.2.1.2 Sample Preparation for MALDI-Biotyping

Isolated colonies from agar plates are suspended in 300 μL water followed by the addition of 900 μL absolute ethanol. After centrifugation at 12,000 rpm for 2 min, the supernatant is removed. The pellet is washed three times by cycles of resuspension and centrifugation in 1 mL water. The final cell pellet is dissolved in 50 μL of 70% formic acid followed by the addition of 50 μL acetonitrile and mixed thoroughly. The suspension is centrifuged at 12,000 rpm for 5 min at room temperature and 0.5 μL of the clear supernatant were mixed with 2,5-dihydroxybenzoic acid (DHB) and spotted in duplicates onto a MALDI target. After air drying, each sample was overlaid with 1.5 μL of MALDI matrix solution and allowed to dry for several minutes before MALDI-TOF mass spectrometry (MS) measurement [36].

97.2.2 Detection Procedures

97.2.2.1 PCR Identification

Based on the 18S rDNA sequences the oligonucleotides shown in Table 97.1 are used for PCR assays specific for pathogenic *Prototheca* species [4,6].

PCR conditions are 35 cycles with 30 s denaturation at 94°C, 30 s annealing at 58°C (63°C for *P. blaschkeae*), and 30 s extension at 72°C. The amplified PCR products and the RFLP fragments were analyzed on a 1% (wt/vol) agarose gel after staining with ethidium bromide.

97.2.2.2 Restriction Fragment Length Polymorphism

Also based on the 18S rDNA sequences the following genotype-specific endonucleases can be used for RFLP analysis to identify both the two genotypes of *P. zopfii* and *P. blaschkeae*: *Kpn*21 (*P. zopfii* genotype 1), *Sma*I (*P. zopfii* genotype 2), and *Bcn*I (*P. blaschkeae*). The RFLP analyses are performed using nonspecific amplicons of the *P. zopfii* and *P. blaschkeae* 18S rDNA. Thereby, the target sequences of the specific restriction enzymes differ from the target sequences of the above described genotype-specific oligonucleotides. The 18S rDNA fragments

TABLE 97.1
Oligonucleotides for Species and Genotype-Specific *Prototheca* PCR

Species	Oligo Name	Sequence	Length of Amplicon (bp)
P. wickerhamii	Proto18S-4f	5′-GACATGGCGAGGATTGACAGA-3′	450
	Proto18S-4r	5′-AGCACACCCAATCGGTAGGA-3′	
	Pwick-F[b]	5′-GTTGGTGGGTTRYCTTGTC-3′	208
	Pwick-R	5′-TTGCATGCATCAGTGTAGCG-3′	
P. zopfii genotype 1	Proto18S-4f[c]	5′-GACATGGCGAGGATTGACAGA-3′	450
	Proto18S-4r[a]	5′-AGCACACCCAATCGGTAGGA-3′	
	Pzopf GT 1r	5′-GCCAAGGCCCCCCGAAG-3′	150
P. zopfii genotype 2	Proto18S-4f[c]	5′-GACATGGCGAGGATTGACAGA-3′	450
	Proto18S-4r[a]	5′-AGCACACCCAATCGGTAGGA-3′	
	Pzopf GT 2r	5′-GTCGGCGGGGCAAAAGC-3′	165
P. blaschkeae	PzGT 3-IKf[c]	5′-CAGGGTTCGATTCCGGAGAG-3′	213
	PzGT 3-IKr[a]	5′-GAATTACCGCGGCTGCTGG-3′	
	Pzopf GT 3r	5′-GTTGGCCCGGCATCGCT-3′	126

[a] Oligonucleotides for internal amplification control (which acts together with the shared forward primer[c])
[b] Wobbles: R = A + GY = C + T.
[c] Forward primer.

were amplified using the oligonucleotides Protol8-2f (5′-CGC GCAAATTACCCAATCC-3′) and Protol8-2r (5′-TGAGAC CGGACAGGGCAA-3′) for *Kpn*21/*Sma*I and Protol8-4f (5′-GACATGGCGAGGATTGACAGA-3′) and Protol8-4r (5′-AGGATGGCTAAC CCACACGA-3′) for *Bcn*I [4,6].

97.2.2.3 MALDI-TOF Biotyping

Identification of *Prototheca* species can also be performed by MALDI-TOF MS of the highly conserved ribosomal proteins. The analyses can be done with a MALDI-TOF or a MALDI-TOF/TOF instrument in combination with a database for biotyping of microorganisms (available, e.g., from Bruker Daltonik Ltd. or Shimadzu Ltd.) that includes the spectrums of ribosomal proteins of all relevant pathogenic fungi and bacteria. The spectrums should be measured in triplicates and are then compared with the data of the database. The results are in general comparable with sequence based techniques basing on the 16S or 18S rDNA [36].

97.3 CONCLUSION

Infections with protothecae are a globally increasing problem for both humans and animals. Thus, it will be advantageous to establish probe-based multiplex real-time PCR assays for simultaneous diagnosis of pathogenic *Prototheca* species. Furthermore, it will be desirable to create significant tools for epidemiological differentiation and clarification of infection chains in future.

REFERENCES

1. Lerche, M., Eine durch Algen (*Prototheca*) hervorgerufene Mastitis der Kuh, *Berl. Münch. Tierärztl. Wochenschr.*, 4, 64, 1952.
2. Krüger, W., Kurze Charakteristik einiger niederer Organismen im Saftflusse der Laubbäume, *Hedwigia*, 33, 16, 1894.
3. Cooke, W.B., Studies in the genus *Prototheca*, *J. Mitch. Soc.*, 84, 213, 1968.
4. Roesler, U. et al., Diversity within the current algal species *Prototheca zopfii*: A proposal for two *Prototheca zopfii* genotypes and description of a novel species, *Prototheca blaschkeae* sp. nov., *Int. J. Syst. Evol. Micr.*, 56, 1419, 2006.
5. Lass-Flörl, C. and Mayr, A., Human protothecosis, *Clin. Microbiol. Rev.*, 20, 230, 2007.
6. Möller, A., Truyen, U., and Roesler U., *Prototheca zopfii* genotype 2: The causative agent of bovine protothecal mastitis? *Vet. Microbiol.*, 120, 370, 2007.
7. Roth, K., Aetiological role of *Prototheca zopfii* genotype 2 in bovine Prototheca mastitis, Doctoral thesis, Free University Berlin, 2009.
8. Janosi, S. et al., Review of the microbiological, pathological, and clinical aspects of bovine mastitis caused by the alga *Prototheca zopfii*, *Vet. Q.*, 23, 58, 2001.
9. Callely, A.G. and Lloyd, D., The metabolism of acetate in the colorless alga, *Prototheca zopfii*, *Biochem. J.*, 90, 483, 1964.
10. Lloyd, D. and Callely, A.G., The assimilation of acetate and propionate by *Prototheca zopfii*. *Biochem. J.*, 97, 176, 1965.
11. Ciferri, O., Thiamine-deficiency of *Prototheca*, a yeast like achloric alga. *Nature*, 178, 1475, 1956.
12. Pore, R.S., *Prototheca* taxonomy, *Mycopathologia*, 90, 129, 1985.
13. Blaschke-Hellmessen, R. and Schuster, H., Untersuchungen zur Überlebensfähigkeit von *Prototheca zopfii* und *Prototheca wickerhamii* unter verschiedenen Umweltbedingungen, *Mh. Vet. Med.*, 39, 852, 1984.
14. Melville, P.A. et al., Evaluation of the susceptibility of *Prototheca zopfii* to milk pasteurization, *Mycopathologia*, 146, 79, 1999.
15. Enders, F. and Weber, A., Pilotstudie zum Vorkommen von Prototheken in Kotproben von Pferden, *Berl. Münch. Tierärztl. Wochenschr.*, 106, 264, 1993.
16. Enders, F. and Weber, A., Untersuchungen zum Vorkommen von Prototheken in Kotproben von Rindern, *Berl. Münch. Tierärztl. Wochenschr.*, 106, 165, 1993.
17. Lass-Flörl, C. et al., Disseminated infection with *Prototheca zopfii* after unrelated stem cell transplantation for leukemia, *J. Clin. Microbiol.*, 42, 4907, 2004.
18. Marques, S. et al., Bovine mastitis associated with *Prototheca blaschkeae*, *J. Clin. Microbiol.*, 46, 1941, 2008.
19. Thompson, G. et al., Algaemia in a dairy cow by *Prototheca blaschkeae*, *Med. Mycol.*, 19, 1, 2009.
20. Thiele, D. and Bergmann, A., Protothecosis in human medicine, *Int. J. Hyg. Envir. Health*, 204, 297, 2002.
21. Tsuji, K. et al., Protothecosis in a patient with systemic lupus erythematosus. *Intern. Med.*, 32, 540, 1993.
22. Osumi, T. et al., *Prototheca zopfii* genotypes isolated from cow barns and bovine mastitis in Japan, *Vet. Microbiol.*, 15, 131, 419, 2008.
23. Bergmann, A., Experimentelle Prototheken-Mastitis beim Rind und Therapie mit Tetramisolhydrochlorid, *Berl. Münch. Tierärztl. Wochenschr.*, 106, 257, 1993.
24. Costa, E.O. et al., Infectious bovine mastitis caused by environmental organisms, *J. Vet. Med. B.*, 45, 65, 1998.
25. Rösler, U. and Hensel, A., Eradication of *Prototheca zopfii* infection in a dairy cattle herd, *Deutsch. Tierärztl. Wochenschr.*, 110, 374, 2003.
26. Cook, J.R. et al., Disseminated protothecosis causing acute blindness and deafness in a dog, *J. Am. Vet. Med. Assoc.*, 184, 1266, 1984.
27. Ginel, P.J. et al., Cutaneous protothecosis in a dog, *Vet. Rec.*, 140, 651, 1997.
28. Rallis, T.S. et al., Protothecal colitis in a German shepherd dog, *Aust. Vet. J.*, 80, 406, 2002.
29. Salvadori, C. et al., Protothecal granulomatous meningoencephalitis in a dog, *J. Small Anim. Pract.*, 49, 531, 2008.
30. Perez, J. et al., Canine cutaneous protothecosis: An immunohistochemical analysis of the inflammatory cellular infiltrate, *J. Comp. Pathol.*, 117, 83, 1997.
31. Rakich, P.M. and Latimer, K.S., Altered immune function in a dog with disseminated protothecosis, *J. Am. Vet. Med. Assoc.*, 185, 681, 1984.
32. Pore, R.S., Selective medium for the isolation of *Prototheca*, *Appl. Microbiol.*, 26, 648, 1973.
33. Espinel-Ingroff, A. et al., Comparison of RapID yeast plus system with API 20C system for identification of common, new, and emerging yeast pathogens, *J. Clin. Microbiol.*, 36, 883, 1998.
34. Laeng, R.H. et al., Protothecosis in an HIV-positive patient, *Am. J. Surg. Pathol.*, 18, 1261, 1994.
35. Rizzi, T.E. et al., More than meets the eye: Subretinal aspirate from an acutely blind dog, *Vet. Clin. Pathol.*, 35, 111, 2006.
36. von Bergen, M. et al., Identification of harmless and pathogenic algae of the genus *Prototheca* by MALDI-MS, *Proteomics Clin. Appl.*, 3, 774, 2009.

98 *Rhinosporidium*

S.N. Arseculeratne

CONTENTS

98.1 INTRODUCTION

Since its first identification in 1900, controversies on the identity, taxonomy, and structure of *Rhinosporidium seeberi* prevailed for many decades. It is a unique pathogen that still, after 118 years, has been unable, because of its *in vitro* uncultivability, to fulfill Koch's classical postulates on the microbial causes of disease. The question *Rhinosporidiosis: what is the cause*? was posed even as recently as 2005.[1] Is it a fungus? If it was, Libero Ajello thought it was a bizarre fungus. Added to this confusion was its later misidentification with the common water cyanobacterium *Microcystis aeruginosa*.[2] It is useful to recapitulate three likely sources of errors that led to this misinterpretation: (i) the failure to delineate the chain of causal relationships, (ii) the inadequate appreciation of the logic of scientific inference, and (iii) the reliance on morphology alone to identify taxonomic relationships.

After refutations of the *Microcystis* theory,[3,4] *R. seeberi* was finally rehabilitated with 10 other saprobic or pathogenic microorganisms, in a new class, the *Mesomycetozoea*. The inability to culture this organism *in vitro* to obtain axenic cultures, contrasting with the majority of bacterial and fungal pathogens, resulted in our ignorance of its biochemical pathways; such biochemical profiles are often used in microbial taxonomy. Recent advances that are described in this chapter justify the hopes[5] that molecular biological methods would ultimately resolve these enigmas of *R. seeberi*.

The role of modern molecular methods in the study of both *R. seeberi* and the disease it causes in humans and in animals, rhinosporidiosis, is best appreciated against a background of the current knowledge on both these topics; a brief review follows that places in perspective, the molecular biological techniques that have been, and could in future be, applied to the resolution of their many enigmas. It will be seen that molecular methods used for research on rhinosporidiosis as a disease, and *R. seeberi* as its causative pathogen are as important as molecular methods for the diagnosis of rhinosporidiosis.

98.1.1 TAXONOMY

A brief summary of the long-held controversies in the taxonomy of *R. seeberi* would serve to illustrate some instances of error that contrast with the specificity that exists in molecular methods that later resolved the controversy:

1892—Malbran in Argentina regarded it as a sporozoon

1900—G.R. Seeber in Argentina described the causative organism

1900—Wernicke named the organism *Rhinosporidium* seeberi

1904—Minchin and Fantham named the organism Rhinosporidium *kinealyi*

1905—Minchin and Fantham found it to have affinities with the Neosporidia

1906—Beattie also claimed affinity with the Neosporidia, allied to the Sarcosporidia

1907—Ridewood and Fantham placed it in the order Haplosporidia, suborder Polysporulea

1912—G.R. Seeber endorsed Wernicke's name Rhinosporidium *seeberi* (Wernicke) and established the identity of *R. kinealyi* with *R. seeberi*

1923—Ashworth considered it a Phycomycete

1973—Vanbreuseghem failed to identify it as an Alga

1999—Herr et al. using molecular methods finally identified it as a Protoctistan and assigned it to a new class, the *Mesomycetozoea*

2000—Fredricks et al. also using molecular methods confirmed its identity as a Protistan

The application of morphological criteria alone, or predominantly as in the *Microcystis aeruginosa* hypothesis of Ahluwalia,[2] is identifiable as a major source of confusion because "Many fungi, which are morphologically similar, may have quite different nutritional demands and habitats" (J.M. Waller, 1995, Personal communication) and because of the "phenomenon of convergence," which is, "in phylogeny, the appearance of similar structures in unrelated organisms"[6]; however, mitochondrial structure has been used in comparative morphology of *R. seeberi* with other *Mesomycetozoea*[7] that were so classified on stricter, specific criteria through molecular biology and phylogeny.

The first report of the application of molecular techniques on *R. seeberi*, and the use of molecular phylogenetic methods in resolving the controversy of *R. seeberi*'s taxonomy was by Herr et al.[8] These authors "amplified and sequenced nearly a full-length 18S small subunit (SSU) ribosomal DNA (rDNA) sequence from *R. seeberi*" in two samples of human nasal rhinosporidial polyps. "Using phylogenetic analysis, by parsimony and distance methods, of *R. seeberi*'s 18S SSU rDNA and that of other eukaryotes," this pathogen that causes rhinosporidiosis in humans and animals clustered "with a novel group of fish parasites referred to as the DRIP Clade (*Dermocystidium*, rosette agent, *Ichthyophonus* and *Psorospermium*), near the animal-fungal divergence." Biologically, members of the genus *Dermocystidium*, and *R. seeberi*, "form spherical structures in infected hosts produce endospores, have not been cultured, and possess mitochondria with flat cristae."

Similar conclusions on its taxonomy were made by an independent group of workers[9] who, using a consensus PCR to amplify a portion of the *R. seeberi* 18S rRNA gene directly from infected tissue, analyzed the phylogenetic relationship of *R. seeberi* to other selected eukaryotes and concluded that *R. seeberi* was "the first known human pathogen from the DRIPs clade, a novel clade of aquatic protistan parasites (Ichthyosporea)." Thus, they confirmed the earlier findings of Herr et al.[8] Fredricks et al.[9] also designed a specific *in situ* fluorescent hybridization probe for the 18S rRNA gene sequence of *R. seeberi*, the specificity of which was such that no labeling was seen on smears of *Candida albicans*, the Rosette agent (of the DRIP Clade to which *R. seeberi* belonged), and tissue sections of *Coccidioides immitis*. As pointed out in the following, the same *in situ* probe was used[10] for the detection of rhinosporidial bodies in deposits from a body of lacustrine ground water (reservoir) in Sri Lanka.

These specific findings with the powerful tools of molecular biology thus laid the former bogeys of mycelial fungi, tapioca starch, and *Microcystis aeruginosa*, as *alter egos* of *Rhinosporidium seeberi*, to rest.

98.1.2 Habitat

Until the question of *R. seeberi*'s natural habitat was specifically resolved by the use of *in situ* hybridization (ISH) with *R. seeberi*-specific probes, there was ample, cogent epidemiological evidence for permanent ("lacustrine") ground waters such as in lakes, ponds, and rivers, as being its natural habitat. The evidence can be summarized as follows: (i) histories from most Asian patients of their prior exposure, as in bathing, to ground waters; (ii) the occurrence of an outbreak in Serbia of rhinosporidiosis in persons who had bathed in the same lake; (iii) the clustering of cases in river-sand workers; (iv) the predominant occurrence of rhinosporidiosis in nasal and ocular sites in humans and animals; (v) reports from the United States of an outbreak of rhinosporidiosis in swans in a particular lake; and (vi) occurrence in animals that have an aquatic habitat, for example, river dolphin, geese, swans, ducks.

It is difficult to differentiate lacustrine (lakes, ponds, reservoirs) waters from soil as being the primary habitat, as these two are in such intimate association in nature, and the differentiation that is sometimes made between ground water and dust as the source of *R. seeberi* is not convincing because dust arises from dry soil, which could, as with true fungi, be provisionally regarded as the ultimate natural habitat of *R. seeberi*. Accordingly, supplementary evidence may be quoted as relevant to soils as the source of *R. seeberi*; (vii) the higher prevalence in males in relation to agricultural occupations over females; (viii) a history of prior trauma from objects associated with agriculture, for example, wood splinters, grass; and (ix) the association of the habit of rubbing stones, probably contaminated with soil-borne *R. seeberi*, to absorb the last drops of urine, with the greater prevalence of urethral rhinosporidiosis in persons with this habit.

Confirmation of lacustrine waters as a habitat of *R. seeberi* was achieved by the second report on the use of molecular methods on *R. seeberi*—the use of ISH with

R. seeberi-specific probes, applied on paraffin-embedded deposits of water from a large reservoir in which many patients with rhinosporidiosis had bathed.[10] This method had the advantage of avoiding inhibitors of the more sensitive technique of PCR, which consistently failed in many hands, to identify rhinosporidial elements in the same water deposits. Possible inhibitors of PCR include silica from sand particles, ions, and humic acid from the organic debris in the reservoir water, which inevitably contaminate the water deposits. Indeed, clinical material, for virological diagnosis through PCR, is not collected on swabs mounted on wooden, but on plastic sticks, lest organic, inhibitory substances leach out from the wood.

In this particular study,[10] endospores from homogenates of human rhinosporidial tissues (after filtration through 25 μm nylon sieves) and a human rhinosporidial polyp embedded in paraffin wax (a 6 μm section) were used as positive controls; and a human non-rhinosporidial polyp served as a negative control. A centrifuged deposit from 40 L of water from a reservoir in which many patients with rhinosporidiosis had bathed, after separation of coarse sediment, was embedded in paraffin wax as with tissues, after preliminary embedding in 1% agarose and dehydration through ethanol gradients. An ISH probe was prepared as an antisense oligonucleotide of a unique region of *R. seeberi*'s 18S rRNA gene sequence (Rhino FISH probe BTGCTGATAGAGTCATTGAATTAACATCTACB) and its complementary control probe (Control FISH probe BACGACTATCTAGTAACTTAATTGTAGATGB) according to the sequences published by Fredricks et al.[9] The paraffin sections after de-waxing and rehydration were treated with lysozyme in EDTA-Tris buffer, pre-hybridized in hybridization buffer, followed by the biotinylated probe in hybridization buffer and hybridized overnight at 45°C. After repeated stringency washes in sodium chloride-EDTA-Tris buffer, the slides were incubated with Vectastatin Elite ABC Complex (Vector Laboratories, the United States) and subjected to tyramide signal amplification (TSA-plus fluorescein system, Perkin-Elmer Life Sciences, the United States).

The results[10] with the water deposit are shown in Figures 98.5 and 98.6 in comparison with diagrams of rhinosporidial endospores (Figures 98.1 and 98.2) and of positive control rhinosporidial bodies (Figures 98.3 and 98.4). The purified endospores (Figure 98.3) and the endospores in the positive

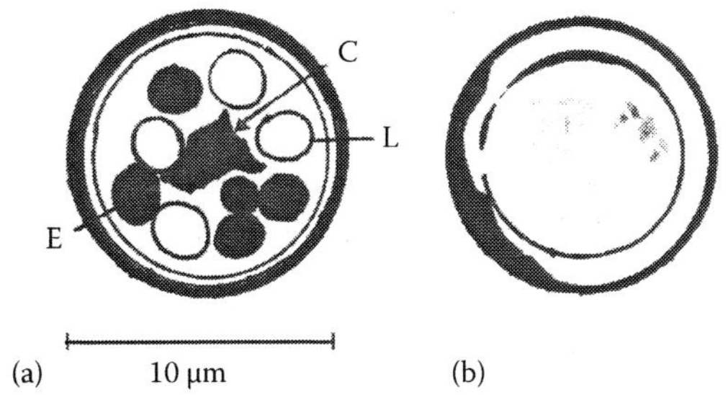

FIGURE 98.1 Diagram of (a) MTT-treated normal rhinosporidial endospore, showing electron-dense bodies (EDBs): (E), lipid body (L), cytoplasm (C); (b) drug-treated endospore, with morphologically abnormal EDB with decreased reduction of MTT by EDB.

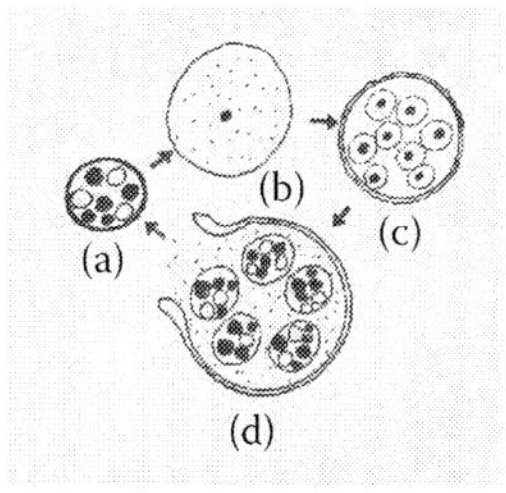

FIGURE 98.2 Diagram (not to scale) of the life cycle of *Rhinosporidium seeberi*, inferred from appearances of the pathogen in rhinosporidial tissues: a, endospore, b, juvenile sporangium, c, immature (intermediate) sporangium, d, mature sporangium releasing endospores to reinitiate the life cycle. (From Kaluarachchi, K. et al., *J. Infect. Dis. Antimicrob. Agents*, 25, 25, 2008. With permission from Infectious Disease Society of Thailand.)

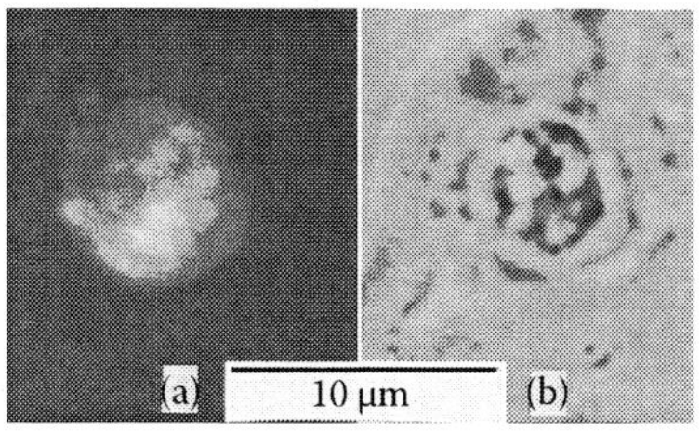

FIGURE 98.3 *In situ* hybridization with a *R. seeberi*-specific molecular (fluorescent) probe of a purified endospore (a) stained with the *R. seeberi*-specific hybridization probe showing deeply fluorescent internal bodies, probably the DNA-containing EDB; (b) the same section under bright light. Initial magnification 400×. (From Kaluarachchi, K. et al., *J. Infect. Dis. Antimicrob. Agents*, 25, 25, 2008. With permission from Infectious Disease Society of Thailand.)

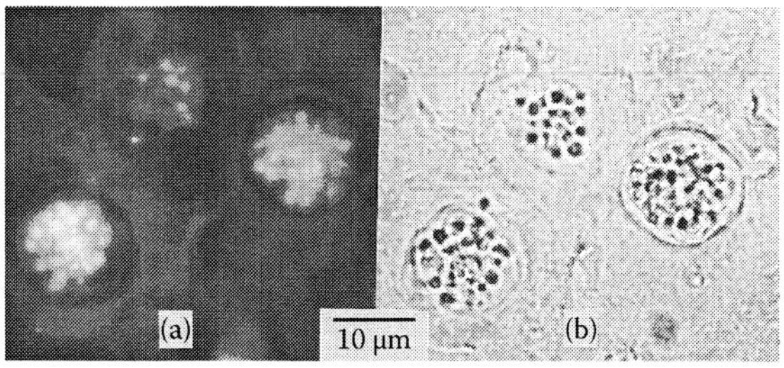

FIGURE 98.4 *In situ* hybridization with a *R. seeberi*-specific molecular (fluorescent) probe of rhinosporidial endospores in rhinosporidial tissue from a human patient, showing (a) mature endospores containing brightly fluorescent EDB and (b) visualized in bright light. Initial magnification 400×. (From Kaluarachchi, K. et al., *J. Infect. Dis. Antimicrob. Agents*, 25, 25, 2008. With permission from Infectious Disease Society of Thailand.)

control rhinosporidial tissue (Figure 98.4) were 8–10 μm spherical bodies with their internal spherical bodies, including the electron dense bodies (EDBs) labeling with the probe. The endospores in the rhinosporidial tissue (Figure 98.4, positive control) were approximately 10 μm spherical entities containing strongly fluorescent spherical bodies, the EDBs, and placed within mature sporangia (Figure 98.4). The negative control non-rhinosporidial polyp, and the non-rhinosporidial areas of the positive (rhinosporidial) polyp did not label with the probe. The labeling of the EDBs within the endospores in Figures 98.3 and 98.4 is added proof that they contain nucleic acids and are thus entitled to their recognition as the true spores of *R. seeberi*, as discussed in the following; this is another instance of the value of molecular techniques in the resolution of the controversies on *R. seeberi*'s morphology and biology. The entities in the water deposit that labeled with the probe were considered, on morphological grounds, to be 10–15 μm endospores (Figure 98.5a) and 15–20 μm juvenile sporangia (Figures 98.5b and 98.6) of *R. seeberi*.

98.1.3 Life Cycle

The life cycle of *R. seeberi* has been logically inferred from the appearance of diverse rhinosporidial elements in diseased tissues of both humans and animals (Figures 98.1 and 98.2); this remains the only approach to this problem because the pathogen has never been cultured *in vitro*. It is convenient to begin the cycle with the classical endospore that most authors have described, rather than the EDB, which is discussed in the following as now meriting its identification as the true spore that initiates the disease. The endospore is at present considered to be the ultimate infective unit of *R. seeberi* that has been implicated by Karunaratne[11] by the term "auto-inoculation" as initiating satellitic rhinosporidial lesions on adjacent mucosae after spillage during traumatization or

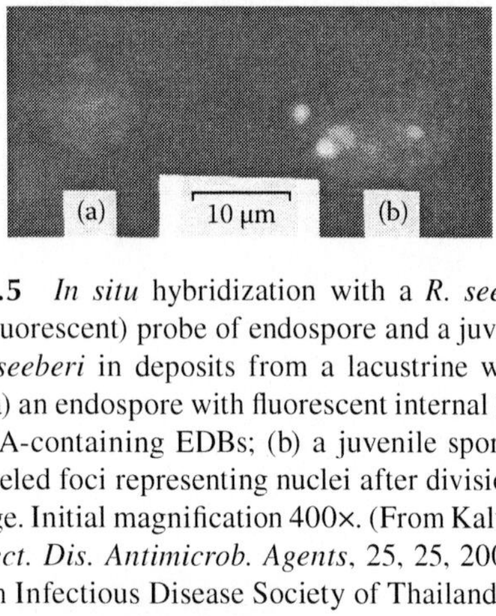

FIGURE 98.5 *In situ* hybridization with a *R. seeberi*-specific molecular (fluorescent) probe of endospore and a juvenile sporangium of *R. seeberi* in deposits from a lacustrine water body in Sri Lanka: (a) an endospore with fluorescent internal bodies, probably the DNA-containing EDBs; (b) a juvenile sporangium with scattered labeled foci representing nuclei after division in the precleavage stage. Initial magnification 400×. (From Kaluarachchi, K. et al., *J. Infect. Dis. Antimicrob. Agents*, 25, 25, 2008. With permission from Infectious Disease Society of Thailand.)

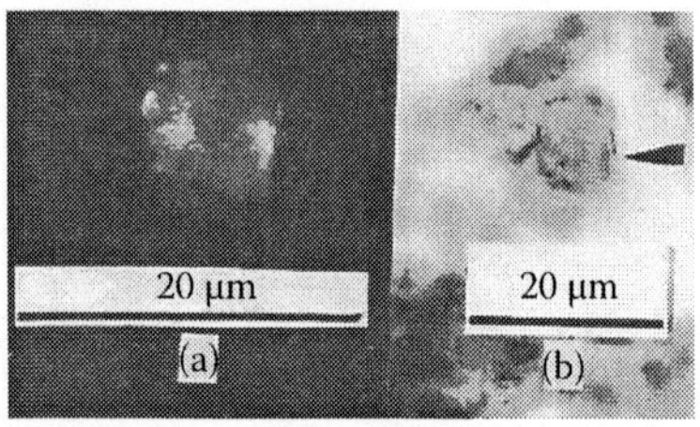

FIGURE 98.6 *In situ* hybridization with a *R. seeberi*-specific molecular (fluorescent) probe of a thin-walled juvenile sporangium of *R. seeberi* in the deposit from a lacustrine water body in Sri Lanka: (a) under UV light showing multiple foci of labeling representing nuclei and (b) under bright light. Initial magnification 400×. (From Kaluarachchi, K. et al., *J. Infect. Dis. Antimicrob. Agents*, 25, 25, 2008. With permission from Infectious Disease Society of Thailand.)

surgery on polyps. It has also been identified in the blood vessels and viscera in a case of disseminated rhinosporidiosis,[12] again implying that it was the endospore that, after release from the primary nasal polyp, was responsible for the systemic lesions.

The rhinosporidial endospore is not comparable with bacterial endospores of *Bacillus* and *Clostridium* spp in structure, chemical composition, and physiological properties. The endospore is produced by endosporulation from the inner surface of the inner cellulose layer of the bilamellar sporangial wall. It is apparently produced by asexual means; noting that a sexual stage(s) have not been described in *R. seeberi*; its maturation ends with its disposition near the sporangial pore through which it is finally extruded. The endospore (10–12 μm, Figures 98.3 and 98.4) can be seen free within rhinosporidial tissues, especially near the pore of the mature sporangia through which it has emerged. It can be seen on smears of discharge from nasal and other sites, both in wet preparations as well as on periodic acid Schiff (PAS)-stained smears. The endospore has a chitin coat and a surface covering of fibrillar material but without flagella that are responsible for motility in bacteria; the endospore has no motility and no flagella-like appendages have been seen in scanning electron micrographs; this feature has been included in taxonomic comparisons. The surface fibrils could mediate the two biological activities shown by the endospores—adherence to mammalian cells and direct hemagglutination; the latter phenomenon could be compared with the direct hemagglutination also mediated by fimbriae on Gram-negative bacilli.

The most controversial element in *R. seeberi* is the spherical refractile granule invariably seen within the mature endospore in human and animal rhinosporidial tissues in histopathological sections and in wet, unstained preparations of the surface material on rhinosporidial polyps. These spherical bodies have also been seen in rhinosporidial tissues of dogs and bovines. This body was also earlier termed the "spherule" or "germinative body." From recent investigations on its nature and significance and its use as an indicator of viability of the endospore, with further work on the endospore's sensitivity to inactivating agents including therapeutic drugs, it has acquired the status as the most significant morphological element in *R. seeberi*; its chequered history over several decades[13] merits a brief recapitulation.

Minchin and Fantham[14] noted a central spot in the refractile granule, which stained with hematoxylin, and concluded that it is a "chromatic nucleus." Beattie suggested[15] that the EDBs were spores and described them as containing a "well-defined nucleus that reacts to the chromatin stains." These bodies stain with acridine orange that identifies nucleic acids, and are Feulgen-positive. Vanbreuseghem et al.[16] having demonstrated DNA in the spherules, which were DNase sensitive (see also Refs. [17,18]), considered them to be "active bodies," the actual spores. Lakshmanan and Kameswaran[19] concluded from histochemical studies that these bodies had a "lipoprotein coat with a protein matrix and a strongly Feulgen-positive centre." The identification of DNA in the spherules[13,20] helped resolve a long-standing controversy on the nature of these bodies—are they nutritive stores, as Ashworth suggested,[21] or are they viable bodies, the ultimate generative units of *R. seeberi*?

The spherules were renamed the EDBs by Kennedy et al.[22] who described them as "irregular membrane-bound, vesicular structures of various sizes containing electron dense material"; the electron micrograph Figure 24.13 in Arseculeratne and Mendoza[13] illustrated this description.

The reduction of the salt MTT (3-[4,5-dimethyl-2-thiazolyl]-2,5-diphenyl tetrazolium bromide) by these bodies[23] supported the conclusion that they were metabolically viable entities, which together with their content of DNA, led to the postulate that they constituted the true spores of *R. seeberi*, confirming the speculation of Lakshmanan and Kameswaran[19] that "... the spherules could be the initial inciting agent of the disease" (see also Ref. [13]). The labeling by *R. seeberi*-specific probes, of the EDBs in the endospores in the control positive rhinosporidial tissues (e.g., Figure 98.4), used by Kaluarachchi et al.[10] is the most cogent evidence that the EDBs contain DNA. These findings lead to the inevitable conclusion[16,24] that the EDBs are indeed the true spores of *R. seeberi* and that the endospore is, therefore, a collection of "spores," the spore-morula, as suggested by Minchin and Fantham,[14] Beattie,[15] and Tirumurti.[25] We endorse that suggestion and propose that this view be accepted with a renaming of the endospore and the EDB, as the "spore-morula" and "spore," respectively.

The resolution of its identity and the central importance of the EDB, the possible ultimate unit of propagation that is now amenable to identification as a target of antimicrobial agents including drugs (see following text), and its role in the propagation of *R. seeberi*, together with the ease with which the endospores can be obtained from rhinosporidial tissues, are features that have enlarged considerably the scope of advances in our knowledge of *R. seeberi*.

The ultrastructure of the endospore reveals a well-marked coat of fibrils that are approximately 0.5–1 μm in length, on the outer surface of its wall.[13] Analogous to the fimbriae on Gram-negative bacilli, these fibrils could be agent that mediates the only biological activity that has been detected in rhinosporidial endospores *in vitro*—adherence to mammalian and nonmammalian cells, including red blood cells. The rhinosporidial fibrils, if analogous with the fimbriae of Gram-negative bacilli, are not organs that confer motility, while flagella that are well known as organs of motility in the bacilli are absent in the nonmotile endospores when seen under high magnifications.

Intra-endosporial elements that have been described[26] include a prominent Golgi apparatus, endoplasmic reticulum, mitochondria, and the spherical bodies that include both EDBs and the lipid bodies. These structures emphasize the invalidity of the conclusions of Ahluwalia and coworkers who regarded the rhinosporidial bodies in rhinosporidial tissues as being sacs of starch[27] and later, spherical forms of the prokaryotic, ubiquitous, aquatic cyanobacterium *Microcystis aeruginosa*[2]; for the latter suggestion, it is surprising that (i) no genetic or immunological homologies between *R. seeberi* and *M. aeruginosa*, were established, for example, by specific PCR or ISH with *R. seeberi*-specific or *Microcystis*-specific probes; the weak precipitation reactions noted with experimental polyclonal sera to *Microcystis aeruginosa* from water and rhinosporidial "round bodies" from patients tissues, could well have been due to *Microcystis* contaminating the nasal polyps from rhinosporidial patients who had bathed in waters containing *Microcystis*; and (ii) the errors from reliance on morphology have not been recognized, as mentioned earlier. Their recommendation that on the alleged *Microcystis aeruginosa* identity of the cause of this disease, the name rhinosporidiosis should be changed is therefore unwarranted.

In smears of the discharge on rhinosporidial polyps, the endospores are readily seen, and they sometimes appear to have a coat that tails off to give the appearance that Beattie[15] graphically described as a "comet-form." This coat is the residue of the mucoid material in which the endospore is embedded inside the mature sporangium. With anti-rhinosporidial antibody tagged with a fluorescent tracer, this coat is brightly stained and this fact might be of use in distinguishing the coated endospore from respiratory transitional epithelial cells in relation to cytodiagnosis of rhinosporidiosis.

The next stage in the growth and development, or what may be termed the "life cycle" of *R. seeberi*, is the *juvenile sporangium* formerly termed the *Trophocyte*.[11] It could be seen in the stroma of rhinosporidial tissues, and within mature sporangia. It is a thin-walled oval body that measures 15–50 μm. The wall is chitinous and encloses chromatin that is either organized into a nucleus with a nucleolus or is fragmented and diffuses in later stages when it becomes an *Intermediate Sporangium*. It could contain a controversial body termed the laminated body (LB) that consists of concentric electron-dense rings surrounding a core of chromatin. The LB, also termed the multi-lamellar *Body*, has also been described in mature endospores that are presumably at the stage of maturing into juvenile sporangia and into intermediate sporangia. Some authors have regarded the LB as the precursor of the endospore, while their presence within the endospore itself is difficult to reconcile with this view. The significance of the LB thus remains controversial but the presence of electron-dense concentric rings in different, early ontogenic stages is well established. The juvenile sporangium has been of recent interest in that they are not infrequently found in smears from the nasal passages of abattoir cattle and buffaloes during a search for concomitant schistosomes (R.P.V.J. Rajapakse and S.N. Arseculeratne 2008, unpublished data). PAS+ bodies in these smears were recognized as juvenile sporangia of *R. seeberi* by their characteristic features of having ridges due to collapse of the thin walls, and cracks that appear as empty sectors in the oval body, features that are readily identified in suspensions of rhinosporidial elements after partial purification from tissue homogenates.

The stages that follow the *juvenile sporangium* are the *intermediate* (*immature*) *sporangium* and then the *mature sporangium* (Figure 98.2). These are also invariably present in rhinosporidial tissues, while the mature sporangia give the characteristic yellow-spotted appearance to the red-colored rhinosporidial polyp that is likened to a strawberry. The mature sporangium is about 150–450 μm in diameter and contains endospores in various stages of maturation. The tri-lamellar wall has an outer chitinous coat with an inner cellulose layer, which give it the characteristic PAS-positive tinctorial character, which is useful in histopathological diagnosis. It is of interest that the possession of chitin is shared with the fungi, a fact that might have erroneously helped in the classification of *R. seeberi* as a fungus. As this sporangium matures, the wall becomes thinner except at a pole termed the annulus that encloses an opening, the pore, through which mature endospores exit. The mature sporangium contains a mucopolysaccharide-containing mucoid matrix in which the mature endospores are embedded. This matrix remains as the coat of the extruded endospore and is probably highly antigenic because anti-rhinosporidial antibody labeled with tracers labels this material very strongly. The matrix also functions in an ingenious manner to absorb water, becoming a swelling gel that features in the extrusion of the endospores; this scene has been captured on a video-recording by Mendoza and colleagues, and refutes the earlier assumption that the endospores are passively released by a rupture of the mature sporangium.

The possession of muco-polysaccharides by the sporangial wall was claimed[28,29] to confer on it, selective permeability to nutrients. This property of permeability is important in relation to the entry of drugs into the sporangium to target the endospores that have been shown to be extremely sensitive to the antimicrobial action of certain drugs including Echinocandin (Caspofungin) and Voriconazole (Arseculeratne 2009, unpublished data) as described in the following. The property of permeability of sporangia is of importance in considerations of drug sensitivity of endospores and sporangia; the failure of chemotherapy has been regarded, probably incorrectly, as having been due to impermeability of the sporangia to drugs.[30]

While the presence of the various stages of rhinosporidial sporangia in rhinosporidial tissues provides that facility of easy histopathological diagnosis, the presence of aberrant appearances of the sporangia and their walls should be remembered in avoiding pitfalls in histopathological diagnosis.[31]

Mitochondria have been recognized in rhinosporidial endospores and their morphology has been invoked in taxonomy; Mendoza et al.[7] described mitochondria with flat cristae, while Fredricks et al.[9] reported tubulo-vesicular mitochondrial cristae. Flat cristae also appear in other members of the Class *Mesomycetozoea* that includes *R. seeberi*. These differences might be attributed to artifacts of processing for electron microscopy. The pitfalls in applying mitochondrial morphology to taxonomy were emphasized by Mendoza et al.[7]

A feature that is of interest in the reproductive physiology of *R. seeberi* is the location of mature sporangia at different levels in the dermis. While they appear to mature in the deeper dermal layers, it can be deduced that as they increase in maturity and size, they "migrate" toward sub-epithelial sites; the word "migrate" does not imply an active motility of the sporangia but merely reflects a progressive translocation of position, probably through the histolytic action of enzymes from rhinosporidial sporangia. This apparent migration implies two associated features; (a) the presence of epidermal elements (probably the *Stratum Granulosum* and the *Stratum Spinosum*) that appear as downgrowths from their usual superficial situation, deeper into the dermis while enclosing the mature sporangia within their "eddies" or whorls. This translocation has been interpreted as illustrating the phenomenon of *Trans-epidermal* (*Epithelial*) *Elimination* of *R. seeberi*[31]; it was pointed out that the translocation is not merely an attempt of the host's tissue reaction in eliminating the pathogen, but is also probably aided by tissue histolysis by enzymes of *R. seeberi* in eroding the subepithelial, dermal tissues to facilitate this translocation to a subepithelial location from where the dissolution and thinning of the epithelium in turn facilitate the expulsion of the mature endospores to the surrounding areas; in these sites, "auto-inoculation" by the free endospores would produce satellite rhinosporidial polyps. The extrusion of the mature endospores is not a passive process, after rupture of the sporangium as has been assumed in the literature; it is an active process based on osmotic action of the mucoid matrix within the sporangium. The process has been described in detail.[32]

Mature sporangia have been isolated from rhinosporidial tissue homogenates or from teased-out polyps by tedious "pipetting-out"; endospores have been obtained from tissue homogenates after filtration through 25 μm nylon sieves to remove tissue debris. Further purification of endospores and sporangia has been achieved on columns of the colloidal silica preparation *Percoll*.[33] Aspects and methods in the study of the viability of endospores have been reviewed.[13]

Culture *in vitro*, as achieved with bacteria and fungi, has never been unequivocally achieved with *R. seeberi*; reports of growth in cell cultures probably refer to the limited retention of viability *in vitro* and limited development of endospores into juvenile sporangia but no evidence is available for serial replication of the putative life cycle *in vitro*, in cell cultures or cell-free systems.

98.1.4 Biological Activity

The most prominent biological activity of *R. seeberi* is, of course, the production of rhinosporidiosis in humans and animals, but there has been no experimental reproduction of this process and hence the mechanisms of rhinosporidial pathogenicity and pathogenesis of rhinosporidiosis cannot be directly explored. On an analogy with other microorganisms, two activities that have been repeatedly demonstrated, that is, adherence to mammalian cells and the ability to cause direct hemagglutination of mammalian and nonmammalian red blood cells (Arseculeratne 2004, unpublished data), could be a route to such studies. A parallel is the direct hemagglutination caused by Gram-negative bacilli through their fimbriae, an activity that has been linked to adhesiveness on epithelia such as that of the urinary tract, in the causation of urinary tract infections.

98.1.4.1 Adherence

Levy et al.[34] reported that endospores were able to adhere strongly to epithelioid cells from a human rectal tumor *in vitro*; they reported that freshly prepared endospores were also able to adhere to HEp2 cells derived from human laryngeal carcinoma cultured in MEM medium with 10% fetal calf serum. Adherence was also shown to HeLa cells and normal human buccal epithelial cells, *in vitro* by similar techniques (Arseculeratne 2004, unpublished data).

98.1.4.2 Hemagglutination

Endospores were also found to be capable of causing direct hemagglutination of mammalian and avian red blood cells. Red cells of the rat and mouse, which have not been reported to be natural hosts of rhinosporidiosis, have shown marked hemagglutinalibility. The reaction is easy to reproduce, at room temperature and is visible in about 1 h. Agitation and dispersal of agglutinated red cells and re-incubation of the mixtures reproduced the hemagglutination,[35] indicating that perhaps an enzymic mechanism that has been postulated with neuraminic acid as the receptor for Gram-negative bacilli, is not involved. Ultrastructural appearances suggest that the endospores in contact with the hemagglutinated red cell develop "spikes" that could conceivably mediate the attachment, in which event they function as do the fimbriae of Gram-negative bacilli.

98.1.4.3 Strain Heterogeneity

The third instance of the application of molecular biological techniques in the study of *R. seeberi*, was the investigation of its strains in respect of genetic heterogeneity.

R. seeberi appears to share with other microorganisms the trait of genetic diversity in its strains. Two reports on this phenomenon exist. Appuhamy et al.[36] used three universal primers REP, ERIC, and BOX on DNA extracted by the *Qiagen* method from partially purified endospores (from homogenates of the human rhinosporidial tissues, purified on *Percoll*), and from normal human spleen and leucocytes as Rhinosporidium-negative controls, after grinding in a bath of liquid nitrogen.[33] Only the BOX primer (BOX AIR 5′ CTACGGCAAGGCGACGCTGACG 3′) gave distinct bands that were analyzed with *Quantity One*® software (BioRad) (Figure 98.7). Six strains of *R. seeberi* from six patients with different clinical presentations of rhinosporidiosis showed that the banding patterns fell into three groups, each containing two strains. There was no correlation of the banding pattern with the characteristics of the clinical disease from which they were obtained, for example, the distribution of the clinical picture in the three groups was as follows:

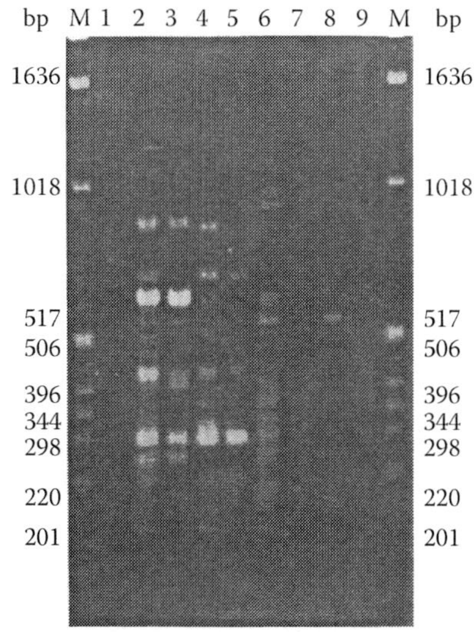

FIGURE 98.7 RAPD analysis by the BOX Universal primer, of six (lanes 2–7) strains of *R. seeberi* from six Sri Lankan patients with different clinical presentations, showing three groups with two strains in each. Lanes 8 and 9 represent human leucocytes and human spleen cells, respectively.

Strain	PCR Group	Site of Rhinosporidiosis
1	a	Disseminated
2	a	Bilateral, nasal
3	b	Oro-pharyngeal
4	b	Disseminated
5	c	Naso-pharyngeal
6	c	Nasal

The second report on genetic heterogeneity in *R. seeberi*, based on molecular methods, was by Silva et al.[37] who examined internal transcribed spacer 1 (ITS1), 5.8S, and ITS2 from rhinosporidial tissue from eight human, two swans, and a dog. "The ITS regions were amplified by PCR using a primer designed from a unique region of *R. seeberi*'s 18S SSU rRNA gene in combination with the ITS4 universal primer." These authors reported that their analysis "suggested that the genus *Rhinosporidium* may possess multiple host-specific strains."

In both reports,[36,37] it was the use of molecular methods that showed the existence of genetic variability of *R. seeberi*, while in the absence of a method for *in vitro* culture, no evidence of phenotypic variation, so often shown in bacteria, is as yet available for *R. seeberi*.

In an ongoing project to investigate strain variation, strains of *R. seeberi* from different human patients are being analyzed with a *R. seeberi*-specific primer[9] for confirmation of identity, and base sequence of the amplification products are being determined to reveal strain variations. It is hoped to include rhinosporidial tissues from animals abroad, noting that rhinosporidiosis in Sri Lankan animals has not yet been detected.

98.1.4.4 Sensitivity to Physical and Antimicrobial Agents

The sensitivity of *R. seeberi* to heat was tested with MTT reduction as the indicator of endospore viability.[23] At a series of temperatures from 37°C to 100°C to determine the thermal death point, it was found that exposure of freshly prepared endospores to 55°C for 10 min totally inactivated the endospores; with 2 min as the thermal death time at 55°C (Arseculeratne 2009, unpublished data).

Investigations on storage conditions that preserved viability were made so that strains of value could be preserved for further investigation, for example, in the event that, epidemiologically, spread within the country might occur, or for comparison with international strains. Storage at 4°C in phosphate buffered saline (pH 7.4) caused a slow loss of viability (MTT-reduction test), while slow freezing and storage at −20°C caused appreciable loss of viability. Partially purified endospores (sieved through 25 μm nylon sieves after homogenization of tissues) suspended in 10% skimmed milk, followed by lyophilization, when resuscitated showed loss of MTT-reducing capacity, but storage of the resuscitated suspensions at 37°C for 2 weeks restored their MTT-reducing capacity and morphology, as in lyophilized and resuscitated bacteria, to approximately pre-lyophilization levels, and thus indicated that lyophilization is a suitable method for preservation of endospores (Arseculeratne 2009, unpublished data).

The MTT-reduction test was then used in a series of trials to investigate the effects of biocides that are commonly used in laboratory and hospital practice, and of antimicrobial drugs in current use. The following generic agents were effective biocides: hydrogen peroxide, glutaraldehyde, chloroxylenol, chlorhexidine, cetrimide, thiomerosal, 70% ethanol, iodine in 70% ethanol, 10% formalin, povidone-iodine, sodium azide, and silver nitrate, while effective brand-named biocides included *Bacillo-floor*, *Sokrena*, *Sterillium*, *Bactolin*, *Bodedex*, *Cutasept*, and *Korsolex*.[38,39]

Of antimicrobial drugs in current use, Amphotericin B, Dapsone, Ketoconazole, Trimethoprim-Sulphadiazine, sodium gluconate, *Berenil* (Diminazine aceturate), and *Imizol* (Imidocarb dipropionate) were found to have anti-rhinosporidial activity; ineffective antimicrobial drugs included, penicillin, streptomycin, gentamicin, pentamidine, metronidazole, and ciprofloxacin.[40] Dapsone, with an *in vitro* inhibitory concentration at the 50% end-point (IC_{50}) of 29 μg/mL) is the commonest antimicrobial drug used on patients with rhinosporidiosis. Recent studies using MTT reduction showed that the anti-tuberculous drug Cycloserine, and the antifungal drugs Echinocandin (Caspofungin) and Voriconazole were even more effective drugs with IC_{50}s of 10 μg/mL and below 0.03 and below 0.02 μg/mL, respectively (Arseculeratne 2009, unpublished data). The high sensitivity of the endospores of *R. seeberi* to Echinocandin that prevents the synthesis of 1,3 β-glucan that is incorporated into the cell walls of fungi, might indicate that 1,3 β-glucan is present in the walls of rhinosporidial endospores though its presence has not been documented in the literature.

With the new *in vitro* test for assessment of viability of rhinosporidial endospores,[22] new light was shed on the relationship of *in vitro* test results of drug sensitivity to responses in drug-treated patients. Dapsone-treated endospores *in vitro* began to show loss of viability on day 1, and were totally inactivated on day 4. On the other hand, Mahakrishnan et al.[41] noted that with oral dapsone *in vivo*, complete regression, clinically, of cutaneous nodules in disseminated rhinosporidiosis, with accentuated granulomatous host's responses histopathologically, and arrest of the maturation and degeneration or absence of endospores, were observed at 18 weeks of Dapsone therapy. Job et al.[42] recorded that, clinically, a reduction in size of the growth at 6 weeks, marked reduction at 36 weeks, and total disappearance after 1 year of treatment, were observed; histopathologically, at 6 weeks no effect on pathogen was observed while in 18 weeks degenerative changes, and in 36 weeks reduction in numbers and size of the pathogen had occurred, with sporangia that had collapsed and been phagocytosed by macrophages. Venkateswaran et al.[43] observed no effect on the pathogen until the sixth week of therapy, and that the patients' tissue responses were augmented after 12 weeks of therapy, with degeneration and disappearance of the pathogen, which were observable after 36 weeks.

Two possible reasons exist for the differences in the time-sequences between *in vitro* and *in vivo* effects, based on (i) the pharmacokinetics of Dapsone *in vivo* (absorption, distribution penetration into tissues, and access to the pathogen), and the pharmacodynamics of its action on the pathogen *in vitro*, through direct confrontation. *In vivo*, the stromal and cellular reactions, with hemorrhage, cystic spaces, oedema, fibrosis, and mainly the existence of the pathogen within downgrowths of squamous epithelium would probably delay the access of the dapsone to the pathogen; and (ii) the presence of a mucoid matrix in which the endospores are embedded, which could inhibit the access of the drug to the endospores, while *in vitro*, the dapsone-pathogen confrontation is immediate. A general correlation, however, between drug susceptibility *in vitro* to clinical outcome of therapy is reported to be present[44,45] and this correlation pertains to the drug sensitivity of the pathogen with drug concentrations. The results summarized above indicate another dimension to the *in vitro–in vivo* correlations exists, which pertains to the pharmacokinetics and the pharmacodynamics of the drug activity in these two situations.

The use of therapeutic antimicrobial drugs in rhinosporidiosis is not extensive yet, to justify the application of molecular techniques, to the problem of drug resistance as in bacterial pathogens; for example, pulsed-field gel electrophoresis of chromosomal DNA of *Salmonella enterica* serovar *typhi* was used to compare strains with multidrug resistance.[46] Bifani et al.,[47] using different molecular biological techniques, investigated streptomycin resistance in strains of *Mycobacterium tuberculosis*. Studies on drug resistance of *R. seeberi* are premature, as the clinical use of antimicrobial drugs in rhinosporidiosis is still in the incipient stages while the analysis of drug resistance *in vitro* of *R. seeberi* from different sources has not been initiated.

98.1.4.5 Immunology

As with other topics, advances in which were hampered by the inability to culture *in vitro*, immunological properties of *R. seeberi* are noteworthy for their scantiness. *In vitro* culture would have yielded the different ontogenic stages of *R. seeberi* for comparative studies on the antigenic composition and antigenic variation of the different stages. Immunological properties of the different stages of *R. seeberi* could illuminate the characteristic clinical features of rhinosporidiosis—chronicity, recurrence, and dissemination.

Indirect methods have provided some advances. Herr et al.[48] demonstrated, by immuno-electron microscopy, the first appearance of a new antigen in the wall in the maturing sporangium. The significance of this finding lay in its relevance to immune evasion as the pathogen's strategy of survival. In this respect, *R. seeberi* has a parallel with Trypanosomes that show antigenic variation. In the same year,[33] partially purified rhinosporidial endospores and early sporangia isolated from human rhinosporidial tissue, further purified on *Percoll* columns, were shown to be immunogenic in experimental rabbits. The resulting antibody, in indirect immunofluorescence tests, labeled endospores and sporangia intensely; the fluorescence in the sporangia was marked in the matrix, especially around the pore, suggesting that the intra-sporangial matrix in which the endospores are embedded is highly immunogenic.

The following strategies could be identified as mechanisms of immune evasion of *R. seeberi*. This statement presupposes that specific immune defenses are protective in rhinosporidiosis, a postulate that, however, lacks experimental confirmation.

(i) *Antigenic variation* in *R. seeberi* as demonstrated by Herr et al.[48]
(ii) *Insensitivity of endospores to specific antibody*. It has been shown that anti-rhinosporidial, polyclonal antibody does not inactivate rhinosporidial endospores *in vitro*,[49] a finding that is similar to the absence of inactivating effects of anti-*Candida* antibody on the viability of *Candida albicans* and *Candida krusei, in vitro*.[50]
(iii) *Immune suppression*[51] involving cell-mediated immune (CMI) responses.
(iv) *Immune deviation* was shown to occur in response to parenteral injections of *R. seeberi* in the foot-pad of mice,[52] a common test object for studies on cell-mediated immunity. Repeated administration of rhinosporidial antigen resulted in a switch from CMI responses (subtype Th 1 to Th 2) to antibody production. The assumption that this deviation results in decreased specific cell-mediated immunity against *R. seeberi* is dependent on the protective role of CMI in rhinosporidiosis, an assumption that is not borne out by the demonstration of the failure of the establishment of progressive rhinosporidiosis in congenitally immunodeficient SCID and NUDE mice, with inocula from human rhinosporidial tissues; there are, however, qualifying factors, apart from immunological, such as devitalization of endospores by slow freezing, which could have contributed to such failure.
(v) *Binding of host immunoglobulins*.[53] When endospores were isolated from homogenates of human rhinosporidial tissues, which were ultrasonicated to release soluble antigens, the extracts on SDS-PAG electrophoresis showed evidence of bound human IgG, IgM, and, to a lesser extent, IgA. These bound immunoglobulins could interfere with immune defenses of the host by masking the antigens of *R. seeberi*. Immunological tests *in vitro* would need to recognize this phenomenon. It is necessary to remove the bound immunoglobulins with Sepharose G prior to the use of rhinosporidial antigens in *in vitro* tests.
(vi) *Immune distraction*.[31] Some areas of rhinosporidial polyps show stromal and cellular infiltration patterns in the absence of rhinosporidial elements. These reactions could be due to the presence of soluble antigens of *R. seeberi*; if this assumption is valid, their presence could result in the binding and removal of anti-rhinosporidial antibody by these antigens, although, as pointed out in (1), anti-rhinosporidial antibody has no adverse effect, at least *in vitro*, on the endospores of *R. seeberi*.
(vii) *Antigen sequestration*. The endospores of *R. seeberi* have no pores as in sporangia, but have thick walls that contain chitin; although they are vegetative bodies that absorb nutrients from the environment, it is probable that the endospore's antigens remain sequestered and not exposed to the immunocompetent cells of the host for the development of an immune response, however inadequate its protective role appears to be.

These mechanisms of immune evasion by *R. seeberi* were summarized earlier.[54]

98.1.5 Experimental Pathogenicity

A major enigma of *R. seeberi* is that, despite clear evidence from histopathology that the lesions of rhinosporidiosis invariably contain rhinosporidial bodies that are viable by the MTT-reduction test, all attempts at reproducing rhinosporidiosis in a wide variety of experimental and farm animals, have failed. These animals include species that are susceptible to the natural disease, such as cattle and buffaloes. Failure to reproduce the disease experimentally in a nonfunctional human eye with optic atrophy, by Chams in Indonesia,[11] is the only experimental attempt on humans.

Living systems such as cell cultures and growing egg embryos too have failed to support the sustained growth of experimental inocula containing *R. seeberi*. The fact that susceptibility factors peculiar to host species, are not involved was demonstrated by the failure to reestablish experimental rhinosporidiosis in bovines with preexisting rhinosporidiosis; the phenomenon of "infection immunity" may be relevant to this failure.

Some explanations for the failure of attempts at producing experimental rhinosporidiosis are as follows:

(i) The use of inocula that had been stored in the frozen state at −20°C. It has been shown that endospores subjected to slow freezing, are nonviable after thawing, as shown on the MTT-reduction test that has been used as a test of viability[23]; this nonviability has been attributed to the rupture of vital intra-endosporial, for example, mitochondrial, cell membranes by ice crystals produced during slow freezing.
(ii) Specific adaptive immune responses have been shown to be probably incapable of conferring anti-rhinosporidial immunity because specific antibody exists in high titer and CMI responses have been demonstrated in patients with rhinosporidiosis, especially of the disseminated variety. *In vitro*, specific antibody has been shown to be incapable of inactivating rhinosporidial endospores.[49] Endospores after injection into dermal tissues have been observed by many workers to have disappeared within 2 weeks of inoculation, a period that probably excludes the development of specific adaptive immune responses. For these reasons, the operation of nonspecific anti-rhinosporidial immunity due to macrophages and neutrophils need to be explored.
(iii) The absence of microorganisms in experimental inocula from homogenized rhinosporidial tissues that, in the natural habitat such as ground waters, could be synergistic with *R. seeberi* in promoting the establishment of sustained infection by the latter. Such synergism has been demonstrated with lactobacilli with *Trichomonas vaginalis*, and *Staphylococcus aureus* with *Candida albicans*.
(iv) Host specificity in strains of *R. seeberi*. It has been shown that strains from animal and human rhinosporidiosis differ in genotype[37]; hence, virulence of strains *R. seeberi* from human patients that have been used in tests of pathogenicity, might not exist for a different species of host.

No information is available on the experimental pathogenicity of the other members of the Class *Mesomycetozoea* to which *R. seeberi* belongs.[7]

98.1.6 Epidemiology and Clinical Features

Rhinosporidiosis was first identified in a human nasal polyp in Argentina in 1892 and its causative pathogen *Rhinosporidium seeberi* was described and named in 1900. Both the disease and *R. seeberi* are unique, and differ from other infective diseases and their causative pathogens, for example, the HIV-AIDS problem that was resolved in 2–3 years and the severe acute respiratory syndrome (SARS) that was elucidated in 2–3 weeks. For many decades, rhinosporidiosis posed enigmas such as the source and mode of infection, the immunity mechanisms, while the pathogen remains uncultivable *in vitro* and unable to establish experimental rhinosporidiosis; unsolved problems were in its biology, life cycle, and its sensitivity to antimicrobial agents. Resolution of some of these problems had to await the development of new laboratory techniques within the last two decades, but yet some enigmas remain. Rhinosporidiosis is a remarkable disease that contains many lessons, both academic and technical, concerning the problems of interpretation of evidence, the derivation of hypotheses on the basis of incomplete evidence, and no less in the philosophy of science that underlies experimentation.

Two terms need definition—*Infectious* and *Infective*. A clear distinction needs to be made in rhinosporidiosis which, while being undoubtedly infective in the sense that all the lesions have demonstrable *R. seeberi*, is non-transmissible between patient and normal persons; it is therefore *Infective* but not *Infectious*. It is regrettable that these two terms that signify two entirely different attributes, which are important aspects of epidemiology, are loosely used in textbooks and journals.

The second most frequent sites of disease are the eye and its adnexae and the term oculosporidiosis has been proposed. We prefer to avoid this term because eye infections can be caused by other spore-bearing microorganisms, and its occurrence in the urinary tract might then prompt a new name Urosporidiosis. We prefer to refer to the disease in these sites as merely ocular rhinosporidiosis and urethral rhinosporidiosis, respectively. Further proposals for modified nomenclature pertaining to *R. seeberi* are made in the following.

Rhinosporidiosis has occurred in about 70 countries worldwide, with a high incidence in sub-Saharan (wet) Africa and south Asia; it is hyperendemic in Sri Lanka (Ceylon) and India. A recent trend has been the occurrence of the disease in emigrant persons and in animals in hitherto rhinosporidiosis-free countries, which is related to international travel and importation of animals.[55] Rhinosporidiosis could therefore be identified as an *emerging infective disease* in both humans and animals. The absence of transmission of *R. seeberi* between diseased humans or animals and uninfected indigenous humans and animals would restrict the propagation of the disease and thus the application of molecular techniques, which otherwise could have characterized the cross-transmitted strains of *R. seeberi*. Molecular techniques have been used in epidemiological studies in *Campylobacter* enteritis (multilocus sequence typing)[56]; in measles (sequence analysis),[57] ribotyping and pulsed-field gel electrophoresis (PFGE) to trace the origin of pandemic strains of *Vibrio parahemolyticus*,[58] Spoligotyping of Beijing strains of *Mycobacterium tuberculosis* emerging in Vietnam,[59] and in epidemiological differentiation of strains of *Flavobacterium meningosepticum* from clinical and environmental sources.[60] Skojen et al.[61] used ribotyping, multilocus enzyme electrophoresis, and PFGE to investigate the resurgence of diphtheria in Russia. With *R. seeberi*, the identification of genotypically different strains from animal and human sources in, especially, the occurrence of rhinosporidiosis as an imported disease might have applications in the future, if more instances of such cases are encountered. As with epidemiological studies of drug sensitivity in *R. seeberi*, epidemiological studies on rhinosporidiosis, using molecular techniques, are not as yet warranted.

It is suspected[13] that factors that determine host susceptibility to infection, after exposure to the pathogen in its natural habitat, exist but the determinants have not been identified, though a male preponderance and lower age groups in patients are widely reported.

Rhinosporidiosis is of insidious onset, slow progression, chronicity, with a not-unusual tendency to disseminate to anatomically unrelated organs and tissues. The upper respiratory tract followed by the eye and adnexae, and the urethra are the commonest sites. The lesions are typical, with strawberry-like polyps that are vascular with a tendency to bleed on handling.

Local spread is by infiltration but regional spread involving lymphatics is very rare; dissemination occurs probably through blood-borne endospores from traumatized polyps or rhinosporidial tissue during surgical resection.

98.1.7 Diagnosis

The urgency of diagnosis of rhinosporidiosis results from its mimicry of malignant disease; rhinosporidial polyps could mimic papillary adenocarcinoma, and squamous carcinoma. Large masses with denuded skin could mimic sarcomas. Benign neoplasms or growths such as fibromas, hemangiomas, benign nasal polyps, angio-fibromas, and viral warts also feature in the differential diagnosis.

98.1.7.1 Clinical Diagnosis

The clinical appearances of rhinosporidial polyps, apart from their site of occurrence, are sufficiently distinctive to provide a tentative diagnosis of rhinosporidiosis. In primary rhinosporidial lesions on the skin, their appearances are atypical and might appear as warts,[13] or in the perineal region, as condylomata. The mucoid material that is invariably present on the surface of nasal and ocular rhinosporidial polyps might be absent on dry–warty polyps, making cytodiagnosis on the latter difficult.

Clinically, rhinosporidial polyps could mimic several infective and noninfective conditions that include benign polyps, rhinoscleroma in the nasal passages; pterygi in the eye; squamous carcinoma on the skin and anterior urethra; and sarcoma, Verruca Vulgaris, and condylomata on the body, caruncles at urethral sites, and in general, hemangiomas. The mimicry of malignant growths emphasizes the urgency of definitive diagnosis of rhinosporidiosis. Actinomycosis (caused by *Actinomyces israelii*) and eumycetoma (caused by many mycelial fungi) could mimic rhinosporidiosis, especially in the feet.

98.1.7.2 Laboratory Diagnosis

98.1.7.2.1 Cytodiagnosis

While clinical appearances do afford a tentative diagnosis of rhinosporidiosis, smears from the surfaces of the polyps when stained with the PAS stain strengthen the diagnosis of rhinosporidiosis. Features that provide that diagnosis are the content of polysaccharide (cellulose) in the endospore and EDB walls that stain deep magenta with the PAS stain. The size (10–20 μm) of the endospore and its content of EDBs, which could be seen as refringent bodies in wet smears and on PAS or MTT staining, affords that certainty. Application of specific anti-*R. seeberi* antisera with fluorescent labeling is more cumbersome and expensive for routine use.

The method of choice for unequivocal diagnosis is histopathology of resected tissue for a retrospective diagnosis, or a biopsy for exploration, given that rhinosporidiosis could mimic infective diseases caused by other pathogens, for example, Rhinoscleroma, or noninfective neoplasms that could be benign or malignant. The invariable presence of *R. seeberi*'s diverse ontogenic stages in rhinosporidial tissues irrespective of site, human, or animal, is the basis of the value of this method. To that extent, the use of molecular diagnostic methods is unnecessary, apart from being expensive and tedious.

Histopathology has pitfalls created by anomalous appearances of the pathogen and to a lesser extent of the host's stromal and cellular reactions; these have been described[31] but may be summarized as comprising degenerate sporangia that have no walls or have walls that are either grossly thickened or attenuated to the point of seeming absence; the intra-sporangial space could be filled with homogeneous eosinophilic material instead of endospores, or be empty; the sporangial walls may be represented only by refractile fragments; the stroma and cellular infiltrate may show unusual appearances consisting of a scarcity of the inflammatory cells, or a predominantly oedematous stroma that lacks the typical cystic spaces, intense vascularity, and hemorrhages. These unusual appearances are rare and histopathological identification of the tissue as rhinosporidial does not normally create a problem.

98.1.7.2.2 Histopathology

Conventional histopathology with the hematoxylin and eosin stain is the mainstay of diagnosis. Selection of the strawberry-like portions of the polyp, especially in the variegated nasopharyngeal growths, is important. Stromal reactions of the host are nonspecific but the diverse ontogenic stages of *R. seeberi*, which are invariably present irrespective of the site of disease in both humans and animals, present no diagnostic problem, given occasional atypical appearances of the parasite.[31] Stromal features characteristic of rhinosporidiosis include cystic spaces, great vascularity, and hemorrhages. Special stains, for example, GMS or mucicarmine, are not required. Molecular methods such as PCR or ISH are also unwarranted in routine diagnosis, except for research purposes.

98.1.7.2.3 Differential Histopathological Diagnosis

Entities both infective and noninfective need to be excluded in the differential diagnosis of rhinosporidiosis. (i) *Coccidiodomycosis*. Rhinosporidial sporangia might be confused with the sporangia of *Coccidioides immitis*. The rhinosporidial sporangia are larger, with thicker walls and with more numerous endospores. They also possess the exit "pore" that might, however, be excluded in a given histological section. The sporangium of *R. seeberi* also differs in staining with muci-carmine.[62] (ii) *Adiaspiromycosis*. This disease is caused by *Emmonsia* spp. especially *E. crescens* in humans and animals. It commonly affects the lungs whereas pulmonary rhinosporidiosis is very rare. Skin lesions, in a single case with AIDS, have been caused by *Emmonsia pasteuriana* while skin lesions are commonly caused by *R. seeberi*. The host responses are similar to those in human and animal rhinosporidiosis and consist of chronic inflammatory cell infiltrates of epithelioid macrophages, lymphocytes, plasma cells, neutrophils, and fewer giant cells and eosinophils; granuloma formation might occur. Embedded in these lesions are the large adiaspores, the characteristics of which, and differentiation from rhinosporidial and coccidioidal bodies, have been described[63]: the adiaspores have a thick capsule that stains with PAS and Gomori Methanamine Silver (GMS) but not with mucicarmine, whereas the rhinosporidial sporangium stains with mucicarmine; the sporangia of *Rhinosporidium* and *Coccidioides immitis* also stain with PAS and GMS, but they are thinner-walled with capsules of thickness ranging from 1 to 2 μm and 3 to 5 μm, respectively, while the *Emmonsia* capsules are 20–70 μm thick. The adiaspores of *E. crescens* are large (50–500 μm) while the mature sporangia *of R. seeberi* could range from 100 to 450 μm in diameter; the rhinosporidial and coccidioidal sporangia have endospores from endosporulation but conidia are not present within the adiaspores of *E. parva*. *In vitro* culture and PCR have also been used in the diagnosis of adiaspiromycosis, while molecular techniques were used to identify genetically strains in different regions of the United States.[63] (iii) *Myospherulosis*. Myospherulosis or Spherulocystic disease[64,65] is a noninfective state that results from alterations by lipids, of erythrocytes or caused by similar alterations of malignant cells or macrophages,[66,67] which occur in cystic spaces, thus resembling endospore-containing sporangia. The GMS or mucicarmine stains are nonreactive with these elements, unlike the rhinosporidial entities.

98.1.7.2.4 Molecular Methods

The molecular techniques enable direct detection of the pathogen or its constituents and not its phenotypic characteristics that are based on its genotype, which are the basis of conventional methods of *in vitro* or *in vivo* culture and of metabolic activity.[68]

The scenarios that benefit most from these new molecular techniques are (i) the presence of the causative pathogen in only very small concentrations that are not revealed by the conventional techniques of microscopy (e.g., tubercle bacilli in cerebrospinal fluid), culture, and animal inoculation; (ii) given the limitations of contemporary methods for culture, the innate uncultivability of the causative pathogen as in leprosy; (iii) the devitalization of the pathogen by therapy with antimicrobial drugs preventing its culture; (iv) fastidious pathogens, the growth requirements of which are either yet unknown or are too exotic for routine use; and (v) the early stages of infective disease when concentrations of the pathogen in the conventional specimens are too low.

In general, the advantages of molecular methods include high sensitivity and specificity contrasting with conventional cultural and immunological techniques, and rapidity in obtaining results, especially when an urgency of diagnosis exists, with patients under life-threatening opportunistic infections in immunocompromised states; disadvantages include the higher cost and greater technical expertise involved than is available in routine diagnostic laboratories in poor countries, where rhinosporidiosis occurs with greater frequency. With rhinosporidiosis, which manifests with single or multiple lesions, a specific histopathological diagnosis could quickly and easily be made on the large majority of rhinosporidial tissues, dispensing with the need for molecular techniques. The only tissue in which rhinosporidial ontogenic stages could be absent, although a host stromal and cellular response are present, is the smooth-surfaced, non-strawberry-like nasopharyngeal polyp[31]; other regions of such polyps are typically strawberry-like with numerous rhinosporidial bodies in growths in this site, and therefore proper sampling of the tissue for histopathology is mandatory. In diseases such as tuberculous meningitis, where the numbers of tubercle bacilli could be very small and could be missed on conventional acid-fast stained smears, PCR has been shown to be useful. Real-time PCR has been shown to improve detection of *Trichomonas vaginalis* infection compared with culture using self-collected vaginal swabs.[69] Such molecular methods have, however, not been used for the diagnosis of rhinosporidiosis, even in portions of tissues that lack visible rhinosporidial bodies because adjacent areas invariably show them.

It is not because these sensitive and specific techniques were not used that serious misidentifications of the causative organism of rhinosporidiosis have been made, for example, that it is not a microorganism at all but sacs of tapioca starch,[27] or the ubiquitous groundwater cyanobacterium *Microcystis aeruginosa*.[2] These errors arose from a lack of understanding of the processes of inference and interpretation of observations for scientific discovery. These red-herrings, however, still mislead, as when it led an author of a paper to write, as late as in 2006. "The causative agent of rhinosporidiosis is a cyanobacterium."[70]

98.1.8 Mechanisms of Anti-Rhinosporidial Immunity

There were no previous reports on the humoral immunity responses (HIR) in patients with rhinosporidiosis, until in 2004 when anti-rhinosporidial antibody levels in patients with rhinosporidiosis and in asymptomatic persons who lived in the same endemic area as patients, were reported.[71] The appreciable titres in asymptomatic persons were attributed to subclinical sensitization to *R. seeberi* in ground waters in which these persons had bathed. The antibody levels in patients bore no relation to the site, extent, and duration of the disease.

CMI responses (CMIR) as assayed by the *in vitro* lymphoproliferative response (LPR) test on patients' (and control asymptomatic persons') peripheral blood lymphocytes, and marker-specific immunohistochemistry on rhinosporidial tissues from patients showed well-marked CMIR[51]; the responses, as with HIR,[52] however, showed no correlation with the site, extent, and duration of the disease. The overall conclusion was that HIR and CMIR are not protective or curative in rhinosporidiosis. On the contrary, suppressor immune responses in rhinosporidiosis were postulated and diverse mechanisms of immune evasion by *R. seeberi* were identified.[13,54]

In the absence of evidence for a role of specific adaptive immunity in protection from the acquisition of the disease and recovery from active disease, a role for immunizing agents (vaccines) is as yet inconceivable. Yet the high sensitivity of *R. seeberi* to physical and chemical agents including antimicrobial drugs augurs well for patients, their contacts, and laboratory workers. The evidence given by Senaratne et al.[72] on the possible mode of initiation of ocular rhinosporidiosis, suggests simple measures for avoiding trauma to the eyes and the respiratory passages, as preventive measures against acquiring rhinosporidiosis.

98.1.9 Rhinosporidiosis in Animals

Rhinosporidiosis has been reported in several genera and species of animals that include farm and domestic animals—horses, mules, dogs, cats, cattle, buffaloes, swans, wild geese, and ducks. In the natural state especially, but not exclusively in poor, tropical countries, these animals have exposure to ground waters that are the habitat of *R. seeberi* and, as with humans, the predominant clinical sites are in the upper respiratory tract. Lesions have also occurred in the eyes. Disseminated cases, which are not uncommon in humans, have not been reported in animals. India and Sri Lanka (Ceylon) have the highest incidence of human rhinosporidiosis, but it is remarkable that rhinosporidiosis in animals in Sri Lanka has not been reported although rhinosporidiosis in animals has been widely reported from India. On the assumption that this disease is easily recognized clinically as in humans, the absence of records of animal rhinosporidiosis in Sri Lanka probably means a true rarity of the disease in animals. Given that buffaloes and cattle are the predominant animals that have had rhinosporidiosis in India, the lack of reports of the disease in Sri Lankan animals is surprising and needs consideration. While there are indigenous varieties of buffaloes and cattle in the rural areas of Sri Lanka, imports of these animals and bovine semen from India have been made by state farms. Cross-breeding between the imported and indigenous varieties could have occurred. No clear claim could therefore be made that Sri Lankan animals are genetically different entailing their insusceptibility to rhinosporidial disease. Rajapakse, Arseculeratne et al. (2008, unpublished data) have found appreciable numbers of abattoir cattle from rural areas that showed rhinosporidial juvenile sporangia in scrapings from their nasal passages, and high serum titres of anti-rhinosporidial IgG; these findings do not necessarily imply active rhinosporidial disease in these animals but could merely reflect the contamination of nasal passages by waterborne *R. seeberi* and subclinical immunization, as these animals constantly wallow in lacustrine bodies of ground water.

The comparison with rhinosporidiosis in humans in Sri Lanka and in India is of interest in terms of the genetics of the host. Sinhalese who form the ethnic majority of the population, and who are the predominant people in the rural areas, are considered to have emigrated from several regions of India, notably Bengal, Gujarat, and Kerala, as well as from other regions in India, while the Tamil populations are considered to have originated from South India. The genetic affinities between the Sri Lankans, especially the Sinhalese who outnumber the other ethnic groups in the occurrence of rhinosporidiosis, and the Indians might underlie the fact that they share the high incidence of rhinosporidiosis, worldwide. What elements of the genotype, for example, HLA types, are involved in susceptibility to rhinosporidosis, remains a topic for investigation.

The histopathology of rhinosporidiosis in animals bears no significant difference from what is seen in human rhinosporidiosis. The immunological status of animals with rhinosporidiosis has not yet been investigated, although an ongoing survey of the seroepidemiology of rhinosporidiosis in Sri Lankan animals indicates that high titres (1/1600 and 1/3200) of serum anti-rhinosporidial IgG are present in cattle (Rajapakse and Arseculeratne 2009, unpublished data); no data is available on the humoral immune status of non-diseased or rhinosporidial animals in India.

An important feature of rhinosporidiosis in animals and of the disease in humans is the close similarity of their clinical presentations, macroscopic and microscopic pathology, although the causative strains are probably genetically different.

98.2 METHODS

Molecular techniques are of significant advantage in the study of both the pathogen and the disease because (i) *R. seeberi* cannot yet be cultured *in vitro*, and (ii) nor has it been possible to establish experimental rhinosporidiosis in experimental animals.

98.2.1 Sample Preparation

For genomic DNA isolation, PCR amplification and sequencing of *R. seeberi*'s 18S SSU rDNA in their phylogenetic analysis of *R. seeberi*, and for electron microscopic investigations, Herr et al.[8] used the following methods.

Biopsied tissues containing *R. seeberi*'s sporangia and endospores were collected from patients with rhinosporidiosis, and transported without fixatives to the laboratory. The sporangia and endospores were physically dissected from the tissues, further purified by centrifugation to remove human cells (~50 *R. seeberi* cells/μL), and then disrupted with glass beads. Alternatively, 100 mg of human tissues containing *R. seeberi*'s sporangia were placed in a mortar and ground under liquid nitrogen. After both of the treatments above, the samples were placed in 1.5 mL Eppendorf tubes, digested with proteinase K in the presence of sodium dodecyl sulfate (SDS), and then extracted with phenol and chloroform.[8]

For electron microscopic investigation, the infected human tissues containing the sporangia and endospores of *R. seeberi* were fixed in 2.5% glutaraldehyde with 0.05 M sodium cacodylate-buffered saline (pH 7.4) at room temperature for 2 h with gentle agitation. The tissues were washed three times with 0.05 M sodium cacodylate-buffered saline for 20 min each. The samples were transferred to a 1% OsO_4 and water solution at room temperature for 4 h with mild agitation. Next, the samples were washed three times with distilled water washes for 20 min each and transferred to 33% followed by 66% Spurr resin in acetone solutions for 30 min in each concentration. The samples were then transferred to 100% Spurr resin for 5 h and then overnight, with resin changes at the end of each period. Ultrathin sections were made by using an ultramicrotome and sectioning the samples into 100 nm sections. The grids with the ultrathin sections were poststained with uranyl acetate for 30 min followed by lead for 3 min. Afterward, a thin layer of carbon was evaporated onto the surfaces of the grids. The ultra-thin-sectioned material was then examined with a Philips CM-10 electron microscope to reveal sporangia at different stages of development characteristic of *R. seeberi*.[8]

Fredricks et al.,[9] who confirmed the Protistan nature of *R. seeberi*, used the following methods for sample preparation for PCR and ISH. The frozen canine nasal polyp was thawed and after a 1 min centrifugation at 500 × g "approximately 0.2 g of the tissue pellet was digested by mechanical disruption and the DNA was purified by adsorption to glass milk in the presence of a chaotropic agent, according to the manufacturer's instructions (Fast Prep. Bio 101, Vista, CA). The DNA was resuspended in 100 μL of 10 mM Tris, 1 mM EDTA buffer at pH 8.5."

For ISH studies, these authors used three blocks of "fixed, paraffin-embedded nasal polyps from unrelated patients with histological confirmed rhinosporidiosis." One tissue sample was a patient in south Asia living in the United States, two samples were from patients in Spain. "Twenty-three blocks of fixed, paraffin-embedded nasal polyps from patients without rhinosporidiosis were obtained from 12 consecutive patients at the Palo Alto Veterans Affairs (VA) hospital who had undergone nasal polypectomy. Two 25-μm sections were cut from each block, and the sections were deparaffinized and digested. DNA from the digests was then purified by the Isoquick method (ORCA, Research Inc., Bothel, WA), and the DNA was resuspended in 25 μL of Tris-EDTA buffer." Control sections included similarly processed, histological normal human lymph node, sections of *Coccidiodes immitis* in bone," and samples of the Rosette Agent and *Dermocystidium salmonis*. Specimens of fungi in culture were transferred to microfuge tubes containing 0.5 mL digestion buffer and 0.1 mL glass beads. "Samples were incubated at 55°C overnight, and the proteinase K was inactivated at 95°C for 10 min and then subjected to two freeze–thaw cycles by immersing tubes in a dry ice–isopropanol bath followed by vortex mixing."

Silva et al.,[37] in analyzing the multiplicity of host-specific strains, used freshly biopsied tissue containing *R. seeberi*'s sporangia and endospores from two Florida swans and a dog from Georgia with rhinosporidiosis. "The tissues were aseptically collected, frozen without fixatives... Portions of each tissue (≈100 mg) were placed in a mortar and ground under liquid nitrogen and their genomic DNA extracted according to the protocol of Herr et al."[8] "Briefly, the samples were placed in Eppendorff tubes and treated with sodium dodecyl sulfate and proteinase K. *R. seeberi*'s genomic DNA separated from the mix after the addition of phenol and chloroform."

98.2.2 Detection Procedures

Herr et al.[8] utilized primers NS1 GTAGTCATATGCTTGT CTC and a degenerate NS8 (5′-TCCGCAGGTTCACC(TA) ACGGA-3′) to amplify nearly a full-length (1790 bp) 18S SSU rDNA sequence from *R. seeberi*. After cloning the amplicons into pCR 2.1-TOPO plasmids (Invitrogen), the nucleotide sequence of the fragment is analyzed by using BigDye Terminator chemistry in an ABI Prism 310 genetic analyzer apparatus (Perkin-Elmer). The 18S SSU rDNA sequence of *R. seeberi* is aligned with those from the GenBank. Phylogenetic analysis is conducted by a distance method (neighbor joining with Kimura's three-parameter multiple-hit correction) and parsimony. Support for internal branches is assessed by using 1000 bootstrapped data sets. Parsimony analyses are performed with the computer software program PAUP (PAUP: Phylogenetic Analysis Using Parsimony, version 3.1; D.L. Swofford, Illinois Natural History Survey, Champaign, IL.). Phylogenetic trees are drawn with stramenopiles and alveolates as the out-groups.

Fredricks et al.[9] used the following detection procedures. "Broad-range PCR primers were designed from a database of >4000 SSU rDNA sequences, with the ARB software package (Technical University, Munich, Germany). Primers were selected that would anneal to most fungal and some protest 18S rDNA but not to 18S rDNA from the chordata (F1-fw CAAGTCTGGTGCCAGCAGCC, F2-rev GATTTCCGTAAGGTGCCGA, F3-rev AATGCTC TATCCCCAMCACG)" as described in their Table.[9] "The broad range of primers F1-fw/F2-rev was tested by using DNA from several diverse fungi."

The PCR primers, and sequencing primers were detailed in Fredricks et al.[9]

> PCR consisted of 40 cycles of amplificationon a Perkin-Elmer GeneAmp 2400 thermal cycler. After an initial activation of Taq gold at 94°C for 10 minutes, each cycle consisted of 30 seconds of melting at 94°C, 30 seconds of annealing

at 56°C, and 30 seconds of extension at 72°C. The last cycle was followed by an extension step at 72°C for 7 minutes. Amplification products were detected by electrophoresis on 2% agarose gels stained with ethidium bromide and visualized with a UV transilluminator.

For fluorescence in situ hybridization (FISH), "Tissue sections on slides were dewaxed by immersion in 99% octane (Sigma, St. Louis, MO). Samples subjected to FISH included *R. seeberi*-infected human nasal polyps, *C. immitis*-infected bone, a Rosette agent-infected cell line, and smears of *Candida albicans*. The *Rhinosporidium* probe was based on the Rhino-rev 18S rDNA primer and was biotinylated at both the 5′ and 3′ ends."

"The control probe, which consisted of the complement of the *Rhinosporidium* probe, was also biotinylated at both ends.

The Rhino FISH probe was BTGCTGATAGAGTCATT GAATTAACATCTACB

The control FISH probe was BACGACTATCTCAGTA ACTTAATTGTAGATGB

To each slide, 50 ng of biotinylated probe in 30 μL of hybridization buffer was added. Cover slips were placed, and the slides were incubated at 40°C overnight in a humid chamber. The hybridization buffer consisted of 10% dextran, 0.2% bovine serum albumin, and 0.01% polyadenosine, in 5× SET buffer, the 25× SET buffer consisted of 3.75 M sodium chloride, 25 mM EDTA, and 0.5M Tris at pH 7.8. Cover slips were removed by immersion in 5× SET buffer at 4°C, and the slides were washed for 10 min per cycle, twice in 0.2× SET buffer at 25°C and once at 40°C. The slides were then subjected to tyramide signal amplification according to the manufacturer's instructions (TSA indirect, NEN Life Sciences, Boston, MA). Cy5-streptavidin (Amersham, Piscataway, NJ) at 1 mg/mL was diluted 1:500 and added to the slides for fluorescence signal detection. Tissue sections were visualized on a Bio-Rad confocal microscope at 200× magnification after the application of 15–20 μL of Vectashield mountant (Sigma) and a cover slip. For electron microscopy, formalin-fixed, paraffin-embedded rhinosporidial nasal polyp was removed from the block, dewaxed with xylene, rehydrated with ethanol, post-stained with 1.5% osmium tetroxide, then dehydrated with ethanol, transferred to propylene oxide, followed by Epon 12 resin, heat-catalyzed at 65°C, and ultrasectioned at 50 nm. The grid-mounted sections were then serially stained with lead hydroxide and uranyl acetate and examined with a Phillips 201 electron microscope at 75 kV."

Silva et al.[37] in their work for demonstrating the existence of multiple host-specific strains, used the following detection procedures. "The ITS1" (TCCGT AGGTG AACCT GCGG), "5.8S and ITS2 sequences of *R. seeberi* were amplified by PCR using the specific primer RhinA2F (5′-TAGTTGCGTGATTTTTCGAA-3′) in combination with the ITS4 universal primer."[73] "These sets of primers were designed to amplify approximately the last 450 bp of 18 SSU rRNA genes, the complete ITS1, 5.8S, and ITS2, and the first 60 bp of 28S large subunit (LSU) rRNA genes. PCR consisted of 40 cycles of amplification on a Perkin Elmer GeneAmp 9700 thermal cycler. After an initial activation of *Taq* Gold (Applied Biosystems) at 95°C for 10 min, each cycle consisted of 1 min of melting at 94°C, 2 min of annealing at 50°C for 18S SSU rRNA genes or 60°C for ITS, and 3 min of extension at 72°C. The last cycle was followed by an extension step at 72°C for 7 min. Amplification products were detected by electrophoresis on 0.8% agarose gels stained with ethidium bromide and visualized by using the Bio-Rad Gel Doc 1000 apparatus with the software Multi-Analyst version 1.0.2 (Bio-Rad, California). Since genomic DNA extraction from paraffin-embedded tissues often results in DNA fragmentation, the ITS sequences from these samples were amplified with the ITS5 and ITS4 universal primers, which amplified smaller fragments than those obtained using the RhinA2F and ITS4 primers.

The PCR amplicons were then cloned into pCR2.1-TOPO plasmids (Invitrogen, Carlsbad, CA) and purified, and at least 10 clones of each sample were then sequenced in both directions by using BigDye Terminator chemistry in an ABI Prism 310 genetic analyzer apparatus (Perkin Elmer, Foster City, CA).

98.3 CONCLUSIONS AND FUTURE PERSPECTIVES

Since 1995 when Kennedy et al.[22] made important proposals for a modified nomenclature of rhinosporidial entities, it seems opportune now to propose a further revision of nomenclature, that of the endospore, which was shown to contain bodies termed the EDBs, the identity and nature of which remained controversial for many decades. With the demonstration of metabolic activity of the EDB and their probable nature as the fundamental germinative unit of *R. seeberi*,[20,23] it would appear reasonable to propose that the endospore, by virtue of its content of EDBs that contain DNA and can therefore be regarded as the ultimate generative bodies of *R. seeberi*,[13] be renamed the "spore-morula," a term first used by Minchin and Fantham,[14] Beattie,[15] and by Tirumurti,[25] while the EDB be renamed the true spore.

The idea of causality, which is one of the central themes in the philosophy of modern science, is pertinent to the problem of misidentification of the nature of *R. seeberi*, firstly as sacs of starch[27] from excessive consumption of tapioca, and then, by the same group of workers, as the ubiquitous groundwater bacterium *Microcystis aeruginosa*.[2] This confusion also illustrates the pitfalls of the sole use of morphological criteria in tracing identities of biological entities, which according to J.M. Waller (Personal communication, 1995) has to reckon with the phenomenon of convergence in fungal taxonomy. The development of ideas regarding the nature and significance of the endospore's EDBs also illustrates the process of scientific advance in terms of revision or modifications, or in Thomas Kuhn's terms, the "falsification" of existing theories.

The idea of causality, and the logic of scientific inference, are illustrated by the inference of the life cycle of *R. seeberi* from its appearances in rhinosporidial tissues, in the absence of direct evidence due to its inability to grow *in vitro*; culture *in vitro* could have been used for this exercise, as has been used with, for example, the life cycles of insect parasites. Jennifer Trusted's important book *The Logic of Scientific Inference* gives the philosophical basis of such an exercise. We consider it a neglect of these basic principles in the philosophy of science, which led to the error of incriminating tapioca starch and then a ubiquitous water bacterium as the cause of this disease. The comments of the distinguished French physiologist Claude Bernard on spurious hypotheses are irresistible: "The man who constructs a hypothesis without supplying an adequate programme for its trial by experiment is a burden to science and to the world."

A further aspect of the philosophy of science, that of the intimate, synergistic relationship between pure (basic, fundamental) research and applied research, is illustrated by the development as recently as 2004, of the basic research on MTT reduction as applied to the endospore,[23] which paved the way for not only the identification of the nature of the EDB, but as importantly, for applied studies on the sensitivity of *R. seeberi* to antimicrobial drugs, information on which was lacking for 118 years.

Molecular techniques have had wide applications for the study of the biological properties of pathogens and the disease they cause: the discrimination between exogenous sources of infection versus endogenous reactivation in tuberculosis, investigation of outbreaks of disease, laboratory cross-contamination, transmission of pathogens, the epidemiological investigations mentioned above, and the sensitivity of pathogens to therapeutic antimicrobial drugs. Despite the high specificity and sensitivity of molecular methods in the diagnosis of diseases caused by microbes, their applicability to rhinosporidiosis is restricted on account of the availability of simpler, less costly, less tedious, but specific methods such as histopathology.

Future applications of molecular methods for rhinosporidiosis and *R. seeberi* could conceivably be in the areas of (i) intercountry transmission of the disease through expanding importation of animals and travel of persons from endemic countries, to determine the mechanisms and routes of the spread of given strains of *R. seeberi*, and (ii) possible development of resistance of *R. seeberi* to drugs and to map the transmission of resistant strains from sources of origin. Molecular methods have been used in the study of drug resistance in bacteria and, though drugs are still in the incipient stages of usage in rhinosporidiosis, mounting evidence of transmission of the disease to non-endemic countries might necessitate such studies on *R. seeberi* in the future.

In view of the spread of *R. seeberi* to hitherto non-endemic countries, strain variation, intercountry genetic variations, and characterization of antigens, are other areas that might benefit from the application of molecular methods, some of which have been listed above as applied to other diseases and their causative pathogens. Problems of the use of molecular techniques in diagnosing, and managing "infectious diseases" in terms of "practicality and costs" have been discussed by Pfaller[74] and are especially relevant in poor, tropical countries in Asia and Africa, where rhinosporidiosis is endemic.

REFERENCES

1. Arseculeratne, S.N., Rhinosporidiosis: What is the cause? *Curr. Opin. Infect. Dis.*, 18, 113, 2005.
2. Ahluwalia, K.B., Maheswari, N., and Deka, R.C., Rhinosporidiosis: A study that resolves aetiologic controversies, *Am. J. Rhinol.*, 11, 479, 1997.
3. Arseculeratne, S.N., *Microcystis aeruginosa* as the causative organism of rhinosporidiosis, *Mycopathologia*, 151, 3, 2000.
4. Mendoza, L., Herr, R.A., and Ajello, L., Authors' reply—Causative agent of rhinosporidiosis, *J. Clin. Microbiol.*, 39, 413, 2001.
5. Kwon-Chung, K.J., Phylogenetic spectrum of fungi that are pathogenic to humans, *Clin. Infect. Dis.*, 19(Suppl. 1), S1, 1994.
6. Steen, E.B., *Dictionary of Biology*, Barnes and Noble Publishers, New York, p. 116.
7. Mendoza, L., Taylor, J.W., and Ajello, L., The class Mesomycetozoaea: A heterogeneous group of microorganisms at the animal-fungal boundary. *Ann. Rev. Microbiol.*, 56, 315, 2002.
8. Herr, R.A. et al., Phylogenetic analysis of *Rhinosporidium seeberi*'s 18S small-subunit ribosomal DNA groups this pathogen among members of the Protoctistan Mesomycetozoa clade, *J. Clin. Microbiol.*, 37, 2750, 1999.
9. Fredricks, D.N. et al., *Rhinosporidium seeberi*: A human pathogen from a novel group of aquatic protistan parasites, *Emerg. Infect. Dis.*, 6, 273, 2000.
10. Kaluarachchi, K. et al., The identification of the natural habitat of *Rhinosporidium seeberi* with *R. seeberi*-specific in situ hybridization probes, *J. Infect. Dis. Antimicrob. Agents*, 25, 25, 2008.
11. Karunaratne, W.A.E., *Rhinosporidiosis in Man*, The Athlone Press, London.
12. Rajam, R.V. et al., Rhinsporidiosis: A study with report of a fatal case with systemic dissemination, *Indian J. Surg.*, 17, 269, 1955.
13. Arseculeratne, S.N. and Mendoza, L., Rhinosporidiosis. In *Topley & Wilson's Microbiology and Microbial Infections. Medical Mycology*, William, G.M. and Roderick, J.H., Eds., 10th edn., Hodder Arnold, London, U.K., 2005, pp. 436–475.
14. Minchin, E.A. and Fantham, H.B., *Rhinosporidium kinealyi* n.g., n.sp., a new sporozoon from the mucous membrane of the septum nasi of man, *Quart. J. Microsc. Sci.*, 49, 521, 1906.
15. Beattie, J.M., *Rhinosporidium kinealyi*: A sporozoon of the nasal mucous membrane, *J. Path. Bact.*, 11, 270, 1906.
16. Vanbreuseghem, R. et al., Troisieme cas Congolais rhinosporidiose. Consideration nouvelles sur la nature des spherules, *Ann. Soc. Belg. Med. Trop.*, 35, 225, 1955.
17. Savino, D.F. and Margo, C.E., Conjunctival rhinosporidiosis: Light and electron microscopic study, *Ophthalmology*, 99, 1482, 1983.
18. Moses, J.S. et al., *Rhinosporidium seeberi*. Light, phase contrast, fluorescent and scanning electron microscopic study, *Mycopathologia*, 114, 17, 1991.

19. Lakshmanan, M. and Kameswaran, S., Development and histochemistry of *Rhinosporidium seeberi*, In Subramanian, C.V., Ed., *Taxonomy of Fungi. Proceedings of the International Symposium on Taxonomy of Fungi*, University of Madras, Madras.
20. Arseculeratne, S.N., Atapattu, D.N., and Wickramaratne, K., Nature and significance of the electron-dense bodies of the endospores of *Rhinosporidium seeberi*; their reactions with MTT (3-[4,5-dimethyl-2-thiazolyl]-2,5-diphenyl tetrazolium bromide) and TMRE (tetramethyl-rhodamine ethyl ester), *Med. Mycol.*, 43, 261, 2005.
21. Ashworth, J.H., On *Rhinosporidium seeberi* (Wernicke 1903) with special reference to its sporulation and affinities, *Trans. Roy. Soc., Edin.*, 53, 301, 1923.
22. Kennedy, F.A., Buggage, R.R., and Ajello, L., Rhinosporidiosis: A description of an unprecedented outbreak in captive swans (*Cygnus* spp.) and a proposal for revision of the ontogenic nomenclature of *Rhinosporidium seeberi*, *J. Med. Vet. Mycol.*, 37, 157, 1995.
23. Arseculeratne, S.N. and Atapattu, D.N., The assessment of the viability of the endospores of *Rhinosporidium seeberi* with MTT (3-[4,5-dimethyl -2- thiazolyl]-2,5-diphenyl tetrazolium bromide), *Mycol. Res.*, 108, 1423, 2004.
24. Teh, E.C. and Kutty, M.K., *Rhinosporidium see*beri: Spherules and their significance, *Pathology*, 7, 133, 1975.
25. Tirumurti, T.S., *Rhinosporidium kinealyi*, *Practitioner*, 93, 704, 1914.
26. Kutty, M.K. and Teh, E.C., *Rhinosporidium seeberi*. An ultrastructural study of its endosporulation phase and trophocyte phase, *Archiv. Pathol.*, 99, 51, 1975.
27. Ahluwalia, K.B., New interpretations in rhinosporidiosis, enigmatic disease of the last nine decades, *J. Submicrosc. Cytol. Pathol.*, 24, 109, 1992.
28. Rao, S.N., *Rhinosporidium seeberi*—A histochemical study, *Indian J. Exp. Biol.*, 4, 10, 1966.
29. Jain, S.N., Aetiology and incidence of rhinosporidiosis. A preliminary report, *Indian J. Otolaryngol.*, 19, 1, 1967.
30. Woodard, B.H. and Hudson, J., Rhinosporidiosis: Ultrastructural study of an infection in South Carolina, *Southern Med. J.*, 77, 1587, 1984.
31. Arseculeratne, S.N., Panabokke, R.G., and Atapattu, D.N., Lymphadenitis, trans-epidermal elimination and unusual histopathology in human rhinosporidiosis, *Mycopathologia*, 153, 57, 2001.
32. Mendoza, L. et al., In vitro studies on the mechanisms of endospore release by *Rhinosporidium seeberi*, *Mycopathologia*, 148, 9, 1999.
33. Atapattu, D.N. et al., Purification of the endospores and sporangia of *Rhinosporidium seeberi* on *Percoll* columns, *Mycopathologia*, 145, 113, 1999.
34. Levy, M.G., Meuten, D.J., and Breitschwerdt, E.B., Cultivation of *Rhinosporidium seeberi* in vitro: Interaction with epithelial cells, *Science*, 234, 474, 1986.
35. Arseculeratne, S.N. and Atapattu, D.N., Biological activities of the endospores of *Rhinosporidium seeberi*, *Proceedings of the 25th Annual Sessions, Kandy Society of Medicine*, Peredemiya, Kandy, Sri Lanka Vol. 22, 2003.
36. Appuhamy, S. et al., Strain variation in *Rhinosporidium seeberi*. *Proceedings of the Sri Lanka College of Microbiologists*, OP7, Colombo, Sri Lanka, 2002.
37. Silva, V. et al., Molecular evidence for multiple host-specific strains in the genus Rhinosporidium, *J. Clin. Microbiol.*, 43, 1865, 2005.
38. Arseculeratne, S.N. et al., The effects of biocides (antiseptics and disinfectants) on the endospores of *Rhinosporidium seeberi*, *Indian J. Med. Microbiol.*, 24, 85, 2006.
39. Eriyagama, N.B. and Arseculeratne, S.N., Further observations on the effects of biocides (Antiseptics and Disinfectants) on the endospores of *Rhinosporidium seeberi*, as assayed by the MTT-reduction test, *J. Infect. Dis. Antimicrob. Agents*, 24, 125, 2007.
40. Arseculeratne, S.N. et al., In vitro susceptibility of the endospores of *Rhinosporidium seeberi* to seven antimicrobial agents, *J. Infect. Dis. Antimicrob. Agents*, 25, 135, 2008.
41. Mahakrishnan A., Rajasekaram V., and Pandian, P.J., Disseminated rhinosporidiosis treated with dapsone, *Trop. Geogr. Med.*, 33, 189, 1981.
42. Job, A. et al., Medical therapy of rhinosporidiosis with dapsone, *J. Laryngol. Otol.*, 107, 809, 1993.
43. Venkateswaran, S. et al., Light and electron microscopic findings in rhinosporidiosis after dapsone therapy, *Trop. Med. Int. Health*, 2, 1128, 1997.
44. Rex, J.H. et al., Antifungal susceptibility testing, *Clin. Microbiol. Rev.*, 6, 367. 1993.
45. Cushion, M.T., Chen, F., and Kloepfer, N., A cytotoxicity assay for evaluation of candidate anti-*Pneumocystis carinii*, *Antimicrob. Agents Chemother.*, 41, 379, 1997.
46. Kariuki, S. et al., Genotypic analysis of Multidrug-Resistant *Salmonella enterica* serovar *typhi*, Kenya, *Emerg. Infect. Dis.*, 6, 649, 2000.
47. Bifani, P. et al., Molecular identification of streptomycin monoresistant *Mycobacterium tuberculosis* related to multidrug-resistant W strain, *Emerg. Infect. Dis.*, 7, 842, 2001.
48. Herr, R.A. et al., Immunolocalization of an endogenous antigenic material of *Rhinosporidium seeberi* expressed only during mature sporangial development, *FEMS Immunol. Med. Microbiol.*, 23, 205, 1999.
49. Arseculeratne, S.N., Atapattu, D.N., and Eriyagama, N.B., Human anti-rhinosporidial antibody does not cause metabolic inactivation or morphological damage in endospores of *Rhinosporidium seeberi, in vitro*, *Indian J. Med. Microbiol.*, 23, 14, 2005.
50. Arseculeratne, S.N. et al., The use of MTT (3-[4,5-dimethyl-2-thiazolyl]-2,5-diphenyl tetrazolium bromide)-reduction as an indicator of the effects of strain-specific, polyclonal rabbit antisera on *Candida albicans* and *C. krusei*, *Indian J. Med. Microbiol.*, 25, 267, 2007.
51. De Silva, N.R. et al., Cell mediated immune responses in human rhinosporidiosis, *Mycopathologia*, 152, 59, 2000.
52. Jayasekera, S. et al., Cell-mediated immune responses (CMIR) to *Rhinosporidium seeberi* in mice, *Mycopathologia*, 152, 69, 2001.
53. Atapattu, D.N. et al., Is human IgG bound by endospores and sporangia of *Rhinosporidium seeberi*? A possible mechanism of immune evasion. *Proc. Kandy Soc. Med.*, 22, 34, 2000.
54. Arseculeratne, S.N., Mechanisms of immune evasion by *Rhinosporidium seeberi*, *Bulletin of the Sri Lanka College of Microbiologists*, Vol. 1, 2003.
55. Leeming, G. et al., Equine rhinosporidiosis in United Kingdom, *Emerg. Infect. Dis.*, 13, 1377, 2007.
56. Dingle, K.E. et al., Molecular characterization of *Campylobacter jejuni* clones: A basis for epidemiologic investigation, *Emerg. Infect. Dis.*, 8, 949, 2002.
57. Rota, P.A. et al., Molecular epidemiology of measles viruses in the United States, 1997–2001, *Emerg. Infect. Dis.*, 8, 902, 2002.

58. Chowdhuri, N.R. et al., Molecular evidence of clonal *Vibrio parahaemolyticus* pandemic strains, *Emerg. Infect. Dis.*, 6, 631, 2000.
59. Anh, D.D. et al., *Mycobacterium tuberculosis* Beijing genotype emerging in Vietnam, *Emerg. Infect. Dis.*, 6, 302, 2000.
60. Colding, H. et al., Ribotyping for differentiating *Flavobacterium meningosepticum* isolates from clinical and environmental sources, *J. Clin. Microbiol.*, 32, 501, 1994.
61. Skojen, V. et al., Molecular characterization of *Corynebacterium diphtheriae* isolates, Russia, 1957–1987. *Emerg. Infect. Dis.*, 8, 516, 2002.
62. Kwon-Chung, K.J. and Bennett, J.E., Rhinosporidiosis. In Kwon-Chung, K.J., and Bennett, J.E., Eds., *Medical Microbiology*, p. 695, Lea & Febiger, Philadelphia, PA, 1992.
63. Sigler, L., Adiaspiromycosis and other infections caused by *Emmonsia* species. In *Topley & Wilson's Microbiology and Microbial Infections. Medical Mycology*, pp. 809–824, William, G.M. and Roderick, J.H., Eds., 10th edn., Hodder Arnold, London, U.K., 2005.
64. McClatchie, S. and Bremner, A.D., Unusual subcutaneous swellings in African patients. *East Afr. Med. J.*, 46, 625, 1969.
65. Hutt, M.S.R., Fernandez, B.J.J., and Templeton, A.C., Myospherulosis (subcutaneous Spherulocystic disease), *Trans. Roy. Soc. Trop. Med. Hyg.*, 65, 182, 1971.
66. Chau, K.Y., Pretorius, J.M., and Stewart, A.W., Myospherulosis in renal carcinoma, *Arch. Path. Lab. Med.*, 124, 1476, 2000.
67. Ali, A. and Flieder, D. et al., Rhinosporidiosis, an unusual affliction, *Arch. Path. Lab. Med.*, 125, 1392, 2001.
68. Speers, D.J., Clinical applications of molecular biology for infectious diseases, *Clin. Biochem. Rev.*, 27, 393, 2006.
69. Caliendo, A.M. et al., Real-time PCR improves detection of *Trichomonas vaginalis* infection compared with culture using self-collected vaginal swabs, *Infect. Dis. Obstet. Gynecol.*, 13, 145, 2005.
70. Dhaulakhandi, D.B. et al., Detection of 16S rRNA gene in round bodies isolated from polyps of rhinosporidiosis. *Infect. Genet. Evol.*, 6, 331, 2006.
71. Arseculeratne, S.N. et al., Anti-rhinosporidial antibody levels in patients with rhinosporidiosis and in asymptomatic persons in Sri Lanka, *Mycopathologia*, 158, 157, 2004.
72. Senaratne, T. et al., Ocular rhinosporidiosis with Staphyloma formation in Sri Lanka. The first report, *J. Infect. Dis. Antimicrob. Agents*, 24, 133, 2007.
73. Gargas, A. and De Priest, P.T., A nomenclature for fungal PCR primers with examples from intron-containing SSU rDNA. *Mycologia*, 88, 745–748, 1996.
74. Pfaller, M.A., Molecular approaches to diagnosing and managing infectious diseases: Practicality and costs, *Emerg. Infect. Dis.*, 7, 312, 2001.

Part VI

Panfungal and Drug Resistance Detection

99 Nucleic Acid-Based Panfungal Detection

Sandra Preuner and Thomas Lion

CONTENTS

99.1 INTRODUCTION

99.1.1 INVASIVE FUNGAL INFECTIONS

Invasive fungal infections (IFIs) are a major cause of severe disease in a variety of critically ill patients. High-risk constellations include primarily immunocompromised individuals with prolonged episodes of profound neutropenia, such as patients with hematologic malignancies[1–3]; recipients of allogeneic bone marrow transplant (BMT),[4,5] peripheral hematopoietic stem cell transplant (HSCT),[6] or solid organ transplant (SOT)[7,8]; patients with AIDS[9]; intensive care patients[10]; patients undergoing long-term corticosteroid therapy[11,12]; and preterm neonates.[13] In addition to the state of immunosuppression, important risk factors of IFI also include long-term exposure to broad-spectrum antibiotic therapy, long presence of central indwelling catheters,[14–18] and the general condition and the geographic location of the patients.[19]

Autopsy studies showed that IFIs may be encountered in as many as 40% of patients with leukemia[20] and accounts for a significant proportion of lethal complications in BMT/HSCT recipients.[21] In hematopoietic stem cell transplantation, the risk of IFI depends on a number of factors including the graft source (i.e., peripheral stem cells, bone marrow, or cord blood), the presence of acute or chronic graft-versus-host disease (GVHD), the administration of steroids, the occurrence of cytomegalovirus (CMV) infection, and the provision or the lack of antifungal prophylaxis.[22,23] It is conceivable that the occurrence of IFI in the context of allogeneic BMT/HSCT is currently underestimated, because in a proportion of patients who apparently die from infectious complications, the pathogen is not identified.[24] Recent observations suggest that the numbers of patients affected by IFI have been increasing because many new therapies induce long-lasting immunosuppression, and improved medical care permits longer survival of critically ill patients.[25]

The diagnosis of IFI, however, can be challenging. Clinical manifestations are mostly nonspecific and frequently overlap with other possible diagnoses, both infectious and noninfectious.[26] The diagnostics of IFI was traditionally based on microbiological, histopathological, and radiological findings. These techniques, however, are often not sensitive or fast enough to permit rational and timely initiation of treatment. In the clinical setting, the onset of antimycotic treatment was and often still is based merely on indirect evidence for systemic fungal infection, i.e., clinical signs of sepsis unresponsive to different antibacterial agents for greater than 72–96h. The observation that early initiation of antifungal treatment greatly improves the outcome in neutropenic patients with IFI has provided the basis for empirical rather than evidence-based treatment of patients with febrile neutropenia.[16,27]

Hence, empirical antimycotic treatment is currently the standard of care in neutropenic patients with recurrent or persisting fever despite broad-spectrum antibacterial treatment.[28] This strategy, however, may lead to significant and possibly deleterious delay of antimycotic therapy in patients with fungal sepsis and, conversely, to a substantial rate of unnecessary treatment in febrile episodes caused by nonfungal pathogens. Another common approach to controlling fungal infections in high-risk clinical situations is prophylactic administration of antifungal agents. Although routine clinical employment of this strategy has apparently been successful in decreasing the incidence of life-threatening fungal disease in the high-risk patient population,[29] a number of problems have emerged. These include not only the considerable costs and the toxic side effects but also changes in the spectrum of fungal pathogens observed in immunosuppressed patients, which are, at least in part, attributable to the widespread administration of broad-range antifungal prophylaxis. In order to overcome the disadvantages inherent in prophylactic and empirical antifungal therapy, which include a high rate of overtreatment and hence unnecessary toxicity on one hand, and the risk of delayed onset of treatment on the other hand, approaches permitting rational, evidence-based antimycotic treatment are urgently needed. Such concepts already have been successfully implemented in patients undergoing allogeneic BMT/HSCT who carry a high risk of severe viral infections. Routine clinical application of highly sensitive and quantitative molecular virus detection assays has provided a basis for efficient treatment and successful prevention of life-threatening invasive viral infections in heavily immunocompromised patients, prior to the onset of clinical symptoms of viral disease.[30,31] The establishment of similar strategies for preemptive antifungal treatment is a major challenge toward the goal of improving the outcome of IFI. To address this task, the development of novel diagnostic tools for rapid and sensitive detection of fungal pathogens is required.

99.1.2 Standard Tests for Fungus Detection

Microbiological cultivation of various clinical specimens including also peripheral blood (PB) has been used routinely for the detection of clinically important fungal pathogens. However, the clinical utility of blood cultures is questionable owing to its limited sensitivity reported to be in the range of 50%.[32,33] In a significant proportion of candidiases, blood cultures were described to fail,[32,34] and the detection of aspergillosis from blood cultures is rarely informative because positive results cannot be readily associated with invasive aspergillosis.[35] Moreover, tests based on the analysis of cultures may take too long to be clinically useful.

Widely used diagnostic tools include serological tests detecting fungal cell wall components. Galactomannan or (1,3)-β-D-glucan polysaccharides constitute the most prominent target molecules. The galactomannan sandwich ELISA (Platelia *Aspergillus*; Bio-Rad, the United States), which is the most commonly used test system, permits the detection of *Aspergillus* (and reportedly also *Penicillium*[36] antigens) but fails to detect other fungal species not containing this polysaccharide in their cell walls. A meta-analysis of the Platelia test revealed a sensitivity of 79% and a specificity of 86% for invasive aspergillosis[37,38] in immunosupressed patients, and the assay was described by other authors to offer good specificity but variable sensitivity.[39,40] This was mainly attributable to the cutoff value initially implemented for the definition of positive results. It is determined by the ratio between the specimen tested and a low-positive control supplied with the kit, and the cutoff value originally recommended by the manufacturer was 1.5.[39,41] After a revision by the U.S. Food and Drug Administration (FDA), the recommended cutoff value was lowered to 0.5 and this change has reportedly resulted in an improved clinical utility of the assay, particularly in adult patients with hematological malignancies.[41,42] However, the European Organization for Research and Treatment of Cancer/Invasive Fungal Infections Cooperative Group (EORTC) and Mycosis Study Group of the National Institute of Allergy and Infectious Diseases (MSG) suggested to follow the recommendations by the manufacturer,[41] and this issue is currently a matter of discussion. The occurrence of false-positive results of galactomannan-based detection assays is a well-documented phenomenon occurring particularly in patients who receive antibiotic treatment with semisynthetic penicillins, e.g., piperacillin–tazobactam or amoxicillin–clavulanate,[43,44] patients with intravenous administration of PLASMA-LYTE® electrolyte solutions (Baxter, Deerfield, Illinois),[37,45] and very commonly in children, apparently in relation to the diet.[46] Other materials and nutrients are described to contain galactomannan, e.g., cotton, cardboard,[47] or soybean,[48] cross-reacting with the Platelia *Aspergillus* assay.

In contrast to galactomannan, the polysaccharide (1,3)-β-D-glucan is present in the cell wall of most fungi, with the exception of *Cryptococcus* spp. and *Zygomycetes*, thereby permitting broader detection of fungal pathogens.[36] The reported sensitivity and specificity of (1,3)-β-D-glucan-based assays for the diagnosis of aspergillosis were in a range between 50%–100% and 44%–98%, respectively,[49–53] and for candidemia between 57%–100% and 44%–92%.[50,51,53–55] False-positive results of (1,3)-β-D-glucan-based detection assays (e.g., Fungitell®; Associates of Cape Cod, the United States) were described in patients treated with immunoglobulin or albumin preparations filtered through cellulose filters, which are known to contain (1,3)-β-D-glucan,[36,56] and in patients suffering from bacterial infections, because some bacteria, e.g., certain streptococci, are known to produce glucan or glucan-like polymers.[57] Tests relying on (1,3)-β-D-glucan detection appear to be more sensitive than galactomannan detection assays in patients with invasive aspergillosis,[53,56] but none of the presently available serological tests provides information on the fungal species present. The intrinsic lack of mycological specificity requires the integration of clinical, radiological, and microbiological data for proper interpretation.[56] Despite the apparent shortcomings, the overall assessment of serological testing by the consensus group of the EORTC/MSG has led to the inclusion of seropositivity in the clinical diagnostic criteria for IFI.[34,41]

Imaging methods play an important role in the diagnosis of IFI, but radiological findings are often nonspecific and may not be conclusive until late in the course of disease.[39] CT scanning can detect invasive mycoses, such as aspergillosis, at an earlier stage of infection. However, the typical signs were described to be less sensitive in nonneutropenic, corticosteroid-treated patients than in patients displaying neutropenia. This observation possibly reflects the different inflammatory response with decreased fungal load and increased inflammatory cellular trafficking in the nonneutropenic lung.[39] Histopathological examination of targeted biopsies of suspicious lesions identified by imaging methods is an important parameter for the diagnosis of proven fungal infection. However, the presence of profound thrombocytopenia in immunocompromised patients often prevents the performance of biopsies owing to the fear of bleeding complications.

99.1.3 Current Approaches to Molecular Fungus Detection

In the early 1990s, molecular detection of fungal infections by polymerase chain reaction (PCR) was introduced and it provided a new option for rapid and sensitive testing. Since then, different molecular platforms were developed, permitting qualitative and/or quantitative detection of fungal pathogens and the identification of fungal species. Standard PCR[58–60] and PCR–ELISA techniques[61–63] were among the first molecular technologies used in the clinical setting. Subsequently, nested PCR systems[64–66] were established to improve the limit of detection. To date, different variations of real-time PCR including SYBR® Green, TaqMan®, HybProbe, and biprobe technologies are available and are used in the clinical setting.[65,67–71] Methods based on SYBR Green-mediated detection of amplified target products display lower intrinsic specificity, because the marker molecule is incorporated into any minor groove of dsDNA, and hence, is more prone to revealing nonspecific signals. Approaches based on the implementation of detection probes in the PCR reaction are inherently more specific, because the specificity of target detection is conferred not only by the amplification primers but also by the hybridization of one or two sequence-specific probes, depending on the format used.

For the identification of fungal species, a plethora of molecular techniques are available including, e.g., multiplex PCR followed by DNA microarray,[72,73] nucleic acid sequence-based amplification (NASBA),[74] fragment-size analysis of variable regions in the fungal genome,[75] DNA sequencing,[60,76] hybridization of PCR amplicons to species-specific capture probes bound to microbeads (Luminex technology, Luminex Corporation),[77–79] or pyrosequencing.[80,81]

Most of the published molecular approaches to the detection of fungal infections focused on important and frequently encountered fungal pathogens. Since the spectrum of pathogens causing IFIs has been changing, as documented by a number of studies,[19,25,82,83] there is an increasing demand for broad-spectrum detection methods covering not only the most commonly occurring fungi but also non-albicans

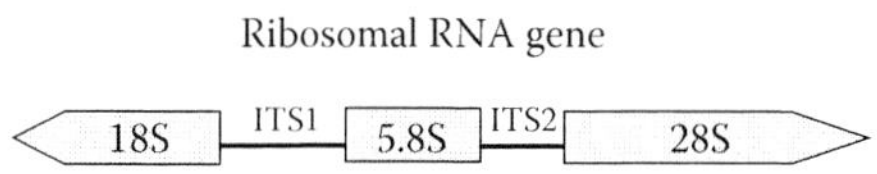

FIGURE 99.1 Schematic representation of the fungal rRNA gene including the highly conserved regions (18S, 5.8S and 28S) and the variable internal transcribed spacer regions (ITS1 and ITS2).

Candida species, *Zygomycetes*, *Fusarium* species, and other emerging fungi. An important prerequisite for reliable and broad-range detection systems is the choice of the appropriate target gene displaying sufficient sequence homology among the fungal species to be covered. Genes exploited for molecular targeting include mitochondrial genes,[84,85] the alkaline protease gene[58] and, most widely, the ribosomal RNA (rRNA) gene (Figure 99.1).[65,67,68,70,71,86,87] The 5.8S rRNA, 18S rRNA, and 28S rRNA regions are highly homologous among most fungal species and are thus regarded as excellent target regions. In addition, rRNA genes are present in multiple copy numbers (≥100)[60] in the fungal genome, thereby permitting highly sensitive molecular detection. The majority of published methods for fungus detection exploiting the multicopy rRNA genes are based on targeting of the 18S rRNA gene.[67,86,87] We and others favor the 28S rRNA gene as target,[65,68,70,71] because it displays a higher level of homology across the fungal genera and species of interest. Fungus detection assays targeting the rRNA genes permit very sensitive diagnosis of a broad range of fungal pathogens and facilitate rapid distinction from other microorganisms causing invasive infections, but they do not permit the identification of specific fungal species. In contrast to the highly homologous regions within the 5.8S rRNA, 18S rRNA, and 28S rRNA genes, the fungal genomes contain the internal transcribed spacer (ITS) 1/2 regions interspersed between the conserved rRNA segments, which display sequences that are highly heterogeneous among fungal species. Molecular analysis of these regions therefore provides a basis for the identification of individual fungal pathogens.[60,72,73,75,76,79]

99.1.3.1 PCR-Based Panfungal Detection

Several groups have developed PCR assays covering varying ranges of fungal specimens.[58–61,63–70,88,89] In the literature, a number of assays permitting the detection of a rather limited number of fungal pathogens have been referred to as panfungal.[90,91] It may conceivably be unrealistic to cover all existing fungal species by a single assay, thus rendering the term panfungal misleading, at least in most instances. Hence, if the term is to be used at all, it should be limited to the designation of assays permitting the detection of a broad range of fungal genera and species with proven or presumed clinical relevance, without apparent restrictions. Most PCR-based broad-range fungus detection assays exploit the homologous regions within the rRNA genes, as outlined in Section 99.1.3. A selection of such assays is shown in Table 99.1. The PCR-based detection assay established in our center, which may deserve the designation panfungal, targets the 28S rRNA[71] (Figure 99.2) and facilitates the detection of ≥80 fungal

TABLE 99.1
Spectrum of Fungal Species Covered by Different Molecular Assays Published as Broad-Range Tests for Fungus Detection. (The Fungal Species Detected are Listed Left from the Respective First Author of the Cited Papers.)

Acremonium spp.	Landlinger et al.[71]	Candida spp.	Leaw et al.[76]	Candida spp.	Einsele et al.[86]
Alternaria spp.		Debaryomyces spp.		Aspergillus spp.	
Aspergillus spp.		Kloeckera spp.		Trichosporon spp.	
Cladosporium spp.		Kluyveromyces spp.		Malassezia spp.	
Histoplasma spp.		Lodderomyces spp.		Fusarium spp.	
Absidia spp.		Saccharomyces spp.		Mucor spp.	
Blastoschizomyces spp.		Zygosaccharomyces spp.		Histoplasma spp.	
Candida spp.		Cryptococcus spp.		Paecilomyces spp.	
Cryptococcus spp.		Rhodotorula spp.		Saccharomyces spp.	
Cunninghamella spp.		Sporobolomyces spp.		Penicillium spp.	
Geotrichum spp.		Trichosporon spp.		Pseudallescheria spp.	
Malassezia spp.		Endomyces spp.			
Mucor spp.		Hansenula spp.			
Rhizopus spp.		Dekkera spp.			
Trichosporon spp.		Arthroascus spp.			
Saccharomyces spp.		Pichia spp.			
Fusarium spp.					
Penicillium spp.					
Scedosporium spp.					
Bipolaris spp.					
Apophysomyces spp.					
Basidiobolus spp.					
Cokeromyces spp.					
Rhizomucor spp.					
Rhodotorula spp.					
Saksenaea spp.					
Aspergillus spp.	Vollmer et al.[70]	Aspergillus spp.	Spiess et al.[73]	Candida spp.	Lau et al.[60]
Candida spp.		Candida spp.		Cryptococcus spp.	
Cryptococcus spp.		Mucor spp.		Fusarium spp.	
Mucor spp.		Fusarium spp.		Scedosporium spp.	
Penicillium spp.		Scedosporium spp.			
Pichia spp.		Rhizopus spp.			
Microsporum spp.		Trichosporon spp.			
Trichophyton spp.					
Scopulariopsis spp.					
Schizosaccharomyces spp.					
Saccharomyces spp.					
Dipodascus spp.					
Aureobasidium spp.					

pathogens. The detection spectrum of the assay was determined experimentally for most species, and the analysis of sequence homologies in databases using BLAST search indicated that the actual detection spectrum may be considerably broader. The assay is divided into two separate reactions, one of which (reaction I) covers clinically important moulds, also including emerging pathogens such as *Fusarium* species and the other (reaction II) primarily yeasts and *Zygomycetes*, including a number of species that have gained importance over the past few years. Despite the high degree of homology of the 28S rRNA target gene across fungal genera and species, it was not possible to develop a single-reaction assay covering all fungal pathogens of interest with adequate efficiency and lacking relevant cross-reactivity with nonfungal eukaryotic genomes, including particularly human sequences. This experience illustrates the complexity of the task and the apparent technical limitations of highly sensitive and specific PCR approaches. However, with regard to the fact that the great majority of IFI are still caused by *Aspergillus* and *Candida* species, the design of the above two reaction assay, permitting rapid distinction between mould and yeast infections, may be helpful in the selection of appropriate first-line antifungal treatment. In the presence of fungal species resistant to standard antimycotic agents,

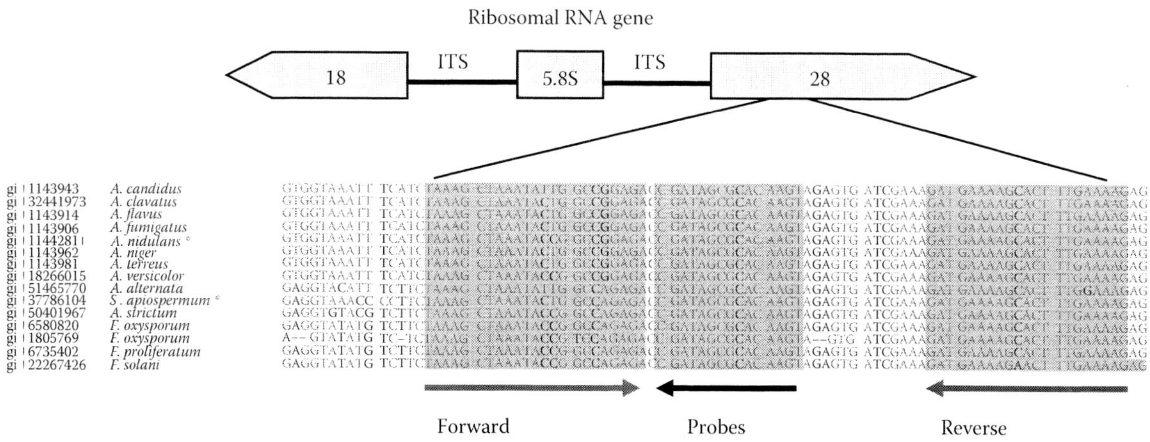

FIGURE 99.2 Representation of the segment from the 28S region of the rRNA gene targeted by our panfungal RQ-PCR assay. The primer and probe positions within sequences highly conserved among different fungal species are indicated.

rapid species identification is required in order to provide a basis for adequate therapy.

99.1.3.2 Current Platforms for the Identification of Fungal Species

The incidence of IFI caused by fungal species resistant to commonly used antifungal agents has been increasing.[25,83,92] Well-documented examples include *Aspergillus terreus* that is known to be resistant to amphotericin B or *Candida krusei* and *Candida glabrata*, which can display intrinsic resistance to fluconazol. Moreover, the rising occurrence of IFI caused by *Zygomycetes* and other hitherto uncommon fungal pathogens, which may be related to the use of antifungal prophylaxis with broad-spectrum antifungal drugs, renders the choice of the most appropriate treatment a challenging task. These observations emphasize the need for the use of clinically applicable tests for rapid and reliable identification of fungal species.

Direct sequencing of a highly variable portion of the fungal genome, such as the ITS2 region, is a possible approach to unambiguous typing of fungal pathogens. However, sequence analysis of clinical isolates from patients with IFI is hampered by the occurrence of mixed fungal infections[75] that may preclude the interpretation of sequencing data. A subcloning step may therefore be required prior to sequencing, which in turn would render this approach too laborious and time consuming to be applicable in routine clinical diagnostics. Identification of fungal species by hybridization to specific probes has been pursued as a diagnostic strategy in patients with IFI for many years. More than a decade ago, Sandhu et al.[93] introduced an oligonucleotide-based hybridization assay for the detection of a variety of fungal pathogens. Thereafter, a number of hybridization-based platforms for the detection and identification of fungal pathogens have been reported, as outlined above.[72,73,79,86] Most of these assays rely on PCR-based target amplification followed by hybridization of the amplicons to probes specific for different fungal genera or species.

An example of a recently published approach to the detection and identification of fungal species is an assay reported by Schabereiter-Gurtner et al.[89] This assay is based on PCR-directed amplification of the highly variable ITS2 region combined with subsequent hybridization to species-specific biprobes (Roche Diagnostics, Basel, Switzerland) facilitating the detection and differentiation between 11 clinically important *Aspergillus* and *Candida* species. Biprobes are sequence-specific probes labeled with Cy5 fluorophore used in combination with SYBR Green. On hybridization of the probes to their specific target sequence, the incorporated SYBR Green excites Cy5 emission by fluorescence resonance energy transfer. Discrimination of different fungal species is facilitated by subsequent melting curve analysis. The PCR amplification is performed by implementing two forward primers specific for various *Hyphomycete* and yeast species, which anneal within the ITS2 region, and a single reverse primer targeting the 28S rRNA gene. After the amplification step, 11 different species-specific biprobes, which are included in the reaction, facilitate the identification of 11 species on the basis of their specific melting temperature profiles on analysis of the melting curve. Owing to the fact that the method relies on specific hybridization of oligonucleotide probes, sequence similarities between fungal species and the occurrence of mutations within the target regions are potential impediments to reliable species identification. This limitation is inherent in all detection assays permitting discrimination between fungal pathogens on the basis of a single oligonucleotide probe for each species to be identified.

The approach to the identification of fungal pathogens developed at our center is based on a microsphere bead hybridization array.[79] The microbeads (Luminex Corporation, Austin, Texas) are prelabeled by different shades of red fluorescent color, resulting in 100 distinct bead classes readily discernable by a fluorescence reader (Figure 99.3a). Individual bead classes are functionalized with covalently bound oligonucleotide capture probes, each specific for a defined DNA sequence derived from highly variable regions of the fungal genome, including primarily the ITS sequences (Figure 99.3b). To account for the possible occurrence of mutations in the targeted fungal gene sequences, which may prevent reliable detection, we have designed a redundant

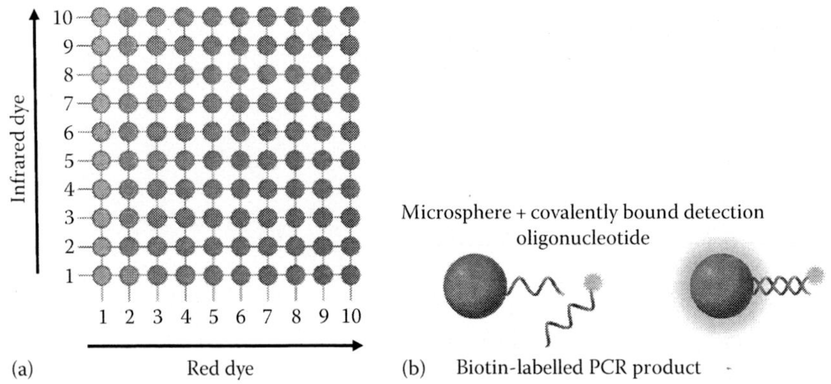

FIGURE 99.3 (a) The Luminex beads (microshperes) are uniquely identifiable by different shades of red/infrared dyes (represented here by different shades of gray) rendering 100 distinct classes of beads. (b) Each class of beads can be functionalized by specific covelently attached oligonucleotide capture probes. In the presence of homologous biotin-labelled PCR products, specific hybridization occurs leading to the presence of dual signals: the (infra)red signal identifying a specific bead class (and thus a specific capture probe) and the green biotin/streptavidin signal reflecting the presence of bound target.

panel of sequence-specific capture oligonucleotides. Each fungal genus or species of interest is covered by up to three independent probes complementary to different regions of the target gene, thus facilitating reliable identification of the fungal pathogen present even in the presence of mutations in one of the targeted sequences. Essential steps of the assay include isolation of DNA from clinical specimens followed by PCR-directed amplification targeting a highly variable region in the fungal genome located in the ITS2 region. To account for the fact that the fungal load in some clinical specimens, particularly in PB, may be very low,[67,94] a semi-nested (two-step) PCR amplification is routinely performed. The use of biotin-labeled primers in the second amplification step facilitates subsequent detection of the amplicons. The amplified fungal sequences are incubated with an array of microbead-bound capture probes specific for a variety of fungal genera and species, and the hybridized PCR products carrying a biotin label are subsequently detected with a streptavidin–phycoerythrin reporter using the Luminex® detection system (Figure 99.4). Currently, the assay includes a panel of 75 probes permitting the identification of 10 fungal genera and 29 fungal species commonly observed in IFI,[19,25,41,82,83] but the detection spectrum can be readily extended and adapted to the individual local requirements. Despite the recent advances in the diagnostic approaches to the identification of fungal pathogens, there is still great potential for improvement, in order to optimally serve the clinical needs.

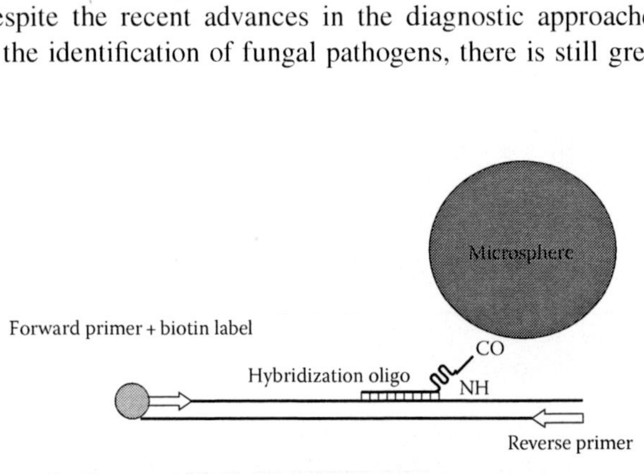

FIGURE 99.4 Principle of the Luminex assay.

99.1.3.3 Panfungal Assays and the Technical Challenges

Screening for a broad range of fungal pathogens is associated with a variety of potential pitfalls. One of the most important impediments to specific detection of fungal sequences is the high degree of homology between the fungal rRNA genes, which are targeted by the great majority of current molecular detection assays, and the rRNA genes within other eukaryotic genomes.[95] This similarity confers a high risk of cross-reactivity with human DNA, which may be a serious limitation in the clinical application of molecular screening for fungal pathogens. In PCR-based detection assays, implementation of short primers and probes of 13–18 nucleotides in length significantly reduces the areas of potential homology and facilitates targeting of unique sequences within individual fungal species. High specificity of detection despite short length can be achieved by using MGB® (minor groove binding) probes (Applied Biosystems) or LNA-(locked nucleic acid) modified nucleotides in primers and probes. MGB probes form extremely stable duplexes with their single-stranded DNA targets and allow the discrimination of target regions differing only by a few bases.[96] Similarly, LNA modifications of oligonucleotides mediate very specific, high-affinity binding to perfectly complementary target sequences,[97] which is sensitive even to single-base mismatches. The use of such modified primer and probe sequences in PCR assays has been shown to permit detection of a very broad range of fungal pathogens without relevant cross-reactivity with other eukaryotic or prokaryotic organisms.[71]

99.1.3.4 Problem of Contamination and the Requirement of Fungus-Grade Reagents

Owing to the ubiquitous presence of fungal organisms, particularly of moulds, in the environment, airborne spores are an important source of contamination, which may lead to false-positive results of fungus detection assays. Contamination may occur during the collection of patient material or during processing of the clinical specimens in the laboratory. Nucleic acid-based detection assays that include amplification of fungal DNA are particularly vulnerable, because even minute traces of contaminating fungal material can lead to false-positive test results. It is essential therefore to perform all processing steps in carefully maintained laminar air flow units. This measure is effective in preventing the occurrence of contamination by airborne spores, but cannot control for the presence of preexisting fungal contaminants. Sterile filtration of all liquid reagents through appropriate filters can efficiently remove fungal spores, if present, but we and others have observed that traces of fungal DNA may occur in a number of commercially available reagents and materials. It is highly advisable therefore to use batches of consumables that were pretested in the laboratory and confirmed to be free of contamination by fungal DNA. The inclusion of multiple negative controls in every assay is imperative to facilitate monitoring of all steps of sample processing.[68] Moreover, the safety and reliability of molecular fungus detection assays could be greatly improved by the introduction of commercially available "fungus-grade" materials and reagents, guaranteed to be completely free of any fungal contaminants.[98]

99.1.3.5 Interpretation of Screening by Broad-Spectrum PCR Assays

PCR-based assays designed to facilitate detection of a wide range of fungal pathogens generally rely on the amplification of gene sequences displaying a high degree of homology among a variety of genera and species. Targeting of highly conserved regions in the fungal genome may also enable the detection of environmental fungal species with supposedly low pathogenicity, such as black moulds including *Alternaria*, *Phialemonium*, *Exophiala*, and *Cladophialophora* spp., whose presence is believed to be of questionable clinical relevance. It appears, however, that virtually any fungal species, including black moulds, may become pathogenic in severely immunocompromised patients.[99] This observation also applies to infections caused by *Penicillium* spp. whose role in high-risk patients has potentially been underestimated.[100] Owing to the variable pathogenicity of different fungal species, positive results of broad-spectrum PCR-mediated fungus detection assays may be difficult to interpret with regard to the imminent risk of invasive fungal disease. It appears reasonable therefore to combine broad-spectrum screening assays with tests permitting identification of the fungal species present to provide a rational basis for clinical management of the infection. Moreover, quantitative assessment of fungal load, which can be readily performed by real-time PCR assays, may be instrumental in differentiating between colonization and impending or ongoing invasive infection. This applies to fungus detection in inherently nonsterile tissues, such as the respiratory tract.[36] PCR analyses of PB were described to provide very high sensitivity and, consequently, very high negative predictive values (NPV) for IFI.[94,101] A negative PCR result in PB therefore correlates with a low risk of IFI and may therefore be regarded as sufficient diagnostic evidence for the absence of imminent fungus-related disease.[75,94,101] It is necessary to bear in mind, however, that the fungal load in PB may be very low in the presence of IFI, and could therefore, at least occasionally, escape detection despite the use of highly sensitive assays. Hence, we suggest that repeated testing at short intervals would increase the reliability of negative test results obtained in high-risk settings, such as febrile neutropenia.

With regard to the interpretation of positive PCR tests in PB, there seems to be a broad consensus: due to the fact that PCR-based fungus detection assays are particularly prone to false-positive results emanating from contamination, a minimum of two positive results obtained in consecutive patient specimens are generally required for molecular evidence of infection. We and others have shown that repeated positivity of PCR testing provides better correlation with the widely accepted EORTC/MSG criteria for proven, probable, and possible IFI.[34,41] It is conceivable that the documentation of changes in the fungal load determined by quantitative PCR monitoring could further improve the clinical interpretation of molecular screening. Currently, molecular methods of fungus detection are not accepted by the EORTC/MSG, mainly on the basis of lacking standardization of the techniques available. Due to the absence of a clearly defined gold standard for the assessment of IFIs, every new method, including also molecular analysis, has to be evaluated against the established EORTC/MSG criteria.[34,41]

99.1.3.6 IFI in Immunocompetent Hospitalized Patients, Healthy Individuals, and Immunocompromised Patients

Although IFIs are infrequent in nonimmunocompromised hosts, a number of cases are documented in the literature. Reported fungal diseases in immunocompetent hospitalized patients include aspergillosis,[102,103] candidiasis,[104] cryptococcosis,[105,106] and also infections with rather uncommon fungal pathogens, which are rarely observed even in immunocompromised patients.[107–111] IFIs with or without fungemia are apparently rare events in healthy individuals, but may occur in specific situations, as recently reviewed by Sipsas and Kontoyiannis.[112] Professions, lifestyles, hobbies, and traveling to endemic areas expose healthy individuals to pathogenic fungi, which occasionally result even in life-threatening IFI.[112] There is evidence that inhaled fungal pathogens can persist for long time periods in macrophages, providing a reservoir for secondary infections when host immunity becomes naturally or artificially weakened.[113] One may speculate therefore that the presence of circulating fungal organisms is an early sign of impending fungal disease in patients with impaired immune response. Moreover, shedding of fungal components into PB from foci localized in individual organs or compartments may facilitate the diagnosis of fungal infection prior to the onset of clinical symptoms or evidence obtained by imaging techniques. Close monitoring of high-risk patients by sensitive molecular screening methods could therefore permit timely detection of IFI and provide a basis for appropriate treatment. Assessment of this notion requires the performance of large clinical trials, which are currently in preparation or ongoing.

99.2 METHODS

99.2.1 Sample Preparation

Different types of clinical specimens are commonly used for the detection and identification of fungal pathogens in patients at high risk for IFI, including, e.g., whole blood, plasma, serum, respiratory secretions, bronchoalveolar lavage specimens, biopsies, and paraffin-embedded tissue specimens from affected sites. Sterile sample collection and adequate preanalytical handling are important measures to minimize the risk of contamination. The selection of appropriate specimens for fungus screening is instrumental for the differentiation between fungal colonization and IFI and is therefore essential for the clinical interpretation. The reported sizes of clinical specimens used for molecular analysis of fungal pathogens are highly variable, ranging from about 200 μL to 10 mL.[101] These differences reflect the fact that the sample size adequate for clinical diagnosis is not well defined and is certainly dependent on the type of specimen investigated and the sensitivity of the methodology used. In addition to appropriate sample size, the efficiency of fungal DNA extraction is a critical parameter for the achievable limit of detection. Hence, suboptimal DNA isolation procedures can lead to false-negative results.[114] The composition of the fungal cell wall, primarily in moulds, renders efficient disruption difficult, and therefore presents a potential impediment to successful isolation of fungal DNA. A plethora of different DNA extraction procedures, both manual and automated, are available.[115] Many of them include pretreatment by enzymatic (e.g., zymolase, lyticase) and/or mechanical approaches (e.g., disruption by glass beads) facilitating fungal DNA release. However, enzymes (e.g., proteinase K that is usually isolated from the fungus *Engyodontium album*) may be contaminated by fungal components.[114] A similar problem may apply to other reagents and materials not produced under completely sterile conditions, which may carry traces of fungal material, and are therefore a potential source of (generally low level) contamination.[116,117] A recent multicenter study showed that a variety of technical approaches to fungal DNA extraction provide satisfactory results,[95] but closed formats, such as the MagNA Pure System (Roche Diagnostics, Penzberg, Germany), may be beneficial by reducing the risk of contamination via airborne spores. Moreover, performing the entire procedure of sample processing in laminar air flow units is essential in order to prevent exogenous contamination from the environment.

We have carefully evaluated a number of extraction procedures for fungal DNA at our center, and in our hands, the combination of enzymatic (lyticase) and mechanical pretreatment (glass beads) followed by automated DNA extraction using the MagNA Pure System provided the best results with regard to efficacy and protection from extraneous contamination.[68,79] It is important to emphasize, however, that the adequacy of any fungal DNA isolation and detection protocol intended for use in the clinical setting has to be assessed in appropriately designed large-scale studies.

99.2.1.1 Fungal DNA Extraction

All reagents used are filtered through 0.2 μm sterile filters (Corning®; Corning Incorporated, Germany) to prevent contamination by fungal spores. A negative control sample (biology grade water replacing the clinical sample) is included in each extraction. PB anticoagulated with EDTA is the type of specimen most commonly investigated in our laboratory. The PB samples are kept at 4°C for at least 4 h to sediment the red blood cells. The entire supernatant, i.e., plasma containing white blood cells, is centrifuged (for 10 min at 15,000 × *g*), and subsequently most of the plasma is removed. The small residual volume and the pellet are resuspended using 430 μL lyticase lysis buffer (LLB; 50 mM Tris pH 7.6, 1 mM EDTA pH 8.0, 0.2% 2-mercaptoethanol, and 1 U/100 μL recombinant lyticase; Sigma) and incubated at 37°C for 1 h. Acid-washed glass beads are added and the solution is vortexed for 3 min to facilitate mechanical cell wall disruption. Four hundred microliters of supernatant are used for DNA extraction on a MagNA Pure compact instrument using the MagNA Pure compact nucleic acid isolation kit I (Roche Diagnostics), according to the manufacturer's recommendations. Protocols for fungal DNA extraction from other clinical specimens are described by Landlinger et al.[79]

99.2.2 Detection Procedures

Genes exploited for molecular targeting of fungal DNA include primarily the rRNA gene (Figure 99.1),[65,67,68,70,71,86,87] and, less commonly, mitochondrial genes,[84,85] the alkaline protease gene,[58] and a small number of other genes.[115] The 5.8S rRNA, 18S rRNA, and 28S rRNA regions are highly homologous among most fungal species and are thus regarded as excellent target regions. In addition, rRNA genes are present in multiple copy numbers (≥100)[60] in the fungal genome, thereby permitting highly sensitive molecular detection. The majority of published methods for fungus detection exploiting the multicopy rRNA genes are based on targeting the 18S rRNA gene.[67,86,87] We and others favor the 28S rRNA gene as the target,[65,68,70,71] because it displays a higher level of homology across the fungal genera and species of interest. Fungus detection assays targeting the rRNA genes permit very sensitive diagnosis of a broad range of fungal pathogens, and facilitate rapid distinction from other microorganisms causing invasive infections, but they do not permit the identification of specific fungal species. In contrast to the highly homologous regions within the 5.8S rRNA, 18S rRNA, and 28S rRNA genes, the fungal genomes contain the ITS 1/2 regions interspersed between the conserved rRNA segments, which display sequences that are highly heterogeneous among fungal species. Molecular analysis of these regions therefore provides a basis for the identification of individual fungal pathogens.[60,72,73,75,76,79]

A broad-range molecular detection assay covering ≥80 different fungal pathogens was recently described by our group.[71] It is based on targeting the 28S rRNA by a real-time PCR approach and comprises the following steps.

99.2.2.1 Panfungal PCR (Patent Pending)

The assay is performed in two separate test tubes, including reaction I permitting the detection and quantification of a variety of moulds and reaction II covering primarily yeasts and *Zygomycetes*. The DNA isolated from clinical specimens is generally analyzed in duplicates, and a number of negative controls (molecular biology grade water replacing the DNA sample) are included in each run. The PCR protocol can be obtained from the authors upon request.

99.2.2.2 Interpretation of PCR Results

For the diagnosis of fungal DNAemia, a minimum of two PCR-positive PB specimens derived at subsequent time points during close follow-up investigation are required.

The panfungal PCR assay has been implemented in a recently performed clinical study including more than 1000 clinical specimens derived from 144 hemato-oncological patients. The observations revealed the expected predominance of mould infections (60%), and also showed a relatively high proportion of mixed infections (>30%), as indicated by the positivity of both reactions of the panfungal PCR assay.[79] The screening results obtained by this assay correlated very well with the EORTC criteria for proven, probable, and possible infections and demonstrated the potential utility of molecular testing for IFIs in clinical diagnostics.

99.3 FUTURE PERSPECTIVES

As shown by a number of studies, including our own observations, the potential value of nucleic acid-based approaches for the detection and identification of fungal pathogens in immunocompromised patients is indisputable. However, PCR-based techniques are currently not regarded as acceptable for clinical diagnostics by the EORTC/MSG, primarily owing to the lack of validation and standardization.[41] For the implementation of reliable and robust molecular methods, various technical challenges have to be addressed, as indicated above. In order to enable controlling the problem of contamination, manufacturers of consumables need to realize the requirement of "fungus-grade" reagents and materials, guaranteed to be free of any traces of fungal components. The development of novel concepts for reliable and highly sensitive identification of a broad spectrum of fungal pathogens, including also the implementation of nanotechnological approaches, is currently underway.

It is reasonable to expect that the optimization of current techniques and ongoing studies on the implementation of molecular screening methods in patient populations at high risk of IFI will satisfy the requirements of the EORTC/MSG in the near future. Collaboration and exchange of experience between diagnostic centers would greatly expedite international standardization of molecular fungus detection assays and could provide a basis for the harmonization of diagnostic approaches facilitating more efficient generation of data by multicenter studies. The exploitation of molecular methods for rapid diagnosis of fungal infections could pave the way to timely initiation of antifungal treatment and selection of the most appropriate antimycotic agents, thereby contributing to improved clinical management and outcome of invasive fungal disease.

REFERENCES

1. Bow, E.J. et al., Invasive fungal disease in adults undergoing remission–induction therapy for acute myeloid leukemia: The pathogenetic role of the antileukemic regimen, *Clin Infect Dis*, 21, 361, 1995.
2. Burch, P.A. et al., Favorable outcome of invasive aspergillosis in patients with acute leukemia, *J Clin Oncol*, 5, 1985, 1987.
3. Morrison, V.A., Haake, R.J. and Weisdorf, D.J., Non-*Candida* fungal infections after bone marrow transplantation: Risk factors and outcome, *Am J Med*, 96, 497, 1994.
4. Baddley, J.W. et al., Invasive mold infections in allogeneic bone marrow transplant recipients, *Clin Infect Dis*, 32, 1319, 2001.
5. Mikulska, M. et al., Risk factors for invasive aspergillosis and related mortality in recipients of allogeneic SCT from alternative donors: An analysis of 306 patients, *Bone Marrow Transplant*, 44, 361, 2009.
6. Marr, K.A., Fungal infections in hematopoietic stem cell transplant recipients, *Med Mycol*, 46, 293, 2008.
7. Singh, N. and Paterson, D.L., *Aspergillus* infections in transplant recipients, *Clin Microbiol Rev*, 18, 44, 2005.
8. Gabardi, S. et al., Invasive fungal infections and antifungal therapies in solid organ transplant recipients, *Transpl Int*, 20, 993, 2007.
9. Selik, R.M., Karon, J.M. and Ward, J.W., Effect of the human immunodeficiency virus epidemic on mortality from opportunistic infections in the United States in 1993, *J Infect Dis*, 176, 632, 1997.
10. Kujath, P. et al., Invasive aspergillosis on a surgical intensive care unit, *Mycoses*, 43(Suppl 2), 23, 2000.
11. Lionakis, M.S. and Kontoyiannis, D.P., Glucocorticoids and invasive fungal infections, *Lancet*, 362, 1828, 2003.
12. Erjavec, Z., Kluin-Nelemans, H. and Verweij, P.E., Trends in invasive fungal infections, with emphasis on invasive aspergillosis, *Clin Microbiol Infect*, 15, 625, 2009.
13. Chapman, R.L. and Faix, R.G., Persistently positive cultures and outcome in invasive neonatal candidiasis, *Pediatr Infect Dis J*, 19, 822, 2000.
14. Hovi, L. et al., Invasive fungal infections in pediatric bone marrow transplant recipients: Single center experience of 10 years, *Bone Marrow Transplant*, 26, 999, 2000.
15. Meunier, F., Aoun, M. and Bitar, N., Candidemia in immunocompromised patients, *Clin Infect Dis*, 14(Suppl 1), S120, 1992.
16. Nucci, M. et al., Risk factors and attributable mortality associated with superinfections in neutropenic patients with cancer, *Clin Infect Dis*, 24, 575, 1997.
17. Rolston, K.V., New trends in patient management: Risk-based therapy for febrile patients with neutropenia, *Clin Infect Dis*, 29, 515, 1999.
18. Rubenstein, E.B. et al., Outpatient treatment of febrile episodes in low-risk neutropenic patients with cancer, *Cancer*, 71, 3640, 1993.

19. Cornely, O.A., *Aspergillus* to *Zygomycetes*: Causes, risk factors, prevention, and treatment of invasive fungal infections, *Infection*, 36, 296, 2008.
20. Bodey, G. et al., Fungal infections in cancer patients: An international autopsy survey, *Eur J Clin Microbiol Infect Dis*, 11, 99, 1992.
21. Castagnola, E. et al., Invasive mycoses in children receiving hemopoietic SCT, *Bone Marrow Transplant*, 41(Suppl 2), S107, 2008.
22. Martino, R. et al., Invasive fungal infections after allogeneic peripheral blood stem cell transplantation: Incidence and risk factors in 395 patients, *Br J Haematol*, 116, 475, 2002.
23. Jantunen, E. et al., Incidence and risk factors for invasive fungal infections in allogeneic BMT recipients, *Bone Marrow Transplant*, 19, 801, 1997.
24. Pagano, L. et al., Fungal infections in recipients of hematopoietic stem cell transplants: Results of the SEIFEM B-2004 study—Sorveglianza Epidemiologica Infezioni Fungine Nelle Emopatie Maligne, *Clin Infect Dis*, 45, 1161, 2007.
25. Richardson, M. and Lass-Florl, C., Changing epidemiology of systemic fungal infections, *Clin Microbiol Infect*, 14(Suppl 4), 5, 2008.
26. Tamura, K. and Drew, R., Antifungal prophylaxis in adult hematopoietic stem cell transplant recipients, *Drugs Today (Barc)*, 44, 515, 2008.
27. Denning, D.W., Invasive aspergillosis, *Clin Infect Dis*, 26, 781, 1998.
28. Cordonnier, C. et al., Empirical versus preemptive antifungal therapy for high-risk, febrile, neutropenic patients: A randomized, controlled trial, *Clin Infect Dis*, 48, 1042, 2009.
29. Murali, S. and Langston, A., Advances in antifungal prophylaxis and empiric therapy in patients with hematologic malignancies, *Transpl Infect Dis*, 11, 480, 2009.
30. Lion, T. et al., Molecular monitoring of adenovirus in peripheral blood after allogeneic bone marrow transplantation permits early diagnosis of disseminated disease, *Blood*, 102, 1114, 2003.
31. Watzinger, F. et al., Real-time quantitative PCR assays for detection and monitoring of pathogenic human viruses in immunosuppressed pediatric patients, *J Clin Microbiol*, 42, 5189, 2004.
32. Ellepola, A.N. and Morrison, C.J., Laboratory diagnosis of invasive candidiasis, *J Microbiol*, 43 Spec No, 65, 2005.
33. Pasqualotto, A.C. and Denning, D.W., Diagnosis of invasive fungal infections—Current limitations of classical and new diagnostic methods, *Eur Oncol Rev*, 2005.
34. Ascioglu, S. et al., Defining opportunistic invasive fungal infections in immunocompromised patients with cancer and hematopoietic stem cell transplants: An international consensus, *Clin Infect Dis*, 34, 7, 2002.
35. Simoneau, E. et al., What is the clinical significance of positive blood cultures with *Aspergillus* sp in hematopoietic stem cell transplant recipients? A 23 year experience, *Bone Marrow Transplant*, 35, 303, 2005.
36. Hope, W.W., Walsh, T.J. and Denning, D.W., Laboratory diagnosis of invasive aspergillosis, *Lancet Infect Dis*, 5, 609, 2005.
37. Wheat, L.J. and Walsh, T.J., Diagnosis of invasive aspergillosis by galactomannan antigenemia detection using an enzyme immunoassay, *Eur J Clin Microbiol Infect Dis*, 27, 245, 2008.
38. Pfeiffer, C.D., Fine, J.P. and Safdar, N., Diagnosis of invasive aspergillosis using a galactomannan assay: A meta-analysis, *Clin Infect Dis*, 42, 1417, 2006.
39. Barnes, R.A., Early diagnosis of fungal infection in immunocompromised patients, *J Antimicrob Chemother*, 61(Suppl 1), i3, 2008.
40. Leeflang, M.M. et al., Galactomannan detection for invasive aspergillosis in immunocompromized patients, *Cochrane Database Syst Rev*, CD007394, 2008.
41. De Pauw, B. et al., Revised definitions of invasive fungal disease from the European Organization for Research and Treatment of Cancer/Invasive Fungal Infections Cooperative Group and the National Institute of Allergy and Infectious Diseases Mycoses Study Group (EORTC/MSG) Consensus Group, *Clin Infect Dis*, 46, 1813, 2008.
42. Maertens, J.A. et al., Optimization of the cutoff value for the Aspergillus double-sandwich enzyme immunoassay, *Clin Infect Dis*, 44, 1329, 2007.
43. Adam, O. et al., Treatment with piperacillin–tazobactam and false-positive *Aspergillus* galactomannan antigen test results for patients with hematological malignancies, *Clin Infect Dis*, 38, 917, 2004.
44. Zandijk, E. et al., False-positive results by the platelia *Aspergillus* galactomannan antigen test for patients treated with amoxicillin–clavulanate, *Clin Vaccine Immunol*, 15, 1132, 2008.
45. Racil, Z. et al., Intravenous PLASMA-LYTE as a major cause of false-positive results of platelia *Aspergillus* test for galactomannan detection in serum, *J Clin Microbiol*, 45, 3141, 2007.
46. Steinbach, W.J. et al., Prospective *Aspergillus* galactomannan antigen testing in pediatric hematopoietic stem cell transplant recipients, *Pediatr Infect Dis J*, 26, 558, 2007.
47. Girmenia, C., Ballaro, D. and Martino, P., Contamination by cardboard particles as a cause of false-positive results for *Aspergillus* galactomannan enzyme immunoassays, *Clin Vaccine Immunol*, 14, 929, 2007.
48. Murashige, N. et al., False-positive results of *Aspergillus* enzyme-linked immunosorbent assays for a patient with gastrointestinal graft-versus-host disease taking a nutrient containing soybean protein, *Clin Infect Dis*, 40, 333, 2005.
49. Kawazu, M. et al., Prospective comparison of the diagnostic potential of real-time PCR, double-sandwich enzyme-linked immunosorbent assay for galactomannan, and a (1 → 3)-beta-D-glucan test in weekly screening for invasive aspergillosis in patients with hematological disorders, *J Clin Microbiol*, 42, 2733, 2004.
50. Odabasi, Z. et al., Beta-D-glucan as a diagnostic adjunct for invasive fungal infections: Validation, cutoff development, and performance in patients with acute myelogenous leukemia and myelodysplastic syndrome, *Clin Infect Dis*, 39, 199, 2004.
51. Pazos, C., Ponton, J. and Del Palacio, A., Contribution of (1 → 3)-beta-D-glucan chromogenic assay to diagnosis and therapeutic monitoring of invasive aspergillosis in neutropenic adult patients: A comparison with serial screening for circulating galactomannan, *J Clin Microbiol*, 43, 299, 2005.
52. Ostrosky-Zeichner, L. et al., International, open-label, noncomparative, clinical trial of micafungin alone and in combination for treatment of newly diagnosed and refractory candidemia, *Eur J Clin Microbiol Infect Dis*, 24, 654, 2005.
53. Persat, F. et al., Contribution of the (1 → 3)-beta-D-glucan assay for diagnosis of invasive fungal infections, *J Clin Microbiol*, 46, 1009, 2008.
54. Akamatsu, N. et al., Preemptive treatment of fungal infection based on plasma (1 → 3) beta-D-glucan levels after liver transplantation, *Infection*, 35, 346, 2007.
55. Digby, J. et al., Serum glucan levels are not specific for presence of fungal infections in intensive care unit patients, *Clin Diagn Lab Immunol*, 10, 882, 2003.

56. Marty, F.M. and Koo, S., Role of (1 → 3)-beta-D-glucan in the diagnosis of invasive aspergillosis, *Med Mycol*, 47(Suppl 1), S233, 2009.
57. Pickering, J.W. et al., Evaluation of a (1 → 3)-beta-D-glucan assay for diagnosis of invasive fungal infections, *J Clin Microbiol*, 43, 5957, 2005.
58. Raad, I. et al., Polymerase chain reaction on blood for the diagnosis of invasive pulmonary aspergillosis in cancer patients, *Cancer*, 94, 1032, 2002.
59. El-Mahallawy, H.A. et al., Evaluation of pan-fungal PCR assay and *Aspergillus* antigen detection in the diagnosis of invasive fungal infections in high risk paediatric cancer patients, *Med Mycol*, 44, 733, 2006.
60. Lau, A. et al., Development and clinical application of a pan-fungal PCR assay to detect and identify fungal DNA in tissue specimens, *J Clin Microbiol*, 45, 380, 2007.
61. Loffler, J. et al., Detection of PCR-amplified fungal DNA by using a PCR–ELISA system, *Med Mycol*, 36, 275, 1998.
62. Lass-Florl, C. et al., Diagnosing invasive aspergillosis during antifungal therapy by PCR analysis of blood samples, *J Clin Microbiol*, 42, 4154, 2004.
63. Florent, M. et al., Prospective evaluation of a polymerase chain reaction–ELISA targeted to *Aspergillus fumigatus* and *Aspergillus flavus* for the early diagnosis of invasive aspergillosis in patients with hematological malignancies, *J Infect Dis*, 193, 741, 2006.
64. Skladny, H. et al., Specific detection of *Aspergillus* species in blood and bronchoalveolar lavage samples of immunocompromised patients by two-step PCR, *J Clin Microbiol*, 37, 3865, 1999.
65. Williamson, E.C. et al., Diagnosis of invasive aspergillosis in bone marrow transplant recipients by polymerase chain reaction, *Br J Haematol*, 108, 132, 2000.
66. Halliday, C. et al., Role of prospective screening of blood for invasive aspergillosis by polymerase chain reaction in febrile neutropenic recipients of haematopoietic stem cell transplants and patients with acute leukaemia, *Br J Haematol*, 132, 478, 2006.
67. Loeffler, J. et al., Quantification of fungal DNA by using fluorescence resonance energy transfer and the light cycler system, *J Clin Microbiol*, 38, 586, 2000.
68. Baskova, L. et al., The Pan-AC assay: A single-reaction real-time PCR test for quantitative detection of a broad range of *Aspergillus* and *Candida* species, *J Med Microbiol*, 56, 1167, 2007.
69. Klingspor, L. and Jalal, S., Molecular detection and identification of *Candida* and *Aspergillus* spp. from clinical samples using real-time PCR, *Clin Microbiol Infect*, 12, 745, 2006.
70. Vollmer, T. et al., Evaluation of novel broad-range real-time PCR assay for rapid detection of human pathogenic fungi in various clinical specimens, *J Clin Microbiol*, 46, 1919, 2008.
71. Landlinger, C. et al., Rapid Detection of Invasive Fungal Infections in Hemato-Oncological Patients, *Blood (ASH Annual Meeting Abstracts)*, 112, 1468, 2008.
72. Leinberger, D.M. et al., Development of a DNA microarray for detection and identification of fungal pathogens involved in invasive mycoses, *J Clin Microbiol*, 43, 4943, 2005.
73. Spiess, B. et al., DNA microarray-based detection and identification of fungal pathogens in clinical samples from neutropenic patients, *J Clin Microbiol*, 45, 3743, 2007.
74. Loeffler, J. et al., Development and evaluation of the nuclisens basic kit NASBA for the detection of RNA from *Candida* species frequently resistant to antifungal drugs, *Diagn Microbiol Infect Dis*, 45, 217, 2003.
75. Landlinger, C. et al., Identification of fungal species by fragment length analysis of the internally transcribed spacer 2 region, *Eur J Clin Microbiol Infect Dis*, 28, 613, 2009.
76. Leaw, S.N. et al., Identification of medically important yeast species by sequence analysis of the internal transcribed spacer regions, *J Clin Microbiol*, 44, 693, 2006.
77. Diaz, M.R. and Fell, J.W., High-throughput detection of pathogenic yeasts of the genus *Trichosporon*, *J Clin Microbiol*, 42, 3696, 2004.
78. Das, S. et al., DNA probes for the rapid identification of medically important *Candida* species using a multianalyte profiling system, *FEMS Immunol Med Microbiol*, 46, 244, 2006.
79. Landlinger, C. et al., Species-specific identification of a wide range of clinically relevant fungal pathogens by the Luminex xMAP™ technology, *J Clin Microbiol*, 47, 1063, 2009.
80. Gharizadeh, B. et al., Identification of medically important fungi by the Pyrosequencing technology, *Mycoses*, 47, 29, 2004.
81. Montero, C.I. et al., Evaluation of pyrosequencing technology for the identification of clinically relevant non-dematiaceous yeasts and related species, *Eur J Clin Microbiol Infect Dis*, 27, 821, 2008.
82. Caston-Osorio, J.J., Rivero, A. and Torre-Cisneros, J., Epidemiology of invasive fungal infection, *Int J Antimicrob Agents*, 32(Suppl 2), S103, 2008.
83. Lass-Florl, C., The changing face of epidemiology of invasive fungal disease in Europe, *Mycoses*, 52, 197, 2009.
84. Costa, C. et al., Development of two real-time quantitative TaqMan PCR assays to detect circulating *Aspergillus fumigatus* DNA in serum, *J Microbiol Methods*, 44, 263, 2001.
85. Ferns, R.B. et al., The prospective evaluation of a nested polymerase chain reaction assay for the early detection of *Aspergillus* infection in patients with leukaemia or undergoing allograft treatment, *Br J Haematol*, 119, 720, 2002.
86. Einsele, H. et al., Detection and identification of fungal pathogens in blood by using molecular probes, *J Clin Microbiol*, 35, 1353, 1997.
87. Jaeger, E.E. et al., Rapid detection and identification of *Candida*, *Aspergillus*, and *Fusarium* species in ocular samples using nested PCR, *J Clin Microbiol*, 38, 2902, 2000.
88. White, P.L. et al., The evolution and evaluation of a whole blood polymerase chain reaction assay for the detection of invasive aspergillosis in hematology patients in a routine clinical setting, *Clin Infect Dis*, 42, 479, 2006.
89. Schabereiter-Gurtner, C. et al., Development of novel real-time PCR assays for detection and differentiation of eleven medically important *Aspergillus* and *Candida* species in clinical specimens, *J Clin Microbiol*, 45, 906, 2007.
90. Dayan, L. et al., *Aspergillus* vertebral osteomyelitis in chronic leukocyte leukemia patient diagnosed by a novel panfungal polymerase chain reaction method, *Spine J*, 7, 615, 2007.
91. Badiee, P. et al., Prospective screening in liver transplant recipients by panfungal PCR–ELISA for early diagnosis of invasive fungal infections, *Liver Transpl*, 13, 1011, 2007.
92. Malani, A.N. and Kauffman, C.A., Changing epidemiology of rare mould infections: Implications for therapy, *Drugs*, 67, 1803, 2007.
93. Sandhu, G.S. et al., Molecular probes for diagnosis of fungal infections, *J Clin Microbiol*, 33, 2913, 1995.
94. Cuenca-Estrella, M. et al., Value of serial quantification of fungal DNA by a real-time PCR-based technique for early diagnosis of invasive *Aspergillosis* in patients with febrile neutropenia, *J Clin Microbiol*, 47, 379, 2009.

95. White, P.L. et al., A consensus on fungal polymerase chain reaction diagnosis?: A United Kingdom–Ireland evaluation of polymerase chain reaction methods for detection of systemic fungal infections, *J Mol Diagn*, 8, 376, 2006.
96. Kutyavin, I.V. et al., 3′-Minor groove binder-DNA probes increase sequence specificity at PCR extension temperatures, *Nucleic Acids Res*, 28, 655, 2000.
97. Simeonov, A. and Nikiforov, T.T., Single nucleotide polymorphism genotyping using short, fluorescently labeled locked nucleic acid (LNA) probes and fluorescence polarization detection, *Nucleic Acids Res*, 30, e91, 2002.
98. Preuner, S. and Lion, T., Towards molecular diagnostics of invasive fungal infections, *Expert Rev Mol Diagn*, 9, 397, 2009.
99. Silveira, F. and Nucci, M., Emergence of black moulds in fungal disease: Epidemiology and therapy, *Curr Opin Infect Dis*, 14, 679, 2001.
100. Lyratzopoulos, G. et al., Invasive infection due to *Penicillium* species other than *P. marneffei*, *J Infect*, 45, 184, 2002.
101. Mengoli, C. et al., Use of PCR for diagnosis of invasive aspergillosis: Systematic review and meta-analysis, *Lancet Infect Dis*, 9, 89, 2009.
102. Karim, M. et al., Chronic invasive aspergillosis in apparently immunocompetent hosts, *Clin Infect Dis*, 24, 723, 1997.
103. Raja, N.S. and Singh, N.N., Disseminated invasive aspergillosis in an apparently immunocompetent host, *J Microbiol Immunol Infect*, 39, 73, 2006.
104. Al-Khuwaitir, T.S. et al., Disseminated *Candida albicans* infection in an immunocompetent host, *Saudi Med J*, 26, 1003, 2005.
105. Capoor, M.R. et al., Disseminated cryptococcosis with necrotizing fasciitis in an apparently immunocompetent host: A case report, *Med Mycol*, 46, 269, 2008.
106. Izumikawa, K. et al., A case of pulmonary cryptococcosis followed by pleuritis in an apparently immunocompetent patient during fluconazole treatment, *Med Mycol*, 46, 595, 2008.
107. Vaideeswar, P. and Pandit, S.P., Primary zygomycosis of the pericardium in an immunocompetent host, *Mycoses*, 50, 232, 2007.
108. Gorman, S.R. et al., *Fusarium oxysporum* pneumonia in an immunocompetent host, *South Med J*, 99, 613, 2006.
109. Purnak, T. et al., A novel fungal pathogen under the spotlight—*Acremonium* spp. associated fungaemia in an immunocompetent host, *Mycoses*, 54, 78, 2011.
110. Abgrall, S. et al., *Scedosporium apiospermum* lung infection with fatal subsequent postoperative outcome in an immunocompetent host, *Clin Infect Dis*, 45, 524, 2007.
111. Borkar, S.A. et al., Brain abscess caused by *Cladophialophora bantiana* in an immunocompetent host: Need for a novel cost-effective antifungal agent, *Indian J Med Microbiol*, 26, 271, 2008.
112. Sipsas, N.V. and Kontoyiannis, D.P., Occupation, lifestyle, diet, and invasive fungal infections, *Infection*, 36, 515, 2008.
113. d'Enfert, C., Hidden killers: Persistence of opportunistic fungal pathogens in the human host, *Curr Opin Microbiol*, 12, 358, 2009.
114. Khot, P.D. and Fredricks, D.N., PCR-based diagnosis of human fungal infections, *Expert Rev Anti Infect Ther*, 7, 1201, 2009.
115. Lau, A. et al., Current status and future perspectives on molecular and serological methods in diagnostic mycology, *Future Microbiol*, 4, 1185, 2009.
116. Loeffler, J. et al., Contaminations occurring in fungal PCR assays, *J Clin Microbiol*, 37, 1200, 1999.
117. Van Burik, J.A. et al., Panfungal PCR assay for detection of fungal infection in human blood specimens, *J Clin Microbiol*, 36, 1169, 1998.

100 Molecular Characterization of Fungal Drug Resistance

Maurizio Sanguinetti, Brunella Posteraro, and Patrizia Posteraro

CONTENTS

100.1 INTRODUCTION

100.1.1 Emergence of Fungal Pathogens and Antifungal Resistance

Medical advances over the last 30 years have led to a significant increase in the incidence of serious life-threatening invasive fungal infections (IFIs) due to an ever-increasing array of pathogens.[1] This paradox seems to be linked to numerous factors such as the immunosuppressive therapies for cancer patients, the rise in the number of individuals undergoing organ transplantation and major surgery, and the AIDS epidemic.[2] Furthermore, patients admitted to the intensive care unit are particularly susceptible to IFI because of the severity of their underlying illness and the excess use of medical and surgical interventions.[3] *Candida* species are the fourth most common pathogen isolated from blood cultures, accounting for 8%–10% of all bloodstream infections acquired in the hospital, with crude mortality rates of 40%.[4] Although *Candida albicans* is the most frequent etiologic agent of invasive candidiasis, *Candida glabrata* has undoubtedly emerged as an important opportunistic fungal pathogen in the United States, followed by *Candida tropicalis*, *Candida parapsilosis*, *Candida krusei*, and other minor *Candida* species, such as *Candida guilliermondii*, *Candida lusitaniae*, *Candida kefyr*, *Candida famata*, and *Candida rugosa*.[4,5] While *C. glabrata* and, to a lesser extent, *C. albicans* remain a focus for the concern regarding antifungal resistance, some of rare *Candida* species, in addition to be responsible for clusters of infection in the hospital setting, exhibit decreased susceptibility to one or more antifungal agents, thus posing a threat of emergence in certain settings.[4] Furthermore, *Aspergillus* infections account for more than 90% of nosocomial mold infections,[6] with *Aspergillus fumigatus* causing the majority of cases of invasive aspergillosis; although less susceptible to antifungal agents, non-*fumigatus* aspergilli are increasingly reported in granulocytopenic and other immunocompromised patients.[7–12] Invasive aspergillosis is also associated with a high mortality rate that ranges from 30% to 90%.[13,14] *Cryptococcus neoformans*, a fungal pathogen with a worldwide distribution, causes disease predominantly in immunocompromised individuals such as human immunodeficiency virus (HIV)-infected patients and those receiving immunosuppressive agents for malignancy, organ transplantation, and other serious medical conditions.[15,16] Beyond *Candida*, *Aspergillus*, and *Cryptococccus* species, opportunistic fungi causing deep fungal infections, however, include less common fungi such as Zygomycetes, *Fusarium* species, *Geotrichum capitatum*, *Scedosporium* species, and dematiaceous fungi, many of which are intrinsically resistant to the available antifungal agents.[17–20]

Microbiological resistance is defined as the ability of a pathogen to continue to reproduce in the presence of an antifungal agent, that is commonly measured as minimum inhibitory concentration (MIC) by an in vitro susceptibility testing assay according to a standard protocol.[21,22] In the presence of a resistant yeast or mold isolate, the MIC of the drug exceeds the susceptibility threshold value or breakpoint for that organism.[21,22] It has yet been suggested that the minimum effective concentration (MEC), defined as the lowest drug concentration that leads to the growth of small, rounded, compact hyphal forms as compared to the hyphal growth seen in the growth control well, should be a more appropriate end point than MIC for testing antifungal susceptibility to echinocandins.[22] Microbiological resistance can be primary (intrinsic) or secondary (acquired). Primary resistance occurs naturally in certain fungi (e.g., *C. krusei* to fluconazole or *C. neoformans* to echinocandins) without prior exposure to the drug, whereas secondary resistance develops following exposure to the antifungal agent and can be either reversible, due to transient adaptation, or be fixed as a result of one or more genetic alterations.[23] This type of resistance, which is paradigmatically exemplified by the development of fluconazole resistance in *C. albicans* isolates in response to long-term or repeated treatment of oropharyngeal candidiasis in AIDS patients, is usually dependent on altered gene expression.[24,25]

Significantly, intrinsic or acquired antifungal resistance is an important contributor to the unacceptably high attributable mortality of systemic fungal infections, since patients infected with a drug-resistant organism undoubtedly experience treatment failure, which in turn influences negatively their clinical outcome.[26] Thus, a better understanding of the mechanisms of antifungal resistance (discerning how fungal pathogens have achieved this) is essential to the prompt and efficient treatment of patients with IFIs and to improving the outcome of such infections. On the other hand, it should be remembered that treatment failure, often referred to as clinical resistance to treatment, could also be the result of the much greater complexity of the interaction between an antimicrobial agent and an infecting microbe in a human host.[27]

In this chapter, we review the molecular mechanisms that underlie the in vitro resistance of yeasts and filamentous fungi to various classes of antifungals, with particular emphasis to those regarding *C. albicans*, *C. glabrata*, *C. neoformans*, and *A. fumigatus*. In addition, we discuss the clinical implications of antifungal drug resistance, as well as how new insights into the physiology of multidrug resistance (MDR) in pathogenic organisms could help in rational drug design.

100.1.2 Antifungal Drugs and Their Targets

Although antifungal drug resistance does not seem to be as much of a problem as resistance to antibacterial agents in bacteria,[27] one long-term concern is that the available therapeutic options are limited, particularly for pathogens that are resistant to more than one class of antifungals (Table 100.1). Fortunately, a number of newly licensed effective antifungal drugs are on hand. After a long period following the release of the first triazole antifungal agents (fluconazole and itraconazole in the early 1990s) and lipid amphotericin B formulations (mid-1990s),[28] physicians of the new millennium can now count on three echinocandins (caspofungin, micafungin, and anidulafungin) and new extended-spectrum azoles such as voriconazole and posaconazole.[29]

Most antifungal agents interfere with biosynthesis or integrity of ergosterol, the major sterol in the fungal cell

TABLE 100.1
In Vitro Susceptibilities of Selected Drug-Resistant Fungi to Various Antifungals

	MIC[a] Range, mg/mL (No. of Isolates)					MEC Range of CAS
Species	AMB	ITC	VRC	PSC	RVC	(No. of Isolates)
Aspergillus terreus	0.25–4 (89)	0.003–1 (89)	0.03–1 (89)	0.03–0.25 (46)	0.03–0.5 (16)	0.015–0.125 (16)
Aspergillus lentulus	0.5–2 (4)	0.25–0.5 (4)	1–4 (4)	0.25 (4)	N/A	N/A
Aspergillus udagawae	1–2 (3)	0.25–2 (3)	0.5–2 (3)	0.125–0.25 (3)	N/A	N/A
Aspergillus ustus	1–2 (8)	1–8 (8)	4–8 (8)	N/A	N/A	2–8 (8)
Aspergillus calidoustus	0.5–1 (11)	>8 (11)	4–8 (11)	>8 (11)	4 (5)	N/A
Zygomycetes	0.06–2 (43)	0.125–>8 (43)	1–>8 (43)	0.5–>8 (14)	0.5–>8 (14)	>8 (8)
Fusarium spp.	0.5–4 (49)	0.5–>8 (49)	0.25–>8 (49)	0.25–>8 (18)	0.25–>8 (18)	>8 (11)
Geotrichum capitatum	0.06–0.25 (23)	0.03–0.5 (23)	0.03–0.5 (23)	N/A	N/A	N/A
Scedosporium apiospermum	0.5–>8 (67)	0.03–4 (67)	0.06–4 (67)	N/A	N/A	N/A
Scedosporium prolificans	4–>8 (11)	4–>8 (11)	2–8 (11)	N/A	N/A	N/A

Sources: Data are compiled from Pfaller, M.A. et al., *Antimicrob. Agents Chemother.*, 46, 1032, 2002; Diekema, D.J. et al., *J. Clin. Microbiol.*, 41, 3623, 2003; Panackal, A.A. et al., *Emerg. Infect. Dis.*, 12, 403, 2006; Baddley, J.W. et al., *J. Clin. Microbiol.*, 47, 3271, 2009; Balajee, S.A. et al., *J. Clin. Microbiol.*, 47, 3138, 2009; Johnson, E.M. et al., *J. Antimicrob. Chemother.*, 42, 741, 1998; Girmenia, C. et al., *Antimicrob. Agents Chemother.*, 47, 3985, 2003; Espinel-Ingroff, A. et al., *J. Antimicrob. Chemother.*, 61, 616, 2008.

Abbreviations: AMB, amphotericin B; ITC, itraconazole; VRC, voriconazole; PSC, posaconazole; RVC, ravuconazole; CAS, caspofungin.

[a] As determined by the CLSI microdilution broth method.

membrane, although other drugs cause disruption of the fungal cell wall.[28,30] Thus, based on their mechanism of action, the antifungal agents currently available for the treatment of invasive mycoses can be grouped into four classes: (i) alteration of membrane function (amphotericin B); (ii) inhibition of DNA or RNA synthesis (flucytosine); (iii) inhibition of ergosterol biosynthesis (azoles [fluconazole, itraconazole, and the newer agents voriconazole, posaconazole, and ravuconazole]); and (iv) inhibition of glucan synthesis (echinocandins [caspofungin, micafungin, and anidulafungin]) (Table 100.2).[31]

Similarly to the mammalian cholesterol, ergosterol plays essential roles in the modulation of membrane fluidity (a nonspecific function) and as a specific signal for cell growth and proliferation. In general, antifungal agents that substantially inhibit the ergosterol biosynthesis are likely to affect also its nonspecific function, thereby causing suppressed cell growth as well.[32]

Polyenes (amphotericin B deoxycholate and its lipid-associated formulations) are incorporated into fungal cell membranes containing ergosterol and less readily into host cell membranes, which contain cholesterol. They, in close association with ergosterol, are thought to form porin channels that allow the leakage of protons, potassium, and other monovalent cations and essential cytoplasmic components, which ultimately results in fungal cell death.[30] It has been suggested, however, that amphotericin B has alternative mechanisms of action, mainly promoting oxidative damage of cell membranes through the generation of reactive oxygen species.[33]

Flucytosine (5-flucytosine), a synthetic fluorinated pyrimidine analogue, is taken up into the cell by a cytosine permease and deaminated into 5-fluorouracil by cytosine deaminase. When converted by cellular pyrimidine-processing enzymes into 5-fluorodeoxyuridine monophosphate and 5-fluoro uridine triphosphate, flucytosine inhibits DNA and RNA synthesis in pathogenic yeasts (*Candida* and *Cryptococcus*)[34] and has been available for many years for use in combination with polyenes or triazoles. Although one of the oldest antifungal agents in clinical practice, the role of this drug is limited to use in association with amphotericin B for treating cryptococcal meningitis.[35] In addition, flucytosine is active against a range of fungal pathogens, including *Candida* species, *Saccharomyces cerevisiae*, and selected dematiaceous molds. In combination therapy, flucytosine might be particularly useful where tissue penetration of the other agent is poor, for example, eye, urine, and meninges. Bone marrow depression and hepatotoxicity may occur; thus, monitoring of serum drug levels is required.[36]

Azoles are the most widely used antifungal agents and exert their action by inhibiting the fungal cytochrome P450 enzyme (Erg11p) involved in the 14α-demethylation of the lanosterol molecule, a step necessary for the ergosterol biosynthesis. This inhibition leads to ergosterol depletion and accumulation of 14α-methylated sterols (lanosterol and 14α-methyl-3-6-diol),[37] both resulting in growth arrest and not in cell death, although specific azoles such as itraconazole have a fungicidal effect in some fungal species (*C. neoformans* and *A. fumigatus*).[38] The use of imidazoles (e.g., ketoconazole) is limited to treating superficial mycoses, whereas the triazoles (e.g., fluconazole) have broad application in therapy of both superficial and IFIs. While first-generation triazoles (fluconazole and itraconazole) are clinically useful "older" agents for a range of IFIs, voriconazole and posaconazole are second-generation triazoles with an extended spectrum of activity against yeasts, *C. neoformans* and molds, including *Aspergillus*, *Scedosporium*, and *Fusarium* species.[39,40] Posaconazole, licensed in May 2007, is the broadest spectrum azole to date and is the only azole drug with activity against zygomycete fungi, thus indicating that diversities among azoles in the spectrum of activity (i.e., in the primary resistance profiles of various fungi) may arise from differences in their affinities to their target.[41] Among the newest investigational triazoles, ravuconazole has an antifungal spectrum narrower than posaconazole and similar to voriconazole.[29] The azoles are relatively safe, but clinicians should be aware of drug–drug interactions (they are metabolized by and also inhibit the hepatic cytochrome P450 enzyme system) and adverse effects, including visual disturbances (with voriconazole), elevations in liver transaminase levels, and skin rashes.[28]

The echinocandins drugs, caspofungin, anidulafungin, and micafungin, represent the first new class of antifungal compounds that inhibit β-1,3-D-glucan synthase, which is responsible for biosynthesis of β-1,3-D-glucan polymers in fungal cell walls.[42] These cyclic lipopeptides were originally

TABLE 100.2
Characteristics of Antifungal Drugs Currently Used for Treatment of Invasive Mycoses

Class	Example Drug	Target, Enzyme	Principal Mechanisms of Resistance Observed in Clinical Isolates
Polyenes	Amphotericin B	Ergosterol	Alterations in specific steps of ergosterol biosynthesis
Pyrimidine analogues (antimetabolites)	Flucytosine	Fungal nucleic acid (RNA and DNA)	Decreased uptake of flucytosine Decreased formation of toxic antimetabolites
Azoles	Fluconazole	Ergosterol biosynthesis, lanosterol 14α-demethylase	Enhanced efflux by upregulation of multidrug transporter genes Target alterations by occurrence of mutations Alterations in specific steps of ergosterol biosynthesis
Echinocandins	Caspofungin	Cell wall, β-1,3-D-glucan synthase	Target alterations by occurrence of mutations

obtained from soil fungi in the 1970s, from which have the three semisynthetic derivatives been developed. The echinocandins show good in vitro activity against *Candida* and *Aspergillus* species but are not active at clinically relevant concentrations against Zygomycetes, *C. neoformans* or *Fusarium* species.[42] The echinocandins are rapidly fungicidal against yeasts but fungistatic against molds, where they block the growing tips of hyphae.[43] Toxicity associated with echinocandins is infrequent because their action is specific to fungal cell walls (glucan is not found in mammalian cells). They are not metabolized via the cytochrome P450 enzymes, so there are minimal drug–drug interactions.[28]

100.1.3 Mechanisms of Resistance to Antifungal Agents

100.1.3.1 Resistance to Polyenes

The occurrence of resistance to polyenes is uncommon and can be caused by a reduction in the amount of plasma membrane ergosterol, to which polyenes bind. Although amphotericin B is active against a wide range of fungi, yet certain fungal species are intrinsically resistant to the drug. Among *Candida* species, primary resistance to amphotericin B has been frequently noted in some isolates of *C. lusitaniae*, *Candida lipolytica*, and *C. guilliermondii*.[4] Intrinsic resistance to polyenes is also extremely rare in *Aspergillus* species, with the exception of isolates of *Aspergillus terreus*.[44] However, identification of amphotericin B-resistant isolates has been problematic due to the very narrow range of MICs obtained, although agar-based methods such as Etest (AB Biodisk, Solna, Sweden) have proven to be the most sensitive and reliable means for detecting resistance to amphotericin B among *Candida* species.[4] Although interpretive breakpoints for amphotericin B have not been established, most clinicians use an MIC of ≥1.0 mg/L to indicate isolates that are possibly "resistant" or, at the very least, may require high doses of amphotericin B for optimal treatment.[4] Using the E-test, Krogh-Madsen et al.[45] documented the emergence of amphotericin B-resistant *C. glabrata* isolates (which were also caspofungin-resistant) during the treatment of a critically ill-liver-transplant recipient, thereby demonstrating the importance of antifungal susceptibility testing in patients who are receiving or who have previously been exposed to antifungal treatment. Molecular epidemiology studies using DNA fingerprinting with the complex Ca3 probe have recently revealed that *C. albicans* can be separated into five major clades, groups I, II, III, SA, and E.[46] These groups exhibit geographical specificity, so group SA is relatively specific (i.e., highly enriched) to South Africa, group E is relatively specific to Europe, and group II is absent in the Southwest United States and South America. Interestingly, a clade-related amphotericin B resistance was documented among South African *C. albicans* isolates, where intrinsically resistant isolates (8.4%) were significantly distributed in all clades, except for clade III.[47]

Mutations in the *ERG3* gene resulting in lack of $\Delta^{5,6}$ desaturase cause depletion of ergosterol in the membrane, thus implying cross-resistance between amphotericin B and azoles in yeasts.[48] Vandeputte et al.[49] found that a missense mutation in the *ERG6* gene causes reduced susceptibility to polyenes in a clinical isolate of *C. glabrata* exhibiting pseudohyphal growth. By contrast, deletions in the *ERG3* gene or extra copies of both genes *ERG3* and *ERG6* do not affect the susceptibility of *A. fumigatus* to amphotericin B.[50] Although mechanisms of the intrinsic resistance of *A. terreus* are not well understood, it is evident that much lower ergosterol content than most other fungal species may account for the poor activity of amphotericin B against this fungus. Alterations in cell wall glucans have been postulated to lead to amphotericin B resistance in *A. flavus*.[51] It has also been suggested that for *A. flavus* (and *C. albicans*), the ability to produce catalase may play a role in amphotericin B resistance. A recent study[52] investigating the basis for intrinsic amphotericin B resistance in *A. terreus* demonstrated that while the ergosterol content, cell wall composition, and lipid peroxidation level had no influence on amphotericin B resistance, the level of catalase production in *A. terreus* was significantly higher than that in *A. fumigatus* ($P < 0.05$). Since oxidative damage has been implicated in amphotericin B action, the authors concluded that the higher-level catalase production might contribute to amphotericin B resistance in *A. terreus*.

Amphotericin-resistant strains obtained in vitro,[53] as well as clinical isolates[54] of *C. neoformans*, have been reported, but amphotericin B-resistant cryptococcal infections remain uncommon.[55]

100.1.3.2 Resistance to Flucytosine

Also for flucytosine, resistance can be intrinsic or acquired and may be due to the deficiency or lack of enzymes implicated in the metabolism of flucytosine or to the deregulation of the pirimidine biosynthetic pathways, whose products can compete with the fluorinated metabolites of flucytosine.[31] Primary resistance in *C. albicans* was observed in around 3% of isolates,[56,57] whereas highest percentages were reported for *C. glabrata*, *C. krusei*, and *C. guilliermondii* isolates.[58] As seen for amphotericin B, the five clades in which clinical isolates of *C. albicans* are clustered differ phenotypically; thus, resistance to flucytosine (MIC, ≥32 mg/L) was found to be restricted to clade I, also suggesting that while intraclade recombination may be common, interclade recombination is rare.[59]

This clade-specific flucytosine resistance is due to a single nucleotide change from cytidylate to thymidylate at position 301 in the uracil phosphoribosyltransferase gene (*FUR1*) of *C. albicans*, thus resulting in the replacement of arginine with cysteine at amino acid position 101 in the enzyme that converts 5-fluorouracil to 5-fluorodeoxyuridine monophosphate.[60,61] This mutation has not been found in flucytosine-resistant isolates within other *C. albicans* clades, and, interestingly, while isolates that are homozygous for this mutation display increased levels of resistance to flucytosine, heterozygous isolates have reduced susceptibility.[61] More recently, a greatest proportion of isolates of *C. albicans* with flucytosine low susceptibility (or resistance) was found in MLST (multilocus enzyme sequence typing) clade I and among isolates homozygous at the locus *MAT*.[62] This study also confirmed the known association between *MAT* homozygosity and azole resistance,[63] thus reinforcing the concept that antifungal resistance is not a property that arises randomly among clinical isolates of *C. albicans* but is found in association with certain definable features of individual strains.[62] The inference that the allele T301C is likely to be the major underlying flucytosine resistance mechanism in *Candida* was supported by the finding that mutations affecting the *FUR1* gene can also confer a very high level of cross-resistance (MICs, ≥512 mg/L) to flucytosine and azole drugs in *C. lusitaniae*.[64] Mutations in the *C. albicans* cytosine deaminase (CaFca1p) may also contribute to flucytosine resistance.[65] Hence, due to the frequent development of resistance in fungi when used as a single agent, flucytosine is best utilized in combination with other antifungal agents, mainly amphotericin B.[66]

100.1.3.3 Resistance to Azole Compounds

It has been known since 1974 that the yeast *Saccharomyces cerevisiae* can acquire resistance to several unrelated antifungal drugs, a phenomenon termed as MDR or pleiotropic drug resistance (PDR). PDR appears to be highly conserved among organisms ranging from bacteria to humans and is caused by the overexpression of membrane-associated protein pumps and, consequently, by the expulsion of a wide range of molecules, including antimicrobial drugs.[32] While in human tumor cells PDR hampers effective chemotherapy, in bacteria and other microorganisms such as fungi, it poses a major obstacle in the prevention of infectious disease. Genes influencing the multidrug-resistant phenotype are typically designated *PDR* loci.

Several investigators have explored at the molecular level the mechanisms responsible for acquisition of azole resistance in clinical isolates of several fungal species, including mainly *C. albicans*, *C. glabrata*, *A. fumigatus*, and, to a lesser extent, *C. neoformans*. These studies have established that different cellular alterations contribute to the decrease of azole susceptibility.[27] Overall, azole resistance can result from (i) enhanced efflux by upregulation of multidrug transporter genes, (ii) amino acid substitutions in Erg11p, (iii) upregulation of the *ERG11* gene, and (iv) alteration of the ergosterol biosynthetic pathway. While the last mechanism was shown to result from an absence of sterol $\Delta^{5,6}$ desaturase activity encoded by the *ERG3* gene mentioned above, leading to the accumulation of ergosta-7, 22-dienol-3β-ol, in azole-resistant *C. albicans* clinical isolates, azole resistance in these cases was coupled with resistance to amphotericin B, as the isolates lacked ergosterol to which this latter antifungal drug binds. By contrast, resistance involving the upregulation of multidrug transporter genes seems to be quite common,[38] even though multiple mechanisms of resistance to azole antifungals can favor the development of stepwise increases in the resistance to these agents.[67] In fact, resistance can also be the result of an increased content of the target enzyme molecules either by gene amplification or by upregulation of the corresponding gene.[38] However, besides a single example of gene amplification of the *ERG11* gene in *C. glabrata*,[68] target enzyme upregulation is thought to have a limited impact in azole resistance of *Candida* species, particularly when resistance must reach high levels.[69] Whereas a weak inhibition of Erg11p seems to be responsible for intrinsic resistance to fluconazole in *C. krusei*,[70] *ERG11* sequences from matched pairs of azole-susceptible and azole-resistant *C. albicans* isolates have shown mutations leading to amino acid changes that result in a decreased affinity for azoles.[38] Functional expression of Ca*ERG11* alleles in *S. cerevisiae* showed the involvement of single mutations in azole resistance, with some CaErg11p alleles containing several mutations with potential additive effects.[71] Common mutations in CaErg11p that confer moderate azole resistance are Y132H, S405F, G464S, and R467K.[25,71,72] A second major mechanism of azole resistance relies on overexpression in the plasma membrane of the proteins that pump the drug out of the cell, thus reducing intracellular azole concentrations to levels at which Erg11p is not inhibited. This mechanism is mediated by two types of multidrug transporters, the major facilitators (MFS, encoded by *MDR* genes) and those of the ATP-binding cassette super families (ABC transporters, encoded by *CDR* genes). These membrane proteins actively translocate compounds across cell membranes using the hydrolysis of ATP (ABC proteins) or the proton-motive force across the plasma membrane (MFS).[73] The ABC proteins share similar molecular architecture with the presence of at least one conserved ABC domain and of membrane-spanning segments (transmembrane segments [TMSs]). The ABC domain typically contains Walker A and Walker B motifs and an ABC signature motif. The ABC domain and TMSs can be arranged in a duplicated forward $(TMS_6\text{-}ABC)_2$ or reverse $(ABC\text{-}TMS_6)_2$ topology; however, "half size" ABC proteins also exist.[74] While *CDR* gene upregulation appears to confer resistance to multiple azoles, *MDR*-encoded efflux pumps have a narrower spectrum specific for fluconazole. Some, but not all, members of the ABC transporter super family are associated with azole resistance in fungi. In *S. cerevisiae*, 31 distinct genes encode ABC proteins of which many are multidrug transporters. By contrast, analysis of the *C. albicans* genome has identified at least 27 ABC proteins,[73] but only Ca*CDR1* and Ca*CDR2* have been shown to play a functional role in MDR,[24,75,76] together with Ca*MDR1*.[67] Indeed, both Ca*CDR1* and Ca*CDR2*, when heterologously expressed in *S. cerevisiae*, confer resistance to azoles and other xenobiotics, including the fluorescent compound and PDR substrate rhodamine 6G,[77] thus suggesting that they are ABC pump substrates.[55]

In order to investigate the molecular epidemiology of azole resistance in *C. albicans* strains displaying high-level fluconazole resistance (MICs, ≥64 mg/L) isolated from HIV-infected patients with oropharyngeal candidiasis, Perea et al.[25] showed that overexpression of genes encoding efflux pumps was detected in 85% of all resistant isolates, whereas alterations in the target enzyme, including Ca*ERG11* mutations and overexpression of the Ca*ERG11* gene, were detected in 65% and 35% of the isolates, respectively. Overall, 75% of the azole-resistant isolates showed combined resistance mechanisms, thus confirming the multifactorial nature of azole resistance in *C. albicans* clinical isolates exhibiting high-level azole resistance, as stated earlier. Accordingly, in a series of azole-resistant *C. albicans* isolates studied by Sanglard[38] that included the isolates of 18 HIV-positive patients, 82% showed upregulation of multidrug transporter genes (mainly Ca*CDR1* and Ca*CDR2*), 63% contained Ca*ERG11* mutations linked to azole resistance, and 50% showed combination of resistance mechanisms. On the other hand, a recent analysis of Cdr protein expression in a collection of *C. albicans* clinical isolates with reduced fluconazole susceptibilities demonstrated that CaCdr1p was expressed in greater amounts than CaCdr2p and that most fluconazole efflux function in these strains was mediated by CaCdr1p rather than CaCdr2p.[78]

C. glabrata is innately less susceptible to azoles than most other *Candida* species and acquires azole resistance rapidly following short-term exposure to triazoles, as recently documented in sequential isolates from fungemic patients.[79–82] Several studies suggested that acquired azole resistance of *C. glabrata* clinical isolates was primarily caused by upregulation of two ABC transporters, Cg*CDR1* and Cg*CDR2*,[79,83–88] although one of the resistant isolates of *C. glabrata* studied by Redding et al.[86] displayed not only overexpression of Cg*CDR1* and Cg*CDR2* but also of Cg*ERG11*. By analyzing the levels of expression of both Cg*CDR1* and Cg*CDR2* in an azole-resistant clinical isolate of *C. glabrata*, Sanglard et al.[85] found that while the upregulation of Cg*CDR1* was significant, Cg*CDR2* was only moderately expressed in the same strain. On the other hand, when high-frequency azole resistance (HFAR; 2×10^{-4} to 4×10^{-4}) was established from an azole-susceptible strain, both Cg*CDR1* and Cg*CDR2* were upregulated to high levels. Interestingly, this HFAR was coupled to the loss of mitochondria. Likewise, the azole resistance of petite mutants of *C. glabrata*, obtained by exposure to fluconazole or induced by ethidium bromide,[89] was linked to the upregulation of nuclear genes encoding efflux proteins.[90] In our study,[79] carried out to examine the expression of Cg*CDR1* and Cg*CDR2* in a collection of fluconazole-resistant *C. glabrata* clinical isolates, we observed that most isolates upregulated both genes and that the high levels of transcripts correlated with their high-level resistance to fluconazole and other azoles, including voriconazole. When levels of expression of Cg*CDR1* were compared with those of Cg*CDR2*, we noticed that Cg*CDR1* upregulation was always clearly manifested in isolates that concomitantly expressed both genes, while Cg*CDR2* was expressed at moderate levels. These findings added support to the idea that Cg*CDR1* was more closely associated with azole resistance than Cg*CDR2*.[84] The importance of Cg*CDR1* and Cg*CDR2* in contributing to azole resistance in *C. glabrata* was indeed corroborated by the fact that both genes appeared to be upregulated at low levels in isolates that exhibited a fluconazole-susceptible dose-dependent phenotype. Conversely, sequencing of Cg*ERG11* and analysis of its expression showed no mutation or upregulation in any *C. glabrata* isolate, suggesting that Cg*ERG11* was not involved in azole resistance. Interestingly, two fluconazole-resistant isolates expressing normal levels of Cg*CDR1* and Cg*CDR2* exhibited increased levels of expression of Cg*SNQ2*, a gene found to be highly similar to Sc*SNQ2* from *S. cerevisiae*. Thus, expression of this gene in an *S. cerevisiae* strain lacking *PDR5*, the archetype of the PDR subfamily of ABC transporters, was shown to mediate not only resistance to azoles but also to 4-nitroquinoline N-oxide, which is an Sc*SNQ2*-specific substrate.[91] We also assessed whether this gene could be implicated in the azole resistance of *C. glabrata* by analyzing its expression in 18 *C. glabrata* isolates representative of various clinical settings. Results of these molecular analyses showed that Cg*SNQ2* was upregulated in some isolates that simultaneously upregulated both Cg*CDR1* and Cg*CDR2*, in support of the concept that several drug efflux pumps can contribute to azole resistance in clinical isolates, but each to different extent. Variations in the structure of azoles are thought to be responsible for the cross-resistance patterns among *Candida* species,[92] so a complete cross-resistance between fluconazole and extended-spectrum triazole antifungals (voriconazole, posaconazole, and itraconazole) has been observed with isolates of *C. glabrata* in association with increased expression of Cg*CDR1* and Cg*CDR2*. This has been manifested clinically where patients with *C. glabrata* fungemia, all with fluconazole exposure, failed treatment with voriconazole.[4,81,93] No such pattern was seen with isolates of *C. krusei*.[92] In this organism, very recent data showed that it is mainly the low affinity of Erg11p for fluconazole, together with the constitutive but low level of expression of the multidrug efflux pump Abc1p, that is responsible for the innate fluconazole resistance of *C. krusei*.[94]

Unlike *Candida* species, primary resistance to azoles (with the exception of fluconazole) does not appear to be a problem, since most *Aspergillus* species are highly susceptible to these drugs.[8] *A. fumigatus* is innately resistant to fluconazole and ketoconazole, so that gene knockout experiments have shown that Cyp51p, the lanosterol 14α-demethylase enzyme, is essential for *A. fumigatus* and that *A. fumigatus* Cyp51Ap, not Cyp51Bp, is responsible for the innate resistance to fluconazole and ketoconazole.[95] However, there are an increasing number of reports of acquired resistance to itraconazole and other triazoles in clinical isolates of *A. fumigatus*,[96–99] although these events remain infrequent to date.[100] So, the statement that acquisition of resistance to azole antifungals among *Aspergillus* species is a relatively rare phenomenon must be qualified by the recognition that analysis of this phenotype has been often complicated by technical difficulties in establishing it,[101] as well as that susceptibility testing of isolates of patients receiving therapy is becoming more frequent.[100] In this context, recommended methods are now promulgated by the CLSI and EUCAST, and work is ongoing to establish epidemiological cutoff and clinical breakpoints.[100] In addition, it should be remembered that accurate speciation of *Aspergillus* species is important especially because rare *Aspergillus* species, such as those belonging to *A. fumigatus* complex (*A. lentulus* and *A. udagawae*) or *A. ustus* complex

(*A. calidoustus*), which are relatively resistant to most antifungal agents (Table 100.1), have recently been reported to cause invasive disease.[102–104]

The availability of the genome sequence of *Aspergillus* has revealed that there is a significant redundancy in genes encoding for efflux transporter, as the genome of *A. fumigatus* is predicted to encode at least 49 ABC and 278 MFS transporters, more than four times the number of such transporters found in *S. cerevisiae*.[9] The first resistance mechanism described in *Aspergillus* involves reduced intracellular concentration of itraconazole caused by overexpression of efflux pumps. Slaven et al.[105] reported the cloning of *atrF*, an ABC transporter gene, whose increased expression was associated with itraconazole resistance in an *A. fumigatus* clinical isolate. Moreover, the overexpression of two other novel genes, Afu*MDR3* (an MFS protein) and Afu*MDR4* (an ABC transporter), was shown to confer a high level of itraconazole resistance of in vitro selected *A. fumigatus* isolates.[106] In spite of the above-mentioned redundancy of efflux transporter-encoding genes, recent comparative studies demonstrated that there is a lack of specificity in their function, which might partially explain the relatively low rate of resistance of *A. fumigatus* to azoles.[9] As stated recently,[107] it is also striking to note that the increased proportion of MFS proteins compared to ABC transporters found in the *Aspergillus* genome with respect to *S. cerevisiae* mimics the situation found in clinically relevant bacterial pathogens, where only MFS proteins are currently seen to participate in evasion of antibacterial therapy.[108] The second well-known resistance mechanism is the modification of the lanosterol 14α-demethylase enzyme that is encoded by the *cyp51A* and *cyp51B* genes,[109] although only mutations in Cyp51A are important for clinical resistance.[96,98,110–112] Mutational hotspots (HS) confirmed to cause resistance have been characterized in the target gene at codons 54, 220, and 98, although other mutations in the *cyp51A* gene have been reported but may not be causal, and additional resistance mechanisms have been postulated (for an overview, see Ref. [100]). To resume, these mutations confer several different susceptibility profiles: (i) resistance to itraconazole and posaconazole associated with amino acid substitutions at Cyp51A glycine 54 (Gly 54); (ii) a pattern of resistance to itraconazole and elevated voriconazole, posaconazole, and ravuconazole MICs associated with amino acid substitutions at methionine Cyp51A 220 (Met 220); (iii) a pattern of azole cross-resistance associated with higher *cyp51A* expression generated by a tandem repeat of a 34 bp sequence in the *cyp51A* gene promoter in combination with an amino acid substitution at Cyp51A leucine 98 (TR-L98H); and (iv) a triazole cross-resistance associated with an amino acid change at Cyp51A glycine 138 to cysteine (G138C). A novel G448S mutation in the Cyp51A gene was detected in a clinical isolate of *A. fumigatus* obtained from a patient with pulmonary aspergilloma treated with voriconazole that was demonstrated to confer acquired resistance to itraconazole and voriconazole.[113] Moreover, of the 34 itraconazole-resistant isolates studied by Howard et al.,[100] 22 (65%) isolates were cross-resistant to voriconazole and 25 (74%) were cross-resistant to posaconazole. Thirteen of the 14 evaluable patients had prior azole exposure, of which 7 (itraconazole-treated) failed to respond to therapy and the remainder failed to improve (i.e., their disease remained stable). A total of 18 amino acid substitutions were found in the Cyp51A enzyme, 4 of which were novel. Interestingly, 3 of 34 isolates had no mutations in their *cyp51A* gene, indicating the presence of other resistance mechanisms. In contrast, in a large investigation on the prevalence of itraconazole resistance in 1912 *A. fumigatus* clinical isolates, collected between 1994 and 2007 in the Netherlands, itraconazole-resistant isolates were found in 32 isolates from 13 patients (1.7%). The itraconazole-resistant isolates also showed elevated MICs of voriconazole, posaconazole, and ravuconazole, with the TR/L98H being shown to be the dominant resistant mechanism (94% of isolates).[112] As specified earlier, the position and type of amino acid substitution within the Cyp51A enzyme determine the pattern of azole cross-resistance, which might reflect the similarities in molecular structures among different triazoles and their interaction with chemically different azole drugs.[41] Therefore, a significant degree of cross-resistance between the structurally similar triazoles, itraconazole and posaconazole, but not between itraconazole and voriconazole had previously been shown by Mosquera and Denning[114] in an vitro study on 17 itraconazole-resistant *A. fumigatus* isolates. Similarly, Meneau and Sanglard,[115] in a large surveillance study that included 400 clinical and 150 environmental *A. fumigatus* isolates, had shown that there was no cross-resistance between itraconazole and voriconazole in 10 (2.5%) clinical and 36 (24%) environmental isolates that exhibited high itraconazole MICs. In contrast, in the recent series of *A. fumigatus* isolates studied by Howard et al.,[100] some isolates exhibited cross-resistance between itraconazole and voriconazole and not posaconazole, probably because of differing breakpoints.

Finally, to test the hypothesis that azole resistance might develop through azole exposure in the environment rather than in azole-treated patients, Snelders et al.[116] compared antifungal susceptibility, resistance mechanisms, and genetic relatedness of itraconazole-resistant *A. fumigatus*, cultured from the indoor hospital environment or the soil in proximity to the hospital, with those of azole-resistant clinical isolates. While cross-resistance was observed for voriconazole, posaconazole, and the azole fungicides metconazole and tebuconazole, a single dominant resistance mechanism was present in both the clinical and 13 of 15 environmental isolates. Interestingly, environmental and clinical isolates were genetically clustered apart from nonresistant isolates, suggesting that patients with azole-resistant aspergillosis might have been colonized with azole-resistant isolates from the environment. In contrast, in the Howard's study,[100] microsatellite typing analysis showed that resistant mutants had evolved from single originally susceptible strains, whereas the presence of genetically identical isolates with different *cyp51A* codon mutations in three patients suggested their independent evolution from the same original strain.

With regard to *C. neoformans*, therapeutic failures related to drug-resistant isolates have been described, particularly to fluconazole, an extensively used drug for the management of cryptococcosis.[117–123] Therefore, relapsed cases of cryptococcosis during prolonged maintenance therapy, in which sequential isolates were found to have increased fluconazole or itraconazole MICs, have increasingly been reported.[124–126] This phenomenon, which is rare in AIDS patients receiving highly active antiretroviral therapy (HAART), has yet become particularly important for those patients with no access to HAART.[55] In *C. neoformans*, the azole resistance mechanisms explored previously have been essentially focused on a reduced azole cellular content[53] or an altered mutation-related activity of Erg11p,[127,128] In an in vitro study conducted by Venkateswarlu et al.,[129] three *C. neoformans* isolates with low-level in vitro fluconazole resistance (their MICs were three- to sixfold higher than those for the initial susceptible isolates) showed changes in the affinity of Erg11p. In contrast, four isolates with MICs 100–200-fold higher than those for the initial susceptible isolates showed decreases in the cellular content of the drug, suggesting that drug efflux transporters may be involved in this high-level fluconazole resistance. Likewise, Joseph-Horne et al.[53,130] found a correlation between azole resistance and reduced cellular content of drug uptake for *C. neoformans* isolates resistant to both azoles and polyenes, which retained high levels of ergosterol and yet displayed cross-resistance to amphotericin B, indicating that MDR mechanism could also exist in *C. neoformans*. It has now been established that in clinical isolates of *C. neoformans*, highest levels of fluconazole resistance are caused by increased drug efflux.[55] Thus, overexpression of the ABC transporter CneAFR1p[131] leads to azole-resistant *C. neoformans* that cannot be treated with azole antifungals, as shown in mouse models using inhalation or intravenous models to produce systemic cryptococcosis.[132] In this study, the overexpression of CneAFR1p led to increased virulence in the mouse models and significantly improved the survival of *C. neoformans* during in vitro macrophage infection. The Cne*AFR1* gene was also shown to play a role in the resistance of *C. neoformans* to microglia-mediated antifungal activity by delaying phagosomal maturation, likely highlighting the potential innate function of CneAFR1p that has been only lately co-opted for antifungal resistance.[133] Consistent with these results, the Kwon-Chung's group[55] demonstrated that the heteroresistant phenotype of aneuploid *C. neoformans* isolates was associated with the increased copy number of the multidrug transporter Cne*AFR1* gene (these strains possessed an extra copy of chromosome A on which Cne*AFR1* is located), and deletion of Cne*AFR1* in a heteroresistant strain led to a reversion of the resistance phenotype. More recently, the same authors analyzed fluconazole-resistant strains isolated from AIDS patients undergoing fluconazole maintenance therapy during 1990–2000.[134] Upon exposure to stepwise increases in the concentration of fluconazole, subpopulations that could grow at higher concentrations emerged. The strains heteroresistant at ≥32 mg/L were significantly more virulent in mice than were those heteroresistant at ≤8 mg/L. During fluconazole treatment of mice infected by strains with low levels of heteroresistance, subpopulations exhibiting higher levels of heteroresistance emerged after a certain period of time. Interestingly, the levels of fluconazole at which the strains manifested heteroresistance generally correlated with the observed degree of virulence in mice. Furthermore, the level of heteroresistance was also associated with the ability of *C. neoformans* yeast cells to survive on media containing toxins and antibiotics produced by soil microorganisms, thereby indicating that the mechanisms that control the level of heteroresistance to azoles *C. neoformans* also contribute to the fungus ability to respond to various other environmental stresses. By the authors' opinion, this intrinsic adaptive mechanism of resistance contributes to relapse of cryptococcosis during azole maintenance therapy, so determination of the heteroresistance level for the strain of each cryptococcosis case will offer helpful insights into the management of long-term fluconazole therapy. Another *C. neoformans* ABC transporter, CneMdrlp, was in 1997 identified by Thornewell et al.[135] Although it belongs, in contrast to CneAFR1p, to the MDR type associated with azole resistance, Cne*MDR1* expression has yet to be associated with azole resistance in clinical isolates or in resistant mutant isolates in vitro. Overexpression of both Cne*AFR1* and Cne*MDR1* homologues from *C. neoformans* strains CDC551 (serotype A) in *S. cerevisiae* ADΔ conferred MDR to a large array of xenobiotics, including all azole antifungals tested, despite their different protein topologies, $(NBD\text{-}TMD)_2$ or $(TMD\text{-}NBD)_2$, respectively.[55]

100.1.3.4 Resistance to Echinocandins

Little is known about the resistance to echinocandin, at least in part due to their more recent approval for clinical use. The most common mechanism of echinocandin resistance is mutation of the drug target,[136,137] the above-mentioned β-1,3-D-glucan synthase, that is an enzyme complex consisting of a regulatory subunit, Rho1p, and a catalytic subunit, Fksp, encoded by three related genes, *FKS1*, *FKS2*, and *FKS3*. It has been shown that in laboratory mutants as well as in clinical isolates of *Candida*, characteristic mutations in Fks1p and Fsk2p were necessary and sufficient to confer reduced susceptibility to echinocandins.[137–140] Mutations, associated with reduced susceptibility to caspofungin, have been observed within two "HS" regions of Fks1p and are located at amino acid positions 640–650 (HS1) and 1345–1365 (HS2) in *C. albicans*.[141,142] Despite the wide range of mutations, the highest frequency of amino acid substitutions were found at Ser645 (S645P, S645F, or S645Y) within the HS1.[139,141] The Fks1p-mediated mechanism extends to other non-*C. albicans* species, including *C. guilliermondii*, *C. krusei*, and *C. tropicalis*,[137] whereas mutations in Fks2p has been linked with echinocandin resistance in *C. glabrata*.[140,143] These mutations, yielding highly elevated MIC values, are genetically dominant and confer cross-resistance to all echinocandin drugs.[137] In addition, they reduce the sensitivity of β-1,3-D-glucan synthase to drug by several 100-fold to a 1000-fold,[141] and strains harboring such mutations may require a concomitant increase in drug to reduce fungal organ burdens in animal infection models.[137,141] With regard to *C. parapsilosis*, Garcia-Effron et al.[144] showed that a naturally occurring proline-to-alanine substitution at position 660 (P660A), immediately distal to the highly conserved HS1 region of Fks1p, accounts for reduced echinocandin susceptibility in this yeast. Kinetic inhibition studies demonstrated that β-1,3-D-glucan synthase from the *C. parapsilosis* group (*C. orthopsilosis*, *C. metapsilosis*, and *C. parapsilosis*) was 1–2 logs less sensitive to echinocandin drugs than the reference enzyme from *C. albicans*. Of note, clinical isolates of *C. albicans* and *C. glabrata*, which harbor mutations at this equivalent position, also showed comparable 2-log decreases in target enzyme sensitivity, which correlated with increased MICs. However, the clinical significance of this reduced susceptibility is unclear, although sporadic treatment failures consistent with clinical resistance have been documented with high MIC isolates. In general, a complex relationship exists between elevated MIC and clinical success,[140,145] probably because fungal cells can induce a variety of adaptive pathways that lead to reduced susceptibility without directly impacting the behavior of the drug on β-1,3-D-glucan synthase. In fact, *C. albicans* may utilize complex stress response pathways to tolerate cell wall stress induced by echinocandins, including upregulation of genes involved in cell wall biosynthesis as well as responses mediated by the cell wall integrity signaling pathways.[146,147] In this context, it should be noted that while echinocandins are thought to be fungicidal against *C. albicans*, this organism has the ability to grow in vitro at concentrations exceeding the MICs of caspofungin.[148] This paradoxical effect, which is referred as the "eagle effect" as such observed in other antimicrobial cell wall-active agents, is strain dependent and has been related to a large increase in chitin content in cell wall, which would appear to compensate for a decrease in β-1,3 glucan and β-1,6 glucan following caspofungin action.[149]

Although breakthrough infections by *A. fumigatus* isolates with elevated MECs have been reported,[150] the mechanisms of resistance in molds, especially *Aspergillus*, have not yet been well characterized. However, the universality of Fks1p modification as a mechanism for fungal resistance to echinocandins has been demonstrated in laboratory strains.[151,152] Two classes of *A. fumigatus* mutant strains exhibiting reduced susceptibility to caspofungin were studied. A site-directed mutation within Afu*FKS1*, the gene encoding for the catalytic subunit of β-1,3-D-glucan synthase in *A. fumigatus*, was introduced and shown to confer low-level (16-fold) reduced susceptibility. In contrast, spontaneous mutants generated by cell wall digestion were susceptible to low levels of drug but displayed nearly normal growth above 0.5 mg/L, suggesting induction of an unknown resistance mechanism.[151] Other laboratory-manipulated strains were found to have mutations in the Afu*ecm33* gene encoding cell wall proteins important for fungal cell wall organization and, in addition to possess caspofungin resistance, were found to be hypervirulent in animal models.[153] Very recently, Arendrup et al.[154] reported the detection of an *A. fumigatus* isolate displaying caspofungin resistance by E-test (MIC, >32 mg/L) obtained from a patient failing caspofungin treatment. As expected, this isolate showed reduced susceptibility to caspofungin in a hematogenous mouse model, but, notably, sequencing of the *FKS1* gene revealed no mutations, whereas gene expression analyses demonstrated that the *FKS1* gene was expressed at a higher level in this isolate than in the susceptible control isolate.

Finally, the mechanism of echinocandin resistance in *C. neoformans* is not fully understood, given that the drug target, β-1,3-D-glucan synthase, is both essential and sensitive to echinocandins in vitro.[155] Possibilities include an echinocandin-resistant β-1,3-D-glucan synthase target, degradation pathways and efflux pumps, although studies indicate that the last are not responsible for echinocandin resistance at least in *C. albicans*.[156]

100.1.4 Transcriptional Control of Antifungal Drug Resistance

As mentioned above, members of the ABC transporter super family are primary efflux transporters that, as their name implies, hydrolyze ATP for export of the substrate. In particular, three subfamilies, referred to as the MDR, multidrug resistance-associated proteins (MRP), and PDR families, are involved in the efflux of toxic compounds; and their members have been extensively studied in *S. cerevisiae*. While the PDR subfamily has the least amount of phylogenetic conservation, which suggests that members of this family are rapidly evolving from outside selective pressures,[157] elements determining the PDR transcriptional control in *S. cerevisiae* appear to be conserved in several well-studied pathogenic fungi.[55] A key determinant in the drug-resistant phenotype of *S. cerevisiae* is provided by the regulation of transactivation capability of regulatory proteins, some of which belong to the well-known family of the zinc cluster proteins.[158] Studies have in fact shown that the Gal4-zinc finger transcription factor ScPdr1p and the paralogue ScPdr3p together confer resistance to several drugs and toxins, through transcriptional activation of ABC transporter genes and members of the major facilitator super family (MFS) of drug efflux pumps, including ScPdr5p, ScSnq2p, and ScYor1p, as well as phospholipid-transfer proteins such as ScPdr16p.[159] ScPdr1p and ScPdr3p both bind to elements referred to as pleiotropic drug response elements (PDREs) located in the promoters of target genes. In vivo footprinting experiments indicated that ScPdr1p and ScPdr3p are likely to be constitutively bound to relevant PDREs, an observation consistent with the constitutive nuclear localization of these proteins.[107] Many point mutations causing resistance to chemically diverse xenobiotics (including azoles) with different targets have been mapped in ScPdr1p and ScPdr3p.[160,161] As a result of single amino acid substitutions, ScPdr1p and ScPdr3p serve as strong, constitutive activators of downstream gene expression, so that hyperactive mutant forms of ScPdr1p drove to the initial identification of the PDR phenotype in *S. cerevisiae*.[160] Sc*PDR3* gene was also shown to be under positive autoregulation of ScPdr3p.[162]

However, while ScPdr1p and ScPdr3p share 36% sequence identity, the regulation of these factors is in turn responsive to different signals.[163] In addition, both ScPdr1p and ScPdr3p can activate the expression of another zinc finger transcription factor-encoding gene called Sc*YRR1*. Increased expression of ScYrr1p can amplify the transcriptional effects of activation of either ScPdr1p or ScPdr3p since this factor recognizes a sequence different from the PDRE, YRRE.[107] The identification of ScYrr1p (initially referred to as ScPdr2p) as an additional regulator of PDR genes provided some of the first evidence of cross talk between regulators of PDR in *S. cerevisiae*.[158]

C. glabrata is evolutionarily closely related to *S. cerevisiae*, and recent studies have identified an ScPdr1 orthologue in *C. glabrata* (CgPdr1p) that controls the expression of its major ABC transporters, CgCdr1p and CgCdr2p,[88,164,165] as well as that of CgSnq2p.[91] Like *S. cerevisiae*, single-point mutations in *CgPDR1* are known to increase the expression of at least Cg*CDR1* and Cg*CDR2*, thus contributing to azole resistance of clinical isolates.[88,165] Very recently, comparison of Cg*PDR1* alleles from azole-susceptible and azole-resistant matched clinical isolates of *C. glabrata* enabled us to identify 57 amino acid substitutions, each positioned in distinct Cg*PDR1* alleles. These substitutions were gain-of-function mutations since they conferred hyperactivity to CgPdr1p as revealed by constitutive high expression of ABC-transporter genes. Interestingly, the major transporters involved in azole *C. glabrata* resistance (Cg*CDR1*, Cg*CDR2*, and Cg*SNQ2*) were not always coordinately expressed in presence of specific Cg*PDR1* gain-of-function mutations, thus suggesting that these are rather *trans*-acting elements (gain-of-function in Cg*PDR1*) than *cis*-acting elements (promoters) that lead to azole resistance by upregulating specific combinations of ABC-transporter genes.[166]

In *S. cerevisiae*, gain-of-function mutations activate over 20 target genes, the major ones being ABC or MFS drug efflux transporters, as mentioned above.[159,167] Similarly, in *C. glabrata*, CgPdr1p coordinately regulate 11 genes that are homologous to ScPdr1p/ScPdr3p targets, but also other genes linked with transport, cell wall biosynthesis, lipid metabolism, and cell stress.[165] Significant differences in the in vitro oxidant, alcohol, and weak acid sensitivities were detected between an azole-resistant strain and its parent, suggesting that the azole resistance-associated expression changes could affect *C. glabrata* tissue-specific virulence.[165] Interestingly, *C. glabrata* isolates with hyperactive Cg*PDR1* alleles that upregulated the expression of ABC transporter genes were shown to be more virulent in mice than isolates with wild-type alleles and gained fitness in the animal model.[166]

Rapid and transient increased expression of drug efflux pumps was recently observed in yeasts in response to antifungal or other xenobiotics.[159,168] Notably, Thakur et al.[159] found that these substances directly bind to a hydrophobic domain xylan-binding domain (XBD) in the Pdr1p orthologues in *S. cerevisiae* and *C. glabrata* and that this binding leads to xenobiotic-dependent activation of drug efflux genes. In particular, the activation domain of ScPdr1p/ScPdr3p at their carboxyl termini was shown to interact with the KIX domain of ScGal11p subunit of the Mediator complex that recruits RNA polymerase II for expression of the ScPdr1p/ScPdr3p-controlled genes. In *C. glabrata*, the interaction of CgPdr1p with antifungal drugs is mediated by one of the two Gal11p homologues. Thus, when treated with xenobiotics, nematode worms infected with *C. glabrata* in which Cg*PDR1* and Cg*GAL11* have been deleted lived longer, thus confirming the importance of the Pdr1p-Gal11p pathway in drug resistance.[169] Interestingly, this is mechanistically similar to regulation of MDR in vertebrates by the PXR (pregnane X receptor) transcription factor, revealing an unexpected functional analogy of fungal and metazoan MDR regulators.[159]

Among the second class of transcription factor that has been associated with regulation of MDR in *S. cerevisiae*, the basic region-leucine zipper, ScYap1p is the better known for its important role in oxidative stress tolerance. Briefly, upon oxidant challenge rapidly accumulates in the nucleus, where it activates antioxidant gene expression. Also, mutant forms of ScYap1p have been described that are constitutively located in the nucleus and are hyper-resistant to certain oxidants. Interestingly, ScYap1p defines a pathway for drug resistance parallel to that of Sc*PDR5*, since activation of ScYap1p controls the expression of different MFS protein-encoding genes (Sc*ATR1* and Sc*FLR1*). Similarly, MDR in *C. glabrata* is also influenced by transcriptional control of MFS protein expression. The Cg*FLR1* gene is induced by the *C. glabrata* Yap1p homologue (Cgap1p) and confers resistance to a range of agents, including fluconazole.[170]

Because of extensive cross-talk between the nucleus and mitochondria, it has long been known that the functional state of mitochondria can influence nuclear gene expression.[171] In *S. cerevisiae*, respiratory deficiency due to loss of the mitochondrial genome (ρ^0 cells) causes a pronounced upregulation of ScPdr3p and a dramatic increase in ScPdr3p-controlled expression of Sc*PDR5*,[163] but not ScPdr1p function. In ρ^0 cells, the elevation in ScPdr5p expression and subsequent increased drug resistance are dependent only on ScPdr3p,[163] thus indicating that the response to mitochondrial deficiency originates directly from Sc*PDR3*. A more recent work demonstrated that the heat shock protein 70 (Hsp70)-derived protein Ssa1p represses ScPdr3p but not ScPdr1p function, and, importantly, ρ^0 cells had less Ssa1p bound to ScPdr3p than ρ^+ cells. As ScPdr3p is autoregulated, this interaction maintains ScPdr3p in a repressed state to ensure that levels of this protein are kept low in the absence of mitochondrial signaling.[172] Likewise, ρ^0 *C. glabrata* cells are highly multidrug resistant by upregulating Cg*CDR1* and, to a lesser extent, Cg*CDR2*,[85,90] as mentioned above. This mitochondrial control of MDR seems to be mediated by the activation of CgPdr1p.[164] In light of the newly discovered role of Hsp70-ScPdr3p association in MDR regulation,[172] as well as high similarities between ScPdr3p and CgPdr1p,[164] we hypothesized that CgPdr1p may be negatively regulated in *C. glabrata* at a posttranslational level, probably through the interaction of Ssa1p. In support of this hypothesis, comparison of proteomic profiles for a matched pair of *C. glabrata* clinical isolates revealed a decreased Ssa1p homologue expression in the azole-resistant petite mutant isolate, as compared to its azole-susceptible parent, which in turn could result in decreased binding of Ssa1p to CgPdr1p.[173]

Although *C. albicans* shares a very similar range of effector genes with *S. cerevisiae*, it seems that the transcriptional mechanisms that regulate the expression of antifungal pumps are quite different from those unraveled in *S. cerevisiae*. In *C. albicans*, resistance to azoles mediated by upregulation of drug efflux pumps is particularly associated with isolates homozygous at the *MAT* gene locus on chromosome 5.[63] This association does not relate directly to mating phenomena but to Ca*TAC1*, located 14 kb away from the *MAT* locus, which encodes the ScPdr1p/Scpdr3p-like zinc finger transcription factor CaTac1p that controls expression of the ABC transporters, CaCdr1p and CaCdr2p.[174] Although CaTac1p and ScPdr1p/ScPdr3p show less than 20% sequence identity, an interesting commonality between Sc*PDR1* and Ca*TAC1* gain-of-function mutations is the spectrum of highly responsive target genes that these two important regulators of MDR induce in their respective organisms. Thus, in *S. cerevisiae*, *PDR1* hyperactive mutants strongly induce *PDR5* and *PDR15*, as well as Rsb1p, a membrane protein that is required for resistance to the long-chain phytosphingosine,[167] whereas in *C. albicans TAC1* hyperactive alleles induce *CDR1* (Pdr5p homologue), *CDR2* (Pdr5p homologue), and *RTA3* (Rsb1p homologue).[107]

Acquired azole resistance in *C. albicans* is often associated with the appearance of aneuploidy and, in particular, with the formation of an isochromosome [i(5L)] containing two left arms of chromosome 5 separated by a centromer.[175] Further studies revealed that *C. albicans* was able to undergo chromosome 5 loss and reduplication in response to azole exposure and thus became homozygous for a hyperactive Ca*TAC1* allele that induced strong CaCdr1p and CaCdr2p expression. This, together with the increase in the Ca*ERG11* copy number, was responsible for the fluconazole resistance phenotype observed in clinical isolates.[176,177] Analogously, mutations in the zinc cluster transcription factor CaMrr1p cause Ca*MDR1*-mediated azole resistance when these mutations are brought to homozygosity by loss of heterozygosity.[178]

Specific regulatory pathways, through which yeast cells respond to different stresses or harmful molecules that make up an ever-changing cellular environment, are broadly characterized into interconnected network, the stress response, and the PDR network, mentioned above.[158] Mechanisms that mitigate drug toxicity are often dependent upon cellular stress responses, which are crucial for the tolerance of the membrane or cell wall stresses exerted by azoles[179] or echinocandins as discussed above.[146,147] In *C. albicans*, Hsp90, an essential chaperone that regulates the form and function of diverse client proteins, was shown to potentiate resistance to the azoles[180] and, more recently, to govern the ability of the fungus to survive exposure to echinocandins, via calcineurin.[181] Impairing Hsp90 function converts the fungistatic azoles into a fungicidal combination and enhances the therapeutic efficacy of azoles in two metazoan models of disseminated *C. albicans* infection.[182] Similarly, inhibitors of Hsp90, which is required for the basal tolerance of *Aspergillus* species to echinocandins, enhances the efficacy of these drugs in an invertebrate model of *A. fumigatus* infection.[182,183] Overall, these studies suggest that compromising Hsp90 function provides a powerful strategy to render existing antifungal drugs more effective in the treatment of life-threatening fungal infections.

100.1.5 Molecular Assessment of Antifungal Resistance

Although conventional mycology will be not replaced in the immediate future, molecular techniques, in particular nucleic acid-based methods, are the fastest growing field in today's clinical laboratories. While the first-generation molecular techniques, by determining only a single parameter like conventional tests, have been focused on the detection and identification of microbial pathogens, newer multiparameter assays using microarrays, multiplex nucleic acid amplification techniques, or mass spectrometry (MS) (see also Section 100.3) paved the way for the parallel determination of multiple drug resistance determinants.[184] On the other hand, the introduction of real-time PCR-based assays allowed the kinetics of the amplification process to be observed and analyzed simultaneously, thus resulting in rapid microbial diagnostics with a reduced contamination risk.[185]

Given the increasing incidence of fungal infections and their attributable impact on mortality and length of hospital stay, early initiation of effective antifungal therapy appears to be crucial. In the case of *Candida* bloodstream infections, administration of empiric antifungal treatment 12 h after a positive blood sample for culture is drawn is associated with greater hospital mortality.[186] Although the MIC of an infecting organism is a presumptive measure for predicting clinical outcome, antifungal susceptibility testing is not routinely performed and yet requires 24–72 h following identification, thus representing an obstacle to effective therapy. In addition, the expanding number of antifungals with both shared and distinct mechanisms of action, as well as the recognition of wide variation in susceptibility due to both intrinsic and acquired antifungal resistance, poses an urgent need for timely and clinically relevant antifungal susceptibility assays.[187]

In this context, molecular methods have the potential to complement and, in some case, replace the growth-based methods for evaluating antifungal susceptibility in routine practice. The development of these methods springs from our increased understanding of the molecular basis of antifungal mechanisms of action and resistance,[187] at least for azole and echinocandin drugs. Thus, these assays cover a range of platforms, including real-time PCR or nucleic acid sequence-based amplification with high fidelity self-reporting probes, microarrays, high-throughput DNA sequencing, and hybridization melt profiles.[188] For instance, Garcia-Effron et al.[13] developed a real-time PCR assay utilizing allele-specific molecular beacons to rapidly and accurately assess

A. fumigatus cyp51A gene mutations conferring resistance to triazole drugs, thereby providing a comprehensive evaluation of drug resistance in this fungal pathogen.

On the other hand, when a database of mutations known to confer resistance to a given antifungal agent is on hand, it becomes possible to predict antifungal drug resistance on the basis of sequence analysis alone.[187]

However, the multifactorial nature of triazole resistance in *Candida* species makes detection complex, even if the recent elucidation of gain-of-function mutations in transcription factors (e.g., Pdr1p) could be exploited for a more specific targeting. In contrast, the limited spectrum of resistance mutations identified for triazole resistance in *A. fumigatus* and echinocandin resistance in *Candida* is ideally suited for molecular detection.[188]

100.2 METHODS

100.2.1 Sample Preparation

The starting material for all nucleic acid-based diagnostic techniques is, of course, the fungal DNA or RNA, depending on the procedure chosen. Fungal isolates, identified at the species level by standard methods, are grown and maintained on Sabouraud dextrose agar (1% peptone, 2% dextrose, 2% agar) medium (Kima, Padua, Italy) or cultivated in yeast extract-peptone-dextrose (YEPD) (1% Bacto Yeast Extract, 2% Bacto Peptone, 2% Bacto Agar) medium, when necessary.

100.2.2 Detection Procedures

100.2.2.1 Molecular Detection of Fungal Resistance to Azole Compounds

Principle

In contrast to *Candida* species, triazole resistance in *A. fumigatus* predominantly involves mutations in *cyp51A* (equivalent to *ERG11*), which consist in substitutions of a specific Gly54 residue in the active site of lanosterol 14α-demethylase enzyme. As mentioned above, the narrow range of resistance-associated mutations facilitated the application of molecular probing technologies for mutation profiling.[188] Molecular beacons are small fluorescent probes that are particularly sensitive and specific for allele discrimination, with the ability to distinguish single nucleotide differences in a DNA sequence.

Thus, by the use of molecular beacons corresponding to the wild-type *cyp51A* gene and seven mutant alleles, a single multiplexed real-time PCR reaction was applied to the analysis of genomic DNA samples from a collection of *A. fumigatus* clinical and laboratory-derived isolates.[189] Out of 48 isolates tested, 23 isolates bearing *cyp51A* alleles with Gly54 mutations were found, and all these isolates exhibited elevated levels of itraconazole resistance. This PCR assay was designed either to identify specific alleles by assigning separate fluorophores to each mutation or overall resistance by assigning a single fluorophore to all mutant *cyp51A* alleles and a separate fluorophore to the wild-type susceptible allele. In this format, high-throughput assessment of drug resistance facilitated a rapid designation of isolates as susceptible or resistant.[188]

Procedure

1. Extract genomic DNA from *A. fumigatus* cells by using the QIAGEN Genomic-tip (Qiagen, Hilden, Germany), following the instructions described in QIAGEN Genomic-tip Protocols.
2. Amplify a 500 bp fragment of *cyp51A* using a 100 μL PCR mixture containing 25 pmol of each of the Af210F and Af709R primers,[189] 2.5 U of i*Taq* DNA polymerase (Bio-Rad Laboratories, Hercules, CA), 0.5 mM deoxynucleoside triphosphates, 50 mM KCl, 4 mM $MgCl_2$, 20 mM Tris–HCl (pH 8.4), and about 100 ng of *A. fumigatus* genomic DNA. The cycling conditions are 1 cycle of 3 min at 95°C; 35 cycles of 30 s at 95°C, 30 s at 55°C, and 1 min at 72°C; and 1 cycle of 3 min at 72°C.
3. Molecular beacons are designed to cover the locus of diversity corresponding to codon 54 of *cyp51A* with Beacon Designer software (version 3.0; Premier Biosoft, Palo Alto, CA) as described.[189] Molecular beacons, one complementary to the wild-type allele and seven complementary to mutant alleles, are labeled, respectively, with fluorophores 5-carboxyfluorescein (FAM) and 6-carboxy-2′,4,4′,5′,7,7′-hexachlorofluorescein (HEX) at the 5′ end, and with the nonfluorescent quencher 4-{[−4-(dimethylamino)-phenyl]-azo}-benzoic acid (dabcyl) at the 3′ end.
4. Perform a multiplex real-time PCR assay using as a template 10 pg of the 500 bp PCR fragment per reaction on a Stratagene Mx4000 multiplex quantitative PCR system (Stratagene, La Jolla, CA) with the "quantitative PCR (multiple standards)" setting. Each 50 μL PCR mixture is composed of 1× Stratagene Core PCR buffer, 20 pmol of each molecular beacon (GGG-FAM and GAG-, GAA-, AAG-, AGG-, CGG-, TGG-, and GTG-HEX), 25 pmol of each of the *cyp51AS* and *cyp51AA* primers,[189] 2.5 U of Stratagene SureStart *Taq* DNA polymerase, 0.4 mM deoxynucleoside triphosphates, and 5 mM $MgCl_2$. PCRs are performed with the following parameters: 1 cycle of 10 min at 95°C and 45 cycles of 30 s at 95°C, 30 s at 55°C, and 30 s at 72°C. The fluorescence is measured three times during the annealing step.

Note. In this single-tube multiplex PCR format, HEX fluorescence (mutant beacons) is observed only when one or more of the mutant templates are present, while the FAM signal from the GGG-FAM molecular beacon is detected only when the wild-type allele is present. Thus, this assay is ideated to give a wild-type (FAM fluorescence) or mutant (HEX fluorescence) output, with the advantage that, in a single reaction, mutant (itraconazole-resistant) and wild-type (itraconazole-susceptible) isolates can be easily distinguished. Although this assay format could not be used to discriminate which mutant allele is present in the DNA template, such an output may be sufficient for clinical laboratories.

100.2.2.2 Molecular Detection of Fungal Resistance to Echinocandins

Principle

As a consequence of mechanism-specific resistance to echinocandins, it is now well recognized that high MIC isolates of *Candida* species from patients failing therapy often contain amino acid substitutions in Fks subunits (Fks1p and/or Fks2p) of the β-1,3-D-glucan synthase target[141] and that the FKS genotype could be a better predictive indicator of resistance than MIC alone.[137] In view of the limited spectrum of mutations conferring (cross-) resistance to echinocandins, real-time PCR with allele-specific beacons has been used to detect prominent mutations associated with echinocandin resistance in *C. albicans*.[139] In that study, mutations in *FKS1* were reliably identified in 85 isolates of *C. albicans* with resistance to caspofungin. Of these isolates, 93% showed changes at Ser645, with 62% displaying a characteristic S645P substitution expressed as either a homozygous or a heterozygous mutation in *FKS1*. Two other prominent amino acid substitutions, S645Y and S645F, were found at frequencies of 22% and 8%, respectively. In addition, new mutations in *FKS1* were identified. Finally, the allele-specific probes were combined in a multiplex real-time PCR assay for rapid detection of echinocandin resistance.

Procedure

1. Extract genomic DNA from *C. albicans* cells grown overnight in liquid YEPD medium with an automated purification system using the EZ1 DNA tissue kit (Qiagen) and the BioRobot EZ1 workstation (Qiagen) that is preceded by the enzymatic treatment of cell pellet derived from the fungal culture with 2 μL zymolyase (4 units/μL; Zymo Research, Orange, CA) in 190 μL of Lysis Buffer (Tris–HCl 50 mM [pH 7.5], EDTA 100 mM, and β-mercaptoethanol 1%).
2. Use published DNA sequences for *FKS1* to design molecular beacons and primers using the Beacon Designer software, as described.[139] Molecular beacons are labeled with fluorophores FAM and HEX at the 5′ end and with dabcyl at the 3′ end.
3. Perform multiplex real-time PCR experiments on a Stratagene MX4000 multiplex quantitative PCR system using the "quantitative PCR (multiple standards)" setting as described above, with a 50 μL PCR mixture containing 1× Stratagene Core PCR buffer, 0.2 μM of each of the four molecular beacons (wild-type-FAM and T1933C-, C1934A-, and C1934T-HEX), 0.25 μM each of primers HS1SN2 and HS1AN2,[139] 2.5 U of SureStart *Taq* DNA polymerase, 0.4 mM deoxynucleoside triphosphates, 4 mM $MgCl_2$, and about 50 ng of *C. albicans* genomic DNA. Cycling conditions are the same as described above, with the single exception of an annealing temperature of 60°C. The fluorescence is measured three times during the annealing step and monitored by both FAM and HEX optical channels of MX4000 hardware.

Note. This multiplex real-time PCR assay focuses on *FKS1* codon 645 mutations T1933C, C1934A, and C1934T, which have been identified in clinical isolates of *C. albicans* showing reduced susceptibility to caspofungin, thus representing a universal closed-tube assay for detection of known *FKS1* mutations conferring resistance to caspofungin. In the assay, the FAM fluorescent signal indicates the presence of the wild-type *FKS1* allele, whereas HEX fluorescence shows the presence of either the T1933C, C1934A, or C1934T *FKS1* mutation in the DNA sample. As demonstrated,[139] DNA extracted from *C. albicans* isolates heterozygous for any of these three mutations yielded both FAM and HEX signals, whereas multiplex real-time PCR with DNA from the *C. albicans* isolate with the two new mutations G1932T and C1934G in *FKS1* yielded neither FAM nor HEX fluorescence. Thus, as *FKS1* new mutations are identified, this assay can be expanded by adding other allele-specific probes and has value for high-throughput screening for routine resistance assessment by using characteristic *FKS1* mutations as a surrogate marker.

100.3 FUTURE PERSPECTIVES AND CONCLUSIONS

The reference assays for antifungal susceptibility testing mentioned above, which are based on the measuring growth under different drug concentrations, are robust and reproducible but remain time-consuming, and end-point reading is somewhat subjective. These limitations pose a challenge for the development of new accurate and reliable testing methodologies with faster turnaround times and where endpoints determination is objective. In this context, Marinach et al.[190] described the application of matrix-assisted laser desorption ionization-time-of-flight mass spectrometry (MALDI-TOF MS), recently emerged as a valid and widely available tool for detection and identification of microbial pathogens,[184] to a novel assay of *C. albicans* susceptibility to fluconazole, based on monitoring modifications of the fingerprint pattern of yeast cells grown in the presence of varying drug concentrations. The method, in addition to show full agreement with the CLSI reference method, allows formulating a new endpoint, the minimal profile change concentration (MPCC), and a value defined as the lowest drug (fluconazole, in this case) concentration at which a mass spectrum profile change can be detected. Notably, the diagnostic profile shift observed with fluconazole did not vary with either the genetic background of the *C. albicans* strain tested (e.g., mating type), or the level or mechanism (*ERG11* mutations, *TAC1* mutation,

CDR1/CDR2, or *MDR1* overexpression) of the resistance to fluconazole.[190] In view of these observations, the authors suggested that a characteristic profile would be associated with each of the different classes (triazoles, echinocandins, polyenes, and antimetabolites) of drugs known to inhibit a given pathogen, for example, *C. albicans*.

In order to individuate putative molecular targets for new drug development, "omic" technologies such as genomic- or proteomic-based approaches leading to a shift from random screens for antifungal agents toward screens against fungus-specific genes or proteins have been extensively used, but to date, their usefulness for antifungal drug discovery remains to be firmly assessed.[191] In contrast, Pukkila-Worley et al.[192] reviewed the use of the *Caenorhabditis elegans* model nematode in drug discovery. In addition to allow high-throughput screens of chemical libraries in vivo, this whole-animal system permits that compounds are simultaneously screened for antifungal efficacy and toxicity to overcome one of the main obstacles in current antimicrobial discovery. Furthermore, the *C. elegans* system may enable the identification of compounds that modulate immune responses or affect fungal virulence that are only expressed during infection. Using this system, a pilot screen for antifungal substances recently identified the caffeic acid phenethyl ester as a compound with both antifungal and host immunomodulatory activity.

With the discovery of the involvement of new molecular domains (e.g., XBD in Pdr1p/Pdr3p and KIX in Gal11p) or novel mediators (e.g., Hsp90) in the pathway leading to antifungal drug resistance, there is now reason to redirect the long-standing search for small molecules that inactivate resistance.[169] For example, these molecules could be used as antagonists that lock Pdr1p orthologues in an inactive conformation and thus prevent activation of efflux pumps resulting in sensitization to standard antifungal therapy, but mainly as blockage agents to prevent drug-Pdr1p/Pdr3p interaction.[159] Alternatively, use of inhibitors to harness Hsp90 function could provide a much-needed therapeutic approach for improving the treatment of fungal disease.[182]

In conclusion, although antifungal drug resistance mechanisms have been elucidated for most important fungal pathogens, further studies are strongly needed for uncommon and emerging pathogenic fungi. Particular renewed enthusiasm should be addressed to understanding regulatory mechanisms that govern antifungal drug resistance. We feel that new insights in these issues will aid in maximizing the clinical utility of current drugs and in developing new antifungal agents and treatment strategies.

REFERENCES

1. Nucci, M. and Marr, K.A., Emerging fungal diseases, *Clin. Infect. Dis.*, 41, 521, 2005.
2. Segal, B.H. et al., Prevention and early treatment of invasive fungal infection in patients with cancer and neutropenia and in stem cell transplant recipients in the era of newer broad-spectrum antifungal agents and diagnostic adjuncts, *Clin. Infect. Dis.*, 44, 402, 2007.
3. Schelenz, S., Management of candidiasis in the intensive care unit, *J. Antimicrob. Chemother.*, 61, i31, 2008.
4. Pfaller, M.A. and Diekema, D.J., Epidemiology of invasive candidiasis: A persistent public health problem, *Clin. Microbiol. Rev.*, 20, 133, 2007.
5. Hachem, R. et al., The changing epidemiology of invasive candidiasis: *Candida glabrata* and *Candida krusei* as the leading causes of candidemia in hematologic malignancy, *Cancer*, 112, 2493, 2008.
6. Pagano, L. et al., The epidemiology of fungal infections in patients with hematologic malignancies: The SEIFEM-2004 study, *Haematologica*, 91, 1068, 2006.
7. Pfaller, M.A. et al., Antifungal activities of posaconazole, ravuconazole, and voriconazole compared to those of itraconazole and amphotericin B against 239 clinical isolates of *Aspergillus* spp. and other filamentous fungi: Report from SENTRY Antimicrobial Surveillance Program, 2000, *Antimicrob. Agents Chemother.*, 46, 1032, 2002.
8. Diekema, D.J. et al., Activities of caspofungin, itraconazole, posaconazole, ravuconazole, voriconazole, and amphotericin B against 448 recent clinical isolates of filamentous fungi, *J. Clin. Microbiol.*, 41, 3623, 2003.
9. Chamilos, G. and Kontoyiannis, D.P., Update on antifungal drug resistance mechanisms of *Aspergillus fumigatus*, *Drug. Resist. Updat.*, 8, 344, 2005.
10. Panackal, A.A. et al., *Aspergillus ustus* infections among transplant recipients, *Emerg. Infect. Dis.*, 12, 403, 2006.
11. Baddley, J.W. et al., Patterns of susceptibility of *Aspergillus* isolates recovered from patients enrolled in the Transplant-Associated Infection Surveillance Network, *J. Clin. Microbiol.*, 47, 3271, 2009.
12. Balajee, S.A. et al., Molecular identification of *Aspergillus* species collected for the Transplant-Associated Infection Surveillance Network, *J. Clin. Microbiol.*, 47, 3138, 2009.
13. Garcia-Effron, G. et al., Rapid detection of triazole antifungal resistance in *Aspergillus fumigatus*, *J. Clin. Microbiol.*, 46, 1200, 2008.
14. Lass-Flörl, C., The changing face of epidemiology of invasive fungal disease in Europe, *Mycoses*, 52, 197, 2009.
15. Mitchell, T.G. and Perfect J.R., Cryptococcosis in the era of AIDS-100 years after the discovery of *Cryptococcus neoformans*, *Clin. Microbiol. Rev.*, 8, 515, 1995.
16. Pukkila-Worley, R. and Mylonakis, E., Epidemiology and management of cryptococcal meningitis: Developments and challenges, *Expert Opin. Pharmacother.*, 9, 551, 2008.
17. Johnson, E.M., Szekely, A., and Warnock, D.W., In-vitro activity of voriconazole, itraconazole and amphotericin B against filamentous fungi, *J. Antimicrob. Chemother.*, 42, 741, 1998.
18. Girmenia, C. et al., In vitro susceptibility testing of *Geotrichum capitatum*: Comparison of the E-test, disk diffusion, and Sensititre colorimetric methods with the NCCLS M27-A2 broth microdilution reference method, *Antimicrob. Agents Chemother.*, 47, 3985, 2003.
19. Pfaller, M.A. and Diekema, D.J., Rare and emerging opportunistic fungal pathogens: Concern for resistance beyond *Candida albicans* and *Aspergillus fumigatus*, *J. Clin. Microbiol.*, 42, 4419, 2004.
20. Espinel-Ingroff, A. et al., Activities of voriconazole, itraconazole and amphotericin B in vitro against 590 moulds from 323 patients in the voriconazole Phase III clinical studies, *J. Antimicrob. Chemother.*, 61, 616, 2008.
21. CLSI. *Reference Method for Broth Dilution Antifungal Susceptibility Testing of Yeasts*; *Approved Standard*, 3rd edn. CLSI, Wayne, PA, 2008.

22. CLSI. Reference *Method for Broth Dilution Antifungal Susceptibility Testing of Filamentous Fungi; Approved Standard*, 2nd edn. CLSI, Wayne, PA, 2008.
23. Anderson, J.B., Evolution of antifungal-drug resistance: Mechanisms and pathogen fitness, *Nat. Rev. Microbiol.*, 3, 547, 2005.
24. Sanglard, D. et al., Mechanisms of resistance to azole antifungal agents in *Candida albicans* isolates from AIDS patients involve specific multidrug transporters, *Antimicrob. Agents Chemother.*, 39, 2378, 1995.
25. Perea, S. et al., Prevalence of molecular mechanisms of resistance to azole antifungal agents in *Candida albicans* strains displaying high-level fluconazole resistance isolated from human immunodeficiency virus-infected patients, *Antimicrob. Agents Chemother.*, 45, 2676, 2001.
26. Tumbarello, M. et al., Fungaemia caused by *Candida glabrata* with reduced susceptibility to fluconazole due to altered gene expression: Risk factors, antifungal treatment and outcome, *J. Antimicrob. Chemother.*, 62, 1379, 2008.
27. Sanglard, D. and Odds, F.C., Resistance of *Candida* species to antifungal agents: Molecular mechanisms and clinical consequences, *Lancet Infect. Dis.*, 2, 73, 2002.
28. Chen, S.C. and Sorrell, T.C., Antifungal agents, *Med. J.*, 187, 404, 2007.
29. Pasqualotto, A.C. and Denning, D.W., New and emerging treatments for fungal infections, *J. Antimicrob. Chemother.*, 61, i19, 2008.
30. Odds, F.C., Brown, A.J., and Gow, N.A., Antifungal agents: Mechanisms of action, *Trends Microbiol.*, 11, 272, 2003.
31. Sanglard, D. and Bille, J., Current understanding of the modes of action and resistance mechanisms to conventional and emerging antifungal agents for treatment of *Candida* infections, in *Candida and Candidiasis*, Calderone, R.A., Ed., ASM Press, Washington, DC, 2002, pp. 349–383.
32. St. Georgiev, V., Membrane transporters and antifungal drug resistance, *Curr. Drug Targets*, 1, 261, 2000.
33. Sokol-Anderson, M.L., Brajtburg, J., and Medoff, G., Amphotericin B-induced oxidative damage and killing of *Candida albicans*, *J. Infect. Dis.*, 154, 76, 1986.
34. White, T.C., Marr, K.A., and Bowden, R.A., Clinical, cellular, and molecular factors that contribute to antifungal drug resistance, *Clin. Microbiol. Rev.*, 11, 382, 1998.
35. Saag, M.S. et al., Practice guidelines for the management of cryptococcal disease. Infectious Diseases Society of America, *Clin. Infect. Dis.*, 30, 710, 2000.
36. Pasqualotto, A.C. et al., Flucytosine therapeutic monitoring: 15 years experience from the UK, *J. Antimicrob. Chemother.*, 59, 791, 2007.
37. Lupetti, A. et al., Molecular basis of resistance to azole antifungals, *Trends Mol. Med.*, 8, 76, 2002.
38. Sanglard, D., Resistance of human fungal pathogens to antifungal drugs, *Curr. Opin. Microbiol.*, 5, 379, 2002.
39. Donnelly, J.P. and De Pauw, B.E., Voriconazole-a new therapeutic agent with an extended spectrum of antifungal activity, *Clin. Microbiol. Infect.*, 10, S107, 2004.
40. Schiller, D.S. and Fung, H.B., Posaconazole: An extended-spectrum triazole antifungal agent, *Clin. Ther.*, 29, 1862, 2007.
41. Xiao, L. et al., Three-dimensional models of wild-type and mutated forms of cytochrome P450 14alpha-sterol demethylases from *Aspergillus fumigatus* and *Candida albicans* provide insights into posaconazole binding, *Antimicrob. Agents Chemother.*, 48, 568, 2004.
42. Denning, D.W., Echinocandin antifungal drugs, *Lancet*, 362, 1142, 2003.
43. Bowman, J.C. et al., The antifungal echinocandin caspofungin acetate kills growing cells of *Aspergillus fumigatus* in vitro, *Antimicrob. Agents Chemother.*, 46, 3001, 2002.
44. Dannaoui, E. et al., Amphotericin B resistance of *Aspergillus terreus* in a murine model of disseminated aspergillosis, *J. Med. Microbiol.*, 49, 601, 1999.
45. Krogh-Madsen, M. et al., Amphotericin B and caspofungin resistance in *Candida glabrata* isolates recovered from a critically ill patient, *Clin. Infect. Dis.*, 42, 938, 2006.
46. Soll, D.R. and Pujol, C., *Candida albicans* clades, *FEMS Immunol. Med. Microbiol.*, 39, 1, 2003.
47. Blignaut, E. et al., Clade-related amphotericin B resistance among South African *Candida albicans* isolates, *Diagn. Microbiol. Infect. Dis.*, 53, 29, 2005.
48. Nolte, F.S. et al., Isolation and characterization of fluconazole- and amphotericin B-resistant *Candida albicans* from blood of two patients with leukemia, *Antimicrob. Agents Chemother.*, 41, 196, 1997.
49. Vandeputte, P. et al., Reduced susceptibility to polyenes associated with a missense mutation in the *ERG6* gene in a clinical isolate of *Candida glabrata* with pseudohyphal growth, *Antimicrob. Agents Chemother.*, 51, 982, 2007.
50. Qiao, J., Liu, W., and Li, R., Antifungal resistance mechanisms of *Aspergillus*, *Jpn. J. Med. Mycol.*, 49, 157, 2008.
51. Loeffler, J. and Stevens, D.A., Antifungal drug resistance, *Clin. Infect. Dis.*, 36, S31, 2003.
52. Blum, G. et al., Potential basis for amphotericin B resistance in *Aspergillus terreus*, *Antimicrob. Agents Chemother.*, 52, 1553, 2008.
53. Joseph-Horne, T. et al., Cross-resistance to polyene and azole drugs in *Cryptococcus neoformans*, *Antimicrob. Agents Chemother.*, 39, 1526, 1995.
54. Kelly, S.L. et al., Resistance to amphotericin B associated with defective sterol delta 8→7 isomerase in a *Cryptococcus neoformans* strain from an AIDS patient, *FEMS Microbiol. Lett.*, 122, 39, 1994.
55. Cannon, R.D. et al., Efflux-mediated antifungal drug resistance, *Clin. Microbiol. Rev.*, 22, 291, 2009.
56. Barchiesi, F. et al., Primary resistance to flucytosine among clinical isolates of *Candida* spp., *J. Antimicrob. Chemother.*, 45, 408, 2000.
57. Pfaller, M.A. et al., In vitro activities of 5-fluorocytosine against 8,803 clinical isolates of *Candida* spp.: Global assessment of primary resistance using National Committee for Clinical Laboratory Standards susceptibility testing methods, *Antimicrob. Agents Chemother.*, 46, 3518, 2002.
58. Perea, S. and Patterson, T.F., Antifungal resistance in pathogenic fungi, *Clin. Infect. Dis.*, 35, 1073, 2002.
59. Pujol, C., Pfaller, M.A., and Soll, D.R., Flucytosine resistance is restricted to a single genetic clade of *Candida albicans*, *Antimicrob. Agents Chemother.*, 48, 262, 2004.
60. Dodgson, A.R. et al., Clade-specific flucytosine resistance is due to a single nucleotide change in the *FUR1* gene of *Candida albicans*, *Antimicrob. Agents Chemother.*, 48, 2223, 2004.
61. Hope, W.W. et al., Molecular mechanisms of primary resistance to flucytosine in *Candida albicans*, *Antimicrob. Agents Chemother.*, 48, 4377, 2004.
62. Odds, F.C., In *Candida albicans*, resistance to flucytosine and terbinafine is linked to MAT locus homozygosity and multilocus sequence typing clade 1, *FEMS Yeast Res.*, 9, 1091, 2009.
63. Rustad, T.R. et al., Homozygosity at the *Candida albicans* MTL locus associated with azole resistance, *Microbiology*, 148, 1061, 2002.

64. Papon, N. et al., Molecular mechanism of flucytosine resistance in *Candida lusitaniae*: Contribution of the *FCY2*, *FCY1*, and *FUR1* genes to 5-fluorouracil and fluconazole cross-resistance, *Antimicrob. Agents Chemother.*, 51, 369, 2007.
65. Kanafani, Z.A. and Perfect, J.R., Antimicrobial resistance: Resistance to antifungal agents: Mechanisms and clinical impact, *Clin. Infect. Dis.*, 46, 120, 2008.
66. Hospenthal, D.R. and Bennett, J.E., Flucytosine monotherapy for cryptococcosis, *Clin. Infect. Dis.*, 27, 260, 1998.
67. White, T.C., Increased mRNA levels of *ERG16*, *CDR*, and *MDR1* correlate with increases in azole resistance in *Candida albicans* isolates from a patient infected with human immunodeficiency virus. *Antimicrob. Agents Chemother.*, 41, 1482, 1997.
68. Marichal, P. et al., Molecular biological characterization of an azole-resistant *Candida glabrata* isolate, *Antimicrob. Agents Chemother.*, 41, 2229, 1997.
69. White, T.C. et al., Resistance mechanisms in clinical isolates of *Candida albicans*, *Antimicrob. Agents Chemother.*, 46, 1704, 2002.
70. Orozco, A.S. et al., Mechanism of fluconazole resistance in *Candida krusei*, *Antimicrob. Agents Chemother.*, 42, 2645, 1998.
71. Sanglard, D. et al., Amino acid substitutions in the cytochrome P-450 lanosterol 14alpha-demethylase (CYP51A1) from azole-resistant *Candida albicans* clinical isolates contribute to resistance to azole antifungal agents, *Antimicrob. Agents Chemother.*, 42, 241, 1998.
72. Marichal, P. et al., Contribution of mutations in the cytochrome P450 14alpha-demethylase (Erg11p, Cyp51p) to azole resistance in *Candida albicans*, *Microbiology*, 145, 2701, 1999.
73. Braun, B.R. et al., A human-curated annotation of the *Candida albicans* genome, *PLoS Genet.*, 1, 36, 2005.
74. Gaur, M., Choudhury, D., and Prasad, R., Complete inventory of ABC proteins in human pathogenic yeast, *Candida albicans*, *J. Mol. Microbiol. Biotechnol.*, 9, 3, 2005.
75. Prasad, R. et al., Molecular cloning and characterization of a novel gene of *Candida albicans*, *CDR1*, conferring multiple resistance to drugs and antifungals, *Curr. Genet.*, 27, 320, 1995.
76. Sanglard, D. et al., Cloning of *Candida albicans* genes conferring resistance to azole antifungal agents: Characterization of *CDR2*, a new multidrug ABC transporter gene, *Microbiology*, 143, 405, 1997.
77. Lamping, E. et al., Characterization of three classes of membrane proteins involved in fungal azole resistance by functional hyperexpression in *Saccharomyces cerevisiae*, *Eukaryot. Cell*, 6, 1150, 2007.
78. Holmes, A.R. et al., ABC transporter Cdr1p contributes more than Cdr2p does to fluconazole efflux in fluconazole-resistant *Candida albicans* clinical isolates, *Antimicrob. Agents Chemother.*, 52, 3851, 2008.
79. Sanguinetti, M. et al., Mechanisms of azole resistance in clinical isolates of *Candida glabrata* collected during a hospital survey of antifungal resistance, *Antimicrob. Agents Chemother.*, 49, 668, 2005.
80. Posteraro, B. et al., Azole resistance of *Candida glabrata* in a case of recurrent fungemia, *J. Clin. Microbiol.*, 44, 3046, 2006.
81. Magill, S.S. et al., Triazole cross-resistance among *Candida* spp.: Case report, occurrence among bloodstream isolates, and implications for antifungal therapy, *J. Clin. Microbiol.*, 44, 529, 2006.
82. Shin, J.H. et al., Changes in karyotype and azole susceptibility of sequential bloodstream isolates from patients with *Candida glabrata* candidemia, *J. Clin. Microbiol.*, 45, 2385, 2007.
83. Miyazaki, H. et al., Fluconazole resistance associated with drug efflux and increased transcription of a drug transporter gene, *PDH1*, in *Candida glabrata*, *Antimicrob. Agents Chemother.*, 42, 1695, 1998.
84. Sanglard, D. et al., The ATP binding cassette transporter gene *CgCDR1* from *Candida glabrata* is involved in the resistance of clinical isolates to azole antifungal agents, *J. Antimicrob. Agents Chemother.*, 43, 2753, 1999.
85. Sanglard, D., Ischer, F., and Bille, J., Role of ATP-binding-cassette transporter genes in high-frequency acquisition of resistance to azole antifungals in *Candida glabrata*, *Antimicrob. Agents Chemother.*, 45, 1174, 2001.
86. Redding, S.W. et al., Multiple patterns of resistance to fluconazole in *Candida glabrata* isolates from a patient with oropharyngeal candidiasis receiving head and neck radiation, *J. Clin. Microbiol.*, 41, 619, 2003.
87. Bennett, J.E., Izumikawa, K., and Marr, K.A., Mechanism of increased fluconazole resistance in *Candida glabrata* during prophylaxis, *Antimicrob. Agents Chemother.*, 48, 1773, 2004.
88. Vermitsky, J.P. and Edlind, T.D., Azole resistance in *Candida glabrata*: Coordinate upregulation of multidrug transporters and evidence for a Pdr1-like transcription factor, *Antimicrob. Agents Chemother.*, 48, 3773, 2004.
89. Brun, S. et al., Relationships between respiration and susceptibility to azole antifungals in *Candida glabrata*, *Antimicrob. Agents Chemother.*, 47, 847, 2003.
90. Brun, S. et al., Mechanisms of azole resistance in petite mutants of *Candida glabrata*, *Antimicrob. Agents Chemother.*, 48, 1788, 2004.
91. Torelli, R. et al., The ATP-binding cassette transporter-encoding gene *CgSNQ2* is contributing to the *CgPDR1*-dependent azole resistance of *Candida glabrata*, *Mol. Microbiol.*, 68, 186, 2008.
92. Pfaller, M.A. et al., Use of fluconazole as a surrogate marker to predict susceptibility and resistance to voriconazole among 13,338 clinical isolates of *Candida* spp. tested by clinical and laboratory standards institute-recommended broth microdilution methods, *J. Clin. Microbiol.*, 45, 70, 2007.
93. Panackal, A.A. et al., Clinical significance of azole antifungal drug cross-resistance in *Candida glabrata*, *J. Clin. Microbiol.*, 44, 1740, 2006.
94. Lamping, E. et al., Abc1p is a multidrug efflux transporter that tips the balance in favor of innate azole resistance in *Candida krusei*, *Antimicrob. Agents Chemother.*, 53, 354, 2009.
95. Mellado, E. et al., Targeted gene disruption of the 14-alpha sterol demethylase (*cyp51A*) in *Aspergillus fumigatus* and its role in azole drug susceptibility, *Antimicrob. Agents Chemother.*, 49, 2536, 2005.
96. Mann, P.A. et al., Mutations in *Aspergillus fumigatus* resulting in reduced susceptibility to posaconazole appear to be restricted to a single amino acid in the cytochrome P450 14alpha-demethylase, *Antimicrob. Agents Chemother.*, 47, 577, 2003.
97. Chen, J. et al., Mutations in the *cyp51A* gene and susceptibility to itraconazole in *Aspergillus fumigatus* serially isolated from a patient with lung aspergilloma, *J. Antimicrob. Chemother.*, 55, 31, 2005.
98. Howard, S.J. et al., Multi-azole resistance in *Aspergillus fumigatus*, *Int. J. Antimicrob. Agents*, 28, 450, 2006.
99. Verweij, P.E., Mellado, E., and Melchers, W.J., Multiple-triazole-resistant aspergillosis, *N. Engl. J. Med.*, 356, 1481, 2007.

100. Howard, S.J. et al., Frequency and evolution of azole resistance in *Aspergillus fumigatus* associated with treatment failure, *Emerg. Infect. Dis.*, 15, 1068, 2009.
101. Moore, C.B. et al., Antifungal drug resistance in *Aspergillus*, *J. Infect.*, 41, 203, 2000.
102. Balajee, S.A. et al., *Aspergillus lentulus* sp. nov., a new sibling species of *A. fumigatus*, *Eukaryot. Cell*, 4, 625, 2005.
103. Pavie, J. et al., Breakthrough disseminated *Aspergillus ustus* infection in allogeneic hematopoietic stem cell transplant recipients receiving voriconazole or caspofungin prophylaxis, *J. Clin. Microbiol.*, 43, 4902, 2005.
104. Varga, J. et al., *Aspergillus calidoustus* sp. nov., causative agent of human infections previously assigned to *Aspergillus ustus*, *Eukaryot. Cell*, 7, 630, 2008.
105. Slaven, J.W. et al., Increased expression of a novel *Aspergillus fumigatus* ABC transporter gene, *atrF*, in the presence of itraconazole in an itraconazole resistant clinical isolate, *Fungal Genet. Biol.*, 36, 199, 2002.
106. Nascimento, A.M. et al., Multiple resistance mechanisms among *Aspergillus fumigatus* mutants with high-level resistance to itraconazole, *Antimicrob. Agents Chemother.*, 47, 1719, 2003.
107. Gulshan, K. and Moye-Rowley, W.S., Multidrug resistance in fungi, *Eukaryot. Cell*, 6, 1933, 2007.
108. Piddock, L.J., Multidrug-resistance efflux pumps—Not just for resistance, *Nat. Rev. Microbiol.*, 4, 629, 2006.
109. Mellado, E. et al., Identification of two different 14-alpha sterol demethylase-related genes (*cyp51A* and *cyp51B*) in *Aspergillus fumigatus* and other *Aspergillus* species, *J. Clin. Microbiol.*, 39, 2431, 2001.
110. Mellado, E. et al., Substitutions at methionine 220 in the 14alpha-sterol demethylase (Cyp51A) of *Aspergillus fumigatus* are responsible for resistance in vitro to azole antifungal drugs, *Antimicrob. Agents Chemother.*, 48, 2747, 2004.
111. Mellado, E. et al., A new *Aspergillus fumigatus* resistance mechanism conferring in vitro cross-resistance to azole antifungals involves a combination of *cyp51A* alterations, *Antimicrob. Agents Chemother.*, 51, 1897, 2007.
112. Snelders, E. et al., Emergence of azole resistance in *Aspergillus fumigatus* and spread of a single resistance mechanism, *PLoS Med.*, 5, e219, 2008.
113. Bellete, B. et al., Acquired resistance to voriconazole and itraconazole in a patient with pulmonary aspergilloma, *Med. Mycol.*, 48, 197, 2010.
114. Mosquera, J. and Denning, D.W., Azole cross-resistance in *Aspergillus fumigatus*, *Antimicrob. Agents Chemother.*, 46, 556, 2002.
115. Meneau, I. and Sanglard, D., Azole and fungicide resistance in clinical and environmental *Aspergillus fumigatus* isolates, *Med. Mycol.*, 43, S307, 2005.
116. Snelders, E. et al., Possible environmental origin of resistance of *Aspergillus fumigatus* to medical triazoles, *Appl. Environ. Microbiol.*, 75, 4053, 2009.
117. Peetermans, W. et al., Fluconazole-resistant *Cryptococcus neoformans* var. *gattii* in an AIDS patient, *Acta Clin. Belg.*, 48, 405, 1993.
118. Birley, H.D. et al., Azole drug resistance as a cause of clinical relapse in AIDS patients with cryptococcal meningitis, *Int. J. STD AIDS*, 6, 353, 1995.
119. Viard, J.P. et al., Fulminant cryptococcal infections in HIV-infected patients on oral fluconazole, *Lancet*, 346, 118, 1995.
120. Armengou, A. et al., Possible development of resistance to fluconazole during suppressive therapy for AIDS-associated cryptococcal meningitis, *Clin. Infect. Dis.*, 23, 1337, 1996.
121. Aller, A.I. et al., Correlation of fluconazole MICs with clinical outcome in cryptococcal infection, *Antimicrob. Agents Chemother.*, 44, 1544, 2000.
122. Saag, M.S. et al., Practice guidelines for the management of cryptococcal disease, Infectious Diseases Society of America, *Clin. Infect Dis.*, 30, 710, 2000.
123. Sar, B. et al., Increasing in vitro resistance to fluconazole in *Cryptococcus neoformans* Cambodian isolates: April 2000 to March 2002, *J. Antimicrob. Chemother.*, 54, 563, 2004.
124. Paugam, A. et al., Increased fluconazole resistance of *Cryptococcus neoformans* isolated from a patient with AIDS and recurrent meningitis, *Clin. Infect. Dis.*, 19, 975, 1994.
125. Davey, K.G. et al., In-vitro susceptibility of *Cryptococcus neoformans* isolates to fluconazole and itraconazole, *J. Antimicrob. Chemother.*, 42, 217, 1998.
126. Friese, G. et al., Development of azole resistance during fluconazole maintenance therapy for AIDS-associated cryptococcal disease, *AIDS*, 15, 2344, 2001.
127. Lamb, D.C. et al., Resistant P45051A1 activity in azole antifungal tolerant *Cryptococcus neoformans* from AIDS patients, *FEBS Lett.*, 368, 326, 1995.
128. Rodero, L. et al., G484S amino acid substitution in lanosterol 14-alpha demethylase (ERG11) is related to fluconazole resistance in a recurrent *Cryptococcus neoformans* clinical isolate, *Antimicrob. Agents Chemother.*, 47, 3653, 2003.
129. Venkateswarlu, K. et al., Fluconazole tolerance in clinical isolates of *Cryptococcus neoformans*, *Antimicrob. Agents Chemother.*, 41, 748, 1997.
130. Joseph-Horne, T. et al., Amphotericin B resistant isolates of *Cryptococcus neoformans* without alteration in sterol biosynthesis, *J. Med. Vet. Mycol.*, 34, 223, 1996.
131. Posteraro, B. et al., Identification and characterization of a *Cryptococcus neoformans* ATP binding cassette (ABC) transporter-encoding gene, Cn*AFR1*, involved in the resistance to fluconazole, *Mol. Microbiol.*, 47, 357, 2003.
132. Sanguinetti, M. et al., Role of *AFR1*, an ABC transporter-encoding gene, in the in vivo response to fluconazole and virulence of *Cryptococcus neoformans*, *Infect. Immun.*, 74, 1352, 2006.
133. Orsi, C.F. et al., The ABC transporter-encoding gene *AFR1* affects the resistance of *Cryptococcus neoformans* to microglia-mediated antifungal activity by delaying phagosomal maturation, *FEMS Yeast Res.*, 9, 301, 2009.
134. Sionov, E. et al., Heteroresistance to fluconazole in *Cryptococcus neoformans* is intrinsic and associated with virulence, *Antimicrob. Agents Chemother.*, 53, 2804, 2009.
135. Thornewell, S.J., Peery, R.B., and Skatrud, P.L., Cloning and characterization of Cne*MDR1*: A *Cryptococcus neoformans* gene encoding a protein related to multidrug resistance proteins, *Gene*, 201, 21, 1997.
136. Douglas, C.M. et al., Identification of the *FKS1* gene of *Candida albicans* as the essential target of 1,3-beta-D-glucan synthase inhibitors, *Antimicrob. Agents Chemother.*, 41, 2471, 1997.
137. Perlin, D.S., Resistance to echinocandin-class antifungal drugs, *Drug. Resist. Updat.*, 10, 121, 2007.
138. Kurtz, M.B. et al., Characterization of echinocandin-resistant mutants of *Candida albicans*: Genetic, biochemical, and virulence studies, *Infect. Immun.*, 64, 3244, 1996.
139. Balashov, S.V., Park, S., and Perlin, D.S., Assessing resistance to the echinocandin antifungal drug caspofungin in *Candida albicans* by profiling mutations in *FKS1*, *Antimicrob. Agents Chemother.*, 50, 2058, 2006.

140. Garcia-Effron, G., Park, S., and Perlin, D.S., Correlating echinocandin MIC and kinetic inhibition of fks1 mutant glucan synthases for *Candida albicans*: Implications for interpretive breakpoints, *Antimicrob. Agents Chemother.*, 53, 112, 2009.
141. Park, S. et al., Specific substitutions in the echinocandin target Fks1p account for reduced susceptibility of rare laboratory and clinical *Candida* sp. isolates, *Antimicrob. Agents Chemother.*, 49, 3264, 2005.
142. Desnos-Ollivier, M. et al., European Committee on Antibiotic Susceptibility Testing. Mutations in the *fks1* gene in *Candida albicans*, *C. tropicalis*, and *C. krusei* correlate with elevated caspofungin MICs uncovered in AM3 medium using the method of the European Committee on Antibiotic Susceptibility Testing, *Antimicrob. Agents Chemother.*, 52, 3092, 2008.
143. Katiyar, S., Pfaller, M., and Edlind, T., *Candida albicans* and *Candida glabrata* clinical isolates exhibiting reduced echinocandin susceptibility, *Antimicrob. Agents Chemother.*, 50, 2892, 2006.
144. Garcia-Effron, G. et al., A naturally occurring proline-to-alanine amino acid change in Fks1p in *Candida parapsilosis*, *Candida orthopsilosis*, and *Candida metapsilosis* accounts for reduced echinocandin susceptibility, *Antimicrob. Agents Chemother.*, 52, 2305, 2008.
145. Kartsonis, N. et al., Caspofungin susceptibility testing of isolates from patients with esophageal candidiasis or invasive candidiasis: Relationship of MIC to treatment outcome, *Antimicrob. Agents Chemother.*, 49, 3616, 2005; *Mol. Microbiol.*, 63, 1399, 2007.
146. Munro, C.A. et al., The PKC, HOG and Ca2+ signalling pathways co-ordinately regulate chitin synthesis in *Candida albicans*, *Mol. Microbiol.*, 63, 1399, 2007.
147. Walker, L.A. et al., Stimulation of chitin synthesis rescues *Candida albicans* from echinocandins, *PLoS Pathog.*, 4, e1000040, 2008.
148. Stevens, D.A., Espiritu, M., and Parmar, R., Paradoxical effect of caspofungin: Reduced activity against *Candida albicans* at high drug concentrations, *Antimicrob. Agents Chemother.*, 48, 3407, 2004.
149. Stevens, D.A. et al., Escape of *Candida* from caspofungin inhibition at concentrations above the MIC (paradoxical effect) accomplished by increased cell wall chitin; evidence for beta-1,6-glucan synthesis inhibition by caspofungin, *Antimicrob. Agents Chemother.*, 50, 3160, 2006.
150. Madureira, A. et al., Breakthrough invasive aspergillosis in allogeneic haematopoietic stem cell transplant recipients treated with caspofungin, *Int. J. Antimicrob. Agents*, 30, 551, 2007.
151. Gardiner, R.E. et al., Characterization of *Aspergillus fumigatus* mutants with reduced susceptibility to caspofungin, *Med. Mycol.*, 43, S299, 2005.
152. Rocha, E.M. et al., A Ser678Pro substitution in Fks1p confers resistance to echinocandin drugs in *Aspergillus fumigates*, *Antimicrob. Agents Chemother.*, 51, 4174, 2007.
153. Romano, J. et al., Disruption of the *Aspergillus fumigatus ECM33* homologue results in rapid conidial germination, antifungal resistance and hypervirulence, *Microbiology*, 152, 1919, 2006.
154. Arendrup, M.C. et al., Establishing in vitro-in vivo correlations for *Aspergillus fumigatus*: The challenge of azoles versus echinocandins, *Antimicrob. Agents Chemother.*, 52, 3504, 2008.
155. Maligie, M.A. and Selitrennikoff, C.P., *Cryptococcus neoformans* resistance to echinocandins: (1,3)beta-glucan synthase activity is sensitive to echinocandins, *Antimicrob. Agents Chemother.*, 49, 2851, 2005.
156. Niimi, K. et al., Overexpression of *Candida albicans CDR1*, *CDR2*, or *MDR1* does not produce significant changes in echinocandin susceptibility, *Antimicrob. Agents Chemother.*, 50, 1148, 2006.
157. Coleman, J.J. and Mylonakis, E., Efflux in fungi: La pièce de résistance, *PLoS Pathog.*, 5, e1000486, 2009.
158. MacPherson, S., Larochelle, M., and Turcotte, B., A fungal family of transcriptional regulators: The zinc cluster proteins, *Microbiol. Mol. Biol. Rev.*, 70, 583, 2006.
159. Thakur, J.K. et al., A nuclear receptor-like pathway regulating multidrug resistance in fungi, *Nature*, 452, 604, 2008.
160. Carvajal, E. et al., Molecular and phenotypic characterization of yeast *PDR1* mutants that show hyperactive transcription of various ABC multidrug transporter genes, *Mol. Gen. Genet.*, 256, 406, 1997.
161. Nourani, A. et al., Clustered amino acid substitutions in the yeast transcription regulator Pdr3p increase pleiotropic drug resistance and identify a new central regulatory domain, *Mol. Gen. Genet.*, 256, 397, 1997.
162. Delahodde, A., Delaveau, T., and Jacq, C., Positive autoregulation of the yeast transcription factor Pdr3p, which is involved in control of drug resistance, *Mol. Cell Biol.*, 15, 4043, 1995.
163. Hallstrom, T.C. and Moye-Rowley, W.S., Multiple signals from dysfunctional mitochondria activate the pleiotropic drug resistance pathway in *Saccharomyces cerevisiae*, *J. Biol. Chem.*, 275, 37347, 2000.
164. Tsai, H.F. et al., *Candida glabrata PDR1*, a transcriptional regulator of a pleiotropic drug resistance network, mediates azole resistance in clinical isolates and petite mutants, *Antimicrob. Agents Chemother.*, 50, 1384, 2006.
165. Vermitsky, J.P. et al., Pdr1 regulates multidrug resistance in *Candida glabrata*: Gene disruption and genome-wide expression studies, *Mol. Microbiol.*, 61, 704, 2006.
166. Ferrari, S. et al., Gain of function mutations in Cg*PDR1* of *Candida glabrata* not only mediate antifungal resistance but also enhance virulence, *PLoS Pathog.*, 5, e1000268, 2009.
167. DeRisi, J., Genome microarray analysis of transcriptional activation in multidrug resistance yeast mutants, *FEBS Lett.*, 470, 156, 2000.
168. Lucau-Danila, A. et al., Early expression of yeast genes affected by chemical stress, *Mol. Cell Biol.*, 25, 1860, 2005.
169. Goffeau, A., Drug resistance: The fight against fungi, *Nature*, 452, 541, 2008.
170. Chen, K.H. et al., The bZip transcription factor Cgap1p is involved in multidrug resistance and required for activation of multidrug transporter gene Cg*FLR1* in *Candida glabrata*, *Gene*, 386, 63, 2007.
171. Traven, A. et al., Interorganellar communication. Altered nuclear gene expression profiles in a yeast mitochondrial DNA mutant, *J. Biol. Chem.*, 276, 4020, 2001.
172. Shahi, P., Gulshan, K., and Moye-Rowley, W.S., Negative transcriptional regulation of multidrug resistance gene expression by an Hsp70 protein, *J. Biol. Chem.*, 282, 26822, 2007.
173. Florio, A., unpublished data, 2007.
174. Coste, A.T. et al., *TAC1*, transcriptional activator of *CDR* genes, is a new transcription factor involved in the regulation of *Candida albicans* ABC transporters *CDR1* and *CDR2*, *Eukaryot. Cell*, 3, 1639, 2004.

175. Selmecki, A., Forche, A., and Berman, J., Aneuploidy and isochromosome formation in drug-resistant *Candida albicans*, *Science*, 313, 367, 2006.
176. Coste, A. et al., A mutation in Tac1p, a transcription factor regulating *CDR1* and *CDR2*, is coupled with loss of heterozygosity at chromosome 5 to mediate antifungal resistance in *Candida albicans*, *Genetics*, 172, 2139, 2006.
177. Selmecki, A. et al., An isochromosome confers drug resistance in vivo by amplification of two genes, *ERG11* and *TAC1*, *Mol. Microbiol.*, 68, 624, 2008.
178. Dunkel, N. et al., Mutations in the multi-drug resistance regulator *MRR1*, followed by loss of heterozygosity, are the main cause of *MDR1* overexpression in fluconazole-resistant *Candida albicans* strains, *Mol. Microbiol.*, 69, 827, 2008.
179. Cowen, L.E. and Steinbach, W.J., Stress, drugs, and evolution: The role of cellular signaling in fungal drug resistance, *Eukaryot. Cell*, 7, 747, 2008.
180. Cowen, L.E. et al., Genetic architecture of Hsp90-dependent drug resistance, *Eukaryot. Cell*, 5, 2184, 2006.
181. Singh, S.D. et al., Hsp90 governs echinocandin resistance in the pathogenic yeast *Candida albicans* via calcineurin, *PLoS Pathog.*, 5, e1000532, 2009.
182. Cowen, L.E. et al., Harnessing Hsp90 function as a powerful, broadly effective therapeutic strategy for fungal infectious disease, *Proc. Natl. Acad. Sci. USA*, 106, 2818, 2009.
183. Cowen, L.E. and Lindquist, S., Hsp90 potentiates the rapid evolution of new traits: Drug resistance in diverse fungi, *Science*, 309, 2185, 2005.
184. Weile, J. and Knabbe, C., Current applications and future trends of molecular diagnostics in clinical bacteriology, *Anal. Bioanal. Chem.*, 394, 731, 2009.
185. Espy, M.J. et al., Real-time PCR in clinical microbiology: Applications for routine laboratory testing, *Clin. Microbiol. Rev.*, 19, 165, 2006.
186. Morrell, M., Fraser, V.J., and Kollef, M.H., Delaying the empiric treatment of *Candida* bloodstream infection until positive blood culture results are obtained: A potential risk factor for hospital mortality, *Antimicrob. Agents Chemother.*, 49, 3640, 2005.
187. Edlind, T.D., Molecular detection of antifungal resistance, in *Molecular Microbiology: Diagnostic Principles and Practice*, Chap. 43, Persing, D.H., Tenover, F.C., Versalovic, J., Tang, Y.W., Unger, E.R., Relman, D.A., White, T.J., Eds., ASM Press, Washington, DC, 2004.
188. Perlin, D.S., Antifungal drug resistance: Do molecular methods provide a way forward? *Curr. Opin. Infect. Dis.*, 22, 568, 2009.
189. Balashov, S.V. et al., Rapid, high-throughput, multiplex, real-time PCR for identification of mutations in the *cyp51A* gene of *Aspergillus fumigatus* that confer resistance to itraconazole, *J. Clin. Microbiol.*, 43, 214, 2005.
190. Marinach, C. et al., MALDI-TOF MS-based drug susceptibility testing of pathogens: The example of *Candida albicans* and fluconazole. *Proteomics*, 9, 4627, 2009.
191. Odds, F.C., Discovery and development of antifungal drugs: A reality check, presented at *4th Trends in Medical Mycology*, Athens, October 18–21, 2009, http://www.TIMM2009.org
192. Pukkila-Worley, R. et al., Antifungal drug discovery through the study of invertebrate model hosts, *Curr. Med. Chem.*, 16, 1588, 2009.

Index

B

C

F

G

N

O

P

T

U

V

W

Z